Use Dynamic Media to Bring Earth Science to Life

Pearson

EARTH SCIENCE

TARBUCK

LUTGENS

Illustrated by TASA

EARTH SCIENCE

FIFTEENTH EDITION

Pearson

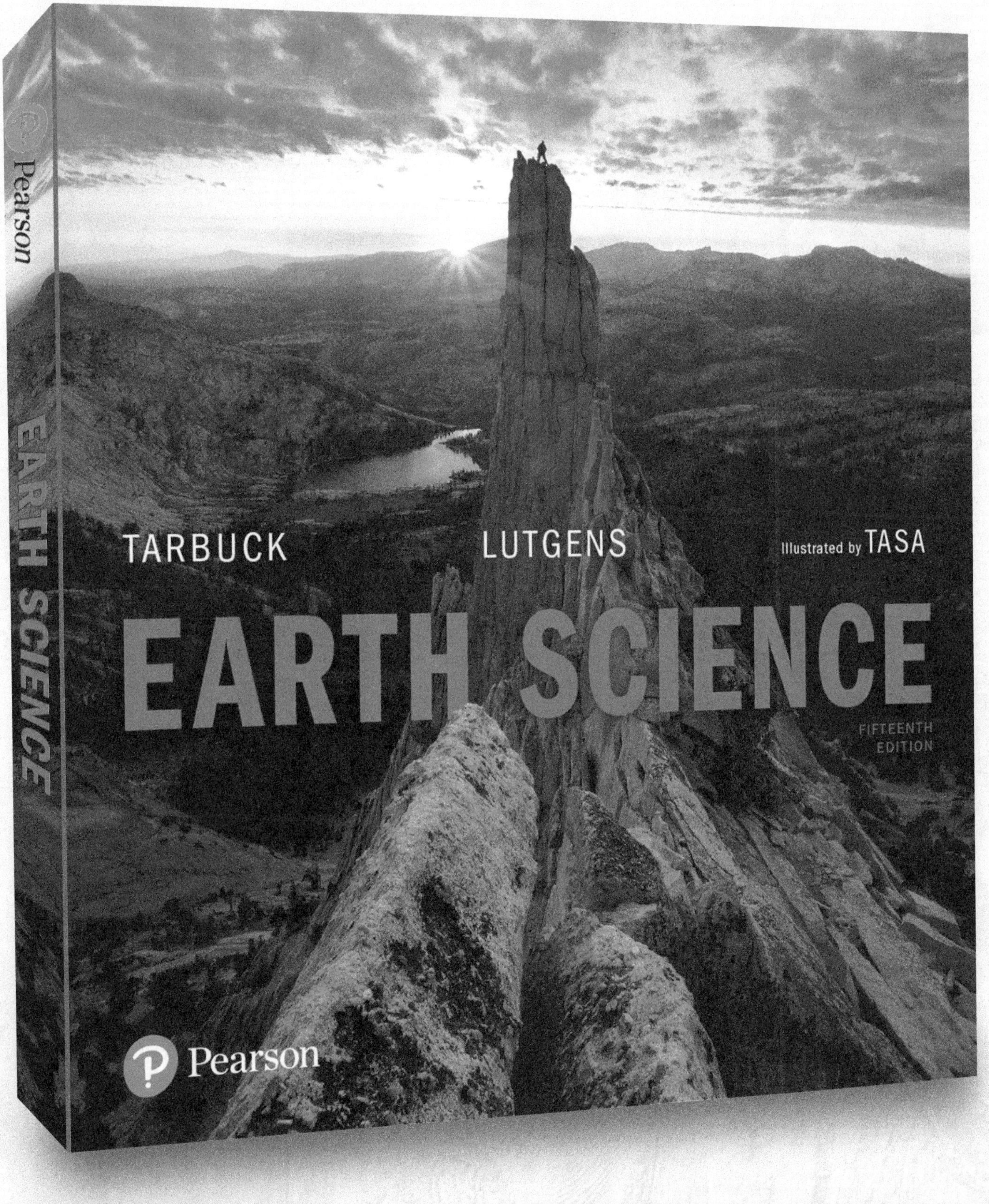

Bring Field Experience to Students' Fingertips...

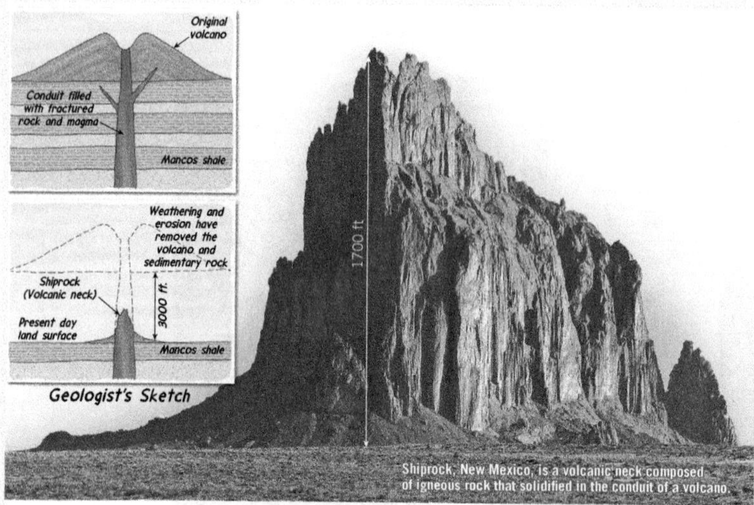

▲ SmartFigure 6.26 **Volcanic neck**
Shiprock, New Mexico, is a volcanic neck that stands about 520 meters (1700 feet) high. It consists of igneous rock that crystallized in the vent of a volcano that has long since been eroded. (Photo by Dennis Tasa)

TUTORIAL
https://goo.gl/TjW5uh

NEW! QR Codes link to SmartFigures

Quick Response (QR) codes link to over 238 videos and animations, giving readers immediate access to five types of dynamic media: Project Condor Quadcopter Videos, Mobile Field Trips, Tutorials, Animations, and Videos to help visualize physical processes and concepts. SmartFigures extend the print book to bring Earth Science to life.

How to download a QR Code Reader

Using a smartphone, students are encouraged to download a QR Code reader app from Google Play or the Apple App Store. Many are available for free. Once downloaded, students open the app and point the camera to a QR Code. Once scanned, they're prompted to open the url to immediately be connected to the digital world and deepen their learning experience with the printed text.

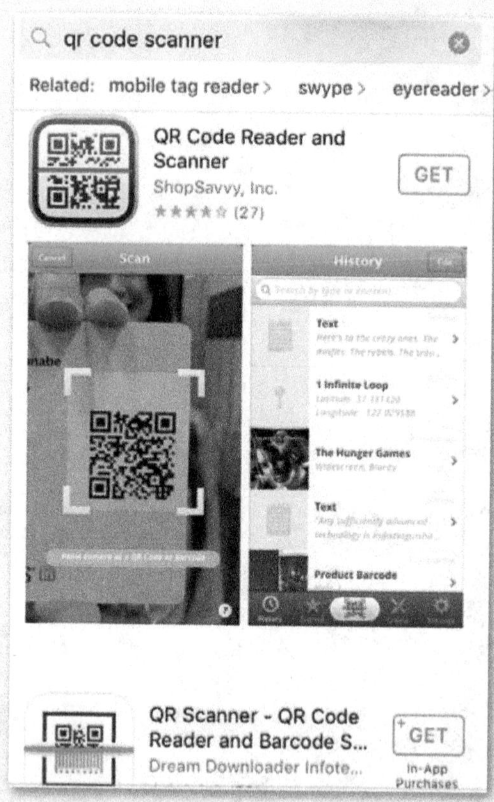

NEW! SmartFigures: Project Condor Quadcopter Videos

Bringing Earth Science to life for students, three geologists, using a quadcopter-mounted GoPro camera, have ventured into the field to film 10 key geologic locations and processes. These process-oriented videos, accessed through QR codes, are designed to bring the field to the classroom and improve the learning experience within the text.

...with SmartFigures

NEW! SmartFigures: Mobile Field Trips
On each trip, students will accompany geologist-pilot-photographer Michael Collier in the air and on the ground to visit and learn about iconic landscapes that relate to discussions in the chapter. These extraordinary field trips are accessed by using QR codes throughout the text. New Mobile Field Trips for the 15th edition include *Formation of a Water Gap*, *Ice Sculpts Yosemite*, *Fire and Ice Land*, *Dendrochronology*, and *Desert Geomorphology*.

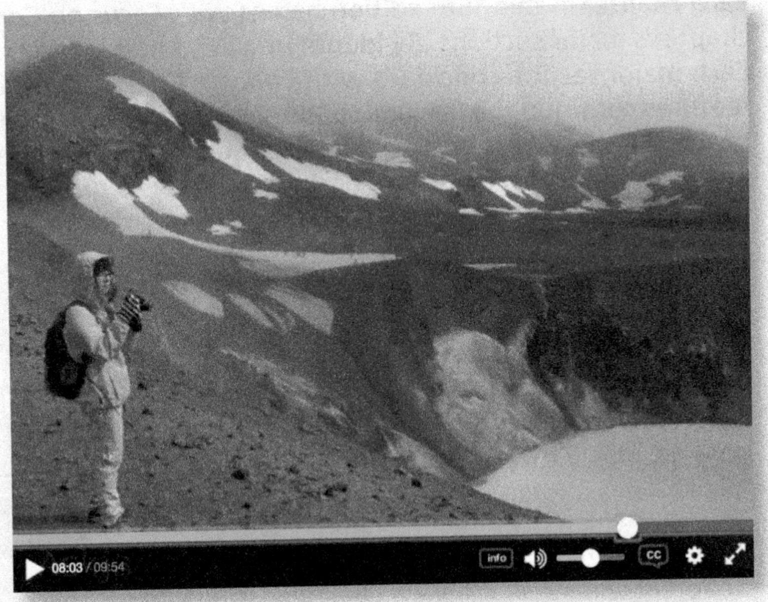

NEW! SmartFigures: Animations
Brief animations created by text illustrator Dennis Tasa animate a process or concept depicted in the textbook's figures. With QR codes, students are given a view of moving figures rather than static art to depict how geologic processes move throughout time.

HALLMARK! SmartFigures: Tutorials
These brief tutorial videos present the student with a 3- to 4-minute feature (mini-lesson), most narrated and annotated by Professor Callan Bentley. Each lesson examines and explains the concepts illustrated by the figure. With over 150 SmartFigure Tutorials inside the text, students have a multitude of ways to enjoy art that teaches.

Clear Learning Path in Each Chapter

Each chapter in this 15th edition begins with *Focus on Concepts*: a set of learning objectives that correspond to the chapter's major sections. By identifying key knowledge and skills, these objectives help students prioritize the material. Each major section concludes with *Concept Checks* so that students can check their learning. Three end-of-chapter features continue the learning path. *Concepts in Review* are coordinated with the *Focus on Concepts* at the beginning of the chapter and with the numbered sections within the chapter, providing a readable and concise overview of key ideas, with photos, diagrams, and questions. The questions and problems in *Give It Some Thought* and *Examining the Earth System* challenge learners by requiring higher order thinking skills to analyze, synthesize, and apply the material.

1) The chapter-opening ***Focus on Concepts*** lists the learning objectives for the chapter. Each section of the chapter is tied to a specific learning objective, providing students with a clear learning path to the chapter content.

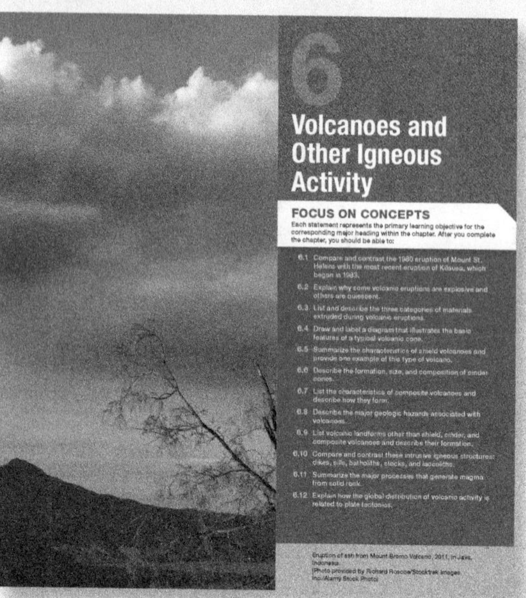

2) Each chapter section concludes with ***Concept Checks***, a set of questions that is tied to the section's learning objectives and allows students to monitor their grasp of significant facts and ideas.

CONCEPT CHECKS 6.8

1. Describe pyroclastic flows and explain why they are capable of traveling great distances.
2. What is a lahar?
3. List at least three volcanic hazards besides pyroclastic flows and lahars.

3) ***Concepts in Review*** provides students with a structured review of the chapter. Consistent with the *Focus on Concepts* and *Concept Checks*, ***Concepts in Review*** is structured around the learning objective for each section.

6 CONCEPTS IN REVIEW

Volcanoes and Other Igneous Activity

6.1 Mount St. Helens Versus Kilauea

Compare and contrast the 1980 eruption of Mount St. Helens with the most recent eruption of Kilauea, which began in 1983.

- Volcanic eruptions cover a broad spectrum from explosive eruptions, like that of Mount St. Helens in 1980, to the quiescent eruptions of Kilauea.

6.2 The Nature of Volcanic Eruptions

Explain why some volcanic eruptions are explosive and others are quiescent.

KEY TERMS: magma, lava, effusive eruption, viscosity, eruption column

- The two primary factors determining the nature of a volcanic eruption are the viscosity (resistance to flow) of the magma and its gas content. In general, magmas that contain more silica are more viscous, while those with lower silica content are more fluid. Temperature also influences viscosity. Hot lavas are more fluid, while cool lavas are more viscous.
- Basaltic magmas, which are fluid and have low gas content, tend to generate effusive (nonexplosive) eruptions. In contrast, silica-rich magmas (andesitic and rhyolitic), which are the most viscous and contain the greatest quantity of gases, are the most explosive.

? Although Kilauea mostly erupts in a gentle manner, what risks might you encounter if you chose to live nearby?

USGS

EXAMINING THE EARTH SYSTEM

1 Speculate about some of the possible consequences that a great and prolonged increase in explosive volcanic activity might have on each of Earth's four spheres.

Jon A Helgason/Yay Micro/AGE Fotostock

GIVE IT SOME THOUGHT

1 Examine the accompanying photo and complete the following:
  a. What type of volcano is shown? What features helped you classify it as such?
  b. What is the eruptive style of such volcanoes? Describe the likely composition and viscosity of its magma.
  c. Which type of plate boundary is the likely setting for this volcano?
  d. Name a city that is vulnerable to the effects of a volcano of this type.

USGS

4) ***Give It Some Thought*** and ***Examining the Earth System*** activities challenge learners by requiring higher-order thinking skills to analyze, synthesize, and apply chapter material.

Exposing Students to Source Data and the Tools of Science

NEW! Each chapter of the 15th edition now concludes with new *Data Analysis* activities. These brief capstone activities send students outside of the book to online science tools and data sets from organizations such as NASA, NOAA, and USGS, empowering students to apply and extend chapter concepts and develop their data analysis and critical thinking skills

DATA ANALYSIS

Recent Volcanic Activity

The Smithsonian Institution Global Volcanism Program and the USGS work together to compile a list of new and changing volcanic activity worldwide. NOAA also uses this information to issue Volcanic Ash Advisories to alert aircraft of volcanic ash in the air.

ACTIVITIES

Go to the Weekly Volcanic Activity Report page at http://volcano.si.edu.

1 What information is displayed on this page?
2 Click on Criteria and Disclaimers. Which volcanoes are not displayed on this map?
3 In what areas is most of the volcanic activity concentrated?
4 Click on Weekly Report. List the new volcanic activity locations. List three ongoing volcanic activity locations.

Click on the name of a volcano under New Activity/Unrest.

5 Where is this volcano located? Be sure to include the city, country, volcanic region name, latitude, and longitude.
6 What is the primary volcanic type?
7 Do some investigating online and in your textbook. What are the key characteristics for this type of volcano?
8 Briefly describe the most recent activity. How was this activity observed?
9 What are the dates for the most recent activity?
10 Click on Eruptive History. What is the earliest date listed for this volcano?
11 Find this volcano on the map on the previous page. Is this volcano near a plate boundary? If so, between which plates? (Use your textbook to determine the location of plate boundaries.)

Go to the Volcanic Ash Advisory Center (VAAC) page at www.ssd.noaa.gov/VAAC/washington.html.

12 List the VAAC locations.
13 Click on Current Volcanic Ash Advisories. When was the most recent Volcanic Ash Advisory issued? What is the location of this advisory?
14 Which of the new volcanic activity locations from question 4 currently have Volcanic Ash Advisories? For each, what is the date of the most recent advisory?

DATA ANALYSIS

The Aral Sea

The Aral Sea was once the fourth-largest lake in the world. This lake has now decreased in size by more than 80%, and the southern Aral Sea has disappeared altogether. This has had devastating effects of the communities around the lake.

ACTIVITIES

Go to NASA's Earth Observatory site at http://earthobservatory.nasa.gov, select World of Change under Special Collections and scroll to select Shrinking Aral Sea. As you step forward in time, you will see the aerial extent of the Aral Sea.

1 When did the Aral Sea begin to shrink? Why did the Aral Sea begin to shrink?
2 How has the shrinking lake affected the quality of the water and farmland in the area?
3 How has the lake's reduction affected summer and winter temperatures?

Step forward in time to see changes in the Aral Sea. The green region is the lake, and the white region around the lake is salt deposits. You may also click on Google Earth to step through time and use the measuring tool to answer some of these questions.

4 What is the east–west distance between the easternmost edge of the Aral Sea in 1960 and the edge of the southern Aral Sea in 2000? 1960 and 2005? 1960 and 2010? 1960 and 2015?
5 What is the distance change between 2000 and 2005? 2005 and 2010? 2005 and 2015?
6 What is the average rate of distance change since 2000? (Remember that rate of change is the distance change divided by the number of years.)
7 Why was there a significant decline in the overall size of the southern Aral Sea after 2005?

Go to "Shrinking Aral Sea" on NASA's Earth Observatory site (https://earthobservatory.nasa.gov/Features/WorldOfChange/aral_sea.php).

8 Compare this image to the Aral Sea images from Earth Observatory. Approximately when was the dust storm image taken? (Giving a range of years is fine.)
9 From which direction is the wind blowing?
10 How long is the dust storm at its longest distance? How wide is the dust storm at its widest distance?
11 Which towns are in the path of this dust storm?

Continuous Learning Before, During, and After Class

BEFORE CLASS

Mobile Media and Reading Assignments Ensure Students Come to Class Prepared

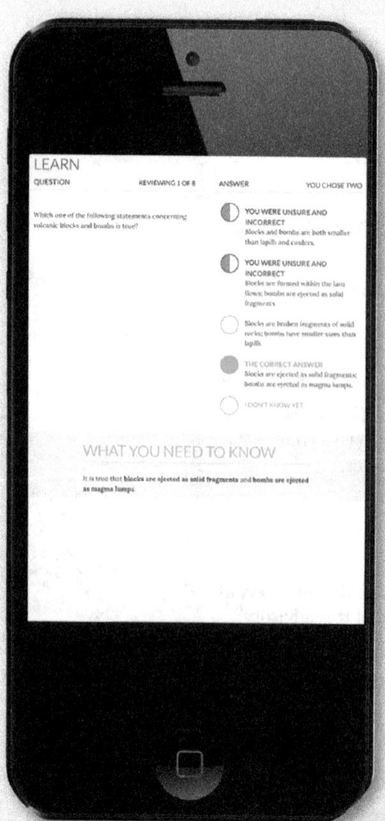

Updated! Dynamic Study Modules help students study effectively by continuously assessing student performance and providing practice in areas where students struggle the most. Each Dynamic Study Module, accessed by computer, smartphone, or tablet, promotes fast learning and long-term retention.

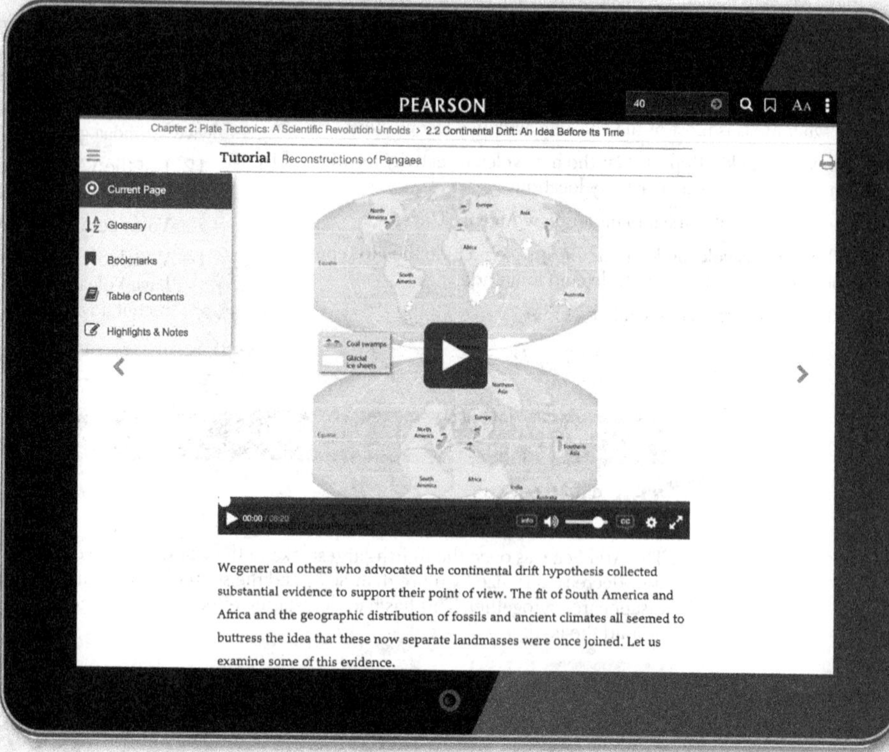

NEW! Interactive eText 2.0 gives students access to the text whenever they can access the internet. eText features include:

- Now available on smartphones and tablets.
- Seamlessly integrated videos and other rich media.
- Accessible (screen-reader ready).
- Configurable reading settings, including resizable type and night reading mode.
- Instructor and student note-taking, highlighting, bookmarking, and search.

Pre-Lecture Reading Quizzes are easy to customize and assign

Reading Quiz Questions ensure that students complete the assigned reading before class and stay on track with reading assignments. Reading Questions are 100% mobile ready and can be completed by students on mobile devices.

with MasteringGeology™

DURING CLASS

Engage students with Learning Catalytics

What has teachers and students excited? Learning Catalytics, a 'bring your own device' student engagement, assessment, and classroom intelligence system, allows students to use their smartphone, tablet, or laptop to respond to questions in class. With Learning Cataltyics, you can:

- Assess students in real time using open-ended question formats to uncover student misconceptions and adjust lecture accordingly.
- Automatically create groups for peer instruction based on student response patterns, to optimize discussion productivity.

> ***"My students are so busy and engaged answering Learning Catalytics questions during lecture that they don't have time for Facebook."***
>
> **Declan De Paor, Old Dominion University**

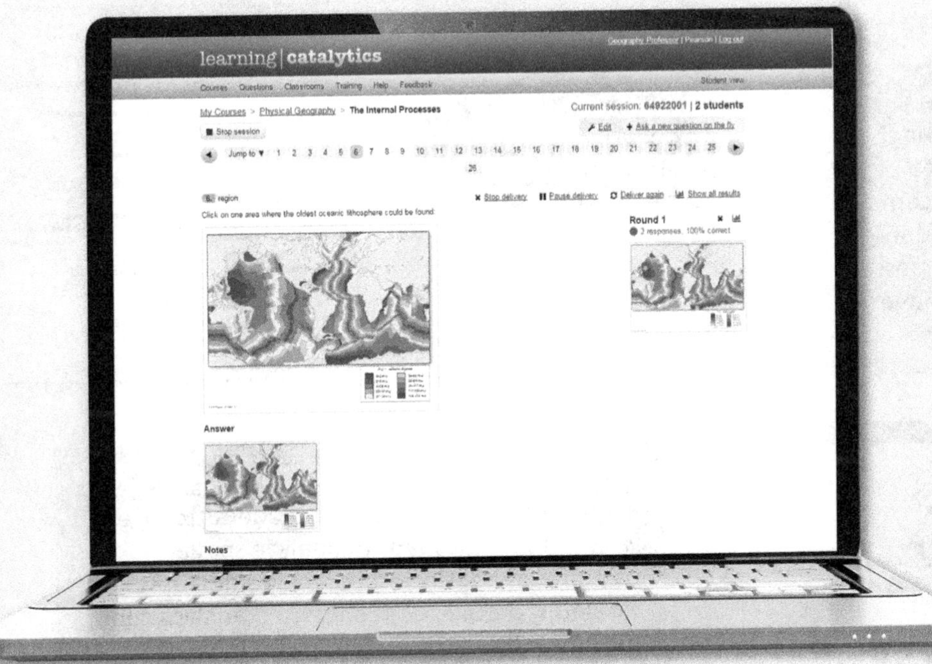

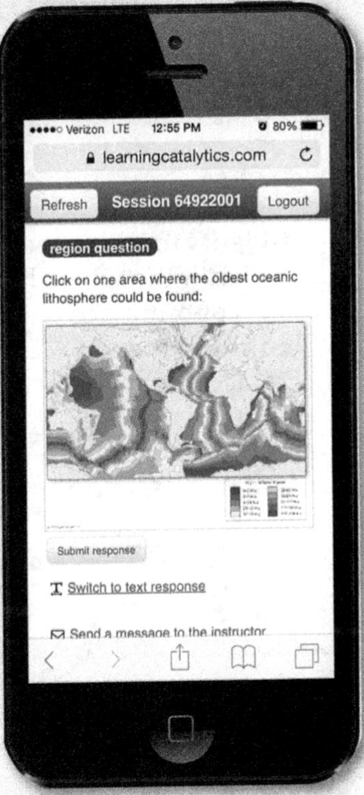

MasteringGeology™

AFTER CLASS

Easy to Assign, Customizable, Media-Rich, and Automatically Graded Assignments

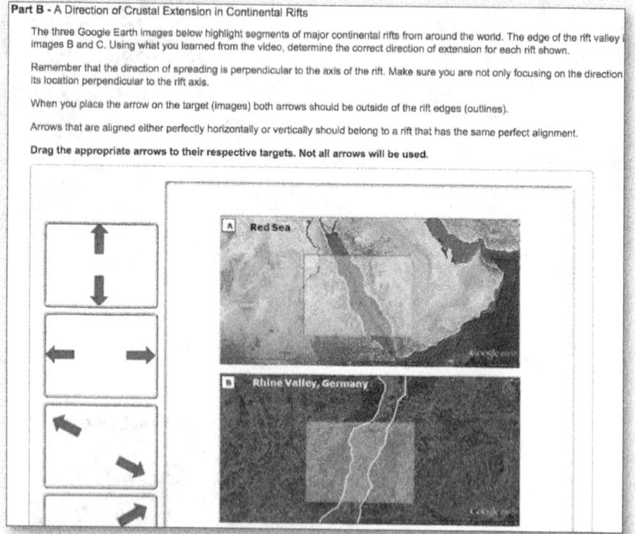

NEW! Project Condor Quadcopter Videos
A series of quadcopter videos with annotations, sketching, and narration help improve the way students learn about monoclines, streams and terraces, and so much more. In MasteringGeology™, these videos are accompanied by assessments to test student understanding.

NEW! 24 Mobile Field Trips take students to iconic locations with Michael Collier in the air and on the ground to learn about places that relate to concepts in the chapter. In Mastering, these videos are accompanied by auto-gradable assessments that will track what students have learned.

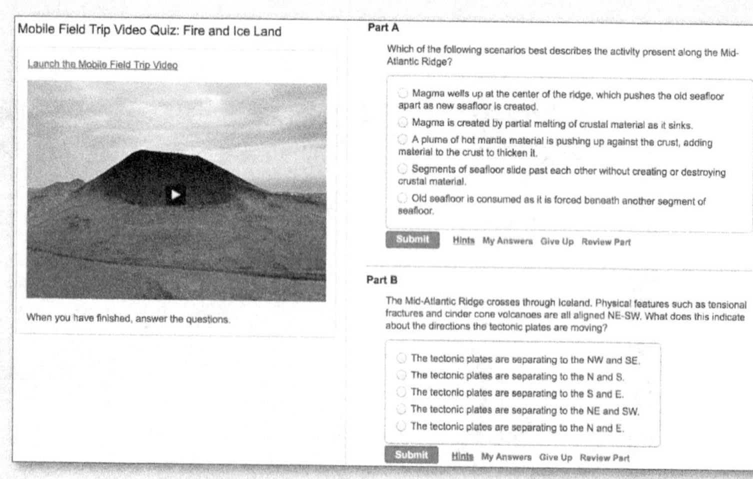

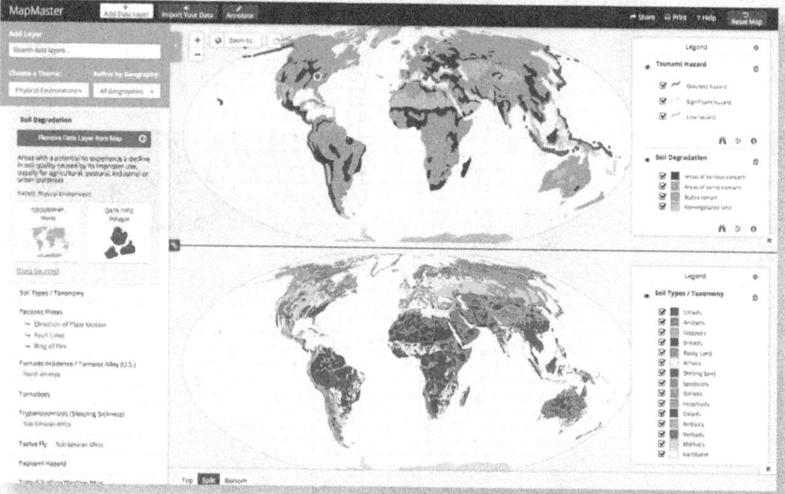

NEW! MapMaster 2.0 Activities are inspired by GIS, allowing students to layer various thematic maps to analyze spatial patterns and data at regional and global scales. Now fully mobile, with enhanced analysis tools, such as split screen, the ability for students to geolocate themselves in the data, and the ability for students to upload their own data for advanced map making. This tool includes zoom, and annotation functionality, with hundreds of map layers leveraging recent data from sources such as NOAA, NASA, USGS, United Nations, CIA, World Bank, UN, PRB, and more.

www.masteringgeology.com

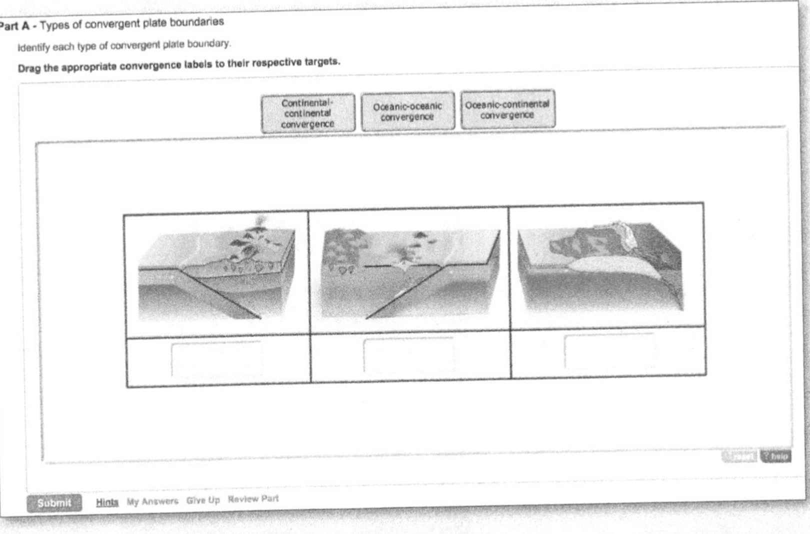

GeoTutors

These coaching activities help students master the most challenging physical geoscience concepts with highly visual, kinesthetic activities focused on critical thinking and application of core geoscience concepts.

GigaPan Activities allow students to take advantage of a virtual field experience with high-resolution imaging technology developed by Carnegie Mellon University in conjunction with NASA.

Encounter Activities

Using Google Earth™ to visualize and explore Earth's physical landscape, Encounter activities provide rich, interactive explorations of geology and Earth Science concepts. Dynamic assessments include questions related to core geoscience concepts. All explorations include corresponding Google Earth KMZ media files, and questions include hints and specific wrong-answer feedback to help coach students toward mastery of the concepts.

Resources for YOU, the Instructor

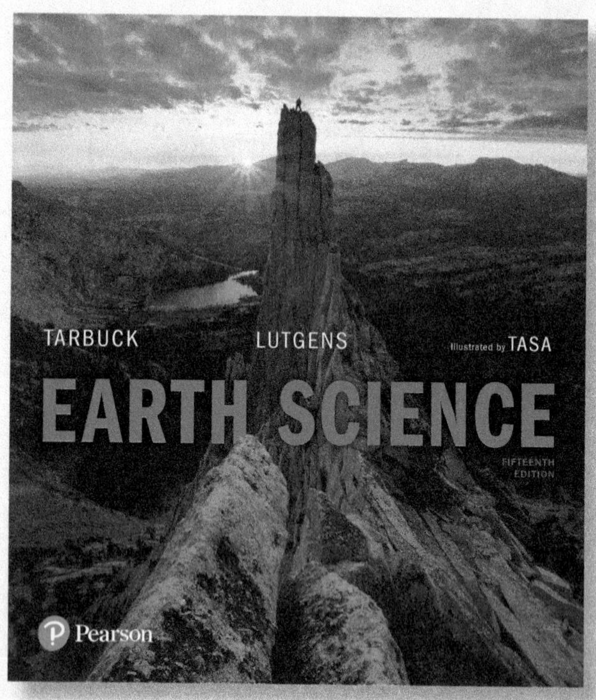

MasteringGeology™ provides everything you need to prep for your course and deliver a dynamic lecture, all in one convenient place. Resources include:

LECTURE PRESENTATION ASSETS FOR EACH CHAPTER

- PowerPoint Lecture Outlines
- PowerPoint Clicker Questions and Jeopardy-style quiz show questions
- All book images and tables in JPEG and PowerPoint formats

TEST BANK

- The Test Bank in Microsoft Word format
- Computerized Test Bank, which includes all the questions from the printed test bank in a format that allows you to easily and intuitively build exams and quizzes.

TEACHING RESOURCES

- *Instructor Resource Manual* in Microsoft Word and PDF formats
- Full access to eText 2.0
- Pearson Community Website (https://communities.pearson.com/northamerica/s/)

Measuring Student Learning Outcomes?

All MasteringGeology assignable content is tagged to learning outcomes from the book, the Earth Science Literacy Initiatives "Big Ideas", and Bloom's Taxonomy. You also have the ability to add your own learning outcomes, helping you track student performance against your learning outcomes. You can view class performance against the specified learning outcomes and share those results quickly and easily by exporting to a spreadsheet.

EARTH SCIENCE

FIFTEENTH
EDITION

ARBUCK LUTGENS Illustrated by TASA

EARTH SCIENCE

FIFTEENTH EDITION

330 Hudson Street, NY NY 10013

Executive Editor, Geosciences Courseware: Christian Botting
Director, Courseware Portfolio Management: Beth Wilbur
Content Producer: William Wenzler, Karen Sanatar
Managing Producer: Mike Early
Courseware Director, Content Development: Ginnie Simione Jutson
Courseware Sr. Analyst: Margot Otway
Editorial Assistant, Geosciences Courseware: Emily Bornhop
Rich Media Content Producers: Mia Sullivan, Libby Reiser
Full Service Vendor: SPi Global
Full Service Project Manager: Patty Donovan
Copyeditor: Kitty Wilson
Design Manager: Mark Ong
Cover and Interior Designer: Jeff Puda
Photo and Illustration Support: Kevin Lear, International Mapping
Rights and Permissions Project Manager: Matthew Perry
Rights and Permissions Management: Ben Ferrini
Manufacturing Buyer: Maura Zaldivar-Garcia
Senior Marketing Manager, Field: Mary Salzman
Marketing Assistant: Ami Sampat
Cover Image Credit: © Grant Ordelheide

Credits and acknowledgments borrowed from other sources and reproduced, with permission, in this textbook appear on the appropriate page within text or are listed below.

Page 10: From J. Bronowski, The Common Sense of Science, p. 148. © 1953 Harvard University Press. **Page 11:** From L. Pasteur, Lecture, University of Lille (7 December 1854). **Page 98:** From R.T. Chamberlain, "Some of the Objections to Wegener's Theory," In: THEORY OF CONTINENTAL DRIFT: A SYMPOSIUM, University of Chicago Press, pp. 83-87, 1928. **Page 349:** From J. Hutton, Theory of Earth, 1700; From J. Hutton, Transactions of the Royal Society of Edinburgh, 1788. **Page 566:** Sir Francis Bacon. **Page 646:** Joseph Louis Lagrange, Oeuvres de Lagrange, 1867.

Library of Congress Cataloging-in-Publication Data

Names: Tarbuck, Edward J., author. | Lutgens, Frederick K., author.
Title: Earth science / Edward J. Tarbuck, Fred K. Lutgens;
illustrated by
Dennis Tasa.
Description: Fifteenth edition. | Hoboken, NJ: Pearson Education, [2017] | Includes index.
Identifiers: LCCN 2017013211 | ISBN 9780134543536 | ISBN 013454353X
Subjects: LCSH: Earth sciences—Textbooks.
Classification: LCC QE26.3.T38 2017 | DDC 550—dc23 LC record
available at https://lccn.loc.gov/2017013211

ISBN-10: 0-134-54353-X
ISBN-13: 978-0-134-54353-6

About Our Sustainability Initiatives

Pearson recognizes the environmental challenges facing this planet, as well as acknowledges our responsibility in making a difference. This book is carefully crafted to minimize environmental impact. The binding, cover, and paper come from facilities that minimize waste, energy consumption, and the use of harmful chemicals. Pearson closes the loop by recycling every out-of-date text returned to our warehouse.

Along with developing and exploring digital solutions to our market's needs, Pearson has a strong commitment to achieving carbonneutrality. As of 2009, Pearson became the first carbon- and climate-neutral publishing company, having reduced our absolute carbon footprint by 22% since then. Pearson has protected over 1,000 hectares of land in Columbia, Costa Rica, the United States, the UK and Canada. In 2015, Pearson formally adopted The Global Goals for Sustainable Development, sponsoring an event at the United Nations General Assembly and other ongoing initiatives. Pearson sources 100% of the electricity we use from green power and invests in renewable energy resources in multiple cities where we have operations, helping make them more sustainable and limiting our environmental impact for local communities.

The future holds great promise for reducing our impact on Earth's environment, and Pearson is proud to be leading the way. We strive to publish the best books with the most up-to-date and accurate content, and to do so in ways that minimize our impact on Earth. To learn more about our initiatives, please visit **https://www.pearson.com/sustainability.html**.

6 2021
www.pearsonhighered.com

BRIEF CONTENTS

CONTENTS

UNIT 2 Forces Within 92

4 Plate Tectonics: A Scientific Revolution Unfolds 93

5 Earthquakes and Earth's Interior 127

6 Volcanoes and Other Igneous Activity 161

19 Weather Patterns and Severe Storms 577

20 World Climates and Global Climate Change 607

UNIT 7 Earth's Place in the Universe 636

21 Origins of Modern Astronomy 637

SmartFigure Media

Use your mobile device and a free Quick Response (QR) code reader app to scan a SmartFigure identified by a QR code, and a video or animation illustrating the SmartFigure's concept launches immediately. No slow websites or hard-to-remember logins required. These mobile media transform textbooks into convenient digital platforms, breathe life into your learning experience, and help you grasp challenging Earth Science concepts.

CONDOR VIDEO
Continental Rifting
https://goo.gl/RXv8qH

Chapter 1

Chapter 2

Chapter 3

Chapter 4

Chapter 5

Chapter 6

Chapter 7

Chapter 8

Chapter 9

Chapter 10

Chapter 11

Chapter 12

Chapter 13

Chapter 14

Chapter 15

Chapter 16

Chapter 17

Chapter 18

Chapter 19

Chapter 20

Chapter 21

Chapter 22

Chapter 23

Chapter 24

PREFACE

Earth Science, 15th edition, is a college-level text designed for an introductory course in Earth Science. It consists of seven units that emphasize broad and up-to-date coverage of basic topics and principles in geology, oceanography, meteorology, and astronomy. The book is intended to be a meaningful, nontechnical survey for undergraduate students who may have a modest science background. Usually these students are taking an Earth Science class to meet a portion of their college's or university's general requirements.

In addition to being informative and up-to-date, *Earth Science,* 15th edition, strives to meet the need of beginning students for a readable and user-friendly text and a highly usable "tool" for learning basic Earth Science principles and concepts.

New and Important Features

This 15th edition is an extensive and thorough revision of *Earth Science* that integrates improved textbook resources with new online features to enhance the learning experience.

- **Significant updating and revision of content.** A basic function of a college science textbook is to present material in a clear, understandable way that is accurate, engaging, and up-to-date. In the long history of this textbook, our number-one goal has always been to keep *Earth Science* current, relevant, and highly readable for beginning students. To that end, every part of this text has been examined carefully. Many discussions, case studies, examples, and illustrations have been updated and revised.
- **Revised organization** In the geology portion of the text, the unit on *Forces Within* now precedes the unit on *Sculpting Earth's Surface.* This was done in response to many users and reviewers of previous editions who wanted the theory of plate tectonics presented earlier in the text because of the unifying role it plays in our understanding of planet Earth. Of course, each unit is basically independent of the others and can be taught in any order desired by the instructor.
- **SmartFigures that make *Earth Science* much more than a traditional textbook.** Through its many editions, an important strength of *Earth Science* has always been clear, logically organized, and well-illustrated explanations. Now, complementing and reinforcing this strength are a series of SmartFigures. Simply by scanning the Quick Response (QR) code next to a SmartFigure with a mobile device, students can follow hundreds of unique and innovative avenues that will increase their insight and understanding of important ideas and concepts. SmartFigures are truly art that teaches! This fifteenth edition of *Earth Science* has more than 220 SmartFigures, of five different types, including many new videos and animations:

  1. **SmartFigure Tutorials.** Each of these 3- to 4-minute features, most prepared and narrated by Professor Callan Bentley, is a mini-lesson that examines and explains the concepts illustrated by the figure.
  2. **SmartFigure Mobile Field Trips.** Scattered throughout this new edition are 24 video field trips that explore classic sites from Iceland to Hawaii. On each trip you will accompany geologist-pilot-photographer Michael Collier in the air and on the ground to see and learn about landscapes that relate to discussions in the chapter.
  3. **SmartFigure Condor Videos.** The 10 *Project Condor* videos take you to locations in the American West. By coupling aerial footage acquired by a drone quadcopter aircraft with ground-level views, effective narratives, annotations, and helpful animations, these videos transport you into the field and engage you in real-life case studies.
  4. **SmartFigure Animations.** These animations and accompanying narrations bring art to life, illustrating and explaining difficult-to-visualize topics and ideas more effectively than static art alone.
  5. **SmartFigure Videos.** Rather than providing a single image to illustrate an idea, these figures include short video clips that help illustrate such diverse subjects as mineral properties and the structure of ice sheets.

- **Revised active learning path.** *Earth Science* is designed for learning. Here is how it is accomplished. Each chapter has been designed to be self-contained so that materials may be taught in a different sequence, according to the preference of the instructor or the needs of the laboratory.

1. Every chapter begins with *Focus on Concepts*. Each numbered learning objective corresponds to a major section in the chapter. The statements identify the knowledge and skills students should master by the end of the chapter and help students prioritize key concepts.
2. Within the chapter, each major section concludes with *Concept Checks* that allow students to check their understanding and comprehension of important ideas and terms before moving on to the next section.
3. *Concepts in Review* is an end-of-chapter feature that coordinates with the *Focus on Concepts* at the start of the chapter and with the numbered sections within the chapter. It is a readable and concise overview of key ideas, with photos, diagrams, and questions that also help students focus on important ideas and test their understanding of key concepts.
4. The questions and problems in *Give It Some Thought* and *Examining the Earth System* challenge learners by involving them in activities that require higher-order thinking skills, such as application, analysis, and synthesis of chapter material. In addition, the activities in *Examining the Earth System* are intended to develop an awareness of and appreciation for some of the Earth system's many interrelationships.

5. The end-of-chapter review material now includes an all-new capstone activity called *Data Analysis* that sends students online to use a variety of interactive science resources and data sets from sources such as USGS, NASA, and NOAA to use various tools to perform data analysis and critical thinking tasks.

- **An unparalleled visual program.** In addition to more than 100 new, high-quality photos and satellite images, dozens of figures are new or have been redrawn by the gifted and highly respected geoscience illustrator Dennis Tasa. Maps and diagrams are frequently paired with photographs for greater effectiveness. Further, many new and revised figures have additional labels that narrate the process being illustrated and guide students as they examine the figures. Overall, the *Earth Science* visual program is clear and easy to understand.
- **MasteringGeology™.** MasteringGeology™ delivers engaging, dynamic learning opportunities—focused on course objectives and responsive to each student's progress—that are proven to help students learn course material and understand difficult concepts. Assignable activities in MasteringGeology™ include SmartFigure (Tutorials, Condor Videos, Animation, Mobile Field Trips, Videos) activities, GigaPan® activities, "Encounter" Earth activities using Google Earth™ activities, GeoTutor activities on the most challenging topics in the geosciences, Geoscience Animation activities, and more. MasteringGeology™ also includes all instructor resources, a robust Study Area with resources for students, and an optional eText version of the textbook.

Digital & Print Resources

MasteringGeology™ with Pearson eText

Used each year by over 3 million science students, the Mastering platform is the most effective and widely used online tutorial, homework, and assessment system for the sciences. Now available with *Earth Science*, 15th edition, **MasteringGeology™** offers tools for use before, during, and after class:

- **Before class:** Assign adaptive Dynamic Study Modules and reading assignments from the eText with Reading Quizzes to ensure that students come prepared for class, having done the reading.
- **During class:** Learning Catalytics, a "bring your own device" student engagement, assessment, and classroom intelligence system, allows students to use smartphones, tablets, or laptops to respond to questions in class. With Learning Catalytics, you can assess students in real-time, using open-ended question formats to uncover student misconceptions and adjust lectures accordingly.
- **After class:** Assign an array of activities such as Mobile Field Trips, Project Condor Quadcopter videos, GigaPan activities, Google Earth Encounter Activities, Geoscience Animations, and much more. Students receive wrong-answer feedback personalized to their answers, which will help them get back on track.

The MasteringGeology Student Study Area also provides students with self-study material including videos, geoscience animations, *In the News* articles, Self Study Quizzes, Web Links, Glossary, and Flashcards.

Pearson eText 2.0 gives students access to the text whenever and wherever they can access the Internet. Features of the Pearson eText include:

- Now available on smartphones and tablets using the Pearson eText 2.0 app
- Seamlessly integrated videos and other rich media
- Fully accessible (screen-reader ready)
- Configurable reading settings, including resizable type and night reading mode
- Instructor and student note-taking, highlighting, bookmarking, and search

For more information or access to MasteringGeology, please visit www.masteringgeology.com.

For Instructors

Instructor Resource Manual (Download Only) The *Instructor Resource Manual* has been designed to help seasoned and new instructors alike, offering the following sections in each chapter: an introduction to the chapter, outline, learning objectives/focus on concepts; teaching strategies; teacher resources; and answers to *Concept Checks* and *Give It Some Thought* questions from the textbook. www.pearsonhighered.com/irc

TestGen Computerized Test Bank (Download Only) TestGen is a computerized test generator that lets instructors view and edit Test Bank questions, transfer questions to tests, and print the test in a variety of customized formats. This Test Bank includes more than 2,000 multiple-choice, matching, and essay questions. Questions are correlated to Bloom's Taxonomy, each chapter's learning objectives, the Earth Science Literacy Initiatives 'Big Ideas', and the Pearson Science Global Outcomes to help instructors better map the assessments against both broad and specific teaching and learning objectives. The Test Bank is also available in Microsoft Word and can be imported into Blackboard, and other LMS. www.pearsonhighered.com/irc

Instructor Resource Materials (Download Only)
All of your lecture resources are now in one easy-to-reach place:

- All of the line art, tables, and photos from the text in JPEG files.
- PowerPoint™ Presentations: three PowerPoint files for each chapter. Cut down on your preparation time, no matter what your lecture needs, by taking advantage of these components of the PowerPoint files:
  - **Exclusive art.** All the photos, art, and tables from the text as JPEG files and PowerPoint slides for each chapter.
  - **Lecture outlines.** This set averages 50 slides per chapter and includes customizable lecture outlines with supporting art.
  - **Classroom Response System (CRS) questions.** Authored for use in conjunction with classroom response systems, these PowerPoint files allow you to electronically poll your class for responses to questions, pop quizzes, attendance, and more.
- Word and PDF versions of the *Instructor Resource Manual.*

For Students

***Applications and Investigations in Earth Science*, 9th Edition (0134746244)**
This manual can be used for any Earth Science lab course, in conjunction with any text. This versatile and adaptable collection of introductory-level laboratory experiences goes beyond traditional offerings to examine the basic principles and concepts of the Earth sciences. With integration of mobile-ready Pre-Lab Videos, the **Ninth Edition** minimizes the need for faculty instruction in the lab, freeing instructors to interact directly with students. Widely praised for its concise coverage and dynamic illustrations by Dennis Tasa, the text contains twenty-three step-by-step exercises that reinforce major topics in geology, oceanography, meteorology, and astronomy.

This edition includes a new lab exercise on Volcanoes, and incorporates MasteringGeology™–the most complete, easy-to-use, and engaging tutorial and assessment tool available. MasteringGeology includes a variety of highly visual, applied, kinesthetic, and automatically-gradable activities to support each lab, as well as a robust Study Study Area with a variety of media and reference resources, and an eText version of the lab manual.

***Laboratory Manual in Physical Geology*, 11th Edition** by the American Geosciences Institute and the National Association of Geoscience Teachers, edited by Vincent Cronin, illustrated by Dennis G. Tasa (0134446607)

This user-friendly, best-selling lab manual examines the basic processes of geology and their applications to everyday life. Featuring contributions from more than 170 highly regarded geologists and geoscience educators, along with an exceptional illustration program by Dennis Tasa, *Laboratory Manual in Physical Geology*, 11th edition, offers an inquiry- and activities-based approach that builds skills and gives students a more complete learning experience in the lab. Pre-lab videos linked from the print labs introduce students to the content, materials, and techniques they will use each lab. These teaching videos help TAs prepare for lab setup and learn new teaching skills. Now with more than 10 new lab activities, the lab manual is also available in MasteringGeology with Pearson eText, allowing teachers to use activity-based exercises to build students' lab skills.

***Dire Predictions: Understanding Global Climate Change*, 2nd Edition** by Michael Mann, Lee R. Kump (0133909778)

Periodic reports from the Intergovernmental Panel on Climate Change (IPCC) evaluate the risk of climate change brought on by humans. But the sheer volume of scientific data remains inscrutable to the general public, particularly to those who may still question the validity of climate change. In just over 200 pages, this practical text presents and expands upon the latest climate change data and scientific consensus of the IPCC's *Fifth Assessment Report* in a visually stunning and undeniably powerful way to the lay reader. Scientific findings that provide validity to the implications of climate change are presented in clear-cut graphic elements, striking images, and understandable analogies. The second edition integrates mobile media links to online media. The text is also available in various eText formats, including an optional eText upgrade option from MasteringGeology courses.

Acknowledgments

Writing a college textbook requires the talents and cooperation of many people. It is truly a team effort, and the authors are fortunate to be part of an extraordinary team at Pearson Education. In addition to being great people to work with, all are committed to producing the best textbooks possible. Special thanks to our Earth Science editor, Christian Botting. We appreciate his enthusiasm, hard work, and quest for excellence. We also appreciate our conscientious Content Producer, Liam Wenzler, whose job it was to keep track of all that was going on—and a lot was going on. As always, our marketing managers, Neena Bali and Mary Salzman, who talk with faculty daily, provide us with helpful advice and many good ideas. The 15th edition of *Earth Science* was certainly improved by the talents of our developmental editor, Margot Otway. Our sincere thanks to Margot for her fine work. It was the job of the production team, led by Patty Donovan at SPi Global, to turn our manuscript into a finished product. The team also included copyeditor Kitty Wilson, proofreader Heather Mann, and photo researcher Kristin Piljay. We think these talented people do great work. All are true professionals, with whom we are very fortunate to be associated.

The authors owe special thanks to three people who were very important contributors to this project.

- Working with Dennis Tasa, who is responsible for all of the text's outstanding illustrations and some excellent animations, is always special for us. He has been part of our team for more than 30 years. We not only value his artistic talents, hard work, patience, and imagination, but his friendship as well.
- As you read this text, you will see dozens of extraordinary photographs by Michael Collier. Most are aerial shots taken from his 60-year-old Cessna 180. Michael was also responsible for preparing the 24 remarkable Mobile Field Trips that are scattered through the text. Among his many awards is the American Geosciences Institute Award for Outstanding contribution to the Public Understanding of Geosciences. We think that Michael's photographs and field trips are the next best thing to being there. We were very fortunate to have had Michael's assistance on *Earth Science*, 15th edition. Thanks, Michael.
- Callan Bentley has been an important addition to the *Earth Science* team. Callan is a professor of geology at Northern Virginia Community College in Annandale, where he has been honored many times as an outstanding teacher. He is a frequent contributor to *Earth* magazine and is author of the popular geology blog *Mountain Beltway*. Callan was responsible for preparing the SmartFigure Tutorials that appear throughout the text. As you take advantage of these outstanding learning aids, you will hear his voice explaining the ideas.

Great thanks also go to those colleagues who prepared in-depth reviews. In particular, we appreciate the valuable input provided by Professor Alan Trujillo at Palomar College, who assisted with the revision of the oceanography chapters, and Professor Redina Herman at Western Illinois University, who helped us strengthen the meteorology chapters. Redina is also responsible for preparing the new end-of-chapter *Data Analysis* feature.

The critical comments and thoughtful input from many other reviewers also helped guide our work and clearly strengthened the text. Special thanks to:

Anna Banda, *El Centro College*
Marianne Caldwell, *Hillsborough Community College - Dale Mabry Campus*
Jennifer Cole, *Western Kentucky University*
Jay Lennartson, *University of North Carolina - Greensboro*
William Meddaugh, *Midwestern State University*
Jeff Niemitz, *Dickinson College*
Mark Peebles, *St. Petersburg College*
Jeff Richardson, *Columbus State Community College*
Sue Riggins, *California State University, Chico*
Kristin Riker-Coleman, *University of Wisconsin - Superior*
Troy Schinkel, *Central Connecticut State University*
Karl Schulze, *Waubonsee Community College*
Christiane Stidham, *Stony Brook University*
John Stimac, *Eastern Illinois University*
Dave Voorhees, *Waubonsee Community College*
Natalie Whitcomb, *Polk State College*

Last but certainly not least, we gratefully acknowledge the support and encouragement of our wives, Joanne Bannon and Nancy Lutgens. Preparation of *Earth Science 15e* would have been far more difficult without their patience and understanding.

Ed Tarbuck
Fred Lutgens

EARTH SCIENCE 15E: MAJOR CHANGES IN THIS EDITION

Global:

- Units 2 and 3 of the book are transposed, so that tectonics and related phenomena are now covered before surface processes.
- Many new SmartFigures are added, including three new types of SmartFigures: *Project Condor* Videos, Animations (many by Dennis Tasa), and Videos.
- Much of the Tasa art is improved with bolder labels or better placement of labels and text.
- New *Data Analysis* activities now conclude each chapter.

Chapter 1:

- The text description of the standard scientific method is replaced with the pictorial version in Figure 1.8.
- In "The Solar System Forms," the description of the collapse of the protosolar nebula is revised and updated.
- Section 1.4, "Earth as a System," now includes the sections on Earth's spheres (hydrosphere, geosphere, biosphere, atmosphere), formerly covered in their own section.
- The 14th edition section that introduced Earth's structure and the basic features of plate tectonics ("A Closer Look at the Geosphere") has been eliminated. In its place, the discussion of the geosphere in Section 1.4 is expanded to introduce Earth's layered structure.
- Section 1.5, "The Face of Earth," is reorganized to cover the ocean basins before the continents, rather than the reverse.
- The 14th edition *GeoGraphics* on world population is eliminated.
- Four new figures are added (Figures 1.1, 1.2, 1.8, 1.18), and three Tasa figures are substantively altered (Figs 1.9. 1.14, 1.16). Six 14th edition figures are deleted (Figures 1.1, 1.3, 1.16–1.19).
- One *Give It Some Thought* question is modified; two 14th edition questions are deleted. One *Examining the Earth System* question is deleted.

Chapter 2:

- In Section 2.4, "Properties of Minerals," the distinction between diagnostic and ambiguous properties is added at the start.
- In Section 2.5, "Mineral Groups," the treatment of the silicate groups is extensively revised. The opening paragraphs of the section "Important Nonsilicate Minerals" are also revised.
- The title of Section 2.6 is changed to "Minerals: A Nonrenewable Resource" from "Natural Resources."
- Two new figures are added: Figure 2.25 and Figure 2.32 (which replaces 14th edition Table 2.1 and Figure 2.31). Six figures are substantively revised: 2.5, 2.8, 2.9, 2.11, 2.12, 2.24; *GeoGraphics* 2.1 is also revised.
- One *Give It Some Thought* question is added and one modified; two 14th edition questions are deleted.

Chapter 3:

- Section "Silica Content as an Indicator of Composition" is removed (in Section 3.2).
- Section "Detrital sedimentary rocks" in Section 3.3 is significantly revised.
- Section "Other Metamorphic Rocks" is added at the end of Section 3.4.
- Section "Nonmetallic Mineral Resources" in Section 3.5 is substantially revised.
- Section "Energy Resources" in Section 3.5 is updated and substantially revised, including the addition of Figure 3.38 to illustrate hydraulic fracturing.
- Three new figures are added: Figures 3.6, 3.22, and 3.38. Eight figures are altered substantively: 3.7, 3.16–3.18, 3.20, 3.21, 3.30, and 3.34.
- One *Give It Some Thought* question is added and one modified; two 14th edition questions are deleted.

Chapter 4:

- The section "Rigid Lithosphere Overlies Weak Asthenosphere" is revised to emphasize the importance of density differences (in Section 4.3).
- The treatment of mantle plumes is updated ("Evidence: Mantle Plumes and Hot Spots" in Section 4.8).
- "Forces that Drive Plate Motion" omits mantle drag (in Section 4.10).
- "Models of Plate–Mantle Convection" in Section 4.10 is updated.
- Two 14th edition figures are deleted (Figures 7.9, 7.11). Fourteen figures are altered substantively: 4.9–4.11, 4.14, 4.15, 4.18, 4.19, 4.21, 4.22, 4.29–4.31, 4.35, and 4.36.
- One *Give It Some Thought* question is added; two are augmented with new question parts.

Chapter 5:

- The chapter introduction describes the 2015 Nepal earthquake.
- The section "Faults and Large Earthquakes" is reorganized to discuss convergent boundaries before transform boundaries, and both discussions are substantially revised. (In Section 5.1.)
- A revised and expanded section "Fault Rupture and Propagation" replaces the 14th edition section "Fault Rupture" (in Section 5.1).
- Section 5.3, "Locating the Source of an Earthquake," is new to the chapter; in the 14th edition, this topic was handled by the *GeoGraphics* "Finding the Epicenter of an Earthquake" (now omitted).
- The section "Intensity Scales" in Section 5.3 now covers the USGS "Did You Feel It?" Community Internet Intensity maps.

- Section 5.6 now covers intraplate as well as plate-boundary earthquakes. (In the 14th edition, intraplate earthquakes were handled in the *GeoGraphics* "Historic Earthquakes East of the Rockies," now omitted.)
- In Section 5.7, the discussion of earthquake prediction and forecasting is extensively revised and updated. A new section "Minimizing Earthquake Hazards" is added, including discussion of earthquake-resistant structures and earthquake warning systems.
- In Section 5.8, the section "Probing Earth's Interior: "Seeing" Seismic Waves" is significantly revised, as are portions of "Earth's Layered Structure."
- Seven new figures are added: 5.13 and 5.14 (which replace the 14th edition *GeoGraphics* "Finding the Epicenter of an Earthquake"); 5.31 and 5.32 (which replace the 14th edition *GeoGraphics* "Historic Earthquakes East of the Rockies"); 5.16, 5.36, and 5.37. Two 14th edition figures are deleted (8.1 and 8.14), in addition to the two *GeoGraphics* just mentioned.
- Six figures are altered substantively: 5.5, 5.18, 5.19, 5.26, 5.35, and 5.38, as well as *GeoGraphics* 5.1.
- Two *Give It Some Thought* questions are added and one is revised; six questions from the 14th edition are deleted.

Chapter 6:

- Considerable editing is done throughout to improve clarity.
- Section 6.2 is substantially rewritten, particularly the sections "Magma: Source Materials for Volcanic Eruptions" and "Effusive Versus Explosive Eruptions."
- More emphasis is put on the fact that most volcanism is submarine (for instance, first paragraph under "Lava Flows" in Section 6.3; the expanded Figure 6.8 on pillow lavas; and the opening paragraph of Section 6.11.)
- Some descriptive text is deleted from the end of "Kilauea: Hawaii's Most Active Volcano" in favor of the *GeoGraphics* on the East Rift Zone (end of Section 6.5)
- 2014 Mount Ontaki incident is added to section on pyroclastic flows, in place of 1991 Mt Unzen flow.
- The section on the destruction of Pompeii is added to Section 6.8; the *GeoGraphics* on this topic is removed.
- The *Eye on Earth* feature on the 1991 Mt Pinatubo eruption is replaced with one about the 2015 eruption of Mount Sinabung.
- The discussion of eruption mechanism for Yellowstone-type caldera eruptions is updated and tightened.
- The discussion of kimberlite and related pipes is deleted from the end of Section 6.9.
- Extensive editing for clarity and readability is done in the section "Decrease in Pressure: Decompression Melting" (in Section 6.11).
- In Section 6.12, volcanism at divergent boundaries is covered before that at convergent boundaries.
- A paragraph on intraplate volcanism associated with mantle plumes is added at the end of Section 6.12.
- Three new figures are added (6.19, 6.31, 6.39); six figures are substantively altered (6.3, 6.8, 6.12, 6.20, 6.21, 6.33, 6.34); two *GeoGraphics* are deleted.
- Three *Give It Some Thought* questions are replaced with new questions; two 14th edition questions are deleted. One *Examining the Earth System* question is deleted.

Chapter 7:

- Section 7.1 is substantially rewritten to improve clarity and effectiveness, including revised treatment of stress and strain, the types of rock deformation, and the factors that affect deformation style.
- The distinction between faults and joints is now covered at the start of Section 7.3.
- The treatment of joints is substantially revised ("Joints" in Section 7.3).
- The description of thrust faulting in the formation of the Himalayas is revised for clarity (paragraph 4 under "The Himalayas" in Section 7.6).
- The description of isostatic balance and its effects is substantially rewritten to improve clarity (Section 7.7).
- More than half of the 35 numbered figures are either substantively revised (19 figures) or new (3 figures). New: 7.4, 7.5, 7.22. Substantively revised: 7.3, 7.6–7.8, 7.12, 7.14, 7.16–7.19, 7.20, 7.21, 7.23–7.25, 7.27, 7.29, 7.30, 7.32. *Eye on Earth* 7.1 and *GeoGraphics* 7.1 are also revised. Three 14th edition figures are omitted: 10.4, 10.18, and 10.20.
- One new *Give It Some Thought* questions is added; three 14th edition questions are deleted.

Chapter 8:

- "Mass movement" is used throughout the chapter in place of "mass wasting."
- The section "Differential Weathering" in Section 4.2 now includes the content of the 14th edition Section 4.3, "Rates of Weathering"; the concept of differential weathering now introduces the section.
- In Section 8.4, "Controls of Soil Formation," the section on climate is revised and is placed second rather than third.
- Section 8.5, "Describing and Classifying Soils," includes the topics of the 14th edition Sections 4.6 ("The Soil Profile") and 4.7 ("Classifying Soils").
- In Section 8.6, erosion by water and by wind are now covered in one section.
- The section "Controls and Triggers of Mass Movement" in Section 8.7 is significantly revised. The Oso, Washington slide is added as an example of water as a trigger.
- Section 8.8, "Types of Mass Movement," includes the topics of the 14th edition Sections 4.11 ("Classifying Mass Wasting Processes"), 4.12 ("Rapid Forms of Mass Wasting"), and 4.13 ("Slow Forms of Mass Wasting").
- Within Section 8.8, the treatment of the mechanism for long-runout landslides is updated (section "Rate of Movement" in Section 8.8);

the section "Debris Flow" provides a more unified treatment of dry versus wet debris flows and omits the Nevado del Ruiz lahars; and the final paragraph on liquefaction is omitted (because it is treated in Chapter 5, which now precedes this chapter).

- One figure is replaced with a new version (Fig. 8.23); four figures are revised substantively (Figs 8.10, 8.19, 8.28, 8.29, and also *Eye on Earth* 8.3); two 14th edition figures are deleted (Figs 4.20, 4.28).
- Two 14th edition *Give It Some Thought* questions are deleted.

Chapter 9:

- "Stream Erosion," now covers corrosion as a means of forming bedrock channels in soluble rocks. Also, in "Suspended Load," Figure 9.14 added to help explain the significance of settling velocity. (Both in Section 9.4.)
- Coverage of stream terraces (including Figure 9.22) is added at the end of Section 9.6.
- Section 9.7 now covers intermittent growth of alluvial fans in dry area
- A discussion of the April 2011 Mississippi flooding us added at the start of "Causes of Floods" in Section 9.8; the description of the1889 dam burst on the Little Conemaugh River is removed.
- The section "Artificial Levees" in Section 9.8 is revised to describe the use of floodways to protect levees.
- Section 9.10 is reorganized to cover wells and artesian systems before springs.
- Seven new figures are added (9.4, 9.8, 9.14, 9.22, 9.26, 9.27, 9.40); three figures are substantively revised (9.2, 9.21, 9.35); five 14th edition figures are deleted (5.1, 5.16, 5.24, 5.25, 5.38). One *GeoGraphics* and one *Eye on Earth* are also deleted.
- One new *Give It Some Thought* question is added; two 14th edition questions are deleted. One *Examining the Earth System* question is added, and four are deleted.

Chapter 10:

- The section on observing and measuring the movement of glacial ice is revised and tightened (in Section 10.2).
- The introduction to "Landforms Created by Glacial Erosion" is rewritten to emphasize the distinction between the effects of valley glaciers and ice sheets (in Section 10.3).
- Section 10.4 is revised to include separate sections on glacial till and stratified drift.
- Section 10.5, "Other Effects of Ice Age Glaciers," is reorganized, and section on sea-level changes are updated.
- Section 10.6 is revised to include Section 10.7 from the previous edition ("Causes of Ice Ages"); it also includes some updating, clarification, and shortening.
- Ten figures are added or substantively altered: 10.4 (photo replaces sketch), 10.8 (new figure part added), 10.9 (new example of retreating glacier), 10.10 (new photo), 10.12 (altered), 10.13 (new figure), 10.17 (new figure), 10.18 (altered), 10.34 (altered), 10.35 (altered).
- One *Give It Some Thought* question is added and one deleted. One *Examining the Earth System* question is deleted.

Chapter 11:

- Section 11.5 is retitled "Numerical Dating with Nuclear Decay" (from "Dating with Radioactivity"), and the text is changed to refer to unstable nuclei and nuclear decay in preference to radioactive nuclei and radioactivity.
- The section "Changes to Atomic Nuclei" (formerly "Radioactivity") is significantly revised for clarity, including revision of Figure 11.19.
- Within the section "Radiometric dating," the description of how daughter nuclei accumulate in a crystal is expanded for clarity.
- Vignettes are added to Figure 11.21 to help convey the concept of half-life.
- The discussion of loss of isotopes as a source of dating error is revised for clarity and no longer refers to closed and open systems (in the section "Using Unstable Isotopes").
- Section 11.7, "The Geologic Time Scale," is moved to the end of the chapter; it no longer comes between the sections "Numerical Dating with Nuclear Decay" and "Determining Numerical Dates for Sedimentary Strata."
- The section "Precambrian Time" within Section 11.7 provides more detail on why the time scale is less detailed for the Precambrian than the Phanerozoic.
- Eight figures are substantively revised (Figures 11.15, 11.16, 11.19–11.22, 11.24, 11.25). One 14th edition *GeoGraphics* is deleted.

Chapter 12:

- The opening paragraphs of Section 12.1 are revised to discuss exoplanet discoveries and the concept of a habitable zone.
- In Section 12.3 ("Origin and Evolution of the Atmosphere and Oceans"), the section "Earth's Primitive Atmosphere" is somewhat expanded, and the section "Oxygen in the Atmosphere" is significantly revised, including an expanded treatment of the Great Oxygenation Event.
- The section "Making Continental Crust" is partially revised and includes mention of the Isua rocks.
- In "Supercontinents and Climate," the discussion of Antarctic glaciation is updated.
- Sections 12.6 through 12.9, on the origin and evolution of life, are significantly updated and revised throughout, and a new section on the end-Cretaceous extinction ("Demise of the Dinosaurs") is added, replacing the former *GeoGraphics* on this topic.
- Nine new figures are added:12.1, 12.2, 12.17, 12.24, 12.28, 12.29 (replacing the 14th edition 12.29), and 12.33–12.35. Six figures are substantively altered (12.3, 12.10, 12.12, 12.16, 12.18, 12.32). Six 14th edition figures are deleted (12.1, 12.2, 12.13, 12.18 12.20, 12.22).
- Three *Give It Some Thought* questions are modified; three 14th edition questions are deleted.

Chapter 13:

- Four figures are substantively altered (13.5, 13.6, 13.13, 13.17); Figure 13.22 now incorporates the photo from a former *Eye on Earth*, which is deleted.

Chapter 14:

- Four figures are substantively altered: Figures 14.2, 14.3, 14.13 (now incorporates the former Table 14.2), and 14.14.
- Three 14th edition *Give It Some Thought* questions are deleted.

Chapter 15:

- The order of Sections 15.7 and 15.8 is reversed: Section 15.7 ("Contrasting America's Coasts") now precedes Section 15.8 (Stabilizing the Shore).
- Section 15.7 is reorganized so that it starts by classifying coasts as emergent and submergent.
- In Section 15.8, the conversion of vulnerable shoreline to parks in Staten Island after Hurricane Sandy is added as an example of coastal land-use change.
- Two figures are substantively altered (15.7, 15.25). Three figures are added: 15.8 (replaces 14th edition 15.8), 15.26, 15.29 (replaces 14th edition 15.26). Two 14th edition figures are deleted (15.30, 15.36).
- One *Examining the Earth System* question is deleted.

Chapter 16:

- In Section 16.2, a paragraph about tropospheric ozone as a pollutant is added.
- Figure 16.18 on the solstices and equinoxes is added, and the corresponding text coverage is made briefer (end of Section 16.4).
- Coverage of thermals is added to the section "Mechanism of Heat Transfer: Convection) (in Section 16.5).
- The description of the greenhouse effect is revised for clarity (end of Section 16.6).
- Two figures are substantively altered (16.6, 16.16). Three figures are added: 16.4, 16.14 (replaces 14th edition 16.14), 16.18, 16.20, 16.23 (replaces 14th edition 16.22). One 14th edition figure is deleted (16.9), as well as figure parts from Figures 16.7 and 16.8. The 14th edition *GeoGraphics* is also deleted.
- One *Examining the Earth System* question is deleted.

Chapter 17:

- In Section 7.2, the sections "Dew Point Temperature" and "How Is Humidity Measured?" are significantly revised for clarity.
- Sections 17.3 ("Adiabatic Temperature Changes and Cloud Formation") and 17.4 ("Processes that Lift Air") are substantially revised to improve clarity.
- Section 17.7, "Types of Fog," is thoroughly rewritten to improve clarity.
- The description of how hail forms is revised for clarity (Section "Hail" in Section 17.9).
- Eleven figures are substantively altered (17.4–17.6, 17.12, 17.14, 17.17–27.19, 17.20, 17.27, 17.34); *GeoGraphics* 17.1 also modified. Two figures are added (17.29, 1731), and also *Eye on Earth* 17.1. One *Eye on Earth* from the 14th edition is deleted.
- One new *Give It Some Thought* question is added; four questions from the 14th edition are deleted. Two *Examining the Earth System* questions are deleted.

Chapter 18:

- Section 18.7 ("El Nino, La Nina, and the Southern Oscillation") is substantially revised.
- Six figures are substantively altered (18.3, 18.14, 18.17, 18.18, 18.24, 18.25). Two figures are added: 18.6 (replaces 14th edition 18.6), 18.8 (replaces 14th edition 18.8). Two 14th edition figures are deleted (18.26, 18.27), as well as the 14th edition *GeoGraphics*.
- One *Examining the Earth System* question is deleted.

Chapter 19:

- The introduction to fronts is revised (beginning of Section 9.2).
- The sections "Tornado Development and Occurrence" and "Tornado Climatology" are significantly revised (in Section 19.5).
- A subsection "The Role of Satellites" is added to the section "Monitoring Hurricanes" in Section 19.6.
- Eight figures are substantively altered (19.2, 19.3, 19.5, 19.14, 19.16, 19.22, 19.28, 19.34). Two figures are added: 19.25 (replaces 14th edition 19.26) and 19.33. One 14th edition figure is deleted (19.15), as well as the 14th edition *GeoGraphics*.
- One new *Give It Some Thought* question is added; one question from the 14th edition is deleted.

Chapter 20:

- Section 20.8, "Human Impact on Global Climate," is revised and brought up to date. This section also now covers aerosols (formerly covered in its own later section).
- Section 20.10, "Some Possible Consequences of Global Warming," revised and brought up to date.
- Two new figures are added: 20.26 (replaces 14th edition Fig 20.27) and 20.28. Three figures are substantively altered (20.19, 20.20, 20.25). Two 14th edition figures are deleted (20.15, 20.29). The *GeoGraphics* and one *Eye on Earth* from the 14th edition are deleted.
- One *Give It Some Thought* question id deleted.

Chapter 21:

- This chapter is edited extensively for conciseness and clarity.
- Section 21.1 is significantly revised for clarity, particularly the section "The Golden Age of Astronomy."
- Section 21.2 is significantly revised for conciseness and, in places, for clarity. Also, Kepler's third law is now expressed mathematically (section "Johannes Kepler"), and, for Newton's law of gravitation, the exact proportionality between mass, distance, and gravitational force is given (section "Isaac Newton").
- Within Section 21.3, "Patterns in the Night Sky" (formerly titled "Positions in the Sky"), star positions are now described in terms of direction and altitude rather than right ascension and declination. Also, the new section "Measurements Using the Celestial Sphere" now covers angular size and angular distance.
- Section 21.4, "The Motions of Earth," no longer describes precession de novo; instead, it reviews the cycles of eccentricity, axial tilt, and precession that were described in Chapter 10.
- Section 21.5, "Motions of the Earth–Moon System," now discusses tidal locking as the reason why one side of the Moon always faces Earth.
- Two new figures are added (21.21, 21.22). Six figures are substantively altered (21.6, 21.11, 21.15, 21.17, 21.18, 21.27). One 14th edition figure is deleted (21.27).
- Two new *Give It Some Thought* questions are modified; two 14th edition questions are deleted. One *Examining the Earth System* question is deleted.

Chapter 22:

- In addition to the sections updated for currency (noted below), the chapter is revised extensively for clarity and conciseness.
- The sections "Nebular Theory: Formation of the Solar System," "Mars: The Red Planet," "Comets: Dirty Snowballs," and "Dwarf Planets" are updated to reflect current research, as is the treatment of cryovolcanism on moons of the outer planets.
- Four new figures are added (22.16, 22.18, 22.33, 22.36). Twelve figures are substantively altered (22.2, 22.4, 22.7, 22.12, 22.17, 22.20, 22.23, 22.28 22.29, 22.30, 22.32, 22.25), as well as is *GeoGraphics* 22.1. Five 14th edition figures are deleted (22.11, 22.22, 22.25, 22.35, 22.37).
- One 14th edition *Give It Some Thought* question is deleted; one *Examining the Earth System* question is deleted.

Chapter 23:

- In addition to the changes discussed below, the text and figures are revised extensively for clarity and conciseness. The chapter now puts more emphasis on current knowledge and less on history.
- Section 23.2, "What Can We Learn from Light?" is thoroughly revised. Line spectra are now referred to as emission and absorption spectra, rather than bright-line and dark-line spectra. The treatment of what spectra tell us about composition and temperature is expanded and put in its own subsections. The explanation of the Doppler effect is revised for clarity. The section now covers the speed of light in vacuum.
- Section 23.3, "Collecting Light Using Optical Telescopes," is almost completely rewritten, with a stronger focus on how observing is done today. The treatment of active optics, adaptive optics, telescope arrays, and astrophotography using film and CCDs is expanded and divided into new subsections.
- Section 23.4, "Radio- and Space-Based Astronomy," is largely rewritten or new, with sections on spaced-based observatories in radio, infrared, x-ray, and gamma-ray wavelengths and on the James Webb Space Telescope.
- Section 23.5, "Our Star: The Sun" is revised, reorganized, and somewhat expanded, with separate sections on the Sun's surface, atmosphere, and interior. It also now covers hydrogen fusion by the p–p process as the source of the Sun's energy.
- Section 23.6, "The Active Sun," is significantly revised, with more coverage of the structure and role of Sun's magnetic field.
- More than half of the figures in the chapter (17 of 30) are revised, replaced, or new. Six figures are substantively altered: 23.1–23.3, 23.20, 23.21, 23.24. Eleven are new: 23.4, 23.6, 23.7, 23.8 (replaces 14th edition 23.6), 23.13, 23.16, 23.18, 23.20, 23.21 (replaces 14th edition 23.17), 23.27, 23.29. Ten 14th edition figures are deleted: 23.4, 23.6, 23.9, 23.13, 23.16, 23.17–23.19, 23.22, 23.25.
- Two 14th edition *Give It Some Thought* questions are deleted.

Chapter 24:

- Section 24.1, "Classifying Stars," now covers stellar luminosity, color, and temperature as well as the H-R diagram and the classes of stars.
- The 14th edition Section 24.2 on the types of nebulae is deleted; retained material is placed elsewhere in this chapter and in Chapter 23.
- Section 24.2, "Stellar Evolution," is extensively revised and updated.
- Sections 24.3 ("Stellar Remnants") and 24.4 ("Galaxies and Galaxy Clusters") are significantly revised for clarity and conciseness, and Section 24.4. includes new information on dwarf galaxies.
- The section on the Universe is moved to the end of the chapter, revised, and combined with the treatment of cosmology.
- Two new figures are added (24.4, 24.9); Figure 24.1 is replaced with a new photo of a different object. Six figures substantively altered (24.2, 24.3, 24.5, 24.8, 24.10, 24.20)). Five 14th edition figures are deleted (24.3–24.7).
- Three 14th edition *Give It Some Thought* questions are deleted; one *Examining the Earth System* question is deleted.

1 Introduction to Earth Science

FOCUS ON CONCEPTS

Each statement represents the primary learning objective for the corresponding major heading within the chapter. After you complete the chapter, you should be able to:

1.1 List and describe the sciences that collectively make up Earth science. Discuss the scales of space and time in Earth science.

1.2 Discuss the nature of scientific inquiry, including the construction of hypotheses and the development of theories.

1.3 Outline the stages in the formation of our solar system.

1.4 List and describe Earth's four major spheres. Define *system* and explain why Earth is considered to be a system.

1.5 List and describe the major features of the ocean basins and continents.

All four of Earth's major spheres are represented in this image from Jasper National Park in the Canadian Rockies. (Photo by Adam Burton/Getty Images)

THE SPECTACULAR ERUPTION OF A VOLCANO, the magnificent scenery of a rocky coast, and the destruction created by a hurricane are all subjects for an Earth scientist. The study of Earth science deals with many fascinating and practical questions about our environment. What forces produce mountains? Why is our daily weather variable? Is climate really changing? How old is Earth, and how is it related to other planets in the solar system? What causes ocean tides? What were the Ice Ages like, and will there be another? Where should we search for water?

The subject of this text is *Earth science*. To understand Earth is not an easy task because our planet is not a static and unchanging mass. Rather, it is a dynamic body with many interacting parts and a long and complex history.

1.1 What Is Earth Science?

List and describe the sciences that collectively make up Earth science. Discuss the scales of space and time in Earth science.

Earth science is the name for all the sciences that collectively seek to understand Earth and its neighbors in space. It includes geology, oceanography, meteorology, and astronomy. Throughout its long existence, Earth has been changing. In fact, it is changing as you read this page and will continue to do so into the foreseeable future. Sometimes the changes are rapid and violent, as when severe storms, landslides, and volcanic eruptions occur. Conversely, many changes take place so gradually that they go unnoticed during a lifetime. Scales of size and space also vary greatly among the phenomena studied in Earth science.

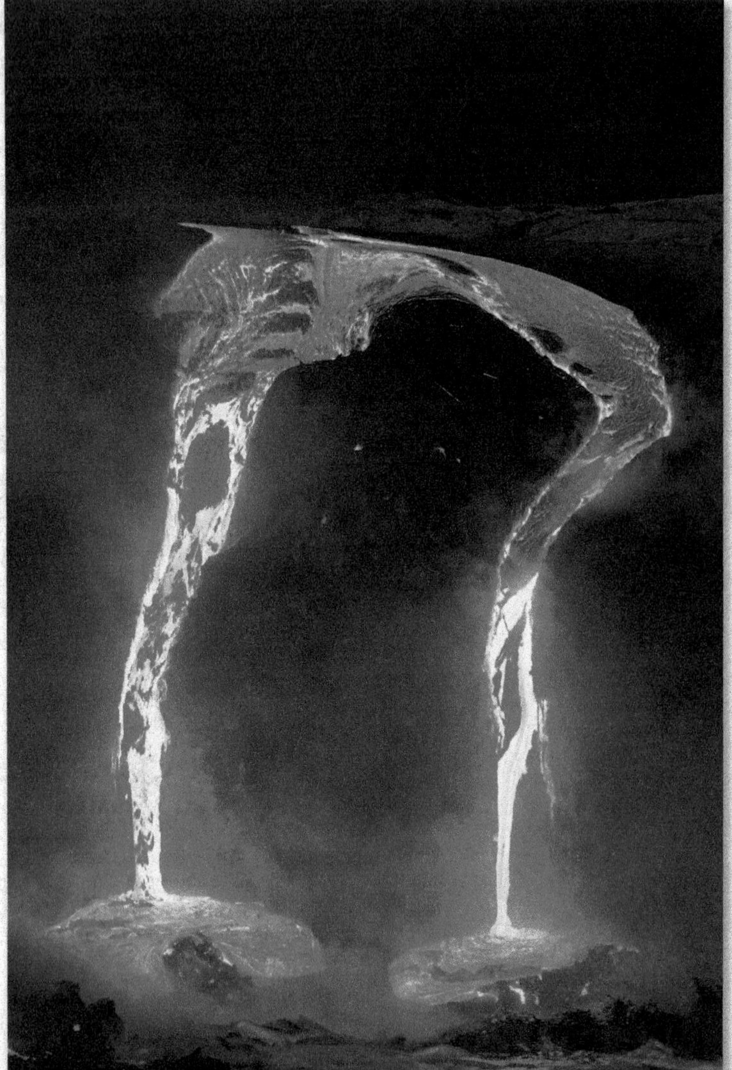

Figure 1.1 **Volcanic eruption** Molten lava from Hawaii's Kilauea Volcano is spilling into the Pacific Ocean. Internal processes are those that occur beneath Earth's surface. Sometimes they lead to the formation of major features at the surface. (Photo by Russ Bishop/Alamy Stock Photo)

Earth science is often perceived as science that is performed outdoors—and rightly so. A great deal of an Earth scientist's study is based on observations and experiments conducted in the field. But Earth science is also conducted in the laboratory, where, for example, the study of various Earth materials provides insights into many basic processes, and the creation of complex computer models allows for the simulation of our planet's complicated climate system. Frequently, Earth scientists require an understanding of and must apply their knowledge about principles from physics, chemistry, and biology. Geology, oceanography, meteorology, and astronomy are sciences that seek to expand our knowledge of the natural world and our place in it.

Geology

In this book, Units 1–4 focus on the science of **geology**, a word that literally means "study of Earth." Geology is traditionally divided into two broad areas: physical and historical.

Physical geology examines the materials composing Earth and seeks to understand the many processes that operate beneath and upon its surface. Earth is a dynamic, ever-changing planet. Internal forces create earthquakes, build mountains, and produce volcanic structures (Figure 1.1). At the surface, external processes break rock apart and sculpt a broad array of landforms. The erosional effects of water, wind, and ice result in a great diversity of landscapes. Because rocks and minerals form in response to Earth's internal and external processes, their interpretation is basic to an understanding of our planet.

In contrast to physical geology, the aim of *historical geology* is to understand the origin of Earth and the development of the planet through its 4.6-billion-year history (**Figure 1.2**). It strives to establish an orderly chronological arrangement of the multitude of physical and biological changes that have occurred in the geologic past. The study of physical geology logically precedes the study of Earth history because we must first understand how Earth works before we attempt to unravel its past.

Oceanography

Earth is often called the "water planet" or the "blue planet." Such terms relate to the fact that more than 70 percent of Earth's surface is covered by the global ocean. If we are to understand Earth, we must learn about its oceans. Unit 5, *The Global Ocean*, is devoted to **oceanography**.

Oceanography is actually not a separate and distinct science. Rather, it involves the application of all sciences in a comprehensive and interrelated study of the oceans in all their aspects and relationships. Oceanography integrates chemistry, physics, geology, and biology. It includes the study of the composition and movements of seawater, as well as coastal processes, seafloor topography, and marine life.

Meteorology

The continents and oceans are surrounded by an atmosphere. Unit 6, *Earth's Dynamic Atmosphere*, examines the mixture of gases that is held to the planet by gravity and thins rapidly with altitude. Acted on by the combined effects of Earth's motions and energy from the Sun, and influenced by Earth's land and sea surface, the formless and invisible atmosphere reacts by producing an infinite variety of weather, which in turn creates the basic pattern of global climates. **Meteorology** is the study of the atmosphere and the processes that produce weather and climate. Like oceanography, meteorology involves the application of other sciences in an integrated study of the thin layer of air that surrounds Earth.

▲ SmartFigure 1.2
Arizona's Grand Canyon The erosional work of the Colorado River along with other external processes created this natural wonder. For someone studying historical geology, hiking down the South Kaibab Trail in Grand Canyon National Park is a trip through time. These rock layers hold clues to millions of years of Earth history. (Photo by Michael Collier)

MOBILE FIELD TRIP
https://goo.gl/kECNV1

▲ Figure 1.3 **Earthquake in Ecuador** On April 16, 2016, a magnitude 7.8 earthquake struck coastal Ecuador. It was the strongest quake in that region in 40 years. There were nearly 700 fatalities and more than 7000 people injured. Natural hazards are *natural processes*. They become hazards only when people try to live where the processes occur. (Photo by Meridth Kohut/Bloomberg via Getty Images)

Astronomy

Unit 7, *Earth's Place in the Universe*, demonstrates that an understanding of Earth requires that we relate our planet to the larger universe. Because Earth is related to all the other objects in space, the science of **astronomy**—the study of the universe—is very useful in probing the origins of our own environment. Because we are so closely acquainted with the planet on which we live, it is easy to forget that Earth is just a tiny object in a vast universe. Indeed, Earth is subject to the same physical laws that govern the many other objects populating the great expanses of space. Thus, to understand explanations of our planet's origin, it is useful to learn something about the other members of our solar system. Moreover, it is helpful to view the solar system as a part of the great assemblage of stars that comprise our galaxy, which is but one of many galaxies.

Earth Science Is Environmental Science

Earth science is an environmental science that explores many important relationships between people and the natural environment. Many of the problems and issues addressed by Earth science are of practical value to people.

Natural Hazards Natural hazards are a part of living on Earth. Every day they adversely affect literally millions of people worldwide and are responsible for staggering damages. Among the hazardous Earth processes studied by Earth scientists are volcanoes, floods, tsunamis, earthquakes, landslides, and hurricanes. Of course, these hazards are *natural* processes. They become hazards only when people try to live where these processes occur.

For most of history, most people lived in rural areas. However, today more people live in cities than in rural areas. This global trend toward urbanization concentrates millions of people into places that are vulnerable to natural hazards. Coastal sites are becoming more vulnerable because development often destroys natural defenses such as wetlands and sand dunes. In addition, there is a growing threat associated with human influences on the Earth system, such as sea level rise that is linked to global warming.* Other urban areas are exposed to seismic (earthquake) and volcanic hazards where inappropriate land use and poor construction practices, coupled with rapid population growth, increase vulnerability (Figure 1.3).

Resources Resources represent another important focus that is of great practical value to people. They include water and soil, a great variety of metallic and nonmetallic minerals, and energy. Together they form the very foundation of modern civilization. Earth science deals with the formation and occurrence of these vital resources and also with maintaining supplies and with the environmental impact of their extraction and use.

People Influence Earth Processes Not only do Earth processes have an impact on people, but we humans can dramatically influence Earth processes as well. Human activities alter the composition of the atmosphere, triggering air pollution episodes and causing global climate change (Figure 1.4). River flooding is natural, but the

*The idea of the Earth system is explored later in the chapter. Global climate change and its effects are a focus of Chapter 20.

▼ Figure 1.4 **People influence the atmosphere** China is plagued by frequent severe air pollution episodes. Fuel combustion from power plants, factories, and motor vehicles provide a high percentage of the pollutants. Meteorological factors determine whether pollutants remain trapped in the city or are dispersed. (Photo by AFP/Stringer/Getty Images)

Galaxy
Solar system
Earth science involves investigations of phenomena that range in size from the atomic level to those that involve large portions of the universe.
Atom
Mineral
Planet
Mountain
Rock

Figure 1.5 **From atoms to galaxies** Earth science studies phenomena on many different scales.

magnitude and frequency of flooding can be changed significantly by human activities such as clearing forests, building cities, and constructing dams. Unfortunately, natural systems do not always adjust to artificial changes in ways that we can anticipate. Thus, an alteration to the environment that was intended to benefit society often has the opposite effect.

At many places throughout this book, you will have opportunities to examine different aspects of our relationship with the physical environment. Moreover, significant parts of some chapters provide the basic knowledge and principles needed to understand environmental problems.

Scales of Space and Time in Earth Science

When we study Earth, we must contend with a broad array of space and time scales (Figure 1.5). Some phenomena are relatively easy for us to imagine, such as the size and duration of an afternoon thunderstorm or the dimensions of a sand dune. Other phenomena are so vast or so small that they are difficult to imagine. The number of stars and distances in our galaxy (and beyond!) or the internal arrangement of atoms in a mineral crystal are examples of such phenomena.

Some of the events we study occur in fractions of a second. Lightning is an example. Other processes extend over spans of tens or hundreds of millions of years. For example, the lofty Himalaya Mountains began forming nearly 50 million years ago, and they continue to develop today.

The concept of **geologic time**, the span of time since the formation of Earth, is new to many nonscientists. People are accustomed to dealing with increments of time that are measured in hours, days, weeks, and years. Our history books often examine events over spans of centuries, but even a century is difficult to appreciate fully. For most of us, someone or something that is 90 years old is *very old*, and a 1000-year-old artifact is *ancient*.

Those who study Earth science must routinely deal with vast time periods—millions or billions (thousands of millions) of years. When viewed in the context of Earth's 4.6-billion-year history, an event that occurred 100 million years ago may be characterized as "recent" by a geologist, and a rock sample that has been dated at 10 million years may be called "young."

An appreciation for the magnitude of geologic time is important in the study of our planet because many processes are so gradual that vast spans of time are needed before significant changes occur. How long

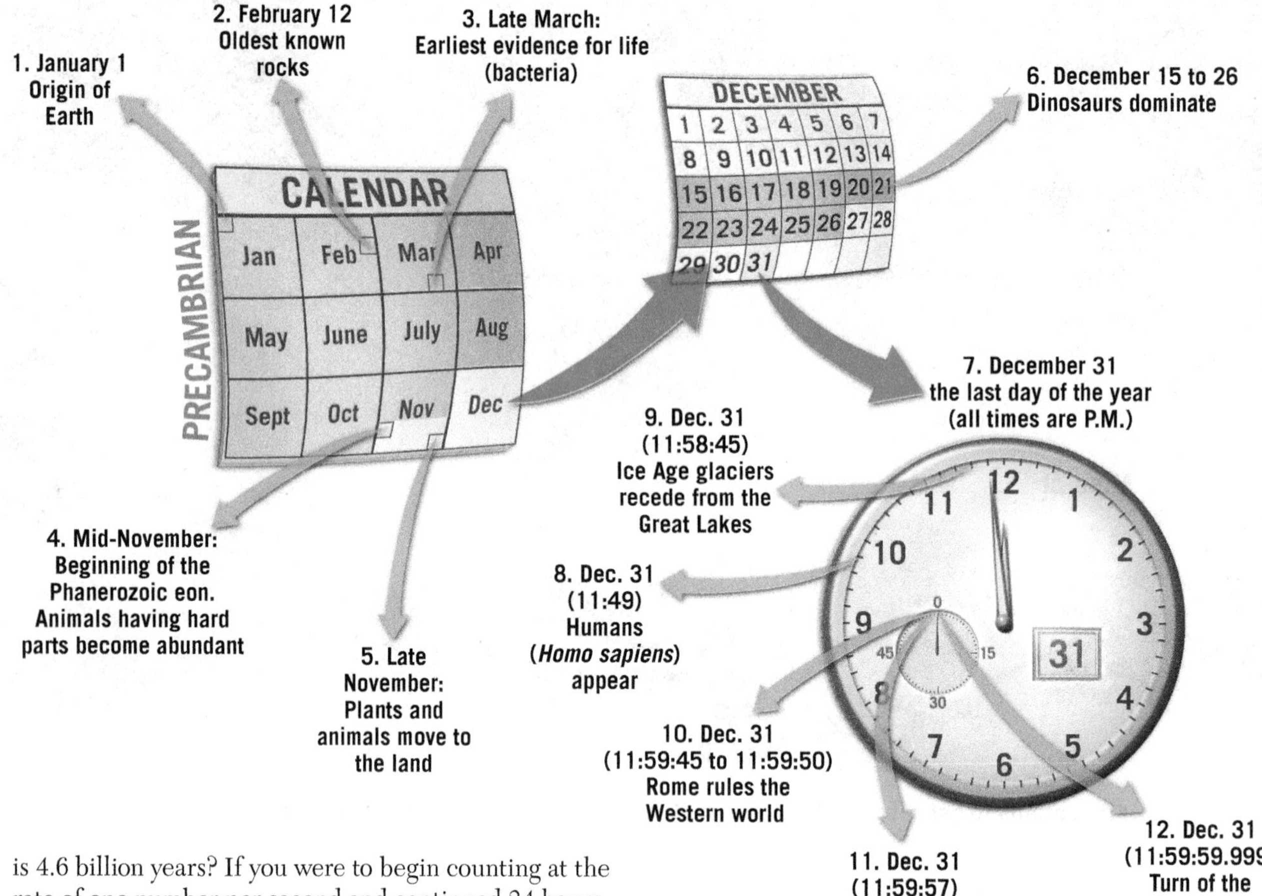
What if we compress the 4.6 billion years of Earth history into a single year?
1. January 1 Origin of Earth
2. February 12 Oldest known rocks
3. Late March: Earliest evidence for life (bacteria)
CALENDAR
PRECAMBRIAN
Jan
Feb
Mar
Apr
May
June
July
Aug
Sept
Oct
Nov
Dec
DECEMBER
6. December 15 to 26 Dinosaurs dominate
7. December 31 the last day of the year (all times are P.M.)
9. Dec. 31 (11:58:45) Ice Age glaciers recede from the Great Lakes
8. Dec. 31 (11:49) Humans (*Homo sapiens*) appear
4. Mid-November: Beginning of the Phanerozoic eon. Animals having hard parts become abundant
5. Late November: Plants and animals move to the land
10. Dec. 31 (11:59:45 to 11:59:50) Rome rules the Western world
11. Dec. 31 (11:59:57) Columbus arrives in the New World
12. Dec. 31 (11:59:59.999) Turn of the millennium

SmartFigure 1.6
Magnitude of geologic time

TUTORIAL
https://goo.gl/V1WFRd

is 4.6 billion years? If you were to begin counting at the rate of one number per second and continued 24 hours a day, 7 days a week and never stopped, it would take about two lifetimes (150 years) to reach 4.6 billion!

The preceding analogy is just one of many that have been conceived in an attempt to convey the magnitude of geologic time. Although helpful, all of them, no matter how clever, only begin to help us comprehend the vast expanse of Earth history. Figure 1.6 provides another interesting way of viewing the age of Earth.

Over the past 200 years or so, Earth scientists have developed the *geologic time scale* of Earth history. It divides the 4.6-billion-year history of Earth into many different units and provides a meaningful time frame within which the events of the geologic past are arranged (see Figure 11.25, page 374). The principles used to develop the geologic time scale are examined in some detail in Chapter 11.

CONCEPT CHECKS 1.1

1. List and briefly describe the sciences that collectively make up Earth science.
2. List at least four different natural hazards. Aside from natural hazards, describe another important connection between people and Earth science.
3. List two examples of size/space scales in Earth science that are at opposite ends of the spectrum.
4. How old is Earth?
5. If you compress geologic time into a single year, how much time has elapsed since Columbus arrived in the New World?

1.2 The Nature of Scientific Inquiry

Discuss the nature of scientific inquiry, including the construction of hypotheses and the development of theories.

In our modern society, we are constantly reminded of the benefits derived from science. But what exactly is the nature of scientific inquiry? Science is a process of investigation that leads to producing knowledge, based on making careful observations and on creating explanations that make sense of the observations. Developing an understanding of how science is performed and how scientists work is an important theme throughout this book. You will explore the difficulties in gathering data and some of the ingenious methods that have been developed to overcome these difficulties. You will also see many examples of how hypotheses are formulated and tested, as well as learn about the evolution and development of some major scientific theories.

All science is based on the assumption that the natural world behaves in a consistent and predictable manner that is comprehensible through careful, systematic study. The overall goal of science is to discover the underlying patterns in nature and then to use that knowledge to make predictions about what should or should not be expected, given certain facts or circumstances. For example, by understanding the circumstances and processes that produce certain cloud types, meteorologists are often able to predict the approximate time and place of their formation and the intensity of the associated weather.

The development of new scientific knowledge involves some basic logical processes that are universally accepted. To determine what is occurring in the natural world, scientists collect data through observation and measurement (Figure 1.7). The data collected often help answer well-defined questions about the natural world. Because some error is inevitable, the accuracy of a particular measurement

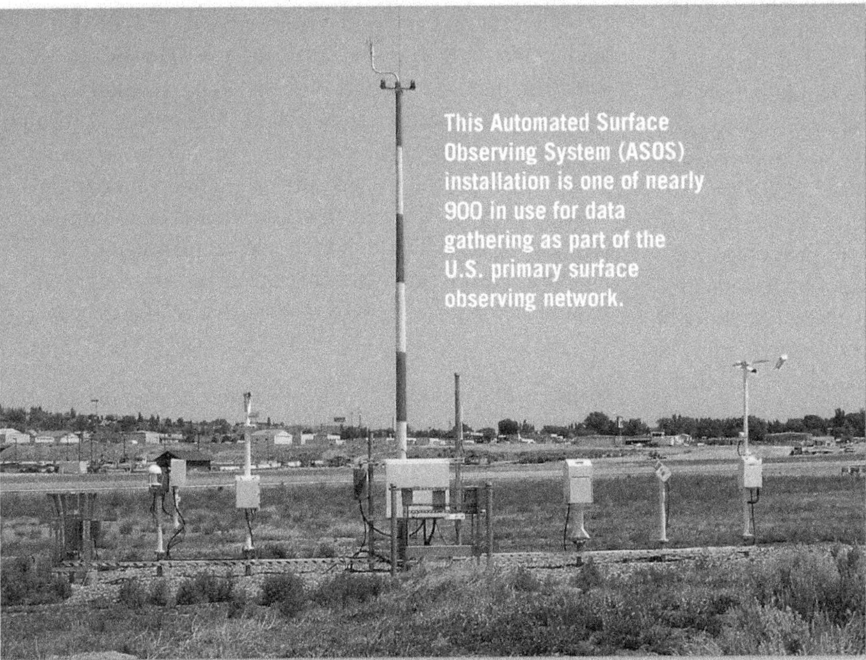

This Automated Surface Observing System (ASOS) installation is one of nearly 900 in use for data gathering as part of the U.S. primary surface observing network.
A.

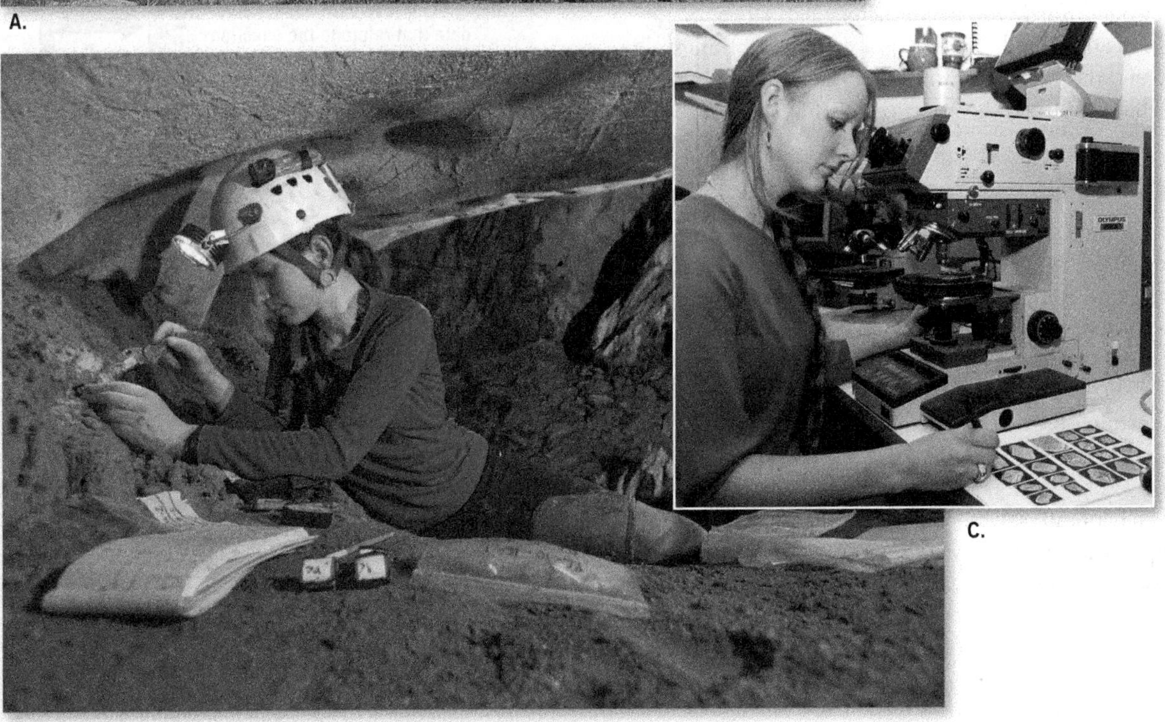

Figure 1.7 Observation and measurement Gathering data and making careful observations are basic parts of scientific inquiry. **A.** This array of instruments automatically records and transmits basic weather data. (Photo by NASA/Science Source) **B.** The Earth sciences frequently involve fieldwork. (Photo by Robbie Shone/Science Source) **C.** In the lab, this researcher is using a special microscope to study the mineral composition of rock samples that were collected during fieldwork. (Photo by Jon Wilson/Science Source)

or observation is always open to question. Nevertheless, these data are essential to science and serve as a springboard for the development of scientific hypotheses and theories.

Hypothesis

Once facts have been gathered and principles have been formulated to describe a natural phenomenon, investigators try to explain how or why things happened in the manner observed. They often do this by constructing a tentative (or untested) explanation, which is called a scientific **hypothesis**. It is best if an investigator can formulate more than one hypothesis to explain a given set of observations. If an individual scientist is unable to devise multiple hypotheses, others in the scientific community will almost always develop alternative explanations. A spirited debate frequently ensues. As a result, extensive research is conducted by proponents of opposing hypotheses, and the results are made available to the wider scientific community in scientific journals.

Before a hypothesis can become an accepted part of scientific knowledge, it must repeatedly pass objective testing and analysis. If a hypothesis cannot be tested, it is not scientifically useful, no matter how interesting it might seem. The verification process requires that *predictions* be made based on the hypothesis being considered and that the predictions be tested by being compared against objective observations of nature. Put another way, hypotheses must fit observations other than those used to formulate them in the first place. Hypotheses that fail rigorous testing are ultimately discarded. The history of science is littered with discarded hypotheses. One of the best known is the Earth-centered model of the universe—a proposal that was supported by the apparent daily motion of the Sun, Moon, and stars around Earth. As the mathematician Jacob Bronowski so ably stated, "Science is a great many things, but in the end they all return to this: Science is the acceptance of what works and the rejection of what does not."

Theory

When a hypothesis has survived extensive scrutiny and when competing hypotheses have been eliminated, it may be elevated to the status of a scientific **theory**. In everyday language, we might say, "That's only a theory." But a scientific theory is a well-tested and widely accepted view that the scientific community agrees best explains certain observable facts.

Some theories that are extensively documented and extremely well supported are comprehensive in scope and may incorporate several well-tested hypotheses. For example, the theory of plate tectonics provides the framework for understanding the origin of mountains, earthquakes, and volcanic activity. In addition, plate tectonics explains the evolution of the continents and the ocean basins through time—ideas that are explored in some detail in Chapters 4 through 7.

Scientific Methods

The process just described, in which researchers gather facts through observations and formulate scientific hypotheses and theories, is called the **scientific method**. Contrary to popular belief, the scientific method is not a standard recipe that scientists apply in a routine manner to unravel the secrets of our natural world. Rather, it is an endeavor that involves creativity and insight. Rutherford and Ahlgren put it this way: "Inventing hypotheses or theories to imagine how the world works and then figuring out how they can be put to the test of reality is as creative as writing poetry, composing music, or designing skyscrapers."†

There is no fixed path for scientists to follow that leads unerringly to scientific knowledge. However, many scientific investigations involve the steps outlined in Figure 1.8. In addition, some scientific discoveries result from purely theoretical ideas that stand up to extensive examination. Some researchers use high-speed computers to create models that simulate what is happening in the "real" world. These models are useful when dealing with natural processes that occur on very long time scales or take place in extreme or inaccessible locations.

†F. James Rutherford and Andrew Ahlgren, *Science for All Americans* (New York: Oxford University Press, 1990), p. 7.

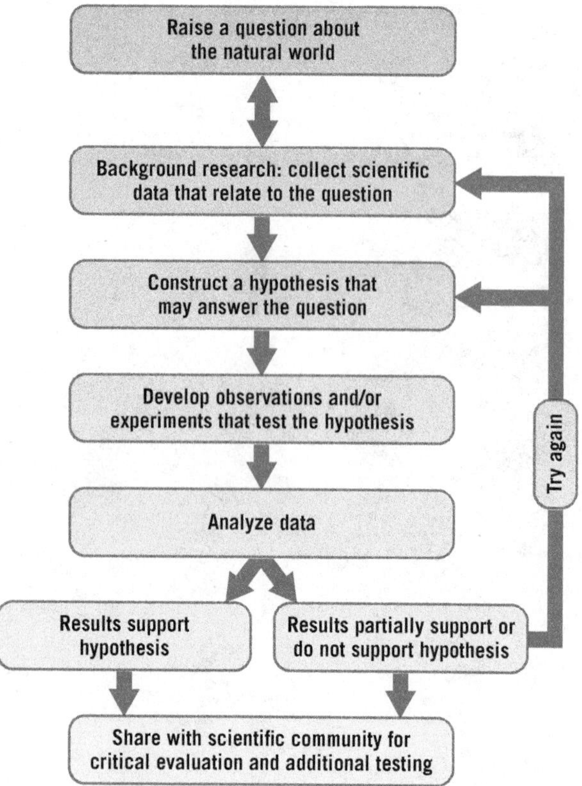

▲ Figure 1.8 **Steps frequently followed in scientific investigations** The diagram depicts the steps involved in the process many refer to as the *scientific method*.

EYE ON EARTH 1.1

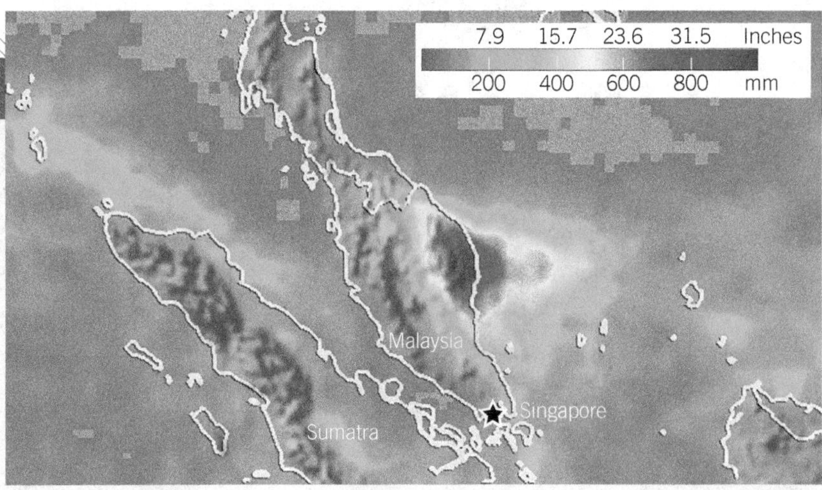

NASA Headquarters

This image shows rainfall data for December 7–13, 2004, in Malaysia. More than 800 millimeters (32 inches) of rain fell along the east coast of the peninsula (darkest red area). The extraordinary rains caused extensive flooding. The data for this image are from NASA's *Tropical Rainfall Measuring Mission* (*TRMM*). This is just one of hundreds of satellites that provide scientists with all kinds of data about our planet.

QUESTION 1 *Suggest some advantages that satellites provide to scientists in terms of gaining information about Earth.*

QUESTION 2 *Gathering data is a basic part of scientific inquiry. Aside from satellites, list at least two ways that Earth scientists gather data.*

Still other scientific advancements are made when a totally unexpected happening occurs during an experiment. These serendipitous discoveries are more than pure luck, for as the nineteenth-century French scientist Louis Pasteur said, "In the field of observation, chance favors only the prepared mind."

Scientific knowledge is acquired through several avenues, so it might be best to describe the nature of scientific inquiry as the *methods of science* rather than as the *scientific method*. In addition, it should always be remembered that even the most compelling scientific theories are still simplified explanations of the natural world.

In this book, you will discover the results of centuries of scientific work. You will see the end product of millions of observations, thousands of hypotheses, and hundreds of theories. We have distilled all of this to give you a "briefing" on Earth science.

Bear in mind that our knowledge of Earth is changing daily, as thousands of scientists worldwide make satellite observations, analyze drill cores from the seafloor, measure earthquakes, develop computer models to predict climate, examine the genetic codes of organisms, and discover new facts about our planet's long history. This new knowledge often updates hypotheses and theories. Expect to see many new discoveries and changes in scientific thinking in your lifetime.

CONCEPT CHECKS 1.2

1. How is a scientific hypothesis different from a scientific theory?
2. Summarize the basic steps followed in many scientific investigations.

1.3 Early Evolution of Earth

Outline the stages in the formation of our solar system.

This section describes the most widely accepted views on the origin of our solar system. The theory summarized here represents the most consistent set of ideas available to explain what we know about our solar system today. **GEOgraphics 1.1** provides a useful perspective on size and scale in the solar system.

The Universe Begins

Our scenario begins about 13.7 billion years ago, with the *Big Bang*, an almost incomprehensible event in which space itself, along with all the matter and energy of the universe, exploded in an instant from tiny to huge dimensions. As the universe continued to expand, subatomic particles condensed to form hydrogen and helium gas, which later cooled and clumped to form the first stars and galaxies. It was in one of these galaxies, the Milky Way, that our solar system, including planet Earth, took form.

The Solar System Forms

Earth is one of eight planets that, along with dozens of moons and numerous smaller bodies, revolve around the Sun. The orderly nature of our solar system helped scientists determine that Earth and the other planets formed at essentially the same time and from the same primordial material as the Sun. The **nebular theory** proposes that the bodies of our solar system evolved

GEOGRAPHICS 1.1

Solar System: Size and Scale

The Sun is the center of a revolving system trillions of miles across, consisting of eight planets, their satellites, and numerous dwarf plan asteroids, comets, and meteoroids.

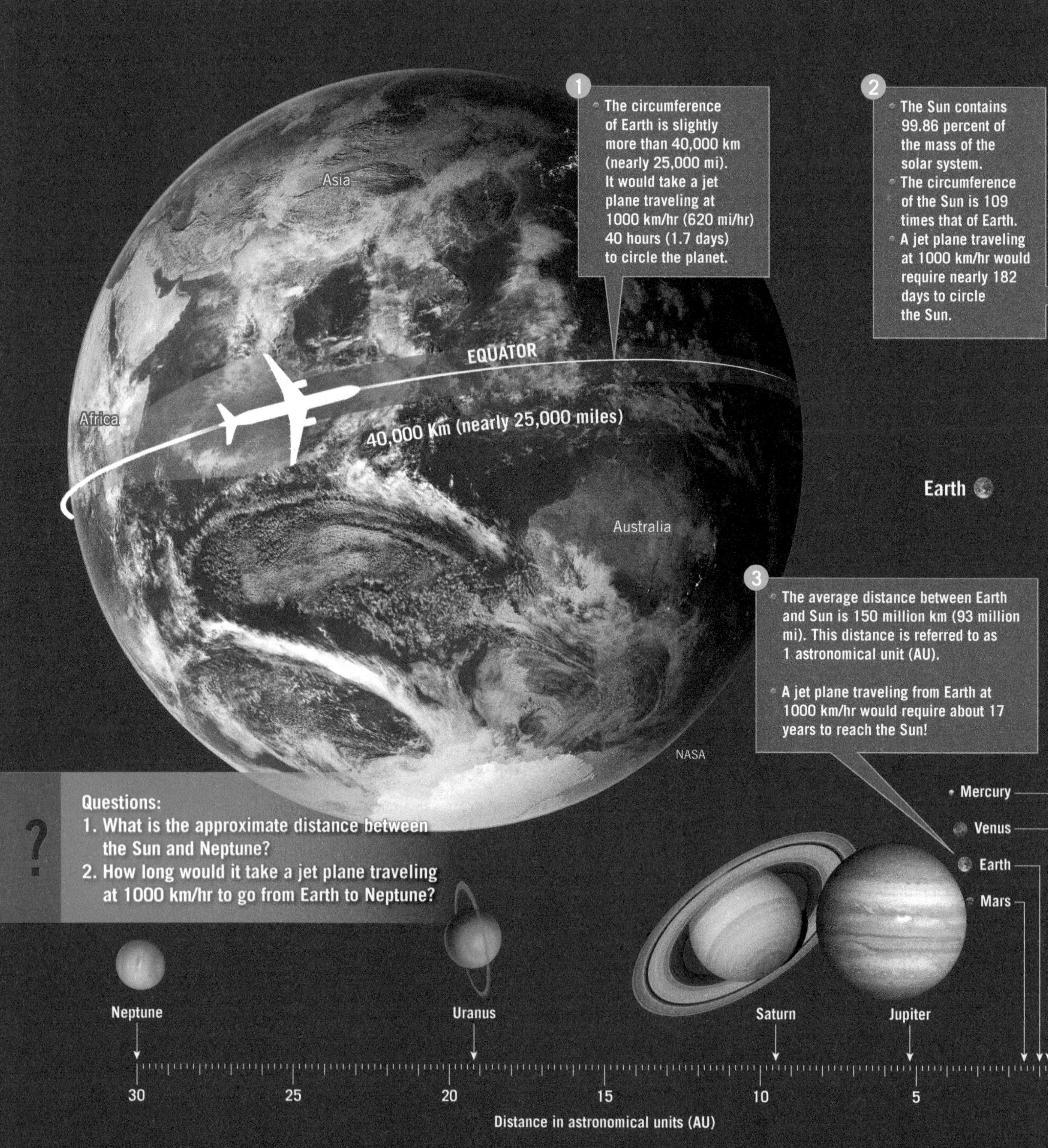

1
The circumference of Earth is slightly more than 40,000 km (nearly 25,000 mi). It would take a jet plane traveling at 1000 km/hr (620 mi/hr) 40 hours (1.7 days) to circle the planet.
2
The Sun contains 99.86 percent of the mass of the solar system.
The circumference of the Sun is 109 times that of Earth.
A jet plane traveling at 1000 km/hr would require nearly 182 days to circle the Sun.
Asia
EQUATOR
Africa
40,000 Km (nearly 25,000 miles)
Australia
NASA
Earth
3
The average distance between Earth and Sun is 150 million km (93 million mi). This distance is referred to as 1 astronomical unit (AU).
A jet plane traveling from Earth at 1000 km/hr would require about 17 years to reach the Sun!
Mercury
Venus
Earth
Mars
Neptune
Uranus
Saturn
Jupiter
30
25
20
15
10
5
Distance in astronomical units (AU)

Questions:

1. What is the approximate distance between the Sun and Neptune?
2. How long would it take a jet plane traveling at 1000 km/hr to go from Earth to Neptune?

from an enormous rotating cloud called the solar nebula (**Figure 1.9**). Besides the hydrogen and helium atoms generated during the Big Bang, the solar nebula consisted of microscopic dust grains and other matter ejected ultimately from long-dead stars. (Nuclear fusion in stars converts hydrogen and helium into the other elements found in the universe.)

Nearly 5 billion years ago, something—perhaps a shock wave from an exploding star (*supernova*)—caused this nebula to start collapsing in response to its own gravitation. As it collapsed, it evolved from a huge, vaguely rotating cloud to a much smaller, fast-spinning disk. The cloud flattened into a disk for the same reason that it is easier to move along with a crowd of circling ice skaters than to cross their path. The orbital plane within the cloud that started out with the largest amount of matter gradually, through collisions and other interactions, incorporated gas and particles that originally had other orbits until all the matter orbited in one plane. The disk spun faster as it shrank, much as ice skaters spin faster when they draw their arms toward their bodies. Most of the cloud's matter ended up in the center of the disk, where it formed the *protosun* (pre-Sun). Astronomers have observed many such disks around newborn stars in neighboring regions of our galaxy.

The protosun and inner disk were heated by the gravitational energy of infalling matter. In the inner disk, temperatures became high enough to cause the dust grains to evaporate. However, at distances beyond the orbit of Mars, the temperatures probably remained

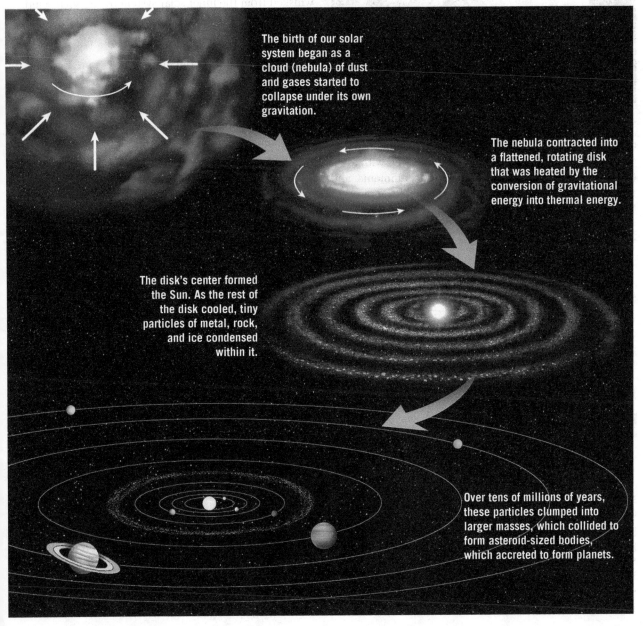

The birth of our solar system began as a cloud (nebula) of dust and gases started to collapse under its own gravitation.
The nebula contracted into a flattened, rotating disk that was heated by the conversion of gravitational energy into thermal energy.
The disk's center formed the Sun. As the rest of the disk cooled, tiny particles of metal, rock, and ice condensed within it.
Over tens of millions of years, these particles clumped into larger masses, which collided to form asteroid-sized bodies, which accreted to form planets.

SmartFigure 1.9
Nebular theory The nebular theory proposes the formation of the solar system.

TUTORIAL
https://goo.gl/HbtC0S

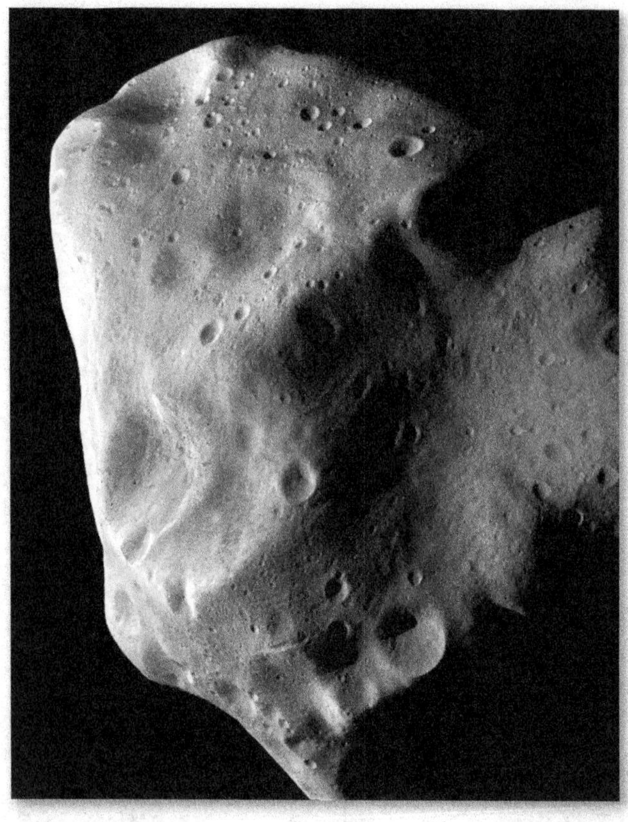

Figure 1.10 **A remnant planetesimal** This image of Asteroid 21 Lutetia was obtained by special cameras aboard the *Rosetta* spacecraft on July 10, 2010. Spacecraft instruments showed that Lutetia is a primitive body (planetesimal) left over from when the solar system formed. (Image courtesy of European Space Agency)

quite low. At −200°C (−328°F), the tiny particles in the outer portion of the nebula were likely covered with a thick layer of frozen water, carbon dioxide, ammonia, and methane. The disk also contained appreciable amounts of the lighter gases hydrogen and helium.

The Inner Planets Form

The formation of the Sun marked the end of the period of contraction and thus the end of gravitational heating. Temperatures in the region where the inner planets now reside began to decline. The decrease in temperature caused substances with high melting points to condense into tiny particles that began to coalesce (join together). Materials such as iron and nickel and the elements of which the rock-forming minerals are composed—silicon, calcium, sodium, and so forth—formed metallic and rocky clumps that orbited the Sun (see Figure 1.9). Repeated collisions caused these masses to coalesce into larger asteroid-size bodies, called *planetesimals*, which in a few tens of millions of years accreted into the four inner planets we call Mercury, Venus, Earth, and Mars (Figure 1.10). Not all of these clumps of matter were incorporated into the planetesimals. Those rocky and metallic pieces that remained in orbit are called asteroids and meteors. Meteors become *meteorites* if they impact Earth's surface.

As more and more material was swept up by these growing planetary bodies, the high-velocity impact of nebular debris caused their temperatures to rise. Because of their relatively high temperatures and weak gravitational fields, the inner planets were unable to accumulate much of the lighter components of the nebular cloud. The lightest of these, hydrogen and helium, were eventually whisked from the inner solar system by the solar wind.

The Outer Planets Develop

At the same time that the inner planets were forming, the larger, outer planets (Jupiter, Saturn, Uranus, and Neptune), along with their extensive satellite systems, were also developing. Because of low temperatures far from the Sun, the material from which these planets formed contained a high percentage of ices—water, carbon dioxide, ammonia, and methane—as well as rocky and metallic debris. The accumulation of ices accounts in part for the large size and low density of the outer planets. The two most massive planets, Jupiter and Saturn, had a surface gravity sufficient to attract and hold large quantities of even the lightest elements—hydrogen and helium.

CONCEPT CHECKS 1.3

1. Name and briefly outline the theory that describes the formation of our solar system.
2. List the inner planets and the outer planets. Describe basic differences in size and composition.

1.4 Earth as a System

List and describe Earth's four major spheres. Define *system* and explain why Earth is considered to be a system.

Anyone who studies Earth soon learns that our planet is a dynamic body with many separate but interacting parts, or *spheres*. The hydrosphere, atmosphere, biosphere, and geosphere and all of their components can be studied separately. However, the parts are *not* isolated. Each is related in some way to the others, producing a complex and continuously interacting whole that we call the **Earth system**.

Earth's Spheres

The images in Figure 1.11 are considered to be classics because they let humanity see Earth differently than ever before. These early views profoundly altered our conceptualizations of Earth and remain powerful images decades after they were first viewed. Seen from space, Earth is breathtaking in its beauty and startling in its solitude. The photos remind us that our home is, after all, a planet—small, self-contained, and in some ways even fragile.

As we look closely at our planet from space, it becomes apparent that Earth is much more than rock and soil. In fact, the most conspicuous features of Earth in Figure 1.11A are swirling clouds suspended above the surface of the vast global ocean. These features emphasize the importance of water on our planet.

The closer view of Earth from space shown in Figure 1.11B helps us appreciate why the physical environment is traditionally divided into three major parts: the water portion of our planet, the *hydrosphere*; Earth's gaseous envelope, the *atmosphere*; and, of course, the solid Earth, or *geosphere*. It needs to be emphasized that our environment is highly integrated and not dominated by rock, water, or air alone. Rather, it is characterized by continuous interactions as air comes in contact with rock, rock with water, and water with air. Moreover, the *biosphere*, which is the totality of all life on our planet, interacts with each of the three physical realms and is an equally integral part of the planet. Thus, Earth can be thought of as consisting of four major spheres: the hydrosphere, atmosphere, geosphere, and biosphere. All four spheres are represented in the chapter-opening photo.

The interactions among Earth's spheres are incalculable. Figure 1.12 provides one easy-to-visualize example. The shoreline is an obvious meeting place for rock, water, and air. In this scene, ocean waves created by the drag of air moving across the water are breaking against the rocky shore.

View called "Earthrise" that greeted *Apollo 8* astronauts as their spacecraft emerged from behind the Moon in December 1968. This classic image let people see Earth differently than ever before.

A.

This image taken from *Apollo 17* in December 1972 is perhaps the first to be called "The Blue Marble." The dark blue ocean and swirling cloud patterns remind us of the importance of the oceans and atmosphere.

B.

SmartFigure 1.11 **Two classic views of Earth from space** The accompanying video commemorates the 45th anniversary of *Apollo 8*'s historic flight by re-creating the moment when the crew first saw and photographed Earth rising from behind the Moon. (NASA)

VIDEO
https://goo.gl/AQKqaa

Hydrosphere

Earth is sometimes called the *blue* planet. Water, more than anything else, makes Earth unique. The **hydrosphere** is a dynamic mass of water that is continually on the move, evaporating from the oceans to the atmosphere, precipitating to the land, and running back to the ocean again. The global ocean is certainly the most prominent feature of the hydrosphere, blanketing nearly 71 percent of Earth's surface to an average depth of about 3800 meters (12,500 feet). It accounts for about

Figure 1.12 **Interactions among Earth's spheres** The shoreline is one obvious interface—a common boundary where different parts of a system interact. In this scene, ocean waves (hydrosphere) that were created by the force of moving air (atmosphere) break against a rocky shore (geosphere). The force of the water can be powerful, and the erosional work that is accomplished can be great. (Photo by Michael Collier)

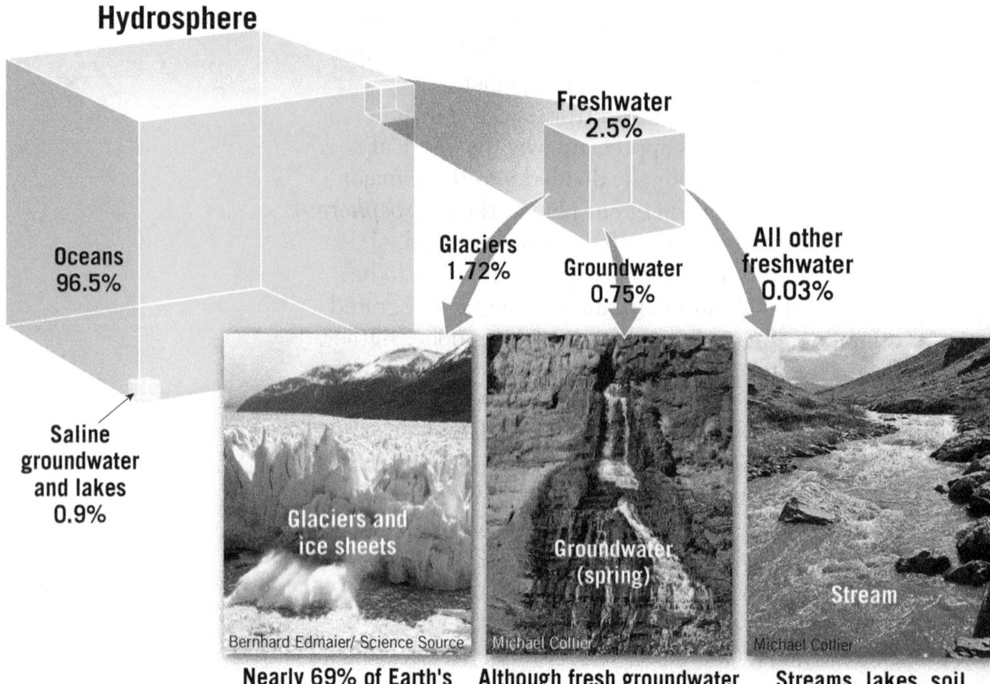

▲ **Figure 1.13 The water planet** Distribution of water in the hydrosphere.

97 percent of Earth's water (**Figure 1.13**). However, the hydrosphere also includes the freshwater found underground and in streams, lakes, and glaciers. Moreover, water is an important component of all living things.

Even though freshwater constitutes only a small fraction of Earth's hydrosphere, it plays an outsized role in Earth's external processes. Streams, glaciers, and groundwater sculpt many of our planet's varied landforms, and freshwater is vital for life on land.

Atmosphere

Earth is surrounded by a life-giving gaseous envelope called the **atmosphere** (**Figure 1.14**). When we watch a high-flying jet plane cross the sky, it seems that the atmosphere extends upward for a great distance. However, when compared to the thickness (radius) of the solid Earth (about 6400 kilometers [4000 miles]), the atmosphere is a very shallow layer. Despite its modest dimensions, this thin blanket of air is an integral part of the planet. It not only provides the air we breathe but also protects us from the Sun's intense heat and dangerous ultraviolet radiation. The energy exchanges that continually

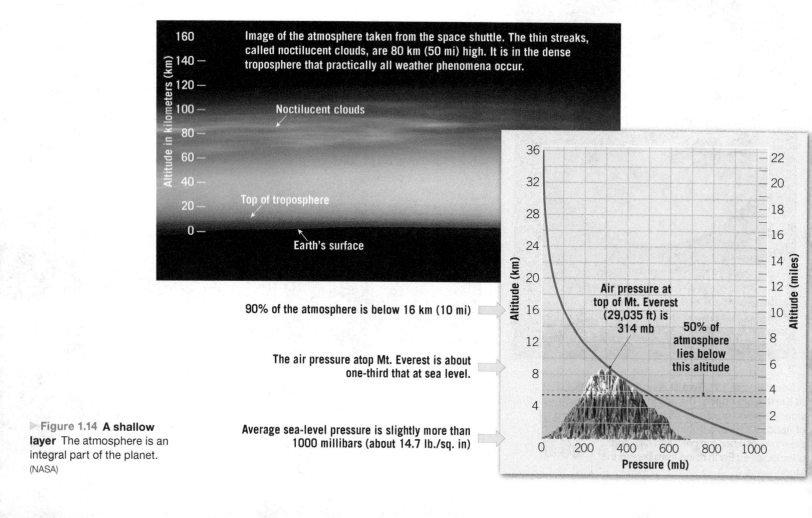

Image of the atmosphere taken from the space shuttle. The thin streaks, called noctilucent clouds, are 80 km (50 mi) high. It is in the dense troposphere that practically all weather phenomena occur.

Altitude in kilometers (km): 160, 140, 120, 100, 80, 60, 40, 20, 0

Noctilucent clouds

Top of troposphere

Earth's surface

90% of the atmosphere is below 16 km (10 mi)

The air pressure atop Mt. Everest is about one-third that at sea level.

Average sea-level pressure is slightly more than 1000 millibars (about 14.7 lb./sq. in)

Air pressure at top of Mt. Everest (29,035 ft) is 314 mb

50% of atmosphere lies below this altitude

Altitude (km): 36, 32, 28, 24, 20, 16, 12, 8, 4

Altitude (miles): 22, 20, 18, 16, 14, 12, 10, 8, 6, 4, 2

Pressure (mb): 0, 200, 400, 600, 800, 1000

► **Figure 1.14 A shallow layer** The atmosphere is an integral part of the planet. (NASA)

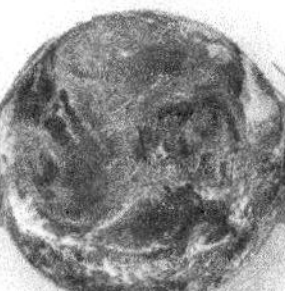

EYE ON EARTH 1.2

This jet is cruising at an altitude of 10 kilometers (6.2 miles).

QUESTION 1 *Refer to the graph in Figure 1.14. What is the approximate air pressure at the altitude where the jet is flying?*

QUESTION 2 *About what percentage of the atmosphere is below the jet (assuming that the pressure at the surface is 1000 millibars)?*

interlight/Shutterstock

occur between the atmosphere and Earth's surface and between the atmosphere and space produce the effects we call *weather* and *climate*. Climate has a strong influence on the nature and intensity of Earth's external processes. When climate changes, these processes respond.

If, like the Moon, Earth had no atmosphere, our planet would be lifeless, and many of the processes and interactions that make the surface such a dynamic place could not operate. Without weathering and erosion, the face of our planet might more closely resemble the lunar surface, which has not changed appreciably in nearly 3 billion years.

Biosphere

The **biosphere** includes all life on Earth (Figure 1.15). Ocean life is concentrated in the sunlit upper waters. Most life on land is also concentrated near the surface, with tree roots and burrowing animals reaching a few meters underground and flying insects and birds reaching a kilometer or so into the atmosphere. A surprising variety of life-forms are also adapted to extreme environments. For example, on the ocean floor, where pressures are extreme and no light penetrates, there are places where vents spew hot, mineral-rich fluids that support communities of exotic life-forms, as in Figure 1.15B. On land, some bacteria thrive in rocks as deep as 4 kilometers (2.5 miles) and in boiling hot springs. Moreover, air currents can carry microorganisms many kilometers into the atmosphere. But even when we consider these extremes, life still must be thought of as being confined to a narrow band very near Earth's surface.

Plants and animals depend on the physical environment for the basics of life. However, organisms do not just respond to their physical environment. Through countless interactions, life-forms help maintain and alter the physical environment. Without life, the makeup and nature of the geosphere, hydrosphere, and atmosphere would be very different.

Geosphere

Lying beneath the atmosphere and the ocean is the solid Earth or **geosphere**, extending from the surface to the center of the planet at a depth of nearly

Figure 1.15 The biosphere The biosphere, one of Earth's four spheres, includes all life. **A.** Tropical rain forests are teeming with life and occur in the vicinity of the equator. (Photo by AGE Fotostock/SuperStock) **B.** Some life occurs in extreme environments such as the absolute darkness of the deep ocean. (Photo by Fisheries and Oceans Canada/Verena Tunnicliffe/Newscom)

A. Tropical rain forests are characterized by hundreds of different species per square kilometer.

B. Microorganisms are nourished by hot, mineral-rich fluids spewing from vents on the deep-ocean floor. The microbes support larger organisms such as tube worms.

6400 kilometers (4000 miles)—by far the largest of Earth's spheres. Much of our study of the solid Earth focuses on the more accessible surface and near-surface features, but it is worth noting that many of these features are linked to the dynamic behavior of Earth's interior. Earth's interior is layered. As **Figure 1.16** shows, we can think of this layering as being due to differences in both *chemical composition* and *physical properties*. On the basis of chemical composition, Earth has three layers: a dense inner sphere called the **core**; the less dense **mantle**; and the **crust**, which is the light and very thin outer skin of Earth. The crust is not a layer of uniform thickness. It is thinnest beneath the oceans and thickest where continents exist. Although the crust may seem insignificant when compared with the other layers of the geosphere, which are much thicker, it was created by the same general processes that formed Earth's present structure. Thus, the crust is important in understanding the history and nature of our planet.

The layering of Earth in terms of physical properties reflects the way Earth's materials behave when various forces and stresses are applied. The term **lithosphere** refers to the rigid outer layer that includes the crust and uppermost mantle. Beneath the rigid rocks that compose the lithosphere, the rocks of the **asthenosphere** are weak and able to slowly flow in response to the uneven distribution of heat deep within Earth.

The two principal divisions of Earth's surface are the continents and the ocean basins. The most obvious difference between these two provinces is their relative vertical levels. The average elevation of the continents above sea level is about 840 meters (2750 feet), whereas the average depth of the oceans is about 3800 meters (12,500 feet). Thus, the continents stand on average 4640 meters (about 4.6 kilometers, or nearly 3 miles) above the level of the ocean floor.

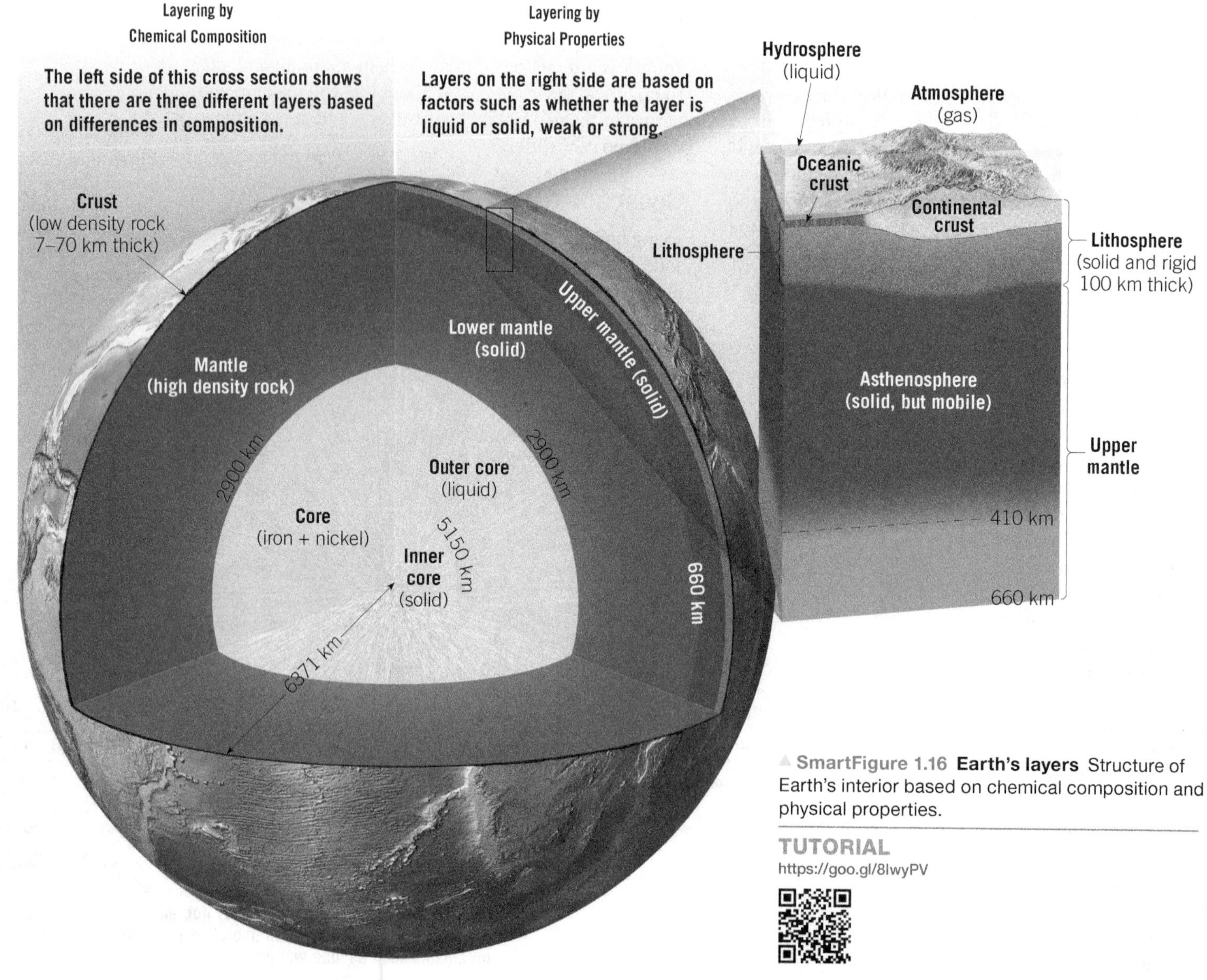

SmartFigure 1.16 **Earth's layers** Structure of Earth's interior based on chemical composition and physical properties.

TUTORIAL
https://goo.gl/8lwyPV

Figure 1.17 **Deadly debris flow** This image provides an example of interactions among different parts of the Earth system. Extraordinary rains triggered this debris flow (popularly called a mudslide) on March 22, 2014, near Oso, Washington. The mass of mud and debris blocked the North Fork of the Stillaguamish River and engulfed an area of about 2.6 square kilometers (1 square mile). Forty-three people perished. (Photo by Michael Collier)

Soil, the thin veneer of material at Earth's surface that supports the growth of plants, may be thought of as part of all four spheres. The solid portion is a mixture of weathered rock debris (geosphere) and organic matter from decayed plant and animal life (biosphere). The decomposed and disintegrated rock debris is the product of weathering processes that require air (atmosphere) and water (hydrosphere). Air and water also occupy the open spaces between solid particles.

Earth System Science

A simple example of the interactions among different parts of the Earth system occurs every winter, as moisture evaporates from the Pacific Ocean and subsequently falls as rain in the mountains of Washington, Oregon, and California, triggering destructive debris flows. The processes that move water from the hydrosphere to the atmosphere and then to the solid Earth have a profound impact on the plants and animals (including humans) that inhabit the affected regions (Figure 1.17).

Scientists have recognized that in order to more fully understand our planet, they must learn how its individual components (land, water, air, and life-forms) are interconnected. This endeavor, called **Earth system science**, aims to study Earth as a *system* composed of numerous interacting parts, or *subsystems*. Rather than look through the limited lens of only one of the traditional sciences—geology, atmospheric science, chemistry, biology, and so on—Earth system science attempts to integrate the knowledge of many academic fields. Using an interdisciplinary approach, those engaged in Earth system science attempt to achieve the level of understanding necessary to comprehend and solve many of our global environmental problems.

A **system** is a group of interacting, or interdependent, parts that form a complex whole. Most of us hear and use the term *system* frequently. We may service our car's cooling *system*, make use of the city's transportation *system*, and be a participant in the political *system*. A news report might inform us of an approaching weather *system*. Further, we know that Earth is just a small part of a larger system known as the *solar system*, which in turn is a subsystem of an even larger system called the Milky Way Galaxy.

The Earth System

The Earth system has a nearly endless array of subsystems in which matter is recycled over and over. One familiar loop, or subsystem, is the *hydrologic cycle*. It represents the unending circulation of Earth's water among the hydrosphere, atmosphere, biosphere, and geosphere (Figure 1.18). Water enters the atmosphere through evaporation from Earth's surface and transpiration from plants. Water vapor condenses in the atmosphere to form clouds, which in turn produce precipitation that falls back to Earth's surface. Some of the rain that falls onto the land sinks in and then is taken up by plants or becomes groundwater, and some flows across the surface toward the ocean.

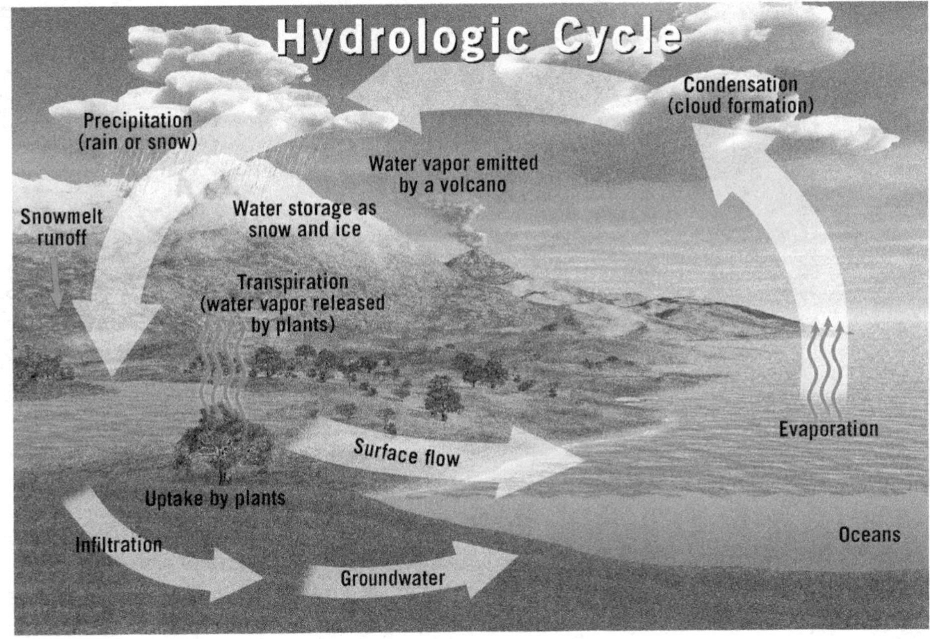

Figure 1.18 **The hydrologic cycle** Water readily changes state from liquid, to gas (vapor), to solid at the temperatures and pressures occurring on Earth. This cycle traces the movements of water among Earth's four spheres. It is one of many subsystems that collectively make up the Earth system.

Viewed over long time spans, the rocks of the geosphere are constantly forming, changing, and re-forming. The loop that involves the processes by which one rock changes to another is called the *rock cycle* and is discussed at some length in Chapter 3. The cycles of the Earth system are not independent of one another. To the contrary, there are many places where the cycles come in contact and interact.

The Parts Are Linked The parts of the Earth system are linked so that a change in one part can produce changes in any or all of the other parts. For example, when a volcano erupts, lava from Earth's interior may flow out at the surface and block a nearby valley. This new obstruction influences the region's drainage system by creating a lake or causing streams to change course. The large quantities of volcanic ash and gases that can be emitted during an eruption might be blown high into the atmosphere and influence the amount of solar energy that can reach Earth's surface. The result could be a drop in air temperatures over the entire hemisphere.

Where the surface is covered by lava flows or a thick layer of volcanic ash, existing soils are buried. This causes the soil-forming processes to begin anew to transform the new surface material into soil (Figure 1.19). The soil that eventually forms will reflect the interactions among many parts of the Earth system—the volcanic parent material, the climate, and the impact of biological activity. Of course, there would also be significant changes in the biosphere. Some organisms and their habitats would be eliminated by the lava and ash, and new settings for life, such as a lake formed by a lava dam, would be created. The potential climate change could also impact sensitive life-forms.

Time and Space Scales The Earth system is characterized by processes that vary on spatial scales from fractions of millimeters to thousands of kilometers. Time scales for

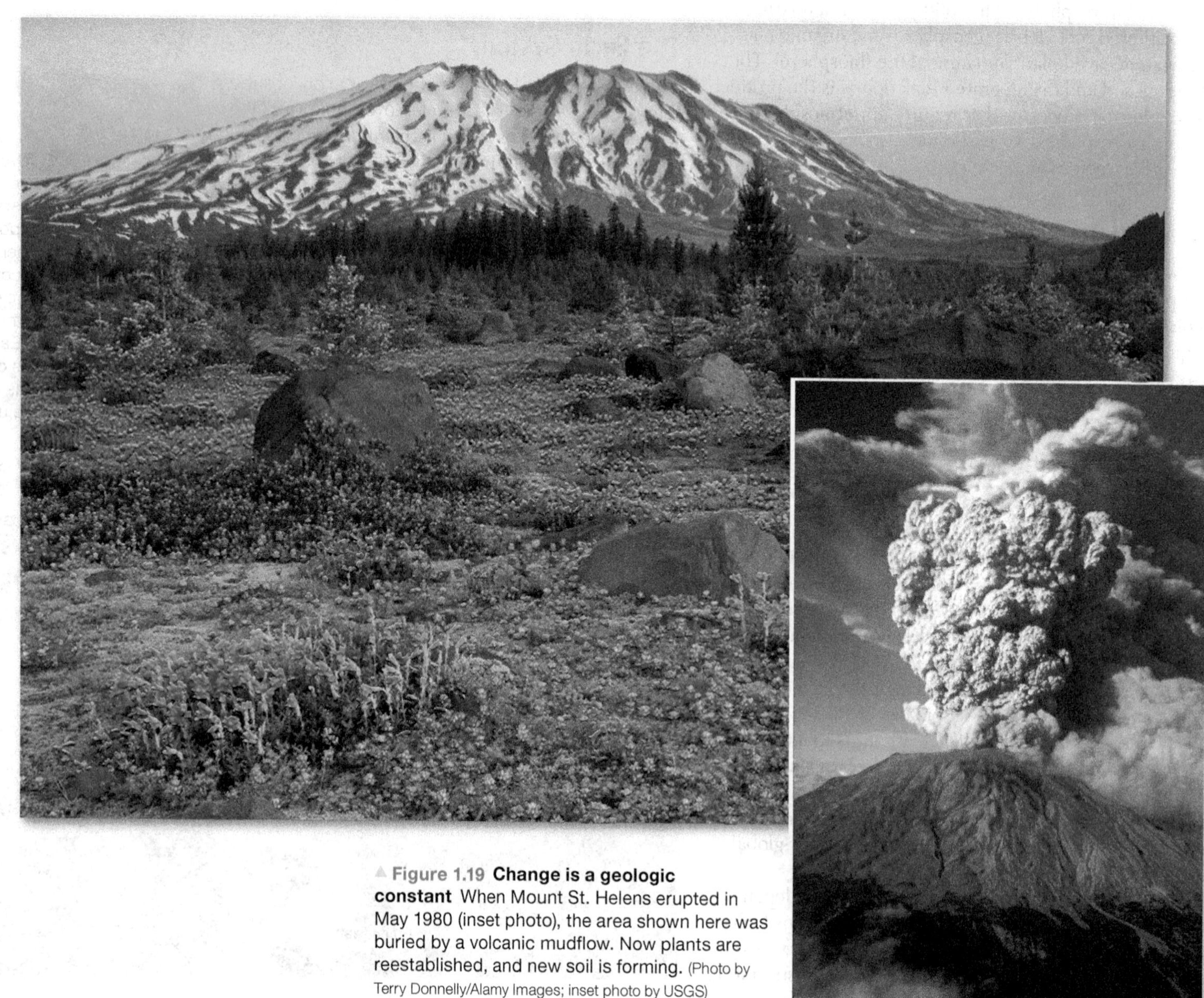

▲ Figure 1.19 **Change is a geologic constant** When Mount St. Helens erupted in May 1980 (inset photo), the area shown here was buried by a volcanic mudflow. Now plants are reestablished, and new soil is forming. (Photo by Terry Donnelly/Alamy Images; inset photo by USGS)

Earth's processes range from milliseconds to billions of years. As we learn about Earth, it becomes increasingly clear that despite significant separations in distance or time, many processes are connected, and a change in one component can influence the entire system.

Energy for the Earth System The Earth system is powered by energy from two sources. The Sun drives external processes that occur in the atmosphere, in the hydrosphere, and at Earth's surface. Energy from the Sun drives weather and climate, ocean circulation, and erosional processes. Earth's interior is the second source of energy. The internal processes that produce volcanoes, earthquakes, and mountains are powered by heat remaining from when our planet formed and heat that is continuously generated by radioactive decay.

People and the Earth System Humans are *part of* the Earth system, a system in which the living and nonliving components are entwined and interconnected. Therefore, our actions produce changes in all the other parts. When we burn gasoline and coal, dispose of our wastes, and clear the land, we cause other parts of the system to respond, often in unforeseen ways. Throughout this book, you will learn about many of Earth's subsystems, including the hydrologic system, the tectonic (mountain-building) system, the rock cycle, and the climate system. Remember that these components *and we humans* are all part of the complex interacting whole we call the Earth system.

The organization of this text involves traditional groupings of chapters that focus on closely related topics. Nevertheless, the theme of *Earth as a system* keeps recurring through *all* major units of this text. It is a thread that weaves through the chapters and helps tie them together.

CONCEPT CHECKS 1.4

1. List and briefly describe the four spheres that constitute the Earth system.
2. Compare the height of the atmosphere to the thickness of the geosphere.
3. How much of Earth's surface do oceans cover? What percentage of Earth's water supply do oceans represent?
4. What is a system? List three examples.
5. What are the two sources of energy for the Earth system?

1.5 The Face of Earth

List and describe the major features of the ocean basins and continents.

The two principal divisions of Earth's surface are the **ocean basins** and the **continents** (Figure 1.20). A significant difference between these two areas is their relative elevation, and it results primarily from differences in their respective densities and thicknesses:

- *Ocean basins.* The average depth of the ocean floor is about 3.8 kilometers (2.4 miles) below sea level, or about 4.5 kilometers (2.8 miles) lower than the average elevation of the continents. The basaltic rocks that comprise the oceanic crust average only 7 kilometers (4.3 miles) thick and have an average density of about 3.0 g/cm^3.
- *Continents.* The continents are remarkably flat features that have the appearance of plateaus protruding above sea level. With an average elevation of about 0.8 kilometer (0.5 mile), continental blocks lie close to sea level, except for limited areas of mountainous terrain. Recall that the continents average about 35 kilometers (22 miles) thick and are composed of granitic rocks that have a density of about 2.7 g/cm^3.

The thicker and less dense continental crust is more buoyant than the oceanic crust. As a result, continental crust floats on top of the deformable rocks of the mantle at a higher level than oceanic crust for the same reason that a large, empty (less dense) cargo ship rides higher than a small, loaded (denser) one.

Major Features of the Ocean Floor

If all water were drained from the ocean basins, a great variety of features would be visible, including chains of volcanoes, deep canyons, plateaus, and large expanses of monotonously flat plains. In fact, the scenery would be nearly as diverse as that on the continents (see Figure 1.20).

During the past 70 years, oceanographers have used modern depth-sounding equipment and satellite technology to map significant portions of the ocean floor. These studies have led them to identify three major regions: *continental margins, deep-ocean basins,* and *oceanic (mid-ocean) ridges.*

Continental Margin The **continental margin** is the portion of the seafloor adjacent to major landmasses. It may include the *continental shelf,* the *continental slope,* and the *continental rise.*

Although land and sea meet at the shoreline, this is *not* the boundary between the continents and the ocean basins. Rather, along most coasts, a gently sloping

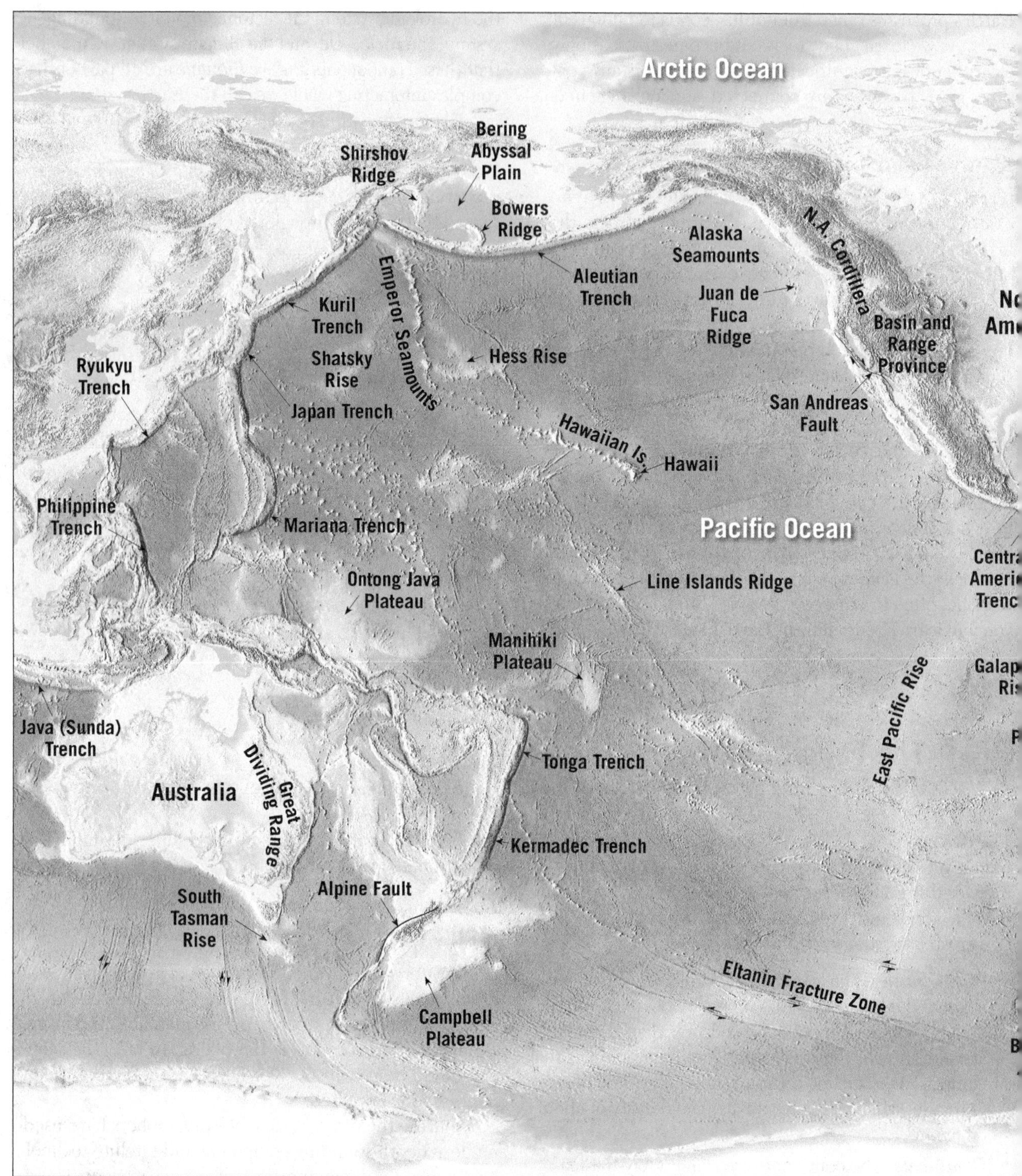

Figure 1.20 **The face of Earth** Major surface features of the geosphere.

platform, called the **continental shelf**, extends seaward from the shore. Because it is underlain by continental crust, it is clearly a flooded extension of the continents. A glance at Figure 1.20 shows that the width of the continental shelf is variable. For example, it is broad along the east and Gulf coasts of the United States but relatively narrow along the Pacific margin of the continent.

The boundary between the continents and the deep-ocean basins lies along the **continental slope**, which is a relatively steep drop-off that extends from the outer edge of the continental shelf, called the *shelf break*, to the floor of the deep ocean (see Figure 1.20). Using this as the dividing line, we find that about 60 percent of Earth's surface is represented by ocean basins and the remaining 40 percent by continents.

In regions where trenches do not exist, the steep continental slope merges into a more gradual incline known as the **continental rise**. The continental rise

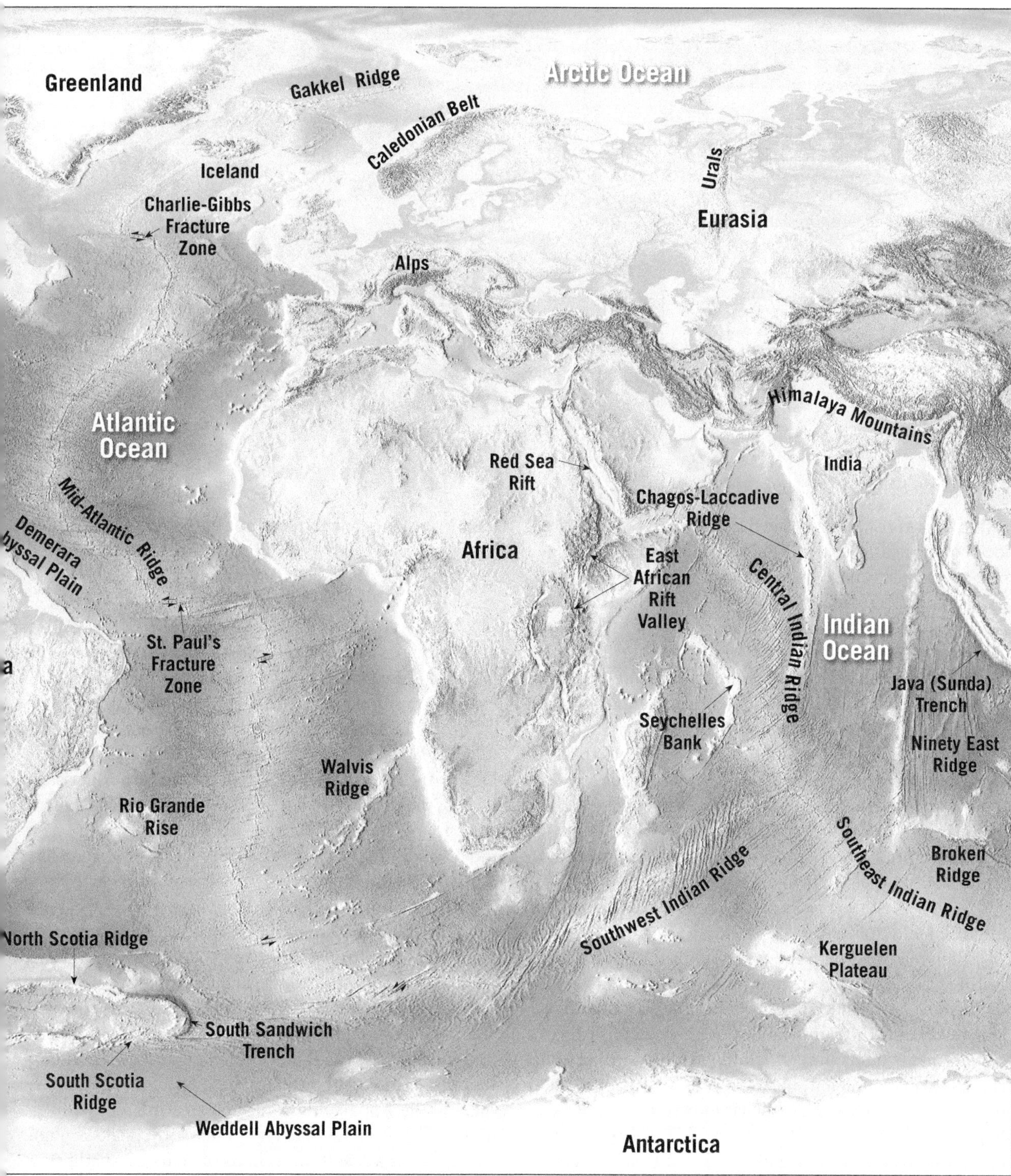

consists of a thick wedge of sediment that moved downslope from the continental shelf and accumulated on the deep-ocean floor.

Deep-Ocean Basins Between the continental margins and oceanic ridges are **deep-ocean basins**. Parts of these regions consist of incredibly flat features called **abyssal plains**. The ocean floor also contains extremely deep depressions that are occasionally more than 11,000 meters (36,000 feet) deep. Although these **deep-ocean trenches** are relatively narrow and represent only a small fraction of the ocean floor, they are nevertheless very significant features. Some trenches are located adjacent to young mountains that flank the continents. For example, in Figure 1.20 the Peru–Chile trench off the west coast of South America parallels the Andes Mountains. Other trenches parallel island chains called *volcanic island arcs*.

Dotting the ocean floor are submerged volcanic structures called **seamounts**, which sometimes form long, narrow chains. Volcanic activity has also produced

EYE ON EARTH 1.3

This photo shows the picturesque coastal bluffs and rocky shoreline along a portion of the California coast, south of San Simeon State Park.

QUESTION 1 *This area, like other shorelines, is described as an interface. What does that mean?*

QUESTION 2 *Does the shoreline represent the boundary between the continent and ocean basin? Explain.*

several large *lava plateaus*, such as the Ontong Java Plateau located northeast of New Guinea. In addition, some submerged plateaus are composed of continental-type crust. Examples include the Campbell Plateau southeast of New Zealand and the Seychelles Bank northeast of Madagascar.

Oceanic Ridges The most prominent feature on the ocean floor is the **oceanic ridge**, or **mid-ocean ridge**. As shown in Figure 1.20, the Mid-Atlantic Ridge and the East Pacific Rise are parts of this system. This broad elevated feature forms a continuous belt that winds for more than 70,000 kilometers (43,000 miles) around the globe, in a manner similar to the seam of a baseball. Rather than consist of highly deformed rock, as do most of the mountains on the continents, the oceanic ridge system consists of layer upon layer of igneous rock that has been fractured and uplifted.

Being familiar with the topographic (or relief) features that comprise the face of Earth is essential to understanding the mechanisms that have shaped our planet. What is the significance of the enormous ridge system that extends through all the world's oceans? What is the connection, if any, between young, active mountain belts and oceanic trenches? What forces crumple rocks to produce majestic mountain ranges? These are a few of the questions that will be addressed in the next chapter, as we begin to investigate the dynamic processes that shaped our planet in the geologic past and will continue to shape it in the future.

Major Features of the Continents

The major features of the continents can be grouped into two distinct categories: uplifted regions of deformed rocks that make up present-day mountain belts and extensive flat, stable areas that have eroded nearly to sea level. Notice in Figure 1.21 that the young mountain belts tend to be long, narrow features at the margins of continents and that the flat, stable areas are typically located in the interiors of the continents.

Mountains The most prominent features of the continents are linear **mountain belts**. Although the distribution of mountains appears to be random, this is not the case. The youngest mountain belts (those less than 100 million years old) are located principally in two major zones. The circum-Pacific belt (the region surrounding the Pacific Ocean) includes the mountains of the western Americas and continues into the western Pacific, in the form of volcanic island arcs (see Figure 1.20). Island arcs are active mountainous regions composed largely of volcanic rocks and deformed sedimentary rocks. Examples include the Aleutian Islands, Japan, the Philippines, and New Guinea.

The other major mountain belt extends eastward from the Alps through Iran and the Himalayas and then dips southward into Indonesia. Careful examination of mountainous terrains reveals that most are places where thick sequences of rocks have been squeezed and highly deformed, as if placed in a gigantic vise. Older mountains are also found on the continents. Examples include the Appalachians in the eastern United States and the Urals in Russia. Their once lofty peaks are now worn low, as a result of millions of years of weathering and erosion.

The Stable Interior Unlike the young mountain belts, which have formed within the past 100 million years, the interiors of the continents, called **cratons**, have been relatively stable (undisturbed) for the past 600 million years or even longer. Typically these regions were involved in mountain-building episodes much earlier in Earth's history.

Within the stable interiors are areas known as **shields**, which are expansive, flat regions composed

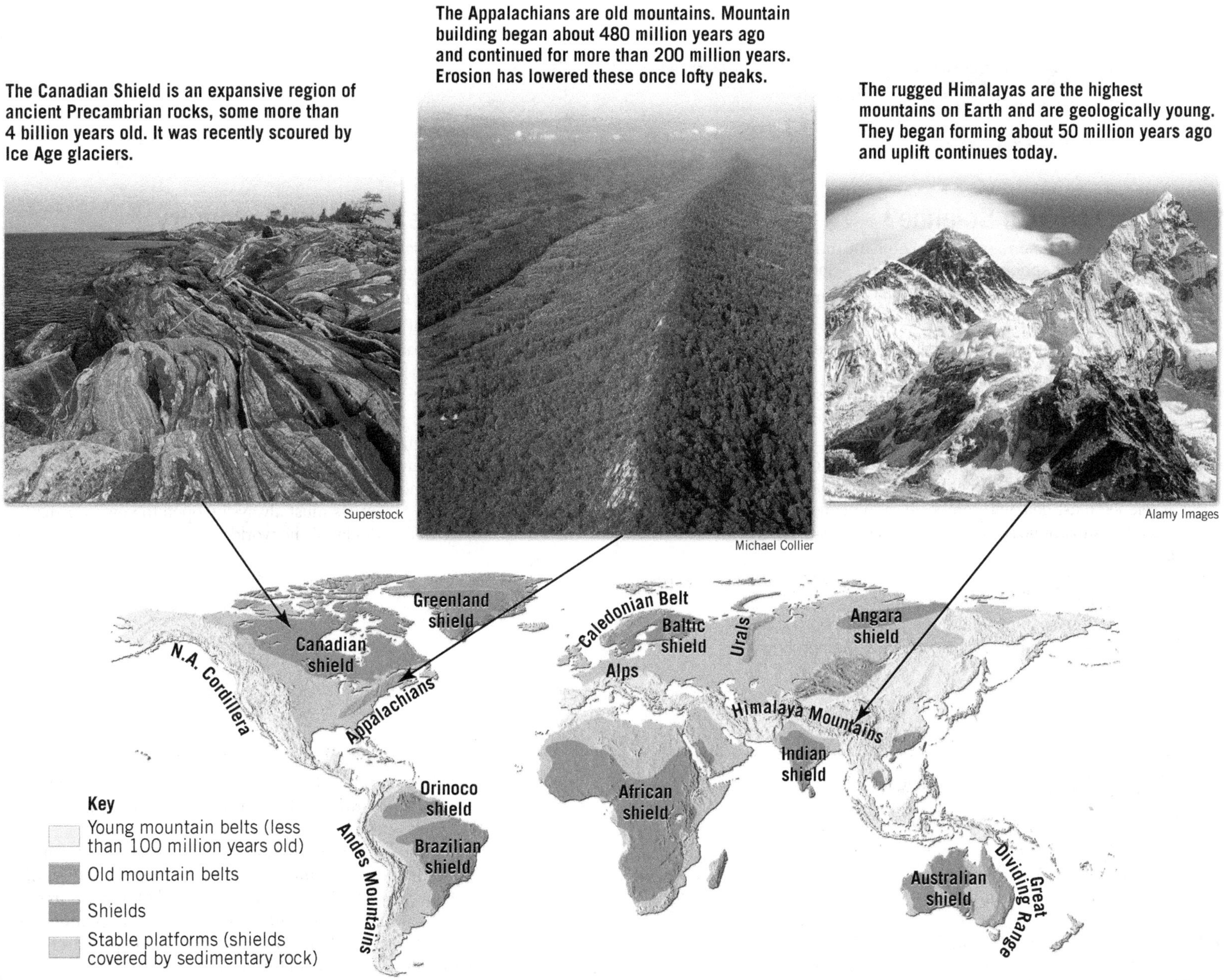

The Canadian Shield is an expansive region of ancient Precambrian rocks, some more than 4 billion years old. It was recently scoured by Ice Age glaciers.

Superstock

The Appalachians are old mountains. Mountain building began about 480 million years ago and continued for more than 200 million years. Erosion has lowered these once lofty peaks.

Michael Collier

The rugged Himalayas are the highest mountains on Earth and are geologically young. They began forming about 50 million years ago and uplift continues today.

Alamy Images

Greenland shield
Canadian shield
N.A. Cordillera
Appalachians
Caledonian Belt
Baltic shield
Urals
Angara shield
Alps
Himalaya Mountains
Indian shield
African shield
Orinoco shield
Brazilian shield
Andes Mountains
Australian shield
Great Dividing Range

Key
- Young mountain belts (less than 100 million years old)
- Old mountain belts
- Shields
- Stable platforms (shields covered by sedimentary rock)

SmartFigure 1.21 **The continents** Distribution of mountain belts, stable platforms, and shields. (Photo by Radius Images/SuperStock, Image Source/Alamy Stock Photo)

TUTORIAL
https://goo.gl/tFkUPM

largely of deformed igneous and metamorphic rocks. Notice in Figure 1.21 that the Canadian Shield is exposed in much of the northeastern part of North America. Radiometric dating of various shields has revealed that they are truly ancient regions. All contain Precambrian-age rocks that are more than 1 billion years old, with some samples approaching 4 billion years in age. Even these oldest-known rocks exhibit evidence of enormous forces that have folded, faulted, and metamorphosed them. Thus, we conclude that these rocks were once part of an ancient mountain system that has since been eroded away to produce these expansive, flat regions.

Other flat areas of the craton exist, in which highly deformed rocks, like those found in the shields, are covered by a relatively thin veneer of sedimentary rocks. These areas are called **stable platforms**. The sedimentary rocks in stable platforms are nearly horizontal, except where they have been warped to form large basins or domes. In North America a major portion of the stable platform is located between the Canadian Shield and the Rocky Mountains.

CONCEPT CHECKS 1.5

1. Compare and contrast continents and ocean basins.
2. Name the three major regions of the ocean floor. What are some physical features associated with each?
3. Describe the general distribution of Earth's youngest mountains.
4. What is the difference between shields and stable platforms?

1 CONCEPTS IN REVIEW
Introduction to Earth Science

1.1 What Is Earth Science?

List and describe the sciences that collectively make up Earth science. Discuss the scales of space and time in Earth science.

KEY TERMS: Earth science, geology, oceanography, meteorology, astronomy, geologic time

- Earth science includes geology, oceanography, meteorology, and astronomy.
- There are two broad subdivisions of geology. Physical geology studies Earth materials and the internal and external processes that create and shape Earth's landscape. Historical geology examines Earth's history.
- The other Earth sciences seek to understand the oceans, the atmosphere's weather and climate, and Earth's place in the universe.
- Earth science must deal with processes and phenomena that vary in size from the subatomic scale of matter to the nearly infinite scale of the universe.
- The time scales of phenomena studied in Earth science range from tiny fractions of a second to many billions of years.
- Geologic time, the span of time since the formation of Earth, is about 4.6 billion years, a number that is difficult to comprehend.

1.2 The Nature of Scientific Inquiry

Discuss the nature of scientific inquiry, including the construction of hypotheses and the development of theories.

KEY TERMS: hypothesis, theory, scientific method

- Scientists make careful observations, construct tentative explanations for those observations (hypotheses), and then test those hypotheses with field investigations, laboratory work, and/or computer modeling.
- In science, a theory is a well-tested and widely accepted explanation that the scientific community agrees best fits certain observable facts.
- As failed hypotheses are discarded, scientific knowledge moves closer to a correct understanding, but we can never be fully confident that we know all the answers. Scientists must always be open to new information that forces change in our model of the world.

1.3 Early Evolution of Earth

Outline the stages in the formation of our solar system.

KEY TERMS: nebular theory

- The nebular theory describes the formation of the solar system. The planets and Sun began forming about 5 billion years ago from a large cloud of dust and gases.
- As the cloud contracted, it rotated faster and assumed a disk shape. Material that was gravitationally pulled toward the center became the protosun. Within the rotating disk, solid matter gradually cohered to form objects called planetesimals, which grew as they swept up more and more of the cloud's debris.
- Because of their high temperatures and weak gravitational fields, the inner planets were unable to accumulate and retain large quantities of the light elements in the disk. Because of the very cold temperatures existing far from the Sun, the large outer planets include huge amounts of lighter materials. These substances account for the comparatively large sizes and low densities of the outer planets.

? Earth is about 4.6 billion years old. If all the planets in our solar system formed at about the same time, how old would you expect Mars to be? Jupiter? The Sun?

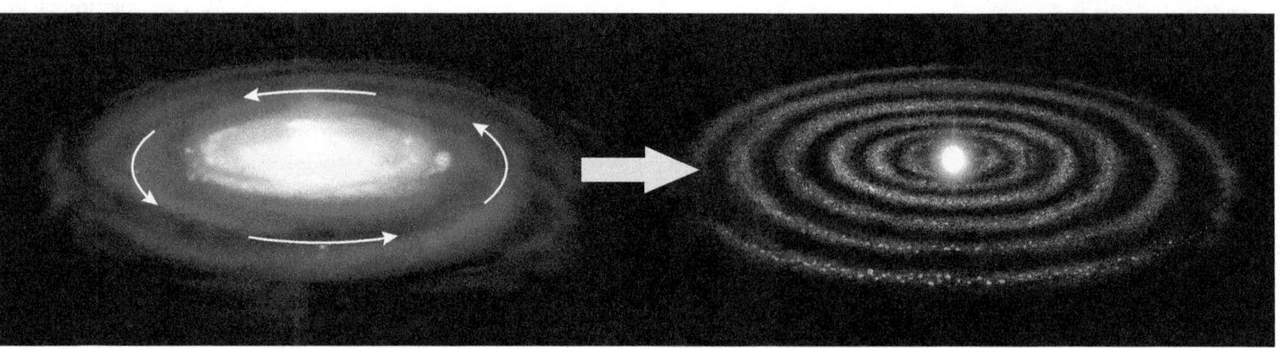

1.4 Earth as a System

List and describe Earth's four major spheres. Define *system* and explain why Earth is considered to be a system.

KEY TERMS: Earth system, hydrosphere, atmosphere, biosphere, geosphere, core, mantle, crust, lithosphere, asthenosphere, Earth system science, system

- Earth's physical environment is traditionally divided into three major parts: the solid Earth, called the geosphere; the water portion of our planet, called the hydrosphere; and Earth's gaseous envelope, called the atmosphere.
- A fourth Earth sphere is the biosphere, the totality of life on Earth. It is concentrated in a relatively thin zone that extends a few kilometers into the hydrosphere and geosphere and a few kilometers up into the atmosphere.
- Of all the water on Earth, about 97 percent is in the oceans, which cover nearly 71 percent of the planet's surface.
- Although each of Earth's four spheres can be studied separately, they are all related in a complex and continuously interacting whole that is called the Earth system.
- Earth system science uses an interdisciplinary approach to integrate the knowledge of several academic fields in the study of our planet and its global environmental problems.
- The two sources of energy that power the Earth system are (1) the Sun, which drives the external processes that occur in the atmosphere, hydrosphere, and at Earth's surface, and (2) heat from Earth's interior, which powers the internal processes that produce volcanoes, earthquakes, and mountains.

? Is glacial ice part of the geosphere, or does it belong to the hydrosphere? Explain your answer.

1.5 The Face of Earth

List and describe the major features of the ocean basins and continents.

KEY TERMS: ocean basin, continent, continental margin, continental shelf, continental slope, continental rise, deep-ocean basin, abyssal plain, deep-ocean trench, seamount, oceanic ridge (mid-ocean ridge), mountain belt, craton, shield, stable platform

- Two principal divisions of Earth's surface are the continents and ocean basins. A significant difference between them, their relative elevations, results primarily from differences in their respective densities and thicknesses.
- Continents have relatively flat, stable core areas called cratons. Where a craton is blanketed by a relatively thin layer of sediment or sedimentary rock, it is called a stable platform. Where a craton is exposed at the surface, it is known as a shield. Wrapping around the edges of some cratons are younger mountain belts, linear zones of intense deformation and metamorphism.
- The ocean basins are rimmed by shallow continental shelves, which are essentially flooded portions of the continents. The deep ocean includes vast abyssal plains and narrow, very deep ocean trenches. Seamounts and lava plateaus interrupt the abyssal plain in some places.

? Put these features of the ocean floor in order from shallowest to deepest: continental slope, deep-ocean trench, continental shelf, abyssal plain, continental rise.

GIVE IT SOME THOUGHT

1 After entering a dark room, you turn on a wall switch, but the light does not come on. Suggest at least three hypotheses that might explain this observation. How would you determine which one of your hypotheses (if any) is correct?

2 The length of recorded history for humankind is about 5000 years. Clearly, most people view this span as being very long. How does it compare to the length of geologic time? Calculate the percentage or fraction of geologic time that is represented by recorded history. To make calculations easier, round the age of Earth to the nearest billion.

3 Refer to the graph in Figure 1.14 to answer the following questions.
   a. If you were to climb to the top of Mount Everest, how many breaths of air would you have to take at that altitude to equal one breath at sea level?
   b. If you are flying in a commercial jet at an altitude of 12 kilometers (about 39,000 feet), about what percentage of the atmosphere's mass is below you?

4 Examine Figure 1.13 to answer these questions.
   a. Where is most of Earth's freshwater stored?
   b. Where is most of Earth's liquid freshwater found?

5 Jupiter, the largest planet in our solar system, is 5.2 astronomical units (AU) from the Sun. How long would it take to go from Earth to Jupiter if you traveled as fast as a jet (1000 kilometers per hour)? Do the same calculation for Neptune, which is 30 AU from the Sun. Referring to GEOgraphics 1.1 on page 12 will be helpful.

6 These rock layers consist of materials such as sand, mud, and gravel that, over a span of millions of years, were deposited by rivers, waves, wind, and glaciers. Each layer was buried by subsequent deposits and eventually compacted and cemented into solid rock. Later, the region was uplifted, and erosion exposed the layers seen here.
   a. Can you establish a relative time scale for these rocks? That is, can you determine which one of the layers shown here is likely oldest and which is probably youngest?
   b. Explain the logic you used.

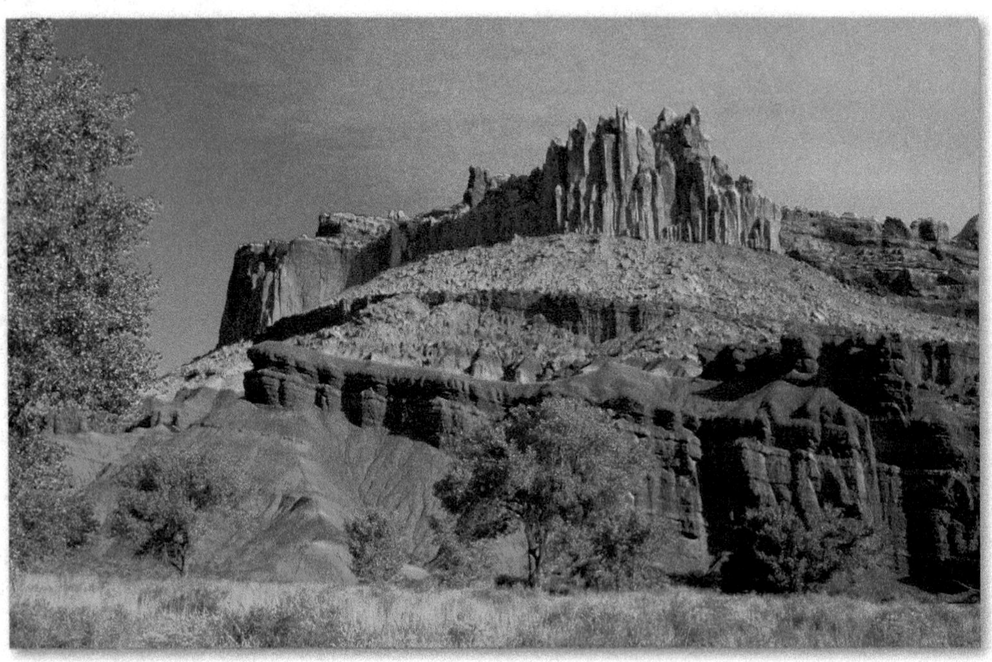

(Photo by M. Timothy O'Keefe/Alamy Stock Photo)

EXAMINING THE EARTH SYSTEM

1 This scene is in British Columbia's Mount Robson Provincial Park. The park is named for the highest peak in the Canadian Rockies.
  a. List as many examples as possible of features associated with each of Earth's four spheres.
  b. Which, if any, of these features was created by internal processes? Describe the role of external processes in this scene.

Michael Wheatley/AGE Fotostock

2 Humans are a part of the Earth system. List at least three examples of how you, in particular, influence one or more of Earth's major spheres.

3 The accompanying photo provides an example of interactions among different parts of the Earth system. It is a view of a debris flow (popularly called a mudslide) that was triggered by extraordinary rains. Which of Earth's four spheres were involved in this natural disaster, which buried a small town on the Philippine island of Leyte? Describe how each might have contributed to or been influenced by the event.

Pat Roque/AP Photo

DATA ANALYSIS

Swift Creek Landslide

The Swift Creek landslide is located in northwestern Washington State, on the west side of Sumas Mountain. The landslide moves on average 3 to 4 meters per year, which is about 1 centimeter per day.

ACTIVITIES

Go to the Western Washington University Swift Creek Landslide Observatory (SCLO) page at http://landslide.geol.wwu.edu and click About SCLO.

1 Why are geologists, government agencies, and local residents interested in the Swift Creek landslide?

Click TimeLapse to open the video generator. Choose a start date of January 1, 2007, and an end date of January 1, 2014. For time of day select between 12pm-3pm for the best lighting. Set the framerate to Medium, for 20 frames per second. Select the camera position and click Generate. It may take about 30 seconds to generate the video.

2 Compare the Upper camera to the Lower camera. Which section of the landslide has a more constant flow rate? How can you tell?

Use the TimeLapse video generator to answer the following question. You may need to watch multiple years and examine each season separately.

3 Is the flowrate of the upper part of the landslide faster during the wet season (October–March) or the dry season (April–September)?

UNIT 1 Earth Materials

2 Matter and Minerals

FOCUS ON CONCEPTS

Each statement represents the primary learning objective for the corresponding major heading within the chapter. After you complete the chapter, you should be able to:

2.1 List the main characteristics that an Earth material must possess to be considered a mineral and describe each characteristic.

2.2 Compare and contrast the three primary particles contained in atoms.

2.3 Distinguish among ionic bonds, covalent bonds, and metallic bonds.

2.4 List and describe the properties used in mineral identification.

2.5 List the common silicate and nonsilicate minerals and describe what characterizes each group.

2.6 Discuss Earth's mineral resources in terms of renewability. Differentiate between mineral resources and ore deposits.

The Cave of Crystals, Chihuahua, Mexico, contains giant gypsum crystals, some of the largest natural crystals ever found. (Photo by Carsten Peter/Speleoresearch & Films/National Geographic/Getty Images)

EARTH'S CRUST AND OCEANS are home to a wide variety of useful and essential minerals. Most people are familiar with the common uses of many basic metals, including aluminum in beverage cans, copper in electrical wiring, and gold and silver in jewelry. However, some people are not aware that pencil "lead" contains the greasy-feeling mineral graphite and that bath powders and many cosmetics contain the mineral talc. Moreover, many do not know that dentists use drill bits impregnated with diamonds to drill through tooth enamel. In fact, practically every manufactured product contains materials obtained from minerals.

In addition to the economic uses of rocks and minerals, every geologic process in some way depends on the properties of these basic Earth materials. Events such as volcanic eruptions, mountain building, weathering and erosion, and even earthquakes involve rocks and minerals. Consequently, a basic knowledge of Earth materials is essential to understanding all geologic phenomena.

2.1 Minerals: Building Blocks of Rocks

List the main characteristics that an Earth material must possess to be considered a mineral and describe each characteristic.

We begin our discussion of Earth materials with an overview of **mineralogy** (*mineral* = mineral, *ology* = study of) because minerals are the building blocks of rocks. Humans have used minerals for both practical and decorative purposes for thousands of years. For example, the common mineral quartz is the source of silicon for computer chips. The first Earth materials mined were flint and chert, which humans fashioned into weapons and cutting tools. As early as 3700 B.C.E. Egyptians began mining gold, silver, and copper. By 2200 B.C.E. humans had discovered how to combine copper with tin to make bronze—a strong, hard alloy. Later, a process was developed to extract iron from minerals such as hematite—a discovery that marked the decline of the Bronze Age. During the Middle Ages, mining of a variety of minerals became common, and the impetus for the formal study of minerals was in place.

In everyday conversation, the term *mineral* is used in several different ways. For example, those concerned with health and fitness extol the benefits of vitamins and minerals. The mining industry typically uses the word *mineral* to refer to anything extracted from Earth, such as coal, iron ore, or sand and gravel. The guessing game *Twenty Questions* usually begins with the question *Is it animal, vegetable, or mineral?* What criteria do geologists use to determine whether something is a mineral (**Figure 2.1**)?

▼ **Figure 2.1 Quartz crystals** A collection of well-developed quartz crystals found near Hot Springs, Arkansas. (Photo by Jeffrey A. Scovil)

Defining a Mineral

Geologists define **mineral** as *any naturally occurring inorganic solid that possesses an orderly crystalline structure and a definite chemical composition that allows for some variation.* Thus, Earth materials that are classified as minerals exhibit the following characteristics:

1. **Naturally occurring.** Minerals form by natural geologic processes. Synthetic materials, meaning those produced in a laboratory or by human intervention, are not considered minerals.

2. **Generally inorganic.** Inorganic crystalline solids, such as ordinary table salt (halite), that are found naturally in the ground are considered minerals. Organic compounds (that is, the kinds of carbon-containing compounds that are made by living things) are generally not considered minerals. Sugar, a crystalline solid like salt but extracted from

sugarcane or sugar beets, is a common example of an organic compound. Many marine animals secrete inorganic compounds, such as calcium carbonate (calcite), in the form of shells and coral reefs. If these materials are buried and become part of the rock record, geologists consider them minerals.

3. **Solid substance.** Only solid crystalline substances are considered minerals. Ice (frozen water) fits this criterion and is considered a mineral, whereas liquid water and water vapor do not.

4. **Orderly crystalline structure.** Minerals are crystalline substances, made up of atoms (or ions) that are arranged in an orderly, repetitive manner (Figure 2.2). This orderly packing of atoms is reflected in regularly shaped objects called *crystals*. Some naturally occurring solids, such as volcanic glass (obsidian), lack a repetitive atomic structure and are not considered minerals.

5. **Definite chemical composition that allows for some variation.** Most minerals are chemical compounds having compositions that can be expressed by a chemical formula. For example, the common mineral quartz has the formula SiO_2, which indicates that quartz consists of silicon (Si) and oxygen (O) atoms, in a 1:2 ratio. This proportion of silicon to oxygen is true for any sample of pure quartz, regardless of its origin. However, the compositions of some minerals can vary *within specific, well-defined limits*. This occurs because certain elements can substitute for others of similar size without changing the mineral's internal structure.

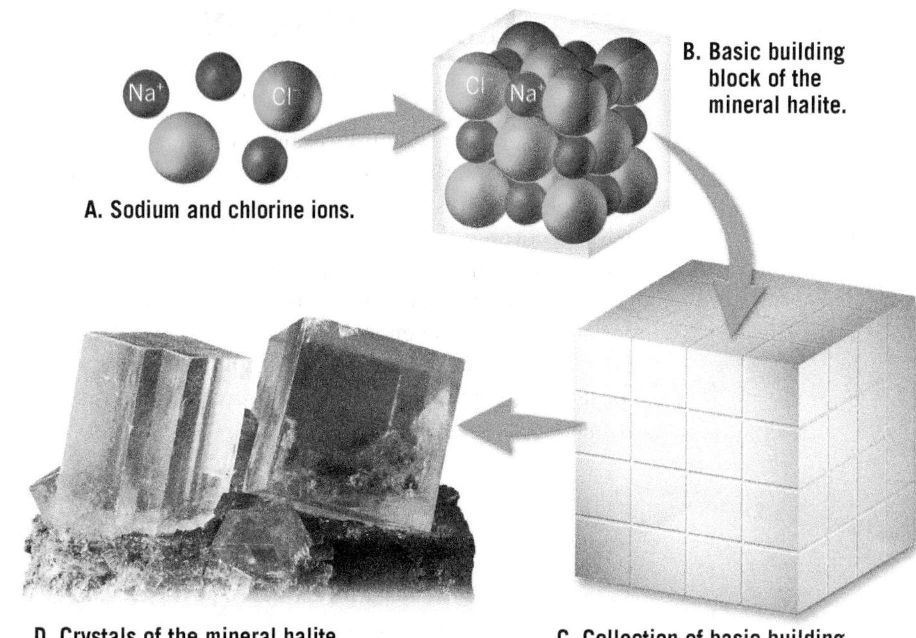

Figure 2.2 Arrangement of sodium and chloride ions in the mineral halite The arrangement of atoms (ions) into basic building blocks that have a cubic shape results in regularly shaped cubic crystals. (Photo by Dennis Tasa)

CONCEPT CHECKS 2.1

1. List five characteristics of a mineral.
2. Based on the definition of mineral, which of the following—gold, liquid water, synthetic diamonds, ice, and wood—are *not* classified as minerals?
3. Define the term *rock*. How do rocks differ from minerals?

What Is a Rock?

In contrast to minerals, rocks are more loosely defined. Simply, a **rock** is any solid mass of mineral or mineral-like matter that occurs naturally as part of our planet. Most rocks, like the sample of granite shown in Figure 2.3, are aggregates of several different minerals. The term *aggregate* implies that the minerals are joined in such a way that their individual properties are retained. Note that the different minerals that make up granite can be easily identified. However, some rocks are composed almost entirely of one mineral. A common example is the sedimentary rock *limestone*, which is an impure mass of the mineral calcite.

In addition, some rocks are composed of nonmineral matter. These include the volcanic rocks *obsidian* and *pumice*, which are noncrystalline glassy substances, and *coal*, which consists of solid organic debris.

Although this chapter deals primarily with the nature of minerals, keep in mind that most rocks are simply aggregates of minerals. Because the properties of rocks are determined largely by the chemical composition and crystalline structure of the minerals contained within them, we will first consider these Earth materials.

SmartFigure 2.3 Most rocks are aggregates of minerals Shown here is a hand sample of the igneous rock granite and three of its major constituent minerals. (Photos by E. J. Tarbuck)

TUTORIAL
https://goo.gl/k5qL1l

2.2 Atoms: Building Blocks of Minerals

Compare and contrast the three primary particles contained in atoms.

When minerals are carefully examined, even under optical microscopes, the innumerable tiny particles of their internal structures are not visible. Nevertheless, scientists have discovered that all matter, including minerals, is composed of minute building blocks called **atoms**—the smallest particles that constitute specific elements and cannot be split by chemical means. Atoms, in turn, contain even smaller particles—*protons* and *neutrons* located in a central **nucleus** that is surrounded by *electrons* (Figure 2.4).

Properties of Protons, Neutrons, and Electrons

Protons and **neutrons** are very dense particles with almost identical masses. By contrast, **electrons** have a negligible mass, about 1/2000 that of a proton. To visualize this difference, imagine a scale on which a proton or neutron has the mass of a baseball, whereas an electron has the mass of a single grain of rice.

Both protons and electrons share a fundamental property called *electrical charge*. Protons have an electrical charge of +1, and electrons have a charge of –1. Neutrons, as the name suggests, have no charge. The charges of protons and electrons are equal in magnitude but opposite in polarity, so when these two particles are paired, the charges cancel each other out. Since matter typically contains equal numbers of positively charged protons and negatively charged electrons, most substances are electrically neutral.

Illustrations sometimes show electrons orbiting the nucleus in a manner that resembles the planets of our solar system orbiting the Sun (see Figure 2.4A). However, electrons do not actually behave this way. A more realistic depiction would show electrons as a cloud of negative charges surrounding the nucleus (see Figure 2.4B). Studies of the arrangements of electrons show that they move about the nucleus in regions called *principal shells*, each with an associated energy level. In addition, each shell can hold a specific number of electrons, with the outermost shell generally containing **valence electrons**. These electrons can be transferred to or shared with other atoms to form chemical bonds.

Most of the atoms in the universe (except hydrogen and helium) were created inside massive stars by nuclear fusion and then released into interstellar space during hot, fiery supernova explosions. As this ejected material cooled, the newly formed nuclei attracted electrons to complete their atomic structure. At the temperatures found at Earth's surface, free atoms (those not bonded to other atoms) generally have a full complement of electrons—one for each proton in the nucleus.

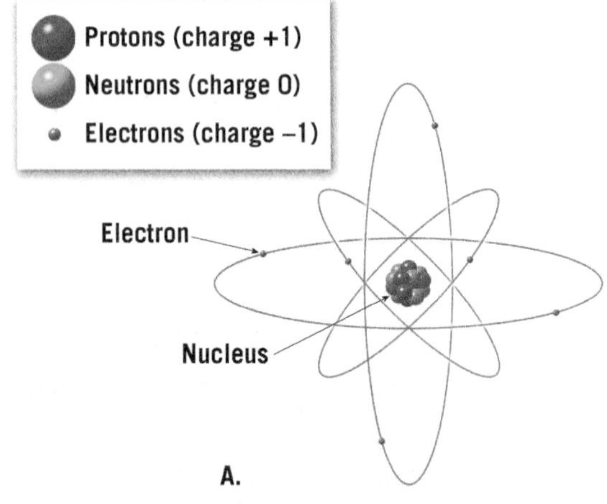

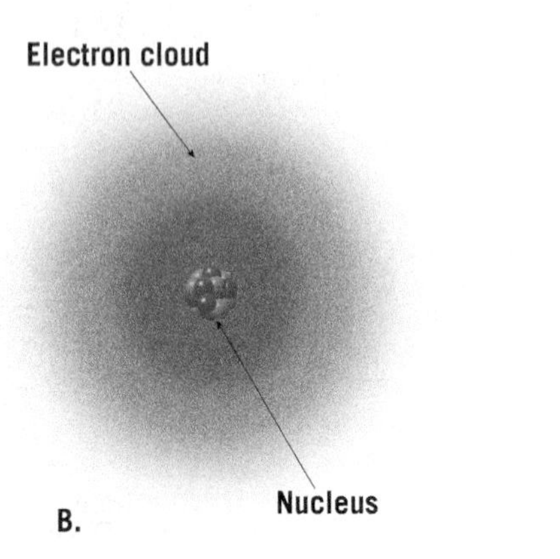

Figure 2.4 **Two models of an atom A.** Simplified view of an atom having a central nucleus composed of protons and neutrons, encircled by high-speed electrons. **B.** This model of an atom shows spherically shaped electron clouds (shells) surrounding a central nucleus. The nucleus contains virtually all of the mass of the atom. The remainder of the atom is the space occupied by negatively charged electrons. (The relative sizes of the nuclei shown are greatly exaggerated.)

Elements: Defined by Their Number of Protons

The simplest atoms have only 1 proton in their nuclei, whereas others have more than 100. The number of protons in the nucleus of an atom, called the **atomic number**, determines the atom's chemical nature. All atoms with the same number of protons have the same chemical and physical properties; collectively they constitute an **element**. There are about 90 naturally occurring elements, and several more have been synthesized in the laboratory. You are probably familiar with the names of many elements, including carbon, nitrogen, and oxygen. All carbon atoms have 6 protons, whereas all nitrogen atoms have 7 protons, and all oxygen atoms have 8.

The **periodic table**, shown in Figure 2.5, is a tool scientists use to organize the known elements. In it, the elements with similar properties line up in columns, referred to as *groups*. Each element is assigned a one- or two-letter symbol. The atomic number and atomic mass for each element are also included in the periodic table.

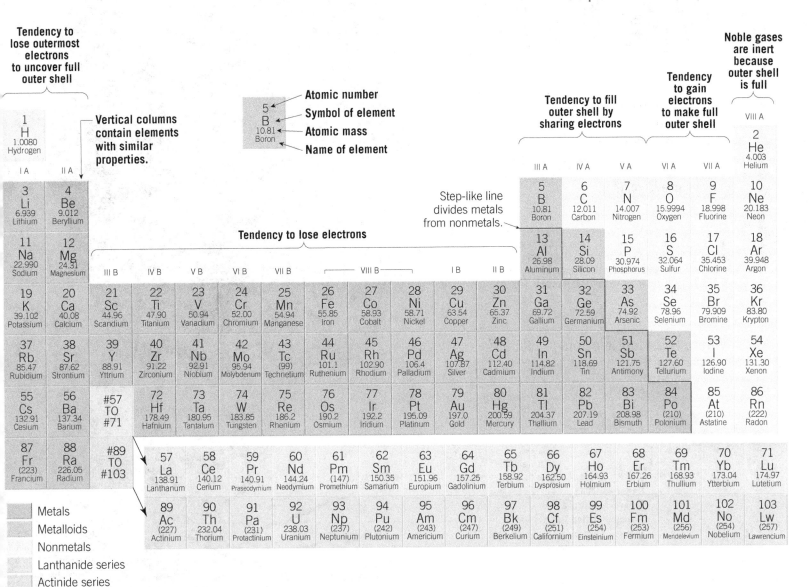

Tendency to lose outermost electrons to uncover full outer shell

Vertical columns contain elements with similar properties.

5 / B / 10.81 / Boron: Atomic number; Symbol of element; Atomic mass; Name of element

Tendency to lose electrons

Step-like line divides metals from nonmetals.

Tendency to fill outer shell by sharing electrons

Tendency to gain electrons to make full outer shell

Noble gases are inert because outer shell is full

I A	II A	III B	IV B	V B	VI B	VII B	VIII B	VIII B	VIII B	I B	II B	III A	IV A	V A	VI A	VII A	VIII A
1 H 1.0080 Hydrogen																	2 He 4.003 Helium
3 Li 6.939 Lithium	4 Be 9.012 Beryllium											5 B 10.81 Boron	6 C 12.011 Carbon	7 N 14.007 Nitrogen	8 O 15.9994 Oxygen	9 F 18.998 Fluorine	10 Ne 20.183 Neon
11 Na 22.990 Sodium	12 Mg 24.31 Magnesium											13 Al 26.98 Aluminum	14 Si 28.09 Silicon	15 P 30.974 Phosphorus	16 S 32.064 Sulfur	17 Cl 35.453 Chlorine	18 Ar 39.948 Argon
19 K 39.102 Potassium	20 Ca 40.08 Calcium	21 Sc 44.96 Scandium	22 Ti 47.90 Titanium	23 V 50.94 Vanadium	24 Cr 52.00 Chromium	25 Mn 54.94 Manganese	26 Fe 55.85 Iron	27 Co 58.93 Cobalt	28 Ni 58.71 Nickel	29 Cu 63.54 Copper	30 Zn 65.37 Zinc	31 Ga 69.72 Gallium	32 Ge 72.59 Germanium	33 As 74.92 Arsenic	34 Se 78.96 Selenium	35 Br 79.909 Bromine	36 Kr 83.80 Krypton
37 Rb 85.47 Rubidium	38 Sr 87.62 Strontium	39 Y 88.91 Yttrium	40 Zr 91.22 Zirconium	41 Nb 92.91 Niobium	42 Mo 95.94 Molybdenum	43 Tc (99) Technetium	44 Ru 101.1 Ruthenium	45 Rh 102.90 Rhodium	46 Pd 106.4 Palladium	47 Ag 107.87 Silver	48 Cd 112.40 Cadmium	49 In 114.82 Indium	50 Sn 118.69 Tin	51 Sb 121.75 Antimony	52 Te 127.60 Tellurium	53 I 126.90 Iodine	54 Xe 131.30 Xenon
55 Cs 132.91 Cesium	56 Ba 137.34 Barium	#57 TO #71	72 Hf 178.49 Hafnium	73 Ta 180.95 Tantalum	74 W 183.85 Tungsten	75 Re 186.2 Rhenium	76 Os 190.2 Osmium	77 Ir 192.2 Iridium	78 Pt 195.09 Platinum	79 Au 197.0 Gold	80 Hg 200.59 Mercury	81 Tl 204.37 Thallium	82 Pb 207.19 Lead	83 Bi 208.98 Bismuth	84 Po (210) Polonium	85 At (210) Astatine	86 Rn (222) Radon
87 Fr (223) Francium	88 Ra 226.05 Radium	#89 TO #103															

57 La 138.91 Lanthanum	58 Ce 140.12 Cerium	59 Pr 140.91 Praseodymium	60 Nd 144.24 Neodymium	61 Pm (147) Promethium	62 Sm 150.35 Samarium	63 Eu 151.96 Europium	64 Gd 157.25 Gadolinium	65 Tb 158.92 Terbium	66 Dy 162.50 Dysprosium	67 Ho 164.93 Holmium	68 Er 167.26 Erbium	69 Tm 168.93 Thullium	70 Yb 173.04 Ytterbium	71 Lu 174.97 Lutetium
89 Ac (227) Actinium	90 Th 232.04 Thorium	91 Pa (231) Protactinium	92 U 238.03 Uranium	93 Np (237) Neptunium	94 Pu (242) Plutonium	95 Am (243) Americium	96 Cm (247) Curium	97 Bk (249) Berkelium	98 Cf (251) Californium	99 Es (254) Einsteinium	100 Fm (253) Fermium	101 Md (256) Mendelevium	102 No (254) Nobelium	103 Lw (257) Lawrencium

Metals
Metalloids
Nonmetals
Lanthanide series
Actinide series

▲ **Figure 2.5 Periodic table of the elements**

▼ **Figure 2.6 Examples of minerals composed of a single element** (Photos by Dennis Tasa)

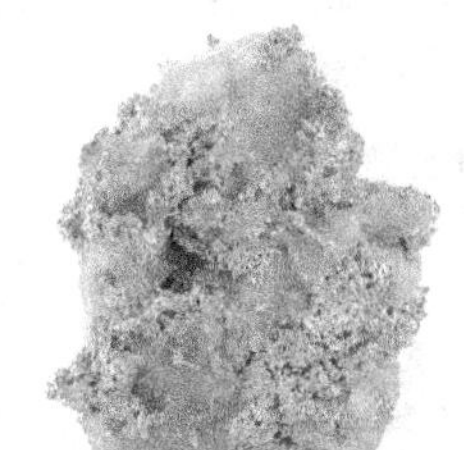

A. Gold on quartz

B. Sulfur

C. Copper

Atoms of the naturally occurring elements are the basic building blocks of Earth's minerals. Most elements join with other elements to form **chemical compounds**. Therefore, most minerals are chemical compounds composed of atoms of two or more elements. These include the minerals quartz (SiO_2), halite (NaCl), and calcite ($CaCO_3$). However, a few minerals, such as diamonds, sulfur, and native gold and copper, are made entirely of atoms of only one element (Figure 2.6). (A metal is called "native" when it is found in its pure form in nature; see GEOgraphics 2.1.)

CONCEPT CHECKS 2.2

1. Make a simple sketch of an atom and label its three main particles. Explain how these particles differ from one another.
2. What is the significance of valence electrons?

GEOGRAPHICS 2.1

Gold

Gold has been treasured since long before recorded history for its beauty.

How valuable is gold?

$41,280

In early 2015, the value of one troy ounce of gold was about US$1,290. Based on that value, a 1000-gram (32-ounce) bar of gold, like the one shown, was worth $41,280. In 1970, the price of gold was less than $40 per troy ounce!

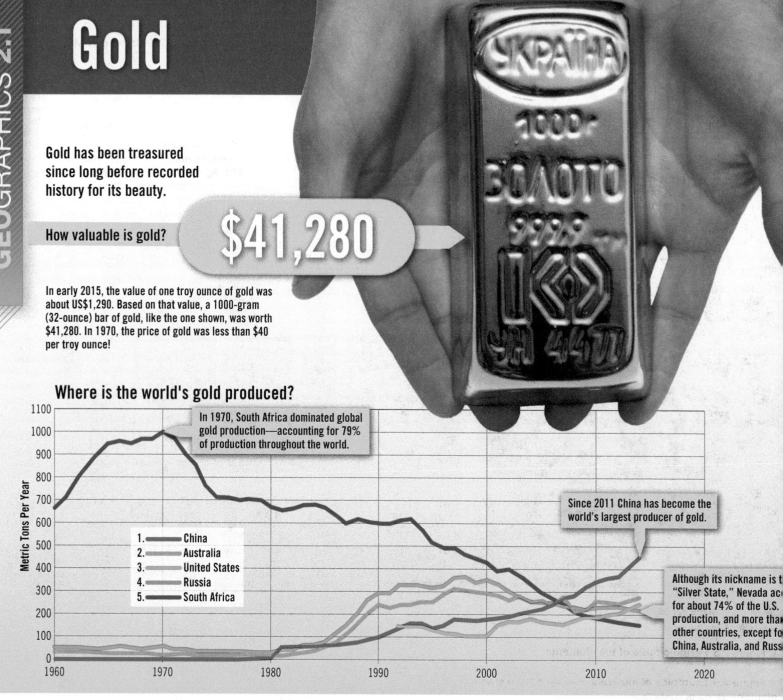

УКРАЇНА
1000г
ЗОЛОТО
999,9

Where is the world's gold produced?

Metric Tons Per Year

1100
1000
900
800
700
600
500
400
300
200
100
0

1960 1970 1980 1990 2000 2010 2020

1. China
2. Australia
3. United States
4. Russia
5. South Africa

In 1970, South Africa dominated global gold production—accounting for 79% of production throughout the world.

Since 2011 China has become the world's largest producer of gold.

Although its nickname is t
"Silver State," Nevada ac
for about 74% of the U.S.
production, and more tha
other countries, except fo
China, Australia, and Russ

(Photo by D7INAMI7S/Shutterstock)

2.3 Why Atoms Bond

Distinguish among ionic bonds, covalent bonds, and metallic bonds.

Under the temperature and pressure conditions found on Earth, most elements do not occur in the form of individual atoms; instead, their atoms bond with other atoms. (A group of elements known as the noble gases are an exception.) Some atoms bond to form *ionic compounds*, some form *molecules*, and still others form *metallic substances*. Why does this happen? Experiments show that electrical forces hold atoms together and bond them to each other. These electrical attractions lower the total energy of the bonded atoms, and this, in turn, generally makes them more stable. Consequently, atoms that are bonded in compounds tend to be more stable than atoms that are free (not bonded).

The Octet Rule and Chemical Bonds

As noted earlier, valence (outer-shell) electrons are generally involved in chemical bonding. Figure 2.7 shows a shorthand way of representing the number of valence electrons for some selected elements. Notice that the elements in Group I have one valence electron each, those in Group II have two valence electrons each, and so on, up to eight valence electron in Group VIII.

The noble gases have very stable electron arrangements with eight valence electrons (except helium, which has two) and, therefore, tend to lack chemical reactivity. Many other

ative gold
ecause gold does not easily react with other elements, often occurs as a native element in nuggets found in ream deposits or as grains in igneous rocks.
Shutterstock
utterstock
What are the uses of gold?
About 50% of gold is used in jewelry. Another 40% is used for currency and investment, and about 10% is used in industry, including electronic devices such as cell phones and televisions. Gold is also used in gourmet foods and cocktails as a decorative ingredient. Because metallic gold is one of the least reactive materials, it has no taste, provides no nutritional value, and leaves the human body unaltered.
Shutterstock
Questions:
1. What is the chemical symbol for gold?
2. What is the term for the property of tenacity, which allows gold to be easily hammered into different shapes?
?
Shutterstock

Electron Dot Diagrams for Some Representative Elements

I	II	III	IV	V	VI	VII	VIII
H•							He:
Li•	•Be•	•B•	•C•	•N•	:O•	:F•	:Ne:
Na•	•Mg•	•Al•	•Si•	•P•	:S•	:Cl•	:Ar:
K•	•Ca•	•Ga•	•Ge•	•As•	:Se•	:Br•	:Kr:

▲ **Figure 2.7 Dot diagrams for certain elements** Each dot represents a valence electron found in the outermost principal shell.

atoms gain, lose, or share electrons during chemical reactions, ending up with electron arrangements of the noble gases. This observation led to a chemical guideline known as the **octet rule**: *Atoms tend to gain, lose, or share electrons until they are surrounded by eight valence electrons.* Although there are exceptions to the octet rule, it is a useful *rule of thumb* for understanding chemical bonding.

When an atom's outer shell does not contain eight electrons, it is likely to chemically bond to other atoms to achieve an octet in its outer shell. A **chemical bond** is a transfer or sharing of electrons that allows each atom to attain a full valence shell of electrons. Some atoms do this by transferring all their valence electrons to other atoms so that an inner shell becomes the full valence shell.

When the valence electrons are transferred between the elements to form ions, the bond is an *ionic bond*. When the electrons are shared between the atoms, the bond is a *covalent bond*. When the valence electrons are shared among all the atoms in a substance, the bonding is *metallic*.

Ionic Bonds: Electrons Transferred

Perhaps the easiest type of bond to visualize is the **ionic bond**, in which one atom gives up one or more valence electrons to another atom to form **ions**—*positively and negatively charged atoms*. The atom that loses electrons becomes a positive ion, and the atom that gains electrons becomes a negative ion. Oppositely charged ions are strongly attracted to one another and join to form *ionic compounds*.

Consider the ionic bonding that occurs between sodium (Na) and chlorine (Cl) to produce the solid ionic compound sodium chloride—the mineral halite (common table salt). Notice in Figure 2.8A that a sodium atom gives up its single valence electron to chlorine and, as a result, becomes a positively charged sodium ion (Na^+). Chlorine, on the other hand, gains one electron and becomes a negatively charged chloride ion (Cl^-). We know that ions having unlike charges attract. Thus, an ionic bond is an attraction of oppositely charged ions to one another that produces an electrically neutral ionic compound.

Figure 2.8B illustrates the arrangement of sodium and chlorine ions in ordinary table salt. Notice that salt consists of alternating sodium and chlorine ions, positioned so that each positive ion is attracted to and surrounded on all sides by negative ions and vice versa. This arrangement maximizes the attraction between ions with opposite charges while minimizing the repulsion between ions with identical charges. Thus, ionic compounds consist of an orderly arrangement of oppositely charged ions assembled in a definite ratio that provides overall electrical neutrality.

The properties of a chemical compound are dramatically different from the properties of the various elements comprising it. For example, sodium is a soft silvery metal that is extremely reactive and poisonous. If you were to consume even a small amount of elemental sodium, you would need immediate medical attention. Chlorine, a green poisonous gas, is so toxic that it was used as a chemical weapon during World War I. Together, however, these elements produce sodium chloride, the edible flavor enhancer that we call table salt. Thus, when elements combine to form compounds, their properties change significantly.

Covalent Bonds: Electron Sharing

Sometimes the forces that hold atoms together cannot be understood on the basis of the attraction of oppositely charged ions. One example is the hydrogen molecule (H_2), in which the two hydrogen atoms are held together tightly and no ions are present. The strong attractive force that holds two hydrogen atoms together results from a **covalent bond**, a chemical bond formed by the *sharing* of one or more valance electrons between a pair of atoms. (Hydrogen is one of the exceptions to the octet rule: Its single shell is full with just two electrons.)

Imagine two hydrogen atoms (each with one proton and one electron) approaching one another, as shown in Figure 2.9. Once they meet, the electron configuration changes so that both electrons primarily occupy the space between the atoms. In other words, the two electrons are shared by both hydrogen atoms and are attracted simultaneously by the positive charge of the proton in the

Figure 2.8 Formation of the ionic compound sodium chloride

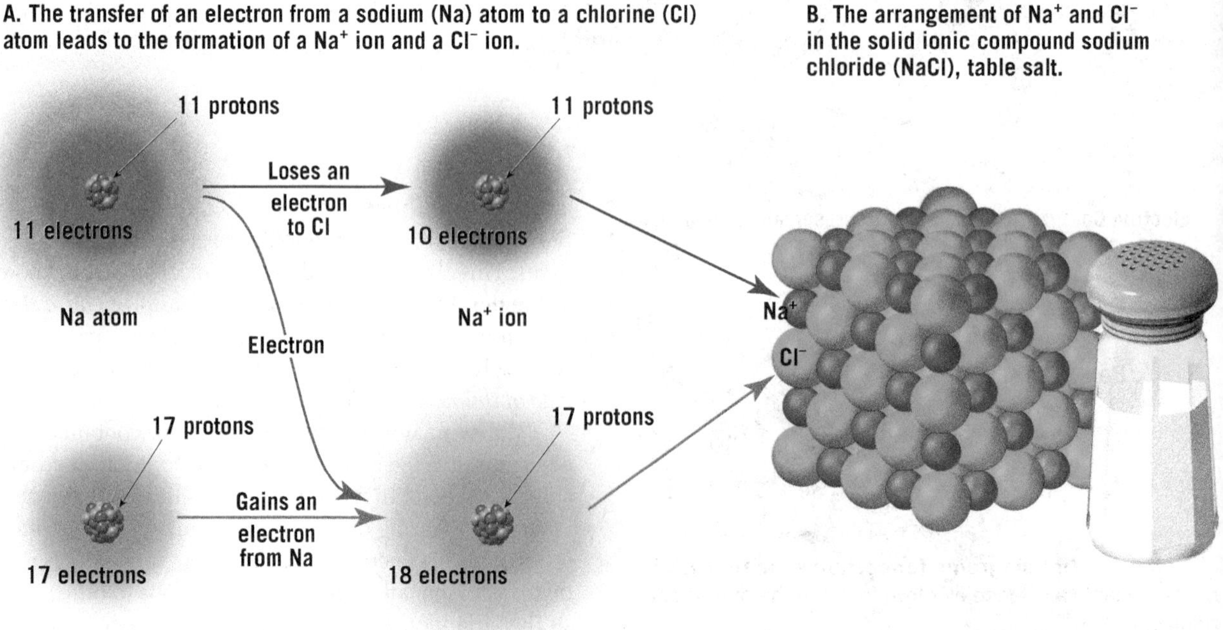
A. The transfer of an electron from a sodium (Na) atom to a chlorine (Cl) atom leads to the formation of a Na^+ ion and a Cl^- ion.
11 protons
11 electrons
Na atom
Loses an electron to Cl
11 protons
10 electrons
Na^+ ion
Electron
17 protons
17 electrons
Cl atom
Gains an electron from Na
17 protons
18 electrons
Cl^- ion
B. The arrangement of Na^+ and Cl^- in the solid ionic compound sodium chloride (NaCl), table salt.
Na^+
Cl^-

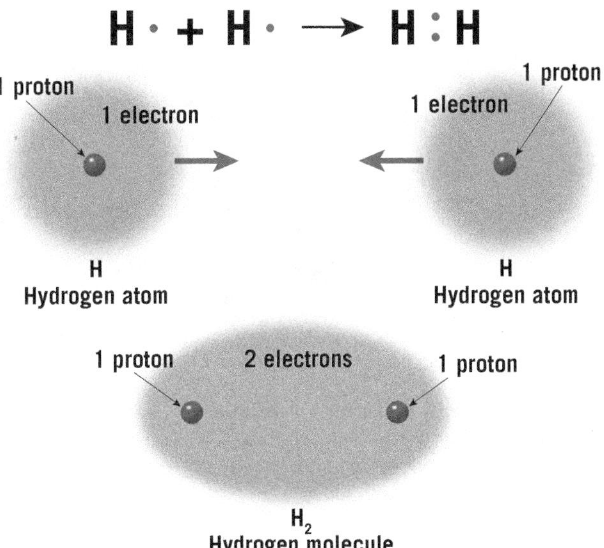

Figure 2.9 **Formation of a covalent bond** When hydrogen atoms bond, the negatively charged electrons are shared by both hydrogen atoms and attracted simultaneously by the positive charge of the proton in the nucleus of each atom.

nucleus of each atom. In this situation the hydrogen atoms do not form ions; instead, the force that holds these atoms together arises from the attraction of oppositely charged particles—positively charged protons in the nuclei and negatively charged electrons that surround these nuclei.

Metallic Bonds: Electrons Free to Move

A few minerals, such as native gold, silver, and copper, are made entirely of metal atoms packed tightly together in an orderly way. The bonding that holds these atoms together results from each atom contributing its valence electrons to a common pool of electrons, which freely move throughout the entire metallic structure. The contribution of one or more valence electrons leaves an array of positive ions immersed in a "sea" of valence electrons, as shown in Figure 2.10.

The attraction between this sea of negatively charged electrons and the positive ions produces the **metallic bonds** that give metals their unique properties. Metals are good conductors of electricity because the valence electrons are free to move from one atom to another. Metals are also *malleable*, which means they can be hammered into thin sheets, and *ductile*, which means they can be drawn into thin wires. By contrast, ionic and covalent solids tend to be *brittle* and fracture when stress is applied. Consider the difference between dropping a metal frying pan and a ceramic plate onto a concrete floor.

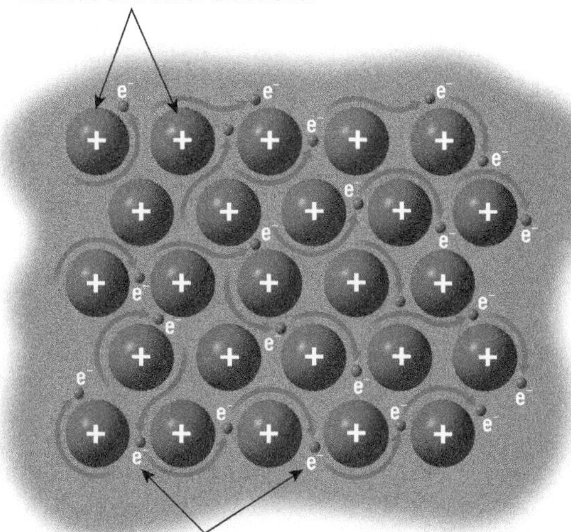

The central core of each metallic atom, which has an overall positive charge, consists of the nucleus and inner electrons.
A "sea" of negatively charged outer electrons, that are free to move throughout the structure, surrounds the positive ions.

Figure 2.10 **Metallic bonding** Metallic bonding is the result of each atom contributing its valence electrons to a common pool of electrons that are free to move throughout the entire metallic structure. The attraction between the "sea" of negatively charged electrons and the positive ions produces the metallic bonds that give metals their unique properties.

CONCEPT CHECKS 2.3

1. What is the difference between an atom and an ion?
2. How does an atom become a positive ion? A negative ion?
3. Briefly distinguish between ionic, covalent, and metallic bonding and discuss the role that electrons play in each.

EYE ON EARTH 2.1

The accompanying image shows one of the world's largest open-pit gold mine, located near Kalgoorlie, Australia. Known as the Super Pit, it originally consisted of a number of small underground mines that were consolidated into a single, open-pit mine. Each year, about 28 metric tons of gold are extracted from the 15 million metric tons of rock shattered by blasting and then transported to the surface.

QUESTION 1 *What is one environmental advantage of underground mining over open-pit mining?*

QUESTION 2 *For those employed at this mine, what change in working conditions would have occurred as it evolved from an underground mine to an open-pit mine?*

(Photo by McPhoto/Blickwinkel/AGE Fotostock)

2.4 Properties of Minerals

List and describe the properties used in mineral identification.

Minerals have definite crystalline structures and chemical compositions that give them unique sets of physical and chemical properties shared by all specimens of that mineral, regardless of when or where they formed. For example, two samples of the mineral quartz will be equally hard and equally dense, and they will break in a similar manner. However, the physical properties of individual samples may vary within specific limits due to ionic substitutions, inclusions of foreign elements (impurities), and defects in the crystalline structure.

Some mineral properties, called **diagnostic properties**, are particularly useful in identifying an unknown mineral. The mineral halite, for example has a salty taste. Because so few minerals share this property, a salty taste is considered a diagnostic property of halite. Other properties of certain minerals, particularly color, vary among different specimens of the same mineral. These properties are referred to as **ambiguous properties**.

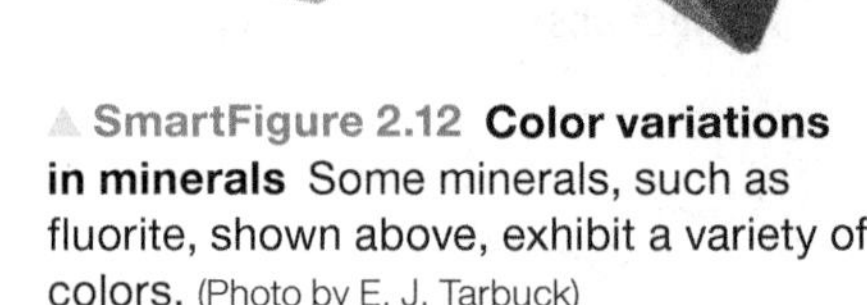

SmartFigure 2.12 **Color variations in minerals** Some minerals, such as fluorite, shown above, exhibit a variety of colors. (Photo by E. J. Tarbuck)

TUTORIAL
https://goo.gl/LWOf

Optical Properties

Of the many diagnostic properties of minerals, their optical characteristics such as luster, color, streak, and ability to transmit light are most frequently used for mineral identification.

A. This freshly broken sample of galena displays a metallic luster.

B. This sample of galena is tarnished and has a submetallic luster.

Figure 2.11 **Metallic versus submetallic luster** (Photos courtesy of E. J. Tarbuck)

Luster The appearance or quality of light reflected from the surface of a mineral is known as **luster**. Minerals that are shiny like a metal, regardless of color, are said to have a *metallic luster* (Figure 2.11A). Some metallic minerals, such as native copper and galena, develop a dull coating or tarnish when exposed to the atmosphere. Because they are not as shiny as samples with freshly broken surfaces, these samples are often said to exhibit a *submetallic luster* (Figure 2.11B).

Most minerals have a *nonmetallic luster* and are described using various adjectives. For example, some minerals are described as being *vitreous*, or *glassy*. Other nonmetallic minerals are described as having a *dull*, or *earthy*, *luster* (a dull appearance like soil) or a *pearly luster* (such as a pearl or the inside of a clamshell). Still others exhibit a *silky luster* (like satin cloth) or a *greasy luster* (as though coated in oil).

Color Although **color** is generally the most conspicuous characteristic of any mineral, it is considered a diagnostic property of only a few minerals. Slight impurities in the common minerals fluorite and quartz, for example, give them a variety of tints, including pink, purple, yellow, white, gray, and even black (Figure 2.12). Other minerals, such as tourmaline, also exhibit a variety of hues, with multiple colors sometimes occurring in the same sample. Thus, the use of color as a means of identification is often ambiguous or even misleading.

Streak The color of a mineral in powdered form, called **streak**, is often useful in identification. A mineral's streak is obtained by rubbing it across a *streak plate* (a piece of unglazed porcelain) and observing the color of the mark it leaves (Figure 2.13). Although a mineral's color may vary

Although the color of a mineral is not always helpful in identification, the streak, which is the color of the powdered mineral, can be very useful.

SmartFigure 2.13 **Streak** (Photo by Dennis Tasa)

VIDEO
https://goo.gl/MdH5j9

from sample to sample, its streak is usually consistent in color. (Note that not all minerals produce a streak when rubbed across a streak plate. Quartz, for example, is harder than a porcelain streak plate and therefore leaves no streak.)

Streak can also help distinguish between minerals with metallic luster and those with nonmetallic luster. Metallic minerals generally have a dense, dark streak, whereas minerals with nonmetallic luster typically have a light-colored streak.

Ability to Transmit Light Another optical property used to identify minerals is the ability to transmit light. When no light is transmitted through a mineral sample, that mineral is described as *opaque*; when light, but not an image, is transmitted, the mineral is said to be *translucent*. When both light and an image are visible through the sample, the mineral is described as *transparent*.

Crystal Shape, or Habit

Mineralogists use the term **crystal shape,** or **habit**, to refer to the common or characteristic shape of individual crystals or aggregates of crystals. Some minerals tend to grow equally in all three dimensions, whereas others tend to be elongated in one direction or flattened if growth in one dimension is suppressed. The crystals of a few minerals can have a regular polygonal shape that is helpful in identification. For example, magnetite crystals sometimes occur as octahedrons, garnets often form dodecahedrons, and halite and fluorite crystals tend to grow as cubes or near-cubes. Most minerals have just one common crystal shape, but a few, such as the pyrite samples shown in Figure 2.14, have two or more characteristic crystal shapes.

In addition, some mineral samples consist of numerous intergrown crystals exhibiting characteristic shapes that are useful for identification. Terms commonly used to describe these and other crystal habits include *equant* (equidimensional), *bladed*, *fibrous*, *tabular*, *cubic*, *prismatic*, *platy*, *blocky*, and *banded*. Some of these habits are pictured in Figure 2.15.

Figure 2.14 **Common crystal shapes of pyrite**

A. Fibrous

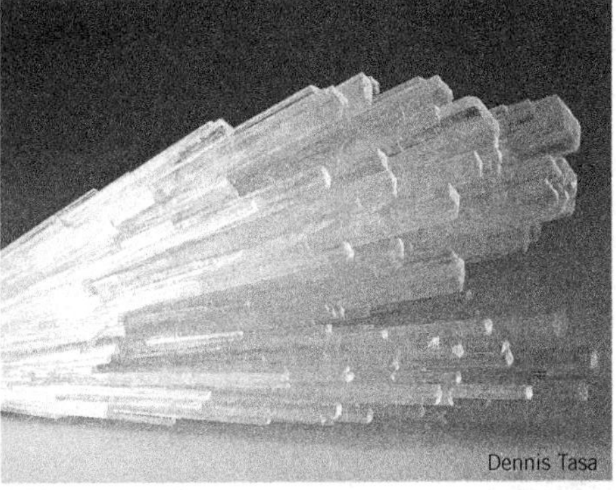

B. Bladed

C. Banded

D. Cubic crystals

SmartFigure 2.15 **Some common crystal habits A.** Thin, rounded crystals that break into fibers. **B.** Elongated crystals that are flattened in one direction. **C.** Minerals that have stripes or bands of different color or texture. **D.** Groups of crystals that are cube shaped.

TUTORIAL
https://goo.gl/vaVDiS

Mineral Strength

How easily minerals break or deform under stress is determined by the type and strength of the chemical bonds that hold the crystals together. Mineralogists use terms including *hardness*, *cleavage*, *fracture*, and *tenacity* to describe mineral strength and how minerals break when stress is applied.

Hardness One of the most useful diagnostic properties is **hardness**, a measure of the resistance of a mineral to abrasion or scratching. This property is determined by rubbing a mineral of unknown hardness against one of known hardness or vice versa. A numerical value of hardness can be obtained by using the **Mohs scale** of hardness, which consists of 10 minerals arranged in order from 1 (softest) to 10 (hardest), as shown in Figure 2.16A. It should be noted that the Mohs scale is a relative ranking and does not imply that a mineral with a hardness of 2, such as gypsum, is twice as hard as mineral with a hardness of 1, like talc. In fact, gypsum is only slightly harder than talc, as Figure 2.16B indicates.

In the laboratory, common objects used to determine the hardness of a mineral can include a human fingernail, which has a hardness of about 2.5, a copper penny (3.5), and a piece of glass (5.5). The mineral gypsum, which has

SmartFigure 2.16
Hardness scales A. The Mohs scale of hardness, with the hardnesses of some common objects. **B.** Relationship between the Mohs relative hardness scale and an absolute hardness scale.

TUTORIAL
https://goo.gl/dvaqIF

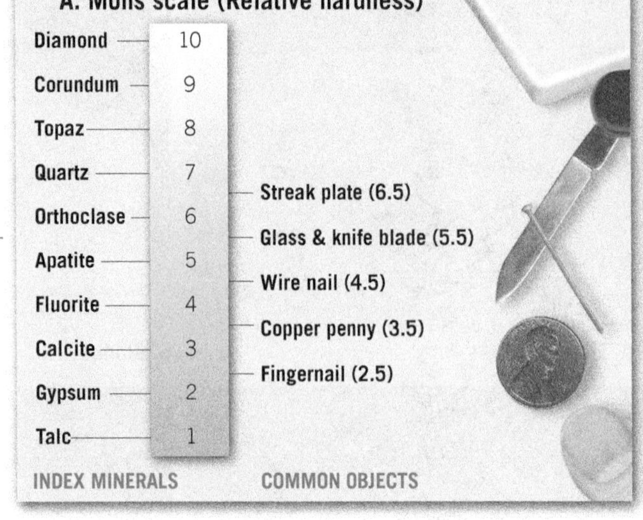

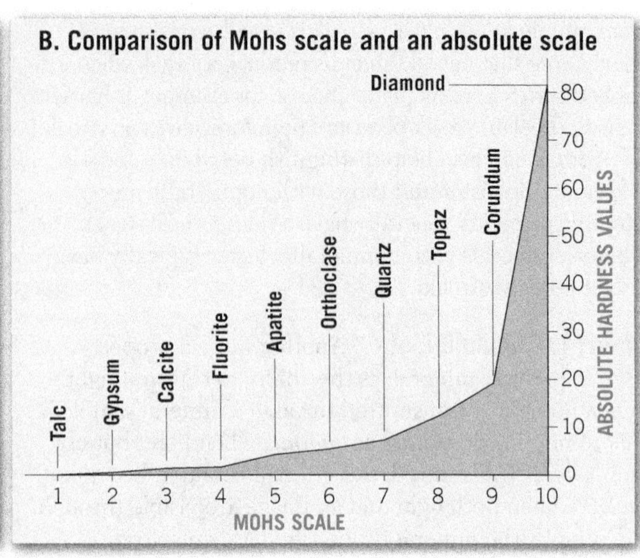

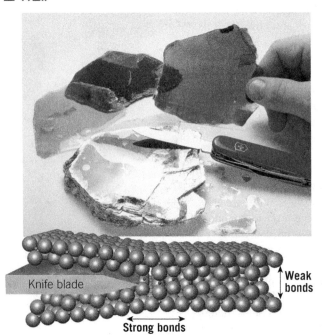

a hardness of 2, can be easily scratched with a fingernail. On the other hand, the mineral calcite, which has a hardness of 3, will scratch a fingernail but will not scratch glass. Quartz, one of the hardest common minerals, will easily scratch glass. Diamonds, hardest of all, scratch anything, including other diamonds.

SmartFigure 2.17
Micas exhibit perfect cleavage The thin sheets shown here exhibit one plane of cleavage. (Photo by Chip Clark/Fundamental Photographs)

ANIMATION
https://goo.gl/LvL2OD

Knife blade
Weak bonds
Strong bonds

Cleavage In the crystal structure of many minerals, some atomic bonds are weaker than others. It is along these weak bonds that minerals tend to break when they are stressed. **Cleavage** (*kleiben* = carve) is the tendency of a mineral to break (cleave) along planes of weak bonding. Not all minerals have cleavage, but those that do can be identified by the relatively smooth, flat surfaces that are produced when the mineral is broken.

The simplest type of cleavage is exhibited by the micas (Figure 2.17). Because these minerals have very weak bonds in one direction, they cleave to form thin, flat sheets. Some minerals have excellent cleavage in one, two, three, or more directions, whereas others exhibit fair or poor cleavage, and still others have no cleavage at all. When minerals break evenly in more than one direction, cleavage is described by *the number of cleavage directions and the angle(s) at which they meet* (Figure 2.18).

Each cleavage surface that has a different orientation is counted as a different direction of cleavage. For example, some minerals, such as halite, cleave to form six-sided cubes. Because a cube is defined by three different sets of parallel planes that intersect at 90-degree angles, cleavage for the mineral halite is described as *three directions of cleavage that meet at 90 degrees*.

Do not confuse cleavage with crystal shape. When a mineral exhibits cleavage, it breaks into pieces that all have the same geometry. By contrast, the smooth-sided quartz crystals shown in Figure 2.1 do not have cleavage. If broken, they fracture into shapes that do not resemble one another or the original crystals.

Fracture Minerals having chemical bonds that are equally, or nearly equally, strong in all directions exhibit a property called **fracture** (Figure 2.19A). When minerals fracture, most produce uneven surfaces and are described as exhibiting *irregular fracture*. However, some minerals, including quartz, sometimes break into smooth, curved surfaces resembling broken glass. Such breaks are called *conchoidal fractures* (Figure 2.19B). Still other minerals exhibit fractures that produce splinters or fibers referred to as *splintery fracture* and *fibrous fracture*, respectively.

Tenacity The term **tenacity** describes how a mineral responds to stress—for instance, whether it tends to break in a brittle fashion or bend elastically. As mentioned earlier, nonmetallic minerals such as quartz and minerals that are ionically bonded, such as fluorite and halite, tend to be *brittle* and fracture or exhibit cleavage when struck. By contrast, native metals, such as copper and gold, are *malleable*, which means they can be hammered without breaking. In addition, minerals that can be cut into thin shavings, including gypsum and talc, are described as *sectile*. Still others, notably the micas, are *elastic* and bend and snap back to their original shape after stress is released.

A. Cleavage in one direction. Example: Muscovite

B. Cleavage in two directions at 90° angles. Example: Feldspar

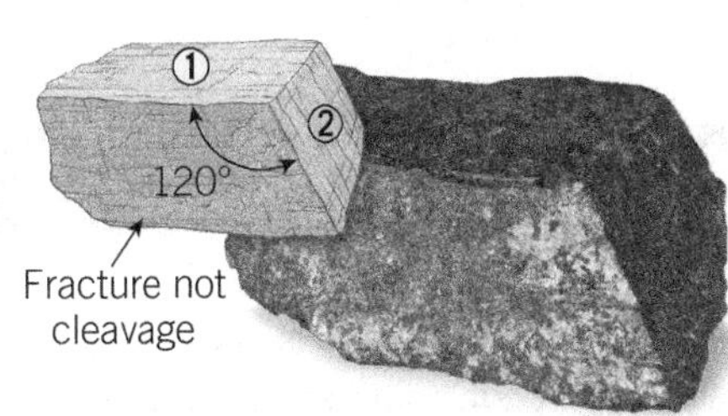

C. Cleavage in two directions not at 90° angles. Example: Hornblende

SmartFigure 2.18 Cleavage directions exhibited by minerals (Photos by E. J. Tarbuck and Dennis Tasa)

TUTORIAL
https://goo.gl/MN1wy7

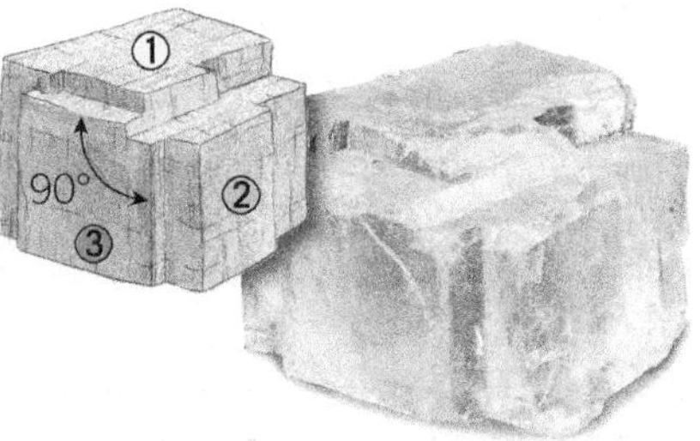

D. Cleavage in three directions at 90° angles. Example: Halite

E. Cleavage in three directions not at 90° angles. Example: Calcite

F. Cleavage in four directions. Example: Fluorite

Density and Specific Gravity

Density, an important property of matter, is defined as mass per unit volume. Mineralogists often use a related measure called **specific gravity** to describe the density of minerals. Specific gravity is a number representing the ratio of a mineral's weight to the weight of an equal volume of water.

Most common minerals have a specific gravity between 2 and 3. For example, quartz has a specific gravity of 2.65. By contrast, some metallic minerals, such as pyrite, native copper, and magnetite, are more than twice as dense and thus have more than twice the specific gravity of quartz. Galena, an ore from which lead is extracted, has a specific gravity of roughly 7.5, whereas 24-karat gold has a specific gravity of approximately 20.

With a little practice, you can estimate the specific gravity of a mineral by hefting it in your hand. Does this mineral feel about as "heavy" as similarly sized rocks you have handled? If the answer is "yes," the specific gravity of the sample will likely be between 2.5 and 3.

Other Properties of Minerals

In addition to the properties discussed thus far, some minerals can be recognized by other distinctive properties. For example, halite is ordinary salt, so it can be quickly identified through taste. Talc and graphite both have distinctive feels: Talc feels soapy, and graphite feels greasy. Further, the streaks of many sulfur-bearing minerals smell like rotten eggs. A few minerals, such as magnetite, have high iron content and can be picked up with a magnet, while some varieties (such as lodestone) are themselves natural magnets and will pick up small iron-based objects such as pins and paper clips (see Figure 2.32F, page 50).

Moreover, some minerals exhibit special optical properties. For example, when a transparent piece of calcite is placed over printed text, the letters appear twice. This optical property is known as *double refraction* (**Figure 2.20**).

One very simple chemical test to detect carbonate minerals involves placing a drop of dilute hydrochloric

A. Irregular fracture (Quartz)

B. Conchoidal fracture (Quartz)

Figure 2.19 Irregular versus conchoidal fracture (Photos by E. J. Tarbuck)

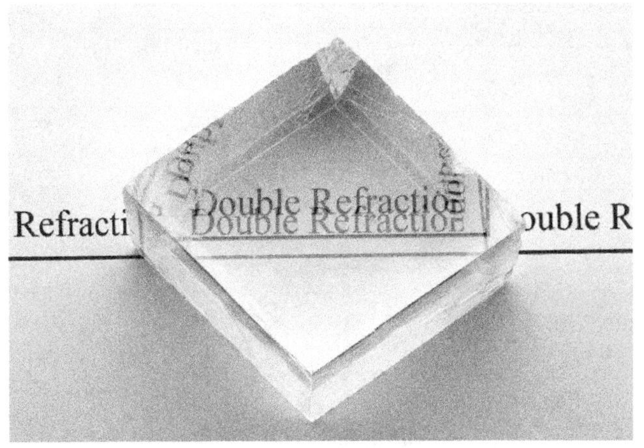

Figure 2.20 Double refraction This sample of calcite exhibits double refraction. (Photo by Chip Clark/Fundamental Photographs)

Glass bottles, like most other manufactured products, contain substances obtained from minerals extracted from Earth's crust and oceans. The primary ingredient in commercially produced glass bottles is the mineral quartz. Glass also contains lesser amounts of the mineral calcite. (Photo by Chris Brignell/Shutterstock)

QUESTION 1 *In what mineral group does quartz belong?*

QUESTION 2 *Glass beer bottles are usually clear, green, or brown. Based on what you know about how the mineral quartz is colored, what do glass manufacturers do to make bottles green and brown?*

QUESTION 3 *Why did some brewers color their glass bottles, rather than use clear glass. (Hint:Search the Internet.)*

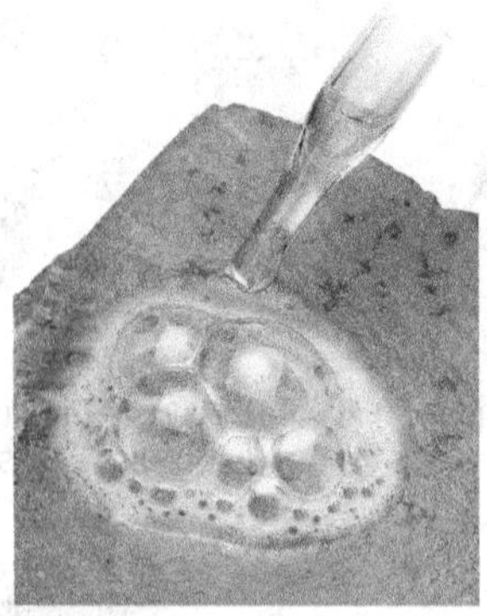

SmartFigure 2.21 **Calcite reacting with a weak acid** (Photo by Chip Clark/ Fundamental Photographs)

VIDEO
https://goo.gl/5L3gns

acid from a dropper bottle onto a freshly broken mineral surface. Samples containing carbonate minerals will effervesce (fizz) as carbon dioxide gas is released (Figure 2.21). This test is especially useful in identifying calcite, a common carbonate mineral.

CONCEPT CHECKS 2.4

1. Define *luster*.
2. Why is color not always a useful property in mineral identification? Give an example of a mineral that supports your answer.
3. What differentiates cleavage from fracture?
4. What is meant by a mineral's *tenacity*? List three terms that describe tenacity.
5. Describe a simple chemical test useful in identifying the mineral calcite.

2.5 Mineral Groups

List the common silicate and nonsilicate minerals and describe what characterizes each group.

More than 4000 minerals have been named, and several new ones are identified each year. Fortunately for students who are beginning to study minerals, no more than a few dozen are abundant. Collectively, these few make up most of the rocks of Earth's crust and are therefore generally known as the **rock-forming minerals**.

Although less abundant, many other minerals are used extensively in the manufacture of products; these are called **economic minerals**. However, rock-forming minerals and economic minerals are not mutually exclusive groups. When found in large deposits, some rock-forming minerals are economically significant. One example is calcite, a mineral that is the primary component of the sedimentary rock limestone. Among calcite's many uses is cement production.

It is worth noting that *only eight elements* make up the vast majority of the rock-forming minerals and represent more than 98 percent (by weight) of the continental crust (Figure 2.22). These elements, in order of most to least abundant, are oxygen (O), silicon (Si), aluminum (Al), iron (Fe), calcium (Ca), sodium (Na), potassium (K), and magnesium (Mg). As shown in Figure 2.22, oxygen and silicon are by far the most common elements in Earth's crust. Furthermore, these two elements readily combine to form the basic "building block" for the most common mineral group, the **silicates**. More than 800 silicate minerals are known, and they account for more than 90 percent of Earth's crust.

Because other mineral groups are far less abundant in Earth's crust than the silicates, they are often grouped together in the category **nonsilicates**. Although not as common as silicates, some nonsilicate minerals are very important economically. They provide us with iron and aluminum to build automobiles, gypsum for plaster and drywall for home construction, and copper wire that carries electricity and connects us to the Internet. In addition to their economic importance, these groups include minerals that are major constituents in sediments and sedimentary rocks.

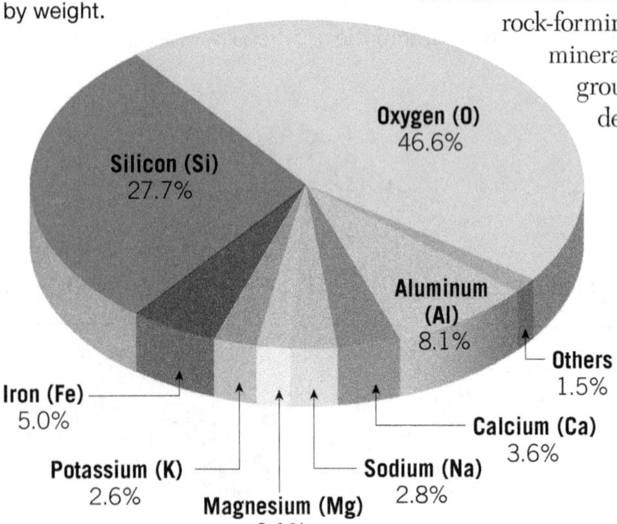

Figure 2.22 **The eight most abundant elements in Earth's continental crust** The numbers represent percentages by weight.

Silicate Minerals

Every silicate mineral contains oxygen and silicon atoms. Except for a few silicate minerals such as quartz, most silicate minerals also contain one or more additional elements in their crystalline structure. These elements give rise to the great variety of silicate minerals and their varied properties.

All silicates have the same fundamental building block, the **silicon–oxygen tetrahedron** (SiO_4^{4-}). This structure consists of four oxygen ions that are covalently bonded to a comparatively smaller silicon ion, forming a tetrahedron—a pyramid shape with four identical faces (Figure 2.23). In some minerals, the tetrahedra are joined into chains, sheets, or three-dimensional networks by sharing oxygen atoms (Figure 2.24). These larger silicate structures are then connected to one another by other elements. The primary elements that join silicate structures are iron (Fe), magnesium (Mg), potassium (K), sodium (Na), and calcium (Ca).

Major groups of silicate minerals and common examples are given in Figure 2.24. The *feldspars* are by far the most plentiful group, comprising about 51 percent of Earth's crust. *Quartz*, the second-most-abundant mineral in the continental crust, is the only common mineral made completely of silicon and oxygen.

Notice in Figure 2.24 that each mineral *group* has a particular silicate *structure*. A relationship exists between this internal structure of a mineral and the *cleavage* it exhibits. Because the silicon–oxygen bonds are strong, silicate minerals tend to cleave between the silicon–oxygen structures rather than across them. For example, the micas have a sheet structure and thus tend to cleave into flat plates (see the muscovite in Figure 2.17). Quartz has equally strong silicon–oxygen bonds in all directions; therefore, it has no cleavage but fractures instead.

How do silicate minerals form? Most of them crystallize from molten rock as it cools. This cooling can occur at or near Earth's surface (low temperature and pressure) or at great depths (high temperature and pressure). The *environment* during crystallization and the *chemical composition of the molten rock* mainly determine which minerals are produced. For example, the silicate mineral olivine crystallizes at high temperatures (about 1200°C [2200°F]), whereas quartz crystallizes at much lower temperatures (about 700°C [1300°F]).

In addition, some silicate minerals form at Earth's surface from the weathered (disintegrated) products of other silicate minerals. Clay minerals are an example. Still other silicate minerals are formed under the extreme pressures associated with mountain building. Each silicate mineral, therefore, has a structure and a chemical composition that *indicate the conditions under which it formed*. Thus, by carefully examining the mineral makeup of rocks, geologists can often determine the circumstances under which the rocks formed.

We will now examine some of the most common silicate minerals, which are divided into two major groups based on their chemical composition.

Common Light Silicate Minerals

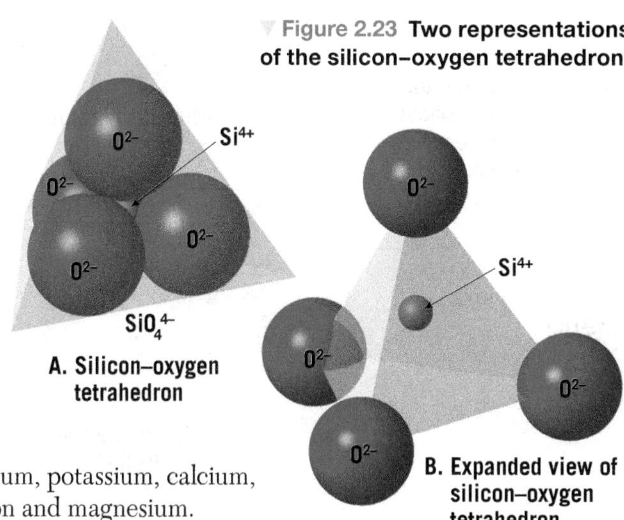

Figure 2.23 **Two representations of the silicon–oxygen tetrahedron**

The **light silicate minerals** are generally light in color and are noticeably less dense than the dark silicates. These differences are mainly attributable to the presence or absence of iron and magnesium, which are "heavy" elements. The light silicates contain varying amounts of aluminum, potassium, calcium, and sodium rather than iron and magnesium.

Feldspar Group

Feldspar minerals are by far the most plentiful silicate group in Earth's crust, comprising about 51 percent of the crust (Figure 2.25). Their abundance can be partially explained by the fact that they can form under a wide range of temperatures and pressures. Two different feldspar structures exist (Figure 2.26). One group of feldspar minerals contains potassium ions in its structure and is therefore termed **potassium feldspar** (Figure 2.26A,B). The other group, called **plagioclase feldspar**, contains both sodium and calcium ions that freely substitute for one another, depending on the environment during crystallization (Figure 2.26C,D). Despite these differences, all feldspar minerals have similar physical properties. They have two planes of cleavage meeting at or near 90-degree angles, are relatively hard (6 on the Mohs scale), and have a luster that ranges from glassy to pearly. As a component in igneous rocks, feldspar crystals can be identified by their rectangular shape and rather smooth, shiny faces.

Potassium feldspar is usually light cream, salmon pink, or occasionally blue-green in color. The plagioclase feldspars, on the other hand, range in color from gray to blue-gray or sometimes black. However, color should not be used to distinguish these groups, as the only way to distinguish the feldspars by looking at them is through the presence of a multitude of fine parallel lines, called *striations*. Striations are found on some cleavage planes of plagioclase feldspar but are not present on potassium feldspar (see Figure 2.26B,D).

Quartz

Quartz (SiO_2) is the second-most-abundant mineral in the continental crust and the only common silicate mineral that consists entirely of silicon and oxygen. In quartz, a three-dimensional framework is developed through the complete sharing of oxygen by adjacent silicon atoms (see Figure 2.24). Thus, all the bonds in quartz are of the strong silicon–oxygen type. Consequently, quartz is hard, resists weathering, and does not have cleavage.

When broken, quartz generally exhibits conchoidal fracture. When pure, quartz is clear and, if allowed to

SmartFigure 2.24
Common silicate minerals Note that the complexity of the silicate structure increases from the top of the chart to the bottom. (Photos by Dennis Tasa and E. J. Tarbuck)

TUTORIAL
https://goo.gl/laowXP

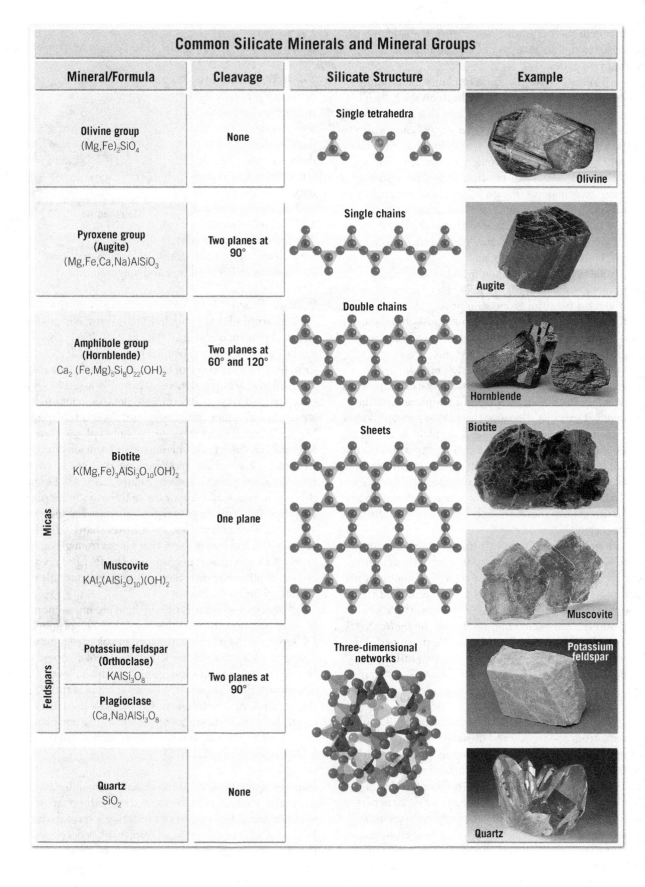

Common Silicate Minerals and Mineral Groups

	Mineral/Formula	Cleavage	Silicate Structure	Example
	Olivine group $(Mg,Fe)_2SiO_4$	None	Single tetrahedra	Olivine
	Pyroxene group (Augite) $(Mg,Fe,Ca,Na)AlSiO_3$	Two planes at 90°	Single chains	Augite
	Amphibole group (Hornblende) $Ca_2(Fe,Mg)_5Si_8O_{22}(OH)_2$	Two planes at 60° and 120°	Double chains	Hornblende
Micas	**Biotite** $K(Mg,Fe)_3AlSi_3O_{10}(OH)_2$	One plane	Sheets	Biotite
Micas	**Muscovite** $KAl_2(AlSi_3O_{10})(OH)_2$	One plane	Sheets	Muscovite
Feldspars	**Potassium feldspar (Orthoclase)** $KAlSi_3O_8$	Two planes at 90°	Three-dimensional networks	Potassium feldspar
Feldspars	**Plagioclase** $(Ca,Na)AlSi_3O_8$	Two planes at 90°	Three-dimensional networks	
	Quartz SiO_2	None	Three-dimensional networks	Quartz

grow without interference, will develop hexagonal crystals that develop pyramid-shaped ends. However, like most other clear minerals, quartz is often colored by inclusions of various ions (impurities) and often forms without developing good crystal faces. The most common varieties of quartz are milky (white), smoky (gray), rose (pink), amethyst (purple), citrine (yellow to brown), and rock crystal (clear) (**Figure 2.27**).

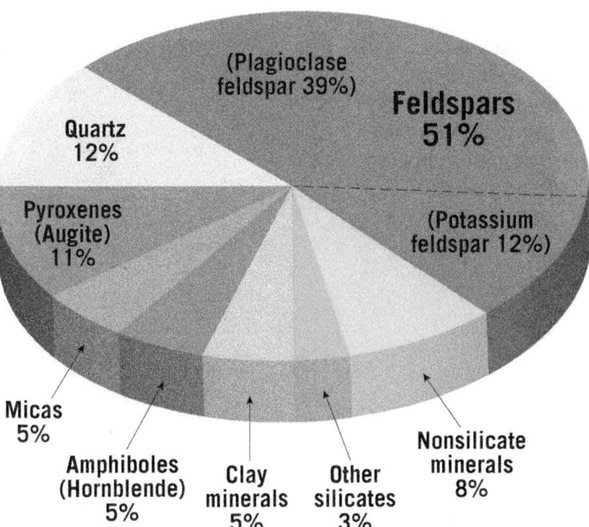

Figure 2.25 **Mineral composition of Earth's crust** Feldspar minerals make up about 51 percent of Earth's crust, and all silicate minerals combined make up about 92 percent of Earth's crust.

Potassium Feldspar

A. Potassium feldspar crystal (orthoclase)

B. Potassium feldspar showing cleavage (orthoclase)

Plagioclase Feldspar

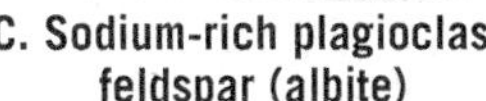

C. Sodium-rich plagioclase feldspar (albite)

D. Plagioclase feldspar showing striations (labradorite)

Figure 2.26 **Some common feldspar minerals A.** Characteristic crystal form of potassium feldspar. **B.** Most salmon-colored feldspar belongs to the potassium feldspar subgroup. (though some are light cream in color). **C.** Sodium-rich plagioclase feldspar tends to be light in color with a pearly luster. **D.** Calcium-rich plagioclase feldspar tends to be gray, blue-gray, or black in color. Labradorite, the sample shown here, exhibits striations on one of its crystal faces. (Photos by Dennis Tasa and E. J. Tarbuck)

Muscovite **Muscovite** is a common member of the mica family. It is light in color and has a pearly luster (see Figure 2.17). Like other micas, muscovite has excellent cleavage in one direction. In thin sheets, muscovite is clear, a property that accounts for its use as window "glass" during the Middle Ages. Because muscovite is very shiny, it can often be identified by the sparkle it gives a rock. If you have ever looked closely at beach sand, you may have seen the glimmering brilliance of the mica flakes scattered among the other sand grains.

Clay Minerals **Clay** is a term used to describe a category of complex minerals that, like the micas, have a sheet structure. Unlike other common silicates, most clay minerals originate as products of the chemical breakdown (chemical weathering) of other silicate minerals. Thus, clay minerals make up a large percentage of the surface material we call soil. (Weathering and soils are discussed in detail in Chapter 8.) Because of soil's importance to agriculture, and because of its role as a supporting material for buildings, clay minerals are extremely important to humans. In addition, clays account for nearly half the volume of sedimentary rocks. Clay minerals are generally very fine grained, which makes them difficult to identify unless they are studied microscopically. Clays are most common in shales, mudstones, and other sedimentary rocks.

One of the most common clay minerals is *kaolinite* (Figure 2.28), which is used in the manufacture of fine china and as a coating for high-gloss paper, such as that used in this textbook. Further, some clay minerals absorb large amounts of water, which allows them to swell to several times their normal size. These clays have been used commercially in a variety of ingenious ways, including as an additive to thicken milkshakes in fast-food restaurants.

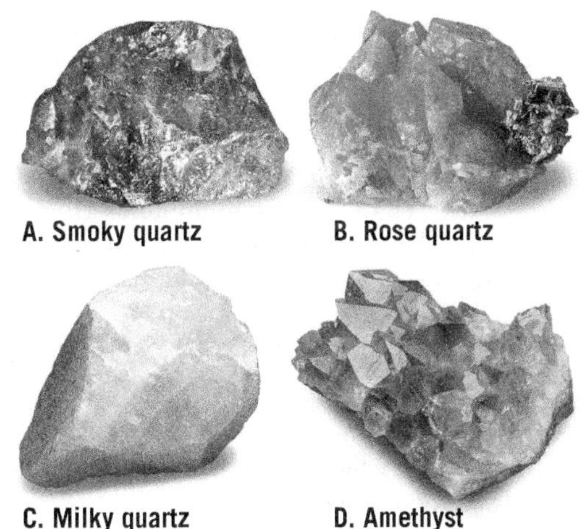

Figure 2.27 **Quartz, the second-most-common mineral in Earth's crust, has many varieties A.** Smoky quartz is commonly found in coarse-grained igneous rocks. **B.** Rose quartz owes its color to small amounts of titanium. **C.** Milky quartz often occurs in veins, which occasionally contain gold. **D.** Amethyst, a purple variety of quartz often used in jewelry, is the birthstone for February. (Photos by Dennis Tasa and E. J. Tarbuck)

Figure 2.28 **Kaolinite** Kaolinite is a common clay mineral formed by weathering of feldspar minerals. (Photo by Dennis Tasa)

Common Dark Silicate Minerals

Dark silicate minerals contain ions of iron and/or magnesium in their structure. Because of their iron content, these silicates are dark in color and have a greater specific gravity than the light silicates.

Olivine Group **Olivine**, a family of high-temperature silicate minerals, are black to olive green in color and have a glassy luster and a conchoidal

Figure 2.29 **Olivine** Commonly black to olive green in color, olivine has a glassy luster and is often granular in appearance. Olivine is commonly found in the igneous rock basalt. (Photo by Dennis Tasa)

fracture (see Figure 2.24, upper-right). Rather than develop large crystals, olivine commonly forms small, rounded crystals that give olivine-rich rocks a granular appearance (Figure 2.29). Olivine and related forms are typically found in basalt, a common igneous rock of the oceanic crust and volcanic areas on the continents; they are thought to constitute up to 50 percent of Earth's upper mantle.

Pyroxene Group The *pyroxenes* are a group of diverse minerals that are important components of dark-colored igneous rocks. The most common member, **augite**, is a black, opaque mineral with two directions of cleavage that meet at nearly a 90-degree angle. Augite is one of the dominant minerals in basalt (Figure 2.30A).

Amphibole Group **Hornblende** is the most common member of a chemically complex group of minerals called *amphiboles* (Figure 2.30B). Hornblende is usually dark green to black in color, and except for its cleavage angles, which are about 60 degrees and 120 degrees, it is very similar in appearance to augite. In a rock, hornblende often forms elongated crystals. This helps distinguish it from pyroxene, which forms rather blocky crystals. Hornblende is found in igneous rocks, where it often makes up the dark portion of an otherwise light-colored rock (see Figure 2.3).

Biotite **Biotite** is a dark, iron-rich member of the mica family (see Figure 2.24). Like other micas, biotite possesses a sheet structure that gives it excellent cleavage in one direction. Biotite also has a shiny black appearance that helps distinguish it from the other dark ferromagnesian minerals. Like hornblende, biotite is a common constituent of igneous rocks, including the rock granite.

Figure 2.30 **Augite and hornblende** These dark-colored silicate minerals are common constituents of a variety of igneous rocks. (Photos by E. J. Tarbuck)

A. Augite

B. Hornblende

Garnet **Garnet** is similar to olivine in that its structure is composed of individual tetrahedra linked by metallic ions. Also like olivine, garnet has a glassy luster, lacks cleavage, and exhibits conchoidal fracture. Although the colors of garnet are varied, this mineral is most often brown to deep red. Well-developed garnet crystals have 12 diamond-shaped faces and are most commonly found in metamorphic rocks (Figure 2.31).

Important Nonsilicate Minerals

Although the nonsilicates make up only about 8 percent of Earth's crust, some nonsilicate minerals, such as gypsum, calcite, and halite, occur as constituents in sedimentary rocks in significant amounts. Many nonsilicates are also economically important.

Figure 2.31 **Well-formed garnet crystal** Garnets come in a variety of colors and are commonly found in mica-rich metamorphic rocks. (Photo by E. J. Tarbuck)

Nonsilicate minerals are typically divided into groups based on the negatively charged ion or complex ion that the members have in common. For example, the *oxides* contain negative oxygen ions (O^{2-}), which bond to one or more kinds of positive ions. Thus, within each mineral group, the basic structure and type of bonding is similar. As a result, the minerals in each group have similar physical properties that are useful in mineral identification. Figure 2.32 lists some of the major nonsilicate mineral groups and includes a few examples of each.

Some of the most common nonsilicate minerals belong to one of three classes of minerals: the carbonates (CO_3^{2-}), the sulfates (SO_4^{2-}), and the halides (Cl^{1-}, F^{1-}, Br^{1-}). The carbonate minerals, which are much simpler structurally than the silicates, are composed of the carbonate ion (CO_3^{2-}) and one or more kinds of positive ions. The two most common carbonate minerals are **calcite**, $CaCO_3$ (calcium carbonate), and **dolomite**, $CaMg(CO_3)_2$ (calcium/magnesium carbonate) (Figure 2.32A,B). Calcite and dolomite are usually found together as the primary constituents in the sedimentary rocks limestone and dolostone. When calcite is the dominant mineral, the rock is called *limestone*, whereas *dolostone* results from a predominance of dolomite. Limestone is used in road aggregate and as a building stone, and it is the main ingredient in Portland cement.

Two other nonsilicate minerals frequently found in sedimentary rocks are **halite** and **gypsum** (Figure 2.32C, I). Both of these minerals are commonly found in thick layers that are the last vestiges of ancient seas that have long since evaporated (Figure 2.33). Like limestone, both halite and gypsum are important nonmetallic resources. Halite is the mineral name for common table salt (NaCl). Gypsum ($CaSO_4 \cdot 2H_2O$), which is calcium sulfate with water bound into the structure, is the mineral from which plaster and other similar building materials are composed.

Most nonsilicate mineral classes contain members that are prized for their economic value. This includes the oxides, whose members *hematite* and *magnetite* are important ores of iron (see Figure 2.32E,F). Also significant are the sulfides, which are basically compounds of sulfur (S) and one or more metals.

Common Nonsilicate Mineral Groups

Mineral Group (key ion(s) or element(s))	Mineral Name	Chemical Formula	Economic Use	Examples
Carbonates (CO_3^{2-})	**Calcite** **Dolomite**	$CaCO_3$ $CaMg(CO_3)_2$	**Portland cement, lime** **Portland cement, lime**	 **A. Calcite** **B. Dolomite**
Halides (Cl^{1-}, F^{1-}, Br^{1-})	**Halite** **Fluorite** **Sylvite**	$NaCl$ CaF_2 KCl	**Common salt** **Used in steel making** **Used as fertilizer**	 **C. Halite** **D. Fluorite**
Oxides (O^{2-})	**Hematite** **Magnetite** **Corundum** **Ice**	Fe_2O_3 Fe_3O_4 Al_2O_3 H_2O	**Ore of iron, pigment** **Ore of iron** **Gemstone, abrasive** **Solid form of water**	 **E. Hematite** **F. Magnetite**
Sulfides (S^{2-})	**Galena** **Sphalerite** **Pyrite** **Chalcopyrite** **Cinnabar**	PbS ZnS FeS_2 $CuFeS_2$ HgS	**Ore of lead** **Ore of zinc** **Sulfuric acid production** **Ore of copper** **Ore of mercury**	 **G. Galena** **H. Chalcopyrite**
Sulfates (SO_4^{2-})	**Gypsum** **Anhydrite** **Barite**	$CaSO_4 \cdot 2H_2O$ $CaSO_4$ $BaSO_4$	**Plaster** **Plaster** **Drilling mud**	 **I. Gypsum** **J. Anhydrite**
Native elements (single elements)	**Gold** **Copper** **Diamond** **Graphite** **Sulfur** **Silver**	Au Cu C C S Ag	**Trade, jewelry** **Electrical conductor** **Gemstone, abrasive** **Pencil lead** **Sulfa drugs, chemicals** **Jewelry, photography**	 **K. Copper** **L. Sulfur**

▲ **Figure 2.32 Important nonsilicate mineral groups** (Photos by Dennis Tasa and E. J. Tarbuck)

Figure 2.33 **Thick bed of halite exposed in an underground mine** In this halite (salt) mine in Grand Saline, Texas, note the person for scale. (Photo by Tom Bochsler/Pearson Education)

Important sulfide minerals include galena (lead), sphalerite (zinc), and chalcopyrite (copper). In addition, native elements—including gold, silver, and carbon (diamonds)—are economically important, as are a host of other nonsilicate minerals—fluorite (flux in making steel), corundum (gemstone, abrasive), and uraninite (a uranium source). See GEOGraphics 2.2 for more information on gemstones.

CONCEPT CHECKS 2.5

1. List the eight most common elements in Earth's crust.
2. Sketch the silicon–oxygen tetrahedron and label its parts.
3. What is the most abundant mineral in Earth's crust?
4. What is the most common carbonate mineral?
5. List six common nonsilicate minerals and their economic uses.

2.6 Minerals: A Nonrenewable Resource

Discuss Earth's mineral resources in terms of renewability. Differentiate between mineral resources and ore deposits.

Earth's crust and oceans are the source of a wide variety of useful and valuable materials. From the first use of rocks such as flint and obsidian to make tools thousands of years ago, the use of Earth materials has expanded, resulting in more complex societies and our modern civilization. The mineral and energy resources we extract from Earth's crust are the raw materials from which we make all the products we use.

Natural resources are typically grouped into broad categories according to (1) their ability to be regenerated (renewable or nonrenewable) or (2) their origin or type. Here we will consider mineral resources. However, other natural resources are indispensable to humans, including air, water, and solar energy.

Figure 2.34 **Solar energy is renewable** The Ivanpah Solar Electric Generating System is a solar thermal plant located in California's Mojave Desert, southwest of Las Vegas. It consists of 173,500 heliostats (mirrors that move so they reflect sunlight at a target), each with two mirrors that focus solar energy on boilers located on one of three centralized towers. The boilers, in turn, generate steam, which turns turbines that generate electricity. (Photo by Steve Proehl/Getty Images/Corbis Documentary)

Renewable Versus Nonrenewable Resources

Resources classified as **renewable** can be replenished over relatively short time spans. Common examples are corn used for food and for making ethanol, natural fibers such as cotton for clothing, and forest products for lumber and paper. Energy from flowing water, wind, and the Sun are also considered renewable (**Figure 2.34**).

By contrast, many other basic resources are classified as **nonrenewable**. Important metals such as iron, aluminum, and copper fall into this category, as do our most widely used fuels: petroleum, natural gas, and coal. Although these and other resources form continuously, the processes that create them are so slow that significant deposits take millions of years to accumulate. Thus, for all practical purposes, Earth contains fixed quantities of these substances. The present supplies will be depleted as they are mined or pumped from the ground. Although some nonrenewable resources, such as the aluminum we use for containers, can be recycled, others, such as the oil burned for fuel, cannot.

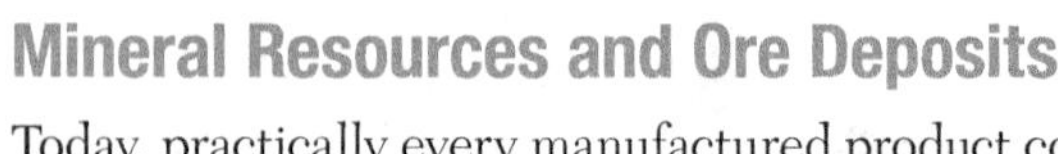

Mineral Resources and Ore Deposits

Today, practically every manufactured product contains materials obtained from minerals. Figure 2.32 lists some of the most economically important mineral groups. **Mineral resources** are occurrences of useful minerals that are formed in such quantities that eventual extraction is reasonably certain. Mineral resources include deposits of metallic minerals that can be presently extracted profitably, as well as known deposits that are not yet economically or technologically recoverable. Materials used for such purposes as building stone, road aggregate, abrasives, ceramics, and fertilizers are not usually called mineral resources; rather, they are classified as *industrial rocks and minerals*.

An **ore deposit** is a naturally occurring concentration of one or more metallic minerals that can be extracted economically. In common usage, the term *ore* is also applied to some nonmetallic minerals such as fluorite and sulfur. Recall that more than 98 percent of Earth's crust is composed of only eight elements, and except for oxygen and silicon, all other elements make up a relatively small fraction of common crustal rocks (see Figure 2.22). Indeed, the natural concentrations of many elements are exceedingly small. A deposit containing the average concentration of an element such as gold has no economic value because the cost of extracting it greatly exceeds the value of the gold that could be recovered.

In order to have economic value, an ore deposit must be highly concentrated. For example, copper makes up about 0.0068 percent of the crust. For a deposit to be considered a copper ore, it must contain a concentration of copper that is about 100 times this amount, or about 0.68 percent. Aluminum, on the other hand, represents about 8.1 percent of the crust and can be extracted profitably when it is found in concentrations 3 or 4 times that amount.

It is important to understand that due to economic or technological changes, a deposit may either become profitable to extract or lose its profitability. If the demand for a metal increases and its value rises sufficiently, the status of a previously unprofitable deposit can be upgraded from a mineral to an ore. Technological advances that allow a resource to be extracted more efficiently and, thus, more profitably than before may also trigger a change of status.

Conversely, changing economic factors can turn what was once a profitable ore deposit into an unprofitable mineral deposit. This situation was illustrated at the copper mining operation located at Bingham Canyon, Utah, one of the largest open-pit mines on Earth (**Figure 2.35**). Mining was halted there in 1985 because outmoded equipment had driven the cost of extracting the copper beyond the current selling price. In 1989 new owners responded by replacing an antiquated 1000-car railroad with modern conveyor belts and large dump trucks for efficiently transporting the ore and waste. The new equipment reduced extraction costs by nearly 30 percent, ultimately returning the copper mine operation to profitability. Today the Bingham Canyon mine produces nearly 18 percent of the refined copper in the United States. In addition to producing about 300,000 metric tons of copper, the Bingham Canyon mine produces about 400,000 troy ounces of gold, 4 million troy ounces of silver, and 25 million pounds of molybdenum.

Over the years, geologists have been keenly interested in learning how natural processes produce localized concentrations of essential minerals. One well-established fact is that occurrences of valuable mineral resources are closely related to the rock cycle. That is, the mechanisms that generate igneous, sedimentary, and metamorphic rocks, including the processes of weathering and erosion, play a major role in producing concentrated accumulations of useful elements.

Moreover, with the development of the theory of plate tectonics, geologists have added another tool for understanding the processes by which one rock is transformed into another. As these rock-forming processes are examined in the following chapters, we consider their role in producing some of our important mineral resources.

CONCEPT CHECKS 2.6

1. List three examples of renewable resources and three examples of nonrenewable resources.
2. Compare and contrast a *mineral resource* and an *ore deposit*.
3. Explain how a mineral deposit that previously could not be mined profitably might be upgraded to an ore deposit.

▼ Figure 2.35 **Aerial view of Bingham Canyon copper mine near Salt Lake City, Utah** Although the amount of copper in the rock is less than 0.5 percent, the huge volume of material removed and processed each day (over 400,000 metric tons) yields enough metal to be profitable. In addition to copper, this mine produces gold, silver, and molybdenum. (Photo by Michael Collier)

Gemstones

Precious stones have been prized since antiquity. Although gemstones are varieties of a particular mineral, misinforma abounds regarding gems and their mineral makeup.

Important Gemstones

Gemstones are classified in one of two categories: precious or semiprecious. Precious gems are rare and generally have hardnesses that exceed 9 on the Mohs scale. Therefore, they are more valuable and thus more expensive than semiprecious gems.

GEM	MINERAL NAME	PRIZED HUES
PRECIOUS		
Diamond	Diamond	Colorless, pinks, blues
Emerald	Beryl	Greens
Ruby	Corundum	Reds
Sapphire	Corundum	Blues
Opal	Opal	Brilliant hues
SEMIPRECIOUS		
Alexandrite	Chrysoberyl	Variable
Amethyst	Quartz	Purples
Cat's-eye	Chrysoberyl	Yellows
Chalcedony	Quartz (agate)	Banded
Citrine	Quartz	Yellows
Garnet	Garnet	Red, greens
Jade	Jadeite or nephrite	Greens
Moonstone	Feldspar	Transparent blues
Peridot	Olivine	Olive greens
Smoky quartz	Quartz	Browns
Spinel	Spinel	Reds
Topaz	Topaz	Purples, reds
Tourmaline	Tourmaline	Reds, blue-greens
Turquoise	Turquoise	Blues
Zircon	Zircon	Reds

Reuters

The Famous Hope Diamond

The deep-blue Hope Diamond is a 45.52-carat gem that is thought to have been cut from a much larger 115-carat stone discovered in India in the mid-1600s. The original 115-carat stone was cut into a smaller gem that became part of the crown jewels of France and was in the possession of King Louis XVI and Marie Antoinette before they attempted to escape France. Stolen during the French Revolution in 1792, the gem is thought to have been recut to its present size and shape. In the 1800s, it became part of the collection of Henry Hope (hence its name) and is on display at the Smithsonian in Washington, DC.

What Constitutes a Gemstone?

When found in their natural state, most gemstones are dull and would be passed over by most people as "just another rock." Gems must be cut and polished by experienced professionals before their true beauty is displayed. Cutting and polishing is accomplished using abrasive material, most often tiny fragments of diamonds that are embedded in a metal disk.

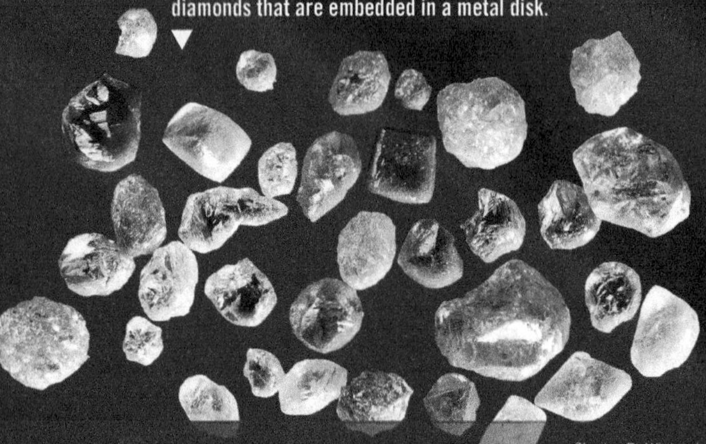

Charles D. Winters/ Photo Researchers, Inc.

Question:
Why are diamonds used as an abrasive material to cut and polish gemstones?

Naming Gemstones

Most precious stones are given names that differ from their parent mineral. For example, sapphire is one of two gems that are varieties of the same mineral, corundum. Trace elements can produce vivid sapphires of nearly every color. Tiny amounts of titanium and iron in corundum produce the most prized blue sapphires. When the mineral corundum contains a sufficient quantity of chromium, it exhibits a brilliant red color. This variety of corundum is called ruby.

Rubies, like sapphires, are varieties of the mineral corundum

Sapphires, showing variations of colors

Dorling Kindersley

Greg C. Grace/ Alamy

CONCEPTS IN REVIEW
Matter and Minerals

2.1 Minerals: Building Blocks of Rocks

List the main characteristics that an Earth material must possess to be considered a mineral and describe each characteristic.

KEY TERMS: mineralogy, mineral, rock

- In Earth science, the word mineral refers to naturally occurring inorganic solids that possess an orderly crystalline structure and a characteristic chemical composition. The study of minerals is called mineralogy.
- Minerals are the building blocks of rocks. Rocks are naturally occurring masses of minerals or mineral-like matter such as natural glass or organic material.

2.2 Atoms: Building Blocks of Minerals

Compare and contrast the three primary particles contained in atoms.

KEY TERMS: atom, nucleus, proton, neutron, electron, valence electron, atomic number, element, periodic table, chemical compound

- Minerals are composed of atoms of one or more elements. All atoms consist of the same three basic components: protons, neutrons, and electrons.
- The atomic number represents the number of protons found in the nucleus of an atom of a particular element. For example, an oxygen atom has eight protons, so its atomic number is eight. Protons and neutrons have approximately the same size and mass, but protons are positively charged, whereas neutrons have no charge.
- Electrons weigh only about 1/2000 as much as protons or neutrons. They occupy the space around the nucleus, where they form what can be thought of as a cloud that is structured into several distinct energy levels called principal shells. The electrons in the outermost principal shell, called valence electrons, are responsible for the bonds that hold atoms together to form chemical compounds.
- Elements that have the same number of valence electrons tend to behave similarly. The periodic table is organized so that elements with the same number of valence electrons form a column, called a group.

? Use the periodic table (see Figure 2.5) to identify the geologically important elements that have the following numbers of protons: (A) 14, (B) 6, (C) 13, (D) 17, and (E) 26.

2.3 Why Atoms Bond

Distinguish among ionic bonds, covalent bonds, and metallic bonds.

KEY TERMS: octet rule, chemical bond, ionic bond, ion, covalent bond, metallic bond

- When atoms are attracted to other atoms, they can form chemical bonds, which generally involve the transfer or sharing of valence electrons. The most stable arrangement for most atoms is to have eight electrons in the outermost principal shell. This concept is called the octet rule.
- To form ionic bonds, atoms of one element give up one or more valence electrons to atoms of another element, forming positively and negatively charged atoms called ions. The ionic bond results from the attraction between oppositely charged ions.
- Covalent bonds form when adjacent atoms share valence electrons.
- In metallic bonds, the sharing is more extensive: The shared valence electrons can move freely through the substance.

? Which of the accompanying diagrams (A, B, or C) best illustrates ionic bonding? What are the distinguishing characteristics of ionic bonding and of covalent bonding?

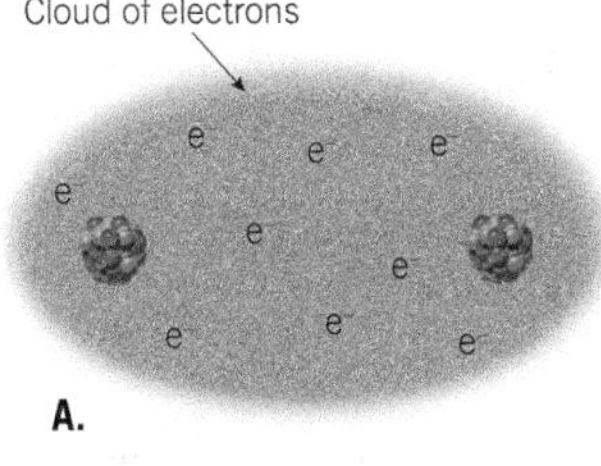

A.

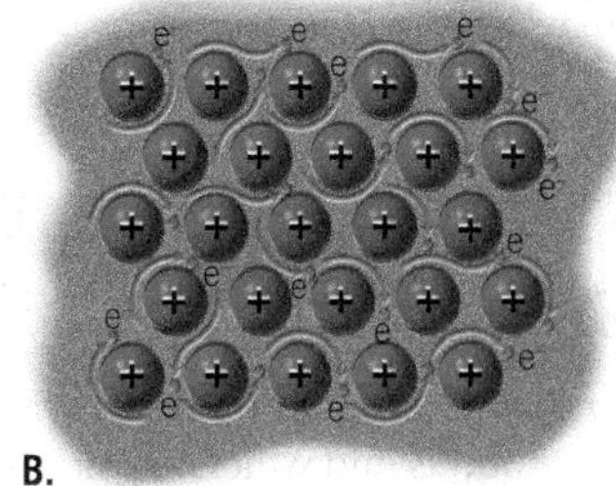
B.

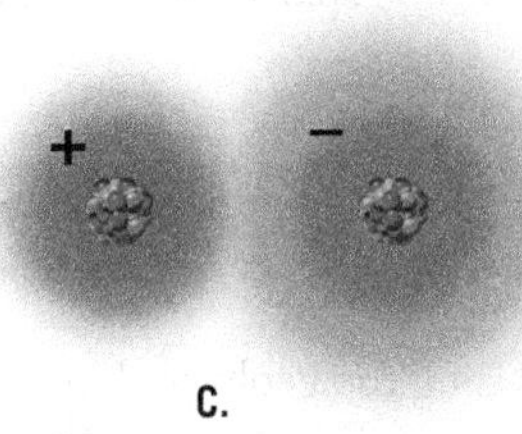

C.

2.4 Properties of Minerals

List and describe the properties used in mineral identification.

KEY TERMS: diagnostic property, ambiguous property, luster, color, streak, crystal shape (habit), hardness, Mohs scale, cleavage, fracture, tenacity, density, specific gravity

- The composition and internal crystalline structure of a mineral give it specific physical properties. Mineral properties useful in identifying minerals are termed *diagnostic properties*.
- Luster is a mineral's ability to reflect light. The terms transparent, translucent, and opaque describe the degree to which a mineral can transmit light. Color can be unreliable for mineral identification, as impurities can "stain" minerals with diverse colors. A more reliable identifier is streak, the color of the powder generated by scraping a mineral against a porcelain streak plate.
- Crystal shape, also called crystal habit, is often useful for mineral identification.

- Variations in the strength of chemical bonds give minerals properties such as hardness (resistance to being scratched) and tenacity (response to deforming stress, such as whether the mineral tends to undergo brittle breakage like quartz, bend elastically like mica, or deform malleably like gold). Cleavage, the preferential breakage of a mineral along planes of weakly bonded atoms, is very useful in identifying minerals.
- The amount of matter packed into a given volume determines a mineral's density. To compare the densities of minerals, mineralogists use a related quantity, known as specific gravity, which is the ratio between a mineral's density and the density of water.
- Other properties are diagnostic for certain minerals but rare in most others; examples include smell, taste, feel, reaction to hydrochloric acid, magnetism, and double refraction.

? Determine which of these specimens exhibit a metallic luster and which have a nonmetallic luster.

2.5 Mineral Groups

List the common silicate and nonsilicate minerals and describe what characterizes each group.

KEY TERMS: rock-forming mineral, economic mineral, silicate, nonsilicate, silicon–oxygen tetrahedron, light silicate mineral, potassium feldspar, plagioclase feldspar, quartz, muscovite, clay, dark silicate mineral, olivine, augite, hornblende, biotite, garnet, calcite, dolomite, halite, gypsum

- Silicate minerals have a basic building block in common: a small pyramid-shaped structure called the silicon–oxygen tetrahedron, which consists of one silicon atom surrounded by four oxygen atoms. Neighboring tetrahedra can share some of their oxygen atoms, causing them to develop long chains, sheet structures, and three-dimensional networks.
- Silicate minerals are the most common mineral class on Earth. They are subdivided into minerals that contain iron and/or magnesium (dark silicates) and those that do not (light silicates). The light silicate minerals are generally light in color and have relatively low specific gravities. Feldspar, quartz, muscovite, and clay minerals are examples. The dark silicate minerals are generally dark in color and relatively dense. Olivine, pyroxene, amphibole, biotite, and garnet are examples.
- Nonsilicate minerals include oxides, which contain oxygen ions that bond to other elements (usually metals); carbonates, which have CO_3^{2-} as a critical part of their crystal structure; sulfates, which have SO_4^{2-} as their basic building block; and halides, which contain a nonmetal ion such as chlorine, bromine, or fluorine that bonds to a metal ion such as sodium or calcium.
- Nonsilicate minerals are often economically important. Hematite is an important source of industrial iron, while calcite is an essential component of cement.

2.6 Minerals: A Nonrenewable Resource

Discuss Earth's mineral resources in terms of renewability. Differentiate between mineral resources and ore deposits.

KEY TERMS: renewable, nonrenewable, mineral resource, ore deposit

- Resources are classified as renewable when they can be replenished over short time spans and nonrenewable when they can't.
- Ore deposits are naturally occurring concentrations of one or more metallic minerals that can be extracted economically using current technology. A mineral resource can be upgraded to an ore deposit if the price of the commodity increases sufficiently or if the cost of extraction decreases. The reverse can also happen.

GIVE IT SOME THOUGHT

1 Using the geologic definition of *mineral* as your guide, determine which of the items in this list are minerals and which are not. If something in this list is not a mineral, explain.

a. Gold nugget
b. Seawater
c. Quartz
d. Cubic zirconia
e. Obsidian
f. Ruby
g. Glacial ice
h. Amber

2 Assume that the number of protons in a neutral atom is 92 and the atomic mass is 238.02. (*Hint:* Refer to the periodic table in Figure 2.5 to answer this question.)

a. What is the name of the element?
b. How many electrons does it have?

3 Research the minerals *quartz* and *calcite*. List five physical characteristics that may be used to distinguish one from another.

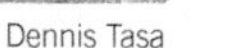

Quartz

Calcite

4 Gold has a specific gravity of almost 20. A 5-gallon bucket of water weighs 40 pounds. How much would a 5-gallon bucket of gold weigh?

5 Examine the accompanying photo of a mineral that has several smooth, flat surfaces that resulted when the specimen was broken.
a. How many flat surfaces are present on this specimen?
b. How many different directions of cleavage does this specimen have?
c. Do the cleavage directions meet at 90-degree angles?

E. J. Tarbuck
Cleaved sample

6 Each of the following statements describes a silicate mineral or mineral group. In each case, provide the appropriate name:
a. The most common member of the amphibole group
b. The most common light-colored member of the mica family
c. The only common silicate mineral made entirely of silicon and oxygen
d. A silicate mineral with a name based on its color
e. A silicate mineral characterized by striations
f. Silicate minerals that originate as a product of chemical weathering

7 What mineral property is illustrated in the accompanying photo?

Dennis Tasa

8 Do an Internet search to determine which minerals are used to manufacture the following products:
a. Stainless steel utensils
b. Cat litter
c. Tums brand antacid tablets
d. Lithium batteries
e. Aluminum beverage cans

EXAMINING THE EARTH SYSTEM

1 Perhaps one of the most significant interrelationships between humans and the Earth system involves the extraction, refinement, and distribution of the planet's mineral wealth. To help you understand these associations, begin by thoroughly researching a mineral commodity that is mined in your local region or state. (You might find useful the information at the U.S. Geological Survey [USGS] website: http://minerals.er.usgs.gov/minerals/pubs/state/.) What products are made from this mineral? Do you use any of these products? Describe the mining and refining of the mineral and the local impact these processes have on each of Earth's spheres (atmosphere, hydrosphere, geosphere, and biosphere). Are any of the effects negative? If so, what, if anything, is being done to end or minimize the damage?

2 Referring to the mineral you described in Question 1, in your opinion does the environmental impact of extracting this mineral outweigh the benefits derived from its products?

DATA ANALYSIS

Global Mineral Resources

Mineral resources are important for the global economy. Some minerals are found only in a few locations, while others are more evenly distributed around the globe. The United States Geological Survey maintains a mineral resource database that can help us explore the distribution of major mineral deposits worldwide.

ACTIVITIES

Go to the Mineral Resource On-Line Spatial Data interactive map at http://mrdata.usgs.gov and click Global. Check the box next to Major Deposits. The following questions relate to the contiguous United States.

1 Return to the map and click on the legend. Scroll down to display the Major Deposits symbology. PGE stands for *platinum-group elements*. What are the two most numerous major mineral deposits? Consult the Internet and list one economic use for each mineral.

2 What are the two least numerous major mineral deposits? List one economic use for each mineral.

3 Which minerals do not have major deposits? List one economic use for each.

4 Where are most of the major gold deposits located? Clay deposits? Iron deposits?

5 According to Figure 1.21 (page 25), what continental feature contains most of the major gold deposits? Clay deposits? Iron deposits?

6 Click on the Map Layers icon and then click on the words Major Deposits for information on this category. What important type of natural resource is not included on this map?

Go to http://mrdata.usgs.gov, select the Mineral Resources tab, and click on the map next to Mineral Resources Data System (MRDS).

7 What information is displayed on this map?

8 Notice that Missouri and Iowa have many mineral resources, but Kansas and Nebraska have few. Why might the mining of mineral deposits change so abruptly at state boundaries?

Zoom in to your region on the map and click on one mineral resource site closest to your location. Underneath Find Features by Clicking the Map Above, click the name of the site to open a new page with site information.

9 Which commodities are available from this site?

MasteringGeology™ Looking for additional review and test prep materials? Visit the Study Area in MasteringGeology to enhance your understanding of this chapter's content by accessing a variety of resources, including Self-Study Quizzes, Geoscience Animations, SmartFigure Tutorials, Mobile Field Trips, *Project Condor* Quadcopter videos, *In the News* articles, flashcards, web links, and an optional Pearson eText.

www.masteringgeology.com

3 Rocks: Materials of the Solid Earth

FOCUS ON CONCEPTS

Each statement represents the primary learning objective for the corresponding major heading within the chapter. After you complete the chapter, you should be able to:

3.1 Sketch, label, and explain the rock cycle.

3.2 Describe the two criteria used to classify igneous rocks and explain how the rate of cooling influences the crystal size of minerals.

3.3 List and describe the different categories of sedimentary rocks and discuss the processes that change sediment into sedimentary rock.

3.4 Define *metamorphism*, explain how metamorphic rocks form, and describe the agents of metamorphism.

3.5 Distinguish between metallic and nonmetallic mineral resources and list at least two examples of each.

Climbers high on El Capitan in California's Yosemite National Park. (Photo by Ron Niebrugge/Alamy Stock Photo)

WHY STUDY ROCKS? You have already learned that some rocks and minerals have great economic value. In addition, all Earth processes depend in some way on the properties of these basic Earth materials. Events such as volcanic eruptions, mountain building, weathering, erosion, and even earthquakes involve rocks and minerals. Consequently, a basic knowledge of Earth materials is essential to understanding most geologic phenomena.

Every rock contains clues about the environment in which it formed. For example, some rocks are composed entirely of small shell fragments of marine organisms, which tells Earth scientists that the rock likely originated in a shallow marine environment. Other rocks contain clues indicating that they formed from a volcanic eruption or deep in Earth during mountain building. Thus, rocks contain a wealth of information about events that have occurred over Earth's long history.

3.1 Earth as a System: The Rock Cycle

Sketch, label, and explain the rock cycle.

Earth as a system is illustrated most vividly when we examine the rock cycle. The **rock cycle** allows us to see many of the interactions among the components and processes of the Earth system (**Figure 3.1**). It helps us understand the origins of igneous, sedimentary, and metamorphic rocks and how these rocks are connected. In addition, the rock cycle demonstrates that any rock type, given the right sequence of events, can be transformed into any other type.

The Basic Cycle

We begin our discussion of the rock cycle with molten rock, called *magma*, which forms by melting that occurs primarily within Earth's crust and upper mantle (see Figure 3.1). Once formed, a magma body rises toward the surface because it is less dense than the surrounding rock. If magma reaches Earth's surface and erupts, we call it *lava*. Eventually, molten rock cools and solidifies, a process called *crystallization* or *solidification*. Molten rock may solidify either beneath the surface or, following a volcanic eruption, at the surface. In either situation, the resulting rocks are called *igneous rocks*.

If igneous rocks are exposed at the surface, they undergo *weathering*, the slow disintegration and decomposition of rocks by the daily influences of the atmosphere. The loose materials that result are often moved downslope by gravity and then picked up and transported by one or more erosional agents—running water, glaciers, wind, or waves. These rock particles and dissolved substances, called *sediment*, are eventually deposited. Although most sediment ultimately comes to rest in the ocean, other sites of deposition include river floodplains, desert basins, lakes, inland seas, and sand dunes.

Next, the sediments undergo *lithification*, a term meaning "conversion into rock." Sediment is usually lithified into *sedimentary rock* when compacted by the weight of overlying materials or when cemented by percolating groundwater that fills the pores with mineral matter.

If the resulting sedimentary rock becomes deeply buried or is involved in the dynamics of mountain building, it will be subjected to great pressures and intense heat. The sedimentary rock may react to the changing environment by turning into the third rock type, *metamorphic rock*. If metamorphic rock is subjected to still higher temperatures, it may melt, creating magma, and the cycle begins again.

Although rocks may appear to be stable, unchanging masses, the rock cycle shows that they are not. The changes, however, take time—sometimes millions or even billions of years. In addition, different locations on Earth are at different stages of the rock cycle. Today, new magma is forming under the island of Hawaii, whereas the rocks that comprise the Colorado Rockies are slowly being worn down by weathering and erosion. Some of this weathered debris will eventually be carried to the Gulf of Mexico, where it will add to the already substantial mass of sediment that has accumulated there.

Alternative Paths

Rocks do not necessarily go through the cycle in the order just described. Other paths are also possible. For example, rather than being exposed to weathering and erosion at Earth's surface, igneous rocks may remain deeply buried (see Figure 3.1). Eventually, these masses may be subjected to the strong compressional forces and high temperatures associated with mountain building. When this occurs, they are transformed directly into metamorphic rocks.

Uplift and erosion may bring even deeply buried rocks of any type to the surface. When this happens, the

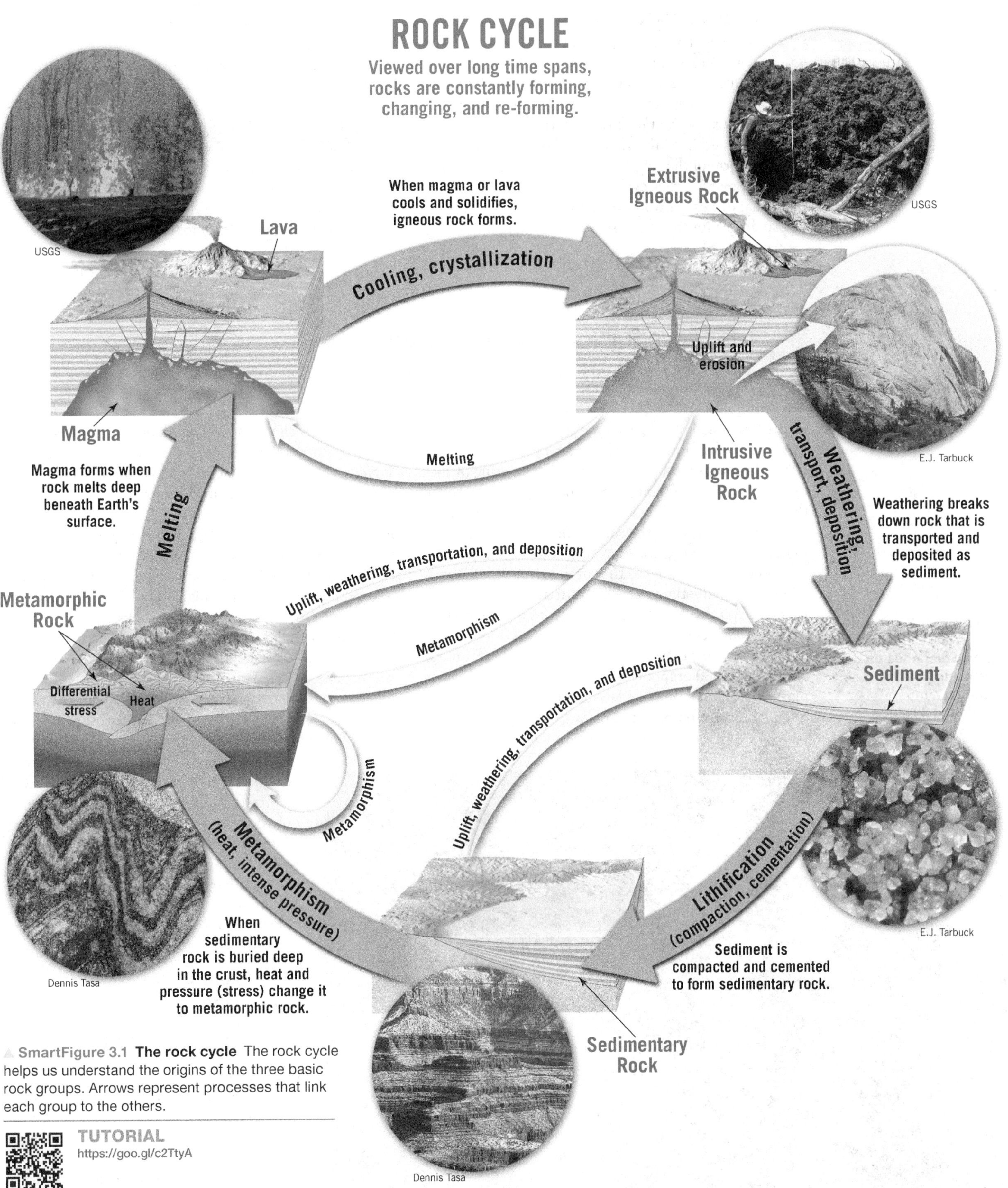
ROCK CYCLE
Viewed over long time spans, rocks are constantly forming, changing, and re-forming.
Lava
When magma or lava cools and solidifies, igneous rock forms.
Extrusive Igneous Rock
USGS
USGS
Cooling, crystallization
Uplift and erosion
Magma
Melting
Intrusive Igneous Rock
Weathering, transport, deposition
E.J. Tarbuck
Magma forms when rock melts deep beneath Earth's surface.
Melting
Weathering breaks down rock that is transported and deposited as sediment.
Uplift, weathering, transportation, and deposition
Metamorphic Rock
Metamorphism
Sediment
Differential stress
Heat
Uplift, weathering, transportation, and deposition
Metamorphism
Metamorphism (heat, intense pressure)
Lithification (compaction, cementation)
When sedimentary rock is buried deep in the crust, heat and pressure (stress) change it to metamorphic rock.
Sediment is compacted and cemented to form sedimentary rock.
E.J. Tarbuck
Dennis Tasa
Sedimentary Rock
Dennis Tasa

SmartFigure 3.1 The rock cycle The rock cycle helps us understand the origins of the three basic rock groups. Arrows represent processes that link each group to the others.

TUTORIAL
https://goo.gl/c2TtyA

material is attacked by weathering processes and turned into new raw materials for sedimentary rocks. Conversely, igneous or metamorphic rocks that form at depth (deep underground) may remain there, where the high temperatures and forces associated with mountain building may metamorphose or even melt them. Over time, rocks may be transformed into any other rock type, or even into a different form of the original type. Rocks may take many paths through the rock cycle.

What drives the rock cycle? Earth's internal heat is responsible for the processes that form igneous and metamorphic rocks. Weathering and the transport of weathered material are external processes, powered by energy from the Sun. External processes produce sedimentary rocks.

CONCEPT CHECKS 3.1

1. Sketch and label the rock cycle. Make sure your sketch includes alternative paths.
2. Use the rock cycle to explain the statement "One rock is the raw material for another."

3.2 Igneous Rocks: "Formed by Fire"

Describe the two criteria used to classify igneous rocks and explain how the rate of cooling influences the crystal size of minerals.

In the discussion of the rock cycle, we pointed out that **igneous rocks** form as *magma* or *lava* cools and crystallizes. **Magma** is most often generated by melting of rocks in Earth's mantle, although some magma originates from the melting of crustal rock. Once formed, a magma body buoyantly rises toward the surface because it is less dense than the surrounding rocks.

When magma reaches the surface, it is called **lava** (Figure 3.2). Sometimes, lava is emitted as fountains, which are produced when escaping gases propel molten rock skyward. On other occasions, magma is ejected explosively from vents, producing a spectacular eruption such as the 1980 eruption of Mount St. Helens. However, most eruptions are not violent; rather, volcanoes most often emit quiet outpourings of lava.

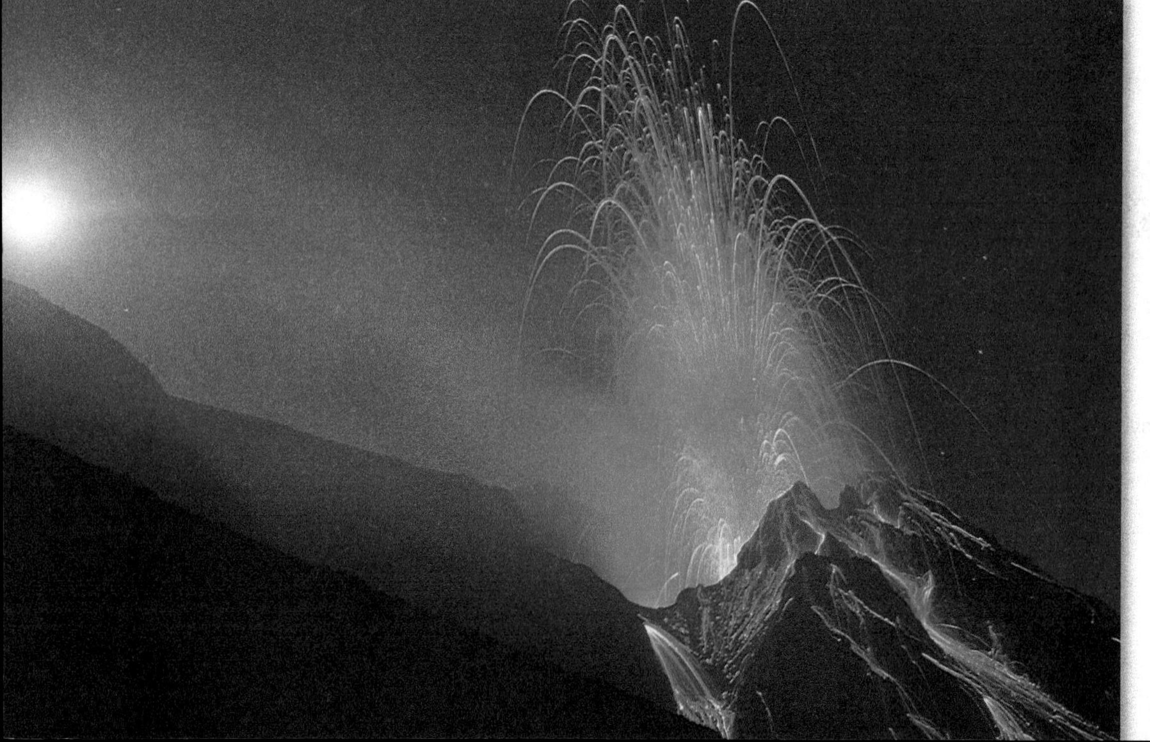

Figure 3.2 **Stromboli, Italy is one of the most active volcanoes on Earth and has been erupting almost continuously since 1932.** (Photo by Zoonar GmbH/Alamy Stock Photo)

Molten rock may solidify at depth, or it may solidify at Earth's surface. When molten rock solidifies *at the surface*, the resulting igneous rocks are classified as **extrusive rocks**, or **volcanic rocks** (after the Roman fire god, Vulcan). Extrusive igneous rocks are abundant in western portions of the Americas, including the volcanic cones of the Cascade Range and the extensive lava flows of the Columbia Plateau. In addition, many oceanic islands, including the Hawaiian Islands, are composed almost entirely of volcanic igneous rocks.

Most magma loses its mobility before reaching Earth's surface and eventually solidifies deep below the surface. When magma solidifies at depth, it forms igneous rocks known as **intrusive rocks**, or **plutonic rocks** (after Pluto, the god of the underworld in classical mythology). Intrusive igneous rocks remain at depth unless portions of the crust are uplifted and the overlying rocks are stripped away by erosion. Exposures of intrusive igneous rocks occur in many places, including Mount Washington, New Hampshire; Stone Mountain, Georgia; Yosemite National Park, California; and Mount Rushmore in the Black Hills of South Dakota (Figure 3.3).

From Magma to Crystalline Rock

Magma is molten rock (or *melt*) composed of ions of the elements found in silicate minerals, mainly silicon and oxygen, that move about freely. Magma also contains gases, particularly water vapor, that are confined within the magma body by the weight (pressure) of the overlying rocks, and it may contain some solids (mineral

Figure 3.3 Mount Rushmore National Memorial, Black Hills of South Dakota This memorial is carved from the intrusive igneous rock granite. This massive igneous body cooled very slowly at depth and has since been uplifted the overlying rocks were stripped away by erosion. (Photo by Barbara A. Harvey/Shutterstock)

crystals). As magma cools, the once-mobile ions begin to arrange themselves into orderly patterns—a process called *crystallization*. As cooling continues, numerous small crystals develop, and ions are systematically added to these centers of crystal growth. When the crystals grow large enough for their edges to meet, their growth ceases. Eventually, all the liquid melt is transformed into a solid mass of interlocking crystals.

The rate of cooling strongly influences crystal size. If magma cools very slowly, ions can migrate over great distances. Consequently, *slow cooling results in the formation of fewer, larger crystals*. On the other hand, if cooling occurs rapidly, the ions lose their motion and quickly combine. This results in a large number of tiny crystals that all compete for the available ions. Therefore, *rapid cooling results in the formation of a solid mass of small intergrown crystals*.

If the molten material is quenched, or cooled almost instantly, there is insufficient time for the ions to arrange themselves into a crystalline network. Solids produced in this manner consist of ions distributed somewhat randomly. Such rocks are called *glass* and are quite similar to ordinary manufactured glass. "Instant" quenching often occurs during violent volcanic eruptions that produce tiny shards of glass called *volcanic ash*.

In addition to the rate of cooling, the composition of a magma and the amount of dissolved gases influence crystallization. Because magmas differ in each of these aspects, the physical appearance and mineral composition of igneous rocks vary widely.

Igneous Compositions

Igneous rocks are composed mainly of silicate minerals. Chemical analysis shows that silicon and oxygen—usually expressed as the silica (SiO_2) content of a magma—are by far the most abundant constituents of igneous rocks. These two elements, plus ions of aluminum (Al), calcium (Ca), sodium (Na), potassium (K), magnesium (Mg), and iron (Fe), make up roughly 98 percent by weight of most magmas. Magma also contains small amounts of many other elements, including titanium and manganese, and trace amounts of much rarer elements, such as gold, silver, and uranium.

As magma cools and solidifies, these elements combine to form two major groups of silicate minerals. The *dark silicates* are rich in iron and/or magnesium and are comparatively low in silica (SiO_2). *Olivine, pyroxene, amphibole*, and *biotite mica* are the common dark silicate minerals of Earth's crust. By contrast, the *light silicates* contain greater amounts of potassium, sodium, and calcium and are richer in silica than dark silicates. Light silicate minerals include *quartz, muscovite mica*, and the most abundant mineral group, the *feldspars*. Feldspars make up at least 40 percent of most igneous rocks. Thus, in addition to feldspar, igneous rocks contain some combination of the other light and/or dark silicates listed earlier.

Granitic (Felsic) Versus Basaltic (Mafic) Compositions Despite their great compositional diversity, igneous rocks (and the magmas from which

SmartFigure 3.4
Composition of common igneous rocks (After Dietrich, Daily, and Larsen)

TUTORIAL
https://goo.gl/goy5yO

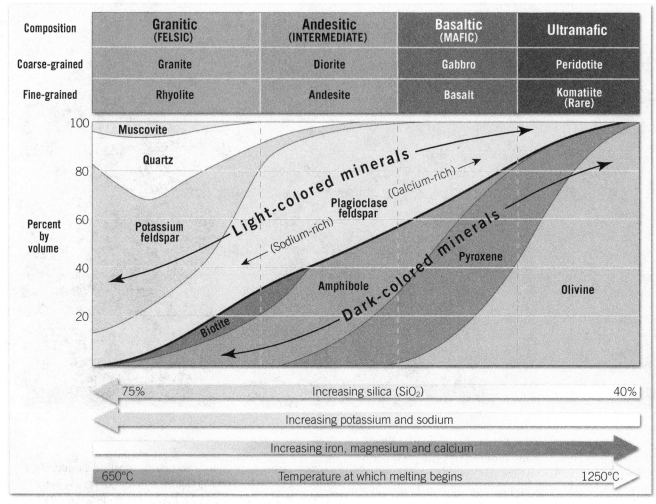

Composition	Granitic (FELSIC)	Andesitic (INTERMEDIATE)	Basaltic (MAFIC)	Ultramafic
Coarse-grained	Granite	Diorite	Gabbro	Peridotite
Fine-grained	Rhyolite	Andesite	Basalt	Komatiite (Rare)

Percent by volume: 100, 80, 60, 40, 20

Muscovite
Quartz
Potassium feldspar
Light-colored minerals
Plagioclase feldspar
(Sodium-rich)
(Calcium-rich)
Biotite
Amphibole
Pyroxene
Olivine
Dark-colored minerals

75% Increasing silica (SiO_2) 40%
Increasing potassium and sodium
Increasing iron, magnesium and calcium
650°C Temperature at which melting begins 1250°C

they form) can be divided into broad groups according to their proportions of light and dark minerals (**Figure 3.4**). Near one end of the continuum are rocks composed almost entirely of light-colored silicates—quartz and potassium feldspar. Igneous rocks in which these are the dominant minerals have a **granitic composition**. Geologists refer to granitic rocks as being **felsic**, a term derived from *feld*spar and *si*lica (quartz). In addition to quartz and feldspar, most granitic rocks contain about 10 percent dark silicate minerals, usually biotite mica and amphibole. Rocks of granitic composition include granite and rhyolite. They are major constituents of the continental crust.

Rocks that contain at least 45 percent dark silicate minerals and calcium-rich plagioclase feldspar (but no quartz) are said to have a **basaltic composition** (see Figure 3.4). Basaltic rocks contain a high percentage of dark silicate minerals, so geologists refer to them as **mafic** (from *ma*gnesium and *fe*rrum, the Latin word for iron). Because of their high iron content, mafic rocks are typically darker and denser than granitic rocks. Rocks of basaltic composition include basalt and gabbro. They make up the ocean floor as well as many of the volcanic islands located within the ocean basins.

Other Compositional Groups As you can see in Figure 3.4, rocks with a composition between granitic and basaltic rocks are said to have an **andesitic composition**, or **intermediate composition**, after the common volcanic rock *andesite*. Intermediate rocks contain at least 25 percent dark silicate minerals, mainly amphibole, pyroxene, and biotite mica, with the other dominant mineral being plagioclase feldspar. This important category of igneous rocks is associated with volcanic activity that is typically confined to the seaward margins of the continents and on volcanic island arcs such as the Aleutian chain.

At the far end of the compositional spectrum are the **ultramafic** igneous rocks, composed almost entirely of dense ferromagnesian minerals (see Figure 3.4). Although ultramafic rocks are rare at Earth's surface, the rock **peridotite**, which is composed mostly of the minerals olivine and pyroxene, is the main constituent of the upper mantle.

What Can Igneous Textures Tell Us?

To describe the *size*, *shape*, and *arrangement* of the mineral grains that make up a rock, geologists use the word **texture**. Texture is an important property because it allows geologists to make inferences about a rock's

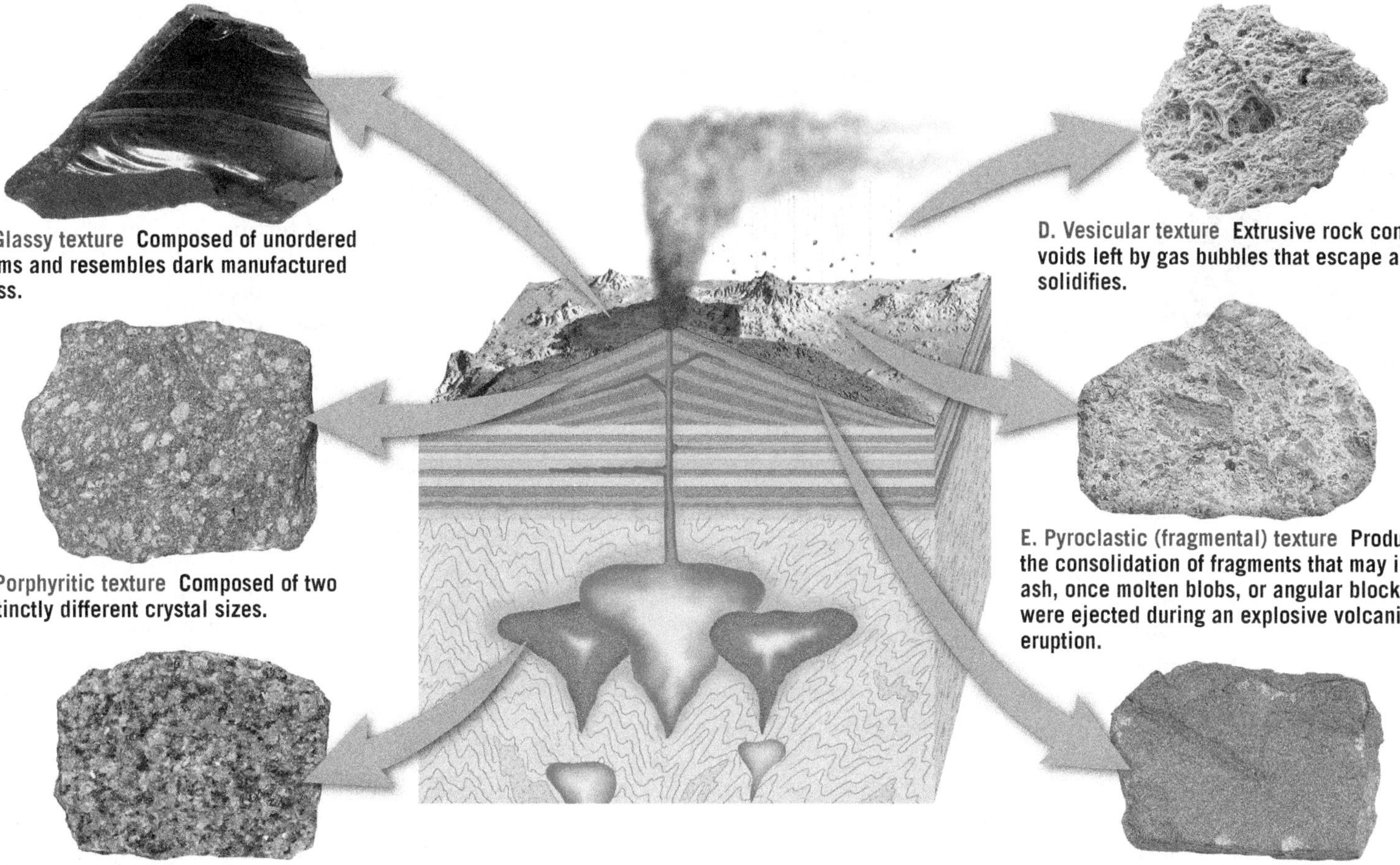

A. Glassy texture **Composed of unordered atoms and resembles dark manufactured glass.**

B. Porphyritic texture **Composed of two distinctly different crystal sizes.**

C. Coarse-grained texture **Mineral grains that are large enough to be identified without a microscope.**

D. Vesicular texture **Extrusive rock containing voids left by gas bubbles that escape as lava solidifies.**

E. Pyroclastic (fragmental) texture **Produced by the consolidation of fragments that may include ash, once molten blobs, or angular blocks that were ejected during an explosive volcanic eruption.**

F. Fine-grained texture **Crystals that are too small for the individual minerals to be identified without a microscope.**

SmartFigure 3.5 **Igneous rock textures** (Photos courtesy of E. J. Tarbuck and Dennis Tasa)

TUTORIAL
https://goo.gl/pYPXWg

origin, based on careful observations of crystal size and other characteristics (Figure 3.5). Rapid cooling produces small crystals, whereas very slow cooling produces much larger crystals. As you might expect, the rate of cooling is slow in magma chambers that lie deep within the crust, whereas a thin layer of lava extruded upon Earth's surface may chill to form solid rock in a matter of hours. Small molten blobs ejected from a volcano during a violent eruption can solidify in mid-air.

Fine-Grained Texture Igneous rocks that form at Earth's surface or as small intrusive masses within the upper crust, where cooling is relatively rapid, exhibit a **fine-grained texture** (see Figure 3.5F). By definition, the crystals that make up fine-grained igneous rocks are so small that individual minerals can be distinguished only with the aid of a microscope or other sophisticated techniques. Therefore, we commonly characterize fine-grained rocks as being light, intermediate, or dark in color.

Coarse-Grained Texture When large masses of magma slowly crystallize at great depth, they form igneous rocks that exhibit a **coarse-grained texture**. Coarse-grained rocks consist of a mass of intergrown crystals that are roughly equal in size and large enough so that the individual minerals can be identified without the aid of a microscope (see Figure 3.5C). Geologists often use a small magnifying lens to aid in identifying minerals in coarse-grained igneous rocks.

Porphyritic Texture A large mass of magma may require thousands or even millions of years to solidify. Because different minerals crystallize under different conditions of temperature and pressure, it is possible for crystals of one mineral to become quite large before others even begin to form. If molten rock containing some large crystals erupts or otherwise moves to a cooler location, the remaining liquid portion of the lava will cool more quickly. The resulting rock, which has large crystals embedded in a matrix of smaller crystals, is said to have a **porphyritic texture** (Figure 3.6). The large crystals in porphyritic rocks are referred to as **phenocrysts**, whereas the matrix of smaller crystals is called **groundmass**.

Vesicular Texture Many extrusive rocks exhibit voids that represent gas bubbles that formed as the lava

Figure 3.6 **Porphyritic texture** The large crystals in porphyritic rocks are called *phenocrysts*, and the small crystals constitute the *groundmass*. (Photos by Dennis Tasa)

solidified. These nearly spherical openings are called *vesicles*, and the rocks that contain them are said to have a **vesicular texture**. Rocks that exhibit a vesicular texture often form in the upper zone of a lava flow, where cooling occurs rapidly enough to preserve the openings produced by the expanding gas bubbles (Figure 3.7). Another common vesicular rock, called *pumice*, forms when silica-rich lava is ejected during an explosive eruption (see Figure 3.5D).

Glassy Texture During some volcanic eruptions, molten rock is ejected into the atmosphere, where it is quenched and becomes a solid. Rapid cooling of this type may generate rocks having a **glassy texture** (see Figure 3.5A). Glass results when unordered ions are "frozen in place" before they are able to unite into an orderly crystalline structure. *Obsidian*, a common type of natural glass, is similar in appearance to dark chunks of manufactured glass.

Figure 3.7 **Vesicular texture** Vesicles form as gas bubbles escape near the top of a lava flow.

Pyroclastic (Fragmental) Texture Another group of igneous rocks is formed from the consolidation of individual rock fragments ejected during explosive volcanic eruptions. The ejected particles might be very fine ash, molten blobs, or large angular blocks torn from the walls of a vent during an eruption. Igneous rocks composed of these rock fragments are said to have a **pyroclastic texture**, or **fragmental texture** (see Figure 3.5E). A common type of pyroclastic rock, called *welded tuff*, is composed of fine fragments of glass that remained hot enough to eventually fuse together.

Common Igneous Rocks

Geologists classify igneous rocks by their texture and mineral composition. The texture of an igneous rock is mainly a result of its cooling history, whereas its mineral composition is largely a result of the chemical makeup of the parent magma (Figure 3.8). Because igneous rocks are classified on the basis of both mineral composition and texture, some rocks have similar mineral constituents but exhibit different textures; therefore, they are given different names.

Granitic (Felsic) Rocks **Granite** is a coarse-grained intrusive igneous rock that forms where large masses of silica-rich magma slowly solidify at depth. Episodes of mountain building may uplift granite and related intrusive rocks, with the processes of weathering and erosion stripping away the overlying crust. Areas where large quantities of granite are exposed at the surface include Pikes Peak in the Rockies, Mount Rushmore in the Black Hills, Stone Mountain in Georgia, and Yosemite National Park in the Sierra Nevada (Figure 3.9).

Granite is perhaps the best-known igneous rock, in part because of its natural beauty, which is enhanced when polished, and partly because of its abundance. Slabs of polished granite are commonly used for tombstones, monuments, and countertops.

Rhyolite is the extrusive equivalent of granite (with the same chemical composition but different texture) and, likewise, is composed essentially of light-colored silicate minerals (see Figure 3.8). This fact accounts for its color, which is usually buff to pink or light gray. Rhyolite is fine grained and frequently contains glass fragments and voids, indicating rapid cooling in a surface environment. In contrast to granite, which is widely distributed as large intrusive masses, rhyolite deposits are less common and generally less voluminous. Yellowstone Park is one well-known exception where extensive rhyolite lava flows are found (along with thick ash deposits of rhyolitic composition).

Obsidian is a common type of volcanic glass. Although dark in color, obsidian usually has a felsic composition; its color results from small amounts of metallic ions in an otherwise clear, glassy substance. Because

IGNEOUS ROCK CLASSIFICATION CHART

	MINERAL COMPOSITION			
	Granitic (FELSIC)	**Andesitic** (INTERMEDIATE)	**Basaltic** (MAFIC)	**Ultramafic**
Dominant Minerals	Quartz Potassium feldspar	Amphibole Plagioclase feldspar	Pyroxene Plagioclase feldspar	Olivine Pyroxene
Accessory Minerals	Plagioclase feldspar Amphibole Muscovite Biotite	Pyroxene Biotite	Amphibole Olivine	Plagioclase feldspar
TEXTURE: **Coarse-grained**	Granite	Diorite	Gabbro	Peridotite
TEXTURE: **Fine-grained**	Rhyolite	Andesite	Basalt	**Komatiite** (rare)
TEXTURE: **Porphyritic** (two distinct grain sizes)	Granite porphyry	Andesite porphyry	Basalt porphyry	Uncommon
TEXTURE: **Glassy**	Obsidian	Less common	Less common	Uncommon
TEXTURE: **Vesicular** (contains voids)	Pumice (also glassy)		Scoria	Uncommon
TEXTURE: **Pyroclastic** (fragmental)	Most fragments < 4mm Tuff or welded tuff	⇄	Most fragments > 4mm Volcanic breccia	Uncommon
Rock Color (based on % of dark minerals)	0% to 25%	25% to 45%	45% to 85%	85% to 100%

SmartFigure 3.8
Classification of igneous rocks, based on their mineral composition and texture Coarse-grained rocks are plutonic, solidifying deep underground. Fine-grained rocks are either volcanic or else solidify as shallow, thin rock bodies. Ultramafic rocks are dark, dense rocks, composed almost entirely of minerals containing iron and magnesium. Although relatively rare on Earth's surface, ultramafic rocks are major constituents of the upper mantle. (Photos by E. J. Tarbuck and Dennis Tasa)

TUTORIAL
https://goo.gl/VOzSR0

Cory Rich/Getty Images

Michael Collier

Granite

Dennis Tasa

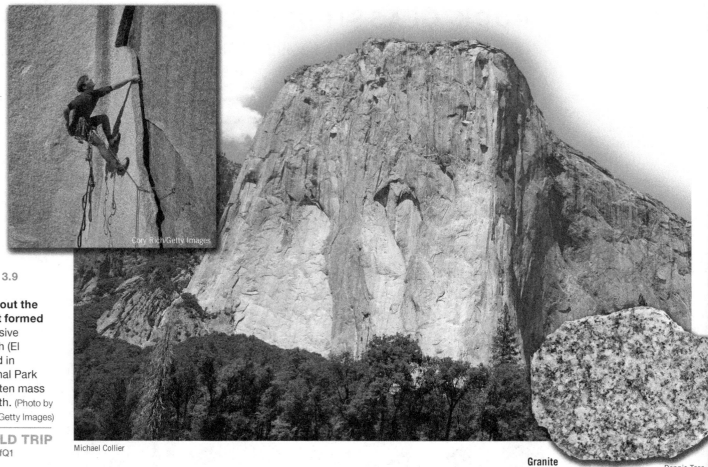

▶ **SmartFigure 3.9 Rocks contain information about the processes that formed them** This massive granitic monolith (El Capitan) located in Yosemite National Park was once a molten mass deep within Earth. (Photo by Corey Rich/Aurora/Getty Images)

MOBILE FIELD TRIP
https://goo.gl/XvMfQ1

of its excellent conchoidal fracture and ability to hold a sharp, hard edge, obsidian was a prized material from which Native Americans chipped arrowheads and cutting tools (Figure 3.10).

Another silica-rich volcanic rock that exhibits a glassy and also vesicular texture is **pumice**. Often found with obsidian, pumice forms when large amounts of gas escape from molten rock to generate a gray, frothy mass (Figure 3.11). In some samples, the vesicles are quite noticeable, whereas in others, the pumice resembles fine shards of intertwined glass. Because of the large volume of air-filled voids, most samples of pumice float in water (see Figure 3.11).

▲ **Figure 3.10 Obsidian, a natural glass** Native Americans used obsidian to make arrowheads and cutting tools. (Photo by Mark Thiessen/National Geographic/Getty Images)

Andesitic (Intermediate) Rocks **Andesite** is a medium-gray extrusive rock. It may be fine grained, or it may have a porphyritic texture (see Figure 3.8), often with phenocrysts of plagioclase feldspar (pale and rectangular) or amphibole (black and elongated). Andesite is a major constituent of many of the volcanoes that are found around the Pacific Rim, including those of the Andes Mountains (after which it is named) and the Cascade Range.

Diorite, the intrusive equivalent of andesite, is a coarse-grained rock that resembles gray granite. However, it can be distinguished from granite because it contains few or no visible quartz crystals and has a higher percentage of dark silicate minerals.

Basaltic (Mafic) Rocks The most common extrusive igneous rock is **basalt**, a very dark green to black, fine-grained volcanic rock composed primarily of pyroxene, olivine, and plagioclase feldspar. Many volcanic islands, such as the Hawaiian Islands and Iceland, are composed

▲ **Figure 3.11 Pumice, a vesicular glassy rock** Pumice is very lightweight because it contains numerous vesicles. (Photo by E. J. Tarbuck; inset photo by Chip Clark/Fundamental Photographs)

Figure 3.12 **Fluid basaltic lava flowing from Kilauea Volcano, Hawaii** (Photo by David Reggie/Perspectives/Getty Images)

mainly of basalt (Figure 3.12). Furthermore, the upper layers of the oceanic crust consist of basalt. In the United States, large portions of central Oregon and Washington were the sites of extensive basaltic outpourings.

The coarse-grained, intrusive equivalent of basalt is **gabbro** (see Figure 3.8). Gabbro is not commonly exposed at the surface, but it makes up a significant percentage of the oceanic crust.

How Igneous Rocks Form

Because igneous rocks exhibit a wide range of compositions, it is logical to assume that they originate from equally diverse magmas. However, geologists have observed that a single volcano, fed by a single magma chamber, may extrude lavas exhibiting quite different compositions. Data of this type led geologists to examine the possibility that magma might change (evolve) and thus become the parent to a variety of igneous rocks. To explore this idea, a pioneering investigation into the crystallization of magma was carried out by N. L. Bowen in the first quarter of the twentieth century.

Bowen's Reaction Series In a laboratory setting, Bowen demonstrated that magma, with its complex chemistry, crystallizes over a temperature range of at least 200°C (360°F), unlike liquids that consist of just a single component (such as water), which solidify at specific temperatures. As magma cools, certain minerals crystallize first at relatively high temperatures. At successively lower temperatures, other minerals begin to crystallize. This succession of minerals, shown in Figure 3.13, became known as **Bowen's reaction series**.

Bowen discovered that the first mineral to crystallize from a body of magma is *olivine*. Further cooling results in the formation of *pyroxene*, as well as *plagioclase feldspar*. At intermediate temperatures, the minerals *amphibole* and *biotite* begin to crystallize.

During the last stage of crystallization, after most of the magma has solidified, the minerals *muscovite* and *potassium feldspar* may form (see Figure 3.13). Finally, *quartz* crystallizes from any remaining liquid. Olivine and quartz are not found in the same igneous rock because quartz crystallizes at much lower temperatures than olivine.

Analysis of igneous rocks provides evidence that this crystallization model approximates what can happen in nature. In particular, we find that minerals that form in the same general range on Bowen's reaction series are

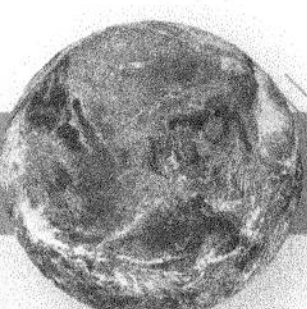

EYE ON EARTH 3.1

These two types of hardened lava, called Pele's hair and Pele's tears, are commonly generated during lava fountaining that occurs at Kilauea Volcano. They are named after Pele, the Hawaiian goddess of fire and volcanoes. Pele's hair consists of goldish strands that are created when tiny blobs of lava are stretched by strong winds. When lava droplets cool quickly, they form Pele's tears, which are sometimes connected to strands of Pele's hair.

QUESTION 1 *What is the texture of Pele's hair and Pele's tears?*

QUESTION 2 *What is the igneous rock name for Pele's tears?*

Pele's hair Marli Miller

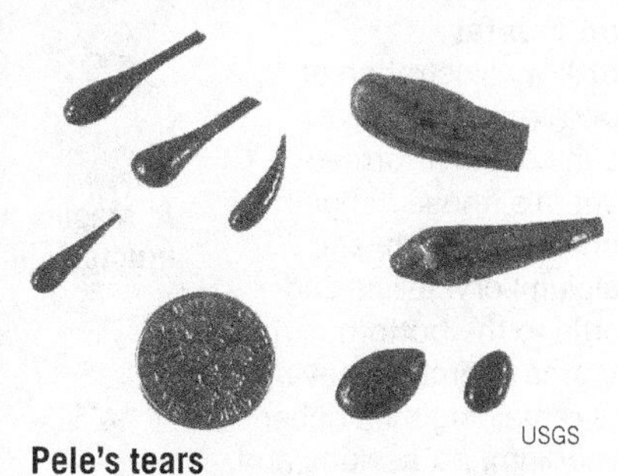

Pele's tears USGS

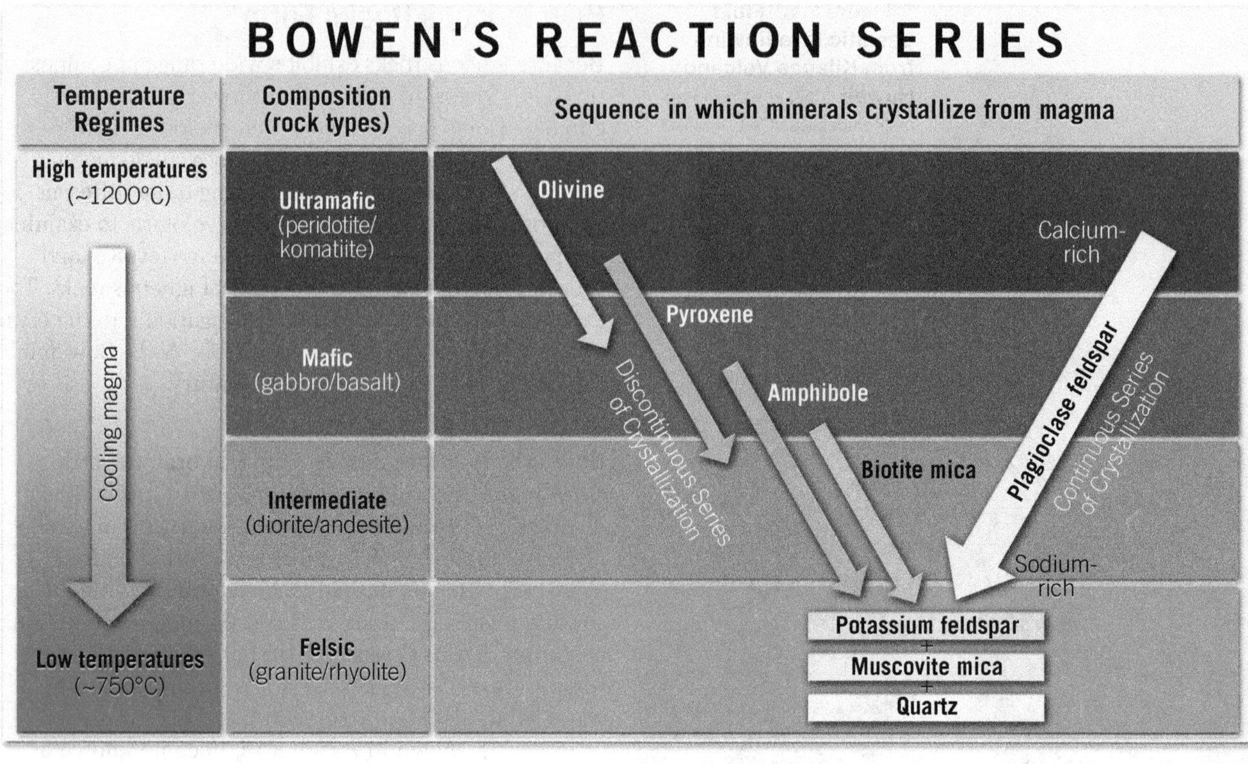

Figure 3.13 **Bowen's reaction series** This diagram shows the sequence in which minerals crystallize from a magma. Compare this figure to the mineral composition of the rock groups in Figure 3.8. Note that each rock group consists of minerals that crystallize in the same temperature range.

found together in the same igneous rocks. For example, notice in Figure 3.13 that the minerals quartz, potassium feldspar, and muscovite, located in the same region of Bowen's diagram, are typically found together as major constituents of the igneous rock *granite*.

Magmatic Differentiation So, how can a single volcano, fed from a single magma chamber, form rocks of quite different composition over its lifetime? First, as Bowen demonstrated, different minerals crystallize from magma according to a predictable pattern. As each mineral forms, it selectively removes certain elements from the melt. For example, crystallization of olivine and pyroxene selectively removes iron and magnesium, leaving the remaining melt more felsic. If some mechanism then physically separates these crystals from the remaining melt, the remaining melt crystallizes to form a different, more felsic rock type.

One such mechanism is **crystal settling**. If early-formed crystals are more dense (heavier) than the remaining melt, they tend to sink toward the bottom of the magma chamber, as shown in Figure 3.14.

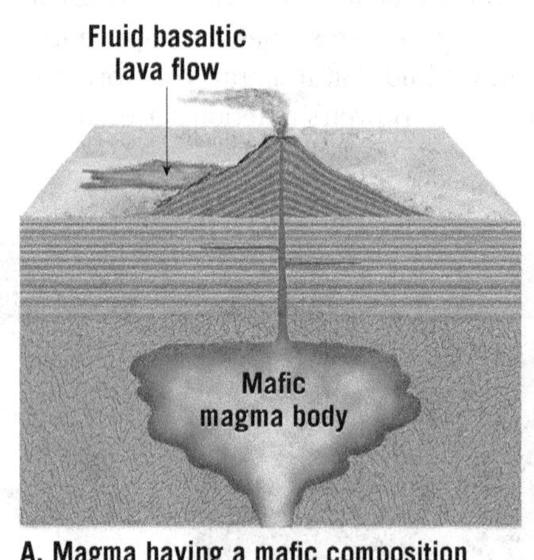

A. Magma having a mafic composition erupts fluid basaltic lavas.

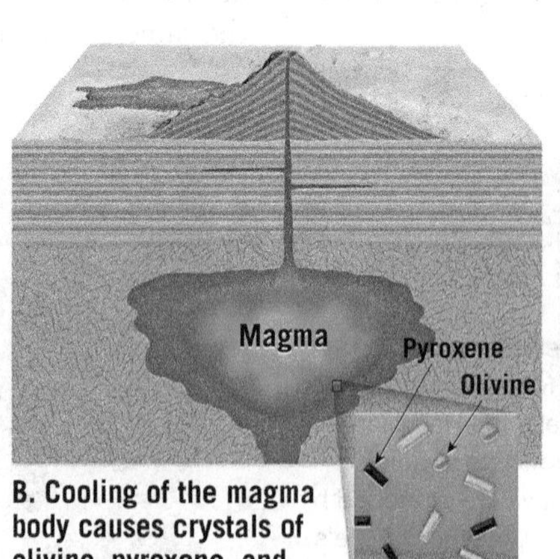

B. Cooling of the magma body causes crystals of olivine, pyroxene, and calcium-rich plagioclase to form and settle out, or crystallize along the magma body's cool margins.

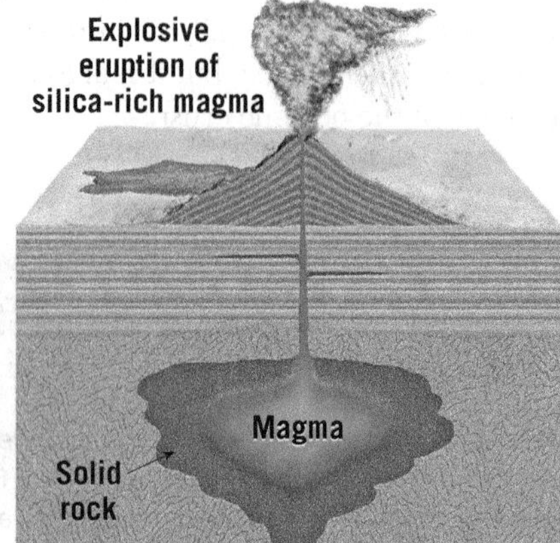

C. The remaining melt will be enriched with silica, and should a subsequent eruption occur, the rocks generated will be more silica-rich and closer to the granitic end of the compositional range than the initial magma.

Figure 3.14 **Magmatic differentiation and crystal settling** Illustration of how a magma evolves as the earliest-formed minerals (those richer in iron, magnesium, and calcium) crystallize and settle to the bottom of the magma chamber, leaving the remaining melt richer in sodium, potassium, and silica (SiO_2).

Consequently, the lower and upper parts of the magma chamber form rocks of differing composition. The formation of one or more secondary magmas from a single parent magma is called **magmatic differentiation**.

At any stage in the evolution of a magma, the solid and liquid components can separate into two chemically distinct units. Furthermore, magmatic differentiation within the secondary magma can generate other chemically distinct masses of molten rock. Consequently, magmatic differentiation and separation of the solid and liquid components at various stages of crystallization can produce several chemically diverse magmas and, ultimately, a variety of igneous rocks.

CONCEPT CHECKS 3.2

1. What is magma? How does magma differ from lava?
2. In what basic settings do intrusive and extrusive igneous rocks originate?
3. How does the rate of cooling influence crystal size? What other factors influence the texture of igneous rocks?
4. What does a porphyritic texture indicate about the history of an igneous rock?
5. List and distinguish among the four basic compositional groups of igneous rocks.
6. How are granite and rhyolite different? In what way are they similar?
7. What is magmatic differentiation? How might this process lead to the formation of several different igneous rocks from a single magma?

3.3 Sedimentary Rocks: Compacted and Cemented Sediment

List and describe the different categories of sedimentary rocks and discuss the processes that change sediment into sedimentary rock.

Recall the rock cycle, which shows the origin of **sedimentary rocks**. Weathering begins the process. Next, gravity and erosional agents remove the products of weathering and carry them to a new location, where they are deposited. Usually, the particles are broken down further during this transport phase. Following deposition, this **sediment** may become lithified, or "turned to rock."

The word *sedimentary* indicates the nature of these rocks, for it is derived from the Latin *sedimentum*, which means "settling," a reference to a solid material settling out of a fluid. Most sediment is deposited in this fashion. Weathered debris is constantly being swept from bedrock and carried away by running water, waves, glacial ice, or wind. Eventually, the material is deposited in lakes, river valleys, seas, and countless other places. The particles in a desert sand dune, the mud on the floor of a swamp, the gravels in a streambed, and even household dust are examples of sediment produced by this never-ending process.

The weathering of bedrock and the transport and deposition of this weathered rock material are continuous processes. Therefore, sediment is found almost everywhere. As piles of sediment accumulate, the materials near the bottom are *compacted* by the weight of the overlying layers. Over long periods, these sediments are *cemented* together by mineral matter deposited from water into the spaces between particles. This forms solid sedimentary rock.

Geologists estimate that sedimentary rocks account for only about 5 percent (by volume) of Earth's outer 16 kilometers (10 miles). However, the importance of this group of rocks is far greater than this percentage implies. If you sampled the rocks exposed at Earth's surface, you would find that the great majority of them are sedimentary (Figure 3.15). Indeed, about 75 percent

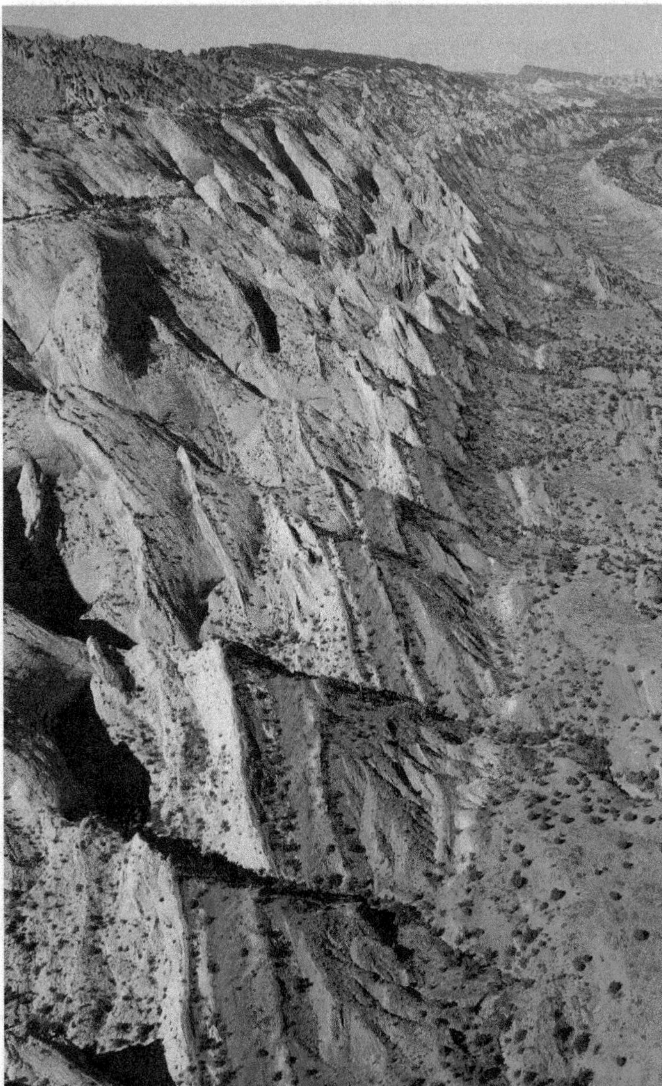

SmartFigure 3.15 **Sedimentary rocks exposed along the Waterpocket Fold, Capitol Reef National Park, Utah** About 75 percent of all rock exposures on continents consist of sedimentary rocks. (Photo by Michael Collier)

MOBILE FIELD TRIP
https://goo.gl/fkQcwr

of all rock outcrops on the continents are sedimentary. Therefore, we can think of sedimentary rocks as comprising a relatively thin and somewhat discontinuous layer in the uppermost portion of the crust. This makes sense because sediment accumulates at the surface.

It is from sedimentary rocks that geologists reconstruct many details of Earth's history. Because sediments are deposited in a variety of different settings at the surface, the rock layers that they eventually form hold many clues to past surface environments. They may also exhibit characteristics that allow geologists to decipher information about the method and distance of sediment transport. Furthermore, sedimentary rocks contain fossils, which are vital evidence in the study of the geologic past.

Finally, many sedimentary rocks are important economically. Bituminous coal, which is burned to provide a significant portion of U.S. electrical energy, is classified as a sedimentary rock. Other major energy resources (such as petroleum and natural gas) occur in pores within sedimentary rocks. Other sedimentary rocks are major sources of iron, aluminum, manganese, and fertilizer, plus numerous materials essential to the construction industry.

Types of Sedimentary Rocks

Materials that accumulate as sediment have two principal sources. First, sediments may originate as solid particles from weathered rocks, such as the igneous rocks described earlier. These particles are called *detritus*, and the sedimentary rocks they form are called **detrital sedimentary rocks**.

Detrital Sedimentary Rocks Though a wide variety of minerals and rock fragments may be found in detrital rocks, clay minerals and quartz dominate. As you learned earlier, clay minerals are the most abundant product of the chemical weathering of silicate minerals, especially the feldspars. Quartz, on the other hand, is abundant because it is extremely durable and very resistant to chemical weathering. Thus, when igneous rocks such as granite are weathered, individual quartz grains are set free.

Particle size is the primary basis for distinguishing among detrital sedimentary rocks. Figure 3.16 presents the size categories for particles making up detrital rocks. When a rock consists of a significant amount of rounded gravel-size particles, it is called **conglomerate**. As Figure 3.16 shows, gravel-size particles can range in size from as large as boulders to particles as small as peas. If the gravel-size particles are angular rather than rounded, the rock is called **breccia**. Angular fragments indicate that the particles were not transported very far from their source prior to deposition and so have not had corners and rough edges abraded. When sand-size grains rather than gravel-size particles prevail, the rock is called **sandstone**.

Shale, which is typically composed of clay minerals, is a name commonly applied to all very fine-grained sedimentary rocks (see Figure 3.16). However, in the strict use of the word, a rock can be classified as shale only if it exhibits the ability to split into thin layers. If a fine-grained rock breaks into chunks or blocks, the name *mudstone*

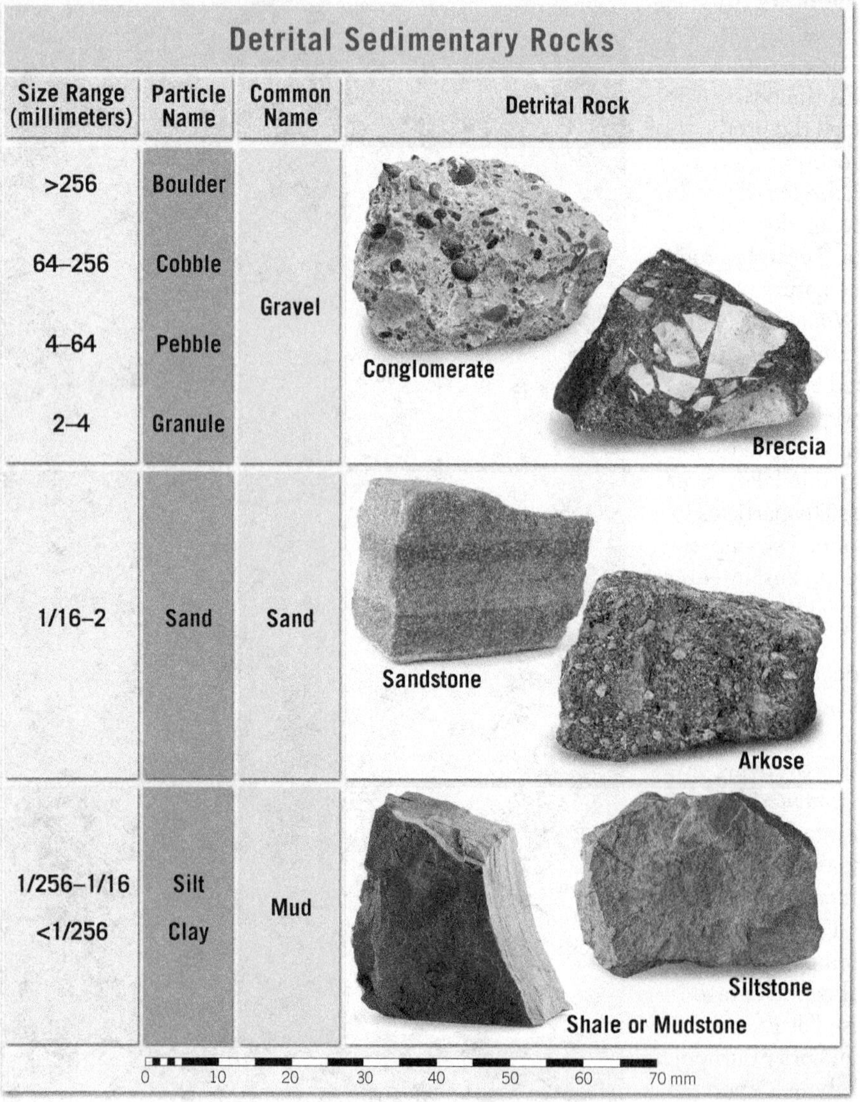

Detrital Sedimentary Rocks

Size Range (millimeters)	Particle Name	Common Name	Detrital Rock
>256	Boulder	Gravel	Conglomerate; Breccia
64–256	Cobble		
4–64	Pebble		
2–4	Granule		
1/16–2	Sand	Sand	Sandstone; Arkose
1/256–1/16	Silt	Mud	Siltstone; Shale or Mudstone
<1/256	Clay		

0 10 20 30 40 50 60 70 mm

Figure 3.16 Detrital sedimentary rocks (Photos by Dennis Tasa and E. J. Tarbuck)

is applied. **Siltstone**, another fine-grained rock, is composed of slightly larger silt-size grains intermixed with clay-sized sediment.

Particle size also provides useful information about the environment in which the sediment was deposited. Currents of water or air sort the particles by size. The stronger the current, the larger the particle size carried. Gravels, for example, are moved by swiftly flowing rivers, rockslides, and glaciers. Less energy is required to transport sand; thus, it is common in windblown dunes, river deposits, and beaches. Because silts and clays settle very slowly, accumulations of these materials are generally associated with the quiet waters of a lake, lagoon, swamp, or marine environment.

Although detrital sedimentary rocks are largely classified by particle size, in certain cases, the mineral composition is also part of the rock name. For example, most sandstones are rich in quartz, so they are referred to as *quartz sandstone*. When sandstone contains more than 25 percent of the mineral feldspar, it is called *arkose*. In addition, rocks consisting of detrital sediments are rarely composed of grains of just one size. Consequently, a rock containing quantities of both sand and silt can be correctly classified as sandy siltstone or silty sandstone, depending on which particle size dominates.

Chemical, Biochemical, and Organic Sedimentary Rocks In contrast to detrital rocks, which are the solid products of weathering, **chemical sedimentary rocks** and **biochemical sedimentary rocks** are derived from material (ions) carried in solution to lakes and seas (Figure 3.17). This material does not remain dissolved in water indefinitely. Under certain conditions and due to physical processes, the dissolved matter precipitates to form *chemical sediments*. An example of chemical sediments resulting from physical processes is the salt left behind as a body of saltwater evaporates.

Precipitation may also occur indirectly through life processes of water-dwelling organisms that form materials called *biochemical sediments*. Many aquatic animals and plants extract the mineral matter dissolved in the water to form shells and other hard parts. After the organisms die, their skeletons may accumulate on the floor of a lake or an ocean.

Limestone, an abundant sedimentary rock, is composed chiefly of the mineral calcite ($CaCO_3$). Nearly 90 percent of limestone is formed from biochemical sediments secreted by marine organisms, and the remaining amount consists of chemical sediments that precipitated directly from seawater.

One easily identified biochemical limestone is **coquina**, a coarse rock composed of loosely cemented shells and shell fragments (Figure 3.18). Another less obvious but familiar example is *chalk*, a soft, porous rock made up almost entirely of the hard parts of microscopic organisms that are no larger than the head of

Chemical, Biochemical, and Organic Sedimentary Rocks

Composition		Texture	Rock Name
Calcite, $CaCO_3$		Fine to coarse crystalline	Crystalline Limestone 
		Very fine-grained crystals	Microcrystalline Limestone 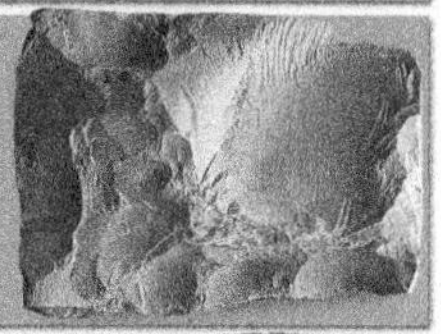
		Fine to coarse crystalline	Travertine 
	Biochemical Limestone	Visible shells and shell fragments loosely cemented	Coquina 
		Various size shells cemented with calcite cement	Fossiliferous Limestone 
		Microscopic shells and clay	Chalk 
Quartz, SiO_2		Very fine crystalline	Chert (light colored) 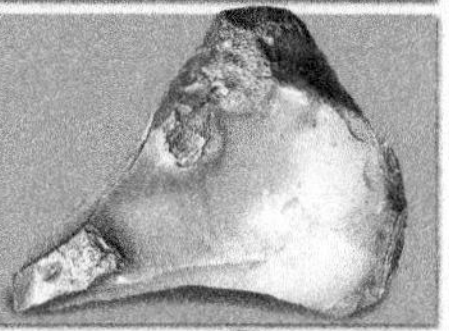
Gypsum $CaSO_4 \bullet 2H_2O$		Fine to coarse crystalline	Rock Gypsum 
Halite, NaCl		Fine to coarse crystalline	Rock Salt 
Altered plant fragments (organic matter)		Fine-grained	Bituminous Coal 

Figure 3.17 **Chemical, biochemical, and organic sedimentary rocks** (Photos by Dennis Tasa and E. J. Tarbuck)

▲ **Figure 3.18 Coquina** This variety of limestone consists of shell fragments; therefore, it has a biochemical origin. (Rock sample photo by E. J. Tarbuck; beach photo by Donald R. Frazier Photolibrary, Inc./Alamy Images)

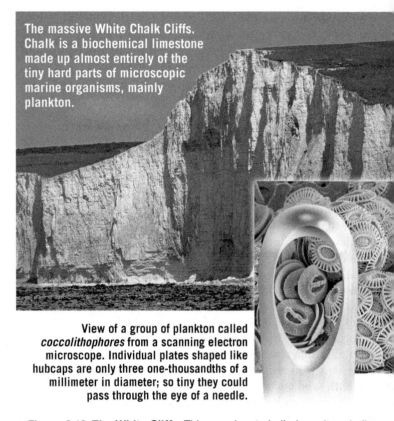

The massive White Chalk Cliffs. Chalk is a biochemical limestone made up almost entirely of the tiny hard parts of microscopic marine organisms, mainly plankton.

View of a group of plankton called *coccolithophores* from a scanning electron microscope. Individual plates shaped like hubcaps are only three one-thousandths of a millimeter in diameter; so tiny they could pass through the eye of a needle.

▲ **Figure 3.19 The White Cliffs** This prominent chalk deposit underlies large portions of southern England as well as parts of northern France. (Photo by David Wall/Alamy; coccolithophores photo by Steve Gschmeissner/SPL/Alamy)

a pin. Among the most famous chalk deposits are the White Chalk Cliffs exposed along the southeast coast of England (Figure 3.19).

Inorganic limestone forms when chemical changes or high water temperatures cause calcium carbonate (calcite) to precipitate. **Travertine**, the type of limestone that decorates caverns, is one example. Groundwater is the source of travertine that is deposited in caves. As water drops reach the air in a cavern, some of the carbon dioxide dissolved in the water escapes, causing calcium carbonate to precipitate.

Dissolved silica (SiO_2) precipitates to form varieties of *microcrystalline rocks*—rocks consisting of very fine-grained quartz crystals (Figure 3.20). Sedimentary rocks composed of microcrystalline quartz include *chert* (light color), *flint* (dark), *jasper* (red), and *petrified wood* (multicolored). These chemical sedimentary rocks may have either an inorganic or biochemical origin, but the mode of origin is usually difficult to determine.

Very often, evaporation causes minerals to precipitate from water. Such minerals include halite, the chief component of *rock salt*, and gypsum, the main ingredient of *rock gypsum*. Both materials have significant commercial importance. Halite is familiar to everyone as the common salt used in cooking and seasoning foods. Of course, it has many other uses and has been considered important enough that people have sought, traded, and fought over it for much of human history. Gypsum is the basic ingredient in plaster of Paris. This material is used most extensively in the construction industry for drywall and plaster.

In the geologic past, many areas that are now dry land were covered by shallow arms of the sea that had

▶ **Figure 3.20 Colorful varieties of chert** Chert is the name applied to a number of dense, hard chemical sedimentary rocks made of microcrystalline quartz. (Photos A and B by E. J. Tarbuck; photo C by Daniel Sambraus/Science Source; photo D by gracious_tiger/Shutterstock)

A. Flint

B. Jasper

C. Chert arrowhead

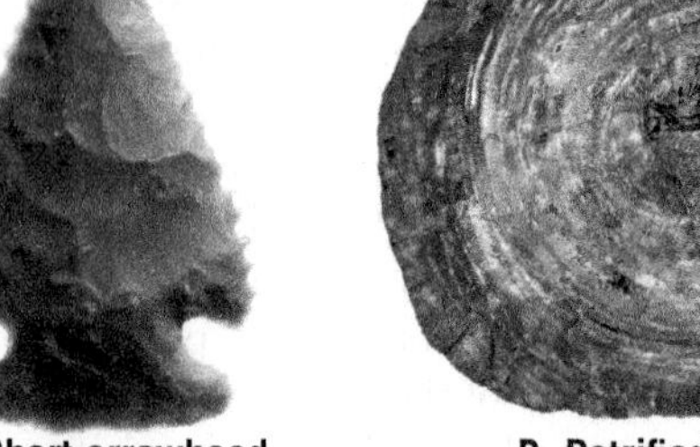
D. Petrified wood

SmartFigure 3.21 **Bonneville Salt Flats** This well-known Utah site was once a large salt lake. (Photo by Jupiter Images/Getty Images; satellite image by NASA)

TUTORIAL
https://goo.gl/eZ5W9p

only narrow connections to the open ocean. Under these conditions, water continually moved into the bay to replace water lost by evaporation. Eventually, the waters of the bay became saturated, and salt deposition began. Today, these arms of the sea are gone, and the remaining deposits are called **evaporite deposits**.

Evaporite deposits are also found in enclosed basins on land. One example is Death Valley, California, where following periods of rainfall or snowmelt in the surrounding mountains, streams carry mineral-rich water into the lowest parts of the valley. As the water in these desert basins evaporates, materials that were dissolved in the water are left behind, forming a white, salt-rich crust on the ground that grows in thickness to generate a *salt flat* (Figure 3.21).

Coal—An Organic Sedimentary Rock In contrast to sedimentary rocks that are calcite or silica rich, **coal** consists mostly of *organic matter*. Because coal is produced by biochemical activity and contains organic matter, it is often classified as a *biochemical*, or *organic, rock*. Examining a piece of *lignite*, also called *brown coal*, under a magnifying glass reveals plant structures such as leaves, bark, and wood that have been chemically altered but remain identifiable (Figure 3.22). This observation supports the conclusion that coal is the end product of the burial of large amounts of plant material over extended periods.

The initial stage in coal formation is the accumulation of large quantities of plant remains. However, special conditions are required for such accumulations because dead plants normally decompose when exposed to the atmosphere. Ideal environments that allow plant matter to accumulate are swamps. Stagnant swamp water is oxygen deficient, which makes it impossible for plant material to completely decay (oxidize). At various times during Earth history, such environments have been common.

Coal forms in a series of stages. With each successive stage, higher temperatures and pressures drive off impurities and volatiles, as shown in Figure 3.22. Lignite and bituminous coals are sedimentary rocks, but anthracite is a metamorphic rock. Anthracite forms when sedimentary layers are subjected to the folding and deformation associated with mountain building, discussed in the final section of this chapter.

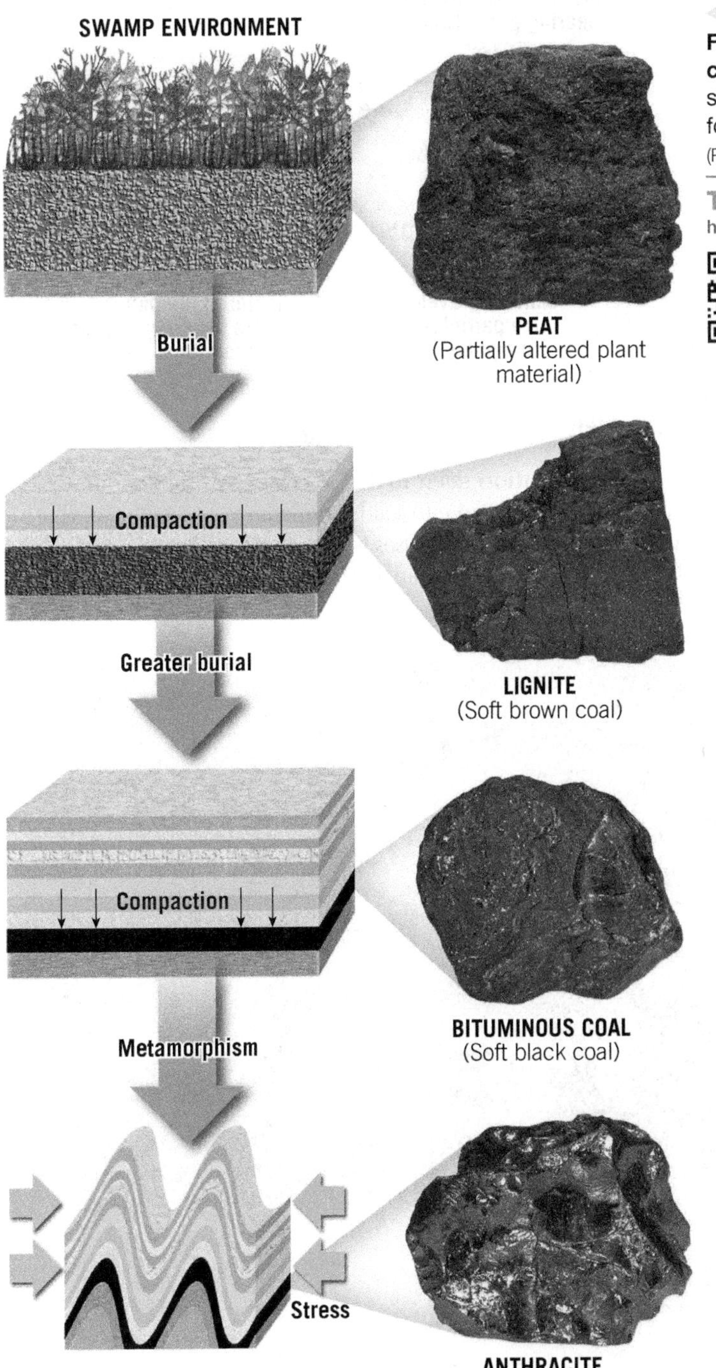

SmartFigure 3.22 **From plants to coal** Successive stages in the formation of coal. (Photos by E. J. Tarbuck)

TUTORIAL
https://goo.gl/KPKKPW

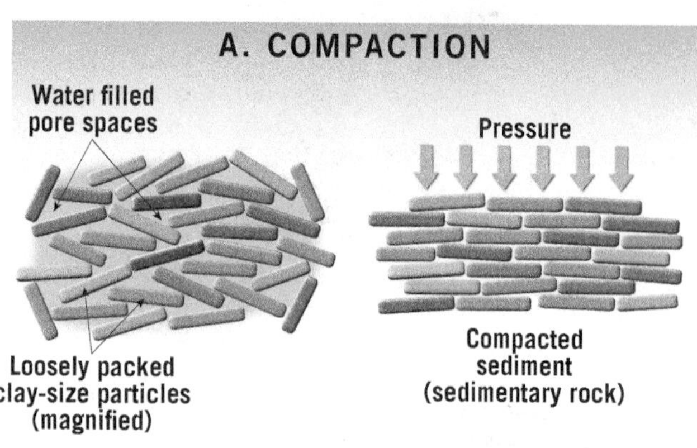

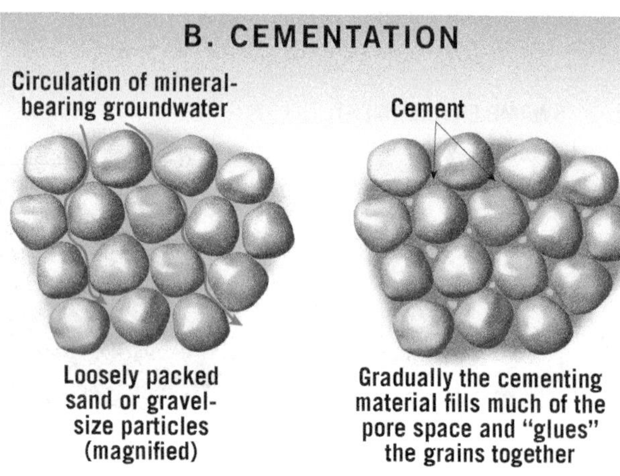

Figure 3.23 Compaction and cementation

Lithification of Sediment

Lithification refers to the processes by which sediments are transformed into solid sedimentary rocks. One of the most common processes is **compaction** (Figure 3.23A). As sediments accumulate through time, the weight of overlying material compresses the deeper sediments. As the grains are pressed closer and closer, pore space is greatly reduced. For example, when clays are buried beneath several thousand meters of material, the volume of the clay may be reduced as much as 40 percent. Compaction is most effective in converting very fine-grained sediments, such as clay-size particles, into sedimentary rocks.

Because sand and coarse sediments (gravel) are not easily compressed, they are generally transformed into sedimentary rocks by the process of **cementation** (Figure 3.23B). The cementing materials are carried in solution by water that percolates through the pore spaces between particles. Over time, the cement precipitates onto the sediment grains, fills the open spaces, and acts like glue, joining the particles together. Calcite, silica, and iron oxide are the most common cements. The cementing material is relatively easy to identify. Calcite cement effervesces (fizzes) in contact with dilute hydrochloric acid. Silica is the hardest cement and thus produces the hardest sedimentary rocks. An orange or red color in a sedimentary rock usually means iron oxide is present.

Figure 3.24 Sedimentary environments A. Ripple marks preserved in sedimentary rocks may indicate a beach or stream channel environment. (Photo by Tim Graham/Alamy) **B.** Mud cracks form when wet mud or clay dries and shrinks, perhaps signifying a tidal flat or desert basin. (Photo by Marli Miller)

Features of Sedimentary Rocks

Sedimentary rocks are particularly important in the study of Earth history. These rocks form at Earth's surface, and as layer upon layer of sediment accumulates, each layer records the nature of the environment at the time the sediment was deposited. These layers, called **strata**, or **beds**, are the *single most characteristic feature of sedimentary rocks* (see Figure 3.15).

Bed thickness ranges from microscopically thin to tens of meters thick. Separating the strata are *bedding planes*, flat surfaces along which rocks tend to separate or break. Generally, each bedding plane marks the end of one episode of sedimentation and the beginning of another.

Sedimentary rocks provide geologists with evidence for deciphering past environments. A conglomerate, for example, indicates a high-energy environment, such as a rushing stream, where only the coarse materials can settle out. By contrast, both coal and black shale (with its high carbon content) are associated with a low-energy, organic-rich environment, such as a swamp or lagoon. Other features found in some sedimentary rocks also give clues to past environments (Figure 3.24).

Fossils, the traces or remains of prehistoric life, are perhaps the most important inclusions found in some sedimentary rocks (Figure 3.25). Knowing the nature of the life-forms that existed at a particular time may help answer many questions about the environment. Was it land or ocean, lake or swamp? Was the climate hot or cold, rainy or dry? Was the ocean water shallow or deep, turbid or clear? Furthermore, fossils are important time indicators and play a key role in matching up rocks from different places that are the same age. Fossils are important tools used in interpreting the geologic past and will be examined in more detail in Chapter 11.

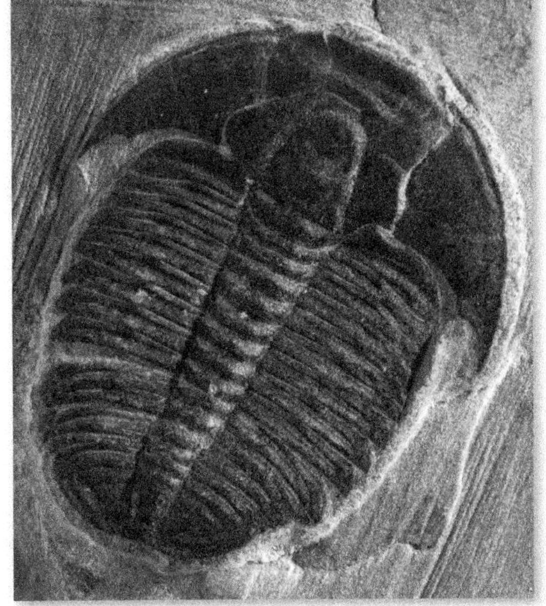

Figure 3.25 **Fossils: Clues to the past** Fossils, the remains or traces of prehistoric life, are primarily associated with sediments and sedimentary rocks. A large variety of trilobites are associated with the Paleozoic era. (Photo by Russell Shively/Shutterstock)

CONCEPT CHECKS 3.3

1. Why are sedimentary rocks important?
2. What minerals are most abundant in detrital sedimentary rocks? In which rocks do these sediments predominate?
3. Distinguish between conglomerate and breccia.
4. What are the two groups of chemical sedimentary rock? Give an example of a rock that belongs to each group.
5. How do evaporites form? Give an example.
6. Describe the two processes by which sediments are transformed into sedimentary rocks. Which is the more effective process in the lithification of sand- and gravel-sized sediments?
7. List three common cements. How might each be identified?
8. What is the single most characteristic feature of sedimentary rocks?

3.4 Metamorphic Rocks: New Rock from Old

Define *metamorphism*, explain how metamorphic rocks form, and describe the agents of metamorphism.

Recall from the discussion of the rock cycle that metamorphism is the transformation of one rock type into another. **Metamorphic rocks** are produced from preexisting igneous, sedimentary, or even other metamorphic rocks. Thus, every metamorphic rock has a *parent rock*—the rock from which it was formed.

Metamorphism, which means "to change form," is a process that leads to changes in the mineralogy, texture (for example, grain size), and sometimes chemical composition of rocks (Figure 3.26). Metamorphism occurs most often when rock is subjected to a significant increase in temperature and/or pressure. In response to

SmartFigure 3.26 **Folded and metamorphosed rocks** This rock outcrop is in Anza-Borrego Desert State Park, California. (Photo by AL Trujillo©APT Photos)

ANIMATION
https://goo.gl/8Lru7x

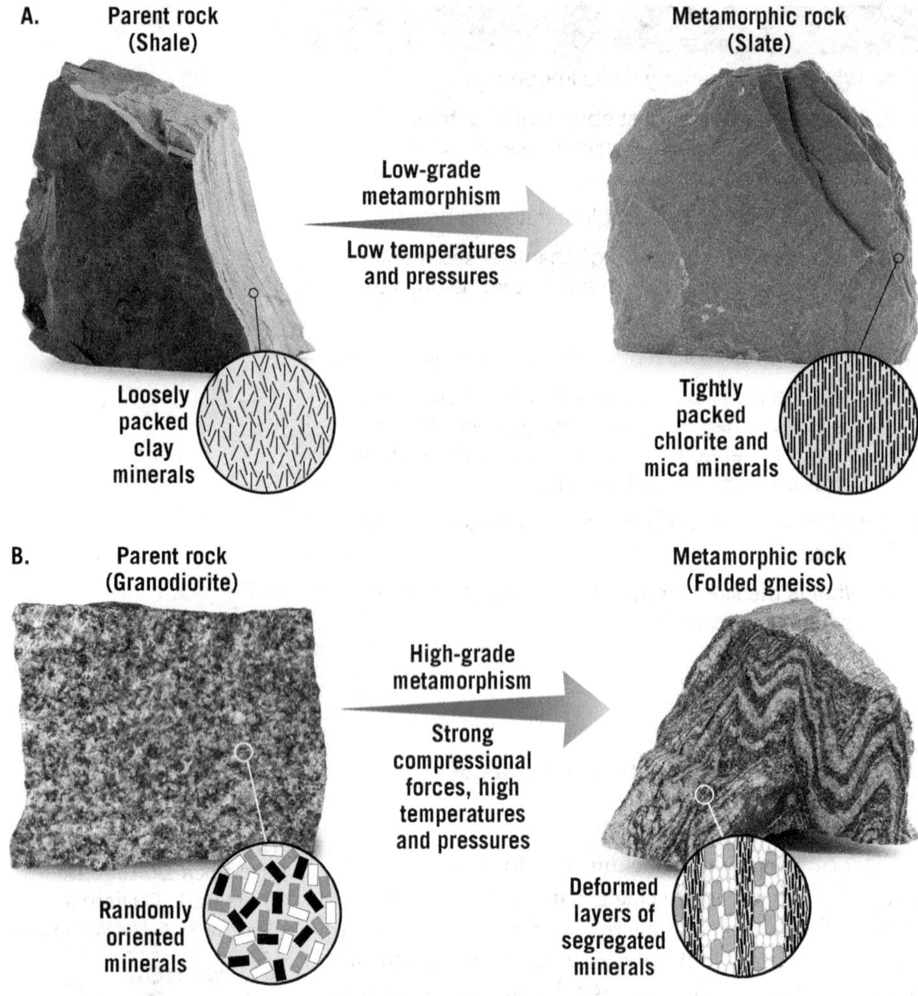

▲ Figure 3.27
Metamorphic grade
A. Low-grade metamorphism, illustrated by the transformation of the common sedimentary rock shale to the more compact metamorphic rock slate. **B.** High-grade metamorphic environments obliterate the existing texture and often change the mineralogy of the parent rock. High-grade metamorphism occurs at temperatures that approach those at which rocks melt. (Photos by Dennis Tasa)

these new conditions, the rock gradually changes until it reaches a state of equilibrium with the new environment. Most metamorphic changes occur at the elevated temperatures and pressures that exist in the zone beginning a few kilometers below Earth's surface and extending into the mantle.

Metamorphism often progresses incrementally, from slight changes (*low-grade metamorphism*) to substantial changes (*high-grade metamorphism*). For example, under low-grade metamorphism, the common sedimentary rock *shale* becomes the more compact metamorphic rock *slate* (Figure 3.27A). Hand samples of shale and slate are sometimes difficult to distinguish, illustrating that the transition from sedimentary to metamorphic rock is often gradual, and the changes can be subtle.

In more extreme environments, metamorphism causes a transformation so complete that the identity of the parent rock cannot be determined. In high-grade metamorphism, such features as bedding planes, fossils, and vesicles that may have existed in the parent rock are obliterated. Furthermore, when rocks deep in the crust (where temperatures are high) are subjected to directed pressure, the entire mass may deform, producing large-scale structures such as folds (Figure 3.27B).

By definition, rock undergoing metamorphism remains essentially solid. In the most extreme metamorphic environments, the temperatures approach those at which rocks melt. However, if appreciable melting occurs, the rocks have entered the realm of igneous activity.

Most metamorphism occurs in one of two settings:

- When rock is intruded by magma, **contact metamorphism** may take place, as the magma heats the adjacent rock to temperatures that cause metamorphic changes.
- During mountain building, great quantities of rock are subjected to pressures and high temperatures associated with large-scale deformation called **regional metamorphism**.

Extensive areas of metamorphic rocks are exposed on every continent. Metamorphic rocks are an important component of many mountain belts, where they make up a large portion of a mountain's crystalline core. Even the stable continental interiors, which are generally covered by sedimentary rocks, are underlain by metamorphic basement rocks. In each of these settings, the metamorphic rocks are usually highly deformed and intruded by igneous masses. Consequently, significant parts of Earth's continental crust are composed of metamorphic and associated igneous rocks.

What Drives Metamorphism?

The agents of metamorphism include *heat, confining pressure, differential stress,* and *chemically active fluids.* During metamorphism, rocks are often subjected to all four metamorphic agents simultaneously. However, the degree of metamorphism and the contribution of each agent vary greatly from one environment to another.

Heat as a Metamorphic Agent *Thermal energy*, commonly referred to as heat, is the most important factor driving metamorphism. It triggers chemical reactions that result in the recrystallization of existing minerals and the formation of new minerals. Thermal energy for metamorphism comes mainly from two sources. Rocks experience a rise in temperature when they are intruded by magma rising from below (contact metamorphism). In this situation, the adjacent host rock is "baked" by the emplaced magma.

By contrast, rocks that formed at Earth's surface experience a gradual increase in temperature and pressure as they are taken to greater depths. In the upper crust, this increase in temperature averages about 25°C per kilometer. When buried to a depth of about 8 kilometers (5 miles), where temperatures are between 150° and 200°C, clay minerals tend to become unstable and begin to recrystallize into other minerals, such as chlorite and muscovite, that are stable in this environment. (Chlorite is a mica-like mineral formed by the metamorphism of iron- and magnesium-rich silicates.) However, many silicate minerals, particularly those found in crystalline igneous rocks—quartz and feldspar, for example—remain stable at these temperatures. Thus, these minerals require much higher temperatures in order to metamorphose and recrystallize.

Confining Pressure and Differential Stress as Metamorphic Agents Pressure, like temperature, increases with depth as the thickness of the overlying rock increases. Buried rocks are subjected to **confining pressure**—similar to water pressure in that the forces are equally applied in all directions (Figure 3.28A). The deeper you go in the ocean, the greater the confining pressure. The same is true for buried rock. Confining pressure causes the spaces between mineral grains to close, producing a more compact rock that has greater density. Further, at great depths, confining pressure may cause minerals to recrystallize into new minerals that display more compact crystalline forms.

During episodes of mountain building, large rock bodies become highly crumpled and metamorphosed (Figure 3.28B). Unlike confining pressure, which "squeezes" rock equally in all directions, the forces that generate mountains are unequal in different directions and are called **differential stress**. As shown in Figure 3.28B, rocks subjected to differential stress are shortened in the direction of greatest stress, and they are elongated, or lengthened, in the direction perpendicular to that stress. The deformation caused by differential stresses plays a major role in developing metamorphic textures.

In surface environments where temperatures are relatively low, rocks are *brittle* and tend to fracture when subjected to differential stress. (Think of a heavy boot crushing a piece of fine crystal.) Continued deformation grinds and pulverizes the mineral grains into small fragments. By contrast, in high-temperature, high-pressure environments deep in Earth's crust, rocks are *ductile* and tend to flow rather than break. (Think of a heavy boot crushing a soda can.) When rocks exhibit ductile behavior, their mineral grains tend to flatten and elongate when subjected to differential stress. This accounts for their ability to generate intricate folds (see Figure 3.26).

Chemically Active Fluids as Metamorphic Agents Ion-rich fluids composed mainly of water and other volatiles (materials that readily change to gases at surface conditions) are believed to play an important role in some types of metamorphism. Fluids that surround mineral grains act as catalysts that promote recrystallization by enhancing ion migration. In progressively hotter environments, these ion-rich fluids become correspondingly more reactive. Chemically active fluids can produce two types of metamorphism, explained below. The first type changes the arrangement and shape of mineral grains within a rock; the second type changes the rock's chemical composition.

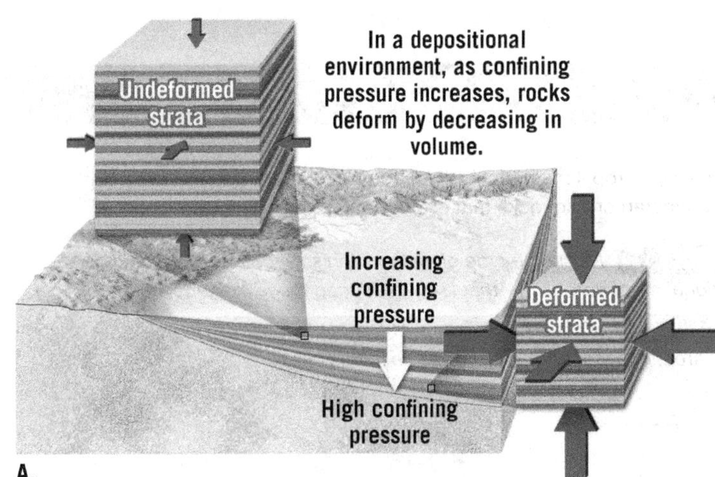
Undeformed strata
In a depositional environment, as confining pressure increases, rocks deform by decreasing in volume.
Increasing confining pressure
Deformed strata
High confining pressure
A.

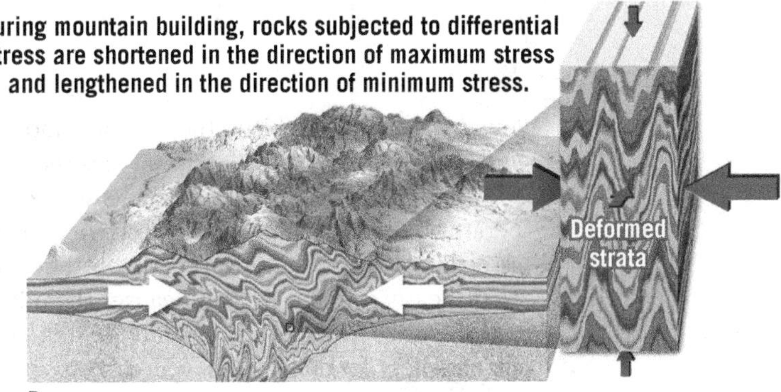
During mountain building, rocks subjected to differential stress are shortened in the direction of maximum stress and lengthened in the direction of minimum stress.
Deformed strata
B.

SmartFigure 3.28
Confining pressure and differential stress

TUTORIAL
https://goo.gl/nBMksQ

When two mineral grains are squeezed together, the parts of their crystalline structures that touch are the most highly stressed. Atoms at these sites are readily dissolved by the hot fluids and move to fill the voids between individual grains. Thus, hot fluids aid in the recrystallization of mineral grains by dissolving material from regions of high stress and then precipitating (depositing) this material in areas of low stress. As a result, *minerals tend to recrystallize and grow longer in a direction perpendicular to compressional stresses.*

When hot fluids circulate freely through rocks, ionic exchange may occur between adjacent rock layers, or ions may migrate great distances before they are finally deposited. The latter situation is particularly common when we consider hot fluids that escape during the crystallization of an intrusive mass of magma. If the rocks surrounding the magma differ markedly in composition from the invading fluids, there may be a substantial exchange of ions between the fluids and host rocks. When this occurs, the overall composition of the surrounding rock changes.

Metamorphic Textures

The degree of metamorphism is reflected in a rock's *texture* and *mineralogy*. (Recall that the term *texture* is used to describe the size, shape, and arrangement of grains within a rock.) When rocks are subjected to

EYE ON EARTH 3.2

This rock outcrop, located in Joshua Tree National Park, consists of dark-colored metamorphic rocks that overlie light-colored igneous rocks.

QUESTION 1 *Name the type of metamorphism—contact (thermal) or regional metamorphism—that likely produced these metamorphic rocks.*

QUESTION 2 *Write a brief statement that describes the geologic history of this area, based on what you observe in this image.*

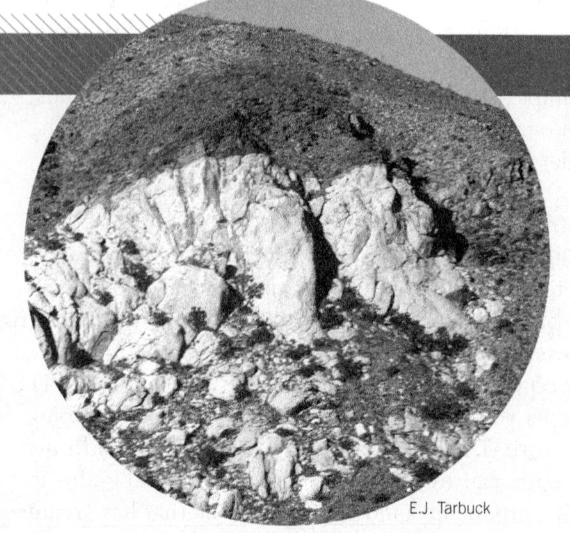

E.J. Tarbuck

low-grade metamorphism, they become more compact and thus denser. A common example is the metamorphic rock slate, which forms when shale is subjected to temperatures and pressures only slightly greater than those associated with the compaction that lithifies sediment. In this case, differential stress causes the microscopic clay minerals in shale to align into the more compact arrangement found in slate.

SmartFigure 3.29
Rotation of platy and elongated mineral grains to produce foliated texture When subjected to differential stress during metamorphism, some mineral grains become reoriented and aligned at right angles to the stress. The resulting orientation of mineral grains gives the rock a foliated (layered) texture. If the coarse-grained igneous rock (granite) on the left underwent intense metamorphism, it could end up closely resembling the metamorphic rock on the right (gneiss). (Photos by E. J. Tarbuck)

ANIMATION
https://goo.gl/ClsjD3

FOLIATION

Before metamorphism (Confining pressure)

Metamorphism

After metamorphism (Differential stress)

Platy and elongated mineral grains having random orientation.

When differential stress causes rocks to flatten, the mineral grains rotate and align roughly perpendicular to the direction of maximum differential stress.

Under more extreme temperature and pressure, stress causes certain minerals to recrystallize. In general, recrystallization encourages the growth of larger crystals. Consequently, many metamorphic rocks consist of visible crystals, much like coarse-grained igneous rocks.

Foliation The term **foliation** refers to any nearly flat arrangement of mineral grains or structural features within a rock. Although foliation may occur in some sedimentary and even a few types of igneous rocks, it is a fundamental characteristic of regionally metamorphosed rocks—that is, rock units that have been strongly deformed, mainly during folding. As shown in Figure 3.29, foliation in metamorphic environments is ultimately driven by compressional stresses that shorten rock units, causing mineral grains in preexisting rocks to develop parallel, or nearly parallel, alignments. Examples of foliation include the *parallel alignment of platy (flat and disk-like) minerals* such as the micas; *elongated* or *flattened pebbles* that are characteristic of metaconglomerates; *compositional banding*, in which dark and light minerals separate, generating a layered appearance; and *rock cleavage*, in which rocks can be easily split into tabular slabs. It is important to note that rock cleavage is not related to the mineral cleavage discussed in Chapter 2.

Nonfoliated Textures Not all metamorphic rocks exhibit a foliated texture. Those that do not are referred to as **nonfoliated** and typically develop in environments where deformation is minimal and the parent rocks are composed of minerals that have a relatively simple chemical composition, such as quartz or calcite. For example, when a fine-grained limestone (made of calcite) is metamorphosed by the intrusion of a hot magma body (contact metamorphism), the small calcite grains recrystallize and form larger interlocking crystals. The resulting rock, *marble*, exhibits large, equidimensional grains that are randomly oriented, similar to those in a coarse-grained igneous rock.

Common Metamorphic Rocks

Figure 3.30 depicts the common rocks produced by metamorphic processes, which are described below.

Foliated Rocks **Slate** is a very fine-grained foliated rock composed of minute mica flakes that are too small to be visible to the unaided eye (see Figure 3.30). A noteworthy characteristic of slate is its excellent rock cleavage, or tendency to break into flat slabs. This property has made slate a useful rock for roof and floor tile, as well as billiard tables (Figure 3.31). Slate is usually generated by the low-grade metamorphism of shale. Less frequently, it is produced when volcanic ash is metamorphosed. Slate's color is variable. Black slate contains organic material, red slate gets its color from iron oxide, and green slate is usually composed of chlorite, a greenish mica-like mineral.

Phyllite represents a degree of metamorphism between slate and schist. Its constituent platy minerals, mainly muscovite and chlorite, are larger than those in slate but not large enough to be readily identifiable with the unaided eye. Although phyllite appears similar to slate, it can be easily distinguished from slate by its glossy sheen and wavy surface (see Figure 3.30).

Schists are moderately to strongly foliated rocks formed by regional metamorphism (see Figure 3.30). They are platy and can be readily split into thin flakes or slabs. Many schists originate from shale parent rock. The term *schist* describes the *texture* of a rock, regardless of composition. For example, schists composed primarily of muscovite and biotite are called *mica schists*.

Gneiss (pronounced "nice") is the term applied to banded metamorphic rocks in which elongated and granular (as opposed to platy) minerals predominate (see Figure 3.30). The most common minerals in gneisses are quartz and feldspar, with lesser amounts of muscovite, biotite, and hornblende. Gneisses exhibit strong segregation of light and dark silicates, giving them a characteristic banded texture. While still deep below the surface where temperatures and pressures are great, banded gneisses can be deformed into intricate folds.

Nonfoliated Rocks **Marble** is a coarse, crystalline rock whose parent rock is limestone (see Figure 3.30). Marble is composed of large interlocking calcite crystals formed from the recrystallization of smaller grains in the parent rock. Because of its color and relative softness (hardness of only 3 on the Mohs scale), marble is a popular building stone. White marble is particularly prized as a stone from which to carve monuments and statues, such as the Lincoln Memorial in Washington, DC, and the Taj Mahal in India (see GEOgraphics 3.1). Marble can also be colored—pink, gray, green, or even black—if the parent rocks from which it formed contain impurities that color the stone.

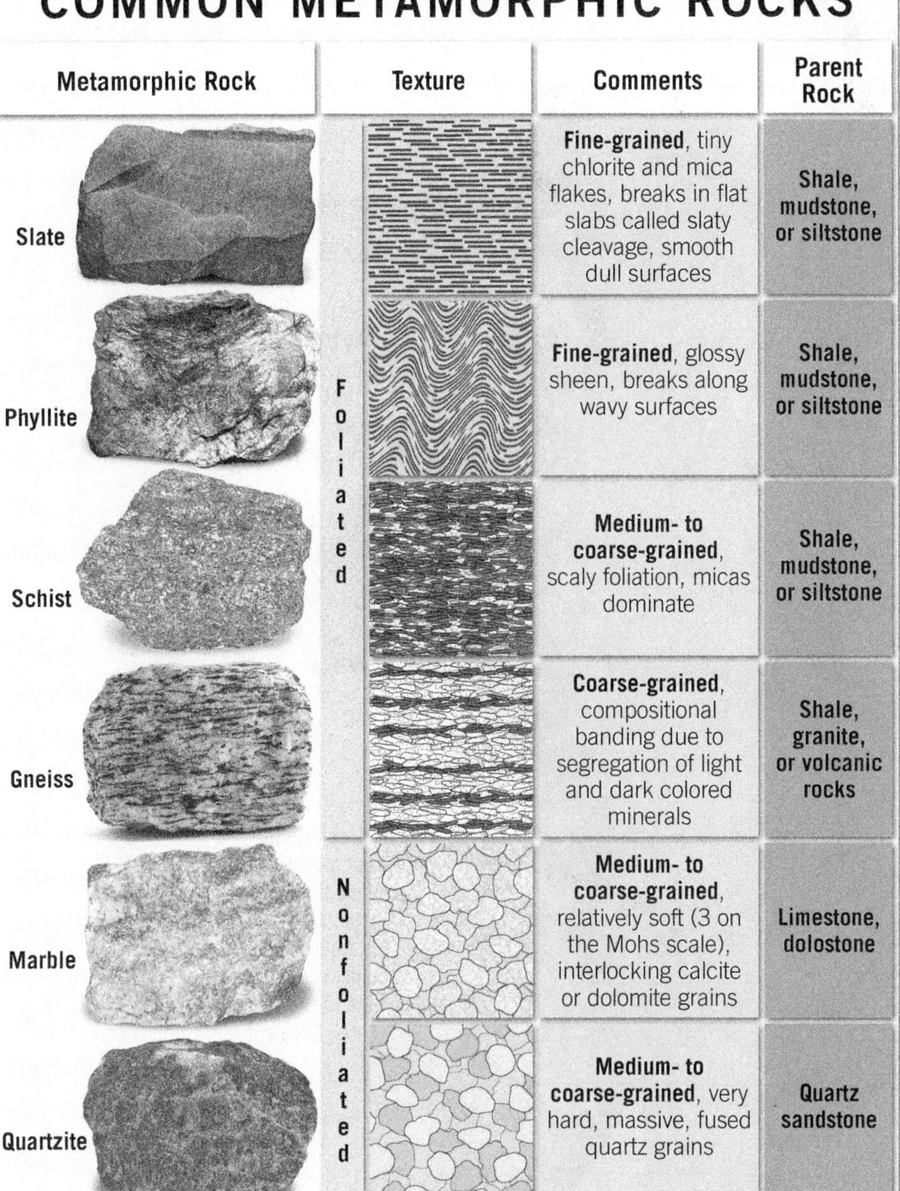

COMMON METAMORPHIC ROCKS

Metamorphic Rock		Texture	Comments	Parent Rock
Slate	Foliated		**Fine-grained**, tiny chlorite and mica flakes, breaks in flat slabs called slaty cleavage, smooth dull surfaces	Shale, mudstone, or siltstone
Phyllite	Foliated		**Fine-grained**, glossy sheen, breaks along wavy surfaces	Shale, mudstone, or siltstone
Schist	Foliated		**Medium- to coarse-grained**, scaly foliation, micas dominate	Shale, mudstone, or siltstone
Gneiss	Foliated		**Coarse-grained**, compositional banding due to segregation of light and dark colored minerals	Shale, granite, or volcanic rocks
Marble	Nonfoliated		**Medium- to coarse-grained**, relatively soft (3 on the Mohs scale), interlocking calcite or dolomite grains	Limestone, dolostone
Quartzite	Nonfoliated		**Medium- to coarse-grained**, very hard, massive, fused quartz grains	Quartz sandstone

Figure 3.30 **Common metamorphic rocks** (Photos by E. J. Tarbuck)

Figure 3.31 **Slate exhibits rock cleavage** Because slate breaks into flat slabs, it has many uses. The larger image shows a quarry near Alta, Norway. (Photo by Fred Bruemmer/Getty Images) In the inset photo, slate is used to roof a house in Switzerland. (Photo by E. J. Tarbuck)

Marble

Marble is crystalline metamorphic rock whose parent rock was limestone or dolostone. Pure marble is white; however, most marble contains impurities such as iron oxide, chlorite, and organic debris which renders the marble pink, green, gray or sometimes black.

Pure white marbles are prized for sculpture

Statue of David carved by Michelangelo from pure Carrara marble between 1501 and 1504. The statue rises over 17 feet above the base.

Venus de Milo is an ancient Greek statue carved out of marble sometime between 130 and 100 B.C. This statue is on permanent display at the Louvre in Paris, France.

Carrara Marble quarries located in the mountains near Carrara, Italy have produced some of the finest marbles for more than two thousand years.

These marble quarries located near Carrara, Italy are so large they can be seen from space.

Circumnavigation/Fotolia

Ray Roberts/Alamy Images

Marble, because of its workability, is a widely used building stone

Taj Mahal Constructed primarily of marble, the Taj Mahal is considered one of the world's most spectacular buildings. Built by Mughal Emperor Shah Jahan in memory of his third wife, the structure is located in northcentral India. Construction began around 1632 and was completed in 1653.

Steve Vider/Superstock

Question:
What is the name of the high-quality marble that has been quarried in Italy for more than 2000 years?

?

Marble is also widely used for floor tile and countertops. The marble in these floor tiles contains impurities that impart a variety of colors and also show evidence of deformation.

The white exterior of the Lincoln Memorial in Washington, DC, is constructed mainly of marble that was quarried near the town of Marble, Colorado. Inside, pink Tennessee "marble" was used for the floors, Alabama marble for the ceilings, and Georgia marble for Lincoln's statue.

d/Shutterstock

Orhan Cam/Shutterstock

SmartFigure 3.32
Garnet-mica schist The dark red garnet crystals are embedded in a matrix of fine-grained micas.
(Photo by E. J. Tarbuck)

MOBILE FIELD TRIP
https://goo.gl/bKnazg

Quartzite is a very hard metamorphic rock most often formed from quartz sandstone (see Figure 3.30). Under moderate- to high-grade metamorphism, the quartz grains in sandstone fuse. Pure quartzite is white, but iron oxide may produce reddish or pinkish stains, and dark minerals may impart a gray color.

Other Metamorphic Rocks

During intermediate- to high-grade metamorphism, recrystallization of existing minerals often produces new minerals that are mainly associated with metamorphic rocks, such as the mineral *garnet*. These newly formed minerals, commonly referred to as *accessory minerals*, tend to form large crystals that are surrounded by smaller crystals of other minerals, such as muscovite and biotite. When naming a metamorphic rock that contains one or more easily recognizable accessory minerals, geologists add a prefix to the appropriate rock name. For example, Figure 3.32 shows a mica schist that contains large dark red garnet crystals embedded in a matrix of fine-grained micas; consequently, this rock is called a *garnet-mica schist*. The metamorphic rock gneiss also frequently contains accessory minerals, including garnet and staurolite. These rocks would be called *garnet gneiss* and *staurolite gneiss*, respectively.

CONCEPT CHECKS 3.4

1. Metamorphism means "to change form." Describe how a rock may change during metamorphism.
2. Explain what is meant by the statement "every metamorphic rock has a parent rock."
3. List the four agents of metamorphism and describe the role of each.
4. Distinguish between regional metamorphism and contact metamorphism.
5. What feature easily distinguishes schist and gneiss from quartzite and marble?
6. In what ways do metamorphic rocks differ from the igneous and sedimentary rocks from which they formed?

3.5 Resources from Rocks and Minerals

Distinguish between metallic and nonmetallic mineral resources and list at least two examples of each.

The outer layer of Earth, which we call the crust, is only as thick when compared to the remainder of the Earth as a peach skin is to a peach, yet it is of supreme importance to us. We depend on it for fossil fuels and as a source of such diverse materials as the iron for automobiles, salt to flavor food, and gold for world trade. In fact, on occasion, the availability or absence of certain Earth materials has altered the course of history. As the world population grows and the material requirements of modern society increase, the need to locate additional supplies of useful minerals becomes more challenging.

Most of the energy and mineral resources used by humans are nonrenewable. Figure 3.33 shows the annual per capita consumption of several important mineral and energy resources. These data reflect an individually prorated share of the materials required by industry to support the needs of our modern society—including a vast array of homes, cars, electronics, cosmetics, packaging, and other products and services.

Metallic Mineral Resources

Some of the most important accumulations of metals, including gold, silver, copper, platinum, and nickel, are produced by igneous and metamorphic processes that concentrate desirable materials to the extent that extraction is economically feasible (Table 3.1). Igneous processes that generate some metal deposits are straightforward. For example, as a large magma body cools, minerals that crystallize early and are heavy tend to settle to the lower portion of the magma chamber. This type of *magmatic differentiation* is particularly important in large basaltic magmas where chromite (ore of chromium), magnetite, and platinum are occasionally generated. Layers of chromite, along with other heavy minerals, are mined from such deposits in the Bushveld Complex in South Africa, which contains over 70 percent of the world's known reserves of platinum.

Igneous processes are also important in generating other types of mineral deposits. For example, as a

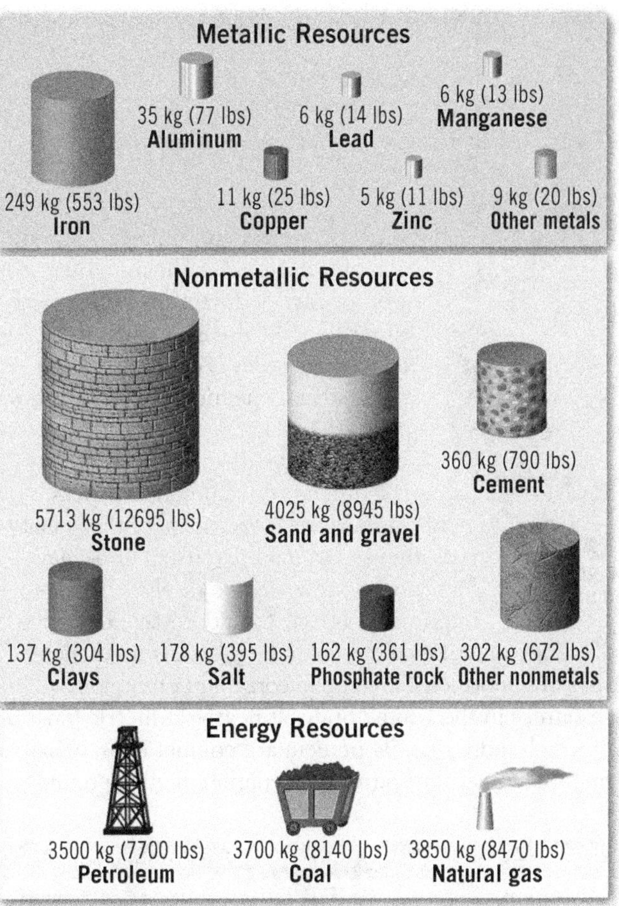

Figure 3.33 How much do each of us use? The annual per capita consumption of metallic and nonmetallic resources for the United States is about 11,000 kilograms (12 tons). About 97 percent of the materials used are nonmetallic. The per capita use of oil, coal, and natural gas exceeds 11,000 kilograms. (U.S. Geological Survey)

Table 3.1 Occurrence of Metallic Minerals

Metal	Principal Ore(s)	Geologic Occurrences
Aluminum	Bauxite	Residual product of weathering
Chromium	Chromite	Magmatic segregation
Copper	Chalcopyrite Bornite Chalcocite	Hydrothermal deposits; contact metamorphism; enrichment by weathering processes
Gold	Native gold	Hydrothermal deposits; placers
Iron	Hematite Magnetite Limonite	Banded sedimentary formations; magmatic segregation
Lead	Galena	Hydrothermal deposits
Magnesium	Magnesite Dolomite	Hydrothermal deposits
Manganese	Pyrolusite	Residual product of weathering
Mercury	Cinnabar	Hydrothermal deposits
Molybdenum	Molybdenite	Hydrothermal deposits
Nickel	Pentlandite	Magmatic segregation
Platinum	Native platinum	Magmatic segregation; placers
Silver	Native silver Argentite	Hydrothermal deposits; enrichment by weathering processes
Tin	Cassiterite	Hydrothermal deposits; placers
Titanium	Ilmenite Rutile	Magmatic segregation; placers
Tungsten	Wolframite Scheelite	Pegmatites; contact metamorphic deposits; placers
Uranium	Uraninite (pitchblende)	Pegmatites; sedimentary deposits
Zinc	Sphalerite	Hydrothermal deposits

granitic magma cools and crystallizes, the residual melt becomes enriched in rare elements and heavy metals, including gold and silver. Furthermore, because water and other volatile substances do not crystallize along with the bulk of the magma body, these fluids make up a high percentage of the melt during the final phase of solidification. Crystallization in a fluid-rich environment enhances the migration of ions and results in the formation of crystals several centimeters, or even a few meters, in length. The resulting rocks, called **pegmatites**, are composed of these unusually large crystals (Figure 3.34).

Most pegmatites are granitic in composition and consist of large crystals of quartz, feldspar, and muscovite. Feldspar is used in the production of ceramics, and muscovite is used for electrical insulation and glitter. In addition to these common silicates, some pegmatites contain semiprecious gems such as beryl, topaz, and tourmaline. Moreover, minerals containing the elements lithium, gold, silver, uranium, and the rare earths are sometimes

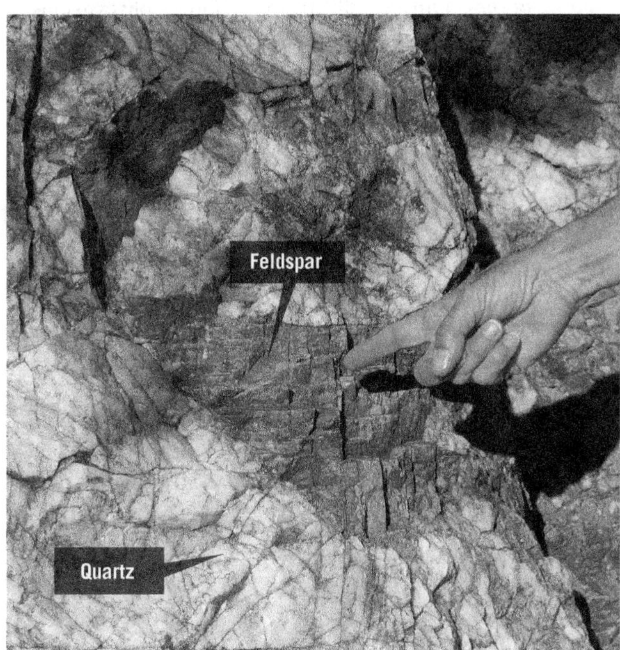

Figure 3.34 Pegmatites This granite pegmatite, found in the inner gorge of the Grand Canyon, is composed mainly of quartz and feldspar. (Photo by Joanne Bannon/E. J. Tarbuck)

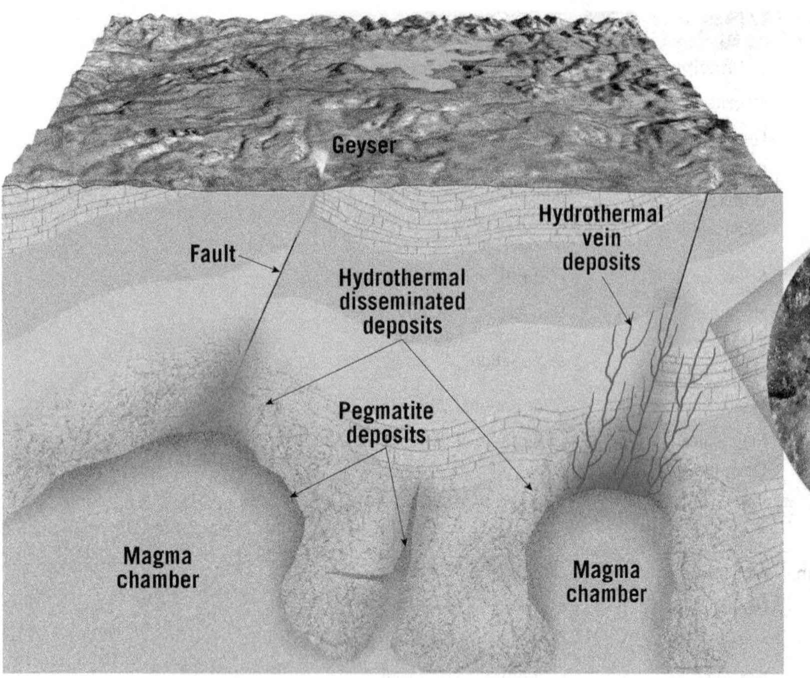

▲ **Figure 3.35 Relationship between an igneous body and associated pegmatites and hydrothermal mineral deposits** (Photo by Greenshoots Communications/Alamy)

found. Most pegmatites are located within large igneous masses or as dikes or veins that cut into the host rock surrounding a magma chamber (Figure 3.35).

Among the most important ore deposits are those generated from hydrothermal (*hydra* = water, *therm* = heat) solutions. Included in this group are the gold deposits of the Homestake Mine in South Dakota; the lead, zinc, and silver ores near Coeur d'Alene, Idaho; the silver deposits of the Comstock Lode in Nevada; and the copper ores of the Keweenaw Peninsula in Michigan.

The majority of hydrothermal deposits originate from hot, metal-rich fluids that are associated with cooling magma bodies. During solidification, liquids plus various metallic ions accumulate near the top of the magma chamber. Because these hot fluids are very mobile, they can migrate great distances through the surrounding host rock before they are eventually deposited. Some of this fluid moves along fractures or bedding planes, where it cools and precipitates the metallic ions to produce **vein deposits** (see Figure 3.35). Many of the most productive deposits of gold, silver, and mercury occur as hydrothermal vein deposits.

Another important type of accumulation generated by hydrothermal activity is called a **disseminated deposit**. Rather than being concentrated in narrow veins and dikes, these ores are distributed as minute masses throughout the entire rock mass (see Figure 3.35). Much of the world's copper is extracted from disseminated deposits, including the huge Bingham Canyon copper mine in Utah (see Figure 2.35, page 51). Because these accumulations contain only 0.4 to 0.8 percent copper, between 125 and 250 metric tons of ore must be mined for every ton of metal recovered. The environmental impact of these large excavations, including the problems of waste disposal, is significant.

Nonmetallic Mineral Resources

Earth materials that are not used as fuels or processed for the metals they contain are referred to as **nonmetallic mineral resources**. Realize that use of the word *mineral* is very broad in this economic context and is quite different from geologists' strict definition of "mineral" found in Chapter 2. Nonmetallic mineral resources are extracted and processed either to make use of the nonmetallic elements they contain or for the physical and chemical properties they possess (Table 3.2). Although these resources have diverse origins, many are sediments or sedimentary rocks. Because a large percentage of these materials are used in manufacturing or to make fertilizers—processes we rarely see—their importance is easy to underestimate.

The quantities of nonmetallic minerals used each year are enormous. Per capita consumption of nonfuel resources in the United States is nearly 11 metric tons, of which more than 94 percent are nonmetallics. Nonmetallic mineral resources are commonly divided into

Table 3.2 Occurrences and Uses of Nonmetallic Minerals

Mineral	Uses	Geologic Occurrences
Apatite	Phosphorus fertilizers	Sedimentary deposits
Asbestos	Incombustible fibers	Metamorphic deposits (chrysotile)
Calcite	Aggregate; steelmaking; soil conditioning; chemicals; cement; building stone	Sedimentary deposits
Clay minerals (kaolinite)	Ceramics; china	Residual product of weathering
Corundum	Gemstones; abrasives	Metamorphic deposits
Diamond	Gemstones; abrasives	Kimberlite pipes; placers
Fluorite	Steelmaking; aluminum refining; glass; chemicals	Hydrothermal deposits
Garnet	Abrasives; gemstones	Metamorphic rocks
Graphite	Pencil lead; lubricant; refractories	Metamorphic rocks
Gypsum	Plaster of Paris	Sedimentary deposits (evaporites)
Halite	Table salt; chemicals; ice control	Sedimentary deposits
Muscovite	Insulators in electrical applications	Igneous intrusions; pegmatites
Quartz	Primary ingredient in glass	Sedimentary deposits (sandstone)
Sulfur	Chemicals; fertilizer manufacture	Sedimentary deposits; hydrothermal deposits
Sylvite	Potassium fertilizers	Sedimentary deposits (evaporites)
Talc	Powder used in paints, cosmetics, etc.	Metamorphic rocks

two broad groups: **building materials** and **industrial minerals**. Some substances fall into both groups. Limestone, a rock formed of the mineral calcite, is perhaps the most versatile and widely used rock of all. As a building material, it is used not only as crushed rock and building stone but also in making cement. As an industrial mineral, limestone is an ingredient in the manufacture of steel and is used in agriculture to neutralize acidic soils.

Building materials also include cut stone, aggregate (sand, gravel, and crushed rock), gypsum for plaster and wallboard, clay for tile and bricks, and cement, which is made from limestone and shale. Cement and aggregate go into the making of concrete, a material that is essential to practically all construction.

A wide variety of resources are classified as industrial minerals. Some are sources of elements or compounds used in the manufacture of chemicals. Others, including fluorite and limestone, are part of the steel-making process; corundum and garnet are used as abrasives; and sylvite, a potassium-rich mineral, is used in the production of fertilizers. Deposits of industrial minerals are generally far less common than those of building materials.

Energy Resources

Earth's tremendous industrialization over the past 2 centuries has been, and continues to be, fueled by burning coal, petroleum, and natural gas. About 81 percent of the energy consumed in the United States today comes from these basic sources. Our reliance on fossil fuels is obvious. Although we are increasing the quantities of energy from alternative sources such as solar, wind, geothermal, and hydroelectric, the U.S. Department of Energy projects that fossil fuels will remain a primary source of energy for decades to come.

Coal In addition to oil and natural gas, coal is commonly called a **fossil fuel**. This designation is appropriate because when coal is burned, energy from the Sun that was stored by plants many millions of years ago is being used, hence the actual burning of "fossils." In the United States, coal fields are widespread and contain supplies that should last for hundreds of years (Figure 3.36).

Coal is plentiful, but its recovery and use present a number of challenges. Surface mining, which is costly, can turn the countryside into a scarred wasteland if careful reclamation is not carried out to restore the land. Currently, all U.S. surface mines are required to reclaim the land. Although underground mining does not scar the landscape to the same degree, it has been costly in terms of human life and health. Strong federal safety regulations have made U.S. mining quite safe. However, collapsing roofs, gas explosions, and the required heavy equipment remain hazards.

In addition, burning coal produces emissions that adversely influence the environment and human health.

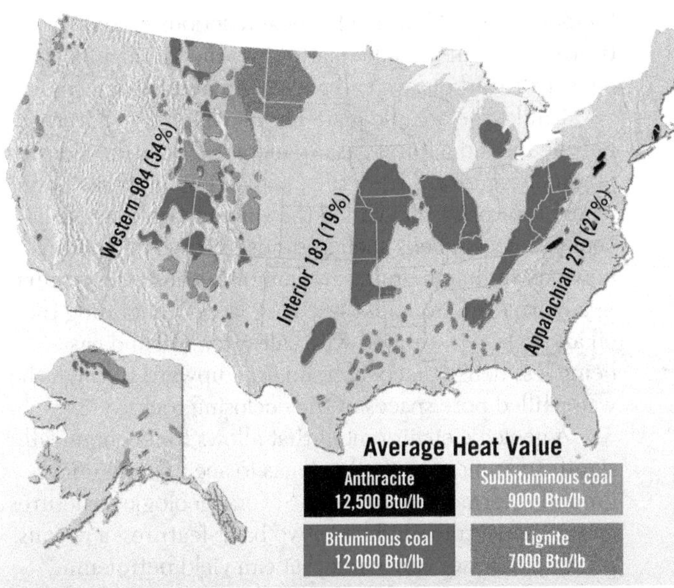

Figure 3.36 **Coal fields of the United States** Most of the coal mined is subbituminous and bituminous. Wyoming and West Virginia are the leading producers of these coals. (Data from U.S. Geological Survey)

Principal emissions resulting from coal combustion include the following:

- Sulfur dioxide (SO_2), which contributes to acid rain and respiratory illnesses
- Nitrogen oxides (NO_x), which contributes to smog and respiratory illnesses
- Particulate matter, which contributes to smog, haze, respiratory illnesses, and lung disease
- Carbon dioxide (CO_2), the primary greenhouse gas produced from the burning of fossil fuels, which plays a significant role in the heating of our atmosphere (Chapter 20 examines this issue in some detail.)

The coal industry has found several ways to reduce sulfur, nitrogen oxides, and other impurities from coal, and it has also developed more effective ways of cleaning coal after it is mined. Although coal consumers have shifted toward greater use of less-polluting low-sulfur coal, significant challenges remain.

Oil and Natural Gas Together, oil and natural gas provided more than 60 percent of the energy consumed in the United States in 2016. In 2011, natural gas surpassed coal for the first time as a source of energy in the United States. An important reason for this is the use of new technologies that have increased production from shale formations. (Hydraulic fracturing is discussed later in this section.)

Like coal, petroleum and natural gas are biological products derived from the remains of organisms. However, while coal formed mostly from plant material that accumulated in swampy environments on land, as shown in Figure 3.22, oil and natural gas originated mainly from microscopic marine plankton in ancient seas millions of years ago. This organic material accumulated in marine sedimentary basins with mud, sand, and other sediments that protected it from oxidation. With increased temperature and burial, chemical reactions gradually transformed this organic matter into the liquid and gaseous hydrocarbons we call petroleum and natural gas.

Unlike the organic matter from which they form, petroleum and natural gas are mobile. These fluids are gradually squeezed from the compacting, mud-rich layers where they originate (called the **source rock**) into adjacent permeable beds such as sandstone, where openings between sediment grains are larger. Because this occurs in a marine environment, the rock layers containing the oil and gas are saturated with saltwater. Oil and gas, being less dense than water, migrate upward through the water-filled pore spaces of the enclosing rocks.

A geologic environment that allows for economically significant amounts of oil and gas to accumulate underground is termed an **oil trap**. Several geologic structures can act as oil traps and have two basic features: a porous, permeable **reservoir rock** that can yield petroleum and natural gas in sufficient quantities to make drilling worthwhile; and a **cap rock**, such as shale, that is virtually impermeable to oil and gas and, hence, traps them in the reservoir rock. Figure 3.37 illustrates the following common oil and natural gas traps:

- **Anticline.** One of the simplest traps is an *anticline*, an up-arched series of sedimentary strata (see Figure 3.37A). The rising oil and gas collect at the top of the fold. Because of its lower density, natural gas collects above the oil. Both rest on the denser water that saturates the reservoir rock.
- **Fault trap.** When strata are displaced so that a dipping reservoir rock abuts an impermeable bed, a *fault trap* forms, as shown in Figure 3.37B. The upward migration of the oil and gas is halted where it encounters the fault.
- **Salt dome.** In the Gulf coastal plain region of the United States, important accumulations of oil occur in association with *salt domes*. Such areas have thick accumulations of sedimentary strata that include layers of rock salt. Salt occurring at great depths has been forced to rise in columns by the pressure of overlying beds. These rising salt columns gradually deform the overlying strata. Because oil and gas migrate to the highest level possible, they accumulate in the upturned sandstone beds adjacent to the salt column (see Figure 3.37C).
- **Stratigraphic (pinchout) trap.** A *stratigraphic trap* results primarily from the original pattern of sedimentation rather than from structural deformation. The stratigraphic trap illustrated in Figure 3.37D exists because a sloping bed of sandstone thins to the point of disappearance.

Regardless of the type of trap, when drilling punctures the lid created by the cap rock, the oil and natural gas, which are under pressure, migrate from the pore spaces of the reservoir rock to the drill pipe. On rare occasions, when fluid pressure is great, it may force oil up the drill hole to the surface, causing a "gusher" at the surface. Usually, however, a pump is required to extract the oil.

Hydraulic Fracturing Some shale deposits contain significant reserves of oil and natural gas that cannot naturally leave because of the rock's low permeability. The practice of **hydraulic fracturing** (often called *fracking*) shatters the shale, opening cracks through which the oil and natural gas can flow into wells and can then be brought to the surface. Figure 3.38 illustrates the process. The fracturing of the shale is initiated by pumping fluids into the rock at very high pressures. The fluid is mostly water but also includes other chemicals that aid in the fracturing process. Some of these chemicals may be toxic, and there are concerns about fracking fluids leaking into freshwater aquifers that supply people with drinking water. The

SmartFigure 3.37
Common oil traps

TUTORIAL
https://goo.gl/a2Ki8m

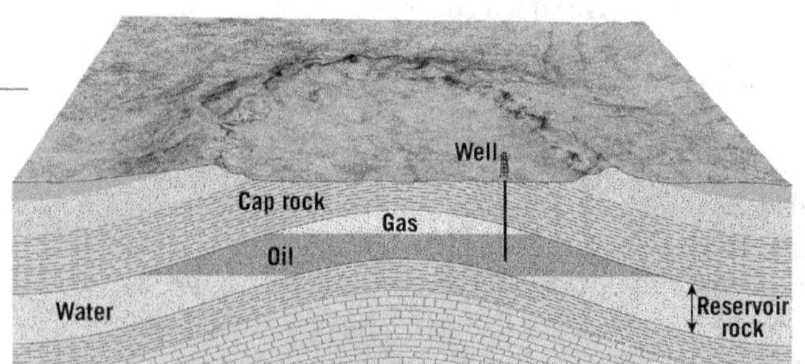

A. Anticline

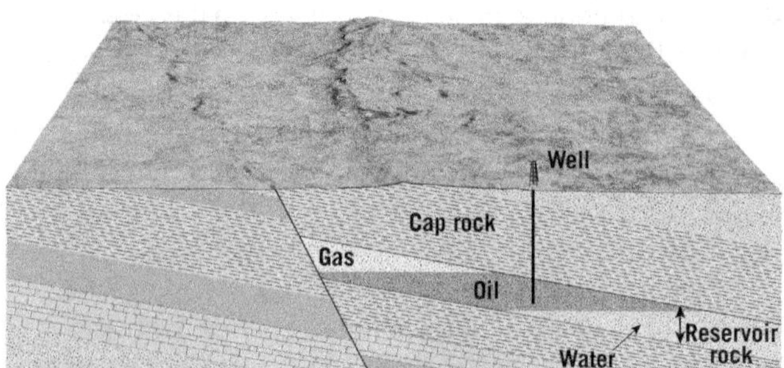

B. Fault trap

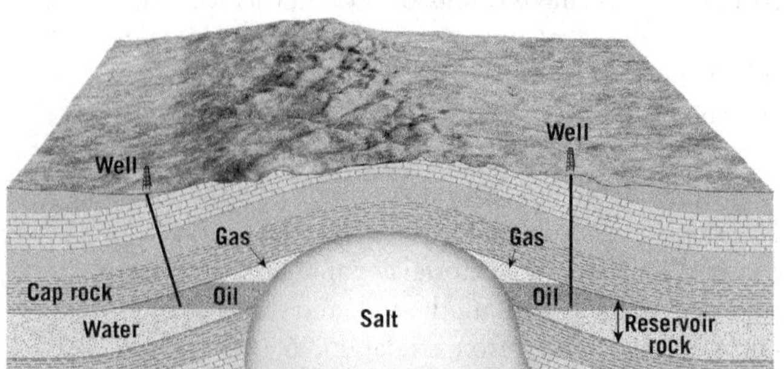

C. Salt dome

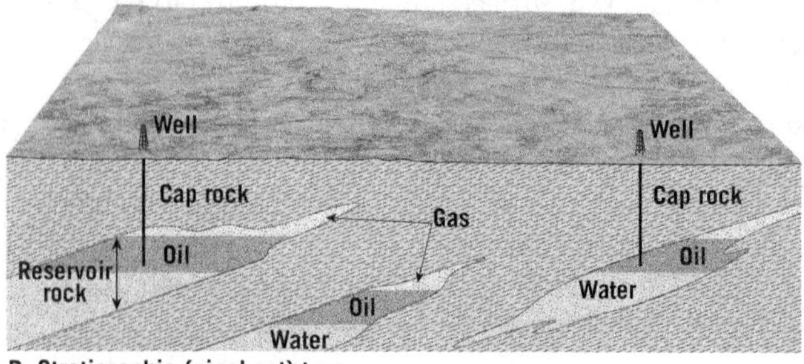

D. Stratigraphic (pinchout) trap

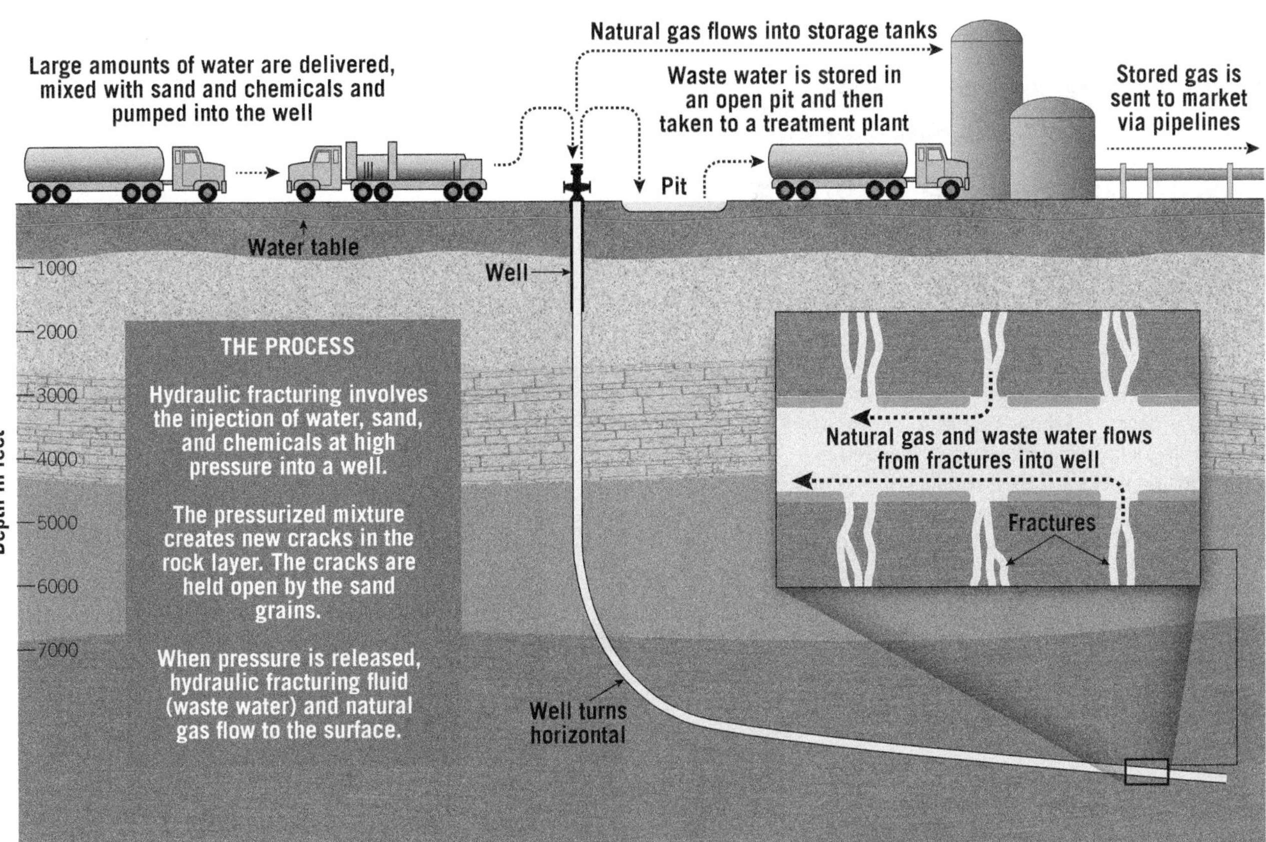

Figure 3.38
Hydraulic fracturing ("fracking") Fracking is commonly used in low-permeability rocks such as shale to increase oil and/or natural gas production.

injection fluid also includes sand, so once fractures open in the shale, the sand grains can keep them propped open and permit the oil and gas to continue to flow.

Once the fracturing has been accomplished, the fracking fluid is brought back to the surface. This wastewater is often injected into deep disposal wells, and in some locations, these injections appear to trigger numerous minor earthquakes. Due to concerns about potential groundwater contamination and induced seismicity, hydraulic fracturing remains a controversial practice. Its environmental effects remain a focus of continuing research.

CONCEPT CHECKS 3.5

1. List two general types of hydrothermal deposits.
2. Nonmetallic resources are commonly divided into two broad groups. What are the two groups, and what are some examples of materials that belong to each?
3. Why are coal, oil, and natural gas called *fossil fuels*?
4. What is an oil trap? Sketch two examples. What do all oil traps have in common?
5. Describe the circumstances in which hydraulic fracturing is used.

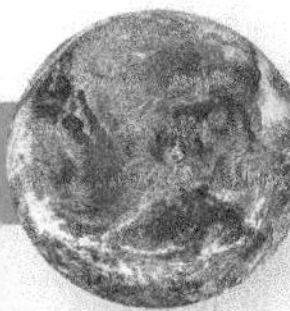

EYE ON EARTH 3.3

The image on the left shows an active landfill where tons of trash and garbage are dumped every day. Eventually this site will be reclaimed to resemble the area shown on the right and become a source of energy. (Left photo by NHPA/SuperStock; right photo by Jim West/Alamy)

QUESTION 1 *Explain how an area filled with trash and waste could become a source of energy. What form of energy will it be? How might it be used?*

QUESTION 2 *Is this type of energy considered renewable or nonrenewable? Explain.*

CONCEPTS IN REVIEW

Rocks: Materials of the Solid Earth

3.1 Earth as a System: The Rock Cycle

Sketch, label, and explain the rock cycle.

KEY TERM: rock cycle

- The rock cycle is a good model for thinking about the transformation of one rock to another due to Earth processes. Igneous rocks form when molten rock solidifies. Sedimentary rocks are made from weathered products of other rocks. Metamorphic rocks are the products of preexisting rocks subjected to conditions of high temperatures and/or pressures. Given the right sequence of conditions, any rock type can be transformed into any other type of rock.

? Name the processes that are represented by each of the letters (A–E) in this rock cycle diagram.

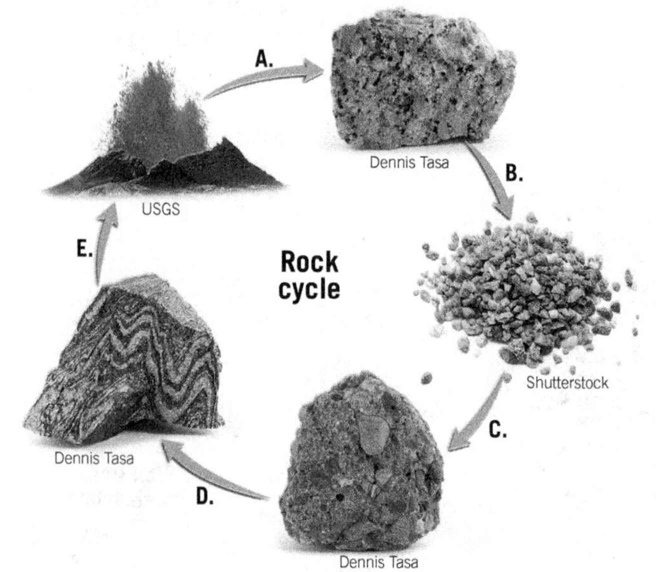

3.2 Igneous Rocks: "Formed by Fire"

Describe the two criteria used to classify igneous rocks and explain how the rate of cooling influences the crystal size of minerals.

KEY TERMS: igneous rock, magma, lava, extrusive (volcanic) rock, intrusive (plutonic) rock, granitic (felsic) composition, basaltic (mafic) composition, andesitic (intermediate) composition, ultramafic, peridotite, texture, fine-grained texture, coarse-grained texture, porphyritic texture, phenocryst, groundmass, vesicular texture, glassy texture, pyroclastic (fragmental) texture, granite, rhyolite, obsidian, pumice, andesite, diorite, basalt, gabbro, Bowen's reaction series, crystal settling, magmatic differentiation

- Completely or partly molten rock is called magma if it is below Earth's surface and lava if it has erupted onto the surface. It consists of a liquid melt that contains gases (volatiles) such as water vapor, and it may contain solids (mineral crystals).
- Magmas that cool at depth produce intrusive igneous rocks, whereas those that erupt onto Earth's surface produce extrusive igneous rocks.
- In geology, texture refers to the size, shape, and arrangement of mineral grains in a rock. Careful observation of the texture of igneous rocks can lead to insights about the conditions under which they formed. Lava on or near the surface cools rapidly, resulting in a large number of very small crystals that gives the rock a fine-grained texture. Magma at depth is insulated by the surrounding rock and cools very slowly. This allows sufficient time for the magma's ions to organize into larger crystals, resulting in a rock with a coarse-grained texture. If crystals begin to form at depth and then the magma rises to a shallow depth or erupts at the surface, it has a two-stage cooling history. The result is a rock with a porphyritic texture.
- Pioneering experimentation by N. L. Bowen revealed that as magma cools, minerals crystallize in a specific order. The dark-colored silicate minerals, such as olivine, crystallize first, at the highest temperatures (1250°C [2300°F]), whereas the light silicates, such as quartz, crystallize last, at the lowest temperatures (650°C [1200°F]). Separation of minerals by mechanisms such as crystal settling results in igneous rocks having a wide variety of chemical compositions.
- Igneous rocks are classified into compositional groups based on the percentage of dark and light silicate minerals they contain. Granitic (or felsic) rocks such as granite and rhyolite are composed mostly of the light-colored silicate minerals potassium feldspar and quartz. Rocks of andesitic (or intermediate) composition such as andesite contain plagioclase feldspar and amphibole. Basaltic (or mafic) rocks such as basalt contain abundant pyroxene and calcium-rich plagioclase feldspar.

3.3 Sedimentary Rocks: Compacted and Cemented Sediment

List and describe the different categories of sedimentary rocks and discuss the processes that change sediment into sedimentary rock.

KEY TERMS: sedimentary rock, sediment, detrital sedimentary rock, conglomerate, breccia, sandstone, shale, siltstone, chemical sedimentary rock, biochemical sedimentary rock, limestone, coquina, travertine, evaporite deposit, coal, lithification, compaction, cementation, strata (beds), fossil

- Although igneous and metamorphic rocks make up most of Earth's crust by volume, sediment and sedimentary rocks are concentrated near the surface.
- Detrital sedimentary rocks are made of solid particles, mostly quartz grains and microscopic clay minerals. Common detrital sedimentary rocks include shale (the most abundant sedimentary rock), sandstone, and conglomerate.
- Chemical and biochemical sedimentary rocks are derived from mineral matter (ions) that is carried in solution to lakes and seas. Under certain conditions, ions in solution precipitate (settle out) to form chemical sediments as a result of physical processes, such as evaporation. Precipitation may also occur indirectly through life processes of water-dwelling organisms that form materials called biochemical sediments. Many water-dwelling animals and plants extract dissolved mineral

matter to form shells and other hard parts. After the organisms die, their skeletons may accumulate on the floor of a lake or an ocean.

- Limestone, an abundant sedimentary rock, is composed chiefly of the mineral calcite ($CaCO_3$). Rock gypsum and rock salt are chemical rocks that form as water evaporates.
- Coal forms from the burial of large amounts of plant matter in low-oxygen depositional environments such as swamps and bogs.
- The transformation of sediment into sedimentary rock is called lithification. The two main processes that contribute to lithification are compaction (a reduction in pore space that results from packing grains more tightly together) and cementation (a reduction in pore space that results from adding new mineral material that acts as a "glue" to bind the grains to each other).

? This photo was taken in the Grand Canyon. What names do geologists use for the characteristic features found in the sedimentary rocks shown in this image?

Dennis Tasa

3.4 Metamorphic Rocks: New Rock from Old

Define *metamorphism*, explain how metamorphic rocks form, and describe the agents of metamorphism.

KEY TERMS: metamorphic rock, metamorphism, contact metamorphism, regional metamorphism, confining pressure, differential stress, foliation, nonfoliated, slate, phyllite, schist, gneiss, marble, quartzite

- When rocks are subjected to elevated temperatures and pressures, they can change form, producing metamorphic rocks. Every metamorphic rock has a parent rock—the rock it used to be prior to metamorphism. When the minerals in parent rocks are subjected to heat and pressure, new minerals can form. Depending on the intensity of alteration, metamorphism ranges from low grade to high grade.
- Heat, confining pressure, differential stress, and chemically active fluids are four agents that drive metamorphic reactions. Any one alone may trigger metamorphism, or all four may act simultaneously.
- Confining pressure results from burial. The force it exerts is the same in all directions, like the pressure exerted by water on a diver. An increase in confining pressure causes rocks to compact into more dense configurations.
- Differential stresses, which occur during mountain building, are greater in one direction than in others. Rocks subjected to differential stress under ductile conditions deep in the crust tend to shorten in the direction of greatest stress and elongate in the direction(s) of least stress, producing flattened or stretched grains. In the shallow crust, most rocks respond to differential stress with brittle deformation, breaking into pieces.
- A common kind of texture is foliation, the planar arrangement of mineral grains. Common foliated metamorphic rocks include (in order of increasing metamorphic grade) slate, phyllite, schist, and gneiss.
- Common nonfoliated metamorphic rocks include quartzite and marble, recrystallized rocks that form from quartz sandstone and limestone, respectively.

? Examine the photograph. Determine whether this rock is foliated or nonfoliated, then determine whether it formed under confining pressure or from differential stress. Which of the pairs of arrows shows the direction of maximum stress?

Dennis Tasa

3.5 Resources from Rocks and Minerals

Distinguish between metallic and nonmetallic mineral resources and list at least two examples of each.

KEY TERMS: pegmatite, vein deposit, disseminated deposit, nonmetallic mineral resource, building material, industrial mineral, fossil fuel, source rock, oil trap, reservoir rock, cap rock, hydraulic fracturing

- Igneous processes concentrate some economically important elements through both magmatic differentiation and the emplacement of pegmatites. Magmas may also release hydrothermal (hot-water) solutions that penetrate surrounding rock, carrying dissolved metals in them. The metal ores may be precipitated as fracture-filling deposits (veins) or may penetrate the surrounding strata, producing countless tiny deposits disseminated throughout the host rock.
- Earth materials that are not used as fuels or processed for the metals they contain are referred to as nonmetallic resources. The two broad groups of nonmetallic resources are building materials (such as gypsum, used for plaster) and industrial minerals (including sylvite, a potassium-rich mineral used to make fertilizers).
- Coal, oil, and natural gas are all fossil fuels. In each, the energy of ancient sunlight, captured by photosynthesis, is stored in the hydrocarbons of plants or other living things buried by sediments.
- Coal is formed from compressed plant fragments, originally deposited in ancient swamps. Coal mining can be risky and environmentally damaging, and burning coal generates several kinds of pollution.
- Oil and natural gas are formed from the heated remains of marine plankton. Together, they account for more than 60 percent of U.S. energy use. Both oil and natural gas leave their source rock (typically shale) and migrate to an oil trap made up of other, more porous rocks, called reservoir rock, covered by a suitable impermeable cap rock.
- Hydraulic fracturing (or "fracking") is a method of opening pore space in otherwise impermeable rocks, permitting natural gas to flow out into wells.

? As you can see here, the Sinclair Oil Company's logo speaks directly to the "fossil" nature of the fuel it sells. However, it is unlikely that any *dinosaur* carbon has ended up in Sinclair's oil. Explain.

Vespasian/Alamy

GIVE IT SOME THOUGHT

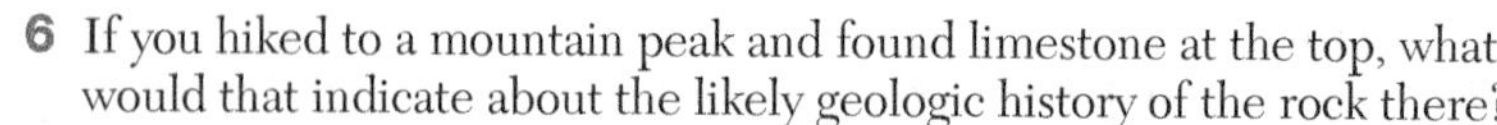

1 Refer to Figure 3.1. How does the rock cycle diagram—in particular, the labeled arrows—support the fact that sedimentary rocks are the most abundant rock type on Earth's surface?

2 Apply your understanding of igneous rock textures to describe the cooling history of each of the igneous rocks labeled A–D. (Photos by E. J. Tarbuck)

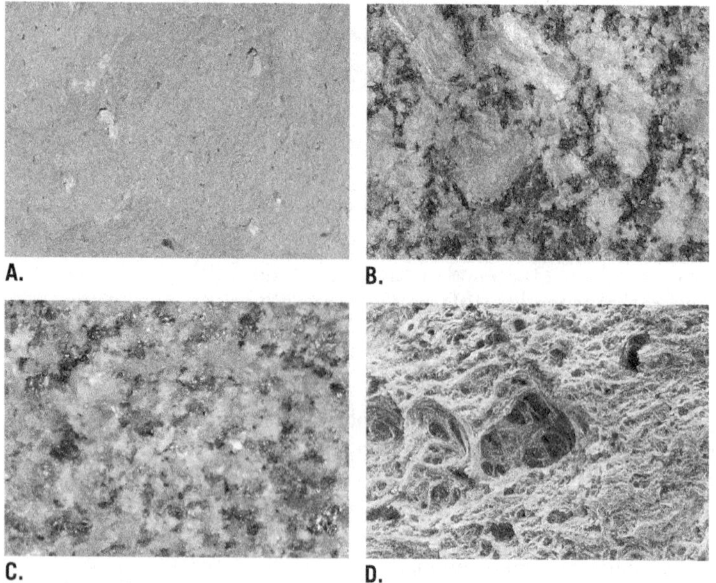

3 Is it possible for two igneous rocks to have the same mineral composition but be different rocks? Support your answer with an example.

4 Would you expect all the crystals in an intrusive igneous rock to be the same size? Explain why or why not.

5 One of the photos below shows an outcrop of metamorphic rock; the other two show igneous and sedimentary outcrops, respectively. Which do you think is the metamorphic rock? Explain why you ruled out the other rock bodies. (Photos by E. J. Tarbuck)

A.

B.

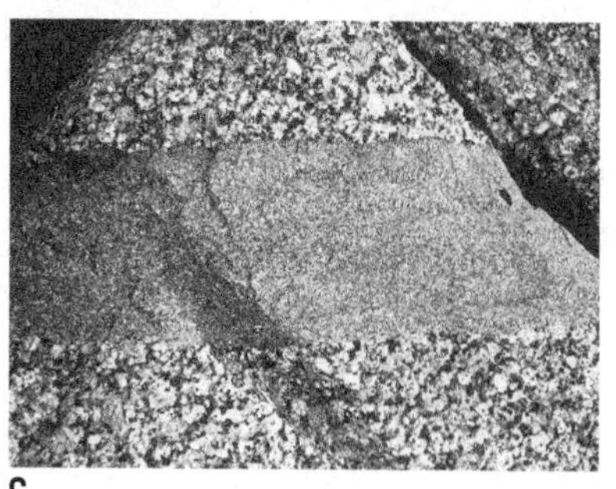

C.

6 If you hiked to a mountain peak and found limestone at the top, what would that indicate about the likely geologic history of the rock there?

7 Give two reasons sedimentary rocks are more likely to contain fossils than igneous rocks.

8 Use your understanding of magmatic differentiation to explain how magmas of different composition can be generated in a cooling magma chamber.

9 Dust collecting on furniture is an everyday example of a sedimentary process. Provide another example of a sedimentary process that might be observed in or around where you live.

10 Examine the accompanying photos, which show the geology of the Grand Canyon. Notice that most of the walls of the canyon consist of layers of sedimentary rocks, but if you were to hike down into what is known as the Inner Gorge, you would encounter the Vishnu Schist, which is metamorphic rock.

a. What process might have been responsible for the formation of the Vishnu Schist? How does this process differ from the processes that formed the sedimentary rocks that are atop the Vishnu Schist?

b. What does the Vishnu Schist tell you about the history of the Grand Canyon prior to the formation of the canyon itself?

c. Why is the Vishnu Schist visible at Earth's surface?

d. Is it likely that rocks similar to the Vishnu Schist exist elsewhere but are not exposed at Earth's surface? Explain.

A. Inner Gorge of the Grand Canyon

B. Close-up of Vishnu Schist (dark color)

EXAMINING THE EARTH SYSTEM

1 The sedimentary rock coquina, shown below, formed in response to interactions among two or more of Earth's spheres. List the spheres associated with the formation of this rock and write a short explanation for each of your choices.

E.J. Tarbuck

2 Of the two main sources of energy that drive the rock cycle—Earth's internal heat and solar energy—which is primarily responsible for each of the three groups of rocks found on and within Earth? Explain your reasoning.

3 Every year about 20,000 pounds of stone, sand, and gravel are mined for each person in the United States.

a. Calculate how many pounds of stone, sand, and gravel will be needed for an individual during an 80-year life span.

b. If 1 cubic yard of rocks weighs roughly 1700 pounds, calculate (in cubic yards) how large a hole must be dug to supply an individual with 80 years' worth of stone, sand, and gravel.

c. A typical pickup truck can carry about a half cubic yard of rock. How many pickup truck loads would be necessary during the 80-year span?

DATA ANALYSIS

Fossil Fuels Near You

Fossil fuels are the primary energy source for the United States. Coal, natural gas, and oil are mined throughout the United States and shipped across the country, sometimes over very long distances, before they make their way to power plants.

ACTIVITIES

Go to the U.S. Energy Information Administration site, at www.eia.gov, and select U.S. Energy Mapping System in the Features box. Click Layers/Legend and click Remove All to remove all symbols from the map. Add the appropriate map layer for each question. Click the + symbol next to a category to see subcategories when needed.

1 Where are the majority of all coal mines located? Are the coal mines in Pennsylvania predominately surface coal mines or underground coal mines?

2 What regions contain the largest concentrations of oil wells? Gas wells?

3 Where are the majority of active off-shore oil and gas platforms located?

Ensure that All Power Plants and all the power plant types under this heading are checked. Deselect all other map layers. Click Find Address, enter your city and state, and click Locate. Click on a power plant location to see information about the plant. Add map layers as needed.

4 Where (in what city, county, and state) is the power plant nearest your location? What is the primary fuel type used by this plant? Approximately how far away is this power plant from your location?

5 Find the nearest natural gas or petroleum power plant. Where (in what city, county, and state) is this power plant located? What is the primary fuel type used by this plant? Approximately how far away is the nearest refining and processing plant for this type of fuel?

6 Find the nearest coal power plant. Where (in what city, county, and state) is this power plant located? Approximately how far away is the nearest coal mine?

MasteringGeology™ Looking for additional review and test prep materials? Visit the Study Area in MasteringGeology to enhance your understanding of this chapter's content by accessing a variety of resources, including Self-Study Quizzes, Geoscience Animations, SmartFigure Tutorials, Mobile Field Trips, *Project Condor* Quadcopter videos, *In the News* articles, flashcards, web links, and an optional Pearson eText.

www.masteringgeology.com

UNIT 2

Forces Within

4 Plate Tectonics: A Scientific Revolution Unfolds

FOCUS ON CONCEPTS

Each statement represents the primary learning objective for the corresponding major heading within the chapter. After you complete the chapter, you should be able to:

4.1 Summarize the view that most geologists held prior to the 1960s regarding the geographic positions of the ocean basins and continents.

4.2 List and explain the evidence Wegener presented to support his continental drift hypothesis.

4.3 List the major differences between Earth's lithosphere and asthenosphere and explain the importance of each in the plate tectonics theory.

4.4 Sketch and describe the movement along a divergent plate boundary that results in the formation of new oceanic lithosphere.

4.5 Compare and contrast the three types of convergent plate boundaries and name a location where each type can be found.

4.6 Describe the relative motion along a transform fault boundary and locate several examples of transform faults on a plate boundary map.

4.7 Explain why plates such as the African and Antarctic plates are increasing in size, while the Pacific plate is decreasing in size.

4.8 List and explain the evidence used to support the plate tectonics theory.

4.9 Describe two methods researchers use to measure relative plate motion.

4.10 Describe plate–mantle convection and explain two of the primary driving forces of plate motion.

Hikers crossing a crevasse in Khumbu Glacier, Mount Everest, Nepal. (Photo by Christian Kober/Robert Harding World Imagery)

PLATE TECTONICS IS THE FIRST THEORY to provide a comprehensive view of the processes that produced Earth's major surface features, including the continents and ocean basins. Within the framework of this model, geologists have found explanations for the basic causes and distribution of earthquakes, volcanoes, and mountain belts. Further, the theory of plate tectonics helps explain the formation and distribution of igneous and metamorphic rocks and their relationship with the rock cycle.

4.1 From Continental Drift to Plate Tectonics

Summarize the view that most geologists held prior to the 1960s regarding the geographic positions of the ocean basins and continents.

Until the late 1960s, most geologists held the view that the ocean basins and continents had fixed geographic positions and were of great antiquity. Over the following decade, scientists came to realize that Earth's continents are not static; instead, they gradually migrate across the globe. These movements cause blocks of continental material to collide, deforming the intervening crust and thereby creating Earth's great mountain chains (Figure 4.1). Furthermore, landmasses occasionally split apart. As continental blocks separate, a new ocean basin emerges between them. Meanwhile, other portions of the seafloor plunge into the mantle. In short, a dramatically different model of Earth's tectonic processes emerged. *Tectonic processes* (*tekto* = to build) are processes that deform Earth's crust to create major structural features, such as mountains, continents, and ocean basins.

This profound reversal in scientific thought has been appropriately called a *scientific revolution*. The revolution began early in the twentieth century as a relatively straightforward proposal termed *continental drift*. For more than 50 years, the scientific community categorically rejected the idea that continents are capable of movement. North American geologists in particular had difficulty accepting continental drift, perhaps because much of the supporting evidence had been gathered from Africa, South America, and Australia, continents with which most North American geologists were unfamiliar.

▼ Figure 4.1 **Himalaya Mountains, as seen from northern India** The tallest mountains on Earth, the Himalayas, were created when the subcontinent of India collided with southeastern Asia. (Photo by Hartmut Postges/Robert Harding)

After World War II, modern instruments replaced rock hammers as the tools of choice for many Earth scientists. Armed with more advanced tools, geologists and a new breed of researchers, including *geophysicists* and *geochemists*, made several surprising discoveries that rekindled interest in the drift hypothesis. By 1968 these developments had led to the unfolding of a far more encompassing explanation, known as the *theory of plate tectonics*.

In this chapter, we will examine the events that led to this dramatic reversal of scientific opinion. We will also briefly trace the development of the *continental drift hypothesis*, examine why it was initially rejected, and consider the evidence that finally led to the acceptance of its direct descendant—the theory of plate tectonics.

CONCEPT CHECKS 4.1

1. Briefly describe the view held by most geologists prior to the 1960s regarding the ocean basins and continents.
2. What group of geologists were the least receptive to the continental drift hypothesis, and why?

4.2 Continental Drift: An Idea Before Its Time

List and explain the evidence Wegener presented to support his continental drift hypothesis.

During the 1600s, as better world maps became available, people noticed that continents, particularly South America and Africa, could be fit together like pieces of a jigsaw puzzle. However, little significance was given to this observation until 1915, when Alfred Wegener (1880–1930), a German meteorologist and geophysicist, wrote *The Origin of Continents and Oceans*. This book outlined Wegener's hypothesis, called **continental drift**, which dared to challenge the long-held assumption that the continents and ocean basins had fixed geographic positions.

Wegener suggested that a single **supercontinent** consisting of all Earth's landmasses once existed.* He named this giant landmass **Pangaea** (pronounced "Pan-jee-ah," meaning "all lands") (Figure 4.2). Wegener further hypothesized that about 200 million years ago, during a time period called the *Mesozoic era* (see Figure 12.3, page 384), this supercontinent began to fragment into smaller landmasses. These continental blocks then "drifted" to their present positions over a span of millions of years.

Wegener and others who advocated the continental drift hypothesis collected substantial evidence to support their point of view. The fit of South America and Africa and the geographic distribution of fossils and ancient climates all seemed to buttress the idea that these now separate landmasses had once been joined. Let us examine some of this evidence.

Evidence: The Continental Jigsaw Puzzle

Like a few others before him, Wegener suspected that the continents might once have been joined when he noticed the remarkable similarity between the coastlines on opposite sides of the Atlantic Ocean. However, other Earth scientists challenged Wegener's use of present-day shorelines to "fit" these continents together. These opponents correctly argued that wave erosion and depositional processes continually modify shorelines. Even if continental displacement had taken place, a good fit today would be unlikely. Because Wegener's original jigsaw fit of the continents was crude, it is assumed that he was aware of this problem (see Figure 4.2).

Scientists later determined that a much better approximation of the outer boundary of a continent is the seaward edge of its continental shelf, which lies submerged a few hundred meters below sea level. In the early 1960s, Sir Edward Bullard and two associates constructed a map that pieced together the edges of the continental shelves of South America and Africa at a depth of about 900 meters (3000 feet) (Figure 4.3). The remarkable fit obtained was more precise than even these researchers had expected.

▼ SmartFigure 4.2 **Reconstructions of Pangaea** The supercontinent Pangaea, as it is thought to have formed in the late Paleozoic and early Mesozoic eras more than 200 million years ago.

TUTORIAL
https://goo.gl/qWfzr2

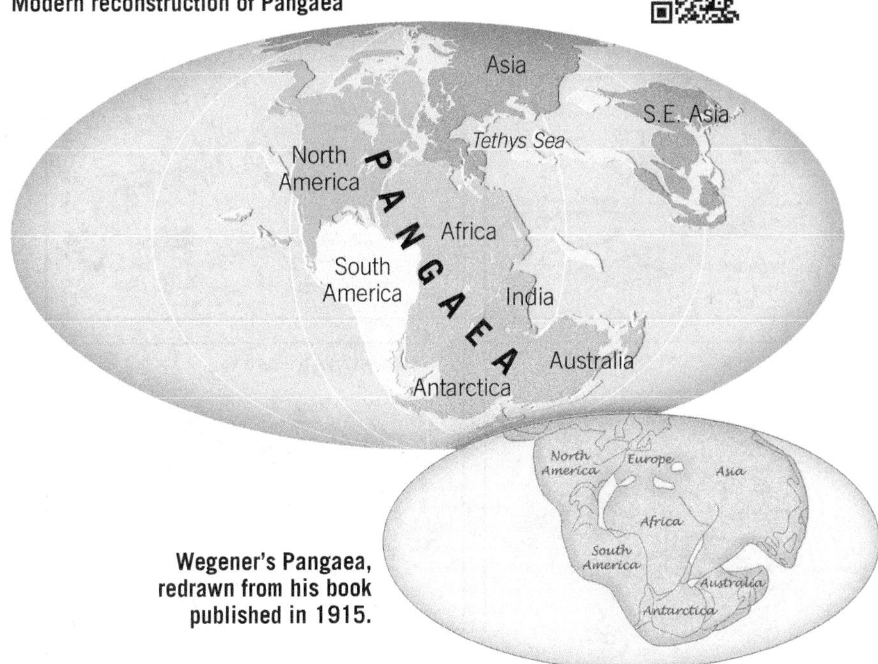

* Wegener was not the first to conceive of a long-vanished supercontinent. Eduard Suess (1831–1914), a distinguished nineteenth-century Austrian geologist, pieced together evidence for a giant landmass comprising South America, Africa, India, and Australia.

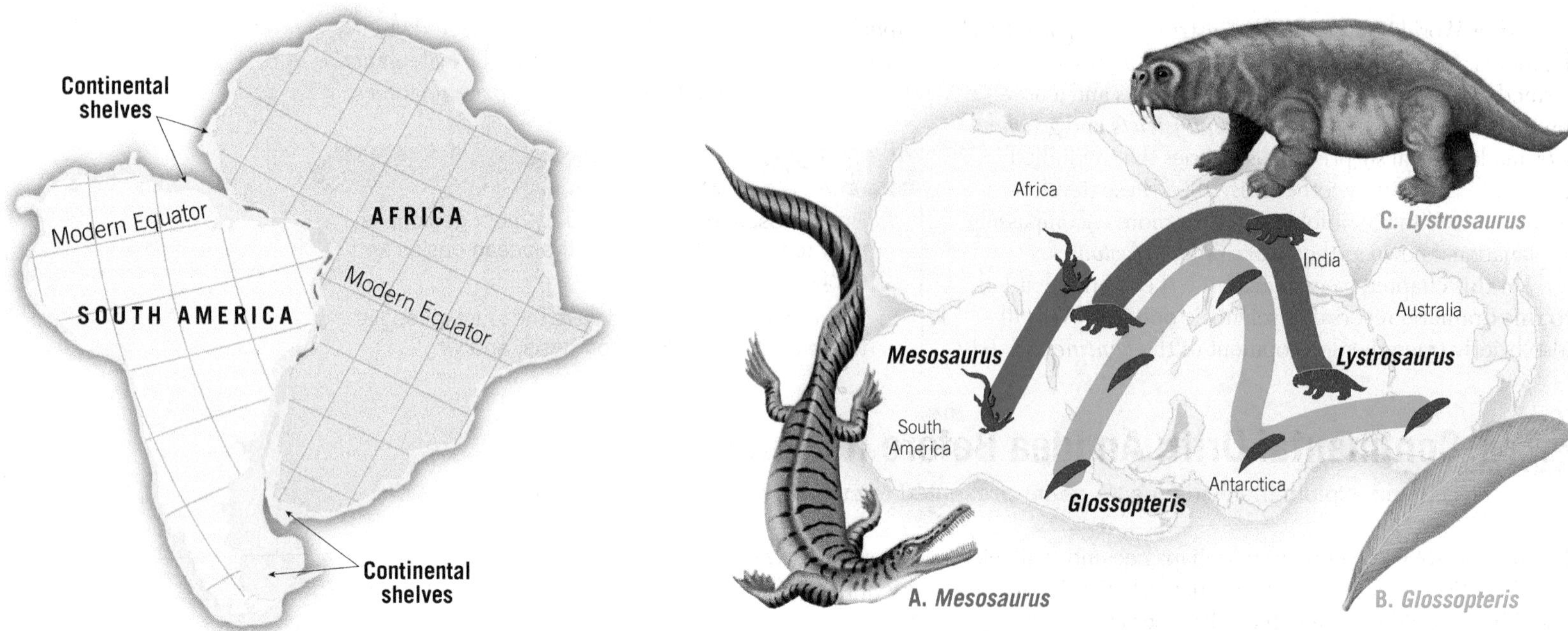

▲ Figure 4.3 **Two of the puzzle pieces** The best fit of South America and Africa occurs along the continental slope at a depth of 500 fathoms (about 900 meters [3000 feet]).

▲ Figure 4.4 **Fossil evidence supporting continental drift** Fossils of identical organisms have been discovered in rocks of similar age in Australia, Africa, South America, Antarctica, and India—continents that are currently widely separated by ocean barriers. Wegener accounted for these occurrences by placing these continents in their pre-drift locations.

Evidence: Fossils Matching Across the Seas

Although the seed for Wegener's hypothesis came from the remarkable similarities of the continental margins on opposite sides of the Atlantic, it was when he learned that identical fossil organisms had been discovered in rocks from both South America and Africa that his pursuit of continental drift became more focused. Wegener learned that most paleontologists (scientists who study the fossilized remains of ancient organisms) agreed that some type of land connection was needed to explain the existence of similar Paleozoic-age life-forms on widely separated landmasses. Just as modern life-forms native to North America are not the same as those of Africa and Australia, Paleozoic-age organisms on widely separated continents should have been distinctly different.

Mesosaurus To add credibility to his argument, Wegener documented several cases in which the same fossil organism is found only on landmasses that are now widely separated, even though it is unlikely that the living organism could have crossed the barrier of a broad ocean (Figure 4.4). A classic example is *Mesosaurus*, a small aquatic freshwater reptile whose fossil remains are mainly found in shales of Permian age (about 260 million years ago) in eastern South America and southwestern Africa. If *Mesosaurus* had been able to make the long journey across the South Atlantic, its remains should be more widely distributed. Because this is not the case, Wegener asserted that South America and Africa must have been joined during that period of Earth history.

How did opponents of continental drift explain the existence of identical fossil organisms in places separated by thousands of kilometers of open ocean? Rafting, transoceanic land bridges (isthmian links), and island stepping stones were the most widely invoked explanations for these migrations (Figure 4.5). We know, for example, that during the Ice Age that ended about

◄ Figure 4.5 **How do land animals cross vast oceans?** These sketches illustrate various early proposals to explain the occurrence of similar species on landmasses now separated by vast oceans. (Used by permission of John C. Holden)

8000 years ago, the lowering of sea level allowed mammals (including humans) to cross the narrow Bering Strait that separates Russia and Alaska. Was it possible that land bridges once connected Africa and South America but later subsided below sea level? Modern maps of the seafloor substantiate Wegner's views and show no such sunken land bridges.

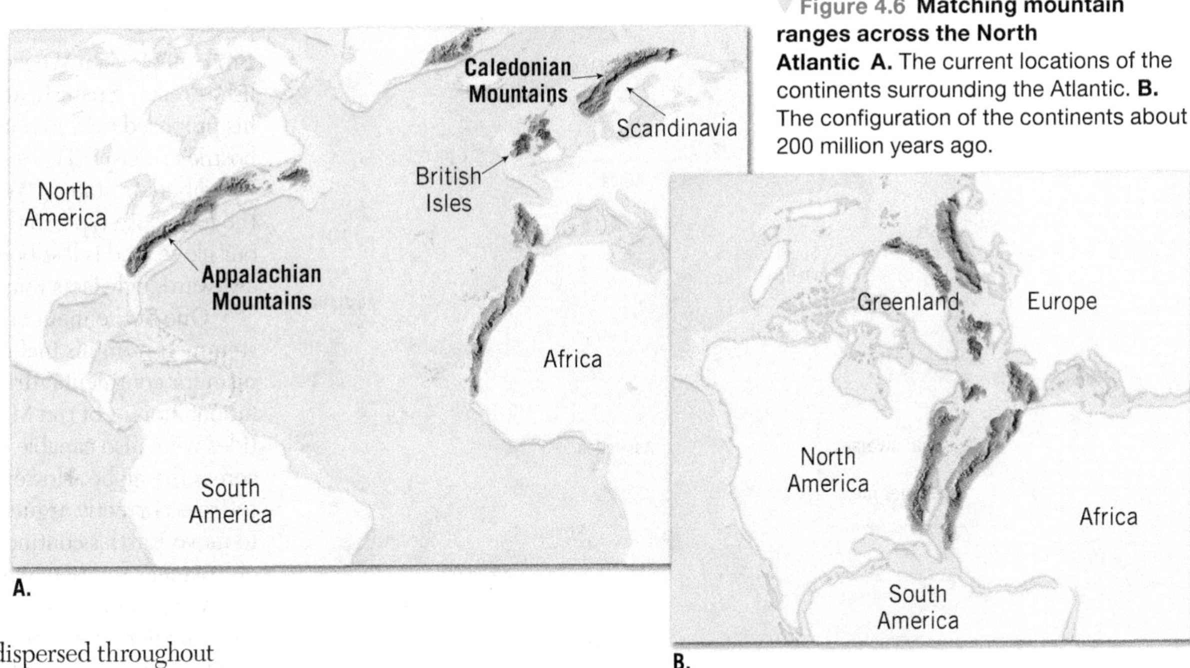

Figure 4.6 **Matching mountain ranges across the North Atlantic A.** The current locations of the continents surrounding the Atlantic. **B.** The configuration of the continents about 200 million years ago.

Glossopteris Wegener also cited the distribution of the fossil "seed fern" *Glossopteris* as evidence for Pangaea's existence (see Figure 4.4). With tongue-shaped leaves and seeds too large to be carried by the wind, this plant was known to be widely dispersed throughout Africa, Australia, India, and South America. Later, fossil remains of *Glossopteris* were also discovered in Antarctica.* Wegener also learned that these seed ferns and associated flora grew only in cool climates—similar to central Canada. Therefore, he concluded that when these landmasses were joined, they were located much closer to the South Pole.

Evidence: Rock Types and Geologic Features

You know that successfully completing a jigsaw puzzle requires maintaining the continuity of the picture while fitting the pieces together. In the case of continental drift, this means that the rocks on either side of the Atlantic that predate the proposed Mesozoic split should match up to form a continuous "picture" when the continents are fitted together as Wegener proposed.

Indeed, Wegener found such "matches" across the Atlantic. For instance, highly deformed igneous rocks in Brazil closely resemble similar rocks of the same age in Africa. Also, the mountain belt that includes the Appalachians trends northeastward through the eastern United States and disappears off the coast of Newfoundland (Figure 4.6A). Mountains of comparable age and structure are found in the British Isles and Scandinavia. When these landmasses are positioned as Wegener proposed (Figure 4.6B), the mountain chains form a nearly continuous belt. As Wegener wrote, "It is just as if we were to refit the torn pieces of a newspaper by matching their edges and then check whether the lines of print run smoothly across. If they do, there is nothing left but to conclude that the pieces were in fact joined in this way."*

Evidence: Ancient Climates

Because Alfred Wegener was a student of world climates, he suspected that paleoclimatic (*paleo* = ancient, *climatic* = climate) data might also support the idea of mobile continents. His assertion was bolstered by the discovery of evidence for a glacial period dating to the late *Paleozoic era* (see Figure 12.3, page 384) in southern Africa, South America, Australia, and India. This meant that about 300 million years ago, vast ice sheets covered extensive portions of the Southern Hemisphere as well as India (Figure 4.7A). Much of the land area that contains evidence of this Paleozoic glaciation presently lies within 30° of the equator, in subtropical or tropical climates.

How could extensive ice sheets form near the equator? One proposal suggested that our planet experienced a period of extreme global cooling. Wegener rejected this explanation because during the same span of geologic time, large tropical swamps existed in several locations in the Northern Hemisphere. The lush vegetation in those swamps was eventually buried and converted to coal (Figure 4.7B). Today these deposits comprise major coal fields in the eastern United States and Northern Europe. Many of the fossils found in these coal-bearing rocks were produced by tree ferns with large fronds—ferns that would have grown in warm, moist climates.*

* In 1912 Captain Robert Scott and two companions froze to death lying beside 16 kilograms (35 pounds) of rock on their return from a failed attempt to be the first to reach the South Pole. These samples, collected on Beardmore Glacier, contained fossil remains of *Glossopteris*.

* Alfred Wegener, *The Origin of Continents and Oceans*, translated from the fourth revised German ed. of 1929 by J. Birman (London: Methuen, 1966).

* It is important to note that coal can form in a variety of climates, provided that large quantities of plant life are buried.

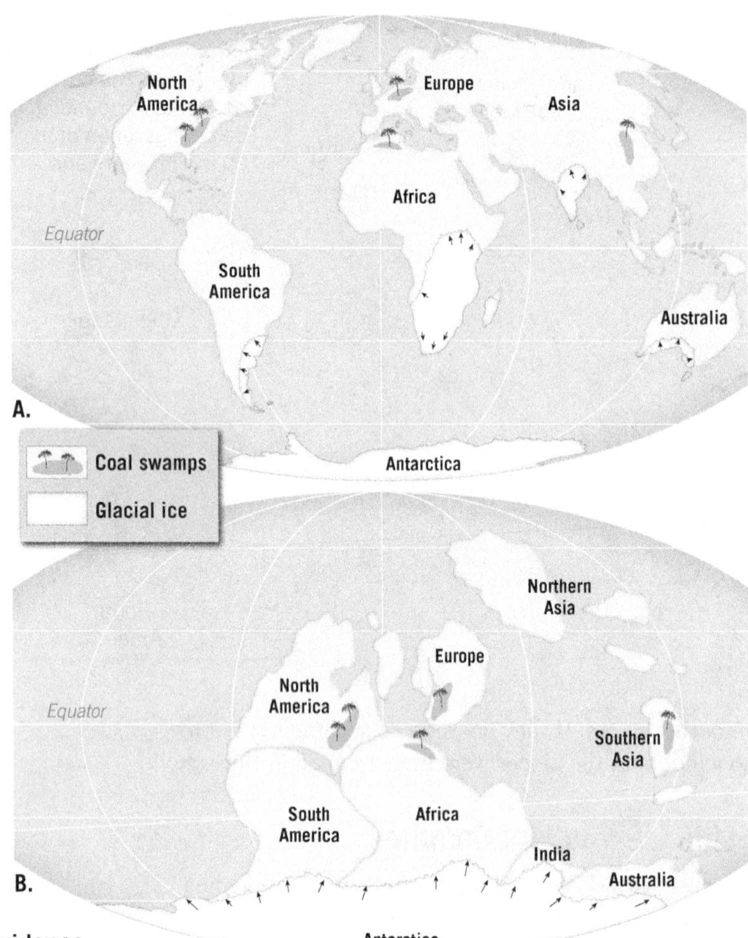

▲ Figure 4.7
Paleoclimatic evidence for continental drift
A. The continents in their current positions show that about 300 million years ago, ice sheets covered extensive areas of the Southern Hemisphere and India. Arrows show the direction of ice movement that can be inferred from the pattern of glacial scratches and grooves found in the bedrock. Tropical coal swamps also existed in areas that are now temperate.
B. Restoring the continents to their pre-drift positions creates a single glaciation centered on the South Pole and puts the coal swamps near the equator.

The existence of these large tropical swamps, Wegener argued, was inconsistent with the proposal that extreme global cooling caused glaciers to form in areas that are currently tropical.

Wegener suggested a more plausible explanation for the late Paleozoic glaciation: The southern continents were joined together in the supercontinent of Pangaea and located near the South Pole (see Figure 4.7B). This would account for the polar conditions required to generate extensive expanses of glacial ice over much of these landmasses. At the same time, this geography places today's northern continents nearer the equator and accounts for the tropical swamps that generated the vast coal deposits.

As compelling as this evidence may have been, 50 years passed before most of the scientific community accepted the concept of continental drift. A number of questions needed answers, including these: How does a glacier develop in hot, arid central Australia? and How do land animals cross what now are broad oceans?

The Great Debate

From 1924 when Wegener's book was translated into English, French, Spanish, and Russian until his death in 1930, his proposed drift hypothesis encountered a great deal of hostile criticism. The respected American geologist R. T. Chamberlain stated, "Wegener's hypothesis in general is of the foot-loose type, in that it takes considerable liberty with our globe, and is less bound by restrictions or tied down by awkward, ugly facts than most of its rival theories."

One of the main objections to Wegener's hypothesis stemmed from his inability to identify a credible mechanism for continental drift. Wegener proposed that gravitational forces of the Moon and Sun that produce Earth's tides were also capable of gradually moving the continents across the globe. However, the prominent physicist Harold Jeffreys correctly argued that tidal forces strong enough to move Earth's continents would have resulted in halting our planet's rotation, which, of course, has not happened.

Wegener also incorrectly suggested that the larger and sturdier continents broke through thinner oceanic crust, much as icebreakers cut through ice. However, no evidence existed to suggest that the ocean floor was weak enough to permit passage of the continents without the continents being appreciably deformed in the process.

In 1930, Wegener made his fourth and final trip to the Greenland Ice Sheet (**Figure 4.8**). Although the primary focus of this expedition was to study this great ice cap and its climate, Wegener continued to test his continental drift hypothesis. While returning from Eismitte, an experimental station located in the center of

▼ Figure 4.8 **Alfred Wegener during an expedition to Greenland** (Photo courtesy of Archive of Alfred Wegener Institute)

Alfred Wegener shown waiting out the 1912–1913 Arctic winter during an expedition to Greenland, where he made a 1200-kilometer traverse across the widest part of the island's ice sheet.

Greenland, Wegener perished along with his Greenland companion. His intriguing idea, however, did not die.

Why was Wegener unable to overturn the established scientific views of his day? Foremost was the fact that, although the central theme of Wegener's drift hypothesis was correct, some details were incorrect. For example, continents do not break through the ocean floor, and tidal energy is much too weak to move continents. Moreover, for any comprehensive scientific hypothesis to gain wide acceptance, it must withstand critical testing from all areas of science. Despite Wegener's great contribution to our understanding of Earth, not *all* of the evidence supported the continental drift hypothesis as he had proposed it. As a result, most of the scientific community (particularly in North America) rejected continental drift or at least treated it with considerable skepticism. However, some scientists recognized the strength of the evidence Wegner had accumulated and continued to pursue the idea.

CONCEPT CHECKS 4.2

1. What was the first line of evidence that led early investigators to suspect that the continents were once connected?
2. Explain why the discovery of the fossil remains of *Mesosaurus* in both South America and Africa, but nowhere else, supports the continental drift hypothesis.
3. Early in the twentieth century, what was the prevailing view of how land animals apparently migrated across vast expanses of open ocean?
4. How did Wegener account for evidence of glaciers in portions of South America, Africa, and India, when areas in North America, Europe, and Asia supported lush tropical swamps?
5. Describe two aspects of Wegener's continental drift hypothesis that were objectionable to most Earth scientists.

4.3 The Theory of Plate Tectonics

List the major differences between Earth's lithosphere and asthenosphere and explain the importance of each in the plate tectonics theory.

Following World War II, oceanographers equipped with new marine tools and ample funding from the U.S. Office of Naval Research embarked on an unprecedented period of oceanographic exploration. Over the next 2 decades, a much better picture of large expanses of the seafloor slowly and painstakingly began to emerge. From this work came the discovery of a global oceanic ridge system that winds through all the major oceans.

In other parts of the ocean, more discoveries were being made. Studies conducted in the western Pacific demonstrated that earthquakes were occurring at great depths beneath deep-ocean trenches. Of equal importance was the fact that dredging of the seafloor did not bring up any oceanic crust that was older than 180 million years. Further, sediment accumulations in the deep-ocean basins were found to be thin, not the thousands of meters that had been predicted. By 1968 these developments, among others, had led to the unfolding of a far more encompassing theory than continental drift, known as the **theory of plate tectonics**.

Rigid Lithosphere Overlies Weak Asthenosphere

According to the plate tectonics model, the crust and the uppermost, and therefore coolest, part of the mantle constitute Earth's strong outer layer, the **lithosphere** (*lithos* = stone). The lithosphere varies in both thickness and density, depending on whether it is oceanic or continental (Figure 4.9). Oceanic lithosphere is about 100 kilometers (60 miles) thick in the deep-ocean basins but is considerably thinner along the crest of the oceanic ridge system—a topic we will consider later. In contrast, continental lithosphere averages about 150 kilometers (90 miles) thick but may extend to depths of 200 kilometers (125 miles) or more beneath the stable interiors of the continents. Further, oceanic and continental crust differ in density. Oceanic crust is composed of basalt, a rock rich in dense iron and magnesium, whereas continental crust is composed largely of less dense granitic rocks. Because of these differences, the overall density of oceanic lithosphere (crust and upper mantle) is greater than the overall density of continental lithosphere. This important difference will be considered in greater detail later in this chapter.

The **asthenosphere** (*asthenos* = weak) is a hotter, weaker region in the mantle that lies below the lithosphere (see Figure 4.9). In the upper asthenosphere (located between 100 and 200 kilometers [60 to 125 miles] depth), the pressure and temperature bring rock very near to

SmartFigure 4.9 **The rigid lithosphere overlies the weak asthenosphere**

TUTORIAL
https://goo.gl/ujkFfZ

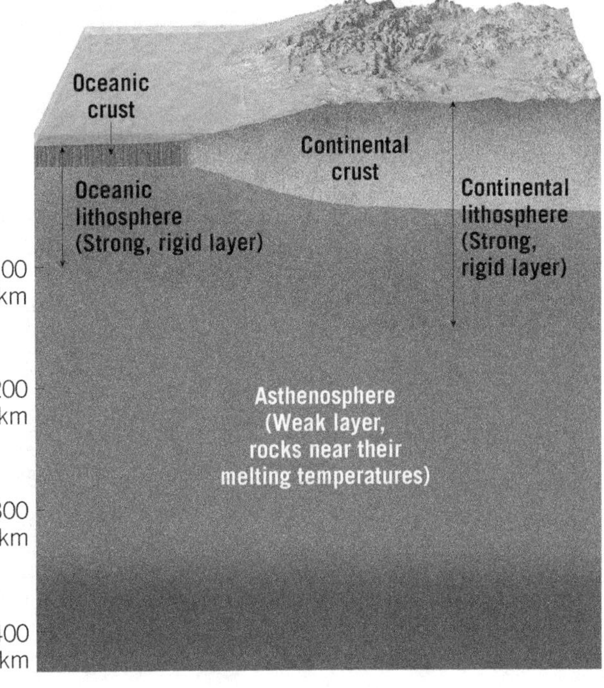

melting. Consequently, although the rock remains largely solid, it responds to forces by *flowing*, similarly to the way clay may deform if you compress it slowly. By contrast, the relatively cool and rigid lithosphere tends to respond to forces acting on it by *bending or breaking but not flowing*. Because of these differences, Earth's rigid outer shell is effectively detached from the asthenosphere, which allows these layers to move independently.

Earth's Major Plates

The lithosphere is broken into numerous segments of irregular size and shape called **lithospheric plates**, or simply **plates**, that are in constant motion with respect to one another (Figure 4.10). Seven major lithospheric plates are recognized and account for 94 percent of Earth's surface area: the *North American*, *South American*, *Pacific*, *African*, *Eurasian*, *Australian-Indian*, and *Antarctic plates*. The largest is the Pacific plate, which encompasses a significant portion of the Pacific basin. Each of the six other large plates consists of an entire continent, as well as a significant amount of oceanic crust. Notice in Figure 4.10 that the South American plate encompasses almost all of South America and about one-half of the floor of the South Atlantic. Note also that none of the plates are defined entirely by the margins of a single continent. This is a major departure from Wegener's continental drift hypothesis, which proposed that the continents move through the ocean floor, not with it.

Intermediate-sized plates include the *Caribbean*, *Nazca*, *Philippine*, *Arabian*, *Cocos*, *Scotia*, and *Juan de Fuca plates*. These plates, with the exception of the Arabian plate, are composed mostly of oceanic lithosphere. In addition, many smaller plates, called *microplates*, have been identified but are not shown in Figure 4.10.

Plate Movement

One of the main tenets of the plate tectonics theory is that plates move as somewhat rigid units relative to all other plates. As plates move, the distance between two locations on different plates, such as New York and London, gradually changes, whereas the distance between sites on the same plate—New York and Denver, for

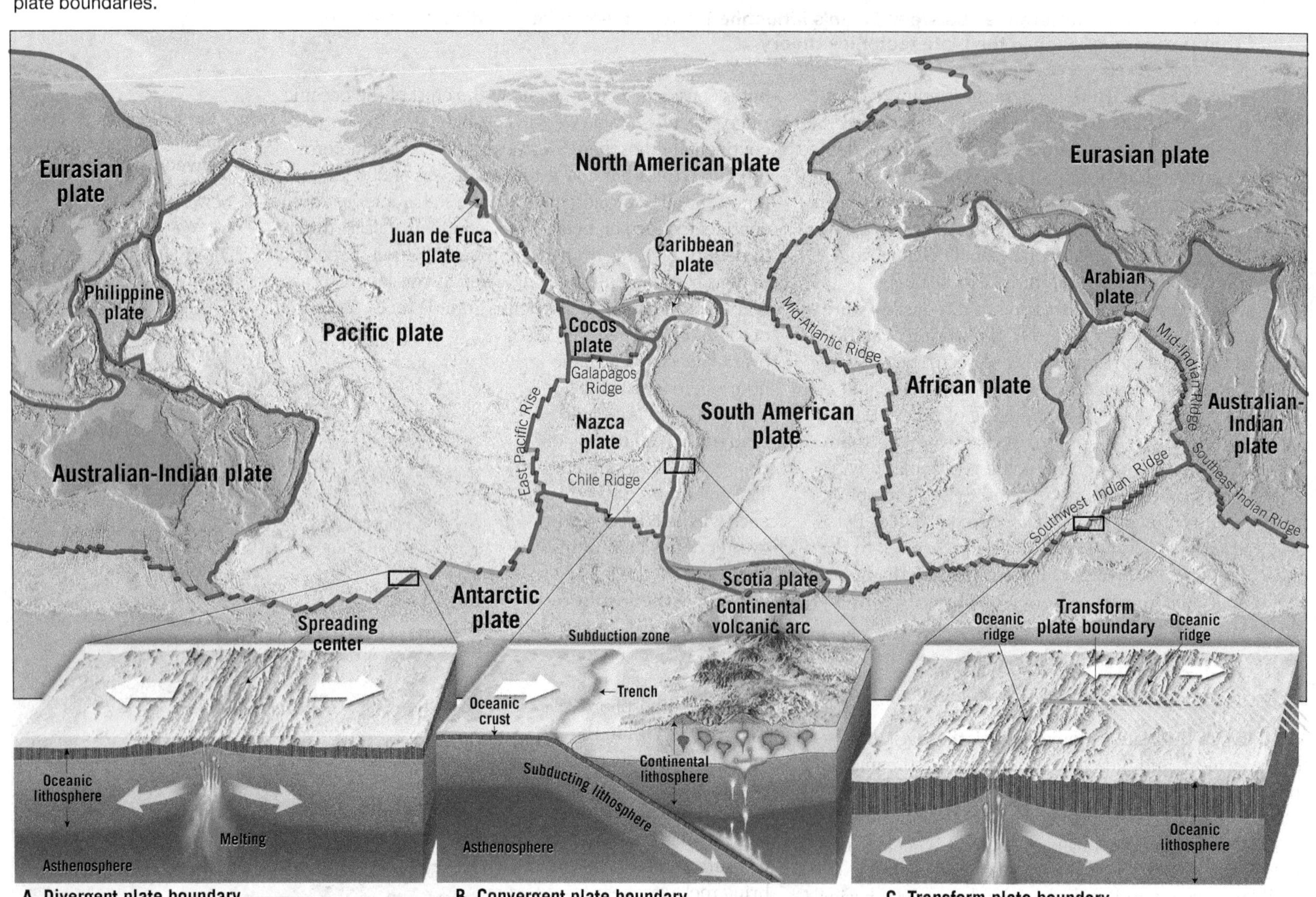

Figure 4.10 Earth's major lithospheric plates The block diagrams below the map illustrate divergent, convergent, and transform plate boundaries.

example—remains relatively constant. However, parts of some plates are comparatively "weak." For example, southern China is literally being squeezed as the Indian subcontinent rams into Asia proper.

Because plates are in constant motion relative to each other, most major interactions among them occur along their *boundaries*, and this is therefore where most deformation occurs. In fact, plate boundaries were first established by plotting the locations of earthquakes and volcanoes. Plates are delimited by three distinct types of boundaries, which are differentiated by the type of movement they exhibit. These boundaries are depicted at the bottom of Figure 4.10 and are briefly described here:

- Divergent plate boundaries—where two plates move apart, resulting in upwelling and partial melting of hot material from the mantle to create new seafloor (see Figure 4.10A)
- Convergent plate boundaries—where two plates move toward each other, resulting either in oceanic lithosphere descending beneath an overriding plate, eventually to be reabsorbed into the mantle, or possibly in the collision of two continental blocks to create a mountain belt (see Figure 4.10B)
- Transform plate boundaries—where two plates grind past each other without the production or destruction of lithosphere (see Figure 4.10C)

Divergent and convergent plate boundaries each account for about 40 percent of all plate boundaries. Transform boundaries account for the remaining 20 percent. In the following sections we will discuss the three types of plate boundaries.

CONCEPT CHECKS 4.3

1. What new findings about the ocean floor did oceanographers discover after World War II?
2. Compare and contrast Earth's lithosphere and asthenosphere.
3. List the seven largest lithospheric plates.
4. List the three types of plate boundaries and describe the relative motion along each.

4.4 Divergent Plate Boundaries and Seafloor Spreading

Sketch and describe the movement along a divergent plate boundary that results in the formation of new oceanic lithosphere.

Most **divergent plate boundaries** (*di* = apart, *vergere* = to move) are located along the crests of oceanic ridges and can be thought of as *constructive plate margins* because this is where new ocean floor is generated (Figure 4.11). Here, two adjacent plates move away from each other, producing long, narrow fractures in the ocean crust. As a result, hot molten rock from the mantle below migrates upward to fill the voids left as the crust is being ripped apart. This molten material gradually cools to produce new slivers of seafloor. In a slow yet unending manner, adjacent plates spread apart, and new oceanic lithosphere forms between them. For this reason, divergent plate boundaries are also called **spreading centers**.

Figure 4.11 **Seafloor spreading** Most divergent plate boundaries are situated along the crests of oceanic ridges—the sites of seafloor spreading.

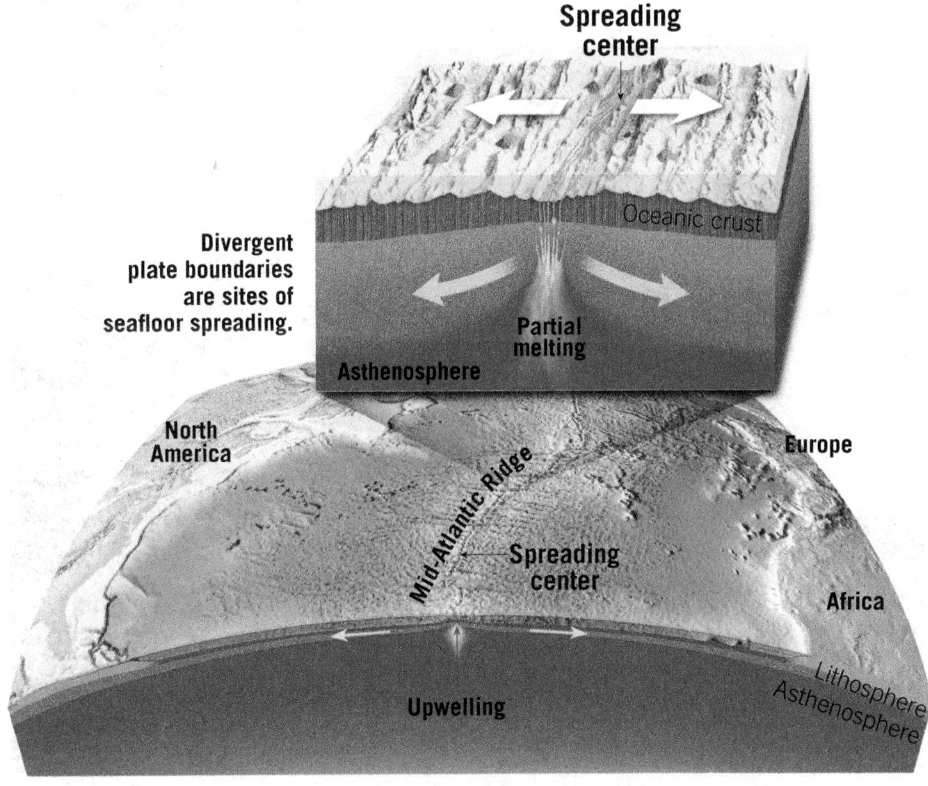

Oceanic Ridges and Seafloor Spreading

The majority of, but not all, divergent plate boundaries are associated with *oceanic ridges*: elevated areas of the seafloor characterized by high heat flow and volcanism. The global **oceanic ridge system** is the longest topographic feature on Earth's surface, exceeding 70,000 kilometers (43,000 miles) in length. As shown in Figure 4.10, various segments of the global ridge system have been named, including the Mid-Atlantic Ridge, East Pacific Rise, and Mid-Indian Ridge.

Representing 20 percent of Earth's surface, the oceanic ridge system winds through all major ocean basins, like the seams on a baseball. Although the crest of the oceanic ridge is commonly 2 to 3 kilometers (1 to 2 miles) higher than the adjacent ocean basins, the term *ridge* may be misleading because it implies "narrow" when, in fact, ridges vary in width from 1000 kilometers (600 miles) to more than 4000 kilometers (2500 miles). Further, along the crest of some ridge segments is a deep canyonlike structure called a **rift valley** (Figure 4.12). This structure is evidence that tensional (pulling apart) forces are actively pulling the ocean crust apart at the ridge crest.

The mechanism that operates along the oceanic ridge system to create new seafloor is appropriately called **seafloor spreading**. Spreading typically averages around 5 centimeters (2 inches) per year, roughly the same rate at which human fingernails grow. Comparatively slow spreading rates of 2 centimeters per year are found along the Mid-Atlantic Ridge, whereas spreading rates exceeding 15 centimeters (6 inches) per year have been measured along sections of the East Pacific Rise. Although these rates of seafloor production are slow on a human time scale, they are rapid enough to have generated all of Earth's current oceanic lithosphere within the past 200 million years.

The primary reason for the elevated position of the oceanic ridge is that newly created oceanic lithosphere is hot and, therefore, less dense than cooler rocks located away from the ridge axis. (Geologists use the term *axis* to refer to a line that follows the general trend of the ridge crest.) As soon as new lithosphere forms, it is slowly yet continually displaced away from the zone of mantle upwelling. Thus, it begins to cool and contract, thereby increasing in density. This thermal contraction accounts for the increase in ocean depth away from the ridge crest. It takes about 80 million years for the temperature of oceanic lithosphere to stabilize and contraction to cease. By this time, rock that was once part of the elevated oceanic ridge system is located in the deep-ocean basin, where it may be buried by substantial accumulations of sediment.

In addition, as the plate moves away from the ridge, cooling of the underlying asthenosphere causes its upper layers to become increasingly rigid and hence to become part of the lithosphere. Stated another way, the thickness of oceanic lithosphere is age dependent. The older (cooler) it is, the greater its thickness. Oceanic lithosphere that exceeds 80 million years in age is about 100 kilometers (60 miles) thick—approximately its maximum thickness.

SmartFigure 4.12
Rift valley in Iceland Thingvellir National Park, Iceland, is located on the western margin of a rift valley roughly 30 kilometers (20 mile) wide. This rift valley is connected to a similar feature that extends along the crest of the Mid-Atlantic Ridge. The cliff in the left half of the image approximates the eastern edge of the North American plate. (Photo by Ragnar Th Sigurdsson/Arctic Images/Alamy Stock Photo)

MOBILE FIELD TRIP
https://goo.gl/D7KNKF

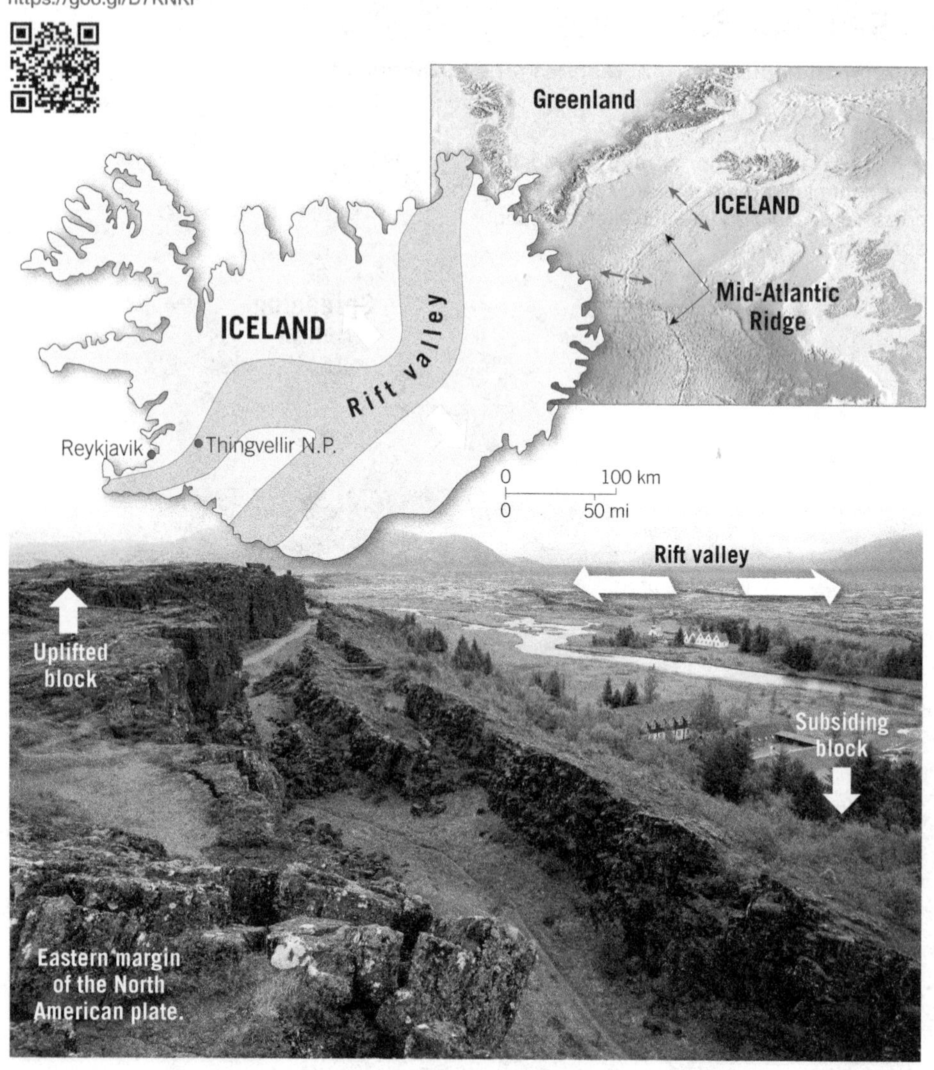

Continental Rifting

Divergent boundaries can develop within a continent and may cause the landmass to split into two or more smaller segments separated by an ocean basin. Continental rifting begins when plate motions produce tensional forces that pull and stretch the lithosphere. This stretching, in turn, promotes mantle upwelling and broad upwarping of the overlying lithosphere (Figure 4.13A). This process thins the lithosphere and breaks the brittle crustal rocks into large blocks. As the tectonic forces continue to pull apart the crust, the broken crustal fragments sink, generating an elongated depression called a **continental rift**, which can widen to form a narrow sea (Figure 4.13B,C) and eventually a new ocean basin (Figure 4.13D).

An example of an active continental rift is the East African Rift (Figure 4.14). Whether this rift will eventually result in the breakup of Africa is a topic of ongoing research. Nevertheless, the East African Rift is an excellent model of the initial stage in the breakup of a continent. Here, tensional forces have stretched and thinned the lithosphere, allowing molten rock to ascend from the mantle. Evidence for this upwelling includes several large volcanic mountains, including Mount Kilimanjaro and Mount Kenya, the tallest peaks in Africa. Research suggests that if rifting continues, the rift valley will lengthen and deepen (see Figure 4.13C).

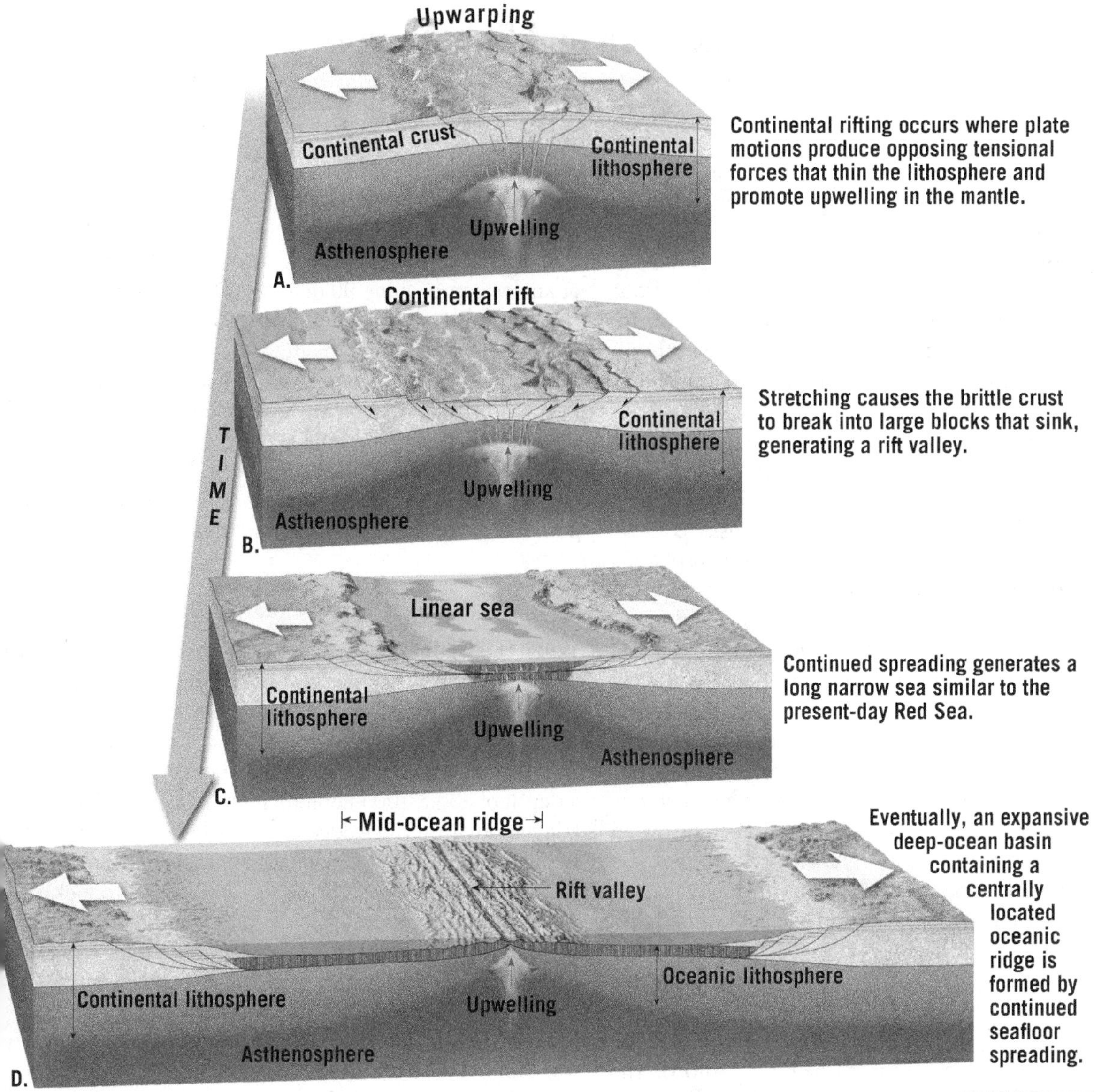

Upwarping

Continental crust

Continental lithosphere

Upwelling

Asthenosphere

A.

Continental rifting occurs where plate motions produce opposing tensional forces that thin the lithosphere and promote upwelling in the mantle.

Continental rift

Continental lithosphere

Upwelling

Asthenosphere

B.

Stretching causes the brittle crust to break into large blocks that sink, generating a rift valley.

TIME

Linear sea

Continental lithosphere

Upwelling

Asthenosphere

C.

Continued spreading generates a long narrow sea similar to the present-day Red Sea.

Mid-ocean ridge

Rift valley

Oceanic lithosphere

Continental lithosphere

Upwelling

Asthenosphere

D.

Eventually, an expansive deep-ocean basin containing a centrally located oceanic ridge is formed by continued seafloor spreading.

SmartFigure 4.13 **Continental rifting: Formation of new ocean basins**

TUTORIAL
https://goo.gl/9CokZD

At some point, the rift valley will become a narrow sea with an outlet to the ocean. The Red Sea, formed when the Arabian Peninsula split from Africa, is a modern example of such a feature and provides us with a view of how the Atlantic Ocean may have looked in its infancy (see Figure 4.13D).

CONCEPT CHECKS 4.4

1. Sketch or describe how two plates move in relation to each other along divergent plate boundaries.
2. What is the average rate of seafloor spreading in modern oceans?
3. List four features that characterize the oceanic ridge system.
4. Briefly describe the process of continental rifting. Name a location where is it occurring today.

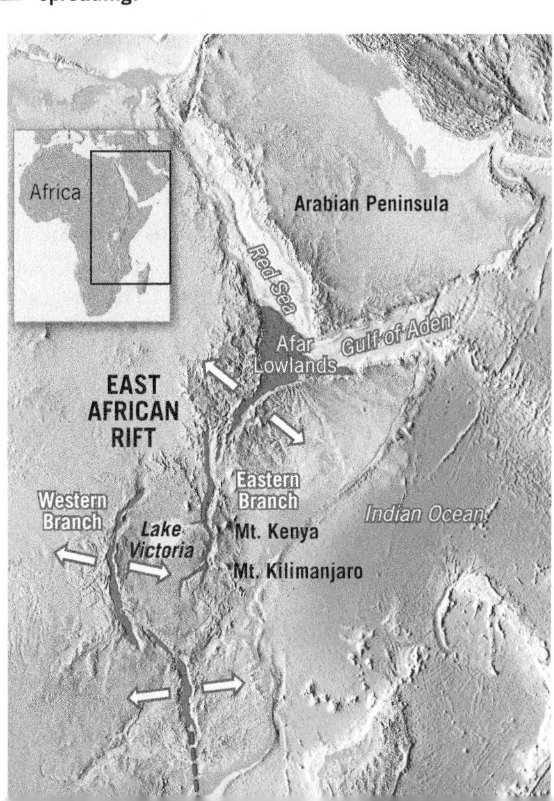

SmartFigure 4.14 **East African Rift valley** The East African Rift valley represents the early stage in the breakup of a continent. Areas shown in red consist of lithosphere that has been stretched and thinned, allowing magma to well up from the mantle.

CONDOR VIDEO
https://goo.gl/RXv8qH

4.5 Convergent Plate Boundaries and Subduction

Compare and contrast the three types of convergent plate boundaries and name a location where each type can be found.

New lithosphere is constantly being produced at the oceanic ridges. However, our planet is not growing larger; its total surface area remains constant. A balance is maintained because older, denser portions of oceanic lithosphere descend into the mantle at a rate equal to seafloor production. This activity occurs along **convergent plate boundaries**, where two plates move toward each other and the leading edge of one is bent downward as it slides beneath the other.

Convergent boundaries are also called **subduction zones** because they are sites where lithosphere is descending (being subducted) into the mantle. Lithospheric plates subduct (sink) because their density is greater than the density of the underlying asthenosphere. (Recall that oceanic crust has a greater density than continental crust because it is largely composed of dense ferromagnesian-rich mineral.) In general, old oceanic lithosphere is about 2 percent more dense than the underlying asthenosphere, and it sinks much like an anchor on a ship. Continental lithosphere, in contrast, is less dense than the underlying asthenosphere and tends to resist subduction. However, there are a few locations where continental lithosphere is thought to have been forced below an overriding plate, albeit to relatively shallow depths.

Deep-ocean trenches are long, linear depressions in the seafloor that are generally located only a few hundred kilometers offshore of either a continent or a chain of volcanic islands such as the Aleutian chain (see Figure 13.6, page 420). These underwater surface features are produced where oceanic lithosphere bends as it descends into the mantle along subduction zones (see Figure 4.15A). An example is the Peru–Chile trench, located along the west coast of South America. It is more than 4500 kilometers (3000 miles) long, and its floor is as much as 8 kilometers (5 miles) below sea level. Western Pacific trenches, including the Mariana and Tonga trenches, are even deeper than those of the eastern Pacific.

Slabs of oceanic lithosphere descend into the mantle at angles that vary from a few degrees to nearly vertical (90 degrees). The angle at which oceanic lithosphere subducts depends largely on its age and, therefore, its density. For example, when seafloor spreading occurs relatively near a subduction zone, as is the case along the coast of Chile (see Figure 4.10B), the subducting lithosphere is young and buoyant, which results in a low angle of descent. As the two plates converge, the overriding plate scrapes over the top of the subducting plate below—a type of forced subduction. Consequently, the region around the Peru–Chile trench experiences great earthquakes, including the 2010 Chile earthquake—one of the 10 largest quakes on record.

As oceanic lithosphere ages (moves farther from the spreading center), it gradually cools, which causes it to thicken and increase in density. In parts of the western Pacific, some oceanic lithosphere is 180 million years old—the thickest and densest in today's oceans. The very dense slabs in this region typically plunge into the mantle at angles approaching 90 degrees. This largely explains why most trenches in the western Pacific are deeper than trenches in the eastern Pacific.

Although all convergent zones have the same basic characteristics, they may vary considerably depending on the type of crustal material involved and the tectonic setting. Convergent boundaries can form *between one oceanic plate and one continental plate*, *between two oceanic plates*, or *between two continental plates* (Figure 4.15).

Oceanic–Continental Convergence

When the leading edge of a plate capped with continental crust converges with a slab of oceanic lithosphere, the buoyant continental block remains "floating," while the denser oceanic slab sinks into the mantle (see Figure 4.15A). When a descending oceanic slab reaches a depth of about 100 kilometers (60 miles), melting is triggered within the wedge of hot asthenosphere that lies above it. But how does the subduction of a cool slab of oceanic lithosphere cause mantle rock to melt? The answer lies in the fact that water contained in the descending plates acts the way salt does to melt ice. That is, "wet" rock in a high-pressure environment melts at substantially lower temperatures than does "dry" rock of the same composition.

Sediments and oceanic crust contain large amounts of water, which is carried to great depths by a subducting plate. As the plate plunges downward, heat and pressure drive out water from the hydrated (water-rich) minerals in the subducting slab. At a depth of roughly 100 kilometers (60 miles), the wedge of mantle rock is sufficiently hot that the introduction of water from the slab below leads to some melting. This process, called **partial melting**, is thought to generate some molten material, which is mixed with unmelted mantle rock. Being less dense than the surrounding mantle, this hot mobile material gradually rises toward the surface. Depending on the environment, these mantle-derived masses of molten rock may ascend through the crust and give rise to a volcanic eruption. However, much of this material never reaches the surface but solidifies at depth—a process that thickens the crust.

The volcanoes of the towering Andes were produced by molten rock generated by the subduction of the Nazca plate beneath the South American continent (see Figure 4.10). Mountain systems like the Andes, which are produced in part by volcanic activity associated

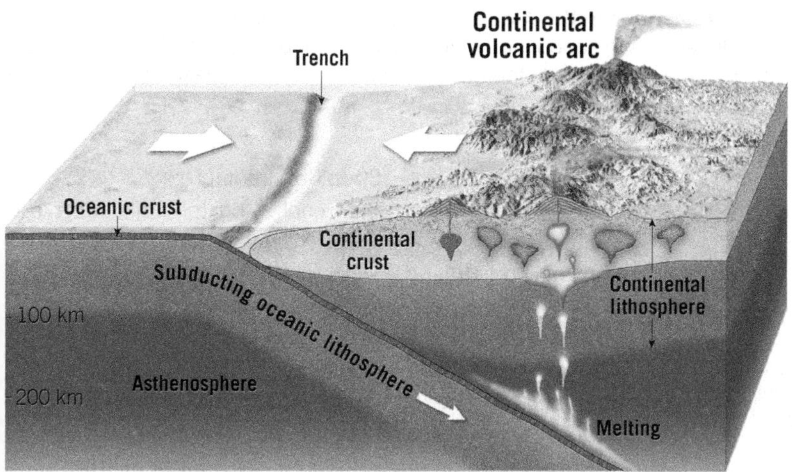

A. Convergent plate boundary where oceanic lithosphere is subducting beneath continental lithosphere.

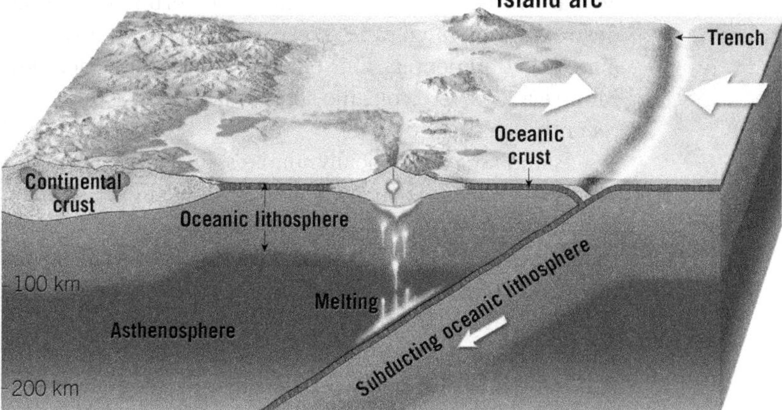

B. Convergent plate boundary involving two slabs of oceanic lithosphere.

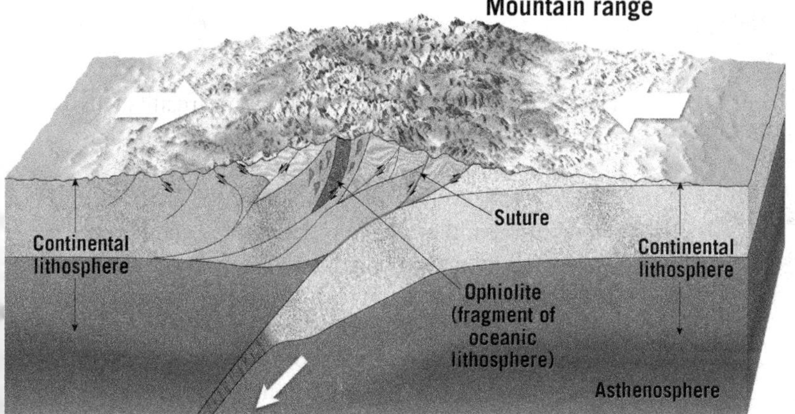

C. Continental collisions occur along convergent plate boundaries when both plates are capped with continental crust.

SmartFigure 4.15 Three types of convergent plate boundaries

TUTORIAL
https://goo.gl/TDOFNu

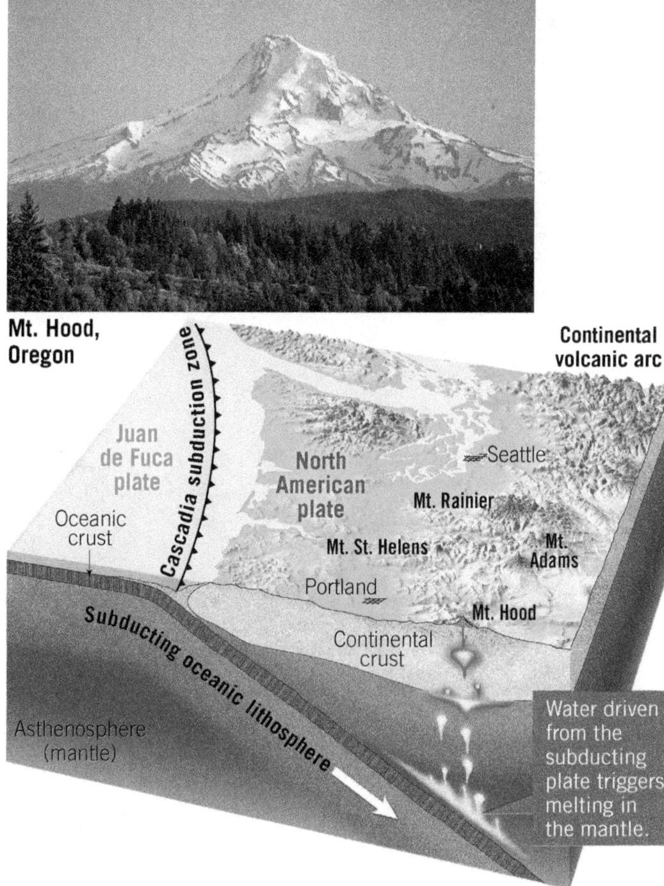
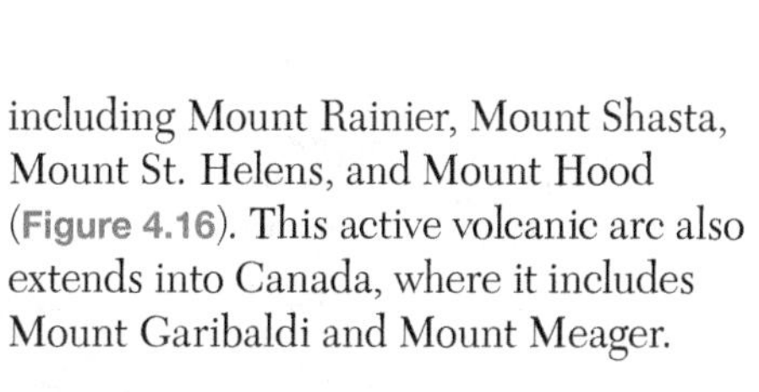

Figure 4.16 Example of an oceanic–continental convergent plate boundary The Cascade Range is a *continental volcanic arc* formed by the subduction of the Juan de Fuca plate beneath the North American plate. Mount Hood, Oregon, is one of more than a dozen large composite volcanoes in the Cascade Range. (Photo by Wallace Garrison/Photolibrary/Getty Images)

with the subduction of oceanic lithosphere, are called **continental volcanic arcs**. The Cascade Range in Washington, Oregon, and California is another mountain system consisting of several well-known volcanoes, including Mount Rainier, Mount Shasta, Mount St. Helens, and Mount Hood (Figure 4.16). This active volcanic arc also extends into Canada, where it includes Mount Garibaldi and Mount Meager.

Oceanic–Oceanic Convergence

An *oceanic–oceanic convergent boundary* has many features in common with oceanic–continental plate margins (see Figure 4.15A,B). Where two oceanic slabs converge, one descends beneath the other, initiating volcanic activity by the same mechanism that operates at all subduction zones (see Figure 4.10). Water released from the subducting slab of oceanic lithosphere triggers melting in the hot wedge of mantle rock above. In this setting, volcanoes grow up from the ocean floor rather than upon a continental platform. Sustained subduction eventually results in a chain of volcanic structures large enough to emerge as islands. The newly formed land, consisting of an arc-shaped chain of volcanic islands, is called a **volcanic island arc**, or simply an **island arc** (Figure 4.17).

The Aleutian, Mariana, and Tonga Islands are examples of relatively young volcanic island arcs. Island arcs

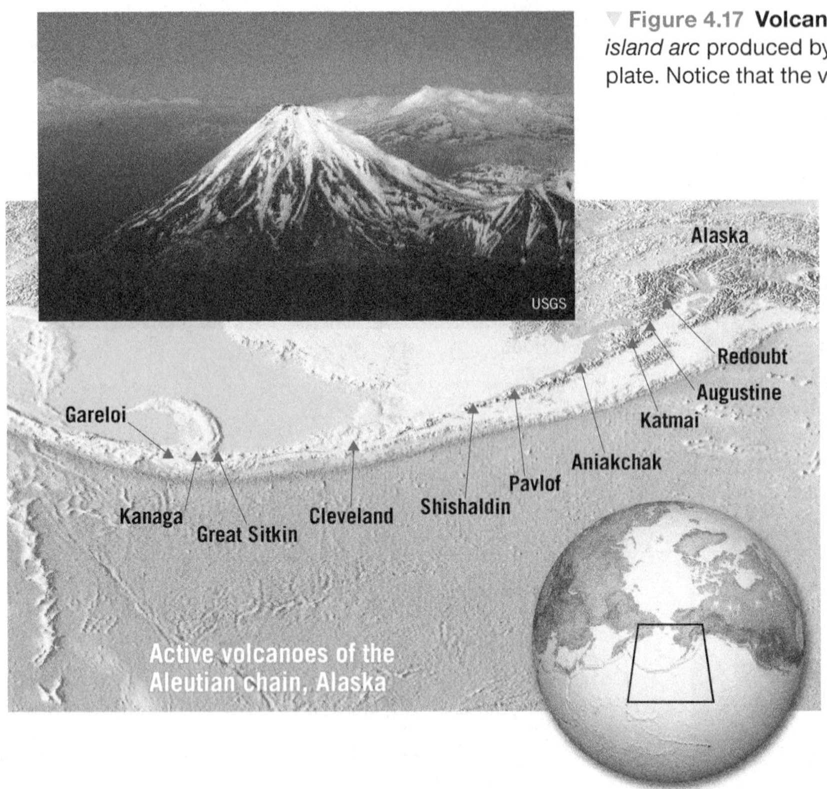

Figure 4.17 Volcanoes in the Aleutian chain The Aleutian Islands are a *volcanic island arc* produced by the subduction of the Pacific plate beneath the North American plate. Notice that the volcanoes of the Aleutian chain extend into Alaska proper.

are generally located 120 to 360 kilometers (75 to 225 miles) from a deep-ocean trench. Located adjacent to the island arcs just mentioned are the Aleutian trench, the Mariana trench, and the Tonga trench.

Most volcanic island arcs are located in the western Pacific. Only two are located in the Atlantic—the Lesser Antilles arc, on the eastern margin of the Caribbean Sea, and the Sandwich Islands, located off the tip of South America. The Lesser Antilles are a product of the subduction of the Atlantic seafloor beneath the Caribbean plate. Located within this volcanic arc are the Virgin Islands of the United States and Britain as well as Martinique, where Mount Pelée erupted in 1902, destroying the town of St. Pierre and killing an estimated 28,000 people. This chain of islands also includes Montserrat, where volcanic activity has occurred as recently as 2010.

Island arcs are typically simple structures made of numerous volcanic cones underlain by oceanic crust that is generally less than 20 kilometers (12 miles) thick. Some island arcs, however, are more complex and are underlain by highly deformed crust that may reach 35 kilometers (22 miles) in thickness. Examples include Japan, Indonesia, and the Alaskan Peninsula. These island arcs are built on material generated by earlier episodes of subduction or on small slivers of continental crust that have rafted away from the mainland.

Continental–Continental Convergence

The third type of convergent boundary results when one landmass moves toward the margin of another because of subduction of the intervening seafloor (Figure 4.18A). Whereas oceanic lithosphere tends to be dense and readily sinks into the mantle, the buoyancy of continental material generally inhibits it from being subducted, at least to any great depth. Consequently, a collision between two converging continental fragments ensues (Figure 4.18B). This process folds and deforms the accumulation of sediments and sedimentary rocks along the continental margins as if they had been placed in a gigantic vise. The result is the formation of a new

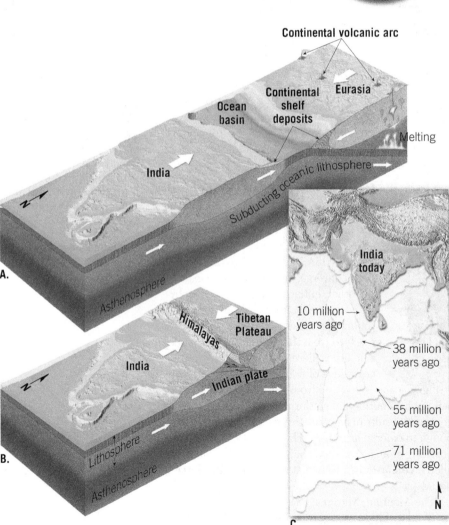

SmartFigure 4.18 The collision of India and Eurasia formed the Himalayas The ongoing collision of the subcontinent of India with Eurasia began about 50 million years ago and produced the majestic Himalayas. Although the map in **C** illustrates only the movement of India, it should be noted that both India and Eurasia were moving as these landmasses collided.

mountain belt composed of deformed sedimentary and metamorphic rocks that often contain slivers of oceanic lithosphere.

Such a collision began about 50 million years ago, when the subcontinent of India "rammed" into Asia, producing the Himalayas—the most spectacular mountain range on Earth (Figure 4.18C). During this collision, the continental crust buckled and fractured and was generally shortened horizontally and thickened vertically. In addition to the Himalayas, several other major mountain systems, including the Alps, Appalachians, and Urals, formed as continental fragments collided. This topic will be considered further in Chapter 7.

CONCEPT CHECKS 4.5

1. Explain why the rate of lithosphere production is roughly equal to the rate of lithosphere destruction.
2. Why does oceanic lithosphere subduct, while continental lithosphere does not?
3. What characteristic of a slab of oceanic lithosphere leads to the formation of a deep oceanic trench as opposed to one that is less deep?
4. What distinguishes a continental volcanic arc from a volcanic island arc?
5. Briefly describe how mountain belts such as the Himalayas form.

4.6 Transform Plate Boundaries

Describe the relative motion along a transform fault boundary and locate several examples of transform faults on a plate boundary map.

Along a **transform plate boundary**, also called a **transform fault**, plates slide horizontally past one another without the production or destruction of lithosphere. The nature of transform faults was discovered in 1965 by Canadian geologist J. Tuzo Wilson, who proposed that these large faults connect two spreading centers (divergent boundaries) or, less commonly, two trenches (convergent boundaries). Most transform faults are found on the ocean floor, where they offset segments of the oceanic ridge system, producing a steplike plate margin (Figure 4.19A). Notice that the zigzag shape of the Mid-Atlantic Ridge in Figure 4.10 (see page 102) roughly reflects the shape of the original rifting that caused the breakup of the supercontinent Pangaea. (Compare the shapes of the continental margins of the landmasses on both sides of the Atlantic with the shape of the Mid-Atlantic Ridge.)

Typically, transform faults are part of prominent linear breaks in the seafloor known as **fracture zones**, which include both active transform faults and their inactive extensions into the plate interior (Figure 4.19B). In a fracture zone, the active transform fault lies *only between* the two offset ridge segments; it is generally defined by weak, shallow earthquakes. On each side of the fault, the seafloor moves away from the corresponding ridge segment. Thus, between the ridge segments, these adjacent slabs of oceanic crust are grinding past each other along

SmartFigure 4.19
Transform plate boundaries Most transform faults offset segments of a spreading center, producing a plate margin that exhibits a zigzag pattern.

TUTORIAL
https://goo.gl/ZT7a9i

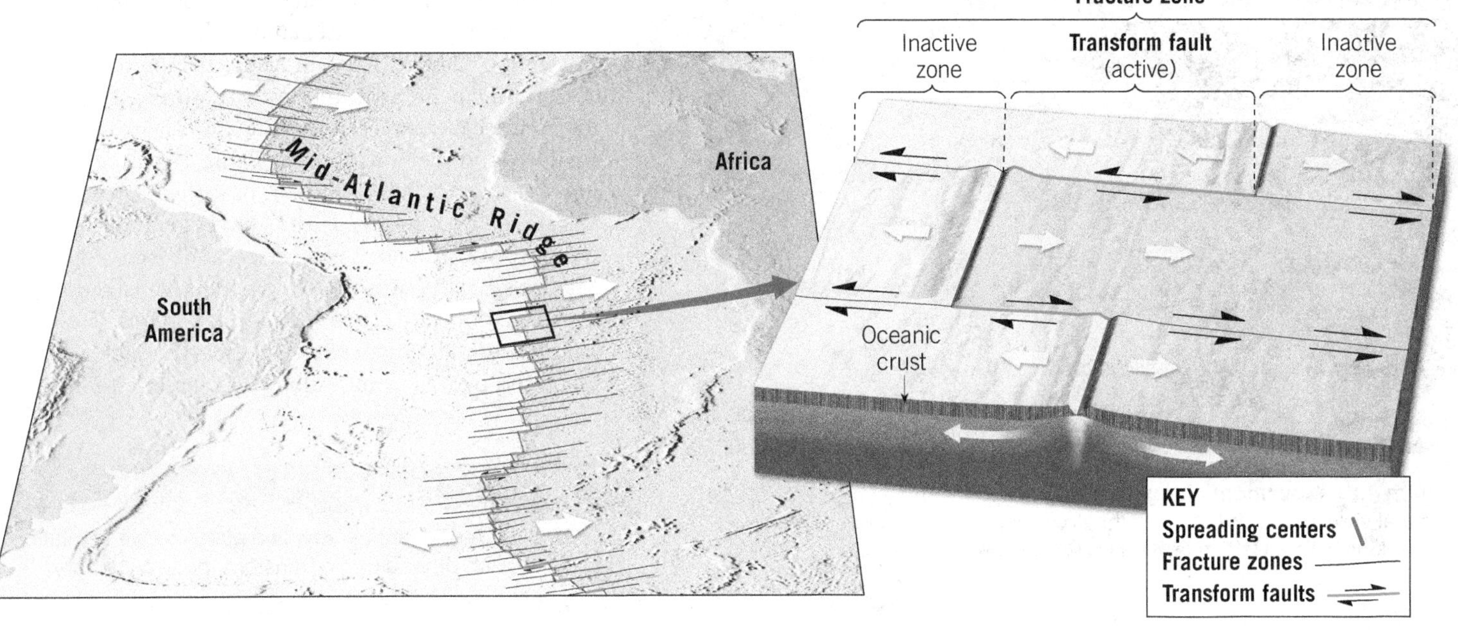

A. The Mid-Atlantic Ridge, with its zigzag pattern, roughly reflects the shape of the zone of rifting that resulted in the breakup of Pangaea.

B. Fracture zones are long, narrow scar-like features in the seafloor that are roughly perpendicular to the offset ridge segments. They include both the active transform fault and its preserved trace.

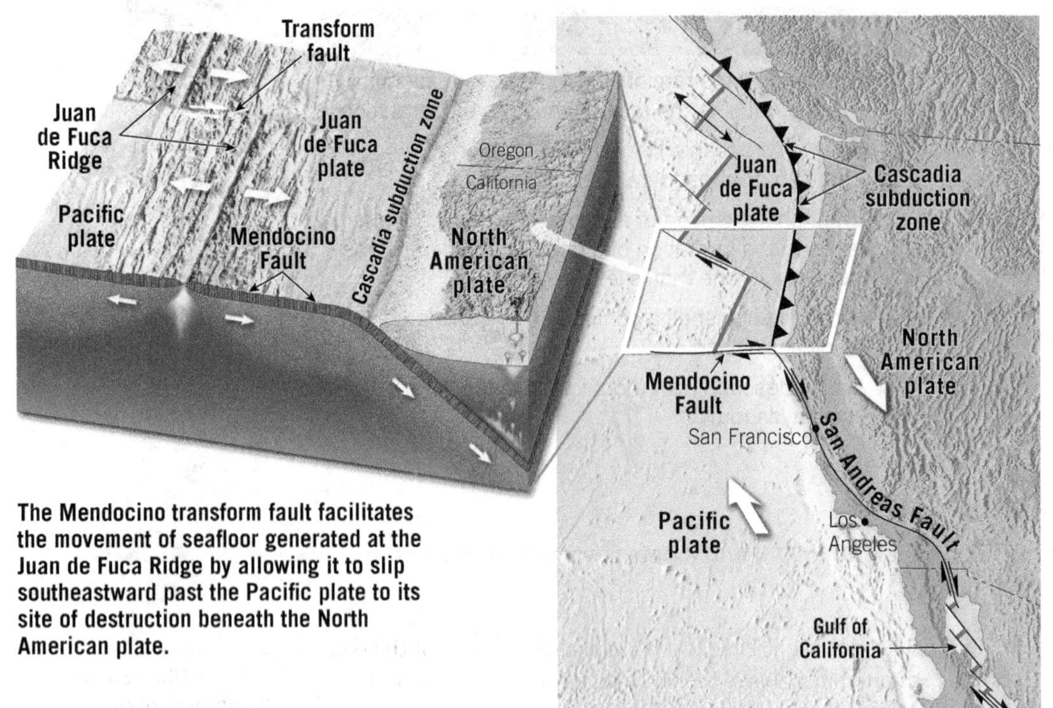

The Mendocino transform fault facilitates the movement of seafloor generated at the Juan de Fuca Ridge by allowing it to slip southeastward past the Pacific plate to its site of destruction beneath the North American plate.

Figure 4.20 Transform faults facilitate plate motion Seafloor generated along the Juan de Fuca Ridge moves southeastward, past the Pacific plate. Eventually it subducts beneath the North American plate. Thus, this transform fault connects a spreading center (divergent boundary) to a subduction zone (convergent boundary). Also shown is the San Andreas Fault, a transform fault connecting a spreading center located in the Gulf of California with the Mendocino Fault.

a transform fault. Beyond the ridge crests, these faults are inactive because the rock on either side moves in the same direction. However, these inactive faults are preserved as linear topographic depressions, which are evidence of past transform fault activity. The trend (orientation) of these fracture zones roughly parallels the direction of plate motion at the time of their formation. Thus, these structures help geologists map the direction of plate motion in the geologic past.

Transform faults also provide the means by which the oceanic crust created at ridge crests can be transported to a site of destruction—the deep-ocean trenches. Figure 4.20 illustrates this situation. Notice that the Juan de Fuca plate moves in a southeasterly direction, eventually being subducted under the west coast of the United States and Canada. The southern end of this plate is bounded by a transform fault called the Mendocino Fault. This transform boundary connects the Juan de Fuca Ridge to the Cascadia subduction zone. Therefore, it facilitates the movement of the crustal material created at the Juan de Fuca Ridge to its destination beneath the North American continent.

Like the Mendocino Fault, most other transform fault boundaries are located within the ocean basins; however, a few cut through continental crust. Two examples are the earthquake-prone San Andreas Fault of California and New Zealand's Alpine Fault. Notice in Figure 4.20 that the San Andreas Fault connects a spreading center located in the Gulf of California to the Cascadia subduction zone and the Mendocino Fault. Along the San Andreas Fault, the Pacific plate is moving toward the northwest, past the North American plate (Figure 4.21). If this movement continues, the part of California west of the fault zone, including Mexico's Baja Peninsula, will become an island off the west coast of the United States and Canada. However, a more immediate concern is the earthquake activity triggered by movements along this fault system.

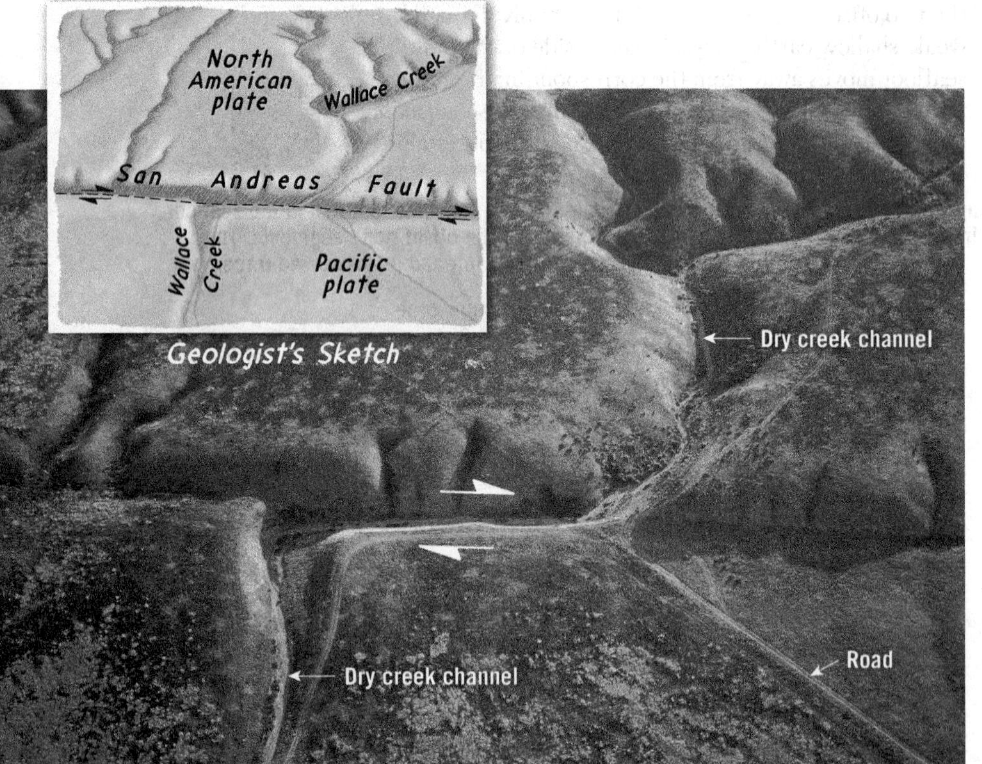

SmartFigure 4.21 **Movement along the San Andreas Fault** This aerial view shows the offset in the dry channel of Wallace Creek near Taft, California. (Photo by Michael Collier)

MOBILE FIELD TRIP
https://goo.gl/fW4cFE

CONCEPT CHECKS 4.6

1. Sketch or describe how two plates move in relation to each other along a transform plate boundary.
2. List two characteristics that differentiate transform faults from the two other types of plate boundaries.

EYE ON EARTH 4.1

Baja California is separated from mainland Mexico by a long narrow sea called the Gulf of California (also known to local residents as the Sea of Cortez). The Gulf of California contains many islands that were created by volcanic activity.

QUESTION 1 *What type of plate boundary is responsible for opening the Gulf of California?*

QUESTION 2 *What major U.S. river originates in the Colorado Rockies and created a large delta at the northern end of the Gulf of California?*

QUESTION 3 *If the material carried by the river in Question 2 had not been deposited, the Gulf of California would extend northward to include the inland sea shown in this satellite image. What is the name of this inland sea?*

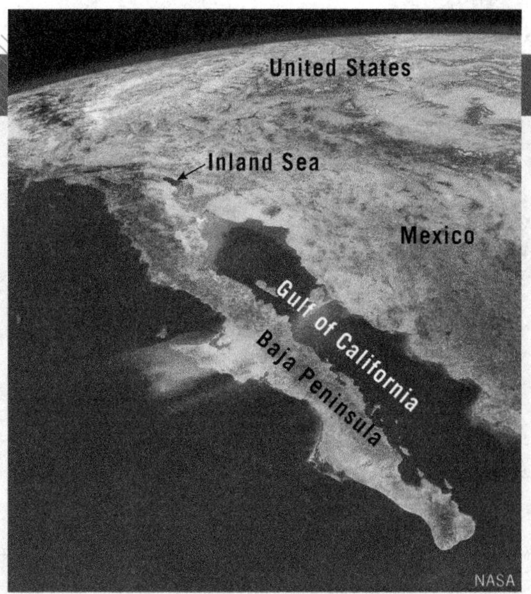

4.7 How Do Plates and Plate Boundaries Change?

Explain why plates such as the African and Antarctic plates are increasing in size, while the Pacific plate is decreasing in size.

Although Earth's total surface area does not change, the size and shape of individual plates are constantly changing. For example, the African and Antarctic plates, which are mainly bounded by divergent boundaries—sites of seafloor production—are continually growing in size as new lithosphere is added to their margins. By contrast, the Pacific plate is being consumed into the mantle along much of its flanks faster than it is being generated along the East Pacific Rise and thus is diminishing in size.

Another result of plate motion is that boundaries migrate. For example, the position of the Peru–Chile trench, which is the result of the Nazca plate being bent downward as it descends beneath the South American plate, has changed over time (see Figure 4.10). Because of the westward drift of the South American plate relative to the Nazca plate, the Peru–Chile trench has migrated in a westerly direction as well.

Plate boundaries can also be created or destroyed in response to changes in the forces acting on the lithosphere. For example, some plates carrying continental crust are presently moving toward one another. In the South Pacific, Australia is moving northward toward southern Asia. If Australia continues its northward migration, the boundary separating it from Asia will eventually become inactive and disappear as these plates become one. Other plates are moving apart. Recall that the Red Sea is the site of a relatively new spreading center that came into existence less than 20 million years ago, when the Arabian Peninsula began to break apart from Africa. The breakup of Pangaea is a classic example of how plate boundaries change through geologic time.

The Breakup of Pangaea

Wegener used evidence from fossils, rock types, and ancient climates to create a jigsaw-puzzle fit of the continents, thereby creating his supercontinent Pangaea. By employing modern tools not available to Wegener, geologists have re-created the steps in the breakup of this supercontinent, an event that began about 180 million years ago. From this work, the dates when individual crustal fragments separated from one another and their relative motions have been well established (Figure 4.22).

An important consequence of Pangaea's breakup was the creation of a "new" ocean basin: the Atlantic. As you can see in Figure 4.22, splitting of the supercontinent did not occur simultaneously along the margins of the Atlantic. The first split developed between North America and Africa. Here, the continental crust was highly fractured, providing pathways for huge quantities of fluid lavas to reach the surface. Today, these lavas are represented by weathered igneous rocks found along the eastern seaboard of the United States—primarily buried beneath the sedimentary rocks that form the continental shelf. Radiometric dating of these solidified lavas indicates that rifting began between 200 million and 190 million years ago. This time span represents the "birth date" for this section of the North Atlantic.

By 130 million years ago, the South Atlantic began to open near the tip of what is now South Africa. As this zone of rifting migrated northward, it gradually opened the South Atlantic (**Figures 4.22B,C**). Continued breakup of

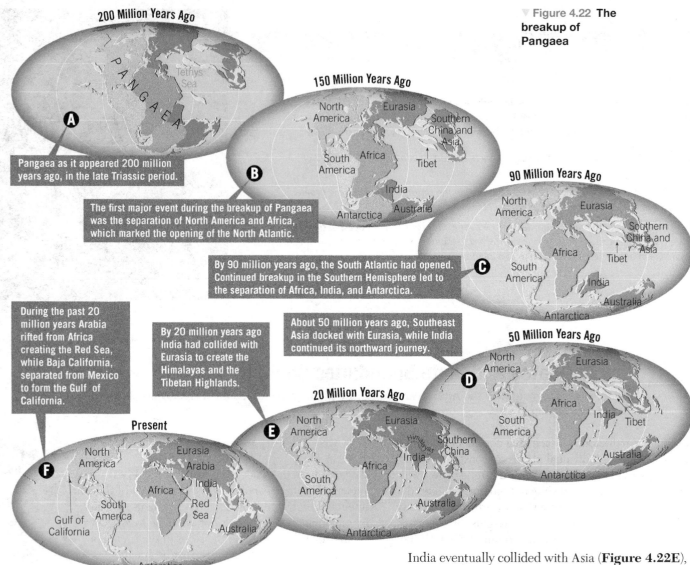

Figure 4.22 **The breakup of Pangaea**

the southern landmass led to the separation of Africa and Antarctica and sent India on a northward journey. By the early Cenozoic era, about 50 million years ago, Australia had separated from Antarctica, and the South Atlantic had become a full-fledged ocean (**Figure 4.22D**).

India eventually collided with Asia (**Figure 4.22E**), an event that began about 50 million years ago and created the Himalayas and the Tibetan Highlands. About the same time, the separation of Greenland from Eurasia completed the breakup of the northern landmass. During the past 20 million years or so of Earth's history, Arabia has rifted from Africa to form the Red Sea, and Baja California has separated from Mexico to form the Gulf of California (**Figure 4.22F**). Meanwhile, the Panama Arc joined North America and South America to produce our globe's familiar modern appearance.

Plate Tectonics in the Future

Geologists have extrapolated present-day plate movements into the future. Figure 4.23 illustrates where Earth's landmasses may be 50 million years from now if present plate movements persist during this time span.

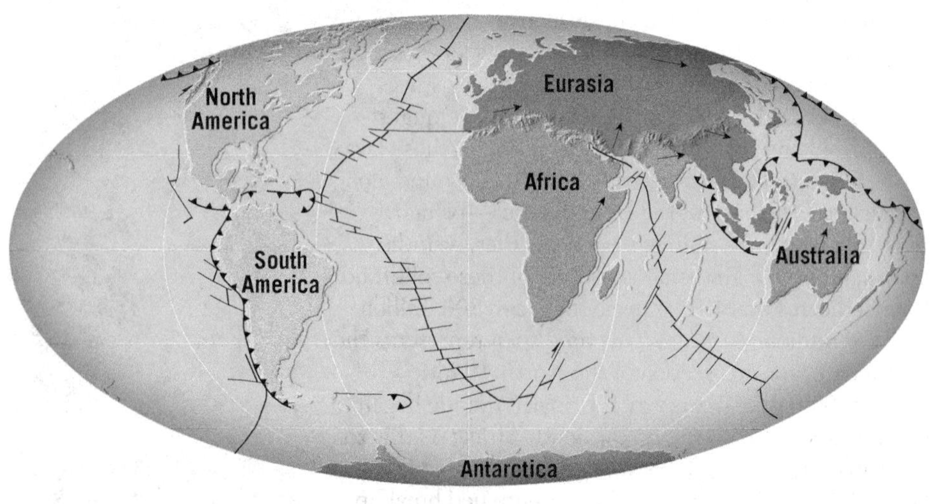

Figure 4.23 **The world as it may look 50 million years from now** This reconstruction is highly idealized and based on the assumption that the processes that caused the breakup and dispersal of the supercontinent of Pangaea will continue to operate. (Based on Robert S. Dietz, John C. Holden, C. Scotese, and others)

In North America we see that the Baja Peninsula and the portion of southern California that lies west of the San Andreas Fault will have slid past the North American plate. If this northward migration continues, Los Angeles and San Francisco will pass each other in about 10 million years, and in about 60 million years the Baja Peninsula will begin to collide with the Aleutian Islands.

If Africa maintains its northward path, it will continue to collide with Eurasia. The result will be the closing of the Mediterranean, the last remnant of a once-vast ocean called the Tethys Ocean, and the initiation of another major mountain-building episode (see Figure 4.23). Australia will be astride the equator and, along with New Guinea, will be on a collision course with Asia. Meanwhile, North and South America will begin to separate, while the Atlantic and Indian Oceans will continue to grow, at the expense of the Pacific Ocean.

A few geologists have even speculated on the nature of the globe 250 million years in the future. In this scenario the Atlantic seafloor will eventually become old and dense enough to form subduction zones around much of its margins, not unlike the present-day Pacific basin. Continued subduction of the Atlantic Ocean floor will result in the closing of the Atlantic basin and the collision of the Americas with the Eurasian–African landmass to form the next supercontinent, shown in Figure 4.24. Support for the possible closing of the Atlantic comes from evidence for a similar event, when an ocean predating the Atlantic closed during Pangaea's formation. Australia is also projected to collide with Southeast Asia by that time. If this scenario is accurate, the dispersal of Pangaea will end when the continents reorganize into the next supercontinent.

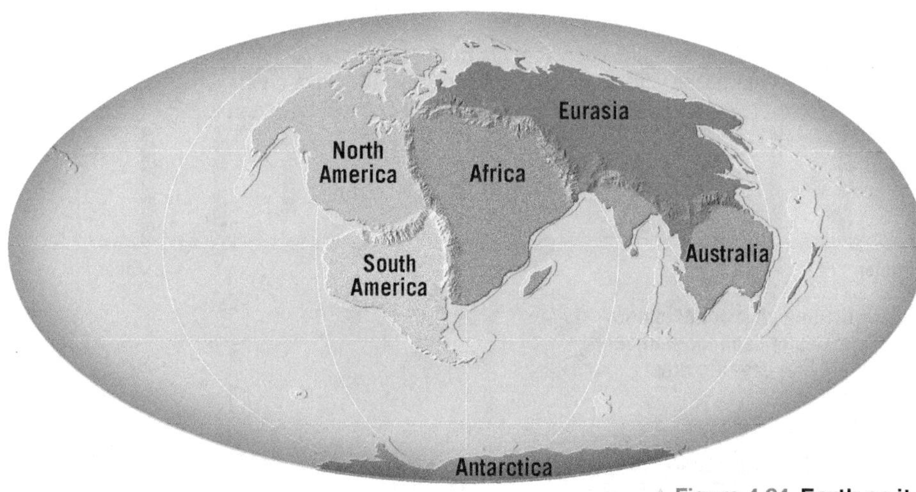

▲ Figure 4.24 **Earth as it may appear 250 million years from now**

Such projections, although interesting, must be viewed with considerable skepticism because many assumptions must be correct for these events to unfold as just described. Nevertheless, changes in the shapes and positions of continents that are equally profound will undoubtedly occur for many hundreds of millions of years to come. Only after much more of Earth's internal heat has been lost will the engine that drives plate motions cease.

CONCEPT CHECKS 4.7

1. Name two plates that are growing in size. Name a plate that is shrinking in size.
2. What new ocean basin was created by the breakup of Pangaea?
3. Briefly describe changes in the positions of the continents if we assume that the plate motions we see today continue 50 million years into the future.

4.8 Testing the Plate Tectonics Model

List and explain the evidence used to support the plate tectonics theory.

Some of the evidence supporting continental drift was presented earlier in this chapter. With the development of plate tectonics theory, researchers began testing this new model of how Earth works. In addition to new supporting data, new interpretations of already existing data often swayed the tide of opinion.

Evidence: Ocean Drilling

Some of the most convincing evidence for seafloor spreading came from the Deep Sea Drilling Project, which operated from 1966 until 1983. One of the early goals of the project was to gather samples of the ocean floor in order to establish its age. To accomplish this, the *Glomar Challenger*, a drilling ship capable of working in water thousands of meters deep, was built. Hundreds of holes were drilled through the layers of sediments that blanket the oceanic crust, as well as into the basaltic rocks below. Rather than use radiometric dating, which can be unreliable on oceanic rocks because of the alteration of basalt by seawater, researchers dated the seafloor by examining the fossil remains of microorganisms found in the sediments resting directly on the crust at each site.

When researchers recorded the age of the sediment from each drill site and its distance from the ridge crest, they found that the sediments increased in age with increasing distance from the ridge. This finding supported the seafloor-spreading hypothesis, which predicted that the youngest oceanic crust would be found at the ridge crest—the site of seafloor production—and the oldest oceanic crust would be located adjacent to the continents.

The distribution and thickness of ocean-floor sediments provided additional verification of seafloor spreading. Drill cores from the *Glomar Challenger* revealed

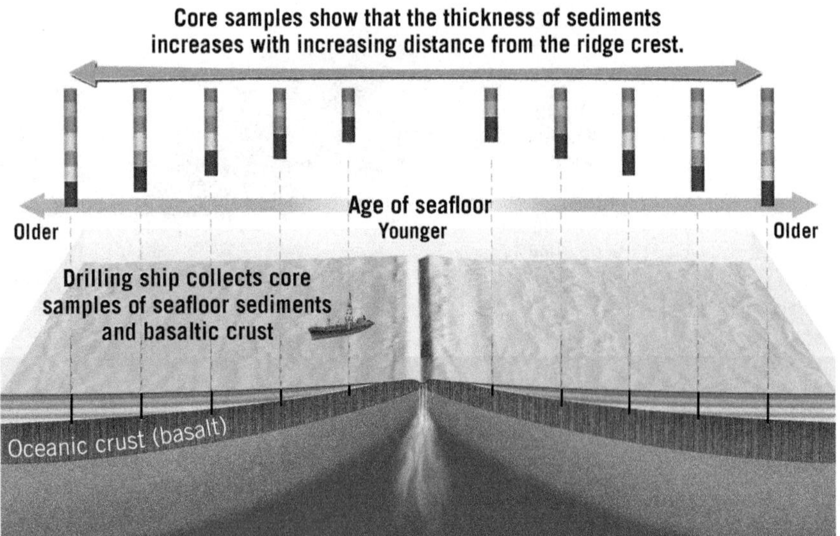
Core samples show that the thickness of sediments increases with increasing distance from the ridge crest.
Age of seafloor
Older
Younger
Older
Drilling ship collects core samples of seafloor sediments and basaltic crust
Oceanic crust (basalt)

B.

▲ Figure 4.25 **Deep-sea drilling A.** Data collected through deep-sea drilling have shown that the ocean floor is indeed youngest at the ridge axis. **B.** The Japanese deep-sea drilling ship *Chikyu*, designed to drill up to 7000 meters (more than 4 miles) below the seafloor, became operational in 2007. (Photo by Itsuo Inouye/AP Images)

that sediments are almost entirely absent on the ridge crest and that sediment thickness increases with increasing distance from the ridge (Figure 4.25A). This pattern of sediment distribution should be expected if the seafloor-spreading hypothesis is correct.

The data collected by the Deep Sea Drilling Project also reinforced the idea that the ocean basins are geologically young because no seafloor older than 180 million years was found. By comparison, most continental crust exceeds several hundred million years in age, and some samples are more than 4 billion years old.

In 1983, a new ocean-drilling program was launched by the Joint Oceanographic Institutions for Deep Earth Sampling (JOIDES). Now the International Ocean Discovery Program (IODP), this ongoing international effort uses multiple vessels for exploration, including the massive 210-meter-long (nearly 690-foot-long) *Chikyu* ("planet Earth" in Japanese), which began operations in 2007 (Figure 4.25B). One of the goals of the IODP is to recover a complete section of the oceanic crust, from top to bottom.

Evidence: Mantle Plumes and Hot Spots

Mapping volcanic islands and *seamounts* (submarine volcanoes) in the Pacific Ocean revealed several linear chains of volcanic structures. One of the most-studied chains consists of at least 129 volcanic structures that extend from the Hawaiian Islands to Midway Island and continue northwestward toward the Aleutian trench (Figure 4.26). Radiometric dating of this linear feature, called the Hawaiian Island–Emperor Seamount chain, showed that the volcanoes increase in age with increasing distance from the Big Island of Hawaii. The youngest volcanic island in the chain (Hawaii) rose from the ocean floor less than 1 million years ago, whereas Midway Island is 27 million years old, and Detroit Seamount, near the Aleutian trench, is about 80 million years old (see Figure 4.26).

One widely accepted hypothesis° proposes that a roughly cylindrical upwelling of hot rock that originates deep in the mantle, called a **mantle plume**, is located beneath the island of Hawaii. As the hot, rocky plume ascends through the mantle, the confining pressure drops, which triggers partial melting. (This process, called *decompression melting*, is discussed in Chapter 6.) The surface manifestation of this activity is a **hot spot**, an area of volcanism, high heat flow, and crustal uplifting that is a few hundred kilometers across. As the Pacific plate moved over the hot spot, which is thought to maintain a relatively fixed position within the mantle, a chain of volcanic structures known as a **hot-spot track** was built. As shown in Figure 4.26, the age of each volcano indicates how much time has elapsed since it was situated over the mantle plume. Of approximately 40 hot spots that are thought to have formed because of upwelling of hot mantle plumes, most, but not all, have hot-spot tracks.

A closer look at the five largest Hawaiian Islands reveals a similar pattern of ages, from the volcanically active island of Hawaii to the inactive volcanoes that make up the oldest island, Kauai (see Figure 4.26). Five million years ago, when Kauai was positioned over the hot spot, it was the *only* modern Hawaiian island in existence. Kauai's age is evident in the island's inactive volcanoes, which have been eroded into jagged peaks and vast canyons. By contrast, the relatively young island of Hawaii exhibits many fresh lava flows, and one of its five major volcanoes, Kilauea, remains active today.

Although the mantle plume hypothesis provides a compelling explanation for volcanism that occurs in the middle of a tectonic plate, the existence of slim mantle plumes that originate near Earth's core–mantle boundary has not been verified by seismic studies. As a result, some geologists have proposed that the source of magma that generated the Hawaiian chain originated from localized melting in the upper mantle.

° Recall from Section 1.2 that a *hypothesis* is a tentative scientific explanation for a given set of observations. Although widely accepted, the validity of the plume hypothesis, unlike the theory of plate tectonics, remains unresolved.

Figure 4.26 **Hot-spot volcanism and the formation of the Hawaiian chain** Radiometric dating of the Hawaiian Islands shows that volcanic activity increases in age moving away from the Big Island of Hawaii.

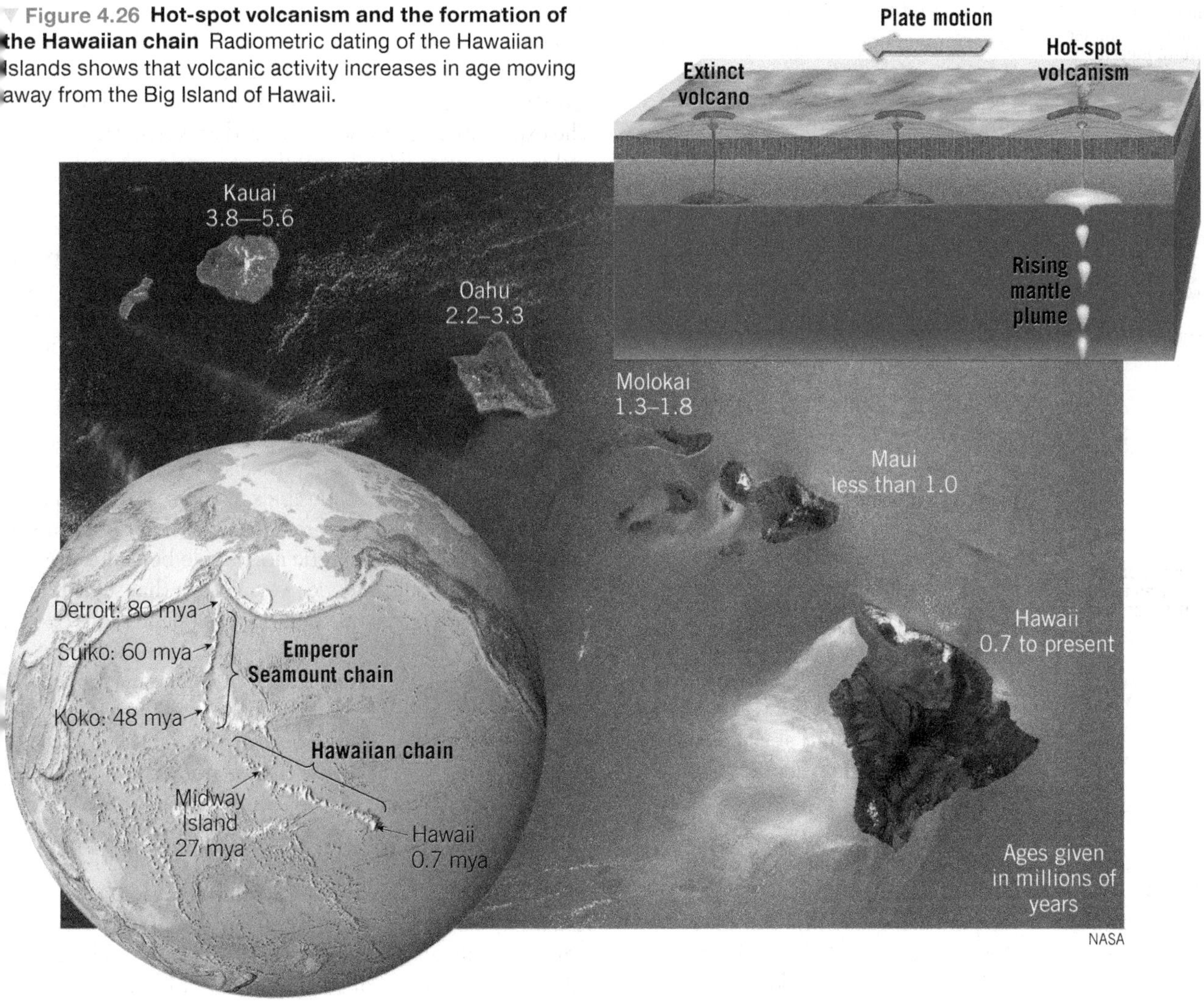

Evidence: Paleomagnetism

You are probably familiar with how a compass operates and know that Earth's magnetic field has north and south magnetic poles. Today these magnetic poles roughly align with the geographic poles that are located where Earth's rotational axis intersects the surface. Earth's magnetic field is similar to that produced by a simple bar magnet. Invisible lines of force pass through the planet and extend from one magnetic pole to the other (Figure 4.27). A compass needle, itself a small magnet free to rotate on an axis, becomes aligned with the magnetic lines of force and points to the magnetic poles.

Earth's magnetic field is less obvious to us than the pull of gravity because we cannot feel it. Movement of a compass needle, however, confirms its presence. In addition, some naturally occurring minerals are magnetic and are influenced by Earth's magnetic field. One of the most common is the iron-rich mineral *magnetite*, which is abundant in lava flows of basaltic composition.* Basaltic lavas erupt at the surface at temperatures greater than 1000°C (1800°F), exceeding a threshold temperature for magnetism known as the **Curie point** (about 585°C [1085°F]).

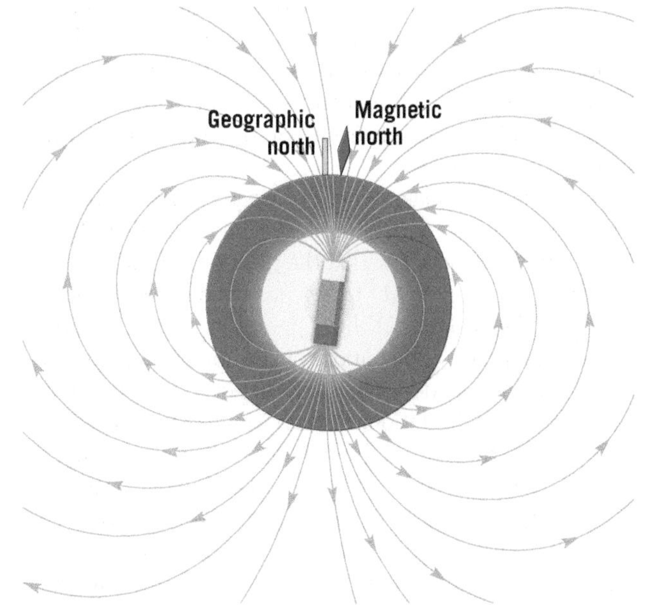

Figure 4.27 **Earth's magnetic field** Earth's magnetic field consists of lines of force much like those a giant bar magnet would produce if placed at the center of Earth.

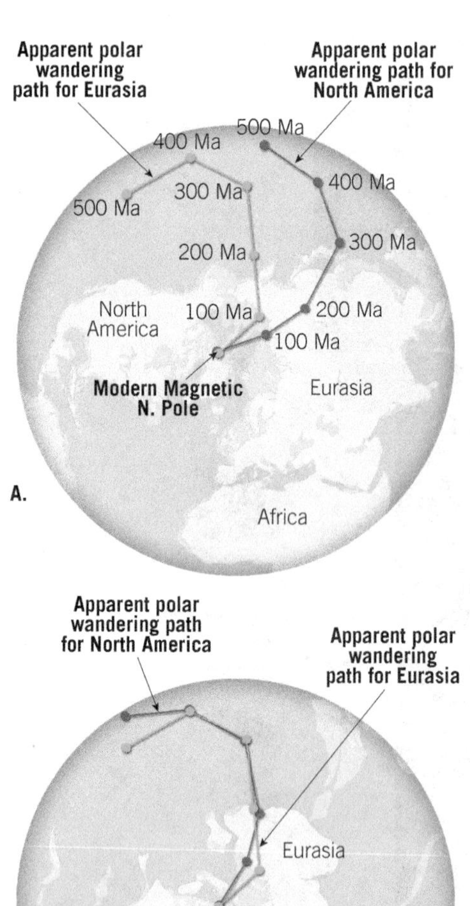

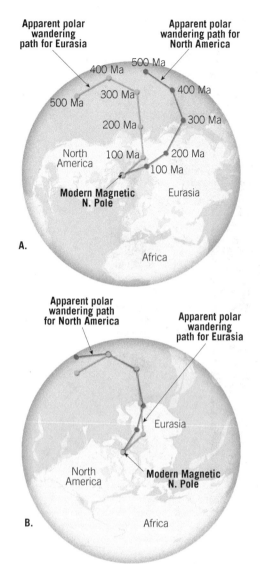

Figure 4.28 Apparent polar-wandering path **A.** Scientists believe that the more westerly path determined from North American data was caused by the westward drift of North America by about 24 degrees from Eurasia. **B.** The positions of the wandering paths when the landmasses are reassembled in their pre-drift locations.

The magnetite grains in molten lava are nonmagnetic, but as the lava cools, these iron-rich grains become magnetized and align themselves in the direction of the existing magnetic lines of force. Once the minerals solidify, the magnetism they possess usually remains "frozen" in this position. Thus, they act like a compass needle because they "point" toward the position of the magnetic poles at the time of their formation. Rocks that formed thousands or millions of years ago and contain a "record" of the direction of the magnetic poles at the time of their formation are said to possess **paleomagnetism**, or **preserved magnetism**.

Apparent Polar Wandering A study of paleomagnetism in ancient lava flows throughout Europe led to an interesting discovery. Taken at face value, the magnetic alignment of iron-rich minerals in lava flows of different ages would indicate that the position of the paleomagnetic poles had changed through time. A plot of the location of the magnetic north pole, as measured from Europe, seemed to indicate that during the past 500 million years, the pole had gradually "wandered" from a location near Hawaii northeastward to its present location over the Arctic Ocean (Figure 4.28). This was strong evidence that either the magnetic north pole had migrated, an idea known as *polar wandering*, or that the poles had remained in place and the continents had drifted beneath them—in other words, Europe had drifted relative to the magnetic north pole.

Although the magnetic poles are known to move in a somewhat erratic path, studies of paleomagnetism from numerous locations show that the positions of the magnetic poles, averaged over thousands of years, correspond closely to the positions of the geographic poles. Therefore, a more acceptable explanation for the apparent polar wandering was provided by Wegener's hypothesis: If the magnetic poles remain stationary, their *apparent movement* is produced by the drift of the seemingly fixed continents.

Further evidence for continental drift came a few years later, when a polar-wandering path was constructed for North America (see Figure 4.28A). For the first 200 million years or so, the paths for North America and Europe were found to be similar in direction—but separated by about 5000 kilometers (3000 miles). Then, during the middle of the Mesozoic era (180 million years ago), they began to converge on the present North Pole. The explanation for these curves is that North America and Europe were joined until the Mesozoic, when the Atlantic began to open. From this time forward, these continents continuously moved apart. When North America and Europe are moved back to their pre-drift positions, as shown in Figure 4.28B, these paths of apparent polar wandering coincide. This is evidence that North America and Europe were once joined and moved relative to the poles as part of the same continent.

Magnetic Reversals and Seafloor Spreading More evidence emerged when geophysicists learned that over periods of hundreds of thousands of years, Earth's magnetic field periodically reverses polarity. During a **magnetic reversal**, the magnetic north pole becomes the magnetic south pole and vice versa. Lava that solidified during a period of reverse polarity has been magnetized with the polarity opposite that of volcanic rocks being formed today. When rocks exhibit the same magnetism as the present magnetic field, they are said to possess **normal polarity**, whereas rocks exhibiting the opposite magnetism are said to have **reverse polarity**.

Once the concept of magnetic reversals was confirmed, researchers set out to establish a time scale for these occurrences. The task was to measure the magnetic polarity of hundreds of lava flows and use radiometric dating techniques to establish the age of each flow. Figure 4.29 shows the **magnetic time scale** established using this technique for the past few million years. The major divisions of the magnetic time scale, *chrons*, last roughly 1 million years each. As more measurements became available, researchers realized that several short-lived reversals (less than 200,000 years long) sometimes occurred during a single chron.

Meanwhile, oceanographers had begun magnetic surveys of the ocean floor in conjunction with their efforts to construct detailed maps of seafloor topography. These magnetic surveys were accomplished by towing very sensitive instruments, called **magnetometers**, behind research vessels (Figure 4.30A). The goal of these geophysical surveys was to map variations in the strength of Earth's magnetic field that arise from differences in the magnetic properties of the underlying crustal rocks.

The first comprehensive study of this type was performed off the Pacific coast of North America and had an unexpected outcome. Researchers discovered alternating stripes of high- and low-intensity magnetism, as shown in Figure 4.30B. This relatively simple pattern of magnetic variation defied explanation until 1963, when Fred Vine and D. H. Matthews demonstrated that the high- and low-intensity stripes supported the concept of seafloor spreading. Vine and Matthews suggested that the stripes of high-intensity magnetism are regions where

° Some sediments and sedimentary rocks also contain enough iron-bearing mineral grains to acquire a measurable amount of magnetization.

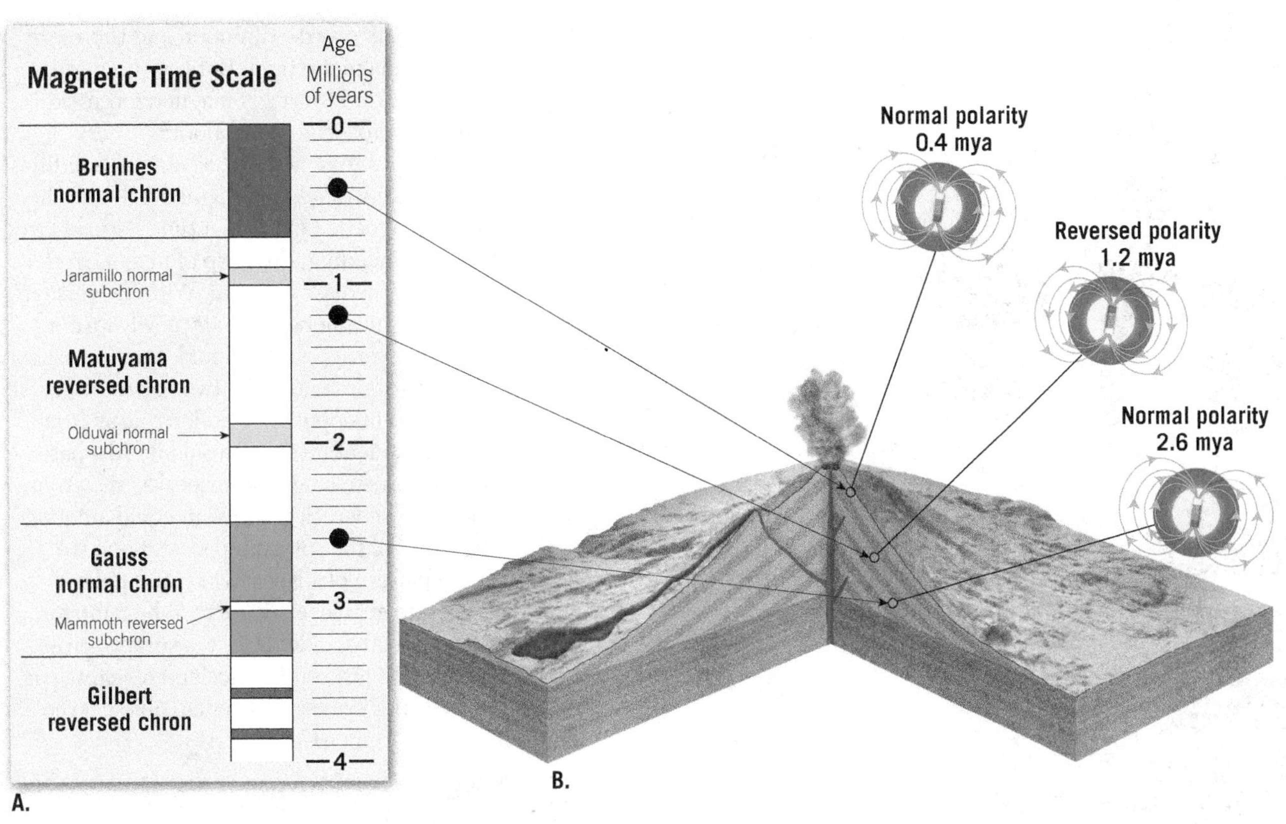

SmartFigure 4.29 **Time scale of magnetic reversals** **A.** Time scale of Earth's magnetic reversals for the past 4 million years. **B.** This time scale was developed by establishing the magnetic polarity for lava flows of known age. (Data from Allen Cox and G. B. Dalrymple)

TUTORIAL
https://goo.gl/KNenyq

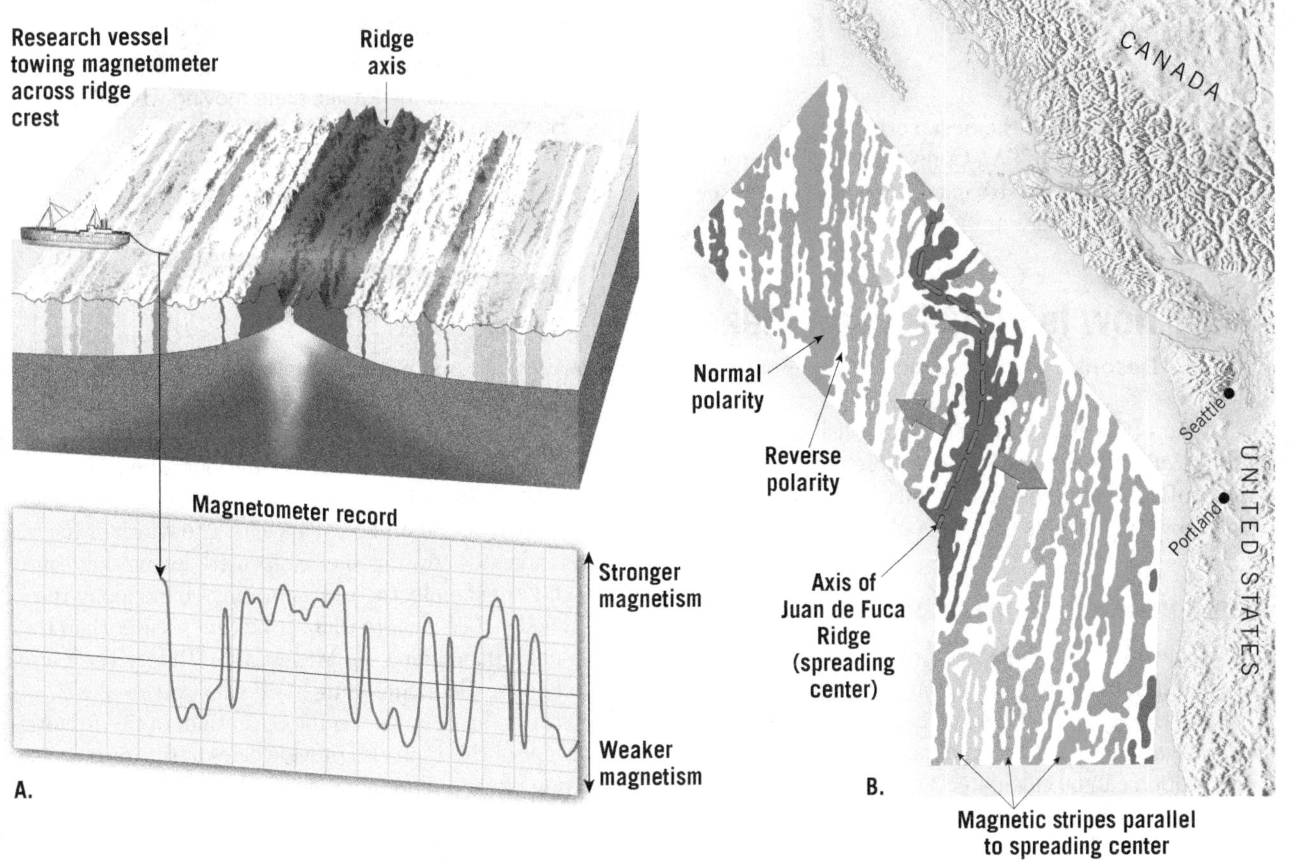

Figure 4.30 **Ocean floor as a magnetic recorder** **A.** Magnetic intensities are recorded when a magnetometer is towed across a segment of the oceanic floor. **B.** Notice the symmetrical stripes of low- and high-intensity magnetism that parallel the axis of the Juan de Fuca Ridge. The colored stripes of high-intensity magnetism occur where normally magnetized oceanic rocks enhance the existing magnetic field. Conversely, the white low-intensity stripes are regions where the crust is polarized in the reverse direction, which weakens the existing magnetic field.

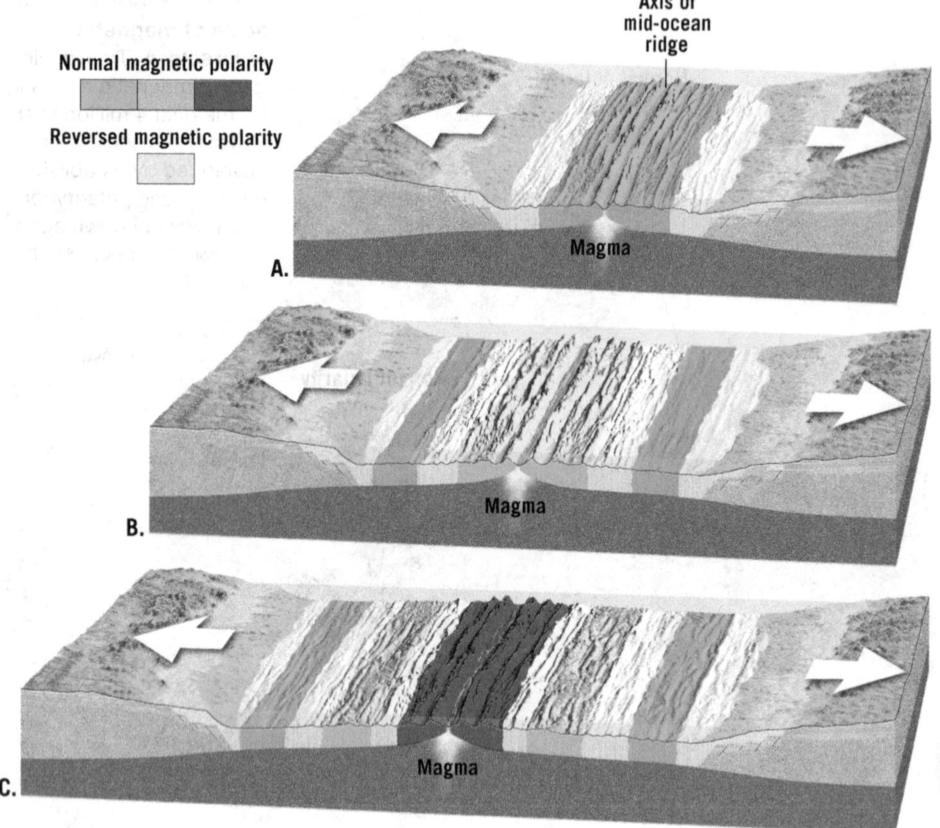

SmartFigure 4.31 **Magnetic reversals and seafloor spreading** When new basaltic rocks form at mid-ocean ridges, they magnetize according to Earth's existing magnetic field. Hence, oceanic crust provides a permanent record of each reversal of our planet's magnetic field over the past 200 million years.

ANIMATION
https://goo.gl/35qNFg

the paleomagnetism of the ocean crust exhibits normal polarity (see Figure 4.29A). Consequently, these rocks *enhance* (reinforce) Earth's magnetic field. Conversely, the low-intensity stripes are regions where the ocean crust is polarized in the reverse direction and therefore *weaken* the existing magnetic field. But how do parallel stripes of normally and reversely magnetized rock become distributed across the ocean floor?

Vine and Matthews reasoned that as magma solidifies at the crest of an oceanic ridge, it is magnetized with the polarity of Earth's magnetic field at that time (Figure 4.31). Because of seafloor spreading, this strip of magnetized crust would gradually increase in width. With a reverse in the polarity of Earth's magnetic field, any newly formed seafloor having this reverse polarity would form in the middle of the old strip. Gradually, the two halves of the old strip would be carried in opposite directions, away from the ridge crest. Subsequent reversals would build a pattern of normal and reverse magnetic stripes, as shown in Figure 4.31. Because new rock is added in equal amounts to both trailing edges of the spreading ocean floor, we should expect the pattern of stripes (width and polarity) found on one side of an oceanic ridge to be a mirror image of those on the other side. In fact, a survey across the Mid-Atlantic Ridge just south of Iceland reveals a pattern of magnetic stripes exhibiting a remarkable degree of symmetry in relation to the ridge axis.

CONCEPT CHECKS 4.8

1. What is the age of the oldest sediments recovered using deep-ocean drilling? How do the ages of these sediments compare to the ages of the oldest continental rocks?
2. How do sedimentary cores from the ocean floor support the concept of seafloor spreading?
3. Assuming that hot spots remain fixed, in what direction was the Pacific plate moving while the Hawaiian Islands were forming?
4. Describe how Fred Vine and D. H. Matthews related the seafloor-spreading hypothesis to magnetic reversals.

4.9 How Is Plate Motion Measured?

Describe two methods researchers use to measure relative plate motion.

A number of methods are used to establish the direction and rate of plate motion. Some of these techniques not only confirm that lithospheric plates move but allow us to trace those movements back in geologic time.

Geologic Measurement of Plate Motion

Using ocean-drilling ships, researchers have obtained dates for hundreds of locations on the ocean floor. By knowing the age of a rock sample and its distance from the ridge axis where it was generated, an average rate of plate motion can be calculated.

Scientists used these data, combined with their knowledge of paleomagnetism stored in hardened lavas on the ocean floor and seafloor topography, to create maps that show the age of the ocean floor. The reddish-orange bands shown in Figure 4.32 range in age from the present to about 30 million years ago. The width of the bands indicates how much crust formed during that time period. For example, the reddish-orange band along the East Pacific Rise is more than three times wider than the same-color band along the Mid-Atlantic Ridge. Therefore, the rate of seafloor spreading has been approximately three times faster in the Pacific basin than in the Atlantic.

Maps of this type also provide clues to the current direction of plate movement. Notice the offsets in the ridges; these are transform faults that connect the spreading centers. Recall that transform faults are aligned parallel

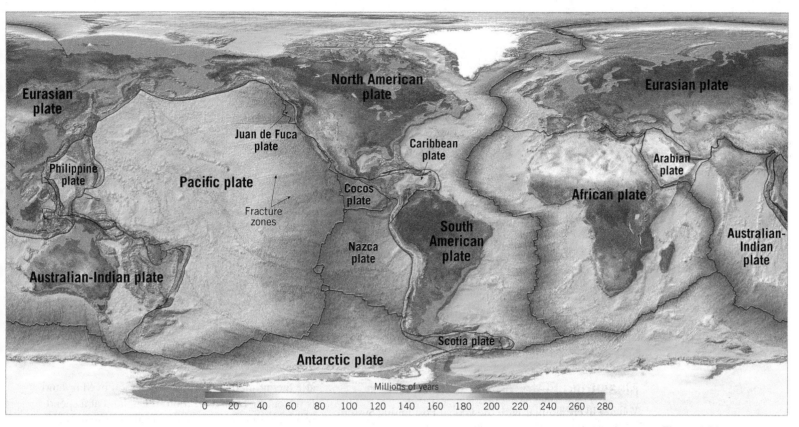

Figure 4.32 **Age of the ocean floor**

to the direction of spreading. Careful measurement of transform faults reveals the direction of plate movement.

To establish the direction of plate motion in the past, geologists can examine the long *fracture zones* that extend for hundreds or even thousands of kilometers from ridge crests. Fracture zones are inactive extensions of transform faults and are therefore a record of past directions of plate motion. Unfortunately, most of the ocean floor is less than 180 million years old, so to look deeper into the past, researchers must rely on paleomagnetic evidence provided by continental rocks.

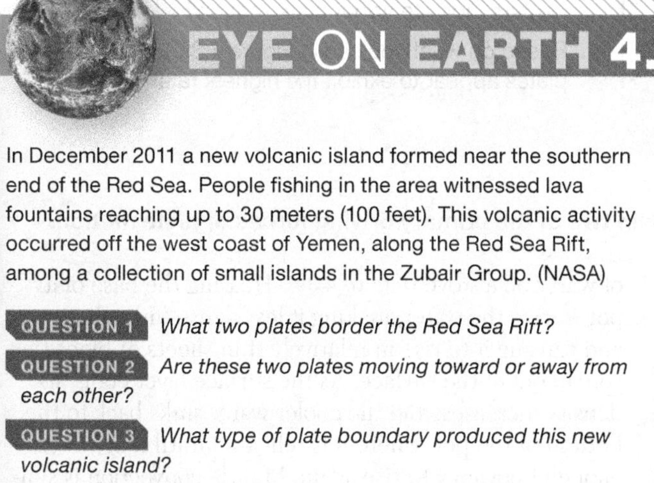

EYE ON EARTH 4.2

In December 2011 a new volcanic island formed near the southern end of the Red Sea. People fishing in the area witnessed lava fountains reaching up to 30 meters (100 feet). This volcanic activity occurred off the west coast of Yemen, along the Red Sea Rift, among a collection of small islands in the Zubair Group. (NASA)

QUESTION 1 *What two plates border the Red Sea Rift?*

QUESTION 2 *Are these two plates moving toward or away from each other?*

QUESTION 3 *What type of plate boundary produced this new volcanic island?*

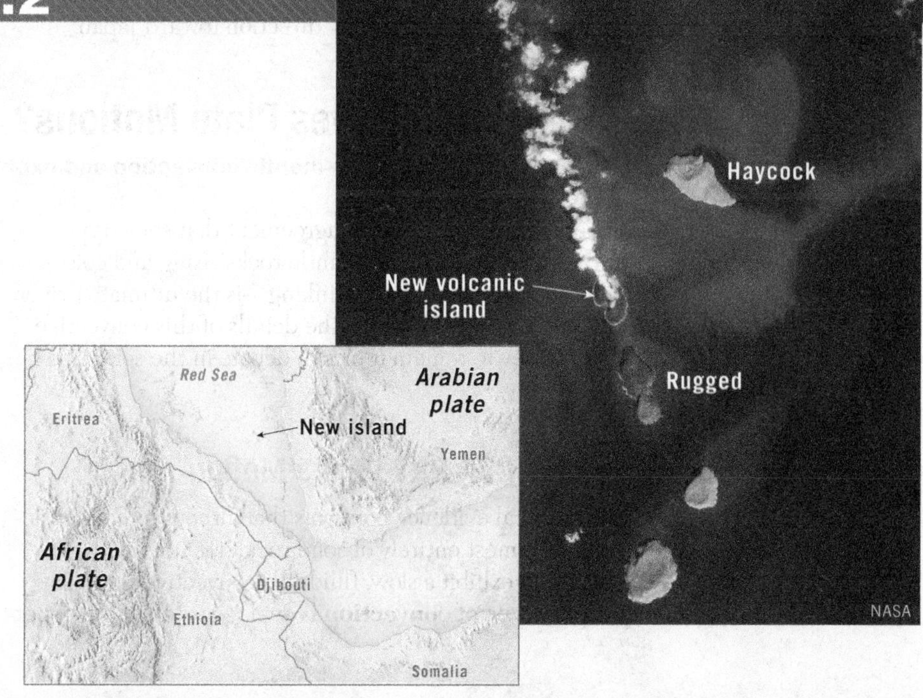

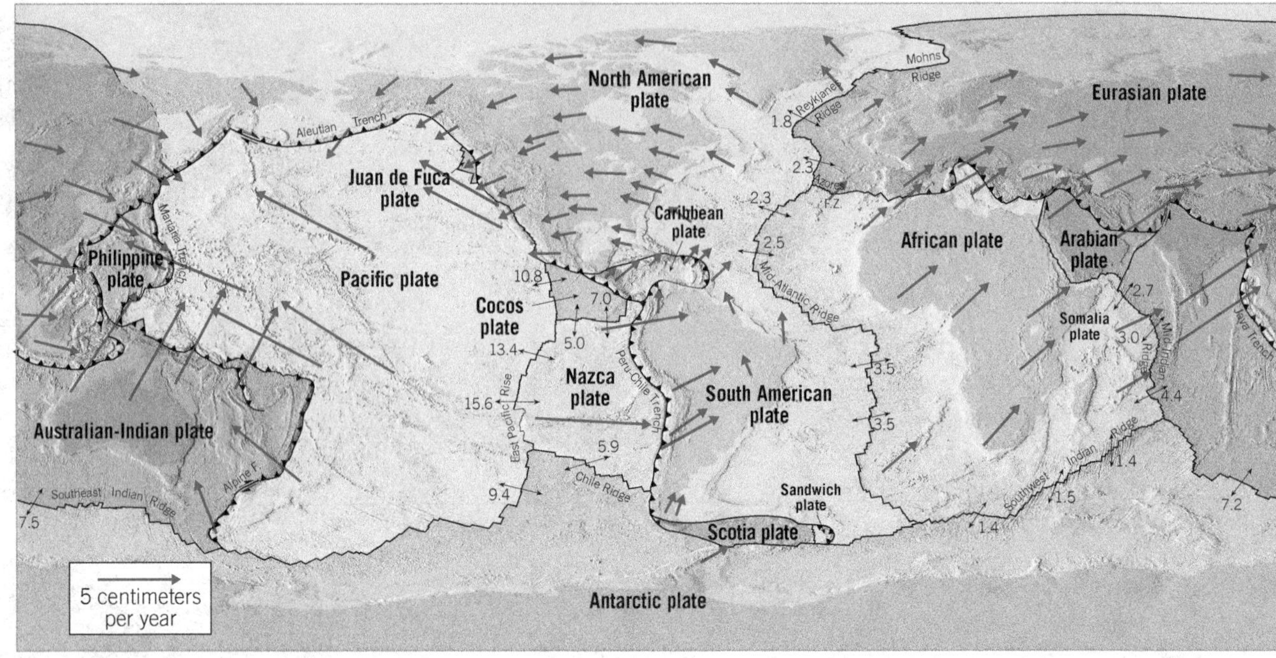

Figure 4.33 **Rates of plate motion** Each red arrows shows plate motion at the location of the arrow's base, determined from GPS data. Longer arrows represent faster spreading rates. The small black arrows and labels show seafloor spreading velocities based mainly on paleomagnetic data. (Seafloor data from DeMets and others; GPS data from Jet Propulsion Laboratory)

Measuring Plate Motion from Space

You are likely familiar with the Global Positioning System (GPS) used to locate one's position in order to provide directions to some other location. In it, satellites send radio signals that are intercepted by GPS receivers located at Earth's surface. The exact position of a site is determined by simultaneously establishing the distance from the receiver to four or more satellites. Researchers use specially designed equipment to locate a point on Earth to within a few millimeters (about the diameter of a small pea). To establish plate motions, GPS data are collected at numerous sites repeatedly over a number of years.

Data obtained from GPS and other techniques are shown in Figure 4.33. Calculations show that Hawaii is moving in a northwesterly direction toward Japan at 8.3 centimeters per year. A location in Maryland is retreating from a location in England at a speed of 1.7 centimeters per year—a value close to the 4.0-centimeters-per-year spreading rate established from paleomagnetic evidence obtained for the North Atlantic. Techniques involving GPS devices have also been useful in confirming small-scale crustal movements, like those occurring along faults in regions known to be tectonically active (for example, the San Andreas Fault).

CONCEPT CHECKS 4.9

1. What do transform faults that connect spreading centers indicate about plate motion?
2. Based on what you see in Figure 4.33, which three plates appear to exhibit the highest rates of motion?

4.10 What Drives Plate Motions?

Describe plate–mantle convection and explain two of the primary driving forces of plate motion.

Researchers are in general agreement that some type of *convection*—with hot mantle rocks rising and cold, dense oceanic lithosphere sinking—is the ultimate driver of plate tectonics. Many of the details of this convective flow, however, remain topics of debate in the scientific community.

Forces That Drive Plate Motion

Geophysical evidence confirms that although the mantle consists almost entirely of solid rock, it is hot and weak enough to exhibit a slow, fluid-like convective flow. The simplest type of **convection** is analogous to heating a pot of water on a stove (Figure 4.34). Heating the base of a pot warms the water, making it less dense (more buoyant) and causing it to rise in relatively thin sheets or blobs that spread out at the surface. As the surface layer cools, its density increases, and the cooler water sinks back to the bottom of the pot, where it is reheated until it achieves enough buoyancy to rise again. Mantle convection is similar to, but considerably more complex than, the model just described.

Geologists generally agree that subduction of cold, dense slabs of oceanic lithosphere is a major driving force of plate motion (Figure 4.35). This phenomenon, called **slab pull**, occurs because cold slabs of oceanic

Figure 4.34 **Convection in a cooking pot** As a stove warms the water in the bottom of a cooking pot, the heated water expands, becomes less dense (more buoyant), and rises. Simultaneously, the cooler, denser water near the top sinks.

Convection is a type of heat transfer that involves the movement of a substance.

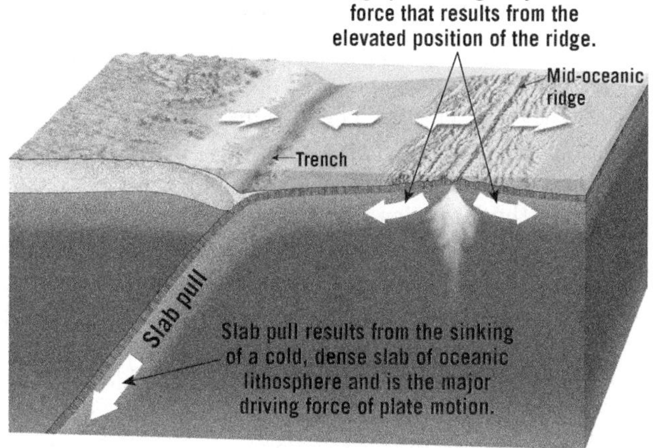

Figure 4.35 **Forces that act on lithospheric plates**

lithosphere are more dense than the underlying warm asthenosphere and hence "sink like a rock"—meaning that they are pulled down into the mantle by gravity.

Another important driving force is **ridge push** (see Figure 4.35). This gravity-driven mechanism results from the elevated position of the oceanic ridge, which causes slabs of lithosphere to "slide" down the flanks of the ridge. Despite its importance, ridge push contributes far less to plate motions than does slab pull. The primary evidence for this is that the fastest-moving plates—the Pacific, Nazca, and Cocos plates—have extensive subduction zones along their margins. By contrast, the spreading rate in the North Atlantic basin, which is nearly devoid of subduction zones, is one of the lowest, at about 4.5 centimeters (about 2 inches) per year.

Models of Plate–Mantle Convection

Although *convection* in the mantle is not yet fully understood, researchers generally agree on the following:

- Convective flow—in which warm, buoyant mantle rocks rise while cool, dense lithospheric plates sink—is the underlying driving force for plate movement.
- Mantle convection and plate tectonics are part of the same system. Subducting oceanic plates drive the cold downward-moving portion of convective flow, while shallow upwelling of hot rock along the oceanic ridge and buoyant mantle plumes are the upward-flowing arms of the convective mechanism.
- Convective flow in the mantle is a major mechanism for transporting heat away from Earth's interior to the surface, where it is eventually radiated into space.

What is not known with certainty is the exact structure of this convective flow. Several models have been proposed for plate–mantle convection, and we will look at two of them.

Whole-Mantle Convection Model One group of researchers favor some type of *whole-mantle convection* model, also called the *plume model*, in which cold oceanic lithosphere sinks to great depths and stirs the entire mantle (Figure 4.36A). The whole-mantle model suggests that the ultimate burial ground for these subducting lithospheric slabs is the core–mantle boundary. The downward flow of these subducting slabs is balanced

Figure 4.36 **Models of mantle convection**

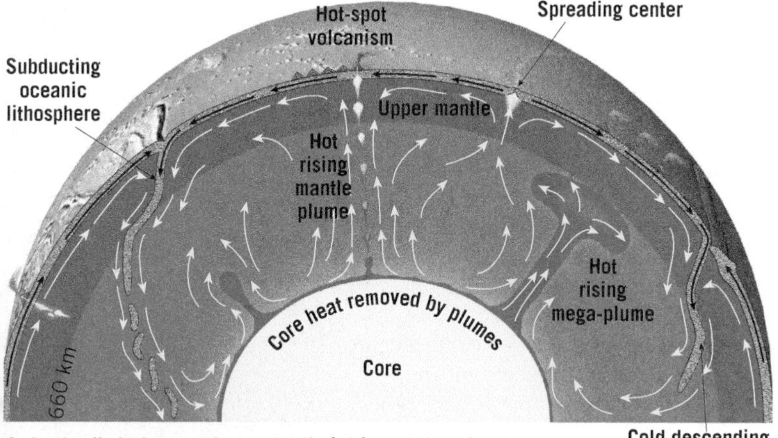

A. In the "whole-mantle model," sinking slabs of cold oceanic lithosphere are the downward limbs of convection cells, while rising mantle plumes carry hot material from the core–mantle boundary toward the surface.

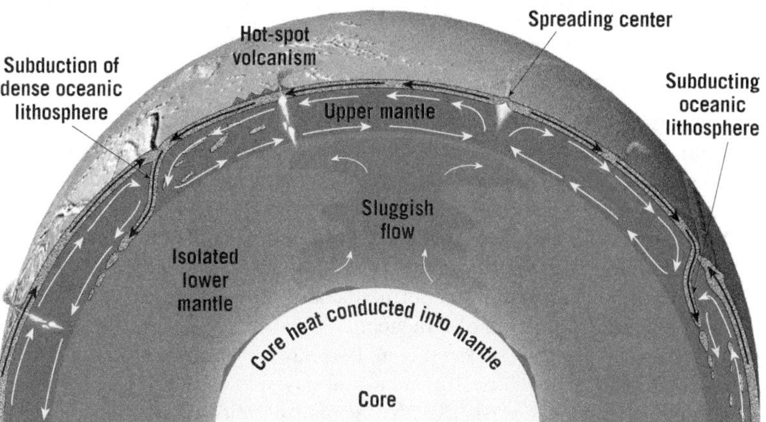

B. The "layer cake model" has two largely disconnected convective layers; a dynamic upper layer driven by descending slabs of cold oceanic lithosphere and a sluggish lower layer that carries heat upward without appreciably mixing with the layer above.

by buoyantly rising mantle plumes that transport hot mantle rock toward the surface.

Two kinds of plumes have been proposed—narrow tube-like plumes and giant upwellings, often referred to as *mega-plumes*. The long, narrow plumes are thought to originate from the core–mantle boundary and produce hot-spot volcanism of the type associated with the Hawaiian Islands, Iceland, and Yellowstone. Scientists believe that areas of large mega-plumes, as shown in Figure 4.36A, occur beneath the Pacific basin and southern Africa. These mega-plumes are thought to explain why southern Africa has an elevation much higher than would be predicted for a stable continental landmass. In the whole-mantle convection model, heat for both the narrow plumes and the mega-plumes is thought to arise mainly from Earth's core, while the deep mantle provides a source for chemically distinct magmas. However, some researchers have questioned that idea and instead propose that the source of magma for most hot-spot volcanism is found in the upper mantle (asthenosphere).

Layer Cake Model Some researchers argue that the mantle resembles a "layer cake" divided at a depth of perhaps 660 kilometers (410 miles) but no deeper than 1000 kilometers (620 miles). As shown in Figure 4.36B, this layered model has two zones of convection—a thin, dynamic layer in the upper mantle and a thick, larger, sluggish one located below. As with the whole-mantle model, the downward convective flow is driven by the subduction of cold, dense oceanic lithosphere. However, rather than reach the lower mantle, these subducting slabs penetrate to depths of no more than 1000 kilometers (620 miles). Notice in Figure 4.36B that the upper layer in the layer cake model is littered with recycled oceanic lithosphere of various ages. According to this model, melting of these fragments would be the source of magma for some of the volcanism that occurs away from plate boundaries, such as the hot-spot volcanism of Hawaii.

In contrast to the active upper mantle, the lower mantle in the layer-cake model is sluggish and does not provide material to support volcanism at the surface. Very slow convection within this layer likely carries heat upward, but very little mixing occurs between these two layers.

Geologists continue to debate the nature of the convective flow in the mantle. As they investigate the possibilities, perhaps a widely accepted hypothesis that combines features from the layer cake model and the whole-mantle convection model will emerge.

CONCEPT CHECKS 4.10

1. Define *slab pull* and *ridge push*. Which of these forces contributes more to plate motion?
2. Briefly describe the two models of plate–mantle convection.
3. What geologic processes are associated with the upward and downward circulation in the mantle?

4 CONCEPTS IN REVIEW

Plate Tectonics: A Scientific Revolution Unfolds

4.1 Continental Drift: An Idea Before Its Time

List and explain the evidence Wegener presented to support his continental drift hypothesis.

KEY TERMS: continental drift, supercontinent, Pangaea

- German meteorologist Alfred Wegener formulated the continental drift hypothesis in 1915. He suggested that Earth's continents are not fixed in place but move slowly over geologic time.
- Wegener proposed a supercontinent called Pangaea that existed about 200 million years ago, during the late Paleozoic and early Mesozoic eras.
- Wegener's evidence that Pangaea existed and later broke into pieces that drifted apart included (1) the shape of the continents, (2) continental fossil organisms that matched across oceans, (3) matching rock types and modern mountain belts on separate continents, and (4) sedimentary rocks that recorded ancient climates, including glaciers on the southern portion of Pangaea.
- Wegener's hypothesis suffered from two flaws: It proposed tidal forces as the mechanism for the motion of continents, and it implied that the continents would have plowed their way through weaker oceanic crust, like boats cutting through a thin layer of sea ice. Most geologists rejected the idea of continental drift when Wegener proposed it, and it wasn't resurrected for another 50 years.

? Why did Wegener choose organisms such as *Glossopteris* and *Mesosaurus* as evidence for continental drift, as opposed to other fossil organisms such as sharks or jellyfish?

John Cancalosi/AGE Fotostock

4.2 The Theory of Plate Tectonics

List the major differences between Earth's lithosphere and asthenosphere and explain the importance of each in the plate tectonics theory.

KEY TERMS: theory of plate tectonics, lithosphere, asthenosphere, lithospheric plate (plate)

- Research conducted after World War II led to new insights that helped revive Wegener's hypothesis of continental drift. Exploration of the seafloor uncovered previously unknown features, including an extremely long mid-ocean ridge system. Sampling of the oceanic crust revealed that it was quite young relative to the continents.
- The lithosphere, Earth's outermost rocky layer, is relatively stiff and deforms by bending or breaking. The lithosphere consists both of crust (either oceanic or continental) and underlying upper mantle. Beneath the lithosphere is the asthenosphere, a relatively weak layer that deforms by flowing.
- The lithosphere consists of numerous segments of irregular size and shape. There are seven large lithospheric plates, another seven intermediate-size plates, and many relatively small microplates. Plates meet along boundaries that may be divergent (moving apart from each other), convergent (moving toward each other), or transform (moving laterally past each other).

4.3 Divergent Plate Boundaries and Seafloor Spreading

Sketch and describe the movement along a divergent plate boundary that results in the formation of new oceanic lithosphere.

KEY TERMS: divergent plate boundary (spreading center), oceanic ridge system, rift valley, seafloor spreading, continental rift

- Seafloor spreading leads to the formation of new oceanic lithosphere at mid-ocean ridge systems. As two plates move apart from one another, tensional forces open cracks in the plates, allowing magma to well up and generate new slivers of seafloor. This process generates new oceanic lithosphere at a rate of 2 to 15 centimeters (1 to 6 inches) each year.
- As it ages, oceanic lithosphere cools and becomes denser. It therefore subsides as it is transported away from the mid-ocean ridge. At the same time, the underlying asthenosphere cools, adding new material to the underside of the plate, which consequently thickens.
- Divergent boundaries are not limited to the seafloor. Continents can break apart, too, starting with a continental rift (as in modern-day east Africa) and potentially producing a new ocean basin between the two sides of the rift.

4.4 Convergent Plate Boundaries and Subduction

Compare and contrast the three types of convergent plate boundaries and name a location where each type can be found.

KEY TERMS: convergent plate boundary (subduction zone), deep-ocean trench, partial melting, continental volcanic arc, volcanic island arc (island arc)

- When plates move toward one another, oceanic lithosphere is subducted into the mantle, where it is recycled. Subduction manifests itself on the ocean floor as a deep linear trench. The subducting slab of oceanic lithosphere can descend at a variety of angles, from nearly horizontal to nearly vertical.
- Aided by the presence of water, the subducted oceanic lithosphere triggers melting in the mantle, which produces magma. The magma is less dense than the surrounding rock and will rise. It may cool at depth, thickening the crust, or it may make it all the way to Earth's surface, where it erupts as a volcano.
- A line of volcanoes that emerge through continental crust is termed a continental volcanic arc, while a line of volcanoes that emerge through an overriding plate of oceanic lithosphere is a volcanic island arc.
- Continental crust resists subduction due to its relatively low density, and so when an intervening ocean basin is completely destroyed through subduction, the continents on either side collide, generating a new mountain range.

? Sketch a typical continental volcanic arc and label the key parts. Then repeat the drawing with an overriding plate made of oceanic lithosphere.

4.5 Transform Plate Boundaries

Describe the relative motion along a transform fault boundary and locate several examples of transform faults on a plate boundary map.

KEY TERMS: transform plate boundary (transform fault), fracture zone

- At a transform boundary, lithospheric plates slide horizontally past one another. No new lithosphere is generated, and no old lithosphere is consumed. Shallow earthquakes signal the movement of these slabs of rock as they grind past their neighbors.
- The San Andreas Fault in California is an example of a transform boundary in continental crust, while the fracture zones between segments of the Mid-Atlantic Ridge are transform faults in oceanic crust.

? On the accompanying tectonic map of the Caribbean, find the Enriquillo Fault. (The location of the 2010 Haiti earthquake is shown as a yellow star.) What kind of plate boundary is shown here? Are there any other faults in the area that show the same type of motion?

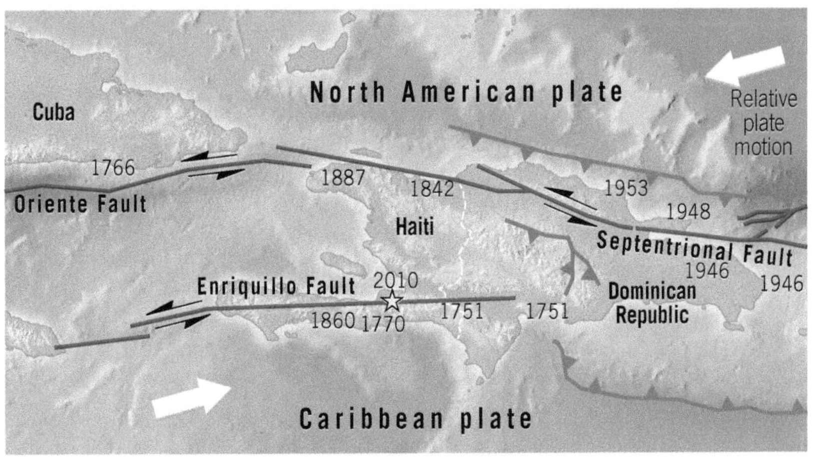

4.6 From Continental Drift to Plate Tectonics

Summarize the view that most geologists held prior to the 1960s regarding the geographic positions of the ocean basins and continents.

- Fifty years ago, most geologists thought that ocean basins were very old and that continents were fixed in place. Those ideas were discarded with a scientific revolution that revitalized geology: the theory of plate tectonics. Supported by multiple kinds of evidence, plate tectonics is the foundation of modern Earth science.

4.7 How Do Plates and Plate Boundaries Change?

Explain why plates such as the African and Antarctic plates are increasing in size, while the Pacific plate is decreasing in size.

- Although the total surface area of Earth does not change, the shape and size of individual plates are constantly changing as a result of subduction and seafloor spreading. Plate boundaries can also be created or destroyed in response to changes in the forces acting on the lithosphere.
- The breakup of Pangaea and the collision of India with Eurasia are two examples of how plates change through geologic time.

4.8 Testing the Plate Tectonics Model

List and explain the evidence used to support the plate tectonics theory.

KEY TERMS: mantle plume, hot spot, hot-spot track, Curie point, paleomagnetism (preserved magnetism), magnetic reversal, normal polarity, reverse polarity, magnetic time scale, magnetometer

- Multiple lines of evidence have verified the plate tectonics model. For instance, the Deep Sea Drilling Project found that the age of the seafloor increases with distance from a mid-ocean ridge. The thickness of sediment atop this seafloor is also proportional to distance from the ridge: Older lithosphere has had more time to accumulate sediment.
- A hot spot is an area of volcanic activity where a mantle plume reaches Earth's surface. Volcanic rocks generated by hot-spot volcanism provide evidence of both the direction and rate of plate movement over time.
- Magnetic minerals such as magnetite align themselves with Earth's magnetic field as rock forms. These preserved magnets are records of the ancient orientation of Earth's magnetic field. This is useful to geologists in two ways: (1) It allows a given stack of rock layers to be interpreted in terms of their orientation relative to the magnetic poles through time, and (2) reversals in the orientation of the magnetic field are preserved as "stripes" of normal and reversed polarity in the oceanic crust. Magnetometers reveal this signature of seafloor spreading as a symmetrical pattern of magnetic stripes parallel to the axis of the mid-ocean ridge.

4.9 How Is Plate Motion Measured?

Describe two methods researchers use to measure relative plate motion.

- Data collected from the ocean floor has established the direction and rate of motion of lithospheric plates. Transform faults point in the direction the plate is moving. Establishing dates for seafloor rocks helps to calibrate the rate of motion.
- GPS satellites can be used to accurately measure the motion of special receivers to within a few millimeters. These "real-time" data support the inferences made from seafloor observations. On average, plates move at about the same rate human fingernails grow: about 5 centimeters (2 inches) per year.

4.10 What Drives Plate Motions?

Describe plate–mantle convection and explain two of the primary driving forces of plate motion.

KEY TERMS: convection, slab pull, ridge push,

- In general, convection (upward movement of less dense material and downward movement of more dense material) appears to drive the motion of plates.
- Slabs of oceanic lithosphere sink at subduction zones because the subducted slab is denser than the underlying asthenosphere. In this process, called slab pull, Earth's gravity tugs at the slab, drawing the rest of the plate toward the subduction zone. As oceanic lithosphere slides down the mid-ocean ridge, it exerts a small additional force, called ridge push.
- Convection may occur throughout the entire mantle, as suggested by the whole-mantle model. Alternatively, it may occur in two layers within the mantle—an active upper mantle and a sluggish lower mantle—as proposed in the layer cake model.

? Compare and contrast mantle convection with the operation of a lava lamp.

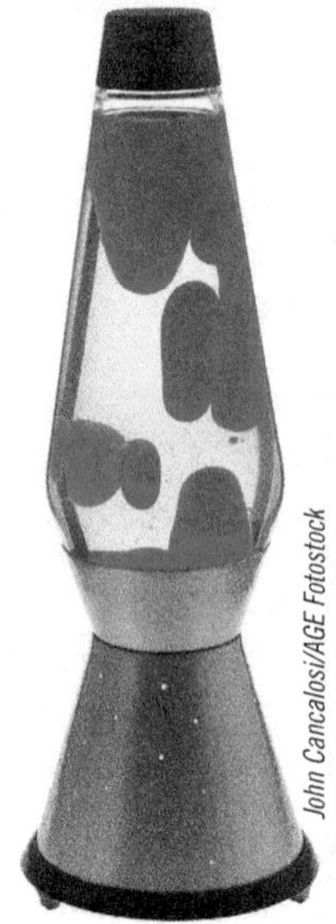

John Cancalosi/AGE Fotostock

GIVE IT SOME THOUGHT

1 Australian marsupials (kangaroos, koalas, etc.) have direct fossil links to marsupial opossums found in the Americas. Yet the modern marsupials in Australia are markedly different from their American relatives. How does the breakup of Pangaea help to explain these differences (*Hint:* See Figure 4.22.)?

2 Refer to the accompanying diagrams illustrating the three types of convergent plate boundaries and complete the following:
  - **a.** Identify each type of convergent boundary.
  - **b.** On what type of crust do volcanic island arcs develop?
  - **c.** Why are volcanoes largely absent where two continental blocks collide?
  - **d.** Describe two ways in which oceanic–oceanic convergent boundaries differ from oceanic–continental boundaries. How are they similar?

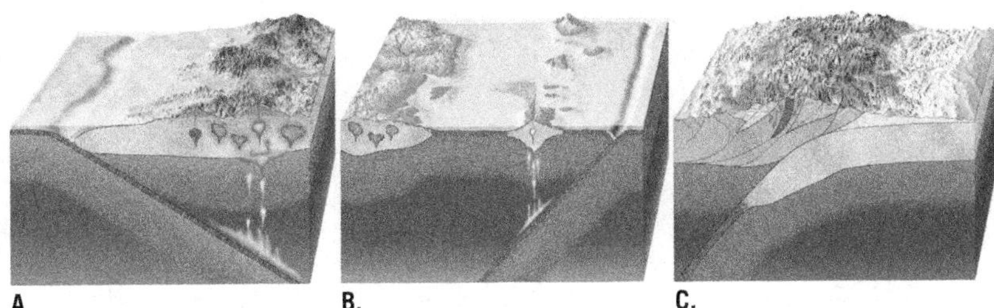

3 Refer to the accompanying hypothetical plate map to answer the following questions:
  - **a.** How many portions of plates are shown?
  - **b.** Explain why active volcanoes are more likely to be found on continents A and B than on continent C.
  - **c.** Provide one scenario in which volcanic activity might be triggered on continent C.

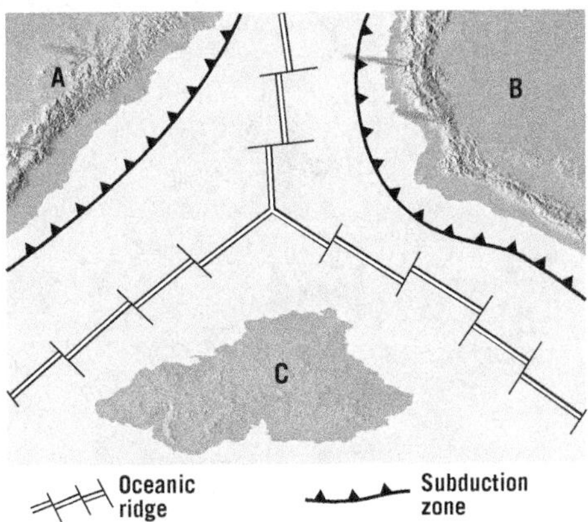

4 Some people predict that California will sink into the ocean. Is this idea consistent with the theory of plate tectonics? Explain.

5 Volcanic islands that form over mantle plumes, such as the Hawaiian chain, are home to some of Earth's largest volcanoes. However, several volcanoes on Mars are gigantic compared to any on Earth. What does this difference tell us about the role of plate motion in shaping the Martian surface?

6 Refer to Section 1.2, "The Nature of Scientific Inquiry," to help answer the following:
  - **a.** What observations led Alfred Wegener to develop his continental drift hypothesis?
  - **b.** Why did most of the scientific community reject the continental drift hypothesis?
  - **c.** Do you think Wegener followed the basic principles of scientific inquiry? Support your answer.

7 Imagine that you are studying seafloor spreading along two different oceanic ridges. Using data from a magnetometer, you produced the two accompanying maps.
  - **a.** From these maps, what can you determine about the relative rates of seafloor spreading along these two ridges? Explain.
  - **b.** Using the scale provided, measure the distance in kilometers from the ridge axis to the beginning (right edge) of the yellowish normal polarity stripe for each ridge. By how many kilometers has each side of these ocean basins spread during the past 1 million years?

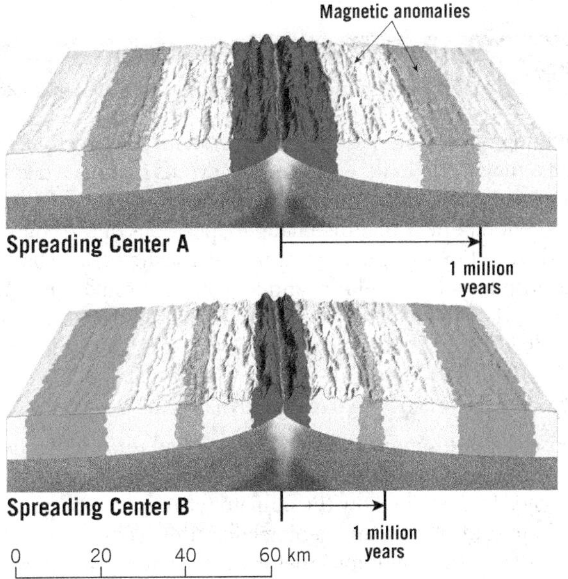

8 Density is a key component in the behavior of Earth materials and is essential to understanding important aspects of the plate tectonics model. Describe three different ways that density and/or density differences play a role in plate tectonics.

9 Explain how the processes that create hot-spot volcanic chains differ from the processes that generate volcanic island arcs.

10 Refer to the accompanying map and the pairs of cities to complete the following: (Boston, Denver), (London, Boston), (Honolulu, Beijing)

a. Which pair of cities is moving apart as a result of plate motion?
b. Which pair of cities is moving closer as a result of plate motion?
c. Which pair of cities is not presently moving relative to each other?

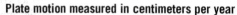

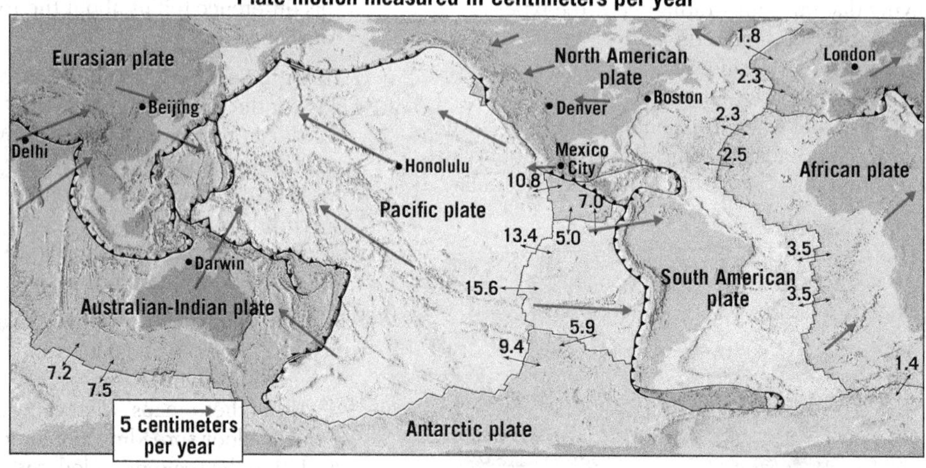

EXAMINING THE EARTH SYSTEM

1 As an integral subsystem of the Earth system, plate tectonics has played a major role in determining the events that have taken place on Earth since its formation about 4.6 billion years ago. Briefly comment on the general effect that the changing positions of the continents and the redistribution of land and water over Earth's surface have had on the atmosphere, hydrosphere, and biosphere through time. You may want to review plate tectonics by investigating the topic on the Internet using a search engine or by visiting the U.S. Geological Survey's (USGS's) This Dynamic Earth: The Story of Plate Tectonics website, at http://pubs.usgs.gov/gip/dynamic/dynamic.html.

2 Suppose that the supercontinent Pangaea had remained intact and in its original location to the present day. Use the accompanying map of Pangaea to describe how the climate (atmosphere), vegetation and animal life (biosphere), and geologic features (geosphere) would be different from the conditions that exist today in the areas currently occupied by the cities of Miami, Florida; Chicago, Illinois; New York, New York; and your college campus location.

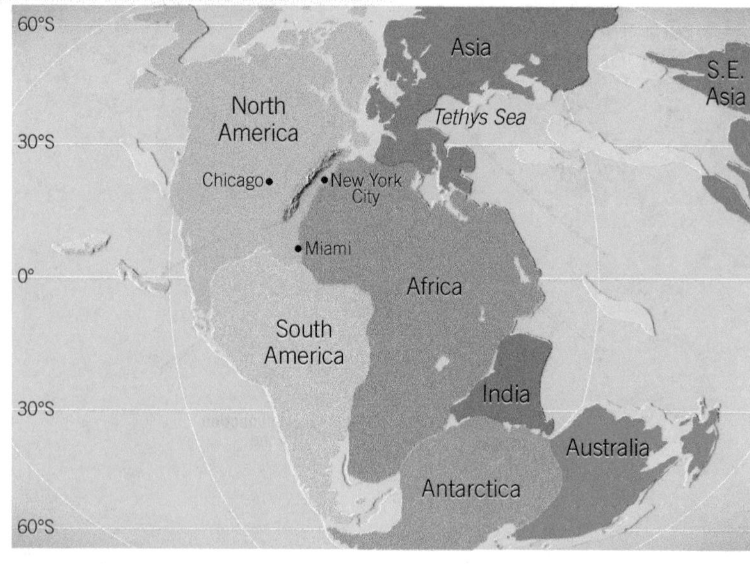

DATA ANALYSIS

Tectonic Plate Movement

Earthquakes and volcanoes provide clues to the type and location of plate boundaries and the movement that happens across them. The Global Positioning System (GPS), a set of 30 satellites used for navigation and geolocation, is also used to track plate movement.

ACTIVITIES

Go to the USGS Earthquake Hazards Program's Latest Earthquakes map, at http://earthquake.usgs.gov/earthquakes/map. Zoom out to the world view and use the settings icon to compare the locations of earthquakes having magnitudes greater than 4.5 with those greater than 2.5 for the past 30 days (you can find these options by clicking the gear icon). Refer to Figure 7.12 for plate boundary locations and types.

1 Along which type of boundary do most strong earthquakes occur?

2 Why might the boundary along the coast of California produce weak earthquakes?

3 Notice that divergent boundaries appear to have the fewest earthquakes. One explanation is that these boundaries produce fewer earthquakes. What might be another explanation?

Zoom in on the convergent boundary on the north side of the Pacific plate.

4 Where are the earthquakes located relative to the plate boundary?

5 What does the location of earthquakes indicate about the movement of plates at this boundary? Which plate is subducting?

Go to the Jet Propulsion Lab's GPS Time Series map, at http://sideshow.jpl.nasa.gov/post/series.html. Read the introduction and then click on a dot near the convergent boundary along the Pacific plate that you analyzed above. Click the graph to open a larger view. Choose a location with at least 10 years of continuous data.

6 What is displayed on this map? Where do the data for this map come from?

7 Is the station you chose moving north (toward higher latitude) or south? On average, how far does this station move (in centimeters) every 10 years?

8 Is this station moving east (toward more positive longitude) or west? On average, how far does this station move (in centimeters) every 10 years?

5 Earthquakes and Earth's Interior

FOCUS ON CONCEPTS

Each statement represents the primary learning objective for the corresponding major heading within the chapter. After you complete the chapter, you should be able to:

5.1 Sketch and describe the mechanism that generates most earthquakes.

5.2 Compare and contrast the types of seismic waves and describe the principle of the seismograph.

5.3 Explain how seismographs are used to locate the epicenter of an earthquake.

5.4 Distinguish between intensity scales and magnitude scales.

5.5 List and describe the major destructive forces that earthquake vibrations can trigger.

5.6 Locate Earth's major earthquake belts on a world map.

5.7 Compare and contrast the goals of short-range earthquake predictions and long-range forecasts.

5.8 Explain how Earth acquired its layered structure and name and describe each of its major layers.

Tsunami striking the coast of Japan on March 11, 2011. (Photo by Sadatsugu Tomizawa/AFP/Getty Images)

ON APRIL 25, 2015, Nepal experienced its worst natural disaster in over 80 years, when an earthquake measuring M 7.8* struck this mountainous region of South Asia. Geologists had anticipated a significant earthquake for decades because of Nepal's location high in the Himalayas, above the collisional boundary where India is being thrust into Asia. The quake killed nearly 9000 people and resulted in injuries to more than 22,000 others. The shallow nature of the 50-second quake resulted in widespread destruction and triggered avalanches on Mt. Everest, claiming the lives of 19 people, including international hikers and Sherpa guides. Numerous landslides throughout the region blocked roads and delayed relief efforts, and Nepal's capital, Kathmandu, reportedly shifted 3 meters (10 feet) to the south during the course of the event.

5.1 What Is an Earthquake?

Sketch and describe the mechanism that generates most earthquakes.

An **earthquake** is ground shaking caused by the sudden and rapid movement of one block of rock slipping past another along fractures in Earth's crust, called **faults**. Most faults are *locked* except for brief, abrupt movements when sudden slippage produces an earthquake. Faults are locked because the confining pressure exerted by the overlying crust is enormous, causing these fractures in the crust to be "squeezed shut."

Earthquakes tend to occur along preexisting faults where internal stresses have caused the crustal rocks to rupture or break into two or more units. The location where slippage begins is called the **hypocenter,** or **focus**. The point on Earth's surface directly above the hypocenter is called the **epicenter** (Figure 5.1).

Large earthquakes release huge amounts of stored-up energy as **seismic waves**—a form of energy that travels through the lithosphere and Earth's interior. The energy carried by these waves causes the material that transmits them to shake. Seismic waves are analogous to waves produced when a stone is dropped into a calm pond. Just as the impact of the stone creates a pattern of circular waves, an earthquake generates waves that radiate outward in all directions from the zone of slippage (see Figure 5.1). Although seismic energy dissipates rapidly as it moves away from the source of an earthquake, sensitive instruments can detect earthquakes even when they occur on the opposite side of Earth.

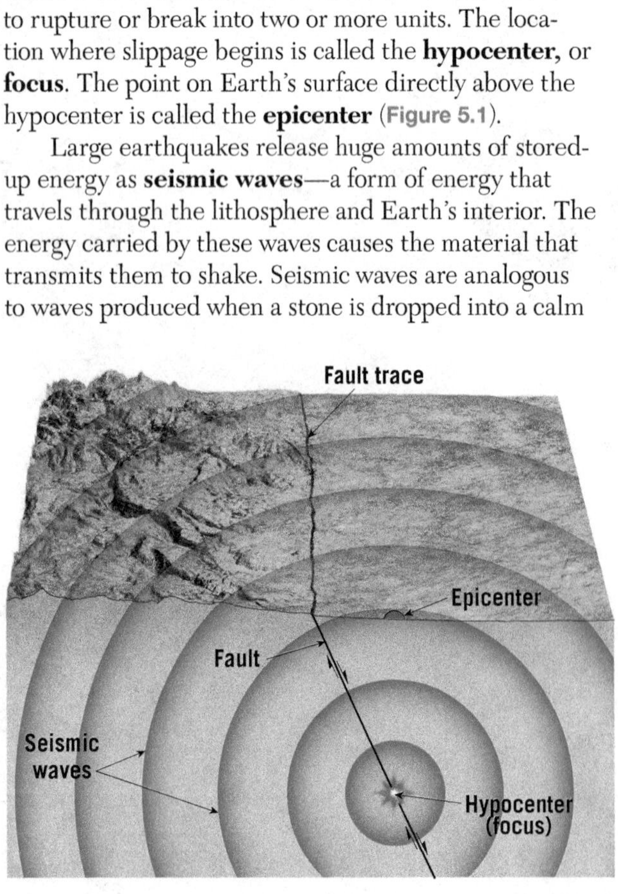

Figure 5.1 Location of an earthquake's hypocenter and epicenter The *hypocenter* is the zone at depth where the initial displacement occurs. The *epicenter* is the surface location directly above the hypocenter.

Thousands of earthquakes occur around the world every day. Fortunately, most are small enough that people cannot detect them. Only about 15 strong earthquakes (magnitude 7 or greater) are recorded each year, many of them occurring in remote regions. The occasional large earthquakes that are triggered near major population centers are among the most destructive natural events on Earth. The shaking of the ground, coupled with the liquefaction of soils, wreaks havoc on buildings, roadways, and other structures. In addition, a quake occurring in a populated area can rupture power and gas lines, causing numerous fires. In the famous 1906 San Francisco earthquake, much of the damage was caused by fires that became uncontrollable when broken water mains left firefighters with only trickles of water (Figure 5.2).

Discovering the Causes of Earthquakes

The energy released by volcanic eruptions, massive landslides, and meteorite impacts can generate earthquake-like waves, but these events are usually weak. What mechanism produces a destructive earthquake? As you have learned, Earth is not a static planet. Because fossils of marine organisms have been discovered thousands of meters above sea level, we know that large sections of Earth's crust have been thrust upward. Other regions, such as California's Death Valley, exhibit evidence of extensive subsidence. In addition to these vertical

*An earthquake's *magnitude* (abbreviation M) is a measure of earthquake strength and is discussed later in this chapter.

n Francisco in flames following the 1906 quake.
oken water mains left firefighters without water.
Fire triggered when a gas line ruptured during the Northridge earthquake in southern California in 1994.

▲ **Figure 5.2 Earthquakes can trigger fires** (Reproduced from the collection of the Library of Congress; inset photo by Hal Garb/AFP/Getty Images)

displacements, offsets in fences, roads, and other structures indicate that horizontal movements between blocks of Earth's crust are also common (Figure 5.3).

The actual mechanism of earthquake generation eluded geologists until H. F. Reid conducted a landmark study following the 1906 San Francisco earthquake. This earthquake was accompanied by horizontal surface displacements of several meters along the northern portion of the San Andreas Fault. Field studies determined that during this single earthquake, the Pacific plate lurched as much as 9.7 meters (32 feet) northward, past the adjacent North American plate. To better visualize this, imagine standing on one side of the fault and watching a person on the other side suddenly slide horizontally about 10 meters to your right.

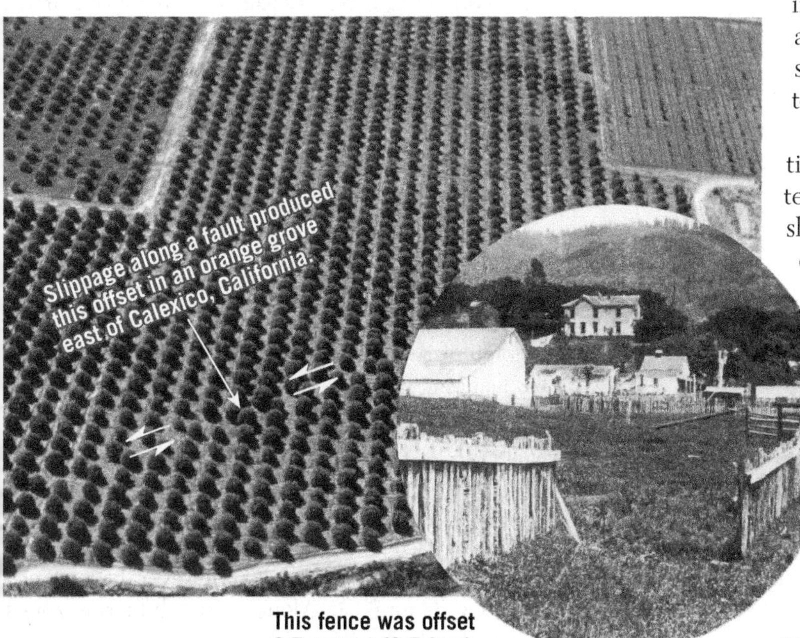

Slippage along a fault produced this offset in an orange grove east of Calexico, California.
This fence was offset 2.5 meters (8.5 feet) during the1906 San Francisco earthquake.

▲ **Figure 5.3 Displacement of human-made structures along a fault** (Color photo by John S. Shelton/ University of Washington Libraries; inset photo by G. K. Gilbert/USGS)

What Reid concluded from his investigations is illustrated in Figure 5.4. Over tens to hundreds of years, differential stress slowly bends the crustal rocks on both sides of a fault. This is much like a person bending a limber wooden stick, as shown in Figure 5.4A,B. Frictional resistance keeps the fault from rupturing and slipping. (Friction inhibits slippage and is enhanced by irregularities that occur along the fault surface.) At some point, the stress along the fault overcomes this frictional resistance, and slip initiates. Slippage allows the deformed (bent) rock to "snap back" to its original, stress-free, shape; a series of earthquake waves radiate outward as it slides (Figure 5.4C,D). Reid termed this "springing back" **elastic rebound** because the rock behaves elastically, much as a stretched rubber band does when it is released.

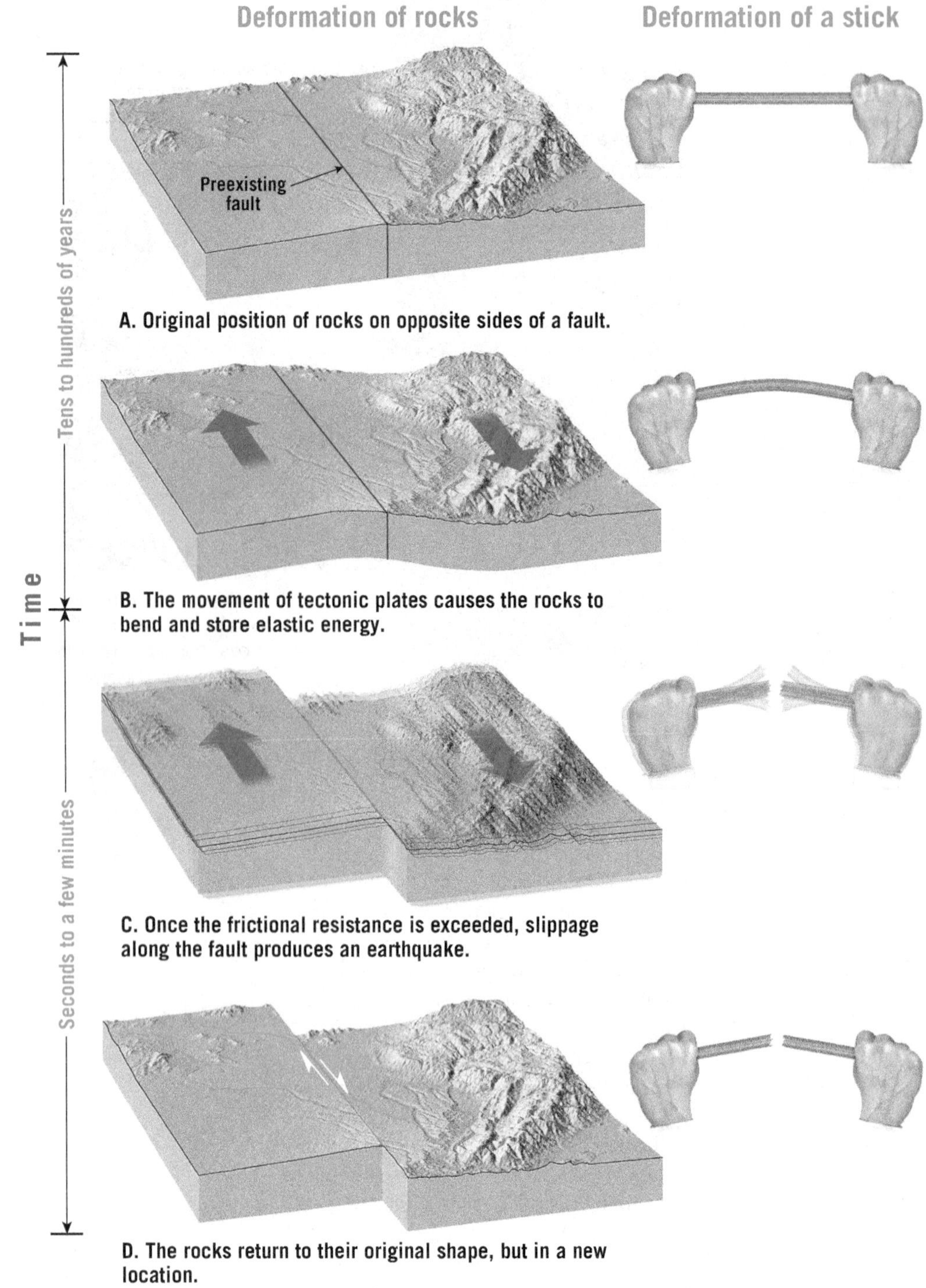

Deformation of rocks

Deformation of a stick

Tens to hundreds of years

Time

Seconds to a few minutes

Preexisting fault

A. Original position of rocks on opposite sides of a fault.

B. The movement of tectonic plates causes the rocks to bend and store elastic energy.

C. Once the frictional resistance is exceeded, slippage along the fault produces an earthquake.

D. The rocks return to their original shape, but in a new location.

▲ SmartFigure 5.4 **Elastic rebound**

TUTORIAL
https://goo.gl/PpfD7W

Aftershocks and Foreshocks

Strong earthquakes are followed by numerous earthquakes of lesser magnitude, called **aftershocks**, which result from crust along the fault surface adjusting to the displacement caused by the main shock. Aftershocks gradually diminish in frequency and intensity over a period of several months following an earthquake. In a little more than a month following the M 7.0 earthquake that devastated Haiti in 2010, the U.S. Geological Survey detected nearly 60 aftershocks with magnitudes of 4.5 or greater. The two largest aftershocks had magnitudes of 6.0 and 5.9, both large enough to inflict further damage. Hundreds of minor tremors were also felt.

Although aftershocks are weaker than the main earthquake, they often trigger the destruction of already weakened structures. For example, in northwestern Armenia in 1988, where many people lived in large apartment buildings constructed of brick and concrete slabs, a moderate earthquake of magnitude 6.9 weakened many structures, and a strong aftershock of magnitude 5.8 completed the demolition.

In contrast to aftershocks, small earthquakes called **foreshocks** often, but not always, precede major earthquakes by days or, in some cases, several years. Monitoring of foreshocks to predict forthcoming earthquakes has been attempted with only limited success.

Faults and Large Earthquakes

The slippage that occurs along faults can be explained by the plate tectonics theory, which states that large slabs of Earth's lithospheric plates are continually grinding past one another. These mobile plates interact with neighboring plates, straining and deforming the rocks along their margins. Faults associated with convergent and transform plate boundaries are the source of most large earthquakes.

Convergent Plate Boundaries Most of Earth's strongest earthquakes occur along large faults associated with convergent plate boundaries. Along convergent boundaries where one continent is colliding with another, the resulting compressional forces slice Earth's crust along numerous large *thrust faults*, discussed in more detail in Section 7.3. The 2015 Nepal earthquake is one example of an earthquake generated along a thrust fault. The epicenter of the quake was located about 80 kilometers (50 miles) north of Kathmandu, where the Indian plate is advancing into the Eurasian plate at a rate of 4.5 centimeters (about 2 inches) per year, driving the uplift of the Himalayas.

When convergence entails the subduction of oceanic lithosphere under another plate, the area of contact between the two plates forms an extensive fault zone, called a **megathrust fault**, that can be several thousand kilometers long (**Figure 5.5**). Along most subduction zones these megathrust faults remain locked for decades or even centuries. As the subducting plate slowly descends, it drags and bends the leading edge of the overlying plate, sometimes producing a bulge on the ocean floor (see Figure 5.26, page 146). Once the frictional forces

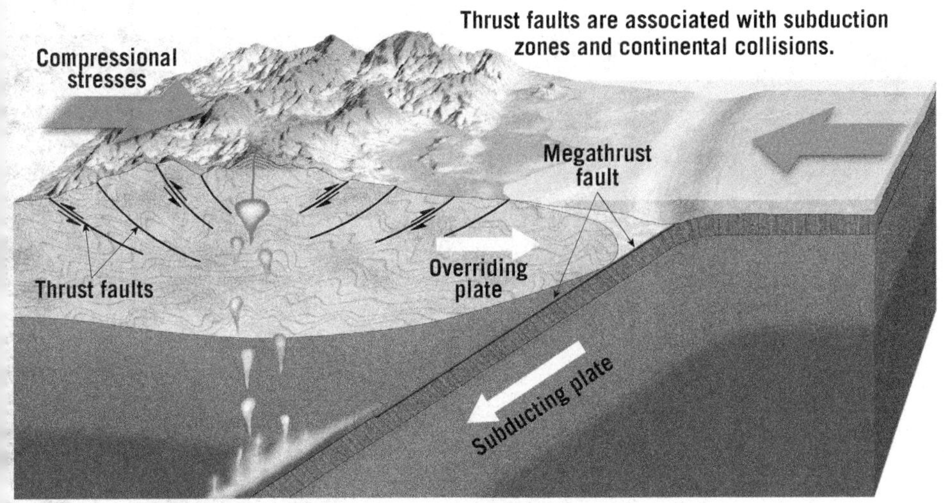

Figure 5.5 **Megathrust faults are the sites of Earth's largest earthquakes** A convergent plate boundary is a site where one plate is subducting beneath another, and the megathrust faults that separate these plates generate most of Earth's largest earthquakes.

between the two stuck plates are exceeded, the overriding plate snaps back to its original shape. This snapping back generates an earthquake whose magnitude depends largely on the size of the zone of slippage.

Megathrust faults have produced the majority of Earth's most powerful and destructive earthquakes, including the 2011 Japan quake (M 9.0), the 2004 Indian Ocean (Sumatra) quake (M 9.1), the 1964 Alaska quake (M 9.2), and the largest earthquake yet recorded, the 1960 Chile quake (M 9.5).

Transform Plate Boundaries Faults in which the dominant displacement is horizontal and parallel to the direction of the fault trace (the line where the fault intersects Earth's surface) are called *strike-slip faults* (discussed in more detail in Section 7.3). Recall from Chapter 4 that *transform plate boundaries* or simply *transform faults* accommodate this type of motion between two tectonic plates. For example, the San Andreas Fault is a large transform fault that lies between the North American plate and the Pacific plate (Figure 5.6). Most large transform faults are not perfectly straight or continuous; instead, they consist of numerous branches and smaller fractures that display kinks and offsets (see Figure 5.6). Earthquakes can occur along any of these branches.

Fault Rupture and Propagation

By studying earthquakes around the globe, geologists have learned that displacement along large faults occurs along discrete fault segments that often behave differently from one another. Some sections of the San Andreas, for example, exhibit slow, gradual displacement known as **fault creep** and produce little seismic shaking. Other segments slip at relatively closely spaced intervals, producing numerous small to moderate earthquakes. Still other segments remain locked and store elastic energy for a few hundred years before they break loose. Ruptures on segments that have been locked for a hundred years or longer usually result in major earthquakes.

Geologists have also discovered that slippage along large faults, such as the San Andreas, does not occur instantaneously. The initial slip occurs at the hypocenter and propagates (travels) along the fault surface. As each section slips, it puts strain on the next section, causing it to slip, as well. Slippage propagates at 2 to 4 kilometers per second—faster than a rifle shot. Rupture of a 100-kilometer (60-mile) fault segment takes about 30 seconds, and rupture of a 300-kilometer (200-mile) segment takes about 90 seconds. As rupturing progresses, it can slow down, speed up, or even jump to a nearby fault segment. Earthquake waves are generated at every point along the fault as that portion of the fault begins to slip.

CONCEPT CHECKS 5.1

1. What is an earthquake? Under what circumstances do most large earthquakes occur?
2. How are faults, hypocenters, and epicenters related?
3. Explain what is meant by *elastic rebound*.
4. What type of fault tends to produce the most destructive earthquakes?

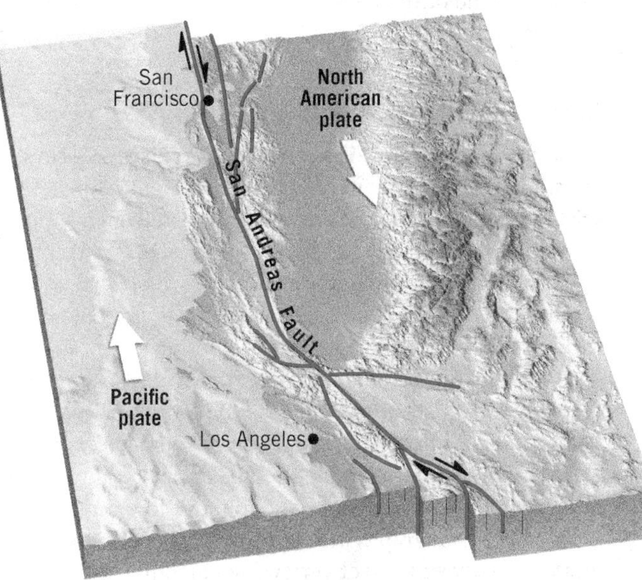

Figure 5.6 **Transform plate boundaries and large earthquakes** The San Andreas Fault is a large fault system separating the Pacific plate from the North American plate. This type of large strike-slip fault, called a transform fault, can generate destructive earthquakes.

EYE ON EARTH 5.1

The Calaveras Fault, a branch of the San Andreas Fault system, cuts directly through the town of Hollister, California. Rather than being "locked," this section of the fault is slowly slipping, producing noticeable offsets and damage to curbs, sidewalks, roads, and buildings. The concrete wall and sidewalk shown here were straight when they were originally constructed.

QUESTION 1 *What term is used to describe the type of slippage observed along the Calaveras Fault in Hollister?*

QUESTION 2 *Are faults that exhibit this type of slippage likely to generate a major earthquake? Explain.*

5.2 Seismology: The Study of Earthquake Waves

Compare and contrast the types of seismic waves and describe the principle of the seismograph.

The study of earthquake waves, **seismology**, dates back to attempts made in China almost 2000 years ago to determine the direction from which these waves originated. The earliest known instrument, invented by Zhang Heng, was a large hollow jar containing a weight suspended from the top (Figure 5.7). The suspended weight (similar to a clock pendulum) was connected to the jaws of several large dragon figurines that encircled the container. The jaws of each dragon held a metal ball. When earthquake waves reached the instrument, the relative motion between the suspended mass and the jar would dislodge some of the metal balls into the waiting mouths of frogs directly below.

Ball dropping

▲ **Figure 5.7 Ancient Chinese seismograph** During an Earth tremor, the dragons located in the direction of the main vibrations would drop a ball into the mouth of a frog below. (Photo from the James E. Patterson Collection, courtesy of F. K. Lutgens)

Instruments That Record Earthquakes

In principle, modern **seismographs,** or **seismometers**, are similar to the instruments used in ancient China. A seismograph has a weight freely suspended from a support that is securely attached to bedrock (Figure 5.8). When vibrations from an earthquake reach the instrument, the **inertia** of the weight keeps it relatively stationary, while Earth and the support move. Inertia can be simply described by this statement: *Objects at rest tend to stay at rest, and objects in motion tend to remain in motion, unless acted upon by an outside force.* You have experienced inertia when you have tried to stop your automobile quickly and your body continued to move forward.

Most seismographs are designed to amplify ground motion in order to detect very weak earthquakes or a

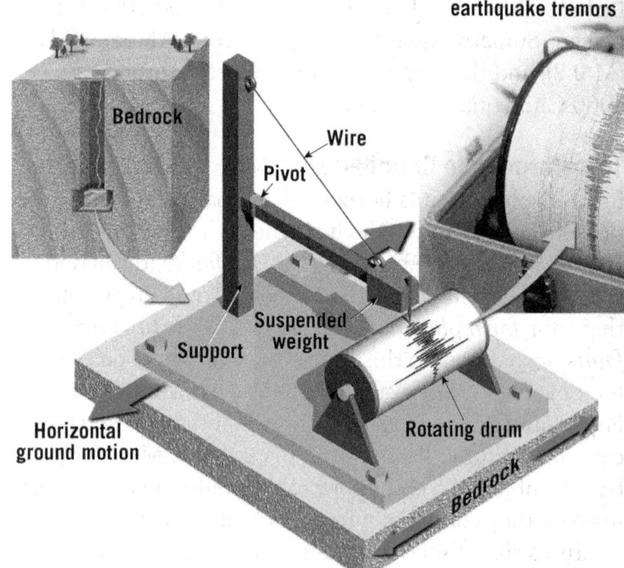

▲ **SmartFigure 5.8 Principle of the seismograph** The inertia of the suspended weight tends to keep it motionless while the recording drum, which is anchored to bedrock, vibrates in response to seismic waves. The stationary weight provides a reference point from which to measure the amount of displacement occurring as a seismic wave passes through the ground. (Photo courtesy of Zephyr/Science Source)

ANIMATION

great earthquake that has occurred in another part of the world. Instruments used in earthquake-prone areas are designed to withstand the violent shaking that can occur near a quake's epicenter.

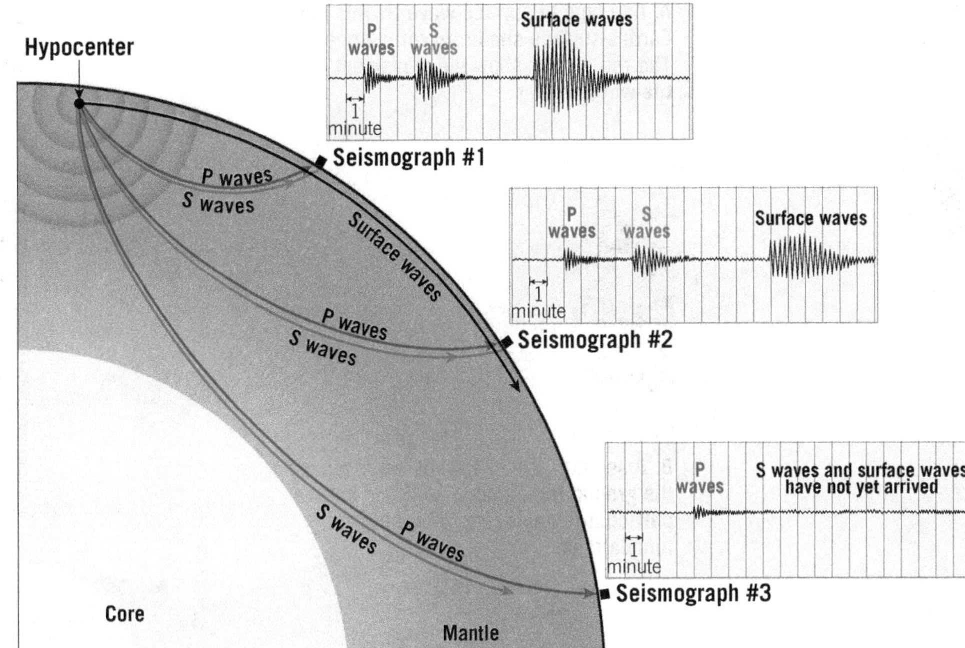

SmartFigure 5.9 **Body waves (P and S waves) versus surface waves** P and S waves travel through Earth's interior, while surface waves travel in the layer directly below the surface. P waves are the first to arrive at a seismic station, followed by S waves and then surface waves.

TUTORIAL
https://goo.gl/PZIJrY

Seismic Waves

The records obtained from seismographs, called **seismograms**, provide useful information about the nature of seismic waves. Seismograms reveal that two main types of seismic waves are generated by the slippage of a rock mass. One of these wave types, called **body waves**, travel through Earth's interior. The other type, called **surface waves**, travel in the rock layers just below Earth's surface (Figure 5.9).

Body Waves Body waves are further divided into two types—**primary waves**, or **P waves**, and **secondary waves**, or **S waves**—and are identified by their mode of travel through intervening materials. P waves are "push/pull" waves; they momentarily push (compress) and pull (stretch) rocks in the direction the waves are traveling (Figure 5.10A). This wave motion is similar to that generated by striking a drum, which moves air back and forth to create sound. Solids, liquids, and gases resist stresses that change their volume when compressed and, therefore, elastically spring back once the stress is removed. Therefore, P waves can travel through all these materials.

By contrast, S waves "shake" the particles at right angles to their direction of travel. This can be illustrated by fastening one end of a rope and shaking the other end, as shown in Figure 5.10B. Unlike P waves, which temporarily change the *volume* of intervening material by alternately squeezing and stretching it, S waves change the *shape* of the material that transmits them. Because fluids (gases and liquids) do not resist stresses that cause changes in shape—meaning fluids do not return to their original shape once the stress is removed—liquids and gases do not transmit S waves.

Surface Waves There are two types of surface waves. One type causes Earth's surface and anything resting on it to move up and down, much as ocean swells toss a ship (Figure 5.11A). The second type of surface wave causes Earth's surface to move from side to side. This motion is particularly damaging to the foundations of structures (Figure 5.11B).

Comparing the Speed and Size of Seismic Waves By examining the seismogram shown in Figure 5.12 you can see that another major difference among the types of seismic waves is their speed of travel. P waves are the first to arrive at a recording station, then S waves, and finally surface waves. Generally, in any solid Earth material, P waves travel about 70 percent faster than S waves, and S waves are roughly 10 percent faster than surface waves.

In addition to the velocity differences in the waves, notice in Figure 5.12 that their seismograph recordings

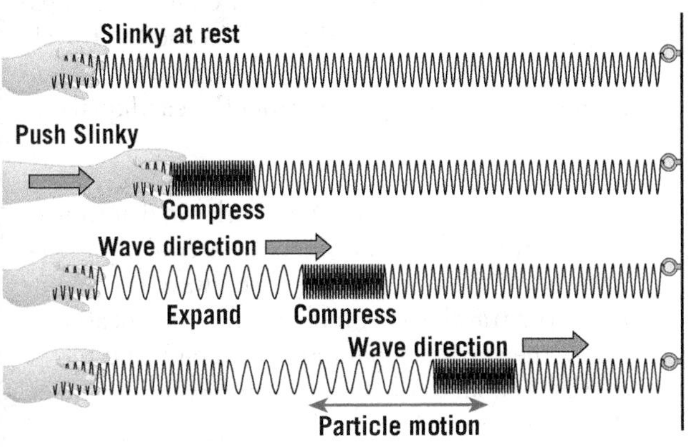

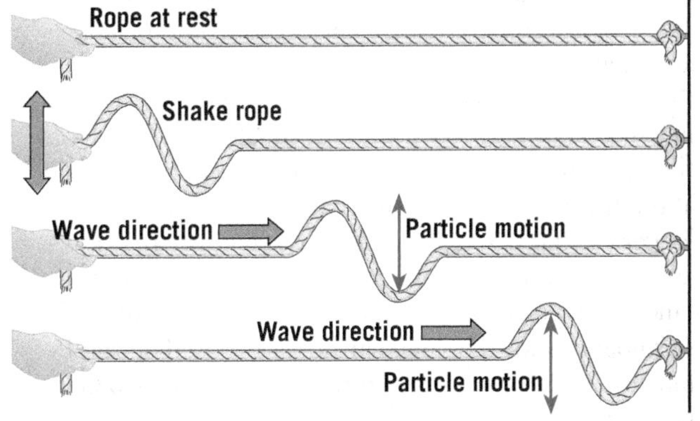

Figure 5.10 **The characteristic motion of P and S waves** During a strong earthquake, ground shaking consists of a combination of various kinds of seismic waves.

SmartFigure 5.11 **Two types of surface waves**

ANIMATION
https://goo.gl/RSV8tN

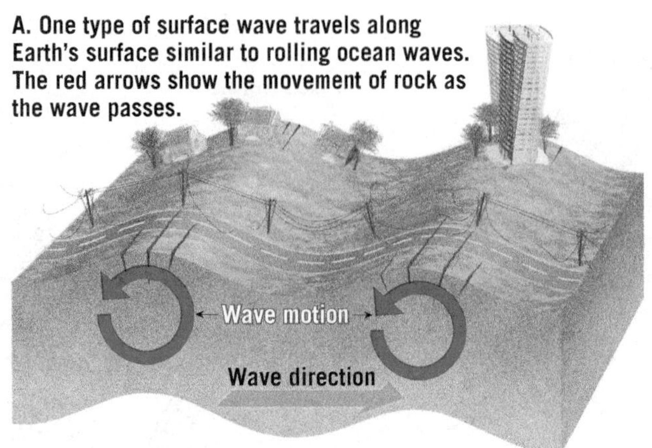

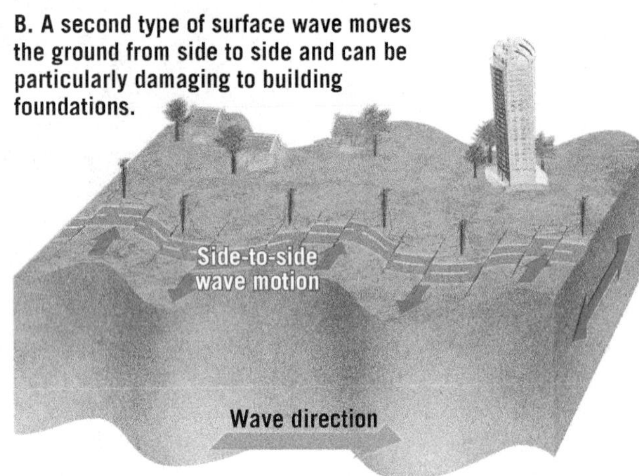

differ in height, or *amplitude*, which reflects the amplitude of shaking they cause. S waves have slightly greater amplitudes than P waves, and surface waves exhibit even greater amplitudes. Surface waves also retain their maximum amplitude longer than P and S waves. As a result, surface waves tend to cause greater ground shaking and, hence, greater property damage, than either P or S waves.

B. A second type of surface wave moves the ground from side to side and can be particularly damaging to building foundations.
Side-to-side wave motion
A. One type of surface wave travels along Earth's surface similar to rolling ocean waves. The red arrows show the movement of rock as the wave passes.
Wave direction

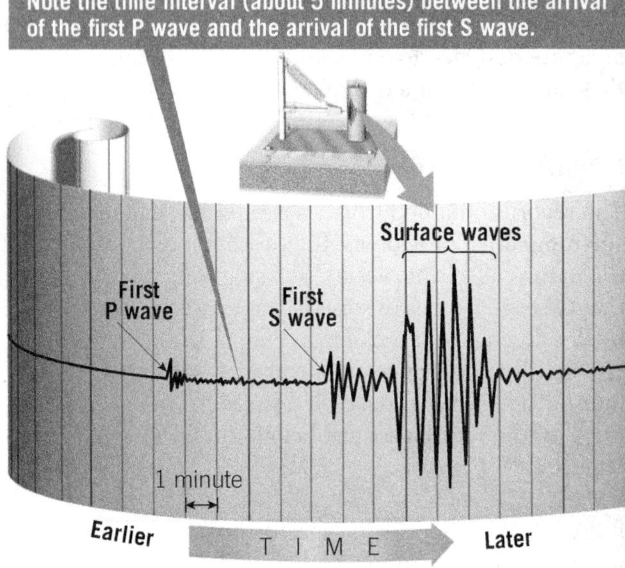

Figure 5.12 **Typical seismogram**

CONCEPT CHECKS 5.2

1. Describe how a seismograph works.
2. List the major differences between P, S, and surface waves.
3. Which type of seismic wave tends to cause the greatest destruction to buildings?

5.3 Locating the Source of an Earthquake

Explain how seismographs are used to locate the epicenter of an earthquake.

When seismologists analyze an earthquake, they first determine its *epicenter*, the point on Earth's surface directly above the hypocenter, or focus (see Figure 5.1). One method used for locating an earthquake's epicenter relies on the fact that P waves travel faster than S waves.

The traveling waves are analogous to two racing automobiles, one faster than the other. The first P wave, like the faster automobile, always wins the race, arriving ahead of the first S wave. The greater the length of the race, the greater the difference in their arrival times at the finish line (the seismic station). Therefore, the longer the interval between the arrival of the first P wave and the arrival of the first S wave, the greater the distance to the epicenter. Figure 5.13 shows three simplified seismograms for the same earthquake. Based on the P–S interval, which city—New York, Nome, or Mexico City—is farthest from the epicenter?

The system for locating earthquake epicenters was developed by using seismograms from earthquakes whose epicenters could be easily pinpointed based on physical evidence. From these seismograms, travel–time graphs were constructed (Figure 5.14). Using the sample seismogram for New York in Figure 5.13 and the travel–time curve in Figure 5.14, we can determine the distance separating the recording station from the earthquake in three steps:

1. Using the seismogram for New York, we determine that the time interval between the arrival of the first P wave and the arrival of the first S wave is 5 minutes.
2. Using the travel–time graph, we find the location where the vertical separation between P and S curves is equal to the P–S time interval (5 minutes in this example).

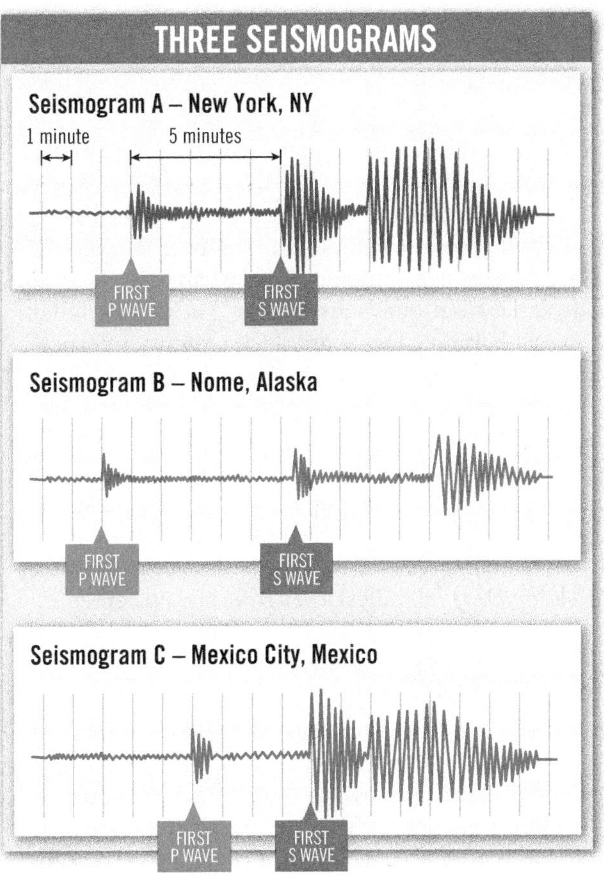

Figure 5.13 Seismograms of the same earthquake recorded at three different locations

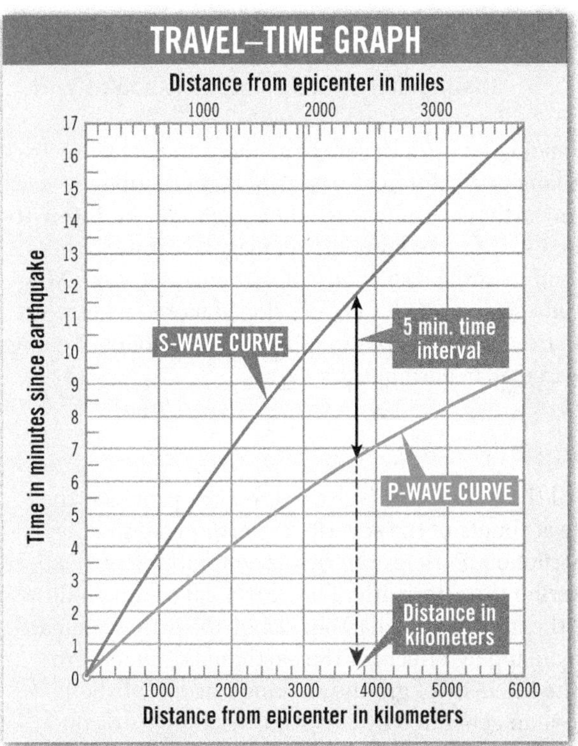

Figure 5.14 **Travel-time graph** A travel-time graph is used to determine the distance to an earthquake's epicenter. The difference in arrival times of the first P and S waves in the example shown is 5 minutes.

3. From the position in step 2, we draw a vertical line to the horizontal axes and read the distance to the epicenter.

Using these steps we determine that the earthquake occurred 3700 kilometers (2300 miles) from the recording instrument in New York City.

Now we know the *distance*, but what about *direction*? The epicenter could be in any direction from the seismic station. Using a method called *triangulation*, we can determine the location of an epicenter if we know the distance to it from two or more additional seismic stations (**Figure 5.15**). On a map or globe, we draw a circle around each seismic station with a radius equal to the distance from that station to the epicenter. The point where the three circles intersect is the approximate epicenter of the quake.

Figure 5.15 **Triangulation to locate an earthquake** This method involves using the distance obtained from three or more seismic stations to establish the location of an earthquake.

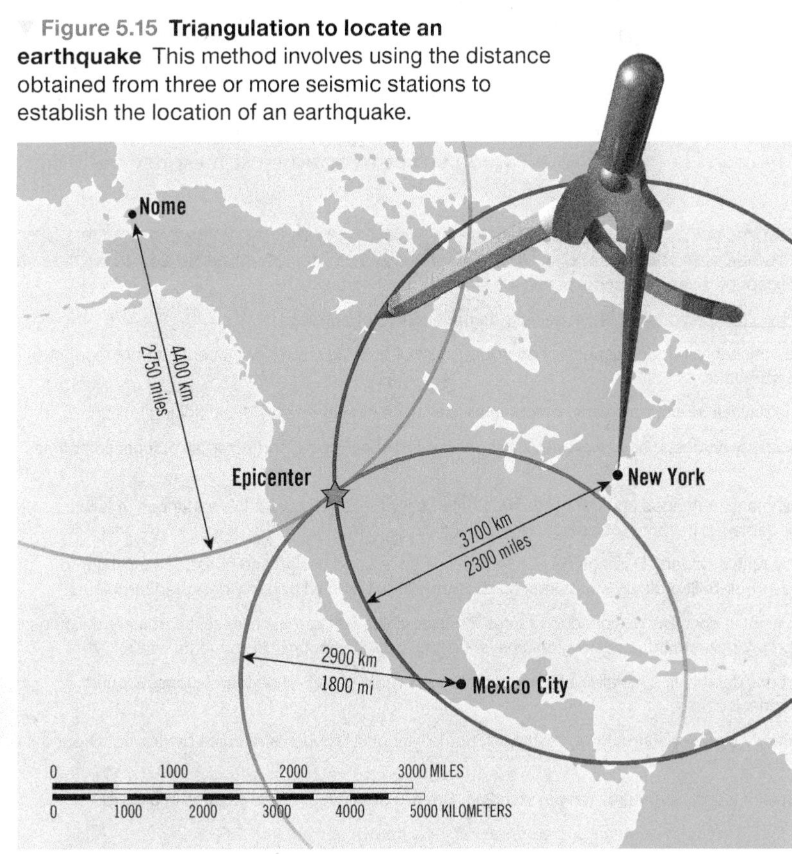

CONCEPT CHECKS 5.3

1. What information does a travel–time graph provide?
2. Briefly describe the *triangulation* method used to locate the epicenter of an earthquake.

5.4 Determining the Size of an Earthquake

Distinguish between intensity scales and magnitude scales.

Seismologists use a variety of methods to determine two fundamentally different measures that describe the size of an earthquake: *intensity* and *magnitude*. An **intensity** scale uses observed property damage to estimate the amount of ground shaking at a particular location. **Magnitude** scales, which were developed more recently, use data from seismographs to estimate the amount of energy released at an earthquake's source.

Intensity Scales

Until the mid-1800s, historical records provided the only accounts of the severity of earthquake shaking and destruction. Perhaps the first attempt to scientifically describe the aftermath of an earthquake came following the great Italian earthquake of 1857. By systematically mapping effects of the earthquake, a measure of the intensity of ground shaking was established. The map generated by this study used lines to connect places of equal damage and hence equal ground shaking. Using this technique, zones of intensity were identified, with the zone of highest intensity representing the location of maximum ground shaking, which is often (but not always) found surrounding the earthquake epicenter.

In 1902, Giuseppe Mercalli developed a more reliable intensity scale, which is still used today in a modified form. The **Modified Mercalli Intensity scale**, shown in Table 5.1, was developed using California buildings as its standard. Based on the 12-point Mercalli Intensity scale, an area in which some well-built wood structures and most masonry buildings are destroyed by an earthquake would be assigned a Roman numeral X.

More recently, the U.S. Geological Survey has developed a website called "Did You Feel It," where Internet users experiencing a quake can enter their zip code and answer questions such as "Did objects fall off shelves?" Within a few hours, a Community Internet Intensity Map, like the one in Figure 5.16 for the 2011 central Virginia earthquake (M 5.8), is generated. As shown in Figure 5.16, shaking strong enough to be felt was reported from Maine to Florida, an area occupied by one-third of the U.S. population. Several national landmarks were damaged, including the Washington Monument and the National Cathedral, located about 130 kilometers (80 miles) away from the epicenter.

Table 5.1 Modified Mercalli Intensity Scale

I	Not felt except by a very few under especially favorable circumstances.
II	Felt only by a few persons at rest, especially on upper floors of buildings.
III	Felt quite noticeably indoors, especially on upper floors of buildings, but many people do not recognize it as an earthquake.
IV	During the day felt indoors by many, outdoors by few. Sensation like heavy truck striking building.
V	Felt by nearly everyone, many awakened. Disturbances of trees, poles, and other tall objects sometimes noticed.
VI	Felt by all; many frightened and run outdoors. Some heavy furniture moved; few instances of fallen plaster or damaged chimneys. Damage slight.
VII	Everybody runs outdoors. Damage negligible in buildings of good design and construction; slight to moderate in well-built ordinary structures; considerable in poorly built or badly designed structures.
VIII	Damage slight in specially designed structures; considerable, with partial collapse, in ordinary buildings; great in poorly built structures (falling chimneys, factory stacks, columns, monuments, walls).
IX	Damage considerable in specially designed structures. Buildings shifted off foundations. Ground cracked conspicuously.
X	Most masonry and frame structures destroyed. Some well-built wooden structures destroyed. Ground badly cracked.
XI	Few, if any, (masonry) structures remain standing. Bridges destroyed. Broad fissures in ground.
XII	Damage total. Waves seen on ground surfaces. Objects thrown upward into air.

Magnitude Scales

To more accurately compare earthquakes around the globe, scientists searched for a way to describe the energy released by earthquakes that did not rely on factors such as building practices, which vary considerably from one part of the world to another. As a result, several magnitude scales were developed.

Richter Magnitude In 1935 Charles Richter of the California Institute of Technology developed the first magnitude scale to use seismic records. As shown in Figure 5.17 (top), the **Richter scale** is calculated by measuring the amplitude of the largest seismic wave (usually an S wave or a surface wave) recorded on a seismogram. Because seismic waves weaken as the distance between the hypocenter and the seismograph increases, Richter developed a method that accounts for the decrease in wave amplitude with increasing distance. Theoretically, as long as equivalent instruments are used, monitoring stations at different locations will obtain the same Richter magnitude for each recorded earthquake. In practice, however, different recording stations often obtain slightly different magnitudes for the same earthquake—a result of the variations in the rock types through which the waves travel.

Earthquakes vary enormously in strength, and great earthquakes produce wave amplitudes thousands of times larger than those generated by weak tremors. To accommodate this wide variation, Richter used a *logarithmic scale* to express magnitude, in which a *10-fold* increase in wave amplitude corresponds to an increase of 1 on the magnitude scale. Thus, the intensity of ground shaking for a magnitude 5 earthquake is 10 times greater than that produced by an earthquake having a Richter magnitude (M_L) of 4 (Figure 5.18).

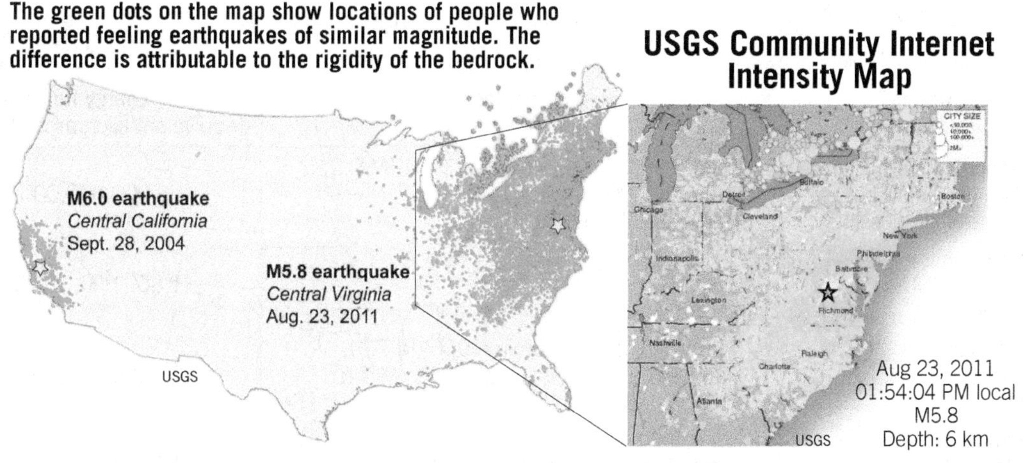

The green dots on the map show locations of people who reported feeling earthquakes of similar magnitude. The difference is attributable to the rigidity of the bedrock.
USGS Community Internet Intensity Map
M6.0 earthquake
Central California
Sept. 28, 2004
M5.8 earthquake
Central Virginia
Aug. 23, 2011
USGS
Aug 23, 2011
01:54:04 PM local
M5.8
Depth: 6 km
USGS

Key for USGS Community Internet Intensity Map

INTENSITY	I	II-III	IV	V	VI	VII	VIII	IX	X
SHAKING	Not felt	Weak	Light	Moderate	Strong	Very strong	Severe	Violent	Extreme
DAMAGE	none	none	none	Very light	Light	Moderate	Moderate/Heavy	Heavy	Very Heavy

◄ SmartFigure 5.16 **USGS Community Internet Intensity Map** Maps like this one are prepared using data collected on the Internet from people responding to questions such as "Did objects fall off shelves?"

TUTORIAL
https://goo.gl/NANACy

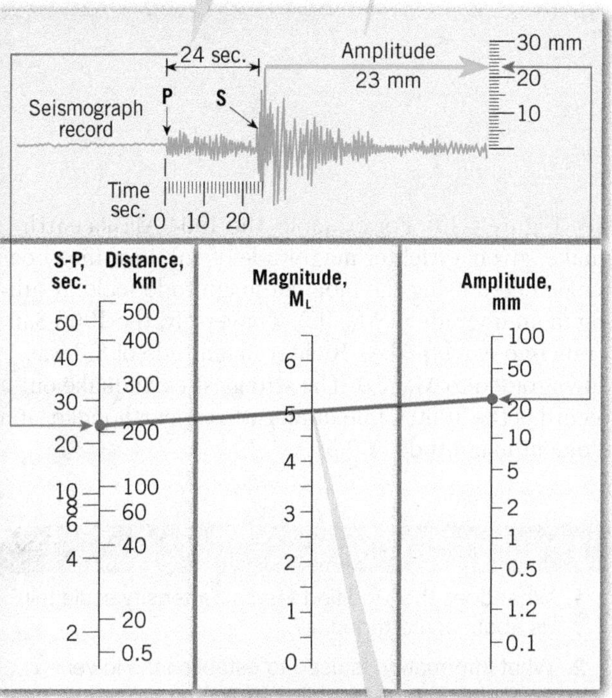

1. Measure the height (amplitude) of the largest wave on the seismogram (23 mm) and plot it on the amplitude scale (right).
2. Determine the distance to the earthquake using the time interval separating the arrival of the first P wave and the arrival of the first S wave (24 seconds) and plot it on the distance scale (left).
24 sec.
Amplitude
23 mm
30 mm
20
10
Seismograph record
P
S
Time sec.
0 10 20
S-P, sec.
Distance, km
Magnitude, M_L
Amplitude, mm
3. Draw a line connecting the two plots and read the Richter magnitude (M_L 5) from the magnitude scale (center).

▲ Figure 5.17 **Determining the Richter magnitude of an earthquake**

In addition, each unit of increase in Richter magnitude equates to roughly a *32-fold increase in the energy released*. Thus, an earthquake with a magnitude of 6.5 releases 32 times more energy than one with a magnitude of 5.5 and roughly 1000 times (32 × 32) more energy than a magnitude 4.5 quake. A major earthquake with a magnitude of 8.5 releases millions of times more energy than the smallest earthquakes felt by humans (Figure 5.19).

The convenience of describing the size of an earthquake by a single number that can be calculated quickly from seismograms makes the Richter scale a powerful tool. Seismologists have since modified Richter's work and developed other Richter-like magnitude scales.

Despite its usefulness, the Richter scale is not adequate for describing very large earthquakes. For example, the 1906 San Francisco earthquake and the 1964 Alaska earthquake have roughly the same Richter magnitudes. However, based on the relative size of the affected areas and the associated tectonic changes, the Alaska earthquake released considerably more energy than the San Francisco quake. Thus, the Richter scale is considered *saturated* for major earthquakes because it

Magnitude vs. Ground Motion and Energy

Difference in Magnitude	Difference in Ground Motion (amplitude)	Difference in Energy Release (approximate)
4.0	10,000 times	1,000,000 times
3.0	1000 times	32,000 times
2.0	100 times	1000 times
1.0	10 times	32 times
0.5	3.2 times	5.5 times
0.1	1.3 times	1.4 times

◄ Figure 5.18 **Magnitude versus ground motion and energy released** An earthquake that is 1 unit of magnitude stronger than another (such as M 6 versus M 5) produces seismic waves that have a maximum amplitude 10 times greater and releases about 32 times more energy than the weaker quake.

Magnitude (Mw)	Average Per Year	Description	Examples	Energy Release (equivalent kilograms of explosive)
9	<1	**Largest recorded earthquakes–** destruction over vast area, massive loss of life possible	Chile, 1960 (M 9.5); Alaska, 1964 (M 9.2); Sumatra, 2004 (M 9.1); Japan, 2011 (M 9.0)	56,000,000,000,000
8	1	**Great earthquakes–** severe economic impact, large loss of life	Chile, 2010 (M 8.8); Mexico City, 1980 (M 8.1)	1,800,000,000,000
7	15	**Major earthquakes–** damage ($ billions), loss of life	San Francisco, California, 1906 (M 7.9); Haiti, 2012 (M 7.0); Nepal, 2015 (M 7.8) Ecuador, 2016 (M 7.8)	56,000,000,000
6	134	**Strong earthquakes–** can be destructive in populated areas, loss of life	Kobe, Japan, 1995 (M 6.9); Loma Prieta, California, 1989 (M 6.9); Northridge, California, 1994 (M 6.7)	1,800,000,000
5	1319	**Moderate earthquakes–** property damage to poorly constructed buildings	Mineral, Virginia, 2011 (M 5.8); Northern New York, 1994 (M 5.8); East of Oklahoma City, Oklahoma, 2011 (M 5.6)	56,000,000
4	13,000	**Light earthquakes–** noticeable shaking of items indoors, some property damage	Western Minnesota, 1975 (M 4.6); Arkansas, 2011 (M 4.7)	1,800,000
3	130,000	**Minor earthquakes–** felt by humans, very light property damage, if any	New Jersey, 2009 (M 3.0); Maine, 2006 (M 3.8); Texas, 2015 (M 3.6)	56,000
2	1,300,000	**Very minor earthquakes–** felt by humans, no property damage		1,800
	Unknown	**Very minor earthquakes–** generally not felt by humans, but may be recorded		56

Frequency and Energy Released by Earthquakes of Different Magnitudes

Data from USGS

Figure 5.19 Frequency of earthquakes with various moment magnitudes

cannot distinguish among them. Despite this shortcoming, Richter-like scales are still used because they allow for quick calculations.

Moment Magnitude For measuring medium and large earthquakes, seismologists now favor a newer scale called **moment magnitude** (M_W), which estimates the total energy released during an earthquake. Moment magnitude is calculated by determining the average amount of slip on the fault, the area of the fault surface that slipped, and the strength of the faulted rock.

Moment magnitude can also be calculated by modeling data obtained from seismograms. The results are converted to a magnitude number, similar to other magnitude scales. As with the Richter scale, each unit increase on the moment magnitude scale equates to roughly a 32-fold increase in the energy released.

Because the moment magnitude scale is better than the Richter scale at estimating the relative size of very large earthquakes, seismologists have used it to recalculate the magnitudes of older strong earthquakes (see Figure 5.19). For example, the 1964 Alaska earthquake, given a Richter magnitude of 8.3, has since been recalculated using the moment magnitude scale, resulting in an upgrade to M_W 9.2. Conversely, the 1906 San Francisco earthquake's Richter magnitude of 8.3 was downgraded to M_W 7.9. The strongest earthquake on record is the 1960 Chilean megathrust earthquake, at a moment magnitude of 9.5.

CONCEPT CHECKS 5.4

1. What does the Modified Mercalli Intensity scale tell us about an earthquake?
2. What information is used to establish the lower numbers on the Mercalli scale?
3. How much more energy does a magnitude 7.0 earthquake release than a magnitude 6.0 earthquake?
4. Why is the moment magnitude scale favored over the Richter scale for large earthquakes?

5.5 Earthquake Destruction

List and describe the major destructive forces that earthquake vibrations can trigger.

The most violent earthquake ever recorded in North America—the 1964 Alaska earthquake—occurred at 5:36 P.M. on March 27, 1964. Felt over most of the state, the earthquake had a moment magnitude of 9.2 and lasted 3 to 4 minutes. This event left 128 people dead and thousands homeless, and it badly disrupted the state's economy. Within 24 hours of the initial shock, 28 aftershocks were recorded, 10 of them exceeding magnitude 6. The epicenter and towns hardest hit by the quake are shown in Figure 5.20.

Many factors determine the degree of destruction that accompanies an earthquake. The most obvious is the *magnitude of the earthquake and its proximity to a populated area.* During an earthquake, the region within 20 to 50 kilometers (12 to 30 miles) of the epicenter tends to experience roughly the same degree of ground shaking, and beyond that limit, vibrations usually diminish rapidly. As described earlier, earthquakes that occur in the stable continental interior, such as the New Madrid, Missouri, earthquakes of 1811–1812, are generally felt over a much larger area than those in earthquake-prone areas such as California.

Destruction from Seismic Vibrations

The 1964 Alaska earthquake provided geologists with insights into the role of ground shaking as a destructive force. As the energy released by an earthquake travels along Earth's surface, it causes the ground to vibrate in a complex manner involving up-and-down as well side-to-side motion. The amount of damage to human-made structures attributable to the vibrations depends on several factors, including (1) the *intensity* and (2) *duration of the vibrations* as well as (3) the *nature of the material on which structures rest* and (4) the *building materials and construction practices of the region.*

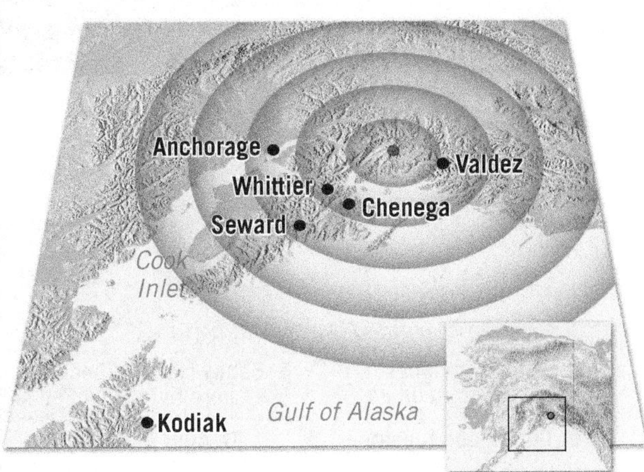

Figure 5.20 **Region most affected by the Alaska earthquake, 1964**

All of the multistory structures in Anchorage were damaged by the vibrations. The more flexible wood-frame residential buildings fared best. A striking example of how construction variations affect earthquake damage is shown in Figure 5.21. You can see that the steel-frame building on the left withstood the vibrations, whereas the poorly designed JCPenney building was badly damaged. Engineers have learned that buildings constructed of blocks and bricks that are not reinforced with steel rods are the most serious safety threats in earthquakes. Unfortunately, most of the structures in the developing world are constructed of unreinforced concrete slabs and bricks made of dried mud—a primary reason the death toll in poor countries such as Haiti and Nepal is usually higher than for earthquakes of similar size in Japan or the United States.

The 1964 Alaska earthquake damaged most large structures in Anchorage, even though they were built according to the earthquake provisions of the Uniform Building Code. Perhaps some of that destruction can be attributed to the unusually long duration of the earthquake. Most quakes involve tremors that last less than a minute. For example, the 1994 Northridge earthquake was felt for about 40 seconds, and the strong vibrations of the 1989 Loma Prieta earthquake lasted less than 15 seconds. But the Alaska quake, which was substantially larger than both of those, reverberated for 3 to 4 minutes.

Amplification of Seismic Waves Although the region near the epicenter experiences seismic waves of about the same energy, destruction may vary considerably in this

Figure 5.21 **Comparing damage to structures** The poorly constructed five-story JCPenney building in Anchorage, Alaska, sustained extensive damage. The steel-frame building (lower left) incurred very little structural damage. (Courtesy of NOAA/Seattle)

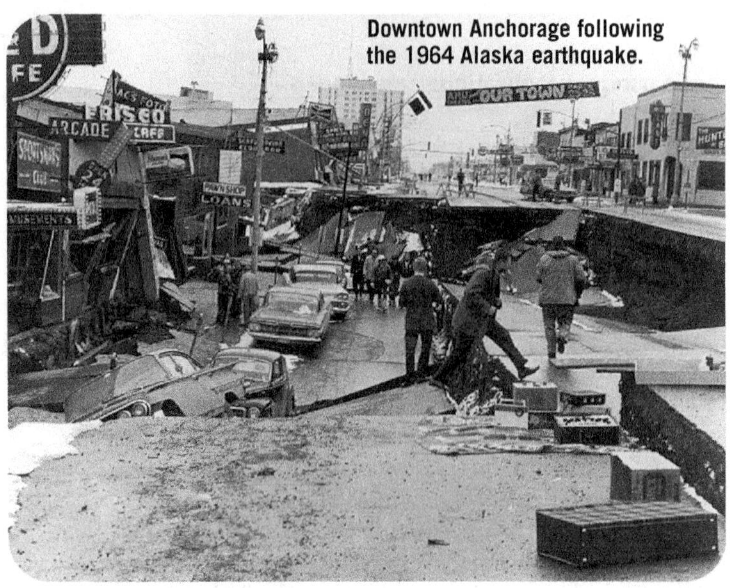

Figure 5.22 **Ground failure caused this street in Anchorage, Alaska, to collapse** (Photo by USGS)

area due to the nature of the ground on which the structures are built. Soft sediments, for example, generally amplify the seismic vibrations more than solid bedrock. Thus, the buildings in Anchorage that were situated on unconsolidated sediments experienced heavy structural damage (Figure 5.22). In contrast, most of the town of Whittier, though much nearer the epicenter, rested on a firm foundation of solid bedrock and, therefore, suffered much less damage from seismic vibrations.

Figure 5.23 **Effects of liquefaction on buildings** (Photo courtesy of USGS)

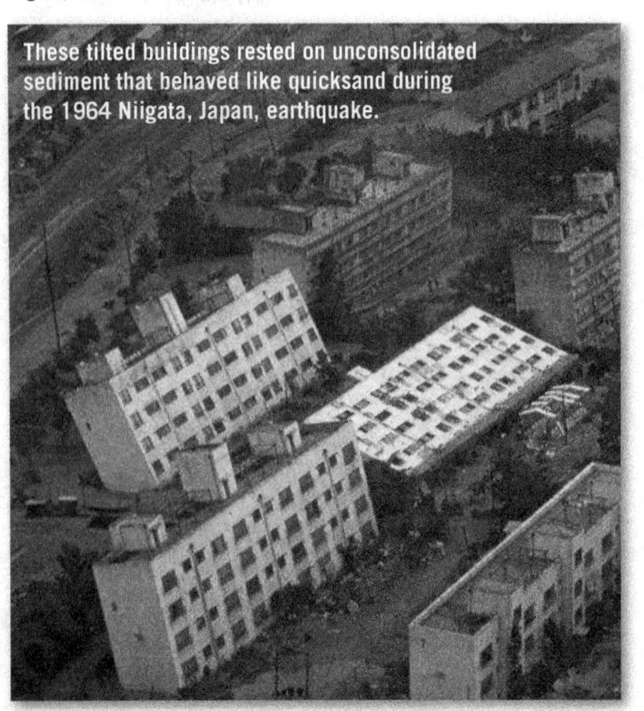

Liquefaction The intense shaking of an earthquake can cause loosely packed water-logged materials, such as sandy stream deposits or fill, to be transformed into a substance that acts like a fluid. The phenomenon of transforming a somewhat stable soil into mobile material capable of rising toward Earth's surface is known as **liquefaction**. When liquefaction occurs, the ground may not be capable of supporting buildings, and underground storage tanks and sewer lines may literally float toward the surface (Figure 5.23).

During the 1989 Loma Prieta earthquake, in San Francisco's Marina District, foundations failed, and geysers of sand and water shot from the ground, evidence that liquefaction had occurred (Figure 5.24). Liquefaction also contributed to the damage inflicted on San Francisco's water system during the 1906 earthquake. During the 2011 Japan earthquake, liquefaction caused entire buildings to sink several feet.

Landslides and Ground Subsidence

The greatest earthquake-related damage to structures is often caused by landslides and ground subsidence triggered by earthquake vibrations. This was the case during the 1964 Alaska earthquake in Valdez and Seward, where the violent shaking caused coastal sediments to slump,

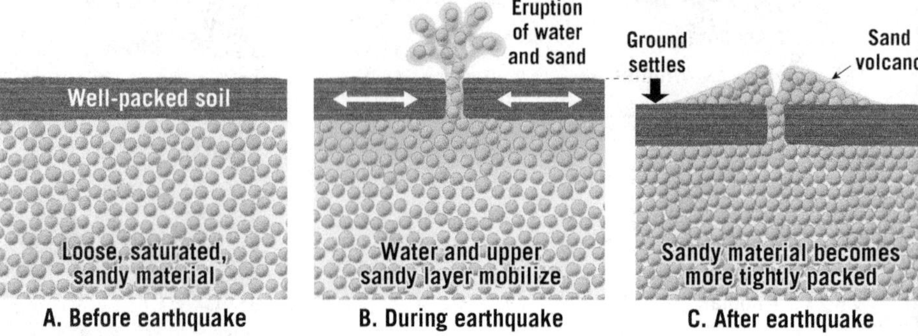

Figure 5.24 **Liquefaction** These "sand volcanoes," produced by the Christchurch, New Zealand, earthquake of 2011, formed when "geysers" of sand and water shot from the ground, an indication that liquefaction occurred. (Photo by Diarmuid/Alamy Stock Photo)

EYE ON EARTH 5.2

Water-saturated sandy soil provides students an opportunity to experience the phenomenon of liquefaction. Liquefaction may occur when ground shaking causes a layer of saturated material to lose strength and act like a fluid.

QUESTION 1 *Describe what you think would happen to a structure built on sandy soil that suddenly experienced liquefaction during an earthquake.*

QUESTION 2 *How might a nearly empty underground storage tank be affected by liquefaction of the surrounding soil?*

carrying away both waterfronts. In Valdez, 31 people died when a dock slid into the sea. Because of the threat of recurrence, the entire town of Valdez was relocated to more stable ground about 7 kilometers (4 miles) away.

Much of the damage in Anchorage was attributed to landslides. Homes in Turnagain Heights were destroyed when a layer of clay lost its strength and over 200 acres of land slid toward the ocean (Figure 5.25). A portion of this spectacular landslide was left in its natural condition as a reminder of this destructive event. The site was appropriately named "Earthquake Park." Downtown Anchorage was also disrupted as sections of the main business district dropped by as much as 3 meters.

Fire

More than a century ago, San Francisco was the economic center of the western United States, largely because of gold and silver mining. Then, at dawn on April 18, 1906, a violent earthquake struck, triggering an enormous firestorm (see Figure 5.2). Much of the city was reduced to ashes and ruins. It is estimated that 3000 people died and more than half of the city's 400,000 residents were left homeless.

The historic San Francisco earthquake reminds us of the formidable threat of fire, which started when the quake severed gas and electrical lines. The initial ground shaking broke the city's water lines into hundreds of disconnected pieces, which made controlling the fires virtually impossible. The fires, which raged out of control for 3 days, were finally contained when expensive houses along Van Ness Avenue were dynamited to provide a fire break, similar to the strategy used in fighting forest fires.

While few deaths were attributed to the San Francisco fires, other earthquake-initiated fires have been more destructive, claiming many more lives. For example, the 1923 earthquake in Japan triggered an estimated 250 fires, devastating the city of Yokohama and destroying more than half the homes in Tokyo. More than 100,000 deaths were attributed to the fires, which were driven by unusually high winds.

Tsunamis

Major undersea earthquakes may set in motion a series of large ocean waves that are known by the Japanese name **tsunami** ("harbor wave"). Most tsunamis are generated by displacement along a megathrust fault that suddenly lifts a large slab of seafloor (Figure 5.26). Once generated, a tsunami resembles a series of ripples formed when a pebble is dropped into a pond. In contrast to ripples, however, tsunamis advance across the ocean at amazing speeds, about 800 kilometers (500 miles) per hour—equivalent to the speed of a commercial airliner. Despite this striking characteristic, a tsunami in the open ocean

SmartFigure 5.25 Turnagain Heights slide caused by the 1964 Alaska earthquake (Photo courtesy of USGS)

TUTORIAL
https://goo.gl/q3Pw2q

A. Vibrations from the Alaska earthquake caused cracks to appear near the edge of the Turnagain Heights bluff.

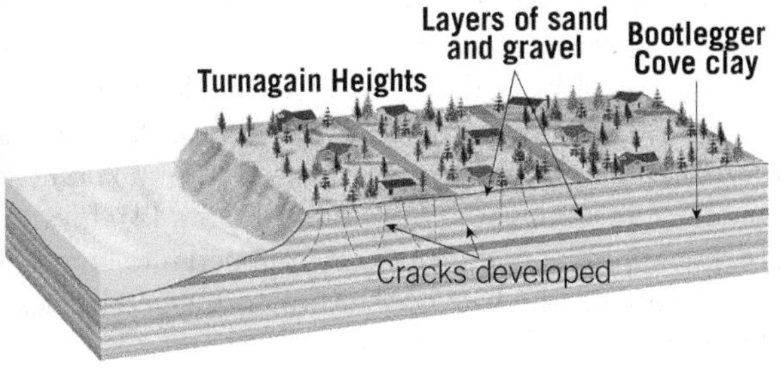

B. Blocks of land began to slide toward the sea on a weak layer called the Bootlegger Cove clay and in less than 5 minutes, as much as 200 meters of the Turnagain Heights bluff area had been destroyed.

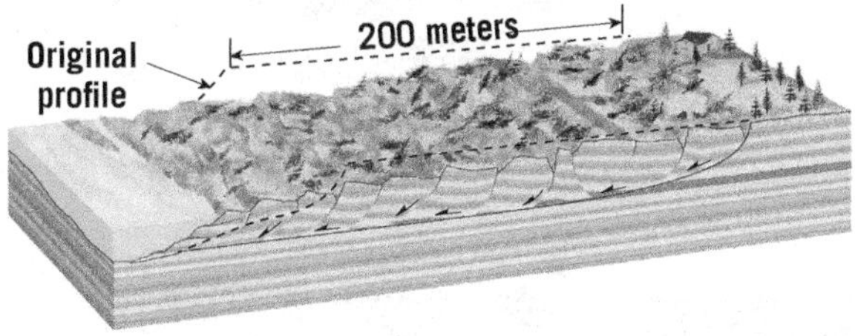

C. Photo of a small area of destruction caused by the Turnagain Heights slide.

A. Before the Earthquake: Along subduction zones the megathrust fault located between the subducting plate and the overriding plate can remain locked for decades or even centuries. As the subducting plate slowly descends it drags and gradually bends the leading edge of overlying plate, sometimes producing a bulge on the ocean floor.

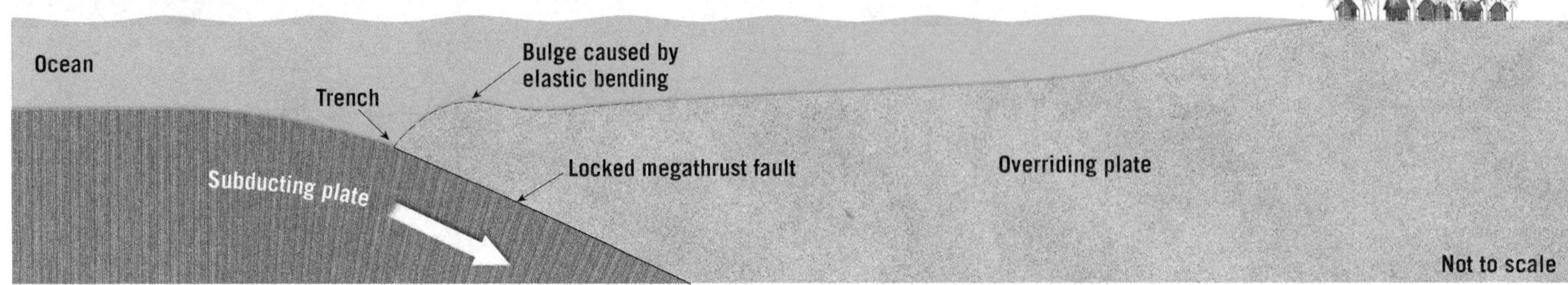

B. During the Earthquake: When the frictional force between the stuck plates is exceeded, the overlying plate lurches seaward and upward. At the same time, the seafloor landward of the zone of displacement stretches and subsides. These rapid up and down vertical motions trigger a tsunami consisting of a train of wave crests and troughs.

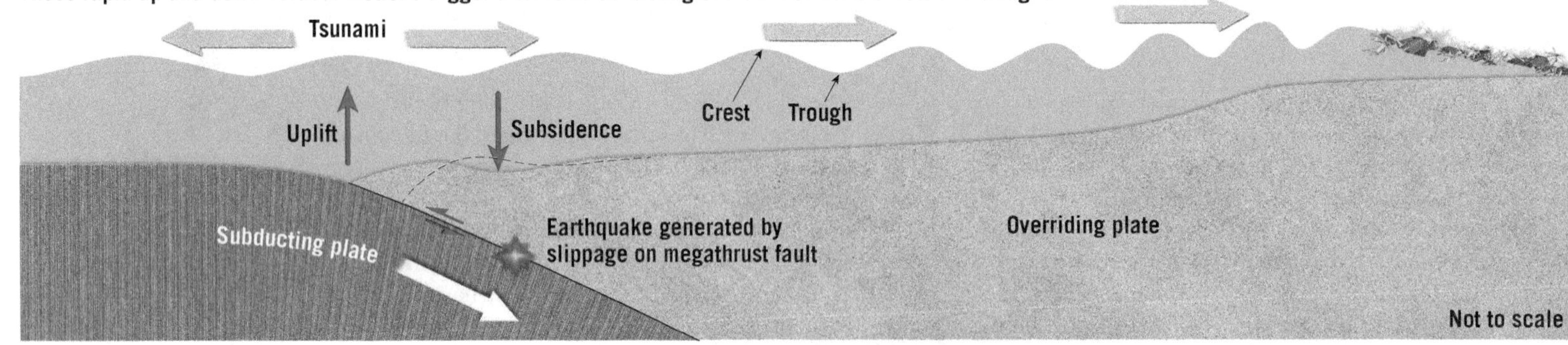

▲ SmartFigure 5.26 **How a tsunami is generated by displacement of the ocean floor during an earthquake** The speed of a tsunami wave correlates with ocean depth. In deep water, these waves can advance at speeds exceeding 800 kilometers (500 miles) per hour. When they enter coastal waters and begin to "feel bottom," they slow down and grow in height. They are still very fast moving, with a speed of 50 kilometers (30 miles) per hour at a depth of 20 meters (65 feet). The size and spacing of the swells in this figure are not to scale.

TUTORIAL
https://goo.gl/glpMvu

can pass undetected because its height (amplitude) is usually less than 1 meter (3 feet), and the distance separating wave crests ranges from 100 to 700 kilometers (60 to 425 miles). However, upon entering shallow coastal waters, these destructive waves "feel bottom" and slow, causing the water to pile up (see Figure 5.26). A few exceptional tsunamis have approached 20 meters (65 feet) in height. As the crest of a tsunami approaches the shore, it appears as a rapid rise in sea level with a turbulent and chaotic surface; it usually does not resemble a breaking wave (Figure 5.27).

▼ SmartFigure 5.27 **Tsunami generated off the coast of Sumatra, 2004** (AFP/Getty Images, Inc.)

ANIMATION
https://goo.gl/WXeRi6

The first warning of an approaching tsunami is often the rapid withdrawal of water from beaches, which is the result of the trough of the first large wave preceding the crest. Some inhabitants of the Pacific basin have learned to heed this warning and quickly move to higher ground. Approximately 5 to 30 minutes after the retreat of water, a surge capable of extending several kilometers inland occurs. In a successive fashion, each surge is followed by a rapid oceanward retreat of the sea. Therefore, people experiencing a tsunami should not return to the shore when the first surge of water retreats.

Tsunami Damage from the 2004 Indonesia Earthquake A massive undersea earthquake of M_W 9.1 occurred near the island of Sumatra on December 26, 2004, sending waves of water racing across the Indian Ocean and Bay of Bengal. It was one of the deadliest natural disasters of any kind in modern times, claiming more than 230,000 lives. As water surged several kilometers inland, cars and trucks were flung around like toys in a bathtub, and fishing boats were rammed into homes. In some locations, the backwash of water dragged bodies and huge amounts of debris out to sea.

Table 5.2 Some Notable Earthquakes

Year	Location	Deaths (est.)	Magnitude*	Comments
856	Iran	200,000		
893	Iran	150,000		
1138	Syria	230,000		
1268	Asia Minor	60,000		
1290	China	100,000		
1556	Shensi, China	830,000		Possibly the greatest natural disaster
1667	Caucasia	80,000		
1727	Iran	77,000		
1755	Lisbon, portugal	70,000		Tsunami damage extensive
1783	Italy	50,000		
1908	Messina, Italy	120,000		
1920	China	200,000	7.5	Landslide buried a village
1923	Tokyo, Japan	143,000	7.9	Fire caused extensive destruction
1948	Turkmenistan	110,000	7.3	Almost all brick buildings near epicenter collapsed
1960	Southern Chile	5700	9.5	The largest-magnitude earthquake ever recorded
1964	Alaska	131	9.2	Greatest-magnitude North american earthquake
1970	Peru	70,000	7.9	Great rockslide
1976	Tangshan, China	242,000	7.5	Estimates for the death toll are as high as 655,000
1985	Mexico City	9500	8.1	Major damage occurred 400 km from epicenter
1988	Armenia	25,000	6.9	Poor construction practices increased the death toll
1990	Iran	50,000	7.4	Landslides and poor construction practices led to great damage
1993	Latur, India	10,000	6.4	Located in stable continental interior
1995	Kobe, Japan	5472	6.9	Damages estimated to exceed $100 billion
1999	Izmit, Turkey	17,127	7.4	Nearly 44,000 injured and more than 250,000 displaced
2001	Gujarat, India	20,000	7.9	Millions homeless
2003	Bam, Iran	31,000	6.6	Ancient city with poor construction
2004	Indian Ocean (Sumatra)	230,000	9.1	Devastating tsunami damage
2005	Pakistan/Kashmir	86,000	7.6	Many landslides; 4 million homeless
2008	Sichuan, China	87,000	7.9	Millions homeless, some towns will not be rebuilt
2010	Port-au-Prince, Haiti	250,000–316,000	7	More than 300,000 injured and 1.3 million homeless
2011	Japan	16,000	9	Majority of the casualties due to a tsunami
2015	Nepal	8000	7.8	About 21,000 injured
2016	Ecuador	700	7.8	More than 28,000 injuries

*Widely differing magnitudes and death tolls have been estimated for some of these earthquakes, particularly those events that occurred prior to the 18th century. When available, moment magnitudes are used.

Source: U.S. Geological Survey.

The destruction was indiscriminate, destroying luxury resorts as well as poor fishing hamlets along the Indian Ocean. Damage was reported as far away as the coast of Somalia in Africa, 4100 kilometers (2500 miles) west of the earthquake epicenter.

Japan Tsunami Because of Japan's location along the circum-Pacific belt and its extensive coastline, it is especially vulnerable to tsunami destruction. The most powerful earthquake to strike Japan in the age of modern seismology was the 2011 Tohoku earthquake (M_W 9.0). This historic earthquake and devastating tsunami resulted in at least 15,890 deaths, more than 3000 people missing, and 6107 injured. Nearly 400,000 buildings, 56 bridges, and 26 railways were destroyed or damaged (Table 5.2).

The majority of human casualties and damage after the Tohoku earthquake were caused by a Pacific-wide tsunami that reached a maximum height of about 8.5 meters (28 feet) and traveled inland 10 kilometers (6 miles) in the region of Sendai, Japan (Figure 5.28). Based on physical evidence, the run-up height, which is the maximum elevation on land that is inundated by water from a tsunami, was almost twice as high—15 meters (50 feet) above sea level. In addition, the tsunami disabled the power supply and cooling mechanisms, which caused the meltdown of three inundated nuclear reactors in Japan's Fukushima Daiichi Nuclear Complex.

Across the Pacific in California, Oregon, Peru, and Chile, some loss of life occurred, and several houses, boats, and docks were destroyed. The tsunami was

▲ **Figure 5.28 Japan tsunami, March 2011** This tsunami breached a seawall and devastated the city of Miyako, Japan, shortly after a magnitude 9.0 earthquake hit northern Japan in 2011. (JIJI PRESS/AFP/Getty Images)

generated when a slab of seafloor located 60 kilometers (37 miles) off the east coast of Japan was suddenly "thrust up" an estimated 5 to 8 meters (16 to 26 feet).

Tsunami Warning System In 1946, a large tsunami struck the Hawaiian Islands without warning. A wave more than 15 meters (50 feet) high left several coastal villages in shambles. This destruction motivated the U.S. Coast and Geodetic Survey to establish a tsunami warning system for coastal areas of the Pacific that today includes 26 countries. Seismic observatories throughout the region report large earthquakes to the Tsunami Warning Center in Honolulu. Scientists at the center use deep-sea buoys equipped with pressure sensors to detect energy released by an earthquake. In addition, tidal gauges measure the rise and fall in sea level that accompany tsunamis, and warnings are issued within an hour. Although tsunamis travel very rapidly, there is sufficient time to warn all except those in the areas nearest the epicenter. For example, a tsunami generated near the Aleutian Islands would take 5 hours to reach Hawaii, and one generated near the coast of Chile would travel 15 hours before reaching the shores of Hawaii (Figure 5.29).

▼ **Figure 5.29 Tsunami travel times** Travel times to Honolulu, Hawaii, from selected locations throughout the Pacific. (Data from NOAA)

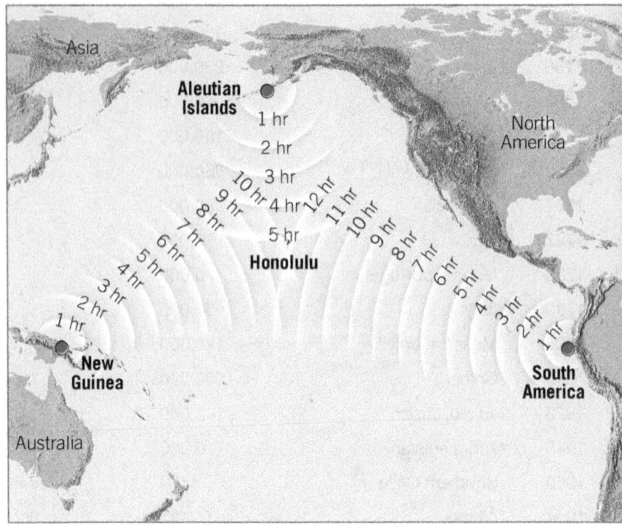

CONCEPT CHECKS 5.5

1. List four factors that influence the amount of destruction that seismic vibrations cause to human-made structures.
2. In addition to the destruction created directly by seismic vibrations, list three other types of destruction associated with earthquakes.
3. What is a tsunami? How are tsunamis generated?
4. List at least three reasons an earthquake with a magnitude of 7.0 might cause more death and destruction than a quake with a magnitude of 8.0.

5.6 Where Do Most Earthquakes Occur?

Locate Earth's major earthquake belts on a world map.

About 95 percent of the energy released by earthquakes originates in a few relatively narrow zones, shown in Figure 5.30. These zones of earthquake activity are located along fault surfaces where tectonic plates interact along one of the three types of plate boundaries—convergent, divergent, and transform plate boundaries.

Earthquakes Associated with Plate Boundaries

The previous section described the relationship between plate tectonics and seismic activity. The zone of greatest seismic activity, called the **circum-Pacific belt**, encompasses the coastal regions of Chile, Central America, Indonesia, Japan, and Alaska, including the Aleutian Islands (see Figure 5.30). Most of the large earthquakes in the circum-Pacific belt occur along convergent plate boundaries where one plate subducts beneath another at a comparatively low angle. Recall that the contacts between the subducting and overlying plates are called *megathrust faults* (see Figure 5.5). There are more than 40,000 kilometers (25,000 miles) of subduction boundaries in the circum-Pacific belt where displacement is dominated by thrust faulting. Ruptures occasionally occur along segments that are nearly 1000 kilometers (600 miles) long, generating catastrophic earthquakes with magnitudes of 8 or greater.

Another major concentration of strong seismic activity, referred to as the *Alpine–Himalayan belt*, runs through the mountainous regions that flank the Mediterranean Sea and extends past the Himalaya Mountains (see Figure 5.30). Tectonic activity in this region is mainly attributed to collisions of the African plate and the Indian subcontinent with the vast Eurasian plate (see Figure 4.18, page 109). These plate interactions created many thrust and strike-slip faults that remain active.

Transform fault boundaries, which are large strike-slip faults, are also a source of strong earthquakes. Examples include California's San Andreas Fault, New Zealand's Alpine Fault, and Turkey's North Anatolian Fault, which produced a deadly earthquake in 1995.

Figure 5.30 shows another continuous earthquake belt that extends thousands of kilometers through the world's oceans. This zone coincides with the oceanic ridge system—a divergent plate boundary—which is an area of frequent but weak seismic activity.

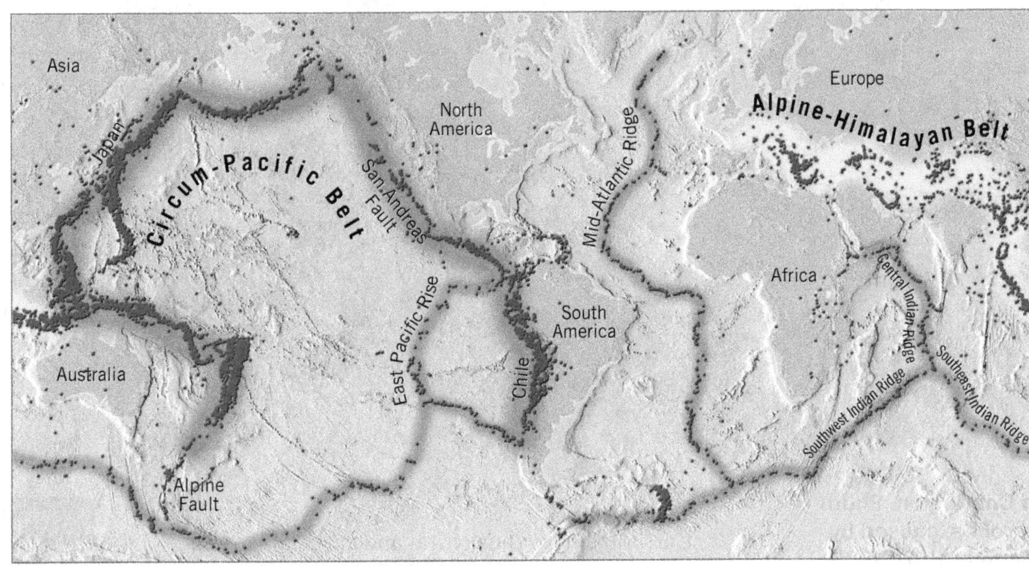

▲ **Figure 5.30 Global earthquake belts** Distribution of nearly 15,000 earthquakes with magnitudes equal to or greater than 5 for a 10-year period, indicated by red dots. (Data from USGS)

Damaging Earthquakes East of the Rockies

When you think "earthquake," you probably think of the western United States or Japan. However, six major earthquakes and several others that inflicted considerable damage have occurred in the central and eastern United States since colonial times (**Figure 5.31**).

Three of these quakes, which occurred as a cluster, had estimated magnitudes of about 7.0 and destroyed what was then the frontier town of New Madrid, Missouri, located in the Mississippi River valley. It has been estimated that if an earthquake the size of the 1811–1812 New Madrid event were to strike in the same location in the next decade, it would result in casualties in the thousands and damages in the tens of billions of dollars.

The greatest historical earthquake in the eastern states occurred on August 31, 1886, in Charleston, South Carolina. This earthquake resulted in 60 deaths, numerous injuries, and great economic loss over an area extending 200 kilometers (120 miles) from Charleston. Within 8 minutes, effects were felt as far away as Chicago and St. Louis, where strong vibrations shook the upper floors of buildings, causing people to rush outdoors. In Charleston alone, more than 100 buildings were destroyed, and 90 percent of the remaining structures were damaged (**Figure 5.32**).

▼ **Figure 5.31 Historical earthquakes east of the Rockies** Large earthquakes are uncommon in the middle of continents, far from the places where plates collide or grind past one another or where one plate slides beneath another. Nevertheless, several damaging earthquakes have occurred in the central and eastern United States since colonial times.

Historic Earthquakes East of the Rockies 1755–2016

	LOCATION	DATE	INTENSITY	MAGNITUDE*	COMMENTS
1	East of Oklahoma City	2011	VII	5.6	Fourteen homes destroyed
2	Mineral, Virginia	2011	VII	5.8	Felt by many
3	Southeastern Illinois	2008	VII	5.4	Occurred along the Wabash Valley Seismic Zone
4	Northeast Kentucky	1980	VII	5.2	Largest earthquake ever recorded in Kentucky
5	Merriman, Nebraska	1964	VII	5.1	Largest earthquake ever recorded in Nebraska
6	Northern New York	1944	VIII	5.8	Left several structures unsafe for occupancy
7	Ossipee Lake, New Hampshire	1947	VII	5.5	Two earthquakes occurred four days apart
8	Western Ohio	1937	VIII	5.4	Extensive damage to plaster walls
9	Valentine, Texas	1931	VIII	5.8	Brick buildings were severely damaged
10	Giles County, Virginia	1897	VIII	5.9	Changed the flow of natural springs
11	Charleston, Missouri	1895	VIII	6.6	Structural damage and liquefaction reported
12	Charleston, South Carolina	1886	X	7.3	Caused 60 deaths, destroyed many buildings
13	New Madrid, Missouri	1811-1812	X	7.0	Three strong earthquakes occurred
14	Cape Ann, Massachusetts	1755	VIII	?	Buildings damaged in Boston

Source: U.S. Geological Survey
*Intensity and magnitudes have been estimated for many of these events.

Figure 5.32 **Damage to Charleston, South Carolina, caused by the August 31, 1886, earthquake** (Photo courtesy of USGS)

Earthquakes in the central and eastern United States occur far less frequently than in California, yet history indicates that the East is vulnerable. Further, these shocks east of the Rockies have generally produced structural damage over a larger area than earthquakes of similar magnitude in California. This is because the underlying bedrock in the central and eastern United States is older and more rigid. As a result, seismic waves can travel greater distances with less attenuation (loss of strength) than in the western United States.

Earthquakes that occur away from plate boundaries are called *intraplate earthquakes*. Intraplate earthquakes can be caused by a variety of factors. For example, stress can rejuvenate ancient fault systems that formed when crustal fragments collided to generate the continent billions of years ago (see Figure 12.12, page 392). In addition, the process called fracking, in which a solution is injected into the ground under high pressure to enhance oil and gas production, has contributed to the recent increase in earthquake activity east of the Rockies. Most of these intraplate earthquakes are weak.

CONCEPT CHECKS 5.6

1. What zone on Earth has the greatest amount of seismic activity?
2. What type of plate boundary is associated with Earth's largest earthquakes?
3. Explain why an earthquake east of the Rockies may produce damage over a larger area than one of similar magnitude in California.

5.7 Earthquakes: Predictions, Forecasts, and Mitigation

Compare and contrast the goals of short-range earthquake predictions and long-range forecasts.

Figure 5.33 **Collapse of the double-decked section of I-880** This section of a double-decked highway, known as the Cypress Viaduct, collapsed during the 1989 Loma Prieta earthquake in California. (Photo by Paul Sakuma/AP Photo)

Strong earthquakes cause damage and loss of life primarily because they usually strike without warning. For example, the vibration that unexpectedly shook the San Francisco area in 1989 caused 63 deaths, heavily damaged the Marina District, and caused the collapse of a double-decked section of I-880 in Oakland, California (Figure 5.33). This level of destruction was the result of an earthquake of moderate magnitude (M_W 6.9). Seismologists warn that other earthquakes of comparable or greater strength can be expected along the San Andreas system, which cuts a nearly 1300-kilometer (800-mile) path through the western one-third of the state (see GEOgraphics 5.1, page 152). An obvious question is: Can earthquakes be predicted?

Short-Range Predictions

It would be extremely desirable if imminent earthquakes could be predicted in the way in which volcanic eruptions sometimes can. Unfortunately, despite substantial efforts in earthquake-prone countries such as Japan, the United States, China, and Russia, no reliable method exists at present for making short-range earthquake predictions.

To be useful, such a method would have to be both accurate and reliable. That is, *it must have a small range of uncertainty with regard to location and timing, and it must produce few failures or false alarms.* Can you imagine the debate that would precede an order to evacuate a large U.S. city, such as Los Angeles or San Francisco, and the expense of doing so?

Research on short-range prediction has concentrated on monitoring possible **precursors**—events or changes that precede an earthquake and thus might provide a warning. In California, for example, seismologists monitor changes in ground elevation and variations in strain levels near active faults. Other researchers measure changes in groundwater levels, while still others try to predict earthquakes based on an increase in the frequency of the foreshocks that precede some, but not all, earthquakes. To date, neither these nor other avenues of research have yielded a reliable method for short-range earthquake prediction.

Long-Range Forecasts

In contrast to short-range predictions, which aim to predict earthquakes within a time frame of hours or days, long-range forecasts are estimates of the likelihood that an earthquake of a certain magnitude will occur in a given place on a time scale of 30 to 100 years or more. Although long-range forecasts are not as informative as we might like, these data provide important guides for building codes so that buildings, dams, and roadways are constructed to withstand expected levels of ground shaking.

Most long-range forecasting strategies are based on the observation that large faults often break in a cyclical manner, producing similar quakes at roughly similar intervals. In other words, as soon as a section of a fault ruptures, the continuing motions of Earth's plates begin to deform (bend) the rocks until they fail (rupture) once more. To determine the characteristic interval for a given fault, seismologists study the historical and geologic record of earthquakes generated by that fault.

Seismic Gaps When seismologists mapped the rupture zones associated with great earthquakes around the globe, they discovered that these zones tend to lie next to each other, without appreciable overlap. A segment that has not experienced an earthquake within the past one to several centuries is called a **seismic gap**. Typically these gaps are accumulating seismic strain and thus represent places where future large earthquakes are likely to occur. Figure 5.34 shows a seismic gap on the megathrust fault that lies offshore of Padang, a low-lying Sumatran city of 800,000 people, which has not ruptured since 1797. Scientists are particularly concerned about this seismic gap because rupture of an adjacent segment of the fault caused the 2004 Indian Ocean (Sumatra) earthquake and tsunami that claimed 230,000 lives.

Paleoseismology **Paleoseismology** (*paleo* = ancient, *seismos* = shake), is the study of the timing, location, and size of prehistoric earthquakes. Paleoseismology studies are often conducted by digging a trench across a suspected fault zone and then looking for evidence of ancient faulting, such as offset sedimentary strata or mud volcanoes. A large vertical offset of the layers of accumulated sediments indicates a large earthquake. Sometimes buried plant debris can be carbon dated, allowing for the timing of recurrence to be established.

One investigation that used this method focused on a segment of the San Andreas Fault that lies north and east of Los Angeles. At this site, the drainage of Pallet Creek has been repeatedly disturbed by successive ruptures along the fault zone (Figure 5.35). Trenches excavated across the creek bed have exposed sediments that have been displaced by several large earthquakes over a span of 1500 years. From these data, researchers determined that strong earthquakes occur an average of once every 135 years. The last major event, the Fort Tejon earthquake, occurred on this segment of the San Andreas Fault in 1857, roughly 150 years ago. Because earthquakes occur on a cyclical basis, a major event in southern California may be imminent.

Paleoseismology has also revealed that powerful earthquakes (magnitude 8 or larger) and associated tsunamis have repeatedly struck the coastal Pacific Northwest over the past several thousand years. Each of these events is attributed to slippage along a section of the megathrust fault associated with the Cascadia subduction zone located off the west coast from southern British Columbia to northern California. The most recent event, which occurred in January 1700, generated a destructive tsunami that caused massive flooding in the coastal lowlands of western North America and was recorded as far away as Japan.

As a result of these findings, public officials have updated building codes and retrofitted some of the region's existing buildings, dams, bridges, and water systems to be more earthquake resistant. One notable example is the recent action taken by the Ocosta

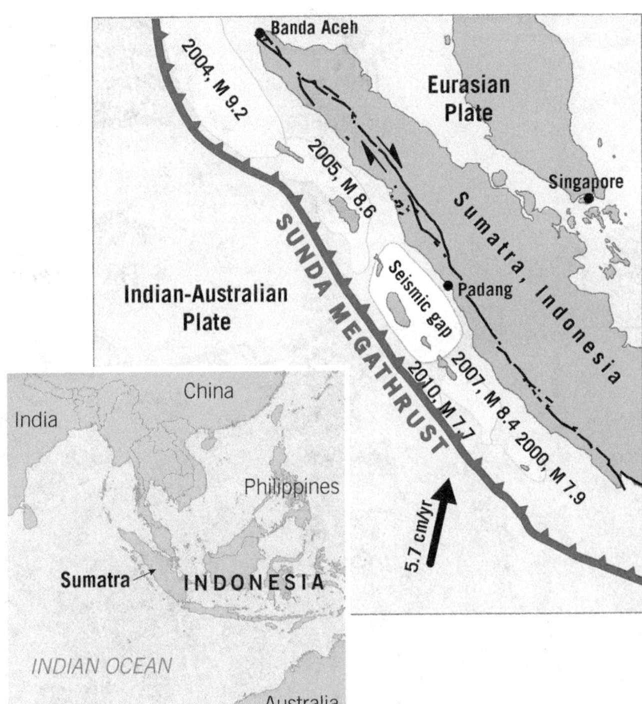

Figure 5.34 Seismic gaps: Tools for forecasting earthquakes Seismic gaps are "quiet zones" that are thought to be storing elastic strain and will eventually produce major earthquakes. This seismic gap is located near Padang, a low-lying coastal city with a population of 800,000 people.

GEOGRAPHICS 5.1

Seismic Risks on the San Andreas Fault System

California's San Andreas Fault runs diagonally from southeast to northwest for nearly 1300 kilometers (800 miles) through much of the western part of the state. For years researchers have been trying to predict the location of the next "Big One"—an earthquake with a magnitude of 8 or greater—along this fault system.

CALIFORNIA

The 1906 San Francisco earthquake caused displacement on the northernmost section of the fault. This event, likely relieved much of the strain that had been building during the previous 200 years or so.

1906 epicenter

San Francisco

Located just south of the 1906 rupture is a section of the San Andreas Fault that exhibits fault creep. When plates gradually slide past each other, less strain accumulates than when the fault is locked.

1857 epicenter

This 300-kilometer-long section of the San Andreas Fault System produced the Fort Tejon earthquake of 1857. Because a portion of the fault has likely accumulated considerable strain since the Fort Tejon quake, the U.S. Geological Survey gives it a 60 percent probability of producing a major earthquake in the next 30 years.

Los Angeles

The next major quake on the San Andreas may well be on its southernmost 200 kilometers—an area that has not produced a large event in about 300 years.

The 1906 San Francisco earthquake was the most devastating in California's history. The quake and resulting fires caused an estimated 3000 deaths and extensively damaged buildings throughout the city.

Tom McHugh/Science Source

Adam Teitelbaum/AFP/Getty Images

On October 17, 1989, millions of television viewers around the world were settling in to watch the third game of the World Series. Instead, they saw their TVs go to black as tremors hit San Francisco's Candlestick Park. Although the Loma Prieta earthquake was centered in a remote section of the Santa Cruz Mountains (100 miles to the south) major damage occurred in the Marina District of San Francisco.

SAN FRANCISCO BAY REGION EARTHQUAKE PROBABILITY

62%

Question:
The section of the San Andreas Fault located south of the San Francisco Bay area exhibits fault creep. Does that section of the fault have a high or low probability of generating a large earthquake? Explain.

?

The U. S. Geological Survey concluded that between 2003 and 2032 there is a 62 percent probability of at least one magnitude 6.7 or greater earthquake striking somewhere in the San Francisco Bay area.

% Probability of magnitude 6.7 or greater quake before 2032 on the indicated faults

Increasing probability along fault segments

Rogers Creek fault
Napa
Concord-Green Valley fault
4%
27%
San Andreas fault
Walnut Creek
Greenville fault
Mt. Diablo thrust fault
Hayward fault
San Francisco
Oakland
3%
21%
San Francisco Bay
3%
Calaveras fault
Palo Alto
10%
San Jose
11%
San Gregorio fault
EXTENT OF RUPTURE IN THE LOMA PRIETA QUAKE
PACIFIC OCEAN
Monterey Bay
Monterey

In mid-January 1994, less than five years after the Loma Prieta event, the Northridge earthquake struck an area slightly north of Los Angeles. This moderate 6.7-magnitude earthquake claimed the lives of 57 people. Nearly 300 schools were severely damaged, and one dozen major roadways buckled. Among these were two of California's major arteries—sections of the Santa Monica Freeway and the Golden State Freeway (I-5) where an overpass collapsed completely and blocked the highway.

Tom McHugh/Photo Researchers, Inc.

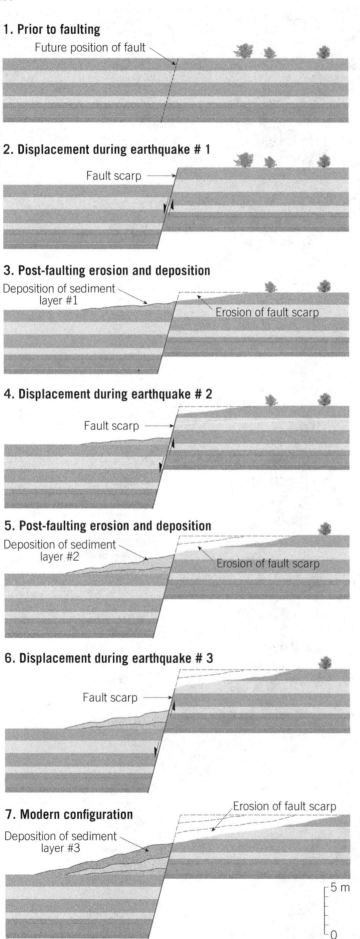

1. Prior to faulting

Future position of fault

2. Displacement during earthquake # 1

Fault scarp

3. Post-faulting erosion and deposition

Deposition of sediment layer #1

Erosion of fault scarp

4. Displacement during earthquake # 2

Fault scarp

5. Post-faulting erosion and deposition

Deposition of sediment layer #2

Erosion of fault scarp

6. Displacement during earthquake # 3

Fault scarp

7. Modern configuration

Erosion of fault scarp

Deposition of sediment layer #3

5 m

0

A.

B. The events depicted in the accompanying diagrams were deciphered by digging a trench (shown here) across the fault zone and studying the displaced sedimentary beds.

Figure 5.35 **Paleoseismology: The study of prehistoric earthquakes** This study was conducted in the Pallet Creek area by digging a trench (B) across a branch of the San Andreas Fault and then looking for evidence of ancient displacements, such as offset sedimentary strata. This simplified diagram shows that vertical displacement occurred on this fault three different times, with each event producing an earthquake. Based on the size of the vertical displacements, these ancient earthquakes had estimated magnitudes between 6.8 and 7.4. (Photo courtesy of USGS)

School District of southwestern Washington State.* This coastal community, located in an area inundated by the 1700 tsunami, took the initiative to build North America's first engineered tsunami refuge atop a new elementary school. The structure consists of a flat roof about 10 meters (30 feet) above ground, accessible from four heavily reinforced stairways. The rooftop has space for roughly 2000 occupants from the school and surrounding residences. This initiative took cues from the 2011 Japan tsunami, but more importantly, was founded on scientific discoveries about Cascadia's earthquake and tsunami risks.

Minimizing Earthquake Hazards

The statement "earthquakes don't kill people; buildings kill people" succinctly expresses the fact that falling structures are by far the greatest cause of casualties

*Based on an article by Marcia McNutt while she was editor-in-chief of *Science*.

▲ Figure 5.36 **Retrofitted building** Webb Tower, located on the campus of the University of Southern California, was retrofitted to help the structure withstand shaking in the event of a strong earthquake. Notice the cross braces on the building's exterior. (Photo by Mel Melcon/Los Angeles Time/Getty Images)

during an earthquake. Thus, in regions where there is a known earthquake hazard, houses, bridges, dams, and other structures should be constructed to withstand at least moderate shaking. In addition, communities need to adopt building codes that require inspection and retrofitting of existing structures to withstand seismic shaking (Figure 5.36). But due to prohibitive costs, the important step of retrofitting existing structures is often not a high priority. For example, a recent study indicated that more than half of the hospitals in Los Angeles County would likely collapse in a strong earthquake.

Earthquake-Resistant Structures The importance of designing new buildings to resist earthquakes and retrofitting older structures is perhaps best illustrated by comparing two earthquakes of similar magnitude—the 1988 Armenian earthquake (M 6.8) and the 1989 San Francisco earthquake (M 6.9). Most of the buildings leveled by the Armenian quake were constructed of unreinforced concrete, which collapsed into rubble—killing an estimated 25,000 people (Figure 5.37). Although the 1989 San Francisco earthquake was very destructive, the death toll (63 lives) was 400 times lower than that of the Armenian quake. The difference was due mainly to building practices; in California, most buildings are either wood framed or built with reinforced concrete that resists collapse, whereas in Armenia the building that collapsed where build of unreinforced concrete. In areas that are not seismically active but have the potential for experiencing a strong earthquake, the challenge is to establish appropriate building codes and educate residents on proactive measures to minimize casualties and property damage in the event of a strong earthquake.

Although the collapse of buildings is the largest cause of earthquake destruction, other hazards, such as ground subsidence caused by liquefaction, landslides, and tsunamis, can also be devastating. The catastrophic 2011 Japan (Tohoku) earthquake (M 9.0) is a sobering reminder of this fact. Because of Japan's strict building codes, the buildings, bridges, and other structures built or retrofitted since 1995 withstood the ground shaking of this powerful earthquake extremely well; only a few buildings collapsed as a direct result of shaking. It was the massive tsunami triggered by the quake that claimed the lives of 93 percent of the estimated 16,000 people who perished in this earthquake (see Table 5.2). Ground subsidence caused by liquefaction and landslides also caused significant damage to buildings, roads, and utilities such as water and gas pipes.

Japan's earthquake preparedness is arguably the best in the world, in large part because it has been struck by 19 major quakes (M 7–7.9) in the past 2 decades. Had the 2011 Japan earthquake occurred in a region where structures are less well engineered, the casualties attributable to collapse caused by shaking would have been considerably higher.

Earthquake Warning Systems Because P waves are less destructive and travel faster than both S waves and surface waves, they can be used to provide a type of earthquake warning system. Japan has operated such a system for over 20 years. One use of this system is to trigger automated shutdown of power-generating facilities and braking of high-speed bullet trains.

In addition, during the 2011 Tohoku earthquake, this system was used to send an automated alert to the nation's television stations as well as to more than 50 million telephones. In the area closest to the epicenter, this alert gave residents about a 10-second chance to take cover. In Tokyo, a city of around 9 million people

▼ Figure 5.37 **Poorly constructed buildings destroyed during the 1988 Armenian earthquake** (Photo by Peter Turnley/Getty Image)

located about 370 kilometers (230 miles) south of the epicenter, citizens had about an 80-second warning before the strong shaking began.

Despite the potential utility of early warning earthquake systems, none exist in North America. Some argue that false alarms could potentially cause more havoc than the benefits derived from what is admittedly a very short warning time.

CONCEPT CHECKS 5.7

1. Are accurate, short-range earthquake predictions currently possible using modern seismic instruments? Explain.
2. What is the value of long-range earthquake forecasts?
3. Which hazard usually causes the most casualties during an earthquake?

5.8 Earth's Interior

Explain how Earth acquired its layered structure and name and describe each of its major layers.

If we could slice Earth in half, the first thing we would notice is that it has distinct layers. The heaviest materials (metals) are in the center. Less dense solids (rocks) are in the middle, and lighter liquids (mainly water) and gases are on top. Within Earth we know these layers as the iron-rich core, the rocky mantle and crust, the liquid ocean, and the gaseous atmosphere.

There are also variations in Earth's composition and temperature with depth, which indicate that the interior of our planet is very dynamic. The rocks of the mantle and crust are in constant, if very slow, motion. In addition to plate tectonics, these motions cause continuous recycling between the surface and the deep interior. Furthermore, it is from Earth's deep interior that the water and air of our oceans and atmosphere are replenished, allowing life to exist at the surface.

Formation of Earth's Layered Structure

According to the *nebular theory* (see Chapter 1, Section 1.3), Earth and the solar system began to form nearly 5 billion years ago due to the gravitational collapse of a huge cloud of dust and gases called a *nebula*. As material accumulated to form Earth (and for a short period afterward), the high-velocity impact of nebular debris and the decay of radioactive elements caused the temperature of our planet to increase steadily. During this time of intense heating, Earth became hot enough to melt iron and nickel. Melting produced dense blobs of liquid metal that sank toward the center of the planet. This process occurred rapidly on the scale of geologic time and produced Earth's dense iron-rich core.

The early period of heating resulted in another process, called *chemical differentiation*, whereby melting formed buoyant masses of molten rock that rose toward Earth's surface and solidified to produce a primitive crust. These rocky materials were rich in oxygen and "oxygen-seeking" elements, particularly silicon and aluminum, along with lesser amounts of calcium, sodium, potassium, iron, and magnesium. In addition, some heavy metals such as gold, lead, and uranium, which have low melting points or were highly soluble in the ascending molten masses, were scavenged from Earth's interior and concentrated in the developing crust. This early period of chemical differentiation established the three basic divisions of Earth's interior: (1) the thin *primitive crust*, (2) the iron-rich *core*, and (3) Earth's largest layer, called the *mantle*, which is located between the crust and core (see Figure 5.39).

Probing Earth's Interior: "Seeing" Seismic Waves

How do we know about the structure and properties of Earth's deep interior? Because light does not travel through rock, we must find other ways to "see" into our planet. The best way to learn about Earth's interior is to dig or drill a hole and examine it directly. Unfortunately, this is possible only at shallow depths. The deepest a drilling rig has ever penetrated is only 12.3 kilometers (7.6 miles), which is about 1/500 of the way to Earth's center. This was an extraordinary accomplishment because temperature and pressure increase rapidly with depth.

Instead, much of what we know about Earth's interior comes from detailed studies of seismic waves generated by large earthquakes. About 3000 earthquakes occur each year that are large enough (about M_W 5.5) to travel all the way through Earth and be recorded by seismographs on the other side of the globe (**Figure 5.38**). The P and S waves from these large earthquakes can be used to "see" into our planet in much the same way that medical technology uses ultrasound waves to generate sonograms.

Despite many recent advances, using the waves recorded on seismograms to visualize Earth's interior structure remains challenging. Seismic waves do not travel along straight paths; instead, they are *reflected*, *refracted*, and *diffracted* as they pass through our planet. They reflect off boundaries between different layers, they refract (change direction) when passing from one layer to another layer, and they diffract (follow a curved path) around some obstacles they encounter. These different wave behaviors have been used to identify the boundaries that exist within Earth.

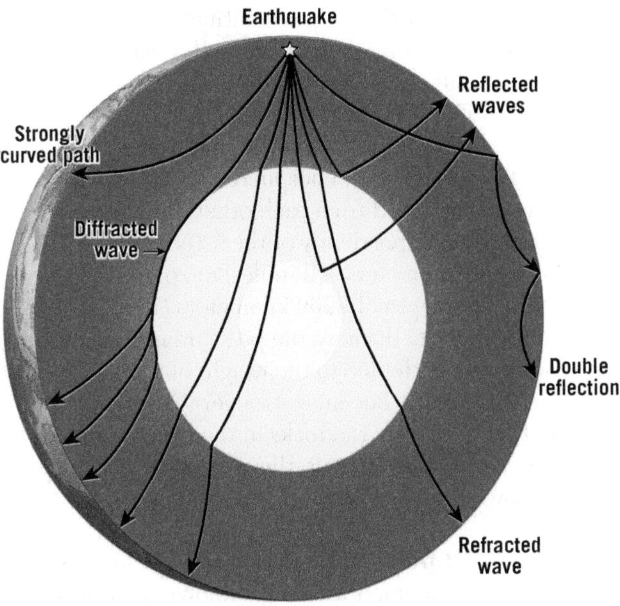

Figure 5.38 **Possible paths that earthquake waves can take** The reason seismic waves follow curved (refracting) paths through Earth's mantle, rather than straight paths, is that the seismic velocity increases with depth due to increasing pressure.

One of the most noticeable behaviors of seismic waves is that they follow strongly curved paths (see Figure 5.38). Seismic waves also travel faster when rock is stiffer or less compressible. These properties of stiffness and compressibility can be used to interpret the composition and temperature of the rock. For instance, when rock is hotter, it becomes less stiff (imagine a chocolate bar left out in the Sun), and waves travel through it more slowly. Waves also travel at different speeds through rocks of different compositions. Thus, the speed at which seismic waves travel can help determine the types of rocks located in Earth's interior, as well as how hot they are.

Earth's Layered Structure

Let's now look in more detail at how Earth's three compositionally distinct layers—the crust, mantle, and core—can be further subdivided into zones based on *physical* properties. The physical properties used to define such regions include whether the layer is solid or liquid and how weak or strong it is. Knowledge of both the chemical composition and physical properties of Earth's layers is essential to our understanding of basic geologic processes, such as volcanism, earthquakes, and mountain building (Figure 5.39).

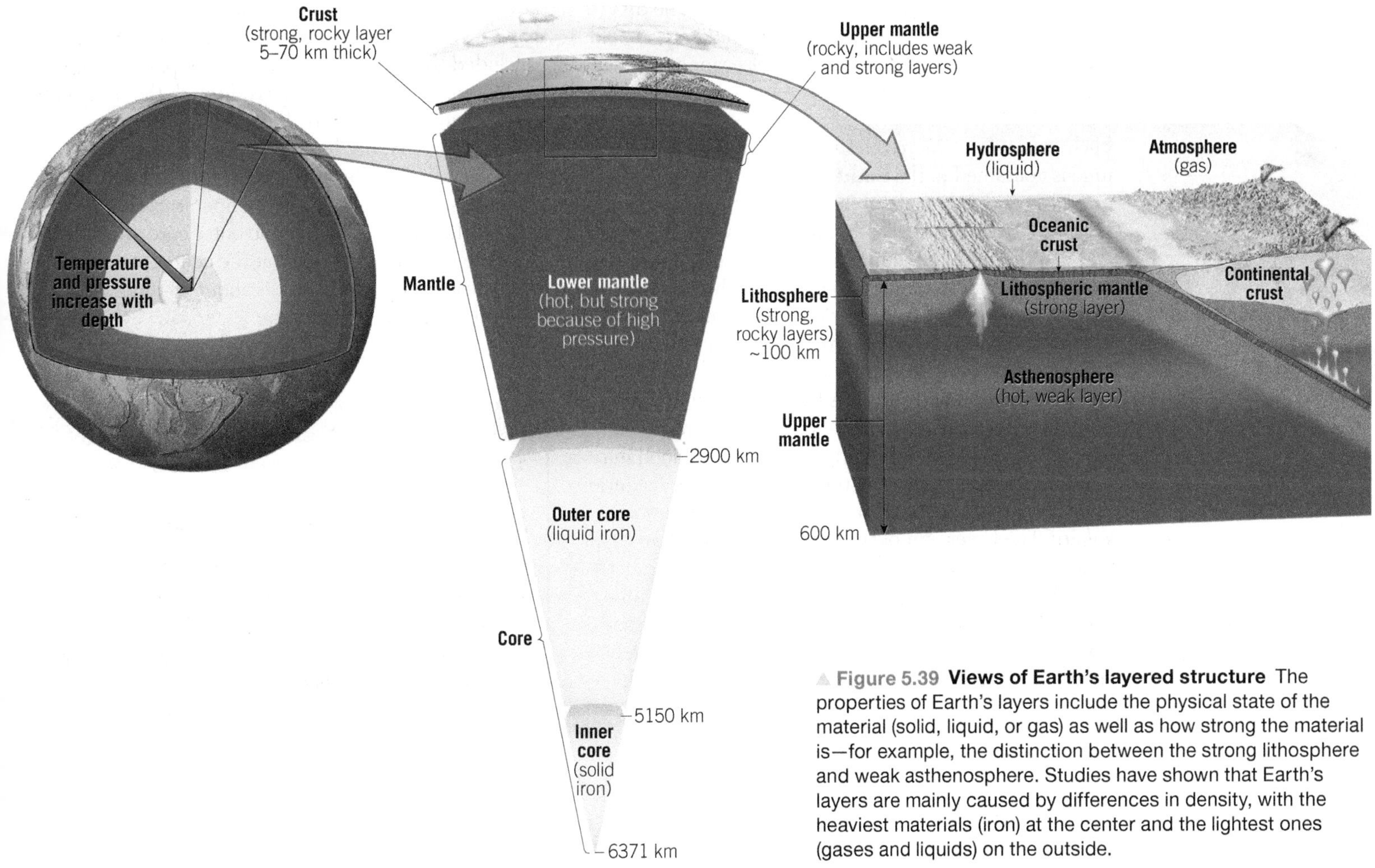

Figure 5.39 **Views of Earth's layered structure** The properties of Earth's layers include the physical state of the material (solid, liquid, or gas) as well as how strong the material is—for example, the distinction between the strong lithosphere and weak asthenosphere. Studies have shown that Earth's layers are mainly caused by differences in density, with the heaviest materials (iron) at the center and the lightest ones (gases and liquids) on the outside.

Earth's Crust The **crust** is Earth's relatively thin, rocky outer skin. The two types of crust—continental and oceanic—have very different compositions, histories, and ages. In fact, oceanic crust is compositionally more similar to Earth's mantle than to its continental crust.

Earth's oceanic crust, which forms along mid-ocean ridges, is about 7 kilometers (4 miles) thick. The rocks of the oceanic crust are younger (about 180 million years old or less) and denser than continental rocks. Ocean crust has a density of about 3.0 g/cm^3 and is composed of the dark igneous rocks *basalt* and *gabbro*.

Unlike oceanic crust, which has a relatively homogeneous chemical composition, continental crust consists of many rock types. Although the upper crust has an average composition of a granitic rock called *granodiorite*, its composition and structure vary considerably from place to place.

Continental crust averages about 40 kilometers (25 miles) thick but can be more than 70 kilometers (40 miles) thick in mountainous regions such as the Himalayas and the Andes. It has an average density of about 2.7 g/cm^3, which is much lower than the density of mantle rock. The low density of the continents relative to the mantle explains why continents are buoyant—acting like giant rafts, floating atop the mantle—and why they cannot be readily subducted into the mantle. The discovery of continental rocks more than 4 billion years old provides evidence that continental rocks cannot be easily recycled into the mantle.

Earth's Mantle More than 82 percent of Earth's volume is contained in the **mantle**, a solid, rocky shell that extends to a depth of about 2900 kilometers (1800 miles) beneath Earth's crust. The boundary between the crust and mantle represents a marked change in chemical composition. The dominant rock type in the uppermost mantle is *peridotite*, which is richer in the metals iron and magnesium than are the rocks found in either the continental or oceanic crust.

The upper mantle extends from the crust–mantle boundary down to a depth of about 660 kilometers (410 miles). Recall that the upper mantle is divided into two different parts. The top portion of the upper mantle, along with the Earth's crust, makes up the stiff *lithosphere*. The lower unit of the upper mantle contains the weak *asthenosphere*.

The **lithosphere** ("sphere of rock") forms Earth's relatively cool, rigid outer shell that is broken into sections of various sizes, called *tectonic plates*. The lithosphere is more than 250 kilometers (155 miles) thick below the oldest portions of the continents and about 100 kilometers (60 miles) thick below the seafloor (see Figure 5.39).

Beneath Earth's lithosphere is a solid, but comparatively weak, layer termed the **asthenosphere** ("weak sphere"). Because the asthenosphere and lithosphere are mechanically detached from each other, the lithosphere is able to move independently of the asthenosphere.

From 660 kilometers (410 miles) deep to the top of the core, at a depth of 2900 kilometers (1800 miles), is the lower mantle. Because the pressure in the mantle increases with depth (due to the weight of the overlying rock), the mantle becomes stronger with depth. Despite their strength, the rocks in the lower mantle are extremely hot, and as a result, they are capable of very gradual flow.

Inner and Outer Core The **core** is thought to consist mainly of iron combined with an unknown quantity of nickel, as well as minor amounts of oxygen, silicon, and sulfur—elements that readily form compounds with iron. Because of the extreme pressure found in the core, this iron-rich material has an average density of more than 10 times the density of water, or 10 g/cm^3.

The **outer core** is a liquid, iron-rich layer 2270 kilometers (1410 miles) thick. The liquid nature of the outer core was discovered when researchers found that S waves do not penetrate the outer core. Scientists concluded that because S waves do not pass through liquids, and S waves do not pass through the outer core, Earth's outer core must be liquid.

At Earth's center lies the **inner core**, a solid dense metallic sphere with a radius of 1216 kilometers (754 miles). Despite its higher temperature, the inner core is solid, not molten like the outer core, due to the immense pressures that exist at the center of our planet. The density at Earth's center is about 13 times that of water, or 13 g/cm^3.

CONCEPT CHECKS 5.8

1. How do seismic waves help scientists describe Earth's interior?
2. How did Earth acquire its layered structure?
3. How do continental crust and oceanic crust differ?
4. Contrast the physical characteristics of the asthenosphere and the lithosphere.
5. How are Earth's inner and outer cores different? How are they similar?

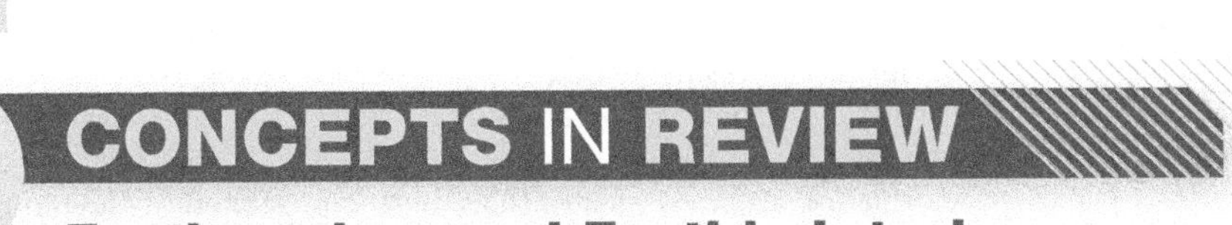

5 CONCEPTS IN REVIEW

Earthquakes and Earth's Interior

5.1 What Is an Earthquake?

Sketch and describe the mechanism that generates most earthquakes.

KEY TERMS: earthquake, fault, hypocenter (focus), epicenter, seismic wave, elastic rebound, aftershock, foreshock, megathrust fault, fault creep

- The sudden movements of large blocks of rock on opposite sides of faults cause most earthquakes. The location where the rock begins to slip is called the hypocenter, or focus. During an earthquake, seismic waves radiate outward from the hypocenter into the surrounding rock. The point on Earth's surface directly above the hypocenter is the epicenter.
- The ultimate cause of earthquakes is differential stress that gradually bends Earth's crust over tens to hundreds of years. Up to a point, frictional resistance along the fault keeps the rock from rupturing and slipping. Once that point is reached, the fault slips, allowing the bent rock to "spring back" to its original shape, generating an earthquake. The springing back is called elastic rebound.
- Convergent plate boundaries and associated subduction zones are marked by megathrust faults. These large faults are responsible for most of the largest earthquakes in recorded history. Megathrust earthquakes may also generate tsunamis.
- The San Andreas Fault in California is an example of a large strike-slip fault that forms a transform plate boundary capable of generating destructive earthquakes.

? Label the blanks on the diagram to show the relationship between earthquakes and faults using the following terms: epicenter, seismic waves, fault, fault trace, and hypocenter.

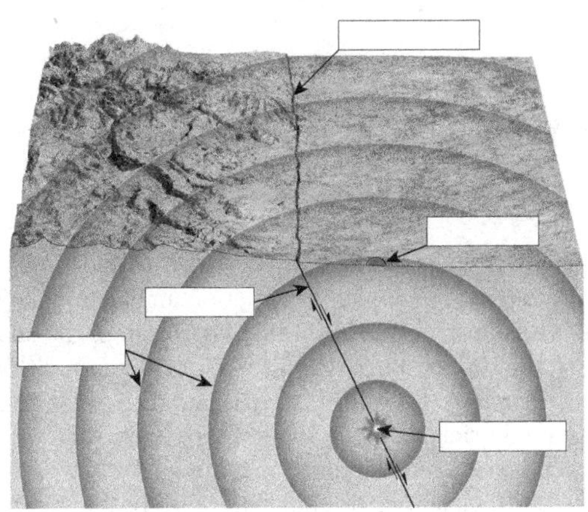

5.2 Seismology: The Study of Earthquake Waves

Compare and contrast the types of seismic waves and describe the principle of the seismograph.

KEY TERMS: seismology, seismograph (seismometer), inertia, seismogram, body waves, surface waves, primary (P) waves, secondary (S) waves

- Seismology is the study of seismic waves. A seismograph measures these waves, using the principle of inertia. While the body of the instrument moves with the waves, the inertia of a suspended weight keeps a sensor stationary to record the displacement between the two.
- A seismogram, a record of seismic waves, reveals two main categories of earthquake waves: body waves (P waves and S waves), which are capable of moving through Earth's interior, and surface waves, which travel only along the upper layers of the crust. P waves are the fastest, S waves are intermediate in speed, and surface waves are the slowest. However, surface waves tend to have the greatest amplitude, S waves are intermediate, and P waves have the lowest amplitude. Large-amplitude waves produce the most shaking, so surface waves usually account for most damage during earthquakes.
- P waves and S waves exhibit different kinds of motion. P waves momentarily push (compress) and pull (stretch) rocks as they travel through a rock body, thereby changing the volume of the rock. S waves impart a shaking motion as they pass through rock, changing the rock's shape but not its volume. Because fluids do not resist forces that change their shape, S waves cannot travel through fluids, whereas P waves can.

? How could you physically demonstrate the difference between P waves and S waves to a friend who hasn't taken a geology course? (*Caution:* Don't hurt your friend!)

5.3 Locating the Source of an Earthquake

Explain how seismographs are used to locate the epicenter of an earthquake.

- The distance separating a recording station from an earthquake's epicenter can be determined by using the difference in arrival times between P and S waves. When the distances are known from three or more seismic stations, the epicenter can be located using a method called triangulation.

5.4 Determining the Size of an Earthquake

Distinguish between intensity scales and magnitude scales.

KEY TERMS: intensity, magnitude, Modified Mercalli Intensity scale, Richter scale, moment magnitude

- Intensity and magnitude are different measures of earthquake strength. Intensity measures the amount of ground shaking at a location due to an earthquake, and magnitude is an estimate of the amount of energy released during an earthquake.
- The Modified Mercalli Intensity scale is a tool for measuring an earthquake's intensity at different locations. The scale is based on verifiable physical evidence that is used to quantify intensity on a 12-point scale.
- The Richter scale takes into account both the maximum amplitude of the seismic waves measured at a given seismograph and that seismograph's distance from the earthquake. The Richter scale is logarithmic, meaning that the next higher number on the scale represents seismic amplitudes that are 10 times greater than those represented by the number below. Furthermore, each larger number on the Richter scale represents the release of about 32 times more energy than the number below it.
- Because the Richter scale does not effectively differentiate between very large earthquakes, the moment magnitude scale was devised. This scale measures the total energy released from an earthquake by considering the strength of the faulted rock, the amount of slippage, and the area of the fault that slipped. Moment magnitude is the modern standard for measuring the size of earthquakes.

5.5 Earthquake Destruction

List and describe the major destructive forces that earthquake vibrations can trigger.

KEY TERMS: liquefaction, tsunami

- Factors influencing how much destruction an earthquake might inflict on a human-made structure include (1) intensity of the shaking, (2) how long shaking persists, (3) the nature of the ground that underlies the structure, and (4) building construction. Buildings constructed of unreinforced bricks and blocks are more likely than other types of structures to be severely damaged in a quake.
- In general, bedrock-supported buildings fare best in an earthquake, as loose sediments amplify seismic shaking.
- Liquefaction may occur when water-logged sediment or soil is severely shaken during an earthquake. Liquefaction can reduce the strength of the ground to the point that it may not support buildings.
- Earthquakes may also trigger landslides or ground subsidence, and they may break gas lines, which can initiate devastating fires.
- Tsunamis are large ocean waves that form when water is displaced, usually by a megathrust fault rupturing on the seafloor. Traveling at the speed of a jet aircraft, a tsunami is hardly noticeable in deep water. However, upon arrival in shallower coastal waters, the tsunami slows down and piles up, producing a wall of water sometimes more than 30 meters (100 feet) in height. Tsunamis cause major destruction in coastal areas if they strike the shoreline. Tsunami warning systems have been established in most of the large ocean basins.

? Of the earthquake hazards discussed earlier, which is (are) the greatest concern in the region where you live? Why?

5.6 Where Do Most Earthquakes Occur?

Locate Earth's major earthquake belts on a world map.

KEY TERM: circum-Pacific belt

- Most earthquake energy is released in the circum-Pacific belt, the ring of megathrust faults rimming the Pacific Ocean. Another earthquake belt is the Alpine–Himalayan belt, which runs along the zone where the Eurasia plate collides with the Indian subcontinent and African plates.
- Earth's oceanic ridge system produces another belt of earthquake activity. Seafloor spreading and active transform faults that separate ridge segments generate many frequent small-magnitude quakes. Transform faults in the continental crust, including the San Andreas Fault, can produce large earthquakes.
- Although most destructive earthquakes are produced along plate boundaries, some occur at considerable distances from plate boundaries. Examples include the 1811–1812 New Madrid, Missouri, earthquakes and the 1886 Charleston, South Carolina, earthquake.

? Outline the circum-Pacific earthquake belt on the accompanying map that has plate boundaries drawn in red. Do the same for the Alpine–Himalayan belt.

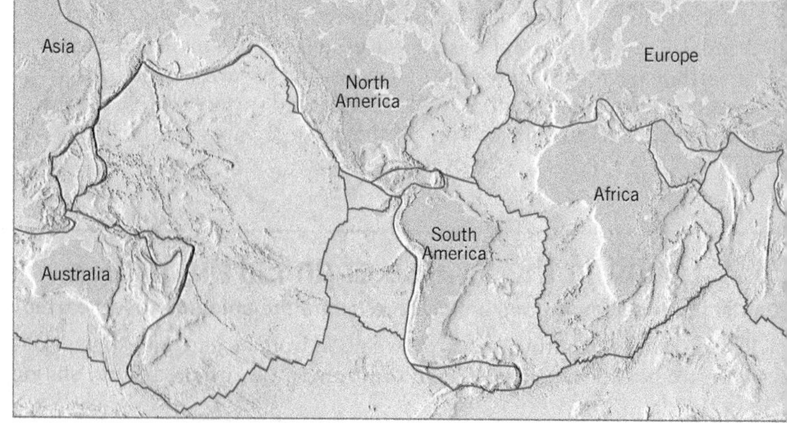

5.7 Earthquakes: Predictions, Forecasts, and Mitigation

Compare and contrast the goals of short-range earthquake predictions and long-range forecasts.

KEY TERMS: precursor, seismic gap, paleoseismology

- Successful earthquake prediction has been an elusive goal of seismology for many years. Attempts at shorter-range predictions (for hours or days) use precursor events such as changes in ground elevation or in strain levels near a fault; they have not been reliable.
- Long-range forecasts (for time scales of 30 to 100 years) are statistical estimates of the likelihood that an earthquake of a given magnitude will occur. Long-range forecasts are useful because they can guide development of building codes and infrastructure.
- Scientists have identified seismic gaps, portions of faults that have been storing strain for a long time, meaning these sites have great potential for experiencing an earthquake in the not-too-distant future. Paleoseismology is another tool used to make long-range forecasts. Because earthquakes occur on a cyclical basis, determining how frequently they have occurred in the past can give some insight into when they are most likely to occur again.

5.8 Earth's Interior

Explain how Earth acquired its layered structure and name and describe each of its major layers.

KEY TERMS: crust, mantle, lithosphere, asthenosphere, core, outer core, inner core

- The layered internal structure of Earth developed due to gravitational sorting of Earth materials early in the history of the planet. The densest material settled to form Earth's core, while the least dense material rose to form Earth's crust, oceans, and atmosphere.
- Seismic waves allow geoscientists to "look into" Earth's interior. Like sonograms used to image human organs, seismic waves generated by large earthquakes reveal details about Earth's layered structure.
- Earth has two distinct kinds of crust: oceanic and continental. Oceanic crust is thinner, denser, and younger than continental crust. Oceanic crust also readily subducts, whereas the less dense continental crust does not.
- The uppermost mantle and crust make up Earth's rigid outer shell, called the lithosphere, which overlies the asthenosphere—a solid but relatively weak layer. The lower mantle is a strong solid layer but capable of very gradual flow.
- Earth's core is very dense and composed of a mixture of iron and nickel, with minor amounts of lighter elements. The outer core is liquid, whereas the inner core is solid.

? Label the layers of Earth's interior shown on the accompanying diagram using the following terms: oceanic crust, continental crust, upper mantle, asthenosphere, and lithosphere.

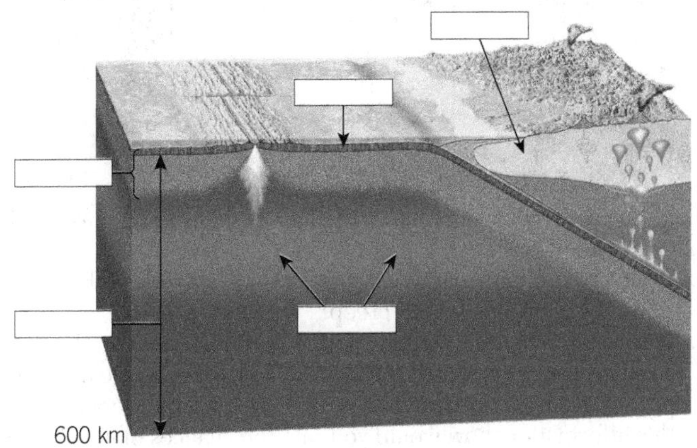

GIVE IT SOME THOUGHT

1 Describe the concept of elastic rebound. Develop an analogy other than a rubber band to illustrate this concept.

2 The accompanying map shows the locations of many of the largest earthquakes in the world since 1900. Refer to the map of Earth's plate boundaries in Figure 4.10, page 103, and determine which type of plate boundary is most often associated with these destructive events.

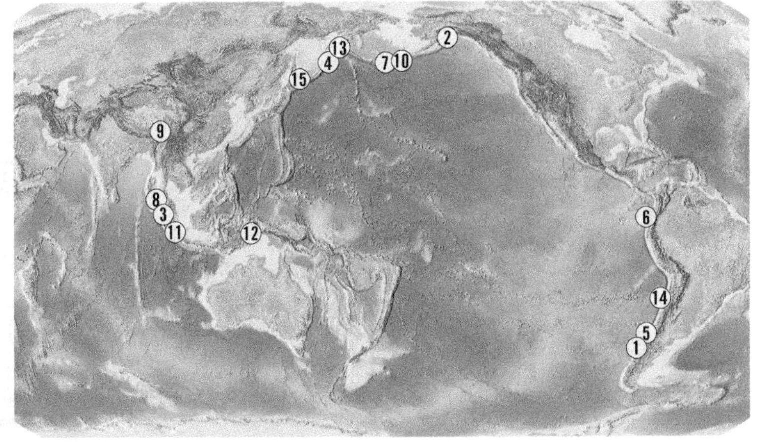

3 Use the accompanying seismogram to answer the following questions:

a. Which of the three types of seismic waves reached the seismograph first?

b. What is the time interval between the arrival of the first P wave and the arrival of the first S wave?

c. Use your answer from Question b and the travel–time graph in Figure 5.14 to determine the distance from the seismic station to the earthquake.

d. Which of the three types of seismic waves had the highest amplitude when it reached the seismic station?

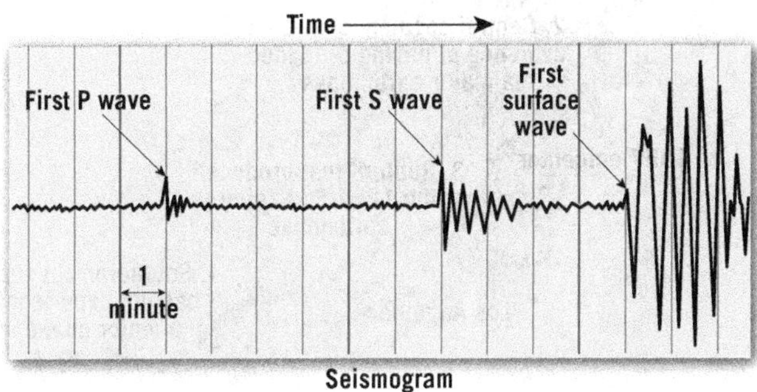

4 You go for a jog on a beach and choose to run near the water, where the sand is saturated. With each step, you notice that your footprint quickly fills with water, but that water is not coming in from the ocean. What is this water's source? For what earthquake-related hazard is this phenomenon a good analogy?

5 On the accompanying Richter scale diagram, first determine the Richter magnitude (M_L) for an earthquake at 400 kilometers distance, with a maximum amplitude of 0.5 millimeter. Second, for this same earthquake (same M_L), determine the amplitude of the biggest waves for a seismograph 40 kilometers from the hypocenter.

6 Using the accompanying map of the San Andreas Fault, answer the following questions:

a. Which of the four segments (1–4) of the San Andreas Fault do you think is experiencing fault creep?

b. Paleoseismology studies have found that the section of the San Andreas Fault that failed during the Fort Tejon quake (segment 3) produces a major earthquake every 135 years, on average. Based on this information, how would you rate the chances of a major earthquake occurring along this section in the next 30 years? Explain.

c. Do you think San Francisco or Los Angeles has the greater risk of experiencing a major earthquake in the near future? Defend your selection.

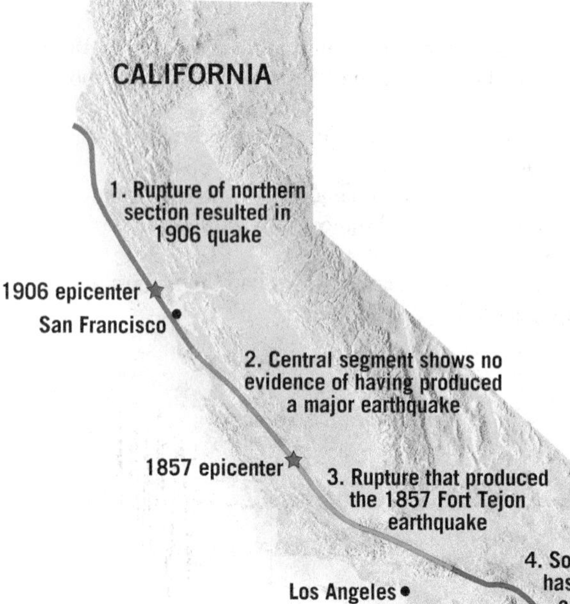

7 The accompanying image shows a double-decked section of Interstate 880 (the Nimitz Freeway) that collapsed during the 1989 Loma Prieta earthquake and caused 42 deaths. About 1.4 kilometers (nearly a mile) of this freeway section, called the Cypress Viaduct, collapsed, while a similar section survived the vibration. Both sections were subsequently demolished and rebuilt as a single-level structure, at a cost of $1.2 billion. Examine the map and seismograms from an aftershock that show the intensity of shaking observed at three nearby locations to answer the following questions:

a. What type of ground material experienced the least amount of shaking during the aftershock?

b. What type of ground materials experienced the greatest amount of ground shaking during the same event?

c. Which of the two sections of the Cypress Viaduct numbered on the map do you think collapsed: #1 or #2? Explain.

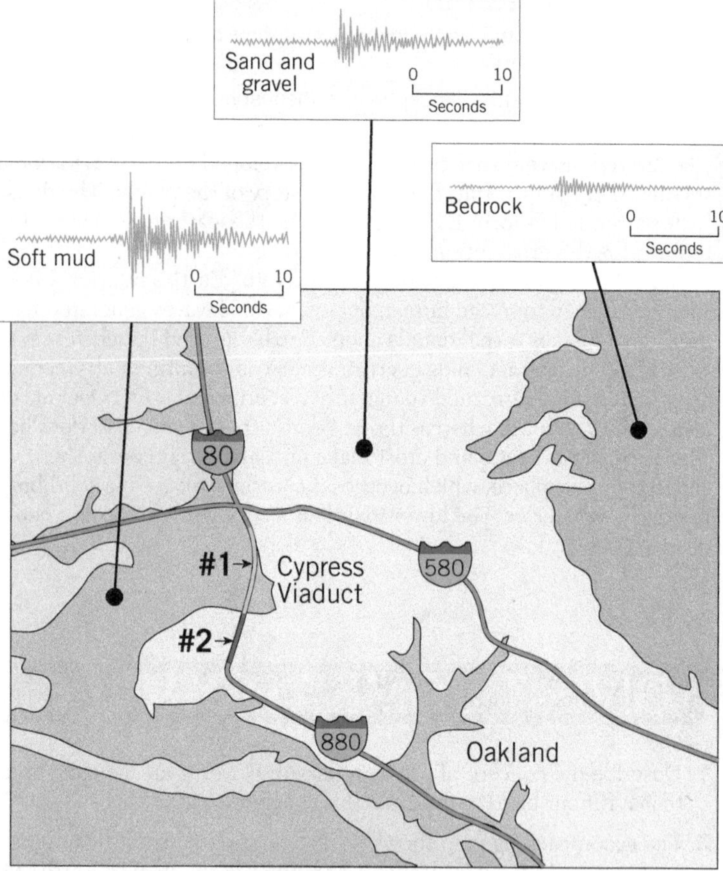

8 Earthquakes below the Yellowstone caldera originate at very shallow depths, about 4 kilometers on average. Below this depth, the rocks are at about 400°C, too hot and weak to store elastic energy. Based on this data, answer the following questions:

a. Calculate the average geothermal gradient in the first 4 kilometers beneath the Yellowstone caldera, assuming an average surface temperature of 0°C (32°F) and a temperature at 4 kilometers of 400°C (752°F). (For this example, the geothermal gradient, which is the increase in temperature with depth, should be measured in degrees centigrade per 100 meters.)

b. At about what depth is the groundwater below the Yellowstone caldera hot enough to "boil" and therefore capable of generating a geyser?

EXAMINING THE EARTH SYSTEM

1 What potentially disastrous phenomenon often occurs when the energy of an earthquake is transferred from the solid earth to the hydrosphere (ocean) at their interface on the floor of the ocean? When the energy from this event is expended along a coast, how might coastal lands and the biosphere be altered?

DATA ANALYSIS

Earthquakes Around the World

Earthquakes happen around the world every day. Many of them are minor and go unnoticed, but some are devastating. The United States Geological Survey (USGS) monitors these earthquakes and the damages they cause.

ACTIVITIES

Go to the USGS Earthquakes page, at http://earthquake.usgs.gov/earthquakes.

1 Where was the most recent significant earthquake in the past 30 days in the United States? What was its magnitude? (*Note:* USGS generally uses moment magnitude [M_W] to indicate magnitude.)

Click on the earthquake's link to bring up more information about the earthquake and then click Pager.

2 What are the estimated fatalities for this earthquake? What are the estimated economic losses?

3 Describe the structures located in the vicinity of this earthquake. Are there any secondary effects?

4 How many people felt moderate and strong shaking? Very strong and severe shaking? Violent and extreme shaking?

Go back to the previous page and click Origin.

5 What was the depth of this earthquake's focus? Be sure to include units and uncertainty.

Go to the USGS Earthquake Hazards Program's Latest Earthquakes map, at http://earthquake.usgs.gov/earthquakes/map. Use the settings icon (which looks like a gear) to find answers to the following questions. Zoom out to see the entire world.

6 How many total earthquakes have occurred worldwide in the past 7 days?

7 How many earthquakes of magnitude 2.5 or higher have occurred worldwide in the past 7 days?

8 How many earthquakes of magnitude 4.5 or higher have occurred worldwide in the past 7 days?

9 What fraction of total earthquakes have had a magnitude of 2.5 or higher? 4.5 or higher?

10 What conclusion can you draw about the relationship between the frequency and strength of earthquakes?

MasteringGeology™ Looking for additional review and test prep materials? Visit the Study Area in MasteringGeology to enhance your understanding of this chapter's content by accessing a variety of resources, including Self-Study Quizzes, Geoscience Animations, SmartFigure Tutorials, Mobile Field Trips, *Project Condor* Quadcopter videos, *In the News* articles, flashcards, web links, and an optional Pearson eText.

www.masteringgeology.com

6 Volcanoes and Other Igneous Activity

FOCUS ON CONCEPTS

Each statement represents the primary learning objective for the corresponding major heading within the chapter. After you complete the chapter, you should be able to:

6.1 Compare and contrast the 1980 eruption of Mount St. Helens with the most recent eruption of Kilauea, which began in 1983.

6.2 Explain why some volcanic eruptions are explosive and others are quiescent.

6.3 List and describe the three categories of materials extruded during volcanic eruptions.

6.4 Draw and label a diagram that illustrates the basic features of a typical volcanic cone.

6.5 Summarize the characteristics of shield volcanoes and provide one example of this type of volcano.

6.6 Describe the formation, size, and composition of cinder cones.

6.7 List the characteristics of composite volcanoes and describe how they form.

6.8 Describe the major geologic hazards associated with volcanoes.

6.9 List volcanic landforms other than shield, cinder, and composite volcanoes and describe their formation.

6.10 Compare and contrast these intrusive igneous structures: dikes, sills, batholiths, stocks, and laccoliths.

6.11 Summarize the major processes that generate magma from solid rock.

6.12 Explain how the global distribution of volcanic activity is related to plate tectonics.

Eruption of ash from Mount Bromo Volcano, 2011, in Java, Indonesia.
(Photo provided by Richard Roscoe/Stocktrek Images, Inc./Alamy Stock Photo)

THE SIGNIFICANCE OF IGNEOUS ACTIVITY may not be obvious at first glance. However, because volcanoes extrude molten rock that formed at great depth, they provide our only means of directly observing processes that occur many kilometers below Earth's surface. Furthermore, Earth's atmosphere and oceans have evolved from gases emitted during volcanic eruptions. Either of these facts is reason enough for igneous activity to warrant our attention.

6.1 Mount St. Helens Versus Kilauea

Compare and contrast the 1980 eruption of Mount St. Helens with the most recent eruption of Kilauea, which began in 1983.

On May 18, 1980, the largest volcanic eruption to occur in North America in historic times transformed a picturesque volcano into a decapitated remnant (Figure 6.1). On that date in southwestern Washington State, Mount St. Helens erupted with tremendous force. The blast blew out the entire north flank of the volcano, leaving a gaping hole. In one brief moment, a prominent volcano whose summit had been more than 2900 meters (9500 feet) above sea level was lowered by more than 400 meters (1350 feet).

The event devastated a wide swath of timber-rich land on the north side of the mountain (Figure 6.2). Trees within a 400-square-kilometer (160-square-mile) area lay intertwined and flattened, stripped of their branches and appearing from the air like toothpicks strewn about. The accompanying mudflows carried ash, trees, and water-saturated rock debris 29 kilometers (18 miles) down the Toutle River. The eruption claimed 59 lives; some died from the intense heat and the suffocating cloud of ash and gases, others from the impact of the blast, and still others from being trapped in mudflows.

The eruption ejected nearly a cubic kilometer of ash and rock debris. Following the devastating explosion, Mount St. Helens continued to emit great quantities of hot gases and ash. The force of the blast was so strong that some ash was propelled more than 18 kilometers (over 11 miles) into the stratosphere. During the next few days, this very fine-grained material was carried around Earth by strong upper-air winds. Crops were damaged in central Montana, and measurable deposits were reported as far away as Oklahoma and Minnesota. Meanwhile, ash fallout in the immediate vicinity exceeded 2 meters (6 feet) in depth. The air over

Figure 6.1 Before-and-after photographs show the transformation of Mount St. Helens The May 18, 1980, eruption of Mount St. Helens occurred in southwestern Washington.

The blast blew out the entire north flank of Mount St. Helens, leaving a gaping hole. In a brief moment, a prominent volcano was lowered by 1350 feet.

Yakima, Washington (130 kilometers [80 miles] to the east), was so filled with ash that residents experienced midnight-like darkness at noon.

Not all volcanic eruptions are as violent as the 1980 Mount St. Helens event. Some volcanoes, such as Hawaii's Kilauea Volcano, generate relatively quiet outpourings of fluid lavas. These quiescent (non-explosive) eruptions are not without some fiery displays; occasionally fountains of incandescent lava spray hundreds of meters into the air (see Figure 6.4), but most lava pours from the vent and flows downslope. During Kilauea's most recent active phase, which began in 1983, more than 180 homes and a national park visitor center have been destroyed by flowing lava igniting material in its path.

Testimony to the quiescent nature of Kilauea's eruptions is the fact that the Hawaiian Volcanoes Observatory has operated on its summit since 1912, despite the fact that Kilauea has had more than 50 eruptive phases since record keeping began in 1823.

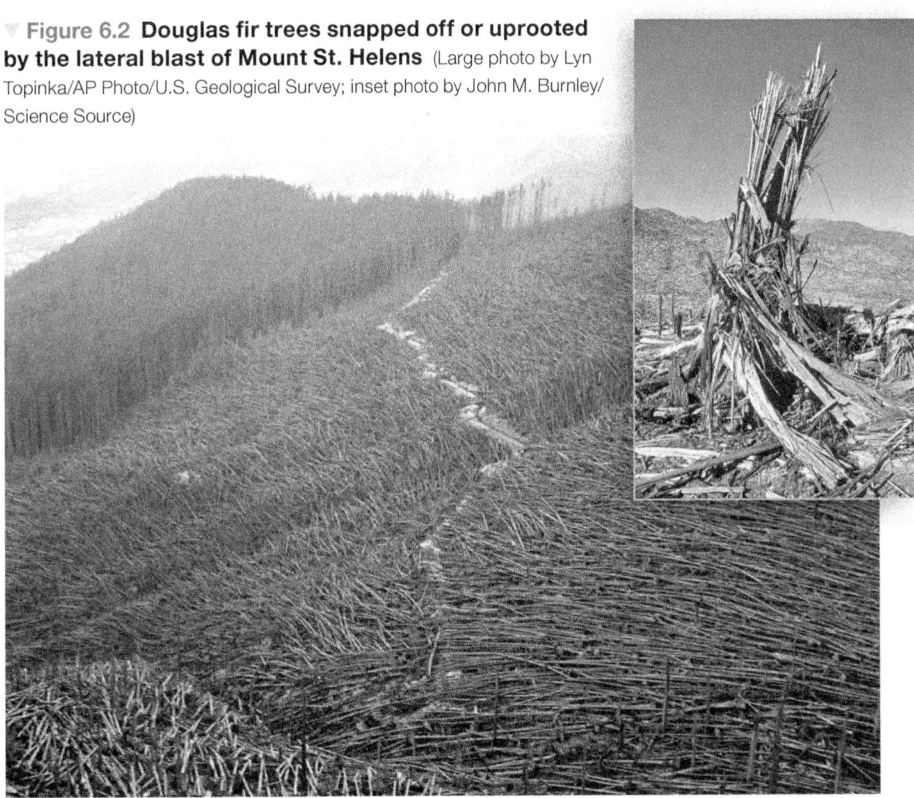

Figure 6.2 **Douglas fir trees snapped off or uprooted by the lateral blast of Mount St. Helens** (Large photo by Lyn Topinka/AP Photo/U.S. Geological Survey; inset photo by John M. Burnley/Science Source)

CONCEPT CHECKS 6.1

1. Briefly compare the 1980 eruption of Mount St. Helens to a typical eruption of Hawaii's Kilauea Volcano.

6.2 The Nature of Volcanic Eruptions

Explain why some volcanic eruptions are explosive and others are quiescent.

Volcanic activity is commonly perceived as a process that produces a picturesque, cone-shaped structure that periodically erupts in a violent manner. However, many eruptions are not explosive, as indicated by Kilauea's activity. What determines the manner in which volcanoes erupt?

Magma: Source Material for Volcanic Eruptions

Recall that **magma**, molten rock that may contain some solid crystalline material and also contains varying amounts of dissolved gas (mainly water vapor and carbon dioxide), is the parent material of igneous rocks. Erupted magma is called **lava**.

Composition of Magma As we discussed in Chapter 3, *basaltic* igneous rocks contain a high percentage of dark silicate minerals and calcium-rich plagioclase feldspar, and as a result, they tend to be dark in color. By contrast, *granite* and its extrusive equivalent *rhyolite* contain mainly light-colored silicate minerals—quartz and potassium feldspar. *Andesitic* igneous rocks have a composition between basaltic and granitic rocks. Correspondingly, basaltic magmas contain a much *lower* percentage of silica (SiO_2) than do granitic magmas. The compositional differences between magmas also affect several other properties, as summarized in Figure 6.3. For example, basaltic (mafic) magmas have the lowest silica content and the lowest gas content, and they erupt at the highest temperatures. By contrast, granitic and rhyolitic (felsic) magmas have the highest silica content and the highest gas content, and they erupt at the lowest temperatures. Andesitic (intermediate) magmas have characteristics between basaltic and granitic magmas.

Where Is Magma Generated? Recall that most magma is generated in Earth's upper mantle by *partial melting of solid rock*. The magmas generated by melting mantle rocks tend to have basaltic (mafic) composition. Once formed, basaltic magma, which is less dense than the surrounding rocks, slowly rises toward Earth's surface. In some settings, this hot molten rock reaches the surface, where it usually produces fluid outflows of basaltic lavas. The largest quantity of basaltic magma erupts on the ocean floor along divergent plate boundaries due to seafloor spreading. Extensive basaltic flows are also thought to result from hot-spot volcanism generated by rising hot mantle plumes (see Figure 6.40).

In continental settings, however, overlying crustal rocks are usually less dense than the ascending basaltic magma, and as a result, the rising molten rock ponds at the crust–mantle boundary. Because the newly formed magma is much hotter than the melting temperature of crustal rocks, the rocks overlying the magma body begin to melt. This process generates a less dense, more silica-rich magma of andesitic or rhyolitic composition, which then continues the journey toward Earth's surface.

Effusive Versus Explosive Eruptions

Volcanic eruptions exhibit a range of behavior from quiescent eruptions that produce outpourings of fluid lava to explosive eruptions. Geologists often refer to quiescent eruptions as **effusive eruptions** (*effus* = pour forth).

The two primary factors that determine how magma erupts are its *viscosity* and *gas content.* **Viscosity** (*viscos* = sticky) is a measure of a fluid's mobility. The more viscous a material, the greater its resistance to flow. For example, syrup is more viscous, and thus more resistant to flow, than water.

Factors Affecting Viscosity Magma's viscosity depends primarily on its temperature and silica content: *The more silica in magma, the greater its viscosity.* Silicon-oxygen tetrahedra begin to link together into long chains early in the crystallization process, which makes the magma more rigid and impedes its flow. Consequently, silica-rich rhyolitic lavas are the most viscous and tend to travel at imperceptibly slow speeds to form comparatively short, thick flows. By contrast, basaltic lavas, which contain much less silica, are relatively fluid and have been known to travel 150 kilometers (90 miles) or more before solidifying. Andesitic magmas, which are intermediate in composition, have flow rates between these extremes.

Temperature affects the viscosity of magma in much the same way it affects the viscosity of pancake syrup: The hotter a magma, the more fluid (less viscous) it will be. As lava cools and begins to congeal, its viscosity increases, and the flow eventually halts.

Role of Gases The nature of volcanic eruptions also depends on the amount of dissolved gases held in the magma body by the pressure exerted by the overlying rock (the confining pressure). The most abundant gases in most magma are water vapor and carbon dioxide. These dissolved gases tend to come out of solution when the confining pressure is reduced. This is analogous to how carbon dioxide behaves in cans and bottles of soft drinks. When you reduce the pressure on a soft drink by opening the cap, the dissolved carbon dioxide quickly separates from the solution to form bubbles that rise and escape.

The viscosity and gas content of magma are directly related to its composition, as shown in Figure 6.3. At one end of the spectrum are basaltic (mafic) magmas, which are very fluid and have a low gas content, sometimes as little as 0.5 percent by weight. At the other extreme are rhyolitic (felsic) magmas, which are highly viscous (sticky) and contain a lot of gas, as much as 8 percent by weight.

Effusive Eruptions

All magmas contain some water vapor and other gases that are kept in solution by the immense pressure of the overlying rock. As magma rises (or the rocks confining the magma fail), the confining pressure drops, causing the dissolved gases to separate from the melt and form large numbers of tiny bubbles. When fluid basaltic magmas erupt, these pressurized gases readily escape. At temperatures that often exceed 1100°C (2000°F), these gases can quickly expand to occupy hundreds of times their original volumes. Occasionally, these expanding gases propel incandescent lava hundreds of meters into the air, producing lava fountains (Figure 6.4). Although spectacular, these fountains are usually harmless and generally not associated with major explosive events that cause great loss of life and property.

Effusive eruptions that involve very fluid basaltic lavas, such as the recent eruptions of Kilauea on Hawaii's

Properties of Magma Bodies with Differing Compositions

Composition	Silica Content (SiO_2)	Gas Content (% by weight)	Eruptive Temperature	Viscosity	Tendency to Form Pyroclastics	Volcanic Landform
Basaltic (MAFIC) High in Fe, Mg, Ca, low in K, Na	**Least** (~50%)	**Least** (0.5–2%)	**Highest** 1000–1250°C	**Least**	**Least**	Shield volcanoes, basalt plateaus, cinder cones
Andesitic (INTERMEDIATE) Varying amounts of Fe, Mg, Ca, K, Na	**Intermediate** (~60%)	**Intermediate** (3–4%)	**Intermediate** 800–1050°C	**Intermediate**	**Intermediate**	Composite cones
Rhyolitic/ Granitic (FELSIC) High in K, Na, low in Fe, Mg, Ca	**Most** (~70%)	**Most** (5–8%)	**Lowest** 650–900°C	**Greatest**	**Greatest**	Pyroclastic flow deposits, lava domes

Figure 6.3 Compositional differences of magma bodies cause their properties to vary.

▲ Figure 6.4 **Lava fountain produced by gases escaping fluid basaltic lava** Mount Etna, located on the island of Sicily, Italy, is one of the most active volcanoes on Earth. Fluid lava is shown erupting from a relatively small cone located on the flanks of the main volcano in 2014. (Photo provided by RealyEasyStar/Rosario Patanè/Alamy Stock Photo)

Big Island, are often triggered by the arrival of a new batch of molten rock, which accumulates in a near-surface magma chamber. Geologists can usually detect such an impending event because the summit of the volcano begins to inflate and rise months or even years before an eruption. The injection of a fresh supply of hot molten rock heats and remobilizes the semi-liquid magma in the chamber. In addition, swelling of the magma chamber fractures the rock above, allowing the fluid magma to move upward along the newly formed fissures, often generating effusions of fluid lava for weeks, months, or possibly years. The eruption of Kilauea that began in 1983 is ongoing.

How Explosive Eruptions Are Triggered

Recall that silica-rich rhyolitic magmas have a relatively high gas content and are quite viscous (sticky) compared to basaltic magmas. As rhyolitic magma rises, the gases remain dissolved until the confining pressure drops sufficiently, at which time tiny bubbles begin to form and increase in size. Because of the high viscosity of rhyolitic magma, gas bubbles tend to remain trapped in the magma, forming a sticky froth.

When the pressure exerted by the expanding magma exceeds the strength of the overlying rock, fracturing occurs. As the frothy magma moves up through the fractures, the resulting drop in confining pressure creates additional gas bubbles. This chain reaction often generates an explosive event in which magma is literally blown into fragments (ash and pumice) that are carried to great heights by the escaping hot gases. (The collapse of a volcano's flank can also greatly reduce the pressure on the magma below, causing an explosive eruption, as exemplified by the 1980 eruption of Mount St. Helens.)

When molten rock in the uppermost portion of the magma chamber is forcefully ejected by the escaping gases, the confining pressure on the magma directly below also drops suddenly. Thus, rather than being a single "bang," an explosive eruption is really a series of violent explosions that can last for a few days.

Because highly gaseous magmas expel fragmented lava at nearly supersonic speeds, they are associated with hot, buoyant **eruption columns** consisting mainly of volcanic ash and gases (Figure 6.5). Eruption columns can rise perhaps 40 kilometers (25 miles) into the atmosphere. It is not uncommon for a portion of an eruption column to collapse, sending hot ash rushing down the volcanic slope at speeds exceeding 100 kilometers (60 miles) per hour. As a result, volcanoes that erupt highly viscous magmas having a high gas content are the most destructive to property and human life.

Following explosive eruptions, partially degassed lava may slowly ooze out of the vent to form thick lava flows or dome-shaped lava bodies that grow over the vent.

CONCEPT CHECKS 6.2

1. List these magmas in order, from the highest to lowest silica content: basaltic (mafic) magma, granitic/rhyolitic (felsic) magma, andesitic (intermediate) magma.
2. List the two primary factors that determine the manner in which magma erupts.
3. Define *viscosity*.
4. Are volcanoes fed by highly viscous magma *more* or *less* likely to be a greater threat to life and property than volcanoes supplied with very fluid magma?

▼ SmartFigure 6.5 **Eruption column generated by viscous, silica-rich magma** Steam and ash eruption column from Mount Tavurvur in eastern Papua New Guinea, 2014. (Photo by Ness Kerton/AFP/Getty Images)

VIDEO
https://goo.gl/RH97D7

Eruptions of highly viscous lavas may produce explosive clouds of hot ash and gases called eruption columns.

6.3 Materials Extruded During an Eruption

List and describe the three categories of materials extruded during volcanic eruptions.

Volcanoes erupt lava, large volumes of gas, and pyroclastic materials (broken rock, lava "bombs," and ash). In this section we will examine each of these materials.

Lava Flows

The vast majority of Earth's lava, more than 90 percent of the total volume, is estimated to be basaltic (mafic) in composition. Most basaltic lavas erupt on the seafloor, via a process termed *submarine volcanism*. Lavas having an andesitic (intermediate) composition account for most of the rest, while rhyolitic (felsic) flows make up as little as 1 percent of the total. Rhyolitic magmas tend to extrude mostly volcanic ash rather than lava.

On land, hot basaltic lavas, which are usually very fluid, generally flow in thin, broad sheets or streamlike ribbons. These fluid lavas have been clocked at speeds exceeding 30 kilometers (19 miles) per hour down steep slopes. However, flow rates of 10 to 300 meters (30 to 1000 feet) per hour are more common. Silica-rich rhyolitic lava, by contrast, often moves too slowly to be perceived. Furthermore, rhyolitic lavas seldom travel more than a few kilometers from their vents. As you might expect, andesitic lavas, which are intermediate in composition, exhibit flow characteristics between these extremes.

Aa and Pahoehoe Flows Fluid basaltic magmas tend to generate two types of lava flows, which are known by their Hawaiian names. The first, called **aa** (pronounced "ah-ah") **flows**, have surfaces of rough jagged blocks with dangerously sharp edges and spiny projections (Figure 6.6A). Crossing a hardened aa flow can be a trying and miserable experience. The second type, **pahoehoe** (pronounced "pah-hoy-hoy") **flows**, exhibit smooth surfaces that sometimes resemble twisted braids of ropes (Figure 6.6B).

Although both lava types can erupt from the same volcano, pahoehoe lavas are hotter and more fluid than aa flows. In addition, pahoehoe lavas can change into aa lava flows, although the reverse (aa to pahoehoe) does not occur. Cooling that occurs as the flow moves away from the vent is one factor that facilitates the change from pahoehoe to aa. The lower temperature increases viscosity and promotes bubble formation. Escaping gas bubbles produce numerous voids (vesicles) and sharp spines in the surface of the congealing lava. As the molten interior advances, the outer crust is broken, transforming the relatively smooth surface of a pahoehoe flow into an aa flow made up of an advancing mass of rough, sharp, broken lava blocks.

Pahoehoe flows often develop cave-like tunnels called **lava tubes** that start as conduits for carrying lava from an active vent to the flow's leading edge (Figure 6.7). Lava tubes form in the interior of a lava flow, where the temperature remains high long after the exposed surface cools and hardens. Because they serve as insulated pathways that allow lava to flow great distances from its source, lava tubes are important features of fluid lava flows.

Pillow Lavas Recall that most of Earth's volcanic output occurs along oceanic ridges (divergent plate boundaries), generating new oceanic crust. When outpourings of lava occur on the ocean floor, the flow's outer skin quickly

Figure 6.6 Lava flows A. A slow-moving, basaltic aa flow advancing over hardened pahoehoe lava. **B.** A typical fluid pahoehoe (ropy) lava. Both of these lava flows erupted from a rift on the flank of Hawaii's Kilauea Volcano. (Photos courtesy of U.S. Geological Survey)

A. Active aa flow overriding an older pahoehoe flow.

B. Pahoehoe flow displaying the characteristic ropy appearance.

A. Lava tubes are cave-like tunnels that once served as conduits carrying lava from an active vent to the flow's leading edge.

Valentine Cave, a lava tube at Lava Beds National Monument, California.

B. Skylights develop where the roofs of lava tubes collapse and reveal the hot lava flowing through the tube.

Figure 6.7 Lava tubes A. A lava flow may develop a solid upper crust, while the molten lava below continues to advance in a conduit called a lava tube. Some lava tubes exhibit extraordinary dimensions. Kazumura Cave, located on the southeastern slope of Hawaii's Mauna Loa Volcano, is a lava tube extending more than 60 kilometers (40 miles). (Photo by Dave Bunnell/Under Earth Images) **B.** The collapsed section of the roof of a lava tunnel results in a skylight. (Photo courtesy of U.S. Geological Survey)

freezes (solidifies) to form volcanic glass. However, the interior lava is able to move forward by breaking through the hardened surface. This process occurs over and over, as molten basalt is extruded like toothpaste from a tightly squeezed tube. The result is a lava flow composed of numerous tube-like structures called **pillow lavas**, stacked one atop the other (Figure 6.8). Pillow lavas are useful when reconstructing geologic history because their presence indicates that the lava flow formed below the surface of a water body.

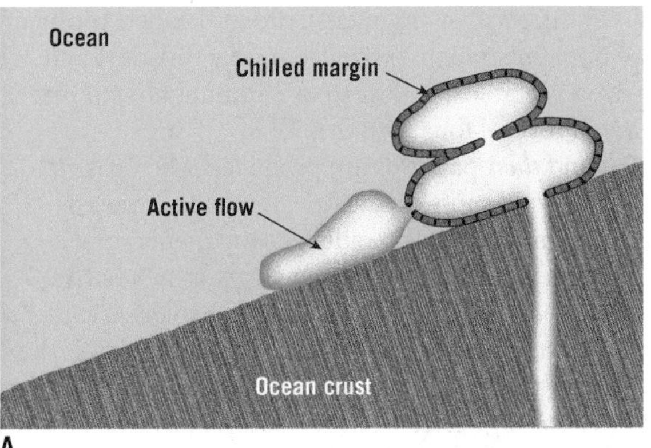

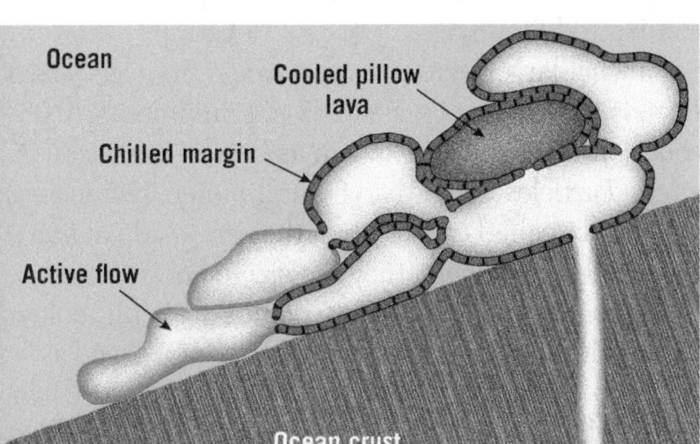

C.

Ocean

Chilled margin

Active flow

Ocean crust

Figure 6.8 Pillow lava These diagrams show the formation of pillow lava. The pillows vary in shape but tend to be elongated tube-like structures. The photo shows an undersea pillow lava flow off the coast of Hawaii. (Photo courtesy of U.S. Geological Survey)

Gases

Recall that magmas contain varying amounts of dissolved gases, called **volatiles**. These gases are held in the molten rock by confining pressure, just as carbon dioxide is held in cans of soft drinks. As with soft drinks, as soon as the pressure is reduced, the gases begin to escape. Obtaining gas samples from an erupting volcano is difficult and dangerous, so geologists usually must estimate the amount of gas originally contained in the magma.

The gaseous portion of most magma bodies ranges from less than 1 percent to about 8 percent of the total weight, with most of this in the form of water vapor. Although the percentage may be small, the actual quantity of emitted gas can exceed thousands of tons per day. Occasionally, eruptions emit colossal amounts of volcanic gases that rise high into the atmosphere, where they may reside for several years.

The composition of volcanic gases is important because these gases contribute significantly to our planet's atmosphere. The most abundant gas typically released into the atmosphere from volcanoes is water vapor (H_2O), followed by carbon dioxide (CO_2) and sulfur dioxide (SO_2), with lesser amounts of hydrogen sulfide (H_2S), carbon monoxide (CO), and nitrogen (N_2). (The relative proportion of each gas varies significantly from one volcanic region to another.) Sulfur compounds are easily recognized by their pungent odor. Volcanoes are also natural sources of air pollution; some emit large quantities of sulfur dioxide (SO_2), which readily combines with atmospheric gases to form toxic sulfuric acid and other sulfate compounds.

Pyroclastic Materials (Tephra)

Particle name	Particle size	Image
Volcanic ash*	Less than 2 mm (0.08 inch)	 USGS; 0 30μ
Lapilli (Cinders)	Between 2 mm and 64 mm (0.08–2.5 inches)	 Dennis Tasa
Volcanic bombs	More than 64 mm (2.5 inches)	 Dennis Tasa; 0 30 mm; 0 1 in.
Volcanic blocks		 USGS

*The term volcanic dust is used for fine volcanic ash less than 0.063 mm (0.0025 inch).

▲ **Figure 6.9 Types of pyroclastic materials** Pyroclastic materials are also commonly referred to as tephra.

Pyroclastic Materials

When volcanoes erupt energetically, they eject pulverized rock and fragments of lava and glass from the vent. The particles produced, **pyroclastic materials** (*pyro* = fire, *clast* = fragment), are also called **tephra**. These fragments range in size from very fine dust and sand-sized volcanic ash (less than 2 millimeters) to pieces that weigh several tons (Figure 6.9).

Ash and *dust* particles are produced when gas-rich viscous magma erupts explosively. As magma moves up in the vent, the gases rapidly expand, generating a melt that resembles the froth that flows from a bottle of champagne. As the hot gases expand explosively, the froth is blown into very fine glassy fragments. When the hot ash falls, the glassy shards often fuse to form a rock called *welded tuff*. Sheets of this material, as well as ash deposits that later consolidate, cover vast portions of the western United States.

Somewhat larger pyroclasts that range from the size of small beads to the size of walnuts (2–64 millimeters [0.08–2.5 inches] in diameter) are known as *lapilli* ("little stones"), or *cinders*. Particles larger than 64 millimeters (2.5 inches) in diameter are called *blocks* when they are made of hardened lava and *bombs* when they are ejected as incandescent lava (see Figure 6.7). Because bombs are semi-molten when ejected, they often take on a streamlined shape as they hurl through the air. Because of their size and weight, bombs and blocks usually fall near the vent; however, they are occasionally propelled great distances. For instance, bombs 6 meters (20 feet) long and weighing about 200 tons were blown 600 meters (2000 feet) from the vent during an eruption of the Japanese volcano Asama.

Pyroclastic materials can be classified by texture and composition as well as by size. For instance, **scoria** is the term for vesicular ejecta produced most often during the eruption of basaltic magmas (Figure 6.10A). These black to reddish-brown fragments are generally found in the size range of lapilli and resemble cinders and clinkers produced by furnaces used to smelt iron.

By contrast, when magmas with andesitic (intermediate) or rhyolitic (felsic) compositions erupt explosively, they emit ash and the vesicular rock **pumice** (Figure 6.10B). Pumice is usually lighter in color and less dense than scoria, and many pumice fragments have so many vesicles that they are light enough to float (see Figure 3.11, page 68).

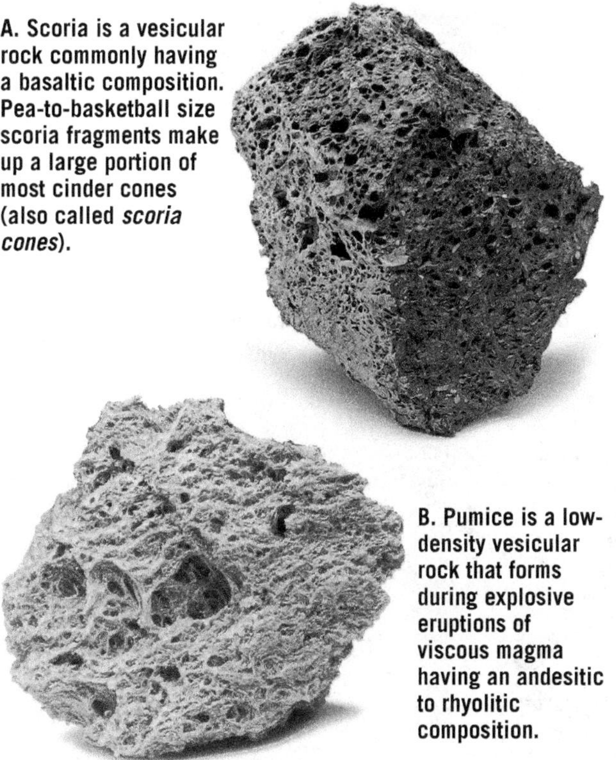

A. Scoria is a vesicular rock commonly having a basaltic composition. Pea-to-basketball size scoria fragments make up a large portion of most cinder cones (also called *scoria cones*).

B. Pumice is a low-density vesicular rock that forms during explosive eruptions of viscous magma having an andesitic to rhyolitic composition.

Figure 6.10 **Common vesicular rocks** Scoria and pumice are volcanic rocks that exhibit a vesicular texture. Vesicles are small holes left by escaping gas bubbles. (Photos by E.J. Tarbuck)

CONCEPT CHECKS 6.3

1. Contrast pahoehoe and aa lava flows.
2. How do lava tubes form?
3. List the main gases released during a volcanic eruption.
4. How do volcanic bombs differ from blocks of pyroclastic debris?
5. What is scoria? How is scoria different from pumice?

6.4 Anatomy of a Volcano

Draw and label a diagram that illustrates the basic features of a typical volcanic cone.

A popular image of a volcano is a solitary, graceful, snow-capped cone, such as Mount Hood in Oregon or Japan's Fujiyama. These picturesque, conical mountains are produced by volcanic activity that occurred intermittently over thousands, or even hundreds of thousands, of years. However, many volcanoes do not fit this image. Cinder cones are quite small and form during a single eruptive phase that lasts a few days to a few years. Alaska's Valley of Ten Thousand Smokes is a flat-topped ash deposit that blanketed a river valley to a depth of 200 meters (600 feet). The eruption that produced it lasted less than 60 hours yet emitted more than 20 times more volcanic material than the 1980 Mount St. Helens eruption.

Volcanic landforms come in a wide variety of shapes and sizes, and each volcano has a unique eruptive history. Nevertheless, volcanologists have been able to classify volcanic landforms and determine their eruptive patterns. In this section we will consider the general anatomy of an idealized volcanic cone. We will follow this discussion by exploring the three major types of volcanic cones—shield volcanoes, cinder cones, and composite volcanoes—as well as their associated hazards.

Volcanic activity frequently begins when a **fissure** (crack) develops in Earth's crust as magma moves forcefully toward the surface. As the gas-rich magma moves up through a fissure, its path is usually localized into a somewhat pipe-shaped **conduit** that terminates at a surface opening called a **vent** (Figure 6.11). The cone-shaped structure we call a **volcanic cone** is often created by successive eruptions of lava, pyroclastic material, or frequently a combination of both, often separated by long periods of inactivity.

Located at the summit of most volcanic cones is a somewhat funnel-shaped depression called a **crater** (*crater* = bowl). Volcanoes built primarily of pyroclastic materials typically have craters that form by gradual accumulation of volcanic debris on the surrounding rim. Other craters form during explosive eruptions, as the rapidly ejected particles erode the crater walls. Craters

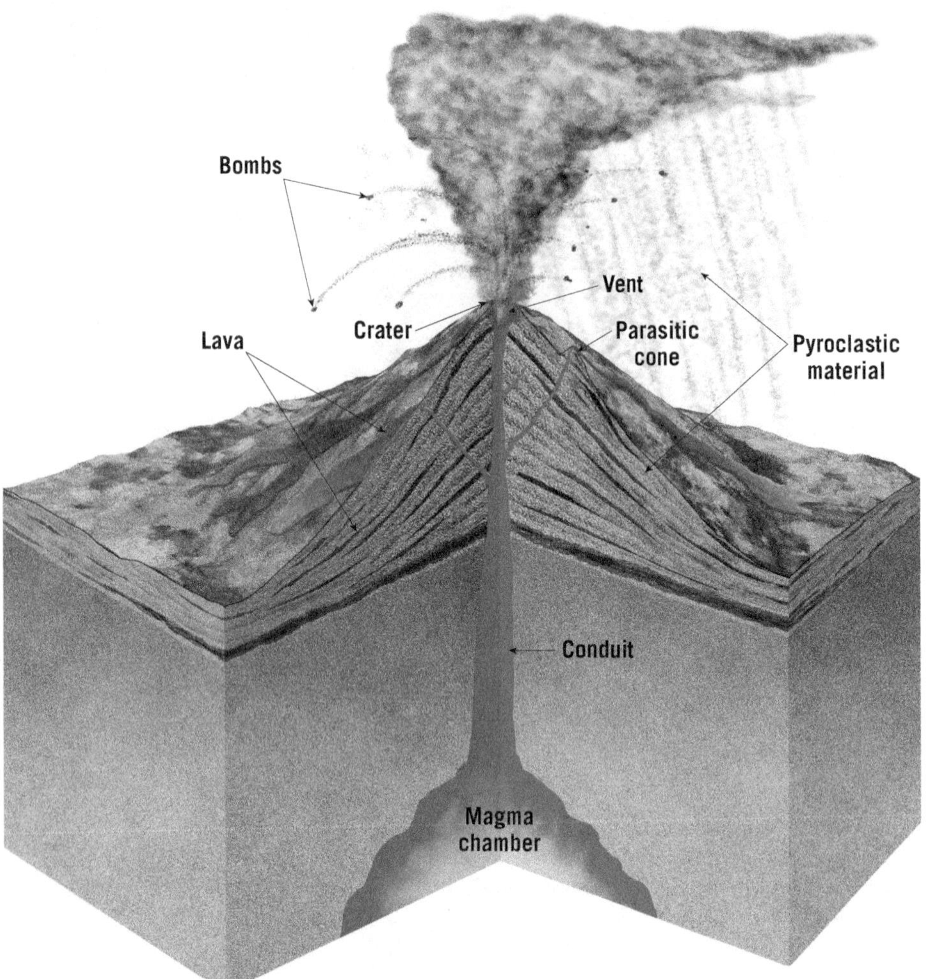

SmartFigure 6.11
Anatomy of a volcano Compare the structure of the "typical" composite cone shown here to that of a shield volcano (Figure 6.12) and a cinder cone (Figure 6.14).

TUTORIAL
https://goo.gl/nbwG5k

also form when the summit area of a volcano collapses following an eruption. Some volcanoes have very large circular depressions, called **calderas**, which have diameters that are greater than 1 kilometer (0.6 mile) and that in rare cases exceed 50 kilometers (30 miles). The formation of various types of calderas will be considered later in this chapter.

During early stages of growth, most volcanic discharges come from a central summit vent. As a volcano matures, material also tends to be emitted from fissures that develop along the flanks (sides) or at the base of the volcano. Continued activity from a flank eruption may produce one or more small **parasitic cones**. Italy's Mount Etna, for example, has more than 200 secondary vents, some of which have built parasitic cones. Many of these vents, however, emit only gases and are appropriately called **fumaroles** (*fumus* = smoke).

CONCEPT CHECKS 6.4

1. Distinguish among a conduit, a vent, and a crater.
2. How is a crater different from a caldera?
3. What is a parasitic cone, and where does it form?
4. What is emitted from a fumarole?

6.5 Shield Volcanoes

Summarize the characteristics of shield volcanoes and provide one example of this type of volcano.

Shield volcanoes are produced by the accumulation of fluid basaltic lavas and exhibit the shape of a broad, slightly domed structure that resembles a warrior's shield (Figure 6.12). Most shield volcanoes begin on the ocean floor as **seamounts** (submarine volcanoes), and a few of them grow large enough to form volcanic islands. In fact, many oceanic islands are either a single shield volcano or, more often, the coalescence of two or more shields built upon massive amounts of pillow lavas. Examples include the Hawaiian Islands, the Canary Islands, Iceland, the Galapagos Islands, and Easter Island. Although less common, some shield volcanoes form on continental crust. Included in this group are Nyamuragira, Africa's most active volcano, and Newberry Volcano, Oregon.

Mauna Loa: Earth's Largest Shield Volcano

Extensive study of the Hawaiian Islands has revealed that they are constructed of a myriad of thin basaltic lava flows, each averaging a few meters thick, intermixed with relatively minor amounts of ejected pyroclastic material. Mauna Loa is the largest of five overlapping shield volcanoes that comprise the Big Island of Hawaii (see Figure 6.12). From its base on the floor of the Pacific Ocean to its summit, Mauna Loa is over 9 kilometers (6 miles) high, exceeding the height of Mount Everest above sea level. The volume of material composing Mauna Loa is roughly 200 times greater than that of the large composite cone Mount Rainier, located in Washington (Figure 6.13).

Like Hawaii's other shield volcanoes, Mauna Loa has flanks with gentle slopes of only a few degrees. This low angle is due to the very hot, fluid lava that traveled "fast and far" from the vent. In addition, most of the lava (perhaps 80 percent) flowed through a well-developed system of lava tubes. Another feature common to active shield volcanoes is one or more large, steep-walled calderas that occupy the summit (see Figure 6.12). Calderas on shield volcanoes usually form when the roof above the magma chamber collapses. This occurs after the magma reservoir empties, either following a large eruption or as magma migrates to the flank of a volcano to feed a fissure eruption.

In their final stage of growth, shield volcanoes erupt more sporadically, and pyroclastic ejections are more common. The lava emitted later tends to be more viscous, resulting in thicker, shorter flows. These eruptions steepen the slope of the summit area, which often becomes capped

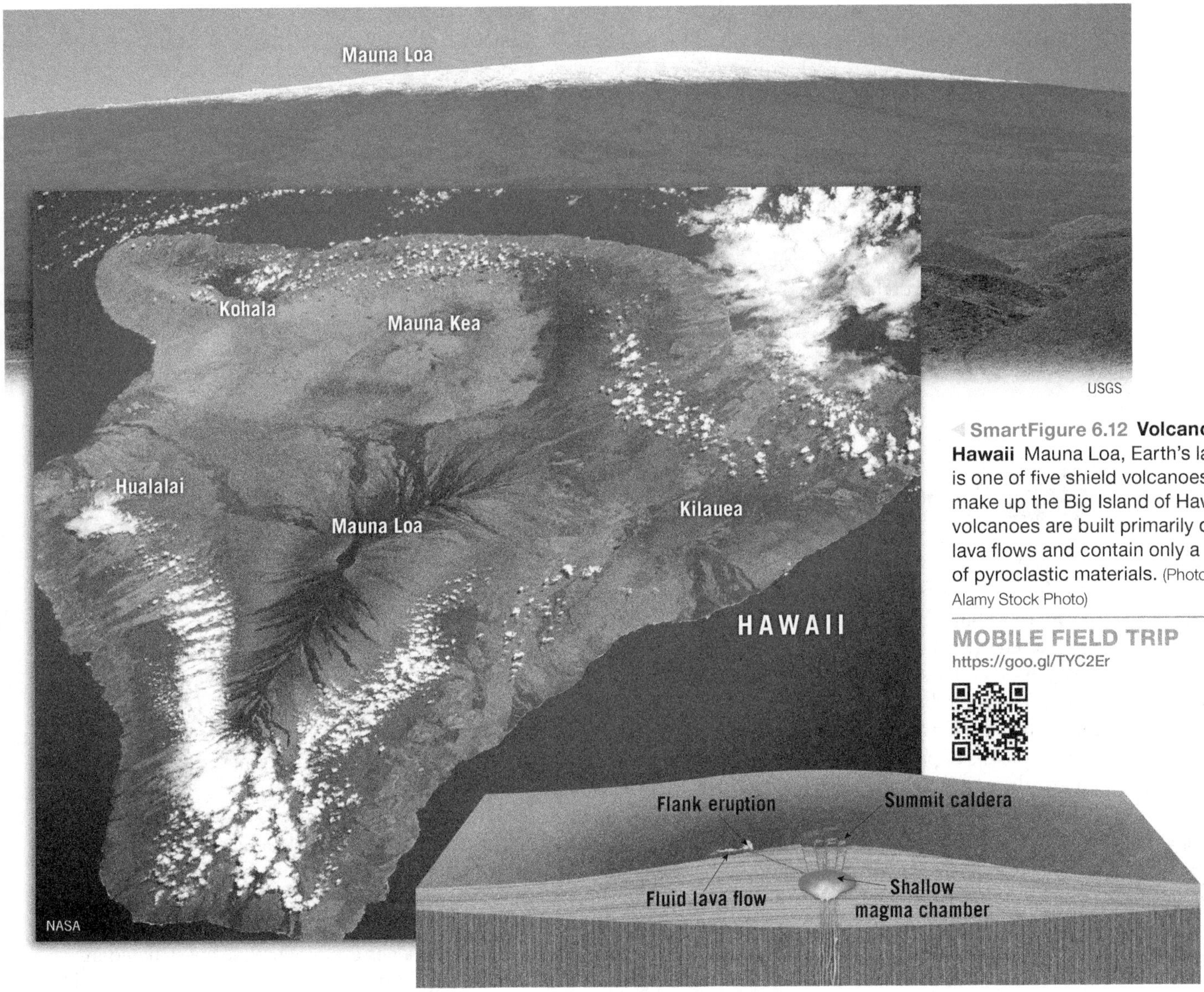

SmartFigure 6.12 Volcanoes of Hawaii Mauna Loa, Earth's largest volcano, is one of five shield volcanoes that collectively make up the Big Island of Hawaii. Shield volcanoes are built primarily of fluid basaltic lava flows and contain only a small percentage of pyroclastic materials. (Photo by Greg Vaughn/ Alamy Stock Photo)

MOBILE FIELD TRIP
https://goo.gl/TYC2Er

with clusters of cinder cones. This explains why Mauna Kea, a more mature volcano that has not erupted in historic times, has a steeper summit than Mauna Loa, which erupted as recently as 1984. Astronomers are so certain that Mauna Kea is "over the hill" that they built an elaborate astronomical observatory on its summit to house some of the world's most advanced and expensive telescopes.

Kilauea: Hawaii's Most Active Volcano

Volcanic activity on the Big Island of Hawaii began on what is now the northwestern flank of the island and has gradually migrated southeastward. It is currently centered on Kilauea Volcano, one of the most active and intensely studied shield volcanoes in the world. Kilauea, located in

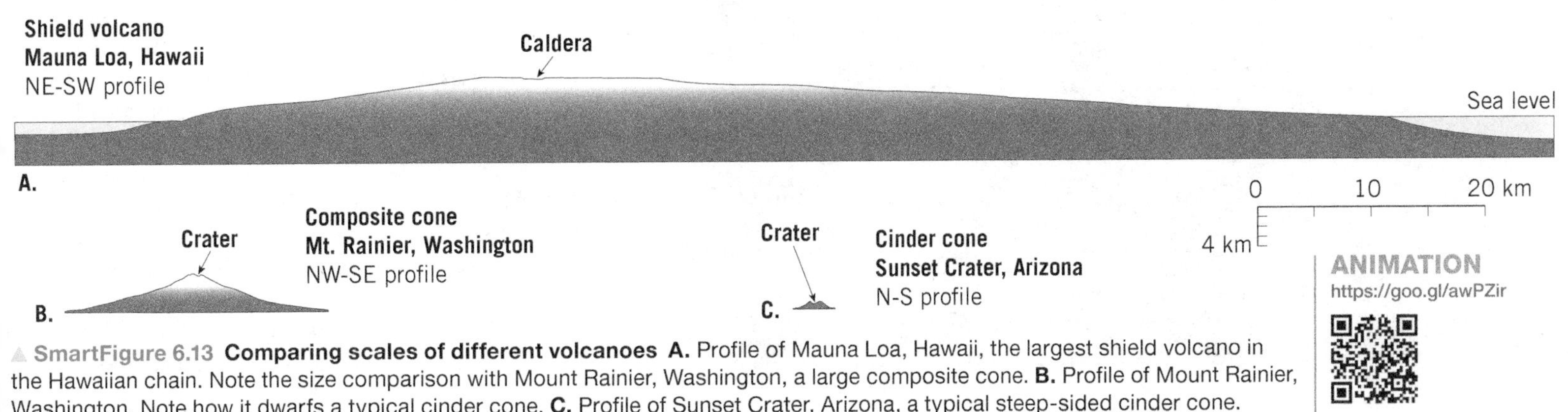

SmartFigure 6.13 Comparing scales of different volcanoes A. Profile of Mauna Loa, Hawaii, the largest shield volcano in the Hawaiian chain. Note the size comparison with Mount Rainier, Washington, a large composite cone. **B.** Profile of Mount Rainier, Washington. Note how it dwarfs a typical cinder cone. **C.** Profile of Sunset Crater, Arizona, a typical steep-sided cinder cone.

ANIMATION
https://goo.gl/awPZir

the shadow of Mauna Loa, has experienced more than 50 eruptions since record keeping began in 1823.

Several months before each eruptive phase, Kilauea inflates as magma gradually migrates upward and accumulates in a central reservoir located a few kilometers below the summit. For up to 24 hours before an eruption, swarms of small earthquakes warn of the impending activity. Most of the recent activity on Kilauea has occurred along the flanks of the volcano, in a region called the *East Rift Zone*. The longest and largest rift eruption ever recorded on Kilauea began in 1983 and continues to this day, with no signs of abating (see GEOgraphics 6.1, page 178).

CONCEPT CHECKS 6.5

1. Describe the composition and viscosity of the lava associated with shield volcanoes.
2. Are pyroclastic materials a significant component of shield volcanoes?
3. Where do most shield volcanoes form—on the ocean floor or on the continents?
4. Where are the best-known shield volcanoes in the United States? Name some examples in other parts of the world.

6.6 Cinder Cones

Describe the formation, size, and composition of cinder cones.

▼ SmartFigure 6.14 **Cinder cones** Cinder cones are built from ejected lava fragments (mostly cinders and bombs) and are relatively small—usually less than 300 meters (1000 feet) in height. This cinder cone, SP Crater, is located north of Flagstone, Arizona. (Photo by Michael Collier)

MOBILE FIELD TRIP
https://goo.gl/X9JvXE

As the name suggests, **cinder cones** (also called **scoria cones**) are built from ejected lava fragments that begin to harden in flight to produce the vesicular rock *scoria* (Figure 6.14). These pyroclastic fragments range in size from fine ash to bombs that may exceed 1 meter (3 feet) in diameter. However, most of the volume of a cinder cone consists of pea- to walnut-sized fragments that are markedly vesicular and have a black to reddish-brown color (see Figure 6.10A). In addition, this pyroclastic material tends to have basaltic composition.

Although cinder cones are composed mostly of loose scoria fragments, some produce extensive lava fields. These lava flows generally form in the final stages of the volcano's life span, when the magma body has lost most of its gas content. Because cinder cones are composed of loose fragments rather than solid rock, the lava usually flows out from the unconsolidated base of the cone rather than from the crater.

Cinder cones have very simple, distinct shapes (see Figure 6.14). Because cinders have a high angle of repose (the steepest angle at which a pile of loose material remains stable), cinder cones are steep-sided, having slopes between 30 and 40 degrees. In addition, cinder cone have large, deep craters relative to the overall size of the structure. Although relatively symmetrical, some cinder cones are elongated and higher on the side that was downwind during the final eruptive phase.

Most cinder cones are produced by a single, short-lived eruptive event. One study found that half of all cinder cones examined were constructed in less than 1 month, and 95 percent of them formed in less than 1 year. Once the event ceases, the magma in the "plumbing" connecting the vent to the magma source solidifies, and the volcano usually does not erupt again. (One exception is Cerro Negro, a cinder cone in Nicaragua, which has erupted more than 20 times since it formed in 1850.) As a result of this typically short life span, cinder cones are small, usually between 30 and 300 meters (100 and 1000 feet) tall. A few rare examples exceed 700 meters (2300 feet) in height.

Cinder cones number in the thousands around the globe. Some occur in groups, such as the volcanic field near Flagstaff, Arizona, which consists of about 600 cones. Others are parasitic cones that are found on the flanks or within the calderas of larger volcanic structures.

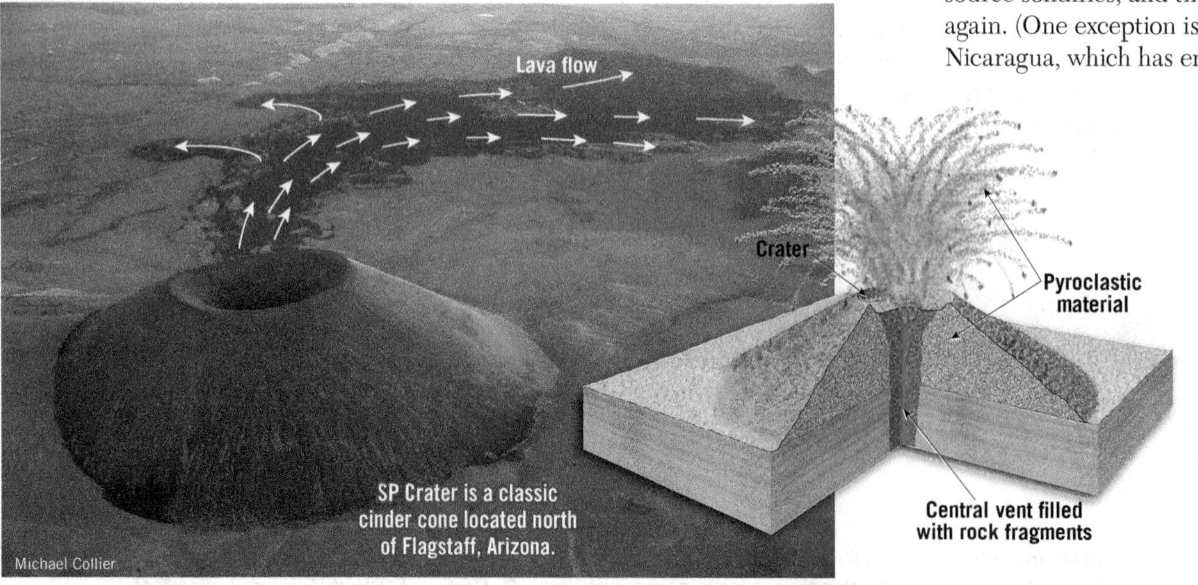

ícutin, a cinder cone located in Mexico, erupted for 9 years.

An aa flow emanating from the base of the cone buried much of the village of San Juan Parangaricutiro, leaving only remnants of the village's church exposed.

SmartFigure 6.15 **Parícutin, a well-known cinder cone** The village of San Juan Parangaricutiro was engulfed by aa lava from Parícutin. Only portions of the church remain. (Photos by Michael Collier)

CONDOR VIDEO
https://goo.gl/Stb3bZ

Parícutin: Life of a Garden-Variety Cinder Cone

One of the very few volcanoes studied by geologists from its very beginning is the cinder cone called Parícutin, located about 320 kilometers (200 miles) west of Mexico City. In 1943, its eruptive phase began in a cornfield owned by Dionisio Pulido, who witnessed the event.

For 2 weeks prior to the first eruption, numerous tremors caused apprehension in the nearby village of Parícutin. Then, on February 20, sulfurous gases began billowing from a small depression that had been in the cornfield for as long as local residents could remember. During the night, hot, glowing rock fragments were ejected from the vent, producing a spectacular fireworks display. Explosive discharges continued, throwing hot fragments and ash occasionally as high as 6000 meters (20,000 feet) into the air. Larger fragments fell near the crater, some remaining incandescent as they rolled down the slope. These materials built an aesthetically pleasing cone, while finer ash fell over a much larger area, burning and eventually covering the village of Parícutin. In the first day, the cone grew to 40 meters (130 feet), and by the fifth day it was more than 100 meters (330 feet) high.

The first lava flow came from a fissure that opened just north of the cone, but after a few months flows began to emerge from the base of the cone. In June 1944, a clinkery aa flow 10 meters (30 feet) thick moved over much of the village of San Juan Parangaricutiro, leaving only remnants of the church exposed (Figure 6.15). After 9 years of intermittent pyroclastic explosions and nearly continuous discharge of lava from vents at its base, the activity ceased almost as quickly as it had begun. Today, Parícutin is just another one of the scores of cinder cones dotting the landscape in this region of Mexico. Like the others, it will not erupt again.

CONCEPT CHECKS 6.6

1. Describe the composition of a cinder cone.
2. How do cinder cones compare in size and steepness of their flanks with shield volcanoes?
3. Over what time span does a typical cinder cone form?

6.7 Composite Volcanoes

List the characteristics of composite volcanoes and describe how they form.

Earth's most picturesque yet potentially dangerous volcanoes are **composite volcanoes**, also known as **stratovolcanoes**. Most are located in a relatively narrow zone that rims the Pacific Ocean, appropriately called the *Ring of Fire* (see Figure 6.36). This active zone includes a chain of continental volcanoes distributed along the west coast of the Americas, including the large cones of the Andes in South America and the Cascade Range of the western United States and Canada.

Classic composite cones are large, nearly symmetrical structures consisting of alternating layers of explosively erupted cinders and ash interbedded with lava flows. A few composite cones, notably Italy's Etna and Stromboli, display very persistent eruption activity, and

GEOGRAPHICS 6.1

Kilauea's East Rift Zone Eruption

Kilauea, one of the most active volcanoes in the world, is located on the island of Hawaii in the shadow of Mauna Loa. Most of the recent activity on Kilauea has occurred along the flanks of the volcano in a region called the East Rift Zone. The longest and largest eruption ever recorded on Kilauea began in 1983, and continues with no signs of abating.
1 Kilauea's most recent eruptive phase began along a 6-kilometer (4 mile) fissure where a 100-meter (300-foot) high "curtain of fire" formed as red-hot basaltic lava was ejected skyward.
Greg Vaughn/Alamy
2 One of many fluid pahoehoe flows that have moved down the flanks of Kilauea since 1983.
3 The activity became localized at a single vent and a series of 44 short-lived episodes of lava fountaining built a cinder and spatter cone—given the Hawaiian name *Puu Oo.*
4 By the summer of 1986 a new vent opened along the rift. Pahoehoe lava flowing from this vent cut off the coastal highway and destroyed more than 180 structures including the National Park Visitor Center.
David Reggie/ Getty Images

1983 to Present

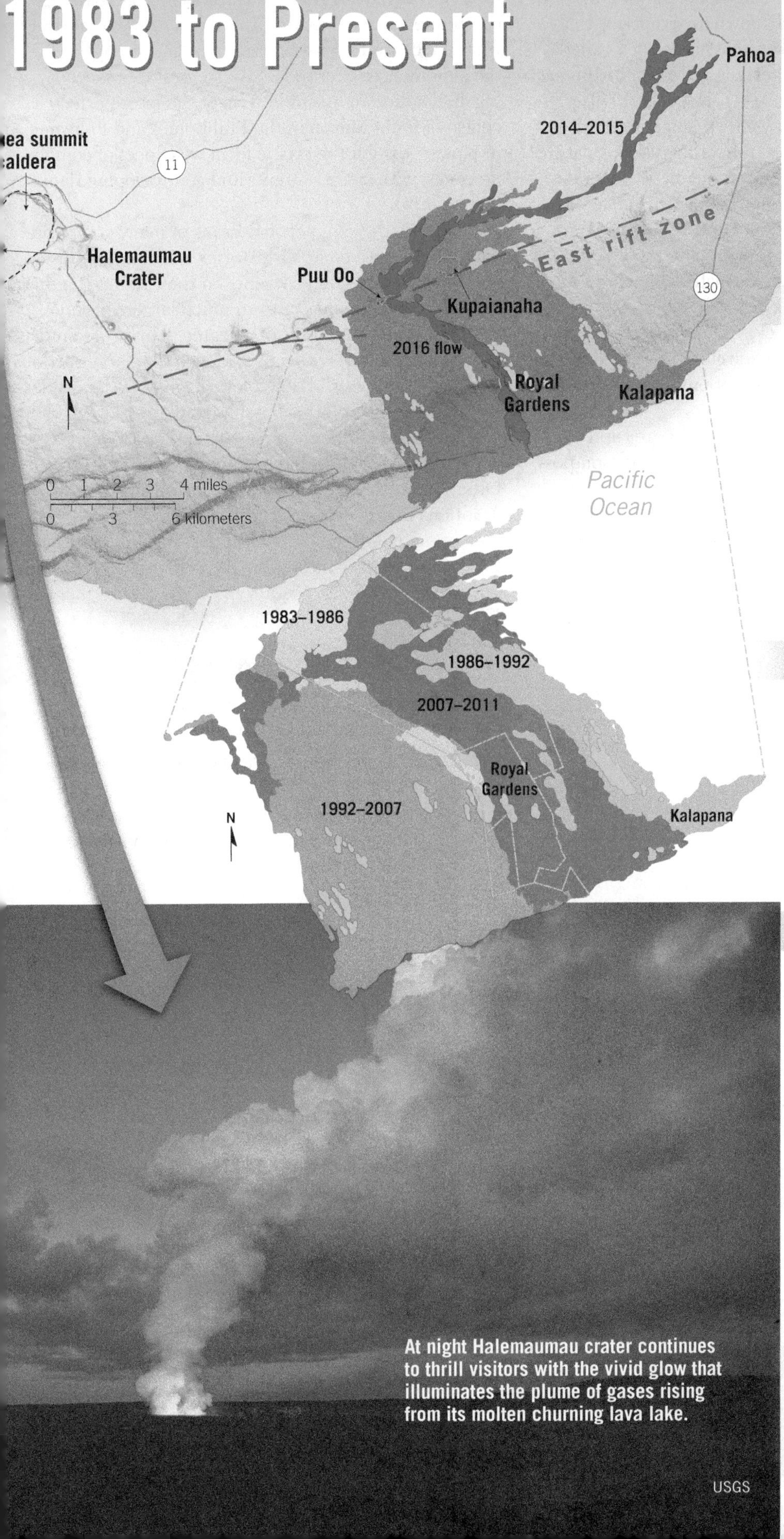

At night Halemaumau crater continues to thrill visitors with the vivid glow that illuminates the plume of gases rising from its molten churning lava lake.

USGS

ERUPTION SUMMARY

1983-1986

The most recent eruption of Kilauea began on January 3, 1983 as a fissure eruption along the East Rift Zone in an area southeast of the summit caldera. These lava fountains quickly built a cinder-and-spatter cone called Puu Oo that produced abundant lava that flowed down the volcano's slope toward the sea.

1986-1992

In 1986 the eruption shifted eastward to new vent that built a small shield volcano called Kupaianaha. Lava tubes fed flows that extended about 12 kilometers (7 miles) to the sea. These flows buried most of the homes in the village of Kalapana under 15-25 meters (50-80 feet) of lava. It also destroyed a section of the Chain of Craters Road, buried Royal Gardens subdivision, and engulfed the famous Black Sand Beach at Kaimu.

1992-2012

Volcanic activity returned to the flanks of Puu Oo in 1992 when lava flowed nearly continuously to the ocean. By 2011, the eruption had added over 500 acres of land to the island, destroyed over 200 structures and buried 14 kilometers (9 miles) of the only major highway through that part of the island.

2014-2016

An eruption that began in 2008 in Halemaumau crater, which resides within Kilauea's summit caldera, remains active. At night, the molten lava lake churns energetically and appears to be spewing fire. In 2016 renewed activity that began on the southeastern flanks of Puu Oo crater generated a lava flow that moved though the abandoned remains of the Royal Gardens subdivision and has reached the Pacific. The flow that began in 2014, which slowly advanced toward and threatened the town of Pahoa and highway 130, has become inactive.

Question:
What is the name of the area on Kilauea where the 1983 eruption began?

molten lava has been observed in their summit craters for decades. Stromboli is so well known for eruptions that eject incandescent blobs of lava that it has been called the "Lighthouse of the Mediterranean." Mount Etna has erupted, on average, once every 2 years since 1979.

Just as shield volcanoes owe their shape to fluid basaltic lavas, composite cones reflect the viscous nature of the material from which they are made. In general, composite cones are the product of silica-rich magma having an andesitic composition. However, many composite cones also emit various amounts of fluid basaltic lava and, occasionally, pyroclastic material having a felsic (rhyolitic) composition. The silica-rich magmas typical of composite cones generate thick, viscous lavas that travel less than a few kilometers. Composite cones are also noted for generating explosive eruptions that eject huge quantities of pyroclastic material.

A conical shape, with a steep summit area and gradually sloping flanks, is typical of most large composite cones. This classic profile, which adorns calendars and postcards, is partially a result of the way viscous lavas and pyroclastic ejected materials contribute to the cone's growth. Coarse fragments ejected from the summit crater tend to accumulate near their source and contribute to the steep slopes around the summit. Finer ejected materials, on the other hand, are deposited as a thin layer over a large area and hence tend to flatten the flank of the cone. In addition, during the early stages of growth, lavas tend to be more abundant and flow greater distances from the vent than they do later in the volcano's history, which contributes to the cone's broad base. As a composite volcano matures, the shorter flows that come from the central vent serve to armor and strengthen the summit area. Consequently, steep slopes exceeding 40 degrees are possible. Two of the most perfect cones—Mount Mayon in the Philippines and Fujiyama in Japan—exhibit the classic form we expect of composite cones, with steep summits and gently sloping flanks (**Figure 6.16**).

Despite the symmetrical forms of many composite cones, most have complex histories. Many composite volcanoes have secondary vents on their flanks that have produced cinder cones or even much larger volcanic structures. Huge mounds of volcanic debris surrounding these structures provide evidence that large sections of these volcanoes slid downslope as massive landslides. Some develop amphitheater-shaped depressions at their summits as a result of explosive lateral eruptions—as occurred during the 1980 eruption of Mount St. Helens. Often, so much rebuilding has occurred since these eruptions that no trace of these amphitheater-shaped scars remain. Other stratovolcanoes, such as Crater Lake, have been truncated by the collapse of their summit (see Figure 6.22).

CONCEPT CHECKS 6.7

1. What name is given to the region having the greatest concentration of composite volcanoes?
2. Describe the materials that compose composite volcanoes.
3. How does the composition and viscosity of lava flows differ between composite volcanoes and shield volcanoes?

▼ Figure 6.16 **Fujiyama, a classic composite volcano** Japan's Fujiyama exhibits the classic form of a composite cone—a steep summit and gently sloping flanks. (Photo by Koji Nakano/Getty Images, Inc-Liaison)

6.8 Volcanic Hazards

Describe the major geologic hazards associated with volcanoes.

Roughly 1500 of Earth's known volcanoes have erupted at least once, and some several times, in the past 10,000 years. Based on historical records and studies of active volcanoes, 70 volcanic eruptions can be expected each year. In addition, 1 large-volume eruption can be expected every decade. These large eruptions account for the vast majority of volcano-related human fatalities.

Today, an estimated 500 million people in places such as Japan, Indonesia, Italy, and Oregon live near active volcanoes. They face a number of volcanic hazards, such as destructive pyroclastic flows, molten lava flows, mudflows called lahars, and falling ash and volcanic bombs.

Pyroclastic Flow: A Deadly Force of Nature

One of the most destructive forces of nature is a **pyroclastic flow**, which consists of hot gases infused with incandescent ash and larger lava fragments. Also known as **nuée ardentes** ("glowing avalanches"), these fiery flows can race down steep volcanic slopes at speeds exceeding 100 kilometers (60 miles) per hour (Figure 6.17). Pyroclastic flows have two components—a low-density cloud of hot expanding gases containing fine ash particles, and a ground-hugging portion composed of pumice and other vesicular pyroclastic material.

Driven by Gravity Pyroclastic flows are propelled by the force of gravity and tend to move in a manner similar to snow avalanches. They are mobilized by expanding volcanic gases released from the lava fragments and by the expansion of heated air that is overtaken and trapped in the moving front. These gases reduce friction between ash and pumice fragments, which gravity propels downslope in a nearly frictionless environment. This is why some pyroclastic flow deposits are found many miles from their source.

Occasionally, powerful hot blasts that carry small amounts of ash separate from the main body of a pyroclastic flow. These low-density clouds, called *surges*, can be deadly but seldom have sufficient force to destroy buildings in their paths. Nevertheless, in 2014, a hot ash cloud from Japan's Mount Ontake killed 47 hikers and injured 69 more.

Pyroclastic flows may originate in a variety of volcanic settings. Some occur when a powerful eruption blasts pyroclastic material out of the side of a volcano. More frequently, however, pyroclastic flows are generated by the collapse of tall eruption columns during an explosive event. When gravity eventually overcomes the initial upward thrust provided by the escaping gases, the ejected materials begin to fall, sending massive amounts of incandescent blocks, ash, and pumice cascading downslope.

The Destruction of St. Pierre In 1902, an infamous pyroclastic flow and associated surge from Mount Pelée, a small volcano on the Caribbean island of Martinique, destroyed the port town of St. Pierre. Although the main pyroclastic flow was largely confined to the valley of Riviere Blanche, a low-density fiery surge spread south of the river and quickly engulfed the entire city. The destruction happened in moments and was so devastating that nearly all of St. Pierre's 28,000 inhabitants were killed. Only 1 person on the outskirts of town—a prisoner protected in a dungeon—and a few people on ships in the harbor were spared (Figure 6.18).

Scientists who arrived on the scene within days found that although St. Pierre was mantled by only a thin layer of volcanic debris, masonry walls nearly 1 meter (3 feet) thick had been knocked over like dominoes, large trees had been uprooted, and cannons had been torn from their mounts.

The Destruction of Pompeii One well-documented event of historic proportions was the C.E. 79 eruption of the Italian volcano we now call Mount Vesuvius. For centuries prior to this eruption, Vesuvius had been dormant, with vineyards adorning its sunny slopes. Yet in less than 24 hours, the entire city of Pompeii (near Naples) and a few thousand of its residents were entombed beneath a layer of volcanic ash and pumice. The city and the victims of the eruption remained buried for nearly 17 centuries.

Figure 6.17 Pyroclastic flows, one of the most destructive volcanic forces A. These pyroclastic flows occurred on Mount Mayon, Philippines, during the 1984 eruption. Pyroclastic flows are composed of hot ash and pumice and/or blocky lava fragments that race down the slope of volcanoes. (Photo courtesy of USGS) **B.** Residents running away from a pyroclastic flow that reached the base of Mound Sinabung, Indonesia, 2014. (Photo by Chaideer Mahyuddin/AFP/Getty Images)

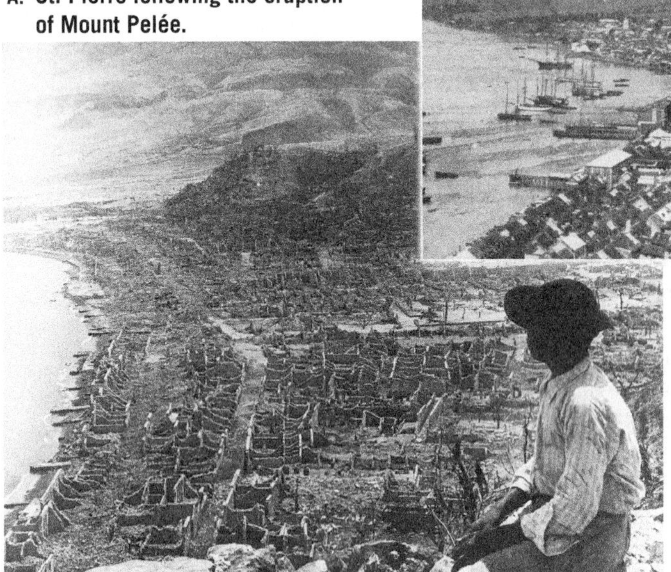

A. St. Pierre following the eruption of Mount Pelée.
B. St. Pierre before the 1902 eruption.

Figure 6.18 Destruction of St. Pierre A. St. Pierre as it appeared shortly after the eruption of Mount Pelée in 1902. (Reproduced from the collection of the Library of Congress) **B.** St. Pierre before the eruption. Many vessels were anchored offshore when this photo was taken, as was the case on the day of the eruption. (Photo by UPPA/Photoshot)

The excavation of Pompeii gave archaeologists a superb picture of ancient Roman life (Figure 6.19A).

By reconciling historic records with detailed scientific studies of the region, volcanologists reconstructed the sequence of events. During the first day of the eruption, a rain of ash and pumice accumulated at a rate of 12 to 15 centimeters (5 to 6 inches) per hour, causing most of the roofs in Pompeii to eventually give way. Then, suddenly, a surge of searing hot ash and gas swept rapidly down the flanks of Vesuvius. This deadly pyroclastic flow killed those who had somehow managed to survive the initial ash and pumice fall. Their remains were quickly buried by falling ash, and subsequent rainfall caused the ash to harden. Over the centuries, the remains decomposed, creating cavities that were discovered by nineteenth-century excavators. Casts were then produced by pouring plaster of Paris into the voids (Figure 6.19B). Mount Vesuvius has had more than two dozen explosive eruptions since C.E. 79, the most recent occurring in 1944. Today, Vesuvius towers over the Naples skyline, a region occupied by roughly 3 million people. Such an image should prompt us to consider how volcanic crises might be managed in the future.

Lahars: Mudflows on Active and Inactive Cones

In addition to violent eruptions, large composite cones may generate a type of fluid mudflow known by its Indonesian name, **lahar**. These destructive flows occur when volcanic debris becomes saturated with water and rapidly moves down steep volcanic slopes, generally following stream valleys. Some lahars are triggered when magma nears the surface of a glacially clad volcano, causing large volumes of ice and snow to melt. Others are generated when heavy rains saturate weathered volcanic deposits. Thus, lahars may occur even when a volcano is *not* erupting.

When Mount St. Helens erupted in 1980, several lahars were generated. These flows and accompanying floodwaters raced down nearby river valleys at speeds exceeding 30 kilometers (20 miles) per hour. These raging rivers of mud destroyed or severely damaged nearly all the homes and bridges along their paths (Figure 6.20). Fortunately, the area was not densely populated.

In 1985, deadly lahars were produced during a small eruption of Nevado del Ruiz, a 5300-meter (17,400-foot) volcano in the Andes Mountains of Colombia. Hot pyroclastic material melted ice and snow that capped the mountain (*nevado* means "snowy" in Spanish) and sent torrents of ash and debris down three major river valleys that flank the volcano. Reaching speeds of 100 kilometers (60 miles) per hour, these mudflows tragically claimed 25,000 lives.

Many consider Mount Rainier, Washington, to be America's most dangerous volcano because, like Nevado del Ruiz, it has a thick, year-round mantle of snow and glacial ice. Adding to the risk is the fact that more than 100,000 people live

Cone produced by the C.E. 79 eruption.

VESUVIUS

Olivier Goujon/Robert Harding

A.

Leonard von Matt/Science Source

B.

Figure 6.19 Pompeii was excavated nearly 17 centuries after the C.E. 79 eruption of Mount Vesuvius A. The ruins of the Roman city of Pompeii as they appear today. In less than 24 hours, Pompeii and all its residents were buried under a layer of volcanic ash and pumice that fell like rain. **B.** Plaster casts of some of the victims of the eruption of Mount Vesuvius.

in the valleys around Rainier, and many homes are built on deposits left by lahars that flowed down the volcano hundreds or thousands of years ago. A future eruption, or perhaps just a period of heavier-than-average rainfall, may produce lahars that could be similarly destructive.

▲ **Figure 6.20 Lahars, mudflows that originate on volcanic slopes A.** This lahar raced down the snow-covered slopes of Mount St. Helens following an eruption on March 19, 1982. **B.** The aftermath of a lahar that formed following the 1982 eruption of Galunggung Volcano in Indonesia.

Other Volcanic Hazards

Volcanoes can be hazardous to human health and property in other ways. Ash and other pyroclastic material can collapse the roofs of buildings or may be drawn into the lungs of humans and other animals or into aircraft engines (Figure 6.21). Volcanic gases, most notably sulfur dioxide, pollute the air and, when mixed with rainwater, can destroy vegetation and reduce the quality of groundwater. Despite the known risks, millions of people live in close proximity to active volcanoes.

Volcano-Related Tsunamis Although **tsunamis** are most often associated with displacement along a fault located on the seafloor (see Chapter 5), some result from the collapse of a volcanic cone. This was dramatically demonstrated during the 1883 eruption on the Indonesian island of Krakatau, when the northern half of a volcano plunged into the Sunda Strait, creating a tsunami that exceeded 30 meters (100 feet) in height. Although Krakatau was uninhabited, an estimated 36,000 people were killed along the coastline of the islands of Java and Sumatra.

Volcanic Ash and Aviation Over the past 2 decades, hundreds of commercial jets have been damaged by inadvertently flying into clouds of volcanic ash. For example, in 1989, a Boeing 747 carrying more than 300 passengers encountered an ash cloud from Alaska's Redoubt Volcano; all four engines clogged with ash and stalled mid-air.

Ash and other pyroclastic materials can collapse roofs, or completely cover buildings.

Lava flows can destroy homes, roads, and other structures in their paths.

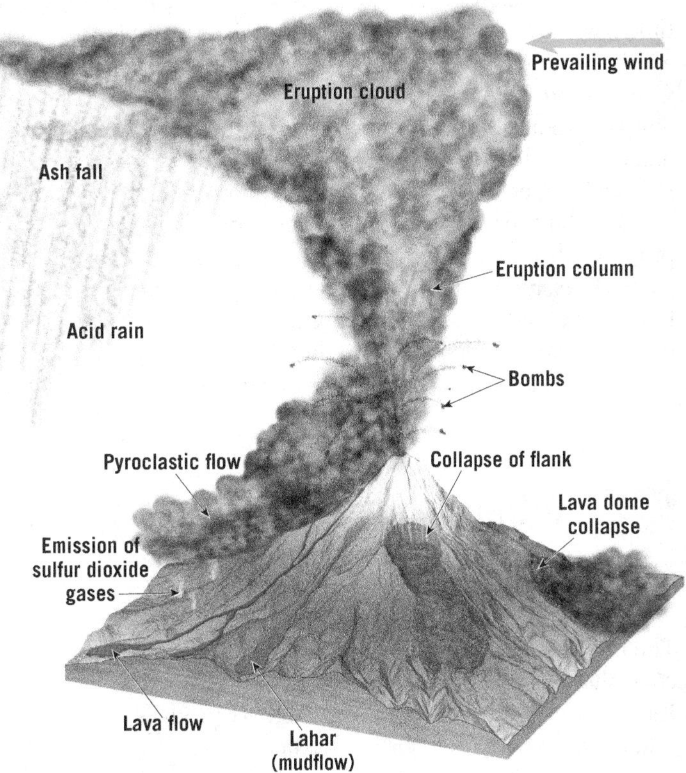

◄ **Figure 6.21 Volcanic hazards** In addition to generating destructive pyroclastic flows and lahars, volcanoes can be hazardous to human health and property in many other ways.

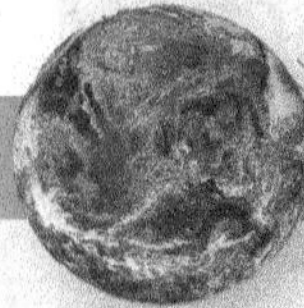

EYE ON EARTH 6.1

This photo shows the February 3, 2015, eruption of Mount Sinabung in North Sumatra, Indonesia. Before awaking in 2010, Mount Sinabung had been dormant since 1600. In recent years it has erupted several times, resulting in the evacuation of more than 30,000 people and at least 23 deaths.

QUESTION 1 *What name is given to the ash- and pumice-laden cloud that is racing down this volcano?*

QUESTION 2 *What type of volcano is associated with these destructive eruptions?*

Fortunately, the pilots were able to dive to a lower altitude and restart the engines, which allowed them to safely land the aircraft in Anchorage.

More recently, the 2010 eruption of Iceland's Eyjafjallajökull Volcano sent ash high into the atmosphere. This thick plume of ash drifted over Europe, causing airlines all across Europe to cancel thousands of flights and leaving hundreds of thousands of travelers stranded. Several weeks passed before air travel resumed its normal schedule.

Engine clogging is not the only problem. Ash can damage an aircraft's fuselage and pit the windshields of the pilot's cabin. In addition, ash has been known to coat a plane to the point that it becomes tail-heavy.

Volcanic Gases and Respiratory Health One of the most destructive volcanic events, called the Laki eruptions, began along a large fissure in southern Iceland in 1783. An estimated 14 cubic kilometers (3.4 cubic miles) of fluid basaltic lavas were released, along with 130 million tons of sulfur dioxide and other poisonous gases. When sulfur dioxide is inhaled, it reacts with moisture in the lungs to produce sulfuric acid, a deadly toxin. More than half of Iceland's livestock died, and the ensuing famine killed 25 percent of the island's human population.

This huge eruption also endangered people and property all across Europe. Crop failure occurred in parts of Western Europe, and thousands of residents perished from lung-related diseases. One report estimated that a similar eruption today would cause more than 140,000 cardiopulmonary fatalities in Europe alone.

Effects of Volcanic Ash and Gases on Weather and Climate Volcanic eruptions can eject dust-sized particles of volcanic ash and sulfur dioxide gas high into the atmosphere. The ash particles reflect sunlight back to space, producing temporary atmospheric cooling. The 1783 Laki eruptions in Iceland appear to have affected atmospheric circulation around the globe. Drought conditions prevailed in the Nile River valley and in India, and the winter of 1784 saw the longest period of below-zero temperatures in New England's history.

Other eruptions that have produced significant effects on climate worldwide include the eruption of Indonesia's Mount Tambora in 1815, which produced the "year without a summer" (1816), and the eruption of El Chichón in Mexico in 1982. El Chichón's eruption, although small, emitted an unusually large quantity of sulfur dioxide that reacted with water vapor in the atmosphere to produce a dense cloud of tiny sulfuric acid droplets. Such particles, called *aerosols*, take several years to settle out of the atmosphere. Like fine ash, these aerosols lower the mean temperature of the atmosphere by reflecting solar radiation back to space.

CONCEPT CHECKS 6.8

1. Describe pyroclastic flows and explain why they are capable of traveling great distances.
2. What is a lahar?
3. List at least three volcanic hazards besides pyroclastic flows and lahars.

6.9 Other Volcanic Landforms

List volcanic landforms other than shield, cinder, and composite volcanoes and describe their formation.

The most widely recognized volcanic structures are the cone-shaped edifices of composite volcanoes that dot Earth's surface. However, volcanic activity produces other distinctive and important landforms.

Calderas

Recall that *calderas* are large steep-sided depressions that have diameters exceeding 1 kilometer (0.6 miles) and have a somewhat circular form. Those less than 1

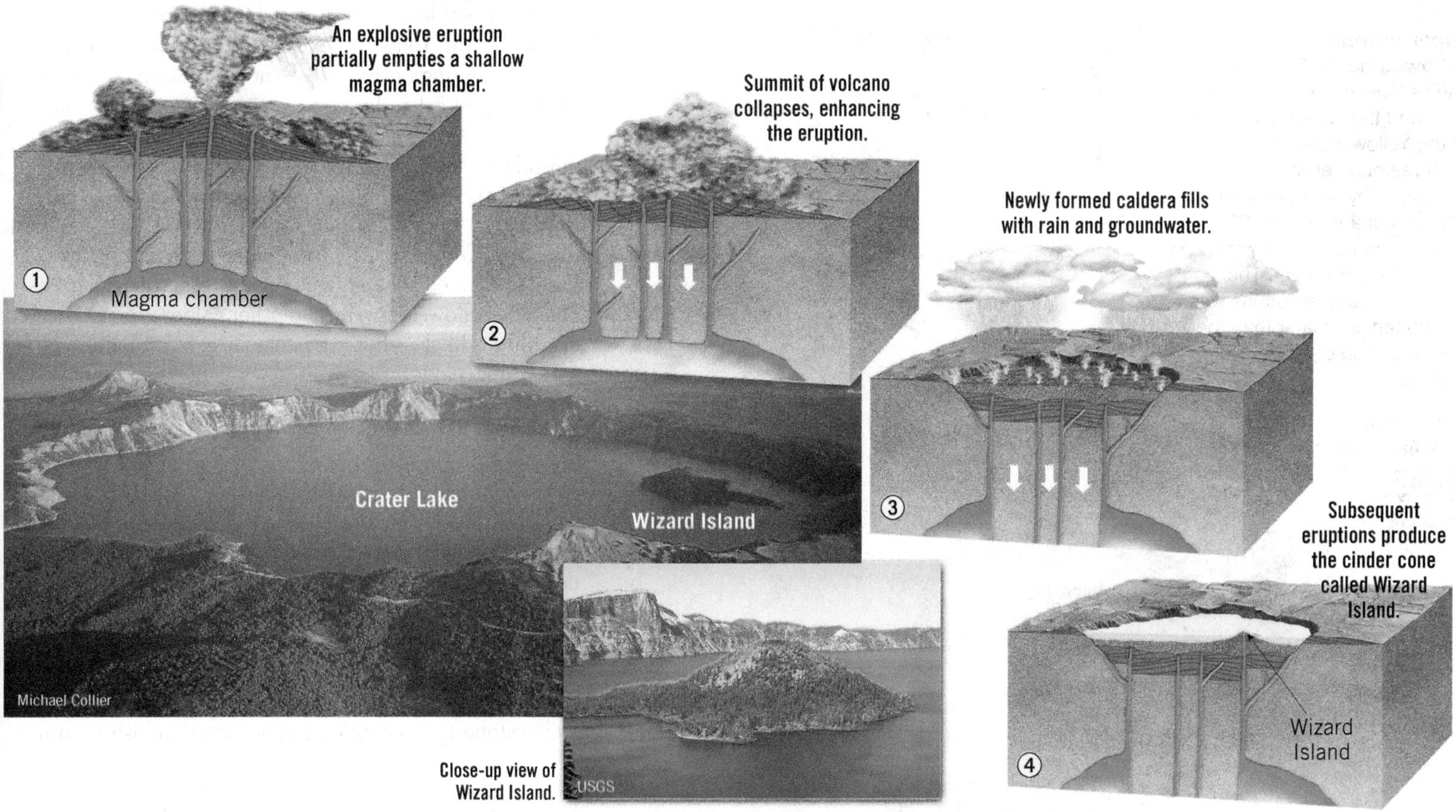

An explosive eruption partially empties a shallow magma chamber.

Summit of volcano collapses, enhancing the eruption.

Newly formed caldera fills with rain and groundwater.

Subsequent eruptions produce the cinder cone called Wizard Island.

① Magma chamber

②

③

④ Wizard Island

Crater Lake

Wizard Island

Michael Collier

Close-up view of Wizard Island.

USGS

▲ SmartFigure 6.22
Formation of Crater Lake–type calderas About 7000 years ago, a violent eruption partly emptied the magma chamber of former Mount Mazama, causing its summit to collapse. Precipitation and groundwater contributed to forming Crater Lake, the deepest lake in the United States—594 meters (1949 feet) deep—and the ninth-deepest lake in the world.

ANIMATION
https://goo.gl/kUCPNB

kilometer across are called *collapse pits* or *craters*. Most calderas are formed by one of the following processes: (1) the collapse of the summit of a large composite volcano following an explosive eruption of silica-rich pumice and ash fragments (*Crater Lake–type calderas*); (2) the collapse of the top of a shield volcano caused by subterranean drainage from a central magma chamber (*Hawaiian-type calderas*); and (3) the collapse of a large area, caused by the discharge of colossal volumes of silica-rich pumice and ash along ring fractures (*Yellowstone-type calderas*).

Crater Lake–Type Calderas Crater Lake, Oregon, is situated in a caldera approximately 10 kilometers (6 miles) wide and 600 meters (more than 1970 feet) deep. This caldera formed about 7000 years ago, when a composite cone named Mount Mazama violently extruded 50 to 70 cubic kilometers of pyroclastic material (Figure 6.22). With the loss of support, 1500 meters (nearly 1 mile) of the summit of this once-prominent cone collapsed, producing a caldera that eventually filled with water. Later, volcanic activity built a small cinder cone in the caldera. Today this cone, called Wizard Island, provides a mute reminder of past activity.

Hawaiian-Type Calderas Unlike Crater Lake–type calderas, many calderas form gradually because of the loss of lava from a shallow magma chamber underlying a volcano's summit. For example, Hawaii's active shield volcanoes, Mauna Loa and Kilauea, both have large calderas at their summits. Kilauea's measures 3.3 by 4.4 kilometers (about 2 by 3 miles) and is 150 meters (500 feet) deep. The walls are almost vertical, and as a result, the caldera looks like a vast, nearly flat-bottomed pit. Kilauea's caldera formed by gradual subsidence as magma slowly drained laterally from the underlying magma chamber, leaving the summit unsupported.

Yellowstone-Type Calderas Historic and destructive eruptions such as that of Mount St. Helens pale in comparison to what happened 630,000 years ago in the region now occupied by Yellowstone National Park, when approximately 1000 cubic kilometers of pyroclastic material erupted. This catastrophic eruption sent showers of ash as far as the Gulf of Mexico and formed a caldera 70 kilometers (43 miles) across (Figure 6.23A). Vestiges of this event are the many hot springs and geysers in the Yellowstone region.

Yellowstone-type eruptions eject huge volumes of pyroclastic materials, mainly in the form of ash and pumice fragments. Typically, these materials are ejected as *pyroclastic flows* that sweep across the landscape, destroying most living things in their paths. Upon coming to rest, the hot fragments of ash and pumice fuse together, forming a welded tuff that closely resembles a solidified lava flow. Despite the immense size of these calderas, the eruptions that produce them are brief, lasting hours to perhaps a few days.

▶ SmartFigure 6.23 **Super-eruptions at Yellowstone** **A.** This map shows Yellowstone National Park and the location and size of the Yellowstone caldera. **B.** Three huge eruptions, separated by relatively regular intervals of about 700,000 years, were responsible for the ash layers shown. The largest of these eruptions was 10,000 times greater than the 1980 eruption of Mount St. Helens.

TUTORIAL
https://goo.gl/y44zXb

Large calderas tend to exhibit a complex eruptive history. In the Yellowstone region, for example, three caldera-forming episodes are known to have occurred over the past 2.1 million years (**Figure 6.23B**). The most recent eruption (630,000 years ago) was followed by episodic outpourings of degassed rhyolitic and basaltic lavas. In the intervening years a slow upheaval of the floor of the caldera produced two elevated regions called *resurgent domes* (see Figure 6.23A). A recent study has determined that a huge magma reservoir still exists beneath Yellowstone; thus, another caldera-forming eruption is likely—but not necessarily imminent.

Unlike calderas associated with shield volcanoes or composite cones, Yellowstone-type calderas are so vast and poorly defined that many were undetected until high-quality aerial and satellite images became available. Other examples of Yellowstone-type calderas are California's Long Valley Caldera; LaGarita Caldera in the San Juan Mountains of southern Colorado; and the Valles Caldera, west of Los Alamos, New Mexico. These and similar calderas found around the globe are among the largest volcanic structures on Earth, hence the name *supervolcanoes*. Volcanologists compare their destructive force to that of the impact of a small asteroid. Fortunately, no Yellowstone-type eruption has occurred in historic times.

Fissure Eruptions and Basalt Plateaus

The greatest volume of volcanic material is extruded from fractures in Earth's crust, called *fissures*. Rather than building cones, **fissure eruptions** usually emit fluid basaltic lavas that blanket wide areas (**Figure 6.24**). In some locations, extraordinary amounts of lava have been extruded along fissures in a relatively short time, geologically speaking. These voluminous accumulations are commonly called **basalt plateaus** because most have a basaltic composition and tend to be rather flat and broad. The Columbia Plateau in

▶ Figure 6.24 **Basaltic fissure eruptions** Lava fountaining from a fissure and formation of fluid lava flows called *flood basalts*. The lower photo shows flood basalt flows near Idaho Falls. (Photo by University of Washington Libraries, Special Collections, KC6673)

the northwestern United States, which consists of the Columbia River basalts, is a product of this type of activity (**Figure 6.25**). Numerous fissure eruptions have buried the landscape, creating a lava plateau nearly 1500 meters (1 mile) thick. Some of the lava remained molten long enough to flow 150 kilometers (90 miles) from its source. The term **flood basalts** appropriately describes these extrusions.

Massive accumulations of basaltic lava, similar to those of the Columbia Plateau, occur elsewhere in the world. One of the largest is known as the Deccan Plateau or Deccan Traps (*traps* = stairs), a thick sequence of flat-lying basalt flows covering nearly 500,000 square kilometers (195,000 square miles) of west-central India. When the Deccan Traps formed about 66 million years ago, nearly 2 million cubic kilometers of lava were extruded over a period of approximately 1 million years. Several other massive accumulations of flood basalts, including the Ontong Java Plateau, have been discovered in the deep-ocean basins (see Figure 6.39).

Volcanic Necks

Most of the lava and materials erupted from a volcano travel through short conduits that connect shallow magma chambers to vents located at the surface. When a volcano becomes inactive, congealed magma is often preserved in the feeding conduit of the volcano as a crudely cylindrical mass. As the volcano succumbs to forces of weathering and erosion, the rock occupying the volcanic conduit, which is highly resistant to weathering, may remain standing above the surrounding terrain long after the cone has been worn away. Shiprock, New Mexico, is a widely recognized and spectacular example of these structures, which geologists call **volcanic necks** (or **plugs**) (**Figure 6.26**). More than 510 meters (1700 feet) high, Shiprock is taller than most skyscrapers and is one of many such landforms that protrude conspicuously from the red desert landscapes of the American Southwest.

CONCEPT CHECKS 6.9

1. Describe the formation of Crater Lake. Compare it to the calderas found on shield volcanoes such as Kilauea.
2. Other than composite volcanoes, what volcanic landform can generate a pyroclastic flow?
3. How do the eruptions that created the Columbia Plateau differ from the eruptions that create large composite volcanoes?
4. What type of volcanic structure is Shiprock, New Mexico, and how did it form?

Figure 6.25 Columbia River basalts A. The Columbia River basalts cover an area of nearly 164,000 square kilometers (63,000 square miles) that is commonly called the Columbia Plateau. Activity here began about 17 million years ago, as lava began to pour out of large fissures, eventually producing a basalt plateau with an average thickness of more than 1 kilometer. **B.** Columbia River basalt flows exposed in the Palouse River Canyon in southwestern Washington State. (Photo by Williamborg)

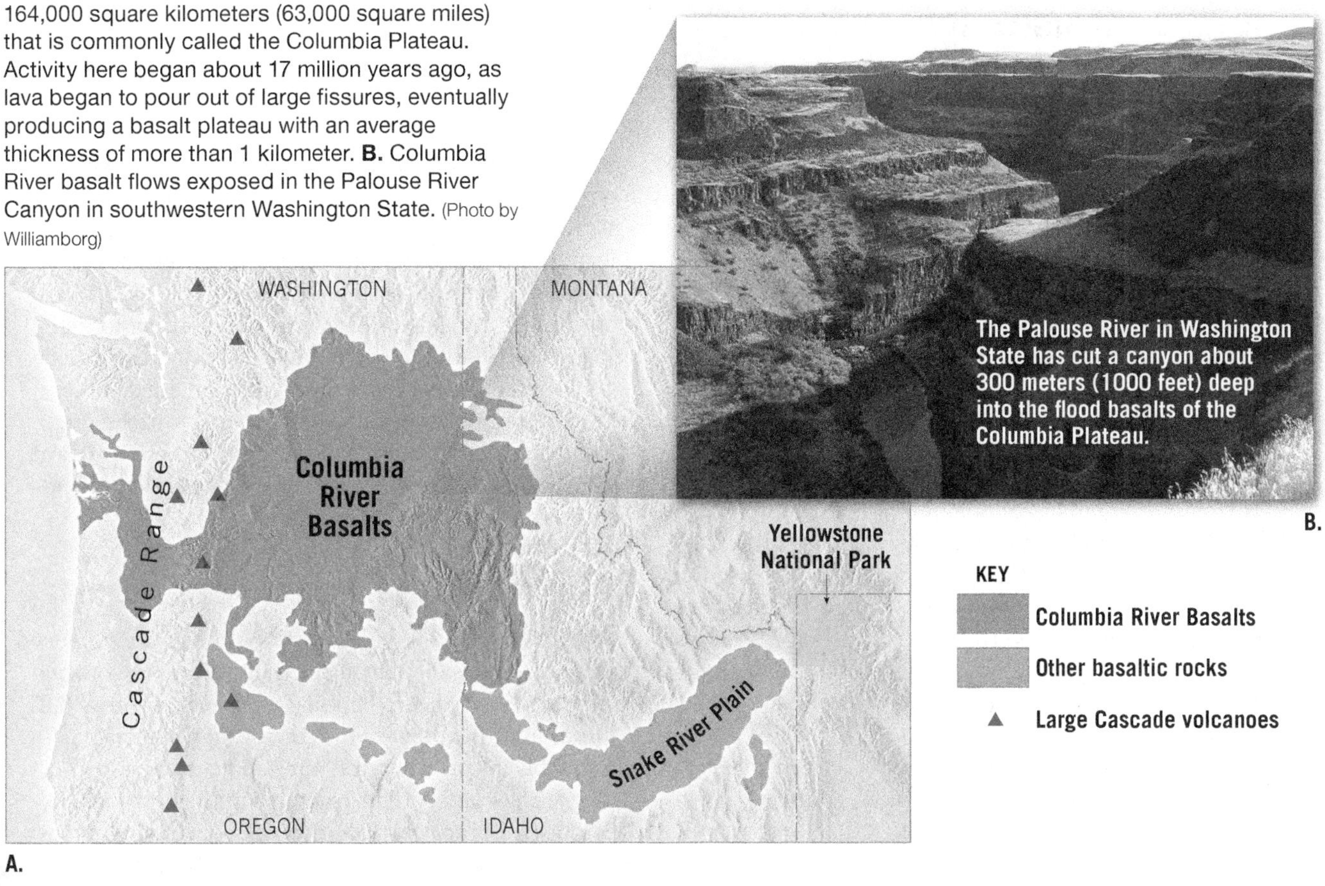

SmartFigure 6.26
Volcanic neck Shiprock, New Mexico, is a volcanic neck that stands about 520 meters (1700 feet) high. It consists of igneous rock that crystallized in the vent of a volcano that has long since been eroded.
(Photo by Dennis Tasa)

TUTORIAL
https://goo.gl/TjW5uh

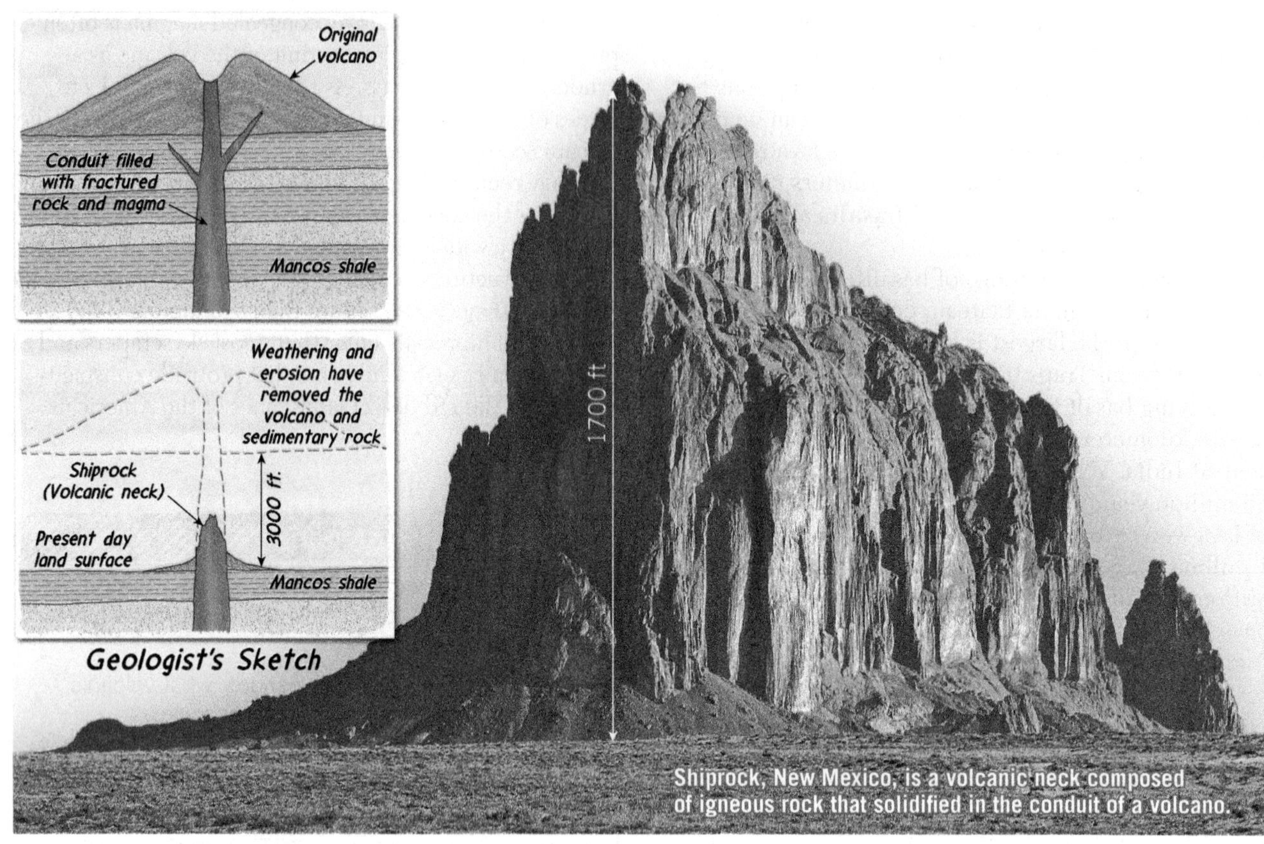

6.10 Intrusive Igneous Activity

Compare and contrast these intrusive igneous structures: dikes, sills, batholiths, stocks, and laccoliths.

Although volcanic eruptions are occasionally violent and spectacular events, most magma crystallizes at depth, without fanfare. Therefore, understanding the igneous processes that occur deep underground is as important to geologists as studying volcanic events.

Nature of Intrusive Bodies

When magma rises through the crust, it forcefully displaces preexisting crustal rocks, termed **host rock**, or **country rock**. The structures that result from the emplacement of magma into preexisting rocks are called **intrusions** or **plutons**. Because all intrusions form far below Earth's surface, they are studied primarily after uplifting and erosion (covered in later chapters) have exposed them. The challenge lies in reconstructing the events that generated these structures in vastly different conditions deep underground, millions of years ago.

Intrusions are known to occur in a great variety of sizes and shapes. Some of the most common types are illustrated in Figure 6.27. Notice that some plutons have a **tabular** (*tabula* = table) shape, whereas others are best described as **massive** (blob shaped). Also, observe that some of these bodies cut across existing structures, such as sedimentary strata, whereas others form when magma is injected between sedimentary layers. Because of these differences, intrusive igneous bodies are generally classified according to their shape as either tabular or massive and by their orientation with respect to the host rock. Igneous bodies are said to be **discordant** (*discordare* = to disagree) if they cut across existing structures and **concordant** (*concordare* = to agree) if they inject parallel to features such as sedimentary strata.

Tabular Intrusive Bodies: Dikes and Sills

Dikes and Sills Tabular intrusive bodies are produced when magma is forcibly injected into a fracture or zone of weakness, such as a bedding surface (see Figure 6.27). **Dikes** are discordant bodies that form when magma is forcibly injected into fractures and cut across bedding surfaces and other structures in the host rock. By contrast, **sills** are nearly horizontal, concordant bodies that form when magma exploits weaknesses between

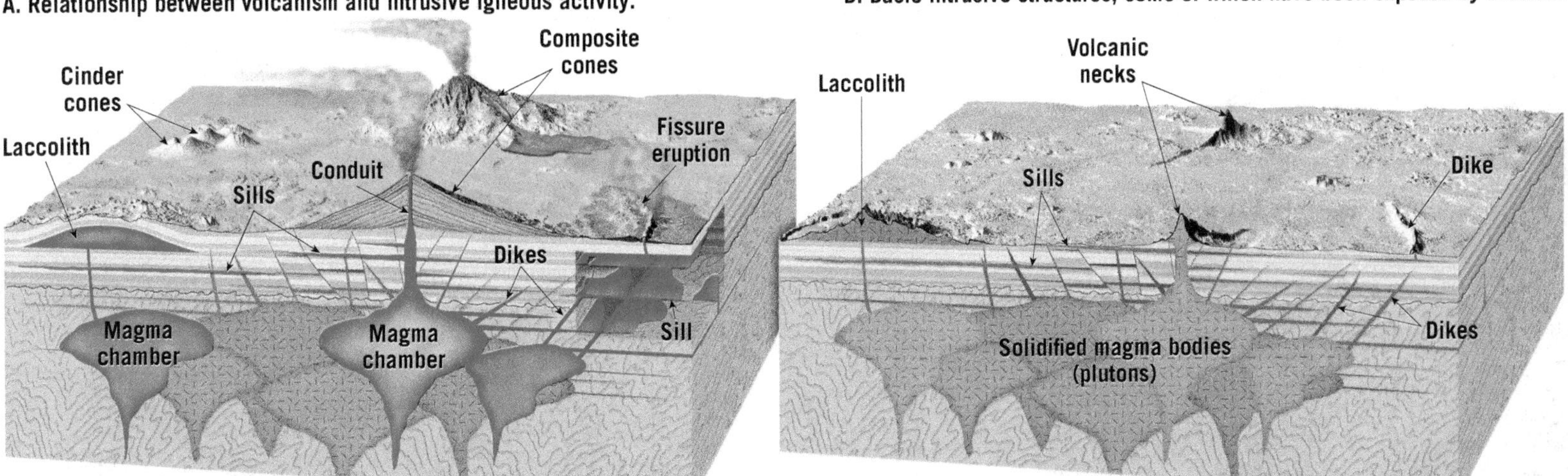

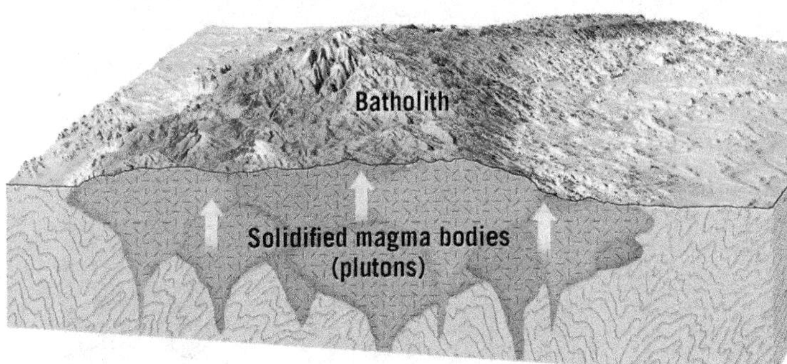

SmartFigure 6.27 **Intrusive igneous structures**

ANIMATION
https://goo.gl/2CGehV

SmartFigure 6.28 **Sill exposed in Sinbad County, Utah** The dark, essentially horizontal band is a sill of basaltic composition that intruded horizontal layers of sedimentary rock. (Photo by Michael Collier)

MOBILE FIELD TRIP
https://goo.gl/qC5DJE

sedimentary beds or other rock structures (**Figure 6.28**). In general, dikes serve as tabular conduits that transport magma upward, whereas sills tend to accumulate magma and increase in thickness.

Dikes and sills are typically shallow features, occurring where the country rocks are sufficiently brittle to fracture. They can range in thickness from less than 1 millimeter to more than 1 kilometer.

While dikes and sills can occur as solitary bodies, dikes tend to form in roughly parallel groups called *dike swarms*. These multiple structures reflect the tendency for fractures to form in sets when tensional forces pull apart brittle country rock. Dikes can also radiate from an eroded volcanic neck, like spokes on a wheel. Where this is the case, the active ascent of rising magma generated fissures in the volcanic cone, out of which lava flowed. Dikes frequently are more resistant and thus weather more slowly than the surrounding rock. Consequently,

▲ SmartFigure 6.29 **Dike exposed in the Spanish Peaks, Colorado** This wall-like dike is composed of igneous rock that is more resistant to weathering than the surrounding material. (Photo by Michael Collier)

CONDOR VIDEO
https://goo.gl/Qm3X6N

▼ Figure 6.30 **Columnar jointing** Giant's Causeway in Northern Ireland is an excellent example of columnar jointing. (Photo by E. J. Tarbuck)

when exposed by erosion, dikes tend to have a wall-like appearance, as shown in Figure 6.29.

Because dikes and sills are relatively uniform in thickness and can extend for many kilometers, they are assumed to be the product of very fluid, and therefore mobile, magmas. One of the largest and most studied sills in the United States is the Palisades Sill. Exposed for 80 kilometers (50 miles) along the west bank of the Hudson River in southeastern New York and northeastern New Jersey, this sill is about 300 meters (1000 feet) thick. Because it is resistant to erosion, the Palisades Sill forms an imposing cliff that can be easily seen from the opposite side of the Hudson.

Columnar Jointing In many respects, sills closely resemble buried lava flows. Both are tabular and can extend over a wide area, and both may exhibit columnar jointing. **Columnar jointing** occurs when igneous rocks cool and develop shrinkage fractures that produce elongated, pillar-like columns that most often have six sides (Figure 6.30). Further, because sills and dikes generally form in near-surface environments and may be only a few meters thick, the emplaced magma often cools quickly enough to generate a fine-grained texture. (Recall that most intrusive igneous bodies have a coarse-grained texture.)

Massive Intrusive Bodies: Batholiths, Stocks, and Laccoliths

Batholiths and Stocks By far the largest intrusive igneous bodies are **batholiths** (*bathos* = depth, *lithos* = stone). Batholiths occur as mammoth linear structures several hundred kilometers long and more than 100 kilometers wide (Figure 6.31). The Sierra Nevada batholith, for example, is a continuous granitic structure that forms much of the "backbone" of the Sierra Nevada in California. An even larger batholith extends for over 1800 kilometers (1100 miles) along the Coast Mountains of western Canada and into southern Alaska. Although batholiths can cover a large area, recent geophysical studies indicate that most are less than 10 kilometers (6 miles) thick. Some are even thinner; the coastal batholith of Peru, for example, is essentially a flat slab with an average thickness of only 2 to 3 kilometers (1 to 2 miles). Batholiths are typically composed of felsic (granitic) and intermediate rock types and are often called "granite batholiths."

Early investigators thought the Sierra Nevada batholith was a huge single body of intrusive igneous rock. Today we know that large batholiths are produced by hundreds of discrete injection of magma that form smaller intrusive bodies (plutons) that intimately crowd against or penetrate one another. These bulbous masses are emplaced over spans of millions of years. The intrusive activity that created the Sierra Nevada batholith, for example, occurred nearly continuously over a 130-million-year period that ended about 80 million years ago (see Figure 6.31).

A batholith is generally defined as a plutonic body having a surface exposure greater than 100 square kilometers (40 square miles). Smaller plutons are termed **stocks**. However, many stocks appear to be portions of much larger intrusive bodies that would be classified as batholiths if they were fully exposed.

Laccoliths A nineteenth-century study by G. K. Gilbert of the U.S. Geological Survey in the Henry Mountains of Utah produced the first clear evidence that igneous intrusions can lift the sedimentary strata they penetrate. Gilbert named the igneous intrusions he observed **laccoliths**, which he envisioned as igneous rock forcibly injected between sedimentary strata, so as to arch the beds above while leaving those below relatively flat. It is now known that the five major peaks of the Henry Mountains are not laccoliths but stocks. However, these central magma bodies are the source material for branching offshoots that are true laccoliths, as Gilbert defined them (Figure 6.32).

Numerous other granitic laccoliths have since been identified in Utah. The largest is a part of the Pine Valley Mountains located north of St. George, Utah. Others are found in the La Sal Mountains near Arches National Park and in the Abajo Mountains directly to the south.

CONCEPT CHECKS 6.10

1. What is meant by the term *country rock*?
2. Describe *dikes* and *sills*, using the appropriate terms from the following list: massive, discordant, tabular, and concordant.
3. Distinguish among batholiths, stocks, and laccoliths in terms of size and shape.

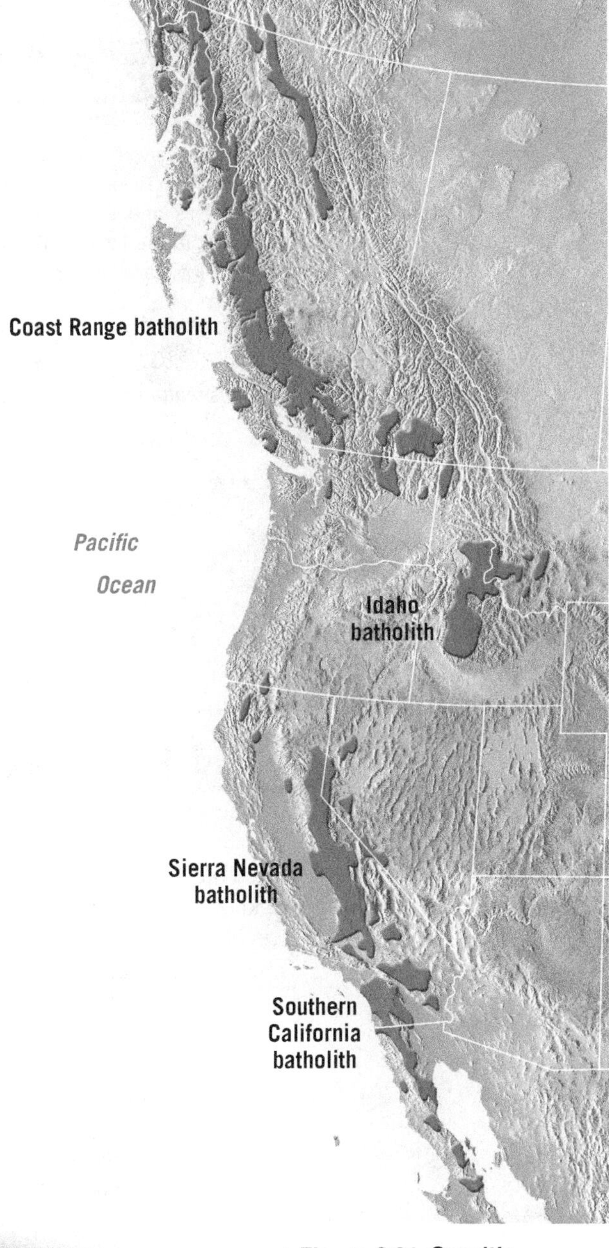

Figure 6.31 **Granitic batholiths along the western margin of North America** These gigantic, elongated bodies consist of numerous plutons that were emplaced beginning about 150 million years ago.

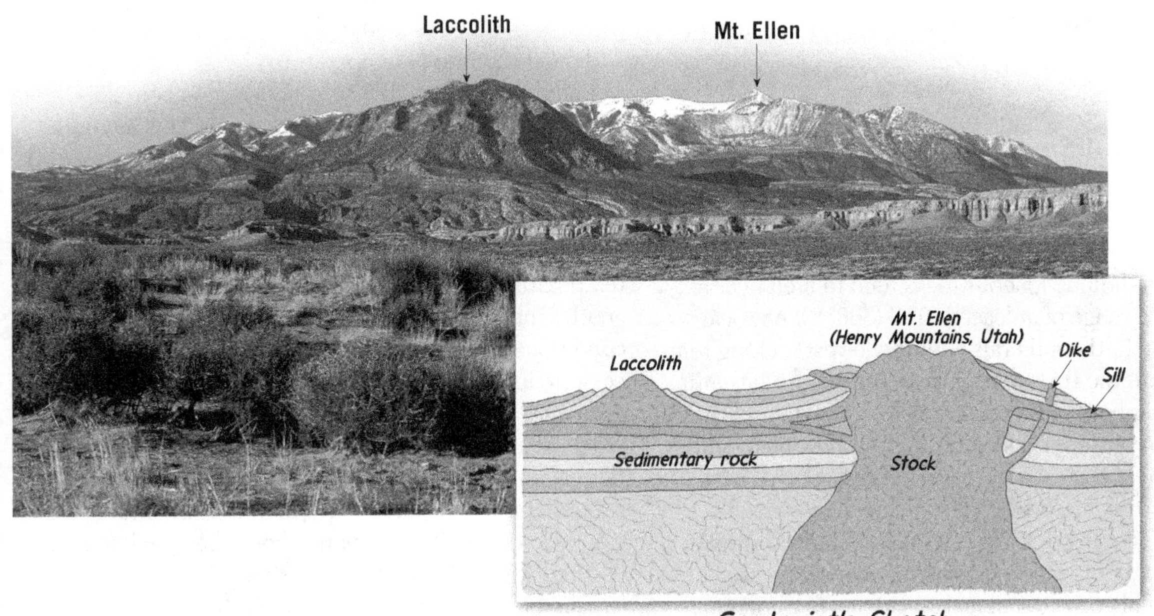

Figure 6.32 **Laccoliths** Mount Ellen in Utah's Henry Mountains is one of five peaks that make up this small mountain range. Although the main intrusions in the Henry Mountains are stocks, numerous laccoliths formed as offshoots of these structures. (Photo by Michael DeFreitas North America/Alamy)

EYE ON EARTH 6.2

The Palisades form impressive cliffs along the western side of the Hudson River for more than 80 kilometers (50 miles). This igneous feature, which is visible from Manhattan, formed when magma was injected between layers of sandstone and shale.

QUESTION 1 *What type of igneous feature constitutes the Palisades?*

QUESTION 2 *Is this feature tabular or massive?*

6.11 Partial Melting and the Origin of Magma

Summarize the major processes that generate magma from solid rock.

Based on evidence from the study of earthquake waves, we know that *Earth's crust and mantle are composed primarily of solid, not molten, rock.* Although the outer core is fluid, this iron-rich material is very dense and remains deep within Earth. So where does magma come from?

Most magma originates in Earth's uppermost mantle. The greatest quantities are produced at divergent plate boundaries, in association with seafloor spreading, with lesser amounts forming at subduction zones, where oceanic lithosphere descends into the mantle. Magma also can be generated when crustal rocks are heated sufficiently to melt.

Partial Melting

Recall that igneous rocks are composed of a mixture of minerals. Since these minerals have different melting points (the temperature at which the mineral changes from solid to liquid), igneous rocks tend to melt over a temperature range of at least 200°C (360°F). As rock begins to melt, the minerals with the lowest melting temperatures are the first to melt. If melting continues, minerals with higher melting points begin to melt, and the composition of the melt steadily approaches the overall composition of the rock from which it was derived.

Most often, however, melting is not complete. The incomplete melting of rocks is known as **partial melting**, a process that produces most magma. Partial melting of rock can be likened to a chocolate chip cookie containing nuts that is left out in the Sun. The chocolate chips represent the minerals with the lowest melting points because they will begin to melt before the other ingredients. Unlike the chocolate chip cookie, when rock partially melts, the molten material separates from the solid components. This molten material is also less dense than the remaining solids, so it rises toward Earth's surface.

Generating Magma from Solid Rock

Workers in underground mines know that temperatures increase as they descend deeper below Earth's surface. Although the rate of temperature change varies considerably from place to place, it *averages* about 25°C (75°F) per kilometer in the *upper* crust. This increase in temperature with depth is known as the **geothermal gradient**. As shown in Figure 6.33, when a typical geothermal gradient is compared to the curve showing the melting points for the mantle rock peridotite, the temperature at which peridotite melts is higher than the geothermal gradient at all depths. Thus, under normal conditions, the mantle is solid. However, tectonic processes trigger melting though various means, including reducing the mantle rock's melting point.

Decrease in Pressure: Decompression Melting

If temperature were the only factor that determined whether rock melts, our planet would be a molten ball covered with a thin, solid outer shell. This is not the case because pressure, which also increases with depth, influences the melting temperatures of rocks.

Melting, which is accompanied by an increase in volume, occurs at progressively higher temperatures with increased depth. This is the result of the steady increase in confining pressure exerted by the weight of overlying rocks. Conversely, *reducing confining pressure lowers a rock's melting temperature*. When confining pressure drops sufficiently, **decompression melting** is triggered. Decompression melting often occurs when hot, solid rock ascends in the upper mantle, thereby moving into regions of lower pressure.

Recall from Chapter 4 that tensional forces along spreading centers promote upwelling where plates diverge. This process is responsible for generating magma along oceanic ridges (divergent plate boundaries) where plates are rifting apart (Figure 6.34). Below the ridge crest, hot mantle rock rises and melts, generating a magma that replaces the material that shifted horizontally away from the ridge axis.

Decompression melting also occurs when ascending mantle plumes reach the uppermost mantle. If this rising magma reaches the surface, it triggers an episode of hot-spot volcanism.

Addition of Water Along with pressure, an important factor affecting the melting temperature of rock is its water content. Water and other volatiles, such as carbon dioxide, act in a similar way to salt melting ice. That is, water causes rock to melt at lower temperatures, just as putting rock salt on an icy sidewalk induces melting.

The introduction of water to generate magma occurs mainly at convergent plate boundaries, where cool slabs of oceanic lithosphere descend into the mantle (Figure 6.35). As an oceanic plate sinks, heat and pressure drive out water from the subducting oceanic crust and overlying sediments. These fluids migrate into the wedge of hot mantle that lies directly above. At a depth of about 100 kilometers (60 miles) this wedge of mantle rock is hot enough that the addition of water leads to some melting. Partial melting of the mantle rock peridotite generates hot basaltic magma whose temperatures may exceed 1250°C (nearly 2300°F).

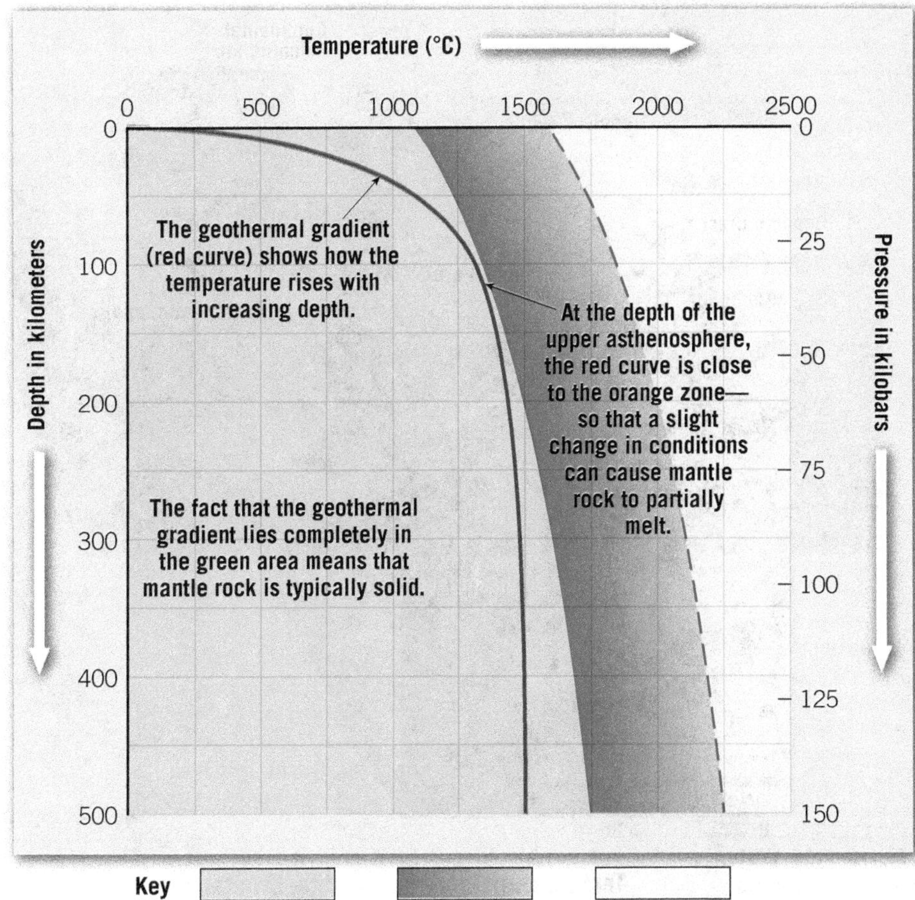

▲ Figure 6.33 **Why the mantle is mainly solid** This diagram shows the geothermal gradient (the increase in temperature with depth) for the crust and upper mantle.

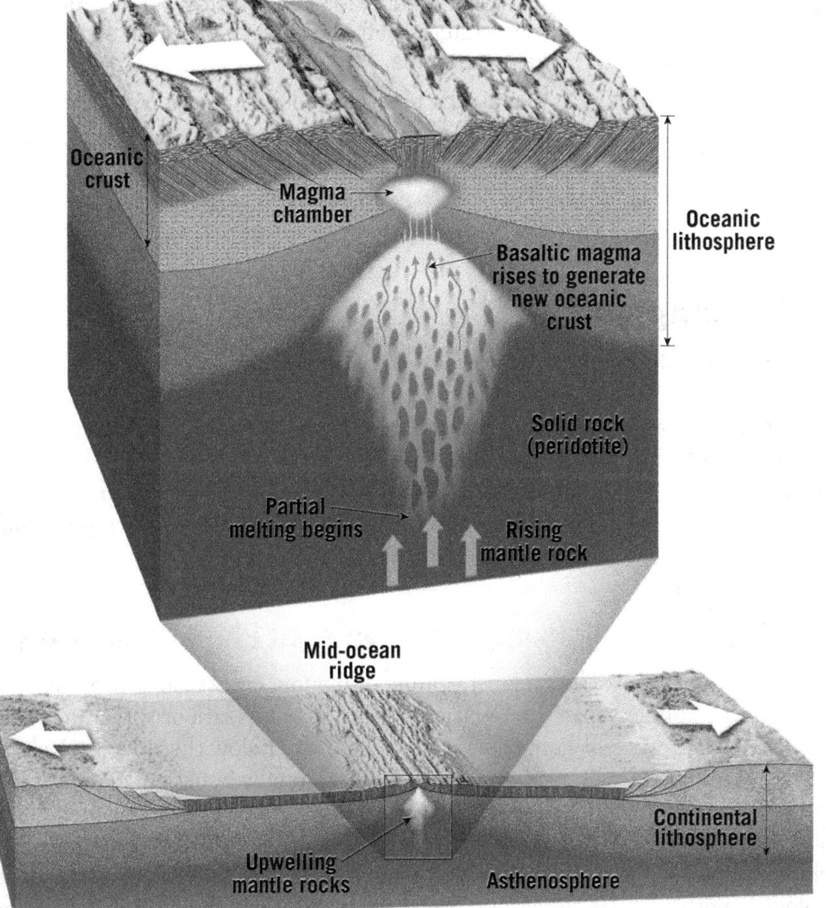

◄ Figure 6.34 **Decompression melting** As hot mantle rock ascends, it experiences continuously decreasing pressure. This drop in confining pressure usually initiates *decompression melting* in the upper mantle.

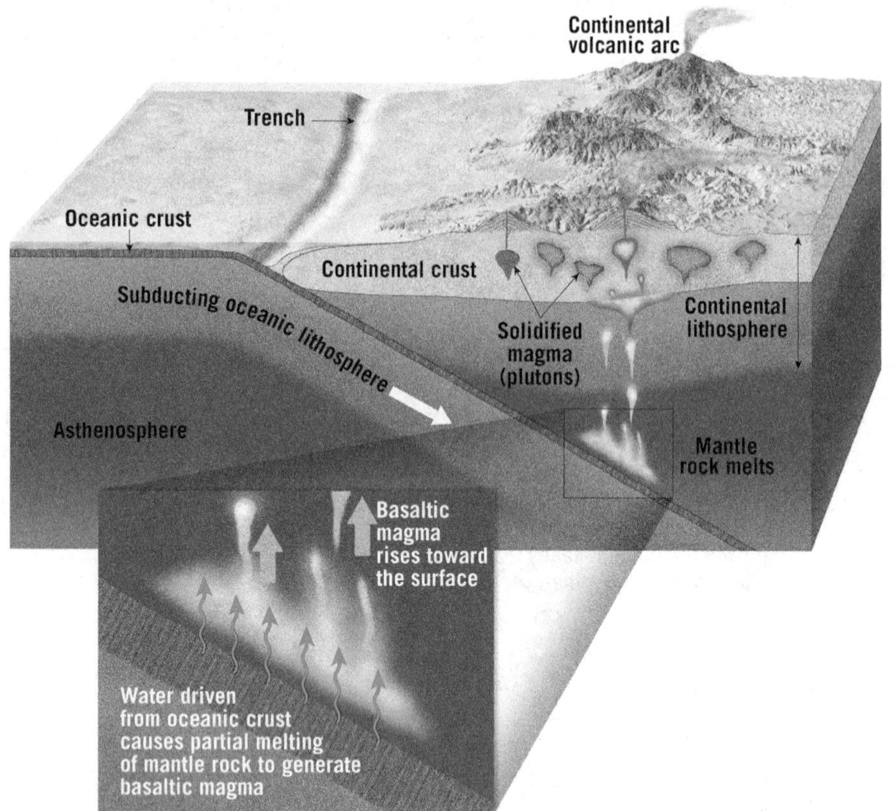

▲ Figure 6.35 **Water lowers the melting temperature of hot mantle rock to trigger partial melting** As an oceanic plate descends into the mantle, water and other volatiles are driven from the subducting crustal rocks into the mantle above.

Temperature Increase: Melting Crustal Rocks Mantle-derived basaltic magma tends to be less dense than the surrounding rocks, which causes the magma to rise buoyantly toward the surface. In oceanic settings, these basaltic magmas often erupt on the ocean floor, generating seamounts, which may grow to form volcanic islands, as exemplified by the Hawaiian Islands. However, in continental settings, basaltic magma often "ponds" beneath low-density crustal rocks. Because the overlying crustal rocks have lower melting temperatures than basaltic magmas, the hot basaltic magma may heat them sufficiently to generate a secondary melt of silica-rich andesitic or granitic magma. If these low-density andesitic or granitic magmas reach the surface, they tend to produce explosive eruptions; such eruptions occur most often at convergent plate boundaries.

Crustal rocks can also melt during continental collisions that result in the formation of a large mountain belt (discussed in detail in Chapter 7). During these events, the crust is greatly thickened, and some crustal rocks are carried to depths where the temperatures are high enough to cause partial melting. The granitic magmas produced in this manner usually solidify before reaching the surface, so volcanism is not typically associated with these collision-type mountain belts.

In summary, magma can be generated by (1) *decompression melting* caused by a decrease in pressure as magma rises; (2) the *introduction of water*, which lowers the melting temperature of hot mantle rock; and (3) *heating of crustal rocks* above their melting temperature.

CONCEPT CHECKS 6.11

1. What is the geothermal gradient? Describe how the geothermal gradient compares with the melting temperatures of the mantle rock peridotite at various depths.
2. Explain the process of decompression melting.
3. What roles do water and other volatiles play in the formation of magma?

6.12 Plate Tectonics and Volcanism

Explain how the global distribution of volcanic activity is related to plate tectonics.

Geologists have known for decades that the global distribution of most of Earth's volcanoes is not random. Most active volcanoes on land are located along the margins of the ocean basins—notably within the circum-Pacific belt known as the **Ring of Fire** (Figure 6.36), where denser oceanic lithosphere subducts under continental lithosphere. Another group of volcanoes includes the innumerable seamounts that form along the crest of the mid-ocean ridges. There are some volcanoes, however, that appear to be randomly distributed around the globe. These volcanic structures comprise most of the islands of the deep-ocean basins, including the Hawaiian Islands, the Galapagos Islands, and Easter Island.

Setting aside for the moment the volcanoes that form in the deep-ocean basins, the development of the theory of plate tectonics provided geologists with a plausible explanation for the distribution of Earth's volcanoes and established the basic connection between plate tectonics and volcanism: *Plate motions provide the mechanisms by which mantle rocks undergo partial melting to generate magma.*

Volcanism at Divergent Plate Boundaries

The greatest volume of magma erupts along divergent plate boundaries associated with seafloor spreading—out of human sight (Figure 6.37B). Below the ridge axis where lithospheric plates are continually being pulled apart, the solid yet mobile mantle rises to fill the rift. Recall that as hot rock rises, it experiences a decrease in confining pressure and may undergo *decompression melting*. This activity continuously adds new basaltic rock to plate

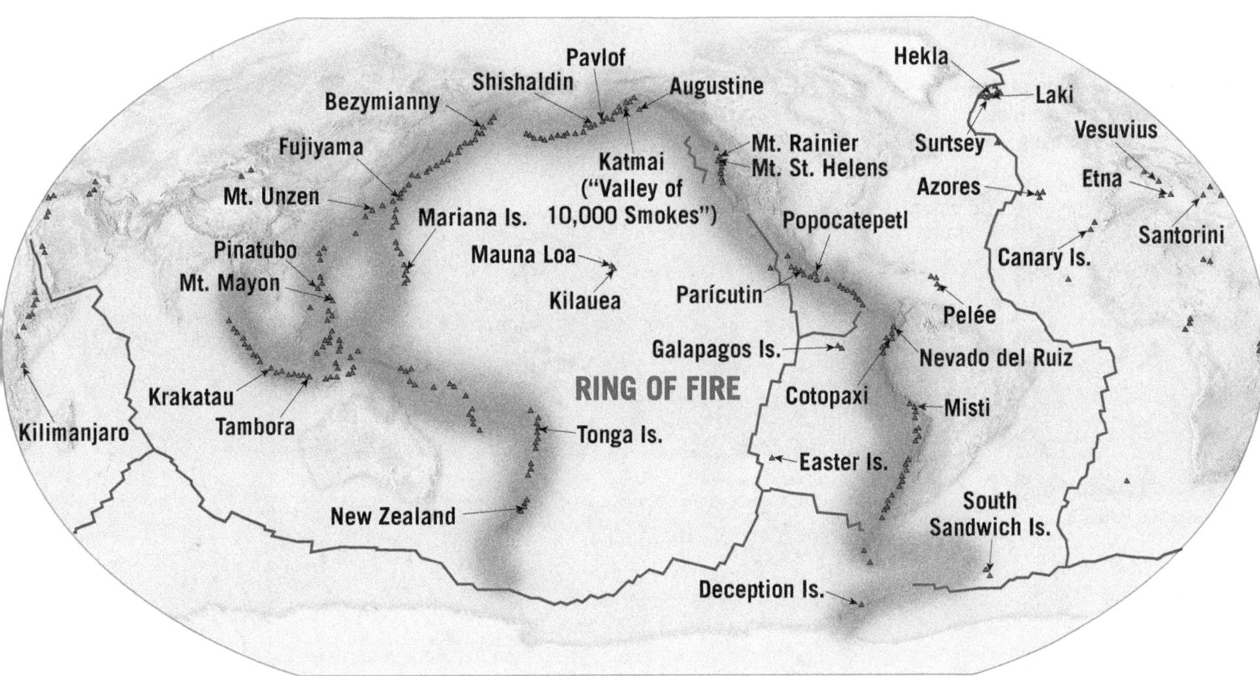

Figure 6.36 Ring of Fire Most of Earth's major volcanoes are located in a zone around the Pacific called the Ring of Fire. Another large group of active volcanoes lies unseen along the mid-ocean ridge system.

margins, temporarily welding them together, only to have them break again as spreading continues. Along some ridge segments, extrusions of pillow lavas build numerous volcanic structures, the largest of which is Iceland.

Although most spreading centers are located along the axis of an oceanic ridge, some are not. In particular, the East African Rift is a site where continental lithosphere is being pulled apart (see Figure 6.37F). Vast outpourings of fluid basaltic lavas as well as several active volcanoes are found in this region of the globe.

Volcanism at Convergent Plate Boundaries

Recall that along convergent plate boundaries, two plates move toward each other, and a slab of dense oceanic lithosphere descends into the mantle. In these settings, water driven from hydrated (water-rich) minerals found in the subducting oceanic crust and overlying sediments triggers partial melting in the hot mantle above (Figure 6.37A).

Volcanism at a convergent plate margin results in the development of a slightly curved chain of volcanoes called a *volcanic arc.* These volcanic chains develop roughly parallel to the associated trench—at distances of 200 to 300 kilometers (100 to 200 miles). Volcanic arcs that develop within the ocean and grow large enough for their tops to rise above the surface are labeled *archipelagos* in most atlases. Geologists prefer the more descriptive term **volcanic island arcs**, or simply **island arcs** (see Figure 6.37A). Several young volcanic island arcs border the western Pacific basin, including the Aleutians, the Tongas, and the Marianas.

Volcanism associated with convergent plate boundaries may also take place where slabs of oceanic lithosphere are subducted under continental lithosphere to produce a **continental volcanic arc** (Figure 6.37E). The mechanisms that generate these mantle-derived magmas are essentially the same as those that create volcanic island arcs. The most significant difference is that continental crust is much thicker and composed of rocks having higher silica content than oceanic crust. Hence, by melting the surrounding silica-rich crustal rocks, mantle-derived magma changes composition as it rises through the crust. The volcanoes of the Cascade Range in the northwestern United States, including Mount Hood, Mount Rainier, Mount Shasta, and Mount St. Helens, are examples of volcanoes generated at a convergent plate boundary along a continental margin (Figure 6.38).

Intraplate Volcanism

We know why igneous activity is initiated along plate boundaries, but why do eruptions occur in the interiors of plates? Hawaii's Kilauea, considered one of the world's most active volcanoes, is situated thousands of kilometers from the nearest plate boundary in the middle of the vast Pacific plate (Figure 6.37C). Sites of **intraplate** (meaning "within the plate") **volcanism** include the large outpourings of fluid basaltic lavas such as those that compose the Columbia Plateau, the Siberian Traps in Russia, India's Deccan Plateau, and several submerged oceanic plateaus, including the Ontong Java Plateau in the western Pacific (Figure 6.39).

One widely accepted hypothesis proposes that most intraplate volcanism occurs when a relatively narrow mass of hot material, called a **mantle plume**, ascends toward the surface (Figure 6.40A).° Although the depth at which mantle plumes originate is a topic of debate, some are

° Recall from Section 1.2 that a *hypothesis* is a tentative scientific explanation for a given set of observations. Although widely accepted, the validity of the mantle plume hypothesis, unlike the theory of plate tectonics, remains unresolved.

A. Convergent Plate Volcanism When an oceanic plate subducts, melting in the mantle produces magma that gives rise to a volcanic island arc on the overlying oceanic crust.

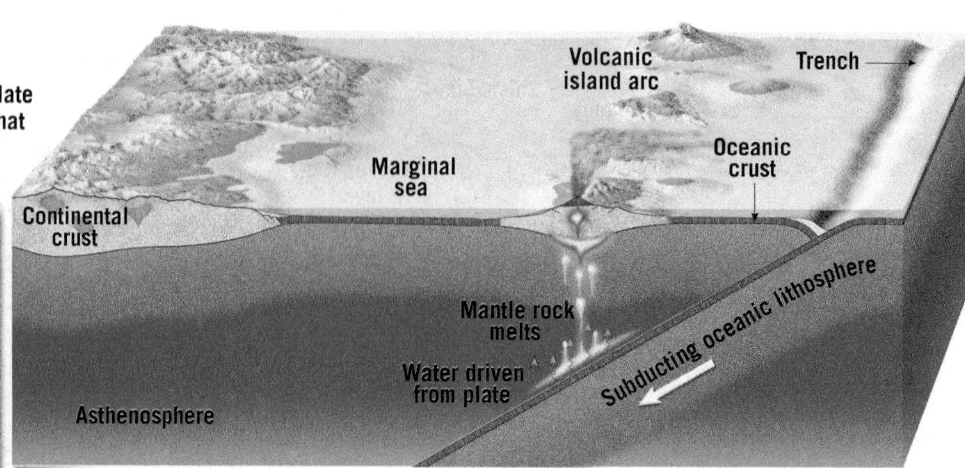

Cleveland Volcano, Aleutian Islands (USGS)

C. Intraplate Volcanism When an oceanic plate moves over a hot spot, a chain of volcanic structures such as the Hawaiian Islands is created.

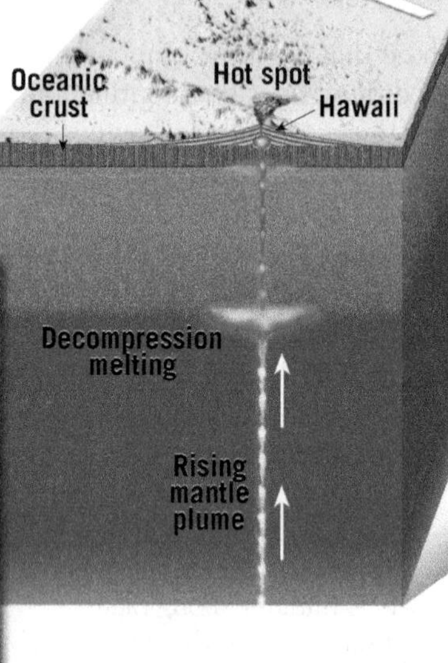

Kilauea, Hawaii (USGS)

E. Convergent Plate Volcanism When oceanic lithosphere descends beneath a continent, magma generated in the mantle rises to form a continental volcanic arc.

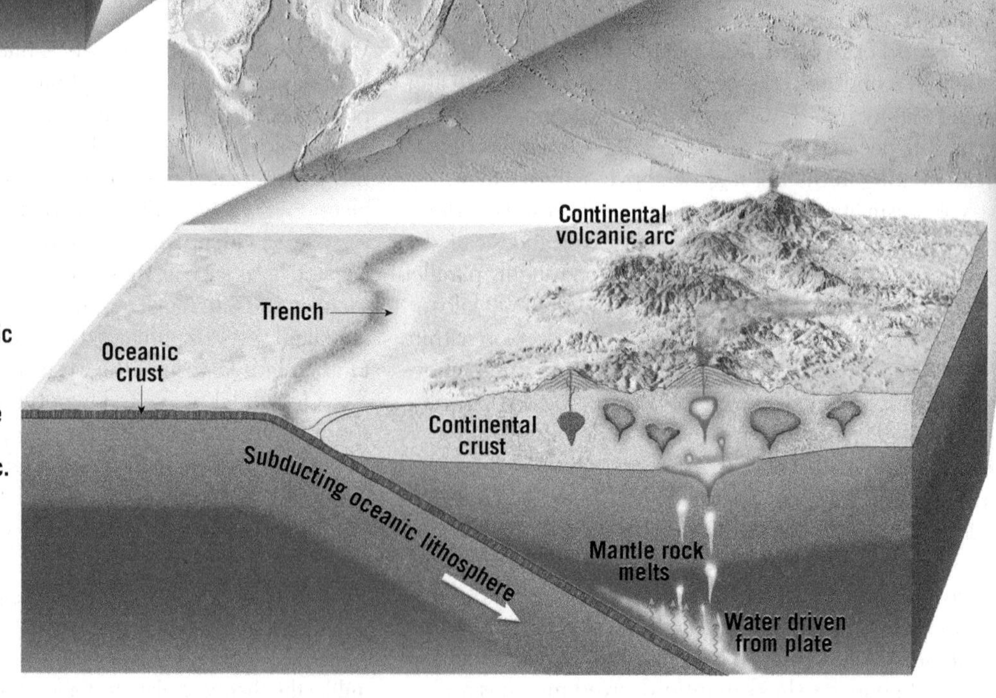

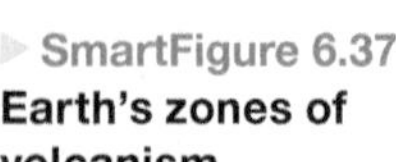

SmartFigure 6.37
Earth's zones of volcanism

TUTORIAL
https://goo.gl/PSN9hc

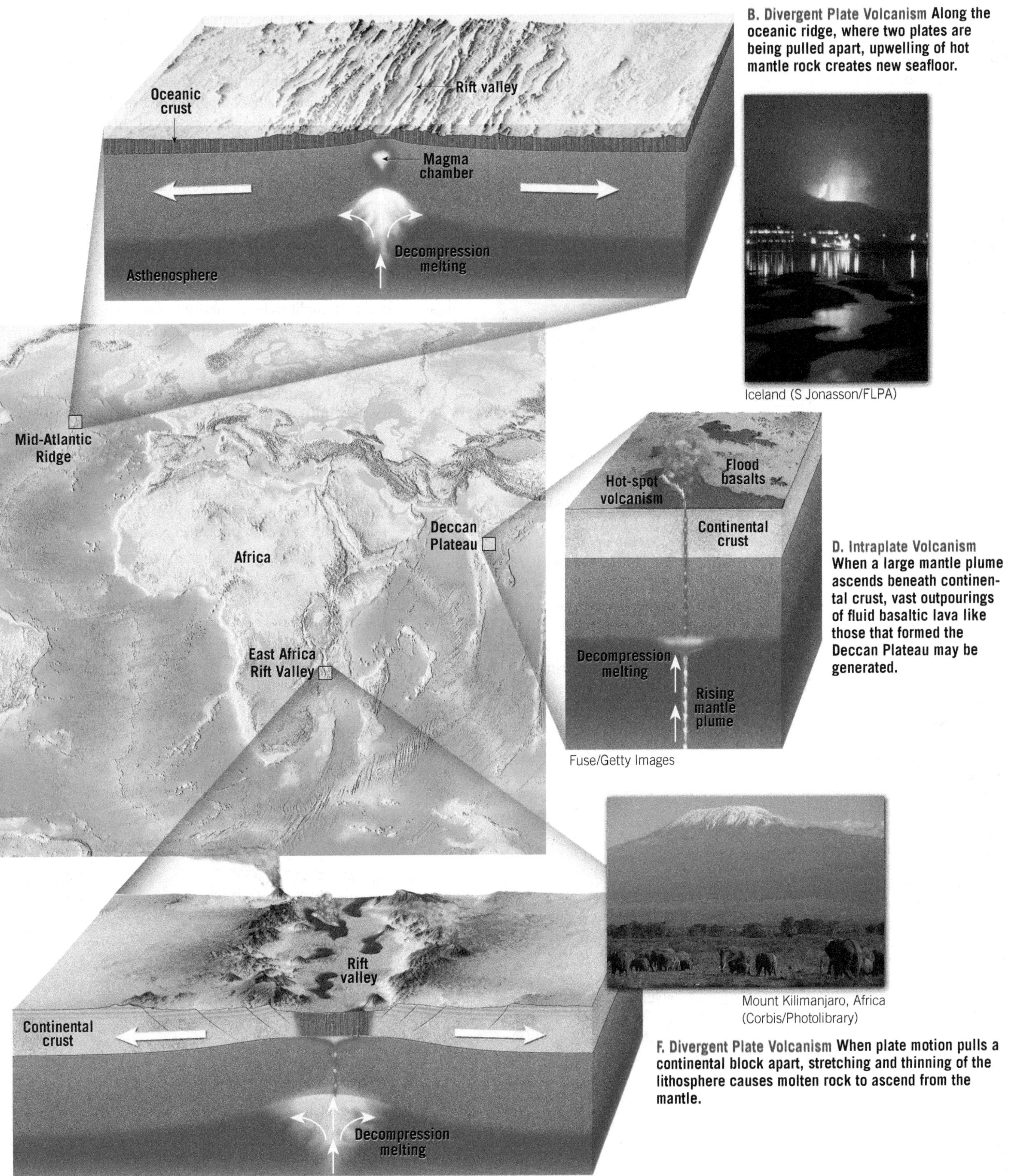

B. Divergent Plate Volcanism Along the oceanic ridge, where two plates are being pulled apart, upwelling of hot mantle rock creates new seafloor.

Iceland (S Jonasson/FLPA)

D. Intraplate Volcanism When a large mantle plume ascends beneath continental crust, vast outpourings of fluid basaltic lava like those that formed the Deccan Plateau may be generated.

Fuse/Getty Images

Mount Kilimanjaro, Africa (Corbis/Photolibrary)

F. Divergent Plate Volcanism When plate motion pulls a continental block apart, stretching and thinning of the lithosphere causes molten rock to ascend from the mantle.

▶ **SmartFigure 6.38**
Subduction-produced Cascade Range volcanoes Subduction of the Juan de Fuca plate along the Cascadia subduction zone produced the Cascade volcanoes.

TUTORIAL
https://goo.gl/gruZZD

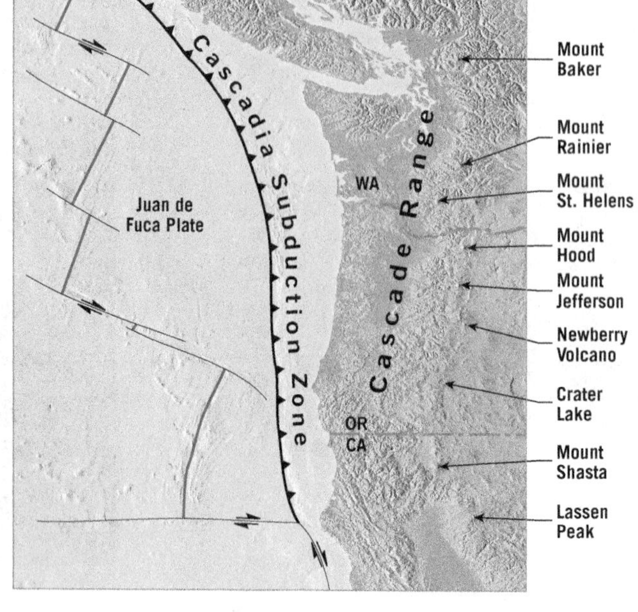

thought to form deep within Earth, at the core–mantle boundary. These plumes of solid yet mobile rock rise toward the surface in a manner similar to the blobs that form within a lava lamp. A lava lamp contains two immiscible liquids in a glass container; as the base of the lamp is heated, the denser liquid at the bottom becomes buoyant and forms blobs that rise to the top. Like the blobs in a lava lamp, a mantle plume has a bulbous head that draws out a narrow stalk or tail beneath it as it rises. The surface manifestation of this activity is called a **hot spot**, an area of volcanism, high heat flow, and crustal uplifting a few hundred kilometers wide.

Large mantle plumes, dubbed **superplumes**, are thought to be responsible for the vast outpourings of basaltic lava that created the large basalt plateaus. When the head of the plume reaches the base of the lithosphere, decompression melting progresses rapidly. This causes the burst of volcanism that emits voluminous flows of lava over a period of 1 million or so years (**Figure 6.40B**). Extreme eruptions of this type would have affected Earth's climate, causing (or at least contributing to) the extinction events recorded in the fossil record.

The comparatively short initial eruptive phase is often followed by millions of years of less voluminous activity, as the plume tail slowly rises to the surface. Extending away from some large flood basalt plateaus is a chain of volcanic structures, similar to the Hawaiian chain (**Figure 6.40C**).

Intraplate volcanism associated with mantle plumes is also thought to be responsible for the massive eruptions of silica-rich pyroclastic material that occurred in continental settings. Perhaps the best known of these hot-spot eruptions are the three caldera-forming eruptions that occurred in the Yellowstone region over the past 2.1 million years (see Figure 6.23).

CONCEPT CHECKS 6.12

1. Are volcanoes in the Ring of Fire generally described as effusive or explosive? Provide an example that supports your answer.
2. How is magma generated along convergent plate boundaries?
3. Volcanism at divergent plate boundaries is most often associated with which magma type? What causes rocks to melt in these settings?
4. What is thought to be the source of magma for most intraplate volcanism?
5. Which type of plate boundary generates the greatest quantity of magma?

▶ **SmartFigure 6.39**
Global distribution of large basalt plateaus The basalt plateaus (shown in red) are thought to be the product of a burst of volcanism generated by partial melting of the bulbous head of a hot mantle plume. The orange dashed lines represent the chain of volcanic structures produced by partial melting of the plume tail. The orange dots are the current surface locations of the hot mantle plumes that likely generated the associated basalt plateaus.

TUTORIAL
https://goo.gl/O2Z6SI

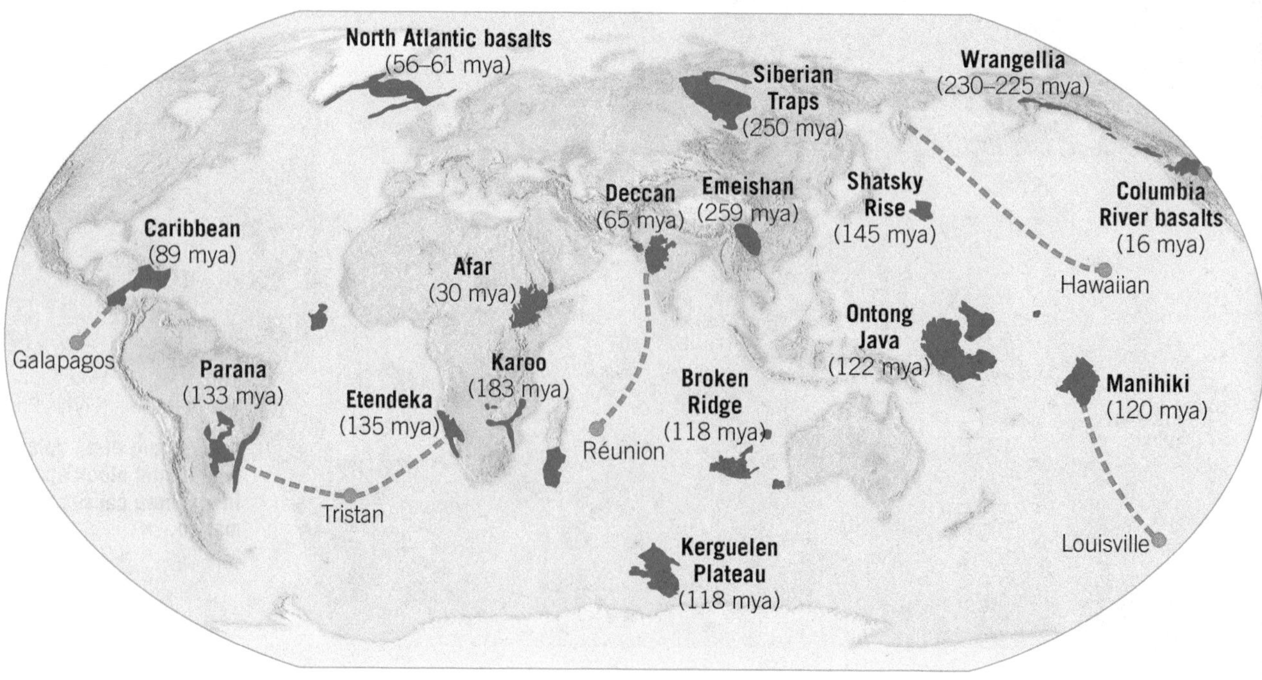

A rising mantle plume with a large bulbous head is thought to generate Earth's large basalt plateaus.

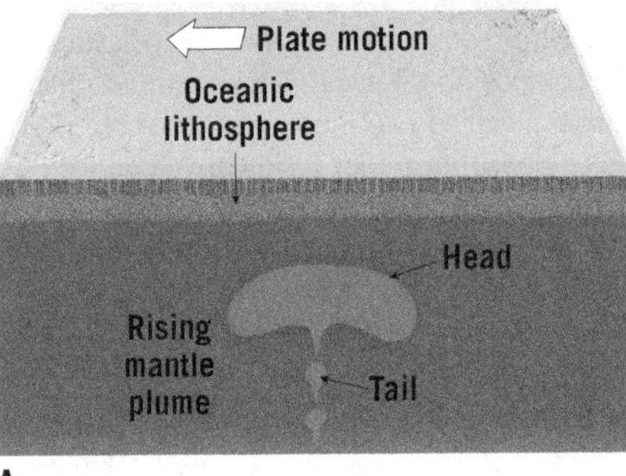

A.

Rapid decompression melting of the plume head produces extensive outpourings of flood basalts over a relatively short time span.

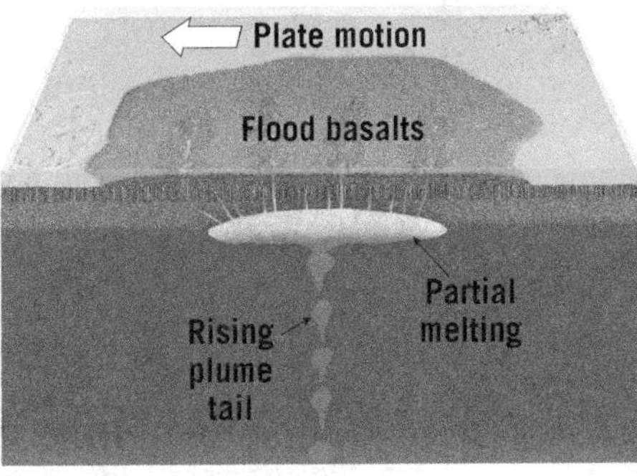

B.

Because of plate movement, volcanic activity from the rising tail of the plume generates a linear chain of smaller volcanic structures.

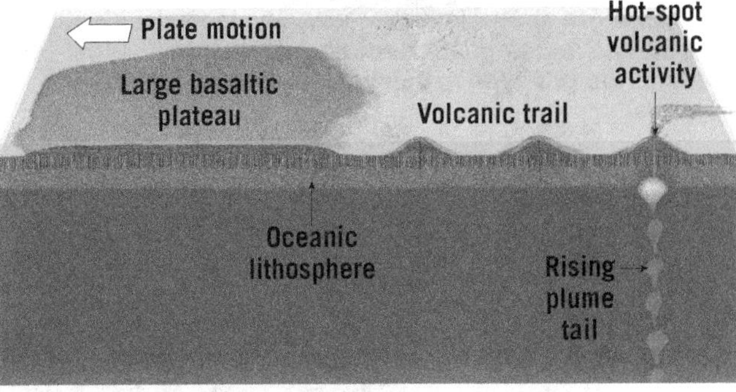

C.

Figure 6.40 **Mantle plumes and large basalt plateaus** Hot-spot volcanism is thought to explain the formation of large basalt plateaus and the chains of volcanic islands associated with these features.

6 CONCEPTS IN REVIEW

Volcanoes and Other Igneous Activity

6.1 Mount St. Helens Versus Kilauea

Compare and contrast the 1980 eruption of Mount St. Helens with the most recent eruption of Kilauea, which began in 1983.

- Volcanic eruptions cover a broad spectrum from explosive eruptions, like that of Mount St. Helens in 1980, to the quiescent eruptions of Kilauea.

6.2 The Nature of Volcanic Eruptions

Explain why some volcanic eruptions are explosive and others are quiescent.

KEY TERMS: magma, lava, effusive eruption, viscosity, eruption column

- The two primary factors determining the nature of a volcanic eruption are the viscosity (resistance to flow) of the magma and its gas content. In general, magmas that contain more silica are more viscous, while those with lower silica content are more fluid. Temperature also influences viscosity. Hot lavas are more fluid, while cool lavas are more viscous.
- Basaltic magmas, which are fluid and have low gas content, tend to generate effusive (nonexplosive) eruptions. In contrast, silica-rich magmas (andesitic and rhyolitic), which are the most viscous and contain the greatest quantity of gases, are the most explosive.

? Although Kilauea mostly erupts in a gentle manner, what risks might you encounter if you chose to live nearby?

USGS

6.3 Materials Extruded During an Eruption

List and describe the three categories of materials extruded during volcanic eruptions.

KEY TERMS: aa flow, pahoehoe flow, lava tube, pillow lava, volatile, pyroclastic material, tephra, scoria, pumice

- Volcanoes erupt molten lava, gases, and solid pyroclastic materials.
- Low-viscosity basaltic lava flows can extend great distances from a volcano. On the surface, they travel as pahoehoe or aa flows. Sometimes the surface of the flow congeals, and lava continues to flow below in tunnels called lava tubes. When lava erupts underwater, the outer surface is chilled instantly to obsidian, while the inside continues to flow, producing pillow lavas.
- The gases most commonly emitted by volcanoes are water vapor and carbon dioxide. Upon reaching the surface, these gases rapidly expand, leading to explosive eruptions that can generate a mass of lava fragments called pyroclastic materials.
- Pyroclastic materials come in several sizes. From smallest to largest, they are ash, lapilli, and blocks or bombs. Blocks exit the volcano as solid fragments, whereas bombs exit as liquid blobs.

- If bubbles of gas in lava don't pop before the lava solidifies, they are preserved as voids called vesicles. Especially frothy, silica-rich lava can cool to make lightweight pumice, while basaltic lava with lots of bubbles cools to make scoria.

? This photo shows layers of volcanic material ejected by a violent eruption and deposited roughly horizontally. What term is used to describe this type of volcanic material?

6.4 Anatomy of a Volcano

Draw and label a diagram that illustrates the basic features of a typical volcanic cone.

KEY TERMS: fissure, conduit, vent, volcanic cone, crater, caldera, parasitic cone, fumarole

- Volcanoes vary in size and form but share a few common features. Most are roughly conical piles of extruded material that collect around a central vent. The vent is usually within a summit crater or caldera. On the flanks of the volcano, there may be smaller vents marked by small parasitic cones, or there may be fumaroles, spots where gas is expelled.

? Label the diagram using the following terms: conduit, vent, lava, parasitic cone, bombs, pyroclastic material.

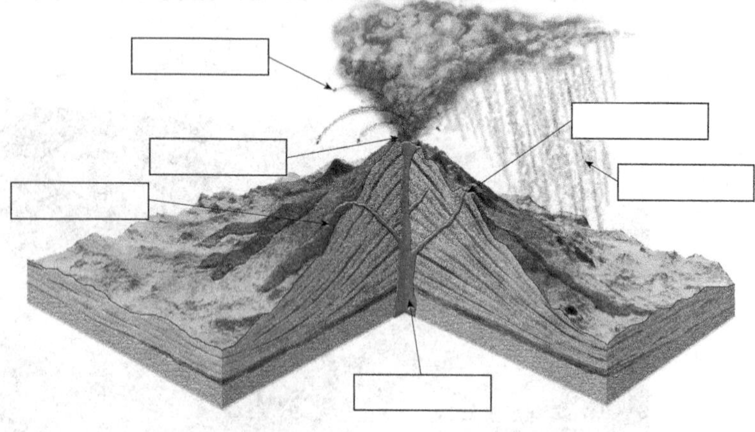

6.5 Shield Volcanoes

Summarize the characteristics of shield volcanoes and provide one example of this type of volcano.

KEY TERMS: shield volcano, seamount

- Shield volcanoes consist of many successive lava flows of low-viscosity basaltic lava but lack significant amounts of pyroclastic debris. Lava tubes help transport lava far from the main vent, resulting in very gentle, shield-like profiles.
- Most shield volcanoes begin as seamounts that grow from Earth's seafloor. Mauna Loa, Mauna Kea, and Kilauea in Hawaii are classic examples of the low, wide form characteristic of shield volcanoes.

6.6 Cinder Cones

Describe the formation, size, and composition of cinder cones.

KEY TERMS: cinder cone, (scoria cone)

- Cinder cones are steep-sided structures composed mainly of pyroclastic debris, typically having a basaltic composition. Lava flows sometimes emerge from the base of a cinder cone but typically do not flow out of the crater.
- Cinder cones are small relative to the other major kinds of volcanoes, reflecting the fact that most form quickly, as single eruptive events. Because they are unconsolidated, cinder cones easily succumb to weathering and erosion.

6.7 Composite Volcanoes

List the characteristics of composite volcanoes and describe how they form.

KEY TERMS: composite volcano, (stratovolcano)

- Composite volcanoes are called "composite" because they consist of both pyroclastic material and lava flows. They typically erupt silica-rich magmas of andesitic or rhyolitic composition. They are much larger than cinder cones and form from multiple eruptions over millions of years.
- Because andesitic and rhyolitic lavas are more viscous than basaltic lava, they accumulate at a steeper angle than does the lava from shield volcanoes. Over time, a composite volcano's combination of lava and cinders produces a towering volcano with a classic symmetrical shape.
- Mount Rainier and the other volcanoes of the Cascade Range in the northwestern United States are good examples of composite volcanoes.

? If your family had to live next to a volcano, would you rather it be a shield volcano, cinder cone, or composite volcano? Explain.

6.8 Volcanic Hazards

Describe the major geologic hazards associated with volcanoes.

KEY TERMS: pyroclastic flow (nuée ardente), lahar, tsunami

- The greatest volcanic hazard to human life is the pyroclastic flow, or nuée ardente. This dense mix of hot gas and pyroclastic fragments races downhill at great speed and incinerates everything in its path. A pyroclastic flow can travel many kilometers from its source volcano. Because pyroclastic flows are hot, their deposits frequently "weld" together into a solid rock called welded tuff.
- Lahars are mudflows that form on volcanoes. These rapidly moving slurries of ash and debris suspended in water tend to follow stream valleys and can result in loss of life and/or significant damage to structures.
- Volcanic ash in the atmosphere can be a risk to air travel when it is sucked into airplane engines. Volcanoes at sea level can generate tsunamis when they erupt or when their flanks collapse into the ocean. Those that spew large amounts of gas such as sulfur dioxide can cause respiratory problems. If volcanic gases reach the stratosphere, they screen out a portion of incoming solar radiation and can trigger short-term cooling at Earth's surface.

6.9 Other Volcanic Landforms

List volcanic landforms other than shield, cinder, and composite volcanoes and describe their formation.

KEY TERMS: fissure eruption, basalt plateau, flood basalt, volcanic neck, (plug)

- Calderas, which can be among the largest volcanic structures, form when the rigid, cold rock above a magma chamber cannot be supported and collapses, creating a broad, roughly circular depression. On shield volcanoes, calderas form slowly as lava drains from the magma chamber beneath the volcano. On a composite volcano, caldera collapse often follows an explosive eruption that can result in significant loss of life and destruction of property.

- Fissure eruptions occasionally produce massive floods of fluid basaltic lava from large cracks, called fissures, in the crust. Layer upon layer of these flood basalts may accumulate to significant thicknesses and blanket a wide area. The Columbia Plateau in the northwestern United States is an example.
- Shiprock, New Mexico, is an example of a volcanic neck where the lava in the "throat" of an ancient volcano congealed to form a plug of solid rock that weathered more slowly than the surrounding volcanic rocks. The surrounding pyroclastic debris eroded, and the resistant neck remains as a distinctive landform.

6.10 Intrusive Igneous Activity

Compare and contrast these intrusive igneous structures: dikes, sills, batholiths, stocks, and laccoliths.

KEY TERMS: host (country) rock, intrusion (pluton), tabular, massive, discordant, concordant, dike, sill, columnar jointing, batholith, stock, laccolith

- When magma intrudes other rocks, it may cool and crystallize before reaching the surface to produce intrusions called plutons. Plutons come in many shapes. They may cut across the host rocks without regard for preexisting structures, or the magma may flow along weak zones in the host rock, such as between the horizontal layers of sedimentary bedding.
- Tabular intrusions may be concordant (sills) or discordant (dikes). Massive plutons may be small (stocks) or very large (batholiths). A blister-like intrusion that lifts the overlying rock layers is a laccolith. As solid igneous rock cools, its volume decreases. Contraction can produce a distinctive fracture pattern called columnar jointing.

? Label the intrusive igneous structures in the accompanying diagram, using the following terms: volcanic neck, sill, batholith, laccolith.

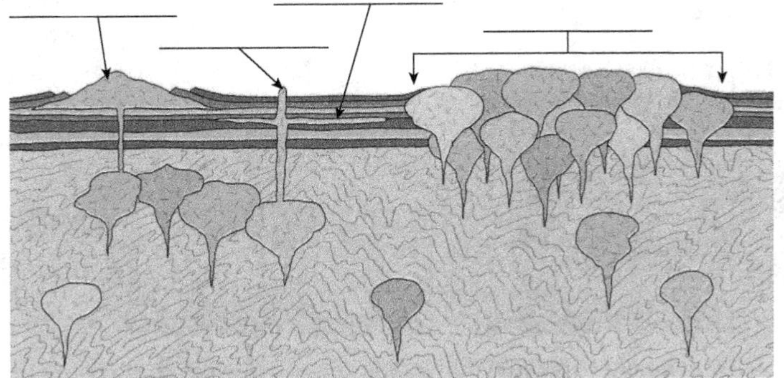

6.11 Partial Melting and the Origin of Magma

Summarize the major processes that generate magma from solid rock.

KEY TERMS: partial melting, geothermal gradient, decompression melting

- Solid rock may melt under three geologic circumstances: when heat is added to the rock, raising its temperature; when already hot rock experiences lower pressures (decompression, as seen at mid-ocean ridges); and when water is added (as occurs at subduction zones).

6.12 Plate Tectonics and Volcanism

Explain how the global distribution of volcanic activity is related to plate tectonics.

KEY TERMS: Ring of Fire, volcanic island arc (island arc), continental volcanic arc, intraplate volcanism, mantle plume, hot spot, superplume

- Volcanoes occur at both convergent and divergent plate boundaries, as well as in intraplate settings.
- At divergent plate boundaries, where lithosphere is being rifted apart, decompression melting is the dominant generator of magma. As warm rock rises, it can begin to melt without the addition of heat.
- Convergent plate boundaries that involve the subduction of oceanic crust are the most common site for explosive volcanoes—most prominently in the Pacific Ring of Fire. The release of water from the subducting plate triggers melting in the overlying mantle. The ascending magma interacts with the lower crust of the overlying plate and can form a volcanic arc at the surface.
- In intraplate settings, the source of magma is a mantle plume—a column of mantle rock that is warmer and more buoyant than the surrounding mantle.

? The accompanying diagram shows one of the tectonic settings where volcanism is a dominant process. Name the tectonic setting and briefly explain how magma is generated in this setting.

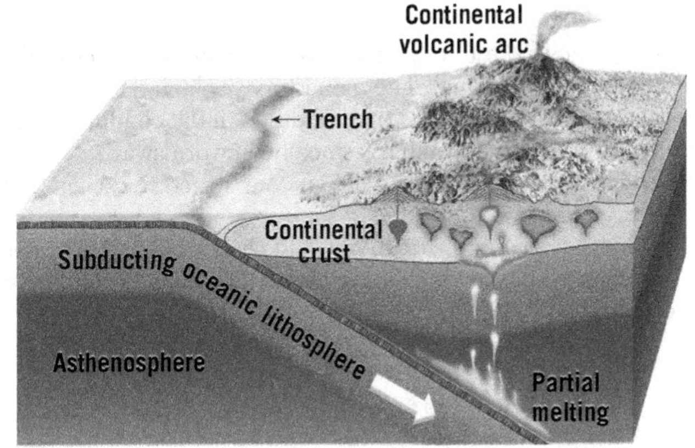

GIVE IT SOME THOUGHT

1 Examine the accompanying photo and complete the following:

a. What type of volcano is shown? What features helped you classify it as such?
b. What is the eruptive style of such volcanoes? Describe the likely composition and viscosity of its magma.
c. Which type of plate boundary is the likely setting for this volcano?
d. Name a city that is vulnerable to the effects of a volcano of this type.

2 Answer the following questions about divergent plate boundaries, such as the Mid-Atlantic Ridge, and their associated lavas:
  a. Divergent boundaries are characterized by eruptions of what type of lava: andesitic, basaltic, or rhyolitic?
  b. What is the main source of the lavas that erupt at divergent plate boundaries?
  c. What process causes the source rocks to melt?

3 For each of the accompanying four sketches, identify the geologic setting (zone of volcanism). Which of these settings will most likely generate explosive eruptions? Which will produce outpourings of fluid basaltic lavas?

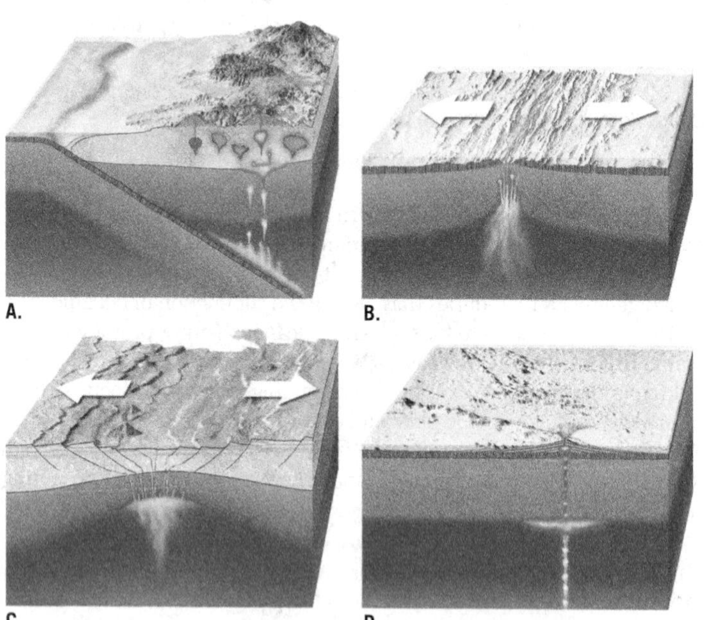

4 This image shows the Buddhist monastery Taung Kalat, located in central Myanmar (Burma). The monastery sits high on a sheer-sided rock made mainly of magmas that solidified in the conduit of an ancient volcano. The volcano has since been worn away.
  a. Based on this information, what igneous structure do you think is shown in this photo?
  b. Would this volcanic structure most likely have been associated with a composite volcano or a cinder cone? Explain how you arrived at your answer.

5 Explain why an eruption of Mount Rainier similar to the 1980 eruption of Mount St. Helens could be considerably more destructive.

6 For each of the volcanoes or volcanic regions listed below, identify whether it is associated with a *convergent* or *divergent* plate boundary or with *intraplate volcanism*.
  a. Crater Lake
  b. Hawaii's Kilauea
  c. Mount St. Helens
  d. East African Rift
  e. Yellowstone
  f. Mount Pelée
  g. Deccan Traps
  h. Fujiyama

7 The formula for the volume of a cone is $V = 1/3\pi r^2 h$ (where V = volume, $\pi = 3.14$, r = radius, and h = height). If Mauna Loa is 9 kilometers high and has a radius of roughly 85 kilometers, what is its approximate total volume?

8 The accompanying image shows a geologist at the end of an unconsolidated flow consisting of lightweight lava blocks that rapidly descended the flank of Mount St. Helens.
  a. What term best describes this type of flow: an aa flow, a pahoehoe flow, or a pyroclastic flow?
  b. What lightweight (vesicular) igneous rock type is likely the main constituent of this flow?

Donald Swanson/USGS

9 Different processes produce magma in different tectonic settings. Consider magma bodies found at locations A, B, and C in the accompanying diagram and describe the process that most likely triggered the melting that produced each.

10 During a field trip with your geology class, you visit an exposure of rock layers similar to the one sketched in the accompanying figure. A fellow student suggests that the layer of basalt is a sill. You disagree. Why do you think the other student is incorrect? What is a more likely explanation for the basalt layer?

Shale
Vesicles
Basalt
Sandstone
Shale
Limestone

EXAMINING THE EARTH SYSTEM

1 Speculate about some of the possible consequences that a great and prolonged increase in explosive volcanic activity might have on each of Earth's four spheres.

Jon A Helgason/Yay Micro/AGE Fotostock

DATA ANALYSIS

Recent Volcanic Activity

The Smithsonian Institution Global Volcanism Program and the USGS work together to compile a list of new and changing volcanic activity worldwide. NOAA also uses this information to issue Volcanic Ash Advisories to alert aircraft of volcanic ash in the air.

ACTIVITIES

Go to the Weekly Volcanic Activity Report page at http://volcano.si.edu.

1 What information is displayed on this page?

2 Click on Criteria and Disclaimers. Which volcanoes are not displayed on this map?

3 In what areas is most of the volcanic activity concentrated?

4 Click on Weekly Report. List the new volcanic activity locations. List three ongoing volcanic activity locations.

Click on the name of a volcano under New Activity/Unrest.

5 Where is this volcano located? Be sure to include the city, country, volcanic region name, latitude, and longitude.

6 What is the primary volcanic type?

7 Do some investigating online and in your textbook. What are the key characteristics for this type of volcano?

8 Briefly describe the most recent activity. How was this activity observed?

9 What are the dates for the most recent activity?

10 Click on Eruptive History. What is the earliest date listed for this volcano?

11 Find this volcano on the map on the previous page. Is this volcano near a plate boundary? If so, between which plates? (Use your textbook to determine the location of plate boundaries.)

Go to the Volcanic Ash Advisory Center (VAAC) page at www.ssd.noaa.gov/VAAC/washington.html.

12 List the VAAC locations.

13 Click on Current Volcanic Ash Advisories. When was the most recent Volcanic Ash Advisory issued? What is the location of this advisory?

14 Which of the new volcanic activity locations from question 4 currently have Volcanic Ash Advisories? For each, what is the date of the most recent advisory?

7 Crustal Deformation and Mountain Building*

FOCUS ON CONCEPTS

Each statement represents the primary learning objective for the corresponding major heading within the chapter. After you complete the chapter, you should be able to:

7.1 Describe the three types of differential stress and identify the tectonic setting most commonly associated with each. Differentiate stress from strain and brittle from ductile deformation.

7.2 List and describe five types of folds.

7.3 Sketch and briefly describe the relative motion of rock bodies located on opposite sides of normal, reverse, and thrust faults as well as both types of strike-slip faults.

7.4 Locate and name Earth's major mountain belts on a world map.

7.5 Sketch a cross section of an Andean-type mountain belt and describe how its major features are generated.

7.6 Summarize the stages in the development of an Alpine-type mountain belt such as the Appalachians.

7.7 Explain the principle of isostasy and how it contributes to the elevated topography of mountain belts.

*This chapter was revised with the assistance of Professor Callan Bentley.

Mount Robson the highest point in the Canadian Rockies, British Columbia, Canada.
(Photo by Alan Majchrowicz/Getty Images)

MOUNTAINS PROVIDE SOME of the most spectacular scenery on our planet. Poets, painters, and songwriters have captured their splendor. Geologists understand that at some time, all continental regions were mountainous masses and that continents grow over time through the addition of mountains to their flanks. Where lithospheric plates collide, the rocks in the collision zone are crumpled and transformed. Even continental interiors hold the remnants of mountain ranges formed from ancient collisions. This chapter explains the forces that create mountains, while the following chapters focus on the processes that weather and carry away rock material to create Earth's varied topography.

7.1 Crustal Deformation

Describe the three types of differential stress and identify the tectonic setting most commonly associated with each. Differentiate stress from strain and brittle from ductile deformation.

Young mountain belts tower above the surrounding landscapes, exhibiting steep slopes and high relief. When the tectonic forces that raised these mountains cease, weathering and erosion will, over long spans of geologic time, grind them down until their roots lie flush with the surrounding topography. Even then, their rocks hold clues to their former grandeur and to the forces that created them.

What Causes Rocks to Deform?

Tectonic forces can cause rocks to *move*, *tilt*, and/or *change shape*. For instance, colliding plates can uplift flat-lying beds of marine limestone so they are exposed at the surface, rotate them so they lie at a steep angle, or crumple the rock layers into folds. Collectively, all of these types of change are called **deformation**. Rock deformation is caused mainly by tectonic forces, and it occurs mostly along plate boundaries—the places where lithospheric plates push together, pull apart, or scrape past each other.

Rocks deform in characteristic ways that can be observed when the rocks are exposed on Earth's surface as an *outcrop* (**Figure 7.1**). For example, the folds in Figure 7.1 represent a typical response to compressional forces at depth. Geologists use the term **tectonic structures**, or **geologic structures**, for the structural

SmartFigure 7.1 **Deformed sedimentary strata** These deformed strata are exposed in a road cut near Palmdale, California. In addition to the obvious folding, light-colored beds are offset along a fault located on the right side of the photograph. (Photo by E. J. Tarbuck)

TUTORIAL
https://goo.gl/BpfvcZ

features that can be observed and that reflect a rock's tectonic history. This chapter will examine three types of tectonic structures: folds (the reshaping of rock layers without breakage), faults (fractures along which one rock sides past another), and joints (cracks in the rock).

To understand rock deformation, we first need to look more closely at the concepts of *stress* and *strain*.

Stress: The Cause of Deformation So far, we've said that tectonic *forces* cause deformation. More precisely, rocks respond to **stress**, which takes into account the area over which a force acts (Stress = Force/Area). Recall from Chapter 3 that a force applied equally in all directions is called **confining pressure**; this type of force compacts mineral grains to reduce the volume of rock bodies (Figure 7.2A). However, confining pressure does not cause deformation; instead, deformation is caused by **differential stress**, in which the force is stronger in one direction and weaker in another.

We will consider three types of differential stress: *compressional*, *tensional*, and *shear*. On a large scale, each type of stress tends to be associated with one type of plate boundary:

1. **Compression.** Differential stress that squeezes a rock mass as if were in a vise is known as **compressional stress** (Figure 7.2B). Compressional stresses are most often associated with convergent plate boundaries. When plates collide, Earth's crust is generally shortened horizontally and thickened vertically. Over millions of years, this deformation produces mountain belts.
2. **Tension.** Differential stress that pulls apart rock bodies is known as **tensional stress** (Figure 7.2C). Along divergent plate boundaries, where plates are moving apart, tensional stresses stretch and lengthen rock bodies horizontally and thin them vertically. For example, in the Basin and Range Province in western North America, tensional forces have fractured and stretched the crust to as much as twice its original width.
3. **Shear.** Differential stress can cause rock to **shear**, which involves the movement of one part of a rock body past another (Figure 7.2D). An everyday example of shear is the slippage that occurs between individual playing cards when the top of the deck is moved relative to the bottom (Figure 7.3). Shear is important at transform plate boundaries, such as the San Andreas Fault, where large segments of Earth's crust slip horizontally past one another.

Strain: A Change in Shape Caused by Stress Recall that differential stress can deform a rock body by causing it to *move*, *tilt*, and/or *change shape*. When differential stress changes a rock's shape, the resulting deformation (distortion) is called **strain**. By observing and measuring the strain imprinted on a rock body, we can infer the type of stress that deformed the rock.

How does a rigid object like a rock change its shape? One way is by undergoing slippage along parallel surfaces of weakness, such as microscopic fractures or foliation surfaces. Like the deck of cards in Figure 7.3, many tiny slips can add up to a significant change in shape.

Mineral grains can also change shape in response to differential stress that does not involve slippage along zones of weakness. Instead, the movement of atoms from a location that is highly stressed to a less-stressed position on the same grain triggers a change in shape—a process called *recrystallization* (see Section 3.4, page 77).

Types of Deformation

Rocks experience three types of deformation that lead to shape changes (strain): *elastic*, *brittle*, and *ductile*.

Elastic Deformation If you open a door that has a spring attached, when you let go of the door, the spring pulls the door shut as it returns to its original shape. We say the spring undergoes **elastic deformation**: It deforms temporarily in response to a stress (the force used to open the door) and returns to its original configuration when the stress is removed. Chemical bonds in a mineral grain act like a spring: When they are elastically deformed, they stretch instead of breaking; when the stress is removed, they snap back to their original length. As you saw in Chapter 5, the energy released by most strong earthquakes comes from stored energy that is suddenly released as rock elastically snaps (elastic rebound) back to its original shape.

Brittle Deformation When a rock's strength is exceeded—that is, when it is deformed beyond its ability to respond elastically—it will either break or be permanently bent. Rocks that break into smaller pieces exhibit **brittle deformation**. From our everyday experience, we know that glass objects, wooden pencils, ceramic plates, and even our bones exhibit brittle deformation when their strength is surpassed. Brittle deformation occurs when stress breaks the chemical bonds that hold a material together.

Ductile Deformation When an object changes shape without breaking, we say that it has undergone **ductile deformation**. When you knead clay or taffy, you are

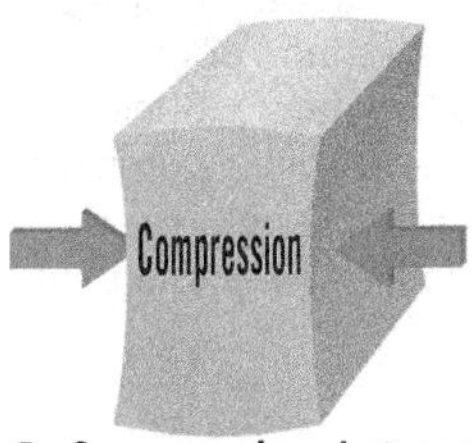

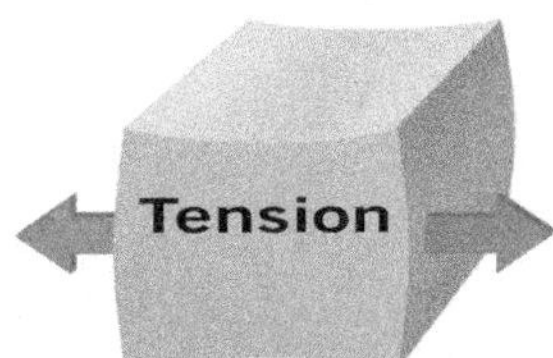

Figure 7.2 Confining pressure and three types of differential stress: Compression, tension, and shear

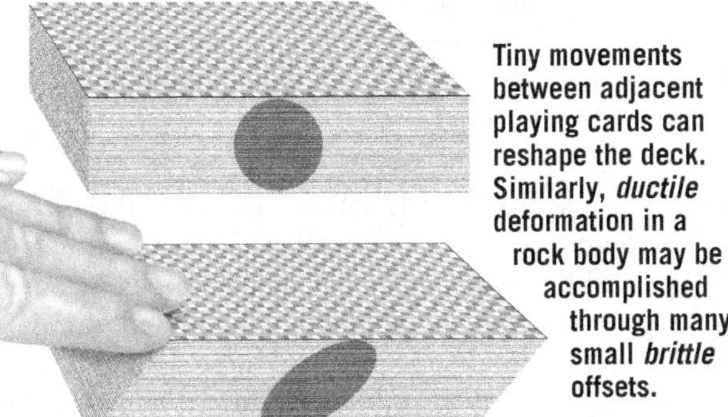

Tiny movements between adjacent playing cards can reshape the deck. Similarly, *ductile* deformation in a rock body may be accomplished through many small *brittle* offsets.

Figure 7.3 Shearing and the resulting deformation (strain) An ordinary deck of playing cards with a circle embossed on its side illustrates shearing and the resulting strain.

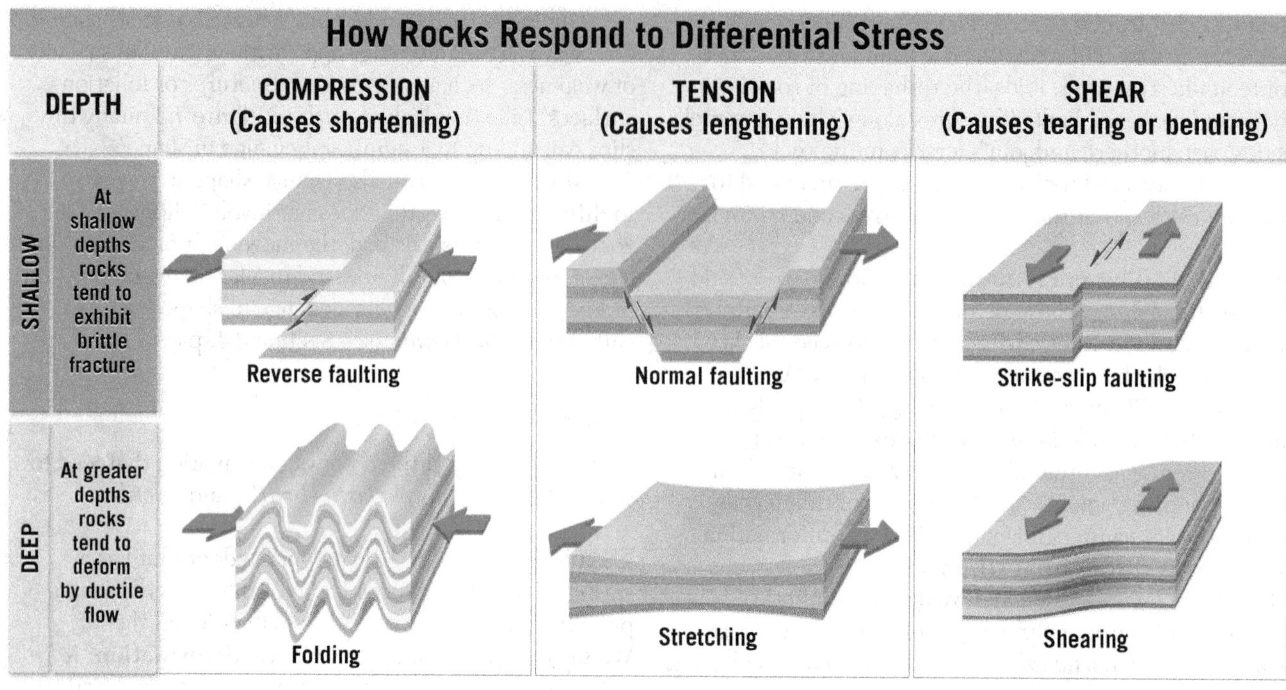

Figure 7.4 Deformation caused by three types of stress Brittle deformation (fracturing and faulting) dominates in the upper crust, where the temperatures are comparatively cool. By contrast, at depths greater than about 10 kilometers (6 miles), where temperatures are high, rock deforms by ductile flow and folding.

deforming it in a ductile way. Ductile deformation in rocks takes place largely through the mechanisms described earlier—slippage along surfaces of weakness within the rock and the gradual reshaping of mineral grains. These processes enable rock to flow very slowly, even though it remains in a solid state. Intricate folds are an example of ductile deformation.

Factors That Affect How Rocks Deform

Figure 7.4 compares the types of brittle and ductile deformation. As the figure indicates, rock deformation tends to be brittle at shallow depths and ductile at greater depths. Four factors influence how a rock deforms; *temperature*, *confining pressure*, the *type of rock*, and *time*.

The Role of Temperature Temperature plays a major role in the behavior of rocks. Where temperatures are high (deep in Earth's crust or adjacent to a heat source such as a magma chamber), rocks are nearer their melting temperatures and are therefore weaker and more capable of ductile deformation (see Figure 7.4). Near the surface or in a comparatively cool environment such as a subduction zone, rocks are more brittle and prone to fracture. This behavior is completely familiar: A cold chocolate bar snaps between your teeth, whereas a warm one bends in your hand.

The Role of Confining Pressure Recall that the confining pressure on rocks increases with depth as the thickness of the overlying rock increases. Because confining pressure squeezes rocks equally from all directions, it tends to make them harder to break and hence less brittle. Thus, at depth, the increase in temperature and the increase in pressure have complementary effects: The increase in temperature enhances ductile behavior, and the increase in pressure tends to keep the rock intact, and it is thus more likely to bend than to fracture.

The Influence of Rock Type The manner in which a particular rock type responds to stress is greatly influenced by its mineral composition and texture. Granite, basalt, and well-cemented quartz sandstones are examples of strong, brittle rocks that tend to fail by breaking (brittle deformation) when subjected to stresses that exceed their strength. By contrast, clay-rich or weakly cemented sedimentary rocks and foliated metamorphic rocks more readily exhibit ductile deformation. Weak rocks that are most likely to behave in a ductile manner (bend or flow) when subjected to differential stress include rock salt, shale, limestone, and schist.

Figure 7.5 shows an example of this type of behavior in a metamorphic rock. The central nonfoliated layer behaved in a brittle manner, breaking into chunks, while the flanking foliated layers responded to the same differential stress in a ductile manner, flowing into the gaps between the chunks. You can create the same effect by biting into a s'more: The graham crackers break, while the warmed marshmallow and chocolate layers ooze.

Rock salt consists of intergrown crystals, so you might think that it would resist ductile deformation. In fact, the opposite is true. Salt crystals readily change shape by recrystallizing in response to differential stresses. That is why salt layers can rise up through other strata to form salt domes. Glacial ice also consists of crystals that deform easily, accommodating internal flow.

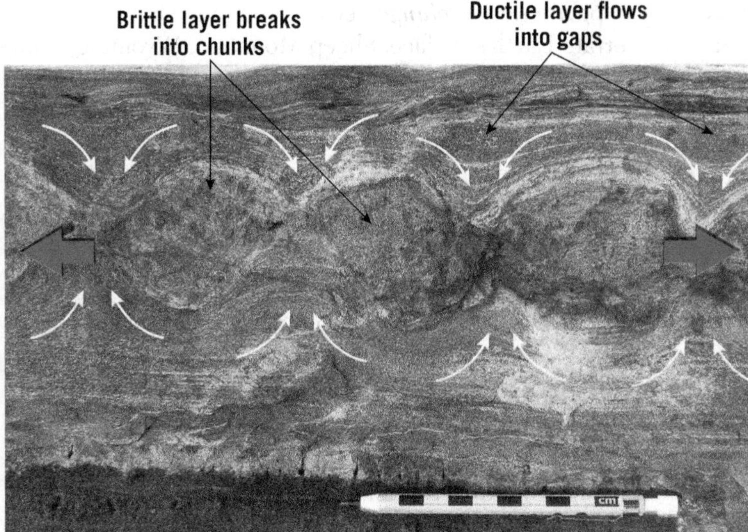

Figure 7.5 **How rock type influences the type of deformation** This structure is an example of *boudinage*, which forms when some portions of a rock body deform in a ductile fashion and others act as brittle units. The central greenish metamorphic layer has broken into a series of chunks, and the surrounding gray sedimentary material has flowed into the gaps between the chunks. (Photo by Callan Bentley)

Time as a Factor The folded rocks of mountain belts show that tens of kilometers of compressional strain (shortening) can be accommodated by ductile deformation. For this to happen, stress has to be applied slowly enough that the sluggish processes of ductile deformation can keep up. If stress is applied to a rock unit too quickly, the rock will deform elastically until its strength is exceeded, and then it will fracture. Taffy exhibits the same behavior on a more familiar time scale: If you hit a bar of taffy against the edge of a table, it will break, but if you put a weight on it and leave it overnight, it will gradually spread and flatten. In practice, rocks that are near Earth's surface tend to accommodate even gradual strain by fracturing; ductile behavior happens mainly at depth.

CONCEPT CHECKS 7.1

1. What is deformation? List several ways in which a rock body might change during deformation.
2. List the three types of differential stress and briefly describe the changes they can impart to rock bodies.
3. What type of plate boundary is most commonly associated with compressional stress?
4. How is strain different from stress?
5. How is brittle deformation different from ductile deformation?
6. List and describe the four factors that affect whether a rock deforms in a brittle or ductile manner.

7.2 Folds: Rock Structures Formed by Ductile Deformation

List and describe five types of folds.

Along convergent plate boundaries, rock strata are often bent into a series of wavelike undulations called **folds**. Folds in flat-lying strata are much like those that would form if you were to hold the ends of a sheet of paper on a flat surface and then push them together. Such folds come in a wide variety of sizes and configurations. Some are broad flexures in which strata hundreds of meters thick have been slightly warped. Others are very tight, even microscopic structures found in metamorphic rocks. Most folds result from *compressional stresses* that result in a lateral shortening and vertical thickening of the crust.

Folds are geologic structures that result when originally planar ("flat") surfaces, such as sedimentary strata, are permanently bent as a result of deformation. Each layer is bent around an imaginary axis called a *hinge line*, or simply a *hinge* (Figure 7.6).

Folds are also described by their *axial plane*, which is a surface that connects all the hinge lines of the folded strata. In simple folds, the axial plane is vertical and divides the fold into two roughly symmetrical *limbs* (see Figure 7.6). However, the axial plane often leans to one side so that one limb is steeper than the other.

Anticlines and Synclines

The two most common types of folds are anticlines and synclines (Figure 7.7). **Anticlines** usually form by the upfolding, or arching, of sedimentary layers.* Typically found in association with anticlines are downfolds, or

*By strict definition, an anticline is a structure in which the oldest strata are found in the center, and a syncline is a structure where the youngest strata are found in the center.

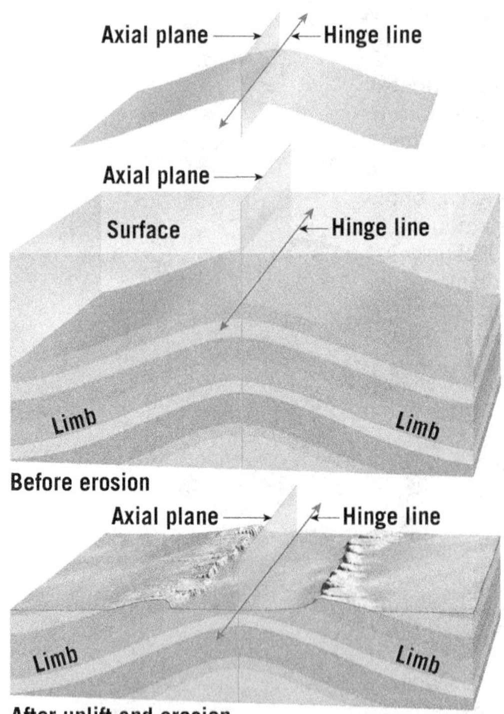

SmartFigure 7.6
Features associated with symmetrical folds The axial plane divides a fold as symmetrically as possible, while the hinge line traces the points of maximum curvature of any layer.

CONDOR VIDEO
https://goo.gl/zWIRuV

troughs, called **synclines**. Notice in Figure 7.7 that the limb of an anticline is also a limb of an adjacent syncline.

Depending on their orientation, these basic folds are described as *symmetrical* when the limbs are mirror images of each other and *asymmetrical* when they are not. The limbs of a symmetrical fold dip at the same angle, resulting in a vertical axial plane. By contrast, the limbs of an asymmetrical fold dip at different angles, which results in an inclined axial plane. An asymmetrical fold is said to be *overturned* if both limbs dip in the same direction, with one limb tilted beyond the vertical (see Figure 7.7). An overturned fold can also "lie on its side" so that the axial plane is horizontal. These *recumbent* folds are common in highly deformed mountain belts such as the Alps.

Folds can also be tilted by tectonic forces so their hinge lines slope downward (Figure 7.8A). Folds of this type are said to *plunge* because the hinge lines penetrate Earth's surface. Sheep Mountain, Wyoming, is an example of a plunging anticline (Figure 7.8C). Notice in Figure 7.8B that a V-shaped pattern is produced when erosion removes the upper layers of a plunging fold and exposes its interior. In the case of an anticline, such as Sheep Mountain, the tip of the V points in the direction of plunge (Figure 7.8C); the opposite is true for a syncline.

A good example of the topography that can result when erosional forces attack folded sedimentary strata is found in the Valley and Ridge Province of the Appalachians (see Figure 7.32). It is important to realize that anticlines typically do not show up as ridges, nor synclines as valleys. Rather, ridges and valleys result from differential weathering and erosion. For example, in the Valley and Ridge Province, resistant sandstone beds remain as imposing ridges separated by valleys that are cut into more easily eroded shale or limestone beds.

Domes and Basins

When a broad upwarping of basement rock deforms the overlying cover of sedimentary strata to produce a circular or slightly elongated bulge, the feature is called a **dome**

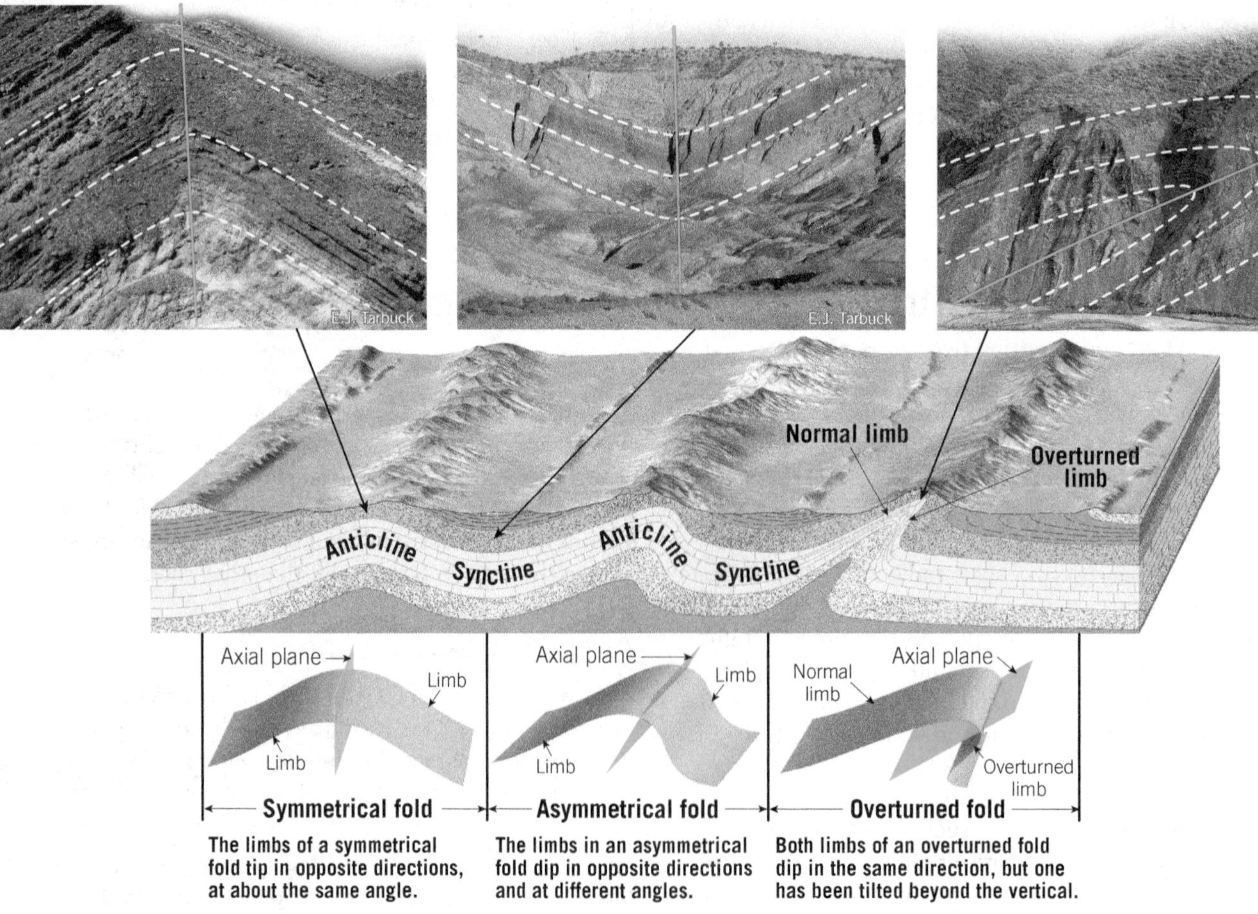

SmartFigure 7.7
Common types of folds The arched structures—folds that go up in the middle—are *anticlines*. Their limbs dip away from each another. Folds that go down in the middle (limbs dip toward each other) are *synclines*.

TUTORIAL
https://goo.gl/EwDKUK

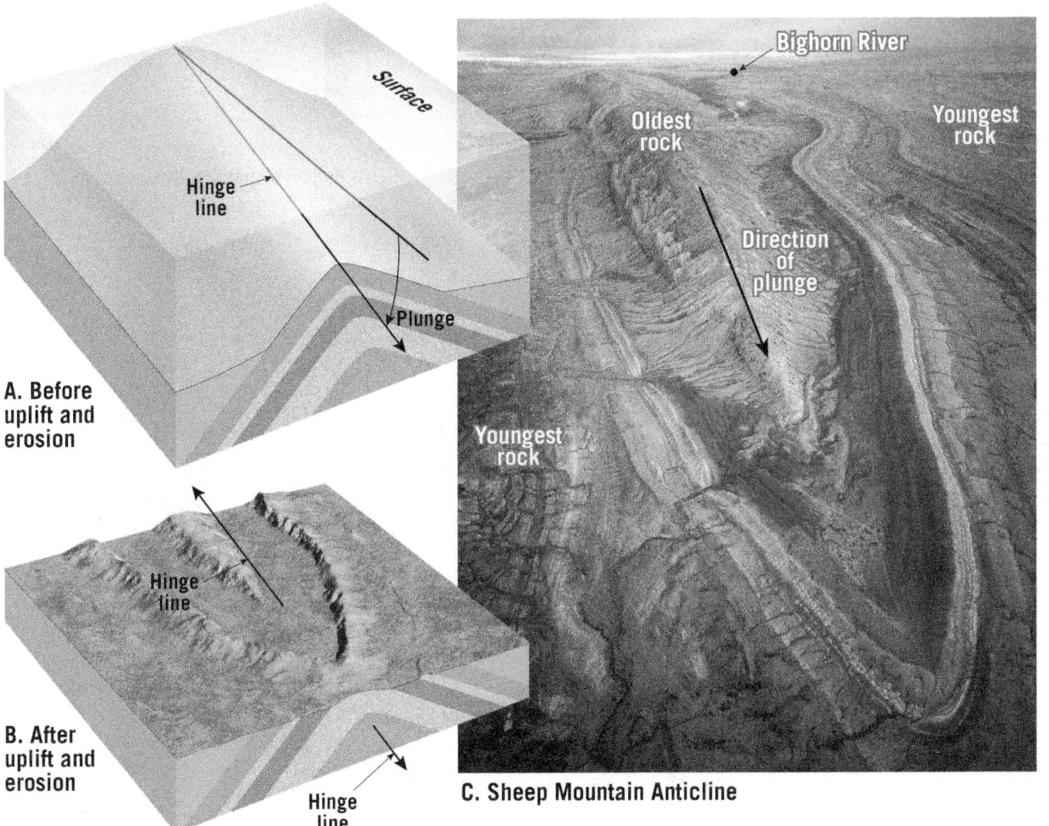

SmartFigure 7.8 **Sheep Mountain, Wyoming** Eroded plunging anticlines, like Sheep Mountain, have an outcrop pattern that "points" in the direction of plunge. (Photo by Michael Collier)

MOBILE FIELD TRIP
https://goo.gl/7LyhfN

(Figure 7.9A). The Black Hills of western South Dakota represent a large structural dome generated by upwarping. Here erosion has stripped away the highest portions of the overlying sedimentary beds, exposing older igneous and metamorphic rocks in the center (Figure 7.10).

Structural domes can also be formed by the intrusion of magma (laccoliths), as shown in Figure 6.27, page 189. In addition, the upward migration of buried salt deposits can produce salt domes like those beneath the Gulf of Mexico. Salt domes are economically important rock structures because when salt migrates upward, the surrounding oil-bearing sedimentary strata deform to form oil reservoirs (see Figure 3.37, page 88).

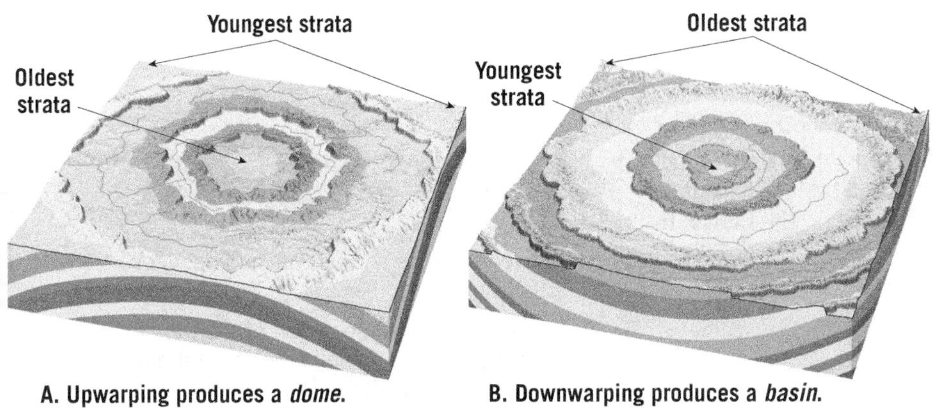

SmartFigure 7.9 **Domes versus basins** Gentle upwarping and downwarping of crustal rocks produce (**A**) domes and (**B**) basins. Erosion of these structures results in an outcrop pattern that ranges from roughly circular to more elongated (elliptical).

TUTORIAL
https://goo.gl/tD2CXw

Figure 7.10 **The Black Hills of South Dakota, a large structural dome** The central core of the Black Hills is composed of resistant Precambrian-age igneous and metamorphic rocks. The surrounding rocks are mainly younger limestones and sandstones.

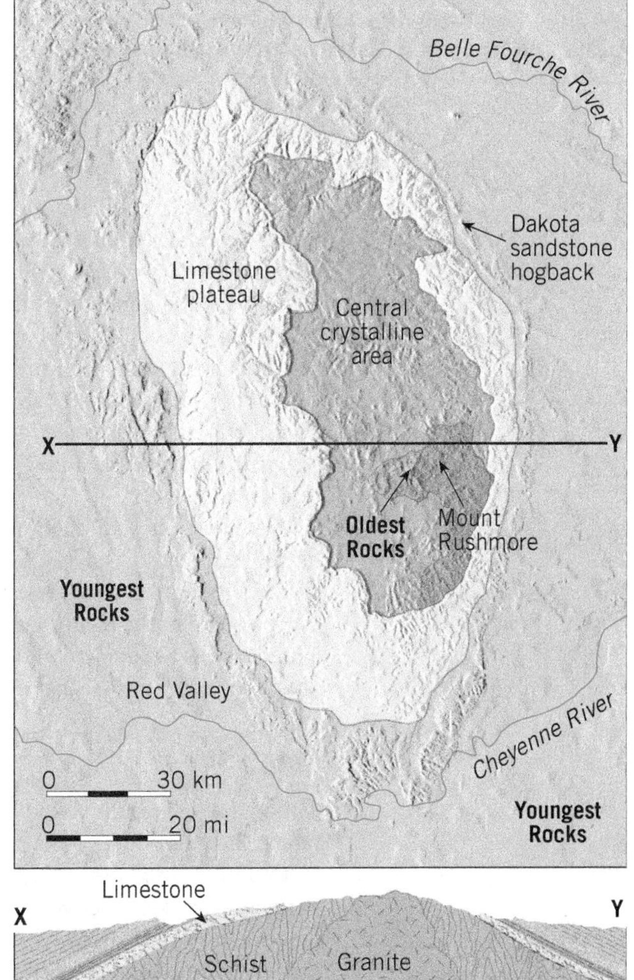

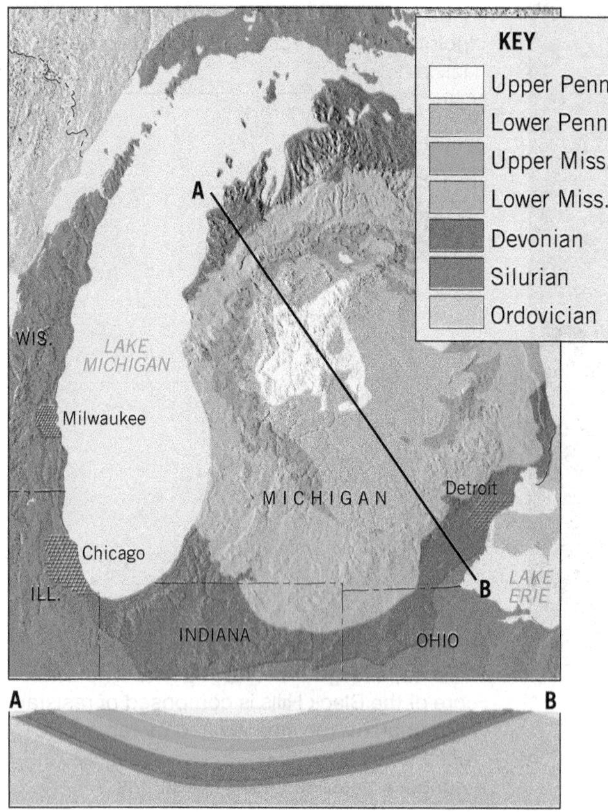

▲ Figure 7.11 **The bedrock geology of the Michigan basin** The youngest rocks are centrally located, and the oldest beds flank this structure.

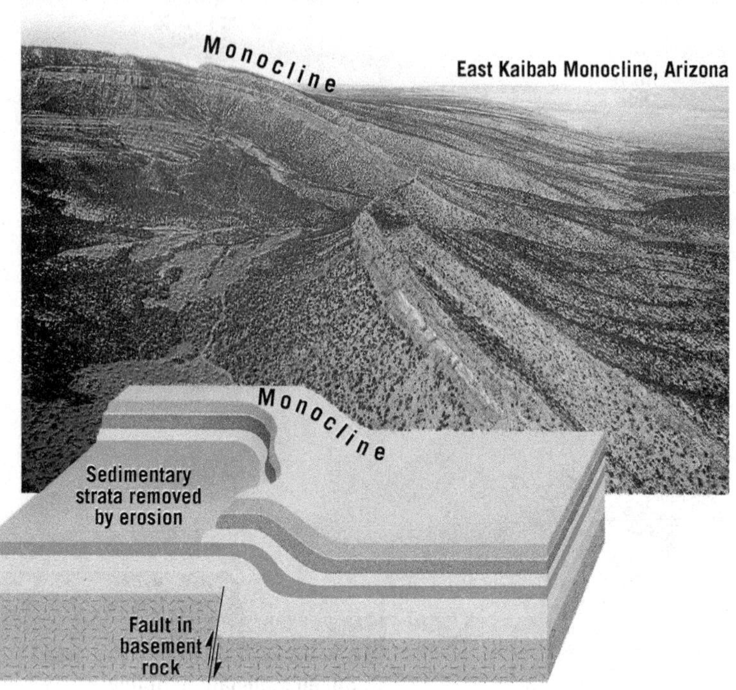

▲ SmartFigure 7.12 **The East Kaibab Monocline, Arizona** This monocline consists of bent sedimentary beds that were deformed by faulting in the bedrock below. The thrust fault does not reach the surface. The inclined strata once extended over the sedimentary layers now exposed at the surface—evidence that a tremendous volume of rock has been eroded from this area. (Photo by Michael Collier)

CONDOR VIDEO
https://goo.gl/dPwXf4

The inverse of a dome is a downwarped structure termed a **basin** (Figure 7.9B). Several large structural basins exist in the United States (Figure 7.11). The basins of Michigan and Illinois have gently sloping beds similar to saucers. These basins are thought to have resulted from large accumulations of sediment, whose weight caused the crust to subside (see the section "The Principle of Isostasy," later in this chapter). A few structural basins may have resulted from giant meteorite impacts.

Because the sedimentary beds in structural basins usually slope at low angles, basins are identified mainly by the age sequence of their strata. In a basin, the youngest rocks are at the center and the oldest on the flanks. In a dome, such as the Black Hills, the reverse is true: The oldest rocks form the core.

Monoclines

Although we have discussed folds and faults separately, they may occur together as a result of the same tectonic stresses. One example can be found in broad regional features called *monoclines*. Particularly prominent features of the Colorado Plateau, **monoclines** (*mono* = one, *kleinen* = incline) are large, steplike folds in otherwise horizontal sedimentary strata (Figure 7.12). These folds appear to have resulted from the reactivation of ancient, steep-dipping reverse faults located in basement rocks beneath the plateau. As large blocks of basement rock were displaced upward, the comparatively ductile sedimentary strata above responded by draping over the fault like clothes hanging over a bench. Displacement along these reactivated faults can exceed 1 kilometer (0.6 miles).

CONCEPT CHECKS 7.2

1. Distinguish between anticlines and synclines, between domes and basins, and between anticlines and domes.
2. Draw a cross-sectional view of a symmetrical anticline. Include a line to represent the axial plane and label both limbs.
3. The Black Hills of South Dakota is a good example of what type of geologic structure?
4. Where are the youngest rocks in a structural basin found: near its center or near its margin?
5. Describe how a monocline forms.

EYE ON EARTH 7.1

This image features a large geologic structure that outcrops in Death Valley National Park, California. (Photo by Michael Collier)

QUESTION 1 *What name would you give to this geologic structure?*

QUESTION 2 *Based on this image, would you describe this fold as symmetrical or asymmetrical?*

QUESTION 3 *Do these rock units mainly display ductile deformation or brittle deformation?*

7.3 Faults and Joints: Rock Structures Formed by Brittle Deformation

Sketch and briefly describe the relative motion of rock bodies located on opposite sides of normal, reverse, and thrust faults as well as both types of strike-slip faults.

Faults and joints are both structures that form where brittle deformation leads to fracturing of Earth's crust. A *joint* is a fracture, whereas a **fault** is a fracture along which motion has occurred, so that the rocks on either side are offset from each other. Figure 7.13 shows a small fault revealed in a road cut, where the sedimentary beds have been offset by a few meters. Faults of this scale usually occur as single discrete breaks. By contrast, large faults, like the San Andreas Fault in California, have displacements of hundreds of kilometers and consist of many interconnecting fault surfaces. These structures, described as *fault zones*, can be several kilometers wide and are often easier to identify from aerial photographs than at ground level. Sudden movements along faults cause most earthquakes. However, the vast majority of faults are remnants of past deformation and are inactive.

Dip-Slip Faults

Faults in which movement is primarily parallel to the slope of the fault surface are called **dip-slip faults**. (In this case, "dip" refers to the angle at which the fault surface is inclined relative to the horizontal.) Geologists identify the rock body that contains the fault's upper surface as the **hanging wall block** and the rock body containing the lower surface as the **footwall block** (Figure 7.14). These names were first used by prospectors and miners who excavated metallic ore deposits such as gold that had precipitated from hydrothermal solutions along inactive fault zones. The miners would walk on the rocks below the mineralized fault zone (the *footwall block*) and hang their lanterns on the rocks above (the *hanging wall block*).

Rapid vertical displacements along dip-slip faults tend to produce long, low cliffs called **fault scarps**, such

SmartFigure 7.13
Faults are fractures where slip has occurred
(Photo by E. J. Tarbuck)

CONDOR VIDEO
https://goo.gl/rCweMh

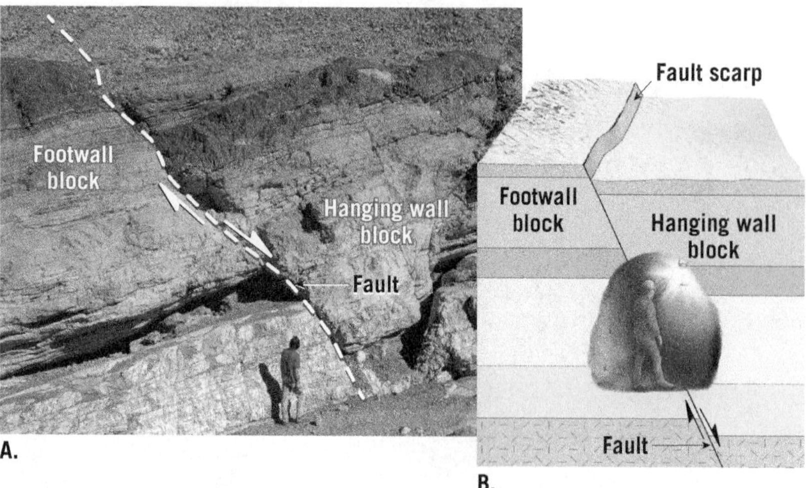

SmartFigure 7.14 **Hanging wall block and footwall block** The rock immediately above a fault surface is the *hanging wall block*, and the one below is called the *footwall block*. These terms were coined by miners who excavated ore deposits that formed along fault zones. The miners hung their lanterns on the rocks above the fault trace (hanging wall block) and walked on the rocks below the fault trace (footwall block). (Photo by Marli Miller)

ANIMATION
https://goo.gl/qkD1uL

as the one shown in Figure 7.15. The movements that form such scarps typically also generate earthquakes.

There are two kinds of dip-slip faults: *normal faults* and *reverse faults*.

Normal Faults When the hanging wall block moves down relative to the footwall block, dip-slip faults are classified as **normal faults** (Figure 7.16). Normal faults are associated with tensional stresses that pull rock units apart, thereby lengthening the crust laterally and thinning it vertically. This "pulling apart" can be accomplished either by uplift that causes the surface to stretch and break or by horizontal forces that have opposing orientations. These faults are called normal faults because it is "normal" for gravity to pull a block of rock down an inclined plane (the fault surface).

Normal faults occur in a variety of sizes. Some are small, having displacements of only a meter or so, like the one shown in the road cut in Figure 7.13. Others, however, extend for tens of kilometers. Most large normal faults have relatively steep dips at the surface but tend to flatten out with depth.

In the western United States, large normal faults are associated with structures called **fault-block mountains**. Excellent examples of fault-block mountains are found in the Basin and Range Province, a region that encompasses Nevada and portions of the surrounding states (Figure 7.17). Here the crust has been elongated and broken to create more than 200 relatively small mountain ranges. Averaging about 80 kilometers (50 miles) in length, the ranges rise 900 to 1500 meters (3000 to 5000 feet) above the adjacent down-faulted topographic basins.

The topography of the Basin and Range Province evolved in association with a system of normal faults trending roughly north–south. Movements along these faults produced alternating uplifted fault blocks called

Figure 7.15 **Fault scarp** This fault scarp was created during the Alaska earthquake of 1964. (Photo courtesy of the USGS)

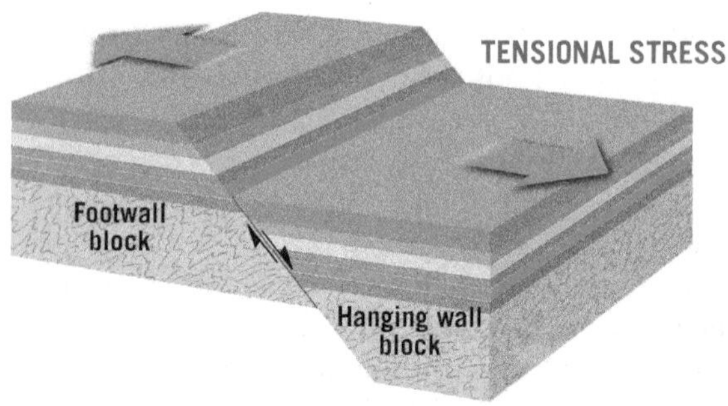

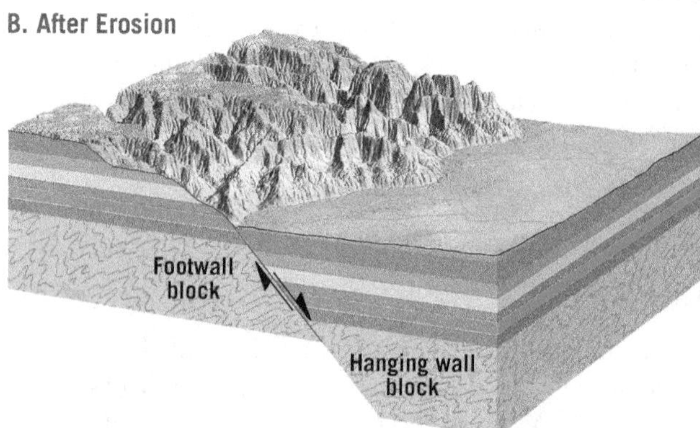

SmartFigure 7.16 **Normal dip-slip fault** The upper diagram illustrates the relative displacement that occurs between the blocks on either side of a fault. The lower diagram shows how erosion may alter the up-faulted blocks.

TUTORIAL
https://goo.gl/eFkg

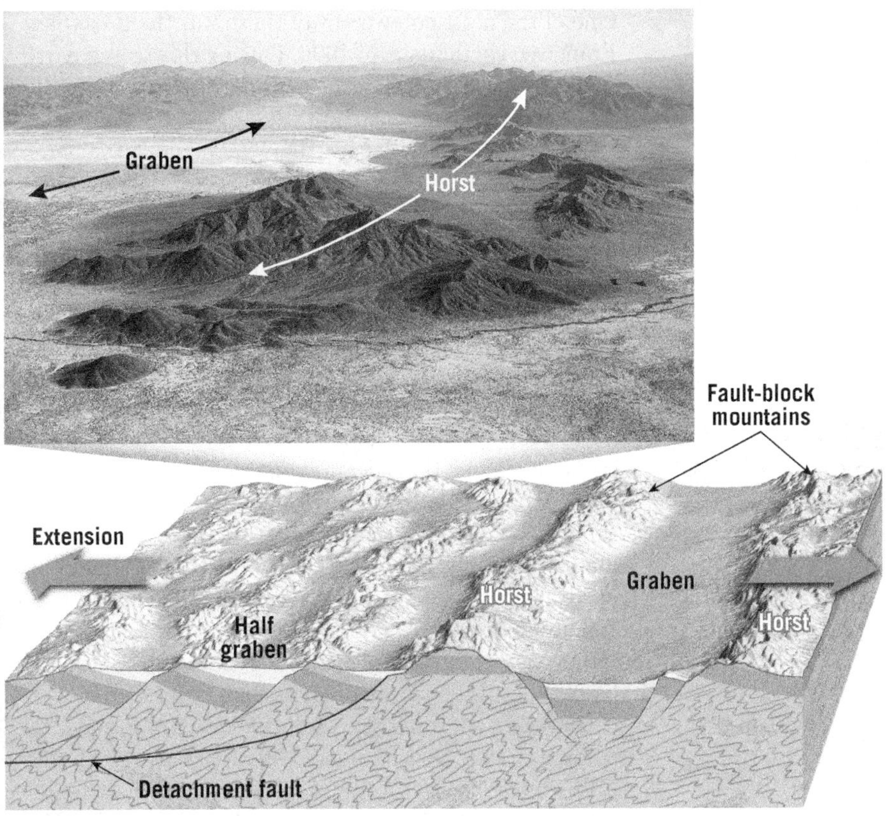

SmartFigure 7.17 **Normal faulting in the Basin and Range Province** Here, tensional stresses have elongated and fractured the crust into numerous blocks. Movement along these faults has tilted the blocks, producing parallel mountain ranges called *fault-block mountains*. The down-faulted blocks (*grabens*) form basins, whereas the up-faulted blocks (*horsts*) erode to form rugged mountainous topography. In addition, numerous tilted blocks (*half-grabens*) form both basins and mountains. (Photo by Michael Collier)

MOBILE FIELD TRIP
https://goo.gl/N7B9bZ

horsts (*horst* = hill) and down-dropped blocks called **grabens** (*graben* = ditch). Horsts form the ranges and are the source of sediments that have accumulated in the topographic basins created by the grabens. As Figure 7.17 illustrates, structures called **half-grabens**, which are tilted fault blocks, also contribute to the alternating topographic highs and lows in the Basin and Range Province.

Also notice in Figure 7.17 that the slopes of many of the large normal faults in the Basin and Range Province decrease with depth and eventually join to form a low-angle, nearly horizontal fault called a **detachment fault**. These faults represent a major boundary between the rocks below, which exhibit ductile deformation, and the rocks above, which exhibit mainly brittle deformation.

Reverse and Thrust Faults Dip-slip faults in which the hanging wall block moves up relative to the footwall block are called **reverse faults** (Figure 7.18). A **thrust fault** is a type of reverse fault in which the fault's angle is less than 45 degrees. Reverse and thrust faults result from compressional stresses that produce horizontal shortening of the crust.

Most high-angle reverse faults are small and accommodate local displacements in regions dominated by other types of faulting. Thrust faults, on the other hand, exist at all scales, with some large thrust faults having displacements ranging from tens to hundreds of kilometers. Movement along a thrust fault can cause the hanging wall block to be thrust nearly horizontally over the footwall block, as shown in Figure 7.19.

Thrust faulting is most pronounced along convergent plate boundaries. Compressional forces associated with colliding plates generally create folds as well as thrust faults that thicken and shorten the crust to produce mountainous topography (see Figure 7.29). Examples of

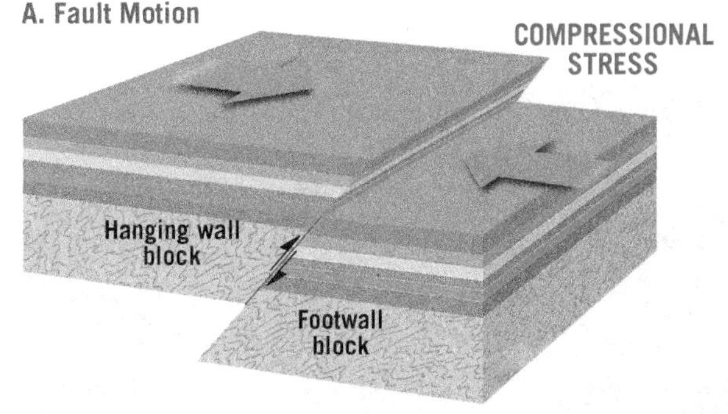

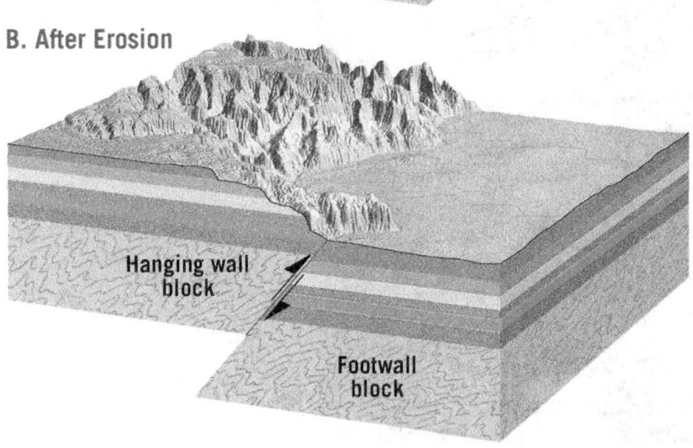

SmartFigure 7.18 **Reverse faults** Reverse faults are generated by compressional stresses that force one block of rock over another.

ANIMATION
https://goo.gl/OvmqP2

SmartFigure 7.19 **Thrust fault** Thrust faults are a type of reverse fault with dips less than 45 degrees.

ANIMATION
https://goo.gl/lejdNe

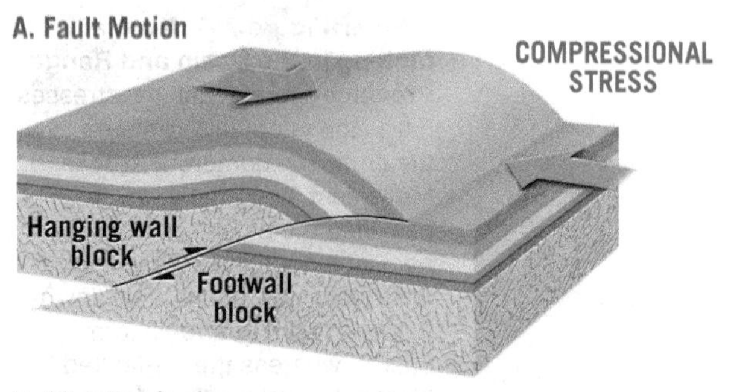

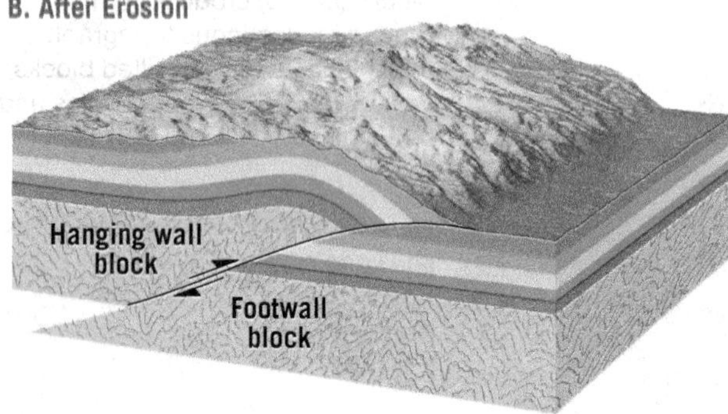

mountainous belts produced by this type of compressional tectonics include the Alps, Northern Rockies, Himalayas, and Appalachians.

Strike-Slip Faults

A fault in which the dominant displacement is horizontal and parallel to the *trend* (direction) of the fault surface is called a **strike-slip fault** (Figure 7.20). The earliest scientific records of strike-slip faulting were made following surface ruptures that produced large earthquakes.

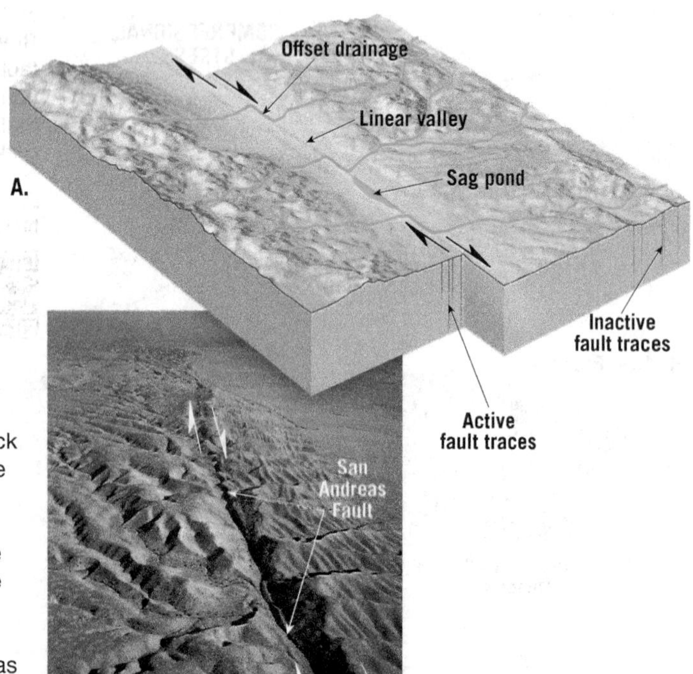

Figure 7.20 **Strike-slip faults A.** The block diagram illustrates the features associated with large strike-slip faults. Notice how the stream channels have been offset by fault movement. **B.** Aerial view of the San Andreas Fault. (Photo by D. Parker/ Science Source)

One of the most noteworthy of these was the great San Francisco earthquake of 1906. During this strong earthquake, structures such as fences and roads that were built across the San Andreas Fault were displaced as much as 4.7 meters (15 feet). Because movement along the San Andreas Fault causes the crustal block on the opposite side of the fault to move to the right as you face the fault, it is called a *right-lateral* strike-slip fault.

The Great Glen Fault in Scotland, which exhibits displacement in the opposite direction, is a well-known example of a *left-lateral* strike-slip fault. Associated with the crushed-up rock along the Great Glen Fault trace are numerous lakes, including the legendary Loch Ness.

As discussed in Chapter 4, **transform faults** are strike-slip faults that accommodate motion between two tectonic plates. (Although all transform faults are strike-slip faults, only those strike-slip faults that form plate boundaries are called transform faults.) Numerous transform faults cut the oceanic lithosphere and link spreading oceanic ridges (see Figure 4.19, page 110). Others accommodate displacement between continental blocks that slip horizontally past each other. Some of the best-known transform faults include California's San Andreas Fault, New Zealand's Alpine Fault, the Middle East's Dead Sea Fault, and Turkey's North Anatolian Fault (see Figure 7.20). Large transform faults like these accommodate relative displacements of up to several hundred kilometers.

Rather than being a single fracture, most continental transform faults consist of a zone of roughly parallel fractures. While this zone may be up to several kilometers wide, the most recent movement is often along a strand only a few meters wide, which may offset features such as stream channels. Crushed and broken rocks produced during faulting are more easily eroded, so depressions such as linear valleys often mark the locations of large strike-slip faults.

Joints

Joints are among the most common geologic structures and can be found in nearly all rock outcrops. As mentioned earlier, joints differ from faults in that no appreciable displacement has occurred along the fracture. Although some joints have random orientations, most occur in roughly parallel groups (Figure 7.21).

Most joints are produced when rocks in Earth's outermost crust are deformed by tensional stresses that stretch the rock layer and cause it to fail by brittle fracture. One way in which rock layers are stretched is in response to relatively subtle regional upwarping and downwarping of Earth's crust. This is illustrated in Figure 7.22, which shows how regional upwarping of the Entrada sandstone in Arches National Park generated a set of nearly parallel joints.

Not all joint are produced by regional tensional stresses, however. Recall from Chapter 6 that *columnar*

Figure 7.21 **Nearly parallel joints in Entrada sandstone, Arches National Park, Utah** These two sets of joints formed by regional upwarping, which caused the rigid Entrada sandstone to fracture. The orientation of the dominant joint set is shown with arrows. The other joint set is oriented perpendicular (90 degrees) to the dominant set. (Photo by Michael Collier)

joints form when igneous rocks cool and develop shrinkage fractures that produce elongated, pillar-like columns (see Figure 6.30, page 191). Another example of jointing can be seen in Figure 8.6, page 245, which illustrates *sheeting*, a process that produces joints that form parallel to the surface of large igneous masses.

Joint patterns profoundly affect the weathering and erosion of bedrock. In particular, joints allow ion-rich water to penetrate to depth and start the weathering process long before the rock is exposed. As a result, joints strongly influence how landforms develop. The iconic landscape in Figure 7.22 formed as weathering and erosion enlarged joints, creating long, narrow walls called *fins*. These narrow rock walls, in turn, are the setting in which differential weathering created the park's famed arches (see Figure 8.2, page 241).

Figure 7.22 **How joints influence the development of landforms** Weathering and erosion along a prominent set of joints in what is now Arches National Park produced a topography called *fins*. **A.** Aerial view of the fins that are located in the Devils Garden area of the park. **B.** Ground view of fins taken near Sand Dune Arch. (Photos by Dennis Tasa)

Because jointing weakens rocks, highly jointed rocks present a risk to construction projects, including bridges, highways, and dams. On June 5, 1976, the Teton Dam in Idaho failed, taking 14 lives and causing nearly $1 billion in property damage. This earthen dam, constructed of easily eroded clays and silts, was situated on highly fractured volcanic rocks. Although attempts were made to fill the voids in the jointed rock, water gradually penetrated the subsurface fractures and undermined the dam's foundation. Eventually the moving water cut a tunnel into the easily erodible clays and silts. Within minutes the dam failed, sending a 20-meter- (65-foot-) high wall of water down the Teton and Snake Rivers.

Jointed rocks also provide physical and economic benefits. For example, highly jointed rocks are often a significant source of groundwater. In addition, some of the world's largest and most important mineral deposits are located along joint systems. Hydrothermal solutions (mineralized fluids) can migrate into fractured host rocks and precipitate economically significant amounts of copper, silver, gold, zinc, lead, and uranium.

CONCEPT CHECKS 7.3

1. Contrast the movements that occur along normal and reverse faults. What type of stress is responsible for each kind of fault?
2. What type of fault is associated with fault-block mountains?
3. How are reverse faults different from thrust faults? In what way are they similar?
4. Describe the relative movement along a strike-slip fault.
5. How are joints different from faults?

7.4 Mountain Building

Locate and name Earth's major mountain belts on a world map.

Mountain building has occurred in the recent geologic past at several locations around the world. Young mountain belts include the American Cordillera (*cordillera* means "spine" or "backbone"), which runs along the western margin of the Americas from southernmost South America to Alaska and includes both the Rockies and the Andes; the Alpine–Himalaya chain, which extends along the margin of the Mediterranean, through Iran to northern India and into Indochina; and the mountainous terrains of the western Pacific, which include volcanic island arcs like Japan, the Philippines, and much of Indonesia. Most of these young mountain belts have formed in the past 100 million years (Figure 7.23). Some, including the Himalayas, began their growth as recently as 50 million years ago.

In addition to these young mountain belts, there are several chains of Paleozoic-age mountains on Earth. Although these older mountain belts are deeply eroded and topographically less prominent, they exhibit the same structural features found in younger mountains. The Appalachians in the eastern United States and the Urals

Figure 7.23 **Earth's major mountain belts** Notice the east–west trend of major mountain belts in Eurasia, in contrast to the north–south trend of the North and South American Cordilleras.

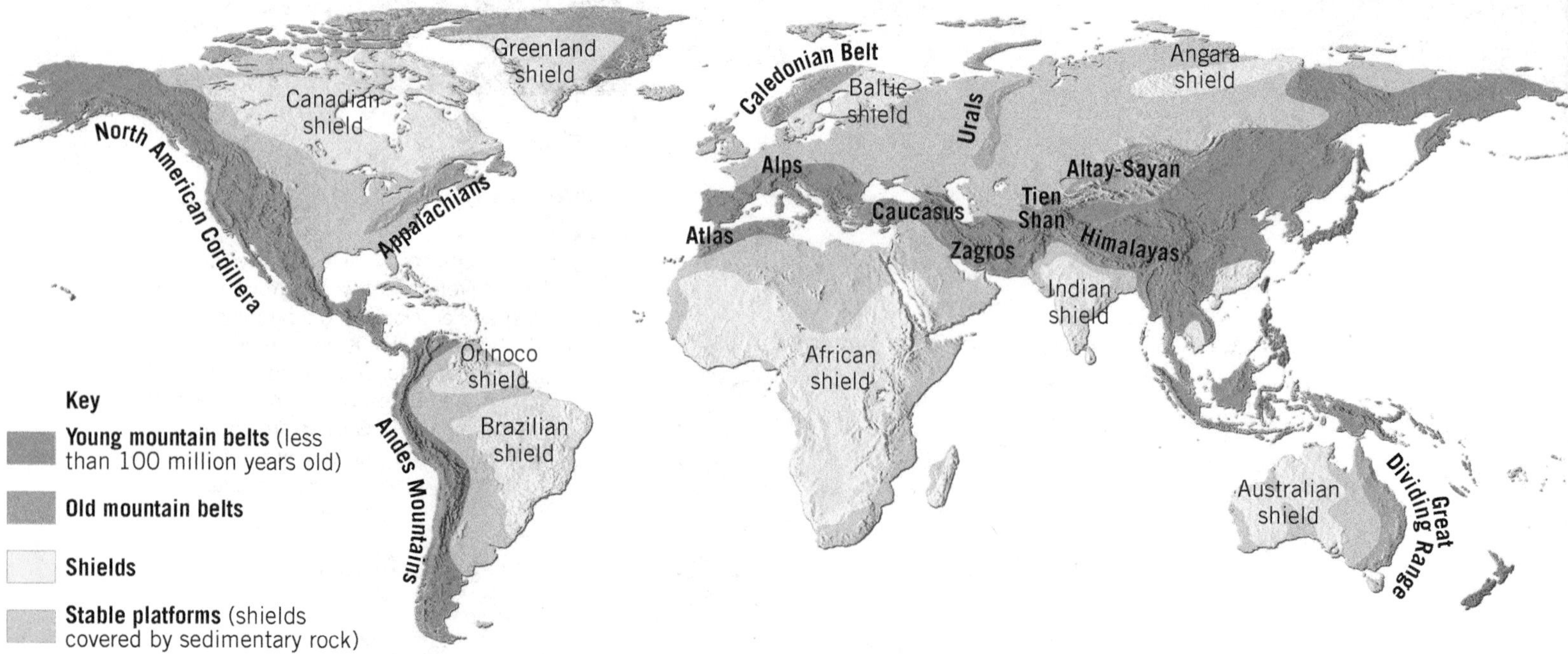

in Russia are classic examples of this group of older, well-worn mountain belts.

The term for the processes that collectively produce a mountain belt is **orogenesis** (*oros* = mountain, *genesis* = to come into being). An episode of mountain building is called an **orogeny**. Most major mountain belts display striking visual evidence of great compressional forces that have shortened the crust horizontally while thickening it vertically. These **collisional mountains** usually result from the collision of one or more small crustal fragments with a continental margin or from the closure of an ocean basin that results in the collision of two major landmasses. As a consequence, collisional mountains contain large quantities of preexisting sediments and sedimentary rocks that at one time lay along the margin of a continent and were subsequently faulted and contorted into a series of folds. Although folding and thrust faulting are often the most conspicuous signs of orogenesis, varying degrees of metamorphism and igneous activity are always present.

The theory of plate tectonics provides a model for orogenesis with excellent explanatory power; it accounts for the origin of virtually all the present mountain belts and most of the ancient ones. According to this model, the tectonic processes that generate Earth's major mountainous terrains occur along convergent plate boundaries. We will next revisit the nature of convergent plate boundaries and then examine how the process of subduction has driven mountain building around the globe.

CONCEPT CHECKS 7.4

1. Define *orogenesis*.
2. Which type of plate boundary is most directly associated with Earth's major mountain belts?

7.5 Subduction and Mountain Building

Sketch a cross section of an Andean-type mountain belt and describe how its major features are generated.

Where oceanic lithosphere subducts beneath oceanic lithosphere, a *volcanic island arc* and related tectonic features develop. Subduction of oceanic lithosphere beneath continental lithosphere, on the other hand, results in the formation of a *continental volcanic arc* and mountainous topography along the margin of a continent. Acting like a conveyor belt, oceanic lithosphere may also bring volcanic island arcs and other crustal fragments to a subduction zone. These crustal elements are generally too buoyant to subduct to any great depth, and they become welded to the overriding plate, which may be another small crustal fragment or a continent. If subduction continues long enough, it can ultimately lead to the closure of an ocean basin and the ensuing collision of two continents.

Island Arc–Type Mountain Building

Island arcs result from the steady subduction of oceanic lithosphere under oceanic lithosphere, which may continue for 200 million years or more. Periodic volcanic activity, the emplacement of igneous plutons at depth, and the accumulation of sediment that is scraped from the subducting plate gradually increase the volume of crustal material capping the upper plate (Figure 7.24). Some large volcanic island arcs, such as Japan, owe their size to having been built on fragments of continental crust that have rifted from a large landmass or to the joining of multiple island arcs over time.

The continued growth of a volcanic island arc can generate mountainous topography consisting of nearly parallel belts of igneous and metamorphic rocks. This activity, however, is viewed as just one phase in the development of Earth's major mountain belts. As you will see later, some volcanic arcs are carried by subducting plates to the margin of large continental blocks, where they become involved in large-scale mountain-building episodes.

Andean-Type Mountain Building

Andean-type mountain building is characterized by subduction beneath a continent rather than oceanic lithosphere, as in the Andes Mountains of South America. Subduction along these active continental margins is associated with long-lasting magmatic activity that builds continental volcanic arcs. The result is crustal thickening, with the crust reaching thicknesses of more than 70 kilometers (45 miles).

The first stage in the development of Andean-type mountain belts occurs along *passive continental margins*.

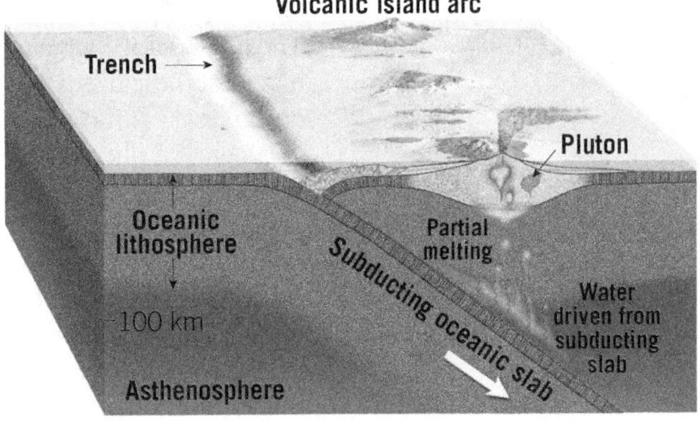

Figure 7.24 **Development of a volcanic island arc** A volcanic island arc forms where one slab of oceanic lithosphere is subducted beneath another slab of the same material.

The east coast of the United States provides a modern example of a passive continental margin where sedimentation has produced a thick platform of shallow-water sandstones, limestones, and shales (Figure 7.25A). At some point, the forces that drive plate motions change, and a subduction zone develops along the margin of the continent. This subduction zone may form because the oceanic lithosphere has become so old and dense that it begins to sink of its own accord. Alternatively, strong compressional forces may help initiate subduction.

▼ Figure 7.25 **Andean-type mountain building**

A. Passive continental margin with an extensive platform of sediments and sedimentary rocks.

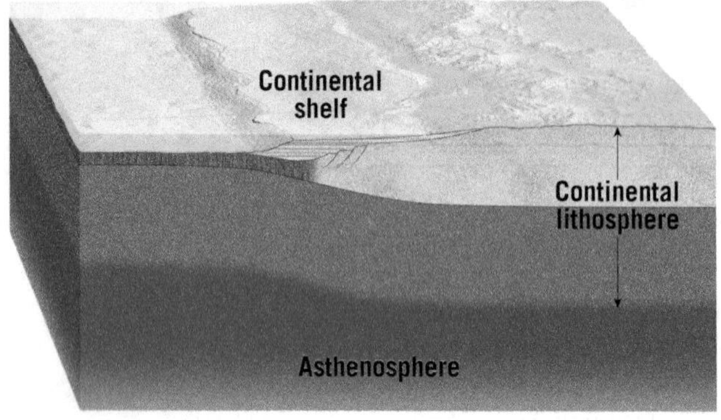

B. Plate convergence generates a subduction zone, and partial melting produces a continental volcanic arc. Compressional forces and igneous activity further deform and thicken the crust, elevating the mountain belt.

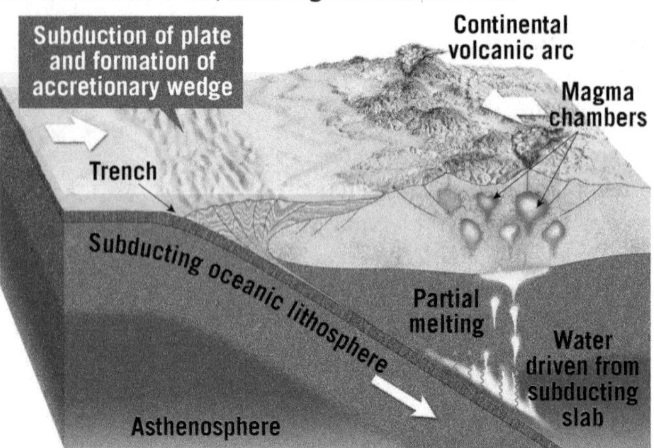

C. Subduction ends and is followed by a period of uplift.

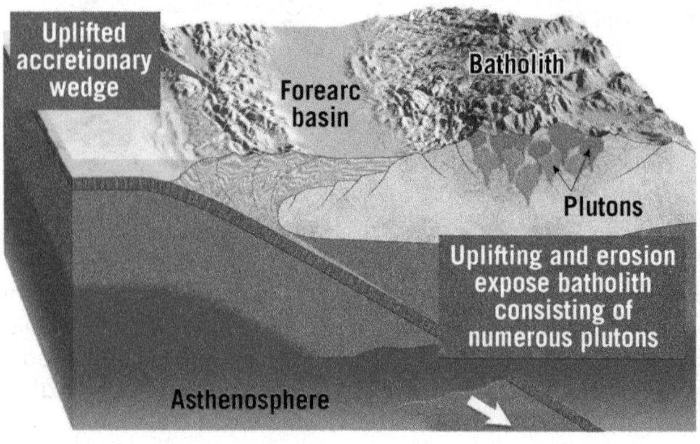

Building Volcanic Arcs Recall that as oceanic lithosphere descends into the mantle, increasing temperatures and pressures drive volatiles (mostly water and carbon dioxide) from the crustal rocks. These mobile fluids migrate upward into the wedge-shaped region of mantle between the subducting slab and upper plate. At a depth of about 100 kilometers (60 miles), these fluids reduce the melting point of hot mantle rock sufficiently to trigger partial melting (Figure 7.25B).

Partial melting of the ultramafic mantle rock peridotite generates magmas with basaltic (mafic) compositions. Because these newly formed basaltic magmas are less dense than the rocks from which they originated, they will rise buoyantly. In continental settings, basaltic magma often ponds beneath the less dense rocks of the crust. The hot basaltic magma may heat these overlying crustal rocks sufficiently to generate a silica-rich magma of intermediate and/or felsic (granitic) composition that can rise to form the continental volcanic arcs characteristic of an Andean-type subduction zone.

Emplacement of Batholiths Because of its low density and great thickness, continental crust significantly impedes the ascent of molten rock. Consequently, much of the magma that intrudes Earth's crust never reaches the surface; instead, it crystallizes at depth to form massive collections of igneous plutons called *batholiths*. The result of this activity is thickening of Earth's crust.

Eventually, uplift and erosion exhume the batholiths. The American Cordillera contains several large batholiths, including the Sierra Nevada batholith of California, the Coast Range batholith of western Canada, and several large igneous bodies in the Andes. Most batholiths consist of intrusive igneous rocks that range in composition from granite to diorite.

Development of an Accretionary Wedge During the development of volcanic arcs, unconsolidated sediments that are carried on the subducting plate, as well as fragments of oceanic crust, may be scraped off and plastered against the edge of the overriding plate, like a wedge of soil scraped up by an advancing bulldozer. The resulting chaotic accumulation of deformed and thrust-faulted sediments and scraps of ocean crust is called an **accretionary wedge** (see Figure 7.25B).

Some of the sediments in an accretionary wedge are muds that accumulated on the ocean floor and were carried to the subduction zone by plate motion. Additional materials are derived from an adjacent continental volcanic arc and consist of volcanic debris and products of weathering and erosion.

Where sediment is plentiful, prolonged subduction may thicken a developing accretionary wedge so that it protrudes above sea level. This has occurred along the southern end of the Puerto Rico trench, where the Orinoco River basin of Venezuela is a major source of sediments. The resulting wedge emerges to form the island of Barbados.

Forearc Basins As an accretionary wedge thickens, it acts as a barrier to the movement of sediment from the volcanic arc to the trench. As a result, sediments begin to collect between the accretionary wedge and the volcanic arc. This region, which is composed of relatively undeformed layers of sediment and sedimentary rocks, is called a **forearc basin** (see Figure 7.25B,C). Subsidence and continued sedimentation in forearc basins can generate a sequence of nearly horizontal sedimentary strata that can attain thicknesses of several kilometers.

Sierra Nevada, Coast Ranges, and Great Valley

California's Sierra Nevada, Coast Ranges, and Great Valley are excellent examples of the tectonic structures that are typically generated along an Andean-type subduction zone. These structures were produced by the subduction of a portion of the Pacific basin (the Farallon plate) under the western margin of California (see Figure 7.25B). The Sierra

Nevada batholith is a remnant of the continental volcanic arc that was produced by many intrusions of magma over a span of more than 100 million years. The Coast Ranges were built from the vast accumulation of sediments (accretionary wedge) that collected along the continental margin.

Beginning about 30 million years ago, subduction gradually ceased along much of the margin of North America, as the spreading center that produced the Farallon plate entered the California trench. The uplifting and erosion that followed removed most of the evidence of past volcanic activity and exposed the core of crystalline igneous and associated metamorphic rocks that make up the Sierra Nevada (Figure 7.25C). The Coast Ranges were uplifted only recently, as evidenced by the young, unconsolidated sediments that currently blanket portions of these highlands.

California's Great Valley is a remnant of the forearc basin that formed between the Sierra Nevada and the accretionary wedge and trench that lay offshore. Throughout much of its history, portions of the Great Valley lay below sea level. This sediment-laden basin contains thick marine deposits and debris eroded from the adjacent continental volcanic arc.

CONCEPT CHECKS 7.5

1. Compare and contrast mountain building at a volcanic island arc with mountain building at an Andean-style continental margin.
2. Describe and give an example of a passive continental margin.
3. In what ways are the Sierra Nevada and the Andes similar?
4. What is an accretionary wedge? Briefly describe its formation.
5. What is a batholith? In what tectonic setting are batholiths being generated?

7.6 Collisional Mountain Belts

Summarize the stages in the development of an Alpine-type mountain belt such as the Appalachians.

Most major mountain belts are generated when one or more buoyant crustal fragments collide with a continental margin as a result of subduction. Whereas oceanic lithosphere, which is relatively dense, readily subducts, continental lithosphere contains significant amounts of low-density crustal rocks and is therefore too buoyant to be subducted deeply or permanently. Consequently, the arrival of a crustal fragment at a trench results in a collision between the two continental blocks.

Cordilleran-Type Mountain Building

A Cordilleran-type orogeny, named after the North American Cordillera, is associated with a Pacific-like ocean—in that, unlike the Atlantic, the Pacific may never close. The rapid rate of seafloor spreading in the Pacific basin is balanced by a high rate of subduction. In this setting, island arcs and small crustal fragments are often carried along until they collide with an active continental margin and accrete (join) onto it. This process of collision and accretion has generated many of the mountainous regions that rim the Pacific. These accreted blocks of crust are called **terranes**. Geologists use this term to describe any crustal fragment that consists of a distinct and recognizable series of rock formations and has been transported and accreted by plate tectonic processes. Notice that *terrane* is a different word from *terrain*; the two are pronounced the same, but *terrain* refers to the shape of the surface topography, or "lay of the land."

The Nature of Terranes What is the nature of the crustal fragments that have become terranes? Some may have been **microcontinents** similar to the modern-day island of Madagascar, located east of Africa in the Indian Ocean. Many others were island arcs similar to Japan, the Philippines, and the Aleutian Islands. Still others may have been submerged oceanic plateaus created by massive submarine outpourings of basaltic lavas (Figure 7.26). More than 100 of these relatively small crustal fragments exist in the modern world.

Accretion and Orogenesis Small structures such as seamounts are generally subducted along with the descending oceanic slab. However, thick sections of oceanic crust, such as the Ontong Java Plateau (which is almost as big as Alaska) or an island arc dominated by low-density andesitic igneous rocks, is too buoyant to subduct. In these situations, a collision between the crustal fragment and the continental margin occurs.

The sequence of events that happen when small crustal fragments reach a Cordilleran-type margin is shown in Figure 7.27. The upper crustal layers are "peeled" from the descending plate and thrust in relatively thin sheets onto the adjacent continental block. Convergence does not generally end with the accretion of a crustal fragment. Rather, new subduction zones typically form seaward of the accreted terrane, and they can carry other island arcs or microcontinents toward a collision with the continental margin. Each collision displaces earlier accreted terranes further inland, adding to the zone of deformation as well as to the thickness and lateral extent of the continental margin.

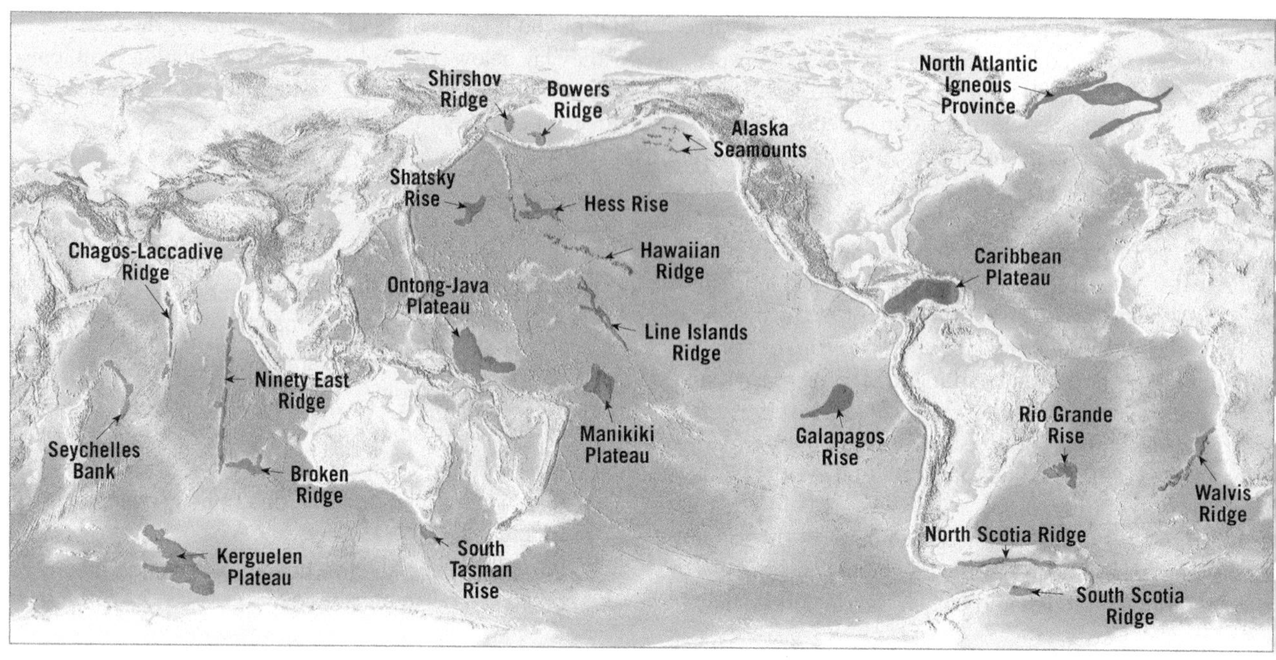

▶ Figure 7.26
Distribution of present-day oceanic plateaus and other submerged crustal fragments These blocks of crust, shaded orange, could someday be accreted to continents as new terranes. (Data from Zvi Ben-Avraham and others)

The North American Cordillera The correlation between mountain building and the accretion of crustal fragments was first developed in studies of the North American Cordillera (**Figure 7.28**). Researchers determined that some of the rocks in the orogenic belts of Alaska and British Columbia contained fossil and paleomagnetic evidence indicating that these strata previously lay much closer to the equator.

It is now known that many of the terranes that make up the North American Cordillera were scattered throughout the Pacific, like the island arcs and oceanic plateaus currently distributed in the western Pacific. During the breakup of Pangaea, the eastern portion of the Pacific basin (the Farallon plate) began to subduct under the western margin of North America. This activity resulted in many additions of crustal fragments along

A. A microcontinent and a volcanic island arc are being carried toward a subduction zone.

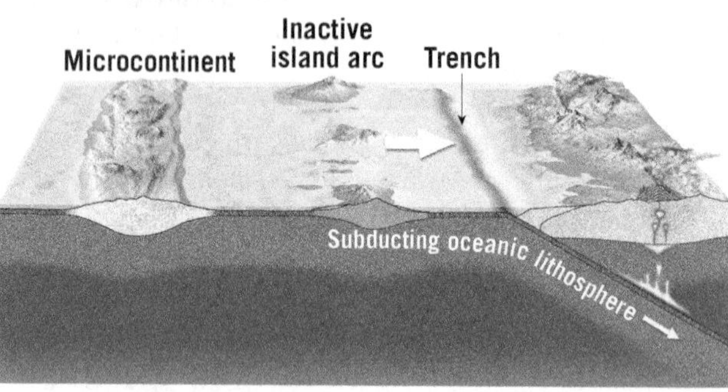

B. The volcanic island arc is sliced off the subducting plate and thrust onto the continent.

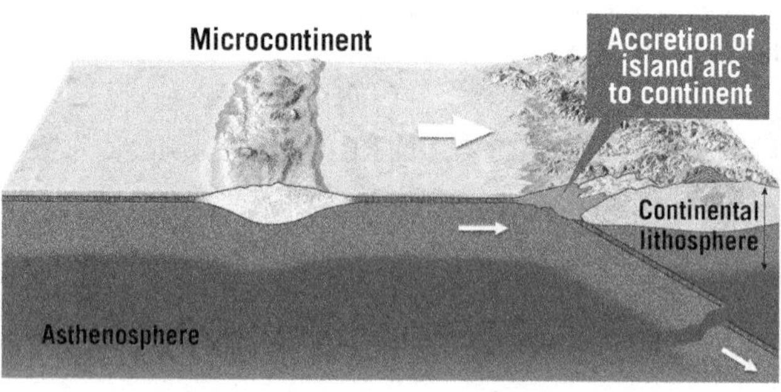

C. A new subduction zone forms seaward of the old subduction zone.

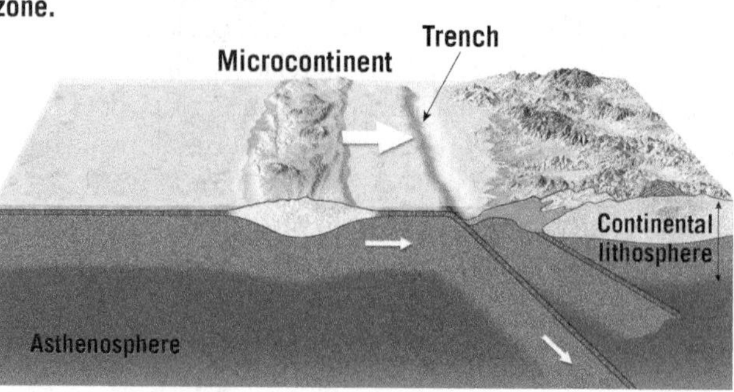

D. The accretion of the microcontinent to the continental margin shoves the remnant island arc further inland and grows the continental margin seaward.

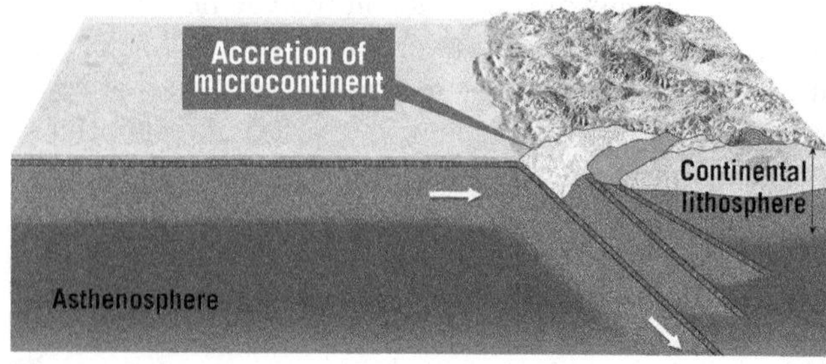

▶ SmartFigure 7.27
Collision and accretion of small crustal fragments to a continental margin

TUTORIAL
https://goo.gl/t8gK91

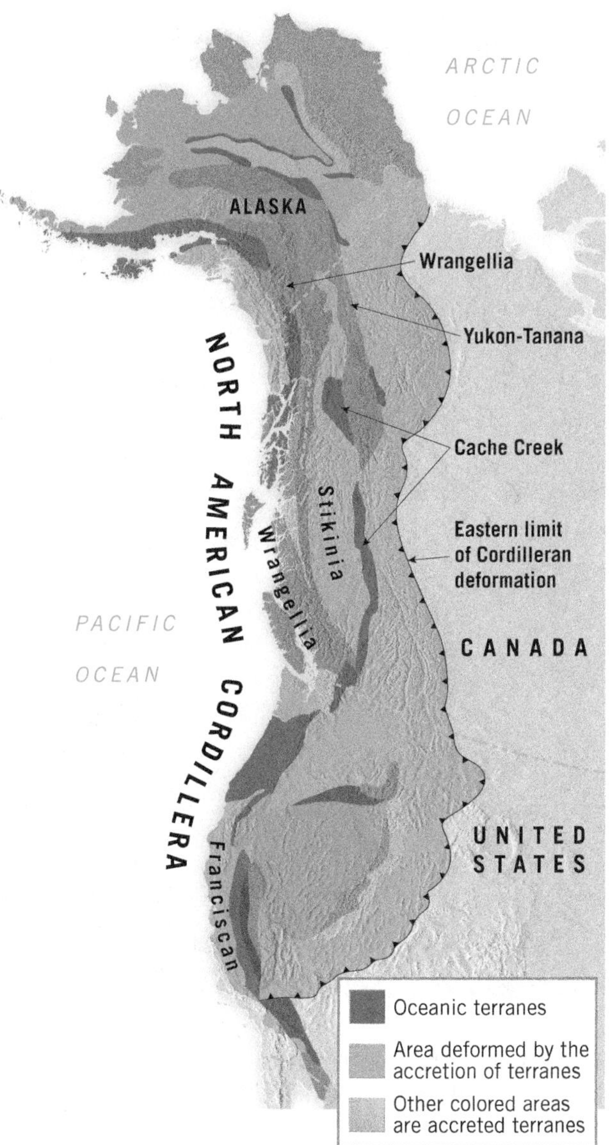

SmartFigure 7.28 **Terranes that have been added to western North America during the past 200 million years** Paleomagnetic studies and fossil evidence indicate that some of these terranes, such as the Cache Creek terrane, originated thousands of kilometers to the south and west of their present locations. (After D. R. Hutchinson and others)

ANIMATION
https://goo.gl/GjDgFr

the entire Pacific margin of the continent—from Mexico's Baja Peninsula to northern Alaska (see Figure 7.27). Geologists expect that many modern microcontinents will likewise be accreted to active continental margins surrounding the Pacific, producing new orogenic belts.

Alpine-Type Mountain Building: Continental Collisions

Alpine-type orogenies are episodes of mountain building that occur where two continental masses collide. They are named after the Alps, which have been intensively studied for more than 200 years. Mountain belts formed by the closure of major ocean basins include the Himalayas, Appalachians, Urals, and Alps. Continental collisions result in the development of mountains characterized by laterally shortened and vertically thickened crust, achieved through deformation such as folding and large-scale thrust faulting. Prior to the collision of the two large landmasses, this type of orogeny may also involve the accretion of smaller continental fragments or island arcs that occupied the ocean basin that once separated the two continental blocks.

The zone where two continents collide and are "welded" together is a **suture**. The same term can be used to describe the boundary between two adjacent accreted terranes. This portion of a mountain belt often preserves slivers of oceanic lithosphere that were trapped between the colliding plates. The unique structure of these pieces of oceanic lithosphere, called *ophiolites*, helps identify the collision boundary.

Next, we will take a closer look at two examples of collisional mountains: the Himalayas and the Appalachians. The Himalayas, Earth's youngest collisional mountains, are still rising. By contrast, the Appalachians are a much older mountain belt, in which active mountain building ceased about 250 million years ago.

The Himalayas

The mountain-building episode that created the Himalayas began between 50 and 30 million years ago, when India began to collide with Asia. Prior to the breakup of Pangaea, India was located between Africa and Antarctica in the Southern Hemisphere. As Pangaea fragmented, India moved rapidly, geologically speaking, a few thousand kilometers in a northward direction.

The subduction zone that facilitated India's northward migration was near the southern margin of Asia (Figure 7.29A). Continued subduction along Asia's margin created an Andean-type plate margin that contained a well-developed continental volcanic arc and an accretionary wedge. India's northern margin, on the other hand, was a passive continental margin consisting of a thick platform of shallow-water sediments and sedimentary rocks.

Geologists have determined that two or perhaps more small crustal fragments were positioned on the subducting plate somewhere between India and Asia. During the closing of the intervening ocean basin, a small crustal fragment, which now forms southern Tibet, reached the trench and was accreted to Asia. This event was followed by the docking of India itself.

As the intervening ocean basin was closing up, the more deformable materials on the continental margins of these landmasses became highly folded and faulted (Figure 7.29B). Two major thrust faults and many smaller ones sliced through the Indian crust. Subsequent motion along these thrust faults caused slices of the Indian crust to be stacked one upon the other.

▶ SmartFigure 7.29 **Continental collision: The formation of the Himalayas** These diagrams illustrate the collision of India with the Eurasian plate that produced the spectacular Himalayas.

ANIMATION
https://goo.gl/x8XSxr

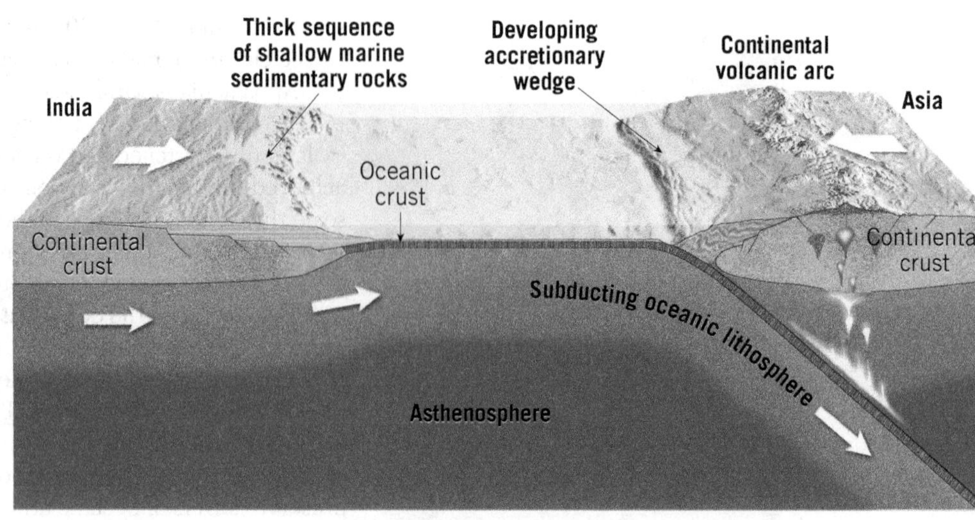

A. Prior to the collision of India and Asia, India's northern margin consisted of a thick platform of continental shelf sediments, whereas Asia's was an active continental margin with a well developed accretionary wedge and volcanic arc.

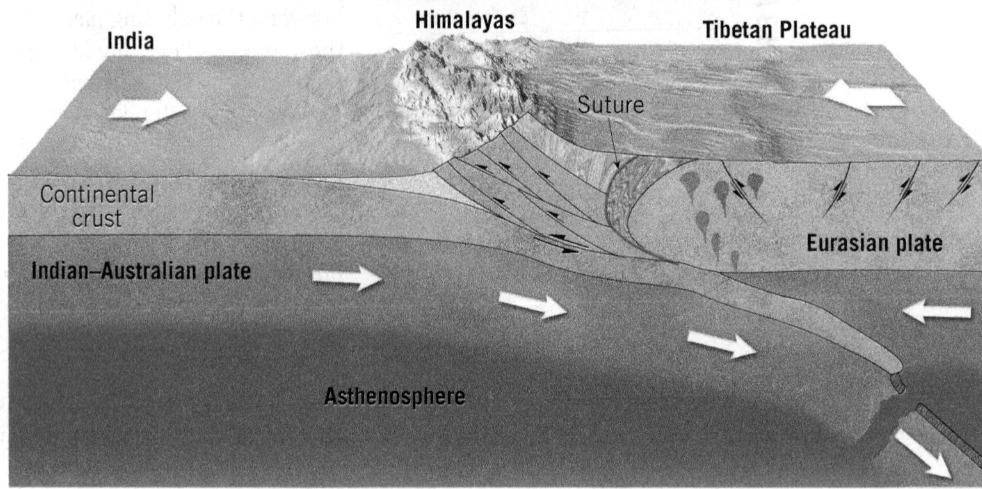

B. The continental collision folded and faulted crustal rocks along the margins of these continents to form the Himalayas. This event was followed by the gradual uplift of the Tibetan Plateau as the subcontinent of India was shoved under Asia.

Today, these slices make up the bulk of the highest peaks in the Himalayas—many of which are capped by tropical marine limestones that formed along what was once the continental shelf.

Map view showing the southeastward displacement of China and the mainland of Southeast Asia as India plowed into Asia.

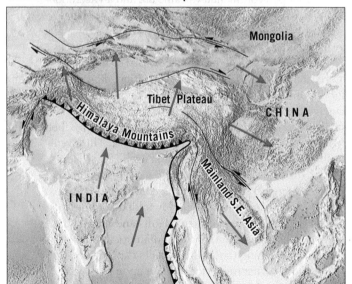

▶ SmartFigure 7.30 **India's continued northward migration severely deformed much of China and Southeast Asia** Global positioning systems now allow scientists to track this deformation in real time.

TUTORIAL
http://goo.gl/6SHx6Y

The formation of the Himalayas was followed by a period of uplift that raised the Tibetan Plateau. Seismic evidence suggests that a portion of the Indian subcontinent was thrust beneath Tibet—a distance of perhaps 400 kilometers (250 miles). If this occurred, the added crustal thickness would account for the lofty landscape of southern Tibet, which has an average elevation of more than 4,500 meters (14,800 feet), higher than the tallest mountain in the contiguous United States.

The collision with Asia slowed but did not stop the northward movement of India, which has since penetrated at least 2000 kilometers (1200 miles) into the mainland of Asia. Crustal shortening and thickening accommodated some of this motion. Much of the remaining penetration into Asia caused lateral displacement of large blocks of the Asian crust by a mechanism described as *escape tectonics*. As shown in Figure 7.30, when India continued its northward trek, parts of Asia were "squeezed" eastward, out of the collision zone. These displaced crustal blocks include much of Southeast Asia (the region between India and China) and sections of China.

Why was the interior of Asia deformed to such a large extent, while India has remained essentially intact?

The answer lies in the nature of these diverse crustal blocks. Much of India is a continental shield composed mainly of old Precambrian rocks (see Figure 7.23). This thick, cold slab of crustal material has been intact for more than 2 billion years and is mechanically strong as a result. By contrast, Southeast Asia was assembled more recently, from the collision of several smaller crustal fragments. Consequently, it is still relatively "warm and weak" from recent periods of mountain building (see Figure 7.30).

The Appalachians

The Appalachian Mountains provide great scenic beauty near the eastern margin of North America, from Alabama to Newfoundland. Mountain belts of similar origin that formed during the same period and were once contiguous are found in the British Isles, Scandinavia, northwestern Africa, and Greenland (see Figure 4.6, page 99). The orogenies that generated this extensive mountain system lasted a few hundred million years and resulted in the assembly of the supercontinent Pangaea. Detailed studies of the Appalachians indicate that this mountain belt was the result of three distinct episodes of mountain building.

Our simplified overview begins roughly 750 million years ago, with the breakup of a supercontinent called Rodinia that predates Pangaea. Much like the breakup of Pangaea, this episode of continental rifting and seafloor spreading generated a new ocean between the rifted continental blocks. Located within this widening ocean basin was a microcontinent near the edge of ancestral Africa.

About 600 million years ago, for reasons geologists do not completely understand, plate motion changed dramatically, and this ancient ocean basin began to close. This led to the development of multiple subduction zones, and the stage was set for the three orogenic events that would lead to the collision of North America and Africa (**Figure 7.31A**).

Taconic Orogeny Around 450 million years ago, the marginal sea between the volcanic island arc and ancestral North America began to close. The collision that ensued, called the *Taconic Orogeny*, caused the volcanic arc along with ocean sediments located on the upper plate to be accreted to the edge of the larger continental block. The remnants of this volcanic arc and oceanic sediments are recognized today as the metamorphic rocks through much of the Appalachian mountain belt (**Figure 7.31B**). For example, schists beneath New York City and Washington, DC, formed at this time. In addition to this pervasive regional metamorphism, numerous magma bodies intruded the crustal rocks along the entire continental margin.

Acadian Orogeny A second episode of mountain building, called the *Acadian Orogeny*, occurred about 350 million years ago. The continued closing of this ancient ocean basin resulted in the collision of a microcontinent with North America (**Figure 7.31C**). This orogeny involved thrust faulting, metamorphism, and the intrusion of many large granite bodies. This event also added substantially to the width of North America, particularly in eastern New England.

Alleghanian Orogeny The final orogeny, called the *Alleghanian Orogeny*, occurred between 250 and 300 million years ago, when Africa collided with North America. This collision displaced material that was accreted earlier by as much as 250 kilometers (155 miles) toward the interior of North America. This event also displaced and further deformed the continental shelf sediments and sedimentary rocks that had once flanked

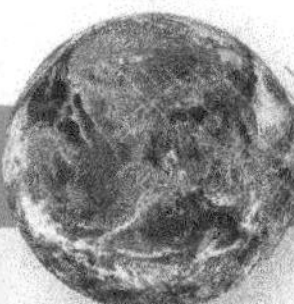

EYE ON EARTH 7.2

These interbedded layers of chert and shale were strongly folded during the growth of an accretionary wedge. A recent period of uplift has exposed these deformed strata near the Marin Headlands, north of San Francisco, California. (Photo by Michael Collier)

QUESTION 1 *What is the nature of the stress that most likely generated these highly folded strata: compressional or tensional?*

QUESTION 2 *Along what type of plate boundaries do accretionary wedges form?*

QUESTION 3 *What type of plate boundary is found today in the San Francisco Bay area?*

SmartFigure 7.31 **Formation of the Appalachian Mountains** The Appalachians formed during the closing of a precursor to the Atlantic Ocean. This event involved three separate stages of mountain building that spanned more than 300 million years. (Based on Zvi Ben-Avraham, Jack Oliver, Larry Brown, and Frederick Cook)

TUTORIAL
https://goo.gl/YZaDyd

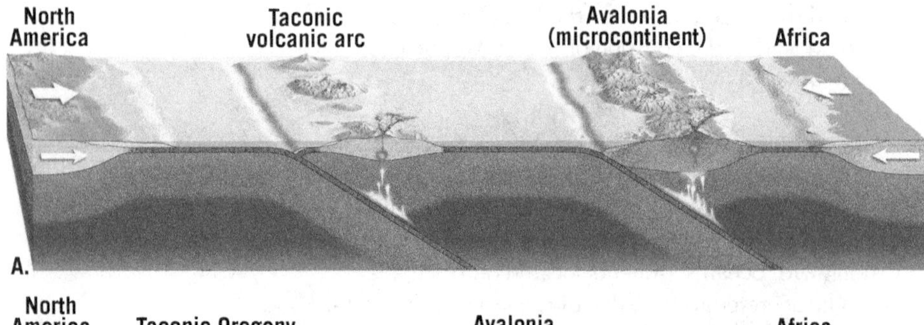

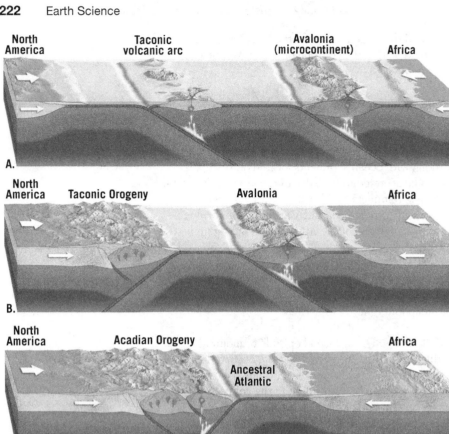

Closing of an Ocean Basin About 600 million years ago, the precursor to the North Atlantic began to close. Located within this ocean basin was an active volcanic island arc off the coast of North America and a microcontinent situated closer to Africa.

North America
Taconic Orogeny
Avalonia
Africa
B.

Taconic Orogeny Around 450 million years ago, the marginal sea between the volcanic island arc and North America closed. The collision, called the Taconic Orogeny, thrust the island arc over the eastern margin of North America.

North America
Acadian Orogeny
Ancestral Atlantic
Africa
C.

Acadian Orogeny A second episode of mountain building, called the Acadian Orogeny, occurred about 350 million years ago and involved the collision of a microcontinent with North America.

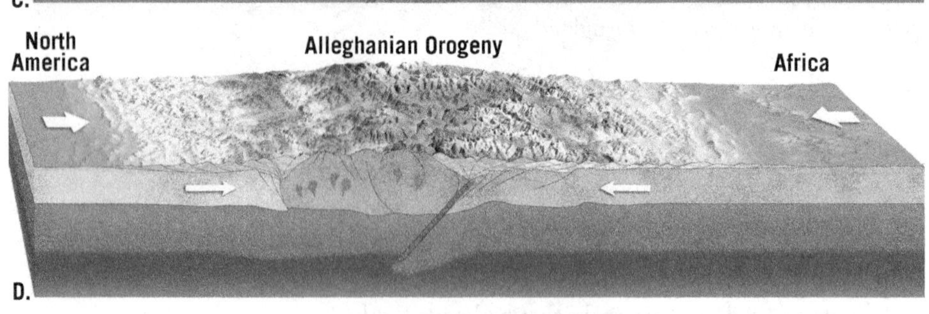

Alleghanian Orogeny The final event, the Alleghanian Orogeny, occurred between 250 and 300 million years ago, when Africa collided with North America. The result was the formation of the Appalachian Mountains.

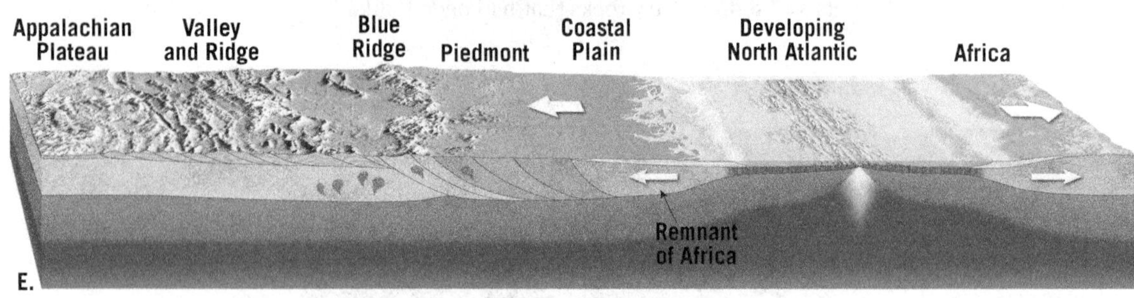

Rifting of Pangaea About 180 million years ago, Pangaea began to break into smaller fragments, a process that ultimately created the modern Atlantic Ocean. Because this new zone of rifting occurred east of the suture that formed when Africa and North America collided, remnants of African crust remain "welded" to the North American plate.

the eastern margin of North America (**Figure 7.31D**). Today these folded and thrust-faulted sandstones, limestones, and shales make up the largely unmetamorphosed rocks of the Valley and Ridge Province (**Figure 7.32**). This structural signature of mountain building can be found as far inland as central Pennsylvania and West Virginia.

With the collision of Africa and North America, the young Appalachians, perhaps as majestic as the Himalayas, lay along the suture, in the interior of Pangaea. The tectonic forces that built the mountains ceased to drive them upward. Then, about 180 million years ago, the new supercontinent began to break into smaller fragments, a process that ultimately created the modern Atlantic Ocean. Because this new zone of rifting occurred east of the suture that formed when Africa and North America collided, remnants of Africa remain stuck

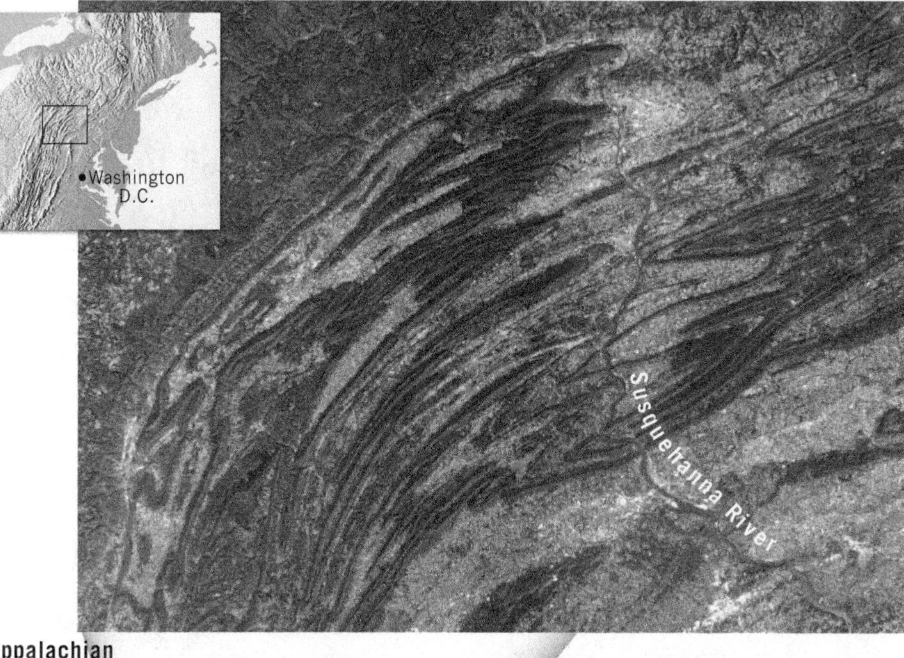

SmartFigure 7.32
The Valley and Ridge Province This false-color image shows the Appalachian Mountains, which consist mainly of folded and faulted sedimentary strata displaced landward along thrust faults as Africa collided with North America. (NASA/GSFC/JPL, MISR Science Team)

MOBILE FIELD TRIP
https://goo.gl/Ihn2PI

to the North American plate (Figure 7.31E). The crust underlying Florida is an example.

Other mountain ranges built from continental collisions include the Alps and the Urals. The Alps formed as Africa and several smaller crustal fragments collided with Europe during the closing of the Tethys Sea. Similarly, the Urals were deformed and uplifted during the assembly of Pangaea, when northern Europe and northern Asia collided, forming a major portion of Eurasia. Unlike the Appalachian belt, however, the Urals did not break apart again after their orogenesis.

CONCEPT CHECKS 7.6

1. Differentiate between *terrane* and *terrain*.
2. Explain why the continental crust of Asia was deformed more than that of the Indian subcontinent during the formation of the Himalayas.
3. Where and how might magma be generated in a newly formed collisional mountain belt?
4. How does the plate tectonics theory help explain the existence of fossil marine life in rocks atop collisional mountains?

7.7 Vertical Motions of the Crust

Explain the principle of isostasy and how it contributes to the elevated topography of mountain belts.

The processes that produced Earth's varied topography are complex (see GEOgraphics 7.1). Beyond the tectonic forces that move rocks laterally and thicken them vertically to produce mountainous topography, additional processes help to shape Earth's surface. As weathering and erosion work to lower mountains, a compensating process called *isostasy* causes them to rise, so they remain mountainous long after the tectonic processes that initially created them have ceased. Also, if tectonic processes raise a mountain belt "too high," the rock at its core will become too weak to support the load, and the mountain will spread.

The Principle of Isostasy

During the 1840s, researchers discovered that Earth's low-density crust "floats" on top of the high-density rocks of the mantle, much as wood floats in water. We will use

GEOGRAPHICS 7.1

The Laramide Rockies

The portion of the Rocky Mountains that extends from southwestern Montana to New Mexico was produced during a period of deformation known as the Laramide Orogeny. This event, which created some of the most picturesque scenery in the United States, peaked about 60 million years ago.

Colorado Rockie
Steamboat Spring
(Photo by Michael Colli

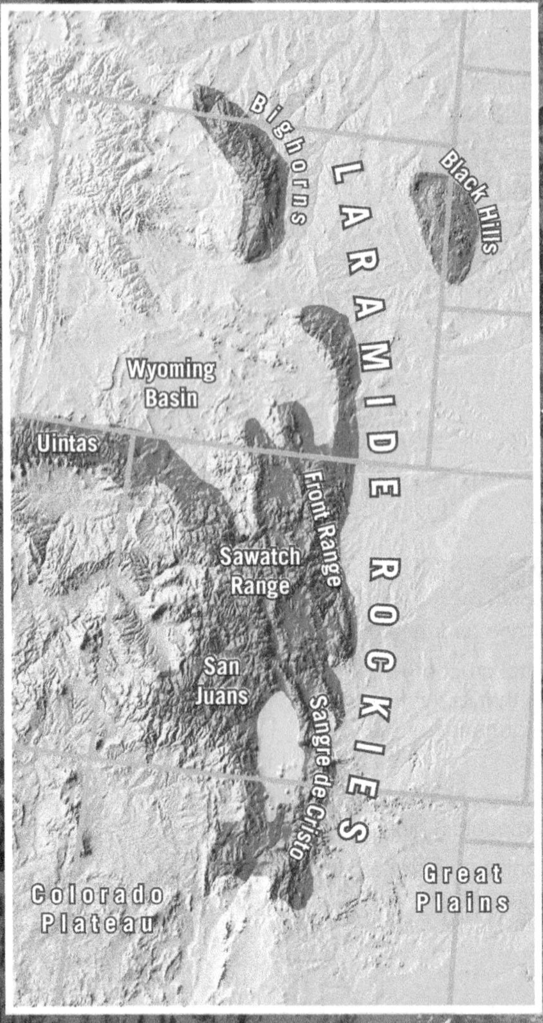

Where are the Laramide Rockies? Sometimes called the Central and Southern Rockies, this mountain belt lies to the east of the Colorado Plateau and includes the Bighorns of Wyoming, the Front Range of Colorado, the Uintas of Utah, and the Sangre de Cristo of southern Colorado and northern New Mexico.

What is the geologic history of the Laramide Rockies? These mountain ra formed when Precambrian age basement roc were uplifted nearly vertically along reverse thrust faults, upwarping the overlying layers younger sedimentary rocks. Uplifting acceler the processes of weathering and erosion, wh removed much of the younger sedimentary c from the highest portions of the uplifted bloc Intrusion of igneous plutons and volcanism occurred simultaneously with this period of mountain building.

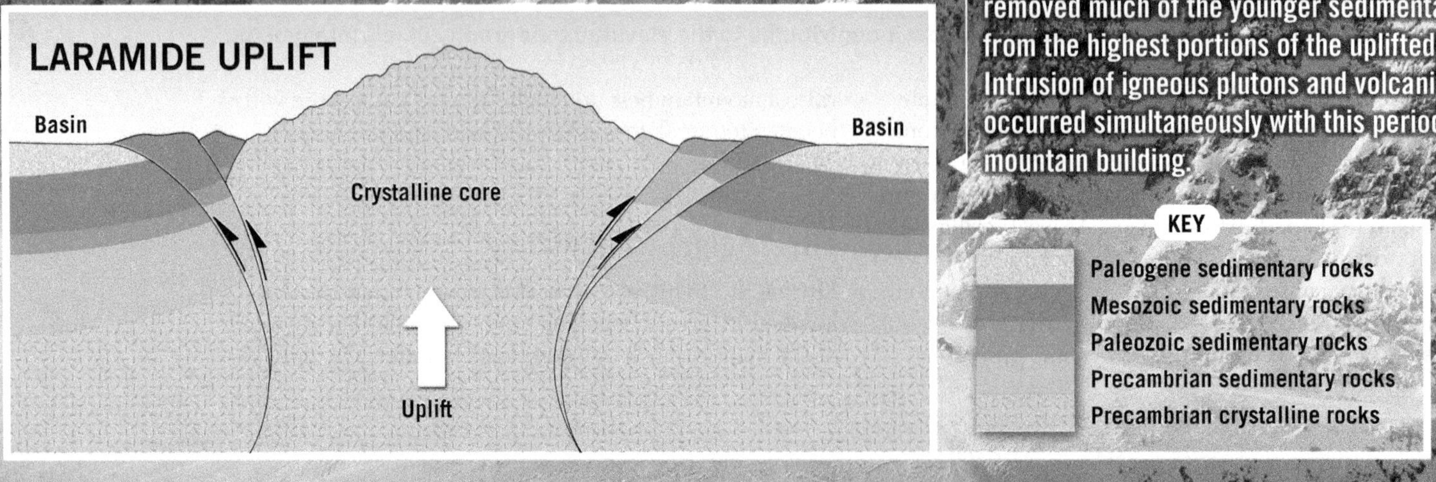

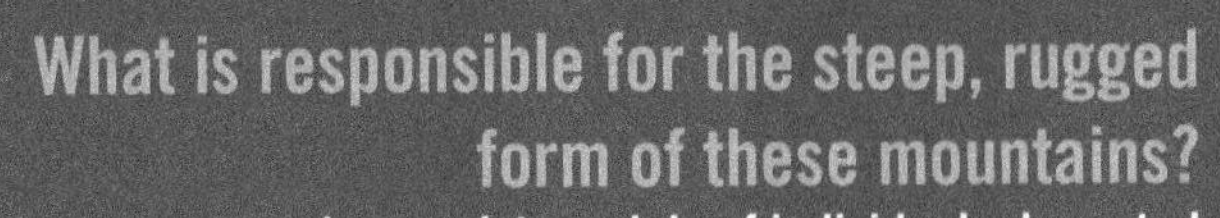

What is responsible for the steep, rugged form of these mountains?

e resulting mountainous topography consists mainly of individual, elongated blocks of igneous and metamorphic rocks flanked by upturned sedimentary strata and separated by sediment filled basins. The cores of these large uplifted blocks include some of the highest and most scenic topography in the West, including Long's Peak, Maroon Bells, and Mount Sneffels, all located in Colorado. Much of the scenic beauty of these mountains can be credited to the work of alpine glaciers.

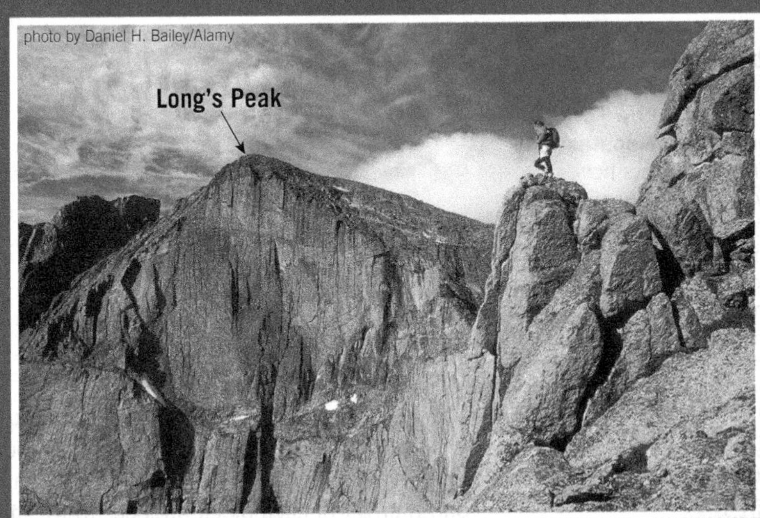

Maroon Bells
photo by Tom Tietz/Shutterstock

How did this mountain belt form?

Because the Laramide Rockies lie more than 1500 kilomers from the plate boundary that produced the Cordilleran Orogeny, it has been difficult to identify the mechanism which produced them. One hypothesis proposes that a thickened slab of the Farallon Plate began to subduct under the west coast of California about 85 million years ago. Because of its thickness, this buoyant slab resisted subduction as it was shoved beneath the continent. As the thick slab continued eastward under the Colorado Plateau it triggered uplift of a least 2 kilometers and built several monoclines. Upon reaching the Rockies, compressional forces squeezed the crust, which responded by developing high angle faults along which igneous and metamorphic basement rock was uplifted. These crystalline blocks form the cores of many of the high peaks that comprise this mountain belt.

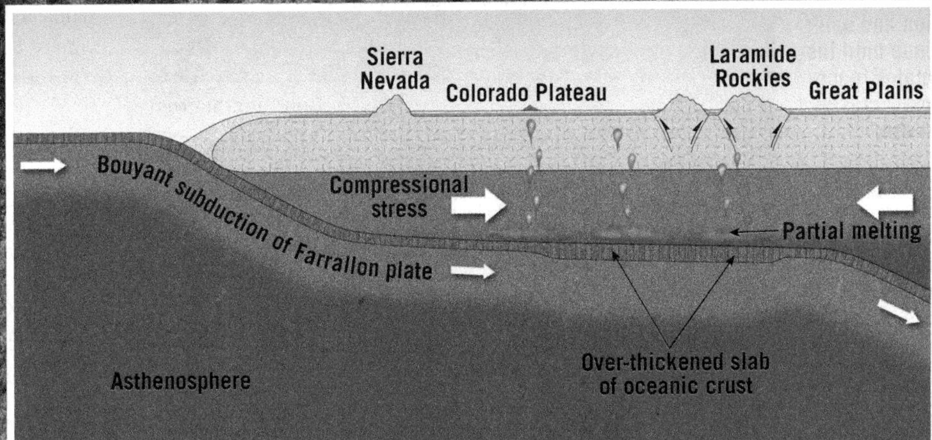

estions:

List the major mountain ranges that comprise the Laramide Rockies.

Why isn't the formation of the Laramide Rockies easily explained using the plate tectonics model?

SmartFigure 7.33 **The principle of isostasy** This drawing shows how wooden blocks of different thicknesses float in water. In a similar manner, thick sections of Earth's crustal material float higher than thinner crustal slabs.

https://goo.gl/zaZa7C

the floating wooden blocks in Figure 7.33 to explore this idea. You can think of these blocks as a model mountain belt, floating in the mantle. Notice that only about one-quarter of each block projects above the water and three-quarters is submerged. This is because wood is about three-quarters as dense as water. Similarly, most of the vertical thickness of a mountain belt forms a buoyant root that is "submerged" in the mantle; the remainder projects above the surrounding crust. This concept—that the crust floats in gravitational balance in the mantle—is called **isostasy**.

Notice that the tallest of the floating blocks in Figure 7.33 stands the highest above the water surface and also sits the deepest. Similarly, the greater the crustal thickness of a mountain belt, the higher it will stand above sea level, and also the deeper its roots will be. Thus, the Himalayas, as the tallest range on Earth, also have the deepest roots.

Now, visualize what would happen if you placed a second small block on top of one of the blocks in Figure 7.33. The combined block would sink until it reached a new isostatic (gravitational) balance, at which point its top would be higher than before, and its bottom would be lower. This process of establishing a new gravitational balance in response to loading or unloading is called **isostatic adjustment**. Notice also that as a block rises or sinks, the surrounding water flows to accommodate it. Similarly, the highly viscous mantle will flow, albeit at an excruciatingly slow rate, when weight is added to or subtracted from the overlying crustal blocks.

Applying the concept of isostatic adjustment, we should expect that when weight is added to the crust, the crust will respond by subsiding, causing the underlying mantle rocks to flow away. When the weight is removed, the crust will rebound, and the mantle rock will flow back underneath. (Visualize what happens to a ship when its cargo is loaded or unloaded.) Scientists have found evidence for crustal subsidence followed by isostatic rebound in areas formerly overlain by Ice Age glaciers. When continental ice sheets covered portions of North America during the Pleistocene epoch, ice masses that averaged about 3 kilometers (2 miles) thick added weight to the crust and caused downwarping by hundreds of meters. In the 8000 years since this ice sheet melted, gradual uplift of as much as 330 meters (1000 feet) has occurred in Canada's Hudson Bay region, where the thickest ice had accumulated.

One of the consequences of isostatic adjustment is that, as erosion cuts into a mountain range, removing mass, the range rises in response to the reduced load (Figure 7.34). In fact, because erosion removes material

When compressional mountains are young they are composed of thick, low density crustal rocks that float on the denser mantle below.

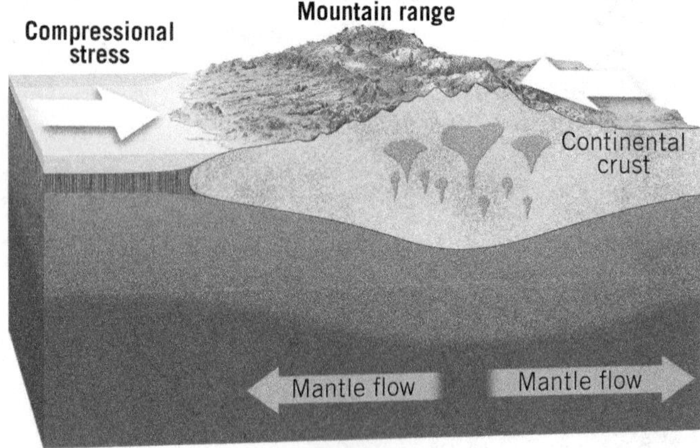

As erosion lowers the mountains, the crust rises in response to the reduced load in order to maintain isostatic balance.

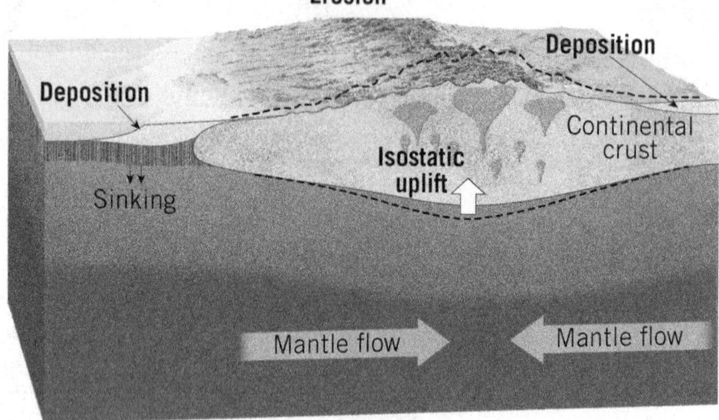

Erosion and uplift continue until the mountains reach "normal" crustal thickness.

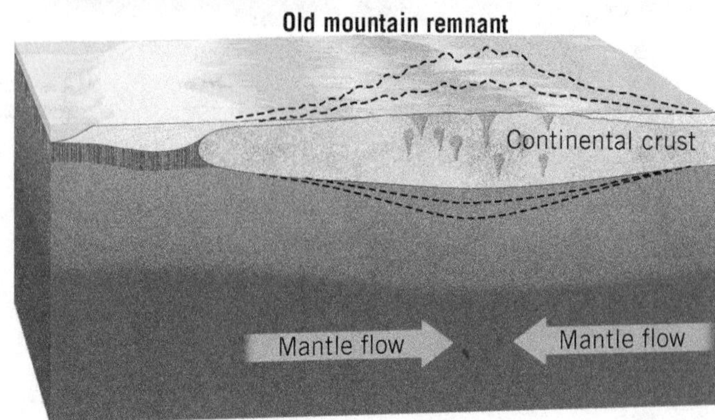

SmartFigure 7.34 **The effects of isostatic adjustment and erosion on mountainous topography** This sequence illustrates how the combined effects of erosion and isostatic adjustment result in a thinning of the crust in mountainous regions.

TUTORIAL
https://goo.gl/kMchrw

mainly by carving canyons and valleys rather than by uniformly wearing down mountain peaks, isostasy may actually "push" the peaks higher than their original height.

The processes of uplift and erosion continue until the mountain block reaches average crustal thickness. When this occurs, these once-elevated structures are near sea level, and the once-deeply buried interior of the mountain is exposed at the surface. In addition, as mountains are worn down, the eroded sediment is deposited on adjacent landscapes, causing these areas to subside (see Figure 7.34).

How High Is Too High?

Where compressional forces are great, such as those driving India into Asia, lofty mountains such as the Himalayas result. Is there a limit on how high a mountain can rise? As mountaintops are elevated, gravity-driven processes such as erosion and mass wasting accelerate, carving the deformed strata into rugged landscapes. Equally important, however, is the fact that gravity also acts on the rocks within the mountain belt. The higher the mountain, the greater the downward force on rocks near the base. Eventually, the rocks deep within the developing mountain, which are relatively warm and weak, begin to flow laterally, as shown in Figure 7.35. This process of **gravitational collapse** is analogous to what happens when a ladle of very thick pancake batter is poured onto a hot griddle. In addition to causing ductile spreading at depth, this process leads to normal faulting and subsidence in the upper, brittle portion of Earth's crust.

Considering these factors, what keeps the Himalayas standing? Simply, the horizontal compressional forces that are driving India into Asia are greater than the vertical force of gravity. However, when India's northward trek ends, the downward pull of gravity, weathering, and erosion will become the dominant forces acting on this mountainous region.

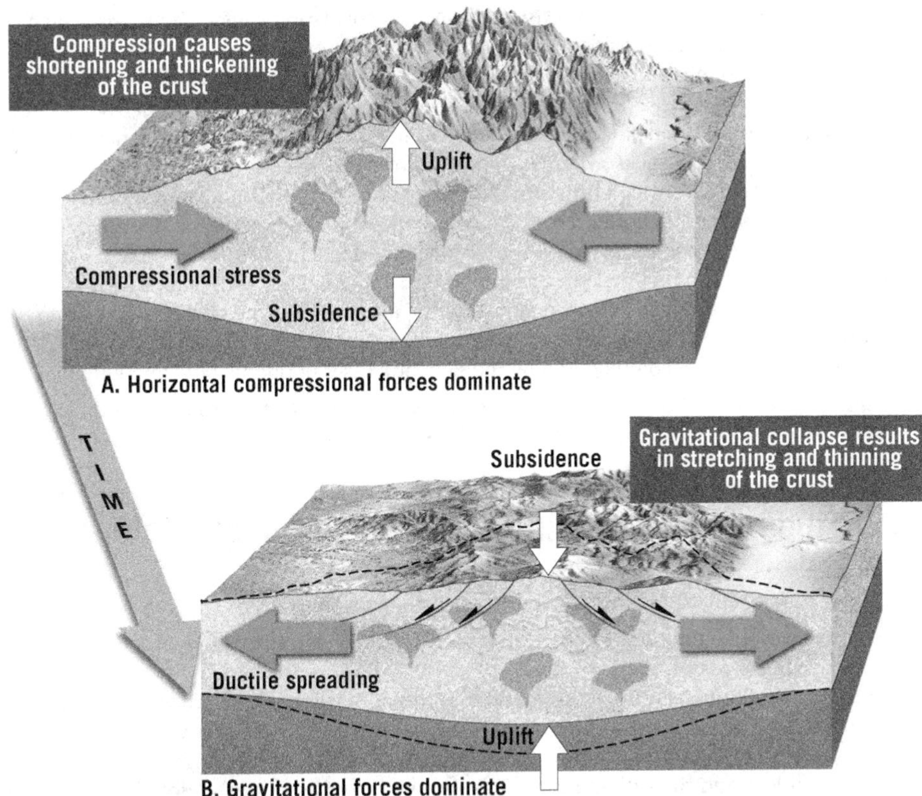

▲ Figure 7.35 **Gravitational collapse** Without compressional forces to support them, mountains gradually collapse under their own weight. Gravitational collapse involves normal faulting in the upper, brittle portion of the crust and ductile spreading in the warm, weak rocks at depth.

CONCEPT CHECKS 7.7

1. Define *isostasy*.
2. Give one kind of evidence that supports the concept of isostatic uplift.
3. What happens to a floating object when weight is added? Subtracted?
4. Briefly describe how the principle of isostatic adjustment applies to changes in the elevations of mountains.
5. Explain the process whereby mountainous regions experience gravitational collapse.

7 CONCEPTS IN REVIEW
Crustal Deformation and Mountain Building

7.1 Crustal Deformation

Describe the three types of differential stress and identify the tectonic setting most commonly associated with each. Differentiate stress from strain and brittle from ductile deformation.

KEY TERMS: deformation, tectonic structure (geologic structure), stress, confining pressure, differential stress, compressional stress, tensional stress, shear, strain, elastic deformation, brittle deformation, ductile deformation

- Tectonic (geologic) structures are structures generated when rocks are deformed by bending or breaking; they include folds, faults, and joints.
- Stress is the force that drives rock deformation. When stress acts equally from all directions, we call it confining pressure. When the stress is greatest in one direction, we call it differential stress. There are three main types of differential stress: compressional, tensional, and shear.

- A rock's strength is its ability to resist permanent deformation. When the stresses on a rock exceed its strength, the rock deforms, usually by folding or faulting.
- Elastic deformation is caused by a temporary stretching of the chemical bonds in a rock. When the stress is released, the rock returns to its original shape. When the rock's strength is exceeded, bonds break, and the rock deforms in either a brittle or ductile fashion. Brittle deformation fractures rocks, whereas ductile deformation changes a rock's shape.
- Whether a rock deforms in a brittle or ductile manner depends on its temperature and its confining pressure. The hotter a rock, the more likely it is to experience ductile deformation. Greater confining pressure makes a rock stronger and less likely to break. Thus, rock deformation tends to be brittle in the shallow crust and ductile at deeper levels.
- Whether deformation is brittle or ductile also depends on the type of rock. For example, shale is weaker than granite, so it is more prone to ductile deformation. If a rock is forced to deform more quickly than can be accommodated by the slow processes of ductile deformation, it will break.

? Did the quarter on the right in the accompanying image experience brittle or ductile deformation?

Anthony Pleva/Alamy

7.2 Folds: Rock Structures Formed by Ductile Deformation

List and describe five types of folds.

KEY TERMS: fold, anticline, syncline, dome, basin, monocline

- Folds are wavelike undulations in layered rocks that develop through ductile deformation caused by compressional stresses.
- Anticlines usually arise by upfolding, or arching, of sedimentary layers, whereas synclines are downfolds, or troughs. Anticlines and synclines may be symmetrical, asymmetrical, overturned, or recumbent.
- When folded rocks erode to form a series of ridges and valleys, the ridges represent resistant beds (not anticlines), and the valleys represent softer beds (not synclines). A fold is said to plunge when its axis penetrates the ground at an angle. This results in a V-shaped outcrop pattern.
- Domes and basins are large bowl- or saucer-shaped folds that produce roughly circular outcrop patterns. When eroded, a dome has the oldest beds in the middle, and a basin has the oldest beds around the margin.
- Monoclines are large steplike folds in otherwise horizontal strata that develop when beds drape over a vertical offset produced by subsurface faulting.

? What name is given to the geologic structure shown in the accompanying image?

E.J. Tarbuck

7.3 Faults and Joints: Rock Structures Formed by Brittle Deformation

Sketch and briefly describe the relative motion of rock bodies located on opposite sides of normal, reverse, and thrust faults as well as both types of strike-slip faults.

KEY TERMS: fault, dip-slip fault, hanging wall block, footwall block, fault scarp, normal fault, fault-block mountain, horst, graben, half-graben, detachment fault, reverse fault, thrust fault, strike-slip fault, transform fault, joint

- Faults and joints are fractures in rock that form through brittle deformation.
- A fault is a fracture along which motion occurs, offsetting the rocks on either side. If the movement is in the direction of the fault's dip (or inclination), the rock above the fault plane is the hanging wall block, and the rock below the fault is the footwall block. If the hanging wall moves *down* relative to the footwall, the fault is a normal fault. If the hanging wall moves *up* relative to the footwall, the fault is a reverse fault. Large normal faults with low dip angles are called detachment faults. Large reverse faults with low dip angles are thrust faults.
- Faults that intersect Earth's surface may produce a "step" in the land known as a fault scarp. Areas of tectonic extension, such as the Basin and Range Province, produce fault-block mountains—horsts separated by neighboring grabens or half-grabens.
- Areas of tectonic compression, such as mountain belts, are dominated by reverse faults that shorten the crust horizontally while thickening it vertically.
- Strike-slip faults have most of their movement in a horizontal direction along the trend of the fault trace. Transform faults are strike-slip faults that serve as tectonic boundaries between lithospheric plates.
- Joints form in the shallow crust when rocks are stressed under brittle conditions. They facilitate groundwater movement and mineralization of economic resources, and they may result in hazards to humans.

? What type of rock structure is shown in each of the accompanying images: faults or joints? Explain how you arrived at your answer.

A.

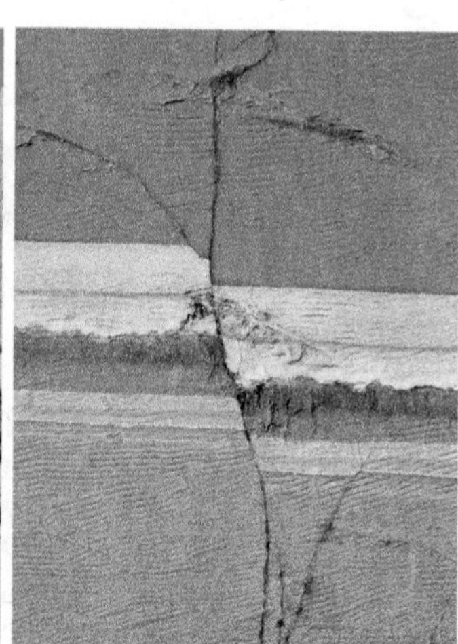

B.

E.J. Tarbuck

7.4 Mountain Building

Locate and name Earth's major mountain belts on a world map.

KEY TERMS: orogenesis, orogeny, collisional mountain

- Orogenesis is the making of mountains. An episode of orogenesis is an orogeny. Most orogenesis occurs along convergent plate boundaries, where compressional forces cause folding and faulting, thickening the crust vertically and shortening it horizontally.

? Look at the South American plate on the map. Explain why the Andes Mountains are located on the western margin of South America rather than the eastern margin.

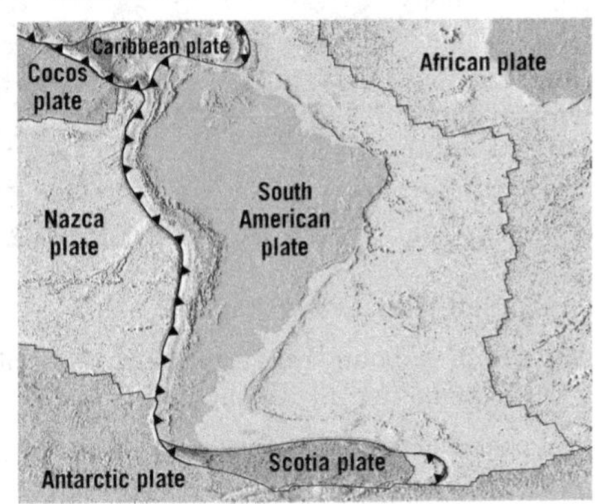

7.5 Subduction and Mountain Building

Sketch a cross section of an Andean-type mountain belt and describe how its major features are generated.

KEY TERMS: accretionary wedge, forearc basin

- The type of convergent margin determines the type of mountains that form. Where one oceanic plate overrides another, a volcanic island arc forms. Where an oceanic plate subducts under a continent, Andean-type mountain building occurs.
- In either case, release of water from the subducted slab triggers melting in the overlying mangle wedge, generating basaltic magmas that rise to the base of the continental crust where they often pond. The hot basaltic magma may heat the overlying crustal rocks sufficiently to generate a silica-rich magma of intermediate or felsic (granitic) composition.
- Sediment scraped off the subducting plate builds an accretionary wedge. Between the accretionary wedge and the volcanic arc is a relatively calm site of sedimentary deposition, the forearc basin.
- The geography of central California preserves an accretionary wedge (Coast Ranges), a forearc basin (Great Valley), and the roots of an Andean-style mountain belt (Sierra Nevada).

7.6 Collisional Mountain Belts

Summarize the stages in the development of an Alpine-type mountain belt such as the Appalachians.

KEY TERMS: terrane, microcontinent, suture

- A terrane is a relatively small crustal fragment (microcontinent, volcanic island arc, or oceanic plateau) that has been carried by an oceanic plate to a continental subduction zone and then accreted onto the continental margin. The North American Cordillera formed by the accretion of many successive terranes.
- The Himalayas and Appalachians were formed by collisions between continents when the intervening ocean basin subducted completely. The Appalachians were caused by the collision of ancestral North America with ancestral Africa more than 250 million years ago. The Himalayas were formed by the collision of India and Eurasia starting around 50 million years ago, and they are still rising.

7.7 Vertical Motions of the Crust

Explain the principle of isostasy and how it contributes to the elevated topography of mountain belts.

KEY TERMS: isostasy, isostatic adjustment, gravitational collapse

- Earth's crust floats in the denser material of the mantle the way wood floats in water. This principle is termed isostasy. If additional weight is placed on the crust (an ice sheet, for example), the crust sinks, and if weight is removed (glacial melting), the crust rebounds. This process of maintaining gravitational equilibrium is called isostatic adjustment. For a mountain belt, isostasy partially offsets the effect of erosion, pushing the mountains up as erosion wears them down.
- When compressional forces raise a mountain belt too high, the rock at the belt's core becomes warm and weak, and the belt spreads, becoming broader and lower.

? Based on the principle of isostasy, predict what will happen to the elevation of the highest peaks in a mountain range if rivers and glaciers erode deep valleys through the mountains, removing large amounts of rock.

GIVE IT SOME THOUGHT

1 Refer to the accompanying diagrams to answer the following:

a. What type of dip-slip fault is shown in Diagram 1? Were the dominant forces during faulting tensional, compressional, or shear?

b. What type of dip-slip fault is shown in Diagram 2? Were the dominant forces during faulting tensional, compressional, or shear?

c. Match the correct pair of arrows in Diagram 3 to the faults in Diagrams 1 and 2.

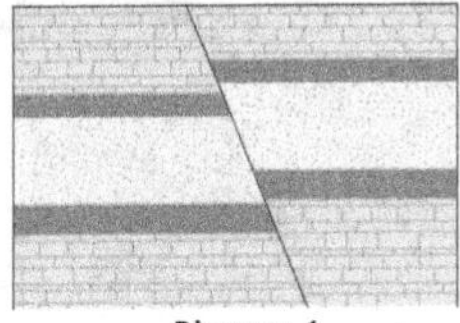

Diagram 1 (cross section)

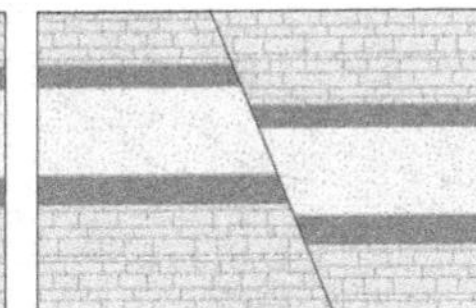

Diagram 2 (cross section)

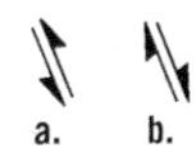

Diagram 3

2 Refer to the accompanying photo to answer the following:

a. The white line shows the approximate location of a fault that displaced these furrows created by a plow. What type of fault caused the offset shown?

USGS

b. Is this a right-lateral or left-lateral fault? Explain.

3 Which of the three types of plate boundaries is primarily associated with normal faulting? Thrust faulting? Strike-slip faulting?

4 Suppose that a sliver of oceanic crust were discovered in the interior of a continent. Would this refute the theory of plate tectonics? Explain.

5 The accompanying photo, taken near the bottom of the Grand Canyon, shows a quartz vein that has been deformed.

a. What type of deformation is exhibited—ductile or brittle?

b. Did this deformation most likely occur near Earth's surface or at great depth?

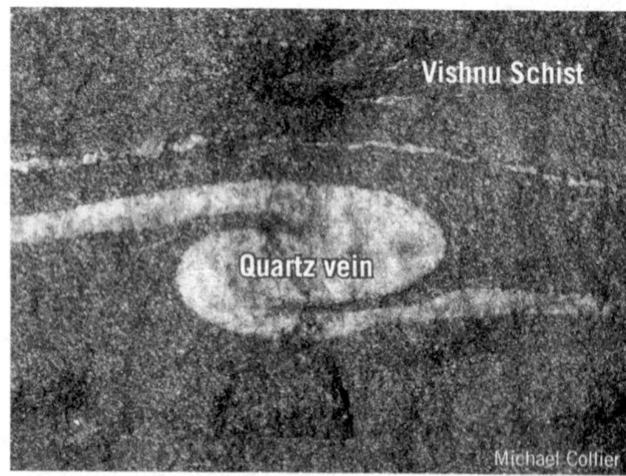

Michael Collier

6 Examine the diagrams depicting the structure of East Africa and the Canadian Rockies.

a. Characterize the type of faulting found at each location and identify the differential stresses that produced these landforms.

b. Along what type of plate boundary did each of these structures form?

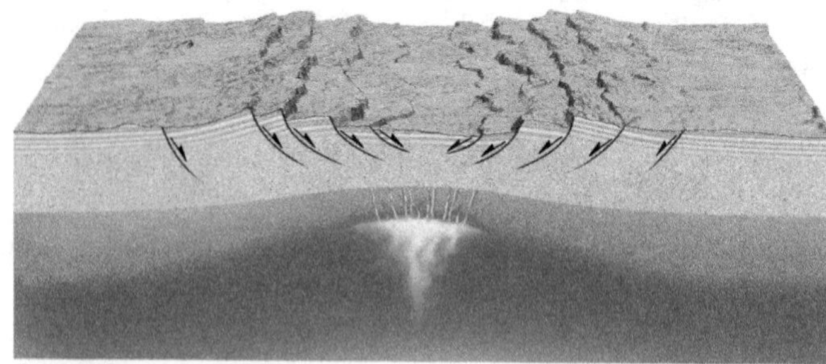
East Africa

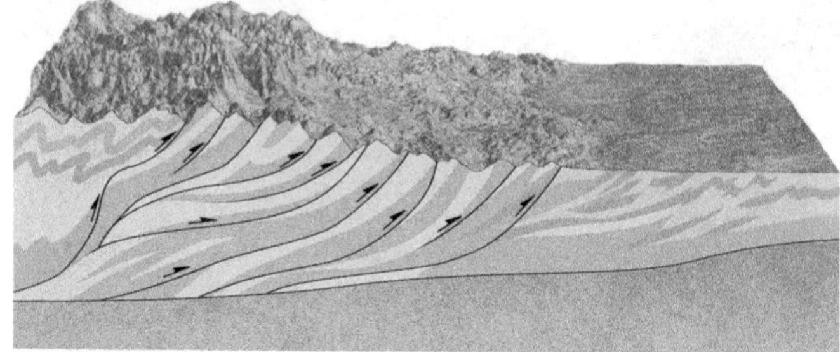
Canadian Rockies

7 Refer to the accompanying map, which shows the location of the Galapagos Rise and the Rio Grande Rise to answer the following questions:

a. Name the type of continental margin found on the west and east coasts of South America. (*Hint:* See Figure 4.10, page 103)

b. Based on your answer to Question a, is the Galapagos Rise or the Rio Grande Rise more likely to end up accreted to South America? Explain your choice.

c. In the distant future, how might a geologist determine that this accreted landmass is distinct from the continental crust to which it accreted?

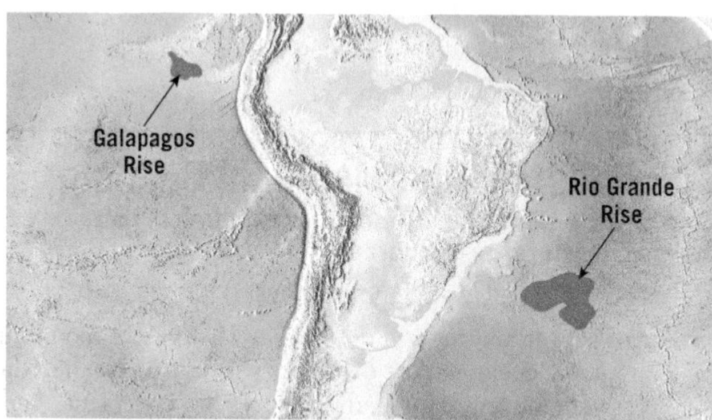

8 The Ural Mountains exhibit a north–south orientation through Eurasia. How does the theory of plate tectonics explain the existence of this mountain belt in the interior of an expansive landmass?

9 Briefly describe the major differences between the evolution of the Appalachian Mountains and the North American Cordillera.

10 Which of the accompanying sketches best illustrates an Andean-type orogeny, a Cordilleran-type orogeny, and an Alpine-type orogeny?

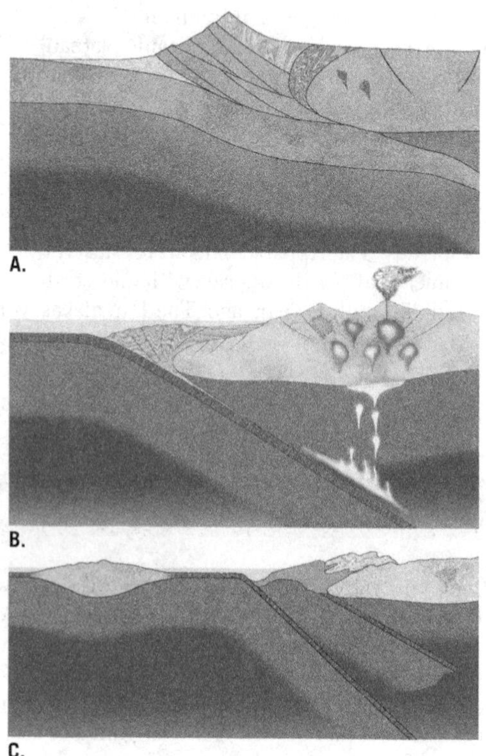

EXAMINING THE EARTH SYSTEM

1 A good example of the interaction among Earth's spheres is the influence of mountains on climate. Examine the accompanying temperature graph for the cities of Seattle and Spokane, Washington. Notice on the inset map that mountains (the Cascades) separate these two cities. The prevailing wind direction in the region is from west to east. (a) Contrast the summer and winter temperatures that occur at each city. Why are they different? *Hint:* Check out the sections "Land and Water" and "Geographic Position" in Chapter 16. The annual rainfall at Spokane (16.6 inches) is less than half that for Seattle (37.1 inches). Can you explain why?

2 The Cascades have had a profound effect on the amount and type of plant and animal life (biosphere) that inhabit the region around Spokane and Seattle. Provide several specific examples to support this statement.

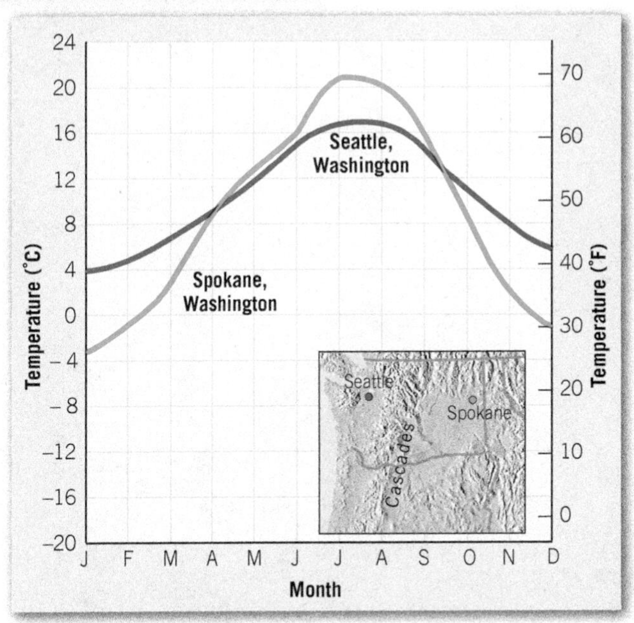

DATA ANALYSIS

Measuring the Movement of Land

As tectonic plates shift, Earth's crust deforms along faults. Scientists measure these movements to investigate the physical processes responsible for fault movement and to develop early warnings for potential earthquakes.

ACTIVITIES

Go to the USGS Earthquake Hazards Program, at http://earthquake.usgs.gov. Click on Monitoring, choose Crustal Deformation Monitoring, and select Fault Creep, Borehole Strain, and Tiltmeter Monitoring Measurements.

1 Click on Monitoring Instruments. Which instruments are used to measure land movement? Describe what each instrument measures.

Go back to the Crustal Deformation Monitoring page and click on Real-Time GPS Data. Click on GPS Real-Time PPP Displacements to display the data regions. Hover over each regions to see its name. Click on the Northern California region. Each triangle represents a data collection location. Click on a location with recent data (less than 30 seconds old) and select View Processed GPS Data.

2 What are the latitude and longitude for this location? Over what time period are data available?

3 Using the North American Fixed graphs, determine the total northward movement from the beginning to the end of the data record. Also determine the total eastward movement. (It may help to use the trend line showing the average motion.) What direction is this location moving toward?

4 What is the range (upward plus downward) of upward motions experienced by this location? What is the average (using the trend line) upward motion?

5 Based on the table below the graph, in which direction is the station you selected moving?

6 Repeat 2–5 for a station in the Southern California region.

Go back to Real-Time GPS Data and click on the GPS Velocities option above the map.

7 Examine the velocities in Northern California, and Southern California. Do most or all of the stations move in the way you determined above? What are the overall directions of movement for these regions?

8 Examine the Northern Rockies region. How would you describe the motion of this region?

UNIT 3

Sculpting Earth's Surface

8 Weathering, Soil, and Mass Movement

FOCUS ON CONCEPTS

Each statement represents the primary learning objective for the corresponding major heading within the chapter. After you complete the chapter, you should be able to:

8.1 List three types of external processes and discuss the role each plays in the rock cycle.

8.2 Define *weathering* and distinguish between the two main categories of weathering. Summarize the factors that influence the type and rate of rock weathering.

8.3 Define *soil* and explain why soil is referred to as an interface.

8.4 List and briefly discuss five controls of soil formation.

8.5 Describe an idealized soil profile. Explain the need for classifying soils.

8.6 Discuss the detrimental impact of human activities on soil and some ways that soil erosion is controlled.

8.7 Discuss the role that mass movements play in the development of landscapes. Summarize the factors that control and trigger mass movement processes.

8.8 List and explain the criteria that are commonly used to classify mass movement processes. Distinguish among six different types of mass movement.

Winter rains saturated old landslide deposits near Oso, Washington, causing a major slide in March 2014. A large mass of mud and debris buried a 2.6-square-kilometer (1-square-mile) area, damming a river, engulfing 49 buildings, and blocking a highway. The event claimed 43 lives.
(Photo by Michael Collier)

EARTH'S SURFACE IS CONSTANTLY CHANGING. Rock is disintegrated and decomposed, moved to lower elevations by gravity, and carried away by water, wind, or ice. In this manner, Earth's physical landscape is sculpted. This chapter focuses on the first two steps of this never-ending process: weathering and mass movements. What causes solid rock to crumble, and why do the type and rate of weathering vary from place to place? What mechanisms act to move weathered debris downslope? Soil, an important product of the weathering process and a vital resource, is also examined.

8.1 Earth's External Processes

List three types of external processes and discuss the role each plays in the rock cycle.

Weathering, mass movements, and erosion are called **external processes** because they occur at or near Earth's surface and are powered by gravity and energy from the Sun. External processes are a basic part of the rock cycle because they are responsible for transforming solid rock into sediment.

To the casual observer, the face of Earth may appear to be unchanging and unaffected by time. In fact, 200 years ago most people believed that mountains, lakes, and deserts were permanent features of an Earth that was thought to be no more than a few thousand years old. Today we know that Earth is 4.6 billion years old and that mountains eventually succumb to weathering and erosion, lakes fill with sediment or are drained by streams, and deserts come and go with changes in climate.

Earth is a dynamic body. Some parts of Earth's surface are gradually elevated by mountain building and volcanic activity. These **internal processes** derive their energy from Earth's interior. Meanwhile, opposing external processes are continually breaking rock apart and moving the debris to lower elevations (Figure 8.1). The latter processes include:

- **Weathering**, which is the physical breakdown (disintegration) and chemical alteration (decomposition) of rocks at or near Earth's surface

SmartFigure 8.1
Excavating the Grand Canyon The walls of the canyon extend far from the channel of the Colorado River. This results primarily from the transfer of weathered debris downslope to the river and its tributaries by mass movement processes. (Photo by Bryan Brazil/Shutterstock)

TUTORIAL
https://goo.gl/KIMCNq

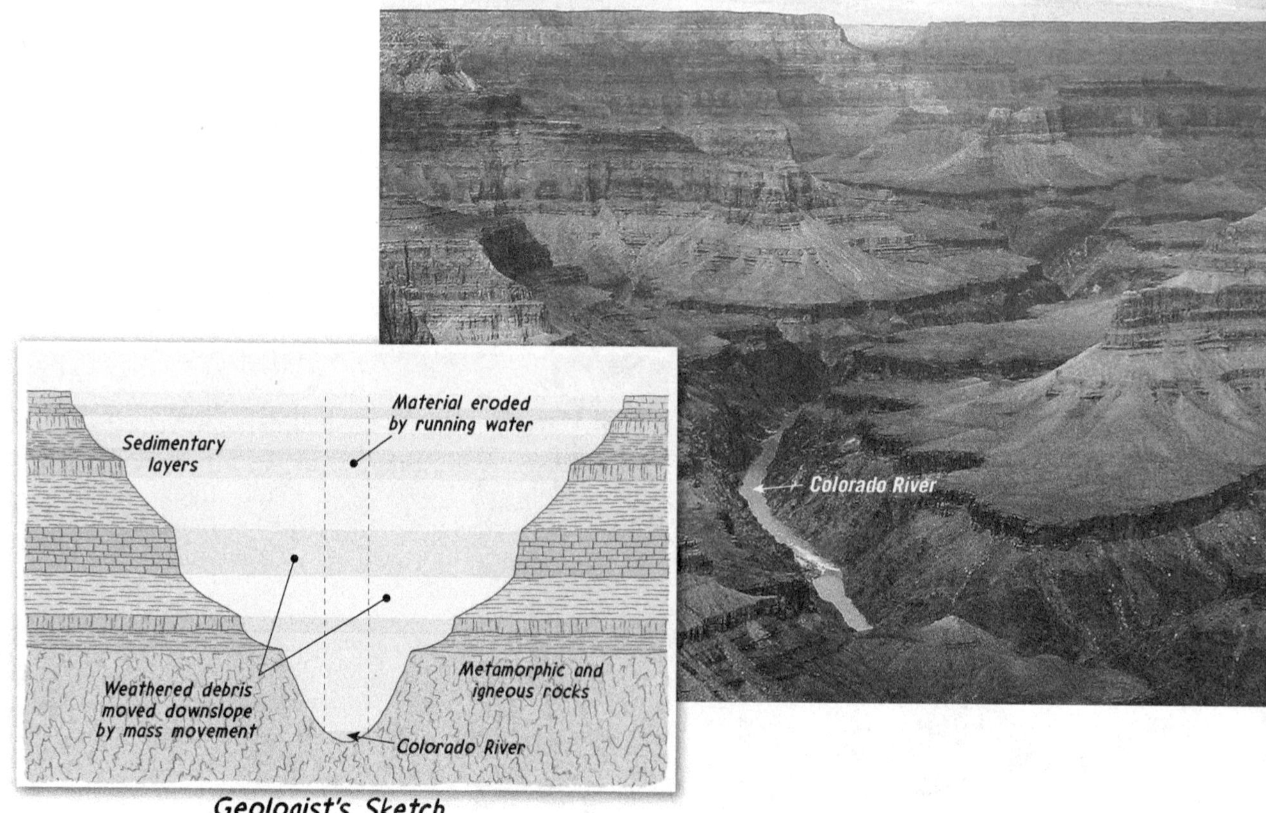

- **Mass movement**, which is the transfer of rock and soil downslope under the influence of gravity
- **Erosion**, which is the physical removal and transport of material by mobile agents such as water, wind, or ice

We first turn our attention to weathering processes and the products generated by these activities. However, weathering cannot be easily separated from the other two processes because, as weathering breaks apart rocks, it facilitates the movement of rock debris by mass movements and erosion. Conversely, the transport of material by mass movements and erosion further disintegrates and decomposes the rock.

CONCEPT CHECKS 8.1

1. List examples of Earth's external and internal processes.
2. From where do these processes derive their energy?

8.2 Weathering

Define *weathering* and distinguish between the two main categories of weathering. Summarize the factors that influence the type and rate of rock weathering.

Weathering goes on all around us, but it seems like such a slow and subtle process that it is easy to underestimate its importance. Yet, it is worth remembering that weathering is a basic part of the rock cycle and thus a key process in the Earth system. Weathering is also important to humans—even to those of us who are not studying geology. For example, many of the life-sustaining minerals and elements found in soil, and ultimately in the food we eat, were freed from solid rock by weathering processes. As Figure 8.2 and many other images in this book illustrate, weathering also contributes to the formation of some of Earth's most spectacular scenery. Of course, these same processes are also responsible for causing the deterioration of many of the structures we build.

All materials are susceptible to weathering. Consider, for example, the fabricated product concrete, which closely resembles the sedimentary rock called conglomerate. A newly poured concrete sidewalk has a smooth, unweathered look. However, not many years later, the same sidewalk will appear chipped, cracked, and rough, with pebbles exposed at the surface. If a tree is nearby, its roots may grow under the concrete, heaving and buckling it. The same natural processes that eventually break apart a concrete sidewalk also act to disintegrate rocks, regardless of their type or strength.

Weathering occurs when rock is mechanically fragmented (disintegrated) and/or chemically altered (decomposed). **Mechanical weathering** is accomplished by physical forces that break rock into smaller and smaller pieces without changing the rock's mineral composition. **Chemical weathering** involves a chemical transformation of rock into one or more new compounds. To understand these two concepts, consider the possible ways to break down a large log. The log disintegrates when it is split into smaller and smaller pieces, whereas decomposition occurs when the log is set afire and burned. GEOgraphics 8.1 provides other examples of weathering.

In the following sections, we discuss the various modes of mechanical and chemical weathering. Although we consider these two categories separately, keep in mind that mechanical and chemical weathering processes usually work simultaneously in nature and reinforce each other.

Mechanical Weathering

When a rock undergoes mechanical weathering, it is broken into smaller and smaller pieces, each retaining the characteristics of the original material. The end result

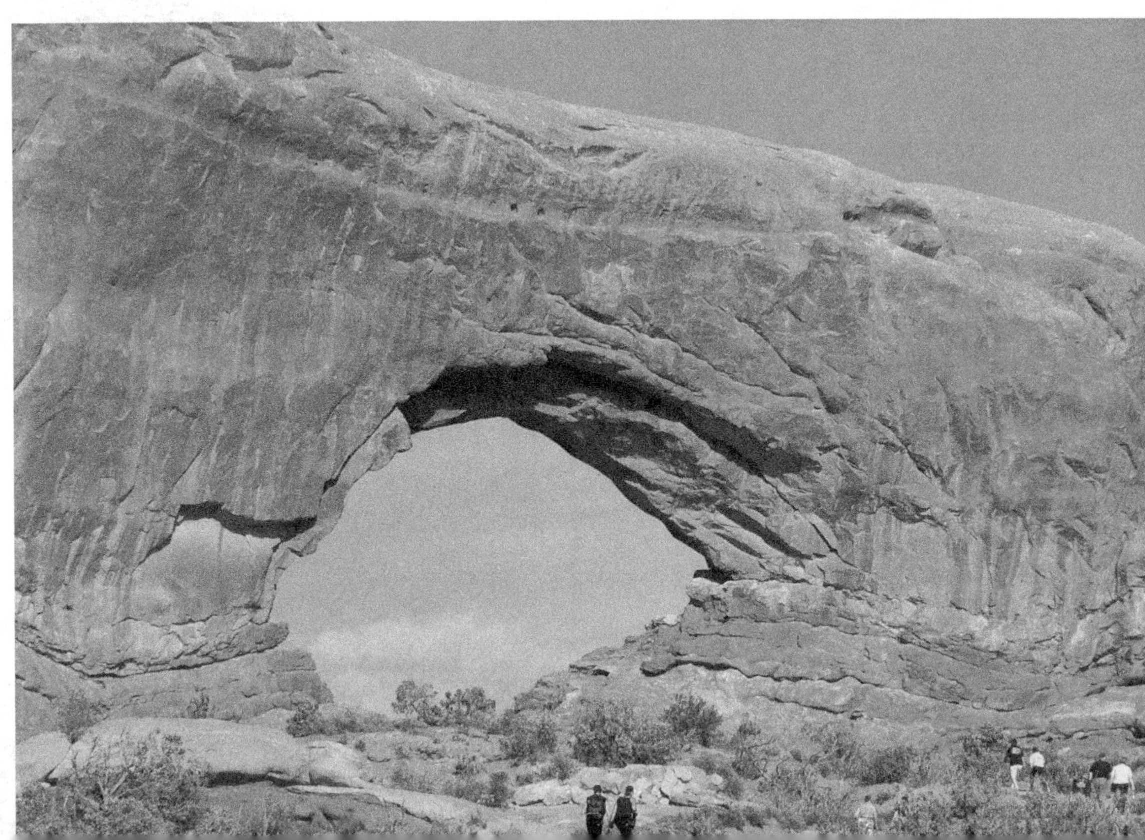

SmartFigure 8.2 **Arches National Park** Mechanical and chemical weathering contributed greatly to the creation of North Window Arch and all of the other arches and rock formations in Utah's Arches National Park. (Photo by Dennis Tasa)

ANIMATION https://goo.gl/bcGvi0

GEOGRAPHICS 8.1

Some Everyday Examples of Weathering

Weathering processes are responsible for the deterioration of objects and materials at Earth's surface. Day to day changes are not obvious, but over time the effects are significant.

Weathering gradually deteriorated the paint protecting this wood siding.

Question:
List one or more additional example of weathering that you have seen.

Even the most "solid" stone structure that people erect gradually succumbs to weathering processes.

Jose Garcia/Alamy Stock Photo

William Britten/iStockphoto

Plant roots can break concrete and rock.

EuroStyle Graphics/Alamy

Potholes are especially common in places that experience wintry conditions. The freeze-thaw cycle coupled with water and road salt contribute to pavement destruction.

Marc Bruxelle/Shutterstock

RATOCA/Fotolia

The chemical weathering process called oxidation is responsible for the rust on this **car.**

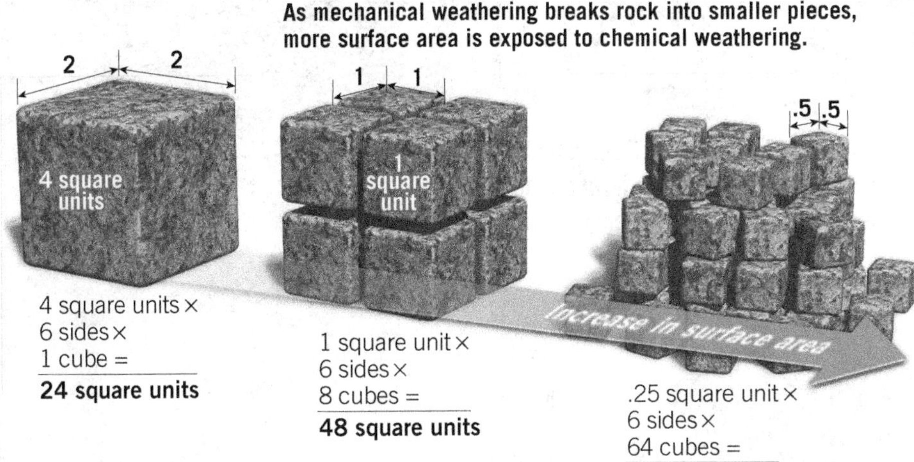

SmartFigure 8.3 **Mechanical weathering increases surface area** Mechanical weathering adds to the effectiveness of chemical weathering because chemical weathering can occur only on exposed surfaces.

TUTORIAL
https://goo.gl/Gqd1iz

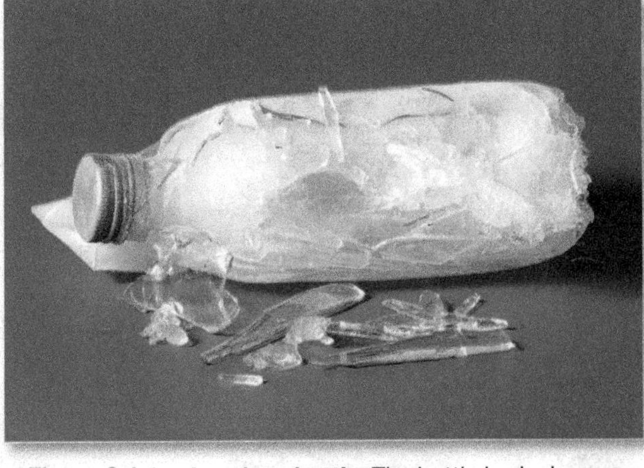

Figure 8.4 **Ice breaks a bottle** The bottle broke because water expands about 9 percent when it freezes. (Photo by Martyn F. Chillmaid/Science Source)

is many small pieces from a single large one. Figure 8.3 shows that breaking a rock into smaller pieces increases the surface area susceptible to chemical attack. Hence, by breaking rocks into smaller pieces, mechanical weathering increases the amount of surface area available for chemical weathering.

In nature, four important physical processes lead to the fragmentation of rock: frost wedging, salt crystal growth, expansion resulting from unloading (sheeting), and biological activity. In addition, although the work of erosional agents such as waves, wind, glacial ice, and running water is usually considered separately from mechanical weathering, it is nevertheless related. As these mobile agents move rock debris, particles continue to be broken and abraded.

Frost Wedging If you leave a glass bottle of water in the freezer a bit too long, the bottle breaks, as shown in Figure 8.4. The bottle breaks because liquid water has the unique property of expanding about 9 percent upon freezing. This is also the reason that poorly insulated or exposed water pipes rupture during frigid weather. You might also expect this same process to fracture rocks in nature. This is, in fact, the basis for the traditional explanation of **frost wedging**. After water works its way into the cracks in rock, the freezing water enlarges the cracks, and angular fragments eventually break off (Figure 8.5 and GEOgraphics 8.2).

For many years, the conventional wisdom was that most frost wedging occurred in the manner just described. However, research has shown that frost wedging can also occur in a different way.* It has long been known that when moist soils freeze, they expand, or *frost heave*, due to the growth of ice lenses. These masses of ice grow larger because they are supplied with water migrating from unfrozen areas as thin liquid films. As more water accumulates and freezes, the soil is heaved upward. A similar process occurs within the cracks and pore spaces of rocks. Lenses of ice grow larger as they attract liquid water from surrounding pores. The growth of these ice masses gradually weakens the rock, causing it to fracture.

Salt Crystal Growth Another expansive force that can split rocks is created by the growth of salt crystals. Rocky shorelines and arid regions are common

*Bernard Hallet, "Why Do Freezing Rocks Break?" *Science* 314 (November 2006): 1092–1099.

SmartFigure 8.5 **Ice breaks rock** In mountainous areas, frost wedging creates angular rock fragments that accumulate to form talus slopes. (Photo by Marli Miller)

TUTORIAL
https://goo.gl/RZRD5W

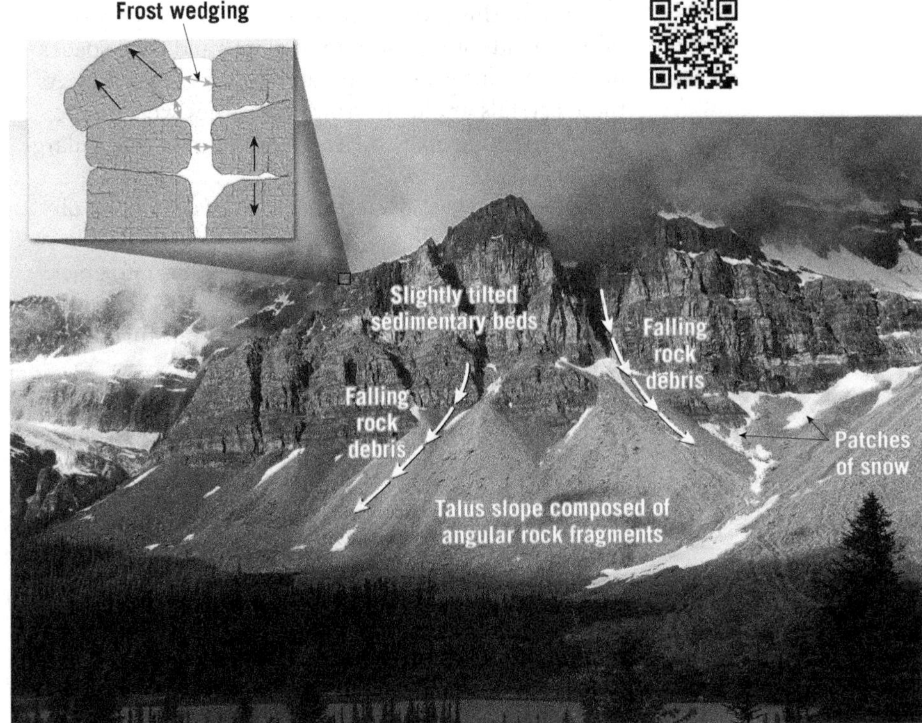

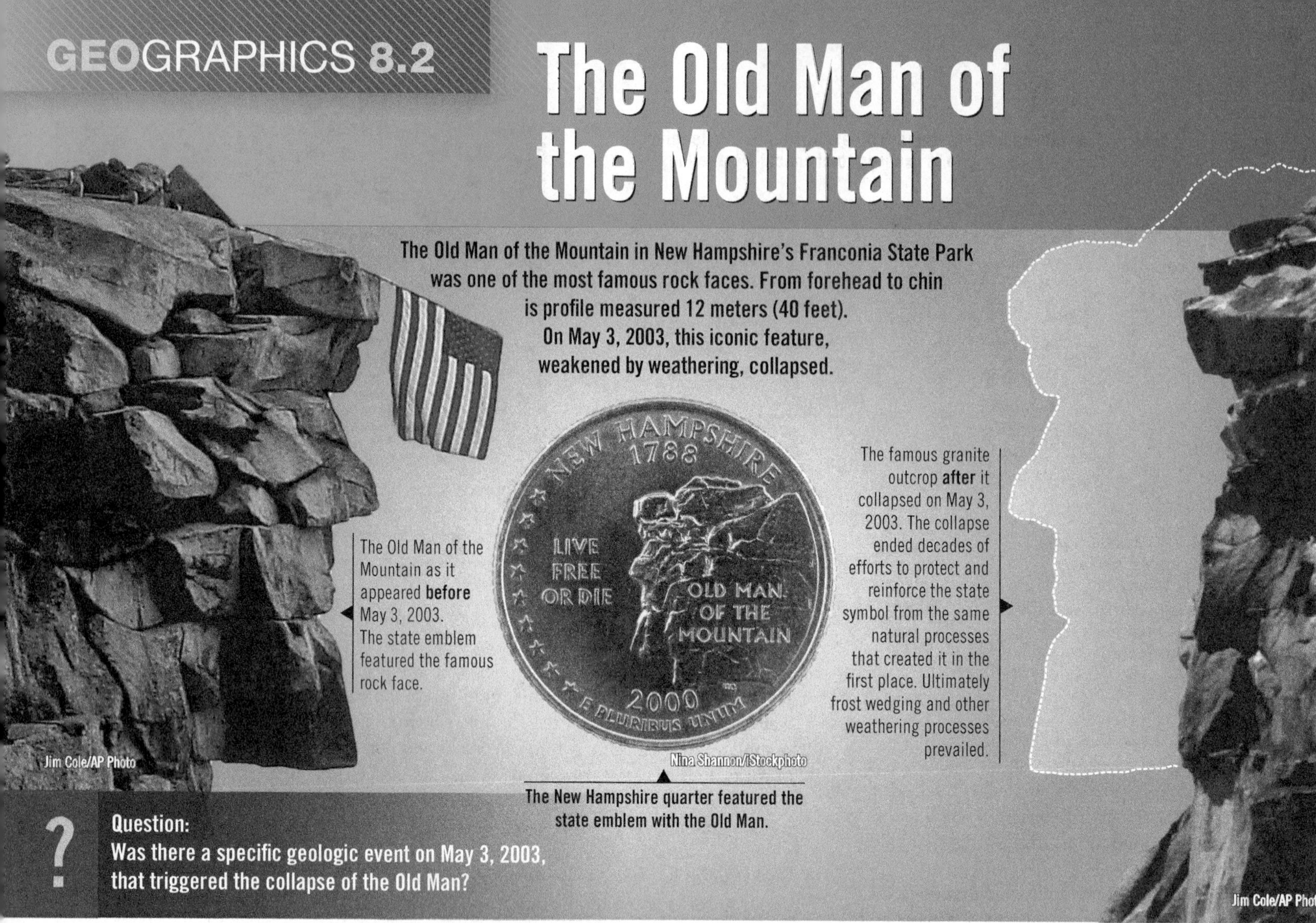

settings for this process. It begins when sea spray or salty groundwater penetrates crevices and pore spaces in rock. As this water evaporates, salt crystals form. As these crystals gradually grow larger, they weaken the rock by pushing apart the surrounding grains or enlarging tiny cracks.

GEOGRAPHICS 8.2
The Old Man of the Mountain
The Old Man of the Mountain in New Hampshire's Franconia State Park was one of the most famous rock faces. From forehead to chin is profile measured 12 meters (40 feet). On May 3, 2003, this iconic feature, weakened by weathering, collapsed.
The Old Man of the Mountain as it appeared before May 3, 2003. The state emblem featured the famous rock face.
NEW HAMPSHIRE
1788
LIVE FREE OR DIE
OLD MAN OF THE MOUNTAIN
2000
E PLURIBUS UNUM
The famous granite outcrop after it collapsed on May 3, 2003. The collapse ended decades of efforts to protect and reinforce the state symbol from the same natural processes that created it in the first place. Ultimately frost wedging and other weathering processes prevailed.
The New Hampshire quarter featured the state emblem with the Old Man.
Question:
Was there a specific geologic event on May 3, 2003, that triggered the collapse of the Old Man?
Jim Cole/AP Photo
Nina Shannon/iStockphoto
Jim Cole/AP Photo

This process also contributes to the crumbling of roadways where salt is spread to melt snow and ice in winter. The salt dissolves in water and seeps into cracks that quite likely originated from frost action. When the water evaporates, the growth of salt crystals further breaks the pavement.

Sheeting When large masses of igneous rock, particularly those composed of granite, are exposed by erosion, concentric slabs begin to break loose. The process generating these onion-like layers is called **sheeting**. It takes place due to the great reduction in pressure that occurs when the overlying rock is eroded away in a process called *unloading*. Figure 8.6 illustrates what happens: As the overburden is removed, the outer parts of the granitic mass expand more than the rock below and separate from the rock body. Continued weathering eventually causes the slabs to separate and peel off, creating **exfoliation domes** (*ex* = off, *folium* = leaf). Excellent examples of exfoliation domes include Stone Mountain in Georgia and Half Dome in Yosemite National Park.

Human activities can also cause sheeting-like fractures as a result of unloading. In deep mines, large rock slabs have been known to explode off the walls of newly cut tunnels. In quarries, fractures can occur parallel to the floor when large blocks of rock are removed.

Fractures can also be caused by contraction as igneous materials cool (see Figure 6.30, page 191), or they can be produced by tectonic forces during mountain building. Fractures produced by these activities often form a definite pattern and are called *joints* (see Figure 8.8 and Figure 7.21, page 219). Joints allow water to penetrate deeply and initiate weathering long before the rock is exposed.

Biological Activity The activities of organisms, including plants, burrowing animals, and humans, can cause weathering. Plant roots in search of minerals and water grow into fractures, and as the roots grow, they wedge

Confining pressure
This large igneous mass formed deep beneath the surface, where confining pressure is great.
Deep pluton
Joints
Expansion and sheeting
Uplift
As erosion removes the overlying bedrock (unloading), the outer parts of the igneous mass expand. Joints form parallel to the surface. Continued weathering causes thin slabs to separate and fall off.
The summit of Half Dome in California's Yosemite National Park is an exfoliation dome and illustrates the onion-like layers created by sheeting.

SmartFigure 8.6 **Unloading leads to sheeting** Sheeting leads to the formation of an exfoliation dome. (Photo by Gary Moon/AGE Fotostock)

TUTORIAL
https://goo.gl/UFWTzX

the rock apart (Figure 8.7). Burrowing animals further break down the rock by moving fresh material to the surface, where physical and chemical processes can more effectively attack it. Where rock has been blasted in search of minerals or for construction projects, the impact of humans is particularly noticeable.

Organisms play many roles in chemical weathering. For example, plant roots, fungi, and lichens that occupy fractures or that may encrust a rock produce acids that promote decomposition. Decaying organisms also produce acids. Moreover, some bacteria are capable of chemically breaking down minerals and harvesting the resulting energy, in essentially the same way that we obtain energy by breaking down the molecules of our food. Such bacteria can live at depths as great as a few kilometers.

Chemical Weathering

In the preceding discussion of mechanical weathering, you learned that breaking rock into smaller pieces aids chemical weathering by increasing the surface area available for chemical attack. It should also be pointed out that chemical weathering contributes to mechanical weathering. It does so by weakening the outer portions of some rocks, which, in turn, makes them more susceptible to being broken by mechanical weathering processes.

Chemical weathering involves the complex processes that alter the internal structures of minerals by removing and/or adding chemical elements. During this transformation, the original rock decomposes into substances that are stable in the surface environment. Consequently,

Plant roots can extend into joints and grow in diameter and length. This process enlarges fractures and breaks rock.

Figure 8.7 **Plants can break rock** Root wedging near Boulder, Colorado. (Photo by Kristin Piljay)

EYE ON EARTH 8.1

This is a close-up view of a massive granite feature in the Sierra Nevada of California. (Photo by Marli Miller)

QUESTION 1 *Relatively thin slabs of granite are separating from this rock mass. Describe the process that caused this to occur.*

QUESTION 2 *What term is applied to the process you described in answering Question 1? What term describes the domelike feature that results from this process?*

the products of chemical weathering will remain essentially unchanged as long as they remain in an environment similar to the one in which they formed.

Water and Carbonic Acid Water is by far the most important agent of chemical weathering. Although pure water is nonreactive, a small amount of dissolved material is generally all that is needed to activate it. Oxygen dissolved in water will *oxidize* some materials. For example, when an iron nail is buried in moist soil, it will develop a coating of rust (iron oxide). Eventually it will rust to the point that it can be broken as easily as a toothpick. When rocks containing iron-rich minerals oxidize, a yellow to reddish-brown rust appears on the surface.

Carbon dioxide (CO_2) dissolves in water (H_2O) to form **carbonic acid** (H_2CO_3)—the same weak acid that makes carbonated soft drinks taste tart. Rain acquires some dissolved carbon dioxide as it falls through the atmosphere, and it picks up more from decaying organic matter as it percolates through soil. Carbonic acid ionizes to form the very reactive hydrogen ion (H^+) and the bicarbonate ion (HCO_3).

Acids such as carbonic acid readily decompose many minerals and produce certain products that are water soluble. For example, the mineral calcite ($CaCO_3$), which composes the common building stones marble and limestone, is easily attacked by even a weakly acidic solution to produce dissolved Ca^{2+} and CO_3^{2-} ions. In nature, over spans of thousands of years, large quantities of limestone are dissolved and carried away by groundwater. This activity is largely responsible for the formation of limestone caverns.

How Granite Weathers To illustrate how a rock rich in silicate minerals chemically weathers when attacked by carbonic acid, we will consider the weathering of granite, an abundant continental rock. Recall that granite consists mainly of quartz and potassium feldspar. The weathering of the potassium feldspar component of granite takes place as follows:

$$\underset{\text{potassium feldspar}}{2\,KAlSi_3O_8} + \underset{\text{carbonic acid}}{2(H^{2+} + CO^{3-})} + \underset{\text{water}}{H_2O} \longrightarrow$$

$$\underset{\text{clay mineral}}{Al_2Si_2O_5(OH)_4} + \underbrace{\underset{\text{potassium ions}}{2\,K^+} + \underset{\text{bicarbonate ions}}{2\,HCO_3}}_{\text{in solution}} + \underset{\text{silica}}{SiO_2}$$

In this reaction, the hydrogen ions (H^+) attack and replace potassium ions (K^+) in the feldspar structure, thereby disrupting the crystalline network. Once removed, the potassium is available as a nutrient for plants or becomes the soluble salt potassium bicarbonate ($KHCO_3$), which may be incorporated into other minerals or carried to the ocean in dissolved form by groundwater and streams.

The most abundant products of the chemical breakdown of feldspar are residual clay minerals. Clay minerals are the end product of weathering and are very stable under surface conditions. Consequently, clay minerals make up a high percentage of the inorganic material in soils. Moreover, the most abundant sedimentary rock, shale, contains a high proportion of clay minerals.

In addition to the formation of clay minerals during this reaction, some silica is removed from the feldspar structure and carried away by groundwater. This dissolved silica eventually precipitates to produce nodules of

Table 8.1 Products of Weathering

Mineral	Residual Products	Material in Solution
Quartz	Quartz grains	Silica
Feldspars	Clay minerals	Silica, K^+, Na^+, Ca^{2+}
Amphibole	Clay minerals Iron oxides	Silica, Ca^{2+}, Mg^{2+}
Olivine	Iron oxides	Silica, Mg^{2+}

chert or flint, fills in the pore spaces between sediment grains, or is carried to the ocean, where microscopic animals remove it to build hard silica shells. To summarize, the weathering of potassium feldspar generates a residual clay mineral, a soluble salt (potassium bicarbonate), and some silica that enters into solution.

Quartz, the other main component of granite, is *very resistant* to chemical weathering; it remains substantially unaltered by weakly acidic solutions. As a result, when granite weathers, the feldspar crystals dull and slowly turn to clay, releasing the once-interlocked quartz grains, which still retain their fresh, glassy appearance. Although some quartz remains in the soil, much is transported to the sea or to other sites of deposition, where it becomes the main constituent of such features as sandy beaches and sand dunes. In time it may become lithified to form the sedimentary rock *sandstone*.

Weathering of Silicate Minerals Table 8.1 lists the weathered products of some of the most common silicate minerals. Remember that silicate minerals make up most of Earth's crust and that these minerals are composed essentially of only eight elements. When chemically weathered, silicate minerals yield sodium, calcium, potassium, and magnesium ions, which form soluble products that may be removed by groundwater. The element iron combines with oxygen, producing relatively insoluble iron oxides, which give soil a reddish-brown or yellowish color. Under most conditions, the three remaining elements—aluminum, silicon, and oxygen—join with water to produce residual clay minerals. However, even the highly insoluble clay minerals are very slowly removed by subsurface water.

Spheroidal Weathering Many rock outcrops have a rounded appearance. This occurs because chemical weathering works inward from exposed surfaces. Figure 8.8 illustrates how angular masses of jointed rock change through time. The process is aptly called **spheroidal weathering**. Because weathering attacks edges from two sides and corners from three sides, these areas wear down faster than does a single flat surface. Gradually, sharp edges and corners become smooth and rounded. Eventually an angular block may evolve into a nearly spherical boulder. Once this occurs, the boulder's shape does not change, but the spherical mass continues to get smaller.

Differential Weathering

Masses of rock do not weather uniformly. Take a moment to look at the photo of Shiprock, New Mexico, in Figure 6.26 (page 188). The durable igneous mass is more resistant to weathering than the surrounding rock

SmartFigure 8.8
The formation of rounded boulders Spheroidal weathering of extensively jointed rock. (Photo by E. J. Tarbuck)

TUTORIAL
https://goo.gl/I7QdFI

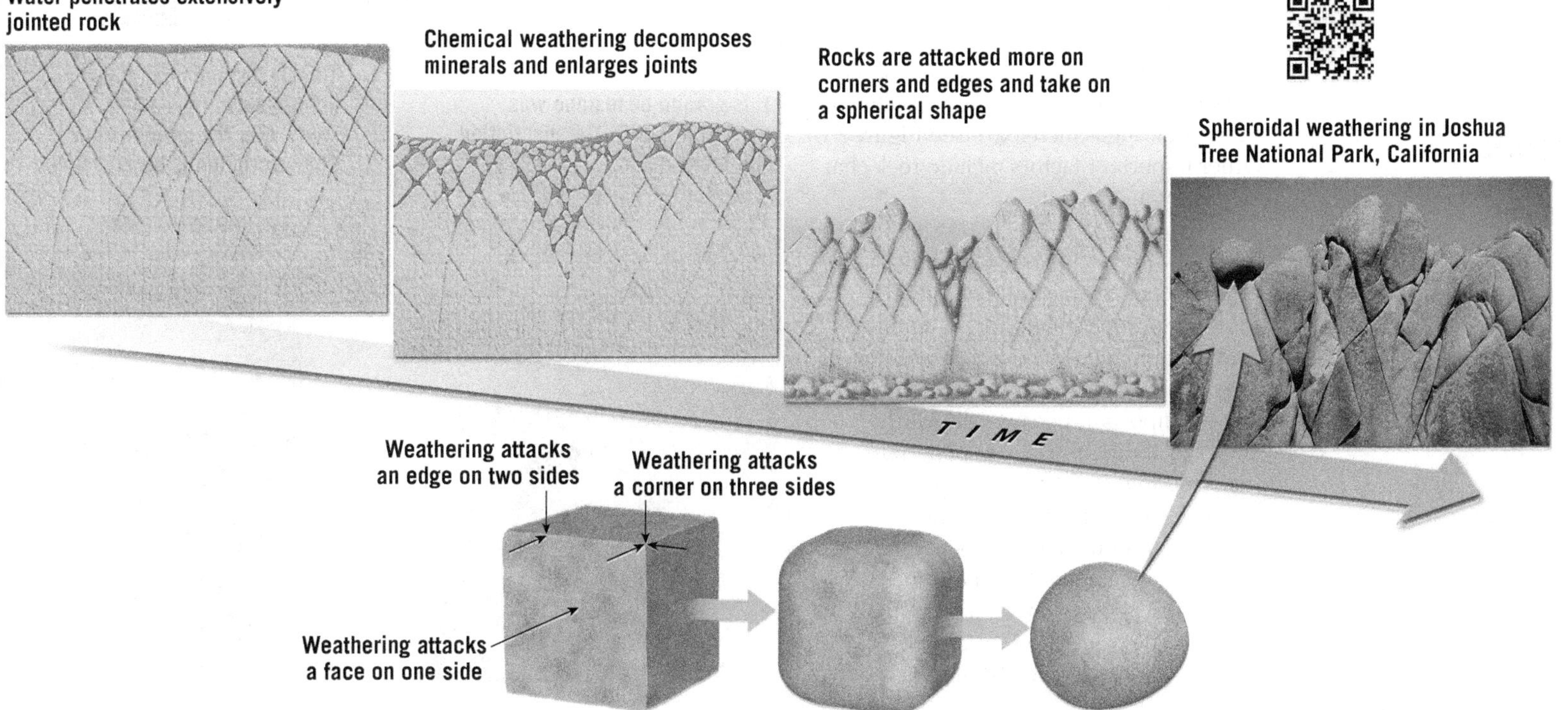
Water penetrates extensively jointed rock
Chemical weathering decomposes minerals and enlarges joints
Rocks are attacked more on corners and edges and take on a spherical shape
Spheroidal weathering in Joshua Tree National Park, California
TIME
Weathering attacks an edge on two sides
Weathering attacks a corner on three sides
Weathering attacks a face on one side

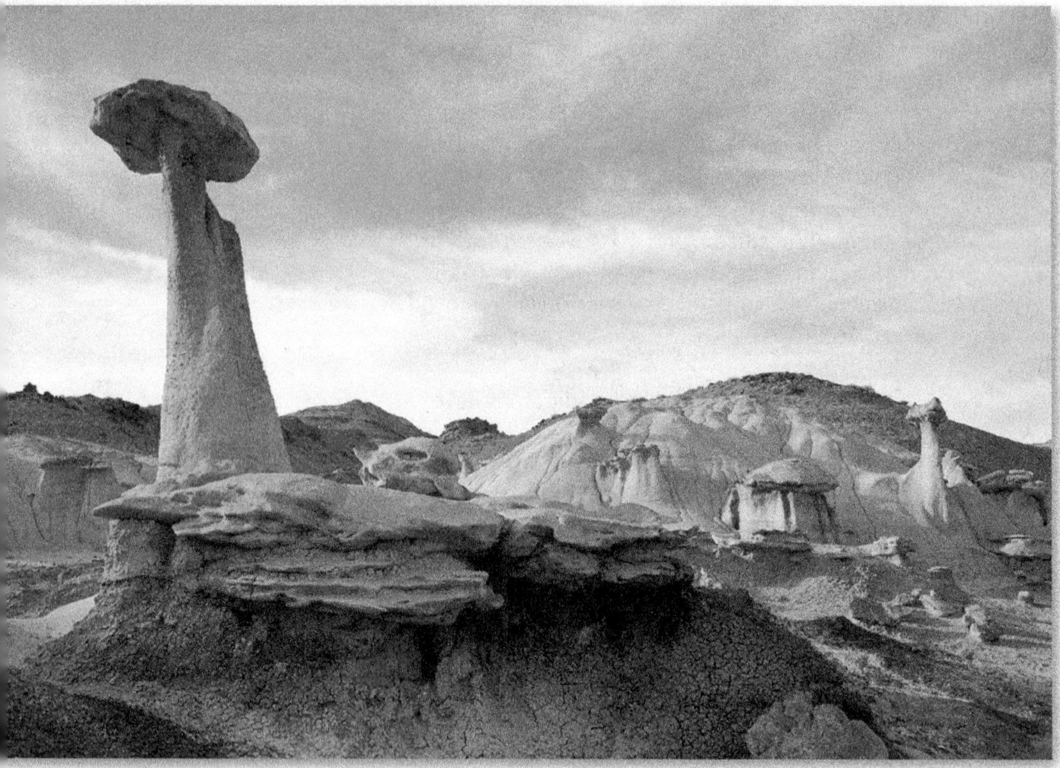

▲ SmartFigure 8.9 **Monuments to weathering** This example of differential weathering is in New Mexico's Bisti Badlands. When weathering accentuates differences in rocks, spectacular landforms are sometimes created. (Photo by Michael Collier)

MOBILE FIELD TRIP
https://goo.gl/MHpjx0

layers and so protrudes high above the surface. A glance at Figure 8.2 shows an additional example of this phenomenon, called **differential weathering**. The results of differential weathering vary in scale from the rough, uneven surface of a concrete sidewalk to the boldly sculpted exposures of bedrock in New Mexico's Bisti Badlands (Figure 8.9).

Several factors influence the type and rate of rock weathering. We have already seen how mechanical weathering affects the rate of weathering. When rocks are broken into smaller pieces, the amount of surface area exposed to chemical weathering is increased (see Figure 8.3). Other important factors include rock characteristics and climate.

Rock Characteristics Rock characteristics encompass all the chemical traits of rocks, including mineral composition and solubility. In addition, physical features such as joints (cracks) can be important because they influence the ability of water to penetrate rock (see Figure 8.8).

The variations in weathering rates due to mineral constituents can be demonstrated by examining the inscriptions on old headstones made from different rock types (Figure 8.10). Headstones of granite, which are composed of silicate minerals, are relatively resistant to chemical weathering. On the other hand, the marble headstone shows signs of extensive chemical alteration over a relatively short period. Marble is composed of calcite (calcium carbonate), which readily dissolves even in a weakly acidic solution.

The silicates, the most abundant mineral group, weather in essentially the same sequence as their order of crystallization. By examining Bowen's reaction series (see Figure 3.13, page 70), you can see that olivine crystallizes first and is therefore the least resistant to chemical weathering, whereas quartz, which crystallizes last, is the most resistant.

Climate Climate elements, particularly temperature and precipitation, are crucial to the rate of rock weathering. For example, the frequency of freeze–thaw cycles greatly affects the amount of frost wedging. Temperature and precipitation also exert a strong influence on the rates of chemical weathering and on the kind and amount of vegetation present. Regions with lush vegetation generally have a thick mantle of soil rich in decayed organic matter from which carbonic and other acids are derived.

▼ SmartFigure 8.10 **Rock type influences weathering** An examination of headstones in the same cemetery shows that the rate of chemical weathering depends on the rock type. (Photos by E. J. Tarbuck)

TUTORIAL
https://goo.gl/BjltRN

This granite headstone was erected in 1868. The inscription still looks fresh.

This headstone of calcite-rich marble dates from 1874, six years after the granite stone. The inscription is barely legible.

The optimal environment for chemical weathering is a combination of warm temperatures and abundant moisture. In polar regions chemical weathering is ineffective because frigid temperatures keep the available moisture locked up as ice, whereas in arid regions there is little moisture to promote chemical weathering.

Human activities influence the composition of the atmosphere, which in turn can impact the rate of chemical weathering. For example, as a consequence of burning large quantities of coal and petroleum, millions of tons of sulfur and nitrogen oxides are released into the atmosphere each year worldwide. Through a series of complex chemical reactions, some of these pollutants are converted into acids that then fall to Earth's surface as *acid precipitation*.

CONCEPT CHECKS 8.2

1. Contrast mechanical and chemical weathering.
2. When a rock is mechanically weathered, how does its surface area change? How does this influence chemical weathering?
3. Distinguish among four types of mechanical weathering.
4. How is carbonic acid formed in nature? What products result when carbonic acid reacts with potassium feldspar?
5. Explain how angular masses of rock often become spherical boulders.
6. List three or more factors that influence the type and rate of weathering.

8.3 Soil: An Indispensable Resource

Define *soil* and explain why soil is referred to as an interface.

Weathering is a key process in the formation of soil. Along with air and water, soil is one of our most indispensable resources. Also, like air and water, soil is often taken for granted. The following quote helps put this vital layer in perspective:

> Science, in recent years, has focused more and more on the Earth as a planet, one that for all we know is unique—where a thin blanket of air, a thinner film of water, and the thinnest veneer of soil combine to support a web of life of wondrous diversity in continuous change†

Soil has accurately been called "the bridge between life and the inanimate world." All life—the entire biosphere—owes its existence to a dozen or so elements that must ultimately come from Earth's crust. After weathering and other processes create soil, plants carry out the intermediary role of assimilating the necessary elements and making them available to animals, including humans.

An Interface in the Earth System

When Earth is viewed as a system, as discussed in Chapter 1, soil is considered an *interface*—a common boundary where different parts of a system interact. This is an appropriate designation because soil forms where the geosphere, the atmosphere, the hydrosphere, and the biosphere meet. Soil develops in response to complex environmental interactions among different parts of the Earth system. Over time, soil gradually evolves to a state of equilibrium, or balance, with the environment. Soil is dynamic and sensitive to almost every aspect of its surroundings. Thus, when environmental changes occur, such as changes in climate, vegetative cover, or animal (including human) activity—soil responds. Any such change gradually alters soil characteristics until a new balance is reached. Although thinly distributed over the land surface, soil functions as a fundamental interface, providing an excellent example of the integration among many parts of the Earth system.

What Is Soil?

With few exceptions, Earth's land surface is covered by **regolith** (*rhegos* = blanket, *lithos* = stone), a layer of rock and mineral fragments produced by weathering. Some would call this material soil, but soil is more than an accumulation of weathered debris. **Soil** is a combination of mineral and organic matter, water, and air—the portion of the regolith that supports the growth of plants. Although the proportions of the major components in soil vary, the same four components are always present to some extent (Figure 8.11). About one-half of the total volume of a good-quality surface soil is a mixture of disintegrated and decomposed rock (mineral matter) and *humus*, the decayed remains of animal and plant life (organic matter). The remaining half consists of pore spaces among the solid particles where air and water circulate.

Figure 8.11 **What is soil?** The pie chart depicts the composition (by volume) of a soil in good condition for plant growth. Although percentages vary, each soil is composed of mineral and organic matter, water, and air. (Photo by Juice Images/Alamy Stock Photo)

†Jack Eddy, "A Fragile Seam of Dark Blue Light," in *Proceedings of the Global Change Research Forum*. U.S. Geological Survey Circular 1086, 1993, p. 15.

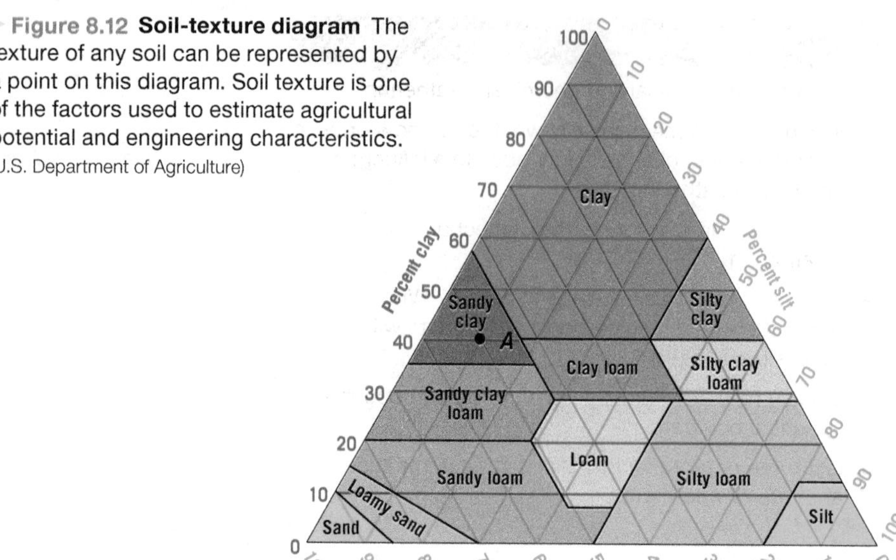

Figure 8.12 **Soil-texture diagram** The texture of any soil can be represented by a point on this diagram. Soil texture is one of the factors used to estimate agricultural potential and engineering characteristics. (U.S. Department of Agriculture)

Although the mineral portion of the soil is usually much greater than the organic portion, humus is an essential component. In addition to being an important source of plant nutrients, humus enhances the soil's ability to retain water. Because plants require air and water to live and grow, the portion of the soil consisting of pore spaces that allow these fluids to circulate is as vital as the solid soil constituents.

Soil water is far from "pure" water; instead, it is a complex solution containing many soluble nutrients. Soil water not only provides the necessary moisture for the chemical reactions that sustain life, it also supplies plants with nutrients in a form they can use. The pore spaces not filled with water contain air. This air is the source of the oxygen and carbon dioxide needed by most microorganisms and plants that live in the soil.

Soil Texture and Structure

Most soils are far from uniform and contain particles of different sizes. **Soil texture** refers to the proportions of different particle sizes. Texture is a very basic soil property because it strongly influences the soil's ability to retain and transmit water and air, both of which are essential to plant growth. Sandy soils may drain too rapidly and dry out quickly. At the opposite extreme, the pore spaces of clay-rich soils may be so small that they inhibit drainage, and long-lasting puddles result. Moreover, when the clay and silt content is very high, plant roots may have difficulty penetrating the soil.

Because soils rarely consist of particles of only one size, *textural categories* have been established, based on the varying proportions of clay, silt, and sand. The standard system of classes used by the U.S. Department of Agriculture is shown in Figure 8.12. For example, point A on this triangular diagram (left center) represents a soil composed of 10 percent silt, 40 percent clay, and 50 percent sand. Such a soil is called a *sandy clay*. The soils called *loam*, which occupy the central portion of the diagram, are those in which no single particle size predominates over the other two. Loam soils are best suited to support plant life because they generally hold moisture and nutrients better than do soils composed predominantly of clay or coarse sand.

Soil particles are seldom completely independent of one another. Rather, they usually form clumps called *peds* that give soils a particular structure. Four basic soil structures are recognized: platy, prismatic, blocky, and spheroidal. Soil structure is important because it influences how easily a soil can be cultivated as well as how susceptible the soil is to erosion. Soil structure also affects a soil's porosity and permeability (that is, the ease with which water can penetrate). This in turn influences the movement of nutrients to plant roots. Prismatic and blocky peds usually allow for moderate water infiltration, whereas platy and spheroidal structures are characterized by slower infiltration rates.

CONCEPT CHECKS 8.3

1. Explain why soil is considered an interface in the Earth system.
2. How is regolith different from soil?
3. Why is texture an important soil property?
4. Using the soil texture diagram in Figure 8.12, name the soil that consists of 60 percent sand, 30 percent silt, and 10 percent clay.

8.4 Controls of Soil Formation

List and briefly discuss five controls of soil formation.

Soil is the product of the complex interplay of several factors. The most important of these are parent material, time, climate, plants and animals, and topography. Although all these factors are interdependent, their roles are examined separately.

Parent Material

The source of the weathered mineral matter from which soils develop is called the **parent material**, and it is a major factor influencing a newly forming soil. Gradually this weathered matter undergoes physical and chemical changes, as soil formation progresses. Parent material may be the underlying bedrock, or it can be a layer of unconsolidated deposits, as in a stream valley. Soils whose parent material is bedrock are termed *residual soils*, while soils that develop on loose sediment are called *transported soils* (Figure 8.13). It should be pointed out that transported soils form *in place* on parent materials that have been carried from elsewhere and deposited by gravity, water, wind, or ice.

Parent material influences soils in two ways. First, the type of parent material influences the rate of weathering and thus the rate of soil formation. (Consider the weathering rates of granite and limestone.) Also, because sediments are already partly weathered and provide more surface area for chemical weathering, soil development on such material usually progresses more rapidly. Second, the chemical makeup of the parent material affects the soil's fertility. This influences the character of the natural vegetation the soil can support.

At one time parent material was thought to be the primary factor causing differences among soils. However, soil scientists came to understand that other factors, especially climate, are more important. In fact, they learned that similar soils often develop from different parent materials and that dissimilar soils can develop from the same parent material. Such discoveries reinforce the importance of the other soil-forming factors.

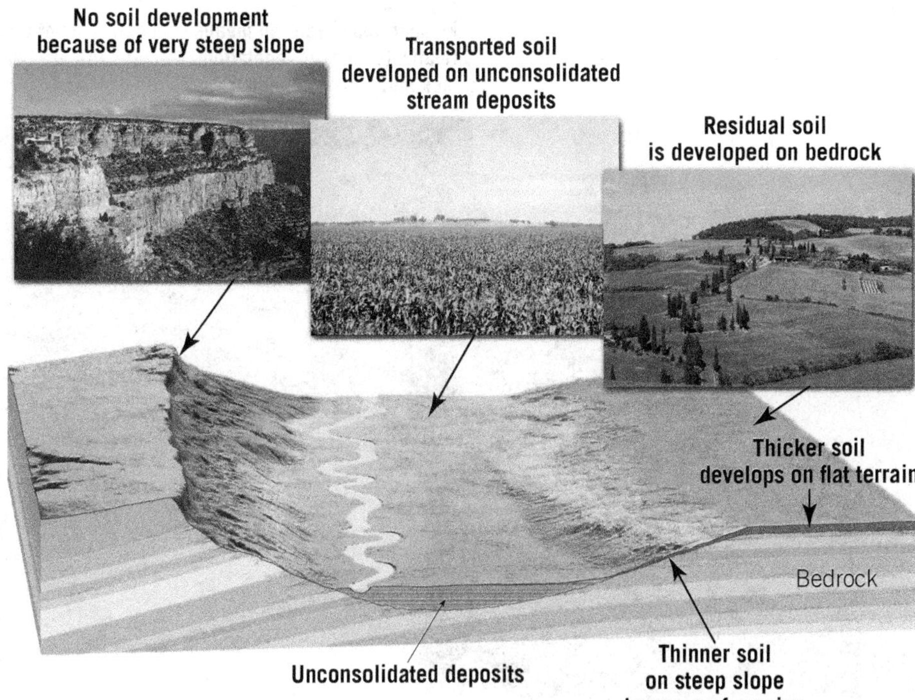
No soil development because of very steep slope
Transported soil developed on unconsolidated stream deposits
Residual soil is developed on bedrock
Thicker soil develops on flat terrain
Bedrock
Unconsolidated deposits
Thinner soil on steep slope because of erosion

Figure 8.13 Slopes and soil development The parent material for residual soils is the underlying bedrock. Transported soils form on unconsolidated deposits. Also note that as slopes become steeper, soil becomes thinner. (Left and center photos by E. J. Tarbuck; right photo by Lucarelli Temistocle/Shutterstock)

Climate

Climate is the most influential factor in soil formation. Temperature and precipitation are the climate elements that exert the strongest impact on soil formation. As noted earlier in this chapter, variations in temperature and precipitation determine whether chemical or mechanical weathering predominates and also greatly influence the rate and depth of weathering. For instance, a hot, wet climate may produce a thick layer of chemically weathered soil in the same amount of time that a cold, dry climate produces a thin mantle of mechanically weathered debris. Also, the amount of precipitation influences the degree to which various materials are removed from the soil by percolating water (a process called *leaching*), thereby affecting soil fertility. Finally, climate conditions are important factors controlling the types of plant and animal life present.

Time

Time is an important component of *every* geologic process, including soil formation. The nature of soil is strongly influenced by the length of time processes have been operating. If weathering has been going on for a comparatively short time, the parent material strongly influences the characteristics of the soil. As weathering continues, the influence of parent material on soil is overshadowed by the other soil-forming factors, especially climate. The amount of time required for various soils to evolve cannot be specified because the soil-forming processes act at varying rates under different circumstances. However, as a rule, the longer a soil has been forming, the thicker it becomes and the less it resembles the parent material.

Plants and Animals

The biosphere plays a vital role in soil formation. The types and abundance of organisms strongly influence the physical and chemical properties of a soil. In fact, for well-developed soils in many regions, the significance of natural vegetation on soil type is frequently implied in the names used by soil scientists, such as *prairie soil*, *forest soil*, and *tundra soil* (Figure 8.14).

Plants and animals furnish organic matter to the soil. Certain bog soils are composed almost entirely of organic matter, whereas desert soils may contain only a tiny percentage. Although the quantity of organic matter varies substantially among soils, almost no soils completely lack organic matter.

The primary source of organic matter is plants, although animals and microorganisms also contribute. Decomposed organic matter supplies important nutrients to plants, as well as to animals and microorganisms living in the soil. Consequently, soil fertility depends in part on the amount of organic matter present. Furthermore, the decay of plant and animal remains causes the

Meager desert rainfall means reduced rates of weathering and relatively meager vegetation. Desert soils are typically thin and lack much organic matter.
In the northern coniferous forest, the organic litter is high in acid resin, which contributes to an accumulation of acid in the soil. As a result, acid leaching is an important soil-forming process.
Soils that develop in well-drained prairie regions typically have a humus-rich surface horizon that is rich in calcium and magnesium. Fertility is usually excellent.

Figure 8.14 Plants influence soil The nature of the vegetation in an area can have a significant influence on soil formation. (Photos by Bill Brooks/Alamy Images; Nickolay Stanev/Shutterstock; and Elizabeth C. Doemer/Shutterstock)

formation of various organic acids. These complex acids hasten the weathering process. Organic matter also has a high water-holding ability and thus aids water retention in a soil.

Microorganisms play an active role in the decay of plant and animal remains. The end product is *humus*, a material that no longer resembles the plants and animals from which it formed. In addition, certain microorganisms aid soil fertility because they have the ability to convert atmospheric nitrogen gas into soil nitrogen compounds.

Earthworms and other burrowing animals mix the mineral and organic portions of a soil. Earthworms, for example, feed on organic matter and thoroughly mix soils in which they live, often moving and enriching many tons per acre each year. Burrows and holes also aid the passage of water and air through the soil.

Topography

The lay of the land can vary greatly over short distances. Such variations in topography can lead to the development of a variety of localized soil types. Many of the differences exist because the length and steepness of slopes significantly affect the amount of erosion and the water content of soil.

On steep slopes, soils are often poorly developed. Due to rapid runoff, the quantity of water soaking in is slight; as a result, the soil's moisture content may not be sufficient for vigorous plant growth. Furthermore, because of accelerated erosion on steep slopes, the soils are thin or nonexistent (see Figure 8.13).

In contrast, waterlogged soils in poorly drained bottomlands have a much different character. Such soils are usually thick and dark. The dark color results from the large quantity of organic matter that accumulates because saturated conditions retard the decay of vegetation. The optimum terrain for soil development is a flat to undulating upland surface. This terrain experiences good drainage, minimum erosion, and sufficient infiltration of water into the soil.

Slope orientation, the direction a slope is facing, also is significant. In the midlatitudes of the Northern Hemisphere, a south-facing slope receives a great deal more sunlight than does a north-facing slope. In fact, a steep north-facing slope may receive no direct sunlight at all. The difference in the amount of solar radiation received causes substantial differences in soil temperature and moisture, which in turn influences the nature of the vegetation and the character of the soil.

Although we have dealt separately with each of the soil-forming factors, remember that *all of them work together* to form soil. No single factor is responsible for a soil being as it is. Rather, the combined influence of parent material, climate, time, plants and animals, and topography determines a soil's character.

CONCEPT CHECKS 8.4

1. List the five basic controls of soil formation.
2. Which factor is most influential in soil formation?
3. How might the direction a slope is facing influence soil formation?

8.5 Describing and Classifying Soils

Describe an idealized soil profile. Explain the need for classifying soils.

The factors controlling soil formation vary greatly from place to place and from time to time, leading to an amazing variety of soil types.

The Soil Profile

Because soil-forming processes operate from the surface downward, soil composition, texture, structure, and color gradually evolve differently at varying depths. These vertical differences, which usually become more pronounced as time passes, divide the soil into zones or layers known as **soil horizons**. If you were to dig a pit in soil, you would see that its walls are layered. Such a vertical section through all of the soil horizons constitutes the **soil profile**.

Figure 8.15 presents an idealized view of a well-developed soil profile in which five horizons are identified. From the surface downward, they are designated as O, A, E, B, and C. These five horizons are common to soils in temperate regions; not all soils have these five layers. The characteristics and extent of horizon development vary in different environments. Thus, different localities exhibit soil profiles that can contrast greatly with one another:

- The *O soil horizon* consists largely of organic material, whereas the layers beneath it consist mainly of mineral matter. The upper portion of the O horizon is primarily plant litter, such as loose leaves and other recognizable organic debris. By contrast, the lower portion of the O horizon is made up of partly decomposed organic matter (humus) in which plant structures can no longer be identified. In addition to plants, the O horizon is teeming with life, including bacteria, fungi, algae, and insects. All these organisms contribute carbon dioxide and organic acids to the developing soil.
- The *A horizon* is largely mineral matter, yet biological activity is high, and humus is generally present—up to 30 percent in some instances. Together the O and A horizons make up what is commonly called the *topsoil*.
- The *E horizon* is a light-colored layer that contains little organic material. As water percolates downward through this zone, finer particles are carried away. This washing out of fine soil components is termed **eluviation**. Water percolating downward also dissolves soluble inorganic soil components and carries them to deeper zones. This depletion of soluble materials from the upper soil is termed **leaching**.
- The *B horizon*, or *subsoil*, is where much of the material removed from the E horizon by eluviation is deposited. Thus, the B horizon is often referred to as the *zone of accumulation*. The accumulation of the fine clay particles enhances water retention in this horizon. In extreme cases clay accumulation can form a very compact, impermeable layer called *hardpan*.
- The O, A, E, and B horizons together constitute the **solum**, or "true soil." It is in the solum that soil-forming processes are active and that living roots and other plant and animal life are largely confined.
- The *C horizon* is characterized by partially altered parent material. Whereas the O, A, E, and B horizons bear little resemblance to the parent material, the parent material is easily identifiable in the C horizon. Although this material is undergoing changes that will eventually transform it into soil, it has not yet crossed the threshold that separates regolith from soil.

The characteristics and extent of development can vary greatly among soils in different environments

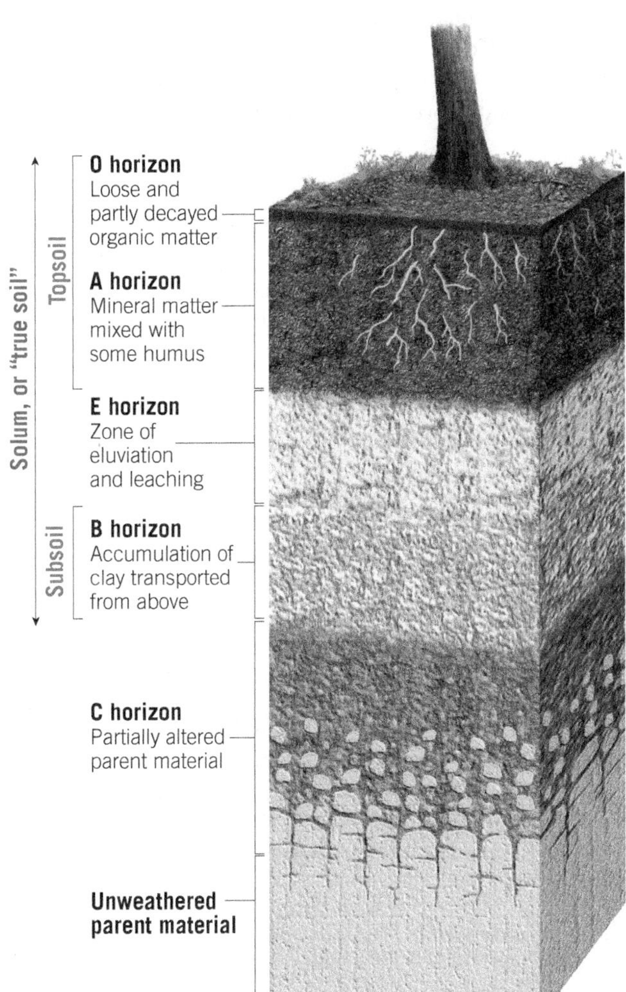

SmartFigure 8.15 **Soil horizons** Idealized soil profile from a humid climate in the middle latitudes.

TUTORIAL
https://goo.gl/ASed3b

Figure 8.16 **Contrasting soil profiles** Soil characteristics and development vary greatly in different environments. (Left photo by USDA; right photo courtesy of E. J. Tarbuck)

Horizons are indistinct in this soil in Puerto Rico, giving it a relatively uniform appearance.

This profile shows a soil in southeastern South Dakota with well-developed horizons.

(Figure 8.16). The boundaries between soil horizons may be sharp, or the horizons may blend gradually from one to another. Consequently, a well-developed soil profile indicates that environmental conditions have been relatively stable over an extended time span and that the soil is *mature*. By contrast, some soils lack horizons altogether. Such soils are called *immature* because soil building has been going on for only a short time. Immature soils are also characteristic of steep slopes, where erosion continually strips away the soil, preventing full development.

Classifying Soils

The great variety of soils on Earth made it essential to devise a means of classifying the vast array of soil data. Establishing categories of items having certain important characteristics in common introduced order and simplicity, which not only aids comprehension and understanding but also facilitates analysis and explanation.

Soil scientists in the United States have devised a system for classifying soils known as the **Soil Taxonomy**. It emphasizes the physical and chemical properties of the soil profile and is organized on the basis of observable soil characteristics. There are six hierarchical categories of classification, ranging from *order*, the broadest category, to *series*, the most specific category. The system recognizes 12 soil orders and more than 19,000 soil series.

The names of the classification units are mostly combinations of Latin or Greek descriptive terms. For example, soils of the order aridosol (from the Latin *aridus* = dry and *solum* = soil) are characteristically dry soils in arid regions. Soils in the order inceptisol

Table 8.2 Basic Soil Orders

Soil Order	Description	Percentage*
Alfisol	Moderately weathered soils formed under boreal forests or broadleaf deciduous forests, rich in iron and aluminum. Clay particles accumulate in a subsurface layer due to leaching in moist environments. Fertile, productive soils because they are neither too wet nor too dry.	9.65
Andisol	Young soils in which the parent material is volcanic ash and cinders, deposited by recent volcanic activity.	0.7
Aridosol	Soils that develop in dry places with insufficient water to remove soluble minerals; may have calcium carbonate, gypsum, or salt accumulation in subsoil; low organic content.	12.02
Entisol	Young soils with limited development and exhibiting properties of the parent material. Productivity ranges from very high for entisols forming on recent river deposits to very low for entisols forming on shifting sand or rocky slopes.	16.16
Gelisol	Young soils with little profile development found in regions with permafrost. Low temperatures and frozen conditions for much of the year slow soil-forming processes.	8.61
Histosol	Organic soils with little or no climatic implications. Found in any climate where organic debris accumulates to form a bog soil. Dark, partially decomposed organic material commonly referred to as *peat*.	1.17
Inceptisol	Weakly developed young soils showing the beginning (inception) of profile development. Most common in humid climates but found from the arctic to the tropics. Native vegetation is most often forest.	9.81
Mollisol	Dark, soft soils developed under grass vegetation, generally found in prairie areas. Humus-rich surface horizon that is rich in calcium and magnesium; excellent fertility. Also found in hardwood forests with significant earthworm activity. Climatic range is boreal or alpine to tropical. Dry seasons are normal.	6.89
Oxisol	Soils formed on old land surfaces unless parent materials were strongly weathered before they were deposited. Generally found in the tropics and subtropical regions. Rich in iron and aluminum oxides, oxisols are heavily leached and hence are poor soils for cultivation.	7.5
Spodosol	Soils found only in humid regions on sandy material. Common in northern coniferous forests and cool humid forests. Beneath the dark upper horizon of weathered organic material lies a light-colored leached horizon, the distinctive property of this soil.	2.56
Ultisol	Soils representing the products of long periods of weathering. Percolating water concentrates clay particles in the lower horizons. Restricted to humid climates in the temperate regions and the tropics, where the growing season is long. Abundant water and a long frost-free period contribute to extensive leaching and poor fertility.	8.45
Vertisol	Soils containing large amounts of clay, which shrink when dry and swell with the addition of water. Found in subhumid to arid climates if sufficient water is available to saturate the soil after periods of drought. Soil expansion and contraction exert stresses on human structures.	2.24

* Percentages refer to the world's ice-free land surface.

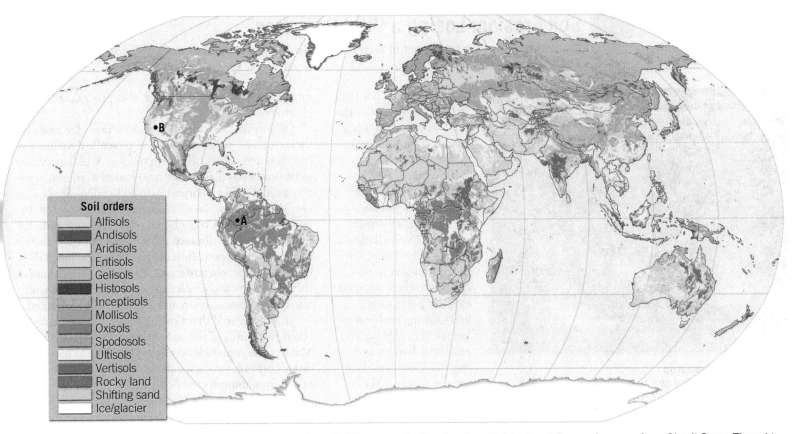

▲ **Figure 8.17 Global soil regions** Worldwide distribution of the Soil Taxonomy's 12 soil orders. Points A and B are references for a *Give It Some Thought* item at the end of the chapter. (Natural Resources Conservation Service/USDA)

(from Latin *inceptum* = beginning and *solum* = soil) are soils with only the beginning, or inception, of profile development.

Brief descriptions of the 12 basic soil orders are provided in **Table 8.2**. **Figure 8.17** shows the complex worldwide distribution pattern of the Soil Taxonomy's 12 soil orders. Like many other classification systems, the Soil Taxonomy is not suitable for every purpose. It is especially useful for agricultural and related land-use purposes, but it is not a useful system for engineers who are preparing evaluations of potential construction sites.

CONCEPT CHECKS 8.5

1. Sketch and label the main soil horizons in a well-developed soil profile.
2. Describe the following features or processes: eluviation, leaching, zone of accumulation, and hardpan.
3. Why are soils classified?
4. Refer to Figure 8.17 and identify three particularly extensive soil orders that occur in the contiguous 48 United States. Describe two soil orders in Alaska.

EYE ON EARTH 8.2

This thick red soil is exposed in one of the states in the United States and is either from the gelisol, mollisol, or oxisol order. (Photo by Sandra A. Dunlap/Shutterstock)

QUESTION 1 *Refer to the descriptions in Table 8.2 and determine the likely soil order shown in this image. Explain your choice.*

QUESTION 2 *Which one of these states is the most likely location of the soil: Alaska, Illinois, or Hawaii?*

8.6 Soil Erosion: Losing a Vital Resource

Discuss the detrimental impact of human activities on soil and some ways that soil erosion is controlled.

Raindrops may strike the surface at velocities approaching 35 km per hour. When a drop strikes an exposed surface, soil particles may splash as high as one meter and land more than a meter away from the point of raindrop impact.

▲ Figure 8.18 **Raindrop impact** Soil dislodged by raindrop impact is more easily moved by sheet erosion. (Photo courtesy U.S. Department of the Navy/Soil Conservation Service/USDA)

Many people do not realize that soil erosion—the removal of topsoil—is a serious environmental problem. Perhaps this is the case because a substantial amount of soil seems to remain even where soil erosion is serious. Nevertheless, although the loss of fertile topsoil may not be obvious to the untrained eye, it is a significant and growing problem as human activities expand and disturb more and more of Earth's surface.

Soil erosion is a natural process; it is part of the constant recycling of Earth materials that we call the *rock cycle*. Once soil forms, erosional forces, especially water and wind, move soil components from one place to another.

Erosion by Water and Wind

Raindrops strike the land with surprising force (Figure 8.18). Each drop acts like a tiny bomb, blasting movable soil particles out of their positions in the soil mass. Water flowing across the surface carries away the dislodged soil particles. Because the soil is moved by thin sheets of water, this process is termed *sheet erosion*.

After this thin, unconfined sheet flows for a relatively short distance, threads of current typically develop, and tiny channels called *rills* begin to form. Still deeper cuts in the soil, known as *gullies*, are created as rills enlarge (Figure 8.19). When normal farm cultivation cannot eliminate the channels, the rills have grown large enough to be called gullies. Although most dislodged soil particles move only a short distance during each rainfall, substantial quantities eventually leave the fields and make their way downslope to a stream. Once in the stream channel, these soil particles, which can now be called *sediment*, are transported downstream and are eventually deposited.

It is estimated that flowing water is responsible for about two-thirds of the soil erosion in the United States. Much of the remainder is caused by wind. When dry conditions prevail, strong winds can remove large quantities of soil from unprotected fields.

Rates of Erosion

Soil erosion is the ultimate fate of practically all soils. In the past, erosion occurred at slower rates than it does today because more of the land surface was covered and protected by trees, shrubs, grasses, and other plants. However, human activities such as farming, logging, and construction, which remove or disrupt the natural vegetation, have greatly accelerated the rate of soil erosion. Without the stabilizing effect of plants, the soil is more easily swept away by the wind or carried downslope by sheet wash.

Severe sheet and rill erosion on an Iowa farm following heavy rains. Just one millimeter of soil from a single acre amounts to about 5 tons.
A.

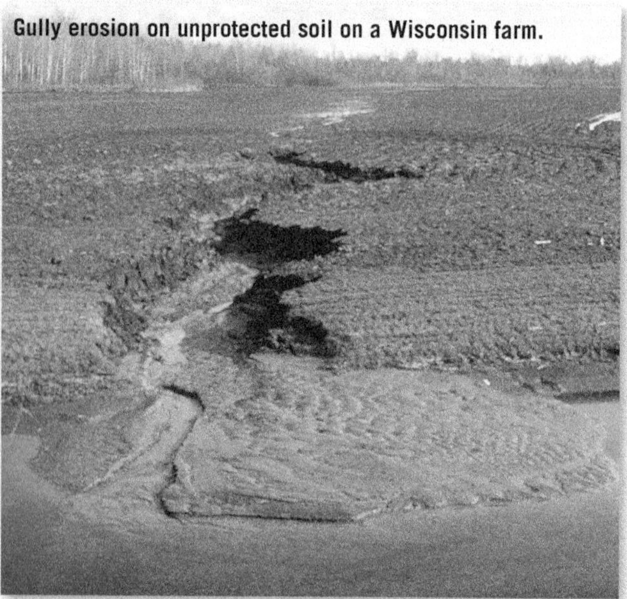

Gully erosion on unprotected soil on a Wisconsin farm.
B.

► Figure 8.19 **Soil erosion A.** Sheetflow and rills. **B.** Rills can grow into deep gullies. (Photo A by Lynn Betts/NRCS; photo B by D. P. Burnside/Science Source)

Natural rates of soil erosion vary greatly from one place to another and depend on soil characteristics as well as factors such as climate, slope, and type of vegetation. Over a broad area, erosion caused by surface runoff may be estimated by determining how much sediment is carried by the streams that drain the region. Studies of this kind made on a global scale indicate that prior to the appearance of humans, sediment transport by rivers to the ocean amounted to just over 9 billion metric tons per year. In contrast, the amount of material currently transported to the sea by rivers is about 24 billion metric tons per year, or more than two and a half times the earlier rate.

▲ **Figure 8.20 Soil conservation** Crops on a farm in northeastern Iowa are planted to decrease water erosion. (Photo courtesy of Erwin C. Cole/USDA/NRCS)

Controlling Soil Erosion

At present, it is estimated that topsoil is eroding faster than it forms on more than one-third of the world's croplands. The results are lower productivity, poorer crop quality, reduced agricultural income, and an ominous future. On every continent, unnecessary soil loss is occurring because appropriate conservation measures are not being taken. Although it is a recognized fact that soil erosion can never be completely eliminated, soil conservation programs can substantially reduce the loss of this basic resource.

Steepness of slope is an important factor in soil erosion. The steeper the slope, the faster water runs off and the greater the erosion. It is best to leave steep slopes undisturbed, but when such slopes are farmed, terraces can be constructed. These nearly flat, steplike surfaces slow runoff and thus decrease soil loss while allowing more water to soak into the ground.

Soil erosion by water also occurs on gentle slopes. Figure 8.20 illustrates one conservation method of planting crops parallel to the contours of the slope. This pattern reduces soil loss by slowing runoff. Strips of grass or cover crops such as hay slow runoff even more, promoting water infiltration and trapping sediment.

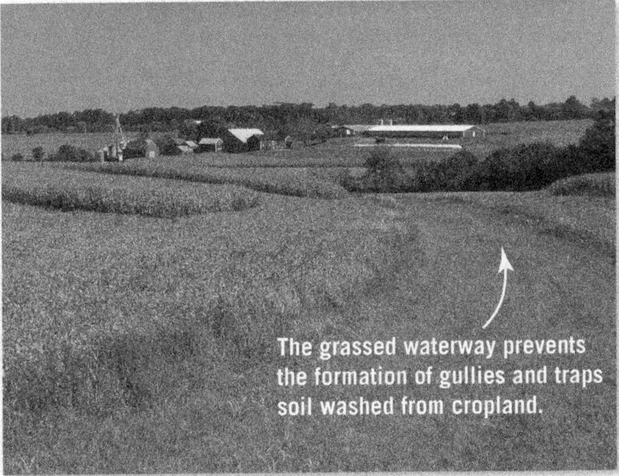

▲ **Figure 8.21 Reducing erosion by water** Grassed waterway on a Pennsylvania farm. (Photo courtesy Bob Nichols/USDA/NRCS)

Creating grassed waterways is another common practice that helps reduce soil erosion (Figure 8.21). Natural drainageways are shaped to form smooth, shallow channels and then planted with grass. The grass prevents the formation of gullies and traps soil washed from cropland. Frequently, crop residues are also left on fields. This debris protects the surface from both water and wind erosion. To protect fields from excessive wind erosion, rows of trees and shrubs are planted to act as windbreaks to slow the wind and deflect it upward (Figure 8.22).

These flat expanses are susceptible to wind erosion, especially when the fields are bare. The rows of trees slow and deflect the wind, which decreases the loss of top soil.

◄ **Figure 8.22 Reducing wind erosion** Windbreaks protecting wheat fields in North Dakota. (Photo courtesy Erwin C. Cole/USDA/NRCS)

CONCEPT CHECKS 8.6

1. Link these phenomena in a description of soil erosion: sheet erosion, gullies, rain drop impact, rills, stream.
2. How have human activities affected the rate of soil erosion?
3. Briefly describe three methods of controlling soil erosion.

8.7 Mass Movement on Slopes: The Work of Gravity

Discuss the role that mass movements play in the development of landscapes. Summarize the factors that control and trigger mass movement processes.

Earth's surface is seldom perfectly flat, instead consisting of slopes of many different varieties. Some are steep and cliff-like; others are moderate or gentle. Slopes can consist of barren rock and rubble or can be mantled with soil and covered by vegetation. Although most slopes appear to be stable and unchanging, they are not static features because the force of gravity causes rock and soil to move downslope by the processes we call *mass movement*. At one extreme, the movement may consist of a roaring debris flow or a thundering rock avalanche. At the other extreme, it may be gradual and practically imperceptible. Landslides are a worldwide natural hazard. When these hazardous processes lead to loss of life and property, they become natural disasters.

Landslides as Geologic Hazards

The chapter-opening photo is a spectacular example of a phenomenon that many people would call a landslide. For most of us, the word *landslide* implies a sudden event in which large quantities of rock and soil plunge down steep slopes. When people and communities are in the way, a natural disaster may result. Landslides constitute major geologic hazards that each year in the United States cause billions of dollars in damages and the loss of 25 to 50 lives. As you will see, many landslides occur in connection with other major natural disasters, including earthquakes, volcanic eruptions, wildfires, and severe storms.

When you view the images in GEOgraphics 8.3, you will likely notice that the term *landslide* seems to refer to several different things—from mudflows to rock avalanches. This variety reflects the fact that although many people, including geologists, frequently use the word *landslide*, the term has no specific definition in geology. Rather, it is a popular, nontechnical word used to describe any or all relatively rapid forms of mass movements.

Landslides are spectacular examples of a basic geologic process called **mass movement**—the downslope movement of rock, regolith, and soil under the direct influence of gravity. Mass movements are distinct from the erosional processes that are examined in subsequent chapters because they do not require a transporting medium such as water, wind, or glacial ice.

The Role of Mass Movement in Landscape Development

In the evolution of most landscapes, mass movement is the step that follows weathering. Once weathering weakens and breaks rock apart, mass movement transfers the debris downslope, where a stream or glacier, acting as a conveyor belt, usually carries it away. Although there may be many intermediate stops along the way, the sediment is eventually transported to its ultimate destination: the sea.

The combined effects of mass movement and running water produce stream valleys, which are the most common and conspicuous of Earth's landforms. If streams alone were responsible for creating the valleys in which they flow, the valleys would be very narrow features. However, the fact that most stream valleys are much wider than they are deep is a strong indication of the significance of mass movement processes in supplying material to streams. The walls of a canyon extend far from the stream because of the transfer of weathered debris downslope to the stream and its tributaries by mass movement. In this manner, streams and mass movements combine to modify and sculpt the surface. Of course, glaciers, groundwater, waves, and wind are also important agents in shaping landforms and developing landscapes.

Slopes Change Through Time

It is clear that if mass movement is to occur, there must be slopes that rock, soil, and regolith can move down. Earth's mountain-building and volcanic processes produce these slopes through sporadic changes in the elevations of landmasses and the ocean floor. If dynamic internal processes did not continually produce regions having higher elevations, the system that moves debris to lower elevations would gradually slow and eventually cease.

Most rapid and spectacular mass movement events occur in areas of rugged, geologically young mountains. Newly formed mountains are rapidly eroded by rivers and glaciers into regions characterized by steep and unstable slopes. It is in such settings that massive destructive landslides occur. As mountain building subsides, mass movement and erosional processes lower the land. Through time, steep and rugged mountain slopes give way to gentler, more subdued terrain. Thus, as a landscape ages, massive and rapid mass movement processes give way to smaller, less dramatic downslope movements that are often imperceptibly slow.

Controls and Triggers of Mass Movement

Whether a slope fails or not depends on the balance between the force pulling material downslope (gravity) and the resisting strength of the rock or soil. While gravity is the driving force of mass movement, several factors play important roles in overcoming friction and material

Landslides as Natural Disasters

lslides don't just occur in remote mountains and yons. People frequently live where rapid mass ement events occur. Here are a few examples.

Even in areas with steep slopes, catastrophic landslides are relatively rare; thus, people living in susceptible areas often do not appreciate the risks of living where they do. This deadly event, **triggered by an earthquake in January 2001, buried 300 homes in Santa Tecla, El Salvador**. Many unsuspecting residents perished.

On **March 1, 2012, this boulder broke loose** from a steep mountain slope in the **French Alps**, crushing a car and then crashing into this house. There was no warning and no obvious trigger for this event.

Question:
List two mass-wasting triggers represented in these images.

?

Weathered rock, precipitous slopes, and the shock of an earthquake combined to produce many **rockfalls and rock avalanches in and around Christchurch, New Zealand on February 22, 2011**.

In January 2011, following a period of torrential rains, this thick slurry of mud, appropriately called a mudflow, buried these cars in **Nova Friburgo, Brazil**. Heavy rains are an important trigger of mass wasting processes.

BUMP

▲ SmartFigure 8.23 **Debris flow in the Colorado Front Range** During the week of September 9–13, 2013, residents in and near Boulder, Colorado, received a harsh reminder of the dangers posed by debris flows. During that 5-day span, nearly continuous rainfall triggered numerous flash floods and more than 1100 debris flows in an area covering more than 3400 square kilometers (1300 square miles). (Photo by Rick Wilking/Reuters)

MOBILE FIELD TRIP
https://goo.gl/kjQUFs

strength to create downslope movements. Long before a landslide occurs, various processes work to reduce the slope's stability, making it gradually more susceptible to the pull of gravity. Eventually, the slope is either weakened or steepened to the point that it crosses the threshold from stabile to unstable. A final event that initiates downslope movement is called a **trigger**. Remember that a trigger is not the sole cause of a mass movement event but just the last of many causes. Among the common factors that trigger mass movement processes are saturation of material with water, oversteepening of slopes, removal of anchoring vegetation, and ground vibrations from earthquakes.

The Role of Water Mass movement is sometimes triggered when heavy rains or melting snow saturate surface materials. The water does not transport the material. Rather, it allows gravity to more easily set the material in motion. This was the case in March 2014, when a massive debris flow (popularly called a *mudslide* in the media) occurred on steep mountain slopes east of Oso, Washington (see the chapter-opening photo and Figure 1.17, page 19). Similarly, in September 2013, heavy rains on steep slopes over a 5-day span triggered more than 1135 debris flows over an area of 3430 square kilometers (1338 square miles) in and near Boulder, Colorado (Figure 8.23).

When the pores in sediment become filled with water, the cohesion among particles is destroyed, and they can slide past one another with relative ease. For example, when sand is slightly moist, it sticks together quite well. However, if enough water is added to fill the openings between the grains, the sand will ooze out in all directions (Figure 8.24). Thus, saturation reduces the internal resistance of materials, which are then easily set in motion by the force of gravity. When clay is wetted, it becomes very slick—another example of the "lubricating" effect of water. Water also adds considerable weight to a mass of material. The added weight in itself may be enough to trigger downslope movement.

▼ Figure 8.24 **Saturation reduces friction** When water saturates sediment, friction among particles is reduced, allowing material to move downslope.

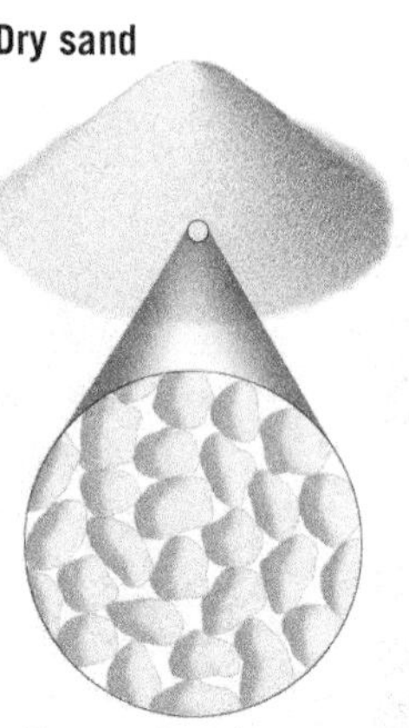
Dry sand
Dry sand grains are bound mainly by friction with one another

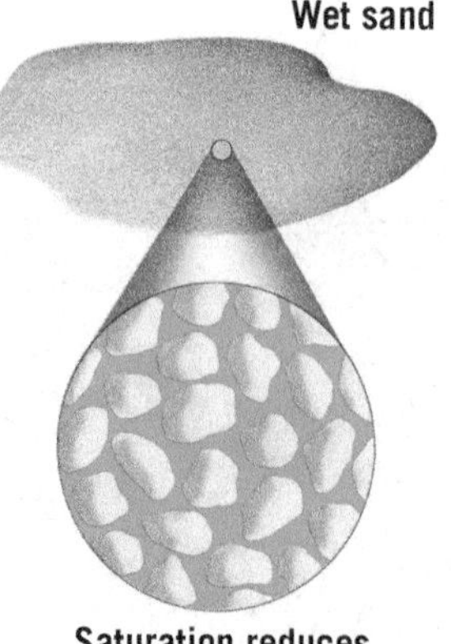
Damp sand
Small amounts of water increase the cohesion among sand grains

Wet sand

Saturation reduces friction and causes the sand to flow

Oversteepened Slopes Oversteepening of slopes is another common trigger of mass movements. As a slope becomes steeper, the forces that resist the downward pull of gravity get weaker. Many situations result in oversteepening. For example, as a stream cuts into a valley wall, it removes material from the base of the wall. Human activities also often create oversteepened slopes that become prime sites for mass movement (Figure 8.25).

Oversteepening has a particularly clear-cut effect on slopes made of loose, granular particles (sand size or coarser). Such materials are stable up to a specific angle called the **angle of repose** (*reposen* = to be at rest) (Figure 8.26). Depending on the size and shape of the particles, the angle of repose varies from 25 to 40 degrees, with larger, more angular particles maintaining steeper slopes. If a slope of loose debris is steepened beyond the angle of repose, the material will adjust by moving downslope until the angle of repose is reestablished.

Cohesive materials such as soil, consolidated regolith, and bedrock do not exhibit a clear-cut angle of repose, so their response to oversteepening is more difficult to predict. Nevertheless, oversteepened slopes of such materials will eventually undergo one or more mass movement processes that will eliminate the oversteepening and restore stability to the slope.

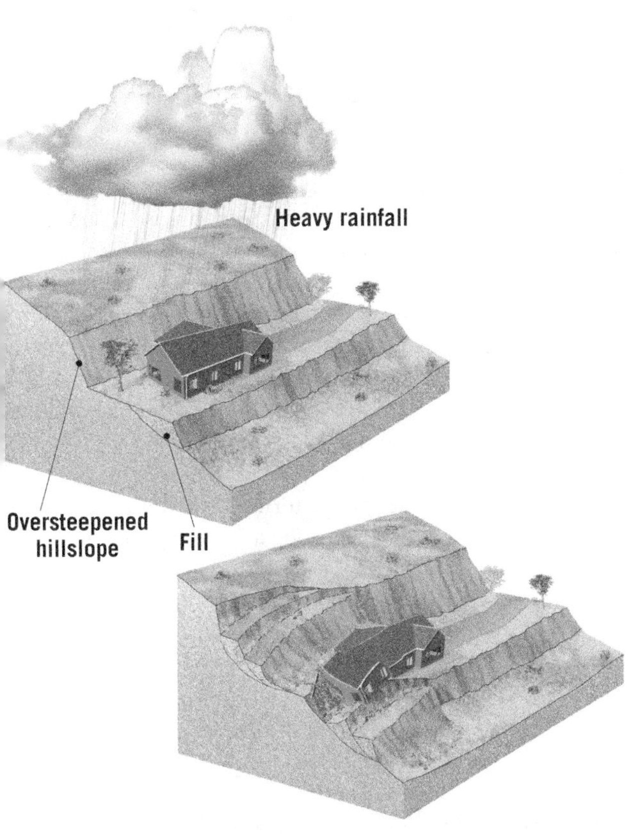

Figure 8.25 **Unstable slopes** Natural processes such as stream and wave erosion can oversteepen slopes. Changing the slope to accommodate a new house or road can also lead to instability and a destructive mass movement event.

Removal of Vegetation Plants protect against erosion and contribute to the stability of slopes because their root systems bind soil and regolith together. In addition, plants shield the soil surface from the erosional effects of raindrop impact (see Figure 8.18). Where plants are lacking, mass movement is enhanced, especially if slopes are steep and water is plentiful. When anchoring vegetation is removed by forest fires or by people (for timber, farming, or development), surface materials frequently move downslope.

Wildfires are inevitable in the western United States, and fast-moving, highly destructive debris flows triggered by intense rainfall are some of the most dangerous postfire hazards. Such events are particularly dangerous because they tend to occur with little warning. Their mass and speed make them particularly destructive. Postfire debris flows are most common in the 2 years after a fire.

Earthquakes as Triggers Conditions favoring mass movement can exist in an area for a long time without movement occurring. An additional factor is sometimes necessary to trigger the movement. Among the most important and dramatic triggers are earthquakes. An earthquake and its aftershocks can dislodge enormous volumes of rock and unconsolidated material. The mass movement event shown in Figure 8.27 was triggered by an earthquake. In many areas that are jolted by earthquakes, it is not ground vibrations directly but landslides and ground subsidence triggered by the vibrations that cause the greatest damage.

Figure 8.26 **Angle of repose** The angle of repose is the steepest angle at which an accumulation of granular particles remains stable. Larger, more angular particles maintain the steepest slopes. (Photo by G. Leavens/Science Source)

CONCEPT CHECKS 8.7

1. What force is responsible for mass movement?
2. Describe the role of mass movement in landscape development.
3. How does water affect mass movement processes?
4. Describe the significance of the angle of repose.
5. How can forest fires and earthquakes influence mass movements?

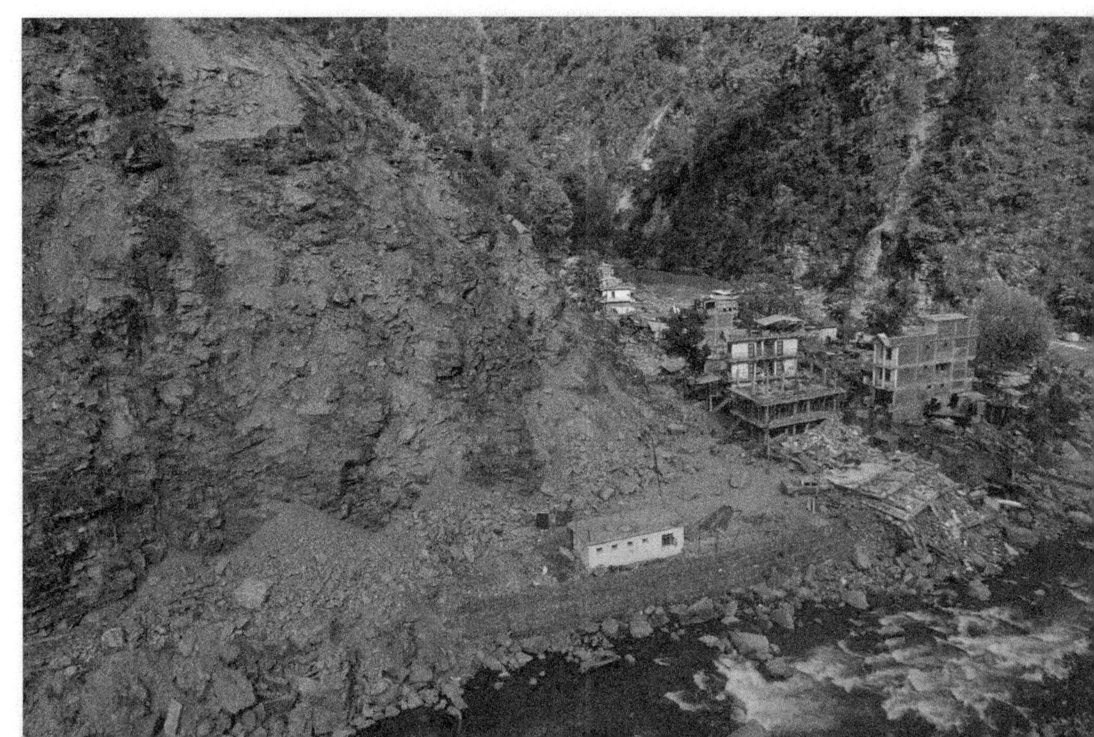

Figure 8.27 **Earthquakes as triggers** A major earthquake in Nepal in April 2015 triggered hundreds of landslides, which destroyed roads and blocked rivers. The damage shown here occurred in the village of Singati in northeastern Nepal. (Photo by Prakash Mathema/AFP/Getty Images)

8.8 Types of Mass Movement

List and explain the criteria that are commonly used to classify mass movement processes. Distinguish among six different types of mass movement.

Geologists use the term *mass movement* to describe a broad array of processes, four of which are illustrated in Figure 8.28.

Classifying Mass Movements

Generally, each type of mass movement is defined by the type of material involved, the kind of motion, and the velocity of the movement.

Type of Material Mass movement processes can be classified based on the material involved. If soil and regolith dominate, terms such as *debris*, *mud*, or *earth* are used in the description. In contrast, when a mass of bedrock breaks loose and moves downslope, the term *rock* may be part of the description.

Type of Motion When mass movement involves the free falling of detached individual pieces of any size, it is termed a **fall**. Falls are common on slopes that are too steep for loose material to remain on the surface. Many falls result when freeze–thaw cycles and/or the action of plant roots loosen rock and then gravity takes over. Rockfall is the primary way in which talus slopes are built and maintained (see Figure 8.5). Sometimes falls may trigger other forms of downslope movement.

Many mass movement processes are **slides**, which occur whenever material remains fairly coherent and moves along a well-defined surface. Sometimes the surface is a joint, a fault, or a bedding plane that is roughly parallel to the slope. However, in the movement called *slump*, the descending material slides en masse along a curved surface of rupture.

A third type of motion is termed **flow**. Flow occurs when material moves downslope as a viscous, often turbulent, fluid. Most flows are saturated with water and typically move as lobes or tongues.

Rate of Movement When mass movement events make the news, a large quantity of material has in all likelihood moved rapidly downslope and has had a disastrous effect on people and property. Indeed, during events called **rock avalanches**, rock and debris can hurtle downslope at speeds exceeding 200 kilometers (125 miles) per hour and may sometimes continue to flow for long distances even after the slope flattens out.

A. Slump: Downward sliding of a mass of rock or unconsolidated material moving as a unit along a curved surface.

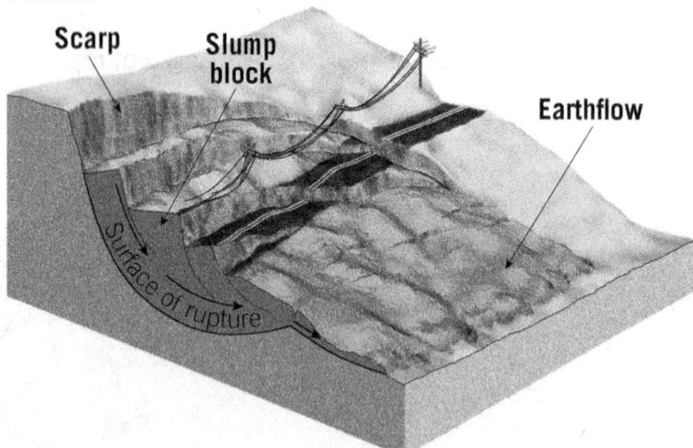

B. Rockslide: Blocks of bedrock break loose and slide very rapidly downslope.

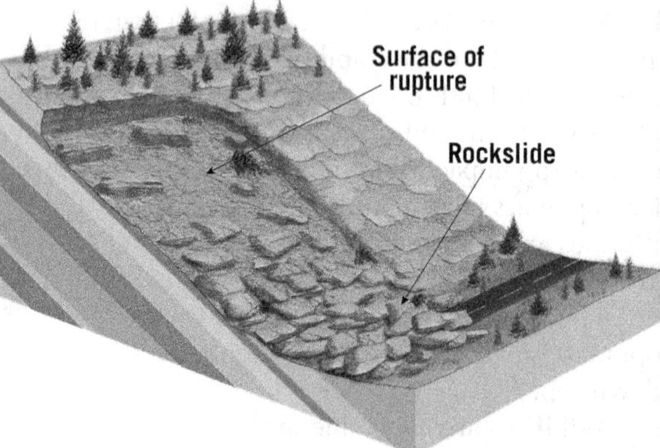

C. Debris flow: A flow of weathered debris containing a large amount of water. Often confined to channels. Sometimes called a mudflow.

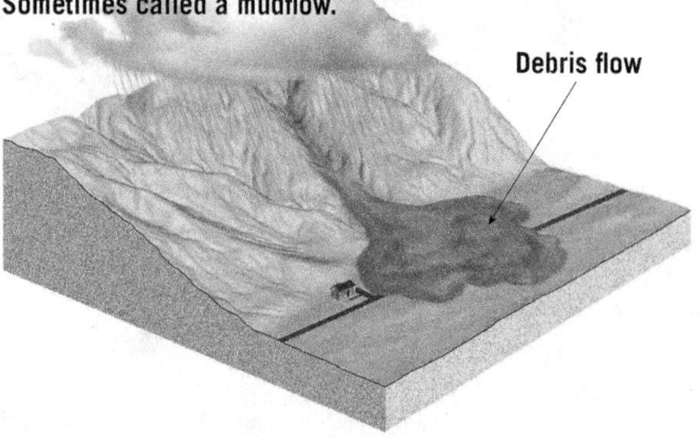

D. Earthflow: A tongue-like flow of water-saturated clay-rich soil on a hillside that breaks away and moves downslope.

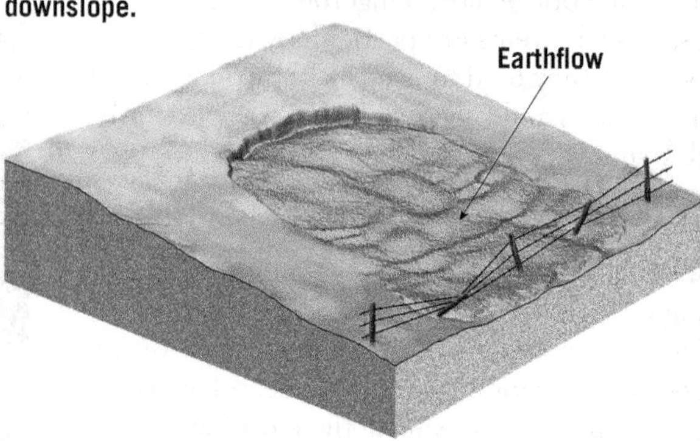

SmartFigure 8.28 **Types of mass movement** The four processes illustrated here are all considered to be relatively rapid forms of mass wasting. Because the material in slumps and rockslides moves along well-defined surfaces, it is said to move by sliding. By contrast, when material moves like a thick fluid, it is described as a flow. Debris flow and earthflow are examples.

ANIMATION
https://goo.gl/7Q2Gmo

EYE ON EARTH 8.3

This photo shows a steep cliff along the shore of Lake Louise in Canada's Banff National Park. Notice the accumulation of angular rock fragments at the base of the sheer rock wall.
(Photo by Marli Miller)

QUESTION 1 *What term is applied to a pile of rock debris such as this?*

QUESTION 2 *Describe the probable combination of weathering and mass movement processes that produced this feature.*

QUESTION 3 *Did the feature probably form rapidly in hours or days or gradually over many years? Explain.*

Geologists used to think that rock avalanches, such as the one that produced the scene in Figure 8.29, must literally "float on air" as they move downslope. That is, their flow was thought to depend on a layer of air that became trapped and compressed beneath the falling mass of debris, allowing the avalanche to move as a buoyant, flexible sheet across the surface. But more recent studies of landslide deposits on other planetary bodies have caused us to question whether this hypothesis explains most or all high-speed, long-runout rock avalanches. Similar deposits from long-runout landslides on Mars cannot have been supported by Mars's tenuous atmosphere and seem instead to have been lubricated by interactions of rock and water.

Most mass movements, however, do not move with the speed of a rock avalanche. In fact, a great deal of mass movement is imperceptibly slow. One process that we will examine later, termed *creep*, results in particle movements that are usually measured in millimeters or centimeters per year. Thus, as you can see, rates of movement can be spectacularly sudden or exceptionally gradual. Although various types of mass movement are often classified as either rapid or slow, such a distinction is highly subjective because a wide range of rates exists between the two extremes. Even the velocity of a single process at a particular site can vary considerably.

Rapid Forms of Mass Movement

The common rapid mass movement processes discussed in this section include slump, rockslide, debris flow, and earthflow. They are most common where slopes are steep. The speed of movement varies from barely perceptible to very rapid.

Slump The downward sliding of a mass of rock or unconsolidated material moving as a unit along a *curved* surface is called **slump** (see Figure 8.28A). Usually the slumped material does not travel spectacularly fast or very far. This is a common form of mass movement, especially in thick accumulations of cohesive materials such as clay. As the movement occurs, a crescent-shaped scarp (cliff) is created at the head, and the block's upper surface is sometimes tilted backward.

Slumping commonly occurs because a slope has been oversteepened. The material on the upper portion of a

Figure 8.29 Blackhawk Rock avalanche This prehistoric event is one of the largest known landslides in North America. (Photo by Michael Collier)

slope is held in place by the material at the bottom of the slope. As this anchoring material is removed, the material above becomes unstable and reacts to the pull of gravity. A common example is a valley wall that becomes oversteepened by the erosional work of a meandering river. Another is a coastal cliff that has been undercut by wave activity at its base.

Rockslide **Rockslides** occur when blocks of bedrock break loose and slide down a slope (see Figure 8.28B). If the material is mostly soil and regolith, the term *debris slide* is used instead. Such events are among the fastest and most destructive mass movements. Usually rockslides take place in a geologic setting where the rock strata are inclined or where joints and fractures exist parallel to the slope. When such a rock unit is undercut at the base of the slope, it loses support, and the rock eventually gives way.

Sometimes an earthquake triggers a rockslide. On other occasions, a rockslide is triggered when rain or melting snow lubricates the underlying surface to the point that friction is no longer sufficient to hold the rock unit in place. As a result, rockslides tend to be most common during spring, when heavy rains and melting snow are most prevalent. The massive Gros Ventre slide shown in Figure 8.30 is a classic example.

SmartFigure 8.30 Gros Ventre rockslide This massive slide occurred on June 23, 1925, just east of the small town of Kelly, Wyoming. (Photo by Michael Collier)

TUTORIAL
https://goo.gl/ek79Ry

The side of the mountain gave way when the tilted sandstone bed, which had been cut through by the river, could no longer maintain its position atop the saturated bed of clay.

Gros Ventre landslide debris

Former land surface

Scar

Lake

Rupture surface

Clay bed

0 0.5 1 Kilometer

Sandstone

Limestone

Even though the Gros Ventre rockslide occurred in 1925, the scar left on the side of Sheep Mountain is still a prominent feature.

Debris Flow A relatively rapid type of mass movement that involves a flow of soil and regolith containing a large amount of water is a **debris flow** (see Figure 8.23 and Figure 8.28C). Debris flows are sometimes called **mudflows** when the material is primarily fine grained. Although they can occur in many different climate settings, debris flows tend to occur most frequently in semiarid mountainous regions.

When a cloudburst creates a sudden flood in a semiarid region, large quantities of soil and regolith are washed into nearby stream channels because there is usually little vegetation to anchor the surface material. The end product is a flowing tongue of well-mixed mud, soil, rock, and water. Its consistency may range from that of wet concrete to a soupy mixture not much thicker than muddy water. The rate of flow therefore depends not only on the slope but also on the water content. When dense, debris flows are capable of carrying or pushing large boulders, trees, and even houses with relative ease. They pose a serious hazard in dry mountainous areas such as southern California, where the construction of homes on canyon hillsides and the removal of anchoring vegetation by brush fires and other means have increased the frequency of these destructive events.

Debris flows called **lahars** are common on the steep slopes of some volcanoes. Historically, lahars have been some of the deadliest volcano hazards. They can occur either during an eruption or when a volcano is quiet. They take place when highly unstable layers of volcanic ash and debris become saturated with water and flow down the steep slopes of a volcano, generally following existing stream channels. Heavy rains often trigger these flows. Other lahars are triggered when large volumes of ice and snow are suddenly melted by heat flowing to the surface from within the volcano or by the hot gases and near-molten debris emitted during a violent eruption.

Earthflow You have seen that debris flows are frequently confined to channels in semiarid regions. In contrast, **earthflows** most often form on hillsides in humid areas during times of heavy precipitation or snowmelt (see Figure 8.28D). When water saturates the soil and regolith on a hillside, the material may break away, leaving a scar on the slope and forming a tongue- or teardrop-shaped mass that flows downslope (Figure 8.31). The materials most commonly involved are rich in clay and silt and contain only small proportions of sand and coarser particles. Earthflows range in size from bodies a few meters long, a few meters wide, and less than 1 meter deep to masses more than 1 kilometer long, several hundred meters wide, and more than 10 meters deep.

Because earthflows are quite viscous, they generally move at slower rates than the more fluid debris flows described in the preceding section. They may move slowly and persistently for periods ranging from days to years. Depending on slope steepness and the material's consistency, velocities range from less than 1 millimeter

Figure 8.31 **Earthflow** This small tongue-shaped earthflow occurred on a newly formed slope along a recently constructed highway in central Illinois. It formed in clay-rich material following a period of heavy rain. Notice the small slump at the head of the earthflow. (Photo by E. J. Tarbuck)

per day up to several meters per day. Movement is typically faster during wet periods. In addition to occurring as isolated hillside phenomena, earthflows commonly take place in association with large slumps. In this situation, they may be seen as tonguelike flows at the base of the slump.

Slow Forms of Mass Movement

Movements such as rockslides, rock avalanches, and debris flows are certainly the most spectacular and catastrophic forms of mass movement. These dangerous events deserve intensive study to enable more effective prediction, timely warnings, and better controls to save lives. However, because of their large size and spectacular nature, they give us a false impression of their importance. Indeed, sudden movements transport less material than does the slow, subtle action of creep. Whereas rapid types of mass movement are characteristic of mountains and steep hillsides, creep takes place on both steep and gentle slopes and is thus much more widespread.

Creep The type of mass movement that involves the gradual downhill movement of soil and regolith is called **creep**. One factor that contributes to creep is the alternate expansion and contraction of surface material caused by freezing and thawing or wetting and drying. As shown in Figure 8.32, freezing or wetting lifts particles at right angles to the slope, and thawing or drying allows the particles to fall back to a slightly lower level. Each cycle therefore moves the material a short distance downhill.

Creep is aided by anything that disturbs the soil. For example, raindrop impact and disturbance by plant roots and burrowing animals may contribute. Creep is also promoted when the ground becomes saturated with water. Following a heavy rain or snowmelt, a waterlogged soil may lose its internal cohesion, allowing gravity to pull the material downslope. Because creep is imperceptibly slow, the process cannot be observed in action. However, the effects of creep can be observed. Creep causes fences and utility poles to tilt and retaining walls to be displaced.

Solifluction When soil is saturated with water, the soggy mass may flow downslope at a rate of a few millimeters or a few centimeters per day or per year. Such a process is called **solifluction** (which means "soil flow"). It is a type of mass movement that is common wherever water cannot escape from the saturated surface layer by infiltrating to deeper levels. A dense clay hardpan in soil or an impermeable bedrock layer can promote solifluction.

Solifluction is also common in regions underlain by **permafrost**. Permafrost refers to the permanently frozen ground that occurs in Earth's harsh tundra and ice-cap climates. Solifluction occurs in a zone above the permafrost called the *active layer*, which thaws in summer and refreezes in winter. During summer, water is unable to percolate into the impervious permafrost layer

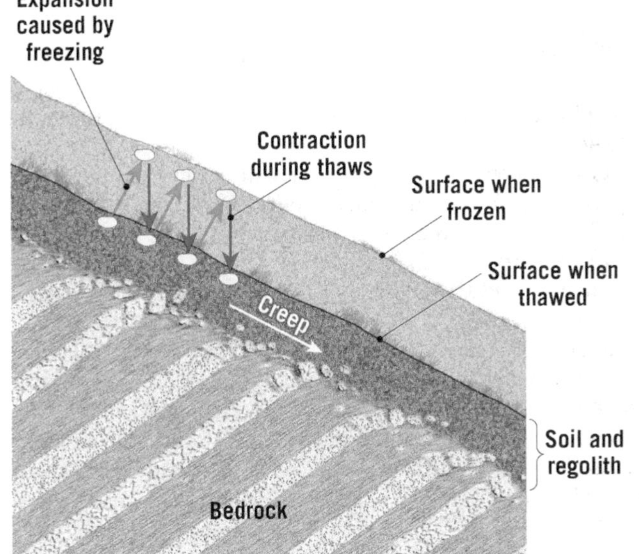

SmartFigure 8.32 **Creep** The repeated expansion and contraction of the surface material causes a net downslope migration of soil and rock particles. (The amount of expansion shown here is exaggerated.)

TUTORIAL
https://goo.gl/DVmxZQ

Figure 8.33 Solifluction lobes near the Arctic Circle in Alaska Solifluction occurs in permafrost regions when the active layer thaws in summer. (Photo by James E. Patterson Collection courtesy of F. K. Lutgens)

below. As a result, the active layer becomes saturated and slowly flows. The process can occur on slopes as gentle as 2 to 3 degrees. Where there is a well-developed mat of vegetation, a solifluction sheet may move in a series of well-defined lobes or overriding folds (Figure 8.33).

CONCEPT CHECKS 8.8

1. What terms are used to classify mass movements by the type of motion they exhibit?
2. Both slumps and rockslides move by sliding. How do these processes differ from one another?
3. What factors contributed to the massive rockslide at Gros Ventre, Wyoming?
4. How is a lahar different from a debris flow that might occur in southern California?
5. Contrast earthflows and debris flows.
6. Describe the basic mechanisms that contribute to creep.

8 CONCEPTS IN REVIEW
Weathering, Soil, and Mass Movement

8.1 Earth's External Processes

List three types of external processes and discuss the role each plays in the rock cycle.

KEY TERMS: external process, internal process, weathering, mass movement, erosion

- Weathering, mass movement, and erosion are responsible for creating, transporting, and depositing sediment. They are called external processes because they occur at or near Earth's surface and are powered by gravity and by energy from the Sun.
- Internal processes lead to volcanic activity and mountain building and derive their energy from Earth's interior.

8.2 Weathering

Define *weathering* and distinguish between the two main categories of weathering. Summarize the factors that influence the type and rate of rock weathering.

KEY TERMS: mechanical weathering, chemical weathering, frost wedging, sheeting, exfoliation dome, carbonic acid, spheroidal weathering, differential weathering

- Mechanical weathering is the physical breaking up of rock into smaller pieces. Rocks can be broken into smaller fragments by frost wedging, salt crystal growth, unloading, and biological activity.
- Chemical weathering alters a rock's chemistry, changing it into different substances.
- Water is by far the most important agent of chemical weathering. Oxygen in water can oxidize some materials, while carbon dioxide (CO_2) dissolved in water forms carbonic acid. The chemical weathering of silicate minerals produces soluble products containing sodium, calcium, potassium, and magnesium, as well as insoluble iron oxides and clay minerals.
- Mechanical weathering aids chemical weathering; the smaller the particles into which rock is broken, the faster the rock will weather.
- The mineral composition of a rock affects the rate of weathering. For example, calcite readily dissolves in mildly acidic solutions. Among the silicate minerals, those that crystallize earlier as magma cools are less resistant to chemical weathering than those that crystallize later.
- Masses of rock do not weather uniformly. Differential weathering refers to the variation in the rate and degree of weathering caused by factors such as rock type, climate, and degree of jointing. Rock weathers most rapidly in an environment with lots of heat to drive reactions and water to facilitate those reactions.

Michael Collier

? How are the two main categories of weathering represented in this photo of human-made objects?

8.3 Soil: An Indispensable Resource

Define *soil* and explain why soil is referred to as an interface.

KEY TERMS: regolith, soil, soil texture

- Soils are vital combinations of organic and nonorganic components found at the interface where the geosphere, atmosphere, hydrosphere, and biosphere meet. This dynamic zone includes the regolith's rocky debris, mixed with humus, water, and air.
- Soil texture refers to the proportions of different particle sizes (clay, silt, and sand) found in soil.

? Label the four components of a soil on this pie chart.

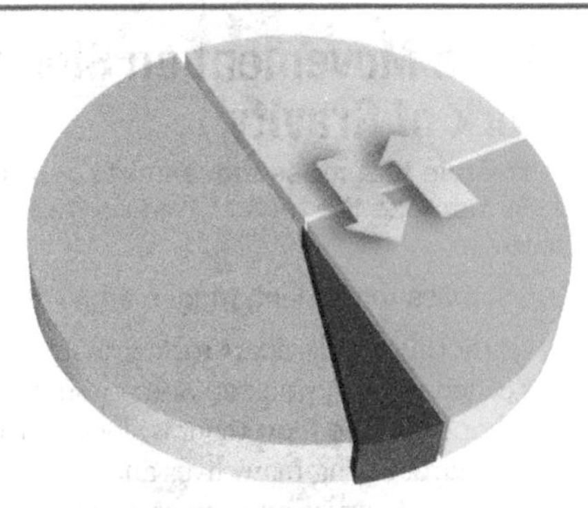

8.4 Controls of Soil Formation

List and briefly discuss five controls of soil formation.

KEY TERMS: parent material

- Residual soils form in place due to the weathering of bedrock, whereas transported soils develop on unconsolidated sediment. Soils form slowly and change in character as they mature.
- The type of parent material influences the rate of weathering and the soil's fertility but only weakly affects the type of mature soil that develops. Climate (particularly temperature and precipitation) exerts the strongest control over soil type.
- The types and abundance of organisms in a locality strongly influence the physical and chemical properties of the soil, particularly by furnishing organic matter and breaking it down to yield nutrients, humus, and organic acids, as well as by mixing and aerating soil.
- The steepness of the slope on which a soil is forming is a key variable, with shallow slopes retaining their soils and steeper slopes shedding them to accumulate elsewhere.

8.5 Describing and Classifying Soils

Describe an idealized soil profile. Explain the need for classifying soils.

KEY TERMS: soil horizon, soil profile, eluviation, leaching, solum, Soil Taxonomy

- Despite the great diversity of soils around the world, there are some broad patterns to the vertical anatomy of soil layers. Organic material, called humus, is added at the top (O horizon), mainly from plant sources. There, it mixes with mineral matter (A horizon). At the bottom, bedrock breaks down and contributes mineral matter (C horizon). In between, some materials are leached out (eluviated) from higher levels (E horizon) and transported to lower levels (B horizon), where they may form an impermeable layer called hardpan.
- The need to bring order to huge quantities of data motivated the establishment of a classification scheme for the world's soils. This Soil Taxonomy features 12 broad orders.

? Which soil order would likely contain a higher proportion of humus: inceptisols or histosols? Which soil order would be more likely found in Brazil: gelisols or oxisols?

8.6 Soil Erosion: Losing a Vital Resource

Discuss the detrimental impact of human activities on soil and some ways that soil erosion is controlled.

- Soil erosion is a natural process; it is part of the constant recycling of Earth materials that we call the rock cycle.
- Because of human activities, soil erosion rates have increased over the past several hundred years. Natural soil production rates are constant, so there is a net loss of soil at a time when a record-breaking number of people live on the planet. Using windbreaks, terracing the land, installing grassed waterways, and plowing the land along horizontal contour lines are all practices that have been shown to reduce soil erosion.

Edwin C. Cole/NCRS

? Why was this row of evergreens planted on this Indiana farm?

8.7 Mass Movement on Slopes: The Work of Gravity

Discuss the role that mass movements play in the development of landscapes. Summarize the factors that control and trigger mass movement processes.

KEY TERMS: mass movement, trigger, angle of repose

- After weathering breaks apart rock, gravity moves the debris downslope, in a process called mass movement. Sometimes this occurs rapidly as a landslide, and at other times the movement is slower. Landslides are a significant geologic hazard, taking many lives and destroying property every year.
- Mass movement serves an important role in landscape development. It widens stream-cut valleys and helps tear down mountains thrust up by internal processes.
- An event that initiates a mass movement process is referred to as a trigger. The addition of water, oversteepening of the slope, removal of vegetation, and shaking due to an earthquake are four important examples. Not all landslides are triggered by one of these four processes, but many are.
- Water added to a slope can expand the pore space between grains, causing them to lose their cohesion. Water also adds a significant amount of mass to a wetted slope.
- Loose granular materials can form a stable slope only up to a specific angle of repose, which is typically between 25 and 40 degrees from horizontal and is steeper for coarser or more angular particles. Cohesive materials do not have a defined angle of repose but eventually respond to oversteepening with mass movement.
- The roots of plants (especially plants with deep roots) act as a three-dimensional "net" that holds soil and regolith particles in place. The removal of vegetation such as by wildfires or various human activities that clear the land on steep slopes can set the stage for significant mass movement.
- Earthquakes are significant triggers that deliver an energetic jolt to slopes poised on the brink of failure.

8.8 Types of Mass Movement

List and explain the criteria that are commonly used to classify mass movement processes. Distinguish among six different types of mass movement.

KEY TERMS: fall, slide, flow, rock avalanche, slump, rockslide, debris flow, mudflow, lahar, earthflow, creep, solifluction, permafrost

- Mass movements can be classified by the type of material involved (e.g., rocks, debris, mud), the type of motion (fall, slide, flow), and the rate of motion (from very fast to centimeters or millimeters per year).
- Rockfalls occur when pieces of bedrock detach and fall freely through the air. Repeated rockfalls are the primary means by which talus slopes are built and maintained.
- Many slides occur when a mass of material moves along a well-defined, flattish surface, such as a bedding plane that tilts into a valley. A rockslide consists of blocks of rock; a debris slide consists mainly of unconsolidated soil and regolith. Motion is generally rapid; rock avalanches can reach speeds exceeding 200 kilometers (125 miles) per hour.
- In a slump, a mass of rock or unconsolidated material (such as clay) slides as a unit along a curved, often spoon-shaped surface. Slumping is a common response to slopes becoming oversteepened.
- In a flow, material moves downslope as a viscous, often turbulent fluid (rather than coherently, as in a slide). Most flows are saturated with water; they range in consistency from very thick to soupy. A debris flow consists of soil and regolith; in a mudflow, the material is mainly fine grained. A lahar is a particularly deadly type of debris flow that develops on steep volcanic slopes. Thick flows may pick up and carry large boulders and trees.
- Earthflows are usually much slower than debris flows and form on hillsides in humid regions. They are associated with silt- and clay-rich materials. Typically, sites of earthflow show an uphill scarp and a lobe of viscous soil on the downhill side.
- Creep is a widespread and important form of mass movement that is very slow. It occurs mainly when freezing (or wetting) causes soil particles to be pushed out away from the slope, only to drop down to a lower position following thawing (or drying).
- Solifluction is the gradual flow of a saturated surface layer that is underlain by an impermeable zone. In arctic regions, the impermeable zone is permafrost.

? What type of motion is illustrated by this cautionary highway sign?

Harris Shiffman/Shutterstock

GIVE IT SOME THOUGHT

1 Describe how plants promote mechanical and chemical weathering but inhibit erosion.

2 Granite and basalt are exposed at Earth's surface in a hot, wet region. Will mechanical weathering or chemical weathering predominate? Which rock will weather more rapidly? Why?

3 The accompanying photo shows Shiprock, a well-known landmark in the northwestern corner of New Mexico. It is a mass of igneous rock that represents the underground "plumbing" of a now-vanished volcano. Extending toward the upper left is a related wall-like igneous structure known as a dike. The igneous features are surrounded by sedimentary rocks. Explain why these once deeply buried igneous features now stand high above the surrounding terrain. What term from Section 8.2 applies to this situation?

Michael Collier

4 What might cause different soils to develop from the same kind of parent material or similar soils to form from different parent materials?

5 Using the map of global soil regions in Figure 8.17, identify the main soil order in the region adjacent to South America's Amazon River (point A on the map) and the predominant soil order in the American Southwest (point B). Briefly contrast these soils. Do they have anything in common? Referring to Table 8.2 might be helpful.

6 Describe at least one situation in which an internal process might cause or contribute to a mass movement process.

7 This soil sample is from a farm in the Midwest. From which horizon was the sample most likely taken—A, E, B, or C? Explain.

Lynn Betts/NRCS

8 The accompanying photo shows a footprint on the Moon left by an *Apollo* astronaut in material popularly called *lunar soil*. Does this material satisfy the definition we use for soil on Earth? Explain why or why not. You may want to refer to Figure 8.11.

NASA

EXAMINING THE EARTH SYSTEM

1 This aerial view shows landslide debris atop Buckskin Glacier in Denali National Park in the rugged Alaska Range. Where Buckskin Glacier ends, its meltwater feeds a river that flows into Cook Inlet, just west of Anchorage. Cook Inlet is an arm of the North Pacific. Many processes have been responsible for creating this scene. Prepare a brief outline or summary that explains the formation and evolution of this landscape. Include internal and external processes and be sure to mention which spheres of the Earth system were involved.

2 Heavy rains in late July 2010 triggered the mass movement that occurred in this mountain valley near Durango, Colorado. Heavy equipment is clearing away material that blocked railroad tracks and significantly narrowed the adjacent stream channel.

a. What type of mass movement likely occurred here? Explain your choice.

b. Most of us are familiar with the phrase "One thing leads to another." It certainly applies to the Earth system. Suppose the material from the Durango mass movement event had completely filled the stream. What other natural hazard might have developed?

3 Because of the burning of fossil fuels such as coal and petroleum, the level of carbon dioxide (CO_2) in the atmosphere has been increasing for more than 150 years. Should this increase tend to accelerate or slow down the rate of chemical weathering of Earth's surface rocks? Explain how you arrived at your conclusion.

4 Discuss the interaction of the atmosphere, geosphere, biosphere, and hydrosphere in the formation of soil.

5 During summer, wildfires are common occurrences in many parts of the western United States. Millions of acres are burned each year. The various parts of Earth's four major spheres interact in uncountable ways. Relate this idea to the situation pictured here.

a. What atmospheric conditions might have preceded and thus set the stage for and/or contributed to this wildfire?

b. What might have ignited the blaze? Suggest a natural possibility and a human possibility.

c. Discuss how wildfires like the one shown here might influence future mass movement events in the area.

DATA ANALYSIS

Soil Types

Soil types have been mapped across the United States. This information is used by farmers and planners to determine how the land can be used.

ACTIVITIES

Go to the USDA's Web Soil Survey, at http://websoilsurvey.sc.egov.usda.gov. Click on the Start WSS button. Select State and County under Quick Navigation; enter your state and county and then click View. Click on the i (information) tool and then click on your location to display location information.

1 What are the latitude and longitude for the location you chose?

2 When was this aerial photograph taken?

Create an area of interest (AOI) around your location by clicking the AOI rectangle tool and dragging the mouse to create a space around the location you chose. This area should be larger than just a few streets in order to capture some variation. When a striped area appears, click the Soil Map tab near the top of the page. Click the Map Unit Legend to see soil types in your area, and then click on a Map Unit Name for details.

3 What type(s) of soil is(are) present at your location?

4 Based on Figure 8.12 (page 250), what is the percentage range in your soil for sand? Silt? Clay?

Click the Soil Data Explorer tab near the top of the page and be sure Suitabilities and Limitations for Use is selected. Select Land Classification and choose Farmland Classification; then click View Rating.

5 Click View Description and determine what information is displayed.

6 What is the predominant rating for this area? What other ratings are available in your area?

7 What percentage is prime farmland? Does this make sense, based on what you know about your area?

Select Land Management and choose Erosion Hazard (Off-Road, Off-Trail); then click View Rating.

8 Click View Description and determine what information is displayed.

9 What is the predominant rating for this area? What other ratings are available in your area? If your area is not rated, find a nearby place that has ratings.

10 Based on what you know about your area's physical features (flat or hilly, urban or rural, etc.), what can you say about the relationship between erosion hazards and the characteristics of the area?

MasteringGeology™ Looking for additional review and test prep materials? Visit the Study Area in MasteringGeology to enhance your understanding of this chapter's content by accessing a variety of resources, including Self-Study Quizzes, Geoscience Animations, SmartFigure Tutorials, Mobile Field Trips, *Project Condor* Quadcopter videos, *In the News* articles, flashcards, web links, and an optional Pearson eText.

www.masteringgeology.com

9 Running Water and Groundwater

FOCUS ON CONCEPTS

Each statement represents the primary learning objective for the corresponding major heading within the chapter. After you complete the chapter, you should be able to:

- 9.1 List the hydrosphere's major reservoirs and describe the different paths that water takes through the hydrologic cycle.
- 9.2 Describe the nature of drainage basins and river systems. Sketch four basic drainage patterns.
- 9.3 Discuss streamflow and the factors that cause it to change.
- 9.4 Summarize the ways in which streams erode, transport, and deposit sediment.
- 9.5 Contrast bedrock and alluvial stream channels. Distinguish between two types of alluvial channels.
- 9.6 Contrast narrow V-shaped valleys, broad valleys with floodplains, and valleys that display incised meanders.
- 9.7 Discuss the formation of deltas, natural levees, and alluvial fans.
- 9.8 Distinguish between regional floods and flash floods. Describe some common flood control measures.
- 9.9 Discuss the importance of groundwater and describe its distribution and movement.
- 9.10 Compare and contrast wells, artesian systems, and springs.
- 9.11 List and discuss three important environmental problems associated with groundwater.
- 9.12 Explain the formation of caverns and the development of karst topography.

The Colorado River winds through Canyonlands National Park in southern Utah. The river once meandered across a relatively flat landscape. Subsequently, the region was gradually lifted upward, while downward erosion lowered the riverbed. The loops, locked within confining walls, are now referred to as incised meanders.
(Photo by Michael Collier)

WATER IS CONTINUALLY ON THE MOVE, from the ocean to the land and back again, in an endless cycle. This chapter deals with the part of the hydrologic cycle that returns water from the land to the sea. Some water travels quickly via rushing streams, and some moves more slowly below the surface. When viewed as part of the Earth system, streams and groundwater represent basic links in the constant cycling of the planet's water. We will examine the factors that influence the distribution and movement of water, as well as look at how water sculpts the landscape. To a great extent, the Grand Canyon, Niagara Falls, Old Faithful, and Mammoth Cave all owe their existence to the action of water on its way to the sea.

9.1 Earth as a System: The Hydrologic Cycle

List the hydrosphere's major reservoirs and describe the different paths that water takes through the hydrologic cycle.

Water is constantly moving among Earth's different spheres—the *hydrosphere*, the *atmosphere*, the *geosphere*, and the *biosphere*. This unending circulation of water is called the **hydrologic cycle**. Earth is the only planet in our solar system that has a global ocean and a hydrologic cycle.

Earth's Water

Water is almost everywhere on Earth—in the oceans, glaciers, rivers, lakes, air, soil, and living tissue. All these reservoirs constitute Earth's hydrosphere. A glance back at Figure 1.13 (page 16) reminds us that the vast bulk of Earth's water, about 96.5 percent, is stored in the global ocean. Ice sheets and glaciers account for most of our planet's freshwater, leaving just a fraction of 1 percent to be divided among lakes, streams, groundwater, and the atmosphere. Although the percentage of Earth's total water found in each of the latter sources is just a small fraction of the total inventory, the absolute quantities are great.

SmartFigure 9.1 The hydrologic cycle The primary movement of water through the cycle is shown by the large arrows. The numbers refer to the annual amount of water taking a particular path.

TUTORIAL
https://goo.gl/ZYRTcf

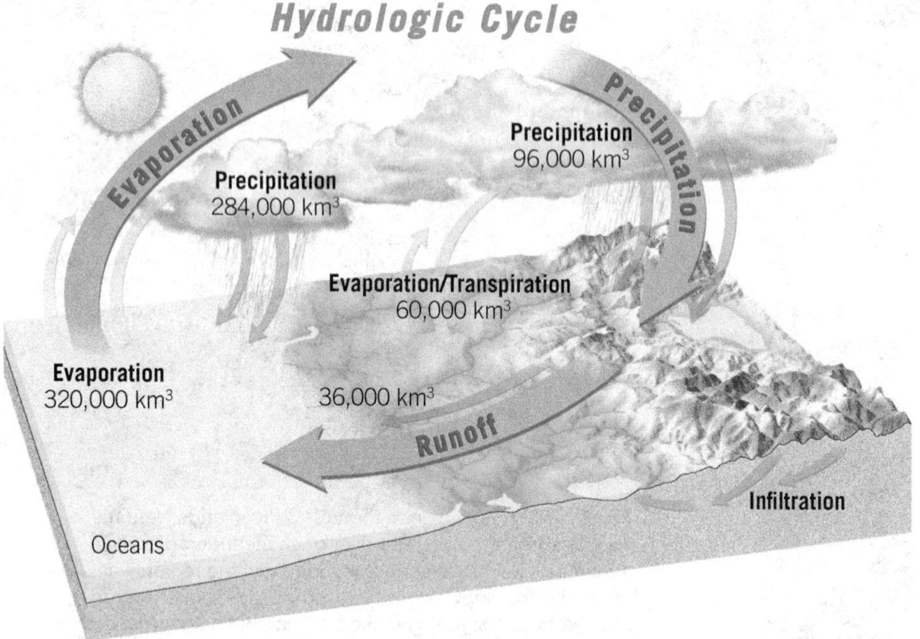

Water's Paths

The hydrologic cycle is a gigantic, worldwide system powered by energy from the Sun, in which the atmosphere provides a vital link between the oceans and continents (Figure 9.1). **Evaporation**, the process by which liquid water changes into water vapor (gas), is how water enters the atmosphere from the ocean and, to a much lesser extent, from the land. Winds often transport moisture-laden air great distances. Complex processes of cloud formation eventually result in precipitation. The precipitation that falls into the ocean has completed its cycle and is ready to begin again. The water that falls on the continents, however, must make its way back to the ocean.

What happens to precipitation once it has fallen on land? A portion of the water soaks into the ground (called **infiltration**), slowly moving downward, then moving laterally, and finally seeping into lakes, streams, or directly into the ocean. When the rate of rainfall exceeds ground's ability to absorb it, the surplus water flows over the surface into lakes and streams, a process called **runoff**. Much of the water that infiltrates or runs off eventually returns to the atmosphere because of evaporation from the soil, lakes, and streams. Also, some of the water that soaks into the ground is absorbed by plants, which then release it into the atmosphere. This process is called **transpiration**. Because both evaporation and transpiration involve the transfer of water from the surface directly to the atmosphere, they are often considered together as the combined process of **evapotranspiration**.

Storage in Glaciers

When precipitation falls in very cold places—at high elevations or high latitudes—the water may not immediately soak in, run off, or evaporate. Instead, it may become part of a snowfield or glacier. In this way, glaciers store large quantities of water on land. If present-day glaciers were to melt and release all their water, sea level would rise by several dozen meters. Such a rise would submerge many heavily populated coastal areas. As you will see in Chapter 10, over the past 2 million years, huge ice sheets have formed and melted on several occasions, each time affecting the balance of the hydrologic cycle.

Water Balance

Figure 9.1 also shows Earth's overall *water balance*, or the volume that passes through each part of the cycle annually. The amount of water vapor in the air at any one time is just a tiny fraction of Earth's total water supply. But the *absolute* quantities that are cycled through the atmosphere over a 1-year period are immense—some 380,000 cubic kilometers (91,000 cubic miles)—enough to cover Earth's entire surface to a depth of about 1 meter (39 inches).

It is important to know that the hydrologic cycle is *balanced*. Because the total amount of water vapor in the atmosphere remains about the same, the average annual precipitation worldwide must be equal to the quantity of water evaporated. However, for all the continents taken together, precipitation exceeds evaporation. Conversely, over the oceans, evaporation exceeds precipitation. Because the level of the world ocean is not dropping, the system must be in balance. Balance is achieved because 36,000 cubic kilometers (8600 cubic miles) of water annually makes its way from the land back to the ocean.

About one-quarter of global precipitation falls on land and flows on and below the surface. This water is *the most important force sculpturing Earth's land surface*. In the rest of this chapter, we will observe the work of water running over the surface, including floods, erosion, and the formation of valleys. Then we will look underground, at the slow labors of groundwater as it forms springs and caverns and provides water for people on its long migration to the sea.

CONCEPT CHECKS 9.1

1. Describe or sketch the movement of water through the hydrologic cycle. Once precipitation has fallen on land, what paths might it take?
2. What is meant by the term *evapotranspiration*?
3. Over the oceans, evaporation exceeds precipitation, yet sea level does not drop. Explain this.

9.2 Running Water

Describe the nature of drainage basins and river systems. Sketch four basic drainage patterns.

Much of the precipitation that falls on land either enters the soil (infiltration) or remains at the surface, moving downslope as runoff. The amount of water that runs off rather than soaking into the ground depends on several factors: (1) the intensity and duration of rainfall, (2) the amount of water already in the soil, (3) the nature of the surface material, (4) the slope of the land, and (5) the extent and type of vegetation. When the surface material is highly impermeable, or when it becomes saturated, runoff is the dominant process. Runoff is also high in urban areas because large areas are covered by impermeable buildings, roads, and parking lots.

Runoff initially flows in broad, thin sheets across hillslopes. This unconfined flow eventually develops threads of current that form tiny channels called *rills*. Where rills merge, the flowing water creates *gullies*, which join to form larger stream channels. At first streams are small, but as one intersects another, larger and larger streams form. Eventually they merge into rivers that carry water from a broad region.

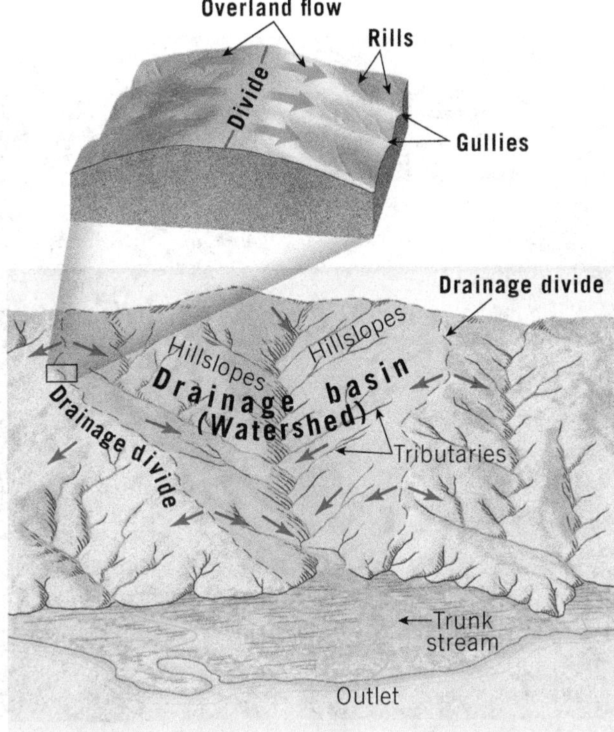

Figure 9.2 Drainage basin and divide A drainage basin or watershed is the area drained by a stream and its tributaries. Boundaries between drainage basins are called divides.

Drainage Basins

Every stream drains an area of land called a **drainage basin**, or **watershed** (Figure 9.2). Each drainage basin is bounded by an imaginary line called a **divide**, something that is clearly visible as a sharp ridge in some

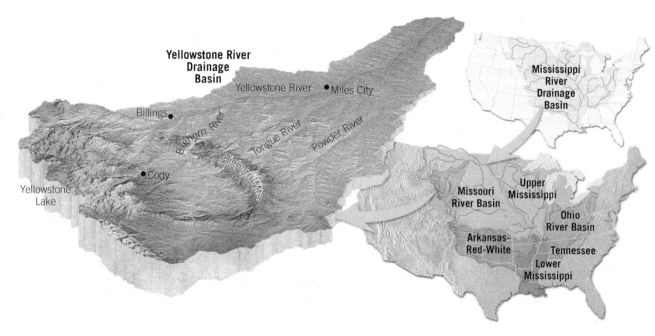

SmartFigure 9.3
Mississippi River drainage basin The drainage basin of the Mississippi River forms a funnel that stretches from Montana and southern Canada in the west to New York State in the east and that runs down to a spout in Louisiana. It consists of many smaller drainage basins. The drainage basin of the Yellowstone River is one of many that contribute water to the Missouri River, which, in turn, is one of many that make up the drainage basin of the Mississippi River.

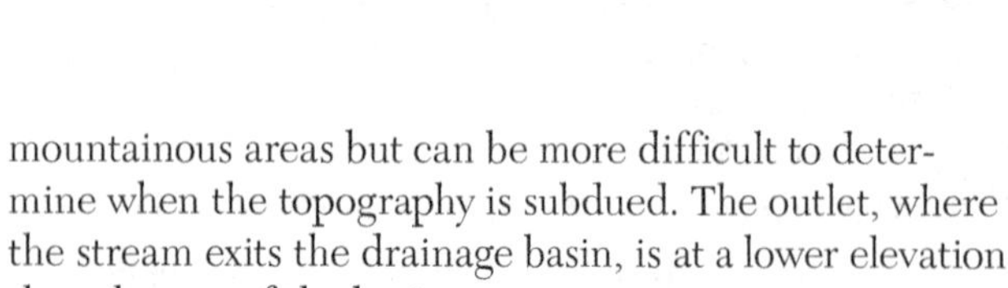

TUTORIAL
https://goo.gl/HU34IU

mountainous areas but can be more difficult to determine when the topography is subdued. The outlet, where the stream exits the drainage basin, is at a lower elevation than the rest of the basin.

Drainage divides range in scale from a small ridge separating two gullies on a hillside to a *continental divide* that splits an entire continent into enormous watersheds. The Mississippi River has the largest watershed in North America, collecting and carrying 40 percent of the flow in the United States (**Figure 9.3**).

By looking at the drainage basin in Figure 9.2, you can see that slopes cover most of the area. Water erosion on the hillsides is aided by the impact of raindrops and by sheet flow, moving downslope as sheets or in rills toward a stream channel (see Figures 8.18, page 256, and 8.19, page 256). Hillslope erosion is the main source of fine particles (clays and fine sand) carried in stream channels.

If you could observe streams in an area similar to that depicted in Figure 9.2 over several years, you would see many of them lengthen by **headward erosion**—that is, by extending the heads of their channels upslope. Headward erosion occurs when the surface flow converging at the head of a channel has enough power to cut the channel deeper (*downcut*). This lowering of the channel creates steeper headward slopes that erode more quickly. Thus, through headward erosion, a valley extends into previously undissected terrain (**Figure 9.4**).

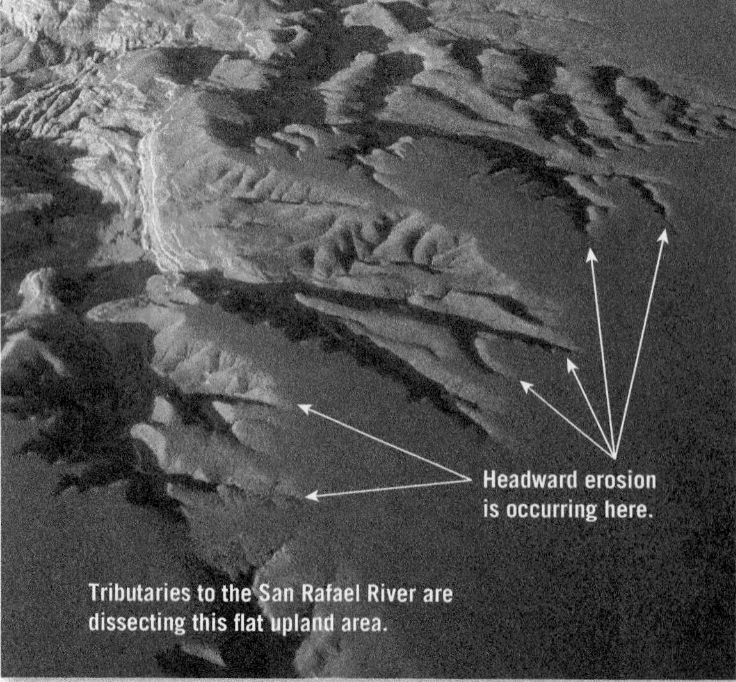

SmartFigure 9.4
Headward erosion A stream lengthens its course by extending the head of its valley upslope into previously undissected terrain. (Photo by Michael Collier)

TUTORIAL
https://goo.gl/J3ulnM

River Systems

A river system includes not only its network of stream channels but its entire drainage basin. It can be divided into three zones, based on the process that dominates in each. These are the zones of *sediment production* (where erosion dominates), *sediment transport*, and *sediment deposition* (**Figure 9.5**). It is important to recognize that sediment is being eroded, transported, and deposited along the entire length of a stream, regardless of which process is dominant within each zone.

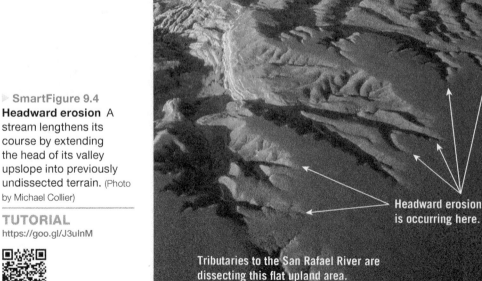

Sediment Production The zone of *sediment production*, where most of the sediment is derived, is located in the headwater region of the river system. Much of the sediment carried by streams begins as bedrock that is subsequently broken down by weathering and then transported downslope

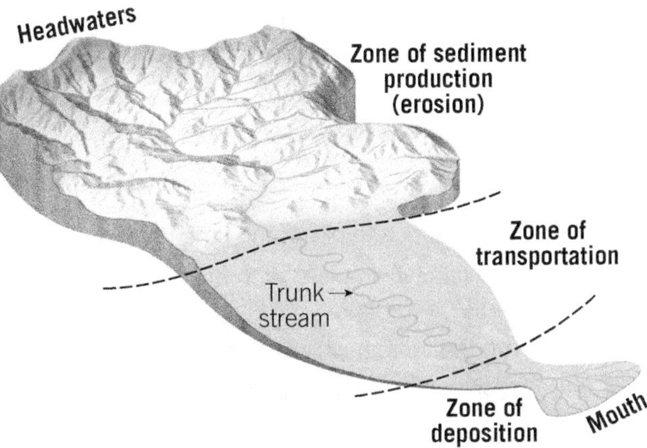
Headwaters
Zone of sediment production (erosion)
Zone of transportation
Trunk stream
Zone of deposition
Mouth

Figure 9.5 **Zones of a river** Each of the three zones is based on the dominant process that is operating in that part of the river system.

by mass wasting and overland flow. Bank erosion can also contribute significant amounts of sediment. In addition, scouring of the channel bed deepens the channel and adds to the stream's sediment load.

Sediment Transport Downstream from the zone of sediment production, material acquired by a stream is transported through the channel network along sections referred to as *trunk streams*. When trunk streams are in balance, the amount of sediment eroded from their banks equals the amount deposited elsewhere in the channel. Although trunk streams rework their channels over time, they are not sources of sediment, nor do they accumulate or store it.

Sediment Deposition When a river reaches the ocean or another large body of water, it slows, and the energy to transport sediment is greatly reduced. Most of the sediment either accumulates at the mouth of the river to form a delta, is reconfigured by wave action to form a variety of coastal features, or is moved far offshore by ocean currents. Because coarse sediment tends to be deposited upstream, it is primarily the fine sediment (clay, silt, and fine sand) that eventually reaches the ocean. Taken together, erosion, transportation, and deposition are the processes by which rivers move Earth's surface materials and sculpt landscapes.

Drainage Patterns

Drainage systems, which are interconnected networks of streams, can exhibit a variety of patterns. The pattern that develops depends primarily on the kind of rock present and/or the structural pattern of joints, faults, and folds. Figure 9.6 illustrates four drainage patterns.

The most commonly encountered drainage pattern is the **dendritic pattern** (see Figure 9.6A). This pattern of irregularly branching tributary streams resembles the branching pattern of a deciduous tree. In fact, the word *dendritic* means "treelike." The dendritic pattern forms where the underlying material is relatively uniform. Because the surface material is essentially uniform in its resistance to erosion, it does not control the pattern of streamflow. Rather, the pattern is determined chiefly by the direction of slope of the land.

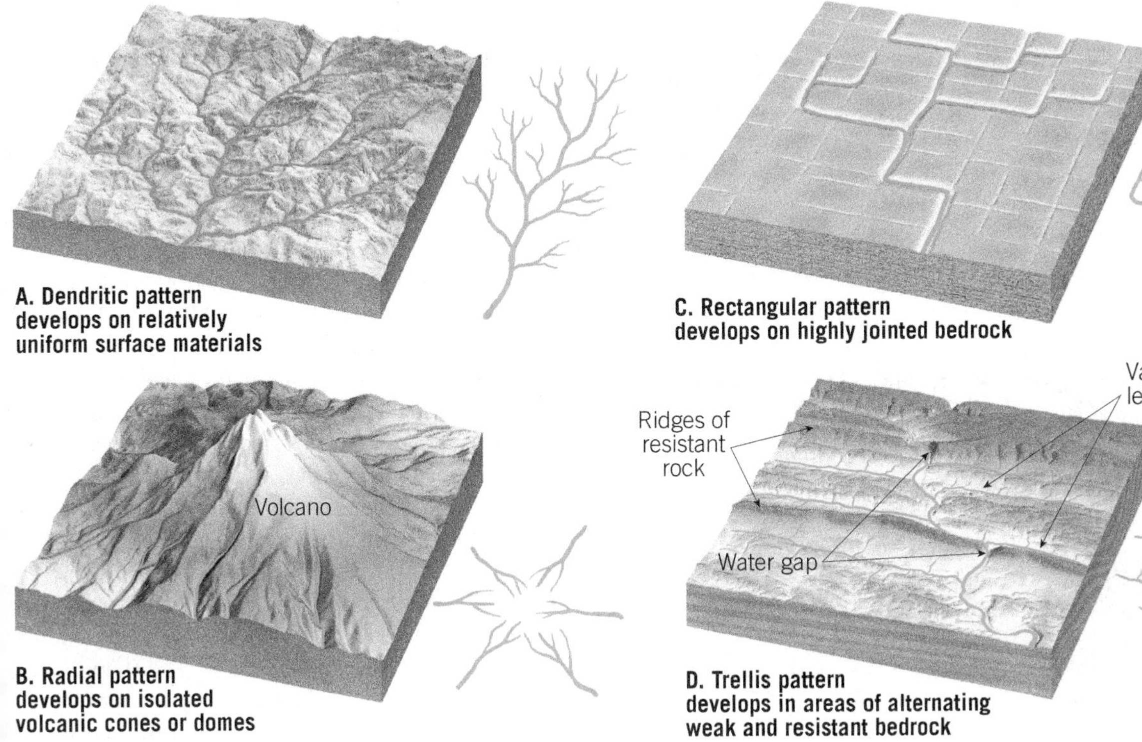
A. Dendritic pattern develops on relatively uniform surface materials
C. Rectangular pattern develops on highly jointed bedrock
B. Radial pattern develops on isolated volcanic cones or domes
Volcano
Ridges of resistant rock
Valleys cut in less-resistant rock
Water gap
D. Trellis pattern develops in areas of alternating weak and resistant bedrock

Figure 9.6 **Drainage patterns** Networks of streams form a variety of patterns.

When streams diverge from a central area like spokes from the hub of a wheel, the pattern is said to be **radial** (see Figure 9.6B). This pattern typically develops on isolated volcanic cones and domal uplifts.

A **rectangular pattern** exhibits many right-angle bends (see Figure 9.6C). This pattern develops when the bedrock is crisscrossed by a series of joints and/or faults. Because these structures are eroded more easily than unbroken rock, their geometric pattern guides the directions of valleys.

A **trellis pattern** is a rectangular drainage pattern in which tributary streams are nearly parallel to one another and have the appearance of a garden trellis (see Figure 9.6D). This pattern forms in areas underlain by alternating bands of resistant and less-resistant rock.

CONCEPT CHECKS 9.2

1. List several factors that cause infiltration and runoff to vary from place to place and time to time.
2. Draw and label a simple sketch of a drainage basin and divide.
3. What are the three main zones of a river system?
4. Prepare a sketch of the four drainage patterns discussed in this section.

9.3 Streamflow Characteristics

Discuss streamflow and the factors that cause it to change.

The water in stream and river channels moves under the influence of gravity. In very slowly flowing streams, water moves in nearly straight-line paths parallel to the stream channel; this is called **laminar flow** (Figure 9.7A). However, streams typically exhibit **turbulent flow** (Figure 9.7B). Strong turbulent behavior occurs in whirlpools and eddies, as well as in roiling whitewater rapids. Even streams that appear smooth on the surface often exhibit turbulent flow near the bottom and sides of the channel, where flow resistance is greatest. Turbulence contributes to a stream's ability to erode its channel because it acts to lift sediment from the streambed.

An important factor influencing stream turbulence is the water's flow velocity. As the velocity of a stream increases, the flow becomes more turbulent. Flow velocities can vary significantly from place to place along a stream, as well as over time, in response to variations in the amount and intensity of precipitation. If you have ever waded into a stream, you may have noticed that the strength of the current increased as you moved into deeper parts of the channel. This is related to the fact that frictional resistance is greatest near the banks and bed of the stream channel.

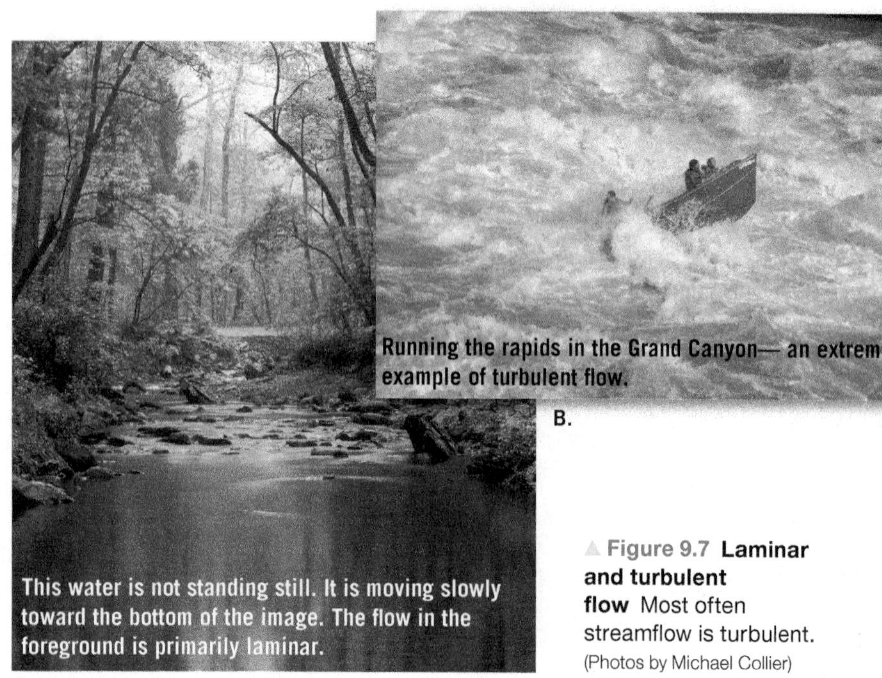

Running the rapids in the Grand Canyon— an extreme example of turbulent flow.
B.
This water is not standing still. It is moving slowly toward the bottom of the image. The flow in the foreground is primarily laminar.
A.

▲ **Figure 9.7 Laminar and turbulent flow** Most often streamflow is turbulent. (Photos by Michael Collier)

Factors Affecting Flow Velocity

The ability of a stream to erode and transport material is directly related to its flow velocity. Even slight variations in flow rate can lead to significant changes in the load of sediment that water can transport. Several factors influence flow velocity and, therefore, control a stream's potential to do "work." These factors include (1) channel slope, or gradient, (2) channel size and cross-sectional shape, (3) channel roughness, and (4) the amount of water flowing in the channel.

Gradient The slope of a stream channel expressed as the vertical drop of a stream over a specified distance is the **gradient**. Portions of the lower Mississippi River have very low gradients of 10 centimeters per kilometer or less. By contrast, some mountain stream channels decrease in elevation at a rate of more than 40 meters per kilometer, a gradient 400 times steeper than the lower Mississippi. Gradient also varies along the length of a particular channel. When the gradient is steeper, more gravitational energy is available to drive channel flow.

Channel Shape, Size, and Roughness A stream's channel is a conduit that guides the flow of water, but the water encounters friction as it flows. The shape, size, and roughness of the channel affect the amount of friction. Larger channels have more efficient flow because a

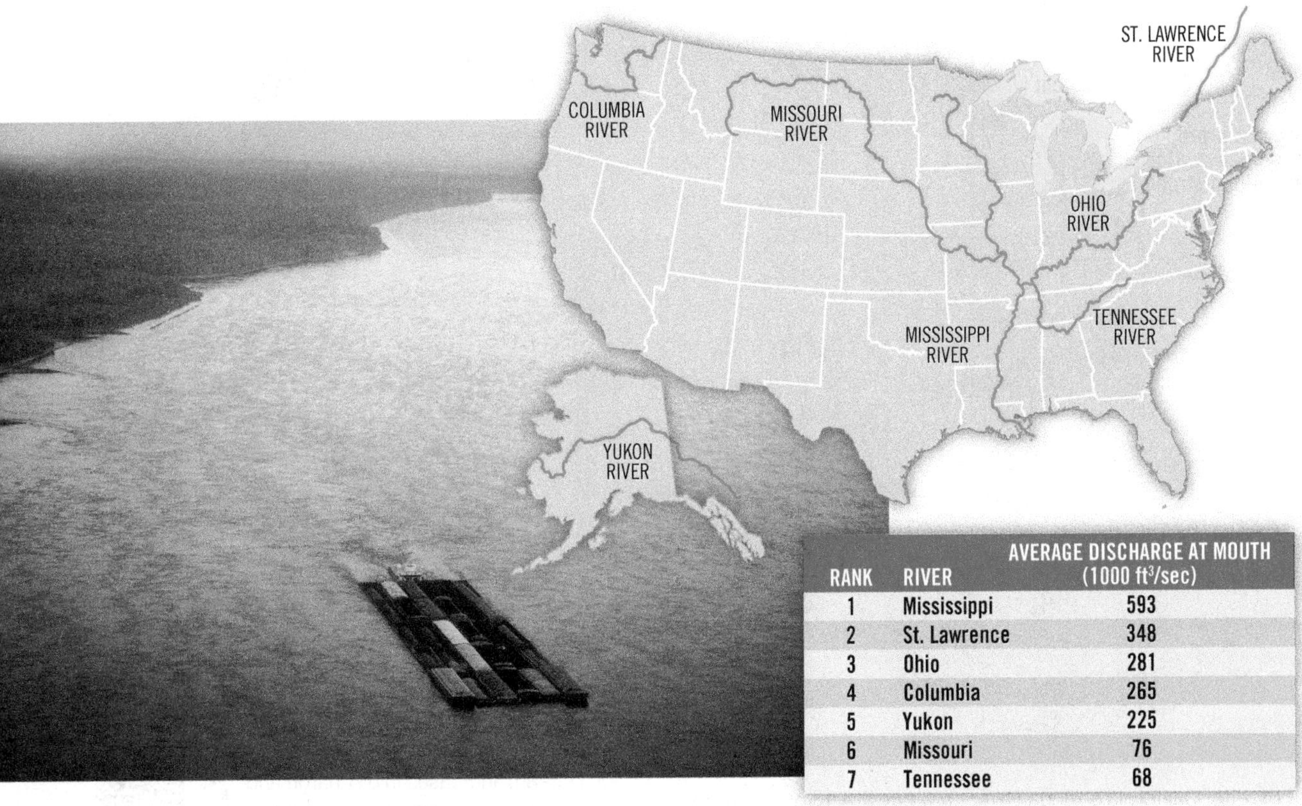

RANK	RIVER	AVERAGE DISCHARGE AT MOUTH (1000 ft³/sec)
1	Mississippi	593
2	St. Lawrence	348
3	Ohio	281
4	Columbia	265
5	Yukon	225
6	Missouri	76
7	Tennessee	68

SmartFigure 9.8 **Largest U.S. rivers** The map shows the seven largest U.S. rivers, based on discharge. The photo shows the Mississippi River near Helena, Arkansas, about 725 kilometers (450 miles) north of the river's mouth. The Mississippi is North America's largest river. From head to mouth, it is nearly 3900 kilometers (2400 miles) long. Its watershed encompasses about 40 percent of the lower 48 states and includes all or parts of 31 states and 2 Canadian provinces. Average discharge at its mouth is about 16,800 cubic meters (593,000 cubic feet) per second. (Photo by Michael Collier)

MOBILE FIELD TRIP
https://goo.gl/UyqOHJ

smaller proportion of water is in contact with the channel. A smooth channel promotes a more uniform flow, whereas an irregular channel filled with boulders creates enough turbulence to slow the stream significantly.

Discharge Streams vary in size from small headwater creeks less than a meter wide to large rivers with widths of several kilometers. The size of a stream channel is largely determined by the amount of water supplied from the watershed. The measure most often used to compare the sizes of streams is **discharge**—the volume of water flowing past a certain point in a given unit of time. Discharge, usually measured in cubic meters per second or cubic feet per second, is determined by multiplying a stream's cross-sectional area by its velocity.

The largest river in North America, the Mississippi, discharges an average of about 16,800 cubic meters (593,000 cubic feet) per second (Figure 9.8). Although this is a huge quantity of water, it is dwarfed by the mighty Amazon in South America, the world's largest river. Fed by a vast rainy region that is nearly three-fourths the size of the conterminous United States, the Amazon discharges about 12 times more water than the Mississippi.

The discharge of a river system changes over time because of variations in the amount of precipitation received by the watershed. Studies show that when discharge increases, the width, depth, and flow velocity of the channel all increase predictably. As we saw earlier, when the size of the channel increases, proportionally less water is in contact with the bed and banks of the channel. Thus, friction, which acts to retard the flow, is reduced, resulting in an increase in the rate of flow.

Streams that exhibit flow only during wet periods are referred to as *intermittent streams*. In arid climates, many streams carry water only occasionally, after a heavy rainstorm, and are called *ephemeral streams*.

Changes from Upstream to Downstream

One useful way of studying a stream is to examine its **longitudinal profile**—a cross-sectional view of a stream from its source area (called the *head* or *headwaters*) to its *mouth*, the point downstream where the river empties into another water body—a river, a lake, or an ocean. As shown in Figure 9.9, the most obvious feature of a typical profile is a constantly decreasing gradient from the head to the mouth. Although many local irregularities may exist, the overall profile is a relatively smooth concave curve.

The change in slope observed on most stream profiles is usually accompanied by an increase in discharge and channel size, as well as a reduction in sediment particle size (Figure 9.10). Along most rivers in humid regions, discharge increases toward the mouth because as we move downstream, more and more tributaries contribute water to the main channel. In order to accommodate the growing volume of water, channel size typically increases downstream as well. Recall that flow velocities are higher in large channels than in small channels. Observations also show a general decline in sediment size downstream, making the channel smoother and more efficient (with less friction).

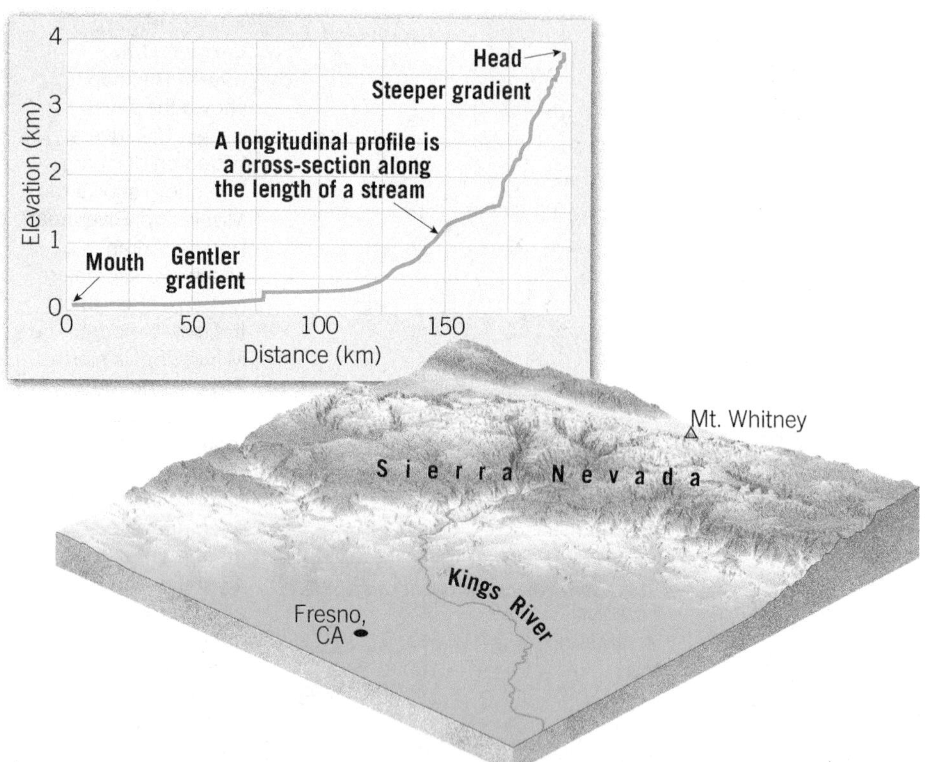

Figure 9.9 **Longitudinal profile** California's Kings River originates high in the Sierra Nevada and flows into the San Joaquin Valley. There is significant vertical exaggeration in this graph.

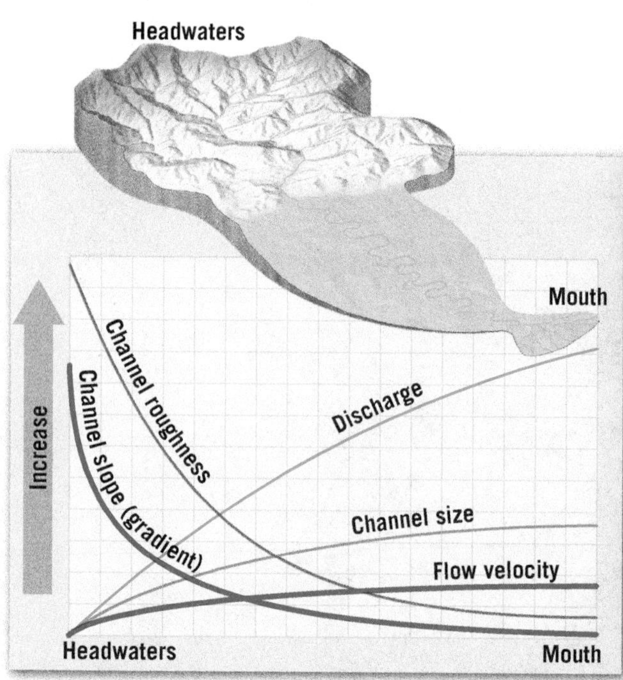

SmartFigure 9.10 **Channel changes from head to mouth** Although the gradient decreases toward the mouth of a stream, increases in discharge and channel size and decreases in roughness more than offset the decrease in slope. Consequently, flow velocity usually increases toward the mouth.

TUTORIAL https://goo.gl/mFXfG4

Although the channel slope decreases—the gradient becomes less steep—toward a stream's mouth, the flow velocity generally increases. This fact contradicts our intuitive assumption that narrow headwater streams are swift, whereas a broad river flowing across subtle topography is slow. Actually, the increase in channel size and discharge and decrease in channel roughness that occur downstream compensate for the decrease in slope, making the stream more efficient (see Figure 9.10). Thus, the average flow velocity is typically lower in headwater streams than in wide rivers that appear to be placid.

CONCEPT CHECKS 9.3

1. Distinguish between laminar flow and turbulent flow.
2. Summarize the factors that influence flow velocity.
3. What is a longitudinal profile of a stream?
4. What typically happens to channel width, channel depth, flow velocity, and discharge between the head and mouth of a stream? Briefly explain why these changes occur.

9.4 The Work of Running Water

Summarize the ways in which streams erode, transport, and deposit sediment.

Streams are Earth's most important erosional agents. In addition to having the ability to deepen and widen their channels, streams also have the capacity to transport the enormous quantities of sediment that are delivered by sheet flow, mass movement, and groundwater. Eventually much of this material is deposited to create a variety of landforms.

Stream Erosion

A stream's ability to accumulate and transport soil and weathered rock is aided by the work of raindrops, which knock sediment particles loose (see Figure 8.18, page 256). When the ground is saturated, rainwater cannot infiltrate, so it flows downslope, transporting some of the material it has dislodged. On barren slopes the sheet flow often erodes small channels, or *rills*, which in time may evolve into larger *gullies* (see Figure 8.19, page 256).

Once flow is confined in a channel, the erosional power of a stream is related to its slope and discharge. When the flow of water is sufficiently strong, it can dislodge particles from the channel and lift them into the moving water. In this manner, the force of running water swiftly erodes poorly consolidated materials on

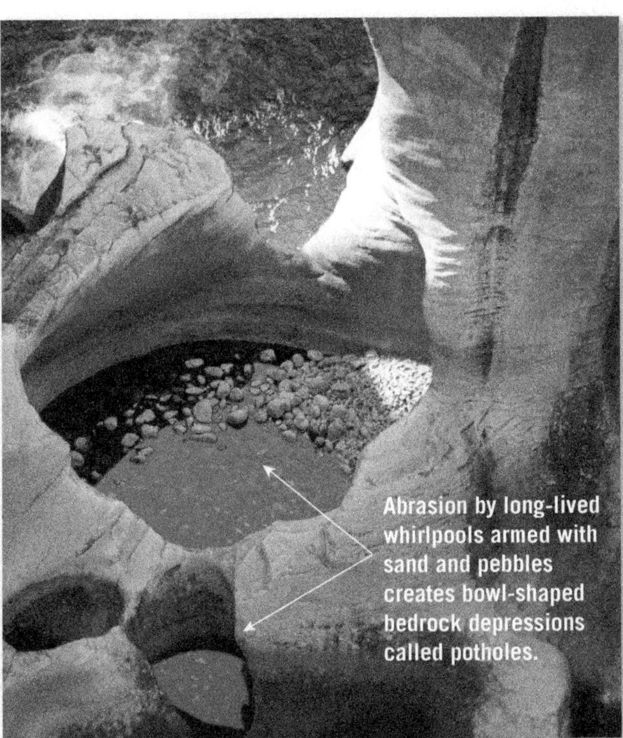

Figure 9.11 **Potholes** The rotational motion of swirling pebbles acts like a drill to create potholes. (Photo by StormStudio/Alamy Stock Photo)

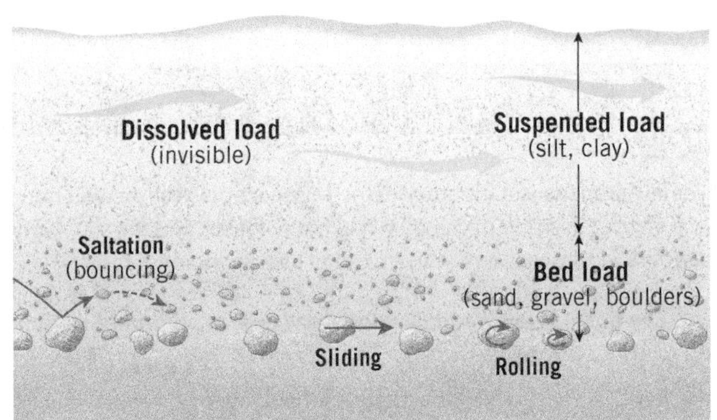

Smart Figure 9.12 **Transport of sediment** Streams transport their load of sediment in three ways. The dissolved and suspended loads are carried in the general flow. The bed load includes coarse sand, gravel, and boulders that move by rolling, sliding, and saltation.

ANIMATION
https://goo.gl/z3OA1T

the bed and sides of a stream channel. On occasion, the banks of the channel may be undercut, dumping even more loose debris into the water to be carried downstream.

In addition to eroding unconsolidated materials, the hydraulic force of streamflow can also cut a channel into solid bedrock. A stream's ability to erode bedrock is greatly enhanced by the particles it carries. These particles can be any size, from large boulders in very fast-flowing waters to sand and gravel-size particles in somewhat slower flow. Just as the particles of grit on sandpaper can wear away a piece of wood, so too can the sand and gravel carried by a stream abrade a bedrock channel. Moreover, pebbles caught in swirling eddies can act like "drills" and bore circular **potholes** into the channel floor (Figure 9.11).

Bedrock channels formed in soluble rock such as limestone are susceptible to *corrosion*—a process in which rock is gradually dissolved by the flowing water. Corrosion is a type of chemical weathering that occurs between the solutions in the stream water and the mineral matter composing the bedrock.

Transportation of Sediment

All streams, regardless of size, transport some weathered rock material (Figure 9.12). Streams also sort the solid sediment they transport because finer, lighter material is carried more readily than larger, heavier particles. Streams transport their load of sediment in three ways: (1) in solution (**dissolved load**), (2) in suspension (**suspended load**), and (3) sliding, rolling, or bouncing along the bottom (**bed load**).

Dissolved Load

Most of the dissolved load is brought to a stream by groundwater and is dispersed throughout the flow. When water percolates through the ground, it acquires soluble soil compounds. Then it seeps through cracks and pores in bedrock, dissolving additional mineral matter. Eventually much of this mineral-rich water finds its way into streams.

The velocity of streamflow has no effect on a stream's ability to carry its dissolved load; material in the solution goes wherever the stream goes. Precipitation of the dissolved mineral matter occurs when the chemistry of the water changes, when organisms create hard parts, or when the water enters an inland "sea," located in an arid climate where the rate of evaporation is high.

Suspended Load

Most streams carry the largest part of their load in *suspension*. Indeed, the muddy appearance created by suspended sediment is the most obvious portion of a stream's load (Figure 9.13). Usually only fine silt- and clay-size particles can be carried this way, but during a flood, larger particles are transported as well. During a flood, the total quantity of material carried in suspension increases dramatically, as people whose homes have been sites for the deposition of this material can attest.

The type and amount of material carried in suspension are controlled by two factors: the stream's flow velocity and the settling velocity of each sediment grain. **Settling velocity** is defined as the speed at which a particle falls

Figure 9.13 **Suspended load** An aerial view of the Colorado River in the Grand Canyon. Heavy rains washed sediment into the river. (Photo by Michael Collier)

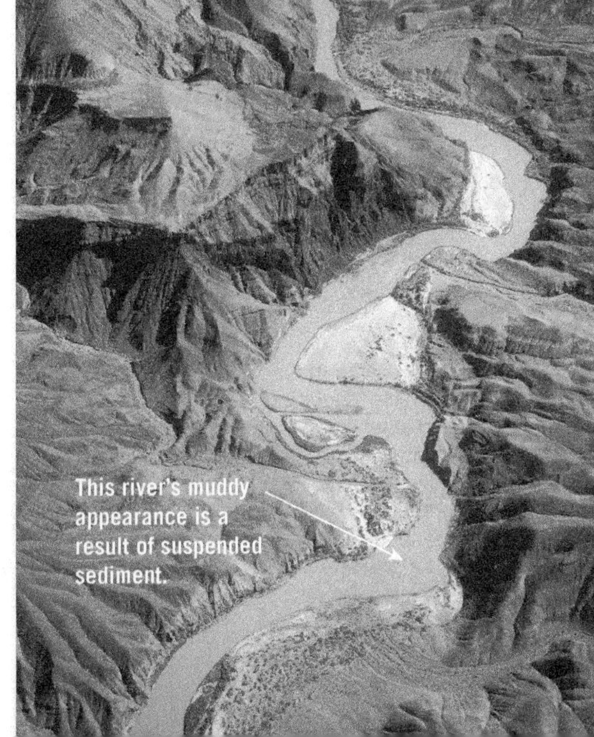

EYE ON EARTH 9.1

The White River in Arkansas is a tributary of the Mississippi River. As you can see in this aerial image, this part of the river has many twists and turns, which are called *meanders* (and described in more detail later in the chapter).

QUESTION 1 *The color of the White River is brown. What part of the stream's load gives it this color?*

QUESTION 2 *If a channel were created across the narrow neck of land shown by the arrow, how would the river's gradient change?*

QUESTION 3 *How would the flow velocity be affected by the formation of such a channel?*

through a still fluid. The larger the particle, the more rapidly it settles toward the streambed (Figure 9.14). In addition to size, the shape and specific gravity of particles also influence settling velocity. Flat grains sink through water more slowly than do spherical grains, and dense particles fall toward the bottom more rapidly than do less dense particles. As long as flow velocity exceeds settling velocity, sediment remains suspended and is transported downstream. Deposition occurs when flow velocity falls below the settling velocity of a particle.

Bed Load A portion of a stream's load of solid material consists of sediment that is too large to be carried in suspension. These coarser particles move along the bottom (bed) of the stream and constitute the *bed load*. In terms of the erosional work accomplished by a downcutting stream, the grinding action of the bed load is of great importance.

The particles that make up the bed load move by rolling, sliding, and saltation. Sediment moving by **saltation** (*saltare* = to leap) appears to jump or skip along the stream bed. This occurs as particles are propelled upward by collisions or lifted by the current and then carried downstream a short distance until gravity pulls them back to the bed of the stream. Particles that are too large or heavy to move by saltation either roll or slide along the bottom, depending on their shapes.

Compared with the movement of suspended load, the movement of bed load through a stream network tends to be less rapid and more localized. A study conducted on a glacially fed river in Norway determined that suspended sediments took only a day to exit the drainage basin, while the bed load required several decades to travel the same distance. Depending on the discharge and slope of the channel, coarse gravels may only be moved during times of high flow, while boulders are moved only during exceptional floods. Once large particles are set in motion, they are usually carried short distances. Along some stretches of a stream, bed load cannot be carried at all until it is broken into smaller particles.

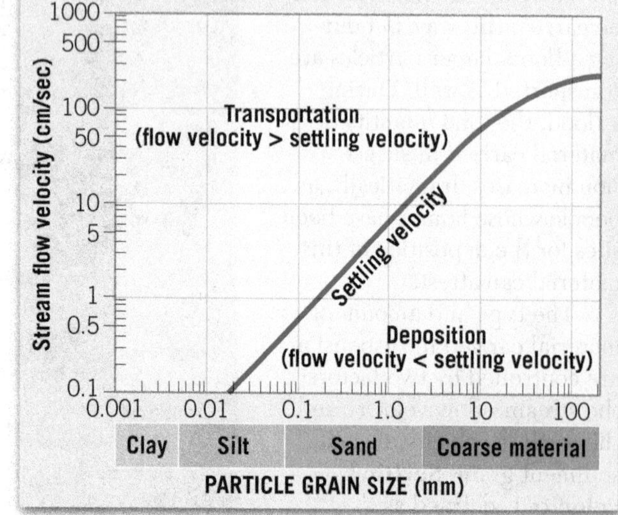

Figure 9.14 Settling velocity Settling velocity is the speed at which a particle falls through still water. Deposition of suspended particles occurs when their settling velocities are greater than the stream's flow velocity.

Capacity and Competence A stream's ability to carry solid particles is described using two criteria: *capacity* and *competence*. **Capacity** is the maximum load of solid particles a stream can transport per unit of time. The greater the discharge, the greater the stream's capacity for hauling sediment. Consequently, large rivers with high flow velocities have large capacities.

Competence is a measure of a stream's ability to transport particles based on size rather than quantity. Flow velocity is the key: Swift streams have greater competencies than slow streams, regardless of channel size. A stream's competence increases proportionately to the square of its velocity. Thus, if the velocity of a stream doubles, the impact force of the water increases four times; if the velocity triples, the force increases nine times, and so forth. Hence, large boulders that are often visible during low water and seem immovable can, in fact, be transported during exceptional floods because of the stream's increased competence.

By now it should be clear why the greatest erosion and transportation of sediment occur during floods. The increase in discharge results in greater capacity, and the increased velocity produces greater competency. Rising velocity makes the water more turbulent, and larger particles are set in motion. In just a few days, or perhaps a few hours, a stream at flood stage can erode and transport more sediment than it does during many months of normal flow.

Deposition of Sediment

Deposition occurs whenever a stream slows, causing a reduction in competence. Put another way, particles are deposited when flow velocity is less than the settling velocity; as a stream's flow velocity decreases, sediment begins to settle, largest particles first. In this manner, stream transport provides a mechanism by which solid particles of various sizes are separated. This process, called **sorting**, explains why particles of similar size are deposited together.

The general term for sediment deposited by streams is **alluvium**. Many different depositional features are composed of alluvium. Some occur within stream channels, some occur on the valley floor adjacent to a channel, and some are found at the mouth of a stream. We will consider the nature of these features later in the chapter.

CONCEPT CHECKS 9.4

1. List two ways in which streams erode their channels.
2. In what three ways does a stream transport its load? Which part of the load moves most slowly?
3. What is the difference between capacity and competence?
4. What is settling velocity? What factors influence settling velocity?

9.5 Stream Channels

Contrast bedrock and alluvial stream channels. Distinguish between two types of alluvial channels.

A basic characteristic of streamflow that distinguishes it from sheet flow is that streamflow is confined to a channel. A stream channel can be thought of as an open conduit that consists of the streambed and banks that act to confine the flow, except during floods.

Although this is somewhat oversimplified, we can divide stream channels into two types. A *bedrock channel* is one in which the stream is actively cutting into solid rock. In contrast, when the bed and banks are composed mainly of unconsolidated sediment, the channel is called an *alluvial channel*.

Bedrock Channels

As the name suggests, bedrock channels are cut into the underlying strata and typically form in the headwaters of river systems where streams have steep slopes. The energetic flow tends to transport coarse particles that actively abrade the bedrock channel. Potholes are often visible evidence of the erosional forces at work.

Steep bedrock channels often develop a sequence of *steps* and *pools*. Steps are steep segments where bedrock is exposed. These steep areas contain rapids or, occasionally, waterfalls. Pools are relatively flat segments where alluvium tends to accumulate.

The channel pattern exhibited by streams cutting into bedrock is controlled by the underlying geologic structure. Even when flowing over rather uniform bedrock, streams tend to exhibit winding or irregular patterns rather than flow in straight channels. Anyone who has gone whitewater rafting has observed the steep, winding nature of a stream flowing in a bedrock channel.

Alluvial Channels

Many stream channels are composed of loosely consolidated sediment (alluvium) and therefore can undergo significant changes in shape because the sediments are continually being eroded, transported, and redeposited. The major factors affecting the shapes of these channels are the average size of the sediment being transported, the channel gradient, and the discharge.

Alluvial channel patterns reflect a stream's ability to transport its load at a uniform rate, while expending the least amount of energy. Thus, the size and type of sediment being carried help determine the nature of the stream channel. Two common types of alluvial channels are *meandering channels* and *braided channels*.

Meandering Streams Streams that transport much of their load in suspension generally move in sweeping bends called **meanders**. These streams flow in relatively deep, smooth channels and transport mainly mud (silt and clay), sand, and occasionally fine gravel. The lower Mississippi River exhibits a channel of this type.

Meandering channels evolve over time as individual meanders migrate across the floodplain. Most of the erosion is focused at the outside of the meander, where velocity and turbulence are greatest. In time, the outside bank is undermined, especially during periods of high water. Because the outside of a meander is a zone of active erosion, it is often referred to as the **cut bank** (Figure 9.15). Debris acquired by the stream at the cut bank moves downstream, where the coarser material is generally deposited as **point bars** on the insides of meanders. In this manner, meanders migrate laterally by eroding the outsides of the bends and depositing sediment on the insides without appreciably changing their shape.

In addition to migrating laterally, the bends in a channel also migrate down the valley. This occurs because erosion is more effective on the downstream (downslope) side of the meander. Sometimes the downstream migration of a meander is slowed when it reaches a more resistant bank material. This allows the next meander upstream to gradually erode the material between the two meanders, as shown in Figure 9.16. Eventually, the river may erode through the narrow neck of land, forming a new, shorter channel segment called a **cutoff**. Because of its shape, the abandoned bend is called an **oxbow lake**.

Braided Streams Some streams consist of a complex network of converging and diverging channels that thread their way among numerous small islands or gravel bars. Because these channels have an interwoven appearance, they are called **braided channels**. Braided channels form where a large proportion of the stream's load consists of coarse material (sand and gravel) and the

SmartFigure 9.15
Formation of cut banks and point bars By eroding its outer bank and depositing material on the inside of the bend, a stream is able to shift its channel.

TUTORIAL
https://goo.gl/eFrBNv

Maximum velocity
Maximum velocity
Erosion of a cut bank along the Newaukum River in southwestern Washington State.
USGS
Deposition of point bar
White River near Vernal, Utah
The cut bank forms on the outside of a meander where flow velocity and turbulence are greatest.
Point bar
Erosion of cut bank
As water slows on the inside of a meander, coarser material is deposited as a point bar.
Michael Collier
Maximum velocity

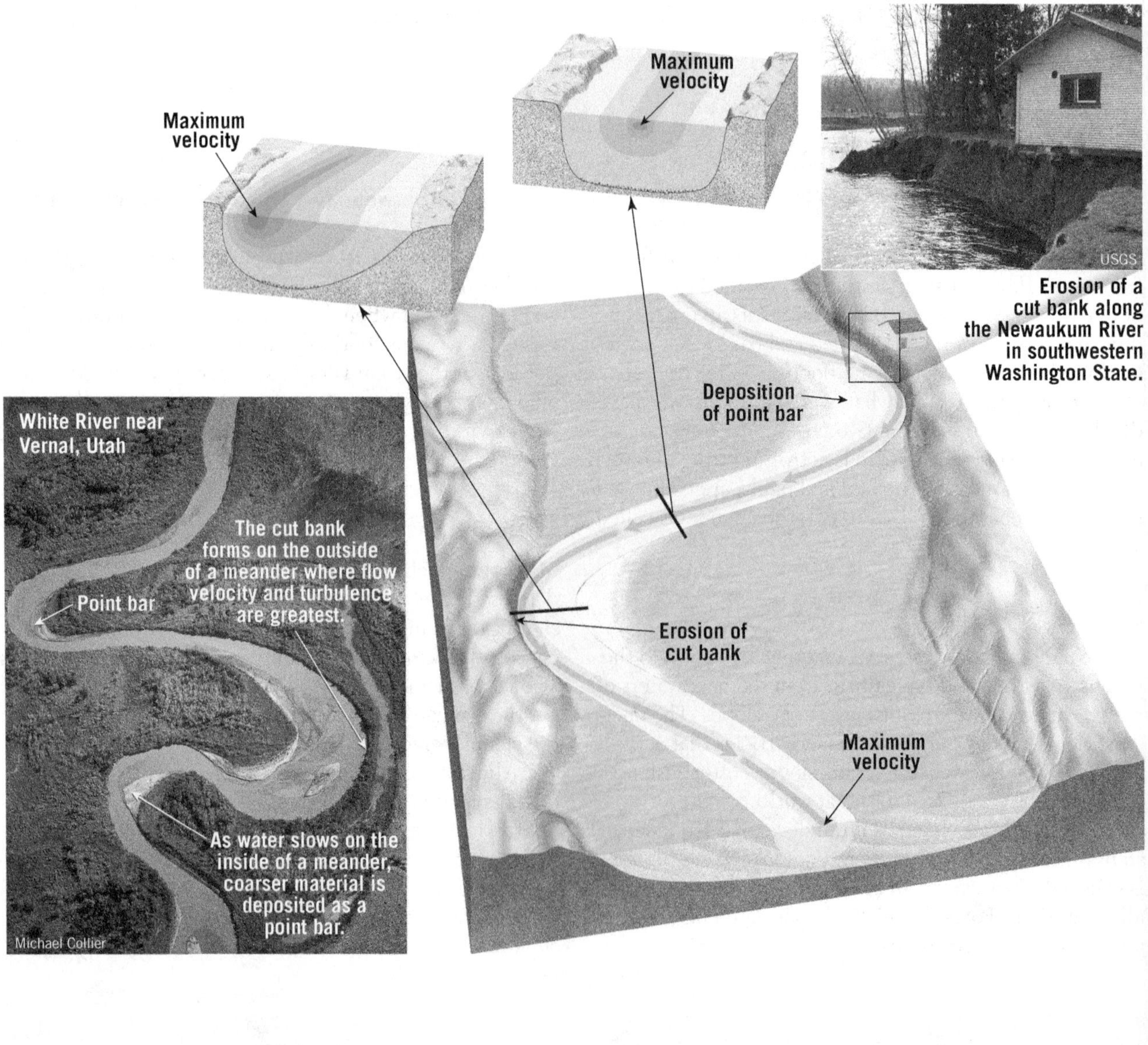

stream has a highly variable discharge. Because the bank material is readily erodible, braided channels are wide and shallow.

One setting in which braided streams form is at the end of glaciers, where there is a large seasonal variation in discharge (Figure 9.17). During the summer, large amounts of ice-eroded sediment are dumped into the meltwater streams flowing away from the glacier. However, when flow is sluggish, the stream deposits the coarsest material as elongated structures called *bars*. The flow then splits into several paths around the bars. During the next period of high flow, the laterally shifting channels erode and redeposit much of this coarse sediment, thereby transforming the entire streambed. In some braided streams, the bars have built semipermanent islands anchored by vegetation.

CONCEPT CHECKS 9.5

1. Are bedrock channels more likely to be found near the head or the mouth of a stream?
2. Describe or sketch the development of a meander, including how an oxbow lake forms.
3. Describe a situation that might cause a stream to become braided.

SmartFigure 9.16 **Formation of an oxbow lake** Oxbow lakes occupy abandoned meanders. This is an aerial view of an oxbow lake created by the meandering Green River near Bronx, Wyoming.
(Photo by Michael Collier)

ANIMATION
https://goo.gl/0lWPw5

Figure 9.17 **Braided stream** The Knik River is a classic braided stream with multiple channels separated by migrating gravel bars. The Knik is choked with sediment from four melting glaciers in the Chugach Mountains north of Anchorage, Alaska.
(Photo by Michael Collier)

9.6 Shaping Stream Valleys

Contrast narrow V-shaped valleys, broad valleys with floodplains, and valleys that display incised meanders.

A **stream valley** consists of a channel and the surrounding terrain that directs water to the stream. It includes the *valley floor*, which is the lower, flatter area that is partially or totally occupied by the stream channel, and the sloping *valley walls* that rise above the valley floor on both sides. Alluvial channels often flow in valleys that have wide valley floors consisting of sand and gravel deposited in the channel and clay and silt deposited by floods. Bedrock channels, on the other hand, tend to be located in narrow V-shaped valleys. In some arid regions, where weathering is slow and rock is particularly resistant, narrow valleys having nearly vertical walls are also found. Such features are called *slot canyons*. Stream valleys exist on a continuum from narrow, steep-sided valleys to valleys that are so flat and wide that the valley walls are not discernible.

Streams, with the aid of weathering and mass movement, shape the landscape through which they flow. As a result, streams continuously modify the valleys they occupy.

Base Level and Stream Erosion

Streams cannot endlessly erode their channels deeper and deeper. There is a lower limit to how deep a stream can erode, a limit called **base level**. Although the idea is relatively straightforward, it is nevertheless a key concept in the study of stream activity. Base level is defined as the lowest elevation to which a stream can erode its channel. Essentially this is the level at which the mouth of a stream enters the ocean, a lake, or another stream. Base level accounts for the fact that most stream profiles have low gradients near their mouths because the streams are approaching the elevation below which they cannot erode their beds.

Two general types of base level are recognized. Sea level is considered the *ultimate base level* because it is the lowest level to which stream erosion could lower the land. *Temporary*, or *local, base levels* include lakes, resistant layers of rock, and main streams that act as base levels for their tributaries. For example, when a stream enters a lake, its velocity quickly approaches zero, and its ability to erode ceases. Thus, the lake prevents the stream from eroding below its level at any point upstream from the lake. However, because the outlet of the lake can cut downward and drain the lake, the lake is only a temporary hindrance to the stream's ability to downcut its channel. In a similar manner, the layer of resistant rock at the lip of a waterfall acts as a temporary base level. Until the ledge of hard rock is eliminated, it limits the amount of downcutting upstream.

Any change in base level causes a corresponding readjustment of stream activities. When a dam is built along a stream, the reservoir that forms behind it raises the base level of the stream (Figure 9.18). Upstream from the dam the gradient is reduced, lowering the stream's velocity and, hence, its sediment-transporting ability. The stream, now having too little energy to transport its entire load, deposits sediment that builds up its channel. Deposition is the dominant process until the stream's gradient increases sufficiently to transport its load.

Figure 9.18 **Building a dam** The base level upstream from the reservoir is raised, which reduces the stream's flow velocity and leads to deposition and a reduced gradient.

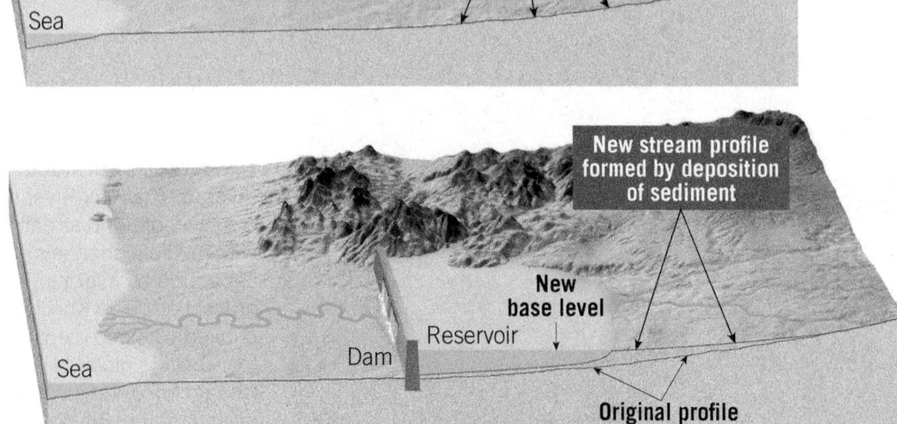

Valley Deepening

When a stream's gradient is steep and the channel is well above base level, downcutting is the dominant activity. Abrasion caused by bed load sliding and rolling along the bottom and the hydraulic power of fast-moving water slowly lower the streambed. The result is usually a V-shaped valley with steep sides. A classic example of a V-shaped valley is the section of the Yellowstone River shown in Figure 9.19.

The most prominent features of a V-shaped valley are *rapids* and *waterfalls*. Both occur where the stream's gradient increases significantly, a situation usually caused by variations in the erodibility of the bedrock into which a stream channel is cutting. Resistant beds create rapids by acting as a temporary base level upstream while allowing downcutting to continue downstream. In time, erosion usually eliminates the resistant rock. Waterfalls are places where the stream makes an abrupt vertical drop.

◀ Figure 9.19 **Yellowstone River** The V-shaped valley, rapids, and waterfalls indicate that the river is vigorously downcutting. (Photo by Charles A. Blakeslee/AGE Fotostock)

Valley Widening

Once a stream has cut its channel closer to base level, downward erosion becomes less dominant. At this point, the stream's channel takes on a meandering pattern, and more of the stream's energy is directed from side to side. The result is a widening of the valley as the river cuts away first at one bank and then the other (Figure 9.20). The continuous lateral erosion caused by shifting of the stream's meanders produces an increasingly broad, flat valley floor covered with alluvium. This feature, called a **floodplain**, is appropriately named because when a river overflows its banks during flood stage, it inundates the floodplain.

Over time the floodplain widens to the point that the stream is actively eroding the valley walls in only a few places. In fact, with large rivers, such as the Mississippi River, the distance from one valley wall to another can exceed 160 kilometers (100 miles).

Changing Base Level and Incised Meanders

We usually expect a stream with a highly meandering course to be on a floodplain in a wide valley. However, certain rivers exhibit meandering channels that flow in steep, narrow valleys. Such meanders are called **incised** (*incisum* = to cut into) **meanders** (see chapter-opening photo).

How do these features form? Originally the meanders probably developed on the floodplain of a stream that was in balance with its base level (Figure 9.21A). Then, a change in base level caused the stream to begin downcutting. Such a change can be caused either by a drop in a downstream base level or by uplift of the land on which the river is flowing. For example, regional uplifting of the Colorado Plateau in the southwestern United States generated incised meanders on several rivers (Figure 9.21B). As the

▼ SmartFigure 9.20
Development of an erosional floodplain Continuous side-to-side erosion by shifting meanders gradually produces a broad, flat valley floor. Alluvium deposited during floods covers the valley floor.

CONDOR VIDEO
https://goo.gl/0sxaWY

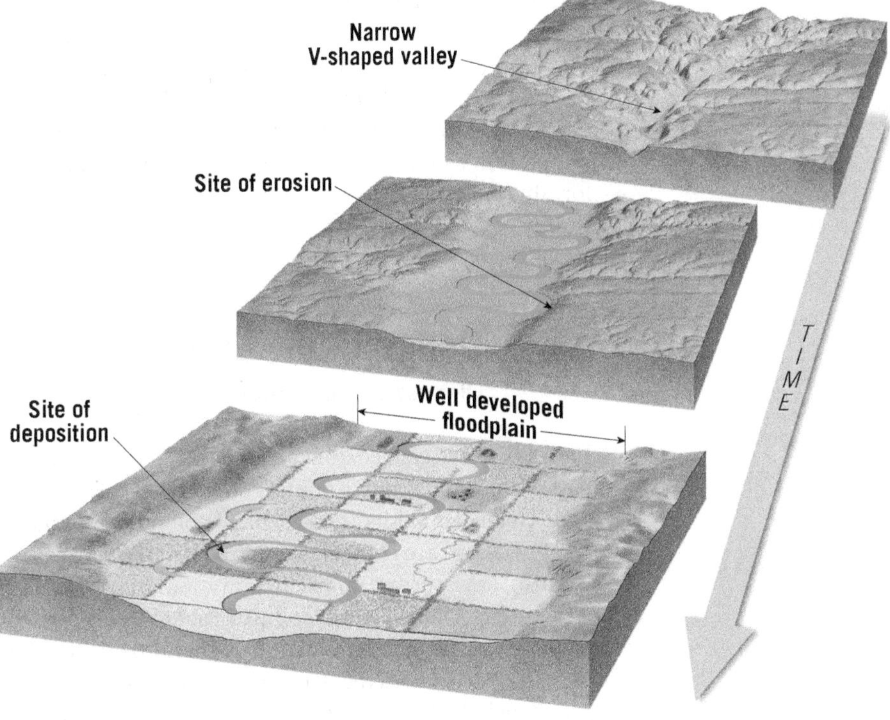

SmartFigure 9.21
Incised meanders These diagrams illustrate the development of incised meanders such as those shown in the chapter-opening photo.

TUTORIAL
https://goo.gl/P3Zlz9

A. Before uplift of the Colorado Plateau, the river was meandering on a floodplain.

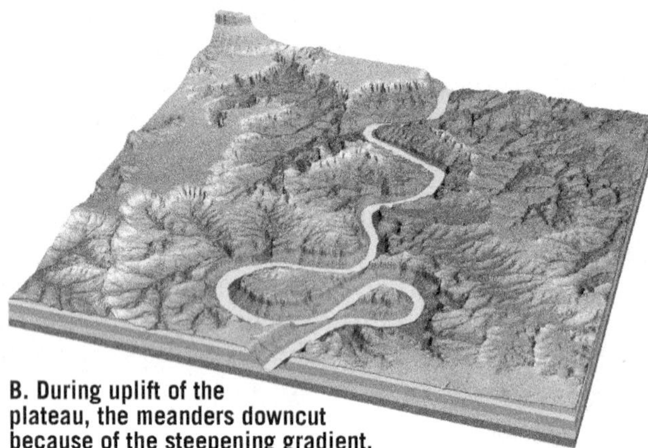

B. During uplift of the plateau, the meanders downcut because of the steepening gradient.

SmartFigure 9.22 **Stream terraces** Terraces result when a stream adjusts to a relative drop in base level.

CONDOR VIDEO
https://goo.gl/YSP8UV

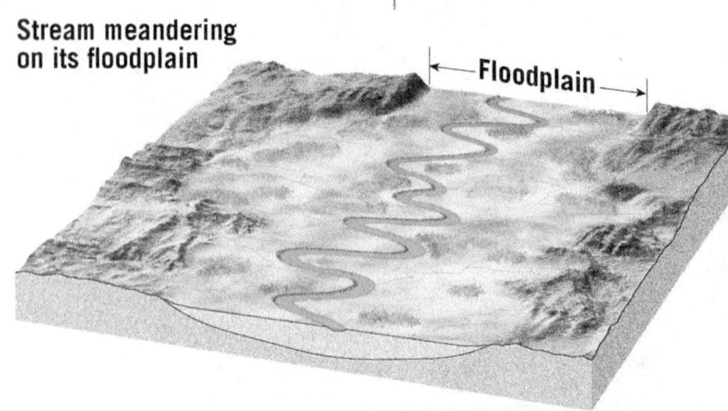

Because of a relative drop in base level, the river erodes downward through previously deposited alluvium. Eventually a new floodplain forms. Terraces represent elevated remnants of the former floodplain.

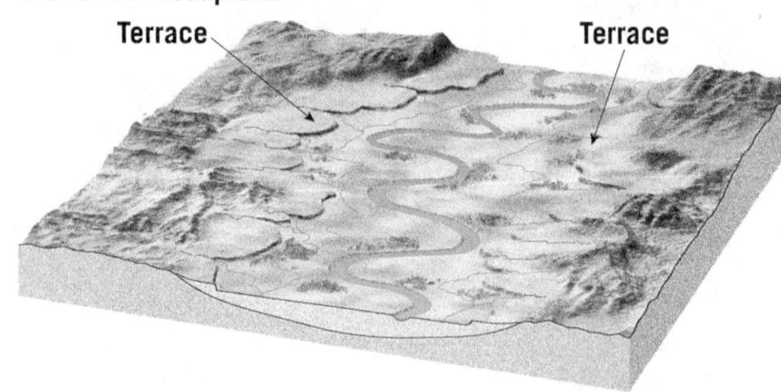

plateau gradually rose, meandering rivers began downcutting because of their steepening gradient.

Other features associated with a relative drop in base level are **stream terraces**. After a river that had been flowing on a floodplain has adjusted to a relative drop in base level, it may once again produce a floodplain at a level below the old one. As shown in Figure 9.22, the remnants of the former floodplain are present as relatively flat surfaces above the newly forming floodplain.

CONCEPT CHECKS 9.6

1. Define *base level* and distinguish between ultimate base level and temporary base level.
2. Explain why V-shaped valleys often contain rapids and waterfalls.
3. Describe or sketch how an erosional floodplain develops.
4. Relate the formation of incised meanders to changes in base level.

9.7 Depositional Landforms

Discuss the formation of deltas, natural levees, and alluvial fans.

Recall that a stream continually picks up sediment in one part of its channel and deposits it downstream. These small-scale channel deposits are most often composed of sand and gravel and are commonly referred to as **bars**. Such features, however, are only temporary, and the material is picked up again and eventually carried to the ocean. In addition to sand and gravel bars, streams also create other depositional features that have somewhat longer life spans. These include *deltas*, *natural levees*, and *alluvial fans*.

Deltas

A **delta** forms where a sediment-laden stream enters the relatively still waters of a lake, an inland sea, or the ocean (Figure 9.23). As the stream's forward motion slows, sediments are deposited by the dying current. As the delta grows outward, the stream's gradient continually lessens. This circumstance eventually causes the channel to become choked with sediment deposited

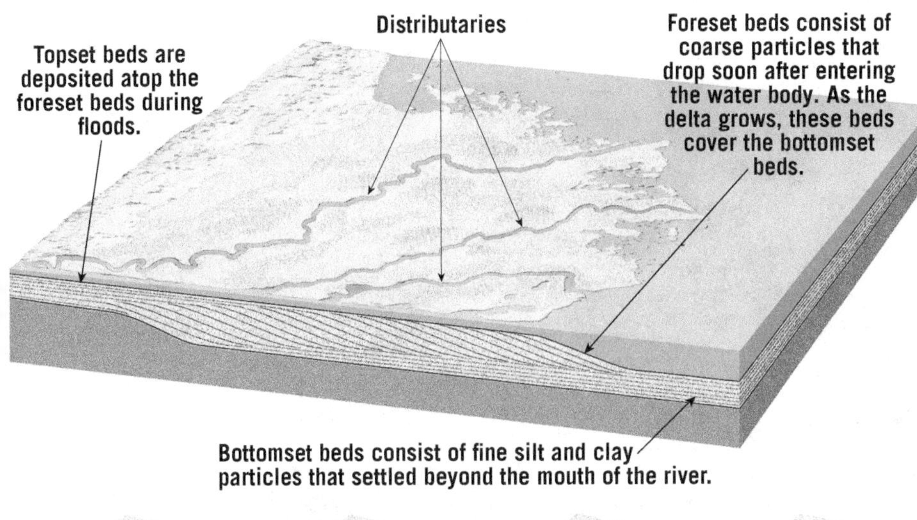

Foreset beds consist of coarse particles that drop soon after entering the water body. As the delta grows, these beds cover the bottomset beds.
Bottomset beds consist of fine silt and clay particles that settled beyond the mouth of the river.

As the stream extends its channel, the gradient is reduced. During flood stage some of the flow is diverted to a shorter, higher-gradient route forming a new distributary.

Figure 9.23 Formation of a simple delta These diagrams illustrate the structure and growth of a simple delta that forms in relatively quiet waters.

from the slowing water. As a consequence, the river seeks a shorter, higher-gradient route to base level, as illustrated in Figure 9.23. This illustration shows the main channel dividing into several smaller ones, called **distributaries**. Most deltas are characterized by these shifting channels that act in an opposite way to that of tributaries.

Rather than carry water into the main channel, distributaries carry water away from the main channel. After numerous shifts of the channel, a delta may grow into a roughly triangular shape like the Greek letter delta (Δ), for which it is named. Not all deltas exhibit this idealized shape, however. Differences in shoreline configuration and in the nature and strength of wave activity result in a diversity of shapes. Many large rivers have deltas extending over thousands of square kilometers. The delta of the Mississippi River is one example. It resulted from the accumulation of huge quantities of sediment derived from the vast region drained by the river and its tributaries (see Figure 9.3). Today, New Orleans rests where there was ocean less than 5000 years ago. Figure 9.24 shows that portion of the Mississippi delta that has been built over the past 6000 years. As you can see, the delta is actually a series of seven coalescing subdeltas. Each formed when the river left its existing channel in favor of a shorter, more direct path to the Gulf of Mexico. The individual subdeltas interfinger and partially cover one another, producing a very complex structure. The present subdelta, called a *bird-foot* delta because of the configuration of its distributaries, has been built by the Mississippi in the past 500 years.

MOBILE FIELD TRIP
https://goo.gl/lkUunm

SmartFigure 9.24 Growth of the Mississippi River delta During the past 6000 years, the river has built a series of seven coalescing subdeltas. The numbers indicate the order in which the subdeltas were deposited. The present bird-foot delta (number 7) represents the activity of the past 500 years. The left inset shows the point where the Mississippi may sometime break through (arrow) and the shorter path it would take to the Gulf of Mexico. The Mississippi delta includes about 12,000 square kilometers (3 million acres) of coastal wetlands—40 percent of all coastal wetlands in the contiguous United States. The Mobile Field Trip explores these wetlands. (Image courtesy of JPL/Cal Tech/NASA)

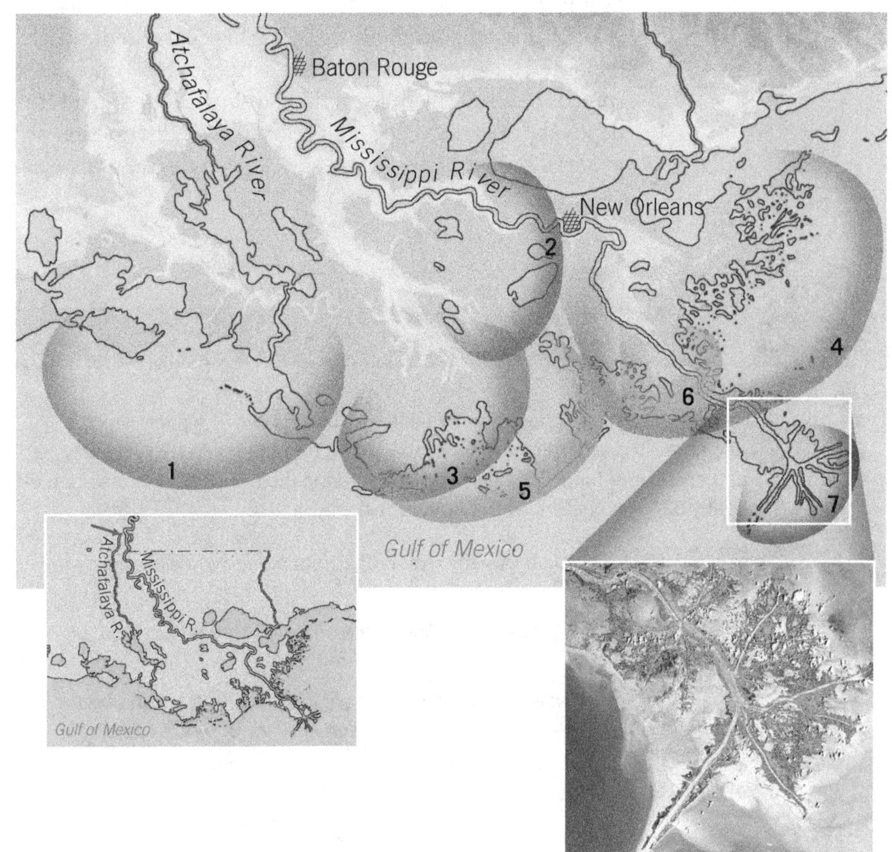

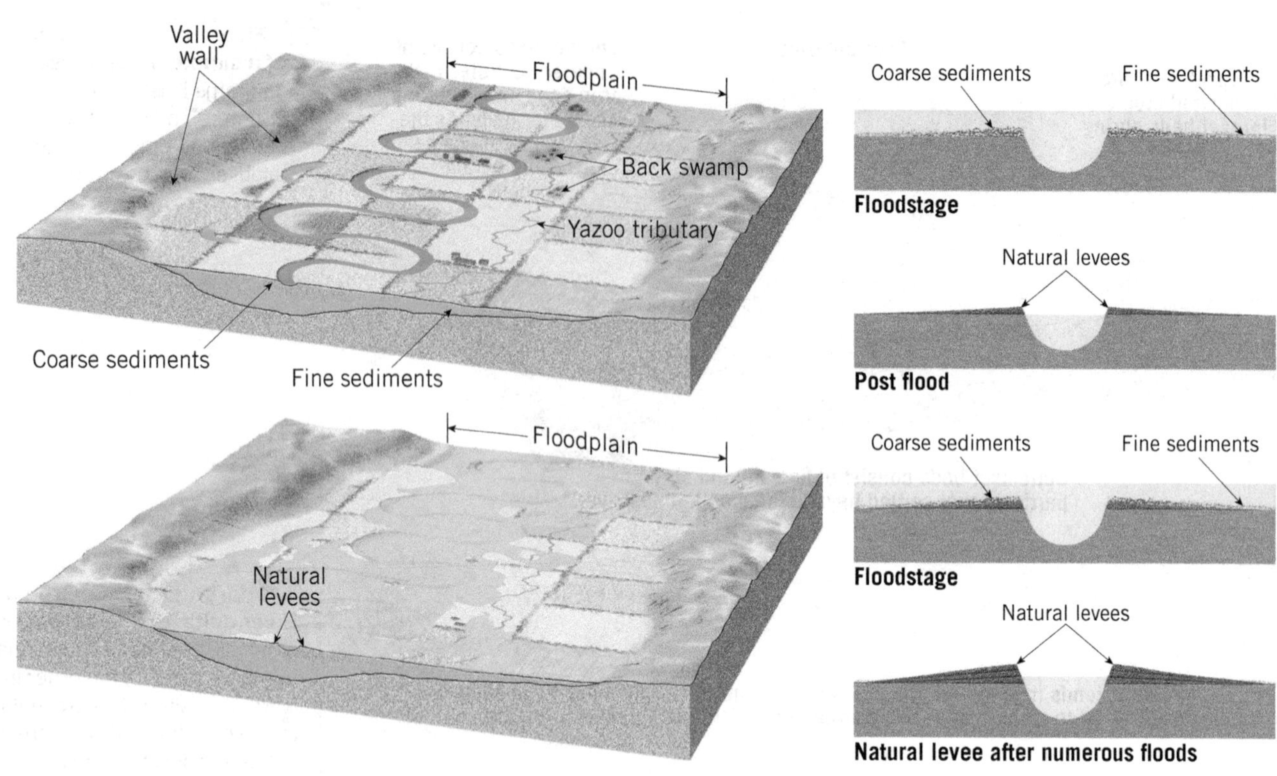

SmartFigure 9.25 **Formation of a natural levee** These gently sloping structures that parallel a river channel are created by repeated floods. Because the ground next to the channel is higher than the adjacent floodplain, back swamps and yazoo tributaries may develop.

ANIMATION
https://goo.gl/UdXGH3

Natural Levees

Some rivers occupy valleys with broad floodplains and build **natural levees** that parallel their channels on both banks (Figure 9.25). Natural levees are built by successive floods over many years. When a stream overflows its banks, its velocity immediately diminishes, leaving coarse sediment deposited in strips bordering the channel. As the water spreads out over the valley, a thin layer of fine sediment is deposited over the valley floor. This uneven distribution of material produces the very gentle slope of the natural levee.

The natural levees of the lower Mississippi rise about 6 meters (20 feet) above the floodplain. The area behind the levee is characteristically poorly drained because water cannot flow up the levee and into the river. Marshes called **back swamps** result. A tributary stream that cannot enter a river because levees block the way often has to flow parallel to the river until it can breach the levee. Such streams are called **yazoo tributaries**, after the Yazoo River, which parallels the Mississippi for more than 300 kilometers (about 190 miles).

EYE ON EARTH 9.2

This satellite image shows the delta of the Yukon River. The river, which originates in northern British Columbia, flows through the Yukon Territory and across the tundra of Alaska before entering the Bering Sea, a distance of nearly 3200 kilometers (about 2000 miles).

QUESTION 1 *Explain why the river breaks into numerous channels as it crosses the delta.*

QUESTION 2 *What term is applied to the channels that radiate across the delta?*

QUESTION 3 *Notice the cloud of sediment in the water surrounding the delta. Are these sediments more likely sand and gravel or silt and clay? Explain.*

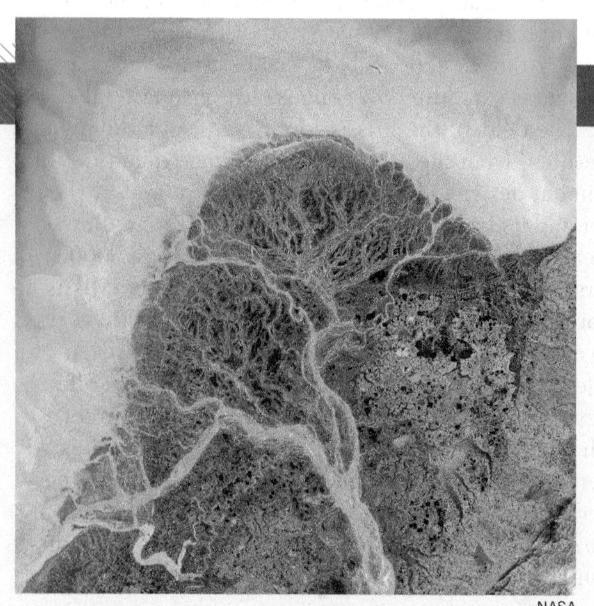

NASA

Alluvial Fans

Alluvial fans typically develop where a high-gradient stream leaves a narrow valley in mountainous terrain and comes out suddenly onto a broad, flat plain or valley floor (see Figure 10.33, page 342). Alluvial fans form in response to the abrupt drop in gradient combined with the change from a narrow channel of a mountain stream to less confined channels at the base of the mountains. The sudden drop in velocity causes the stream to dump its load of sediment quickly in a distinctive cone- or fan-shaped accumulation. As illustrated in Figure 10.33 (page 342), the surface of the fan slopes outward in a broad arc from an apex at the mouth of the steep valley. Usually, coarser material is dropped near the apex of the fan, while finer material is carried toward the base of the deposit.

Between rainy periods in deserts, little or no water flows across an alluvial fan, which is evident in the many dry channels that cross its surface. Thus, fans in dry regions grow intermittently, receiving considerable water and sediment only during wet periods. Because steep canyons in dry regions are prime locations for debris flows, many alluvial fans have debris-flow deposits interbedded with the coarse alluvium.

CONCEPT CHECKS 9.7

1. What feature may form where a stream enters the relatively still waters of a lake, an inland sea, or the ocean?
2. What are distributaries, and why do they form?
3. Briefly describe the formation of a natural levee. How is this feature related to back swamps and yazoo tributaries?
4. Describe the formation of an alluvial fan.

9.8 Floods and Flood Control

Distinguish between regional floods and flash floods. Describe some common flood control measures.

When the discharge of a stream becomes so great that it exceeds the capacity of its channel, it overflows its banks as a **flood**. Floods are among the most common and most destructive of all natural hazards. They are, nevertheless, simply part of the *natural* behavior of streams.

Causes of Floods

Rivers flood because of the weather. Rapid melting of snow and/or major storms that bring heavy rains over a large areas cause most *regional floods*. In April 2011, unrelenting storms brought record rains to the Mississippi watershed. The Ohio Valley, which makes up the eastern portion of the Mississippi's drainage basin, received nearly 300 percent of its normal springtime precipitation. When that rainfall combined with water from the past winter's extensive and rapidly melting snowpack, the Mississippi River and many of its tributaries began to swell to record levels by early May. The resulting floods were among the largest and most damaging in nearly a century. Like most other regional floods, these were associated with weather phenomena that could be forecast with a good deal of accuracy. This allowed adequate time to warn and evacuate thousands of people who were in harm's way. Although economic losses approached $4 billion, loss of life was small.

Flash floods often occur with little warning and are potentially deadly because they produce rapid rises in water levels and can have devastating flow velocities (see GEOgraphics 9.1). Rainfall intensity and duration, surface conditions, and topography are among the factors that influence flash flooding. Mountainous areas are susceptible to flash floods because steep slopes can quickly funnel runoff into narrow canyons. Urban areas are also susceptible to them because a high percentage of the surface area is composed of impervious surfaces, such as roofs, streets, and parking lots, where runoff is very rapid.

Human interference with a stream system can worsen or even cause floods. A prime example is the failure of a dam or an artificial levee. These structures are built for flood protection and are designed to contain floods of a certain magnitude. If a dam or levee fails or is washed out, the water behind it is released and becomes a flash flood.

Flood Control

Several strategies have been devised to eliminate or reduce the catastrophic effects of floods. Engineering efforts include the construction of artificial levees, the building of flood-control dams, and river channelization.

Artificial Levees *Artificial levees* are earthen mounds built on the banks of a river to increase the volume of water the channel can hold. These most common of stream-containment structures have been used since ancient times and continue to be used today. Artificial levees are usually easy to distinguish from natural levees because their slopes are much steeper. When exceptional floods threaten to overwhelm levees in densely populated areas, water is sometimes intentionally diverted

GEOGRAPHICS 9.1

Flash Floods

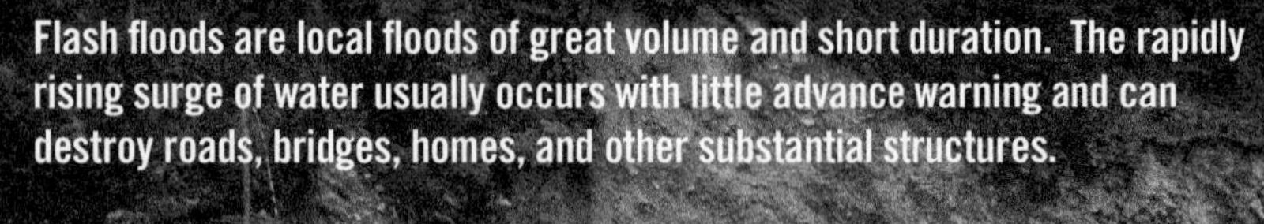

Flash floods are local floods of great volume and short duration. The rapidly rising surge of water usually occurs with little advance warning and can destroy roads, bridges, homes, and other substantial structures.

USGS

The power of a flash flood is illustrated by the Big Thompson River flood of July 31, 1976, in Colorado. During a four-hour span more than 30 centimeters (12 inches) of rain fell on portions of the river's small drainage basin. This amounted to nearly three-quarters of the average yearly total. The flash flood in the narrow canyon lasted only a few hours, but cost 139 people their lives.

Urban development increases runoff. As a result, peak discharge and flood frequency increase. A recent study indicated that the area of impervious surfaces in the 48 contiguous United States is roughly equal to the area of the state of Ohio (44,000 mi^2).

Most people do not appreciate the power of moving water. Many automobiles will float and be swept away in a strong current that is only 2 feet deep. More than half of all U. S. flash-flood fatalities are auto related!

Question:
Briefly explain how urban development influences the peak discharge of a strea

Average Annual Storm-Related Deaths in the U.S. (1986–2015)

In most years floods are responsible for the greatest number of storm-related deaths. The average number of hurricane deaths was dramatically affected by Hurricane Katrina in 2005 (more than 1000). For all other years on this graph, hurricane fatalities numbered fewer than 20.

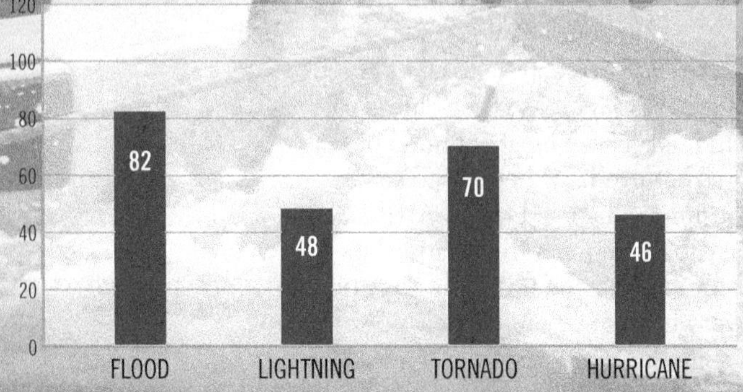

Effect of Urban Development on Flooding

Streamflow in Mercer Creek, an urban stream in western Washington, increases more quickly, reaches a higher peak discharge, and has a larger volume during a one-day storm on February 1, 2000, than streamflow in Newaukum Creek, a nearby rural stream. Streamflow during the following week, however, was greater in Newaukum Creek.

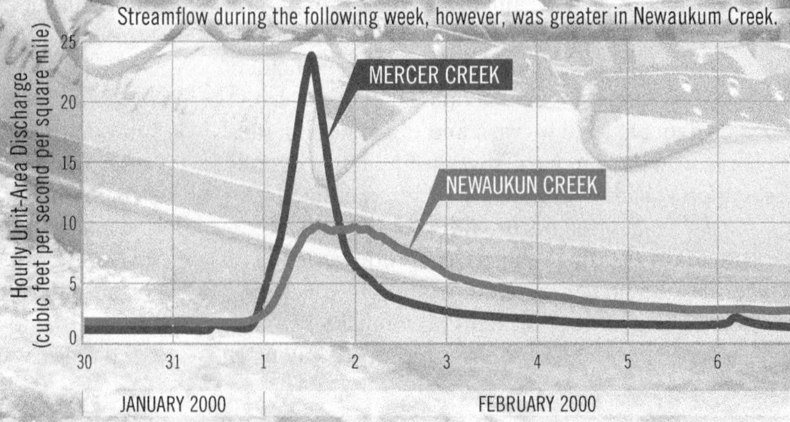

from a river by creating openings in artificial levees. The purpose is to spare vulnerable urban areas by allowing water to flood sparsely populated rural areas. The areas that are intentionally flooded are called *floodways* (Figure 9.26).

Flood-Control Dams *Flood-control dams* are built to store floodwater and then let it out slowly. This action lowers the flood crest by spreading it out over a longer time span. Since the 1920s, thousands of dams have been built on nearly every major river in the United States. Many dams have significant non-flood-related functions, such as providing water for irrigated agriculture and for hydroelectric power generation. Many reservoirs are also major regional recreational facilities.

Although dams may reduce flooding and provide other benefits, building these structures also has significant costs and consequences. For example, reservoirs created by dams may cover fertile farmland, useful forests, historic sites, and scenic valleys. Of course, dams trap sediment. Therefore, deltas and floodplains downstream erode because they are no longer replenished with silt during floods. Large dams can also cause significant ecological damage to river environments that took thousands of years to establish.

Building a dam is not a permanent solution to flooding. Sedimentation behind a dam causes the volume of its reservoir to gradually diminish, reducing the effectiveness of this flood-control measure.

Channelization *Channelization* involves altering a stream channel in order to speed the flow of water and prevent it from reaching flood height. This may simply involve clearing a channel of obstructions or dredging a channel to make it wider and deeper.

Another alteration involves straightening a channel by creating *artificial cutoffs*. The idea is that by shortening the stream, the gradient and the flow velocity are both increased. By increasing velocity, the larger discharge associated with flooding can be dispersed more rapidly.

Since the early 1930s, the U.S. Army Corps of Engineers has created many artificial cutoffs on the Mississippi for the purpose of increasing the efficiency of the channel and reducing the threat of flooding. In all, the river has been shortened more than 240 kilometers (150 miles). These efforts have been somewhat successful in reducing the height of the river in flood stage. However, channel shortening led to higher gradients and accelerated erosion of riverbank material, both of which necessitated further intervention. Following the creation of artificial cutoffs, massive riverbank protection to reduce erosion was installed along several stretches of the lower Mississippi.

A Nonstructural Approach All of the flood-control measures described so far have involved structural solutions aimed at "controlling" a river. These solutions are expensive and often give people residing on the floodplain a false sense of security.

Today, many scientists and engineers advocate a nonstructural approach to flood control. They suggest that an alternative to artificial levees, dams, and channelization is sound floodplain management. By identifying high-risk areas, appropriate zoning regulations can be implemented to minimize development and promote more appropriate land use.

CONCEPT CHECKS 9.8

1. Contrast regional floods and flash floods.
2. List and briefly describe three basic flood-control strategies.
3. What is meant by a *nonstructural approach* to flood control?

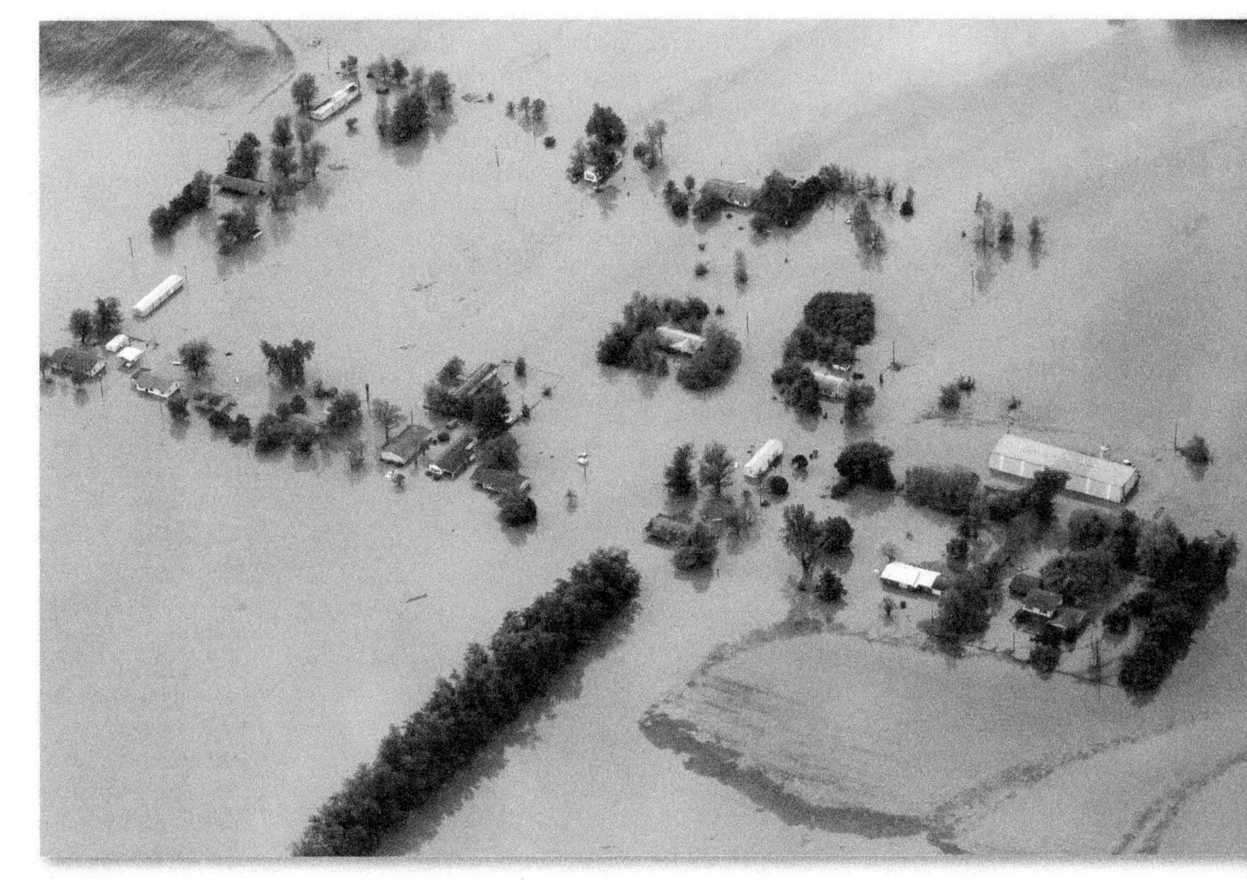

Figure 9.26 **Birds Point–New Madrid Floodway** In early May 2011 the intentional demolition of portions of an artificial levee on the west side of the Mississippi River immersed this floodway. A total of 520 square kilometers (200 square miles) of Missouri farmland were flooded. This rare action prevented the inundation of the small town of Cairo, Illinois. (Photo by David Carson/Zuma Press/Newscom)

9.9 Groundwater: Water Beneath the Surface

Discuss the importance of groundwater and describe its distribution and movement.

Groundwater is one of our most important and widely available resources, yet people's perceptions of the subsurface environment from which it comes are often unclear and incorrect. This is because the groundwater environment is hidden from view except in caves and mines, and the impressions people gain from these subsurface openings are often misleading. Observations on the land surface give an impression that Earth is "solid." This view is not changed very much when we enter a cave and see water flowing in a channel that appears to have been cut into solid rock.

Because of such observations, many people believe that groundwater occurs only in underground "rivers." However, actual rivers underground are extremely rare. In reality, much of the subsurface environment is not solid at all. Rather, it includes countless tiny *pore spaces* between grains of soil and sediment plus narrow joints and fractures in bedrock. Together, these spaces add up to an immense volume. Where these subsurface pore spaces are saturated with water, that stored water is called **groundwater**.

The Importance of Groundwater

Only a tiny percentage of Earth's water occurs underground. Nevertheless, this small percentage, stored in the rocks and sediments beneath Earth's surface, is a vast quantity. When the oceans are excluded and only sources of freshwater are considered, the significance of groundwater becomes more apparent.

Figure 9.27 shows an estimate of the distribution of freshwater in the hydrosphere. Clearly, the largest volume occurs as glacial ice. Groundwater is ranked second, with slightly more than 30 percent of the total. However, when glacial ice is excluded and just *liquid* water is considered, about 96 percent is groundwater. Without question, *groundwater represents the largest reservoir of freshwater that is readily available to humans.* Its value in terms of economics and human well-being is incalculable.

Worldwide, wells and springs provide water for cities, crops, livestock, and industry. In the United States, groundwater is the source of about 40 percent of the water used for all purposes (except hydroelectric power generation and power plant cooling). Groundwater is the drinking water for about 44 percent of the population and provides 40 percent of the water used for irrigation. In some areas, however, overuse of this basic resource has caused serious problems, including streamflow depletion, land subsidence, and increased pumping costs. In addition, groundwater contamination resulting from human activities is a real and growing threat in many places.

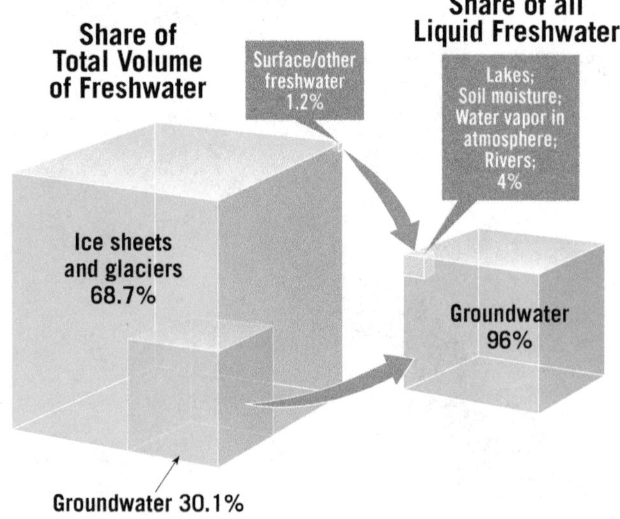

Figure 9.27 **Earth's freshwater** Groundwater is the major reservoir of liquid freshwater.

Geologic Importance of Groundwater

Geologically, groundwater is important as an erosional agent. The dissolving action of groundwater slowly removes soluble rock such as limestone, causing surface depressions known as sinkholes to form and creating subterranean caverns (Figure 9.28). The final section of this chapter describes the landforms associated with groundwater. Groundwater is also an equalizer of streamflow. Much of the water that flows in rivers is not direct runoff from rain and snowmelt. Rather, a large percentage of precipitation soaks into the ground and then moves slowly to stream channels. Groundwater is thus a form of storage that sustains streams during periods when rain does not fall. When we see water flowing in a river during a dry period, it is water from rain that fell at some earlier time and was stored underground.

Distribution of Groundwater

When rain falls, some of the water runs off, some returns to the atmosphere through evaporation and transpiration, and the remainder soaks into the ground. This last path is the primary source of practically all groundwater. The amount of water that takes each of these paths, however, varies greatly from time to time and place to place. Influential factors include the steepness of the slope, the nature of the surface material, the intensity of the rainfall, and the type and amount of vegetation. Heavy rains falling on steep slopes underlain by impervious materials will obviously result in a high percentage of the water running off. Conversely, if rain falls steadily and gently on more gradual slopes composed of materials that are more easily penetrated by water, a much larger percentage of the water soaks into the ground.

Underground Zones Some of the water that soaks in does not travel far because it is held by molecular attraction as a surface film on soil particles. This near-surface zone is called the *belt of soil moisture*. It is crisscrossed by roots, voids left by decayed roots, and animal and worm burrows that enhance the infiltration of rainwater into the soil. Soil water is used by plants for life functions and transpiration. Some of this water also evaporates directly back into the atmosphere.

Water that is not held as soil moisture penetrates downward until it reaches a zone where all the open spaces in sediment and rock are completely filled with water. This is the **zone of saturation**. Water within it is called *groundwater*. The upper limit of this zone is known as the **water table**. The area above the water table where the soil, sediment, and rock are not saturated is called the **unsaturated zone** (Figure 9.29). Although a considerable amount of water can be present in the unsaturated zone, this water cannot be pumped by wells because it clings too tightly to rock and soil particles. By contrast, below the water table, the water pressure is great enough to allow water to enter wells, thus permitting groundwater to be withdrawn for use. We will examine wells more closely later in the chapter.

Water Table The water table is rarely level, as we might expect something called a *table* to be. Instead, its shape is usually a subdued replica of the land's surface, reaching its highest elevations beneath hills and decreasing in height toward valleys. The water table of a wetland (swamp) is right at the surface. Lakes and streams generally occupy areas low enough that the water table is above the land surface.

Several factors contribute to the irregular surface of the water table. One important influence is the fact that groundwater moves very slowly. Because of this, water tends to "pile up" beneath high areas between stream valleys. If rainfall were to cease completely, these water "hills" would slowly subside and gradually approach the level of the adjacent valleys. However, new supplies of rainwater are usually added often enough to prevent this. Nevertheless, in times of extended drought, the water table may drop enough to dry up shallow wells. Other causes of the uneven water table are variations in rainfall and in the permeability of Earth materials from place to place.

Storage and Movement of Groundwater

The nature of subsurface materials strongly influences the rate of groundwater movement and the amount of groundwater that can be stored. Two factors are especially important: porosity and permeability.

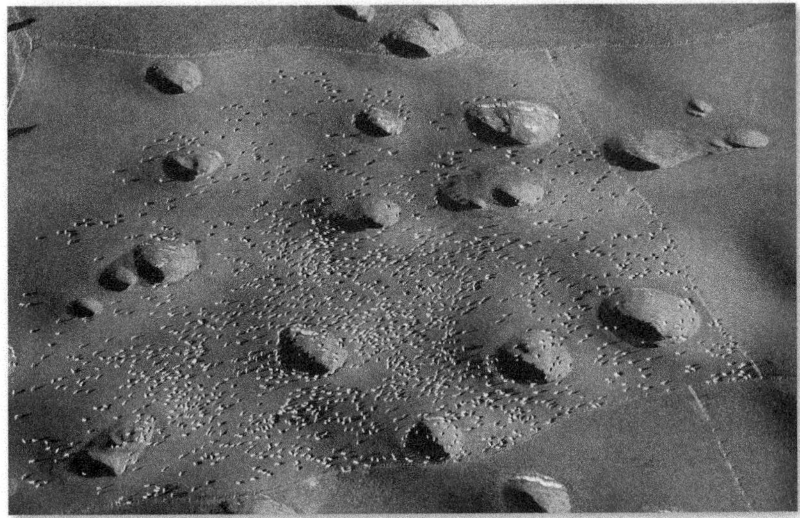

Figure 9.28 **Caverns and sinkholes A.** This is a view of the interior of New Mexico's Carlsbad Caverns. The dissolving action of acidic groundwater created the caverns. Later, groundwater deposited the limestone decorations. (Photo by Clint Farlinger/Alamy) **B.** Groundwater was responsible for creating these depressions, called sinkholes, west of Timaru on New Zealand's South Island. The white dots in this photo are grazing sheep. (Photo by David Wall/Alamy)

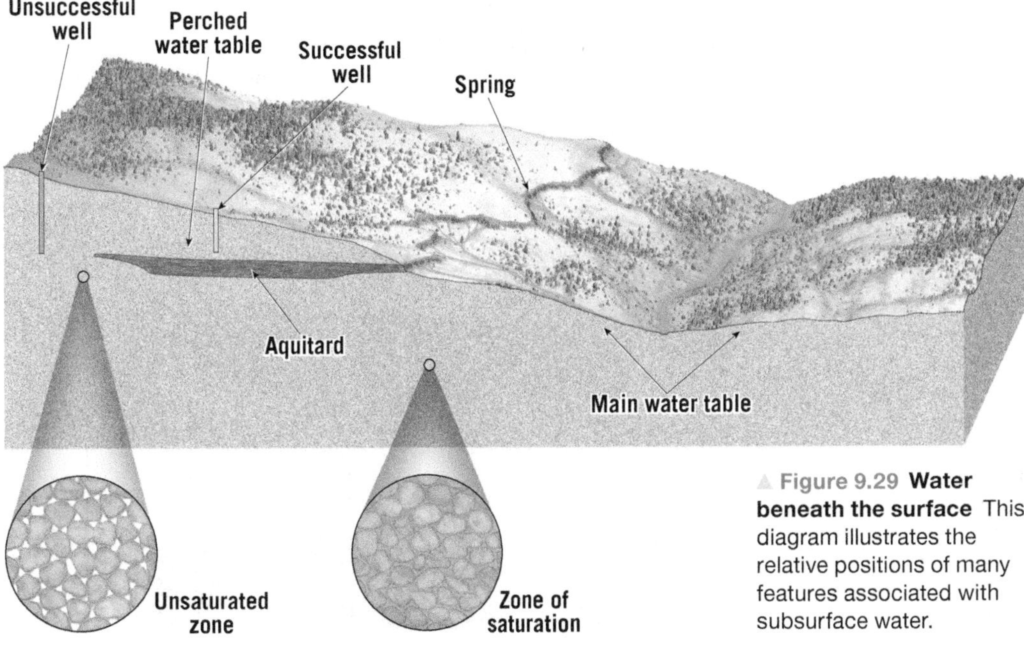

Figure 9.29 **Water beneath the surface** This diagram illustrates the relative positions of many features associated with subsurface water.

The beaker on the left is filled with 1000 ml of sediment. The beaker on the right is filled with 1000 ml of water.

The sediment-filled beaker now contains 500 ml of water. Pore spaces (porosity) must represent 50 percent of the volume of the sediment.

▲ Figure 9.30 **Porosity demonstration** Porosity is the percentage of the total volume of rock or sediment that consists of pore spaces.

Porosity Water soaks into the ground because bedrock, sediment, and soil contain countless voids, or openings. These openings, which are similar to those of a sponge, are called *pore spaces*. The quantity of groundwater that can be stored depends on the **porosity** of the material, which is the percentage of the total volume of rock or sediment that consists of pore spaces (Figure 9.30). Voids most often are spaces between sedimentary particles; also common are joints, faults, cavities formed by the dissolving of soluble rock such as limestone, and vesicles (voids left by gases escaping from lava).

Variations in porosity can be great. Sediment is commonly quite porous, and open spaces may occupy 10 percent to 50 percent of the sediment's total volume. Pore space depends on the size and shape of the grains, how they are packed together, the degree of sorting, and, in sedimentary rocks, the amount of cementing material. Most igneous and metamorphic rocks, as well as some sedimentary rocks, are composed of tightly interlocking crystals, so the voids between grains may be negligible. In these rocks, fractures must provide the voids.

Permeability Porosity alone cannot measure a material's capacity to yield groundwater. Rock or sediment may be very porous and still prohibit water from moving through it. The **permeability** of a material indicates its ability to *transmit* a fluid. Groundwater moves by twisting and turning through interconnected small openings. The smaller the pore spaces, the slower the groundwater moves. If the spaces between particles are too small, or if these spaces are not connected to each other, water cannot move at all. For example, clay's ability to store water can be great due to its high porosity, but its pore spaces are so small that water is unable to move through it. Thus, we say that clay is *impermeable*.

Aquitards and Aquifers Impermeable layers such as clay that hinder or prevent water movement are termed **aquitards** (*aqua* = water, *tard* = slow). In contrast, larger particles, such as sand or gravel, have larger pore spaces. Therefore, water moves with relative ease. Permeable rock strata or sediments that transmit groundwater freely are called **aquifers** ("water carriers"). Aquifers are important because they are the water-bearing layers sought by well drillers.

Groundwater Movement

The movement of most groundwater is exceedingly slow, from pore to pore. A typical rate is a few centimeters per day. The energy that makes the water move is provided by the force of gravity. In response to gravity, water moves from areas where the water table is high to zones where the water table is lower. This means that water usually gravitates toward a stream channel, lake, or spring. Although some water takes the most direct path down the slope of the water table, much of the water follows long, curving paths toward the area where it is discharged.

Figure 9.31 shows how water percolates into a stream from all possible directions. Some paths clearly turn upward, apparently against the force of gravity, and enter through the bottom of the channel. This is easily explained: The deeper you go into the zone of saturation, the greater the water pressure. Thus, the looping curves followed by water in the saturated zone may be thought of as a compromise between the downward pull of gravity and the tendency of water to move toward areas of reduced pressure.

Recharge area
Recharge area
Discharge area
Discharge area
Water table

The looping flow lines result from the downward pull of gravity and the tendency of groundwater to move toward areas of reduced pressure.

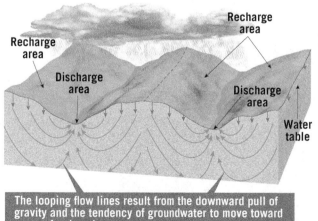

▶ Figure 9.31 **Groundwater movement** Arrows show paths of groundwater movement through uniformly permeable material.

CONCEPT CHECKS 9.9

1. About what percentage of freshwater is groundwater? How does this figure change if glacial ice is excluded?
2. What are two geologic roles for groundwater?
3. When it rains, what factors influence the amount of water that soaks in?
4. Define *groundwater* and relate it to the water table.
5. Distinguish between porosity and permeability. Contrast aquifer and aquitard.
6. What factors cause water to follow the paths shown in Figure 9.31?

9.10 Wells, Artesian Systems, and Springs

Compare and contrast wells, artesian systems, and springs.

A great deal of groundwater eventually makes its way to the surface. Sometimes this occurs as a naturally flowing spring or as a spectacularly erupting geyser. We bring much of the groundwater we use to the surface by pumping it from a well. To understand these phenomena, it is necessary to understand Earth's sometimes complex underground "plumbing."

Wells and Artesian Systems

According to the National Groundwater Association, there are more than 16 million water wells for all purposes in the United States. Private household wells constitute the largest share—more than 13 million. About 500,000 new residential wells are drilled each year.

Wells The most common method for removing groundwater is to use a **well**, a hole drilled into the zone of saturation. Wells serve as small reservoirs into which groundwater migrates and from which it can be pumped to the surface.

The water table level can fluctuate considerably during the course of a year, dropping during dry seasons and rising following periods of precipitation. Therefore, to ensure a continuous supply of water, a well must penetrate below the water table. Whenever a substantial amount of water is withdrawn from a well, the water table around the well is lowered. This effect, termed **drawdown**, decreases with increasing distance from the well. The result is a depression in the water table, roughly conical in shape, known as a **cone of depression** (Figure 9.32). For most small domestic wells, the cone of depression is negligible. However, when wells are used for irrigation or for industrial purposes, the withdrawal of water can be great enough to create a very wide and steep cone of depression that may substantially lower the water table in an area and cause nearby shallow wells to become dry. The lower diagram in Figure 9.32 illustrates this situation.

Artesian Systems In most wells, water cannot rise on its own. If water is first encountered at a depth of 30 meters (100 feet), it remains at that level, fluctuating perhaps 1 or 2 meters with seasonal wet and dry periods. However, in some wells, water rises, sometimes overflowing at the surface.

Artesian system refers to a situation in which groundwater rises in a well above the level where it was initially encountered. For such a situation to occur, two conditions must exist (Figure 9.33): (1) Water must be confined to an aquifer that is inclined so that one end is exposed at the surface, where it can receive water; and (2) aquitards both above and below the aquifer must be present to prevent the water from escaping. Such an aquifer is called a **confined aquifer**. When such a layer is tapped, the pressure created by the weight of the water above forces the water to rise. If there were no friction, the water in the well would rise to the level of the water at the top of the aquifer. However, friction reduces the height of this pressure surface. The greater the distance from the recharge area (the area where water enters the inclined aquifer), the greater the friction and the smaller the rise of water.

In Figure 9.33, Well 1 is a *nonflowing artesian well* because at this location, the pressure surface is below ground level. When the pressure surface is above the ground and a well is drilled into the aquifer, a *flowing artesian well* is created (Well 2 in Figure 9.33). Not all artesian systems are wells. Groundwater may reach the surface by rising along a natural fracture such as a fault rather than through an artificially produced hole; this is called an *artesian spring*. In deserts, artesian springs are sometimes responsible for creating oases.

Artesian systems act as "natural pipelines," transmitting water from remote areas of recharge great distances to the points of discharge. In this manner, water that fell in central Wisconsin years ago is now taken from the ground and used by communities many kilometers to the south, in Illinois. In South Dakota, such a system brings water from the western Black Hills eastward across the state.

On a different scale, city water systems may be considered examples of artificial artesian systems

SmartFigure 9.32
Cone of depression
A. For most small domestic wells, the cone of depression is negligible. **B.** When wells are heavily pumped, the cone of depression can be large and may lower the water table such that nearby shallower wells may be left dry.

ANIMATION
https://goo.gl/QidrnV

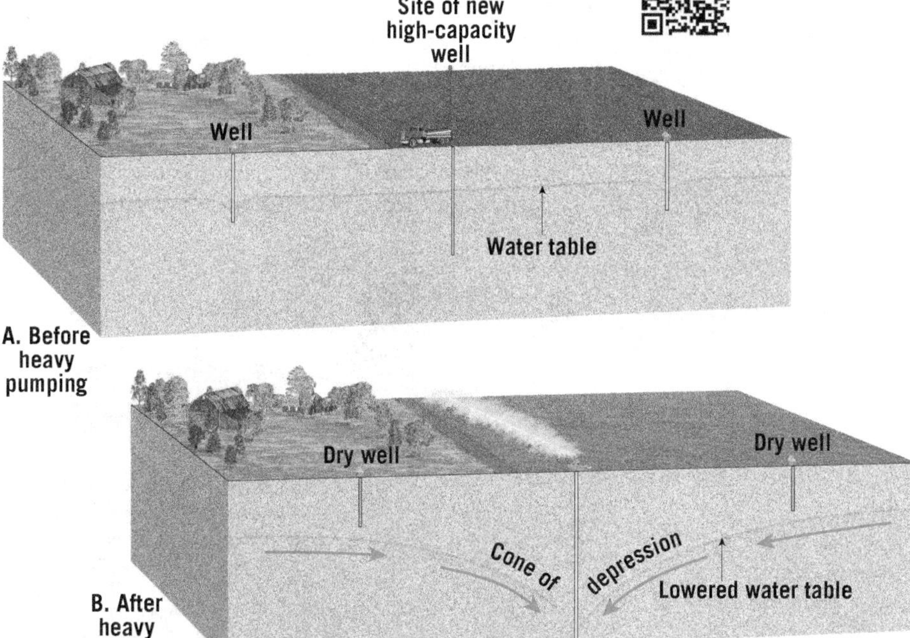

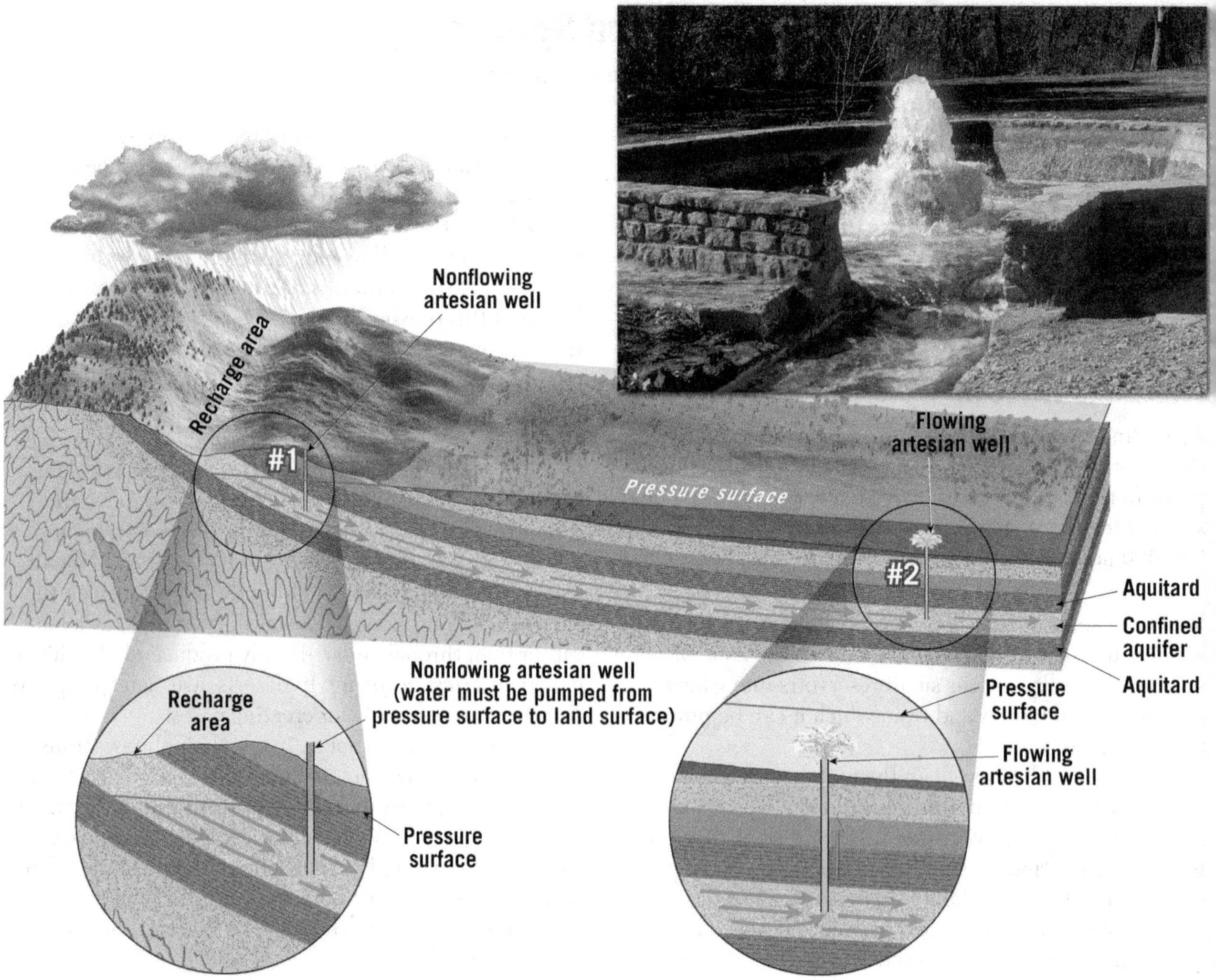

SmartFigure 9.33 **Artesian systems** These groundwater systems occur where an inclined aquifer is surrounded by impermeable beds (aquitards). Such aquifers are called *confined aquifers*. The photo shows a flowing artesian well. (Photo from the James E. Patterson Collection, courtesy of F. K. Lutgens)

TUTORIAL
https://goo.gl/Jk0ALo

(Figure 9.34). A water tower, into which water is pumped, may be considered the area of recharge, the pipes the confined aquifer, and the faucets in homes the flowing artesian wells.

Springs

Springs have aroused the curiosity and wonder of people for thousands of years. The fact that springs were (and to some people still are) rather mysterious phenomena is not difficult to understand because water is flowing freely from the ground in all kinds of weather, in seemingly inexhaustible supply but with no obvious source. Today, we know that the source of springs is water from the zone of saturation and that the ultimate source of this water is precipitation.

Whenever the water table intersects the ground surface, a natural flow of groundwater results, which we call a **spring** (Figure 9.35). Many springs form when an aquitard blocks the downward movement of groundwater and forces it to move laterally. Where the permeable bed (aquifer) outcrops in a valley, one or more springs result.

Another situation that can produce a spring is illustrated in Figure 9.29. Here an aquitard is situated above the main water table. As water percolates downward, a portion accumulates above the aquitard to create a localized zone of saturation and a **perched water table**. Springs, however, are not confined to places where a perched water table creates a flow at the surface. Many geologic situations lead to the formation of springs because subsurface conditions vary greatly from place to place. Even in areas underlain by impermeable crystalline rocks, permeable zones may exist in the form of fractures or solution channels. If these openings fill with water and intersect the ground surface along a slope, a spring results.

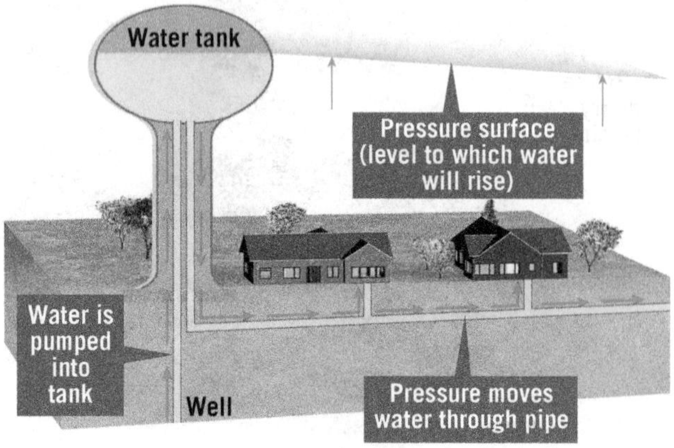

Figure 9.34 **City water systems** City water systems can be considered artificial artesian systems.

Figure 9.35 **Vasey's Paradise** A spring is a natural outflow of groundwater that occurs when the water table intersects the surface. This spring creates a waterfall after emerging from the steep rock wall of the Grand Canyon. (Photo by Michael Collier)

Hot Springs There is no universally accepted definition of **hot spring**. One frequently used definition is that the water in a hot spring is 6° to 9°C (10° to 15°F) warmer than the average annual air temperature for the locality where it occurs. In the United States alone, there are well over 1000 such springs.

Temperatures in deep mines and oil wells usually rise with increasing depth, an average of about 2°C per 100 meters (1°F per 100 feet), a figure known as the *geothermal gradient*. Therefore, when groundwater circulates at great depths, it becomes heated. If the hot water rises rapidly to the surface, it may emerge as a hot spring. The water of some hot springs in the eastern United States is heated in this manner. The springs at Hot Springs National Park in Arkansas are one example. Water temperatures of these springs average about 60°C (140°F).

The great majority (more than 95 percent) of the hot springs (and geysers) in the United States are found in the West. The reason for this distribution is that the sources of heat for most hot springs are magma bodies and hot igneous rocks, and it is in the West that igneous activity has occurred most recently. The hot springs and geysers of the Yellowstone region are well-known examples.

Geysers Intermittent fountains in which columns of hot water and steam are ejected with great force, often rising 30 to 60 meters (100 to 200 feet) into the air, are called **geysers**. After the jet of water ceases, a column of steam rushes out, often with a thunderous roar. Perhaps the most famous geyser in the world is Old Faithful in Yellowstone National Park (Figure 9.36). The great abundance, diversity, and spectacular nature of Yellowstone's geysers and other thermal features undoubtedly was the primary reason for its becoming the first national park in

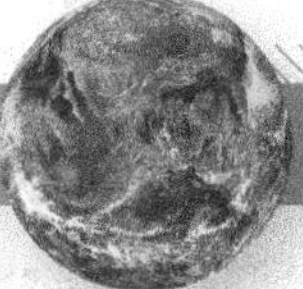

EYE ON EARTH 9.3

In 1900, when this well was drilled near Woonsocket in eastern South Dakota, a "gusher" of water resulted. The stream of water from a 3-inch pipe reached a height of nearly 30 meters (100 feet). Thousands of additional wells now tap the same aquifer.

QUESTION 1 *Describe or sketch the subsurface geologic situation that was responsible for this fountain of water.*

QUESTION 2 *What term is applied to a well such as this?*

QUESTION 3 *Today wells that tap this aquifer do not flow freely at the surface but must be pumped. Suggest a likely reason.*

(Photo by N.H. Darton/USGS)

Figure 9.36 **Old Faithful** This geyser in Wyoming's Yellowstone National Park, is one of the most famous in the world. (Photo by Jeff Vanuga/Getty Images)

the United States. Geysers are also found in other parts of the world, notably New Zealand and Iceland. In fact, the Icelandic word *geysa*, meaning "to gush," gives us the name *geyser.*

Geysers occur where extensive underground chambers exist within hot igneous rocks. As relatively cool groundwater enters the chambers, it is heated by the surrounding rock. At the bottom of the chamber, the water is under great pressure because of the weight of the overlying water. This great pressure prevents the water from boiling at the normal surface temperature of 100°C (212°F). For example, at the bottom of a 300-meter (1000-foot) water-filled chamber, water must attain a temperature of nearly 230°C (450°F) before it will boil. Heating causes the water to expand, and as a result, some of the water is forced out at the surface. This loss of water reduces the pressure on the remaining water in the chamber, which lowers the boiling point. As a result, a portion of the water deep within the chamber quickly turns to an expanding mass of steam, which causes the geyser to erupt. Following the eruption, cool groundwater again seeps into the chamber, and the cycle begins anew.

CONCEPT CHECKS 9.10

1. Relate drawdown to cone of depression.
2. In Figure 9.29, two wells are at the same level. Why is one successful and the other not?
3. Sketch a simple cross section of an artesian system with a flowing artesian well. Label aquitards, the aquifer, and the pressure surface.
4. Describe the circumstances that created the spring in Figure 9.29.
5. What is the source of heat for most hot springs and geysers? Describe what occurs to cause a geyser to erupt.

9.11 Environmental Problems Related to Groundwater

List and discuss three important environmental problems associated with groundwater.

Like many of our other valuable natural resources, groundwater is being exploited. In some areas, overuse threatens the groundwater supply. In other places, groundwater withdrawal has caused the ground and everything resting on it to sink. Still other localities are concerned with the possible contamination of the groundwater supply.

Treating Groundwater as a Nonrenewable Resource

For many, groundwater appears to be an endlessly renewable resource, for it is continually replenished by rainfall and melting snow. In contrast, groundwater in some regions has been and continues to be treated as a *nonrenewable* resource. Where this occurs, the amount of water available to recharge the aquifer is significantly less than the amount being withdrawn.

The High Plains, a relatively dry region that extends from South Dakota to western Texas, is one example of an extensive agricultural economy that is largely dependent on irrigation using groundwater (**Figure 9.37**). Underlying about 111 million acres (450,000 square kilometers [174,000 square miles]) in parts of eight states, the High Plains aquifer is one of the largest and most agriculturally significant aquifers in the United States. It accounts for about 30 percent of all groundwater withdrawn for irrigation in the country. In the southern part of this region, which includes the Texas panhandle, the natural recharge of the aquifer is very slow, and the problem of declining groundwater levels is acute. In fact, in years of average or below-average precipitation, recharge is negligible because all or nearly all of the meager rainfall is returned to the atmosphere by evaporation and transpiration.

Therefore, where intense irrigation has been practiced for an extended period, depletion of groundwater

Figure 9.37 Mining groundwater The High Plains aquifer is one of the largest aquifers in the United States. **B.** In parts of the High Plains aquifer, water is pumped from the ground faster than it is replenished. In such instances, groundwater is being treated as a nonrenewable resource. The U.S. Geological Survey estimates that during the past 60 years, water in storage in the High Plains aquifer declined about 267 acre feet (about 87 trillion gallons), with 60 percent of the total decline occurring in west Texas. (Photo by James L. Amos/Getty Images). **C.** Groundwater provides more than 54 billion gallons per day in support of agriculture in the United States. (Photo by Michael Collier)

can be severe. Declines in the water table at rates as great as 1 meter (3 feet) per year have led to an overall drop of between 15 and 60 meters (50 and 200 feet) in some areas. Under these circumstances, it can be said that the groundwater is literally being "mined." Even if pumping were to cease immediately, it would take thousands of years for the groundwater to be fully replenished.

Groundwater depletion has been a concern in the High Plains and other areas of the West for many years, but it is worth pointing out that the problem is not confined to that part of the country. Increased demands on groundwater resources have overstressed aquifers in many areas, not just in arid and semiarid regions.

Land Subsidence Caused by Groundwater Withdrawal

As you will see later in this chapter, surface subsidence can result from natural processes related to groundwater. However, the ground may also sink when water is pumped from wells faster than natural recharge processes can replace it. This effect is particularly pronounced in areas underlain by thick layers of loose sediments. As water is withdrawn, the water pressure drops, and the weight of the overburden is transferred to the sediment. The greater pressure packs the sediment grains more tightly together, and the ground subsides.

Many areas can be used to illustrate such land subsidence. A classic example in the United States occurred in the San Joaquin Valley of California (**Figure 9.38**). Other well-known cases of land subsidence resulting from groundwater pumping in the United States include Las Vegas, Nevada; New Orleans and Baton Rouge, Louisiana; portions of southern Arizona; and the Houston–Galveston area of Texas. In the low-lying coastal area between Houston and Galveston, land subsidence ranges from 1.5 to 3 meters (5 to 10 feet). The result is that about 78 square kilometers (30 square miles) are permanently flooded.

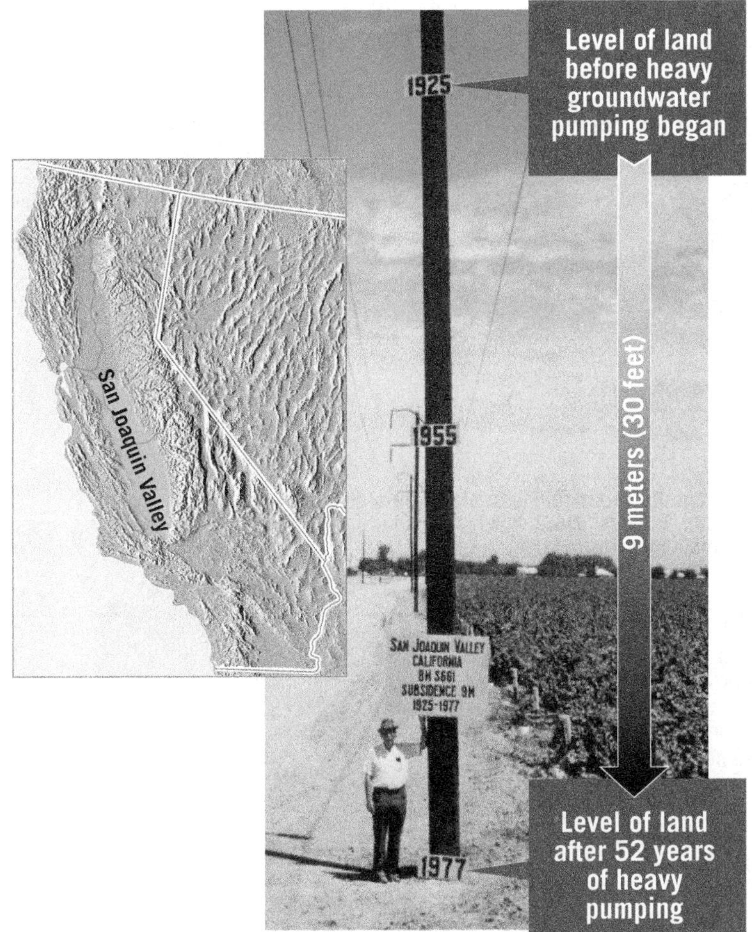

Figure 9.38 That sinking feeling! The San Joaquin Valley, an important agricultural area, relies heavily on irrigation. Between 1925 and 1975, this part of the valley subsided almost 9 meters (30 feet) because of the withdrawal of groundwater and the resulting compaction of sediments. (Photo courtesy of U.S. Geological Survey)

Outside the United States, one of the most spectacular examples of subsidence occurred in Mexico City, a portion of which is built on a former lake bed. In the first half of the twentieth century, thousands of wells were sunk into the water-saturated sediments beneath the city. As water was withdrawn, portions of the city subsided 6 meters (20 feet) or more.

Groundwater Contamination

The pollution of groundwater is a serious matter, particularly in areas where aquifers provide a large part of the water supply. One common source of groundwater pollution is sewage. Its sources include an ever-increasing number of septic tanks, as well as farm wastes and inadequate or broken sewer systems.

If sewage water that is contaminated with bacteria enters the groundwater system, it may become purified through natural processes. The harmful bacteria can be mechanically filtered by the sediment through which the water percolates, destroyed by chemical oxidation, and/or assimilated by other organisms. For purification to occur, however, the aquifer must be of the correct composition. For example, extremely permeable aquifers (such as highly fractured crystalline rock, coarse gravel, or cavernous limestone) have such large openings that contaminated groundwater may travel long distances without being cleansed. In this case, the water flows too rapidly and is not in contact with the surrounding material long enough for purification to occur. This is the problem at Well 1 in Figure 9.39A.

Conversely, when the aquifer is composed of sand or permeable sandstone, the water can sometimes be purified after traveling only a few dozen meters through it. The openings between sand grains are large enough to permit water movement, yet the movement of the water is slow enough to allow ample time for its purification (Well 2, Figure 9.39B).

Other sources and types of contamination also threaten groundwater supplies. These include widely used substances such as highway salt, fertilizers that are spread across the land surface, and pesticides. In addition, a wide array of chemicals and industrial materials may leak from pipelines, storage tanks, landfills, and holding ponds. Some of these pollutants are classified as *hazardous*, meaning that they are either flammable, corrosive, explosive, or toxic. As rainwater oozes through the refuse, it may dissolve a variety of potential contaminants. If the leached material reaches the water table, it mixes with the groundwater and contaminates the supply.

Because groundwater movement is usually slow, polluted water might go undetected for a long time. In fact, contamination is sometimes discovered only after drinking water has been affected and people become ill. By this time, the volume of polluted water might be very large, and even if the source of contamination is removed immediately, the problem is not solved. Although the sources of groundwater contamination are numerous, there are relatively few solutions.

Once the source of the problem has been identified and eliminated, the most common practice is simply to abandon the water supply and allow the pollutants to be flushed away gradually. This is the least costly and easiest solution, but the aquifer must remain unused for many years. To accelerate this process, polluted water is sometimes pumped out and treated. Following removal of the tainted water, the aquifer is allowed to recharge naturally or, in some cases, the treated water or other freshwater is pumped back in. This process is costly and time-consuming, and it may be risky because there is no way to be certain that all the contamination has been removed. Clearly, the most effective solution to groundwater contamination is prevention.

Figure 9.39 Comparing two aquifers In this example, the limestone aquifer allowed the contamination to reach a well, but the sandstone aquifer did not.

Although the contaminated water has traveled more than 100 meters before reaching Well 1, the water moves too rapidly through the cavernous limestone to be purified.

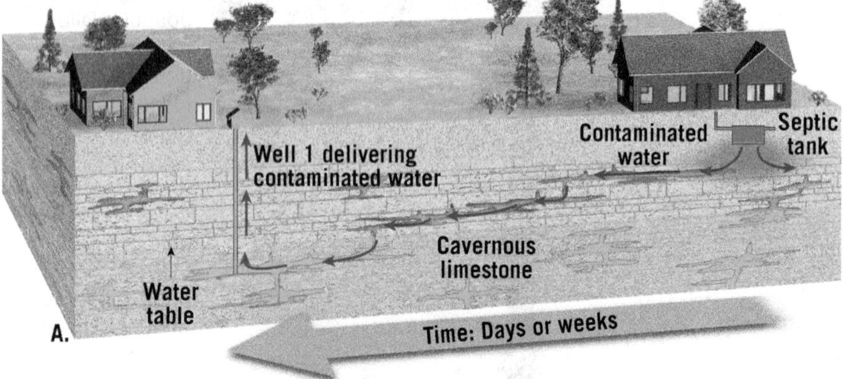

As the discharge from the septic tank percolates through the permeable sandstone, it moves more slowly and is purified in a relatively short distance.

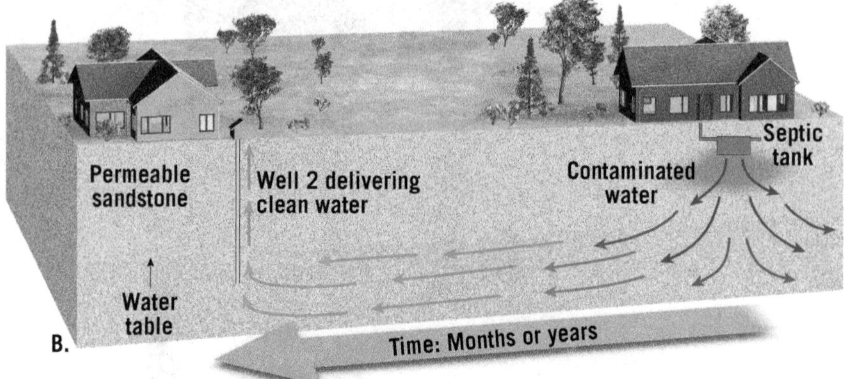

CONCEPT CHECKS 9.11

1. Describe the problem associated with pumping groundwater for irrigation in parts of the High Plains.
2. Explain why ground may subside after groundwater is pumped to the surface.
3. Which aquifer would be most effective in purifying polluted groundwater: coarse gravel, sand, or cavernous limestone?

9.12 The Geologic Work of Groundwater

Explain the formation of caverns and the development of karst topography.

Groundwater can dissolve rock. This fact is key to understanding how caverns and sinkholes form. Soluble rocks, especially limestone, underlie millions of square kilometers of Earth's surface, and in these rocks, groundwater carries on its important role as an erosional agent (Figure 9.40). Limestone is nearly insoluble in pure water but is quite easily dissolved by water containing small quantities of carbonic acid, and most groundwater contains this acid. It forms because rainwater readily dissolves carbon dioxide from the air and from decaying plants. Therefore, when groundwater comes in contact with limestone, the carbonic acid reacts with calcite (calcium carbonate) in the rocks to form calcium bicarbonate, a soluble material that is then carried away in solution.

Caverns

The most spectacular results of groundwater's erosional handiwork are limestone **caverns**. In the United States alone, about 17,000 caves have been discovered. Although most are relatively small, some have spectacular dimensions. Carlsbad Caverns in southeastern New Mexico and Mammoth Cave in Kentucky are famous examples. One chamber in Carlsbad Caverns has an area equivalent to 14 football fields and enough height to accommodate the U.S. Capitol Building. At Mammoth Cave, the total length of interconnected caverns extends for more than 540 kilometers (340 miles).

Most caverns form at or below the water table, in the zone of saturation. Here acidic groundwater follows lines of weakness in the rock, such as joints and bedding planes. As time passes, the dissolving process slowly creates cavities and gradually enlarges them into caverns. Material removed by the groundwater is eventually discharged into streams and carried to the ocean.

The features that arouse the greatest curiosity for most cavern visitors are the stone formations that give some caverns a wonderland appearance. These are not erosional features, like the caverns in which they reside, but depositional features. They are created by the seemingly endless dripping of water over great spans of time. The calcium carbonate that is left behind produces the limestone we call *travertine*. These cave deposits, however, are also commonly called *dripstone*, an obvious reference to their mode of origin.

Although the formation of caverns takes place in the zone of saturation, the deposition of dripstone is not possible until the caverns are above the water table, in the unsaturated zone. This commonly occurs as nearby streams cut their valleys deeper, lowering the water table as the elevation of the rivers drops. As soon as the chamber is filled with air, the conditions are right for the decoration phase of cavern building to begin.

Of the various dripstone features found in caverns, perhaps the most familiar are **stalactites**. These icicle-like pendants hang from the ceiling of a cavern and form

SmartFigure 9.40 **Kentucky's Mammoth Cave area** Portions of Kentucky are underlain by limestone. Dissolution by groundwater has created a landscape characterized by caves and sinkholes. (Photo by Michael Collier)

MOBILE FIELD TRIP
https://goo.gl/jt13CE

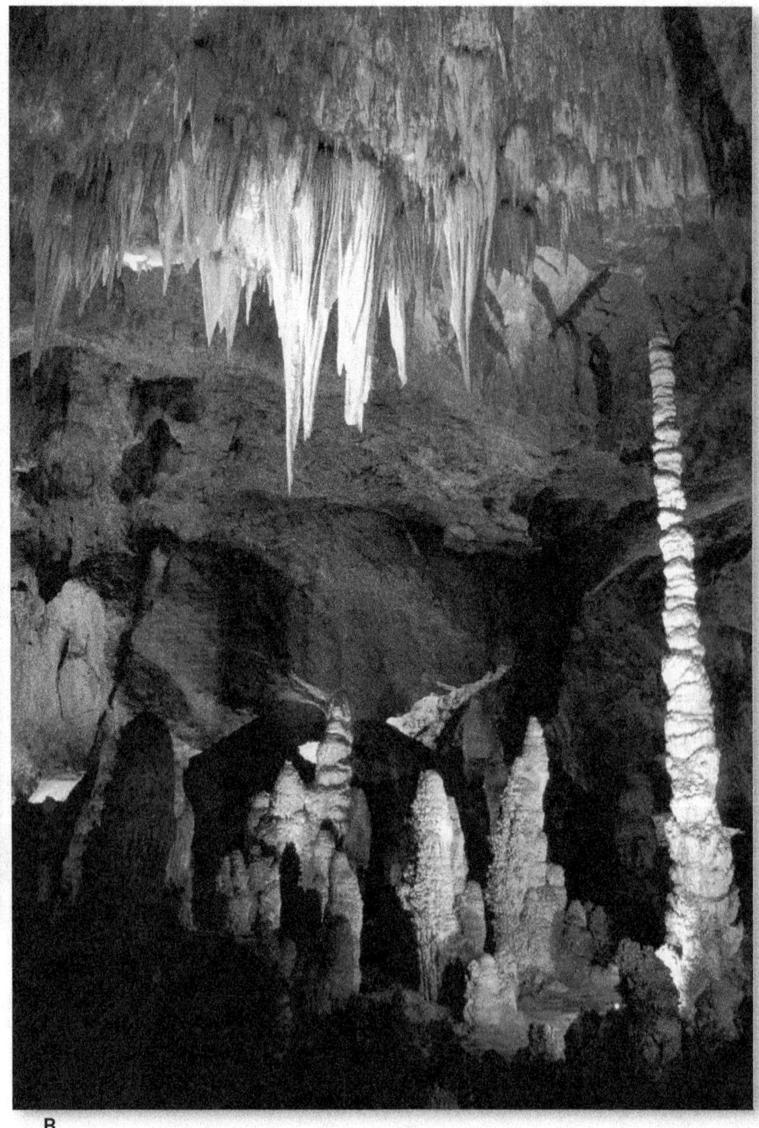

Figure 9.41 **Cave decorations** Dripstone features are of many types, including stalactites, stalagmites, and columns. **A.** Close-up of a delicate live soda-straw stalactite in Chinn Springs Cave, Independence County, Arkansas. (Photo by Dante Fenolio/Science Source) **B.** Stalagmites and stalactites in New Mexico's Carlsbad Caverns National Park. (Photo by Fritz Poelking/Glow Images)

where water seeps through cracks above. When water reaches air in the cave, some of the dissolved carbon dioxide escapes from the drop, and calcite begins to precipitate. Deposition occurs as a ring around the edge of the water drop. As drop after drop follows, each leaves an infinitesimal trace of calcite behind, and a hollow limestone tube is created. Water then moves through the tube, remains suspended momentarily at the end, contributes a tiny ring of calcite, and falls to the cavern floor. The stalactite just described is appropriately called a *soda straw* (Figure 9.41A). Often the hollow tube of the soda straw becomes plugged, or its supply of water increases. In either case, the water is forced to flow and deposit along the outside of the tube. As deposition continues, the stalactite takes on the more common conical shape.

Formations that develop on the floor of a cavern and reach upward toward the ceiling are called **stalagmites** (Figure 9.41B). The water supplying the calcite for stalagmite growth falls from the ceiling and splatters over the surface. As a result, stalagmites do not have a central tube and are usually more massive in appearance and more rounded on their upper ends than stalactites. Given enough time, a downward-growing stalactite and an upward-growing stalagmite may join to form a *column*.

Karst Topography

Many areas of the world have landscapes that, to a large extent, have been shaped by the dissolving power of groundwater. Such areas are said to exhibit **karst topography**, named for the *Krs* region in the border area between Slovenia and Italy, where such topography is strikingly developed. In the United States, karst landscapes occur in many areas that are underlain by limestone, including portions of Kentucky, Tennessee, Alabama, southern Indiana, and central and northern Florida (Figure 9.42). Generally, arid and semiarid areas do not develop karst topography because there is insufficient groundwater. When karst features exist in such regions, they are likely to be remnants of a time when rainier conditions prevailed.

Sinkholes Karst areas typically have irregular terrain punctuated with many depressions called **sinkholes** or, simply, **sinks** (see Figure 9.28B). In the limestone areas of Florida, Kentucky, and southern Indiana, literally tens

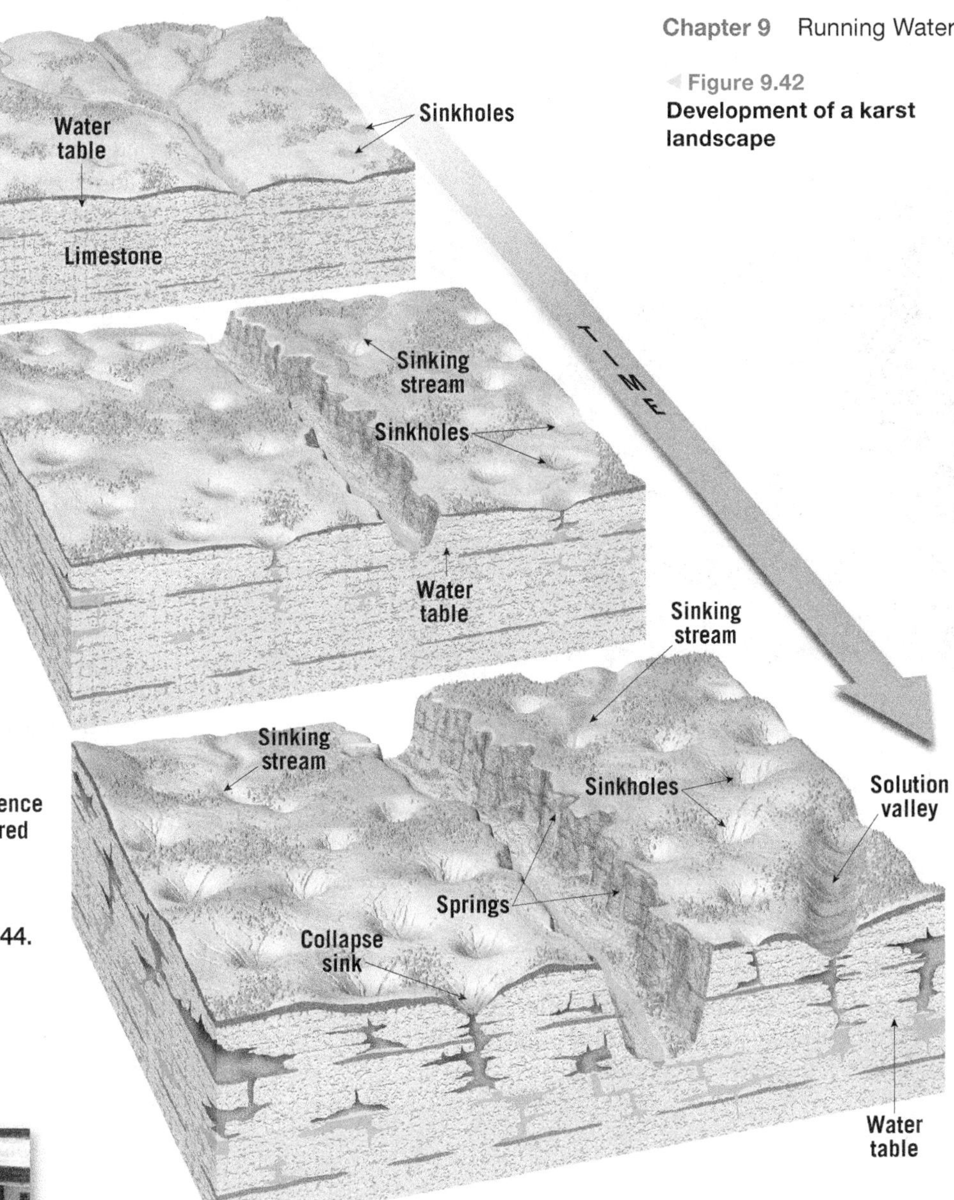

During early stages, groundwater percolates through limestone along joints and bedding planes. Solution activity creates and enlarges caverns at and below the water table.

With time, caverns grow larger and the number and size of sinkholes increase. Surface drainage is frequently funneled below ground.

Collapse of caverns and coalescence of sinkholes form larger, flat-floored depressions. Eventually solution activity may remove most of the limestone from the area, leaving isolated remnants as in Figure 9.44.

Figure 9.42
Development of a karst landscape

of thousands of these depressions vary in depth from just 1 to 2 meters (3 to 6 feet) to a maximum of more than 50 meters (165 feet).

Sinkholes commonly form in one of two ways. Some develop gradually over many years, without any physical disturbance to the rock. In these situations, the limestone immediately below the soil is dissolved by downward-seeping rainwater that is freshly charged with carbon dioxide. These depressions are usually not deep and are characterized by relatively gentle slopes. By contrast, sinkholes can also form suddenly and without warning when the roof of a cavern collapses under its own weight. Typically, the depressions created in this manner are steep sided and deep. When they form in populous areas, they may represent a serious geologic hazard. Such a situation is clearly the case in **Figure 9.43**.

Figure 9.43 Sinkholes can be geologic hazards This sinkhole formed suddenly in the backyard of a home in Lake City, Florida. Sinkholes such as this one form when the roof of a cavern collapses. (Photo by Jon M. Fletcher/The Florida Times-Union/AP Images)

▲ **Figure 9.44 Tower karst landscape in China** One of the best-known and most distinctive regions of tower karst development is along the Li River in the Guilin District of southeastern China. (Photo by Philippe Michel/AGE/Fotostock)

In addition to a surface pockmarked by sinkholes, karst regions characteristically show a striking lack of surface drainage (streams). Following a rainfall, runoff is quickly funneled below ground, through sinks. It then flows through caverns until it finally reaches the water table. Where streams exist at the surface, their paths are usually short. The names of such streams often give a clue to their fate. The Mammoth Cave area of Kentucky, for example, is home to Sinking Creek, Little Sinking Creek, and Sinking Branch. Some sinkholes become plugged with clay and debris, creating small lakes or ponds.

Tower Karst Landscapes Some regions of karst development exhibit landscapes that look very different from the sinkhole-studded terrain depicted in Figure 9.43. One striking example is an extensive region in southern China that is described as exhibiting *tower karst*. As **Figure 9.44** shows, the term *tower* is appropriate because the landscape consists of a maze of isolated steep-sided hills that rise abruptly from the ground. Each is riddled with interconnected caves and passageways. This type of karst topography forms in wet tropical and subtropical regions having thick beds of highly jointed limestone. In such settings, groundwater dissolves large volumes of limestone, leaving only these residual towers. Karst development occurs more rapidly in tropical climates due to the abundant rainfall and greater availability of carbon dioxide from the decay of lush tropical vegetation. The extra carbon dioxide in the soil means there is more carbonic acid for dissolving limestone. Other tropical areas of advanced karst development include portions of Puerto Rico, western Cuba, and northern Vietnam.

CONCEPT CHECKS 9.12

1. How does groundwater create caverns?
2. How do stalactites and stalagmites form?
3. Describe two ways in which sinkholes form.

9 CONCEPTS IN REVIEW

Running Water and Groundwater

9.1 Earth as a System: The Hydrologic Cycle

List the hydrosphere's major reservoirs and describe the different paths that water takes through the hydrologic cycle.

KEY TERMS: hydrologic cycle, evaporation, infiltration, runoff, transpiration, evapotranspiration

- Water moves through the hydrosphere's many reservoirs by evaporating, condensing into clouds, and falling as precipitation. Once it reaches the ground, rain can either soak in, evaporate, be returned to the atmosphere by plant transpiration, or run off. Running water is the most important agent sculpting Earth's varied landscapes.

9.2 Running Water

Describe the nature of drainage basins and river systems. Sketch four basic drainage patterns.

KEY TERMS: drainage basin (watershed), divide, headward erosion, dendritic pattern, radial pattern, rectangular pattern, trellis pattern

- The land area that contributes water to a stream is its drainage basin. Drainage basins are separated by imaginary lines called divides.
- As a generalization, river systems tend to erode at the upstream end, transport sediment through the middle section, and deposit sediment at the downstream end.
- A stream erodes most effectively in a headward direction, thereby lengthening its course.

? Identify each of the drainage patterns depicted in the accompanying sketch.

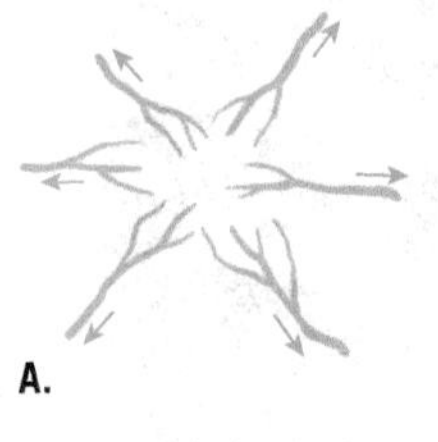

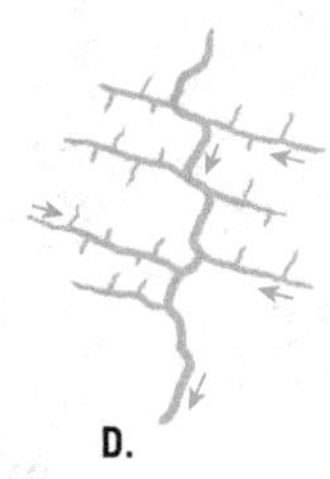

9.3 Streamflow Characteristics

Discuss streamflow and the factors that cause it to change.

KEY TERMS: laminar flow, turbulent flow, gradient, discharge, longitudinal profile

- The flow of water in a stream may be laminar or turbulent. A stream's flow velocity is influenced by the channel's gradient; the size, shape, and roughness of the channel; and the stream's discharge.
- A cross-sectional view of a stream from head to mouth is a longitudinal profile. Usually the gradient and roughness of the stream channel decrease going downstream, whereas the size of the channel, stream discharge, and flow velocity increase in the downstream direction.

? Sketch a typical longitudinal profile. Where does most erosion happen? Where is sediment transport the dominant process?

9.4 The Work of Running Water

Summarize the ways in which streams erode, transport, and deposit sediment.

KEY TERMS: pothole, dissolved load, suspended load, bed load, settling velocity, saltation, capacity, competence, sorting, alluvium

- Streams erode when turbulent water lifts loose particles from the streambed. The focused "drilling" of the stream armed with swirling particles also creates potholes in solid rock.
- Streams transport their load of sediment dissolved in water, in suspension, and along the bottom (bed) of the channel.
- A stream's ability to transport solid particles is described using two criteria: Capacity refers to how much sediment a stream is transporting, and competence refers to the particle sizes the stream is capable of moving.
- Streams deposit sediment when velocity slows and competence is reduced. This results in sorting, the process by which like-size particles are deposited together.

9.5 Stream Channels

Contrast bedrock and alluvial stream channels. Distinguish between two types of alluvial channels.

KEY TERMS: meander, cut bank, point bar, cutoff, oxbow lake, braided channel

- Bedrock channels are cut into solid rock and are most common in headwaters areas where gradients are steep. Rapids and waterfalls are common features.
- Alluvial channels are dominated by streamflow through alluvium previously deposited by the stream. A floodplain usually covers the valley floor, with the river meandering or moving through braided channels.
- Meanders change shape through erosion at the cut bank (the outer edge of the meander) and deposition of sediment on point bars (the inside of a meander). A meander may become cut off and form an oxbow lake.

? The town of Carter Lake is the *only* portion of the state of Iowa that lies on the west side of the Missouri River. It is bounded on the north by its namesake, Carter Lake, on the south by the Missouri River, and on the east and west by Nebraska. After examining the map, prepare a hypothesis that explains how this unusual situation could have developed.

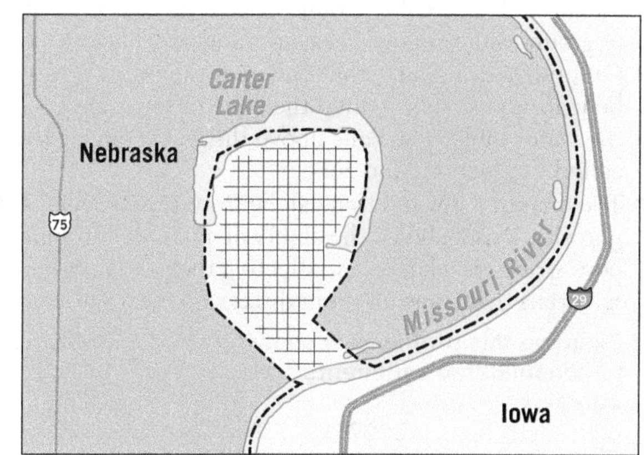

9.6 Shaping Stream Valleys

Contrast narrow V-shaped valleys, broad valleys with floodplains, and valleys that display incised meanders.

KEY TERMS: stream valley, base level, floodplain, incised meander, stream terrace

- A stream valley includes the channel itself, the adjacent floodplain, and the relatively steep valley walls. Streams erode downward until they approach base level, the lowest point to which a stream can erode its channel. A river flowing toward the ocean (the ultimate base level) may encounter several local base levels along its route. These could be lakes or resistant rock layers that retard downcutting by the stream.
- A stream valley is widened through the meandering action of the stream, which erodes the valley walls and widens the floodplain. If base level drops or if the land is uplifted, a stream downcuts. If it is underlain by bedrock, the stream may develop incised meanders. Streams underlain by deep alluvium are likely to develop terraces.

? Meanders are associated with a river that is eroding from side to side, whereas narrow canyons are associated with rivers that are vigorously downcutting. The river in this image is confined to a narrow canyon but is also meandering. Explain.

Michael Collier

9.7 Depositional Landforms

Discuss the formation of deltas, natural levees, and alluvial fans.

KEY TERMS: bar, delta, distributary, natural levee, back swamp, yazoo tributary, alluvial fan

- A delta may form where a river deposits sediment in another water body at its mouth. The partitioning of streamflow into multiple distributaries spreads sediment in different directions.
- Natural levees result from sediment deposited along the margins of a stream channel by many flooding events. Because the levees slope gently away from the channel, the adjacent floodplain is poorly drained, resulting in back swamps and yazoo tributaries flowing parallel to the main river.
- Alluvial fans are fan-shaped deposits of alluvium that form where steep mountain fronts drop down into adjacent valleys.

9.8 Floods and Flood Control

Distinguish between regional floods and flash floods. Describe some common flood control measures.

KEY TERMS: flood

- Floods are usually triggered by heavy rains and/or snowmelt. Sometimes human interference can worsen or even cause floods. Flood control measures include the building of artificial levees and dams. Channelization may involve creating artificial cutoffs. Many scientists and engineers advocate a nonstructural approach to flood control that involves more appropriate land use.

? Artificial levees are constructed to protect property from floods. Sometimes artificial levees in rural areas are intentionally opened up to protect a city from experiencing flooding. How does this work?

9.9 Groundwater: Water Beneath the Surface

Discuss the importance of groundwater and describe its distribution and movement.

KEY TERMS: groundwater, zone of saturation, water table, unsaturated zone, porosity, permeability, aquitard, aquifer

- Groundwater represents the largest reservoir of freshwater that is readily available to humans. Geologically, groundwater is an equalizer of streamflow, and the dissolving action of groundwater produces caverns and sinkholes.
- Groundwater is water that occupies the pore spaces in sediment and rock in a zone beneath the surface called the zone of saturation. The upper limit of this zone is called the water table. The zone above the water table where the material is not saturated is called the unsaturated zone.
- The quantity of water that can be stored in the open spaces in rock or sediment is termed porosity. Permeability, the ability of a material to transmit a fluid through interconnected pore spaces, is a key factor affecting the movement of groundwater. Aquifers are permeable materials that transmit groundwater freely, whereas aquitards are impermeable materials.

? Examine this profile view showing the distribution of water in relatively uniform unconsolidated sediments and label the various portions of the groundwater complex.

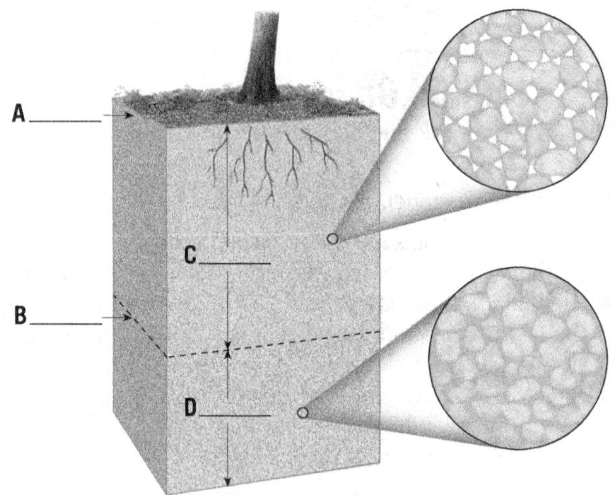

9.10 Wells, Artesian Systems, and Springs

Compare and contrast wells, artesian systems, and springs.

KEY TERMS: well, drawdown, cone of depression, artesian system, confined aquifer, spring, perched water table, hot spring, geyser

- Wells, which are openings bored into the zone of saturation, withdraw groundwater and may create roughly conical depressions in the water table known as cones of depression.
- Artesian wells tap into inclined aquifers bounded above and below by aquitards. For a system to qualify as artesian, the water in the well must be under sufficient pressure for the water to rise above the top of the confined aquifer. Artesian wells may be flowing or nonflowing, depending on whether the pressure surface is above or below the ground surface.
- Springs occur where the water table intersects the land surface and a natural flow of groundwater results. When groundwater circulates deep below the surface, it may become heated and emerge at the surface as a hot spring. Geysers occur when groundwater is heated in underground chambers and expands, with some water quickly changing to steam and causing the geyser to erupt. The source of heat for most hot springs and geysers is hot igneous rock.

ifong/Shutterstock

? Relate this image to an artesian system.

9.11 Environmental Problems Related to Groundwater

List and discuss three important environmental problems associated with groundwater.

- Groundwater can be "mined" by being extracted at a rate that is greater than the rate of replenishment. When groundwater is treated as a nonrenewable resource, as it is in parts of the High Plains aquifer, the water table drops, in some cases by 60 meters (200 feet).
- The extraction of groundwater can cause pore space to decrease in volume and the grains of loose Earth materials to pack more closely together. This overall compaction of sediment volume results in the subsidence of the land surface.
- Contamination of groundwater with sewage, highway salt, fertilizer, or industrial chemicals is another issue of critical concern. Once groundwater is contaminated, the problem is very difficult to solve, requiring expensive remediation or even abandonment of the aquifer.

9.12 The Geologic Work of Groundwater

Explain the formation of caverns and the development of karst topography.

KEY TERMS: cavern, stalactite, stalagmite, karst topography, sinkhole (sink),

- Groundwater dissolves rock, in particular limestone, leaving behind void spaces in the rock. Caverns form at the zone of saturation, but later dropping of the water table may leave them open and dry—and available for people to explore.
- Dripstone is rock deposited by dripping of water containing dissolved calcium carbonate inside caverns. Features made of dripstone include stalactites, stalagmites, and columns.
- Karst topography develops in limestone regions and exhibits irregular terrain punctuated with many depressions called sinkholes. Some sinkholes form when the cavern roofs collapse.

? Identify the three cavern deposits labeled in this photograph.

Miroslav/AGE Fotostock

GIVE IT SOME THOUGHT

1 A river system consists of three zones, based on the dominant process operating in each part of the river system. On the accompanying illustration, match each process with one of the three zones:
   a. Sediment production (erosion)
   b. Sediment deposition
   c. Sediment transport

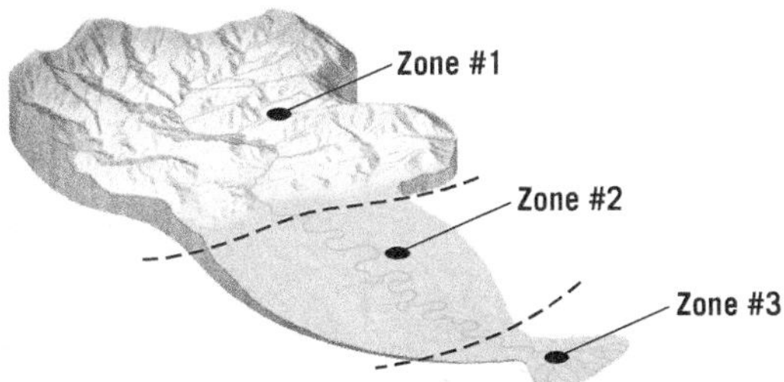

2 This image, taken from the International Space Station, shows a plateau and canyons in central Saudi Arabia. Over a long time span, how will the canyons likely change? How will the plateau be affected? Is there a term that describes the process?

NASA

3 What factors influence how much rain will soak into the ground compared to how much will run off?

4 If you collect a jar of water from a stream, what part of its load will settle to the bottom of the jar? What portion will remain in the water indefinitely? What part of the stream's load would probably not be represented in your sample?

5 The Middle Fork of the Salmon River flows for about 175 kilometers (110 miles) through a rugged wilderness area in central Idaho.
   a. Is the river flowing in an alluvial channel or a bedrock channel? Explain.
   b. What process is dominant here: valley deepening or valley widening?
   c. Is the area shown in this image more likely near the mouth of the river or the head?

Michael Collier

6 What is the likely difference between an intermittent stream (one that flows off and on) and a stream that flows all the time, even during extended dry periods?

7 The cemetery in this photo is located in New Orleans, Louisiana. As in other cemeteries in the area, all of the burial plots are above ground. Based on what you have learned in this chapter, suggest a reason for this rather unusual practice.

Zack Frank/Shutterstock

8 During a trip to the grocery store, your friend wants to buy some bottled water. Some brands promote the fact that their product is artesian. Other brands boast that their water comes from a spring. Your friend asks, "Is artesian water or spring water necessarily better than water from other sources?" How would you answer?

9 This black-and-white photo from the 1930s shows Franklin Roosevelt enjoying the hot springs at the presidential retreat at Warm Springs, Georgia. The temperature of these hot springs is always near 32°C (90°F). This area has no history of recent volcanic activity. What is the likely reason these springs are so warm?

New York Daily News/Getty Images

10 Imagine that you are an environmental scientist who has been hired to solve a groundwater contamination problem. Several homeowners have noticed that their well water has a funny smell and taste. Some think the contamination is coming from a landfill, but others think it might be a nearby cattle feedlot or chemical plant. Your first step is t gather data from wells in the area and prepare the map of the water table shown here.

a. Based on your map, can any of the three potential sources of contamination be eliminated? If so, explain.

b. What other steps would you take to determine the source of the contamination?

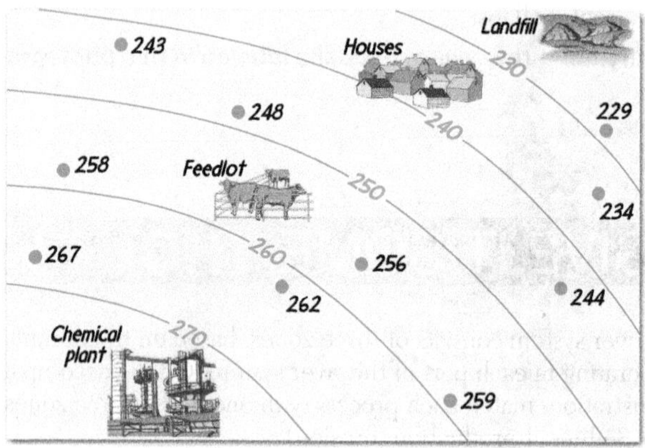

EXAMINING THE EARTH SYSTEM

1 Building a dam is one method of regulating the flow of a river to control flooding. Dams and their reservoirs may also provide recreational opportunities and water for irrigation and hydroelectric power generation. This image, from near Page, Arizona, shows Glen Canyon Dam on the Colorado River upstream from the Grand Canyon and a portion of Lake Powell, the reservoir it created.

Michael Collier

a. How did the behavior of the stream likely change upstream from Lake Powell?

b. How might the behavior of the Colorado River downstream from the dam have been affected?

c. Given enough time, how might the reservoir change?

d. Speculate on the possible environmental impacts of building a dam such as this one.

2 The map shows that the drainage basin of the Republican River occupies portions of Colorado, Nebraska, and Kansas. A significant part of the basin is considered semiarid. In 1943, the three states made a legal agreement to share the river's water. In 1998, Kansas went to court to force farmers in southern Nebraska to substantially reduce the amount of groundwater they used for irrigation. Nebraska officials claimed that the farmers were not taking water from the Republican River and thus were not violating the 1943 agreement. The court ruled in favor of Kansas.

a. Explain why the court ruled that groundwater in southern Nebraska should be considered part of the Republican River system.

b. How might heavy irrigation in a drainage basin influence the flow of a river?

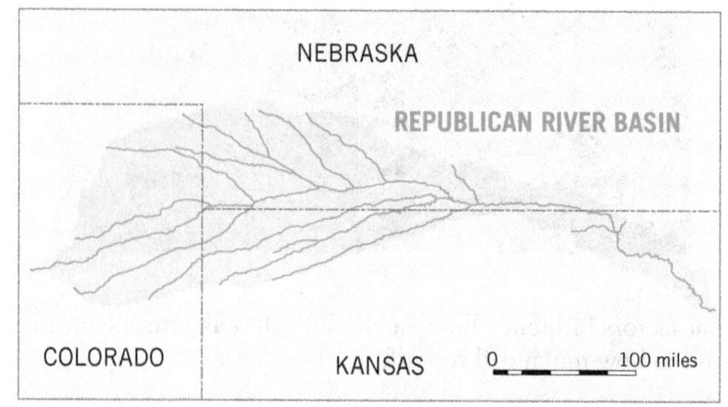

3 This satellite image shows a portion of the desert in northern Saudi Arabia, a region known for its abundant sunshine, high temperatures, and meager rainfall. The green circles are agricultural fields that are about 1 kilometer (0.62 mile) in diameter. Water for irrigation is pumped from deep aquifers and distributed around a center point within each field—a technique known as center-pivot irrigation. The deep aquifers contain water that dates to the Ice Age, about 20,000 years ago, a time when the climate in this region was wetter and milder.

a. Is it likely that agricultural activity in this region is sustainable indefinitely? Explain.

b. A significant portion of the water placed on these fields is "lost" (not available to the crops). Suggest a reason for the loss of water.

c. Relate what is likely occurring to the water table in the region pictured here to an example of a similar situation in the United States.

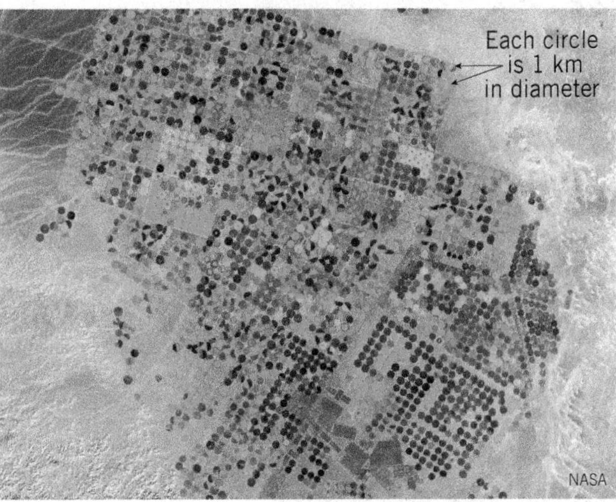
Each circle is 1 km in diameter
NASA

DATA ANALYSIS

Streamflow Rates Near You

Rivers are an important resource for communities, but they can cause dangerous flooding. River water levels are monitored across the United States to determine flow rates and flood stage (the level at which the water surface has risen enough to be a hazard.)

ACTIVITIES

Go to http://waterwatch.usgs.gov. Click on the Current Streamflow map.

1 Click on the word Map to bring up information about the map. What do the map dots display? For how long must data have been recorded at a streamgage for the gage to appear on this map?

2 Why might some states have few data points in the winter? What are the effects of ice?

3 Click anywhere in the Explanation—Percentile Classes table below the map to bring up information about the classes. What do the percentile classes mean?

4 Based on the Current Streamflow map, which areas of the United States are experiencing higher-than-normal streamflow? Which areas are experiencing lower-than-normal streamflow?

5 Explore the different water-resource regions by clicking on the Water-Resources Regions drop-down menu. Which water-resource region are you located in?

Click on a ranked (colored) stream gage near your current location that has Hydrograph, Peak, and Forecast graphs. Be sure the Peak graph also shows the National Weather Service Flood Stage level. You can click on each of these graphs to bring up a larger version.

6 Where is this streamflow site located?

7 What is the flood stage water height for this location? What is the current water height for this location?

8 What is the drainage area (in square miles)? How does this compare to nearby waterways?

9 What is the current discharge rate (in cubic feet per second)? How does this compare to nearby waterways?

10 Click on Hydrograph. Is the current discharge rate higher or lower than the median daily discharge rate? Does this make sense, given what you know about the local conditions in your area and upstream? (You may need to do some investigating online to answer this.)

11 Click on Peak. When was the most recent flood stage for this waterway? When was the largest flood stage?

MasteringGeology™ Looking for additional review and test prep materials? Visit the Study Area in MasteringGeology to enhance your understanding of this chapter's content by accessing a variety of resources, including Self-Study Quizzes, Geoscience Animations, SmartFigure Tutorials, Mobile Field Trips, *Project Condor* Quadcopter videos, *In the News* articles, flashcards, web links, and an optional Pearson eText.

www.masteringgeology.com

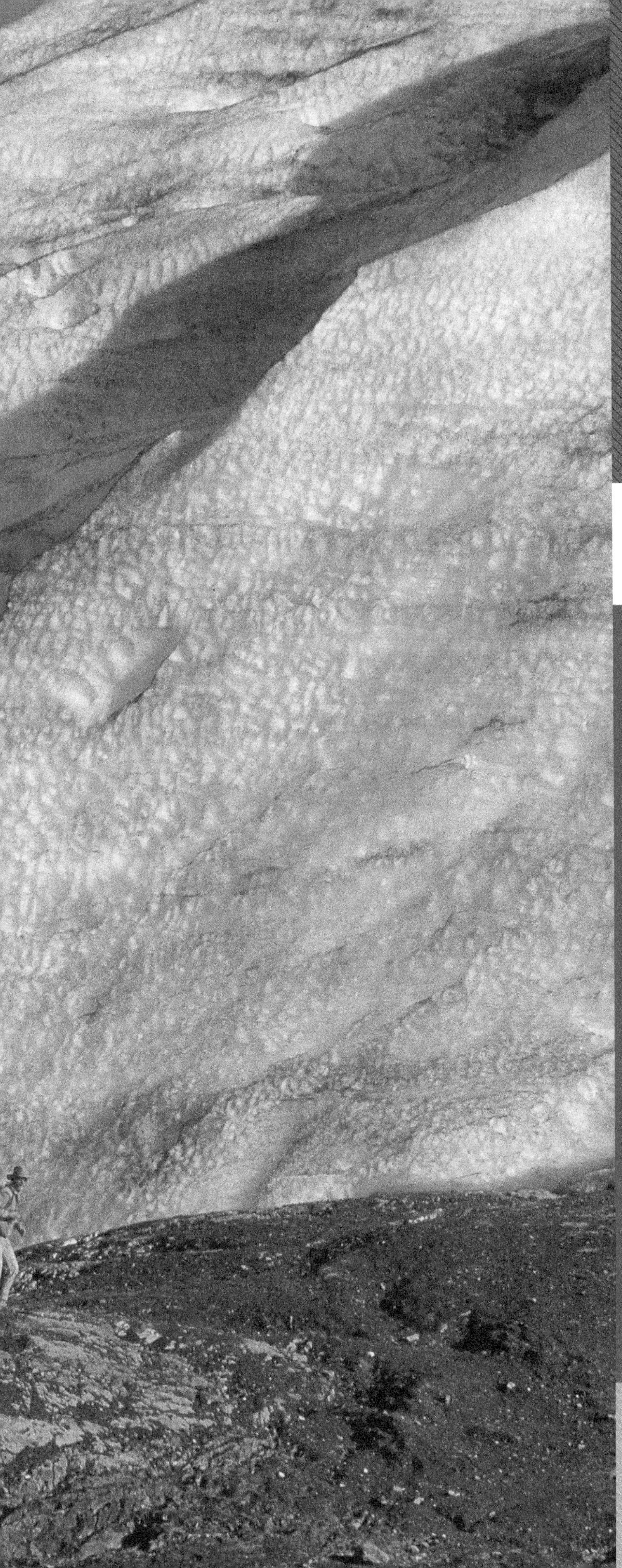

10

Glaciers, Deserts, and Wind

FOCUS ON CONCEPTS

Each statement represents the primary learning objective for the corresponding major heading within the chapter. After you complete the chapter, you should be able to:

10.1 Explain the role of glaciers in the hydrologic and rock cycles. Describe the different types of glaciers, their characteristics, and their present-day distribution.

10.2 Describe how glaciers move, the rates at which they move, and the significance of the glacial budget.

10.3 Discuss the processes of glacial erosion. Identify and describe the major topographic features sculpted by glacial erosion.

10.4 Distinguish between the two basic types of glacial deposits and briefly describe the features associated with each type.

10.5 Describe and explain several important effects of Ice Age glaciers other than the formation of erosional and depositional landforms.

10.6 Discuss the extent of glaciation and climate variability during the Quaternary Ice Age. Summarize some of the current ideas about the causes of ice ages.

10.7 Describe the general distribution and extent of Earth's dry lands and the role that water plays in modifying desert landscapes.

10.8 Discuss the stages of landscape evolution in the Basin and Range region of the western United States.

10.9 Describe the ways that wind transports sediment and the features created by wind erosion.

10.10 Explain how loess deposits differ from deposits of sand. Discuss the movement of dunes and distinguish among different dune types.

Hiker at the terminus of Exit Glacier in Kenai Fiords National Park near Seward, Alaska. (Photo by Michael Collier)

LIKE THE RUNNING SURFACE WATER AND GROUNDWATER that were the focus of Chapter 9, glaciers and wind are significant erosional agents. They are responsible for creating many different landforms and are part of an important link in the rock cycle, in which the products of weathering are transported and deposited as sediment.

Climate has a strong influence on the nature and intensity of Earth's external processes, a fact is dramatically illustrated in this chapter. The existence and extent of glaciers are largely controlled by Earth's changing climate. The development of arid landscapes also illustrates the strong link between climate and geology.

Today, glaciers cover nearly 10 percent of Earth's land surface; however, in the recent geologic past, ice sheets were three times more extensive, covering vast areas with ice thousands of meters thick. Many regions still bear the mark of these glaciers. The first part of this chapter examines glaciers and the erosional and depositional features they create. The second part explores dry lands and the geologic work of wind. Because desert and near-desert conditions prevail over an area as large as that affected by the massive glaciers of the Ice Age, the nature of such landscapes is indeed worth investigating.

10.1 Glaciers and the Earth System

Explain the role of glaciers in the hydrologic and rock cycles. Describe the different types of glaciers, their characteristics, and their present-day distribution.

A **glacier** is a thick ice mass that forms over hundreds or thousands of years. It originates on land from the accumulation, compaction, and recrystallization of snow. A glacier appears to be motionless, but it is not; glaciers move very slowly. Although glaciers are found in many parts of the world, most are located in remote areas, near Earth's poles or in high mountains.

Many present-day landscapes were modified by the widespread glaciers of the most recent Ice Age, and they still strongly reflect the handiwork of ice. The basic features of such diverse places as the Alps, Cape Cod, and Yosemite Valley were fashioned by now-vanished masses of glacial ice. Moreover, Long Island, the Great Lakes, and the fiords of Norway and Alaska all owe their existence to glaciers.

Glaciers: A Part of Two Basic Cycles

Glaciers are part of two important cycles in the Earth system—the hydrologic cycle and the rock cycle. In Chapter 9 you learned that the water of the hydrosphere is constantly cycled through the atmosphere, biosphere, and geosphere. Over and over again, water evaporates from the oceans into the atmosphere, precipitates on the land, and flows in rivers and underground back to the sea. However, when precipitation falls at high elevations or high latitudes, the water may not immediately make its way toward the sea. Instead, it may become part of a glacier. Although the ice will eventually melt, allowing the water to continue its path to the sea, water can be stored as glacial ice for many tens, hundreds, or even thousands of years. During the time that the water is part of a glacier, the moving mass of ice can do enormous amounts of work. Like running water, groundwater, wind, and waves, glaciers are dynamic erosional agents that accumulate, transport, and deposit sediment. This activity is a basic part of the rock cycle.

Valley (Alpine) Glaciers

Literally thousands of relatively small glaciers exist in lofty mountain areas, where they usually follow valleys originally occupied by streams. Unlike the rivers that previously flowed in these valleys, the glaciers advance slowly, perhaps only a few centimeters each day. Because of their setting, these moving ice masses are termed **valley glaciers**, or **alpine glaciers** (Figure 10.1). Each glacier is a stream of ice, bounded by precipitous rock walls, that flows downvalley from a snow accumulation center near its head. Like rivers, valley glaciers can be long or short, wide or narrow, single or with branching tributaries. Generally, alpine glaciers are longer than they are wide; some extend for just a fraction of a kilometer, whereas others go on for many dozens of kilometers. The west branch of the Hubbard Glacier, for example, runs through 112 kilometers (nearly 70 miles) of mountainous terrain in Alaska and the Yukon Territory.

Ice Sheets

Ice sheets exist on a much larger scale than valley glaciers. These enormous masses flow out in all directions from one or more snow-accumulation centers and completely obscure all but the highest areas of underlying terrain. The low total annual solar energy reaching the poles makes these regions hospitable to great ice accumulations. Presently, each of Earth's polar regions supports an ice sheet: on Greenland in the Northern Hemisphere and on Antarctica in the Southern Hemisphere (Figure 10.2).

SmartFigure 10.1 **Valley glacier** This tongue of ice, also called an *alpine glacier*, is still eroding the Alaskan landscape. Dark stripes of sediment within these glaciers are called medial moraines. This is Johns Hopkins Glacier in Alaska's Glacier Bay National Park. (Photo by Michael Collier)

MOBILE FIELD TRIP
https://goo.gl/CWZNtK

Greenland's ice sheet occupies 1.7 million square kilometers (663,000 square miles), about 80 percent of the island.

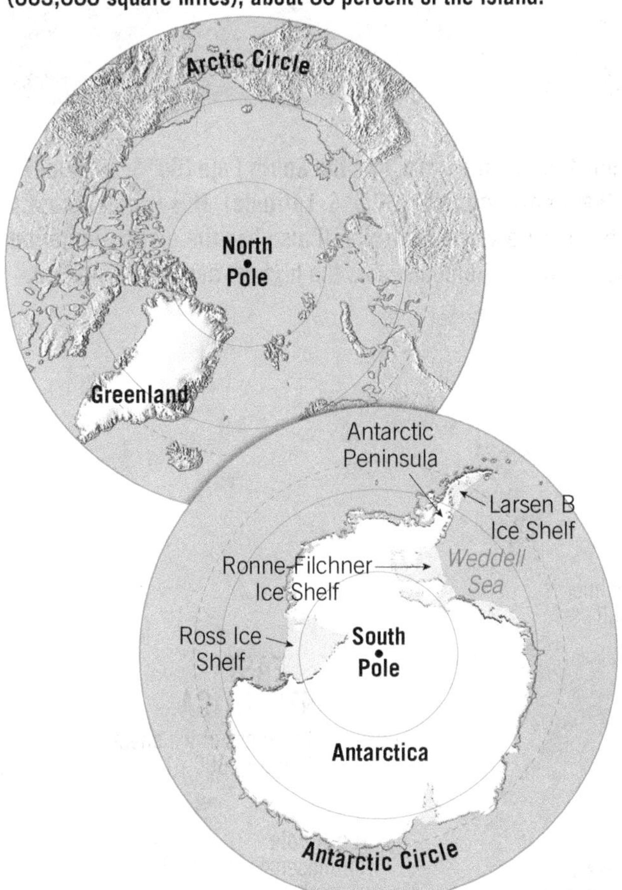

The area of the Antarctic Ice Sheet is almost 14 million square kilometers (5,460,000 square miles). Ice shelves occupy an additional 1.4 million square kilometers (546,000 square miles).

SmartFigure 10.2 **Ice sheets** The only present-day ice sheets are those covering Greenland and Antarctica. Their combined areas represent almost 10 percent of Earth's land area.

VIDEO
https://goo.gl/x7UVyL

Ice Age Ice Sheets

About 18,000 years ago, glacial ice covered not only Greenland and Antarctica but also large portions of North America, Europe, and Siberia. This date in Earth history is appropriately known as the *Last Glacial Maximum.* The term implies that there were other glacial maximums, and this is indeed the case. Throughout the Quaternary period, which began about 2.6 million years ago and extends to the present, ice sheets formed, advanced over broad areas, and then wasted away. These alternating glacial and interglacial periods have occurred over and over again.

Greenland and Antarctica

Some people mistakenly think that the North Pole is covered by glacial ice, but this is not the case. The ice that covers the Arctic Ocean is **sea ice**—frozen seawater. Sea ice floats because ice is less dense than liquid water. Although sea ice never completely disappears from the Arctic, the area covered expands and contracts with the seasons. The thickness of sea ice ranges from a few centimeters for new ice to 4 meters for sea ice that has survived for years. By contrast, glaciers can be hundreds or thousands of meters thick.

GEOGRAPHICS 10.1

Antarctica Fact File

Earth's southernmost continent surrounds the South Pole (90° S. Latitude) and is almost entirely south of the Antarctic Circle (66.5° S. Latitude). This icy landmass is the fifth largest continent and is twice as large as Australia. It also has the distinction of being the coldest, driest, and windiest continent and also has the highest average elevation.

Linda Martel/ANSMET

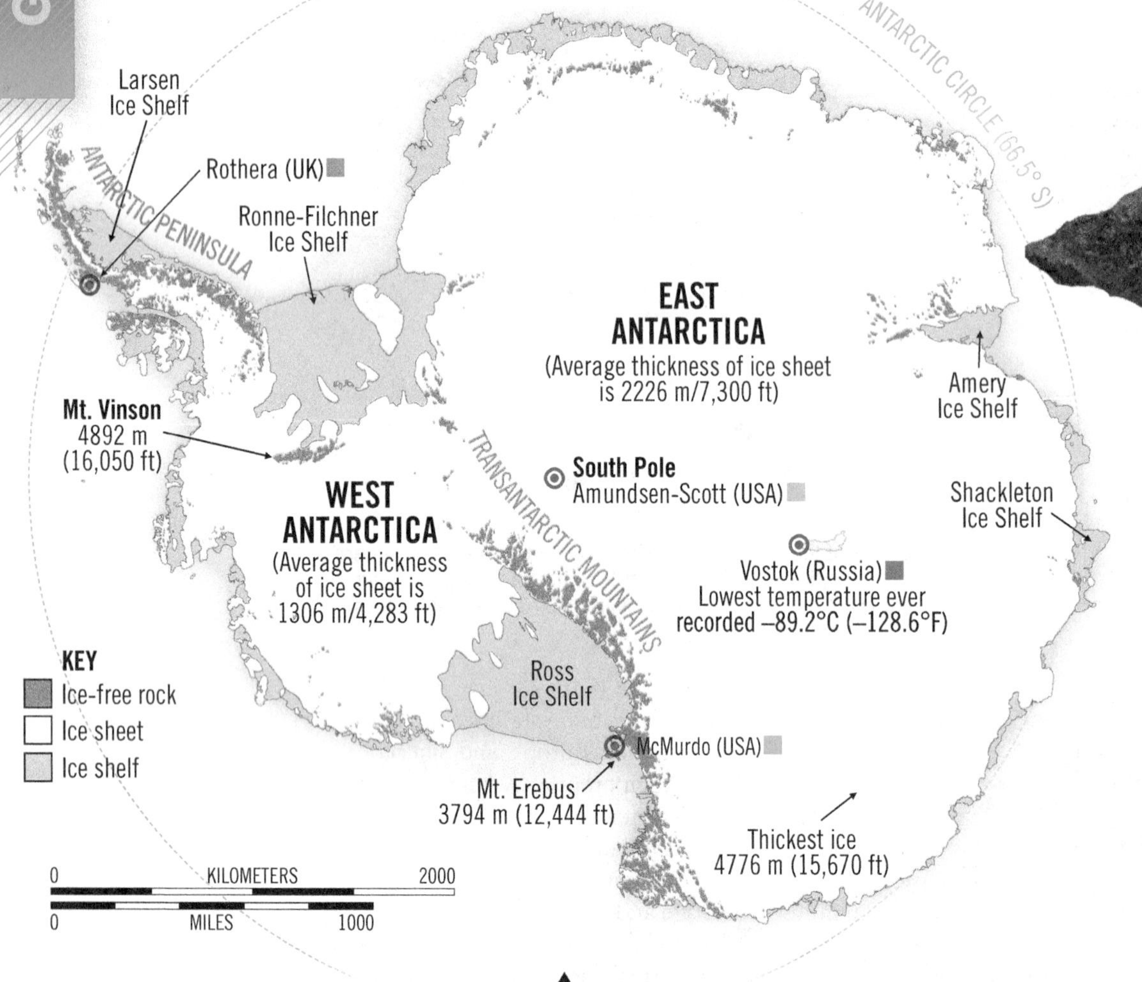

Antarctica is 1.4 times larger than the United States and about 58 times bigger than the United Kingdom. The continent is almost completely ice covered. The ice-free area amounts to only 44,890 square kilometers (17,330 square miles) or just 0.32 percent (32/100 of 1 percent) of the continent.

1 CM

Hand Out/ZUMAPRESS/Newscom

Meteorites are rocky or metallic particles from space that have fallen on Earth. These ancient fragments provide clues to the origin and history of our solar system. Antarctica is an especially good place to collect these dark masses from space because even small ones are relatively easy to spot. In addition ice flow patterns tend to concentrate them in certain areas.

Practically all of the continent belongs to the same climate classification, aptly termed ice cap climate, in which the average temperature of the warmest month is 0° C (32° F) or below. Take a look at the graph and you will see that some areas are much colder than others.

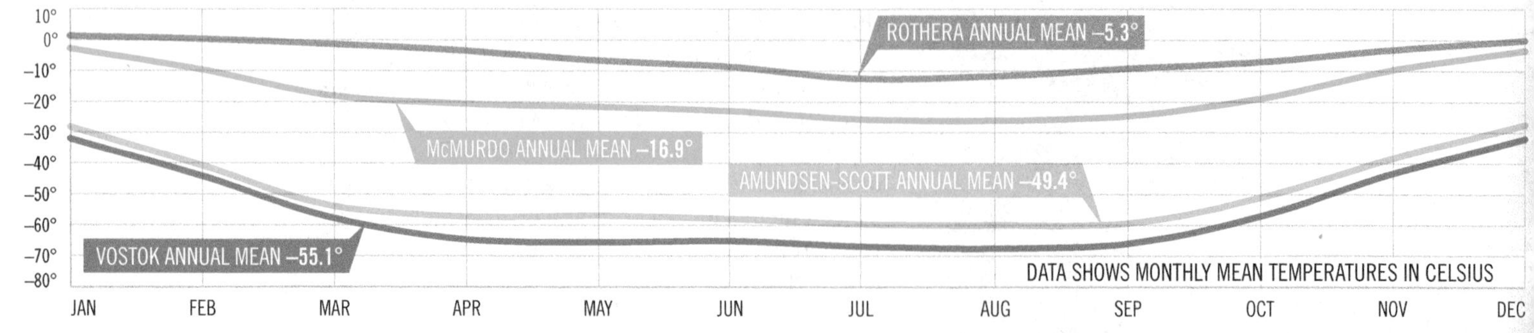

John Goodge/NSF

Rick Price/Getty Images

What if the ice melted? The discharge at the mouth of the Mississippi River is 17,300 cubic meters (593,000 cubic feet) per second. If Antarctica's ice sheets melted at a suitable rate, they could maintain the flow of the Mississippi River for more than 50,000 years! If all of the continent's ice were to melt, sea level would rise by an estimated 56 meters (more than 180 feet). Antarctica's ice represents about 65 percent of Earth's entire supply of freshwater.

The Transantarctic Mountains are a 3300-kilometer- (2600-mile-) long range that separates the West Antarctic Ice Sheet and the East Antarctic Ice Sheet. Vinson Massif is the highest peak at 4892 meters (16,050 feet). Most of the mountains are buried beneath the continent's huge ice sheets.

About half of the continent's coastal areas are characterized by ice shelves. The Ross Ice Shelf is about the size of France, whereas the Ronne-Filchner Ice Shelf has an area similar to that of Spain.

uestions:

Is the glacial ice thickest in East Antarctica or West Antarctica?

If all of Antarctica's ice were to melt, about how much would sea level rise?

McMurdo Station is the main U.S. scientific research station. It is the largest installation on the continent, capable of supporting more than 1200 residents. The total population at all research stations is about 4000 in summer and 1000 in winter. There are no permanent (indigenous) human residents on the continent.

ZUMA Wire Service/Alamy

Dan Leeth/Alamy

Glaciers form exclusively on land, and in the Northern Hemisphere, Greenland supports an ice sheet. Greenland extends between about 60° and 80° north latitude. This largest island on Earth is covered by an imposing ice sheet that occupies 1.7 million square kilometers (663,000 square miles), or about 80 percent of the island. Averaging nearly 1500 meters (5000 feet) thick, the ice extends 3000 meters (10,000 feet) above the island's bedrock floor in some places.

In the Southern Hemisphere, the huge Antarctic Ice Sheet attains a maximum thickness of about 4300 meters (14,000 feet) and covers an area of more than 13.9 million square kilometers (5.4 million square miles)—nearly the entire continent. Because of the proportions of these huge features, they are often called *continental ice sheets*. There is more about Antarctica and its ice sheets in **GEOgraphics 10.1**. The combined areas of present-day continental ice sheets represent almost 10 percent of Earth's land area.

Ice Shelves Along portions of the Antarctic coast, glacial ice flows into the adjacent ocean, creating features called **ice shelves**. These are large, relatively flat masses of floating ice that extend seaward from the coast but remain attached to the land along one or more sides. The shelves are thickest on their landward sides, and they become thinner seaward. They are sustained by ice from the adjacent ice sheet and are also nourished by snowfall and the freezing of seawater to their bases. Antarctica's ice shelves extend over approximately 1.4 million square kilometers (600,000 square miles). The Ross and Ronne-Filchner Ice Shelves are the largest; the Ross Ice Shelf covers an area approximately the size of Texas (see Figure 10.2).

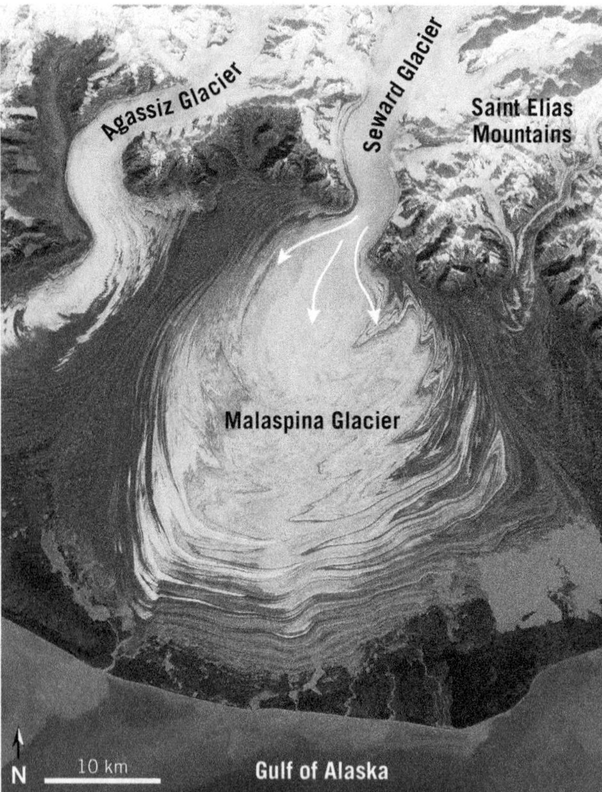

▲ **Figure 10.4 Piedmont glacier** The ice of a piedmont glacier spills from a steep valley onto a relatively flat plain, where it spreads out. Malaspina Glacier, in southeastern Alaska, fills most of this image. Covering roughly 3880 square kilometers (1500 square miles), it extends nearly 45 kilometers (28 miles) from the mountain front nearly to the sea. (NASA)

Ice caps completely bury the underlying terrain but are much smaller than ice sheets.

▶ **SmartFigure 10.3 Iceland's Vatnajökull ice cap** In 1996 the Grímsvötn Volcano erupted beneath this ice cap, an event that triggered melting and floods. (NASA)

MOBILE FIELD TRIP
https://goo.gl/SEhxye

Other Types of Glaciers

In addition to valley glaciers and ice sheets, other types of glaciers are also recognized. Covering some uplands and plateaus are masses of glacial ice called **ice caps**. Like ice sheets, ice caps completely bury the underlying landscape, but they are much smaller than the continental-scale features. Ice caps occur in many places, including Iceland and several of the large islands in the Arctic Ocean (**Figure 10.3**).

Another type, known as **piedmont glaciers**, occupy broad lowlands at the bases of steep mountains and form when one or more valley glaciers emerge from the confining walls of mountain valleys. The advancing ice spreads out to form a broad sheet. The size of individual piedmont glaciers varies greatly. Among the largest is the broad Malaspina Glacier, along the coast of southern Alaska. It covers more than 5000 square kilometers (2000 square miles) of the flat coastal plain at the foot of the lofty St. Elias range (**Figure 10.4**).

Often ice caps and ice sheets feed **outlet glaciers**. These tongues of ice flow down valleys, extending outward from the margins of these larger ice masses. The tongues are essentially valley glaciers that are avenues for ice movement from an ice cap or ice sheet through mountainous terrain to the sea. Where they encounter the ocean, some outlet glaciers spread out as floating ice shelves. Often large numbers of icebergs are produced.

CONCEPT CHECKS 10.1

1. Where are glaciers found on Earth today, and what percentage of Earth's land surface do they cover?
2. Describe how glaciers fit into the hydrologic cycle. What role do they play in the rock cycle?
3. List and briefly distinguish among four types of glaciers.
4. What is the difference between an ice sheet, sea ice, and an ice shelf?

10.2 How Glaciers Move

Describe how glaciers move, the rates at which they move, and the significance of the glacial budget.

The way in which ice moves is complex, and there are two basic types of ice movement. The first of these, *plastic flow*, involves movement *within* the ice. Ice behaves as a brittle solid until the pressure on it is equivalent to the weight of about 50 meters (165 feet) of ice. Once that load is surpassed, ice behaves as a plastic material, and flow begins. Such flow occurs due to the molecular structure of ice. Glacial ice consists of layers of molecules stacked one upon the other. The bonds between layers are weaker than those within each layer. Therefore, when a stress exceeds the strength of the bonds between the layers, the layers remain intact and slide over one another. A second and often equally important mechanism of glacial movement occurs when an entire ice mass slips along the ground. The lowest portions of most glaciers probably move by this sliding process.

The uppermost 50 meters (165 feet) of a glacier is appropriately referred to as the **zone of fracture**. Because there is not enough overlying ice to cause plastic flow, this upper part of the glacier consists of brittle ice. Consequently, the ice in this zone is carried along piggyback-style by the ice below. When the glacier moves over irregular terrain, the zone of fracture is subjected to tension, resulting in cracks called **crevasses** (Figure 10.5). These gaping cracks, which often make travel across glaciers dangerous, may extend to depths of 50 meters (165 feet). Below this depth, plastic flow seals them off.

Figure 10.5 **Crevasses** As a glacier moves, internal stresses cause large cracks to develop in the brittle upper portion of the glacier, called the zone of fracture. Crevasses, which can extend to depths of 50 meters (165 feet), can make travel across glaciers dangerous. (Photo by Wave/Glow Images)

Observing and Measuring Movement

Unlike the movement of water in streams, the movement of glacial ice is not obvious. If we could watch a valley glacier move, we would see that as with the water in a river, all of the ice does not move downstream at the same rate. Flow is greatest in the center of the glacier because the drag created by the walls and floor of the valley slow the base and sides.

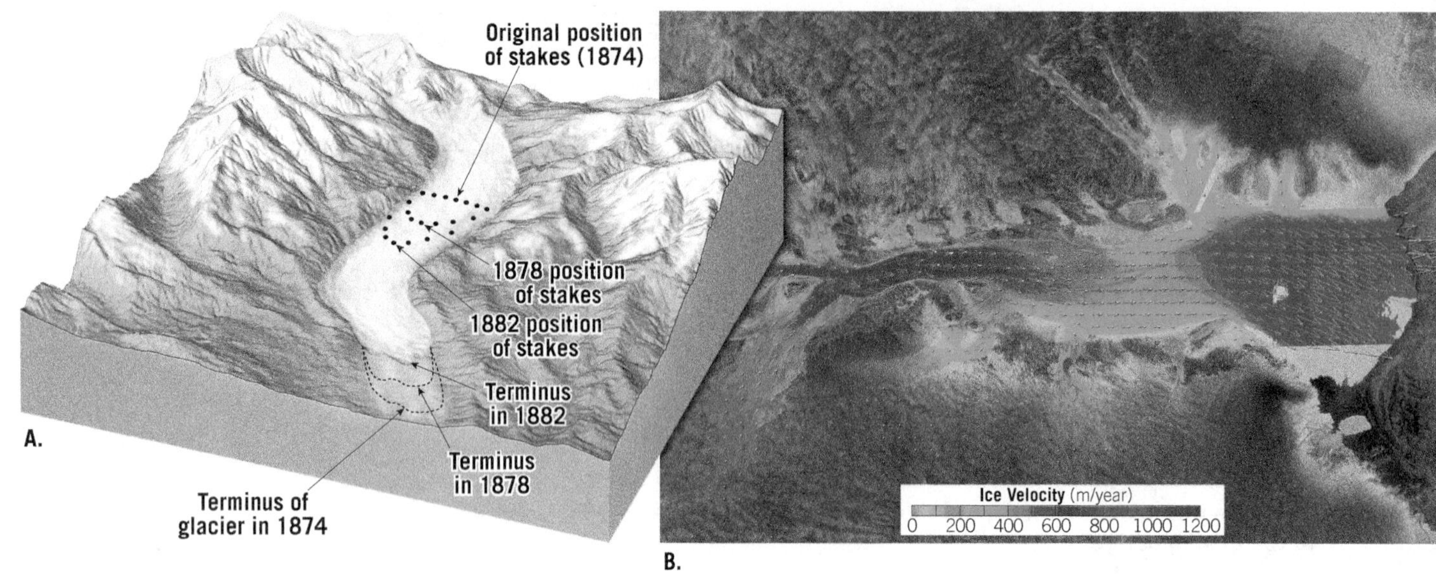

Figure 10.6 **Measuring the movement of a glacier A.** Ice movement and changes were measured in the terminus of Rhône Glacier, Switzerland. In this classic study of a valley glacier, the movement of stakes clearly shows that glacial ice moves and that movement along the sides of the glacier is slower than movement in the center. Also notice that even though the ice front was retreating, the ice within the glacier was advancing. **B.** This satellite image provides detailed information about the movement of Antarctica's Lambert Glacier. Ice velocities are determined from pairs of images obtained 24 days apart using radar data. (NASA)

Figure 10.6A shows how glacial movement was first investigated during the nineteenth century. Markers were placed in a straight line across an alpine glacier (in this example, Rhône Glacier in the Swiss Alps), and the original position of the line was marked on the valley walls. The positions of the markers were noted periodically, revealing the nature of the movement described above. In this particular study, investigators also mapped the position of the glacier's terminus, demonstrating that the terminus could retreat even as the ice within the glacier moved forward.

How rapidly does glacial ice move? Average rates vary considerably from one glacier to another. Some glaciers move so slowly that trees and other vegetation may become well established in the debris that accumulates on the glacier's surface. Others advance up to several meters each day. Recent satellite radar imaging provided insights into movements within the Antarctic Ice Sheet. Portions of some outlet glaciers move at rates greater than 800 meters (2600 feet) per year; on the other hand, ice in some interior regions creeps along at less than 2 meters (6.5 feet) per year (Figure 10.6B). Movement of some glaciers is characterized by occasional periods of extremely rapid advance called *surges*, followed by periods during which movement is much slower.

SmartFigure 10.7 **Zones of a glacier** The snowline separates the zone of accumulation and the zone of wastage. Whether the ice front advances, retreats, or remains stationary depends on the balance or lack of balance between accumulation and wastage (ablation).

TUTORIAL https://goo.gl/y3g5ig

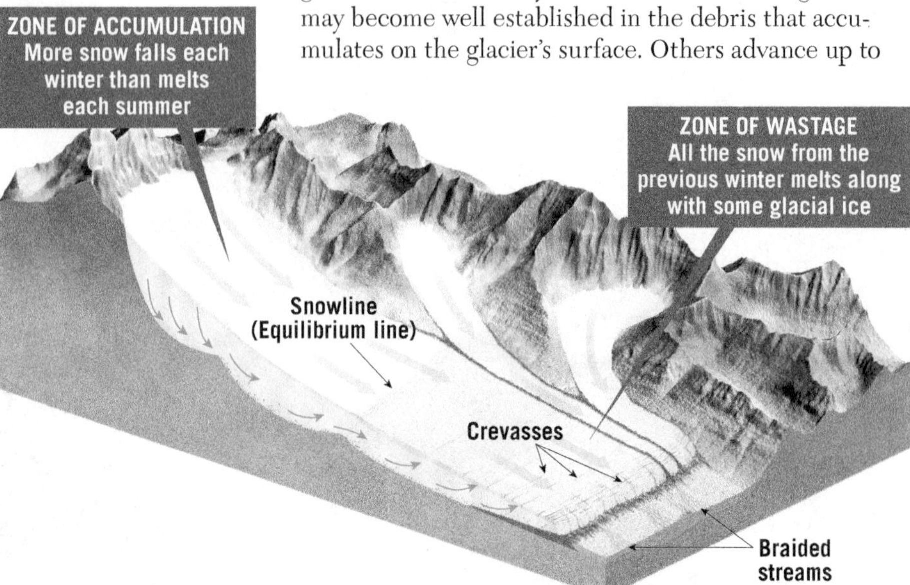

Budget of a Glacier: Accumulation Versus Wastage

Snow is the raw material from which glacial ice originates; therefore, glaciers form in areas where more snow falls in winter than melts during summer. Glaciers are constantly gaining and losing ice.

Glacial Zones

Snow accumulation and ice formation occur in the **zone of accumulation**. Its outer limits are defined by the **snowline**, or **equilibrium line**—the elevation at which the accumulation and wasting of glacial ice is equal. The elevation of this boundary varies greatly, from sea level in polar regions to altitudes approaching 5000 meters (16,000 feet) near the equator. Above the snowline, in the zone of accumulation, the addition of snow thickens the glacier and promotes movement. Below the snowline is the **zone of wastage** (ablation). Here there is a net loss to the glacier as all of the snow from the previous winter melts, as does some of the glacial ice (Figure 10.7).

Figure 10.8 Icebergs Icebergs form when large masses of ice break off from the front of a glacier after it reaches a water body, in a process known as calving. **A.** Calving of one of Greenland's outlet glaciers created this iceberg. As the sketch illustrates, the photo shows only the "tip of the iceberg." (Photo by Melissa King/Shutterstock) **B.** This satellite image documents the breakup of the Larsen B Ice Shelf, adjacent to the Antarctic Peninsula. Thousands of very large "flat-topped" icebergs were created in the process. (NASA)

Geologist's Sketch

A. Only about 20 percent or less of an iceberg protrudes above the waterline.

Icebergs produced by the breakup of ice shelf

March 7, 2002

50km

B.

In addition to melting, glaciers also waste as large pieces of ice break off the front of a glacier, in a process called **calving**. Calving creates **icebergs** in places where glaciers reach the sea (Figure 10.8). Because icebergs are just slightly less dense than seawater, they float very low in the water, with more than 80 percent of their mass submerged. Along the margins of Antarctica's ice shelves, calving is the primary means by which the ice shelves lose mass. The relatively flat icebergs produced here can be several kilometers across and 600 meters (2000 feet) thick. By comparison, thousands of irregularly shaped icebergs are produced by outlet glaciers flowing from the margins of the Greenland Ice Sheet. Many drift southward and find their way into the North Atlantic, where they can be hazardous to navigation.

Glacial Budget Whether the margin of a glacier advances, retreats, or remains stationary depends on the *budget* of the glacier. The **glacial budget** refers to the balance or lack of balance between accumulation at the upper end of a glacier and loss at the lower end. If ice accumulation exceeds wastage, the glacial front advances until the two factors balance. At this point, the terminus of the glacier becomes stationary.

If a warming trend increases wastage and/or if a drop in snowfall decreases accumulation, the ice front will retreat. As the terminus of the glacier retreats, the extent of the zone of wastage diminishes. Therefore, in time a new balance will be reached between accumulation and wastage, and the ice front will again become stationary.

Whether the margin of a glacier is advancing, retreating, or stationary, the ice within the glacier continues to flow forward. In the case of a receding glacier,

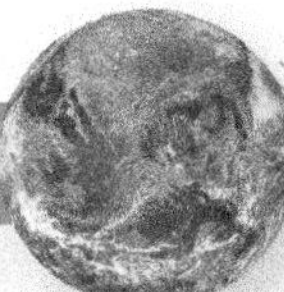

EYE ON EARTH 10.1

This photo shows an iceberg floating in the ocean near the coast of Greenland.

QUESTION 1 *How do icebergs form? What term applies to this process?*

QUESTION 2 *Using the knowledge you have gained about these features, explain the common phrase "It's only the tip of the iceberg."*

QUESTION 3 *Is an iceberg the same as sea ice? Explain.*

(Photo by GybasDigiPhoto/Shutterstock)

Figure 10.9 **Retreating glaciers** These two images were taken 78 years apart from about the same vantage point, along the southwest coast of Greenland. Between 1935 and 2013 the outlet glacier that is the primary focus of these photos retreated about 3 kilometers (about 2 miles). (NASA/Earth Observatory)

the ice still flows forward but not rapidly enough to offset wastage. This point is illustrated in Figure 10.6A. As the line of stakes within the Rhône Glacier continued to move downvalley, the terminus of the glacier slowly retreated upvalley.

Glaciers in Retreat: Unbalanced Glacial Budgets

Because glaciers are sensitive to changes in temperature and precipitation, they provide clues about changes in climate. With few exceptions, valley glaciers around the world have been retreating at unprecedented rates over the past century. Many valley glaciers have disappeared altogether. For example, 150 years ago, there were 147 glaciers in Montana's Glacier National Park. Today only 37 remain. Greenland's ice sheet and portions of Antarctica's ice are also shrinking. The photos in Figure 10.9 provide an example.

CONCEPT CHECKS 10.2

1. Describe two components of glacial movement.
2. How rapidly does glacial ice move? Provide some examples.
3. What are crevasses, and where do they form?
4. Under what circumstances will the front of a glacier advance? Retreat? Remain stationary?

10.3 Glacial Erosion

Discuss the processes of glacial erosion. Identify and describe the major topographic features sculpted by glacial erosion.

Glaciers erode tremendous volumes of rock. For anyone who has observed the terminus of an alpine glacier, the evidence of its erosive force is clear (Figure 10.10). You can witness firsthand the release of rock fragments of various sizes from the ice as it melts. All signs lead to the conclusion that the ice has scraped, scoured, and torn rock debris from the floor and walls of the valley and carried it downvalley. In addition, in mountainous regions, mass-movement processes also make substantial contributions to the sediment load of a glacier. The photo that accompanies *Examining the Earth System* Problem 3 in Chapter 8 (page 270) provides an excellent example.

Once a glacier acquires rock debris, that debris cannot settle out as does the load carried by a stream

Figure 10.10 **Evidence of glacial erosion** As the terminus of this glacier wastes away, it deposits large quantities of sediment. This image near the terminus of Exit Glacier in Alaska's Kenai Fiords National Park shows that the rock debris dropped by the melting ice is a jumbled mixture of different-size sediments. (Photo by Michael Collier)

or by the wind. Consequently, glaciers can carry huge blocks that no other erosional agent could possibly budge. Although today's glaciers are of limited importance as erosional agents, many landscapes that were modified by the widespread glaciers of the recent Ice Age still reflect to a high degree the work of ice.

How Glaciers Erode

Glaciers erode land primarily in two ways: plucking and abrasion. First, as a glacier flows over a fractured bedrock surface, it loosens and lifts blocks of rock and incorporates them into the ice. This process, known as **plucking**, occurs when meltwater penetrates the cracks and joints along the rock floor of the glacier and freezes. Because water expands when it freezes, it exerts tremendous leverage that pries the rock loose. In this manner, sediment of all sizes becomes part of the glacier's load.

The second major erosional process is **abrasion**. As the ice and its load of rock fragments slide over bedrock, they function like sandpaper, smoothing and polishing the surface below. The pulverized rock produced by the glacial gristmill is appropriately called **rock flour**. So much rock flour may be produced that meltwater streams leaving a glacier often have the grayish appearance of skim milk—visible evidence of the grinding power of the ice.

When the ice at the bottom of a glacier contains large rock fragments, long scratches and grooves called **glacial striations** may be gouged into the bedrock (Figure 10.11A). These linear scratches on the bedrock surface provide clues to the direction of glacial movement. By mapping the striations over large areas, glacial flow patterns can often be reconstructed.

Not all abrasive action produces striations. The rock surface over which the glacier moves may also become highly polished by the ice and its load of finer particles. The broad expanses of smoothly polished granite in California's Yosemite National Park provide an excellent example (Figure 10.11B).

As is the case with other agents of erosion, the rate of glacial erosion is highly variable. This rate is largely controlled by four factors: (1) speed of glacier movement; (2) ice thickness; (3) shape, abundance, and hardness of the rock fragments in the ice at the base of the glacier; and (4) erodibility of the surface beneath the glacier. These factors can all vary from place to place and from one time to another, with resulting variation in the degree of landscape modification.

Landforms Created by Glacial Erosion

The erosional effects of valley glaciers and ice sheets are quite different. A visitor to a glaciated mountain region is likely to see sharp and angular topography. This is because alpine glaciers tend to accentuate the irregularities of the mountain landscape by creating steeper canyon walls and making bold peaks even more jagged. By contrast, continental ice sheets generally override the terrain and hence subdue rather than accentuate the irregularities they encounter. Although the erosional potential of ice sheets is enormous, landforms carved by these huge ice masses usually do not inspire the same awe as do the erosional features

Glacial abrasion created the scratches and grooves in this bedrock.
A.
Glacially polished granite in California's Yosemite National Park.
B.

Figure 10.11 Glacial abrasion Moving glacial ice, armed with sediment, acts like sandpaper, scratching and polishing rock. (Photos by Michael Collier)

created by valley glaciers. Figure 10.12 shows a hypothetical mountain area before, during, and after glaciation. You will refer to this figure often in the following discussion.

Glaciated Valleys A hike up a glaciated valley reveals a number of striking ice-sculpted features. The valley itself is often a dramatic sight. Unlike streams, which create their own valleys, glaciers take the path of least resistance, following the paths of existing stream valleys. Prior to glaciation, mountain valleys are characteristically narrow and V-shaped because streams are well above base level and are therefore downcutting. However, during glaciation, these narrow valleys are transformed as the glacier widens and deepens them, creating a U-shaped **glacial trough** (see Figure 10.12C and Figure 10.13). In addition to producing a broader and deeper valley, the glacier also straightens the valley. As ice flows around sharp curves, its great erosional force removes the spurs of land that extend into the valley.

The amount of glacial erosion depends in part on the thickness of the ice. Consequently, main glaciers, also called *trunk glaciers*, cut their valleys deeper than do their smaller tributary glaciers. Thus, after the ice has receded, the valleys of tributary glaciers are left standing above the main glacial trough and are termed **hanging valleys**. Rivers flowing through hanging valleys may produce spectacular waterfalls, such as those in Yosemite National Park, California (see Figure 10.12C).

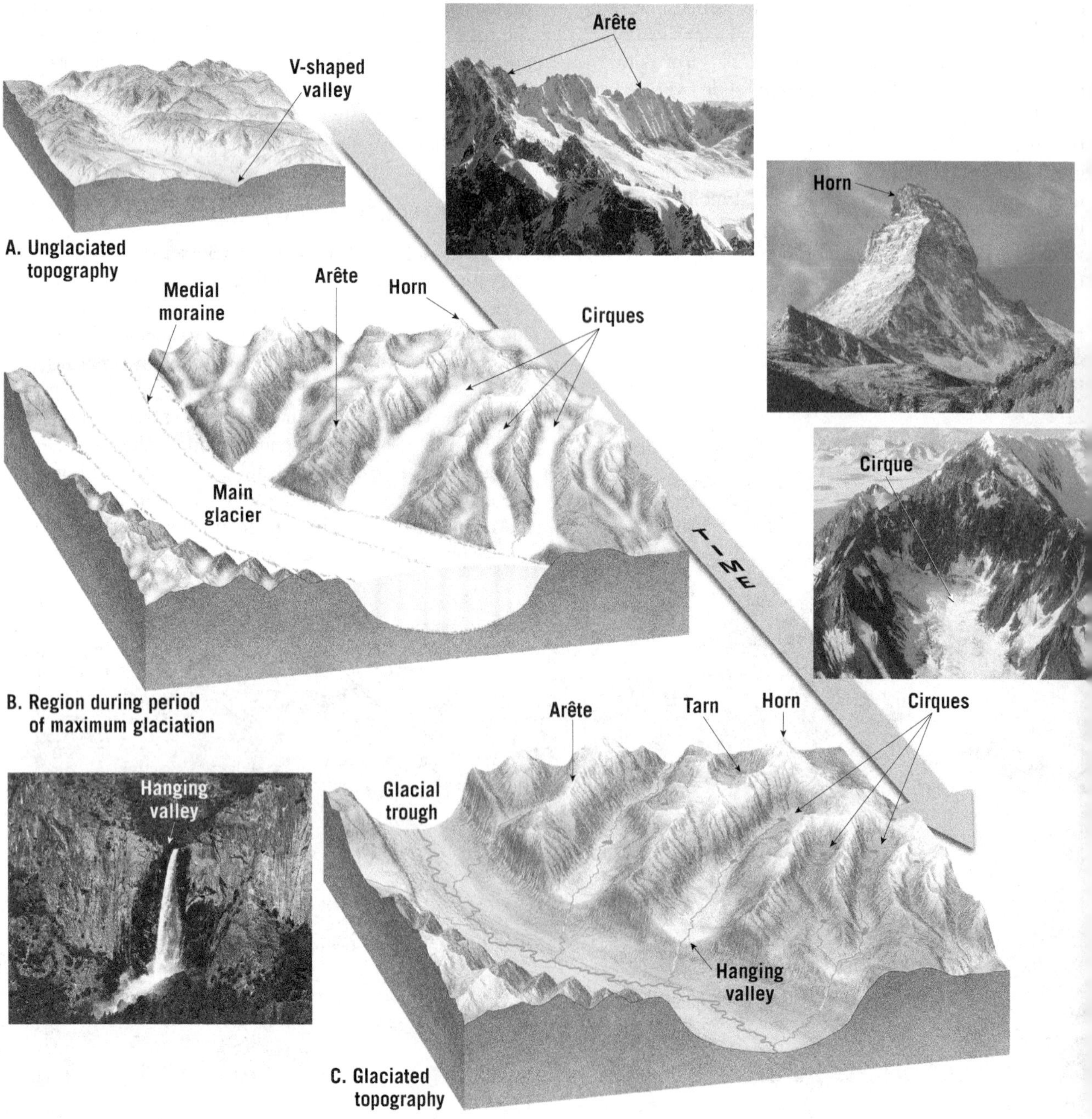

SmartFigure 10.12 **Erosional landforms created by alpine glaciers** The unglaciated landscape (**A**) is modified by valley glaciers (**B**). After the ice recedes (**C**), the terrain looks very different than it looked before glaciation. (Photo from the James E. Patterson Collection, courtesy of F. K. Lutgens; horn photo by Andy Selinger/AGE fotostock; cirque photo by Marli Miller; hanging valley photo by James Mott/Alamy Stock Photo)

TUTORIAL
https://goo.gl/VRAu3y

Cirques At the head of a glacial valley is a characteristic and often imposing feature associated with an alpine glacier—a **cirque**. As Figure 10.12C illustrates, these bowl-shaped depressions have precipitous walls on three sides but are open on the downvalley side. The cirque is the focal point of the glacier's growth because it is the area of snow accumulation and ice formation. Cirques begin as irregularities in the mountainside that are subsequently enlarged by frost wedging and plucking along the sides and bottom of the glacier. The glacier in turn acts as a conveyor belt that carries away the debris. After the glacier has melted away, the cirque basin is sometimes occupied by a small lake called a *tarn* (see Figure 10.12C).

Arêtes and Horns The Alps, Northern Rockies, and many other mountain landscapes sculpted by valley glaciers reveal more than glacial troughs and cirques. In addition, sinuous, knife-edged ridges called **arêtes** and sharp, pyramid-like peaks termed **horns** project above the surroundings (see Figure 10.12C). Both features can originate from the same basic process: the enlargement of cirques produced by plucking and frost action. Several cirques around a single high mountain create the spires of rock called *horns*. As the cirques enlarge and converge, an isolated horn is produced. A famous example is the Matterhorn in the Swiss Alps.

Arêtes can form in a similar manner except that the cirques are not clustered around a point but rather exist on opposite sides of a divide. As the cirques grow, the divide separating them is reduced to a very narrow, knifelike partition. An arête can also be created when glaciers that flow in parallel valleys narrow the intervening ridge as they scour and widen their valleys.

Fiords **Fiords** are deep, often spectacular, steep-sided inlets of the sea that exist in many high-latitude areas of the world where mountains are adjacent to the ocean (Figure 10.14). Norway, British Columbia, Greenland, New Zealand, Chile, and Alaska all have coastlines characterized by fiords. They are glacial troughs that became submerged as the ice left the valleys and sea level rose following the Ice Age.

The depths of some fiords can exceed 1000 meters (3300 feet). However, the great depths of these flooded troughs are only partly explained by the post–Ice Age rise in sea level. Unlike the situation

SmartFigure 10.13 **A U-shaped glacial trough** Prior to glaciation, a mountain valley is typically narrow and V-shaped. During glaciation, an alpine glacier widens, deepens, and straightens the valley, creating the classic U-shape shown here in the Sierra Nevada, west of Bishop, California. (Photo by Michael Collier)

ANIMATION
https://goo.gl/mzSk4k

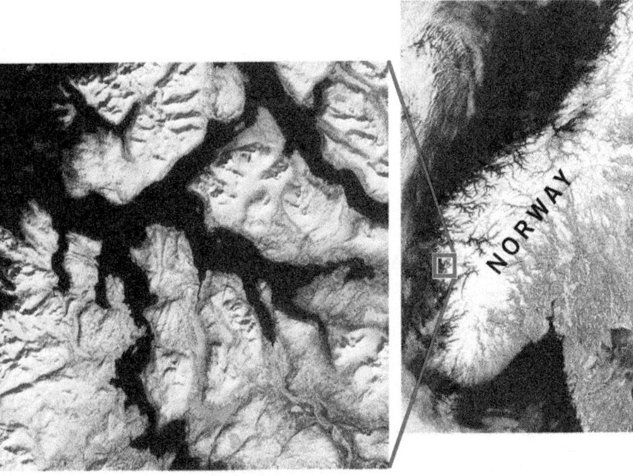

Figure 10.14 **Fiords** The coast of Norway is known for its many fiords. Frequently these ice-sculpted inlets of the sea are hundreds of meters deep. (Satellite images courtesy of NASA; photo by Inger Yoshio Tomii/SuperStock)

governing the downward erosional work of rivers, sea level does not act as a base level for glaciers. As a consequence, glaciers are capable of eroding their beds far below the surface of the sea. For example, a valley glacier 300 meters (1000 feet) thick can carve its valley floor more than 250 meters (800 feet) below sea level before downward erosion ceases and the ice begins to float.

CONCEPT CHECKS 10.3

1. How do glaciers acquire their load of sediment?
2. How does a mountain valley that has been glaciated differ from one that has not? Describe the erosional features created by valley glaciers.
3. Relate fiords to glacial troughs.

10.4 Glacial Deposits

Distinguish between the two basic types of glacial deposits and briefly describe the features associated with each type.

A glacier picks up and transports a huge load of rock debris as it slowly advances across the land. Where the ice melts, these materials are deposited. Such deposits can play a significant role in forming the physical landscape. For example, in many areas once covered by the ice sheets of the recent Ice Age, the bedrock is rarely exposed because the terrain is completely mantled by glacial deposits that are dozens or even hundreds of meters thick. The general effect of these deposits is to reduce the local relief and thus level the topography. Indeed, rural country scenes familiar to many of us—rocky pastures in New England, wheat fields in the Dakotas, rolling farmland in the Midwest—result directly from glacial deposition.

Glacial till is an unsorted mixture of many different sediment sizes.

A close examination of glacial till often reveals cobbles that have been scratched as they were dragged along by the ice.

Figure 10.15 Glacial till Unlike sediment deposited by running water and wind, material deposited directly by a glacier is not sorted. Figure 15.12 provides another good example of till. (Top photo by Michael Collier; bottom photo by E. J. Tarbuck)

Types of Glacial Drift

Long before the theory of an extensive Ice Age was proposed, much of the soil and rock debris covering portions of Europe was recognized as having come from elsewhere. At the time, these foreign materials were believed to have been "drifted" into their present positions by floating ice during an ancient flood. As a consequence, the term *drift* was applied to this sediment. Although rooted in a concept that was not correct, this term was so well established by the time the true glacial origin of the debris became widely recognized that it remained part of the glacial vocabulary. Today, **glacial drift** is an all-embracing term for sediments of glacial origin, no matter how, where, or in what form they were deposited.

Geologists divide glacial drift into two distinct types: (1) materials deposited directly by the glacier, which are known as *till*, and (2) sediments laid down by glacial meltwater, called *stratified drift*.

Glacial Till As glacial ice melts and drops its load of rock fragments, **till** is deposited. Unlike moving water and wind, ice cannot sort the sediment it carries; therefore, deposits of till are characteristically unsorted mixtures of many particle sizes (Figure 10.15). A close examination of this sediment shows that many of the pieces are scratched and polished as a result of being dragged along by the glacier. Such pieces help distinguish till from other deposits that are a mixture of different sediment sizes, such as material from a debris flow or a rockslide.

Boulders found in the till or lying free on the surface are called **glacial erratics** if they are different from the bedrock below (Figure 10.16). Of course, this means that they must have been derived from a source outside the area where they are found. By studying glacial erratics as well as the mineral composition of the remaining till, geologists are sometimes able to trace the path of a lobe of ice. In portions of New England and other areas, erratics dot pastures and farm fields.

In some places, these large rocks were cleared from fields and piled to make walls.

Stratified Drift As the name implies, **stratified drift** is sorted according to the size and weight of the particles. Ice is not capable of sorting the way running water can, and therefore these materials are not deposited directly by the glacier as till is but instead reflect the sorting action of glacial meltwater.

Some deposits of stratified drift are made by streams issuing directly from the glacier. Other stratified deposits involve sediment that was originally laid down as till and later picked up, transported, and redeposited by meltwater beyond the margin of the ice. Accumulations of stratified drift often consist largely of sand and gravel because the meltwater is not capable of moving larger material and because the finer rock flour remains suspended and is commonly carried far from the glacier. In consequence, these deposits may be actively mined as aggregate for road work and other construction projects.

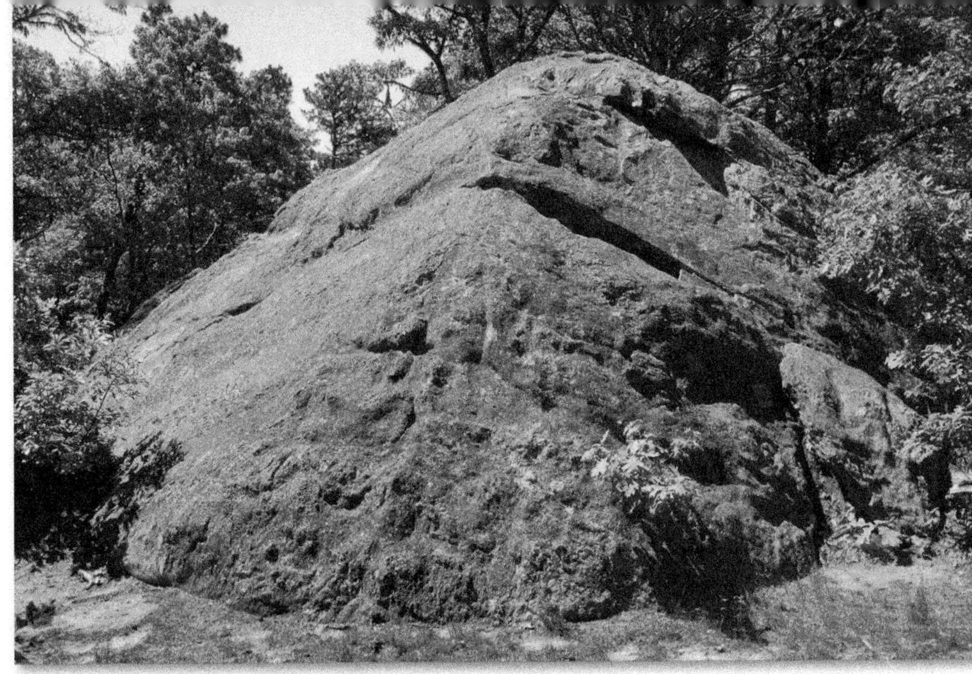

SmartFigure 10.16 **Glacial erratic** This large glacially transported boulder, called Doane Rock, is a prominent feature near Nauset Bay on Cape Cod. Such boulders are called *glacial erratics.* (Photo by Michael Collier)

MOBILE FIELD TRIP
https://goo.gl/fz5L6k

Moraines, Outwash Plains, and Kettles

Perhaps the most widespread features created by glacial deposition are *moraines*, which are simply layers or ridges of till. Lateral and medial moraines are found only in mountain valleys, whereas end moraines and ground moraines are associated with areas affected by either ice sheets or valley glaciers.

Lateral and Medial Moraines The sides of a valley glacier accumulate large quantities of debris from the valley walls. When the glacier wastes away, these materials are left as ridges, called **lateral moraines**, along the sides of the valley. **Medial moraines** are formed when two valley glaciers coalesce to form a single ice stream (Figure 10.17). The till that was once carried along the edge of each glacier joins to form a single dark stripe of debris within the newly enlarged glacier. Creation of these dark stripes within the ice stream is obvious proof that glacial ice moves because the medial moraine could not form if the ice did not flow downvalley. A large alpine glacier may have several medial moraines, each forming where a tributary glacier joins the main valley.

End Moraines and Ground Moraines Sometimes a glacier is compared to a conveyor belt. No matter whether the front of a glacier or ice sheet is advancing, retreating, or stationary, it is constantly moving sediment forward and dropping it at its terminus.

An **end moraine** is a ridge of till that forms at the terminus of a glacier or ice sheet whenever the terminus is stationary. That is, the end moraine forms when the ice is wasting away near the end of the glacier at a rate equal to the forward advance of the glacier. Although the terminus of the glacier is stationary, the ice continues to flow forward, delivering a continuous supply of sediment in the same manner a conveyor belt delivers goods to the end of a production line. As the ice melts, the till is dropped, and the end moraine grows. The longer the ice front remains stationary, the larger

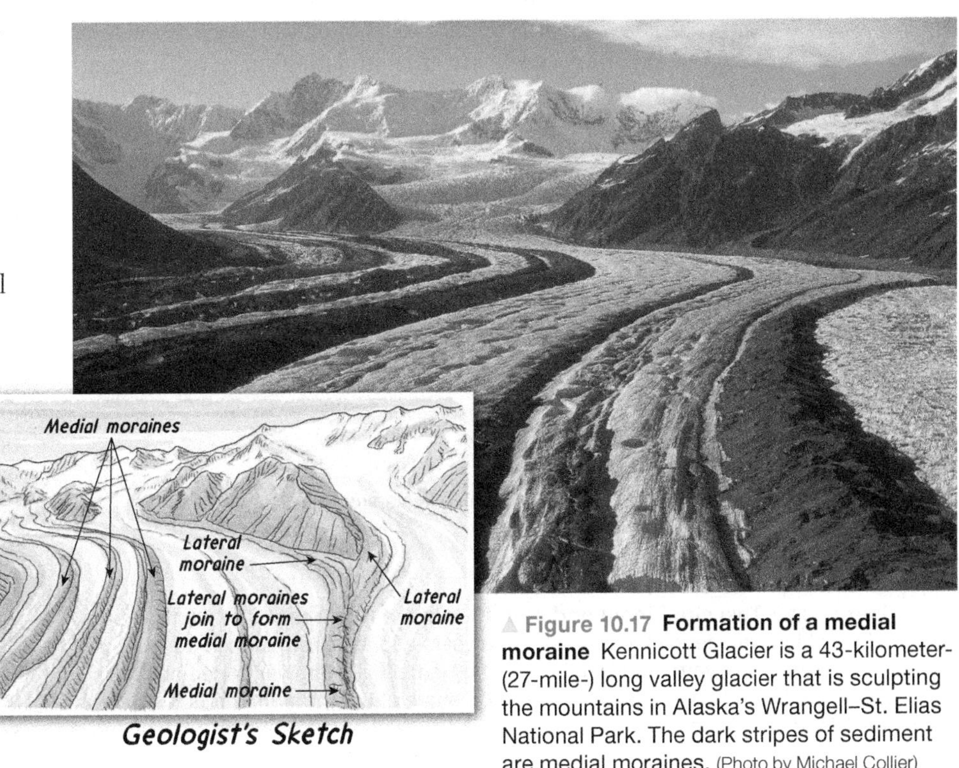

Figure 10.17 **Formation of a medial moraine** Kennicott Glacier is a 43-kilometer- (27-mile-) long valley glacier that is sculpting the mountains in Alaska's Wrangell–St. Elias National Park. The dark stripes of sediment are medial moraines. (Photo by Michael Collier)

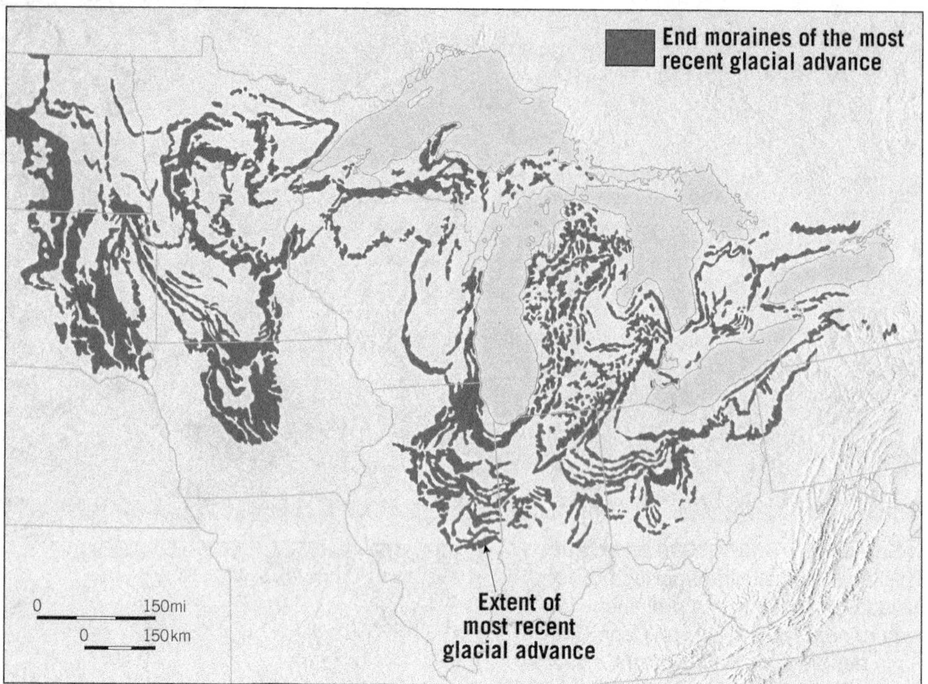

Figure 10.18 **End moraines of the Great Lakes region** End moraines deposited during the most recent stage of glaciation are the most prominent features in many parts of this region.

the ridge of till becomes. End moraines are common features of both valley glaciers and ice sheets. Figure 10.10 shows a portion of an end moraine at the terminus of Alaska's Exit Glacier.

Eventually, wastage exceeds nourishment. At this point, the front of the glacier begins to recede in the direction from which it originally advanced. However, as the ice front retreats, the conveyor-belt action of the glacier continues to provide fresh supplies of sediment to the terminus. In this manner, a large quantity of till is deposited as the ice melts away, creating a rock-strewn, undulating surface. This gently rolling layer of till deposited as the ice front recedes is termed **ground moraine**. Ground moraine has a leveling effect, filling in low spots and clogging old stream channels, often leading to a derangement of the existing drainage system. In areas where this layer of till is still relatively fresh, such as the northern Great Lakes region, poorly drained swampy lands are quite common. Periodically, a glacier will retreat to a point where wastage and nourishment once again balance. When this happens, the ice front stabilizes, and a new end moraine forms.

The pattern of end moraine formation and ground moraine deposition may be repeated many times before the glacier has completely vanished. Such a pattern is illustrated in Figure 10.18. The very first end moraine to form marks the farthest advance of the glacier and is called the *terminal end moraine*. End moraines that form as the ice front occasionally stabilizes during retreat are termed *recessional end moraines*. Terminal and recessional moraines are essentially alike; the only difference between them is their relative positions.

End moraines deposited by the most recent major stage of Ice Age glaciation are prominent features in many parts of the Midwest and Northeast. In Wisconsin, the wooded, hilly terrain of the Kettle Moraine near Milwaukee is a particularly picturesque example. A well-known example in the Northeast is Long Island. This linear strip of glacial sediment that extends northeastward from New York City is part of an end moraine complex that stretches from eastern Pennsylvania to Cape Cod, Massachusetts (Figure 10.19).

Figure 10.20 represents a hypothetical area during and following glaciation. This figure depicts landscape features, such as the end moraines just described, as well as depositional landforms similar to what might be encountered if you were traveling in the upper Midwest or New England. You will be referred to this figure several times as you read the following paragraphs on glacial deposits.

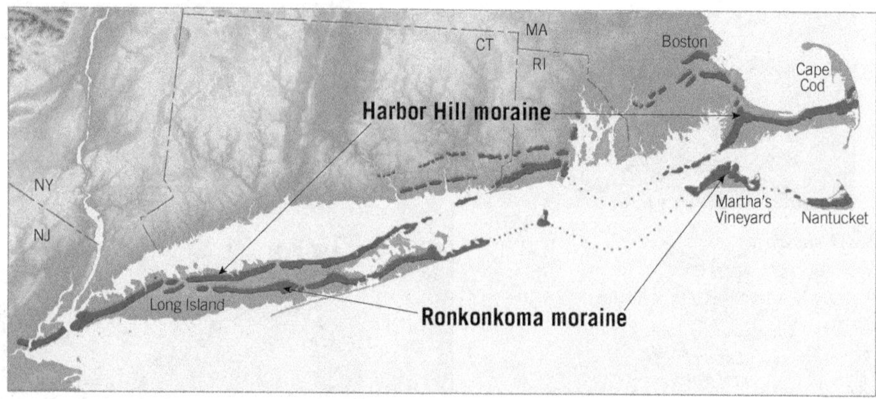

Figure 10.19 **Two significant end moraines in the Northeast** The Ronkonkoma moraine, which was deposited about 20,000 years ago, extends through central Long Island, Martha's Vineyard, and Nantucket. The Harbor Hill moraine formed about 14,000 years ago and extends along the north shore of Long Island, through southern Rhode Island and Cape Cod.

Outwash Plains and Valley Trains At the same time that an end moraine is forming, water from the melting glacier cascades over and through the till, sweeping some of it out in front of the growing ridge of unsorted debris. Meltwater generally emerges from the ice in rapidly moving streams that are often choked with suspended material and carry a substantial bed load as well. Water leaving the glacier moves onto the relatively flat surface beyond and rapidly loses velocity. As a consequence, much of its bed load is dropped, and the meltwater begins weaving a complex pattern of braided channels. Such a situation is shown in Figure 10.20. In this way, a broad, ramplike surface composed of stratified drift is built adjacent to the downstream edge of most end moraines. When the feature is formed in association with an ice sheet, it is termed an **outwash plain**; when it is largely confined to a mountain valley, it is usually called a **valley train**.

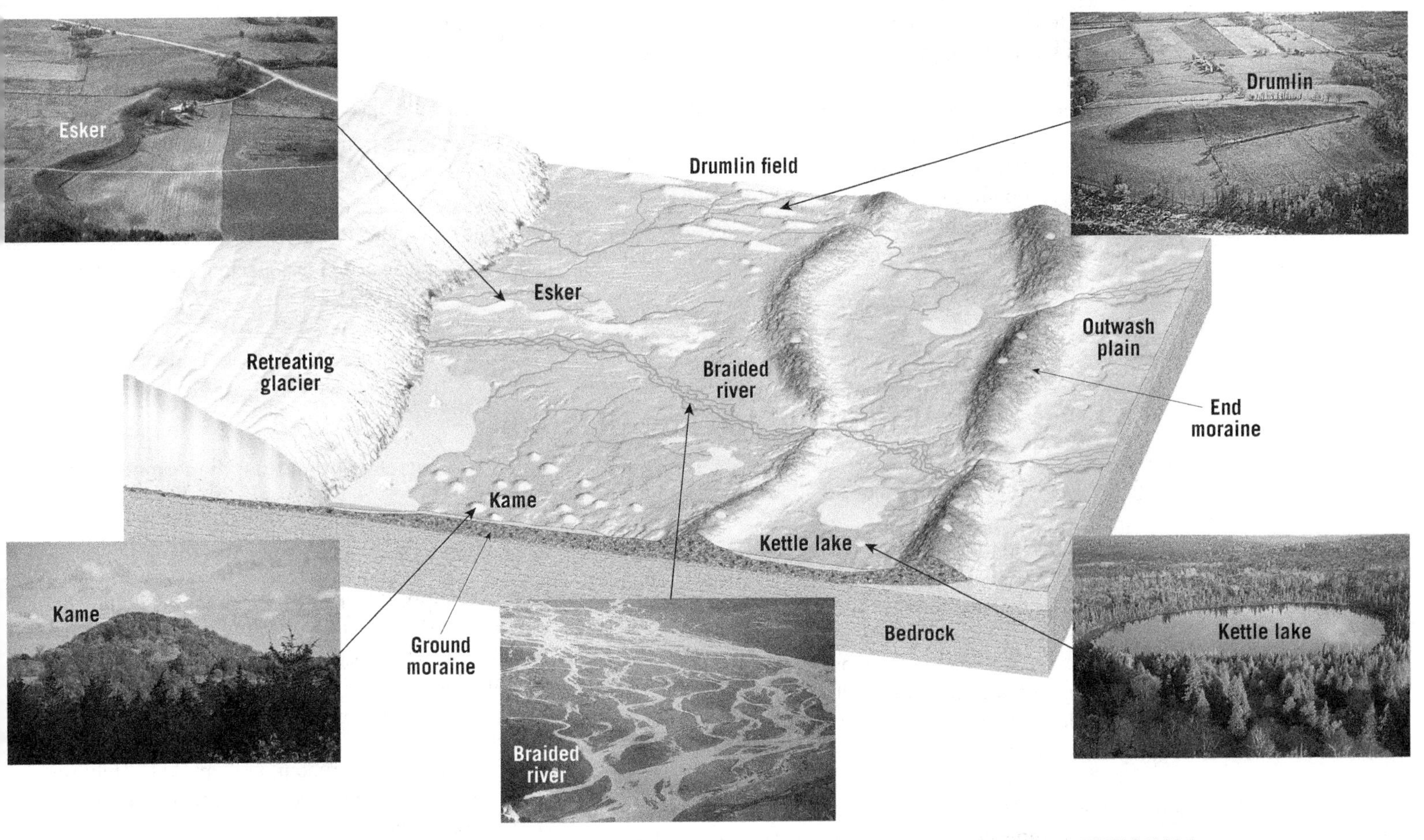

SmartFigure 10.20 **Common depositional landforms** This diagram depicts a hypothetical area affected by ice sheets in the recent geologic past. (Drumlin photo courtesy of Ward's Natural Science Establishment; esker photo by Richard P. Jacobs/JLM Visuals; kame photo by John Dankwardt; kettle lake photo by Carlyn Iverson/Science Source; braided river photo by Michael Collier)

TUTORIAL https://goo.gl/76mWl1

Kettles End moraines, outwash plains, and valley trains are often pockmarked with basins, or depressions, known as **kettles** (see Figure 10.20). Kettles form when blocks of stagnant ice become buried in drift and eventually melt, leaving pits in the glacial sediment. Most kettles do not exceed 2 kilometers in diameter, and the typical depth of most kettles is less than 10 meters (33 feet). Water often fills the depression and forms a pond or lake. One well-known example is Walden Pond near Concord, Massachusetts. It is here that Henry David Thoreau lived alone for 2 years in the 1840s and about which he wrote *Walden*, his classic of American literature.

Drumlins, Eskers, and Kames

Moraines are not the only landforms deposited by glaciers. Some landscapes are characterized by numerous elongate parallel hills made of till. Other areas exhibit conical hills and relatively narrow winding ridges composed mainly of stratified drift.

Drumlins **Drumlins** are streamlined asymmetrical hills composed of till (see Figure 10.20). They range in height from 15 to 60 meters (50–200 feet) and average 0.4 to 0.8 kilometer (0.25–0.50 mile) in length. The steep side of the hill faces the direction *from* which the ice advanced, whereas the gentler slope points in the direction the ice moved. Drumlins are not found singly but rather occur in clusters, called *drumlin fields*. One such cluster, east of Rochester, New York, is estimated to contain about 10,000 drumlins. Their streamlined shape indicates that they were molded in the zone of flow within an active glacier. It is thought that drumlins originate when glaciers advance over previously deposited drift and reshape the material.

Eskers and Kames In some areas that were once occupied by glaciers, sinuous ridges composed largely of sand and gravel may be found. These ridges, called **eskers**, are deposits made by streams flowing in tunnels beneath the ice, near the terminus of a glacier (see Figure 10.20). They may be several meters high and extend for many kilometers. In some areas they are mined for sand and gravel, and for this reason, eskers are disappearing in some localities.

Kames are steep-sided hills that, like eskers, are composed of sand and gravel (see Figure 10.20). Kames originate when glacial meltwater washes sediment into openings and depressions in the stagnant wasting terminus of a glacier. When the ice eventually melts away, the stratified drift is left behind as mounds or hills.

CONCEPT CHECKS 10.4

1. What term can be applied to any glacial deposit? What is the difference between till and stratified drift?
2. Describe the formation of a medial moraine.
3. Distinguish between terminal end moraine, recessional end moraine, and ground moraine. Relate these moraines to the budget of a glacier.
4. List four glacial depositional features other than moraines.

10.5 Other Effects of Ice Age Glaciers

Describe and explain several important effects of Ice Age glaciers other than the formation of erosional and depositional landforms.

In addition to their massive erosional and depositional work, the ice sheets and glaciers of the Ice Age glaciers had other effects, sometimes profound, on the landscape. For example, as the ice advanced and retreated, animals and plants were forced to migrate or perish; a number of plants and animals became extinct. Other effects of Ice Age glaciers that are described in this section involve adjustments in Earth's crust due to the addition and removal of ice and sea-level changes associated with the formation and melting of ice sheets. The advance and retreat of ice sheets also led to significant changes in the routes taken by rivers. In some regions, glaciers acted as dams that created large lakes. When these ice dams failed, the effects on the landscape were profound. In some areas that today are deserts, lakes of another type, called pluvial lakes, formed.

During the Last Glacial Maximum, about 18,000 years ago, sea level was nearly 100 meters (330 feet) lower than it is today.

Present level
Mean sea level (m)
-20
-40
-60
-80
-100
Last glacial maximum
20 18 16 14 12 10 8 6 4 2 0
Thousands of years ago

Maximum extent of glaciation
Coastline 18,000 years ago

During the Last Glacial Maximum, the shoreline extended out onto the present-day continental shelf.

SmartFigure 10.21
Changing sea level As ice sheets form and then melt away, sea level falls and rises, causing the shoreline to shift.

ANIMATION
https://goo.gl/QL4wnv

Crustal Subsidence and Rebound

In areas that were major centers of ice accumulation, such as Scandinavia and northern Canada, the land has been slowly rising for the past several thousand years. The land had downwarped under the tremendous weight of 3-kilometer- (almost 2-mile-) thick ice sheets. Since the removal of this immense load, the crust has been adjusting by gradually rebounding upward.* Uplift of nearly 300 meters (1000 feet) has occurred in the Hudson Bay region.

Sea-Level Changes

One of the most interesting and perhaps dramatic effects of the Ice Age was the fall and rise of sea level that accompanied the advance and retreat of the glaciers. Although the total volume of glacial ice today is great, exceeding 25 million cubic kilometers, during the Last Glacial Maximum the volume of glacial ice amounted to about 70 million cubic kilometers, or 45 million cubic kilometers more than at present. Because we know that the snow from which glaciers are made ultimately comes from the evaporation of ocean water, the growth of ice sheets must have caused a worldwide drop in sea level (**Figure 10.21**). Indeed, estimates suggest that sea level was as much as 100 meters (330 feet) lower than it is today. Thus, land that is presently flooded by the oceans was dry. The Atlantic coast of the United States lay more than 100 kilometers (60 miles) to the east of New York City, France and Britain were joined where the famous English Channel

*For a more complete discussion of this concept, termed *isostatic adjustment*, see the section "The Principle of Isostasy" in Chapter 7, page 229.

s today, Alaska and Siberia were connected across the Bering Strait, and Southeast Asia was tied by dry land to the islands of Indonesia.

Changing Rivers

Many present-day stream courses bear little resemblance to their preglacial routes. The Missouri River once flowed northward toward Hudson Bay in Canada. The Mississippi River followed a path through central Illinois, and the head of the Ohio River reached only as far as Indiana (**Figure 10.22**). A comparison of the two parts of Figure 10.22 shows that the Great Lakes were created by glacial erosion during the Ice Age. Prior to the Quaternary period, the basins occupied by these huge lakes were lowlands with rivers that ran eastward to the Gulf of St. Lawrence.

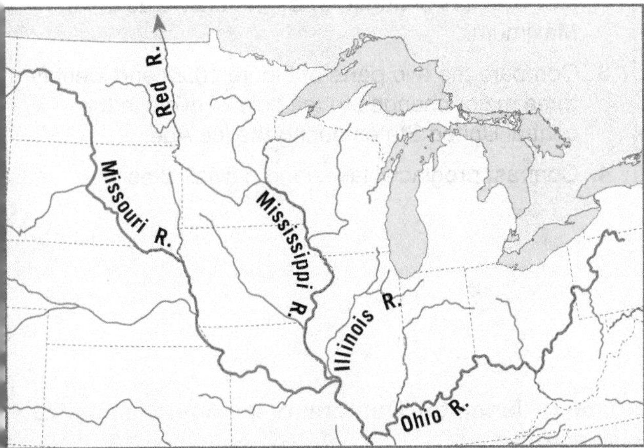

A. This map shows the Great Lakes and the familiar present-day pattern of rivers. Quaternary ice sheets played a major role in creating this pattern.

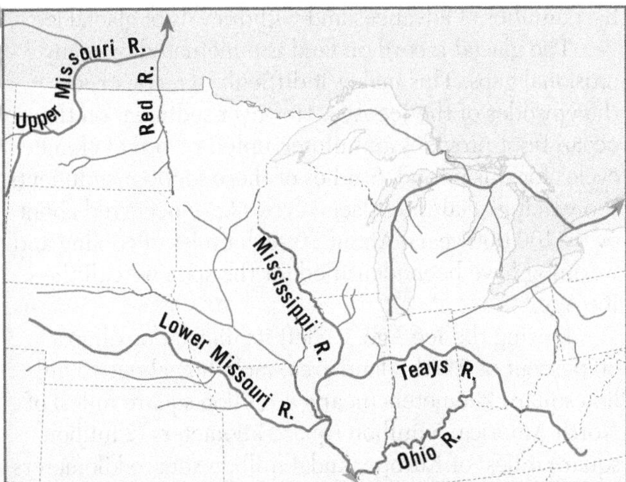

B. Reconstruction of drainage systems prior to the Ice Age. The pattern was very different from today, and the Great Lakes did not exist.

Figure 10.22 Changing rivers The advance and retreat of ice sheets caused major changes in the routes followed by rivers in the central United States.

Ice Dams Create Proglacial Lakes

Ice sheets and alpine glaciers can act as dams and create lakes by trapping glacial meltwater and blocking the flow of rivers. Some of these lakes are relatively small, short-lived impoundments. Others can be large and exist for hundreds or thousands of years.

Figure 10.23 is a map of Lake Agassiz—the largest lake to form during the Ice Age in North America. It came into existence about 12,000 years ago and lasted for about 4500 years. With the retreat of the ice sheet came enormous volumes of meltwater. The Great Plains generally slope upward to the west. As the terminus of the ice sheet receded northeastward, meltwater was trapped between the ice on one side and the sloping land on the other, causing Lake Agassiz to deepen and spread across the landscape. Such water bodies are termed **proglacial lakes**, referring to their position just beyond the outer limits of a glacier or ice sheet. Research shows that the shifting of glaciers and the failure of ice dams can cause the rapid release of huge volumes of water. Such events occurred during the history of Lake Agassiz.

Pluvial Lakes

While the formation and growth of ice sheets was an obvious response to significant changes in climate, the existence of the glaciers themselves triggered important climatic changes in the regions beyond their margins.

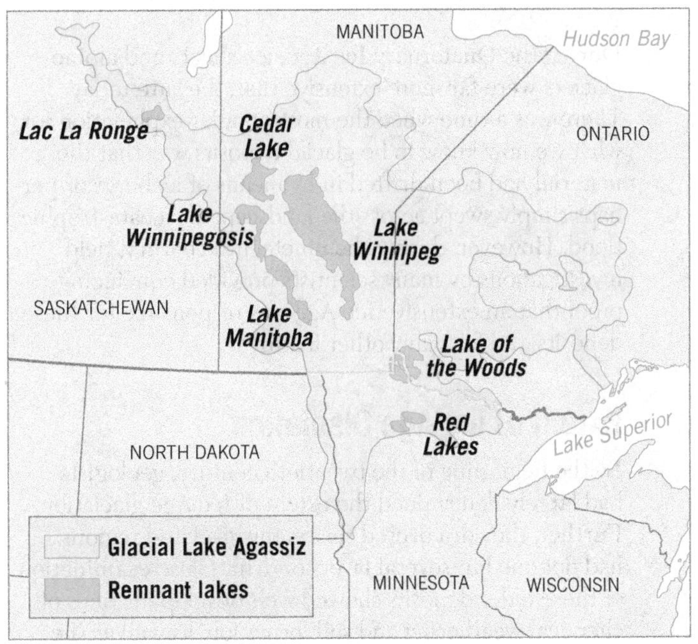

Figure 10.23 Glacial Lake Agassiz This lake was an immense feature—bigger than all of the present-day Great Lakes combined. Modern-day remnants of this proglacial water body are still major landscape features.

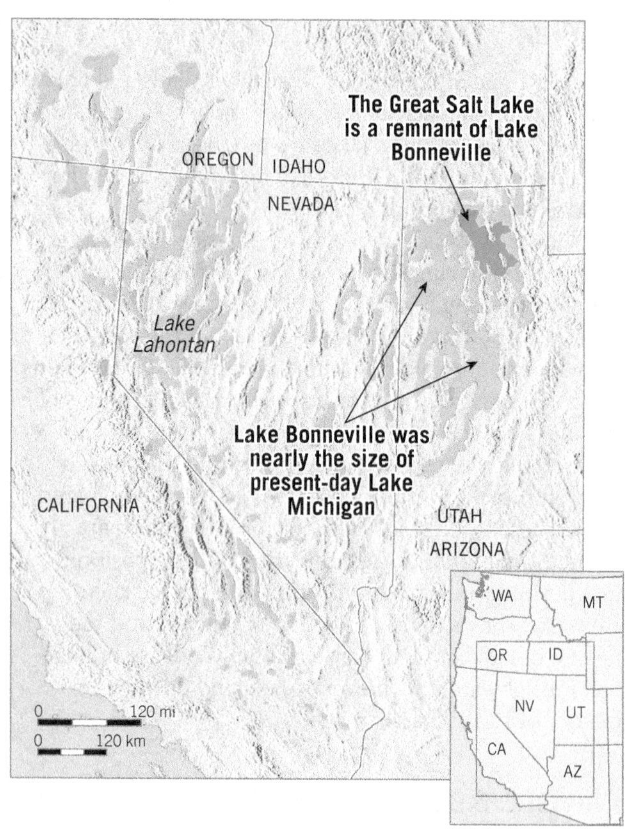

Figure 10.24 Pluvial lakes During the Ice Age, the Basin and Range region experienced a wetter climate than it has today. Many basins turned into large lakes.

In arid and semiarid areas on all the continents, temperatures were lower and thus evaporation rates were lower, but at the same time, precipitation totals were moderate. This cooler, wetter climate formed many **pluvial lakes** (*pluvia* = rain). In North America, pluvial lakes were concentrated in the vast Basin and Range region of Nevada and Utah (Figure 10.24). Although most pluvial lakes are now gone, a few remnants remain, the largest being Utah's Great Salt Lake.

CONCEPT CHECKS 10.5

1. List four effects of Ice Age glaciers, aside from the formation of major erosional and depositional features.
2. Examine Figure 10.21 and determine how much sea level has changed since the Last Glacial Maximum.
3. Compare the two parts of Figure 10.22 and identify three major changes to the flow of rivers in the central United States during the Ice Age.
4. Contrast proglacial lakes and pluvial lakes.

10.6 The Ice Age

Discuss the extent of glaciation and climate variability during the Quaternary Ice Age. Summarize some of the current ideas about the causes of ice ages.

During the Quaternary Ice Age, ice sheets and alpine glaciers were far more extensive than they are today. There was a time when the most popular explanation for what we now know to be glacial deposits was that the material had been drifted in by means of icebergs or perhaps simply swept across the landscape by a catastrophic flood. However, during the nineteenth century, field investigations by many scientists provided convincing proof that an extensive Ice Age was responsible for these deposits and for many other features.

Extent of Ice Age Glaciation

By the beginning of the twentieth century, geologists had largely determined the extent of Ice Age glaciation. Further, they discovered that many glaciated regions had not one but several layers of drift. Close examination of these older deposits showed well-developed zones of chemical weathering and soil formation, as well as the remains of plants that require warm temperatures. The evidence was clear: There had not been just one glacial advance but several, each separated by an extended period when climates were as warm as or warmer than at present. The Ice Age was not simply a time when the ice advanced over the land, lingered for a while, and then receded. Rather, it was a complex period characterized by a number of advances and withdrawals of glacial ice.

The glacial record on land is punctuated by many erosional gaps. This makes it difficult to reconstruct the episodes of the Ice Age. However sediment on the ocean floor provides an uninterrupted record of climate cycles for this period. Studies of these seafloor sediments show that glacial/interglacial cycles have occurred about every 100,000 years. About 20 such cycles of cooling and warming have been identified for the span we call the Ice Age.

During the Ice Age, ice left its imprint on almost 30 percent of Earth's land area, including about 10 million square kilometers (nearly 4 million square miles) of North America, 5 million square kilometers (2 million square miles) of Europe, and 4 million square kilometers (1.6 million square miles) of Siberia (Figure 10.25). The amount of glacial ice in the Northern Hemisphere was roughly twice that in the Southern Hemisphere. The primary reason is that the southern polar ice could not spread far beyond the margins of Antarctica. By contrast, North America and Eurasia provided great expanses of land for the spread of ice sheets.

Today we now know that the Ice Age began between 2 million and 3 million years ago. This means that most of the major glacial episodes occurred during a division of the geologic time scale called the **Quaternary period**. However, the glacial/interglacial cycles that characterize the Ice Age constitute only the most recent and extreme manifestation of a longer period of relatively cool climate, with glaciation beginning in Antarctica around 40 million years ago.

Causes of Ice Ages

A great deal is known about glaciers and glaciation. Much has been learned about glacier formation and movement, the extent of glaciers past and present, and the features created by glaciers, both erosional and depositional. However, the causes of glacial ages are not completely understood.

Although widespread glaciation has been rare in Earth's history, the Quaternary Ice Age is not the only glacial period for which a record exists. Earlier glaciations are indicated by rock layers called *tillite*, a sedimentary rock formed when glacial till becomes lithified. Such deposits, found in strata of several different ages, usually contain striated rock fragments, and some overlie grooved and polished bedrock surfaces or are associated with sandstones and conglomerates that show features of outwash deposits. For example, our Chapter 4 discussion of evidence supporting the continental drift hypothesis mentioned a glacial period that occurred in late Paleozoic time (see Figure 4.7, page 100). In addition, two Precambrian glacial episodes have been identified in the geologic record, the first approximately 2 billion years ago and the second about 600 million years ago. Furthermore, a well-documented record of a earlier glacial age is found in late Paleozoic rocks that are about 250 million years old and that exist on several landmasses.[†]

Any theory that attempts to explain the causes of ice ages must successfully answer two basic questions:

- *What causes the onset of glacial conditions?* For continental ice sheets to have formed, average temperature must have been somewhat lower than at present and perhaps substantially lower than throughout much of geologic time. Thus, a successful theory would have to account for the cooling that finally leads to glacial conditions.
- *What caused the alternating glacial and interglacial stages that have been documented for the Quaternary period?* Whereas the first question deals with long-term trends in temperature on a scale of millions of years, this question relates to much shorter-term changes.

[†]The terms *Precambrian* and *Paleozoic* refer to time spans on the geologic time scale of Earth history. For more on the geologic time scale, see Chapter 11 and Figure 11.25.

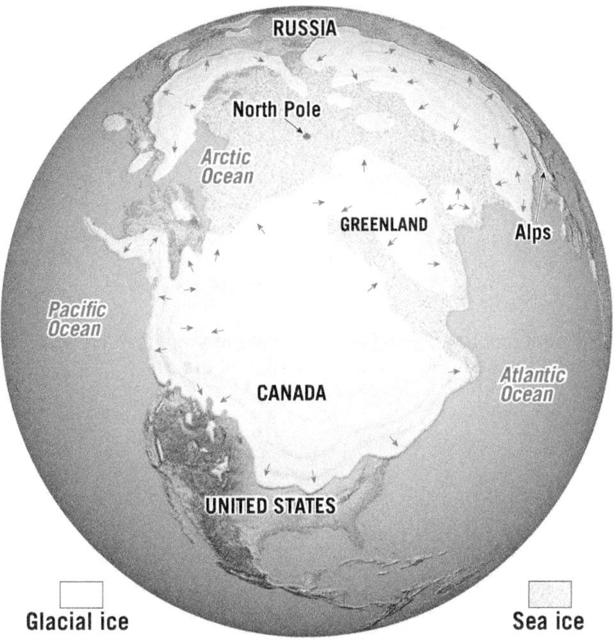

Figure 10.25 Where was the ice? This map shows the maximum extent of ice sheets in the Northern Hemisphere during the Ice Age.

Although the scientific literature contains many hypotheses related to the possible causes of glacial periods, we will discuss only a few major ideas to summarize current thought.

Plate Tectonics

Probably the most attractive proposal for explaining the fact that extensive glaciations have occurred only a few times in the geologic past comes from the theory of plate tectonics.[‡] Because glaciers can form only on land, we know that landmasses must exist somewhere in the higher latitudes before an ice age can commence. Many scientists suggest that ice ages have occurred only when Earth's shifting crustal plates have carried the continents from tropical latitudes to more poleward positions.

Glacial features in present-day Africa, Australia, South America, and India indicate that these regions, which are now tropical or subtropical, experienced an ice age near the end of the Paleozoic era, about 250 million years ago. However, there is no evidence that ice sheets existed during this same period in what are today the higher latitudes of North America and Eurasia. For many years, this puzzled scientists. Was the climate in these relatively tropical latitudes once like it is today in Greenland and Antarctica? Why did glaciers not form in North America and Eurasia? Until the plate tectonics theory was formulated, there had been no reasonable explanation.

Today, scientists realize that the areas containing these ancient glacial features were joined together to form part of a single supercontinent (Pangaea) and were located

[‡]A complete discussion of plate tectonics is presented in Chapter 4.

Figure 10.26 A late Paleozoic ice age Shifting tectonic plates sometimes move landmasses to high latitudes, where the formation of ice sheets is possible.

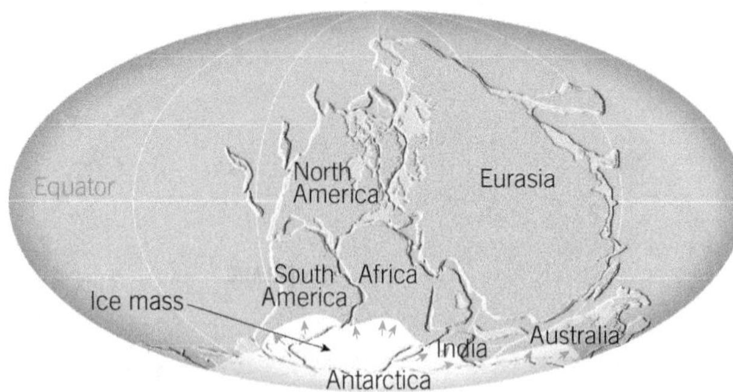

The supercontinent Pangaea showing the area covered by glacial ice near the end of the Paleozoic era.

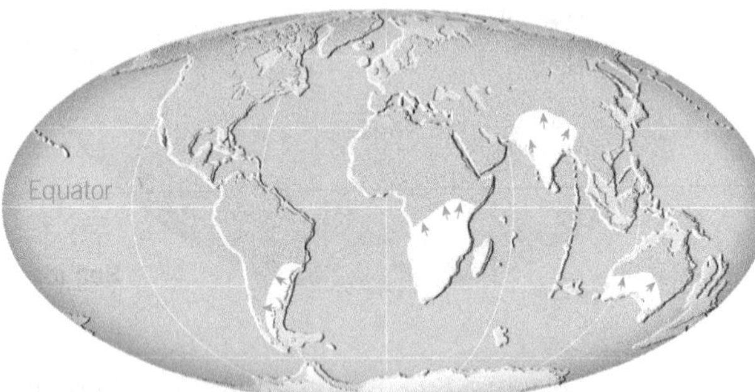

The continents as they appear today. The white areas indicate where evidence of the late Paleozoic ice sheets exists.

at latitudes far to the south of their present positions. Later, this landmass broke apart, and its pieces, each moving on a different plate, migrated toward their present locations (Figure 10.26). Now we know that during the geologic past, plate movements accounted for many dramatic climatic changes as landmasses shifted in relation to one another and moved to different latitudinal positions. Shifting landmasses also caused changes in oceanic circulation, altering the transport of heat and moisture and consequently the climate as well. Because the rate of plate movement is very slow—a few centimeters per year—appreciable changes in the positions of the continents occur only over great spans of geologic time. Thus, climate changes brought about by shifting plates are extremely gradual and occur on a scale of millions of years.

Variations in Earth's Orbit

Because climate changes brought about by moving plates are extremely gradual, the plate tectonics theory cannot be used to explain the alternating glacial and interglacial climates that occurred during the Quaternary period. Therefore, we must look to some other triggering mechanism that may cause climate change on a scale of thousands rather than millions of years. Today, many scientists strongly suspect that the climatic oscillations that characterized the Quaternary may be linked to changes in Earth's orbit. This hypothesis was first developed and strongly advocated by Serbian scientist Milutin Milankovitch and is based on the premise that variations in incoming solar radiation are a principal factor in controlling Earth's climate.

Milankovitch formulated a comprehensive mathematical model based on the following elements (Figure 10.27):

- Variations in the shape (*eccentricity*) of Earth's orbit about the Sun
- Changes in *obliquity*—that is, changes in the angle that Earth's axis makes with the plane of its orbit
- The wobbling of Earth's axis, called *precession*

Using these factors, Milankovitch calculated variations in the receipt of solar energy and the corresponding surface temperature of Earth back into time, in an attempt to correlate these changes with the climate fluctuations of the Quaternary. In explaining climate changes that result from these three variables, note that they cause little or no variation in the *total* solar energy reaching the ground. Instead, their impact is felt because they change the degree of contrast between the seasons. Somewhat milder winters in the middle to high latitudes means greater snowfall totals, whereas cooler summers bring a reduction in snowmelt.

Among the studies that added considerable credibility to this astronomical hypothesis is one in which deep-sea sediments containing certain climatically sensitive microorganisms were analyzed to establish a chronology of temperature changes going back nearly 500,000 years.[§] This time scale of climatic change was then compared to astronomical calculations of eccentricity, obliquity, and precession to determine whether a correlation did indeed exist. Although the study was very involved and mathematically complex, the conclusions were straightforward. The researchers found that major variations in climate over the past several hundred thousand years were closely associated with changes in the geometry of Earth's orbit; that is, cycles of climate change were shown to correspond closely with the periods of obliquity, precession, and orbital eccentricity. More specifically, the researchers stated: "It is concluded that changes in the earth's orbital geometry are the fundamental cause of the succession of Quaternary ice ages."[**]

Let us briefly summarize the ideas that were just described. The plate tectonics theory helps explain the widely spaced and nonperiodic onset of glacial conditions at various times in the geologic past, whereas the astronomical model proposed by Milankovitch and supported by the work of J. D. Hays and his colleagues furnishes an explanation for the alternating glacial and interglacial episodes of the Pleistocene.

§J. D. Hays, John Imbrie, and N. J. Shackelton, "Variations in the Earth's Orbit: Pacemaker of the Ice Ages," *Science* 194 (1976): 1121–1132.

**J. D. Hays et al., p. 1131.

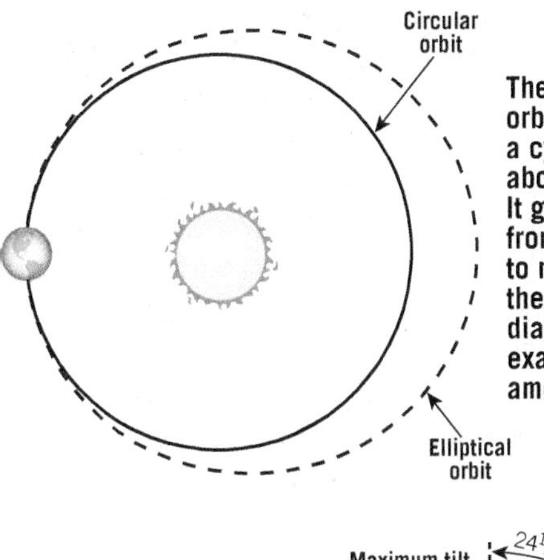

Circular orbit
The shape of Earth's orbit changes during a cycle that spans about 100,000 years. It gradually changes from nearly circular to more elliptical and then back again. This diagram greatly exaggerates the amount of change.
Elliptical orbit

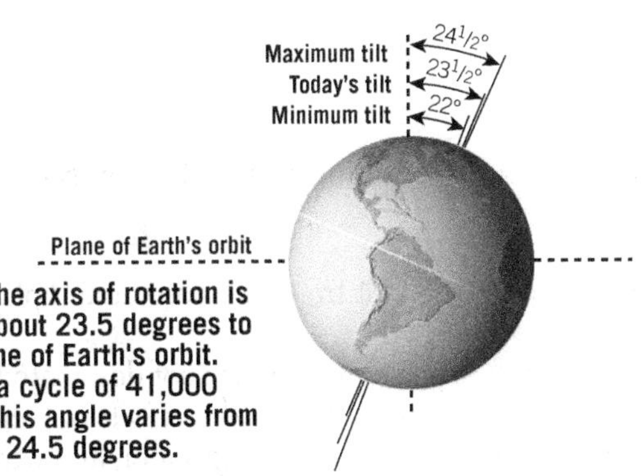

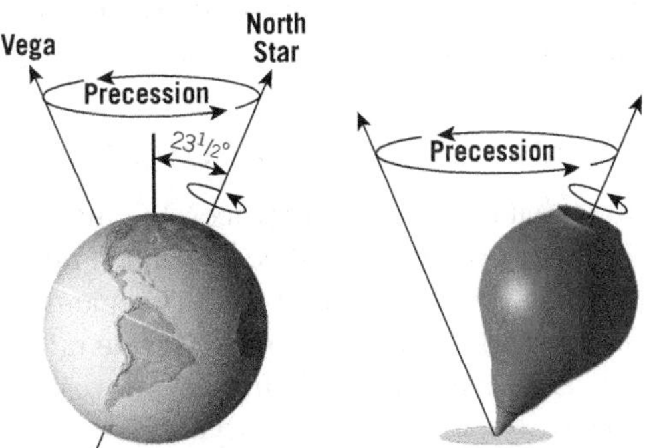

Maximum tilt
Today's tilt
Minimum tilt

Today the axis of rotation is tilted about 23.5 degrees to the plane of Earth's orbit. During a cycle of 41,000 years, this angle varies from 22.1 to 24.5 degrees.

Earth's axis wobbles like a spinning top. Consequently, the axis points to different spots in the sky during a cycle of about 26,000 years.

SmartFigure 10.27 **Orbital variations** Periodic variations in Earth's orbit are linked to alternating glacial and interglacial conditions during the Ice Age.

TUTORIAL
https://goo.gl/KE2Rx7

Other Factors

Variations in Earth's orbit correlate closely with the timing of glacial/interglacial cycles. However, the variations in solar energy reaching Earth's surface caused by these orbital changes do not adequately explain the magnitude of the temperature changes that occurred during the most recent ice age. Other factors must also have contributed. One factor involves variations in the chemical composition of the atmosphere. Other influences involve changes in the reflectivity of Earth's surface and in ocean circulation. Let's take a brief look at these factors.

Chemical analyses of air bubbles that become trapped in glacial ice at the time of ice formation indicate that during glacial episodes, the atmosphere contained less of the gases carbon dioxide and methane than during interglacials (Figure 10.28). Carbon dioxide and methane are important "greenhouse" gases, which means they trap radiation emitted by Earth and contribute to the heating of the atmosphere.[††] When the amount of carbon dioxide and methane in the atmosphere increases, global temperatures rise, and when there is a reduction in these gases, as occurred during glacial episodes, temperatures fall. Therefore, reductions in the concentrations of greenhouse gases help explain the magnitude of the temperature drop that occurred during glacial times. Although scientists know that concentrations of carbon dioxide and methane dropped, they do not know what caused the drop. As often occurs in science, observations gathered during one investigation yield information and raise questions that require further analysis and explanation.

Whenever Earth enters a glacial period, extensive areas of land that were once ice free are covered with ice and snow. In addition, a colder climate causes the area covered by sea ice (frozen surface sea water) to expand as well. Ice and snow reflect a large portion of incoming solar energy back to space. Thus, energy that would have warmed Earth's surface and the air above is lost, and global cooling is reinforced.

Yet another factor that influences climate during glacial times relates to ocean currents, which, as you will learn in Chapter 15, are a complex matter. Research has

[††]For more on this idea, see the section "Heating the Atmosphere: The Greenhouse Effect" in Chapter 16 and the section "Carbon Dioxide, Trace Gases, and Global Warming" in Chapter 20.

Figure 10.28 **Ice cores contain clues to shifts in climate** This scientist is slicing an ice core from Antarctica for analysis. He is wearing protective clothing and a mask to minimize contamination of the sample. Chemical analyses of ice cores can provide important data about past climates. (Photo by British Antarctic Survey/Science Source)

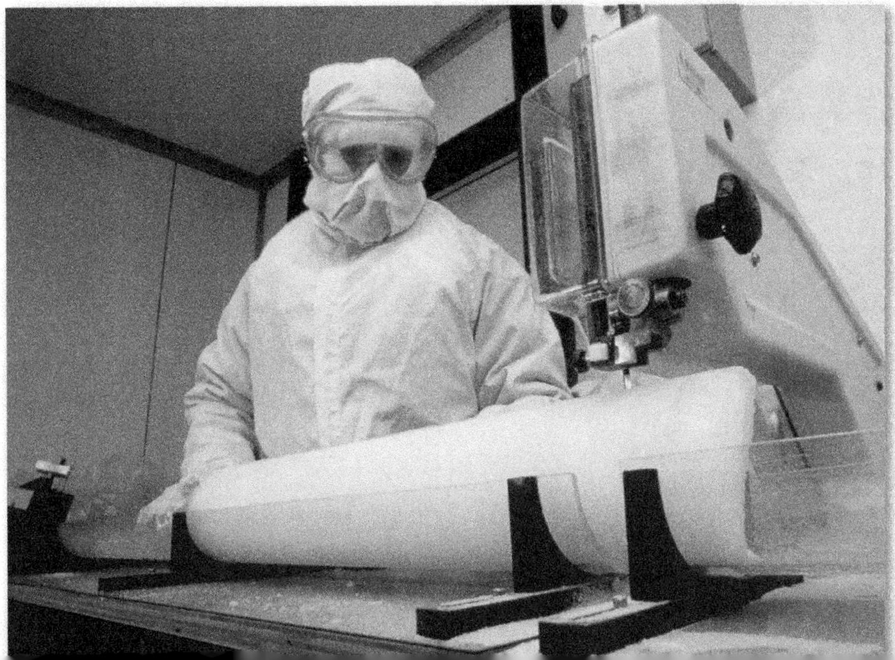

shown that ocean circulation changes during ice ages. For example, studies suggest that the warm current that transports large amounts of heat from the tropics toward higher latitudes in the North Atlantic was significantly weaker during the Ice Age. This would lead to a colder climate in Europe, amplifying the cooling that can be attributed to orbital variations.

In conclusion, it should be noted that our understanding of the causes of glacial episodes is not complete. The ideas that were just discussed do not represent all the possible explanations. Additional factors may be, and probably are, involved.

CONCEPT CHECKS 10.6

1. About what percentage of Earth's land surface was affected by glaciers during the Quaternary period?
2. Where were ice sheets more extensive during the Ice Age: the Northern Hemisphere or the Southern Hemisphere? Why?
3. How does the theory of plate tectonics help us understand the causes of ice ages?
4. Does the theory of plate tectonics alone explain alternating glacial/interglacial climates during the Quaternary period? Why or why not?

10.7 Deserts

Describe the general distribution and extent of Earth's dry lands and the role that water plays in modifying desert landscapes.

The dry regions of the world encompass about 42 million square kilometers (about 16 million square miles)—a surprising 30 percent of Earth's land surface. No other climate group covers so large a land area. The word *desert* literally means "deserted," or "unoccupied." For many dry regions, this is a very appropriate description. Yet where water is available in deserts, plants and animals thrive. Nevertheless, the world's dry regions are among the least familiar land areas on Earth outside the polar realm.

Desert landscapes frequently appear stark. Their profiles are not softened by a carpet of soil and abundant plant life. Instead, barren rocky outcrops with steep, angular slopes are common. At some places the rocks are tinted orange and red. At others they are gray and brown and streaked with black. For many visitors desert scenery exhibits a striking beauty; to others the terrain seems bleak. No matter which feeling is elicited, it is clear that deserts are very different from the more humid places where most people live.

As you will see, arid regions are not dominated by a single geologic process. Rather, the effects of tectonic (mountain-building) forces, running water, and wind are all apparent. Because these processes combine in different ways from place to place, the appearance of desert landscapes varies a great deal as well (Figure 10.29).

▼ Figure 10.29 **Nevada's Great Basin Desert** Mountains separate this area from Pacific moisture and thus contribute to its aridity. When rare storms occur, the sparse vegetation does little to protect the surface from erosion. The appearance of desert landscapes varies a great deal from place to place. (Photo by Dennis Tasa)

Distribution and Causes of Dry Lands

We all recognize that deserts are dry places, but just what is meant by the word *dry*? That is, how much rain defines the boundary between humid and dry regions?

Sometimes, it is arbitrarily defined by a single rainfall figure, such as 25 centimeters (10 inches) per year of precipitation. However, the concept of dryness is relative; it refers to *any situation in which a water deficiency exists*. Climatologists define **dry climate** as a climate in which yearly precipitation is less than the potential loss of water by evaporation.

Within these water-deficient regions, two climatic types are commonly recognized: **desert**, or arid, and **steppe**, or semiarid. The two categories have many features in common; their differences are primarily a matter of degree. The steppe is a marginal and more humid variant of the desert and represents a transition zone that surrounds the desert and separates it from bordering humid climates. Maps showing the distribution of desert and steppe regions reveal that dry lands are concentrated in the subtropics and in the middle latitudes (Figure 10.30).

Deserts in places such as Africa, Arabia, and Australia primarily result from the prevailing global distribution of air pressure and winds. Coinciding with dry regions in the lower latitudes are zones of high air pressure known as the *subtropical highs*. These pressure systems are characterized by subsiding air currents.

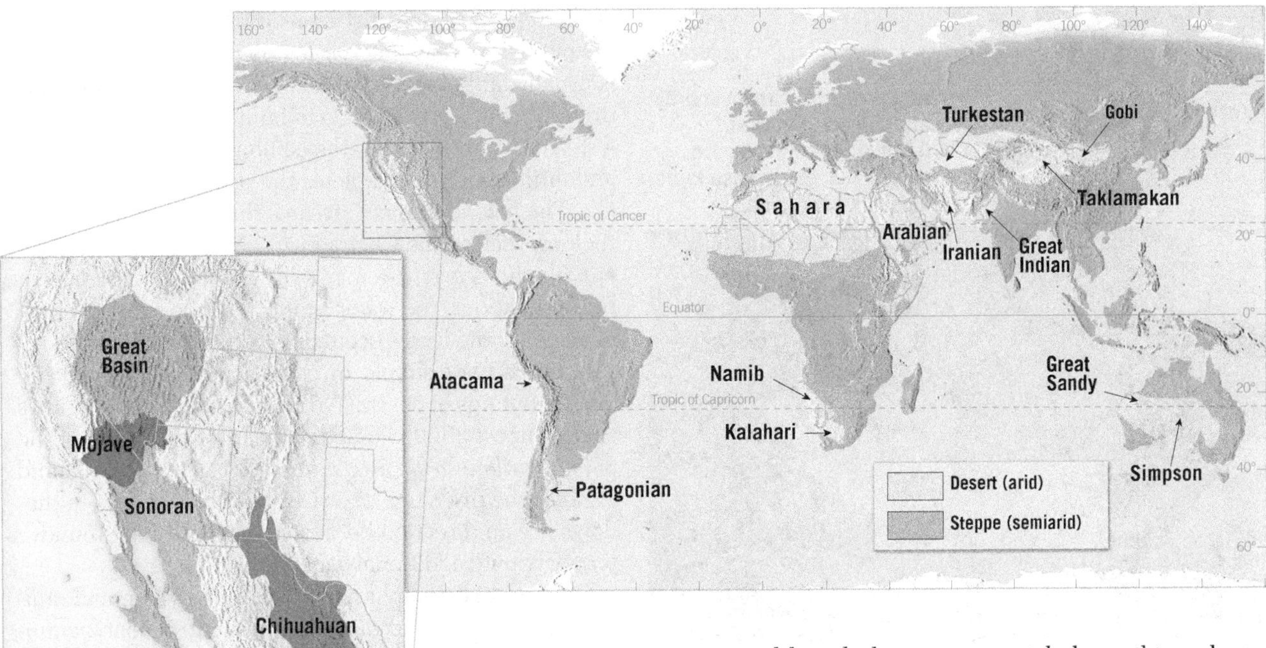

SmartFigure 10.30 Dry climates Arid and semiarid climates cover about 30 percent of Earth's land surface. The dry region of the American West is commonly divided into four deserts, two of which extend into Mexico.

TUTORIAL
https://goo.gl/mCl6C8

When air sinks, it is compressed and warmed. Such conditions are just the opposite of what is needed to produce clouds and precipitation. Consequently, these regions are known for their clear skies, sunshine, and dryness. There is more about pressure systems and the distribution of precipitation in Chapter 18.

Middle-latitude deserts and steppes exist principally because they are sheltered in the deep interiors of large landmasses. They are far removed from the ocean, which is the ultimate source of moisture for cloud formation and precipitation. In addition, the presence of high mountains across the paths of prevailing winds further acts to separate these areas from water-bearing maritime air masses (see Figure 10.29). In North America, the Coast Ranges, Sierra Nevada, and Cascades are the foremost mountain barriers to moisture from the Pacific (see Figure 17.11, page 534). In their rainshadow lies the dry and expansive Basin and Range region of the American West.

Middle-latitude deserts provide an example of how mountain-building processes affect climate. Without mountains, wetter climates would prevail where dry regions exist today.

Geologic Processes in Arid Climates

The angular rock exposures, the sheer canyon walls, and the rocky and pebble- or sand-covered surfaces of deserts contrast sharply with the rounded hills and curving slopes of more humid places. To a visitor from a humid region, a desert landscape may seem to have been shaped by forces different from those operating in wetter areas. However, although the contrasts might be striking, they do not reflect different processes. They merely disclose the differing effects of the same processes that operate under contrasting climate conditions.

Dry-Region Weathering Recall from Chapter 8 that water plays an important role in chemical weathering. Consequently, chemical weathering processes are not as prominent in dry regions as in humid regions. In humid regions, relatively well-developed soils support an almost continuous cover of vegetation. Here the slopes and rock edges are rounded. Such a landscape reflects the strong influence of chemical weathering in a humid climate. By contrast, much of the weathered debris in deserts consists of unaltered rock and mineral fragments—the result of mechanical weathering processes. In dry lands rock weathering of any type is greatly reduced because of the lack of moisture and the scarcity of organic acids from decaying plants. Chemical weathering, however, is not completely lacking in deserts. Over long spans of time, clays and thin soils do form, and many iron-bearing silicate minerals oxidize, producing the rust-colored stain that tints some desert landscapes.

The Role of Water Deserts have scant precipitation and few major rivers. Nevertheless, water plays an important role in shaping landscapes in dry regions. Permanent streams are normal in humid regions, but almost all desert streams are dry most of the time. Deserts have intermittent streams, or **ephemeral streams**, which means they carry water only in response to specific episodes of rainfall. A typical ephemeral stream might flow only a few days or perhaps just a few hours during the year. In some years the channel may carry no water at all.

This fact is obvious even to a casual observer who, while traveling in a dry region, notices how often the road crosses a dry channel. However, when the rare

An ephemeral stream shortly after a heavy shower. Although such floods are short-lived, they cause large amounts of erosion.
Most of the time desert stream channels are dry.
A familiar sign in desert areas. Roads dip into washes which can rapidly fill with water following a heavy rain.
POTENTIAL FLASH FLOOD AREAS
NEXT 21 MILES

▲ Figure 10.31
Ephemeral stream This example is near Arches National Park in southern Utah. (Photos by Demetrio Carrasco/DK Images)

heavy showers do occur, so much rain falls in such a short time that all of it cannot soak in. Because the vegetative cover is sparse, runoff is largely unhindered and consequently rapid, often creating flash floods along valley floors (**Figure 10.31**). Such floods, however, are quite unlike floods in humid regions. Whereas a flood on a river such as the Mississippi may take many days to reach its crest and then subside, a desert flood arrives suddenly and subsides quickly. Because much of the surface material is not anchored by vegetation, the amount of erosional work that occurs during a single short-lived rain event is impressive. The muddy torrent in Figure 10.31 illustrates this fact.

Throughout the world, a number of names are used for ephemeral stream channels. Two of the most common in the dry western United States are *wash* and *arroyo*. In other parts of the world, a dry desert stream channel may be called a *wadi* (Arabian Peninsula and North Africa), a *donga* (South America), or a *nullah* (India).

Humid regions are notable for their integrated drainage systems. But in arid regions, streams usually lack extensive systems of tributaries. In fact, a basic characteristic of desert streams is that they are small and die out before reaching the sea. Because the water table is usually far below the surface, few desert streams can draw upon it, as streams do in humid regions. Without a steady supply of water, the combination of evaporation and infiltration soon depletes the stream.

The few permanent streams that do cross arid regions, such as the Colorado and Nile Rivers, originate *outside* the desert, often in well-watered mountains. In these situations, the water supply must be great to compensate for the losses occurring as the stream crosses the desert. For example, after the Nile leaves the lakes and mountains of central Africa that are its source, it traverses almost 3000 kilometers (nearly 1900 miles) of the Sahara *without a single tributary*. By contrast, in humid regions, the discharge of a river usually increases in the downstream direction because tributaries and groundwater contribute additional water along the way.

It should be emphasized that, although annual rainfall totals are low and rain events are infrequent, *running water does most of the erosional work in deserts*. This is contrary to the common misconception that wind is the most important erosional agent sculpting desert landscapes. Although wind erosion is indeed more significant in dry areas than elsewhere, most desert landforms are carved by running water. As you will see in Section 10.9, the main role of wind is in the transportation and deposition of sediment, which creates and shapes the ridges and mounds we call dunes.

CONCEPT CHECKS 10.7

1. Define *dry climate*. How extensive are the desert and steppe regions of Earth?
2. How does the rate of rock weathering in dry climates compare to the rate in humid regions?
3. What is an ephemeral stream?
4. When a permanent stream such as the Nile River crosses a desert, does discharge increase or decrease as you move downstream? How does this compare to a river in a humid area?
5. What is the most important agent of erosion in deserts?

10.8 Basin and Range: The Evolution of a Mountainous Desert Landscape

Discuss the stages of landscape evolution in the Basin and Range region of the western United States.

Dry regions typically lack permanent streams and often have **interior drainage**. This means they have a discontinuous pattern of ephemeral streams that do not flow out of the desert to the ocean. In the United States, the dry Basin and Range region provides an excellent example. The region includes southern Oregon, all of Nevada, western Utah, southeastern California, southern Arizona, and southern New Mexico. The name *Basin and Range* is an apt description for this almost 800,000-square-kilometer (312,000-square-mile) region, which is characterized by more than 200 relatively small mountain ranges that rise 900 to 1500 meters (3000–5000 feet) above the basins that separate them. The origin of these fault-block mountains is examined in Chapter 7. In this section, we look at how surface processes change the landscape.

In the Basin and Range region, as in other regions like it around the world, most erosion occurs without reference to the ocean (ultimate base level) because the interior drainage never reaches the sea. Even where permanent streams flow to the ocean, few tributaries exist, and thus only a narrow strip of land adjacent to the stream has sea level as its ultimate level of land reduction.

The block diagrams in Figure 10.32 depict how the landscape has evolved in the Basin and Range region and illustrate the landforms described in the following paragraphs. During and following uplift of the mountains, mass movement and running water carve the elevated masses and deposit large quantities of sediment in adjacent basins. Relief is greatest during the early stage, and elevation differences gradually diminish as the mountains are lowered and sediment fills the basins.

When the occasional torrents of water produced by sporadic rains or periods of snowmelt high in the mountains move down the mountain canyons, they are heavily loaded with sediment. Emerging from the confines of the canyon, the runoff spreads over the gentler slopes at the base of the mountains and quickly loses velocity. Consequently, most of the sediment load is dumped within a short distance. The result is a cone of debris known as an **alluvial fan** at the mouth of a canyon. Over the years, a fan enlarges, eventually coalescing with fans from adjacent canyons to produce an apron of sediment, called a **bajada**, along the mountain front.

On the rare occasions of abundant rainfall, streams may flow across the alluvial fans to the center of the basin, converting the basin floor into a shallow **playa lake**. Playa lakes last only a few days or weeks before evaporation and infiltration remove the water. The dry, flat lake bed that remains is termed a *playa*. Playas occasionally become encrusted with salts (*salt flats*) that are left behind when the water in which they were dissolved evaporates. Figure 10.33 includes a satellite view (larger image) and an aerial view of a portion of California's Death Valley, a classic Basin and Range landscape. Many of the features just described are prominent, including a bajada (left side of valley), alluvial fans, a playa lake, and extensive salt flats.

With the ongoing erosion of the mountain mass and the accompanying sedimentation, the local relief continues to diminish. Eventually nearly the entire mountain mass is gone. Thus, by the late stages of erosion, the mountain areas are reduced to a few large bedrock knobs (called *inselbergs*) projecting above the sediment-filled basin.

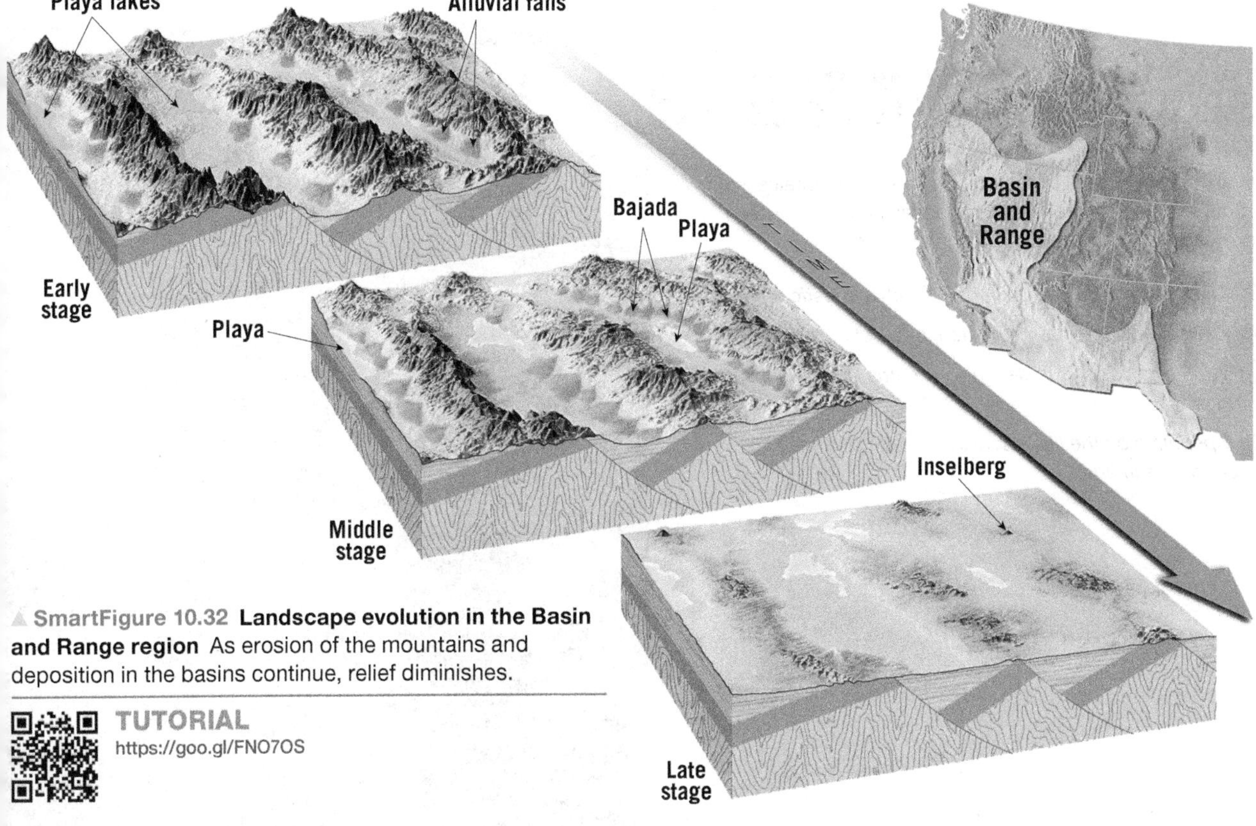

▲ SmartFigure 10.32 **Landscape evolution in the Basin and Range region** As erosion of the mountains and deposition in the basins continue, relief diminishes.

TUTORIAL
https://goo.gl/FNO7OS

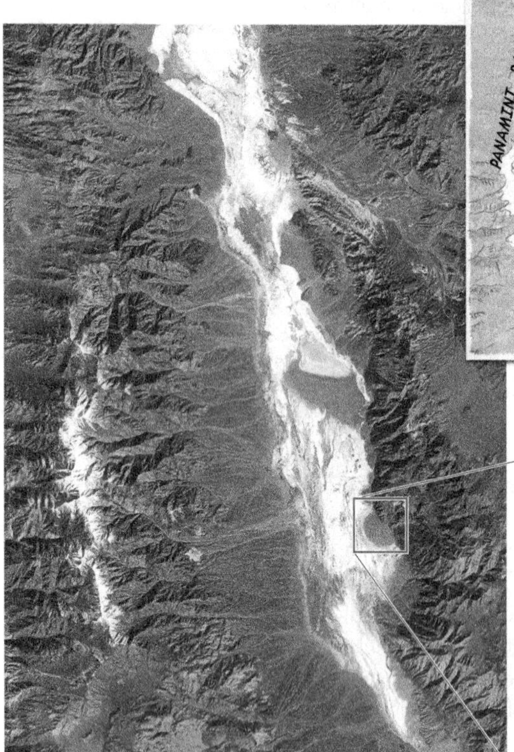
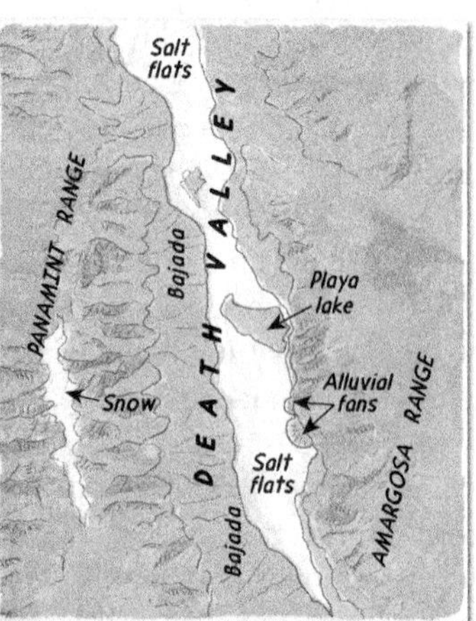

SmartFigure 10.33 **Death Valley: A classic Basin and Range landscape** Shortly before the satellite image (left) was taken in February 2005, heavy rains led to the formation of a small playa lake—the pool of greenish water on the basin floor. By May 2005, the lake had reverted to a salt-covered playa. (NASA)
The small photo is a closer view of one of Death Valley's many alluvial fans. (Photo by Michael Collier)

MOBILE FIELD TRIP
https://goo.gl/GurOxw

Each of the stages of landscape evolution in an arid climate depicted in Figure 10.32 can be observed in the Basin and Range region. Recently uplifted mountains in an early stage of erosion are found in southern Oregon and northern Nevada. Death Valley, California, and southern Nevada fit into the more advanced middle stage, whereas the late stage, with its inselbergs, can be seen in southern Arizona.

CONCEPT CHECKS 10.8

1. What is meant by *interior drainage*?
2. Describe the features and characteristics associated with each stage in the evolution of a mountainous desert.
3. Where in the United States can each stage of desert landscape evolution be observed?

EYE ON EARTH 10.2

This satellite image shows a small portion of the Zagros Mountains in dry southern Iran. Streams in this region flow only occasionally. The green tones on the image identify productive agricultural areas.

QUESTION 1 *Identify the large feature labeled with a question mark.*

QUESTION 2 *Explain how the feature named in Question 1 formed.*

QUESTION 3 *What term is used to describe streams like the ones that occur in this region?*

QUESTION 4 *Speculate on the likely source of water for the agricultural areas in this image.*

10.9 Wind Erosion

Describe the ways that wind transports sediment and the features created by wind erosion.

Moving air, like moving water, is turbulent and able to pick up loose debris and transport it to other locations. Just as in a stream, the velocity of wind increases with height above the surface. Also as with a stream, wind transports fine particles in suspension, while heavier ones are carried as bed load. However, the transport of sediment by wind differs from transport by running water in two significant ways. First, wind's lower density compared to water renders it less capable of picking up and transporting coarse materials. Second, because wind is not confined to channels, it can spread sediment over large areas, as well as high into the atmosphere.

Compared to running water and glaciers, wind is a relatively modest force in sculpting landforms. Recall that even in deserts, most erosion is performed by running water, not by wind. It is also important to point out that wind erosion is more effective in arid lands than in humid areas because in humid places, moisture binds particles together, and vegetation anchors the soil. For wind to be an effective erosional force, dryness and scanty vegetation are essential. When such circumstances exist, wind may pick up, transport, and deposit great quantities of fine sediment. During the 1930s, parts of the Great Plains experienced vast dust storms. The plowing under of the natural vegetative cover for farming, followed by severe drought, exposed the land to wind erosion and led to the area being labeled the Dust Bowl.

Red Sea
SAUDI ARABIA
SUDAN
Movement of dust
Sahara Desert
Satellite image showing thick plumes of dust from the Sahara Desert blowing across the Red Sea.
ERITREA
YEMEN

SmartFigure 10.34 **Wind's suspended load** Dust storms can cover huge areas, and dust can be transported great distances. (NASA)

VIDEO
https://goo.gl/P4bXft

Deflation, Blowouts, and Desert Pavement

One way that wind erodes is by **deflation**—the lifting and removal of loose material. Moving air has low competence (ability to transport different-sized particles) and can suspend only fine sediment, such as clay and silt (Figure 10.34). Larger grains of sand are rolled or skipped along the surface (a process called *saltation*) and comprise the bed load (Figure 10.35). Particles larger than sand are usually not transported by wind. The effects of deflation are sometimes difficult to notice because the entire surface is being lowered at the same time, but the impact of this process can be significant.

Blowouts The most noticeable results of deflation in some places are shallow depressions called **blowouts** (Figure 10.36). In the Great Plains region, from Texas north to Montana, thousands of blowouts can be seen. They range from small dimples less than 1 meter (3 feet) deep and 3 meters (10 feet) wide to depressions that are over 45 meters (50 feet) deep and several kilometers across.

Desert Pavement In portions of many deserts, the surface is characterized by a layer of coarse pebbles and cobbles that are too large to be moved by the wind. This stony veneer, called **desert pavement**, may form as deflation lowers the surface by removing sand and silt from poorly sorted materials. As Figure 10.37A illustrates, the concentration of larger particles at the surface gradually increases as the finer particles are blown away. Eventually, a continuous cover of coarse particles remains.

Wind
Saltating sand grains.

SmartFigure 10.35 **Transporting sand** The bed load that wind carries consists of sand grains that move by bouncing along the surface. Sand never travels far from the surface, even when winds are very strong. (Photo by Bernd Zoller/Photolibrary)

ANIMATION
https://goo.gl/wnc9t6

The man is pointing to where the ground surface was when the grasses began to grow. Wind erosion lowered the land surface to the level of his feet.
Clumps of anchored soil
Unanchored soil
Sand dune
1.2 meters

Figure 10.36 **Blowouts** Deflation is especially effective in creating these depressions when the land is dry and largely unprotected by anchoring vegetation. (Photo courtesy of USDA/Natural Resources Conservation Service)

Studies have shown that the process depicted in Figure 10.37A does not adequately explain all occurrences of desert pavement. As a result, an alternate explanation was formulated and is illustrated in Figure 10.37B. This hypothesis suggests that pavement develops on a surface that initially consists of coarse pebbles. Over time, protruding cobbles trap fine wind-blown grains that settle and sift downward through the spaces between the larger surface stones. The process is aided by infiltrating rainwater.

Once desert pavement becomes established, a process that might take hundreds of years, the surface is effectively protected from further deflation if left undisturbed. However, because the layer is only one or two stones thick, the passage of vehicles or animals can dislodge the pavement and expose the fine-grained material below. If this happens, the surface is no longer protected from deflation.

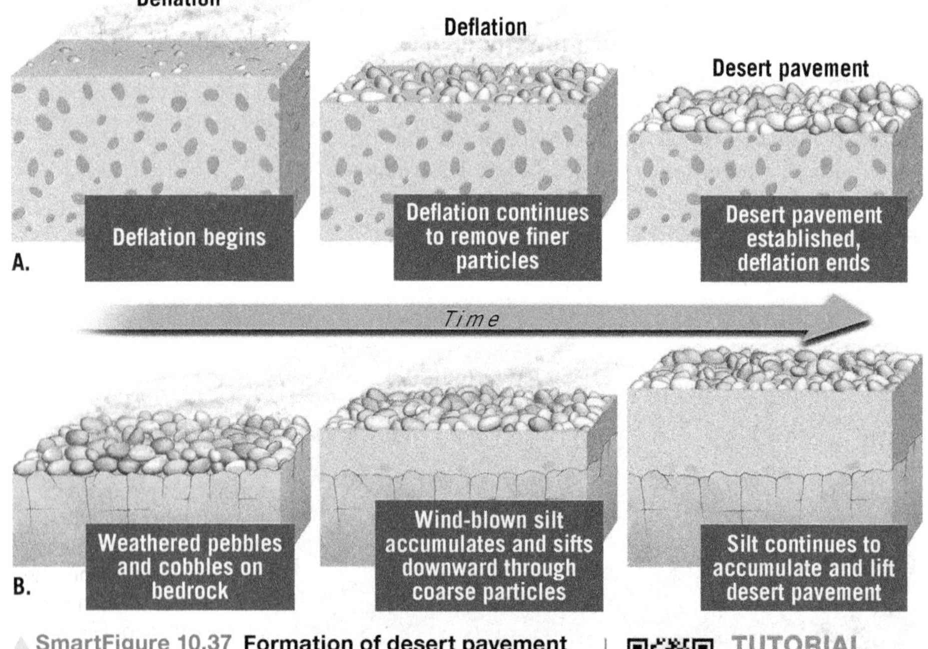
Deflation
Deflation
Desert pavement
A.
Deflation begins
Deflation continues to remove finer particles
Desert pavement established, deflation ends
Time
B.
Weathered pebbles and cobbles on bedrock
Wind-blown silt accumulates and sifts downward through coarse particles
Silt continues to accumulate and lift desert pavement

SmartFigure 10.37 **Formation of desert pavement** **A.** This model shows an area with poorly sorted surface deposits. Over time, deflation lowers the surface, and coarse particles become concentrated. **B.** In this model, the surface is initially covered with cobbles and pebbles. Over time, windblown dust accumulates at the surface and gradually sifts downward.

TUTORIAL
https://goo.gl/Q9CZw8

Wind Abrasion

Like glaciers, streams, and waves, wind erodes in part by *abrasion*. In dry regions as well as along some beaches, windblown sand cuts and polishes exposed rock surfaces. However, abrasion is often credited for accomplishments far beyond its actual capabilities. Such features as balanced rocks that stand high atop narrow pedestals and intricate detailing on tall pinnacles are *not* the results of wind abrasion. Sand seldom travels more than 1 meter above the surface, so the wind's sandblasting effect is obviously limited in vertical extent. However, in areas prone to such activity, telephone poles have actually been cut through near their bases. For this reason, collars may be fitted on the poles to protect them from being "sawed" down.

CONCEPT CHECKS 10.9

1. Why is wind erosion more effective in arid regions than in humid areas?
2. What are blowouts? What term describes the process that creates these features?
3. Briefly describe two hypotheses used to explain the formation of desert pavement.

10.10 Wind Deposits

Explain how loess deposits differ from deposits of sand. Discuss the movement of dunes and distinguish among different dune types.

Although wind is relatively unimportant in producing *erosional* landforms, significant *depositional* landforms are created by wind in some regions. Accumulations of windblown sediment are particularly conspicuous in the world's dry lands and along many sandy coasts. Wind deposits are of two distinctive types: (1) extensive blankets of silt, called *loess*, which once were carried in suspension, and (2) mounds and ridges of sand from the wind's bed load, which we call *dunes*.

Loess

In some parts of the world, the surface topography is mantled with deposits of windblown silt called **loess**. Dust storms deposited this material over thousands of years. When loess is breached by streams or road cuts, it tends to maintain vertical cliffs and lacks any visible layers, as you can see in Figure 10.38.

The distribution of loess worldwide indicates that there are two primary sources for this sediment: deserts and glacial deposits of stratified drift. The thickest and most extensive deposits of loess on Earth occur in western and northern China. They were blown there from the extensive desert basins of central Asia. Accumulations of 30 meters (100 feet) are not uncommon, and thicknesses of more than 100 meters (325 feet) have been measured. It is this fine, buff-colored sediment that gives China's Yellow River (Huang He) its name.

In the United States, deposits of loess are significant in many areas, including South Dakota, Nebraska, Iowa, Missouri, and Illinois, as well as portions of the Columbia Plateau in the Pacific Northwest. Unlike the deposits in China, the loess in the United States, as well as in Europe, is an indirect product of glaciation. Its source is deposits of stratified drift. During the retreat of the ice sheets, many river valleys were choked with sediment deposited by meltwater. Strong westerly winds sweeping across the barren floodplains picked up the finer sediment and dropped it as a blanket on areas adjacent to the valleys.

Sand Dunes

Like running water, wind drops its load of sediment when velocity falls and the energy available for transport diminishes. Thus, sand begins to accumulate wherever an obstruction across the path of the wind slows its movement. Unlike deposits of loess, which form blanketlike layers over broad areas, winds commonly deposit sand in mounds or ridges called **dunes**.

Moving air encountering an object, such as a clump of vegetation or a rock, sweeps around and over the

This vertical bluff near the Mississippi River in southern Illinois is about 3 meters (10 feet) high.

Figure 10.38
Loess In some regions, the surface is mantled with deposits of windblown silt.
(Photo from the James E. Patterson Collection, courtesy of F. K. Lutgens)

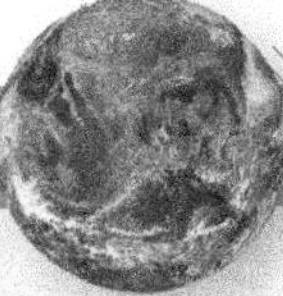

EYE ON EARTH 10.3

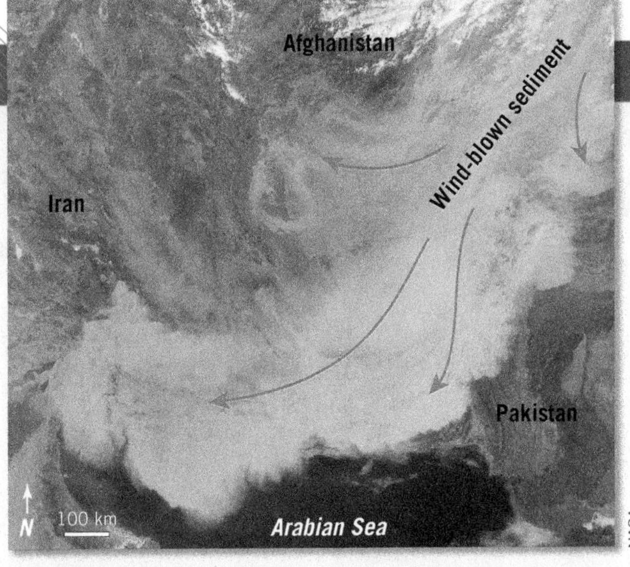

This satellite image shows a large plume of windblown sediment covering large portions of Iran, Afghanistan, and Pakistan in March 2012. The airborne material is thick enough to completely hide the area beneath it. On either side of the plume, skies are mostly clear.

QUESTION 1 *What term is applied to the erosional process that was responsible for producing this plume?*

QUESTION 2 *Is the wind-transported material in the image more likely bed load or suspended load?*

QUESTION 3 *People sometimes refer to events like the one pictured here as "sandstorms." Is that an appropriate description? Why or why not?*

Wind
Strong winds move sand up the relatively gentle windward slope.
As sand accumulates at the dune crest, the slope steepens and some of the sand slides down the steep *slip face.*

▶ SmartFigure 10.39 **White Sands National Monument** The dunes at this landmark in southeastern New Mexico are composed of gypsum. The dunes slowly migrate with the wind. (Photos by Michael Collier)

MOBILE FIELD TRIP
https://goo.gl/UFKrAJ

object, leaving a "shadow" of slower-moving air behind the obstacle and a smaller zone of quieter air just in front of the obstacle. Some of the saltating sand grains moving with the wind come to rest in these wind shadows. As the accumulation of sand continues, it forms an increasingly efficient wind barrier, trapping even more sand. If there is a sufficient supply of sand and the wind blows steadily long enough, the mound of sand grows into a dune.

Many dunes have an asymmetrical profile, with the leeward (sheltered) slope being steep and the windward slope being more gently inclined. The dunes in Figure 10.39 are a good example. Sand moves up the gentler slope on the windward side by saltation. Just beyond the crest of the dune, where wind velocity is reduced, the sand accumulates. As more sand collects, the slope steepens, and eventually some of it slides or slumps under the pull of gravity. In this way, the leeward slope of the dune, called the **slip face**, maintains an angle of about 34 degrees, the angle of repose for loose dry sand. (Recall from Chapter 8 that the angle of repose is the steepest angle at which loose material remains stable.) Continued sand accumulation, coupled with periodic slides down the slip face, results in the slow migration of the dune in the direction of air movement.

As sand is deposited on the slip face, layers form that are inclined in the direction the wind is blowing. These sloping layers are called **cross beds** (Figure 10.40). When the dunes are eventually buried under other layers of sediment and become part of the sedimentary rock

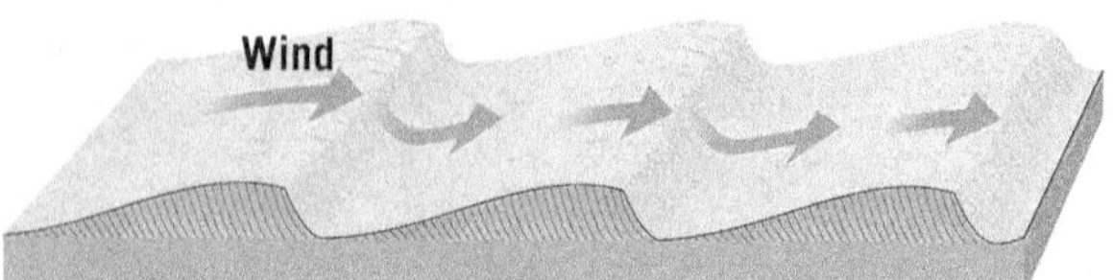
Wind

Dunes commonly have an asymmetrical shape and migrate with the wind.

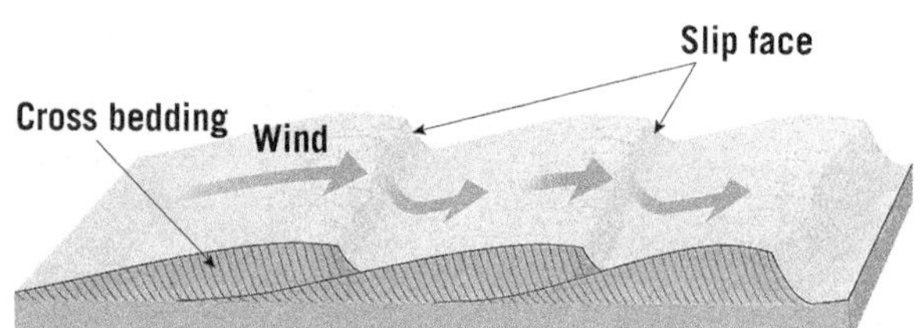
Slip face
Cross bedding
Wind

Sand grains deposited on the slip face at the angle of repose create the cross bedding of dunes.

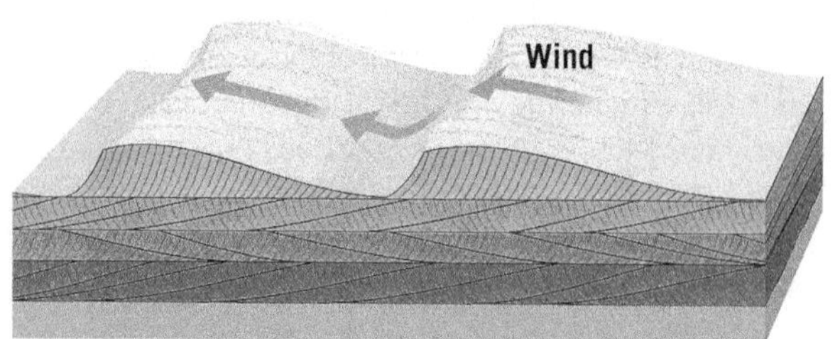
Wind

When dunes are buried and become part of the sedimentary rock record, the cross bedding is preserved.

Cross bedding is an obvious characteristic of the Navajo Sandstone in Zion National Park, Utah.

▲ SmartFigure 10.40 **Cross bedding** As sand is deposited on the slip face, layers form that are inclined in the direction the wind is blowing. With time, complex patterns develop in response to changes in wind direction. (Photo by Dennis Tasa)

TUTORIAL
https://goo.gl/ltqAKo

record, their asymmetrical shape is destroyed, but the cross beds remain as a testimony to their origin. Nowhere is cross bedding more prominent than in the sandstone walls of Zion Canyon in Utah, shown on the right side of Figure 10.40.

Types of Sand Dunes

Dunes are not just random heaps of windblown sediment. Rather, they are accumulations that usually assume patterns that are surprisingly consistent. Dunes come in a wide assortment of forms, and some irregular dunes do not fit easily into any category. For simplicity, we will consider just the six basic dune types shown in Figure 10.41. Several factors influence the form and size that dunes ultimately assume. These include wind direction and velocity, availability of sand, and the amount of vegetation present.

Barchan Dunes Solitary sand dunes shaped like crescents and with their tips pointing downwind are called **barchan dunes** (see Figure 10.41A). These dunes form where supplies of sand are limited and the surface is relatively flat, hard, and lacking vegetation. They migrate slowly with the wind at a rate of up to 15 meters (50 feet) annually. Their size is usually modest, with the largest barchans dunes reaching heights of about 30 meters (100 feet) and the maximum spread between their tips approaching 300 meters (nearly 1000 feet). When the wind direction is nearly constant, the crescent form of these dunes is nearly symmetrical. However, when the wind direction is not perfectly fixed, one tip becomes larger than the other.

Transverse Dunes In regions where the prevailing winds are steady, sand is plentiful, and vegetation is sparse or absent, the dunes form a series of long ridges that are separated by troughs and oriented at right angles to the prevailing wind. Because of this orientation, they are termed **transverse dunes** (see Figure 10.41B). Typically, many coastal dunes are of this type. In addition, transverse dunes are common in many arid regions where the extensive surface of wavy sand is sometimes called a *sand sea*. In some parts of the Sahara and Arabian Deserts, transverse dunes reach heights of 200 meters (650 feet), are 1 to 3 kilometers (0.5 to 2 miles) across, and can extend for distances of 100 kilometers (60 miles) or more.

Barchanoid Dunes There is a relatively common dune form that is intermediate between isolated barchans and extensive waves of transverse dunes. Such dunes, called **barchanoid dunes**, form scalloped rows of sand oriented at right angles to the wind (see Figure 10.41C). The rows resemble a series of barchans that have been positioned side by side. Visitors exploring the gypsum dunes at White Sands National Monument in New Mexico will recognize this form (see Figure 10.39).

Longitudinal Dunes **Longitudinal dunes** are long ridges of sand that form more or less parallel to the prevailing wind and where sand supplies are moderate

SmartFigure 10.41
Types of sand dunes
Factors that influence the form and size of dunes include wind direction and velocity, the availability of sand, and the amount of vegetation.

TUTORIAL
https://goo.gl/zeh51r

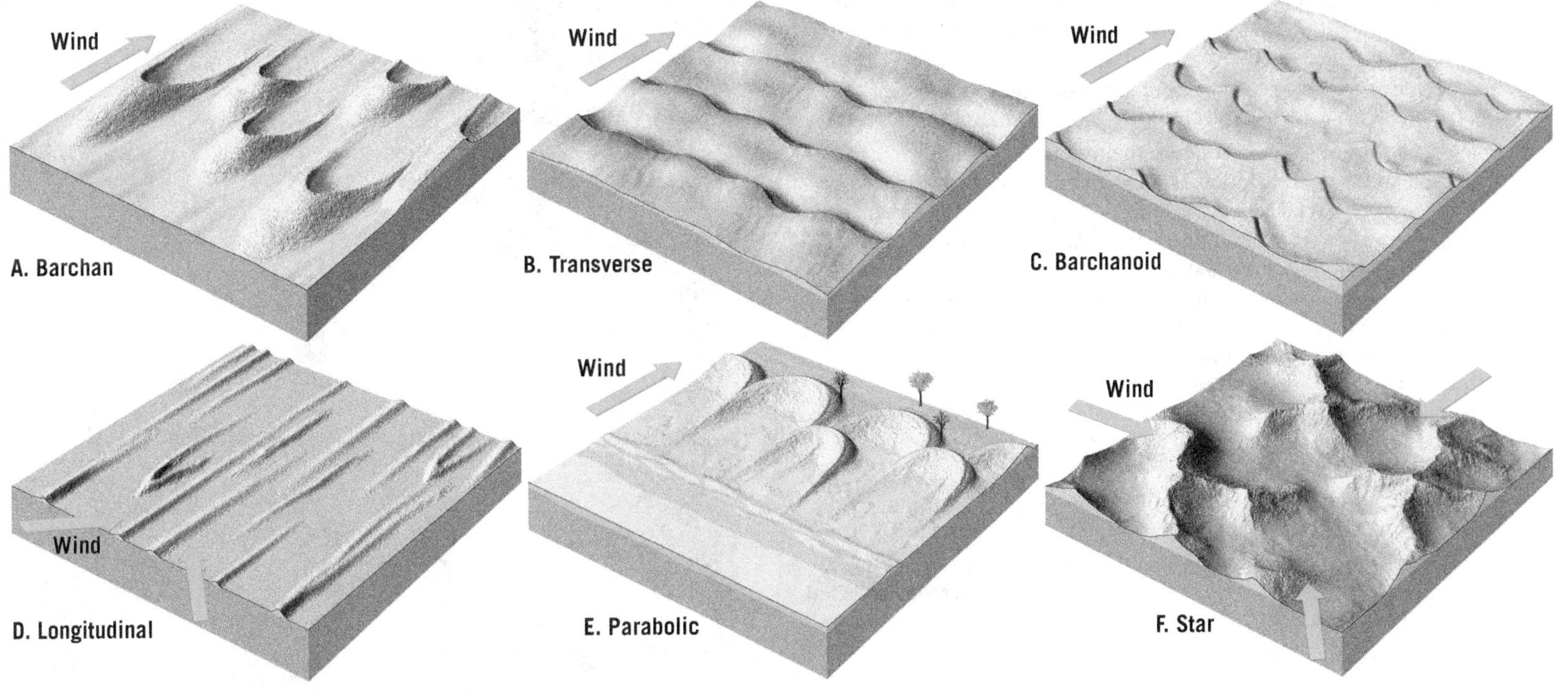

(see Figure 10.41D). Apparently the prevailing wind direction varies somewhat but remains in the same quadrant of the compass. Although the smaller types are only 3 or 4 meters high and several tens of meters long, in some large deserts, longitudinal dunes can reach great size. For example, in portions of North Africa, Arabia, and central Australia, these dunes may approach a height of 100 meters and extend for distances of more than 100 kilometers (62 miles).

Parabolic Dunes Unlike the other dune types described thus far, **parabolic dunes** form where vegetation partially covers the sand. The shape of these dunes resembles the shape of barchans except that their tips point into the wind rather than downwind (see Figure 10.41E). Parabolic dunes often form along coasts where there are strong onshore winds and abundant sand. If the sand's sparse vegetative cover is disturbed at some spot, deflation creates a blowout. Sand is then transported out of the depression and deposited as a curved rim that grows higher as deflation enlarges the blowout.

Star Dunes Confined largely to parts of the Sahara and Arabian Deserts, **star dunes** are isolated hills of sand that exhibit a complex form (see Figure 10.41F). Their name is derived from the fact that the bases of these dunes resemble multipointed stars. Usually three or four sharp-crested ridges diverge from a central high point that in some cases may approach a height of 90 meters (300 feet). As their form suggests, star dunes develop where wind directions are variable.

CONCEPT CHECKS 10.10

1. Contrast loess and sand dunes in terms of composition and how they form.
2. How are some loess deposits related to glaciers?
3. Describe how sand dunes migrate.
4. What is cross bedding?
5. List and briefly distinguish among basic dune types.

10 CONCEPTS IN REVIEW
Glaciers, Deserts, and Wind

10.1 Glaciers and the Earth System

Explain the role of glaciers in the hydrologic and rock cycles. Describe the different types of glaciers, their characteristics, and their present-day distribution.

KEY TERMS: glacier, valley, (alpine) glacier, ice sheet, sea ice, ice shelf, ice cap, piedmont glacier, outlet glacier

- A glacier is a thick mass of ice that originates on land from the compaction and recrystallization of snow and that shows evidence of past or present flow. Glaciers are part of both the hydrologic cycle and the rock cycle. They store and release freshwater, and they transport and deposit large quantities of sediment.
- Valley glaciers flow down mountain valleys, whereas ice sheets are very large masses, such as those that cover Greenland and Antarctica. During the Last Glacial Maximum, around 18,000 years ago, large areas of Earth were covered by glacial ice.
- When valley glaciers exit confining mountains, they may spread out into broad lobes called piedmont glaciers. Similarly, ice shelves form when glaciers flow into the ocean, producing a layer of floating ice.
- Ice caps are like small ice sheets. Both ice sheets and ice caps may be drained by outlet glaciers, which often resemble valley glaciers.

? This satellite image shows ice in the high latitudes of the Northern Hemisphere. What term is applied to the ice at the North Pole? What term best describes Greenland's ice? Are both considered glaciers? Explain.

10.2 How Glaciers Move

Describe how glaciers move, the rates at which they move, and the significance of the glacial budget.

KEY TERMS: zone of fracture, crevasse, zone of accumulation, snowline, (equilibrium line), zone of wastage, calving, iceberg, glacial budget

- Glaciers move in part by flowing under pressure. On the surface of a glacier, ice is brittle. Below about 50 meters (165 feet), pressure is great, and ice flows like a plastic material. In addition, the bottom of a glacier may slide along its bed.
- Fast glaciers may move 800 meters (2600 feet) per year, while slow glaciers may move only 2 meters (6.5 feet) per year. Some glaciers experience periodic surges of rapid movement.
- A glacier's budget is the balance between formation of new ice from snow in the zone of accumulation and loss of ice in the zone of wastage. When the budget is positive, the glacier's terminus advances; when the budget is negative, the terminus retreats.

? This image shows that melting is one way that glacial ice wastes away. What is another way that ice is lost from a glacier?

Robbie Shone/Photo Researchers, Inc

10.3 Glacial Erosion

Discuss the processes of glacial erosion. Identify and describe the major topographic features sculpted by glacial erosion.

KEY TERMS: plucking, abrasion, rock flour, glacial striations, glacial trough, hanging valley, cirque, arête, horn, fiord

- Glaciers acquire sediment through plucking from the bedrock beneath the glacier, by abrasion of the bedrock using sediment already in the ice, and when mass-wasting processes drop debris on top of the glacier. Grinding of the bedrock produces grooves and scratches called glacial striations.
- Erosional features produced by valley glaciers include glacial troughs, hanging valleys, cirques, arêtes, horns, and fiords.

? Examine the illustration of a mountainous landscape after glaciation. Identify the landforms that resulted from glacial erosion.

10.4 Glacial Deposits

Distinguish between the two basic types of glacial deposits and briefly describe the features associated with each type.

KEY TERMS: glacial drift, till, glacial erratic, stratified drift, lateral moraine, medial moraine, end moraine, ground moraine, outwash plain, valley train, kettle, drumlin, esker, kame

- Any sediment of glacial origin is called drift. The two distinct types of glacial drift are till, which is unsorted material deposited directly by the ice, and stratified drift, which is sediment sorted and deposited by meltwater from a glacier.
- The most widespread features created by glacial deposition are layers or ridges of till, called moraines. Associated with valley glaciers are lateral moraines, formed along the sides of the valley, and medial moraines, formed between two valley glaciers that have merged. End moraines, which mark the former position of the front of a glacier, and ground moraines, undulating layers of till deposited as the ice front retreats, are common to both valley glaciers and ice sheets.

? Examine the illustration of depositional features left behind by a retreating ice sheet. Identify the features and indicate which landforms are composed of till and which are composed of stratified drift.

10.5 Other Effects of Ice Age Glaciers

Describe and explain several important effects of Ice Age glaciers other than the formation of erosional and depositional landforms.

KEY TERMS: proglacial lake, pluvial lake

- In addition to erosional and depositional features, other effects of Ice Age glaciers include the forced migration of organisms and adjustments of the crust by rebounding upward after removal of the immense load of ice.
- Ice sheets are nourished by water that ultimately comes from the ocean, so when ice sheets grow, sea level falls, and when they melt, sea level rises.
- Advance and retreat of ice sheets caused significant changes to the paths followed by rivers. Proglacial lakes formed when glaciers acted as dams to create lakes by trapping glacial meltwater or blocking rivers. In response to the cooler and wetter glacial climate, pluvial lakes formed in areas such as present-day Nevada.

10.6 The Ice Age

Discuss the extent of glaciation and climate variability during the Quaternary Ice Age. Summarize some of the current ideas about the causes of ice ages.

KEY TERMS: Quaternary period

- The Ice Age that began between 2 and 3 million years ago is a complex period characterized by numerous advances and withdrawals of glacial ice. Most of the major glacial episodes occurred during a span on the geologic time scale called the Quaternary period, which continues today. The existence of multiple layers of drift on land and an uninterrupted record of climate cycles preserved in seafloor sediments provide evidence of the occurrence of several glacial advances during the Ice Age.
- While rare, Ice Ages have occurred in Earth history prior to the recent glaciations we call the Ice Age. Lithified till, called tillite, is a major line of evidence for these ancient ice ages. There are several reasons glacial ice might accumulate globally, including the position of the continents, which is driven by plate tectonics. Antarctica's position over the South Pole is doubtless a key reason for its massive ice sheets, for instance.
- The Quaternary period is marked by not only glacial advances but also intervening episodes of glacial retreat. One way to explain these oscillations is through variations in Earth's orbit, which lead to seasonal variations in the distribution of solar radiation. The orbit's shape varies (eccentricity), the tilt of the planet's rotational axis varies (obliquity), and the axis slowly "wobbles" over time (precession). These three effects, which occur on different time scales, collectively account for alternating colder and warmer periods during the Quaternary.
- Additional factors that may be important for initiating or ending glaciations include rising or falling levels of greenhouse gases, changes in the reflectivity of Earth's surface, and variations in the ocean currents that redistribute heat energy from warmer to colder regions.

? About 250 million years ago, parts of India, Africa, and Australia were covered by ice sheets, while Greenland, Siberia, and Canada were ice free. Explain how this could be the case.

10.7 Deserts

Describe the general distribution and extent of Earth's dry lands and the role that water plays in modifying desert landscapes.

KEY TERMS: dry climate, desert, steppe, ephemeral stream

- Dry climates cover about 30 percent of Earth's land area. These regions have yearly precipitation totals that are less than the potential loss of water through evaporation. Deserts are drier than steppes, but both climate types are considered water deficient.
- Dry regions in the lower latitudes coincide with the zones of subsiding air and high air pressure known as subtropical highs. Middle-latitude deserts exist because of their positions in the deep interiors of large continents far removed from oceans. Mountains also act to shield these regions from humid marine air masses.
- Practically all desert streams are dry most of the time (ephemeral). Nevertheless, running water is responsible for most of the erosional work in a desert. Although wind erosion is more significant in dry areas than elsewhere, the main role of wind in a desert is to transport and deposit sediment.

10.8 Basin and Range: The Evolution of a Mountainous Desert Landscape

Discuss the stages of landscape evolution in the Basin and Range region of the western United States.

KEY TERMS: interior drainage, alluvial fan, bajada, playa lake

- The Basin and Range region of the western United States is characterized by interior drainage, with streams eroding uplifted mountain blocks and depositing sediment in interior basins. Alluvial fans, bajadas, playas, playa lakes, salt flats, and inselbergs are features often associated with these landscapes.

? Identify the lettered features in this photo. How did they form?

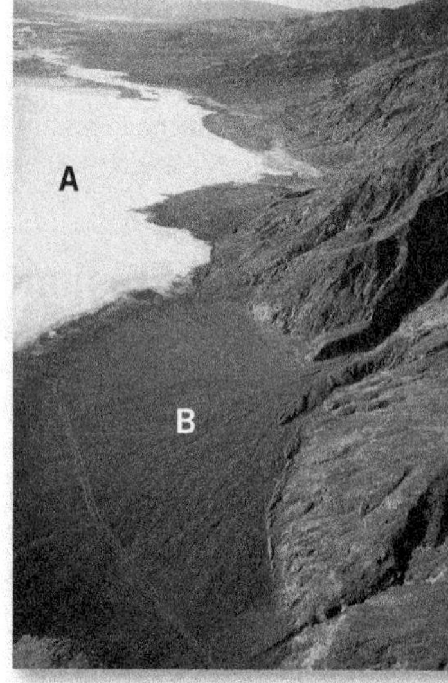

10.9 Wind Erosion

Describe the ways that wind transports sediment and the features created by wind erosion.

KEY TERMS: deflation, blowout, desert pavement

- For wind erosion to be effective, dryness and scant vegetation are essential. Deflation, the lifting and removal of loose material, often produces shallow depressions called blowouts.
- Abrasion, the sandblasting effect of wind, is often given too much credit for producing desert features. However, abrasion does cut and polish rock near the surface.

? What term is applied to the layer of coarse pebbles covering this desert surface? How might it have formed?

Michael Collier

10.10 Wind Deposits

Explain how loess deposits differ from deposits of sand. Discuss the movement of dunes and distinguish among different dune types.

KEY TERMS: loess, dune, slip face, cross bed, barchan dunes, transverse dunes, barchanoid dunes, longitudinal dunes, parabolic dunes, star dunes

- Wind deposits are of two distinct types: extensive blankets of silt, called loess, carried by wind in suspension, and mounds and ridges of sand, called dunes, which are formed from sediment that is carried as part of the wind's bed load.
- Most loess is derived from either deserts or areas that have recently been glaciated. In the latter case, wind blowing across stratified drift picks up silt-size grains.
- Dunes accumulate due to the difference in wind energy on the upwind and downwind sides of the dune (or an object that initiates dune formation). Sand that is blown up the gently sloping upwind side settles out on the downwind slip face, where it periodically avalanches down to maintain the angle of repose. These processes cause the dune to migrate downwind. Inside the dune, the buried slip faces may be preserved as cross beds.
- There are six major kinds of dunes. Their shapes result from the pattern of prevailing winds, the amount of available sand, and the presence of vegetation.

? This close-up photo shows a small portion of one side of a barchan dune. What term is applied to this side of the dune? Is the prevailing wind direction "coming out" of the photo or "going into" the photo? Explain. Why did some of the sand break away and slide?

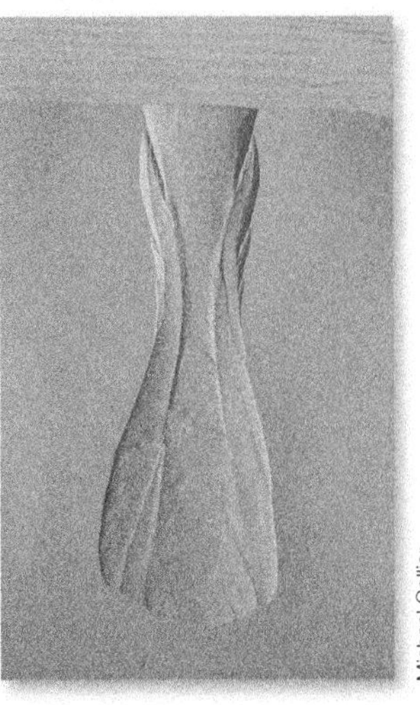
Michael Collier

GIVE IT SOME THOUGHT

1 The accompanying diagram shows the results of a classic experiment used to determine how glacial ice moves in a mountain valley. The experiment was carried out over an 8-year span. Refer to this diagram and answer the following:

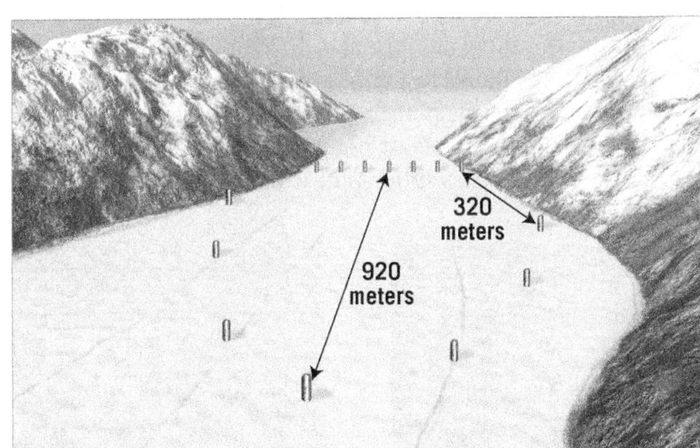

a. What was the average yearly rate of ice advance in the center of the glacier?
b. About how fast was the center of the glacier advancing *per day*?
c. What was the average rate at which ice advanced along the sides of the glacier?
d. Why was the rate at the center different than the rate along the sides?

2 Studies have shown that during the Ice Age, the margins of some ice sheets advanced southward from the Hudson Bay region at rates ranging from about 50 to 320 meters per year.
a. Determine the maximum amount of time required for an ice sheet to move from the southern end of Hudson Bay to the south shore of present-day Lake Erie, a distance of 1600 kilometers.
b. Calculate the minimum number of years required for an ice sheet to move this distance.

3 If the budget of a valley glacier were balanced for an extended span of time, what feature would you expect to find at the terminus of the glacier? Now assume that the glacier's budget changes so that wastage exceeds accumulation. How would the terminus of the glacier change? Describe the deposit you would expect to form under these conditions.

4 Is the glacial deposit shown in the photo an example of till or stratified drift? Is it more likely part of an end moraine or an esker?

E. J. Tarbuck

5 Assume that you and a nongeologist friend are visiting the glacier shown in Figure 10.1. After studying the glacier for quite a long time, your friend asks, "Do these things really move?" How would you convince your companion that this glacier does indeed move, using evidence that is clearly visible in this image? What other figure in this chapter would aid your explanation?

6 If Earth were to experience another glacial cycle, one hemisphere would have substantially more expansive ice sheets than the other. Would it be the Northern Hemisphere or the Southern Hemisphere? What is the reason for the large disparity?

7 Is either of the following statements true? Are they both true? Explain.
a. Wind does its most effective erosional work in dry places.
b. Wind is the most important agent of erosion in deserts.

8 This is an aerial view of the Preston Mesa dunes in northern Arizona.

Michael Collier

a. Which one of the basic dune types is shown here?
b. Sketch a simple profile (side view) of one of these dunes. Add an arrow to show the prevailing wind direction and label the dune's slip face.
c. These dunes gradually migrate across the surface. Describe the processes responsible for this movement.

9 Bryce Canyon National Park, shown in the accompanying photo, is in dry southern Utah. It is carved into the eastern edge of the Paunsaugunt Plateau. Erosion has sculpted the colorful limestone into bizarre shapes, including spires called "hoodoos." As you and a companion (who has not studied geology) are viewing the scenery in Bryce Canyon, your friend says, "It's amazing how wind has created this incredible scenery!" Now that you have studied arid landscapes, how would you respond to your companion's statement?

ozoptimest/Shutterstock

10 Compare the sediment deposited by a stream, the wind, and a glacier. Which deposit should have the most uniform grain size? Which one would exhibit the poorest sorting? Explain your choices.

EXAMINING THE EARTH SYSTEM

1 Assume that you are teaching an introductory Earth science class and that you have just assigned this chapter to your students. A student in the class asks why glaciers, deserts, and wind are treated in the same chapter. Formulate a response that connects these topics.

2 This image is a close-up of Surprise Glacier in Alaska's Prince William Sound. Prepare two brief descriptions that relate to Surprise Glacier (and to all other glaciers as well): one that describes how Surprise Glacier fits into the hydrologic cycle and another that explains how Surprise Glacier fits into the rock cycle. Is Surprise Glacier a part of the hydrosphere? Is it a part of the geosphere? Some scientists think that ice should be a separate sphere of the Earth system, called the *cryosphere*. Does such an idea have merit? Explain.

Michael Collier

3 These two images show an ephemeral stream in Niger, a country in North Africa's Sahara Desert. What is the local term for an ephemeral stream in this part of the world? Prepare a brief story that would explain the contrast between the two images. Try to include all four of Earth's spheres in your story.

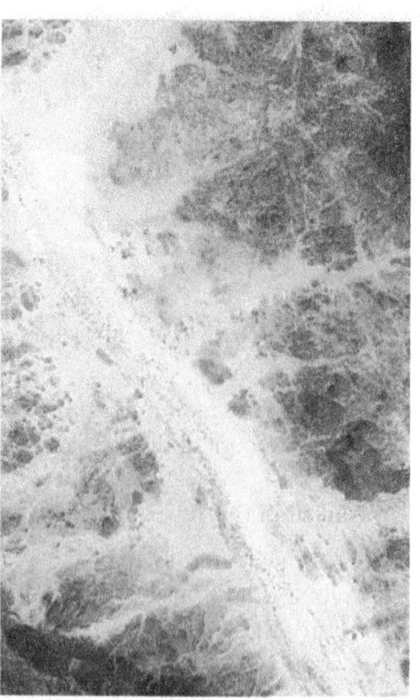

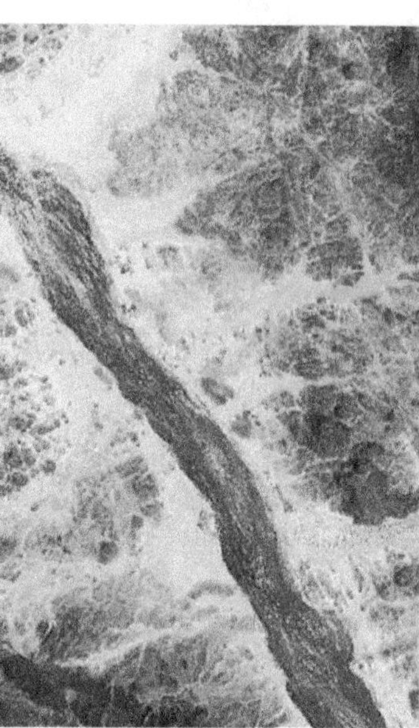

DATA ANALYSIS

The Aral Sea

The Aral Sea was once the fourth-largest lake in the world. This lake has now decreased in size by more than 80%, and the southern Aral Sea has disappeared altogether. This has had devastating effects of the communities around the lake.

ACTIVITIES

Go to NASA's Earth Observatory site at http://earthobservatory.nasa.gov, select World of Change under Special Collections and scroll to select Shrinking Aral Sea. As you step forward in time, you will see the aerial extent of the Aral Sea.

1 When did the Aral Sea begin to shrink? Why did the Aral Sea begin to shrink?

2 How has the shrinking lake affected the quality of the water and farmland in the area?

3 How has the lake's reduction affected summer and winter temperatures?

Step forward in time to see changes in the Aral Sea. The green region is the lake, and the white region around the lake is salt deposits. You may also click on Google Earth to step through time and use the measuring tool to answer some of these questions.

4 What is the east–west distance between the easternmost edge of the Aral Sea in 1960 and the edge of the southern Aral Sea in 2000? 1960 and 2005? 1960 and 2010? 1960 and 2015?

5 What is the distance change between 2000 and 2005? 2005 and 2010? 2005 and 2015?

6 What is the average rate of distance change since 2000? (Remember that rate of change is the distance change divided by the number of years.)

7 Why was there a significant decline in the overall size of the southern Aral Sea after 2005?

Go to "Shrinking Aral Sea" on NASA's Earth Observatory site (https://earthobservatory.nasa.gov/Features/WorldOfChange/aral_sea.php).

8 Compare this image to the Aral Sea images from Earth Observatory. Approximately when was the dust storm image taken? (Giving a range of years is fine.)

9 From which direction is the wind blowing?

10 How long is the dust storm at its longest distance? How wide is the dust storm at its widest distance?

11 Which towns are in the path of this dust storm?

MasteringGeology™ Looking for additional review and test prep materials? Visit the Study Area in MasteringGeology to enhance your understanding of this chapter's content by accessing a variety of resources, including Self-Study Quizzes, Geoscience Animations, SmartFigure Tutorials, Mobile Field Trips, *Project Condor* Quadcopter videos, *In the News* articles, flashcards, web links, and an optional Pearson eText.

www.masteringgeology.com

UNIT 4 Deciphering Earth's History

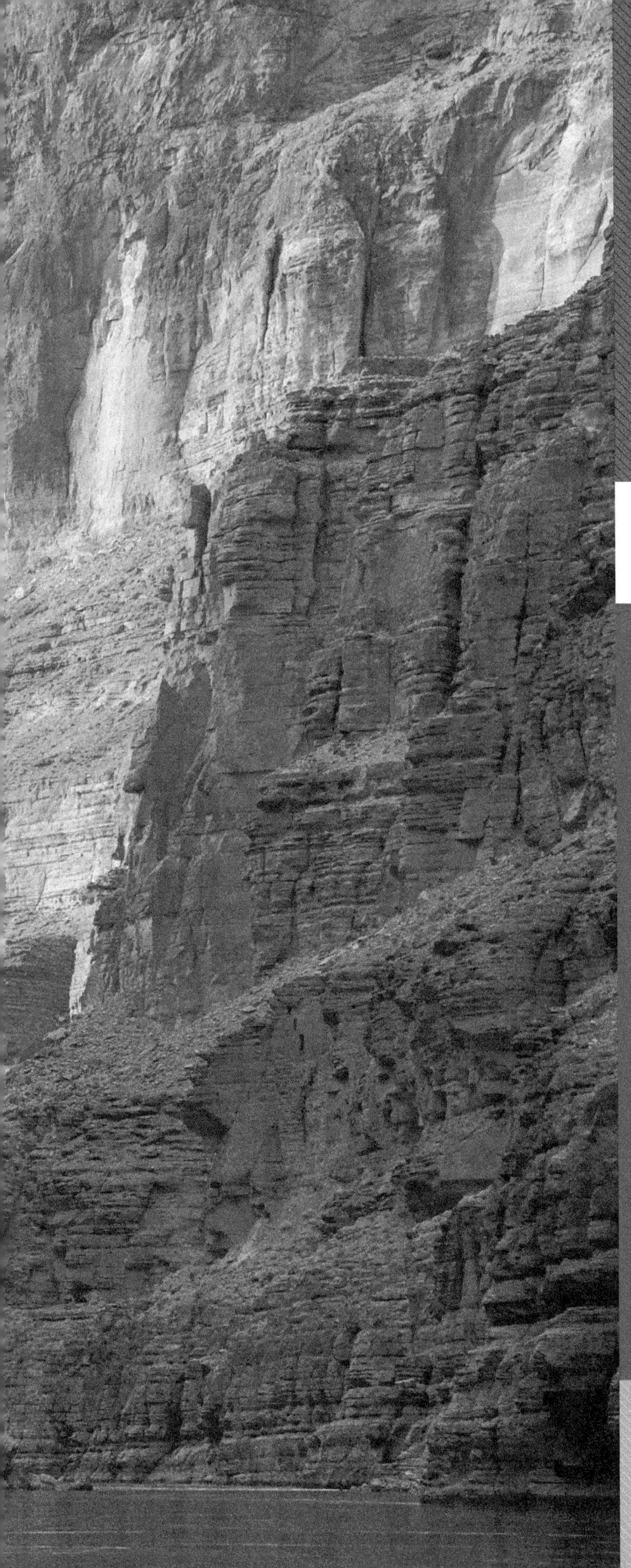

11 Geologic Time

FOCUS ON CONCEPTS

Each statement represents the primary learning objective for the corresponding major heading within the chapter. After you complete the chapter, you should be able to:

- **11.1** Explain the principle of uniformitarianism and discuss how it differs from catastrophism.
- **11.2** Distinguish between numerical and relative dating and apply relative dating principles to determine a time sequence of geologic events.
- **11.3** Define *fossil* and discuss the conditions that favor the preservation of organisms as fossils. List and describe various types of fossilization.
- **11.4** Explain how rocks of similar age that are in different places can be matched up.
- **11.5** Discuss three ways that atomic nuclei change and explain how unstable isotopes are used to determine numerical dates.
- **11.6** Explain how reliable numerical dates are determined for layers of sedimentary rock.
- **11.7** Distinguish among the four basic time units that make up the geologic time scale and explain why the time scale is considered to be a dynamic tool.

The Colorado River above Havasu Creek in Grand Canyon National Park. Millions of years of Earth history are exposed in the canyon's rock walls. (Photo by Michael Collier)

IN THE LATE EIGHTEENTH CENTURY, James Hutton recognized the immensity of Earth history and the importance of time as a component in all geologic processes. In the nineteenth century, Sir Charles Lyell and others effectively demonstrated that Earth had experienced many episodes of mountain building and erosion, which must have required great spans of geologic time. Although these pioneering scientists understood that Earth was very old, they had no way of determining its age in years. Was it tens of millions, hundreds of millions, or even billions of years old? Long before geologists could establish a geologic calendar that included numerical dates in years, they gradually assembled a time scale using relative dating principles. What are these principles? What part do fossils play? With the discovery of radioactivity and radiometric dating techniques, geologists can now assign quite accurate dates to many of the events in Earth history. What is radioactivity? Why is it a good "clock" for dating the geologic past?

11.1 A Brief History of Geology

Explain the principle of uniformitarianism and discuss how it differs from catastrophism.

The hiker in **Figure 11.1** is perched on the rim of the Grand Canyon. Today we know that the strata beneath the hiker represent hundreds of millions of years of Earth history. We also have the knowledge and skills that are necessary to unravel the complex story contained in these rocks. Such an understanding is a relatively recent accomplishment.

The nature of our Earth—its materials and processes—has been a focus of study for centuries. However, the late 1700s is generally regarded as the beginning of modern geology. It was during this time that James Hutton published his important work *Theory of the Earth*. Prior to that time, a great many explanations about Earth history relied on supernatural events.

Figure 11.1 **Contemplating geologic time** This hiker is resting atop the Kaibab Formation, the uppermost layer in the Grand Canyon. (Photo by Michael Collier)

Catastrophism

In the mid-1600s, James Ussher, Anglican Archbishop of Armagh, Primate of All Ireland, published a work that had immediate and profound influence. A respected scholar of the Bible, Ussher constructed a chronology of human and Earth history in which he determined that Earth was only a few thousand years old, having been created in 4004 B.C.E. Ussher's treatise earned widespread acceptance among Europe's scientific and religious leaders, and his chronology was soon printed in the margins of the Bible itself.

During the 1600s and 1700s, the doctrine of **catastrophism** strongly influenced people's thinking about Earth. Briefly stated, catastrophists believed that Earth's varied landscapes had been fashioned primarily by great catastrophes. Features such as mountains and canyons, which today we know take great periods of time to form, were explained as having been produced by sudden and often worldwide disasters of unknowable causes that no longer operate. This philosophy was an attempt to fit the rate of Earth's processes to the prevailing ideas about Earth's age.

The Birth of Modern Geology

Modern geology began in the late 1700s, when James Hutton, a Scottish physician and gentleman farmer, published his *Theory of the Earth*. In this work, Hutton put forth a fundamental principle that is a pillar of geology today: **uniformitarianism**. It simply states that *the physical, chemical, and biological laws that operate today have also operated in the geologic past.* This means that the forces and processes that we observe presently shaping our planet have been at work for a very long time. Thus, to understand ancient rocks, we must first understand present-day processes and their results because, as the saying goes, "the present is the key to the past."

Prior to Hutton's *Theory of the Earth*, no one had effectively demonstrated that geologic processes occur over extremely long periods of time. However, Hutton persuasively argued that processes that appear to be weak and slow acting can, over long spans of time, produce effects that are just as great as those resulting from sudden catastrophic events. Unlike his predecessors, Hutton cited verifiable observations to support his ideas.

For example, when he argued that mountains are sculpted and ultimately destroyed by weathering and the work of running water and that their wastes are carried to the oceans by processes that can be observed, Hutton said, "We have a chain of facts which clearly demonstrates that the materials of the wasted mountains have traveled through the rivers"; and, further, "There is not one step in all this progress that is not to be actually perceived." He went on to summarize these thoughts by asking a question and immediately providing the answer: "What more can we require? Nothing but time."

Geology Today

The basic tenets of uniformitarianism are just as viable today as in Hutton's day. We realize more strongly than ever before that the present gives us insight into the past and that the physical, chemical, and biological laws that govern geologic processes remain unchanging through time. However, we also understand that the doctrine should not be taken too literally. To say that geologic processes in the past were the same as those occurring today is not to suggest that they always had the same relative magnitude or that they operated at precisely the same rate. Moreover, some important geologic processes are not currently observable, but evidence that they occur is well established. For example, we know that Earth has experienced impacts from large meteorites even though we have no human witnesses. Such events altered Earth's crust, modified its climate, and strongly influenced life on the planet.

The acceptance of the concept of uniformitarianism, however, meant the acceptance of a very long history for Earth. Although Earth's processes vary in their intensity, they still take a long time to create or destroy major landscape features. For example, geologists have established that mountains once existed in portions of present-day Minnesota, Wisconsin, Michigan, and Manitoba—a region that today consists of low hills and plains. Erosion (processes that wear away land) gradually destroyed those peaks. The rock record contains evidence showing that Earth has experienced *many cycles* of mountain building and erosion. Concerning the ever-changing nature of Earth through great expanses of geologic time, Hutton made a statement that was to become his most famous. In concluding his classic 1788 paper published in the *Transactions of the Royal Society of Edinburgh*, he stated, "The results, therefore, of our present enquiry is, that we find no vestige of a beginning—no prospect of an end."

It is important to remember that although many features of our physical landscape may seem to be unchanging over the decades we observe them, they are nevertheless changing—but on time scales of hundreds, thousands, or even many millions of years.

CONCEPT CHECKS 11.1

1. Contrast catastrophism and uniformitarianism.
2. How did each philosophy view the age of Earth?

11.2 Creating a Time Scale: Relative Dating Principles

Distinguish between numerical and relative dating and apply relative dating principles to determine a time sequence of geologic events.

Beneath the hiker in Figure 11.1 are thousands of meters of sedimentary strata that go as far back more than 540 million years ago. These strata rest atop even older sedimentary, metamorphic, and igneous rocks from a span known as the Precambrian. Some of these rocks are 2 billion years old. Although the Grand Canyon's rock record has numerous interruptions, the rocks beneath the hiker contain clues to great spans of Earth history. Earth's long and complicated history is recorded in the structure, sequence, and properties of its rocks, sediments, and fossils.

The Importance of a Time Scale

Like the pages in a long and complicated history book, rocks record the geologic events and changing life-forms of the past. The book, however, is not complete. Many pages, especially in the early chapters, are missing. Others are tattered, torn, or smudged. Yet enough of the book remains to allow much of the story to be deciphered.

Interpreting Earth history is a prime goal of the science of geology. Like a modern-day sleuth, a geologist must interpret the clues found preserved in the rocks. By studying rocks, especially sedimentary rocks, and the features they contain, geologists can unravel the complexities of the past.

Geologic events by themselves, however, have little meaning until they are put into a time perspective. Studying history, whether it is the Civil War or the age of dinosaurs, requires a calendar. Among geology's major contributions to human knowledge are the *geologic time scale* and the discovery that Earth history is exceedingly long.

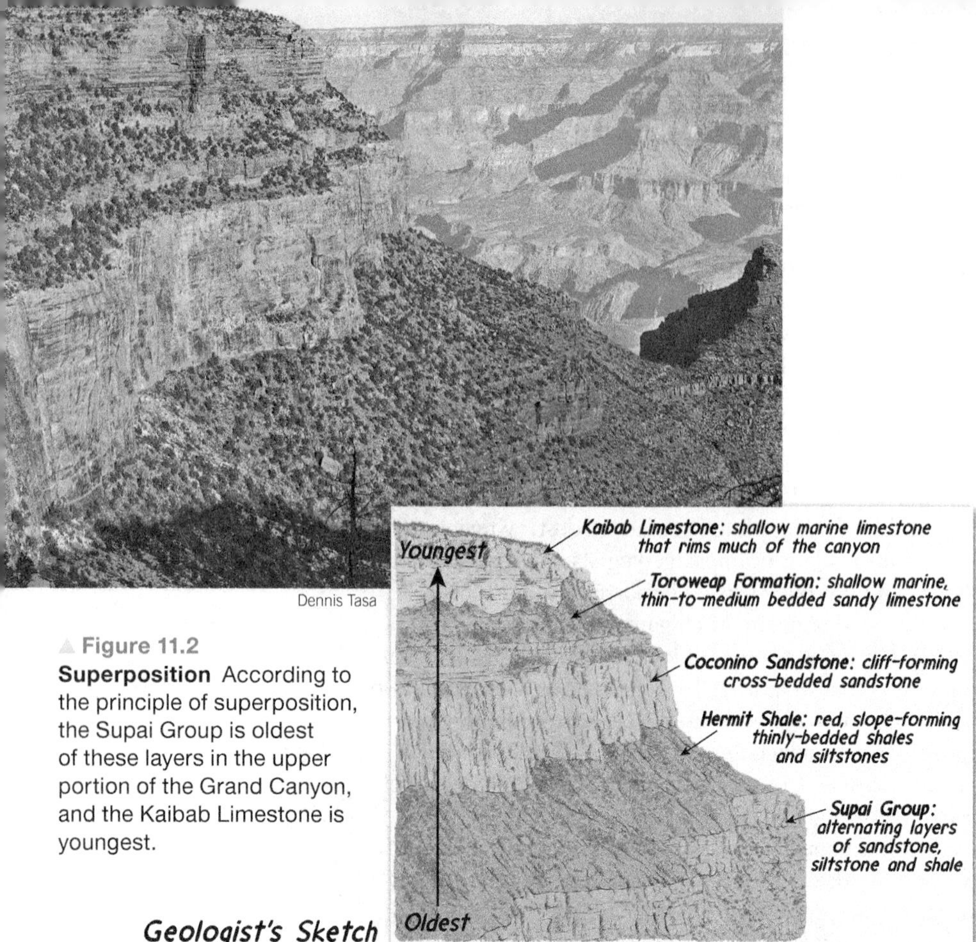

Dennis Tasa

▲ Figure 11.2 **Superposition** According to the principle of superposition, the Supai Group is oldest of these layers in the upper portion of the Grand Canyon, and the Kaibab Limestone is youngest.

Numerical and Relative Dates

The geologists who developed the geologic time scale revolutionized the way people think about time and perceive our planet. They learned that Earth is much older than anyone had previously imagined and that its surface and interior have been changed over and over again by the same geologic processes that operate today.

Numerical Dates During the late 1800s and early 1900s, attempts were made to determine Earth's age. Although some of the methods appeared promising at the time, none of those early efforts proved to be reliable. What those scientists were seeking was a **numerical date**. Such dates specify the actual number of years that have passed since an event occurred. Today, our understanding of radioactivity allows us to accurately determine numerical dates for rocks that represent important events in Earth's distant past. We will study radioactivity later in this chapter. Prior to the discovery of radioactivity, geologists had no reliable method of numerical dating and had to rely solely on relative dating.

Relative Dates When we place rocks in their proper *sequence of formation*—indicating which formed first, second, third, and so on—we are establishing **relative dates**. Such dates cannot tell us how long ago something took place, only that it followed one event and preceded another. The relative dating techniques that were developed are valuable and still widely used. Numerical dating methods do not replace these techniques; they simply supplement them. To establish a relative time scale, a few basic principles or rules had to be discovered and applied. They were major breakthroughs in thinking at the time, and their discovery was an important scientific achievement.

Principle of Superposition

Nicolas Steno, a Danish anatomist, geologist, and priest (1638–1686), was the first to recognize a sequence of historical events in an outcrop of sedimentary rock layers. Working in the mountains of western Italy, Steno applied a very simple rule that has become the most basic principle of relative dating—the **principle of superposition** (*super* = above; *positum* = to place). This principle simply states that in an undeformed sequence of sedimentary rocks, each bed is older than the one above and younger than the one below. Although it may seem obvious that a rock layer could not be deposited with nothing beneath it for support, it was not until 1669 that Steno clearly stated this principle.

This rule also applies to other surface-deposited materials, such as lava flows and beds of ash from volcanic eruptions. Applying the principle of superposition to the beds exposed in the upper portion of the Grand Canyon, we can easily place the layers in their proper order. Among those that are pictured in Figure 11.2, the sedimentary rocks in the Supai Group are the oldest, followed in order by the Hermit Shale, Coconino Sandstone, Toroweap Formation, and Kaibab Limestone.

Principle of Original Horizontality

Steno is also credited with recognizing the importance of another basic rule, the **principle of original horizontality**, which states that layers of sediment are generally deposited in a horizontal position. Thus, if we observe rock layers that are flat, we know that they have not been disturbed and still have their *original* horizontality. The layers in the Grand Canyon illustrate this in Figures 11.1 and 11.2. But if they are folded or inclined at a steep angle, they must have been moved into that position by crustal disturbances sometime *after* their deposition (Figure 11.3).

▼ Figure 11.3 **Original horizontality** Most layers of sediment are deposited in a nearly horizontal position. When we see strata that are folded or tilted, we can assume that they were moved into that position by crustal disturbances *after* their deposition. (Photo by Marco Simoni/Robert Harding World Imagery)

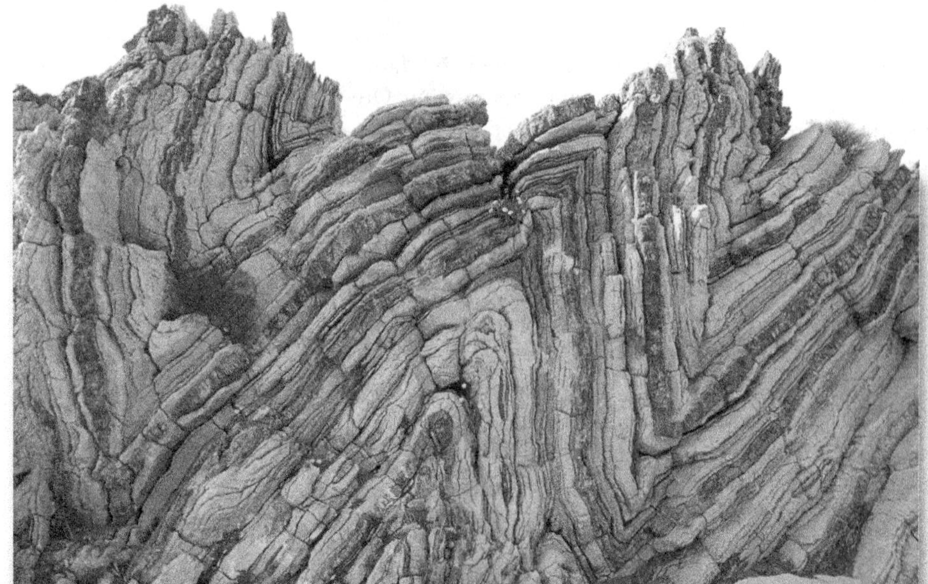

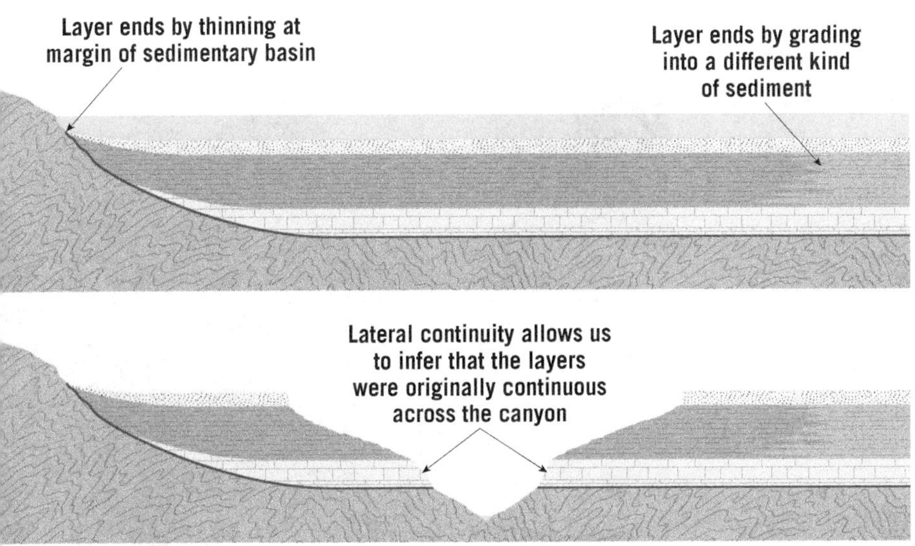

Layer ends by thinning at margin of sedimentary basin
Layer ends by grading into a different kind of sediment
Lateral continuity allows us to infer that the layers were originally continuous across the canyon

Figure 11.4 Lateral continuity Sediments are deposited over a large area in a continuous sheet. Sedimentary strata extend continuously in all directions until they thin out at the edge of a depositional basin or grade into a different type of sediment.

Principle of Lateral Continuity

The **principle of lateral continuity** refers to the fact that sedimentary beds originate as continuous layers that extend in all directions until they eventually grade into a different type of sediment or until they thin out at the edge of the basin of deposition (Figure 11.4). For example, when a river creates a canyon, we can assume that identical or similar strata on opposite sides once spanned the canyon. Although rock outcrops may be separated by a considerable distance, the principle of lateral continuity tells us that those outcrops once formed a continuous layer. This principle allows geologists to relate rocks in isolated outcrops to one another. Combining the principles of lateral continuity and superposition lets us extend relative age relationships over broad areas. This process, called *correlation*, is examined in Section 11.4.

Principle of Cross-Cutting Relationships

Figure 11.5 shows a mass of rock that is offset by a fault, a fracture in rock along which displacement occurs. It is clear that the rocks must be older than the fault that broke them. The **principle of cross-cutting relationships** states that geologic features that cut across rocks must have formed *after* the rocks they cut through. Igneous intrusions provide another example. The dike shown in Figure 11.6 is a tabular mass of igneous rock that cuts through the surrounding rocks. The magmatic heat from igneous intrusions often creates a narrow "baked" zone of contact metamorphism on the adjacent rock, also indicating that the intrusion occurred after the surrounding rocks were in place.

Principle of Inclusions

Sometimes inclusions can aid in the relative dating process. *Inclusions* are fragments of one rock unit that have been enclosed within another. The **principle of inclusions** is logical and straightforward: The rock mass adjacent to the one containing the inclusions must have been there first in order to provide the rock fragments. Therefore, the rock mass that contains inclusions is the younger of the two. For example, when magma intrudes into surrounding rock, blocks of the surrounding rock may become dislodged and incorporated into the magma. If these pieces do not melt, they remain as inclusions, known as *xenoliths*. In another example, when sediment is deposited atop a weathered mass of bedrock, pieces of the weathered rock become incorporated into the younger sedimentary layer (Figure 11.7).

l/gLu1IR

SmartFigure 11.5 Cross-cutting fault The rocks are older than the fault that displaced them. (Morley Read/Alamy)

Fault

Figure 11.6 Cross-cutting dike An igneous intrusion is younger than the rocks that are intruded. (Photo by Jonathan.s.kt)

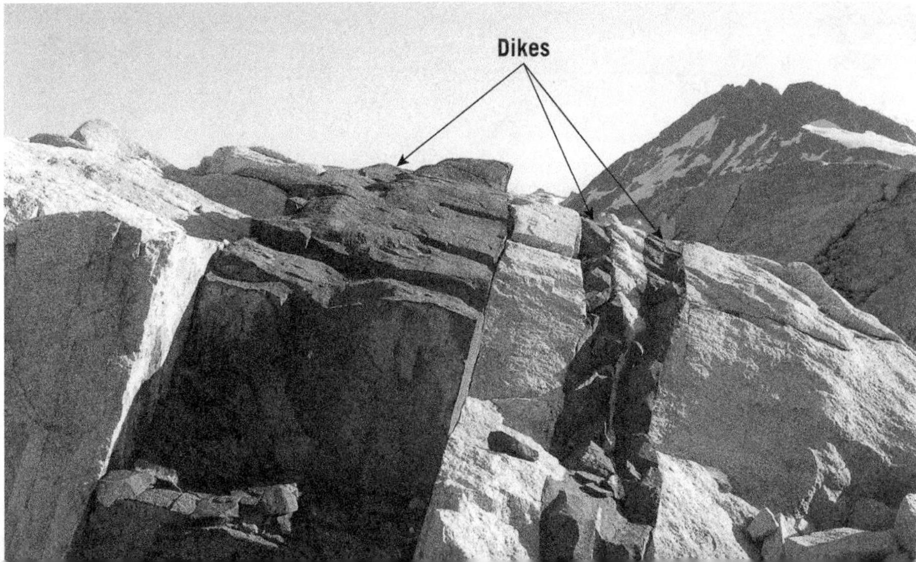

Dikes

EYE ON EARTH 11.1

This image shows West Cedar Mountain in southern Utah. The gray rocks are shale that originated as muddy river delta deposits. The sediments composing the orange sandstone were deposited by a river.

QUESTION 1 *Place the events related to the geologic history of this area in proper sequence. Explain your logic. Include the following: uplift, sandstone, erosion, and shale.*

QUESTION 2 *What term is applied to the type of dates you established for this site?*

Shale
Sandstone

Michae

Unconformities

When we observe layers of rock that have been deposited essentially without interruption, we call them **conformable**. Particular sites exhibit conformable beds representing certain spans of geologic time. However, no place on Earth has a complete set of conformable strata that represents the entire history of Earth.

Throughout Earth history, the deposition of sediment has been interrupted over and over again. All such breaks in the rock record are termed *unconformities*. An **unconformity** represents a long period during which deposition ceased, erosion removed previously formed rocks, and then deposition resumed. In each case, uplift and erosion are followed by subsidence and renewed sedimentation. Unconformities are important features because they represent significant geologic events in Earth history. There are three basic types of unconformities, and their recognition helps geologists identify what intervals of time are not represented by strata and thus are missing from the geologic record.

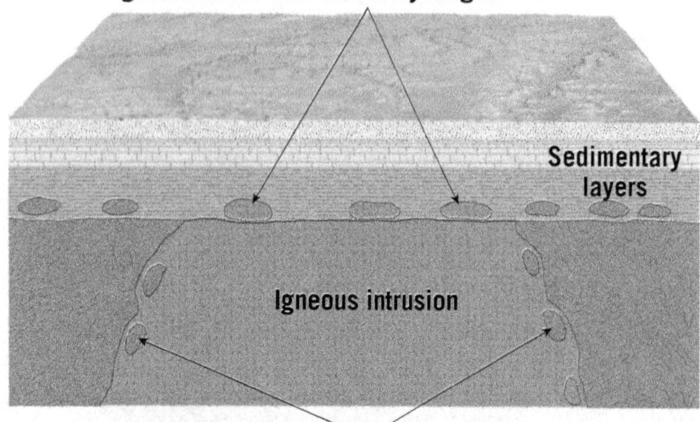
These inclusions of igneous rock contained in the adjacent sedimentary layer indicate that the sediments were deposited atop the weathered igneous mass and thus are younger.
Sedimentary layers
Igneous intrusion
Xenoliths are inclusions in an igneous intrusion that form when pieces of surrounding rock are incorporated into magma.

SmartFigure 11.7 **Inclusions** The rock containing inclusions is younger than the inclusions.

TUTORIAL
https://goo.gl/s8USDC

Angular Unconformity

Perhaps the most easily recognized unconformity is an **angular unconformity**. It consists of tilted or folded sedimentary rocks that are overlain by younger, more flat-lying strata. An angular unconformity indicates that during a pause in deposition, a period of deformation (folding or tilting) and erosion occurred (Figure 11.8).

When James Hutton studied an angular unconformity in Scotland more than 200 years ago, he understood that it represented a major episode of geologic activity (Figure 11.9). He and his colleagues also appreciated the immense time span implied by such relationships. When a companion later wrote of their visit to this site, he stated that "the mind seemed to grow giddy by looking so far into the abyss of time."

Disconformity

A **disconformity** is a gap in the rock record that represents a period during which erosion rather than deposition occurred. Imagine that a series of sedimentary layers is deposited in a shallow marine setting. Following this period of deposition, sea level falls or the land rises, exposing some the sedimentary layers. During this span, when the sedimentary beds are above sea level, no new sediment accumulates, and some of the existing layers are eroded away. Later, sea level rises or the land subsides, submerging the landscape. Now the surface is again below sea level, and a new series of sedimentary beds is deposited. The boundary separating the two sets of beds is a disconformity—a span for which there is no rock record

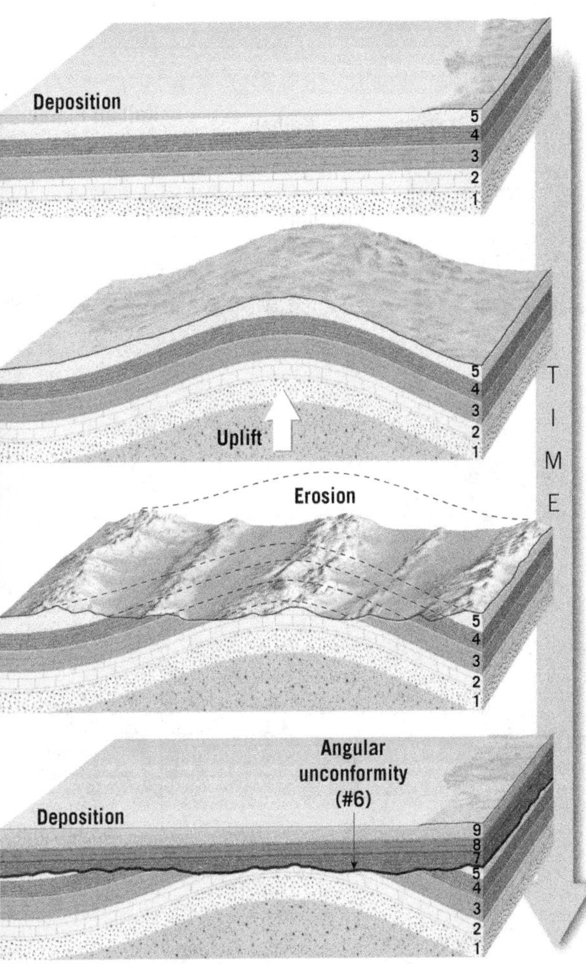

Deposition
5
4
3
2
1
Uplift
T
I
M
E
Erosion
Angular
unconformity
(#6)
Deposition
9
8
7

SmartFigure 11.8 **Formation of an angular unconformity** An angular unconformity represents an extended period during which deformation and erosion occurred. Numbers on the diagram indicate the order in which events occurred.

TUTORIAL
https://goo.gl/fkUmV5

Above the unconformity lie gently dipping beds of reddish sandstone and conglomerate
Angular unconformity
Rock hammer
Below the unconformity lie nearly vertical beds of sandstone and shale

Figure 11.9 **Siccar Point, Scotland** James Hutton studied this famous unconformity in the late 1700s. (Photo by Marli Miller)

(Figure 11.10). Because the layers above and below a disconformity are parallel, these features are sometimes difficult to identify unless you notice evidence of erosion such as a buried stream channel.

Nonconformity The third basic type of unconformity is a **nonconformity**, in which younger sedimentary strata overlie older metamorphic or intrusive igneous rocks (Figure 11.11). Just as angular unconformities and some disconformities imply crustal movements, so too do nonconformities. Intrusive igneous masses and metamorphic rocks originate far below the surface. Thus, for

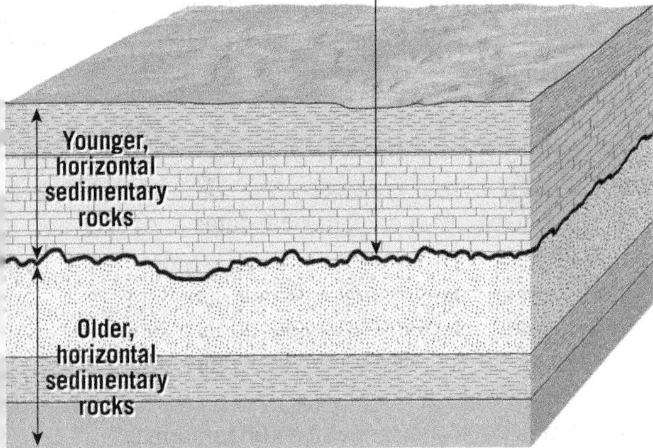

Disconformity
Gap in the rock record represents a period of nondeposition and erosion
Younger, horizontal sedimentary rocks
Older, horizontal sedimentary rocks

Figure 11.10 **Disconformity** The layers on both sides of this gap in the rock record are essentially parallel.

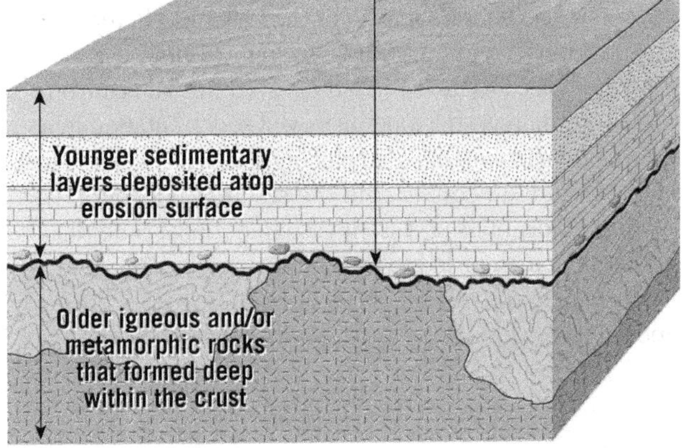

Nonconformity
Period of uplift and erosion that exposed the deep rocks at the surface
Younger sedimentary layers deposited atop erosion surface
Older igneous and/or metamorphic rocks that formed deep within the crust

Figure 11.11 **Nonconformity** Younger sedimentary rocks rest atop older metamorphic or igneous rocks.

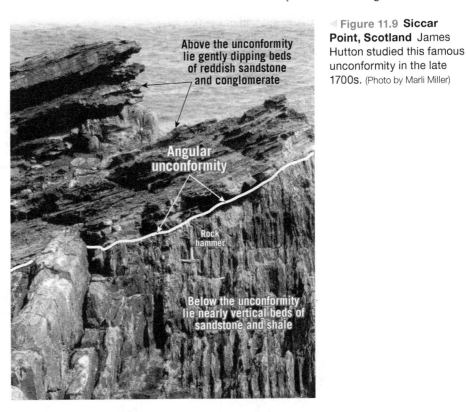

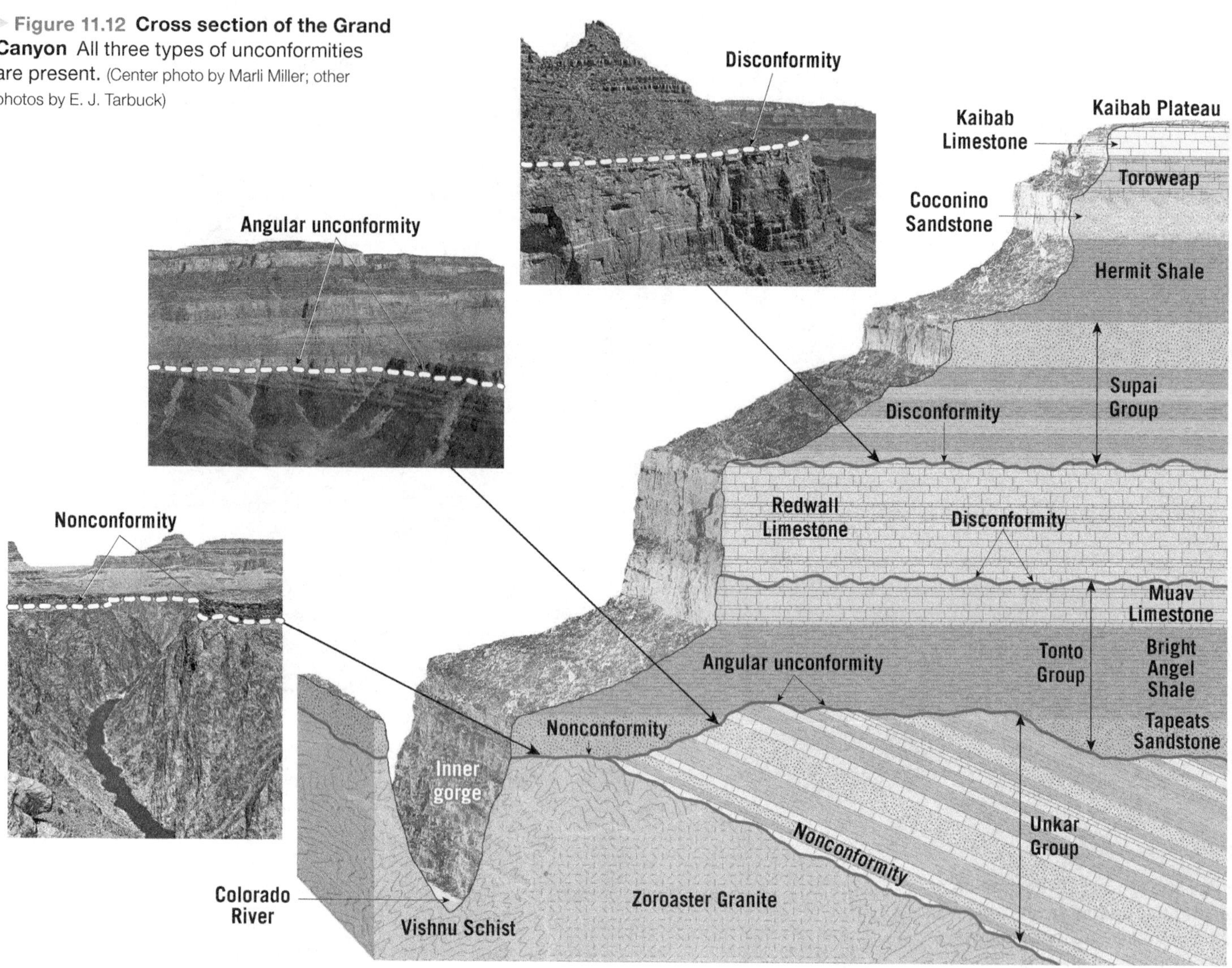

Figure 11.12 Cross section of the Grand Canyon All three types of unconformities are present. (Center photo by Marli Miller; other photos by E. J. Tarbuck)

a nonconformity to develop, there must be a period of uplift and erosion of overlying rocks. Once exposed at the surface, the igneous or metamorphic rocks are subjected to weathering and erosion and then undergo subsidence and renewed sedimentation.

Unconformities in the Grand Canyon The rocks exposed in the Grand Canyon of the Colorado River represent a tremendous span of geologic history. It is a wonderful place to take a trip through time. The canyon's colorful strata record a long history of sedimentation in a variety of environments—advancing seas, rivers and deltas, tidal flats and sand dunes. But the record is not continuous. Unconformities represent vast amounts of time that have not been recorded in the canyon's layers. Figure 11.12 is a geologic cross section of the Grand Canyon. All three types of unconformities can be seen in the canyon walls.

Applying Relative Dating Principles

If you apply the principles of relative dating to the hypothetical geologic cross section in Figure 11.13, you can place in proper sequence the rocks and the events they represent. The statements within the figure summarize the logic used to interpret the cross section.

In this example, we establish a relative time scale for the rocks and events in the area of the cross section. Remember that this method gives us no idea how many years of Earth history are represented, for we have no numerical dates. Nor do we know how this area compares to any other.

CONCEPT CHECKS 11.2

1. Distinguish between numerical dates and relative dates.
2. Sketch and label four simple diagrams that illustrate each of the following: superposition, original horizontality, lateral continuity, and cross-cutting relationships.
3. What is the significance of an unconformity?
4. Distinguish among angular unconformity, disconformity, and nonconformity.

Working out the geologic history of a hypothetical region

Interpretation:

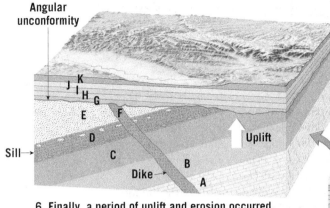

1. Beneath the ocean, beds A, B, C, and E were deposited in that order (law of superposition).

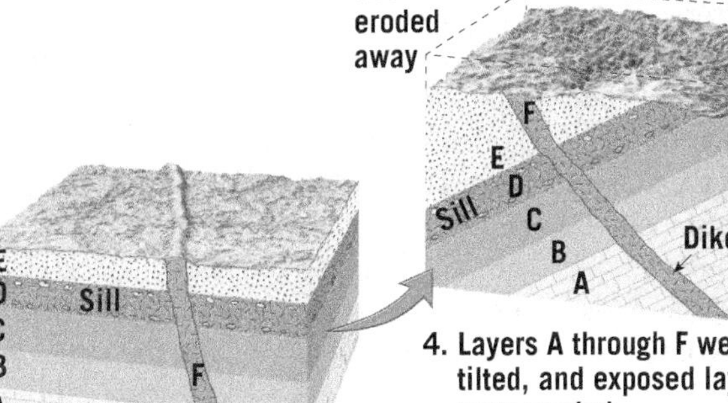

2. Uplift and intrusion of a sill (layer D). We know that sill D is younger than beds C and E because of the inclusions in the sill of fragments from beds C and E.

3. Next is the intrusion of dike F. Because the dike cuts through layers A through E, it must be younger (principle of cross-cutting relationships).

4. Layers A through F were tilted, and exposed layers were eroded.

5. Next, beds G, H, I, J, and K were deposited, in that order, atop the erosion surface to produce an angular unconformity.

6. Finally, a period of uplift and erosion occurred. The irregular surface and stream valley indicate that another gap in the rock record is being created by erosion.

SmartFigure 11.13
Applying principles of relative dating

TUTORIAL
https://goo.gl/sHKpNK

EYE ON EARTH 11.2

This close-up shows pieces of diorite (darkest rock) in granite. The thin white line is a vein of quartz. Think of a vein as a tiny dike.

QUESTION 1 *What term is applied to the pieces of diorite?*

QUESTION 2 *Place the quartz vein, diorite, and granite in order from oldest to youngest.*

Marli Miller

11.3 Fossils: Evidence of Past Life

Define *fossil* and discuss the conditions that favor the preservation of organisms as fossils. List and describe various types of fossilization.

Fossils, the remains or traces of prehistoric life, are important inclusions in sediment and sedimentary rocks. They are basic and important tools for interpreting the geologic past. The scientific study of fossils is called **paleontology**. It is an interdisciplinary science that blends geology and biology in an attempt to understand all aspects of the evolution of life over the vast expanse of geologic time (see GEOgraphics 11.1). Knowing the nature of the life-forms that existed at a particular time helps researchers understand past environmental conditions. Further, fossils are important time indicators and play a key role in correlating rocks of similar ages that are from different places.

Types of Fossils

Fossils are preserved in many ways. The remains of relatively recent organisms may not have been altered at all. Objects such as teeth, bones, and shells are common examples (Figure 11.14). Far less common are entire animals, flesh included, that have been preserved because of rather unusual circumstances. Remains of prehistoric elephant relatives called mammoths that were frozen in the Arctic tundra of Siberia and Alaska are examples, as are the mummified remains of sloths preserved in a dry cave in Nevada.

Permineralization When mineral-rich groundwater permeates porous tissue such as bone or wood, minerals precipitate out of solution and fill pores and empty spaces, a process called *permineralization*. The formation of *petrified wood* involves permineralization with silica, often from a volcanic source such as a surrounding layer of volcanic ash. The wood is gradually transformed into chert, sometimes with colorful bands from impurities such as iron or carbon (Figure 11.15A). The word *petrified* literally means "turned into stone." Sometimes the microscopic details of the petrified structure are faithfully retained.

Skeleton of a mammoth, a prehistoric relative of modern elephant, from the La Brea tar pits.
Excavating bones from pit 91. It is a site rich in unaltered Ice Age organisms. Scientists have been excavating here since 1915.

Figure 11.14 **La Brea tar pits, Los Angeles** The fossils here are actual (unaltered) remains. (Excavation photo by Reed Saxon/AP Wide World Photo; skeleton photo by Martin Shields/Alamy)

Molds and Casts Another common mode of fossilization is *molds* and *casts*. When a shell or another structure is buried in sediment and then dissolved by underground water, a *mold* is created. The mold faithfully reflects the shape and surface marking of the organism; however, it does not reveal any information concerning its internal structure. If these hollow spaces are subsequently filled with mineral matter, *casts* are created (Figure 11.15B).

Carbonization and Impressions A type of fossilization called *carbonization* is particularly effective at preserving leaves and delicate animal forms. It occurs when fine sediment encases the remains of an organism. As time passes, pressure squeezes out the liquid and gaseous components and leaves behind a thin residue of carbon (Figure 11.15C). Black shale deposited as organic-rich mud in oxygen-poor environments often contains abundant carbonized remains. If the film of carbon is lost from a fossil preserved in fine-grained sediment, a replica of the surface, called an *impression*, may still show considerable detail (Figure 11.15D).

Amber Delicate organisms, such as insects, are difficult to preserve, and consequently they are relatively rare in the fossil record. However, *amber*—the hardened resin of ancient trees—can preserve them in exquisite three-dimensional detail. The spider in Figure 11.15E was preserved after being trapped in a drop of sticky resin. Resin sealed off the insect from the atmosphere and protected the remains from damage by water and air. As the resin hardened, a protective pressure-resistant case was formed.

Trace Fossils In addition to the fossils already mentioned, there are numerous other types, many of them only traces of prehistoric life. Examples of such indirect evidence include:

- Tracks—animal footprints made in soft sediment that later turned into sedimentary rock.
- Burrows—tubes in sediment, wood, or rock made by an animal. These holes may later become filled with mineral matter and preserved. Some of the oldest-known fossils are believed to be worm burrows.

How is paleontology different from archaeology?

People frequently confuse these two areas of study because a common perception of both paleontologists and archaeologists is of scientists carefully extracting important clues about the past from layers of rock or sediment. While it is true that scientists in both disciplines "dig" a lot, the focus of each is different.

Archaeology

Archaeologists help us learn about how our human ancestors met the challenges of life in the past.

Archaeologists focus on the material remains of past human life. These remains include both the objects used by people long ago, called *artifacts*, and the buildings and other structures associated with where people lived, called *sites*.

Paleontology

Paleontologists study fossils and are concerned with all life forms in the geologic past. These scientists are excavating the fossil remains of *Albertasaurus*, a carnivore similar to *Tyrannosaurus rex*.

Question:
Briefly distinguish between paleontology and archaeology.

?

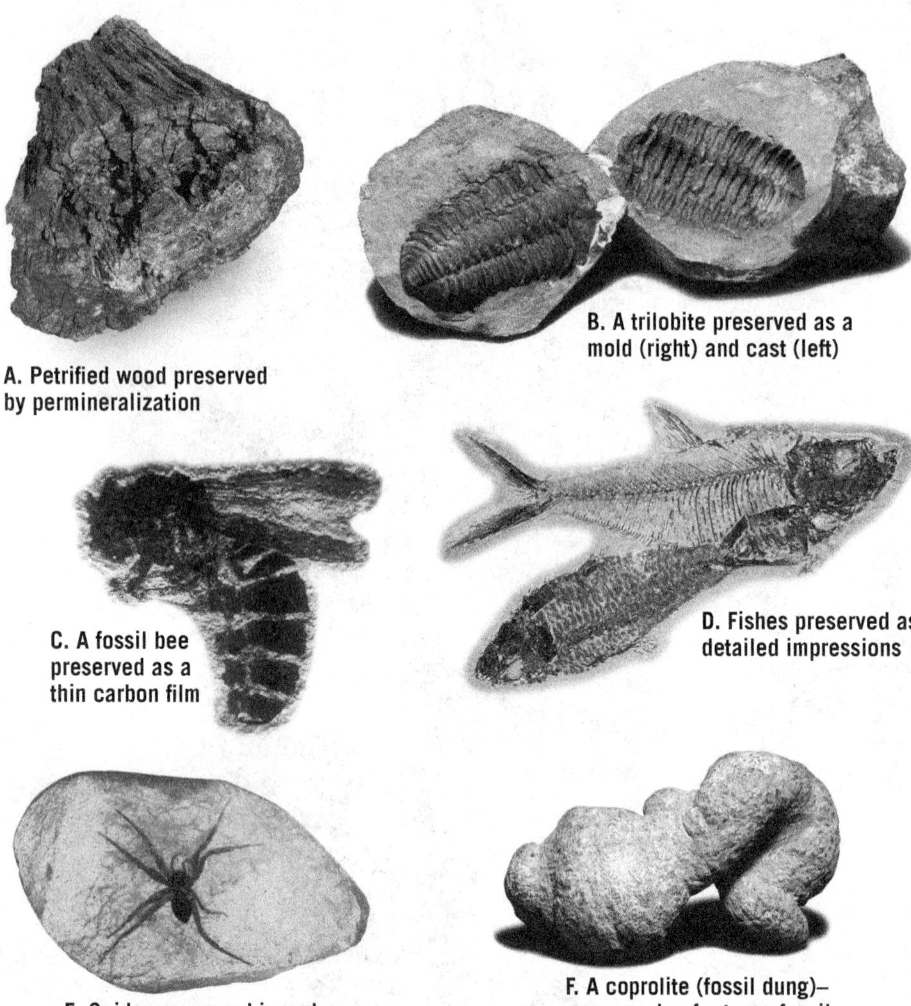

A. Petrified wood preserved by permineralization
B. A trilobite preserved as a mold (right) and cast (left)
C. A fossil bee preserved as a thin carbon film
D. Fishes preserved as detailed impressions
E. Spider preserved in amber
F. A coprolite (fossil dung)– an example of a trace fossil

▲ **Figure 11.15 Types of fossilization** (Photo A by Bernhard Edmaier/Science Source; photo B by E. J. Tarbuck; photo C by Florissant Fossil Beds National Monument; photo D by E. J. Tarbuck; photo E by Colin Keates/Dorling Kindersley Media Library; photo F by E. J. Tarbuck)

- Coprolites—fossil dung can provide useful information pertaining to the size and food habits of organisms (Figure 11.15F).
- Gastroliths—highly polished stomach stones that were used in the grinding of food by some dinosaurs and other organisms.

Conditions Favoring Preservation

Only a tiny fraction of the organisms that have lived during the geologic past have been preserved as fossils. Normally, the remains of an animal or a plant are destroyed. Under what circumstances are they preserved? Two special conditions favor fossilization: rapid burial and the possession of hard parts.

When an organism perishes, its soft parts usually are quickly eaten by scavengers or decomposed by bacteria. Occasionally, however, the remains are buried by sediment. When this occurs, the remains are protected from the surface environment, where destructive processes operate. Rapid burial, therefore, is an important condition favoring preservation.

In addition, animals and plants have a much better chance of being preserved as part of the fossil record if they have hard parts. Although traces and imprints of soft-bodied animals such as jellyfish, worms, and insects exist, they are not common. Flesh usually decays so rapidly that preservation is exceedingly unlikely. Hard parts such as shells, bones, and teeth predominate in the record of past life.

Because preservation is contingent on special conditions, the record of life in the geologic past is biased. The fossil record of those organisms with hard parts that lived in areas of sedimentation is quite abundant. However, we get only an occasional glimpse of the vast array of other life-forms that did not meet the special conditions favoring preservation.

CONCEPT CHECKS 11.3

1. Describe several ways that an animal or a plant can be preserved as a fossil.
2. List three examples of trace fossils.
3. What conditions favor the preservation of an organism as a fossil?

11.4 Correlation of Rock Layers

Explain how rocks of similar age that are in different places can be matched up.

To develop a geologic time scale that is applicable to the entire Earth, rocks of similar age in different regions must be matched up. Such a task is called **correlation**. Correlating the rocks from one place to another makes possible a more comprehensive view of the geologic history of a region. Figure 11.16, for example, shows the correlation of strata at three sites on the Colorado Plateau in southern Utah and northern Arizona. No single locale exhibits the entire sequence, but correlation reveals a more complete picture of the sedimentary rock record.

Correlation Within Limited Areas

Within a limited area, geologists can correlate rocks of one locality with those of another by simply walking along the outcropping edges, but this may not be possible when the rocks are mostly concealed by soil and vegetation. Correlation over short distances is often achieved by noting the position of a bed in a sequence of strata. Or a layer may be identified in another location if it is composed of distinctive or uncommon minerals. However, when correlation between widely separated areas or between continents is the objective, geologists must rely on fossils.

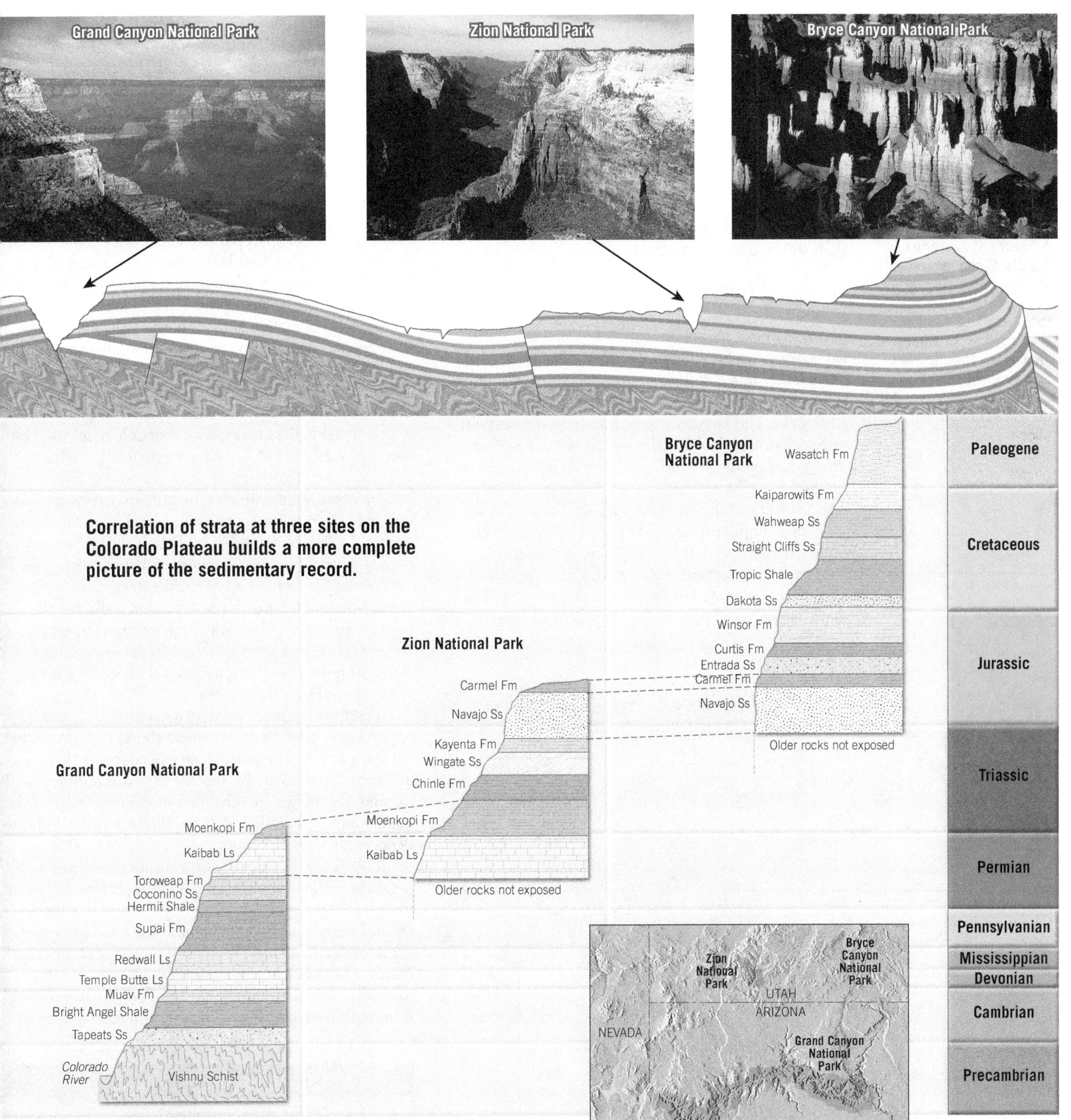

▲ **Figure 11.16 Correlation** Matching strata at three locations on the Colorado Plateau. Dashed red lines show correlation between sites. (Photos by E. J. Tarbuck)

▲ **Figure 11.17 Index fossils** Since microfossils are often very abundant, widespread, and quick to appear and become extinct, they constitute ideal index fossils. This scanning electron micrograph shows marine microfossils from the Miocene epoch. (Photo by Biophoto Associates/Science Source)

Fossils and Correlation

The existence of fossils had been known for centuries, yet it was not until the late 1700s and early 1800s that their significance as geologic tools was made evident. During this period, an English engineer and canal builder, William Smith, discovered that each rock formation in the canals he worked on contained fossils unlike those in the beds either above or below. Further, he noted that sedimentary strata in widely separated areas could be identified—and correlated—based on their distinctive fossil content.

Principle of Fossil Succession Based on Smith's classic observations and the findings of many later geologists, one of the most important and basic principles in historical geology was formulated: *Fossil organisms succeed one another in a definite and determinable order, and therefore any time period can be recognized by its fossil content.* This has come to be known as the **principle of fossil succession**. In other words, when fossils are arranged according to their age, they do not present a random or haphazard picture. To the contrary, fossils document the evolution of life through time.

For example, an Age of Trilobites is recognized quite early in the fossil record. Then, in succession, paleontologists recognize an Age of Fishes, an Age of Coal Swamps, an Age of Reptiles, and an Age of Mammals. These "ages" pertain to groups that were especially plentiful and characteristic during particular time periods. Within each of the "ages" are many subdivisions, based, for example, on certain species of trilobites and certain types of fish, reptiles, and so on. This same succession of dominant organisms, never out of order, is found on every continent.

Index Fossils and Fossil Assemblages When fossils were found to be time indicators, they became the most useful means of correlating rocks of similar age in different regions. Geologists pay particular attention to certain fossils called **index fossils** (Figure 11.17). These fossils are widespread geographically but limited to a short span of geologic time, so their presence provides an important method of matching rocks of the same age. Rock formations, however, do not always contain a specific index fossil. In such situations, a group of fossils, called a **fossil assemblage**, is used to establish the age of the bed. Figure 11.18 illustrates how an assemblage of fossils may be used to date rocks more precisely than could be accomplished by the use of any single fossil.

▲ SmartFigure 11.18 **Fossil assemblage** Overlapping ranges of fossils help date rocks more exactly than using a single fossil.

TUTORIAL
https://goo.gl/lpkEBT

Environmental Indicators In addition to being important, and often essential, tools for correlation, fossils are important environmental indicators. Although we can deduce much about past environments by studying the nature and characteristics of sedimentary rocks, a close examination of the fossils present can usually provide a great deal more information. For example, when the remains of certain clam shells are found in limestone, a geologist quite reasonably assumes that the region was once covered by a shallow sea.

Fossils can also at times be used to identify the approximate position of an ancient shoreline. Given what we know of living organisms, we can conclude that fossil animals with thick shells, capable of withstanding pounding and surging waves, inhabited shorelines. On the other hand, animals with thin, delicate shells probably indicate deep, calm offshore waters.

Fossils also can be used to indicate the former temperature of the water. Certain kinds of present-day corals must live in warm and shallow tropical seas like those around Florida and The Bahamas. When similar types of coral are found in ancient limestones, they indicate the marine environment that must have existed when they were alive. These examples illustrate how fossils can help unravel the complex story of Earth history.

CONCEPT CHECKS 11.4

1. What is the goal of correlation?
2. State the principle of fossil succession in your own words.
3. Contrast index fossil and fossil assemblage.
4. Along with their value for correlation, how else are fossils useful to geologists?

11.5 Numerical Dating with Nuclear Decay

Discuss three ways that atomic nuclei change and explain how unstable isotopes are used to determine numerical dates.

In addition to establishing relative dates by using the principles described in the preceding sections, scientists can also obtain reliable numerical dates for events in the geologic past. For example, we know that Earth is about 4.6 billion years old and that the dinosaurs became extinct about 66 million years ago. Dates that are expressed in millions and billions of years truly stretch our imagination because our personal calendars involve time measured in hours, weeks, and years. In this section you will learn about radioactivity and its application in radiometric dating. Our understanding of changes in the nuclei of atoms has allowed us to determine that geologic time is vast. This immense span is often referred to as *deep time*. Radiometric dating allows us to measure it quantitatively.

Reviewing Basic Atomic Structure

Recall from Chapter 2 that each atom has a *nucleus* that contains protons and neutrons and that the nucleus is orbited by electrons. *Electrons* have a negative electrical charge, and *protons* have a positive charge. A *neutron* has no charge (it is electrically neutral), but it can be converted to a positively charged proton plus a negatively charged electron.

The *atomic number* (each element's identifying number) is the number of protons in the nucleus. Every element has a different number of protons and thus a different identifying atomic number (hydrogen = 1, carbon = 6, oxygen = 8, uranium = 92, etc.). Atoms of the same element always have the same number of protons, so the atomic number stays constant.

Practically all of an atom's mass (99.9 percent) is in the nucleus, indicating that electrons have virtually no mass at all. So, by adding the protons and neutrons in an atom's nucleus, we derive the atom's *mass number*. The number of neutrons can vary, and these variants, or *isotopes*, have different mass numbers.

For example, uranium's nucleus always has 92 protons, so its atomic number is always 92. But its neutron population varies, so uranium has three isotopes: uranium-234 (protons + neutrons = 234), uranium-235, and uranium-238. All three isotopes are mixed in nature. They look the same and behave the same in chemical reactions.

Changes to Atomic Nuclei

Usually, the forces that stabilize atomic nuclei are strong. However, in some isotopes, the forces that bind protons and neutrons are not strong enough to keep them together forever. Such nuclei are *unstable* and spontaneously break apart in a process called **nuclear decay** (also called **radioactive decay**). As time goes by, more and more of the unstable atoms decay, producing an ever-growing number of stable isotopes. Not all isotopes are unstable—there are stable isotopes, too—but here we focus on the unstable isotopes and the stable isotopes they produce.

What happens when unstable atoms break apart? Three common types of nuclear decay are illustrated in Figure 11.19:

- Alpha particles (α particles) may be emitted from the nucleus. An alpha particle is composed of 2 protons

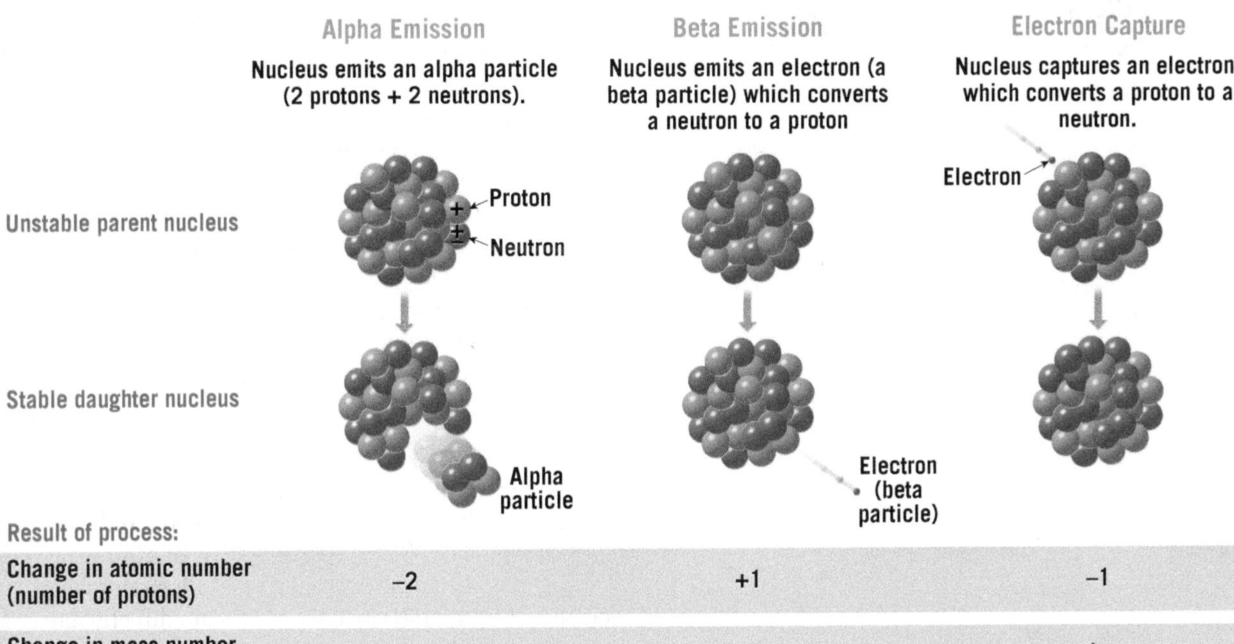
Alpha Emission
Nucleus emits an alpha particle (2 protons + 2 neutrons).
Beta Emission
Nucleus emits an electron (a beta particle) which converts a neutron to a proton
Electron Capture
Nucleus captures an electron, which converts a proton to a neutron.
Unstable parent nucleus
Proton
Neutron
Electron
Stable daughter nucleus
Alpha particle
Electron (beta particle)
Result of process:

Result of process:	Alpha Emission	Beta Emission	Electron Capture
Change in atomic number (number of protons)	−2	+1	−1
Change in mass number (number of protons + neutrons)	−4	no change	no change

Figure 11.19 Changing atomic nuclei Notice that in each example, the number of protons (atomic number) in the nucleus changes, thus producing a different element.

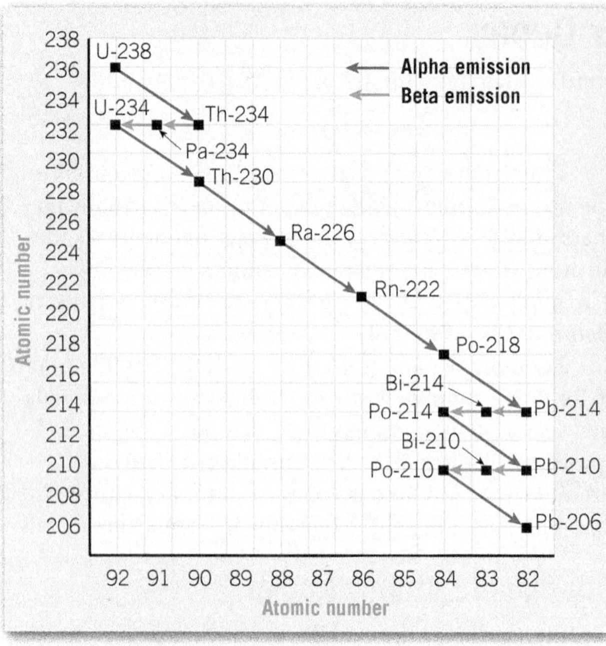

Figure 11.20 Decay of U-238 Uranium-238 is an example of a nuclear decay series. Before the stable end product (Pb-206) is reached, many different isotopes are produced as intermediate steps.

SmartFigure 11.21 Changing parent/daughter ratios Change is exponential. Half of the unstable parent atoms remain after one half-life. After a second half-life, one-quarter of the parent atoms remain, and so forth.

TUTORIAL
https://goo.gl/o58WCb

and 2 neutrons. Thus, the emission of an alpha particle means that the mass number of the isotope is reduced by 4, and the atomic number is lowered by 2.

- When an electron (often confusingly referred to as a "beta particle," or β particle), is emitted from a nucleus, the mass number remains unchanged because electrons have practically no mass. However, the electron is produced when a neutron (which has no charge) decays to produce the electron plus a proton. Because the nucleus now contains one more proton than before, the atomic number increases by 1. It's no longer the same element!
- Sometimes an electron is captured by the nucleus. The electron combines with a proton and forms an additional neutron. As in the last example, the mass number remains unchanged. However, because the nucleus now contains one fewer proton, the atomic number decreases by 1.

An unstable (radioactive) isotope is referred to as the *parent*, and the isotopes resulting from the decay of the parent are termed the *daughter products*. But the path from parent to daughter isn't always direct. Uranium-238, one of the most important isotopes for geologic dating, provides an example of the complexity (**Figure 11.20**). When the radioactive parent, uranium-238 (atomic number 92, mass number 238) decays, it follows a number of steps, emitting a total of 8 alpha particles and 6 electrons before finally becoming the stable daughter product lead-206 (atomic number 82, mass number 206).

Radiometric Dating

Nuclear decay provides a reliable way of calculating the ages of rocks and minerals that contain particular unstable isotopes. The procedure is called **radiometric dating**. Radiometric dating is reliable because the rates of decay for many isotopes have been precisely measured and do not vary under the physical conditions that exist in Earth's outer layers. Therefore, each unstable isotope used for dating has been decaying at a fixed rate since the formation of the mineral crystals in which we find it, and the products of its decay have been accumulating in that crystal at a corresponding rate. For example, some minerals are able to incorporate uranium atoms in their crystal lattice. When such a mineral crystallizes from magma, it contains no lead (the stable daughter product) from previous decay. The radiometric "clock" starts at this point. As the uranium in this newly formed mineral decays, atoms of the daughter product accumulate, trapped in the crystal, and eventually build up to measurable levels. Similarly, when a crystal of feldspar forms, some of the potassium atoms incorporated into its lattice will be the unstable isotope potassium-40. These atoms will decay at a steady rate by electron capture to produce the daughter argon-40. Over time, there is less and less of the parent potassium and more and more of the daughter argon.

Half-Life

The time required for half of the nuclei in a sample of a given unstable isotope to decay is called the **half-life** of that isotope. Half-life is a common way of expressing the rate of radioactive decay. **Figure 11.21** illustrates what occurs when a radioactive parent decays directly into its stable daughter product. When the quantities of parent and daughter are equal (ratio 1:1), we know that one half-life has transpired. When one-quarter of the original parent atoms remain and three-quarters have decayed to the daughter product, the parent/daughter ratio is 1:3, and we know that two half-lives have passed. After three half-lives, the ratio of parent atoms to daughter atoms is 1:7 (one parent atom for every seven daughter atoms).

If the half-life of a radioactive isotope is known and the parent/daughter ratio can be determined, the age of the sample can be calculated. For example, assume that the half-life of a hypothetical unstable isotope is 1 million years, and the parent/daughter ratio in a sample is 1:15. This ratio indicates that four half-lives have passed and that the sample must be 4 million years old.

Notice that the *percentage* of radioactive atoms that decay during one half-life is always the same: 50 percent. However, the *actual number* of atoms that decay with the passing of each half-life continually decreases. Thus, as the percentage of radioactive parent atoms declines, the proportion of stable daughter atoms rises, with the increase in daughter atoms just matching the drop in parent atoms. This fact is the key to radiometric dating.

Using Unstable Isotopes

Of the many radioactive isotopes that exist in nature, five have proved particularly useful in providing radiometric ages for ancient rocks (**Table 11.1**). Rubidium-87, thorium-232, and the two listed isotopes of uranium are used only for dating rocks that are millions of years old, but potassium-40 is more versatile. Although the half-life of potassium-40 is 1.3 billion years, analytical techniques make it possible to detect tiny amounts of its stable daughter product, argon-40, in some rocks that are younger than 100,000 years. Another important reason for its frequent use is that potassium is an abundant constituent of many common minerals, particularly micas and feldspars.

A Complex Process Although the basic principle of radiometric dating is simple, the actual procedure is quite complex. The chemical analysis that determines the quantities of parent and daughter must be painstakingly precise. In addition, some radioactive materials do not decay directly into the stable daughter product, and this fact may further complicate the analysis. In the case of uranium-238, there are 13 intermediate unstable daughter products formed before the 14th and last daughter product, the stable isotope lead-206, is produced (see Figure 11.20).

Table 11.1 Isotopes Frequently Used in Radiometric Dating of Rocks

Radioactive Parent	Stable Daughter Product	Currently Accepted Half-Life Values
Uranium-238	Lead-206	4.5 billion years
Uranium-235	Lead-207	704 million years
Thorium-232	Lead-208	14.1 billion years
Rubidium-87	Strontium-87	47.0 billion years
Potassium-40	Argon-40	1.3 billion years

Sources of Error It is important to understand that an accurate radiometric date can be obtained only if there has been no leakage of parent or daughter isotopes between the mineral crystal and its surroundings in the time since the mineral formed. This is not always the case. In fact, a limitation of the potassium-argon method arises from the fact that argon is a gas, and it may leak from minerals, resulting in a radiometric age that is lower than the actual age. Indeed, losses can be significant if the rock is subjected to high temperatures. If the rock is heated to the point where *all* of the argon in its minerals escapes, then its radiometric clock will be reset, and radiometric dating will give the time of thermal resetting, not the true age of the rock.

For other radiometric clocks, a loss of daughter atoms can occur if the rock has been subjected to weathering or leaching. To avoid such a problem, one simple safeguard is to use only fresh, unweathered material and not samples that exhibit signs of chemical alteration.

To guard against error in radiometric dating, scientists often use cross-checks, subjecting a sample to two different methods. If the results agree, the likelihood is high that the date is reliable. If the results are appreciably different, other cross-checks must be employed to determine which, if either, is correct.

Earth's Oldest Rocks Radiometric dating has produced literally thousands of dates for events in Earth history. Rocks exceeding 3.5 billion years in age are found on all of the continents. Earth's oldest rocks (so far) may be as old as 4.28 billion years (b.y.). Discovered in northern Quebec, Canada, on the shores of Hudson Bay, these rocks may be remnants of Earth's earliest crust. Rocks from western Greenland have been dated at 3.7 to 3.8 b.y., and rocks nearly as old are found in the Minnesota River valley and northern Michigan (3.5 to 3.7 b.y.), in southern Africa (3.4 to 3.5 b.y.), and in western Australia (3.4 to 3.6 b.y.). Tiny crystals of the mineral zircon having radiometric ages as old as 4.3 b.y. have been found in younger sedimentary rocks in western Australia. The source rocks for these tiny durable grains either no longer exist or have not yet been found.

Radiometric dating has vindicated the ideas of Hutton, Darwin, and others, who more than 150 years ago inferred that geologic time must be immense. Indeed, modern dating methods have proved that there has been enough time for the processes we observe to have accomplished tremendous tasks.

Dating with Carbon-14

It is possible to date some relatively recent events by using carbon-14. Carbon-14 is the radioactive isotope of carbon. The process is often called **radiocarbon dating**. Because the half-life of carbon-14 is only 5730 years, radiocarbon dating can be used for dating events from the historic past as well as those from very recent

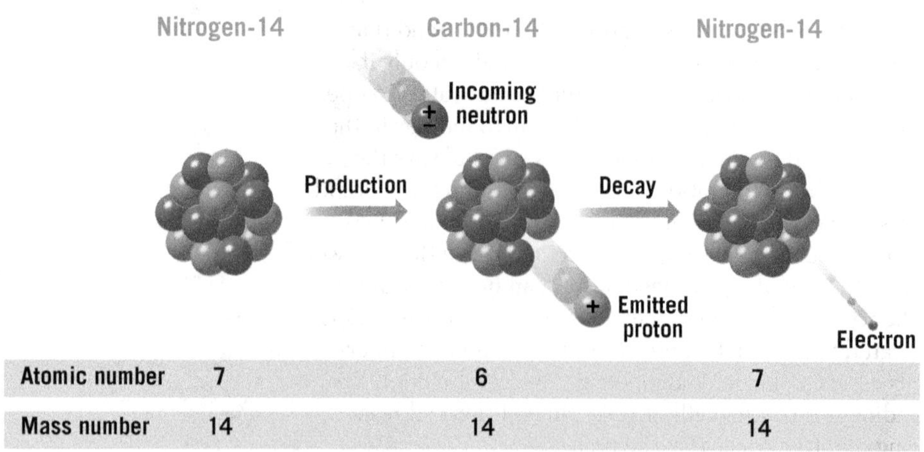

▲ Figure 11.22
Carbon-14 Production and decay of radiocarbon. These sketches represent the nuclei of the respective atoms.

geologic history. In some cases carbon-14 can be used to date events as far back as 70,000 years.

Carbon-14 (^{14}C) is continuously produced in the upper atmosphere as a result of cosmic-ray bombardment. Cosmic rays (high-energy particles) shatter the nuclei of gas atoms, releasing neutrons. Some of the neutrons are absorbed by nitrogen atoms (atomic number 7, mass number 14), causing each nucleus to emit a proton. As a result, the atomic number decreases by 1 (to 6), and a different element, carbon-14, is created (Figure 11.22). This isotope of carbon quickly becomes incorporated into carbon dioxide, which circulates in the atmosphere and is absorbed by living matter. As a result, all organisms—including you—contain a small amount of carbon-14. You "top off" your ^{14}C levels every time you eat something.

As long as an organism is alive, the decaying radiocarbon is continually replaced, and the proportions of carbon-14 and carbon-12 remain constant. Carbon-12 is the stable and most common isotope of carbon. However, when any plant or animal dies, the amount of carbon-14 gradually decreases as it decays to nitrogen-14 by beta emission. By comparing the proportions of carbon-14 and carbon-12 in a sample, radiocarbon dates can be determined. It is important to emphasize that carbon-14 can only be used to date organic materials, such as wood, charcoal, bones, flesh, and cloth.

▲ Figure 11.23 **Cave art** Chauvet Cave in southern France, discovered in 1994, contains some of the earliest-known cave paintings. Radiocarbon dating indicates that most of the images were drawn between 30,000 and 32,000 years ago. (Photo by Javier Trueba/MSF/Science Source)

Although carbon-14 is only useful in dating the last small fraction of geologic time, it is a valuable tool for anthropologists, archaeologists, and historians, as well as for geologists who study very recent Earth history (Figure 11.23). In fact, the development of radiocarbon dating was considered so important that the chemist who discovered this application, Willard F. Libby, received a Nobel Prize in 1960.

CONCEPT CHECKS 11.5

1. List three ways that unstable nuclei change. For each type, describe how the atomic number and atomic mass change.
2. Sketch a simple diagram that explains the idea of half-life.
3. Why is radiometric dating a reliable method for determining numerical dates?
4. For what time span does radiocarbon dating apply?

11.6 Determining Numerical Dates for Sedimentary Strata

Explain how reliable numerical dates are determined for layers of sedimentary rock.

Although reasonably accurate numerical dates have been worked out for the periods of the geologic time scale, the task is not without difficulty. The primary difficulty in assigning numerical dates to units of time is that not all rocks can be dated by using radiometric methods. For a radiometric date to be useful, all the minerals in the rock must have formed at about the same time. For this reason, unstable isotopes can be used to determine when minerals in an igneous rock crystallized and when pressure and heat created new minerals in a metamorphic rock.

However, samples of sedimentary rock can only rarely be dated directly by radiometric means. Although a detrital sedimentary rock may include particles that contain unstable isotopes, the rock's age cannot be accurately determined because the grains composing the rock are not the same age as the rock in which they occur. Rather, the sediments have been weathered from rocks of diverse ages.

Radiometric dates obtained from metamorphic rocks may also be difficult to interpret because the age of a

EYE ON EARTH 11.3

This is a close-up view of the detrital sedimentary rock conglomerate. This rock contains radioactive isotopes that will yield numerical dates.

QUESTION 1 *Although radioactive isotopes are present, a reliable numerical date for this conglomerate cannot be accurately determined. Explain.*

QUESTION 2 *How might a numerical age range be established for the conglomerate layer?*

particular mineral in a metamorphic rock does not necessarily represent the time when the rock initially formed. Instead, the date might indicate any one of a number of subsequent metamorphic phases.

If samples of sedimentary rocks rarely yield reliable radiometric ages, how can numerical dates be assigned to sedimentary layers? Usually geologists must relate the strata to datable igneous masses, as in Figure 11.24. In this example, radiometric dating has determined the ages of the volcanic ash bed in the Morrison Formation and the dike cutting the Mancos Shale and Mesaverde Formation. The sedimentary beds below the ash are obviously older than the ash, and all the layers above the ash are younger (based on the principle of superposition). The dike is younger than the Mancos Shale and the Mesaverde Formation but older than the Wasatch Formation because the dike does not intrude this topmost layer (based on the principle of cross-cutting relationships).

From this kind of evidence, geologists estimate that the Morrison Formation was deposited more than 160 million years ago, as indicated by the ash bed. Further, they conclude that deposition of the Wasatch Formation began after the intrusion of the dike, 66 million years ago. This is one example of literally thousands that illustrate how datable materials are used to "bracket" the various episodes in Earth history within specific time periods. It shows the necessity of combining laboratory dating methods with field observations of rocks.

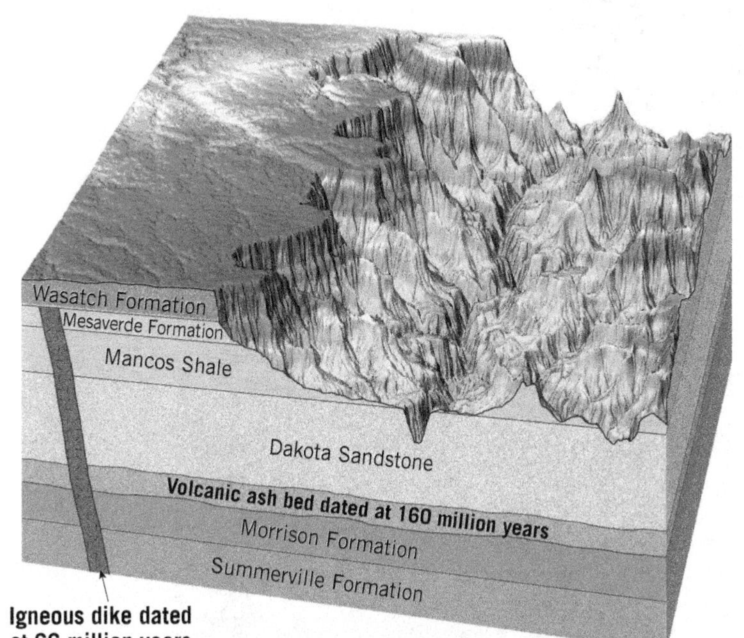

▲ **Figure 11.24 Dating sedimentary strata** Numerical dates for sedimentary layers are usually determined by examining their relationship to igneous rocks.

CONCEPT CHECKS 11.6

1. Briefly explain why it is often difficult to assign a reliable numerical date to a sample of sedimentary rock.
2. How might a numerical date for a layer of sedimentary rock be determined?

11.7 The Geologic Time Scale

Distinguish among the four basic time units that make up the geologic time scale and explain why the time scale is considered to be a dynamic tool.

Geologists have divided the whole of geologic history into units of varying length. Together, they compose the **geologic time scale** of Earth history (Figure 11.25). The major units of the time scale were delineated during the nineteenth century, principally by scientists in Western Europe. Because radiometric dating was unavailable at that time, the entire time scale was created using methods of relative dating. It was only in the twentieth century that radiometric methods permitted numerical dates to be added.

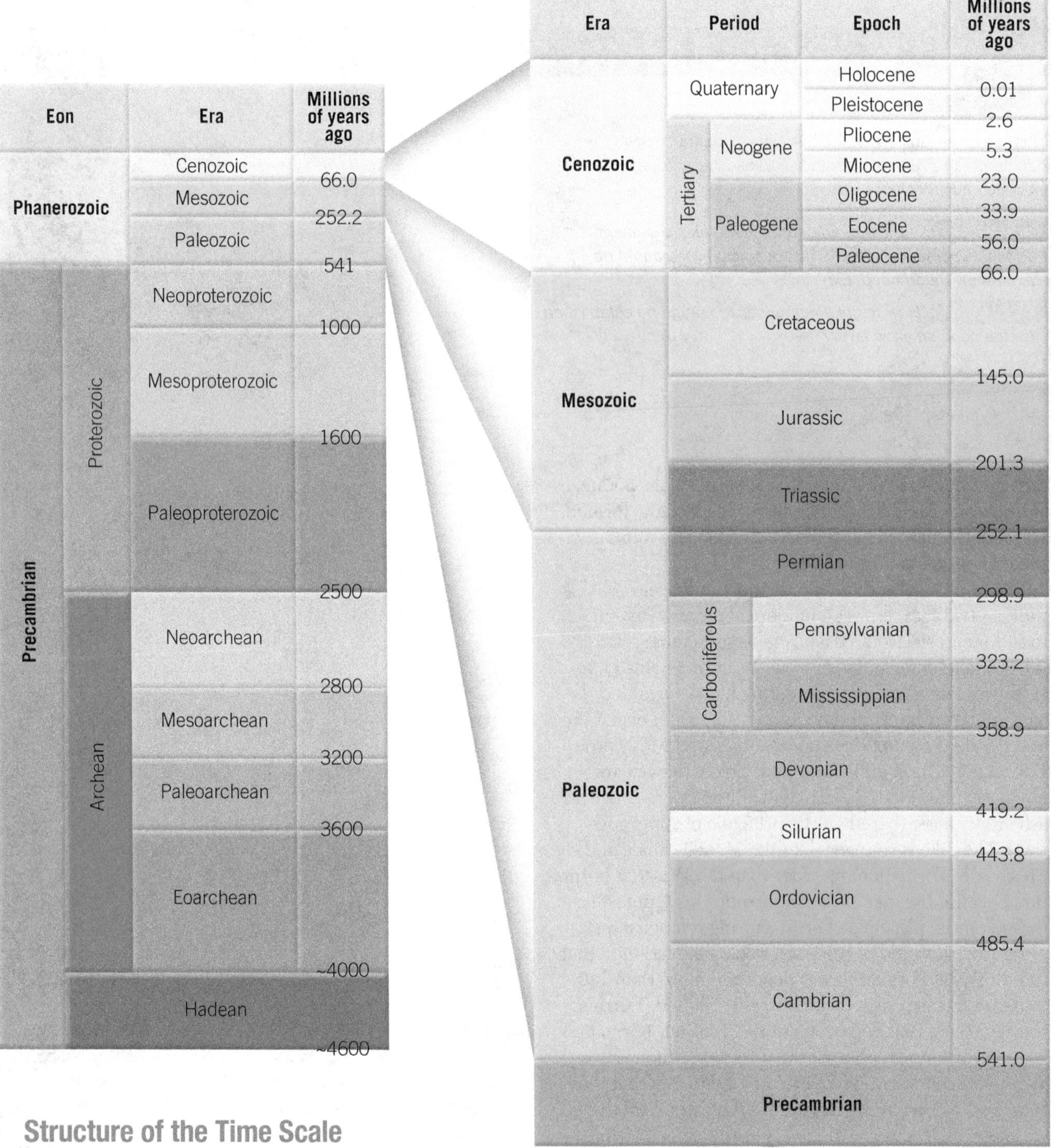

Eon
Era
Millions of years ago
Phanerozoic
Cenozoic
66.0
Mesozoic
252.2
Paleozoic
541
Precambrian
Proterozoic
Neoproterozoic
1000
Mesoproterozoic
1600
Paleoproterozoic
2500
Archean
Neoarchean
2800
Mesoarchean
3200
Paleoarchean
3600
Eoarchean
~4000
Hadean
~4600
Era
Period
Epoch
Millions of years ago
Cenozoic
Quaternary
Holocene
0.01
Pleistocene
2.6
Tertiary
Neogene
Pliocene
5.3
Miocene
23.0
Paleogene
Oligocene
33.9
Eocene
56.0
Paleocene
66.0
Mesozoic
Cretaceous
145.0
Jurassic
201.3
Triassic
252.1
Paleozoic
Permian
298.9
Carboniferous
Pennsylvanian
323.2
Mississippian
358.9
Devonian
419.2
Silurian
443.8
Ordovician
485.4
Cambrian
541.0
Precambrian

Figure 11.25 Geologic time scale: A basic reference The time scale divides the vast 4.6-billion-year history of Earth into eons, eras, periods, and epochs. Numbers on the time scale represent time in millions of years before the present. The Precambrian accounts for more than 88 percent of geologic time. Numerical dates were added long after the time scale was established using relative dating techniques. The dates appearing on this time scale are those currently accepted by the International Commission on Stratigraphy (ICS) in 2015. The color scheme used on this chart was selected because it is similar to that used by the ICS.

Structure of the Time Scale

The geologic time scale subdivides the 4.6-billion-year history of Earth into many different units and provides a meaningful time frame within which the events of the geologic past are arranged. As shown in Figure 11.25, **eons** represent the greatest expanses of time. The eon that began about 541 million years ago is the **Phanerozoic**, a term derived from Greek words meaning "visible life." It is an appropriate description because the rocks and deposits of the Phanerozoic eon contain abundant fossils that document major evolutionary trends.

Another glance at the time scale reveals that eons are divided into **eras**. The Phanerozoic eon consists of the **Paleozoic era** (*paleo* = ancient, *zoe* = life), the **Mesozoic era** (*meso* = middle, *zoe* = life), and the **Cenozoic era** (*ceno* = recent, *zoe* = life). As the names imply, these eras are bounded by profound worldwide changes in life-forms.*

Each era of the Phanerozoic eon is further divided into time units known as **periods**. The Paleozoic has seven, and the Mesozoic and Cenozoic each have three. Each of these periods is characterized by a somewhat less profound change in life-forms compared with the eras.

* Major changes in life-forms are discussed in Chapter 12.

Each of the periods is divided into still smaller units called **epochs**. As you can see in Figure 11.25, seven epochs have been named for the periods of the Cenozoic. The epochs of other periods usually are simply termed *early*, *middle*, and *late*.

Precambrian Time

Notice that the geologic time scale is considerably less detailed prior to the beginning of the Cambrian period, 541 million years ago. The nearly 4 billion years that precede the Cambrian are divided into two eons, the **Archean** (*archaios* = ancient) and the **Proterozoic** (*proteros* = before, *zoe* = life). It is also common for this vast expanse of time to simply be referred to as the **Precambrian**.

Why is the huge expanse of Precambrian time, which represents about 88 percent of Earth history, not divided into numerous eras, periods, and epochs? The reason is that Precambrian history is not known in great enough detail. In geology, as in human history, the farther back we go, the less we know. We know much more about the past decade than we know about the first century C.E. Similarly, in Earth history, the more recent past has the freshest, least disturbed, and most observable record. The further back in time a geologist goes, the more fragmented the record and clues become. There are also other reasons to explain our lack of a detailed time scale for this vast segment of Earth history:

- The first abundant fossil evidence does not appear in the geologic record until the beginning of the Cambrian period. Prior to the Cambrian, simple life-forms such as algae, bacteria, fungi, and worms predominated. All of these organisms lack hard parts, an important condition favoring preservation. For this reason, there is only a relatively meager Precambrian fossil record. Many exposures of Precambrian rocks have been studied in some detail, but correlation is often difficult when fossils are lacking.
- Because Precambrian rocks are very old, most have been subjected to a great many changes. Much of the Precambrian rock record is composed of highly distorted metamorphic rocks. This makes the interpretation of past environments difficult because many of the clues present in the original sedimentary rocks have been destroyed.

Radiometric dating has provided a partial solution to the troublesome task of dating and correlating Precambrian rocks. But untangling the complex Precambrian record still remains a daunting task.

Terminology and the Geologic Time Scale

Some terms are associated with the geologic time scale but are not officially recognized as being a part of it. The best known, and most common, example is *Precambrian*—the informal name for the eons that came before the current Phanerozoic eon. Although the term *Precambrian* has no formal status on the geologic time scale, it has been traditionally used as though it does.

Hadean is another informal term that is found on some versions of the geologic time scale and is used by many geologists. It refers to the earliest interval (eon) of Earth history—before the oldest-known rocks. When the term was coined in 1972, the age of Earth's oldest-known rocks was about 3.8 billion years. Today that number stands at slightly greater than 4 billion, and, of course, is subject to revision. The name *Hadean* derives from *Hades*, Greek for "underworld"—a reference to the "hellish" conditions that prevailed on Earth early in its history.

Effective communication in the geosciences requires that the geologic time scale consist of standardized divisions and dates. So, who determines which names and dates on the geologic time scale are "official"? The organization that is largely responsible for maintaining and updating this important document is the International Commission on Stratigraphy (ICS), a committee of the International Union of Geological Sciences. Advances in the geosciences require that the scale be periodically updated to include changes in unit names and boundary age estimates.

For example, the geologic time scale shown in Figure 11.25 was updated in 2015. After considerable dialogue among geologists who focus on very recent Earth history, the ICS changed the date for the start of the Quaternary period and the Pleistocene epoch from 1.8 million to 2.6 million years ago. Perhaps by the time you read this, other changes will have been made.

If you were to examine a geologic time scale from just a few years ago, it is quite possible that you would see the Cenozoic era divided into the Tertiary and Quaternary periods. However, on more recent versions, the space formerly designated as Tertiary is divided into the Paleogene and Neogene periods. As our understanding of this time span has changed, so too has its designation on the geologic time scale. Today, the Tertiary period is considered a "historic" name and is given no official status on the ICS version of the time scale. Many time scales still contain references to the Tertiary period, though, including Figure 11.25. One reason for this is that a great deal of past (and some current) geologic literature uses this name.

For those who study historical geology, it is important to realize that the geologic time scale is a dynamic tool that continues to be refined as our knowledge and understanding of Earth history evolve.

CONCEPT CHECKS 11.7

1. List the four basic units that make up the geologic time scale.
2. Why is *zoic* part of so many names on the geologic time scale?
3. What term applies to *all* of geologic time prior to the Phanerozoic eon? Why is this span *not* divided into as many smaller time units as the Phanerozoic eon?
4. To what does the term *Hadean* apply? Is it an "official" part of the geologic time scale?

CONCEPTS IN REVIEW
Geologic Time

11.1 A Brief History of Geology

Explain the principle of uniformitarianism and discuss how it differs from catastrophism.

KEY TERMS: catastrophism, uniformitarianism

- Early ideas about the nature of Earth were based on religious traditions and notions of great catastrophes.
- In the late 1700s, James Hutton emphasized that the same slow processes have acted over great spans of time and are responsible for Earth's rocks, mountains, and landforms. This similarity of processes over vast spans of time led to this principle being called uniformitarianism.

11.2 Creating a Time Scale: Relative Dating Principles

Distinguish between numerical and relative dating and apply relative dating principles to determine a time sequence of geologic events.

KEY TERMS: numerical date, relative date, principle of superposition, principle of original horizontality, principle of lateral continuity, principle of cross-cutting relationships, principle of inclusions, conformable, unconformity, angular unconformity, disconformity, nonconformity

- The two types of dates that geologists use to interpret Earth history are (1) relative dates, which put events in their proper sequence of formation, and (2) numerical dates, which pinpoint the time in years when an event took place.
- Relative dates can be established using the principles of superposition, original horizontality, cross-cutting relationships, and inclusions. Unconformities, gaps in the geologic record, may be identified during the relative dating process.

? The accompanying diagram is a cross section of a hypothetical area. Place the lettered features in the proper sequence, from oldest to youngest. Where in the sequence can you identify an unconformity? Which principles did you use to establish the sequence?

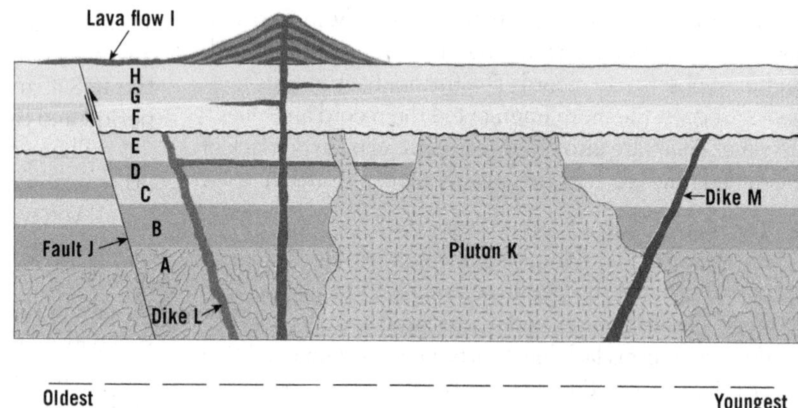

11.3 Fossils: Evidence of Past Life

Define *fossil* and discuss the conditions that favor the preservation of organisms as fossils. List and describe various types of fossils.

KEY TERMS: fossil, paleontology

- Fossils are remains or traces of ancient life. Paleontology is the branch of science that studies fossils.
- Fossils can form through many processes. For an organism to be preserved as a fossil, it usually needs to be buried rapidly. Also, an organism's hard parts are most likely to be preserved because soft tissue decomposes rapidly in most circumstances.

? What term is used to describe the type of fossil that is shown here? Briefly describe how it formed.

E.J. Tarbuck

11.4 Correlation of Rock Layers

Explain how rocks of similar age that are in different places can be matched up.

KEY TERMS: correlation, principle of fossil succession, index fossil, fossil assemblage

- Matching up exposures of rock that are the same age but are in different places is called correlation. By correlating rocks from around the world, geologists developed the geologic time scale and obtained a fuller perspective on Earth history.
- Fossils can be used to correlate sedimentary rocks in widely separated places by using the rocks' distinctive fossil content and applying the principle of fossil succession. This principle states that fossil organisms succeed one another in a definite and determinable order, and, therefore, a time period can be recognized by examining its fossil content.
- Index fossils are particularly useful in correlation because they are widespread and associated with a relatively narrow time span. The overlapping ranges of fossils in an assemblage may be used to establish an age for a rock layer that contains multiple fossils.
- Fossils may be used to establish ancient environmental conditions that existed when sediment was deposited.

11.5 Numerical Dating with Nuclear Decay

Discuss three ways that atomic nuclei change and explain how unstable isotopes are used to determine numerical dates.

KEY TERMS: nuclear, (radioactive) decay, radiometric dating, half-life, radiocarbon dating

- Nuclear decay is the spontaneous breaking apart of certain unstable atomic nuclei. Three common forms of nuclear decay are (1) emission of an alpha particle from the nucleus, (2) emission of a beta particle (electron) from the nucleus, and (3) capture of an electron by the nucleus.
- Radiometric dating refers to the procedure by which unstable isotopes are used to determine numerical ages of rocks and minerals. It is reliable because the rates of decay for the isotopes that are used have been precisely measured and do not vary.
- The length of time it takes for one-half of the nuclei of an unstable parent isotope to change into its stable daughter product is called the half-life of that isotope. If the half-life is known, and the parent/daughter ratio can be measured, the age of the sample can be calculated.

? Measurements of zircon crystals containing trace amounts of uranium from a specimen of granite yield parent/daughter ratios of 25 percent parent (uranium-235) and 75 percent daughter (lead-206). The half-life of uranium-235 is 704 million years. How old is the granite?

11.6 Determining Numerical Dates for Sedimentary Strata

Explain how reliable numerical dates are determined for layers of sedimentary rock.

- Sedimentary strata are usually not directly datable using radiometric techniques because they consist of the material produced by the weathering of other rocks. A particle in a sedimentary rock comes from some older source rock. If you were to date the particle using unstable isotopes, you would get the age of the source rock, not the age of the sedimentary rock.
- One way geologists assign numerical dates to sedimentary rocks is to use relative dating principles to relate them to datable igneous masses, such as dikes and volcanic ash beds. A layer may be older than one igneous feature and younger than another.

? Express the numerical age of the sandstone layer in the diagram as accurately as possible.

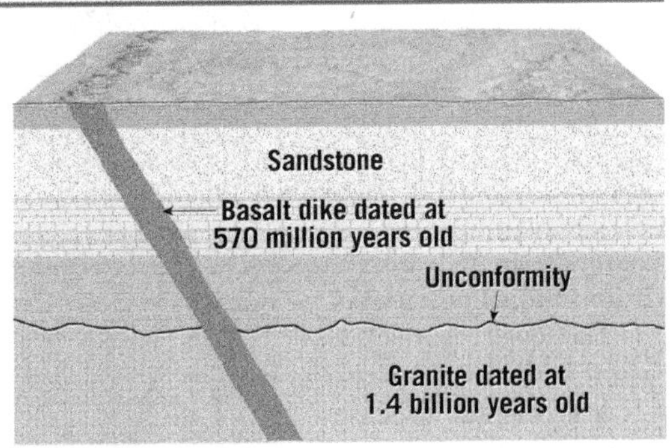

11.7 The Geologic Time Scale

Distinguish among the four basic time units that make up the geologic time scale and explain why the time scale is considered to be a dynamic tool.

KEY TERMS: geologic time scale, eon, Phanerozoic eon, era, Paleozoic era, Mesozoic era, Cenozoic era, period, epoch, Archean, Proterozoic, Precambrian

- Earth history is divided into units of time on the geologic time scale. Eons are divided into eras, which each contain multiple periods. Periods are divided into epochs.
- Precambrian time includes the Archean and Proterozoic eons. It is followed by the Phanerozoic eon, which is well documented by abundant fossil evidence, resulting in many subdivisions.
- The geologic time scale is a work in progress, continually being refined as new information becomes available.

? Is the Mesozoic an example of an eon, an era, a period, or an epoch? What about the Jurassic?

GIVE IT SOME THOUGHT

1. The accompanying image shows the metamorphic rock gneiss, a basaltic dike, and a fault. Place these three features in their proper sequence (which came first, second, and third) and explain your logic.

Marli Miller

2. A mass of granite is in contact with a layer of sandstone. Using a principle described in this chapter, explain how you might determine whether the sandstone was deposited on top of the granite or whether the magma that formed the granite was intruded after the sandstone was deposited.

3. A hypothetical unstable isotope has a half-life of 10,000 years. If the ratio of radioactive parent to stable daughter product is 1:3, how old is the rock that contains the radioactive material?

4. Solve the problems below that relate to the magnitude of Earth history. To make calculations easier, round Earth's age to 5 billion years.
   - **a.** What percentage of geologic time is represented by recorded history? (Assume 5000 years for the length of recorded history.)
   - **b.** Humanlike ancestors (hominins) have been around for roughly 5 million years. What percentage of geologic time is represented by these ancestors?
   - **c.** The first abundant fossil evidence for multicellular organisms does not appear until the beginning of the Cambrian period, about 540 million years ago. What percentage of geologic time is represented by this abundant fossil evidence?

5 This scenic image is from Monument Valley in the northeastern corner of Arizona. The bedrock in this region consists of layers of sedimentary rocks. Although the prominent rock exposures ("monuments") in this photo are widely separated, we can infer that they represent a once-continuous layer. Discuss the principle that allows us to make this inference.

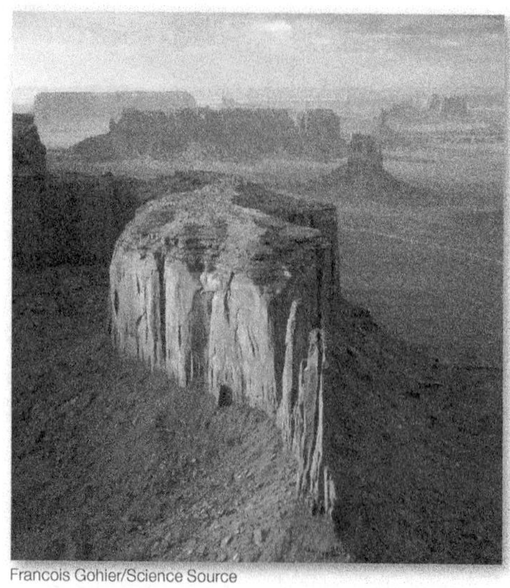
Francois Gohier/Science Source

6 The accompanying photo shows two layers of sedimentary rock. The lower layer is shale from the late Mesozoic era. Note the old river channel that was carved into the shale after it was deposited. Above is a younger layer of boulder-rich breccia. Are these layers conformable? Explain why or why not. What term from relative dating applies to the line separating the two layers?

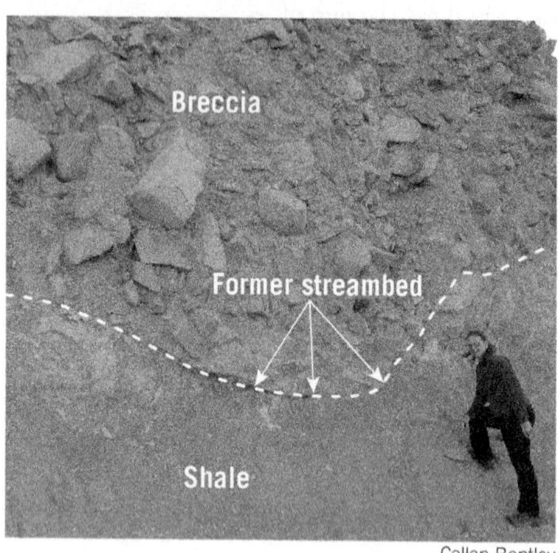

Callan Bentley

7 If an unstable isotope of thorium (atomic number 90, mass number 232) emits 6 alpha particles and 4 beta particles during the course of radioactive decay, what are the atomic number and mass number of the stable daughter product?

8 A portion of a popular college text in historical geology includes 10 chapters (281 pages) in a unit titled "The Story of Earth." Two chapters (49 pages) are devoted to Precambrian time. By contrast, the last two chapters (67 pages) focus on the most recent 23 million years, with 25 of those pages devoted to the Holocene Epoch, which began 10,000 years ago.

- **a.** Compare the percentage of pages devoted to the Precambrian to the percentage of geologic time that this span represents.
- **b.** How does the number of pages about the Holocene compare to its percentage of geologic time?
- **c.** Suggest some reasons the text seems to have such an unequal treatment of Earth history.

9 These polished stones are called *gastroliths*. Explain how such objects can be considered fossils. What category of fossil are they? Name another example of a fossil in this category.

Francois Gohier/Photo Researchers, Inc.

EXAMINING THE EARTH SYSTEM

1 The accompanying photo shows a large petrified log in Arizona's Petrified Forest National Park. Describe the transition of this tree from being part of the biosphere to being a component of the geosphere. How might the hydrosphere and/or atmosphere have played a role in the transition?

Buddy Mays/Alamy

2 The famous angular unconformity at Scotland's Siccar Point shown in this photo was originally studied by James Hutton in the late 1700s.

- **a.** Describe in a general way what occurred to produce this feature.
- **b.** Suggest ways in which all four spheres of the Earth system could have been involved in producing Siccar Point.
- **c.** The Earth system is powered by energy from two sources. How are both sources represented in the Siccar Point unconformity?

Marli Miller

3 This scene in Montana's Glacier National Park shows layers of Precambrian sedimentary rocks. The darker layer contained within the sedimentary layers is igneous. The narrow, light-colored areas adjacent to the igneous rock were created when molten material that formed the igneous rock baked the adjacent rock.

Marli Miller

a. Is the igneous layer more likely a lava flow that was laid down at the surface prior to the deposition of the layers above it or a sill that was intruded after all the sedimentary layers were deposited? Explain.

b. Is it likely that the igneous layer will exhibit a vesicular texture? Explain.

c. To which group (igneous, sedimentary, or metamorphic) does the light-colored rock belong? Relate your explanation to the rock cycle.

DATA ANALYSIS

Fossils and Geologic Time

Fossils are extremely useful for stratigraphy. In the Phanerozoic, time unit boundaries typically represent widespread changes in fossil assemblages. Often, these changes represent mass extinctions.

ACTIVITIES

Go to the Paleobiology Database at https://paleobiodb.org/ and click Explore. Click Help to learn how to navigate this page. Click X to exit the Help page. Click on the taxa browser (image of an insect) and type Life in the search box. Click "Life (unranked)" to see the location of all fossils.

1 What type of organism is most prevalent in this record? What percent of all listed fossils does it represent?

2 What percent of the listed fossils are mammals (Mammalia)?

3 What is the total number of occurrences displayed? What is the total number of collections (dig sites)?

4 One explanation for the prevalence of the organism you named for Question 1 is that it was the dominant species at the time. What might be another explanation?

Foraminifera (forams) are small marine organisms that are used widely for stratigraphy. In the taxa browser search box, type Foraminifera, and click Foraminifera (phylum). Click on the Stats button (image of a line graph) to display the diversity of foram fossils over time. Mouse over the colored blocks to see the name of the time period.

5 Identify points on this graph that could represent a mass extinction among forams, followed by the appearance of new types? Would such a feature make a good stratigraphic marker?

6 Identify three places where a mass extinction occurs at the boundary between geologic time units.

Double-click on the black Geologic Time band at the bottom to see the full range of geologic time. Double-click on a time unit to zoom and to see numerical ages. Mouse over a time unit to see its full name.

7 Single-click successively on the Mesoproterozoic, Neoproterozoic, and Paleozoic eras. Judging from these data, when do fossils first become abundant?

8 Why is the geologic time scale much more finely divided in the Phanerozoic than in earlier eons?

9 Within the Phanerozoic, roughly how long does a typical period last? What are the longest and shortest periods?

12 Earth's Evolution Through Geologic Time

FOCUS ON CONCEPTS

Each statement represents the primary learning objective for the corresponding major heading within the chapter. After you complete the chapter, you should be able to:

12.1 List the principal characteristics that make Earth habitable.

12.2 Outline the major stages in Earth's evolution, from the Big Bang to the formation of our planet's layered internal structure.

12.3 Describe how Earth's atmosphere and oceans formed and evolved through time.

12.4 Explain the formation of continental crust, how continental crust becomes assembled into continents, and the role that the supercontinent cycle has played in this process.

12.5 List and discuss the major geologic events in the Paleozoic, Mesozoic, and Cenozoic eras.

12.6 Describe some of the hypotheses on the origin of life and the characteristics of early prokaryotes, eukaryotes, and multicellular organisms.

12.7 List the major developments in the history of life during the Paleozoic era.

12.8 Briefly explain the major developments in the history of life during the Mesozoic era.

12.9 Discuss the major developments in the history of life during the Cenozoic era.

Multiple views of Earth as seen from space. (Photos provided by NASA)

EARTH HAS A LONG AND COMPLEX HISTORY. Time and again, the splitting and colliding of continents has led to the formation of new ocean basins, the creation of great mountain ranges, and the evolution of Earth's continents. Furthermore, the degree to which life on our planet has evolved cannot be overstated. In this chapter, we will first explore the characteristics that make Earth habitable and then briefly examine Earth's physical and biological evolution though geologic time.

12.1 What Makes Earth Habitable?

List the principal characteristics that make Earth habitable.

Not since Galileo first pointed his telescope toward the sky has there been as much excitement and popular interest as there is today in the discovery of planets that orbit other stars, called **exoplanets**. Much of this interest undoubtedly stems from human curiosity—more specifically, our desire to know if we are alone in the universe.

The discoveries of most of the exoplanets in recent years have been made using NASA's *Kepler* telescope (Figure 12.1). As of 2016, this space telescope had identified more than 2000 exoplanets and numerous planetary candidates. One goal of the *Kepler* mission is to survey nearby sections of the Milky Way in search of Earth-size planets orbiting in or near the habitable zone of planetary systems. The **habitable zone** can be broadly defined as the region around a host star where a planet with sufficient atmospheric pressure can maintain liquid water on its surface.

Figure 12.1 shows some of the small exoplanets recently discovered using the *Kepler* space telescope. All of these planets are located in the habitable zone of one of three types of stars. G-type stars are Sun-like, yellow stars having masses about 0.8 to 1.2 times the mass of the Sun, with surface temperatures of about 5800°C (about 10,000°F). Somewhat smaller, the K-type stars are orange stars with masses between 0.45 and 0.8 that of the Sun and surface temperatures of about 4000°C (about 7000°F). These stars are thought to be likely candidates to host habitable planets because they have longer life spans than Sun-like stars and emit less ultraviolet radiation, which is harmful to an organism's DNA. M-type stars are reddish in color and less than half the size of the Sun. For various reasons, they were once thought to be unlikely candidates to host habitable planets. However, because they are by far the most common type of star, they might actually be a common home for life.

▼ **Figure 12.1 Small exoplanets found in the habitable zones of their host star** (Photo courtesy of NASA)

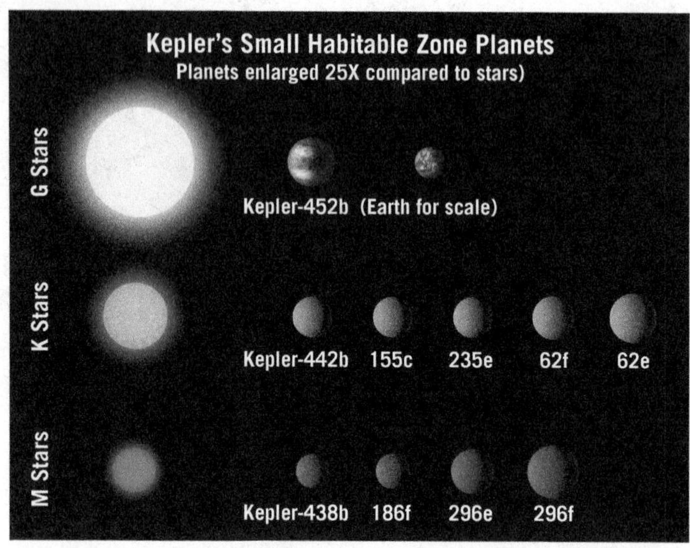

Astronomers generally consider habitable planets to be *terrestrial planets* (Earth-like planets) that orbit within the habitable zone, with conditions roughly comparable to those of Earth and, thus, favorable to supporting Earth-like life. The planet Venus offers a sobering view of the importance of the habitability zone. In terms of size, mass, and composition, Venus is nearly Earth's twin. However, it orbits a little closer to the Sun than does Earth—and its atmosphere has evolved into a thick blanket that keeps temperatures at the surface hot enough to melt lead and, therefore, most certainly too hot to sustain life.

What fortuitous events produced Earth, a planet so hospitable to life? Earth was not always as we find it today. During its formative years, our planet supported a magma ocean. It also survived a several-hundred-million-year period of extreme bombardment by asteroids. The oxygen-rich atmosphere that makes many modern life-forms possible developed long after the solar system was born. Serendipitously, Earth seems to be the right planet, in the right location, at the right time.

The Right Planet

What are some of the characteristics that make Earth unique among the planets of our solar system? Consider the following:

- If Earth were considerably larger (more massive), its force of gravity would be proportionately greater. Like the giant planets, Earth might have retained a thick, hostile atmosphere consisting of ammonia and methane, and possibly hydrogen and helium.
- If Earth were significantly smaller, oxygen, water vapor, and other volatiles would escape to space and be lost forever. Thus, like the Moon and Mercury, both of which lack appreciable atmospheres, Earth would be devoid of life.
- If Earth did not have a rigid lithosphere overlaying a weak asthenosphere, plate tectonics would not operate. The continental crust (Earth's "highlands") would not have formed without the recycling of plates. Consequently, the entire planet would likely be covered

by an ocean a few kilometers deep. As author Bill Bryson so aptly stated, "There might be life in that lonesome ocean, but there certainly wouldn't be baseball."*

- Most surprising, perhaps, is the fact that if our planet did not have a molten metallic outer core, most of the life-forms on Earth would not exist. Fundamentally, without the flow of iron in the core, Earth could not support a magnetic field. It is the magnetic field that prevents lethal cosmic rays from showering Earth's surface and stripping away our atmosphere.

The Right Location

A primary factor that determines whether a planet is habitable is its location with respect to its host star. The following scenarios substantiate Earth's advantageous position:

- If Earth were about 10 percent closer to the Sun, our atmosphere would be more like that of Venus and consist mainly of the greenhouse gas carbon dioxide. Earth's surface temperature would then be too hot to support higher life-forms.
- If Earth were about 10 percent farther from the Sun, the problem would be reversed: It would be too cold. The oceans would freeze over, and Earth's active water cycle would not exist. Without liquid water, all life would perish.
- Earth is near a star of modest size. Stars like the Sun have a life span of roughly 10 billion years and emit radiant energy at a fairly constant level during most of this time. Giant stars, on the other hand, consume their nuclear fuel at very high rates and "burn out" in a few hundred million years. Therefore, Earth's proximity to a modest-sized star allowed enough time for the evolution of humans, who first appeared on this planet only a few million years ago.

The Right Time

The last, but certainly not the least, fortuitous factor for Earth is timing. The first organisms to inhabit Earth came into existence roughly 3.8 billion years ago. From that point in Earth's history, innumerable changes occurred: Life-forms came and went, and the physical environment of our planet was transformed in many ways. Consider two of the many timely Earth-altering events:

- Earth's atmosphere has developed over time. Earth's primitive atmosphere is thought to have been composed mostly of nitrogen, water vapor, methane, and carbon dioxide—but no free oxygen, that is, oxygen not combined with other elements. Fortunately, microorganisms evolved that released oxygen into the atmosphere through the process of *photosynthesis*. About 2.5 billion years ago, an atmosphere with free oxygen came into existence. The result was the evolution of the ancestors of the vast array of multicellular organisms that we find on Earth today.
- About 66 million years ago, our planet was struck by an asteroid 10 kilometers (6 miles) in diameter. This impact contributed to a mass extinction that obliterated nearly three-quarters of all plant and animal species—including dinosaurs (Figure 12.2).† Although it may not seem lucky, the extinction of dinosaurs opened new habitats for small mammals that survived the impact. These habitats, along with evolutionary forces, led to the development of the many large mammals that occupy our modern world. Without this event, mammals might have remained mostly small and inconspicuous.

As various observers have noted, Earth developed under "just right" conditions to support higher life-forms. Astronomers refer to this as the *Goldilocks scenario*. Like the classic "Goldilocks and the Three Bears" fable, Venus is too hot (Papa Bear's porridge), and Mars is too cold (Mama Bear's porridge), but Earth is just right (Baby Bear's porridge).

Viewing Earth's History

The remainder of this chapter focuses on the origin and evolution of planet Earth—the one place in the universe we know fosters life. As you learned in Chapter 11,

A Short History of Nearly Everything (Broadway Books, 2003).

† We use the term *dinosaurs* here to refer to all members of this group except birds.

Figure 12.2 **Digging into the past** An asteroid impact about 66 million years ago contributed to the extinction of the dinosaurs, which opened new habitats for mammals that survived the event. (Photo by Kristi Curry Rogers/ Macalester College)

Figure 12.3 The geologic time scale Numbers represent time in millions of years before the present. The Precambrian accounts for about 88 percent of geologic time.

Relative Time Span		Boundary (millions of years ago)
Phanerozoic	Cenozoic	
	Mesozoic	
	Paleozoic	
Precambrian	Proterozoic	2500
	Archean	~4000
	Hadean*	~4600

Era	Period		Epoch	Millions of years ago	Development of Plants and Animals
Cenozoic	Quaternary		Holocene	0.01	Humans develop
			Pleistocene	2.6	
	Tertiary	Neogene	Pliocene	5.3	
			Miocene	23.0	Large mammals flourish
		Paleogene	Oligocene	33.9	
			Eocene	56.0	
			Paleocene	66.0	Extinction of dinosaurs and many other species
Mesozoic	Cretaceous			145.0	First known flowering plants
	Jurassic			201.3	First known birds
	Triassic			252.1	Dinosaurs flourish First known mammals
Paleozoic	Permian			298.9	Extinction of trilobites and many other marine animals
	Carboniferous	Pennsylvanian		323.2	First known reptiles Large coal swamps
		Mississippian		358.9	Amphibians abundant
	Devonian			419.2	First insect fossils Fishes dominant
	Silurian			443.8	First land plants
	Ordovician			485.4	First known fishes Cephalopods abundant
	Cambrian			541.0	Trilobites abundant First organisms with shells
Precambrian				~4600	First multicelled organisms First one-celled organisms Origin of Earth

* Hadean is the informal name for the span that begins at Earth's formation and ends with Earth's earliest-known rocks.

researchers utilize many tools to interpret clues about Earth's past. Using these tools, as well as clues contained in the rock record, scientists continue to unravel many complex events of the geologic past. This chapter provides a brief overview of the history of our planet and its life-forms—a journey that takes us back about 4.6 billion years, to the formation of Earth. Later, we will consider how our physical world assumed its present state and how Earth's inhabitants changed through time. As you read this chapter, refer to the *geologic time scale* presented in Figure 12.3.

CONCEPT CHECKS 12.1

1. Explain why Earth is just the right size.
2. In what way does Earth's molten, metallic core help protect Earth's life-forms?
3. Why is Earth's location in the solar system ideal for the development of complex life-forms like humans?

12.2 Birth of a Planet

Outline the major stages in Earth's evolution, from the Big Bang to the formation of our planet's layered internal structure.

The universe began about 13.8 billion years ago with the *Big Bang*, when all matter and space came into existence. Shortly thereafter, the two simplest elements, hydrogen and helium, formed. These basic elements were the ingredients for the first star systems. Several billion years later, our home galaxy, the Milky Way, came into existence. It was within a band of stars and nebular debris in an arm of this spiral galaxy that the Sun and planets took form nearly 4.6 billion years ago.

From the Big Bang to Heavy Elements

One of the products of the Big Bang was an array of subatomic particles, including protons, neutrons, and electrons (Figure 12.4). Later, as the debris cooled, these subatomic particles combined to generate atoms of hydrogen and helium, the two lightest elements. Within a few hundred million years of the Big Bang, the primordial hydrogen and helium had condensed and coalesced to form the first stars and galaxies (see Figure 12.4C). Within the cores of these early stars, heating triggered the process of *nuclear fusion*, in which hydrogen nuclei combine to form helium nuclei, releasing enormous amounts of radiant energy (heat, light, and cosmic rays). As these stars aged and died (in some cases via cataclysmic **supernova** explosions), other nuclear reactions generated all the elements on the periodic table, seeding them into interstellar space. It is from this material, as well as primordial hydrogen and helium, that our Sun and solar system formed. Based on this scenario, all the atoms in your body except for hydrogen were produced billions of years ago, in now-defunct stars, and the gold in your jewelry was produced mainly during supernova explosions.

From Planetesimals to Protoplanets

Recall that the solar system, including Earth, formed about 4.6 billion years ago from the **solar nebula**, a large rotating cloud of interstellar dust and gas (see Figure 12.4E). As the solar nebula contracted, most of the matter collected in the center to create the hot *protosun*. The remaining materials formed a thick, flattened, rotating disk, within which matter gradually cooled and condensed into grains and clumps of icy, rocky, and metallic material. Repeated collisions resulted in most of this solid material eventually collecting into asteroid-sized objects called **planetesimals**.

The composition of planetesimals was largely determined by their proximity to the protosun. As you might expect, temperatures were highest in the inner solar system and decreased toward the outer edge of the disk. Therefore, between the present orbits of Mercury and Mars, the planetesimals were composed mainly of materials with high melting temperatures—metals and rocky substances. The planetesimals that formed beyond the orbit of Mars, where temperatures are low, contained high percentages of ices—water, carbon dioxide, ammonia, and methane—as well as smaller amounts of rocky and metallic debris.

Through repeated collisions and accretion (sticking together), these planetesimals grew into eight **protoplanets**, as well as dwarf planets and some larger moons (see Figure 12.4G). During this process, matter was concentrated into fewer and fewer bodies, each having greater and greater masses.

At some point in Earth's early evolution, a giant impact occurred between a Mars-sized object and a young, semimolten Earth. This collision ejected huge amounts of debris into space, some of which coalesced to form the Moon (see Figure 12.4J,K,L).

Earth's Early Evolution

As material continued to collide and accumulate, the high-velocity impacts of interplanetary debris (planetesimals) and the decay of radioactive elements caused the temperature of our planet to steadily increase. This early period of heating resulted in a magma ocean that was perhaps several hundred kilometers deep. Within the magma ocean, buoyant masses of molten rock rose toward the surface and eventually solidified to produce thin rafts of crustal rocks. Geologists call this early period of Earth's history the **Hadean**, which began with Earth's formation about 4.6 billion years ago and ended roughly 4 billion years ago (Figure 12.5). The name *Hadean* is derived from the Greek word *Hades*, meaning "the underworld," referring to the "hellish" conditions on Earth at the time.

During this period of intense heating, Earth became so hot that iron and nickel began to melt. Melting produced liquid blobs of heavy metal that gravitationally sank toward the center of Earth. This process occurred rapidly on the scale of geologic time and produced Earth's dense iron-rich core. As you learned in Chapter 5, the formation of a molten iron core was the first of many stages of chemical differentiation in which Earth converted from a homogeneous body, with roughly the same matter at all depths, to a layered planet with material sorted by density (see Figure 12.4I).

This period of chemical differentiation established the three major divisions of Earth's interior—the iron-rich *core*; the thin *primitive crust*; and Earth's thickest

SmartFigure 12.4
Major events that led to the formation of early Earth Ages are in billions of years (Ga).

TUTORIAL
http://goo.gl/aqiQ0Z

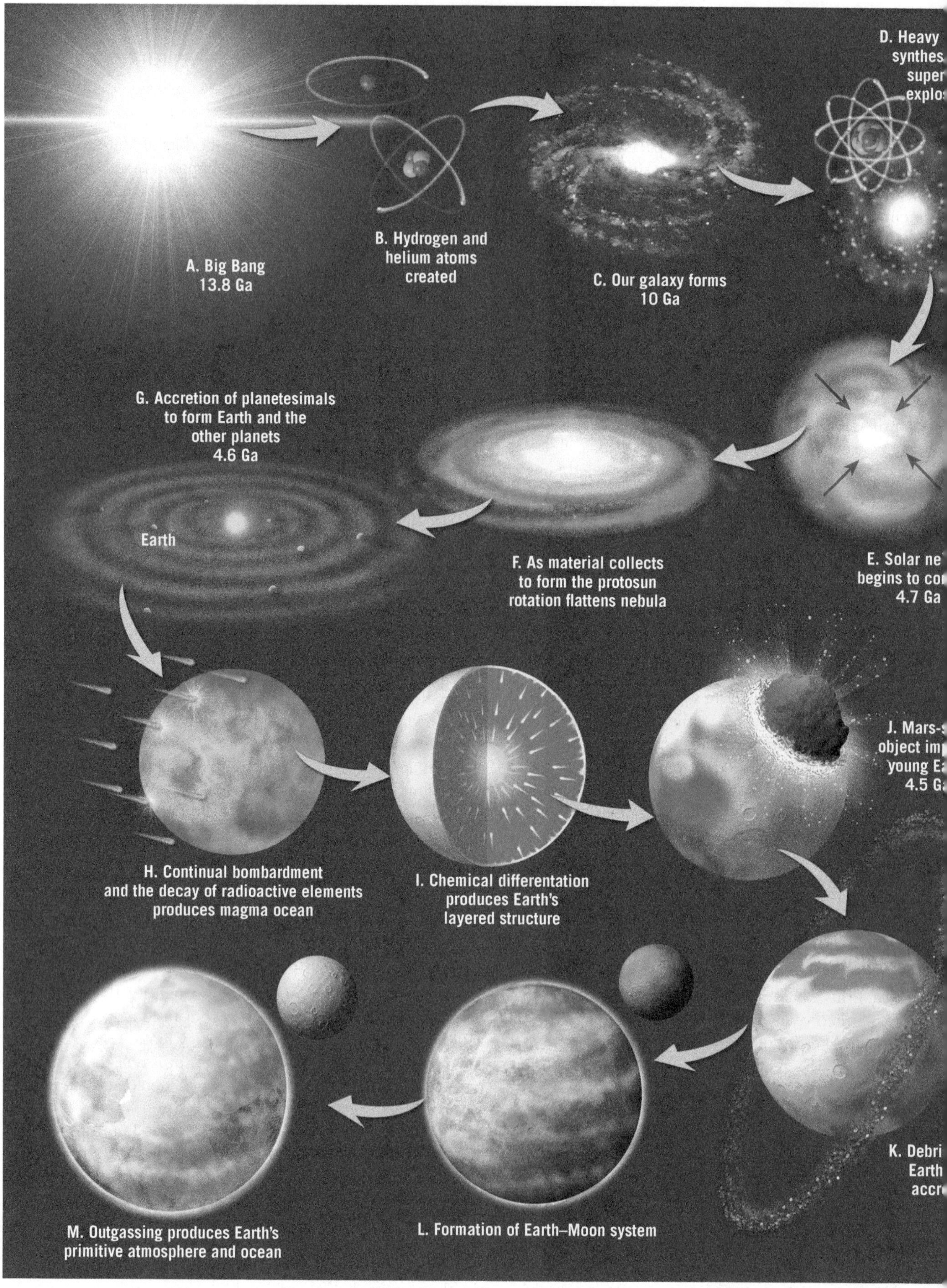

A. Big Bang
13.8 Ga

B. Hydrogen and
helium atoms
created

C. Our galaxy forms
10 Ga

D. Heavy
synthes
super
explo

E. Solar ne
begins to co
4.7 Ga

F. As material collects
to form the protosun
rotation flattens nebula

G. Accretion of planetesimals
to form Earth and the
other planets
4.6 Ga

Earth

H. Continual bombardment
and the decay of radioactive elements
produces magma ocean

I. Chemical differentation
produces Earth's
layered structure

J. Mars-
object im
young E
4.5 G

K. Debri
Earth
accr

L. Formation of Earth–Moon system

M. Outgassing produces Earth's
primitive atmosphere and ocean

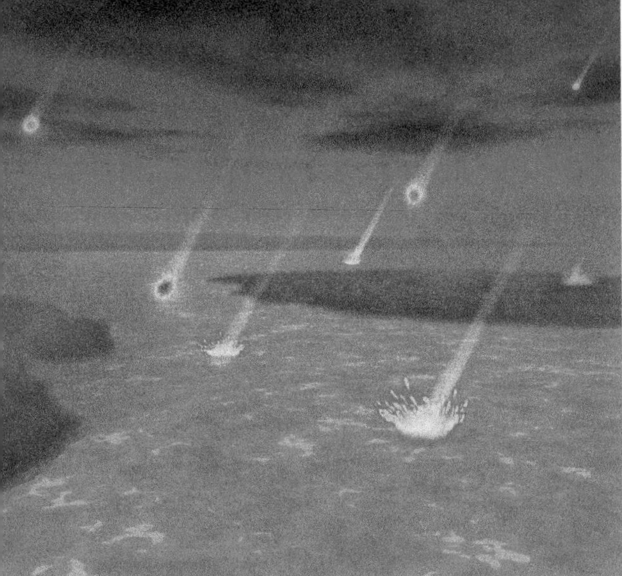

Figure 12.5 **Artistic depiction of Earth during the Hadean** The Hadean is an unofficial eon of geologic time that occurred before the Archean. Its name refers to the "hellish" conditions on Earth. During the early Hadean, Earth had a magma ocean and experienced intense bombardment by nebular debris.

layer, the *mantle*, located between the core and the crust. In addition, the lightest materials—including water vapor, carbon dioxide, and other gases—escaped to form a primitive atmosphere and, shortly thereafter, the oceans.

CONCEPT CHECKS 12.2

1. What two elements made up most of the very early universe?
2. Name the cataclysmic event in which an exploding massive star produces elements heavier than iron.
3. Briefly describe the formation of the planets from the solar nebula.
4. Describe the conditions on Earth during the Hadean.

12.3 Origin and Evolution of the Atmosphere and Oceans

Describe how Earth's atmosphere and oceans formed and evolved through time.

We can be thankful for our atmosphere; without it, there would be no greenhouse effect, and Earth would be nearly 60°F colder. Earth's water bodies would be frozen nearly solid, making the hydrologic cycle nonexistent.

The air we breathe is a relatively stable mixture of 78 percent nitrogen, 21 percent oxygen, about 1 percent argon (an inert gas), and small amounts of other gases such as carbon dioxide and water vapor. However, our planet's original atmosphere was substantially different.

Earth's Primitive Atmosphere

Late in Earth's formative period, its atmosphere probably consisted of gases most common in the early solar system: hydrogen, helium, methane, ammonia, carbon dioxide, and water vapor. The lightest of these—hydrogen and helium—likely escaped into space because Earth's gravity was too weak to hold them. The gases that remained—methane, ammonia, carbon dioxide, and water vapor—contain the basic ingredients of life: carbon, hydrogen, oxygen, and nitrogen.

Over time, Earth's atmosphere was enhanced by a process called **outgassing**, by which gases trapped in the planet's interior are released. Outgassing from hundreds of active volcanoes continues to be an important planetary function worldwide (Figure 12.6). These eruptions release mainly water vapor, carbon dioxide, and sulfur dioxide, with minor amounts of other gases. As a result, Earth's atmosphere gradually became enriched in carbon dioxide (most of the water vapor condensed to form liquid water) and was probably similar in this respect to the atmospheres of Venus and Mars.

Equally important, molecular oxygen (O_2) was not present in Earth's atmosphere in appreciable amounts for at least the first 2 billion years of Earth history. Molecular oxygen is often called "free oxygen" because it consists of oxygen atoms that are not bound to other elements, such as hydrogen (in water molecules, H_2O) or carbon (in carbon dioxide, CO_2).

Figure 12.6 **Outgassing produced Earth's first enduring atmosphere** Outgassing continues today from hundreds of active volcanoes worldwide. (Photo by Lilja Kristjansdo/NordicPhotos/Getty Images)

Oxygen in the Atmosphere

As Earth's surface cooled, water vapor condensed to form clouds, and torrential rains began to fill low-lying areas that eventually became the oceans. In those oceans, nearly 3.5 billion years ago, primitive bacteria known as *cyanobacteria* (once called blue-green algae), developed the ability to carry out photosynthesis and began to release oxygen into the water. **Photosynthesis** is the production of energy-rich molecules of sugar from molecules of carbon dioxide (CO_2) and water (H_2O), using sunlight as the energy source. The sugars (glucose and other sugars) generated by photosynthesis are used in metabolic processes by living things, and the by-product of photosynthesis is molecular oxygen.

Initially, the newly released molecular oxygen likely combined with other elements through processes that included the chemical weathering of rocks. Scientists also found evidence that appreciable quantities of oxygen were "soaked up" by iron that was dissolved in the young ocean. Apparently, large quantities of iron were released into the young ocean by hydrothermal vents that spewed hot water solutions containing iron and other metals.

Iron has tremendous affinity for oxygen. When these two elements join, they become iron oxide (rust). These early iron oxide accumulations on the seafloor created alternating layers of iron-rich rocks and chert, called **banded iron formations**. Most banded iron deposits accumulated in the Precambrian eon, between 3.5 and 2 billion years ago, and represent the world's most important reservoirs of iron ore.

As photosynthesizing organisms proliferated, oxygen began to build in the atmosphere. Chemical analysis of rock suggests that molecular oxygen began to appear in significant amounts in the atmosphere around 2.3 billion years ago, a phenomenon termed the **Great Oxygenation Event**. One positive benefit of the Great Oxygenation Event is that, when struck by sunlight, oxygen molecules form a compound called *ozone* (O_3), a type of oxygen molecule composed of three oxygen atoms. Ozone, which absorbs much of the Sun's harmful ultraviolet radiation before it reaches Earth's surface, is concentrated between 10 and 50 kilometers (6 and 30 miles) above Earth's surface, in a layer called the *stratosphere*. Thus, as a result of the Great Oxygenation Event, Earth's landmasses were protected from ultraviolet radiation, which is particularly harmful to DNA—the genetic blueprints for living organisms. Marine organisms had always been shielded from harmful ultraviolet radiation by seawater, but the development of the atmosphere's protective ozone layer made the continents more hospitable as well.

During the billion years following the Great Oxygenation Event, oxygen levels in the atmosphere probably fluctuated but remained below current levels. Then, just prior to the start of the Cambrian period 541 million years ago, the level of free oxygen in the atmosphere began to increase. The availability of abundant oxygen in the atmosphere contributed to the proliferation of aerobic life-forms (oxygen-consuming organisms). On the other hand, it likely wiped out huge portions of Earth's anaerobic organisms (organisms that do not require oxygen for respiration), for which oxygen is poisonous.

One apparent spike in oxygen levels occurred during the Pennsylvanian period (about 300 million years ago), when oxygen made up as much as 35 percent of the atmosphere, compared to today's level of 21 percent. One possible effect of this increase in oxygen is the occurrence of unusually large insects from that time period. (Studies have shown that, in some insects at least, higher oxygen levels promote larger size.) A fossil dragonfly found in 1979 had a wingspan of 50 centimeters (20 inches), and an even larger specimen had a 75-centimeter (30-inch) wingspan and was named *Meganeura* (*mega* = large). One hypothesis proposes that the climate during the Pennsylvanian period was ideal for plant growth, both in the extensive swampy areas on land and in the oceans (see Figure 12.26). With all the trees and plankton in the sea producing oxygen via photosynthesis, the environment was favorable for the development of large insects.

Evolution of Earth's Oceans

When Earth cooled sufficiently to allow water vapor to condense, rainwater fell and collected in low-lying areas. By 4 billion years ago, scientists estimate that as much as 90 percent of the current volume of seawater was contained in the developing ocean basins. Because volcanic eruptions released into the atmosphere large quantities of sulfur dioxide, which readily combines with water to form sulfuric acid, the earliest rainwater was highly acidic. The level of acidity was even greater than the acid rain that damaged lakes and streams in eastern North America during the latter part of the twentieth century. Consequently, Earth's rocky surface weathered at an accelerated rate. The products released by chemical weathering included atoms and molecules of various substances—including sodium, calcium, potassium, and silica—that were carried by running water into the newly formed oceans. Some of these dissolved substances precipitated to become chemical sediment that mantled the ocean floor. Other substances formed soluble salts, which increased the salinity of seawater. Research suggests that the salinity of the oceans initially increased rapidly, but it has remained relatively constant over the past 2 billion years.

Earth's oceans also serve as a repository for tremendous volumes of carbon dioxide, a major constituent of the primitive atmosphere. This is significant because carbon dioxide is a greenhouse gas that strongly influences the heating of the atmosphere. Venus, once thought to be very similar to Earth, has an atmosphere composed of 97 percent carbon dioxide, which produces an extreme greenhouse effect. As a result, Venus's surface temperature is 475°C (nearly 900°F).

Carbon dioxide is readily soluble in seawater, where it often combines with other atoms or molecules to produce various chemical precipitates. One of the most

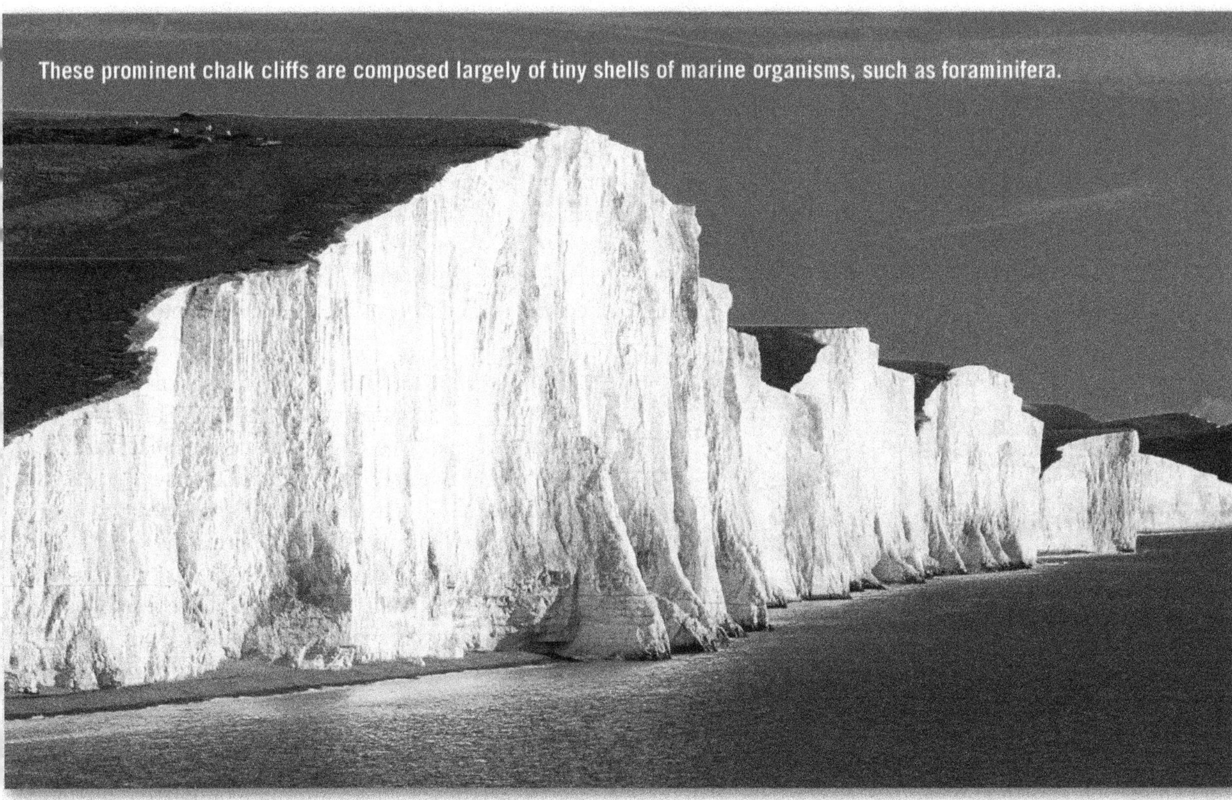

These prominent chalk cliffs are composed largely of tiny shells of marine organisms, such as foraminifera.

Figure 12.7 White Cliffs of Dover, England Similar chalk deposits are also found in northern France. (Photo by Imagesources/Glow Images)

common compounds generated by mineral precipitation is calcium carbonate ($CaCO_3$). Crystalline calcium carbonate is the mineral calcite, the main component of the sedimentary rock limestone. About 541 million years ago, marine organisms began to extract large quantities of calcium carbonate from seawater to make their shells and other hard parts. Trillions of tiny marine organisms, such as foraminifera, deposited their shells on the seafloor at the end of their life cycle. Some of these deposits can be observed today in the chalk beds exposed along the White Cliffs of Dover, England (Figure 12.7). By "locking up" carbon dioxide, these limestone deposits store massive amounts of this greenhouse gas so that it cannot easily reenter the atmosphere. Thus, the evolution of life-forms that secrete calcium carbonate shells aided in the removal of this greenhouse gas from the atmosphere.

CONCEPT CHECKS 12.3

1. What is meant by *outgassing*, and what modern phenomenon serves that role today?
2. List the most abundant gases that were added to Earth's early atmosphere through the process of outgassing.
3. Why was the evolution of photosynthesizing bacteria important for the evolution of large, oxygen-consuming organisms like ourselves?
4. Why was rainwater highly acidic early in Earth's history?
5. How does the ocean remove carbon dioxide from Earth's atmosphere? What role do tiny marine organisms, such as foraminifera, play in the removal of carbon dioxide?

12.4 Precambrian History: The Formation of Earth's Continents

Explain the formation of continental crust, how continental crust becomes assembled into continents, and the role that the supercontinent cycle has played in this process.

Earth's first 4 billion years are encompassed in the time span called the *Precambrian*. Representing nearly 90 percent of Earth's history, the Precambrian is divided into the *Archean eon* ("ancient age"), the *Proterozoic eon* ("early life age"), and an informal time span referred to as the Hadean. Our knowledge of this ancient time is limited because much of the early rock record has been obscured by the very Earth processes you have been studying, especially plate tectonics, erosion, and deposition. Most Precambrian rocks lack fossils, which hinders correlation of rock units (see Chapter 11). In addition, rocks this old are often metamorphosed and deformed, extensively eroded, and frequently concealed by younger strata. Indeed, Precambrian history is written in scattered, speculative episodes, like a long book with many missing chapters.

The crust covering this lava lake is continually being replaced with fresh lava from below, much like the way Earth's crust was recycled early in its history.

▲ **Figure 12.8 Earth's early crust was continually recycled** (Photo courtesy of USGS)

Earth's First Continents

Geologists have discovered tiny crystals of the mineral zircon in continental rocks that formed 4.4 billion years ago—evidence that the continents began to form early in Earth's history. By contrast, the oldest rocks found in the ocean basins are generally less than 200 million years old.

What differentiates continental crust from oceanic crust? Recall that oceanic crust is a relatively dense (3.0 g/cm^3) homogeneous layer of basaltic rocks derived from partial melting of the rocky upper mantle. In addition, oceanic crust is thin, averaging only 7 kilometers (4 miles) thick. Continental crust, on the other hand, is composed of a variety of rock types, has an average thickness of nearly 40 kilometers (25 miles), and contains a large percentage of low-density (2.7 g/cm^3), silica-rich rocks such as granite.

▼ **Figure 12.9 Earth's oldest preserved continental rocks are more than 3.8 billion years old** (Photo courtesy of James L. Amos/Corbis Documentary/Getty Images)

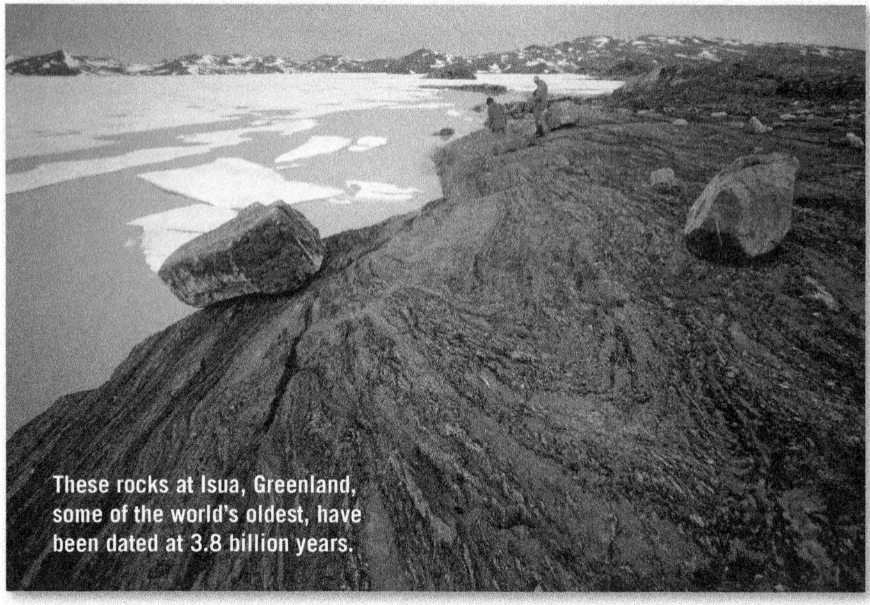

These rocks at Isua, Greenland, some of the world's oldest, have been dated at 3.8 billion years.

The significance of the differences between continental crust and oceanic crust cannot be overstated in a review of Earth's geologic evolution. Oceanic crust, because it is relatively thin and dense, is found several kilometers below sea level—unless of course it has been pushed onto a landmass by tectonic forces. Continental crust, because of its great thickness and lower density, may extend well above sea level. Also, recall that dense oceanic crust of normal thickness readily subducts, whereas thick, buoyant blocks of continental crust resist being recycled into the mantle.

Making Continental Crust The formation of continental crust is a continuation of the gravitational segregation of Earth materials that began during the final stage of our planet's formation. Dense metallic material, mainly iron and nickel, sank to form Earth's core, leaving behind the less dense rocky material that forms the mantle. It is from Earth's rocky mantle that low-density, silica-rich minerals were gradually distilled to form continental crust. This process is analogous to making sour mash whiskey. In the production of whiskeys, various grains such as corn are fermented to generate alcohol. This mixture is then heated or distilled, which drives off the lighter material (alcohol) and leaves behind the sour mash as the by-product. In a similar manner, partial melting of mantle rocks generates low-density, silica-rich melts that buoyantly rise to the surface to form Earth's crust, leaving behind the dense mantle rocks (see Chapter 6). However, little is known about the details of the mechanisms that generated these silica-rich melts during the Archean eon.

Earth's first crust was probably ultramafic in composition, but because physical evidence no longer exists, we are not certain. The hot, turbulent mantle that most likely existed during the Archean eon recycled most of this crustal material back into the mantle. In fact, it may have been continuously recycled, in much the same way that the "crust" that forms on a lava lake is repeatedly replaced with fresh lava from below (Figure 12.8).

The oldest preserved continental rocks occur as small, highly deformed *terranes*, which are incorporated within somewhat younger blocks of continental crust (Figure 12.9). One of these is a 3.8-billion-year-old terrane located near Isua, Greenland. Slightly older crustal rocks, called the Acasta Gneiss, have been discovered in Canada's Northwest Territories.

Some geologists think that a type of plate-like motion operated early in Earth's history. In addition, hot-spot volcanism was likely active during this time. However, because the mantle was hotter in the Archean than it is today, both of these phenomena would have progressed at faster rates than their modern counterparts. Hot-spot volcanism, due to mantle plumes, is thought to have created immense shield volcanoes as well as oceanic plateaus. Simultaneously, subduction of oceanic crust generated volcanic island arcs. Collectively, these relatively small crustal fragments represent the first phase in creating stable, continent-size landmasses.

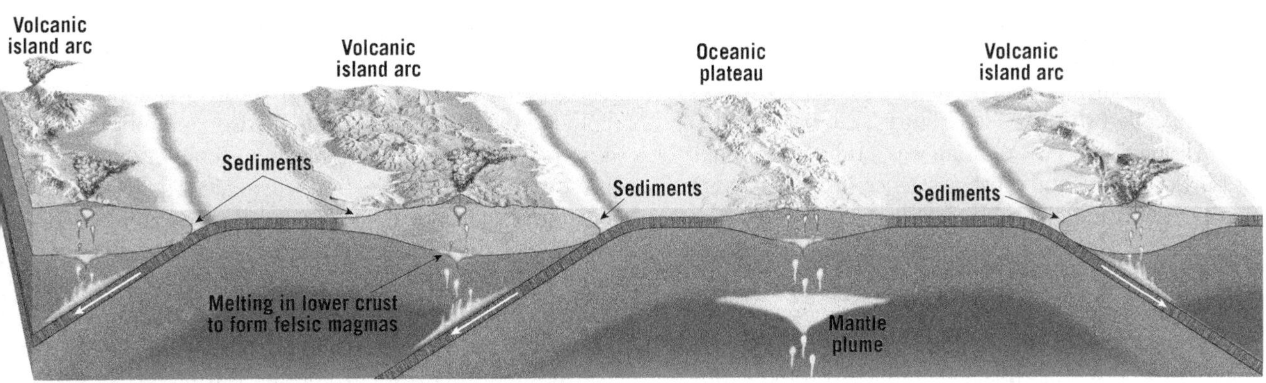

A. Scattered crustal fragments separated by ocean basins

SmartFigure 12.10
The formation of continents The growth of large continental masses occurs through the collision and accretion of smaller crustal fragments.

TUTORIAL
http://goo.gl/7y4htd

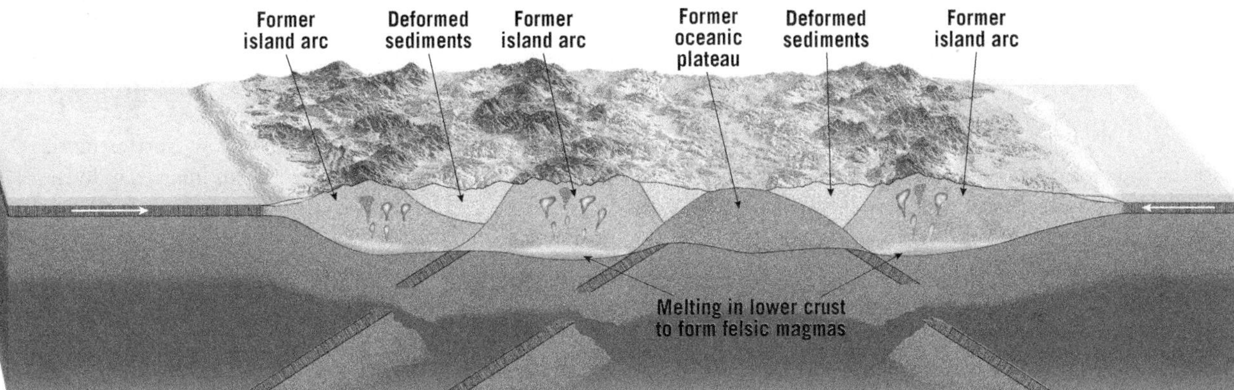

B. Collision of volcanic island arcs and oceanic plateau to form a larger crustal block

From Continental Crust to Continents The growth of larger continental masses was accomplished through collision and accretion of many thin, highly mobile crustal fragments, as illustrated in Figure 12.10. This type of collisional tectonics deformed and metamorphosed sediments caught between converging crustal fragments, thereby shortening and thickening the developing crust. In the deepest regions of these collision zones, partial melting of the thickened crust generated silica-rich magmas that ascended and intruded the rocks above. This led to the formation of large crustal provinces that, in turn, accreted with others to form even larger crustal blocks called **cratons**.

The assembly of a large craton involves the accretion of several crustal blocks that cause major mountain-building episodes similar to India's collision with Asia. Figure 12.11 shows the extent of crustal material that was

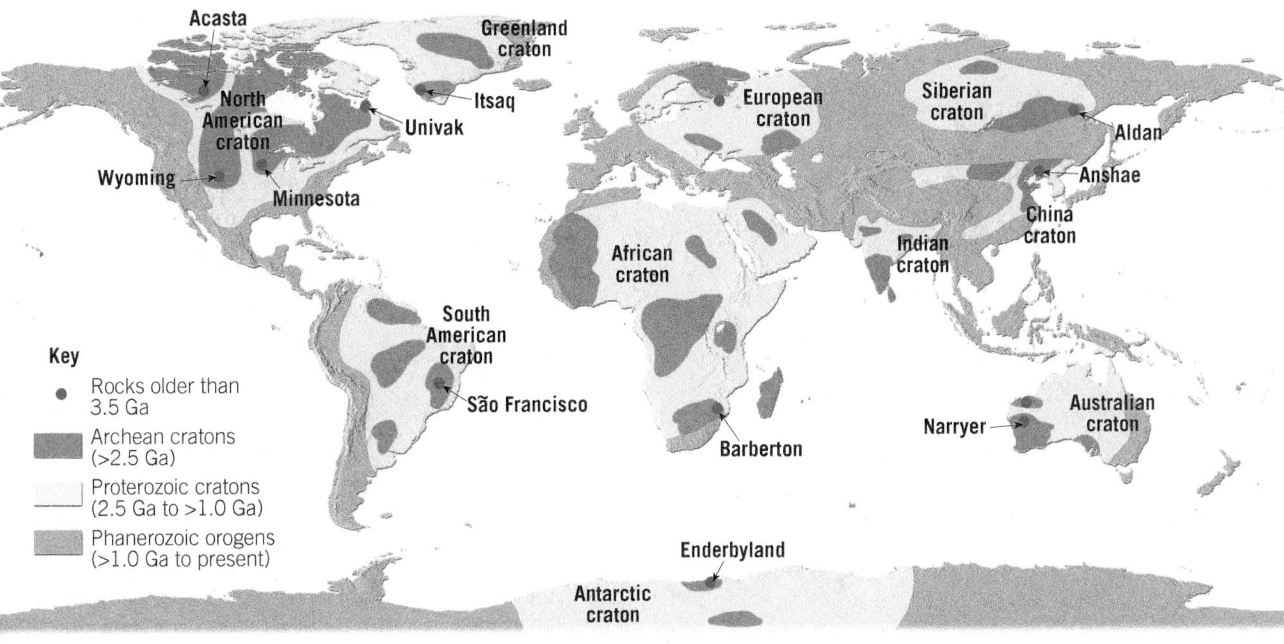

Figure 12.11
Distribution of crustal material remaining from the Archean and Proterozoic eons Ages are in billions of years (Ga).

produced during the Archean and Proterozoic eons and remains today. The regions within a modern continent where these ancient cratons are exposed at Earth's surface are called **shields** (see Figure 7.23, page 220).

Although the Precambrian was a time when much of Earth's continental crust was generated, a substantial amount of crustal material was destroyed as well. Some was lost via weathering and erosion. In addition, during much of the Archean, it appears that thin slabs of continental crust were subducted into the mantle. However, by about 3 billion years ago, cratons grew sufficiently large and thick to resist subduction. After that time, weathering and erosion became the primary processes of crustal destruction. By the close of the Precambrian, an estimated 85 percent of the modern continental crust had formed.

The Making of North America

North America provides an excellent example of the development of continental crust and its piecemeal assembly into a continent. With the exception of a few locations noted in Figure 12.11, very little continental crust older than 3.5 billion years remains. In the late Archean, between 3 and 2.5 billion years ago, there was a period of major continental growth, which is shown in purple in Figure 12.12. During this span, the accretion of numerous island arcs and other fragments generated several large crustal provinces. North America contains some of these crustal units, including the Superior and Hearne-Rae cratons shown in Figure 12.12, but just where these ancient continental blocks formed is unknown.

About 1.9 billion years ago, these crustal provinces collided to produce the Trans-Hudson mountain belt (see Figure 12.12). (Such mountain-building episodes were not restricted to North America; ancient deformed strata of similar age are also found on other continents.) This event built the North American craton, around which several large and numerous small crustal fragments were later added. One of these late arrivals is the Appalachian province. In addition, several terranes were added to the western margin of North America during the Mesozoic and Cenozoic eras to generate the mountainous North American Cordillera (see Figure 7.28, page 225).

Supercontinents of the Precambrian

At different times, parts of what is now North America combined with other continental landmasses to form a supercontinent. **Supercontinents** are large landmasses that contain all, or nearly all, the existing continents. Pangaea was the most recent, but certainly not the only, supercontinent to exist in the geologic past. The earliest well-documented supercontinent, *Rodinia*, formed during the Proterozoic eon, about 1.1 billion years ago (Figure 12.13). Although geologists are still studying its construction, it is clear that Rodinia's configuration was quite different from Pangaea's. One obvious distinction is North America's position near the center of this ancient landmass.

Between 800 and 600 million years ago, Rodinia gradually split apart. By the end of the Precambrian many of

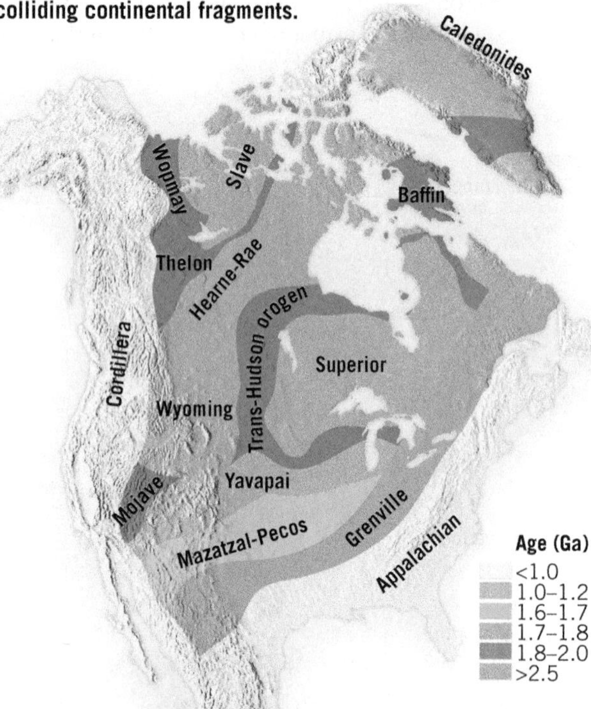

SmartFigure 12.12 **The major geologic provinces of North America** The age of each province is in billions of years (Ga).

TUTORIAL
http://goo.gl/QGu7RI

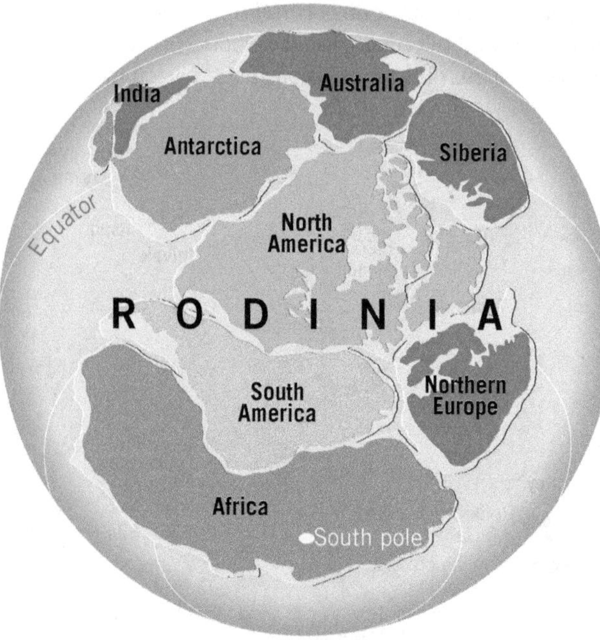

Figure 12.13 **Possible configuration of the supercontinent Rodinia** For clarity, the continents are drawn with somewhat modern shapes, not their actual shapes from 1 billion years ago. (After P. Hoffman, J. Rogers, and others)

the fragments had reassembled, producing a large landmass in the Southern Hemisphere called *Gondwana*, composed mainly of present-day South America, Africa, India, Australia, and Antarctica (Figure 12.14). Other continental fragments also developed—North America, Siberia, and Northern Europe. We consider the fate of these Precambrian landmasses in the next section.

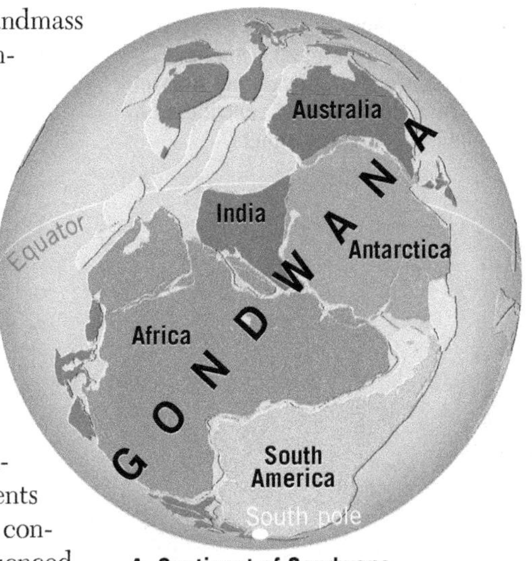

A. Continent of Gondwana **B. Continents not a part of Gondwana**

Figure 12.14 **Reconstruction of Earth as it may have appeared in late Precambrian time** The southern continents were joined into a single landmass called Gondwana. Other landmasses that were not part of Gondwana include North America, northwestern Europe, and northern Asia. (After P. Hoffman, J. Rogers, and others)

Supercontinent Cycle The **supercontinent cycle** involves rifting and dispersal of one supercontinent followed by a long period during which the fragments are gradually reassembled into a new supercontinent with a different configuration. The assembly and dispersal of supercontinents had a profound impact on the evolution of Earth's continents. In addition, this phenomenon greatly influenced global climates and contributed to periodic episodes of rising and falling sea level.

Supercontinents and Climate The movement of continents changes the patterns of ocean currents and global winds, which influences the global distribution of temperature and precipitation. The formation of the Antarctic's vast ice sheet is one example of how the movement of a continent is thought to have contributed to climate change. Although eastern Antarctica remained over the South Pole for more than 100 million years, Antarctica was not covered by a stable continental-scale ice sheet until about 34 million years ago. Prior to this period of glaciation, South America and Antarctica were connected. As shown in Figure 12.15A, this arrangement of landmasses helped maintain a circulation pattern in which warm ocean currents reached the coast of Antarctica and aided in keeping Antarctica mainly ice free. This is similar to the way in which the modern Gulf Stream helps keep Iceland mostly ice free, despite its name.

As South America separated from Antarctica and moved northward, a pattern of ocean circulation developed that flowed from west to east around the entire continent of Antarctica (Figure 12.15B). This cold current, called the West Wind Drift, effectively isolated the entire Antarctic coast from the warm, poleward-directed currents in the southern oceans. This change in circulation, along with a period of global cooling, probably resulted in the growth of Antarctica's massive ice sheet.

Local and regional climates are also influenced by large mountain systems created by the collision of large cratons. One example is the collision of the Indian subcontinent with southern Asia that generated the Himalayas. Because of their high elevations, mountains exhibit markedly lower average temperatures than surrounding lowlands. In addition, air rising over these lofty structures promotes condensation and precipitation, leaving the region downwind relatively dry. A modern analogy

EYE ON EARTH 12.1

The oldest-known sample of Earth is a 4.4-billion-year-old zircon crystal found in a metaconglomerate in the Jack Hills area of western Australia. Zircon is a silicate mineral that occurs in trace amounts in most granitic rocks. (Photo by John W. Valley/NSF)

QUESTION 1 *What is the parent rock of a metaconglomerate?*

QUESTION 2 *Assuming that this zircon crystal originated as part of a granite intrusion, briefly describe its journey from the time of its formation until it was discovered in the Jack Hills.*

QUESTION 3 *Is this zircon crystal younger or older than the metaconglomerate in which it was found? Explain.*

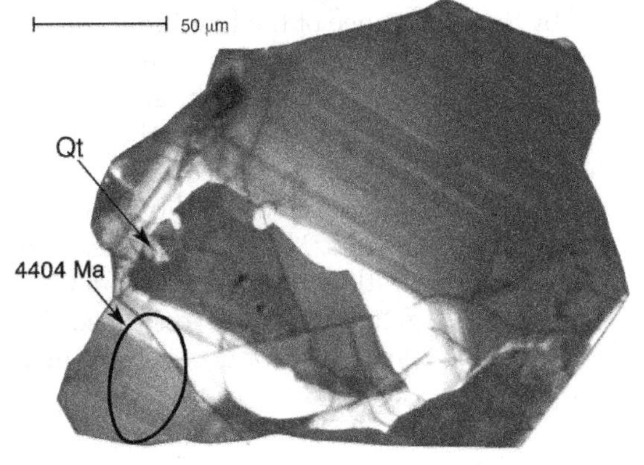

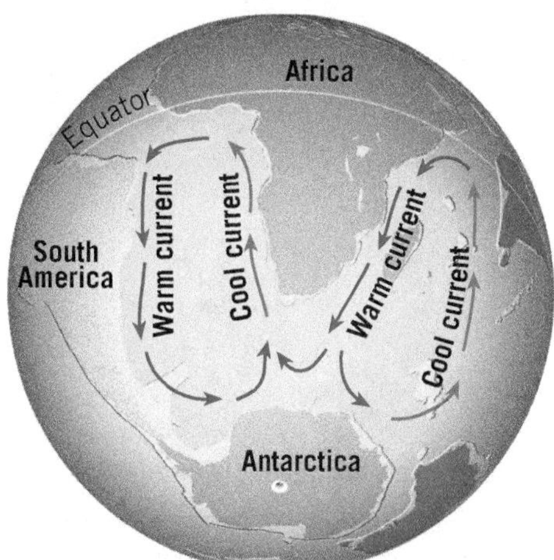
50 million years ago warm ocean currents kept Antarctica nearly ice free.
Africa
Equator
South America
Warm current
Cool current
Warm current
Cool current
Antarctica

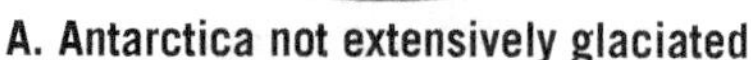
A. Antarctica not extensively glaciated

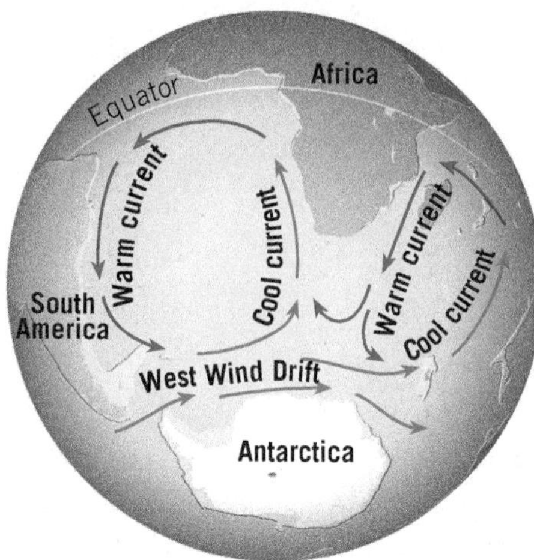
As South America separated from Antarctica, the West Wind Drift developed. This newly formed ocean current effectively cut Antarctica off from warm currents and contributed to the formation of its vast ice sheets.
Africa
Equator
South America
Warm current
Cool current
Warm current
Cool current
West Wind Drift
Antarctica
B. Antarctica covered by continental-size ice sheets

SmartFigure 12.15
Connection between ocean circulation and the climate in Antarctica

TUTORIAL
http://goo.gl/0rwB5K

is the wet, heavily forested western slopes of the Sierra Nevada compared to the dry climate of the Great Basin Desert that lies directly to the east.

Supercontinents and Sea-Level Changes Significant and numerous sea-level changes have been documented in geologic history, and many of them appear to have been related to the assembly and dispersal of supercontinents. If sea level rises, shallow seas advance onto the continents. Evidence for periods when the seas advanced onto the continents include thick sequences of ancient marine sedimentary rocks that blanket large areas of modern landmasses—including much of the eastern two-thirds of the United States.

The supercontinent cycle and sea-level changes are directly related to rates of *seafloor spreading*. When the rate of spreading is rapid, as it is along the East Pacific Rise today, the production of warm oceanic crust is also high. Because new, warm oceanic crust is less dense (takes up more space) than cold crust, fast-spreading ridges occupy more volume in the ocean basins than do slow-spreading centers. (Think of getting into a tub filled with water.) As a result, when rates of seafloor spreading increase, more seawater is displaced, which results in the sea level rising. This, in turn, causes shallow seas to advance onto the low-lying portions of the continents.

CONCEPT CHECKS 12.4

1. Briefly explain how low-density continental crust was produced from Earth's rocky mantle.
2. Describe how cratons came into being.
3. What is the supercontinent cycle? What supercontinent preceded Pangaea?
4. Give an example of how the movement of a continent can trigger climate change.
5. Explain how the rate of seafloor spreading is related to changes in sea level.

12.5 Geologic History of the Phanerozoic: The Formation of Earth's Modern Continents

List and discuss the major geologic events in the Paleozoic, Mesozoic, and Cenozoic eras.

The time span since the close of the Precambrian, called the *Phanerozoic eon*, encompasses 541 million years and is divided into three eras: *Paleozoic*, *Mesozoic*, and *Cenozoic*. The beginning of the Phanerozoic is marked by the appearance of the first life-forms with hard parts such as shells, scales, bones, or teeth—all of which greatly enhance the chances for an organism to be preserved in the fossil record. Thus, the study of Phanerozoic crustal history was aided by the availability of fossils, which improved our ability to date and correlate geologic events. Moreover, because every organism is associated with its own particular environmental niche, the greatly improved fossil record provided invaluable information for deciphering ancient environments.

Paleozoic History

As the Paleozoic era opened, what is now North America hosted no plants or animals large enough to be seen—just tiny microorganisms such as bacteria. There were no Appalachian or Rocky Mountains; the continent was largely a barren lowland. Several times during the early Paleozoic, shallow seas moved inland and then receded from the continental interior, leaving behind the thick deposits of limestone, shale, and clean sandstone that mark the shorelines of these previously midcontinent shallow seas.

Formation of Pangaea One of the major events of the Paleozoic era was the formation of the supercontinent **Pangaea**. This event began with a series of collisions that over millions of years joined North America, Europe, Siberia, and other smaller crustal fragments to form a large continent called **Laurasia** (Figure 12.16). Located south of Laurasia was the vast southern continent called **Gondwana**, which encompassed five modern landmasses—South America, Africa, Australia, Antarctica, and India—and perhaps portions of China. Evidence of extensive continental glaciation places this landmass near the South Pole. By the late Paleozoic, Gondwana had migrated northward and collided with Laurasia to begin the final stage in Pangaea's assembly.

The accretion of all of Earth's major landmasses to form Pangaea spans more than 300 million years and resulted in the formation of several mountain belts. The collision of northern Europe (mainly Norway) with Greenland produced the Caledonian Mountains, whereas the joining of northern Asia (Siberia) and Europe created the Ural Mountains. Northern China is also thought to have accreted to Asia by the end of the Paleozoic, whereas southern China may not have become part of Asia until after Pangaea had begun to rift. (Recall that India did not begin to accrete to Asia until about 50 million years ago.)

Pangaea reached its maximum size between 300 and 250 million years ago, as Africa collided with North America (see Figure 12.16D). This event marked the final and most intense period of mountain building in the long history of the Appalachian Mountains (see Figure 7.31, page 228). This mountain-building event produced the Central Appalachians of the Atlantic states, as well as New England's northern Appalachians and mountainous structures that extend into Canada (Figure 12.17).

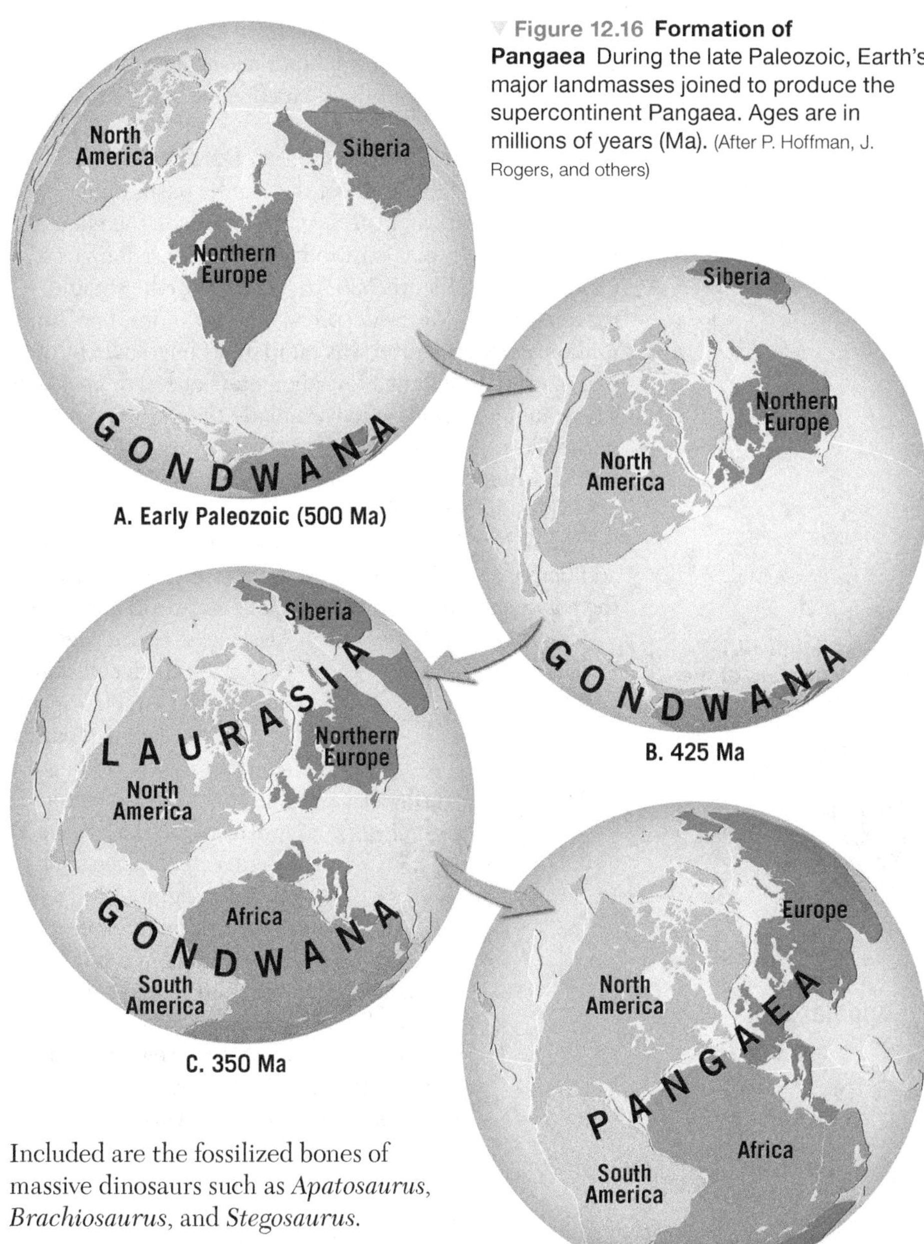

Figure 12.16 **Formation of Pangaea** During the late Paleozoic, Earth's major landmasses joined to produce the supercontinent Pangaea. Ages are in millions of years (Ma). (After P. Hoffman, J. Rogers, and others)

Mesozoic History

Spanning about 186 million years, the Mesozoic era is divided into three periods: the *Triassic*, *Jurassic*, and *Cretaceous*. Major geologic events of the Mesozoic include the breakup of Pangaea and the evolution of our modern ocean basins.

Changes in Sea Levels The Mesozoic era began with much of the world's continents above sea level. The exposed Triassic strata are primarily red sandstones and mudstones that lack marine fossils, features that indicate a terrestrial environment. (The red color in sandstone comes from the oxidation of iron.)

As the Jurassic period opened, the sea invaded western North America. Adjacent to this shallow sea, extensive continental sediments were deposited on what is now the Colorado Plateau. The most prominent is the Navajo Sandstone, a cross-bedded, quartz-rich layer that in some places approaches 300 meters (1000 feet) thick. These remnants of massive dunes indicate that an enormous desert occupied much of the American Southwest during early Jurassic times (Figure 12.18). Another well-known Jurassic deposit is the Morrison Formation—one of the world's richest storehouses of dinosaur fossils. Included are the fossilized bones of massive dinosaurs such as *Apatosaurus*, *Brachiosaurus*, and *Stegosaurus*.

Coal Formation in Western North America As the Jurassic period gave way to the Cretaceous, shallow seas again encroached upon much of western North America, as well as the Atlantic and Gulf coastal regions. This led to the formation of "coal swamps" (see Chapter 3) similar to those of the Paleozoic era. Today, the Cretaceous coal deposits in the western United States and Canada are economically important. For example, the Crow Native American reservation in Montana holds nearly 20 billion tons of high-quality, Cretaceous-age coal.

The Breakup of Pangaea Another major event of the Mesozoic era was the breakup of Pangaea. About 185 million years ago, a rift developed between what is now North America and western Africa, marking the birth of the Atlantic Ocean. As Pangaea gradually broke apart, the westward-moving North American plate began to override the Pacific basin. This tectonic event triggered a continuous wave of deformation that moved inland along the entire western margin of North America.

Formation of the North American Cordillera By Jurassic times, subduction of the Pacific basin under the North American plate began to produce the chaotic mixture of rocks that exist today in the Coast Ranges of California (see Figure 7.25, page 222). Further inland, igneous activity was widespread, and for more than 100 million years volcanism was rampant as huge masses of magma rose to within a few kilometers of Earth's surface. The remnants of this activity include the granitic plutons of the Sierra Nevada, as well as the Idaho batholith and British Columbia's Coast Range batholith.

The subduction of the Pacific basin under the western margin of North America also resulted in the piecemeal addition of crustal fragments to the entire Pacific margin of the continent—from Mexico's Baja Peninsula to northern Alaska (see Figure 7.28, page 225). Each collision displaced crustal fragments (terranes) accreted earlier farther inland, adding to the zone of deformation as well as to the thickness and lateral extent of the continental margin.

Compressional forces moved huge rock units in a shingle-like fashion toward the east. Across much of North America's western margin, older rocks were thrust eastward over younger strata, for distances exceeding 150 kilometers (90 miles). Ultimately, this activity was responsible for generating a vast portion of the North American Cordillera that extends from Wyoming to Alaska.

Toward the end of the Mesozoic, the southern portions of the Rocky Mountains developed. This mountain-building event, called the *Laramide Orogeny*, occurred when large blocks of deeply buried Precambrian rocks were lifted nearly vertically along steeply dipping faults, upwarping the overlying younger sedimentary strata. The mountain ranges produced by the Laramide Orogeny include Colorado's Front Range, the Sangre de Cristo of New Mexico and Colorado, and the Bighorns of Wyoming.

SmartFigure 12.17
Major provinces of the Appalachian Mountains

TUTORIAL
http://goo.gl/Y3wezV

Northern Appalachians

Central Appalachians

Appalachian Plateau

Valley and Ridge

Blue Ridge Mountains

Piedmont

Coastal Plain

A.

B.

Appalachian Plateau (Underlain by nearly flat-lying sedimentary strata of Paleozoic age.)

Valley and Ridge (Highly folded and thrust-faulted sedimentary rocks of Paleozoic age.)

Blue Ridge (Hilly to mountainous terrain consisting of slices of basement rock of Precambrian age.)

Piedmont (Crustal fragments of metamorphosed sedimentary and igneous rocks that were added to North America.)

Coastal Plain (Area of low relief underlain by gradualy sloping sedimentary strata and unlithified sediments.)

Cenozoic History

The Cenozoic era, or "era of recent life," encompasses the past 66 million years of Earth history. It was during this span that the physical landscapes and life-forms of our modern world came into existence. The Cenozoic era represents a considerably smaller fraction of geologic time than either the Paleozoic or the Mesozoic, but we know much more about this time span because the rock formations are more widespread and less disturbed than those of any preceding era.

Most of North America was above sea level during the Cenozoic era. However, the eastern and western margins of the continent experienced markedly dissimilar events because of their different plate boundary relationships. The Atlantic and Gulf coastal regions, far removed from an active plate boundary, were tectonically stable. By contrast, western North America was the leading edge of the North American plate, and the plate interactions during the Cenozoic account for many events of mountain building, volcanism, and earthquakes.

Eastern North America The stable continental margin of eastern North America was the site of abundant marine sedimentation. The most extensive deposits surrounded the Gulf of Mexico, from the Yucatan Peninsula to Florida, where a massive buildup of sediment caused the crust to downwarp. In many instances, faulting created structures in which oil and natural gas accumulated. Today, these and other petroleum traps (see Figure 3.37, page 88) are the Gulf coast's most economically important resource, as evidenced by numerous offshore drilling platforms.

Early in the Cenozoic, the Appalachians had eroded to create a low plain. Later, isostatic adjustments again raised the region and rejuvenated its rivers. Streams eroded with renewed vigor, gradually sculpting the surface into its present-day topography. Sediments from this erosion were deposited along the eastern continental margin, where they accumulated to a thickness of many kilometers. Today, portions of the strata deposited during the Cenozoic are exposed as the gently sloping Atlantic and Gulf coastal plains, where a large percentage of the eastern and southeastern United States population resides.

Western North America In the West, the Laramide Orogeny responsible for building the southern Rocky Mountains was coming to an end. As erosional forces lowered the mountains, the basins between uplifted ranges began to fill with sediment. East of the Rockies, a large wedge of sediment from the eroding mountains created the gently sloping Great Plains.

Beginning in the Miocene epoch, about 20 million years ago, a broad region from northern Nevada into Mexico experienced crustal extension that created more than 100 fault-block mountain ranges. Today, they rise abruptly above the adjacent basins, forming the Basin and Range Province (see Figure 7.17, page 217).

During the development of the Basin and Range Province, the entire western interior of the continent gradually uplifted. This event elevated the Rockies and rejuvenated many of the West's major rivers. As the rivers became incised, many spectacular gorges were created, including the Grand Canyon of the Colorado River, the Grand Canyon of the Snake River, and the Black Canyon of the Gunnison River.

Volcanic activity was also common in the West during much of the Cenozoic. Beginning in the Miocene epoch, great volumes of fluid basaltic lava flowed from fissures in portions of present-day Washington, Oregon, and Idaho. These eruptions built the 3.4-million-square-kilometer (1.3-million-square-mile) Columbia Plateau. Immediately west of the vast Columbia Plateau, volcanic activity was different in character. Here, more viscous magmas with higher silica content erupted explosively, creating the Cascades, a chain of stratovolcanoes extending from northern California into Canada, some of which are still active (see Figure 6.38, page 198).

As the Cenozoic was drawing to a close, the effects of mountain building, volcanic activity, isostatic adjustments, and extensive erosion and sedimentation created the physical landscape we know today. All that remained of the Cenozoic era was the final 2.6-million-year episode called the Quaternary period. During this most recent, and ongoing, phase of Earth's history, humans evolved and the action of glacial ice, wind, and running water added to our planet's long, complex geologic history.

Close-up view of cross bedding in the Navajo Sandstone, Zion National Park

▲ Figure 12.18 **Massive, cross bedded sandstone cliffs in Zion National Park** These sandstone cliffs are the remnants of ancient sand dunes that were part of an enormous desert during the Jurassic period. (Photo by Michael Collier; inset photo by Dennis Tasa)

CONCEPT CHECKS 12.5

1. During which period of geologic history did the supercontinent Pangaea come into existence? During which period did it begin to break apart?
2. Describe the climate of the present-day American Southwest during early Jurassic time.
3. Where is most Cretaceous age coal found today in the United States?
4. Compare and contrast eastern and western North America's geology during the Cenozoic era.

12.6 Earth's First Life

Describe some of the hypotheses on the origin of life and the characteristics of early prokaryotes, eukaryotes, and multicellular organisms.

The oldest fossils provide evidence that life on Earth was established at least 3.5 billion years ago (**Figure 12.19**). Microscopic fossils similar to modern cyanobacteria have been found in silica-rich chert deposits worldwide. Notable examples include southern Africa, where rocks date to more than 3.1 billion years ago, and the Lake Superior region of western Ontario and northern Minnesota, where the Gunflint Chert contains some fossils older than 2 billion years. Chemical traces of organic matter in even older rocks have led paleontologists to conclude that life may have existed much earlier.

Origin of Life

How did life begin? This question sparks considerable debate, and hypotheses abound. Requirements for life, in addition to a hospitable environment, include the chemical raw materials, principally complex organic compounds that are essential for life. The organic molecules that provide the primary structural material for life and contribute to its functioning are **proteins** (from the Greek *proteiso*, which means primary, hence "primary substance"). Proteins consist of long chains

Evolution of Life Through Geologic Time

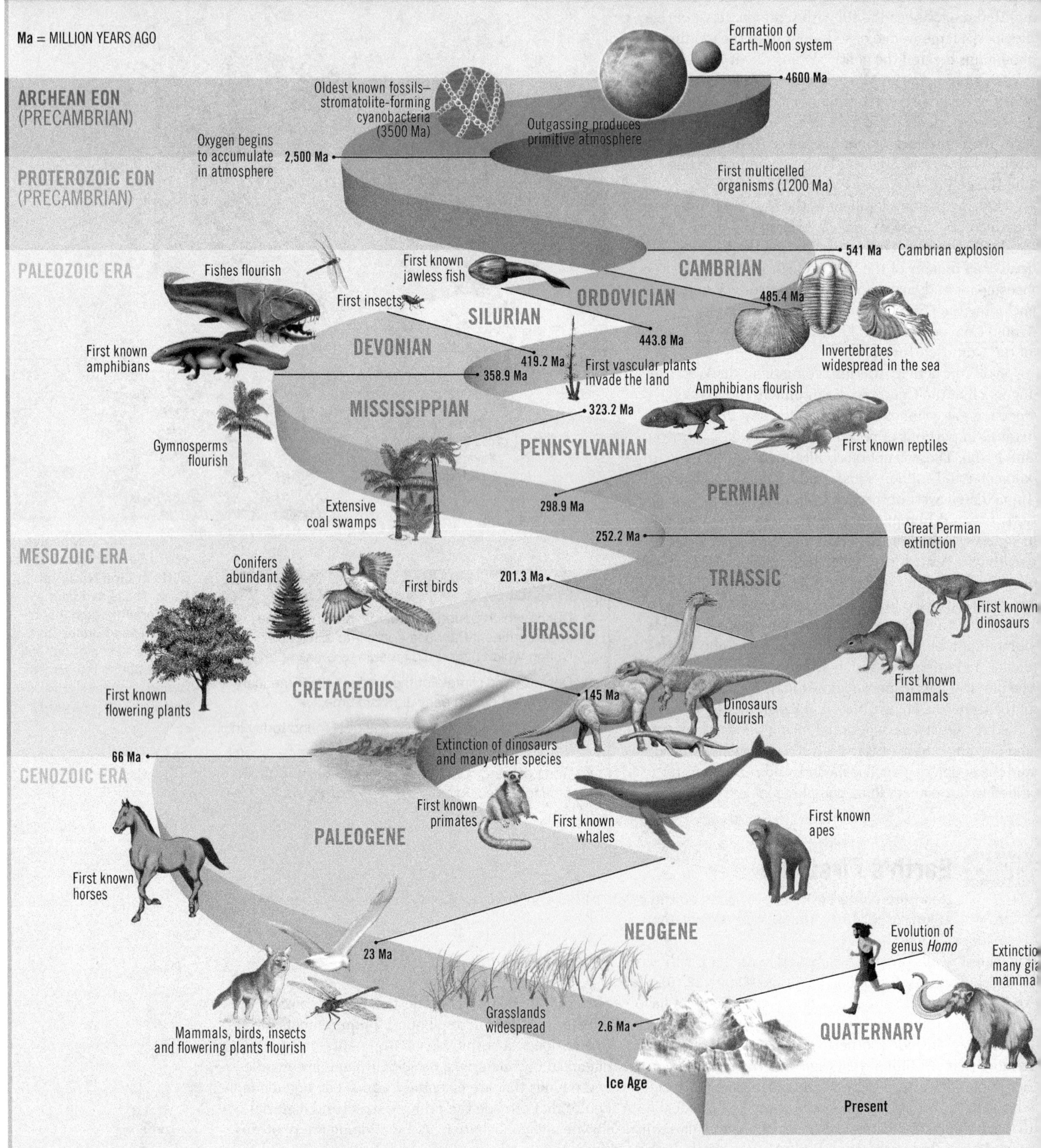

▲ **Figure 12.19 Evolution of life though geologic time** Ages are in millions of years (Ma).

made up of comparatively small molecular units called *amino acids*. Because proteins cannot make copies of themselves—a necessary condition for the proliferation of life—the production of proteins from simpler amino acids requires a template. These information-carrying, self-replicating (or *genetic*) components of life are types of *nucleic acids* with which you are likely familiar—DNA and RNA.

What was the source of the basic organic molecules that became Earth's first life? The first organic molecules may have been synthesized from carbon dioxide and nitrogen, both of which were plentiful in Earth's primitive atmosphere. Some scientists suggest that these gases could have been easily reorganized into amino acids by ultraviolet light. Others consider lightning to have been the impetus, as the groundbreaking experiments conducted by biochemists Stanley Miller and Harold Urey attempted to demonstrate.

Still other researchers suggest that amino acids arrived "ready-made," delivered by asteroids or comets that collided with a young Earth. Evidence for this hypothesis comes from a group of meteorites, called *carbonaceous chondrites*, which contain amino acid–like organic compounds.

Yet another hypothesis proposes that the organic material needed for life came from the methane and hydrogen sulfide that spews from deep-sea hydrothermal vents (black smokers). It is also possible that life originated in hot springs similar to those in Yellowstone National Park.

Earth's First Life: Prokaryotes

Regardless of where or how life originated, it is clear that the journey from "then" to "now" involved change (see Figure 12.19). The first known organisms were simple, single-cell **prokaryotes** (bacteria and similar microbes), in which DNA is not segregated from the rest of the cell in a nucleus.

A major triumph in the history of prokaryote life was the evolution of a primitive type of photosynthesis, which provided the energy that allowed life to proliferate. Recall that photosynthesis by *cyanobacteria*, which began to inhabit Earth about 3.5 billion years ago, contributed to the gradual rise in the level of oxygen in the atmosphere. Fossil evidence for the existence of these microscopic bacteria are distinctively layered mats, called **stromatolites**, which are composed of slimy material secreted by these organisms, along with trapped sediments (Figure 12.20A). What is known about these ancient fossils comes mainly from the study of modern stromatolite structures found in Shark Bay, Australia (Figure 12.20B). Today's stromatolites look like stubby pillars built as these microbes slowly move upward to avoid being buried by the sediment that is continually deposited on them.

Evolution of Eukaryotes

The oldest fossils of larger and more complex organisms, called **eukaryotes**, are about 2.1 billion years old. Eukaryotic cells have their genetic material segregated into a nucleus, and they are more complex in other ways than their prokaryotic precursors. While the first eukaryotes were single-celled, all the multicellular organisms that now inhabit our planet—trees, birds, fish, reptiles, and humans—are eukaryotes.

During much of the Precambrian, life consisted exclusively of single-celled organisms. It wasn't until about 1.2 billion years ago that multicellular eukaryotes evolved. Green algae, one of the first multicellular organisms, contained chloroplasts (used in photosynthesis) and were the likely ancestors of modern plants. The first

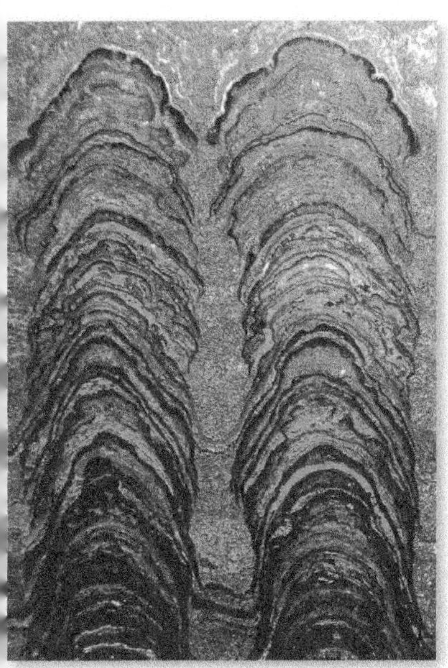
A.

B.

Figure 12.20 **Stromatolites are among the most common Precambrian fossils A.** Cross-section though fossil stromatolites deposited by cyanobacteria. (Photo by Sinclair Stammers/Science Source) **B.** Modern stromatolites exposed at low tide in western Australia. (Photo by Bill Bachman/Science Source)

marine multicellular animals did not appear until somewhat later, perhaps 600 million years ago.

Fossil evidence suggests that life underwent evolutionary change at an excruciatingly slow pace until nearly the end of the Precambrian. At that time, Earth's continents were largely barren, and the oceans were populated mainly with tiny organisms, many too small to be seen with the naked eye. Nevertheless, the stage was set for the evolution of much more diverse plants and animals.

CONCEPT CHECKS 12.6

1. What group of organic compounds combine to form proteins and is therefore necessary for life as we know it?
2. What are stromatolites? What group of organisms is thought to have produced them?
3. Compare *prokaryotes* with *eukaryotes*. To which group do all modern multicellular organisms belong?

12.7 Paleozoic Era: Life Explodes

List the major developments in the history of life during the Paleozoic era.

The Cambrian period marks the beginning of the Paleozoic era, a time span that saw the emergence of a spectacular variety of new life-forms. All major **invertebrate** (animals lacking backbones) groups became widespread during the Cambrian, including jellyfish, sponges, worms, mollusks (such as clams and snails), and arthropods (such as insects and crabs). This expansion in biodiversity, which began about 541 million years ago, is known as the **Cambrian explosion**.

Early Paleozoic Life-Forms

The Cambrian explosion, which lasted about 20 to 30 million years, resulted in an immense variety of invertebrate animals that inhabited Earth's oceans. The development of hard shells and skeletons resulted in predators to develop sharp claws and modified mouth parts to capture and break apart their prey. Other animals developed defense mechanisms, including spikes or armor.

Figure 12.21 **Fossil of a trilobite** Trilobites, which were abundant in the early Paleozoic, found food on the ocean bottom. (Photo by Sarkao/Shutterstock)

The Cambrian period was the golden age of *trilobites* (Figure 12.21). Like modern crabs and lobsters, trilobites had a jointed external skeleton, which enabled them to be mobile and obtain food. More than 600 genera of these mud-burrowing scavengers and grazers flourished worldwide.

The Ordovician period marked the appearance of abundant cephalopods—mobile, highly developed mollusks that became the major predators of their time (Figure 12.22). Descendants of these cephalopods include the squid, octopus, and chambered nautilus that inhabit our modern oceans. Cephalopods were the first truly large organisms on Earth, including one species that reached a length of nearly 10 meters (more than 30 feet).

The early diversification of animals was driven, in part, by the emergence of predatory lifestyles. The larger mobile cephalopods preyed on trilobites that were typically smaller than a child's hand. The evolution of efficient movement was often associated with the development of greater sensory capabilities and more complex nervous systems. These early animals elaborated sensory devices for detecting light, odor, and touch.

Mid-Paleozoic Life

Approximately 450 million years ago, green algae that had adapted to survive at the water's edge gave rise to the first multicellular land plants. Some of the adaptations needed for sustaining plant life on land were the ability to transport water and minerals internally as well as stay upright, despite gravity and winds. By 420 million years ago, plants stood upright and possessed a primitive *vascular system* to transport ion-rich water. These leafless vertical spikes were about the size of a human index finger, about 10 centimeters (4 inches) tall. By the beginning of the Mississippian period, there were forests with treelike plants tens of meters tall (Figure 12.23).

In the ocean, the **vertebrates** (animals with backbones), began to thrive—primarily the ancestors of modern fish. An important derived characteristic of these early vertebrates was jaws, which enabled them to grab, pry, or bite off a chunk of flesh. One example was the armor-plated fishes called *placoderms* (meaning "plate-skinned") that evolved during the Ordovician (Figure 12.24).

Other fish evolved during the Devonian, including sharks with cartilage skeletons and bony fish with hard internal skeletons—the two groups to which most modern fish belong. The first large vertebrates, fish

#1
#2
#3
#4
#5
▲ Figure 12.22 **Artistic depiction of a shallow Ordovician sea** During the Ordovician period (488–444 million years ago), the shallow waters of an inland sea over central North America contained an abundance of marine invertebrates. Shown in this reconstruction are (1) corals, (2) a trilobite, (3) a snail, (4) brachiopods, and (5) a straight-shelled cephalopod. (Photo by Ron Testa/Field Museum Library/Getty Images)

proved to be faster swimmers than most invertebrates and possessed acute senses and large brains. They became dominant predators of the sea, which is why the Devonian period is sometimes referred to as the "Age of the Fishes."

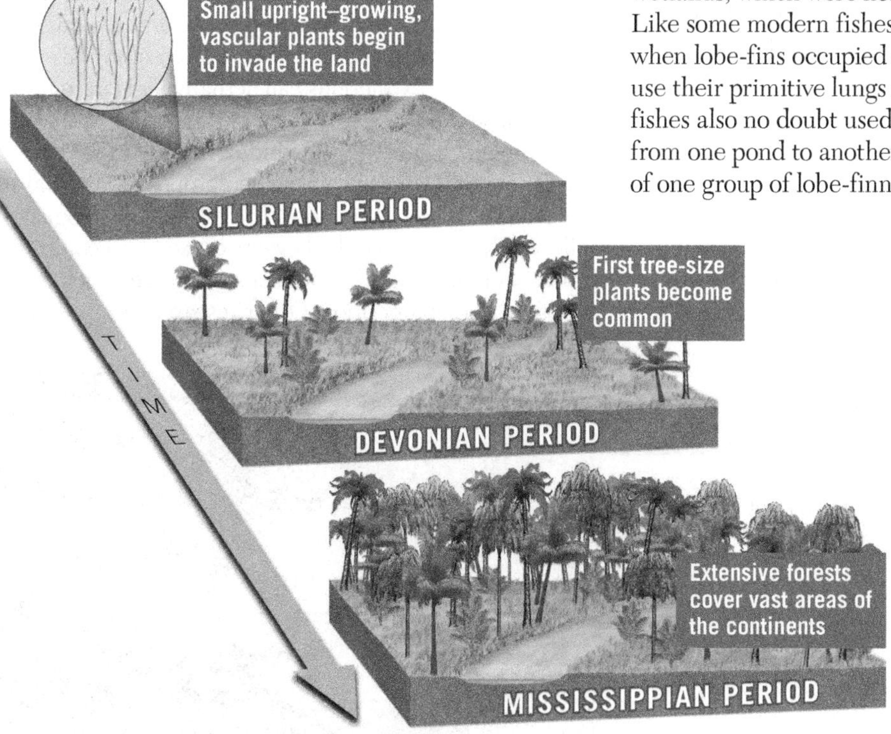
Small upright-growing, vascular plants begin to invade the land
SILURIAN PERIOD
First tree-size plants become common
DEVONIAN PERIOD
Extensive forests cover vast areas of the continents
MISSISSIPPIAN PERIOD
TIME
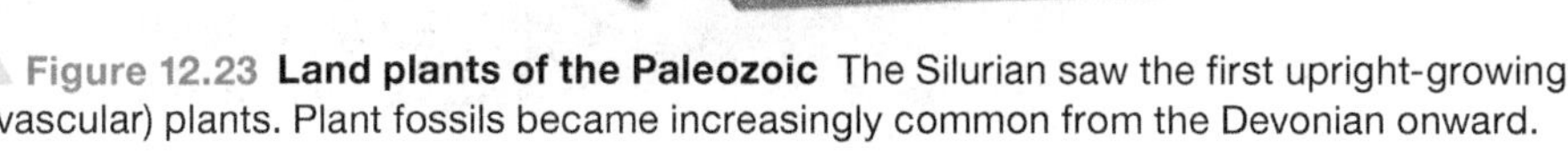
▲ Figure 12.23 **Land plants of the Paleozoic** The Silurian saw the first upright-growing (vascular) plants. Plant fossils became increasingly common from the Devonian onward.

Vertebrates Move to Land

At the beginning of the Devonian period all vertebrates shared the same fish-like anatomy. Shortly thereafter, a group of fishes called *lobe-finned fishes* or simply *lobe-fins* began to adapt to terrestrial environments (Figure 12.25A). The Devonian landscape included abundant coastal wetlands, which were home to a wide range of lobe-fins. Like some modern fishes, such as the African lungfish, when lobe-fins occupied oxygen-poor water, they would use their primitive lungs for breathing. Some lobe-finned fishes also no doubt used their stout bony fins to move from one pond to another. By the late Devonian, the fins of one group of lobe-finned fishes had evolved into four

▼ Figure 12.24 **Fossil of an armored fish belonging to a group called the placoderms** This particular placoderm was a formidable predator that grew up to 10 meters (30 feet) in length, although most were much smaller. (Photo courtesy of Field Museum LibraryPremium Archive/Getty Images)

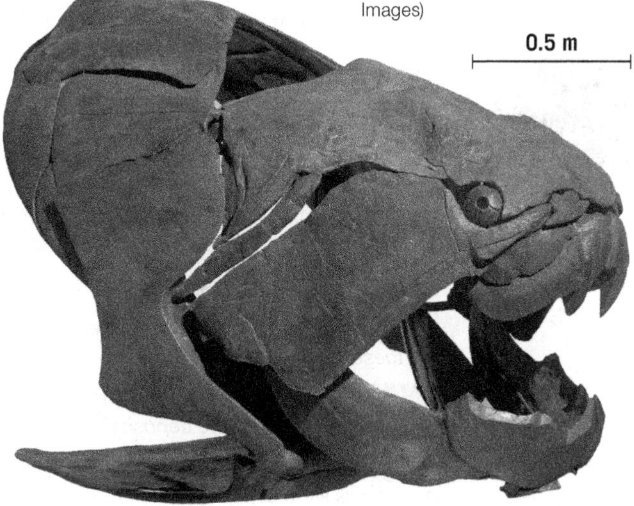
0.5 m

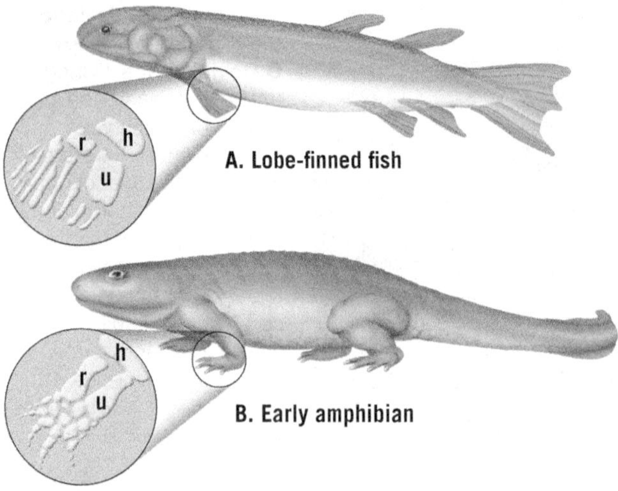

Figure 12.25
Comparison of the anatomical features of a lobe-finned fish and an early amphibian A. The fins on the lobe-finned fish contained the same basic elements (*h*, humerus, or upper arm; *r*, radius, and *u*, ulna, which correspond to the lower arm) as those of the amphibians. **B.** This amphibian is shown with the standard five toes, but early amphibians had as many as eight toes, as well as other characteristics that made them quite different from modern amphibians. Eventually the amphibians evolved to have a standard toe count of five.

limbs and feet with digits (Figure 12.25B). The limbs were used to support the weight of the animal on land, and the feet aided in walking. By about 365 million years ago, this group of lobe-fins had given rise to the four-legged ancestors of the first amphibians and eventually to other land-dwelling vertebrate groups, including reptiles and mammals (Figure 12.26).

Modern amphibians, such as frogs, toads, and salamanders, are small and occupy mainly ponds or damp habitats such as swamps and rainforests. Conditions during the late Paleozoic were ideal for these newcomers to land. Large tropical swamps teeming with insects and millipedes extended across North America, Europe, and Siberia (Figure 12.27). With an abundance of ideal habitats, amphibians diversified rapidly. Some even took on lifestyles and forms similar to those of modern reptiles such as crocodiles.

Despite their success on land, amphibians are not fully adapted to drier habitats. Amphibians must lay their eggs in water or moist environments on land because their eggs lack a shell and dehydrate in dry air. In fact, **amphibian** means "both ways of life." Frogs, for instance, lay their eggs in water and develop as aquatic tadpoles with gills and tails (fish characteristics) before maturing into air-breathing adults with four legs.

Reptiles: The First True Terrestrial Vertebrates

Fossil evidence indicates that about 310 million years ago, the first reptiles evolved from an ancestor common to both reptiles and amphibians. The earliest known reptiles resembled lizards with small sharp teeth. The **reptile** group includes snakes, turtles, lizards, and crocodiles, as well as extinct groups such as dinosaurs, ichthyosaurs, and plesiosaurs. (Birds, which are descendants of one group of dinosaurs, can also be considered reptiles, depending on the classification system used.)

Reptiles share several derived characteristics that are beneficial for life on land. For example, unlike amphibians, most reptiles have scales made of the protein keratin (also found in human fingernails) that help prevent the loss of body fluids and resist abrasion. More importantly, most reptiles lay shell-covered eggs, called **amniotic eggs** that contain a fluid that bathes the embryo—a significant evolutionary step (Figure 12.28). Because the reptile embryo matures in this watery environment, the shelled egg has been characterized as a "private aquarium" in which the embryos of these land vertebrates spend their water-dwelling stage of life. With this amniotic egg, the remaining ties to a watery environment were finally broken. Thus, unlike amphibians, the first reptiles were able to occupy a wider range of terrestrial habitats—most notably arid regions.

The Great Permian Extinction

A **mass extinction** occurred at the close of the Permian period, and in it a large number of Earth's species became extinct. During this mass extinction, 70 percent of all land-dwelling vertebrate species and perhaps 90 percent of all marine organisms were obliterated; it was the most severe of five mass extinctions to occur over the past 500 million years. Each extinction wreaked havoc with the existing biosphere, wiping out large numbers of species. In each case, however, survivors created new biological communities, which were often more diverse. Therefore, mass extinctions can actually invigorate life on Earth, as the few hardy survivors eventually filled the environmental niches left behind by the victims.

Several mechanisms have been proposed to explain these ancient mass extinctions. Initially, paleontologists believed these were gradual events caused by a combination of climate change and biological forces, such as predation and competition. Other research groups have

EYE ON EARTH 12.2

The rocks shown here are Cambrian-age stromatolites of the Hoyt Limestone, exposed at Lester Park, near Saratoga Springs, New York. (Photo by Michael C. Rygel)

Michael C. Rygel via Wikimedia Commons

QUESTION 1 *Using Figure 12.3, determine approximately how many years ago these rocks were deposited.*

QUESTION 2 *What is the name of the group of organisms that likely produced these limestone deposits?*

QUESTION 3 *What was the environment like in this part of New York when these rocks were deposited?*

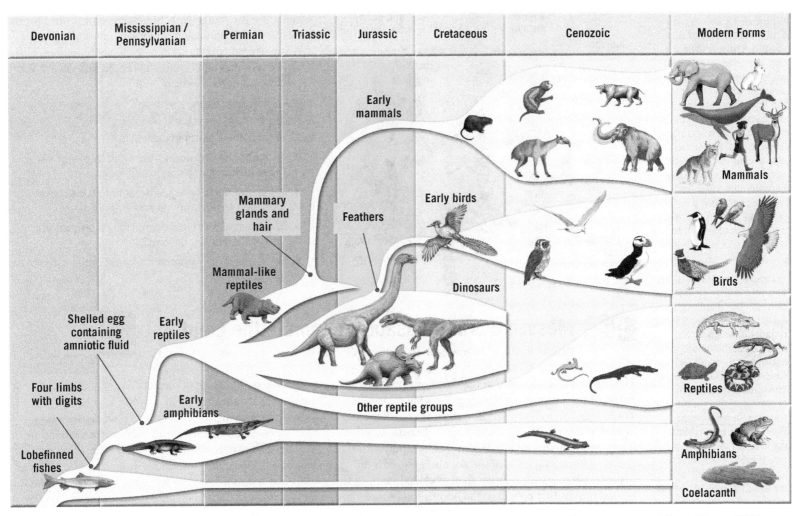

Devonian
Mississippian / Pennsylvanian
Permian
Triassic
Jurassic
Cretaceous
Cenozoic
Modern Forms

Early mammals
Mammary glands and hair
Feathers
Early birds
Mammal-like reptiles
Dinosaurs
Shelled egg containing amniotic fluid
Early reptiles
Four limbs with digits
Early amphibians
Other reptile groups
Lobefinned fishes
Mammals
Birds
Reptiles
Amphibians
Coelacanth

SmartFigure 12.26 **Relationships of major land-dwelling vertebrate groups and their divergence from lobefin fish**

TUTORIAL
http://goo.gl/q6ltN0

attempted to link certain mass extinctions to the explosive impact of a large asteroid striking Earth's surface.

The most widely held view is that the Permian mass extinction was driven mainly by volcanic activity because it coincided with a period of voluminous eruptions of flood basalts that blanketed about 1.6 million square kilometers (624,000 square miles), an area nearly the size of Alaska. This series of eruptions, which lasted roughly 1 million years, occurred in northern Russia, in an area called the Siberian Traps. It was the largest volcanic eruption in the past 500 million years. The release of huge amounts of carbon dioxide likely generated a period of accelerated

Figure 12.27 **Artistic depiction of a Pennsylvanian-age coal swamp** Shown are scale trees (left), seed ferns (lower left), and horsetails (right). Also note the large dragonfly. (Photo by John Weinstein/Field Museum Library/Premium Archive/Getty Images)

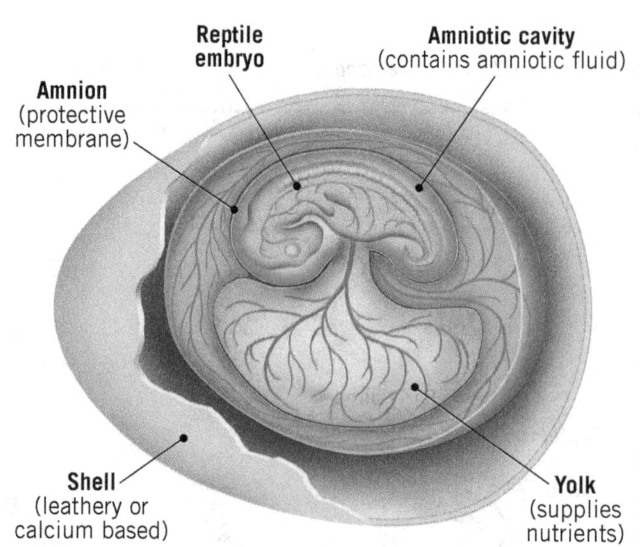

Figure 12.28 **The shelled egg of a reptile** The *amniotic egg* contains specialized membranes, including the *amnion*, which encloses the fluid that bathes the embryo and acts as a hydraulic shock absorber to protect it. Mammals also produce an amniotic egg, but in most groups of mammals the outer membranes form a placenta and an umbilical cord that is attached to the uterus of a pregnant adult.

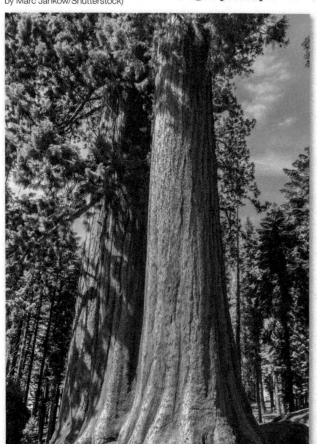

greenhouse warming, while the emission of sulfur dioxide is credited with producing copious amounts of acid rain and low-oxygen conditions in marine environments. These drastic environmental changes likely put excessive stress on many of Earth's life-forms.

CONCEPT CHECKS 12.7

1. What is the Cambrian explosion?
2. Describe the obstacles plants had to overcome in order to inhabit the continents.
3. What group of animals is thought to have moved onto land to become the first amphibians?
4. What features of typical amphibians prevent them from living entirely on land?
5. What major developments allowed reptiles to move inland?

12.8 Mesozoic Era: Dinosaurs Dominate the Land

Briefly explain the major developments in the history of life during the Mesozoic era.

The life-forms that existed at the dawn of the Mesozoic era were the survivors of the great Permian extinction. These organisms diversified in many ways to fill the biological voids created at the close of the Paleozoic. While life on land underwent a radical transformation with the rise of the dinosaurs, life in the sea also entered a dramatic phase of transformation that produced many of the animal groups that prevail in the oceans today, including modern groups of fish, crustaceans, mollusks, and starfish and their relatives.

Figure 12.29 **The giant sequoia is a cone-bearing conifer** The giant sequoia is perhaps the largest living organism, weighing as much as 2,500 metric tons, equivalent to about 24 blue whales or 40,000 humans. The natural distribution of giant sequoias is the western Sierra Nevada, California. A close relative is the coast redwood, the tallest living trees, which can grow to about 115 meters (380 feet) tall. (Photo by Marc Jankow/Shutterstock)

Gymnosperms: The Dominant Mesozoic Trees

Conditions on land favored organisms that could adapt to drier climates. One useful evolutionary adaptation was the **seed**, an embryo packaged with a supply of nutrients inside a protective coating. Two groups of seed plants exist, *gymnosperms* and *angiosperms*. **Gymnosperms** produce "naked seeds" that develop on modified leaves, usually scale-like structures that form a cone. Think of the rather flat structures that make up a pinecone. Gymnosperm seeds are referred to as naked because they are not enclosed in a structure, such as those found in flowers that eventually become fruits—apple seeds, for example. Once gymnosperm seeds mature, the scales of the cone separate, and the seeds are released. Unlike the first plants to invade the land, the more primitive ferns, seed-bearing gymnosperms no longer had to depend on a water body for fertilization. Consequently, they readily adapted to drier habitats.

The gymnosperms dominated terrestrial ecosystems throughout much of the Mesozoic era, which lasted from 252 to 66 million years ago. Examples of this group include cycads, which resemble large pineapple plants; ginkgo trees, which have fan-shaped leaves; and the largest plants of the time, the *conifers*, whose modern descendants include pines, junipers, and redwoods (Figure 12.29). The best-known fossil occurrence of these ancient trees is in northern Arizona's Petrified Forest National Park. Here, huge petrified logs lie exposed at the surface, exhumed by the weathering of rocks of the Triassic Chinle Formation (Figure 12.30).

Reptiles Take Over the Land, Sea, and Sky

Among the animals, reptiles readily adapted to the drier Permian and Triassic environment. On land, the dinosaurs ranged in size from small bird-sized bipeds (animals that move on two feet) to 40-meter (130 feet) long quadrupeds (animals with four feet) with necks long enough to allow them to graze from the tops of tall trees.

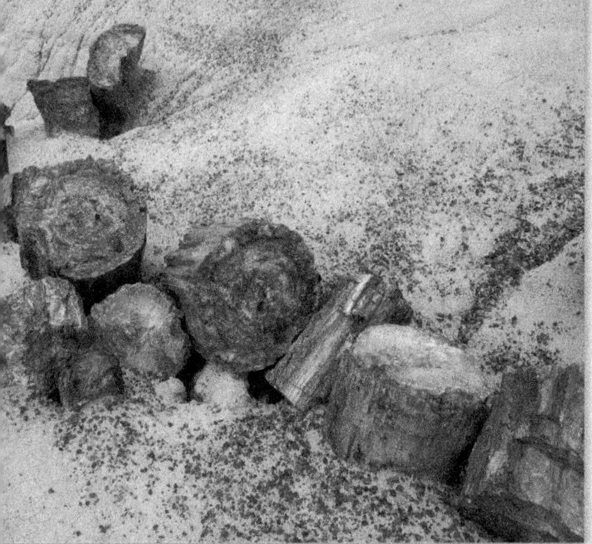

Figure 12.30 **Petrified logs of Triassic age, Arizona's Petrified Forest National Park** (Photo by Bernd Siering/Premium/AGE Fotostock)

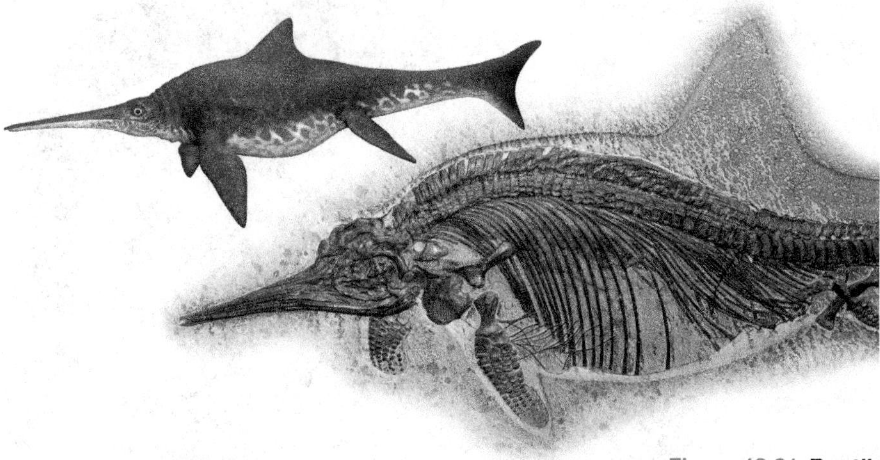

Figure 12.31 **Reptiles returned to the sea** *Ichthyosaurs* are one of the groups of reptiles that returned to the sea during the Mesozoic. (Photo by Blickwinkel/Koenig/Alamy Stock Photo)

Other large dinosaurs were carnivorous, such as the infamous bipedal *Tyrannosaurus.*

Some reptiles evolved with specialized characteristics that allowed them to inhabit drastically different habitats. One group, the pterosaurs, became airborne. How the largest pterosaurs—some of which had wing spans greater than 11 meters (35 feet) and weighed more than 90 kilograms (200 pounds)—took flight is still unknown. Other reptiles returned to the sea, including fish-eating *plesiosaurs* and *ichthyosaurs* (Figure 12.31). These reptiles became proficient swimmers, breathing by means of lungs rather than gills, much like modern sea-going mammals such as whales and dolphins.

By about 160 million years ago, a group of feathered dinosaurs gave rise to modern birds (Figure 12.32). Many researches consider *Archaeopteryx* to be the first known bird. *Archaeopteryx* had feathered wings but retained teeth, clawed digits in its wings, and a long tail with many vertebrae. A recent study concluded that *Archaeopteryx* flew well at high speeds, but unlike most modern birds, it could not take off from a standing position. Rather, these descendants of bird-like dinosaurs took flight by running and leaping into the air. Other researchers disagree and envision them as climbing animals that glided down to the ground, following the idea that birds evolved from tree-dwelling gliders. Whether the first birds took to the air from the ground *up* or from the trees *down* or by *both mechanisms* is a question scientists continue to debate.

Dinosaurs flourished for nearly 160 million years. However, by the close of the Mesozoic, all dinosaurs (except for the ancestors of modern birds) and many related reptile groups became extinct. The huge land-dwelling dinosaurs, the marine plesiosaurs, and the flying pterosaurs are known only through the fossil record. What caused this great extinction?

Demise of the Dinosaurs

The boundaries between divisions on the geologic time scale represent times of significant geologic and/or biological change. Of special interest is the boundary between the Mesozoic era ("middle life") and the Cenozoic era ("recent life"), about 66 million years ago. During this transition, roughly three-quarters of

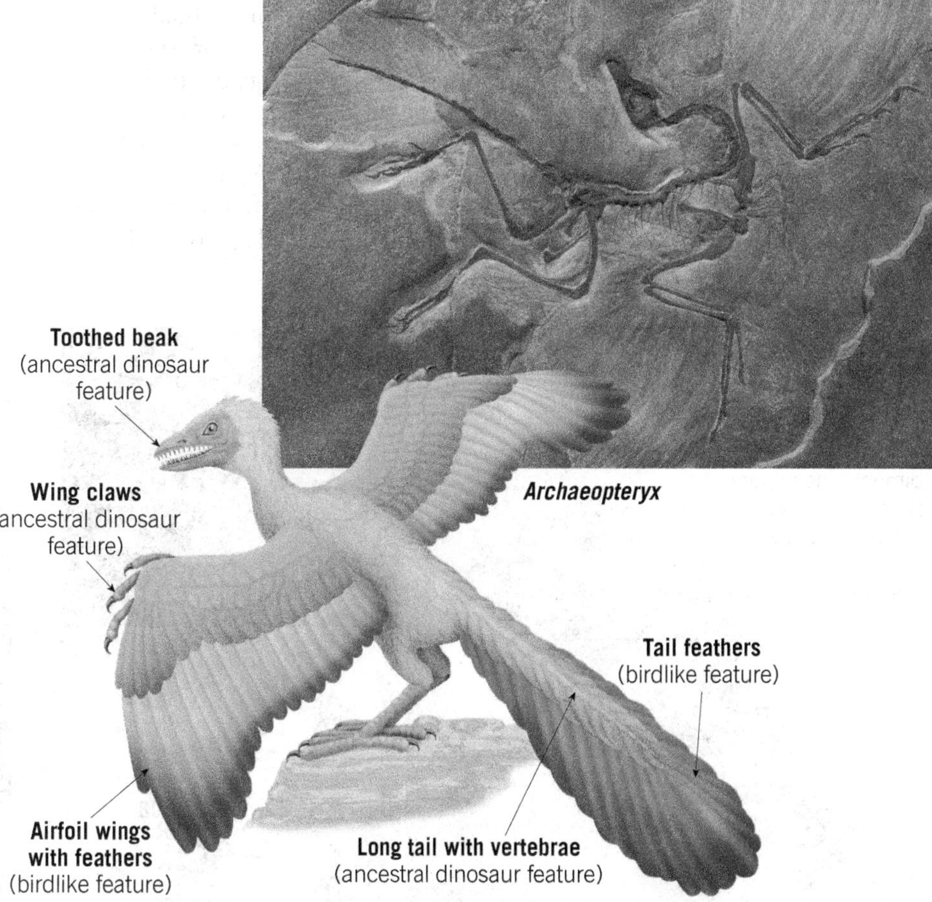

Figure 12.32 ***Archaeopteryx*, a transitional form related to modern birds** *Archaeopteryx* had feathered wings and a feathered tail that appear intended for flight, but in many ways it resembled ancestral dinosaurs. The sketch shows an artist's reconstruction of *Archaeopteryx.* (Photo by Michael Collier)

▲ **Figure 12.33 Artist's rendering of *Allosaurus*** *Allosaurus* was a large carnivorous dinosaur that lived in the late Jurassic period (155–145 million years ago). (Image by Roger Harris/Science Source)

all plant and animal species died out in another mass extinction. This boundary marks the end of the era in which dinosaurs and other large reptiles dominated the landscape (**Figure 12.33**) and the beginning of the era when mammals assumed that role.

What could have triggered the extinction of one of the most successful groups of land animals? An increasing number of researchers support the view that the dinosaurs fell victim to a "one–two punch." The first blow occurred during the last few million years of the Mesozoic era; climate data indicates that during this time, average temperature over the land increased by more than 20°C (40°F) in a few tens of thousands of years—a blink of the eye in geologic time. This episode of global warming is thought to have coincided with massive basaltic eruptions, which produced the Deccan Plateau, located in what is now India. The Deccan eruptions released massive amounts of carbon dioxide, which caused a period of greenhouse warming that resulted in a dramatic rise in temperatures. Presumably, this period of global warming snuffed out some species and hobbled others.

The final blow came about 66 million years ago, when our planet was struck by a stony meteorite, a relic from the formation of the solar system. The errant mass of rock was approximately 10 kilometers (6 miles) in diameter and was traveling at about 90,000 kilometers per hour at the time of impact. It collided with the southern portion of North America in a shallow tropical sea—now Mexico's Yucatan Peninsula (**Figure 12.34**).

Following the impact, suspended dust greatly reduced the amount of sunlight reaching Earth's surface, which resulted in global cooling ("impact winter") and inhibited photosynthesis, disrupting food production. Long after the dust settled, the sulfur compounds added to the atmosphere by the blast remained suspended and reflected solar radiation back to space, perpetuating the unusually cold temperatures.

One piece of evidence that points to a catastrophic collision 66 million years ago is a thin layer of sediment, less than 1 centimeter thick, discovered in several places around the globe. This sediment contains a high level of the element *iridium*, rare in Earth's crust but found in high proportions in stony meteorites (**Figure 12.35**). This layer presumably contains scattered remains of the meteorite responsible for the environmental changes that

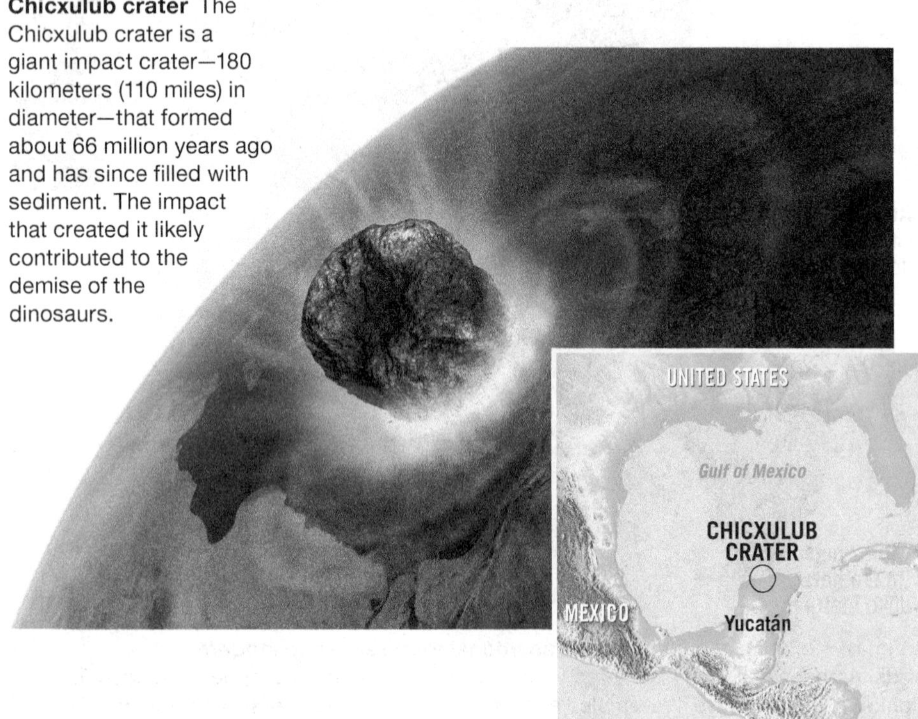

▶ **Figure 12.34 Chicxulub crater** The Chicxulub crater is a giant impact crater—180 kilometers (110 miles) in diameter—that formed about 66 million years ago and has since filled with sediment. The impact that created it likely contributed to the demise of the dinosaurs.

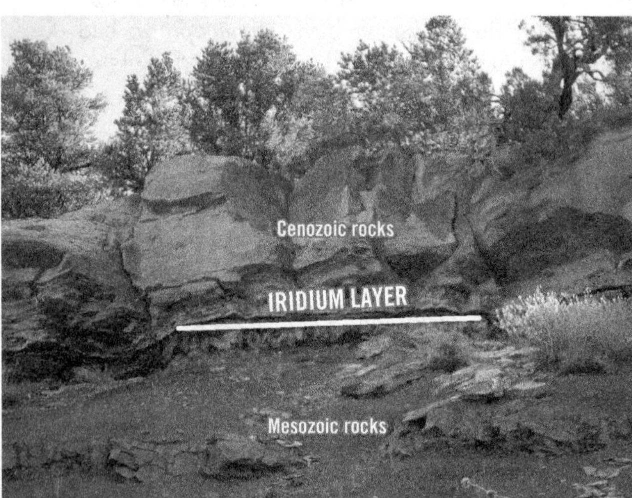

▲ **Figure 12.35 Iridium layer: Evidence for a catastrophic impact 66 million years ago** A thin layer of sediment has been discovered worldwide at Earth's physical boundary separating the Mesozoic and Cenozoic eras. This sediment contains a high level of iridium, an element that is rare in Earth's crust but found in much higher concentrations in stony meteorites. (Photo courtesy of National Parks)

provided the second and final blow that led to the demise of many reptile groups.

Regardless of what caused this massive extinction, its outcome provides a valuable lesson in understanding the role that catastrophic events play in shaping our planet's physical landscape and biosphere. The extinction of the dinosaurs opened habitats for the mammals that survived. These new habitats, along with evolutionary forces, led to the development of the great diversity of mammals that occupy our modern world.

CONCEPT CHECKS 12.8

1. What group of plants became the dominant trees during the Mesozoic era? Name a modern descendant of this group.
2. What was the dominant reptile group on land during the Mesozoic?
3. What group of reptiles gave rise to modern birds?
4. Identify two groups of reptiles that returned to life in the sea.

12.9 Cenozoic Era: Mammals Diversify

Discuss the major developments in the history of life during the Cenozoic era.

During the Cenozoic era, from about 65.5 million years ago to the present, mammals replaced dinosaurs as the dominant vertebrates on land. Among plants, **angiosperms** (flowering plants), which appeared about 140 million years ago, came to occupy most terrestrial environments. Today, angiosperms make up roughly 90 percent of all plant species.

The development of flowering plants strongly influenced the evolution of both birds and mammals that feed on seeds and fruits, as well as many insect groups. During the middle of the Cenozoic, one type of angiosperm, grasses, spread rapidly to form grasslands—a completely novel type of landscape. At the same time, herbivorous mammal groups gave rise to a great variety of grazers, and carnivores evolved to prey on them (Figure 12.36).

During the Cenozoic era, the ocean teemed with modern fish such as tuna, swordfish, and barracuda. In addition, some mammals, including seals, whales, and walruses, took up life in the sea.

From Dinosaurs to Mammals

Mammals coexisted with dinosaurs for nearly 100 million years but remained mostly small and probably mostly nocturnal. Then, about 66 million years ago, fate intervened when the large meteorite collided with Earth and dealt the final crashing blow to the reign of the dinosaurs. This transition from one prominent group to another is clearly visible in the fossil record.

Mammals are named for their *mammary glands* that produce milk for their offspring. Another distinctive

B.

A.

Figure 12.36 **Angiosperms became the dominant plants during the Cenozoic** Angiosperms, commonly known as flowering plants, consist of a group of seed plants that have reproductive structures called flowers and fruits. **A.** The most diverse and widespread of modern plants, many angiosperms display easily recognizable flowers. (Photo by WDG Photo/Shutterstock) **B.** Some angiosperms, including grasses, have very tiny flowers. The expansion of the grasslands during the Cenozoic era greatly increased the diversity of grazing mammals and the predators that feed on them. (Photo A by Torleif Svensson/Corbis/Getty Images)

characteristic of mammals is hair. Also, like birds, mammals are **endothermic**, which means they are capable of maintaining a constant body temperature thorough metabolic activity, and they have large brains for their size compared to other vertebrate groups.

With the demise of the large Mesozoic dinosaurs, Cenozoic mammals diversified rapidly. The many forms that exist today evolved from small primitive mammals that were characterized by short legs; flat, five-toed feet; and small brains. Their development and specialization took four principal directions: increase in size, increase in brain capacity, specialization of teeth to accommodate more diverse diets, and specialization of limbs to be better equipped for a particular lifestyle or habitat.

Mammal Groups

The three major lineages of living mammals, the *monotremes* (egg-laying mammals), *marsupials* (mammals with a pouch), and *placentals* (the group to which humans belong), emerged during the Mesozoic. The groups differ principally in their modes of reproduction. The monotremes lay hard-shelled eggs, an ancestral characteristic retained in this group as well as by most modern reptiles. The platypus, which is native to Australia and Tasmania, is one of only a handful of monotreme species remaining.

Young marsupials, on the other hand, are born live at a very early stage of development. At birth, the tiny and immature young enter the mother's pouch to suckle as they complete their development. Today, marsupials are found primarily in Australia, where they took a separate evolutionary path largely isolated from placental mammals. Modern marsupials include kangaroos, opossums, and koalas (**Figure 12.37**).

Figure 12.37 **Kangaroos, examples of marsupial mammals** After the breakup of Pangaea, the Australian marsupials evolved independently. (Photo by Martin Harvey/Photolibrary/Getty Images)

Placental mammals, also called *eutherians*, complete their embryonic development in their mother's uterus and are nourished via a placenta, to which they are connected by an umbilical cord. Most mammals, including humans, are placental. Other members of this group include wolves, elephants, bats, manatees, and monkeys.

Humans: Mammals with Large Brains and Bipedal Locomotion

Both fossil and genetic evidence suggest that, by 6.5 million years ago, our own group, the hominins, had branched off from the line leading to modern chimpanzees. Scientists have a good record of this evolution in fossils found in several sedimentary basins in Africa, including the Rift Valley system in East Africa. To date, anthropologists have unearthed roughly 20 extinct species of primates that are more closely related to humans than to chimpanzees.

The genus *Australopithecus*, which came into existence about 4.2 million years ago, showed skeletal characteristics that were intermediate between our apelike ancestors and modern humans. In particular, *Australopithecus* walked upright. Evidence for this bipedal stride includes footprints preserved in 3.2-million-year-old ash deposits at Laetoli, Tanzania (**Figure 12.38**). This new way of moving around made it possible for our human ancestors to leave forested habitats and to travel long distances for hunting and gathering food.

The earliest fossils of our genus, *Homo*, include the remains of *Homo habilis*, nicknamed "handy man" because they were often found with sharp stone tools in sedimentary deposits from 2.4 to 1.6 million years ago. *Homo habilis* had a shorter jaw and a larger brain than its predecessor.

During the next 1.3 million years, our ancestors developed substantially larger brains and long slender legs with hip joints adapted for long-distance walking. These species (including *Homo erectus*) ultimately gave rise to our species, *Homo sapiens*, as well as to some extinct related species, including the Neanderthals (*Homo neanderthalis*). Despite having larger brains than present-day humans and being able to fashion hunting tools from wood and stone, Neanderthals became extinct about 28,000 years ago. At one

Figure 12.38 **Footprints of *Australopithecus* in ash deposits at Laetoli, Tanzania** (Photo by John Reader/Science Source)

time, Neanderthals were considered a stage in the evolution of *Homo sapiens*, but that view has largely been abandoned.

Based on our current understanding, *Homo sapiens* originated in Africa about 200,000 years ago and began to spread around the globe. The oldest-known *Homo sapiens* fossils outside Africa were found in the Middle East and date to 115,000 years ago. Humans are known to have coexisted with Neanderthals and other prehistoric populations, with remains found in Siberia, China, and Indonesia. Further, there is mounting genetic evidence that our ancestors interbred with members of some of these groups.

Figure 12.39 Cave painting of animals by early humans (Photo courtesy of Sisse Brimberg/National Geographic/Getty Images)

By 36,000 years ago, humans were producing spectacular cave paintings in Europe (**Figure 12.39**). About 28,000 years ago, all prehistoric hominin populations except for *Homo sapiens* died out.

Large Mammals and Extinction

During the rapid mammal diversification of the Cenozoic era, some species became very large. For example, by the Oligocene epoch (about 23 million years ago), a hornless rhinoceros evolved that stood nearly 5 meters (16 feet) high. It is the largest land mammal known to have existed. As time passed, many other mammals evolved to large forms—more, in fact, than now exist. Many of these large forms were common as recently as 11,000 years ago. However, a wave of late Pleistocene extinctions rapidly eliminated many of these animals from the landscape.

North America experienced the extinction of mastodons and mammoths, both large relatives of the modern elephants (**Figure 12.40**). In addition, saber-toothed cats, giant beavers, large ground sloths, horses, giant bison, and others died out. In Europe, late Pleistocene extinctions included woolly rhinos, large cave bears, and Irish elk. Scientists remain puzzled about the reasons for these more recent extinctions of large mammals. Because these large animals survived several major glacial advances and interglacial periods, it is difficult to ascribe their extinction to climate change. Some scientists hypothesize that early humans hastened the decline of these mammals by selectively hunting large forms.

CONCEPT CHECKS 12.9

1. What animal group became the prominent land animals of the Cenozoic era?
2. Explain how the demise of the dinosaurs impacted the development of mammals.
3. Where did researchers discover most of the evidence for the early evolution of our hominin ancestors?
4. What two characteristics best separate humans from other mammals?
5. Describe one hypothesis that explains the extinction of large mammals in the late Pleistocene.

Figure 12.40 Artist's rendition of mammoths These relatives of modern elephants were among the large mammals that became extinct at the close of the Ice Age. (Image courtesy of INTERFOTO/Alamy)

12 CONCEPTS IN REVIEW
Earth's Evolution Through Geologic Time

12.1 What Makes Earth Habitable?

List the principal characteristics that make Earth habitable.

KEY TERMS: exoplanet, habitable zone

- As far as we know, Earth is unique among planets in hosting life. The planet's size, composition, and distance from the Sun all contribute to conditions that support life.

12.2 Birth of a Planet

Outline the major stages in Earth's evolution, from the Big Bang to the formation of our planet's layered internal structure.

KEY TERMS: supernova, solar nebula, planetesimal, protoplanet, Hadean

- The universe is thought to have formed about 13.8 billion years ago, with the Big Bang, which generated space, time, energy, and matter, including the elements hydrogen and helium. Elements heavier than hydrogen and helium were synthesized by nuclear reactions that occur in stars.
- Earth and the solar system formed around 4.6 billion years ago, with the contraction of a solar nebula. Collisions between clumps of matter in this spinning disk resulted in the growth of planetesimals and then protoplanets. Over time, the matter of the solar nebula was concentrated into a smaller number of larger bodies: the Sun, the rocky inner planets, the icy outer planets, moons, comets, and asteroids.
- The early Earth was hot enough for rock and iron to melt, thanks to the kinetic energy of impacting asteroids and planetesimals as well as the decay of radioactive isotopes. This allowed iron to sink to form Earth's core and rocky material to rise to form the mantle and crust.

? The accompanying image shows Comet Shoemaker-Levy 9 impacting Jupiter in 1994. After this event, what happened to Jupiter's total mass? How was the number of objects in the solar system affected?

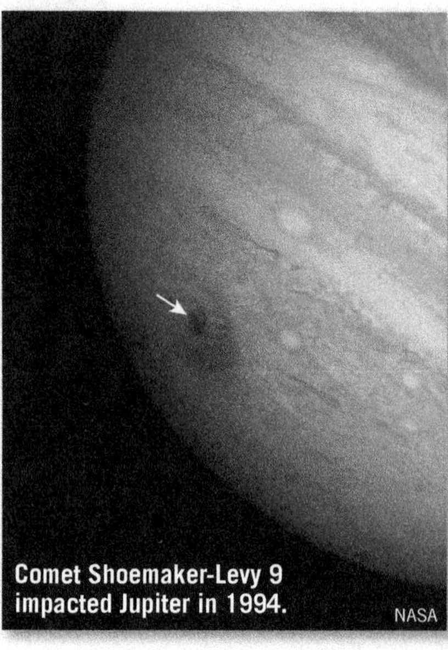
Comet Shoemaker-Levy 9 impacted Jupiter in 1994. NASA

12.3 Origin and Evolution of the Atmosphere and Oceans

Describe how Earth's atmosphere and oceans formed and evolved through time.

KEY TERMS: outgassing, photosynthesis, banded iron formation, Great Oxygenation Event

- Earth's atmosphere formed as volcanic outgassing added mainly water vapor and carbon dioxide to the primordial atmosphere of gases common in the early solar system: methane and ammonia.
- Free oxygen began to accumulate through photosynthesis by cyanobacteria, which released oxygen as a waste product. Much of this early oxygen immediately reacted with iron dissolved in seawater and settled to the ocean floor as chemical sediments called banded iron formations. The Great Oxygenation Event of 2.5 billion years ago marks the first evidence of significant amounts of free oxygen in the atmosphere.
- Earth's oceans formed after the planet's surface had cooled. Soluble ions weathered from the crust were carried to the ocean, making it salty. The oceans also absorbed tremendous amounts of carbon dioxide from the atmosphere.

12.4 Precambrian History: The Formation of Earth's Continents

Explain the formation of continental crust, how continental crust becomes assembled into continents, and the role that the supercontinent cycle has played in this process.

KEY TERMS: craton, shield, supercontinent, supercontinent cycle

- The Precambrian includes the Archean and Proterozoic eons. Our knowledge of these eons is limited because erosion has destroyed much of the rock record.
- Continental crust was produced over time through the recycling of ultramafic and mafic crust in an early version of plate tectonics. Small crustal fragments formed and amalgamated into large crustal provinces called cratons. Over time, North America and other continents grew through the accretion of new terranes around the edges of this central "nucleus" of crust.

(12.4 continued)

- Early cratons not only merged but sometimes rifted apart. The supercontinent Rodinia formed around 1.1 billion years ago and then rifted apart, opening new ocean basins. Over time, these ocean basins also closed to form a new supercontinent called Pangaea around 250 million years ago. Like Rodinia before it, Pangaea broke up as part of the ongoing supercontinent cycle.
- The formation of elevated oceanic ridges following the breakup of a supercontinent displaced enough water that sea level rose, and shallow seas flooded the continents. The breakup of continents can also influence the direction of ocean currents, with important consequences for climate.

? Consult Figure 12.12 and briefly summarize the history of the assembly of North America over the past 3.5 billion years.

12.5 Geologic History of the Phanerozoic: The Formation of Earth's Modern Continents

List and discuss the major geologic events in the Paleozoic, Mesozoic, and Cenozoic eras.

KEY TERMS: Pangaea, Laurasia, Gondwana

- The Phanerozoic eon began 545 million years ago and is divided into the Paleozoic, Mesozoic, and Cenozoic eras.
- In the Paleozoic era, North America experienced a series of collisions that resulted in the rise of the young Appalachian mountain belt, as part of the assembly of Pangaea. High sea levels caused the ocean to cover vast areas of the continent and resulted in a thick sequence of sedimentary strata.
- During the Mesozoic, Pangaea broke up, and the Atlantic Ocean began to form. As the North American continent moved westward, the Cordillera began to rise due to subduction and the accretion of terranes along the west coast. In the Southwest, vast deserts accumulated thick layers of dune sand, while environments in the East were conducive to the formation and subsequent burial of coal swamps.
- In the Cenozoic era, a thick sequence of sediments was deposited along North America's Atlantic margin and the Gulf of Mexico. Meanwhile, western North America experienced an extraordinary episode of crustal extension; the Basin and Range Province resulted.

? Contrast the tectonics of eastern and western North America during the Mesozoic era.

12.6 Earth's First Life

Describe some of the hypotheses on the origin of life and the characteristics of early prokaryotes, eukaryotes, and multicellular organisms.

KEY TERMS: protein, prokaryote, stromatolite, eukaryote

- Life began from nonlife. Amino acids, a necessary building block for proteins, may have been assembled with energy from ultraviolet light or lightning, or in a hot spring, or may have been delivered later to Earth via meteorites.
- The first organisms were relatively simple single-celled prokaryotes that thrived in the absence of oxygen. They may have formed by 3.8 billion years ago. The advent of photosynthesis allowed microbial mats to build up and form stromatolites.
- Eukaryotes have larger, more complex cells than prokaryotes. The oldest-known eukaryotic cells date from around 2.1 billion years ago. Eukaryotic cells gave rise to the great diversity of multicellular organisms.

? The accompanying image shows fossil stromatolites. In what way did the ancient cyanobacteria that built these structures contribute to the evolution of Earth's atmosphere?

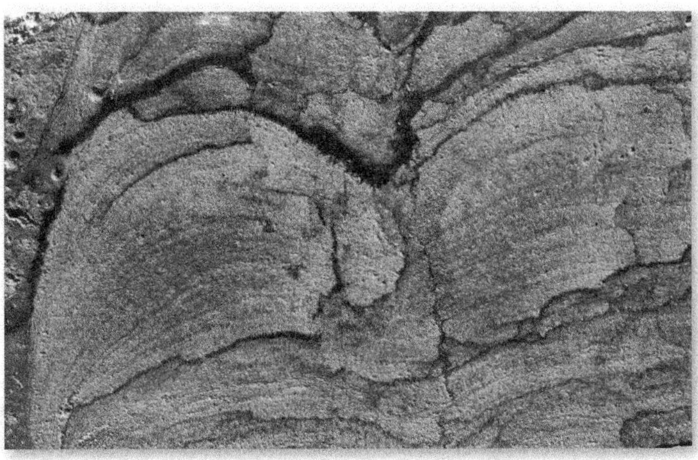

Biophoto Associates/Science Source

12.7 Paleozoic Era: Life Explodes

List the major developments in the history of life during the Paleozoic era.

KEY TERMS: invertebrate, Cambrian explosion, vertebrate, amphibian, reptile, amniotic egg, mass extinction

- Abundant fossil hard parts appear in sedimentary rocks at the beginning of the Cambrian period. These shells and other skeletal material came from a profusion of new animals, including trilobites and cephalopods.
- Plants colonized the land around 400 million years ago and soon diversified into forests.
- In the Devonian, some lobe-finned fishes gradually evolved into the first amphibians. A subset of the amphibian population evolved waterproof skin and shelled eggs and split off to become the reptile line.
- The Paleozoic era ended with the largest mass extinction in the geologic record. This deadly event may have been related to the eruption of the Siberian Traps flood basalts.

? What advantages do reptiles have over amphibians for life on dry land?

12.8 Mesozoic Era: Dinosaurs Dominate the Land

Briefly explain the major developments in the history of life during the Mesozoic era.

KEY TERMS: seed, gymnosperm

- Plants diversified during the Mesozoic. The flora of that time was dominated by gymnosperms, the first plants with seeds.
- The dinosaurs came to dominate the land, pterosaurs took to the air, and a suite of marine reptiles swam the seas. The first birds evolved during the Mesozoic, as evidenced by *Archaeopteryx*, a transitional fossil.
- Like the Paleozoic, the Mesozoic ended with a mass extinction. This extinction was due to a massive meteorite impact and a period of extensive volcanism, both of which released particulate matter into the atmosphere and dramatically altered Earth's climate and disrupted its food chain.

12.9 Cenozoic Era: Mammals Diversify

Discuss the major developments in the history of life during the Cenozoic era.

KEY TERMS: angiosperm, mammal, endothermic

- Flowering plants, called angiosperms, diversified and spread around the world through the Cenozoic era.
- Once the giant Mesozoic reptiles were extinct, mammals were able to diversify on the land, in the air, and in the oceans. Mammals are endothermic (control their body temperature metabolically), have hair, and nurse their young with milk. Marsupial mammals are born very young and then move to a pouch on the mother, while placental mammals spend a longer time *in utero* and are born in a relatively mature state compared to marsupials.
- Humans evolved from primate ancestors in Africa over a period of about 6.5 million years. They are distinguished from their ape ancestors by an upright, bipedal posture and large brains, as well as the use of elaborate tools. The oldest anatomically modern human fossils are 200,000 years old. Some of these humans migrated out of Africa and coexisted with Neanderthals and other human-like populations.

GIVE IT SOME THOUGHT

1 Referring to Figure 12.4, write a brief summary of the events that led to the formation of Earth.

2 The accompanying photograph shows layered iron-rich rocks called banded iron formations. What does the existence of these 2.5-billion-year-old rocks tell us about the evolution of Earth's atmosphere?

Graeme Churchard (Bristol, UK)

3 Describe how the sudden appearance of oxygen in the atmosphere about 2.5 billion years ago influenced the development of modern life-forms.

4 Five mass extinctions, in which 50 percent or more of Earth's marine species became extinct, are documented in the fossil record. Use the accompanying graph, which depicts the time and extent of each mass extinction, to answer the following:

a. Which of the five mass extinctions was the *most extreme*? Identify this extinction by name and when it occurred.
b. When did the *most recent* mass extinction occur?
c. During the most recent mass extinction, what prominent group of land animals was eliminated?
d. What group of terrestrial animals experienced a major period of diversification following the most recent mass extinction?

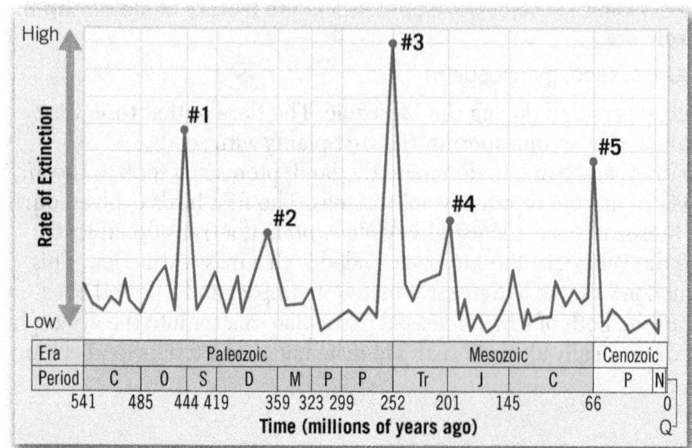

5 Currently, oceans cover about 71 percent of Earth's surface. However, early in Earth history the oceans covered a greater percentage of the surface. Explain.

6 Contrast the eastern and western margins of North America during the Cenozoic era in terms of their relationships to plate boundaries.

7 Between 300 and 250 million years ago, plate movement assembled all the previously separated landmasses together to form the supercontinent Pangaea. Pangaea's formation resulted in deeper ocean basins and a drop in sea level, causing shallow coastal areas to dry up. Thus, in addition to rearranging the geography of our planet, continental drift had a major impact on life on Earth. Use the accompanying diagram showing the movement of hypothetical landmasses and the information above to answer the following:

a. Which of the following types of habitats would likely diminish in size during the formation of a supercontinent: deep-ocean habitats, wetlands, shallow marine environments, or terrestrial (land) habitats? Explain.
b. During the breakup of a supercontinent, would sea level remain the same, rise, or fall?
c. Explain how and why the development of an extensive oceanic ridge system that forms during the breakup of a supercontinent affects sea level.

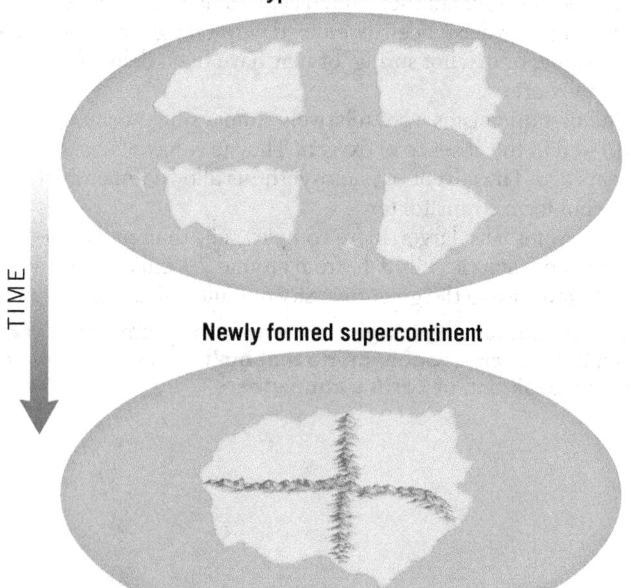

EXAMINING THE EARTH SYSTEM

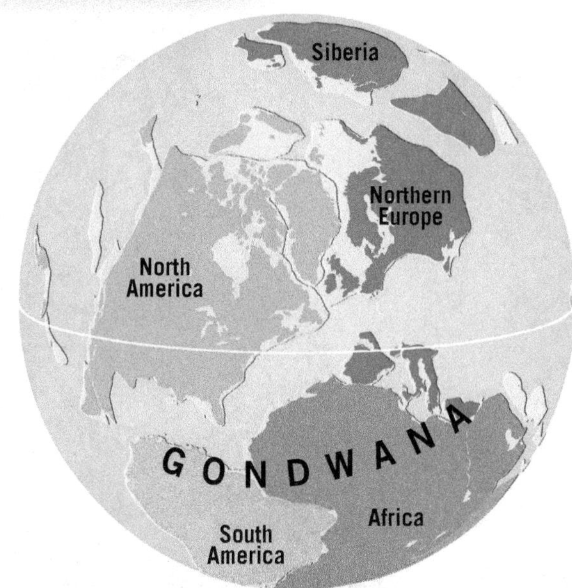

1 The Earth system has been responsible for both the conditions that favored the evolution of life on this planet and the mass extinctions that have occurred throughout geologic time. Describe the role of the biosphere, hydrosphere, and solid Earth in forming the current level of atmospheric oxygen. How did Earth's outer-space environment interact with the atmosphere and biosphere to contribute to the great mass extinction that marked the end of the dinosaurs?

2 Most of the vast North American coal resources from Pennsylvania to Illinois began forming during the Pennsylvanian and Mississippian periods of Earth history. (This time period is also referred to as the Carboniferous period.) Using Figure 12.27, an artistic depiction of a Pennsylvania period coal swamp, describe the climatic and biological conditions associated with this unique environment. Next, examine the accompanying diagram that illustrates the geographic position of North America during the period of coal formation. Where, relative to the equator, was North America located during the time of coal formation? Why is it unlikely that a similar coal-forming environment will repeat itself in North America in the near future? (You may find it helpful to visit the University of California Time Machine Exhibit at www.ucmp.berkeley.edu/carboniferous/carboniferous.html.)

DATA ANALYSIS

Fossils in Your Area

Fossils have been found around the world and are used as evidence of plate tectonics and evolution. Fossils come from many different periods of geologic time and provide clues about what the planet was like.

ACTIVITIES

Go to the Paleobiology Database at https://paleobiodb.org and click the blue Explore button. Click Help to learn how to navigate this page, and click x to exit the Help page. Zoom in to your state. Hovering over a circle will highlight the relevant time period on the geologic timeline. Hovering over the colored blocks on the timeline will display the name of the period.

1 Which geologic time periods are represented? Which time period has the most sites?

2 Zoom to the fossil site closest your location. Which eon does this fossil find belong to? Which era? Period? Epoch? Stage (time period shown at the bottom of the timeline), if available?

3 Click on the fossil site. If the site has multiple fossil collections, select one of them. The number in parentheses gives the number of kinds of fossils found.

4 What is the lithology of this site (if reported)? Click on the Occurrences tab. What group or groups of organisms (at the class or family level) are listed? In common language, what kinds of creatures are these? (Use the Internet as needed.)

5 Zoom out until you can see several fossil sites in your region. What other eras and periods are shown in your area, if any?

6 Investigate the fossil sites by clicking on them. What other types of fossils have been found in your region?

7 Based on the fossils in the area, describe the likely environmental conditions at that time. Information in the chapter will help you answer this question.

Zoom in to a region were fossils come from a different era (shown by different colors) than the ones in your region.

8 What geologic time periods are shown?

9 What types of fossils have been found in this region?

MasteringGeology™ Looking for additional review and test prep materials? Visit the Study Area in MasteringGeology to enhance your understanding of this chapter's content by accessing a variety of resources, including Self-Study Quizzes, Geoscience Animations, SmartFigure Tutorials, Mobile Field Trips, *Project Condor* Quadcopter videos, *In the News* articles, flashcards, web links, and an optional Pearson eText.

www.masteringgeology.com

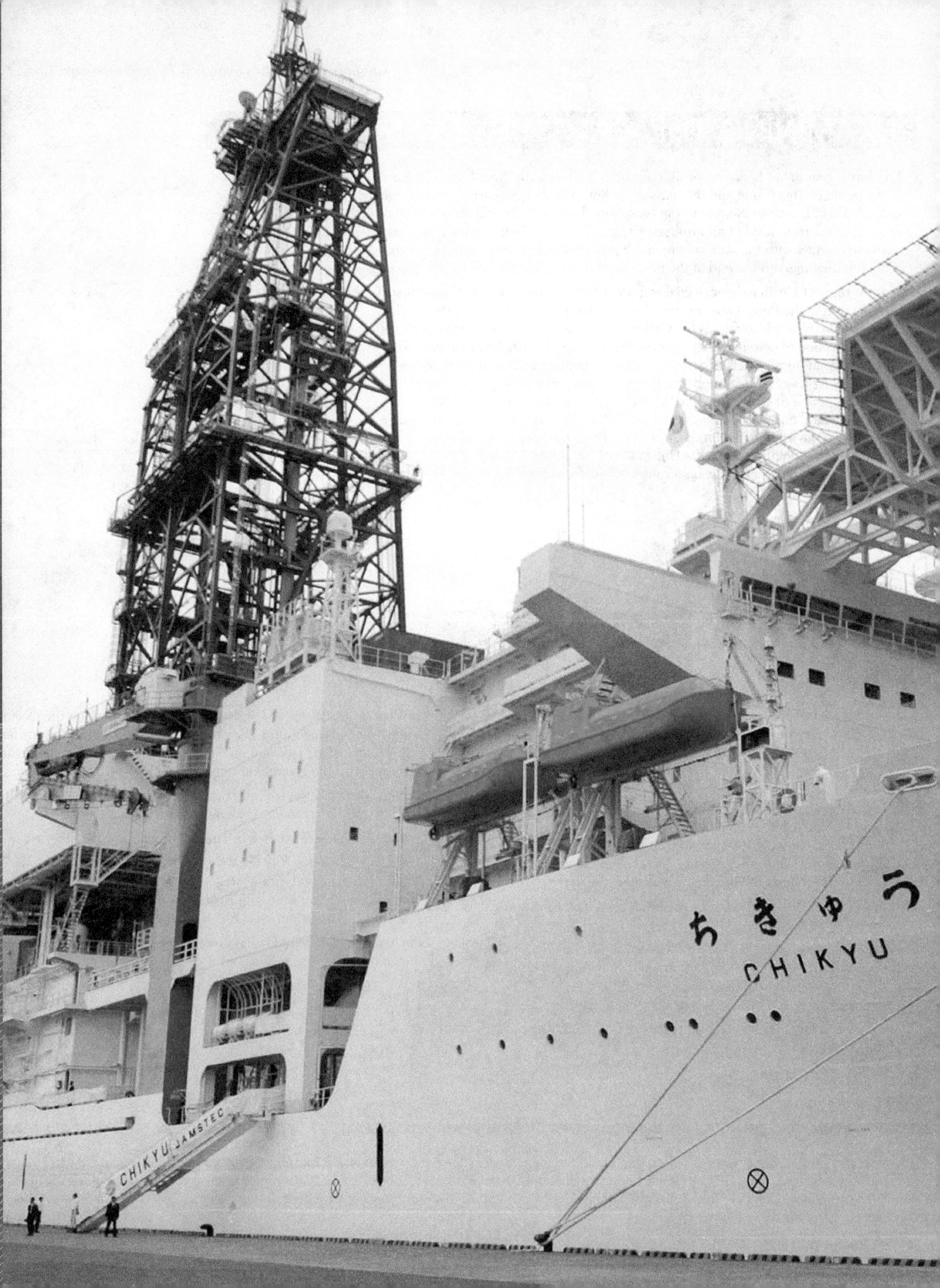
ちきゅう
CHIKYU
CHIKYU JAMSTEC

13 The Ocean Floor*

FOCUS ON CONCEPTS

Each statement represents the primary learning objective for the corresponding major heading within the chapter. After you complete the chapter, you should be able to:

13.1 Discuss the extent and distribution of oceans and continents on Earth. Identify Earth's four main ocean basins.

13.2 Define *bathymetry* and summarize the various techniques used to map the ocean floor.

13.3 Compare a passive continental margin with an active continental margin and list the major features of each.

13.4 List and describe the major features associated with deep-ocean basins.

13.5 Summarize the basic characteristics of oceanic ridges.

13.6 Distinguish among three categories of seafloor sediment and explain how biogenous sediment can be used to study climate change.

13.7 Specify the important resources and potential resources associated with the ocean floor.

*This chapter was prepared with the assistance of Professor Alan P. Trujillo, Palomar College.

The *Chikyu* (meaning "Earth" in Japanese) is one of the most advanced scientific drilling vessels. It is part of the Integrated Ocean Drilling Program (IODP).
(Photo by Itsuo Inouye/AP Photo)

HOW DEEP IS THE OCEAN? How much of Earth is covered by seawater? What does the seafloor look like? Answers to these and other questions about the oceans and the basins they occupy are sometimes elusive and often difficult to determine. Suppose that all the water were drained from the ocean. What would be seen? Plains? Mountains? Canyons? Plateaus? Indeed, the ocean conceals all these features—and more. And what about the carpet of sediment that covers much of the seafloor? Where did it come from, and what can we learn by studying it? This chapter provides answers to these questions and much more about our watery planet.

13.1 The Vast World Ocean

Discuss the extent and distribution of oceans and continents on Earth. Identify Earth's four main ocean basins.

The ocean is one of the most prominent features of our planet. The fact that our planet has so much liquid water at its surface is why Earth is frequently called the *water planet* or the *blue planet*. There is a good reason for these names: Nearly 71 percent of Earth's surface is covered by the global ocean (Figure 13.1). The focus of this chapter and Chapters 14 and 15 is **oceanography**, the scientific study of the oceans. It is an interdisciplinary science that draws on the methods and knowledge of geology, chemistry, physics, and biology to study all aspects of the world ocean.

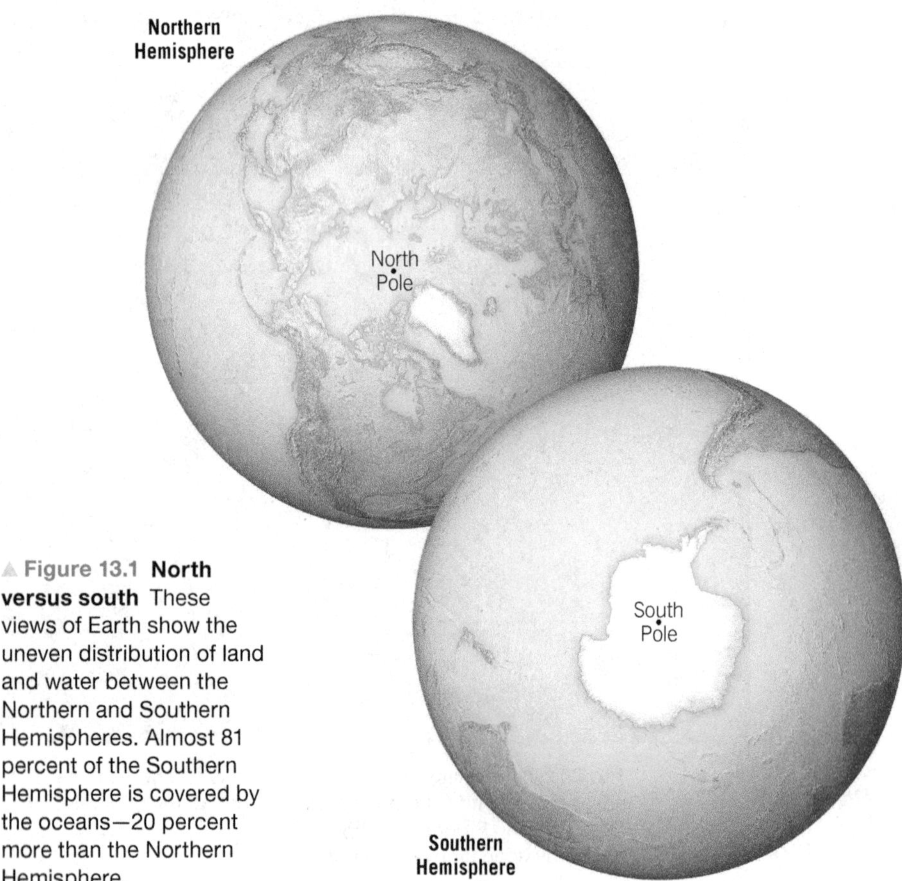

▲ Figure 13.1 **North versus south** These views of Earth show the uneven distribution of land and water between the Northern and Southern Hemispheres. Almost 81 percent of the Southern Hemisphere is covered by the oceans—20 percent more than the Northern Hemisphere.

Geography of the Oceans

The area of Earth is about 510 million square kilometers (197 million square miles). Of this total, approximately 360 million square kilometers (140 million square miles), or 71 percent, is represented by oceans and marginal seas (meaning seas around the ocean's margin, like the Mediterranean Sea and Caribbean Sea). Continents and islands comprise the remaining 29 percent, or 150 million square kilometers (58 million square miles).

By studying a globe or world map, we can easily see that the continents and oceans are not evenly divided between the Northern and Southern Hemispheres (see Figure 13.1). In the Northern Hemisphere, nearly 61 percent of the surface is water, and about 39 percent is land. In the Southern Hemisphere, on the other hand, almost 81 percent of the surface is water, and only 19 percent is land. It is no wonder then that the Northern Hemisphere is called the *land hemisphere* and the Southern Hemisphere is called the *water hemisphere*.

Figure 13.2A shows the distribution of land and water in the Northern and Southern Hemispheres. Between latitudes 45° north and 70° north, there is actually more land than water, whereas between 40° south and 65° south, there is almost no land to interrupt oceanic and atmospheric circulation.

The world ocean can be divided into four main ocean basins (Figure 13.2B):

- The *Pacific Ocean*, which is the largest ocean and the largest single geographic feature on the planet, accounts for more than one-third of all surface area on Earth and over half of the ocean surface area of Earth. In fact, the Pacific Ocean is so large that all the continents could fit into the space occupied by it—and have room left over! It is also the world's deepest ocean, with an average depth of 3940 meters (13,000 feet, or about 2.5 miles).
- The *Atlantic Ocean* is about half the size of the Pacific Ocean and is not quite as deep. It is a relatively

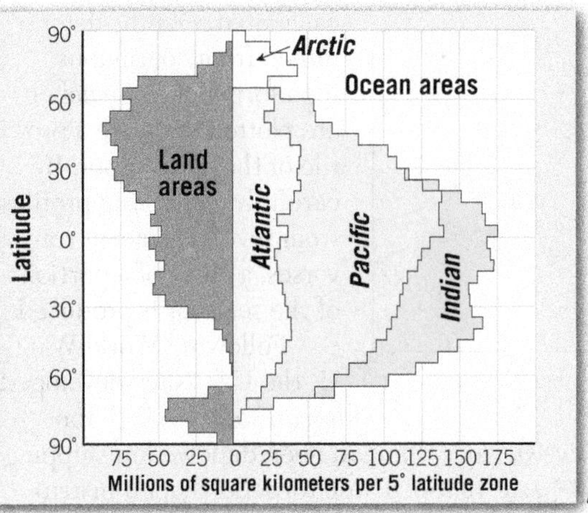

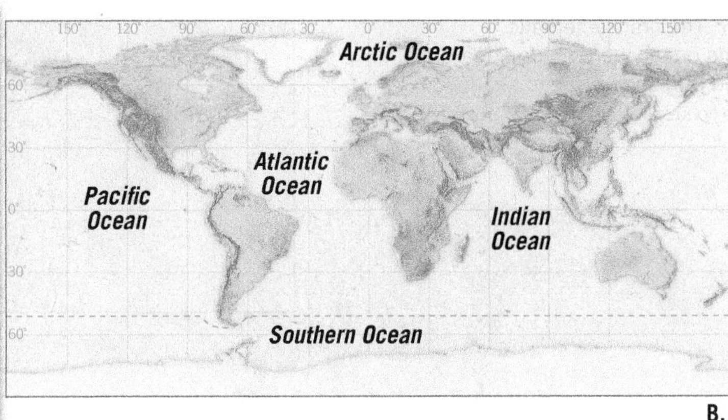

SmartFigure 13.2
Distribution of land and water A. The graph shows the amount of land and water in each 5° latitude belt. **B.** The world map provides a more familiar view.

TUTORIAL
https://goo.gl/aBd6Wg

narrow ocean compared to the Pacific and is bounded by nearly parallel continental margins.

- The *Indian Ocean* is slightly smaller than the Atlantic Ocean but has about the same average depth. Unlike the Pacific and Atlantic Oceans, it is largely a Southern Hemisphere water body.
- The *Arctic Ocean* is about 7 percent the size of the Pacific Ocean and is only a little more than one-quarter as deep as the rest of the oceans.

Oceanographers also recognize an additional ocean near the continent of Antarctica in the Southern Hemisphere. Defined by the meeting of currents near Antarctica called the Antarctic Convergence, the *Southern Ocean*, or *Antarctic Ocean*, is actually those portions of the Pacific, Atlantic, and Indian Oceans south of about 50° south latitude.

Comparing the Oceans to the Continents

A major difference between continents and ocean basins is their relative levels. The average elevation of the continents above sea level is about 840 meters (2756 feet), whereas the average depth of the oceans is nearly four and a half times this amount—3729 meters (12,234 feet, or about 2.5 miles). The volume of ocean water is so large that if Earth's solid mass were perfectly smooth (level) and spherical, the oceans would cover Earth's entire rocky surface to a uniform depth of more than 2000 meters (1.2 miles)!

CONCEPT CHECKS 13.1

1. How does the area of Earth's surface covered by the oceans compare with that of the continents?
2. Contrast the distribution of land and water in the Northern Hemisphere and the Southern Hemisphere.
3. Excluding the Southern Ocean, name the four main ocean basins. Contrast them in terms of area and depth.
4. How does the average depth of the oceans compare to the average elevation of the continents?

13.2 An Emerging Picture of the Ocean Floor

Define *bathymetry* and summarize the various techniques used to map the ocean floor.

If all water were removed from the ocean basins, a great variety of features would be seen, including broad volcanic peaks, deep trenches, extensive plains, linear mountain chains, and large plateaus. In fact, the scenery would be very diverse and different from that on the continents.

Mapping the Seafloor

The complex nature of ocean-floor topography did not begin to become known until the historic 3½-year voyage of the British scientific expedition aboard the research vessel HMS *Challenger* (Figure 13.3). From December

Figure 13.3 **HMS *Challenger*** The first systematic bathymetric measurements of the ocean were made aboard the HMS *Challenger*, which departed England in December 1872 and returned in May 1876. (Image courtesy of the Library of Congress)

1872 to May 1876, the *Challenger* expedition made the first comprehensive study of the global ocean. During the 127,500-kilometer (79,200-mile) voyage, the ship and its crew of scientists traveled to every ocean except the Arctic. Throughout the voyage, they sampled a multitude of ocean properties, including water depth, which was accomplished by laboriously lowering a long, weighted line overboard and then retrieving it. Using this process, the *Challenger* made the first recording of the deepest-known point on the ocean floor in 1875. This spot, on the floor of the western Pacific, was later named the *Challenger Deep*.

Modern Bathymetric Techniques The measurement of ocean depths and the charting of the shape (topography) of the ocean floor is known as **bathymetry** (*bathos* = depth, *metry* = measurement). Today, sound energy is used to measure water depths. The basic approach employs **sonar**, an acronym for *so*und *na*vigation and *r*anging. The first devices that used sound to measure water depth, called **echo sounders**, were developed early in the twentieth century. Echo sounders work by transmitting a pulse of sound (which is an acoustic wave called a *ping*) into the water in order to produce an echo when it bounces off any object, such as a large marine organism or the ocean floor (Figure 13.4). A sensitive receiver intercepts the echo reflected from the bottom, and a clock precisely measures the travel time to fractions of a second. By knowing the speed of sound waves in water—about 1500 meters (4900 feet) per second—and the time required for the energy pulse to reach the ocean floor and return, depth can be calculated. Depths determined from continuous monitoring of these echoes are plotted to create a profile of the ocean floor. By carefully combining profiles from several adjacent traverses, a chart of a portion of the seafloor is produced.

Following World War II, the U.S. Navy developed *sidescan sonar* to look for explosive devices that had been deployed in shipping lanes (Figure 13.5A). These torpedo-shaped instruments, called *towfish*, can be towed behind a ship, where they send out a fan of sound extending to either side of the ship's track. By combining swaths of sidescan sonar data, researchers produced the first photograph-like images of the seafloor. Although early sidescan sonar devices provided valuable views of the seafloor, they did not provide accurate bathymetric (water depth) data.

This drawback was resolved in the 1990s, with the development of *high-resolution multibeam sonar* instruments (see Figure 13.5A). These systems use hull-mounted sound sources that send out a fan of sound and then record reflections from the seafloor through a set of narrowly focused receivers aimed at different

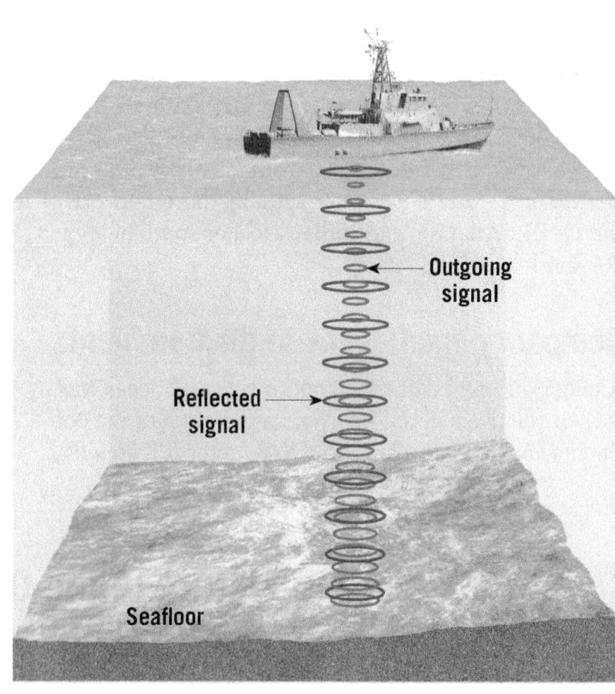

Figure 13.4 **Echo sounder** An echo sounder determines water depth by measuring the time interval required for an acoustic wave to travel from a ship to the seafloor and back. The speed of sound in water is 1500 meters per second. Therefore, depth = ½ (1500 m/sec × echo travel time).

A. Sidescan sonar and multibeam sonar operating from the same research vessel.

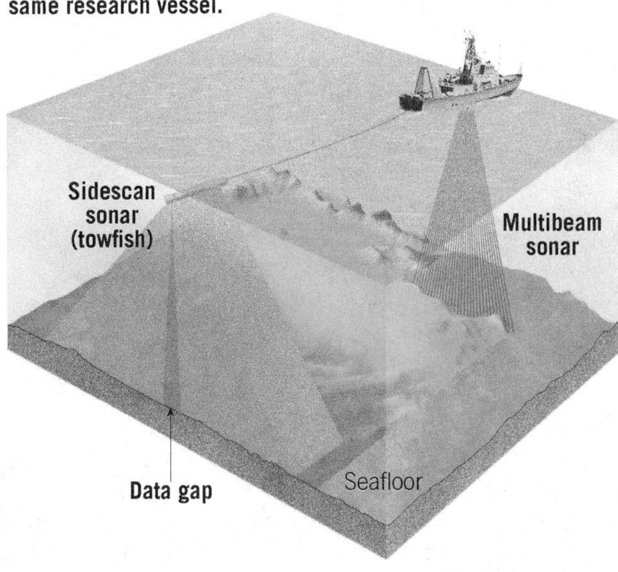

B. Color-enhanced perspective map of the seafloor and coastal landforms in the Los Angeles area of California.

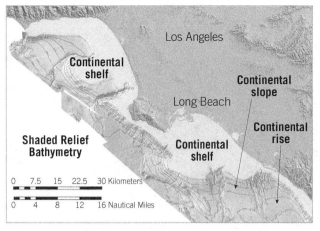

Los Angeles
Continental shelf
Continental slope
Long Beach
Continental rise
Shaded Relief Bathymetry
Continental shelf
0 7.5 15 22.5 30 Kilometers
0 4 8 12 16 Nautical Miles

Figure 13.5 Sidescan and multibeam sonar

angles. Rather than obtain the depth of a single point every few seconds, this technique makes it possible for a survey ship to map the features of the ocean floor along a strip tens of kilometers wide. These systems can collect bathymetric data of such high resolution that they can distinguish depths that differ by less than a meter (Figure 13.5B). When multibeam sonar is used to make a map of a section of seafloor, the ship travels through the area in a regularly spaced back-and-forth pattern known as "mowing the lawn."

Despite their greater efficiency and enhanced detail, research vessels equipped with multibeam sonar travel at a mere 10–20 kilometers (6 to 12 miles) per hour. It would take at least 100 vessels outfitted with this equipment hundreds of years to map the entire seafloor. This explains why only about 5 percent of the seafloor has been mapped in detail—and why large portions of the seafloor have not yet been mapped with sonar. In fact, large, remote areas of the seafloor as big as the state of Colorado have not been surveyed by ships at all.

Mapping the Ocean Floor from Space Breakthroughs in technology have led to more detailed ocean floor maps. One breakthrough that enhanced our understanding of the seafloor involves measuring the shape of the ocean surface from space. After compensating for waves, tides, currents, and atmospheric effects, it was discovered that the ocean surface is not perfectly "flat." Because massive seafloor features exert stronger-than-average gravitational attraction, they produce elevated areas on the ocean surface. Conversely, canyons and trenches create slight depressions.

Satellites equipped with *radar altimeters* are able to measure these subtle differences by bouncing microwaves off the sea surface (Figure 13.6). These devices can measure variations as small as a few centimeters. Along with traditional sonar depth measurements, these data are used to produce detailed ocean-floor maps, such as the one shown in Figure 13.7.

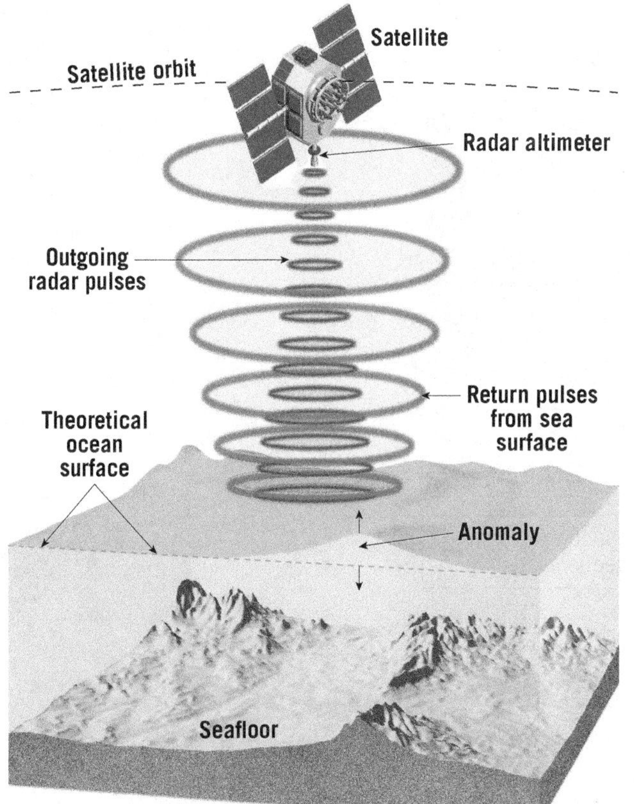

SmartFigure 13.6 Satellite altimeter A satellite altimeter measures the variation in sea-surface elevation, which is caused by gravitational attraction and mimics the shape of the seafloor. The sea-surface anomaly is the difference between the measured and theoretical ocean surface.

TUTORIAL
https://goo.gl/n66Rra

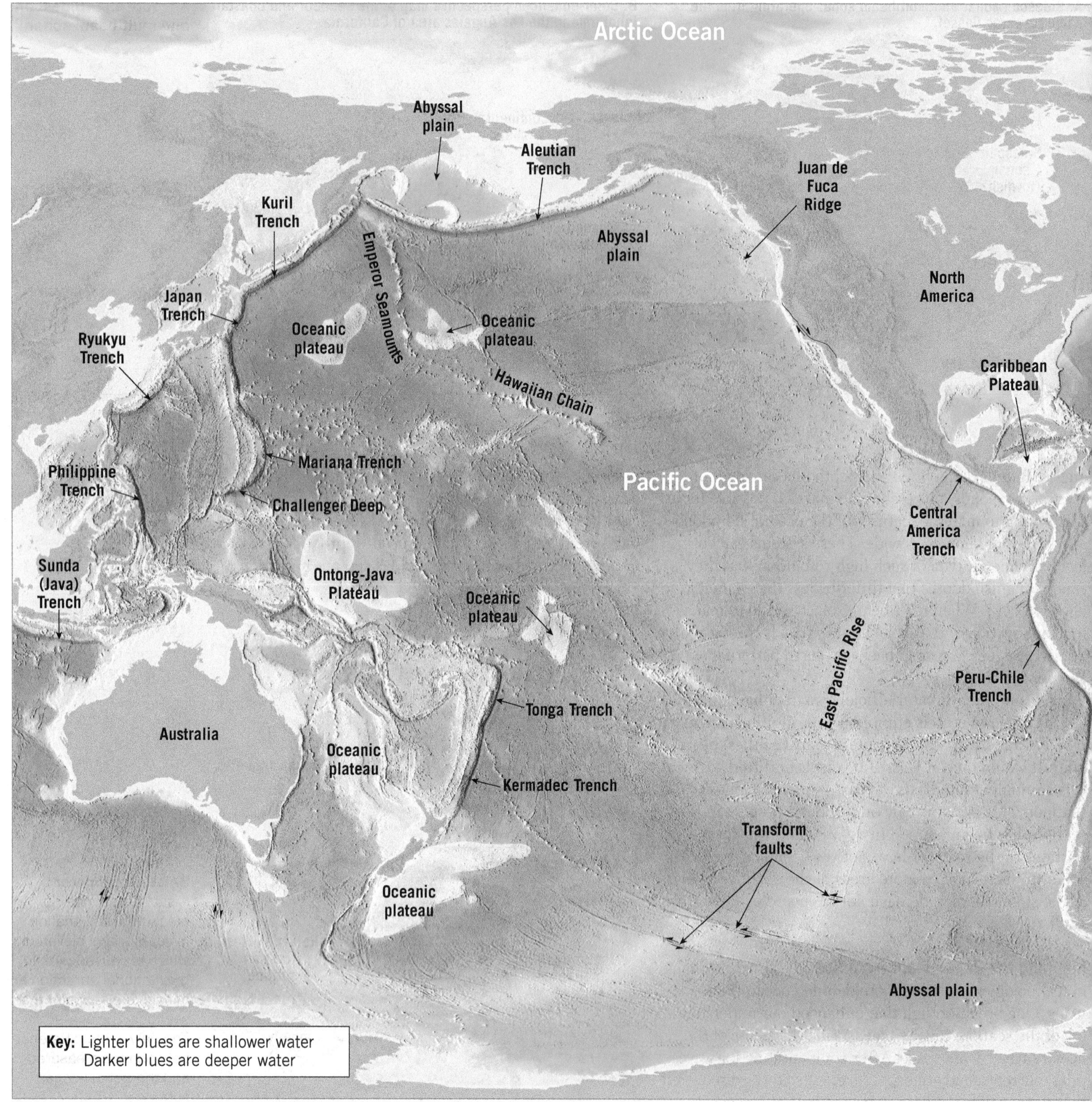

Figure 13.7 **Major features of the seafloor**

Provinces of the Ocean Floor

Oceanographers studying the topography of the ocean floor recognize three major units: *continental margins*, the *deep-ocean basin*, and the *oceanic (mid-ocean) ridge*. The map in Figure 13.8 outlines these provinces for the North Atlantic Ocean, and the profile at the bottom of the illustration shows the varied topography. Such profiles usually have their vertical dimension exaggerated many times—40 times in this case—to make topographic

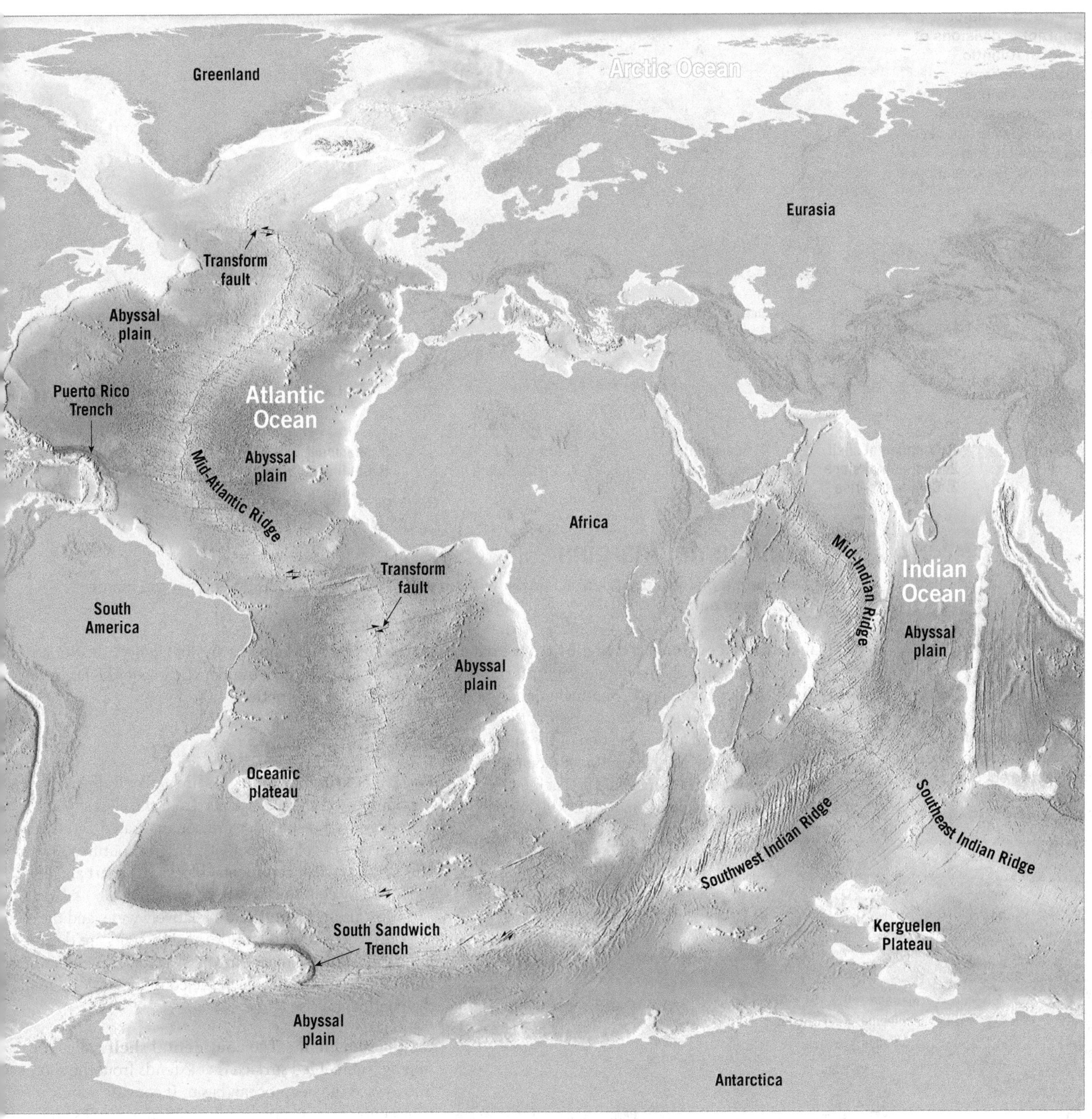

features more conspicuous. Vertical exaggeration, however, makes slopes shown in seafloor profiles appear to be *much* steeper than they actually are.

CONCEPT CHECKS 13.2

1. Define *bathymetry*.
2. Describe how satellites orbiting Earth can determine features on the seafloor without being able to directly observe them beneath several kilometers of seawater.
3. List the three major provinces of the ocean floor.

Figure 13.8 **Major topographic divisions of the North Atlantic** Map view (*top*) and corresponding profile view (*bottom*). On the profile, the vertical scale has been exaggerated to make topographic features more conspicuous.

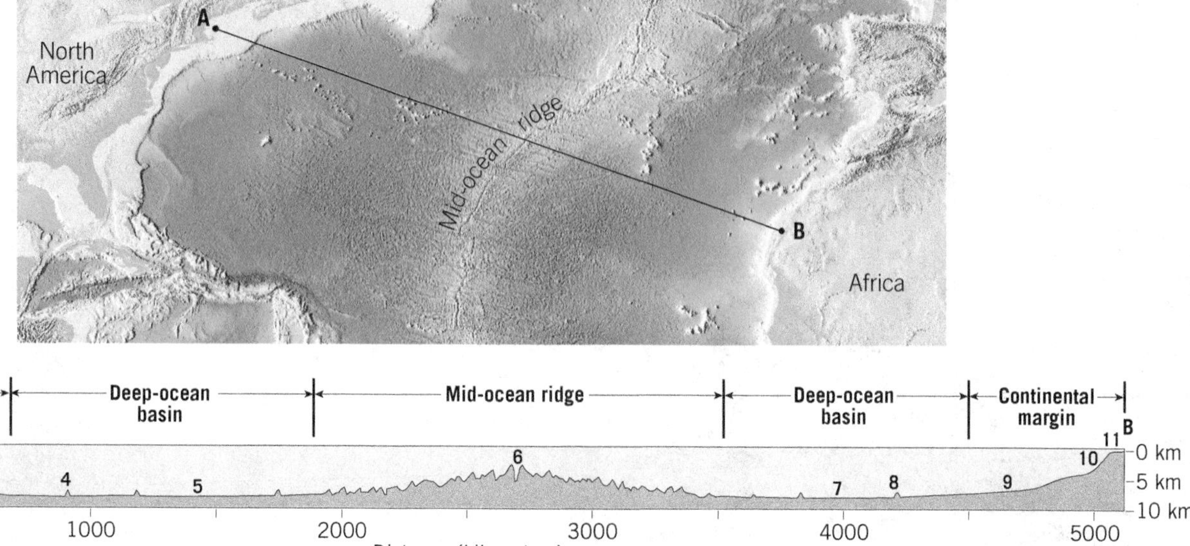

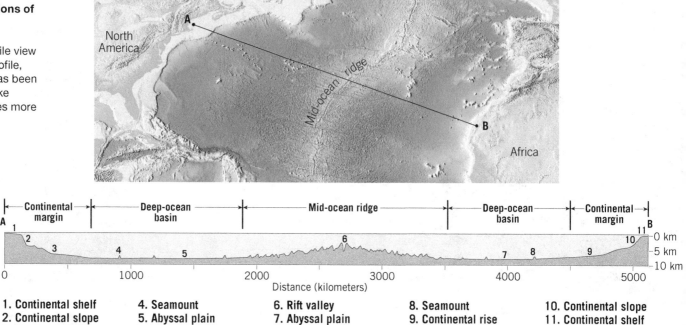

1. Continental shelf
2. Continental slope
3. Continental rise
4. Seamount
5. Abyssal plain
6. Rift valley
7. Abyssal plain
8. Seamount
9. Continental rise
10. Continental slope
11. Continental shelf

13.3 Continental Margins

Compare a passive continental margin with an active continental margin and list the major features of each.

As the name implies, **continental margins** are the outer margins of the continents where continental crust transitions to oceanic crust. Two types of continental margin have been identified: *passive* and *active*. Nearly the entire Atlantic Ocean and a large portion of the Indian Ocean are surrounded by passive continental margins (see Figure 13.6). By contrast, most of the Pacific Ocean is bordered by active continental margins, which are marked by the subduction zones in Figure 13.9. Notice that many of those active subduction zones lie far beyond the margins of the continents.

Figure 13.9 **Distribution of Earth's subduction zones** Most of the active subduction zones surround the Pacific basin.

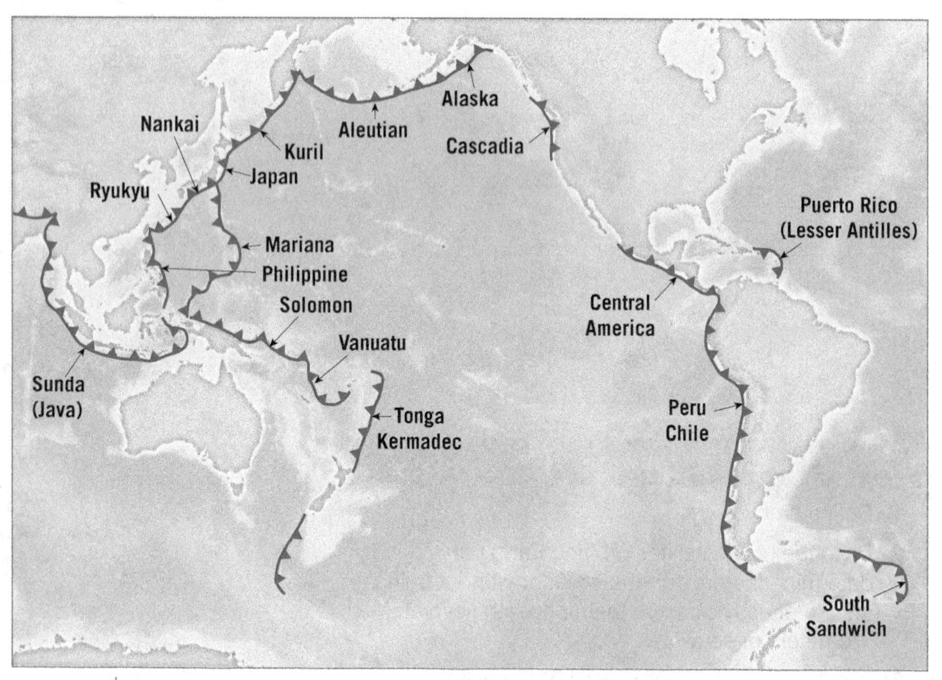

Passive Continental Margins

Passive continental margins are geologically inactive regions located a great distance from the closest plate boundary. As a result, they are not associated with strong earthquakes or volcanic activity. Passive continental margins develop when continental blocks rift apart and are separated by continued seafloor spreading.

Most passive margins are relatively wide, and they are sites where large quantities of sediments are deposited. The features comprising passive continental margins include the continental shelf, the continental slope, and the continental rise (Figure 13.10).

Continental Shelf The **continental shelf** is a gently sloping, submerged surface that extends from the shoreline toward the deep-ocean basin. It consists mainly of continental crust capped with sedimentary rocks and sediments eroded from adjacent landmasses.

The width of the continental shelf varies greatly. The shelf is very narrow along portions of some continents, whereas along others it extends seaward more than 1500 kilometers (930 miles). The average inclination of the continental shelf is only about one-tenth of 1 degree, a slope so slight that it would appear to an observer to be a horizontal surface.

The continental shelf tends to be relatively featureless; however, some areas are mantled by extensive glacial deposits and thus are quite rugged. In addition, some continental shelves are dissected by large valleys that run from the coastline into deeper waters. Many of these *shelf valleys* are the seaward extensions of river valleys on the adjacent landmass. They were eroded during the most recent Ice Age (Quaternary period), when enormous quantities of water were stored in vast ice sheets on the continents, causing sea level to drop at least 100 meters (330 feet). Because of this sea-level drop, rivers extended their valleys, and land-dwelling plants and animals migrated to the newly exposed portions of the continents. Dredging off the coast of North America has retrieved the ancient remains of numerous land dwellers, such as mammoths, mastodons, and horses, providing further evidence that portions of the continental shelves were once above sea level.

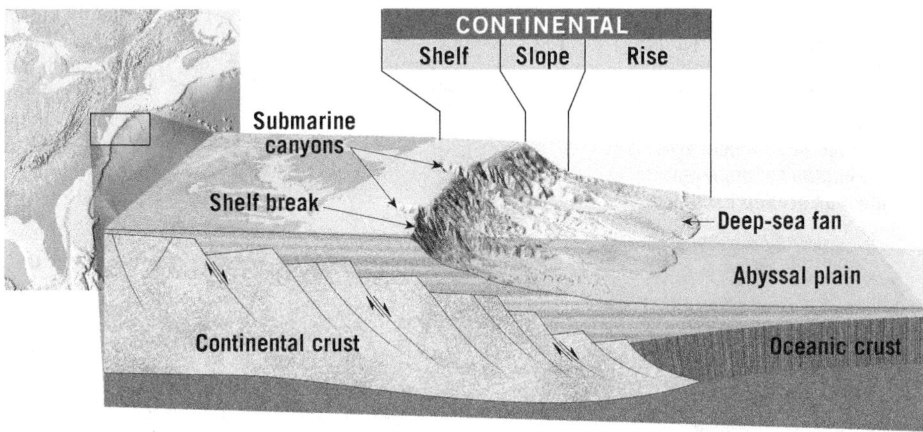

SmartFigure 13.10 **Passive continental margin** The slopes shown for the continental shelf and continental slope are greatly exaggerated. The continental shelf has an average slope of one-tenth of 1 degree, whereas the continental slope has an average slope of about 5 degrees.

TUTORIAL
http://goo.gl/jM81Sn

Although continental shelves represent only 7.5 percent of the total ocean area, they have economic and political significance because they contain important reservoirs of oil and natural gas, and they support important fishing grounds.

Continental Slope Marking the seaward edge of the continental shelf is the **continental slope**, a relatively steep zone that marks the boundary between continental crust and oceanic crust. Although the inclination of the continental slope varies greatly from place to place, it averages about 5 degrees and in places exceeds 25 degrees.

Continental Rise The continental slope merges into a more gradual incline known as the **continental rise** that may extend seaward for hundreds of kilometers. The continental rise consists of a thick accumulation of sediment that has moved down the continental slope and onto deep-ocean floor. Most of the sediments are delivered to the seafloor by *turbidity currents* that periodically flow down *submarine canyons* (discussed below). When these muddy slurries emerge from the mouth of a canyon onto the relatively flat ocean floor, they deposit sediment that forms a **deep-sea fan**. As fans from adjacent submarine canyons grow, they merge to produce a continuous wedge of sediment at the base of the continental slope, forming the continental rise.

Submarine Canyons and Turbidity Currents Deep, steep-sided valleys known as **submarine canyons** are cut into the continental slope and may extend across the entire continental rise to the deep-ocean basin (Figure 13.11). Although some of these canyons appear to be the seaward extensions of river valleys, many others do not line up in this manner. In addition, submarine canyons extend to depths far below the maximum lowering of sea level during the Ice Age, so their formation cannot be related to stream erosion.

Most submarine canyons have been excavated by turbidity currents. **Turbidity currents** are episodic

This image shows a perspective view of the continental margin, looking southwest toward the Palos Verdes Peninsula near Los Angeles.

QUESTION 1 *Match the features labeled 1 through 4 with the following terms: continental shelf, continental slope, continental rise, and submarine canyon.*

QUESTION 2 *Based on the features in this image, what type of continental margin is shown here?*

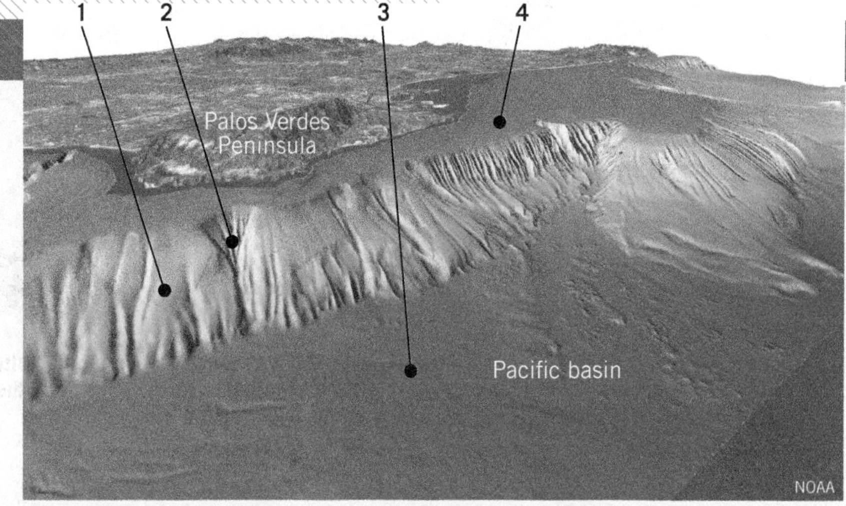

Figure 13.11 Turbidity currents and submarine canyons Turbidity currents are an important factor in the formation of submarine canyons. (Photo by Marli Miller)

Beds deposited by turbidity currents are called turbidites. Each event produces a single bed characterized by a decrease in sediment size from bottom to top, a feature known as graded bedding.

Turbidity currents are downslope movements of dense, sediment-laden water. They are created when sand and mud on the continental shelf and/or slope are dislodged and thrown into suspension. Because the mud-choked water is denser than normal seawater, it flows downslope, eroding and accumulating more sediment.

Turbidity current

Turbidite deposits

Fine particles settle out last

Graded beds

Coarse particles settle out first

Submarine canyons

Turbidity current

Deep-sea fans

SmartFigure 13.12 Active continental margins

TUTORIAL
https://goo.gl/koxEsk

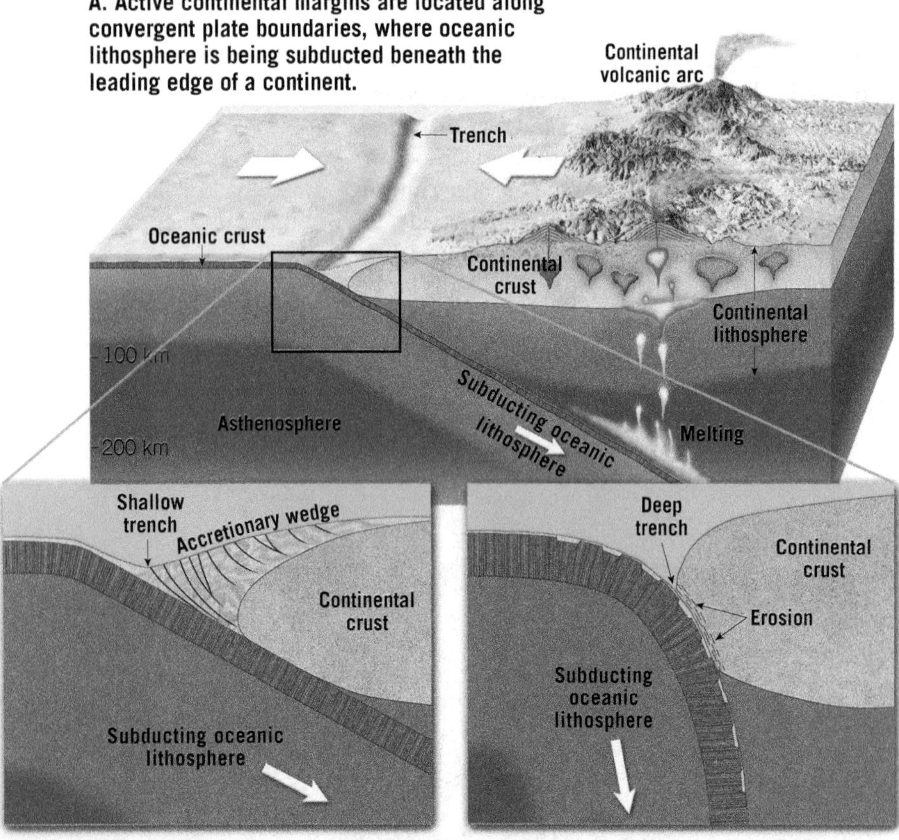

A. Active continental margins are located along convergent plate boundaries, where oceanic lithosphere is being subducted beneath the leading edge of a continent.

B. Accretionary wedges develop along subduction zones, where sediments from the ocean floor are scraped from the descending oceanic plate and pressed against the edge of the overriding plate.

C. Subduction erosion occurs where sediment and rock scraped off the bottom of the overriding plate are carried into the mantle by the subducting plate.

downslope movements of dense, sediment-laden water. They are created when rock debris, sand, and mud on the continental shelf and slope are dislodged and form a slurry. The mud-thickened mass is denser than seawater and moves downslope, eroding and accumulating more sediment as it goes, similar to a flash flood in stream channels on land. The erosional work repeatedly carried on by these muddy torrents is the major force in the excavation of most submarine canyons.

Turbidity currents usually originate along the continental slope and continue across the continental rise, still cutting channels. Eventually, they lose momentum and come to rest along the floor of the deep-ocean basin. As these currents slow, suspended sediments begin to settle out. First, the coarser rock debris is dropped, followed by successively finer accumulations of sand, silt, and then clay. These deposits, called *turbidites*, display a decrease in sediment grain size from bottom to top, a feature known as *graded bedding*. Turbidity currents are an important mechanism of sediment transport in the ocean. By the action of turbidity currents, submarine canyons are excavated, and sediments are carried to the deep-ocean floor.

Active Continental Margins

Most **active continental margins** are located along convergent plate boundaries where oceanic lithosphere is being subducted beneath the leading edge of a continent (Figure 13.12). Active continental margins are characterized by having a narrow continental shelf, a steep continental slope, and a deep ocean trench just offshore.

Deep-ocean trenches are the major topographic expression at convergent plate boundaries. These deep, narrow furrows surround most of the Pacific Rim.

Along some subduction zones, sediments from the ocean floor and pieces of oceanic crust are scraped from the descending oceanic plate and plastered against the edge of the overriding plate. This chaotic accumulation of deformed sediment and scraps of oceanic crust is called an **accretionary wedge** (*ad* = toward, *crescere* = to grow). Prolonged plate subduction can produce massive accumulations of sediment along active continental margins.

The opposite process, known as **subduction erosion**, characterizes many other active continental margins. Rather than sediment accumulating along the front of the overriding plate, sediment and rock are scraped off the bottom of the overriding plate and transported into the mantle by the subducting plate. Subduction erosion is particularly effective when the angle of descent is steep. Sharp bending of the subducting plate causes faulting in the ocean crust and a rough surface, as shown in Figure 13.12.

CONCEPT CHECKS 13.3

1. List the three major features of a passive continental margin.
2. Describe the differences between active and passive continental margins. Where is each type found?
3. Discuss the process that is responsible for creating most submarine canyons.
4. How are active continental margins related to plate tectonics?
5. Briefly explain how an accretionary wedge forms. What is meant by *subduction erosion*?

13.4 Features of Deep-Ocean Basins

List and describe the major features associated with deep-ocean basins.

Between the continental margin and the oceanic ridge lies the **deep-ocean basin** (see Figure 13.8). The size of this region—almost 30 percent of Earth's surface—is roughly comparable to the percentage of land that presently projects above sea level. This region includes *deep-ocean trenches*, which are extremely deep linear depressions in the ocean floor; remarkably flat areas known as *abyssal plains*; tall volcanic peaks called *seamounts* and *guyots*; and extensive areas of lava flows piled one atop the other called *oceanic plateaus*.

Deep-Ocean Trenches

Deep-ocean trenches are long, relatively narrow troughs that are the deepest parts of the ocean. Most trenches are located along the margins of the Pacific Ocean, where many exceed 10 kilometers (6 miles) in depth (see Figure 13.7). A portion of one—the Challenger Deep in the Mariana trench—has been measured at a record 10,994 meters (36,070 feet) below sea level, making it the deepest-known part of the world ocean (Figure 13.13). Only two trenches are located in the Atlantic: the Puerto Rico trench and the South Sandwich trench.

Although deep-ocean trenches represent only a very small portion of the area of the ocean floor, they are nevertheless significant geologic features. Trenches are sites of plate convergence where slabs of oceanic lithosphere subduct and plunge back into the mantle. In addition to earthquakes being created as one plate "scrapes" against another, volcanic activity is also

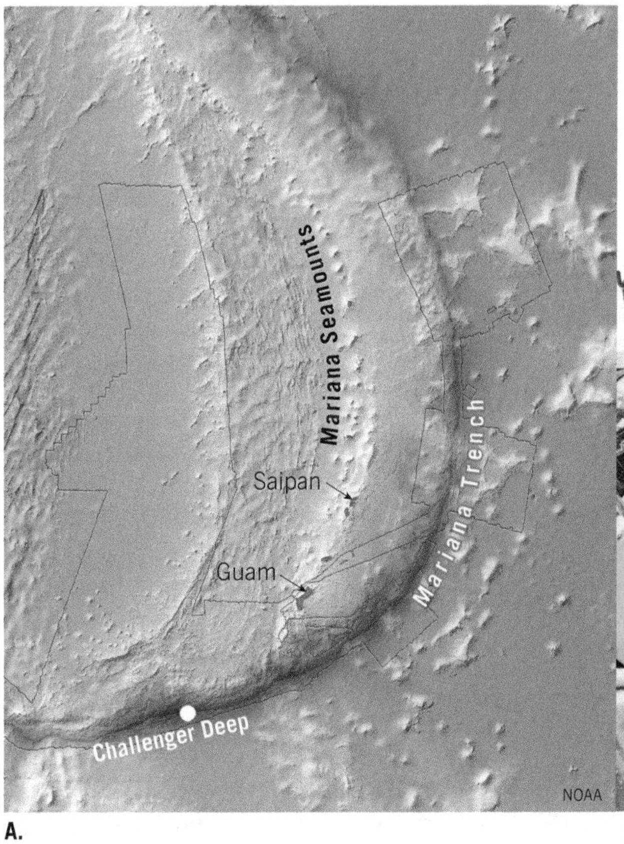

Mariana Seamounts
Mariana Trench
Saipan
Guam
Challenger Deep
NOAA
A.

B.

Figure 13.13 The Challenger Deep
A. Located near the southern end of the Mariana trench, the Challenger Deep is the deepest place in the global ocean, about 10,994 meters (36,070 feet) deep. **B.** Film director James Cameron (*Titanic* and *Avatar*) made news in March 2012 as the first person to take a deep-diving submersible to the bottom of the Challenger Deep in more than 50 years. (Photo by Mark Thiessen/AFP/Getty Images/Newscom)

triggered by plate subduction. As a result, some trenches run parallel to an arc-shaped row of active volcanoes called a **volcanic island arc**. In Figure 13.13, the Mariana Seamounts are such a feature. Furthermore, **continental volcanic arcs**, such as those making up portions of the Andes and Cascades, are located parallel to trenches that lie adjacent to continental margins. The volcanic activity associated with the trenches that surround the Pacific Ocean explains why the region is called the *Ring of Fire*.

Abyssal Plains

Abyssal (*a* = without, *byssus* = bottom) **plains** are deep, incredibly flat features; in fact, these regions are likely the most level places on Earth. The abyssal plain found off the coast of Argentina, for example, has less than 3 meters (10 feet) of relief over a distance exceeding 1300 kilometers (800 miles). The monotonous topography of abyssal plains is occasionally interrupted by the protruding summit of a partially buried volcanic peak (seamount).

Using *seismic reflection profilers*, instruments that generate signals designed to penetrate far below the ocean floor, researchers have determined that abyssal plains owe their relatively featureless topography to thick accumulations of sediment that have buried an otherwise rugged ocean floor (**Figure 13.14**). The nature of the sediment indicates that these plains consist primarily of three types of sediment: (1) fine sediments from land transported far out to sea by winds, ocean currents, and turbidity currents, (2) mineral matter that has precipitated out of seawater, and (3) shells and skeletons of microscopic marine organisms.

Abyssal plains occur in all oceans. However, the floor of the Atlantic has the most extensive abyssal plains because it has few trenches to act as traps for sediment carried down the continental slope.

Figure 13.14 Seismic profile This seismic cross section and matching sketch across a portion of the Madeira Abyssal Plain in the eastern Atlantic show the irregular oceanic crust buried by sediments. (Image courtesy Woods Hole Oceanographic Institution, Charles Hollister, translator. Used with permission of the Woods Hole Oceanographic Institution.)

Seismic reflection profile

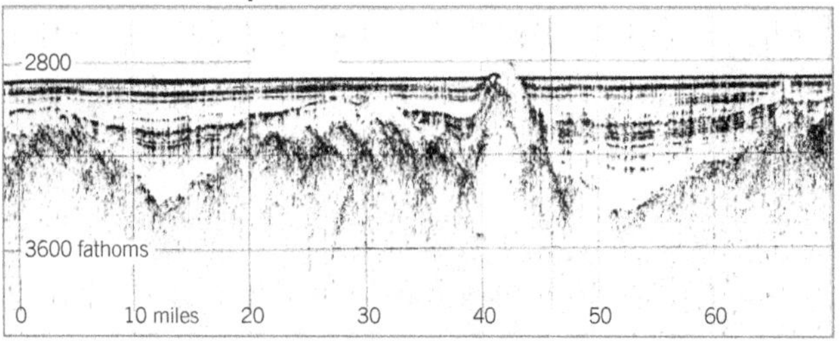

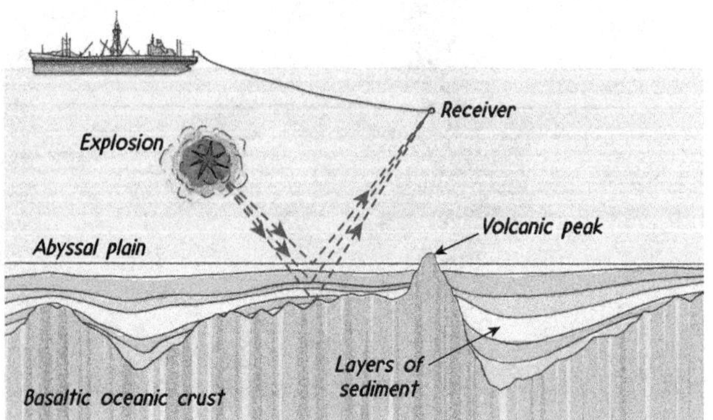

Volcanic Structures on the Ocean Floor

Dotting the seafloor are numerous volcanic structures of various sizes. Many occur as isolated features that resemble volcanoes on land. Others occur as long, narrow chains that stretch for thousands of kilometers, while still others are massive structures that cover an area the size of Texas.

Seamounts and Volcanic Islands

Submarine volcanoes, called **seamounts**, may rise hundreds of meters above the surrounding seafloor. It is estimated that more than one million seamounts exist worldwide. Some grow large enough to become oceanic islands, but most do not have a sufficiently long eruptive history to create a volcano so tall that it emerges above sea level. Although seamounts are found on the floors of all the oceans, they are most common in the Pacific.

Some, like the Hawaiian Island–Emperor seamount chain, which stretches from the Hawaiian islands to the Aleutian trench, form over volcanic hot spots (see Figure 13.6 and Figure 4.26, page 115). Others are born near oceanic ridges.

If a volcano grows large enough before it is carried from its magma source by plate motion, the structure may emerge as a *volcanic island*. Examples of volcanic islands include Easter Island, Tahiti, Bora Bora, the Galapagos Islands, and the Canary Islands. Some islands are sites of substantial coral reefs. GEOgraphics 13.1 explores the development of coral reefs associated with volcanic islands.

Guyots During their existence, inactive volcanic islands are gradually but inevitably lowered to near sea level by the forces of weathering and erosion. As a moving plate slowly carries inactive volcanic islands away from the elevated oceanic ridge or hot spot over which they formed, they gradually sink and disappear below the water surface. Submerged, flat-topped seamounts that formed in this manner are called **guyots**.*

Oceanic Plateaus The ocean floor contains several massive **oceanic plateaus**, which resemble lava plateaus composed of flood basalts found on the continents. Oceanic plateaus, which can be more than 30 kilometers (20 miles) thick, are generated from vast outpourings of fluid basaltic lavas.

Some oceanic plateaus appear to have formed quickly in geologic terms. Examples include the Ontong Java Plateau, which formed in less than 3 million years, and the Kerguelen Plateau, which formed in 4.5 million years (see Figure 13.6). Like basalt plateaus on land, oceanic plateaus are thought to form when the bulbous head of a rising mantle plume melts and produces vast underwater outpourings of basalt.

CONCEPT CHECKS 13.4

1. Explain how deep-ocean trenches are related to convergent plate boundaries.
2. Why are abyssal plains more extensive on the floor of the Atlantic than on the floor of the Pacific?
3. How does a flat-topped seamount, called a *guyot*, form?
4. What features on the ocean floor most resemble basalt plateaus on the continents? How are they formed?

*The term *guyot* comes from the name of Arnold Guyot, Princeton University's first geology professor. It is pronounced "GEE-oh," with a hard g, as in *give*.

13.5 The Oceanic Ridge System

Summarize the basic characteristics of oceanic ridges.

Along well-developed divergent plate boundaries, the seafloor is elevated, forming a broad linear mountain range on the seafloor called the **oceanic ridge** or **rise,** or **mid-ocean ridge**. Knowledge of the oceanic ridge system comes from soundings of the ocean floor, core samples from deep-sea drilling, visual inspection using deep-diving submersibles, and firsthand inspection of slices of ocean floor that have been thrust onto dry land during continental collisions (Figure 13.15). Oceanic ridges are characterized by extensive normal and strike-slip faulting, earthquakes, high heat flow, and recent volcanism.

▲ Figure 13.15 **The deep-diving submersible *Alvin*** Human Occupied Vehicle (HOV) *Alvin* enables data collection and observation by two scientists to depths reaching 4,500 meters (15,000 feet), during dives lasting up to 10 hours. Commissioned in 1964 as one of the world's first deep-ocean submersibles, *Alvin* has remained state-of-the-art as a result of numerous overhauls and upgrades. The most recent was completed in 2013. With seven reversible thrusters, it can hover in the water, maneuver over rugged topography, or rest on the seafloor. It can collect data throughout the water column, produce a variety of maps, and perform photographic surveys. *Alvin* also has two robotic arms that can manipulate instruments and obtain samples. *Alvin* gives researchers in-person access to about two-thirds of the ocean floor. (Photo by Rod Catanach KRT/Newscom)

Anatomy of the Oceanic Ridge System

The oceanic ridge system winds through all major oceans in a manner similar to the seam on a baseball. It is the longest topographic feature on Earth, at more than 70,000 kilometers (43,000 miles) in length (Figure 13.16). The crest of the ridge typically stands 2 to 3 kilometers above the adjacent deep-ocean floor and is associated with a divergent plate boundary where new oceanic crust is created.

Notice in Figure 13.16 that large sections of the oceanic ridge system have been named based on their locations within the various ocean basins. Some ridges run through the middle of ocean basins, where they are appropriately called *mid-ocean* ridges. The Mid-Atlantic Ridge and the Mid-Indian Ridge are examples. By contrast, the East Pacific Rise is *not* a "mid-ocean" feature. Rather, as its name implies, it is located in the eastern Pacific, far from the center of the ocean.

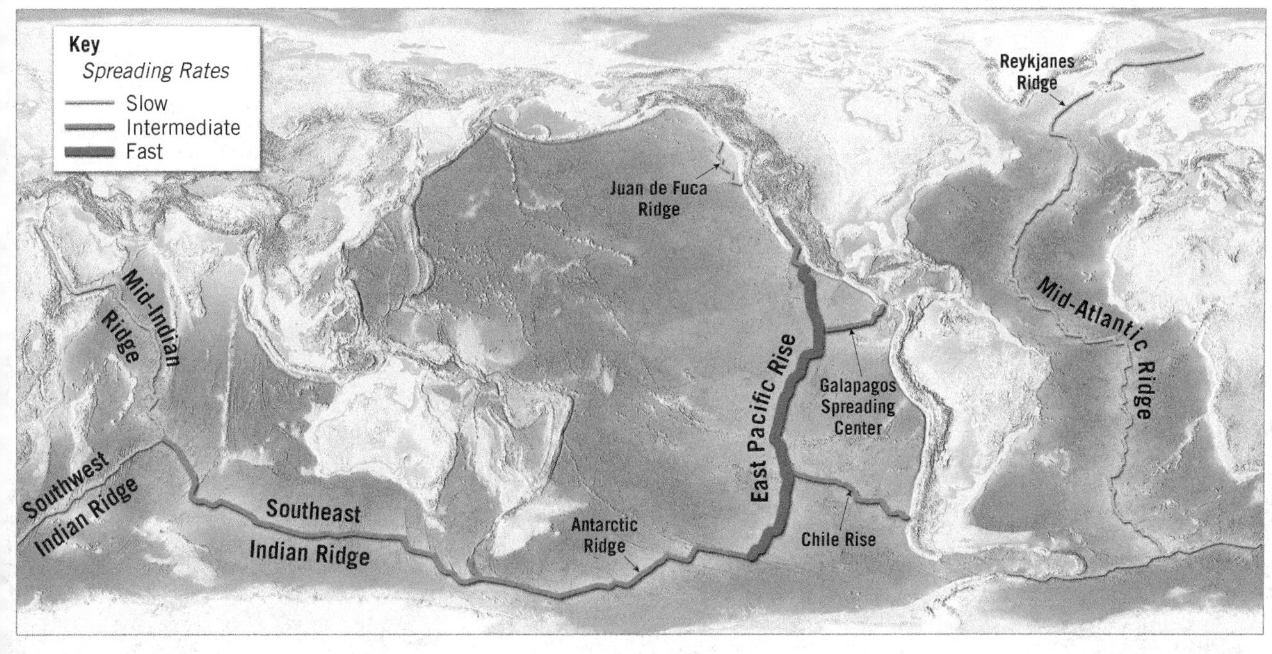

◄ Figure 13.16 **Distribution of the oceanic ridge system** The map shows ridge segments that exhibit slow, intermediate, and fast spreading rates.

Explaining Coral Atolls: Darwin's Hypothesis

Douglas Peeples Photograpl

Coral atolls are ring-shaped structures that often extend from slightly above sea level to depths of several thousand meters. What causes atolls to form, and how do they attain such great thicknesses?

Reef-building corals only grow in warm, clear sunlit water. The depth of most active reef growth is limite to warm tropical water no more than about 45 meters (150 feet) deep. The strict environmental conditio required for coral growth create an interesting paradox: How can corals—which require warm, shallov sunlit water no deeper than a few dozen meters—create massive structures such as coral atolls that extend to great depths?

Anatomy of an individual coral polyp

TENTACLE

MOUTH

STOMACH

CALCIUM CARBONATE SUPPORT STRUCTURE

Most corals secrete a hard external skeleton made of calcium carbonate. Colonies of these corals build large calcium carbonate structures, called reefs, where new colonies grow atop the strong skeletons of previous colonies. Sponges and algae may attach to the reef, further enlarging the structure.

2.0 cm

Corals: Tiny colonial animals Corals are marine animals that typically live in compact colonies of many identical organisms called polyps.

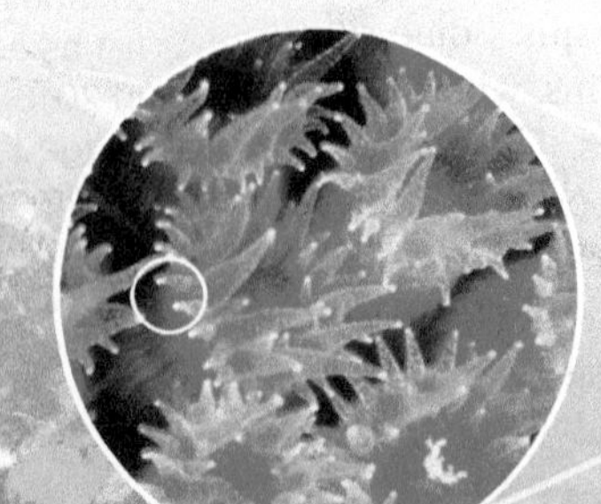

Reinhard Dirsherl/AGE Fotostock

What Darwin observed Naturalist Charles Darwin was one of the first to formulate a hypothesis on the origin of these ringed-shaped atolls. Aboard the British ship HMS *Beagle* during its famous global circumnavigation, Darwin noticed a progression of stages in coral reef development from (1) a fringing reef along the margins of a volcano to (2) a barrier reef with a volcano in the middle to (3) an atoll, consisting of a continuous or broken ring of coral reef surrounding a central lagoon.

Darwin's hypothesis Darwin's hypothesis asserts that, in addition to being lowered by erosional forces, many volcanic islands gradually sink. Darwin also suggested that corals respond to the gradual change in water depth caused by the subsiding volcano by building the reef complex upward. During Darwin's time, however, there was no plausible mechanism to account for how or why so many volcanic islands sink.

Plate tectonics and coral atolls The plate tectonics theory provides the most current scientific explanation regarding how volcanic islands become extinct and sink to great depths over long periods of time. Some volcanic islands form over relatively stationary mantle plumes, causing the lithosphere to be buoyantly uplifted. Over spans of millions of years, these volcanic islands gradually sink as moving plates carry them away from the region of hot-spot volcanism.

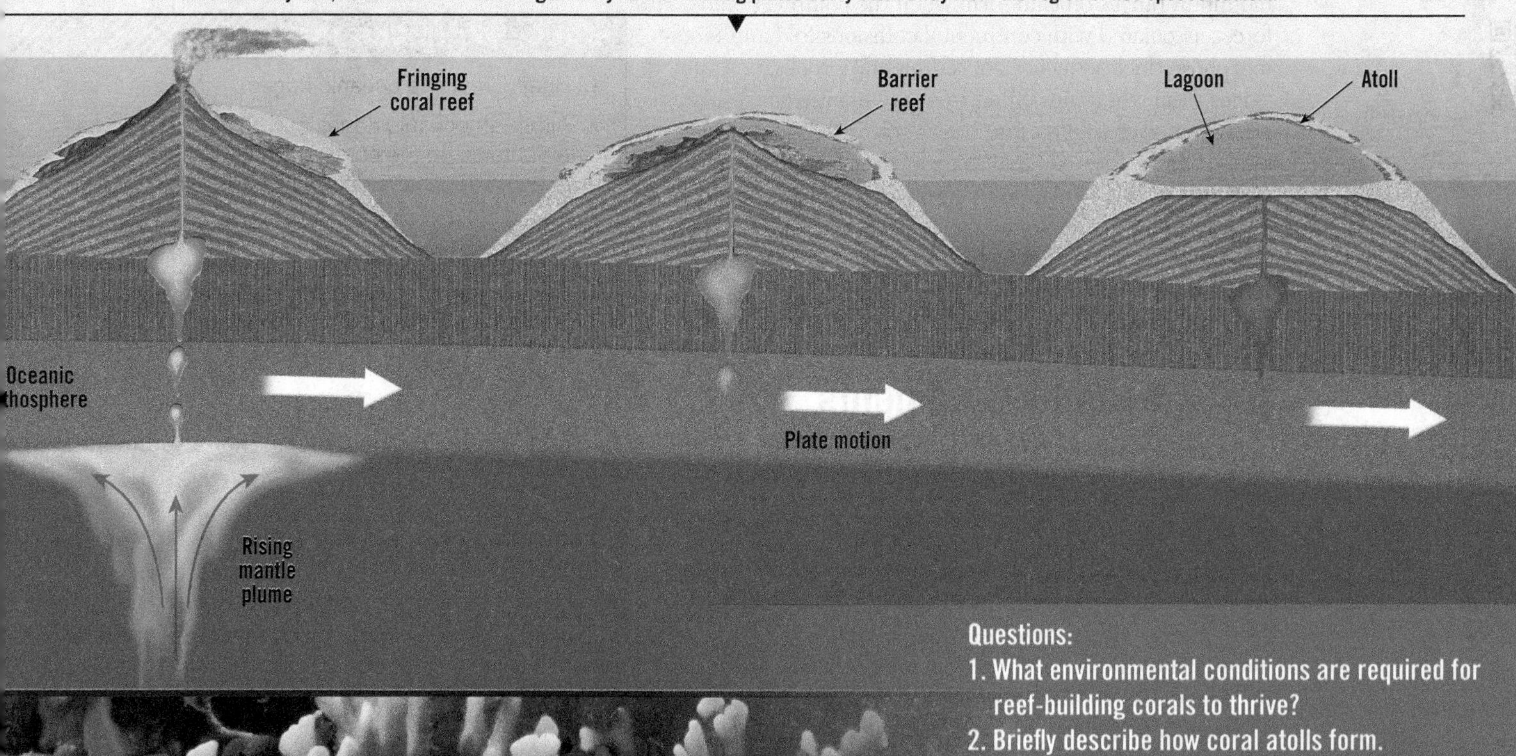

Questions:

1. What environmental conditions are required for reef-building corals to thrive?
2. Briefly describe how coral atolls form.

?

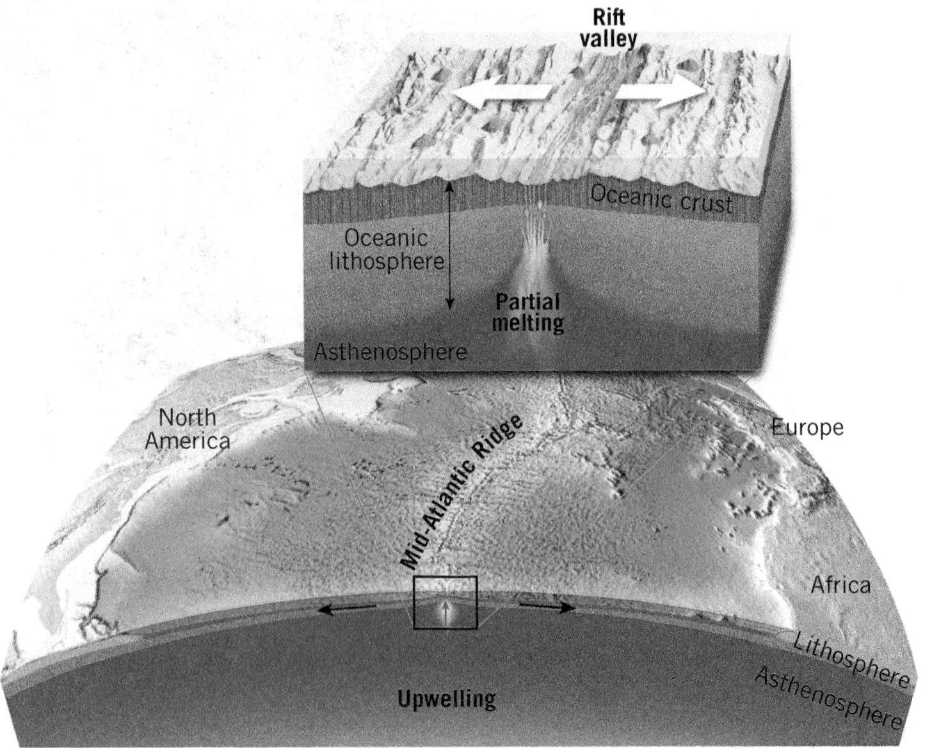

SmartFigure 13.17 **Rift valleys** The axes of some segments of the oceanic ridge system contain deep down-faulted structures called *rift valleys*. Some may exceed 50 kilometers (31 miles) in width and 2000 meters (6600 feet) in depth.

TUTORIAL
http://goo.gl/4KcuA1

The term *ridge* is somewhat misleading because these features are not narrow and steep as the term implies but have widths of 1000 to 4000 kilometers (600 to 2500 miles) and the appearance of broad, elongated swells that exhibit varying degrees of ruggedness. Furthermore, the ridge system is broken into segments that range from a few tens to hundreds of kilometers in length. Each segment is offset from the adjacent segment by a transform fault.

Oceanic ridges are as high as some mountains on the continents, but the similarities end there. Whereas most mountain ranges on land form when the compressional forces associated with continental collisions fold and metamorphose thick sequences of sedimentary rocks, oceanic ridges form where upwelling from the mantle generates new oceanic crust. Oceanic ridges consist of layers and piles of newly formed basaltic rocks that are buoyantly uplifted by the hot mantle rocks from which they formed.

Along the axes of some segments of the oceanic ridge system are deep, down-faulted structures that are called **rift valleys** because of their striking similarity to the continental rift valleys in East Africa (Figure 13.17). Some rift valleys, including those along the rugged Mid-Atlantic Ridge, are typically 30 to 50 kilometers (20 to 30 miles) wide and have walls that tower 500 to 2500 meters (1640 to 8200 feet) above the valley floor. This makes them comparable to the deepest and widest part of Arizona's Grand Canyon.

Why Is the Oceanic Ridge Elevated?

The primary reason for the elevated position of the ridge system is that newly created oceanic lithosphere is hot and therefore less dense than cooler rocks of the deep-ocean basin. As the newly formed basaltic crust travels away from the ridge crest, it is cooled from above as seawater circulates through the pore spaces and fractures in the rock. In addition, it cools because it gets farther and farther from the zone of hot mantle upwelling. As a result, the lithosphere gradually cools, contracts, and becomes denser. This thermal contraction accounts for the greater ocean depths that occur away from the ridge. It takes as much as 80 million years of cooling and contraction for rock that was once part of an elevated ocean ridge system to relocate to the deep-ocean basin.

As lithosphere is displaced away from the ridge crest, cooling also causes a gradual increase in lithospheric thickness. This happens because the boundary between the lithosphere and asthenosphere is a thermal (temperature) boundary. Recall that the lithosphere is Earth's cool, stiff outer layer, whereas the asthenosphere is a comparatively hot and weak layer. As material in the uppermost asthenosphere ages and cools, it becomes stiff and rigid. Thus, the upper portion of the asthenosphere is gradually converted to lithosphere simply by cooling. Oceanic lithosphere continues to thicken until it is about 80 to 100 kilometers (50 to 60 miles) thick. Thereafter, its thickness remains relatively unchanged until it is subducted.

CONCEPT CHECKS 13.5

1. Briefly describe oceanic ridges.
2. Although oceanic ridges can be as tall as some mountains on the continents, list some ways that oceanic ridges are different from mountains.
3. Where do rift valleys form along the oceanic ridge system?
4. What is the primary reason for the elevated position of the oceanic ridge system?

13.6 Seafloor Sediments

Distinguish among three categories of seafloor sediment and explain how biogenous sediment can be used to study climate change.

The vast majority of the ocean floor is covered with sediment. Part of this material is deposited by turbidity currents, and the rest slowly settles to the seafloor from above and is sometimes referred to as "marine snow." The thickness of this carpet of debris varies greatly. In some trenches, which act as traps for sediments originating on the continental margin, accumulations may approach 10 kilometers (6 miles). In general, however, sediment accumulations on the deep ocean floor are considerably less. In the Pacific Ocean, for example, uncompacted sediment measures about 600 meters (2000 feet) or less, whereas on the floor of the Atlantic, the thickness varies from 500 to 1000 meters (1600 to 3300 feet).

Types of Seafloor Sediments

Seafloor sediments can be classified according to their origin into three broad categories: **terrigenous** ("derived from land"), **biogenous** ("derived from organisms"), and **hydrogenous** ("derived from water"). Although each category is discussed separately, remember that all seafloor sediments are mixtures. No marine sediments come entirely from a single source.

Terrigenous Sediment Terrigenous sediment consists primarily of mineral grains that were weathered from continental rocks and transported to the ocean. Larger particles (sand and gravel) usually settle rapidly near shore, whereas the very smallest particles take years to settle to the ocean floor and may be carried thousands of kilometers by ocean currents. As a result, virtually every area of the ocean receives some terrigenous sediment. The rate at which this sediment accumulates on the deep-ocean floor, though, is very slow. The time required to form a 1-centimeter (0.4-inch) abyssal clay layer, for example, can be as much as 10,000 years. Conversely, on the continental margins near the mouths of large rivers, terrigenous sediment accumulates rapidly and forms thick deposits.

Biogenous Sediment In the deep ocean, biogenous sediment consists of microscopic shells and skeletons of marine animals and algae (Figure 13.18). This debris is produced by free-floating microscopic marine organisms called *plankton* that live in the sunlit waters near the ocean surface. Once these organisms die, their hard *tests* (*testa* = shell) continually "rain" down and accumulate on the seafloor below.

The most common biogenous sediment is *calcareous* ($CaCO_3$) *ooze*, which, as the name implies, has the consistency of thick mud. This sediment is produced from the tests of organisms that inhabit warm surface waters. When the organisms die, calcareous hard parts slowly sink through a cool layer of water; when they reach a certain depth, they begin to dissolve. This occurs because the deeper cold seawater is richer in carbon dioxide, which forms carbonic acid that dissolves calcium carbonate and is thus more acidic than warm water. In seawater deeper than about 4500 meters (15,000 feet), calcareous tests completely dissolve before they reach the bottom. Consequently, calcareous ooze does not accumulate if the water depth is deeper than 4500 meters (15,000 feet).

Other biogenous sediments include *siliceous* (SiO_2) *ooze* and phosphate-rich material. The former is composed primarily of tests of diatoms (single-celled algae) and radiolaria (single-celled animals), whereas the latter is derived from the bones, teeth, and scales of fish and other marine organisms.

Hydrogenous Sediment Hydrogenous sediment consists of minerals that crystallize directly from seawater through various chemical reactions. For example, some limestones are formed when calcium carbonate precipitates directly from the water; however, most limestone is composed of biogenous sediment.

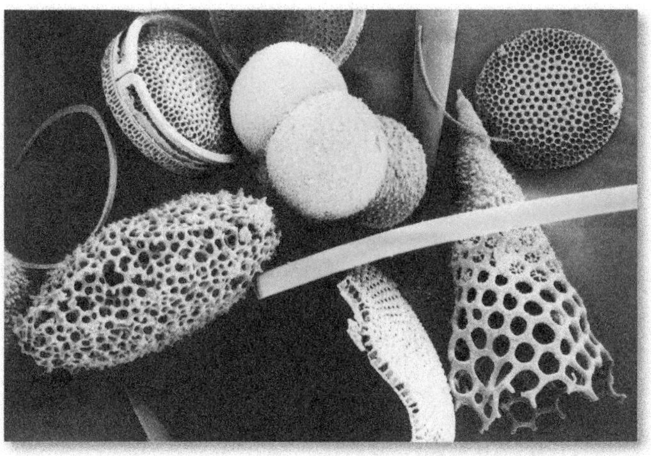

Figure 13.18 Marine microfossils: An example of biogenous sediment These tiny, single-celled organisms are sensitive to even small fluctuations in temperature. Seafloor sediments containing these fossils are useful sources of data on climate change. This is a false-color image. (Photo by Biophoto Associates/Science Source)

Some of the most common types of hydrogenous sediment include the following:

- *Manganese nodules* are rounded, hard lumps of manganese, iron, and other metals that precipitate in concentric layers around a central object (such as a volcanic pebble or a grain of sand). The nodules can be up to 20 centimeters (8 inches) in diameter and are often littered across large areas of the deep seafloor (Figure 13.19A).
- *Calcium carbonates* form by precipitating directly from seawater in warm climates. If this material is buried and hardened, it forms limestone. Most limestone, however, is composed of biogenous sediment.
- *Metal sulfides* are usually precipitated as coatings on rocks near black smokers associated with the crest of a mid-ocean ridge (Figure 13.19B). These deposits contain iron, nickel, copper, zinc, silver, and other metals in varying proportions.
- *Evaporites* form where evaporation rates are high and there is restricted open-ocean circulation. As water evaporates from such areas, the remaining seawater

SmartFigure 13.19 Examples of hydrogenous sediment A. Manganese nodules. (Photo by Charles A. Winter/Science Source) **B.** This black smoker is spewing hot, mineral-rich water. When the heated solutions meet cold seawater, metal sulfides precipitate and form mounds of minerals around these hydrothermal vents. (Photo by Verena Tunnicliffe/Uvic/Fisheries and Oceans Canada/Newscom)

TUTORIAL
https://goo.gl/2BMRcv

A.

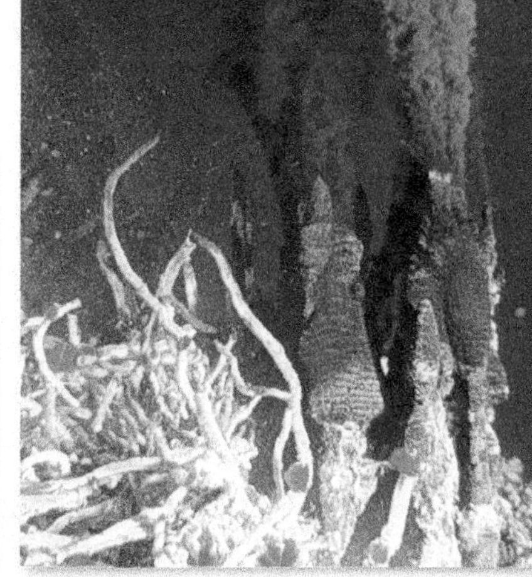

B.

▲ Figure 13.20 **Data from the seafloor** This scientist is holding a sediment core obtained by an oceanographic research vessel. Cores of sediment provide data that allow for a more complete understanding of past climates. Notice the "library" of sediment cores in the background. (Photo by INGO WAGNER/EPA/Newscom)

becomes saturated with dissolved minerals, which then begin to precipitate. Heavier than seawater, these crystals sink to the bottom or form a characteristic white crust of evaporite minerals around the edges of these areas. Evaporites are collectively termed *salts*; some evaporite minerals taste salty, such as *halite* (common table salt, NaCl), and some do not, such as the calcium sulfate minerals *anhydrite* ($CaSO_4$) and *gypsum* ($CaSO_4 \cdot 2H_2O$).

Seafloor Sediment—A Storehouse of Climate Data

Reliable human climate records go back only a couple hundred years, at best. How do scientists learn about climates and climate change prior to that time? They must reconstruct past climates from *proxy data*—that is, data from sources that respond to and reflect changing climate conditions. An interesting and important source of proxy data are biogenous seafloor sediments.

We know that the parts of the Earth system are linked so that a change in one part can produce changes in any or all of the other parts. For example, changes in atmospheric and oceanic temperatures are reflected in the nature of life in the sea.

Most seafloor sediments contain the remains of organisms that once lived near the sea surface (the ocean–atmosphere interface). When such near-surface organisms die, their shells slowly settle to the floor of the ocean, where they become part of the sedimentary record. Because the numbers and types of organisms living near the sea surface change with the climate, the sediments on the seafloor faithfully record changes in climate dating back millions of years. These seafloor sediments are a "library" of worldwide climate change, unique in its long-term record of history and unlike anything on land.

Thus, in seeking to understand climate change as well as other environmental transformations, scientists are tapping the huge reservoir of data in seafloor sediments (Figure 13.20). The sediment cores gathered by drilling ships such as the one in the chapter-opening photo have provided invaluable data that have significantly expanded our knowledge and understanding of past climates.

One notable example of the importance of seafloor sediments to our understanding of climate change relates to unraveling the fluctuating atmospheric and oceanic conditions of the Quaternary Ice Age. The records of temperature changes contained in cores of sediment from the ocean floor have been critical to our present understanding of this recent span of Earth history.

CONCEPT CHECKS 13.6

1. Distinguish among the three basic types of seafloor sediments. Give an example of each.
2. Explain why biogenous sediments are useful in studying past climates.

13.7 Resources from the Seafloor

Specify the important resources and potential resources associated with the ocean floor.

The seafloor is rich in mineral and organic resources. Most, however, are not easily accessible, and recovery involves technological challenges and high cost. Nevertheless, certain resources have high value and thus make appealing exploration targets.

Energy Resources

Among the nonliving resources extracted from the oceans, more than 95 percent of the economic value comes from energy products. The main energy products are oil and natural gas, which are currently being extracted, and gas hydrates, which are not yet being utilized but have significant potential.

Oil and Natural Gas The ancient remains of microscopic organisms, buried within marine sediments before they could completely decompose, are the source of today's deposits of oil and natural gas. The percentage of world oil produced from offshore regions has increased from trace amounts in the 1930s to more than 30 percent today. Most of this increase results from continuing technological advancements employed by offshore drilling platforms.

Major offshore reserves exist in the Persian Gulf, in the Gulf of Mexico, off the coast of Southern California, in the North Sea, and in the East Indies. Additional reserves are located off the northern coast of Alaska and in the Canadian Arctic, Asian seas, Africa, and Brazil. Because the likelihood of finding major new reserves on land is small, future offshore exploration will continue to be important, especially in deeper waters of the continental margins. A major environmental concern about offshore petroleum exploration is the possibility of oil spills caused by inadvertent leaks or blowouts during the drilling process (Figure 13.21).

▲ **Figure 13.21 Disaster in the Gulf of Mexico** On April 20, 2010, the mobile floating drilling platform named *Deepwater Horizon* exploded while working on a 10,680-meter (35,000-foot) well in approximately 1600 meters (5200 feet) of water, causing a devastating oil spill. (Photo by John Mosier/ZUMA Press/Newscom)

Gas Hydrates **Gas hydrates** are unusually compact chemical structures made of water and natural gas. The most common type of natural gas is methane, which produces *methane hydrate*. Gas hydrates occur beneath permafrost areas on land and under the ocean floor at depths below 525 meters (1720 feet).

Most oceanic gas hydrates are created when bacteria break down organic matter trapped in seafloor sediments, producing methane gas with minor amounts of ethane and propane. These gases combine with water in deep-ocean sediments (where pressures are high and temperatures are low) in such a way that the gas is trapped inside a lattice-like cage of water molecules.

Vessels that have drilled into gas hydrates have retrieved cores of mud mixed with chunks or layers of gas hydrates (Figure 13.22A) that fizzle and evaporate quickly when they are exposed to the relatively warm, low-pressure conditions at the ocean surface. Gas hydrates resemble chunks of ice but ignite when lit by a flame because methane and other flammable gases are released as the hydrate melts (Figure 13.22B).

Some estimates indicate that as much as 20 quadrillion cubic meters (700 quadrillion cubic feet) of methane are locked up in sediments containing gas hydrates. This is equivalent to about *twice* as much carbon as Earth's coal, oil, and natural gas reserves combined, so gas hydrates have great potential. A major drawback in exploiting reserves of gas hydrates is that they rapidly decompose at surface temperatures and pressures. In the future, however, these vast seafloor reserves of energy may help power modern society.

Other Resources

Other major resources from the seafloor include sand and gravel, evaporative salts, and manganese nodules.

Sand and Gravel The offshore sand-and-gravel industry is second in economic value only to the petroleum industry. Sand and gravel, which includes rock fragments that are washed out to sea and shells of marine organisms, are mined by offshore barges using suction dredges. Sand and gravel are primarily used as aggregate in concrete, as fill material in grading projects, and on recreational beaches.

In some cases, materials of high economic value are associated with offshore sand and gravel deposits. Gem-quality diamonds, for example, are recovered from gravels on the continental shelf offshore of South Africa and Australia. Sediments rich in tin have been mined from some offshore areas of Southeast Asia. Platinum and gold have been found in deposits in gold-mining areas throughout the world, and some Florida beach sands are rich in titanium.

Evaporative Salts When seawater evaporates, the salts increase in concentration until they can no longer remain dissolved; they then precipitate out of solution and form salt deposits, which can then be harvested. The most economically important salt is *halite* (common table salt). Halite is widely used for seasoning, curing, and preserving foods. It is also used in water conditioners, in agriculture, in the clothing industry for dying fabric, and to de-ice roads. Since ancient times, the ocean has been an important source of salt for human consumption, and the sea remains a significant supplier (Figure 13.23).

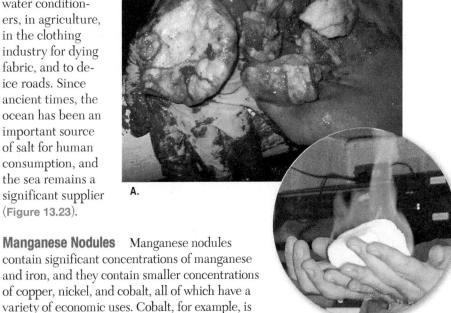

Manganese Nodules Manganese nodules contain significant concentrations of manganese and iron, and they contain smaller concentrations of copper, nickel, and cobalt, all of which have a variety of economic uses. Cobalt, for example, is deemed "strategic" (essential to U.S. national security)

▼ **Figure 13.22 Gas hydrates A.** This sample was retrieved from the deep-ocean floor and shows white ice-like gas hydrate mixed with mud. (Photo by USGS National Center) **B.** Gas hydrates evaporate when exposed to surface conditions and release natural gas, which can be ignited. They are called "ice that burns." (Photo by Yonhap News/Newscom)

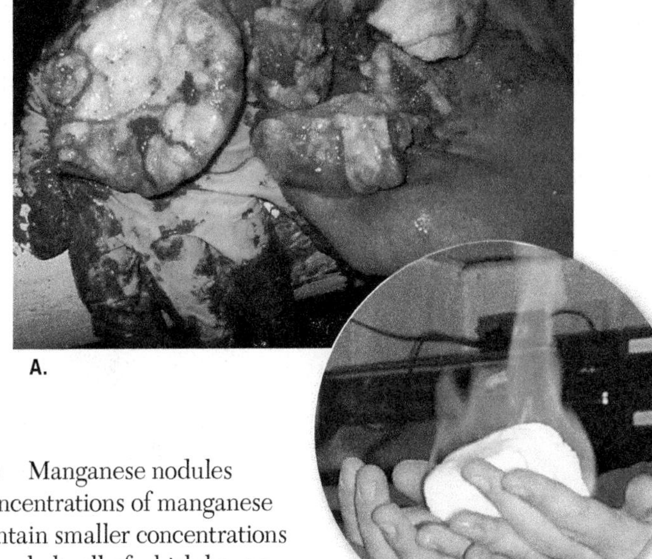

▲ Figure 13.23 **Harvesting sea salt** Seawater is one important source of common salt (sodium chloride). Saltwater is held in shallow ponds, where solar energy evaporates the water. The nearly pure salt deposits that eventually form are essentially artificial evaporate deposits. (Photo by Geof Kirby/Alamy Stock Photo)

because it is required to produce dense, strong alloys with other metals and is used in high-speed cutting tools, powerful permanent magnets, and jet engine parts. Technologically, mining the deep-ocean floor for manganese nodules is possible but not yet economically profitable.

Manganese nodules are widely distributed, but not all regions have the same potential for mining. Good locations have abundant nodules that contain the economically optimum mix of copper, nickel, and cobalt. Sites meeting these criteria, however, are relatively limited. In addition, there are political problems of establishing mining rights far from land and environmental concerns about disturbing large portions of the deep-ocean floor.

CONCEPT CHECKS 13.7

1. Which type of seafloor resource is presently most valuable?
2. What are gas hydrates?
3. What nonenergy seafloor resource is most valuable?

CONCEPTS IN REVIEW

The Ocean Floor

13.1 The Vast World Ocean

Discuss the extent and distribution of oceans and continents on Earth. Identify Earth's four main ocean basins.

KEY TERM: oceanography

- Oceanography is an interdisciplinary science that draws on the methods and knowledge of biology, chemistry, physics, and geology to study all aspects of the world ocean.
- Earth's surface is dominated by oceans. Nearly 71 percent of the planet's surface area is oceans and marginal seas. In the Southern Hemisphere, about 81 percent of the surface is water.
- Of the three major oceans—the Pacific, Atlantic, and Indian—the Pacific Ocean is the largest, contains slightly more than half of the water in the world ocean, and has the greatest average depth—3940 meters (12,927 feet).

? Name the four principal oceans identified with letters on this world map.

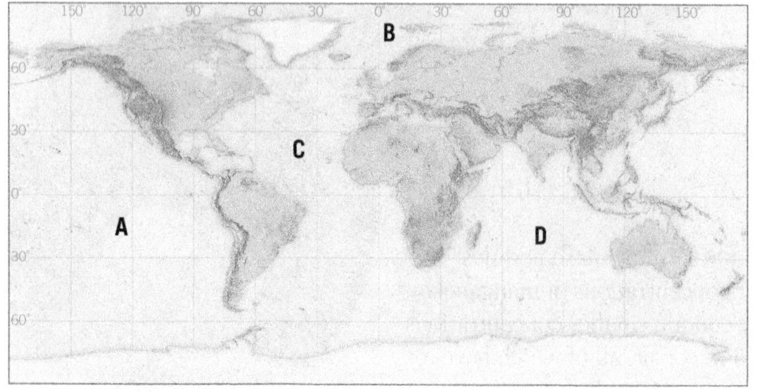

13.2 An Emerging Picture of the Ocean Floor

Define *bathymetry* and summarize the various techniques used to map the ocean floor.

KEY TERMS: bathymetry, sonar, echo sounder

- Seafloor mapping is done with sonar—shipboard instruments that emit pulses of sound that "echo" off the bottom. Satellites are also used to map the ocean floor. Their instruments measure slight variations in sea level that result from differences in the gravitational pull of features on the seafloor. Accurate maps of seafloor topography can be made using these data.
- Mapping efforts have revealed three major areas of the ocean floor: continental margins, deep-ocean basins, and oceanic ridges.

? Do the math: How many seconds would it take an echo sounder's ping to make the trip from a ship to the Challenger Deep (11,022 meters [36,161 feet]) and back? Recall that depth = ½ (1500 m/sec × echo travel time).

13.3 Continental Margins

Compare a passive continental margin with an active continental margin and list the major features of each.

KEY TERMS: continental margin, passive continental margin, continental shelf, continental slope, continental rise, deep-sea fan, submarine canyon, turbidity current, active continental margin, accretionary wedge, subduction erosion

- Continental margins are transition zones between continental and oceanic crust. Active continental margins occur where a plate boundary and the edge of a continent coincide, usually on the leading edge of a plate. Passive continental margins are on the trailing edges of continents, far from plate boundaries.

- Heading offshore from the shoreline of a passive margin, a submarine traveler would first encounter the gently sloping continental shelf and then the steeper continental slope, marking the end of the continental crust and the beginning of the oceanic crust. Beyond the continental slope is another gently sloping section, the continental rise, made of sediment transported by turbidity currents through submarine canyons and piled up in deep-sea fans atop the oceanic crust.
- Submarine canyons are deep, steep-sided valleys that originate on the continental slope and may extend to the deep-ocean basin. Many submarine canyons have been excavated by turbidity currents (downslope movements of dense, sediment-laden water).
- At an active continental margin, material may be added to the leading edge of a continent in the form of an accretionary wedge (common at shallow-angle subduction zones), or material may be scraped off the edge of a continent by subduction erosion (common at steeply dipping subduction zones).

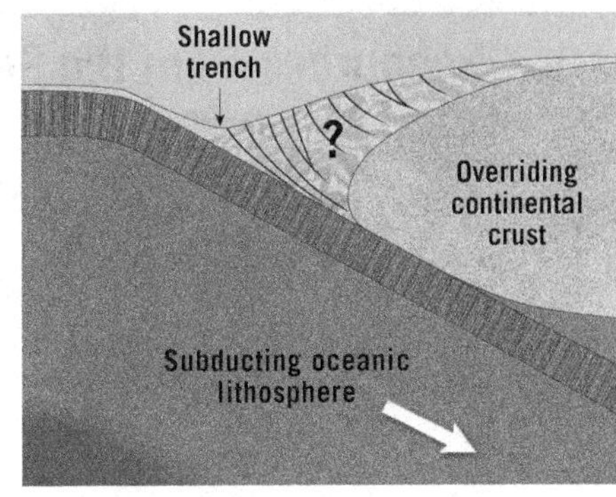

? What type of continental margin is depicted by this diagram? Be as specific as possible. Name the feature indicated by the question mark.

13.4 Features of Deep-Ocean Basins

List and describe the major features associated with deep-ocean basins.

KEY TERMS: deep-ocean basin, deep-ocean trench, volcanic island arc, continental volcanic arc, abyssal plain, seamount, guyot, oceanic plateau

- The deep-ocean basin makes up about half of the ocean floor's area. Much of it is abyssal plain (deep, featureless sediment-draped crust). Subduction zones and deep-ocean trenches also occur in deep-ocean basins. Paralleling trenches are volcanic island arcs (if the subduction goes underneath oceanic lithosphere) or continental volcanic arcs (if the overriding plate has continental lithosphere on its leading edge).
- There are a variety of volcanic structures on the deep-ocean floor. Seamounts are submarine volcanoes; if they pierce the surface of the ocean, we call them volcanic islands. Guyots are old volcanic islands that have had their tops eroded off before they sink below sea level. Oceanic plateaus are unusually thick sections of oceanic crust formed by massive underwater eruptions of lava.

? On this cross-sectional view of a deep-ocean basin, label the following features: a seamount, a guyot, a volcanic island, an oceanic plateau, and an abyssal plain.

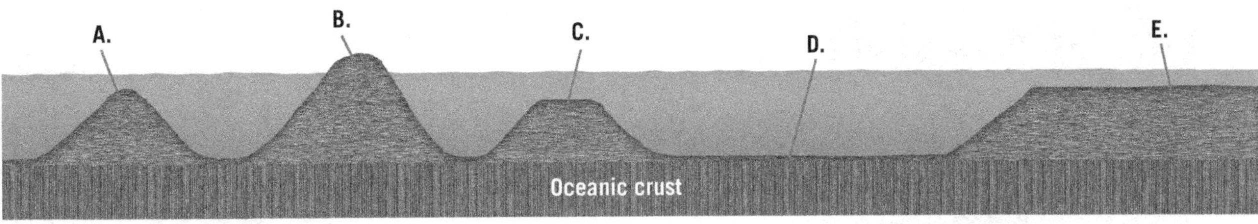

13.5 The Oceanic Ridge System

Summarize the basic characteristics of oceanic ridges.

KEY TERMS: oceanic ridge or rise (mid-ocean ridge), rift valley

- The oceanic ridge system is the longest topographic feature on Earth, wrapping around the world through all major ocean basins. It is a few kilometers tall, a few thousand kilometers wide, and a few tens of thousands of kilometers long. The summit is the place where new oceanic crust is generated, often marked by a rift valley.
- Oceanic ridges are elevated features because they are warm and therefore less dense than older, colder oceanic lithosphere. As oceanic crust moves away from the ridge crest, heat loss causes the oceanic crust to become denser and subside. After 80 million years, crust that was once part of an oceanic ridge is in the deep-ocean basin, far from the ridge.

13.6 Seafloor Sediments

Distinguish among three categories of seafloor sediment and explain how biogenous sediment can be used to study climate change.

KEY TERMS: terrigenous sediment, biogenous sediment, hydrogenous sediment

- There are three broad categories of seafloor sediments. Terrigenous sediment consists primarily of mineral grains that were weathered from continental rocks and transported to the ocean; biogenous sediment consists of shells and skeletons of marine animals and plants; and hydrogenous sediment includes minerals that crystallize directly from seawater through various chemical reactions.
- Seafloor sediments are helpful in studying worldwide climate change because they often contain the remains of organisms that once lived near the sea surface. The numbers and types of these organisms change as the climate changes, and their remains in seafloor sediments record these changes.

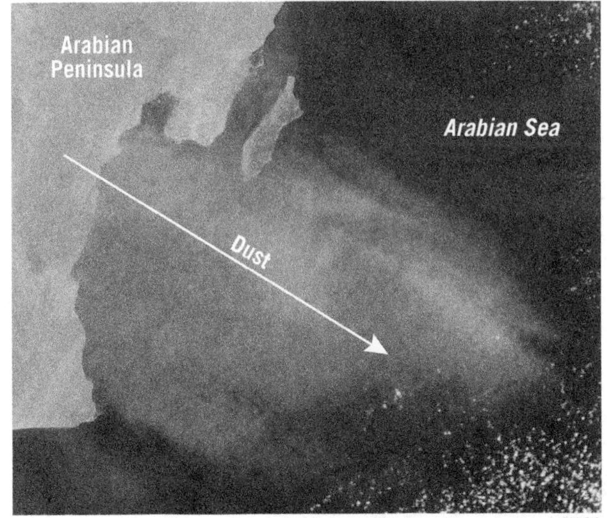

NASA

? This satellite image from February 4, 2013, shows a plume of dust moving from the southern Arabian Peninsula over the Arabian Sea, the name given to a portion of the Indian Ocean. The wind will subside, and the dust will settle onto the water and slowly sink, eventually reaching the ocean floor. To which category of seafloor sediment will this material belong?

13.7 Resources from the Seafloor

Specify the important resources and potential resources associated with the ocean floor.

KEY TERM: gas hydrate

- Oil and natural gas represent more than 95 percent of the value of the nonliving resources extracted from the oceans.
- About 30 percent of world oil production is from offshore areas. Environmental concerns include leaks and blowouts.
- Gas hydrates are complex molecules made of natural gas and water. Some form under high pressure and low temperature in marine sediments. Gas hydrates also form in permafrost environments. When brought to Earth's surface, they rapidly decompose into methane and water.
- Significant nonenergy resources and potential resources include sand and gravel, salt, and manganese nodules. About 30 percent of the world's supply of salt is extracted from seawater.

? Name two *potential* resources from the seafloor—one energy and one nonenergy.

GIVE IT SOME THOUGHT

1 Refer to the accompanying diagram of the eastern seaboard of the United States to complete the following:
   a. Associate each of the following with a letter on the map: continental rise, continental shelf, continental slope, and shelf break.
   b. How does the size of the continental shelf surrounding Florida compare to the size of the Florida peninsula?
   c. Why are there no deep-ocean trenches on this map?

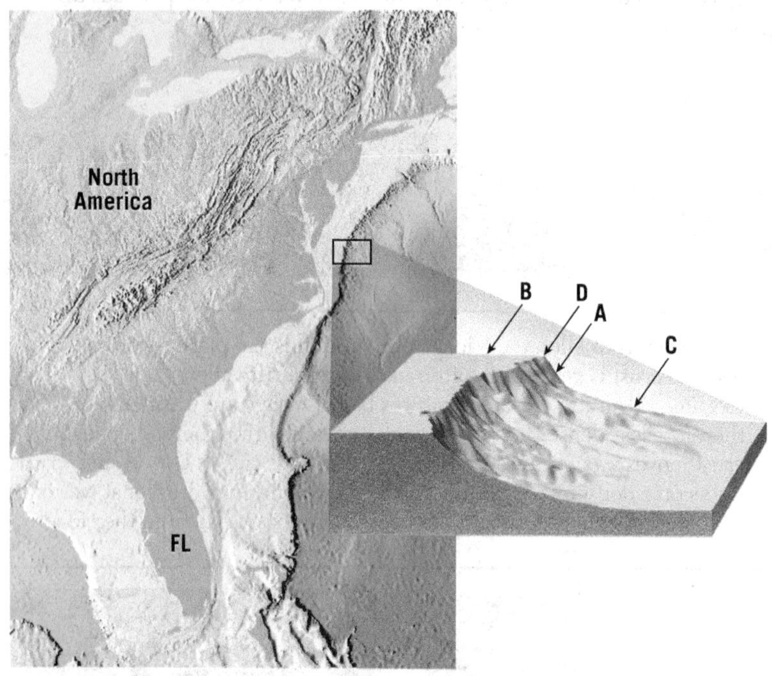

2 Assuming that the average speed of sound waves in water is 1500 meters per second, determine the water depth if a signal sent out by an echo sounder on a research vessel requires 6 seconds to strike bottom and return to the recorder aboard the ship.

3 Refer to Figure 13.2A to answer the following questions:
   a. Water dominates Earth's surface, but not everywhere. In what Northern Hemisphere latitude belt is there more land than water?
   b. In what latitude belt is there no land at all?

4 Are the continental margins surrounding the Atlantic Ocean primarily active or passive? How about the margins surrounding the Pacific Ocean? Based on your response to the foregoing questions, indicate whether each ocean basin is getting larger, shrinking, or staying the same size. Explain your answer.

5 Imagine that you and a passenger are in a deep-diving submersible such as the one shown here in the North Pacific Ocean near Alaska's Aleutian Islands. During a typical 6- to 10-hour dive, you encounter a long, narrow depression on the ocean floor. Your passenger asks whether you think it is a submarine canyon, a rift valley, or a deep-ocean trench. How would you respond? Explain your choice.

Jeff Rotman/Alamy

6 Examine the accompanying sketch, which shows three sediment layers on the ocean floor. What term is applied to such layers? What process was responsible for creating these layers? Are these layers more likely part of a deep-sea fan or an accretionary wedge?

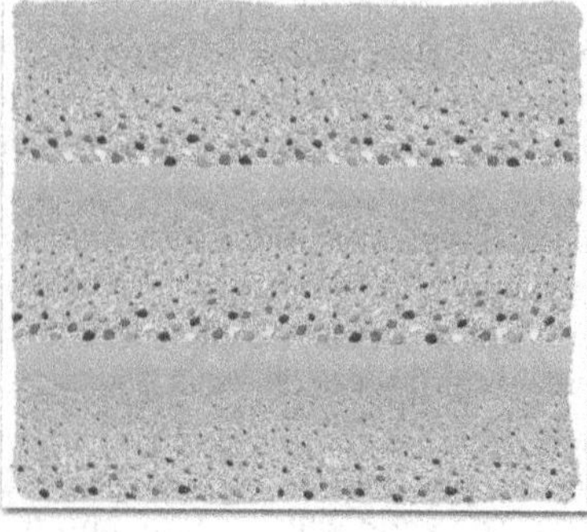

EXAMINING THE EARTH SYSTEM

1 Describe some of the exchanges of matter and energy that take place at the following interfaces:
a. ocean surface and atmosphere
b. ocean water and ocean floor
c. ocean biosphere and ocean water

2 This image from a scanning electron microscope shows the shell (test) of a tiny organism called a foraminiferan that was collected from the ocean floor by a research vessel. When alive, the organism that created this shell lived in the shallow surface waters of the ocean. Relate this tiny shell to each of Earth's four major spheres. Why are microscopic fossils such as this one useful in the study of climate change?

Dee Breger/Science Source

3 Sediment on the seafloor often leaves clues about various conditions that existed during deposition. What do the following layers in a seafloor core indicate about the environment in which each layer was deposited?
- Layer 5 (top): A layer of fine clays
- Layer 4: Siliceous ooze
- Layer 3: Calcareous ooze
- Layer 2: Fragments of coral reef
- Layer 1 (bottom): Rocks of basaltic composition with some metal sulfide coatings

Explain how one area of the seafloor could experience such varied conditions of deposition.

4 Reef-building corals are responsible for creating atolls—ring-shaped structures that extend from the surface of the ocean to depths of thousands of meters. These corals, however, can only live in warm, sunlit water no more than about 45 meters (150 feet) deep. This presents a paradox: How can corals, which require warm, sunlit water, create structures that extend to great depths? Explain the apparent contradiction.

Reinhard Dirscherl/Getty Images

DATA ANALYSIS

Exploring the Ocean Surface

One of the goals of the National Oceanic and Atmospheric Administration (NOAA) is to learn more about Earth's oceans, including ecosystems and surface topography. Research expeditions are conducted to explore various aspects of the ocean world.

ACTIVITIES

Go to the NOAA's Ocean Explorer at http://oceanexplorer.noaa.gov. Click on Explorations.

1 What is the purpose of the Explorations page?

Click on Digital Atlas.

2 Where have most of the research cruises been conducted?

3 How many missions have occurred during all years displayed on the map?

4 How many missions have occurred in the current year?

5 Which year had the largest number of missions? How many missions were conducted during that year?

Click on Search by Theme. Click No Themes and then select Canyons. Click on an expedition location on the map to learn more about the expedition. Find an expedition that has the GIS Tools tab.

6 What was the name of the expedition? When did this expedition take place?

7 What landmass is nearest to this research site?

Click on Expedition Web Site on the Summary tab. Find a mission statement or detailed description of the mission. (You may have to search for the expedition title.)

8 What are the goals of this expedition? Why is this research mission needed?

Click the GIS Tools tab and choose Ship Track and Plot on Map.

9 What are the approximate length and width of the area being explored?

10 Describe the ship's exploration pattern.

11 What named ocean features are near the ship's track?

14

Ocean Water and Ocean Life*

FOCUS ON CONCEPTS

Each statement represents the primary learning objective for the corresponding major heading within the chapter. After you complete the chapter, you should be able to:

14.1 Define *salinity* and list the main elements that contribute to the ocean's salinity. Describe the sources of dissolved substances in seawater and causes of variations in salinity.

14.2 Discuss temperature, salinity, and density changes with depth in the open ocean.

14.3 Distinguish among plankton, nekton, and benthos. Summarize the factors used to divide the ocean into marine life zones.

14.4 Contrast ocean productivity in polar, midlatitude, and tropical settings.

14.5 Define *trophic level* and discuss the efficiency of energy transfer between different trophic levels.

*This chapter was prepared and revised with the assistance of Professor Alan P. Trujillo, Palomar College.

Modern coral reefs are unique ecosystems of animals, plants, and their associated geologic framework. They are home to about 25 percent of all marine species. Because of this great diversity, they are sometimes referred to as the ocean equivalent of rain forests. (Photo by Mark Conlin/V&W/Image Quest Marine)

WHAT IS THE DIFFERENCE BETWEEN PURE WATER AND SEAWATER? One of the most obvious differences is that seawater contains dissolved substances that give it a distinctly salty taste. These dissolved substances are not simply sodium chloride (table salt); they include various other salts, metals, and even dissolved gases. In fact, every known naturally occurring element is found dissolved in at least trace amounts in seawater. Unfortunately, the salt content of seawater makes it unsuitable for drinking or for irrigating most crops and causes it to be highly corrosive to many materials. Yet many parts of the ocean are teeming with life that is superbly adapted to the marine environment.

There is an amazing variety of life in the ocean, from microscopic bacteria and algae to the largest organism alive today, the blue whale. Water is the major component of nearly every life-form on Earth, and our own body fluid chemistry is remarkably similar to the chemistry of seawater.

14.1 Composition of Seawater

Define *salinity* and list the main elements that contribute to the ocean's salinity. Describe the sources of dissolved substances in seawater and causes of variations in salinity.

Seawater consists of about 3.5 percent (by weight) dissolved mineral substances that are collectively termed "salts." Although the percentage of dissolved components may seem small, the actual quantity is huge because the ocean is so vast. For example, the oceans contain enough salt to cover the entire planet with a layer more than 150 meters (500 feet) thick (about the height of a 50-story skyscraper).

Salinity

Salinity (*salinus* = salt) is the total amount of solid material dissolved in water. More specifically, it is the ratio of the mass of dissolved substances to the mass of the water sample. Many common quantities are expressed in percent (%), which is really *parts per hundred.* Because the proportion of dissolved substances in seawater is such a small number, oceanographers typically express salinity in *parts per thousand* (‰). Thus, the average salinity of seawater is 3.5%, or 35‰.

Figure 14.1 shows the principal elements that contribute to the ocean's salinity. If one wanted to make artificial seawater, it could be approximated by following the recipe shown in Table 14.1. From this table, it is evident that most of the salt in seawater is sodium chloride—common table salt. Sodium chloride together with the next four most abundant salts comprise more than 99 percent of all dissolved substances in the sea. Although only eight elements make up these five most abundant salts, seawater contains all of Earth's other naturally occurring elements. Despite their presence in minute quantities, many of these elements are very important in maintaining the necessary chemical environment for life in the sea.

Sources of Sea Salts

What are the primary sources for the vast quantities of dissolved substances in the ocean? Chemical weathering of rocks on the continents is one important source.

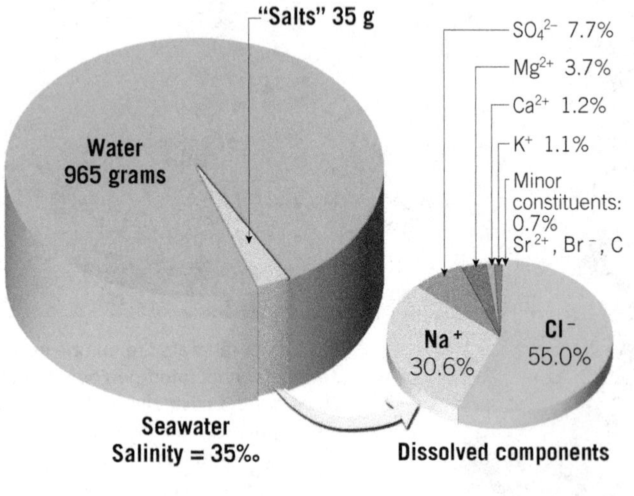

"Salts" 35 g
Water 965 grams
Seawater Salinity = 35‰
SO_4^{2-} 7.7%
Mg^{2+} 3.7%
Ca^{2+} 1.2%
K^+ 1.1%
Minor constituents: 0.7% Sr^{2+}, Br^-, C
Na^+ 30.6%
Cl^- 55.0%
Dissolved components

▶ **Figure 14.1 Composition of seawater** The diagram shows the relative proportions of water and dissolved components in a typical sample of seawater.

Table 14.1 Recipe for Artificial Seawater

To Make Seawater, Combine:	Amount (grams)
Sodium chloride (NaCl)	23.48
Magnesium chloride ($MgCl_2$)	4.98
Sodium sulfate (Na_2SO_4)	3.92
Calcium chloride ($CaCl_2$)	1.10
Potassium chloride (KCl)	0.66
Sodium bicarbonate ($NaHCO_3$)	0.192
Potassium bromide (KBr)	0.096
Hydrogen borate (H_3BO_3)	0.026
Strontium chloride ($SrCl_2$)	0.024
Sodium fluoride (NaF)	0.003

Then Add: Pure water (H_2O) to form 1000 grams of solution.

These dissolved materials are delivered to the oceans by streams at an estimated rate of more than 2.3 billion metric tons (5 trillion pounds) annually. The second major source of elements found in seawater is from Earth's interior. Through volcanic eruptions, large quantities of water vapor and other gases have been emitted during much of geologic time. This process, called *outgassing*, is the principal source of water in the oceans. Certain elements—notably chlorine, bromine, sulfur, and boron—were outgassed along with water and exist in the ocean in much greater abundance than could be explained by weathering of continental rocks alone.

Although rivers and volcanic activity continually contribute dissolved substances to the oceans, the salinity of seawater is not increasing. In fact, evidence suggests that the composition of seawater has been relatively stable for millions of years. Why doesn't the sea get saltier? The answer is because material is being removed just as rapidly as it is added. For example, some dissolved components are withdrawn from seawater by organisms as they build hard parts. Other components are removed when they chemically precipitate out of the water as sediment. Still others are exchanged at the oceanic ridge by *hydrothermal* (*hydro* = water, *thermos* = hot) *activity*. The net effect is that the overall makeup of seawater has remained relatively constant through time.

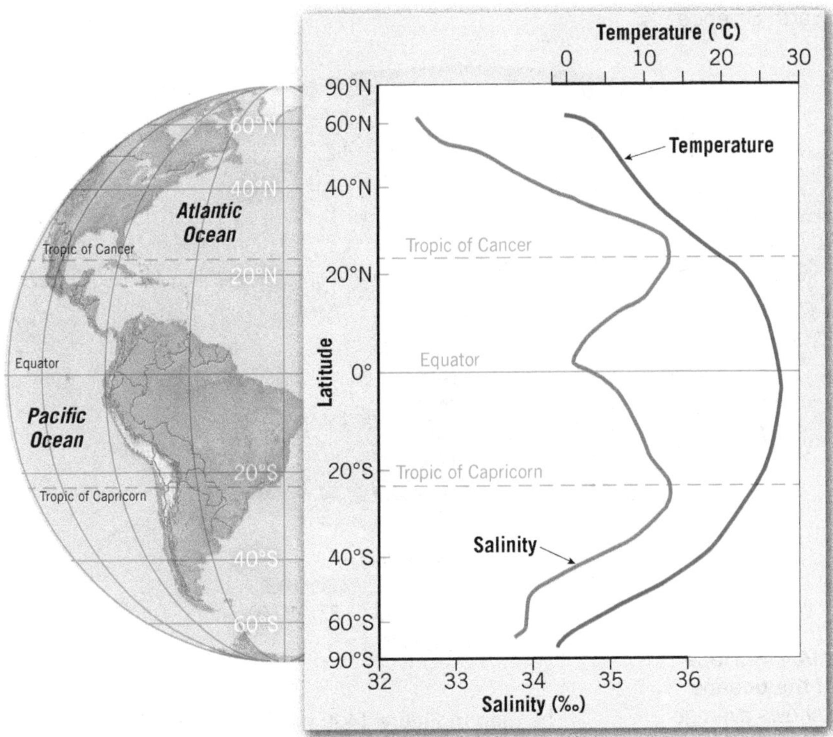

SmartFigure 14.2 **Variations in surface temperature and salinity with latitude** Average temperatures are highest near the equator and decrease toward the poles. Important factors influencing salinity are variations in rainfall and rates of evaporation. For example, in the dry subtropics in the vicinity of the Tropics of Cancer and Capricorn, evaporation removes more water than is replaced by the meager rainfall, resulting in high surface salinities. In the wet equatorial region, abundant rainfall reduces surface salinities. (After Trujillo, Alan P.; Thurman, Harold V., Essentials of Oceanography, 12th Ed., ©2017, p.159. Reprinted and electronically reproduced by permission of Pearson Education, Inc.)

TUTORIAL
https://goo.gl/WAzHDu

Processes Affecting Seawater Salinity

Because the ocean is well mixed, the relative abundances of the major components in seawater are essentially constant, no matter where the ocean is sampled. Variations in salinity, therefore, primarily result from changes in the water content of the solution.

Various surface processes alter the amount of water in seawater, thereby affecting salinity. The processes that add large amounts of freshwater to seawater—and thereby decrease salinity—are rain- and snowfall, river discharge, and the melting of icebergs and sea ice. Conversely, large amounts of freshwater can be removed from seawater—thereby increasing seawater salinity—by evaporation (in which liquid water becomes a gas and enters the atmosphere) and the formation of sea ice. High salinities, for example, are found where evaporation rates are high, as is the case in the dry subtropical regions (roughly between 25° and 35° north or south latitude). Conversely, where large amounts of precipitation dilute ocean waters, as in the midlatitudes (between 35° and 60° north or south latitude) and near the equator, lower salinities prevail. The graph in Figure 14.2 shows this pattern.

Surface salinity in polar regions varies seasonally due to the formation and melting of sea ice. When seawater freezes in winter, only a small proportion of sea salts become part of the ice. Therefore, the salinity of the remaining seawater increases. In summer when sea ice melts, the addition of the freshwater dilutes the solution, and seawater salinity decreases (Figure 14.3).

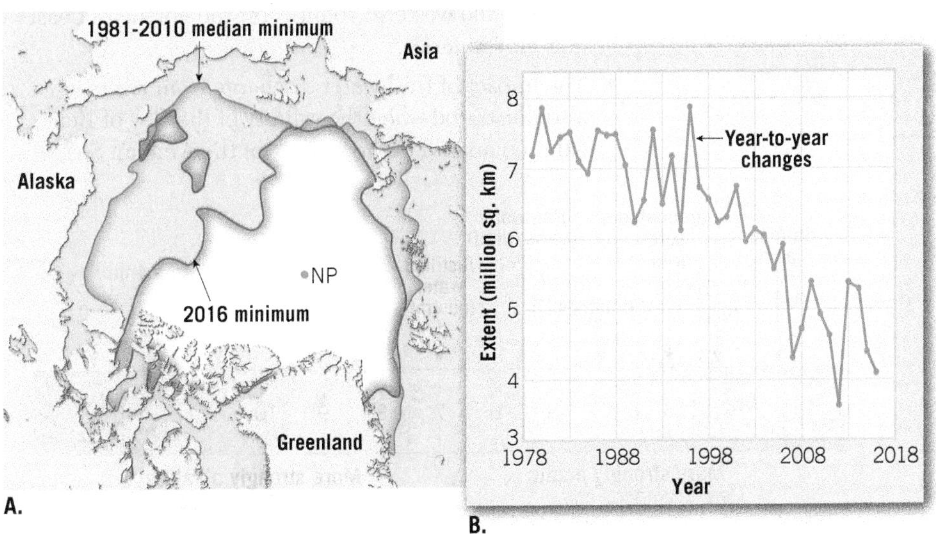

SmartFigure 14.3 **Tracking sea ice changes** Sea ice is frozen seawater. The formation and melting of sea ice influence the surface salinity of the ocean. In winter the Arctic Ocean is completely ice covered. In summer a portion of the ice melts. **A.** This map shows the extent of sea ice in early September 2016 compared to the average extent for the period 1981 to 2010. The sea ice that does not completely melt in summer is getting thinner. **B.** The graph clearly depicts the trend of reduced sea ice cover at the end of the summer melt period. The trend is related to global warming. (NASA)

VIDEO
http://goo.gl/bVornN

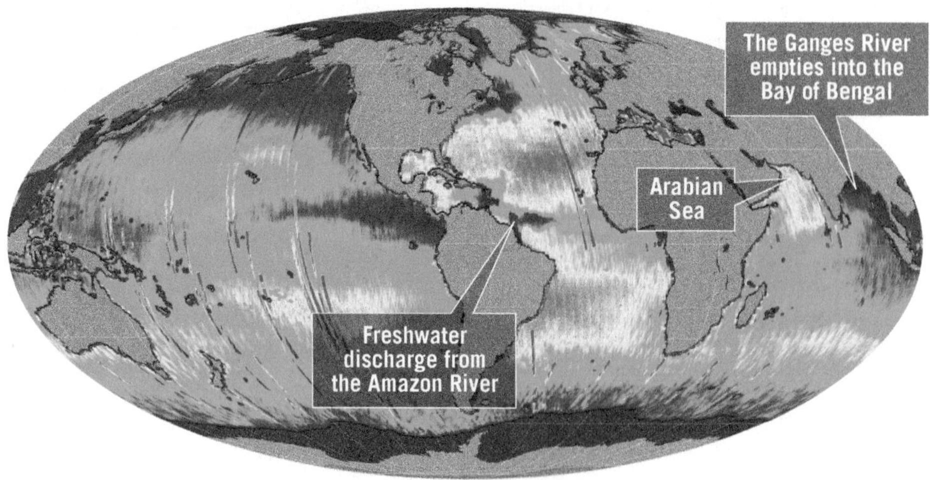

Figure 14.4 Surface salinity of the oceans This map shows average surface salinity for September 2012. It was created using data from the *Aquarius* satellite. (NASA)

The map in Figure 14.4, produced with data from the *Aquarius* satellite, shows how ocean surface salinity varies worldwide. Notice that the overall pattern on the map matches the graph in Figure 14.2. Also notice the following:

- The Atlantic Ocean is the saltiest ocean, with salinity greater than 37‰ in some places. This is because loss of freshwater by evaporation exceeds input from rainfall and river runoff.
- Notice that the huge discharge of freshwater from the Amazon River produces a zone of lower-salinity seawater in the western Atlantic, off the northeast coast of South America.
- The impact of freshwater discharge from rivers is also illustrated when the salinity of the Bay of Bengal is compared to the salinity of the Arabian Sea.

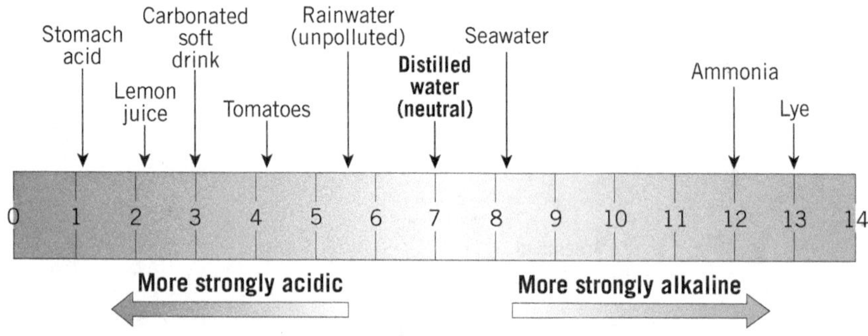

SmartFigure 14.5 The pH scale This scale is the common measure of the degree of acidity or alkalinity of a solution. The scale ranges from 0 to 14, with a value of 7 indicating a solution that is neutral. Values below 7 indicate greater acidity, whereas numbers above 7 indicate greater alkalinity. It is important to note that the pH scale is logarithmic; that is, each whole number increment indicates a tenfold difference. Thus, pH 4 is 10 times more acidic than pH 5 and 100 times (10 × 10) more acidic than pH 6.

TUTORIAL
https://goo.gl/qvhBKt

Freshwater from the Ganges River causes the surface salinity of the Bay of Bengal to be lower than the salinity of the Arabian Sea.

Recent Increase in Ocean Acidity

Human activities, especially burning fossil fuels and deforestation, have been adding ever-increasing quantities of carbon dioxide to the atmosphere for many decades. This change in atmospheric composition is linked in a cause-and-effect way to global climate change, a topic addressed in Chapter 20. Rising levels of atmospheric carbon dioxide also have some serious implications for ocean chemistry and for marine life. About one-third of the human-generated carbon dioxide ends up dissolved in the oceans. This causes the ocean's pH to drop, making seawater more acidic. The pH scale is shown and briefly described in Figure 14.5.

How does dissolved carbon dioxide affect seawater pH? When carbon dioxide (CO_2) from the atmosphere dissolves in seawater (H_2O), it forms carbonic acid (H_2CO_3). This lowers the ocean's pH and changes the balance of certain chemicals found naturally in seawater. In fact, the oceans have already absorbed enough carbon dioxide for surface waters to have experienced a pH decrease of 0.1 unit since preindustrial times, and an additional pH decrease is likely in the future. Moreover, if the current trend of human-induced carbon dioxide emissions continues, by the year 2100, the ocean will experience a pH decrease of at least 0.3 unit, which represents a change in ocean chemistry that has not occurred for millions of years. Such a shift would make it more difficult for certain marine creatures to build hard parts out of calcium carbonate because acidic water can dissolve calcium carbonate. The decline in pH would thus threaten calcite-secreting organisms as diverse as microbes and corals. That concerns marine scientists because of the potential consequences for other sea life that depends on the health and availability of these organisms.

CONCEPT CHECKS 14.1

1. What is salinity, and how is it usually expressed? What is the average salinity of the ocean?
2. What are the six most abundant elements dissolved in seawater? What is produced when the two most abundant elements combine?
3. What are the two primary sources for the elements comprising the dissolved components in seawater?
4. List several factors that cause salinity to vary from place to place and from time to time.
5. Is the pH of the ocean increasing or decreasing? What is responsible for the changing pH?

14.2 Variations in Temperature and Density with Depth

Discuss temperature, salinity, and density changes with depth in the open ocean.

Temperature and density are basic ocean water properties that influence such things as deep-ocean circulation and the distribution and types of life-forms. A sampling of the open ocean from the surface to the seafloor shows that these basic properties change with depth and that the changes are not the same everywhere. How and why seawater temperature and density change with depth is the focus of this section.

Temperature Variations

If a thermometer were lowered from the surface of the ocean into deeper water, what temperature pattern would be found? Surface waters are warmed by the Sun, so they are generally warmer than deeper waters. Surface seawater is also warmer in the tropics than toward the poles.

Figure 14.6 shows two graphs of temperature versus depth: one for high-latitude regions such as the Gulf of Alaska, and one for low-latitude regions, such as the waters near Hawaii. At low latitudes, the curve shows that the temperature at the surface is high but then decreases rapidly with depth because of the inability of the Sun's rays to penetrate very far into the ocean. Below a depth of about 1000 meters (3300 feet), the temperature is just a few degrees above freezing and remains relatively constant down to the ocean floor. The layer of ocean water between about 300 meters (980 feet) and 1000 meters (3300 feet), where temperature changes rapidly with depth, is called the **thermocline** (*thermo* = heat, *cline* = slope). The thermocline is very important because it inhibits mixing between the warm surface waters and the colder, denser waters below. The high-latitude curve in Figure 14.6 displays a quite different pattern: the temperature is nearly the same—just a few degrees above freezing—all the way from the surface to the ocean floor. In other words, there is no thermocline; instead, the water column is *isothermal* (*iso* = same, *thermal* = heat).

Midlatitude waters display an intermediate pattern. During the summer, they develop a strong thermocline; as winter chills the surface waters, this thermocline breaks down, and the temperature curve adopts the high-latitude pattern. (Some high-latitude waters also develop an extremely weak summer thermocline.)

Density Variations

Density is defined as mass per unit volume; it can be thought of as a measure of *how heavy something is for its size*. For instance, an object that has low density is lightweight for its size—for example, a dry sponge, foam packing, or a surfboard. Conversely, an object that has high density is heavy for its size—for example, cement and many metals.

Density is an important property of ocean water because it determines the water's vertical position in the ocean. For example, freshwater has lower density than saltwater, so when a river discharges into the ocean, its freshwater tends to spread out on top of the higher-density seawater rather than mix with it.

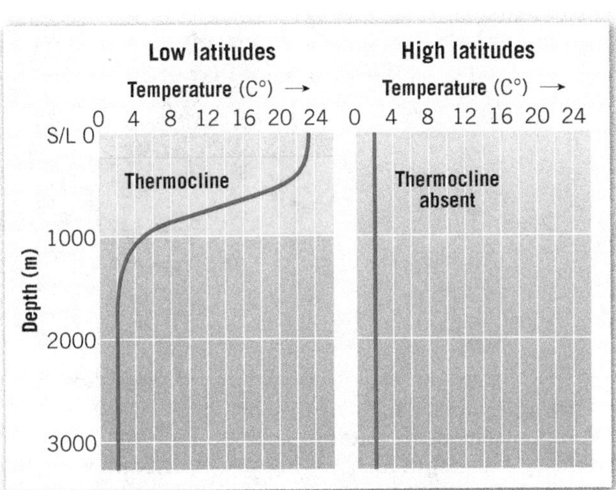

Figure 14.6 Variations in ocean water temperature with depth for low- and high-latitude regions The layer of rapidly changing temperature called the *thermocline* is not present in high latitudes. S/L is the standard abbreviation for *sea level*.

Factors Affecting Seawater Density

Seawater density is influenced by two main factors: *salinity* and *temperature*. The dissolved substances that make water salty also increase its density; the higher the salinity of water, the greater its density (Figure 14.7). An increase in temperature, on the other hand, causes water to expand and decreases its density. Such a relationship, where one variable decreases as a result of another variable's increase, is known as an *inverse relationship*, where one variable is *inversely proportional* with the other.

Temperature has the greatest influence on surface seawater density because variations in surface seawater temperature are greater than salinity variations. In fact, only in the extreme polar areas of the ocean, where temperatures are low and remain relatively constant, does salinity significantly affect density. Cold water that also has high salinity is some of the highest-density water in the world.

Density Variation with Depth By extensively sampling ocean waters, oceanographers have mapped how temperature and salinity—and the water's resulting density—vary with depth. Figure 14.8 shows two graphs of density versus depth: one for high-latitude regions and one for low-latitude regions.

At low latitudes, surface waters are low in density because they are warm. Density then increases rapidly to about 1000 meters (3300 feet) as temperature decreases, after which it remains high and constant to the ocean floor. The layer between about 300 meters (980 feet) and 1000 meters (3300 feet), where density changes rapidly with depth, is called the **pycnocline** (*pycno* = density, *cline* = slope). The pycnocline is important

Figure 14.7 The Dead Sea Because of its salinity of 330‰ (almost 10 times the average salinity of seawater), Dead Sea water has high density. As a result, it also has high buoyancy that allows swimmers to float easily. (Photo by Peter Guttman/Corbis)

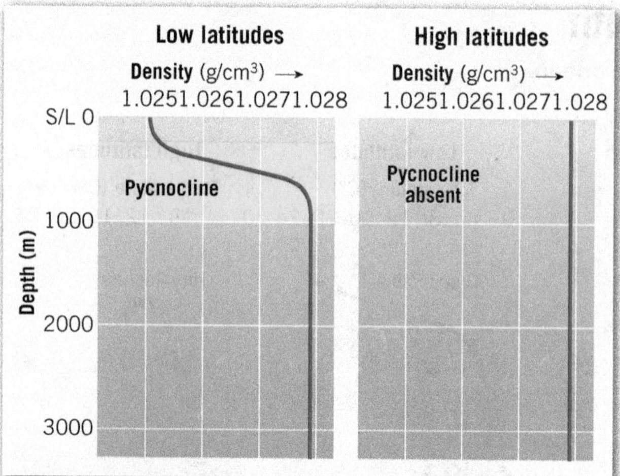

SmartFigure 14.8
Variations in ocean water density with depth for low- and high-latitude regions The layer of rapidly changing density, called the *pycnocline*, is present in the low latitudes but absent in the high latitudes.

TUTORIAL
https://goo.gl/rS9tqR

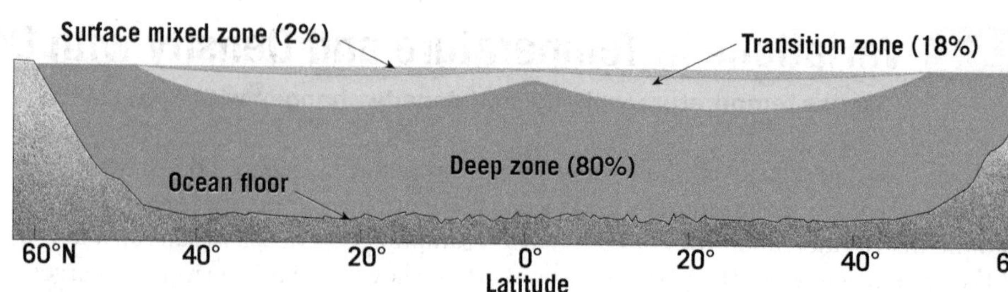

Figure 14.9 **The ocean's layers** Oceanographers recognize three main layers in the ocean, based on water density, which varies with temperature and salinity. The warm *surface mixed zone* accounts for only 2 percent of ocean water; the *transition zone* includes the thermocline and the pycnocline and accounts for 18 percent of ocean water; and the *deep zone* contains cold, high-density water that accounts for 80 percent of ocean water.

because gravity prevents higher-density water from rising spontaneously to mix with overlying less dense water. Hence, the pycnocline constitutes a barrier to mixing between surface water and deeper water.

As you may now be able to predict, the high-latitude curve in Figure 14.8 is almost vertical: the water is cold and correspondingly dense from the surface to the bottom. Because there is no pycnocline, we say that the water column is *isopycnal* (*iso* = same, *pycno* = density).

Ocean Layering

As we have just seen, much of the ocean is layered according to density, just like Earth's interior, with low-density surface water overlying higher-density water below. Except for some shallow inland seas with a high rate of evaporation, the highest-density water is found at the greatest ocean depths. Oceanographers generally recognize a three-layered structure in most parts of the open ocean: a shallow surface mixed zone, a transition zone, and a deep zone (Figure 14.9).

Because solar energy is received at the ocean surface, it is here that water temperatures are warmest. The mixing of these waters by waves as well as turbulence from currents and tides distributes heat gained at the surface through a shallow layer. Consequently, this *surface mixed zone* has nearly uniform temperatures. The thickness and temperature of this layer vary, depending on latitude and season. The zone usually extends to about 300 meters (980 feet) but may attain a thickness of 450 meters (1500 feet). The surface mixed zone accounts for only about 2 percent of ocean water.

Below the Sun-warmed zone of mixing, the temperature falls abruptly with depth (see Figure 14.6). Here, a distinct layer called the *transition zone* exists between the warm surface layer above and the deep zone of cold water below. The transition zone includes a prominent thermocline and associated pycnocline and accounts for about 18 percent of ocean water.

Below the transition zone is the *deep zone*, where sunlight never reaches and water temperatures are just a few degrees above freezing. As a result, water

EYE ON EARTH 14.1

Icebergs such as this one are common in high-latitude oceans. The iceberg pictured here occurred in the Southern Ocean near Antarctica.

QUESTION 1 *Where do icebergs originate?*

QUESTION 2 *If the iceberg melted, would the water be fresh or salty?*

QUESTION 3 *When icebergs melt, how is surface salinity affected, if at all?*

(Photo by Steve Bloom Images/Alamy)

density remains constant and high. Remarkably, the deep zone includes about 80 percent of ocean water, indicating the immense depth of the ocean. (The average depth of the ocean is more than 3700 meters [12,200 feet].)

In high latitudes, the three-layer structure does not exist because the water column is isothermal and isopycnal, which means there is no rapid change in temperature or density with depth. Consequently, good vertical mixing between surface and deep waters occurs mostly in high-latitude regions. Here, cold, high-density water forms at the surface, sinks, and initiates deep-ocean currents, which are discussed in Chapter 15.

CONCEPT CHECKS 14.2

1. Contrast temperature variations with depth in the high and low latitudes. Why do high-latitude waters generally lack a thermocline?
2. What two factors influence seawater density? Which one has the greater influence on surface seawater density?
3. Contrast density variations with depth in the high and low latitudes. Why do high-latitude waters generally lack a pycnocline?
4. Describe the ocean's layered structure. Why does the three-layer structure not exist in high latitudes?

14.3 The Diversity of Ocean Life

Distinguish among plankton, nekton, and benthos. Summarize the factors used to divide the ocean into marine life zones.

A wide variety of organisms inhabit the marine environment. These organisms range in size from microscopic bacteria and algae to blue whales, which are as long as three buses lined up end to end. Of the 2.3 million known species worldwide, biologists have identified more than 228,000 marine species, a number that is constantly increasing as new organisms are discovered and classified.

Most marine organisms live within the sunlit surface waters of the ocean. Strong sunlight supports **photosynthesis** (*photo* = light, *syn* = with, *thesis* = an arranging) by marine algae, which either directly or indirectly provide food for the vast majority of marine organisms. All marine algae live near the surface because they need sunlight; most marine animals also live near the surface because this is where food can be obtained. In shallow-water areas close to land, sunlight reaches all the way to the bottom, resulting in an abundance of marine life on the ocean floor.

There are advantages and disadvantages to living in the marine environment. One advantage is that there is an abundance of water available, which all types of life need. One disadvantage is that maneuvering in water, which has high density and impedes movement, can be difficult. The success of marine organisms depends on their ability to avoid predators, find food, and cope with the physical challenges of their environment.

Classification of Marine Organisms

Marine organisms can be classified according to where they live (their habitat) and how they move (their mobility). Organisms that inhabit the water column can be classified as either *plankton* (floaters) or *nekton* (swimmers). All other organisms are *benthos* (bottom dwellers).

Plankton (Floaters) All organisms—algae, animals, and bacteria—that drift with ocean currents are **plankton** (*planktos* = wandering). Just because plankton drift does not mean they are unable to swim. Many plankton can swim but either move very weakly or move only vertically within the water column.

Among plankton, algae (photosynthetic cells, most of which are microscopic) are called **phytoplankton** (*phyto* = plant, *planktos* = wandering), and animals are called **zooplankton** (*zoo* = animal, *planktos* = wandering). Representative members of each group are shown in Figure 14.10.

Figure 14.10 Phytoplankton and zooplankton (floaters) Schematic drawings of various phytoplankton and zooplankton. (After Trujillo, Alan P.; Thurman, Harold V., Essentials of Oceanography, 12th Ed., ©2017, p.380. Reprinted and electronically reproduced by permission of Pearson Education, Inc., Upper Saddle River, New Jersey.)

Phytoplankton:
(1) Coccolithophores; (2–4) Diatoms; (5–7) Dinoflagellates.

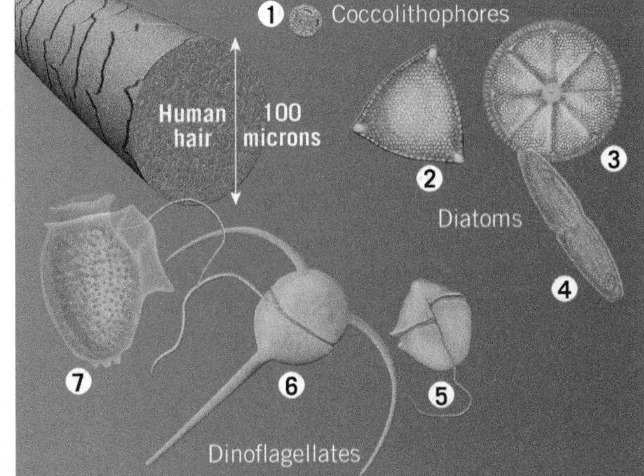

Zooplankton:
(1) Squid larva; (2) Copepod; (3) Snail larva; (4) Fish larva; (5) Arrowworm; (6) Foraminifers; (7) Radiolarian.

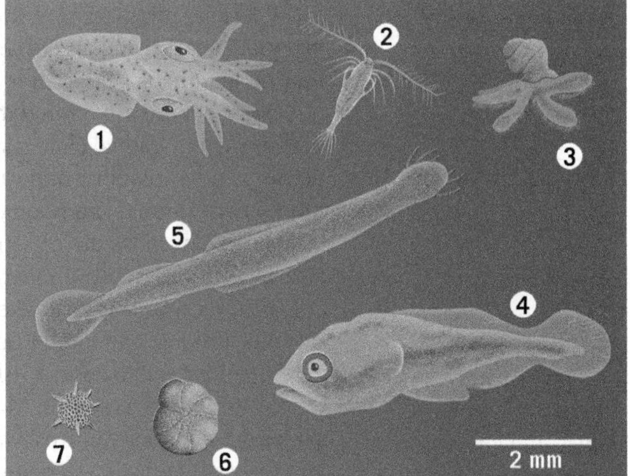

Figure 14.11 **Nekton** All animals capable of moving independently of ocean currents are nekton. **A.** Gray reef shark, Bikini Atoll. **B.** California market squid. **C.** School of grunts, Florida Keys. **D.** Pygmy killer whales in shallow surface waters along the Kona Coast, Big Island, Hawaii. (Photo A by Yann Hubert/Science Source; Photo B by Tom McHugh/Science Source; Photo C by Georgie Holland/AGE fotostock; Photo D by Masa Ushioda/Image Quest Marine)

Plankton are extremely abundant and very important within the marine environment. In fact, most of Earth's **biomass**—the mass of all living organisms—consists of plankton adrift in the oceans. Even though a very high percentage of marine species are bottom dwelling, the vast majority of the ocean's biomass is planktonic.

Nekton (Swimmers) All animals capable of moving independently of ocean currents, by swimming or other means of propulsion, are called **nekton** (*nektos* = swimming). They are capable not only of determining their position within the ocean but also, in many cases, of long migrations. Nekton include most adult fish, including sharks, and also other species, such as squid, marine mammals, and even marine reptiles (Figure 14.11).

Most nekton are limited to certain environments in the ocean. Gradual changes in temperature, salinity, density, and availability of nutrients effectively limit their range. For example, temporary shifts of water masses in the ocean can cause die-offs among fish. High water pressure at depth normally limits the vertical range of nekton.

Fish may appear to exist everywhere in the oceans, but they are most abundant near continents and islands and in colder waters. Some fish, such as salmon, ascend freshwater rivers to spawn. Many eels do just the reverse, growing to maturity in freshwater and then descending streams to breed in the great depths of the ocean.

Benthos (Bottom Dwellers) The term **benthos** (*benthos* = bottom) describes organisms living on or in the ocean bottom. *Epifauna* (*epi* = upon, *fauna* = animal) live on the surface of the seafloor, either attached to rocks or moving along the bottom. *Infauna* (*in* = inside, *fauna* = animal) live buried in the sand or mud. Some benthos, called *nektobenthos*, live on the bottom but also swim or crawl through the water above the ocean floor. Examples of benthos are shown in Figure 14.12.

The shallow coastal ocean floor contains a wide variety of physical conditions and nutrient levels, both of which have allowed a great number of species to evolve. Moving across the bottom from the shore into deeper water, the number of benthos species may remain relatively constant, but the biomass of benthos organisms decreases. In addition, shallow coastal areas are the only locations where large marine algae (often called "seaweeds") are found attached to the bottom. This is the case because these are the only areas of the seafloor that receive sufficient sunlight to support photosynthesis.

Throughout most of the deeper parts of the seafloor, animals live in perpetual darkness, where photosynthesis cannot occur. They must feed on each other, or on whatever nutrients fall from the productive surface waters above. The deep-sea bottom is an environment of coldness, stillness, and darkness. Under these conditions, life progresses slowly,

A.

B.

C.

D.

SmartFigure 14.12
Benthos Organisms living on or in the ocean bottom are classified as benthos. **A.** Sea star. **B.** Yellow tube sponge. **C.** Green sea urchin. **D.** Coral crab. (Photo A by David Hall/Science Source; Photo B by Andrew J. Martinez/Science Source; Photo C by Andrew J. Martinez/Science Source; Photo D by Images & Stories/Alamy)

TUTORIAL
https://goo.gl/70D0QD

and organisms that live in the deep sea usually are widely distributed because food sources that come from surface waters arrive in sparse amounts on the deep-sea floor.

There are exceptions to the situation just described. Instead of sparse life, some spots on the deep-ocean floor exhibit abundant life-forms. These places, called *hydrothermal vents*, are found along portions of the oceanic ridge system. There is more about these unique features in GEOgraphics 14.1.

Marine Life Zones

The distribution of marine organisms is affected by the chemistry, physics, and geology of the oceans. Marine organisms are influenced by a variety of physical oceanographic factors. Some of these factors—such as availability of sunlight, distance from shore, and water depth—are used to divide the ocean into distinct marine life zones (Figure 14.13).

Availability of Sunlight

The upper part of the ocean into which sunlight penetrates is called the **photic** (*photos* = light) **zone**. The clarity of seawater is affected by many factors, such as the amount of plankton, suspended sediment, and decaying organic particles in the water. In addition, the amount of sunlight varies with atmospheric conditions, time of day, season of the year, and latitude.

The **euphotic** (*eu* = good, *photos* = light) **zone** is the portion of the photic zone near the surface where light is strong enough for photosynthesis to occur. In the open ocean, this zone can reach a depth of 100 meters (330 feet), but the zone is much shallower close to shore, where water is typically less clear. In the euphotic zone, phytoplankton use sunlight to produce food molecules, which makes them the basis of most oceanic food webs.

Seawater absorbs longer-wavelength light (the red end of the spectrum) more effectively than shorter wavelengths (the blue and violet end); this is why underwater shots look blue. Red and orange light penetrate only a short way below the surface. Yellow and green light make it farther, and blue and violet light penetrate the farthest. In fact, faint traces of blue and violet light can still be measured in extremely clear water at depths of 1000 meters (3300 feet).

Although photosynthesis cannot occur much below 100 meters (330 feet), there is enough light in the lower photic zone for marine animals to avoid predators, find food, recognize their species, and locate mates. Even deeper is the **aphotic** (*a* = without, *photos* = light) **zone**, where there is no sunlight.

Distance from Shore

Marine life zones are also subdivided based on distance from shore. The area where the land and ocean meet and overlap is called the

Figure 14.13 Marine life zones Criteria for determining life zones include availability of light, distance from shore, and water depth.

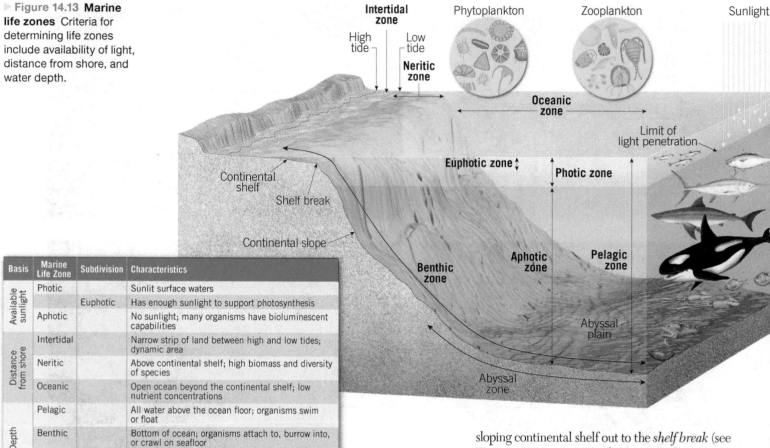

Basis	Marine Life Zone	Subdivision	Characteristics
Available sunlight	Photic		Sunlit surface waters
		Euphotic	Has enough sunlight to support photosynthesis
	Aphotic		No sunlight; many organisms have bioluminescent capabilities
Distance from shore	Intertidal		Narrow strip of land between high and low tides; dynamic area
	Neritic		Above continental shelf; high biomass and diversity of species
	Oceanic		Open ocean beyond the continental shelf; low nutrient concentrations
Depth	Pelagic		All water above the ocean floor; organisms swim or float
	Benthic		Bottom of ocean; organisms attach to, burrow into, or crawl on seafloor
		Abyssal	Deep-sea bottom; dark,cold, high pressure; sparse life

intertidal zone. This narrow strip of land between high and low tides is alternately covered and uncovered by seawater with each tidal change. Even though it appears to be a harsh place to live, with crashing waves, periodic drying out, and rapid changes in temperature, salinity, and oxygen concentrations, many species live here and are superbly adapted to the dramatic environmental changes.

Seaward from the low-tide line is the **neritic** (*neritos* = of the coast) **zone**, which covers the gently sloping continental shelf out to the *shelf break* (see Figure 13.10, page 423). This zone can be very narrow or may extend hundreds of kilometers from shore. The neritic zone is often shallow enough for sunlight to reach all the way to the ocean floor, putting it entirely within the photic zone.

Although the neritic (near-shore) zone covers only about 5 percent of the world's oceans, it is rich in both biomass and number of species. Many organisms find the conditions here ideal because photosynthesis occurs readily. In addition, nutrients wash in from the land and well up from the depths in areas where there is coastal upwelling. The bottom also provides shelter and habitat.

EYE ON EARTH 14.2

These dolphins were photographed in the relatively shallow waters near the coast of southern California.

QUESTION 1 *Which of these terms best fits the organisms shown here: plankton, nekton, or benthos? Explain.*

QUESTION 2 *Is the marine life zone shown in this photo intertidal, neritic, or oceanic?*

QUESTION 3 *Which term, benthic or pelagic, best fits the situation shown in the photo?*

(Photo by Danny Frank/Fotosearch RM/AGE Fotostock)

This zone is so rich, in fact, that it supports 90 percent of the world's commercial fisheries.

Beyond the continental shelf is the **oceanic zone**. The open ocean reaches great depths, and as a result, surface waters typically have lower nutrient concentrations because nutrients tend to sink out of the photic zone. This low nutrient concentration usually results in much smaller populations than the more productive neritic zone.

Water Depth A third method of classifying marine habitats is based on water depth. Open ocean of *any* depth is called the **pelagic** (*pelagios* = of the sea) **zone**. Animals in this zone swim or float freely. The photic part of the pelagic zone is home to phytoplankton, zooplankton, and nekton, such as tuna, sea turtles, and dolphins. The aphotic part has strange species like viperfish and giant squid that are adapted to life in deep water.

Benthos organisms such as giant kelp, sponges, crabs, sea anemones, sea stars, and marine worms that attach to, crawl upon, or burrow into the seafloor occupy parts of the **benthic** (*benthos* = bottom) **zone**. The benthic zone includes any sea-bottom surface, regardless of its distance from shore, and is mostly inhabited by benthos organisms.

The **abyssal** (*a* = without, *byssus* = bottom) **zone** is a subdivision of the benthic zone and includes the deep-ocean floor, such as *abyssal plains*. This zone is characterized by extremely high water pressure, consistently low temperature, no sunlight, and sparse life. Three food sources exist at abyssal depths: (1) tiny decaying particles steadily "raining" down from above, which provide food for filter-feeders such as brittle stars and burrowing worms; (2) large fragments or entire dead bodies falling at scattered sites, which supply meals for actively searching fish, invertebrate communities, and bacteria; and (3) the hydrothermal vents described in GEOgraphics 14.1.

CONCEPT CHECKS 14.3

1. Describe the lifestyles of plankton, nekton, and benthos and give examples of each. Which group comprises the largest biomass?
2. List three physical factors that are used to divide the ocean into marine life zones. How does each factor influence the abundance and distribution of marine life?
3. Why are there greater numbers and types of organisms in the neritic zone than in the oceanic zone?

14.4 Ocean Productivity

Contrast ocean productivity in polar, midlatitude, and tropical settings.

Why do some regions of the ocean teem with life, while others are relatively barren? The answer is related to the amount of primary productivity in various parts of the ocean. **Primary productivity** is the rate at which organisms store energy through the formation of organic matter (carbon-based compounds) using energy derived from solar radiation (*photosynthesis*) or chemical reactions (*chemosynthesis*). Although chemosynthesis supports hydrothermal vent biocommunities along the oceanic ridge, it is much less significant than photosynthesis in worldwide oceanic productivity.

Two factors influence a region's photosynthetic productivity: the *availability of nutrients* (such as nitrates, phosphorus, iron, and silica) and the *amount of solar radiation* (sunlight). Thus, marine life is most abundant in places with ample nutrients and good sunlight. Oceanic productivity, however, varies dramatically because of the uneven distribution of nutrients throughout the photic zone and seasonal changes in the availability of solar energy.

As explained earlier, many regions of the ocean have a permanent or seasonal *thermocline* (and resulting *pycnocline*). This layer forms a barrier to vertical mixing and prevents nutrients from being resupplied to surface waters as they are consumed by phytoplankton. In the midlatitudes, a thermocline develops only during the summer season, and in polar regions a thermocline does not usually develop at all. The degree to which waters develop a thermocline profoundly affects the productivity observed at different latitudes.

Productivity in Polar Oceans

Polar regions such as the Arctic Ocean's Barents Sea, which is off the northern coast of Europe, experience continuous darkness for about 3 months of winter and continuous illumination for about 3 months during summer. Productivity of phytoplankton—mostly single-celled algae called *diatoms*—peaks there during May (**Figure 14.14**), when the Sun rises high enough in the sky that there is deep penetration of sunlight into the water. As soon as the diatoms develop, zooplankton—mostly small crustaceans called *copepods* and larger *krill*—begin feeding on them. The zooplankton biomass peaks in June and continues at a relatively high level until winter darkness begins in October.

Recall that temperature and density change very little with depth in polar regions (see Figures 14.6 and 14.8), so these waters are *isothermal* and *isopycnal*, and there is no barrier to mixing between surface waters and deeper, nutrient-rich waters. In the summer, however, a slight, seasonal thermocline develops, creating a layering of surface and deeper waters.

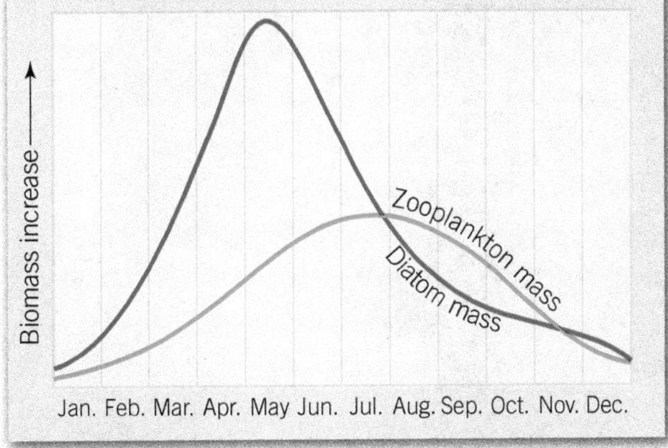

Figure 14.14 One example of productivity in polar oceans illustrated by the Barents Sea A springtime increase in diatom mass is followed closely by an increase in zooplankton abundance.

Deep-Sea Hydrothermal Vents

Deep-sea hydrothermal vents are openings in the oceanic crust from which geothermally heated water rise They are found mainly along the oceanic ridge system where tectonic plates rift apart, resulting in the production of new seafloor by upwelling magma.

When these hot, mineral-rich fluids reach the seafloor their temperatures can exceed 350 °C, but because of the extremely high pressures exerted by the water column above, they do not boil. When this hydrothermal fluid comes into contact with the much colder seawater, mineral matter rapidly precipitates to form shimmering smoke-like clouds called "*black smokers.*" The particles that compose the black smokers eventually settle out of the seawater. These deposits may contain economically significant amounts of iron, copper, zinc, lead, and occasionally silver and gold.

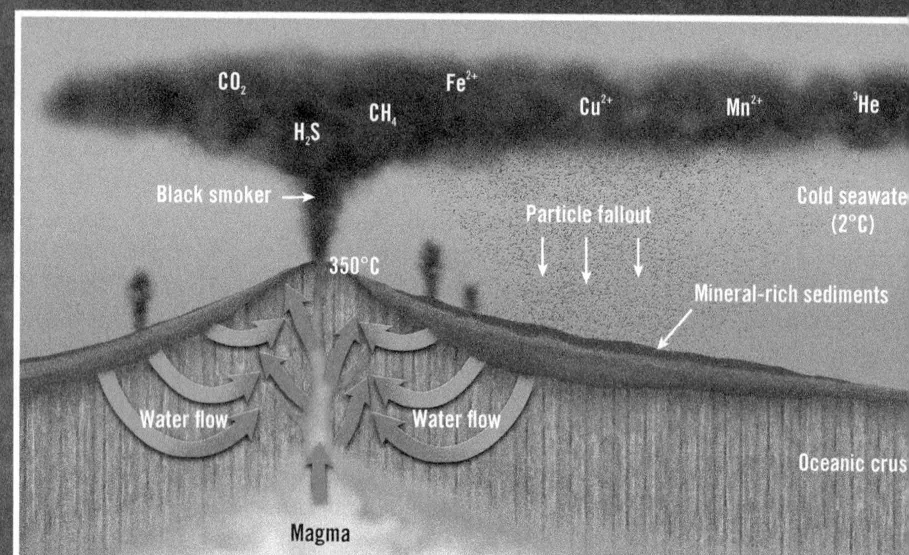

At oceanic ridges, cold seawater circulates hundreds of meters down into the highly fractured basaltic crust, where it is heated by magmatic sources. Along the way, the hot water strips metals and other elements such as sulfur from surrounding rock. This heated fluid eventually becomes hot and buoyant enough to rise along conduits and fractures toward the surface.

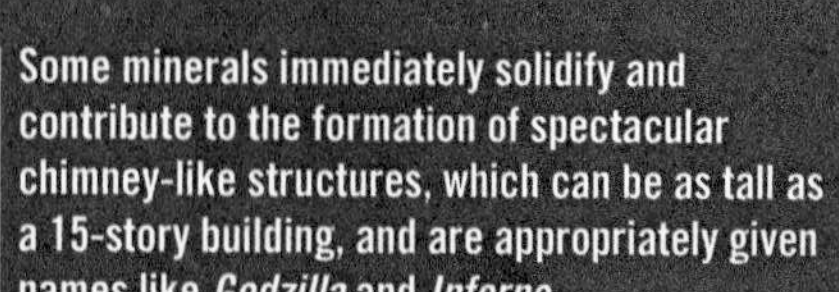

Some minerals immediately solidify and contribute to the formation of spectacular chimney-like structures, which can be as tall as a 15-story building, and are appropriately given names like *Godzilla* and *Inferno.*

NOAA

Tube worms
Verena Tunnicliffe/Uvic/Fisheries and Oceans Canada/News.com

B. Murton/Southampton Oceanography Centre/Science Source

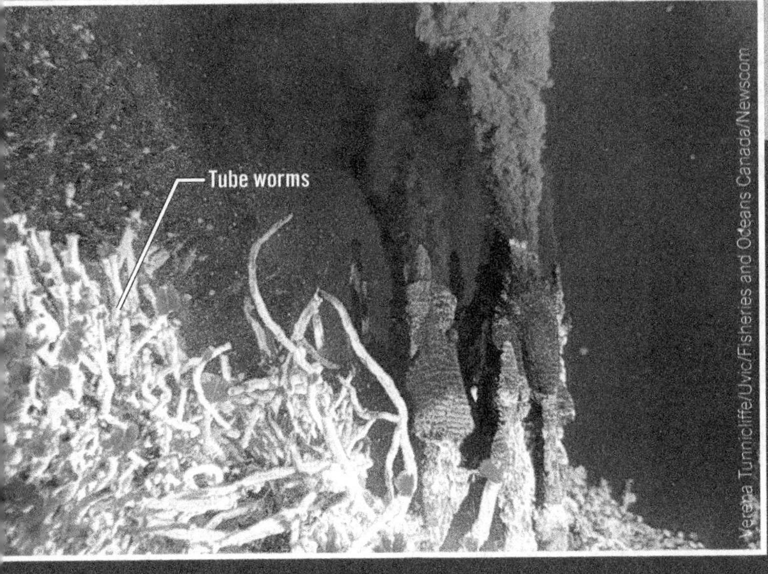

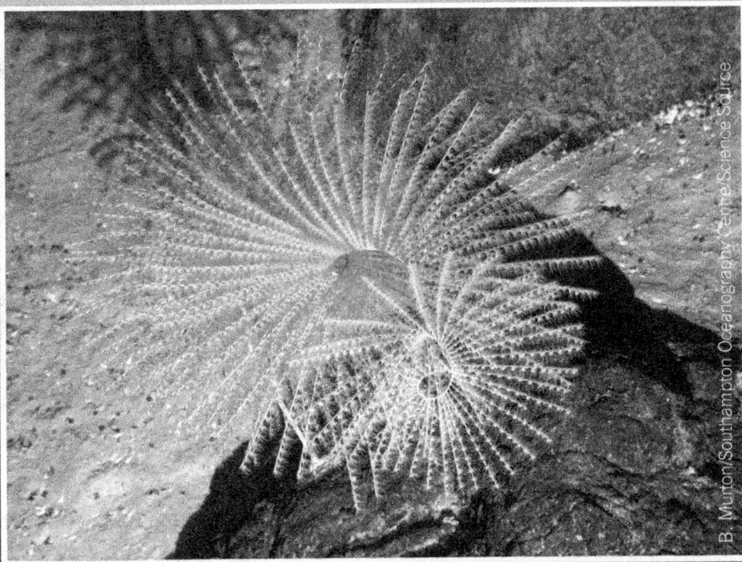

rothermal vents are remarkable for the unique types of marine life they support. In these extreme environments, completely devoid of light, microorganisms utilize mineral-rich hermal fluids to perform chemosynthesis—the conversion of carbon into organic compounds without sunlight for energy. The microbial communities, in turn, support larger, more plex animals such as fish, crabs, mussels, clams, and perhaps the most conspicuous creatures, tube worms, which can be up to 3 meters (10 feet) long. With their white chitinous s and bright red plumes, tube worms rely entirely on bacteria growing in their trophosome, an internal organ designed for harvesting bacteria. The symbiotic bacteria rely on tube worms to provide them with a suitable habitat and, in return, they use chemosynthesis to provide carbon-based nutrients to the tube worms.

Question:
Where are hydrothermal vents found?

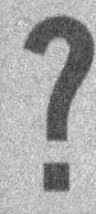

Most hydrothermal vents are found around the oceanic ridge system including some small spreading centers such as the Juan de Fuca Ridge and Galapagos Rift, as well as in the back arc basins that lie behind subduction zones.

KEY
Active
Unconfirmed

GLOBAL DISTRIBUTION OF HYDROTHERMAL VENT FIELDS

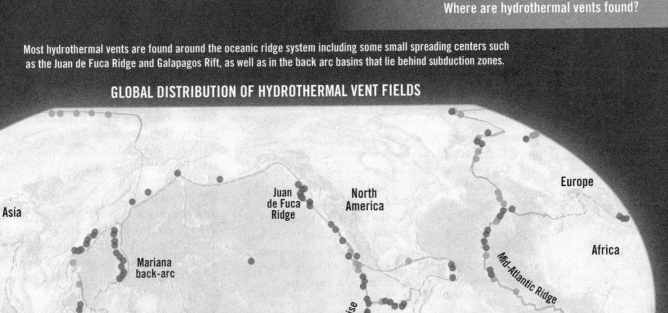

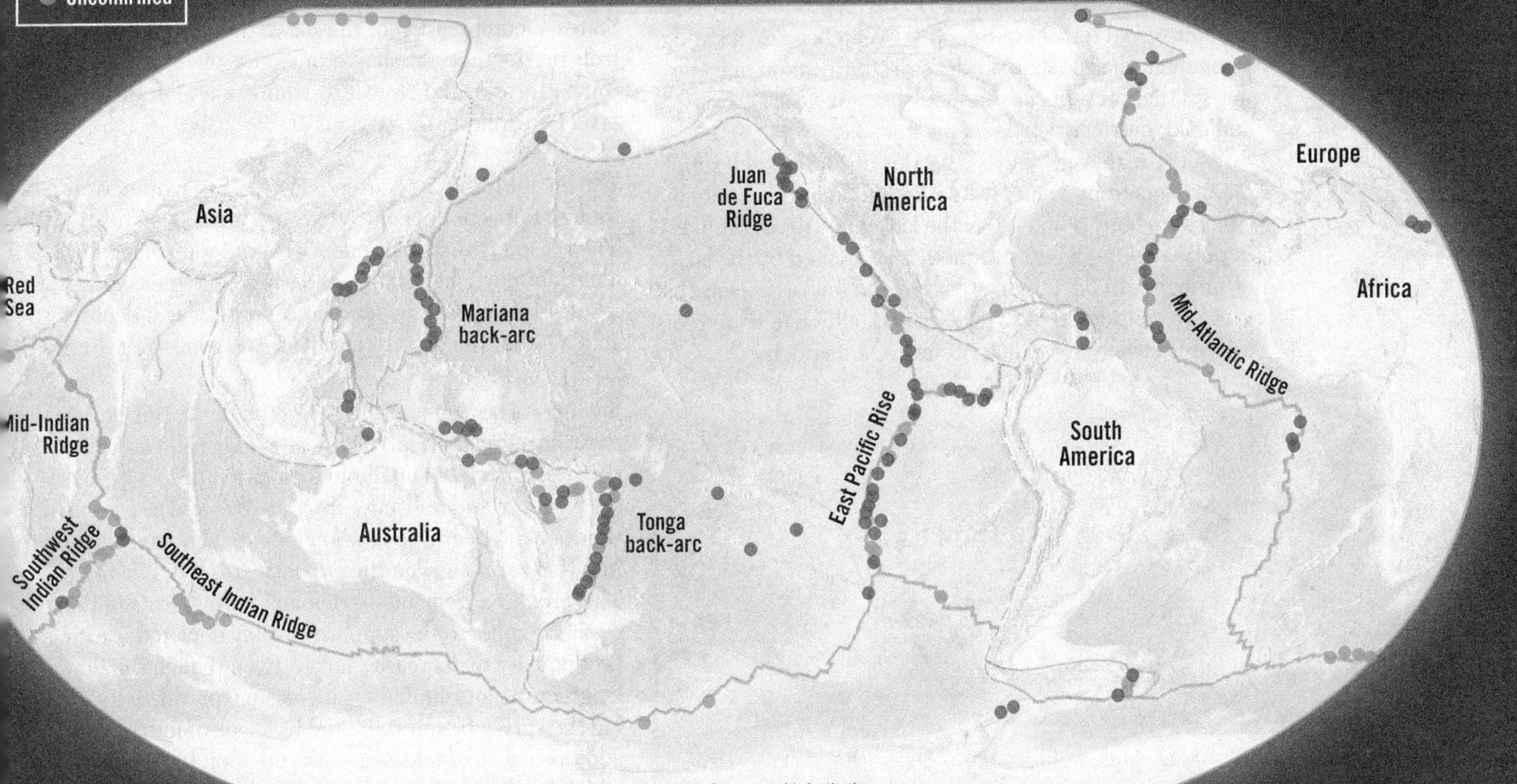

Asia
Red Sea
Mid-Indian Ridge
Southwest Indian Ridge
Southeast Indian Ridge
Australia
Mariana back-arc
Tonga back-arc
Juan de Fuca Ridge
North America
East Pacific Rise
South America
Mid-Atlantic Ridge
Europe
Africa

Data from Woods Hole Oceanographic Institution

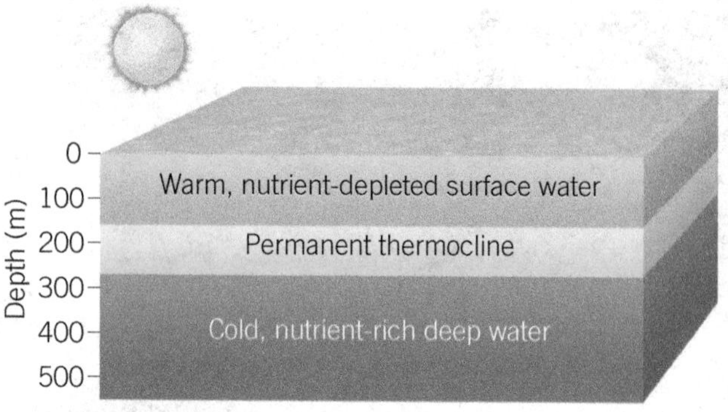

▲ Figure 14.15
Productivity in tropical oceans Although tropical regions receive adequate sunlight year-round, a permanent thermocline prevents the mixing of surface and deep water. As phytoplankton consume nutrients in the surface layer, productivity is limited because the thermocline prevents replenishment of nutrients from deeper water. Thus, productivity remains at a steady, low level.

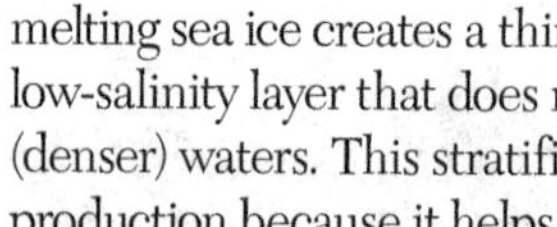

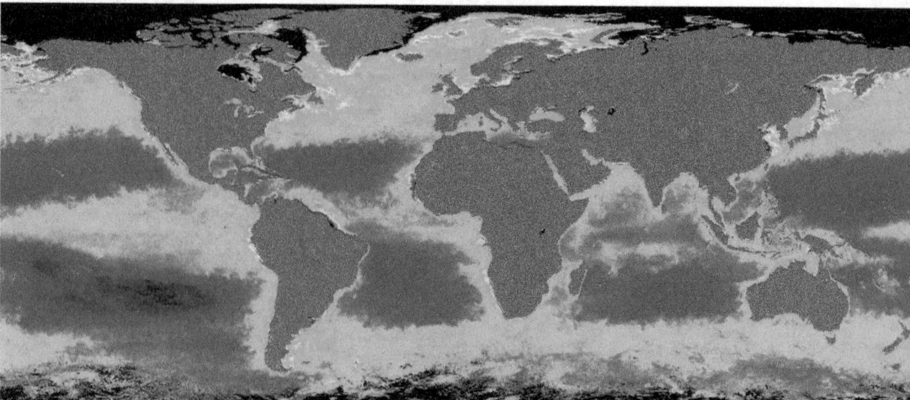

▲ Figure 14.17 **Chlorophyll concentrations measured by satellite instruments** Bright greens, yellows, and reds indicate that the northern oceans were alive with plant life in the spring of 2006. High chlorophyll concentrations indicate high amounts of photosynthesis. Observations of global chlorophyll patterns tell scientists where ocean surface plants (phytoplankton) are growing, which is an indicator of where marine ecosystems are thriving. Such global maps also give scientists an idea of how much carbon the plants are soaking up, which is important in understanding the global carbon budget. (NASA)

Moreover, in some areas, melting sea ice creates a thin, low-salinity layer that does not readily mix with deeper (denser) waters. This stratification is crucial to summer production because it helps prevent phytoplankton from being carried into deeper, darker waters. Instead, phytoplankton are concentrated in the sunlit surface waters, where they reproduce continuously during near-24-hour sunlight.

Because of the constant supply of nutrients brought up from deeper waters, high-latitude surface waters typically have high nutrient concentrations. The lack of available solar energy during the winter, however, limits photosynthetic productivity in these areas.

Productivity in Tropical Oceans

You may be surprised to learn that productivity is low in tropical regions of the open ocean such as near Hawaii or in the Caribbean Sea. Because the Sun is more directly overhead, light penetrates much deeper into tropical oceans than in midlatitude and polar waters, and solar energy is available year-round. However, productivity is low in tropical regions of the open ocean because a strong, permanent thermocline produces a stratification of water masses that prevents mixing between warm surface waters and cold, nutrient-rich deeper waters (Figure 14.15). In essence, the thermocline is a barrier that blocks the supply of nutrients from deeper waters below. So, productivity in tropical regions is limited by the lack of nutrients (unlike in polar regions, where productivity is limited by the lack of sunlight). In fact, these areas have so few organisms that they are considered biological deserts, although coral reefs and areas where upwelling occurs are exceptions.

▼ SmartFigure 14.16
Productivity in midlatitude oceans (Northern Hemisphere)
The graph shows the relationship among phytoplankton, zooplankton, amount of sunshine, and nutrient levels for surface waters.

TUTORIAL
https://goo.gl/r65pgA

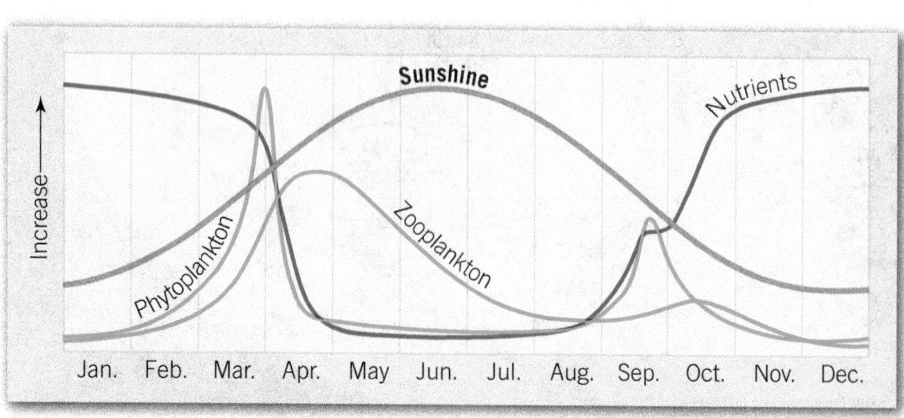

Productivity in Midlatitude Oceans

As we have seen, productivity is limited by available sunlight during the winter in polar regions and by a lack of nutrient supply in the tropics. In midlatitude regions, such as the coastal regions of the contiguous United States, a combination of these two limiting factors controls productivity, as shown in Figure 14.16 (which shows the pattern for the Northern Hemisphere; in the Southern Hemisphere, the seasons are reversed).

Winter Productivity is very low during winter, even though nutrient concentrations are highest at this time. The reason is that solar energy is limited because the length of daylight is short, and the Sun angle is low. As a result, the depth at which photosynthesis can occur is so shallow that the growth of phytoplankton is very limited.

Spring The Sun rises higher in the sky during spring, creating a greater depth at which photosynthesis can occur. A *spring bloom* of phytoplankton occurs because solar energy and nutrients are available, and a seasonal thermocline begins to develop (due to increased solar heating) that traps algae in the euphotic zone (Figure 14.17). This creates a tremendous demand for nutrients in the euphotic zone, so the supply is quickly depleted, causing productivity to decrease sharply. Even though the days are lengthening and sunlight is increasing, productivity during the spring bloom is limited by a lack of nutrients. In addition, phytoplankton are preyed upon by zooplankton, which further decreases their abundance.

Summer The Sun rises even higher in the summer, so surface waters continue to warm. A strong thermocline develops that prevents vertical mixing, so nutrients depleted from surface waters cannot be replaced by those from deeper waters. Throughout summer, the phytoplankton population remains relatively low (see Figure 14.16).

Fall Solar radiation diminishes in the fall as the Sun moves lower in the sky, so sea-surface temperatures drop, and the summer thermocline breaks down. Nutrients return to surface waters as stronger winds cause the mixing of surface and deeper waters. These conditions create a *fall bloom* of phytoplankton, which is much smaller than the spring bloom (see Figure 14.16). The fall bloom is very short-lived because sunlight (not nutrient supply, as in the spring bloom) becomes the limiting factor as winter approaches to repeat the seasonal cycle.

Figure 14.18 compares the seasonal variation in phytoplankton biomass of tropical, polar, and midlatitude regions. The total area under each curve represents photosynthetic productivity. The graph shows the dramatic peak in productivity in polar oceans during the summer; the steady, low rate of productivity year-round in the tropical oceans; and the pattern of seasonal productivity that occurs in midlatitude oceans. It also shows that the highest overall productivity occurs in the midlatitudes.

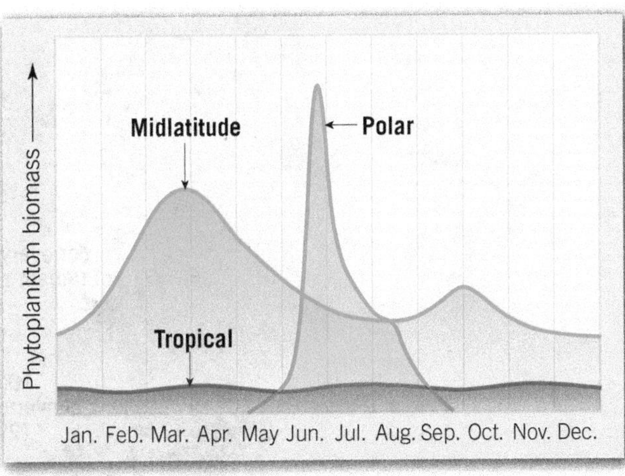

Figure 14.18 **Comparing productivity** This comparison of tropical, midlatitude, and polar oceans in the Northern Hemisphere shows seasonal variations in phytoplankton biomass. The total (shaded) area under each curve represents annual photosynthetic productivity.

CONCEPT CHECKS 14.4

1. List two methods by which primary productivity is accomplished in the ocean. Which one is most significant? What two factors influence it?
2. Compare the biological productivity of polar, midlatitude, and tropical regions of the ocean.

14.5 Oceanic Feeding Relationships

Define *trophic level* and discuss the efficiency of energy transfer between different trophic levels.

Marine algae, plants, bacteria, and bacteria-like archaea are the main oceanic producers. As these producers make food (organic matter) available to the consuming animals of the ocean, the food passes from one feeding population to the next. Only a small percentage of the energy taken in at any level is passed on to the next because energy is consumed and lost at each level. As a result, the producers' biomass in the ocean is many times greater than the mass of the top consumers, such as sharks or whales.

Trophic Levels

Chemical energy stored in the mass of the ocean's algae (the "grass of the sea") is transferred to the animal community mostly through feeding. Many types of zooplankton are *herbivores* (*herba* = grass, *vora* = eat), eating diatoms and other microscopic marine algae. Larger herbivores feed on the larger algae and marine plants that grow attached to the ocean bottom near shore.

The herbivores (grazers) are then eaten by larger animals, the *carnivores* (*carni* = meat, *vora* = eat). They in turn are eaten by another population of larger carnivores, and so on. Each of these feeding stages is called a **trophic** (*tropho* = nourishment) **level**.

As on land, most organisms in the ocean are larger than their prey. However, in the oceans, many organisms live by filter-feeding—that is, they strain much tinier organisms out of the water. Clams, barnacles, anchovies, and herring make their living in this way, and so do blue whales. Up to 30 meters (100 feet) long, the blue whale is possibly the largest animal that has ever existed on Earth, yet it feeds mostly on krill, which have a maximum length of only 6 centimeters (2.4 inches).

Transfer Efficiency

The transfer of energy between trophic levels is very inefficient. For instance, only about 2 *percent*, on average, of the light energy absorbed by algae is ultimately synthesized into food and made available to herbivores. We say that the algal trophic level is about 2 percent efficient.

Figure 14.19 shows the passage of energy between trophic levels through an entire ecosystem, from the solar energy assimilated by phytoplankton through all trophic levels to the ultimate consumer—humans. Because energy is lost at each trophic level, it takes thousands of smaller marine organisms to produce a single fish that is so easily consumed during a meal!

Food Chains and Food Webs

A **food chain** is a sequence of organisms through which energy is transferred, starting with an organism that is the primary producer, then an herbivore, then one or more carnivores, and finally culminating with the "top carnivore," which is not usually preyed upon by any other organism.

Because energy transfer between trophic levels is inefficient, it is advantageous for fishers to choose a population that feeds as close to the primary producing population as possible. This increases the biomass

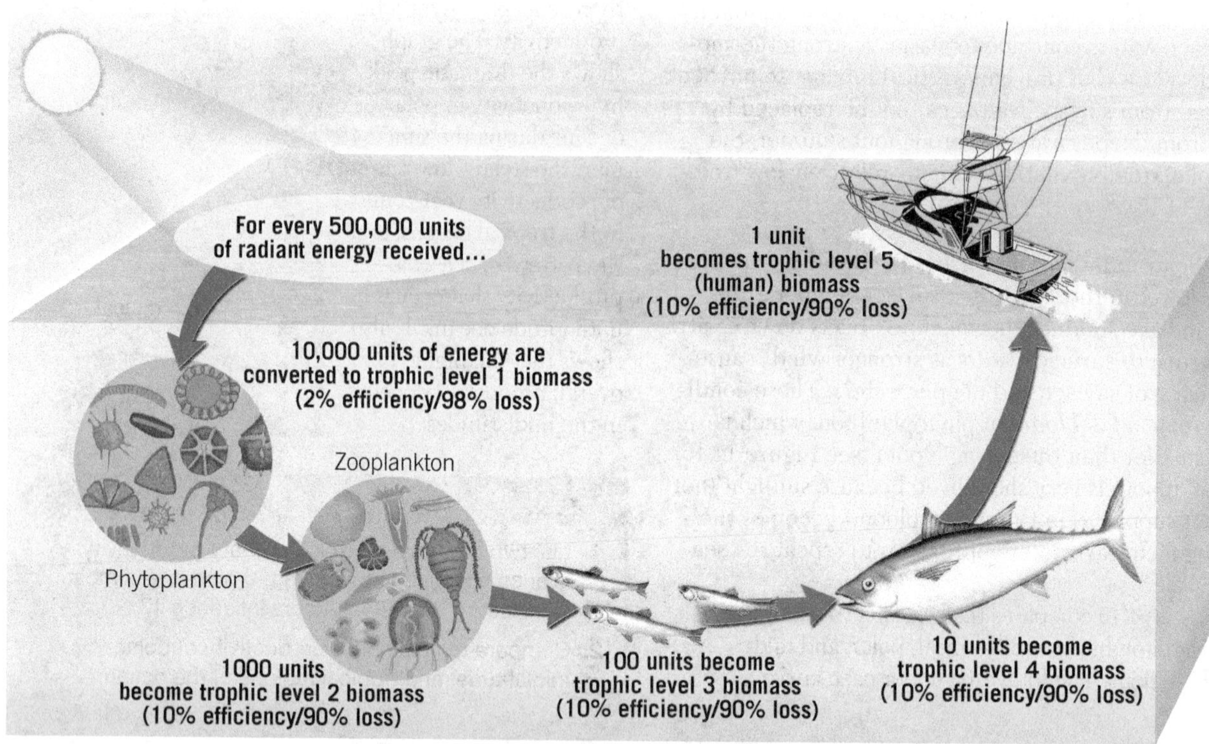

▲ SmartFigure 14.19 **Ecosystem energy flow and efficiency** For every 500,000 units of radiant energy available to the producers (phytoplankton), only 1 unit of mass is added to the fifth trophic level (humans). Average phytoplankton transfer efficiency is 2 percent (98 percent loss), and all other trophic levels average 10 percent efficiency (90 percent loss). The ultimate effect of energy transfer between trophic levels is that the number of individuals and the total biomass decrease at successive trophic levels because the amount of available energy decreases.

TUTORIAL
https://goo.gl/0YhP6R

▼ Figure 14.20 **Comparing a food chain and a food web** **A.** A food chain is the passage of energy along a single path, such as from diatoms to copepods to Newfoundland herring. Feeding relationships are rarely this simple. **B.** A food web showing multiple paths for food sources of the North Sea herring.

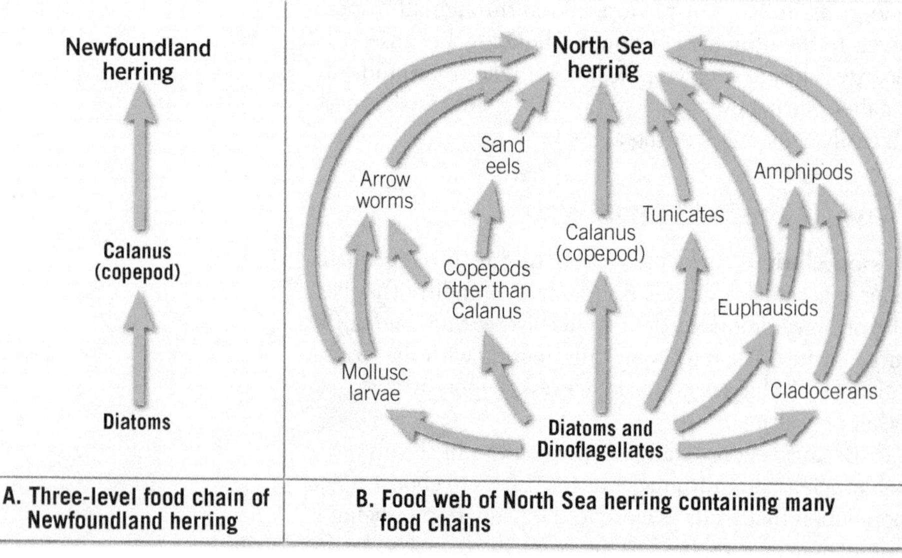

available for food and the number of fish available to be taken by a fishery. Newfoundland herring, for example, are an important fishery that usually represents the third trophic level in a food chain. They feed primarily on small copepods that feed, in turn, on diatoms (Figure 14.20A).

Feeding relationships are rarely as simple as that of the Newfoundland herring. More often, top carnivores in a food chain feed on a number of different animals, each of which has its own simple or complex feeding relationships. This constitutes a **food web**, as shown in Figure 14.20B for North Sea herring.

Animals that feed through a food web rather than a food chain are more likely to survive because they have alternative foods to eat should one of their food sources diminish in quantity or even disappear. Newfoundland herring, on the other hand, eat only copepods, so the disappearance of copepods would catastrophically affect their population.

CONCEPT CHECKS 14.5

1. Discuss energy transfer between trophic levels. Which transfer is most inefficient?
2. Describe the advantage that a top carnivore gains by eating from a food web rather than a food chain.

14 CONCEPTS IN REVIEW
Ocean Water and Ocean Life

14.1 Composition of Seawater

Define *salinity* and list the main elements that contribute to the ocean's salinity. Describe the sources of dissolved substances in seawater and causes of variations in salinity.

KEY TERM: salinity

- Salinity is the proportion of dissolved salts to pure water, usually expressed in parts per thousand (‰). The average salinity in the open ocean is about 35‰. The principal elements that contribute to the ocean's salinity are chlorine (55 percent) and sodium (31 percent). The primary sources for the elements in sea salt are chemical weathering of rocks on the continents and volcanic outgassing on the ocean floor.
- Variations in seawater salinity are primarily caused by gain or loss of water. The main causes of reduced salinity are precipitation, runoff from land, and the melting of icebergs and sea ice. The main causes of increased salinity are evaporation and the formation of sea ice.

? How are emissions from this coal-burning power plant influencing the acidity of seawater? Explain.

Michael Collier

14.2 Variations in Temperature and Density with Depth

Discuss temperature, salinity, and density changes with depth in the open ocean.

KEY TERMS: thermocline, density, pycnocline

- The ocean's surface temperature is related to the amount of solar energy received and varies as a function of latitude. Low-latitude regions have relatively warm surface water and distinctly colder water at depth, creating a thermocline, which is a layer of rapid temperature change. No thermocline exists in high-latitude regions because there is little temperature difference between the top and bottom of the water column (that is, the water column is isothermal).
- The density of seawater depends mainly on its temperature and secondarily on salinity. Cold, high-salinity water is densest. Low-latitude regions have distinctly denser (colder) water at depth than at the surface, creating a pycnocline, which is a layer of rapidly changing density. No pycnocline exists in high-latitude regions (that is, the water column is isopycnal).
- At low latitudes, most regions of the open ocean exhibit a three-layered structure based on water density. The shallow surface mixed zone has warm and nearly uniform temperatures. The transition zone includes a prominent thermocline and associated pycnocline. The deep zone is continually dark and cold and accounts for 80 percent of the water in the ocean. In high latitudes, the three-layered structure does not exist; in temperate latitudes, it is present during the summer.

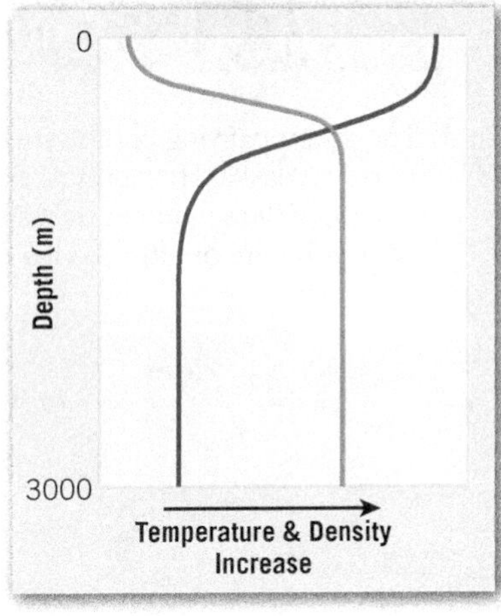

? The accompanying graph depicts variations in ocean water density and temperature with depth for a location near the equator. Which line represents temperature, and which represents density? Explain.

14.3 The Diversity of Ocean Life

Distinguish among plankton, nekton, and benthos. Summarize the factors used to divide the ocean into marine life zones.

KEY TERMS: photosynthesis, plankton, phytoplankton, zooplankton, biomass, nekton, benthos, photic zone, euphotic zone, aphotic zone, intertidal zone, neritic zone, oceanic zone, pelagic zone, benthic zone, abyssal zone

- Marine organisms can be classified into one of three groups, based on habitat and mobility. Plankton are free-floating forms with little power of locomotion, nekton are swimmers, and benthos are bottom dwellers. Most of the ocean's biomass is planktonic.
- Three criteria are frequently used to establish marine life zones. Based on availability of sunlight, the ocean can be divided into the photic zone (which includes the euphotic zone) and the aphotic zone. Based on distance from shore, the ocean can be divided into the intertidal zone, the neritic zone, and the oceanic zone. Based on water depth, the ocean can be divided into the pelagic zone and the benthic zone (which includes the abyssal zone).

De Meester Johan/Arterra Picture Library/Alamy Stock Photo

? If you were a commercial fisher, in which of these marine life zones would you concentrate your efforts—intertidal, neritic, or oceanic? Explain.

14.4 Oceanic Productivity

Contrast ocean productivity in polar, midlatitude, and tropical settings.

KEY TERM: primary productivity

- Primary productivity is the amount of carbon fixed by organisms through the synthesis of organic matter using energy derived from solar radiation (photosynthesis) or chemical reactions (chemosynthesis). Chemosynthesis is much less significant than photosynthesis in worldwide oceanic productivity. Photosynthetic productivity in the ocean varies due to the availability of nutrients and amount of solar radiation.
- Oceanic photosynthetic productivity varies at different latitudes because of seasonal changes and the development of a thermocline. In polar oceans, the availability of solar radiation limits productivity even though nutrient levels are high. In tropical oceans, a strong thermocline exists year-round, so the lack of nutrients generally limits productivity. In midlatitude oceans, productivity peaks in the spring and fall and is limited by the lack of solar radiation in winter and by the lack of nutrients in summer.

14.5 Oceanic Feeding Relationships

Define *trophic level* and discuss the efficiency of energy transfer between different trophic levels.

KEY TERMS: trophic level, food chain, food web

- The Sun's energy is utilized by phytoplankton and converted to chemical energy, which is passed through different trophic levels. On average, only about 10 percent of the mass taken in at one trophic level is passed on to the next. As a result, the size of individuals increases but the number of individuals decreases with each trophic level of a food chain or food web. Overall, the total biomass of populations decreases at successive trophic levels.

? What trophic level is represented by the person in this photo?

GIVE IT SOME THOUGHT

1 The accompanying photo shows sea ice in the Beaufort Sea near Barrow, Alaska. How do seasonal changes in the amount of sea ice influence the salinity of the remaining surface water? Is water density greater before or after sea ice forms? Explain.

2 Say that someone brings several water samples to your laboratory. His problem is that the labels are incomplete. He knows samples A and B are from the Atlantic Ocean and that one came from near the equator and the other from near the Tropic of Cancer. But he does not know which one is which. He has a similar problem with samples C and D. One is from the Red Sea, and the other is from the Baltic Sea. Applying your knowledge of ocean salinity, how would you identify the location of each sample? How were you able to figure this out?

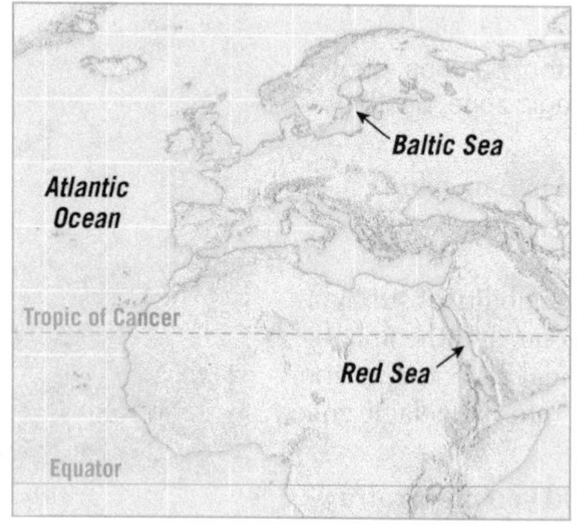

3 After sampling a column of water from the surface to a depth of 3000 meters (nearly 10,000 feet), a colleague aboard an oceanographic research vessel tells you that the water column is *isopycnal*. What does this mean? What conditions create such a situation? What would have to happen in order to create a pycnocline?

4 Tropical environments on land are well known for their abundant life; rain forests are an example. By contrast, biological productivity in tropical oceans is meager. Why is this the case?

5 The accompanying graph relates to the abundance of ocean life (productivity) in a polar region of the Northern Hemisphere. Which line represents phytoplankton, and which represents zooplankton? How did you figure this out? Why are the curves so low from November through February?

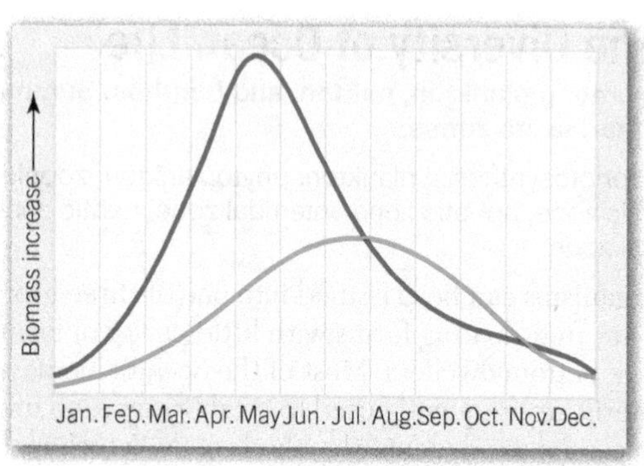

6 Refer to Figure 14.19. What is the average efficiency of energy transfer between trophic levels? Use this efficiency to determine how much phytoplankton mass is required to add 1 gram of mass to a killer whale, which is a third-level carnivore.

EXAMINING THE EARTH SYSTEM

1 Reef-building corals are tiny invertebrate colonial animals that live in warm, sunlit marine environments. They extract calcium carbonate from seawater and secrete an external skeleton. Although individuals are tiny, colonies are capable of creating massive reefs. Many other organisms also make the reef structure their home. Corals are a part of the biosphere that inhabit the hydrosphere. The solid calcium carbonate reefs they build may ultimately become the sedimentary rock limestone, a part of the geosphere. How are coral reefs related to the atmosphere? Can you come up with more than one connection? Explain.

Dirscherl Reinhard/Prisma/AGE Fotostock

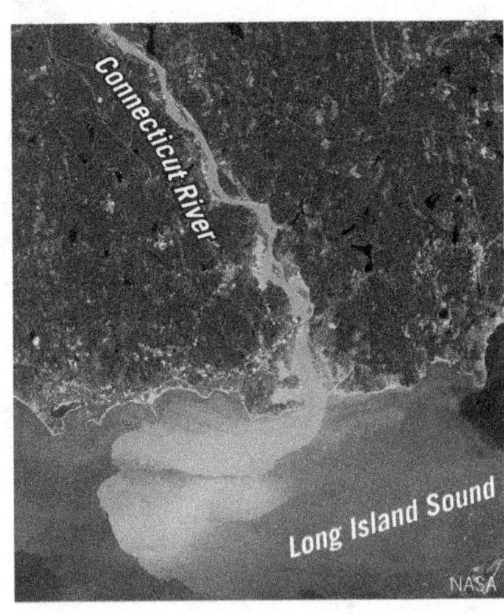

2 A storm on the Atlantic coast produces sediment-rich runoff that causes water in the euphotic zone to become cloudy. Describe how this would affect the plankton, nekton, and benthos.

3 Hydrothermal vents, such as the one shown here, occur on the ocean floor along mid-ocean ridges. GEOgraphics 14.1 examines these features, where very hot, mineral-rich water is emitted into the cold water of the deep-ocean environment.

a. What category of sediment is created by the process described above—hydrogenous, biogenous, or terrigenous?

b. How are hydrothermal vents related to the biosphere? How can there be life in these deep, cold, and dark environments? Explain.

IFE, URI-IAO, UW, Lost City Science Party; NOAA/OAR/OER; The Lost City 2005 Expedition

DATA ANALYSIS

Chlorophyll Concentrations in the Ocean

Tiny floating plants called phytoplankton underpin most marine food chains. The MODIS (Moderate Resolution Imaging Spectroradiometer) sensor on NASA's *Aqua* satellite measures phytoplankton abundance by monitoring the concentration of chlorophyll, a green pigment that is essential for photosynthesis.

ACTIVITIES

Go to the NASA Earth Observations at http://neo.sci.gsfc.nasa.gov. Click on Ocean and choose Chlorophyll Concentration. Read both the Basic and Intermediate versions of the About This Dataset description.

1 How is the chlorophyll concentration determined?

2 Why do scientists measure the chlorophyll concentration in the ocean?

3 Why is chlorophyll important to Earth's climate system?

Be sure that 1 Mo is selected under View by Date and select the most recent year that has 12 months of data.

4 Examine each month's map to see trends in chlorophyll concentration for the Northern Hemisphere in the Atlantic Ocean. How does the location of the peak chlorophyll change throughout the course of the year?

5 Examine the Southern Hemisphere seasonal trend in chlorophyll concentration. During which months would you predict a large phytoplankton population?

6 There is relatively high chlorophyll concentration near the equatorial region for every month. What does this imply about nutrient availability in the equatorial region?

Go back to the February map, then under Downloads, click on 3600 × 1800 to display a 0.1 degree resolution map. The image will open in a new window.

7 Notice that there are often high chlorophyll concentrations near the coasts. What is one plausible explanation for this?

MasteringGeology™ Looking for additional review and test prep materials? Visit the Study Area in MasteringGeology to enhance your understanding of this chapter's content by accessing a variety of resources, including Self-Study Quizzes, Geoscience Animations, SmartFigure Tutorials, Mobile Field Trips, *Project Condor* Quadcopter videos, *In the News* articles, flashcards, web links, and an optional Pearson eText.

www.masteringgeology.com

15 The Dynamic Ocean

FOCUS ON CONCEPTS

Each statement represents the primary learning objective for the corresponding major heading within the chapter. After you complete the chapter, you should be able to:

15.1 Discuss the factors that create and influence surface-ocean currents and describe the effect these currents have on climate.

15.2 Explain the processes that produce coastal upwelling and the ocean's deep circulation.

15.3 Explain why the shoreline is considered a dynamic interface and identify the basic parts of the coastal zone.

15.4 List and discuss the factors that influence the height, length, and period of a wave and describe the motion of water within a wave.

15.5 Describe how waves erode and move sediment along the shore.

15.6 Describe the features typically created by wave erosion and those resulting from sediment deposited by longshore transport processes.

15.7 Distinguish between emergent and submergent coasts. Contrast the erosion problems faced on the Atlantic and Gulf coasts with those along the Pacific coast.

15.8 Summarize the ways in which people deal with shoreline erosion problems.

15.9 Explain the cause of tides and their monthly cycles. Describe the horizontal flow of water that accompanies the rise and fall of tides.

In contrast to the gently sloping coastal plains of the Atlantic and Gulf coasts, the Pacific Coast is characterized by relatively narrow beaches that are often backed by steep cliffs and mountain ranges. These crashing waves and sea stacks are at Soberanes Point along the California coast.
(Photo by Jamie Pham/Zoonar/AGE Fotostock)

THE RESTLESS WATERS OF THE OCEAN ARE CONSTANTLY IN MOTION, powered by many different forces. Winds, for example, generate surface currents, which influence coastal climate and provide nutrients that affect the abundance of algae and other marine life in surface waters. Winds also produce waves that carry energy from storms to distant shores, where their impact erodes the land. In some regions, density differences create deep-ocean circulation, which is important for ocean mixing and recycling of nutrients. In addition, the Moon and the Sun produce tides, which periodically raise and lower average sea level. This chapter examines these movements of ocean waters and their effects on coastal regions.

15.1 The Ocean's Surface Circulation

Discuss the factors that create and influence surface-ocean currents and describe the effect these currents have on climate.

You may have heard of the Gulf Stream, an important and well-studied surface current in the Atlantic Ocean that flows northward along the East Coast of the United States (Figure 15.1). Surface currents like this one are set in motion by the wind. At the water surface, where the atmosphere and ocean meet, energy is passed from moving air to the water through friction. The drag exerted by winds blowing steadily across the ocean causes the surface layer of water to move. Thus, major horizontal movements of surface waters are closely related to the global pattern of prevailing winds.* As an example, the small map in Figure 15.2 shows how the wind belts known as the *trade winds* and the *prevailing westerlies* create large, circular-moving loops of water in the Atlantic Ocean. The same wind belts influence the other oceans as well, so that a similar pattern of currents can also be seen in the Pacific and Indian Oceans. Essentially, the pattern of surface-ocean circulation closely matches the pattern of global winds but is also strongly influenced by the distribution of major landmasses and by the spinning of Earth on its axis.

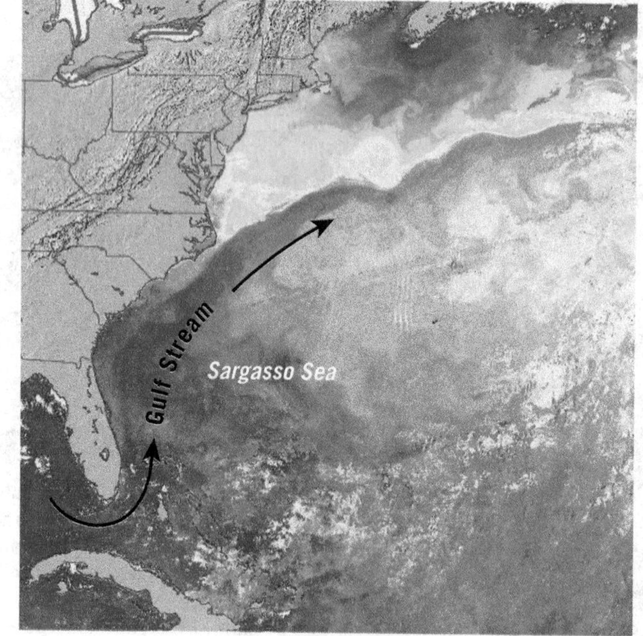

▶ SmartFigure 15.1
The Gulf Stream In this satellite image off the East Coast of the United States, orange and yellow represent higher water temperatures, and blue indicates cooler water temperatures. The Gulf Stream transports heat from the subtropics far into the North Atlantic. (NOAA)

ANIMATION
https://goo.gl/Jnpvl2

The Pattern of Surface-Ocean Currents

Huge, circular-moving current systems dominate the surfaces of the oceans. These large whirls of water within an ocean basin are called **gyres** (*gyros* = circle). The large map in Figure 15.2 shows the world's five main gyres: the *North Pacific Gyre*, the *South Pacific Gyre*, the *North Atlantic Gyre*, the *South Atlantic Gyre*, and the *Indian Ocean Gyre* (which exists mostly within the Southern Hemisphere). The center of each gyre coincides with the subtropics at about 30° north or south latitude, so they are often called *subtropical gyres*. Four main currents generally exist within each subtropical gyre (see Figure 15.2).

Coriolis Effect As shown in Figure 15.2, subtropical gyres rotate clockwise in the Northern Hemisphere and counterclockwise in the Southern Hemisphere. Why do the gyres flow in different directions in the two hemispheres? Although wind is the force that generates surface currents, other factors also influence the movement of ocean waters. The most significant of these is the **Coriolis Effect**. Because of Earth's rotation, currents are deflected to the *right* in the Northern Hemisphere and to the *left* in the Southern Hemisphere.† As a consequence, gyres flow in opposite directions in the two different hemispheres.

North Pacific Currents The four main currents that comprise the North Pacific Gyre are the North Equatorial Current, the Kuroshio Current, the North Pacific Current, and the California Current. Tracking of floating

*Details about the global pattern of winds appear in Chapter 18.
†The Coriolis effect is more fully explained in Chapter 18.

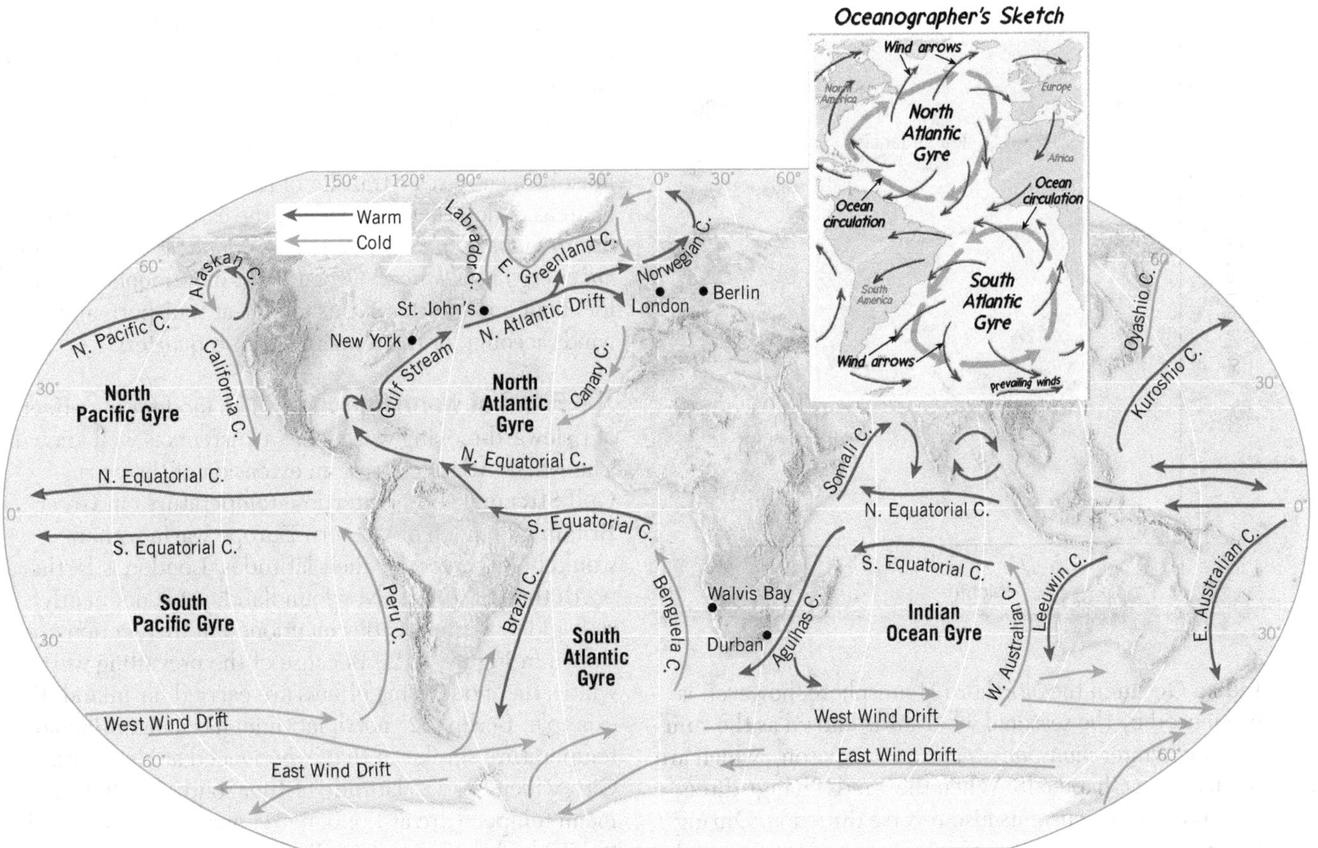

SmartFigure 15.2 **Major surface-ocean currents** The ocean's surface circulation is organized into five major gyres. Poleward-moving currents are warm, and equatorward-moving currents are cold. Ocean currents play an important role in redistributing heat around the globe. Cities mentioned in the text discussion are shown on this map. In the smaller inset map, broad arrows show the idealized surface circulation for the Atlantic, and the thin arrows show prevailing winds. Winds provide the energy that drives the ocean's surface circulation. (NOAA)

TUTORIAL
https://goo.gl/G4t7Yn

objects that are released intentionally or accidentally into the ocean reveals that it takes an average of about 6 years for the objects to go all the way around the gyre.

North Atlantic Currents The North Atlantic Ocean has four main currents, too (see Figure 15.2). Beginning near the equator, the North Equatorial Current is deflected northward through the Caribbean, where it becomes the Gulf Stream. As the Gulf Stream moves along the East Coast of the United States, it is strengthened by the prevailing westerly winds and is deflected to the east (to the right) offshore of North Carolina into the North Atlantic. As it continues northeastward, it gradually widens and slows until it becomes a vast, slowly moving current known as the North Atlantic Current, which, because of its sluggish nature, is also known as the North Atlantic Drift.

As the North Atlantic Current approaches Western Europe, it splits, part of it moving northward past Great Britain, Norway, and Iceland, carrying heat to these otherwise chilly areas. The other part is deflected southward as the cool Canary Current. As the Canary Current moves southward, it eventually merges into the North Equatorial Current, completing the gyre. The North Atlantic Ocean basin is about half the size of the North Pacific, and it takes floating objects about 3 years to go completely around this gyre.

The circular motion of gyres leaves a large central area that has no well-defined currents. In the North Atlantic, this zone of calmer waters is known as the Sargasso Sea, named for the large quantities of *Sargassum*, a type of floating seaweed encountered there (see Figure 15.1).

Southern Hemisphere Currents The ocean basins in the Southern Hemisphere exhibit a similar pattern of flow as in the Northern Hemisphere basins, with surface currents that are influenced by wind belts, the positions of continents, and the Coriolis effect. In the South Atlantic and South Pacific, for example, surface-ocean circulation is very much the same as in their Northern Hemisphere counterparts except that the direction of flow is counterclockwise (see Figure 15.2).

Indian Ocean Currents The Indian Ocean exists mostly in the Southern Hemisphere, so it follows a surface circulation pattern similar to other Southern Hemisphere ocean basins (see Figure 15.2). The small portion of the

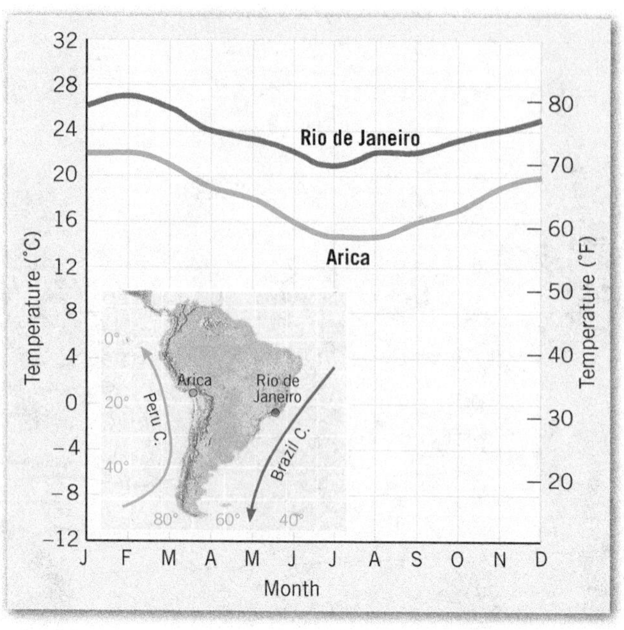

Figure 15.3 The chilling effect of a cold current Monthly mean temperatures for Rio de Janeiro, Brazil, and Arica, Chile, both of which are coastal cities near sea level. Even though Arica is closer to the equator, its temperatures are cooler than Rio de Janeiro's. Arica is influenced by the cold Peru Current, whereas Rio de Janeiro is adjacent to the warm Brazil Current.

Figure 15.4 Chile's Atacama Desert This is the driest desert on Earth. Average rainfall at the wettest locations in this desert is not more than 3 millimeters (0.12 inch) per year. Stretching nearly 1000 kilometers (600 miles), the Atacama is situated between the Pacific Ocean and the towering Andes Mountains. The cold Peru Current makes this slender zone cooler and drier than it would otherwise be. (Photo by Jacques Jangoux/Science Source)

Indian Ocean in the Northern Hemisphere, however, is influenced by the seasonal wind shifts known as the summer and winter *monsoons* (*mausim* = season), which are discussed in Chapter 18. When the winds change direction, the surface currents also reverse direction. During the summer, winds blow from the Indian Ocean toward the Asian landmass. In the winter, the winds reverse and blow out from Asia over the Indian Ocean. You can see this reversal when you compare the January and July windsin Figure 18.18. When the winds change direction, the surface currents also reverse direction.

West Wind Drift The only current that completely encircles Earth is the *West Wind Drift* (see Figure 15.2). It flows around the ice-covered continent of Antarctica, where no large landmasses are in the way, so its cold surface waters circulate in a continuous loop. It moves in response to the Southern Hemisphere prevailing westerly winds, and portions of it split off into the adjoining southern ocean basins. Its strong flow also helps define the Southern Ocean or Antarctic Ocean, which is really the portions of the Pacific, Atlantic, and Indian Oceans south of about 50° south latitude.

Ocean Currents Influence Climate

Surface-ocean currents have an important effect on climate. When Earth is considered as a whole, the energy gained from incoming solar radiation is equal to the energy radiated back out to space. However, that is not true for most individual latitudes. Low latitudes gain more solar energy than they radiate to space; the reverse is true for high latitudes. Because the tropics are not heating up, nor the polar regions getting colder, there must be a large-scale transfer of heat from areas of excess to areas of deficit. This is indeed the case. *The transfer of heat by winds and ocean currents equalizes these latitudinal energy imbalances.* Ocean water movements account for about one-quarter of this total heat transport, and winds account for the remaining three-quarters.

The Effect of Warm Currents The moderating effect of poleward-moving warm ocean currents is well known. The North Atlantic Drift, an extension of the warm Gulf Stream, keeps wintertime temperatures in Great Britain and much of Western Europe warmer than would be expected for their latitudes. London is farther north than St. John's, Newfoundland, yet is not nearly so frigid in winter. (Cities mentioned in this section are shown in Figure 15.2.) Because of the prevailing westerly winds, the moderating effects are carried far inland. For example, Berlin (52° north latitude) has a mean January temperature similar to that experienced at New York City, which lies 12° latitude farther south. The January mean temperature at London (51° north latitude) is 4.5°C (8.1°F) higher than at New York City.

Cold Currents Chill the Air In contrast to warm ocean currents like the Gulf Stream, the effects of which are felt most during the winter, cold currents exert their greatest influence in the tropics and during the summer months in the middle latitudes. For example, the cool Benguela Current off the western coast of southern Africa moderates the tropical heat along this coast. Walvis Bay (23° south latitude), a town adjacent to the Benguela Current, is 5°C (9°F) cooler in summer than Durban, which is 6° latitude farther poleward but on the eastern side of South Africa, away from the influence of the cold current. The east and west coasts of South America provide another example. Figure 15.3 shows monthly mean temperatures for Rio de Janeiro, Brazil, which is influenced by the warm Brazil Current, and Arica, Chile, which is adjacent to the cold Peru Current. Closer to home, because of the cold California Current, summer temperatures in subtropical coastal southern California are lower by 6°C (10.8°F) or more compared to East Coast stations.

Cold Currents Increase Aridity In addition to influencing temperatures of adjacent land areas, cold currents have other climatic influences. For example, where tropical deserts exist along the west coasts of continents, cold ocean currents have a dramatic impact. The principal west coast deserts are the Atacama in Peru and Chile and the Namib in southwestern Africa (Figure 15.4). The aridity along these coasts is intensified because the lower atmosphere is chilled by cold offshore waters. When this occurs, the air becomes very stable and resists the upward movement necessary to create precipitation-producing clouds. In addition, the presence of a cold offshore current causes temperatures to approach and often reach the dew point, defined as the temperature at which water vapor condenses. As a result, these areas are characterized by high relative humidities and much fog. Thus, not all subtropical deserts are hot with low humidities and clear skies. Rather, the presence of cold currents transforms some subtropical deserts into relatively cool, damp places that are often shrouded in fog.

CONCEPT CHECKS 15.1

1. What is the primary driving force of surface-ocean currents?
2. How does the Coriolis effect influence ocean currents?
3. Name the five subtropical gyres and identify the main surface currents in each.
4. How do ocean currents influence climate? Provide at least three examples.

15.2 Upwelling and Deep-Ocean Circulation

Explain the processes that produce coastal upwelling and the ocean's deep circulation.

The preceding discussion focused mainly on the horizontal movements of the ocean's surface waters. In this section, you will learn that the ocean also exhibits significant vertical movements and a slow-moving, multilayered deep-ocean circulation. Some vertical movements are related to wind-driven surface currents, whereas deep-ocean circulation is strongly influenced by seawater density differences.

Coastal Upwelling

In addition to producing horizontal-moving surface currents, winds can also cause *vertical* water movements. **Upwelling**, the rising of cold water from deeper layers to replace warmer surface water, is a common wind-induced vertical movement. One type of upwelling, called *coastal upwelling*, is most characteristic along the west coasts of continents, most notably along California, western South America, and West Africa.

Coastal upwelling occurs in these areas when winds blow toward the equator and parallel to the coast (Figure 15.5). Coastal winds combined with the Coriolis effect cause surface water to move *away* from shore. As the surface layer moves away from the coast, it is replaced by water that "upwells" from below the surface. This slow upward movement of water from depths of 50 to 300 meters (165 to 1000 feet) brings water that is cooler than the original surface water and results in lower surface-water temperatures near the shore.

For swimmers who are accustomed to the warm waters along the mid-Atlantic shore of the United States, a swim in the Pacific off the coast of central California can be a chilling surprise. In August, when temperatures in the Atlantic are 21°C (70°F) or higher, central California's surf is only about 15°C (60°F).

Upwelling brings greater concentrations of dissolved nutrients, such as nitrates and phosphates, to the ocean surface. These nutrient-enriched waters from below promote the growth of microscopic plankton, which in turn support extensive populations of fish and other marine organisms. Figure 15.5 includes a satellite image that shows high productivity due to coastal upwelling off the southwest coast of Africa.

Deep-Ocean Circulation

Deep-ocean circulation has a significant vertical component and accounts for the thorough mixing of deep-water masses. This component of ocean circulation is caused by density differences among water masses that cause denser water to sink and slowly spread out beneath the surface. Because the density variations that cause deep-ocean circulation are caused by differences in temperature and salinity, deep-ocean circulation is also referred

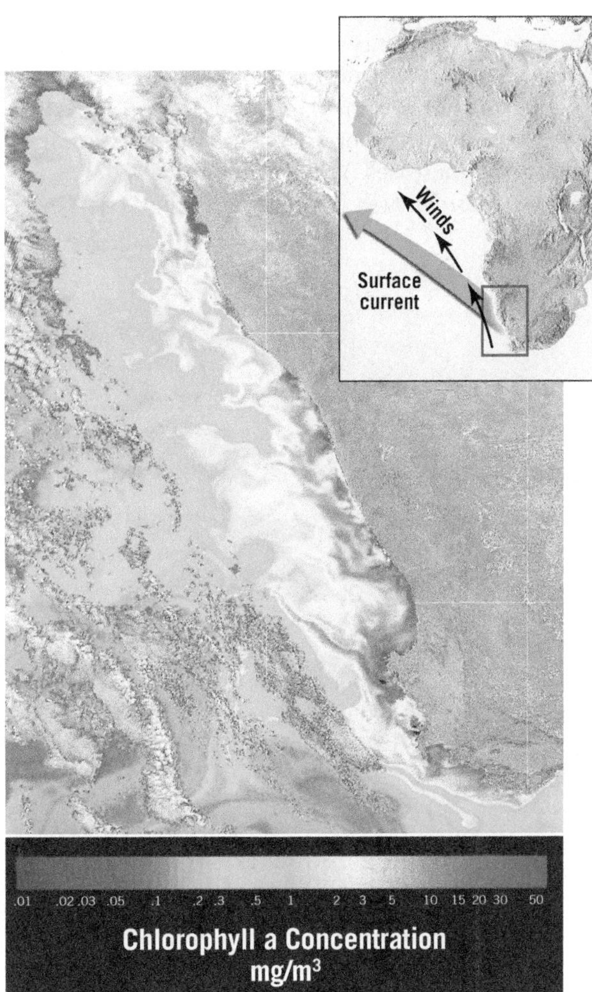

SmartFigure 15.5
Coastal upwelling Coastal upwelling occurs along the west coasts of continents, where winds blow toward the equator and parallel to the coast. The Coriolis effect (deflection to the left in the Southern Hemisphere) causes surface water to move away from the shore, which brings cold, nutrient-rich water to the surface. This satellite image shows chlorophyll concentrations along the southwest coast of Africa (February 21, 2001). An instrument aboard the satellite detects changes in seawater color caused by changing concentrations of chlorophyll. High chlorophyll concentrations indicate high rates of photosynthesis, which is linked to the upwelling nutrients. Red indicates high concentrations, and blue indicates low concentrations. (Provided by from the SeaWIFS Project, NASA/Goddard Space Flight Center and ORBIMAGE)

TUTORIAL
https://goo.gl/lgsqs6

▲ Figure 15.6 **Sea ice near Antarctica** When seawater freezes, sea salts do not become part of the ice. Consequently, the surface salinity of the remaining seawater increases, which makes it denser and prone to sink. (Photo by John Higdon/AGE Fotostock)

to as **thermohaline** (*thermo* = heat, *haline* = salt) **circulation**.

An increase in seawater density can be caused by either a *decrease* in temperature or an *increase* in salinity. Density changes due to salinity variations are important in very high latitudes, where water temperature remains low and relatively constant.

Most water involved in deep-ocean currents (thermohaline circulation) begins in high latitudes at the surface. In these regions, where surface waters are cold, salinity increases when sea ice forms (Figure 15.6). When seawater freezes to form sea ice, most salts do not become part of the ice. As a result, the salinity (and therefore the density) of the remaining seawater increases. When this surface water becomes dense enough, it sinks, initiating deep-ocean currents. Once this water sinks, it is removed from the physical processes that increased its density in the first place, and so its temperature and salinity remain largely unchanged for the duration of the time it spends in the deep ocean.

Near Antarctica, surface conditions create some of the highest-density water in the world. This cold saline brine slowly sinks to the seafloor, spreads out, and moves throughout the ocean basins in sluggish currents. After sinking from the surface of the ocean, deep waters will not reappear at the surface for an average of 500 to 2000 years.

A simplified model of ocean circulation is similar to a conveyor belt that travels from the Atlantic Ocean through the Indian and Pacific Oceans and back again (Figure 15.7). In this model, warm water in the ocean's upper layers flows poleward, becomes chilled and is converted to dense water, sinks and flows into the deep ocean, and finally returns toward the equator and upwells as cold deep water that eventually completes the circuit. As this "conveyor belt" moves around the globe, it influences global climate by converting warm water to cold and liberating heat to the atmosphere.

CONCEPT CHECKS 15.2

1. Describe the process of coastal upwelling. Why is an abundance of marine life associated with these areas?
2. Why is deep-ocean circulation referred to as *thermohaline circulation*?
3. Describe or make a simple sketch of the ocean's conveyor-belt circulation.

▼ SmartFigure 15.7 **Idealized conveyor-belt circulation** Source areas for deep water (purple ovals) occur in high-latitude areas where surface water cools, becomes denser, and sinks. These source areas feed the flow of deep, high-density water (blue bands), which slowly drifts through all the oceans. Deep water returns to the surface in localized areas of upwelling and also as gradual, uniform upwelling throughout the ocean basins. Surface currents (red bands) complete the conveyor by returning water to the source areas. (After Trujillo, Alan P.; Thurman, Harold, *Essentials of Oceanography*, 12th ed., ©2017, p. 238.)

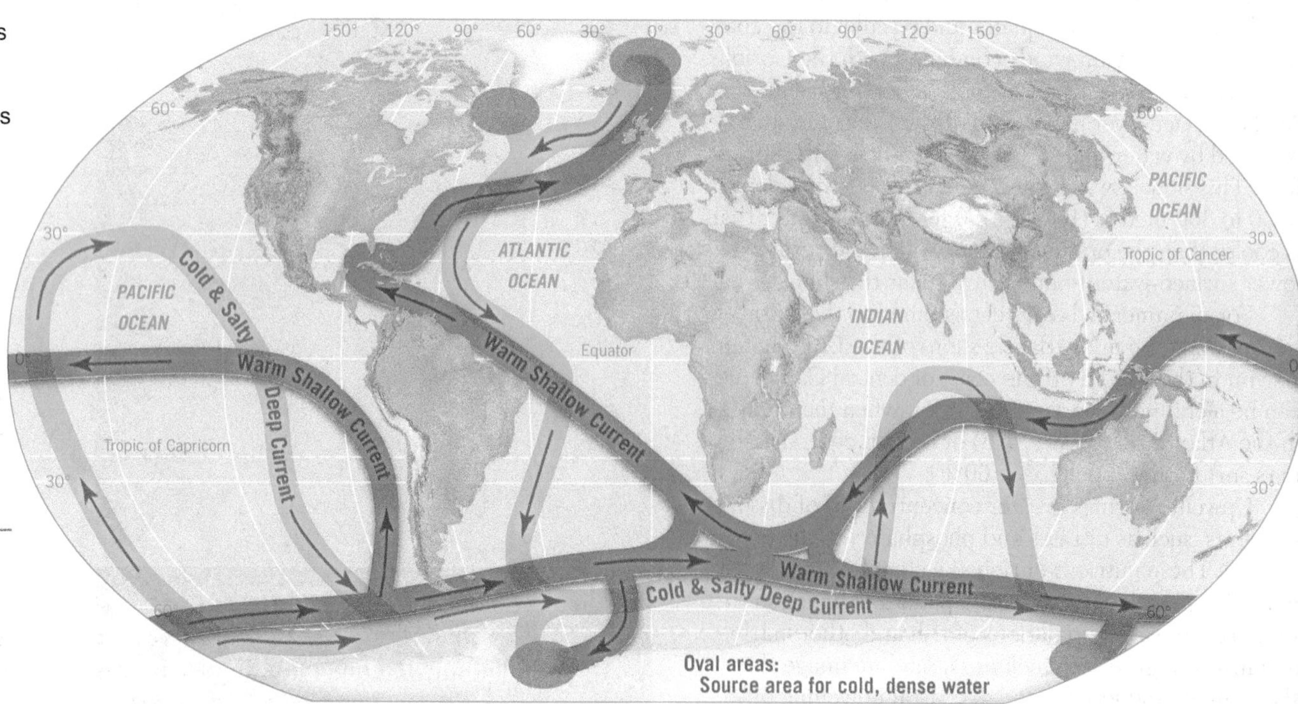

ANIMATION
https://goo.gl/GWNJnl

15.3 The Shoreline: A Dynamic Interface

Explain why the shoreline is considered a dynamic interface and identify the basic parts of the coastal zone.

Shorelines are dynamic environments. Their topography, geologic makeup, tidal conditions, sea-level fluctuations, and climate vary greatly from place to place. Continental and oceanic processes converge along coasts to create landscapes that frequently undergo rapid change. When it comes to the deposition of sediment, coasts are transition zones between marine and continental environments.

Nowhere is the restless nature of the ocean's water more noticeable than along the shore—the dynamic interface among air, land, and sea. An *interface* is a common boundary where different parts of a system interact. This is certainly an appropriate designation for the coastal zone. Here we can see the rhythmic rise and fall of tides and observe waves rolling in and breaking. Sometimes the waves are low and gentle. At other times they pound the shore with awesome fury.

▲ Figure 15.8 **Teetering on the edge** Bluff failure caused by storm waves in January 2016 resulted in these apartments in Pacifica, California, being condemned. When these buildings were erected in the 1970s, they were safely away from the cliffs. Over the years, several measures were attempted to reduce erosion of the sandstone cliffs. All proved to be inadequate. (Photo by Terry Chea/AP Images)

The Coastal Zone

Although it may not be obvious, the shoreline is constantly being modified by waves. Crashing surf is very effective at eroding the adjacent land. Wave activity also moves sediment toward and away from the shore, as well as along it. Such activity sometimes produces narrow sandbars that frequently change size and shape as powerful storm waves come and go.

Present-Day Shorelines The nature of present-day shorelines is not just the result of the relentless attack of the land by the sea. The shore has a complex character that results from multiple geologic processes. For example, practically all coastal areas were affected by the worldwide rise in sea level that accompanied the melting of glaciers at the close of the Pleistocene epoch. As the sea encroached landward, the shoreline retreated, becoming superimposed upon existing landscapes that had resulted from such diverse processes as stream erosion, glaciation, volcanic activity, and the forces of mountain building.

Human Activity Today, the coastal zone is experiencing intensive human activity. Unfortunately, people often treat the shoreline as if it were a stable platform on which structures can safely be built. This attitude inevitably leads to conflicts between people and nature. Figure 15.8 provides a recent example from the coast of California. Many coastal landforms are relatively fragile, short-lived features and are inappropriate sites for development.

Coastal Features and Terminology

In general conversation, several terms are used when referring to the boundary between land and sea. In the preceding paragraphs, the terms *shore, shoreline, coastal zone*, and *coast* were all used. Moreover, when many think of the land–sea interface, the word *beach* comes to mind. Let's take a moment to clarify these terms and introduce some other terminology used by those who study the land–sea boundary zone. You will find it helpful to refer to Figure 15.9, which is an idealized profile of the coastal zone.

The **shoreline** is the line that marks the contact between land and sea. Each day, as tides rise and fall, the position of the shoreline migrates. Over longer time spans, the average position of the shoreline gradually shifts as sea level rises or falls.

The **shore** is the area that extends between the lowest tide level and the highest elevation on land that is affected by storm waves. By contrast, the **coast** extends inland from the shore as far as ocean-related features can be found. The **coastline** marks the coast's seaward edge, whereas the inland boundary is not always obvious or easy to determine.

As Figure 15.9 illustrates, the shore is divided into the *foreshore* and the *backshore*. The **foreshore** is the area exposed when the tide is out (low tide) and submerged when the tide is in (high tide). The **backshore** is landward of the high-tide shoreline. It is usually dry, affected by waves only during storms. Two other zones are commonly identified. The **near-shore zone** lies between the low-tide shoreline and the line where waves break at low tide. Seaward of the near-shore zone is the **offshore zone**.

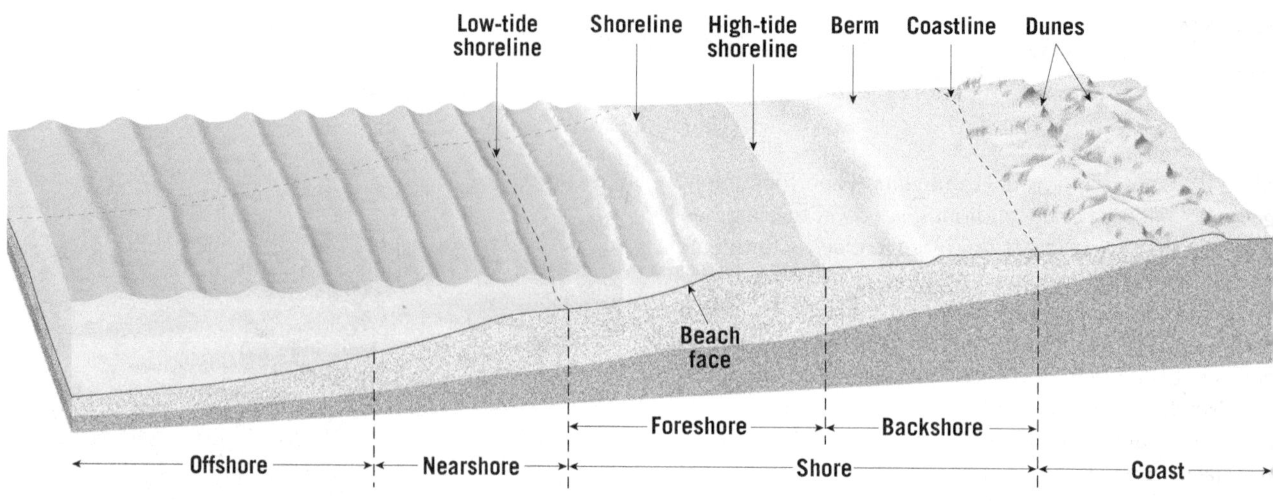

Figure 15.9 The coastal zone The transition zone between land and sea consists of several parts.

Beaches

For many, a beach is the sandy area where people lie in the sunshine and walk along the water's edge. Technically, a **beach** is an accumulation of sediment found along the landward margin of an ocean or a lake. Along straight coasts, beaches may extend for tens or hundreds of kilometers. Where coasts are irregular, beach formation may be confined to the relatively quiet waters of bays.

Beaches consist of one or more **berms**, which are relatively flat platforms often composed of sand that are adjacent to coastal dunes or cliffs and marked by a change in slope at the seaward edge. Another part of the beach is the **beach face**, which is the wet sloping surface that extends from the berm to the shoreline. Where beaches are sandy, sunbathers usually prefer the berm, whereas joggers prefer the wet, hard-packed sand of the beach face.

Beaches are composed of whatever material is locally abundant. The sediment for some beaches is derived from the erosion of adjacent cliffs or nearby coastal mountains. Other beaches are built from sediment delivered to the coast by rivers.

Although the mineral makeup of many beaches is dominated by durable quartz grains, other minerals may be dominant. For example, in areas such as southern Florida, where there are no mountains or other sources of rock-forming minerals nearby, most beaches are composed of shell fragments and the remains of organisms that live in coastal waters (**Figure 15.10A**). Some beaches

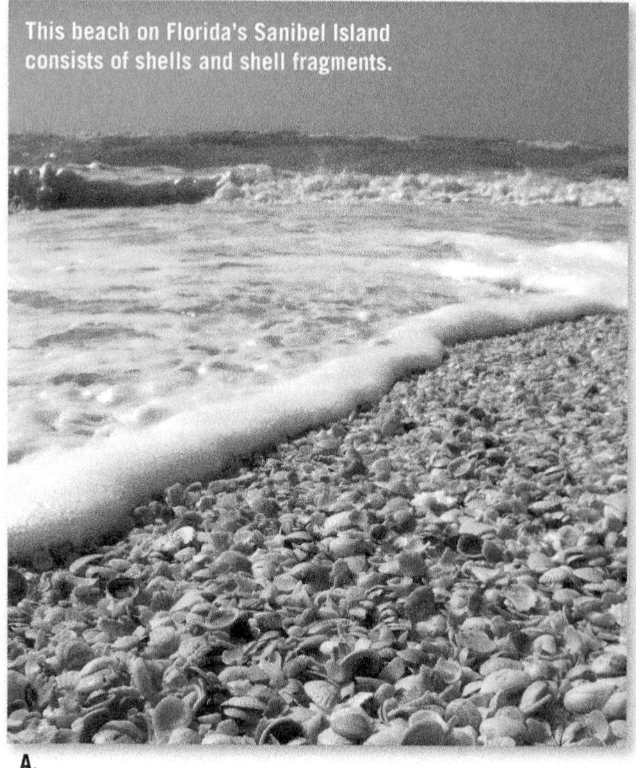

This beach on Florida's Sanibel Island consists of shells and shell fragments.

A.

The black sands on this beach in Hawaii were derived from the weathering of nearby basaltic lava flows.

B.

Figure 15.10 Beaches A beach is an accumulation of sediment on the landward margin of an ocean or a lake and can be thought of as material in transit along the shore. Beaches are composed of whatever material is locally available. (Photo A by David R. Frazier Photolibrary, Inc./Alamy; photo B by E. J. Tarbuck)

on volcanic islands in the open ocean are composed of weathered grains of the basaltic lava that makes up the islands or of coarse debris eroded from coral reefs that develop around islands in low latitudes (Figure 15.10B).

Regardless of the composition, the material that comprises the beach does not stay in one place. Instead, crashing waves are constantly moving it. Thus, beaches can be thought of as material in transit along the shore.

CONCEPT CHECKS 15.3

1. Why is the shoreline considered an *interface*?
2. Distinguish among shore, shoreline, coast, and coastline.
3. What is a *beach*? Distinguish between *beach face* and *berm*.

15.4 Ocean Waves

List and discuss the factors that influence the height, length, and period of a wave and describe the motion of water within a wave.

Ocean waves are energy traveling along the interface between ocean and atmosphere. They can carry energy from a storm far out at sea over distances of several thousand kilometers. That's why even on calm days, the ocean still has waves that travel across its surface. When observing waves, always remember that you are watching *energy* travel through a medium (water). If you make waves by tossing a pebble into a pond, splashing in a pool, or blowing across the surface of a cup of coffee, you are imparting *energy* to the water, and the waves you see are the visible evidence of the energy passing through.

Wind-generated waves provide most of the energy that shapes and modifies shorelines. Where the land and sea meet, waves that may have traveled unimpeded for hundreds or thousands of kilometers suddenly encounter a barrier that will not allow them to advance farther and must absorb their energy. Stated another way, the shore is the location where a practically irresistible force confronts an almost immovable object. The conflict that results is never-ending and sometimes very dramatic.

Wave Characteristics

Most ocean waves derive their energy and motion from the wind. When a breeze has a speed less than 3 kilometers (2 miles) per hour, only wavelets appear. At greater wind speeds, more stable waves gradually form and advance with the wind.

Characteristics of ocean waves are illustrated in Figure 15.11, which shows a simple, nonbreaking waveform. The tops of the waves are the *crests*, which are separated by *troughs*. Halfway between the crests and troughs is the *still water level*, which is the level that the water would occupy if there were no waves. The vertical distance between trough and crest is called the **wave height**, and the horizontal distance between successive crests or successive troughs is the **wavelength**. The time it takes one full wave—one wavelength—to pass a fixed position is the **wave period**.

The height, length, and period that are eventually achieved by a wave depend on three factors: (1) wind speed, (2) length of time the wind has blown (*duration*), and (3) *fetch*, the distance that wind has traveled across open water. As the quantity of energy transferred from the wind to the water increases, both the height and steepness of the waves increase. Eventually, a critical point is reached where waves grow so tall that they topple over, forming ocean breakers called *whitecaps*.

For a particular wind speed, there is a maximum fetch and duration of wind beyond which waves will no longer increase in size. When the maximum fetch and duration are reached for a given wind velocity, the waves are said to be "fully developed." The reason that waves can grow no further is that they are losing as much energy through the breaking of whitecaps as they are receiving energy from the wind.

When the wind stops or changes direction, or the waves leave the storm area where they were created, the waves continue on without relation to local winds. The waves also undergo a gradual change to *swells*, which describes any wave that has traveled out if its area of origin. Swells are lower in height and longer in length and are capable of carrying a storm's energy to distant shores. Because many independent wave systems exist at the same time, the sea surface acquires a complex and irregular pattern, sometimes producing very large waves. The sea waves that are seen from shore are usually a mixture of swells from faraway storms and waves created by local winds.

SmartFigure 15.11
Wave basics An idealized drawing of a nonbreaking wave, showing its basic parts and the movement of water with increasing depth.

ANIMATION
https://goo.gl/h0NxLb

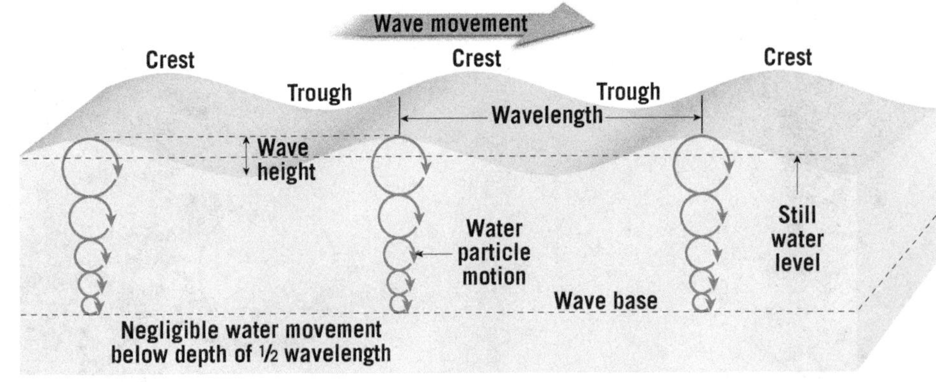

EYE ON EARTH 15.1

This surfer is enjoying a ride on a large wave along the coast of Maui.

QUESTION 1 *What was the source of energy that created this wave?*

QUESTION 2 *How was the wavelength changing just prior to the time when this photo was taken?*

QUESTION 3 *Why was the wavelength changing?*

QUESTION 4 *Many waves exhibit circular orbital motion. Is that true of the wave in this photo? Explain*

Ron Dahlquist/Getty Images

Circular Orbital Motion

Waves can travel great distances across ocean basins. In one study, waves generated near Antarctica were tracked as they traveled through the Pacific Ocean basin. After more than 10,000 kilometers (more than 6,000 miles), the waves finally expended their energy a week later, along the shoreline of Alaska's Aleutian Islands. It is important to realize that what crossed the Pacific was the waveform, not the water. As a wave passes, each bit of water moves in a near-circle and returns to about where it started. This movement, which transfers wave energy, is called **circular orbital motion**.

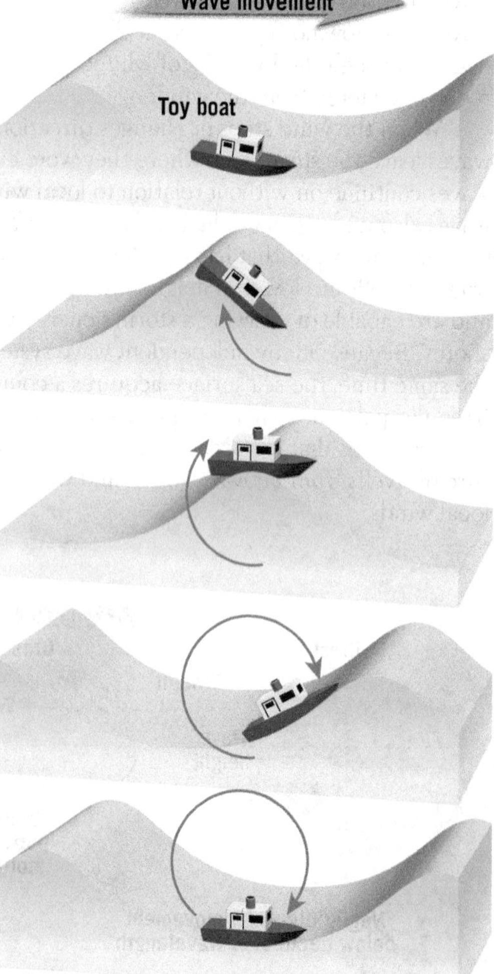

▶ SmartFigure 15.12 **Passage of a wave** The movements of the toy boat show that the waveform advances, but the water does not advance appreciably from the original position. In this sequence, the wave moves from left to right as the boat (and the water in which it is floating) rotates in an imaginary circle, without moving significantly to the right along with the wave.

TUTORIAL
https://goo.gl/uH2qhW

Observation of an object floating in waves shows that it moves not only up and down but also slightly forward and backward with each successive wave. When the movement of the floating toy boat shown in **Figure 15.12** is traced as a wave passes, it can be seen that the boat moves in a circle and returns to essentially the same place. Circular orbital motion allows a waveform (the wave's shape) to move forward *through the water* while the individual water particles that transmit the wave move around in a circle. Wind moving across a field of wheat causes a similar phenomenon: The wheat itself doesn't travel across the field, but the waves do.

The energy contributed by the wind to the water is transmitted not only along the surface of the sea but also downward. However, beneath the surface, the circular motion rapidly diminishes until, at a depth equal to one-half the wavelength measured from still water level, the movement of water particles becomes negligible. This depth is known as the **wave base**. The dramatic decrease of wave energy with depth is shown by the rapidly diminishing diameters of water-particle orbits in Figure 15.11.

Waves in the Surf Zone

As long as a wave is in deep water, it is unaffected by water depth (**Figure 15.13**, left). However, when a wave approaches the shore, the water becomes shallower and influences wave behavior. The wave begins to "feel

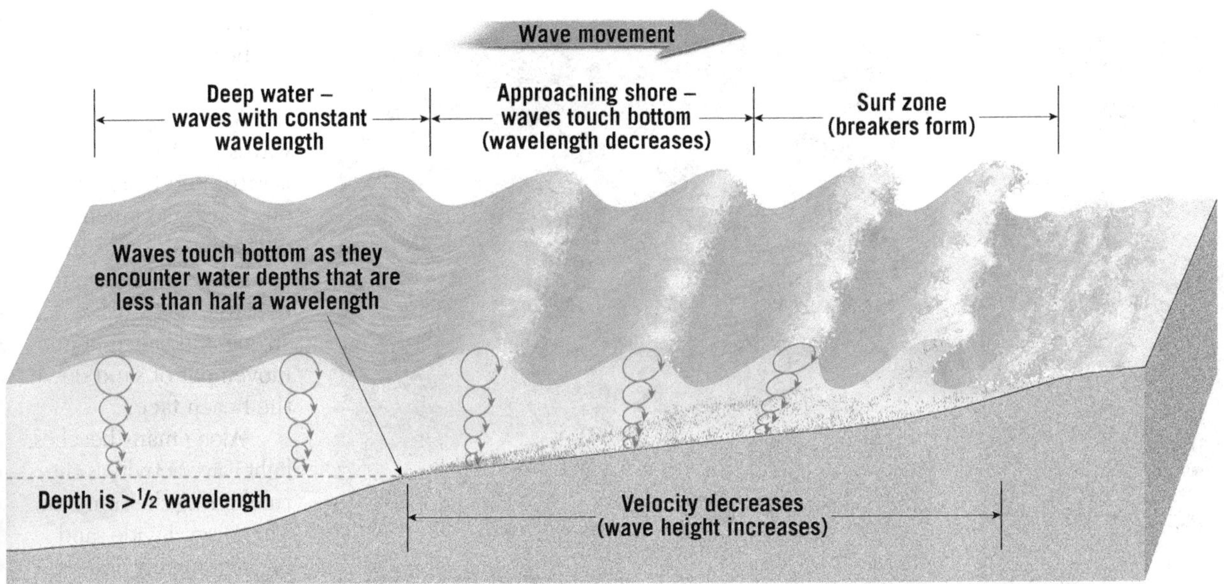

SmartFigure 15.13 **Waves approaching the shore** Waves touch bottom as they encounter water depths that are less than half a wavelength. As a result, the wave speed decreases, and the faster-moving waves farther from shore begin to catch up, which causes the distance between waves (the wavelength) to decrease. This causes an increase in wave height to the point where the waves finally pitch forward and break in the surf zone.

ANIMATION
https://goo.gl/4PiqxA

bottom" at a water depth equal to its wave base. Such depths interfere with water movement at the base of the wave and slow its advance (see Figure 15.13, center).

As a wave advances toward the shore, the slightly faster waves farther out to sea catch up, decreasing the wavelength. As the speed and length of the wave diminish, the wave steadily grows higher. Finally, a critical point is reached when the wave is too steep to support itself, and the wave front collapses, or *breaks* (see Figure 15.13, right), causing water to advance up the shore.

The turbulent water created by breaking waves is called **surf**. On the landward margin of the surf zone, the turbulent sheet of water from collapsing breakers, called *swash*, moves up the slope of the beach face. When the energy of the swash has been expended, the water flows back down the beach face toward the surf zone as *backwash*.

CONCEPT CHECKS 15.4

1. List three factors that determine the height, length, and period of a wave.
2. Describe the motion of a floating object as a wave passes.
3. How do the speed, length, and height of a wave change as the wave moves into shallow water and breaks?

15.5 The Work of Waves

Describe how waves erode and move sediment along the shore.

Along most coasts, waves break along the shoreline about 10,000 times each day. During calm weather, wave action is minimal. However, just as streams do most of their work during floods, waves accomplish most of their work during storms. The impact of large, powerful, storm-induced waves against the shore can be awesome in its violence (**Figure 15.14**).

Figure 15.14 **Storm waves along the Portugal coast** When large waves break against the shore, the force of the water can be powerful, and the erosional work that is accomplished can be great. (Photo by Zacarias da Mata/Fotolia)

Wave Erosion

Each breaking wave may hurl thousands of tons of water against the land, sometimes causing the ground to literally tremble. The pressures exerted by Atlantic waves in wintertime, for example, average nearly 10,000 kilograms per square meter (more than 2000 pounds per square foot). The force during storms is even greater. It is no wonder that cracks and crevices are quickly opened in cliffs, seawalls, breakwaters, and anything else that is

A.

Smooth, rounded rocks along the shore are an obvious reminder that abrasion can be intense in the surf zone.

B.

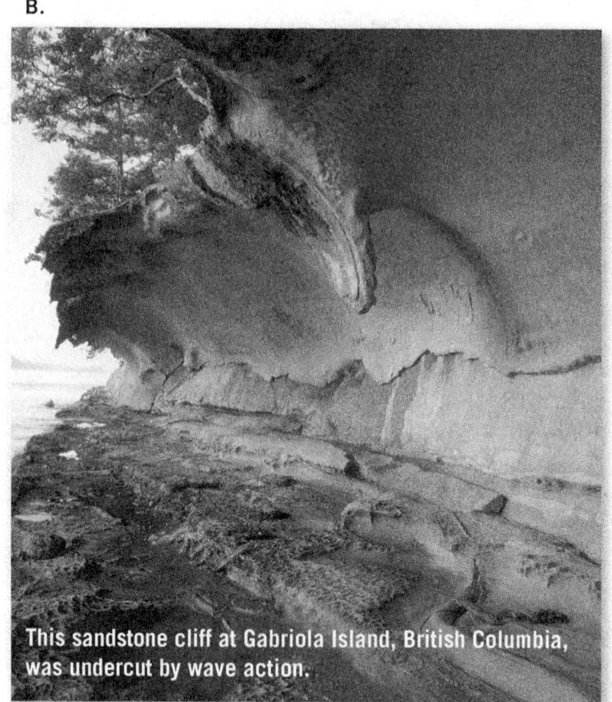

This sandstone cliff at Gabriola Island, British Columbia, was undercut by wave action.

▲ **Figure 15.15**
Abrasion—Sawing and grinding Breaking waves armed with rock debris can do a great deal of erosional work. (Photo A by Michael Collier; photo B by Fletcher & Baylis/Science Source)

subjected to these enormous shocks. Water is forced into every opening, causing air in the cracks to become highly compressed by the thrust of crashing waves. When the wave subsides, the air expands rapidly, dislodging rock fragments and enlarging and extending fractures.

In addition to the erosion caused by wave impact and pressure, **abrasion**—the sawing and grinding action of the water armed with rock fragments—is also important. In fact, abrasion is probably more intense in the surf zone than in any other environment. Smooth, rounded stones and pebbles along the shore are obvious reminders of the relentless grinding action of rock against rock in the surf zone (Figure 15.15A). Further, large waves can pick up and use these fragments as "tools" as they cut horizontally into the land (Figure 15.15B).

Sand Movement on the Beach

Beaches are sometimes called "rivers of sand." The reason is that the energy from breaking waves often causes large quantities of sand to move roughly parallel to the shoreline, both along the beach face and in the surf zone. Wave energy also causes sand to move perpendicular to (toward and away from) the shoreline.

Movement Perpendicular to the Shoreline If you stand ankle deep in water at the beach, you will see that swash and backwash move sand toward and away from the shoreline. Whether there is a net loss or addition of sand depends on the level of wave activity. When wave activity is relatively light (less energetic waves), much of the swash soaks into the beach, which reduces the backwash. Consequently, the swash dominates and causes a net movement of sand up the beach face toward the berm.

When high-energy waves prevail, the beach is saturated from previous waves, so much less of the swash soaks in. As a result, erosion occurs because backwash is strong and causes a net movement of sand down the beach face.

Along many beaches, light wave activity is the rule during the summer. Therefore, a wide sand berm gradually develops. During winter, when storms are frequent and more powerful, strong wave activity erodes and narrows the berm. A wide berm that may have taken months to build can be dramatically narrowed in just a few hours by high-energy waves created by strong winter storms.

Wave Refraction The bending of waves, called **wave refraction**, plays an important part in shoreline processes (Figure 15.16). It affects the distribution of energy along the shore and thus strongly influences where and to what degree erosion, sediment transport, and deposition will take place.

The shore is seldom oriented exactly parallel to approaching ocean waves. Rather, most waves move toward the shore at a slight angle. However, when they reach the shallow water of a smoothly sloping bottom, they bend and tend to become parallel to the shore. Such bending occurs because the part of the wave nearest the shore reaches shallow water and slows first, whereas the part that is still in deep water continues forward at its full speed. The net result is a wave front that may approach nearly parallel to the shore, regardless of the original direction of the wave.

Because of refraction, wave impact is concentrated against the sides and ends of headlands that project into the water, whereas wave attack is weakened in bays. This differential wave attack along irregular coastlines is illustrated in Figure 15.16. The shallow water near a headland causes waves to refract toward the headland, focusing their energy and also causing them to attack the headland from all three sides. By contrast, refraction in the bays causes waves to diverge and expend less energy. In these zones of weakened wave activity, sediments can accumulate and form sandy beaches. Over a long period, erosion of the headlands and deposition in the bays tends to straighten irregular shorelines.

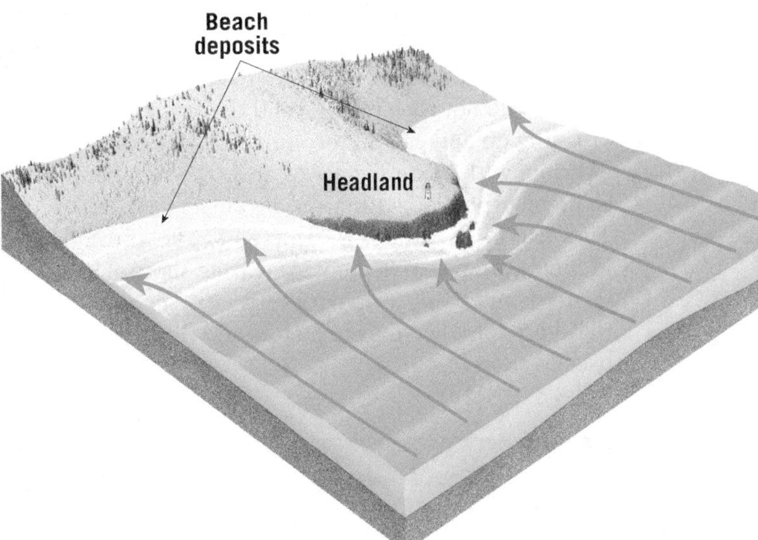
As these waves approach nearly straight on, refraction causes the wave energy to be concentrated at headlands (resulting in erosion) and dispersed in bays (resulting in deposition).
Beach deposits
Headland
Waves travel at original speed in deep water
Waves "feel bottom" and slow down in surf zone
Shoreline
Result: waves bend so that they strike the shore more directly
Wave refraction at Rincon Point, California

SmartFigure 15.16 **Wave refraction** As waves first touch bottom in the shallows along an irregular coast, they are slowed, causing them to bend (refract) and align nearly parallel to the shoreline. (Photo by Rich Reid/National Geographic/Getty Images)

TUTORIAL
https://goo.gl/3m2GHC

Movement Parallel to Shore Although waves are refracted, most still reach the shore at some angle, however slight. Consequently, the uprush of water from each breaking wave (the swash) is at an angle to the shoreline. However, the backwash is straight down the slope of the beach. The effect of this pattern of water movement is to transport sediment in a zigzag pattern along the beach face (Figure 15.17). This movement is called **beach drift**, and it can transport sand and pebbles hundreds or even thousands of meters each day.

Waves that approach the shore at an angle also produce currents within the surf zone that flow parallel to the shore. These currents move substantially more sediment than beach drift. Because the water in the surf zone is turbulent, these **longshore currents** easily move the fine suspended sand and roll larger sand and gravel along the bottom. When the sediment transported by longshore currents is added to the quantity moved by beach drift, the total amount can be very impressive. At Sandy Hook, New Jersey, for example, the quantity of sand transported along the shore over a 48-year period averaged almost 750,000 tons annually. For a 10-year period in Oxnard, California, more than 1.5 million tons of sediment moved along the shore each year.

Both rivers and coastal zones move water and sediment from one area (*upstream*) to another (*downstream*). The beach has therefore often been characterized as a "river of sand." Beach drift and longshore currents, however, move in a zigzag pattern, whereas rivers flow mostly in a turbulent, swirling fashion. In addition, the direction of flow of longshore currents along a shoreline can change if the direction that waves approach the beach changes, whereas rivers always flow in the same direction (downhill). Nevertheless, longshore currents generally

SmartFigure 15.17 **Movement parallel to shore** The two components of this sediment-moving system, beach drift and longshore currents, are created by breaking waves that approach the shoreline at an angle. These processes move large quantities of material along the beach and in the surf zone. (Photo by University of Washington Libraries, Special Collections, John Shelton Collection, KC14461)

TUTORIAL
https://goo.gl/FmkYbu

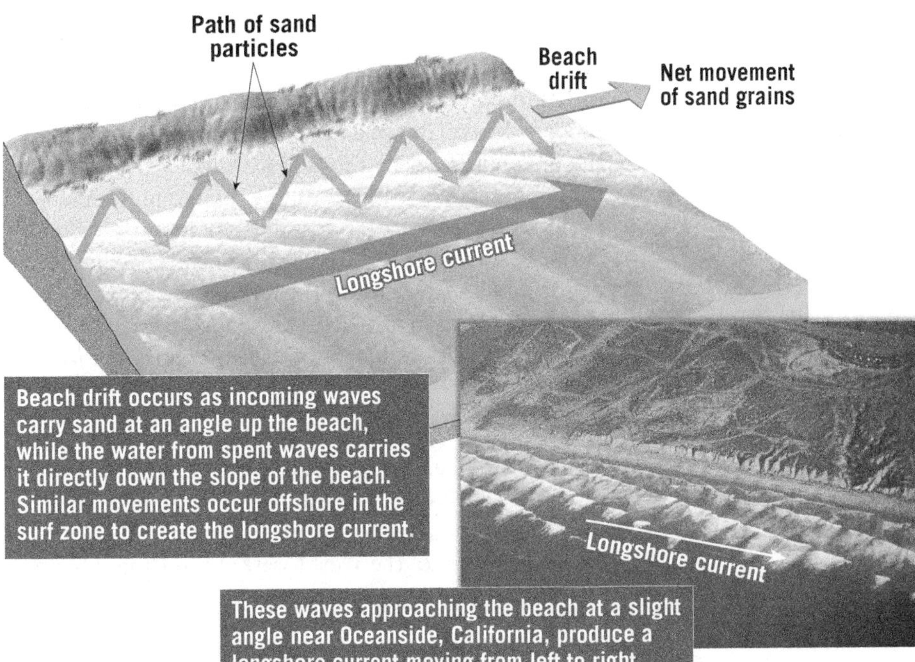
Path of sand particles
Beach drift
Net movement of sand grains
Longshore current
Beach drift occurs as incoming waves carry sand at an angle up the beach, while the water from spent waves carries it directly down the slope of the beach. Similar movements occur offshore in the surf zone to create the longshore current.
Longshore current
These waves approaching the beach at a slight angle near Oceanside, California, produce a longshore current moving from left to right.

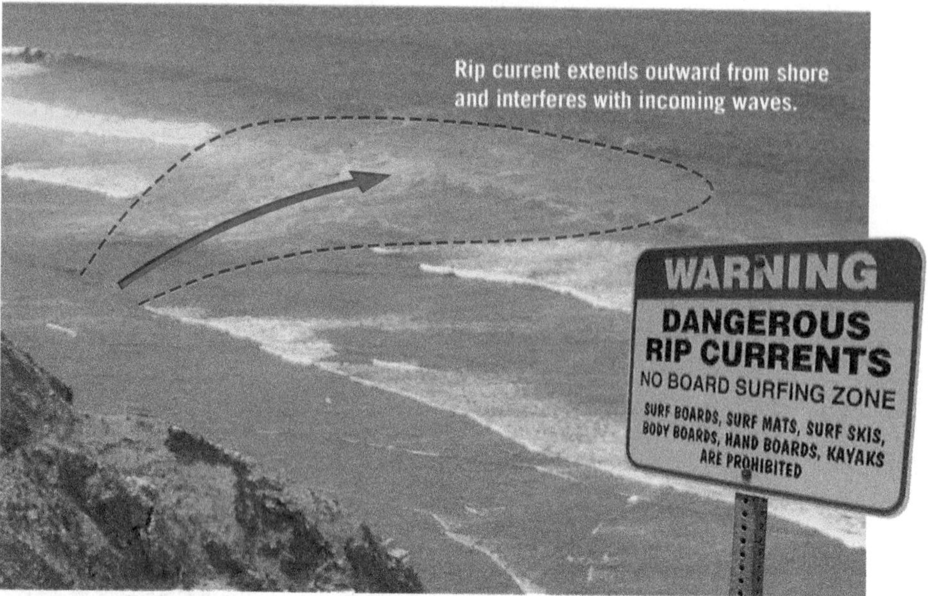

Rip current extends outward from shore and interferes with incoming waves.
WARNING
DANGEROUS RIP CURRENTS
NO BOARD SURFING ZONE
SURF BOARDS, SURF MATS, SURF SKIS, BODY BOARDS, HAND BOARDS, KAYAKS ARE PROHIBITED

▲ **Figure 15.18**
Rip current These concentrated movements of water flow opposite the direction of breaking waves. (Photos by AP Trujillo © APT Photos)

flow southward along both the Atlantic and Pacific shores of the United States.

Rip Currents Concentrated movements of water that flow *opposite* the direction of breaking waves are called **rip currents**. (Sometimes rip currents are incorrectly called *rip tides*, although they are unrelated to tidal phenomena.) Most of the backwash from spent waves finds its way back to the open ocean as an unconfined flow across the ocean bottom called *sheet flow*. However, sometimes a portion of the returning water moves seaward in the form of surface rip currents. These currents do not travel far beyond the surf zone before breaking up and can be recognized by the way they interfere with incoming waves or by the sediment that is often suspended within the rip current (Figure 15.18). They can be hazardous to swimmers, who, if caught in them, can be carried out away from shore. The best strategy for exiting a rip current is to swim *parallel* to the shore for a few tens of meters and then ride a wave back to shore.

CONCEPT CHECKS 15.5

1. Describe two ways in which waves cause erosion.
2. Why do waves that are approaching the shoreline often bend?
3. What is the effect of wave refraction along an irregular coastline?
4. Describe the two processes that contribute to longshore transport.

15.6 Shoreline Features

Describe the features typically created by wave erosion and those resulting from sediment deposited by longshore transport processes.

A fascinating assortment of shoreline features can be observed along the world's coastal regions. Although the same processes cause change along every coast, not all coasts respond in the same way. Interactions among different processes and the relative importance of each process depend on local factors. The factors include (1) the proximity of a coast to sediment-laden rivers, (2) the degree of tectonic activity, (3) the topography and composition of the land, (4) prevailing winds and weather patterns, and (5) the configuration of the coastline and near-shore areas. Features that originate primarily because of erosion are called *erosional features*, whereas accumulations of sediment produce *depositional features*.

Erosional Features

Many coastal landforms owe their origin to erosional processes. Such erosional features are common along the rugged and irregular New England coast and along the steep shorelines of the West Coast of the United States.

Wave-Cut Cliffs, Wave-Cut Platforms, and Marine Terraces As the name implies, **wave-cut cliffs** originate in the cutting action of the surf against the base of coastal land. As erosion progresses, rocks overhanging the notch at the base of the cliff crumble into the surf, and the cliff retreats. A relatively flat, benchlike surface, called a **wave-cut platform**, is left behind by the receding cliff (Figure 15.19, left). The platform broadens as wave attack continues. Some debris produced by the breaking waves may remain along the water's edge as sediment on the beach, and the remainder is transported farther seaward. If a wave-cut platform is uplifted above sea level by tectonic forces, it becomes a **marine terrace** (Figure 15.19, right). Marine terraces are easily recognized by their gentle seaward-sloping shape and are often desirable sites for coastal roads, buildings, or agriculture.

Sea Arches and Sea Stacks Because of refraction, waves vigorously attack headlands that extend into the sea. The surf erodes the rock selectively, wearing away the softer or more highly fractured rock at the fastest rate. At first, sea caves may form. When caves on opposite sides of a headland unite, a **sea arch** results (Figure 15.20). Eventually, the arch falls in, leaving an isolated remnant, or **sea stack**, on the wave-cut platform (see Figure 15.20). In time, it too will be consumed by the action of waves.

Figure 15.19 **Wave-cut platform and marine terrace** This wave-cut platform is exposed at low tide along the California coast at Bolinas Point near San Francisco. A wave-cut platform was uplifted to create the marine terrace. (Photo by University of Washington Libraries, Special Collections, John Shelton Collection, KC5902)

Depositional Features

Sediment that is transported along the shore tends to be deposited in areas where wave energy is low. Such processes produce a variety of depositional features.

Spits, Bars, and Tombolos Where beach drift and longshore currents are active, several features related to the movement of sediment along the shore may develop. A **spit** (*spit* = spine) is an elongated ridge of sand that projects from the land into the mouth of an adjacent bay. Often the end in the water hooks landward in response to the dominant direction of the longshore current. Both images in Figure 15.21 show spits. The term **baymouth bar** is applied to a sandbar that completely crosses a bay, sealing it off from the open ocean (see Figure 15.21A). Such a feature tends to form across bays where currents are weak, allowing a spit to extend to the other side. A **tombolo** (*tombolo* = mound) is a ridge of sand or gravel that connects an island to the mainland or to another island and forms in the wave shadow of a sea stack.

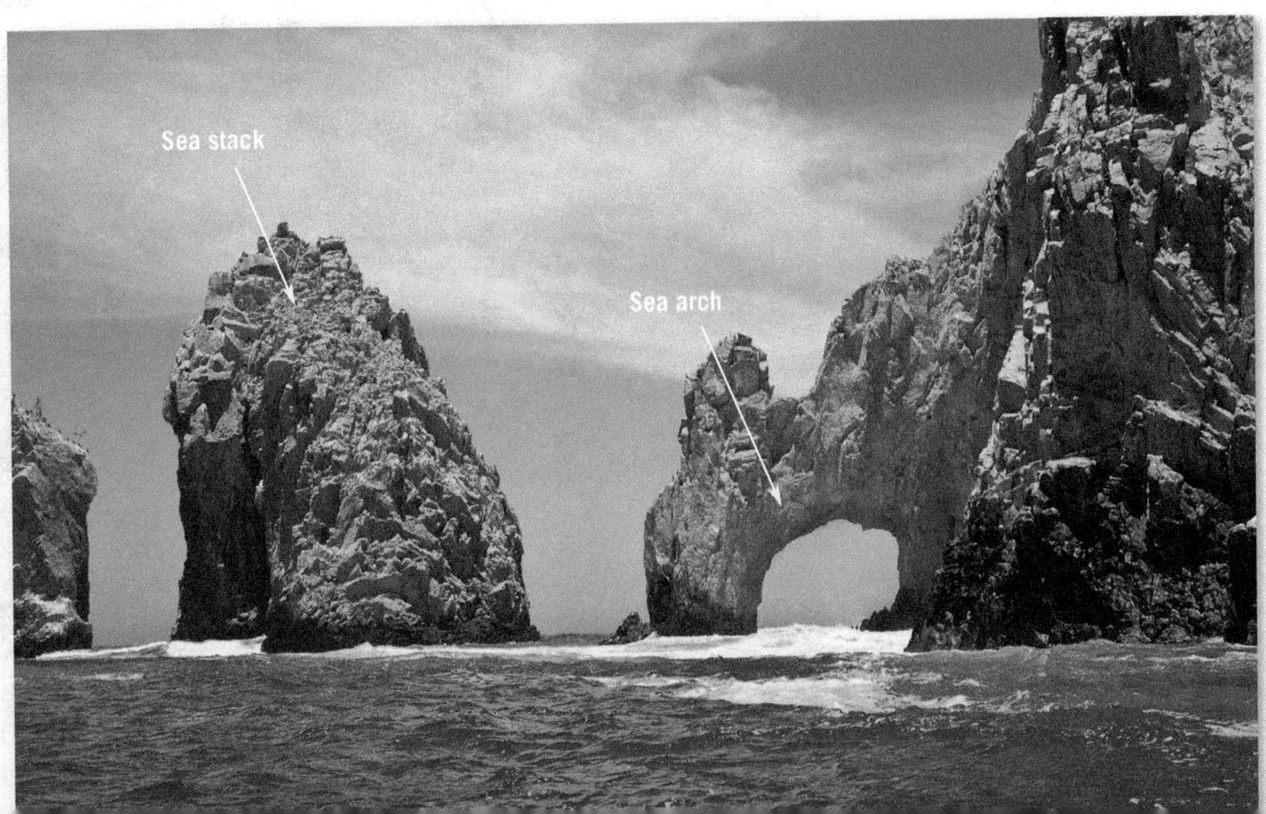

Figure 15.20 **Sea arch and sea stack** These features at the tip of Mexico's Baja Peninsula resulted from the vigorous wave attack of a headland. (Photo by Lew Robertson/Brand X Pictures/Getty Images)

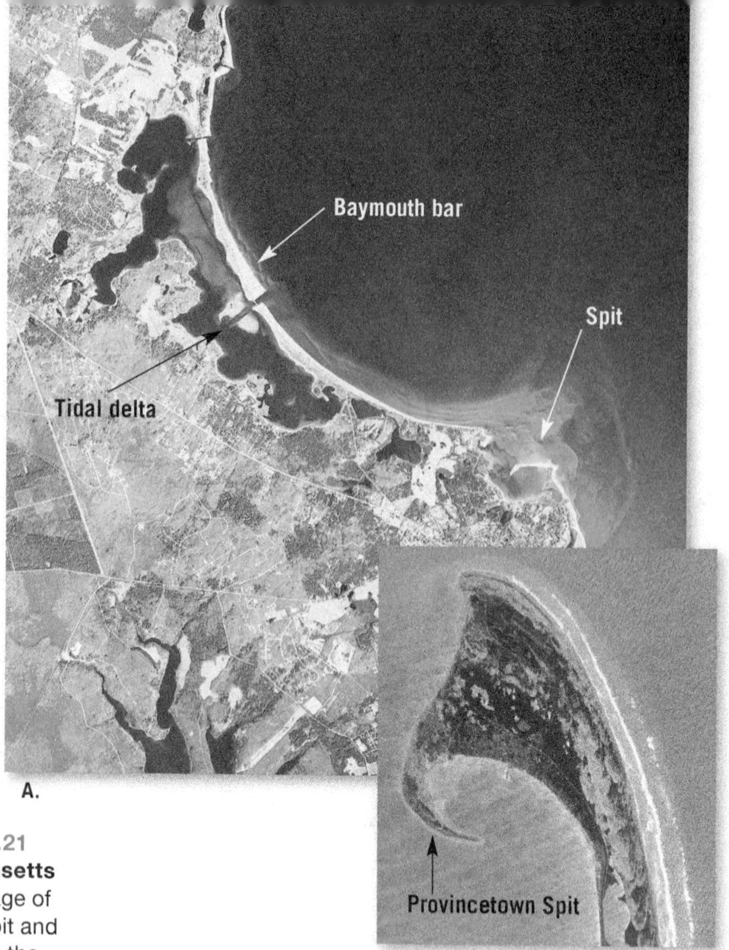

SmartFigure 15.21
Coastal Massachusetts
A. High-altitude image of a well-developed spit and baymouth bar along the coast of Martha's Vineyard. (Image courtesy of ASCS/USDA)
B. This photograph, taken from the International Space Station, shows Provincetown Spit at the tip of Cape Cod. (NASA Earth Observing System)

MOBILE FIELD TRIP
https://goo.gl/fz5L6k

Barrier Islands The Atlantic and Gulf coastal plains are relatively flat and slope gently seaward. The shore zone is characterized by **barrier islands**. These low ridges of sand parallel the coast at distances of 3 to 30 kilometers (2 to 19 miles) offshore. From Cape Cod, Massachusetts, to Padre Island, Texas, nearly 300 barrier islands rim the coast. Figure 15.22 shows an example from North Carolina.

Most barrier islands are 1 to 5 kilometers (0.6 to 3 miles) wide and 15 to 30 kilometers (9 to 18 miles) long. The tallest features are sand dunes, which usually reach heights of 5 to 10 meters (16 to 33 feet); in a few areas, unvegetated dunes are more than 30 meters (100 feet) high. The lagoons separating these narrow islands from the shore are zones of relatively quiet water that allow small craft traveling between New York and northern Florida to avoid the rough waters of the North Atlantic.

Barrier islands form in several ways. Some originate as spits that were subsequently severed from the mainland by wave erosion or by the general rise in sea level following the last episode of glaciation. Others were created when turbulent waters in the line of breakers heaped up sand that had been scoured from the ocean floor. Finally, some barrier islands may be former sand dune ridges that originated along the shore during the most recent glacial period, when sea level was lower. As the ice sheets melted, sea level rose and flooded the area behind the beach–dune complex.

The Evolving Shore

A shoreline continually undergoes modification, regardless of its initial configuration. Along a coastline that is characterized by varied geology, the pounding surf may initially increase its irregularity because the waves erode the weaker rocks more easily than the stronger ones. However, if a shoreline remains stable, marine erosion and deposition eventually produce a straighter, more regular coast.

Figure 15.23 illustrates the evolution of an initially irregular coast that remains relatively stable and shows many of the coastal features discussed in this section. As headlands are eroded and erosional features such as wave-cut cliffs and wave-cut platforms are created,

Figure 15.22 **Barrier islands** Nearly 300 barrier islands line the Gulf and Atlantic coasts. The islands along the coast of North Carolina are excellent examples. (Photo by Michael Collier)

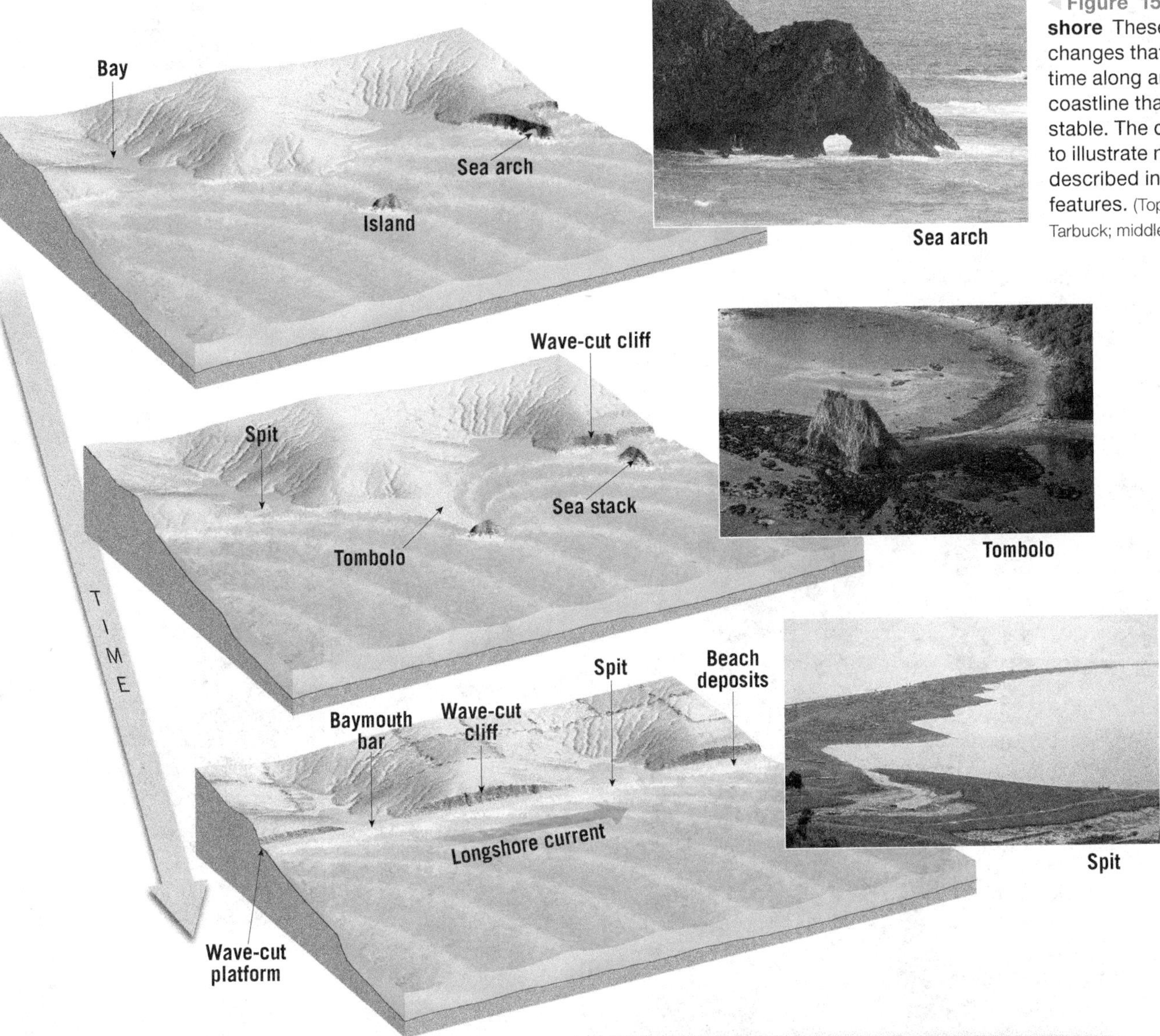

Figure 15.23 The evolving shore These diagrams illustrate changes that can take place through time along an initially irregular coastline that remains tectonically stable. The diagrams also serve to illustrate many of the features described in the section on shoreline features. (Top and bottom photos by E. J. Tarbuck; middle photo by Michael Collier)

sediment is produced that is carried along the shore by beach drift and longshore currents. Some material is deposited in the bays, while other debris is formed into depositional features such as spits and baymouth bars. At the same time, rivers fill the bays with sediment. Ultimately, a smooth coast results.

CONCEPT CHECKS 15.6

1. How is a marine terrace related to a wave-cut platform?
2. Describe the formation of the features labeled in Figures 15.20 and 15.21.
3. List three ways that barrier islands form.

15.7 Contrasting America's Coasts

Distinguish between emergent and submergent coasts. Contrast the erosion problems faced on the Atlantic and Gulf coasts with those along the Pacific coast.

As GEOgraphics 15.1 illustrates, the Pacific Coast of the United States is strikingly different from the Atlantic and Gulf coast regions. Some of the differences are related to plate tectonics. The West Coast represents the leading edge of the North American plate, and thus, it experiences active uplift and deformation. By contrast, the East Coast is far from any active plate boundary and relatively quiet tectonically. Because of this basic geologic difference, the nature of shoreline erosion problems along America's opposite coasts is different.

Coastal Classification

The great variety of shorelines demonstrates their complexity. Indeed, to understand any particular coastal area, many factors must be considered, including rock types, size and direction of waves, frequency of storms, tidal range, and offshore topography. In addition, recall from Chapter 10 that practically all coastal areas were affected by the worldwide rise in sea level that accompanied the melting of ice sheets following the Last Glacial Maximum.

GEOGRAPHICS 15.1

A Brief Tour of America's Coasts*

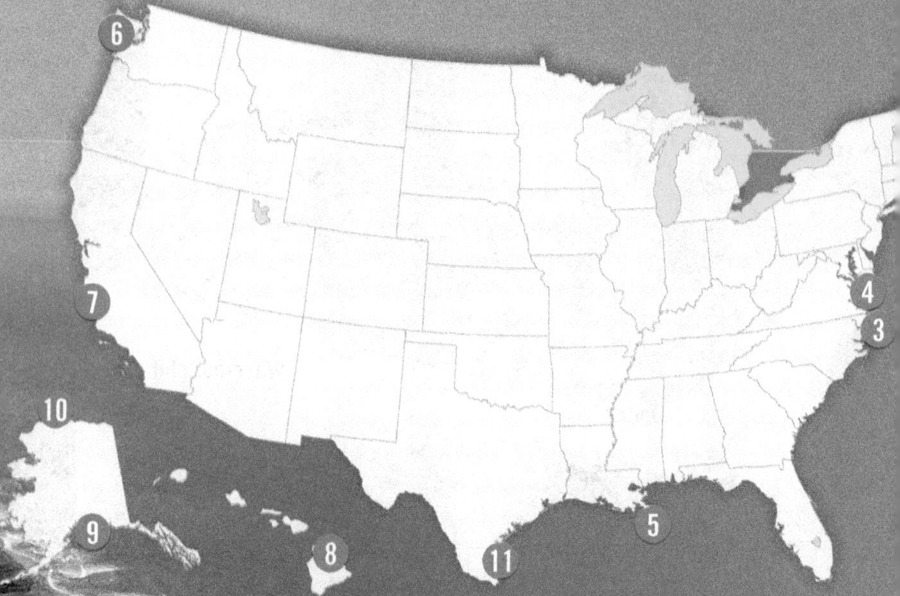

1 A small portion of the Cape Cod coast shows the entrance to Nauset Bay. Depending on the whims of recent storms and the strength of coastal currents, the opening into the bay may only be a few hundred feet wide. Tidal currents have created an underwater sandbar just inside the harbor.

10 Sea ice hugs Alaska's north slope near Barrow. The Arctic shore is locked in ice for much of the year.

7 This highway clings to the California Coast south of Big Sur. Uplift is occurring in this area near the boundary separating the Pacific and North American plates.

3 North Carolina's Outer Banks. Oregon Inlet Bridge connecting Bodie Island (foreground) and Hatteras Island. These narrow wisps of sand are part of an extensive barrier island system. The beach and dunes on the left face the Atlantic Ocean. On the right are the quieter waters of Pamlico Sound.

*Prepared with the assistance of Michael Collier. All photos by Michael Collier.

delta of the Mississippi r is a major feature in Gulf of Mexico. This -lying coastal zone is a e of low soggy islands are barely above sea l with a myriad of ural distributries and ficial channels.

Coasts are among Earth's most dynamic landscapes. Waves, tides, and currents continuously shape this interface between land and sea. Coasts may also exhibit the effects of mountain building, sea level changes, rivers, glaciers, and people. Here is a very small glimpse at the diversity and beauty of America's costs.

9 A small glacier flows down a steep mountain slope into the Cook Inlet southwest of Anchorage, Alaska. The total length of Alaska's irregular coastline is nearly 71,000 kilometers (44,000 miles).

2 Spruce and fir cover Turtle Island, part of Maine's Acadia National Park. This region was sculpted by Ice Age glaciers, then flooded when sea level rose as the ice sheets melted.

4 Mobjack Bay near Gloucester, Virginia, is part of the much larger Chesapeake Bay. These bays, called estuaries, are actually river valleys that were drowned when sea level rose at the end of the Ice Age.

6 Waves have aggressively attacked the coast north of La Push Harbor, Washington. In places, remnants of the cliff remain as sea stacks.

8 Cliffs and waterfalls along the north coast of Hawaii's Big Island. This portion of the island receives abundant rainfall. There are still active volcanoes elsewhere on the island.

11 Padre Island is a barrier island that protects the coast of southern Texas from storms. Winds off the Gulf of Mexico create dunes, which offer a temporary footing for vegetation that is occasionally stripped away by hurricanes.

Question:
Use these images to identify at least two shoreline features described in section 20.4.

?

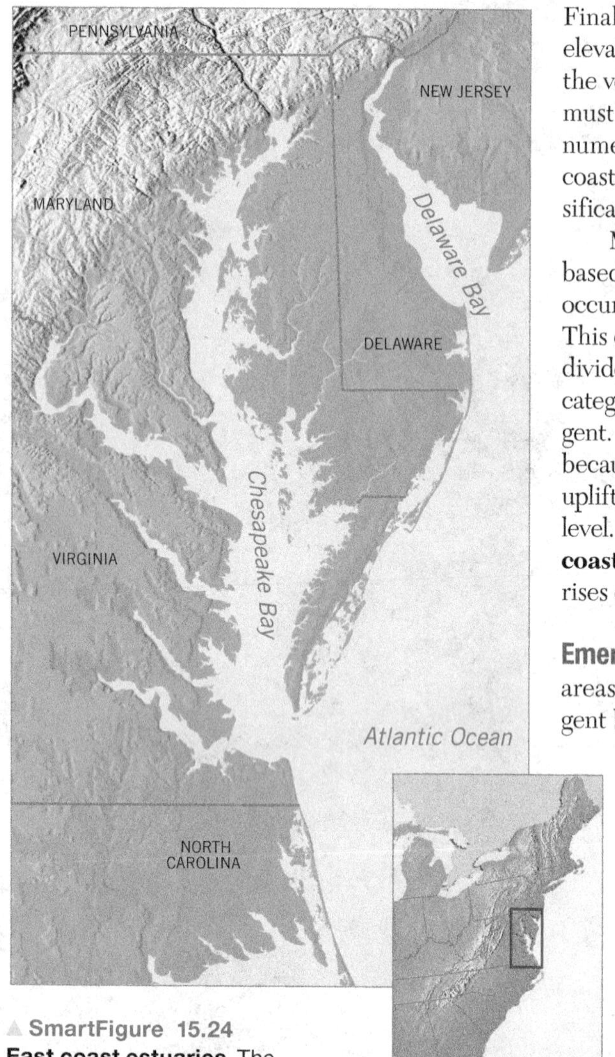

▲ SmartFigure 15.24
East coast estuaries The lower portions of many river valleys were flooded by the rise in sea level that followed the end of the Quaternary Ice Age, creating large estuaries such as the Chesapeake and Delaware Bays.

TUTORIAL
https://goo.gl/CkLRFL

Finally, tectonic events that elevate or drop the land or change the volume of ocean basins must be taken into account. The numerous factors that influence coastal areas make shoreline classification difficult.

Many geologists classify coasts based on the changes that have occurred with respect to sea level. This commonly used classification divides coasts into two general categories: emergent and submergent. **Emergent coasts** develop because an area experiences either uplift of the land or a drop in sea level. Conversely, **submergent coasts** are created when sea level rises or the land subsides.

Emergent Coasts In some areas, the coast is clearly emergent because rising land or a falling sea level exposes wave-cut cliffs and platforms. Excellent examples include portions of coastal California, where uplift has occurred in the recent geologic past. The marine terrace shown in Figure 15.19 illustrates this situation. In the case of the Palos Verdes Hills, south of Los Angeles, seven different terrace levels exist, indicating seven episodes of uplift. The ever-persistent sea is now cutting a new platform at the base of the cliff. If uplift follows, it too will become an elevated marine terrace.

Other examples of emergent coasts include regions that were once buried beneath great ice sheets. When glaciers were present, their weight depressed the crust, and when the ice melted, the crust began gradually to spring back. Consequently, prehistoric shoreline features may now be found high above sea level. The Hudson Bay region of Canada is such an area; portions of it are still rising at a rate of more than 1 centimeter per year.

Submergent Coasts In contrast to the preceding examples, other coastal areas show definite signs of submergence. Shorelines that have been submerged in the relatively recent past are often highly irregular because the sea typically floods the lower reaches of river valleys flowing into the ocean. The ridges separating the valleys, however, remain above sea level and project into the sea as headlands. These drowned river mouths, which are called **estuaries**, characterize many coasts today. Along the Atlantic coastline, the Chesapeake and Delaware Bays are examples of large estuaries created by submergence (Figure 15.24). The picturesque coast of Maine, particularly in the vicinity of Acadia National Park, is another excellent example of an area that was flooded by the postglacial rise in sea level and transformed into a highly irregular coastline.

Keep in mind that most coasts have complicated geologic histories. With respect to sea level, at various times many coasts have emerged and then submerged. Each time, they may retain some of the features created during the previous situation.

Atlantic and Gulf Coasts

Much of the coastal development along the Atlantic and Gulf coasts has occurred on barrier islands. Typically, a barrier island, also termed a *barrier beach* or *coastal barrier*, consists of a wide beach that is backed by dunes and separated from the mainland by marshy lagoons. The broad expanses of sand and exposure to the ocean have made barrier islands exceedingly attractive sites for development. Unfortunately, development has taken place more rapidly than our understanding of barrier island dynamics has increased. Figure 15.25, shows how the historic 21-story lighthouse at Cape Hatteras National Seashore was threatened by erosion and had to be relocated.

Because barrier islands face the open ocean, they receive the full force of major storms that strike the coast. When a storm occurs, the barriers absorb the energy of the waves primarily through the movement of sand. A single major storm can move huge quantities of sand, leaving a barrier island looking very different than it did prior to the storm. Just as a flexible sapling may survive winds that destroy an oak tree, so too do barrier islands survive the force of huge waves—not with unyielding strength but by yielding to the storm.

Pacific Coast

In contrast to the broad, gently sloping coastal plains of the Atlantic and Gulf coasts, much of the Pacific coast is characterized by relatively narrow beaches that are backed by steep cliffs and mountain ranges. The chapter-opening photo provides a good example. Recall that America's western margin is a more rugged and tectonically active region than the eastern margin. Because uplift continues, a rise in sea level in the west is not so readily apparent. Nevertheless, like the shoreline erosion problems facing the Atlantic coast's barrier islands, west coast difficulties also stem largely from the alteration of a natural system by people.

A major problem facing the Pacific shoreline, and especially portions of southern California, is a significant narrowing of many beaches. The bulk of the sand on many of these beaches is supplied by rivers that transport it from the mountains to the coast. Over the years, this natural flow of material to the coast has been interrupted by dams built for irrigation and flood control. The reservoirs effectively trap the sand that would otherwise nourish

Various attempts to protect the lighthouse failed. They included building groins and beach nourishment. By 1999, when this photo was taken, the lighthouse was only 36 meters (120 ft.) from the water.
New location of lighthouse
To save the famous candy-striped landmark, the National Park Service authorized moving the structure. After the $12 million move, it is expected to be safe for 50 years or more.
884 meters (2900 ft.)

Figure 15.25 Relocating the Cape Hatteras lighthouse After the failure of a number of efforts to protect this 21-story lighthouse, the nation's tallest, from being destroyed due to an eroding shoreline, the structure finally had to be moved. (Photos by Drew.C. Wilson/Virginian-Pilot/AP Images)

Figure 15.26 Pacoima Dam and Reservoir Dams such as this one in the San Gabriel Mountains near Los Angeles, trap sediment that otherwise would have nourished beaches along the nearby coast. (Photo by Michael Collier)

Dam

the beach environment (Figure 15.26). When the beaches were wider, they protected the cliffs behind them from the force of storm waves. Now, however, the waves move across the narrowed beaches without losing much energy and cause more rapid erosion of the sea cliffs. Figure 15.8 provides a dramatic example. In efforts over the years to halt cliff retreat at Pacifica, California, the city piled rocks along the beach, drilled reinforcement rods into the bluffs, and coated the face of the cliffs with reinforced concrete. Ultimately, the Pacific Ocean won the battle.

Shoreline erosion along the Pacific coast varies considerably from one year to the next, largely because of the sporadic occurrence of storms. As a consequence, when the infrequent but serious episodes of erosion occur, the damage is often blamed on the unusual storms and not on coastal development or the sediment-trapping dams that may be great distances away. As sea level rises at an increasing rate in the years to come, increased shoreline erosion and sea-cliff retreat should be expected along many parts of the Pacific coast. Coastal vulnerability to sea-level rise is examined in more detail as part of a discussion of the possible consequences of global warming in Chapter 20.

CONCEPT CHECKS 15.7

1. Are estuaries associated with submergent coasts or emergent coasts? Explain.
2. What observable features would lead you to classify a coastal area as emergent?
3. Briefly describe what happens when storm waves strike an undeveloped barrier island.
4. How might building a dam on a river that flows to the sea affect a coastal beach?

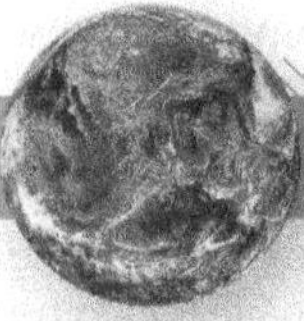

EYE ON EARTH 15.2

This is an aerial view of a small portion of a coastal area in the United States.

QUESTION 1 *Is this an emergent coast or a submergent coast?*

QUESTION 2 *Provide an easily seen line of evidence to support your answer to Question 1.*

QUESTION 3 *Is the location more likely along the coast of North Carolina or California? Explain.*

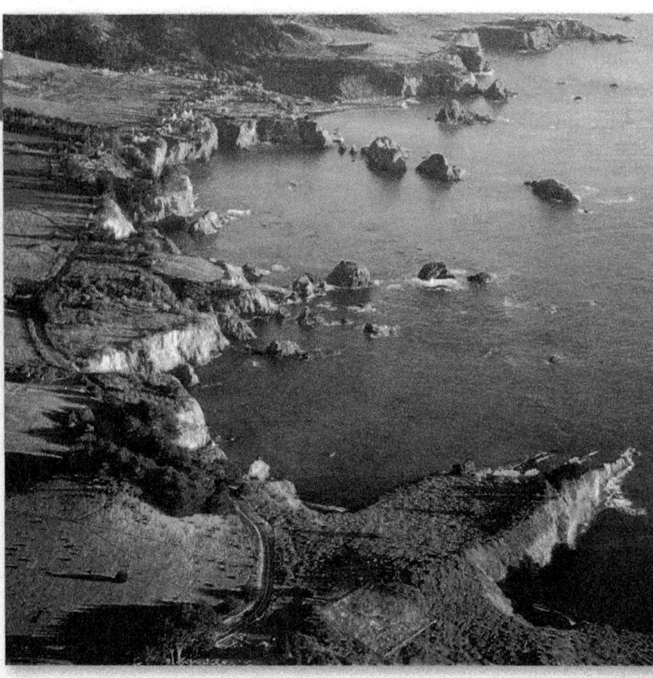

15.8 Stabilizing the Shore

Summarize the ways in which people deal with shoreline erosion problems.

The coastal zone teems with human activity. Unfortunately, people often treat the shoreline as if it were a stable platform on which structures can be built safely. Such development jeopardizes both people and the shoreline because many coastal landforms are relatively fragile, short-lived features that are easily damaged. As anyone who has endured a tsunami or a strong coastal storm knows, the shoreline is not always a safe place to live. Figure 15.8 illustrates this point.

Compared with natural hazards such as earthquakes, volcanic eruptions, and landslides, shoreline erosion appears to be a more continuous and predictable process that causes relatively modest damage to limited areas. In reality, the shoreline is one of Earth's most dynamic places and can change rapidly in response to natural forces. Storms, for example, are capable of eroding beaches and cliffs at rates that far exceed the long-term average. Such bursts of accelerated erosion not only have a significant impact on the natural evolution of a coast but can also have a profound impact on people who reside in the coastal zone. Erosion along the coast causes significant property damage. Huge sums are spent annually not only to repair damage but also in an attempt to prevent or control erosion. Already a problem at many sites, shoreline erosion is certain to become increasingly serious as extensive coastal development continues.

During the past 100 years, growing affluence and increasing demands for recreation have brought unprecedented development to many coastal areas. As both the number and the value of buildings have increased, so too have efforts to protect property from storm waves by stabilizing the shore. Also, controlling the natural migration of sand is an ongoing struggle in many coastal areas. Such interference can result in unwanted changes that are difficult and expensive to correct.

Hard Stabilization

Structures built to protect a coast from erosion or to prevent the movement of sand along a beach are known as **hard stabilization**. Hard stabilization can take many forms and often results in predictable yet unwanted outcomes. Hard stabilization includes jetties, groins, breakwaters, and seawalls.

Jetties Since relatively early in America's history, a principal goal in coastal areas has been the development and maintenance of harbors. In many cases, this has involved the construction of jetty systems. **Jetties** are usually built in pairs and extend into the ocean at the entrances to rivers and harbors. With the flow of water confined to a narrow zone, the ebb and flow caused by the rise and fall of the tides keep the sand in motion and prevent deposition in the channel. However, as illustrated in Figure 15.27, a jetty may act as a dam against which the longshore current and beach drift deposit

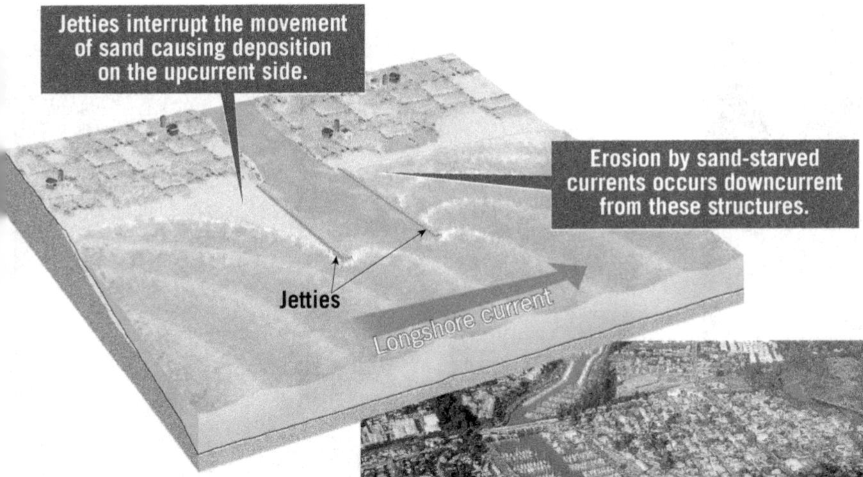

▲ **Figure 15.27 Jetties** These structures, built at entrances to rivers and harbors, are intended to prevent deposition in the navigation channel. The photo is an aerial view at Santa Cruz Harbor, California. (Photo by U.S. Army Corps of Engineers, Washington)

sand. At the same time, wave activity removes sand on the other side. Because the other side is not receiving any new sand, there is soon no beach at all.

Groins To maintain or widen beaches that are losing sand, groins are sometimes constructed. A **groin** (*groin* = ground) is a barrier built at a right angle to the beach to trap sand that is moving parallel to the shore. Groins are usually constructed of large rocks but may also be composed of wood. These structures often do their job so effectively that the longshore current beyond the groin becomes sand-starved. As a result, the current erodes sand from the beach on the downstream side of the groin.

To offset this effect, property owners downstream from the structure may erect a groin on their property. In this manner, the number of groins multiplies, resulting in a *groin field* (Figure 15.28). An example of such proliferation is the shoreline of New Jersey, where hundreds of these structures have been built. Because it has been shown that groins often do not provide a satisfactory solution, building them is no longer the preferred method of keeping beach erosion in check.

Breakwaters and Seawalls Hard stabilization can also be built parallel to the shoreline. One such structure, a **breakwater**, may be used to protect boats from the force of large breaking waves by creating a quiet water zone near the shore. However, when this is done, the reduced wave activity along the shore behind the structure may allow sand to accumulate. If this happens, the boat anchorage will eventually fill with sand, while the downstream beach erodes and retreats. At Santa Monica, California, the building of a breakwater created such a problem (Figure 15.29).

Another type of hard stabilization built parallel to the shore is a **seawall**, which is designed to armor the coast and defend property from the force of breaking waves. Waves expend much of their energy as they move across an open beach. Seawalls cut this process short by reflecting the force of unspent waves seaward. As a consequence, the beach to the seaward side of the seawall experiences significant erosion and may, in some instances, be eliminated entirely (Figure 15.30). Once the width of the beach is reduced, the seawall is subjected to even greater pounding by the waves. Eventually this battering takes such a toll on the seawall that it will fail, and a larger, more expensive structure must be built to take its place.

The wisdom of building temporary protective structures along shorelines is increasingly questioned. The opinions of many coastal scientists and engineers is that halting an eroding shoreline with protective structures benefits only a few and seriously degrades or destroys the natural beach and the value it holds for the majority. Protective structures divert the ocean's energy temporarily from private properties but usually refocus that energy on the adjacent beaches. Many structures interrupt the natural sand flow in coastal currents, robbing affected beaches of vital sand replacement.

▼ **Figure 15.28 Groins** These wall-like structures trap sand that is moving parallel to the shore. This series of groins is along the shoreline north of Ship Bottom, New Jersey. The view is toward the north, and the primary direction of the longshore current is toward the bottom of the photo (toward the south). (Photo by University of Washington Libraries, Special Collections, John Shelton Collection, KCS2-31)

Figure 15.29 Breakwater at Santa Monica, California These paired aerial photos show the beach and pier at Santa Monica **A.** before the breakwater was built, and (Photo by University of California, Santa Barbara Library) **B.** after construction. The breakwater disrupted longshore transport and caused the seaward growth of the beach. After the breakwater was destroyed by storm waves in 1983, the bulge disappeared, and the shoreline returned to its pre-breakwater appearance. (Photo by University of California, Santa Barbara Library)

A. The shoreline and pier at Santa Monica as it appeared in September 1931, before the breakwater was constructed in 1933. Note that the pier is on stilts and thus does not affect longshore transport.

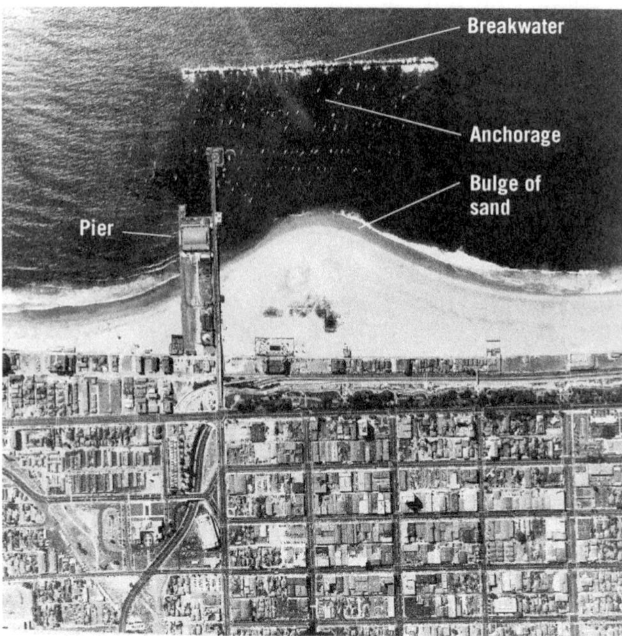

B. The same area in 1949. Construction of the breakwater to create a boat anchorage disrupted the longshore transport of sand and caused a bulge of sand on the beach in the wave shadow behind the breakwater.

Alternatives to Hard Stabilization

Armoring the coast with hard stabilization has several potential drawbacks, including the cost of the structure and the loss of sand on the beach. Alternatives to hard stabilization include beach nourishment and relocation.

Beach Nourishment One approach to stabilizing shoreline sands without hard stabilization is **beach nourishment**. As the term implies, this practice involves adding large quantities of sand to the beach system (**Figure 15.31**). Extending beaches seaward makes buildings along the shoreline less vulnerable to destruction by storm waves and enhances recreational uses.

The process of beach nourishment is straightforward. Sand is pumped by dredges from offshore or trucked from inland locations. The "new" beach, however, will not be the same as the former beach. Because replenishment sand is from somewhere else, typically not another beach, it is new to the beach environment. The new sand is usually different in size, shape, sorting, and composition. These differences pose problems in terms of erodibility and the kinds of life the new sand will support.

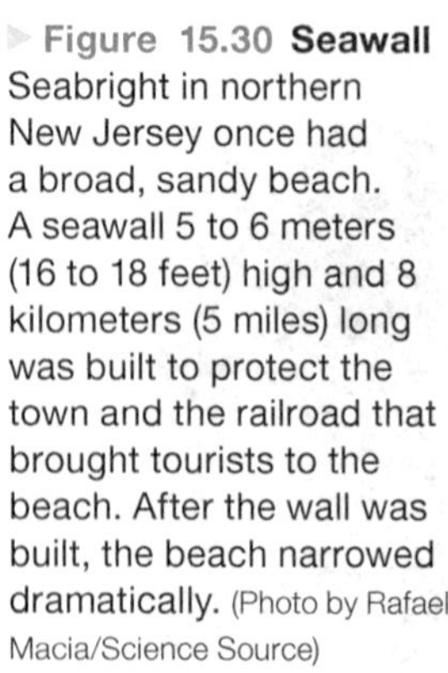

Figure 15.30 Seawall Seabright in northern New Jersey once had a broad, sandy beach. A seawall 5 to 6 meters (16 to 18 feet) high and 8 kilometers (5 miles) long was built to protect the town and the railroad that brought tourists to the beach. After the wall was built, the beach narrowed dramatically. (Photo by Rafael Macia/Science Source)

Beach nourishment is not a permanent solution to the problem of shrinking beaches. The same processes that removed the sand in the first place will eventually remove the replacement sand as well. Nevertheless, the number of nourishment projects has increased in recent years, and many beaches, especially along the Atlantic coast, have had their sand replenished many times. Virginia Beach, Virginia, has been nourished more than 50 times.

Beach nourishment is costly. For example, a modest project might involve 50,000 cubic yards of sand distributed across a half mile of shoreline. A good-sized dump truck holds about 10 cubic yards of sand. So this small project would require about 5000 dump-truck loads. Many projects extend for many miles. Nourishing beaches typically costs millions of dollars per mile.

Figure 15.31 **Beach nourishment** If you visit a beach along the Atlantic coast, it is more and more likely that you will walk into the surf zone atop a beach composed of sand from somewhere else. In this image, an offshore dredge is pumping sand to a beach. (Photo by imagebroker/Alamy Stock Photo)

Changing Land Use Instead of building structures such as groins and seawalls to hold the beach in place or adding sand to replenish eroding beaches, another option is available. Many coastal scientists and planners are calling for a policy shift from defending and rebuilding beaches and coastal property in high-hazard areas to *relocating* storm-damaged buildings in those places and letting nature reclaim the beach. This option is similar to an approach the federal government adopted for river floodplains following the devastating 1993 Mississippi River floods, in which vulnerable structures are either abandoned or relocated on higher, safer ground.

A recent example of changing land use occurred on New York's Staten Island, following Hurricane Sandy in 2012. The state turned some vulnerable shoreline areas of the island into waterfront parks. The parks act as buffers to protect inland homes and businesses from strong storms while providing the community with needed open space and access to recreational opportunities.

Land use changes can be controversial. People with significant near-shore investments want to rebuild and defend coastal developments from the erosional wrath of the sea. Others, however, argue that with sea level rising, the impact of coastal storms will get worse in the decades to come, and oft-damaged structures should be abandoned or relocated to improve personal safety and reduce costs. Such ideas will no doubt be the focus of much study and debate as states and communities evaluate and revise coastal land-use policies.

CONCEPT CHECKS 15.8

1. List three examples of *hard stabilization* and describe what each is intended to do. How does each affect sand distribution on a beach?
2. What are two alternatives to hard stabilization, and what potential problems are associated with each?

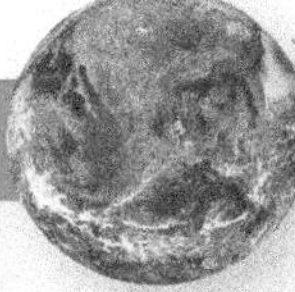

EYE ON EARTH 15.3

This structure along the eastern shore of Lake Michigan was built at the entrance to Port Shelton, Michigan.

QUESTION 1 *What is the purpose of the artificial structure pictured here?*

QUESTION 2 *What term is applied to structures such as this?*

QUESTION 3 *Suggest a reason there is a greater accumulation of sand on one side of the structure than the other.*

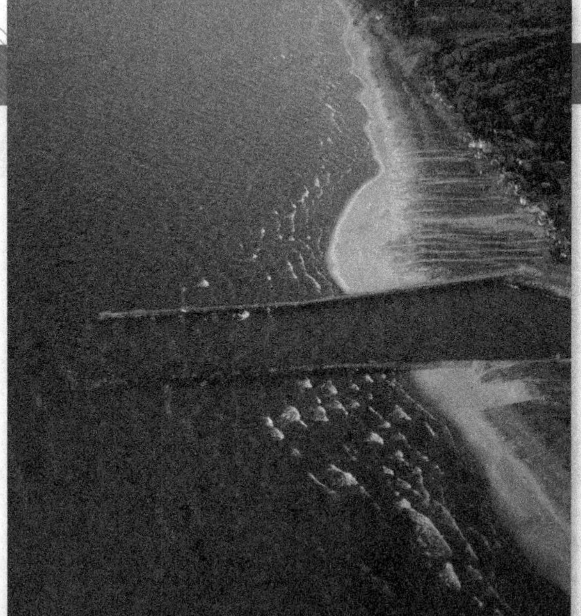

15.9 Tides

Explain the cause of tides and their monthly cycles. Describe the horizontal flow of water that accompanies the rise and fall of tides.

Tides are daily changes in the elevation of the ocean surface. Their rhythmic rise and fall along coastlines have been known since antiquity. Other than waves, they are the easiest ocean movements to observe (**Figure 15.32**).

Although known for centuries, tides were not explained satisfactorily until Sir Isaac Newton applied the law of gravitation to them in 1687. Newton showed that there is a mutual attractive force between two bodies, such as between Earth and the Moon. Because the atmosphere and the ocean are fluids and are free to move, they are deformed by tidal forces. Hence, ocean tides result from the gravitational attraction exerted upon Earth by the Moon and, to a lesser extent, by the Sun.

Causes of Tides

To illustrate how tides are produced, consider an idealized case in which Earth is a rotating sphere covered to a uniform depth with water (**Figure 15.33**). Furthermore, ignore the effect of the Sun for now. It is easy to see how the Moon's gravitational force can cause the water to bulge on the side of Earth nearer the Moon because this side is physically closer. In addition, however, an equally large tidal bulge is produced on the side of Earth directly *opposite* the Moon.

Both tidal bulges are caused, as Newton discovered, by the pull of gravity. Gravity is inversely proportional to the square of the distance between two objects, meaning simply that it quickly weakens with distance. In this case, the two objects are the Moon and Earth. Because the force of gravity decreases with distance, the Moon's gravitational pull on Earth is slightly greater on the near side of Earth than on the far side. The result of this differential pulling is to stretch (elongate) the "solid" Earth very slightly. In contrast, the world ocean, which is mobile, is deformed quite dramatically by this effect, producing the two opposing tidal bulges of similar size.

Because the position of the Moon changes only moderately in a single day, the tidal bulges remain in place while Earth rotates "through" them. For this reason, if you stand on the seashore for 24 hours, Earth will rotate you through alternating areas of higher and lower water. As you are carried into each tidal bulge, the tide rises, and as you are carried into the intervening troughs between the tidal bulges, the tide falls. Therefore, most places on Earth experience two high tides and two low tides each day.

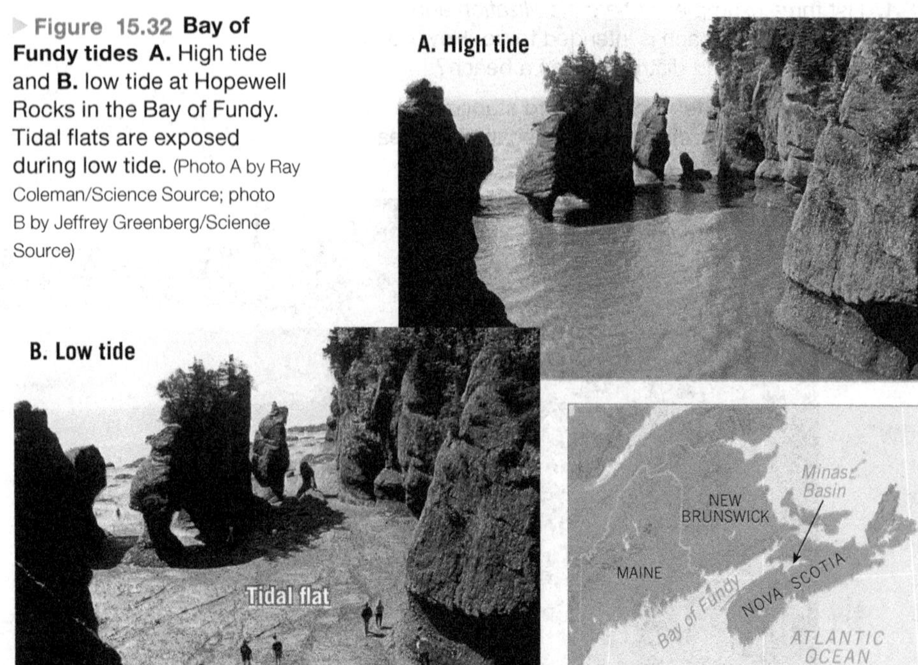

Figure 15.32 Bay of Fundy tides A. High tide and **B.** low tide at Hopewell Rocks in the Bay of Fundy. Tidal flats are exposed during low tide. (Photo A by Ray Coleman/Science Source; photo B by Jeffrey Greenberg/Science Source)

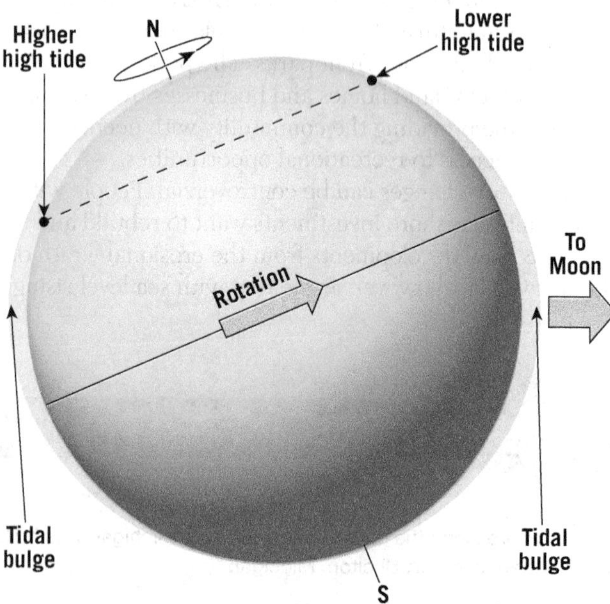

Figure 15.33 Idealized tidal bulges caused by the Moon If Earth were covered to a uniform depth with water, there would be two tidal bulges: one on the side of Earth facing the Moon (right) and the other on the opposite side of Earth (left). Depending on the Moon's position, tidal bulges may be inclined relative to Earth's equator. In this situation, Earth's rotation causes an observer to experience two unequal high tides in a day.

In addition, the tidal bulges migrate as the Moon revolves around Earth about every 29½ days. As a result, the tides, like the time of moonrise, shift about 50 minutes later each day. In essence, the tidal bulges exist in fixed positions relative to the Moon, which slowly moves progressively eastward as it orbits Earth. After 29½ daysthe cycle is complete, and a new one begins.

Many locations may show an inequality between the high tides during a given day. Depending on the Moon's position, the tidal bulges may be inclined relative to the equator, as in Figure 15.33. This figure shows how one high tide experienced by an observer in the Northern Hemisphere is considerably higher than the high tide that occurs half a day later. A Southern Hemisphere observer would experience the opposite effect.

Monthly Tidal Cycle

The primary body that influences the tides is the Moon, which makes one complete revolution around Earth every 29½ days. The Sun, however, also influences the tides. It is far larger than the Moon, but because it is much farther away, its effect is considerably less. In fact, the Sun's tide-generating effect is only about half that of the Moon.

Near the times of new and full moons, the Sun and Moon are aligned, and their forces are added together (Figure 15.34A). The combined gravity of these two tide-producing bodies causes larger tidal bulges (higher high tides) and larger tidal troughs (lower low tides), producing a large tidal range. These are called the **spring** (*springen* = to rise up) **tides**, which have no connection with the spring season but occur twice a month, during the time when the Earth–Moon–Sun system is aligned. Conversely, at about the time of the first and third quarters of the Moon, the gravitational forces of the Moon and Sun act on Earth at right angles, and each partially offsets the influence of the other (Figure 15.34B). As a result, the daily tidal range is less. These are called **neap** (*nep* = scarcely, or barely, touching) **tides**, and they also occur twice each month. Each month, then, there are two spring tides and two neap tides, each about 1 week apart.

Tidal Patterns

In the previous section, the basic causes of tides and the monthly tidal cycle were explained for a uniform, water-covered Earth. Keep in mind, however, that these idealized considerations cannot be used to predict either the height or the time of actual tides at a particular place. Many factors—including the shape of the coastline, the configuration of ocean basins, the Coriolis effect, and water depth—greatly influence the tides. Consequently, tides at various locations respond differently to the tide-producing forces. Thus, the nature of the tides at any coastal location can be determined most accurately by observation. The predictions in tidal tables and tidal data on nautical charts are based on the patterns identified in such observations.

Three main tidal patterns exist worldwide. A **diurnal** (*diurnal* = daily) **tidal pattern** is characterized by a single high tide and a single low tide each tidal day (Figure 15.35). Tides of this type occur along the northern shore of the Gulf of Mexico, among other locations. A **semidiurnal** (*semi* = twice, *diurnal* = daily) **tidal pattern** exhibits two high tides and two low tides each tidal day, with the two highs about the same height and the two lows about the same height (see Figure 15.35). This type of tidal pattern is common along the Atlantic Coast of the United States. A **mixed tidal pattern** is similar to a semidiurnal pattern except that it is characterized by a large inequality in high water heights, low water heights, or both (see Figure 15.35). In this case, there are usually

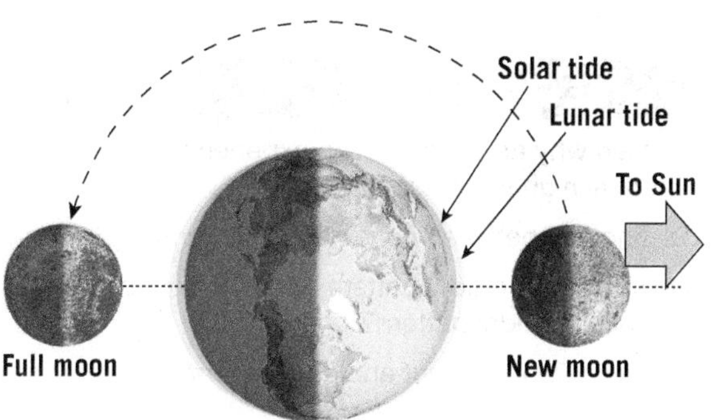

A. Spring Tide When the Moon is in the full or new position, the tidal bulges created by the Sun and Moon are aligned and there is a large tidal range.

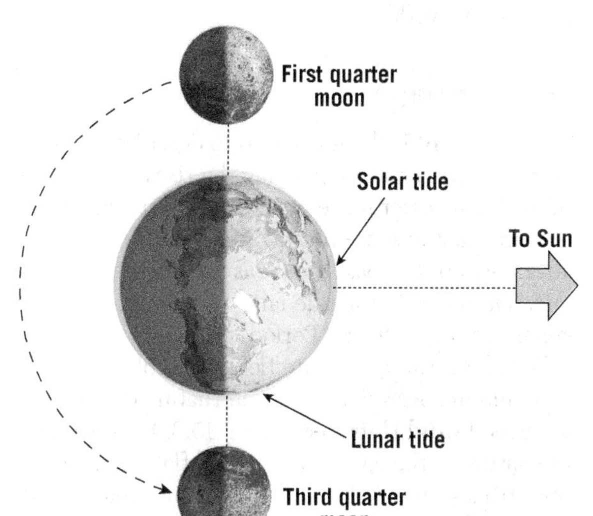

B. Neap Tide When the Moon is in the first-or third-quarter position, the tidal bulges produced by the Moon are at right angles to the bulges created by the Sun and the tidal range is smaller.

SmartFigure 15.34
Spring and neap tides Earth–Moon–Sun positions influence the tides.

ANIMATION
http://goo.gl/4rUIGf

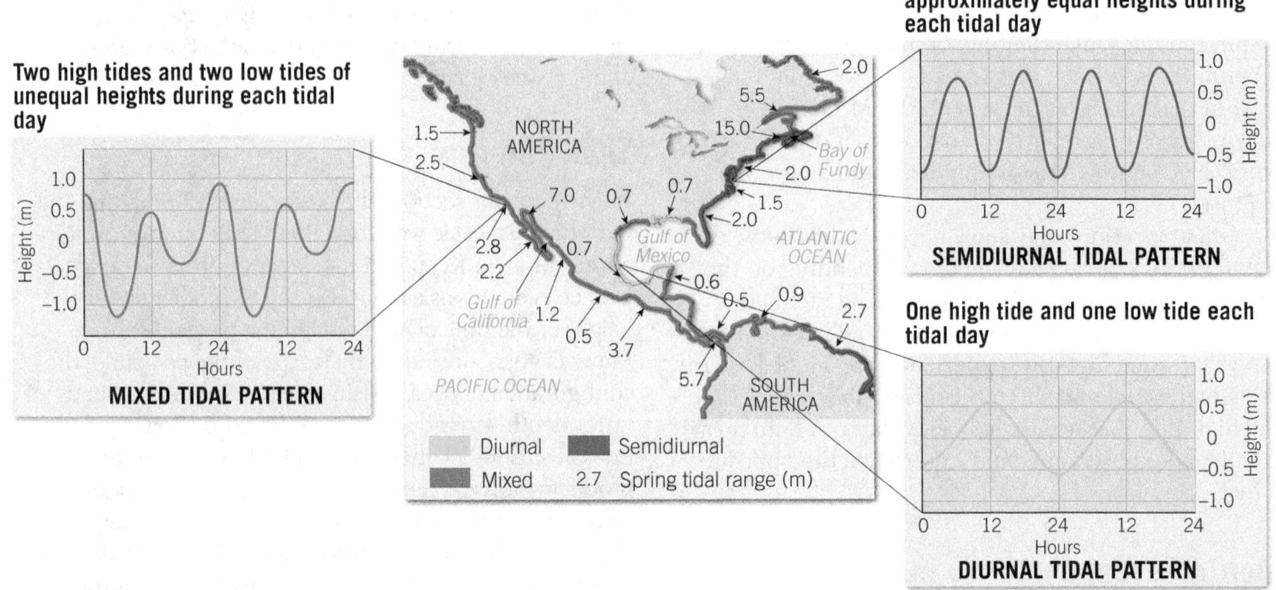

SmartFigure 15.35

Tidal patterns A diurnal tidal pattern (lower right) shows one high tide and one low tide each tidal day. A semidiurnal pattern (upper right) shows two high tides and two low tides of approximately equal heights during each tidal day. A mixed tidal pattern (left) shows two high tides and two low tides of unequal heights during each tidal day.

TUTORIAL
https://goo.gl/26JTPe

two high and two low tides each tidal day, with two high tides of different heights and two low tides of different heights each tidal day. Such tides are prevalent along the Pacific Coast of the United States and in many other parts of the world.

Tidal Currents

Tidal current is the term used to describe the *horizontal* flow of water accompanying the rise and fall of the tides. These water movements induced by tidal forces can be important in some coastal areas. Tidal currents that advance into the coastal zone as the tide rises are called *flood currents*. As the tide falls, seaward-moving water generates *ebb currents*. Periods of little or no current, called *slack water*, separate flood and ebb. The areas covered and uncovered by these alternating tidal currents are called **tidal flats** (see Figure 15.32). Depending on the nature of the coastal zone, tidal flats vary from narrow strips seaward of the beach to zones that may extend several kilometers.

Although tidal currents are not important in the open sea, they can be rapid in bays, river estuaries, straits, and other narrow places. Off the coast of Brittany in France, for example, tidal currents that accompany a high tide of 12 meters (40 feet) may attain a speed of 20 kilometers (12 miles) per hour. Tidal currents are not generally considered to be major agents of erosion and sediment transport, but notable exceptions occur where tides move through narrow inlets. Here they scour the narrow entrances to many harbors that would otherwise become choked with sediment.

CONCEPT CHECKS 15.9

1. Explain why an observer can experience two unequal high tides during one day.
2. Distinguish between *neap tides* and *spring tides*.
3. How is a mixed tidal pattern different from a semidiurnal tidal pattern?
4. Contrast *flood current* and *ebb current*.

The Dynamic Ocean

15.1 The Ocean's Surface Circulation

Discuss the factors that create and influence surface-ocean currents and describe the effect these ocean currents have on climate.

KEY TERMS: gyre, Coriolis effect

- The ocean's surface pattern of the world's major wind belts. Surface currents are parts of huge, slowly moving loops of water called gyres that are centered in the subtropics of each ocean basin. The positions of the continents and the Coriolis effect also influence the movement of ocean water within gyres. Because of the Coriolis effect, subtropical gyres move clockwise in the Northern Hemisphere and counterclockwise in the Southern Hemisphere. Generally, four main currents comprise each subtropical gyre.
- Ocean currents can have a significant effect on climate. Poleward-moving warm ocean currents moderate winter temperatures in the middle latitudes. Cold currents exert their greatest influence during summer in middle latitudes and year-round in the tropics. In addition to cooler temperatures, cold currents are associated with greater fog frequency and drought.

? Assume that arrow A represents prevailing winds in a Northern Hemisphere ocean. Which arrow—A, B, or C—best represents the surface-ocean current in this region? Explain.

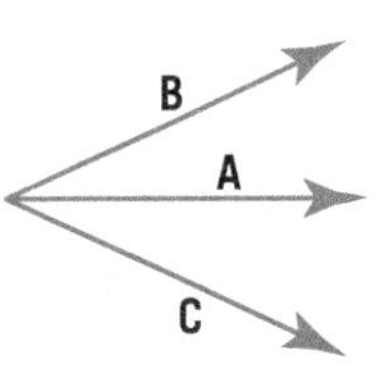

15.2 Upwelling and Deep-Ocean Circulation

Explain the processes that produce coastal upwelling and the ocean's deep circulation.

KEY TERMS: upwelling, thermohaline circulation

- Upwelling, the rising of colder water from deeper layers, is a wind-induced movement that brings cold, nutrient-rich water to the surface. Coastal upwelling is most characteristic along the west coasts of continents.
- In contrast to surface currents, deep-ocean circulation is governed by gravity and driven by density differences. The two factors that can create a dense mass of water are temperature and salinity, so the movement of deep-ocean water is often termed thermohaline circulation. Most water involved in thermohaline circulation begins in high latitudes at the surface, when the salinity of the cold water increases as a result of sea ice formation. This dense water sinks, initiating deep-ocean currents.

15.3 The Shoreline: A Dynamic Interface

Explain why the shoreline is considered a dynamic interface and identify the basic parts of the coastal zone.

KEY TERMS: shoreline, shore, coast, coastline, foreshore, backshore, near-shore zone, offshore zone, beach, berm, beach face

- The shore is the area extending between the lowest tide level and the highest elevation on land that is affected by storm waves. The coast extends inland from the shore as far as ocean-related features can be found. The shore is divided into the foreshore and backshore. Seaward of the foreshore are the near-shore and offshore zones.
- A beach is an accumulation of sediment along the landward margin of the ocean or a lake. Among its parts are one or more berms and the beach face. Beaches are composed of whatever material is locally abundant and can be thought of as material in transit along the shore.

? Assume that you took this photo and that the photo was taken at high tide. On which part of the shore are you standing?

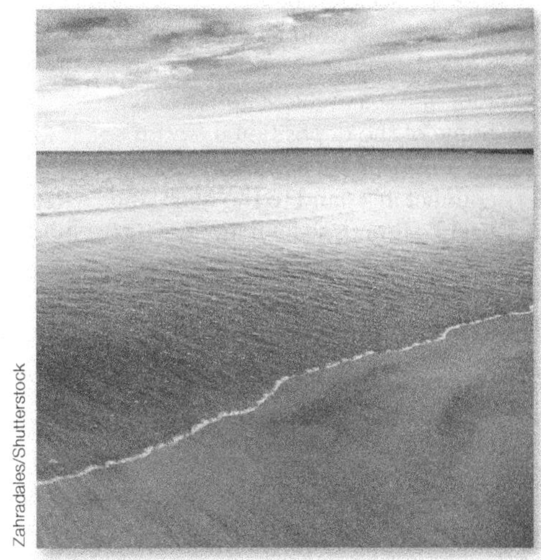
Zahradales/Shutterstock

15.4 Ocean Waves

List and discuss the factors that influence the height, length, and period of a wave and describe the motion of water within a wave.

KEY TERMS: wave height, wavelength, wave period, circular orbital motion, wave base, surf

- Waves are moving energy, and most ocean waves are initiated by wind. The three factors that influence the height, wavelength, and period of a wave are (1) wind speed, (2) length of time the wind has blown (duration), and (3) fetch, the distance that the wind has traveled across open water. Once waves leave a storm area, they are termed swells, which are symmetrical, longer-wavelength waves.
- As waves travel, water particles transmit energy by circular orbital motion, which extends to a depth equal to one-half the wavelength (the wave base). When a wave enters water that is shallower than the wave base, it slows down, which allows waves farther from shore to catch up. As a result, wavelength decreases and wave height increases. Eventually the wave breaks, creating turbulent surf in which water rushes toward the shore.

15.5 The Work of Waves

Describe how waves erode and move sediment along the shore.

KEY TERMS: abrasion, wave refraction, beach drift, longshore current, rip current

- Waves provide most of the energy that modifies shorelines. Wave erosion is caused by wave impact pressure and abrasion (the sawing and grinding action of water armed with rock fragments).
- As they approach the shore, waves refract (bend) to align nearly parallel to the shore. Refraction occurs because a wave travels more slowly in shallower water, allowing the part still in deeper water to catch up. Wave refraction causes wave erosion to be concentrated against the sides and ends of headlands and dispersed in bays.
- Waves that approach the shore at an angle transport sediment parallel to the shoreline. On the beach face, this movement of sediment, called beach drift, is due to the fact that the incoming swash pushes sediment obliquely upward, whereas the backwash pulls it directly downslope. Longshore currents are a similar phenomenon in the surf zone, capable of transporting very large quantities of sediment parallel to a shoreline.

? What process is causing wave energy to be concentrated on the headland? Predict how this area will appear in the future.

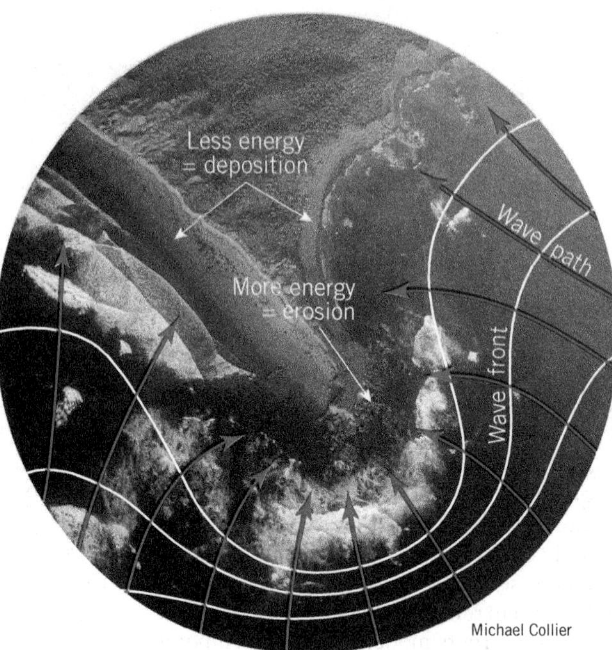

Michael Collier

15.6 Shoreline Features

Describe the features typically created by wave erosion and those resulting from sediment deposited by longshore transport processes.

KEY TERMS: wave-cut cliff, wave-cut platform, marine terrace, sea arch, sea stack, spit, baymouth bar, tombolo, barrier island

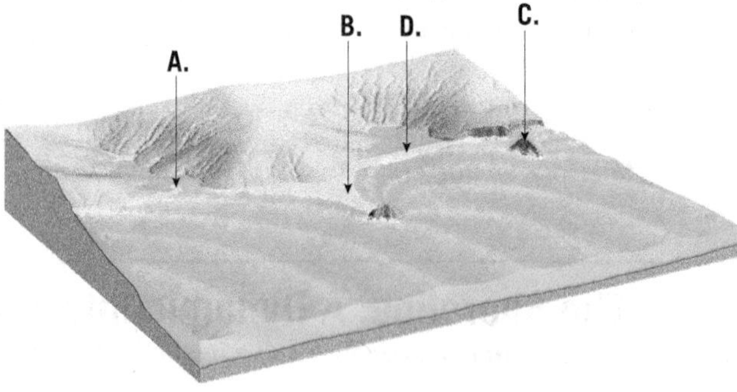

- Erosional features include wave-cut cliffs (which originate from the cutting action of the surf against the base of coastal land), wave-cut platforms (relatively flat, bench-like surfaces left behind by receding cliffs), and marine terraces (uplifted wave-cut platforms). Erosional features also include sea arches (formed when a headland is eroded and two sea caves from opposite sides unite) and sea stacks (formed when the roof of a sea arch collapses).
- Some of the depositional features that form when sediment is moved by beach drift and longshore currents are spits (elongated ridges of sand that project from the land into the mouth of an adjacent bay), baymouth bars (sandbars that completely cross a bay), and tombolos (ridges of sand that connect an island to the mainland or to another island). Along the Atlantic and Gulf coastal plains, the coastal region is characterized by offshore barrier islands, which are low ridges of sand that parallel the coast.

? Identify the lettered features in this diagram.

15.7 Contrasting America's Coasts

Distinguish between emergent and submergent coasts. Contrast the erosion problems faced on the Atlantic and Gulf coasts with those along the Pacific Coast.

KEY TERMS: emergent coast, submergent coast, estuary

- Coasts may be classified by their changes relative to sea level. Emergent coasts are sites of either land uplift or sea-level fall. Marine terraces are features of emergent coasts. Submergent coasts are sites of land subsidence or sea-level rise. One characteristic of submergent coasts is drowned river valleys called estuaries. In the United States, the Pacific coast is emergent, and the Atlantic and Gulf coasts are submergent.
- The Atlantic and Gulf coasts of the United States are lined in many places by barrier islands—dynamic expanses of sand that see a lot of change during storm events. Many of these low and narrow islands have also been prime sites for real estate development.
- The Pacific coast's big issue is the narrowing of beaches due to sediment starvation. Rivers that drain to the coast (bringing it sand) have been dammed, resulting in reservoirs that trap sand before it can make it to the coast. Narrower beaches offer less resistance to incoming waves, often leading to erosion of bluffs behind the beach.

? What term is applied to the masses of rock protruding from the water in this photo? How did they form? Is the location more likely along the Gulf Coast or the coast of California? Explain.

15.8 Stabilizing the Shore

Summarize the ways in which people deal with shoreline erosion problems.

KEY TERMS: hard stabilization, jetty, groin, breakwater, seawall, beach nourishment

- Hard stabilization refers to any structures built along the coastline to prevent movement of sand or shoreline erosion. Jetties project out from the coast, with the goal of keeping inlets open. Groins are also oriented perpendicular to the coast, but with the goal of slowing beach erosion by longshore currents. Breakwaters are parallel to the coast but located some distance offshore. Their goal is to blunt the force of incoming ocean waves, often to protect boats. Like breakwaters, seawalls are parallel to the coast, but they are built on the shoreline itself. Often the installation of hard stabilization results in increased erosion elsewhere.
- Beach nourishment is an expensive alternative to hard stabilization. Sand is pumped onto a beach from some other area, temporarily replenishing the sediment supply. Another possibility is relocating buildings away from high-risk areas and leaving the beach to be shaped by natural processes.

? Based on their position and orientation, identify the four kinds of hard stabilization illustrated in this diagram.

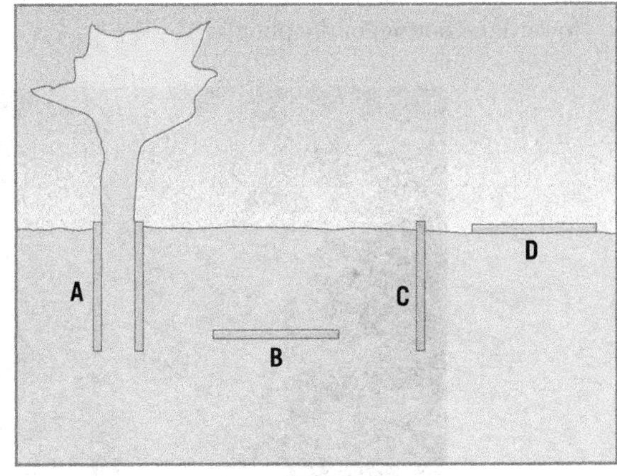

15.9 Tides

Explain the cause of tides and their monthly cycles. Describe the horizontal flow of water that accompanies the rise and fall of tides.

KEY TERMS: , tide, spring tide, neap tide, diurnal tidal pattern, semidiurnal tidal pattern, mixed tidal pattern, tidal current, tidal flat

- Tides are daily changes in ocean-surface elevation. They are caused by gravitational pull on ocean water by the Moon and, to a lesser extent, the Sun. When the Sun, Earth, and Moon all line up about every 2 weeks (full moon or new moon), the tides are most exaggerated. When a quarter moon is in the sky, the Moon is pulling on Earth's water at a right angle relative to the Sun, and the daily tidal range is minimized as the two forces partially counteract one another.
- Tides are strongly influenced by local conditions, including the shape of the local coastline and the depth of the ocean basin. Tidal patterns may be diurnal (one high tide per day), semidiurnal (two high tides per day), or mixed (similar to semidiurnal but with significant inequality between high tides).
- A flood current is the landward movement of water during the shift between low tide and high tide. When high tide transitions to low tide again, the movement of water away from the land is an ebb current. Ebb currents may expose tidal flats to the air. If a tide passes through an inlet, the current may carry sediment that gets deposited as a tidal delta.

? Would spring tides and neap tides occur on an Earth-like planet that had no moon? Explain.

GIVE IT SOME THOUGHT

1 In this chapter you learned that global winds are the force that drives surface-ocean currents. A glance at the accompanying map, however, shows a surface current that does not exactly coincide with the prevailing wind. Provide an explanation.

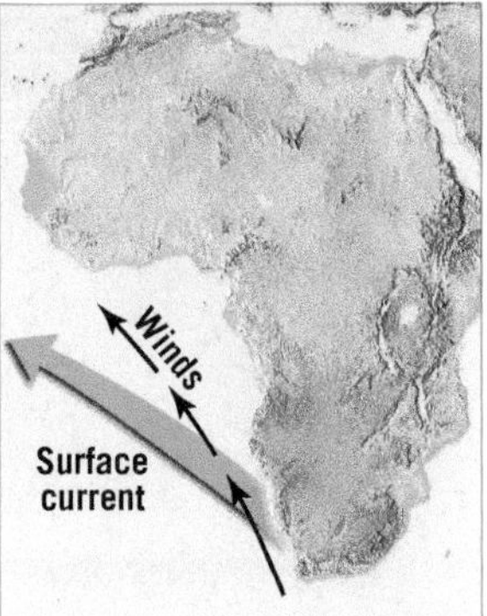

2 During a visit to the beach, you get in a small rubber raft and paddle out beyond the surf zone. Tiring, you stop and take a rest. Describe the movement of your raft during your rest. How does this movement differ, if at all, from what you would have experienced if you had stopped paddling while in the surf zone?

3 You and a friend set up an umbrella and chairs at a beach. Your friend then goes into the surf zone to play Frisbee with another person. Several minutes later, your friend looks back toward the beach and is surprised to see that she is no longer near where the umbrella and chairs were set up. Although she is still in the surf zone, she is 30 or 40 meters away from where she started. How would you explain to your friend why she moved along the shore?

4 A friend wants to purchase a vacation home on a barrier island. If consulted, what advice would you give your friend?

5 The force of gravity plays a critical role in creating ocean tides. The more massive an object, the stronger the pull of gravity. Explain why the Sun's influence is only about half that of the Moon, even though the Sun is much more massive than the Moon.

6 Examine the accompanying aerial photo that shows a portion of the New Jersey shoreline. What term is applied to the wall-like structures that extend into the water? What is their purpose? In what direction are beach drift and longshore currents moving sand: toward the top or toward the bottom of the photo?

Photo by University of Washington Libraries, Special Collections, John Shelton Collection, KCS2-26

7 This photo shows a portion of the Maine coast. The brown muddy area in the foreground is influenced by tidal currents. What term is applied to this muddy area? Name the type of tidal current this area will experience in the hours to come.

Marli Miller

EXAMINING THE EARTH SYSTEM

1 Palm trees in Scotland? Yes, in the 1850s and 1860s, amateur gardeners planted palm trees on the western shore of Scotland. The latitude here is 57° north, about the same as the northern portion of Labrador, across the Atlantic in Canada. Surprisingly, these exotic plants flourished. Suggest a possible explanation for how these palms can survive at such a high latitude.

South West Images Scotland/Alamy

2 If wastage (melting and calving) of the Greenland Ice Sheet were to dramatically increase, how would the salinity of the adjacent North Atlantic be affected? Speculate about how this might influence thermohaline circulation.

Paul Souders/Getty Images

3 In this chapter, the shoreline was described as a "dynamic interface." What is an interface? List and briefly describe some other interfaces in the Earth system. You need not confine yourself to examples from this chapter.

DATA ANALYSIS

Tides and Sea-Level Rise

Tides account for short-term changes in coastal water levels. Long-term changes in water levels are also observed as the level of the ocean itself changes.

ACTIVITIES

Go to the Tides and Currents page, at http://tidesandcurrents.noaa.gov. Click on Tides and Currents Map and zoom in to view Miami, Florida. Click on the Virginia Key station next to Miami. You can click on Help in the top right corner to learn how to use this map.

1 What time is the next tide? Is this a high or low tide?

2 What will be the height (in feet) of this tide?

3 How long does it take to go from high tide to low tide? High tide to high tide?

4 Is this a diurnal, semidiurnal, or mixed tidal pattern?

Examine high tide for stations in the New Jersey, Delaware, and Maryland region.

5 Notice that the time of high tide changes by location. What factors would control the time of high tide?

Click a station on the East Coast that allows you click on Sea Level Trends.

6 How many years are shown on the graph? How long is the data record for this station?

7 Use the Linear Mean Sea Level Trend line to determine the total sea level change over the data record.

Go back to the NOAA Tides and Currents page. Click on Sea Level Trends and zoom out to see the entire world.

8 In general, which regions are experiencing higher sea levels? Lower sea levels?

Go to the Sea Level Rise and Coastal Flooding Impacts site, at https://coast.noaa.gov/slr/. Zoom in to a state along the East Coast. Use the Sea Level Rise slider to raise the sea level.

9 What is the effect of rising sea level? What percentage of the coastline is affected?

Click on several Visualization Locations and use the slider to see the effect of rising sea level in each location.

10 Which locations did you visualize, and what effects did you see?

UNIT 6

Earth's Dynamic Atmosphere

16

The Atmosphere: Composition, Structure, and Temperature

FOCUS ON CONCEPTS

Each statement represents the primary learning objective for the corresponding major heading within the chapter. After you complete the chapter, you should be able to:

16.1 Distinguish between weather and climate and name the basic elements of weather and climate.

16.2 List the major gases composing Earth's atmosphere and identify the components that are most important to understanding weather and climate.

16.3 Interpret a graph that shows changes in air pressure from Earth's surface to the top of the atmosphere. Sketch and label a graph that shows atmospheric layers based on temperature.

16.4 Explain what causes the Sun angle and length of daylight to change during the year and describe how these changes produce the seasons.

16.5 Distinguish between heat and temperature. List and describe the three mechanisms of heat transfer.

16.6 Sketch and label a diagram that shows the paths taken by incoming solar radiation. Summarize the greenhouse effect.

16.7 Calculate five commonly used types of temperature data and interpret a map that depicts temperature data using isotherms.

16.8 Discuss the principal controls of temperature and use examples to describe their effects.

16.9 Interpret the patterns depicted on world maps of January and July temperatures.

This snow-capped mountain in Alaska's Denali National Park reminds us that temperatures decrease with an increase in altitude in the lowest layer of the atmosphere. Altitude is one of several factors that cause air temperatures to vary from place to place. (Photo by Arterra Picture Library/Alamy)

EARTH'S ATMOSPHERE IS UNIQUE. No other planet in our solar system has an atmosphere with the exact mixture of gases or the heat and moisture conditions necessary to sustain life as we know it. The gases that make up Earth's atmosphere and the controls to which they are subject are vital to our existence. In this chapter we begin our examination of the ocean of air in which we all must live. We will answer a number of basic questions: What is the composition of the atmosphere? Where does the atmosphere end, and where does outer space begin? What causes the seasons? How is air heated? What factors control temperature variations around the globe?

16.1 Focus on the Atmosphere

Distinguish between weather and climate and name the basic elements of weather and climate.

Weather influences our everyday activities, our jobs, and our health and comfort. Many of us pay little attention to the weather unless we are inconvenienced by it or it adds to our enjoyment outdoors. Nevertheless, there are few other aspects of our physical environment that affect our lives more than the phenomena we collectively call the weather.

Weather in the United States

The United States occupies an area that stretches from the tropics of Hawaii to beyond the Arctic Circle in Alaska. It has thousands of miles of coastline and extensive regions that are far from the influence of the ocean. Some landscapes are mountainous, and others are dominated by plains. It is a place where Pacific storms strike the West Coast, and the East is sometimes influenced by events in the Atlantic and the Gulf of Mexico. Those in the center of the country commonly experience weather events triggered when frigid southward-bound Canadian air masses clash with northward-moving ones from the Gulf of Mexico.

Stories about weather are a routine part of the daily news. Articles and reports about the effects of heat, cold, floods, drought, fog, snow, ice, and strong winds are commonplace. Of course, storms of all kinds are frequently front-page news (Figure 16.1). Beyond its direct impact on the lives of individuals, the weather has a strong effect on the world economy, influencing agriculture, energy use, water resources, transportation, and industry.

Weather clearly influences our lives a great deal. Yet, it is important to realize that people influence the atmosphere and its behavior as well (Figure 16.2). Significant political and scientific decisions involve these impacts. Important examples are air pollution control and the effects of human activities on global climate and the atmosphere's protective ozone layer. So there is a need for increased awareness and understanding of our atmosphere and its behavior.

Weather and Climate

Acted on by the combined effects of Earth's motions and energy from the Sun, our planet's formless and invisible envelope of air reacts by producing an infinite variety of

▼ Figure 16.1 **Memorable weather event** Few aspects of our physical environment influence our daily lives more than the weather. During the winter of 2014–2015, Boston experienced record-breaking snows that at times paralyzed the city. A total of 280.9 centimeters (110.6 inches) fell, the most in 120 years of record keeping. (Photo by Brian Snyder/Reuters)

▼ Figure 16.2 **People influence the atmosphere** China is plagued by air quality issues. Major contributors are coal-burning power plants. (Photo by AFP/Stringer/Getty Images)

weather, which in turn creates the basic pattern of global climates. Although not identical, weather and climate have much in common.

Weather is constantly changing, sometimes from day to day and sometimes from hour to hour. It is a term that refers to the state of the atmosphere at a given time and place. Whereas changes in the weather are continuous and sometimes seemingly erratic, it is nevertheless possible to arrive at a generalization of these variations. Such a description of aggregate weather conditions is termed **climate**. It is based on observations that have been accumulated over many years. Climate is often defined simply as "average weather," but this is an inadequate definition. To more accurately portray the character of an area, variations and extremes must also be included, as well as the probabilities that such departures will take place. For example, farmers need to know the average rainfall during the growing season, and they also need to know the frequency of extremely wet and extremely dry years. Thus, climate is the sum of all statistical weather information that helps describe a place or region.

Suppose you were planning a vacation in an unfamiliar place. You would probably want to know what kind of weather to expect. Such information would help you select clothes to pack and could influence decisions regarding activities you might engage in during your stay. Unfortunately, weather forecasts that go beyond a few days are not very dependable. Therefore, you might ask someone who is familiar with the area about what kind of weather to expect. "Are thunderstorms common?" "Does it get cold at night?" "Are the afternoons sunny?" What you are seeking is information about the climate, the conditions that are typical for that place. Another useful source of such information is the great variety of climate tables, maps, and graphs that are available. For example, the graph in Figure 16.3 shows average daily high and low temperatures for each month, as well as extremes for New York City.

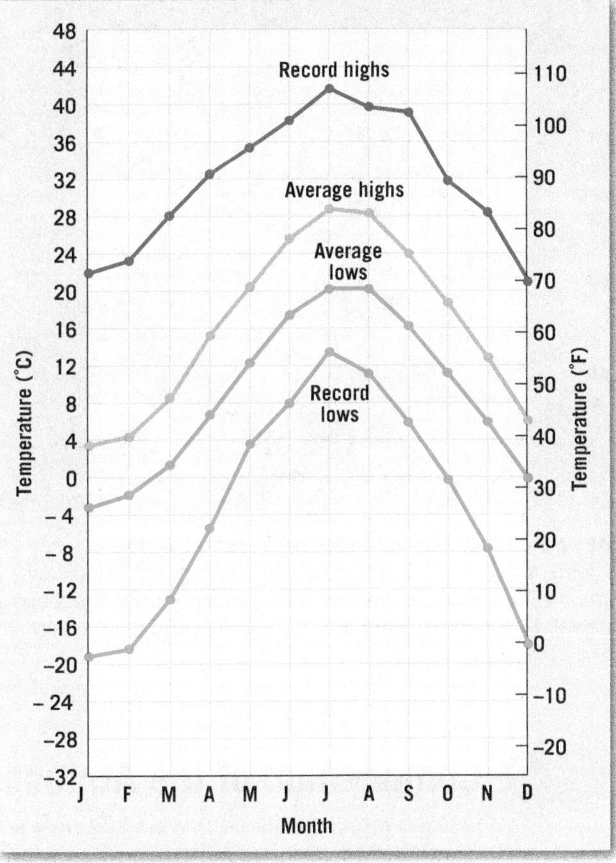

Figure 16.3 Graphs can display climate data This graph shows daily temperature data for New York City. In addition to the average daily maximum and minimum temperatures for each month, extremes are also shown. As this graph shows, there can be significant departures from the average.

Such information could no doubt help as you planned your trip. But it is important to realize that *climate data cannot predict the weather.* Although a place may usually (climatically) be warm, sunny, and dry during the time of your planned vacation, you may actually experience cool, overcast, and rainy weather. There is a well-known saying that summarizes this idea: "Climate is what you

EYE ON EARTH 16.1

This is a scene on a summer day in southern Utah's Sinbad Country. Hebes Mountain is in the center of the image.

QUESTION 1 *Write two brief statements about this place—one that clearly relates to weather and another that might be part of a description of climate.*

QUESTION 2 *Explain the reasoning associated with each statement.*

expect, but weather is what you get." Figure 16.4 provides an example.

Weather and climate are expressed in terms of the same basic **elements**—quantities or properties that are measured regularly. The most important elements are (1) air temperature, (2) humidity, (3) type and amount of cloudiness, (4) type and amount of precipitation, (5) air pressure, and (6) the speed and direction of the wind. These elements are the major variables from which weather patterns and climate types are deciphered. Although you will study these elements separately at first, keep in mind that they are very much interrelated. A change in any one of the elements will often bring about changes in the others.

▲ Figure 16.4 **November tornado** Although "tornado season" in central Illinois is considered to be April through June, a devastating tornado struck Washington, Illinois, on November 17, 2013. With maximum winds of 306 kilometers (190 miles) per hour, the storm caused complete destruction of well-built homes. More than 630 homes and 2500 vehicles were destroyed. (Photo by Brian Kersey/UPI/Newscom)

CONCEPT CHECKS 16.1

1. Distinguish between weather and climate.
2. Write two brief statements about your current location: one that relates to weather and one that relates to climate.
3. What is an *element*?
4. List the basic elements of weather and climate.

16.2 Composition of the Atmosphere

List the major gases composing Earth's atmosphere and identify the components that are most important to understanding weather and climate.

Sometimes the term *air* is used as if it were a specific gas, but it is not. Rather, **air** is a *mixture* of many discrete gases, each with its own physical properties, in which varying quantities of tiny solid and liquid particles are suspended.

Major Components

The composition of air is not constant; it varies from time to time and from place to place. However, if the water vapor, dust, and other variable components were removed from the atmosphere, we would find that its makeup is very stable worldwide up to an altitude of about 80 kilometers (50 miles).

As you can see in Figure 16.5, two gases—nitrogen and oxygen—make up 99 percent of the volume of clean, dry air. Although these gases are the most plentiful components of air and are of great significance to life on Earth, they are of minor importance in affecting weather phenomena. The remaining 1 percent of dry air is mostly the inert gas argon (0.93 percent) plus tiny quantities of a number of other gases.

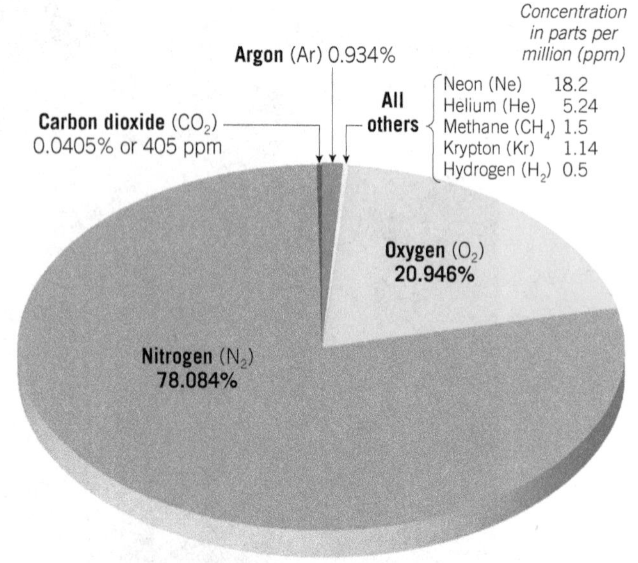

► Figure 16.5 **Composition of the atmosphere** Proportional volume of gases composing dry air. Nitrogen and oxygen clearly dominate.

Carbon Dioxide (CO_2)

Carbon dioxide, although present in only minute amounts (0.0405 percent, or 405 parts per million [ppm]), is nevertheless an important constituent of air. Carbon dioxide is of great interest to meteorologists because it is an efficient absorber of energy emitted by Earth and thus influences the heating of the atmosphere. Although the proportion of carbon dioxide in the atmosphere is relatively uniform, its percentage has been rising steadily for 200 years. Figure 16.6 is a graph that shows the growth in atmospheric CO_2 since 1958. Much of this rise is attributed to the burning of ever-increasing quantities of fossil fuels, such as coal and oil. Some of this additional carbon dioxide is absorbed by the ocean or is used by plants, but about 45 percent remains in the air. Estimates project that by sometime n the second half of the twenty-first century, CO_2 levels will be twice as high as the pre-industrial level.

Most atmospheric scientists agree that increased carbon dioxide concentrations have contributed to a warming of Earth's atmosphere over the past several decades and will continue to do so in the decades to come. The magnitude of such temperature changes is uncertain and depends partly on the quantities of CO_2 contributed by human activities in the years ahead. The role of carbon dioxide in the atmosphere and its possible effects on climate are examined in Chapter 20.

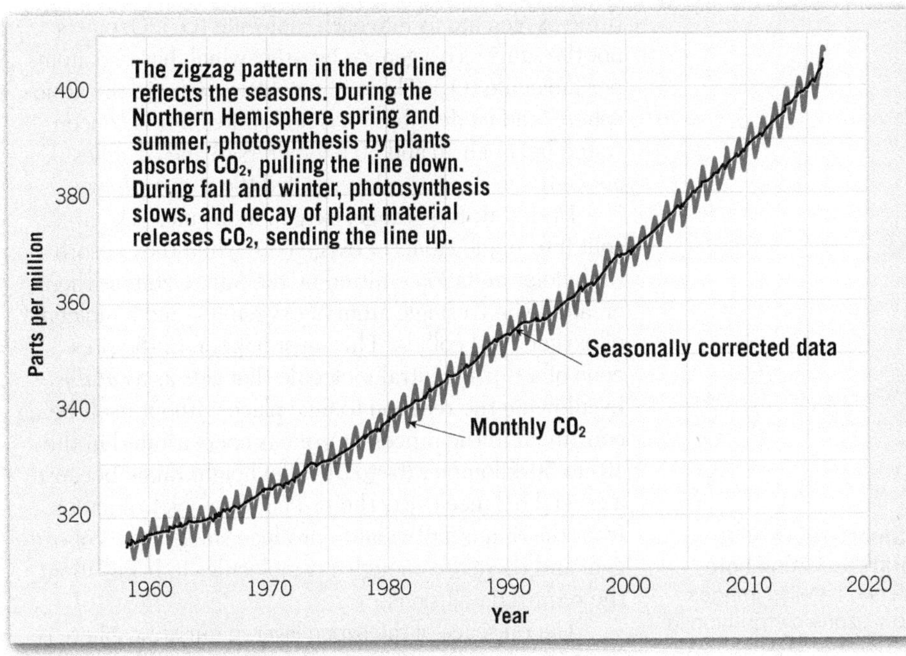
The zigzag pattern in the red line reflects the seasons. During the Northern Hemisphere spring and summer, photosynthesis by plants absorbs CO_2, pulling the line down. During fall and winter, photosynthesis slows, and decay of plant material releases CO_2, sending the line up.
Seasonally corrected data
Monthly CO_2
Parts per million
400
380
360
340
320
1960
1970
1980
1990
2000
2010
2020
Year

SmartFigure 16.6 **Monthly CO_2 concentrations** Atmospheric CO_2 has been measured at Mauna Loa Observatory, Hawaii, since 1958. There has been a consistent increase since monitoring began. This graphic portrayal is known as the *Keeling Curve*, in honor of the scientist who originated the measurements. (NOAA)

TUTORIAL
http://goo.gl/Y479oq

Variable Components

Air also includes many gases and particles whose quantities vary significantly in different times and places. Important examples include water vapor, dust particles, and ozone. Although usually present in small percentages, they can have significant effects on weather and climate.

Water Vapor You are probably familiar with the term *humidity* from watching weather reports on TV. Humidity is a reference to water vapor in the air. As you will learn in Chapter 17, there are several ways to express humidity. The amount of water vapor in the air varies considerably, from practically none at all up to about 4 percent by volume. Why is such a small fraction of the atmosphere so significant? The fact that water vapor is the source of all clouds and precipitation would be enough to explain its importance. However, water vapor has other roles. Like carbon dioxide, water vapor absorbs heat given off by Earth as well as some solar energy. It is therefore important when we examine the heating of the atmosphere.

When water changes from one state to another (see Figure 17.2 on page 527), it absorbs or releases heat. This energy is termed *latent heat*, which means "hidden heat." As we shall see in later chapters, water vapor in the atmosphere transports this latent heat from one region to another, and it is the energy source that helps drive many storms.

Aerosols The movements of the atmosphere are sufficient to keep a large quantity of solid and liquid particles suspended within it. Although visible dust sometimes clouds the sky, these relatively large particles are too heavy to stay in the air very long. Many other particles are microscopic and remain suspended for considerable periods of time. They may originate from many sources, both natural and human made, and include sea salts from breaking waves, fine soil blown into the air, smoke and soot from fires, pollen and microorganisms lifted by the wind, ash and dust from volcanic eruptions, and more (Figure 16.7). Collectively, these tiny solid and liquid particles are called **aerosols**.

From a meteorological standpoint, these tiny, often invisible particles can be significant. First, many act as surfaces on which water vapor can condense, an important function in the formation of clouds and fog. Second, aerosols can absorb, reflect, and scatter incoming solar radiation. Thus, when an air-pollution episode is occurring or when ash fills the sky following a volcanic eruption, the amount of sunlight reaching Earth's surface can be measurably reduced. Finally, aerosols contribute to an optical phenomenon we have all observed—the varied hues of red and orange at sunrise and sunset.

Ozone Another important component of the atmosphere is **ozone**, which is a form of oxygen that combines

Dust storm
Air pollution

SmartFigure 16.7 **Aerosols** This satellite image shows two examples of aerosols. First, a dust storm is blowing across northeastern China toward the Korean Peninsula. Second, a dense haze toward the south (bottom center) is human-generated air pollution. (NASA)

VIDEO
http://goo.gl/t8anlg

three oxygen atoms into each molecule (O_3). Ozone is not the same as oxygen we breathe, which has two atoms per molecule (O_2). There is very little ozone in the atmosphere, and its distribution is not uniform. It is concentrated between 10 and 50 kilometers (6 and 31 miles) above the surface, in a layer called the *stratosphere*.

In this altitude range, oxygen molecules (O_2) are split into single atoms of oxygen (O) when they absorb ultraviolet radiation emitted by the Sun. Ozone is then created when a single atom of oxygen (O) and a molecule of oxygen (O_2) collide. This must happen in the presence of a third, neutral molecule that acts as a *catalyst* by allowing the reaction to take place without itself being consumed in the process. Ozone is concentrated in the 10- to 50-kilometer (6- to 31-mile) height range because a crucial balance exists there: The ultraviolet radiation from the Sun is sufficient to produce single atoms of oxygen, and there are enough gas molecules to bring about the required collisions.

The presence of the ozone layer in our atmosphere is crucial to those of us who dwell on Earth. The reason is that ozone absorbs much of the potentially harmful ultraviolet (UV) radiation from the Sun. If ozone did not filter a great deal of the ultraviolet radiation, and if the Sun's UV rays reached the surface of Earth undiminished, our planet would be uninhabitable for most life as we know it. Thus, anything that reduces the atmosphere's naturally occurring ozone could affect the well-being of life on Earth. Just such a problem exists and is described in the next section.

Although naturally occurring ozone in the stratosphere is critical to life on Earth, ozone is regarded as a pollutant when produced at ground level because it can damage vegetation and can be harmful to human health. Ozone is a major component in a noxious mixture of gases and particles called *photochemical smog*, which forms when strong sunlight triggers reactions among pollutants from car exhaust and industrial sources.

SmartFigure 16.8
Antarctic ozone hole The two satellite images show ozone distribution in the Southern Hemisphere on the days in September 1979 and October 2015 when the ozone hole was largest. The dark blue shades over Antarctica correspond to the region with the sparsest ozone. The ozone hole is not technically a "hole" where no ozone is present but is actually a region of exceptionally depleted ozone in the stratosphere over the Antarctic that occurs in the spring. (NASA)

TUTORIAL
https://goo.gl/Y9RVfo

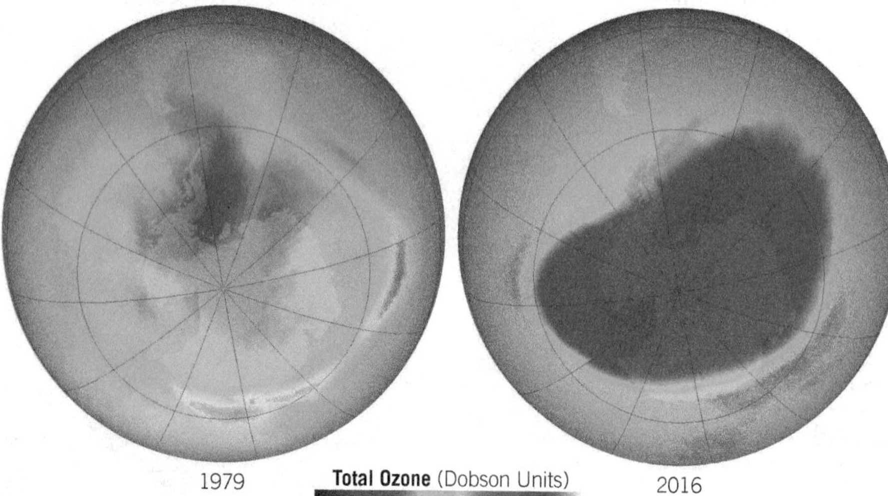
The ozone hole on October 1, 2016, reached 23 million km²
(8.9 million mi²)—an area nearly as large as North America.
1979
2016
Total Ozone (Dobson Units)
0 100 200 300 400 500 600 700

Ozone Depletion: A Global Issue

Although stratospheric ozone is 10 to 50 kilometers (6 to 31 miles) above Earth's surface, it is vulnerable to human activities. Chemicals produced by people are breaking up ozone molecules in the stratosphere, weakening our shield against UV rays. This loss of ozone is a serious global-scale environmental problem. Measurements over the past three decades confirm that ozone depletion is occurring worldwide and is especially pronounced above Earth's poles You can see this effect over the South Pole in Figure 16.8.

Over the past 70 years, people have unintentionally placed the ozone layer in jeopardy by polluting the atmosphere. The most significant of the offending chemicals are known as *chlorofluorocarbons* (CFCs for short). Over the decades, many uses were developed for CFCs: They were used as coolants for air-conditioning and refrigeration equipment, cleaning solvents for electronic components, and propellants for aerosol sprays, and they were used in the production of certain plastic foams.

Because CFCs are practically inert (that is, not chemically active) in the lower atmosphere, some of these gases gradually make their way to the ozone layer, where sunlight separates the chemicals into their constituent atoms. The chlorine atoms released this way break up some of the ozone molecules.

Because ozone filters out most of the UV radiation from the Sun, a decrease in its concentration permits more of these harmful wavelengths to reach Earth's surface. The most serious threat to human health is an increased risk of skin cancer. An increase in damaging UV radiation also can impair the human immune system as well as promote cataracts, a clouding of the eye lens that reduces vision and may cause blindness if not treated.

In response to this problem, an international agreement known as the *Montreal Protocol* was developed under the sponsorship of the United Nations to eliminate the production and use of CFCs. More than 190 nations eventually ratified the treaty. Although strong action was taken, CFC levels in the atmosphere will not drop rapidly. Once in the atmosphere, CFC molecules can take many years to reach the ozone layer, and once there, they can remain active for decades. Not until between 2060 and 2075 is the abundance of ozone-depleting gases projected to fall to values that existed before the ozone hole began to form in the 1980s.

CONCEPT CHECKS 16.2

1. Is *air* a specific gas? Explain.
2. What are the two major components of clean, dry air? What proportion does each represent?
3. Why are water vapor and aerosols important constituents of Earth's atmosphere?
4. What is ozone? Why is ozone important to life on Earth? What are CFCs, and what is their connection to the ozone problem?

16.3 Vertical Structure of the Atmosphere

Interpret a graph that shows changes in air pressure from Earth's surface to the top of the atmosphere. Sketch and label a graph that shows atmospheric layers based on temperature.

To say that the atmosphere begins at Earth's surface and extends upward is obvious. However, where does the atmosphere end, and where does outer space begin? There is no sharp boundary; the atmosphere rapidly thins as you travel away from Earth, until there are too few gas molecules to detect.

Pressure Changes

To understand the vertical extent of the atmosphere, let us examine the changes in atmospheric pressure with height. Atmospheric pressure is simply the weight of the air above. At sea level, the average pressure is slightly more than 1000 millibars (mb). This corresponds to a weight of slightly more than 1 kilogram per square centimeter (14.7 pounds per square inch). The pressure at higher altitudes is less (Figure 16.9).

One-half of the atmosphere lies below an altitude of 5.6 kilometers (3.5 miles). At about 16 kilometers (10 miles), 90 percent of the atmosphere has been traversed, and above 100 kilometers (62 miles), only 0.00003 percent of all the gases making up the atmosphere remains. Even so, traces of our atmosphere extend far beyond this altitude, gradually merging with the emptiness of space.

Figure 16.9 Air pressure changes with altitude The rate of pressure decrease with increase in altitude is not constant. Pressure decreases rapidly near Earth's surface and more gradually at greater heights. Put another way, the graph shows that the vast bulk of the gases making up the atmosphere are very near Earth's surface and that the gases gradually merge with the emptiness of space.

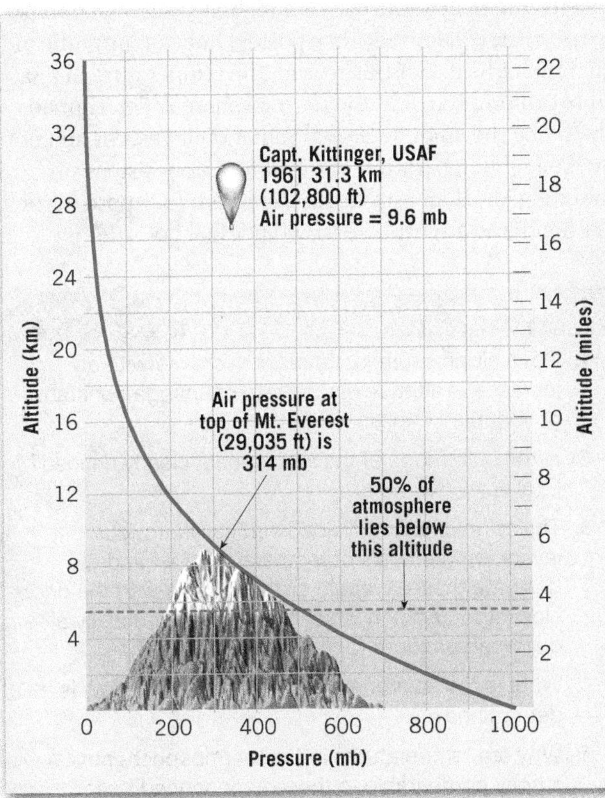

Temperature Changes

By the early twentieth century, much had been learned about the lower atmosphere. The upper atmosphere was partly known from indirect methods. Data from balloons and kites showed that near Earth's surface, air temperature drops with increasing height. This phenomenon is felt by anyone who has climbed a high mountain and is obvious in pictures of snow-capped mountaintops rising above snow-free lowlands, as in the chapter-opening photo. We divide the atmosphere vertically into four layers on the basis of temperature (Figure 16.10).

Troposphere The lowermost layer, in which we live, where temperature decreases with an increase in altitude, is the **troposphere**. The term literally means the region

Figure 16.10 Thermal structure of the atmosphere
Earth's atmosphere is traditionally divided into four layers, based on temperature.

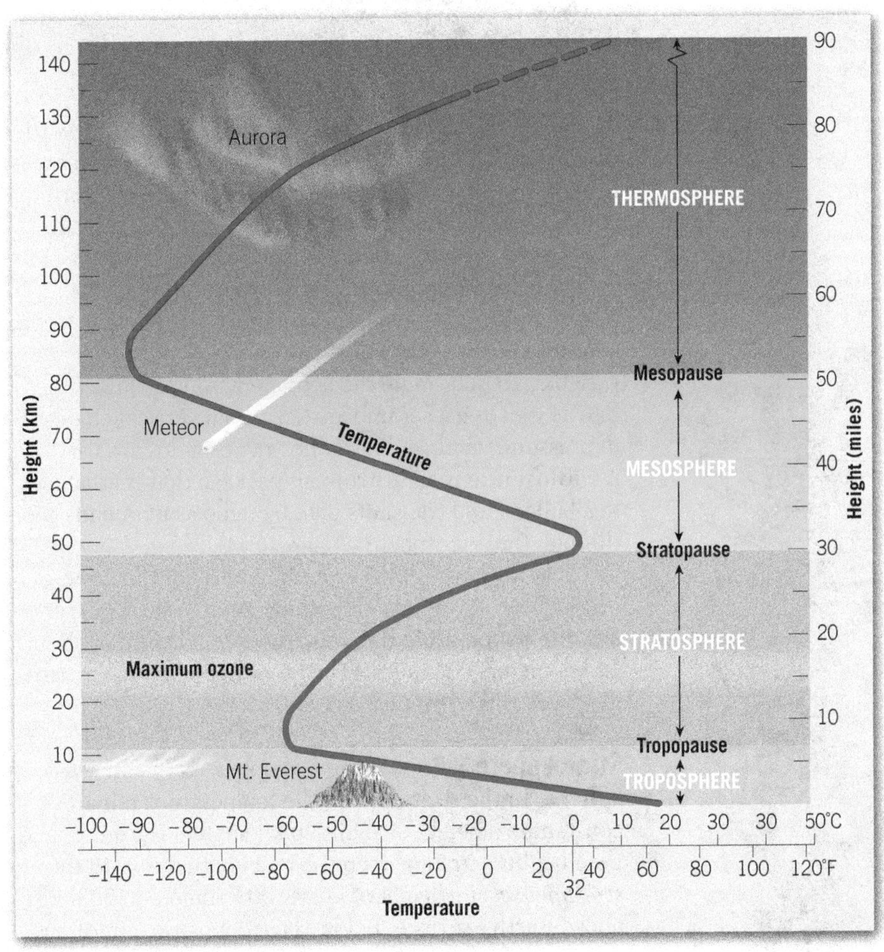

▶ Figure 16.11
Radiosonde A radiosonde is a lightweight package of instruments that is carried aloft by a small weather balloon. It transmits data on vertical changes in temperature, pressure, and humidity in the troposphere. The troposphere is where practically all weather phenomena occur, so it is very important to have frequent measurements.
(Photo by David R. Frazier/ Danita Delimont/Newscom)

where air "turns over," a reference to the appreciable vertical mixing of air in this lowermost zone. The troposphere is the chief focus of meteorologists because it is in this layer that essentially all important weather phenomena occur.

The temperature decrease in the troposphere is called the **environmental lapse rate**. Although its average value is 6.5°C per kilometer (3.5°F per 1000 feet), a figure known as the *normal lapse rate*, its value is variable. To determine the actual environmental lapse rate as well as gather information about vertical changes in pressure, wind, and humidity, radiosondes are used. A **radiosonde** is an instrument package that is attached to a balloon and transmits data by radio as it ascends through the atmosphere (Figure 16.11).

The thickness of the troposphere is not the same everywhere; it varies with latitude and season. On average, the temperature drop continues to a height of about 12 kilometers (7.4 miles). The outer boundary of the troposphere is the *tropopause*.

Stratosphere Beyond the tropopause is the **stratosphere.** In the stratosphere, the temperature remains constant to a height of about 20 kilometers (12 miles) and then begins a gradual increase that continues until the *stratopause*, at a height of nearly 50 kilometers (30 miles) above Earth's surface. Below the tropopause, atmospheric properties such as temperature and humidity are readily transferred by large-scale turbulence and mixing. Above the tropopause, in the stratosphere, they are not. Temperatures increase in the stratosphere because it is in this layer that the atmosphere's ozone is concentrated. Recall that ozone absorbs ultraviolet radiation from the Sun. As a consequence, the stratosphere is heated.

Mesosphere In the third layer, the **mesosphere**, temperatures again decrease with height until, at the *mesopause*—approximately 80 kilometers (50 miles) above the surface—the temperature approaches −90°C (−130°F). The coldest temperatures anywhere in the atmosphere occur at the mesopause. Because accessibility is difficult, the mesosphere is one of the least explored regions of the atmosphere. It cannot be reached by the highest research balloons, nor is it accessible to the lowest orbiting satellites. Recent technological developments are just beginning to fill this knowledge gap.

Thermosphere The fourth layer extends outward from the mesopause and has no well-defined upper limit. It is the **thermosphere**, a layer that contains only a tiny fraction of the atmosphere's mass. In the extremely rarefied air of this outermost layer, temperatures again increase, due to the absorption of very short-wave, high-energy solar radiation by atoms of oxygen and nitrogen.

Temperatures rise to extremely high levels of more than 1000°C (1800°F) in the thermosphere. But such temperatures are not comparable to those experienced near Earth's surface. Temperature is defined in terms of the average speed at which molecules move. Because the gases of the thermosphere are moving at very high speeds, the temperature is very high. But the gases are so sparse that, collectively, they possess only an insignificant quantity of heat. For this reason, the temperature of a satellite orbiting Earth in the thermosphere is determined chiefly by the amount of solar radiation it absorbs and not by the high temperature of the almost nonexistent surrounding air. If an astronaut inside were to expose his or her hand, the atmosphere would not feel hot.

CONCEPT CHECKS 16.3

1. Does air pressure increase or decrease with an increase in altitude? Is the rate of change constant or variable? Explain.
2. Is the outer edge of the atmosphere clearly defined? Explain.
3. The atmosphere is divided vertically into four layers, on the basis of temperature. List and describe these layers in order, starting with the one closest to Earth. In which layer does practically all our weather occur?
4. What is the environmental lapse rate, and how is it determined?
5. Why are temperatures in the thermosphere not strictly comparable to those experienced near Earth's surface?

16.4 Earth–Sun Relationships

Explain what causes the Sun angle and length of daylight to change during the year and describe how these changes produce the seasons.

Nearly all the energy that drives Earth's variable weather and climate comes from the Sun. Earth intercepts only a minute percentage of the energy given off by the Sun—less than 1 two-billionth. This may seem to be an insignificant amount, until we realize that it is several hundred thousand times the electrical-generating capacity of the United States.

Solar energy is not distributed evenly over Earth's land–sea surface. The amount of energy received varies with latitude, time of day, and season of the year. Contrasting images of polar bears on ice rafts and palm trees along a remote tropical beach serve to illustrate the extremes. It is the unequal heating of Earth that creates winds and drives the ocean's currents. These movements of air and water, in turn, transport heat from the tropics toward the poles, in an unending attempt to balance energy inequalities. The consequences of these processes are the phenomena we call weather.

If the Sun were "turned off," global winds and ocean currents would quickly cease. Yet as long as the Sun shines, the winds *will* blow and weather *will* persist. So to understand how the atmosphere's dynamic weather machine works, we must first know why different latitudes receive varying quantities of solar energy and why the amount of solar energy changes to produce the seasons. As you will see, the variations in solar heating are caused by the motions of Earth relative to the Sun and by variations in Earth's land–sea surface.

Earth's Motions

Earth has two principal motions—its **rotation** about its axis and its *orbital motion* around the Sun. The axis is an imaginary line running through the poles. Our planet rotates on its axis once every 24 hours, producing the daily cycle of daylight and darkness. At any moment, half of Earth is experiencing daylight and the other half darkness. The line separating the dark half of Earth from the lighted half is called the **circle of illumination**.

Each year, Earth makes one slightly elliptical orbit around the Sun. The distance between Earth and Sun averages about 150 million kilometers (93 million miles). Because Earth's orbit is not perfectly circular, however, the distance varies during the course of a year. Each year, on about January 3, our planet is about 147.3 million kilometers (91.5 million miles) from the Sun, which is closer than at any other time—a position known as *perihelion*. About 6 months later, on July 4, Earth is about 152 million kilometers (94.5 million miles) from the Sun, farther away than at any other time—a position called *aphelion*. Although Earth is closest to the Sun and receives up to 7 percent more energy in January than in July, this difference plays only a minor role in producing seasonal temperature variations, as evidenced by the fact that Earth is closest to the Sun during the Northern Hemisphere winter.

What Causes the Seasons?

If variations in the distance between the Sun and Earth are not responsible for seasonal temperature changes, what is? The gradual but significant change in the length of daylight certainly accounts for some of the difference we notice between summer and winter. Furthermore, a gradual change in the angle (altitude) of the Sun above the horizon is also a contributing factor (Figure 16.12). For example, someone living in Chicago, Illinois, experiences the noon Sun highest in the sky in late June. But as summer gives way to autumn, the noon Sun appears lower in the sky, and sunset occurs earlier each evening.

The seasonal variation in the angle of the Sun above the horizon affects the amount of energy received at Earth's surface in two ways. First, when the Sun is directly overhead (at a 90-degree angle), the solar rays are most concentrated and thus most intense. The lower the angle, the more spread out and less intense is the solar radiation that reaches the

SmartFigure 16.12
The changing Sun angle Daily paths of the Sun for a place located at 40° north latitude for **A.** summer solstice, **B.** spring or fall equinox, and **C.** winter solstice. As we move from summer to winter, the angle of the noon Sun decreases from 73½ to 26½ degrees—a difference of 47 degrees. Notice also how the location of sunrise (east) and sunset (west) changes during a year.

ANIMATION
https://goo.gl/PbzOgf

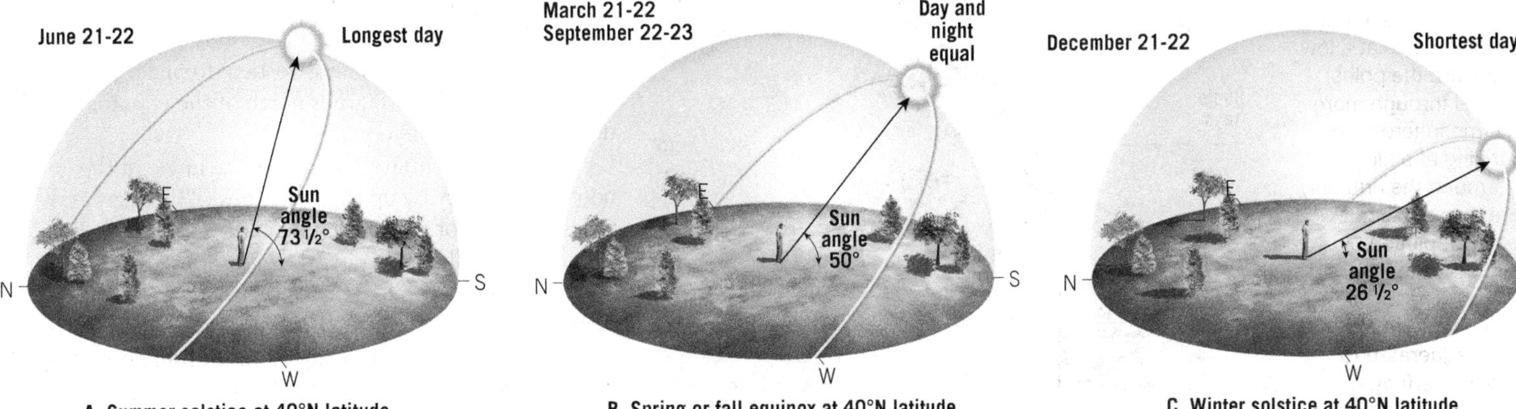

A. Summer solstice at 40°N latitude

B. Spring or fall equinox at 40°N latitude

C. Winter solstice at 40°N latitude

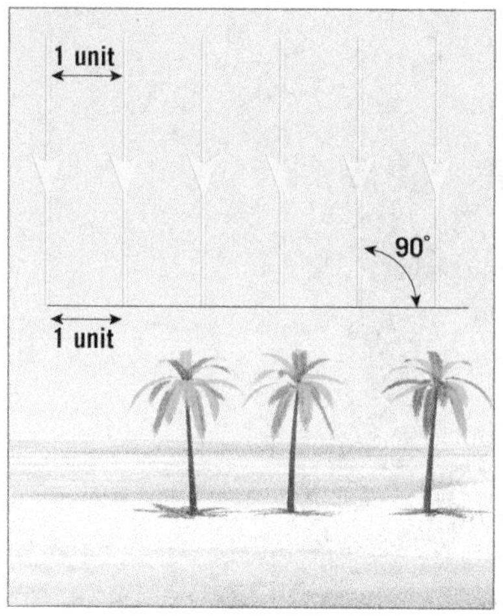

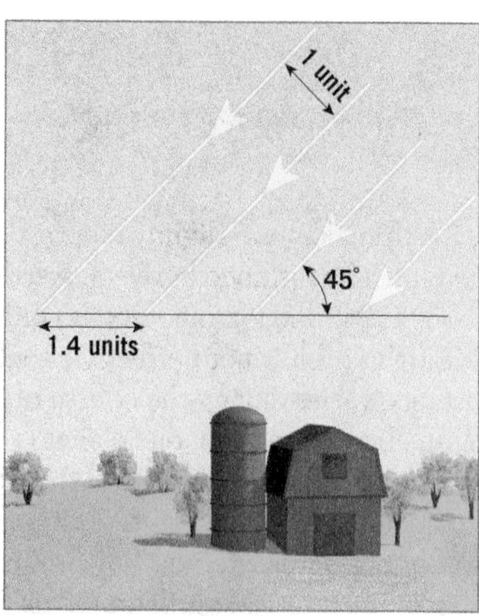

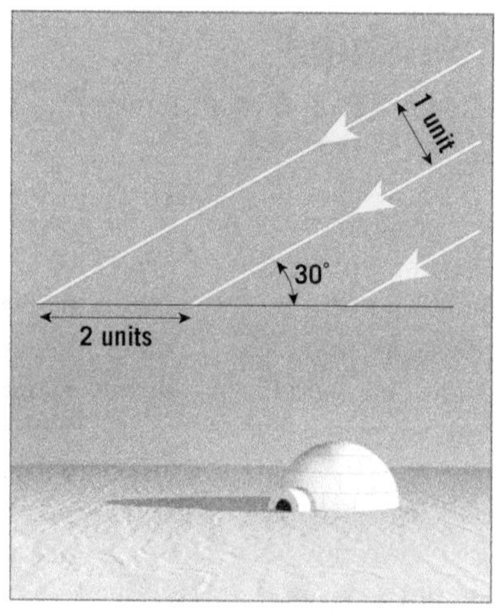

A. B.

SmartFigure 16.13 **Changes in the Sun's angle cause variations in the amount of solar energy that reaches Earth's surface A.** The higher the angle, the more intense the solar radiation reaching the surface. **B.** If a flashlight beam strikes a surface at a 90-degree angle, a small intense spot is produced; however, if it strikes at any other angle, the area illuminated is larger but noticeably dimmer.

ANIMATION
https://goo.gl/3eiSHB

surface (Figure 16.13A). To illustrate this principle, hold a flashlight at a right angle to a surface and then change the angle (Figure 16.13B).

Second, but of lesser importance, the angle of the Sun determines the path solar rays take as they pass through the atmosphere (Figure 16.14). When the Sun is directly overhead, the rays strike the atmosphere at a 90-degree angle and travel the shortest possible route to the surface. This distance is referred to as *1 atmosphere*. However, rays entering at a 30-degree angle travel through twice this distance before reaching the surface, while rays at a 5-degree angle travel through a distance roughly equal to the thickness of 11 atmospheres. The longer the path, the greater the chance that sunlight will be dispersed by the atmosphere, which reduces the intensity at the surface. These conditions account for the fact that we cannot look directly at the midday Sun, but we can enjoy gazing at a sunset.

It is important to remember that Earth's shape is spherical. On any given day, the only places that will receive vertical (90-degree) rays from the Sun are located along one particular line of latitude. As we move either north or south of this location, the Sun's rays strike at ever-decreasing angles. The nearer a place is to the latitude receiving vertical rays of the Sun, the higher will be its noon Sun and the more concentrated will be the radiation it receives (see Figure 16.14).

Figure 16.14 **The Sun angle affects the path of sunlight through the atmosphere** Notice that rays striking Earth at a low angle (toward the poles) must travel through more of the atmosphere than rays striking at a high angle (around the equator) and are thus subject to greater depletion by reflection, scattering, and absorption. The thickness of the atmosphere is greatly exaggerated to illustrate this effect.

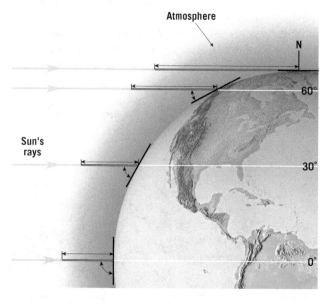

Earth's Orientation

What causes fluctuations in Sun angle and length of daylight that occur during the course of a year? Variations occur because Earth's orientation to the Sun continually changes as it travels along its orbit. Earth's axis (the imaginary line through the poles around which Earth rotates) is not perpendicular to the plane of its orbit around the Sun. Instead, it is tilted 23½ degrees from the perpendicular, as shown in Figure 16.15. This is called the **inclination of the axis**. If the axis were not inclined, Earth would lack seasons. Because the axis remains pointed in the same direction (toward the North Star), the orientation of Earth's axis to the Sun's rays is constantly changing (see Figure 16.15).

For example, on one day in June each year, the axis is such that the Northern Hemisphere is "leaning" 23½ degrees *toward* the Sun (see Figure 16.15, left). Six months later, in December, when Earth has moved to the opposite side of its orbit, the Northern Hemisphere "leans" 23½ degrees *away from* the Sun (see Figure 16.15, right). On days between these extremes, Earth's axis is leaning at amounts less than 23½ degrees to the rays of the Sun. This change in orientation causes the spot where the Sun's rays are vertical to make an annual migration from 23½ degrees north of the equator to 23½ degrees south of the equator.

In turn, this migration causes the angle of the noon Sun to vary by up to 47 degrees (23½ degrees + 23½ degrees) for many locations during the year. For example, a midlatitude city like New York (about 40° north latitude) has a maximum noon Sun angle of 73½ degrees when the Sun's vertical rays reach their farthest northward location in June and a minimum noon Sun angle of 26½ degrees 6 months later.

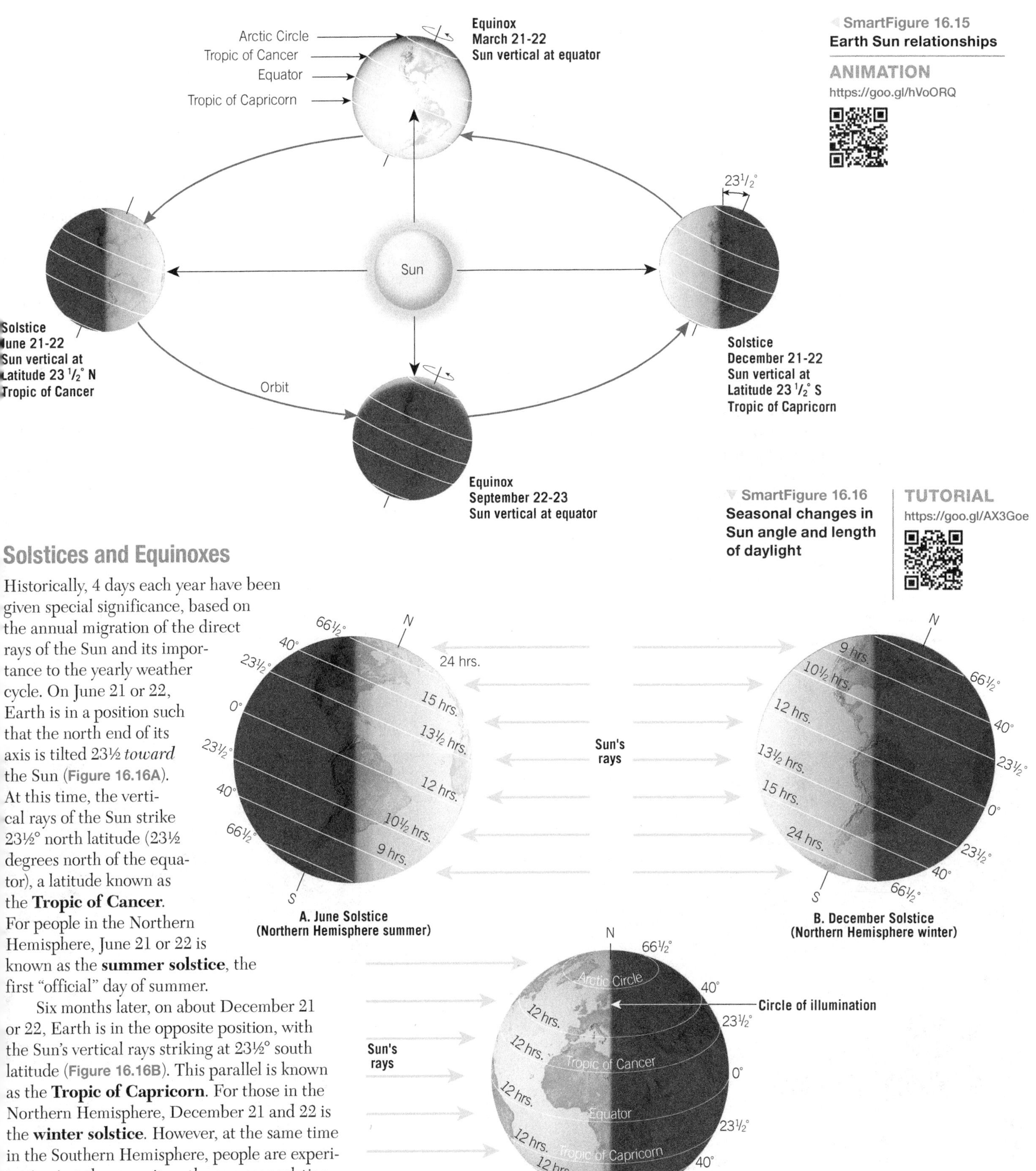

SmartFigure 16.15
Earth Sun relationships

ANIMATION
https://goo.gl/hVoORQ

SmartFigure 16.16
Seasonal changes in Sun angle and length of daylight

TUTORIAL
https://goo.gl/AX3Goe

Solstices and Equinoxes

Historically, 4 days each year have been given special significance, based on the annual migration of the direct rays of the Sun and its importance to the yearly weather cycle. On June 21 or 22, Earth is in a position such that the north end of its axis is tilted 23½ *toward* the Sun (Figure 16.16A). At this time, the vertical rays of the Sun strike 23½° north latitude (23½ degrees north of the equator), a latitude known as the **Tropic of Cancer**. For people in the Northern Hemisphere, June 21 or 22 is known as the **summer solstice**, the first "official" day of summer.

Six months later, on about December 21 or 22, Earth is in the opposite position, with the Sun's vertical rays striking at 23½° south latitude (Figure 16.16B). This parallel is known as the **Tropic of Capricorn**. For those in the Northern Hemisphere, December 21 and 22 is the **winter solstice**. However, at the same time in the Southern Hemisphere, people are experiencing just the opposite—the summer solstice.

Midway between the solstices are the equinoxes. September 22 or 23 is the date of the

Table 16.1 Length of Daylight

Latitude (degrees)	Summer Solstice	Winter Solstice	Equinoxes
0	12 h	12 h	12 h
10	12 h 35 min	11 h 25 min	12 h
20	13 h 12 min	10 h 48 min	12 h
30	13 h 56 min	10 h 04 min	12 h
40	14 h 52 min	9 h 08 min	12 h
50	16 h 18 min	7 h 42 min	12 h
60	18 h 27 min	5 h 33 min	12 h
70	24 h (for 2 mo)	0 h 00 min	12 h
80	24 h (for 4 mo)	0 h 00 min	12 h
90	24 h (for 6 mo)	0 h 00 min	12 h

autumnal (fall) equinox in the Northern Hemisphere, and March 21 or 22 is the date of the **spring equinox**. On these dates, the vertical rays of the Sun strike the equator (0° latitude) because Earth is in such a position in its orbit that the axis is tilted neither toward nor away from the Sun (**Figure 16.16C**).

The length of daylight versus darkness is also determined by Earth's position in orbit. The length of daylight on June 21 or 22, the summer solstice in the Northern Hemisphere, is greater than the length of night. This fact can be established from Figure 16.16A by comparing the fraction of a given latitude that is on the "day" side of the circle of illumination with the fraction on the "night" side. The opposite is true for the winter solstice, when the nights are longer than the days. Again for comparison, let us consider New York City, which has about 15 hours of daylight on June 21 and only about 9 hours on December 21. (You can see this in Figure 16.16 and **Table 16.1**.) Also note from Table 16.1 that the farther north of the equator you are on June 21, the longer the period of daylight. When you reach the Arctic Circle (66½° north latitude), the length of daylight is 24 hours. This is the land of the "midnight Sun," which does not set for about 6 months at the North Pole (**Figure 16.17**).

Figure 16.18 summarizes the characteristics of the solstices and equinoxes for a location in the Northern Hemisphere. An examination of Figure 16.18 should make it apparent why a midlatitude location is warmest in the summer—when the days are longest and the angle of the Sun above the horizon is highest. The winter solstice facts are the reverse: The days are shortest, and the Sun angle is lowest. During an equinox (meaning "equal night"), the length of daylight is 12 hours *everywhere* on Earth because the circle of illumination passes directly through the poles, thus dividing the lines of latitude in half.

All locations situated at the same latitude have identical Sun angles and lengths of daylight. If the Earth–Sun relationships were the only controls of temperature, we would expect these places to have identical temperatures as well. Obviously, such is not the case. Although the angle of the Sun above the horizon and the length of daylight are very important controls of temperature, other factors must be considered—a topic that is addressed in Section 16.7.

In summary, seasonal fluctuations in the amount of solar energy reaching various places on Earth's surface are caused by the tilt of Earth's axis relative to the Sun and the resulting variations in Sun angle and length of daylight.

▼ **Figure 16.17 Midnight Sun** Multiple exposures of the Sun, representative of midsummer in the high latitudes. This example shows the midnight Sun in Norway. (Photo by Martin Woike/AGE Fotostock)

EYE ON EARTH 16.2

This image shows the first sunrise of 2008 at Amundsen–Scott Station, a research facility at the South Pole. At the moment the Sun cleared the horizon, the weathered U.S. flag was seen whipping in the wind above a sign marking the location of the geographic South Pole.

QUESTION 1 *What was the approximate date that this photograph was taken?*

QUESTION 2 *How long after this photo was taken did the Sun set at the South Pole?*

QUESTION 3 *On what date would you expect Amundsen–Scott Station to experience its highest noon Sun angle? Explain.*

(NASA)

Characteristics of the Solstices and Equinoxes for the Northern Hemisphere			
Characteristics	Summer Solstice	Winter Solstice	Equinoxes
Date of occurrence	June 21-22	December 21-22	Spring: March 21-22 Fall: September 22-23
Vertical rays of the sun	Tropic of Cancer (23½° N)	Tropic of Capricorn (23½° S)	Equator
Length of daylight	Longest period of daylight	Shortest period of daylight	Equal days and nights
Angle of noon sun	At its highest point above horizon	At its lowest point above horizon	At an intermediate position above horizon

Figure 16.18 Characteristics of solstices and equinoxes for the Northern Hemisphere

CONCEPT CHECKS 16.4

1. Do the annual variations in Earth–Sun distance adequately account for seasonal temperature changes? Explain.
2. Create a simple sketch that shows why the intensity of solar radiation striking a spot on Earth's surface changes when the Sun angle changes.
3. Briefly explain the primary cause of the seasons.
4. Why are the Tropic of Cancer and the Tropic of Capricorn significant?
5. After examining Table 16.1, write a general statement that relates the season, latitude, and length of daylight.

16.5 Energy, Heat, and Temperature

Distinguish between heat and temperature. List and describe the three mechanisms of heat transfer.

The universe is made up of a combination of matter and energy. The concept of matter is easy to grasp because it is the "stuff" we can see, smell, and touch. Energy, on the other hand, is abstract and therefore more difficult to describe. For our purposes, we define energy simply as *the capacity to do work*. We can think of work as being accomplished whenever matter is moved. You are likely familiar with some of the common forms of energy, such as thermal, chemical, nuclear, radiant (light), and gravitational energy. One type of energy is described as *kinetic energy*, which is energy of motion. Recall that matter is composed of atoms or molecules that are constantly in motion and therefore possesses kinetic energy.

Heat is a term that is commonly used synonymously with *thermal energy*. In this usage, heat is energy possessed by a material arising from the internal motions of its atoms or molecules. Whenever a substance is heated, its atoms move faster and faster, which leads to an increase in its heat content. **Temperature**, on the other hand, is related to the average kinetic energy of a material's atoms or molecules. Stated another way, the term *heat* generally refers to the quantity of energy present, whereas the word *temperature* refers to the intensity—that is, the degree of "hotness."

Heat and temperature are closely related concepts. Heat is the energy that flows because of temperature differences. In all situations, *heat is transferred from warmer to cooler objects*. Thus, if two objects of different temperature are in contact, the warmer object will become cooler and the cooler object will become warmer until they both reach the same temperature.

The flow of energy can occur in three ways: *conduction*, *convection*, and *radiation* (Figure 16.19). Although they are presented separately, all three mechanisms of heat transfer can operate simultaneously. In addition, working in tandem, these processes may transfer heat between the Sun and Earth and between Earth's surface, the atmosphere, and outer space.

SmartFigure 16.19
The three mechanisms of heat transfer: Conduction, convection, and radiation

TUTORIAL
https://goo.gl/ch9TfH

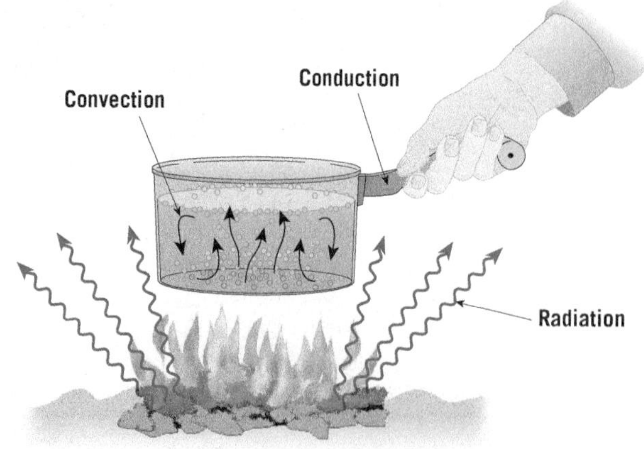

Mechanism of Heat Transfer: Conduction

Conduction is familiar to all of us. Anyone who has touched a metal spoon that was left in a hot pan has discovered that heat was conducted through the spoon. **Conduction** is *the transfer of heat through matter by molecular activity*. The energy of molecules is transferred through collisions from one molecule to another, with the heat flowing from the area of higher temperature to that of lower temperature.

The ability of substances to conduct heat varies considerably. Metals are good conductors, as those of us who have touched hot metal have quickly learned. Air, conversely, is a very poor conductor of heat. Consequently, conduction is important only between Earth's surface and the air directly in contact with the surface. As a means of heat transfer for the atmosphere as a whole, conduction is the least significant.

Mechanism of Heat Transfer: Convection

Much of the heat transport that occurs in the atmosphere occurs via convection. **Convection** is *the transfer of heat by mass movement or circulation within a substance*. It takes place in fluids (e.g., liquids like the ocean and gases like air) where the atoms and molecules are free to move about.

The pan of water in Figure 16.19 illustrates the nature of simple convective circulation. Radiation from the fire warms the bottom of the pan, which conducts heat to the water near the bottom of the container. As the water is heated, it expands and becomes less dense than the water above. Because of this new buoyancy, the warmer water rises. At the same time, cooler, denser water near the top of the pan sinks to the bottom, where it becomes heated. As long as the water is heated unequally—that is, from the bottom up—the water will continue to "turn over," producing a *convective circulation*.

In a similar manner, most of the heat acquired in the lowest portion of the atmosphere by way of radiation and conduction is transferred upward by convection. For example, on a hot, sunny day the air above a plowed field will be heated more than the air above the surrounding croplands. As warm, less dense air above the plowed field buoys upward, it is replaced by the cooler air above the croplands (Figure 16.20). In this way, a convective flow is established. The warm parcels of rising air, called *thermals*, are what hang-glider pilots use to keep their crafts soaring. Convection of this type not only transfers heat but also transports moisture (water vapor) aloft, which may condense into clouds (at the *condensation level* labeled in the figure). The result is the increase in cloudiness that frequently can be observed on warm summer afternoons.

On a much larger scale, the global convective circulation of the atmosphere is driven by the unequal heating of Earth's surface. These complex movements are responsible for the redistribution of heat between hot equatorial regions and frigid polar latitudes and will be discussed in detail in Chapter 18.

EYE ON EARTH 16.3

This infrared (IR) image produced by the *GOES-14* satellite displays cold objects as bright white and hot objects as black. The hottest (blackest) features shown are land surfaces, and the coldest (whitest) features are the tops of towering storm clouds. Recall that we cannot see infrared (thermal) radiation, but we have developed instruments that are capable of extending our vision into the long-wavelength portion of the electromagnetic spectrum.

QUESTION 1 *Several areas of cloud development and potential storms are shown on this IR image. One is a well-developed tropical storm named Hurricane Bill. Can you locate this storm?*

QUESTION 2 *What is an advantage of IR images over visible images?*

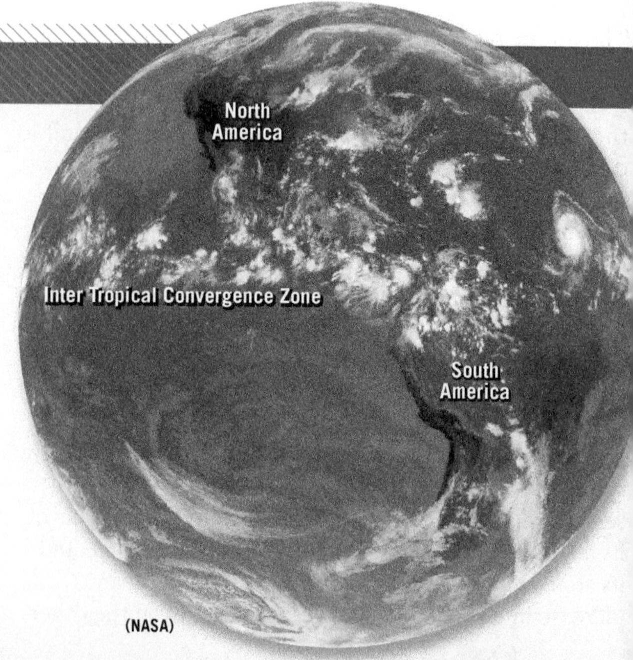

(NASA)

Mechanism of Heat Transfer: Radiation

The third mechanism of heat transfer is **radiation**. As shown in Figure 16.19, radiation travels out in all directions from its source. Unlike conduction and convection, which need a medium to travel through, radiant energy readily travels through the vacuum of space. Thus, radiation is the heat-transfer mechanism by which solar energy reaches our planet.

Solar Radiation From our everyday experience, we know that the Sun emits light and heat as well as the ultraviolet rays that cause suntan. Although these forms of energy comprise a major portion of the total energy that radiates from the Sun, they are only part of a large array of energy called radiation, or **electromagnetic radiation**. This array or spectrum of electromagnetic energy is shown in Figure 16.21. All radiation, whether x-rays, radio waves, or heat waves, travels through the vacuum of space at 300,000 kilometers (186,000 miles) per second and only slightly more slowly through our atmosphere.

Nineteenth-century physicists were so puzzled by the seemingly impossible phenomenon of energy traveling through the vacuum of space without a medium to transmit it that they assumed that a material, which they named *ether*, existed between the Sun and Earth. This medium was thought to transmit radiant energy in much the same way that air transmits sound waves. Of course, this was incorrect. We now know that, like gravity, radiation requires no material for transmission.

In some respects, the transmission of radiant energy parallels the motion of the gentle swells in the open ocean. Like ocean swells, electromagnetic waves come in various sizes. For our purpose, the most important characteristic is their *wavelength*, or the distance from one crest to the next. Radio waves have the longest wavelengths, ranging to tens of kilometers, whereas gamma waves are the shortest, being less than one-billionth of a centimeter long.

Visible light, as the name implies, is the only portion of the spectrum we can see. We often refer to visible light as "white" light because it appears "white" in color. However, it is easy to show that white light is really a mixture of colors, each corresponding to a specific wavelength. Using a prism, white light can be divided into a rainbow color array. Figure 16.21 shows that violet has the shortest wavelength of visible light—0.4 micrometer (1 micrometer is 0.0001 centimeter)—and red has the longest wavelength—0.7 micrometer.

Located adjacent to red, and having a longer wavelength, is **infrared** radiation, which we cannot see but can detect as heat. The closest invisible waves to violet are called **ultraviolet (UV)** rays. They are responsible for the sunburn that can occur after intense exposure to the Sun. Although we divide radiant energy into groups based on our ability to perceive the different types, all forms of radiation are basically the same. When any form of radiant energy is absorbed by an object, the result is an increase in molecular motion, which causes a corresponding increase in temperature.

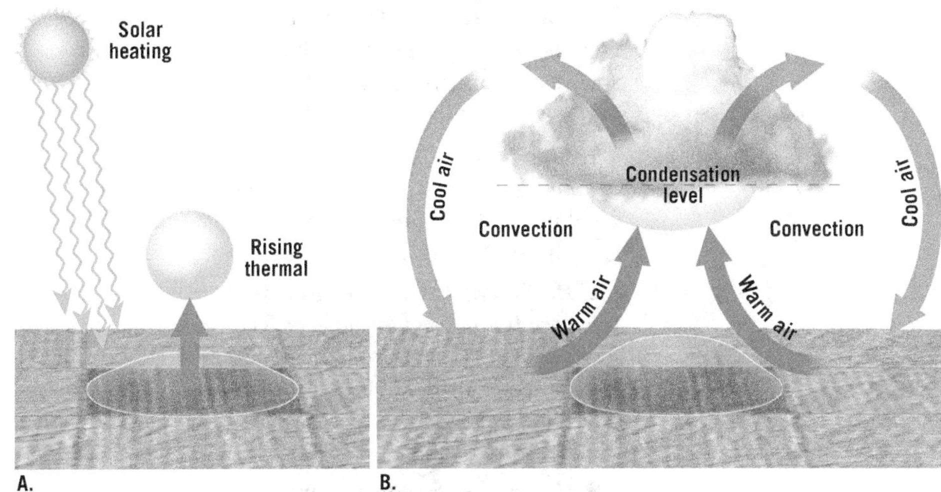

Figure 16.20 **Rising warmer air and descending cooler air are examples of convective circulation A.** Heating of Earth's surface produces thermals of rising air that transport heat and moisture aloft. **B.** The rising air cools, and if it reaches the condensation level, clouds form.

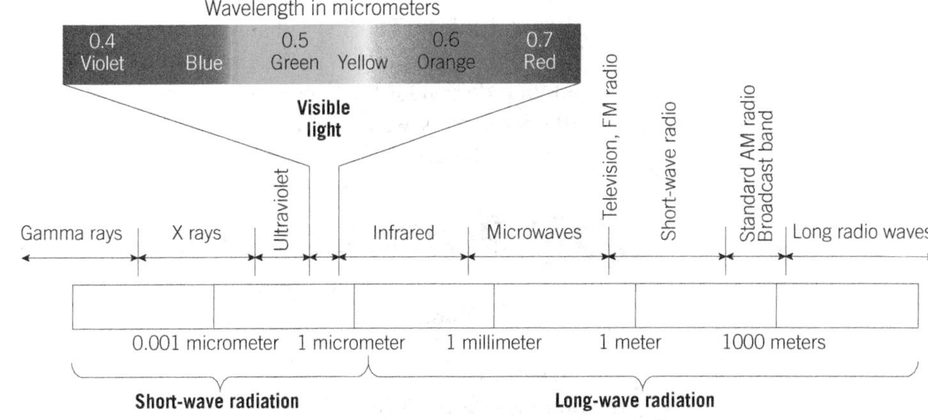

Figure 16.21 **The electromagnetic spectrum** The diagram above illustrates the wavelengths and names of various types of radiation. Visible light consists of an array of colors we commonly call the "colors of the rainbow."

Dennis Tasa

Laws of Radiation To obtain a better understanding of how the Sun's radiant energy interacts with Earth's atmosphere and land–sea surface, it is helpful to have a general understanding of the basic laws governing radiation:

1. *All objects, at whatever temperature, emit radiant energy.* Hence, not only hot objects such as the Sun but also Earth, including its polar ice caps, continually emit energy.
2. *Hotter objects radiate more total energy per unit area than do colder objects.* The Sun, which has a surface temperature of nearly 6000°C (10,000°F), emits about 160,000 times more energy per unit area than does Earth, which has an average surface temperature of about 15°C (59°F).
3. *Hotter objects radiate more energy in the form of short-wavelength radiation than do cooler objects.* We can visualize this law by imagining a metal bar heated to white hot in a forge and allowed to cool. As it cools, it gradually dims, and its color changes from white through yellow to red as progressively more of its energy is radiated at longer wavelengths. Even when it no longer glows visibly, if you place your hand near the metal, the still-longer infrared radiation will be detected as heat. The Sun radiates maximum energy at 0.5 micrometer, which is in the visible range. The maximum radiation for Earth occurs at a wavelength of 10 micrometers, well within the infrared (heat) range. Because the maximum Earth radiation is roughly 20 times longer than the maximum solar radiation, Earth radiation is often called *long-wave radiation*, and solar radiation is called *short-wave radiation*.
4. *Objects that are good absorbers of radiation are good emitters as well.* Earth's surface and the Sun approach being perfect radiators because they absorb and radiate with nearly 100 percent efficiency for their respective temperatures. On the other hand, *gases are selective absorbers and radiators*. Thus, the atmosphere, which is nearly transparent (does not absorb) to certain wavelengths of radiation, is nearly opaque (a good absorber) to others. Our experience tells us that the atmosphere is transparent to visible light; that is why visible light readily reaches Earth's surface. The atmosphere is much less transparent to the longer-wavelength radiation emitted by Earth.

CONCEPT CHECKS 16.5

1. Distinguish between heat and temperature.
2. Describe the three basic mechanisms of heat transfer. Which mechanism is *least* important as a means of heat transfer in the atmosphere?
3. In what part of the electromagnetic spectrum does the Sun radiate maximum energy? How does this compare to Earth?
4. Describe the relationship between the temperature of a radiating body and the wavelengths it emits.

16.6 Heating the Atmosphere

Sketch and label a diagram that shows the paths taken by incoming solar radiation. Summarize the greenhouse effect.

The goal of this section is to describe how energy from the Sun heats Earth's surface and atmosphere. It is important to know the paths taken by incoming solar radiation and the factors that cause the amount of solar radiation taking each path to vary.

Solar radiation 100%
30% lost to space by reflection and scattering
5% backscattered to space by the atmosphere
20% reflected from clouds
20% of radiation absorbed by atmosphere and clouds
50% of direct and diffused radiation absorbed by land and sea
5% reflected from land-sea surface

SmartFigure 16.22 **Paths taken by solar radiation** This diagram shows the average distribution of incoming solar radiation, by percentage. More solar radiation is absorbed by Earth's surface than by the atmosphere.

TUTORIAL
http://goo.gl/IqkUlj

What Happens to Incoming Solar Radiation?

When radiation strikes an object, three different results usually occur. First, some of the energy is *absorbed* by the object. Recall that when radiant energy is absorbed, it is converted to heat, which causes an increase in temperature. Second, substances such as water and air are transparent to certain wavelengths of radiation. Such materials simply *transmit* this energy. Radiation that is transmitted does not contribute energy to the object. Third, some radiation may "bounce off" the object without being absorbed or transmitted. *Reflection* and *scattering* are responsible for redirecting incoming solar radiation. In summary, *radiation may be absorbed, transmitted, or redirected (reflected or scattered).*

Figure 16.22 shows the fate of incoming solar radiation averaged for the entire globe. Notice that the atmosphere is quite transparent to incoming solar radiation.

On average, about 50 percent of the solar energy that reaches the top of the atmosphere is absorbed at Earth's surface. Another 30 percent is reflected back to space by the atmosphere, clouds, and reflective surfaces. The remaining 20 percent is absorbed by clouds and the atmosphere's gases. What determines whether solar radiation will be transmitted to the surface, scattered, reflected outward, or absorbed by the atmosphere? As you will see, the answer depends greatly on the wavelength of the energy being transmitted, as well as on the nature of the intervening material.

Reflection and Scattering

Reflection is the process whereby light bounces back from an object at the same angle at which it encounters a surface and with the same intensity. By contrast, **scattering** is a general process in which radiation is forced to deviate from a straight trajectory. When a beam of light strikes an atom, a molecule, or a tiny particle in the atmosphere, it can spread out in all directions (Figure 16.23). Scattering disperses light both forward and backward. Whether solar radiation is reflected or scattered depends largely on the size of the intervening particles and the wavelength of the light.

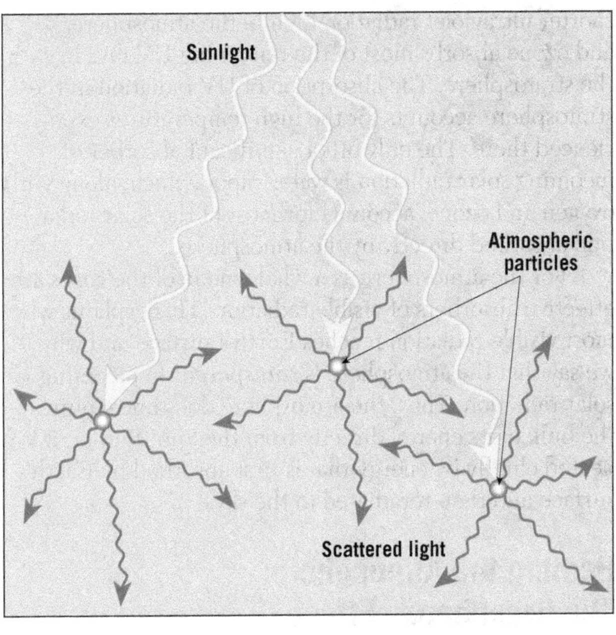

Figure 16.23 Scattering by atmospheric particles When sunlight is scattered, the rays travel in different directions. Usually more energy is scattered in the forward direction than is backscattered.

Reflection and Earth's Albedo

Energy is returned to space from Earth in two ways: reflection and emission of radiant energy. The portion of solar energy that is reflected back to space leaves in the same short wavelengths in which it came to Earth. About 30 percent of the solar energy that reaches the outer atmosphere is reflected back to space. Included in this figure is the amount sent skyward by backscattering. This energy is lost to Earth and does not play a role in heating the atmosphere.

The fraction of the total radiation that is reflected by a surface is called its **albedo**. Thus, the albedo for Earth as a whole (the *planetary albedo*) is 30 percent. However, the albedo varies considerably both from place to place and from one time to another, depending on the amount of cloud cover and particulate matter in the air, on the angle of the Sun's rays, and on the nature of the surface. A lower Sun angle means that more atmosphere must be penetrated, thus making the "obstacle course" longer and the loss of solar radiation greater (see Figure 16.14). Figure 16.24 shows the albedos for various surfaces. Note that the angle at which the Sun's rays strike a water surface greatly affects its albedo.

Diffused Light

Although incoming solar radiation travels in a straight line, small dust particles and gas molecules in the atmosphere scatter some of this energy in all directions. The result, called **diffused light**, explains how light reaches into the area beneath a shade tree and how a room is lit in the absence of direct sunlight. Further, scattering accounts for the brightness and even the blue color of the daytime sky. In contrast, bodies such as the Moon and Mercury, which are without atmospheres, have dark skies and "pitch-black" shadows, even during daylight hours. Overall, about half of the solar radiation that is absorbed at Earth's surface arrives as diffused (scattered) light.

Absorption

As stated earlier, gases are **selective absorbers**, meaning that they absorb some wavelengths strongly, some moderately, and some only slightly. When a gas molecule absorbs radiation, the energy is transformed into internal molecular motion, which is detectable as a rise in temperature.

Nitrogen, the most abundant constituent in the atmosphere, is a poor absorber of all types of incoming solar radiation. Oxygen and ozone are efficient absorbers of ultraviolet radiation. Oxygen removes most of the

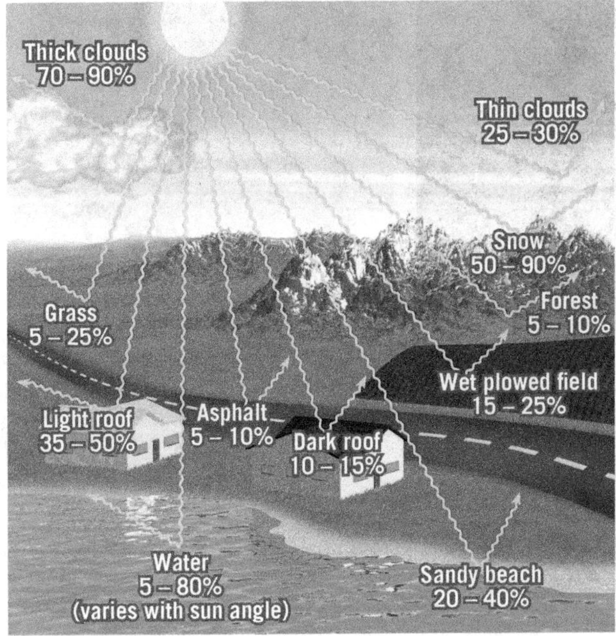

Figure 16.24 Albedo (reflectivity) of various surfaces In general, light-colored surfaces tend to reflect more sunlight than dark-colored surfaces and thus have higher albedos.

shorter ultraviolet radiation high in the atmosphere, and ozone absorbs most of the remaining UV rays in the stratosphere. The absorption of UV radiation in the stratosphere accounts for the high temperatures experienced there. The only other significant absorber of incoming solar radiation is water vapor, which, along with oxygen and ozone, accounts for most of the solar radiation absorbed directly by the atmosphere.

For the atmosphere as a whole, none of the gases are effective absorbers of visible radiation. This explains why most visible radiation reaches Earth's surface and why we say that the atmosphere is *transparent* to incoming solar radiation. Thus, the atmosphere does not acquire the bulk of its energy directly from the Sun. Rather, it is heated chiefly by energy that is first absorbed by Earth's surface and then reradiated to the sky.

Heating the Atmosphere: The Greenhouse Effect

Approximately 50 percent of the solar energy that strikes the top of the atmosphere reaches Earth's surface and is absorbed. Most of this energy is then reradiated skyward. Because Earth has a much lower surface temperature than the Sun, the radiation that it emits has longer wavelengths than solar radiation.

The atmosphere as a whole is an efficient absorber of the longer wavelengths emitted by Earth (*terrestrial radiation*). Water vapor and carbon dioxide are the principal absorbing gases. Water vapor absorbs roughly five times more terrestrial radiation than do all the other gases combined and accounts for the warm temperatures found in the lower troposphere, where it is most highly concentrated. Because the atmosphere is quite transparent to shorter-wavelength solar radiation and more readily absorbs longer-wavelength radiation emitted by Earth, the atmosphere is heated from the ground up rather than vice versa. This explains the general drop in temperature with increasing altitude experienced in the troposphere. The farther from the "radiator," the colder it becomes.

When the gases in the atmosphere absorb radiation emitted by Earth, they warm, but they eventually radiate this energy away. Some energy travels skyward, where it may be reabsorbed by other gas molecules, although that happens progressively less often with increasing height as the atmosphere thins and the concentration of water vapor decreases. The remainder travels Earthward and is again absorbed by Earth. For this reason, Earth's surface is continually being supplied with heat from the atmosphere as well as from the Sun. Without these absorptive gases in our atmosphere, Earth would be a truly frigid place.

This very important phenomenon, illustrated in Figure 16.25, received the name **greenhouse effect** because people at the time thought it was the main way greenhouses work. Like Earth's atmosphere, the glass

SmartFigure 16.25
The greenhouse effect Earth's greenhouse effect is compared with two of our close solar system neighbors.

TUTORIAL
http://goo.gl/2wxon2

Airless bodies like the Moon All incoming solar radiation reaches the surface. Some is reflected back to space. The rest is absorbed by the surface and radiated directly back to space. As a result the lunar surface has a much lower average surface temperature than Earth.

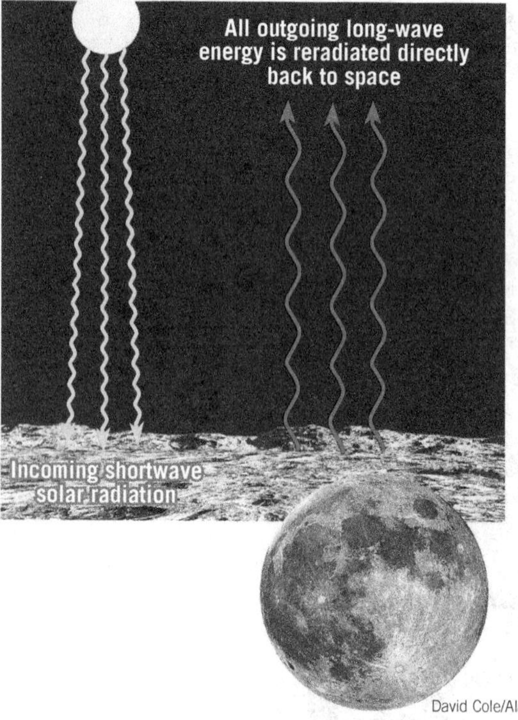
All outgoing long-wave energy is reradiated directly back to space
Incoming shortwave solar radiation
David Cole/Alamy

Bodies with modest amounts of greenhouse gases like Earth The atmosphere absorbs some of the long-wave radiation emitted by the surface. A portion of this energy is radiated back to the surface and is responsible for keeping Earth's surface 33°C (59°F) warmer than it would otherwise be.

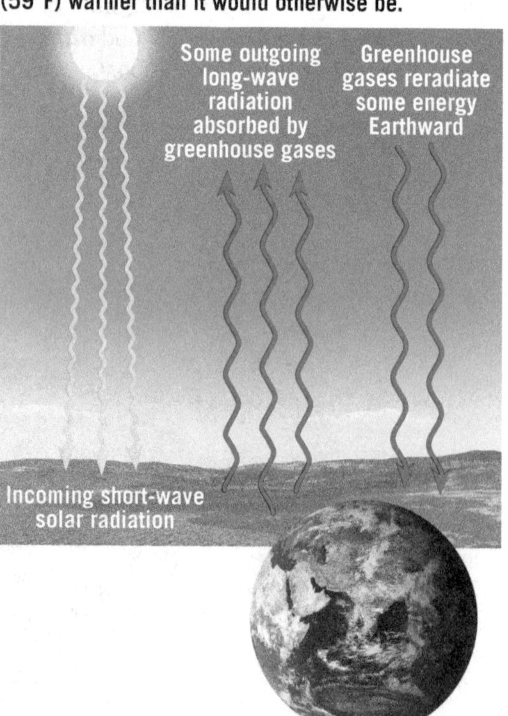
Some outgoing long-wave radiation absorbed by greenhouse gases
Greenhouse gases reradiate some energy Earthward
Incoming short-wave solar radiation
NASA

Bodies with abundant greenhouse gases like Venus Venus experiences extraordinary greenhouse warming, which is estimated to raise its surface temperature by 523°C (941°F).

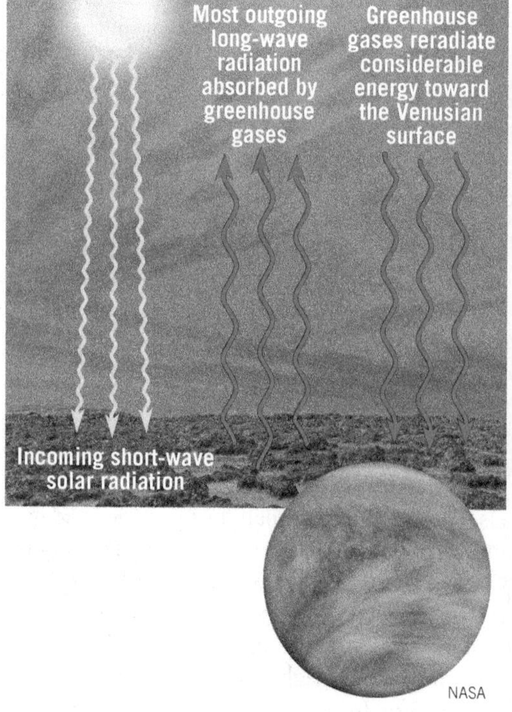
Most outgoing long-wave radiation absorbed by greenhouse gases
Greenhouse gases reradiate considerable energy toward the Venusian surface
Incoming short-wave solar radiation
NASA

of a greenhouse is largely transparent to visible light but somewhat opaque to the longer-wavelength radiation emitted by materials in the greenhouse. Greenhouse glass allows short-wavelength light to enter and heat plants and soil but prevents much of the longer-wavelength radiation these objects emit from leaving. In the case of greenhouses, we now know that this effect is much less important than the simple fact that the greenhouse prevents the warmer interior air from mixing with cooler outside air. Nevertheless, in the case of the atmosphere, the term *greenhouse effect* is still used.

CONCEPT CHECKS 16.6

1. What are the three paths taken by incoming solar radiation?
2. What factors cause albedo to vary from time to time and from place to place on Earth?
3. Explain why the atmosphere is heated chiefly by radiation emitted from Earth's surface rather than by direct solar radiation.
4. Prepare and label a sketch that explains the greenhouse effect for Earth's atmosphere.

16.7 For the Record: Air Temperature Data

Calculate five commonly used types of temperature data and interpret a map that depicts temperature data using isotherms.

People probably notice changes in air temperature more often than they notice changes in any other element of weather. At a weather station, the temperature is monitored on a regular basis from instruments mounted in an instrument shelter (Figure 16.26). The shelter protects the instruments from direct sunlight and allows a free flow of air.

The daily maximum and minimum temperatures underlie much of the basic temperature data compiled by meteorologists:

- By adding the maximum and minimum temperatures and then dividing by two, the **daily mean temperature** is calculated.
- The **daily range** of temperature is computed by finding the difference between the maximum and minimum temperatures for a given day.
- The **monthly mean** is calculated by adding together the daily means for each day of the month and dividing by the number of days in the month.
- The **annual mean** is an average of the 12 monthly means.
- The **annual temperature range** is computed by finding the difference between the highest and lowest monthly means.

Mean temperatures are particularly useful for making comparisons, whether on a daily, monthly, or annual basis. It is quite common to hear weather reporters make statements such as "Last month was the hottest July on record" or "Today, Chicago was 10 degrees warmer than Miami." Temperature ranges are also useful statistics because they give an indication of extremes.

To examine the distribution of air temperatures over large areas, isotherms are commonly used. An **isotherm** is a line that connects points on a map that have the same temperature (*iso* = equal, *therm* = temperature). Therefore, all points through which an isotherm passes have identical temperatures for the time period indicated. Generally, isotherms representing 5° or 10° temperature differences are used, but any interval may be chosen. Figure 16.27 illustrates how isotherms are drawn on a map. Notice that most isotherms do not pass directly through the observing stations because the station readings may not coincide with the values chosen for the isotherms. Only an occasional station temperature will be exactly the same as the value of the isotherm, so it is usually necessary to draw the lines by estimating the proper position between stations.

Maps with isotherms are valuable tools because they make the temperature distribution visible at a glance. Areas of low and high temperatures are easy to pick out. In addition, the amount of temperature change per unit of distance, called the **temperature gradient**, is easy to visualize. Closely spaced isotherms indicate a rapid rate of temperature change, whereas more widely spaced lines indicate a more gradual rate of change. You can see this in Figure 16.27. The isotherms are closer in Colorado and Utah (steeper temperature gradient), whereas the isotherms are spread farther in Texas (gentler temperature gradient). Without isotherms, a map would be covered with numbers representing temperatures at dozens or hundreds of places, which would make patterns difficult to see.

Figure 16.26 Measuring temperature electrically This modern shelter contains an electrical thermometer called a *thermistor.* A shelter protects instruments from direct sunlight and allows for the free flow of air. (Photo by GIPhotoStock/Science Source)

▶ SmartFigure 16.27 **Isotherms** Map showing high temperatures for a spring day. Isotherms are lines that connect points of equal temperature. The temperatures on this map are in degrees Fahrenheit. Showing temperature distribution in this way makes patterns easier to see. Notice that most isotherms do not pass directly through the observing stations. It is usually necessary to draw isotherms by estimating their proper position between stations. On television and in many newspapers, temperature maps are in color. Rather than label isotherms, maps label the area *between* isotherms. For example, the zone between the 60° and 70° isotherms is labeled "60s."

TUTORIAL

https://goo.gl/O2XCrh

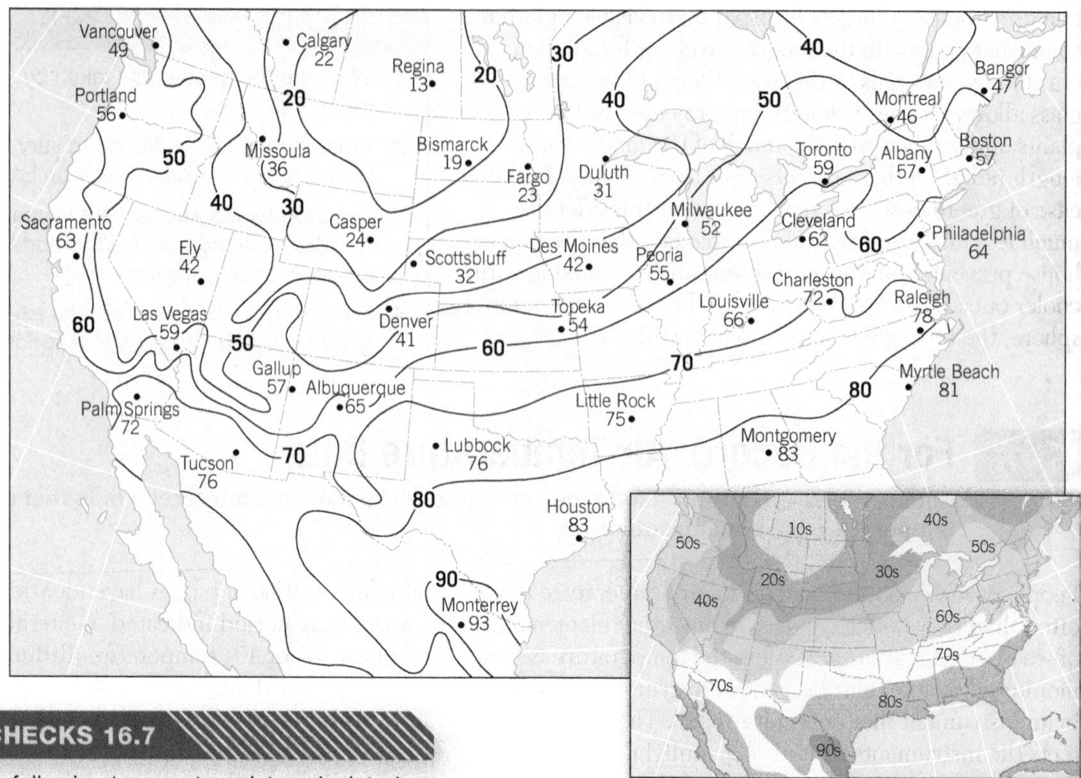

CONCEPT CHECKS 16.7

1. How are the following temperature data calculated: daily mean, daily range, monthly mean, annual mean, and annual range?
2. What are isotherms, and what is their purpose?

16.8 Why Temperatures Vary: The Controls of Temperature

Discuss the principal controls of temperature and use examples to describe their effects.

A **temperature control** is any factor that causes temperature to vary from place to place and from time to time. Earlier in this chapter we examined the most important cause for temperature variations—differences in the receipt of solar radiation. Because variations in Sun angle and length of daylight depend on latitude, they are responsible for warm temperatures in the tropics and colder temperatures at more poleward locations. Of course, seasonal temperature changes at a given latitude occur as the Sun's vertical rays migrate toward and away from a place during the year. Figure 16.28 reminds us of the importance of latitude as a control of temperature.

But latitude is not the only control of temperature; if it were, we would expect all places along the same parallel of latitude to have identical temperatures. This is clearly not the case. For example, Eureka, California, and New York City are both coastal cities at about the same latitude, and both have an annual mean temperature of 11°C (52°F). However, New York City is 9°C (16°F) warmer than Eureka in July and 10°C (18°F) cooler in January. In another example, two cities in Ecuador—Quito and Guayaquil—are relatively close to each other, yet the annual mean temperatures of these two cities differ by 12°C (21°F). To explain these situations and countless others, we must realize that factors other than latitude also exert a strong influence on temperature. In the next sections, we examine these other controls, which include differential heating of land and water, altitude, geographic position, cloud cover and albedo, and ocean currents.*

Land and Water

The heating of Earth's surface directly influences the heating of the air above it. Therefore, to understand variations in air temperature, we must understand the variations in heating properties of the different surfaces that Earth presents to the Sun—soil, water, trees, ice, and so on. Different land surfaces absorb varying amounts of incoming solar energy, which in turn cause variations in the temperature of the air above. The greatest contrast, however, is not between different land surfaces but between land and water. Figure 16.29 illustrates this idea nicely. This satellite image shows surface temperatures in

*For a discussion of the effects of ocean currents on temperature, see Chapter 15.

portions of Nevada, California, and the adjacent Pacific Ocean on the afternoon of May 2, 2004, during a spring heat wave. Land-surface temperatures are clearly much higher than water-surface temperatures. The image shows the extreme high surface temperatures in southern California and Nevada in dark red.[†] Surface temperatures in the Pacific Ocean are much lower. The peaks of the Sierra Nevada, still capped with snow, form a cool blue line down the eastern side of California.

In side-by-side areas of land and water, such as those shown in Figure 16.29, *land heats more rapidly and to higher temperatures than water, and it cools more rapidly and to lower temperatures than water.* Variations in air temperatures, therefore, are much greater over land than over water.

Why do land and water heat and cool differently? Several factors are responsible:

- The **specific heat** (the amount of energy needed to raise the temperature of 1 gram of a substance 1°C) is far greater for water than for land. Thus, water requires a great deal more heat to raise its temperature the same amount than does an equal quantity of land.
- Land surfaces are opaque, so heat is absorbed only at the surface. Water, being more transparent, allows heat to penetrate to a depth of many meters.
- The water that is heated often mixes with water below, thus distributing the heat through an even larger mass.
- Evaporation (a cooling process) from water bodies is greater than that from land surfaces.

All these factors collectively cause water to warm more slowly, store greater quantities of heat, and cool more slowly than land.

Monthly temperature data for two cities will demonstrate the moderating influence of a large water body and the extremes associated with land (Figure 16.30). Vancouver, British Columbia, is located along the windward Pacific coast, whereas Winnipeg, Manitoba, is in a continental position far from the influence of water. The two cities are at about the same latitude and thus experience similar Sun angles and lengths of daylight. Winnipeg, however, has a mean January temperature that is 20°C lower than Vancouver's. Conversely, Winnipeg's July mean is 2.6°C higher than Vancouver's. Although their latitudes are nearly the same, Winnipeg, which has no water influence, experiences much greater temperature extremes than does Vancouver. The key to Vancouver's moderate year-round climate is the Pacific Ocean.

On a different scale, the moderating influence of water may also be demonstrated when temperature variations in the Northern and Southern Hemispheres are compared.

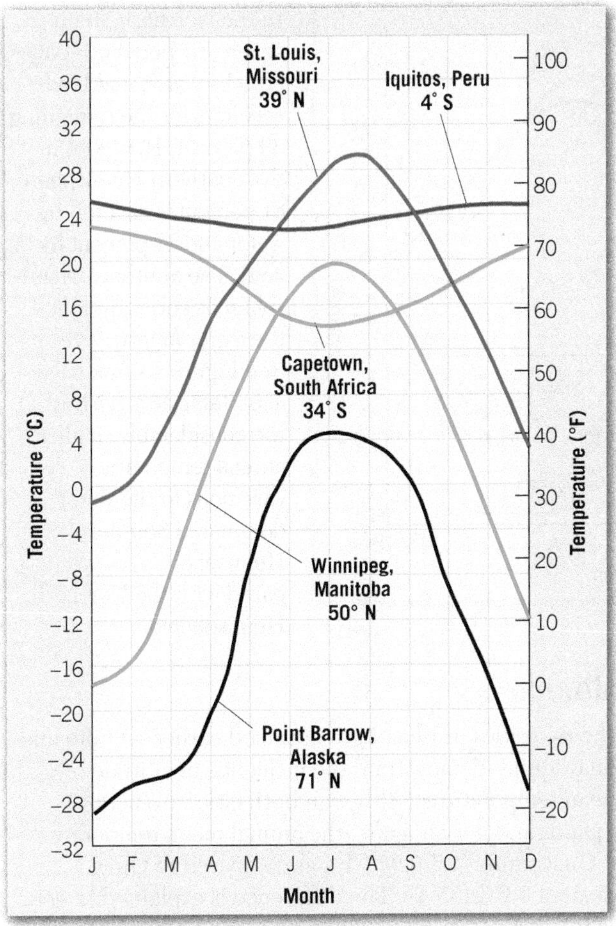

Figure 16.28 Latitude is a major control of temperature Data for these five cities remind us that latitude (Earth–Sun relationships) is a significant factor influencing temperatures. Places located at higher latitudes experience larger temperature differences between summer and winter. Note that Cape Town, South Africa, experiences winter in June, July, and August.

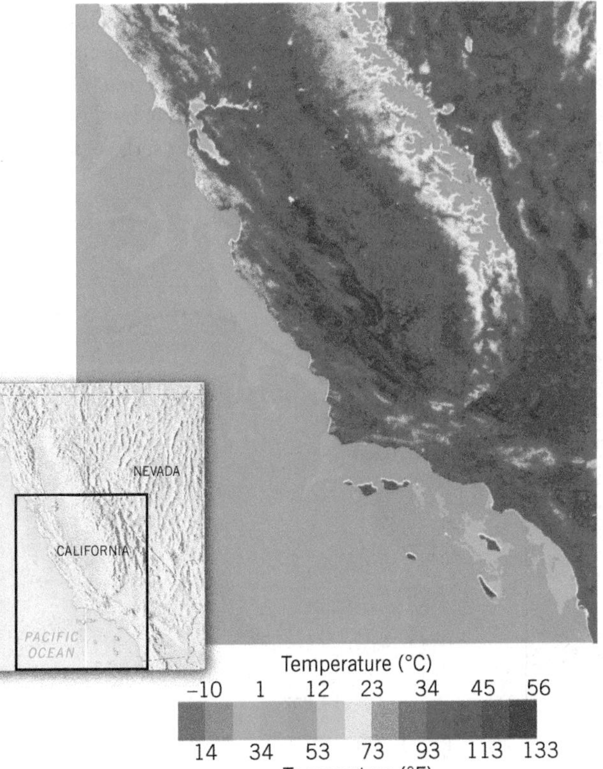

Figure 16.29 Differential heating of land and water This satellite image from the afternoon of May 2, 2004, illustrates an important control of air temperature. Water-surface temperatures in the Pacific Ocean are much lower than land-surface temperatures in California and Nevada. The narrow band of cool temperatures in the center of the image is associated with mountains (the Sierra Nevada). The cooler water temperatures immediately offshore are associated with the California Current. (NASA)

[†]Realize that when a land surface is hot, the air above is cooler. For example, while the surface of a sandy beach can be painfully hot, the air temperature above the surface is more comfortable.

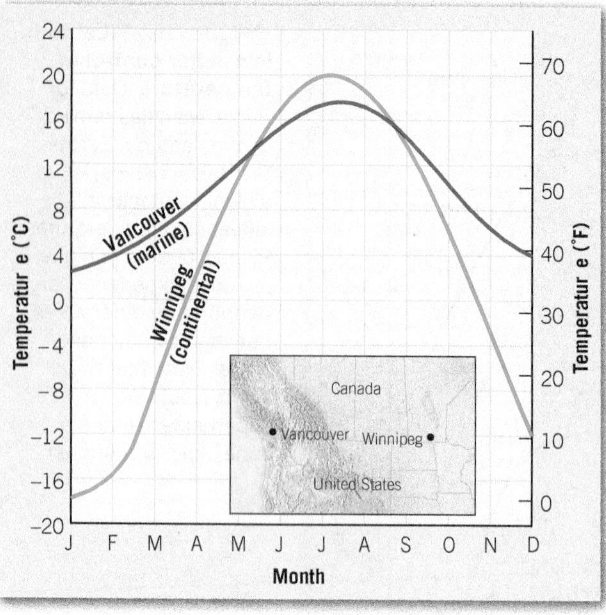

SmartFigure 16.30 **Monthly mean temperatures for Vancouver, British Columbia, and Winnipeg, Manitoba** Vancouver has a much smaller annual temperature range, due to the strong marine influence of the Pacific Ocean. Winnipeg illustrates the greater extremes associated with an interior location.

TUTORIAL

https://goo.gl/g4jOhk

In the Northern Hemisphere, 61 percent is covered by water, and land accounts for the remaining 39 percent. However, in the Southern Hemisphere, 81 percent is covered by water and 19 percent by land. The Southern Hemisphere is correctly called the *water hemisphere* (see Figure 13.1 on page 416). **Table 16.2** portrays the considerably smaller annual temperature variations in the water-dominated Southern Hemisphere as compared with the Northern Hemisphere.

Altitude

The two cities in Ecuador mentioned earlier—Quito and Guayaquil—demonstrate the influence of altitude on mean temperatures. Although both cities are near the equator and not far apart, the annual mean temperature at Guayaquil is 25°C (77°F), as compared to Quito's mean of 13°C (55°F). The difference is explained largely by the difference in the cities' elevations: Guayaquil is only 12 meters (40 feet) above sea level, whereas Quito is high in the Andes Mountains, at 2800 meters (9200 feet). **Figure 16.31** provides another example.

Recall that temperatures drop an average of 6.5°C per kilometer in the troposphere; thus, cooler temperatures are to be expected at greater heights. Yet the magnitude of the difference is not explained completely by the normal lapse rate. If the normal lapse rate is used, we would expect Quito to be about 18°C cooler than Guayaquil, but the difference is only 12°C. The fact that high-altitude places such as Quito are warmer than the value calculated using the normal lapse rate results from the absorption and reradiation of solar energy by the ground surface.

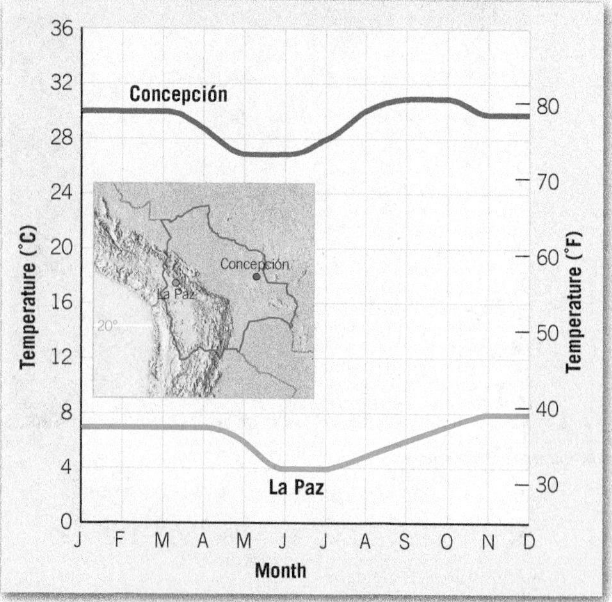

Figure 16.31 **Monthly mean temperatures for Concepción and La Paz, Bolivia** Both cities have nearly the same latitude (about 16° degrees south). However, because La Paz is high in the Andes, at 4103 meters (13,461 feet), it experiences much cooler temperatures than Concepción, which is at an elevation of 490 meters (1608 feet).

Table 16.2 Variation in Annual Mean Temperature Range (°C) with Latitude

Latitude	Northern Hemisphere	Southern Hemisphere
0	0	0
15	3	4
30	13	7
45	23	6
60	30	11
75	32	26
90	40	31

Geographic Position

The geographic setting can greatly influence the temperatures experienced at a specific location. A coastal location where prevailing winds blow from the ocean onto the shore (a **windward coast**) experiences considerably different temperatures than does a coastal location where the prevailing winds blow from the land toward the ocean (a **leeward coast**). In the first situation, the windward coast will experience the full moderating influence of the ocean—cool summers and mild winters—compared to an inland station at the same latitude.

A leeward coast, on the other hand, will have a more continental temperature pattern because the winds do not carry the ocean's influence onshore. Eureka, California, and New York City, the two cities mentioned earlier, illustrate this aspect of geographic position. The annual temperature range at New York City is 19°C (34°F) greater than Eureka's (**Figure 16.32**).

Seattle and Spokane, both in the state of Washington, illustrate a second aspect of geographic position: mountains that act as barriers. Although Spokane is only about 360 kilometers (220 miles) east of Seattle, the towering Cascade Range separates the cities. Consequently, Seattle's temperatures show a marked marine influence, but Spokane's are more typically continental (**Figure 16.33**). Spokane is 7°C (13°F) cooler than Seattle in January and 4°C (7°F) warmer than Seattle in July. The annual range at Spokane is 11°C (20°F) greater than at Seattle. The Cascade Range effectively cuts off Spokane from the moderating influence of the Pacific Ocean.

Cloud Cover and Albedo

You may have noticed that clear days are often warmer than cloudy ones and that clear nights are usually cooler than cloudy ones. Cloud cover is another factor that influences temperature in the lower atmosphere. Studies using satellite images show that at any particular

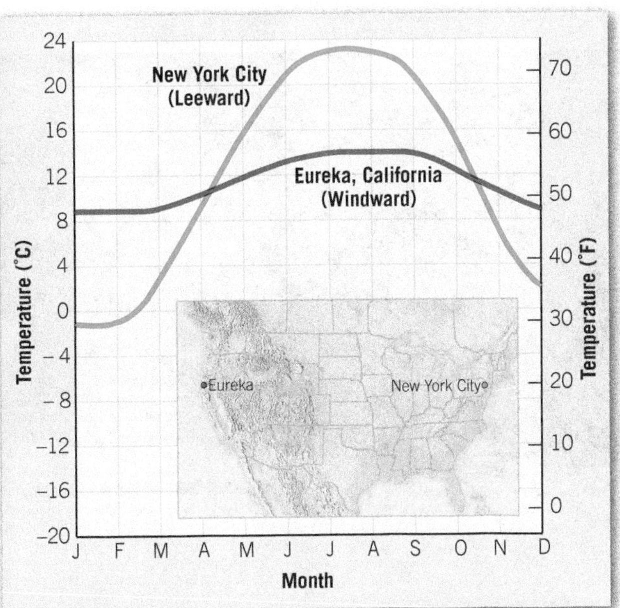

Figure 16.32 **Monthly mean temperatures for Eureka, California, and New York City** These two cities are coastal and located at about the same latitude. Because Eureka is strongly influenced by prevailing winds from the ocean and New York City is not, the annual temperature range at Eureka is much smaller.

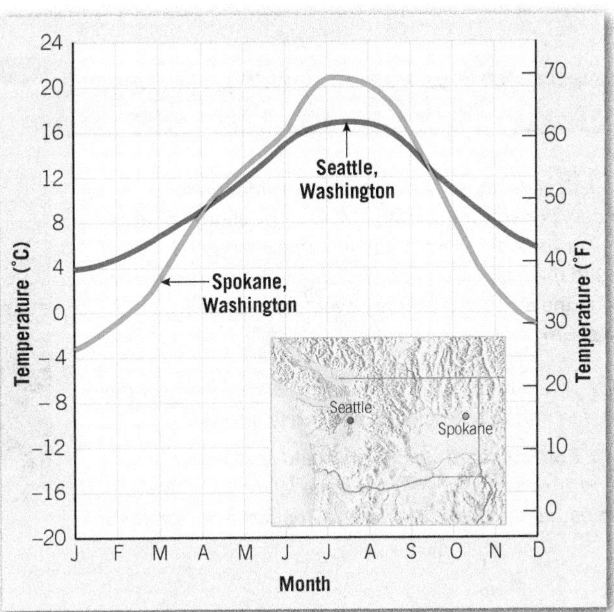

Figure 16.33 **Monthly mean temperatures for Seattle and Spokane, Washington** Because the Cascade Mountains cut off Spokane from the moderating influence of the Pacific Ocean, Spokane's annual temperature range is greater than Seattle's.

time, about half of our planet is covered by clouds. Cloud cover is important because many clouds have a high albedo; therefore, clouds reflect a significant portion of the sunlight that strikes them back into space (see Figure 16.24). By reducing the amount of incoming solar radiation, clouds reduce daytime temperatures.

At night, clouds have the opposite effect as during daylight: They act as a blanket, absorbing radiation emitted by Earth's surface and reradiating a portion of it back to the surface. Consequently, some of the heat that otherwise would have been lost remains near the ground. Thus, nighttime air temperatures do not drop as low as they would on a clear night. The effect of cloud cover is to reduce the daily temperature range by lowering the daytime maximum and raising the nighttime minimum (Figure 16.34).

Clouds are not the only phenomenon that increase albedo and thereby reduces air temperatures. We also recognize that snow- and ice-covered surfaces have high albedos. This is one reason mountain glaciers do not melt away in the summer, and it is why snow may still be present on a mild spring day. In addition, during the winter, when snow covers the ground, daytime maximums on a sunny day are less than they otherwise would be because energy that the land would have absorbed and used to heat the air is reflected and lost.

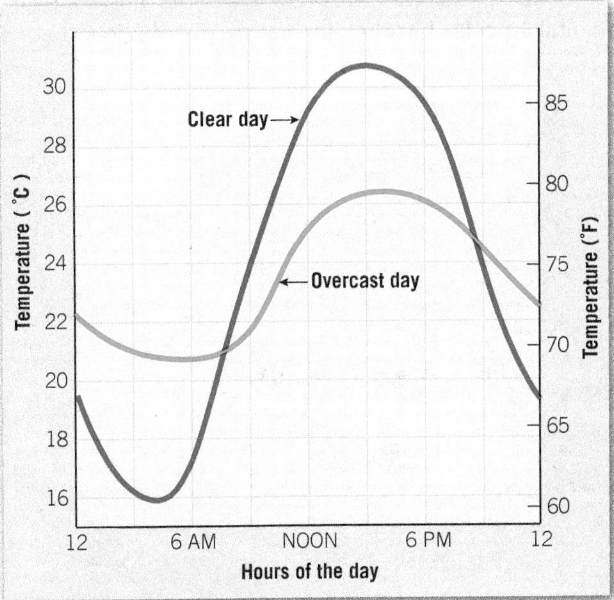

SmartFigure 16.34 **The daily cycle of temperature at Peoria, Illinois, for two July days** Clouds reduce the daily temperature range. During daylight hours, clouds reflect solar radiation back to space. Therefore, the maximum temperature is lower than if the sky were clear. At night, the minimum temperature will not fall as low because clouds retard the loss of heat.

TUTORIAL
https://goo.gl/TjgQQK

CONCEPT CHECKS 16.8

1. List the factors that cause land and water to heat and cool differently.
2. Quito, Ecuador, is located on the equator and is *not* a coastal city. It has an average annual temperature of only 13°C (55°F). What is the likely cause for this low average temperature?
3. In what ways can geographic position be considered a control of temperature?
4. How does cloud cover influence the maximum temperature on an overcast day? How is the nighttime minimum influenced by clouds?

EYE ON EARTH 16.4

For more than a decade, scientists have used the Moderate Resolution Imaging Spectroradiometer (MODIS, for short) aboard NASA's *Aqua* and *Terra* satellites to gather surface temperature data from around the globe. This image shows average land-surface temperatures for the month of February over a 10-year span (2001–2010). (NASA)

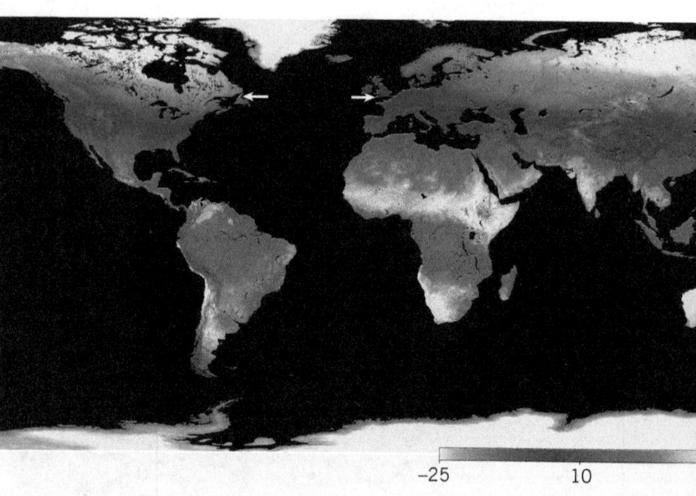

QUESTION 1 *What are the approximate temperatures for southern Great Britain and northern Newfoundland (white arrows)?*

QUESTION 2 *Both southern Great Britain and northern Newfoundland are coastal areas at the same latitude, yet average February temperatures are quite different. Suggest a reason for this disparity.*

16.9 World Distribution of Temperature

Interpret the patterns depicted on world maps of January and July temperatures.

Take a moment to study the two world isothermal maps in **Figures 16.35** and **16.36**. From hot colors near the equator to cool colors toward the poles, these maps portray sea-level temperatures in the seasonally extreme months of January and July. Temperature distribution is shown by using isotherms. On these maps you can study global temperature patterns and the effects of the controlling factors of temperature, especially latitude, the distribution of land and water, and ocean currents. As with most other isothermal maps of large regions, all temperatures on these world maps have been reduced to sea level to eliminate the complications caused by differences in altitude.

On both maps, the isotherms generally trend east and west and show a decrease in temperatures poleward from the tropics. They illustrate one of the most fundamental aspects of world temperature distribution: that

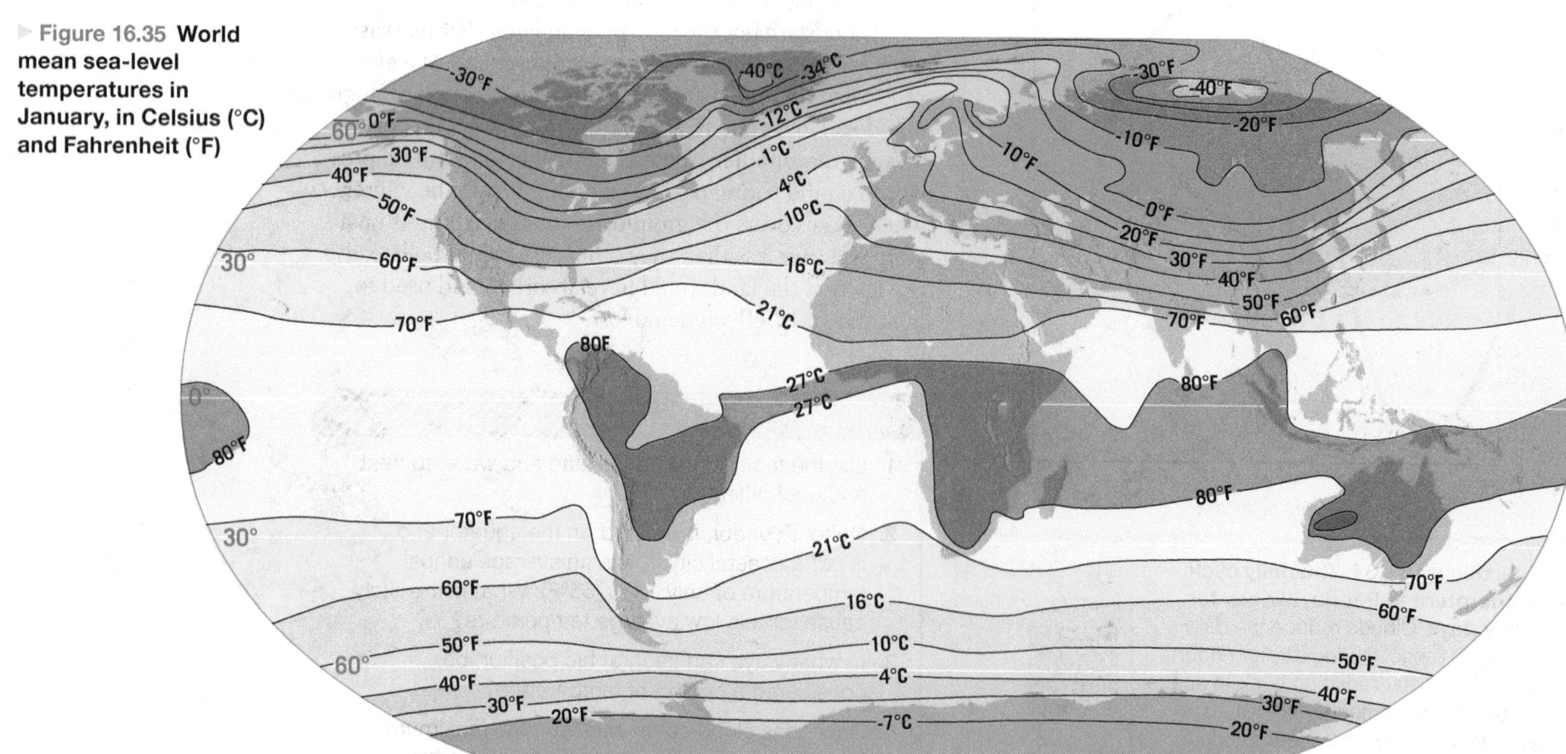

Figure 16.35 World mean sea-level temperatures in January, in Celsius (°C) and Fahrenheit (°F)

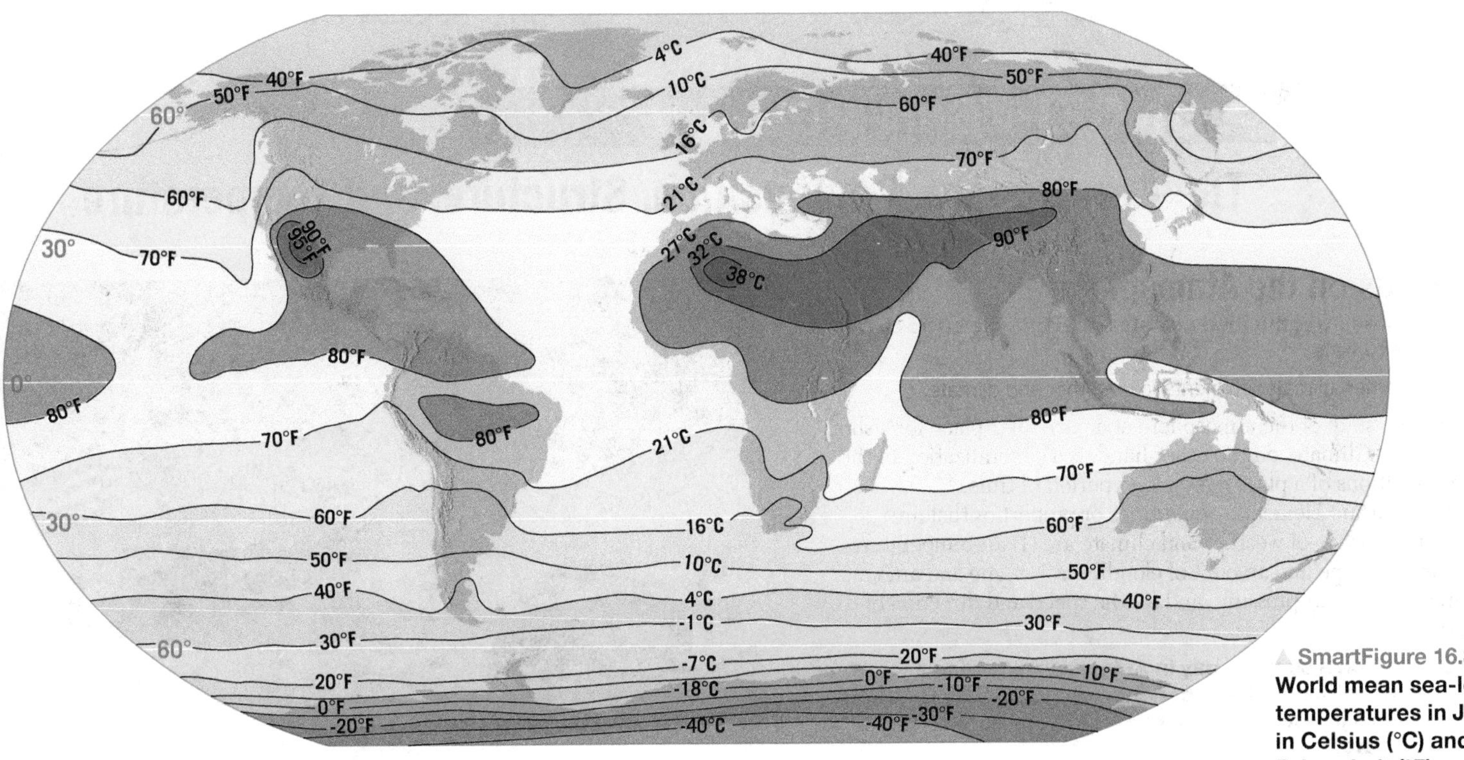

▲ SmartFigure 16.36 **World mean sea-level temperatures in July, in Celsius (°C) and Fahrenheit (°F)**

TUTORIAL
https://goo.gl/9nS328

the effectiveness of incoming solar radiation in heating Earth's surface and the atmosphere above it is largely a function of latitude.

Moreover, there is a latitudinal shifting of temperatures caused by the seasonal migration of the Sun's vertical rays. To see this, compare the color bands by latitude on the two maps. On the January map, the "hot spots" of 30°C are *south* of the equator, but in July they have shifted *north* of the equator.

If latitude were the only control of temperature distribution, our analysis could end here, but that is not the case. The added effect of the differential heating of land and water is also reflected on the January and July temperature maps. The warmest and coldest temperatures are found over land; note the coldest area, a purple oval in Siberia, and the hottest areas, the deep orange ovals, all over land. Because temperatures do not fluctuate as much over water as over land, the north–south migration of isotherms is greater over the continents than over the oceans.

In addition, it is clear that the isotherms in the Southern Hemisphere, where there is little land and where the oceans predominate, are much more regular than in the Northern Hemisphere, where they bend sharply northward in July and southward in January over the continents.

Isotherms also show the presence of ocean currents. Warm currents cause isotherms to be deflected toward the poles, whereas cold currents cause an equatorward bending. The horizontal transport of water poleward warms the overlying air and results in air temperatures that are higher than would otherwise be expected for the latitude. Conversely, currents moving toward the equator produce cooler-than-expected air temperatures.

Because Figures 16.35 and 16.36 show the seasonal extremes of temperature, they can be used to evaluate variations in the annual range of temperature from place to place. A comparison of the two maps shows that a station near the equator has a very small annual range because it experiences little variation in the length of daylight, and it always has a relatively high Sun angle. A station in the middle latitudes, however, experiences wide variations in Sun angle and length of daylight and hence large variations in temperature. Therefore, we can state that the annual temperature range increases with an increase in latitude.

Moreover, land and water also affect seasonal temperature variations, especially outside the tropics. A continental location must endure hotter summers and colder winters than a coastal location. Consequently, outside the tropics, the annual temperature range will increase with an increase in continentality.

CONCEPT CHECKS 16.9

1. Why do isotherms generally trend east–west?
2. Why do isotherms shift north and south from season to season?
3. Where do isotherms shift most: over land or water? Explain.
4. Which area on Earth experiences the highest annual temperature range?

The Atmosphere: Composition, Structure, and Temperature

16.1 Focus on the Atmosphere

Distinguish between weather and climate and name the basic elements of weather and climate.

KEY TERMS: weather, climate, elements (of weather and climate)

- Weather is the state of the atmosphere at a particular place for a short period of time. Climate, on the other hand, is a generalization of the weather conditions of a place over a long period of time.
- The most important elements—quantities or properties that are measured regularly—of weather and climate are (1) air temperature, (2) humidity, (3) type and amount of cloudiness, (4) type and amount of precipitation, (5) air pressure, and (6) the speed and direction of the wind.

? This is a scene on a summer day in Antarctica, showing a joint British–American research team. Write two brief statements about the locale in this image—one that relates to weather and one that relates to climate.

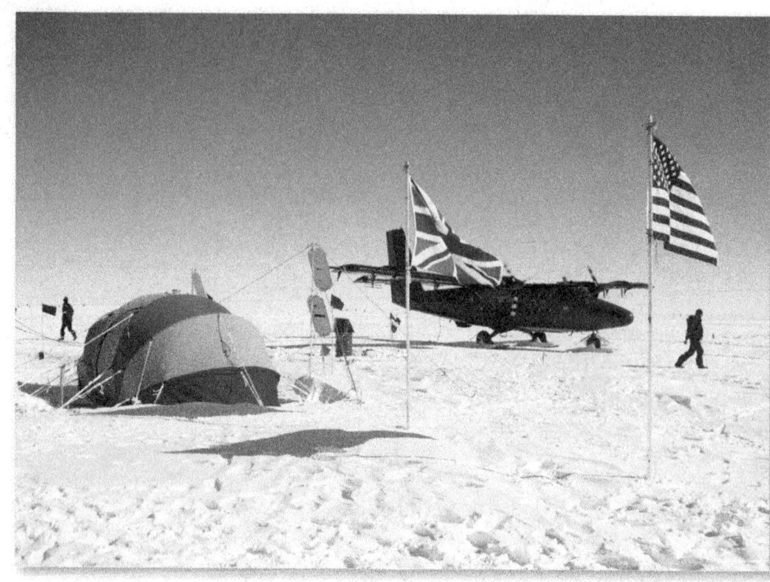
David Vaughan/Science Source

16.2 Composition of the Atmosphere

List the major gases composing Earth's atmosphere and identify the components that are most important to understanding weather and climate.

KEY TERMS: air, aerosols, ozone

- Air is a mixture of many discrete gases, and its composition varies from time to time and from place to place. If water vapor, dust, and other variable components of the atmosphere are removed, clean, dry air is composed almost entirely of nitrogen (N_2) and oxygen (O_2). Carbon dioxide (CO_2), although present only in minute amounts, is important because it has the ability to absorb heat radiated by Earth and thus helps keep the atmosphere warm.
- Among the variable components of air, water vapor is important because it is the source of all clouds and precipitation. Like carbon dioxide, water vapor can absorb heat emitted by Earth. When water changes from one state to another, it absorbs or releases heat. In the atmosphere, water vapor transports this latent ("hidden") heat from place to place; latent heat provides the energy that helps drive many storms.
- Aerosols are tiny solid and liquid particles that are important because they may act as surfaces on which water vapor can condense and are also absorbers and reflectors of incoming solar radiation.
- Ozone, a form of oxygen that combines three oxygen atoms into each molecule (O_3), is concentrated in the 10- to 50-kilometer (6- to 31-mile) height range in the atmosphere. This gas is important to life because of its ability to absorb potentially harmful ultraviolet radiation from the Sun.

? This graph shows changes in one atmospheric component between January 2014 and mid-2016. Which gas is it? How did you figure it out? Why is the line so wavy?

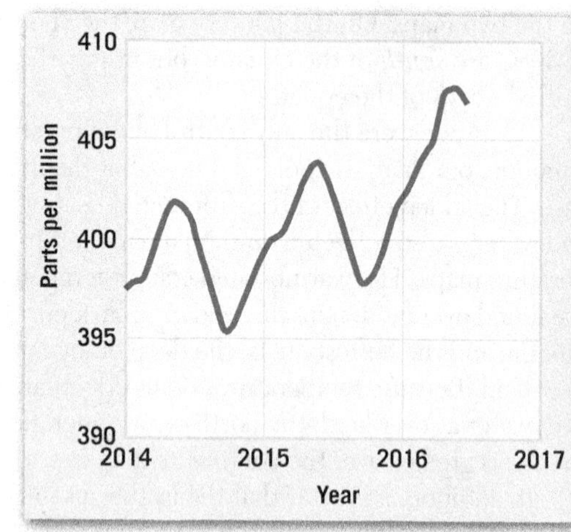

16.3 Vertical Structure of the Atmosphere

Interpret a graph that shows changes in air pressure from Earth's surface to the top of the atmosphere. Sketch and label a graph that shows atmospheric layers based on temperature.

KEY TERMS: troposphere, environmental lapse rate, radiosonde, stratosphere, mesosphere, thermosphere

- Because the atmosphere gradually thins with increasing altitude, it has no sharp upper boundary but simply blends into outer space.
- Based on temperature, the atmosphere is divided vertically into four layers. The troposphere is the lowermost layer. In the troposphere, temperature usually decreases with increasing altitude. This environmental lapse rate is variable but averages about 6.5°C per kilometer (3.5°F per 1000 feet). Essentially, all important weather phenomena occur in the troposphere.
- Beyond the troposphere is the stratosphere, which warms with increasing altitude because of absorption of UV radiation by ozone. In the mesosphere, temperatures again decrease with increasing altitude. Above the mesosphere is the thermosphere, a layer with only a tiny fraction of the atmosphere's mass and no well-defined upper limit.

? When the weather balloon in this photo was launched, the surface temperature was 17°C. The balloon is now at an altitude of 1 kilometer. What term is applied to the instrument package being carried aloft by the balloon? In what layer of the atmosphere is the balloon? If average conditions prevail, what is the air temperature at this altitude? How did you figure this out?

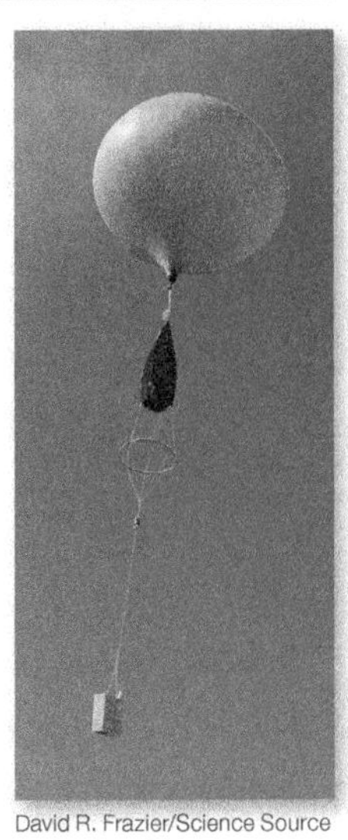
David R. Frazier/Science Source

16.4 Earth–Sun Relationships

Explain what causes the Sun angle and length of daylight to change during the year and describe how these changes produce the seasons.

KEY TERMS: rotation, circle of illumination, inclination of the axis, Tropic of Cancer, summer solstice, Tropic of Capricorn, winter solstice, autumnal (fall) equinox, spring equinox

- The two principal motions of Earth are (1) rotation about its axis, which produces the daily cycle of daylight and darkness, and (2) orbital motion around the Sun, which produces yearly variations.
- The seasons are caused by changes in the angle at which the Sun's rays strike Earth's surface and the changes in the length of daylight at each latitude. These seasonal changes are the result of the tilt of Earth's axis as it orbits the Sun.

? Assume that the date is December 22. At what latitude do the Sun's rays strike the ground vertically at noon? Is this date an equinox or a solstice? What is the season in Australia?

16.5 Energy, Heat, and Temperature

Distinguish between heat and temperature. List and describe the three mechanisms of heat transfer.

KEY TERMS: heat, temperature, conduction, convection, radiation, electromagnetic radiation, visible light, infrared, ultraviolet (UV)

- Heat refers to the quantity of energy present in a material, whereas temperature refers to intensity, or the degree of "hotness."
- The three mechanisms of heat transfer are (1) conduction, the transfer of heat through matter by molecular activity; (2) convection, the transfer of heat by the movement of a mass or substance from one place to another; and (3) radiation, the transfer of heat by electromagnetic waves.
- Electromagnetic radiation is energy emitted in the form of rays, or waves, called electromagnetic waves. All radiation is capable of transmitting energy through the vacuum of space. One of the most important differences between electromagnetic waves is their wavelengths, which range from very long for radio waves to very short for gamma rays. Visible light is the only portion of the electromagnetic spectrum we can see.

Johner Images RF/AGE Fotostock

- Some basic laws that relate to radiation are (1) all objects emit radiant energy; (2) hotter objects radiate more total energy than do colder objects; (3) the hotter the radiating body, the shorter the wavelengths of maximum radiation; and (4) objects that are good absorbers of radiation are good emitters as well.

? Describe how each of the three basic mechanisms of heat transfer are illustrated in this image.

16.6 Heating the Atmosphere

Sketch and label a diagram that shows the paths taken by incoming solar radiation. Summarize the greenhouse effect.

KEY TERMS: reflection, scattering, albedo, diffused light, selective absorbers, greenhouse effect

- About 50 percent of the solar radiation that strikes the atmosphere reaches Earth's surface. About 30 percent is reflected back to space. The remaining 20 percent of incoming solar energy is absorbed by clouds and the atmosphere's gases. The fraction of radiation reflected by a surface is called the albedo of that surface.
- Radiant energy absorbed at Earth's surface is eventually radiated skyward. Because Earth has a much lower surface temperature than the Sun, its radiation is in the form of long-wave infrared radiation. Because atmospheric gases, primarily water vapor and carbon dioxide, are more efficient absorbers of long-wave radiation than of short-wave radiation, the atmosphere is heated from the ground up.
- Greenhouse effect is the term for the selective absorption of Earth's long-wave radiation by water vapor and carbon dioxide, which results in Earth's average temperature being warmer than it would be otherwise.

16.7 For the Record: Air Temperature Data

Calculate five commonly used types of temperature data and interpret a map that depicts temperature data using isotherms.

KEY TERMS: daily mean temperature, daily range, monthly mean, annual mean, annual temperature range, isotherm, temperature gradient

- Daily mean temperature is an average of the daily maximum and daily minimum temperatures, whereas the daily range is the difference between the daily maximum and daily minimum temperatures. The monthly mean is determined by averaging the daily means for a particular month. The annual mean is an average of the 12 monthly means, whereas the annual temperature range is the difference between the highest and lowest monthly means.
- Temperature distribution is shown on a map by using isotherms, which are lines of equal temperature. Temperature gradient is the amount of temperature change per unit of distance. Closely spaced isotherms indicate a rapid rate of change.

16.8 Why Temperatures Vary: The Controls of Temperature

Discuss the principal controls of temperature and use examples to describe their effects.

KEY TERMS: temperature control, specific heat, windward coast, leeward coast

- Controls of temperature are factors that cause temperature to vary from place to place and from time to time. Latitude (Earth–Sun relationships) is one example. Ocean currents (discussed in Chapter 10) provide another example.
- Unequal heating of land and water is a temperature control. Because land and water heat and cool differently, land areas experience greater temperature extremes than do water-dominated areas.
- Altitude is an easy-to-visualize control: The higher up you go, the colder it gets; therefore, mountains are cooler than adjacent lowlands.
- Geographic position as a temperature control involves factors such as mountains acting as barriers to marine influence and a place being on a windward coast or a leeward coast.

? The graph shows monthly high temperatures for Urbana, Illinois, and San Francisco, California. Although the two cities are located at about the same latitude, the temperatures they experience are quite different. Which line on the graph represents Urbana, and which represents San Francisco? How did you figure this out?

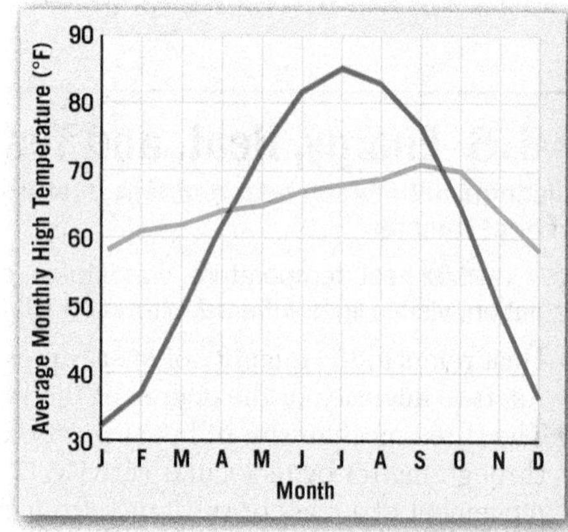

16.9 World Distribution of Temperature

Interpret the patterns depicted on world maps of January and July temperatures.

- On world maps showing January and July mean temperatures, isotherms generally trend east–west and show a decrease in temperature from the equator to the poles. When the two maps are compared, a latitudinal shifting of temperatures is seen. Bending isotherms reveal the locations of ocean currents.
- Annual temperature range is small near the equator and increases with an increase in latitude. Outside the tropics, annual temperature range also increases as marine influence diminishes.

? Refer to Figure 16.35. What causes the isotherms to bend in the North Atlantic?

GIVE IT SOME THOUGHT

1 Determine which statements refer to weather and which refer to climate. (*Note:* One statement includes aspects of *both* weather and climate.)
   a. The baseball game was rained out today.
   b. January is Omaha's coldest month.
   c. North Africa is a desert.
   d. The high this afternoon was 25°C.
   e. Last evening a tornado ripped through central Oklahoma.
   f. I am moving to southern Arizona because it is warm and sunny.
   g. Thursday's low of −20°C is the coldest temperature ever recorded for that city.
   h. It is partly cloudy.

2 This map shows the mean percentage of possible sunshine received in the month of November across the 48 contiguous United States.
   a. Does this map relate more to climate or to weather?
   b. If you visited Yuma, Arizona, in November, would you *expect* to experience a sunny day or an overcast day?
   c. Might what you actually experience during your visit be different from what you expected? Explain.

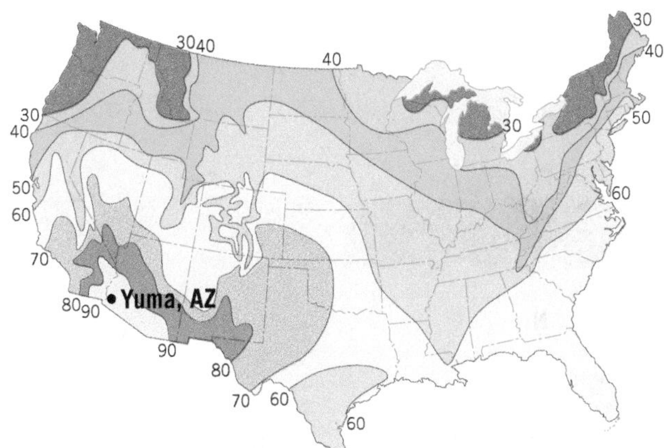

3 Refer to the graph in Figure 16.3 to answer the following questions about temperatures in New York City:
   a. What is the approximate average daily high temperature in January? In July?
   b. Approximately what are the highest and lowest temperatures ever recorded?

4 Which of the three mechanisms of heat transfer is clearly illustrated in each of the following situations?
   a. Driving a car with the seat heater turned on
   b. Sitting in an outdoor hot tub
   c. Lying inside a tanning bed
   d. Driving a car with the air conditioning turned on

5 The circumference of Earth at the equator is 24,900 miles. Calculate how fast someone at the equator is rotating in miles per hour. If the rotational speed of Earth were to slow down, how might this impact daytime highs and nighttime lows?

6 Rank the following according to the wavelengths of radiant energy each emits, from the shortest wavelengths to the longest:
   a. A light bulb with a filament glowing at 4000°C
   b. A rock at room temperature
   c. A car engine at 140°C

7 Imagine being at the beach in this photo on a sunny summer afternoon.
   a. Describe the temperatures you would expect to measure at the surface of the beach and at a depth of 12 inches in dry sand.
   b. If you stood waist deep in the water and measured the water's surface temperature and its temperature at a depth of 12 inches, how would these measurements compare to those taken on the beach?

Tequilab/Shutterstock

8 On which summer day would you expect the *greatest* temperature range? Which would have the *smallest* range in temperature? Explain your choices.
   a. Cloudy skies during the day and clear skies at night
   b. Clear skies during the day and cloudy skies at night
   c. Clear skies during the day and clear skies at night
   d. Cloudy skies during the day and cloudy skies at night

9 The accompanying sketch map represents a hypothetical continent in the Northern Hemisphere. One isotherm has been placed on the map.
   a. Is the temperature higher at City A or City B?
   b. Is the season winter or summer? How are you able to determine this?
   c. Describe (or sketch) the position of this isotherm 6 months later.

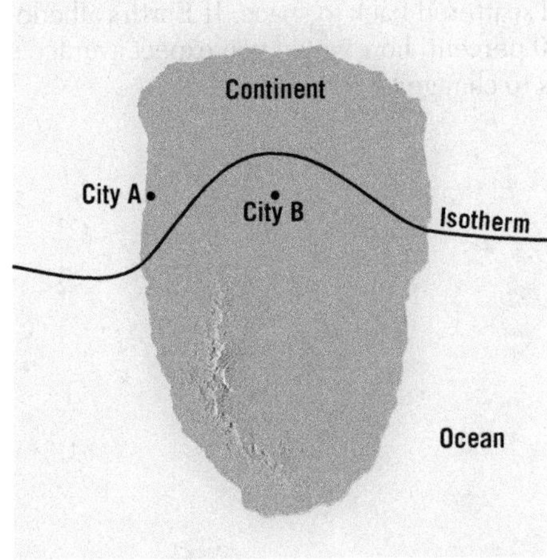

10 This photo shows a snow-covered area in the middle latitudes on a sunny day in late winter. Assume that 1 week after this photo was taken, conditions were essentially identical, except that the snow was gone. Would you expect the air temperatures to be different on the two days? If so, which day would be warmer? Suggest an explanation.

CoolR/Shutterstock

11 The Sun shines continually at the North Pole for 6 months, from the spring equinox until the fall equinox, yet temperatures never get very warm. Explain why this is the case.

12 The data below are mean monthly temperatures in degrees Celsius for an inland location that lacks any significant ocean influence. *Based on annual temperature range*, what is the approximate latitude of this place? Are these temperatures what you would normally expect for this latitude? If not, what control would explain these temperatures?

J	F	M	A	M	J	J	A	S	O	N	D
6.1	6.6	6.6	6.6	6.6	6.1	6.1	6.1	6.1	6.1	6.6	6.6

EXAMINING THE EARTH SYSTEM

1 Earth's axis is inclined 23½ degrees to the plane of its orbit. What if the inclination of the axis changed? Answer the following questions that address this possibility:

a. How would seasons be affected if Earth's axis were perpendicular to the plane of its orbit?

b. Describe the seasons if Earth's axis were inclined 40°. Where would the Tropics of Cancer and Capricorn be located? How about the Arctic and Antarctic Circles?

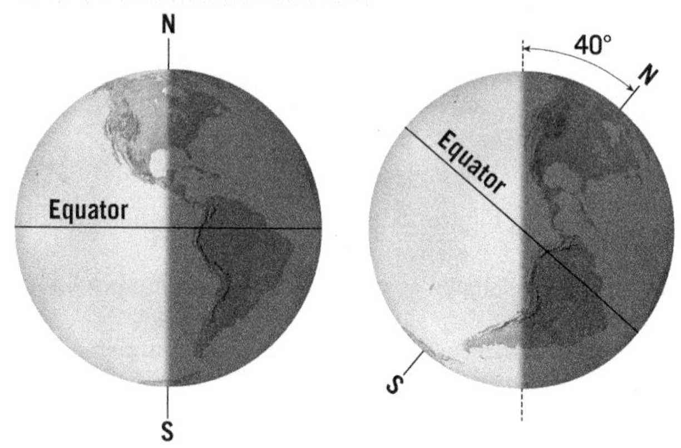

2 Speculate on the changes in global temperatures that might occur if Earth had substantially more land area and less ocean area than at present. How might such changes influence the biosphere?

3 Figure 16.22 shows that about 30 percent of the Sun's energy is reflected and scattered back to space. If Earth's albedo were to increase to 50 percent, how would you expect average surface temperatures to change? Explain.

DATA ANALYSIS

Ozone Hole Trends

The National Aeronautics and Space Administration monitors ozone concentrations in the stratosphere. Data from satellite instruments are used to create maps showing the amount of ozone and the extent of the ozone hole.

ACTIVITIES

Go to Ozone Hole Watch site, at http://ozonewatch.gsfc.nasa.gov/SH.html.

1 What is a Dobson unit? What is significant about 220 Dobson units?

2 How is the area of the ozone hole determined?

3 For the most recent date, was the ozone hole present over Antarctica? How did you determine this?

In the Animations section, click on the daily animation for the most recent year.

4 During which months is the ozone hole most prominent? What season is this in the Southern Hemisphere?

5 Describe the variation in shape and location of the ozone hole.

Go back to the Ozone Hole Watch page and examine the Annual Records graph for average ozone hole area and minimum ozone.

6 What year was the largest average ozone hole area? What was the area of the ozone hole during that year?

7 What is the area of the ozone hole for the most recent full year?

8 What year was the minimum average ozone measured? What was the amount of ozone, in Dobson units?

9 What is the minimum ozone amount for the current year?

Click on the Meteorology tab and examine the most recent plots for ozone hole area and ozone minimum.

10 How does the most recent full year's ozone hole area compare to the largest ozone hole area (thin black line)? Discuss both the area and timing of the ozone hole area.

11 Based on the animations, graphs, and Meteorology tab readings, what three factors control the size of the ozone hole?

12 How does the most recent full year's minimum ozone value compare to the lowest ozone values measured (bottom thin black line)? Discuss both the ozone amount and the timing of the lowest ozone values.

17

Moisture, Clouds, and Precipitation

FOCUS ON CONCEPTS

Each statement represents the primary learning objective for the corresponding major heading within the chapter. After you complete the chapter, you should be able to:

17.1 Summarize the six processes by which water changes from one state of matter to another. For each, indicate whether energy is absorbed or released.

17.2 Write a generalization relating air temperature and the amount of water vapor needed to saturate air.

17.3 Describe adiabatic temperature changes and explain why the wet adiabatic rate of cooling is less than the dry adiabatic rate.

17.4 List and describe the four mechanisms that cause air to rise.

17.5 Describe how atmospheric stability is determined and compare conditional instability with absolute instability.

17.6 Name and describe the 10 basic cloud types, based on form and height. Contrast nimbostratus and cumulonimbus clouds and their associated weather.

17.7 Identify the basic types of fog and describe how each forms.

17.8 Describe the Bergeron process and explain how it differs from the collision–coalescence process.

17.9 Describe the atmospheric conditions that produce sleet, freezing rain (glaze), and hail.

17.10 List the advantages of using weather radar versus a standard rain gauge to measure precipitation.

Large lenticular cloud forming downwind of Mt. Drum, Wrangell Mountains, Alaska.
(Photo by Michael De Young/Getty Images)

WATER VAPOR IS AN ODORLESS, colorless gas that mixes freely with the other gases of the atmosphere. Unlike oxygen and nitrogen—the two most abundant components of the atmosphere—water can change from one state of matter to another (solid, liquid, or gas) at the temperatures and pressures experienced on Earth (**Figure 17.1**). Because of this unique property, water leaves the oceans as a gas and returns to the oceans as a liquid.

As you observe day-to-day weather changes, you might ask: Why is it generally more humid in the summer than in the winter? Why do clouds form on some occasions but not on others? Why do some clouds look thin and harmless, whereas others form dark and ominous towers? Answers to these questions involve the role of water vapor in the atmosphere, the central theme of this chapter.

17.1 Water's Changes of State

Summarize the six processes by which water changes from one state of matter to another. For each, indicate whether energy is absorbed or released.

Water is the only substance that naturally exists on Earth as a solid (ice), liquid, and gas (water vapor). Because all forms of water are composed of water molecules (H_2O), each consisting of one oxygen atom bonded to two hydrogen atoms, the primary difference among water's three phases is the arrangement of these water molecules.

Ice, Liquid Water, and Water Vapor

Ice is composed of water molecules that form a tight, orderly network held together by mutual molecular attractions, as shown in **Figure 17.2**. As a consequence, the water molecules in ice are not free to move relative to each other but rather vibrate about fixed sites. When ice is heated, the molecules oscillate more rapidly. When the rate of molecular movement increases sufficiently, the bonds between the water molecules begin to break, resulting in melting.

In the liquid state, water molecules are still tightly packed but are moving fast enough that they are able to slide past one another. As a result, liquid water is fluid and takes the shape of its container.

As liquid water gains heat from its environment, some of the molecules acquire enough energy to break the remaining molecular attractions and escape from the surface, becoming water vapor. Water-vapor molecules are widely spaced compared to liquid water and exhibit very energetic random motion. Unlike a liquid, a gas will expand to occupy a container of any size, and it can also be compressed. For example, you can easily put more and more air into a tire and increase its volume only slightly. However, you can't put 10 gallons of gasoline into a 5-gallon can.

▼ SmartFigure 17.1
Heavy rain and hail at a ballpark in Wichita, Kansas (Photo by Fernando Salazar/AP Image)

ANIMATION
https://goo.gl/HCd3wX

Latent Heat

Whenever water changes state, heat is exchanged between water and its surroundings. Heat is absorbed, for example, when water evaporates (see Figure 17.2). Meteorologists often measure heat energy in calories. One **calorie** is the amount of heat required to raise the temperature of 1 gram of liquid water 1°C (1.8°F). Thus, when 10 calories of heat are added to 1 gram of water, the water molecules vibrate faster, and a 10°C (18°F) temperature rise occurs.

Under certain conditions, a substance can absorb heat without an accompanying rise in temperature. For example, when a glass of water with ice is warmed, the temperature of the ice–water mixture remains a constant 0°C (32°F) until all the ice has melted. If adding heat does not raise the temperature, where does this energy go? In this case, the added energy goes into breaking the molecular attractions between the water molecules in the ice cubes.

Because the heat used to melt ice does not produce a temperature change, it is referred to as **latent heat**.

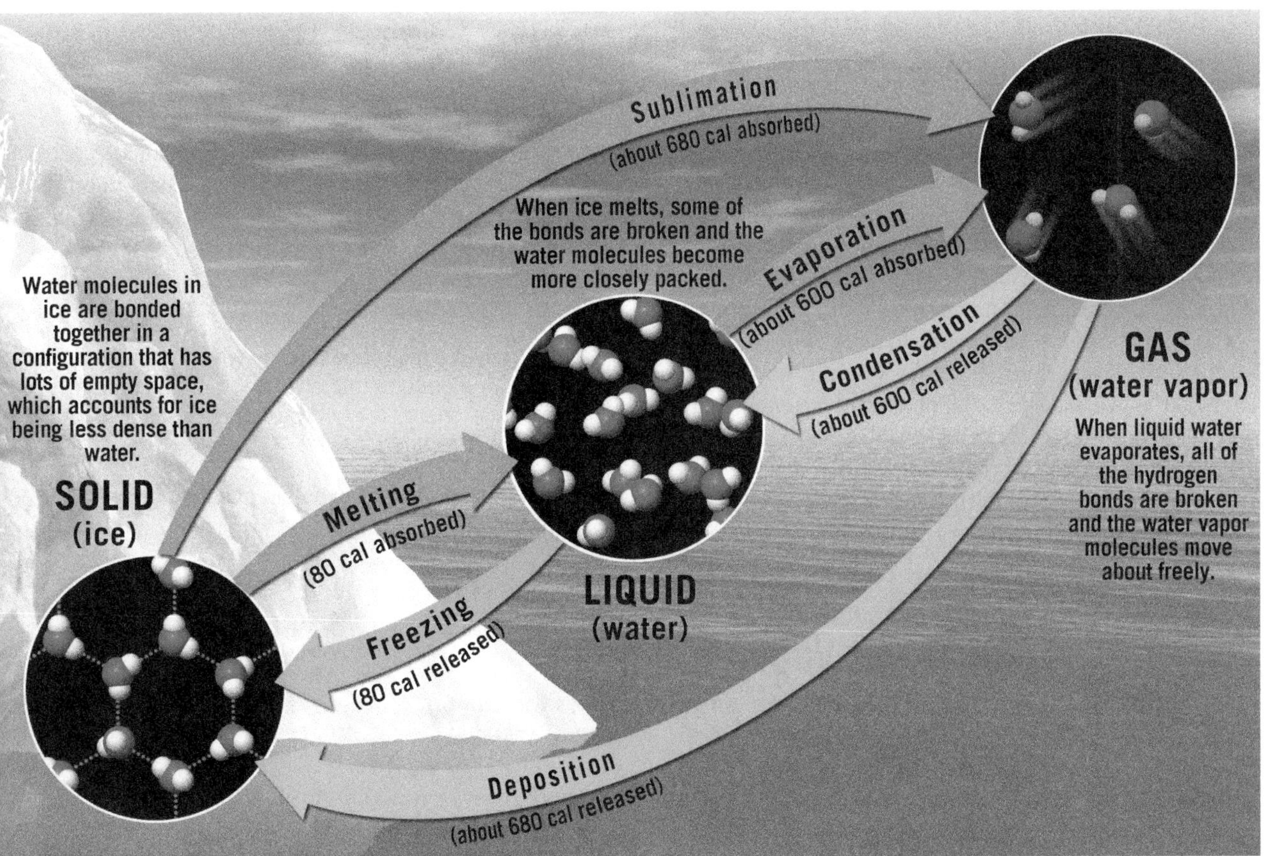

SmartFigure 17.2
Changes of state involve an exchange of heat The amounts shown here are the approximate numbers of calories absorbed or released when 1 gram of water changes from one state of matter to another.

ANIMATION
https://goo.gl/1ANliR

(*Latent* means "hidden," like the latent fingerprints hidden at a crime scene.) This energy can be thought of as being *stored in liquid water*, and it is not released to its surroundings as heat until the liquid returns to the solid state.

Melting 1 gram of ice requires 80 calories, an amount referred to as *latent heat of melting*. *Freezing*, the reverse process, releases these 80 calories per gram to its surroundings as *latent heat of freezing*.

Evaporation and Condensation We saw that heat is absorbed when ice is converted to liquid water. Heat is also absorbed during **evaporation**, the process of converting a liquid to a gas (water vapor). The energy absorbed by water molecules during evaporation is used to give them the motion needed to escape the surface of the liquid and become a gas. This energy is referred to as the *latent heat of vaporization*. During the process of evaporation, the higher-temperature (faster-moving) molecules escape the surface. As a result, the average molecular motion (temperature) of the remaining water is reduced—hence the common expression "evaporation is a cooling process." You have undoubtedly experienced this cooling effect when stepping dripping wet from a swimming pool or bathtub. In such a situation, the energy used to evaporate water comes from your skin, and you feel cool.

The reverse process, **condensation**, occurs when water vapor changes to the liquid state. During condensation, water-vapor molecules release energy (*latent heat of condensation*) in an amount equivalent to what was absorbed during evaporation. When condensation occurs in the atmosphere, it results in the formation of such phenomena as fog and clouds (Figure 17.3A).

Condensation of water vapor generates phenomena such as dew, clouds, and fog.
A.

Frost on a windowpane, an example of deposition.
B.

Figure 17.3
Examples of condensation and deposition (Photo A by NaturePL/SuperStock; photo B by elen_studio/Fotolia)

As you will see, latent heat plays an important role in many atmospheric processes. In particular, when water vapor condenses to form cloud droplets, latent heat of condensation is released, warming the surrounding air and giving it buoyancy. When the moisture content of air is high, this process can spur the growth of towering storm clouds.

Sublimation and Deposition You are probably least familiar with the last two processes illustrated in Figure 17.2—sublimation and deposition. **Sublimation** is the conversion of a solid directly to a gas, without passing through the liquid state. Examples you may have observed include the gradual shrinking of unused ice cubes in the freezer and the rapid conversion of dry ice (frozen carbon dioxide) to wispy clouds that quickly disappear.

Deposition refers to the reverse process: the conversion of a vapor directly to a solid. This change occurs, for example, when water vapor is deposited as ice on solid objects such as grass or windows (**Figure 17.3B**). These deposits are called *white frost* or *hoar frost*, or simply *frost*. A household example of the process of deposition is the "frost" that accumulates in a freezer. As shown in Figure 17.2, deposition releases an amount of energy equal to the total amount released by condensation and freezing.

CONCEPT CHECKS 17.1

1. Summarize the processes by which water changes from one state of matter to another. Indicate whether energy is absorbed or released.
2. What is *latent heat*?
3. What is a common example of sublimation?
4. How does frost form?

17.2 Humidity: Water Vapor in the Air

Write a generalization relating air temperature and the amount of water vapor needed to saturate air.

Water vapor constitutes only a small fraction of the atmosphere, varying from as little as one-tenth of 1 percent up to about 4 percent by volume. But the importance of water in the air is far greater than these small percentages would indicate. Indeed, scientists agree that *water vapor* is the most important gas in the atmosphere when it comes to understanding atmospheric processes.

Humidity is the amount of water vapor in air. Meteorologists employ several methods to express the water-vapor content of the air. Here we examine three: mixing ratio, relative humidity, and dew-point temperature.

Saturation

Before we consider these humidity measures further, it is important to understand the concept of **saturation**. Imagine a closed jar that contains water overlain by dry air, both at the same temperature. As the water begins to evaporate from the water surface, a small increase in pressure can be detected in the air above. This increase is the result of the motion of the water-vapor molecules that were added to the air through evaporation. In the open atmosphere, this pressure is termed **vapor pressure** and is defined as the part of the total atmospheric pressure that can be attributed to the water-vapor content.

In the closed container, as more and more molecules escape from the water surface, the steadily increasing vapor pressure in the air above forces more and more of these molecules to return to the liquid. Eventually the number of vapor molecules returning to the surface will balance the number leaving. At that point, the air is *saturated*: It can hold no more water vapor. However, if we add heat to the container, which would increase the temperature of the water and air, more water will evaporate before a balance is reached. Consequently, at higher temperatures, more moisture is required to reach saturation. The amount of water vapor required for saturation at various temperatures is shown in **Table 17.1**.

Mixing Ratio

Not all air is saturated, of course. Thus, we need ways to express how humid a parcel of air is. One method is to specify the amount of water vapor contained in a unit

Table 17.1 Amount of Water Vapor Needed to Saturate 1 Kilogram of Dry Air at Various Temperatures

Temperature °C (°F)	Water-Vapor Content at Saturation (grams)
−40 (−40)	0.1
−30 (−22)	0.3
−20 (−4)	0.75
−10 (14)	2
0 (32)	3.5
5 (41)	5
10 (50)	7
15 (59)	10
20 (68)	14
25 (77)	20
30 (86)	26.5
35 (95)	35
40 (104)	47

of air. The **mixing ratio** is *the mass of water vapor in a unit of air compared to the remaining mass of dry air*:

$$\text{mixing ratio} = \frac{\text{mass of water vapor (grams)}}{\text{mass of dry air (kilograms)}}$$

Because the mixing ratio is expressed in units of mass (usually in grams per kilogram), it is not affected by changes in pressure or temperature—the quantity of water vapor in the air remains the same. However, measuring the mixing ratio by direct sampling is time-consuming. Thus, meteorologists commonly use other methods to express the moisture content of the air. These include relative humidity and dew-point temperature.

Relative Humidity

The most familiar and, unfortunately, the most misunderstood term used to describe the moisture content of air is relative humidity. **Relative humidity** *is the ratio of the air's actual water-vapor content to the amount of water vapor required for saturation at that temperature (and pressure)*. Thus, unlike the mixing ratio, relative humidity indicates how near the air is to saturation, rather than the actual quantity of water vapor in the air.

To illustrate, we see from Table 17.1 that at 25°C (77°F), air is saturated when it contains 20 grams of water vapor per kilogram of dry air. Thus, if the air contains 10 grams of water vapor per kilogram of dry air on a 25°C day, the relative humidity is expressed as 10/20, or 50 percent. If air with a temperature of 25°C has a water-vapor content of 20 grams per kilogram, the relative humidity would be expressed as 20/20, or 100 percent. When the relative humidity reaches 100 percent, the air is saturated.

Because relative humidity depends both on the air's water-vapor content and on the amount of moisture required for saturation, it can be changed in either of two ways. First, relative humidity can be changed by adding or removing water vapor. Second, because the amount of moisture required for saturation is a function of air temperature (the warmer the air, the more water is required to saturate it), relative humidity varies with temperature.

How Changes in Moisture Affect Relative Humidity

In nature, moisture is added to the air mainly via evaporation from the oceans. However, plants, soil, and smaller bodies of water also make substantial contributions.

Notice in Figure 17.4 that when water vapor is added to a parcel of air, the relative humidity of the parcel increases until saturation occurs (100 percent relative humidity). What if even more moisture is added to this parcel of saturated air? Does the relative humidity exceed 100 percent? Normally, this situation does not occur. Instead, as in the closed container described previously, the excess water vapor condenses to form liquid water.

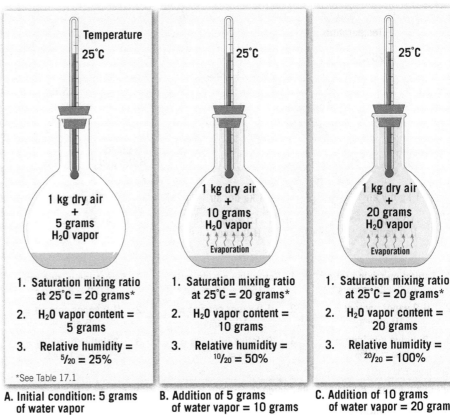

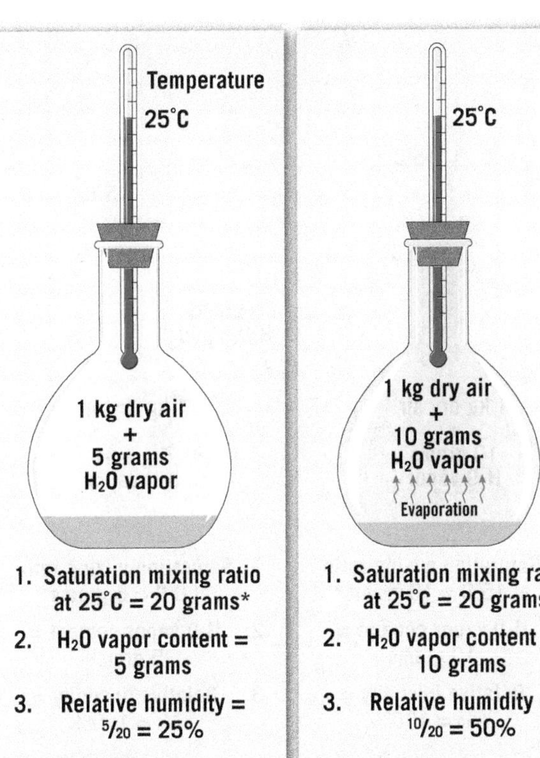

▲ Figure 17.4 **At a constant temperature, the relative humidity increases as water vapor is added to the air** The saturation mixing ratio for air at 25°C is 20 g/kg (see Table 17.1). As the water-vapor content in the flask increases, the relative humidity rises from 25 percent in **A.** to 100 percent in **C.**

You may have experienced such a situation while taking a hot shower. The water is composed of very energetic (hot) molecules, which means that the rate of evaporation is high. As long as you run the shower, the process of evaporation continually adds water vapor to the unsaturated air in the bathroom. If you stay in a hot shower long enough, the air eventually becomes saturated, and the excess water vapor begins to condense on the mirror, window, tile, and other cool surfaces in the room.

How Changes in Temperature Affect Relative Humidity

The second condition that affects relative humidity is air temperature. Examine Figure 17.5 carefully. Note in Figure 17.5A that when air at 25°C contains 10 grams of water vapor per kilogram of air, it has a relative humidity of 50 percent. This can be verified by referring to Table 17.1. Here we can see that at 25°C, air is saturated when it contains 20 grams of water vapor per kilogram of air. Because the air in Figure 17.5A contains 10 grams of water vapor, its relative humidity is 10/20, or 50 percent.

When the air in the flask is cooled from 25°C to 15°C, as shown in Figure 17.5B, the relative humidity increases from 50 to 100 percent. We can conclude that when the water-vapor content remains constant, *a decrease in temperature results in an increase in relative humidity*.

But there is no reason to assume that cooling would cease the moment the air reached saturation. What happens when the air is cooled below the temperature at which saturation occurs? Figure 17.5C illustrates this

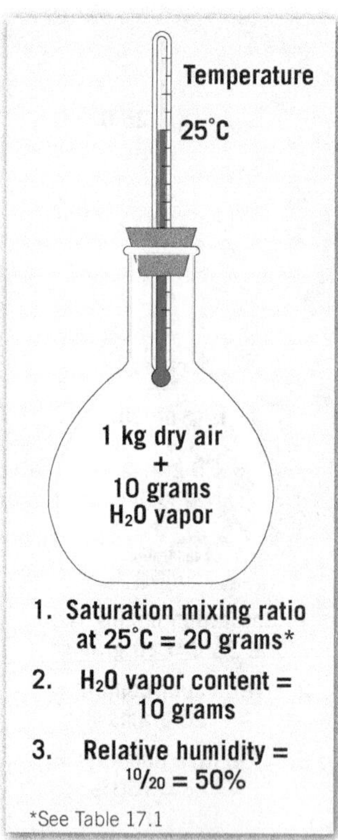

A. Initial condition: 25°C

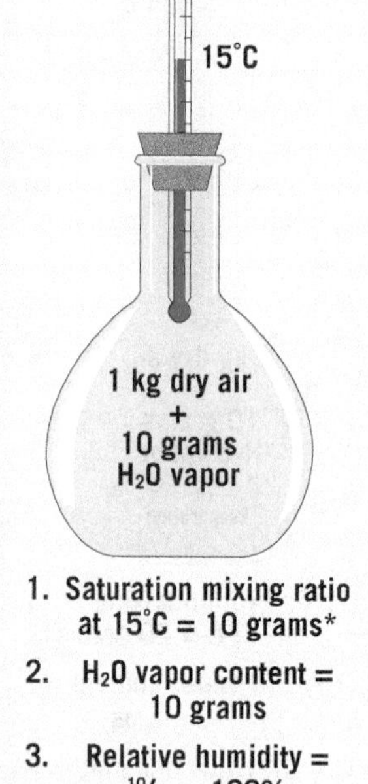

B. Cooled to 15°C

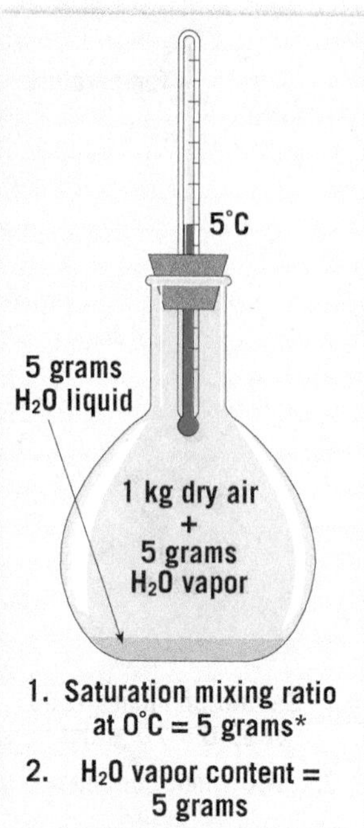

C. Cooled to 5°C

Figure 17.5 Relative humidity varies with temperature When the water-vapor content (mixing ratio) remains constant, the relative humidity changes when the air temperature increases or decreases. In this example, when the temperature of the air in the flask is lowered from 25°C in **A.** to 15°C in **B.**, the relative humidity increases from 50 to 100 percent. Further cooling from 15°C in B to 5°C in **C.** causes one-half of the water vapor to condense. In nature, when saturated air cools, the result is condensation in the form of clouds, dew, or fog.

situation. Notice from Table 17.1 that when the flask is cooled to 5°C, the air is saturated at 5 grams of water vapor per kilogram of air. Because this flask originally contained 10 grams of water vapor, 5 grams of water vapor will condense to form liquid droplets that collect at the bottom of the container. In the meantime, the relative humidity of the air inside remains at 100 percent.

Similarly, when rising air reaches an elevation where it is cooled below its dew-point temperature, some of the water vapor condenses to form clouds. Because clouds are made of tiny liquid droplets (or ice crystals), this moisture is no longer part of the water-vapor content of the air.

We can summarize the effects of temperature on relative humidity as follows: When the water-vapor content of air remains at a constant level, a decrease in air temperature results in an increase in relative humidity, and an increase in temperature causes a decrease in relative humidity. Figure 17.6 illustrates the variations in temperature and relative humidity during a typical day and the relationship described above.

Dew-Point Temperature

The **dew-point temperature**, or simply the **dew point**, of a given parcel of air is *the temperature at which water vapor begins to condense*. The term *dew point* stems from the fact that at night, objects near the ground often cool below the dew-point temperature and become coated with dew. You have undoubtedly seen "dew" form on an ice-cold drink on a humid summer day (Figure 17.7). In nature, cooling air below its dew-point temperature typically generates dew, fog, or clouds when the dew point is above freezing and frost when it is below freezing (0°C [32°F]).

Dew point can also be defined as the *temperature at which a parcel of air reaches saturation* and, hence, is directly related to the *actual moisture content* of that parcel. Recall that the saturation vapor pressure is

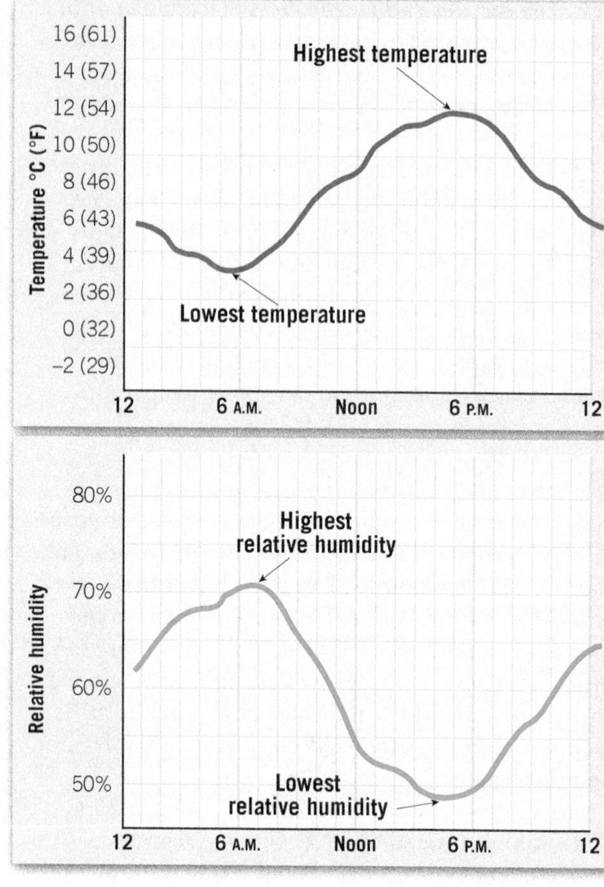

Figure 17.6 Daily variation in temperature and relative humidity during a typical spring day in Washington, DC

Figure 17.7 Condensation and dew-point temperature Condensation, or "dew," occurs when a cold drinking glass chills the surrounding layer of air below the dew-point temperature. (Photo by Nitr/Fotolia)

Table 17.2 Dew-Point Thresholds

Dew-Point Temperature	Threshold
≤ 10°F	Significant snowfall is inhibited
≥ 55°F	Minimum for severe thunderstorms to form
≥ 65°F	Considered humid by most people
≥ 70°F	Typical of the rainy tropics
≥ 75°F	Considered oppressive by most

temperature dependent. In fact, for every 10°C (18°F) increase in temperature, the amount of water vapor needed for saturation approximately doubles. Therefore, *saturated air* at 0°C (32°F) contains about half the water vapor of *saturated air* at 10°C (50°F) and roughly one-fourth that of *saturated air* at 20°C (68°F). Because the dew point is the temperature at which saturation occurs, we can conclude that high dew-point temperatures indicate moist air and, conversely, low dew-point temperatures indicate dry air (**Table 17.2**).

More precisely, based on what we have learned about vapor pressure and saturation, we can state that for every 10°C (18°F) increase in the dew-point temperature, air contains about twice as much water vapor. Therefore, we know that when air over Fort Myers, Florida, has a dew-point temperature of 25°C (77°F), it contains about twice the water vapor as the air over St. Louis, Missouri, with a dew point of 15°C (59°F) and four times that of air over Tucson, Arizona, with a dew point of 5°C (41°F).

Because the dew-point temperature is a good measure of the amount of water vapor in the air, it commonly appears on weather maps. When the dew point exceeds 65°F (18°C), most people consider the air to feel humid; air with a dew point of 75°F (24°C) or higher is considered oppressive. Notice on the map in **Figure 17.8** that much of the southeastern United States has dew-point temperatures that exceed 65°F (18°C). Also notice in Figure 17.8 that although the Southeast is dominated by humid conditions, most of the remainder of the country is experiencing drier air.

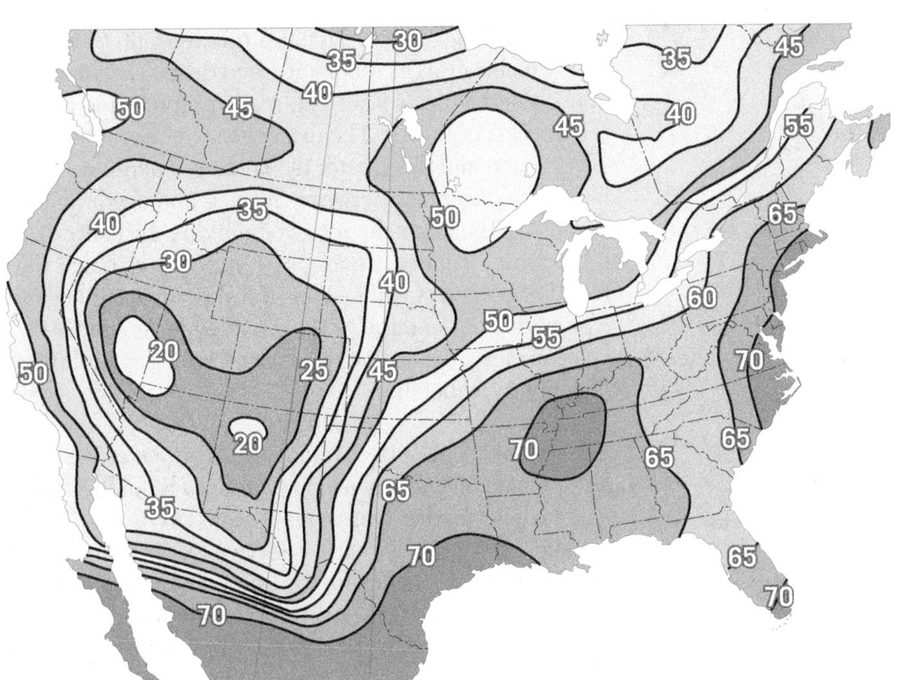

SmartFigure 17.8 **Surface map showing dew-point temperatures on a typical September day** Dew-point temperatures above 60°F dominate the southeastern United States, indicating that this region is blanketed with humid air.

TUTORIAL
https://goo.gl/re8J8P

How Is Humidity Measured?

Instruments called **hygrometers** (*hygro* = moisture, *metron* = measuring instrument) are used to measure the moisture content of the air.

Psychrometers One of the simplest hygrometers, a **psychrometer** (called a *sling psychrometer* when connected to a handle and spun), consists of two identical thermometers mounted side by side (**Figure 17.9A**). One thermometer, called the *dry bulb*, measures air temperature, and the other, called the *wet bulb*, has a thin cloth wick tied at the bottom. This cloth wick is saturated with water, and a continuous current of air is passed over the wick, either by swinging the psychrometer or by using an electric fan to move air past the instrument (**Figure 17.9B,C**). As a result, water evaporates from the

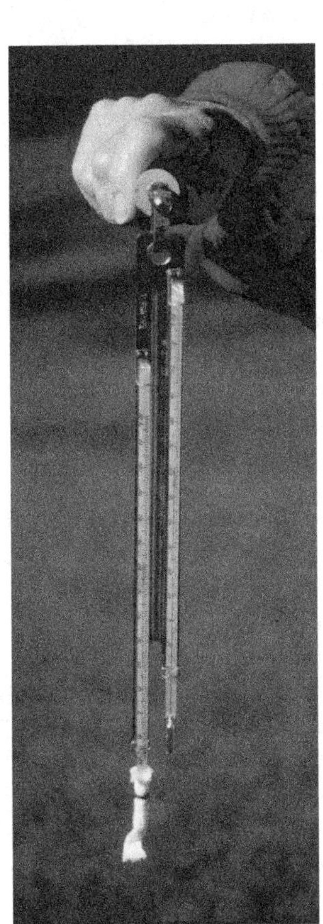

A. The dry-bulb thermometer gives the current air temperature.

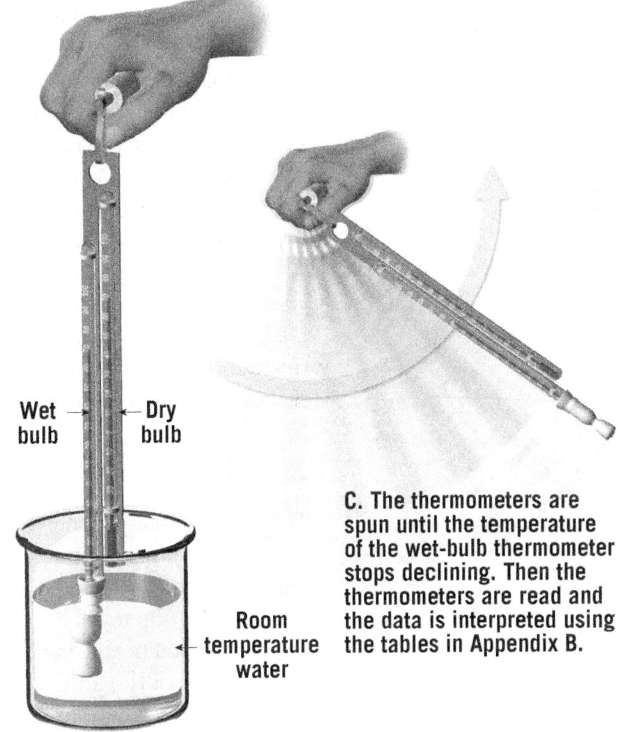

B. The wet-bulb thermometer is covered with a cloth wick that is dipped in water.

C. The thermometers are spun until the temperature of the wet-bulb thermometer stops declining. Then the thermometers are read and the data is interpreted using the tables in Appendix B.

Figure 17.9 **Sling psychrometer used to determine both relative humidity and dew point** (Photo by E. J. Tarbuck)

wick, absorbing heat energy from the wet-bulb thermometer, which causes its temperature to drop. The amount of cooling that takes place is directly proportional to the dryness of the air: The drier the air, the greater the evaporation and the greater the cooling. Therefore, the larger the difference between the wet- and dry-bulb temperatures, the lower the relative humidity. By contrast, if the air is saturated, no evaporation will occur, and the two thermometers will have identical readings. By using a psychrometer and the tables provided in Appendix B, you can easily determine the relative humidity and the dew-point temperature.

Electric Hygrometers Today, a variety of *electric hygrometers* are widely used to measure humidity. The Automated Weather Observing System (AWOS) operated by the National Weather Service (NWS) employs an electric hygrometer that works on the principle of *capacitance*—a material's ability to store an electrical charge. The sensor consists of a thin hygroscopic (water-absorbent) film that is connected to an electric current. As the film absorbs or releases water, the capacitance of the sensor changes at a rate proportional to the relative humidity of the surrounding air. Thus, relative humidity can be measured by monitoring the change in the film's capacitance. Higher capacitance means higher relative humidity.

CONCEPT CHECKS 17.2

1. List three measures used to express humidity.
2. If the amount of water vapor in the air remains unchanged, how does a decrease in temperature affect relative humidity?
3. Define *dew-point temperature*.
4. Which measure of humidity—relative humidity or dew point—best describes the actual quantity of water vapor in a mass of air?
5. Briefly describe the principle of a psychrometer.

17.3 Adiabatic Temperature Changes and Cloud Formation

Describe adiabatic temperature changes and explain why the wet adiabatic rate of cooling is less than the dry adiabatic rate.

Recall that condensation occurs when sufficient water vapor is added to the air or, more commonly, when the air is cooled to its dew-point temperature. Condensation may produce dew, fog, or clouds. Heat near Earth's surface is readily exchanged between the ground and the air directly above. As the ground loses heat in the evening (radiation cooling), dew may condense on the grass, while fog may form slightly above Earth's surface. Thus, surface cooling that occurs after sunset produces some condensation. Cloud formation, however, often takes place during the warmest part of the day—an indication that another mechanism must operate aloft that cools air sufficiently to generate clouds.

Adiabatic Temperature Changes

The process that generates most clouds is easily visualized. Have you ever pumped up a bicycle tire with a hand pump and noticed that the pump barrel became very warm? When you applied energy to *compress* the air, the motion of the gas molecules increased, and the temperature of the air rose. Conversely, if you allow air to escape from a bicycle tire, the air *expands*; the gas molecules move less rapidly, and the air cools. You have probably felt the same effect while applying hair spray or spray deodorant: The propellant gas cools as it expands out of the nozzle. Temperature changes of this type, in which heat energy is neither added nor subtracted, are called **adiabatic temperature changes**. In the cases described, the change in temperature is caused by a change in pressure. When air is compressed, it warms, and when air is allowed to expand, it cools.

Adiabatic Cooling and Condensation

To simplify the discussion of adiabatic cooling, imagine a volume of air enclosed in a thin expandable membrane. Meteorologists call this imaginary volume of air a **parcel**. Typically, we consider a parcel to be a few hundred cubic meters in volume, and we assume that it acts independently of the surrounding air. We can also assume that no heat is transferred into or out of the parcel. Although this model is highly idealized, over short time spans, actual parcels of air moving up or down in the atmosphere behave in much this way. (In nature, a rising or descending column of air is sometimes infiltrated by the surrounding air, a process called *entrainment*. For the following discussion, however, we assume that no mixing of this type occurs.)

Dry Adiabatic Rate Recall from Chapter 16 that atmospheric pressure decreases with height. Any time a parcel of air moves upward, therefore, the pressure surrounding it gradually decreases. As a result, it expands and cools adiabatically. Unsaturated air cools at a constant rate of 10°C for every 1000 meters of ascent

(17.5°F per 1000 feet). Conversely, descending air comes under increasing pressure and is compressed and heated by 10°C for every 1000 meters of descent (Figure 17.10). This rate of cooling or heating applies only to *unsaturated air* and is known as the **dry adiabatic rate** ("dry" because the air is unsaturated).

Wet Adiabatic Rate If an air parcel rises high enough, it will eventually cool to its dew point, triggering the process of condensation. The altitude at which a parcel reaches saturation and begins to form clouds is called the **lifting condensation level**, or simply **condensation level**. At the lifting condensation level, an important change occurs: The *latent heat* that was absorbed by the water vapor when it evaporated is released as **sensible heat**—energy that can be measured with a thermometer. Although the parcel will continue to cool adiabatically as it rises, the release of latent heat slows the rate of cooling. In other words, when a parcel of air ascends above the lifting condensation level, the rate at which it cools is reduced. This slower rate of cooling is called the **wet adiabatic rate**, also commonly referred to as the *moist*, or *saturated, adiabatic rate* (see Figure 17.10).

Because the amount of latent heat released depends on the amount of moisture present in the rising air (generally between 0 and 4 percent), the wet adiabatic rate varies from about 5°C per 1000 meters for air with a high moisture content to 9°C per 1000 meters for air with a low moisture content.

To summarize, rising air expands and cools at the dry adiabatic rate from the surface up to the lifting condensation level, after which it cools at the slower wet adiabatic rate.

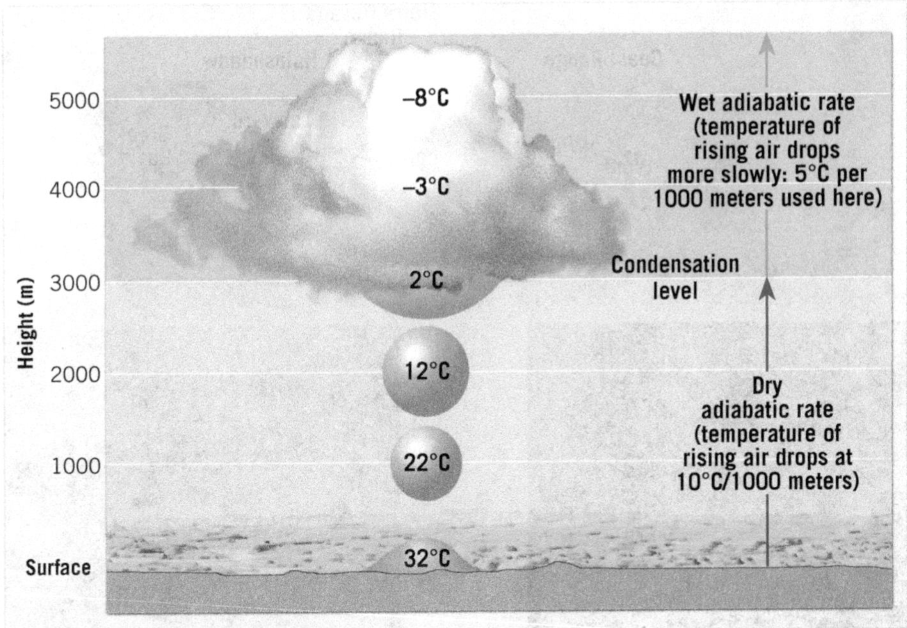

▲ Figure 17.10 **Dry versus wet adiabatic rates of cooling** Rising air cools at the relatively constant dry adiabatic rate of 10°C per 1000 meters until the air reaches the dew point and condensation (cloud formation) begins. As air continues to rise, the latent heat released by condensation reduces the rate of cooling. Because the amount of latent heat released depends on the amount of moisture present in the rising air, the wet adiabatic rate varies from about 5°C per 1000 meters for air with high water vapor content to 9°C per 1000 meters for dry air.

CONCEPT CHECKS 17.3

1. What name is given to the processes whereby the temperature of air changes without the addition or subtraction of heat?
2. Why does air expand as it moves upward through the atmosphere?
3. At what rate does unsaturated air cool when it rises through the atmosphere?
4. Why does the adiabatic rate of cooling change when condensation begins?
5. Why does the wet adiabatic rate not have a constant value?

17.4 Processes That Lift Air

List and describe the four mechanisms that cause air to rise.

Why does air rise on some occasions to produce clouds, but not on others? Generally, the tendency is for air to resist vertical movement; air near the surface tends to stay near the surface, and air aloft tends to remain aloft. However, the following four processes can cause air to rise, thereby generating clouds:

1. *Orographic lifting*, in which air is forced to rise over a mountainous barrier
2. *Frontal lifting*, in which warmer, less-dense air is forced over cooler, denser air
3. *Convergence*, which is a pileup of horizontal airflow that results in upward movement
4. *Localized convective lifting*, in which unequal surface heating causes localized pockets of air to rise because of their buoyancy

Orographic Lifting

Orographic lifting occurs when elevated terrain, such as a mountain range, act as a barrier to the flow of air (Figure 17.11). As air ascends a mountain slope, adiabatic cooling often generates clouds and copious precipitation. In fact, many of the rainiest places in the world are located on windward mountain slopes.

By the time air reaches the leeward side of a mountain, much of its moisture has been lost. If the air descends, it warms adiabatically, making condensation and precipitation even less likely. As shown in Figure 17.11, the result can be a **rainshadow desert** like the Great Basin Desert of the western United States, which lies only a few hundred kilometers from the Pacific Ocean but is effectively cut off from the ocean's moisture

A. Orographic lifting leads to precipitation on windward slopes.

B. By the time air reaches the leeward side of the mountains, much of the moisture has been lost, resulting in a *rainshadow desert*.

▲ Figure 17.11 **Orographic lifting and precipitation** (Photo A by Dean Pennala/Shutterstock; photo B by Dennis Tasa)

by the imposing Sierra Nevada. The Gobi Desert of Mongolia, the Takla Makan of China, and the Patagonia Desert of Argentina are other examples of deserts that exist because they are on the leeward sides of large mountain systems.

Frontal Lifting

If orographic lifting were the only mechanism that forced air aloft, the relatively flat central portion of North America would be an expansive desert rather than the area known as "the nation's breadbasket." Fortunately, this is not the case.

In central North America, warm and cold air masses often collide, producing boundaries called **fronts**. Rather than mixing, the cooler, denser air mass acts as a barrier over which the warmer, less-dense air rises. This process, called **frontal lifting**, or **frontal wedging**, is illustrated in Figure 17.12.

In the middle latitudes, most precipitation is due to weather-producing fronts associated with storm systems called *midlatitude cyclones*. Therefore, we will examine fronts in detail in Chapter 19.

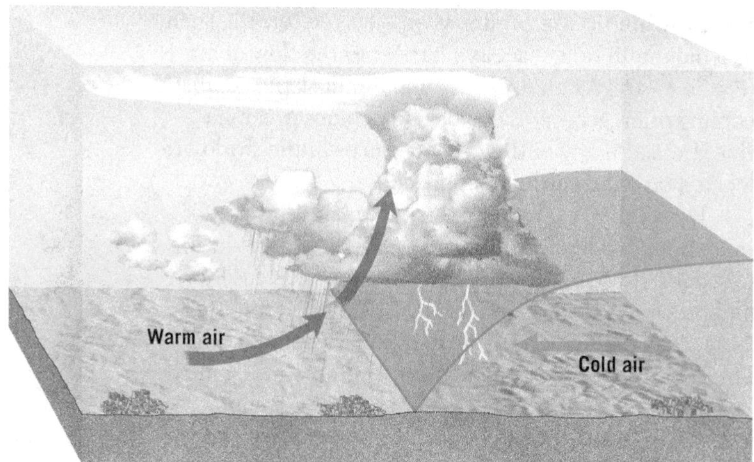

► Figure 17.12 **Frontal lifting** Colder, denser air acts as a barrier over which warmer, less-dense air is forced to rise.

Convergence

When the wind pattern near Earth's surface is such that more air is entering an area than is leaving—a phenomenon called **convergence**—lifting occurs (Figure 17.13). When air is compressed by convergence, it escapes by moving upward. One situation in which convergence causes lifting is when air is drawn into a center of *low pressure*, such as a midlatitude cyclone or hurricane. The inward flow of air at the bottom of these systems is balanced by rising air, cloud formation, and usually precipitation.

Convergence can also occur when an obstacle slows or restricts horizontal airflow (wind). For example, when air moves from a relatively smooth surface, such as the ocean, onto an irregular landscape, increased

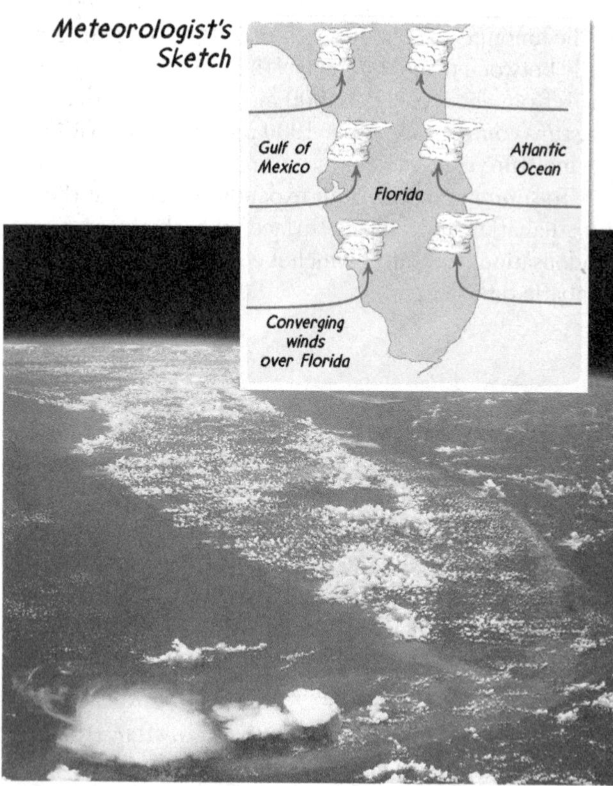

▲ SmartFigure 17.13 **Convergence over the Florida peninsula** When surface air converges, it is forced to rise. Florida provides a good example. On warm days, airflow from the Atlantic Ocean and Gulf of Mexico onto the Florida peninsula generates many mid-afternoon thunderstorms. (Courtesy of NASA)

TUTORIAL
https://goo.gl/uyi6Aw

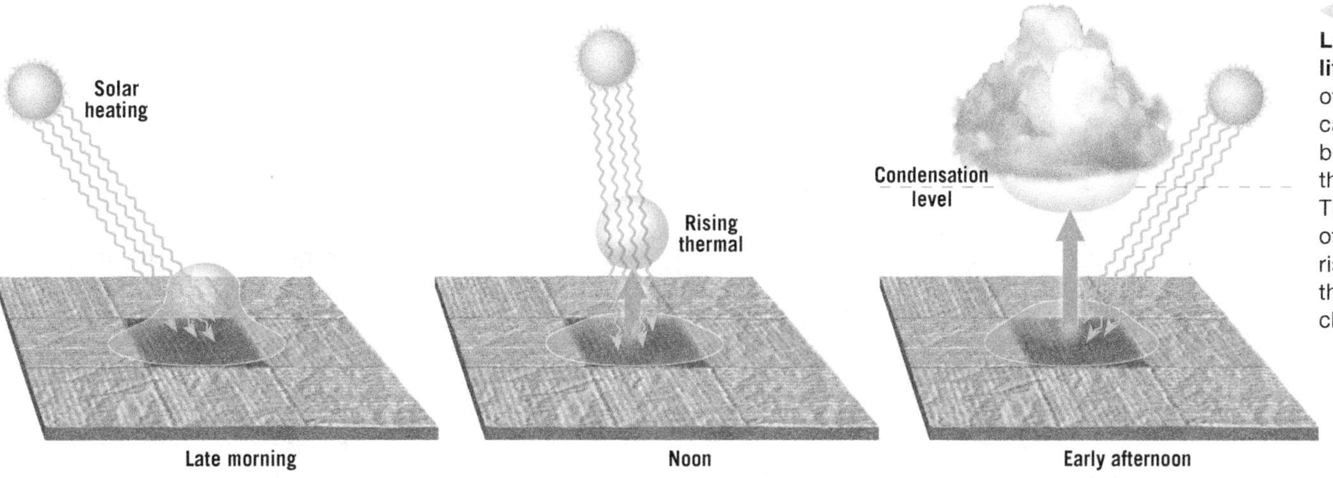

Figure 17.14 **Localized convective lifting** Unequal heating of Earth's surface causes pockets of air to be warmed more than the surrounding air. These buoyant parcels of hot air (thermals) rise, and if they reach the condensation level, clouds form.

friction reduces its speed. The result is a pileup of air (convergence).

The Florida peninsula provides an excellent example of the role that convergence can play in initiating cloud development and precipitation (see Figure 17.13). On warm days, the airflow is from the ocean to the land along both coasts of Florida. This leads to a pileup of air along the coasts and general convergence over the peninsula. This pattern of convergence and uplift is aided by intense solar heating of the land. As a result, Florida's peninsula experiences the greatest frequency of mid-afternoon thunderstorms in the United States.

Localized Convective Lifting

On warm summer days, unequal heating of Earth's surface may cause some pockets of air to be warmed more than the surrounding air (Figure 17.14). For instance, a plowed field or parking lot will absorb more radiation and hence impart more warmth to the overlying air than will adjacent fields of crops. Consequently, the overlying parcel of air, being warmer (less dense) than the surrounding air, will be buoyed upward. Such rising parcels of warmer air are called *thermals*. Thermals can carry birds such as hawks and eagles—and also human hang gliders—to great heights.

The phenomenon that produces rising thermals is called **localized convective lifting**, or simply **convective lifting**. When these warm parcels of air rise above the lifting condensation level, clouds form and occasionally produce mid-afternoon rain showers. The height of clouds produced in this fashion is somewhat limited because the buoyancy caused solely by unequal surface heating is confined to, at most, the first few kilometers of the atmosphere. Also, the accompanying rains, although occasionally heavy, are of short duration and widely scattered, a phenomenon called *sun showers*.

CONCEPT CHECKS 17.4

1. Explain why the Great Basin of the western United States is dry. What term is applied to this type of desert?
2. How does frontal lifting cause air to rise?
3. Define *convergence*. Identify two types of weather systems associated with convergence in the lower atmosphere.
4. Why does Florida have abundant mid-afternoon thunderstorms?
5. Describe convective lifting.

17.5 The Critical Weathermaker: Atmospheric Stability

Describe how atmospheric stability is determined and compare conditional instability with absolute instability.

When air rises, it cools and usually produces clouds. Why do clouds vary so much in size, and why does the resulting precipitation vary so much? The answers are closely related to the *stability* of the air.

Recall that a parcel of air can be thought of as having a thin, flexible cover that allows it to expand but prevents it from mixing with the surrounding air (picture a hot-air balloon). Imagine that you find such a parcel, with some given starting temperature, and forcibly lift it to a higher location. Its temperature will decrease because of expansion; its final temperature will depend on how warm it was to begin with and how high you lift it up.

When you let go of the parcel, what happens? If it is cooler (and hence denser) than the surrounding air, it will sink back down to its original location. Air of this type, called **stable air**, resists upward movement. But if it is *warmer* than the surrounding air (and hence less dense), it will continue to rise. Specifically, it will rise

Figure 17.15 **Hot air rises** If air is warmer and therefore less dense than its surroundings, it rises. Hot-air balloons rise through the atmosphere for this reason. (Photo by Steve Vidler/SuperStock)

until it reaches an altitude where its temperature equals that of its surroundings. This is exactly how a hot-air balloon works, rising as long as it is warmer and less dense than the surrounding air (Figure 17.15). This type of air is termed **unstable air**.

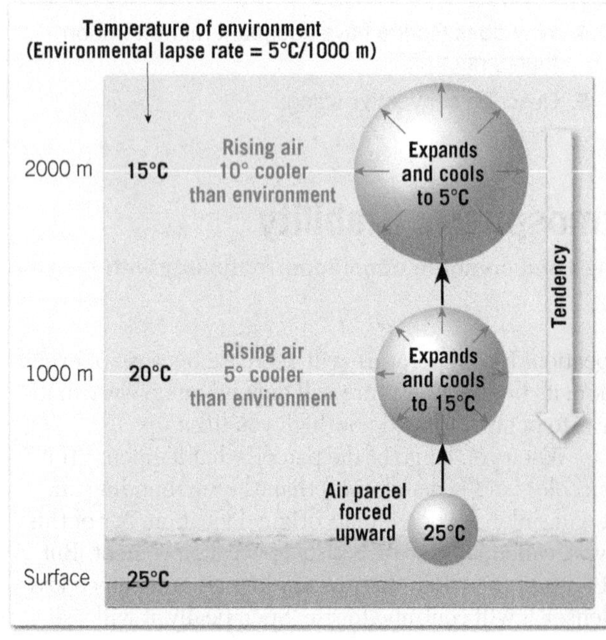

SmartFigure 17.16 **How the stability of the air is determined** When an unsaturated parcel of air is forced to rise, it expands and cools at the dry adiabatic rate of 10°C per 1000 meters. In this example, the temperature of the rising parcel of air is lower than that of the surrounding environment; therefore, the parcel is heavier than the surrounding air and, if allowed to do so, will sink to its original position. Air of this type is referred to as *stable*.

ANIMATION
https://goo.gl/oBRjlp

Types of Stability

Stability is a property of air that describes whether it resists rising (is stable) or may rise spontaneously (is unstable). To determine the stability of a given parcel of air, we first need to know how the temperature of the atmosphere above the parcel changes with height. Recall from Chapter 16 that this measure, determined from observations made by radiosondes and aircraft, is called the **environmental lapse rate**. It is important not to confuse this with *adiabatic temperature changes*, which are changes in the temperature of a rising or sinking parcel of air caused by expansion or compression.

To illustrate, we examine a situation in which the environmental lapse rate is 5°C per 1000 meters (Figure 17.16). Under this condition, when air at the surface has a temperature of 25°C, the air at 1000 meters will be 5° cooler, or 20°C, the air at 2000 meters will have a temperature of 15°C, and so forth. At first glance, it appears that the air at the surface is less dense than the air at 1000 meters because it is 5° warmer. However, if the air near the surface were unsaturated and were to rise to 1000 meters, it would expand and cool at the dry adiabatic rate of 10°C per 1000 meters. Therefore, upon reaching 1000 meters, its temperature would have dropped 10°C. Being 5° cooler than its environment, it would be denser and tend to sink to its original position. Hence, we say that the air near the surface is potentially cooler than the air aloft and therefore will not rise on its own. The air just described is *stable* and resists vertical movement.

Absolute Stability Stated quantitatively, **absolute stability** prevails when the environmental lapse rate is less than the wet adiabatic rate. Figure 17.17 depicts this situation using an environmental lapse rate of 5°C per 1000 meters and a wet adiabatic rate of 6°C per 1000 meters. Note that at 1000 meters, the temperature of the surrounding air is 15°C, while the rising parcel of air has cooled to 10°C and is therefore the denser air. Even if this stable air were to be forced above the condensation level, it would remain cooler and denser than its environment, and thus it would tend to return to the surface.

The most stable conditions occur when the temperature in a layer of air actually increases with altitude rather than decreases. When such a reversal occurs, a *temperature inversion* is said to exist. Temperature inversions frequently occur on clear nights as a result of radiation cooling of Earth's surface. Under these conditions, an inversion is created because the ground and the air immediately above will cool more rapidly than the air aloft. When warm air overlies cooler air, it acts as a lid and prevents appreciable vertical mixing. Because of this, temperature inversions are responsible for trapping pollutants in a narrow zone near Earth's surface.

Absolute Instability At the other extreme from absolute stability, air is said to exhibit **absolute instability** when

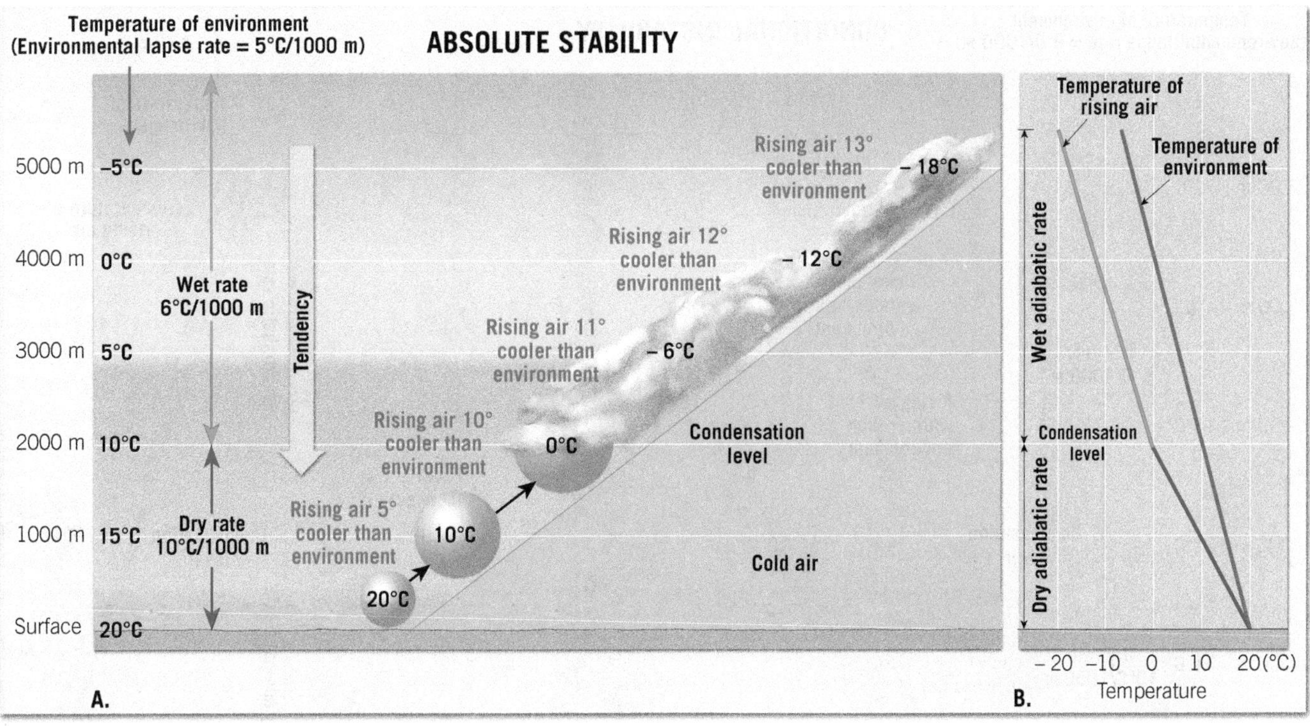

SmartFigure 17.17 **Atmospheric conditions that result in absolute stability** Absolute stability prevails when the environmental lapse rate is less than the wet adiabatic rate. **A.** The rising parcel of air is always cooler and heavier than the surrounding air, producing stability. **B.** Graphical representation of the conditions shown in part A.

TUTORIAL
https://goo.gl/oYR2ga

the environmental lapse rate is greater than the dry adiabatic rate. As shown in Figure 17.18, the ascending parcel of air is always warmer than its environment and will continue to rise because of its own buoyancy. However, the conditions needed to render the air absolutely unstable mainly occur near Earth's surface. On hot, sunny days the air above some surfaces, such as shopping center parking lots, is heated more than the air over adjacent surfaces. These invisible pockets of more intensely heated air, being less dense than the air aloft, will rise like a hot-air balloon. This phenomenon produces the small, fluffy clouds we associate with fair weather. Occasionally, when the surface air is considerably warmer than the air aloft, clouds with considerable vertical development can form.

Conditional Instability A more common type of atmospheric instability is called **conditional instability**. This occurs when moist air has an environmental lapse rate

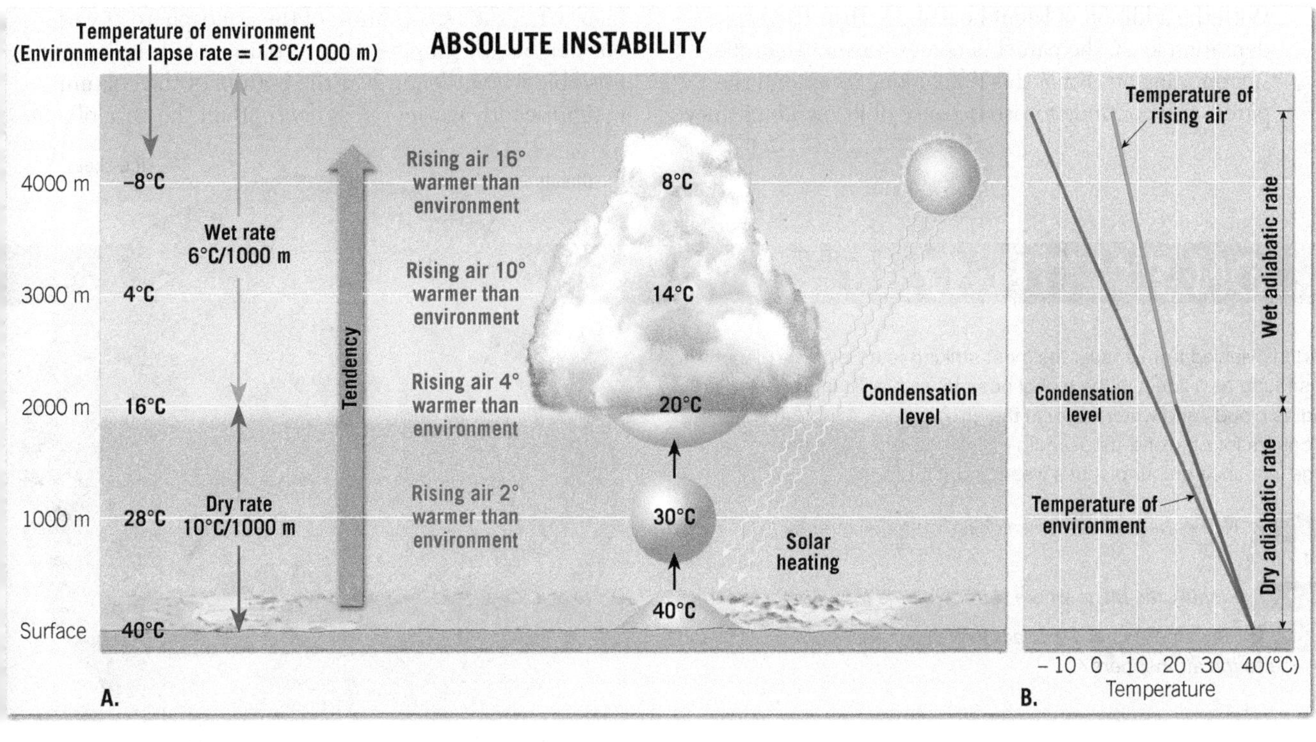

Figure 17.18 **Atmospheric conditions that result in absolute instability** Absolute instability can develop when solar heating causes the lowermost layer of the atmosphere to be warmed to a much higher temperature than the air aloft. **A.** The result is a steep environmental lapse rate that renders the atmosphere unstable. **B.** Graphical representation of the conditions shown in part A.

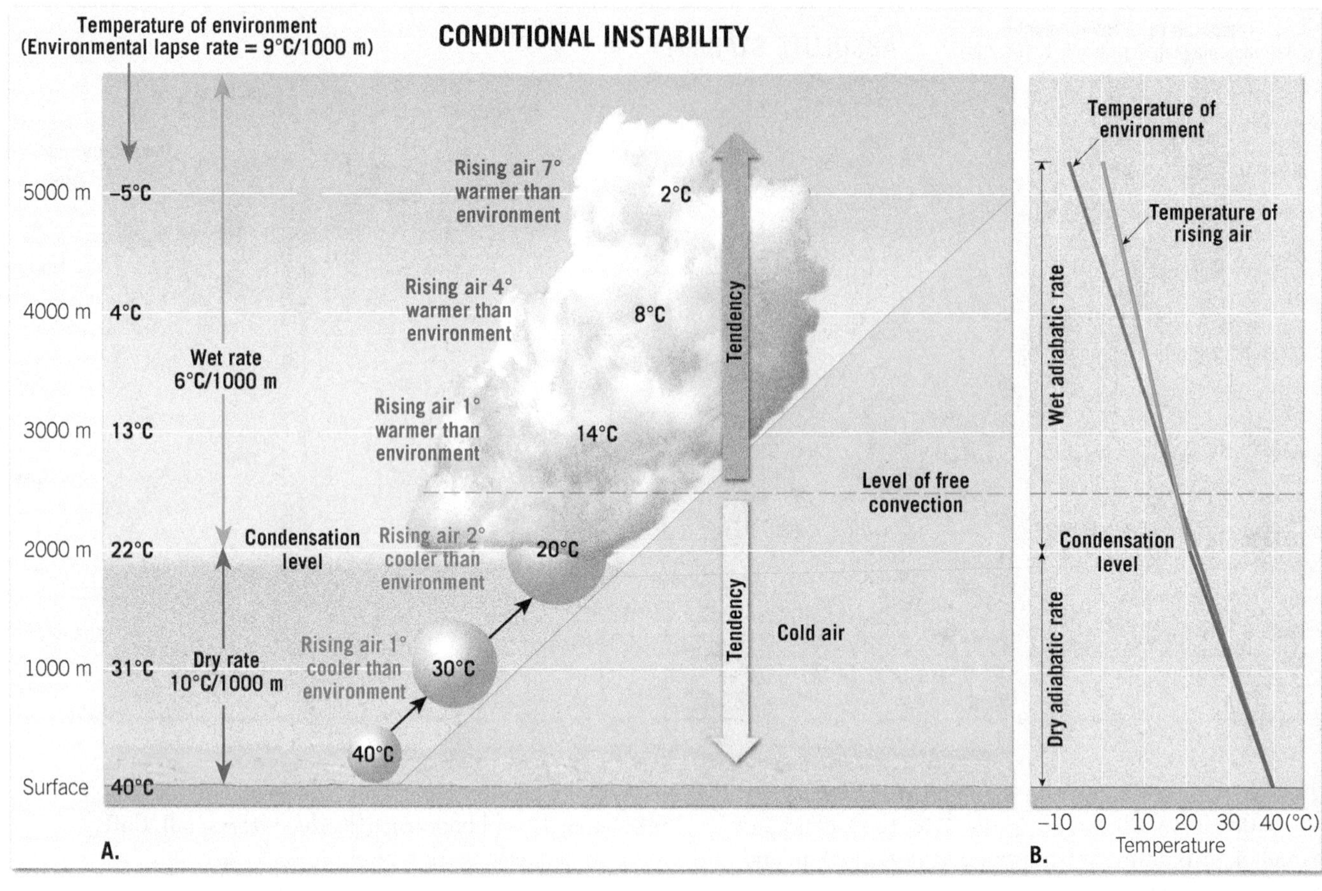

Figure 17.19 Atmospheric conditions that result in conditional instability Conditional instability may result when warm air is forced to rise along a frontal boundary. Note that the environmental lapse rate of 9°C per 1000 meters lies between the dry and wet adiabatic rates. **A.** The parcel of air is cooler than the surrounding air up to nearly 3000 meters, so in this range its tendency is to sink toward the surface (it is stable). Above this level, however, condensation causes the parcel to be warmer than its environment, so it rises because of its own buoyancy (it is unstable). Thus, when conditionally unstable air is forced to rise, the result can be towering cumulus clouds. **B.** Graphical representation of the conditions shown in part A.

between the dry and wet adiabatic rates (between 5°C and 10°C per 1000 meters). Simply, the atmosphere is said to be conditionally unstable when it is *stable* for an *unsaturated* parcel of air but *unstable* for a *saturated* parcel of air. Notice in **Figure 17.19** that the rising parcel of air is cooler than the surrounding air for nearly 3000 meters. With the addition of latent heat above the lifting condensation level, the parcel becomes warmer than the surrounding air. From this point along its ascent, the parcel will continue to rise because of its own buoyancy, without an outside lifting force. Thus, conditional instability depends on whether the rising air is saturated. The word *conditional* is used because the air must be forced upward, such as over mountainous terrain, before it becomes unstable and rises because of its own buoyancy.

In summary, the stability of air is determined by measuring the temperature of the atmosphere at various heights. In simple terms, a column of air is deemed unstable when the air near the bottom of the column is significantly warmer (less dense) than the air aloft,

EYE ON EARTH 17.1

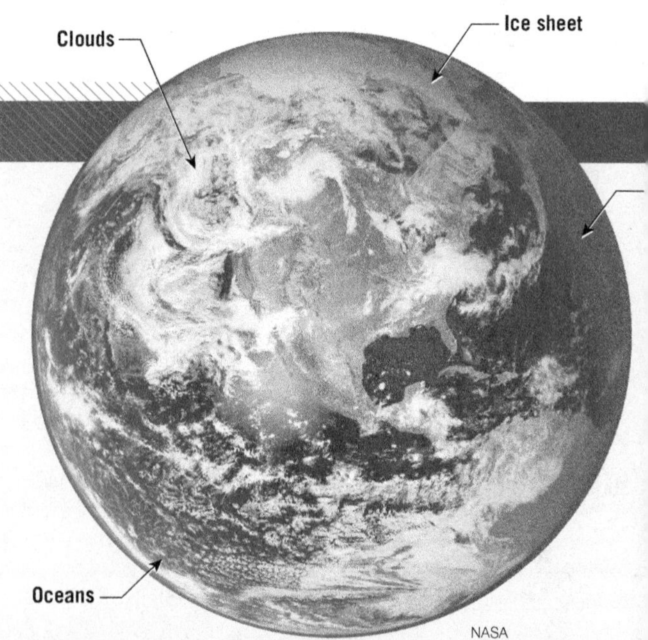

When Earth is viewed from space, the most striking feature of the planet is *water*. It is found as a liquid in the global oceans, as a solid in the polar ice caps, and as clouds and water vapor in the atmosphere. Although water vapor accounts for only one-thousandth of 1 percent of the water on Earth, it has a huge influence on our planet's weather and climate.

QUESTION 1 *Name the process by which ice changes directly from a solid to water vapor.*

QUESTION 2 *Identify the place where water vapor is found in this image.*

QUESTION 3 *How does water vapor act to transfer heat from Earth's land–sea surface to the atmosphere?*

indicating a steep environmental lapse rate. Under these conditions, the air actually turns over, as the warm air below rises and displaces the colder air aloft. Conversely, the air column is considered to be stable when the temperature drops relatively slowly with increasing altitude. The most stable conditions occur during a temperature inversion, when the temperature actually increases with height. Under these conditions, there is very little vertical air movement.

Stability and Daily Weather

From the previous discussion, we can conclude that stable air resists upward movement, whereas unstable air ascends freely because of its own buoyancy. But how do these facts manifest themselves in our daily weather?

Because stable air resists upward movement, we might conclude that clouds will not form when stable conditions prevail in the atmosphere. Although this seems reasonable, recall that processes exist that *force* air aloft. These processes include orographic lifting, frontal wedging, and convergence. When stable air is forced aloft, the clouds that form are widespread and have little vertical thickness when compared to their horizontal dimension, and precipitation, if any, is light to moderate.

By contrast, clouds associated with the lifting of unstable air are towering and often generate thunderstorms and occasionally even tornadoes. For this reason, we can conclude that on a dreary, overcast day with light drizzle, stable air has been forced aloft. On the other hand, during a day when cauliflower-shaped clouds appear to be growing as if bubbles of hot air are surging upward, we can be fairly certain that the ascending air is unstable.

In summary, stability plays an important role in determining our daily weather. To a large degree, stability determines the type of clouds that develop and whether precipitation will come as a gentle shower or a heavy downpour.

CONCEPT CHECKS 17.5

1. Explain the difference between the environmental lapse rate and adiabatic cooling.
2. How is the stability of air determined?
3. Write a statement relating the environmental lapse rate to stability.
4. What types of clouds and precipitation, if any, form when stable air is forced aloft?
5. Describe the weather associated with unstable air.

17.6 Condensation and Cloud Formation

Name and describe the 10 basic cloud types, based on form and height. Contrast nimbostratus and cumulonimbus clouds and their associated weather.

To review briefly, condensation occurs when water vapor in the air changes to a liquid. The result of this process may be dew, fog, or clouds. For any of these forms of condensation to occur, the air must be saturated. Saturation occurs most commonly when air is cooled to its dew point, or, less often, when water vapor is added to the air.

Condensation Nuclei and Cloud Formation

Under normal atmospheric conditions, condensation occurs only when a *surface* exists on which the water vapor can condense. When dew forms, objects at or near the ground, such as grass and car windows, serve this purpose. But when condensation occurs high above the ground, tiny bits of particulate matter, known as **cloud condensation nuclei**, serve as surfaces for water-vapor condensation. These nuclei are very important, for in their absence, a relative humidity well in excess of 100 percent is needed to produce clouds.

Condensation nuclei such as microscopic dust, smoke, pollen, and salt particles (from the ocean) are profuse in the lower atmosphere. Because of this abundance of particles, relative humidity rarely exceeds 101 percent. Some particles, such as ocean salt, are particularly good nuclei because they absorb water. These particles are termed **hygroscopic** (*hygro* = moisture, *scopic* = to seek) **nuclei**. When condensation takes place, the tiny cloud droplets grow quickly at first, but their growth slows as they use up the excess water vapor. The result is a cloud consisting of millions upon millions of tiny water droplets, all so fine that they remain suspended in air. When cloud formation occurs at below-freezing temperatures, tiny ice crystals form. Thus, a cloud might consist of water droplets, ice crystals, or both.

Cloud Classification

Clouds are among the most conspicuous and observable aspects of the atmosphere and its weather. **Clouds** are a form of condensation best described as *visible aggregates of minute droplets of water or tiny crystals of ice.* In addition to being prominent and sometimes spectacular features in the sky, clouds are of continual interest to meteorologists because they indicate what is going on in the atmosphere.

In 1803, English naturalist Luke Howard published a cloud classification scheme that serves as the basis of our present-day system. According to Howard's system, clouds are classified on the basis of two criteria: *form* and *height* (Figure 17.20). We will look at the basic cloud forms (shapes) first and then examine cloud height.

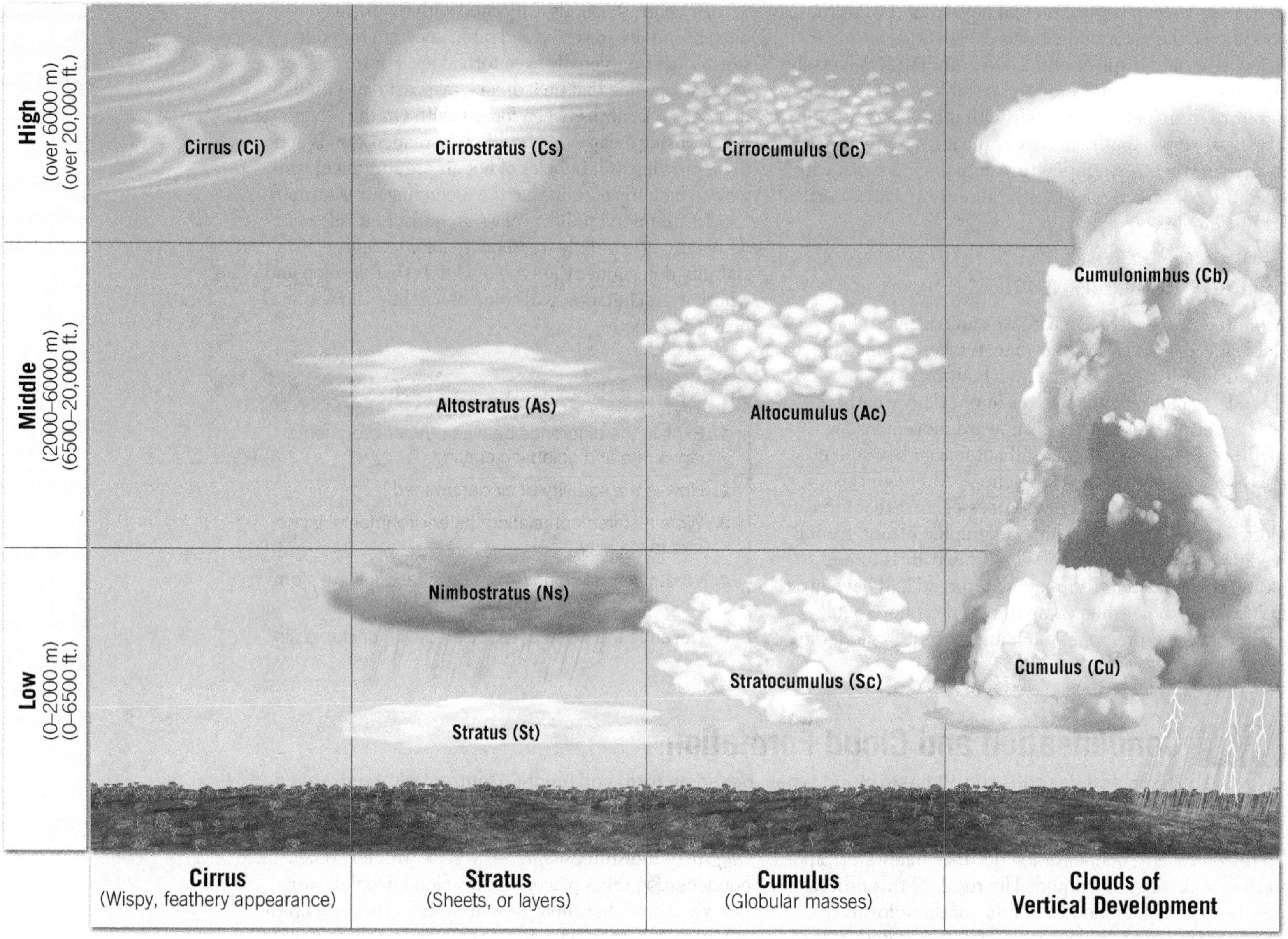

High (over 6000 m) (over 20,000 ft.)

Cirrus (Ci)

Cirrostratus (Cs)

Cirrocumulus (Cc)

Middle (2000–6000 m) (6500–20,000 ft.)

Cumulonimbus (Cb)

Altostratus (As)

Altocumulus (Ac)

Low (0–2000 m) (0–6500 ft.)

Nimbostratus (Ns)

Stratocumulus (Sc)

Cumulus (Cu)

Stratus (St)

Cirrus (Wispy, feathery appearance)

Stratus (Sheets, or layers)

Cumulus (Globular masses)

Clouds of Vertical Development

▲ SmartFigure 17.20
Classification of clouds based on height and form

TUTORIAL
https://goo.gl/K7XrSO

Cloud Forms Clouds are classified based on how they appear when viewed from Earth's surface. The basic forms, or shapes, are:

- **Cirrus** (*cirriform*) clouds are high, white, and thin. They form delicate veil-like patches or wisplike strands and often have a feathery appearance. (*Cirrus* is Latin for "curl" or "filament.")
- **Stratus** (*stratiform*) clouds consist of sheets or layers (*strata*) that cover much or all of the sky.
- **Cumulus** (*cumuliform*) clouds consist of globular cloud masses that are often described as cottonlike in appearance. Normally cumulus clouds exhibit a flat base and appear as rising domes or towers. (*Cumulus* means "heap" or "pile" in Latin.) Cumulus clouds form within a layer of the atmosphere where there is some convection and rising air.

All clouds have at least one of these three basic forms, and some are a combination of two of them; for example, stratocumulus clouds are mostly sheetlike structures composed of long parallel rolls or broken globular patches. In addition, the term **nimbus** (Latin for "raincloud") is added to the name of a cloud that is a major producer of precipitation. Thus, *nimbostratus* denotes a flat-lying rain cloud.

Cloud Heights In terms of their height, clouds are classified as high, middle, low, and clouds of vertical development (see Figure 17.20). **High clouds** form in the highest and coldest region of the troposphere and normally have bases above 6000 meters (20,000 feet). Temperatures at these altitudes are usually below freezing, so the high clouds are generally composed of ice crystals or supercooled water droplets. **Middle clouds** occupy heights from 2000 to 6000 meters (6500 to 20,000 feet) and may be composed of water droplets or ice crystals, depending on the time of year and temperature profile of the atmosphere. **Low clouds** form nearer Earth's surface—up to an altitude of about 2000 meters (6500 feet)—and are generally composed of water droplets. Clouds that extend upward to span more than one height range are called

clouds of vertical development. The altitudes at which high, middle, and low clouds occur can vary somewhat with season and latitude. For example, at high (poleward) latitudes and during cold winter months, high clouds generally occur at lower altitudes.

The internationally recognized cloud types are described in the sections that follow.

High Clouds Three cloud types make up the family of high clouds (above 6000 meters [20,000 feet]): *cirrus*, *cirrostratus*, and *cirrocumulus*. *Cirrus* (Ci) clouds are thin and delicate and sometimes appear as hooked filaments called "mares' tails" (Figure 17.21A). As the names suggest, **cirrocumulus** (Cc) clouds consist of fluffy masses (Figure 17.21B), whereas **cirrostratus** (Cs) clouds are flat

Figure 17.21 Common forms of different cloud types (Photos A, B, D, E, F, and G by E. J. Tarbuck; photo C by Jung-Pang Wu/Moment/Getty Images; photo H by Doug Millar/Science Source)

A. Cirrus

B. Cirrocumulus

C. Cirrostratus

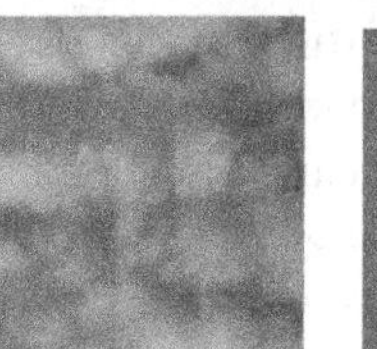

D. Altocumulus

E. Altostratus

F. Nimbostratus

G. Cumulus

H. Cumulonimbus

layers (Figure 17.21C). Because of the low temperatures and small quantities of water vapor present at high altitudes, all high clouds are thin and white and are made up of ice crystals. Furthermore, high clouds do not produce precipitation that reaches the surface. However, when cirrus clouds are followed by cirrocumulus clouds and increased sky coverage, they may warn of impending stormy weather.

Middle Clouds Clouds that appear in the middle range (2000–6000 meters [6500–20,000 feet]) have the prefix *alto* as part of their name. **Altocumulus** (Ac) clouds are composed of globular masses that differ from cirrocumulus clouds in being thicker and denser (Figure 17.21D). **Altostratus** (As) clouds create a uniform white to grayish sheet covering the sky, with the Sun or Moon visible as a bright spot (Figure 17.21E). Infrequent light snow or drizzle may accompany these clouds.

Low Clouds There are three members in the family of low clouds (below 2000 meters [6500 feet]): *stratus*, *stratocumulus*, and *nimbostratus*. *Stratus* (St) is a uniform foglike layer of clouds that frequently covers much of the sky. These clouds sometimes produce light precipitation. When stratus clouds develop a scalloped bottom that appears as long parallel rolls or broken globular patches, they are called **stratocumulus** (Sc) clouds.

Nimbostratus (Ns) clouds derive their name from the Latin *nimbus*, which means "raincloud," and *stratus*, "to cover with a layer" (Figure 17.21F). Nimbostratus clouds tend to produce constant precipitation and low visibility. These clouds normally form under stable conditions when air is forced to rise, as along a front (discussed in Chapter 19). Such forced ascent of stable air leads to the formation of a stratified cloud deck that is widespread and that may grow into the middle level of the troposphere. Precipitation associated with nimbostratus clouds is generally light to moderate (but can be heavy) and is usually of long duration, covering a large area.

Clouds of Vertical Development Some clouds have their bases in the low height range but extend upward into the middle or high altitudes. Consequently, clouds in this category are called *clouds of vertical development*. They tend to be associated with unstable air, which rises because of its buoyancy. Although *cumulus* clouds are usually connected with fair weather (Figure 17.21G), they may grow dramatically under the proper circumstances. Once upward movement is triggered, acceleration can be powerful, and clouds with great vertical extent may form. The end result is often a towering cloud, called a **cumulonimbus** (Cb), which usually produces moderate to heavy rain showers or a thunderstorm (Figure 17.21H).

Definite weather patterns can often be associated with particular clouds or certain combinations of cloud types, so it is important to become familiar with cloud descriptions and characteristics. Table 17.3 lists the 10 basic cloud types that are recognized internationally and gives some characteristics of each.

CONCEPT CHECKS 17.6

1. Explain why a glass containing an ice-cold drink often becomes wet when it sits out at room temperature.
2. What role do condensation nuclei play in the formation of clouds?
3. What are the two criteria by which clouds are classified?
4. Why are high clouds always thin in comparison to low and middle clouds?
5. List the basic cloud types and describe each based on its form (shape) and height (altitude).

Table 17.3 Cloud Types and Characteristics

Cloud Family and Height	Cloud Type	Characteristics
High clouds: above 6000 meters (20,000 feet)	Cirrus	Thin, delicate, fibrous, ice-crystal clouds. Sometimes appear as hooked filaments called "mares' tails" (see Figure 17.21A).
	Cirrocumulus	Thin, white, ice-crystal clouds in the form of ripples, waves, or globular masses, all in a row. May produce a "mackerel sky." Least common of the high clouds (see Figure 17.21B).
	Cirrostratus	Thin sheet of white, ice-crystal clouds that may give the sky a milky look. Sometimes produce halos around the Sun or Moon (see Figure 17.21C).
Middle clouds: 2000–6000 meters (6500–20,000 feet)	Altocumulus	White to gray clouds often composed of separate globules; "sheep-back" clouds (see Figure 17.21D).
	Altostratus	Stratified veil of clouds that are generally thin and may produce very light precipitation. When thin, the Sun or Moon may be visible as a bright spot, but no halos are produced (see Figure 17.21E).
Low clouds: below 2000 meters (6500 feet)	Stratocumulus	Soft gray clouds in globular patches or rolls. Rolls may join together to make a continuous cloud.
	Stratus	Low uniform layer resembling fog but not resting on the ground. May produce drizzle.
	Nimbostratus	Amorphous layer of dark gray clouds. Some of the chief precipitation-producing clouds (see Figure 17.21F).
Clouds of vertical development: 500–18,000 meters (1600–60,000 feet)	Cumulus	Dense, billowy clouds, often characterized by flat bases. May occur as isolated clouds or closely packed (see Figure 17.21G).
	Cumulonimbus	Towering cloud sometimes spreading out on top to form an "anvil head." Associated with heavy rainfall, thunder, lightning, hail, and tornadoes (see Figure 17.21H).

17.7 Types of Fog

Identify the basic types of fog and describe how each forms.

Fog is defined as *a cloud with its base at or very near the ground*. Physically, there are no differences between fog and a cloud; their appearances and structures are the same. The essential difference is the method and place of formation. While clouds mainly form when air rises and cools adiabatically, fog occurs because air either is cooled below its dew-point temperature or gains water through evaporation until it becomes saturated (evaporation fog).

Fog is considered an atmospheric hazard because it reduces visibility (**Figure 17.22**). Official weather stations report fog only when it is thick enough to reduce visibility to 1 kilometer (0.6 mile) or less. Dense fog can cut visibility to a few dozen meters or less, making travel by any mode difficult and dangerous.

Fogs Caused by Cooling

When the temperature of a layer of air in contact with the ground falls below its dew point, condensation produces fog. Depending on the prevailing conditions, fogs formed by cooling are called either *radiation fog*, *advection fog*, or *upslope fog*.

Radiation Fog As the name implies, **radiation fog** results when ground cooled by radiation cools the overlying layer of air. It is a nighttime phenomenon that requires clear skies and a high relative humidity. As the night progresses, a thin layer of air near the ground is cooled below its dew point, resulting in the formation of fog.

Because the air containing the fog is relatively cold and dense, it flows downslope in hilly terrain. As a result, radiation fog is thickest in valleys, whereas the surrounding hills may remain clear (see Figure 17.22A). Normally, radiation fog dissipates within 1 to 3 hours after sunrise. The common phrase that fog "lifts" is misleading; actually, as the Sun warms the ground, the fog evaporates from the bottom up.

Advection Fog When warm, moist air blows over a cold surface, it becomes chilled by contact with the cold surface below. If cooling is sufficient, the result is a blanket of fog called **advection fog**. (The term *advection* refers to air moving horizontally.) A classic example is the frequent advection fog around San Francisco's Golden Gate Bridge (**Figure 17.23**). The fog experienced in San Francisco, California, as well as many other west coast locations, is produced when warm, moist air from the Pacific Ocean moves over the cold California Current.

Advection fog is also a common wintertime phenomenon in the Southeast and Midwest when relatively warm, moist air from the Gulf of Mexico and Atlantic moves over cold and occasionally snow-covered surfaces to produce widespread foggy conditions. This type of advection fog tends to be thick and produce hazardous driving conditions.

Upslope Fog When relatively humid air moves up a gradually sloping landform or, in some cases, up the steep slopes of a mountain, **upslope fog** can form.

SmartFigure 17.22
Radiation fog A. Satellite image of dense fog in California's San Joaquin Valley on November 20, 2002. This early morning fog was caused by radiation cooling that occurred on a clear, cool evening. It was responsible for several accidents in the region, including a 14-car pileup. (Courtesy of NASA) **B.** Radiation fog can make the morning commute hazardous. (Photo by Tim Gainey/Alamy)

VIDEO
https://goo.gl/yiN8iz

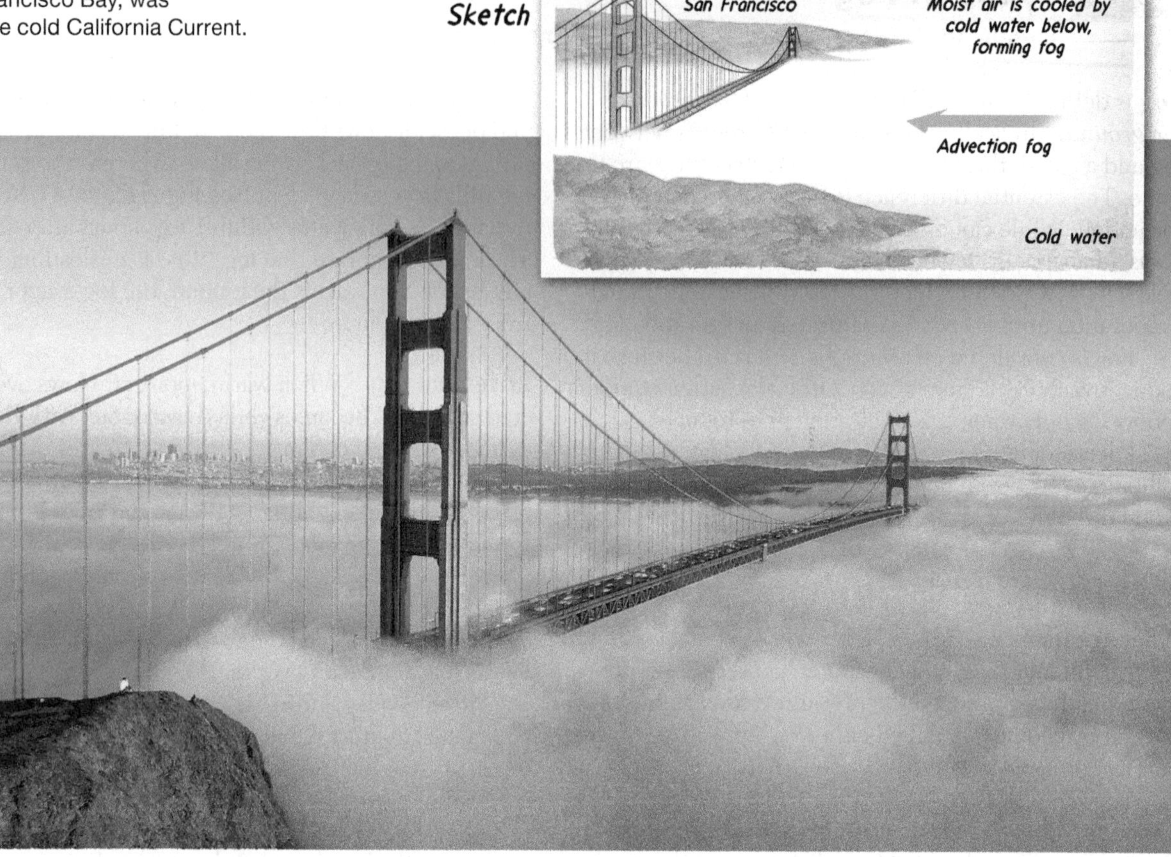

Figure 17.23 **Advection fog rolling into San Francisco Bay** This fog bank, rolling into San Francisco Bay, was generated as moist air passed over the cold California Current. (Photo by Ed Pritchard/Stone/Getty Images)

Because of the upward movement, air expands and cools adiabatically. If the dew point is reached, an extensive layer of fog will form.

It is easy to visualize how upslope fog might form in mountainous terrain. However, in the United States, upslope fog also occurs in the Great Plains, when humid air moves from the Gulf of Mexico toward the Rocky Mountains. (Recall that Denver, Colorado, is called the "mile-high city," and the Gulf of Mexico is at sea level.) Air flowing "up" the Great Plains expands and cools adiabatically by as much as 12°C (22°F), which can result in extensive upslope fog in the western plains.

Evaporation Fogs

When air forms fog mainly because of the addition of water vapor, the fog is called an *evaporation fog*. Two types of evaporation fogs are recognized: *steam fog* and *frontal (precipitation) fog*.

Steam Fog When cool, unsaturated air moves over a warm water body, enough moisture may evaporate to saturate the air directly above, generating a layer of fog. The added moisture and energy often makes the saturated air buoyant enough to rise. Because the foggy air looks like the "steam" that forms above a hot cup of coffee, the phenomenon is called **steam fog** (Figure 17.24). Steam fog is a fairly common occurrence over lakes and rivers on clear, crisp autumn mornings when the water is still relatively warm but the air is comparatively cold.

Figure 17.24 **Steam fog rising from Sierra Lake, Blanca, Arizona** (Photo by Michael Collier)

Frontal (Precipitation) Fog Frontal boundaries where a warm, moist air mass is forced to rise over cooler, dryer air below generate **frontal (precipitation) fog**. Fog develops because the raindrops falling from relatively warm air above the frontal surface evaporate in the cooler air below, causing it to become saturated. Frontal fog, which can be quite thick, is most common on cool days during extended periods of light rainfall.

Where Is Fog Most Common? The frequency of dense fog varies considerably from place to place (Figure 17.25). As might be expected, fog incidence is highest in coastal areas, especially where cold currents prevail, as along the Pacific and New England coasts. Relatively high frequencies are also found in the Great Lakes region and in the humid Appalachian Mountains of the eastern United States.

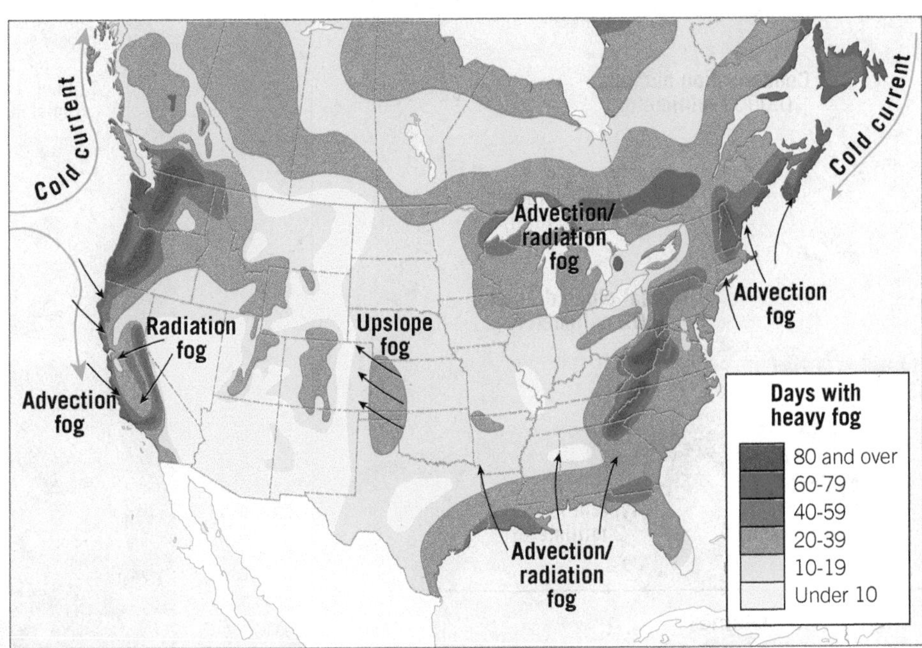

SmartFigure 17.25 **Map showing the average number of days per year with heavy fog** Coastal areas, particularly the Pacific Northwest and New England, where cold ocean currents prevail, have high occurrences of dense fog.

TUTORIAL
https://goo.gl/qVgLja

CONCEPT CHECKS 17.7

1. Distinguish between clouds and fog.
2. List five main types of fog and describe how each type forms.

17.8 How Precipitation Forms

Describe the Bergeron process and explain how it differs from the collision–coalescence process.

If all clouds contain water, why do some produce precipitation while others drift placidly overhead? This seemingly simple question perplexed meteorologists for many years.

Typical cloud droplets are miniscule—0.02 millimeter (20 micrometers) in diameter (Figure 17.26), or about a quarter the width of a human hair (about 0.075 millimeter [75 micrometers]). Because of their small size, cloud droplets in still air fall incredibly slowly. An average cloud droplet falling from a cloud base would require several hours to reach the ground—except that it would actually evaporate

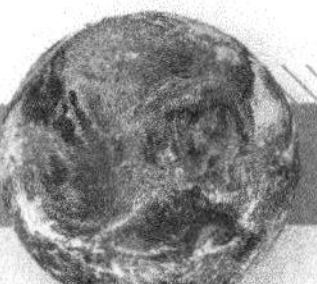

EYE ON EARTH 17.2

The cloud perched on the top of this volcano is called a cap cloud, and it may remain in place for hours. Cap clouds belong to a group of clouds called orographic clouds.

QUESTION 1 *Describe how cap clouds form, based on the fact that they are also called orographic clouds.*

QUESTION 2 *Explain why clouds in this group have relatively flat bases.*

QUESTION 3 *Cap clouds are related to another cloud type that has a very similar shape and forms in mountainous regions. Do an Internet search to find the name of the related cloud type.*

Adam Jones/Science Source

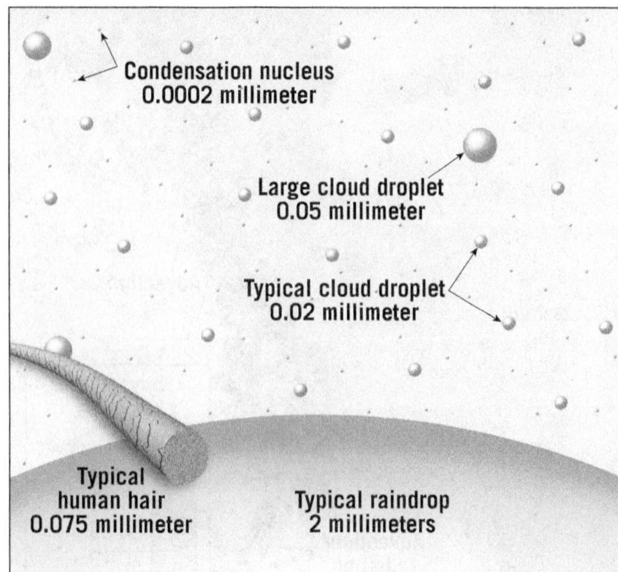

Figure 17.26 **Diameters of particles involved in condensation and precipitation processes**

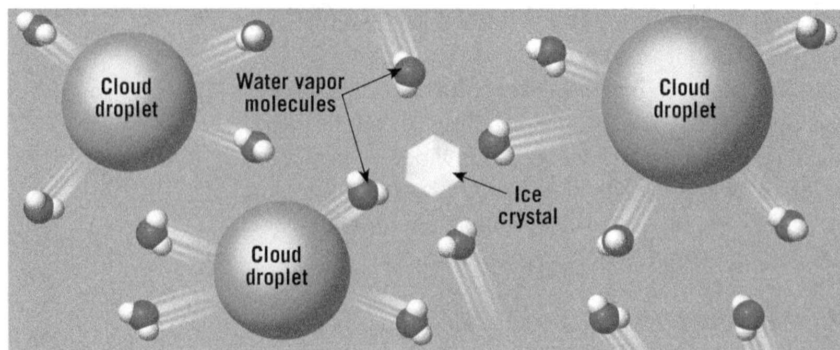

Figure 17.27 **The Bergeron process** Ice crystals grow at the expense of cloud droplets until they are large enough to fall.

within a few meters in the unsaturated air below the cloud.

How large must a cloud droplet grow in order to fall as precipitation? A typical raindrop has a diameter of about 2 millimeters, or 100 times that of the average cloud droplet (see Figure 17.26). However, the *volume* of a typical raindrop is 1 million times that of a cloud droplet. Thus, for precipitation to form, cloud droplets must grow in volume by roughly 1 million times.

Two processes are responsible for the formation of precipitation: the *Bergeron process* and the *collision–coalescence process*.

Precipitation from Cold Clouds: The Bergeron Process

You have probably watched a TV documentary in which mountain climbers braved intense cold and ferocious snow storms to scale ice-covered peaks. Although it is hard to imagine, similar conditions exist in the upper portions of towering cumulonimbus clouds, even on sweltering summer days. It is within these cold clouds that a mechanism called the **Bergeron process** generates much of the precipitation that occurs in the middle and high latitudes.

The Bergeron process is based on the fact that cloud droplets remain liquid at temperatures as low as −40°C (−40°F). Liquid water at temperatures below freezing is termed **supercooled**, and it becomes solid, or freezes, upon impact with a surface. This explains why airplanes collect ice when they pass through a cloud composed of subzero droplets, a condition called *icing*. Supercooled water droplets also freeze upon contact with particles in the atmosphere known as **ice nuclei**, or **freezing nuclei**. Because ice nuclei are relatively sparse, cold clouds primarily consist of supercooled droplets intermixed with a lesser amount of ice crystals.

When ice crystals and supercooled water droplets coexist in a cloud, the conditions are ideal for generating precipitation. Because ice crystals have a greater affinity for water vapor than does liquid water, they collect the available water vapor at a much faster rate. In turn, the water droplets evaporate to maintain saturation and replenish the diminishing water vapor, thereby providing a continual source of moisture for the growth of ice crystals. As shown in Figure 17.27, the result is that the ice crystals grow larger—at the expense of the water droplets, which shrink in size.

Eventually, this process generates ice crystals large enough to fall as snowflakes. During their descent, these ice crystals become larger as they intercept supercooled cloud droplets that freeze on them. When Earth's surface temperature is about 4°C (39°F) or higher, snowflakes usually melt before they reach the ground and continue their descent as rain.

Precipitation from Warm Clouds: The Collision–Coalescence Process

A few decades ago, meteorologists believed that the Bergeron process was responsible for the formation of most precipitation. However, it was discovered that copious rainfall may be produced within clouds located well below the freezing level (*warm clouds*), particularly in the tropics. This led to the proposal of a second mechanism thought to produce precipitation—the **collision–coalescence process**.

Research has shown that clouds composed entirely of liquid droplets must contain some droplets larger than 20 micrometers (0.02 millimeter) for precipitation to form. These large droplets usually form when *hygroscopic particles* (particles that attract water), such as sea salt, are abundant in the atmosphere. Hygroscopic particles begin to remove water vapor from the air when the relative humidity is under 100 percent, and the cloud droplets that form on them can grow quite large. Because the rate at which drops fall is size dependent, these "giant" droplets fall most rapidly. As they plummet, they collide with smaller, slower droplets (Figure 17.28). After many such collisions, these droplets may grow large enough to fall to the surface without evaporating. Updrafts also aid this process because the airflow propels the droplets so they remain in the cloud longer, allowing for additional collisions.

Raindrops can grow to a maximum size of 5 millimeters, at which point they fall at a rate of 33 kilometers (20 miles) per hour. Beyond this size and speed, the water's surface tension, which holds the drop together, is overcome by the drag imposed by the air, causing the drops to break apart (see Figure 17.28). The resulting breakup of a large raindrop produces numerous smaller drops that begin anew the task of sweeping up cloud droplets.

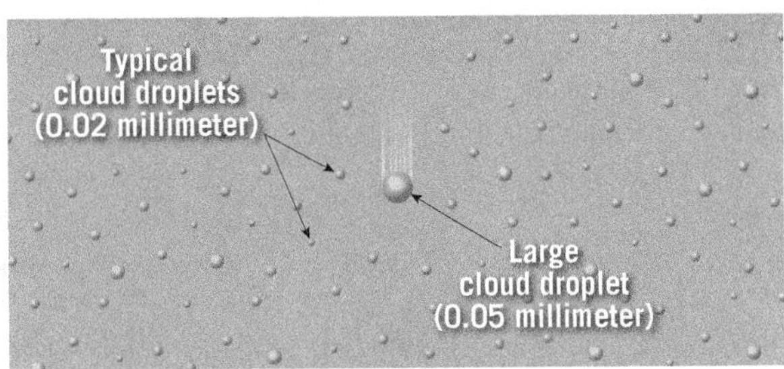

A. Because large cloud droplets fall more rapidly than smaller droplets, they are able to sweep up the smaller ones in their path and grow.

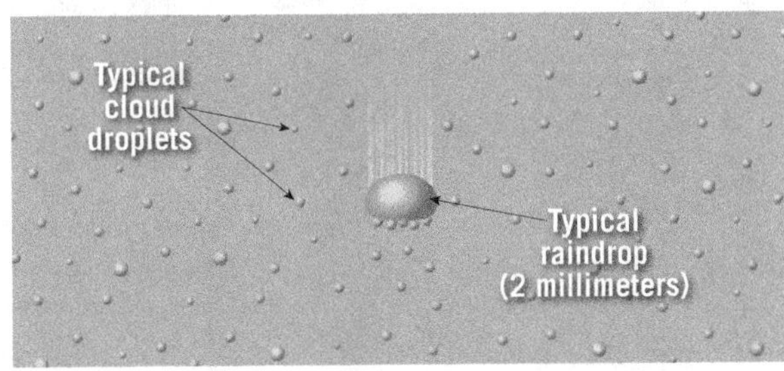

B. As drops increase in size, their fall velocity increases, resulting in increased air resistance, which causes the raindrop to flatten.

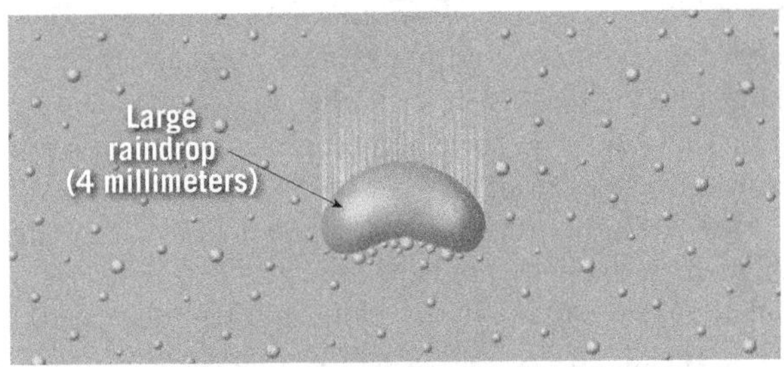

C. As the raindrop approaches 4 millimeters in size, it develops a depression in the bottom.

D. Finally, when the diameter exceeds about 5 millimeters, the depression grows upward almost explosively, forming a donut-like ring of water that immediately breaks into smaller drops.

Figure 17.28 The collision–coalescence process The collision–coalescence process involves multiple collisions of tiny cloud droplets that stick together (coalesce) to form raindrops large enough to reach the ground before evaporating.

CONCEPT CHECKS 17.8

1. Describe the temperature conditions in clouds that are required to form precipitation by the Bergeron process.
2. Explain how snow that formed high in a towering cloud might produce rain.
3. Briefly summarize the collision–coalescence process.

17.9 Forms of Precipitation

Describe the atmospheric conditions that produce sleet, freezing rain (glaze), and hail.

Atmospheric conditions vary greatly both geographically and seasonally, resulting in several different types of precipitation. Rain and snow are the most common and familiar forms, but others, listed in Table 17.4, are important as well. Sleet, freezing rain (glaze), and hail often produce hazardous weather and occasionally inflict considerable damage.

Rain, Drizzle, and Mist

In meteorology, the term **rain** is restricted to drops of water that fall from a cloud and have a diameter of at least 0.5 millimeter (0.02 inch). Most rain originates either in nimbostratus clouds or in towering cumulonimbus clouds; the latter are capable of producing unusually

Table 17.4 Forms of Precipitation

Type	Approximate Size	State of Matter	Description
Mist	0.005–0.05 mm	Liquid	Droplets large enough to be felt on the face when air is moving 1 meter/second. Associated with stratus clouds.
Drizzle	0.05–0.5 mm	Liquid	Small uniform drops that fall from stratus clouds, generally for several hours.
Rain	0.5–5 mm	Liquid	Generally produced by nimbostratus or cumulonimbus clouds. When heavy, rain can be highly variable from one place to another.
Sleet	0.5–5 mm	Solid	Small, spherical to lumpy ice particles that form when raindrops freeze while falling through a layer of subfreezing air. Because the ice particles are small, damage, if any, is generally minor. Sleet can make travel hazardous.
Freezing rain (glaze)	Layers 1 mm–2 cm thick	Solid	Produced when supercooled raindrops freeze on contact with solid objects. Glaze can form a thick coating of ice heavy enough to seriously damage trees and power lines.
Rime	Variable accumulations	Solid	Deposits usually consisting of ice feathers that point into the wind. These delicate, frostlike accumulations form as supercooled cloud or fog droplets encounter objects and freeze on contact.
Snow	1 mm–2 cm	Solid	The crystalline nature of snow allows it to assume many shapes, including six-sided crystals, plates, and needles. Snow is produced in supercooled clouds, where water vapor is deposited as ice crystals that remain frozen during their descent.
Hail	5 mm–10 cm or larger	Solid	Occurs as hard, rounded pellets or irregular lumps of ice. Hail is produced in large cumulonimbus clouds, where frozen ice particles and supercooled water coexist.
Graupel	2–5 mm	Solid	"Soft hail" that forms when rime collects on snow crystals to produce irregular masses of "soft" ice. Because these particles are softer than hailstones, they normally flatten out upon impact.

heavy rainfalls known as *cloudbursts*. As mentioned in the previous section, raindrops rarely exceed about 5 millimeters (0.2 inch) in diameter because the surface tension that holds the drops together is exceeded by the frictional drag of the air.

Fine, uniform drops of water less than 0.5 millimeter (0.02 inch) in diameter are called **drizzle** and require about 10 minutes to fall from a cloud 1000 meters (3300 feet) overhead. Drizzle can be so fine that the tiny drops appear to float, and their impact is almost imperceptible. Precipitation containing the very smallest droplets able to reach the ground is called **mist**.

Snow

Snow is precipitation in the form of ice crystals (snowflakes) or, more often, aggregates of crystals. The size, shape, and concentration of snowflakes depend to a great extent on the temperature at which they form.

Recall that at very low temperatures, the moisture content of air is low. The result is the formation of very light, fluffy snow made up of individual six-sided ice crystals. This is the "powder" that downhill skiers love so much. By contrast, at temperatures warmer than about −5°C (23°F), the ice crystals join together into larger clumps consisting of tangled aggregates of crystals. Snowfalls composed of these composite snowflakes are generally heavy and have high moisture content, which makes them ideal for making snowballs.

Sleet and Freezing Rain (Glaze)

Sleet consists of clear to translucent ice pellets. Depending on intensity and duration, sleet can cover the ground much like a thin blanket of snow. **Freezing rain**, or **glaze**, on the other hand, falls as supercooled raindrops that freeze on contact with roads, power lines, and other surfaces.

As shown in Figure 17.29, both sleet and freezing rain occur in the winter, and they most often form along a warm front where a mass of relatively warm air is forced over a layer of subfreezing air near the ground. Both begin as snow, which melts to form raindrops as it falls though the layer of warm air below. When the newly formed raindrops encounter a thick cold layer of air below the frontal boundary, *sleet* results. In this setting, as raindrops fall through the subfreezing air, they refreeze and reach the ground as small pellets of ice roughly the size of the raindrops from which they formed.

If, however, the layer of cold air near the ground is not thick enough to cause the raindrops to refreeze, they instead become supercooled—that is, they remain liquid at temperatures below freezing (see Figure 17.29). Upon striking subfreezing objects on Earth's surface, these supercooled raindrops instantly turn to ice, creating a coating of *freezing rain*. Freezing rain makes walking and driving extremely hazardous, and when the ice grows thick, it can break tree limbs and down power lines (Figure 17.30).

Hail

Hail is precipitation in the form of hard, rounded pellets or irregular lumps of ice with diameters of 5 millimeters (0.20 inches) or more. Hail is produced in the middle to upper reaches of tall cumulonimbus clouds, where updrafts can sometimes exceed speeds of 160 kilometers (100 miles) per hour and where the air temperature is below freezing. Hailstones begin as small embryonic ice pellets or *graupel* that coexist with supercooled droplets. The ice pellets grow by collecting supercooled water

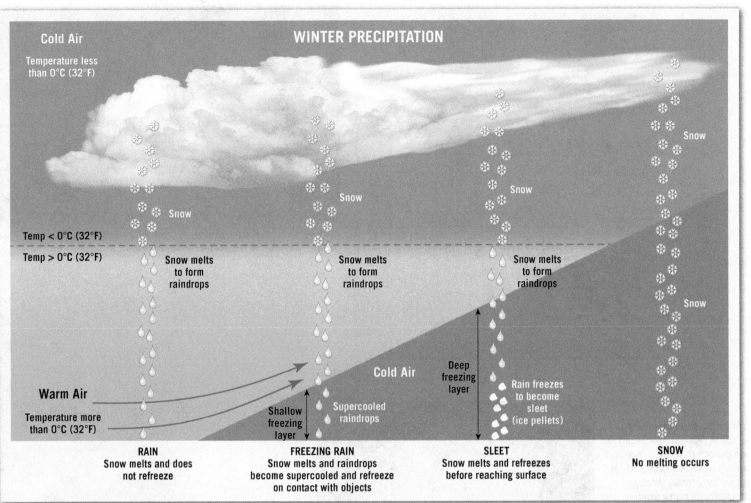

Figure 17.29 Formation of sleet versus freezing rain These forms of precipitation occur in winter, when warm air (along a warm front) is forced over a layer of subfreezing air. When rain passes through a cold layer of air and freezes, the resulting ice pellets are called sleet. Freezing rain forms when the cold layer of air is not deep enough to refreeze the raindrops, resulting in supercooled droplets freezing on contact with objects at the surface.

droplets and, sometimes, other small pieces of hail as they are lifted by updrafts within the cloud.

Cumulonimbus clouds that produce hail have a complex system of updrafts and downdrafts. As shown in Figure 17.31A, a region of intense updrafts suspends rain and hail aloft, producing a rain-free region surrounded by an area of downdrafts and heavy precipitation. The largest hailstones are generated around the core of the most intense zone of updraft, where they rise slowly enough to collect appreciable amounts of supercooled water. The process continues until a hailstone grows too heavy to be supported by the updraft or encounters a downdraft and falls to the surface.

Large hailstones often show alternating layers of clear and milky ice (Figure 17.31B). These layers reflect two different processes by which a hailstone can grow, termed *wet growth* and *dry growth*. Wet growth occurs when hail collects supercooled droplets quickly, causing

Figure 17.30 Freezing rain In January 1998, an ice storm of historic proportions caused enormous damage in New England and southeastern Canada. Nearly 5 days of freezing rain (glaze) left millions without electricity—some for as long as a month. (Photo by Dick Blume/Syracuse Newspapers/The Image Works)

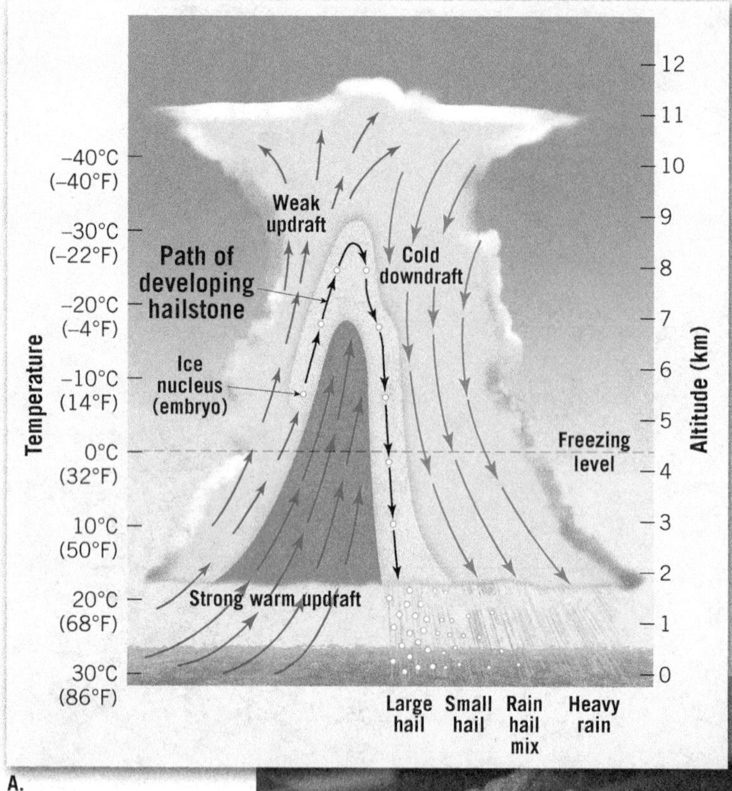

▲ Figure 17.31 **Formation of hailstones A.** Hailstones begin as small ice pellets that grow through the addition of supercooled water droplets as they move through a cloud. Updrafts carry stones upward, increasing the size of the hail by adding layers of ice. Eventually, the hailstones either grow too large to be supported by the updraft or else encounter a downdraft. **B.** This cut hailstone, which fell over Coffeyville, Kansas, in 1970, originally weighed 766 grams (1.69 pounds). (Courtesy of University Corporation for Atmospheric Research)

▲ Figure 17.32 **Hail damage** A strong hail storm destroyed property in its path in eastern Nebraska. (Photo by Mike Hollingshead/Getty Images)

▲ Figure 17.33 **Rime** Rime consists of delicate ice crystals that form when supercooled fog or cloud droplets freeze on contact with objects. (Photo by Marcus Siebert Image Broker/Age Fotostock)

the surface of the hailstone to remain wet. As this wet layer gradually freezes, it produces clear, bubble-free ice. Dry growth, on the other hand, occurs when the accumulation of droplets occurs at a much slower rate. The supercooled droplets freeze immediately on contact, trapping tiny air bubbles that result in milky-looking ice.

The record for the largest hailstone ever found in the United States was set on July 23, 2010, in Vivian, South Dakota. The stone was over 20 centimeters (8 inches) in diameter and weighed nearly 900 grams (2 pounds). The stone that held the previous record of 766 grams (1.69 pounds) fell in Coffeyville, Kansas, in 1970 (see Figure 17.31B). The diameter of the stone found in South Dakota also surpassed the previous record of a 17.8-centimeter (7-inch) stone that fell in Aurora, Nebraska, in 2003. Even larger hailstones have reportedly been recorded in Bangladesh, where a 1987 hailstorm killed more than 90 people.

The destructive effects of large hailstones are well known, especially to farmers whose crops have been devastated in a few minutes and to people whose windows, roofs, and cars have been damaged (**Figure 17.32**). In the United States, annual hail damage can run into the hundreds of millions of dollars.

Rime

Rime is a deposit of ice crystals formed by the freezing of supercooled fog or cloud droplets on objects whose surface temperature is below freezing. When rime forms on trees, it adorns them with its characteristic ice feathers, which can be spectacular to observe (**Figure 17.33**).

In these situations, objects such as pine needles act as ice nuclei, causing the supercooled droplets to freeze on contact. When a wind is blowing, the windward surfaces of objects tend to accumulate rime.

CONCEPT CHECKS 17.9

1. Compare and contrast rain, drizzle, and mist.
2. Describe sleet and freezing rain. Why does freezing rain result on some occasions and sleet on others?
3. How does hail form? What factors govern the ultimate size of hailstones?

17.10 Measuring Precipitation

List the advantages of using weather radar versus a standard rain gauge to measure precipitation.

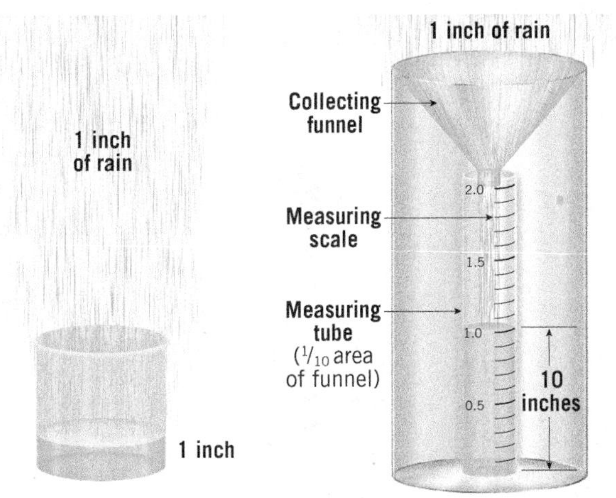

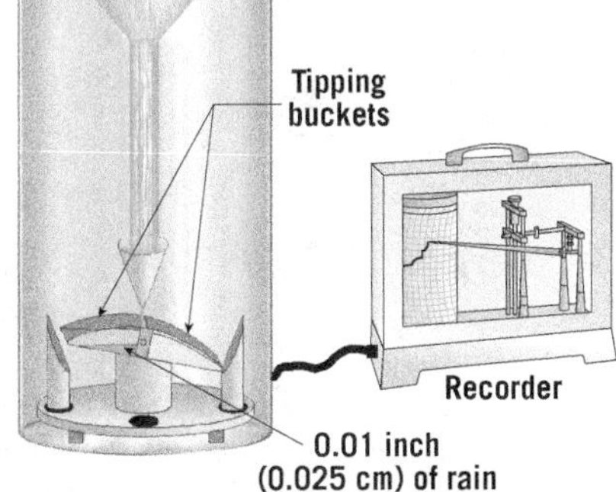

Figure 17.34 **Precipitation measurement** **A.** The simplest gauge is any straight-sided container left in the rain. **B.** The standard rain gauge increases the height of water collected by a factor of 10, allowing for accurate rainfall measurement to the nearest 0.025 centimeter (0.01 inch). **C.** The tipping-bucket rain gauge contains two "buckets," each holding the equivalent of 0.025 centimeter (0.01 inch) of liquid precipitation. When one bucket fills, it tips, and the other bucket takes its place.

The most common form of precipitation, rain, is the easiest to measure. Any open container having a consistent cross section throughout can be used as a rain gauge (Figure 17.34A). In practice, rain gauges are designed to measure small amounts of rainfall accurately and to reduce loss from evaporation.

A **standard rain gauge** has a diameter of about 20 centimeters (8 inches) at the top (Figure 17.34B). When the water is caught, a funnel conducts the rain into a cylindrical measuring tube that has a cross-sectional area only one-tenth as large as the receiver. Consequently, rainfall depth is magnified 10 times, which allows for accurate measurements to the nearest 0.025 centimeter (0.01 inch). When the amount of rain is less than 0.025 centimeter, it is reported as a *trace of precipitation*.

As Figure 17.34C illustrates, the **tipping-bucket gauge** consists of two compartments, each one capable of holding 0.025 centimeter (0.01 inch) of rain, situated at the base of a funnel. When one "bucket" fills, it tips and empties its water. Meanwhile, the other "bucket" takes its place at the mouth of the funnel. Each time a compartment tips, an electrical circuit is closed, and 0.025 centimeter (0.01 inch) of precipitation is automatically recorded on a graph.

Measuring Snowfall

Snowfall is typically measured by depth and water equivalent. One way to measure the depth of snow is by using a calibrated stick. The actual measurement is not difficult, but choosing a representative spot can be. Even when winds are light or moderate, snow drifts freely. As a rule, it is best to take several measurements in an open place, away from trees and obstructions, and then average them. To obtain the water equivalent, a core taken from the snow may be weighed, or snowfall captured in a gauge may be melted and then measured as though it were rain.

The quantity of water in a given volume of snow is not constant. You may have heard media weathercasters say, "Every 10 inches of snow equals 1 inch of rain." But the actual water content of snow may deviate widely from this figure. It may take as much as 30 inches of light and fluffy dry snow (30:1) or as little as 4 inches of wet snow (4:1) to produce 1 inch of water.

Precipitation Measurement by Weather Radar

Using **weather radar**, the National Weather Service (NWS) produces maps like the one in Figure 17.35, in which different colors illustrate precipitation intensity. The development of weather radar has given meteorologists an important tool to track storm systems and the precipitation patterns they produce, even when the storms are as far as a few hundred kilometers away.

Radar units have transmitters that send out short pulses of radio waves toward a storm system. The specific wavelengths selected depend on the objects being detected. Wavelengths between 3 and 10 centimeters are employed when monitoring precipitation. Radio waves at these wavelengths can penetrate clouds composed of small droplets, but they are reflected

GEOGRAPHICS 17.1

Our Water Supply

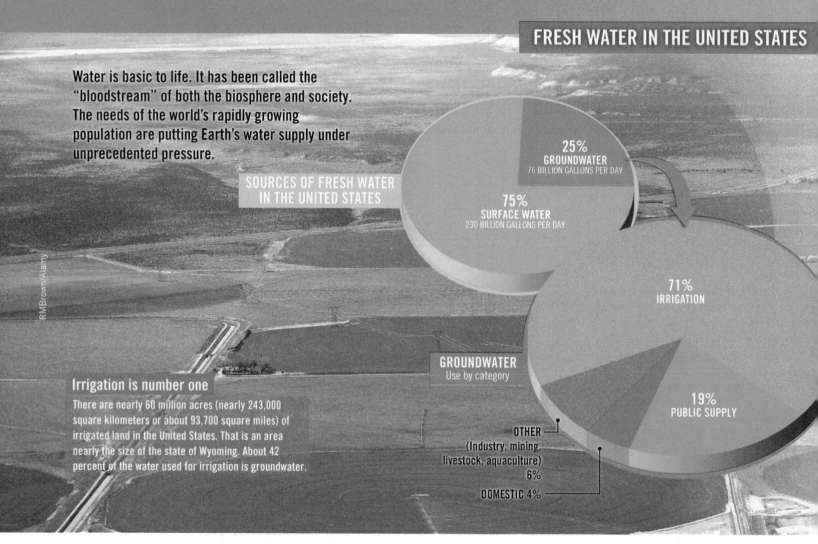

FRESH WATER IN THE UNITED STATES

Water is basic to life. It has been called the "bloodstream" of both the biosphere and society. The needs of the world's rapidly growing population are putting Earth's water supply under unprecedented pressure.

SOURCES OF FRESH WATER IN THE UNITED STATES

25% GROUNDWATER
76 BILLION GALLONS PER DAY

75% SURFACE WATER
230 BILLION GALLONS PER DAY

GROUNDWATER
Use by category

71% IRRIGATION

19% PUBLIC SUPPLY

OTHER (Industry, mining, livestock, aquaculture) 6%

DOMESTIC 4%

Irrigation is number one

There are nearly 60 million acres (nearly 243,000 square kilometers or about 93,700 square miles) of irrigated land in the United States. That is an area nearly the size of the state of Wyoming. About 42 percent of the water used for irrigation is groundwater.

RMBrown/Alamy

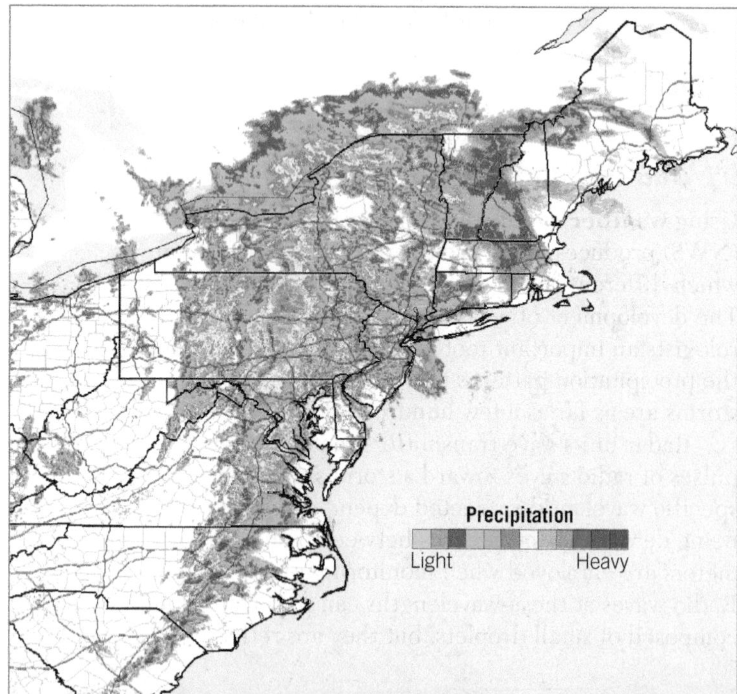

Figure 17.35 Doppler radar display produced by the National Weather Service Colors indicate different intensities of precipitation. Note the band of heavy precipitation (orange and red colors) along the eastern seaboard. (Courtesy of NOAA)

by larger raindrops, ice crystals, and hailstones. The reflected signal, called an *echo*, is received and displayed on a screen. Because the echo is "brighter" when the precipitation is more intense, modern weather radar is able to depict both the regional extent and the rate of precipitation. Also, because the measurements are in real time, they are particularly useful in short-term forecasting.

CONCEPT CHECKS 17.10

1. Although any open container can serve as a rain gauge, what advantages does a standard rain gauge provide?
2. Identify one advantage that weather radar has over a standard rain gauge.

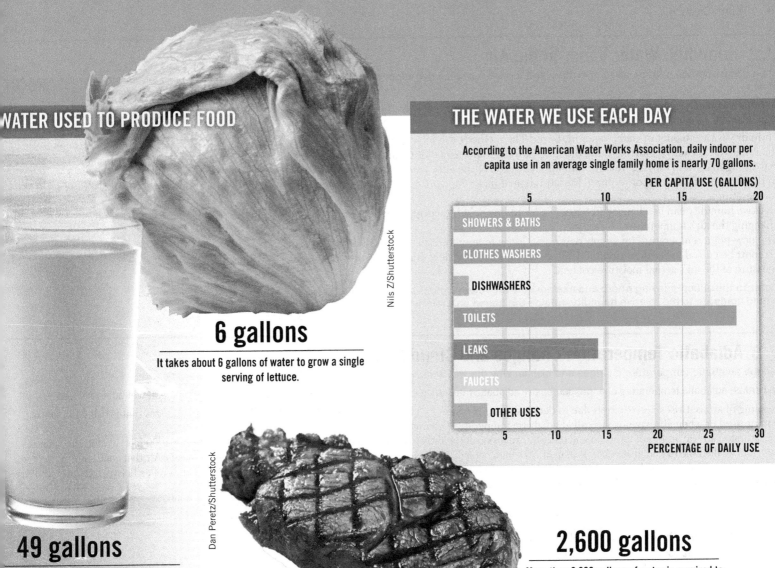

WATER USED TO PRODUCE FOOD

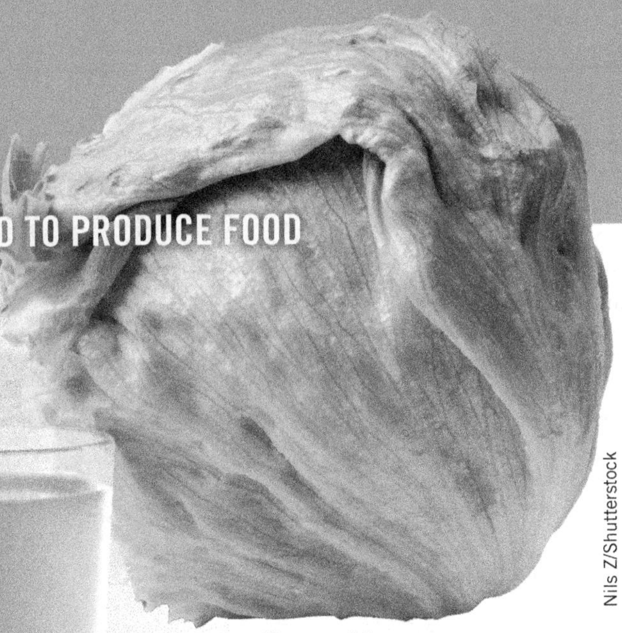

Nils Z/Shutterstock

6 gallons

It takes about 6 gallons of water to grow a single serving of lettuce.

49 gallons

almost 49 gallons of water to produce st one eight ounce glass of milk.

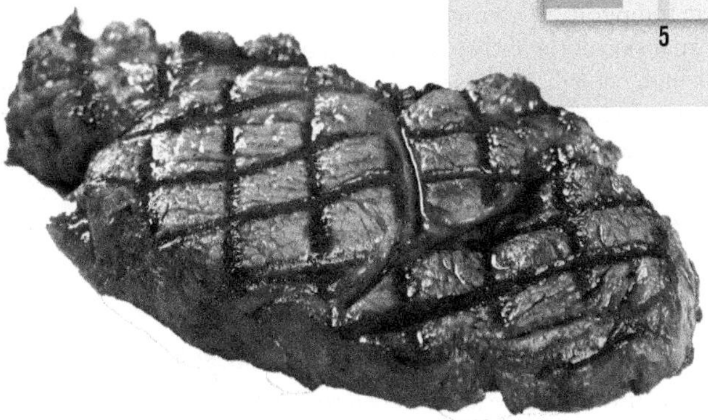

Dan Peretz/Shutterstock

2,600 gallons

More than 2,600 gallons of water is required to produce a single serving of steak.

THE WATER WE USE EACH DAY

According to the American Water Works Association, daily indoor per capita use in an average single family home is nearly 70 gallons.

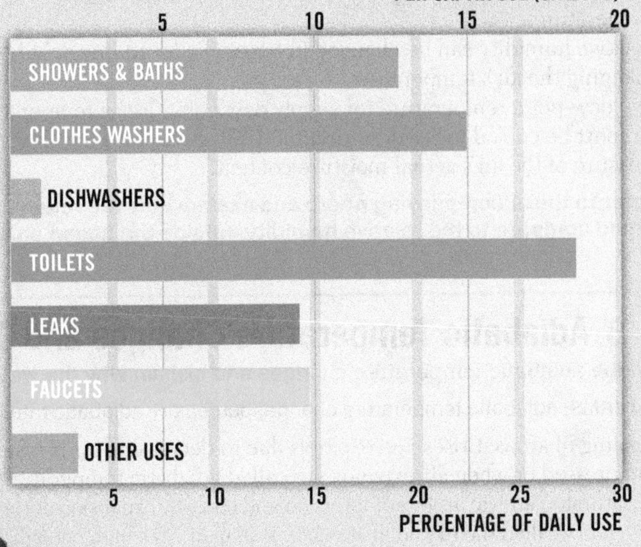

PER CAPITA USE (GALLONS)

5 10 15 20

SHOWERS & BATHS

CLOTHES WASHERS

DISHWASHERS

TOILETS

LEAKS

FAUCETS

OTHER USES

5 10 15 20 25 30

PERCENTAGE OF DAILY USE

17 CONCEPTS IN REVIEW

Moisture, Clouds, and Precipitation

17.1 Water's Changes of State

Summarize the six processes by which water changes from one state of matter to another. For each, indicate whether energy is absorbed or released.

KEY TERMS: calorie, latent heat, evaporation, condensation, sublimation, deposition

- Water exists in all three states of matter (solid, liquid, or gas) at the temperatures and pressures near Earth's surface. The gaseous form of water is water vapor.
- The processes by which matter changes state are evaporation (liquid to gas), condensation (gas to liquid), melting (solid to liquid), freezing (liquid to solid), sublimation (solid to gas), and deposition (gas to solid). During each change, latent (hidden, or stored) heat is either absorbed or released.

? Label the accompanying diagram with the appropriate terms for the changes of state that are shown.

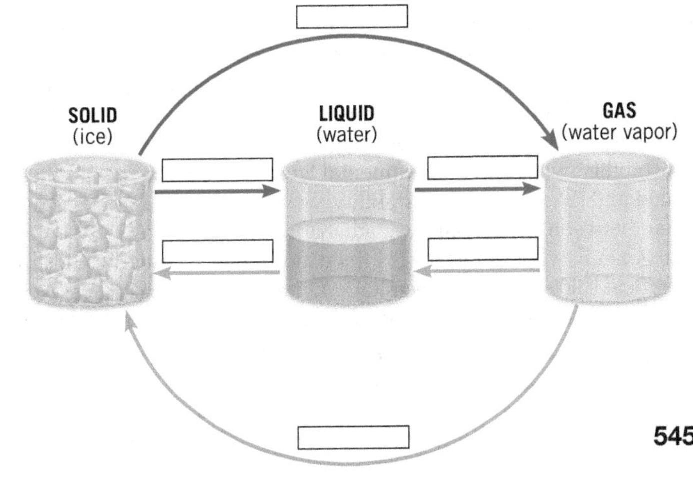

17.2 Humidity: Water Vapor in the Air

Write a generalization relating air temperature and the amount of water vapor needed to saturate air.

KEY TERMS: humidity, saturation, vapor pressure, mixing ratio, relative humidity, dew-point temperature (dew point), hygrometer, psychrometer

- Humidity is the amount of water vapor in the air. The methods used to express humidity quantitatively include (1) mixing ratio, the mass of water vapor in a unit of air compared to the remaining mass of dry air; (2) relative humidity, the ratio of the air's actual water-vapor content to the amount of water vapor required for saturation at that temperature; and (3) dew-point temperature.
- Relative humidity can be changed in two ways: by adding or subtracting water vapor or by changing the air's temperature.
- The dew-point temperature (or simply dew point) is the temperature to which a parcel of air must be cooled to reach saturation. Unlike relative humidity, dew-point temperature is a measure of the air's actual moisture content.

? Refer to the accompanying photo and explain how the relative humidity inside this house would compare to the relative humidity outside the house on a winter day.

Clynt Garnham Housing/Alamy

17.3 Adiabatic Temperature Changes and Cloud Formation

Describe adiabatic temperature changes and explain why the wet adiabatic rate of cooling is less than the dry adiabatic rate.

KEY TERMS: adiabatic temperature change, parcel, dry adiabatic rate, lifting condensation level (condensation level), sensible heat, wet adiabatic rate

- Cooling of air as it rises and expands due to decreasing air pressure is the basic cloud-forming process. Temperature changes that result when air is compressed or when air expands are called adiabatic temperature changes.
- Unsaturated air warms by compression and cools by expansion at the rather constant rate of 10°C per 1000 meters (5.5°F per 1000 feet) of altitude change, a quantity called the dry adiabatic rate. When air rises high enough, it cools sufficiently to cause condensation and form clouds. Air that continues to rise above the condensation level cools at the wet adiabatic rate, which varies from 5°C to 9°C per 1000 meters of ascent. The difference between the wet and dry adiabatic rates is due to the latent heat released by condensation, which slows the rate at which air cools as it ascends.

17.4 Processes That Lift Air

List and describe the four mechanisms that cause air to rise.

KEY TERMS: orographic lifting, rainshadow desert, front, frontal lifting (wedging), convergence, localized convective lifting (convective lifting)

- Four mechanisms that cause air to rise are (1) orographic lifting, where air is forced to rise over elevated terrain such as a mountain barrier; (2) frontal lifting, where warmer, less-dense air is forced over cooler, denser air along a front; (3) convergence, a pileup of horizontal airflow resulting in an upward flow; and (4) localized convective lifting, where unequal surface heating causes localized pockets of air to rise because of their buoyancy.

17.5 The Critical Weathermaker: Atmospheric Stability

Describe how atmospheric stability is determined and compare conditional instability with absolute instability.

KEY TERMS: stable air, unstable air, environmental lapse rate, absolute stability, absolute instability, conditional instability

- Stable air resists vertical movement, whereas unstable air rises because of its buoyancy. The stability of a parcel of air is determined by the local environmental lapse rate (the temperature of the atmosphere at various heights). The three fundamental conditions of the atmosphere are (1) absolute stability, when the environmental lapse rate is less than the wet adiabatic rate; (2) absolute instability, when the environmental lapse rate is greater than the dry adiabatic rate; and (3) conditional instability, when moist air has an environmental lapse rate between the dry and wet adiabatic rates.
- In general, when stable air is forced aloft, the associated clouds have little vertical thickness, and precipitation, if any, is light. In contrast, clouds associated with unstable air are towering and can produce heavy precipitation.

? Describe the atmospheric conditions that were likely associated with the development of the towering cloud shown in the accompanying photo.

Rolf Nussbaumer/Alamy Stock Photo

17.6 Condensation and Cloud Formation

Name and describe the 10 basic cloud types, based on form and height. Contrast nimbostratus and cumulonimbus clouds and their associated weather.

KEY TERMS: cloud condensation nuclei, hygroscopic nuclei, cloud, cirrus, stratus, cumulus, nimbus, high clouds, middle clouds, low clouds, clouds of vertical development, cirrocumulus, cirrostratus, altocumulus, altostratus, stratocumulus, nimbostratus, cumulonimbus

- For water vapor to condense into cloud droplets, the air must reach saturation, and there must be a surface on which the water vapor can condense. The resulting cloud droplets are tiny and are held aloft by the slightest updrafts.
- Clouds are classified on the basis of their form and height. The three basic cloud forms are cirrus (high, white, thin wisps or sheets), cumulus (globular, individual cloud masses), and stratus (sheets or layers).
- Cloud heights can be high, with bases above 6000 meters (20,000 feet); middle, from 2000 (6500 feet) to 6000 meters; or low, below 2000 meters. Clouds of vertical development have bases in the low height range and extend upward into the middle or high range.

? Which of the three basic cloud forms (cirrus, cumulus, or stratus) is illustrated by each of the accompanying images, A–C?

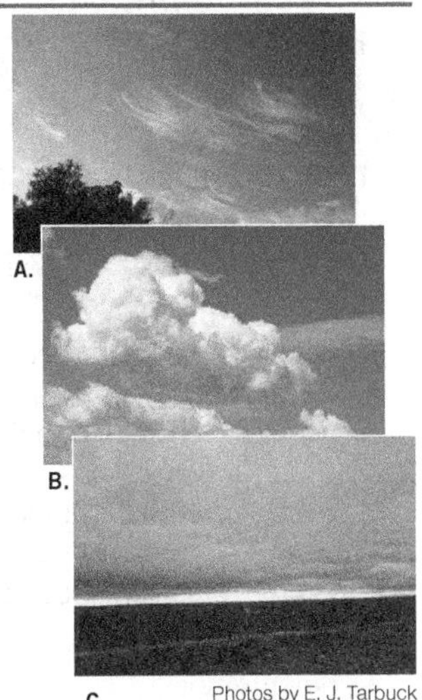

Photos by E. J. Tarbuck

17.7 Types of Fog

Identify the basic types of fog and describe how each forms.

KEY TERMS: fog, radiation fog, advection fog, upslope fog, steam fog, frontal (precipitation) fog

- Fog is a cloud with its base at or very near the ground. Fogs form when air is cooled below its dew point or when enough water vapor is added to the air to cause saturation.
- Fogs formed by cooling include radiation fog, advection fog, and upslope fog. Fogs formed by the addition of water vapor are steam fog and frontal fog.

? Identify the fog type shown in the accompanying image and describe the mechanism that generated it.

Pat and Chuck Blackley/Alamy

17.8 How Precipitation Forms

Describe the Bergeron process and explain how it differs from the collision–coalescence process.

KEY TERMS: Bergeron process, supercooled, ice (freezing) nuclei, collision–coalescence process

- For precipitation to form, millions of cloud droplets must join together into drops that are large enough to reach the ground before evaporating.
- The two mechanisms that generate precipitation are the Bergeron process, which produces precipitation from cold clouds primarily in the middle and high latitudes, and the collision–coalescence process, which occurs in warm clouds and primarily in the tropics.

17.9 Forms of Precipitation

Describe the atmospheric conditions that produce sleet, freezing rain (glaze), and hail.

KEY TERMS: rain, drizzle, mist, snow, sleet, freezing rain (glaze), hail, rime

- The two most common and familiar forms of precipitation are rain and snow. Rain can form in either warm or cold clouds. When it falls from cold clouds, it begins as snow that melts before reaching the ground.
- Sleet consists of spherical to lumpy ice particles that form when raindrops freeze while falling through a thick layer of subfreezing air. Freezing rain results when supercooled raindrops freeze upon contact with cold objects. Rime consists of delicate frostlike accumulations that form as supercooled fog droplets encounter objects and freeze on contact. Hail consists of hard, rounded pellets or irregular lumps of ice produced in towering cumulonimbus clouds, where frozen ice particles and supercooled water coexist.

17.10 Measuring Precipitation

List the advantages of using weather radar versus a standard rain gauge to measure precipitation.

KEY TERMS: standard rain gauge, tipping-bucket gauge, weather radar

- Two instruments commonly used to measure rain are the standard rain gauge and the automated tipping-bucket gauge. The two most common measurements of snow are depth and water equivalent.
- Modern weather radar has given meteorologists an important tool to track storm systems and precipitation patterns, even when the storms are as far as a few hundred kilometers away.

GIVE IT SOME THOUGHT

1 In what state of matter is the "steam" rising from the liquid in the accompanying photo? (*Hint:* Is it possible to see water vapor?)

Photo by Dmitry Kolmakov/Shutterstock

2 Refer to Figure 17.2 to complete the following:

a. In which state of matter is water the most dense?

b. In which state of matter are water molecules most energetic?

c. In which state of matter is water highly compressible?

3 The primary mechanism by which the human body cools itself is perspiration.

a. Explain how perspiring cools the skin.

b. Refer to the data for Phoenix, Arizona, and Tampa, Florida, in Table A. In which city would it be easier to stay cool by perspiring? Explain your choice.

TABLE A

City	Temperature	Dew-point Temperature
Phoenix, AZ	101°F	47°F
Tampa, FL	101°F	77°F

4 The accompanying graph shows how air temperature and relative humidity change on a typical summer day in the Midwest. Assuming that the dew-point temperature remained constant, what would be the best time of day to water a lawn to minimize evaporation of the water spread on the grass?

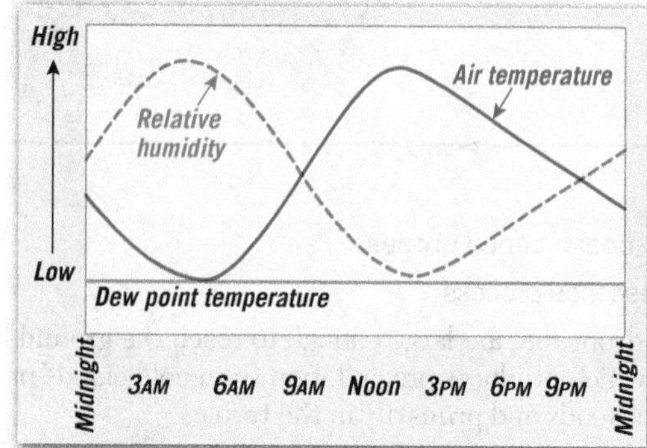

5 Refer to Table 17.1. How much more water is contained in saturated air at a tropical location with a temperature of 40°C compared to a polar location with a temperature of −10°C?

6 Use the data in Table B, to complete the following:

a. Which city has a higher relative humidity?

b. Which city has the greater quantity of water vapor in the air?

c. In which city is the air closest to its saturation point with respect to water vapor?

TABLE B

City	Temperature	Dew-point Temperature
Phoenix, AZ	101°F	47°F
Bismark, ND	39°F	38°F

7 Large cumulonimbus clouds like the one shown in Figure 17.21H are roughly 12 kilometers tall and 8 kilometers in diameter. Assume that the droplets in each cubic meter of the cloud total 0.5 cubic centimeter. How much liquid would a cloud of that size contain? How many gallons is this? (*Note:* 3785 cm^3 = 1 gallon.)

8 The accompanying diagram shows air flowing from the ocean over a coastal mountain range. Assume that the dew-point temperature remains constant in dry air (air having a relative humidity less than 100 percent). If the air parcel becomes saturated, the dew-point temperature will cool at the wet adiabatic rate as it ascends, but it will not change as the air parcel descends. Use this information to complete the following:

a. Determine the air temperature and dew-point temperature for the air parcel at each location (B–G) shown on the diagram.

b. At what elevation will clouds begin to form (with relative humidity = 100 percent)?

c. Compare the air temperatures at points A and G. Why are they different?

d. How did the water vapor content of the air change as the parcel of air traversed the mountain? (*Hint:* Compare dew-point temperatures.)

e. On which side of the mountain might you expect lush vegetation, and on which side would you expect desert-like conditions?

f. Where in the United States might you find a situation like what is pictured here?

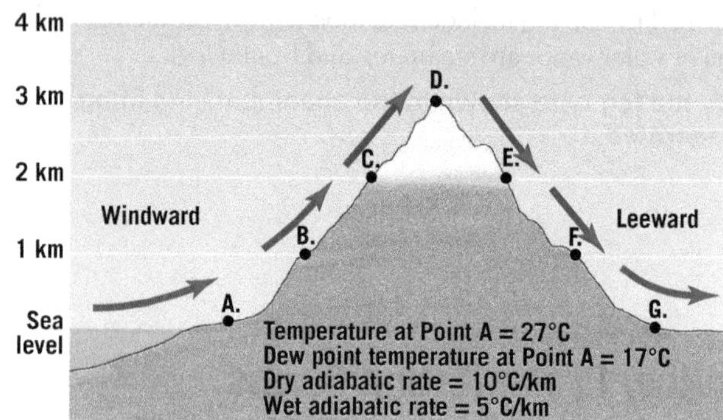

9 Weather radar provides information on the intensity of precipitation in addition to the total amount of precipitation that falls over a given time period. Table C shows the relationship between radar reflectivity (echo intensity) values and rainfall rates. If radar measured a reflectivity value of 47 dBZ for 2½ hours over a location, how much rain will have fallen there?

Table C
Conversion of radar reflectivity to rainfall rate

Radar Reflectivity (dBZ)	Rainfall Rate (inches/hr)
65	16+
60	8.0
55	4.0
52	2.5
47	1.3
41	0.5
36	0.3
30	0.1
20	trace

10 The photo shows the largest recorded hailstone in the United States, which fell on July 23, 2010, in Vivian, South Dakota. The stone was 8 inches in diameter and 18.62 inches in circumference, and it weighed nearly 2 pounds. The maximum fall speed of a spherical hailstone of diameter d can be calculated using $V = k2d$, where $k = 20$ if d is in centimeters and V is in meters per second. Estimate the strength (speed) of the updrafts in this storm that were necessary to support the stone just before it began to fall. Convert your answer from meters per second to feet per second and then to miles per hour.

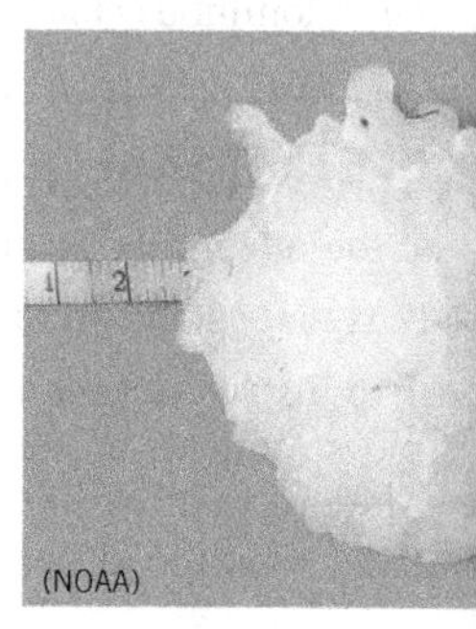
(NOAA)

EXAMINING THE EARTH SYSTEM

1 The amount of precipitation that falls at any particular place and time is controlled by the quantity of moisture in the air and many other factors. How might each of the following alter the precipitation at a particular locale?
  a. An increase in the elevation of the land
  b. A decrease in the area covered by forests and other types of vegetation
  c. Lowering of average ocean-surface temperatures
  d. An increase in the percentage of time that the winds blow from an adjacent body of water
  e. A major episode of global volcanism lasting a decade

2 Phoenix and Flagstaff, Arizona, are both located in the southwestern United States, less than a 2-hour drive apart. Using the accompanying climate diagram (which gives the elevations of the two cities), describe the impact that elevation has on the precipitation and temperature of each city. Use the Internet to compare and explain the natural vegetation of these locations. Next, check the current weather conditions for Phoenix and Flagstaff, Arizona, by using The Weather Channel website, at www.weather.com. Do the current conditions seem to fit climate data? Explain.

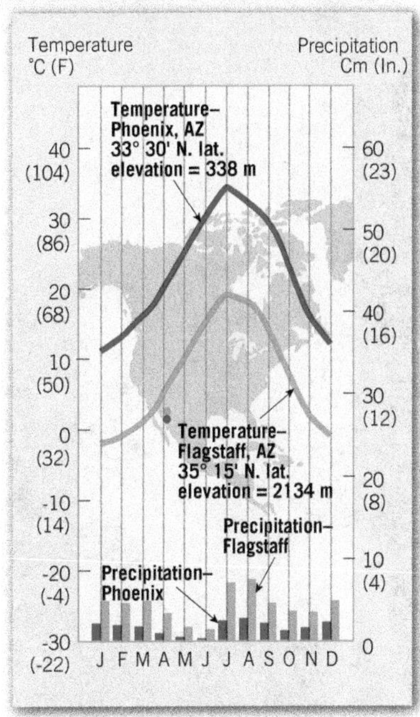

DATA ANALYSIS

Cloud Climatology

The International Satellite Cloud Climatology Project (ISCCP) collects cloud data from several satellites. The role of clouds in climate is an area of great interest and uncertainty.

ACTIVITIES

Go to the ISCCP Monthly Means and Climatology page, at http://isccp.giss.nasa.gov/products/browsed2.html. Retain the variable Total Cloud Amount (%) and time period Mean Annual and click View.

1 The map indicates the average annual percentage of cloud-covered sky. What is the range of cloud cover amounts for the United States?

2 What is the annual average cloud cover amount (%) for your approximate location?

Go back and compare the variables VIS-IR Low Cloud Amount (%), VIS-IR Low Middle Cloud Amount (%), and VIS-IR High Cloud Amount (%) to view the cloud amounts derived from visible and infrared satellite data.

3 Look at the mean data for the current month. Are the greatest number of clouds in the upper, middle, or lower atmosphere?

4 Examine the seasonally averaged cloud amounts for the United States. During what season(s) are high clouds most common? Middle clouds? Low clouds?

5 Examine the annual mean high clouds globally. Where are most of these clouds located? What causes this pattern?

6 Examine the annual mean low clouds globally. Where are most of these clouds?

Go back and view the Total Cloud Amount (%) for the current month. Also look outside and examine the actual sky.

7 According to the ISCCP data, what is the mean cloud amount for your location during the current month?

8 How does the actual amount of cloud cover in the sky compare to the monthly average?

9 What types of clouds are currently in the sky? Do the clouds indicate that the atmosphere is stable or unstable?

10 What is the most likely lifting mechanism responsible for the current clouds, if any?

18 Air Pressure and Wind

FOCUS ON CONCEPTS

Each statement represents the primary learning objective for the corresponding major heading within the chapter. After you complete the chapter, you should be able to:

18.1 Define *air pressure* and describe the instruments used to measure this weather element.

18.2 Discuss the three forces that act on the atmosphere to either create or alter winds.

18.3 Contrast the weather associated with low-pressure centers (cyclones) and high-pressure centers (anticyclones).

18.4 Summarize Earth's idealized global circulation. Describe how continents and seasonal temperature changes complicate the idealized pattern.

18.5 List three types of local winds and describe their formation.

18.6 Describe the instruments used to measure wind. Explain how wind direction is expressed using compass directions.

18.7 Describe the Southern Oscillation and its relationship to El Niño and La Niña. List the climate impacts of El Niño and La Niña on North America.

18.8 Discuss the major factors that influence the global distribution of precipitation.

Wind energy is a fast-growing type of renewable energy generated worldwide.
(Photo by Planetpix/Alamy Stock Photo)

OF THE VARIOUS ELEMENTS OF WEATHER AND CLIMATE, changes in air pressure are the least noticeable. When listening to a weather report, we are generally interested in moisture conditions (humidity and precipitation), temperature, and perhaps wind. Rarely do people wonder about air pressure. Although people do not generally notice the hour-to-hour and day-to-day variations in air pressure, such changes are very important factors in producing changes in our weather. Variations in air pressure from place to place cause the movement of air we call wind and are a significant factor in weather forecasting. As we will see, air pressure is closely tied to the other elements of weather in a cause-and-effect relationship.

18.1 Understanding Air Pressure

Define *air pressure* and describe the instruments used to measure this weather element.

In Chapter 16 we noted that **air pressure** is simply the pressure exerted by the weight of air above. Average air pressure at sea level is about 1 kilogram per square centimeter, or 14.7 pounds per square inch—also called *1 atmosphere*. Specifically, a column of air 1 square inch in cross section, measured from sea level to the top of the atmosphere, would weigh about 14.7 pounds (Figure 18.1). This is roughly the same pressure that is produced by a 1-square-inch column of water 10 meters (33 feet) in height. With some simple arithmetic, you can calculate that the air pressure exerted on the top of a small (50 centimeter-by-100 centimeter [20 inch-by-40 inch]) school desk exceeds 5000 kilograms (11,000 pounds), or about the weight of a 50-passenger school bus. Why doesn't the desk collapse under the weight of the ocean of air above? Simply, air pressure is exerted in all directions—down, up, and sideways. Thus, the air pressure pushing down on the desk exactly balances the air pressure pushing up on the desk.

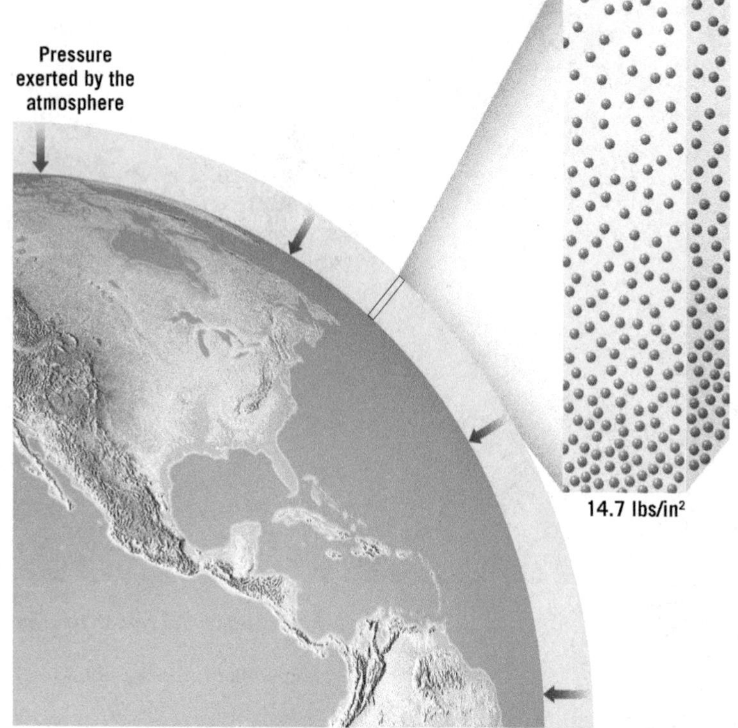

▶ Figure 18.1 **sea-level pressure** Air pressure can be thought of as the force exerted by the weight of the atmosphere above. A column of air 1 square inch in cross section extending from sea level to the top of the atmosphere would weigh about 14.7 pounds.

Visualizing Air Pressure

Imagine a tall aquarium that has the same dimensions as the small desk mentioned in the preceding paragraph. When this aquarium is filled to a height of 10 meters (33 feet), the water pressure at the bottom equals 1 atmosphere (1 kilogram per square centimeter [14.7 pounds per square inch]). Now, imagine what will happen if this aquarium is placed on top of our student desk so that all the force is directed downward. Compare this to what results when the desk is placed inside the aquarium and allowed to sink to the bottom. In the latter example, the desk survives because the water pressure is exerted in all directions, not just downward, as in our earlier example. The desk, like your body, is "built" to withstand the pressure of 1 atmosphere. It is important to note that although we do not generally notice the pressure exerted by the ocean of air around us, except when ascending or descending in an airplane or a tall elevator, it is nonetheless substantial. The pressurized suits that astronauts use on space walks are designed to duplicate the atmospheric pressure experienced at Earth's surface. Without these protective suits to keep body fluids from boiling away, astronauts would perish in minutes.

The concept of air pressure can also be understood if we examine the behavior of gas molecules. Gas molecules, unlike molecules in a liquid or solid, are not bound to one another but move freely throughout the space available to them. When two gas molecules collide, which happens frequently under normal conditions, they bounce off each other like elastic balls. If a gas is confined to a container, this motion is restricted by its sides, much as the walls of a handball court redirect the motion of the handball. The continuous bombardment of gas molecules against the sides of the container exerts an outward push that we call air pressure. Although the atmosphere is without walls, it is confined from below by Earth's surface and effectively from above because the force of gravity prevents its escape. Here, we can define *air pressure* as the force exerted against a surface by the continuous collision of gas molecules.

Measuring Air Pressure

When meteorologists measure atmospheric pressure, it is expressed in units called *millibars*. Standard sea-level pressure is 1013.2 millibars (mb). Although the millibar has been the unit of measure on all U.S. weather maps since January 1940, the media use "inches of mercury" to describe atmospheric pressure. In the United States, the National Weather Service converts millibar values to inches of mercury for public and aviation use (Figure 18.2).

Inches of mercury is easy to understand. The use of mercury for measuring air pressure dates from 1643, when Torricelli, a student of the famous Italian scientist Galileo, invented the **mercury barometer**. Torricelli correctly described the atmosphere as a vast ocean of air that exerts pressure on us and all objects around us. To measure this force, he filled a glass tube, which was closed at one end, with mercury. He then inverted the tube into a dish of mercury (Figure 18.3). Torricelli found that the mercury flowed out of the tube until the weight of the column was balanced by the pressure that the atmosphere exerted on the surface of the mercury in the dish. In other words, the weight of mercury in the column equaled the weight of the same diameter column of air that extended from the ground to the top of the atmosphere.

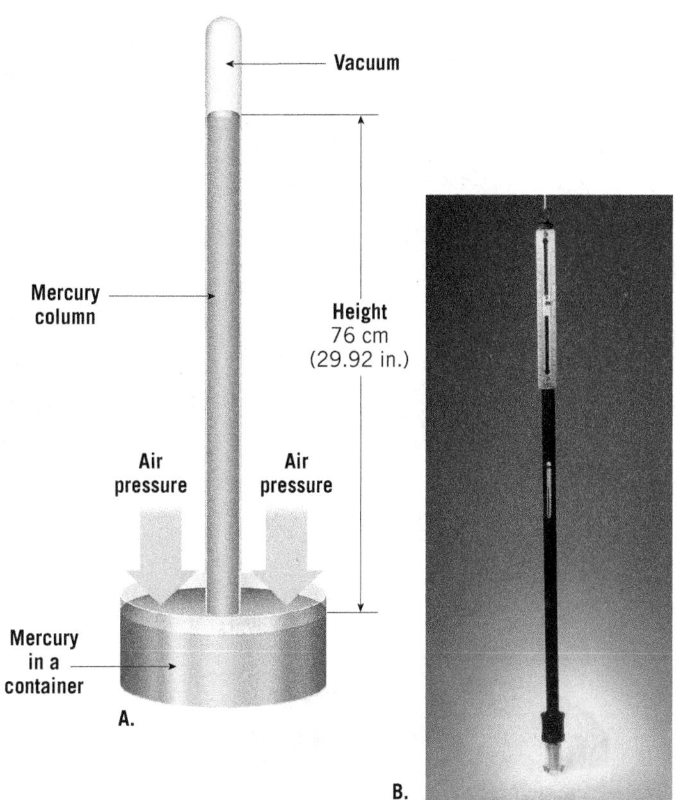

Figure 18.3 **Simple mercury barometer**
A. The weight of the column of mercury is balanced by the pressure exerted on the dish of mercury by the air above. If the pressure decreases, the column of mercury falls; if the pressure increases, the column rises.
B. A mercury barometer. (Photo by Charles E. Winters/Science Source)

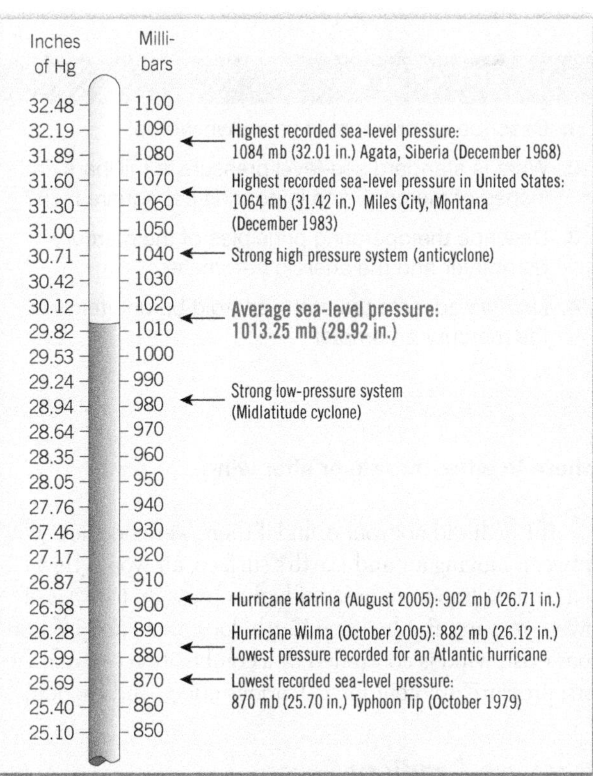

SmartFigure 18.2 **Inches and millibars** A comparison of two units commonly used to express air pressure.

TUTORIAL
https://goo.gl/IkGr8B

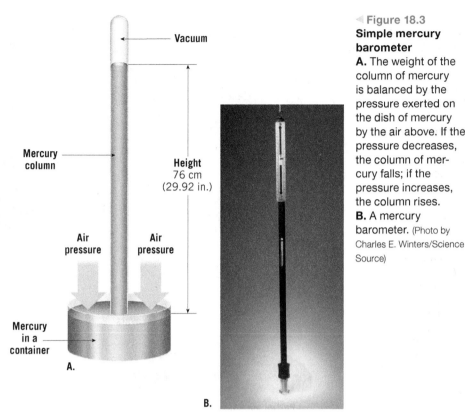

When air pressure increases, the mercury in the tube rises. Conversely, when air pressure decreases, so does the height of the mercury column. With some refinements, the mercury barometer invented by Torricelli is still the standard pressure-measuring instrument used today. Standard atmospheric pressure at sea level equals 29.92 inches of mercury.

The need for a smaller and more portable instrument for measuring air pressure led to the development of the **aneroid barometer** (*aneroid* means "without liquid"). Instead of having a mercury column held up by air pressure, an aneroid barometer uses a partially evacuated metal chamber (Figure 18.4). The chamber is extremely sensitive to variations in air pressure and changes shape, compressing as the pressure increases and expanding as the pressure decreases. A series of levers transmit the movements of the chamber to a pointer on a dial that is calibrated to indicate inches of mercury and/or millibars.

As shown in Figure 18.4, the face of an aneroid barometer intended for home use is inscribed with words such as *fair, change, rain,* and *stormy*. Notice that "fair"

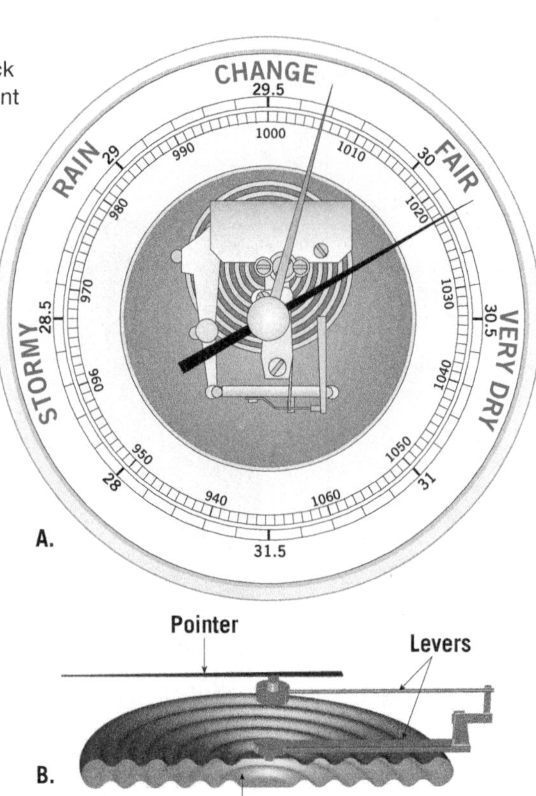

▶ Figure 18.4 **Aneroid barometer A.** The black pointer shows the current air pressure. When the barometer is read, the observer moves the other pointer to coincide with the current air pressure. Later, when the barometer is checked, the observer can see whether the air pressure has risen, fallen, or remained steady. **B.** An aneroid barometer has a partially evacuated chamber that changes shape, compressing as air pressure increases and expanding as pressure decreases.

corresponds with high-pressure readings, whereas "rain" is associated with low pressures. Barometric readings, however, may not always indicate the weather. The dial may point to "fair" on a rainy day or to "rain" on a fair day. To "predict" the local weather, the change in air pressure over the past few hours is more important than the current pressure reading. Falling pressure is often associated with increasing cloudiness and the possibility of precipitation, whereas rising air pressure generally indicates clearing conditions. It is useful to remember, however, that particular barometer readings or trends do not always correspond to specific types of weather.

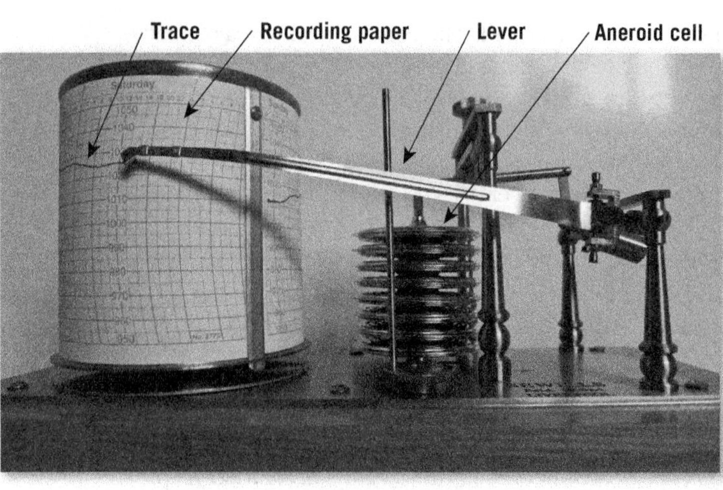

▲ Figure 18.5 **Aneroid barograph** This instrument makes a continuous record of air pressure. The cylinder with the graph is a clock that turns once per day or once per week. (Photo courtesy of Stuart Aylmer/Alamy)

Another advantage of an aneroid barometer is that it can easily be connected to a recording mechanism. The resulting instrument, a **barograph**, provides a continuous record of pressure changes with the passage of time (Figure 18.5). Another important adaptation of the aneroid barometer is its use to indicate altitude for aircraft, mountain climbers, and mapmakers.

CONCEPT CHECKS 18.1

1. Describe air pressure in your own words.
2. What is standard sea-level pressure in millibars, in inches of mercury, and in pounds per square inch?
3. Describe the operating principles of the mercury barometer and the aneroid barometer.
4. List two advantages of the aneroid barometer over the mercury barometer.

18.2 Factors Affecting Wind

Discuss the three forces that act on the atmosphere to either create or alter winds.

In Chapter 17, we examined the upward movement of air and its role in cloud formation. As important as vertical motion is, far more air moves horizontally, the phenomenon we call **wind**. What causes wind?

Simply stated, wind is the result of horizontal differences in air pressure. *Air flows from areas of higher pressure to areas of lower pressure.* You may have experienced this when opening something that is vacuum packed. The noise you hear is caused by air rushing from the higher pressure outside the can or jar to the lower pressure inside. Wind is nature's attempt to balance such inequalities in air pressure. Because unequal heating of Earth's surface generates these pressure differences, *solar radiation is the ultimate energy source for most wind.*

If Earth did not rotate, and if there were no friction between moving air and Earth's surface, air would flow in a straight line from areas of higher pressure to areas of lower pressure. But because Earth does rotate and friction does exist, wind is controlled by a combination of three factors: pressure gradient force, Coriolis effect, and friction.

Pressure Gradient Force

The force that generates winds results from horizontal pressure differences. When air is subjected to greater pressure on one side than on another, the imbalance produces a force that is directed from the region of higher pressure toward the area of lower pressure. Thus, pressure differences cause the wind to blow, and the greater these differences, the greater the wind speed.

Variations in air pressure over Earth's surface are determined from barometric readings taken at thousands of weather stations. These pressure measurements are shown on surface weather maps using **isobars** (*iso* = equal, *bar* = pressure), or lines connecting places of equal air pressure (**Figure 18.6**). The *spacing* of isobars indicates the amount of pressure change occurring over a given distance, which is called the **pressure gradient force**. Pressure gradient is analogous to gravity acting on a ball rolling down a hill. A steep pressure gradient, like a steep hill, causes greater acceleration of a parcel of air than does a weak pressure gradient (a gentle hill). Thus, the relationship between wind speed and the pressure gradient is straightforward: *Closely spaced isobars indicate a steep pressure gradient and strong winds; widely spaced isobars indicate a weak pressure gradient and light winds.* Figure 18.6 illustrates the relationship between the spacing of isobars and wind speed. Note also that the pressure gradient force is always directed at *right angles* to the isobars.

In order to draw isobars on a weather map to show air pressure patterns, meteorologists must compensate for the *elevation* of each station. Otherwise, high-elevation locations, such as Denver, Colorado, would always be mapped as having low pressure. This compensation is accomplished by converting all pressure measurements to sea-level equivalents.

Figure 18.7 is a surface weather map that shows isobars (representing corrected sea-level air pressure) and winds. Wind *direction* is shown as wind arrow shafts, and *speed* is shown as wind bars (see the key accompanying the map in the figure). The isobars on such maps are rarely straight or evenly spaced. Consequently, wind generated by the pressure gradient force typically changes speed and direction as it flows.

The area of somewhat circular closed isobars in eastern North America represented by the red letter *L* is a *low-pressure system*. In western Canada, a *high-pressure system*, denoted by the blue letter *H*, can also be seen. We will discuss *highs* and *lows* in the next section.

In summary, the *horizontal pressure gradient is the driving force of wind.* The magnitude of the pressure gradient force is shown by the spacing of isobars. The direction of force is always from areas of higher pressure toward areas of lower pressure and at right angles to the isobars.

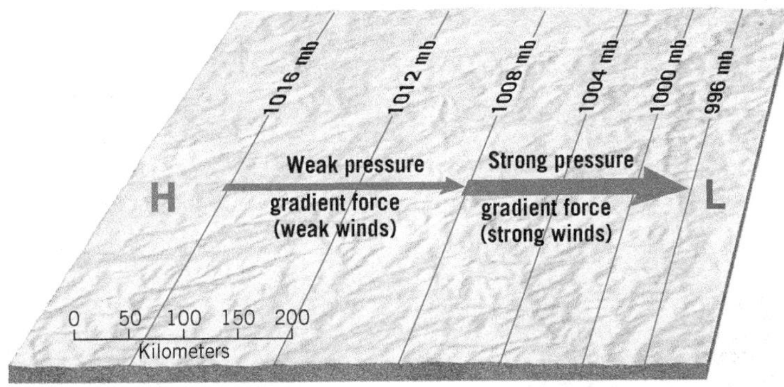

Figure 18.6 Isobars show the pressure gradient Isobars are lines connecting places of equal air pressure. The spacing of isobars indicates the amount of pressure change occurring over a given distance—called the *pressure gradient*. Closely spaced isobars indicate a strong pressure gradient and high wind speeds, whereas widely spaced isobars indicate a weak pressure gradient and low wind speeds.

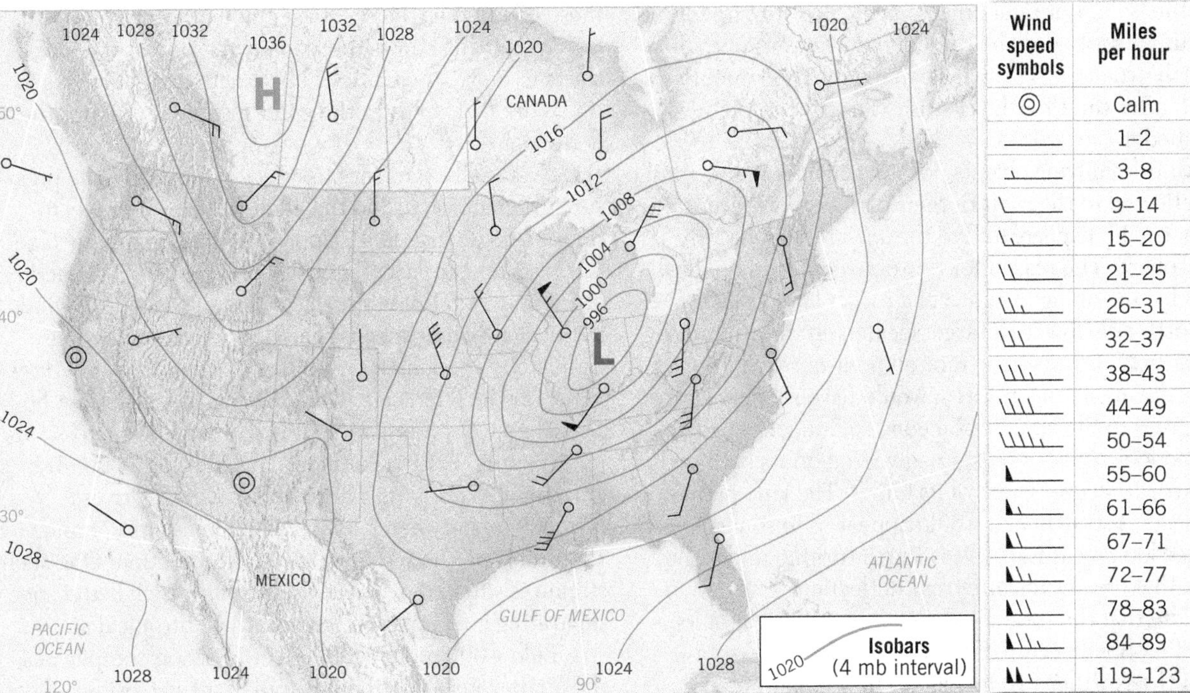

SmartFigure 18.7 Isobars on a weather map Isobars are used to show the distribution of pressure on daily weather maps. Isobars are seldom straight but usually form broad curves. Concentric isobars indicate cells of high and low pressure. The "wind flags" indicate the expected airflow surrounding pressure cells and are plotted as "flying" with the wind (that is, the wind blows toward the station circle). Notice on this map that the isobars are more closely spaced, and the wind speed is faster around the low-pressure center than around the high.

TUTORIAL
https://goo.gl/fuiOOB

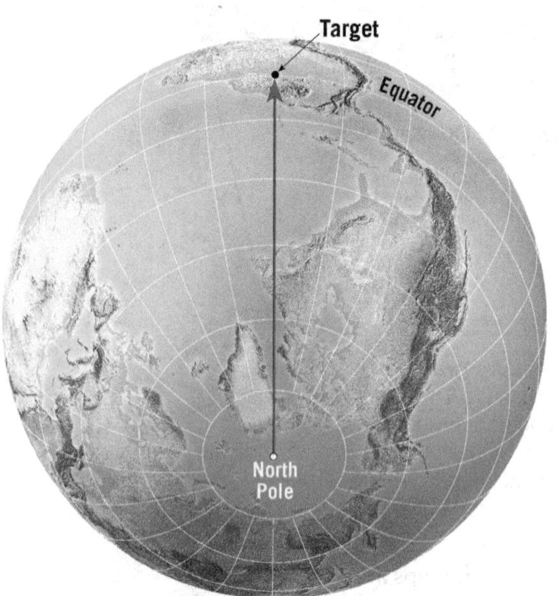

SmartFigure 18.8 **Coriolis effect** The deflection caused by Earth's rotation is illustrated by a rocket traveling for 1 hour from the North Pole to a location on the equator. **A.** On a nonrotating Earth, the rocket would fly straight to its target. **B.** However, Earth rotates 15° each hour. Thus, although the rocket travels in a straight line, when we plot the path of the rocket on Earth's surface, it follows a curved path that veers to the right of the target. The video illustrates the Coriolis effect by using a playground merry-go-round.

VIDEO
https://goo.gl/0bDeRk

Coriolis Effect

Figure 18.7 shows the typical air movements associated with high- and low-pressure systems. As expected, the air moves out of the regions of higher pressure and into the regions of lower pressure. However, the wind does not cross the isobars at right angles, as the pressure gradient force directs it to do. The direction deviates as a result of Earth's rotation. This has been named the **Coriolis effect**, after the French scientist who first thoroughly described it.

All free-moving objects or fluids, including the wind, are deflected to the *right* of their path of motion in the Northern Hemisphere and to the *left* in the Southern Hemisphere. The reason for this deflection can be illustrated by imagining the path of a rocket launched from the North Pole toward a target located on the equator (Figure 18.8). If the rocket took an hour to reach its target, during its flight, Earth would have rotated 15 degrees to the east. To someone standing on Earth, it would look as if the rocket had veered off its path and hit Earth 15 degrees west of its target. The true path of the rocket is straight and would appear so to someone out in space looking at Earth. It is Earth turning under the rocket that creates the *apparent* deflection.

Note that the rocket is deflected to the right of its path of motion because of the counterclockwise rotation of the Northern Hemisphere. In the Southern Hemisphere, the effect is reversed. Clockwise rotation produces a similar deflection but to the *left* of the path of motion. The same deflection is experienced by wind regardless of the direction it is moving.

We attribute the apparent shift in wind direction to the Coriolis effect. This deflection (1) is always directed at right angles to the direction of airflow; (2) affects only wind direction, not wind speed; (3) is affected by wind speed (the stronger the wind, the greater the deflection); and (4) is strongest at the poles and weakens equatorward, becoming nonexistent at the equator.

Note that any free-moving object will experience a deflection caused by the Coriolis effect. The U.S. Navy dramatically discovered this fact in World War II. During target practice, long-range guns on battleships continually missed their targets by as much as several hundred yards until ballistic corrections were made for the changing position of a seemingly stationary target. Over a short distance, however, the Coriolis effect is small.

Friction with Earth's Surface

The effect of friction on wind is important only within a few kilometers of Earth's surface. Friction acts to slow air movement and, as a consequence, alters wind direction. To illustrate friction's effect on wind direction, let us look at a situation in which friction has no role. Above the friction layer, the pressure gradient force and Coriolis effect work together to direct the flow of air. Under these conditions, the pressure gradient force causes air to start moving across the isobars. As soon as the air starts to move, the Coriolis effect acts at right angles to this motion. The faster the wind speed, the greater the deflection.

Eventually, the Coriolis effect will balance the pressure gradient force, and the wind will blow parallel to the isobars (Figure 18.9). Upper-air winds generally take this path and are called **geostrophic winds**. The lack of friction with Earth's surface allows geostrophic winds to travel at higher speeds than do surface winds. This can be observed in Figure 18.10 by noting the wind flags, many of which indicate winds of 50 to 100 miles per hour.

The most prominent features of upper-level flow are **jet streams**. First encountered by high-flying bombers during World War II, these fast-moving "rivers" of air travel between 120 and 240 kilometers (75 and 150 miles) per hour in a west-to-east direction. One such stream is situated over the polar front, which is the zone separating cool polar air from warm subtropical air.

Below 600 meters (2000 feet), friction complicates the airflow just described. Recall that the Coriolis effect

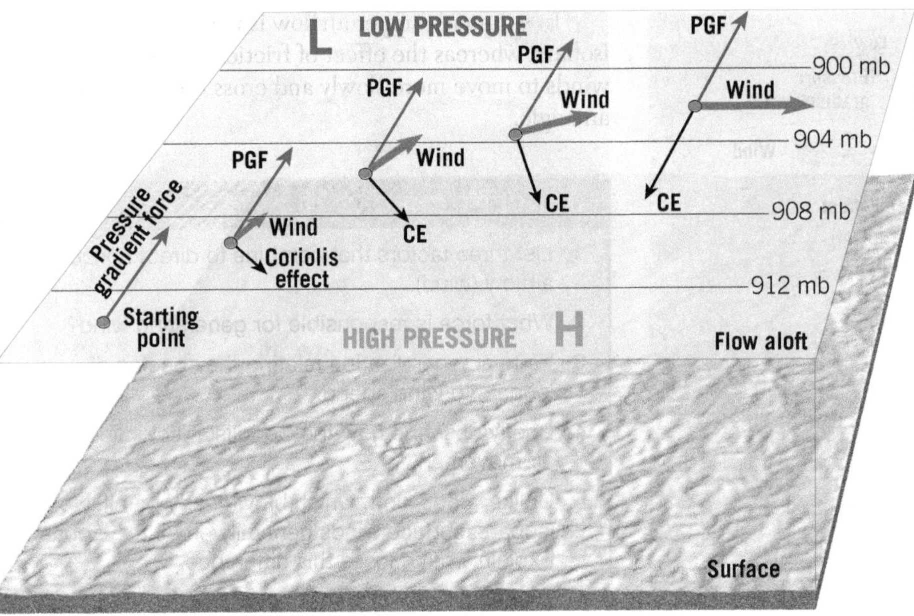

Figure 18.9 Geostrophic wind The only force acting on a stationary parcel of air is the pressure gradient force. Once the air begins to accelerate, the Coriolis effect deflects it to the right in the Northern Hemisphere. Greater wind speeds result in a stronger Coriolis effect (deflection) until the flow is parallel to the isobars. At this point, the pressure gradient force and Coriolis effect are in balance, and the flow is called a *geostrophic wind.* It is important to note that in the "real" atmosphere, airflow is continually adjusting for variations in the pressure field. As a result, the adjustment to geostrophic equilibrium is much more irregular than shown.

is proportional to wind speed. Friction lowers the wind speed, so it reduces the Coriolis effect. Because the pressure gradient force is not affected by wind speed, it wins the tug of war, as shown in Figure 18.11. The result is a movement of air at an angle across the isobars, toward the area of lower pressure.

The roughness of the terrain determines the angle of airflow across the isobars. Over the smooth ocean surface, friction is low, and the angle is small. Over rugged terrain, where friction is higher, the angle that air makes as it flows across the isobars can be as great as 45 degrees.

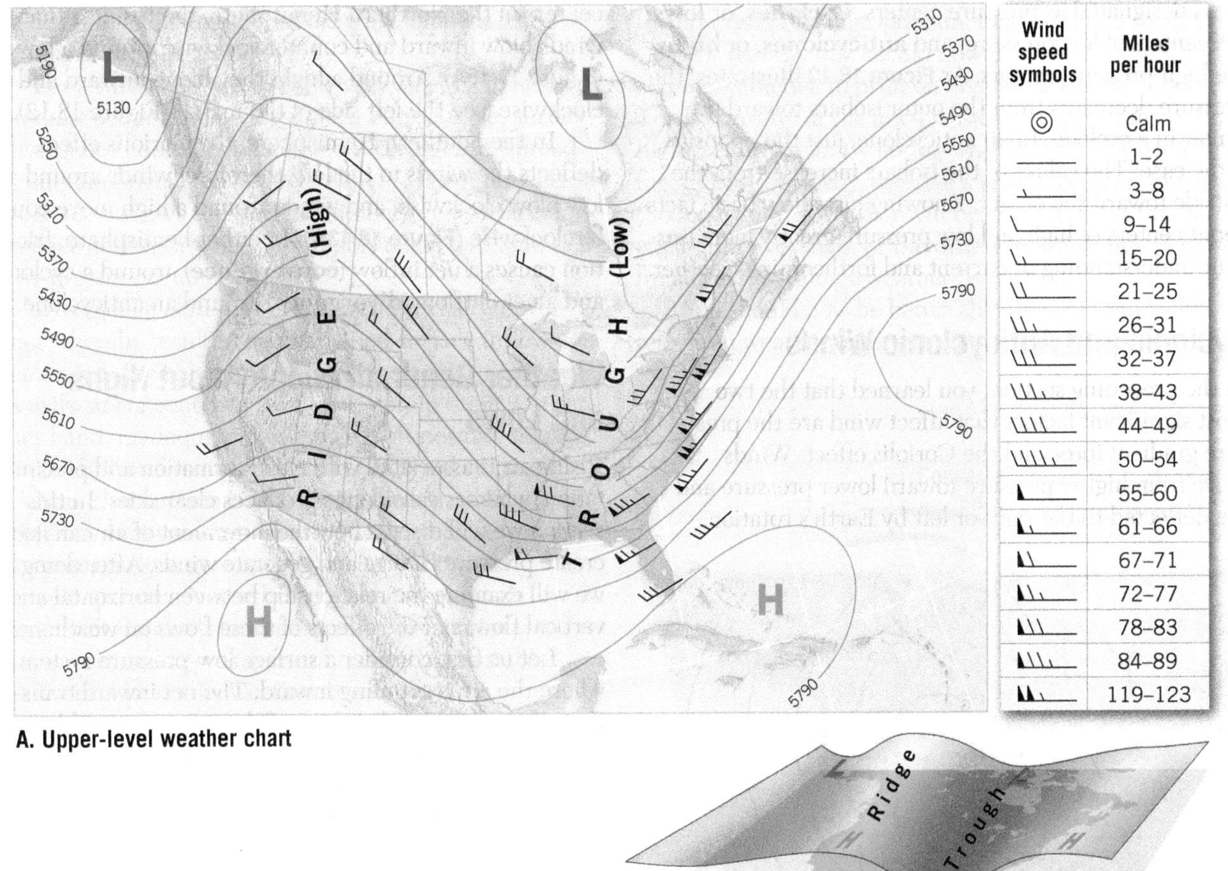

A. Upper-level weather chart

B. Representation of upper-level chart

Figure 18.10 Simplified upper-air weather chart A. This simplified weather chart shows the direction and speed of the upper-air winds. Note from the flags that the airflow is almost parallel to the contours. Like most other upper-air charts, this one shows variations in the height (in meters) at which a selected pressure (500 millibars) is found, rather than showing variations in pressure at a fixed height, like surface maps. Do not let this confuse you because there is a simple relationship between height contours and pressure. **B.** Places experiencing 500-millibar pressure at higher altitudes (toward the south on this map) are experiencing higher pressures than are places where the height contours indicate lower altitudes. Thus, *higher-elevation* contours indicate *higher* surface pressures, and *lower-elevation* contours indicate *lower* surface pressures.

A.

B.

Figure 18.15 Weather generalizations related to pressure centers
A. A rainy day in London. Low-pressure systems are frequently associated with cloudy conditions and precipitation. (Photo by Lourens Smak/Alamy) **B.** Clear skies and "fair" weather may be expected when an area is under the influence of high pressure. (Photo by Vidler Steve/Prisma Bildagentur/Alamy Stock Photo)

projected paths of cyclones and anticyclones. The "villain" on these weather programs is always the low-pressure center, which produces "bad" weather in any season. Lows move in roughly a west-to-east direction across the United States due to prevailing upper air winds, and require a few days to more than a week for the journey. Because their paths can be somewhat erratic, accurate prediction of their migration is difficult, although it is essential for short-range forecasting.

Meteorologists must also determine whether the flow aloft will intensify an embryo storm or act to suppress its development. Because of the close tie between conditions at the surface and those aloft, a great deal of emphasis has been placed on the importance and understanding of the total atmospheric circulation, particularly in the mid-latitudes. We will now examine the workings of Earth's general atmospheric circulation and then again consider the structure of the cyclone in light of this knowledge.

CONCEPT CHECKS 18.3

1. Prepare a diagram with isobars and wind arrows that shows the winds associated with surface cyclones and anticyclones in both the Northern and Southern Hemispheres.
2. For a surface low-pressure center to exist for an extended period, what condition must exist aloft?
3. What general weather conditions are to be expected when the pressure tendency is rising? When the pressure tendency is falling?

18.4 General Circulation of the Atmosphere

Summarize Earth's idealized global circulation. Describe how continents and seasonal temperature changes complicate the idealized pattern.

The underlying cause of wind is unequal heating of Earth's surface. In tropical regions, more solar radiation is received than is radiated back to space. In polar regions, the opposite is true: less solar energy is received than is lost. The atmosphere acts as a giant heat-transfer system, moving warm air poleward and cool air equatorward, in an attempt to balance these differences. On a smaller scale, but for the same reason, ocean currents also contribute to this global heat transfer. The general circulation is complex, and a great deal has yet to be explained. We can, however, develop a general understanding by first considering the circulation that would occur on a nonrotating Earth having a uniform surface. We will then modify this system to fit observed patterns.

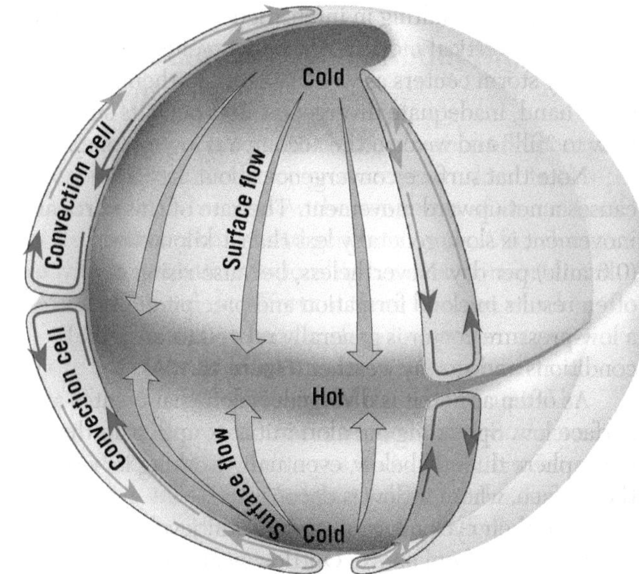

Figure 18.16 Global circulation on a nonrotating Earth A simple convection system is produced by unequal heating of the atmosphere.

Circulation on a Nonrotating Earth

On a hypothetical nonrotating planet with a smooth surface of either all land or all water, two large thermally produced cells would form (**Figure 18.16**). The heated equatorial air would rise until it reached the tropopause, which acts like a lid and deflects the air poleward. Eventually, this upper-level airflow would reach the poles, sink, spread out in all directions at the surface, and move back toward the equator. Once there, it would be reheated and start its journey over again. This hypothetical circulation system has upper-level air flowing poleward and surface air flowing equatorward.

If we add the effect of rotation, this simple convection system will break down into smaller cells. Figure 18.17 illustrates the three pairs of cells proposed to carry on the task of heat redistribution on a rotating planet. The polar and tropical cells retain the characteristics of the thermally generated convection described earlier. The nature of the midlatitude circulation is more complex and will be discussed in more detail later in this chapter.

Idealized Global Circulation

Near the equator, the rising air is associated with the pressure zone known as the **equatorial low**. This region of ascending moist, hot air is marked by abundant precipitation. Because this region of low pressure is a zone where winds from the north and south converge, it is also referred to as the **intertropical convergence zone (ITCZ)**. As the diverging upper-level flow from the equatorial low reaches 20° to 30° latitude, north or south, it sinks back toward the surface. This subsidence and associated adiabatic heating produce hot, arid conditions. The center of this zone of subsiding dry air is the **subtropical high**, which encircles the globe near 30° latitude, north and south (see Figure 18.17). The great deserts of Australia, Arabia, and Africa exist because of the stable, dry condition caused by the subtropical highs.

At the surface, airflow is outward from the center of the subtropical high. Some of the air travels equatorward and is deflected by the Coriolis effect, producing the reliable **trade winds**. The remainder travels poleward and is also deflected, generating the prevailing **westerlies**

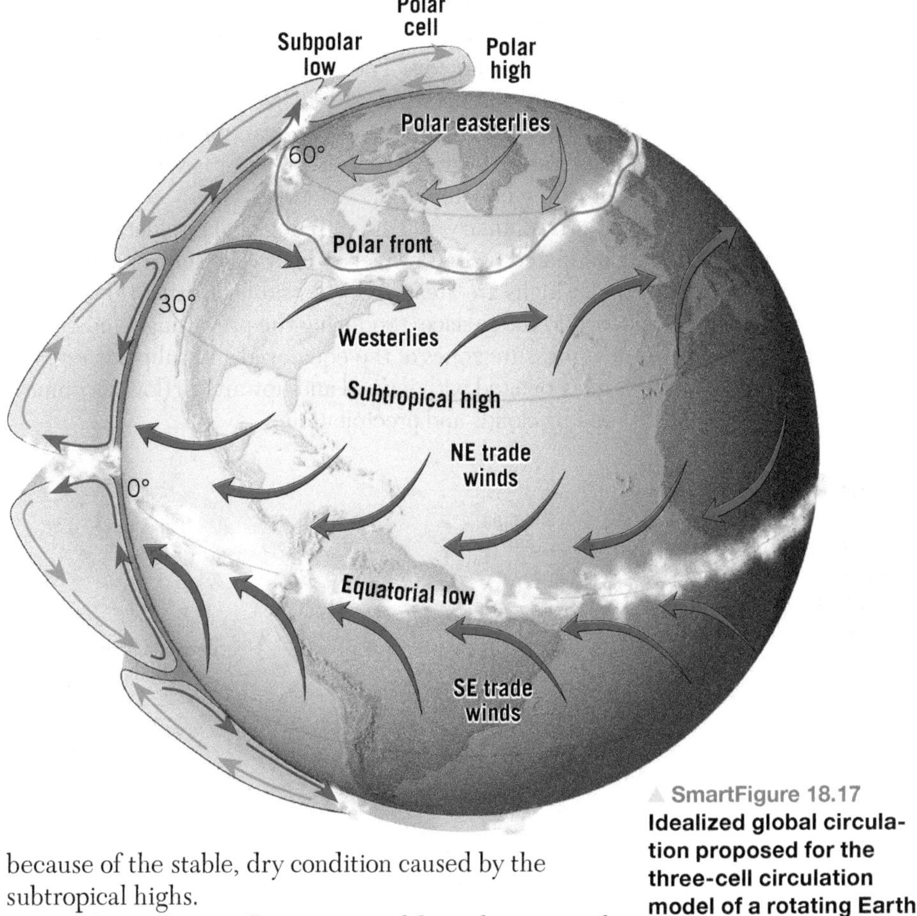

SmartFigure 18.17
Idealized global circulation proposed for the three-cell circulation model of a rotating Earth

TUTORIAL
https://goo.gl/8lhbvd

EYE ON EARTH 18.1

This composite satellite image shows a January dust storm in Africa that produced a dust plume that reached all the way to the northeast coast of South America. It has been estimated that plumes such as this transport about 40 million tons of dust from the Sahara Desert to the Amazon basin each year. The minerals carried by these dust plumes help replenish nutrients in rain forest soils that are continually being washed out of the soils by heavy tropical rains.

QUESTION 1 *Notice that the dust plume is following a curved path. Does the atmospheric circulation carrying this dust plume exhibit a clockwise or counterclockwise rotation?*

QUESTION 2 *What global pressure system was responsible for transporting this dust from Africa to South America? Referring to Figure 18.18 might be helpful.*

NASA

of the midlatitudes. As the westerlies move poleward, they encounter the cool **polar easterlies** in the region of the **subpolar low**. The interaction of these warm and cool winds produces the stormy belt known as the **polar front**. The source region for the variable polar easterlies is the **polar high**. Here, cold polar air is subsiding and spreading equatorward.

In summary, this simplified global circulation is dominated by four pressure zones. The subtropical and polar highs are areas of dry subsiding air that flows outward at the surface, producing the prevailing winds. The low-pressure zones of the equatorial and subpolar regions are associated with inward and upward airflow accompanied by clouds and precipitation.

Influence of Continents

Up to this point, we have described the surface pressure and associated winds as continuous belts around Earth. However, the only truly continuous pressure belt is the subpolar low in the Southern Hemisphere, where the ocean is uninterrupted by landmasses. At other latitudes, particularly in the Northern Hemisphere, where landmasses break up the ocean surface, large seasonal temperature differences disrupt the pattern. Figure 18.18 shows the resulting pressure and wind patterns for January and July. The circulation over the oceans is dominated by semipermanent cells of high pressure in the subtropics and cells of low pressure over the subpolar

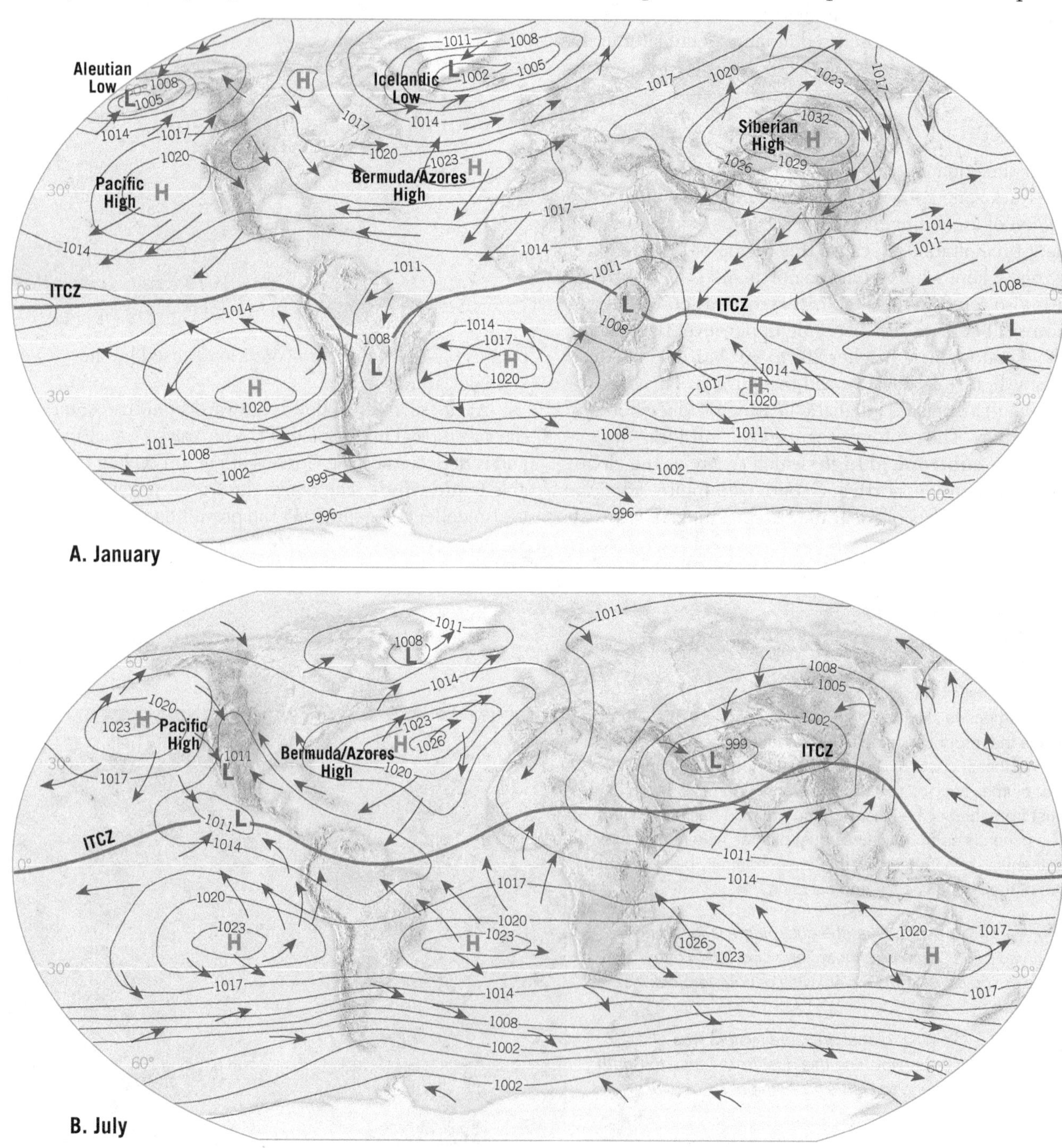

Figure 18.18
Average surface air pressure These maps show the average surface air pressure, in millibars, for **A.** January and **B.** July, with associated winds.

regions. The subtropical highs are responsible for the trade winds and westerlies, as mentioned earlier.

The large landmasses, on the other hand, particularly Asia, become cold in the winter and develop a seasonal high-pressure system from which surface flow is directed off the land (see Figure 18.18A). In the summer, the opposite occurs: The landmasses are heated and develop a low-pressure cell, which permits air to flow onto the land (18.18B). These seasonal changes in wind direction are known as the **monsoons**. During warm months, areas such as India experience a flow of warm, water-laden air from the Indian Ocean, which produces the rainy summer monsoon. The winter monsoon is dominated by dry continental air. A similar situation exists, but to a lesser extent, over North America.

In summary, the general circulation is produced by semipermanent cells of high and low pressure over the oceans and is complicated by seasonal pressure changes over land.

The Westerlies

The circulation in the midlatitudes, the zone of the westerlies, is complex and does not fit the convection system proposed for the tropics. Between 30° and 60° latitude, the general west-to-east flow is interrupted by the migration of cyclones and anticyclones. In the Northern Hemisphere, these cells move from west to east around the globe, creating an anticyclonic (clockwise) flow or a cyclonic (counterclockwise) flow in their area of influence. A close correlation exists between the paths taken by these surface pressure systems and the position of the upper-level airflow, indicating that the upper air steers the movement of cyclonic and anticyclonic systems.

Among the most obvious features of the flow aloft are the seasonal changes. The steep temperature gradient across the middle latitudes in the winter months corresponds to a stronger flow aloft. In addition, the polar jet stream fluctuates seasonally such that its average position migrates southward with the approach of winter and northward as summer nears. By midwinter, the jet core may penetrate as far south as central Florida.

Because the paths of low-pressure centers are guided by the flow aloft, we can expect the southern tier of states to experience more of their stormy weather in the winter season. During the hot summer months, the storm track is across the northern states, and some cyclones never leave Canada. The northerly storm track associated with summer also applies to Pacific storms, which move toward Alaska during the warm months, thus producing an extended dry season for much of the west coast. The number of cyclones generated is seasonal as well, with the largest number occurring in the cooler months, when the temperature gradients are greatest. This fact is in agreement with the role of cyclonic storms in the distribution of heat across the midlatitudes.

CONCEPT CHECKS 18.4

1. Referring to the idealized model of atmospheric circulation, in which belt of prevailing winds is most of the United States?
2. The trade winds diverge from which pressure belt?
3. Which prevailing wind belts converge in the stormy region known as the polar front?
4. Which pressure belt is associated with the equator?
5. Explain the seasonal change in winds associated with India. What term is applied to this seasonal wind shift?

18.5 Local Winds

List three types of local winds and describe their formation.

Now that we have examined Earth's large-scale circulation, let us turn briefly to winds that influence much smaller areas. Remember that all winds are produced for the same reason: pressure differences that arise because of temperature differences caused by unequal heating of Earth's surface. **Local winds** are small-scale winds produced by a locally generated pressure gradient. Those described here are caused either by topographic effects or by variations in surface composition in the immediate area.

Land and Sea Breezes

In coastal areas during the warm summer months, the land is heated more intensely during the daylight hours than is the adjacent body of water. As a result, the air above the land surface heats, expands, and rises, creating an area of lower pressure. A **sea breeze** then develops because cooler air over the water (higher pressure) moves toward the warmer land (lower pressure) (Figure 18.19A). The sea breeze begins to develop shortly before noon and generally reaches its greatest intensity during the mid- to late afternoon. These relatively cool winds can be a significant moderating influence on afternoon temperatures in coastal areas. Small-scale sea breezes can also develop along the shores of large lakes. People who live in a city near the Great Lakes, such as Chicago, recognize this lake effect, especially in the summer. They are reminded daily by weather reports of the cooler temperatures near the lake as compared to warmer outlying areas.

SmartFigure 18.19 **Sea and land breezes**

TUTORIAL
https://goo.gl/LBVzDB

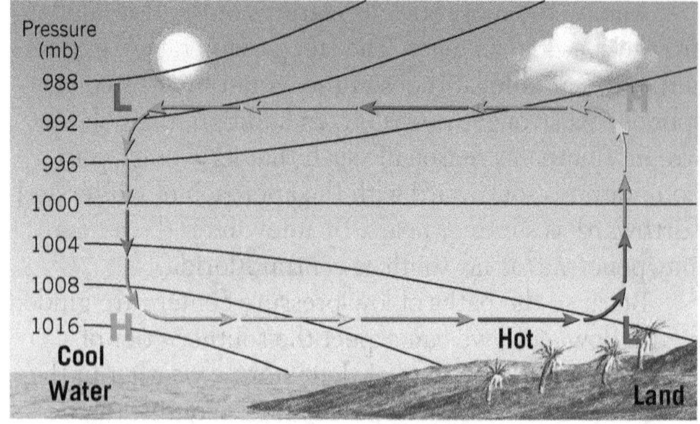

A. During daylight hours, cooler and denser air over the water moves onto the land, generating a sea breeze.

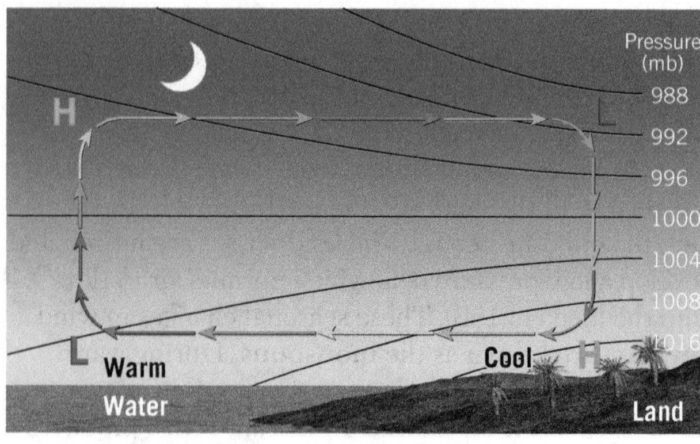

B. At night the land cools more rapidly than the sea, generating an offshore flow called a land breeze.

At night, the reverse may take place: The land cools more rapidly than the sea, and the **land breeze** develops (Figure 18.19B).

Mountain and Valley Breezes

A daily wind similar to land and sea breezes occurs in many mountainous regions. During daylight hours, the air along the slopes of the mountains is heated more intensely than the air at the same elevation over the valley floor. Because this warmer air is less dense, it glides up along the slope and generates a **valley breeze** (Figure 18.20A). The occurrence of these daytime upslope breezes can often be identified by the cumulus clouds that develop on adjacent mountain peaks.

After sunset, the pattern may reverse. Rapid radiation cooling along the mountain slopes produces a layer of cooler air next to the ground. Because cool air is denser than warm air, it drains downslope into the valley. This movement of air is called a **mountain breeze** (Figure 18.20B). The same type of cool air drainage can occur in places that have very modest slopes. The result is that the coldest pockets of air are usually in the lowest spots. Like many other winds, mountain and valley breezes have seasonal tendencies. Although valley breezes are most common during the warm season, when solar heating is most intense, mountain breezes tend to be more dominant in the cold season.

Chinook and Santa Ana Winds

Warm, dry winds sometimes move down the eastern slopes of the Rockies, where they are called **chinooks**. Such winds are often created when a strong pressure gradient develops in a mountainous region. As the air descends the leeward slopes of the mountains, it is heated adiabatically (by compression). Because condensation may have occurred as the air ascended the windward side, releasing latent heat, the air descending the leeward slope will be warmer and drier than it was at a similar elevation on the windward side. Although the temperature of these winds is generally less than 10°C (50°F), which is not particularly warm, the winds occur mostly in the winter and spring, when the affected areas may be experiencing below-freezing temperatures. Thus, by comparison, these dry, warm winds often bring a drastic change. When the ground has a snow cover, these winds are known to melt it in short order.

Figure 18.20 **Valley and mountain breezes**

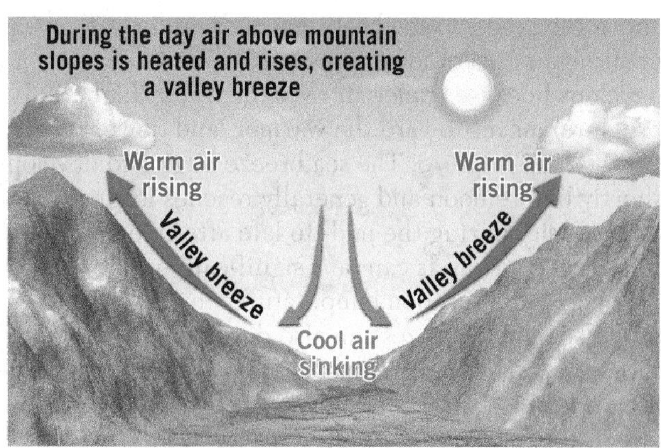

A. Valley breeze

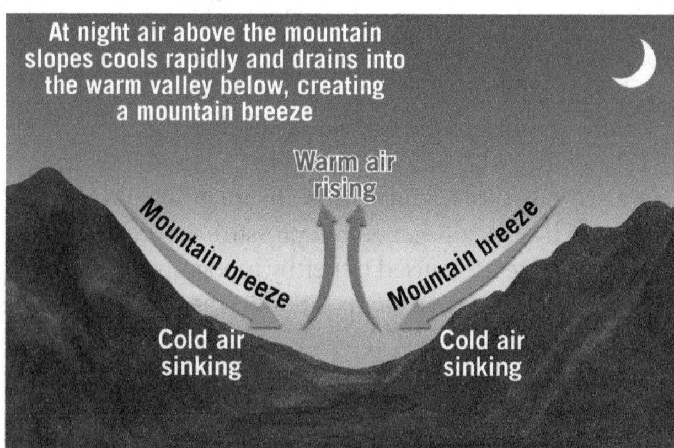

B. Mountain breeze

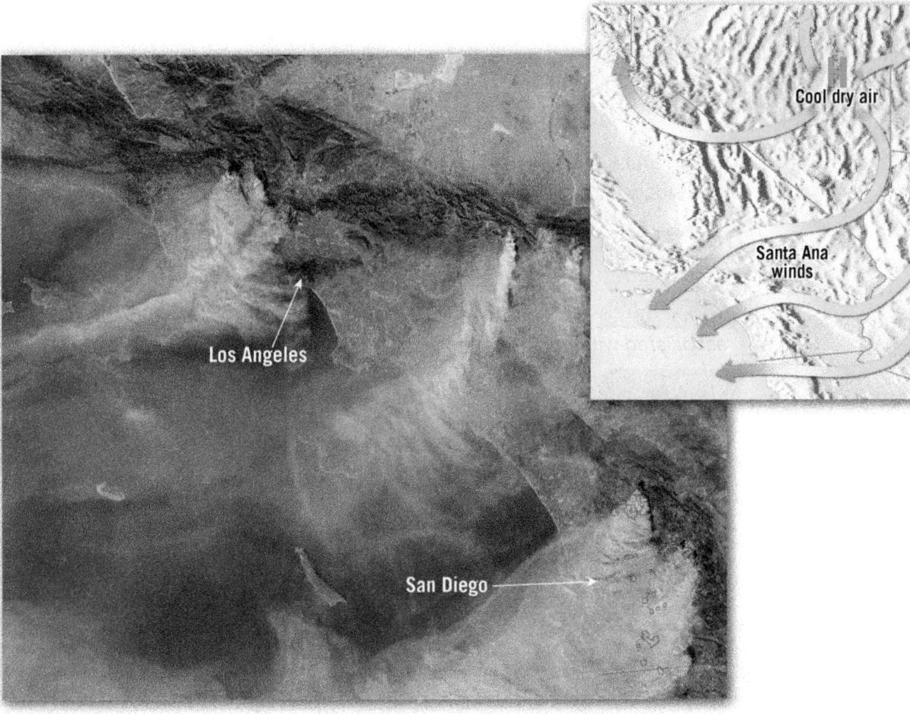

Figure 18.21 **Santa Ana winds** Wildfires fanned by Santa Ana winds raged in southern California in October 2003. These fires scorched more than 740,000 acres and destroyed more than 3000 homes. The inset shows an idealized high-pressure center composed of cool, dry air that drives Santa Ana winds. Adiabatic heating causes the air temperature to increase and the relative humidity to decrease. (NASA image)

A chinooklike wind that occurs in southern California is the **Santa Ana**. This hot, desiccating wind greatly increases the threat of fire in this already dry area (Figure 18.21).

CONCEPT CHECKS 18.5

1. What is a local wind?
2. Describe the formation of a sea breeze.
3. Does a land breeze blow toward or away from the shore?
4. During what time of day would you expect to experience a well-developed valley breeze—midnight, late morning, or late afternoon?

18.6 Measuring Wind

Describe the instruments used to measure wind. Explain how wind direction is expressed using compass directions.

Two basic wind measurements, direction and speed, are particularly significant to weather observers. One simple device for determining both measurements, a *wind sock*, is a common sight at small airports and landing strips (Figure 18.22A). The cone-shaped bag is open at both ends and is free to change position with shifts in wind direction. The degree to which the sock is inflated is an indication of wind speed.

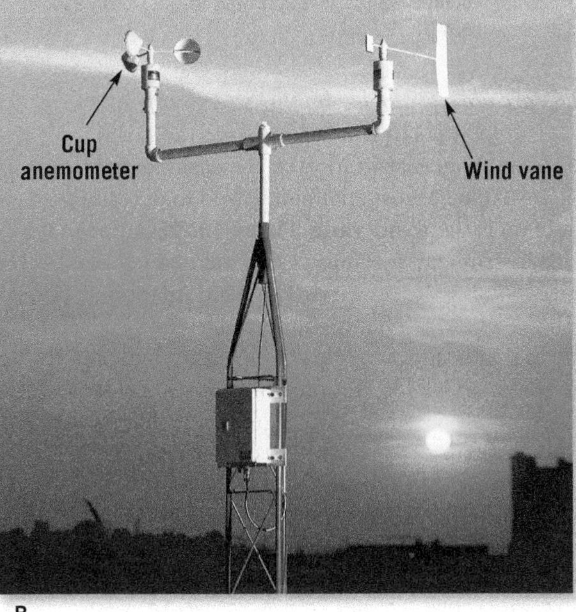

Figure 18.22 **Measuring the wind** **A.** A wind sock is a device for determining wind direction and estimating wind speed. Wind socks are common sights at small airports and landing strips. (Photo by Lourens Smak/Alamy Images) **B.** Wind vane (right) and cup anemometer (left). The wind vane shows wind direction, and the anemometer measures wind speed. (Photo by Belfort Instrument Company)

EYE ON EARTH 18.2

This mountain area was cloud free as this summer day began. By afternoon these clouds had formed. The clouds were associated with a local wind.

QUESTION 1 *With which local wind are the clouds in this photo most likely associated?*

QUESTION 2 *Describe the process that created the local wind that was associated with the formation of these clouds.*

QUESTION 3 *Would you expect clouds such as these to form at night?*

Herbert Koeppel/Alamy Images

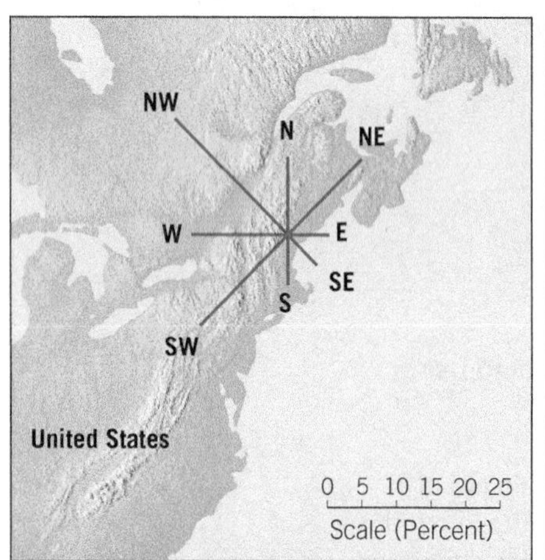

A. Wind frequency for winter in the northeastern United States.

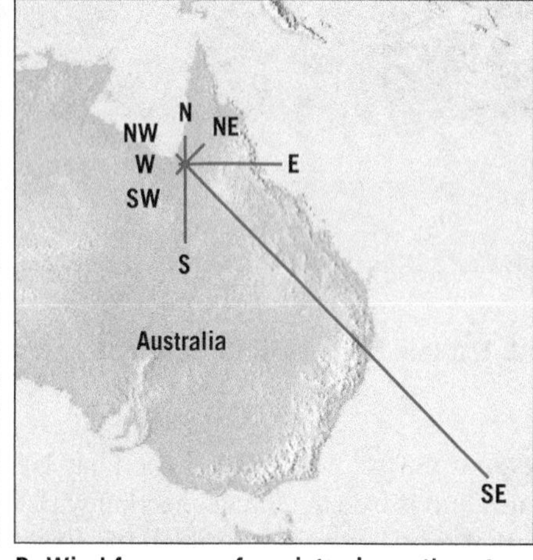

B. Wind frequency for winter in northeastern Australia. Note the reliability of the southeast trade winds in Australia as compared to the westerlies in the northeastern United States.

▲ Figure 18.23 **Wind roses** These graphs show the percentages of time winds come from various directions.

Winds are always labeled by the direction from which they blow. A north wind blows *from* the north *toward* the south, an east wind *from* the east *toward* the west. The instrument most commonly used to determine wind direction is the **wind vane** (Figure 18.22B, upper right). This instrument, a common sight on many buildings, always points *into* the wind. The wind direction is often shown on a dial that is connected to the wind vane. The dial indicates wind direction, either by points of the compass (N, NE, E, SE, etc.) or by a scale of 0 to 360 degrees. On the latter scale, 0 degrees and 360 degrees are both north, 90 degrees is east, 180 degrees is south, and 270 degrees is west.

Wind speed is commonly measured using a **cup anemometer** (see Figure 18.22B, upper left). The wind speed is read from a dial much like the speedometer of an automobile. Places where winds are steady and speeds are relatively high are potential sites for tapping wind energy.

When the wind consistently blows more often from one direction than from any other, it is called a **prevailing wind**. You may be familiar with the prevailing westerlies that dominate the circulation in the midlatitudes. In the United States, for example, these winds consistently move the "weather" from west to east across the continent. Embedded within this general eastward flow are cells of high and low pressure, with their characteristic clockwise and counterclockwise flow. As a result, the winds associated with the westerlies, as measured at the surface, often vary considerably from day to day and from place to place. By contrast, the direction of airflow associated with the belt of trade winds is much more consistent, as can be seen in Figure 18.23.

By knowing the locations of cyclones and anticyclones in relation to where you are, you can predict the changes in wind direction that will occur as a pressure center moves past. Because changes in wind direction often bring changes in temperature and moisture conditions, the ability to predict the winds can be very useful. In the Midwest, for example, a north wind may bring cool, dry air from Canada, whereas a south wind may bring warm, humid air from the Gulf of Mexico. Sir Francis Bacon summed it up nicely when he wrote, "Every wind has its weather."

CONCEPT CHECKS 18.6

1. What are the two basic wind measurements? What instruments are used to make these measurements?
2. *From* what direction does a northeast wind blow? *Toward* what direction does a south wind blow?

18.7 El Niño, La Niña, and the Southern Oscillation

Describe the Southern Oscillation and its relationship to El Niño and La Niña. List the climate impacts of El Niño and La Niña on North America.

El Niño was first recognized by fishermen from Ecuador and Peru, who noticed that the waters of the eastern Pacific sometimes got warmer in December or January. Because the warming usually occurred near the Christmas season, the event was named El Niño—"little boy," or "Christ child," in Spanish. These periods of abnormal warming happen at irregular intervals of 2 to 7 years and usually persist for spans of 9 months to 2 years.

El Niño is noted for its potentially catastrophic impact on the weather and economies of Peru, Chile, and Australia, among other countries. As shown in Figure 18.24A, during an El Niño, strong equatorial countercurrents amass large quantities of *warmer-than-normal* water along the west coast of South America. The unusually warm water and associated low pressure cause arid areas of Peru and Chile to receive unusually heavy rains that can lead to major flooding. In addition, the warm surface water blocks the upwelling of colder, nutrient-filled water—the primary food source for millions of small feeder fish (mainly anchovies).

Within 1 or 2 years, the circulation associated with El Niño is usually replaced by a La Niña event (Figure 18.24B). **La Niña**, which means "little girl," is essentially the opposite of El Niño and refers to *colder-than-normal sea-surface temperatures* in the central and eastern Pacific. As Figure 18.24B illustrates, the atmospheric circulation in the equatorial Pacific during La Niña is dominated by strong trade winds. These wind systems, in turn, generate a strong equatorial current that flows westward from South America toward Australia and Indonesia. This circulation pattern is often associated with flooding in northeastern Australia and Indonesia, whereas especially dry conditions prevail along the west coast of South America.

During La Niña, the cold Peru Current that flows equatorward along the coast of Chile and Peru intensifies, which encourages upwelling of deep nutrient-rich water. As a result, La Niña has come to be known as the "gift giver" in that region because the added nutrients are a boon to marine life, making fishing particularly good.

Global Impact of El Niño

The climate fluctuations associated with El Niño and La Niña were known for years but were considered local phenomena. Eventually, scientists discovered that El Niño and La Niña are part of the global atmospheric circulation pattern that affects the weather at great distances from the equatorial Pacific. Although the effects are variable, some locales appear to be affected more consistently.

During an El Niño event, winter temperatures are warmer than normal in the north-central United States and parts of Canada (Figure 18.25A). In addition, significantly wetter winters are experienced in the southwestern United States and northwestern Mexico, while the southeastern United States experiences wetter and cooler conditions. One major benefit of El Niño is a lower-than-average number of Atlantic hurricanes.

Global Impact of La Niña

La Niña was once thought to represent the normal conditions that occur between El Niño events. However, meteorologists now consider La Niña an important

Figure 18.24 **The relationship between El Niño, La Niña, and the Southern Oscillation A.** During an El Niño event, the pressure over the eastern Pacific drops, while the pressure over the western Pacific rises. This causes the trade winds to diminish, leading to an eastward movement of warm water along the equator. This strengthens the equatorial countercurrent, causing surface waters of the central and eastern Pacific to warm, with far-reaching consequences for weather patterns. **B.** During a La Niña event, strong trade winds drive the equatorial currents toward the west. At the same time, the strong Peru Current causes upwelling of cold water along the west coast of South America.

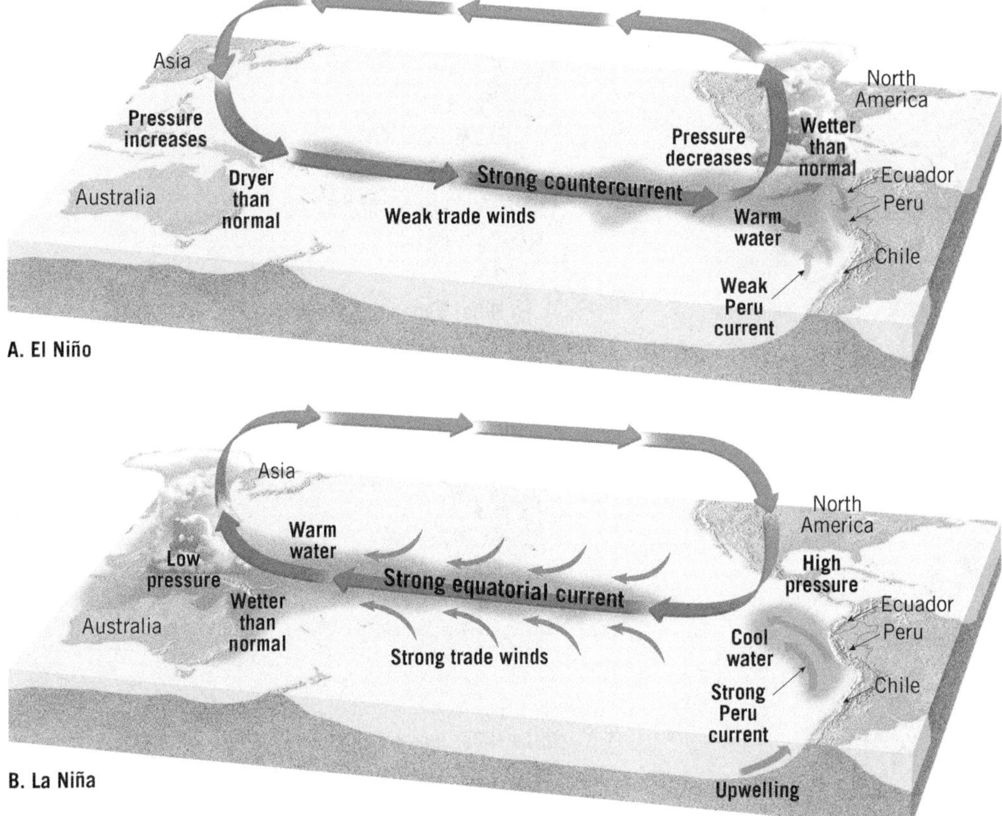

Figure 18.25 Climate impacts of El Niño and La Niña El Niño has the most significant impact on the climate of North America during the winter. In addition, El Niño affects areas around the tropical Pacific in both winter and summer. Likewise, La Niña has its greatest impact on North America's weather in the winter but influences other areas throughout the year.

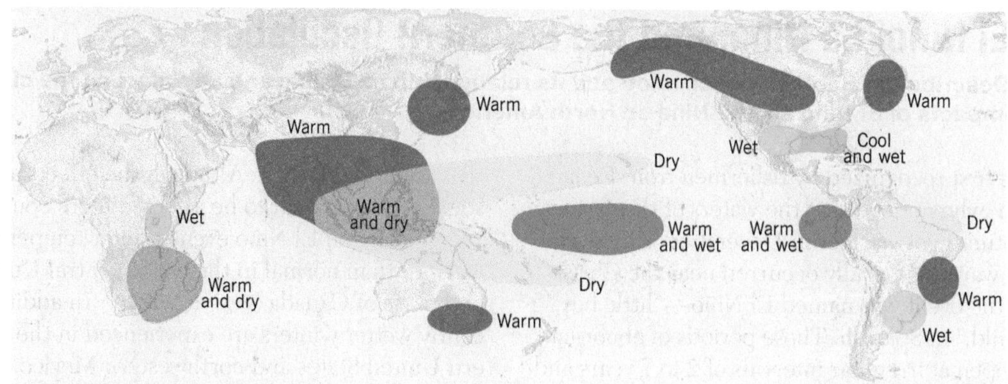

A. El Niño: December to February

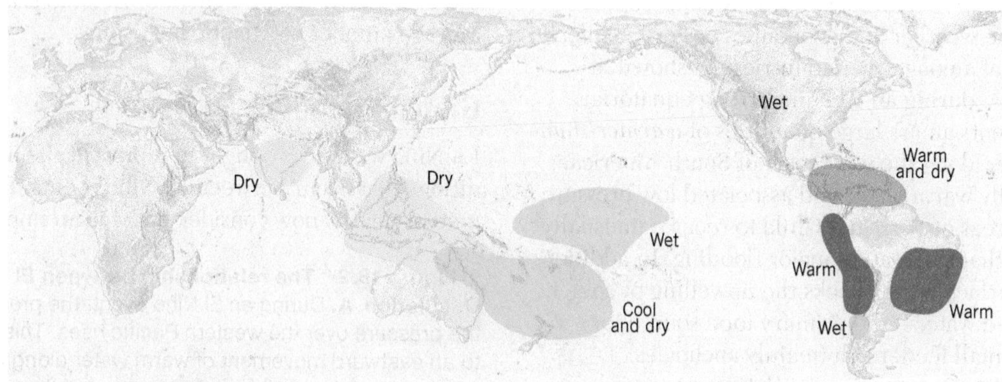

B. El Niño: June to August

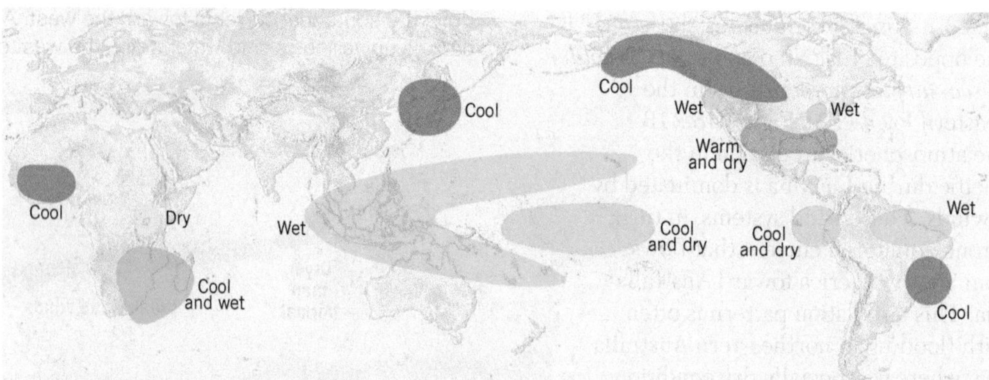

C. La Niña: December to February

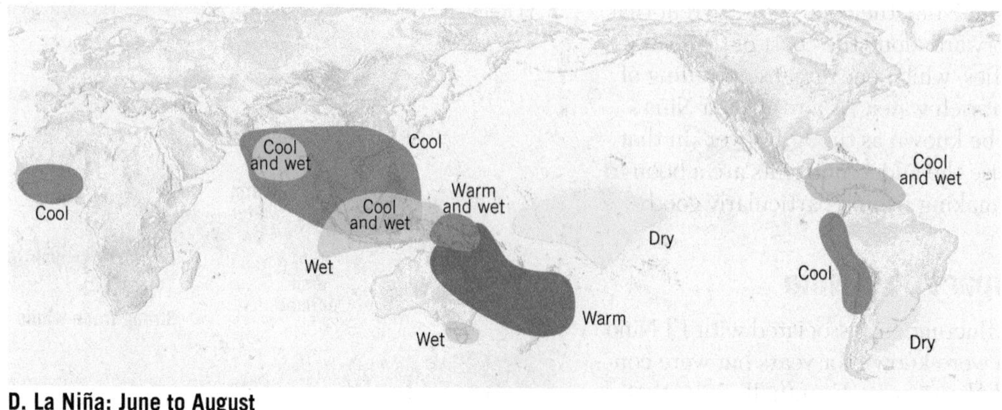

D. La Niña: June to August

atmospheric phenomenon in its own right. Researchers learned that colder-than-average surface temperatures in the central and eastern Pacific can trigger a La Niña event that exhibits a distinctive set of weather patterns.

Typical La Niña winter weather includes cooler and wetter conditions over the northwestern United States and especially cold winter temperatures in the Northern Plains states, while unusually warm conditions occur in the Southwest and Southeast (Figure 18.25C). In the western Pacific, La Niña events are associated with wetter-than-normal conditions. Another La Niña impact is more frequent hurricane activity in the Atlantic. A recent study concluded that the cost of hurricane damages in the United States is 20 times greater in La Niña years than in El Niño years.

Southern Oscillation

Although the triggers are not fully understood, we do know that El Niño and La Niña events are intimately related to changes in the global pressure patterns. Each time an El Niño occurs, the barometric pressure drops over large portions of the eastern Pacific and rises in the tropical portions of the western Pacific (see Figure 18.24A). Then, as a major El Niño event comes to an end, the pressure difference between these two regions swings back in the opposite direction, triggering a La Niña event (see Figure 18.24B).

This seesaw pattern of atmospheric pressure between the eastern and western Pacific is called the **Southern Oscillation**. Low pressure in the western Pacific strengthens the trade winds, which move warm, tropical water toward Australia and Indonesia. By contrast, a drop in pressure in the eastern Pacific weakens the trade winds and strengthens the equatorial countercurrents, which amass large quantities of warm water along the coasts of Peru and Chile. The changes in atmospheric pressure create the wind patterns that produce the weather associated with El Niño and La Niña events.

The link between the weather occurring in widely separated regions of the globe is called a *teleconnection*. For example, the warming of the eastern Pacific associated with an El Niño event is linked to winter flooding in southern California. In addition, there appears to be a teleconnection between a strong El Niño and a weak Asian monsoon.

Because sea-surface temperatures are often predictable months in advance, understanding teleconnection patterns allows meteorologists to make climate predictions for distant locations. For instance, predicting the start of an El Niño enables prediction of North American precipitation and temperature patterns weeks or even months in advance. National Weather Service predictions for periods from 1 to 13 months into the future are called *climate outlooks*.

CONCEPT CHECKS 18.7

1. Describe how a major El Niño event tends to affect the weather in Peru and Chile as compared to Indonesia and Australia.
2. Describe the sea-surface temperatures on both sides of the tropical Pacific during a La Niña event.
3. How does a major La Niña event influence the hurricane season in the North Atlantic?
4. Briefly describe the Southern Oscillation and how it is related to El Niño and La Niña.

18.8 Global Distribution of Precipitation

Discuss the major factors that influence the global distribution of precipitation.

A casual glance at Figure 18.26 shows a relatively complex pattern for the distribution of precipitation. Although the map appears to be complicated, the general features of the map can be explained by applying our knowledge of global winds and pressure systems.

The Influence of Pressure and Wind Belts

In general, regions influenced by high pressure, with its associated subsidence and diverging winds, experience relatively dry conditions. On the other hand, regions under the influence of low pressure and its converging winds and ascending air receive ample precipitation. This pattern is illustrated by the fact that the tropical regions dominated by the equatorial low are the rainiest regions on Earth. It is here that we find the rain forests of the Amazon basin in South America and the Congo basin in Africa. The warm, humid trade winds converge to yield abundant rainfall throughout the year. By contrast, areas dominated by subtropical high-pressure cells clearly receive much smaller amounts of precipitation. These are the regions of extensive subtropical deserts. In the Northern Hemisphere, the largest desert is the Sahara. Examples in the Southern Hemisphere include the Kalahari in southern Africa and the dry lands of Australia.

Other Factors

If Earth's pressure and wind belts were the only factors controlling precipitation distribution, the pattern shown in Figure 18.26 would be simpler. The inherent nature of the air is also an important factor in determining precipitation potential. Because cold air has a low capacity for moisture compared with warm air, we would expect

Centimeters	Inches
200 and over	80 and over
150-199	60-79
100-149	40-59
50-99	20-39
25-49	10-19
Under 25	Under 10

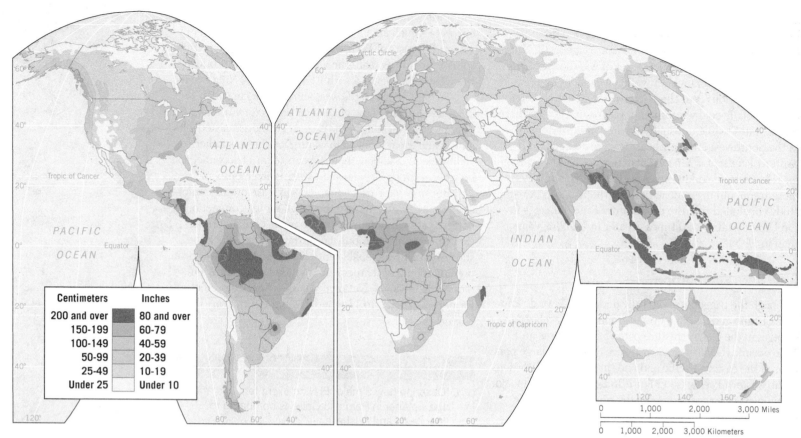

Figure 18.26 Global distribution of average annual precipitation

a latitudinal variation in precipitation, with low latitudes receiving the greatest amounts of precipitation and high latitudes receiving the smallest amounts. Figure 18.26 indeed shows heavy rainfall in equatorial regions and meager precipitation in high-latitude areas. Recall that the dry region in the warm subtropics is explained by the presence of the subtropical high.

The distribution of land and water also complicates the precipitation pattern. Large landmasses in the middle latitudes commonly experience decreased precipitation toward their interiors. For example, central North America and central Eurasia receive considerably less precipitation than do coastal regions at the same latitude.

Mountain barriers also alter precipitation patterns. Windward mountain slopes receive abundant precipitation, whereas leeward slopes and adjacent lowlands are often deficient in moisture.

CONCEPT CHECKS 18.8

1. With which global pressure belt are the rain forests of Africa's Congo basin associated? Which pressure system is linked to the Sahara Desert?
2. What factors, in addition to the distribution of wind and pressure, influence the global distribution of precipitation?

EYE ON EARTH 18.3

This satellite image was produced with data from the *Tropical Rainfall Measuring Mission* (*TRMM*). Notice the band of heavy rainfall shown in reds and yellows that extends east–west across the image.

QUESTION 1 *With which pressure zone is this band of rainy weather associated?*

QUESTION 2 *Is it more likely that this image was acquired in July or January? Explain.*

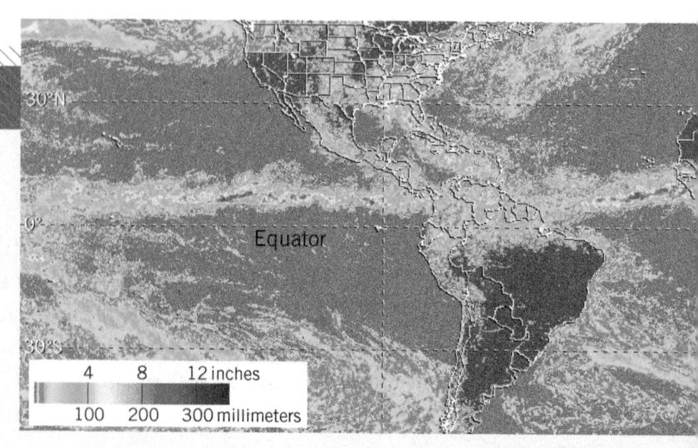

CONCEPTS IN REVIEW
Air Pressure and Wind

18.1 Understanding Air Pressure

Define *air pressure* and describe the instruments used to measure this weather element.

KEY TERMS: air pressure, mercury barometer, aneroid barometer, barograph

- Air has weight: At sea level, it exerts a pressure of 1 kilogram per square centimeter (14.7 pounds per square inch), or 1 atmosphere.
- Air pressure is the force exerted by the weight of air above. With increasing altitude, there is less air above to exert a force, and thus air pressure decreases with altitude—rapidly at first and then much more slowly.
- The unit meteorologists use to measure atmospheric pressure is the millibar. Standard sea-level pressure is expressed as 1013.2 millibars. Isobars are lines on a weather map that connect places of equal air pressures.
- A mercury barometer measures air pressure using a column of mercury in a glass tube sealed at one end and inverted in a dish of mercury. It measures atmospheric pressure based on the height of the column of mercury in the barometer. Standard atmospheric pressure at sea level equals 29.92 inches of mercury. As air pressure increases, the mercury in the tube rises, and when air pressure decreases, so does the height of the column of mercury.
- Aneroid ("without liquid") barometers consist of partially evacuated metal chambers that compress as air pressure increases and expand as pressure decreases.

18.2 Factors Affecting Wind

Discuss the three forces that act on the atmosphere to either create or alter winds.

KEY TERMS: wind, isobar, pressure gradient force, Coriolis effect, geostrophic wind, jet stream

- Wind is controlled by a combination of (1) the pressure gradient force, (2) the Coriolis effect, and (3) friction. The pressure gradient force, which results from pressure differences, is the primary force that drives wind. It is depicted by the spacing of isobars on a map. Closely spaced isobars indicate a steep pressure gradient and strong winds; widely spaced isobars indicate a weak pressure gradient and light winds.
- The Coriolis effect, which is due to Earth's rotation, produces deviation in the path of wind due to Earth's rotation (to the right in the Northern Hemisphere and to the left in the Southern Hemisphere). Friction, which significantly influences airflow near Earth's surface, is negligible above a height of a few kilometers.
- Above a height of a few kilometers, the Coriolis effect is equal to and opposite the pressure gradient force, which results in geostrophic winds. Geostrophic winds follow a path parallel to the isobars, with velocities proportional to the pressure gradient force.

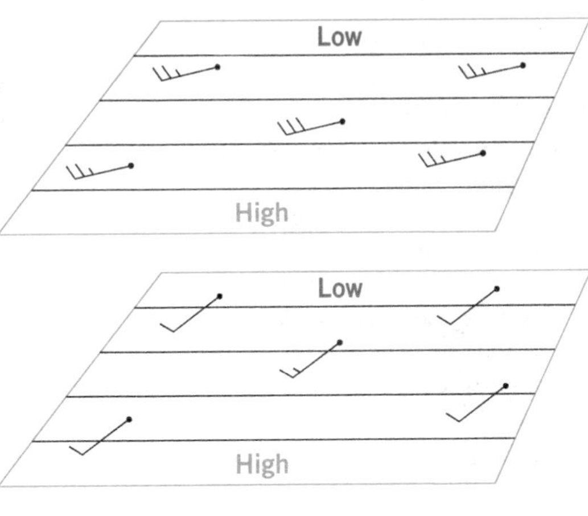

? These diagrams show surface winds at two locations. All factors in both situations are identical except that one surface is land and the other is water. Which diagram represents winds over the land? Explain your choice.

18.3 Highs and Lows

Contrast the weather associated with low-pressure centers (cyclones) and high-pressure centers (anticyclones).

KEY TERMS: cyclone (low), anticyclone (high), convergence, divergence, pressure (barometric) tendency

- The two types of pressure centers are (1) cyclones, or lows (centers of low pressure), and (2) anticyclones, or highs (centers of high pressure). In the Northern Hemisphere, winds around a low (cyclone) are counterclockwise and inward. Around a high (anticyclone), they are clockwise and outward. In the Southern Hemisphere, the Coriolis effect causes winds to be clockwise around a low and counterclockwise around a high.
- Because air rises and cools adiabatically in a low-pressure center, cloudy conditions and precipitation are often associated with their passage. In a high-pressure center, descending air is compressed and warmed; therefore, cloud formation and precipitation are unlikely in an anticyclone, and "fair" weather is usually expected.

? Assume that you are an observer checking this aneroid barometer several hours after it was last checked. What is the pressure tendency? How did you figure this out? What does the tendency shown on the barometer indicate about forthcoming weather?

18.4 General Circulation of the Atmosphere

Summarize Earth's idealized global circulation. Describe how continents and seasonal temperature changes complicate the idealized pattern.

KEY TERMS: equatorial low, intertropical convergence zone (ITCZ), subtropical high, trade winds, westerlies, polar easterlies, subpolar low, polar front, polar high, monsoon

- If Earth's surface were uniform, four belts of pressure oriented east to west would exist in each hemisphere. Beginning at the equator, the four belts would be the (1) equatorial low, also referred to as the intertropical convergence zone (ITCZ), (2) subtropical high at about 25° to 35° on either side of the equator, (3) subpolar low, situated at about 50° to 60° latitude, and (4) polar high, near Earth's poles.
- Particularly in the Northern Hemisphere, large seasonal temperature differences over continents disrupt the idealized, or zonal, global patterns of pressure and wind. In winter, large, cold landmasses develop a seasonal high-pressure system from which surface airflow is directed off the land. In summer, landmasses are heated, and a low-pressure system develops over them, which permits air to flow onto the land. These seasonal changes in wind direction are known as monsoons.
- In the middle latitudes, between 30° and 60° latitude, the general west-to-east flow of the westerlies is interrupted by the migration of cyclones and anticyclones. The paths taken by these cyclonic and anticyclonic systems is closely correlated to upper-level airflow and the polar jet stream. The average position of the polar jet stream, and hence the paths followed by cyclones, migrates southward with the approach of winter and northward as summer nears.

? The accompanying sketch shows a cross section of the idealized circulation in the Northern Hemisphere. Match the appropriate number on the sketch to each of the following features: (a) equatorial low, (b) polar front, (c) subtropical high, (d) polar high.

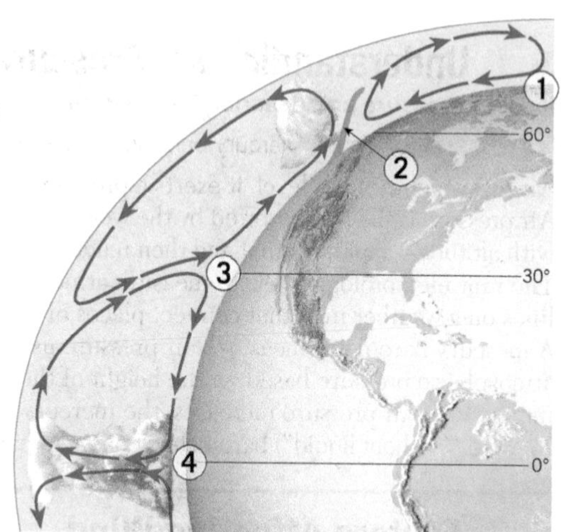

18.5 Local Winds

List three types of local winds and describe their formation.

KEY TERMS: local wind, sea breeze, land breeze, valley breeze, mountain breeze, chinook, Santa Ana

- Local winds are small-scale winds produced by a locally generated pressure gradient. Sea and land breezes form along coasts and are brought about by temperature contrasts between land and water. Valley and mountain breezes occur in mountainous areas where the air along slopes heats differently than does the air at the same elevation over the valley floor. Chinook and Santa Ana winds are warm, dry winds created when air descends the leeward side of a mountain and warms by compression.

? It is late afternoon on a warm summer day, and you are enjoying some time at the beach. Until the last hour or two, winds were calm. Then a breeze began to develop. Is it more likely a cool breeze from the water or a warm breeze from the adjacent land area? Explain.

18.6 Measuring Wind

Describe the instruments used to measure wind. Explain how wind direction is expressed using compass directions.

KEY TERMS: wind vane, cup anemometer, prevailing wind

- The two basic wind measurements are direction and speed. Winds are always labeled by the direction from which they blow. Wind direction is measured with a wind vane, and wind speed is measured using a cup anemometer.

? When designing an airport, it is important to have planes take off *into* the wind. Refer to the accompanying wind rose and describe the orientation of the runway and the direction planes would usually travel when taking off. Where on Earth might you find a wind rose like this?

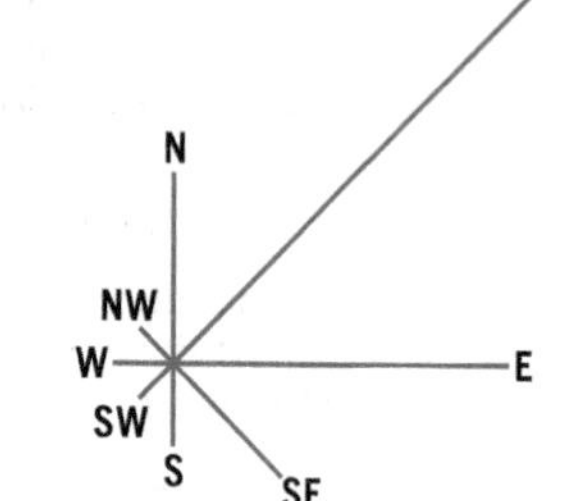

18.7 El Niño, La Niña, and the Southern Oscillation

Describe the Southern Oscillation and its relationship to El Niño and La Niña. List the climate impacts of El Niño and La Niña on North America.

KEY TERMS: El Niño, La Niña, Southern Oscillation

- El Niño refers to episodes of ocean warming in the eastern Pacific along the coasts of Ecuador and Peru. It is associated with weak trade winds, a strong eastward-moving equatorial countercurrent, a weakened Peru Current, and diminished upwelling along the western margin of South America.
- A La Niña event is associated with colder-than-average surface temperatures in the eastern Pacific. La Niña is linked to strong trade winds, a strong westward-moving equatorial current, and a strong Peru Current with significant coastal upwelling.
- El Niño and La Niña events are part of the global circulation and are related to a seesaw pattern of atmospheric pressure between the eastern and western Pacific called the Southern Oscillation. El Niño and La Niña events influence weather on both sides of the tropical Pacific Ocean as well as weather in the United States.

? This image shows floods occurring in eastern Australia. Is this region more likely being influenced by La Niña or El Niño?

Torsten Blackwood/AFP/Getty Images

18.8 Global Distribution of Precipitation

Discuss the major factors that influence the global distribution of precipitation.

- The general features of the global distribution of precipitation can be explained by global winds and pressure systems. In general, regions influenced by high pressure, with its associated subsidence and divergent winds, experience dry conditions. Regions under the influence of low pressure, with its converging winds and ascending air, receive ample precipitation.
- Air temperature, the distribution of continents and oceans, and the location of mountains also influence the distribution of precipitation.

? Why do most high-latitude regions receive relatively meager precipitation?

GIVE IT SOME THOUGHT

1. Mercury is 13.5 times denser (heavier) than water. If you built a barometer using water rather than mercury, how tall, in inches, would it have to be to record standard sea-level pressure?

2. This satellite image shows a tropical cyclone (hurricane).
   a. Examine the cloud pattern and determine whether the flow is clockwise or counterclockwise.
   b. In which hemisphere is the storm located?
   c. What factor determines whether the flow is clockwise or counterclockwise?

NASA

3. Use the accompanying drawing to answer the following:
   a. Identify the arrows labeled A and B as representing either the pressure gradient force, the Coriolis effect, or friction.
   b. Is the wind in this drawing occurring near Earth's surface or aloft?

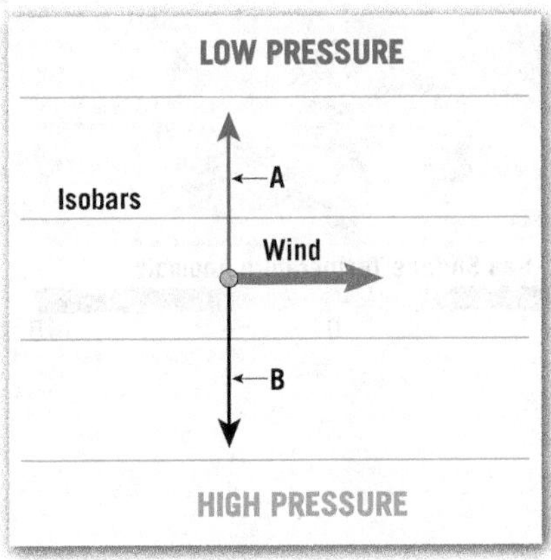

4. If divergence in the jet stream above a surface low-pressure center exceeds convergence at the surface, will surface winds likely get stronger or weaker? Explain.

5. You and a friend are watching TV on a rainy day, when the weather reporter says, "The barometric pressure is 28.8 inches and rising." Hearing this, you say, "It looks like fair weather is on its way." Your friend responds with the following questions: "I thought air pressure had something to do with the weight of air. How does inches relate to weight? And why do you think the weather is going to improve?" How would you respond to your friend's queries?

6. If you live in the Northern Hemisphere and are directly west of the center of a cyclone, what is the probable wind direction at your location? What if you were west of an anticyclone?

7. The accompanying map is a simplified surface weather map for April 2, 2011, on which three pressure cells are numbered.
   a. Identify which of the pressure cells are anticyclones (highs) and which are cyclones (lows).
   b. Which pressure cell has the steepest pressure gradient and therefore the strongest winds?
   c. Refer to Figure 18.2 to determine whether pressure cell 3 should be considered strong or weak.

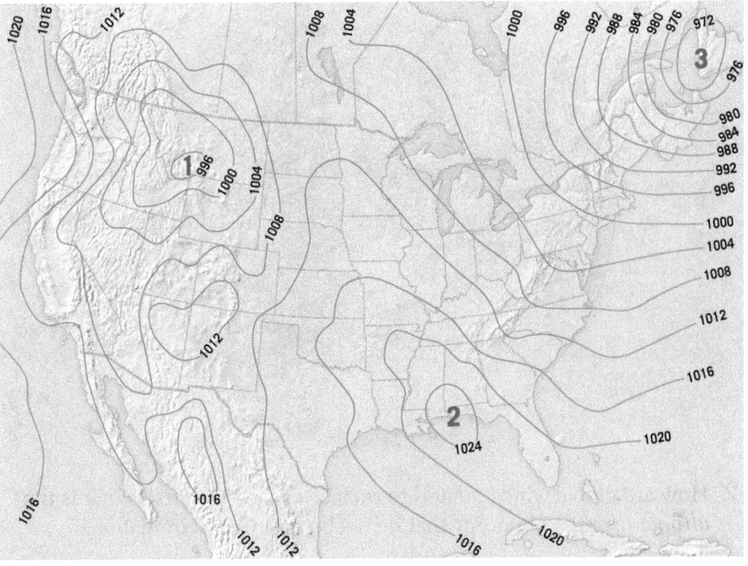

8 If Earth did not rotate on its axis and if its surface were completely covered with water, what direction would a boat drift if it started its journey in the middle latitudes of the Northern Hemisphere? (*Hint:* What would the global circulation pattern be like for a nonrotating Earth?)

9 The accompanying maps of Africa show the distribution of precipitation for July and January. Which map represents July, and which represents January? How were you able to figure this out?

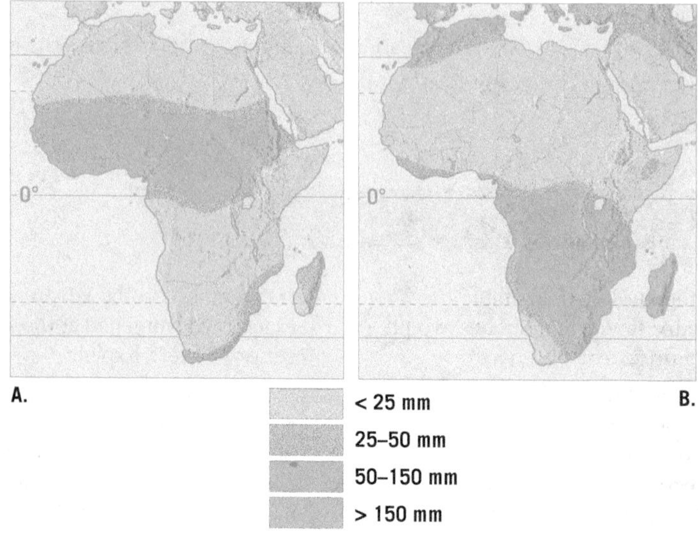

EXAMINING THE EARTH SYSTEM

1 Examine this classic image of Africa from space and pick out the region dominated by the equatorial low and the areas influenced by the subtropical highs in each hemisphere. What clue(s) did you use? Speculate on the differences in the biosphere between the regions dominated by high pressure and the zone influenced by low pressure.

2 How are global winds related to surface ocean currents? What is the *ultimate* source of energy that drives both of these circulations?

3 The accompanying image shows wintertime sea-surface temperature anomalies (differences from average) over the equatorial Pacific Ocean. Based on this map, answer the following questions:

a. In what phase was the Southern Oscillation (El Niño or La Niña) when this image was made?

b. Would the trade winds be strong or weak at this time?

c. If you lived in Australia during this event, what weather conditions would you expect?

d. If you were attending college in the southeastern United States during winter months, what type of weather conditions would you expect? (*Hint:* See Figure 18.25.)

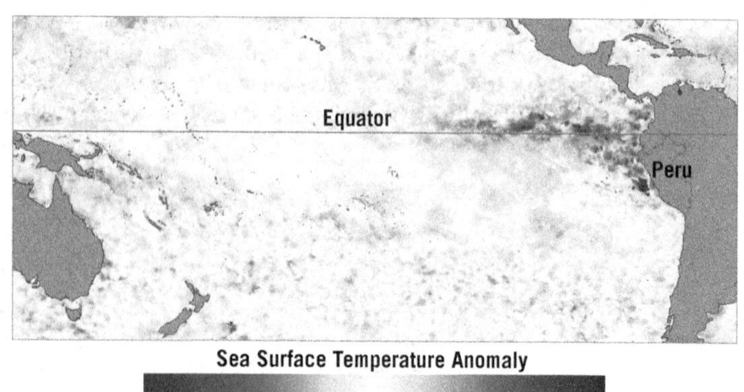

DATA ANALYSIS

Sea-Surface Temperatures

Satellites observe sea-surface temperatures (SST) in the equatorial Pacific Ocean. These data are used to predict and track the progress of El Niño and La Niña events.

ACTIVITIES

Go to the Unisys Weather page, at http://weather.unisys.com. Click on Surface Data under Analyses and then click on Daily Plots under Other Pages. Click on SST Data under Plots to bring up a current sea surface temperature map.

1 Where are the warmest waters? Coldest?

2 Based on sea-surface temperatures, where do you think warm ocean currents are located? Cold ocean currents? Compare your answers to the ocean current map in Figure 15.2.

3 Are temperatures off the west coast of equatorial South America slightly warmer or slightly colder than water farther to the west? Based on the temperature pattern, would you expect the trade winds in the equatorial Pacific to be strong or weak? How can you tell?

Click on SST Anom at the top of the map to display the sea-surface temperature anomalies map. These anomalies are the difference between the current temperature and the 1981–2010 average temperature.

4 Are the temperatures in the western part of the equatorial Pacific warmer or colder than average? What about the temperatures in the central part? In the eastern part?

5 Estimate the difference in degrees Celsius between the current and average temperatures for the western, central, and eastern regions.

6 Based on what you know about the Southern Oscillation, is the eastern equatorial Pacific currently exhibiting El Niño, La Niña, or normal characteristics?

Go to NOAA's Southern Oscillation Index (SOI) page, at www.ncdc.noaa.gov/teleconnections/enso/indicators/soi/.

7 What is the SOI?

8 How would a difference in sea level pressure cause the temperature patterns associated with El Niño and La Niña?

9 According to the graph has the SOI been predominately positive or negative in the past few months? Does this indicate El Niño or La Niña?

10 Zoom the graph according to the instructions just above it. Is the SOI positive or negative for the most recent month? How many months has the SOI been in this phase?

11 Does the answer you found in Question 10 match your prediction from Question 6? Why or why not?

19

Weather Patterns and Severe Storms*

FOCUS ON CONCEPTS

Each statement represents the primary learning objective for the corresponding major heading within the chapter. After you complete the chapter, you should be able to:

19.1 Discuss air masses, their classification, and associated weather.

19.2 Compare and contrast typical weather associated with a warm front and a cold front. Describe an occluded front and a stationary front.

19.3 Summarize the weather associated with the passage of a mature midlatitude cyclone. Describe how airflow aloft is related to cyclones and anticyclones at the surface.

19.4 List the basic requirements for thunderstorm formation and locate places on a map that exhibit frequent thunderstorm activity. Describe the stages in the development of a thunderstorm.

19.5 Summarize the atmospheric conditions and locations that are favorable to the formation of tornadoes. Discuss tornado destruction and tornado forecasting.

19.6 Identify areas of hurricane formation on a world map and discuss the conditions that promote hurricane formation. List the three broad categories of hurricane destruction.

*This chapter was revised with the assistance of Professor Redina Herman, Western Illinois University.

A storm is classified as a thunderstorm only after thunder is heard. Because thunder is produced by lightning, lightning must also occur. (Photo by Saverio Maria Gallotti/Alamy)

TORNADOES AND HURRICANES RANK among nature's most destructive forces. Each spring, newspapers report the death and destruction left in the wake of bands of tornadoes. During late summer and fall, we hear occasional news reports about hurricanes. Storms with names such as Katrina, Matthew, Sandy, and Ike make front-page headlines. Thunderstorms, although less intense and far more common than tornadoes and hurricanes, are also part of our discussion on severe weather in this chapter. Before looking at violent weather, however, we will study the atmospheric phenomena that most often affect our day-to-day weather: air masses, fronts, and traveling midlatitude cyclones. We will see the interplay of the elements of weather discussed in Chapters 16, 17, and 18.

19.1 Air Masses

Discuss air masses, their classification, and associated weather.

For many people who live in the middle latitudes, which include much of the United States, summer heat waves and winter cold spells are familiar experiences. In the first instance, several days of high temperatures and oppressive humidity may finally end when a series of thunderstorms pass through the area, followed by a few days of relatively cool relief. By contrast, the clear skies that often accompany a span of frigid subzero days may be replaced by thick gray clouds and a period of snow as temperatures rise to levels that seem mild compared to those that existed just a day earlier. In both examples, what was experienced was a period of generally constant weather conditions followed by a relatively short period of change and then the reestablishment of a new set of weather conditions that remained for perhaps several days before changing again.

What Is an Air Mass?

The weather patterns just described result from movements of large bodies of air, called air masses. An **air mass**, as the term implies, is an immense body of air, usually 1,600 kilometers (1,000 miles) or more across and perhaps several kilometers thick, that is characterized by a similarity of temperature and moisture across a given altitude. When this air moves out of its region of origin, it will carry these temperatures and moisture conditions with it, eventually affecting a large portion of a continent (Figure 19.1).

An excellent example of the influence of an air mass is illustrated in Figure 19.2, which shows a cold, dry mass from northern Canada moving southward. With a beginning temperature of −46°C (−51°F), the air mass warms to −33°C (−27°F) by the time it reaches Winnipeg. It continues to warm as it moves southward through the Great Plains and into Mexico. Throughout its southward journey, the air mass becomes warmer. But it also brings some of the coldest weather of the winter to the places in its path. Thus, the air mass is modified, but it also modifies the weather in the areas over which it moves.

The horizontal uniformity of an air mass is not perfect, of course. Because air masses extend over large areas, small differences occur in temperature and humidity from place to place. Still, the differences observed within an air mass are small compared to the rapid changes experienced across air-mass boundaries.

Since it may take several days for an air mass to move across an area and be modified, the region under its influence will likely experience fairly constant weather, a situation called **air-mass weather**. Certainly there are some day-to-day variations, but the weather will be relatively consistent until an air-mass boundary comes through and a different air mass influences the area. The air-mass concept is an important one because it is closely related to the study of atmospheric disturbances. Most disturbances in the middle latitudes originate along the boundary zones that separate different air masses.

▼ **Figure 19.1 Lake-effect snow storm** This satellite image shows a cold, dry air mass moving from its source region in Canada across Lake Superior. It illustrates the process that leads to lake-effect snow storms. (NASA)

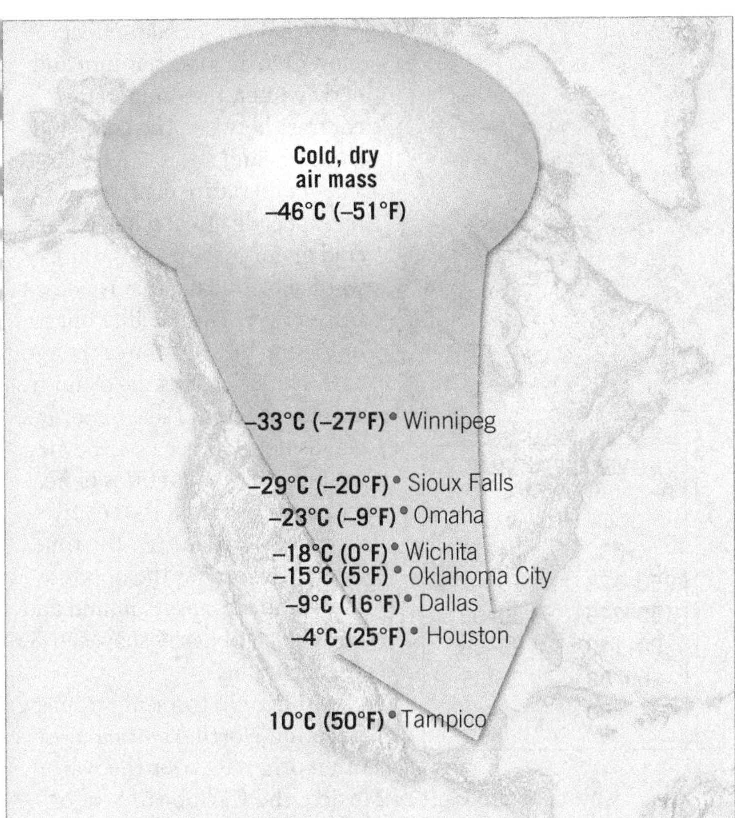

Figure 19.2 **An invasion of frigid air** As this very cold air mass moved southward from Canada, it brought some of the coldest weather of the winter to the areas in its path. As it advanced into the United States, the air mass slowly got warmer. Thus, the air mass was gradually modified at the same time that it modified the weather in the areas over which it moved. (From *Physical Geography: A Landscape Appreciation*, 9th edition, by Tom L. McKnight and Darrell Hess, ©2008. Reprinted and electronically reproduced by permission of Pearson Education, Inc., Upper Saddle River, NJ.)

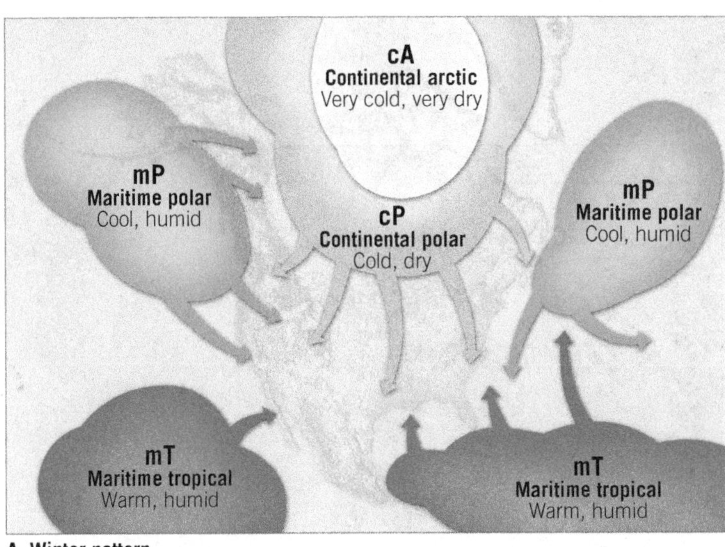

A. Winter pattern

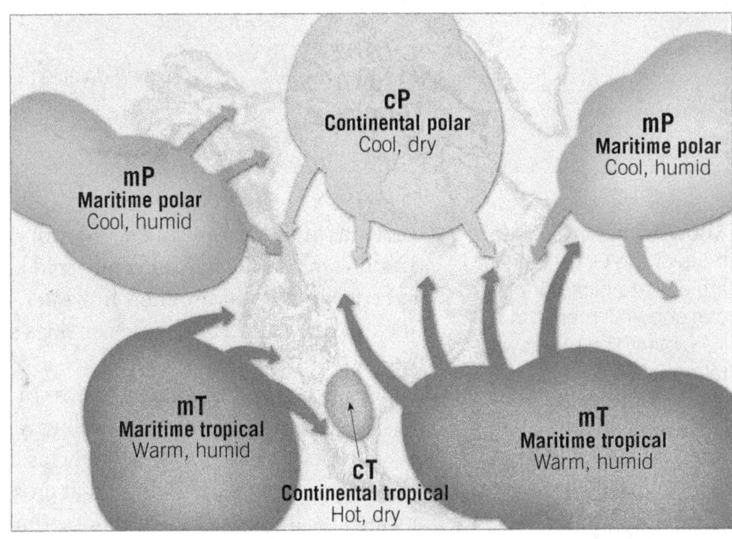

B. Summer pattern

Figure 19.3 **Air-mass source regions for North America** Source regions are largely confined to subtropical and subpolar locations. The fact that the middle latitudes are where cold and warm air masses clash, often because the converging winds of a traveling cyclone draw them together, means that this zone lacks the conditions necessary to be a source region. The differences between polar and arctic are relatively small and serve to indicate the degree of coldness of the respective air masses. By comparing the **A.** winter and **B.** summer maps, it is clear that the movement and temperature characteristics fluctuate.

Source Regions

When a portion of the lower atmosphere moves slowly or stagnates over a relatively uniform surface, the air assumes the distinguishing features of that area, particularly with regard to temperature and moisture conditions.

The area where an air mass acquires its characteristic properties of temperature and moisture is called its **source region**. The source regions that produce air masses influencing North America are shown in Figure 19.3.

Air masses are classified according to their source region. **Polar (P) air masses** and **arctic (A) air masses** originate in high latitudes toward Earth's poles, whereas those that form in low latitudes are called **tropical (T) air masses**. The designation *polar, arctic*, or *tropical* gives an indication of the temperature characteristics of an air mass. *Polar* and *arctic* indicate cold, and *tropical* indicates warm.

In addition, air masses are classified according to the nature of the surface in the source region. **Continental (c) air masses** form over land, and **maritime (m) air masses** originate over water. The designation *continental* or *maritime* thus suggests the moisture characteristics of the air mass. Continental air is likely to be dry, and maritime air is likely to be humid.

The basic types of air masses according to this scheme of classification are continental polar (cP), continental arctic (cA), continental tropical (cT), maritime polar (mP), and maritime tropical (mT).

Weather Associated with Air Masses

Continental polar and maritime tropical air masses influence the weather of North America most, especially east of the Rocky Mountains. Continental polar air masses

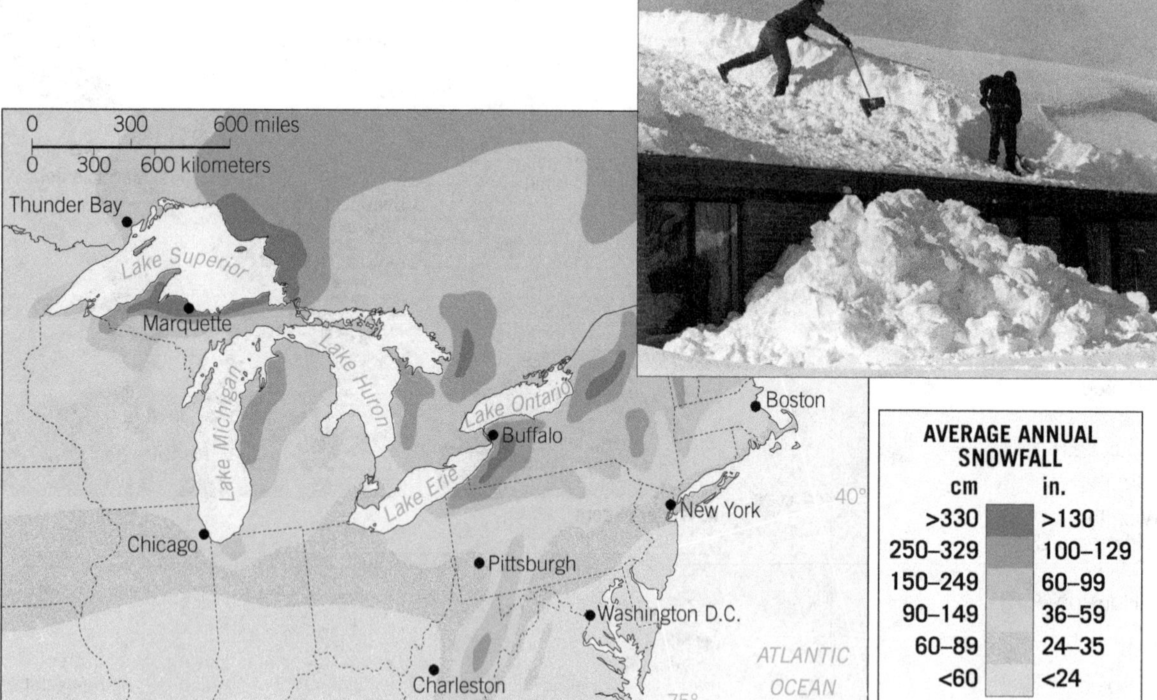

SmartFigure 19.4 **Snowfall map** The snow belts of the Great Lakes are easy to pick out on this snowfall map. (Data from NOAA) The photo was taken following a 6-day lake-effect snowstorm in November 1996 that dropped 175 centimeters (nearly 69 inches) of snow on Chardon, Ohio, setting a new state record. (Photo by Tony Dejak/AP Images)

TUTORIAL
https://goo.gl/BXIF6T

originate in northern Canada, interior Alaska, and the Arctic—areas that are uniformly cold and dry in winter and cool and dry in summer. In winter, an invasion of continental polar air brings the clear skies and cold temperatures we associate with a cold wave as it moves southward from Canada into the United States. In summer, this air mass may bring a few days of cooling relief.

Although cP air masses are not, as a rule, associated with heavy precipitation, those that cross the Great Lakes during late autumn and winter sometimes bring snow to the leeward shores. These localized storms often form when the surface weather map indicates no apparent cause for a snowstorm. These are known as **lake-effect snows**, and they make Buffalo and Rochester, New York, among the snowiest cities in the United States (Figure 19.4).

What causes lake-effect snow? During late autumn and early winter, the temperature contrast between the lakes and adjacent land areas can be large.* The temperature contrast can be especially great when a very cold cP air mass pushes southward across a lake that is not yet frozen over. The satellite image in Figure 19.1 illustrates the process. Notice that as the cloud-free air moves across Lake Superior, clouds develop because the air acquires large quantities of heat and moisture from the relatively warm lake surface. By the time the cP air reaches the opposite shore, the air mass is humid and unstable, and heavy snow showers are occurring.

Maritime tropical air masses affecting North America most often originate over the warm waters of the Gulf of Mexico, the Caribbean Sea, or the adjacent Atlantic Ocean. As you might expect, these air masses are warm, moisture laden, and usually unstable. Maritime tropical air is the source of much, if not most, of the precipitation in the eastern two-thirds of the United States. In summer, when an mT air mass invades the central and eastern United States, and occasionally southern Canada, it brings the high temperatures and oppressive humidity typically associated with its source region.

Of the two remaining air masses, maritime polar and continental tropical, the latter has the least influence on the weather of North America. Hot, dry continental tropical air, originating in the Southwest and Mexico during the summer, only occasionally affects the weather outside its source region.

During the winter, maritime polar air masses coming from the North Pacific often originate as continental polar air masses in Siberia. The cold, dry cP air is transformed into relatively mild, humid, unstable mP air during its long journey across the North Pacific (Figure 19.5). As this mP air arrives at the western shore of North America, it is often accompanied by low clouds and shower activity. When this air advances inland against the western mountains, orographic uplift produces heavy rain or snow on the windward slopes of the mountains. Maritime polar air also originates in the North Atlantic, off the coast of eastern Canada, and occasionally influences the weather of the northeastern United States.

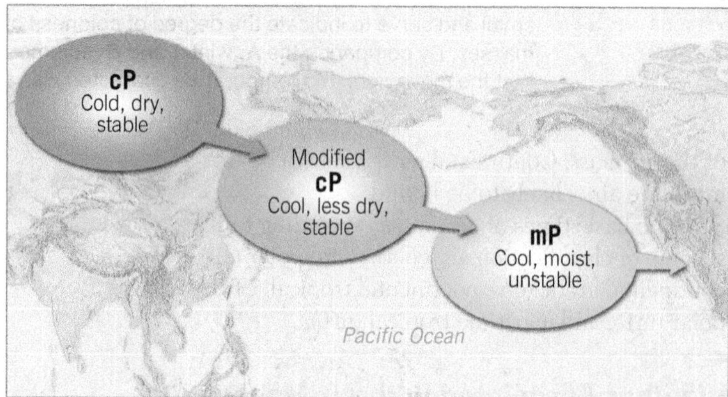

Figure 19.5 **Air-mass modification** During winter, maritime polar (mP) air masses in the North Pacific usually begin as continental polar (cP) air masses in Siberia. The cP air is modified to mP as it slowly crosses the ocean.

*Recall that land cools more rapidly and to lower temperatures than water. See the discussion of land and water in the section "Why Temperatures Vary: The Controls of Temperature" in Chapter 16.

EYE ON EARTH 19.1

This satellite image from December 27, 2010, shows a strong winter storm off the east coast of the United States.

QUESTION 1 *Can you identify the very center of the storm?*

QUESTION 2 *What air mass is being drawn into the storm to produce the dense clouds in the upper right?*

QUESTION 3 *What term is applied to a storm such as this?*

QUESTION 4 *Farther south, a cold air mass over the southeastern states is cloud free. What is its likely classification? Explain how it is being modified as it moves over the Atlantic.*

Figure 19.6 **Classic nor'easter** The satellite image shows a strong winter storm called a *nor'easter* along the coast of New England on January 12, 2011. In winter, a nor'easter exhibits a weather pattern in which strong northeast winds carry cold, humid mP air from the North Atlantic into New England and the middle Atlantic states. The ground-level view of the storm in Boston shows that the combination of ample moisture and strong convergence can result in heavy snow. (Satellite image by NASA; photo by Michael Dwyer/Alamy Images)

In winter, when New England is on the northern or northwestern side of a passing low-pressure center, the counterclockwise cyclonic winds draw in maritime polar air. The result is a storm characterized by snow and cold temperatures, known locally as a **nor'easter** (Figure 19.6).

CONCEPT CHECKS 19.1

1. Define *air mass*. What is air-mass weather?
2. On what basis are air masses classified?
3. Compare the temperature and moisture characteristics of the following air masses: cP, cA, mP, mT, and cT.
4. Which air mass is associated with lake-effect snow? What causes lake-effect snow?

19.2 Fronts

Compare and contrast typical weather associated with a warm front and a cold front. Describe an occluded front and a stationary front.

One prominent feature of middle-latitude weather is how suddenly and dramatically it can change. Most of these sudden changes are associated with the passage of weather fronts. **Fronts** are boundary surfaces that separate air masses of different densities—one of which is usually warmer and contains more moisture than the other. However, fronts can form between any two contrasting air masses. When the vast sizes of air masses are considered, the zones (fronts) that separate them are relatively narrow and are shown as lines on weather maps.

Generally, the air mass located on one side of a front moves faster than the air mass on the other side. Thus, one air mass actively advances into the region occupied by another and collides with it. During World War I, Norwegian meteorologists visualized these zones of air-mass interactions as analogous to battle lines and tagged them "fronts," as in battlefronts. It is along these zones of "conflict" that storms develop and produce much of the precipitation and severe weather in the belt of the westerlies.

As one air mass moves into a region occupied by another, minimal mixing occurs along the frontal surface. Instead, the air masses retain their identity as one is displaced upward over the other. No matter which air mass is advancing, it is always the warmer, less-dense air that is forced aloft, whereas the cooler, denser air acts as a wedge on which lifting occurs. The process of warm air gliding up and over a cold air mass is termed **overrunning**.

Warm Fronts

When the surface position of a front moves so that warm air occupies territory formerly covered by cooler air, it is called a **warm front** (Figure 19.7). On a weather map, the surface position of a warm front is shown by a red line with red semicircles protruding into the cooler air. The direction the symbols are pointing is the direction the front is moving.

East of the Rockies, warm tropical air often enters the United States from the Gulf of Mexico and overruns receding cool air. As the cool air retreats, friction with the ground slows the advance of the surface position of the front more so than its position aloft. Stated another way, less dense, warm air has a hard time displacing denser, cold air. For this reason, the boundary separating these air masses acquires a very gradual slope. The average slope of a warm front is about 1:200, which means that if you are 200 kilometers (120 miles) ahead of the surface location of a warm front, you will find the frontal surface at a height of 1 kilometer (0.6 mile).

As warm air ascends the retreating wedge of cold air, it expands and cools adiabatically to produce clouds and, frequently, precipitation. The sequence of clouds shown in Figure 19.7 typically precedes a warm front. The first sign of the approach of a warm front is the appearance of cirrus clouds overhead. These high clouds form 1000 kilometers (600 miles) or more ahead of the surface front, where the overrunning warm air has ascended high up the wedge of cold air.

As the front nears, cirrus clouds grade into cirrostratus, which blend into denser sheets of altostratus. About 300 kilometers (180 miles) ahead of the front, thicker stratus and nimbostratus clouds appear, and rain or snow begins. Because of their slow rate of advance and very low slope, warm fronts usually produce light to moderate precipitation over a large area for an extended period. This precipitation is often in the form of snow in the winter. During late fall and early spring, the precipitation often changes from snow to sleet to freezing rain and finally to rain as the warm front approaches. Warm fronts are occasionally associated with thunderstorms in the spring and fall. This occurs when the overrunning air is unstable and the temperatures on opposite sides of the front contrast sharply. At the other extreme, a warm front associated with a dry air mass could pass without creating clouds or precipitation.

A gradual increase in temperature occurs with the passage of a warm front. The increase is most noticeable when there is a large temperature difference between the adjacent air masses. The moisture content and stability of the encroaching warm air mass largely determine when clear skies will return. During summer, cumulus, and occasionally cumulonimbus, clouds may be embedded in the warm unstable air mass that follows the front. Precipitation from these clouds can be heavy but is usually scattered and of short duration.

SmartFigure 19.7
Warm front This diagram shows the idealized clouds and weather associated with a warm front. During most of the year, warm fronts produce light to moderate precipitation over a wide area.

ANIMATION
https://goo.gl/r6qTf0

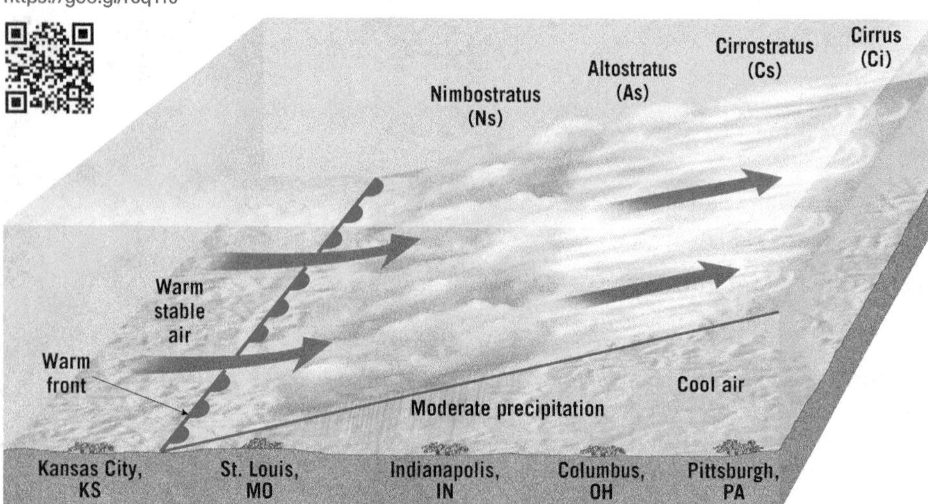

Cold Fronts

When dense cold air is actively advancing into a region occupied by warmer air, the boundary is called a **cold front** (Figure 19.8). As with warm fronts, friction tends to slow the surface position of a cold front more so than its position aloft. However, because of the relative positions of the adjacent air masses, the cold front steepens as it moves. On average, cold fronts are about twice as steep as warm fronts, having a slope of perhaps 1:100. In addition, cold fronts advance at speeds around 35 to 50 kilometers (20 to 35 miles) per hour compared to 25 to 35 kilometers (15 to 20 miles) per hour for warm fronts. These two differences—rate of movement and steepness of slope—largely account for the more violent nature of cold-front weather compared to the weather generally accompanying a warm front (Figure 19.9).

As a cold front approaches, commonly from the west or northwest, towering clouds can often be seen in the distance. Near the front, a dark band of ominous clouds foretells the coming weather. The forceful lifting of air along a cold front is often so rapid that the latent heat released when water vapor condenses appreciably

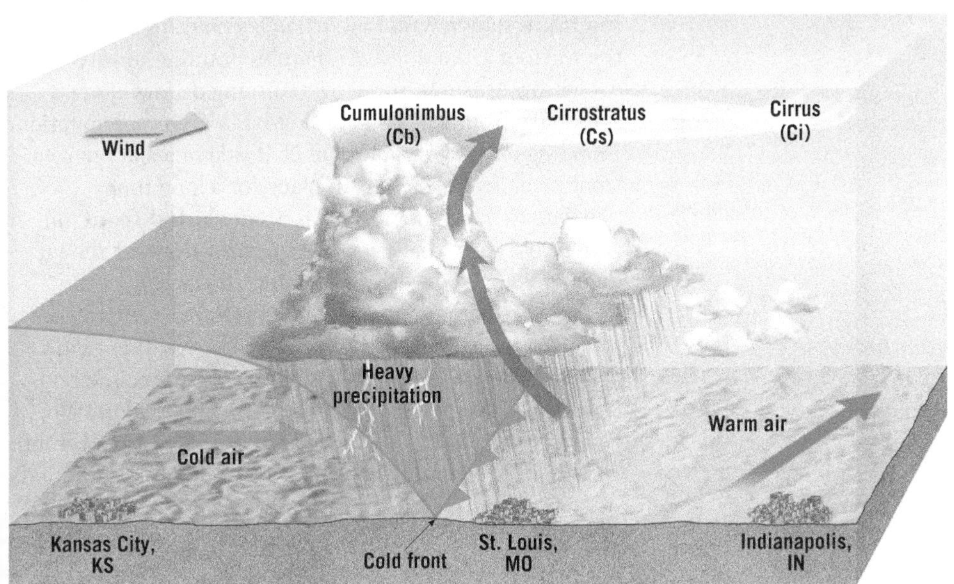

SmartFigure 19.8 **Cold front** Fast-moving cold front and cumulonimbus clouds. Thunderstorms may occur if the warm air is unstable.

TUTORIAL
https://goo.gl/ieWWUS

increases the air's buoyancy. The heavy downpours and vigorous wind gusts associated with mature cumulonimbus clouds frequently result. A cold front produces roughly the same amount of lifting as a warm front but over a shorter distance. As a result, the intensity of precipitation is greater, but the duration is shorter. In addition, a marked temperature drop and a wind shift from the south to west or northwest accompany the passage of the front. The sometimes violent weather and sharp temperature contrast along the cold front are symbolized on a weather map by a blue line with blue triangle-shaped points that extend into the warmer air mass (see Figure 19.8).

The weather behind a cold front is dominated by a subsiding and relatively cold air mass. Thus, clearing usually begins soon after the front passes. Although the compression of air due to subsidence causes some adiabatic heating, the effect on surface temperatures is minor. In winter, the long, cloudless nights that often follow the passage of a cold front allow for abundant radiation cooling that reduces surface temperatures. When a cold front moves over a relatively warm area, surface heating can produce shallow convection. This, in turn, may generate low cumulus or stratocumulus clouds behind the front.

Stationary Fronts and Occluded Fronts

Occasionally, the flow on both sides of a front is neither toward the cold air mass nor toward the warm air mass but almost parallel to the line of the front. Thus, the surface position of the front does not move. This condition is called a **stationary front**. On a weather map, stationary

Figure 19.9 **Hailstones litter the ground** Cumulonimbus clouds along a cold front produced hail and heavy rain in central Mississippi in March 2013. (Photo by T. Coraghessan Boyle/Rogelio V. Solis/AP Images)

Figure 19.10 **Stages in the formation of an occluded front**

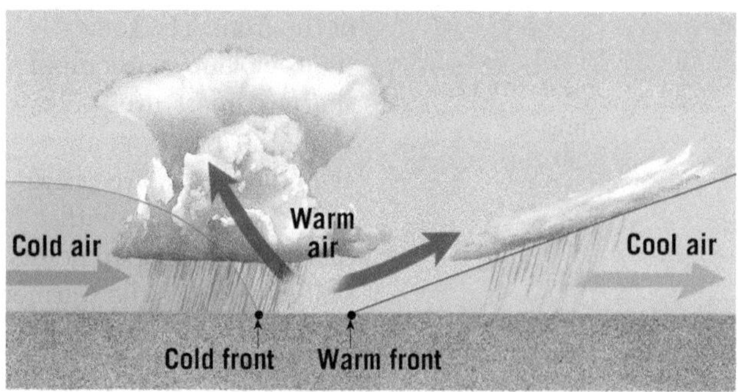

In this example, the air behind the cold front is colder and denser than the air ahead of the warm front.

The surface cold front moves faster than the surface warm front and overtakes it to form an occluded front.

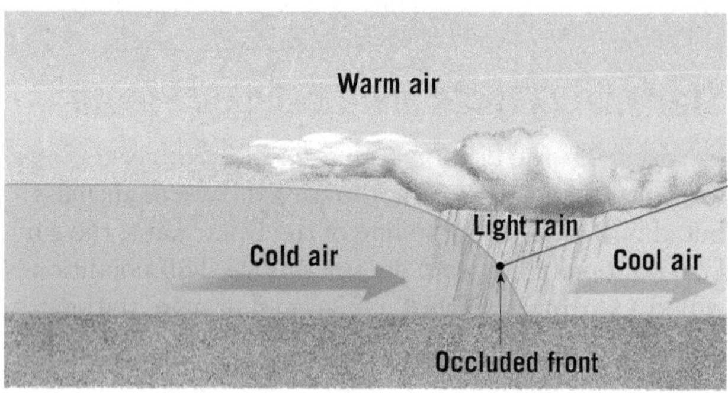

The denser cold air lifts the warm air and advances into and displaces the cool air.

fronts are shown with blue triangles pointing toward the warmer air and red semicircles pointing toward the cooler air. At times, some overrunning occurs along a stationary front, causing gentle to moderate precipitation. Flooding can occur along the cold side of a stationary front if the front remains in place for a long time.

The fourth type of front is an **occluded front**, an active cold front that overtakes a warm front, as shown in Figure 19.10. As the advancing cold air wedges the warm air upward, a new front emerges between the advancing cold air and the cool air. The weather of an occluded front is generally complex. Most precipitation is associated with the warm air being forced aloft. When conditions are suitable, however, the newly formed front is capable of initiating precipitation of its own. Occluded fronts may result in heavy rain in the warmer months or a blizzard in the winter. On a weather map, occluded fronts are shown with purple triangles and semicircles both pointing in the direction of the front's advance.

A word of caution is in order concerning the weather associated with various fronts. Although the preceding discussion will help you recognize the weather patterns associated with fronts, remember that these descriptions are generalizations. The weather generated along any individual front may or may not conform fully to this idealized picture. Fronts, like all other aspects of nature, do not lend themselves to classification as easily as we would like.

CONCEPT CHECKS 19.2

1. Compare the weather of a typical warm front with that of a typical cold front.
2. Why is cold-front weather usually more severe than warm-front weather?
3. Describe a stationary front and an occluded front.

19.3 Midlatitude Cyclones

Summarize the weather associated with the passage of a mature midlatitude cyclone. Describe how airflow aloft is related to cyclones and anticyclones at the surface.

So far, we have examined the basic elements of weather as well as the dynamics of atmospheric motions. We are now ready to apply our knowledge of these diverse phenomena to an understanding of day-to-day weather patterns in the middle latitudes. For our purposes, *middle latitudes* refers to the region between southern Florida and Alaska. The primary weather producers here are **midlatitude, or middle-latitude, cyclones**. On weather maps they are shown by an *L*, meaning *low-pressure system*. Figure 19.11 shows two views of a large idealized midlatitude cyclone with probable air masses, fronts, and surface wind patterns.

Midlatitude cyclones are large centers of low pressure that generally travel from west to east. Lasting from a few days to more than a week, these weather systems have a counterclockwise circulation, with an airflow inward toward their centers. Most midlatitude cyclones also have a cold front extending from the central area of low pressure and, frequently, a warm front as well. Convergence and forceful lifting initiate cloud development and often cause abundant precipitation in the vicinity of the low-pressure center.

As early as the 1800s, it was known that midlatitude cyclones were the bearers of precipitation and severe

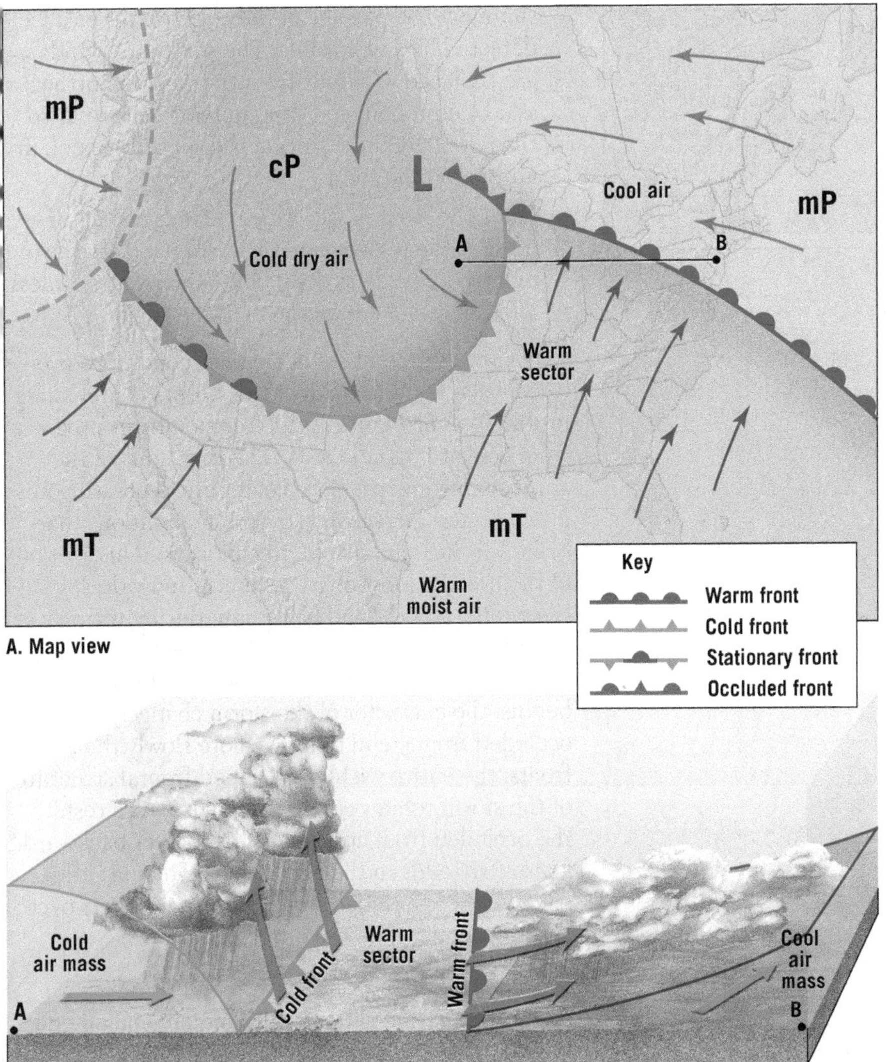

SmartFigure 19.11 **Idealized structure of a large, mature midlatitude cyclone A.** This map view shows fronts, air masses, and surface winds. **B.** The three-dimensional view is a cross section through warm and cold fronts along a line from point A to point B.

TUTORIAL
https://goo.gl/Qs5VVo

weather. But it was not until the early part of the 1900s that a model was developed to explain how cyclones form. A group of Norwegian scientists formulated and published this model, created primarily from near-surface observations, in 1918.

Years later, as data from the middle and upper troposphere and from satellite images became available, modifications were necessary. However, this model is still a useful working tool for interpreting the weather. If you keep this model in mind when you observe changes in the weather, the changes will no longer come as a surprise. You should begin to see some order in what once appeared to be disorder, and you might even occasionally "predict" the impending weather.

Idealized Weather of a Midlatitude Cyclone

The midlatitude cyclone model provides a useful tool for examining the weather patterns of the middle latitudes. Figure 19.12 illustrates the distribution of clouds and thus the regions of possible precipitation associated with a mature system. Compare this drawing to the satellite image shown in Figure 19.13. It is easy to see why we often refer to the cloud pattern of a midlatitude cyclone as having a "comma" shape.

Guided by the westerlies aloft, cyclones generally move eastward across the United States, so we can expect the first signs of their arrival in the west. However, often in the region of the Mississippi River valley, cyclones begin a more northeasterly path and occasionally move directly northward. A midlatitude cyclone typically requires 2 to 4 days to move completely across a region. During that brief period, abrupt changes in atmospheric conditions may be experienced. This is particularly true in the winter and spring, when the largest temperature contrasts occur across the middle latitudes.

Using Figure 19.12 as a guide, we will now consider these weather producers and what we should expect from them as they move over an area. To facilitate our discussion, Figure 19.12 includes two profiles along lines A–E and F–G:

- Imagine the change in weather as you move along profile A–E. At point A, the sighting of high cirrus clouds would be the first sign of the approaching cyclone. These high clouds can precede the surface front by 1000 kilometers (600 miles) or more, and they generally are accompanied by falling pressure. As the warm front advances, a lowering and thickening of the cloud deck is noticed.
- Usually within 12 to 24 hours after the first sighting of cirrus clouds, light precipitation begins (point B). As the front nears, the rate of precipitation increases, a rise in temperature is noticed, and winds begin to change from east or southeast to south or southwest.

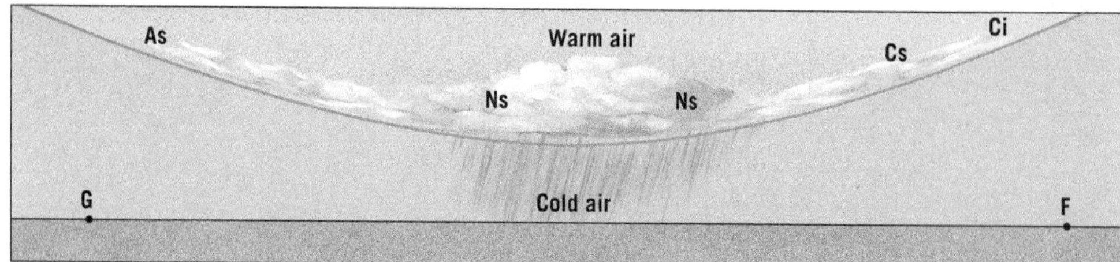

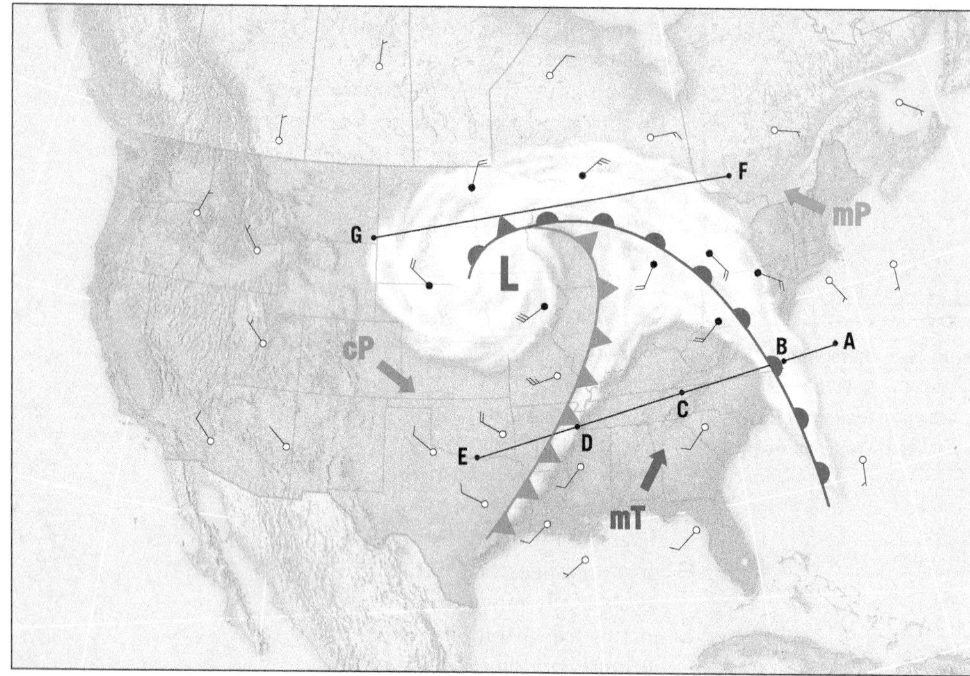

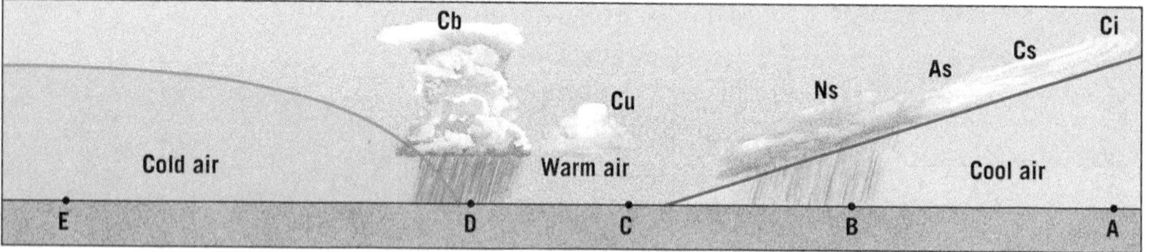

▲ SmartFigure 19.12
Cloud patterns typically associated with a mature midlatitude cyclone The middle section is a map view. Note the cross-sectional lines (F–G, A–E). Above the map is a vertical cross section along line F–G. Below the map is a section along A–E. For cloud abbreviations, refer to Figures 19.7 and 19.8.

TUTORIAL
https://goo.gl/D7ItJG

- With the passage of the warm front, the area is under the influence of a maritime tropical air mass (point C). Generally, the region affected by this sector of the cyclone experiences warm to hot temperatures, southwesterly winds, fairly high humidity, and skies that may be clear or contain cumulus clouds.
- The relatively warm, humid weather of the warm sector passes quickly and is replaced by gusty winds and precipitation generated along the cold front. The approach of a rapidly advancing cold front is marked by a wall of dark clouds (point D). Severe weather accompanied by heavy precipitation, hail, and an occasional tornado is a definite possibility, especially during spring and summer. The passage of the cold front is easily detected by a wind shift: The southwest winds are replaced by winds from the west to northwest and by a pronounced drop in temperature. Also, the rising pressure hints at the subsiding cool, dry air behind the front.
- Once the front passes, skies clear as cooler air invades the region (point E). Often a day or two of almost cloudless deep blue skies occurs, unless another cyclone is edging into the region.

A very different set of weather conditions prevails in the regions north of the storm's center along profile F–G of Figure 19.12. Temperatures progress from cool at F to cold at G. The first hints of the approaching low-pressure center are decreasing air pressure and increasingly overcast conditions that bring varying amounts of precipitation. This section of the cyclone most often generates snow during the winter months and heavy rain during warmer months.

Once the formation of an occluded front begins, the character of the storm changes. Because occluded fronts tend to move more slowly than other fronts, the entire wishbone-shaped frontal structure of the storm rotates counterclockwise. As a result, the occluded front appears to "bend over backward." This effect adds to the misery of the region influenced by the occluded front because it lingers over the area longer than the other fronts.

▼ **Figure 19.13 Satellite view of a mature midlatitude cyclone** This storm swept across the central United States and produced strong wind gusts (up to 125 kilometers [78 miles] per hour), rain, hail, and snow. It also spawned 61 tornadoes on October 26, 2010. This cyclone set a record for the lowest pressure not associated with a hurricane ever recorded over land in the continental United States: 28.21 inches of mercury. It is easy to see why we often refer to the cloud pattern of a cyclone as having a "comma" shape. (NASA)

The Role of Airflow Aloft

When the earliest studies of midlatitude cyclones were made, little was known about the nature of the airflow in the middle and upper troposphere. Since then, a close relationship has been established between surface disturbances and the flow aloft. Airflow aloft plays an important role in maintaining cyclonic and anticyclonic circulation. In fact, more often than not, these rotating surface wind systems are actually generated by upper-level flow.

Recall that airflow around a surface cyclone (low-pressure system) is inward, a fact that leads to mass convergence, or coming together (Figure 19.14). The resulting accumulation of air must be accompanied by a corresponding increase in surface pressure. Consequently, we might expect a low-pressure system to "fill" rapidly and be eliminated. However, this does not occur. On the contrary, cyclones often exist for a week or longer. For this to happen, surface convergence must be offset by a mass outflow at some level aloft (see Figure 19.14). As long as divergence (spreading out) aloft is equal to or greater than surface inflow, the low pressure and its accompanying convergence can be sustained.

Because cyclones are bearers of stormy weather, they have received far more attention than anticyclones. Anticyclones on a weather map represent the air masses that were discussed earlier. For example, the surface high-pressure system shown in Figure 19.14 is likely a continental polar air mass. Because a close relationship exists between surface highs and lows, it is difficult to separate any discussion of these two types of pressure systems. The surface air that feeds a cyclone, for example, generally originates as air flowing out of an anticyclone. Consequently, cyclones and anticyclones typically are found adjacent to each other. Like a cyclone, an anticyclone depends on the flow far above to maintain its circulation. Divergence at the surface is balanced by convergence aloft and general subsidence of the air column (see Figure 19.14).

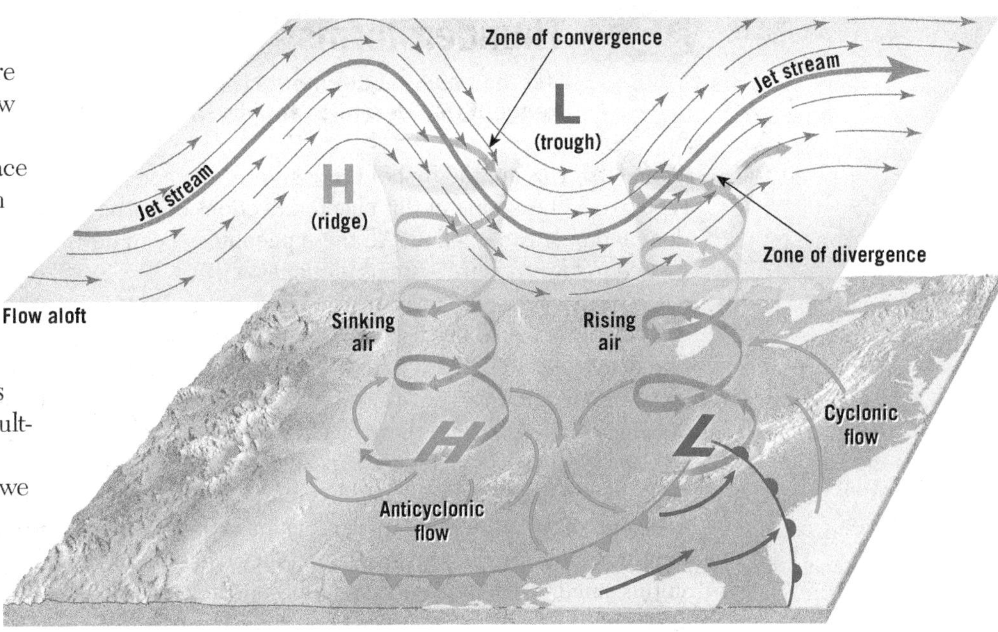

Figure 19.14 **Flow aloft influences surface winds and pressure** Idealized depiction showing the support that divergence and convergence aloft provide to cyclonic and anticyclonic circulation at the surface. Divergence aloft initiates upward air movement, reduced surface pressure, and cyclonic flow. On the other hand, convergence along the jet stream results in general subsidence of the air column, increased surface pressure, and anticyclonic surface winds.

CONCEPT CHECKS 19.3

1. Briefly describe the weather associated with the passage of a mature midlatitude cyclone when the center of low pressure is about 200 to 300 kilometers (125 to 200 miles) north of your location.
2. If the midlatitude cyclone described in Question 1 took 3 days to pass your location, on which day would temperatures likely be warmest? On which day would they likely be coldest?
3. What winter weather might be expected with the passage of a mature midlatitude cyclone when the center of low pressure is located about 100 to 200 kilometers (60 to 125 miles) south of your location?
4. Briefly explain how flow aloft aids the formation of cyclones at the surface.

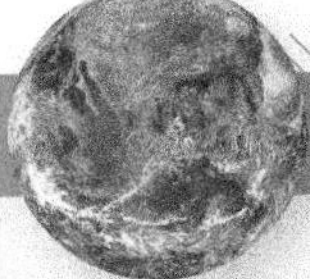

EYE ON EARTH 19.2

This image of a line of clouds was taken by astronauts aboard the International Space Station. The dashed line on the image shows the approximate surface position of the front responsible for the cloud development. Assume that this front is located over the central United States as you answer the following questions.

NASA

QUESTION 1 *What is the cloud type of the tallest clouds in the image?*

QUESTION 2 *Are these clouds more typical of a cold front or a warm front?*

QUESTION 3 *Is the front most likely moving toward the southeast or toward the northwest?*

QUESTION 4 *Is the air mass located to the southeast of the front more likely continental polar (cP) or maritime tropical (mT)?*

19.4 Thunderstorms

List the basic requirements for thunderstorm formation and locate places on a map that exhibit frequent thunderstorm activity. Describe the stages in the development of a thunderstorm.

Thunderstorms are the first of three severe weather types we will examine in this chapter. Sections on tornadoes and hurricanes follow. All these phenomena can be related to low-pressure systems (cyclones).

Severe weather is more fascinating than everyday weather phenomena. The lightning display and booming thunder generated by a severe thunderstorm can be a spectacular event that elicits both awe and fear (see the chapter-opening photo). Of course, hurricanes and tornadoes also attract a great deal of much-deserved attention. A single tornado outbreak or hurricane can cause many deaths as well as billions of dollars in property damage. In a typical year, the United States experiences thousands of violent thunderstorms, hundreds of tornadoes, and a few hurricanes.

What's in a Name?

Up to now we have examined midlatitude cyclones, which play an important role in causing day-to-day weather changes. Yet the use of the term *cyclone* is often confusing. To many people, the term implies only an intense storm, such as a tornado or a hurricane. When a hurricane unleashes its fury on India or Bangladesh, for example, it is usually reported in the media as a *cyclone* (the term denoting a hurricane in that part of the world).

Similarly, tornadoes are referred to as *cyclones* in some places. This custom is particularly common in portions of the Great Plains of the United States. Recall that in *The Wizard of Oz*, Dorothy's house was carried from her Kansas farm to the land of Oz by a cyclone. Indeed, the nickname for the athletic teams at Iowa State University is the *Cyclones* (Figure 19.15). Although hurricanes and tornadoes are, in fact, cyclones, the vast majority of cyclones are *not* hurricanes or tornadoes. The term *cyclone* simply refers to the circulation around any low-pressure center, no matter how large or intense it is.

Tornadoes and hurricanes are both smaller and more violent than midlatitude cyclones. Midlatitude cyclones can have a diameter of 1600 kilometers (1000 miles) or more. By contrast, hurricanes average only 600 kilometers (375 miles) across, and tornadoes, with a typical diameter of just 0.25 kilometer (0.16 mile), are much too small to show up on a weather map.

The thunderstorm, a much more familiar weather event, hardly needs to be distinguished from tornadoes, hurricanes, and midlatitude cyclones. Unlike the flow of air about these latter storms, the circulation associated with thunderstorms is characterized by strong up-and-down movements. Winds in the vicinity of a thunderstorm do not follow the inward spiral of a cyclone, but they are typically variable and gusty.

Thunderstorms can form "on their own," away from cyclonic storms, and they can also form in conjunction with cyclones. For instance, thunderstorms are frequently spawned along the cold front of a midlatitude cyclone, where on rare occasions a tornado may descend from the thunderstorm's cumulonimbus tower. Hurricanes also generate widespread thunderstorm activity. Thus, thunderstorms are related in some manner to all three types of cyclones mentioned here.

Figure 19.15 The term *cyclone* Sometimes the use of the term *cyclone* can be confusing. (Satellite image courtesy of NASA; logo courtesy of Iowa State University)

In parts of the Great Plains, *cyclone* is a synonym for *tornado*. The nickname for the athletic teams at Iowa State University is the *Cyclones*.*

In southern Asia and Australia, the term *cyclone* is applied to storms that are called *hurricanes* in the United States. This image shows Cyclone Yasi, which struck eastern Australia in February 2011.

* Iowa State University is the only Division I school to use cyclones as its team name. The pictured logo was created to better communicate the school's image by combining the mascot, a cardinal bird named Cy, and the Cyclone team name.

Thunderstorm Occurrence

Almost everyone has observed various small-scale phenomena that result from the vertical movements of relatively warm, unstable air. Perhaps you have seen a dust devil over an open field on a hot day, whirling its dusty load to great heights. Or maybe you have noticed a bird glide effortlessly skyward on an invisible thermal of hot air. These examples illustrate the dynamic thermal instability that occurs during the development of a thunderstorm.

A **thunderstorm** is a storm that generates lightning and thunder. Thunderstorms frequently produce gusty winds, heavy rain, and hail. A thunderstorm may be produced by a single cumulonimbus cloud and influence only a small area, or it may be associated with clusters of cumulonimbus clouds covering a large area.

Thunderstorms form when warm, humid air rises in an unstable environment. Various mechanisms can trigger the upward air movement needed to create thunderstorm-producing cumulonimbus clouds. One mechanism, the unequal heating of Earth's surface, significantly contributes to the formation of ordinary thunderstorms, which are also called *air-mass thunderstorms*. These storms are associated with the scattered puffy cumulonimbus clouds that commonly form *within* maritime tropical air masses and produce scattered thunderstorms on summer days. Such storms are usually short-lived and seldom produce strong winds or hail.

Another type of thunderstorm not only benefits from uneven surface heating but is associated with the lifting of warm air, as occurs along a front or a mountain slope. Moreover, diverging winds aloft frequently contribute to the formation of these storms because they tend to draw air from lower levels upward beneath them. Some of the thunderstorms of this type may produce high winds, damaging hail, flash floods, and tornadoes. Such storms are described as *severe*.

At any given time, an estimated 2000 thunderstorms are in progress on Earth. As we would expect, the greatest number occur in the tropics, where warmth, plentiful moisture, and instability are always present. About 45,000 thunderstorms take place each day, and more than 16 million occur annually around the world. The lightning from these storms strikes Earth 100 times each second (Figure 19.16A). Annually, the United States experiences about 100,000 thunderstorms and millions of lightning strikes. A glance at Figure 19.16B shows that thunderstorms are most frequent in Florida and the eastern Gulf coast region, where such activity is recorded between 70 and 100 days each year. The region on the eastern side of the Rockies in Colorado and New Mexico is next, with thunderstorms occurring on 60 to 70 days each year. Most of the rest of the nation experiences thunderstorms on 30 to 50 days annually. The western margin of the United States has little thunderstorm activity because air from the adjacent Pacific Ocean tends to be cool and stable. Thunderstorms are also relatively uncommon in the northern tier of states and in Canada, because warm, moist, unstable mT air seldom penetrates these areas.

Stages of Thunderstorm Development

All thunderstorms require warm, moist air, which, when lifted, releases sufficient latent heat to provide the buoyancy necessary to maintain its upward flight. This instability and associated buoyancy are triggered by a number of different processes, yet most thunderstorms have a similar life history.

Because instability and buoyancy are enhanced by high surface temperatures, thunderstorms are most common in the afternoon and early evening (Figure 19.17A). However, surface heating alone is not sufficient for the growth of towering cumulonimbus clouds. A solitary cell of rising hot air produced by surface heating could, at best, produce a small cumulus cloud, which would evaporate within 10 to 15 minutes.

The development of 12,000-meter (40,000-foot) (or, on rare occasions, 18,000-meter [60,000-foot]) cumulonimbus towers requires a continual supply of moist

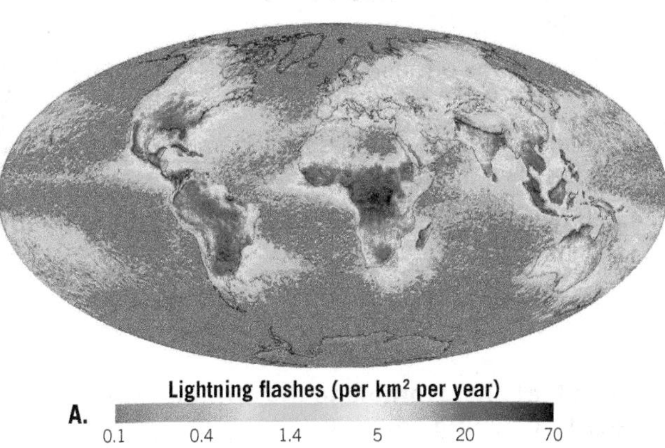
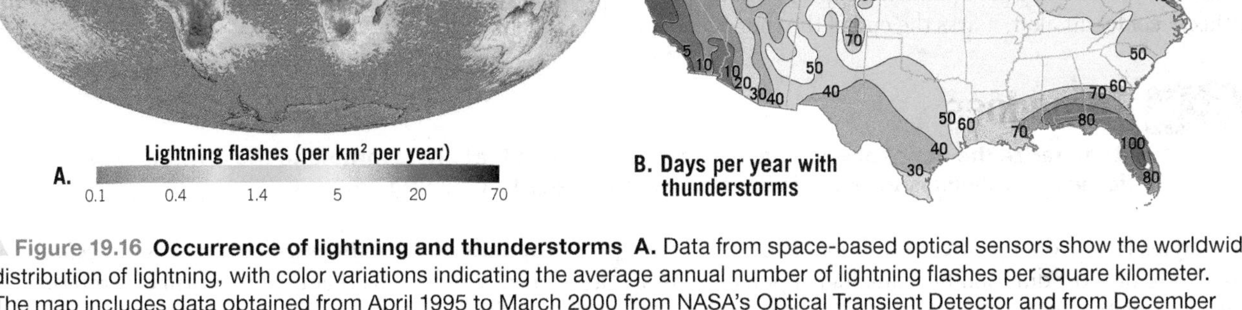

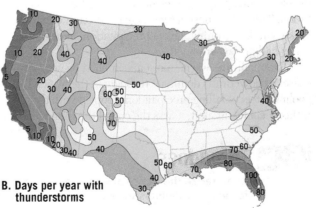

Figure 19.16 **Occurrence of lightning and thunderstorms A.** Data from space-based optical sensors show the worldwide distribution of lightning, with color variations indicating the average annual number of lightning flashes per square kilometer. The map includes data obtained from April 1995 to March 2000 from NASA's Optical Transient Detector and from December 1997 to November 2000 from NASA's Lightning Imaging Sensor. Both of these satellite-based sensors use high-speed cameras capable of detecting brief lightning flashes even under daytime conditions. (NASA) **B.** Average number of days each year with thunderstorms. The humid subtropical climate that dominates the southeastern United States receives much of its precipitation in the form of thunderstorms. Most of the Southeast averages 50 or more days each year with thunderstorms. (Environmental Data Service, NOAA)

A.

B.

Figure 19.17 **Cumulus development A.** Buoyant thermals often produce fair-weather cumulus clouds that soon evaporate into the surrounding air, making it more humid. As this process of cumulus development and evaporation continues, the air eventually becomes sufficiently humid so that newly forming clouds do not evaporate but continue to grow. **B.** This developing cumulonimbus cloud became a towering August thunderstorm over central Illinois. (Photos by E. J. Tarbuck)

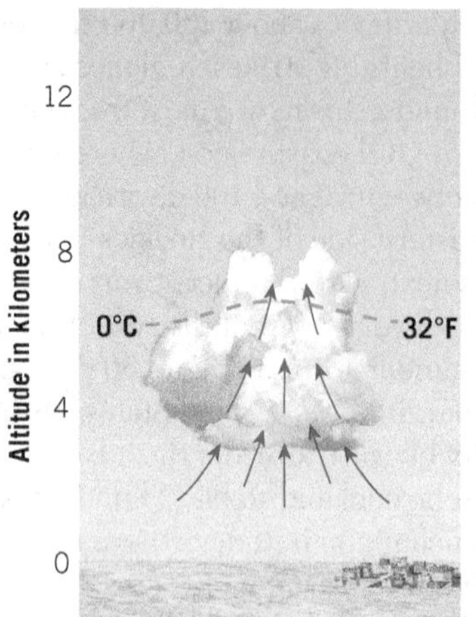
12
8
4
0
Altitude in kilometers
0°C
32°F
During the cumulus stage, strong updrafts provide moisture that condenses and builds the cloud.

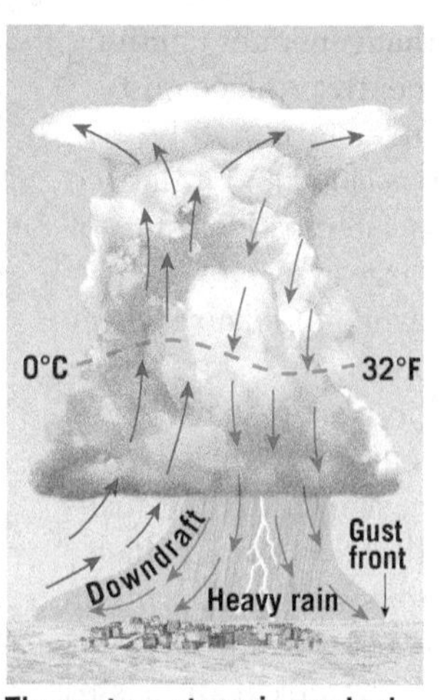
0°C
32°F
Downdraft
Gust front
Heavy rain
The mature stage is marked by heavy preciptation. Updrafts exist side by side with downdrafts and continue to enlarge the cloud.

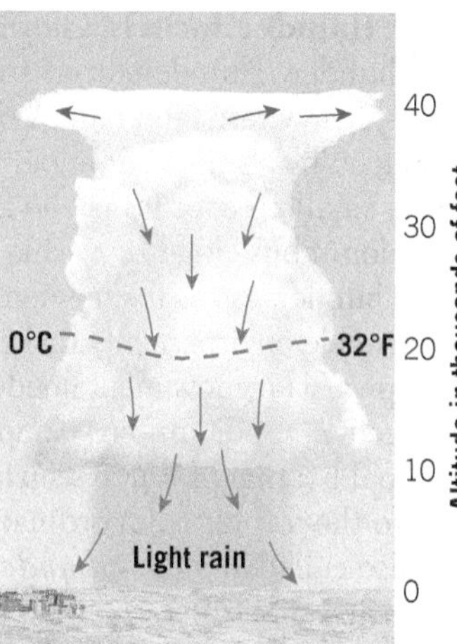
0°C
32°F
Light rain
40
30
20
10
0
Altitude in thousands of feet
When updrafts disappear, precipitation becomes light and then stops. Without a supply of moisture from updrafts, the cloud evaporates.

SmartFigure 19.18 **Development of an ordinary thunderstorm** Once a cloud passes beyond the freezing level, the Bergeron process begins producing precipitation. Eventually, the accumulation of precipitation in the cloud is too great for the updraft to support. The falling precipitation causes drag on the air and initiates a downdraft. Once downdrafts dominate, rainfall diminishes, and the cloud starts to dissipate.

TUTORIAL
https://goo.gl/nrOfL2

air (Figure 19.17B). Each new surge of warm air rises higher than the last, adding to the height of the cloud (Figure 19.18). These updrafts occasionally reach speeds greater than 100 kilometers (60 miles) per hour, based on the size of hailstones they are capable of carrying upward. Usually within an hour, the amount and size of precipitation that has accumulated is too much for the updrafts to support, and consequently downdrafts develop in one part of the cloud, releasing heavy precipitation. This is the most active stage of the thunderstorm. Gusty winds, lightning, heavy precipitation, and sometimes hail are experienced.

Eventually the warm, moist air supplied by updrafts ceases as downdrafts dominate throughout the cloud. The cooling effect of falling precipitation, coupled with the influx of colder air aloft, marks the end of the thunderstorm activity. The life span of a typical cumulonimbus cell within a thunderstorm complex is only about an hour, but as the storm moves, fresh supplies of warm, water-laden air generate new cells to replace those that are dissipating.

CONCEPT CHECKS 19.4

1. Briefly compare and contrast midlatitude cyclones, hurricanes, and tornadoes. How are thunderstorms related to each?
2. What are the basic requirements for the formation of a thunderstorm?
3. Where are thunderstorms most common on Earth? In the United States?
4. Summarize the stages in the development of a thunderstorm.

19.5 Tornadoes

Summarize the atmospheric conditions and locations that are favorable to the formation of tornadoes. Discuss tornado destruction and tornado forecasting.

Tornadoes form from the strongest thunderstorms. These events are of short duration but rank high among nature's most destructive forces (Figure 19.19). Their sporadic occurrence and violent winds cause many deaths each year. The nearly total destruction in some stricken areas has led many to liken their passage to bombing raids during war (Figure 19.20).

Tornadoes, sometimes called *twisters* or *cyclones*, are violent windstorms that take the form of a rotating column of air, or *vortex*. Pressures within some tornadoes have been estimated to be as much as 10 percent lower than pressures immediately outside the tornado. Drawn by the much lower pressure in the center of the vortex, air near the ground rushes into the tornado from all

▲ Figure 19.19 **Condensation and debris make tornadoes visible** A tornado is a violently rotating column of air in contact with the ground. The air column is visible when it contains condensation or when it contains dust and debris. Often the appearance is a result of both. When the column of air is aloft and does not produce rotation at the ground, the visible portion is properly called a *funnel cloud*. (Photo by Jason Persoff Stormdoctor/Cultura RM Exclusive/Getty Images)

▲ SmartFigure 19.20 **Tornado destruction at Moore, Oklahoma** On May 20, 2013, central Oklahoma was devastated by an EF-5 tornado, the most severe category. It took 24 lives, injured 377, and caused damages in excess of $2 billion. At least 13,000 structures were destroyed or damaged. The tornado was on the ground for 39 minutes, and its path extended for 27 kilometers (17 miles). At its peak, the tornado was 2.1 kilometers (1.3 miles) wide and had winds of 340 kilometers (210 miles) per hour. (Photo by Jewel Samad/Getty Images)

VIDEO
https://goo.gl/xTPT5X

directions. As the air streams inward, it spirals upward around the core until it eventually merges with the airflow of the parent thunderstorm deep in the cumulonimbus tower. Because of the tremendous pressure gradient associated with a strong tornado, maximum winds can sometimes approach 480 kilometers (300 miles) per hour.

A tornado may consist of a single vortex, but within many stronger tornadoes are smaller whirls called *suction vortices* that rotate within the main vortex (**Figure 19.21**). Suction vortices have diameters of only about 10 meters (33 feet) and rotate very rapidly. This structure accounts for occasional observations of virtually total destruction of one building while another one, just 10 meters (33 feet) away, suffers little damage.

Tornado Development and Occurrence

Tornadoes can form in any situation that produces severe weather, including cold fronts, squall lines, and tropical cyclones (hurricanes). Usually the most intense tornadoes are those that form in association with huge thunderstorms called *supercells*. An important precondition linked to tornado formation in severe thunderstorms is the development of a mesocyclone. A **mesocyclone** is a vertical cylinder of rotating air, typically about 3 to 10 kilometers (2 to 6 miles) across, that develops in the updraft of a severe thunderstorm. The formation of this large vortex often precedes tornado formation by 30 minutes or so.

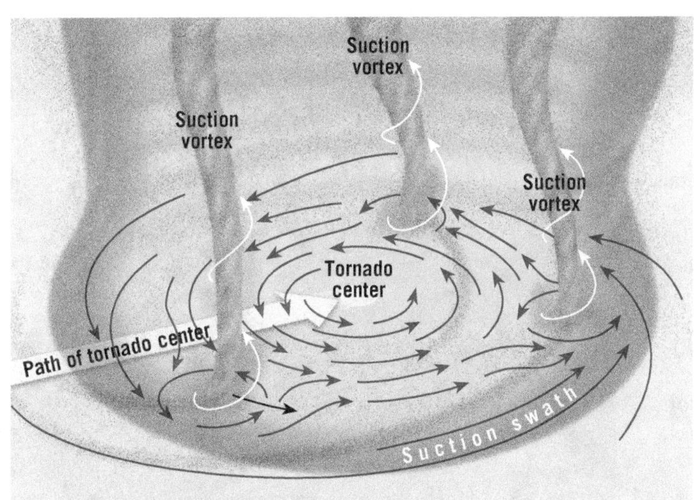

▲ SmartFigure 19.21 **Multiple-vortex tornado** Some tornadoes have multiple suction vortices. These small and very intense vortices are roughly 10 meters (30 feet) across and move in a counterclockwise path around the tornado center. Because of this multiple-vortex structure, one building might be heavily damaged and another one, just 10 meters away, might suffer little damage.

ANIMATION
https://goo.gl/x3QT9U

We have learned that in order to create a thunderstorm, there must be moisture, instability, and a lifting mechanism; but to get a storm that produces a strong tornado, there needs to be one additional condition—wind shear. **Wind shear** is a change in wind speed and/or direction with height. Mesocyclone formation depends on the presence of wind shear. Moving upward from the surface, winds change direction from southerly to westerly, and wind speed increases. The *speed wind shear* (that is, stronger winds aloft and weaker winds near the surface) produces a rolling motion about a horizontal axis, as shown in Figure 19.22A. If conditions are right, strong updrafts in the storm tilt the horizontally rotating air to a nearly vertical alignment (Figure 19.22B,C). This produces the initial rotation within the cloud interior.

At first, the mesocyclone is wider, shorter, and more slowly rotating than will be the case in later stages. Subsequently, the mesocyclone is stretched vertically and narrowed horizontally, causing wind speeds to accelerate in an inward vortex (just as spinning ice skaters accelerate by pulling in their arms). Next, the narrowing column of rotating air stretches downward until a portion of the cloud protrudes below the cloud base to produce a very dark, slowly rotating *wall cloud*. Finally, a slender and rapidly spinning vortex emerges from the base of the wall cloud to form a *funnel cloud*. If the funnel cloud makes contact with the surface, it is then classified as a *tornado*.

The formation of a mesocyclone does not necessarily mean that tornado formation will follow. Only about half of all mesocyclones produce tornadoes. Forecasters cannot determine in advance which mesocyclones will spawn tornadoes.

Tornado Climatology

Recall that severe thunderstorms—and hence tornadoes—are most often spawned along the cold front of a midlatitude cyclone or in association with supercell thunderstorms. Such severe storms are likeliest to form during the spring, when the air masses associated with midlatitude cyclones are most likely to have greatly contrasting conditions. Continental polar air from Canada may still be very cold and dry, whereas maritime tropical air from the Gulf of Mexico is warm, humid, and unstable. The greater the contrast, the more intense the storm tends to be.

These two contrasting air masses are most likely to meet in the central United States because there is no significant natural barrier separating the center of the country from the arctic or the Gulf of Mexico. Consequently, this region generates more tornadoes than any other part of the country or, in fact, the world. Figure 19.23, which depicts the average annual tornado incidence in the United States over a 27-year period, readily substantiates this fact.

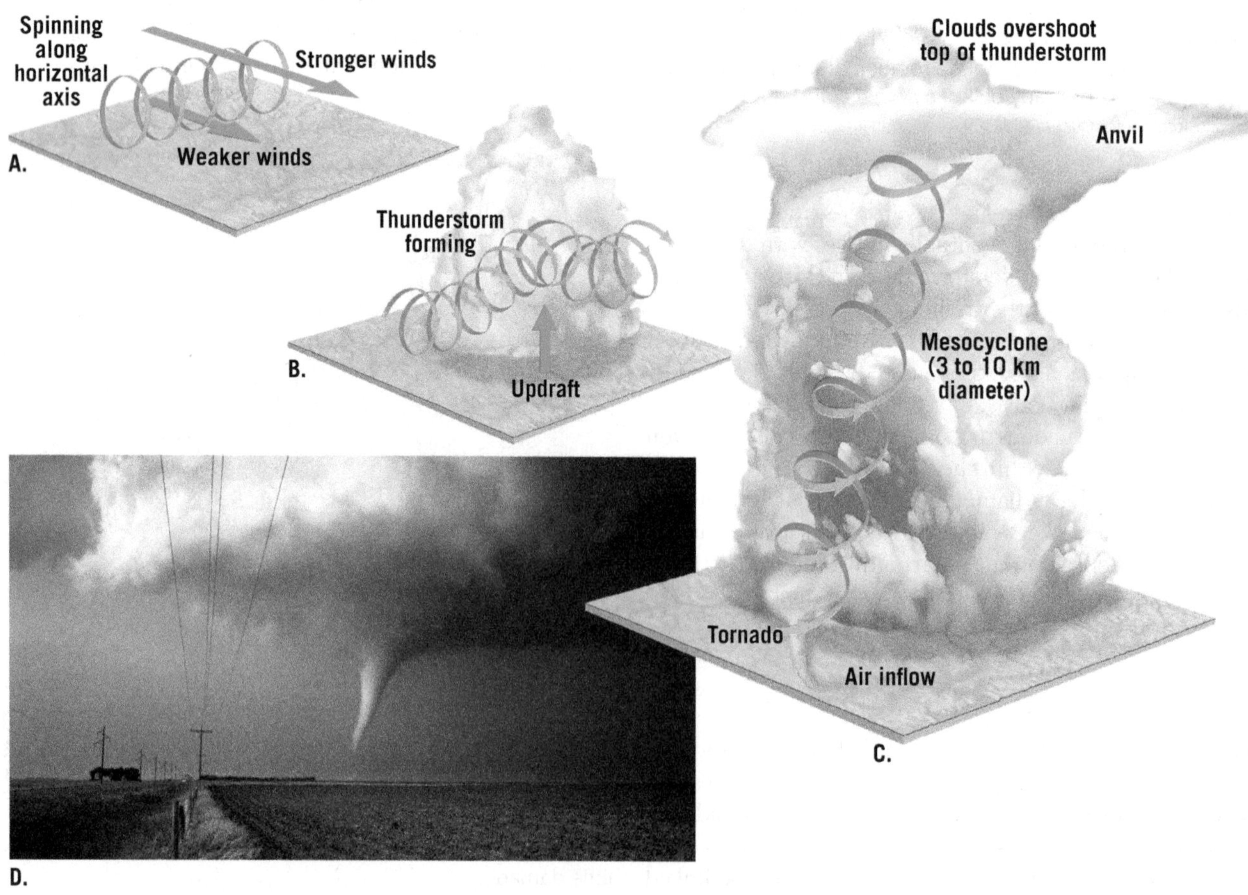

SmartFigure 19.22 **Mesocyclone formation often precedes tornado formation A.** Winds are stronger aloft than at the surface (called *speed wind shear*), producing a rolling motion about a horizontal axis. **B.** Strong thunderstorm updrafts tilt the horizontally rotating air to a nearly vertical alignment. **C.** The mesocyclone, a vertical cylinder of rotating air, is established. **D.** If a tornado develops, it will descend from a slowly rotating wall cloud in the lower portion of the mesocyclone. (Photo by Gene Rhoden/Weatherpix/Getty Images)

TUTORIAL
https://goo.gl/4O8MXL

On average, about 1300 tornadoes are reported annually in the United States. However, the actual number that occur from one year to the next varies greatly. During the 15-year period from 2001 through 2015, for example, yearly totals ranged from a low of 938 in 2002 to a high of 1894 in 2011. Tornadoes occur during every month of the year. April through June is the period of greatest tornado frequency in the United States, and December and January are the months of lowest activity (see Figure 19.23, graph insert).

The preceding paragraphs describe tornado climatology. Recall from the discussion of weather and climate in Chapter 16 that climate provides a statistical perspective of atmospheric behavior; remember the well-known saying: "Climate is what you expect, but weather is what you get." However, as the saying warns, weather events do not always occur when statistical probabilities suggest. Figure 16.4 (page 496) provides an example from November 2013.

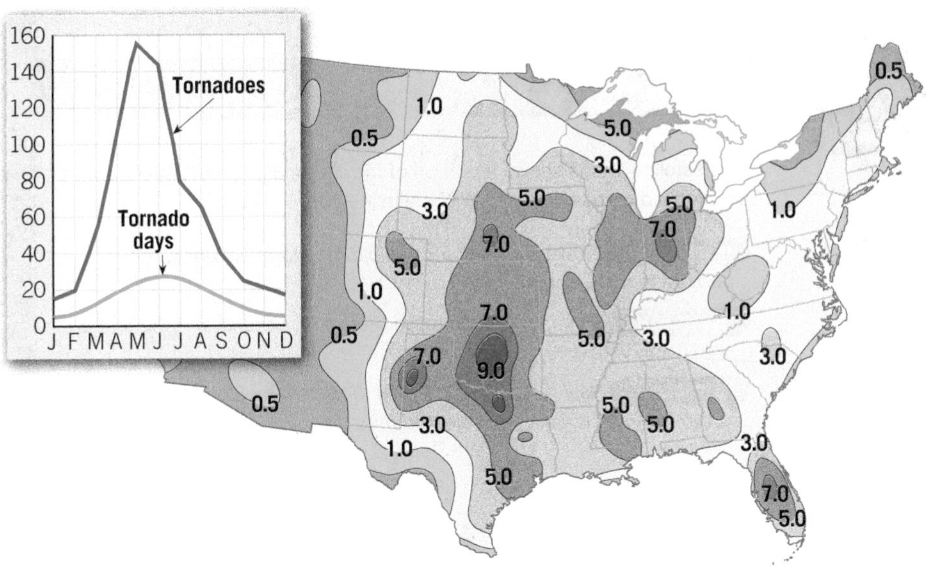

Figure 19.23 **Tornado occurrence** The map shows average annual tornado incidence per 26,000 square kilometers (10,000 square miles) for a 27-year period. The graph shows the average number of tornadoes and tornado days each month in the United States for the same period.

Tornado Destruction and Loss of Life

The potential for tornado destruction depends largely on the strength of the winds generated by the storm. Because tornadoes generate the strongest winds in nature, they have accomplished many seemingly impossible tasks, such as the ones shown in Figure 19.24. Although it may seem impossible for winds to cause some of the extensive damage attributed to tornadoes, tests in engineering facilities have repeatedly demonstrated that winds in excess of 320 kilometers (200 miles) per hour are capable of incredible feats.

Most tornado losses are associated with the rare storms that strike urban areas or devastate entire small communities. The amount of destruction caused by such storms depends to a significant degree (but not completely) on the strength of the winds. A wide spectrum of tornado strengths, sizes, and lifetimes are observed. The commonly used guide to tornado intensity is the **Enhanced Fujita intensity scale, or EF-scale** for short (Table 19.1). Because tornado winds cannot be measured directly, a rating on the EF-scale is determined by assessing the worst damage produced by a storm. Although widely used, the EF-scale is not perfect. Unlike the original Fujita scale, the EF-scale takes into consideration the structural integrity of buildings, but it does not do a good job of classifying tornadoes that pass through regions with no structures because there is no

Figure 19.24 **Tornado winds: The strongest in nature A.** The force of the wind during a tornado near Wichita, Kansas, in April 1991 was enough to drive this piece of metal into a utility pole. (Photo by John Sokich/NOAA) **B.** The remains of a truck wrapped around a tree in Bridge Creek, Oklahoma, on May 4, 1999, following a major tornado outbreak. (LM Otero/AP Photo)

A.

B.

EYE ON EARTH 19.3

This satellite image shows a portion of the diagonal path left by a tornado as it moved across northern Wisconsin in 2007.

QUESTION 1 *Toward what direction did the storm advance: the northeast or the southwest?*

QUESTION 2 *Did the tornado more likely occur ahead of or behind a cold front? Explain.*

QUESTION 3 *Is it more probable that the storm took place in March or June? Why is the date you selected more likely?*

NASA

damage to assess. These tornadoes get rated as weak, if at all, no matter how strong the wind speed.

Although the greatest part of tornado damage is caused by violent winds, most tornado injuries and deaths result from flying debris. The proportion of tornadoes that result in loss of life is small. In most years, slightly fewer than 2 percent of all reported tornadoes in the United States are "killers." Although the percentage of tornadoes resulting in death is small, each tornado is potentially lethal. When tornado fatalities and storm intensities are compared, the results are quite interesting: The majority (63 percent) of tornadoes are weak (EF-0 and EF-1), and the number of storms decreases as tornado intensity increases. The distribution of tornado fatalities, however, is just the opposite. Although only 2 percent of tornadoes are classified as violent (EF-4 and EF-5), they account for nearly 70 percent of tornado deaths.

Tornado Forecasting

Because severe thunderstorms and tornadoes are small and relatively short-lived phenomena, they are among the most difficult weather features to forecast precisely. Nevertheless, the prediction, detection, and monitoring of such storms are some of the most important services provided by professional meteorologists. Both the timely issuance and dissemination of watches and warnings are critical to the protection of life and property.

The Storm Prediction Center (SPC) located in Norman, Oklahoma, is part of the National Weather Service (NWS) and the National Centers for Environmental Prediction (NCEP). The mission of the SPC is to provide timely and accurate forecasts and watches for severe thunderstorms and tornadoes.

Severe thunderstorm outlooks are issued several times daily. *Day 1* outlooks identify the areas that are likely to be affected by severe thunderstorms during the next 6 to 30 hours, and *day 2* outlooks extend the forecast through the following day. Both outlooks describe the type, coverage, and intensity of the severe weather expected. Many local NWS field offices also issue severe weather outlooks that provide more local descriptions of the severe weather potential for the next 12 to 24 hours.

Tornado Watches and Warnings

Tornado watches alert the public to the possibility of tornadoes over a specified area for a particular time interval. Watches serve to fine-tune forecast areas already identified in severe weather outlooks. A typical watch covers an area of about 65,000 square kilometers (25,000 square miles) for a 4- to 6-hour period. A tornado watch is an important part of the tornado alert system because it sets in motion the procedures necessary to deal adequately with detection, tracking, warning, and response. Watches are generally reserved for organized severe weather events where the tornado threat will affect at least 26,000 square kilometers (10,000 square miles) and/or persist for at least 3 hours. Watches typically are not issued when the threat is thought to be isolated and/or short-lived.

Whereas a tornado watch is designed to alert people to the possibility of tornadoes, a **tornado warning** is issued by local offices of the NWS when a tornado has actually been sighted in an area or is indicated by weather radar. It warns of a high probability of imminent danger.

Table 19.1 Enhanced Fujita Intensity Scale*

Scale	Wind Speed		Damage
	Km/Hr	Mi/Hr	
EF-0	105–137	65–85	*Light.* Some damage to siding and shingles.
EF-1	138–177	86–110	*Moderate.* Considerable roof damage. Winds can uproot trees and overturn single-wide mobile homes. Flagpoles bend.
EF-2	178–217	111–135	*Considerable.* Most single-wide mobile homes destroyed. Permanent homes can shift off foundations. Flagpoles collapse. Softwood trees debarked.
EF-3	218–265	136–165	*Severe.* Hardwood trees debarked. All but small portions of houses destroyed.
EF-4	266–322	166–200	*Devastating.* Complete destruction of well-built residences, large sections of school buildings.
EF-5	>322	>200	*Incredible.* Significant structural deformation of mid- and high-rise buildings.

*The original Fujita scale was developed by T. Theodore Fujita in 1971 and put into use in 1973. The Enhanced Fujita intensity scale is a revision that was put into use in February 2007. Winds speeds are estimates (not measurements) based on damage and represent 3-second gusts at the point of damage.

SmartFigure 19.25
Doppler radar
A. Doppler radar sites in the United States. Go to http://radar.weather.gov to see a similar map. You can click on any site to view the current National Weather Service Doppler radar display. **B.** Doppler on Wheels is a portable unit that researchers use in field studies of severe weather events. (Photo by University Corporation for Atmospheric Research)

VIDEO
https://goo.gl/CzmyY5

Warnings are issued for much smaller areas than are watches, usually covering portions of a county or counties. In addition, they are in effect for much shorter periods, typically 30 to 60 minutes. Because a tornado warning may be based on an actual sighting, warnings are occasionally issued after a tornado has already developed. However, most warnings are issued prior to tornado formation, sometimes by several tens of minutes, based on Doppler radar data and/or spotter reports of funnel clouds.

If the direction and the approximate speed of a storm are known, an estimate of its most probable path can be made. Because tornadoes often move erratically, the warning area is fan-shaped downwind from the point where the tornado has been spotted. Improved forecasts and advances in technology have contributed to a significant decline in tornado deaths over the past 50 years.

Doppler Radar Many of the difficulties that once limited the accuracy of tornado warnings have been reduced or eliminated by an advancement in radar technology called **Doppler radar** (Figure 19.25). Doppler radar not only performs the same tasks as conventional radar but also has the ability to detect motion directly. Doppler radar can detect the initial formation and subsequent development of a mesocyclone, the intense rotating wind system in the lower part of a thunderstorm that frequently precedes tornado development. Almost all mesocyclones produce damaging hail, severe winds, or tornadoes. Those that produce tornadoes (about 50 percent) can sometimes be distinguished by their stronger wind speeds and their sharper gradients of wind speeds. The tornado itself is too small to be detected directly by the radar.

It should also be pointed out that not all tornado-bearing storms have clear-cut radar signatures and that other storms can give false signatures. Detection, therefore, is sometimes a subjective process, and a given display could be interpreted in several ways. Consequently, trained observers, called *storm spotters*, continue to be an important part of the warning system.

The benefits of Doppler radar are many. As a research tool, it not only provides data on the formation of tornadoes but also helps meteorologists gain new insights into thunderstorm development, the structure and dynamics of hurricanes, and air-turbulence hazards that plague aircraft. As a practical tool for tornado detection, Doppler radar has significantly improved our ability to track thunderstorms and issue warnings.

CONCEPT CHECKS 19.5

1. Why do tornadoes have such high wind speeds?
2. What general atmospheric conditions are most conducive to the formation of tornadoes?
3. During what months is tornado activity most pronounced in the United States?
4. What scale is commonly used to rate tornado intensity? How is a rating on this scale determined?
5. Distinguish between a tornado watch and a tornado warning.

19.6 Hurricanes

Identify areas of hurricane formation on a world map and discuss the conditions that promote hurricane formation. List the three broad categories of hurricane destruction.

Most of us view the weather in the tropics with favor. Places such as the islands of the Caribbean are known for their lack of significant day-to-day variations. Warm breezes, steady temperatures, and rains that come as heavy but brief tropical showers are often the rule. It is ironic that these relatively tranquil regions produce some of the most violent storms on Earth.

Hurricanes are intense centers of low pressure that form over tropical oceans and are characterized by intense convective (thunderstorm) activity and strong

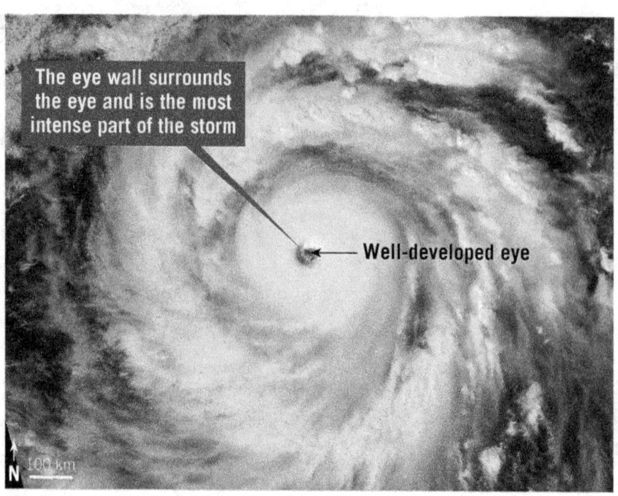

Figure 19.26 **Super Typhoon Jangmi** In the western Pacific, hurricanes are called *typhoons*. This storm struck portions of Taiwan, China, and Japan in late September 2008. It was the strongest storm worldwide that year, with sustained winds that reached 270 kilometers (165 miles) per hour. The counterclockwise spiral of the clouds indicates that it is a Northern Hemisphere storm. (NASA)

cyclonic circulation (Figure 19.26). Sustained winds must equal or exceed 119 kilometers (74 miles) per hour. Unlike midlatitude cyclones, hurricanes lack contrasting air masses and fronts. Rather, the source of energy that produces and maintains hurricane-force winds is the huge quantity of latent heat liberated during the formation of the storm's cumulonimbus towers.

The vast majority of hurricane-related deaths and damage are caused by relatively infrequent, yet powerful, storms. Hurricane Sandy devastated parts of the Caribbean and the coastal areas of New Jersey, New York, and Connecticut, causing billions of dollars in damages in late October 2012. The storm that pounded an unsuspecting Galveston, Texas, in 1900 was not just the deadliest U.S. hurricane ever, but the deadliest natural disaster of *any kind* to affect the United States. The deadliest and most costly storm in recent memory occurred in August 2005, when Hurricane Katrina devastated the Gulf Coast of Louisiana, Mississippi, and Alabama and took an estimated 1800 lives. Although hundreds of thousands fled before the storm made landfall, thousands of others were caught by the storm. In addition to the human suffering and tragic loss of life that were left in the wake of Hurricane Katrina, the financial losses caused by the storm were practically incalculable.

Profile of a Hurricane

Hurricanes form mostly between the latitudes of 5° and 20° over all the tropical oceans except the South Atlantic and the eastern South Pacific (Figure 19.27). The North Pacific has the greatest number of storms, averaging 20 each year. Fortunately for those living in the coastal regions of the southern and eastern United States, fewer than 5 hurricanes, on average, develop annually in the warm sector of the North Atlantic.

These intense tropical storms are known in various parts of the world by different names. In the western Pacific, they are called *typhoons*, and in the Indian Ocean, including the Bay of Bengal and Arabian Sea, they are simply called *cyclones*. In the following discussion, these storms will be referred to as hurricanes. The term *hurricane* is derived from the name *Huracan*, a Carib god of evil.

Although many tropical disturbances develop each year, only a few reach hurricane status. By international agreement, a hurricane has

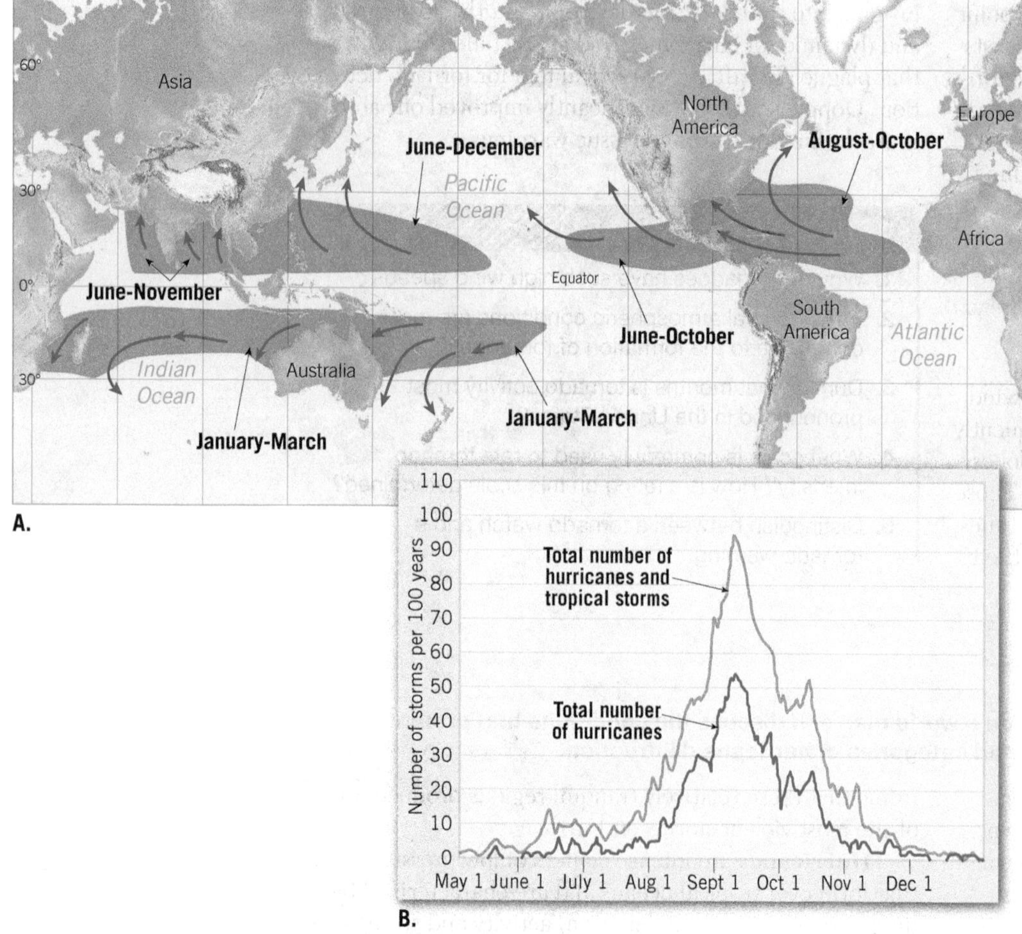

SmartFigure 19.27 **The where and when of hurricanes A.** The world map shows the regions where most hurricanes form as well as their principal months of occurrence and the tracks they most commonly follow. Hurricanes do not develop within about 5° of the equator because the Coriolis effect (a force related to Earth's rotation that gives storms their "spin") there is too weak. Because warm ocean-surface temperatures are necessary for hurricane formation, hurricanes seldom form poleward of 20° latitude or over the cool waters of the south Atlantic and the eastern south Pacific. **B.** The graph shows the frequency of tropical storms and hurricanes from May 1 through December 31 in the Atlantic basin. It shows the number of storms to be expected over a span of 100 years. The period from late August through October is clearly the most active. (Data from National Hurricane Center/NOAA)

VIDEO
https://goo.gl/JPdwbo

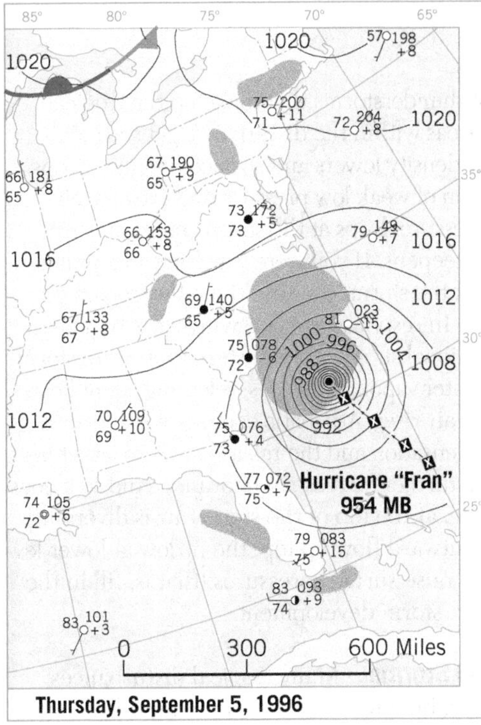

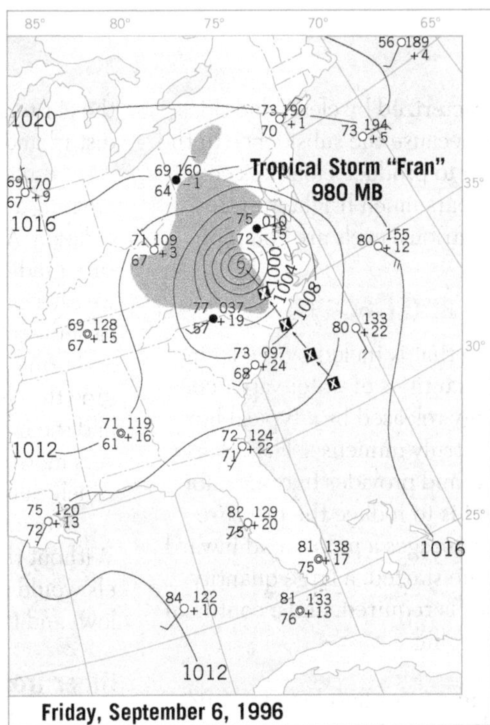

Figure 19.28 Hurricane Fran These weather maps show Hurricane Fran at 7:00 A.M. EST on two successive days, September 5 and 6, 1996. On September 5, winds exceeded 190 kilometers (118 miles) per hour. As the storm moved inland, heavy rains caused flash floods, killed 30 people, and caused more than $3 billion in damages. The station information plotted off the Gulf and Atlantic coasts is from data buoys, which are remote floating instrument packages. The small boxes extending southeast from the storm's center show the position of the eye at 6-hour intervals. Numerals and symbols on the maps refer to temperature, dew point, and wind speed and direction at select weather stations.

wind speeds in excess of 119 kilometers (74 miles) per hour and a rotary circulation. Mature hurricanes average 600 kilometers (375 miles) across, although they can range in diameter from 100 kilometers (60 miles) up to about 1500 kilometers (930 miles). From the outer edge to the center, the barometric pressure has on occasion dropped 60 millibars, from 1010 millibars to 950 millibars or less. The lowest pressures ever recorded in the Western Hemisphere are associated with these storms. (Pressures inside tornadoes are often lower but are usually estimated; obtaining actual measurements from tornadoes is both difficult and dangerous.)

A steep pressure gradient generates the rapid, inward-spiraling winds of a hurricane (Figure 19.28). As the air rushes toward the center of the storm, its velocity increases. This occurs for the same reason that skaters with their arms extended spin faster as they pull their arms in close to their bodies.

As the inward rush of warm, moist surface air approaches the core of the storm, it turns upward and ascends in a ring of cumulonimbus towers (Figure 19.29A). This doughnut-shaped wall of intense convective activity surrounding the center of the storm is called the **eye wall**. It is here that the greatest wind speeds and heaviest rainfall occur (Figure 19.29B). Surrounding the eye wall are curved bands of clouds that trail away in a spiral fashion. Near the top of the hurricane, the airflow is outward, carrying the rising air away from the storm center, thereby providing room for more inward flow at the surface.

At the very center of the storm is the **eye** of the hurricane. This well-known feature is a zone about 20 kilometers (12.5 miles) in diameter where precipitation ceases and winds subside. It offers a brief but deceptive break from the extreme weather in the enormous curving wall clouds that surround it. The air within the eye gradually descends and heats by compression, making it the warmest part of the storm. Although many people

SmartFigure 19.29 Conditions inside a hurricane (Data from World Meteorological Organization)

VIDEO
https://goo.gl/9XuBl4

Outflow of air at the top of the hurricane is important because it prevents the convergent flow at lower levels from "filling in" the storm.

Eye

Sinking air in the eye warms by compression.

Eye wall, the zone where winds and rain are most intense.

Tropical moisture spiraling inward creates rain bands that pinwheel around the storm center.

A. Cross section of a hurricane. Note that the vertical dimension is greatly exaggerated. (After NOAA)

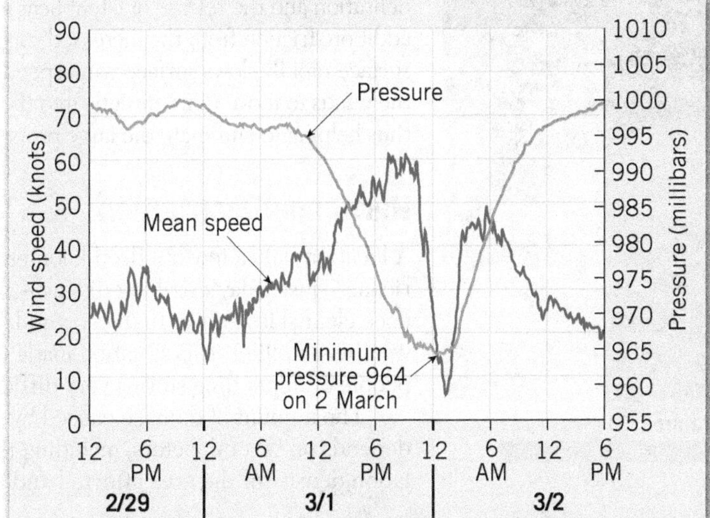

B. Measurements of surface pressure and wind speed during the passage of Cyclone Monty at Mardie Station, Western Australia, between February 29 and March 2, 2004. (Hurricanes are called "cyclones" in this part of the world.)

believe that the eye is characterized by clear blue skies, this is usually not the case because the subsidence in the eye is seldom strong enough to produce cloudless conditions. Although the sky appears much brighter in this region, scattered clouds at various levels are common.

Hurricane Formation and Decay

A hurricane is a heat engine that is fueled by the latent heat liberated when huge quantities of water vapor condense. The amount of energy released by a typical hurricane in just a single day is truly immense. The release of latent heat warms the air and provides buoyancy for its upward flight. The result is to reduce the pressure near the surface, which encourages a more rapid inward flow of air. To get this engine started, a large quantity of warm, moisture-laden air is required, and a continual supply is needed to keep it going.

Hurricane Formation Hurricanes develop most often in the late summer, when ocean waters have reached temperatures of 27°C (80°F) or higher and thus are able to provide the necessary heat and moisture to the air (Figure 19.30). This ocean-water temperature requirement accounts for the fact that hurricanes do not form over the relatively cool waters of the South Atlantic and the eastern South Pacific. For the same reason, few hurricanes form poleward of 20° latitude. Although water temperatures are sufficiently high, hurricanes do not form within 5° of the equator because the Coriolis effect is too weak to initiate the necessary rotary motion.

Many tropical storms begin as disorganized arrays of clouds and thunderstorms that develop weak pressure gradients but exhibit little or no rotation. Such areas of low-level convergence and lifting are called *tropical disturbances*. Most of the time, these zones of convective activity die out. However, tropical disturbances occasionally grow larger and develop a strong cyclonic rotation.

What happens on occasions when conditions favor hurricane development? As latent heat is released from the clusters of thunderstorms that make up the tropical disturbance, areas within the disturbance get warmer. As a result, air density lowers and surface pressure drops, creating a region of weak low pressure and cyclonic circulation. As pressure drops at the storm center, the pressure gradient steepens. If you were watching an animated weather map of the storm, you would see the isobars get closer together. In response, surface wind speeds increase and bring additional supplies of moisture to nurture storm growth. The water vapor condenses, releasing latent heat, and the heated air rises. Adiabatic cooling of rising air triggers more condensation and the release of more latent heat, which causes a further increase in buoyancy. And so it goes.

Meanwhile, at the top of the storm, air is diverging. Without this outward flow up top, the inflow at lower levels would soon raise surface pressures (that is, fill in the low) and thwart storm development.

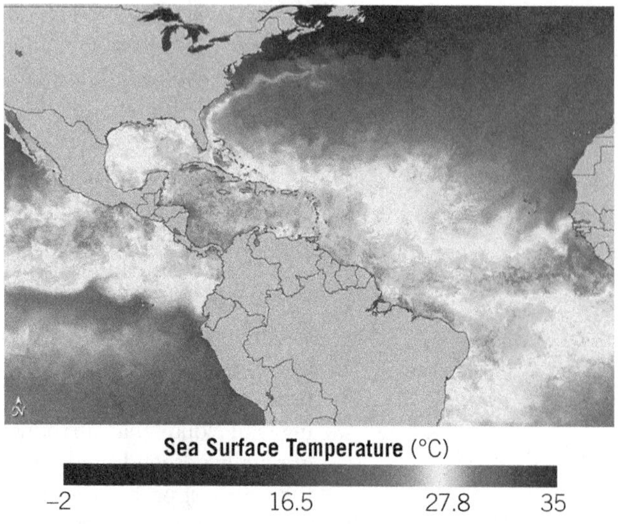

Sea Surface Temperature (°C)
−2 16.5 27.8 35

Figure 19.30 **Sea-surface temperatures** Among the necessary ingredients for a hurricane is warm ocean temperatures above 27°C (80°F). This color-coded satellite image from June 1, 2010, shows sea-surface temperatures at the beginning of hurricane season. (NASA)

Other Tropical Storms Many tropical disturbances occur each year, but only a few develop into full-fledged hurricanes. By international agreement, lesser tropical cyclones are placed in different categories, based on wind strength. When a cyclone's strongest winds do not exceed 61 kilometers (38 miles) per hour, it is called a **tropical depression**. When winds are between 61 and 119 kilometers (38 and 74 miles) per hour, the cyclone is termed a **tropical storm** and given a name (Andrew, Katrina, Sandy, etc.). If the tropical storm becomes a hurricane (more than119 kilometers/74 miles per hour), the name remains the same. Each year, between 80 and 100 tropical storms develop around the world. Of these, usually half or more eventually become hurricanes.

Hurricane Decay Hurricanes diminish in intensity whenever they (1) move over ocean waters that cannot supply warm, moist tropical air; (2) move onto land; or (3) reach a location where the large-scale flow aloft is unfavorable. When a hurricane moves onto land, it loses its punch rapidly. The most important reason for this rapid demise is the fact that the storm's source of warm, moist air is cut off. When an adequate supply of water vapor does not exist, condensation and the release of latent heat must diminish. In addition, friction from the increased roughness of the land surface rapidly slows surface wind speeds. This factor causes the winds to move more directly into the center of the low, thus helping to eliminate the large pressure differences.

Hurricane Destruction

A location only a few hundred kilometers from a hurricane—just 1 day's striking distance away—may experience clear skies and virtually no wind. Prior to the age of weather satellites, this situation made the task of warning people of impending storms very difficult.

The amount of damage caused by a hurricane depends on several factors, including the size and population density of the area affected and the shape of the

A. B.

Figure 19.31 **Storm surge destruction A.** This classic image shows the aftermath of the historic Galveston hurricane. The storm struck an unsuspecting and unprepared city on September 8, 1900. It was the worst natural disaster in U.S. history, taking the lives of 8000 people. Entire blocks were swept clean, and mountains of debris accumulated around the few remaining buildings. (Photo by Associated Press) **B.** This is Crystal Beach, Texas, on September 16, 2008, 3 days after Hurricane Ike came ashore. At landfall the storm had sustained winds of 165 kilometers (105 miles) per hour. The extraordinary storm surge caused most of the damage shown here. (Photo by Smiley N. Pool/Rapport Press/Newscom)

ocean bottom near the shore. The most significant factor, of course, is the strength of the storm. By studying past storms, a scale has been established to rank the relative intensities of hurricanes. As Table 19.2 indicates, a *category 5* storm is the worst possible, whereas a *category 1* hurricane is least severe. (In the Pacific, a super typhoon is equivalent to a strong category 4 or 5 storm.)

During hurricane season, it is common to hear scientists and reporters use the numbers from the **Saffir–Simpson hurricane scale**. When Hurricane Katrina made landfall, sustained winds were 225 kilometers (140 miles) per hour, making it a strong category 4 storm. Storms that fall into category 5 are rare. Damage caused by hurricanes can be divided into three categories: (1) storm surge, (2) wind damage, and (3) heavy rains and inland flooding.

Storm Surge The most devastating damage in the coastal zone is usually caused by storm surge (Figure 19.31). It not only accounts for a large share of coastal property losses but also is responsible for a high percentage of all hurricane-caused deaths. A **storm surge** is a dome of water 65 to 80 kilometers (40 to 50 miles) wide that sweeps across the coast near the point where the eye makes landfall. If all wave activity were smoothed out, the storm surge would be the height of the water above normal tide level. In addition, tremendous wave activity is superimposed on the surge. The worst surges occur in places like the Gulf of Mexico, where the continental shelf is very shallow and gently sloping. In addition, local features such as bays and rivers can cause the surge height to double and increase in speed, and high tide conditions during a surge can compound the water volume and damage.

As a hurricane advances toward the coast in the Northern Hemisphere, storm surge is always most intense where winds are blowing *toward* the shore. In addition, on this side of the storm, the forward movement of the hurricane contributes to the storm surge. In Figure 19.32,

Table 19.2 Saffir–Simpson Hurricane Scale

Scale Number (category)	Central Pressure (millibars)	Winds (km/hr)	Storm Surge (meters)	Damage
1	980	119–153	1.2–1.5	Minimal
2	965–979	154–177	1.6–2.4	Moderate
3	945–964	178–209	2.5–3.6	Extensive
4	920–944	210–250	3.7–5.4	Extreme
5	<920	>250	>5.4	Catastrophic

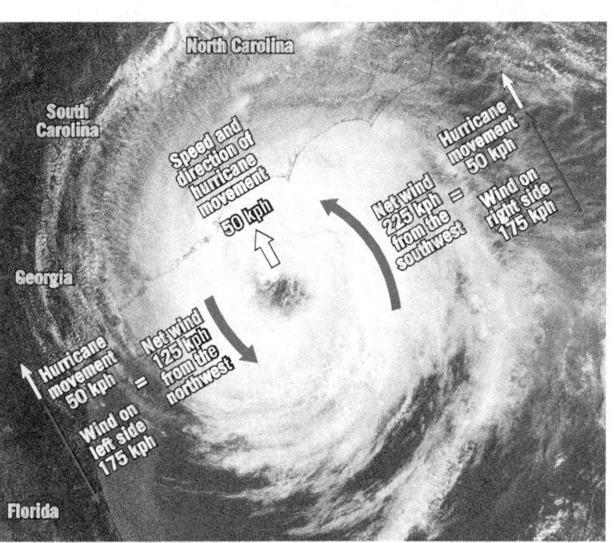

Figure 19.32 **An approaching hurricane** This hypothetical storm, with peak winds of 175 kilometers (109 miles) per hour, is moving toward the coast at 50 kilometers (31 miles) per hour. On the right side of the eye when looking in the direction the hurricane is moving, the 175-kilometer-per-hour winds are in the same direction as the movement of the storm (50 kilometers per hour). Therefore, the *net* wind speed on the right side of the storm is 225 kilometers (140 miles) per hour. On the left side, the hurricane's winds are blowing opposite the direction of storm movement, so the *net* winds of 125 kilometers (78 miles) per hour are away from the coast. Storm surge will be greatest along the part of the coast hit by the right side of the advancing hurricane. (NASA)

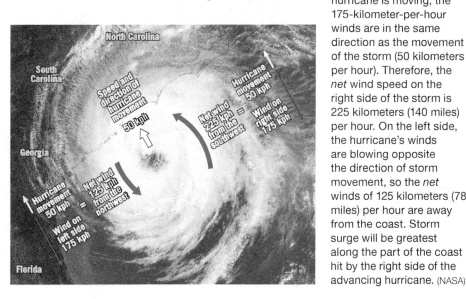

assume that a hurricane with peak winds of 175 kilometers (109 miles) per hour is moving toward the shore at 50 kilometers (31 miles) per hour. In this case, the net wind speed on the right side of the advancing storm is 225 kilometers (140 miles) per hour. On the left side, the hurricane's winds are blowing opposite the direction of storm movement, so the net winds are *away* from the coast at 125 kilometers (78 miles) per hour. Along the shore facing the left side of the oncoming hurricane, the water level may actually decrease as the storm makes landfall.

Wind Damage Destruction caused by wind is perhaps the most obvious of the classes of hurricane damage. Debris such as signs, roofing materials, and small items left outside become dangerous flying missiles in hurricanes. For some structures, the force of the wind is sufficient to cause total ruin. Mobile homes are particularly vulnerable. High-rise buildings are also susceptible to hurricane-force winds. Upper floors are most vulnerable because wind speeds usually increase with height. Recent research suggests that people should stay below the tenth floor but remain above any floors at risk for flooding. In regions with good building codes, wind damage is usually not as catastrophic as storm-surge damage. However, hurricane-force winds affect a much larger area than storm surge and can cause huge economic losses. For example, in 1992 it was largely the winds associated with Hurricane Andrew that produced more than $25 billion of damage in southern Florida and Louisiana.

A hurricane may produce tornadoes that contribute to the storm's destructive power. Studies have shown that more than half of the hurricanes that make landfall produce at least one tornado. In 2004 the number of tornadoes associated with tropical storms and hurricanes was extraordinary. Tropical Storm Bonnie and five land-falling hurricanes—Charley, Frances, Gaston, Ivan, and Jeanne—produced nearly 300 tornadoes that affected the southeastern and mid-Atlantic states.

Heavy Rains and Inland Flooding The torrential rains that accompany most hurricanes pose a third significant threat: flooding. Whereas the effects of storm surge and strong winds are concentrated in coastal areas, heavy rains may affect places hundreds of kilometers from the coast for up to several days after the storm has lost its hurricane-force winds.

The severity of flooding is not only influenced by the rate at which rain is falling but also by how quickly or slowly the storm is moving. A slowly moving hurricane or tropical storm can increase the risk of flooding along the coast or far inland. For example, Tropical Storm Alison (2001) is among the costliest and deadliest storms in U.S. history because it was moved very slowly.

Monitoring Hurricanes

Today we have the benefit of numerous observational tools for monitoring tropical storms and hurricanes. Using input from satellites, aircraft reconnaissance, coastal radar, and remote data buoys in conjunction with sophisticated computer models, meteorologists monitor and forecast storm movements and intensity. The goal is to issue timely watches and warnings.

The Role of Satellites The greatest single advancement in tools used for observing hurricanes has been the development of meteorological satellites. Vast areas of open ocean must be observed in order to detect a hurricane. Before satellites, this was an impossible task. Today instruments aboard satellites can detect a potential storm even before it develops its characteristic circular cloud pattern.

In recent years two methods of using satellite-acquired data to monitor hurricane intensity have been developed. One technique uses instruments aboard a satellite to estimate wind speeds within a storm. A second method uses satellites to identify areas of extraordinary cloud development, called *hot towers*, in the eye wall of an approaching hurricane (Figure 19.33).

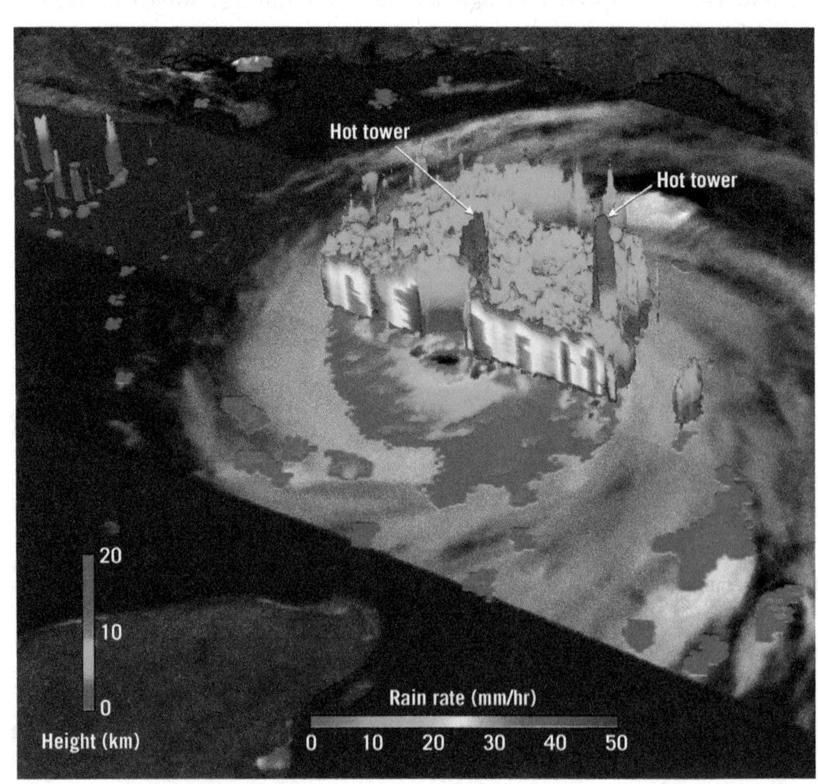

▼ SmartFigure 19.33 **Hot towers** This *Tropical Rainfall Measuring Mission* (*TRMM*) satellite image of Hurricane Katrina was acquired early on August 28, 2005. The cutaway view of the inner portion of the storm shows cloud height on one side and rainfall rates on the other. Two hot towers (in red) are visible: one in an outer rain band and the other in the eye wall. The eye wall tower rises 16 kilometers (10 miles) above the ocean surface and is associated with an area of intense rainfall. Towers this tall near the core often indicate that a storm is intensifying. Katrina grew from a category 3 to a category 4 storm soon after this image was received. (NASA)

VIDEO
https://goo.gl/jW4kai

Track Forecasts The predicted path of a hurricane is called the *track forecast*. The track forecast is probably the most basic information because accurate prediction of other storm characteristics (winds and rainfall) is of little value if there is significant uncertainty about where the storm is going. Accurate track forecasts are important because they can lead to timely evacuations from the surge zone, where the greatest number of deaths usually occur. Track forecasts have been steadily improving. Current 5-day track forecasts are now as accurate as the 3-day forecasts of 20 years ago (**Figure 19.34**). Despite improvements in accuracy, forecast uncertainty still requires that hurricane warnings be issued for relatively large coastal areas. Only about one-quarter of an average warning area experiences hurricane conditions.

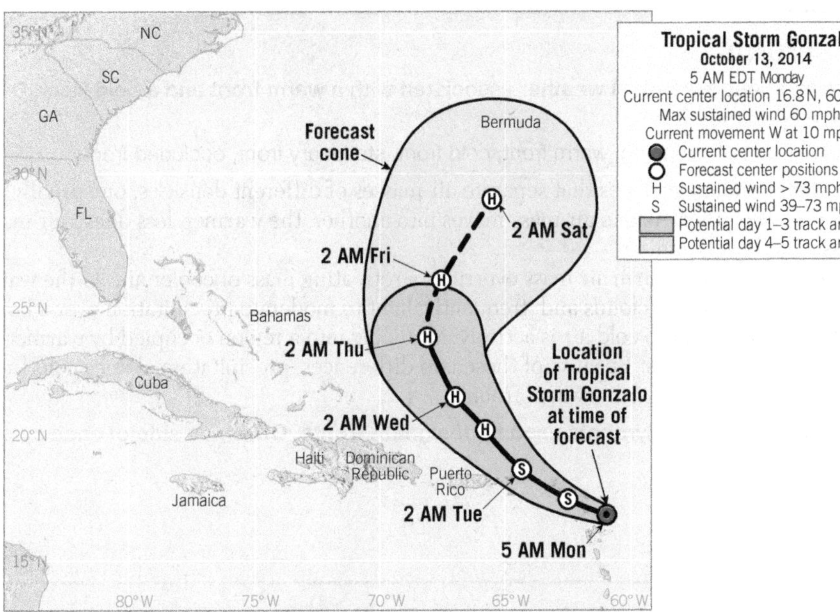

SmartFigure 19.34 **Five-day track forecast for Tropical Storm Gonzalo, issued at 5 A.M. EDT, Monday, October 13, 2014** When a hurricane track forecast is issued by the National Hurricane Center, it is termed a *forecast cone*. The cone represents the probable track of the center of the storm and is formed by enclosing the area swept out by a set of circles along the forecast track (at 12 hours, 24 hours, 36 hours, etc.). The circles get larger as they extend further into the future. The entire track of an Atlantic tropical cyclone can be expected to remain entirely within the cone roughly 60 to 70 percent of the time. (National Weather Service/National Hurricane Center)

VIDEO
https://goo.gl/2h0Oyd

CONCEPT CHECKS 19.6

1. Define *hurricane*. What names are used for this type of storm in other parts of the world?
2. In what latitude zone do hurricanes develop?
3. Distinguish between the eye and the eye wall of a hurricane. How do conditions differ in these zones?
4. What is the source of energy that drives a hurricane?
5. Why do hurricanes *not* form near the equator? Explain the lack of hurricanes in the South Atlantic and eastern South Pacific.
6. When do most hurricanes in the North Atlantic and Caribbean occur? Why are these months the most common times for hurricanes?
7. Why does the intensity of a hurricane diminish rapidly when it moves over land?
8. What are the three broad categories of hurricane damage?

19 CONCEPTS IN REVIEW
Weather Patterns and Severe Storms

19.1 Air Masses

Discuss air masses, their classification, and associated weather.

KEY TERMS: air mass, air-mass weather, source region, polar (P) air mass, arctic (A) air mass, tropical (T) air mass, continental (c) air mass, maritime (m) air mass, lake-effect snow, nor'easter

- An air mass is a large body of air, usually 1600 kilometers (1000 miles) or more across, that is characterized by a sameness of temperature and moisture at any given altitude. When this air moves out of its region of origin, called the source region, it carries these temperatures and moisture conditions elsewhere, perhaps eventually affecting a large portion of a continent.
- Air masses are classified according to the nature of the surface in the source region and the latitude of the source region. Continental (c) designates an air mass of land origin, with the air likely to be dry; a maritime (m) air mass originates over water and, therefore, will be relatively humid. Polar (P) and arctic (A) air masses originate in high latitudes and are cold. Tropical (T) air masses form in low latitudes and are warm. According to this classification scheme, the four main types of air masses are continental polar (cP), continental tropical (cT), maritime polar (mP), and maritime tropical (mT).
- Continental polar (cP) and maritime tropical (mT) air masses influence the weather of North America most, especially east of the Rocky Mountains. Maritime tropical air is the source of much, if not most, of the precipitation received in the eastern two-thirds of the United States.

? Identify the source region associated with each letter on this map. One letter *is not* associated with a source region. Which one is it?

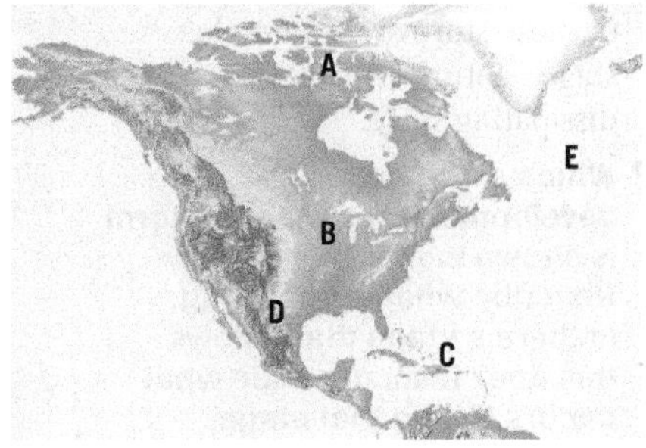

19.2 Fronts

Compare and contrast typical weather associated with a warm front and a cold front. Describe an occluded front and a stationary front.

KEY TERMS: front, overrunning, warm front, cold front, stationary front, occluded front

- Fronts are boundary surfaces that separate air masses of different densities, one usually warmer and more humid than the other. As one air mass moves into another, the warmer, less dense air mass is forced aloft in a process called overrunning.
- Along a warm front, a warm air mass overrides a retreating mass of cooler air. As the warm air ascends, it cools adiabatically to produce clouds and, frequently, light to moderate precipitation over a large area.
- A cold front forms where cold air is actively advancing into a region occupied by warmer air. Cold fronts are about twice as steep as and move more rapidly than warm fronts. Because of these two differences, precipitation along a cold front is generally more intense and of shorter duration than precipitation associated with a warm front.

? Identify each of these symbols used to designate fronts. On which side of each symbol are the warmer air and the cooler air?

A. B. C. D.

19.3 Midlatitude Cyclones

Summarize the weather associated with the passage of a mature mid-latitude cyclone. Describe how airflow aloft is related to cyclones and anticyclones at the surface.

KEY TERMS: midlatitude (middle-latitude) cyclone

- The primary weather producers in the middle latitudes are large centers of low pressure that generally travel from west to east, called midlatitude cyclones. These bearers of stormy weather, which last from a few days to a week, have a counterclockwise circulation pattern in the Northern Hemisphere, with an inward flow of air toward their centers.
- Most midlatitude cyclones have a cold front and frequently a warm front extending from the central area of low pressure. Convergence and forceful lifting along the fronts initiate cloud development and frequently cause precipitation. The particular weather experienced by an area depends on the path of the cyclone.
- Guided by west-to-east-moving jet streams, cyclones generally move eastward across the United States. Airflow aloft (divergence and convergence) plays an important role in maintaining cyclonic and anticyclonic circulation. In cyclones, divergence aloft supports the inward flow at the surface.

19.4 Thunderstorms

List the basic requirements for thunderstorm formation and locate places on a map that exhibit frequent thunderstorm activity. Describe the stages in the development of a thunderstorm.

KEY TERMS: thunderstorm

- Thunderstorms are caused by the upward movement of warm, moist, unstable air. They are associated with cumulonimbus clouds that generate heavy rainfall, lightning, thunder, and occasionally hail and tornadoes.
- Air-mass thunderstorms frequently occur in maritime tropical (mT) air during spring and summer in the middle latitudes. Generally, three stages are involved in the development of these storms: the cumulus stage, mature stage, and dissipating stage.

? Which stage in the development of a thunderstorm is shown in this sketch? Describe what is occurring. Is there a stage that follows this one? If so, describe what occurs during that stage.

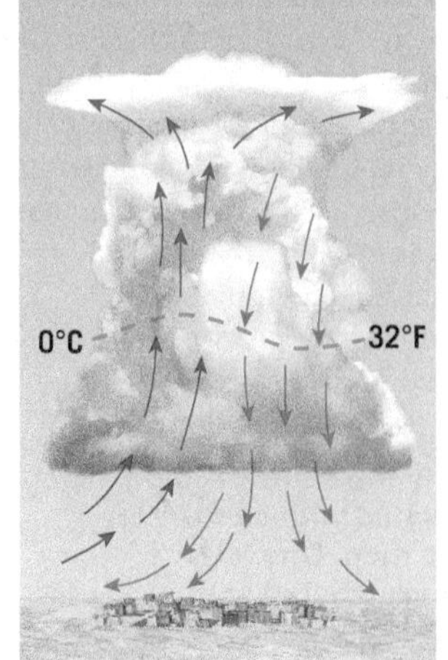

19.5 Tornadoes

Summarize the atmospheric conditions and locations that are favorable to the formation of tornadoes. Discuss tornado destruction and tornado forecasting.

KEY TERMS: tornado, mesocyclone, wind shear, Enhanced Fujita intensity scale (EF-scale), tornado watch, tornado warning, Doppler radar

- A tornado is a violent windstorm that takes the form of a rotating column of air called a vortex that extends downward from a cumulonimbus cloud. Many strong tornadoes contain smaller internal vortices. Because of the tremendous pressure gradient associated with a strong tornado, maximum winds can approach 480 kilometers (300 miles) per hour.
- Tornadoes are most often spawned along the cold front of a midlatitude cyclone or in association with a supercell thunderstorm. Tornadoes also form in association with tropical cyclones (hurricanes). In the United States, April through June is the period of greatest tornado activity, but tornadoes can occur during any month of the year.
- Most tornado damage is caused by the tremendously strong winds. One commonly used guide to tornado intensity is the Enhanced Fujita intensity scale (EF-scale). A rating on the EF-scale is determined by assessing damage produced by the storm.
- Because severe thunderstorms and tornadoes are small and short-lived phenomena, they are among the most difficult weather features to forecast precisely. When weather conditions favor the formation of tornadoes, a tornado watch is issued. The National Weather Service issues a tornado warning when a tornado has been sighted in an area or is indicated on Doppler radar.

19.6 Hurricanes

Identify areas of hurricane formation on a world map and discuss the conditions that promote hurricane formation. List the three broad categories of hurricane destruction.

KEY TERMS: hurricane, eye wall, eye, tropical depression, tropical storm, Saffir–Simpson hurricane scale, storm surge

- Hurricanes, the greatest storms on Earth, are tropical cyclones with wind speeds in excess of 119 kilometers (74 miles) per hour. These complex tropical disturbances develop over tropical ocean waters and are fueled by the latent heat that is liberated when huge quantities of water vapor condense.
- Hurricanes form most often in late summer, when ocean-surface temperatures reach 27°C (80°F) or higher and thus are able to provide the necessary heat and moisture to the air. Hurricanes diminish in intensity when they move over cool ocean water that cannot supply adequate heat and moisture, move onto land, or reach a location where large-scale flow aloft is unfavorable.
- The Saffir–Simpson scale ranks the relative intensities of hurricanes. A 5 on the scale represents the strongest storm possible, and a 1 indicates the lowest severity. Damage caused by hurricanes is divided into three categories: (1) storm surge, (2) wind damage, and (3) heavy rains and inland flooding.

? This graph shows the number of hurricanes in the North Atlantic between May and December over a 100-year span. Why is the occurrence of hurricanes low in early summer?

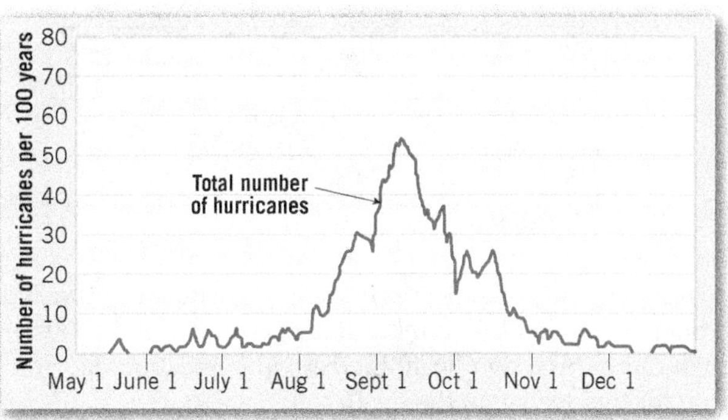

GIVE IT SOME THOUGHT

1. Refer to Figure 19.4 to answer these questions:
   - **a.** Thunder Bay and Marquette are both on the shore of Lake Superior, yet Marquette gets much more snow than Thunder Bay. Why is this the case?
   - **b.** Notice the narrow, north–south zone of relatively heavy snow east of Pittsburgh and Charleston. This region is too far from the Great Lakes to receive lake-effect snowfall. Speculate on a likely reason for the higher snowfalls here. Does your answer explain the shape of this snowy zone?
2. Refer to the accompanying weather map to answer the following questions:
   - **a.** What is a likely wind direction at each city?
   - **b.** Identify the likely air mass that is influencing each city.
   - **c.** Identify the cold front, warm front, and occluded front.
   - **d.** What are the barometric tendencies at city A and city C?
   - **e.** Which one of the three cities is probably coldest? Which one is probably warmest?

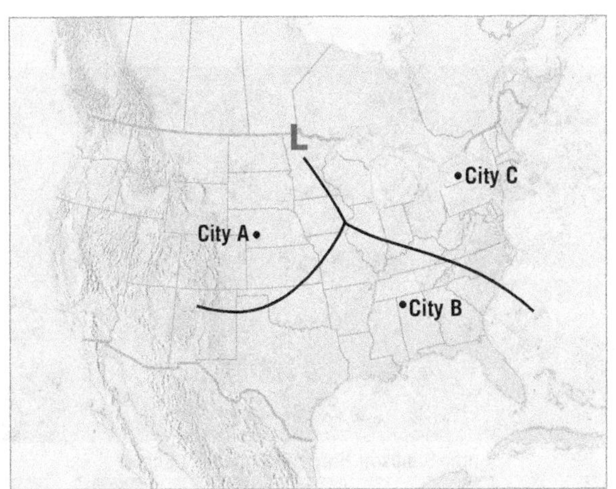

3. Apply your knowledge of fronts to explain the following weather proverb:
   Rain long foretold, long last;
   Short notice, soon past.
4. If you hear that a cyclone is approaching, should you immediately seek shelter? Why or why not?
5. The accompanying diagrams show surface temperatures with isotherms labeled in degrees Fahrenheit for noon and 6 P.M. on January 29, 2008. On this day, a powerful front moved through Missouri and Illinois.
   - **a.** What type of front passed through the Midwest?
   - **b.** Describe how the temperature changed in St. Louis, Missouri, over the 6-hour period.
   - **c.** Describe the likely shift in wind direction in St. Louis during this time span.

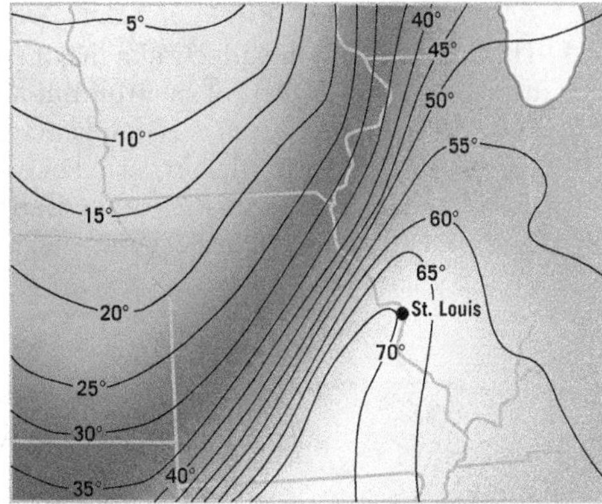

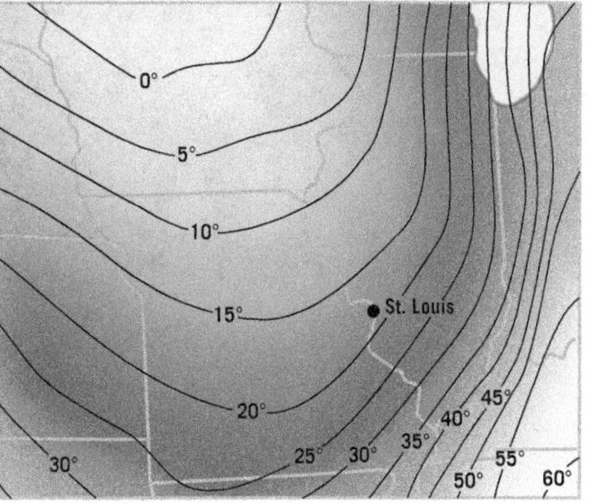

6 If you were located 400 kilometers ahead of the surface position of a typical warm front that had a slope of 1:200, how high would the frontal surface be above you?

7 The accompanying table lists the number of tornadoes reported in the United States by decade. Propose a reason to explain why the totals for the 1990s and 2000s are so much higher than for the 1950s and 1960s.

Number of U.S. Tornadoes Reported, by Decade	
Decade	Number of Tornadoes Reported
1950–1959	4796
1960–1969	6613
1970–1979	8579
1980–1989	8196
1990–1999	12,138
2000–2009	12,914

8 The number of tornado deaths in the United States in the 2000s was less than 40 percent the number that occurred in the 1950s, even though there was a significant increase in population. Suggest a likely reason for the decline in the death toll.

9 A television meteorologist is able to inform viewers about the intensity of an approaching hurricane. However, the meteorologist can report the intensity of a tornado only after it has occurred. Why is this the case?

10 Figure 19.28 shows Hurricane Fran on successive days. Did the storm get stronger or weaker on September 6 compared to September 5? What feature on the weather map (other than the label change from one day to the next) is most useful in figuring this out at a glance? Suggest a reason that the storm's intensity changed.

11 Refer to the graph in Figure 19.29. Explain why wind speeds are greatest when the slope of the pressure curve is steepest.

12 Assume that it is late September 2021, and that the eye of Hurricane Fred, a category 5 storm, is projected to follow the path shown on the accompanying map of Texas. Answer the following questions:

a. Name the stages of development that Fred must have gone through to become a hurricane. At what stage did the storm receive its name?

b. If the storm follows the projected path, will the city of Houston experience Fred's fastest winds and greatest storm surge? Explain why or why not.

c. What is the greatest threat to life and property if this storm approaches the Dallas–Fort Worth area?

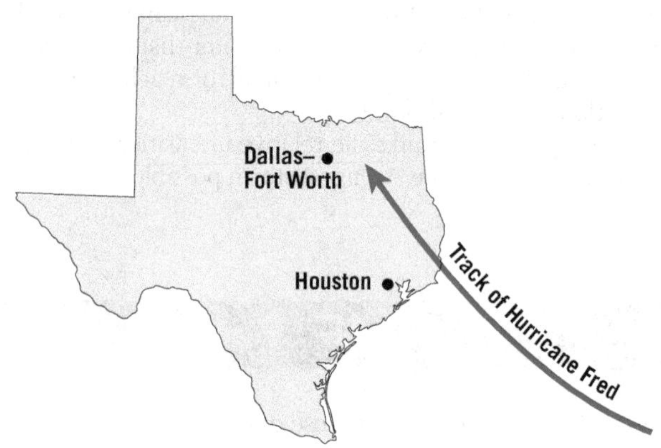

EXAMINING THE EARTH SYSTEM

1 This image shows the effects of a major snowstorm that dropped nearly 2 meters (7 feet) of snow on Buffalo, New York, in December 2001. This weather event was unrelated to a midlatitude cyclone. Places not far from Buffalo received only modest amounts of snow or no snow at all. Which spheres of the Earth system interacted in the Great Lakes region to produce this snowstorm? What term is applied to heavy snows such as this?

Mike Groll/Stringer/Getty Images

2 This world map shows the tracks and intensities of thousands of hurricanes and other tropical cyclones. It was put together by the National Hurricane Center and the Joint Typhoon Warning Center.

a. What area has experienced the greatest number of category 4 and 5 storms?

b. Why do hurricanes not form in the very heart of the tropics, astride the equator?

c. Explain the absence of storms in the South Atlantic and the eastern South Pacific.

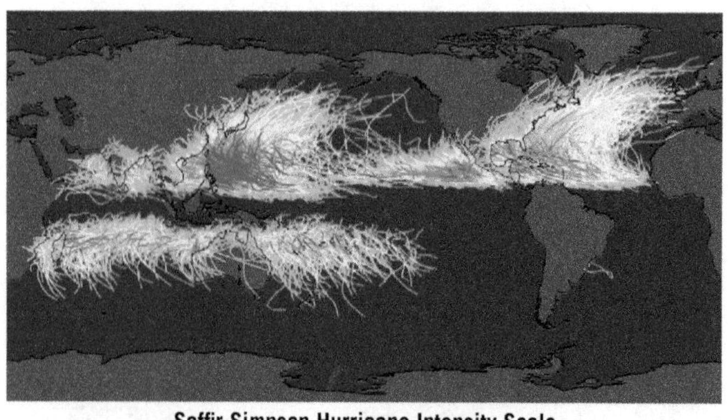

3 The situations described below involve interactions between the atmosphere and Earth's surface. In each case, indicate whether the air mass is being made more stable or more unstable. Briefly explain each choice.

a. An mT air mass moving northward from the Gulf of Mexico over the southeastern United States in winter

b. An mT air mass from the Gulf of Mexico moving northward over the southeastern United States in summer

c. A wintertime cP air mass from Siberia moving eastward from Asia across the North Pacific

4 This satellite image shows Tropical Cyclone Favia as it came ashore along the coast of Mozambique, Africa, on February 22, 2007. This powerful storm was moving from east to west. Portions of the storm had sustained winds of 203 kilometers (126 miles) per hour as it made landfall. Letters A–D relate to Question c.

a. Identify the eye and the eye wall of the cyclone.

b. Based on wind speed, classify the storm using the Saffir–Simpson hurricane scale.

c. Which one of the lettered sites should experience the strongest storm surge? Explain.

d. Describe the possible effects of the storm on coastal lands (geosphere), drainage networks (hydrosphere), and plant and animal life (biosphere).

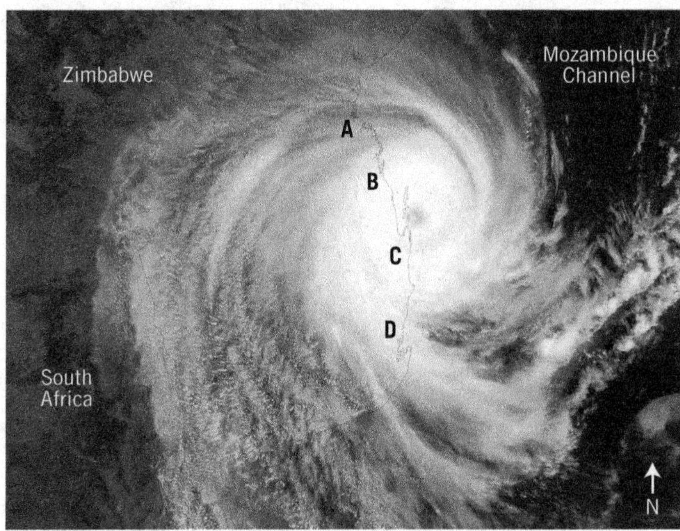

NASA

DATA ANALYSIS

Current Weather Conditions

Weather data are collected worldwide. These data keep people informed about what to wear and whether to bring an umbrella. More importantly, these data are used by forecasters to predict upcoming weather changes and hazards.

ACTIVITIES

Go to the National Weather Service page, at www.weather.gov, to display a map of current weather-related watches and warnings for the United States.

1 List the two current weather alerts that cover the largest area.

2 Under the map click on the name of the alert next to its color square. In one to two sentences, describe the current weather situation for each of your chosen alerts.

3 Go back to the map and click on your current location to bring up a more zoomed-in map. What weather advisories, if any, are currently active in your region? Click on each and give a brief (one to two sentences each) description.

4 Click on your exact location on the zoomed-in map of your region map to bring up a local forecast. What weather advisories, if any, are currently active at your location?

5 How is your weather expected to change over the next 2 days?

Go back to the page with the map of your region. Then, depending on your region's forecast office page, either scroll down and click on the Weather Map, or click on the green Forecast Maps button to bring up the current surface analysis showing fronts and regions of precipitation. The default map is Today's Forecast, which is the first tab above the map.

6 Which of the fronts discussed in the chapter are shown on the map? Where is each located? What type of precipitation is near each of the fronts?

7 Where are the midlatitude cyclones, if any, located on the map? How are the midlatitude cyclones on the map similar to and different from the idealized cyclone shown in Figure 19.12? Be sure to compare the types of fronts, frontal orientation, and any precipitation.

8 Click on the Tomorrow's Forecast tab. How has each of the fronts moved?

9 Compare the weather shown on the map to tomorrow's forecasted weather from Question 5. Does the weather forecast for tomorrow make sense, given the location and movement of fronts? Why or why not?

10 Click on the Day 3 Forecast tab. How has each of the fronts moved?

11 Compare the weather shown on the map to the day after tomorrow's forecasted weather from Question 5. Does the weather forecast for day 3 make sense, given the location and movement of fronts? Why or why not?

MasteringGeology™ Looking for additional review and test prep materials? Visit the Study Area in MasteringGeology to enhance your understanding of this chapter's content by accessing a variety of resources, including Self-Study Quizzes, Geoscience Animations, SmartFigure Tutorials, Mobile Field Trips, *Project Condor* Quadcopter videos, *In the News* articles, flashcards, web links, and an optional Pearson eText.

www.masteringgeology.com

20 World Climates and Global Climate Change

FOCUS ON CONCEPTS

Each statement represents the primary learning objective for the corresponding major heading within the chapter. After you complete the chapter, you should be able to:

20.1 List the five parts of the climate system and provide examples of each.

20.2 Explain why classification is a necessary process when studying world climates. Discuss the criteria used in the Köppen system of climate classification.

20.3 Compare the two broad categories of tropical climates.

20.4 Contrast low-latitude dry climates and middle-latitude dry climates.

20.5 Distinguish among five different humid middle-latitude climates.

20.6 Contrast ice cap and tundra climates.

20.7 Summarize the characteristics associated with highland climates.

20.8 Summarize the nature and cause of the atmosphere's changing composition since about 1750. Describe the climate's response.

20.9 Contrast positive- and negative-feedback mechanisms and provide examples of each.

20.10 Summarize some of the possible consequences of global warming.

Glaciers are sensitive to changes in temperature and precipitation and therefore provide clues about changes in climate. Like most other glaciers and ice sheets worldwide, Margerie Glacier in Alaska's Glacier Bay National Park is losing mass and retreating. (Photo by Don Paulson/AGE Fotostock)

THE FOCUS OF THIS CHAPTER IS *CLIMATE*, the long-term aggregate of weather. Climate is more than just an expression of average atmospheric conditions. To accurately portray the character of a place or an area, variations and extremes must also be included. Climate strongly influences the nature of plant and animal life, the soil, and many external geologic processes. Climate influences people as well.

Although climate has a significant impact on people, we are learning that people also have a strong influence on climate. In fact, today global climate change caused by humans is an important global environmental issue. Unlike changes in the geologic past, which represented natural variations, modern climate change is dominated by human effects that are sufficiently large that they exceed the bounds of natural variability. Moreover, these changes are likely to continue for many centuries. The outcomes of this venture into the unknown with climate could be very disruptive not only to humans but to many other life-forms as well.

20.1 The Climate System

List the five parts of the climate system and provide examples of each.

Throughout this book, you have been reminded that Earth is a complex system that consists of many interacting parts. A change in any one part can produce changes in any or all of the other parts—often in ways that are neither easy to foretell nor immediately apparent. This fact is certainly true when it comes to the study of climate and climate change.

To understand and appreciate climate, it is important to realize that climate involves more than just the atmosphere. Indeed, we must recognize that there is a **climate system** that includes the atmosphere, hydrosphere, geosphere, biosphere, and cryosphere. The first four spheres were discussed in Chapter 1; the **cryosphere** refers to the portion of Earth's surface where water is in solid form. This includes snow, glaciers, sea ice, freshwater ice, and frozen ground (permafrost). The climate system *involves the exchanges of energy and moisture that occur among the five spheres.* These exchanges link the atmosphere to the other spheres to form an extremely complex, interactive whole. Changes to the climate system do not occur in isolation. Rather, when one part of it changes, the other components also react. The major components of the climate system are shown in **Figure 20.1**.

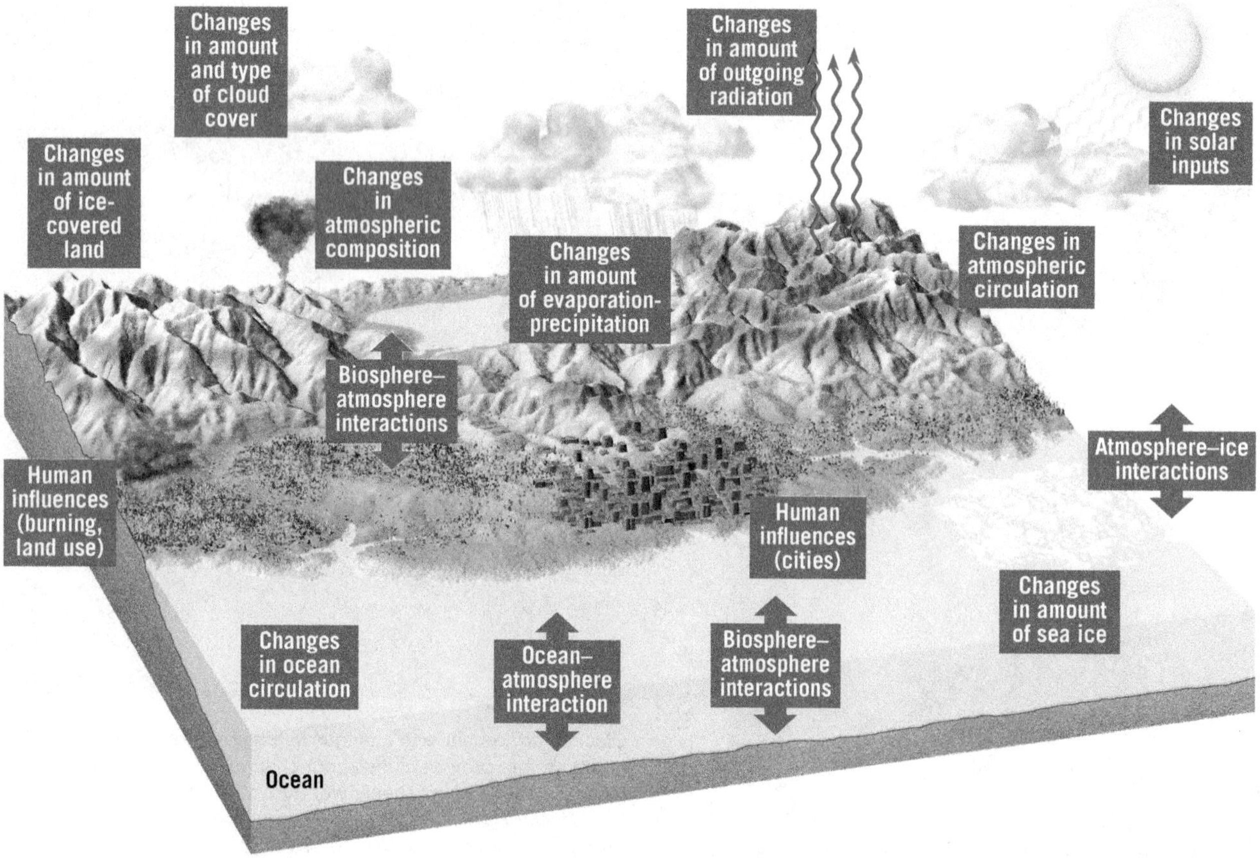

Figure 20.1 Earth's climate system Schematic view showing several components of Earth's climate system. Many interactions occur among the various components on a wide range of space and time scales, making the system extremely complex.

Climate has a profound impact on many of Earth's external processes. When climate changes, these processes respond. A glance back at the rock cycle in Chapter 3 reminds us about many of the connections. Of course, rock weathering has an obvious climate connection, as do processes that operate in arid, tropical, and glacial landscapes. Phenomena such as debris flows and river flooding are often triggered by atmospheric events such as periods of extraordinary rainfall. Clearly, the atmosphere is a basic link in the hydrologic cycle. Other connections involve the impact of internal processes on the atmosphere. For example, the particles and gases emitted by volcanoes can change the composition of the atmosphere, and mountain building can have a significant impact on regional temperature, precipitation, and wind patterns.

The study of sediments, sedimentary rocks, and fossils clearly demonstrates that, through the ages, practically every place on our planet has experienced wide swings in climate, from ice ages to conditions associated with subtropical coal swamps or desert dunes. Chapter 12 reinforces this fact. Time scales for climate change vary from decades to millions of years.

CONCEPT CHECKS 20.1

1. What are the five major parts of the climate system?
2. List at least five connections between climate and Earth's external and internal processes.

20.2 World Climates

Explain why classification is a necessary process when studying world climates. Discuss the criteria used in the Köppen system of climate classification.

Previous chapters have already presented the spatial and seasonal variations of the major elements of weather and climate. Chapter 16 examined the controls of temperature and the world distribution of temperature. In Chapter 18, you studied the general circulation of the atmosphere and the global distribution of precipitation. You are now ready to investigate the *combined* effects of these variations in different parts of the world. The varied nature of Earth's surface and the many interactions that occur among atmospheric processes give every location on our planet a distinctive, even unique, climate. However, we are not going to describe the unique climatic character of countless different locales. Instead, we introduce the major climate regions of the world. The discussion in this chapter examines large areas and uses particular places only to illustrate the characteristics of these major climate regions.

Temperature and precipitation are the most important elements in a climate description because they have the greatest influence on people and their activities, and they also have an important impact on the distribution of such phenomena as vegetation and soils. Nevertheless, other factors are also important for a complete climatic description. When possible, some of these factors are introduced in our discussion of world climates.

Climate Classification

The complex distribution of temperature, precipitation, pressure, and wind across Earth's surface creates a huge diversity of local climates. Thus, like other

EYE ON EARTH 20.1

This image shows a portion of the Alaska Range and surrounding landscape in Denali National Park.

QUESTION 1 *Which of the five parts of the climate system are represented in this photo?*

QUESTION 2 *Speculate on how climate change in the coming decades might cause changes to the area shown here.*

Photo by Arterra Picture Library/Alamy

scientists, climate researchers must make sense of a vast array of complex data. (Consider astronomy, which deals with billions of stars, and biology, which studies millions of complex organisms.) To cope, we must devise some means of *classifying* the data—that is, establishing groups of items that have common characteristics. A good classification scheme makes the data easier to understand and also facilitates analysis and explanation.

In an early attempt at climate classification, the ancient Greeks divided each hemisphere into three zones: *torrid*, *temperate*, and *frigid* (Figure 20.2). The basis of this simple scheme was Earth–Sun relationships. The boundaries were the four astronomically important parallels of latitude: the Tropic of Cancer (23.5° north), the Tropic of Capricorn (23.5° south), the Arctic Circle (66.5° north), and the Antarctic Circle (66.5° south). Thus, the globe was divided into winterless climates and summerless climates and an intermediate type that had features of the other two.

Few other attempts were made until the beginning of the twentieth century. Since then, many climate-classification schemes have been devised. Remember that the classification of climates (or of anything else) is not a natural phenomenon but the product of human ingenuity. The value of any particular classification system is determined largely by its *intended use*. A system designed for one purpose may not work well for another.

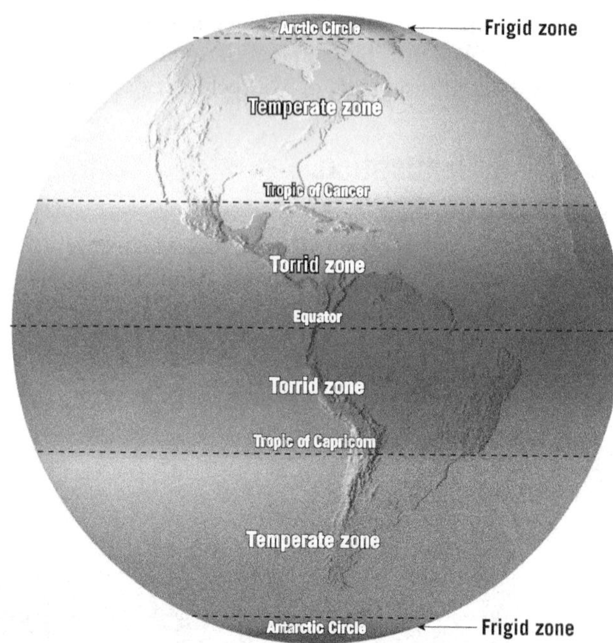

Figure 20.2 **An early climate classification** Among the first attempts at climate classification was one made by the ancient Greeks. They divided each hemisphere into three zones. The winterless *torrid* zone was separated from the summerless *frigid* zone by the *temperate* zone, which had features of the other two.

The Köppen Classification

In this chapter, we use a classification devised by Russian-born German climatologist Wladimir Köppen (1846–1940). As a tool for presenting the general world pattern of climates, the **Köppen classification** has been the best-known and most used system for decades. It is widely accepted for many reasons. For example, it uses only easily obtained data: mean monthly and annual values of temperature and precipitation. Furthermore, the criteria are unambiguous, are relatively simple to apply, and divide the world into climate regions in a realistic way.

Köppen believed that the distribution of natural vegetation is an excellent expression of the totality of climate. Consequently, the boundaries he chose were largely based on the limits of certain plant associations. Five principal groups were recognized, and each group was designated by a capital letter, as follows:

A. **Humid tropical (A).** Winterless climates; all months have a mean temperature above 18°C (64°F).
B. **Dry (B).** Climates where evaporation exceeds precipitation; there is a constant water deficiency.
C. **Humid middle-latitude, mild winters (C).** The average temperature of the coldest month is below 18°C (64°F) but above −3°C (27°F).
D. **Humid middle-latitude, severe winters (D).** The average temperature of the coldest month is below −3°C (27°F), and the warmest monthly mean exceeds 10°C (50°F).
E. **Polar (E).** Summerless climates; the average temperature of the warmest month is below 10°C (50°F).

Notice that four of the major groups (A, C, D, and E) are defined on the basis of temperature characteristics, and the fifth, the B group, has precipitation as its primary criterion. Each of the five groups is further subdivided by using the criteria and symbols presented in Figure 20.3.

A strength of the Köppen classification is the relative ease with which boundaries are determined. However, these boundaries cannot be viewed as fixed. On the contrary, all climate boundaries shift from year to year (Figure 20.4). The boundaries shown on climate maps are simply average locations based on data collected over many years. Thus, a climate boundary should be regarded as a broad transition zone and not a sharp line.

Letter Symbol 1st	2nd	3rd	
A			Average temperature of the coldest month is 18°C or higher.
	f		Every month has 6 cm of precipitation or more.
	m		Short dry season; precipitation in driest month less than 6 cm but equal to or greater than 10 – $R/25$ (R is annual rainfall in cm).
	w		Well-defined winter dry season; precipitation in driest month less than 10 – $R/25$.
	s		Well-defined summer dry season (rare).
B			Potential evaporation exceeds precipitation. The dry–humid boundary is defined by the following formulas: (Note: R is the average annual precipitation in cm, and T is the average annual temperature in °C.) When R is less than the calculated value the climate is dry. $R < 2T + 28$ when 70% or more of rain falls in warmer 6 months. $R < 2T$ when 70% or more of rain falls in cooler 6 months. $R < 2T + 14$ when neither half year has 70% or more of rain.
	S		Steppe — The BS–BW boundary is 1/2 the dry–humid boundary.
	W		Desert — The BS–BW boundary is 1/2 the dry–humid boundary.
		h	Average annual temperature is 18°C or greater.
		k	Average annual temperature is less than 18°C.
C			Average temperature of the coldest month is under 18°C and above –3°C.
	w		At least 10 times as much precipitation in a summer month as in the driest winter month.
	s		At least three times as much precipitation in a winter month as in the driest summer month; precipitation in driest summer month less than 4 cm.
	f		Criteria for w and s cannot be met.
		a	Warmest month is over 22°C; at least 4 months over 10°C.
		b	No month above 22°C; at least 4 months over 10°C.
		c	One to 3 months above 10°C.
D			Average temperature of coldest month is –3°C or below; average temperature of warmest month is greater than 10°C.
	w		Same as under C.
	s		Same as under C.
	f		Same as under C.
		a	Same as under C.
		b	Same as under C.
		c	Same as under C.
		d	Average temperature of the coldest month is –38°C or below.
E			Average temperature of the warmest month is below 10°C.
	T		Average temperature of the warmest month is greater than 0°C and less than 10°C.
	F		Average temperature of the warmest month is 0°C or below.

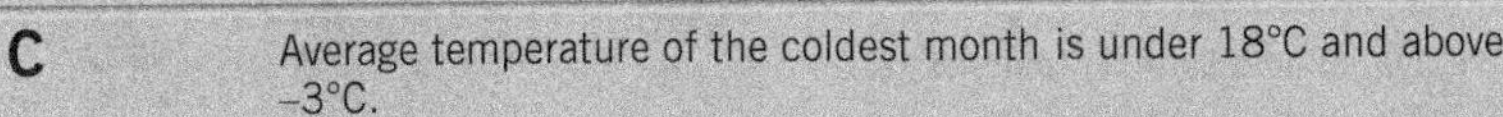

Figure 20.3 The Köppen system of climate classification This system uses easily obtained data: mean monthly and annual values of temperature and precipitation. When using this figure to classify climate data, first determine whether the data meet the criteria for the E climates. If the climate is not polar, proceed to the criteria for B climates. If the data do not fit into either the E or B groups, check the data against the criteria for A, C, and D climates, in that order. (Photos, in order from A–E, are by earlytwenties/Shutterstock, Witold Skrypczak/Alamy Stock Photo, Ed Reschke/Getty Images, Martin Shields/Alamy Stock Photo, and J.G. Paren/Science Source)

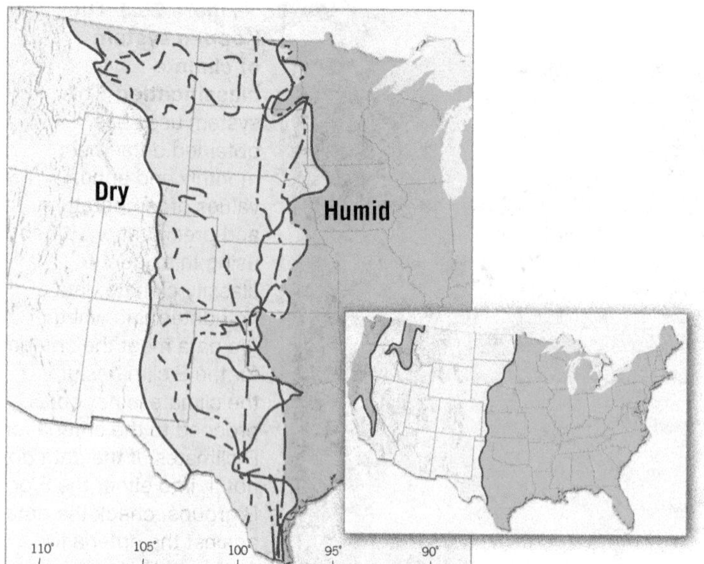

▲ Figure 20.4 **Conditions change from year to year** Yearly fluctuations in the dry–humid boundary during a 5-year period. The small inset shows the average position of the dry–humid boundary.

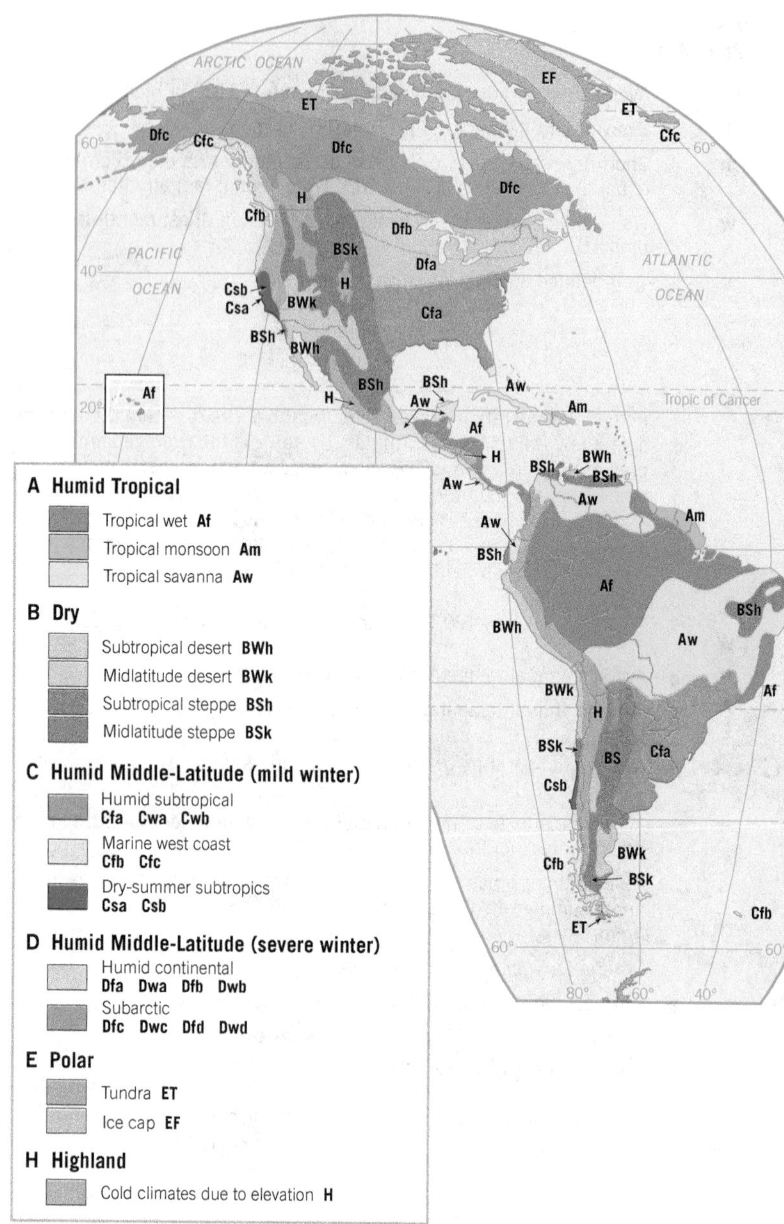

▲ Figure 20.5 **Climates of the world** This map is based on the Köppen classification.

The world distribution of climates according to the Köppen classification is shown in Figure 20.5. You will refer to this map several times as Earth's climates are discussed in the following pages.

CONCEPT CHECKS 20.2

1. Why is classification often a necessary task in science?
2. What climate data are needed to classify a climate using the Köppen system?
3. Should climate boundaries, such as those shown on the world map in Figure 20.5, be regarded as fixed? Explain.

20.3 Humid Tropical (A) Climates

Compare the two broad categories of tropical climates.

Within the A group of climates, two main types are recognized: wet tropical climates (Af and Am) and tropical wet and dry (Aw).

The Wet Tropics

The constantly high temperatures and year-round rainfall in the wet tropics combine to produce the most luxuriant vegetation found in any climatic realm: the **tropical rain forest** (Figure 20.6).

The environment of the wet tropics characterizes almost 10 percent of Earth's land area. An examination of Figure 20.5 shows that Af and Am climates form a discontinuous belt astride the equator that typically extends 5° to 10° into each hemisphere. The poleward margins are most often marked by diminishing rainfall, but occasionally decreasing temperatures mark the boundary. Because of the general decrease in temperature with height in the troposphere, this climate region is restricted to elevations below 1000 meters (nearly 3300 feet).

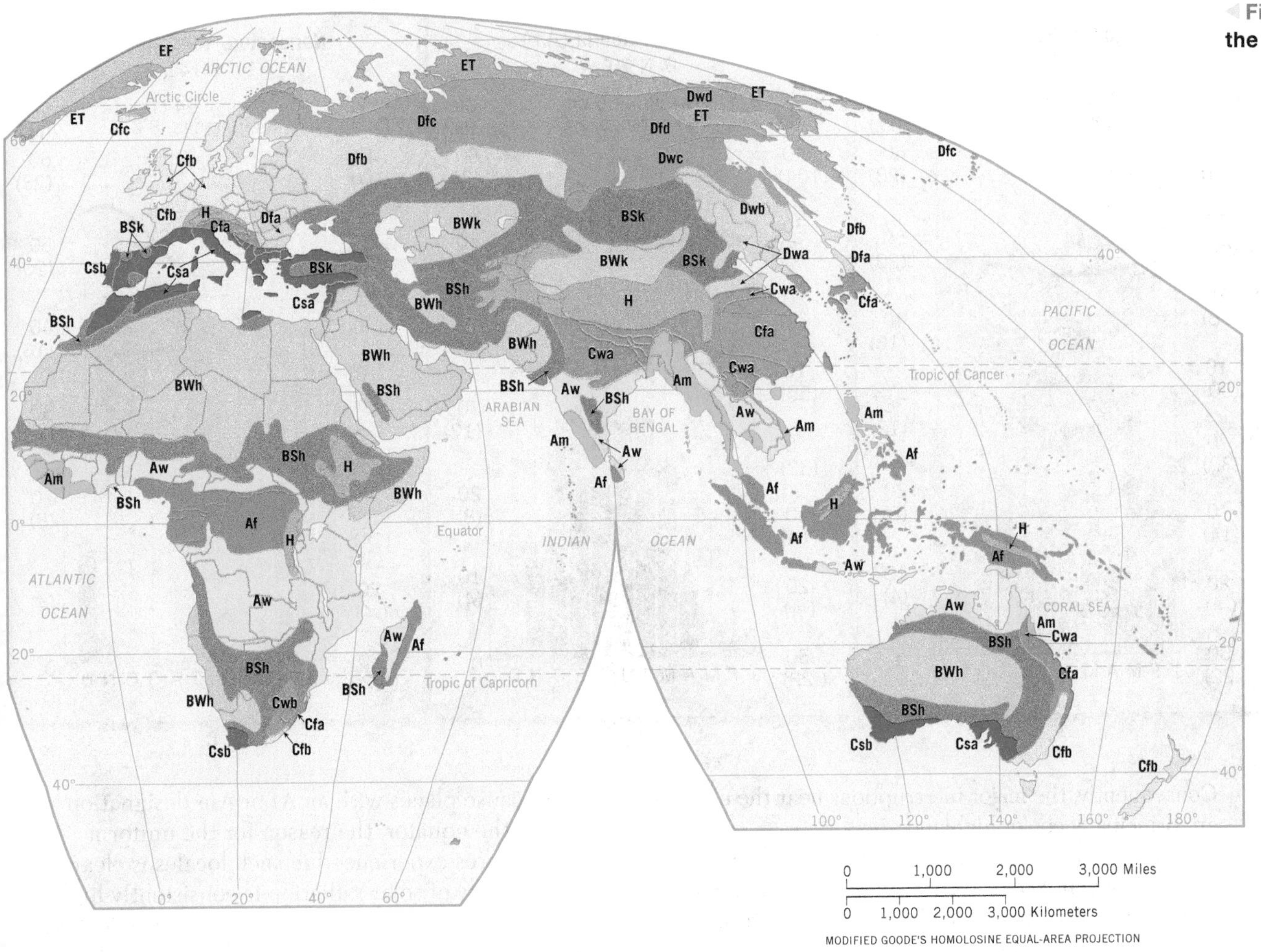

Figure 20.5 Climates of the world (*continued*)

0 1,000 2,000 3,000 Miles
0 1,000 2,000 3,000 Kilometers
MODIFIED GOODE'S HOMOLOSINE EQUAL-AREA PROJECTION

SmartFigure 20.6 **Tropical rain forest** Unexcelled in luxuriance and characterized by hundreds of different species per square kilometer, the tropical rain forest is a broadleaf evergreen forest that dominates the wet tropics. This image shows Borneo's Segama River passing through virgin tropical rain forest. (Photo by Peter Lilja/AGE Fotostock)

TUTORIAL
https://goo.gl/7wixyY

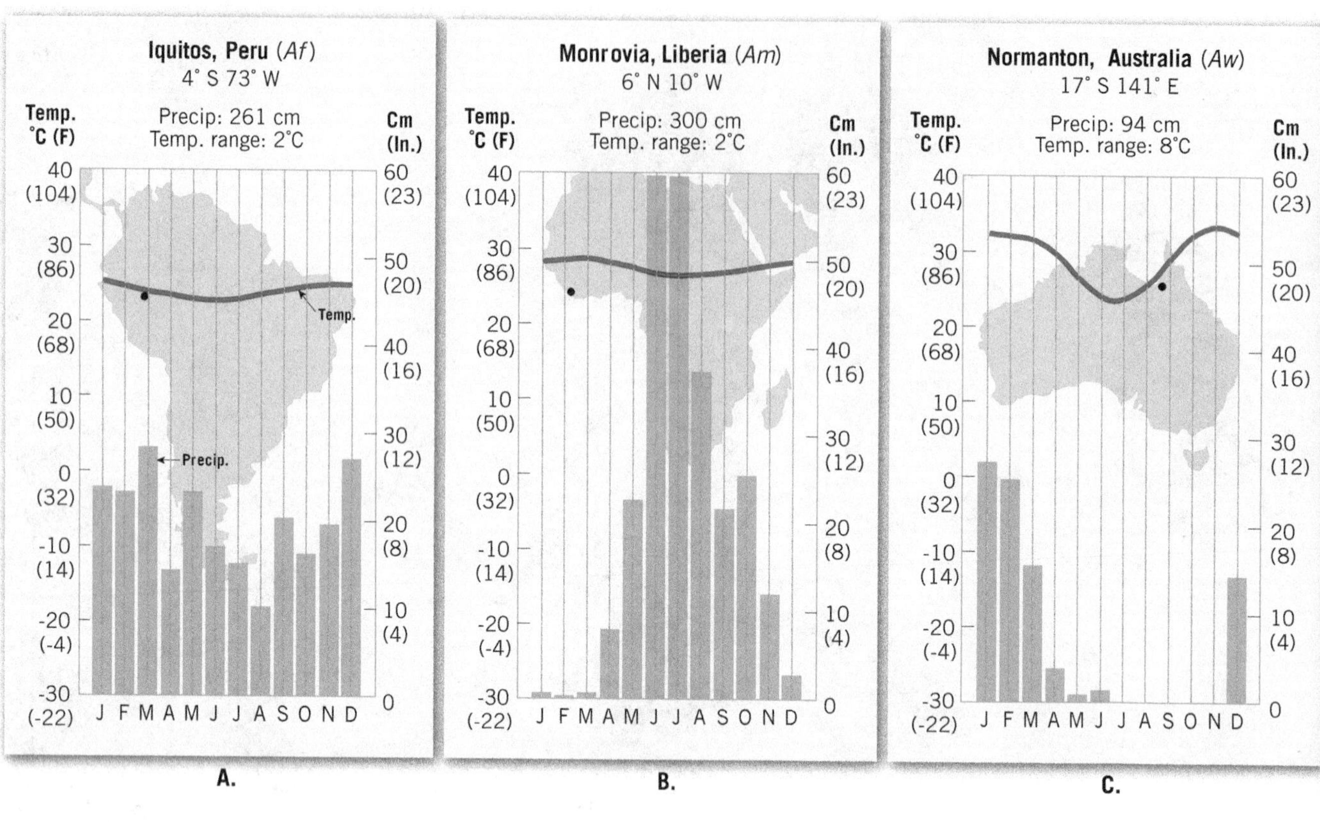

Figure 20.7 Humid tropical climates By comparing these three climatic diagrams, the primary differences among the A climates can be seen. **A.** Iquitos, the Af station, is wet throughout the year. **B.** Monrovia, the Am station, has a short dry season. **C.** As is true for all Aw stations, Normanton has an extended dry season and a higher annual temperature range than the others.

Consequently, the major interruptions near the equator are principally cooler highland areas.

Data for two representative stations in the wet tropics are shown in Figure 20.7A,B. A brief examination reveals the most obvious features that characterize the climate in these areas:

- Temperatures usually average 25°C (77°F) or more each month. Consequently, not only is the annual mean temperature high, but the annual temperature range is very small.
- The total precipitation for the year is high, often exceeding 200 centimeters (80 inches).
- Although rainfall is not evenly distributed throughout the year, tropical rain forest stations are generally wet in all months. If a dry season exists, it is very short.

Because places with an Af or Am designation lie near the equator, the reason for the uniform temperatures experienced in such locales is clear: The intensity of solar radiation is consistently high. The rays of the Sun are always relatively vertical, and changes in the length of daylight throughout the year are slight; therefore, seasonal temperature variations are minimal.

The region is strongly influenced by the equatorial low. Its converging trade winds and the accompanying ascent of warm, humid, unstable air produce conditions that are ideal for the formation of precipitation.

Figure 20.8 Tropical savanna grassland This tropical savanna in Tanzania's Serengeti National Park, with its stunted, drought-resistant trees, was probably strongly influenced by seasonal burnings carried out by native human populations. (Photo by Kondrachov Vladimir/Shutterstock)

Tropical Wet and Dry

In the latitude zone poleward of the wet tropics and equatorward of the subtropical deserts lies the transitional **tropical wet and dry climate** (Aw). Here the rain forest gives way to the **savanna**, a tropical grassland with scattered drought-tolerant trees (Figure 20.8). Because temperature characteristics among all A climates are quite similar, the primary factor that distinguishes the Aw climate from Af and Am is precipitation. Although the overall amount of precipitation in the tropical wet and dry realm is often considerably less than in the wet tropics, the most distinctive feature of this climate is not the annual rainfall total but the markedly seasonal character of the rainfall. The climate diagram for Normanton, Australia (Figure 20.7C), clearly illustrates this trait. As the equatorial low advances poleward in summer, the rainy season commences and features weather patterns typical of the wet tropics. Later, with the retreat of the equatorial low, the subtropical high advances into the region and brings with it pronounced dryness. In some Aw regions, such as India, Southeast Asia, and portions of Australia, the alternating periods of rainfall and dryness are associated with a well-established monsoon circulation (see Chapter 18).

CONCEPT CHECKS 20.3

1. What is the main factor that distinguishes Aw climates from Af and Am? How is this difference reflected in the vegetation of these climate regions?
2. How do the equatorial low and the subtropical high influence the seasonal distribution of rainfall in the Aw climate?

20.4 Dry (B) Climates

Contrast low-latitude dry climates and middle-latitude dry climates.

It is important to realize that the concept of dryness is a relative one and refers to any situation in which a water deficiency exists. Climatologists define a dry climate as one in which the yearly precipitation is not as great as the potential loss of water by evaporation. Thus, dryness is not only related to annual rainfall totals but is also a function of evaporation, which in turn depends closely on temperature.

To establish the boundary between dry and humid climates, the Köppen classification uses formulas that involve three variables: average annual precipitation, average annual temperature, and seasonal distribution of precipitation. The use of average annual temperature reflects its importance as an index of evaporation. The amount of rainfall defining the humid–dry boundary increases as the annual mean temperature increases. The use of seasonal precipitation as a variable is also related to this idea. If rain is concentrated in the warmest months, loss to evaporation is greater than if the precipitation is concentrated in the cooler months.

Within the regions defined by a general water deficiency are two climatic types: **arid**, or **desert** (BW), and **semiarid**, or **steppe** (BS). These two groups have many features in common; their differences are primarily a matter of degree. The semiarid is a marginal and more humid variant of the arid climate type and represents a transition zone that surrounds the desert and separates it from the bordering humid climates (see Figure 10.30, page 339).

Low-Latitude Deserts and Steppes

The heart of low-latitude dry climates lies in the vicinities of the Tropics of Cancer and Capricorn. A glance at Figure 20.5 shows a virtually unbroken desert environment stretching for more than 9300 kilometers (nearly 6000 miles) from the Atlantic coast of North Africa to the dry lands of northwestern India. In addition to this single great expanse, the Northern Hemisphere contains another, much smaller area of subtropical desert and steppe in northern Mexico and the southwestern United States. In the Southern Hemisphere, dry climates dominate Australia. Almost 40 percent of the continent is desert, and much of the remainder is steppe. In addition, arid and semiarid areas are found in southern Africa and make a limited appearance in coastal Chile and Peru.

The existence of this dry subtropical realm is primarily the result of the prevailing global distribution of air pressure and winds. Earth's low-latitude deserts and steppes coincide with the subtropical high-pressure belts where air is subsiding (see Figure 18.17, page 569). When air sinks, it is compressed and warmed. Such conditions are the opposite of what is needed for cloud formation and precipitation. Therefore, clear skies, a maximum of sunshine, and dryness are to be expected. The classic view of Africa and the Arabian Peninsula from space in Figure 20.9 reinforces this idea. The climate diagrams for Cairo, Egypt, and Monterrey, Mexico (Figure 20.10A,B), illustrate the characteristics of low-latitude dry climates.

Middle-Latitude Deserts and Steppes

Unlike their low-latitude counterparts, middle-latitude deserts and steppes are not controlled by the subsiding air masses associated with high pressure. Instead, these dry lands exist principally because of their positions in the deep interiors of large landmasses far removed from the oceans, which are the ultimate source of moisture for cloud formation and precipitation. The presence of high mountains

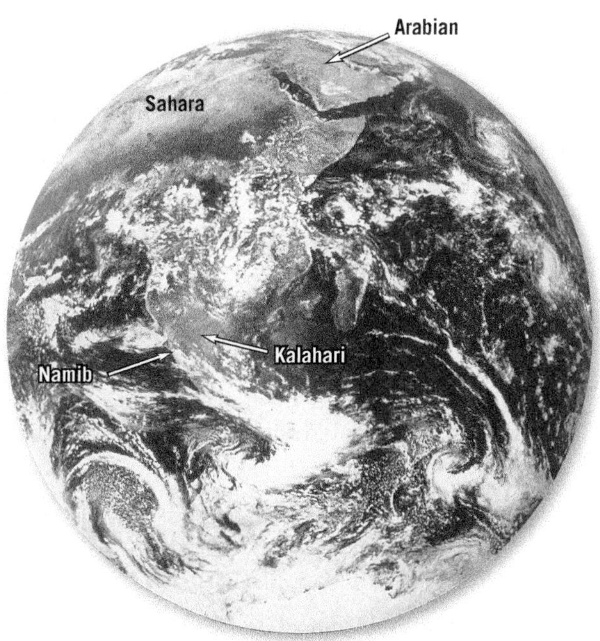

Figure 20.9 **Deserts of Africa and the Arabian Peninsula** In this view of Earth from space, North Africa's Sahara Desert, the adjacent Arabian Desert, and the Kalahari and Namib Deserts in southern Africa are clearly visible as tan-colored, cloud-free zones. These low-latitude deserts are dominated by the dry, subsiding air associated with pressure belts known as the *subtropical highs*. By contrast, the band of clouds that extends across central Africa and the adjacent oceans coincides with the equatorial low-pressure belt, the rainiest region on Earth. (NASA)

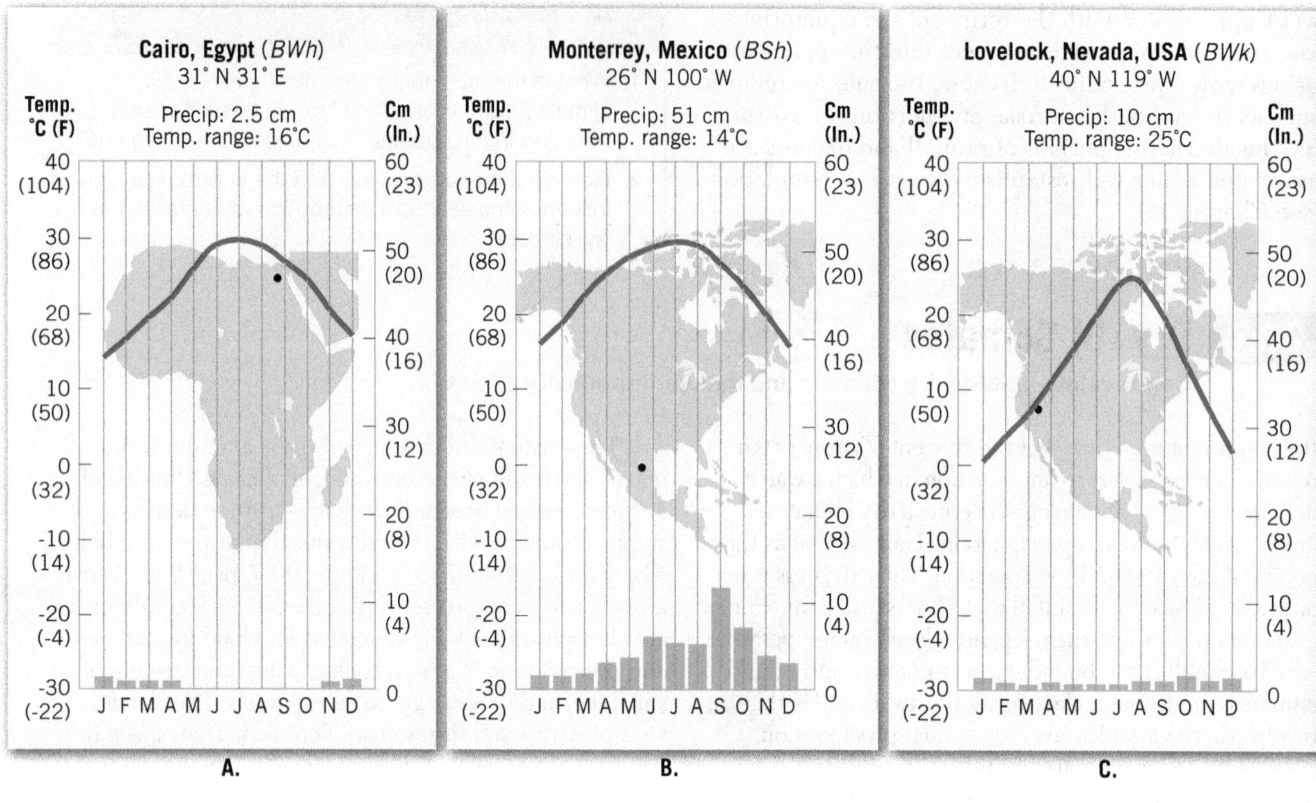

▶ **Figure 20.10**
Climate diagrams for arid and semiarid stations Stations **A** and **B** are in the subtropics, whereas **C** is in the middle latitudes. Cairo and Lovelock are classified as deserts; Monterrey is a steppe. Lovelock, Nevada, may also be called a rainshadow desert.

across the paths of prevailing winds further acts to separate these areas from water-bearing maritime air masses.

Windward sides of mountains are often wet. As prevailing winds meet mountain barriers, the air is forced to ascend, producing clouds and precipitation. By contrast, the leeward sides of mountains are usually much dryer and are often arid enough to be referred to as rainshadow deserts (see Figure 17.11, page 534). Because many middle-latitude deserts occupy sites on the leeward sides of the mountains, they can also be classified as rainshadow deserts (**Figure 20.10C**). In North America, the Coast Ranges, Sierra Nevada, and Cascades are the foremost mountain barriers. In Asia, the great Himalayan chain prevents the summertime monsoon flow of moist Indian Ocean air from reaching the interior. Because the Southern Hemisphere lacks extensive land areas in the middle latitudes, only a small area of desert and steppe is found in this latitude range, existing primarily in the rain shadow of the towering Andes.

In the case of middle-latitude deserts, we have an example of the impact of tectonic processes on climate. Rainshadow deserts exist because of the mountains produced at convergent plate boundaries. Without such mountain-building episodes, wetter climates would prevail where many dry regions exist today.

CONCEPT CHECKS 20.4

1. Why is the amount of precipitation that defines the boundary between humid and dry climates variable?
2. What is the primary reason (control) for the existence of the dry subtropical realm (BWh and BSh)?
3. What factors contribute to the existence of middle-latitude deserts and steppes?

20.5 Humid Middle-Latitude Climates (C and D Climates)

Distinguish among five different humid middle-latitude climates.

The humid middle-latitude climates dominate large portions of North America, Europe, and Asia. Humid middle-latitude climates are divided into two categories: C climates, which have mild winters, and D climates, which have severe winters.

Humid Middle-Latitude Climates with Mild Winters (C Climates)

Although the term *subtropical* is often used for the C climates, it can be misleading. Although many areas with C climates do indeed possess some near-tropical characteristics, other regions do not. For example, we would be stretching the use of the term *subtropical* to describe the climates of coastal Alaska and Norway, which belong to the C group. Within the C group of climates, several subgroups are recognized.

Humid Subtropics Located on the eastern sides of the continents, in the 25° to 40° latitude range, the **humid subtropical climate** dominates the southeastern United States, as well as other similarly situated areas around the

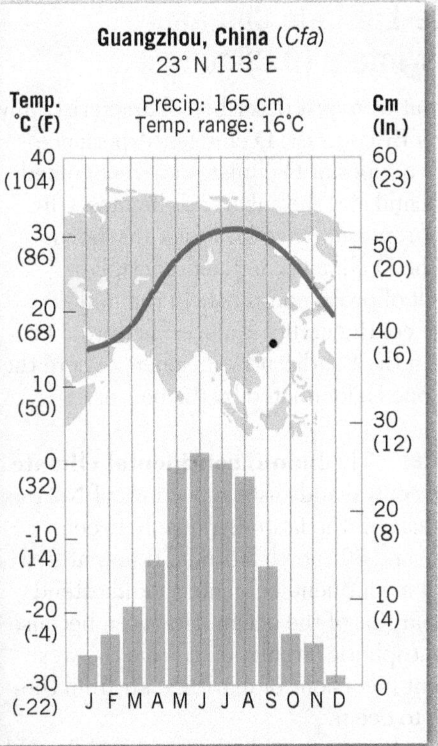

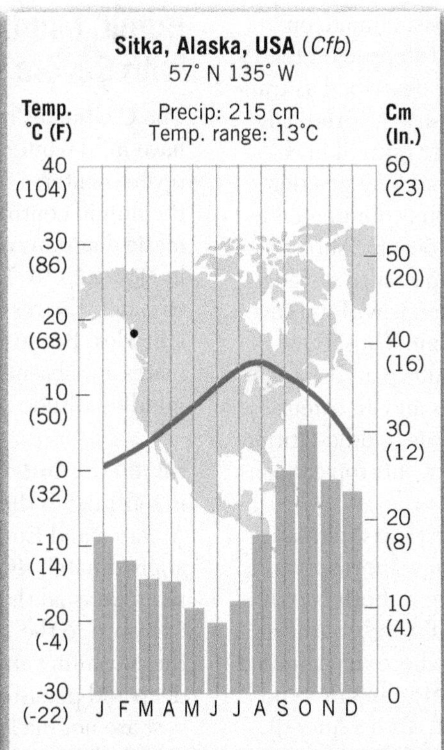

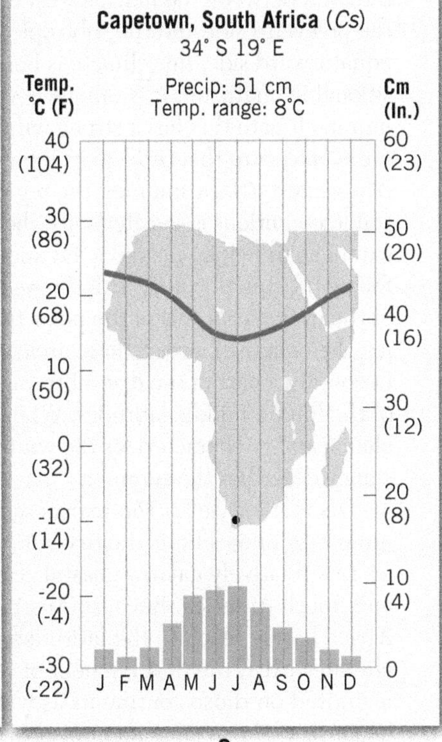

Figure 20.11 **Examples of C climates A.** Humid subtropical, **B.** marine west coast, and **C.** dry-summer subtropical.

world (see Figure 20.5). The climate diagram for Guangzhou, China (Figure 20.11A), shows a typical example. In summer, the humid subtropics experience hot, sultry weather of the type one expects to find in the rainy tropics. Daytime temperatures are generally high, and because both mixing ratio and relative humidity are high, the night brings little relief. An afternoon or evening thunderstorm is also possible, for these areas experience such storms on an average of 40 to 100 days each year, the majority during the summer months.

As summer turns to autumn, the humid subtropics lose their similarity to the rainy tropics. Although winters are mild, frosts are common in the higher-latitude Cfa areas and occasionally plague the tropical margins as well. The winter precipitation is also different in character from the summer. Some is in the form of snow, and most is generated along fronts of the frequent middle-latitude cyclones that sweep over these regions.

Marine West Coast Situated on the western (windward) side of continents, from about 40° to 65° north and south latitude, is a climate region dominated by the onshore flow of oceanic air (Figure 20.12). In North America, the **marine west coast climate** extends from near the U.S.–Canadian border northward as a narrow belt into southern Alaska (Figure 20.11B). The largest area of Cfb climate is found in Europe because there are no mountain barriers blocking the movement of cool maritime air from the North Atlantic.

The prevalence of maritime air masses means that mild winters and cool summers are the rule, as is an ample amount of rainfall throughout the year. Although there is no pronounced dry period, there is a drop in monthly precipitation totals during the summer. The reason for the reduced summer rainfall is the poleward migration of the oceanic subtropical highs. Although the areas of marine west coast climate are situated too far poleward to be dominated by these dry anticyclones, their influence is sufficient to cause a decrease in warm season rainfall.

Dry-Summer Subtropics The **dry-summer subtropical climate** is typically located along the west sides of continents between latitudes 30° and 45°.

Figure 20.12 **Marine west coast climate** Fog is common along the rocky Pacific coastline at Olympic National Park, Washington. As the name of this climate implies, the ocean exerts a strong influence. (Photo by Richard J. Green/Science Source)

Situated between the marine west coast climate on the poleward side and the subtropical steppes on the equatorward side, this climate is best described as transitional in character. It is unique because it is the only humid climate that has a strong winter rainfall maximum, a feature that reflects its intermediate position (Figure 20.11C). In summer, the region is dominated by stable conditions associated with the oceanic subtropical highs. In winter, as the wind and pressure systems follow the Sun equatorward, the region is within range of the cyclonic storms of the polar front. Thus, during the course of a year, these areas alternate between becoming a part of the dry subtropics and an extension of the humid middle latitudes. Whereas middle-latitude changeability characterizes the winter, subtropical constancy describes the summer.

As is the case for the marine west coast climate, mountain ranges limit the dry-summer subtropics to a relatively narrow coastal zone in both North and South America. Because Australia and southern Africa barely reach to the latitudes where dry-summer climates exist, the development of this climatic type is limited on those continents as well. Consequently, because of the arrangement of the continents and their mountain ranges, inland development occurs only in the Mediterranean basin. Here the zone of subsidence extends far to the east in summer; in winter, the sea is a major route of cyclonic disturbances. Because the dry-summer climate is particularly extensive in this region, the name *Mediterranean climate* is often used as a synonym.

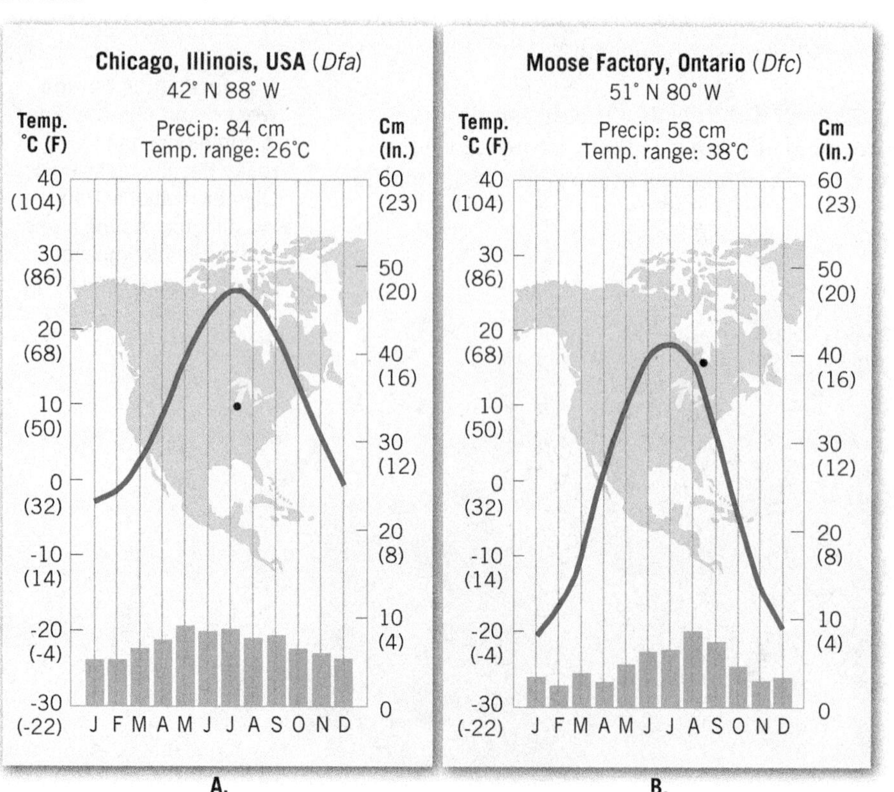

Figure 20.13
Examples of D climates Climates in this category are associated with the interiors of large landmasses in the mid- to high latitudes of the Northern Hemisphere. Winters can be harsh in Chicago's humid continental (Dfa) climate, and the subarctic environment (Dfc) of Moose Factory is more extreme.

Humid Middle-Latitude Climates with Severe Winters (D Climates)

The C climates that were just described characteristically have mild winters. By contrast, D climates experience severe winters. Two types of D climates are recognized: the humid continental and the subarctic climates. Climatic diagrams of representative locations are shown in Figure 20.13. The D climates are land-controlled climates, the result of broad continents in the middle latitudes. Because continentality is a basic feature, D climates are absent in the Southern Hemisphere, where the middle-latitude zone is dominated by the oceans.

Humid Continental The **humid continental climate** is confined to the central and eastern portions of North America and Eurasia, in the latitude range between approximately 40° and 50° north latitude. It may at first seem unusual that a continental climate should extend eastward to the margins of the ocean. However, because the prevailing atmospheric circulation is from the west, deep and persistent incursions of maritime air from the east are not likely to occur.

Both winter and summer temperatures in the humid continental climate can be characterized as relatively severe. Consequently, annual temperature ranges are high throughout the climate.

Precipitation is generally greater in summer than in winter. Precipitation totals generally decrease toward the interior of the continents, as well as from south to north, primarily because of increasing distance from the sources of maritime tropical (mT) air. Furthermore, the more northerly stations are also influenced for a greater part of the year by drier polar air masses.

Wintertime precipitation in humid continental climates is chiefly associated with the passage of fronts connected with traveling midlatitude cyclones. Part of this precipitation is in the form of snow, and the proportion of snow increases with latitude. Although precipitation is often considerably less during the cold season, it is usually more conspicuous than the greater amounts that fall during summer. An obvious reason is that snow remains on the ground, often for extended periods.

Subarctic Situated north of the humid continental climate and south of the polar tundra is an extensive **subarctic climate** region covering broad, uninterrupted expanses from western Alaska to Newfoundland in North America and from Norway to the Pacific coast of Russia in Eurasia (see Figure 20.5). It is often referred to as the **taiga** climate, for its extent closely corresponds to the northern coniferous forest region of the same name (Figure 20.14). Although scrawny, the spruce, fir, larch, and birch trees in the taiga represent the largest stretch of continuous forest on the surface of Earth.

Here in the source regions of continental polar air masses, the outstanding feature is certainly the

Figure 20.14 **Subarctic climate** The northern coniferous forest, also called the *taiga*, is associated with subarctic climates. This climate typically experiences the highest annual temperature ranges on Earth. This scene is in Quebec's Gaspé Peninsula. (Photo by Michael P. Gadomski/Science Source)

dominance of winter. Winter is long, and temperatures are bitterly cold. Winter minimum temperatures are among the lowest ever recorded outside the ice sheets of Greenland and Antarctica. In fact, for many years, the world's coldest temperature was attributed to Verkhoyansk in east-central Siberia, where the temperature dropped to −68°C (−90°F) on February 5 and 7, 1892. Over a 23-year period, this same station had an average monthly minimum of −62°C (−80°F) during January. Although exceptional temperatures, they illustrate the extreme cold that envelops the taiga in winter.

By contrast, summers in the subarctic are remarkably warm, despite their short duration. However, when compared with regions farther south, this short season must be characterized as cool. The extremely cold winters and relatively warm summers combine to produce the highest annual temperature ranges on Earth. Because these far northerly continental interiors are the source regions for cP air masses, there is very limited moisture available throughout the year. Precipitation totals are therefore small, with a maximum occurring during the warmer summer months.

CONCEPT CHECKS 20.5

1. Describe and explain the differences between summertime and wintertime precipitation in the humid subtropical climate (Cfa).
2. Why is the marine west coast climate (Cfb) represented by only slender strips of land in North and South America, and why is it very extensive in Western Europe?
3. The dry-summer subtropics are described as transitional. Explain why this is true.
4. Why is the humid continental climate confined to the Northern Hemisphere?
5. Describe and explain the annual temperature range you should expect in the realm of the taiga.

20.6 Polar (E) Climates

Contrast ice cap and tundra climates.

Polar climates are those in which the mean temperature of the warmest month is below 10°C (50°F). Thus, just as the tropics are defined by their year-round warmth, the polar realm is known for its enduring cold. As winters are periods of perpetual night, or nearly so, winter temperatures at most polar locations are understandably bitter. During the summer months temperatures remain cool despite the long days; the Sun is low in the sky, and its oblique rays give limited warmth. Although polar climates are classified as humid, precipitation is generally meager. Evaporation, of course, is also limited. The scanty precipitation totals are easily understood in view of the temperature characteristics of the region. The amount of water vapor in the air is always small because low mixing ratios must accompany low temperatures. Usually precipitation is most abundant during the warmer summer months, when the moisture content of the air is highest.

Two types of polar climates are recognized. The **tundra climate** (ET) is a treeless climate found almost exclusively in the Northern Hemisphere. Because of the combination of high latitude and continentality, winters are severe, summers are cool, and annual temperature ranges are high (**Figure 20.15A**). Furthermore, yearly precipitation is small, with a modest summer maximum.

The **ice cap climate** (EF) does not have a single monthly mean above 0°C (32°F) (**Figure 20.15B**). Consequently, because the average temperature for all months is below freezing, the growth of vegetation is prohibited, and the landscape is one of permanent ice and snow. This climate of perpetual frost covers a

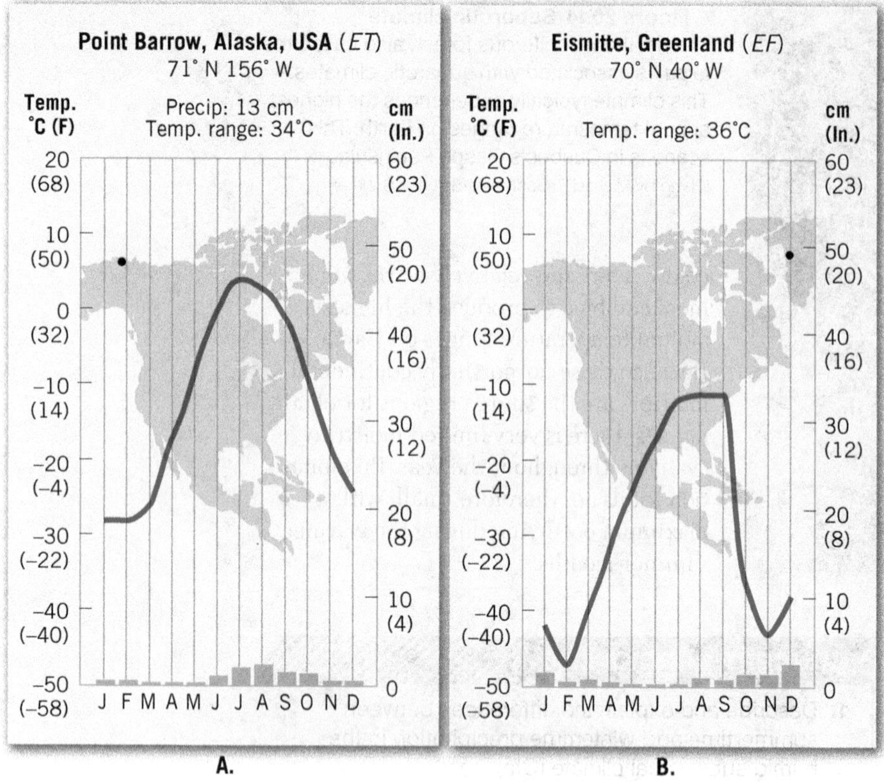

◀ SmartFigure 20.15 **Examples of E climates** These climatic diagrams represent the two basic types of polar climates. **A.** Barrow, Alaska, exhibits a tundra (ET) climate. **B.** Eismitte, Greenland, a station located on a massive ice sheet, is classified as an ice cap (EF) climate.

TUTORIAL
https://goo.gl/3WsHsP

surprisingly large area of more than 15.5 million square kilometers (6 million square miles), or about 9 percent of Earth's land area. Aside from scattered occurrences in high mountain areas, it is confined to the ice sheets of Greenland and Antarctica (see Figure 10.2, page 317).

CONCEPT CHECKS 20.6

1. Although polar regions experience extended periods of sunlight in the summer, temperatures remain cool. Explain.
2. Why are precipitation totals low in polar climates? Which season has the most precipitation? Why?
3. Where are EF climates most extensively developed?

20.7 Highland Climates

Summarize the characteristics associated with highland climates.

It is a well-known fact that mountains have climate conditions that are distinctly different from those found in adjacent lowlands. Compared to nearby places at lower elevations, sites with **highland climates** are cooler and usually wetter. Unlike the world climate types already discussed, which consist of large, relatively homogeneous regions, the outstanding characteristic of highland climates is the great diversity of climatic conditions that occur.

The best-known climatic effect of increased altitude is lower temperatures. In addition, an increase in precipitation due to orographic lifting usually occurs at higher elevations. Despite the fact that mountain stations are colder and often wetter than locations at lower elevations, highland climates are often very similar to those in adjacent lowlands in terms of seasonal temperature cycles and precipitation distribution. Figure 20.16 illustrates this relationship.

Phoenix, at an elevation of 338 meters (1109 feet), lies in the desert lowlands of southern Arizona. By contrast, Flagstaff is located at an altitude of 2100 meters (7000 feet), on the Colorado Plateau in northern Arizona. When summer averages climb to 34°C (93°F) in Phoenix, Flagstaff is experiencing a pleasant 19°C (66°F), which is a full 15°C (27°F) cooler. Although the temperatures at each city are quite different, the pattern of monthly temperature changes for each place is similar. Both experience their minimum and maximum monthly means in the same months. When precipitation data are examined, both places have a similar seasonal pattern, but the amounts at Flagstaff are higher in every month. In addition, because of its higher altitude, much of Flagstaff's winter precipitation is in the form of snow. By contrast, all of the precipitation at Phoenix is rain.

Because topographic variations are pronounced in mountains, every change in slope with respect to the Sun's rays produces a different microclimate. In the Northern Hemisphere, south-facing slopes are warmer and dryer because they receive more direct sunlight than north-facing slopes and deep valleys. Wind direction and speed in mountains can be highly variable. Mountains create various obstacles to winds. Locally, winds may be funneled through valleys or forced over ridges and around mountain peaks. When weather conditions are fair, mountain and valley breezes are created by the topography itself.* Perhaps the terms *variety* and *changeability* best describe mountain climates.

We know that climate strongly influences vegetation, which is the basis for the Köppen system. Thus, where there are vertical differences in climate, we should expect a vertical zonation of vegetation as well. By

*Mountain and valley breezes are described in the section "Local Winds" in Chapter 18. Also see Figure 18.20.

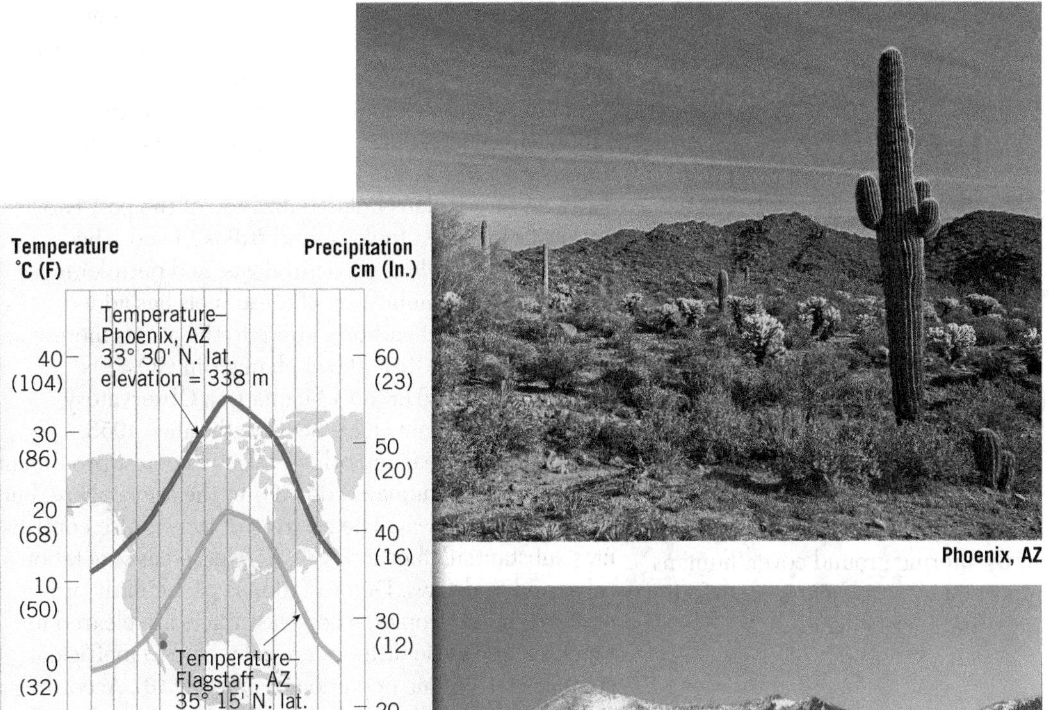

SmartFigure 20.16 Highland climate Diagrams for two stations in Arizona illustrate the general influence of elevation on climate. Flagstaff is cooler and wetter because of its position on the Colorado Plateau, nearly 1800 meters (6000 feet) higher than Phoenix. Only scanty drought-tolerant natural vegetation can survive in the hot, dry climate of southern Arizona, near Phoenix. (Photo by Design Pics/SuperStock) The natural vegetation associated with the cooler, wetter highlands near Flagstaff, Arizona, is much different from the desert lowlands. (Photo by Jim Cole/Alamy Stock Photo)

TUTORIAL
https://goo.gl/CNQwpV

ascending a mountain, we can view dramatic vegetation changes that otherwise might require a poleward journey of thousands of kilometers (see Figure 20.16). This occurs because altitude duplicates, in some respects, the influence of latitude on the distribution of vegetation.

CONCEPT CHECKS 20.7

1. The Arizona cities of Flagstaff and Phoenix are relatively close to one another yet have contrasting climates. Briefly describe the differences and why they occur.

20.8 Human Impact on Global Climate

Summarize the nature and cause of the atmosphere's changing composition since about 1750. Describe the climate's response.

Many processes can change Earth's climate. In Chapter 10, we examined the main factors that are thought to have caused ice ages, including tectonic factors and variations in Earth's orbit. Chapter 6 discussed how explosive volcanic eruptions can modify the atmosphere. Natural processes such as these will continue to affect Earth's climate in the future.

When relatively recent and future changes in our climate are considered, we must also examine the impact of human beings. In this section, we examine how humans contribute to global climate change. One impact largely results from the addition of carbon dioxide and other greenhouse gases to the atmosphere. A second impact is related to the addition of human-generated aerosols to the atmosphere.

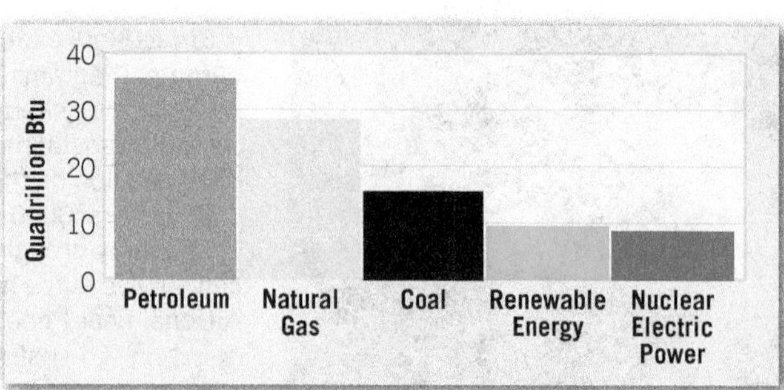

Figure 20.17 U.S. energy consumption The graph shows energy consumption in 2015. The total was 97.65 quadrillion Btu. A quadrillion is 10 raised to the 15th power and is equal to about the amount of energy in 45 million tons of coal, or 1 trillion cubic feet of natural gas, or 170 million barrels of crude oil. The burning of fossil fuels represents about 81.3 percent of the total. (Data from U.S. Energy Information Administration)

Human influence on regional and global climate did not just begin with the onset of the modern industrial period. People have been modifying the environment over extensive areas for thousands of years. The use of fire and the overgrazing of marginal lands by domesticated animals have both reduced the abundance and distribution of vegetation. By altering ground cover, humans have modified such important climate factors as surface albedo, evaporation rates, and surface winds.

Rising CO_2 Levels

In Chapter 16 you learned that carbon dioxide (CO_2) represents only about 0.0405 percent (405 parts per million) of the gases that make up clean, dry air. Nevertheless, it is a very significant component meteorologically. Carbon dioxide is influential because it is transparent to incoming short-wavelength solar radiation, but it is not transparent to some of the longer-wavelength outgoing Earth radiation. A portion of the energy leaving the ground is absorbed by atmospheric CO_2. This energy is subsequently re-emitted, part of it back toward the surface, thereby keeping the air near the ground warmer than it would be without CO_2. Thus, along with water vapor, carbon dioxide is largely responsible for the *greenhouse effect* of the atmosphere.

The tremendous industrialization of the past two centuries has been fueled—and still is fueled—by burning fossil fuels: coal, natural gas, and petroleum (**Figure 20.17**). Combustion of these fuels has added great quantities of carbon dioxide to the atmosphere. Figure 16.6 (page 497) shows changes in CO_2 concentrations at Hawaii's Mauna Loa Observatory, where measurements have been made since 1958.

The use of coal and other fuels is the most prominent means by which humans add CO_2 to the atmosphere, but it is not the only way. The clearing of forests also contributes substantially because CO_2 is released as vegetation is burned or decays. Deforestation is particularly pronounced in the tropics, where vast tracts are cleared for ranching and agriculture or are subjected to inefficient commercial logging operations (**Figure 20.18**). According to United Nations estimates, nearly 10.2 million hectares (25.1 million acres) of tropical forest were permanently destroyed each year during the 1990s. Between the years 2000 and 2005, the average figure increased to 10.4 million hectares (25.7 million acres) per year.

Some of the excess CO_2 is taken up by plants or is dissolved in the ocean. It is estimated that about 45 percent remains in the atmosphere. **Figure 20.19** is a graphic record of changes in atmospheric CO_2 extending

EYE ON EARTH 20.2

Amundsen–Scott South Pole Station is an American research facility. Among the phenomena that are monitored there are variations in atmospheric composition. The accompanying graph shows changes in the air's CO_2 content at South Pole Station (90° south latitude) and at a similar facility at Barrow, Alaska (71° north latitude).

QUESTION 1 *Describe how the two lines on the graph differ.*

QUESTION 2 *Which line on the graph represents the South Pole, and which represents Barrow, Alaska?*

QUESTION 3 *Explain how you were able to determine which line is which.*

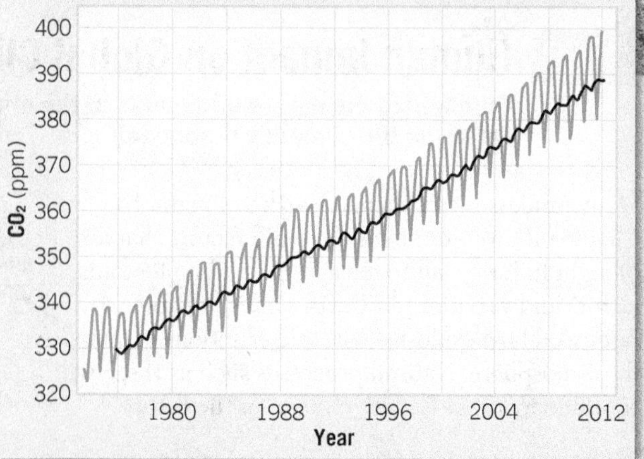

Figure 20.18 **Tropical deforestation** Clearing of the tropical rain forest is a serious environmental issue. In addition to causing a loss of biodiversity, it is a significant source of carbon dioxide. Fires are frequently used to clear the land. This scene is in Brazil's Amazon basin. (Photo by Nigel Dickinson/Alamy)

back 800,000 years. Over this long span, natural fluctuations have varied from about 180 to 300 ppm. As a result of human activities, the present CO_2 level is more than 30 percent higher than its highest level over the past 800,000 years. The rapid increase in CO_2 concentrations since the onset of industrialization is obvious. The annual rate at which atmospheric CO_2 concentrations are growing has been increasing over the past several decades.

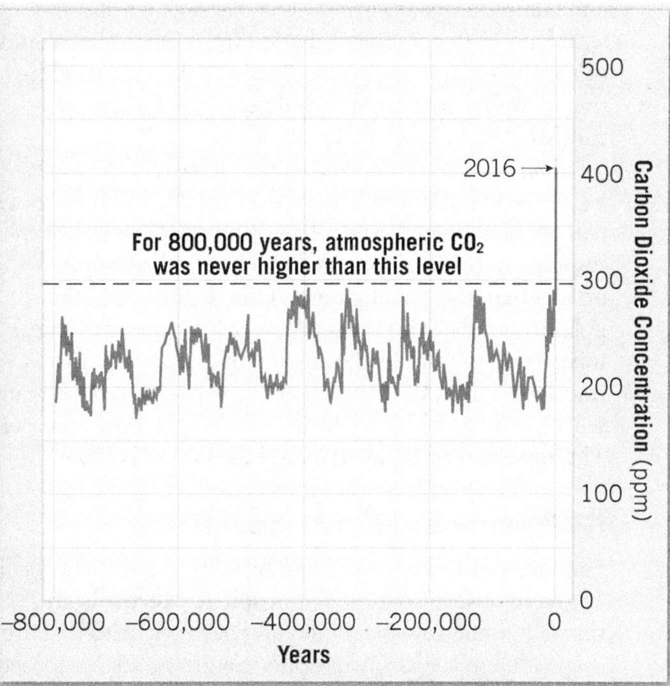

Figure 20.19 **CO_2 concentrations over the past 800,000 years** Most of these data come from the analysis of air bubbles trapped in ice cores. The record since 1958 comes from direct measurements at Mauna Loa Observatory, Hawaii. The rapid increase in CO_2 concentrations since the onset of the Industrial Revolution is obvious. (Based on data from NOAA)

The Atmosphere's Response

Given the increase in the atmosphere's CO_2 content, have global temperatures actually increased? The answer is yes. According to a 2013 report by the Intergovernmental Panel on Climate Change (IPCC), "Warming of the climate system is unequivocal, as is now evident from observations of increases in global average air and ocean temperatures, widespread melting of snow and ice, and rising global sea level."* Most of the observed increase in global average temperatures since the mid-twentieth century is *extremely likely* due to the observed increase in human-generated greenhouse gas concentrations. As used by the IPCC, *extremely likely* indicates a probability of 95 to 100 percent. The planet's average surface air temperature has risen about 1°C (1.8°F) since the late nineteenth century, with most of the warming occurring in the past 35 years. All but 1 of the 16 warmest years on record have occurred since 2001. Surface air temperatures in 2015 were the warmest since record keeping began in 1880, shattering the mark set in 2014 (Figure 20.20A).

Weather patterns and other natural cycles cause fluctuations in average temperatures from year to year. This is especially true on regional and local levels. For example, while the globe experienced notably warm temperatures in 2014, parts of the continental United States were cooler than normal. By contrast, 2014 was the warmest year on record for much of Europe and parts of Russia, and ocean temperatures were at a record high. Regardless of regional differences in any year, increases in greenhouse gas levels are causing a long-term rise in global temperatures. Although each calendar year will not necessarily be warmer than the one before, scientists expect each decade to be warmer than the previous one. An examination of the decade-by-decade temperature trend in Figure 20.20B bears this out.

What about the future? Projections for the years ahead depend in part on the quantities of greenhouse gases that are emitted. Figure 20.21 shows the best estimates of global warming for several different scenarios. The 2013 IPCC report states that if there is a doubling of the pre-industrial level of carbon dioxide (280 ppm) to 560 ppm, the *likely* temperature increase will be in the range of 2° to 4.5°C (3.5° to 8.1°F). The increase is *very unlikely* (1 to 10 percent probability) to be less than 1.5°C (2.7°F), and values higher than 4.5°C (8.1°F) are possible.

*IPCC, "Summary for Policymakers," in *Climate Change 2013: The Physical Science Basis*. The Intergovernmental Panel on Climate Change is an authoritative group of scientists that provides advice to the world community through periodic reports that assess the state of knowledge of the causes and effects of climate change.

Figure 20.20 **Global Temperatures A.** The world map shows how average temperatures in 2015 deviated from the mean for the 1951–1980 base period. The high latitudes in the Northern Hemisphere clearly stand out. (NASA/GISS) **B.** The decade-by-decade temperature trend since 1950 is illustrated in this bar graph. Continued increases in the atmosphere's greenhouse gas levels are driving a long-term increase in global temperatures. Each calendar year is not necessarily warmer than the year before, but, since 1950, each decade has been warmer than the previous one. (Data from NASA)

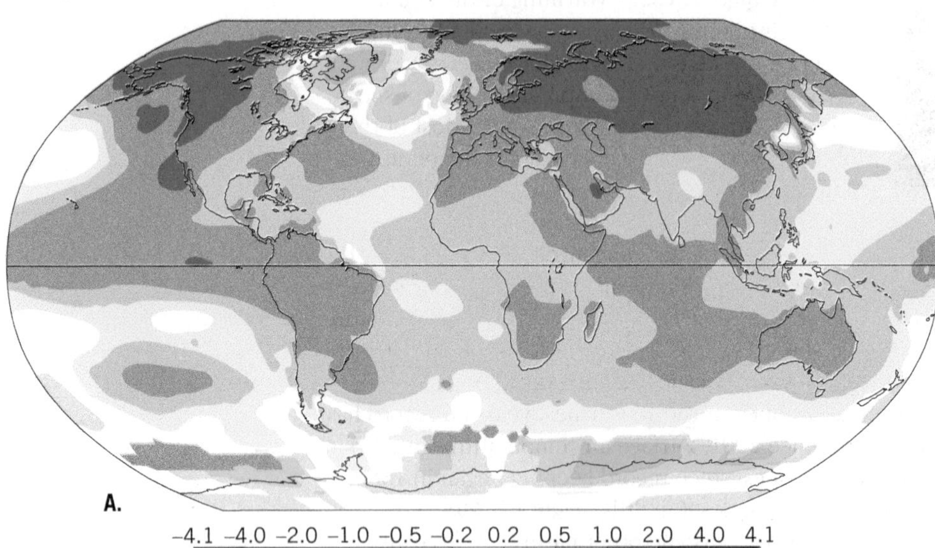

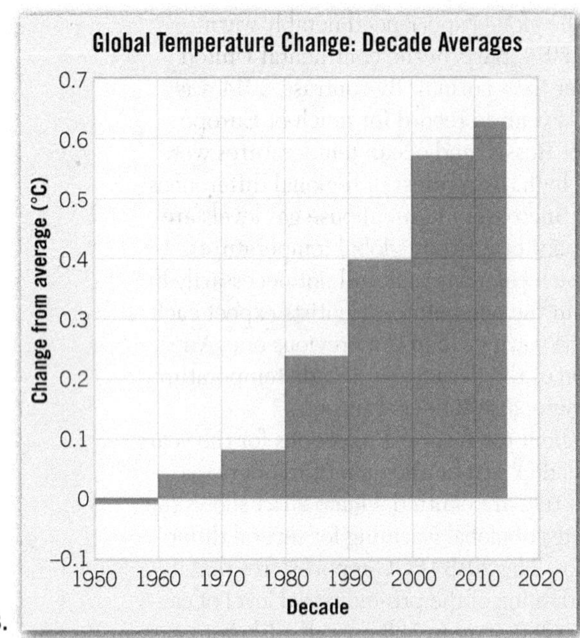

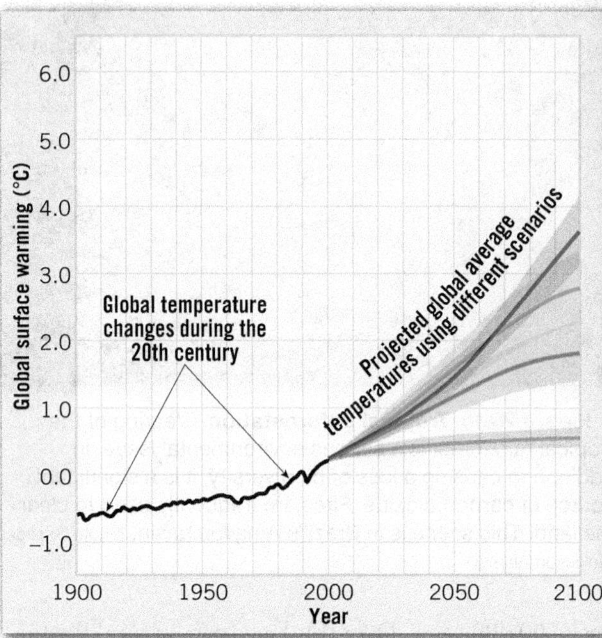

Figure 20.21 **Temperature projections to 2100** The right half of the graph shows projected global warming based on different emissions scenarios. The shaded zone adjacent to each colored line shows the uncertainty range for each scenario. The basis for comparison (0.0 on the vertical axis) is the global average for the period 1980 to 1999. The orange line represents the scenario in which CO_2 concentrations were held constant at values for the year 2000. (NOAA)

The Role of Trace Gases

Carbon dioxide is not the only gas that contributes to the global increase in temperature. Industrial and agricultural activities of people are causing a buildup of several trace gases that also play significant roles. The substances are called *trace gases* because their concentrations are much lower than the concentration of carbon dioxide. The trace gases that are most important are methane (CH_4), nitrous oxide (N_2O), hydrofluorocarbons (HFCs), and chlorofluorocarbons (CFCs). These gases absorb wavelengths of outgoing radiation from Earth that would otherwise escape into space (**Figure 20.22**). These additional human-generated greenhouse gases add significantly to the impact of CO_2 alone.

Sophisticated computer models show that the warming of the lower atmosphere caused by CO_2 and trace gases will not be the same everywhere. Rather, the temperature response in polar regions could be two to three times greater than the global average. One reason is that the polar troposphere is very stable, which suppresses vertical mixing and thus limits the amount of surface heat that is transferred upward. In addition, the expected reduction in sea ice will contribute to the greater temperature increase. This topic will be explored more fully in Section 20.9.

How Aerosols Influence Climate

Humans can influence climate by adding aerosols as well as greenhouse gases to the atmosphere. **Aerosols** are tiny, often microscopic, liquid and solid particles that are suspended in the air. Atmospheric aerosols are composed of many different materials, including soil, smoke, sea salt, and sulfuric acid. Natural sources are numerous and include such phenomena as dust storms and volcanoes.

Most human-generated aerosols come from the sulfur dioxide emitted during the combustion of fossil fuels and as a consequence of burning vegetation to clear agricultural land. Chemical reactions in the atmosphere

▲ Figure 20.22 **Methane** Methane is produced by anaerobic bacteria in wet places, where oxygen is scarce. (*Anaerobic* means "without air," specifically oxygen.) Such places include swamps, bogs, wetlands, and the guts of termites and grazing animals such as cattle and sheep. Methane is also generated in flooded paddy fields ("artificial swamps") used for growing rice. Coal mining and drilling for oil and natural gas are other sources of methane. (Photo by Robert Harding/SuperStock)

convert the sulfur dioxide into sulfate aerosols, the same material that produces acid precipitation. The satellite images in Figure 20.23 provide an example.

How do aerosols affect climate? Aerosols act directly by reflecting sunlight back to space and indirectly by making clouds "brighter" reflectors. The second effect relates to the fact that many aerosols (such as those composed of salt or sulfuric acid) attract water and thus are especially effective as cloud condensation nuclei. The large quantity of aerosols produced by human activities (especially industrial emissions) trigger an increase in the number of cloud droplets that form within a cloud. A greater number of small droplets increases the cloud's brightness, causing more sunlight to be reflected back to space.

One category of aerosols, called **black carbon**, is soot generated by combustion processes and fires. Unlike most other aerosols, black carbon warms the atmosphere because it is an effective absorber of incoming solar radiation. In addition, when deposited on snow and ice, black carbon reduces surface albedo, thus increasing the amount of light absorbed. Nevertheless, despite the warming effect of black carbon, the overall effect of atmospheric aerosols is to cool Earth.

Studies indicate that the cooling effect of human-generated aerosols offsets a portion of the global warming caused by the growing quantities of greenhouse gases in the atmosphere. The magnitude and extent of the cooling effect of aerosols is uncertain. This uncertainty is a significant hurdle in advancing our understanding of how humans alter Earth's climate.

It is important to point out some significant differences between global warming by greenhouse gases and aerosol cooling. After being emitted, greenhouse gases, such as carbon dioxide, remain in the atmosphere for many decades. By contrast, aerosols released into the troposphere remain there for only a few days or, at most, a few weeks before they are "washed out" by precipitation. Because of their short lifetime in the troposphere, aerosols are distributed unevenly over the globe. As expected, human-generated aerosols are concentrated near the areas that produce them—namely industrialized regions that burn fossil fuels and land areas where vegetation is burned.

CONCEPT CHECKS 20.8

1. Why has the CO_2 level of the atmosphere been increasing over the past 200 years?
2. How has the atmosphere responded to the growing CO_2 levels? How are temperatures in the lower atmosphere likely to change as CO_2 levels continue to increase?
3. Aside from CO_2, what trace gases are contributing to global temperature change?
4. List the main sources of human-generated aerosols and describe their net effect on atmospheric temperatures.

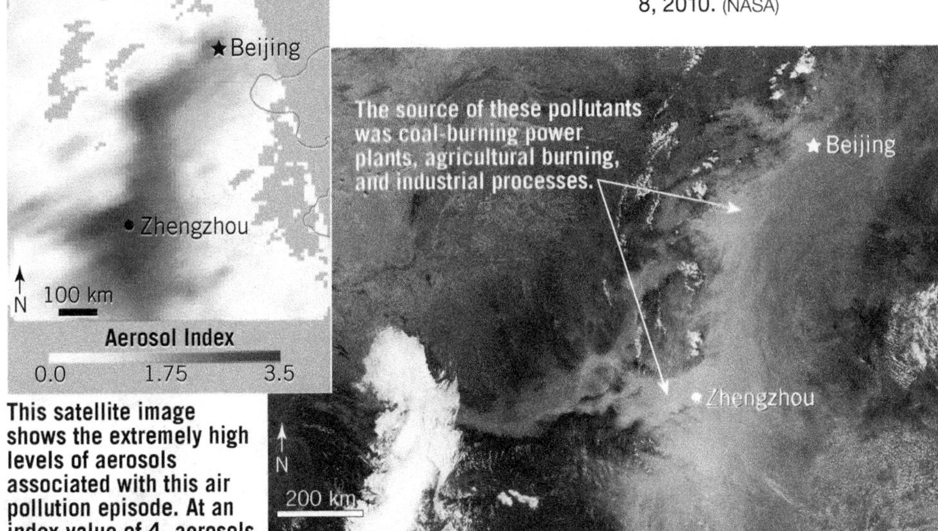

▼ Figure 20.23 **Human-generated aerosols** These satellite images show a serious air pollution episode that plagued China on October 8, 2010. (NASA)

20.9 Climate-Feedback Mechanisms

Contrast positive- and negative-feedback mechanisms and provide examples of each.

Climate is a very complex interactive physical system. Thus, when any component of the climate system is altered, scientists must consider many possible responses, some of which amplify the initial effect and some of which counterbalance it. These possible responses are called **climate-feedback mechanisms**. They complicate climate-modeling efforts and add greater uncertainty to climate predictions.

Types of Feedback Mechanisms

What climate-feedback mechanisms are related to carbon dioxide and other greenhouse gases? One important mechanism is that warmer surface temperatures increase evaporation rates. This in turn increases the atmosphere's water vapor content. Remember that water vapor is a powerful absorber of radiation emitted by Earth. Therefore, an increase in atmospheric water vapor tends to reinforce the temperature increase caused by carbon dioxide and the trace gases.

Recall that the future temperature increase at high latitudes may be two to three times greater than the global average. This projection is based in part on the likelihood that the area covered by sea ice will decrease as surface temperatures rise. Because ice reflects a much larger percentage of incoming solar radiation (has a higher albedo) than does open water, its melting results in a substantial increase in the solar energy absorbed at the surface (**Figure 20.24**). This in turn magnifies the temperature increase created by higher levels of greenhouse gases.

The climate-feedback mechanisms just described magnify the temperature rise caused by the buildup of carbon dioxide. Because these effects reinforce the initial change, they are called **positive-feedback mechanisms**. Conversely, **negative-feedback mechanisms** produce results that are just the opposite of the initial change and tend to offset it.

Clouds present an interesting example because they have both negative and positive feedback effects on temperature. One probable result of a global temperature rise is an increase in cloud cover due to the higher moisture content of the atmosphere. Most clouds are good reflectors of solar radiation, so increasing cloudiness should reduce the amount of solar energy available to heat the atmosphere—a negative-feedback effect. However, clouds act as a positive-feedback mechanism by absorbing and emitting radiation from Earth's surface that would otherwise be lost from the troposphere.

Which effect, if either, is stronger? Observations and modeling demonstrate that the overall effect of clouds is to slightly cool the planet. This means that the warming caused by the buildup of greenhouse gases is slightly offset by the resulting increase in cloud cover.

Global warming caused by human-induced changes in atmospheric composition continues to be one of the most-studied aspects of climate change. Although no models yet incorporate the full range of potential factors and feedbacks, there is a strong scientific consensus that the increasing levels of atmospheric carbon dioxide and trace gases is creating a warmer planet with a different distribution of climate regimes.

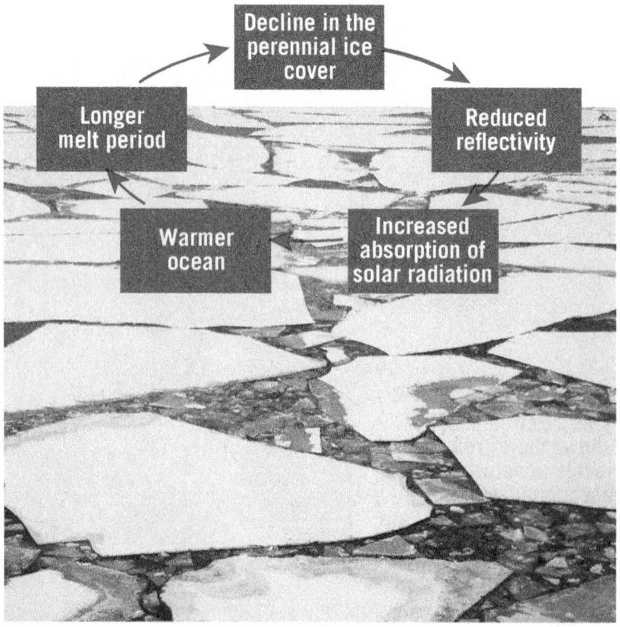

Figure 20.24 Sea ice as a feedback mechanism The image shows the springtime breakup of sea ice near Antarctica. The diagram shows a likely feedback loop. A reduction in sea ice acts as a positive-feedback mechanism because surface albedo decreases, and the amount of energy absorbed at the surface increases. (Photo by Radius Images/Alamy Stock Photo)

Computer Models of Climate: Important yet Imperfect Tools

Earth's climate system is amazingly complex. Comprehensive state-of-the-science climate simulation models are among the basic tools used to develop possible climate-change scenarios. Called *general circulation models* (*GCMs*), they are based on fundamental laws of physics and chemistry and incorporate human and biological interactions. GCMs are used to simulate many variables, including temperature, rainfall, snow cover, soil moisture, winds, clouds, sea ice, and ocean circulation over the entire globe through the seasons and over spans of decades.

In many other fields of study, hypotheses can be tested by direct experimentation in the laboratory or by field observations and measurements. However, this is often not possible in the study of climate. Rather, scientists must construct computer models of how our planet's climate system works. If we understand the climate system correctly and construct a model appropriately, then the behavior of the model climate system should mimic the behavior of Earth's climate system (**Figure 20.25**).

What factors influence the accuracy of climate models? Clearly, mathematical models are *simplified* versions of the real Earth and cannot capture its full complexity, especially at smaller geographic scales. Moreover, computer models used to simulate future climate change must make many assumptions that significantly influence predictions. They must consider a wide range of possible changes in population, economic growth, fossil fuel consumption, technological development, improvements in energy efficiency, and more.

Despite many obstacles, our ability to use supercomputers to simulate climate continues to improve. Although today's models are far from infallible, they are powerful tools for understanding what Earth's future climate might be like.

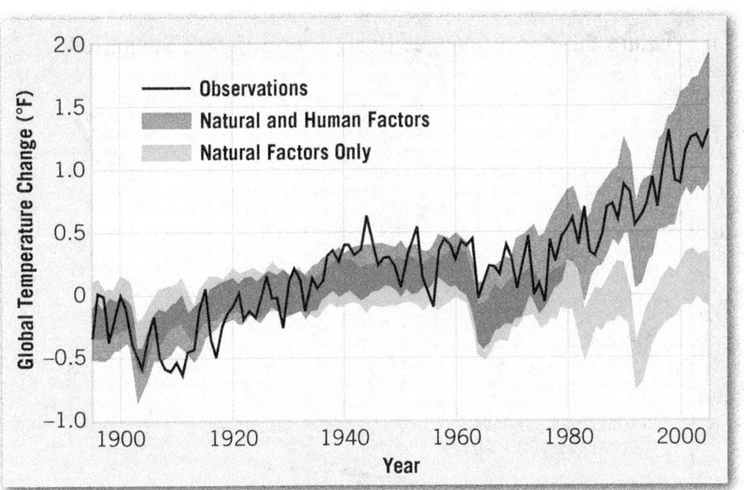

Figure 20.25 Separating human and natural influences on climate The blue band shows how global average temperatures would have changed over the past century due to natural forces only, as simulated by computer models of climate. The red band is the model's simulation of the effects of human and natural forces combined. The black line shows actual observed global average temperatures. As the blue band indicates, without human influences, temperatures over the past century would actually have first warmed and then cooled slightly over recent decades. Bands of color are used to express the range of uncertainty. (U.S. Global Change Research Program)

CONCEPT CHECKS 20.9

1. Distinguish between positive and negative climate-feedback mechanisms.
2. Provide at least one example of each type of feedback mechanism.
3. List some factors that influence the accuracy of computer models of climate.

20.10 Some Possible Consequences of Global Warming

Summarize some of the possible consequences of global warming.

What consequences can be expected as the carbon dioxide content of the atmosphere reaches a level that is twice what it was early in the twentieth century? Because the climate system is complex, predicting the occurrence of specific effects in particular places is speculative. It is not yet possible to pinpoint such changes. Nevertheless, there are plausible scenarios for larger scales of space and time.

As noted, the magnitude of the temperature increase will not be the same everywhere. The temperature rise will probably be smallest in the tropics and increase toward the poles. As for precipitation, the models indicate that some regions will experience significantly more precipitation and runoff. However, others will experience a decrease in runoff due to reduced precipitation or greater evaporation caused by higher temperatures.

Table 20.1 lists possible effects of global warming based on the IPCC's projections for the late twenty-first century, ranked in decreasing order of certainty. Probabilities are based on the quality, volume, and consistency of the evidence and the extent of agreement among scientists. The risk associated with each of the projections in Table 20.1 is a combination of the probability of occurrence and the severity of the damage if it were to occur. This means that even projections labeled "unlikely" should not be ignored because if any of those events were to happen, the consequences would be extremely serious.

Sea-Level Rise

A significant impact of human-induced global warming is a rise in sea level. As this occurs, coastal cities, wetlands, and low-lying islands will be threatened with more frequent flooding, increased shoreline erosion, and saltwater encroachment into coastal rivers and aquifers.

How is a warmer atmosphere related to a rise in sea level? One significant factor is thermal expansion. Higher air temperatures warm the adjacent upper layers of the ocean, which in turn causes the water to expand and sea level to rise.

A second factor contributing to global sea-level rise is melting glaciers. With few exceptions, glaciers around the world have been retreating at unprecedented rates over the past century. Some mountain glaciers have disappeared altogether. A satellite study spanning 20 years showed that the mass of the Greenland and Antarctic Ice Sheets dropped an average of 475 gigatons per year. (A gigaton is 1 billion metric tons.) That is enough water to raise sea level 1.5 millimeters (0.05 inch) per year. The loss of ice was not steady but was occurring at an accelerating rate during the study period. During the same span, mountain glaciers and ice caps lost an average of slightly more than 400 gigatons per year.

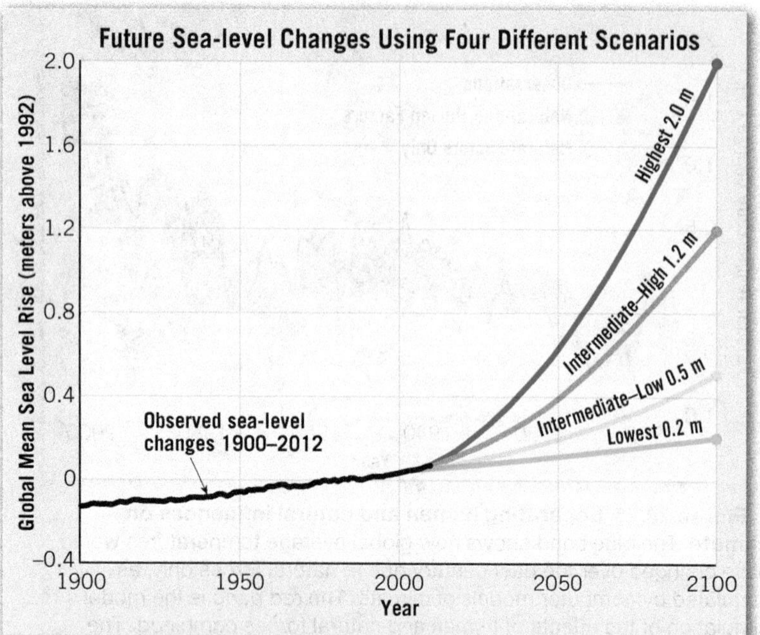

▲ **SmartFigure 20.26 Rising sea level** This graph shows changes in sea level between 1900 and 2012 and projections to 2100, using four different scenarios. Currently the highest and lowest projections are considered to be extremely unlikely. The greatest uncertainty surrounding estimates is the rate and magnitude of ice sheet loss from Greenland and Antarctica. Zero on the graph represents mean sea level in 1992. (NASA)

VIDEO
https://goo.gl/vADY1r

Research indicates that sea level has risen about 25 centimeters (9.75 inches) since 1870, with the rate of sea-level rise accelerating in recent years. What about future changes in sea level? As **Figure 20.26** indicates, the estimates of future sea-level rise are uncertain. The four scenarios depicted on the graph represent estimates based on different degrees of ocean warming and ice sheet loss and range from 0.2 meter (8 inches) to 2 meters (6.6 feet). The lowest scenario is an extrapolation of the annual rate of sea-level rise that occurred between 1870 and 2000 (1.7 millimeter/year). However, when the rate of sea-level rise for the period 1993 to 2012 is examined, the annual change is 3.17 millimeters/year. Such data show that there is a reasonable chance that sea level will rise considerably more than the lowest scenario indicates.

Scientists realize that even modest rises in sea level along a gently sloping shoreline, such as the Atlantic and Gulf coasts of the United States, will lead to significant erosion and severe permanent inland flooding (**Figure 20.27**). If this happens, many beaches and wetlands will disappear, and coastal civilization will be severely disrupted. Low-lying and densely populated places such as Bangladesh and the small island nation of the Maldives are especially vulnerable. The average elevation in the Maldives is 1.5 meters (less than 5 feet), and its highest point is just 2.4 meters (less than 8 feet) above sea level.

TABLE 20.1 IPCC Projections for the Late Twenty-First Century

Projection	Probability
• Cold days and nights will be warmer and less frequent over most land areas • Hot days and nights will be warmer and more frequent over most land areas • The extent of permafrost will decline • Ocean acidification will increase as the atmosphere accumulates CO_2 • Northern Hemisphere glaciation will not initiate before the year 3000 • Global mean sea level will rise and continue to do so for many centuries	Virtually certain (99–100%)
• Arctic sea ice cover will continue to shrink and thin, and Northern Hemisphere spring snow cover will decrease • The dissolved oxygen content of the ocean will decrease by a few percent • The rate of increase in atmospheric CO_2, methane, and nitrous oxide will reach levels unprecedented in the past 10,000 years • The frequency of warm spells and heat waves will increase • The frequency of heavy precipitation events will increase • Precipitation amounts will increase in high latitudes • The ocean's conveyor-belt circulation will weaken • The rate of sea-level rise will exceed that of the late twentieth century • Extreme high sea-level events will increase, as will ocean wave heights of midlatitude storms	Very likely (90–100%)
• If the atmospheric CO_2 level stabilizes at double the present level, global temperatures will rise by between 1.5°C (2.7°F) and 4.5°C (8.1°F) • Areas affected by drought will increase • Precipitation amounts will decline in the subtropics • The loss of glaciers will accelerate in the next few decades	Likely (66–100%)
• Intense tropical cyclone activity will increase • The West Antarctic Ice Sheet will pass the melting point if global warming exceeds 5°C (9°F)—this is relative, not absolute	About as likely as not (33–66%)
• Antarctic and Greenland Ice Sheets will collapse due to surface warming	← Not likely (0–33%)

Probability (%): 0 10 20 30 40 50 60 70 80 90

Where the slope is gentle a small rise in sea level causes a substantial shift
Original shoreline
Shoreline shift
Sea level rise

SmartFigure 20.27 **Slope of the shoreline** The slope of a shoreline is critical to determining the degree to which sea-level changes will affect it. As sea level gradually rises, the shoreline retreats, and structures that were once thought to be safe from wave attack become vulnerable.

TUTORIAL
http://goo.gl/u5fVXR

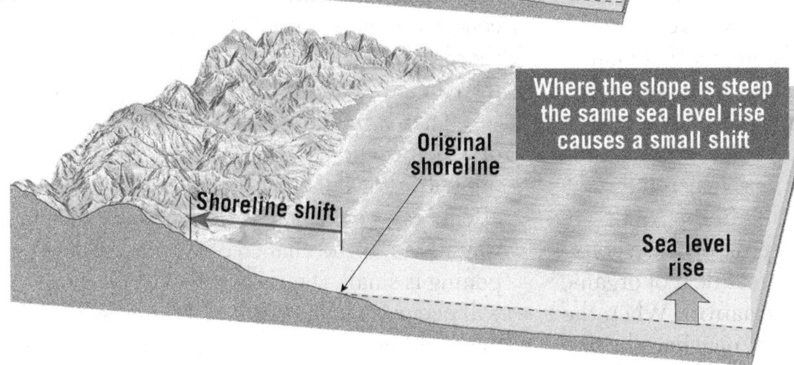

Where the slope is steep the same sea level rise causes a small shift
Original shoreline
Shoreline shift
Sea level rise

Because rising sea level is a gradual phenomenon, coastal residents may overlook it as an important contributor to shoreline flooding and erosion problems. Rather, the blame may be assigned to other forces, especially storm activity. Although a given storm may be the immediate cause, the magnitude of its destruction may result from the relatively small sea-level rise that allowed the storm's power to cross a much greater land area.

The Changing Arctic

The effects of global warming are most pronounced in the high latitudes of the Northern Hemisphere. For more than 30 years, the extent and thickness of sea ice have been rapidly declining. In addition, permafrost temperatures have been rapidly rising, and the area affected by permafrost has been decreasing. Meanwhile, alpine glaciers and the Greenland Ice Sheet have been shrinking. Another sign that the Arctic is rapidly warming is related to plant growth (Figure 20.28). A 2013 study showed that vegetation growth at northern latitudes now resembles that which characterized areas 4° to 6° of latitude farther south as recently as 1982. That is a distance of 400 to 700 kilometers (250 to 430 miles). One researcher characterized the finding this way: "It's like Winnipeg, Manitoba, moving to Minneapolis-St. Paul in only 30 years."

Arctic Sea Ice Climate models are in general agreement that one of the strongest signals of global warming should be a loss of sea ice in the Arctic. This is indeed occurring. The map in Figure 14.3A (page 441) compares the average sea ice extent for September 2015 to the long-term September average for the period 1981–2010. (September represents the end of the melt period, when the area covered by sea ice is at a minimum.) On that date the extent was about 4.4 million square kilometers (1.7 million square miles)—the fourth lowest minimum of the satellite era, which began in 1979. Figure 14.3B (page 441), which shows year-to-year changes, clearly depicts the trend. Not only is the area covered by sea ice declining but the remaining sea ice has become thinner, making it more vulnerable to further melting.

Models that best match historical trends project that Arctic waters may be virtually ice-free in the late summer by the 2030s. As was noted earlier in this chapter, a reduction in sea ice represents a positive-feedback mechanism that reinforces global warming.

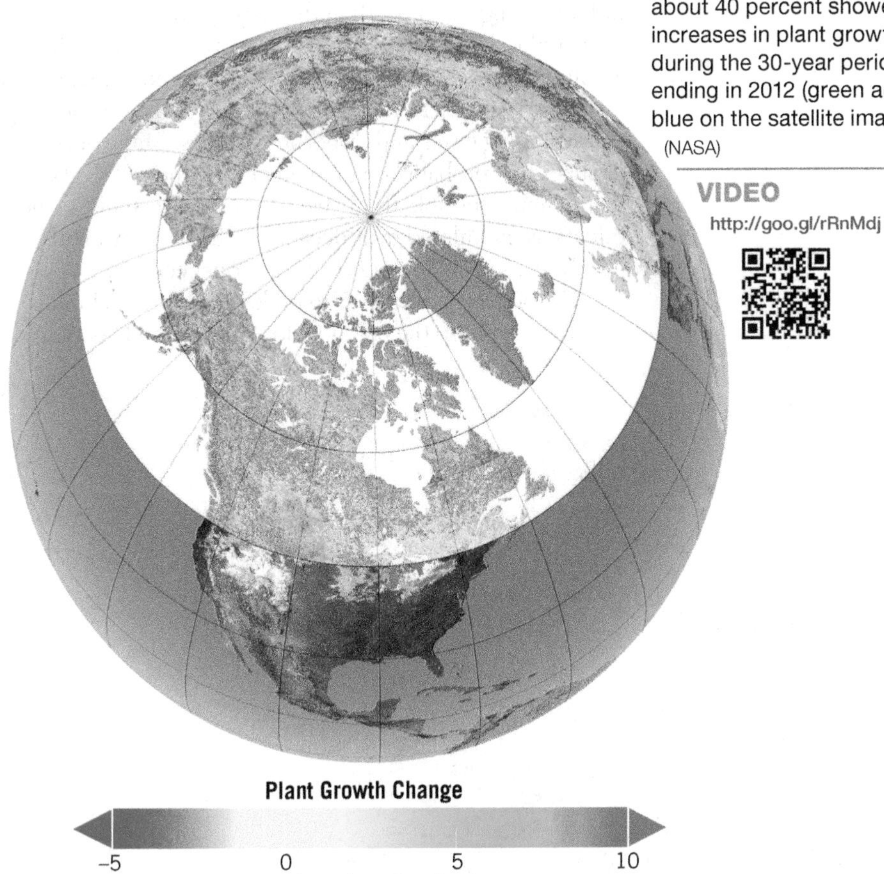

Plant Growth Change
-5
0
5
10
Percent per decade

SmartFigure 20.28 **Climate change spurs plant growth beyond 45° north** Of the 26 million square kilometers (10 million square miles) of northern vegetated lands, about 40 percent showed increases in plant growth during the 30-year period ending in 2012 (green and blue on the satellite image). (NASA)

VIDEO
http://goo.gl/rRnMdj

Permafrost Chapter 8 included a brief discussion of permanently frozen ground called *permafrost* that occurs in large portions of the high latitudes of the Northern Hemisphere. Evidence indicates that the extent of permafrost in the Northern Hemisphere has decreased over the past decade, as would be expected under long-term warming conditions.

Studies in Alaska show that thawing is occurring in interior and southern parts of the state, where permafrost temperatures are near the thaw point. As Arctic temperatures continue to rise, some models project that near-surface permafrost may be lost entirely from large parts of Alaska by the end of the century.

Thawing permafrost represents a potentially significant positive-feedback mechanism that may reinforce global warming. When vegetation dies in the Arctic, cold temperatures inhibit its decomposition. As a consequence, over thousands of years, a great deal of organic matter has become stored in the permafrost. When the permafrost thaws, organic matter that may have been frozen for millennia comes out of "cold storage" and decomposes. The result is the release of carbon dioxide and methane—greenhouse gases that contribute to global warming. Thus, like decreasing sea ice, thawing permafrost is a positive-feedback mechanism.

The Potential for "Surprises"

You have seen that climate in the twenty-first century, unlike during the preceding thousand years, is not expected to be stable. Rather, change is occurring. The amount and rate of future climate shifts depend primarily on current and future human-caused emissions of heat-trapping gases and airborne particles. Many of the changes will probably be gradual environmental shifts, nearly imperceptible from year to year. Nevertheless, the effects, accumulated over decades, will have powerful economic, social, and political consequences.

Despite our best efforts to understand future climate shifts, there is also the potential for "surprises." This simply means that due to the complexity of Earth's climate system, we might experience relatively sudden, unexpected changes or see some aspects of climate shift in unanticipated ways. Many future climate scenarios predict steadily changing conditions, giving the impression that humanity will have time to adapt. However, the scientific community has indicated that at least some changes will be abrupt, perhaps crossing a threshold, or "tipping point," so quickly that there will be little time to react. This is a reasonable concern because abrupt changes occurring over periods as short as decades or even years have been a natural part of the climate system throughout Earth history.

There are many examples of potential surprises, each of which would have large consequences. We simply do not know how far the climate system or other systems it affects can be pushed before they respond in unexpected ways. Even if the chance of any particular surprise happening is small, the chance that at least one such surprise will occur is much greater. In other words, although we may not know which of these events will occur, it is likely that one or more will eventually occur.

The impact on climate of an increase in atmospheric CO_2 and trace gases is obscured by some uncertainties. Yet climate scientists continue to improve our understanding of the climate system and the potential impacts and effects of global climate change. Policymakers are confronted with responding to the risks posed by greenhouse gas emissions, knowing that our understanding is imperfect. They must also face the fact that climate-induced environmental changes cannot be reversed quickly, if at all, due to the lengthy time scales associated with the climate system.

CONCEPT CHECKS 20.10

1. Describe the factors that are causing sea level to rise.
2. Is global warming greater near the equator or near the poles? Explain.
3. Based on Table 20.1, what projected changes relate to something other than temperature?

CONCEPTS IN REVIEW

World Climates and Global Climate Change

20.1 The Climate System

List the five parts of the climate system and provide examples of each.

KEY TERMS: climate system, cryosphere

- Climate is the aggregate of weather conditions for a place or region over a long period of time.
- Earth's climate system involves the exchanges of energy and moisture that occur among the atmosphere, hydrosphere, solid Earth, biosphere, and cryosphere (the ice and snow that exist at Earth's surface).

20.2 World Climates

Explain why classification is a necessary process when studying world climates. Discuss the criteria used in the Köppen system of climate classification.

KEY TERM: Köppen classification

- Climate classification brings order to large quantities of information, which aids comprehension and understanding and facilitates analysis and explanation.
- The most important elements in climate descriptions are temperature and precipitation because they have the greatest influence on people and their activities, and they also have an important impact on the distribution of vegetation and the development of soils.
- An early attempt at climate classification by the Greeks divided each hemisphere into three zones: torrid, temperate, and frigid. Many climate classifications have been devised, with the value of each determined by its intended use.
- The Köppen classification, which uses mean monthly and annual values of temperature and precipitation, is a widely used system. The boundaries Köppen chose were largely based on the limits of certain plant associations.
- The Köppen classification uses five principal climate groups, each with subdivisions. Four of the climate groups—A, C, D, and E—are defined on the basis of temperature characteristics, and the fifth, the B group, has precipitation as its primary criterion.

? Why are three different formulas used to determine whether a climate is considered dry?

20.3 Humid Tropical (A) Climates

Compare the two broad categories of tropical climates.

KEY TERMS: tropical rain forest, tropical wet and dry, savanna

- Humid tropical (A) climates are winterless, with all months having a mean temperature above 18°C (64°F).
- Wet tropical climates (Af and Am), which lie near the equator, have constantly high temperatures and enough rainfall to support the most luxuriant vegetation (tropical rain forest) found in any climatic realm.
- Tropical wet and dry climates (Aw) are found poleward of the wet tropics and equatorward of the subtropical deserts, where the rain forest gives way to the tropical grasslands and scattered drought-tolerant trees of the savanna. The most distinctive feature of this climate is the seasonal character of the rainfall.

? When does the rainy season occur in a tropical wet and dry climate: winter or summer? Explain.

20.4 Dry (B) Climates

Contrast low-latitude dry climates and middle-latitude dry climates.

KEY TERMS: arid (desert), semiarid (steppe)

- Dry (B) climates, in which the yearly precipitation is less than the potential loss of water by evaporation, are subdivided into two types: arid or desert (BW) and semiarid or steppe (BS).
- Differences between desert and steppe are primarily a matter of degree, with semiarid being a marginal and more humid variant of arid.
- Low-latitude deserts and steppes coincide with the clear skies caused by subsiding air beneath the subtropical high-pressure belts.
- Middle-latitude deserts and steppes exist principally because of their position in the deep interiors of large landmasses far removed from the ocean. Because many middle-latitude deserts occupy sites on the leeward sides of mountains, they can also be classified as rainshadow deserts.

Dennis Tasa

? This photo was taken in Nevada's Great Basin Desert, looking west toward the Sierra Nevada. What is the basic cause of the arid conditions in this region?

20.5 Humid Middle-Latitude Climates (C and D Climates)

Distinguish among five different humid middle-latitude climates.

KEY TERMS: humid subtropical climate, marine west coast climate, dry-summer subtropical climate, humid continental climate, subarctic climate, taiga

- Middle-latitude climates with mild winters (C climates) occur where the average temperature of the coldest month is below 18°C (64°F) but above −3°C (27°F). Three C climate subgroups exist.
- Humid subtropical climates (Cfa) are located on the eastern sides of the continents, in the 25° to 40° latitude range. Summer weather is hot and sultry, and winters are mild. In North America, the marine west coast climate (Cfb, Cfc) extends from near the U.S.–Canada border northward as a narrow belt into southern Alaska. The prevalence of maritime air masses means that mild winters and cool summers are the rule. Dry-summer subtropical climates (Csa, Csb) are typically located along the west sides of continents between latitudes 30° and 45°. In summer, the regions are dominated by stable, dry conditions associated with the oceanic subtropical highs. In winter, they are within range of the cyclonic storms of the polar front.
- Humid middle-latitude climates with severe winters (D climates) are land-controlled climates that are absent in the Southern Hemisphere. The average temperature of the coldest month is −3°C (27°F) or below, and the warmest monthly mean exceeds 10°C (50°F).
- Humid continental climates (Dfa, Dfb, Dwa, Dwb) are confined to the eastern portions of North America and Eurasia in the latitude range between approximately 40° and 50° north latitude. Both winter and summer temperatures are relatively severe. Precipitation is generally greater in summer than in winter. Subarctic climates (Dfc, Dfd, Dwc, Dwd) are situated north of the humid continental climates and south of the polar tundras. The outstanding feature of subarctic climates is the dominance of winter; summers are short but remarkably warm. The highest annual temperature ranges on Earth occur here.

20.6 Polar (E) Climates

Contrast ice cap and tundra climates.

KEY TERMS: polar climate, tundra climate, ice cap climate

- Polar climates (ET, EF) are those in which the mean temperature of the warmest month is below 10°C (50°F). Annual temperature ranges are extreme, with the lowest annual means on the planet. Although polar climates are classified as humid, precipitation is generally meager, with many nonmarine stations receiving less than 25 centimeters (10 inches) annually. Two types of polar climates are recognized.
- The tundra climate (ET) is found almost exclusively in the Northern Hemisphere. The 10°C (50°F) summer isotherm represents its equatorward limit. It is a treeless region of grasses, sedges, mosses, and lichens with permanently frozen subsoil, called permafrost.
- The ice cap climate (EF) does not have a single monthly mean above 0°C (32°F). Consequently, the growth of vegetation is prohibited, and the landscape is one of permanent ice and snow. The ice sheets of Greenland and Antarctica are important examples.

20.7 Highland Climates

Summarize the characteristics associated with highland climates.

KEY TERM: highland climate

- Highland climates are characterized by a great diversity of climatic conditions over a small area. Although the best-known climatic effect of increased altitude is lower temperatures, greater precipitation due to orographic lifting is also common. Variety and changeability best describe highland climates. Because atmospheric conditions fluctuate with altitude and exposure to the Sun's rays, a nearly limitless variety of local climates occur in mountainous regions.

? Which of the local winds discussed in Chapter 18 are likely associated with highland climates?

20.8 Human Impact on Global Climate

Summarize the nature and cause of the atmosphere's changing composition since about 1750. Describe the climate's response.

KEY TERMS: aerosols, black carbon

- Humans have been modifying the environment for thousands of years. By burning vegetation and allowing domestic animals to overgraze the land, people have modified such important climatic factors as surface albedo, evaporation rates, and surface winds.
- Human activities produce climate change in large part by releasing carbon dioxide (CO_2) and trace gases. Humans release CO_2 when they cut down forests and when they burn fossil fuels such as coal, oil, and natural gas. A steady rise in atmospheric CO_2 levels has been documented at Mauna Loa, Hawaii, and other locations around the world.
- More than half of the carbon released by humans is absorbed by new plant matter or dissolved in the oceans. About 45 percent remains in the atmosphere, where it can influence climate for decades. Air bubbles trapped in glacial ice reveal that there is currently about 30 percent more CO_2 than the atmosphere has contained in the past 800,000 years.
- As a result of the extra heat retained by added CO_2, Earth's atmosphere has warmed by about 0.8°C (1.4°F) in the past 100 years, most of it since 1980. Temperatures are projected to increase by another 2°C to 4.5°C (3.6°F to 8.1°F) in the future.
- Trace gases such as methane, nitrous oxide, and CFCs also play a significant role in increasing global temperature.
- Aerosols—tiny liquid and solid particles suspended in the air—are produced by human and natural sources and affect global climate. Many aerosols reflect a portion of incoming solar radiation back to space and therefore have a cooling effect. Some aerosols, called black carbon, absorb incoming solar radiation and warm the atmosphere. When black carbon is deposited on snow and ice, it reduces surface albedo and increases the amount of light absorbed at the surface.

? Do aerosols spend more or less time in the atmosphere than greenhouse gases such as carbon dioxide? What is the significance of this difference in residence time? Explain.

20.9 Climate-Feedback Mechanisms

Contrast positive- and negative-feedback mechanisms and provide examples of each.

KEY TERMS: climate-feedback mechanism, positive-feedback mechanism, negative-feedback mechanism

- A change in one part of the climate system may trigger changes in other parts of the climate system that amplify or diminish the initial effect. These climate-feedback mechanisms are called positive-feedback mechanisms if they reinforce the initial change and negative-feedback mechanisms if they counteract the initial effect.
- The melting of sea ice due to global warming (decreasing albedo and hence increasing the initial effect of warming) is one example of a positive-feedback mechanism. The production of more clouds (blotting out incoming solar radiation, leading to cooling) is an example of a negative-feedback mechanism.
- Computer models of climate give scientists a tool for testing hypotheses about climate change. Although these models are far simpler than the real climate system, they are useful tools for predicting the future climate.

? Changes in precipitation and temperature due to climate change can increase the risk of forest fires. Describe two ways that the event shown in this photo could contribute to global warming.

Michael Collier

20.10 Some Possible Consequences of Global Warming

Summarize some of the possible consequences of global warming.

- In the future, Earth's surface temperature is likely to continue to rise. The temperature increase will likely be greatest in the polar regions and least in the tropics. Some areas will get drier, and other areas will get wetter.
- Sea level is predicted to rise for several reasons, including the melting of glacial ice and thermal expansion (a given mass of seawater takes up more volume when it is warm than when it is cool). Low-lying, gently sloped coastal areas (which are often highly populated) are most at risk.
- The extent and thickness of sea ice in the Arctic have been declining since satellite observations began in 1979.
- Because of the warming of the Arctic, permafrost is melting, releasing CO_2 and methane to the atmosphere in a positive-feedback loop.
- Because the climate system is complicated, dynamic, and imperfectly understood, it could produce sudden, unexpected changes with little warning.

GIVE IT SOME THOUGHT

1 Refer to Figure 20.1, which illustrates various components of Earth's climate system. Boxes represent interactions or changes that occur in the climate system. Select three boxes and provide an example of an interaction or change associated with each. Explain how these interactions may influence temperature.

2 Refer to Figure 20.5, which shows climates of the world. Humid continental (Dfb and Dwb) and subarctic (Dfc) climates are usually described as being "land controlled"—that is, they lack marine influence. Nevertheless, these climates are found along the margins of the North Atlantic and the North Pacific oceans. Explain why this occurs.

3 It has been suggested that global warming over the past several decades likely would have been greater were it not for the effect of certain types of air pollution. Explain how this could be true.

4 Refer to the monthly rainfall data (in millimeters) for three cities in Africa and match each city to the correct location (1, 2, or 3) on the map. How were you able to figure this out? *Bonus:* Which figure in Chapter 18 would be especially useful in explaining or illustrating why these places have rainfall maximums and minimums when they do?

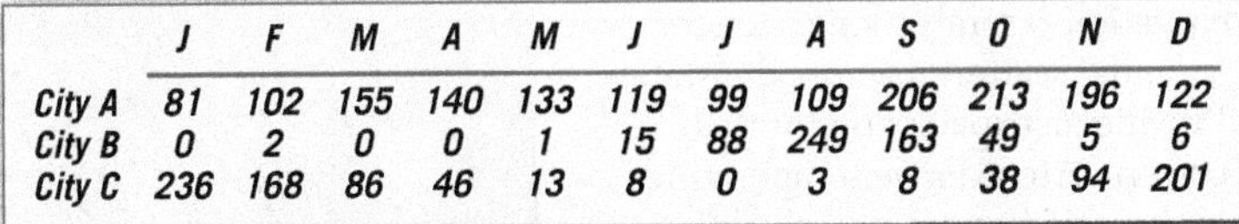

	J	F	M	A	M	J	J	A	S	O	N	D
City A	81	102	155	140	133	119	99	109	206	213	196	122
City B	0	2	0	0	1	15	88	249	163	49	5	6
City C	236	168	86	46	13	8	0	3	8	38	94	201

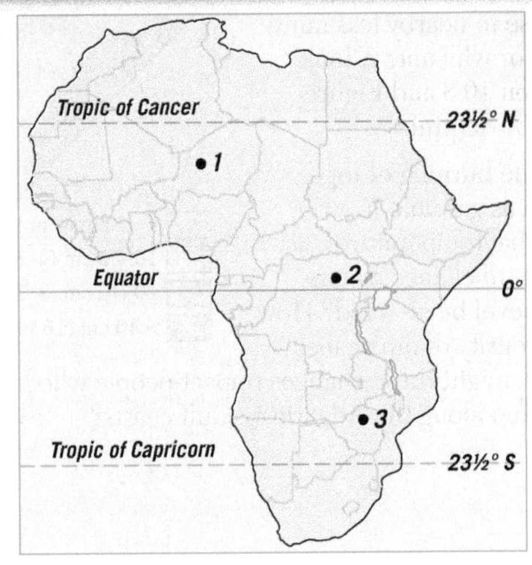

5 Motor vehicles are a significant source of CO_2. Using electric cars, such as the one pictured here, is one way to reduce emissions from this source. Although these vehicles emit little or no CO_2 or other air pollutants directly into the air, can they still be connected to such emissions? If so, explain.

David Pearson/Alamy Stock Photo

6 If a fellow student who, unlike you, had not studied climate were to ask, "Isn't the greenhouse effect a bad thing because it's responsible for global warming?" How would you respond?

7 During a conversation, an acquaintance indicates that he is skeptical about global warming. When you ask why he feels that way, he says, "The past couple of years in this area have been among the coolest I can remember." While you assure this person that it is useful to question scientific findings, you suggest to him that his reasoning in this case may be flawed. Use your understanding of the definition of *climate* along with one or more graphs in the chapter to persuade this person to reevaluate his reasoning.

EXAMINING THE EARTH SYSTEM

1 The Köppen climate classification is based on the fact that there is an excellent association between natural vegetation (biosphere) and climate (atmosphere). Briefly describe the climate conditions (temperature and precipitation) and natural vegetation associated with each of the following Köppen climates: Af, BWh, Dfc, and ET.

2 Examine the precipitation map for the state of Nevada. Notice that the areas receiving the most precipitation resemble long, slender "islands" scattered across the state. Provide an explanation for this pattern. Are average temperatures in these wetter areas likely different from those in nearby less rainy places? Why or why not? A look back at Section 10.8 and Figure 10.32 might be helpful.

< 10 cm (4 in.)
10–20 cm (4–8 in.)
20–40 cm (8–16 in.)
> 40 cm (16 in.)

3 How might the burning of fossil fuels, such as gasoline in a car, influence global temperature? As this temperature change occurs, how will sea level be affected? How might the intensity of hurricanes change? How might these changes impact people who live on a beach or barrier island along the Atlantic or Gulf coasts?

4 This satellite image from August 2007 shows the effects of tropical deforestation in a portion of the Amazon basin in western Brazil. Intact forest is dark green, whereas cleared areas are tan (bare ground) or light green (crops and pasture). Notice the relatively dense smoke in the left center of the image. How does deforestation of tropical forests change the composition of the atmosphere? Describe the effect that tropical deforestation has on global warming.

NASA

DATA ANALYSIS

Climate Classification and Climate Change

As climate changes, the world climate classification boundaries will also change. Predictions of how these will shift are made for various climate change scenarios.

ACTIVITIES

At http://koeppen-geiger.vu-wien.ac.at/shifts.htm, go to the Observed and Projected Climate Shifts page. Compare the 1976–2000 map to the A1FI 2076–2100 map by opening each map in a separate window. The A1FI climate change scenario includes rapid economic growth and intense use of fossil fuels. Zoom in to the United States.

1 What are the climate classification regions based on for the Koppen-Geiger climate classification map?

2 What is the Köppen-Geiger climate classification code in your approximate location according to the 1976–2000 map? Use the information on the map to describe your climate.

3 What is the Köppen-Geiger climate classification code in your approximate location according to the 2076–2100 map? Describe your projected climate.

4 In what direction do climate zones shift in the Eastern United States? Western United States?

5 Compare the 2076–2100 map to the 1976-2000 map. How is the climate in the Great Lakes region projected to change? Discuss both precipitation and temperature in your answer.

A shift in growing season has already been observed in the United States. Go to the Planting your spring garden? article at https://www.climate.gov/news-features/featured-images/planting-your-spring-garden-consider-climate%E2%80%99s-%E2%80%98new-normal%E2%80%99.

6 What is plotted on the map?

7 What data are used to determine the old Normals? New Normals?

8 When was the planting zone map updated?

9 What are the implications of the new climate normal for gardeners and landscapers?

10 Compare the "Updated Normals" and the "Previous Normals" maps. What is your previous climate-related planting zone? What is your updated zone?

11 According to the Zone Changes map, how have climate zones shifted in the Eastern United States? Western United States?

UNIT 7 Earth's Place in the Universe

21

Origins of Modern Astronomy*

FOCUS ON CONCEPTS

Each statement represents the primary learning objective for the corresponding major heading within the chapter. After you complete the chapter, you should be able to:

21.1 Explain the geocentric view of the solar system and describe how it differs from the heliocentric view.

21.2 List and describe the contributions to modern astronomy of Nicolaus Copernicus, Tycho Brahe, Johannes Kepler, Galileo Galilei, and Isaac Newton.

21.3 Describe how constellations are used in modern astronomy.

21.4 Describe the two primary motions of Earth and explain the difference between a solar day and a sidereal day.

21.5 Sketch the changing configuration of the Earth–Moon system that produces the regular cycle we call the phases of the Moon.

21.6 Sketch the configuration of the Earth–Moon–Sun system that produces a lunar eclipse and the configuration that produces a solar eclipse.

*This chapter was revised with the assistance of Professors Teresa Tarbuck and Mark Watry.

The Milky Way and Venus are visible behind a dish antenna. (Photo by Babak Tafreshi/Getty Images)

THE SCIENCE OF ASTRONOMY PROVIDES A RATIONAL WAY of knowing and understanding the origins of Earth, the solar system, and the universe. Earth was once thought to be unique, different in every way from everything else in the universe. However, through the science of astronomy, we have discovered that our planet is similar to other objects in the universe and that the physical laws that apply on Earth also apply everywhere else in the universe.

In this chapter we examine the transformation from the ancient view of the universe, which focused on the *positions and movements of celestial objects* (mainly planets), to the modern perspective, which focuses on understanding *how* these celestial objects came to be and *why* they move the way they do.

21.1 Ancient Astronomy

Explain the geocentric view of the solar system and describe how it differs from the heliocentric view.

Long before recorded history, people were aware of the close relationship between events on Earth and the positions of heavenly bodies. They realized that changes in the seasons and floods of great rivers such as the Nile in Egypt occurred when certain celestial bodies, including the Sun, Moon, planets, and stars, appeared in particular places in the heavens. Early agrarian cultures, whose survival depended on seasonal change, believed that if these heavenly objects could control the seasons, they could also strongly influence all Earthly events. These beliefs undoubtedly encouraged early civilizations to begin keeping records of the positions of celestial objects.

▼ **Figure 21.1 The Bayeux Tapestry that hangs in Bayeux, France** This tapestry shows the apprehension caused by Halley's Comet in 1066 C.E. This event preceded the defeat of King Harold by William the Conqueror. (Photo by DEA/G DAGLI ORTI/De Agostini Editore/AGE Fotostock)

The ancient Chinese, Egyptians, and Babylonians are well known for their record keeping. These cultures recorded the positions of the Sun, the Moon, and the five visible planets as these objects moved slowly against the background of "fixed" stars. Eventually, it was not enough to track the motions of celestial objects; predicting their future positions (to avoid getting married at an unfavorable time, for example) became important.

The Chinese recorded every appearance of the famous Halley's Comet for at least 10 centuries. However, because this comet appears only once every 76 years, they were unable to link these appearances to establish that what they saw was the same object multiple times. Like most other ancients, the Chinese considered comets to be mystical. Generally, comets were seen as bad omens and were blamed for a variety of disasters, from wars to plagues (**Figure 21.1**).

The ancient Chinese also kept quite accurate records of "guest stars." Today we know that a "guest star" is a normal star, usually too faint to be visible, which increases suddenly in brightness as it undergoes an explosion. We call this phenomenon a *nova* (*novus* = new) or *supernova* (**Figure 21.2**).

The Golden Age of Astronomy

The "Golden Age" of early astronomy (600 B.C.E.–150 C.E.) was centered in Greece. Although the early Greeks have been criticized for using purely philosophical arguments to explain natural phenomena, they employed observational data as well. The Greeks developed geometry and trigonometry and used the basics of these disciplines to measure the sizes of and distances to the largest-appearing bodies in the heavens—the Sun and the Moon.

When watching the Sun, Moon, and stars rise along the eastern horizon and move across the sky, it seems reasonable to think that Earth is stationary and everything in the sky is moving around you. If you

Figure 21.2 **The sudden appearance of a "guest star"** The Chinese recorded the sudden appearance of a "guest star" in 1054 C.E. The scattered remains of that supernova is the Crab Nebula in the constellation Taurus. This image comes from the Hubble Space Telescope. (Photo by J Hester/A Loll/NASA)

observe the skies as ancient astronomers did, you will also find that the stars seem fixed in position relative to each other, whereas seven heavenly bodies—the Sun, the Moon, and the five visible planets (Mercury, Venus, Mars, Jupiter, and Saturn)—shift position from night to night against the stars. The Greeks imagined that the universe consisted of eight transparent concentric spheres that revolved around a central, motionless Earth. The stars were fixed to the outermost sphere, and the seven moving bodies (*planetai* in Greek, meaning "wanderers") each occupied an inner sphere. This conceptual model, in which the heavens circle a motionless Earth, is called the **geocentric** (*geo* = Earth, *centric* = centered) view of the universe. The idea was largely accepted because people can't sense Earth's motion, and Earth seems too large to spin. However, modern science has taught us that Earth moves, causing visible distant objects to appear to move across the sky. Earth's rotation was not confirmed experimentally until 1851.

Determining Earth's Shape Contrary to common belief, the explorer Columbus did not prove that the world is spherical rather than flat. The Greeks had assembled convincing evidence that Earth is a sphere. For example, the famous Greek philosopher Aristotle (384–322 B.C.E.) pointed out that Earth must be spherical because it always casts a curved shadow when it eclipses the moon (see Section 21.6).

Measuring Earth's Circumference The first successful attempt to establish the size of Earth is credited to Eratosthenes (276–194 B.C.E.). Eratosthenes observed the angles of the noonday Sun in two Egyptian cities that were roughly north and south of each other—Syene (presently Aswan) and Alexandria (Figure 21.3). Finding that the angles of the noonday Sun differed by 7 degrees, or 1/50 of a complete circle, he concluded that the circumference of Earth must be 50 times the distance between these two cities. The cities were 5000 *stadia* apart, so Eratosthenes came up with a measurement of 250,000 *stadia*.

Many historians believe the *stadia* was 157.6 meters (517 feet), which would make Eratosthenes's calculation of Earth's circumference—39,400 kilometers (24,428 miles)—very close to the modern value of 40,075 kilometers (24,902 miles).

A Sun-Centered Universe? The first Greek to profess a *Sun-centered*, or **heliocentric** (*helios* = Sun, *centric* = centered), universe was Aristarchus (312–230 B.C.E.). Aristarchus also used simple geometric relations to calculate the relative distances from Earth to the Sun and the Moon. He later used these data to calculate their sizes. As a result of an observational error beyond his control, he came up with measurements that were much too small. However, he did discover that the Sun was many times more distant than the Moon and many times larger than Earth. The latter fact may have prompted him to suggest a Sun-centered universe. Nevertheless, partially because of the strong influence of Aristotle's writings, the Earth-centered view dominated Western thought for nearly 2000 years.

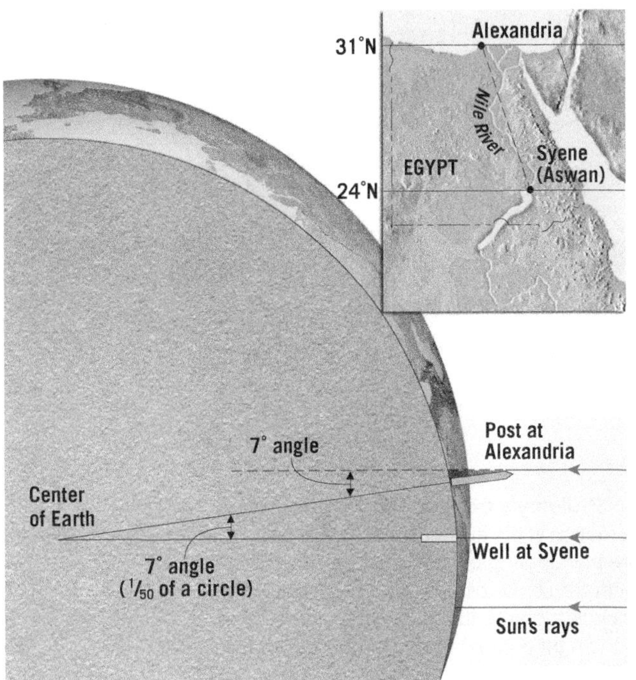

SmartFigure 21.3 **Orientation of the Sun's rays at Syene (Aswan) and Alexandria, Egypt, on June 21** From these data, Eratosthenes calculated Earth's circumference.

TUTORIAL
https://goo.gl/eiOiGe

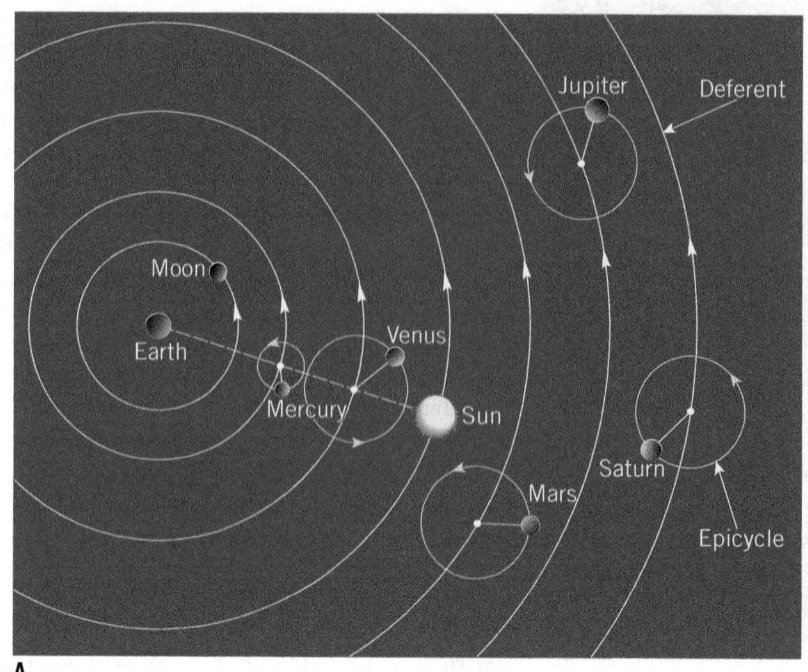

▶ **Figure 21.4 The universe according to Ptolemy, second century C.E. A.** Ptolemy believed that the star-studded celestial sphere made a daily trip around a motionless Earth. In addition, he proposed that the Sun, Moon, and planets made trips of various lengths along individual orbits. **B.** A three-dimensional model of an Earth-centered system. Ptolemy likely utilized something similar to this to calculate the motions of the heavens. (Photo by Science Museum, London, UK/Bridgeman Images)

Mapping the Stars Possibly the greatest of the early Greek astronomers was Hipparchus (second century B.C.E.), best known for his star catalog. Hipparchus determined the locations of almost 850 stars, which he divided into six groups, according to their brightness—a system that is still used today. He measured the length of the year to within minutes of the modern value and developed a method for predicting the times of lunar eclipses to within a few hours.

Although many of the Greek discoveries were lost during the Middle Ages, the Earth-centered view that the Greeks proposed became entrenched in Europe.

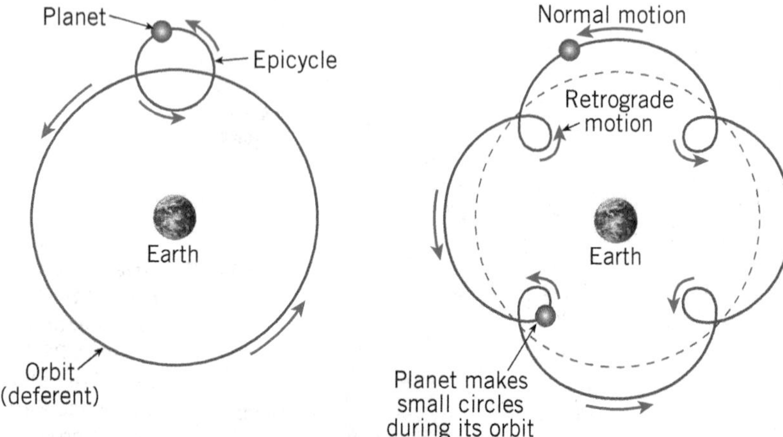

▲ **SmartFigure 21.5 Ptolemy's explanation of retrograde motion** Retrograde motion is the apparent backward motion of planets against the background of fixed stars. In Ptolemy's model, the planets move on small circles (epicycles) while they orbit Earth on larger circles (deferents). Through trial and error, Ptolemy discovered the right combination of circles to produce the retrograde motion observed for each planet.

https://goo.gl/5bgOeZ

Ptolemy's Model of the Universe

Much of our knowledge of Greek astronomy comes from a 13-volume treatise compiled by Claudius Ptolemy in 141 C.E. and known to later scholars as the *Almagest*. In addition to presenting a summary of Greek astronomical knowledge, Ptolemy is credited with developing a detailed geocentric model of the universe, known as the **Ptolemaic system** (Figure 21.4).

The Greeks believed that heavenly things must be made of perfect circles, which they regarded as the purest geometric shape. And indeed, the Sun, the Moon, and the stars can be modeled as following circular paths around Earth. But the motion of the five visible planets against the background stars is more complex. Periodically, each planet appears to stop, reverse direction for weeks or months at a time, and then resume its forward motion through the sky. These apparent reversals are called **retrograde** (*retro* = to go back, *gradus* = walking) **motion**.

To account for these motions, Ptolemy devised a complex system whereby planets move along small circles (*epicycles*) while orbiting Earth on large circles (*deferents*), as shown in Figure 21.5. His model predicted the motions of the planets with remarkable accuracy for at least a century. When it fell too far out of step with the observed positions, it was simply recalibrated, using the new observed positions as a starting point.

Ptolemy's model is, of course, not correct. Figure 21.6 shows the actual cause of retrograde motion for the planet Mars. Because Earth takes less time to orbit the Sun than Mars, it overtakes its neighbor. While Earth is "passing" Mars, Mars *appears* to be moving backward, in retrograde motion. Similarly, when you pass another

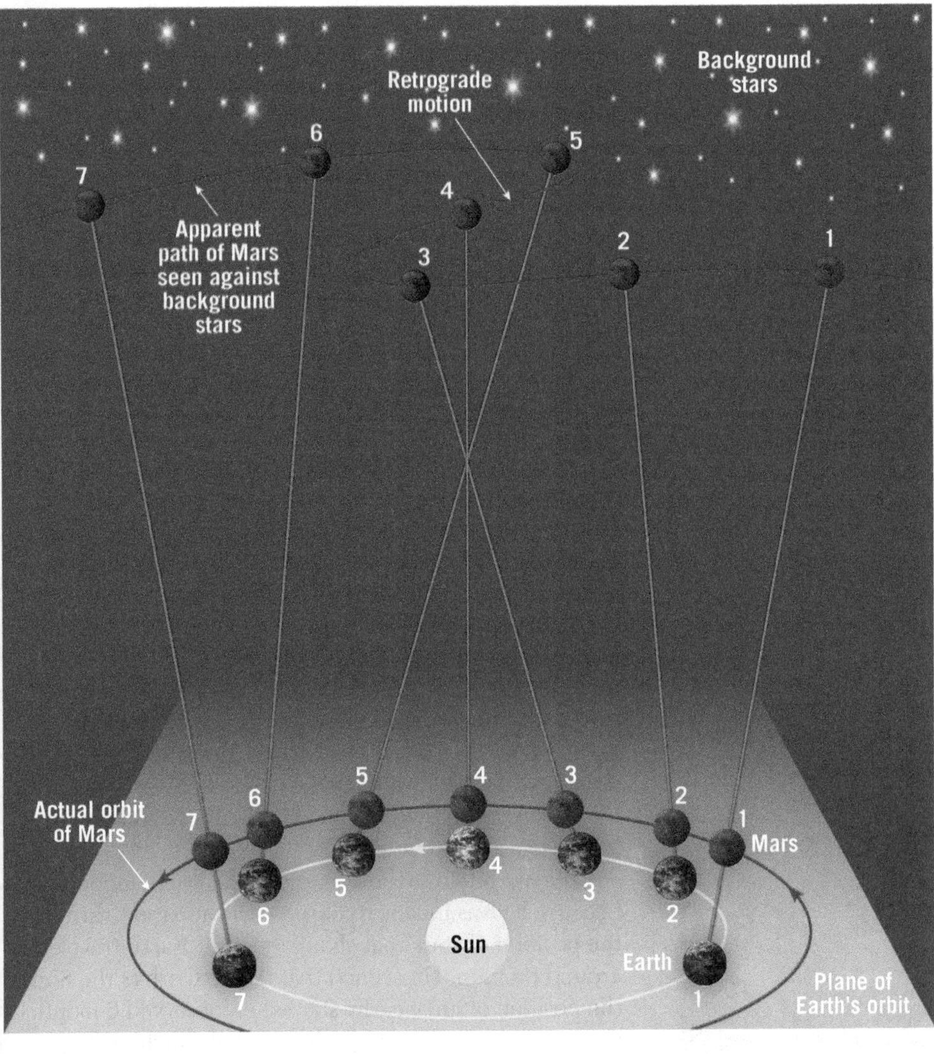

SmartFigure 21.6 **Retrograde motion of Mars, as seen against the background of distant stars** Mars takes nearly twice as long to orbit the Sun as Earth does, so Earth periodically overtakes and passes it. During each passing, Mars appears to move "backward" relative to the background stars (it exhibits retrograde motion).

TUTORIAL
https://goo.gl/K9bze0

car, the other car appears to move backward relative to yourself.

Much accumulated knowledge disappeared as libraries were destroyed after the decline of Greek and Roman civilizations, and the center of astronomical study moved east, to Baghdad, where, fortunately, Ptolemy's work was translated into Arabic. When it was later reintroduced into Europe, it became fused with the teachings of the Church—a problem for those who would later question it.

CONCEPT CHECKS 21.1

1. Why did the ancients believe that celestial objects had some influence over their lives?
2. What is the modern explanation for "guest stars" suddenly appearing in the night sky?
3. Explain the *geocentric* view of the universe.
4. Describe what produces the retrograde motion of Mars. What geometric arrangements did Ptolemy use to explain this motion?

21.2 The Birth of Modern Astronomy

List and describe the contributions to modern astronomy of Nicolaus Copernicus, Tycho Brahe, Johannes Kepler, Galileo Galilei, and Isaac Newton.

Modern astronomy's development was more than a scientific endeavor; it required a break from deeply entrenched Western philosophical and religious views that had been held for thousands of years. This transition was brought about by the discovery of a new and much larger universe, governed by discernible laws.

We examine the work of five noted scientists involved in this progression from an astronomy that merely describes what is observed to one that tries to explain why the universe behaves as it does: Nicolaus Copernicus, Tycho Brahe, Johannes Kepler, Galileo Galilei, and Sir Isaac Newton.

▲ Figure 21.7 **Painting of Polish astronomer Nicolaus Copernicus (1473–1543)** Copernicus's claim that the Sun rather than Earth was the center of the cosmos was very controversial for more than 100 years after his death. (Detlev van Ravenswaay/Science Source)

Nicolaus Copernicus

For almost 13 centuries after the time of Ptolemy, very few astronomical advances were made in Europe. The first great astronomer to emerge after the Middle Ages was Nicolaus Copernicus (1473–1543) from Poland (**Figure 21.7**). After discovering Aristarchus's writings, Copernicus became convinced that Earth is a planet, just like the other five known planets of that time, and that the planets orbit the Sun. This *heliocentric* (Sun-centered) model was a major break from the ancient and prevailing idea that a motionless Earth lies at the center of all movement in the universe. However, Copernicus retained a link to the past and used circles to represent the orbits of the planets. Because of this, he was unable to accurately predict the future locations of the planets and found it necessary to add smaller circles (epicycles) like those Ptolemy had used. The discovery that the planets actually have *elliptical* orbits was made a century later by Johannes Kepler.

Copernicus's monumental work, *De Revolutionibus Orbium Coelestium* (*On the Revolution of the Heavenly Spheres*), which set forth his controversial Sun-centered solar system, was published as he lay on his deathbed. Hence, he never suffered the criticisms that fell on many of his followers for deposing Earth from its central and unique place in the universe.

▶ Figure 21.8 **Tycho Brahe (1546–1601) in his observatory in Uraniborg, on the then-Danish island Hven** Tycho (central figure) and the background are painted on the wall of the observatory within the arc of the sighting instrument called a quadrant. On the far right, Tycho can be seen "sighting" a celestial object through the "hole" in the wall. Tycho's accurate measurements of Mars enabled Johannes Kepler to formulate his three laws of planetary motion. (Photo by © Royal Geographical Society, London, UK/Bridgeman Images)

Tycho Brahe

Tycho Brahe (1546–1601), a Danish nobleman, was born 3 years after the death of Copernicus. Tycho reportedly became interested in astronomy while viewing a solar eclipse that astronomers had predicted. He persuaded King Frederick II to establish an observatory near Copenhagen, which Tycho headed. There he designed and built pointers (the telescope would not be invented for a few more decades), which he used for 20 years to systematically measure the locations of the heavenly bodies in an effort to disprove the Copernican theory (**Figure 21.8**). His observations, particularly of Mars, were far more precise than any made previously and are his legacy to astronomy.

Tycho did not believe in the Copernican model because he was unable to observe the apparent shift in the position of stars that should result if Earth traveled around the Sun. He argued that if Earth orbits the Sun, the position of any nearby star, when observed 6 months apart and hence from opposite sides of Earth's orbit, should shift with respect to the more distant stars—a phenomenon called *stellar parallax*.

Parallax is easy to experience. Hold your index finger up in front of your face. Close one eye and then the other, and observe how your finger seems to shift position against background objects. Now hold your finger farther away and notice that the apparent shift is smaller. Therein lay the flaw in Tycho's argument. He was right about parallax, but the distance to even the nearest stars is so huge compared to the width of Earth's orbit that he was unable to observe it. Today, space-based telescopes use parallax to measure distances to stars in our galaxy.

With the death of his patron, the king of Denmark, Tycho was forced to leave his observatory. Known for his arrogance and extravagant nature, Tycho was unable to continue his work under Denmark's new ruler and moved to Prague, in the present-day Czech Republic. There, in the last year of his life, he acquired an able assistant, Johannes Kepler. Kepler retained most of Tycho's observations and put them to exceptional use. Ironically, the data Tycho collected to refute the Copernican view of the solar system would later be used by Kepler to support it.

Figure 21.9 **German Astronomer Johannes Kepler (1571–1630)** Kepler's contribution to modern astronomy was the derivation of his three laws of planetary motion. (Photo by Imagno/Getty Images)

Johannes Kepler

If Copernicus ushered out the old astronomy, Johannes Kepler (1571–1630) ushered in the new (Figure 21.9). Armed with Tycho's data, a good mathematical mind, and, of much importance, a strong belief in the accuracy of Tycho's work, Kepler derived three basic laws of planetary motion. The first two laws resulted from his inability to fit Tycho's observations of Mars to a circular orbit. Unwilling to concede that the discrepancies were due to observational error, he searched for another solution. This endeavor led him to discover that the orbit of Mars is not a perfect circle but is slightly elliptical (Figure 21.10). Kepler also realized that the orbital speed of Mars varies in a predictable way: Mars speeds up as it approaches the Sun and slows down as it moves away.

In 1609, after nearly a decade of work, Kepler proposed his first two laws of planetary motion:

1. *The path of each planet around the Sun is an ellipse, with the Sun at one focus rather than at the center* (see Figure 21.10). This law implies that a planet's distance from the Sun varies during its orbit. By using elliptical orbits, rather than circles, Kepler created a Sun-centered model that predicted planetary motions with great accuracy.

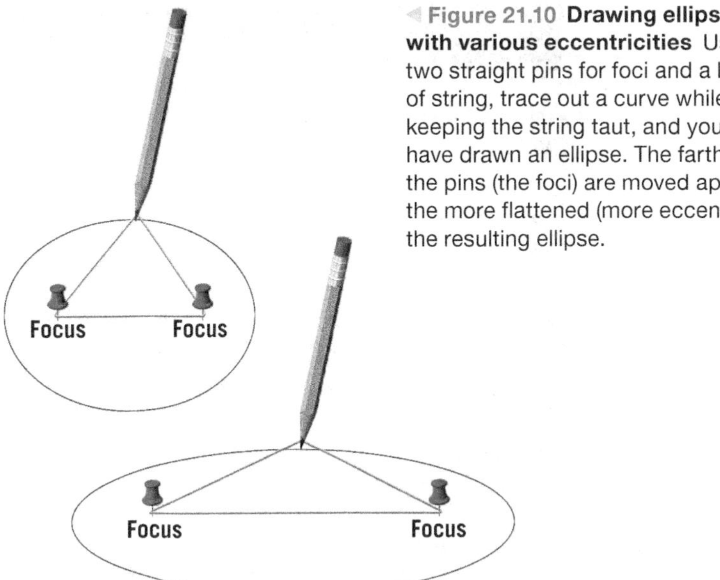

Figure 21.10 **Drawing ellipses with various eccentricities** Using two straight pins for foci and a loop of string, trace out a curve while keeping the string taut, and you will have drawn an ellipse. The farther the pins (the foci) are moved apart, the more flattened (more eccentric) is the resulting ellipse.

2. *A planet moves faster when closest to the Sun and slower when farthest from the Sun* (Figure 21.11). More precisely, Kepler's second law tells us that a planet's orbital speed varies in such a way that a line connecting it to the Sun sweeps out equal areas in equal time intervals, as shown in Figure 21.11.

Kepler was religiously devout and believed that "the Creator" had made a universe with order that would be reflected in the positions and motions of the planets. The uniformity he tried to find eluded him for nearly a decade. Then, in 1619, Kepler published his third law, in *The Harmony of the Worlds*:

3. *Distant planets orbit the Sun at slower average speeds than planets located closer to the Sun.* Mathematically, Kepler's third law states that a planet's orbital period squared is equal to its mean solar distance cubed, written $p^2 = a^3$, where p is a planet's orbital period and a is its mean distance from the Sun. In its simplest form, the orbital period (the time needed to make a complete orbit around the Sun) is measured in Earth years, and the planet's distance to the Sun is expressed in terms of Earth's mean distance to the Sun. The latter "yardstick" is called the **astronomical unit (AU)**, which is equal to about 150 million kilometers (93 million miles). Using these units, the solar distances of the planets can

Figure 21.11 **Illustration of Kepler's second law** This law states that a planet (Earth in this illustration) travels slower when it is farther from the Sun (aphelion) and faster when it is closest (perihelion). The eccentricity of Earth's orbit is greatly exaggerated in this diagram.

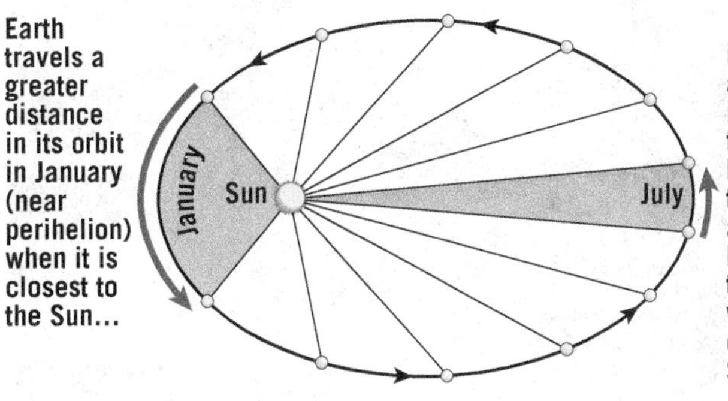

Earth travels a greater distance in its orbit in January (near perihelion) when it is closest to the Sun...
January
Sun
July
...than it does in July (near aphelion) when it is farthest from the Sun. Because the time intervals are the same, Earth must be traveling faster when it is closest to the Sun.

Table 21.1 Periods of Orbits and Solar Distances of Planets

Planet	Mean Solar Distance (AU)*	Period (years)	Eccentricity**
Mercury	0.39	0.24	0.206
Venus	0.72	0.62	0.007
Earth	1.00	1.00	0.017
Mars	1.52	1.88	0.093
Jupiter	5.20	11.86	0.048
Saturn	9.54	29.46	0.0565
Uranus	19.18	84.01	0.046
Neptune	30.06	164.80	0.011

* AU = astronomical unit

** The eccentricity of an orbit is the degree to which it is elliptical (longer than wide). A circle has an eccentricity of zero.

be calculated when their orbital periods are known and vice versa. For example, Mars has an orbital period of 1.88 years, and 1.88 squared equals 3.54. The cube root of 3.54 is 1.52, and that is the mean distance from Mars to the Sun, in astronomical units (Table 21.1).

Kepler's findings provide strong support for the Copernican theory that the planets orbit the Sun. However, the task of determining *why* planets move in this way would remain for Galileo Galilei and Sir Isaac Newton.

Galileo Galilei

Galileo Galilei (1564–1642) was the greatest Italian scientist of the Renaissance (Figure 21.12). A contemporary of Kepler, Galileo also strongly supported the Copernican theory of a heliocentric system. Galileo's greatest contributions to science were his descriptions of the behavior of moving objects, which he derived from experimentation. The method of using experiments to determine natural laws had essentially been lost since the time of the early Greeks.

All astronomical discoveries before Galileo's time were made without the aid of a telescope. In 1609, Galileo heard that a Dutch lens maker had devised a system of lenses that magnified objects. Apparently without ever having seen a telescope, Galileo constructed his own, which magnified distant objects to three times the size seen by the unaided eye. He immediately made others, the best having a magnification of about 30 (Figure 21.13).

With the telescope, Galileo was able to view the universe in a new way. He made many important discoveries that supported the Copernican view of the universe, including the following:

1. *Galileo observed Jupiter's four largest satellites, or moons* (Figure 21.14). This finding dispelled the old idea that Earth was the sole center of motion in the universe, for here, plainly visible, was another center of motion—Jupiter. This finding also countered the frequently used argument that the Moon would be left behind if Earth traveled around the Sun.
2. *Planets are spheres rather than just points of light, as had previously been thought.* This indicated that the planets could be Earth-like as opposed to star-like.

Figure 21.12 **Italian scientist Galileo Galilei (1564–1642)** Galileo was the first scientist to use a new invention, the telescope, to observe the Sun, Moon, and planets in more detail than ever before. (Photo by Nimatallah/Art Resource, NY)

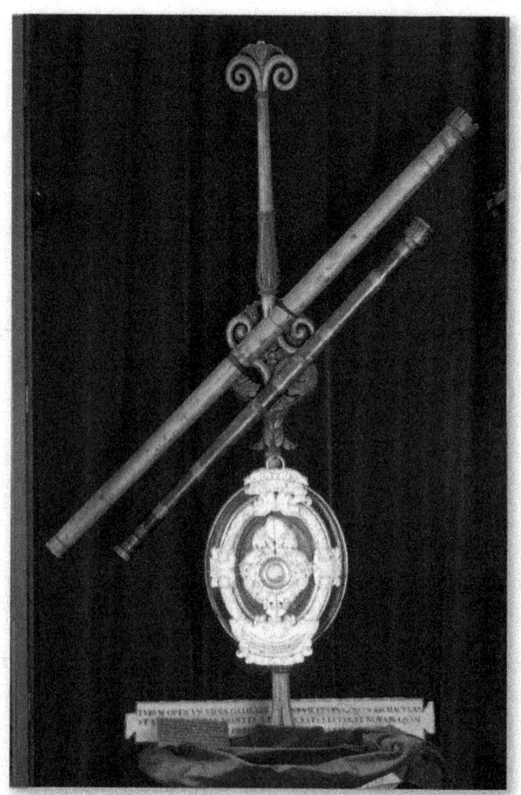

Figure 21.13 **One of Galileo's many telescopes** Although Galileo did not invent the telescope, he built several—the largest of which had a magnification of 30. (Photo by Gianni Tortoli/Science Source)

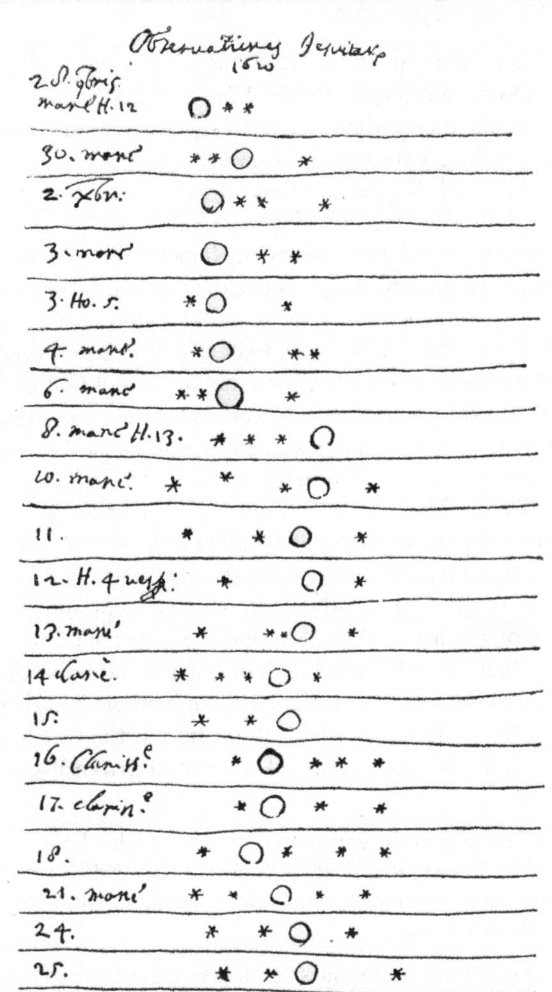

Figure 21.14 **Sketch by Galileo of Jupiter and its four largest satellites** Using a telescope, Galileo discovered Jupiter's four largest moons (drawn as stars) and noted that their positions change nightly. You can observe these same changes by using binoculars. (Yerkes Observatory Photograph/ University of Chicago)

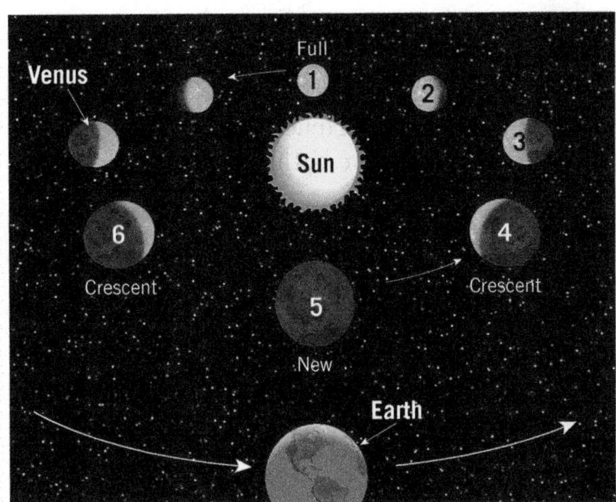

A. In the Copernican (Sun-centered) system, Venus orbits the Sun and hence all of the phases of Venus should be visible from Earth. The numbers corrispond to the numbered images in part B.

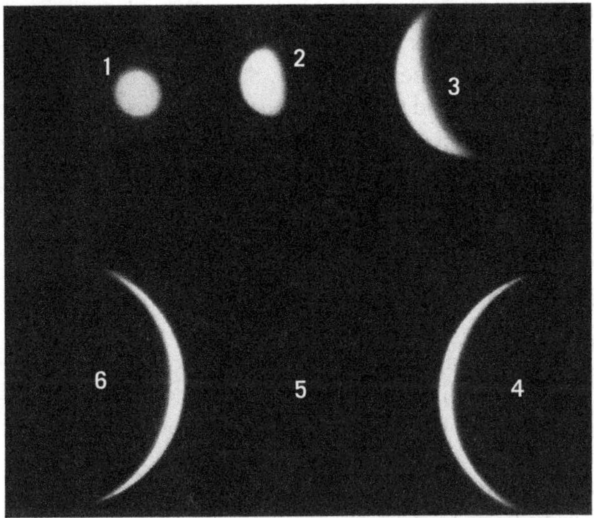

B. As Galileo observed, Venus goes through a series of Moonlike phases. Venus appears smallest during the full phase when it is farthest from Earth and largest in the crescent phase when it is closest to Earth. This led Galileo to conclude that the Sun was the center of the solar system.

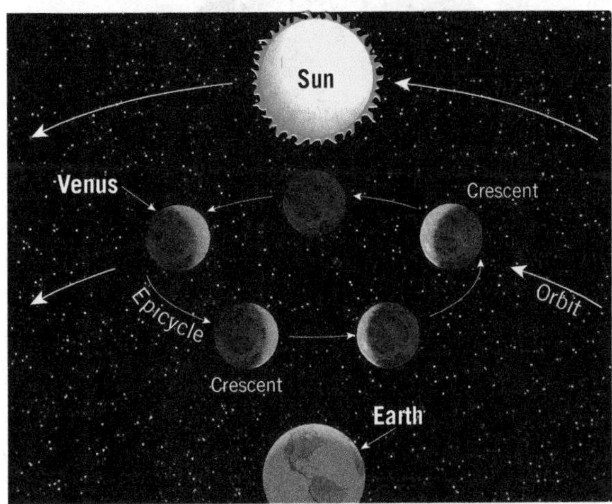

C. In the Ptolemaic (Earth-centered) system, Venus and the Sun both orbit the Earth, and Venus moves around an epicycle while following its larger orbit. In this model, only the crescent and new-Moon phases of Venus would be visible from Earth.

SmartFigure 21.15 **Using a telescope, Galileo discovered that Venus has phases like Earth's Moon** Because Venus exhibits full as well as crescent phases, and it appears smallest when full and largest in its crescent phase, Galileo concluded that the Sun is the center of the solar system. (Photo by Lowell Observatory Archives)

TUTORIAL
https://goo.gl/iMX5d2

3. *Venus exhibits phases just as the Moon does, and Venus appears smallest when it is in full phase and thus is farthest from Earth* (Figure 21.15A,B). This observation demonstrates that Venus orbits its source of light—the Sun. In the Ptolemaic system, shown in Figure 21.15C, the orbit of Venus lies between Earth and the Sun, which means that only the crescent phases of Venus should ever be seen from Earth.
4. *The Moon's surface is not smooth, as the ancients had proclaimed.* Rather, Galileo saw mountains, craters, and plains. These features indicated that the Moon is similar to Earth and not a perfect celestial sphere. He thought the plains might be bodies of water, and this idea was strongly promoted by others, as we can tell from the names given to these features (Sea of Tranquility, Sea of Storms, etc.).
5. *The Sun has sunspots—dark regions caused by slightly lower temperatures.* Galileo tracked

the movement of these spots (which may have caused the eye damage that later blinded him) and estimated the rotational period of the Sun at just under a month. Hence, another heavenly body was found to have both "blemishes" and rotational motion.

Each of these observations eroded a bedrock principle held by the prevailing view on the nature of the universe.

In 1616, the Roman Catholic Church condemned Copernican theory as contrary to biblical scripture because it did not put humans in their rightful place at the center of creation, and Galileo was told to abandon this theory. Undeterred, Galileo began writing his most famous work, *Dialogue of the Great World Systems*, and in 1630 he went to Rome, seeking permission from Pope Urban VIII to publish it. Because the book was a dialogue that expounded both the Ptolemaic and Copernican systems, publication was allowed. However, Galileo's detractors were quick to realize that he was promoting the Copernican view. Sale of the book was quickly halted, and Galileo was called before the Inquisition. Convicted of proclaiming doctrines contrary to religious teachings, he was sentenced to permanent house arrest, under which he remained for the last 10 years of his life.

Despite this restriction, and his grief following the death of his eldest daughter, Galileo continued to work. In 1637 he became totally blind yet completed his finest scientific work, a book on the study of motion in which he stated that the natural tendency of an object in motion is to remain in motion. Later, as more scientific evidence supporting the Copernican system was discovered, the Church allowed Galileo's works to be published.

Figure 21.16 **Prominent English scientist Sir Isaac Newton (1642–1727)** Newton discovered that gravity is the force that holds planets in orbit around the Sun. (Photo by DEA/G. NIMATALLAH/Getty Images)

Sir Isaac Newton

Sir Isaac Newton (1642–1727) was born in the year of Galileo's death (Figure 21.16). His many accomplishments in mathematics and physics led a successor to say, "Newton was the greatest genius that ever existed."

Although Kepler and those who followed attempted to explain what caused planetary motion, their explanations were less than satisfactory. Kepler believed that some force pushed the planets along in their orbits. Galileo, however, correctly reasoned that no force is required to keep an object in motion. Instead, Galileo proposed that the natural tendency for a moving object that is unaffected by an outside force is to continue moving at a uniform speed and in a straight line. Newton later formalized this concept, **inertia**, as his first law of motion.

The problem, then, was not explaining the force that keeps the planets moving but rather determining the force that *keeps them from going in a straight line out to space.* It was to this end that Newton conceptualized the force of gravity. At the young age of 23, he envisioned a force that extends from Earth into space and holds the Moon in orbit around Earth. Although others had theorized the existence of such a force, he was the first to formulate and test the **law of universal gravitation**. It states:

> Every body in the universe attracts every other body with a force that is directly proportional to their masses and inversely proportional to the square of the distance between them.

Therefore, more massive objects exert a greater gravitational attraction than do less massive objects. More precisely, if the mass of either of two objects is doubled, then the force of gravity between them is doubled. If the mass of one of the two objects is tripled, then the force of gravity between them is tripled. If the mass of both of the objects is doubled, then the force of gravity between them is quadrupled; and so on.

The law of gravitation also states that the gravitational force decreases with distance. For example, two objects 3000 kilometers apart exert 3^2, or 9 times, less gravitational attraction on each other than if they were 1000 kilometer apart.

Newton proved that this law of gravity, combined with the tendency of a planet to remain in straight-line motion, mathematically yielded the elliptical orbits that Kepler had established on the basis of observations. Earth, for example, moves forward in its orbit about 30 kilometers (18.5 miles) each second, and during the same second, the force of gravity pulls it toward the Sun about 0.5 centimeter (1/8 inch). Therefore, as Newton concluded, it is the combination of Earth's forward motion and its "falling" motion that defines its orbit (Figure 21.17). If gravity were eliminated, Earth would move in a straight line out into space. Conversely, if Earth's forward motion suddenly stopped, gravitational attraction would cause it to crash into the Sun.

CONCEPT CHECKS 21.2

1. What major change from the Ptolemaic system did Copernicus propose? Why was this change a philosophical departure?
2. What data did Tycho Brahe collect that was useful to Johannes Kepler in his quest to describe planetary motion?
3. Who discovered that planetary orbits are ellipses rather than circles?
4. Explain how Galileo's discovery of the phases of Venus supported the Copernican view of a Sun-centered universe.
5. In your own words, describe the law of universal gravitation. Who discovered this important concept?

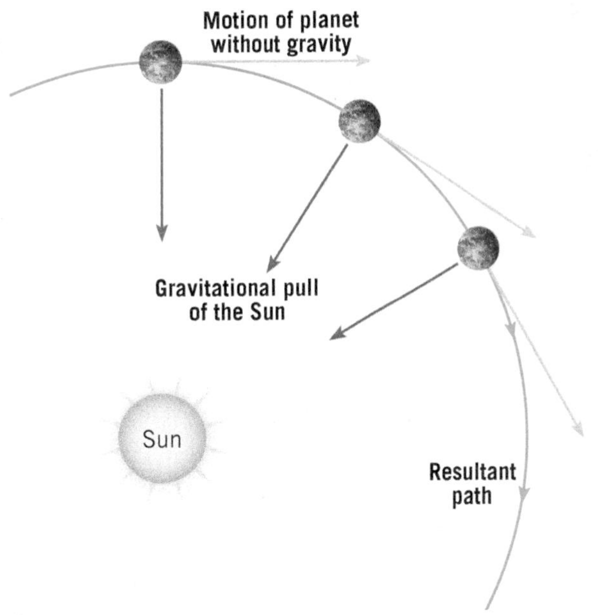

SmartFigure 21.17 **Orbital motion of Earth and other planets**

TUTORIAL
https://goo.gl/wuUWzU

21.3 Patterns in the Night Sky

Describe how constellations are used in modern astronomy.

Have you ever experienced the awe inspired by a clear dark sky illuminated with countless stars? In this section we will discuss some of the basic patterns seen on a clear night.

Constellations

The natural fascination people have with the star-studded skies led them to name the patterns they saw. Western culture has inherited the **constellations** (*con* = with, *stella* = star) that the Greeks named in honor of mythological characters or great heroes, such as Orion the hunter (see GEOgraphics 21.1). Sometimes it takes a bit of imagination to identify the intended subjects, as most constellations were probably not originally thought of as likenesses. The Greeks actually inherited many of their constellations from the Babylonians, Egyptians, and Mesopotamians. Other cultures, including Native Americans and Chinese, had their own systems of constellations.

It's important to realize that two stars that appear close together in the sky may actually be far apart because one may be much more distant than the other. Thus, most often, the stars that form a constellation have no actual relationship to each other.

Because the planets and moons of the solar system all orbit the Sun in roughly the same plane (see Section 21.4), the planets, Sun, and Moon all appear to move along a band around the sky known as the *zodiac*. Because Earth's Moon cycles through its phases about 12 times each year, the Babylonians divided the zodiac into 12 constellations. The dozen constellations of the zodiac are Aries, Taurus, Gemini, Cancer, Leo, Virgo, Libra, Scorpio, Sagittarius, Capricorn, Aquarius, and Pisces. These names may be familiar to you as the astrological signs of the zodiac.

Today, astronomers recognize 88 constellations and use them to divide the sky into regions, just as state boundaries divide the United States. Every star in the sky is within the boundaries of one of these constellations. Astronomers use constellations to roughly identify the area of the heavens they are observing. For a student, constellations provide a good way to become familiar with the night sky.

Some bright stars have acquired proper names, such as Sirius, Arcturus, and Betelgeuse. In addition, the brightest stars in a constellation are generally named in order of their brightness by the letters of the Greek alphabet—alpha (α), beta (β), and so on—followed by the name of the parent constellation. For example, Sirius, the brightest star in the constellation Canis Major (Larger Dog), is also called Alpha (α) Canis Majoris.

The Celestial Sphere

Seen in the night sky, the stars all look equally far away. Therefore, it is easy to understand why the early Greeks regarded the stars as being fixed to a **celestial sphere** that surrounds Earth (Figure 21.18).

Figure 21.18 **Basic features of the celestial sphere**

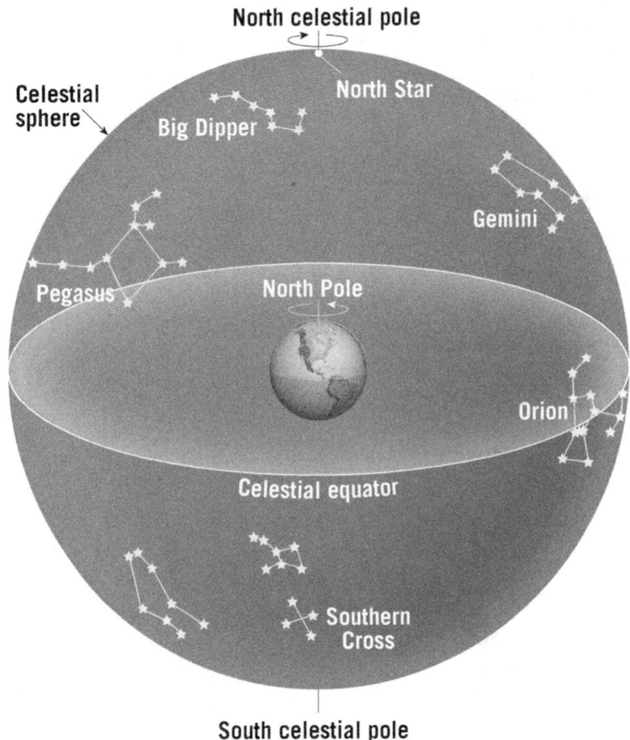

GEOGRAPHICS 21.1

Orion the Hunter

Orion, a prominent constellation located on the celestial equator, is visible around the world. It is one of the most conspicuous constellations to decorate the winter skies in the Northern Hemisphere. Named after a hunter in Greek mythology, its brightest stars are Rigel, a luminous blue-white supergiant, and Betelgeuse, a red supergiant. Many of the other brightest stars in the constellation are hot, massive blue stars.

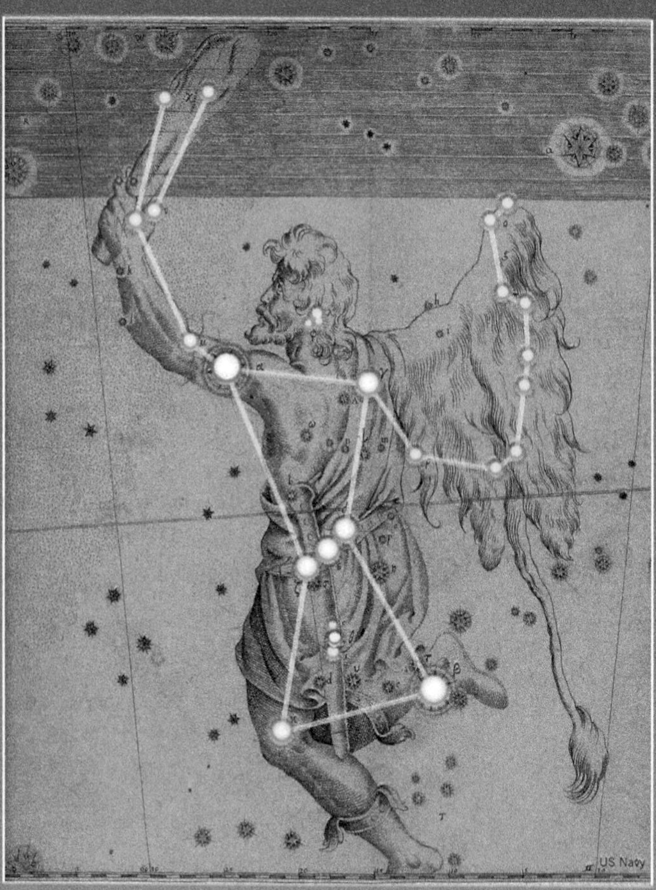
US Navy

▲ Artist's depiction of Orion based on Greek mythology. To ancient Chinese observers, these stars formed part of the mythologicl creature called the White Tiger of the West.

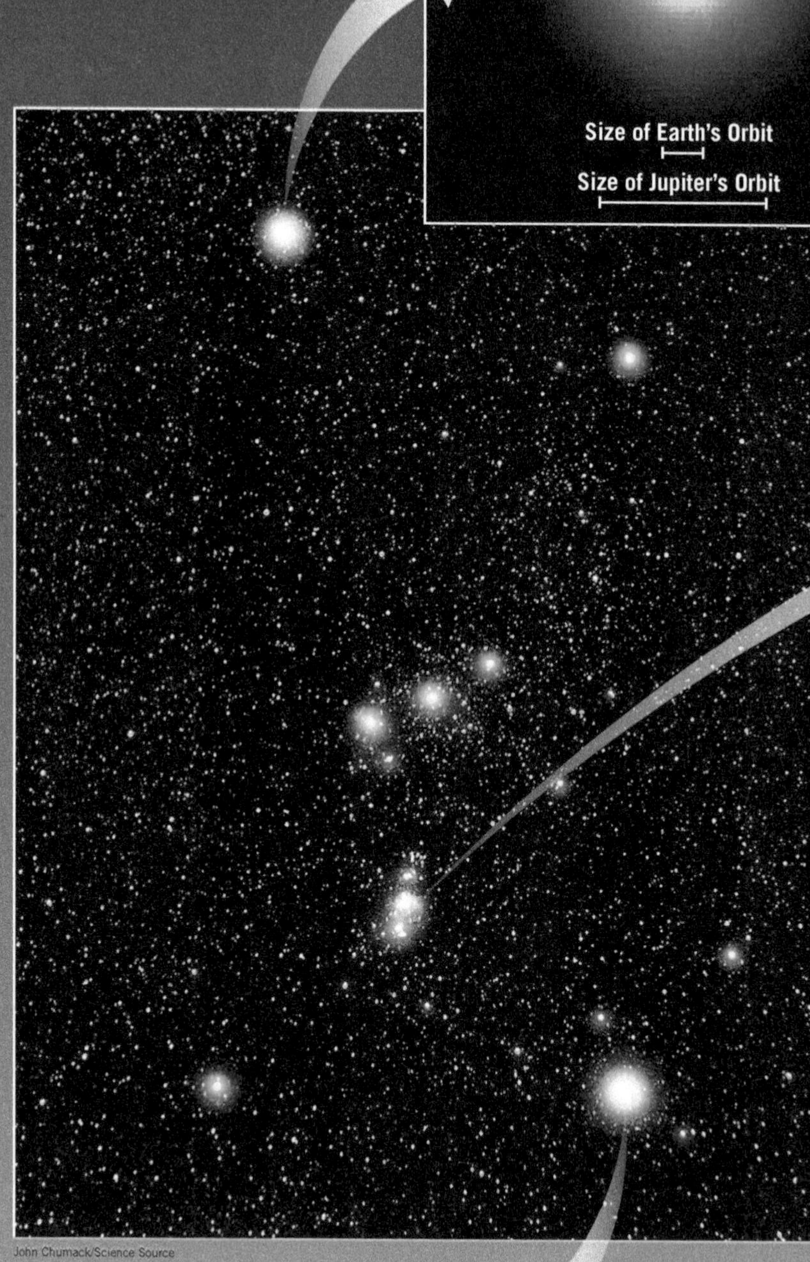

John Chumack/Science Source

Rigel, also known as Beta Orionis (β Orionis), is the brightest star in the constellation (it is sometimes outshone by Betelgeuse) and is the seventh brightest star in the night sky. Rigel is actually a triple star system. The primary star (Rigel A) of the three-star system is a blue-white massive star that is about 130,000 times more luminous than the Sun.

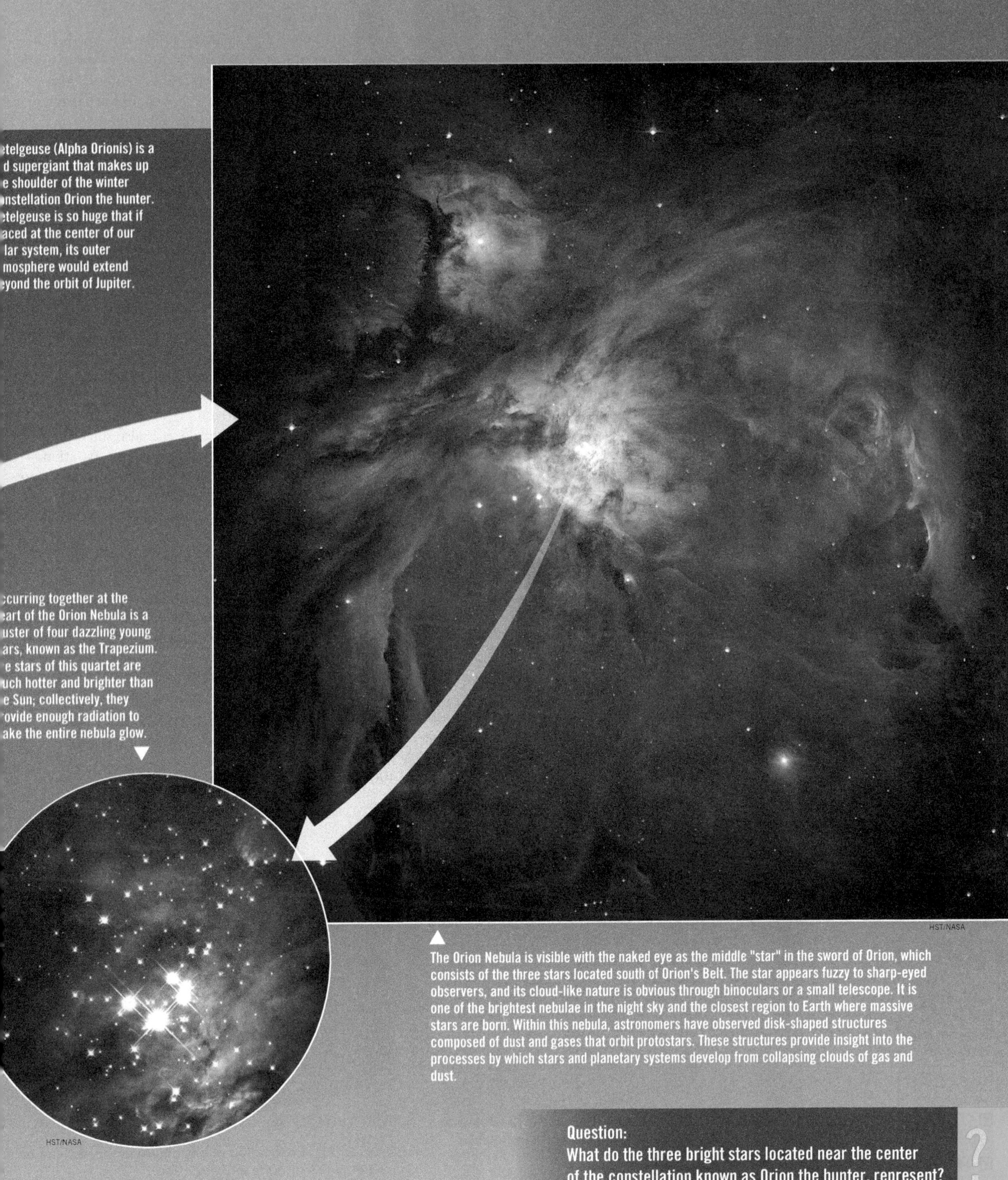

etelgeuse (Alpha Orionis) is a d supergiant that makes up e shoulder of the winter nstellation Orion the hunter. etelgeuse is so huge that if aced at the center of our lar system, its outer mosphere would extend eyond the orbit of Jupiter.

ccurring together at the eart of the Orion Nebula is a uster of four dazzling young ars, known as the Trapezium. e stars of this quartet are uch hotter and brighter than e Sun; collectively, they ovide enough radiation to ake the entire nebula glow.

HST/NASA

HST/NASA

The Orion Nebula is visible with the naked eye as the middle "star" in the sword of Orion, which consists of the three stars located south of Orion's Belt. The star appears fuzzy to sharp-eyed observers, and its cloud-like nature is obvious through binoculars or a small telescope. It is one of the brightest nebulae in the night sky and the closest region to Earth where massive stars are born. Within this nebula, astronomers have observed disk-shaped structures composed of dust and gases that orbit protostars. These structures provide insight into the processes by which stars and planetary systems develop from collapsing clouds of gas and dust.

Question:
What do the three bright stars located near the center of the constellation known as Orion the hunter, represent?

▲ **Figure 21.19 Star trails in the region of Polaris (north celestial pole), taken using a time exposure.** (Photo by Ed Lackert/Getty Image)

Although no such sphere exists, we still use this convenient concept to locate the stars and other celestial objects.

If you watch the sky over several hours, you will see that the stars remain fixed relative to each other while sweeping from east to west across the sky, as though they rotate around an axis that extends through Earth's North and South Poles. Although this apparent motion is actually due to Earth spinning on its axis, we designate a *north celestial pole* and the *south celestial pole* where an imaginary line through Earth's axis of rotation intersects the (imaginary) celestial sphere (see Figure 21.18). Now, imagine a plane passing through Earth's equator that extends outward until it intersects the celestial sphere. The intersection of this plane with the celestial sphere is called the *celestial equator*.

The north celestial pole happens to be very near the bright star whose various names reflect its location: "pole star," Polaris, and North Star. As shown in Figure 21.19, to an observer in the Northern Hemisphere, the stars appear to circle Polaris because it, like the North Pole, is the center of motion. Figure 21.20 shows how to locate the North Star using two stars, called pointer stars, located on the outside of the "dipper" in the easily located group of stars named the Big Dipper.

Measurements Using the Celestial Sphere

The celestial sphere is a useful tool for describing the location of objects in the night sky. However, when you go outside, you can see only one-half of the celestial sphere from any location, the other half being blocked from view by the ground. As shown in Figure 21.21, the location of any object in the sky can be described by its **direction** along the horizon and its **altitude** above the horizon. The direction is usually measured in degrees clockwise from due north. In Figure 21.21 the location of the object being described is located at 90 degrees along the horizon, or simply due east, and has an altitude of 50 degrees above the horizon.

When viewing the night sky, it is also sometimes useful to describe the apparent size of, or separation between, the objects we are viewing. The **angular size**, or **angular diameter**, of an object is the arc it appears to span in your field of view. To estimate the angular size of objects in the night sky, recall that a circle is 360 degrees. So if you go outside and point to the horizon and then gradually raise your arm so it is pointing directly overhead, your arm would have traversed an arc of 90 degrees, one-fourth of a circle. The Sun and full Moon have angular diameters of about ½ of a degree when viewed from Earth.

It is important to note that angular size does not indicate an object's true size because angular size also depends on distance. The Sun, which is 400 times larger in diameter than the Moon, has the same angular diameter as the Moon to observers on Earth because it is about 400 times farther away.

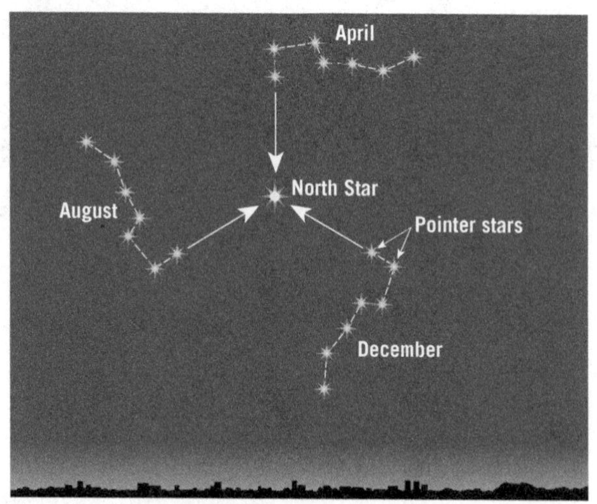

▶ **SmartFigure 21.20 Locating the North Star (Polaris) from the pointer stars in the Big Dipper**

TUTORIAL
https://goo.gl/Cnq9Pg

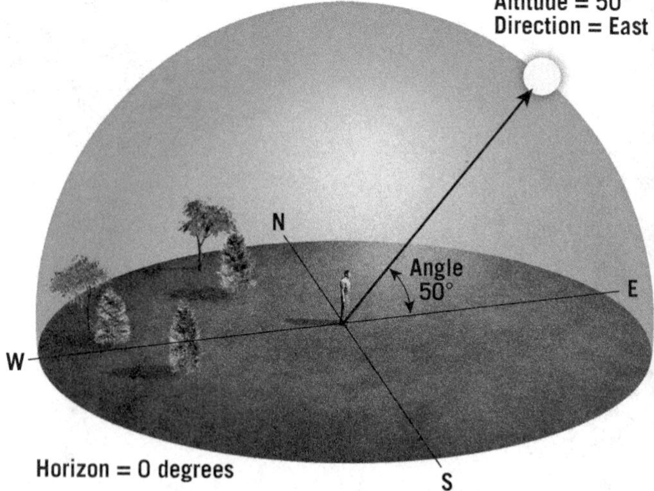

▲ **Figure 21.21 The location of any object in the sky can be described by its altitude and direction.**

The **angular distance** refers to the angle that appears to separate two objects in the sky. For example, the angular distance between the two pointer stars in the Big Dipper shown in Figure 21.20 is about 5 degrees. You can use your outstretched hand to estimate angular sizes and distances in the sky. Your hand with fingers fully spread spans about 20 degrees, your index finger about 1 degree, and the back of your clenched fist about 10 degrees (Figure 21.22).

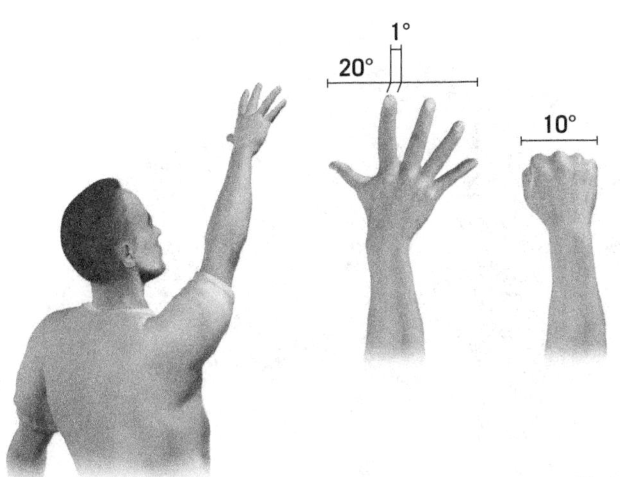

Figure 21.22 **Estimating angular sizes or distances with your outstretched hand**

CONCEPT CHECKS 21.3

1. How are constellations used in modern astronomy?
2. How many constellations are currently recognized?
3. Describe the *celestial sphere*.
4. Describe how you can use your outstretched hand to estimate angular distances.

21.4 The Motions of Earth

Describe the two primary motions of Earth and explain the difference between a solar day and a sidereal day.

The most basic motions of Earth are its daily **rotation** (spin) and its yearly **orbit**, or **revolution**, around the Sun. Earth rotates once each day on its axis, which is an imaginary line that connects the North Pole to the South Pole. A lesser motion is called *axial precession*.

Earth's Rotation: Spinning on Its Axis

The main consequences of Earth's rotation are day and night. Earth's rotation has become a standard tool for measuring time because it is so dependable and easy to use. You may be surprised to learn that Earth's rotation is measured in two ways, making two kinds of days. Most familiar is the **mean solar day**, the time interval from noon on one day until noon on the next, which averages about 24 hours. Noon is when the Sun has reached its highest point in the sky.

The **sidereal** (*sider* = star, *at* = pertaining to) **day**, on the other hand, is the time it takes for Earth to make one complete rotation (360 degrees) with respect to a star other than our Sun. The sidereal day is measured by the time required for a star to reappear at the identical position in the sky. The sidereal day has a period of 23 hours, 56 minutes, and 4 seconds (measured in solar time), which is almost 4 minutes shorter than the mean solar day. This difference results because the direction to distant stars changes only infinitesimally, whereas the direction to the Sun changes by almost 1 degree each day as Earth advances in its orbit around the Sun. This difference is shown in Figure 21.23.

In daily life, we use the mean solar day rather than the sidereal day to measure time because we want "noon" to occur when the Sun is overhead. In sidereal time, "noon" occurs 4 minutes earlier each day, so after 6 months, it occurs in the middle of the night.

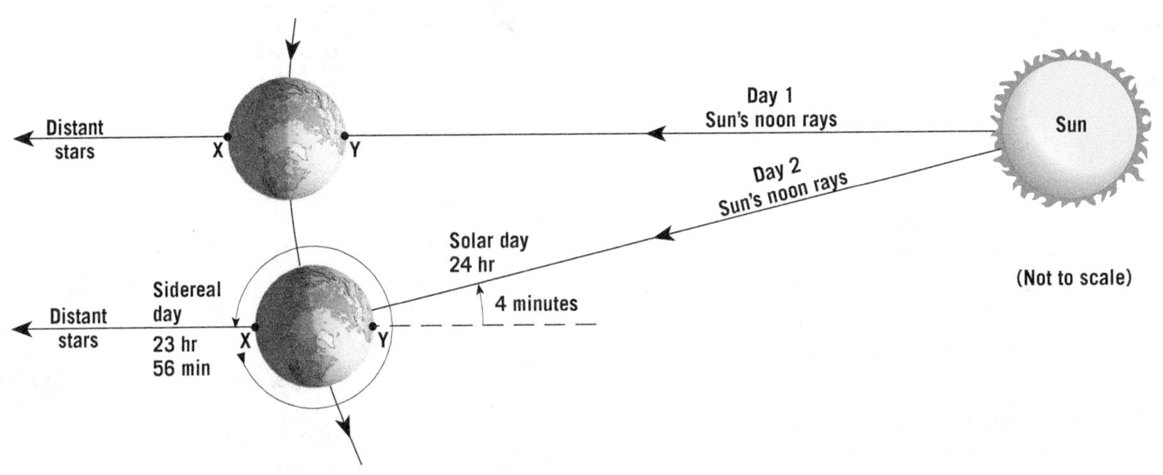

Figure 21.23 **The difference between a solar day and a sidereal day** Locations X and Y are directly opposite each other. It takes Earth 23 hours and 56 minutes to make one rotation with respect to the stars (sidereal day). However, notice that after Earth has rotated once with respect to the stars, point Y has not yet returned to the "noon position" with respect to the Sun. Earth has to rotate another 4 minutes to complete the solar day.

▲ **Figure 21.24 The ecliptic and the plane of the ecliptic** Earth's orbital motion causes the apparent position of the Sun to shift about 1 degree each day on the celestial sphere. The ecliptic is the path the Sun appears to trace through the stars. The plane of the ecliptic is an imaginary plane that connects the points on the ecliptic.

Observatories, however, use clocks that keep sidereal time. If a star is sighted directly south of an observatory at 9:00 P.M. (sidereal time), it will appear in the same direction at that time every (sidereal) day.

Earth's Orbit Around the Sun

Earth travels around the Sun in an elliptical orbit at an average speed of 107,000 kilometers (66,000 miles) per hour. Its average distance from the Sun is 150 million kilometers (93 million miles), but because its orbit is an ellipse, Earth's distance from the Sun varies. At **perihelion** (*peri* = near, *helios* = sun), it is 147 million kilometers (91.5 million miles) distant, which occurs about January 3 each year. At **aphelion** (*apo* = away, *helios* = sun), Earth is 152 million kilometers (94.5 million miles) distant, which occurs about July 4.

Because of Earth's orbital movement, the Sun appears to move relative to the constellations. Each day, this apparent movement amounts to an angular distance equal to about 1 degree, or about twice the Sun's apparent angular size. The apparent annual path of the Sun against the backdrop of the celestial sphere is called the **ecliptic** (Figure 21.24). The planets and the Moon travel in nearly the same plane as Earth. Hence, their paths on the celestial sphere also lie near the ecliptic.

The imaginary plane that connects points along the ecliptic is called the **plane of the ecliptic**. As measured from this imaginary plane, Earth's axis is tilted about 23½ degrees (see Figure 21.24). This angle is very important to Earth's inhabitants because the inclination of Earth's axis causes the yearly cycle of seasons, a topic discussed in detail in Chapter 16.

EYE ON THE UNIVERSE 21.1

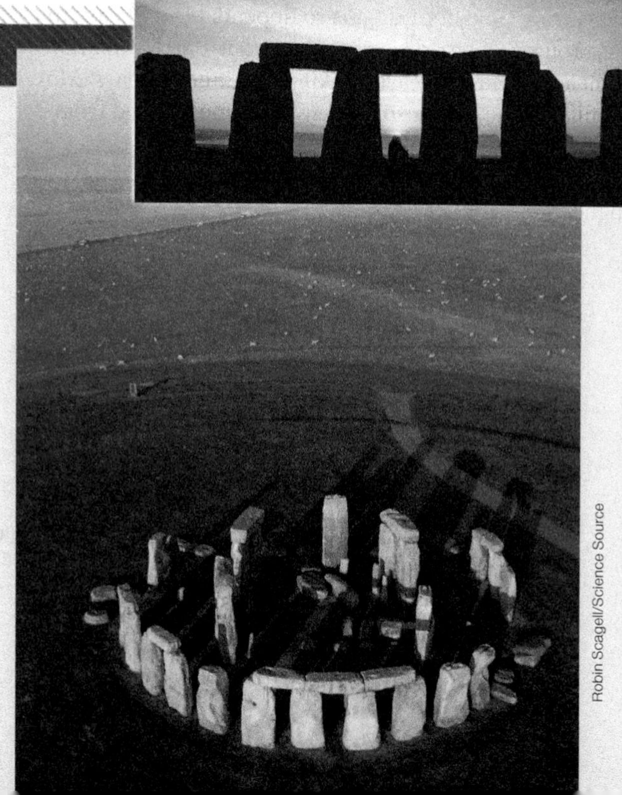

Stonehenge, a prehistoric monument whose construction began about 5000 years ago, is located north of the modern city of Salisbury, England. It is the remains of a ring of massive stones; the largest stands 9 meters (30 feet) tall and weighs 25 tons. Some of the smaller stones appear to have been transported 250 kilometers (150 miles), from southwestern Wales. The arrangement of the stones indicates that Stonehenge must have had an astronomical connection. Each year on June 21 or 22, an observer standing within the stone circle, looking though the entrance, would see the Sun rising above the heel stone, as shown in the inset image. (To answer the following questions, you may want to refer to the Section 16.4 on Earth–Sun relationships.)

QUESTION 1 *What is the significance of June 21–22? Does it mark the occurrence of the spring equinox, the summer equinox, the spring solstice, or the summer solstice?*

QUESTION 2 *On June 21–22, at what latitude do the rays of the Sun strike Earth's surface vertically?*

QUESTION 3 *Most world maps and globes label the latitude that is the answer to Question 2. What is this line of latitude called?*

Other Motions of Earth

In Chapter 10, we described three ways in which Earth's orbital motion and axial spin change cyclically over time (see Figure 10.27, page 337). The eccentricity of Earth's orbit changes slightly over a cycle that spans about 100,000 years; the tilt of its axis increases and decreases by a few degrees on a cycle of 41,000 years; and the axis precesses on a cycle of about 26,000 years.

Finally, Earth accompanies the Sun on its journey through space. The Sun orbits the galactic center, moving at approximately 250 kilometers (150 miles) per second and making a circuit every 230 million years. Meanwhile, our galaxy is engaged in a gravitational dance with its neighbor the Andromeda Galaxy; in the very far future, the two will merge.

CONCEPT CHECKS 21.4

1. Describe the three primary motions of Earth.
2. Explain the difference between the mean solar day and the sidereal day.
3. Define *ecliptic*.

21.5 Motions of the Earth–Moon System

Sketch the changing configuration of the Earth–Moon system that produces the regular cycle we call the phases of the Moon.

Earth has one natural satellite, the Moon. In addition to accompanying Earth in its annual trek around the Sun, our Moon orbits Earth about once each month. When viewed from a Northern Hemisphere perspective, the Moon moves counterclockwise (eastward) around Earth. The Moon's orbit is elliptical, causing the Earth–Moon distance to vary by about 6 percent, averaging 384,401 kilometers (238,329 miles).

The motions of the Earth–Moon system constantly change the relative positions of the Sun, Earth, and Moon. The results are some of the most noticeable astronomical phenomena: the *phases of the Moon* and the occasional *eclipses of the Sun and Moon*.

Lunar Motions

The cycle of the Moon through its phases requires 29½ days—a time span called the **synodic month**. This cycle was the basis for the first Roman calendar. However, this is the *apparent period* of the Moon's orbit around Earth and not the true period, which takes only 27½ days and is known as the **sidereal month**. The reason for the difference of nearly 2 days each cycle is shown in Figure 21.25. Notice that as the Moon orbits Earth, the Earth–Moon system also moves in an orbit around the Sun. Consequently, even after the Moon has made a complete orbit around Earth, it has not yet reached its starting position with respect to the Sun, which is directly between the Sun and Earth (*new-Moon phase*). This motion takes an additional 2 days.

Because the Moon rotates on its axis only once every 27⅓ days, any location on its surface experiences periods of daylight and darkness lasting about 2 weeks. This, along with the absence of an atmosphere, accounts for the high surface temperature of 127°C (261°F) on the day side of the Moon and the low surface temperature of −173°C (−280°F) on its night side.

An interesting fact is that tidal forces exerted by Earth on the Moon over geologic time have "locked" the Moon's rotation so that the same side of the Moon always faces Earth. That is, from Earth, we always see just one lunar hemisphere; the "far side" never rotates into view. As you can see from Figure 21.24, this means that the Moon rotates exactly once on its axis for every orbit around Earth. One consequence is that all the staffed *Apollo* Moon landings took place on the Earth-facing side, from which radio signals can reach Earth.

Phases of the Moon

The first astronomical phenomenon to be understood by ancient observers was the regular cycle of the **phases of the Moon**. On a monthly basis, we observe the phases

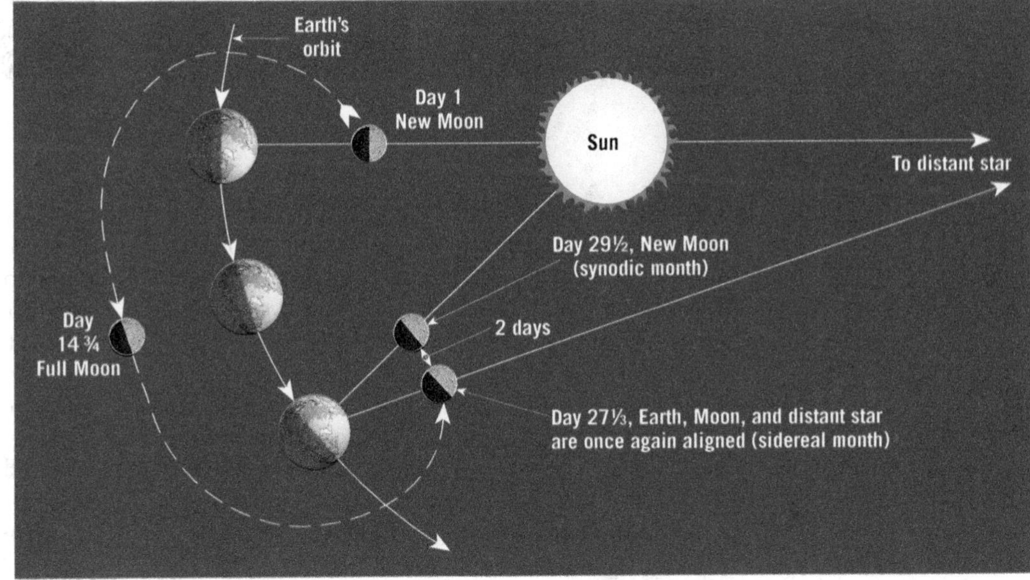

Figure 21.25 **The difference between a sidereal month (27⅓ days) and a synodic month (29½ days)** Distances and angles are not shown to scale.

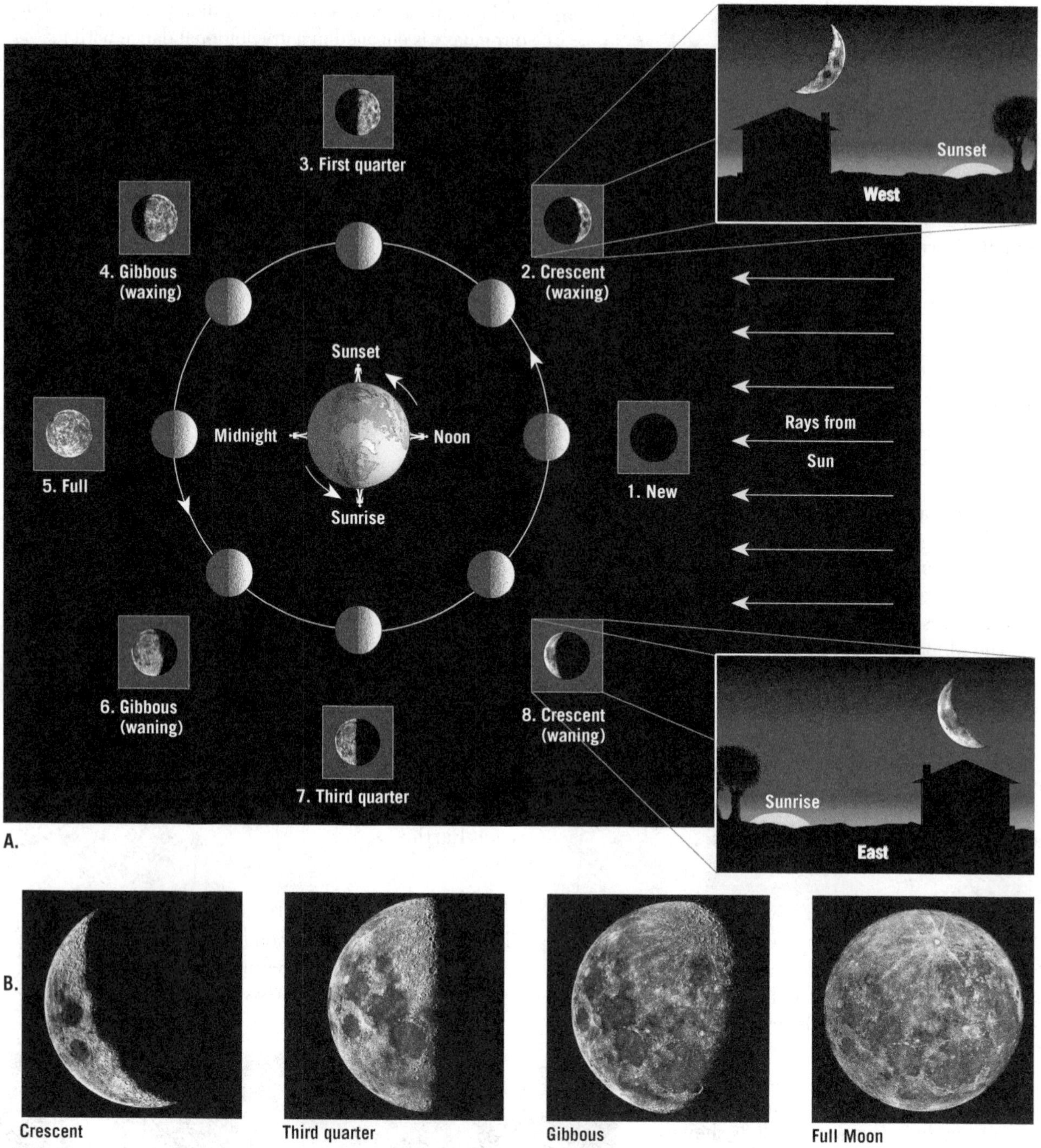

▶ SmartFigure 21.26
Phases of the Moon
A. The outer figures show the phases as seen from Earth. **B.** Compare these photographs with the diagram. (Photos © UC Regents/Lick Observatory)

TUTORIAL
https://goo.gl/q1xLsW

3. First quarter
4. Gibbous (waxing)
2. Crescent (waxing)
Sunset
West
Sunset
Midnight
Noon
5. Full
1. New
Rays from Sun
Sunrise
6. Gibbous (waning)
8. Crescent (waning)
7. Third quarter
Sunrise
East
A.

B.
Crescent
Third quarter
Gibbous
Full Moon

as a systematic change in the amount of the Moon that appears illuminated (**Figure 21.26**). We will choose the "new-Moon" position in the cycle as a starting point. About 2 days after the new Moon, a thin sliver (*crescent phase*) can be seen with the naked eye low in the western sky just after sunset. During the following week, the illuminated portion of the Moon that is visible from Earth increases (*waxing*) to a half-circle (*first-quarter phase*) that can be seen from about noon to midnight. In another week, the complete disk (*full-Moon phase*) can be seen rising in the east as the Sun sinks in the west. During the next 2 weeks, the percentage of the Moon that can be seen steadily declines (*waning*), until the Moon disappears altogether (*new-Moon phase*). The cycle soon begins anew, with the reappearance of the crescent Moon.

The lunar phases are a result of the motion of the Moon and the sunlight that is reflected from its surface (see Figure 21.26B). Half of the Moon is illuminated at all times (note the inner group of Moon sketches in Figure 21.26A). But to an earthbound observer, the percentage of the bright side that is visible depends on the location of the Moon with respect to the Sun and Earth. When the Moon lies *between* the Sun and Earth, none of its bright side faces Earth, so we see the new-Moon ("no-Moon") phase. Conversely, when the Moon lies on the side of Earth opposite the Sun, all of its lighted side faces Earth, so we see the full Moon. At all positions between these extremes, an intermediate amount of the Moon's illuminated side is visible from Earth.

CONCEPT CHECKS 21.5

1. Compare the *synodic month* with the *sidereal month*.
2. The Moon rotates on its axis very slowly (once in 27⅓ days). How does this affect the lunar surface temperature?
3. What phenomenon results from the fact that the Moon rotates once on its axis for every orbit that it makes around Earth?

21.6 Eclipses of the Sun and Moon

Sketch the configuration of the Earth–Moon–Sun system that produces a lunar eclipse and the configuration that produces a solar eclipse.

Along with understanding the Moon's phases, the early Greeks also realized that eclipses are simply shadow effects. When the Moon passes directly between Earth and the Sun, which can occur only during the new-Moon phase, it casts a dark shadow on Earth, producing a **solar eclipse** (*eclipsis* = failure to appear) (Figure 21.27). Conversely, the Moon is eclipsed (**lunar eclipse**) when it moves within Earth's shadow, a situation that is possible only during the full-Moon phase (Figure 21.28).

Why does a solar eclipse not occur with every new-Moon phase and a lunar eclipse with every full Moon? They would if the orbit of the Moon lay exactly along the plane of Earth's orbit. However, the Moon's orbit is inclined about 5 degrees to the plane of the ecliptic. Thus, during most new-Moon phases, the shadow of the Moon passes either above or below Earth; during most full-Moon phases, the shadow of Earth misses the Moon. An eclipse can take place only when a new- or full-Moon phase occurs while the Moon's orbit crosses the plane of the ecliptic.

Because these conditions are normally met only twice a year, the usual number of eclipses per year is four. These occur as a set of one solar and one lunar eclipse, followed 6 months later with another set. Occasionally the alignment is such that three eclipses can occur in a 1-month period—at the beginning, middle, and end. These occur as a solar eclipse flanked by two lunar eclipses or vice versa. Furthermore, it occasionally happens that the first set of eclipses for the year occurs at the very beginning of a year, the second set in the middle, and a third set before the calendar year ends, resulting in six eclipses in that year. More rarely, if one of these sets consists of three eclipses, the total number of eclipses in a year can reach seven, which is the maximum.

During a total lunar eclipse, Earth's circular shadow moves slowly across the disk of the full Moon. When totally eclipsed, the Moon is completely within Earth's shadow but is still visible as a coppery disk because Earth's atmosphere bends some long-wavelength light (red) into its shadow. Some of this light reflects off the Moon and back to us. A total eclipse of the Moon can last up to 4 hours and is visible to anyone on the side of Earth facing the Moon.

During a solar eclipse, the Moon *completely* hides the Sun (producing a total solar eclipse)

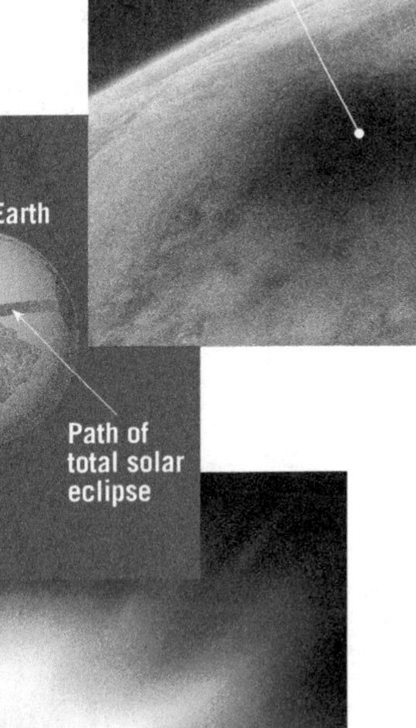

SmartFigure 21.27
Solar eclipse A. Observers in the zone of the umbral shadow see a total solar eclipse. Those located in the penumbra see only a partial eclipse. The path of the solar eclipse moves eastward across Earth. **B.** During a total solar eclipse, the blotted-out solar disk is surrounded by an irregularly shaped halo called the *corona*. (Photo by Maksim Nikalayenka/Shutterstock) **C.** Image of Moons shadow as seen from space. (NASA)

TUTORIAL
https://goo.gl/gUGTAq

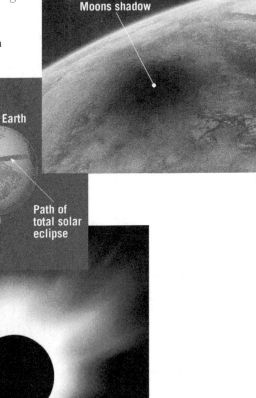

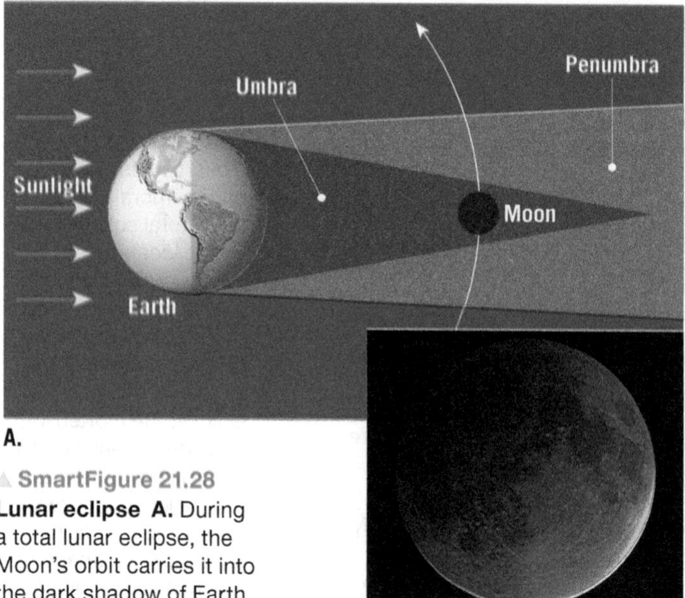

SmartFigure 21.28
Lunar eclipse A. During a total lunar eclipse, the Moon's orbit carries it into the dark shadow of Earth (umbra). During a partial eclipse, only a portion of the Moon enters the umbra. **B.** During a total lunar eclipse, a dark, copper-colored Moon is observed. The Moon appears this way because a small amount of light reaches it after being bent (refracted) and reddened by Earth's atmosphere. The light is red for the same reason that sunsets are red. (Photo by Eckhard Slawik/Science Source)

TUTORIAL
https://goo.gl/KSG5Mu

Figure 21.29 **Montage of images showing a complete solar eclipse** This sequence of photos moving from the left to the right shows the stages of a total solar eclipse. (Photo by Dr. Fred Espenak/Science Source)

along only a narrow track on Earth's surface—never wider than 275 kilometers (170 miles), about the size of South Carolina. This zone of total eclipse is called the *umbra* (see Figure 21.27). Outside this track, a much broader swath experiences a partial eclipse, in which the Moon covers just part of the Sun's disk, casting a partial shadow called the *penumbra*. Although solar eclipses occur several times a year, any given location on Earth experiences one only rarely. In the contiguous United States, a total solar eclipse will occur on August 21, 2017, followed by another on April 8, 2024.

An observer lucky enough to experience a total solar eclipse will see the Moon slowly block the Sun from view and the sky darken (**Figure 21.29**). Near totality, a sharp drop in temperature of a few degrees is experienced. The solar disk is completely blocked for a maximum of only 7 minutes because the Moon's shadow (the umbra) is so small. At totality, the dark Moon is seen covering the complete solar disk, and only the Sun's brilliant white outer atmosphere (corona) is visible (see Figure 21.27B).

CONCEPT CHECKS 21.6

1. Sketch the locations of the Sun, Moon, and Earth during a solar eclipse and during a lunar eclipse.
2. How many eclipses normally occur each year?
3. Solar eclipses are slightly more common than lunar eclipses. Why, then, is it more likely that your region of the country will experience a lunar eclipse than a solar eclipse?
4. How long can a total eclipse of the Moon last? How about a total eclipse of the Sun?

21 CONCEPTS IN REVIEW
Origins of Modern Astronomy

21.1 Ancient Astronomy

Explain the geocentric view of the solar system and describe how it differs from the heliocentric view.

KEY TERMS: geocentric, heliocentric, Ptolemaic system, retrograde motion

- The early Greeks held a geocentric ("Earth-centered") view of the universe, in which Earth is a motionless sphere at the center of the universe, orbited by the Moon, the Sun, and the planets that were known at the time—Mercury, Venus, Mars, Jupiter, and Saturn.
- The early Greeks believed that the stars traveled daily around Earth on a transparent, hollow celestial sphere. In 141 C.E. Claudius Ptolemy documented this geocentric view, now called the Ptolemaic system, which became the dominant view of the solar system for over 15 centuries.

21.2 The Birth of Modern Astronomy

List and describe the contributions to modern astronomy of Nicolaus Copernicus, Tycho Brahe, Johannes Kepler, Galileo Galilei, and Isaac Newton.

KEY TERMS: astronomical unit (AU), inertia, law of universal gravitation

- Modern astronomy evolved during the 1500s and 1600s, facilitated by scientists going beyond merely describing what is observed to explaining why the universe behaves as it does.
- Nicolaus Copernicus (1473–1543) proposed that the Sun, rather than Earth, is the center of the solar system, although he perpetuated the erroneous view that orbits must be circular. His Sun-centered view was rejected by the establishment of his day.
- Tycho Brahe's (1546–1601) observations of the planets were far more precise than any made previously and are his legacy to astronomy.
- Johannes Kepler (1571–1630) used Tycho Brahe's observations to usher in a new astronomy with the formulation of his three laws of planetary motion.
- After constructing his own telescope, Galileo Galilei (1564–1642) made many important discoveries that supported the Copernican view of a Sun-centered solar system. This included discovering moons around Jupiter, thus proving that Earth was not the center of all planetary motion.
- Sir Isaac Newton (1642–1727) demonstrated that the orbit of a planet is a result of the planet's inertia (its tendency to move in a straight line) and the Sun's gravitational attraction, which bends the planet's path into an elliptical orbit.

21.3 Patterns in the Night Sky

Describe how constellations are used in modern astronomy.

KEY TERMS: constellation, celestial sphere, direction, altitude, angular size (angular diameter), angular distance

- As early as 5000 years ago, people began naming the configurations of stars, called *constellations*, in honor of mythological characters or great heroes. Today, 88 constellations are recognized that divide the sky into regions, just as state boundaries divide the United States.
- Although we realize that the stars are not fixed to a celestial sphere that surrounds Earth, we use this convenient idea to describe the locations of the stars and other celestial objects.
- The location of any object in the sky can be described by its direction along the horizon and its altitude above the horizon. The direction is usually measured in degrees clockwise from due north. The apparent size of a celestial object can be described as the angle of view it spans (angular size); the same measure is used for apparent distances.

21.4 The Motions of Earth

Describe the two primary motions of Earth and explain the difference between a solar day and a sidereal day.

KEY TERMS: rotation, orbit (revolution), mean solar day, sidereal day, perihelion, aphelion, ecliptic, plane of the ecliptic

- The most basic motions of Earth are its daily rotation on its axis (spin) and its yearly orbit, or revolution, around the Sun.
- Earth's rotation can be measured in two ways, making two kinds of days. The mean solar day is the time interval from one noon (the time of day when the Sun is highest in the sky) to the next, which averages about 24 hours. In contrast, the sidereal day is the time it takes for Earth to make one complete rotation with respect to a star other than the Sun, a period of 23 hours, 56 minutes, and 4 seconds.
- Earth travels around the Sun in an elliptical orbit at an average distance from the Sun of 150 million kilometers (93 million miles). At perihelion (closest to the Sun), which occurs in January, Earth is 147 million kilometers (91.5 million miles) from the Sun. At aphelion (farthest from the Sun), which occurs in July, Earth is 152 million kilometers (94.5 million miles) distant. The imaginary plane that connects Earth's orbit with the celestial sphere is called the plane of the ecliptic.

21.5 Motions of the Earth–Moon System

Sketch the changing configuration of the Earth–Moon system that produces the regular cycle we call the phases of the Moon.

KEY TERMS: synodic month, sidereal month, phases of the Moon

- One of the first astronomical phenomena to be understood was the regular cycle of the phases of the Moon. The phases of the Moon are a result of the motion of the Moon around Earth and the portion of the bright side of the Moon that is visible to an observer on Earth.
- The cycle of the Moon through its phases requires 29½ days, a time span called the synodic month. However, the true period of the Moon's orbit around Earth is 27⅓ days and is known as the sidereal month. The difference of nearly 2 days is due to the fact that as the Moon orbits Earth, the Earth–Moon system also advances in its orbit around the Sun.

NASA's Goddard Space Flight Center

21.6 Eclipses of the Sun and Moon

Sketch the configuration of the Earth–Moon–Sun system that produces a lunar eclipse and the configuration that produces a solar eclipse.

KEY TERMS: solar eclipse, lunar eclipse

- In addition to understanding the Moon's phases, the early Greeks also realized that eclipses are simply shadow effects. When the Moon passes directly between Earth and the Sun, which can occur only during the new-Moon phase, it casts a dark shadow on Earth, producing a solar eclipse.
- A lunar eclipse takes place when the Moon moves within the shadow of Earth during the full-Moon phase.
- Because the Moon's orbit is inclined about 5 degrees to the plane that contains Earth and the Sun (the plane of the ecliptic), during most new- and full-Moon phases, no eclipse occurs. Only if a new- or full-Moon phase occurs as the Moon crosses the plane of the ecliptic can an eclipse take place. The usual number of eclipses is four per year.

? Does the accompanying image show a total eclipse of the Sun or a total eclipse of the Moon? Explain your reasoning.

NASA

GIVE IT SOME THOUGHT

1 Refer to Figure 21.3 and imagine that Eratosthenes had measured the difference in the angles of the noonday Sun between Syene and Alexandria to be 10 degrees instead of 7 degrees. Consider how this measurement would have affected his calculation of Earth's circumference as you answer the following questions.
   a. Would this new measurement lead to a more accurate calculation?
   b. Would this new measurement lead to an estimate for the circumference of Earth that is larger or smaller than Eratosthenes's original estimate?

2 The accompanying diagram shows two imaginary planets orbiting the Sun, both adhering to Kepler's laws and Newton's law of universal gravitation. Use the diagram and the fact that Planet A has a nearly circular orbit while Planet B has a highly elliptical orbit to answer the following questions.
   a. Does Planet B travel faster or slower when it is closest to the Sun than at other times?
   b. Which planet takes longer to orbit the Sun?
   c. Does the gravitational attraction between these planets increase or decrease as their orbits move them closer together?
   d. Which planet has the highest average orbital speed?
   e. Which planet travels at about the same speed throughout its orbit?

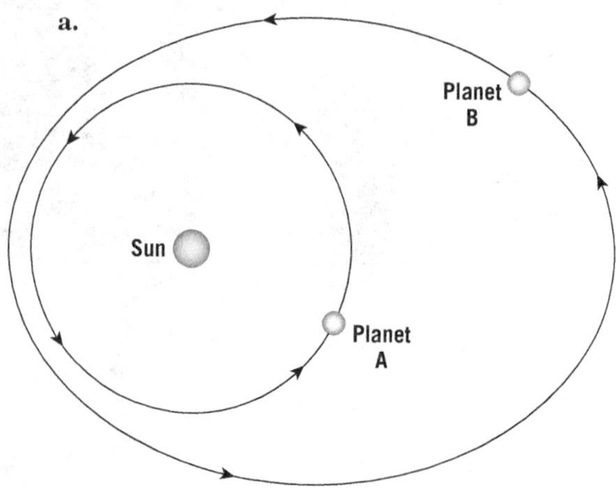

3 Galileo used his telescope to observe the planets and moons in our solar system. These observations allowed him to determine the positions and relative motions of the Sun, Earth, and other objects in the solar system. Refer to Figure 21.15A, which shows an Earth-centered solar system, and Figure 21.15B, which shows a Sun-centered solar system, to complete the following:
   a. Describe the phases of Venus that an observer on Earth would see for the Earth-centered model of the solar system.
   b. Describe the phases of Venus that an observer on Earth would see for the Sun-centered model of the solar system.
   c. Explain how Galileo used observations of the phases of Venus to determine whether Venus orbits Earth or the Sun.

4 Refer to the accompanying diagram, which shows three asteroids (A, B, and C), each of which feels the gravitational pull of its partner asteroid. How does the strength of the gravitational force felt by each asteroid (A, B, and C) compare? (Assume that all these asteroids are composed of the same material.)

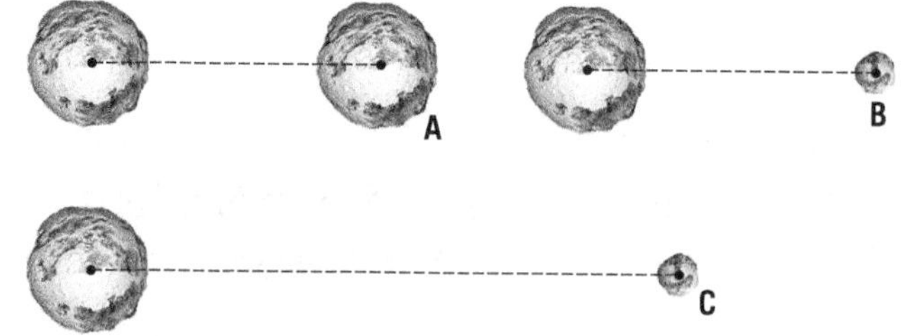

5 Refer to the accompanying diagram, which shows two pairs of asteroids, Pair D and Pair E. Is it possible for the asteroids in Pair D to be experiencing the same degree of gravitational force as the asteroids in Pair E? Explain your answer.

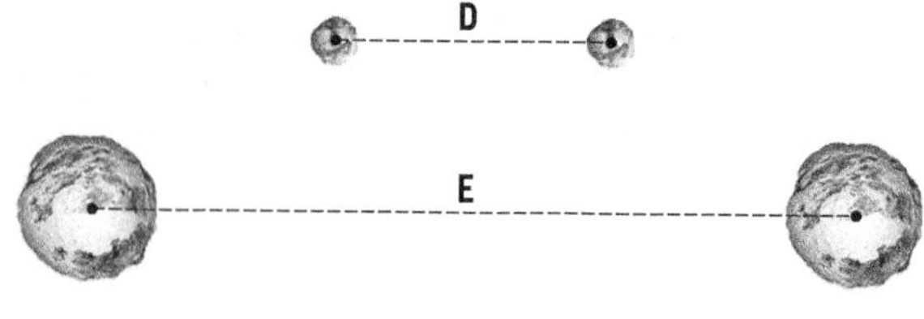

6 Imagine that Earth rotates on its axis at half its current rate. How much time would be required to capture the time-exposure photo shown in Figure 21.19?

7 If we were able to reverse the direction of Earth's rotation, would the solar day be longer, shorter, or the same?

8 Refer to Figure 21.26 to complete the following:
 a. What is different about the crescent phase that precedes the new-Moon phase and the one that follows the new-Moon phase?
 b. What phase of the Moon occurs approximately 1 week after the new Moon? 2 weeks?

9 If the Moon's orbit were precisely aligned with the plane of Earth's orbit, how many eclipses (solar and lunar) would occur in a 6-month period?

10 Refer to the accompanying photo of the Moon to complete the following:
 a. When you observe the phase of the Moon shown, is the Moon waxing or waning?
 b. What time of day can this phase of the Moon be observed?

©UC Regents/Lick Oservatory

EXAMINING THE EARTH SYSTEM

1 Currently, Earth is closest to the Sun in January (147 million kilometers [91.5 million miles]) and farthest from the Sun in July (152 million kilometers [94.5 million miles]). As a result of the precession of Earth's axis, 12,000 years from now perihelion (closest approach) will occur in July, and aphelion (most distant position) will take place in January. Assuming no other changes, how might this change *average* summer temperatures for your location? What about *average* winter temperatures? What might the impact be on the biosphere and hydrosphere? (To aid your understanding of the effect of Earth's orbital parameters on the seasons, you may want to review the section "Variations in Earth's Orbit" in Section 10.6, page 336.)

DATA ANALYSIS

Observing the Night Sky

Astronomers of the past kept detailed records of the changing positions of the Sun, moon, planets, and stars. These observations were used to monitor the passage of time and guide travelers to their destinations. In addition to aiding navigation, tracking constellations is also a fun hobby.

ACTIVITIES

Go to the Harvard-Smithsonian Center for Astrophysics Sky Report page at https://www.cfa.harvard.edu/skyreport to determine what is in the night sky in the current month.

1 Which planets can be seen in the sky this month? In which direction should you look for each planet? Is each planet visible after sunset, or before sunrise?

2 Are there any comets visible this month? If so, which one(s)? When are they visible?

3 Are there any meteor showers visible this month? If so, which one(s)? When are they most visible? Are there any limitations to visibility? How many meteors per hour may be seen?

Go to http://www.beckstromobservatory.com/ (the Beckstrom Observatory page) and read about the observatory.

4 Where is Beckstrom Observatory located? What is the size of the observatory's main telescope?

5 Click on Tonight's Sky and read the information about the Evening and Morning Star Maps. How should you use these maps to find constellations?

Download and print the current month's Evening or Morning Star Map, depending on when you are observing the sky. Bring this map with you when you go outside to find constellations. Try to find a dark location with as little light pollution as possible.

6 Based on the information from Question 1, which planets are you able to locate? Depending on your observation date, which comet(s) or meteor shower(s) are you able to see?

7 Using the Star Map, which constellations are you able to locate? (Note: It will take a few minutes for your eyes to adjust to the dark.) If you have difficulty locating the celestial objects listed for the current month, suggest reasons why these may not be visible to you.

MasteringGeology™ Looking for additional review and test prep materials? Visit the Study Area in MasteringGeology to enhance your understanding of this chapter's content by accessing a variety of resources, including Self-Study Quizzes, Geoscience Animations, SmartFigure Tutorials, Mobile Field Trips, *Project Condor* Quadcopter videos, *In the News* articles, flashcards, web links, and an optional Pearson eText.

www.masteringgeology.com

22

Touring Our Solar System*

FOCUS ON CONCEPTS

Each statement represents the primary learning objective for the corresponding major heading within the chapter. After you complete the chapter, you should be able to:

22.1 Describe the formation of the solar system according to the nebular theory. Compare and contrast the terrestrial and Jovian planets.

22.2 List and describe the major features of Earth's Moon and explain how maria basins were formed.

22.3 Outline the principal characteristics of Mercury, Venus, and Mars. Describe their similarities to and differences from Earth.

22.4 Summarize and compare the features of Jupiter, Saturn, Uranus, and Neptune, including their ring systems.

22.5 List and describe the principal characteristics of the small bodies that inhabit the solar system.

*This chapter was revised with the assistance of Professors Teresa Tarbuck and Mark Watry.

This selfie of the *Curiosity* rover as it moves up the slopes of Mount Sharp on Mars is a composite of several images captured by a camera located at the end of the rover's robotic arm. (Photo courtesy of NASA)

PLANETARY SCIENTISTS STUDY THE FORMATION AND EVOLUTION

of the bodies in our solar system and beyond—including the eight planets and myriad smaller objects, such as moons, dwarf planets, asteroids, comets, and meteoroids. Studying these objects provides valuable insights into the dynamic processes that operate on Earth. Understanding how other atmospheres evolve helps scientists build better models for predicting climate change. In addition, seeing how erosional forces work on other bodies allows us to observe the many ways landscapes are created. Finally, the uniqueness of Earth, a body that harbors life, is revealed through the exploration of other planetary bodies.

22.1 Our Solar System: An Overview

Describe the formation of the solar system according to the nebular theory. Compare and contrast the terrestrial and Jovian planets.

The Sun is the center of a revolving system, trillions of miles wide, consisting of eight planets, their satellites, and numerous smaller bodies—dwarf planets, asteroids, comets, and meteoroids. An estimated 99.85 percent of the mass of our solar system is contained within the Sun. Collectively, the planets account for most of the remaining 0.15 percent. Starting from the Sun, the planets are Mercury, Venus, Earth, Mars, Jupiter, Saturn, Uranus, and Neptune (**Figure 22.1**).

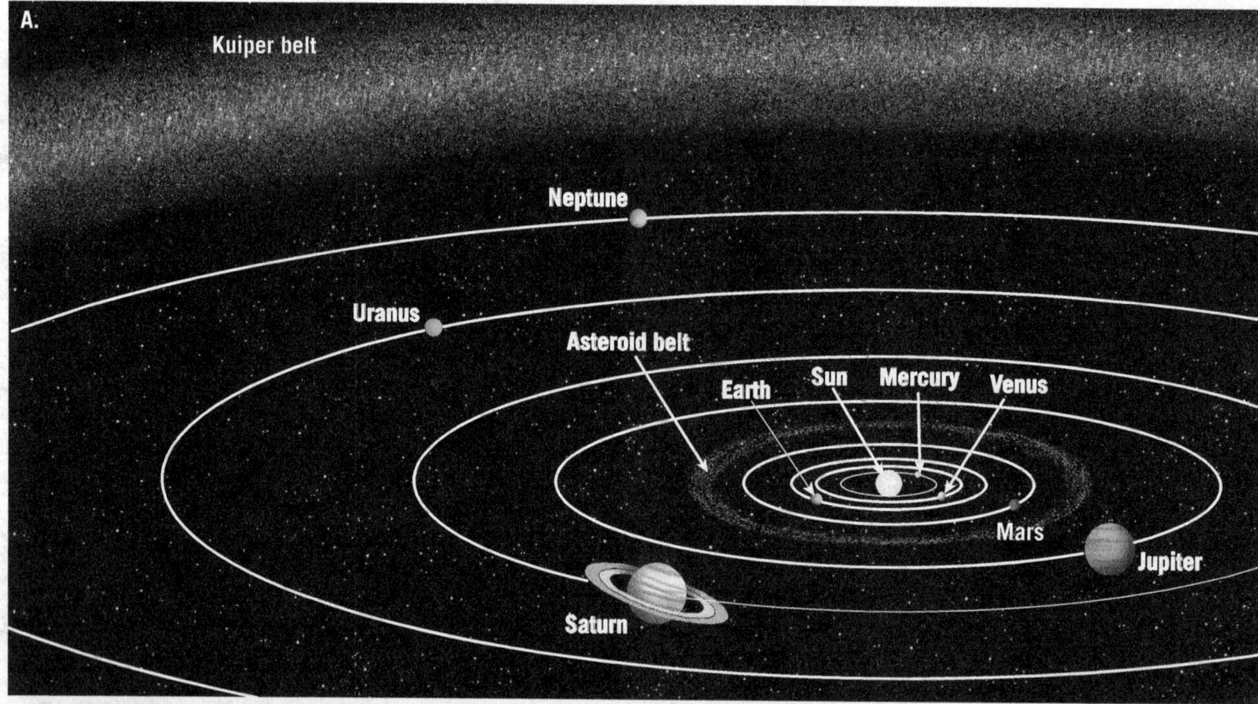

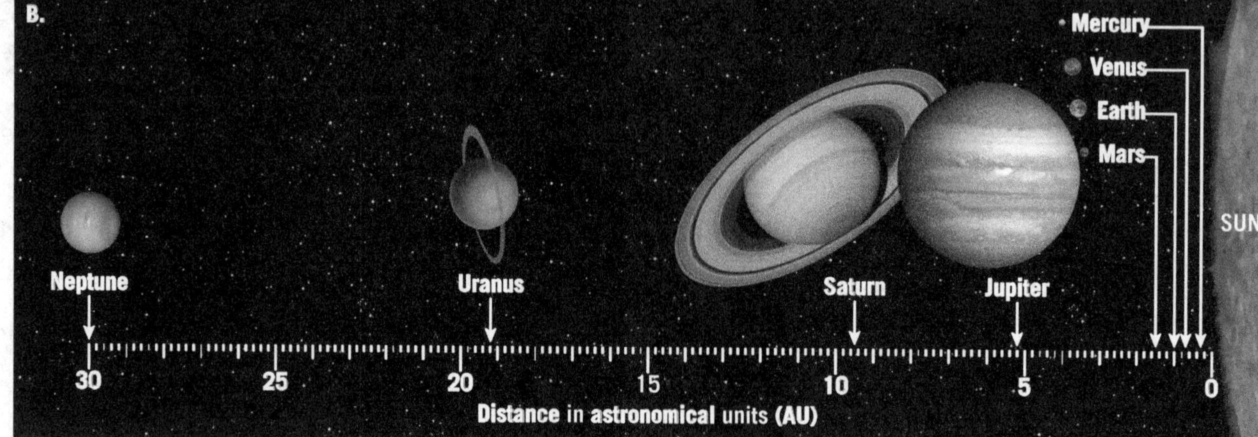

▶ SmartFigure 22.1
Planetary orbits
A. Artistic view of the solar system; the planets are not drawn to scale. **B.** The distances of the planets from the Sun and relative sizes of the planets shown using two different scales. The distances are given in astronomical units (AU), where 1 AU is equal to the mean distance from Earth to the Sun—150 million kilometers (93 million miles). If the Sun and planets were shown at the scale used for the distances, they would be about 5000 times smaller than illustrated here.

TUTORIAL
https://goo.gl/AflR5d

Tethered to the Sun by gravity, the planets travel in the same direction, on slightly elliptical orbits; those nearest the Sun travel fastest (Table 22.1). Mercury has the highest orbital velocity, 48 kilometers (30 miles) per second, and the shortest period of revolution around the Sun, 88 Earth days. By contrast, Neptune has an orbital speed of just 5.3 kilometers (3.3 miles) per second and requires 165 Earth-years to complete one revolution. Most large bodies orbit the Sun approximately in the same plane. The planets' inclination with respect to the Earth–Sun orbital plane, known as the *ecliptic*, is shown in Table 22.1.

Nebular Theory: Formation of the Solar System

The **nebular theory**, which describes the formation of the solar system, proposes that the Sun and the planets formed from the same rotating cloud of interstellar gases, called the *solar nebula*. Much of what we know about the formation of the Sun and other stars comes from observing regions in which stars are forming currently. In all cases, star formation occurs within expansive interstellar clouds (see Section 24.2, page 721). Based on this observation, we conclude that the **solar nebula** began as a slowly rotating cloud of cold, mostly gaseous material composed of over 98 percent hydrogen and helium; all other elements combined made up less than 2 percent of the total. The solid material consisted of tiny grains often referred to as *interstellar dust*. As this enormous cloud of dust and gases collapsed, gravitational energy was converted into energy of motion and into thermal energy, causing the contracting gases to greatly increase in temperature. Most of the material in the solar nebula collected in the center, where temperatures were hottest, forming the *protosun*.

As the solar nebula collapsed, it also spun faster and faster, as happens when ice skaters pull their arms in as they spin. The rotational motion caused some of the material to form a flattened, rotating disk that surrounded the hot protosun. Within this rotating disk, matter gradually cooled and condensed into tiny metallic grains and clumps of icy and rocky material. Repeated collisions resulted in most of that material clumping together into increasingly larger chunks that eventually became boulder-sized objects called **planetesimals**, which means "pieces of planets." The largest planetesimals grew rapidly because of their stronger gravitational attraction and quickly became hundreds of kilometers in diameter.

The compositions of the planetesimals were determined largely by their proximity to the protosun. As you might expect, temperatures were highest in the inner solar system and decreased toward the outer edge of the rotating disk. Between the 88-day orbit of Mercury and the 687-day orbit of Mars, the planetesimals were composed mainly of materials with high melting temperatures—metals and rocky substances. In this region,

Table 22.1 Planetary Data

Planet	Symbol	Mean Distance from Sun: AU*	Mean Distance from Sun: Millions of Miles	Mean Distance from Sun: Millions of Kilometers	Orbital Period	Inclination of Orbit	Orbital Velocity: mi/s	Orbital Velocity: km/s
Mercury	☿	0.39	36	58	88 days	7°00′	29.5	47.5
Venus	♀	0.72	67	108	245 days	3°24′	21.8	35.0
Earth	⊕	1.00	93	150	365.25 days	0°00′	18.5	29.8
Mars	♂	1.52	142	248	687 days	1°51′	14.9	24.1
Jupiter	♃	5.20	483	778	12 years	1°18′	8.1	13.1
Saturn	♄	9.54	886	1427	30 years	2°29′	6.0	9.6
Uranus	⛢	19.18	1783	2870	84 years	0°46′	4.2	6.8
Neptune	♆	30.06	2794	4497	165 years	1°46′	3.3	5.3

Planet	Period of Rotation Around Axis	Diameter: Miles	Diameter: Kilometers	Relative Mass (Earth = 1)	Average Density (g/cm³)	Polar Flattening (%)	Eccentricity**	Number of Known Satellites†
Mercury	59 days	3015	4878	0.06	5.4	0.0	0.206	0
Venus	243 days	7526	12,104	0.82	5.2	0.0	0.007	0
Earth	$23^h56^m04^s$	7920	12,756	1.00	5.5	0.3	0.017	1
Mars	$24^h37^m23^s$	4216	6794	0.11	3.9	0.5	0.093	2
Jupiter	9^h56^m	88,700	143,884	317.87	1.3	6.7	0.048	67
Saturn	10^h30^m	75,000	120,536	95.14	0.7	10.4	0.056	62
Uranus	17^h14^m	29,000	51,118	14.56	1.2	2.3	0.047	27
Neptune	16^h07^m	28,900	50,530	17.21	1.7	1.8	0.009	14

*AU = astronomical unit, Earth's mean distance from the Sun

**Eccentricity is a measure of the amount an orbit deviates from a circular shape. The larger the number, the less circular the orbit.

†Includes all satellites discovered as of July 2016. Satellites are celestial bodies that orbit a planet rather than orbiting a star like the Sun.

through repeated collisions and accretion, these asteroid-sized rocky bodies combined to form four **protoplanets** that eventually became the terrestrial planets Mercury, Venus, Earth, and Mars.

By contrast, the gas giants—Jupiter, Saturn, Uranus, and Neptune—have orbital periods that range from about 12 to 165 years and contain more than 150 times the mass of the terrestrial planets. These planets accreted from planetesimals that originated beyond the orbit of Mars (beyond the *frost line*), where temperatures were low enough so compounds that remained gases in the inner solar system condensed to form ices. As a result, these planetesimals contained high percentages of ices—mainly ices of water, carbon dioxide, ammonia, and methane—as well as smaller amounts of rocky and metallic debris. The fact that the outer reaches of the solar system contained much larger quantities of ices than metallic and rocky material accounts in part for the large sizes and low densities of the outer planets. The two most massive planets, Jupiter and Saturn, also had surface gravities sufficient to attract and retain even large quantities of hydrogen and helium gas, the lightest elements.

During the first few hundred million years of its existence, the solar system had a dynamic and violent history. This was a period of intense bombardment as the planets cleared their orbits by collecting much of the remaining leftover material. The "scars" of this period are still evident on the Moon's surface. Because of the gravitational effect of the planets, particularly Jupiter, small bodies were flung into planet-crossing orbits or interstellar space. The small fraction of interplanetary matter that survived this violent period became either asteroids or comets, the latter residing mainly in the outer reaches of the solar system.

Despite the fact that the nebular theory successfully accounts for nearly all of the major features of the solar system, recent discoveries have led some planetary scientists to conclude that the nebular model may need some tweaking. In particular, based on thousands of *exoplanets* (worlds that orbit other stars; see Section 12.1, page 374) discovered by surveys, such as NASA's *Kepler* mission, planetary scientists began to realize that solar systems that resemble ours are relatively rare. Many of the first-known exoplanets were "hot Jupiters," gas giants speeding around their stars with orbital periods of a few days. Furthermore, of the planetary systems discovered thus far, most contain one or more "super Earths" (planets a few times larger than Earth) with orbital periods of less than 100 days. Recall that Mercury, the runt of the solar system and the innermost planet, has an orbit of 88 days.

How do giant planets end up in scorching proximity to a star, where ices can't possibly exist? Why does the inner solar system lack super Earths that orbit close to the Sun? Finding answers to these and other questions related to the development of our solar system will give planetary scientists much to contemplate.

The Planets: Internal Structures and Atmospheres

The planets fall into two groups, based on location, size, and density: the **terrestrial (Earth-like) planets** (Mercury, Venus, Earth, and Mars) and the **Jovian (Jupiter-like) planets** (Jupiter, Saturn, Uranus, and Neptune). Because of their locations relative to the Sun, the four terrestrial planets are also known as *inner planets*, and the four Jovian planets are known as *outer planets*. A correlation exists between planetary locations and sizes: The inner planets are substantially smaller than the outer planets, also called *gas giants*. For example, Neptune (the smallest Jovian planet) has nearly 4 times the diameter and 17 times the mass of Earth or Venus.

Other properties that differ among the planets include density, chemical composition, orbital period, and number of satellites (see Table 22.1). Variations in the chemical compositions of planets are largely responsible for their density differences. The average density of the terrestrial planets, which consist mainly of rock and metal, is about 5 times the density of water, whereas the average density of the Jovian planets, with a high proportion of hydrogen, helium, and low-density compounds, is only 1.5 times that of water. In fact, Saturn has a density only 0.7 times that of water, which means it would float in a sufficiently large tank of water. The outer planets are also characterized by long orbital periods and numerous satellites.

Recall from Chapter 5 that shortly after Earth formed, segregation of material formed three major layers, defined by their chemical composition—the crust, mantle, and core. This type of chemical differentiation also occurred in the other planets, but because the terrestrial planets are compositionally different from the Jovian planets, the nature of these layers differs as well (Figure 22.2).

Interiors of Terrestrial Planets

The terrestrial planets are dense, having relatively large cores composed mainly of iron and nickel. Based on seismological evidence, we know that Earth's outer core is molten. We also know that Earth's strong magnetic field is generated by convection within the molten outer core, aided by moderately rapid planetary rotation.

Mars's core is thought to be partially molten but not hot enough to support convection; as a result, Mars lacks a magnetic field. Venus does not have a magnetic field either, even though its core is thought to have a molten metallic layer much like Earth's. Presumably, it lacks a magnetic field either because the core is not hot enough to drive convection or because the planet's 243-day rotation period is too slow to generate a magnetic field. Researchers were surprised to discover that Mercury, despite its small size and slow 59-day period of rotation, possesses a measurable magnetic field, albeit only 1 percent the strength of Earth's. This may be the result of Mercury's partially molten metallic core, which is unusually large compared to the size of the planet.

Silicate minerals and other lighter compounds make up the mantles of the terrestrial planets. Finally, the silicate crusts of terrestrial planets are relatively thin compared to their mantles.

Interiors of the Jovian Planets The two largest Jovian planets, Jupiter and Saturn, likely have small, solid cores consisting of iron compounds, like the cores of the terrestrial planets, and rocky material similar to Earth's mantle. Progressing outward, the layer above the core consists of liquid hydrogen that is under extremely high temperatures and pressures. Substantial evidence indicates that under these conditions, hydrogen behaves like a metal in that its electrons move freely about and efficiently conduct both heat and electricity. Jupiter's intense magnetic field is thought to be the result of electric currents flowing within a spinning layer of liquid metallic hydrogen. Saturn's magnetic field is much weaker than Jupiter's, due to its smaller shell of liquid metallic hydrogen. Above this metallic layer, scientists believe both Jupiter and Saturn are composed of molecular liquid hydrogen intermixed with helium. The outermost layer is a gaseous atmosphere consisting mainly of hydrogen and helium, along with small amounts of water, ammonia, methane, and other substances.

Uranus and Neptune also have small iron-rich rocky cores, but their mantles are likely hot, dense water, methane, and ammonia. Like Jupiter and Saturn, they have atmospheres dominated by hydrogen and helium.

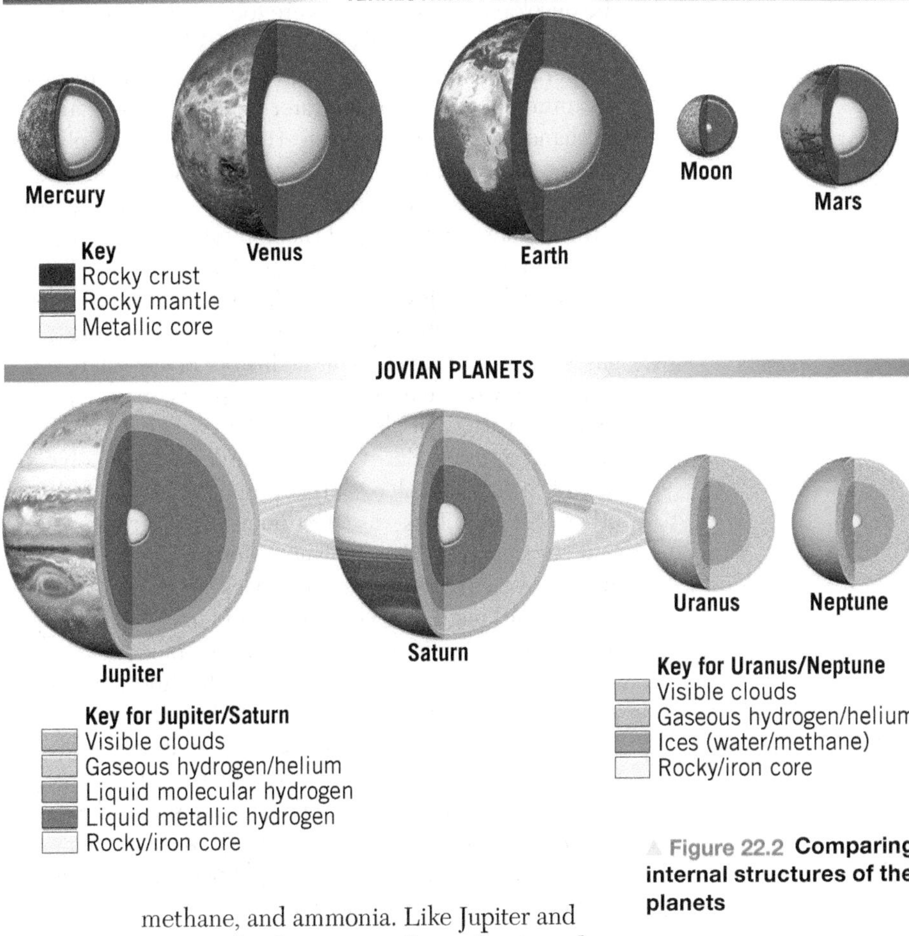

Figure 22.2 **Comparing internal structures of the planets**

The Atmospheres of the Planets The Jovian planets have very thick atmospheres composed mainly of hydrogen and helium, with lesser amounts of water, methane, ammonia, and other hydrogen compounds. By contrast, the terrestrial planets, including Earth, have relatively meager atmospheres that are typically dominated by carbon dioxide or nitrogen, and most small solar system bodies are airless. Two factors explain these significant differences: solar heating (temperature) and gravity (Figure 22.3). These variables determine what planetary gases, if any, were captured by the planets during the formation of the solar system and which were ultimately retained.

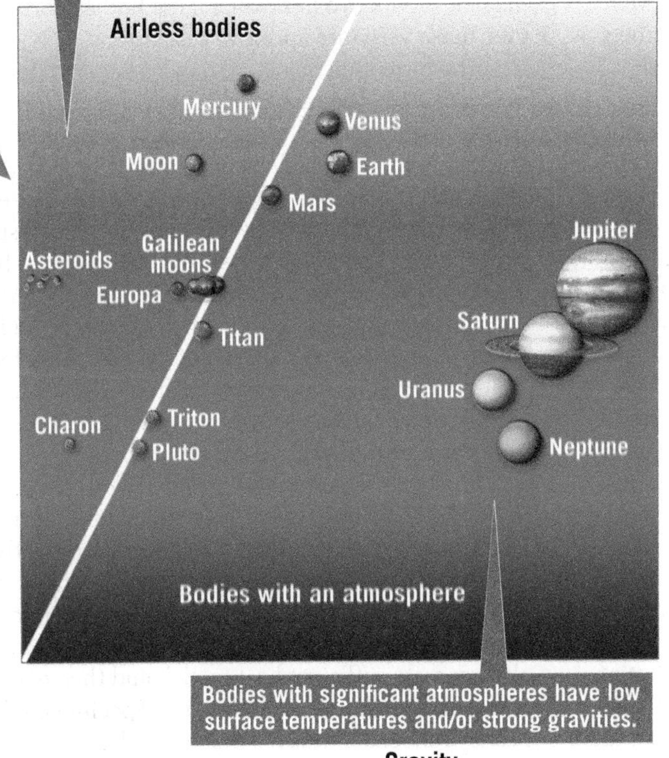

SmartFigure 22.3 **Bodies with atmospheres versus airless bodies** Airless worlds have relatively warm surface temperatures and/or weak gravities. Bodies with significant atmospheres have low surface temperatures and/or strong gravities.

TUTORIAL
https://goo.gl/g4Snrw

During planetary formation, the inner regions of the developing solar system were too hot for ices and gases to condense, but the outer planets formed where temperatures were low and solar heating of planetesimals was minimal. This allowed water vapor, ammonia, and methane to condense into ices. Hence, the Jovian planets contain large amounts of these volatiles. As the planets grew, the largest Jovian planets, Jupiter and Saturn, also attracted large quantities of the lightest gases, hydrogen and helium, due to their strong gravitational fields. This explains why the Jovian planets have thick atmospheres.

How did Earth acquire water and other volatile gases? It seems that early in the history of the solar system, gravitational tugs by the gas giants, mainly Jupiter and Saturn, sent planetesimals into very eccentric orbits. As a result, Earth was bombarded with icy objects (mainly comet-like objects) that brought in water and other elements. This was a fortuitous event for organisms that currently inhabit our planet. Mercury, our Moon, and numerous other small bodies lack significant atmospheres, although they certainly would have been bombarded by icy bodies early in their development.

Airless bodies develop where solar heating is strong and/or gravities are weak. Simply stated, *small warm bodies* have a better chance of losing their atmospheres: Gas molecules are more energetic (and hence faster-moving) on a warm body, and they need less speed to escape the weak gravity of a small body. Mercury, the smallest and least massive of the eight planets, has a low surface gravity, which makes holding on to an atmosphere challenging. In addition, because it is the planet closest to the Sun, it is constantly bombarded by solar wind. Mercury therefore has the thinnest atmosphere of all the planets.

The somewhat larger terrestrial planets, Earth, Venus, and Mars, retain some heavy gases, including nitrogen, carbon dioxide, oxygen, and even water vapor. However, their atmospheres are miniscule compared to their total mass. Early in their development, the terrestrial planets may have had thicker atmospheres. Over time, however, these primitive atmospheres gradually changed as light gases trickled away into space. For example, Earth's atmosphere continues to leak hydrogen and helium (the two lightest gases) into space. This phenomenon occurs near the top of Earth's atmosphere, where air is so tenuous that nothing stops the fastest-moving particles from flying off into space. The speed required to escape a planet's gravity is called **escape velocity**. Because hydrogen is the lightest gas, it most easily reaches the speed needed to overcome Earth's gravity. Billions of years in the future, the loss of hydrogen (one of the components of water) will eventually "dry out" Earth's oceans, ending its hydrologic cycle.

The massive Jovian planets have strong gravitational fields, which partially explains their thick atmospheres. Furthermore, because of their great distances from the Sun, solar heating is minimal. Because the molecular motion of a gas is temperature dependent, even hydrogen and helium move too slowly to escape the gravitational pull of the Jovian planets. Cold temperatures also explain why Saturn's moon Titan, which is small compared to Earth but much farther from the Sun and therefore colder, retains an atmosphere.

Planetary Impacts

Impacts between solar system bodies have occurred throughout the history of the solar system. On bodies that have little or no atmosphere (like the Moon) and, therefore, no air resistance, even the smallest pieces of interplanetary debris (meteorites) can reach the surface. At high enough velocities, this debris can produce microscopic cavities on individual mineral grains. By contrast, large **impact craters** result from collisions with massive bodies, such as asteroids and comets.

Planetary impacts were considerably more common in the early history of the solar system than they are today. Following that early period of intense bombardment, the rate of cratering diminished dramatically, and it now remains essentially constant. Because weathering and erosion are almost nonexistent on the Moon and Mercury, evidence of their cratered past is clearly evident.

On larger bodies, the presence of an atmosphere may cause the impacting objects to break up and/or decelerate. For example, Earth's atmosphere causes meteoroids with masses of less than 10 kilograms (22 pounds) to lose up to 90 percent of their speed as they penetrate the atmosphere. Therefore, impacts of low-mass bodies produce only small craters on Earth. Our atmosphere is much less effective in slowing large bodies; fortunately, they make very rare appearances.

The formation of a large impact crater is illustrated in Figure 22.4. The meteoroid's high-speed impact compresses the material it strikes, causing an almost instantaneous rebound, which ejects material from the surface. On Earth, impacts can occur at speeds that exceed 50 kilometers (30 miles) per second. Impacts at such high speeds produce shock waves that compress both the impactor and the material being impacted. Almost instantaneously, the over-compressed material rebounds and explosively ejects material out of the newly formed crater. In addition, large craters often exhibit a central peak, such as the one shown in Figure 22.5. These central peaks also result from crustal rebound.

Much of the material expelled, called *ejecta*, lands in or near the crater, where it accumulates to form a rim. The remaining material forms a blanket around the crater. Upon impact, large meteoroids may eject large projectiles that strike the surrounding landscape to generate smaller structures called *secondary craters*. Large meteoroids also generate sufficient heat to melt and then eject some of the impacted rock as glass beads. Specimens of glass beads produced in this manner, as well as melt breccia consisting of broken fragments welded by the heat of impact, have been collected on Earth and on the Moon.

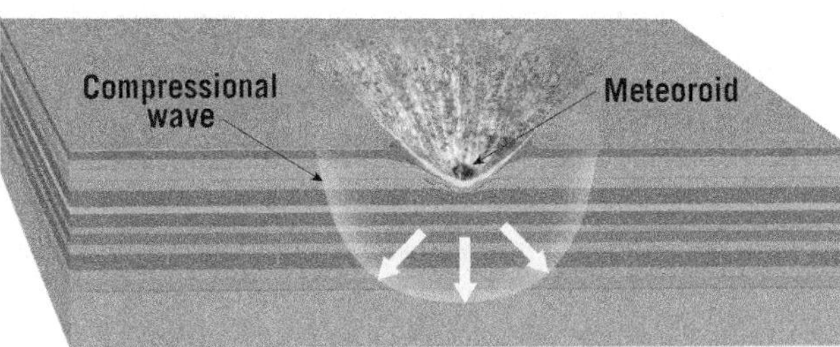

A. The energy of a rapidly moving body is transformed into heat and shock waves.

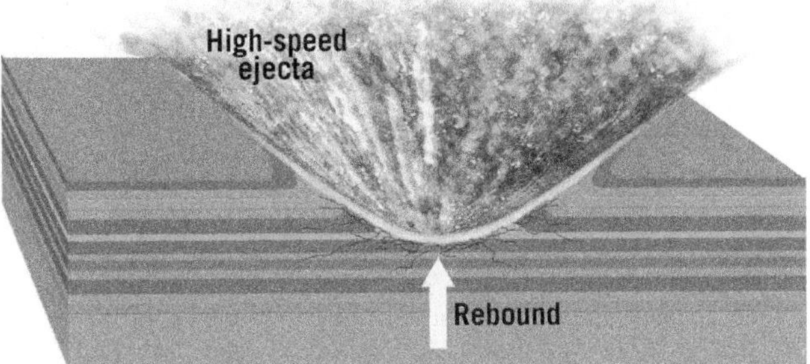

B. The rebound of over-compressed rock causes debris to be explosively ejected from the crater. Some of this material may melt and be deposited as glass beads.

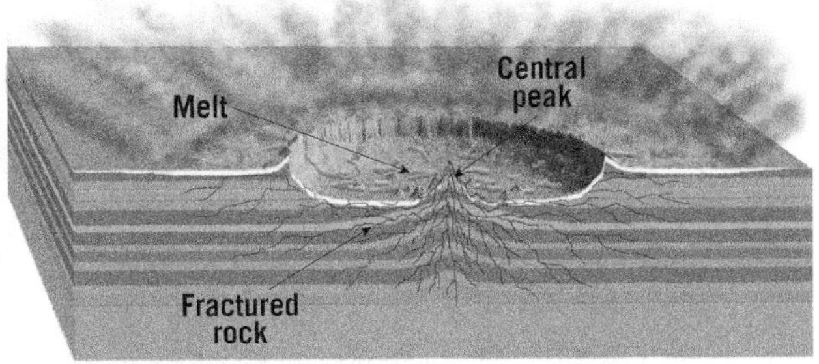

C. Large craters may contain areas of rock that was melted by the impact and a rebounded central peak.

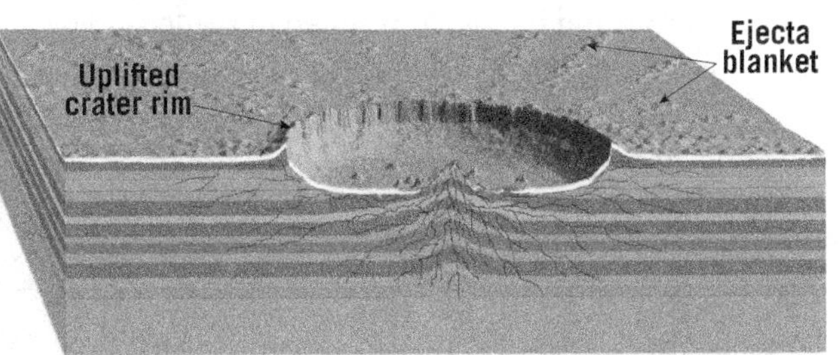

D. Ejected material forms a "blanket" around the crater.

▲ Figure 22.4 **Formation of an impact crater**

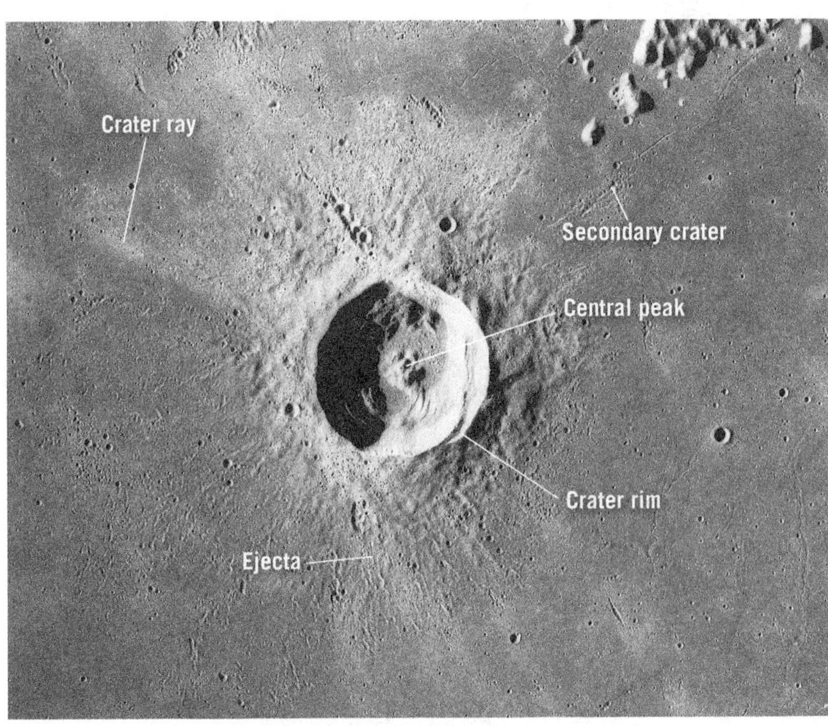

▲ Figure 22.5 **Lunar crater Euler** This 20-kilometer-wide (12-mile-wide) crater is located in the southwestern part of Mare Imbrium. Clearly visible are the bright rays, central peak, secondary craters, and large accumulation of ejecta near the crater rim. (Courtesy of NASA)

CONCEPT CHECKS 22.1

1. Briefly outline the steps in the formation of our solar system, according to the nebular theory.
2. By what criteria are planets considered either terrestrial or Jovian?
3. What accounts for the large density differences between the terrestrial and Jovian planets?
4. Explain why the terrestrial planets have meager atmospheres compared to the Jovian planets.

22.2 Earth's Moon: A Chip Off the Old Block

List and describe the major features of Earth's Moon and explain how maria basins were formed.

The Earth–Moon system is unique partly because of the Moon's large size relative to other bodies in the inner solar system. The Moon's diameter is 3475 kilometers (2160 miles), about one-fourth of Earth's 12,756 kilometers (7926 miles), and its surface temperature averages about 107°C (225°F) during daylight hours and −153°C (−243°F) at night. Because its period of rotation on its axis equals its period of revolution around Earth, the same lunar hemisphere always faces Earth. All of the landings of staffed *Apollo* missions were confined to the side of the Moon that faces Earth.

The Moon's density is 3.3 times that of water, comparable to that of mantle rocks on Earth but considerably less than Earth's average density (5.5 times that of water).

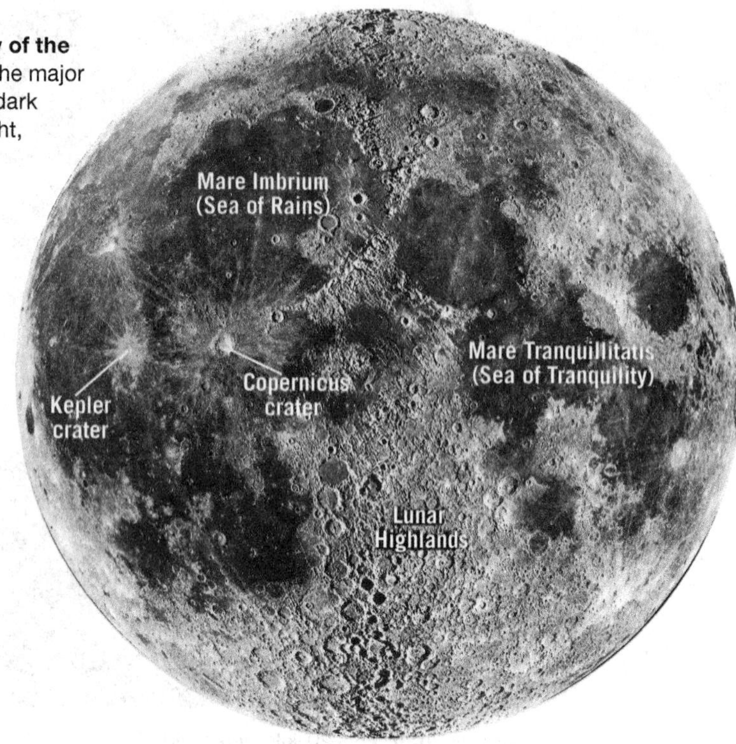

Figure 22.6
Telescopic view of the lunar surface The major features are the dark maria and the light, highly cratered highlands. (Lick Observatory Publications Office)

The Moon's low mass relative to Earth results in a lunar gravitational attraction that is one-sixth that of Earth. The Moon's small mass (and low gravity) is the primary reason it was not able to retain an atmosphere.

How Did the Moon Form?

Current models show that Earth is too small to have formed with a moon, particularly one so large. Furthermore, a captured moon would likely have an eccentric orbit similar to the captured moons that orbit the Jovian planets.

Astronomers generally agree that the Moon formed as a result of a collision between a Mars-sized body and a youthful, semimolten Earth about 4.5 billion years ago. During this collision, some of the ejected debris was thrown into orbit around Earth and gradually coalesced to form the Moon. Computer simulations show that most of the ejected material would have come from the rocky mantle of the impactor, while its core was assimilated into the growing Earth. This *impact model* is consistent with the Moon having a proportionately smaller core than Earth's and, hence, a lower density.

The Lunar Surface

When Galileo first pointed his telescope toward the Moon, he observed two different types of terrain: dark lowlands and brighter, highly cratered highlands (Figure 22.6). Because the dark regions appeared to be smooth, resembling seas on Earth, they were called **maria** (*mar* = sea, singular *mare*). The *Apollo 11* mission showed conclusively that the maria are exceedingly smooth plains composed of basaltic lavas. These vast plains are strongly concentrated on the side of the Moon facing Earth and cover about 16 percent of the lunar surface. The lack of large volcanic cones on these surfaces is evidence of high eruption rates of very fluid basaltic lavas similar to the Columbia Plateau flood basalts on Earth.

By contrast, the Moon's light-colored areas resemble Earth's continents, so the first observers dubbed them *terrae* (Latin for "lands"). These areas are now called **lunar highlands** because they are elevated several kilometers above the maria. Rocks retrieved from the highlands are mainly breccias, pulverized by massive bombardment early in the Moon's history. The arrangement of terrae and maria has resulted in the legendary "face" of the "man in the Moon."

EYE ON THE UNIVERSE 22.1

Mariner 9 obtained this image of Phobos, one of two tiny moons of Mars. Phobos has a diameter of only 24 kilometers (15 miles). The two moons of Mars were not discovered until the late 1800s because their size made them nearly impossible to see with earlier telescopes.

QUESTION 1 *In what way is Phobos similar to Earth's Moon?*

QUESTION 2 *List characteristics of Phobos that make it different from Earth's Moon.*

QUESTION 3 *Do an Internet search to learn how Phobos and its companion moon, Deimos, got their names.*

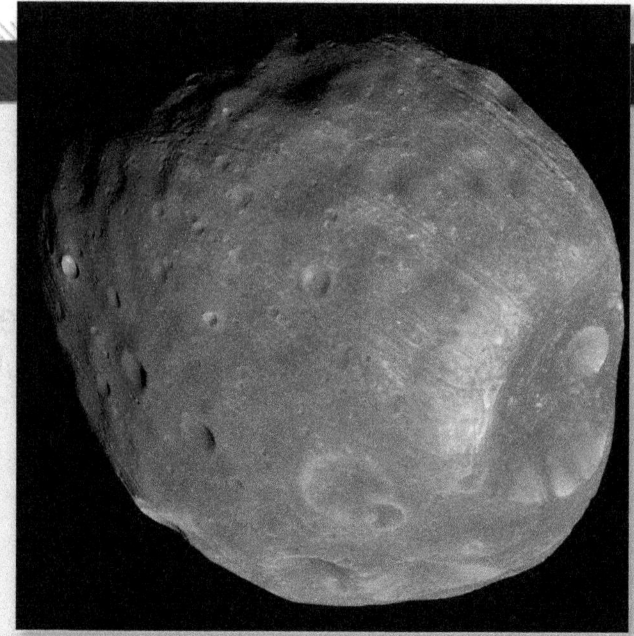

Some of the most obvious lunar features are impact craters. A meteoroid 3 meters (10 feet) in diameter can blast out a crater 50 times larger, or about 150 meters (500 feet) in diameter. The larger craters shown in Figure 22.6, such as Kepler and Copernicus (32 and 93 kilometers [20 and 58 miles] in diameter, respectively), were created from bombardment by bodies 1 kilometer (0.62 mile) or more in diameter.

History of the Lunar Surface The evidence used to unravel the history of the lunar surface comes primarily from radiometric dating of rocks returned from *Apollo* missions and studies of crater densities—counting the number of craters per unit area. Because planets and moons have continually been struck by meteoroids throughout their history, the greater the density of craters on a surface feature, the older the feature is inferred to be. Such evidence suggests that, after the Moon coalesced, it passed through four phases: (1) formation of the original crust, (2) excavation of the large impact basins, (3) filling of maria basins, and (4) formation of rayed craters.

During the late stages of its accretion, the Moon's outer shell was most likely completely melted—literally a magma ocean. Then, about 4.4 billion years ago, the magma ocean began to cool and underwent magmatic differentiation (see Chapter 2). Most of the dense minerals, olivine and pyroxene, sank, while less-dense silicate minerals floated to form the Moon's crust. The highlands are made of these igneous rocks, which rose buoyantly like "scum" from the crystallizing magma. The most common highland rock type is *anorthosite*, composed mainly of calcium-rich plagioclase feldspar.

Once formed, the lunar crust was continually impacted as the Moon swept up debris from the solar nebula. During this time, several large impact basins were created. Then, about 3.8 billion years ago, the Moon, as well as the rest of the solar system, experienced a sudden drop in the rate of meteoritic bombardment.

The Moon's next major event was the filling of the large impact basins that had been created at least 300 million years earlier (Figure 22.7). Radiometric dating of the maria basalts puts their age between 3.0 billion and 3.5 billion years—considerably younger than the initial lunar crust.

The maria basalts are thought to have originated at depths between 200 and 400 kilometers (125 and 250 miles). They were likely generated by a slow rise in temperature attributed to the decay of radioactive elements. Partial melting probably occurred in several isolated pockets, as indicated by the diverse chemical makeup of the rocks retrieved during the *Apollo* missions. Recent evidence suggests that some mare-forming eruptions may have occurred as recently as 1 billion years ago.

Other lunar surface features related to this period of volcanism include small shield volcanoes (8 to 12 kilometers [5 to 7.5 miles] in diameter), evidence of pyroclastic eruptions, narrow winding valleys (*rilles*) thought to be collapsed lava tubes, and long linear depressions similar to down-faulted valleys (*grabens*) on Earth.

The last prominent features to form were rayed craters, as exemplified by the roughly 93-kilometer-wide (58-mile-wide) Copernicus crater shown in Figure 22.6. Light-colored material ejected from these craters, called *rays* because they radiate outward, blankets the maria surfaces and many older, rayless craters. The relatively young Copernicus crater is thought to be about 1 billion years old. Had it formed on Earth, weathering and erosion would have long since obliterated it.

Impact of an asteroid-size body produced a huge crater hundreds of kilometers in diameter and disturbed the lunar crust far beyond the crater.

Filling of the impact crater with fluid basalts, perhaps derived from partial melting deep within the lunar mantle.

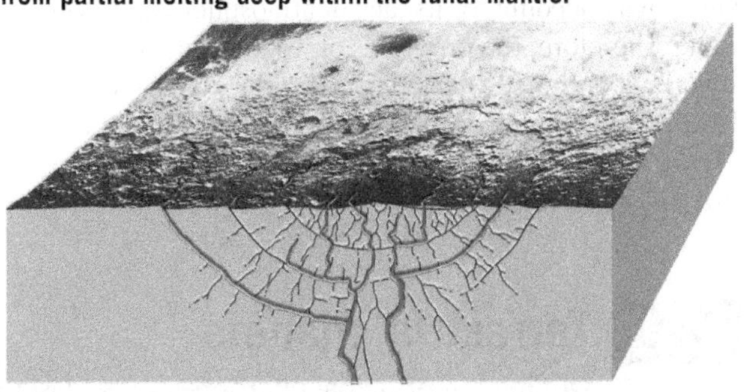

Today these lava-filled basins make up the lunar maria and similar large structures on Mercury.

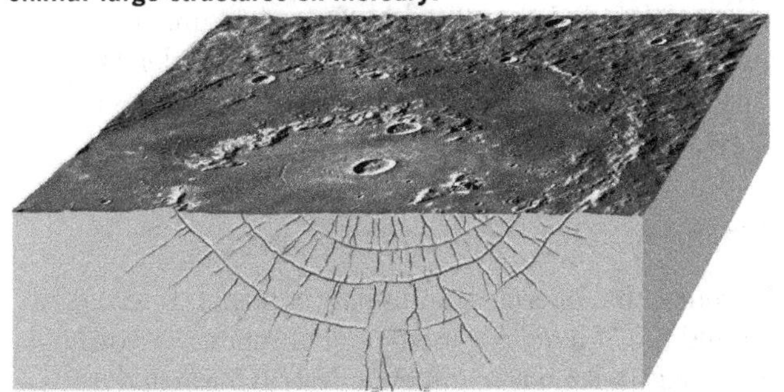

SmartFigure 22.7 **Formation and filling of large impact basins** (Photo courtesy of NASA)

TUTORIAL
https://goo.gl/NgHPCL

▲ **Figure 22.8 Astronaut Harrison Schmitt, sampling the lunar surface** Notice the footprint (inset) in the lunar "soil," called regolith, which lacks organic material and is therefore not a true soil. (Courtesy of NASA)

Today's Lunar Surface The Moon's small mass and low gravity account for its lack of atmosphere and flowing water; therefore, the processes of weathering and erosion that continually modify Earth's surface are absent. In addition, tectonic forces are no longer active on the Moon, so quakes and volcanic eruptions have ceased. Because the Moon is unprotected by an atmosphere, erosion is dominated by the impact of tiny particles from space (*micrometeorites*) that continually bombard its surface and gradually smooth the landscape. This activity has crushed and repeatedly mixed the upper portions of the lunar crust.

Both the maria and terrae are mantled with a layer of gray, unconsolidated debris derived from a few billion years of meteoric bombardment (Figure 22.8). This soil-like layer, properly called **lunar regolith** (*rhegos* = blanket, *lithos* = stone), is anywhere from 2 to 20 meters (6.6 to 66 feet) thick, composed of igneous rocks, breccia, glass beads, and fine *lunar dust*.

CONCEPT CHECKS 22.2

1. Briefly describe the origin of the Moon.
2. Compare and contrast the Moon's maria and highlands.
3. How are maria on the Moon similar to the Columbia Plateau in the Pacific Northwest?
4. How is crater density used in the relative dating of the Moon's surface features?
5. Summarize the major stages in the development of the modern lunar surface.

22.3 Terrestrial Planets

Outline the principal characteristics of Mercury, Venus, and Mars. Describe their similarities to and differences from Earth.

The terrestrial planets, in order from the Sun, are Mercury, Venus, Earth, and Mars. Here we consider Mercury, Venus, and Mars and compare their features to those of Earth.

Mercury: The Innermost Planet

Mercury, the innermost and smallest planet, revolves around the Sun quickly (88 days) but rotates so slowly on its axis that a day–night cycle lasts 176 Earth days. Thus, 1 "night" on Mercury is roughly equivalent to 3 months on Earth and is followed by the same duration of daylight. Mercury has the greatest extremes of surface temperature among the planets, from as low as −173°C (−280°F) at night to noontime temperatures exceeding 427°C (800°F), hot enough to melt tin and lead, and making life as we know it impossible.

Mercury absorbs most of the sunlight that strikes it, reflecting only 6 percent into space, a characteristic of terrestrial bodies with little or no atmosphere. The minuscule amount of gas that makes up Mercury's nearly nonexistent atmosphere may have originated from several sources, including ionized gas from the Sun, ices that vaporized during a relatively recent comet impact, and outgassing of the planet's interior.

Although Mercury is small and scientists initially expected the planet's interior to have already cooled, NASA's *Messenger* spacecraft found in 2012 that Mercury has a magnetic field, although it is about 100 times weaker than Earth's. This suggests that an outer layer of Mercury's large core remains molten and capable of convection—a requirement for generating a magnetic field.

Mercury resembles Earth's Moon in that it has very low reflectivity, no sustained atmosphere, numerous volcanic features, and a heavily cratered terrain (Figure 22.9). The largest-known impact crater on Mercury is Caloris Basin (1300 kilometers [800 miles] in diameter). Like our Moon, Mercury has extensive smooth plains; they cover nearly 40 percent of the area imaged by *Mariner 10*. Most of these smooth areas are associated with large impact basins, including Caloris Basin, where lava partially filled the basins and surrounding lowlands. Consequently, they appear to be similar in origin to lunar maria. Recently, *Messenger* found evidence of other types of volcanism on Mercury, including a huge flood basalt province reminiscent of, but much larger than, Earth's Columbia Plateau. Researchers also confirmed the presence of substantial deposits of ice within perpetually shadowed polar craters.

Venus: The Veiled Planet

Venus, second only to the Moon in brilliance in the night sky, is named for the Roman goddess of love and beauty. It orbits the Sun in a nearly perfect circle once every 225 Earth days. However, Venus rotates in the opposite direction of the other planets (*retrograde motion*) at an agonizingly slow pace: 1 Venus day is equivalent to about 243 Earth days. Venus has the densest atmosphere of the terrestrial planets, consisting mostly of carbon dioxide (97 percent)—and it is the prototype for an extreme *greenhouse effect*. As a consequence, the surface temperature of Venus averages more than 450°C (900°F) both day and night. Temperature variations at the surface are generally minimal because of the intense mixing within the planet's dense atmosphere. Investigations of the planet's extreme and uniform surface temperature led scientists to more fully understand how the greenhouse effect operates on Earth.

Figure 22.9 **Two views of Mercury** On the left is a monochromatic image, while the image on the right is color enhanced. These are high-resolution mosaics constructed from thousands of images obtained by the *Messenger* orbiter. (Courtesy of NASA)

The composition of the Venusian interior is probably similar to Earth's. Although Venus, like Earth, probably has a molten outer core, Venus essentially lacks a magnetic field, which indicates that the core does not convect. That could be due to some combination of the planet's slow rotation and an insufficient thermal gradient. However, scientists do think that mantle convection operates on Venus, but the processes of plate tectonics do not appear to have contributed to the present Venusian topography.

The surface of Venus is completely hidden from view by a thick cloud layer composed mainly of tiny sulfuric acid droplets. Between 1961 and 1984, despite extreme temperatures and pressures, 10 Russian spacecraft landed successfully and transmitted data, including surface images. As expected, however, all the probes were crushed by the planet's immense atmospheric pressure, approximately 90 times that on Earth, within an hour of landing. Using radar imaging, the unstaffed spacecraft *Magellan* mapped Venus's surface in stunning detail (Figure 22.10).

A few thousand impact craters have been identified on Venus—far fewer than on Mercury and Mars but more than on Earth. Researchers expected that Venus would show evidence of extensive cratering from the heavy bombardment period but found instead that a period of extensive volcanism was responsible for resurfacing Venus. The planet's thick atmosphere also limits the number of impacts by breaking up large incoming meteoroids and incinerating most of the small debris.

About 80 percent of the Venusian surface consists of low-lying plains covered by lava flows, some of which traveled along lava channels that extend hundreds of kilometers. Venus's Baltis Vallis, the longest-known lava channel in the solar system, meanders 6800 kilometers (4255 miles) across the planet. More than 1000 volcanoes with diameters greater than 20 kilometers (12 miles) have been identified on Venus. However, high surface pressures keep the gaseous components in lava from escaping and limit production of pyroclastic material and lava fountaining, phenomena that tend to steepen volcanic cones.

Figure 22.10 **Global view of the surface of Venus** This computer-generated, false-color image of Venus was constructed from years of investigations, culminating with the *Magellan* mission. The twisting bright features that cross the globe are highly fractured ridges and canyons of the eastern Aphrodite highland. (Courtesy of NASA)

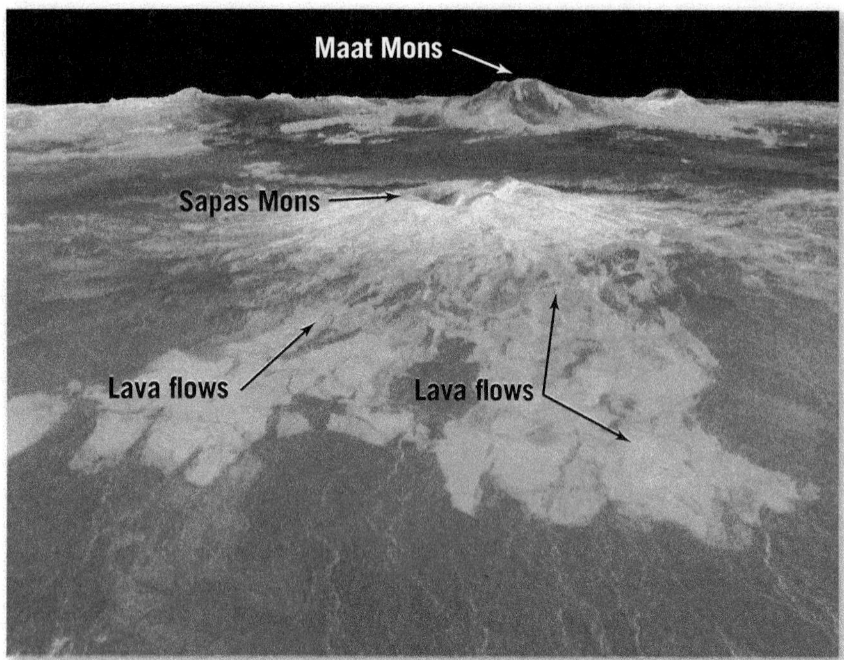

▲ **Figure 22.11 Volcanoes on Venus** Sapas Mons is a broad volcano, 400 kilometers (250 miles) wide. The bright areas in the foreground are lava flows. Another large volcano, Maat Mons, is in the background. (This image has considerable vertical exaggeration.) (Courtesy of NASA)

In addition, Venus's high temperatures allow lava to remain mobile longer and, thus, flow far from the vent. Both of these factors result in volcanoes that tend to be flatter and wider than those on Earth or Mars (**Figure 22.11**). Maat Mons, the largest volcano on Venus, is about 8.5 kilometers (5 miles) high and 400 kilometers (250 miles) wide. By comparison, Mauna Loa, Earth's largest volcano, is about 9 kilometers high (5.5 miles) and only 120 kilometers (75 miles) wide.

Venus also has major highlands consisting of plateaus, ridges, and topographic rises that stand above the plains. The rises are thought to have formed where hot mantle plumes encountered the base of the planet's crust, causing uplift. Much like mantle plumes on Earth, abundant volcanism is associated with mantle upwelling on Venus. Recent data collected by the European Space Agency's *Venus Express* suggest that Venus's highlands contain silica-rich granitic rock. These elevated landmasses resemble Earth's continents, albeit on a smaller scale.

Mars: The Red Planet

Mars, approximately one-half the diameter of Earth, revolves around the Sun in 687 Earth days. Mean surface temperatures range from lows of −140°C (−220°F) at the poles in the winter to highs of 20°C (68°F) at the equator in the summer. Although seasonal temperature variations are similar to Earth's, daily temperature variations are greater due to the very thin atmosphere of Mars (only 1 percent as dense as Earth's). The tenuous Martian atmosphere consists primarily of carbon dioxide (95 percent), with small amounts of nitrogen, oxygen, and water vapor.

Martian Topography Mars, like the Moon, is pitted with impact craters. The smaller craters are usually filled with wind-blown dust—confirming that Mars is a dry, desert world. The reddish color of the Martian landscape is due to iron oxide (rust). Large impact craters provide information about the nature of the Martian surface. For example, where the Martian surface is composed of dry rocky debris, ejecta similar in size and shape to that surrounding lunar craters is found. However, some Martian craters feature ejecta that looks like muddy slurry was splashed from the crater. Planetary geologists infer that a layer of permafrost (frozen, icy soil) lies below portions of the Martian surface and that the heat of impact melted the ice to produce the fluid-like appearance of these ejecta.

About two-thirds of the surface of Mars consists of heavily cratered highlands, concentrated mostly in its southern hemisphere (**Figure 22.12**). The period of extreme cratering occurred early in the planet's history and ended about 3.8 billion years ago, as it did in the rest

Olympus Mons
Tharsis bulge
Equator
Valles Marineris
Argyre
Northern lava plains
Elysium Mons
Equator
Cratered highlands
Hellas
−8 Low −4 0 4 Altitude (km) 8 12 High

◄ **Figure 22.12 Two hemispheres of Mars** Color represents height above (or below) the mean planetary radius: White is about 12 kilometers above average, and dark blue is 8 kilometers below average. (Courtesy of NASA)

of the solar system. Thus, Martian highlands are similar in age to the lunar highlands.

The remaining one-third of the planet, located in the northern hemisphere, is covered by low plains. Based on their relatively low crater counts, these northern plains are younger than the highlands. Their flat topography, possibly the smoothest surface in the solar system, is consistent with vast outpourings of fluid basaltic lavas. Visible on the plains are volcanic cones, some with summit pits (craters) and lava flows with wrinkled edges. If Mars once had abundant water, it would have flowed to the north, which is lower in elevation, possibly forming an expansive ocean.

Located along the Martian equator is an enormous elevated region about the size of North America, called the *Tharsis bulge*. This feature, about 10 kilometers (6 miles) high, appears to have been uplifted and capped with a massive accumulation of volcanic rock that includes the solar system's largest volcanoes.

The tectonic forces that created the Tharsis region also produced fractures that radiate from its center, like spokes on a bicycle wheel. Along the eastern flanks of the bulge, a series of vast canyons called *Valles Marineris* (Mariner Valleys) developed (see Figure 22.12). This canyon network was largely created by down-faulting rather than the stream erosion that carved Arizona's Grand Canyon. Thus, it consists of graben-like valleys similar to the East African Rift Valley. Once formed, Valles Marineris grew thanks to water erosion and collapse of the rift walls. The main canyon is more than 5000 kilometers (3000 miles) long, 7 kilometers (4 miles) deep, and 100 kilometers (60 miles) wide.

Other prominent features on the Martian landscape are large impact basins. Hellas, the largest visible impact basin on the planet, is about 2300 kilometers (1400 miles) in diameter and has the planet's lowest elevation. Debris ejected from this basin contributed to the elevation of the adjacent highlands. Other buried crater basins even larger than Hellas exist, including Utopia Basin, where *Viking 2* landed.

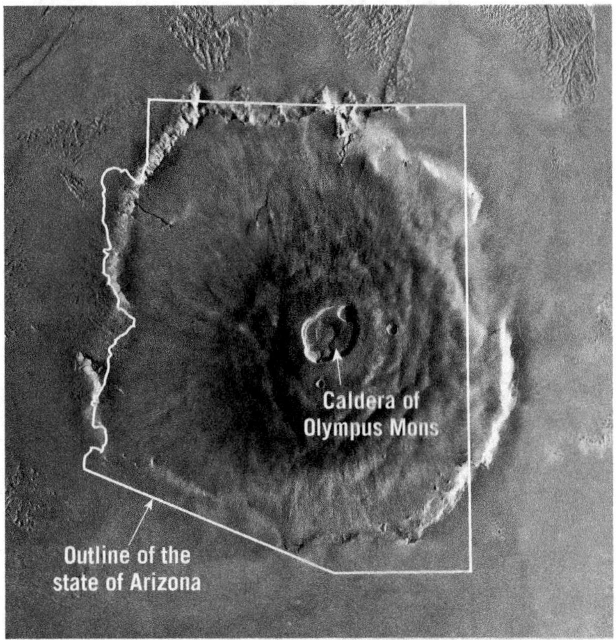

SmartFigure 22.13 **Olympus Mons** This massive inactive shield volcano on Mars covers an area about the size of the state of Arizona. (Courtesy of NASA)

TUTORIAL
https://goo.gl/XxMzhS

Volcanoes on Mars Volcanism has been prevalent on Mars during most of its history. The scarcity of impact craters on some volcanic surfaces suggests that the planet is still active. Mars has several of the solar system's largest known volcanoes, including the largest, Olympus Mons, which is about the size of Arizona and stands nearly three times higher than Mount Everest. This enormous volcano was active as recently as a few million years ago and resembles Earth's Hawaiian shield volcanoes (Figure 22.13).

How did the volcanoes on Mars grow so much larger than similar structures on Earth? The largest volcanoes on the terrestrial planets tend to form where plumes of hot rock rise from deep within their interiors. On Earth, moving plates keep the crust in constant motion. Consequently, mantle plumes tend to produce a chain of volcanic structures, like the Hawaiian Islands. By contrast, plate movement on Mars is absent, so successive eruptions accumulate in the same location, creating enormous volcanoes rather than a string of smaller ones.

Wind Erosion on Mars The dominant force currently shaping the Martian surface is wind erosion. Extensive dust storms with winds up to 270 kilometers (170 miles) per hour can persist for weeks. Dust devils have also been photographed. Most of the Martian landscape resembles Earth's rocky deserts, with abundant dunes and low areas partially filled with dust.

Water on Mars in the Past Considerable evidence indicates that in the first 1 billion years of the planet's history, liquid water flowed on Mars's surface, creating stream valleys and related features (Figure 22.14). One location where running water was involved in carving

Figure 22.14 **Similar rock outcrops on Mars and Earth** This set of images compares a rock outcrop on Mars (left) with similar rocks on Earth. The rock outcrop on Earth formed in a streambed, and the similarity of the Martian rocks suggests that they formed in a similar environment. Based on this finding, scientist John Grotzinger concluded that there was once "a vigorous flow on the surface of Mars." (Courtesy of NASA)

This NASA image obtained by *Curiosity* rover shows rounded gravel fragments within a rock outcrop consistent with the sedimentary rock conglomerate. Weathered rock fragments can be seen below.

A typical sample of the sedimentary rock conglomerate that contains rounded gravel fragments deposited in a stream bed on Earth.

Mars Exploration

Since the first close-up picture of Mars was obtained in 1965, spacecraft voyages to the fourth planet from the Sun have revealed a world that is strangely familiar. Mars has a thin atmosphere, polar ice caps, volcanoes, lava plains, sand dunes, and seasons. Unlike Earth, Mars appears to lack any appreciable amount of liquid water on its surface. However many Martian landscapes suggest that, in the past, running water was an effective erosional agent. The defining question for Mars exploration is "Has Mars ever harbored life?"

NASA's *Phoenix* lander dug into the Martian surface to uncover water ice in a northern region of the planet. Whether ice becomes available as liquid water to support microbial life remains unanswered.

MARS LANDING SITES

NASA

The U.S. has successfully landed seven spacecrafts on the surface of Mars. The most recent was NASA's *Curiosity*, which landed in Gale Crater in August, 2012.

NASA's *Curiosity* rover, the size of a car, gracefully landed on Mars after decelerating from 13,000 miles per hour to a complete stop. The landing, described as "seven minutes of terror," began a two-year mission in and around Gale Crater to discover signs of past or present microbial life.

NASA

NASA

The layers in the background, located at the base of Mount Sharp, are thought to be surviving remnants of extensive deposits laid down in a lake long ago, or possibly wind-delivered sediments subsequently cemented together by groundwater. *Curiosity* uses 10 instruments to investigate whether Mars had ever provided a water-rich environment.

This image captured by *Curiosity* shows rounded sand grains, which suggests the grains were carried long distances, possibly by running water.

CAPE ST. VINCENT

VICTORIA CRATER

NASA

This false-color image obtained by Rover *Opportunity* shows Cape St. Vincent, one of many promontories that jut out from the walls of Victoria Crater. Below the loose, jumbled rocks, layering in the crater walls shows evidence of ancient wind-blown dunes.

	Spacecraft*	Type	Landed or entered orbit	Years Active***
1	*Curiosity*	Rover	August 2012	Remains in operation
2	*Phoenix*	Lander	May 2008	Ran out of power during its first Martian winter
3	*Mars Reconnaissance*	Orbiter	March 2006	Planned 2-year mission, remains in operation
4	*Spirit*	Rover	January 2004	Planned 4-year mission, operated for more than 6 years
5	*Opportunity*	Rover	January 2004	Planned 90-day mission, remains in operation
6	*Odyssey*	Orbiter	October 2001	Remains active, longest active spacecraft in orbit around another planet
7	*Viking I*	Lander**	July 1976	Operational for more than 6 years
8	*Viking II*	Lander**	September 1976	Operational for more than 3 years
9	*Maven*	Orbiter	September 2014	Remains in operation

* Selected NASA missions; ** Also had an orbiter; *** As of 2017

Question:
What evidence indicates that Mars may have been habitable in the distant past?

?

NASA

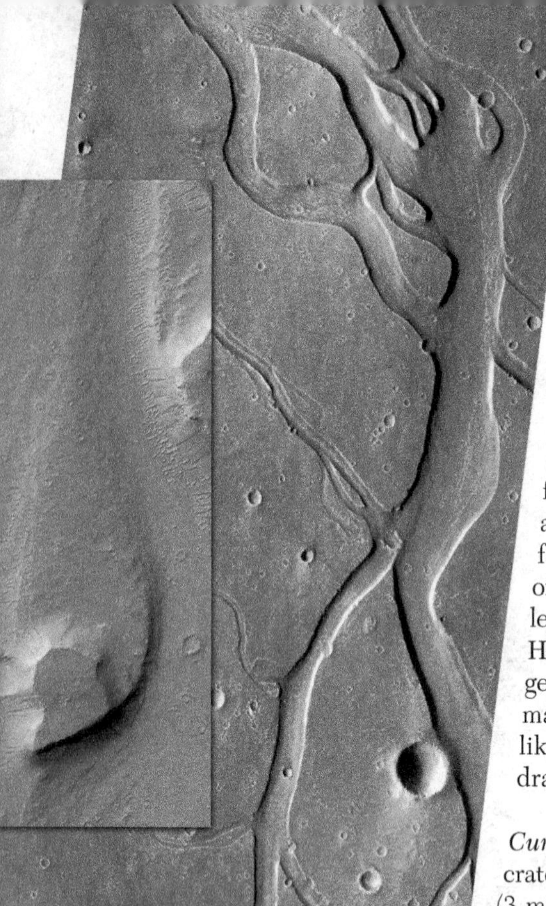

▲ Figure 22.15 **Earth-like stream channels** These streamlike channels are strong evidence that Mars once had flowing water. Inset shows a close-up of a streamlined island where running water encountered resistant material along its channel. (Courtesy of NASA)

valleys can be seen in the *Mars Reconnaissance Orbiter* image in Figure 22.15. Notice the streamlike banks that contain numerous teardrop-shaped islands. These valleys appear to have been cut by catastrophic floods with discharge rates more than 1000 times greater than those of the Mississippi River. Most of these large flood channels emerge from areas of chaotic topography that appear to have formed when the surface collapsed. The most likely source of water for these flood-created valleys was the melting of subsurface ice. However, not all Martian valleys were generated from water released in this manner. Some exhibit branching, treelike patterns resembling dendritic stream drainage networks on Earth.

On August 6, 2012, the Mars rover *Curiosity* landed in Gale Crater, an impact crater that contains a 5-kilometer-high (3-mile-high) accumulation of sediment called Mount Sharp. As of October 2016, *Curiosity* had traveled over 10 kilometers (6 miles), examining the lower slopes of this layered mountain (see GEOgraphics 22.1). At a target zone NASA calls "Big Arm," the Mars Hand Lens Imager (MAHLI) took images of sediment containing rounded grains, which must have traveled long distances before being deposited. Analysis of these sediments indicates that Gale Crater periodically filled with water, forming a lake that lasted hundreds or possibly even thousands of years. This is strong evidence that Mars must have had a much thicker atmosphere that supported a hydrologic cycle similar to Earth's. It also means that other craters most likely supported long-lived lakes that may have provided suitable habitat for microbial life.

Does Liquid Water Exist on Present-Day Mars? We have learned from NASA's *Phoenix Mars Lander*, which dug into the Martian surface, that poleward of about 30° latitude, ice can be found within 1 meter (3 feet) of the surface. Furthermore, Mars's permanent polar ice caps are composed of mainly water ice, blanketed by a thin layer of carbon dioxide ice during the cold season. Current estimates place the maximum amount of water ice held by the Martian polar ice caps at about 1.5 times the amount covering Greenland. However, this water appears to remain frozen year-round.

Recent images from a high-resolution camera aboard NASA's *Mars Reconnaissance Orbiter* show dark streaks on Mars, called *recurring slope lineae* (Figure 22.16). Researchers believe that these streaks, which appear seasonally on steep, relatively warm Martian slopes, may be caused by the flow of briny (salty) liquid water. Although these dark streaks are just 0.5 to 5 meters (1.6 to 16 feet) wide, they can extend for hundreds of meters downslope. In addition, these features appear during warm weather but fade away when the temperatures drop, providing further evidence that liquid water is involved in their formation. Salts, which are widespread on the Martian surface, would lower the freezing point of water from 0°C (32°F) to −70°C (−94°F). The discovery of these dark streaks has implications for future human exploration of Mars. In the late 2030s, when NASA plans to send astronauts to the red planet, the presence of liquid water—even very salty water—would aid that ambitious effort.

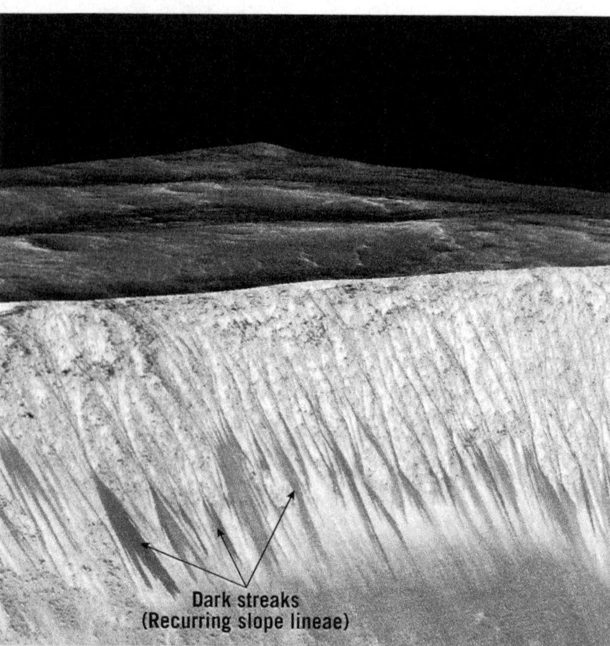

▲ Figure 22.16 **Dark streaks on Mars, thought to be caused by the flow of briny (salty) water** These streaks, called *recurring slope lineae*, are found on steep, warm Martian slopes and disappear during the cold season. (Photo courtesy of NASA)

CONCEPT CHECKS 22.3

1. What body in our solar system is most like Mercury?
2. Venus was once referred to as "Earth's twin." How are these two planets similar? How do they differ from one another?
3. What surface features do Mars and Earth have in common?
4. Why are the largest volcanoes on Earth so much smaller than the largest ones on Mars?
5. What evidence suggests that Mars had an active hydrologic cycle in the past?

22.4 Jovian Planets

Summarize and compare the features of Jupiter, Saturn, Uranus, and Neptune, including their ring systems.

The four Jovian planets, in order from the Sun, are Jupiter, Saturn, Uranus, and Neptune. Because of their location within the solar system and their size and composition, they are also commonly called the *outer planets* and the *gas giants*.

Jupiter: Lord of the Heavens

The giant among planets, Jupiter has a mass 2.5 times greater than the combined mass of all other planets, satellites, and asteroids in the solar system. However, it pales in comparison to the Sun, with only 1/800 of the Sun's mass.

Jupiter orbits the Sun once every 12 Earth years, and it rotates more rapidly than any other planet, completing one rotation in slightly less than 10 hours. When viewed telescopically, the effect of this fast spin is noticeable. The bulge of the equatorial region and the slight flattening at the poles are evident (see the "Polar Flattening" column in Table 22.1).

Jupiter's appearance is mainly attributable to the colors of light reflected from its three main cloud layers (**Figure 22.17**). The warmest, and lowest, layer is composed mainly of water ice and appears blue-gray, while the middle layer, where temperatures are lower, consists of brown to orange-brown clouds of ammonium hydrosulfide droplets. These colors are thought to be by-products of chemical reactions occurring in Jupiter's atmosphere. Near the top of its atmosphere lie white wispy clouds of ammonia ice.

Because of its immense gravity, Jupiter is shrinking a few centimeters each year. This contraction generates most of the heat that drives Jupiter's atmospheric circulation. Thus, unlike winds on Earth, which are driven by solar energy, the heat emanating from Jupiter's interior produces the huge convection currents observed in its atmosphere.

Jupiter's convective flow produces alternating dark-colored *belts* and light-colored *zones*, as shown in Figure 22.17. The light clouds (*zones*) are regions where warm material is ascending and cooling, whereas the dark belts represent cool material that is sinking and warming. This convective circulation, along with Jupiter's rapid rotation, generates the high-speed, east–west flow observed between the belts and zones.

The largest storm on the planet is the Great Red Spot. This enormous anticyclonic storm, twice the size of Earth, has been known for 300 years. In addition to the Great Red Spot, there are various white and brown oval-shaped storms. The white ovals are the cold cloud tops of huge storms many times larger than hurricanes on Earth. The brown storm clouds reside at lower levels in the atmosphere. Lightning in the white oval storms has been photographed by the *Cassini* spacecraft, but the strikes appear to be less frequent than on Earth.

In 2016 *Juno* spacecraft send back the first images of Jupiter's north polar region (**Figure 22.18**). This region of Jupiter is bluer than the rest of the planet and exhibits numerous storm systems, unlike anything previously seen on any of the gas giants in our solar system.

Jupiter's magnetic field, the strongest in the solar system, is probably generated by a rapidly rotating, liquid metallic hydrogen layer surrounding its core. Bright *auroras* associated with the magnetic field have been photographed over Jupiter's poles (**Figure 22.19**). Unlike Earth's auroras, which occur only in conjunction with heightened solar activity, Jupiter's auroras are continuous.

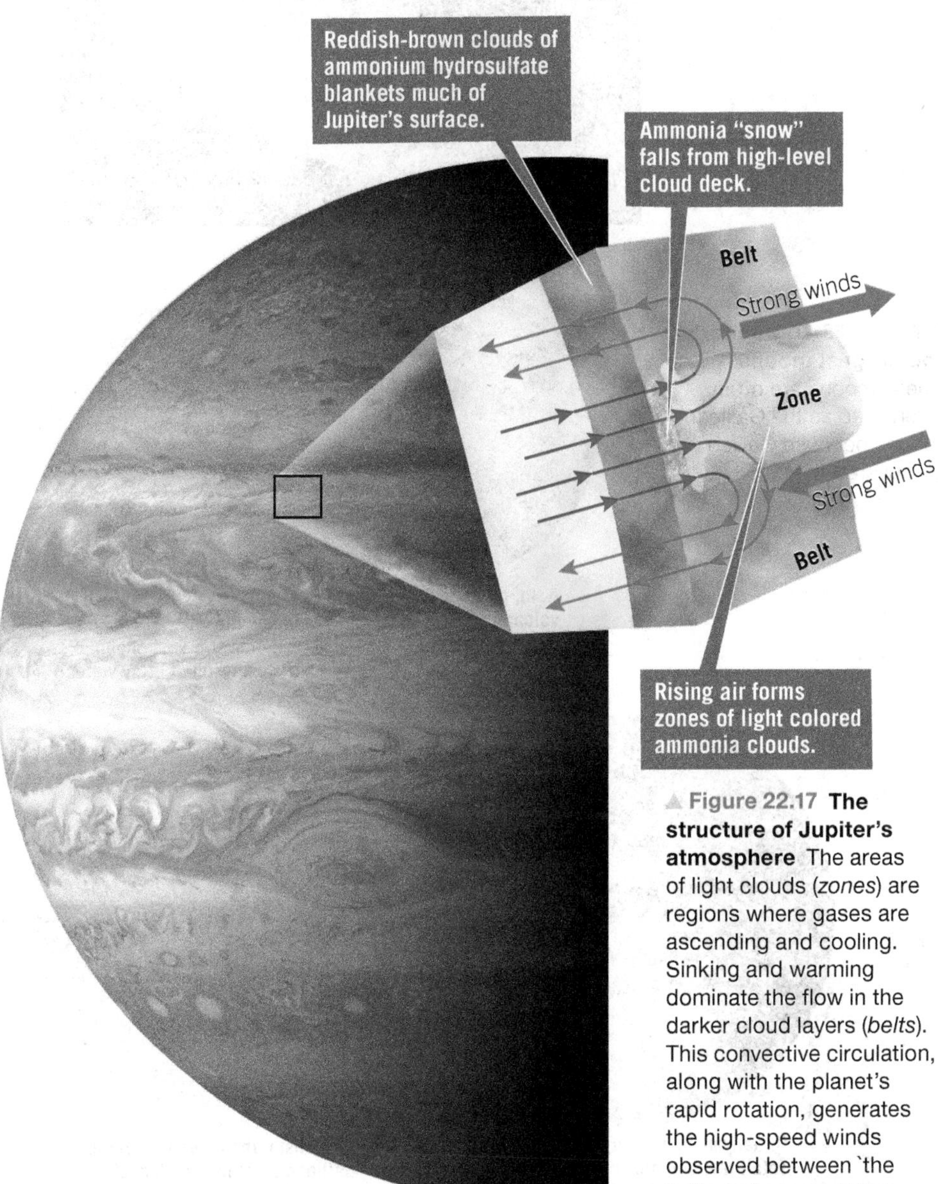
Reddish-brown clouds of ammonium hydrosulfate blankets much of Jupiter's surface.
Ammonia "snow" falls from high-level cloud deck.
Belt
Strong winds
Zone
Strong winds
Belt
Rising air forms zones of light colored ammonia clouds.

▲ **Figure 22.17 The structure of Jupiter's atmosphere** The areas of light clouds (*zones*) are regions where gases are ascending and cooling. Sinking and warming dominate the flow in the darker cloud layers (*belts*). This convective circulation, along with the planet's rapid rotation, generates the high-speed winds observed between `the belts and zones. (NASA)

Figure 22.18 **First image taken of Jupiter's north polar region** This image was captured by NASA's *Juno* spacecraft in 2016, when it flew about 4200 kilometers (2500 miles) from Jupiter's swirling clouds. (Courtesy of NASA)

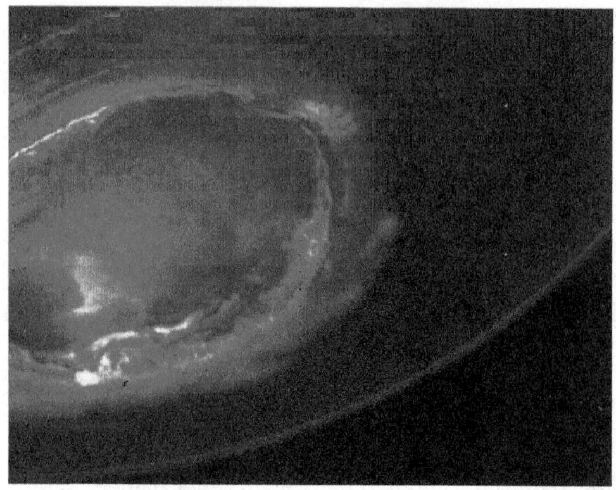

Figure 22.19 **Infrared image of the southern aurora of Jupiter** This image, captured by *Juno* in 2016, shows the wavelength of light emitted by excited hydrogen ions found in the planet's polar regions. (Courtesy of NASA)

Jupiter's Moons Jupiter's satellite system, consisting of 67 moons discovered thus far, resembles a miniature solar system. Galileo discovered the 4 largest satellites, called Galilean satellites, in 1610 (Figure 22.20). The two largest, Ganymede and Callisto, are roughly the size of Mercury, whereas the two smaller ones, Europa and Io, are about the size of Earth's Moon. The eight largest moons appear to have formed around Jupiter as the solar system condensed.

Jupiter also has many very small satellites, most of which revolve in the opposite direction (retrograde motion) of the largest moons and have eccentric (elongated) orbits steeply inclined to the Jovian equator. These satellites appear to be asteroids or comets that passed near enough to be either gravitationally captured by Jupiter or remnants of the collisions of larger bodies.

The Galilean moons can be observed with binoculars or a small telescope and are interesting in their own right. Images from *Voyagers 1* and *2* revealed, to the surprise of most geoscientists, that each of the four Galilean satellites is a unique world (Figure 22.20). The *Galileo* mission also unexpectedly revealed that the composition of each satellite is strikingly different, implying a different evolution for each. For example, Ganymede has a dynamic core that generates a strong magnetic field not observed in other satellites.

The innermost Galilean moon, Io, is perhaps the most volcanically active body in our solar system. More than 80 active, sulfurous volcanic centers have been discovered. Umbrella-shaped plumes have been observed rising from Io's surface to heights exceeding 100 kilometers (60 miles) (Figure 22.21A). The heat source for volcanic activity is tidal energy generated by a relentless "tug of war" between Jupiter and the other Galilean satellites—with Io as the rope. The gravitational field of Jupiter and the other nearby satellites pull and push on Io's tidal bulge as its slightly eccentric orbit takes it

Figure 22.20 **Jupiter's four largest moons** These moons are often referred to as the Galilean moons because Galileo discovered them. (Courtesy of NASA)

A. Io, perhaps the most volcanically active body in our solar system, has more than 80 active, sulfurous volcanic structures.

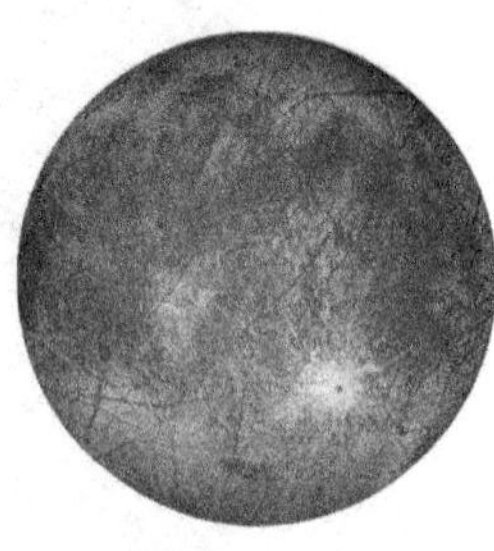

B. Europa's icy surface is quite flat and thought to cover a vast ocean composed of briny water.

C. Ganymede, the largest of the Jovian satellites, contains both smooth as well as cratered regions, which suggest this body is still active.

D. Callisto, the outermost of the Galilean satellites, is densely cratered, much like Earth's Moon.

alternately closer to and farther from Jupiter. This gravitational flexing of Io is transformed into heat (similar to the back-and-forth bending of a piece of sheet metal) and results in Io's spectacular sulfurous volcanic eruptions. Lava, thought to be mainly composed of silicate minerals, regularly erupts on its surface (Figure 22.21B).

One of the best prospects of finding liquid water within our solar system lies beneath the icy surfaces of some of Jupiter's moons. For example, detailed images from *Galileo* have revealed that Europa's icy surface is quite young and exhibits cracks apparently filled with dark fluid from below. This suggests that under its icy shell, Europa must have a warm, mobile interior—perhaps an ocean. Because liquid water is a necessity for life as we know it, there is considerable interest in sending an orbiter to Europa—and, eventually, a lander capable of launching a robotic submarine—to determine whether it harbors life.

Jupiter's Rings One of the surprising aspects of the *Voyager 1* mission was the discovery of Jupiter's ring system. More recently, the ring system was thoroughly investigated by the *Galileo* mission. By analyzing how these rings scatter light, researchers determined that the rings are composed of fine, dark particles similar in size to smoke particles. Furthermore, the faint nature of the rings indicates that these minute particles are widely dispersed. The main ring is composed of particles believed to be fragments blasted from the surfaces of Metis and Adrastea, two small moons of Jupiter. Impacts on Jupiter's moons Amalthea and Thebe are believed to be the source of the debris from which the outer gossamer ring formed.

This plume of volcanic gases and debris is rising more than 100 kilometers (60 miles) above Io's surface.
A.

The bright red area on the left side of the image (see arrow) is newly erupted lava.
B.

Figure 22.21 A volcanic eruption on Jupiter's moon Io (Photo by University of Arizona/JPL/NASA)

Saturn: The Elegant Planet

Requiring more than 29 Earth years to make one revolution, Saturn is almost twice as far from the Sun as Jupiter, yet their atmospheres, compositions, and internal structures are remarkably similar. The most striking feature of Saturn is its system of rings, first observed by Galileo in 1610 (Figure 22.22). Through his primitive telescope, the rings appeared as two small bodies adjacent to the planet. Their ring nature was determined 50 years later by Dutch astronomer Christiaan Huygens.

Saturn's atmosphere, like Jupiter's, is dynamic. Although the bands of clouds are fainter and wider near the equator, rotating "storms" similar to Jupiter's Great Red Spot occur in Saturn's atmosphere, as does intense lightning. Although the atmosphere is about 93 percent hydrogen and 3 percent helium by volume, the clouds (or condensed gases) are composed mainly of ammonia, ammonium hydrosulfide, and water, each segregated by temperature. Also, much like Jupiter, Saturn emits roughly twice as much energy as it receives from the Sun. This implies that it must have

Figure 22.22 Saturn's major rings The two bright rings, called A ring (outer) and B ring (inner), are separated by the Cassini division. A second small gap (Encke gap) is also visible as a thin line in the outer portion of the A ring. (Courtesy of NASA)

an internal heat source, which may come from chemical differentiation in its interior.

Saturn's Moons The Saturnian satellite system consists of 62 known moons, of which 53 have been named. The moons vary significantly in size, shape, surface age, and origin. Some of the moons are "original" satellites that formed in tandem with their parent planet. Most of Saturn's smallest moons have irregular shapes and are only a few tens of kilometers in diameter.

Saturn's largest moon, Titan, is larger than Mercury and is the second-largest satellite in the solar system. Titan and Neptune's Triton are the only satellites in the solar system known to have substantial atmospheres. Titan was visited and photographed by the *Cassini-Huygens* probe in 2005. The atmospheric pressure at Titan's surface is about 1.5 times that at Earth's surface, and the atmospheric composition is about 98 percent nitrogen and 2 percent methane, with trace organic compounds. Titan has Earth-like landforms and geologic processes, such as dune formation and streamlike erosion caused by methane "rain." In addition, the northern latitudes appear to have lakes of liquid methane.

Enceladus is another unique satellite of Saturn—one of a few icy moons that erupt "fluid" ice containing minor amounts of other debris. This amazing manifestation of volcanism, called **cryovolcanism** (from the Greek *kryos*, meaning "frost") describes the eruption of magmas derived from the partial melting of ice instead of silicate rocks (Figure 22.23). In the south polar region, areas called "tiger stripes," consisting of large fractures with ridges on either side, erupt geyserlike plumes. The material ejected by these eruptions is thought to be the source of material that replenishes Saturn's E ring.

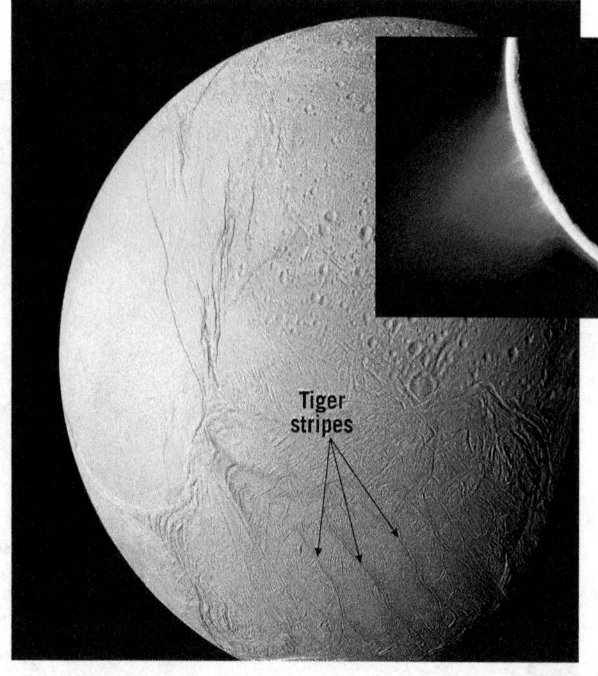

Figure 22.23
Enceladus, one of Saturn's tectonically active, icy satellites Enceladus has active linear features, called tiger stripes, which are a source region for cryovolcanic activity. Inset image shows jets spurting ice particles, water, and organic compounds from the area of the tiger stripes. (Courtesy of NASA)

Saturn's Ring System In the early 1980s, the nuclear-powered *Voyagers 1* and *2* explored Saturn to within 160,000 kilometers (100,000 miles) of its cloud tops. More information was collected about Saturn in that short time than had been acquired since Galileo first viewed this "elegant planet" in the early 1600s. More recently, observations from ground-based telescopes, the Hubble Space Telescope, and the *Cassini-Huygens* spacecraft, have added to our knowledge of Saturn's ring system. In 1995 and 1996, when the positions of Earth and Saturn allowed the rings to be viewed edge-on, Saturn's faintest rings and satellites became visible. (The rings were visible edge-on again in 2009.)

Saturn's ring system is more like a large rotating disk of varying density and brightness than a series of independent ringlets. Each ring is composed of individual particles—mainly water ice, with lesser amounts of rocky debris—that circle the planet while regularly impacting one another. There are only a few gaps; most of the areas that look like empty space contain either fine dust particles or coated ice particles that are inefficient light reflectors.

Most of Saturn's rings fall into one of two categories, based on density. Saturn's main (bright) rings, designated A and B, are tightly packed and contain particles ranging in size from a few centimeters (pebble-size) to tens of meters (house-size), with most of the particles being roughly the size of a large snowball (see Figure 22.22). In these dense rings, particles collide frequently as they orbit the planet. Although Saturn's main rings (A and B) are 40,000 kilometers (25,000 miles) wide, they are very thin, only 10 to 30 meters (30 to 100 feet) from top to bottom.

At the other extreme are Saturn's faint rings. Saturn's outermost ring (E ring), not visible in Figure 22.22, is composed of widely dispersed, tiny particles. Recall that cryovolcanism (eruption of a water-ice mixture) on Saturn's satellite Enceladus is thought to be the source of material for the E ring.

Studies have shown that the gravitational tugs of nearby moons tend to "shepherd" the ring particles by gravitationally altering their orbits (Figure 22.24). For example, the F ring, which is very narrow, appears to be the work of satellites located on either side that confine the ring by pulling back particles that try to escape. On the other hand, the Cassini division, a clearly visible gap in Figure 22.22, arises from the gravitational pull of Mimas, one of Saturn's moons.

Some of the ring particles are believed to be debris ejected from the moons embedded in the rings. It is also possible that material is continually recycled between the rings and the ring moons. The ring moons gradually sweep up particles, which are subsequently ejected by collisions with large chunks of ring material, or perhaps by energetic collisions with other moons. It seems, then, that planetary rings are not the timeless features that we once thought; rather, they are continually recycled.

The origin of planetary ring systems is still being debated. Saturn's rings probably formed when objects

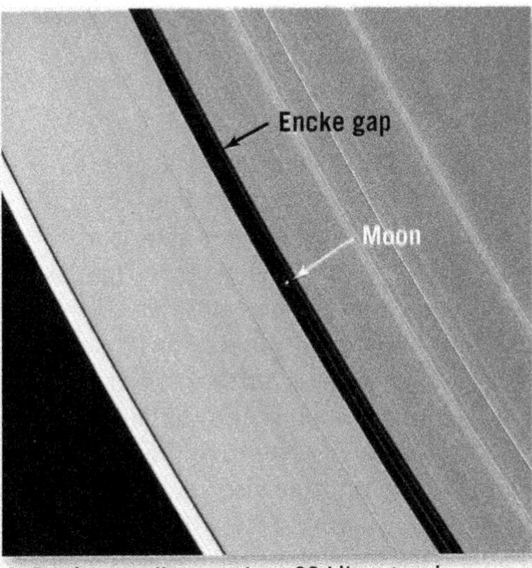

A. Pan is a small moon about 30 kilometers in diameter that orbits in the Encke gap, located in the A ring. It is responsible for keeping the Encke gap open by sweeping up any stray material that may enter.

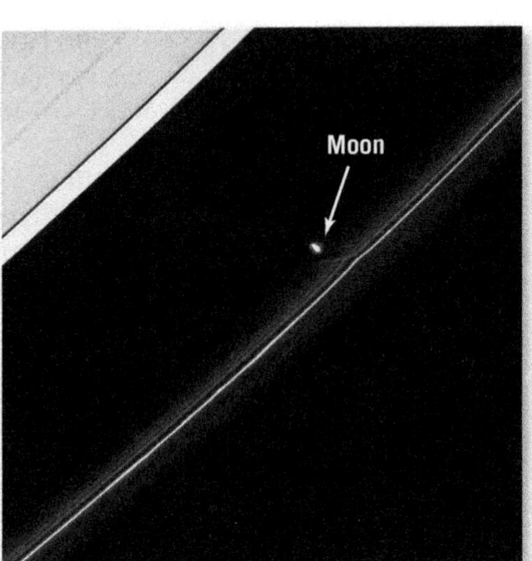

B. Prometheus, a potato-shaped moon, acts as a ring shepherd. Its gravity helps confine the particles that make up Saturn's thin Fring.

Figure 22.24 **Two of Saturn's ring moons** (Courtesy of NASA)

like comets, asteroids, or perhaps even moons were pulled apart by Saturn's strong gravity. Pieces of these objects would have collided with each other, breaking into even smaller pieces. Collisions among these fragments would tend to jostle one another and cause them to spread out to form the flat, thin ring system we observe today. Saturn's rings, as well as those of the other planets, are thought to be short-lived compared to the age of our solar system. This means that Saturn probably lacked rings early in its history, and its existing ring system will likely dissipate in the distant future.

Uranus and Neptune: Twins

Although Earth and Venus have many similar traits, Uranus and Neptune are perhaps more deserving of being called "twins." They are nearly equal in diameter (both about four times the size of Earth), and they are both bluish in appearance, as a result of methane in their atmospheres. Their days are nearly the same length, and their cores are made of rocky silicates and iron—similar to the other Jovian planets. Their mantles, made mainly of water, ammonia, and methane, are thought to be very different from those of Jupiter and Saturn. One of the most pronounced differences between Uranus and Neptune is the time they take to complete one revolution around the Sun—84 and 165 Earth years, respectively.

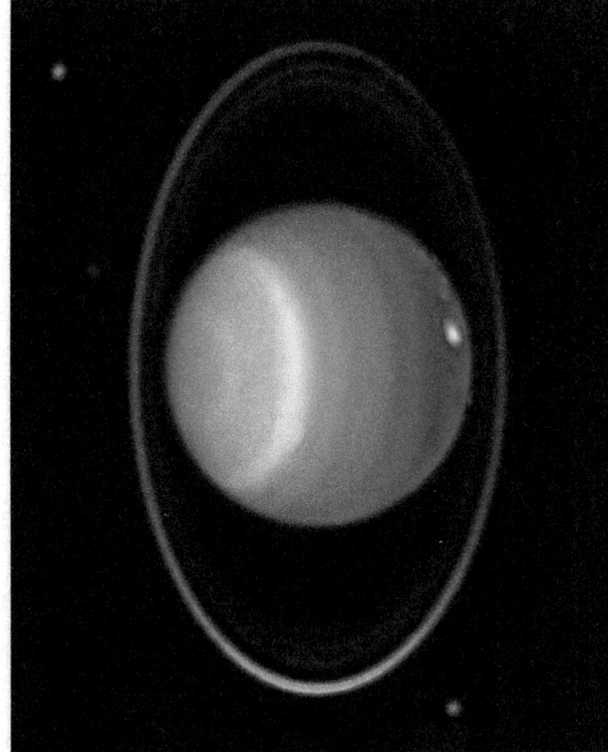

Figure 22.25 **Uranus, surrounded by its major rings and a few of its known moons** Also visible in this image are cloud patterns and several oval storm systems. This false-color image was generated from data obtained by Hubble's Near Infrared Camera. (Courtesy of NASA)

Uranus: The Sideways Planet Unique to Uranus is the orientation of its axis of rotation. Whereas the other planets resemble spinning toy tops as they circle the Sun, Uranus is like a top that has been knocked on its side but remains spinning (Figure 22.25). This unusual characteristic of Uranus is likely due to one or more impacts that

EYE ON THE UNIVERSE 22.2

In 2012, the Hubble Space Telescope captured these images of auroras above the giant planet Uranus. These light shows on Uranus appear to last for only a few minutes and consist of faint glowing dots. These are unlike auroras on Earth, which can color the sky shades of green, red, or purple for hours.

QUESTION 1 *What is unusual about the location of the auroras on Uranus?*

QUESTION 2 *What does this indicate about the locations of Uranus's "north" and "south" magnetic poles?*

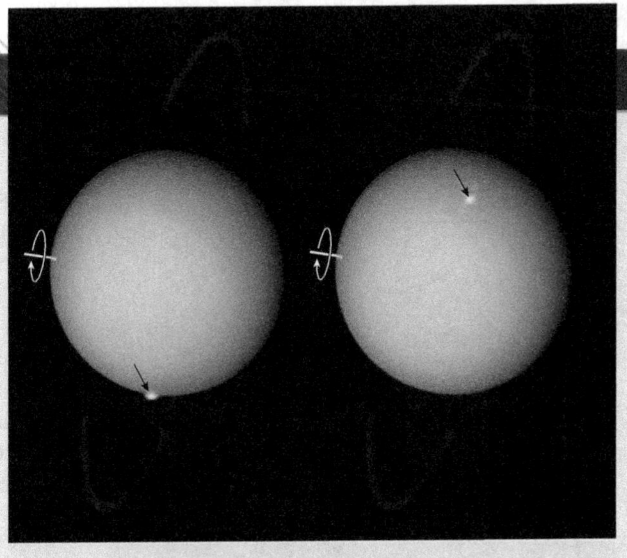

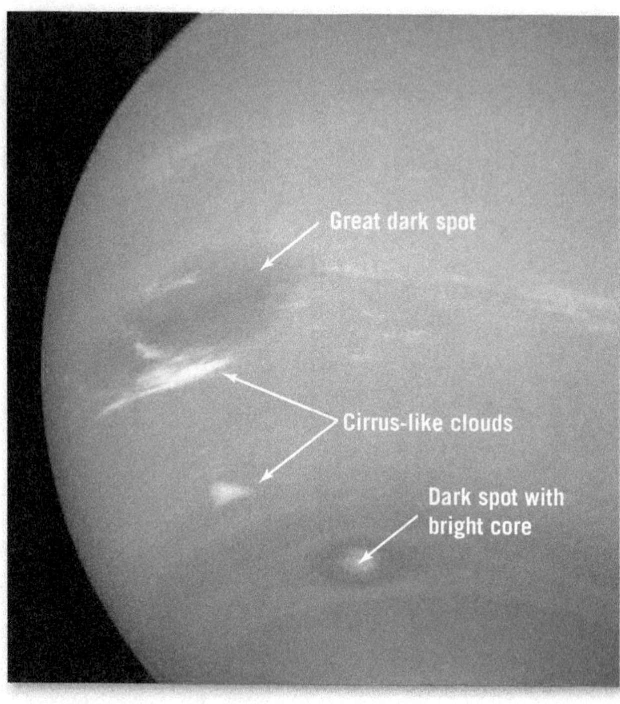

Figure 22.26 **Neptune's dynamic atmosphere** (Courtesy of NASA)

essentially knocked the planet sideways from its original orientation early in its evolution.

Uranus shows evidence of huge storm systems larger than Earth's continents. Recent photographs from the Hubble Space Telescope also reveal banded clouds composed mainly of ammonia and methane ice—similar to the cloud systems of the other Jovian planets.

Uranus's Moons Spectacular views from *Voyager 2* showed that Uranus's five largest moons have varied terrains. Some have long, deep canyons and linear scars, whereas others possess large, smooth areas on otherwise crater-riddled surfaces. Studies conducted at NASA's Jet Propulsion Laboratory suggest that Miranda, the innermost of the five largest moons, was recently geologically active—most likely driven by gravitational heating, as occurs on Io.

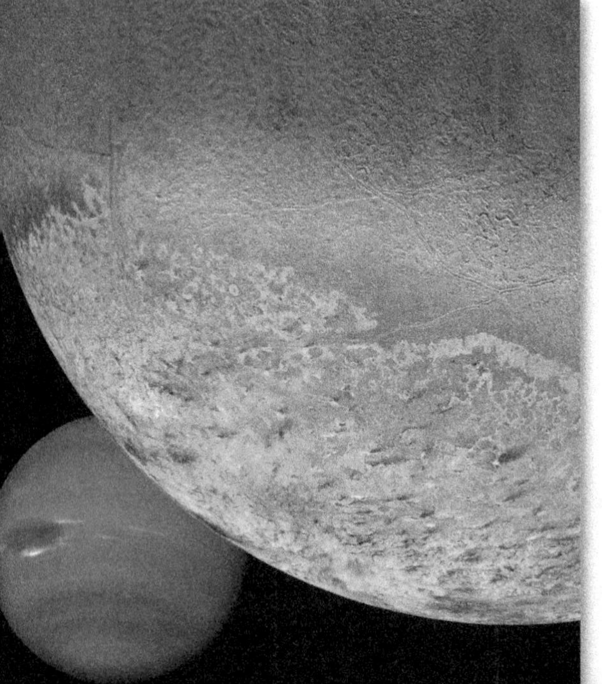

Figure 22.27 **Triton, Neptune's largest moon** The bottom of the image shows Triton's wind- and sublimation-eroded south polar cap. Sublimation is the process whereby a solid (ice) changes directly to a gas. (Courtesy of NASA)

Uranus's Rings A surprise discovery in 1977 showed that Uranus has a ring system. The discovery was made as Uranus passed in front of a distant star and blocked its view, a process called *occultation* (*occult* = hidden). Observers saw the star "wink" briefly five times (meaning five rings) before the primary occultation and again five times afterward. More recent ground- and space-based observations indicate that Uranus has at least 10 sharp-edged, distinct rings orbiting its equatorial region. Interspersed among these distinct structures are broad sheets of dust.

Neptune: The Windy Planet Because of Neptune's great distance from Earth, astronomers knew very little about this planet until 1989. Twelve years and nearly 3 billion miles of *Voyager 2* travel provided investigators an amazing opportunity to view the outermost planet.

Neptune has a dynamic atmosphere, much like that of the other Jovian planets (Figure 22.26). Record wind speeds approaching 2400 kilometers (1500 miles) per hour encircle the planet, making Neptune the windiest places in the solar system. Neptune also exhibits large dark spots, thought to be rotating storms similar to Jupiter's Great Red Spot. However, Neptune's storms appear to have comparatively short life spans—usually only a few years. Another feature that Neptune has in common with Uranus is layers of white, cirrus-like clouds (probably frozen methane) about 50 kilometers (30 miles) above the main cloud deck.

Neptune's Moons Neptune has 14 known satellites, the largest of which is Triton; the remaining 13 are small, irregularly shaped bodies (Figure 22.27). Like Enceladus, Triton exhibits cryovolcanism. Triton's icy magma is a mixture of water ice, methane, and probably ammonia. When partially melted, this mixture behaves as molten rock does on Earth. In fact, upon reaching the surface, these magmas can generate quiet outpourings of ice lavas that can flow great distances from their source—similar to the fluid basaltic flows on Hawaii. They also occasionally produce explosive eruptions that can generate the ice equivalent of volcanic ash. In 1989, *Voyager 2* detected active plumes on Triton that rose 8 kilometers (5 miles) above the surface and were blown downwind for more than 100 kilometers (60 miles).

Neptune's Rings Neptune has five named rings; two of them are broad, and three are narrow, perhaps no more than 100 kilometers (60 miles) wide. The outermost ring appears to be partially confined by the satellite Galatea. Neptune's rings, like Jupiter's, appear faint, which suggests that they are composed mostly of dust-size particles. Neptune's rings also display red colors, indicating that the dust is composed of organic compounds.

CONCEPT CHECKS 22.4

1. What is the nature of Jupiter's Great Red Spot?
2. What is distinctive about Jupiter's satellite Io?
3. How are Jupiter and Saturn similar to one another?
4. What two roles do ring moons play in the nature of planetary ring systems?
5. How are Saturn's satellite Titan and Neptune's satellite Triton similar to one another?

22.5 Small Solar System Bodies

List and describe the principal characteristics of the small bodies that inhabit the solar system.

There are countless chunks of debris in the vast spaces separating the eight planets and in the outer reaches of the solar system. In 2006, the International Astronomical Union organized solar system objects not classified as planets or moons into two broad categories: (1) **small solar system bodies**, including *asteroids*, *comets*, and *meteoroids*, and (2) **dwarf planets**. The newest grouping, dwarf planets, includes Ceres, a body about 1000 kilometers (600 miles) in diameter and the largest-known object in the asteroid belt, and Pluto, formerly considered a planet.

Asteroids and meteoroids are compositionally quite similar, both being composed of rocky and/or metallic material similar to that which makes up the terrestrial planets. They are usually distinguished by size, with asteroids being much larger than meteoroids, although the exact size difference is not well defined. Comets, on the other hand, are collections of ices, with lesser amounts of dust and small rocky particles. Comets mainly inhabit the outer reaches of the solar system.

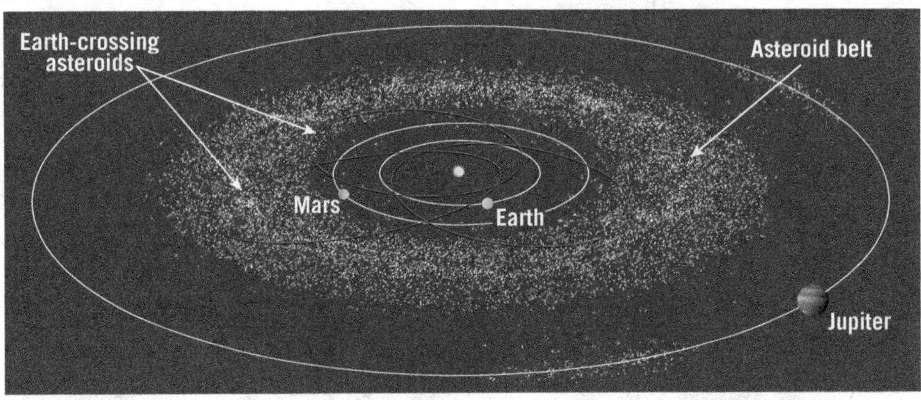

▲ **Figure 22.28 The asteroid belt** The orbits of most asteroids lie between Mars and Jupiter. Also shown in red are the orbits of a few known near-Earth asteroids.

Asteroids: Leftover Planetesimals

Asteroids are small bodies (planetesimals) that remain from the formation of the solar system, which means they are about 4.6 billion years old. The orbits of more than 100,000 asteroids have been accurately measured, and thousands more have orbits that are incompletely known, keeping them off the "official" list.

Most asteroids orbit the Sun between Mars and Jupiter, in the region known as the **asteroid belt** (Figure 22.28). There are only about 2 dozen asteroids that are more than 200 kilometers (125 miles) across. However, our solar system hosts an estimated 1 to 2 million asteroids larger than 1 kilometer (0.6 mile) across and many millions that are smaller.

A smaller number of asteroids travel along eccentric orbits that take them near the Sun, and about 1300 of these, called *Earth-crossing asteroids*, will eventually collide with Earth. Many of the recent large impact craters discovered on Earth resulted from collisions with asteroids. Although these events are rare, they merit our attention because of their potential destruction. As a result, observational initiatives that aim to measure asteroid orbits with great accuracy are ongoing.

Asteroid Structure and Composition The largest asteroids are roughly spherical because, as for planets and large moons, gravity determines their shape. Indirect evidence from meteorites suggests that the largest asteroids were heated by impact events early in their history, which caused them to melt. This resulted in an early period of chemical differentiation that produced their dense iron-rich cores and rocky mantles.

Most asteroids, however, are small and have irregular shapes, which led planetary geologists to conclude that they are leftover debris from the solar nebula. In addition, these small asteroids have densities lower than scientists originally predicted, indicating that they are relatively porous bodies, like "piles of rubble," loosely bound together by their weak gravitational fields (Figure 22.29).

In February 2001, an American spacecraft became the first visitor to an asteroid. Although it was not designed for landing, *NEAR Shoemaker* landed successfully on Eros and collected information that has planetary geologists both intrigued and perplexed. Images obtained as the spacecraft drifted toward the surface of Eros revealed a barren, rocky surface composed of particles ranging in size from fine dust to boulders up to 10 meters (30 feet) across. Researchers unexpectedly discovered that fine debris tends to concentrate in the

▼ **Figure 22.29 Asteroid Itokawa** The barren rocky surface of asteroid Itokawa appears to be a pile of rubble held together by the asteroid's weak gravitational field. This potato-shaped asteroid orbits between Mars and Jupiter and is only about 0.5 kilometers across—about the size of five football fields. (Courtesy of Japan Aerospace Exploration Agency)

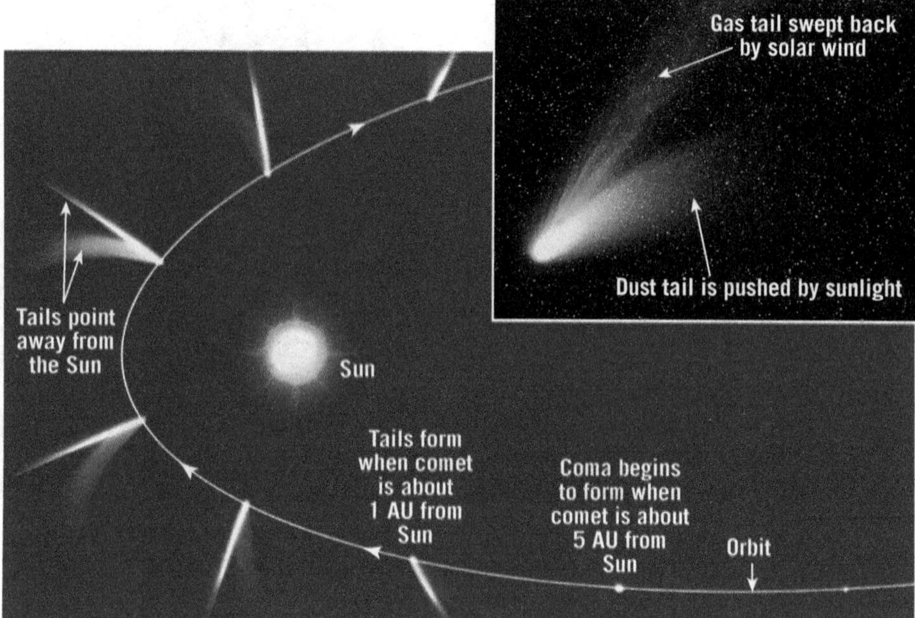
Gas tail swept back by solar wind
Dust tail is pushed by sunlight
Tails point away from the Sun
Sun
Tails form when comet is about 1 AU from Sun
Coma begins to form when comet is about 5 AU from Sun
Orbit

▲ **Figure 22.30 Changing orientation of a comet's tail as it orbits the Sun** (Photo by Dan Schechter/ Science Source)

low areas, where it forms flat deposits resembling ponds. Surrounding the low areas, the landscape is marked by an abundance of large boulders. One of several hypotheses to explain the boulder-strewn topography is seismic shaking, which would cause the boulders to move upward as the finer materials sink. This is analogous to what happens when a jar of sand and various-sized pebbles is shaken: The larger pebbles rise to the top, while the smaller sand grains settle to the bottom (sometimes referred to as the Brazil-nut effect).

Exploring Asteroids In November 2005, the Japanese probe *Hayabusa* made a soft landing on a small near-Earth asteroid named Itokawa and picked up some rocky debris before returning to Earth in June 2010 (see Figure 22.29). Chemical analyses of samples from this mission show that the surface of this asteroid is nearly identical in composition to rocky meteorites. This finding strongly supports the idea that asteroids are the source of most meteoroids large enough to reach Earth's surface.

Hayabusa 2, launched in 2014, is expected to land on its target (asteroid 1999 JU3) in July 2018. After exploring this asteroid by digging into its surface to extract fresh samples, it is scheduled to return to Earth in 2020.

▼ **Figure 22.31 Coma and nucleus of Comet Holmes** The nucleus of Comet Holmes is the bright yellow spot within the reddish-orange coma. (Courtesy of NASA)

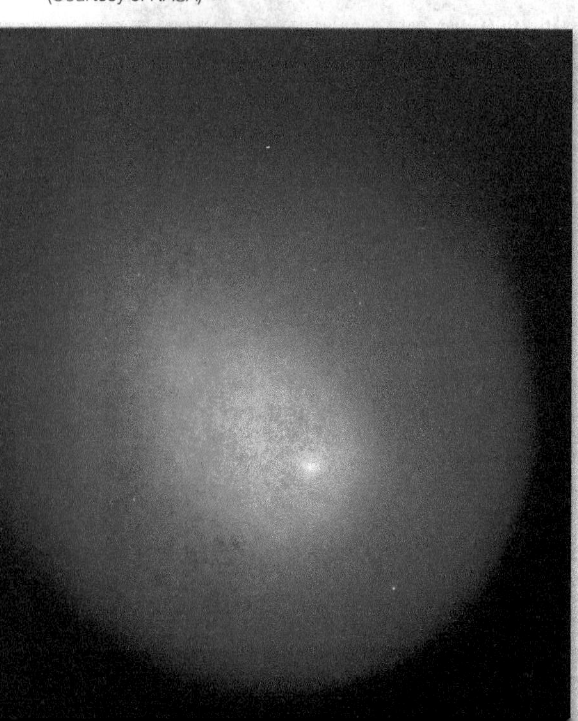

Comets: Dirty Snowballs

Comets, like asteroids, are leftover material from the formation of the solar system. They are loose collections of rocky material, dust, water ice, and frozen gases (ammonia, methane, and carbon dioxide), thus the nickname "dirty snowballs." Recent space missions to comets have shown their surfaces to be dry and dusty, which indicates that their ices are hidden beneath a layer of rocky debris.

Most comets reside in the outer reaches of the solar system and take hundreds of thousands of years to complete a single orbit around the Sun. However, a smaller number of *short-period comets* (those having orbital periods of less than 200 years), such as the famous Halley's Comet, make regular encounters with the inner solar system (Figure 22.30). The shortest-period comet (Encke's Comet) orbits the Sun once every 3 years.

Structure and Composition of Comets The phenomena associated with comets come from a small central body called the **nucleus**. These structures are typically 1 to 10 kilometers in diameter, but comet nuclei 40 kilometers across have been observed. When a comet comes within about 5 AU of the Sun, solar energy heats its surface sufficiently to cause its icy components to begin to vaporize into gas. The escaping gases carry dust from the comet's surface, producing a huge dusty atmosphere called a **coma** (Figure 22.31). As a comet approaches the inner solar system, the coma grows, and some of the dust and gas is pushed away from the Sun to form the comet's tails, which can grow to hundreds of millions of kilometers in length.

Bright comets have two visible tails: a dark blue tail that points straight away from the Sun and a brighter tail that points away from the Sun but also curves slightly in the direction from which the comet came. Scientists have determined the mechanisms that account for the formation of these tails. The faint, straight *gas tail* consists of ionized cometary gases that are pushed away from the coma by the pressure of the *solar wind* of charged particles emitted by the Sun. The brighter, curved *dust tail* consists of dust particles that are pushed away from the coma by the much weaker pressure of sunlight (*radiation pressure*). The dust tail is curved because its bright, relatively slow-moving particles record how the direction toward the Sun changes as the comet moves along its orbit.

As a comet's orbit carries it away from the Sun, the gases forming the coma dissipate, the tails disappear, and the comet returns to cold storage. Material that was blown from the coma to form the tails is lost forever. When all the gases are expelled, the inactive comet continues its orbit without a coma or tail. Sometimes the comet literally disintegrates into small fragments that continue to orbit the Sun. Scientists believe that few comets remain active for more than a few hundred close orbits of the Sun.

In 2015 the *Rosetta* spacecraft, launched by the European Space Agency, obtained a close-up image of the nucleus of Comet 67P/Churyumov–Gerasimenko that offered a new perspective on the extent of the comet's activity. Figure 22.32 shows jets of gas and dust emanating from the central region of the nucleus and

Figure 22.32 **Jets of gas and dust erupting from the nucleus of Comet 67P** This comet's full name is Comet 67P/Churyumov–Gerasimenko; like most other comets, it is named after its discoverers. It is a regular visitor to the inner solar system, orbiting the Sun every 6.5 years. (Courtesy of European Space Agency)

extending toward the upper right of the image. This image also shows the nebulous glow of material escaping the bright sunlit surface of the comet.

The Realm of Comets: The Kuiper Belt and Oort Cloud

Most comets originate in one of two regions: the *Kuiper belt* or the *Oort cloud*. Named in honor of astronomer Gerald Kuiper, who predicted its existence, the **Kuiper belt** hosts a large group of icy objects that reside in the outer solar system, beyond the orbit of Neptune. Pluto's orbit lies within the Kuiper belt, and in 2005 scientists discovered Eris, a Kuiper belt body more massive than Pluto.

Like most bodies located in the inner solar system, Kuiper belt comets orbit the Sun in the same direction and along roughly the same plane as the planets (Figure 22.33A). This disc-shaped structure is thought to contain about 100,000 bodies more than 100 kilometers (60 miles) across, as well as many smaller objects. The largest Kuiper belt objects are much larger than the comets we observe in the inner solar system, probably because relatively small objects are more likely to have their orbits altered than are larger bodies.

Kuiper belt objects are thought to be leftover planetesimals that formed and remain in the frigid outer reaches of the solar system. Interactions with other nearby bodies or the gravitational influence of one of the Jovian planets occasionally alters their orbits sufficiently to send them into our view.

Named for Dutch astronomer Jan Oort, the **Oort cloud** consists of icy planetesimals that form a roughly spherical shell around the outer reaches of the solar system (Figure 22.33B). Oort cloud objects have random orbits at distances often greater than 50,000 times the Earth–Sun distance. This places them at nearly half the distance to Proxima Centauri, the star nearest the Sun. These icy objects are so loosely bound to the solar system that the gravitational effect of a distant passing star may send an occasional Oort cloud body into a highly eccentric orbit that carries it toward the Sun. Because most comets that enter the inner solar system have random orbits, these are assumed to have come from the Oort cloud.

The Oort cloud is estimated to contain at least a trillion comets. How did so many comets end up in the outer reaches of the solar system? The most widely accepted hypothesis is that these bodies formed early in our solar system's history, in the region occupied by the still-developing Jovian planets. Rather than being captured by one of these growing planets, the comets were gravitationally flung in all directions.

Meteors, Meteoroids, and Meteorites

Nearly everyone has seen **meteors**, commonly (but inaccurately) called "shooting stars." These streaks of light can be observed in as little as the blink of an eye or can last

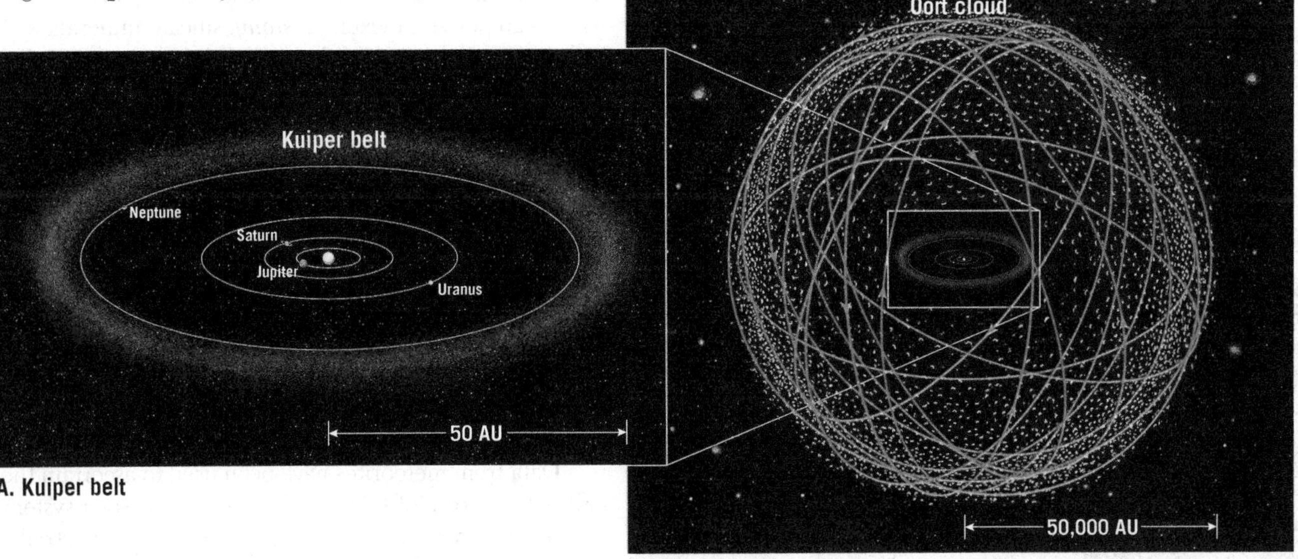

Figure 22.33 **The realm of the comets: The Oort cloud and Kuiper belt** Most comets reside in one of two places. **A.** Located beyond the orbit of Neptune is the Kuiper belt, a disk of icy objects with roughly circular orbits that travel in the same direction as the planets. **B.** The Oort cloud is a spherical cloud containing icy planetesimals located roughly 50,000 times farther from the Sun than Earth.

as "long" as a few seconds. They occur when a small solid particle, a **meteoroid**, enters Earth's atmosphere from interplanetary space. Heat, created by friction between the meteoroid and Earth's atmosphere, produces the streak of light we see trailing across the sky. Most meteoroids originate from one of three sources: (1) interplanetary debris missed by the gravitational sweep of the planets during solar system formation, (2) material ejected from the asteroid belt, or (3) the rocky remains of comets that once passed through Earth's orbit.

Meteoroids smaller than about 1 meter (3 feet) in diameter generally vaporize before reaching Earth's surface. Some, called *micrometeorites*, are so tiny and their rate of fall so slow that they drift to Earth continually as space dust. Researchers estimate that thousands of meteoroids enter Earth's atmosphere every day. After sunset on a clear, dark night, many are bright enough to be seen with the naked eye.

Meteor Showers Occasionally, meteor sightings increase dramatically to 60 or more per hour. Such displays, called **meteor showers**, result when Earth encounters a swarm of meteoroids traveling in the same direction at nearly the same speed as Earth. The close association of these swarms to the orbits of some short-term comets strongly suggests that they represent material lost by these comets (Table 22.2). Some swarms, not associated with the orbits of known comets, are probably the scattered remains of the nucleus of a long-defunct comet. The notable *Perseid meteor shower* that occurs each year around August 12 is likely material ejected from the comet Swift–Tuttle on previous approaches to the Sun.

Table 22.2 Major Meteor Showers

Shower	Approximate Dates	Associated Comet
Quadrantids	January 4–6	Unknown
Lyrids	April 20–23	Comet Thatcher
Eta Aquarids	May 3–5	Halley's Comet
Delta Aquarids	July 30	Unknown
Perseids	August 12	Comet Swift–Tuttle
Draconids	October 7–10	Comet Giacobini–Zinner
Orionids	October 20	Halley's Comet
Taurids	November 3–13	Comet Encke
Andromedids	November 14	Comet Biela
Leonids	November 18	Comet Tempel–Tuttle
Geminids	December 4–16	Unknown

Meteorites: Visitors to Earth The remains of meteoroids that have impacted Earth's surface are called **meteorites** (Figure 22.34). Most meteoroids large enough to survive passage through the atmosphere likely started out as asteroids; a chance collision or gravitational interactions with Jupiter modifies the orbit and sends the asteroid toward Earth. Earth's gravity does the rest. A few meteorites are fragments of the Moon, Mars, or possibly even Mercury, ejected by a violent asteroid impact. Before *Apollo* astronauts brought Moon rocks back to Earth, meteorites were the only extraterrestrial materials that could be studied in laboratories.

A few large meteorites have blasted craters on Earth's surface that strongly resemble craters on the Moon. More than 40 terrestrial craters with diameters larger than 20 kilometers (12 miles) exist. These craters exhibit features that could only have been produced by an explosive impact of an asteroid or perhaps even a comet nucleus. More than 250 smaller craters are also thought to have impact origins. Notable among them is Arizona's Meteor Crater, a cavity more than 1 kilometer (0.6 mile) wide and 170 meters (560 feet) deep, with an upturned rim that rises above the surrounding countryside (Figure 22.35). More than 30 tons of iron fragments have been found in the immediate area, but attempts to locate the main body have been unsuccessful. Based on the amount of erosion observed on the crater rim, the impact likely occurred within the past 50,000 years.

Types of Meteorites Classified by their composition, meteorites are either (1) *irons*, mostly aggregates of iron with 5–20 percent nickel; (2) *stony*, silicate minerals with inclusions of other minerals; or (3) *stony–irons*, mixtures of the two. Although stony meteorites are the most common, irons are found in large numbers because metallic meteorites withstand impacts better, weather more slowly, and are easily distinguished from terrestrial rocks. Iron meteorites are probably fragments of once-molten cores of large asteroids or small planets.

One type of stony meteorite, called a *carbonaceous chondrite*, contains organic compounds and occasionally simple amino acids, which are some of the basic building blocks of life. This discovery confirms similar findings in observational astronomy, which indicate that numerous organic compounds exist in interstellar space.

Data from meteorites have been used to ascertain the internal structure of Earth and the age of the solar system. If meteorites represent the composition of the terrestrial planets, as some planetary geologists suggest, our planet

▼ SmartFigure 22.34 **Iron meteorite found near Meteor Crater, Arizona** (Photo by M2 Photography/Alamy Stock Photo)

TUTORIAL
https://goo.gl/yQzIDJ

SmartFigure 22.35 **Meteor Crater, near Winslow, Arizona** This cavity is about 1.2 kilometers (0.75 mile) across and 170 meters (560 feet) deep. The solar system is cluttered with asteroids and comets that can strike Earth with explosive force. (Photo by Michael Collier)

TUTORIAL
https://goo.gl/qvD1wu

must contain a much larger percentage of iron than is indicated by surface rocks. This is one reason that geologists think Earth's core is mostly iron and nickel. In addition, radiometric dating of meteorites indicates that the age of our solar system is about 4.6 billion years. This "old age" has been confirmed by data obtained from lunar samples.

Dwarf Planets

Pluto, once considered the ninth planet, was discovered in 1930 by Clyde Tombaugh, who was searching for a yet-undiscovered planet in order to explain irregularities in the orbit of Uranus. Astronomers quickly realized that Pluto was too tiny and too distant to account for such irregularities. Pluto has a diameter of about 2370 kilometers (1470 miles), or about one-fifth that of Earth and less than half that of Mercury—long considered the solar system's "runt."

More attention was given to Pluto's status as a planet when astronomers began to discover other large Kuiper belt bodies. Clearly, Pluto and these other large Kuiper belt bodies were completely different from either the terrestrial planets or the Jovian planets. In 2006, the International Astronomical Union, the group responsible for naming and classifying celestial objects, voted to designate a new class of solar system objects called *dwarf planets*. To be classified a dwarf planet, a celestial body must orbit the Sun and be essentially spherical due to its own gravity but not large enough to sweep its orbits clear of other debris. By this definition, Pluto is recognized as a dwarf planet and the prototype for this new category of planetary objects. Other dwarf planets include Eris, Makemake, and Haumea, many Kuiper belt objects, and Ceres, the largest-known asteroid.

Exploration of Pluto In July 2015, NASA's *New Horizons* spacecraft shot past Pluto, after a 9-year journey that brought it within 12,500 kilometers (7800 miles) of Pluto's surface. Images transmitted from *New Horizons* show Pluto to be an active body with several distinct terrains, including mountainous areas consisting of blocks of water-ice, flat ice plains composed of nitrogen-ice, and rugged areas indicating a long history of impacts.

One of the most interesting regions, named Sputnik Planitia (*planitia* = plain, or flat area), is the left part of Pluto's iconic "heart" feature (Figure 22.36A). Sputnik Planitia is a large ice field about 1050 kilometers (650 miles) in diameter, with a maximum depth of about 4 kilometers (2.5 miles). Found in Pluto's northern hemisphere, it contains lobes of ice flowing around its edges, similar to glaciers on Earth. With surface temperatures of about −235°C (just shy of −400°F), Pluto is too cold for Sputnik Planitia's ice field to be made of water. Instead, it is composed mainly of frozen nitrogen, with lesser amounts of carbon monoxide and methane ices, which flow at these frigid temperatures.

As shown in Figure 22.36B, much of the surface of Sputnik Planitia consists of irregular polygons separated by linear troughs. This structure led researchers to conclude that heat from the planet's interior has generated a

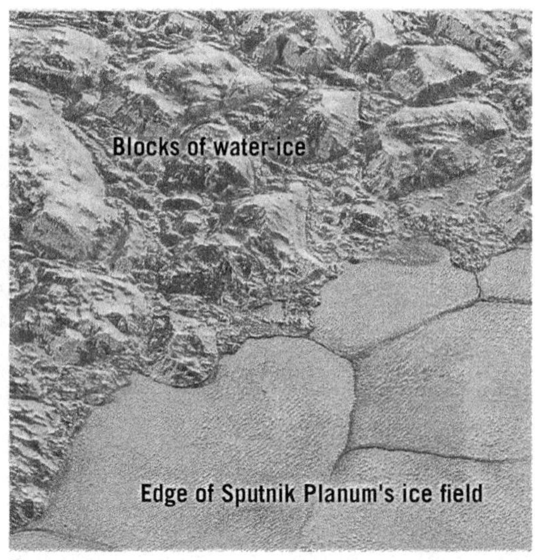

Figure 22.36 **Images of Pluto obtained from NASA's *New Horizons* spacecraft A.** This enhanced color image was generated to detect differences in the composition and texture of Pluto's surface. The bright area in the lower central region, informally named Sputnik Planitia, consists mainly of soft, nitrogen-rich ices that, in some locations, appear to flow like glaciers on Earth. (Photo by Johns Hopkins University Applied Physics Laboratory/Southwest Research Institute/NASA) **B.** This close-up view shows Sputnik Planitia in the lower right, which in this region forms a nearly level plane broken into cellular units. The area in the upper left consists of blocks of water-ice, some standing almost 2.5 kilometers (about 1.5 miles) above the surrounding plains. (Courtesy of NASA)

type of convective flow, whereby upwelling of nitrogen ice occurs in the center of the cells, which spreads out along the surface and eventually sinks at the linear troughs. This convective model is supported by the nearly 100-meter (300-foot) height difference between the centers of these cells and their lower margins. Because no craters were discovered on the surface of Sputnik Planitia, researchers conclude that its surface is relatively young—less than 100 million years old.

One hypothesis proposes that Sputnik Planitia began as a large impact basin, much like the large impact basins, called maria, on the Moon (see Figure 22.7). Once formed, the impact basin on Pluto began to fill with nitrogen ice, rather than basaltic magma, like the maria on the Moon. Pluto's atmosphere, a tenuous layer of gases consisting mainly of nitrogen, is thought to be the source of nitrogen ice that blankets Sputnik Planitia. Two processes are involved: *sublimation* (the change of a solid directly to a gas) of nitrogen ice in regions where the planet receives the most direct sunlight, and *deposition* of nitrogen gas to form ice in the much colder impact basin.

What's Next Having completed its flyby of Pluto, *New Horizons* changed course and was aimed at another icy Kuiper belt object about 40 kilometers (25 miles) in diameter. A close encounter with this tiny icy object is expected to occur January 1, 2019.

CONCEPT CHECKS 22.5

1. Compare and contrast asteroids and comets. Where are most asteroids found?
2. Where are most comets thought to reside? What eventually becomes of comets that orbit close to the Sun?
3. Differentiate among the following solar system bodies: meteoroid, meteor, and meteorite.
4. What are the three main sources of meteoroids?
5. What characteristic does a celestial body possess to be classified as a dwarf planet?

22 CONCEPTS IN REVIEW

Touring Our Solar System

22.1 Our Solar System: An Overview

Describe the formation of the solar system according to the nebular theory. Compare and contrast the terrestrial and Jovian planets.

KEY TERMS: nebular theory, solar nebula, planetesimal, protoplanet, terrestrial (Earth-like) planet, Jovian (Jupiter-like) planet, escape velocity, impact crater

- Our Sun is the most massive body in our solar system, which includes planets, dwarf planets, moons, and other small bodies. The planets orbit in the same direction and at speeds proportional to their distance from the Sun, with inner planets moving faster and outer planets moving more slowly.
- The solar system began as a solar nebula before condensing due to gravity. While most of the matter ended up in the Sun, some material formed a thick disk around the early Sun and later clumped together into larger and larger bodies. Planetesimals collided to form protoplanets, and protoplanets grew into planets.
- The four terrestrial planets are enriched in rocky and metallic materials, whereas the Jovian planets have a higher proportion of ice and gas. The terrestrial planets are relatively dense, with thin atmospheres, while the Jovian planets are less dense and have thick atmospheres.
- Smaller planets have less gravity to retain gases in their atmosphere. Lightweight gases such as hydrogen and helium more easily reach escape velocity, so the atmospheres of the terrestrial planets tend to be enriched in heavier gases, such as water vapor, carbon dioxide, and nitrogen.

22.2 Earth's Moon: A Chip Off the Old Block

List and describe the major features of Earth's Moon and explain how maria basins were formed.

KEY TERMS: maria, lunar highlands, lunar regolith

- The Moon has a composition that is approximately the same as that of Earth's mantle. The Moon likely formed from a collision between a Mars-sized protoplanet and the early Earth.
- The lunar surface is dominated by light-colored lunar highlands (or terrae) and darker lowlands called maria, the latter formed primarily from flood basalts. Both terrae and maria are partially covered by lunar regolith produced by micrometeorite bombardment.

? Briefly describe how our Moon formed and how its formation accounts for its low density compared to that of Earth.

22.3 Terrestrial Planets

Outline the principal characteristics of Mercury, Venus, and Mars. Describe their similarities to and differences from Earth.

- Mercury has a very thin atmosphere and a weak magnetic field. Like Earth's moon, Mercury has both heavily cratered areas and smooth plains; the smooth plains are similar to lunar maria.
- Venus has a very dense atmosphere, dominated by carbon dioxide. The resulting extreme greenhouse effect produces surface temperatures around 450°C (900°F). The topography of Venus has been resurfaced by active volcanism.
- Mars has about 1 percent as much atmosphere as Earth, so it is relatively cold (−140°C to 20°C [−220°F to 68°F]). Mars appears to be the closest planetary analog to Earth, showing surface evidence of rifting, volcanism, and modification by flowing water. Volcanoes on Mars are much bigger than volcanoes on Earth because of the lack of plate motion on Mars.

? As you can see from this graph, Mercury's temperature varies significantly from "day" to "night," but Venus's temperature is relatively constant "around the clock." Suggest a reason for this difference.

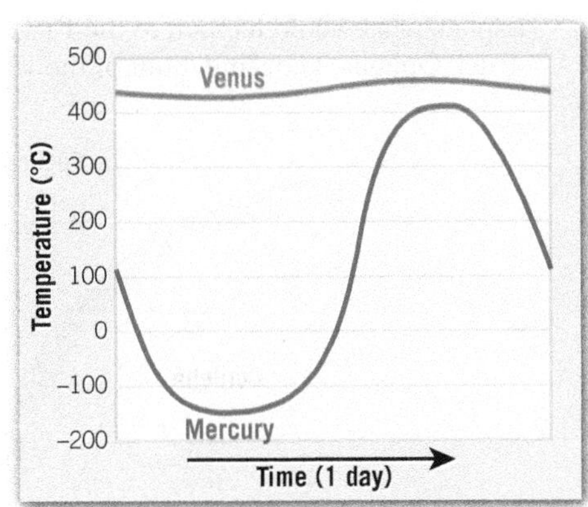

22.4 Jovian Planets

Summarize and compare the features of Jupiter, Saturn, Uranus, and Neptune, including their ring systems.

KEY TERMS: cryovolcanism

- Jupiter's mass is several times larger than the combined mass of everything else in the solar system except for the Sun. Convective flow, combined with its three cloud layers, produces its characteristic banded appearance. Persistent, giant rotating storms exist between these bands. Many moons orbit Jupiter, including Io, which shows active volcanism, and Europa, which is believed to have a liquid ocean under its icy shell.
- Saturn, like Jupiter, is big, gaseous, and endowed with dozens of moons. Some moons show evidence of tectonics, while Titan has its own atmosphere. Saturn's well-developed rings are made of many particles of water ice and rocky debris.

(22.4 continued)

- Uranus, like its "twin" Neptune, has a blue atmosphere dominated by methane, and its diameter is about four times greater than Earth's. Uranus rotates sideways relative to the plane of the solar system. It has a relatively thin ring system and at least five moons.
- Neptune has an active atmosphere, with fierce wind speeds and giant storms. It has 1 large moon, Triton, which shows evidence of cryovolcanism, as well as 13 smaller moons and a ring system.

22.5 Small Solar System Bodies

List and describe the principal characteristics of the small bodies that inhabit the solar system.

KEY TERMS: small solar system body, dwarf planet, asteroid, asteroid belt, comet, nucleus, coma, Kuiper belt, Oort cloud, meteor, meteoroid, meteor shower, meteorite

- Small solar system bodies include rocky asteroids and icy comets. Both are basically scraps left over from the formation of the solar system or fragments from later impacts.
- Most asteroids are concentrated in a wide belt between the orbits of Mars and Jupiter. Some are rocky, some are metallic, and some are basically "piles of rubble," loosely held together by their own weak gravity.
- Comets are dominated by ices, "dirtied" by rocky material and dust. Most originate in either the Kuiper belt beyond Neptune or the Oort cloud. When a comet's orbit brings it through the inner solar system, solar radiation causes its ices to vaporize, generating the coma and its characteristic "tail."
- A meteoroid is a small rocky or metallic body traveling through space. When it enters Earth's atmosphere, it flares briefly as a meteor before either burning up or striking Earth's surface to become a meteorite. Asteroids and material lost from comets as they travel through the inner solar system are the most common sources of meteoroids.
- Bodies massive enough to have a spherical shape but not so massive as to have cleared their orbits of debris are classified as dwarf planets. They include the rocky asteroid Ceres as well as the icy worlds Pluto and Eris, which are located in the Kuiper belt.

? Shown here are four small solar system bodies. Identify each and explain the differences among them.

Muellek Josef/Shutterstock
National Science Foundation
Jerry Schad/Photo Researchers, Inc.
NASA/JPL-Caltech/UCAL/MPS/DLR/IDA

GIVE IT SOME THOUGHT

1 Assume that a solar system was discovered in a nearby region of the Milky Way Galaxy. The accompanying table shows data that have been gathered about three of the planets orbiting the central star of this newly discovered solar system. Using Table 22.1 as a guide, classify each planet as Jovian, terrestrial, or neither. Explain your reasoning.

	Planet 1	Planet 2	Planet 3
Relative Mass (Earth = 1)	1.2	15	0.1
Diameter (km)	11,000	52,000	2200
Mean Distance from Star (AU)	1.4	17	35
Density (g/cm3)	4.8	1.22	1.8
Orbital Eccentricity	0.01	0.05	0.23

2 In order to conceptualize the size and scale of Earth and Moon as they relate to the rest of the solar system, complete the following:

a. Approximately how many Moons (diameter 3475 kilometers [2160 miles]) would fit side by side across the diameter of Earth (diameter 12,756 kilometers [7926 miles])?

b. Given that the Moon's orbital radius is 384,798 kilometers, approximately how many Earths would fit side by side between Earth and the Moon?

c. Approximately how many Earths would fit side by side across the Sun, whose diameter is about 1,390,000 kilometers?

d. Approximately how many Suns would fit side by side between Earth and the Sun, a distance of about 150,000,000 kilometers?

3 The accompanying graph shows the temperatures at various distances from the Sun during the formation of our solar system. (You can assume that the planets formed at roughly their current distances from the Sun, although that may not be strictly true.) Use it to answer the following:

a. Which planet or planets formed at locations in the solar system where the temperature was hotter than the boiling point of water?

b. Which planet or planets formed at locations in the solar system where the temperature was cooler than the freezing point of water?

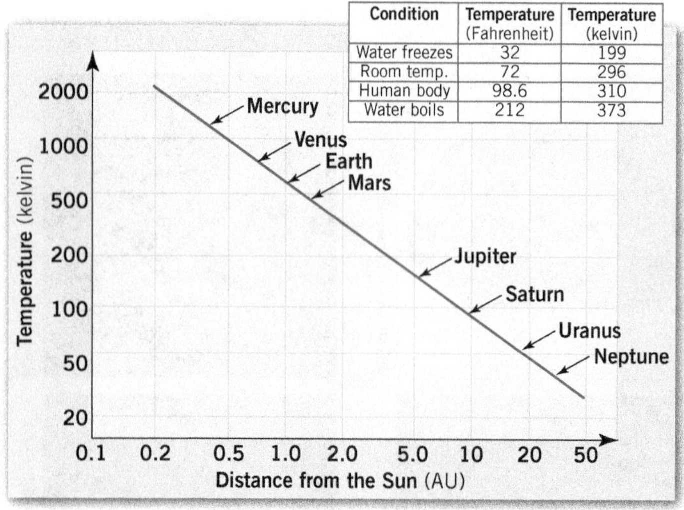

4 This sketch shows four primary craters (A, B, C, and D). The impact that produced Crater A produced two secondary craters (labeled a) and three rays. Crater D has one secondary crater (labeled d). Rank the four primary craters from oldest to youngest and explain your ranking.

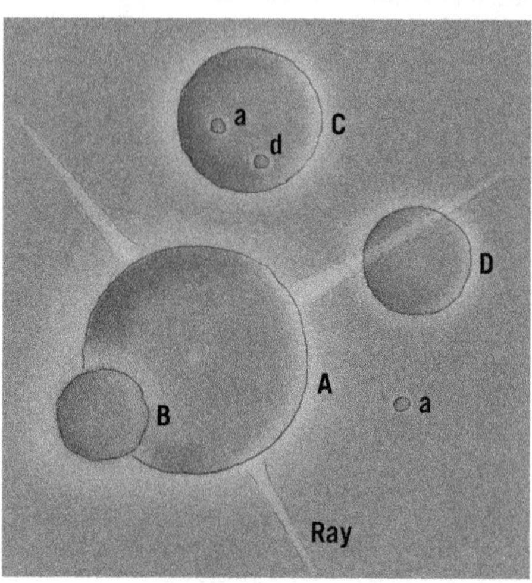

5 Halley's Comet has a mass estimated at 100 billion tons. Furthermore, it is estimated to lose about 100 million tons of material when its orbit brings it close to the Sun. With an orbital period of 76 years, calculate the maximum remaining life span of Halley's Comet.

6 The accompanying diagram shows two of Uranus's moons, Ophelia and Cordelia, which act as shepherd moons for the Epsilon ring. Explain what would happen to the Epsilon ring if a large asteroid struck Ophelia, knocking it out of the Uranian system.

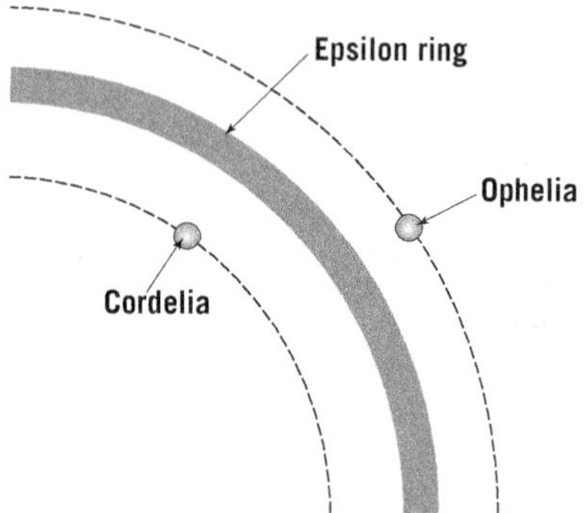

7 Assume that three irregularly shaped planet-like objects, each smaller than our Moon, have just been discovered orbiting the Sun at a distance of 35 AU. One of your friends argues that the objects should be classified as planets because they are large and orbit the Sun. Another friend argues that the objects should be classified as dwarf planets, like Pluto. State whether you agree or disagree with either or both of your friends. Explain your reasoning.

EXAMINING THE EARTH SYSTEM

1 On Earth the four major spheres (atmosphere, hydrosphere, geosphere, and biosphere) interact as a system with occasional influences from our near-space neighbors. Which of these spheres are absent, or nearly absent, on the Moon? Because the Moon lacks these spheres, list at least five processes that operate on Earth but are absent on the Moon.

2 Among the planets in our solar system, Earth is unique because water exists in all three states (solid, liquid, and gas) on and near its surface. In what state(s) of matter is water found on Mars?

a. How would Earth's hydrologic cycle be different if its orbit were inside the orbit of Venus?

b. How might Earth's hydrologic cycle be different if its orbit were outside the orbit of Mars?

DATA ANALYSIS

Topography of Venus

Rocky planets and natural satellites exhibit geology only recently photographed by passing spacecraft. Impact craters are present on every rocky planet, but these features are not preserved in the same way on all planets.

ACTIVITIES

Go to the USGS Astrogeology Science Center at https://astrogeology.usgs.gov/ and click on Planetary Nomenclature at the bottom of the page. Click on Nomenclature and select Venus. To navigate these data, hover over a feature type in the list to see a short description of the feature. Click on the feature type to see a table of named features of this type. You can reorder column data by clicking on the up and down arrows at the top of selected columns, such as name or diameter. (Note that reordering the table may take a minute or two.)

1 Click on Crater, Craters to examine the craters of Venus. How many of the planet's craters have been named? What is the name of the largest approved crater on Venus? What is its diameter?

2 Note that the largest crater in the list has been disallowed. In fact, it has been reclassified as a corona. The exact origin of coronae is unknown, but they appear to be unique to Venus. Go back to the Venus: Search by Feature Types page and click on Corona, Coronae. How many coronae have been named on Venus?

3 What is the name of the largest corona? What is its diameter?

Return to the Venus: Search by Feature Types page, then click on Venus Map under Interactive Images & Maps. Click on the Venus Global GIS Viewer. Check the box next to Venus Global GIS and add only the Nomenclature layer. To display the legend, click the + next to the Nomenclature box, then click the + next to the Venus Nomenclature label.

4 Based on the colors on the map, what is the most commonly named feature on Venus?

5 Check the box next to Volcanic Catalogs, and click the + next to that box. Then check only the box next to USGS Venus Volcano Catalog 1999 layer. How common are volcanoes on Venus, compared to other features?

Now uncheck the Venus Global GIS box. Without closing the Venus map tab, go back to the Planetary Names tab, and in a second window, open the Lunar Global GIS map by choosing The Moon from the Nomenclature list and clicking Lunar Map under Interactive Images & Maps. Then open the Moon Global GIS Viewer. Do the same for Mars: in a third window, open the Mars Global GIS map by choosing Mars from the Nomenclature list, then select Mars to display a list of its features. Click on Map of Mars under Interactive Images & Maps, and open the Mars Global GIS Viewer. Compare the Venus, Mars, and Lunar maps. You will need to zoom in and pan around to see surface features clearly.

6 Rank these bodies from most densely packed craters to least densely packed craters. On which body are craters most easily distinguished?

7 Based on what you have learned from this chapter, suggest reasons why craters on Venus differ in size, shape, and density compared to craters on Mars and Earth's moon.

23

Light, Telescopes, and the Sun*

FOCUS ON CONCEPTS

Each statement represents the primary learning objective for the corresponding major heading within the chapter. After you complete the chapter, you should be able to:

23.1 Compare and contrast the wave properties of light with the particle properties of light.

23.2 Explain how the three types of spectra are generated and what they tell astronomers about the radiating body that produced them.

23.3 Describe the two properties that make telescopes with large mirrors more useful than those with small mirrors.

23.4 List the advantages of radio-wave and orbiting observatories over ground-based optical telescopes.

23.5 Sketch the Sun's structure and describe each of its major layers. Summarize the process called the proton–proton chain reaction.

23.6 List and describe four types of solar storms that occur on the Sun.

*This chapter was revised with the assistance of Professor Teresa Tarbuck.

Comet Hale-Bopp, above the observatories on the summit of Mauna Kea, Hawaii.
(Photo by David Nunuk/Science Source)

BECAUSE ASTRONOMERS CANNOT STUDY THE UNIVERSE by bringing it into the laboratory, and the vast majority of celestial objects are too far away to visit, astronomers collect and study the things that come to Earth from space. Overwhelmingly, this means collecting and studying light, emitted or reflected by objects found in the universe (**Figure 23.1**). In fact, almost everything that is known about the universe beyond the solar system comes from the analysis of light from distant sources. This chapter examines the properties and utility of light, some of the tools astronomers use to collect and study light, and what is known about the nearest source of light, the Sun.

23.1 Light: Messenger from Space

Compare and contrast the wave properties of light with the particle properties of light.

Although visible light is most familiar to us, it constitutes only a tiny sliver of an array of light known as **electromagnetic radiation**. Included in this array are *gamma rays*, *x-rays*, *ultraviolet light*, *visible light*, *infrared light* (heat), and *radio waves* (**Figure 23.2**). All forms of electromagnetic energy (light) travel through the vacuum of space at a rate of 300,000 kilometers (186,000 miles) per second—a staggering 26 billion kilometers per day. Light tells us about the processes that created it and about the matter that exists between its source and Earth.

Nature of Light

Experiments have demonstrated that light can be described in two ways. In some instances, light behaves like waves, and in others, it behaves like tiny particles.

Wave Properties of Light Light waves are analogous to swells in the ocean and are characterized by **wavelength**—the distance from one wave crest to the next. Electromagnetic wavelengths range from several kilometers for long radio waves to less than one-billionth of a centimeter for gamma rays (see Figure 23.2). Another important wave property of light is **frequency**—the number of waves that pass a point during 1 second, expressed in *hertz*. The frequency of visible light ranges from 430 trillion hertz for red light to 750 trillion hertz for violet light.

Most light occurs at wavelengths that are either too long or too short for our eyes to detect; however, the primary characteristics of all electromagnetic radiation can be described using visible light as an example. The extremely narrow band of electromagnetic radiation that humans can see (labeled *visible light* in Figure 23.2) astronomers often call *white light*. In the 1600s Newton used clear glass prisms to demonstrate that white light consists of an array of colors. As white light passes through a prism, it spreads into the familiar rainbow of colors, ranging from short-wavelength violet through long-wavelength red.

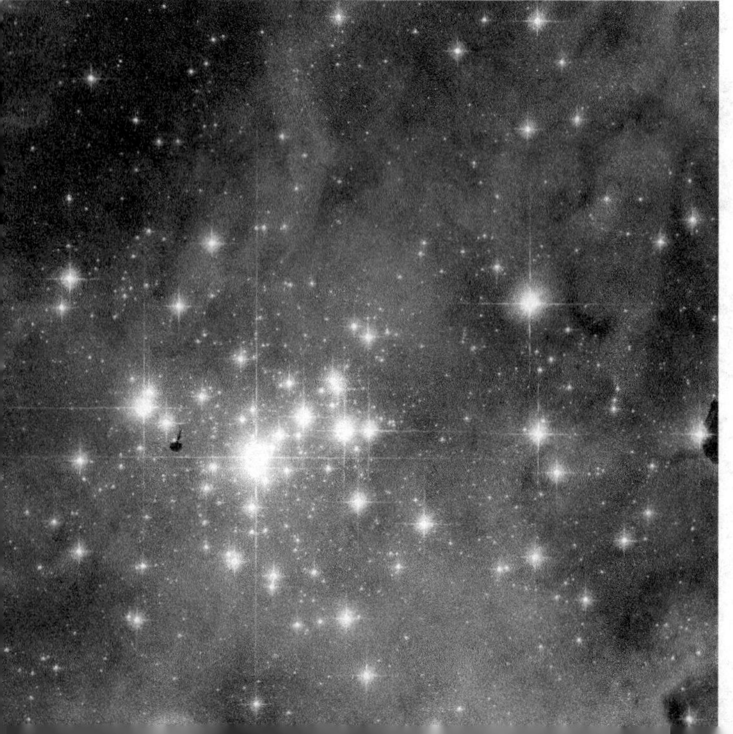

Figure 23.1 Star cluster that contains some of the brightest stars in the Milky Way Galaxy Called Trumpler 14, this cluster is located 8000 light-years away, in the Carina Nebula, a large star-forming region. (Courtesy of NASA)

Particle Properties of Light Wave theory, however, cannot explain some of the observed characteristics of light. In these cases, light acts like a stream of particles, analogous to infinitesimally small bullets fired from a machine gun. These particles, called **photons**, can exert pressure (push) on matter, which is called **radiation pressure**. Recall from Chapter 22 that photons from the Sun are responsible for pushing material away from a comet to produce its dust tail.

Each photon has a specific amount of energy, which is related to its wavelength in a simple way: *Shorter wavelengths* correspond to *more energetic photons*. Thus, violet light consists of more energetic photons than red light, and ultraviolet light is more energetic than violet light—while gamma rays are the most energetic form of electromagnetic radiation and radio waves are the least energetic (see Figure 23.2). High-energy photons, such as x-rays, tend to plow through matter before they are absorbed. That is why x-rays are used in medical imaging: They can readily pass though soft tissue but are absorbed by bones, which are denser.

Which theory of light—the wave theory or the particle theory—is correct? The answer is that both are correct because each predicts the behavior of light for certain phenomena. As George Abell, a prominent astronomer, stated about all scientific laws, "The mistake is only to apply them to situations that are outside their range of validity."

Why Study Light?

Light is truly a messenger from space that allows us explore the universe around us. Light energy from all parts of the electromagnetic spectrum continually

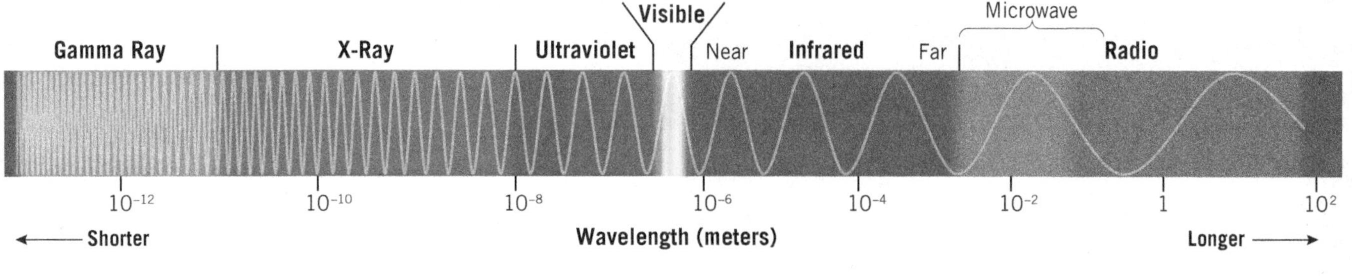

Figure 23.2 Electromagnetic radiation The electromagnetic spectrum ranges from short-wavelength gamma radiation to long-wavelength radio waves.

bombard Earth—from stars, interstellar matter, and galaxies throughout the universe. For example, matter that is being drawn into a black hole is heated to unimaginable temperatures, causing it to emit high-energy x-rays before it is engulfed. By contrast, when less violent processes occur, smaller amounts of low-energy radiation are released. This may occur when a star heats a nearby cloud of dust and gases and raises the nebula's temperature, causing it to emit visible light and glow.

In addition, the intensity of the light emitted and its wavelength tell us about the type of process that is generating the light. This information can often be used to support or refute scientific hypotheses. For example, theoretical studies predicted the existence of black holes long before observational evidence existed. The concept of black holes gained considerable support when x-rays matching the wavelengths predicted by this hypothesis were detected around objects suspected of being black holes.

CONCEPT CHECKS 23.1

1. What term is used to collectively describe gamma rays, x-rays, ultraviolet light, visible light, infrared radiation, and radio waves?
2. Which color of the visible spectrum has the longest wavelength? The shortest?
3. How does the amount of energy contained in a photon relate to its wavelength?

23.2 What Can We Learn from Light?

Explain how the three types of spectra are generated and what they tell astronomers about the radiating body that produced them.

When Sir Isaac Newton used a prism to spread white light into its component colors, he unknowingly initiated the field of **spectroscopy**—the study of the properties of light and the way light interacts with matter. Spectroscopy employs an instrument called a **spectroscope**, which contains a prism (or other material) that spreads out the various wavelengths of light so they can be accurately measured. Using a spectroscope, light collected telescopically from a distant object can be analyzed to determine a vast array of information, including the object's temperature, its composition, and how fast it is moving.

Three Types of Spectra

The rainbow of colors Newton produced is called a *continuous spectrum*, which provides a view of all wavelengths of visible light without interruption. Following Newton's discovery, scientists learned that two other types of spectra, *emission line spectrum* and *absorption line spectrum*, exist.

Continuous Spectrum A **continuous spectrum** consists of a continuous band of wavelengths that looks like the rainbow of colors that Newton observed. It is produced by an incandescent object (*incandescent* means "to emit bright light when hot"). For example, the filament in an old-fashioned incandescent light bulb produces a continuous spectrum when heated to thousands of degrees by an electric current (Figure 23.3A).

SmartFigure 23.3 Formation of the three types of spectra
A. Continuous spectrum.
B. Emission line spectrum.
C. Absorption line spectrum.

TUTORIAL
https://goo.gl/0vvn95

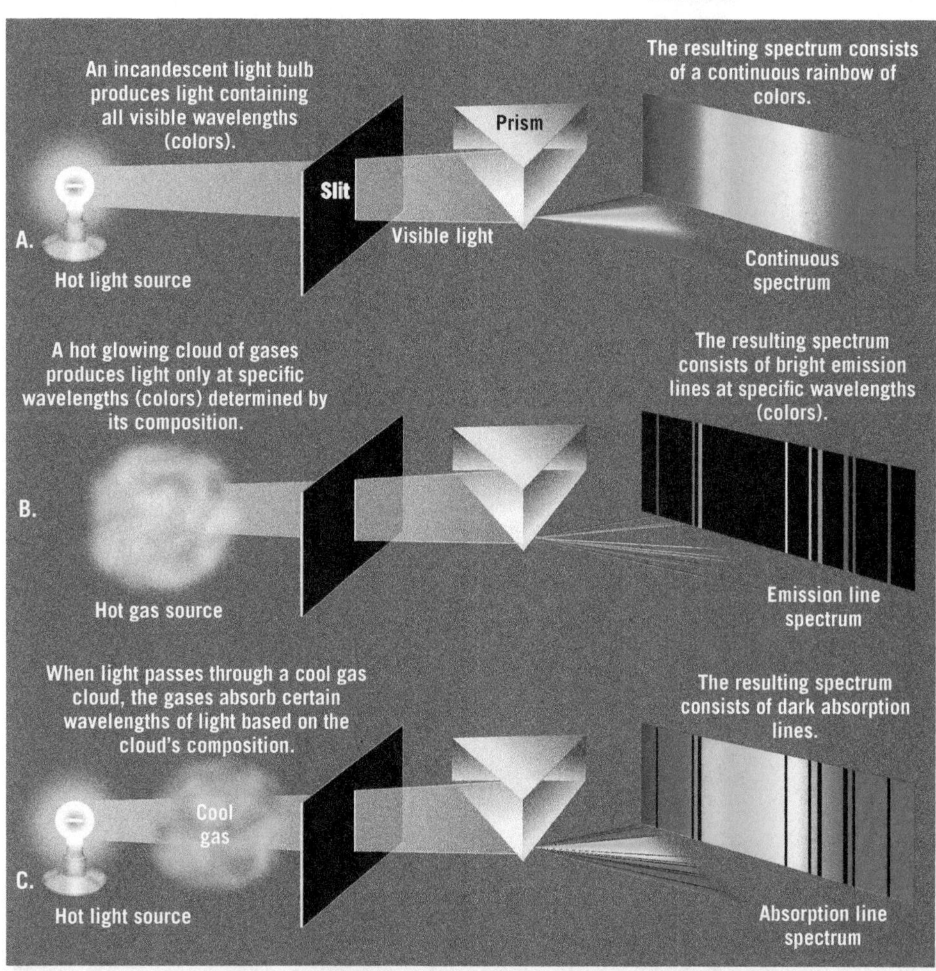

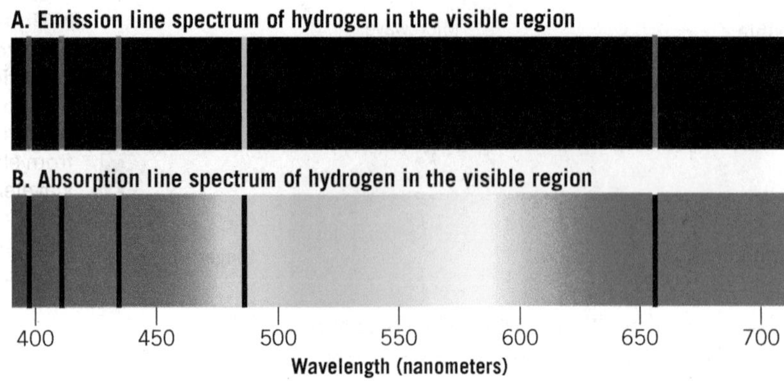

▲ Figure 23.4 **Emission and absorption spectra of hydrogen in the visible region** Notice that the bright emission lines align with the dark absorption lines.

Emission Line Spectrum An **emission line spectrum** consists of a series of narrow, discrete lines at specific wavelengths. (Figure 23.3B). Emission spectra are produced by hot, glowing gases at low pressure. Figure 23.4A shows the emission spectra produced by hot hydrogen gas, the most abundant element in the universe.

The Orion Nebula, shown in Figure 23.5, produces an emission spectrum because the hydrogen gases that comprise this nebula are heated sufficiently, by the young hot stars forming within, that they emit light. The brightest emission line produced by hydrogen is red, and the Orion Nebula has a red glow that is characteristic of excited hydrogen gas. In a similar manner, fluorescent light bulbs produce an emission line spectrum when the gases within them are excited by an electric current.

▼ Figure 23.5 **The Orion Nebula, an emission nebula** The red color of Orion Nebula is produced by hydrogen gas that is heated by nearby hot stars. Bright enough to be seen by the naked eye, the Orion Nebula is located in the sword of the hunter in the constellation of the same name. (Photo by National Optical Astronomy Observatory/Association of Universities for Research in Astronomy/National Science Foundation (NOAO/AURA/NSF))

Absorption Line Spectrum An **absorption line spectrum** consists of a series of dark lines subtracted from a continuous spectrum. It is generated when light passes through a comparatively cool gas at low pressure (Figure 23.3C). For example, if we take a glass container filled with cool hydrogen gas and pass through it light from a continuous source, an absorption spectrum is produced. As shown in Figure 23.4B, the dark lines on an absorption line spectrum for hydrogen are in exactly the same location as the bright lines found on its emission spectrum.

What Does Light Tell Us About Composition?

Both emission line and absorption line spectra allow us to determine the composition of distant objects in the universe. All ions, atoms, and molecules can be identified by their unique set of absorption lines, much as humans are identified by their unique sets of fingerprints.

Most stars produce an absorption line spectrum because the light they emit passes through their gaseous outer atmosphere, which absorbs specific wavelengths of light. Elements that exist in the gaseous state in the Sun's atmosphere, such as iron, calcium, and sodium, have been identified by studying the Sun's absorption line spectrum (Figure 23.6). Organic molecules have also been discovered in distant interstellar clouds of dust and gases using this technique.

What Does Light Tell Us About Temperature?

Every object radiates a *continuous thermal spectrum* that depends on its temperature. A human body, for example, emits infrared light (commonly called thermal radiation or simply heat), which can be seen using night-vision goggles. A continuous thermal spectrum contains two important pieces of information that help us determine the temperature of the radiating object.

First, a continuous thermal spectrum provides information about the object's total energy output. As Figure 23.7 shows, *hotter objects radiate more total energy per unit area than colder objects*. Specifically, Figure 23.7 shows the relative amount, or **intensity**, of the light emitted at each wavelength for objects of different temperatures. A star hotter than the Sun emits more light (has higher intensity) than the Sun at *every* wavelength. (Astronomers use the Kelvin scale to measure temperature; this scale starts at the coldest possible temperature, known as *absolute zero* [0 K]. See Appendix A.)

Second, a continuous thermal spectrum contains information about the surface temperature of the radiating body. Specifically, *hotter objects radiate a larger proportion of their energy at shorter wavelengths (higher energy)*. For instance, if you heat steel in a forge until it glows blue-white and then allow it to cool, its color will change from white to yellow to red as it emits its

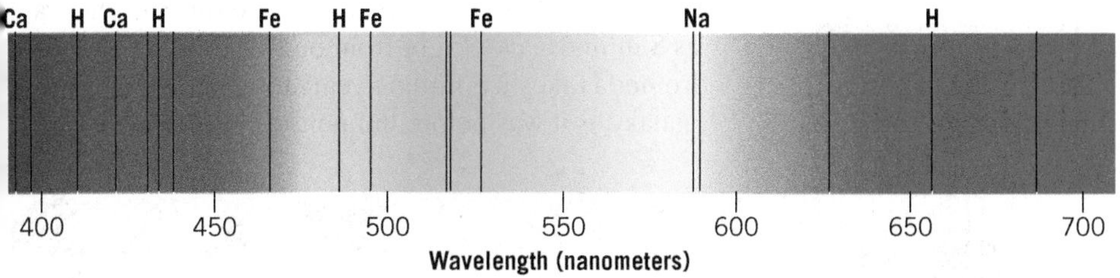

Figure 23.6 A low-resolution view of the absorption line spectrum of the Sun in the visible-light region Because each element absorbs light at characteristic frequencies, this spectrum makes it possible to identify the elements present in the Sun's atmosphere, including hydrogen (H), iron (Fe), sodium (Na), and calcium (Ca). A high-resolution view of the Sun's spectrum would have several hundred absorption lines, revealing that the Sun contains most of the elements found on Earth, although most of them occur in small amounts compared to hydrogen and helium.

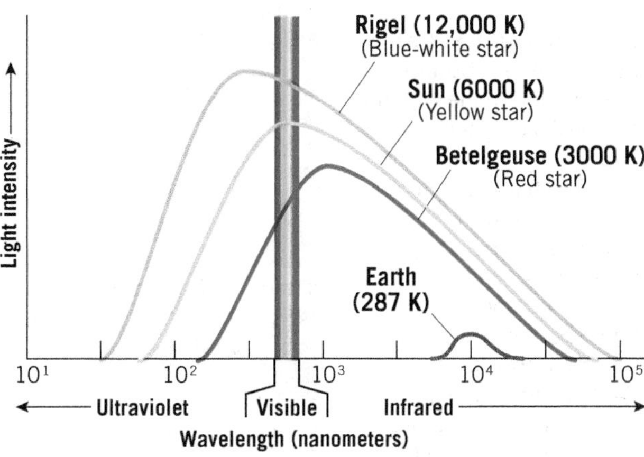

Figure 23.7 Idealized graphs showing the relative intensity of light emitted by objects of different surface temperatures Rigel, a hot blue-white star, radiates more energy at all wavelengths than the much cooler Sun, and it radiates most intensely at shorter wavelengths. (Its curve peaks in the ultraviolet, whereas the Sun's peak is in the visible light.)

maximum energy at progressively longer wavelengths. Even once it no longer glows visibly, you will be able to feel the heat (infrared radiation) it emits. All hot glowing objects show this behavior, so it follows that blue-white stars are hotter than yellow stars (like the Sun), which are hotter than red stars (see Figure 23.7).

Also notice in Figure 23.7 that Earth radiates only infrared light, while the Sun's maximum output occurs in visible wavelengths. Objects much hotter than the Sun radiate much of their light at even shorter wavelengths, such as ultraviolet light, x-rays, or even gamma rays.

To sum up: *Hotter objects emit more light per unit area and emit more of their light at shorter wavelengths (higher energy).*

What Does Light Tell Us About the Motion of Distant Objects?

You may have heard the way an ambulance siren seems to drop in pitch as it passes you. This is called the **Doppler effect**. As the siren approaches, its motion causes its sound waves to bunch together, resulting in a shorter wavelength and hence a higher pitch (Figure 23.8). The opposite happens as the siren recedes: Its motion stretches the sound waves, producing a lower pitch. Stated another way, as a siren approaches, its sound waves are shortened, producing a higher frequency (more waves arriving per unit of time), resulting in a higher pitch. Similarly, the pitch of a receding sound source is lowered because the wavelengths are stretched.

In the case of light, when a source is moving away, the light appears redder than it actually is because its waves are lengthened (see Figure 23.8). Conversely, approaching objects have their light waves shifted toward the blue (shorter-wavelength) end of the spectrum. The same effect is produced if you are moving and the light remains stationary. The Doppler effect is observed for all types of light waves.

Doppler shifts for distant stars are generally established by comparing the absorption (dark) lines in the spectrum of a star with spectra produced in the laboratory (Figure 23.9). When the spectral lines are shifted toward the long end of the electromagnetic spectrum, called a **redshift**, the star is moving away from the observer. By contrast, when the spectral lines are moving toward the short end (blue/violet end) of the spectrum, the object is approaching Earth. The amount of the shift indicates the speed at which the object is approaching or receding; larger shifts indicate faster speeds. For example, in Figure 23.9, star C (large redshift) is moving away from us faster than star B (smaller redshift).

Two types of Doppler shifts are important in astronomy: those caused by local motions and those caused by the expansion of the universe. Doppler shifts due to local motions can be used to measure how fast one star orbits another in a binary (two-star) system or how fast a pulsating star expands and contracts. However, most galaxies exhibit redshifts, indicating that most of the universe is moving away from Earth. These shifts are caused by the expansion of the universe (a topic that will be covered in Chapter 24).

A final important point about light is that it has a finite speed. In a vacuum, light travels at a constant 300,000,000 meters per second (671,000,000 miles per hour). Light takes about 8 minutes to travel from the Sun

SmartFigure 23.8 The Doppler effect for a moving sound source

TUTORIAL
https://goo.gl/BTZ8cy

A receding source produces sound waves that have longer wavelengths and thus lower frequencies, which results in a lower pitch.

An approaching source produces sound waves that have shorter wavelengths and thus higher frequencies, which results in a higher pitch.

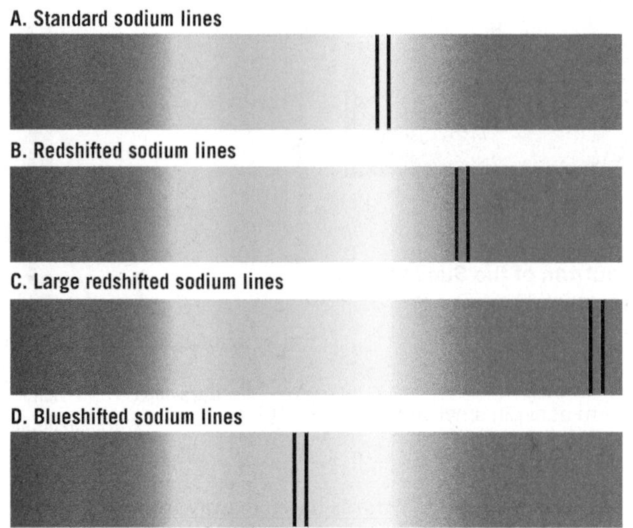

Figure 23.9
Determining the relative motion of two bodies We can determine whether a celestial object is approaching or receding from Earth by comparing its spectrum with that of motionless laboratory samples. **A.** Standard absorption spectrum for sodium produced in the laboratory. **B.** and **C.** Redshifted sodium lines produced by stars moving away from us. **D.** Blueshifted sodium lines produced by an approaching star.

to Earth. This means that we always see the Sun the way it was 8 minutes ago. Light from our neighboring galaxy Andromeda takes 2.5 million years to reach us, so we see this galaxy as it was before humans evolved.

CONCEPT CHECKS 23.2

1. What is *spectroscopy*?
2. What can be learned about a star (or other radiating objects) from its absorption line spectrum?
3. Describe the relationship between the temperature of a radiating body and the wavelengths of light it emits.
4. Briefly describe the Doppler effect and explain how astronomers determine whether a star is moving toward or away from Earth.

23.3 Collecting Light Using Optical Telescopes

Describe the two properties that make telescopes with large mirrors more useful than those with small mirrors.

The earliest tools used to observe the heavens were human eyes—the only mechanisms early astronomers like Tycho Brahe had available. However, the human eye is a poor instrument for astronomical observation because it cannot collect much light, is not very sensitive to faint colors, and collects only visible light. The invention of the telescope greatly improved our ability to explore the universe. Optical telescopes collect light with visible (or nearly visible) wavelengths and come in two basic types—*refracting* and *reflecting* telescopes.

Refracting Telescopes

Much like the one used by Galileo, **refracting telescopes** employ lenses, similar to those used in ordinary eyeglasses, to collect and focus light (Figure 23.10). The light coming from a distant star can be thought of as a ray or beam by the time it reaches Earth. A telescope lens intercepts some portion of this incoming light. To collect more light, a larger lens can be used. However, lenses can only be supported around their edges, so if they are too large, their weight will deform them and cause the image to be distorted.

The world's largest refracting telescope is at Yerkes Observatory in Williams Bay, Wisconsin. Completed in 1897, the Yerkes telescope has a lens that is a little more than 1 meter (40 inches) in diameter, while its tube is an amazing 18 meters (60 feet) long. With the completion of the Yerkes telescope, refracting telescopes had essentially reached their maximum size limit.

Figure 23.10
Refracting telescope A refracting telescope uses a lens to collect and focus light.

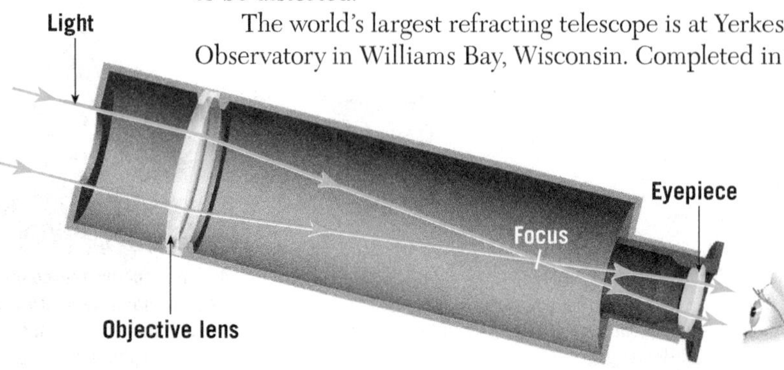

Reflecting Telescopes

Newton invented the first reflecting telescope. Unlike a refracting telescope, in which the light passes through a lens, a **reflecting telescope** uses mirrors to collect and focus light (Figure 23.11). Because mirrors can be made thinner than lenses of similar size, and they can be supported from the back, all large telescopes constructed today are reflecting telescopes.

In a reflecting telescope, the *primary mirror* collects the light; the larger the primary mirror, the greater the light-collecting capacity. Light gathered by the primary mirror is reflected to a *secondary mirror* located in front of it. The secondary mirror then reflects the light to a point, called the *focus*, where the eye or an instrument can observe it (See Figure 23.11).

Most reflecting telescopes have a single primary mirror ground to a nearly perfect parabolic shape (Figure 23.12A). The Hubble Space Telescope has a 2.4-meter (94.5-inch) mirror that is ground to within about one-millionth of an inch of being a perfect paraboloid (Figure 23.12B). Once ground, the surface of the mirror is coated with a highly reflective material.

For more than 40 years, the largest reflecting telescope on Earth was the Hale Telescope on Mount Palomar near San Diego, California, with a 5-meter (200-inch) primary mirror. Larger mirrors were too difficult to make and too prone to warping. Today,

Figure 23.11 **Newton's reflecting telescope** (Photo by Dave King © Dorling Kindersley, Courtesy of The Science Museum, London)

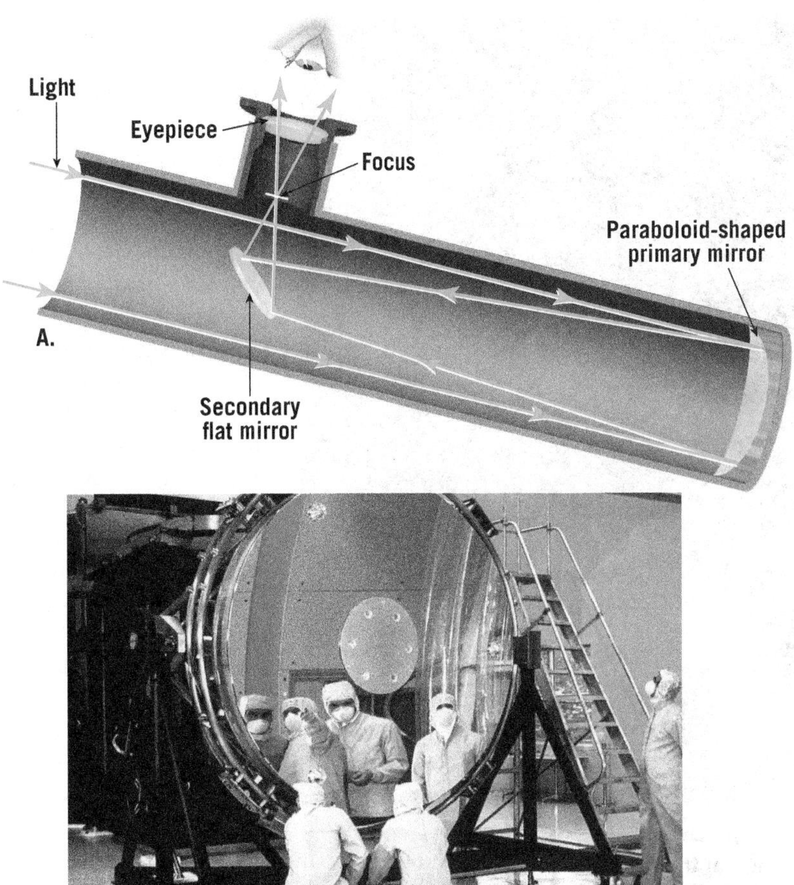

SmartFigure 23.12
Reflecting telescope
A. Diagram illustrating how paraboloid-shaped mirrors, like those used in reflecting telescopes, gather light. **B.** Preparation of the 2.4-meter (94.5-inch) mirror for the Hubble Space Telescope. (Courtesy of NASA

TUTORIAL
https://goo.gl/z6qMb5

with the aid of large rotating furnaces, larger and thinner mirrors are being produced. The furnaces spin so that molten glass near the center of the mold moves gradually outward to the edges, producing the desired curved shape. After casting, the mirror is gradually cooled for several months and then polished to within a millionth of a centimeter of a perfect parabolic shape. Using this process, mirrors that exceed 8 meters (more than 26 feet) in diameter have been constructed.

Some of the largest telescopes operating today are housed at the Paranal Observatory, located in the extremely dry Atacama Desert of Chile. Operated by the European Southern Observatory (ESO), Paranal has four large telescopes (called collectively the Very Large Telescope), each with an 8.2-meter (almost 27-foot) mirror—built and polished in Europe and transported to the observatory in Chile.

Even larger optical telescopes use segmented, rather than single mirrors. Examples include the twin Keck Observatory telescopes located atop Mauna Kea, on the Big Island of Hawaii. Each Keck Observatory telescope has a primary mirror about 10 meters (33 feet) in diameter and consisting of 36 relatively thin segments. A few much larger telescopes are in the planning or development stages, including the Giant Magellan Telescope (GMT), which will consist of 7 mirrors, each 8.4 meters (about 27 feet) in diameter. The mirrors for this GMT are currently being cast under the east wing of the University of Arizona football stadium. The GMT will have 10 times the resolution of the Hubble, allowing us to see fainter objects at much greater distances. The largest optical telescope yet envisioned is the European Extremely Large Telescope (ELT), which will have a 39-meter (nearly 130-foot) primary mirror, composed of 798 hexagonal mirror segments.

Why Build Large Optical Telescopes?

If you observe the sky on a clear night without light pollution, your pupils will gradually enlarge so that more light reaches your retinas. When your pupils are fully open, you will see roughly 3000 stars. No matter how long you look up, no additional stars will be visible. A telescope, because of its much larger **light-gathering area**, can see more stars than the unaided eye. Telescopes, therefore, can be considered giant eyes that collect far more photons of light. As a result, larger telescopes can see fainter objects.

Because large telescopes can see extremely distant objects, they can also see farther back in time, to when the universe was young. Recall that the finite speed of light means that we see objects as they were when the light left them, so that the farther away we look, the younger the objects we see. Large telescopes allow us to see back to the time when galaxies were just forming. By comparing these newborn galaxies with less distant (and therefore older) galaxies, we can see how galaxies have evolved over time.

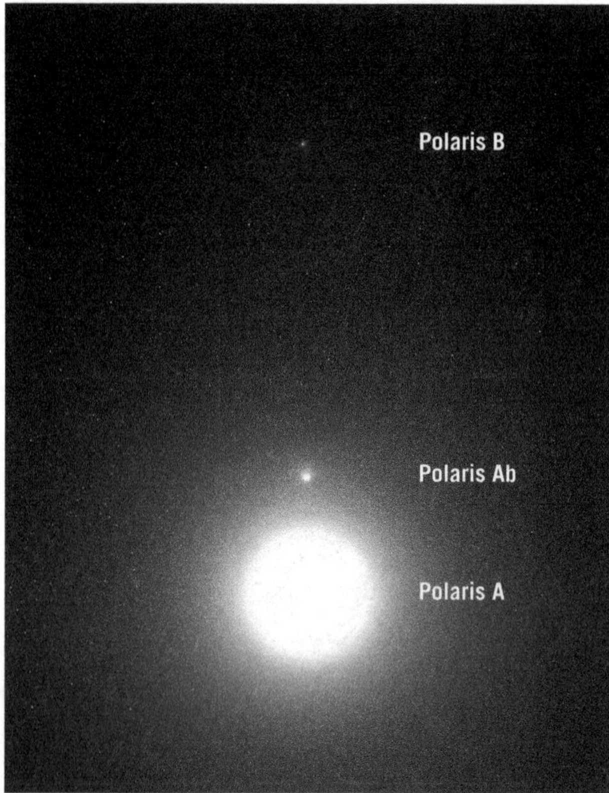

▲ Figure 23.13 **High-resolution image of the North Star (Polaris)** With the unaided eye, the North Star appears as a single star, but use of a telescope (which collects more light) shows it to be three stars. (NASA)

Another reason to build larger telescopes is that they can produce sharper, more detailed images, a property called **resolution**. For example, when we view Polaris (the North Star) with the unaided eye, we see a single point of light. However, using a telescope with greater resolution, we can see three stars—a supergiant about six times as massive as the Sun and two small companion stars (Figure 23.13). The larger the primary lens or mirror of a telescope, the greater its potential resolution (Figure 23.14).

For large ground-based optical telescopes, resolution is limited by atmospheric **turbulence**—the blurring and twinkling of astronomic objects such as stars caused by the movement of air in Earth's atmosphere. To minimize this problem, modern optical observatories are placed on mountain summits (Figure 23.15), such as the Mauna Kea Observatory on the Big Island of Hawaii and the Paranal observatory on a mountain in the Atacama Desert of northern Chile.

Advances in Light Collection

Early astronomers observed the skies with their eyes and recorded what they saw in drawings. Starting with Galileo, telescopes supplanted eyes as the main tool of astronomers. We will look now at additional technologies that help astronomers to collect and record light.

Photographic Film and CCDs Photographic film was a revolutionary improvement over drawings. It is not distorted by human biases, it records reasonably accurate light intensities, and it records faint objects more accurately than can the human eye. Most importantly, photographic film can be exposed for hours on end, enabling it to gather enough photons to record very faint objects.

Photography in general and in astronomy in particular has largely abandoned film for CCDs (charge-coupled devices). The large CCD cameras that astronomers use can collect almost all the photons of incoming light that strike the chip, which greatly reduces exposure times. Using CCD cameras with remotely controlled telescopes, the entire night sky can be imaged in as little as a few nights, giving astronomers a view of the continually changing universe that has never before been possible. Light can also be collected by more than one telescope and synchronized by computers to make a single image. In addition, CCD chips can be designed to detect wavelengths of infrared and ultraviolet light. Some astronomical objects are more visible at these wavelengths than in visible light. The stunning image in Figure 23.16 was created by digitally combining images obtained using visible, infrared, and ultraviolet light.

Active Optics In conjunction with increasingly thin mirrors, telescope designers have developed a sophisticated support system to control mirror shape. This system, called **active optics**, uses motors to drive numerous precisely controlled supports that push or pull on the mirror to compensate for the distortions produced by gravity, wind, and other forces. Active optics has made very large, thin mirrors possible. The twin Keck Observatory telescopes, for example, use active optics to focus incoming starlight.

Arrays of Telescopes Astronomers have found ways to connect telescopes together so that, in effect, they function as one larger telescope. This type of array is called an **interferometer**. Connecting telescopes into an array increases their resolution even more than their light-gathering area. One of the most advanced optical interferometers is located at the Paranal Observatory in Chile. This array consists of the four large telescopes, with main mirrors 8.2 meters (almost 27 feet) in diameter, and four movable 1.8-meter (6-feet) auxiliary telescopes. When the telescopes are working together, their light beams are combined using a complex system of mirrors located in tunnels. This interferometer is capable of constructing images with a resolution equivalent to distinguishing two headlights of a car at the distance from Earth to the Moon.

▼ Figure 23.14 **Comparing the resolutions of two telescopes** Appearance of the galaxy in the constellation Andromeda, using telescopes of different resolutions. (Images Copyright © The Association of Universities for Research in Astronomy, Inc. All Rights Reserved)

A.

B.

EYE ON THE UNIVERSE 23.1

The accompanying image shows one of the twin telescopes located at Keck Observatory atop Hawaii's Mauna Kea volcano, at an elevation of 4200 meters (13,800 feet). The 10-meter (33-foot) mirrors consists of 36 distinct segments. These telescopes can work independently or in tandem to double their light-gathering capacity.

QUESTION 1 *Examine the primary mirror located near the center of the image. Based on what you see, describe the shape of the 36 segments that make up this mirror.*

QUESTION 2 *Assume that these mirrors have a perfectly circular shape and a diameter of 10 meters (33 feet); also assume that they are working in tandem (gathering light from the same source). How much more light-gathering area do they have compared to the Hale Telescope, which has a mirror with a 5-meter (16-foot) diameter? (Hint: Compare the surface areas of these instruments.)*

QUESTION 3 *The Keck Observatory telescopes are located at an elevation of more than 4000 meters (13,500 feet), far out in the Pacific on Mauna Kea. The Hale Telescope is located 80 kilometers (50 miles) northeast of San Diego, California, at an altitude of 1700 meters (5600 feet). What advantage do the Keck Observatory telescopes have compared to the Hale Telescope?*

(Photo by Enrico Sacchetti/Science Source)

Un-twinkling the Stars: Adaptive Optics Even when located on mountaintops, advanced telescopes still face the distorting effects of atmospheric turbulence. A technology called **adaptive optics** can eliminate much of this distortion. An adaptive optics system uses computer-controlled deformable mirrors to cancel out in real time the effects of atmospheric turbulence. To do this, the computer needs a way to determine how the atmosphere is distorting the incoming light. If a bright star is present near the object being studied, the computer can simply manipulate the deformable mirrors so that this reference star's image stays sharp. Astronomers can also create artificial stars by shining a powerful laser beam into Earth's upper atmosphere to produce a point of light. Adaptive optics has allowed astronomers to observe finer details than otherwise would be possible from Earth's surface.

CONCEPT CHECKS 23.3

1. What two properties make telescopes with large mirrors more useful than those with small mirrors?
2. List two advantages of charge-coupled devices (CCDs) over photographic film.
3. Name the technique devised by astronomers to greatly eliminate the blurring caused by atmospheric turbulence.

Figure 23.15 Kitt Peak National Observatory on a starlit night (Photo by Susan E. Degginger/Alamy Stock Photo)

Figure 23.16 Detailed view of the Cat's Eye Nebula This nebula, which consists of several concentric shells, formed when a Sun-like star ejected its outer gaseous layers to form this complex structure. Hubble's Advanced Camera for Surveys produced this stunning image by collecting light in the visible as well as the infrared and ultraviolet wavelengths. (Photo by ESA, HEIC,/The Hubble Heritage Team (STScI/AURA)/NASA)

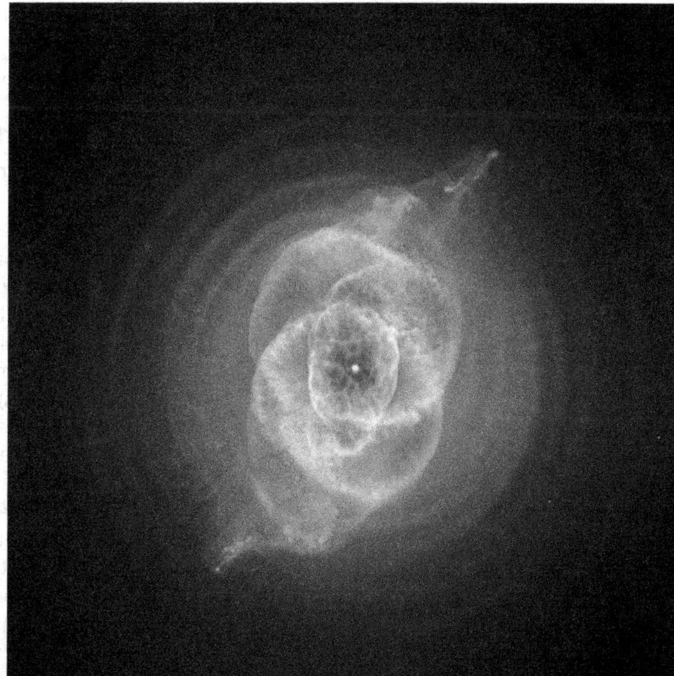

23.4 Radio- and Space-Based Astronomy

List the advantages of radio-wave and orbiting observatories over ground-based optical telescopes.

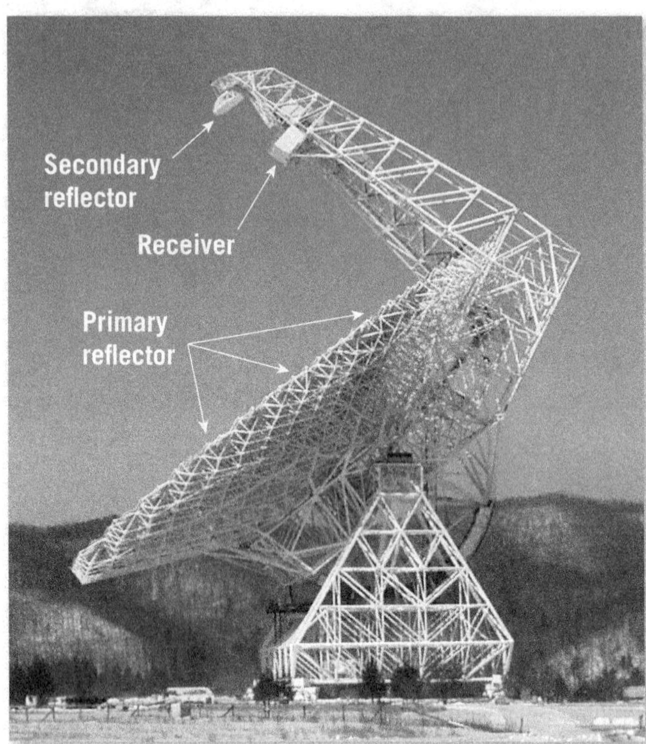

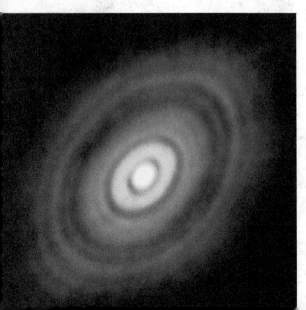

▲ Figure 23.17 **The 100-meter (330-foot) steerable Robert C. Byrd radio telescope at Green Bank, West Virginia** The dish acts like the mirror of a reflecting optical telescope to focus radio waves onto a secondary reflector, which transmits it to a receiver. (Photo by National Radio Astronomy Observatory)

Much of the radiation produced by celestial objects either cannot penetrate our atmosphere or cannot be detected by optical telescopes. As a result, astronomers have developed other observational techniques to cover the remaining portions of the electromagnetic spectrum.

Radio Telescopes: Observing the Invisible

Radio waves are detected using "big dishes" called **radio telescopes** (Figure 23.17). The parabolic dish of a radio telescope operates much like the mirror of an optical telescope, focusing the incoming radio waves onto a secondary reflector, which transmits the signals to a receiver. Because radio waves are roughly 100,000 times longer than waves of visible light, the surface of a dish need not be as smooth as a mirror. In fact, except with the shortest radio waves, wire mesh is an adequate reflector.

Radio telescopes are used to study objects that include pulsars, which are rapidly rotating corpses of dying stars that emit short pulses of radio waves; energetic quasars (*quasi-stellar radio sources*); and massive black holes that reside at the centers of galaxies, hidden from the view of optical telescopes. Other less energetic objects can also be studied, including cold clouds of gas found in interstellar space. One of the sharpest images ever taken by an array of radio telescopes shows the disk surrounding the young star HL Tauri (Figure 23.18). The gaps in this disk may mark where planets are forming.

Radio dishes must be large because radio signals from celestial sources tend to be very weak. The very largest, which sacrifices steerability for size, is the Five-hundred-meter Aperture Spherical Telescope (FAST), built into a bowl-shaped depression in China's southeast province of Guizhou. Its dish is the size of 30 U.S. football fields.

The relatively poor resolution of radio telescopes can be improved by linking them into an array. The Very Large Array, located in the desert near Socorro, New Mexico, is composed of 27 radio telescopes in a Y-shaped configuration (Figure 23.19). When working as an interferometer, this array provides the resolution of a radio telescope up to 36 kilometers (22 miles) in diameter. It is possible to link together radio telescopes that are located far apart to form a **very-long-baseline interferometer (VLBI)**. One example, called the Very Large Baseline Array, consists of 10 radio observatories with locations from Hawaii to the U.S. Virgin Islands, nearly a third of the way around the globe.

◄ Figure 23.18 **Protoplanetary disk surrounding the young star HL Tauri** This image of a protoplanetary disk, taken by Atacama Large Millimeter Array (ALMA), is sharper than those obtained in visible light by the Hubble Space Telescope. ALMA is an array of radio telescopes located in the Atacama Desert of Chile at an altitude of about 5000 meters (more than 16,000 feet). (Photo by The European Southern Observatorg (ESO))

▲ Figure 23.19 **Twenty-seven identical radio telescopes operate together to form the Very Large Array** This large array of radio telescopes is located near Socorro, New Mexico. (Photo A. by Joe McNally/Getty Images; B. by denerzel21/Fotolia)

Orbiting Observatories: Detecting All Forms of Light

There are several reasons so many telescopes have been placed in space. Earth-based telescopes can only observe on cloudless nights and are subject to light pollution from city lights. More importantly, Earth's atmosphere blocks out much of the electromagnetic spectrum (**Figure 23.20**). All radiation shorter than the longest ultraviolet waves is blocked, and water vapor in the lower atmosphere absorbs most infrared light. Orbiting observatories circumvent many of these problems.

Some space-based telescopes detect optical light, and some are designed to work in other regions of the electromagnetic spectrum. Some orbiting observatories are fitted with multiple instruments so that a single object can be studied simultaneously in more than one region of the spectrum. The Hubble Space Telescope, for example can observe celestial objects in parts of the infrared and ultraviolet regions of the spectrum as well as in visible light. More than 100 space observatories have been launched. We will look how these telescopes have aided our understanding of the universe.

Radio-Wave Space Observatories Because all but the longest radio waves readily penetrate Earth's atmosphere, most radio astronomy is done on the ground. However, space-based radio telescopes offer the promise of doing *very* long-baseline interferometry, which could yield extraordinary resolution.

Space-based telescopes also offer advantages for detecting microwaves, the shortest type of radio waves and the electromagnetic radiation that is used in a microwave oven. The Planck space observatory successfully mapped the *cosmic microwave background*—radiation omitted from the Big Bang—with high resolution. This energy, emitted at the very beginning of the universe, fills the sky.

Infrared Space Observatories Because infrared light has less energy than visible light, it brings us information about cool celestial objects that are invisible to optical telescopes. These include small cool stars that are too faint to be detected at visible wavelengths, planets that lay outside our solar system, and cold molecular clouds located in interstellar space. Infrared light can also pass through clouds of interstellar gas that block visible light, allowing us to see what is within and behind them.

One of the most advanced infrared observatories is the Spitzer Space Telescope (SST), launched in 2003. Moving in a unique Earth-trailing orbit around the Sun, the SST sees an optically invisible universe composed of cool dust and gas, as well as cool stars and exoplanets.

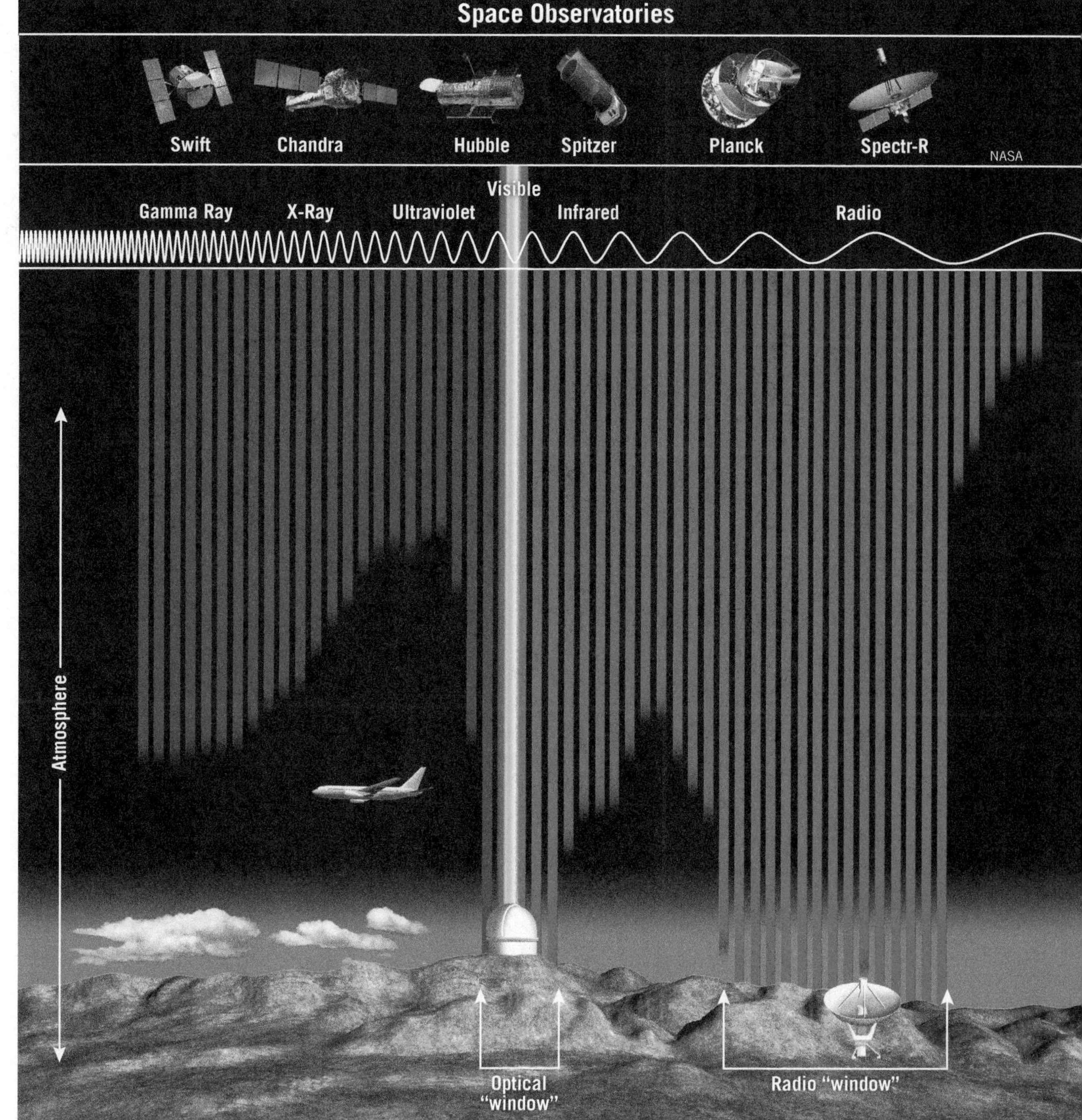

▼ **Figure 23.20 Diagram showing the approximate depths to which various wavelengths of light can penetrate Earth's atmosphere**

Hubble Space Telescope

In 1990, the Space Shuttle Atlantis carried the Hubble Space Telescope (HST) into an orbit 557 kilometers (347 miles) above Earth's surface. At this altitude, the HST avoids many of the problems experienced by Earth-bound telescopes, which are affected by the atmosphere that distorts and blocks much of the light reaching our planet. The thousands of images beamed back to Earth have helped scientists solve many great mysteries. For example, data from the HST allowed scientists to determine that the age of the universe is between 13 and 14 billion years, significantly narrowing the previous range of 10 to 20 billion years.

These pillar-likes structures are columns of cool hydrogen gas and dust: the incubators of new stars. Captured in stunning detail by the HST, they are part of the Eagle Nebula, a nearby star-forming area about 6500 light-years away.

HST/NASA

HST/N

The Hubble Space telescope produced images of remote galaxies with unprecedented clarity and sharpness, including small galaxies dubbed "tadpoles" that existed when the universe was young. This image, called the Hubble Deep Field, was acquired in what appears to be an empty part of the sky located near the Big Dipper. Thousands of galaxies in various stages of evolution are visible in the image, captured by the HST focusing entirely on the same tiny patch of sky for ten consecutive days.

HST/NASA

This striking view of The Ring Nebula was obtained by the Hubble Space Telescope. Barrel-shaped clouds, called planetary nebula because they appeared spherical to early observers, are actually composed of ejected material from a dying sun-like star. The HST is also credited with imaging several supernova remnants, including the Crab Nebula, which formed just 900 years ago.

From Hubble images, astronomers deduced that galaxies grow by the accretion of smaller galaxies. Despite many successes, they cannot yet explain the variety of galaxies that exist, from large spiral galaxies such as the one shown in the accompanying image, to round, featureless elliptical galaxies.

HST/NASA

HST/NASA

Ultraviolet Space Observatories Ultraviolet light is the signature of hot young massive stars as well as of very old stars and highly evolved galaxies. Ultraviolet observatories in space are able to detect chemical composition, densities, and temperatures of interstellar dust and gas clouds and the temperature and composition of hot young stars, which has helped astronomers better understand star formation. The Hubble Space Telescope has some ultraviolet capabilities.

X-ray Space Observatories Like ultraviolet light, x-rays are blocked by Earth's atmosphere. High-energy x-rays do not reflect from a telescopic mirror like visible light but simply pass through. However, just as a bullet ricochets when it hits a wall at a steep angle, x-rays ricochet when they barely graze the surface of a mirror. The telescope system used to detect x-rays employs a group of mirrors (shaped like a tube) that are oriented nearly parallel to the path of the incoming x-rays to focus them toward a detector.

The Chandra X-ray Observatory (CXO) is able to observe objects such as black holes, quasars, and high-temperature gases at x-ray wavelengths. The CXO has recorded a black hole pulling in matter and two black holes merging into one. It has also provided an independent measurement of the age of the universe, reinforcing the estimated age of 13.8 billion years. The CXO has also shown what galaxies were like when the universe was only a few billion years old.

Combining data from Chandra X-ray Observatory with that of other observatories has made possible a more complete picture of the universe. For example, the image of the Crab Nebula shown in **Figure 23.21** is a composite made using data from the CXO, the Hubble Space Telescope, and the Spitzer Space Telescope. The Chandra x-ray image, shown in blue, shows that the hot star left behind by the explosion that produced the Crab Nebula is spewing out a blizzard of high-energy particles (x-rays).

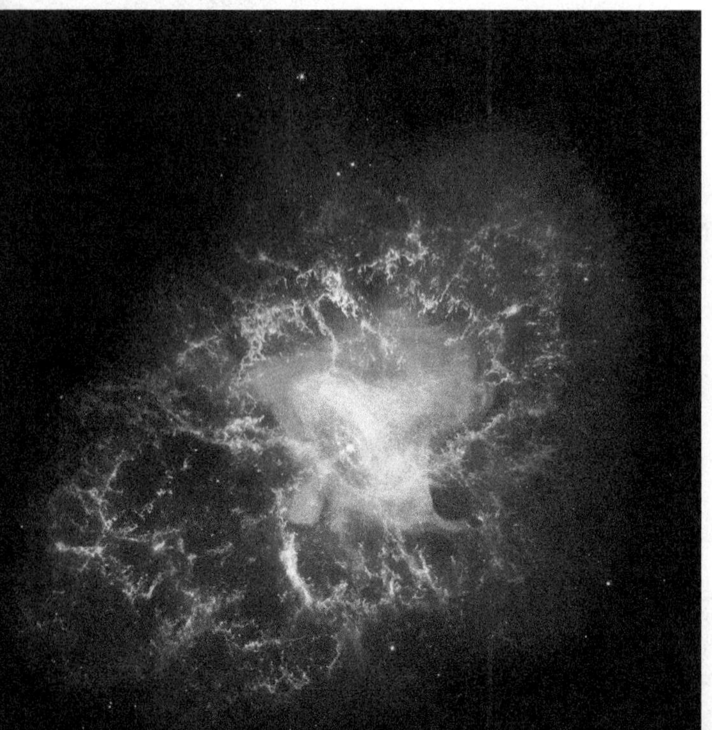

▼ Figure 23.21 **Composite image of the Crab Nebula in x-rays, visible light, and infrared light** This composite was made using data from the CXO, the Hubble Space Telescope, and the Spitzer Space Telescope. The Chandra x-ray image, in blue, shows that the hot star left behind by the explosion that produced the Crab Nebula is spewing out a blizzard of x-ray-emitting high-energy particles. This important structure of the Crab Nebula is not seen in visible light. (NASA/Marshall Space Flight Center)

Gamma Ray Space Observatories Gamma rays are the highest-energy form of light, which makes it very difficult to produce "telescopes" that can collect them. A detector is unable to collect gamma rays, but it emits a signal whenever it is struck by a gamma ray.

Gamma-ray space telescopes are designed to study light from some of the most violent physical processes in the universe. These instruments primarily study *gamma ray bursts* and matter being engulfed by black holes. Gamma ray bursts are flashes of gamma rays that come from seemingly random places deep in the universe. They are probably the most luminous and, therefore, the most energetic events occurring in the universe since the Big Bang. It is quite likely that observed gamma ray bursts are emitted by a massive, rapidly rotating star as it collapses to form a black hole (see Section 24.4, page 735).

The Hubble Space Telescope and Beyond

The Hubble Space Telescope, launched in 1990, must be considered one of the most important instruments in the history of astronomy. Its 2.4-meter (8-foot) primary mirror has produced images with a sensitivity and resolution that are only now being matched by much larger (10-meter [33-foot]) ground-based telescopes.

Discoveries Made by Hubble An enormous range of discoveries have been made with the Hubble Space Telescope. For example, it has shown that disks of dust are common around young stars, providing support for the nebular hypothesis of solar system formation. Hubble also discovered decisive evidence that supermassive black holes reside at the centers of many large galaxies by imaging the movements of dust and gas in their interiors. In addition, Hubble has allowed us to look farther out into the universe (and, thus, further back in time) than ever before, in the process of producing the most "elusive" astronomical image ever taken, the *Ultra-Deep Field* (see **GEOgraphics 23.1**). This image was acquired by looking at a patch of "empty" sky for a total of 1 million seconds; the faintest objects put only one photon per minute into the exposure.

James Webb Space Telescope The scientific successor to the Hubble Space Telescope, called the *James Webb Space Telescope*, is scheduled for launch in 2018. The Webb is an infrared telescope whose primary mirror is 6.5 meters (over 20 feet) in diameter, providing more than six times the light-gathering area of the Hubble telescope. In addition, the Webb will be located about 150 million kilometers away from Earth and will move with Earth as it orbits the Sun, allowing its large sunshield to more effectively block heat and light from both Earth and the Sun.

The James Webb Space Telescope is designed to observe the first stars and galaxies formed after the Big Bang, to peer into nebular clouds where stars and planetary systems are forming today, and to measure the physical and chemical properties of planetary systems around distant stars. It will even be able to record spectra from the atmospheres of distant planets, revealing whether they contain the water vapor needed for Earth-like life. These observations are all more easily made in the far-red and infrared wavelengths for which the Webb is optimized than in visible light.

The Webb will also use its sensitive instrument package to observe objects in the solar system that lie beyond Earth's orbit. It will be particularly good at detecting faint icy and rocky objects in the outer solar system, including dwarf planets and small Kuiper belt objects. These studies will help test the nebular theory of the solar system's formation.

CONCEPT CHECKS 23.4

1. Why are radio telescopes much larger than optical telescopes?
2. What are some of the advantages of radio telescopes over optical telescopes?
3. Explain why space makes a good site for an optical observatory.

23.5 Our Star: The Sun

Sketch the Sun's structure and describe each of its major layers. Summarize the process called the proton–proton chain reaction.

The Sun is one of an estimated 100 to 400 billion stars that make up the Milky Way Galaxy. Although the Sun is of little significance to the universe as a whole, to those of us who inhabit Earth, it is the primary source of energy. Everything from the food we eat to the fossil fuels we burn in our automobiles is ultimately derived from solar energy. The Sun is also important in astronomy, since it is the only star we can study close up. However, because the Sun is so bright and emits eye-damaging radiation, observing it directly is unsafe. To circumvent this issue, astronomers project the Sun's image onto a surface where it can be safely studied.

For a star, the Sun is fairly average. However, on the scale of our solar system, it is truly gigantic, exceeding Earth's diameter by more than 100 times and Earth's volume by 1.25 million times. Yet because of its gaseous nature, the Sun's average density is only one-quarter that of Earth (a little greater than the density of water).

The Sun's Surface

The Sun's visible surface, called the **photosphere** (*photos* = light, *sphere* = ball), is aptly named because it radiates most of the observable sunlight that produces the bright disk of the Sun (Figure 23.22). Although we call it a "surface," it is actually a layer of incandescent gas about 500 kilometers (300 miles) thick, with a pressure less than 1/100 that of our atmosphere.

The photosphere is neither smooth nor uniformly bright; instead, it has a granular appearance. The Sun's **granules** are relatively small (Texas-sized) bright spots surrounded by narrow dark regions and are the result of convective flow (Figure 23.23). Hot, glowing gas rises at the center of a granule and then cools and darkens as it spreads laterally, eventually sinking back into the interior. Granules survive for only 10 to 20 minutes and are continuously replaced. This convective movement of gas—in effect, "boiling"—transfers energy outward from the Sun's interior to the photosphere, where it is radiated to space (see Figure 23.22).

The Sun's Atmosphere

The Sun's atmosphere consists primarily of hydrogen and helium gases, with about 2 percent other elements. These gases are *ionized*, meaning that the atoms have dissociated into ions and free electrons. Such an ionized gas is called a *plasma*.

Chromosphere Just above the photosphere lies the **chromosphere** ("color sphere"), a tenuous layer of hot, incandescent gases a few thousand kilometers thick. The chromosphere is normally visible for a few moments during a total solar eclipse. However, astronomers can always observe it by using an instrument to block the light from the photosphere. Under such conditions, it appears as a

SmartFigure 23.22
Diagram of the Sun's structure

TUTORIAL
https://goo.gl/D8yZRd

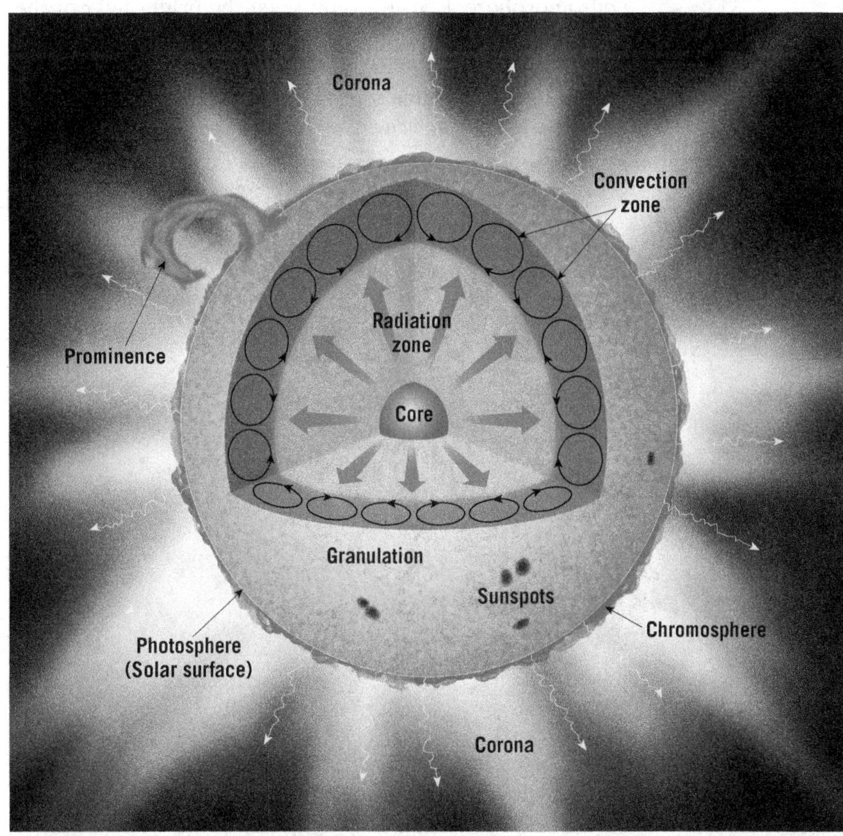

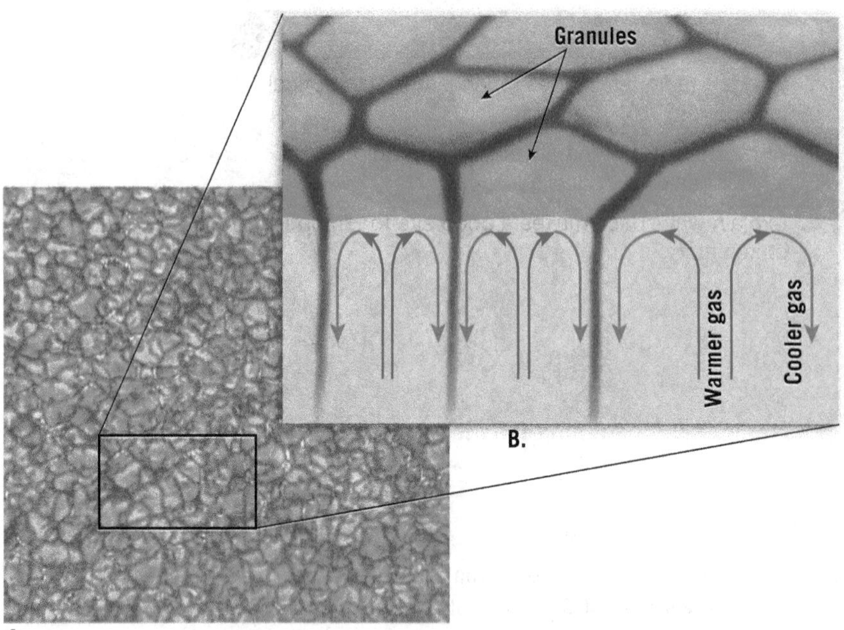

Figure 23.23 **Granules of the Sun's surface A.** Granules appear as yellowish-orange patches about the size of Texas; they last for only 10 to 20 minutes before being replaced by new granules. **B.** Granules owe their brightness to hotter gases that are rising from below. As this gas spreads laterally, cooling causes it to darken and sink back into the interior. (Courtesy of NASA/BBSO/NSF)

thin red rim around the Sun. Because the chromosphere consists of incandescent gases at low pressure, it produces an emission spectrum. One of the bright lines of hydrogen contributes a good portion of the chromosphere's total output and accounts for its red color. Because of its temperature, about 10,000 K, the chromosphere is also the layer that emits most of the Sun's ultraviolet light.

Corona The outermost portion of the solar atmosphere, the **corona** (*corona* = crown) is tenuous and, like the chromosphere, is visible only when the bright light of the photosphere is blocked (Figure 23.24). This envelope of ionized gases normally extends 1 million kilometers or so from the Sun and produces a glow about half as bright as that of the full Moon. The corona is very hot, about 1 million degrees Kelvin, which explains why it is the source of most of the x-rays emitted by the Sun. Despite its high temperatures, the gases of the corona are so tenuous that they emit very little thermal energy (heat).

At the outer fringe of the corona, the ionized gases have speeds great enough to escape the gravitational pull of the Sun. This stream of escaping charged particles (protons and electrons) constitutes the **solar wind**, which travels outward through the solar system. The solar wind interacts with the bodies of the solar system; for instance, it continually bombards lunar rocks and alters their appearance. Earth's magnetic field channels these charged particles toward the poles, where they produce the auroral displays we call the Northern and Southern Lights (see Figure 23.30).

Figure 23.24 **Solar corona** This image was obtained during a total solar eclipse. (Photo courtesy of NASA)

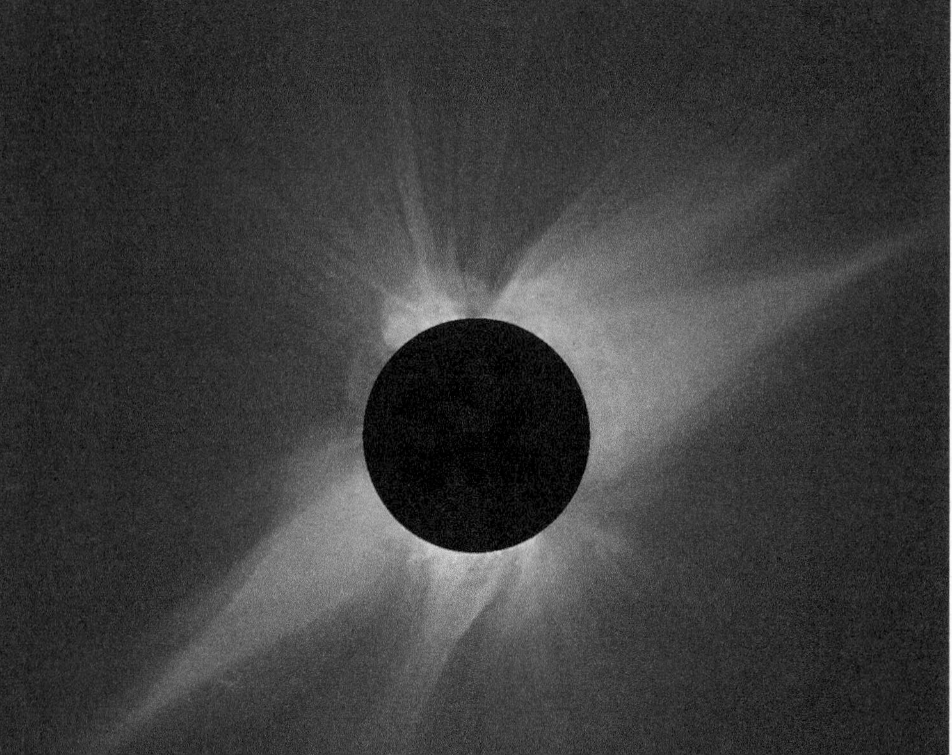

The Sun's Interior

The Sun's interior, which contains almost all its mass, cannot be observed directly. However, the movement of gases within the Sun's interior produces sound waves that can be detected, much as the seismic waves produced by earthquakes can be recorded and studied. Based on these observations and theoretical studies, we know the Sun's interior consists of three distinct layers (see Figure 23.22). The **core** is the region where the Sun's energy is generated. Above the central core is the **radiation zone**. In this zone, the energy generated in the core moves outward mainly as high-energy photons of light. Matter is so dense in the radiation zone that the photons of light are continually absorbed and reradiated. As a result, it takes photons more than 100,000 years to navigate through this layer. Located between the Sun's radiation zone and the photosphere is the Sun's **convection zone**. In this zone, energy is transported outward mainly by the convective motion of hot gases.

The Source of Solar Energy

The source of the Sun's energy, **nuclear fusion**, was not discovered until the late 1930s. In the Sun's core, a multistep nuclear reaction called the **proton–proton chain reaction** converts four hydrogen nuclei (protons) into the nucleus of a helium nucleeus (Figure 23.25). The energy

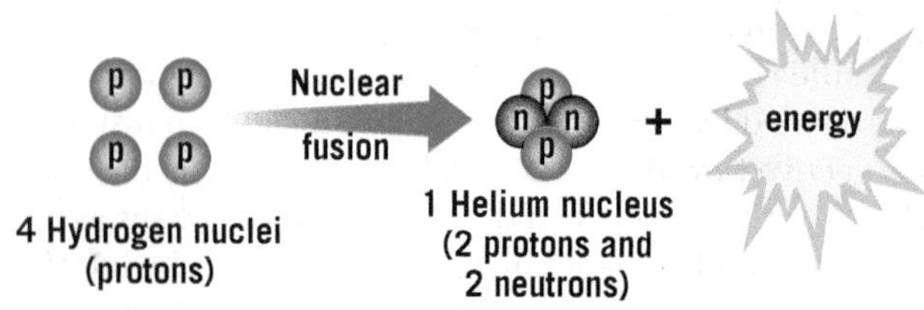

Figure 23.25 **Deep in the Sun's interior, a nuclear reaction called the proton–proton chain converts four hydrogen nuclei (protons) into the nucleus of a helium atom** The energy released from the proton–proton reaction represents matter being converted to radiant energy. (The proton–proton chain also releases neutrinos, not discussed here.)

released from this proton–proton reaction is a result of matter being converted to radiant energy. Specifically, one helium nucleus has only 99.3% as much mass as the four hydrogen nuclei that combine to produce it. The missing mass is emitted as energy in accordance with Einstein's formula $E = mc^2$, where E equals energy, m equals mass, and c equals the speed of light. The energy released from the conversion of even a small amount of mass to energy is enormous. The Sun converts roughly 600 million tons of hydrogen to 596 tons of helium each second. This means that the Sun, through nuclear fusion, converts 4 million tons of matter to energy each second. Although this amount seems huge, it is minuscule in relationship to the total mass of the Sun.

How long can the Sun produce energy at its current rate before all of the hydrogen fuel in its core has been consumed? Even at its enormous rate of consumption, the Sun has enough fuel to sustain itself in its current state for approximately another 5 billion years. After that, the Sun will grow dramatically, becoming a red giant and engulfing Earth before finally reaching the end of its life.

CONCEPT CHECKS 23.5

1. Why is the Sun significant to the study of astronomy?
2. Why is the photosphere considered the Sun's "surface"?
3. What "fuel" does the Sun consume?

23.6 The Active Sun

List and describe four types of solar storms that occur on the Sun.

Most of the solar energy that reaches Earth is a result of a rather steady, continuous emission of radiation from the Sun's photosphere. In addition to this predictable aspect of our Sun's energy output, there is a much more irregular component, characterized by solar activity, also referred to as *solar storms*, that includes *sunspots*, *prominences*, *solar flares*, and *coronal mass ejections*. We know that the Sun's strong magnetic field is the ultimate cause of solar storms; however, what remains unknown is exactly how and where the Sun's magnetic field is generated.

Sunspots

One if the most conspicuous features on the Sun's surface is the dark blotches called **sunspots** (Figure 23.26A). In 1610, Galileo, using one of his early telescopes, recorded the location of sunspots over a period of several days. From their motion, he estimated that the Sun rotates on its axis about once a month. Later observations indicated that the time required for one rotation varied slightly depending on solar latitude. A sunspot located near the Sun's equator rotates once in 25 days, whereas one located 70° from the solar equator requires 33 days for one rotation. This nonuniform rotation is evidence of the Sun's gaseous nature.

Sunspots begin as small, dark spots about 1600 kilometers (1000 miles) in diameter that last for only a few hours; however, some grow into blemishes larger than the diameter of Earth and may last for a month or more. The largest sunspots often occur in pairs surrounded by several smaller sunspots (Figure 23.26B).

An individual spot contains a black center, the *umbra* (*umbra* = shadow), which is rimmed by a lighter region, the *penumbra* (*paene* = almost, *umbra* = shadow) (Figure 23.26C). Sunspots appear dark only by contrast with the brilliant photosphere, a fact accounted for by their temperature, which is roughly 1500 degrees cooler than the solar surface. If these dark spots could be observed away from the Sun, they would appear many times brighter than the full Moon.

Like all other solar storms, sunspots are governed by the Sun's magnetic field, which is perhaps best illustrated using *magnetic field lines*, as shown in Figure 23.27A. When the Sun's surface is particularly active, sunspots

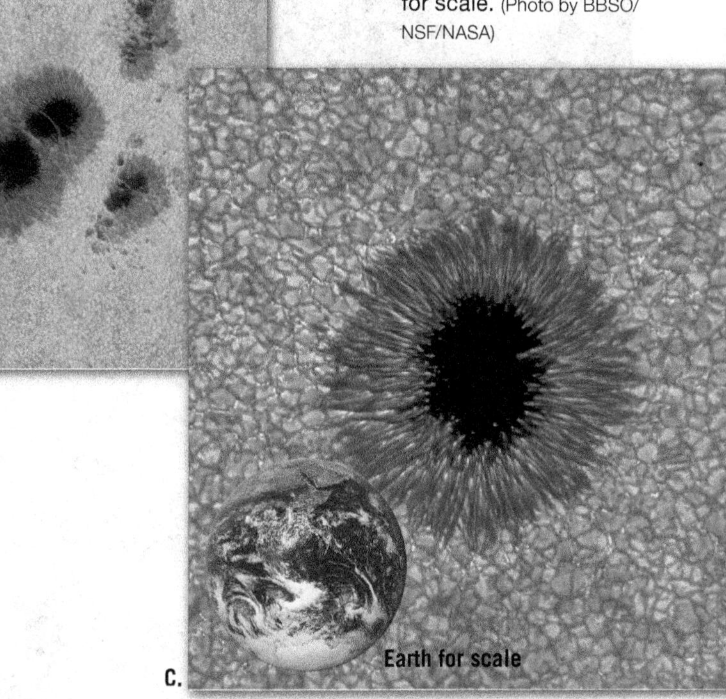

Figure 23.26 Sunspots **A.** Large sunspot group on the solar disk. (Photo by SDO/NASA) **B.** The largest sunspots often occur in pairs surrounded by several smaller sunspots. (Photo by © Alan Friedman/ avertedimagination.com) **C.** An individual sunspot has a dark central umbra surrounded by a lighter penumbra. Earth is shown for scale. (Photo by BBSO/ NSF/NASA)

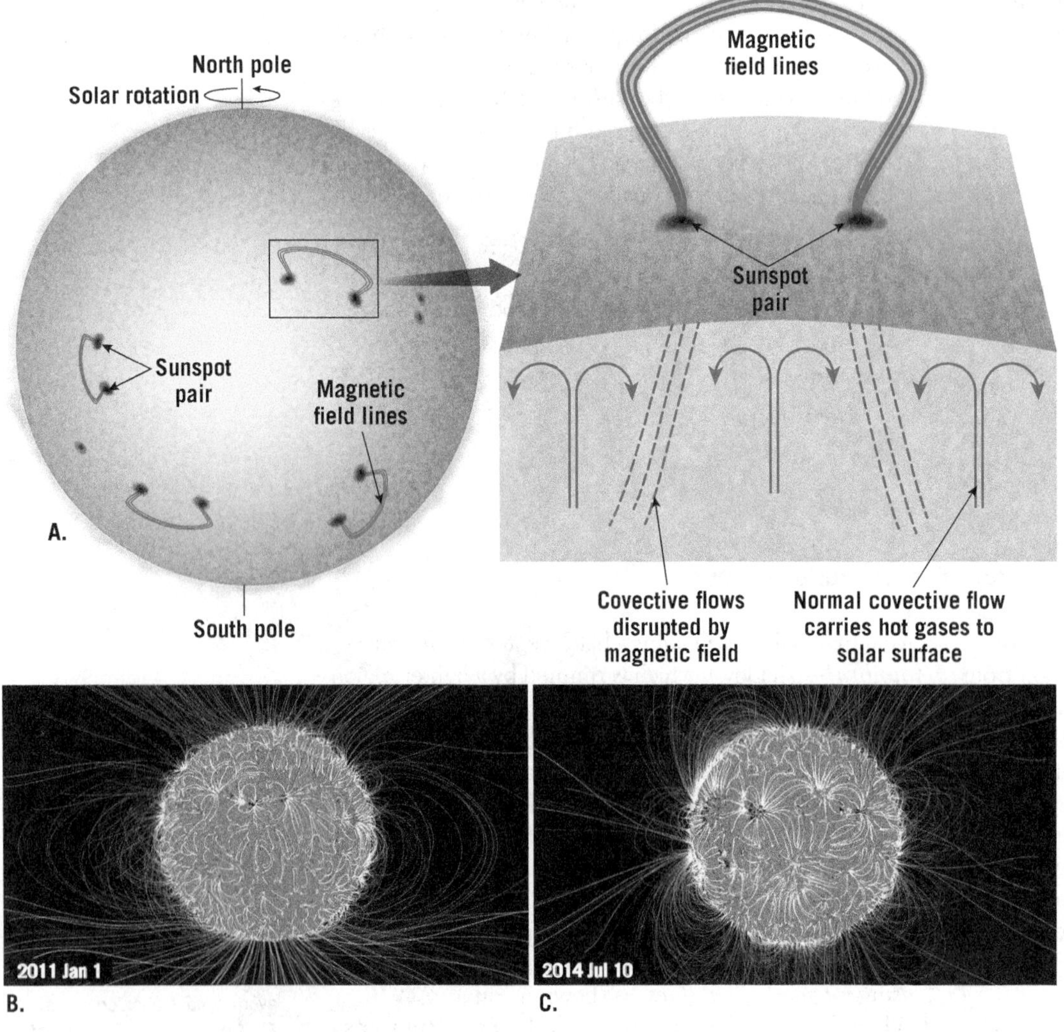

Figure 23.27 **Sunspots and the Sun's magnetic field** **A.** Magnetic field lines connecting a pair of sunspots. **B.** Side-by side comparison of how the Sun's magnetic field changed from January 2011 to July 2014. Notice how the magnetic field is more concentrated near the poles in 2011, 3 years after a solar minimum. **C.** By 2014, the magnetic field has become more twisted and disorderly, making conditions ripe for solar storms, such as sunspots, solar flares, and coronal mass ejections. (Courtesy of NASA)

are common, and the Sun's magnetic field lines connect pairs of sunspots. In addition, the magnetic field emanating from a typical sunspot is roughly 1000 times stronger than the magnetic field in the surrounding, undisturbed region of the photosphere. Astronomers have concluded that sunspots are cooler than the neighboring solar surface because the magnetic field disrupts the convective flow below the sunspot, which normally carries hot gases to the Sun's surface (see Figure 23.27A).

During the early nineteenth century, it was thought that a tiny planet named Vulcan orbited the Sun inside the orbit of Mercury. In the search for Vulcan, an accurate record of sunspot occurrences was kept, and it showed that the number of sunspots varies in a roughly 11-year cycle. At its *maximum*, the number of sunspots reaches perhaps 100 or more at a given time. Then, over a period of 5 to 7 years, their numbers decline to a *minimum*, when only a few or even none are visible. Like other solar disturbances, we know that the Sun's magnetic field drives the roughly 11-year sunspot cycle.

At a sunspot minimum, the Sun's magnetic field is weak and comparatively smooth, and the magnetic field lines are concentrated at the poles (**Figure 23.27B**). As the solar maximum approaches, the magnetic field becomes more complicated, with numerous small structures emanating throughout the solar surface (**Figure 23.27C**). Sometimes the magnetic field lines become twisted and knotted to the point that they "snap" and reorganize themselves into more simplified structures. This process releases huge amounts of energy that generates even more energetic solar storms than sunspots.

Figure 23.28 **Solar prominence.** (NASA)

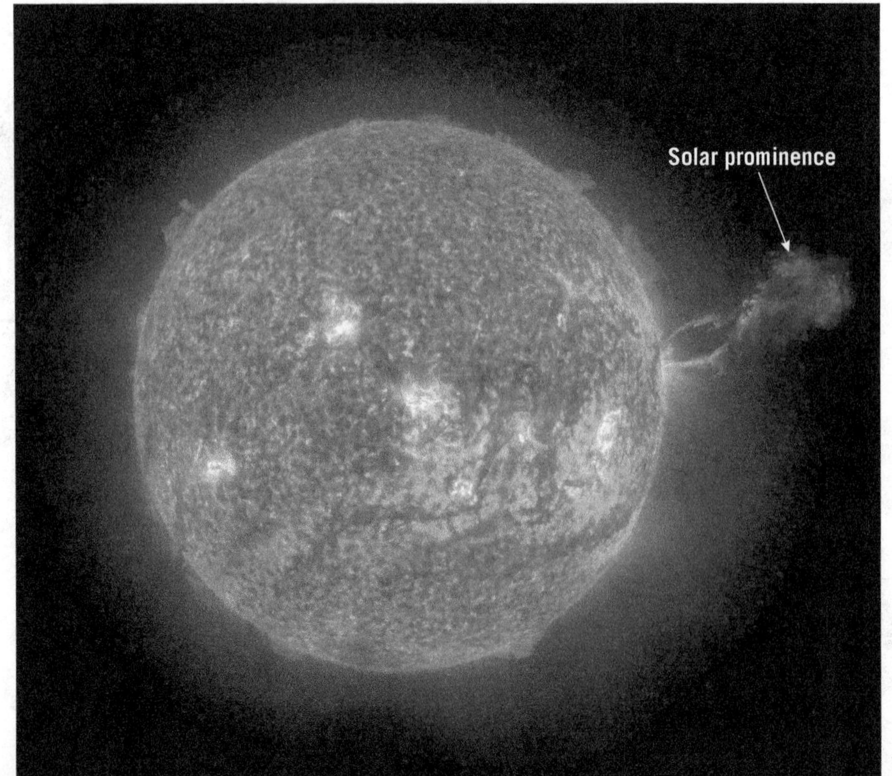

Prominences

Among the most spectacular features of the active Sun are **prominences** (*prominere* = to jut out). These huge cloudlike structures, consisting of concentrations of chromospheric gases, are best observed when they are on the edge, or limb, of the Sun, where they often appear as bright arches that extend well into the corona (**Figure 23.28**). *Quiescent prominences* have

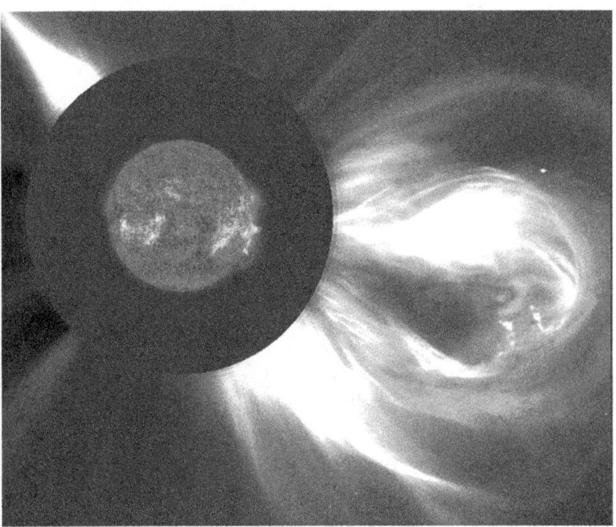

▲ Figure 23.29 **Coronal mass ejection, hurling a large cloud of particles into space** This is a composite image taken by two instruments on the Solar and Heliospheric Observatory (SOHO). The red disk extending around the Sun blocks sunlight to allow details in the corona to be observed. (NASA)

▲ Figure 23.30 **Aurora Borealis (Northern Lights)** This phenomenon, caused by particles ejected during a solar flare, also occurs near the South Pole, where it is called the Aurora Australis (Southern Lights). (Photo by Andy Farrer/Getty Images)

the appearance of a fine tapestry and seem to hang motionless for days at a time, but pictures reveal that the material within them is continually falling like luminescent rain.

By contrast, *eruptive prominences* rise almost explosively away from the Sun. These active prominences reach velocities up to 1000 kilometers (620 miles) per second and may leave the Sun entirely, producing a *coronal mass ejection*. Both eruptive and quiescent prominences consist of ionized chromospheric gases trapped by magnetic fields that extend from sunspots or other regions of intense solar activity.

Solar Flares and Coronal Mass Ejections

Solar flares are brief outbursts that normally last an hour or so. A solar flare appears as a sudden brightening of the region above a sunspot cluster. During a solar flare enormous quantities of energy are released across the entire electromagnetic spectrum, much of it in the form of ultraviolet and x-ray radiation. Simultaneously, fast-moving atomic particles are ejected, causing the *solar wind* to intensify noticeably.

Sometimes solar flares, or other large solar storms, eject huge quantities of charged particles from the Sun's corona. These particles travel outward from the Sun as huge masses called **coronal mass ejections**, which, if traveling in our direction, can reach Earth in a couple of days (Figure 23.29). Upon reaching Earth, a coronal mass ejection creates a magnetic storm in our upper atmosphere, which can hamper radio communications, damage electronic components on orbiting satellites, and even disrupt the transmission of electrical power.

One positive effect of these large solar storms is unusually strong **auroras**, also called the Northern and Southern Lights (Figure 23.30). Following a coronal mass ejection, the upper atmosphere near Earth's magnetic poles is set aglow for several nights. Auroral displays, because they are caused by solar activity, vary in intensity with the 11-year sunspot cycle.

CONCEPT CHECKS 23.6

1. What did Galileo learn about the Sun from his observations of sunspots?
2. Briefly describe the 11-year sunspot cycle.
3. Why are sunspots cooler than the surrounding photosphere?

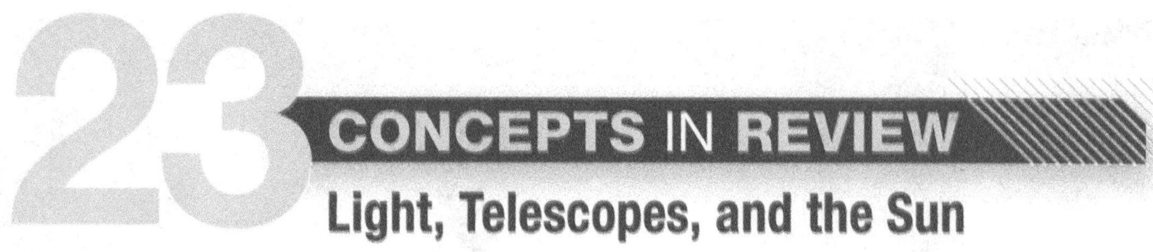

23 CONCEPTS IN REVIEW
Light, Telescopes, and the Sun

23.1 Light: Messenger from Space

Compare and contrast the wave properties of light with the particle properties of light.

KEY TERMS: electromagnetic radiation, wavelength, frequency, photon, radiation pressure

- Electromagnetic radiation occurs in a spectrum of wavelengths. From short to long, these are gamma rays, x-rays, ultraviolet light, visible light, infrared radiation (heat), and radio waves. The shorter the wavelength, the more energy the radiation carries.
- Electromagnetic radiation can be described as behaving either like waves or like a stream of particles known as photons. Both descriptions are valid; they apply in different circumstances.

23.2 What Can We Learn from Light?

Explain how the three types of spectra are generated and what they tell astronomers about the radiating object that produced them.

KEY TERMS: spectroscopy, spectroscope, continuous spectrum, emission line spectrum, absorption line spectrum, intensity, Doppler effect, redshift

- Spectroscopy is the study of the interaction of matter and light.
- The three types of spectra are called (1) continuous spectra, (2) absorption line spectra, and (3) emission line spectra. A continuous spectrum provides information about the radiating object's energy output and surface temperature. Absorption and emission spectra provide information about the object's composition. Emission spectra are produced by incandescent (glowing) gases; absorption spectra are produced when light passes through a gas. The spectra of most stars are absorption spectra.
- Spectroscopy can also be used to determine the motion of an object by measuring the Doppler effect.

? Use the accompanying diagram, which shows two absorption spectra, to determine whether the object is moving toward or away from the observer. Explain your reasoning.

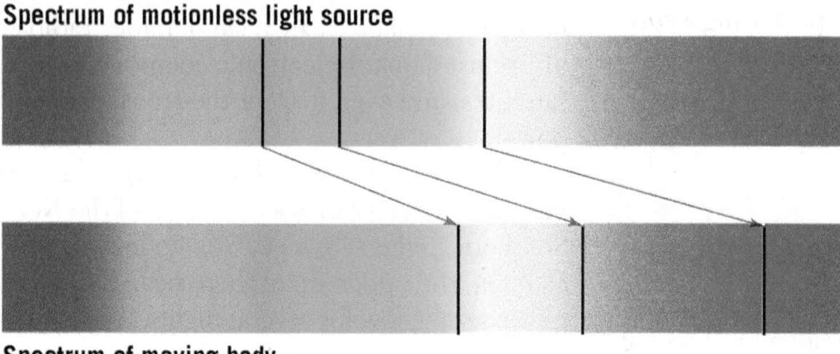

23.3 Collecting Light Using Optical Telescopes

Describe the two properties that make telescopes with large mirrors more useful than those with small mirrors.

KEY TERMS: refracting telescope, reflecting telescope, light-gathering area, resolution, turbulence, active optics, interferometer, adaptive optics

- There are two types of optical telescopes: (1) the refracting telescope, which uses a lens to bend or refract light, and (2) the reflecting telescope, which uses a curved mirror to focus light. Most large modern telescopes are reflectors.
- Telescopes simply collect light. When analyzed, the collected light can be used to determine the temperature, composition, relative motion, and distance to a celestial object.
- Historically, astronomers relied on their eyes to collect light. Then photographic film was developed. Now, light is collected using charge-coupled devices (CCDs).
- Advances such as active optics, interferometry, and adaptive optics have great improved the resolution of images acquired by modern telescopes. Still, Earth-bound optical telescopes can only "view" a tiny portion of the electromagnetic spectrum.

23.4 Radio- and Space-Based Astronomy

List the advantages of radio-wave and orbiting observatories over ground-based optical telescopes.

KEY TERMS: radio telescope, very-long-baseline interferometer (VLBI)

- Much of the radiation produced by celestial objects cannot penetrate our atmosphere, so other types of observatories have been developed.
- The detection of radio waves is accomplished by "big dishes" known as radio telescopes. A parabolic-shaped dish, often made of wire mesh, operates in a manner similar to the mirror of a reflecting telescope.
- Orbiting observatories, like the Hubble Space Telescope, circumvent many of the problems caused by Earth's atmosphere and have led to numerous significant discoveries in astronomy.

23.5 Our Star: The Sun

Sketch the Sun's structure and describe each of its major layers. Summarize the process called the proton–proton chain reaction.

KEY TERMS: photosphere, granule, chromosphere, corona, solar wind, core, radiation zone, convection zone, nuclear fusion, proton–proton chain reaction

- The photosphere (visible surface) of the Sun radiates most of the light we see. Unlike most surfaces to which we are accustomed, it consists of a layer of incandescent gas less than 500 kilometers (300 miles) thick and has a grainy texture consisting of numerous relatively small, bright markings called granules.
- Just above the photosphere lies the chromosphere, a relatively thin layer of hot incandescent gases a few thousand kilometers thick.
- At the uppermost edge of the solar atmosphere, called the corona, ionized gases escape the gravitational pull of the Sun and stream toward Earth at high speeds, producing the solar wind.
- Deep in the solar interior, a type of nuclear fusion called the proton–proton chain reaction converts four hydrogen nuclei into the nucleus of a helium atom. During the reaction, some of the matter is converted into the energy that the Sun ultimately radiates into space.

23.6 The Active Sun

List and describe four types of solar storms that occur on the Sun.

KEY TERMS: sunspot, prominence, solar flare, coronal mass ejection, aurora

- Sunspots are the most common type of solar activity. The number of sunspots observable on the solar disk varies in an 11-year cycle.
- Prominences, huge cloudlike structures best observed when they are on the Sun's edge, or limb, are produced by ionized chromospheric gases trapped by magnetic fields that extend from regions of intense solar activity.
- One of the most explosive events associated with sunspots is large solar flares, which are brief outbursts that release enormous quantities of energy.
- Solar flares and other solar storms can sometimes produce coronal mass ejections, which consist of radiation and fast-moving particles that cause the solar wind to intensify. If the ejected particles reach Earth, they can disrupt radio communication and produce the auroras, also called the Northern and Southern Lights.

? The accompanying image is a close-up view of the Sun's surface. What name is given to the Sun's surface? Name the labeled features.

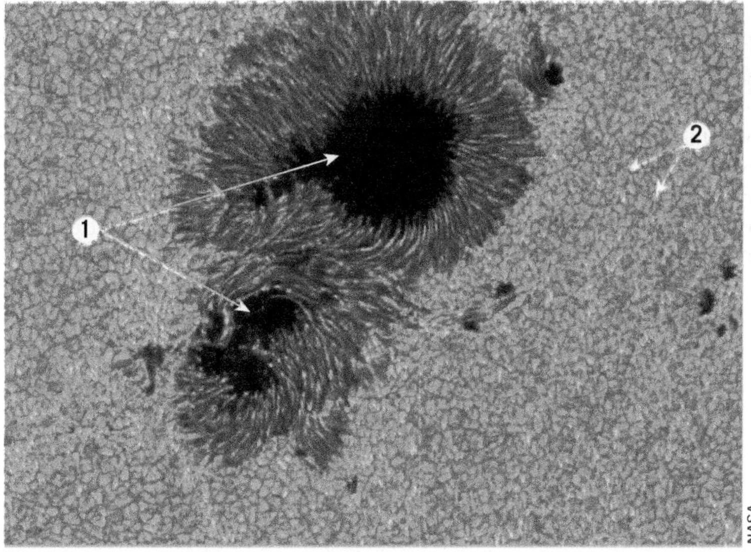

GIVE IT SOME THOUGHT

1 Refer to the accompanying spectra, which represent four identical stars in our galaxy. One star is not moving relative to you, another is moving away from you, and two stars are moving toward you. Determine which star is which and explain how you reached your conclusion.

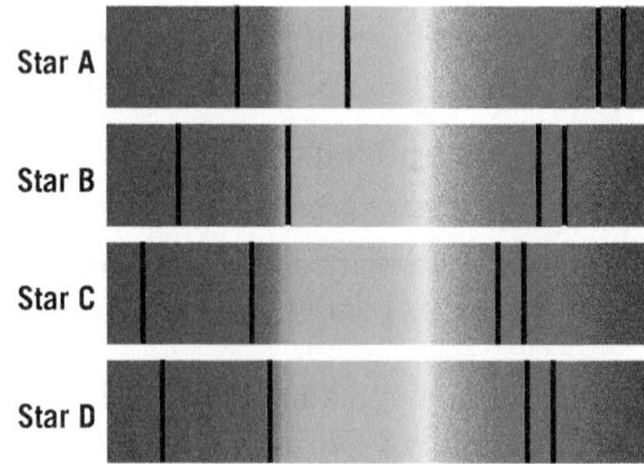

2 Imagine that you are responsible for funding the construction of observatories. After considering the four proposals listed below, state whether you would or would not recommend funding for each proposal and explain your reasoning.

Proposal A: A ground-based x-ray telescope on top of a mountain in Arizona, designed to observe supernovae in distant galaxies.

Proposal B: A space-based 6-meter reflecting infrared telescope designed to observe very distant galaxies.

Proposal C: A ground-based 8-meter refracting optical telescope located on the top of Mauna Kea in Hawaii, designed to measure the spectra of stars in our galaxy.

3 An important absorption line in the spectrum of stars occurs at a wavelength of 656 nm for stars not moving toward or away from Earth. Imagine that you observe four stars in our galaxy and discover that this absorption line is at the wavelength shown in the accompanying diagram. Using this information, answer the following questions and explain the reasoning behind your answers. If an answer cannot be determined from the given information, explain.

STAR	*Wavelength of Absorption Line*
A	*649 nm*
B	*656 nm*
C	*658 nm*
D	*647 nm*

a. Which of these stars is moving toward Earth the fastest?
b. Which of these stars is closest to Earth?
c. Which of these stars is moving away from Earth?

4 Consider the following discussion among three of your classmates regarding why telescopes are put in space. Support or refute each statement.

Student 1: "I think it is because the atmosphere distorts and magnifies light, which causes objects to look larger than they actually are."

Student 2: "I thought it was because some of the wavelengths of light being sent out from the telescopes can be blocked by Earth's atmosphere, so the telescopes need to be above the atmosphere."

Student 3: "Wait, I thought it was because by moving the telescope above the atmosphere, the telescope is closer to the objects, which makes them appear brighter."

5 Refer to Figure 23.20 to answer the following questions:
a. Is the atmosphere mostly transparent or mostly opaque to visible light?
b. Is the atmosphere mostly transparent or mostly opaque to radio waves with a wavelength of 1 meter (3 feet)?
c. Is the atmosphere mostly transparent or mostly opaque to gamma rays?

EXAMINING THE EARTH SYSTEM

1 Of the two sources of energy that power the Earth system, the Sun is the main driver of Earth's external processes. If the Sun increased its energy output by 10 percent, what would happen to global temperatures? What effect would this temperature change have on the percentage of water that exists as ice? What would be the impact on the position of the ocean shoreline? Speculate about whether the change in temperature might produce an increase or a decrease in the amount of surface vegetation. In turn, what impact might this change in vegetation have on the level of atmospheric carbon dioxide? How would such a change in the amount of carbon dioxide in the atmosphere affect global temperatures?

DATA ANALYSIS

Observing the Sun

While it is unsafe to look directly at the Sun with the naked eye, there are many ground and satellite-based instruments that continuously observe the Sun. This information is used to create solar forecasts, including forecasts for the Northern Lights.

ACTIVITIES

Go to the Solar and Heliospheric Observatory (SOHO) page at https://sohowww.nascom.nasa.gov/ and click on the image under The Sun Now to show the latest SOHO images. Click on About These Images and scroll down.

1 What is EIT an acronym for? What is displayed on an EIT image?

2 What is the temperature of the material shown on the 304 image? 171? 195? 284?

3 Which EIT image shows solar material farthest from the sun's surface? Closest to the Sun's surface? How do you know?

4 What is MDI an acronym for? What is displayed on an MDI continuum image?

5 What is displayed on an MDI magnetogram image? What do black and white areas indicate?

Use the back arrow to return to the latest SOHO images. Clicking on an image will open a larger version.

6 What is the date and time for the most recent imagery?

7 What features on the Sun's surface are responsible for the uneven brightness shown in EIT 284? (Refer to Figure 23.20.)

8 Based on EIT 284 imagery, where are strong prominences located on the Sun?

9 How many sunspots do you observe in the most recent image? If there are sunspots, can you see umbra? Penumbra?

10 What is the relationship between strong magnetic field regions in the MDI (SDO/HMI) Magnetogram and prominences show on the EIT 284 image?

Go to NOAA's Space Weather Prediction Center page at http://www.swpc.noaa.gov/ and click on the image below The Aurora to see the current forecast for the Northern Lights.

11 Is the aurora likely to be visible in any region of the United States? If so, which region(s)? Explain the connection between solar activity and the Northern Lights.

24 Beyond Our Solar System

FOCUS ON CONCEPTS

Each statement represents the primary learning objective for the corresponding major heading within the chapter. After you complete the chapter, you should be able to:

24.1 Explain the criteria used to classify stars and define *main-sequence star.*

24.2 List and describe the stages in the evolution of a typical Sun-like star.

24.3 Compare and contrast the final state of Sun-like stars to the remnants of the most massive stars.

24.4 List the three major types of galaxies. Explain the formation of large elliptical galaxies.

24.5 Describe Edwin Hubble's discoveries about the nature of the universe and summarize the Big Bang theory.

The night sky over Grand Teton National Park, Wyoming. (Photo by Babak Tafreshi/Getty Images)

ASTRONOMERS AND COSMOLOGISTS STUDY THE NATURE OF OUR VAST UNIVERSE, attempting to answer questions such as these: Is our Sun a typical star? How do stars form, and how do they differ from one another? What happens when stars exhaust their fuel? How large is the universe? Did it have a beginning? Will it have an end? This chapter explores these questions and others.

24.1 Classifying Stars

Explain the criteria used to classify stars and define *main-sequence star.*

When you look up on a clear night, you might get the impression that all stars are basically alike. But if you look more closely, you will notice some subtle differences. Although most stars appear similar in brightness, a few look brighter. Careful examination of the brightest stars will reveal that some are reddish, while others glow bluish-white. For example, of the two brightest stars in the constellation Orion, Rigel appears bluish-white, whereas Betelgeuse has a red hue (**Figure 24.1**). A pair of binoculars or a telescope reveals yet more stars and makes their variations in brightness and colors easier to see.

Figure 24.1 **Bright, colorful stars in the constellation Orion the Hunter** The brightest stars in this constellation are the red supergiant Betelgeuse, on the upper left, and the blue supergiant Rigel, on the lower right. (Photo by John Chumack/Science Source)

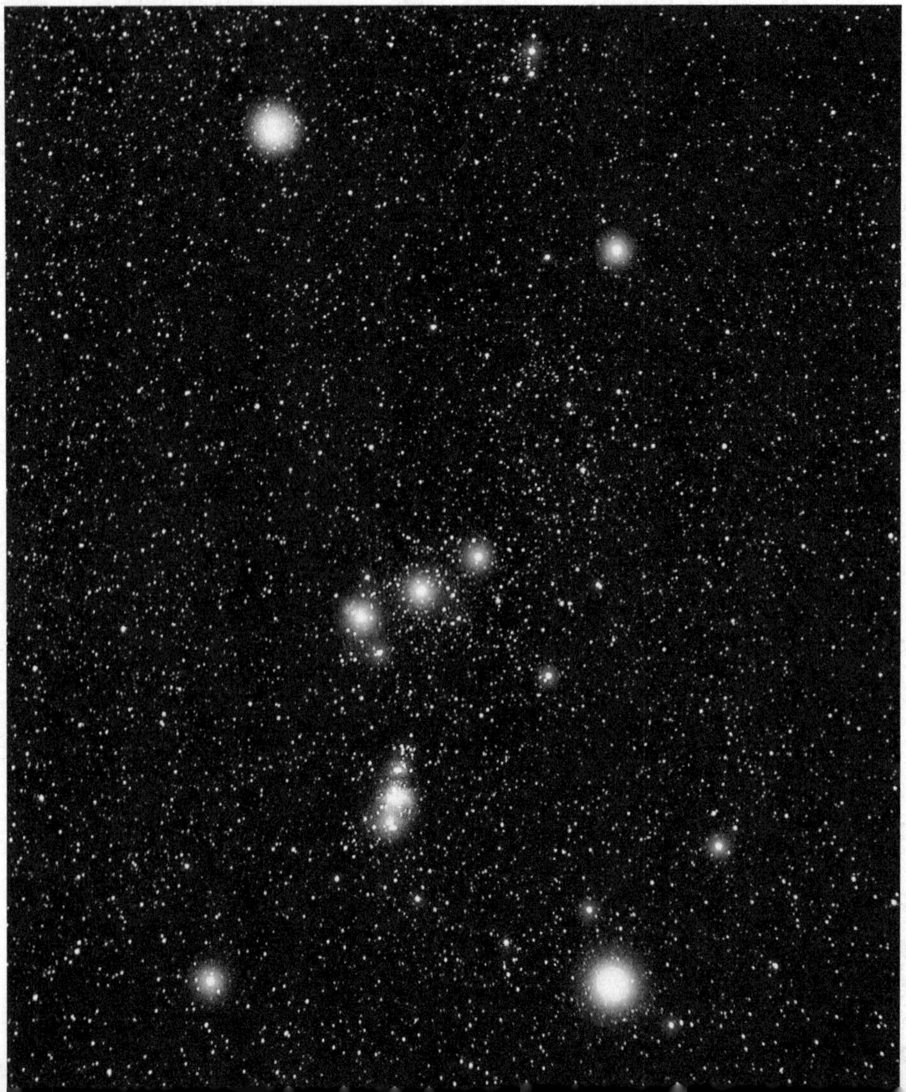

Stellar Luminosity

The oldest means of classifying stars is based on their *brightness*, also called *luminosity* or *magnitude*. Three factors control the brightness of a star as seen from Earth: how large it is, how hot it is, and its distance from Earth. The stars in the night sky come in a grand assortment of sizes, temperatures, and distances, and their brightnesses vary widely.

Apparent Magnitude Some stars appear dimmer than others because they are farther away. A star's brightness, *as it appears when viewed from Earth*, is called its **apparent magnitude**. The early Greek observer Hipparchus developed the first system to classify stars according to their apparent brightness. Because he could only reliably see six different brightness levels, he created six categories. These categories, which are still employed, are called *magnitudes*, with first magnitude being the brightest and sixth magnitude the dimmest. With the invention of the telescope, many stars fainter than the sixth magnitude were discovered.

In the mid-1800s, a method was developed to standardize the magnitude scale by comparing the light coming from stars of the first magnitude to those of the sixth magnitude. It was determined that a first-magnitude star is about 100 times brighter than a sixth-magnitude star. Based on this scale, any two stars that differ by five magnitudes have a ratio in brightness of 100 to 1. Hence, a third-magnitude star is 100 times brighter than an eighth-magnitude star. It follows, then, that the brightness ratio of two stars differing by one magnitude is about 2.5,* such that a star of the first magnitude is about 2.5 times brighter than a star of the second magnitude.

Because some celestial bodies are brighter than first-magnitude stars, zero and negative magnitudes were introduced. On this scale, the Sun has an apparent magnitude of −26.7. At its brightest, Venus has an apparent magnitude of −4.3. At the other end of the scale, the Hubble Space Telescope can view stars with an apparent magnitude of 30, more than 1 billion times dimmer than stars that are visible to the unaided eye.

*Calculations: $2.512 \times 2.512 \times 2.512 \times 2.512 \times 2.512$, or 2.512 raised to the fifth power, equals 100.

Absolute Magnitude (Luminosity) Apparent magnitudes seemed like good approximations of the true brightness of stars when we thought the universe was very small—containing no more than a few thousand stars that were all very close to Earth. However, we now know that the universe is unimaginably large and contains innumerable stars at wildly varying distances. Astronomers are interested in the "true" brightness of stars, so they devised a measure called **absolute magnitude**, or **luminosity**.

To establish the absolute magnitude of a star, astronomers must account for the *distance* separating the observer and the star. This is done by determining the luminosity (magnitude) that a star would have if it were located at a standard distance—about 32.6 light-years from Earth. (The distance light travels in 1 year is called a **light-year**—slightly less than 10 trillion kilometers [6 trillion miles].) For example, if the Sun, which has an apparent magnitude of −26.7, were located 32.6 light-years from Earth, it would have an absolute magnitude of about 5. Thus, stars with absolute magnitudes greater than 5 (smaller numerical value) are intrinsically brighter than the Sun but appear much dimmer only because of their vast distance from Earth. Table 24.1 lists the absolute and apparent magnitudes of some stars, as well as their distances from Earth. Most stars have an absolute magnitude between −10 (very bright) and 20 (very dim). The Sun is near the midpoint of this range.

Stellar Color and Temperature

Because color is primarily a manifestation of a star's surface temperatures, astronomers use color to classify stars. Very hot stars with surface temperatures above 30,000K emit most of their energy in the form of short-wavelength light and, therefore, appear bluish in color. (Stellar temperatures are expressed in kelvins [K]; see Appendix A.) On the other hand, red stars are cooler, having surface temperatures generally less than 3000K and emitting most of their energy as longer-wavelength red light. Stars such as the Sun with surface temperatures between 5000K and 6000K appear yellow-white.

Hertzsprung–Russell Diagrams (H-R Diagrams)

Early in the twentieth century, Einar Hertzsprung and Henry Russell independently studied the relationship between a star's luminosity (absolute magnitude) and its surface temperature. Their work resulted in the development of a graph called a **Hertzsprung–Russell diagram (H-R diagram)**. To produce an H-R diagram, astronomers survey a portion of the sky and plot each star according to its *absolute magnitude (luminosity)* and *temperature* (Figure 24.2). This allows astronomers to classify stars and, as discussed later, understand their evolution.

Table 24.1 Distance, Apparent Magnitude, and Absolute Magnitude of Some Stars

Star	Distance from Earth (light-years)	Apparent Magnitude	Absolute Magnitude
Sun	0	−26.7	5.0
Alpha Centauri	4.27	0.0	4.4
Sirius	8.70	−1.4	1.5
Arcturus	36	−0.1	−0.3
Betelgeuse	520	0.8	−5.5
Deneb	1600	1.3	−6.9

Notice that the stars plotted on the H-R diagram in Figure 24.2 are not uniformly distributed. Rather, about 90 percent of all stars fall along a band that runs from the upper-left corner to the lower-right corner of the diagram. The stars found along this band are called **main-sequence stars**. As you can see in Figure 24.2, the hottest (blue) main-sequence stars are intrinsically the brightest, and, conversely, the coolest (red) are the dimmest. Main-sequence stars are comparatively stable stars that produce energy by fusing hydrogen to form helium in their cores.

Astronomers also discovered that the absolute magnitude of main-sequence stars is directly related to their mass. The hottest (blue) stars can be up to 200 times more massive than the Sun, whereas the coolest (red) stars can be less than one-tenth as massive. Therefore, on the H-R diagram, the main-sequence stars appear in decreasing order, from hotter, more massive blue stars to cooler, less massive red stars.

Note the location of the Sun in Figure 24.2. The Sun is a yellow main-sequence star with an absolute magnitude, or "true" brightness, of about 5. Because the vast majority of main-sequence stars have magnitudes between −10 (very bright) and 20 (very dim), the Sun's midpoint position in this range results in its classification as an "average size star." Recall that stellar magnitudes are measured so that the lower the number, the brighter the star and vice versa.

Nearly 90 percent of all stars have luminosities and temperatures that place them on the band of main-sequence stars, but a few stars have properties that do not follow the main-sequence pattern. Above the right-hand portion of the main sequence on the H-R diagram is a group of very luminous stars called **giants** or, on the basis of their color, **red giants** (see Figure 24.2). The size of a star can be estimated by comparing it with stars of known size that have the same surface temperature. Recall that stars having equal surface temperatures radiate the same amount of energy per unit area. Any difference in the brightness of two stars having the same surface temperature can be attributed to their relative sizes. Therefore, if one red star is 100 times more luminous than another red star, it must have a surface area that is 100 times larger. Therefore, stars with large radiating surfaces are appropriately called *giants*.

SmartFigure 24.2 **Hertzsprung–Russell diagram** Astronomers use these diagrams to study stellar evolution by plotting stars according to their temperatures and luminosities (absolute magnitudes). Stellar temperatures are given in kelvins (K); to convert to degrees Celsius (°C), subtract 273.

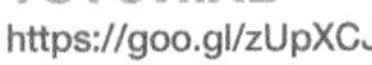

TUTORIAL
https://goo.gl/zUpXCJ

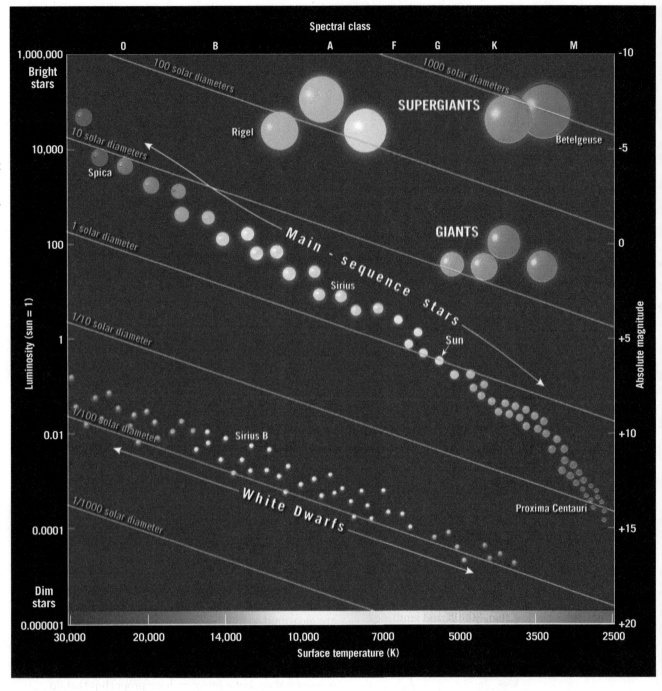

Some stars are so immense that they are called **supergiants**. Betelgeuse, the bright red supergiant in the constellation Orion, has a radius about 800 times that of the Sun. If this star were at the center of our solar system, it would extend beyond the orbit of Mars, and Earth would find itself buried inside!

In the lower portion of the H-R diagram, the opposite condition is observed. These stars are much fainter than main-sequence stars of the same temperature and, by the same reasoning, are much smaller. Some approximate Earth in size (see Figure 24.2). These tiny stars are called **white dwarfs**. Despite their name, the hottest white dwarfs are blue in color, but as they cool, they gradually become white and then reddish in color, before eventually going dark.

H-R diagrams are an important tool for interpreting stellar evolution—the stages in which stars, similar to living things, are born, age, and die. Because the vast majority of stars lie on the main sequence, we can be relatively certain that most stars spend most of their active years as main-sequence stars. Only a few percent are giants, and perhaps 10 percent are white dwarfs.

CONCEPT CHECKS 24.1

1. Define *absolute magnitude*.
2. How does the Sun compare in size and brightness to other main-sequence stars?
3. Describe how the H-R diagram is used to determine which stars are "giants."

24.2 Stellar Evolution

List and describe the stages in the evolution of a typical Sun-like star.

The idea of describing how a star is born, ages, and dies may seem a bit presumptuous, for most stars have life spans in the billions of years. However, by studying stars of different ages at different points in their life cycles, astronomers have been able to assemble a model for stellar evolution.

The method used to create this model is analogous to how an alien, upon reaching Earth, might determine the developmental stages of human life. By observing large numbers of humans, this stranger would witness the onset of life, the progression of life in children and adults, and the death of the elderly. From this information, the alien could put the stages of human development into their natural sequence. Based on the relative abundance of humans in each stage of development, it would even be possible to conclude that humans spend more of their lives as adults than as toddlers. Similarly, astronomers have pieced together the life story of stars.

The stages in the evolution of a typical Sun-like star are illustrated in Figure 24.3.

Stellar Birth

The birthplaces of stars are interstellar clumps of matter, rich in dust and gases, called **nebulae** (meaning "clouds"; the singular of *nebulae* is *nebula*). Most of the nebulae that generate stars are **molecular clouds**—that is, clouds cold and dense enough for hydrogen atoms to join together into hydrogen molecules. In the Milky Way, these clouds consist of about 92 percent hydrogen and 7 percent helium. The remainder consists of heavier atoms, molecules such as carbon monoxide, and tiny grains of *interstellar dust*. Although most molecular clouds are vast in size and their total mass is hundreds of times greater than the Sun, their particles are widely dispersed.

Figure 24.3 H-R diagram illustrating the evolution of an average Sun-like star

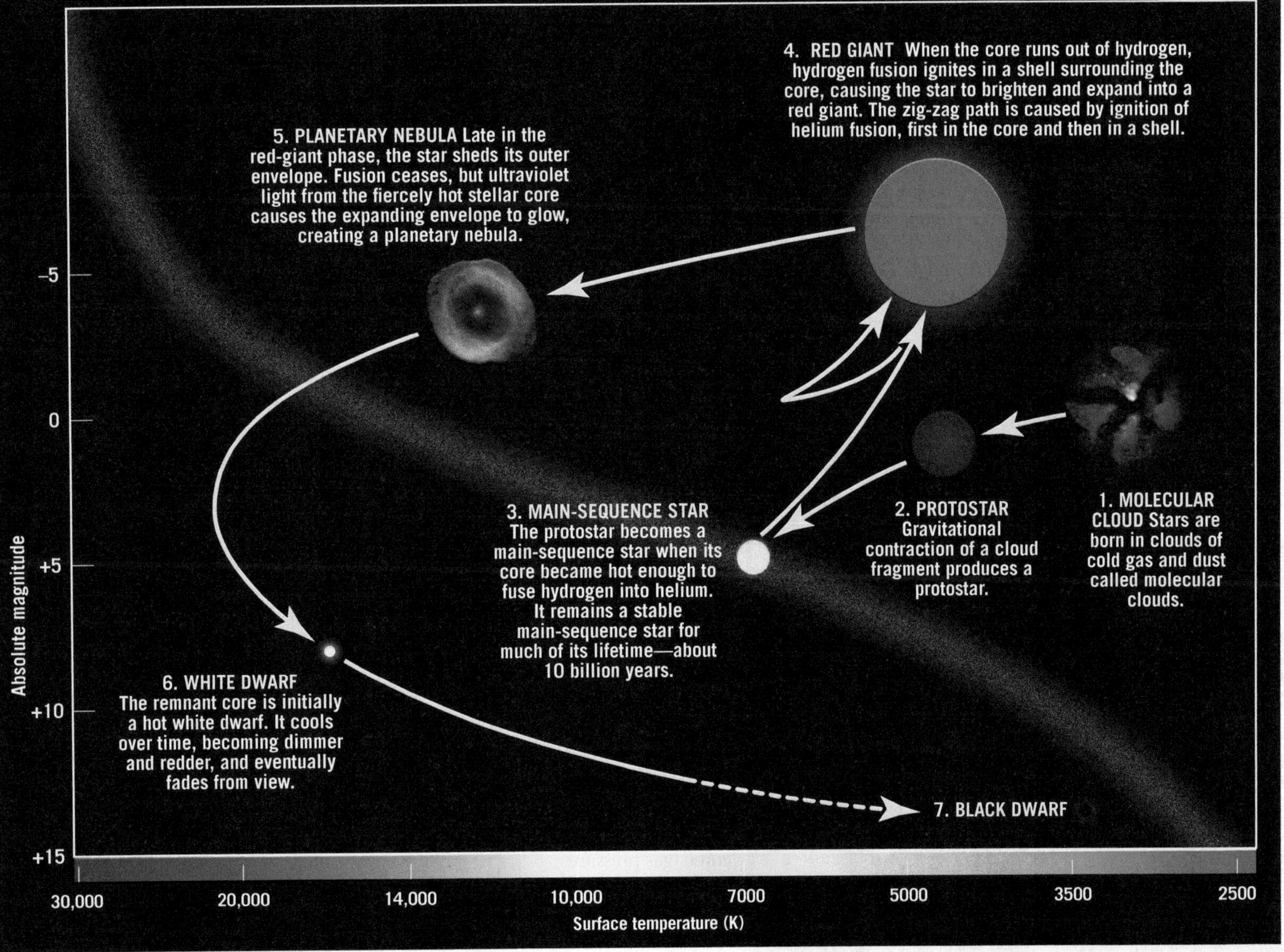

4. RED GIANT When the core runs out of hydrogen, hydrogen fusion ignites in a shell surrounding the core, causing the star to brighten and expand into a red giant. The zig-zag path is caused by ignition of helium fusion, first in the core and then in a shell.
5. PLANETARY NEBULA Late in the red-giant phase, the star sheds its outer envelope. Fusion ceases, but ultraviolet light from the fiercely hot stellar core causes the expanding envelope to glow, creating a planetary nebula.
3. MAIN-SEQUENCE STAR The protostar becomes a main-sequence star when its core became hot enough to fuse hydrogen into helium. It remains a stable main-sequence star for much of its lifetime—about 10 billion years.
2. PROTOSTAR Gravitational contraction of a cloud fragment produces a protostar.
1. MOLECULAR CLOUD Stars are born in clouds of cold gas and dust called molecular clouds.
6. WHITE DWARF The remnant core is initially a hot white dwarf. It cools over time, becoming dimmer and redder, and eventually fades from view.
7. BLACK DWARF
Absolute magnitude
-5
0
+5
+10
+15
30,000
20,000
14,000
10,000
7000
5000
3500
2500
Surface temperature (K)

Nevertheless, some clouds contain somewhat denser star-forming regions such as the "Pillars of Creation" in the Eagle Nebula (Figure 24.4). In these regions, gases and dust clump together to form even denser masses that attract more matter until they are massive enough to contract gravitationally to form stars.

Every stage of a star's life is ruled by gravity. The mutual gravitational attraction of particles in a thin, gaseous cloud causes the cloud to collapse on itself. As the cloud's center is squeezed by unimaginable pressures, its temperature increases; eventually its nuclear furnace ignites, and a star is born. A star is a ball of very hot gases, caught between the opposing forces of gravity trying to contract it and thermal nuclear energy trying to expand it. Eventually, a star's nuclear fuel is exhausted, gravity prevails, and the star collapses into a small, dense stellar remnant.

Protostar Stage

As these massive clouds of dust and gases collapse, gravitational energy is converted into thermal energy, causing the contracting gases to gradually increase in temperature. As they become warmer, these gaseous bodies begin to radiate energy in the form of long-wavelength red light. Because these large red objects are not hot enough to engage in nuclear fusion, they are not yet stars. The name **protostar** is applied to these bodies.

During the protostar stage, gravitational contraction continues, slowly at first and then much more rapidly. This collapse causes the core of the developing star to heat much faster than its outer envelope. When the core reaches a temperature of 10 million K, the temperature and pressure are so extreme that groups of four hydrogen nuclei (through a several-step process) fuse into a single helium nucleus, in a process called **hydrogen fusion**. Recall from Chapter 23 that this nuclear reaction, in which hydrogen nuclei are fused into helium nuclei, is called the *proton–proton chain reaction*.

Figure 24.4 Star-forming regions in the Eagle Nebula These dark, eerie pillar-like structures are columns of cool interstellar dust, hydrogen, and helium, which are incubators for new stars. (NASA)

Main-Sequence Stage

The immense amount of heat released by hydrogen fusion causes the gas particles inside stars to move with increased vigor, raising the internal *gas pressure*—also called *thermal pressure*. At some point, the increased motion produces an outward force (gas pressure) that balances the inward-directed force of gravity. Upon reaching this balance, a star become a *stable main-sequence star* (see Figure 24.3). In other words, a main-sequence star is one in which the force of gravity, in an effort to squeeze the star into the smallest possible ball, is precisely balanced by the gas pressure created by nuclear fusion in the star's interior.[†]

During the main-sequence stage, stars experience minimal changes in size or energy output. Hydrogen is continually being converted into helium, and the energy released keeps the gas pressure sufficiently high to prevent gravitational collapse.

How long can stars maintain this balance? Hot, massive blue stars radiate energy at such an enormous rate that they substantially deplete their nuclear fuel in only a few million years, approaching the end of their main-sequence stage relatively rapidly. By contrast, the smallest (red) main-sequence stars may take hundreds of billions of years to burn their hydrogen; they live practically forever. A yellow star, such as the Sun, remains a main-sequence star for about 10 billion years. Because the solar system is about 5 billion years old, the Sun is expected to remain a stable main-sequence star for about another 5 billion years.

Red Giant Stage

A Sun-like star spends 90 percent of its life as a hydrogen-burning main-sequence star. Once the hydrogen fuel in the star's core is depleted, it evolves over a period of about a billion years to become a *red giant* (see Figure 24.3). More massive stars evolve at a much faster rate. Evolution to the red giant stage begins when the hydrogen in a star's core is consumed, leaving behind a central core enriched in helium. Without hydrogen fusion supplying thermal energy, the star's core no longer has the gas pressure necessary to support itself against the inward force of gravity. As a result, the star's core begins to contract.

Collapse of the core causes the star's internal temperature to rise rapidly as gravitational energy is converted into thermal energy. Eventually the hydrogen-rich shell surrounding the star's inert helium core becomes hot enough to initiate even more vigorous hydrogen fusion than occurred in the core. This accelerated rate of hydrogen fusion causes the star's outer gaseous shell to heat and expand. Sun-like stars become bloated *red giants*, which are roughly 100 times larger than their main-sequence size. More massive stars become *supergiants*, which can be thousands of times larger than their main-sequence size.

As a star expands, its surface cools, which explains the star's red color. The core, however, continues to slowly contract, raising its internal temperature and the rate of hydrogen fusion in the overlying shell. Eventually, the star's internal temperature reaches 100 million K. This astonishing temperature triggers another nuclear reaction in the core, in which three helium nuclei are converted into one carbon nucleus, plus energy. Beyond this point,

[†]In very massive stars, radiation pressure—the force produced by photons as they escape a star's interior—also plays a major role in supporting the star against gravitational collapse.

the red giant consumes both hydrogen and helium to produce energy. In stars more than about eight times as massive as the Sun, other thermonuclear reactions occur that generate elements even heavier than carbon.

The transition to core helium fusion causes the star to become somewhat smaller, bluer, and less luminous. But this is just a temporary retreat. When the helium in the core has all been converted to carbon, core fusion again ceases, and the core again contracts and heats up. This event triggers helium fusion in the shell surrounding its spent carbon core. This helium fusion in the shell around the core, coupled with the continued hydrogen fusion in the shell above, causes the star to expand to become a red giant for a second time. This furious rate of hydrogen and helium fusion, called *double-shell fusion*, is thought to last for about 30 million years in a star with a mass near that of the Sun.

Stars more than about eight times the mass of the Sun have interior temperatures many times greater than those of Sun-like stars. As a result, they are able to fuse carbon in their cores to produce even heavier elements. Once all the carbon in a massive star's core is fused, the core once again shrinks, and because of the star's immense gravity, it is heated sufficiently to trigger other nuclear reactions. Through these various complex nuclear reactions, massive stars are able to fuse elements heavier than carbon, including oxygen, magnesium, and silicon. The final product of these thermal nuclear reactions is iron, which accumulates in the core. But iron is unique: It is the one element from which nuclear energy cannot be generated. Once the iron in the core piles up sufficiently, the star is without a source of energy to resist the crush of gravity—a topic we will consider next.

Burnout and Death

What happens to stars in the final phase of their lives? We know that stars, regardless of their size, must eventually exhaust their usable nuclear fuel and collapse in response to their immense gravity (Figure 24.5). Because the mass of a star determines how hot it can get and thus

SmartFigure 24.5
Evolutionary stages of stars having various masses

TUTORIAL
https://goo.gl/wWcMue

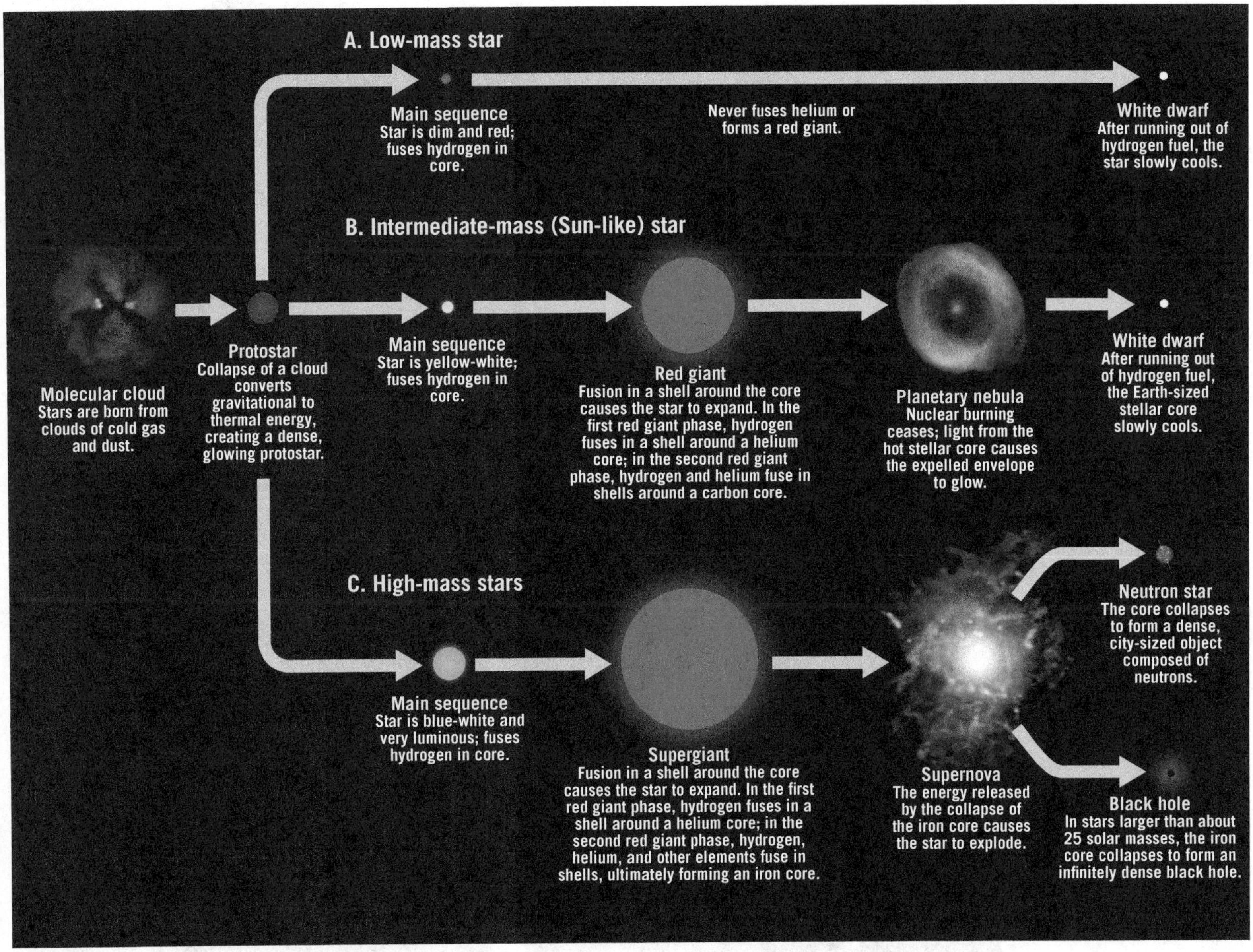

A. Low-mass star

Main sequence
Star is dim and red; fuses hydrogen in core.

Never fuses helium or forms a red giant.

White dwarf
After running out of hydrogen fuel, the star slowly cools.

B. Intermediate-mass (Sun-like) star

Molecular cloud
Stars are born from clouds of cold gas and dust.

Protostar
Collapse of a cloud converts gravitational to thermal energy, creating a dense, glowing protostar.

Main sequence
Star is yellow-white; fuses hydrogen in core.

Red giant
Fusion in a shell around the core causes the star to expand. In the first red giant phase, hydrogen fuses in a shell around a helium core; in the second red giant phase, hydrogen and helium fuse in shells around a carbon core.

Planetary nebula
Nuclear burning ceases; light from the hot stellar core causes the expelled envelope to glow.

White dwarf
After running out of hydrogen fuel, the Earth-sized stellar core slowly cools.

C. High-mass stars

Main sequence
Star is blue-white and very luminous; fuses hydrogen in core.

Supergiant
Fusion in a shell around the core causes the star to expand. In the first red giant phase, hydrogen fuses in a shell around a helium core; in the second red giant phase, hydrogen, helium, and other elements fuse in shells, ultimately forming an iron core.

Supernova
The energy released by the collapse of the iron core causes the star to explode.

Neutron star
The core collapses to form a dense, city-sized object composed of neutrons.

Black hole
In stars larger than about 25 solar masses, the iron core collapses to form an infinitely dense black hole.

the fusion reactions it can sustain, low-mass stars and high-mass stars have different fates. To simplify the discussion of the final stage in stellar lives, we divide stars into three groups, based on mass: low-mass stars, intermediate-mass (Sun-like) stars, and high-mass stars.

Death of Low-Mass Stars Stars less than about one-half the mass of the Sun (0.5 solar mass) have low surface temperatures (red color), and because of their small size are frequently called *red dwarfs* (see Figure 24.2). Red dwarfs are the most common stars in the universe and are thought to remain stable for at least 100 billion years, or possibly much longer. Because the interiors of low-mass stars are convective, hydrogen and helium mix continuously throughout much of the star's lifetime. Thus, these stars fuse all the hydrogen they contain, rather than just the hydrogen in their core region, as is the case for more massive stars. Because these stars never become hot enough to ignite helium, they do not bloat into red giants (Figure 24.5A). Instead, at the end of their long main-sequence lifetime, they gradually shrink into hot, dense white dwarfs—a topic we will consider later.

Death of Intermediate-Mass (Sun-Like) Stars Stars with masses ranging between one-half and eight times that of the Sun all have a similar evolutionary history (Figure 24.5B). During their red giant phase, Sun-like stars fuse hydrogen and, eventually, helium fuel at accelerated rates. Once this fuel is exhausted, these stars (like low-mass stars) evolve into Earth-size bodies of great density—white dwarfs.

Toward the end of their red giant phase, Sun-like stars cast off their bloated outer atmospheres, creating an expanding cloud of gas and revealing their fiercely hot core—a white dwarf. This white dwarf heats the expanding gas cloud, causing it to glow. One example of these picturesque clouds, called **planetary nebulae**, is shown in Figure 24.6.

Death of Massive Stars In contrast to Sun-like stars, which expire rather nonviolently, stars that are more than about eight times the mass of the Sun have relatively short life spans and terminate in brilliant explosions, called **supernovas** (Figure 24.5C). During supernova events, these stars become millions of times brighter than they were in prenova stages. If a star located in close proximity to Earth produced such an outburst, its brilliance would surpass that of the Sun. Fortunately for us, supernovae are relatively rare events in the Milky Way Galaxy; none have been observed in our galaxy since the advent of the telescope, though Tycho Brahe and Johannes Kepler each recorded one, about 30 years apart, late in the sixteenth century. Recall from Chapter 21 that Chinese astronomers recorded an even brighter supernova in 1054 C.E. Today, the remnant of that great outburst is the Crab Nebula, shown in Figure 24.7.

Supernova events are triggered when a massive star has consumed its nuclear fuel (though they can be produced by mechanisms other than the collapse of a massive star). Without the heat-producing fusion needed to balance its immense gravitational field, the star collapses, releasing a huge amount of thermal energy, which ejects the star's outer shell into space as a fiery supernova event. The scattered debris from the supernova carries with it the elements produced by nuclear fusion in the star, as well as elements produced during the catastrophic supernova explosion. Millions or billions of years later, this debris may be incorporated into new generations of stars and the Earth-like planets that may surround them. To quote Carl Sagan, "We are made of star stuff."

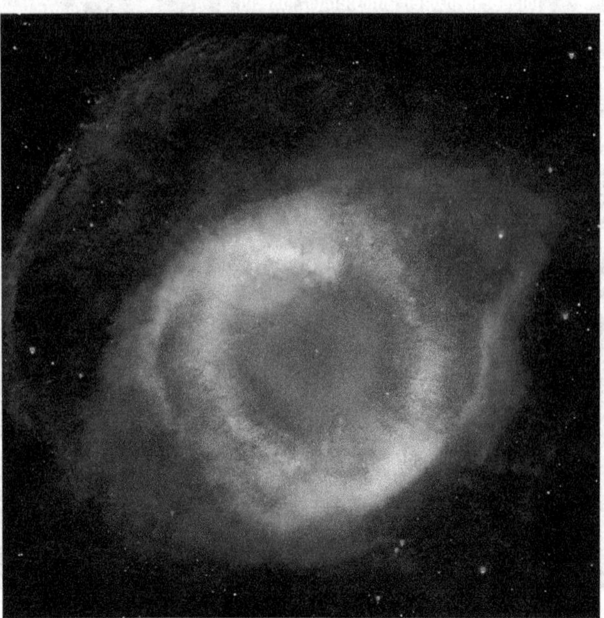

Figure 24.6 **Planetary nebula** The Helix Nebula, a planetary nebula, is the ejected outer envelope of a Sun-like star that collapsed from a red giant to a white dwarf. (Photo by ESA, and C.R. O'Dell (Vanderbilt University)/NASA)

Figure 24.7 **Crab Nebula in the constellation Taurus** This spectacular nebula is thought to be the remains of the supernova of 1054 C.E. (NASA)

Researchers theorize that during this type of supernova, the star's interior collapses into an incredibly hot object, possibly no larger than 20 kilometers (12 miles) in diameter. These incomprehensibly dense bodies have been named *neutron stars*. Some supernova events are thought to produce even smaller and more intriguing objects called *black holes*. We consider the nature of white dwarfs, neutron stars, and black holes in the following section.

CONCEPT CHECKS 24.2

1. Describe how main-sequence stars become giants.
2. Why are less massive stars thought to age more slowly than more massive stars, despite the fact they have much less "fuel"?
3. List the steps thought to be involved in the evolution of Sun-like stars.

24.3 Stellar Remnants

Compare and contrast the final state of Sun-like stars to the remnants of the most massive stars.

Eventually, all stars consume their nuclear fuel and collapse into one of three types of celestial object: *white dwarfs*, *neutron stars*, or *black holes*. How a star's life ends and what final form it takes depend largely on the star's mass during its main-sequence stage (Table 24.2).

White Dwarfs

All that is left of a Sun-like star after its outer layers are ejected into space is its core, which has collapsed to become a *white dwarf*. These objects, typically the size of Earth, have masses roughly equal to the Sun and are composed of super-compressed gases. One teaspoon of matter from a white dwarf would weigh about 5 tons on Earth. Matter of this type forms when electrons are displaced closer and closer together until the gas particles can be thought of as having no "elbow room" and, unlike ordinary gases, they resist being compressed further. This phenomenon, called *degeneracy pressure*, inhibits electrons from getting too close together and is sufficient to support the core of a Sun-like star against further gravitational collapse.

The surface of a young white dwarf is extremely hot, sometimes exceeding 25,000K, which accounts for its white or bluish-white color. Without an internal source of energy, these stars slowly cool as they continually radiate thermal energy into space, and they eventually become cold, burned-out embers called *black dwarfs*. However, current estimates of the cooling rates of white dwarfs indicates that our galaxy is not yet old enough for any black dwarfs to have formed.

Table 24.2 Summary of Evolution for Stars of Various Masses

Initial Mass* of Main-Sequence Star†	Main-Sequence Color	Giant Phase	Evolution After Giant Phase	Terminal State (final mass)
0.001	None (planet)	No	Not applicable	Planet (0.001)
0.1	Red	No	Not applicable	White dwarf (0.1)
1–8	Yellow	Yes	Planetary nebula	White dwarf (<1.4)
10	White	Yes	Supernova	Neutron star (1.4–3)
25	Blue	Yes (Supergiant)	Supernova	Black hole (>3)

*Masses in this table are *solar masses*, where 1 = the Sun's mass, 0.1 = one-tenth the Sun's mass, etc.

†These masses are estimates.

Neutron Stars

A study of white dwarfs produced a surprising conclusion: The smallest white dwarfs are the *most massive*, and the largest are the *least massive* (Figure 24.8). This occurs because white dwarfs are compressed by their immense gravity. The more massive a white dwarf, the more tightly it is able to squeeze its own matter, and hence the smaller it is.

This conclusion led to the prediction that stellar remnants even *smaller* and *more massive* than white dwarfs must exist. Named **neutron stars**, these objects are the remnants (cores) of massive stars (originally more than eight times as massive as the Sun). Neutron stars are the product of an explosive supernova event in which the outer envelope of a star is violently ejected by the energy released as the dense core of a massive star collapses into a very hot star less than 20 kilometers (12 miles) in diameter.

The immense gravitational force of a massive star's core is able to overcome degeneracy pressure, which in less massive stars inhibits electrons from getting too close together. When the core of a massive star collapses, the electrons are forced to combine with the protons located *inside* the nucleus to produce neutrons (hence the name *neutron star*). A pea-size sample of this matter would weigh 100 million tons. This approximates the density of an atomic nucleus; thus, a neutron star can be thought of as a large nucleus composed mainly of neutrons that are packed together almost seamlessly.

Although neutron stars have high surface temperatures, their small size greatly limits their luminosity, making them difficult to locate visually. However, newly formed neutron stars have a very strong magnetic field and a high rate of rotation. As stars collapse, they rotate

▼ **Figure 24.8 Comparing the sizes of white dwarfs of different masses to Earth** Counterintuitively, the greater the mass of a white dwarf, the smaller it will be. (NASA)

faster—similar to the way ice skaters rotate faster as they pull their arms in as they spin. Radio waves generated by the rapidly rotating magnetic fields of neutron stars are concentrated into two narrow zones that align with the star's magnetic poles. Consequently, these stars resemble rapidly rotating beacons emitting strong radio waves. If Earth happened to be in the path of these beacons, the star would appear to blink on and off, or *pulsate*, as the radio waves swept past.

In the early 1970s, a source that radiates short pulses of radio energy, named a **pulsar** (*pulsating radio source*), was discovered in the Crab Nebula (Figure 24.9). Visual inspection of this radio source indicated that it was coming from a small, hot star centered in the nebula. The Crab pulsar is the remains of the supernova of 1054 C.E. (see Figure 24.9). Although most pulsars discovered thus far emit radio waves, others emit ultraviolet light, x-rays, and even gamma rays.

▲ Figure 24.9 **Crab Pulsar: A young neutron star centered in the Crab Nebula** This is the first pulsar to have been associated with a known supernova. The energy emitted from this star illuminates the Crab Nebula. (Photo by NASA EOS Earth Observing System)

Black Holes

Although neutron stars are extremely dense, they are not the densest objects in the universe. Stellar evolutionary theory predicts that neutron stars cannot exceed three times the mass of the Sun. Above this mass, not even tightly packed neutrons can withstand the star's gravitational pull. Following a supernova explosion, if the core of the remaining star exceeds about three solar masses, gravity prevails, and the stellar remnant collapses into an object even more dense than a neutron star. (The mass of such a star prior to the supernova event was likely 25 times that of the Sun.) The incredible objects created by such a collapse are called **black holes**.

Einstein's theory of general relativity predicts that even though black holes are extremely hot, their surface gravity is so immense that even light cannot escape. Consequently, they literally disappear from sight. Anything that moves too close to a black hole will be swept in and devoured by its immense gravitational field.

How do astronomers find objects whose gravitational field prevents the escape of all matter and energy? These scientists theorize that as matter is pulled into a black hole, it should become extremely hot and emit a flood of x-rays before it is engulfed. Because *isolated* black holes do not have a source of matter to engulf, astronomers decided to look at binary-star (two-star) systems for evidence of matter emitting x-rays while being rapidly swept into a region of apparent nothingness.

X-rays cannot penetrate our atmosphere; therefore, the existence of black holes was not confirmed until the advent of orbiting observatories. The first black hole to be identified, Cygnus X-1, orbits a massive supergiant companion once every 5.6 days. The gases that are pulled from the companion form an *accretion disk* that spirals around a "void" while emitting a steady stream of x-rays (Figure 24.10). Recent observations have determined that pairs of jets extend outward from these accretion disks and are thought to return some of this material back to space (see Figure 24.10).

Cygnus X-1, which is about eight or nine times as massive as our Sun, probably formed from a star of approximately 40 solar masses. Since the discovery of Cygnus X-1, many other x-ray sources have been discovered and are thought to be black holes.

Astronomers have established that black holes are common objects in the universe and vary considerably in size. Small black holes have masses approximately 10 times that of our Sun but are only about 32 kilometers (20 miles) across, less than the distance of a marathon course. Intermediate black holes have masses 1000 times that of our Sun, and the largest black holes (*supermassive black holes*), found in the centers of galaxies, are estimated to be millions of solar masses. The earliest stars are thought to have been massive, and their deaths could have provided the seeds that eventually formed the supermassive black holes at the centers of galaxies.

▼ Figure 24.10 **Artist's view of a black hole and a giant companion star** (NASA)

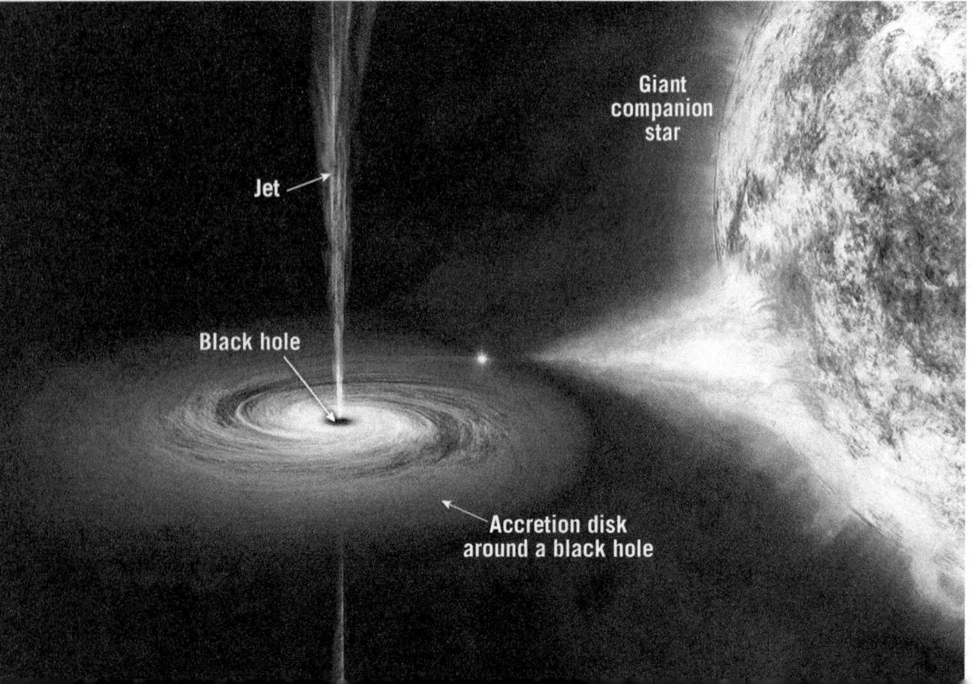

CONCEPT CHECKS 24.3

1. Explain how it is possible for the *smallest* white dwarfs to be the *most* massive.
2. What is the final state of a medium-mass (Sun-like) star?
3. How do the "lives" of the most massive stars end? What are the two possible products of this event?

EYE ON THE UNIVERSE 24.1

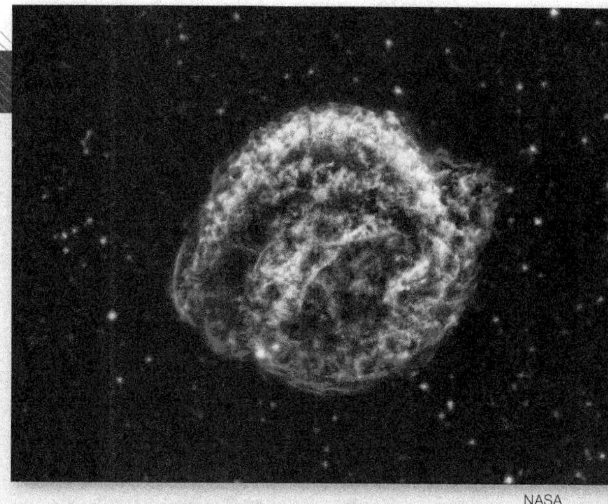

NASA

The accompanying image shows the remnant of a supernova that Johannes Kepler observed in 1604. The red, green, and blue show low-, intermediate-, and high-energy x-rays observed by NASA's Chandra X-ray Observatory.

QUESTION 1 *If a supernova explosion were to occur within the immediate vicinity of our solar system, what might be some possible consequences of the intense x-ray and gamma radiation that would reach Earth?*

QUESTION 2 *When the collapse of the core of a massive star results in a supernova, what type of stellar remnant remains?*

24.4 Galaxies and Galaxy Clusters

List the three major types of galaxies. Explain the formation of large elliptical galaxies.

On a clear and moonless night away from city lights, you can see a truly marvelous sight—a band of light stretching from horizon to horizon. With his telescope, Galileo discovered that this band of light was composed of countless individual stars. Today, we know that the Sun is actually a part of this vast system of stars—the Milky Way Galaxy (see GEOgraphics 24.1)

Galaxies (*galaxias* = milky), including the Milky Way, are collections of interstellar matter (dust and gases), stars, and stellar remnants that are gravitationally bound together (Figure 24.11A). Recent observational data indicate that supermassive black holes may exist at the centers of most galaxies. In addition, spherical halos of very tenuous gas and numerous star clusters, called *globular clusters*, surround most galaxies (Figure 24.11B,C).

A. Oblique view

B. Edge-on view

C. Globular cluster

SmartFigure 24.11 **Galaxies are collections of stars and interstellar matter that are gravitationally bound** **A.** Oblique view of a large spiral galaxy. Spiral galaxies typically have a greater concentration of older stars near their center, which gives the central bulge its yellowish color. By contrast, the arms of spiral galaxies contain numerous hot, young stars that give these structures a bluish or violet tint. **B.** Edge-on view showing the central bulge. **C.** Surrounding most large galaxies is a spherical halo that includes groups of stars called globular clusters. (Image A, NASA; images B and C, European Southern Observatory)

TUTORIAL
https://goo.gl/30Ousy

The Milky Way

The Milky Way derives it name from its appearance as a dim, "milky" glowing band that arches across the night sky. In this magnificent 360-degree panoramic image we see the plane of our galaxy, edge-on from our perspective on Earth. From this vantage point, the components of the Milky Way come into view. We can see the galaxy's bright central bulge, its disk that contains both dark and glowing nebulae that harbor bright, young stars. Also visible adjacent to the Milky Way are a few of its satellite galaxies.

ESO

This profile view shows that, in addition to its dense central bulge and galactic disl containing spiral arms, the Milky Way is surrounded by a spherical halo that exten far beyond the galactic disk. Although the halo lacks star formation, recent eviden indicates that it contains a large amount of hot gas. The halo also contains old sta and numerous large stellar groups called globular clusters.

The stars in globular cluster NGC 6397 are more than 13 billion years old, which is somewhat close to the age of the universe.

An artist's conception of the Milky Way Galaxy showing its dense central bulge and flat disk consisting of curved spiral arms. Our galaxy is a barred-spiral type that contains more than 200 billion stars. Its diameter exceeds 100,000 light-years and it rotates such that our solar system makes one complete trip around the galactic center every 250 million years. The galactic center of our galaxy houses a supermassive black hole with a mass of at least 40,000 Suns. Like the black holes found in the center of most large galaxies this behemoth will consume anything that happens to wander by.

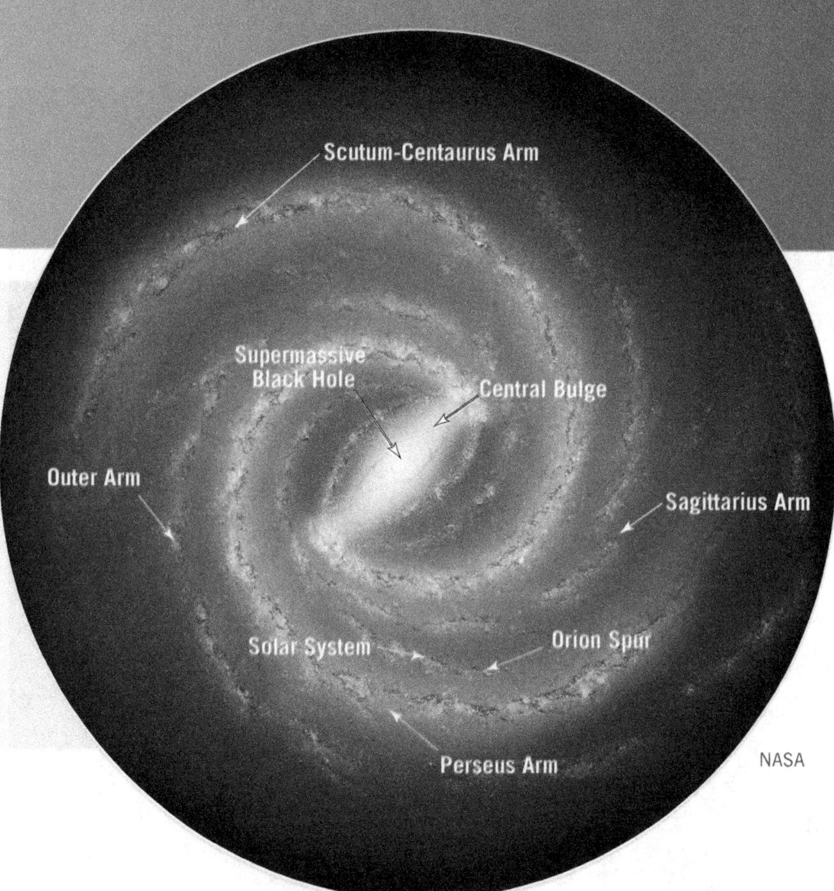

NASA

Our galaxy is just one of an assemblage of more than 50 galaxies that collectively make up the Local Group. Members include Andromeda Galaxy, which is an even larger spiral galaxy than the Milky Way. For many years astronomers thought that the Magellanic Clouds, which can be seen as fuzzy patches in the southern sky, were our closest galactic neighbors. But, in fact, the closest galaxy so far discovered, named Canis Major Dwarf Galaxy, lies within our galaxy. Astronomers have recently concluded that the Milky Way grew to its current size by "eating up" dwarf galaxies like Canis Major.

Large Magellanic Cloud

Small Magellanic Cloud

ESO

▲ Figure 24.12 **Dramatic image of the spiral galaxy Messier 83** Messier 83 is thought to be very similar in structure to the Milky Way Galaxy, although smaller. (Photo by European Southern Observatory)

▼ Figure 24.13 **Barred spiral galaxy** (NASA)

The first galaxies were small and composed mainly of young massive stars and abundant interstellar matter. These galaxies grew quickly by accreting nearby dust and gases and colliding and merging with other galaxies. In fact, our Milky Way Galaxy is currently absorbing at least two tiny satellite galaxies.

Types of Galaxies

Among the hundreds of billions of galaxies in the universe, astronomers classify galaxies into three major categories: *spiral*, *elliptical*, and *irregular*. Within each category are many variations.

Spiral Galaxies Our Milky Way Galaxy is an example of a large **spiral galaxy**. Spiral galaxies have a thin flat *disk*, consisting of stars, gas, and dust, and a central concentration of stars called a *bulge*. As shown in Figure 24.12, spiral galaxies get their name from the spiral arms that extend from the central bulge into the galactic disk. Surrounding these structures is the nearly invisible *halo*, which is composed of hot gases, old stars that tend to occur in clusters, and dark matter—a topic we will consider later. The nearly spheroid-shaped halo blends almost seamlessly with the disk and bulge.

The disk component of a spiral galaxy consists of a large population of young hot stars and dust-laden clouds that follow orderly orbits around the galactic center. Groups of these young hot stars appear as bright patches of blue and violet light in Figure 24.12. By contrast, the central bulge contains a large population of older stars, which gives the galaxy's center a yellowish glow. In addition, the stars in the central bulge, as well as those found in the halo, travel along orbits having many different inclinations.

About two-thirds of spiral galaxies have a band of stars extending outward from the central bulge that merges with the spiral arms. These are called **barred spiral galaxies** (Figure 24.13). Recent investigations have found evidence that our galaxy has a bar structure. What produces these bar-shaped structures is a matter of ongoing research. Astronomers estimate that over 70 percent of the largest galaxies in the universe are spiral galaxies.

Elliptical Galaxies As the name implies, **elliptical galaxies** have ellipsoid shapes, but they can be nearly spherical (Figure 24.14). Elliptical galaxies, like spiral galaxies, have a central bulge and halo but lack a well-defined disk component.

Elliptical galaxies tend to be composed of older, low-mass stars (red) and have minimal amounts of dust and cool gases. Thus, unlike the arms of spiral galaxies, they have low rates of star formation. As a result, elliptical galaxies appear yellow to red in color, as compared to the bluish tint emanating from the young, hot stars in the arms of spiral galaxies.

The very largest and smallest galaxies are elliptical. All small galaxies, regardless of type, are known as **dwarf galaxies**. The Milky Way's two small companions, the Large and Small Magellanic Clouds, are dwarf galaxies (see GEOgraphics 24.1). In fact, many of the galaxies located nearby are dwarf elliptical galaxies. Their small size and lack of brightness make dwarf galaxies almost impossible to detect at distances greater than a few million light-years. Nevertheless, based on the large number of nearby dwarf galaxies, astronomers suspect that these small galaxies may be the most common type of galaxy in the universe.

The very largest known galaxies (1 million light-years in diameter) are also elliptical, but these are rare. For comparison, the Milky Way, a large spiral galaxy, is about one-tenth the diameter of a massive elliptical galaxy. Large elliptical galaxies are believed to result from the merger of numerous smaller galaxies.

Irregular Galaxies Approximately 25 percent of known galaxies show no symmetry and are classified as **irregular galaxies**. Some were once spiral or elliptical galaxies that were subsequently distorted by the gravity of a neighboring galaxy. The Milky Way's two small companion galaxies, the Large and Small Magellanic Clouds, are irregular galaxies. They are named for explorer Ferdinand Magellan, who observed them when he circumnavigated the globe in 1520.

Recent images of the Large Magellanic Cloud reveal a central bar-shaped structure. Thus, the Large Magellanic Cloud was once a barred spiral galaxy that was subsequently distorted, probably by the gravitation of the Milky Way.

Galaxy Clusters

Once astronomers discovered that stars occur in groups (galaxies), they set out to determine whether galaxies themselves also occur in groups or whether they are randomly distributed. They found that galaxies are grouped into gravitationally bound clusters (Figure 24.15). Some large **galaxy clusters** contain thousands of galaxies. Our own galaxy cluster, called the **Local Group**, consists of more than 40 galaxies and may contain many undiscovered dwarf galaxies. Of the Local Group galaxies, three are large spiral galaxies, including the Milky Way and the Andromeda Galaxy.

Galaxy clusters also reside in huge groups called *superclusters*. There are possibly 10 million superclusters; our Local Group is located in the Virgo Supercluster. From visual observations, it appears that superclusters are the largest entities in the universe.

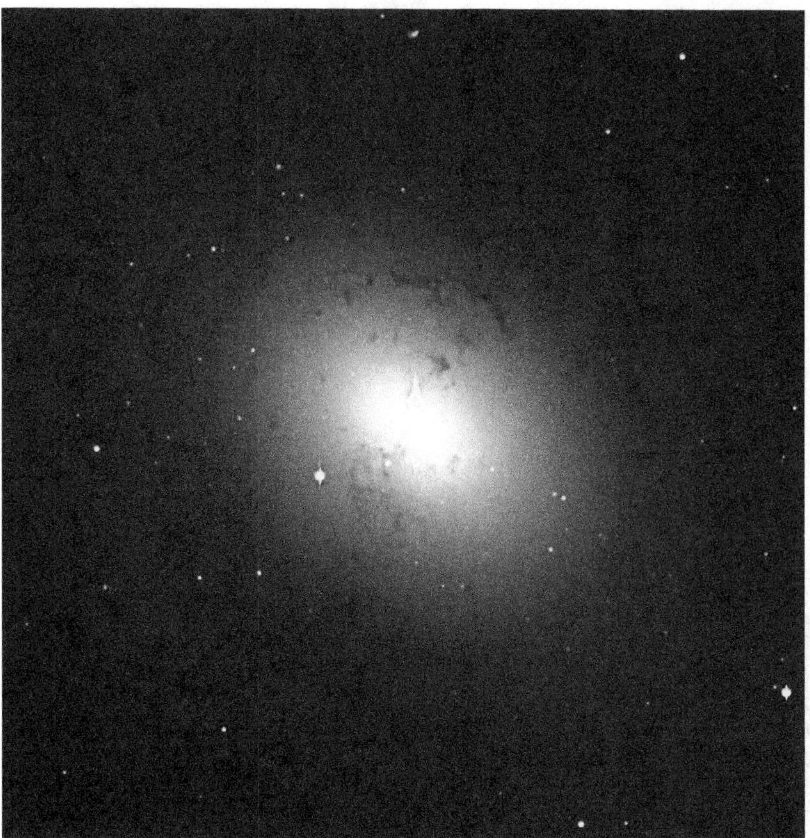

▲ **Figure 24.14 Large elliptical galaxy** This large elliptical galaxy belongs to the Fornax Cluster. Dark clouds of interstellar matter are visible within the central nucleus of this galaxy. Some of the star-like objects in this image are large groups of stars (globular clusters) that belong to the galaxy. (European Southern Observatory)

▼ **Figure 24.15 The Fornax Galaxy Cluster** This is one of the nearest groupings of galaxies to our Local Group. Although many of the galaxies shown are elliptical, an elegant barred spiral galaxy is visible in the lower right. (European Southern Observatory/J. Emerson/VISTA)

Galactic Collisions

Within galaxy clusters, galaxies frequently interact gravitationally with one another. For example, a large galaxy may engulf a dwarf satellite galaxy. In this case, the larger galaxy will retain its form, while the smaller galaxy will be torn apart and assimilated into the larger galaxy.

Galactic interactions may also involve two galaxies of similar size passing through one another without merging. It is unlikely that the individual stars within these galaxies will collide because they are so widely dispersed. However, the interstellar matter will likely interact, triggering an intense period of star formation.

In an extreme case, two large galaxies may collide and merge into a single large galaxy (Figure 24.16). Many of the largest elliptical galaxies were likely produced by the merger of two or more large spiral galaxies. Some studies have predicted that in 2 to 4 billion years, there is a 50 percent probability that the Milky Way and Andromeda Galaxies will collide and merge.

CONCEPT CHECKS 24.4

1. Compare the three main types of galaxies.
2. What type of galaxy is our Milky Way?
3. Describe a possible scenario for the formation of a large elliptical galaxy.

Figure 24.16 **The collision of the Antennae Galaxies** When two galaxies collide, their clouds of dust and gas become compressed, triggering a wave of star formation, as shown by the bright regions in this image. Actual stars rarely collide. (NASA)

24.5 The Universe

Describe Edwin Hubble's discoveries about the nature of the universe and summarize the Big Bang theory.

For most of human existence, our universe was thought to be Earth-centered, containing only the Sun, Moon, five planets, and the roughly 6000 stars visible to the naked eye. Even after the Copernican view of a Sun-centered universe became widely accepted, the entire universe was believed to consist of a single galaxy—the Milky Way, composed of innumerable stars, along with many faint "fuzzy patches," thought to be clouds of dust and gases.

How Large Is It?

In the mid-1700s, German philosopher Immanuel Kant proposed that many of the telescopically visible fuzzy patches of light scattered among the stars were actually distant galaxies similar to the Milky Way. Kant described them as "island universes." Each galaxy, he believed, contained billions of stars and was a universe in itself. In Kant's time, however, the weight of opinion favored the hypothesis that the faint patches of light occurred within our galaxy. Admitting otherwise would have implied a vastly larger universe, thereby diminishing the status of Earth and, likewise, humankind.

Edwin Hubble's Discovery In 1919 Edwin Hubble arrived at the observatory at Mount Wilson, California, to conduct research using a 2.5-meter (100-inch) telescope, then the world's largest and most advanced astronomical instrument. Armed with this modern tool, Hubble set out to solve the mystery of the "fuzzy patches." At that time, the debate was still raging as to whether the fuzzy patches were "island universes," as Kant had proposed more than 150 years earlier, or clouds of dust and gases (nebulae).

To accomplish this task, Hubble studied a group of pulsating stars known as *Cepheid variables*—extremely bright variable stars that increase and decrease in brightness in a repetitive cycle. These stars are significant because their absolute magnitude can be determined by knowing the rate at which they pulsate. When the absolute magnitude of a star is compared to its observed brightness, astronomers can determine a reliable approximation of its distance from Earth. This is similar to how we judge the distance of an oncoming vehicle when driving at night. Thus, Cepheid variables are important because they can be used to measure large astronomical distances.

Hubble found several Cepheid variables embedded in one of the fuzzy patches. Because these intrinsically bright stars appeared faint, Hubble concluded that they must lie outside the Milky Way. Indeed, he concluded that this fuzzy patch lies more than 2 million light-years away. This bright patch in the night sky is now known as the *Andromeda Galaxy* (Figure 24.17).

Based on his observations, Hubble determined that the universe extended far beyond the limits of our imagination. Today, we know that there are hundreds of billions of galaxies. For example, researchers estimate that a million galaxies exist in the portion of the sky bounded by the cup of the Big Dipper. There are literally more stars in the heavens than grains of sand in all the beaches on Earth.

What Does Light Tell Us About the Size of the Universe? As discussed in Chapter 23, telescopes actually "look back in time," which accounts for much of the knowledge astronomers have acquired about the history of the universe. Light from celestial objects that are great distances from Earth require millions or even billions of years to reach Earth. Therefore, the farther out telescopes can "see," the farther back in time astronomers can study. Even Andromeda, the closest large galaxy, is a staggering 2.5 million light-years away. Light that left the Andromeda Galaxy 2.5 million years ago is just now reaching Earth, allowing scientists to observe this galaxy as it was 2.5 million years ago. We have now observed the almost unimaginably faint light from primordial galaxies as they existed almost 13 billion years ago.

A Brief History of the Universe

Cosmology is the study of the universe, including its properties, structure, and evolution. Over the years, cosmologists have developed a comprehensive theory

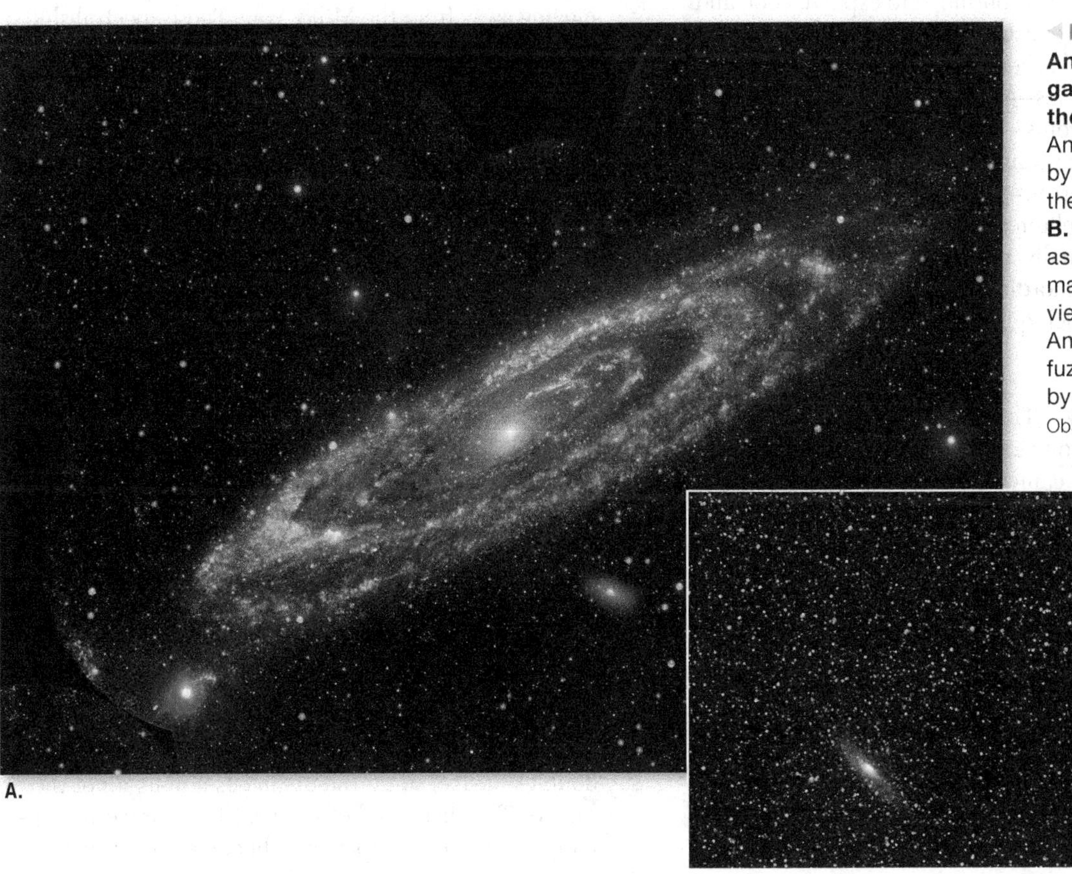

Figure 24.17
Andromeda: A nearby galaxy that is larger than the Milky Way A. Photo of Andromeda Galaxy, taken by a telescope aboard the GALEX Orbiter. (NASA) **B.** Andromeda Galaxy, as it appears under low magnification. When viewed with the naked eye, Andromeda appears as a fuzzy patch surrounded by stars. (European Southern Observatory)

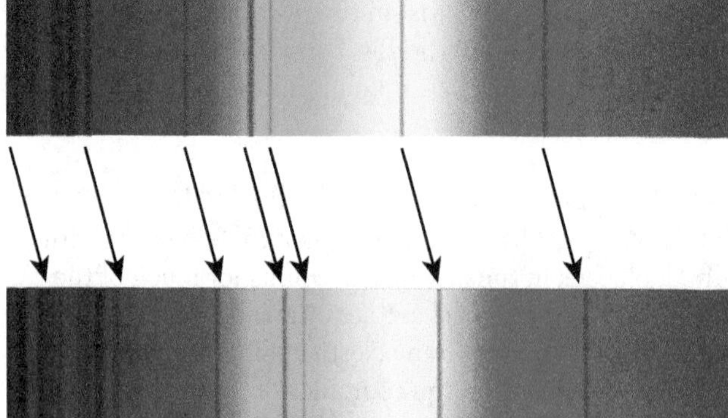

Figure 24.18
Cosmological redshifts Illustration of the shift in spectral lines toward the red end of the spectrum, which occurs when an object emitting light is receding from an observer.

that describes the structure and evolution of the universe in order to answer questions such as these: How did the universe evolve to its present state? How long has it existed? How will it end? Modern cosmology addresses these important issues and helps us understand the universe we inhabit.

The Big Bang Theory The model that most accurately describes the birth and current state of the universe is the **Big Bang theory**.† According to this theory, all of the energy and matter of the universe originally existed in an incomprehensibly hot and dense state. About 13.8 billion years ago, our universe began as a cataclysmic explosion, and it has continued to expand, cool, and evolve to its current state.

During the earliest stages of this expansion, only subatomic particles—the building blocks of protons, neutrons, and atoms—existed. Not until 380,000 years after the initial expansion did the universe cool sufficiently for electrons and protons to combine to form hydrogen and helium atoms—the lightest elements in the universe. Eventually, temperatures decreased sufficiently to allow this primordial gas to condense gravitationally into clumps, which quickly evolved into the first stars and galaxies.

The First Stars The first stars probably formed about 200 million years after the "Big Bang," when clouds of gas became dense enough to collapse under their own gravitation. Like the primordial gas from which they formed, these *first-generation stars* consisted primarily of hydrogen and small amounts of helium. The first stars were monsters, perhaps 30 to 300 times more massive than the Sun and shining millions of times brighter. Massive stars, however, have relatively short lifetimes ending in violent, explosive deaths. During their lives and even during their explosive deaths, these massive stars synthesized heavier elements and expelled them into space. Some of this ejected matter was incorporated into subsequent generations of stars such as our Sun. Our Sun and planetary system, having formed about 5 billion years ago (nearly 9 billion years after the Big Bang), is a latecomer to the universe.

†The name "Big Bang" was originally coined by cosmologist Fred Hoyle as a sarcastic comment on the believability of the theory.

Evidence for an Expanding Universe

In 1912 Vesto Slipher, while working at the Lowell Observatory in Flagstaff, Arizona, was the first to discover that galaxies exhibit motion. The motions he detected were twofold: Galaxies rotate, and galaxies move relative to each other. Slipher's efforts focused on the shifts in the spectral lines of the light emanating from galaxies. (As discussed in Chapter 23, when an object is moving toward or away from an observer, the Doppler effect shifts the position of its spectral lines.) When a galaxy is moving away from Earth, its spectral lines shift toward the red end of the spectrum (longer wavelengths), and when it is moving toward Earth, its spectral lines shift toward the blue end of the spectrum (shorter wavelengths).

In 1929, Edwin Hubble's study of galaxies expanded the groundwork established by Slipher. Hubble noticed that most galaxies, except those of the Local Group, have spectral shifts toward the red end of the spectrum (Figure 24.18). Therefore, most galaxies appear to be moving away from the Milky Way. These spectral shifts were later named **cosmological redshifts** because the movement they revealed resulted from the expansion of the universe.

Recall that Hubble had also found a way to measure galactic distances. By comparing the distance to a galaxy with Vesto Slipher's measurements of its redshift, Hubble made an unexpected discovery: the redshifts of galaxies increase with distance, and the most distant galaxies are receding from the Milky Way at the fastest rate. This concept, now called **Hubble's law**, states that *galaxies recede at speeds proportional to their distances from the observer.*

This discovery surprised Hubble because conventional wisdom was that the universe was unchanging and would likely remain unchanged. What cosmological theory could explain Hubble's findings? Researchers concluded that an expanding universe could account for the redshifts Hubble observed.

To help visualize why Hubble's law implies an expanding universe, imagine a batch of raisin bread dough that has been set out to rise for a few hours (Figure 24.19). In this analogy, the raisins represent galaxies, and the dough represents space. As the dough doubles in size, so do the distances among all of the raisins. The distance between raisins that were originally 2 centimeters apart will become 4 centimeters,

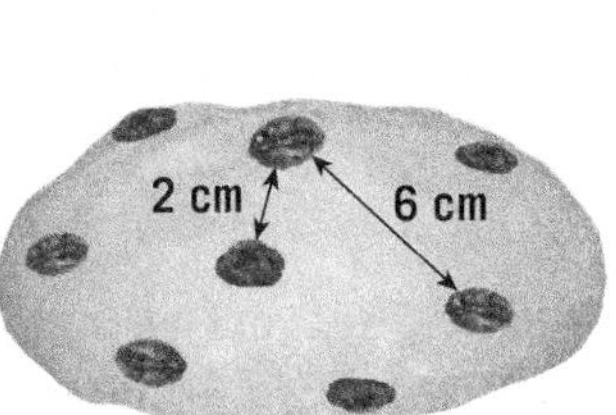
2 cm
6 cm

A. Raisin bread dough before it rises.

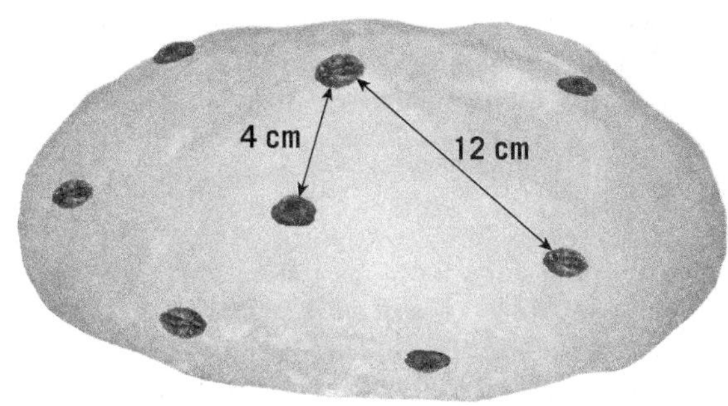
4 cm
12 cm

B. Raisin bread dough a few hours later.

SmartFigure 24.19 Raisin bread analogy for an expanding universe As the dough rises, raisins that were originally farthest apart travel greater distances than those located closer together. Thus, in an expanding universe (as with the raisins), more space is created between two objects that are farther apart than between two objects that are closer together.

TUTORIAL
https://goo.gl/NN4vFb

while the distance between raisins originally 6 centimeters apart will increase to 12 centimeters. The raisins that were originally farthest apart traveled greater distances than those located closer together. Therefore, in an expanding universe, as in our analogy, more space is created between two objects located farther apart than between two objects that are closer together.

Another feature of the expanding universe can be demonstrated using the bread analogy. Regardless of which raisin you look at, it will move away from all the other raisins. Likewise, at any point in the universe, all other galaxies (except those in the same cluster) are receding from an observer at that location. Hubble's law implies a centerless universe that is expanding uniformly. The Hubble Space Telescope is named in honor of Edwin Hubble's invaluable contributions to the scientific understanding of the universe.

Predictions of the Big Bang Theory

Recall from "The Nature of Scientific Inquiry" in Chapter 1 that in order for a hypothesis to become an accepted component of scientific knowledge (a theory), it must incorporate predictions that can be tested. One prediction of the Big Bang model is that the universe was initially unimaginably hot, and researchers should be able to detect the remnant of that heat. The electromagnetic radiation (light) emitted by a white-hot universe would have extremely high energy and short wavelengths. However, according to the Big Bang theory, the continued expansion of the universe would have stretched the waves so that by now they should be detectable in the microwave region of the spectrum (at the short-wavelength end of the radio spectrum). Scientists began to search for this "missing" radiation, which they named *cosmic microwave background radiation*. As predicted, in 1965, this microwave radiation was detected and found to fill the entire visible universe.

Detailed observations of the cosmic microwave background radiation have confirmed many details of the Big Bang theory, including the order and timing of important events in the early history of the universe.

What Is the Fate of the Universe?

Cosmologists have developed different scenarios for the ultimate fate of the universe (Figure 24.20). In one scenario, the stars will slowly burn out, and galaxies will become ever more widely separated in an endless dark, cold universe. This scenario is sometimes called

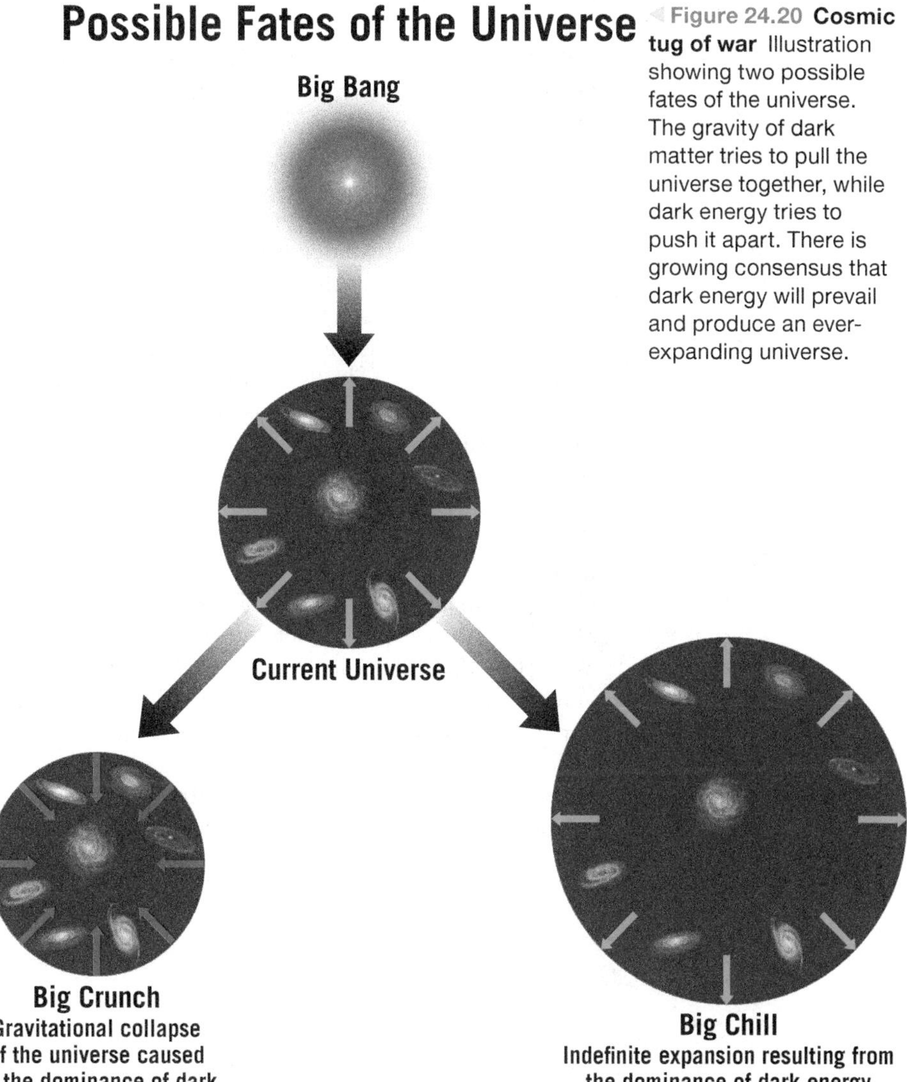
Possible Fates of the Universe
Big Bang
Current Universe
Big Crunch
Gravitational collapse of the universe caused by the dominance of dark matter pulling it together
Big Chill
Indefinite expansion resulting from the dominance of dark energy pushing the universe apart

Figure 24.20 Cosmic tug of war Illustration showing two possible fates of the universe. The gravity of dark matter tries to pull the universe together, while dark energy tries to push it apart. There is growing consensus that dark energy will prevail and produce an ever-expanding universe.

the "Big Chill" because the universe will slowly cool as it expands, to the point that it will be unable to sustain life. Another possibility is that the outward flight of the galaxies will slow and eventually stop. Gravitational contraction would follow, causing all matter to eventually collide and coalesce into the high-energy, high-density state from which the universe began. This fiery death of the universe, the Big Bang operating in reverse, has been called the "Big Crunch."

Whether the universe will expand forever or eventually collapse upon itself is partially contingent on its density. If the average density of the universe is greater than an amount known as its *critical density* (equivalent to five atoms for every cubic meter), gravitational attraction should be sufficient to stop the outward expansion and cause the universe to collapse. On the other hand, if the density of the universe is less than the critical value, the universe will expand forever. Complicating the possibilities for the fate of the universe are two other constituents that are thought to exist: *dark matter* and *dark energy*.

Dark Matter The universe contains perhaps 100 billion galaxies, many with billions of stars, massive clouds of gas and dust, and large numbers of planets, moons, and other debris. Yet everything we see is like the tip of the cosmic iceberg; it accounts for a small percentage of the total mass of the universe. Astronomers came to this conclusion after studying the rotational periods of stars as they orbit the center of the Milky Way Galaxy. The law of gravity states that the stars closest to the galactic center should travel faster than those near the galaxy's outer edge. (This is the reason Mercury travels around the Sun at a much faster speed than Neptune.) Yet these researchers found that all stars orbit the galactic center at roughly the same speed. This implies that something within our galaxy is tugging on the stars. This yet-undetected material is called **dark matter**.

Approximately one-quarter of the universe consists of dark matter, which does not absorb or emit light but interacts with "visible" matter in the universe through gravity. Thus, dark matter exerts a force that helps hold our galaxy together and at the same time works to slow the overall expansion of the universe.

Although the concept of dark matter may sound foreboding, it simply allows for the possible existence of matter that does not interact with electromagnetic radiation, such as visible light. Because most of our knowledge about the universe comes to us via light, if there is a form of matter that does not interact with light, we will not be able to "see" it—hence the term *dark matter*.

Dark Energy In the early 1990s most cosmologists held the view that gravity was certain to slow the expansion of the universe over time, resulting eventually in the Big Crunch. However, in 1998 Hubble Space Telescope observations of very distant galaxies showed that the universe is actually expanding faster today than it was early in its history. Therefore, the expansion of the universe was not slowing due to gravity as scientists had thought but instead was speeding up. To explain this unexpected result, researchers concluded that some unusual material, generally referred to as **dark energy**, must exist. Unlike dark matter, which works to slow the expansion of the universe, dark energy exerts a force that pushes matter apart, causing the expansion to accelerate.

It has not been determined whether dark matter and dark energy are related—or even exactly what they are. Most researchers think that dark matter consists of a type of subatomic particle that has not yet been detected. Dark energy may have its own particle, but there is no evidence of the existence of this particle.

There is growing consensus among cosmologists that dark energy, which is propelling the universe outward, is the dominant force. If dark energy is, in fact, the driving force behind the fate of the universe, the universe will expand forever (see Figure 24.20). Consider the following as astronomers search for dark energy: "Absence of evidence is not evidence of absence."

CONCEPT CHECKS 24.5

1. Explain how Edwin Hubble used Cepheid variables to change our view of the structure of the universe.
2. In your own words, explain how astronomers determined that the universe is expanding.
3. Which view of the fate of the universe is currently favored: the Big Crunch or the Big Chill?

CONCEPTS IN REVIEW
Beyond Our Solar System

24.1 Classifying Stars

Explain the criteria used to classify stars and define *main-sequence star*.

KEY TERMS: apparent magnitude, absolute magnitude (luminosity), light-year, Hertzsprung–Russell diagram (H-R diagram), main-sequence star, giant, red giant, supergiant, white dwarf

- Hertzsprung–Russell diagrams are constructed by plotting the absolute magnitudes (luminosity) and temperatures of stars on graphs.
- Stars are positioned within H-R diagrams as follows: (1) Main-sequence stars, 90 percent of all stars, are in the band that runs from the upper-left corner (massive, hot blue stars) to the lower-right corner (low-mass, red stars); (2) red giants and supergiants, very luminous stars with large diameters, are located in the upper-right position; and (3) white dwarfs, which are small stars, are located in the lower portion.

? Identify the type of stars located in the positions labeled A, B, and C on the accompanying H-R diagram.

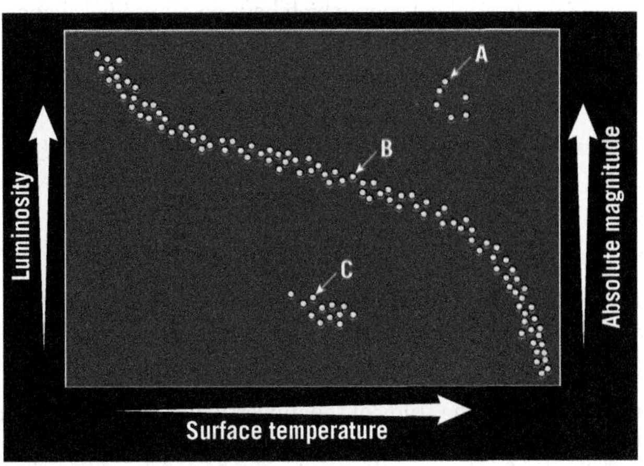

24.2 Stellar Evolution

List and describe the stages in the evolution of a typical Sun-like star.

KEY TERMS: nebula, molecular cloud, protostar, hydrogen fusion, planetary nebula, supernova

- Stars originate from the gravitational collapse of a molecular cloud. The energy released by collapse causes the cloud's center to become a hot, luminous protostar.
- The protostar becomes a star when its core reaches a temperature of about 10 million K, igniting hydrogen fusion by the proton–proton chain. Hydrogen fusion involves the conversion of four hydrogen nuclei into a single helium nucleus with the release of thermal nuclear energy.
- Two opposing forces act on a star: gravity, which tries to contract it into the smallest possible ball, and gas pressure (created by thermal nuclear energy), which tries to expand it. When the forces come into balance, the star becomes a stable main-sequence star.
- A star's main-sequence lifetime ends when the hydrogen fuel in the core is exhausted. In all but low-mass (red dwarf) stars, other types of nuclear fusion then cause the outer envelope to expand enormously (hundreds to thousands of times), making the star a red giant or supergiant. When a star exhausts all of its usable nuclear fuel, gravity takes over, and the stellar remnant collapses into a small, dense body.

24.3 Stellar Remnants

Compare and contrast the final state of Sun-like stars to the remnants of the most massive stars.

KEY TERMS: neutron star, pulsar, black hole

- The final fate of a star is determined primarily by its mass.
- Low-mass and intermediate-mass stars end up as white dwarfs. In the case of intermediate-mass stars such as our Sun, the newborn white dwarf is typically surrounded by an expanding clouds of glowing gas called a planetary nebula.
- Massive stars terminate in a brilliant explosion called a supernova. Supernova events can produce small, extremely dense neutron stars composed largely of neutrons, or smaller, even denser black holes—objects that have such immense gravity that light cannot escape from them.

24.4 Galaxies and Galaxy Clusters

List the three major types of galaxies. Explain the formation of large elliptical galaxies.

KEY TERMS: galaxy, spiral galaxy, barred spiral galaxy, elliptical galaxy, dwarf galaxy, irregular galaxy, galaxy cluster, Local Group

- The various types of galaxies include (1) irregular galaxies, which lack symmetry and account for about 25 percent of the known galaxies; (2) spiral galaxies, which are disk-shaped and have a greater concentration of stars near their centers and arms extending from their central nucleus; and (3) elliptical galaxies, which have an ellipsoidal shape and may be nearly spherical. Some spiral galaxies, called barred spirals, have a central bar connecting the spiral arms.
- Galaxies can be grouped into galaxy clusters, some containing thousands of galaxies. Our own, called the Local Group, contains at least 40 galaxies.

? What type of galaxy is shown in the accompanying image?

NASA

24.5 The Universe

Describe Edwin Hubble's discoveries about the nature of the universe and summarize the Big Bang theory.

KEY TERMS: cosmology, Big Bang theory, cosmological redshift, Hubble's law, dark matter, dark energy

- Cosmology is the study of the universe, including its properties, structure, and evolution.
- The universe consists of hundreds of billions of galaxies, most containing billions of stars.
- Evidence for an expanding universe came from the study of redshifts in the spectra of galaxies. Edwin Hubble concluded that the observed redshifts result from the expansion of space.
- The model that most accurately describes the birth and current state of the universe is the Big Bang theory. According to this model, the universe began about 13.8 billion years ago, in a cataclysmic explosion, and then it continued to expand, cool, and evolve to its current state.
- One question that remains is whether the universe will expand forever in a Big Chill or gravitationally contract in a Big Crunch. Dark matter works to slow the expansion of the universe, while dark energy exerts a force that pushes matter outward and causes the expansion to speed up. Most cosmologists favor an endless, ever-expanding universe.

GIVE IT SOME THOUGHT

1 Assume that NASA is sending space probes to each of the locations listed below. Arrange the order in which each probe will encounter its destination, *from nearest Earth to farthest*.
  a. Polaris (the North Star)
  b. A comet near the outer edge of our solar system
  c. Jupiter
  d. The far edge of the Milky Way Galaxy
  e. The Andromeda Galaxy
  f. The Sun

2 How a star evolves is closely related to its mass as a main-sequence star. Complete the accompanying diagram by labeling the evolutionary stages for the three groups of main-sequence stars shown.

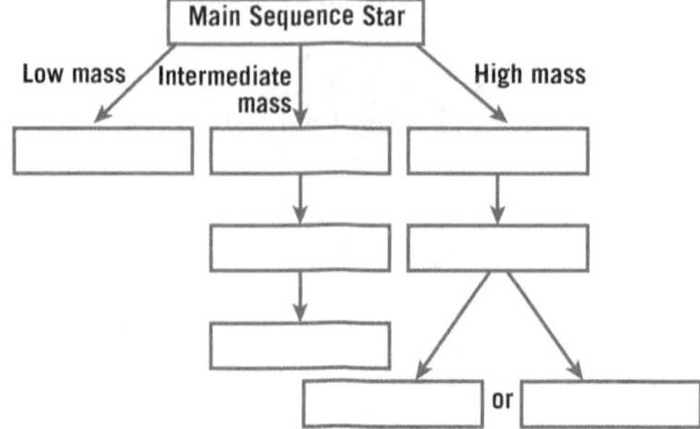

3 Use the information provided below about three main-sequence stars (A, B, and C) and Figure 24.2 to complete and explain your reasoning for the following:
- Star A has a main-sequence life span of 5 billion years.
- Star B has the same luminosity (absolute magnitude) as the Sun.
- Star C has a surface temperature of 3000K.

  a. Rank the mass of these stars from *greatest to least*.
  b. Rank the energy output of these stars from *greatest to least*.
  c. Rank the main-sequence life span of these stars from *longest to shortest*.

4 The masses of three clouds of gas and dust (nebulae) are provided below:
- Cloud A is 60 times the mass of the Sun.
- Cloud B is 7 times the mass of the Sun.
- Cloud C is 2 times the mass of the Sun.

Imagine that all the material in each cloud will collapse to form a single star. Use this information to answer the questions and explain your reasoning.
  a. Which cloud or clouds, if any, will evolve into a red main-sequence star?
  b. Which of the stars that will form from these clouds, if any, will reach the giant stage?
  c. Which of the stars that will form from these clouds, if any, will go through the supernova stage?

5 Refer to the accompanying photos of an elliptical galaxy and a spiral galaxy to answer the following:
   a. Which image (A or B) is an elliptical galaxy?
   b. Which of these galaxies appears to contain more young, hot, massive stars? How did you determine your answer?
   c. When stars are born from a cloud of dust and gases, high- and low-mass stars form at about the same time. Which group of stars, high-mass or low-mass, will die out first? Over time, how will this affect the color of the light we observe coming from this group of stars?

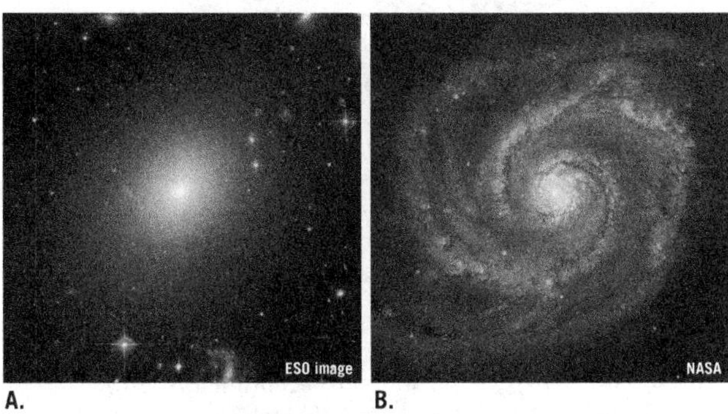

A. B.

EXAMINING THE EARTH SYSTEM

1 Briefly describe how the atmosphere, hydrosphere, geosphere, and biosphere are each related to the death of stars that occurred billions of years ago.

2 Scientists are continuously searching the Milky Way Galaxy for other stars that may have planets. What types of stars would most likely have a planet or planets suitable for life as we know it? (*Hint:* Do an Internet search.)

DATA ANALYSIS

Searching for Planets Around Other Stars

Our Sun has four rocky planets, four giant gas planets, five dwarf planets, more than 150 moons, and thousands of asteroids. There are billions of stars in the night sky, many of which have planetary systems of their own. A planet rotating around a distant star is called an exoplanet.

ACTIVITIES

Go to the NASA Exoplanet Exploration page at https://exoplanets.nasa.gov/. Hover over Explore and select 5 Ways to Find a Planet.

1 Describe each of the five ways astronomers use to detect an exoplanet.

Go to the Exoplanets Data Explorer page at http://exoplanets.org/. Read the information and then click on Table. Clicking on a column title will reorganize the data. Clicking on the units label just below the column title will change units for some columns, such as orbital period.

2 How many total confirmed planets are there? How many unconfirmed planets (Kepler candidates) are there?

3 Earth has a semi-major axis orbit of 1.0 AU. Which exoplanet has a semi-major axis orbit closest to Earth's orbital size?

4 Earth has an orbital period around our Sun of 365.24 days. Which exoplanet has an orbital period around its star most similar to Earth's orbital period?

The column labeled Msin(i) lists an estimate for the mass of each exoplanet in terms of its size compared to Jupiter ($m_{jupiter}$.) Click on $m_{jupiter}$ and change the mass unit to Earth Mass (m_{Earth}). A m_{earth} of 5.0 means that the exoplanet is 5 times more massive than Earth.

5 Which planet in the database has a mass most similar to that of Earth?

6 Find the orbital period and semi-major axis for this exoplanet. Is its orbital period longer or shorter than Earth's orbital period? Is its semi-major axis orbital size larger or smaller than Earth's orbital size?

7 This planet orbits a star with a temperature similar to Earth's. Would this planet be habitable for humans? Why or why not?

APPENDIX A

Metric and English Units Compared

Units

1 kilometer (km)	= 1000 meters (m)
1 meter (m)	= 100 centimeters (cm)
1 centimeter (cm)	= 0.39 inch (in.)
1 mile (mi)	= 5280 feet (ft)
1 foot (ft)	= 12 inches (in.)
1 inch (in.)	= 2.54 centimeters (cm)
1 square mile (mi^2)	= 640 acres (a)
1 kilogram (kg)	= 1000 grams (g)
1 pound (lb)	= 16 ounces (oz)
1 fathom	= 6 feet (ft)

Conversions

Length

When you want to convert:	multiply by:	to find:
inches	2.54	centimeters
centimeters	0.39	inches
feet	0.30	meters
meters	3.28	feet
yards	0.91	meters
meters	1.09	yards
miles	1.61	kilometers
kilometers	0.62	miles

Area

When you want to convert:	multiply by:	to find:
square inches	6.45	square centimeters
square centimeters	0.15	square inches
square feet	0.09	square meters
square meters	10.76	square feet
square miles	2.59	square kilometers
square kilometers	0.39	square miles

Volume

When you want to convert:	multiply by:	to find:
cubic inches	16.38	cubic centimeters
cubic centimeters	0.06	cubic inches
cubic feet	0.028	cubic meters
cubic meters	35.3	cubic feet
cubic miles	4.17	cubic kilometers
cubic kilometers	0.24	cubic miles
liters	1.06	quarts
liters	0.26	gallons
gallons	3.78	liters

Masses and Weights

When you want to convert:	multiply by:	to find:
ounces	28.35	grams
grams	0.035	ounces
pounds	0.45	kilograms
kilograms	2.205	pounds

Temperature

When you want to convert degrees Fahrenheit (°F) to degrees Celsius (°C), subtract 32 degrees and divide by 1.8.

When you want to convert degrees Celsius (°C) to degrees Fahrenheit (°F), multiply by 1.8 and add 32 degrees.

When you want to convert degrees Celsius (°C) to kelvins (K), delete the degree symbol and add 273. When you want to convert kelvins (°C), add the degree symbol and subtract 273.

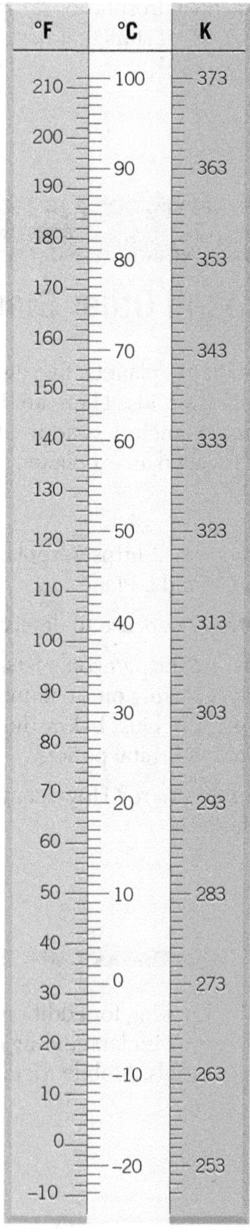

Figure A.1
Temperature scales

APPENDIX B

Relative Humidity and Dew-Point Tables

TABLE B.1 Relative Humidity (Percent)*

Dry bulb (°C)	Depression of Web-Bulb (Dry-bulb temperature − Wet-bulb temperature = Depression of the wet bulb)																					
	1	2	3	4	5	6	7	8	9	10	11	12	13	14	15	16	17	18	19	20	21	22
−20	28																					
−18	40																					
−16	48	0																				
−14	55	11																				
−12	61	23																				
−10	66	33	0																			
−8	71	41	13																			
−6	73	48	20	0																		
−4	77	54	32	11																		
−2	79	58	37	20	1																	
0	81	63	45	28	11																	
2	83	67	51	36	20	6																
4	85	70	56	42	27	14																
6	86	72	59	46	35	22	10	0														
8	87	74	62	51	39	28	17	6														
10	88	76	65	54	43	38	24	13	4													
12	88	78	67	57	48	38	28	19	10	2												
14	89	79	69	60	50	41	33	25	16	8	1											
16	90	80	77	62	54	45	37	29	21	14	7	1										
18	91	81	72	64	56	48	40	33	26	19	12	6	0									
20	91	82	74	66	58	51	44	36	30	23	17	11	5									
22	92	83	75	68	60	53	46	40	33	27	21	15	10	4	0							
24	92	84	76	69	62	55	49	42	36	30	25	20	14	9	4	0						
26	92	85	77	70	64	57	51	45	39	34	28	23	18	13	9	5						
28	93	86	78	71	65	59	53	45	42	36	31	26	21	17	12	8	4					
30	93	86	79	72	66	61	55	49	44	39	34	29	25	20	16	12	8	4				
32	93	86	80	73	68	62	56	51	46	41	36	32	27	22	19	14	11	8	4			
34	93	86	81	74	69	63	58	52	48	43	38	34	30	26	22	18	14	11	8	5		
36	94	87	81	75	69	64	59	54	50	44	40	36	32	28	24	21	17	13	10	7	4	
38	94	87	82	76	70	66	60	55	51	46	42	38	34	30	26	23	20	16	13	10	7	5
40	94	89	82	76	71	67	61	57	52	48	44	40	36	33	29	25	22	19	16	13	10	7

Dry-Bulb (Air) Temperature

Relative Humidity Values

* To determine the relative humidity, find the air (dry-bulb) temperature on the vertical axis (far left) and the depression of the wet bulb on the horizontal axis (top). Where the two meet, the relative humidity is found. For example, when the dry-bulb temperature is 20°C and a wet-bulb temperature is 14°C, then the depression of the wet bulb is 6°C (20°C−14°C). From Table B.1, the relative humidity is 51 percent and from Table B.2, the dew point is 10°C.

TABLE B.2 Dew-Point Temperature (°C)

Dry bulb (°C)	Depression of Web-Bulb (Dry-bulb temperature − Wet-bulb temperature = Depression of the wet bulb)																					
	1	2	3	4	5	6	7	8	9	10	11	12	13	14	15	16	17	18	19	20	21	22
−20	−33																					
−18	−28																					
−16	−24																					
−14	−21	−36																				
−12	−18	−28																				
−10	−14	−22																				
−8	−12	−18	−29																			
−6	−10	−14	−22																			
−4	−7	−12	−17	−29																		
−2	−5	−8	−13	−20																		
0	−3	−6	−9	−15	−24																	
2	−1	−3	−6	−11	−17																	
4	1	−1	−4	−7	−11	−19																
6	4	1	−1	−4	−7	−13	−21															
8	6	3	1	2	−5	−9	−14															
10	8	6	4	1	−2	−5	−9	−14	−18													
12	10	8	6	4	1	−2	−5	−9	−16													
14	12	11	9	6	4	1	−2	−5	−10	−17												
16	14	13	11	9	7	4	1	−1	−6	−10	−17											
18	16	15	13	11	9	7	4	2	−2	5	10	−19										
20	19	17	15	14	12	10	7	4	2	−2	−5	−10	−19									
22	21	19	17	16	14	12	10	8	5	3	−1	−5	−10	−19								
24	23	21	20	18	16	14	12	10	8	6	2	−1	−5	−10	−18							
26	25	23	22	20	18	17	15	13	11	9	6	3	0	−4	−9	−18						
28	27	25	24	22	27	19	17	16	14	11	9	7	4	1	−3	−9	16					
30	29	27	26	24	23	21	19	18	16	14	12	10	8	5	1	−2	−8	−15				
32	31	29	28	27	25	24	22	21	19	17	15	13	11	8	5	2	−2	−7	−14			
34	33	31	30	29	27	26	24	23	21	20	18	16	14	12	9	6	3	−1	−5	−12	−29	
36	35	33	32	31	29	28	27	25	24	22	20	19	17	15	13	10	7	4	0	−4	−10	
38	37	35	34	33	32	30	29	28	26	25	23	21	19	17	15	13	11	8	5	1	−3	9
40	39	37	36	35	34	32	31	30	28	27	25	24	22	20	18	16	14	12	9	6	2	−2

Dry-Bulb (Air) Temperature

Dew-Point Values

GLOSSARY

A

Aa flow A type of lava flow that has a jagged, blocky surface.

Abrasion The grinding and scraping of a rock surface by the friction and impact of rock particles carried by water, wind, or ice.

Absolute instability Air that has a lapse rate greater than the dry adiabatic rate.

Absolute magnitude The apparent brightness of a star if it were viewed from a distance of 10 parsecs (32.6 light-years). Used to compare the true brightness of stars (also called their *luminosity*).

Absolute stability Refers to air with a lapse rate less than the wet adiabatic rate.

Absorption line spectrum A continuous spectrum with dark absorption lines superimposed.

Abyssal plain A very level area of the deep-ocean floor, usually lying at the foot of the continental rise.

Abyssal zone A subdivision of the benthic zone characterized by extremely high pressures, low temperatures, low oxygen, few nutrients, and no sunlight.

Accretionary wedge A large wedge-shaped mass of sediment that accumulates in subduction zones. Here, sediment is scraped from the subducting oceanic plate and accreted to the overriding crustal block.

Active continental margin A portion of the seafloor adjacent to the continents that is usually narrow and consisting of highly deformed sediments. These margins occur where oceanic lithosphere is being subducted beneath the margin of a continent.

Active optics On a telescope, a system that uses many computer-controlled motors to compensate for mirror distortions caused by gravity, wind, etc.

Adaptive optics On a telescope, a system that uses many computer-controlled motors to change the shape of the telescope mirror so as to cancel out, in real time, the effects of atmospheric turbulence.

Adiabatic temperature change Cooling or warming of air caused when air is allowed to expand or is compressed, not because heat is added or subtracted.

Advection fog A fog formed when warm, moist air blows over a cool surface.

Aerosols Tiny solid and liquid particles suspended in the atmosphere.

Aftershocks Smaller earthquakes that follow a main earthquake.

Air A mixture of many discrete gases, of which nitrogen and oxygen are most abundant and in which varying quantities of tiny solid and liquid particles are suspended.

Air mass A large body of air that is characterized by relatively homogeneous temperature and humidity.

Air pressure The force exerted by the weight of a column of air above a given point.

Air-mass weather The conditions experienced in an area as an air mass passes over it. Because air masses are large and fairly homogenous, air-mass weather is fairly constant and may last for several days.

Albedo The reflectivity of a substance, usually expressed as a percentage of the incident radiation reflected.

Alluvial fan A fan-shaped deposit of sediment formed when a stream's slope is abruptly reduced.

Alluvium Unconsolidated sediment deposited by a stream.

Alpine glacier A glacier confined to a mountain valley, which in most instances had previously been a stream valley. Also called a *valley glacier*.

Altitude In measuring the position of objects in the sky (such as a star or planet), the object's height above the horizon, measured in degrees. Used with the direction to the object along the horizon.

Altocumulus A type of cumulus clouds that form at medium altitude (2000–6000 meters [6500–20,000 feet]). Altocumulus clouds are white to gray "sheep-back" clouds, often composed of separate globules.

Altostratus A type of stratus clouds that form at medium altitude (2000–6000 meters [6500–20,000 feet]). Altostratus clouds are generally thin and may produce light precipitation.

Ambiguous property A property of a mineral that is not diagnostic because it varies among different specimens of the mineral.

Amniotic egg A fluid-filled, shelled egg that can be laid on land. Amniotic eggs are among the adaptations that enabled early reptiles to occupy environments not open to amphibians.

Amphibian A vertebrate such as a frog or salamander that has legs but must lay its eggs in water.

Andesite A gray, fine-grained igneous rock, primarily of volcanic origin and commonly exhibiting a porphyritic texture.

Andesitic composition A compositional group of igneous rocks that contains at least 25 percent dark silicate minerals. The other dominant mineral is plagioclase feldspar. Also called *intermediate composition*.

Aneroid barometer An instrument for measuring air pressure that consists of evacuated metal chambers that are very sensitive to variations in air pressure.

Angiosperm A flowering plant in which fruits contain the seeds.

Angle of repose The steepest angle at which loose material remains stationary, without sliding downslope.

Angular diameter The angle that an object appears to span in your field of view. Used to describe the apparent size of astronomical objects. Also called *angular size*.

Angular distance The angle that appears to separate two objects in the sky. Used to describe the apparent separation of astronomical objects.

Angular size The angle that an object appears to span in your field of view. Used to describe the apparent size of astronomical objects. Also called *angular diameter*.

Angular unconformity An unconformity in which the strata below dip at an angle different from that of the beds above.

Annual mean With respect to air temperature, the average of the 12 monthly temperature means.

Annual temperature range The difference between the highest and lowest monthly temperature means.

Anticline A fold in sedimentary strata that resembles an arch; the opposite of *syncline*.

Anticyclone A high-pressure center characterized by a clockwise flow of air in the Northern Hemisphere. Also called a *high*.

Aphelion The place in the orbit of a planet where the planet is farthest from the Sun.

Aphotic zone The portion of the ocean where there is no sunlight.

Apparent magnitude The brightness of a star when viewed from Earth.

Aquifer Rock or soil through which groundwater moves easily.

Aquitard Impermeable beds that hinder or prevent groundwater movement.

Archean The second eon of Precambrian time, following the Hadean and preceding the Proterozoic. It extends between 3.8 billion and 2.5 billion years before the present.

Arctic (A) air mass A bitterly cold air mass that forms over the frozen Arctic Ocean.

Arête A narrow knifelike ridge separating two adjacent glaciated valleys.

Arid One of the two types of dry climate; the driest of the dry climates. Also called *desert*.

Artesian system A system in which groundwater under pressure rises above the level of the aquifer.

Asteroid belt The region in which most asteroids orbit the Sun between Mars and Jupiter.

Asteroids Thousands of small planetlike bodies, ranging in size from a few hundred kilometers to less than a kilometer, whose orbits lie mainly between those of Mars and Jupiter.

Asthenosphere A subdivision of the mantle situated below the lithosphere. This zone of weak material exists below a depth of about 100 kilometers (60 miles) and in some regions extends as deep as 700 kilometers (430 miles). The rock within this zone is easily deformed.

Astronomical unit (AU) The average distance from Earth to the Sun; 1.5×10^8 kilometers (93×10^6 miles).

Astronomy The scientific study of the universe, which includes the observation and interpretation of celestial bodies and phenomena.

Atmosphere The gaseous portion of a planet; the planet's envelope of air. One of the traditional subdivisions of Earth's physical environment.

Atom The smallest particle that exists as an element.

Atomic number The number of protons in the nucleus of an atom.

Augite A black, opaque silicate mineral of the pyroxene group that is a dominant component of basalt.

Aurora A bright display of ever-changing light caused by solar radiation interacting with the upper atmosphere in the region of the poles.

Autumnal equinox The equinox that occurs on September 21–23 in the Northern Hemisphere and on March 21–22 in the Southern Hemisphere. Also called the *fall equinox*.

B

Backshore The inner portion of the shore, lying landward of the high-tide shoreline. It is usually dry, being affected by waves only during storms.

Back swamp A poorly drained area on a floodplain that results when natural levees are present.

Bajada An apron of sediment along a mountain front created by the coalescence of alluvial fans.

Banded iron formations A finely layered iron and silica-rich (chert) layer deposited mainly during the Precambrian.

Bar The common term for sand and gravel deposits in a stream channel.

Barchan dune A solitary sand dune shaped like a crescent with its tips pointing downward.

Barchanoid dune A type of dune in which the dunes form scalloped rows of sand oriented at right angles to the wind. This form is intermediate between isolated barchans and extensive waves of transverse dunes.

Barograph A recording barometer.

Barometric tendency The nature of the change in atmospheric pressure over the past several hours. It can be a useful aid in short-range weather prediction. Also called *pressure tendency*.

Barred spiral galaxy A galaxy that has straight arms extending from its nucleus.

Barrier island A low, elongate ridge of sand that parallels the coast.

Basalt A fine-grained igneous rock of mafic composition.

Basalt plateau The broad and extensive accumulation of lava from a succession of flows emanating from fissure eruptions.

Basaltic composition A compositional group of igneous rocks indicating that the rock contains substantial dark silicate minerals and calcium-rich plagioclase feldspar. Also called *mafic composition*.

Base level The level below which a stream cannot erode.

Basin A circular downfolded structure.

Batholith A large mass of igneous rock that formed when magma was emplaced at depth, crystallized, and subsequently exposed by erosion.

Bathymetry The measurement of ocean depths and the charting of the shape or topography of the ocean floor.

Baymouth bar A sandbar that completely crosses a bay, sealing it off from the open ocean.

Beach An accumulation of sediment found along the landward margin of the ocean or a lake.

Beach drift The transport of sediment in a zigzag pattern along a beach caused by the uprush of water from obliquely breaking waves.

Beach face The wet, sloping surface that extends from the berm to the shoreline.

Beach nourishment The process by which large quantities of sand are added to the beach system to offset losses caused by wave erosion.

Bed load Sediment that is carried by a stream along the bottom of its channel.

Beds Parallel layers of sedimentary rock. Also called *strata*.

Benthic zone The marine life zone that includes *any* seafloor surface, regardless of its distance from shore.

Benthos The forms of marine life that live on or in the ocean bottom.

Bergeron process A theory that relates the formation of precipitation to supercooled clouds, freezing nuclei, and the different saturation levels of ice and liquid water.

Berm The dry, gently sloping zone on the backshore of a beach at the foot of the coastal cliffs or dunes.

Big Bang theory The theory which proposes that the universe originated as a single mass, which subsequently exploded.

Biochemical sedimentary rock Sediment that forms when material dissolved in water is precipitated by water-dwelling organisms. Shells are common examples.

Biogenous sediment Seafloor sediments consisting of material of marine-organic origin.

Biomass The total mass of a defined organism or group of organisms in a particular area or ecosystem.

Biosphere The totality of life on Earth; the parts of the solid Earth, hydrosphere, and atmosphere in which living organisms can be found.

Biotite A dark iron-rich mineral and a member of the mica family that has excellent cleavage.

Black carbon Soot generated by combustion processes and fires.

Black hole A massive star that has collapsed to such a small volume that its gravity prevents the escape of all radiation.

Blowout A depression excavated by the wind in easily eroded deposits.

Body waves Seismic waves that travel through Earth's interior.

Bowen's reaction series A concept proposed by N. L. Bowen that illustrates the relationships between magma and the minerals crystallizing from it during the formation of igneous rocks.

Braided channel A stream channel consisting of numerous intertwining channels.

Breakwater A structure that protects a near-shore area from breaking waves.

Breccia A sedimentary rock composed of angular fragments that were lithified.

Brittle deformation Deformation that involves the fracturing of rock. It is associated with rocks near the surface.

Building material A category of nonsilicate mineral resource used in building.

C

Cactolith A quasi-horizontal chonolith composed of anastomosing ductoliths, whose distal ends curl like a harpolith, thin like a sphenolith, or bulge discordantly like an akmolith or ethmolith.

Calcite Calcium carbonate ($CaCO_3$), one of the two most common carbonate minerals.

Caldera A large depression typically caused by collapse or ejection of the summit area of a volcano.

Calorie The amount of heat required to raise the temperature of 1 gram of water 1°C.

Calving Wastage of a glacier that occurs when large pieces of ice break off into water.

Cambrian explosion The huge expansion in biodiversity that occurred at the beginning of the Paleozoic era.

Cap rock A necessary part of an oil trap which is impermeable and hence keeps upwardly mobile oil and gas from escaping at the surface.

Capacity The total amount of sediment a stream is able to transport.

Carbonic acid A weak acid formed when carbon dioxide is dissolved in water. It plays an important role in chemical weathering.

Catastrophism The concept that Earth was shaped by catastrophic events of a short-term nature.

Celestial sphere An imaginary hollow sphere on which the ancients believed the stars were hung and carried around Earth.

Cementation One way in which sedimentary rocks are lithified. As material precipitates from water that percolates through the sediment, open spaces are filled, and particles are joined into a solid mass.

Cenozoic era A span on the geologic time scale beginning about 65 million years ago following the Mesozoic era.

Chemical bond A strong attractive force that exists between atoms in a substance. It involves the transfer or sharing of electrons that allows each atom to attain a full valence shell.

Chemical compound A substance formed by the chemical combination of two or more elements in definite proportions and usually having properties different from those of its constituent elements.

Chemical sedimentary rock Sedimentary rock consisting of material that was precipitated from water by either inorganic or organic means.

Chemical weathering The processes by which the internal structure of a mineral is altered by the removal or addition of elements.

Chinook A wind blowing down the leeward side of a mountain and warming by compression.

Chromosphere The first layer of the solar atmosphere, directly above the photosphere.

Cinder cone A rather small volcano built primarily of pyroclastics ejected from a single vent. Also called a *scoria cone*.

Circle of illumination The great circle that separates daylight from darkness.

Circular orbital motion A reference to the movement of water in a wave. As a wave travels, energy is passed along by moving in a circle. The waveform advances, but the water does not advance appreciably.

Circum-Pacific belt An area approximately 40,000 kilometers (24,000 miles) in length surrounding the basin of the Pacific Ocean where oceanic lithosphere is continually subducted beneath the surrounding continental plates, causing most of Earth's largest earthquakes.

Cirque An amphitheater-shaped basin at the head of a glaciated valley produced by frost wedging and plucking.

Cirrocumulus A type of cumulus clouds that form at high altitude (above 6000 meters [20,000 feet]). Cirrocumulus clouds are thin sheets of white ice-crystal cloud that may give the sky a milky look.

Cirrostratus A type of stratus cloud that forms at high altitude (above 6000 meters [20,000 feet]). Cirrostratus clouds are thin, delicate, fibrous ice-crystal clouds, such as mares' tails.

Cirrus One of three basic cloud forms; also one of the three high cloud types. Cirrus clouds are thin, delicate ice-crystal clouds often appearing as veil-like patches or thin, wispy fibers.

Clay A group of light-colored silicates that typically form as products of chemical weathering of igneous rocks. It is a major component of soil and sedimentary rocks. Kaolinite is a common clay mineral derived from the weathering of feldspar.

Cleavage The tendency of a mineral to break along planes of weak bonding.

Climate A description of aggregate weather conditions; the sum of all statistical weather information that helps describe a place or region.

Climate system The exchanges of energy and moisture that occur among the atmosphere, hydrosphere, solid Earth, biosphere, and cryosphere.

Climate-feedback mechanism Several different possible outcomes that may result when one of the atmosphere's elements is altered.

Cloud A form of condensation best described as a dense concentration of suspended water droplets or tiny ice crystals.

Cloud condensation nuclei Microscopic particles that serve as surfaces on which water vapor condenses.

Clouds of vertical development Clouds that have their bases in the low-height range but extend upward into the middle or high altitudes.

Coal A sedimentary rock consisting primarily of organic matter, formed in stages from accumulations of large quantities of undecayed plant material. It is used as a fossil fuel.

Coarse-grained texture An igneous rock texture in which the crystals are roughly equal in size and large enough that individual minerals can be identified with the unaided eye.

Coast A strip of land that extends inland from the coastline as far as ocean-related features can be found.

Coastline The coast's seaward edge. The landward limit of the effect of the highest storm waves on the shore.

Cold front A front along which a cold air mass thrusts beneath a warmer air mass.

Collisional mountains Mountains in which compressive horizontal forces have shortened and thickened the crust. Most major mountain belts are of this type.

Collision–coalescence process A theory of raindrop formation in warm clouds (above 0°C) in which large cloud droplets (giants) collide and join together with smaller droplets to form a raindrop. Opposite electrical charges may bind the cloud droplets together.

Color A phenomenon of light by which otherwise identical objects may be differentiated.

Columnar jointing A pattern of cracks that form during cooling of molten rock to generate columns that are generally six sided.

Coma The fuzzy, gaseous component of a comet's head.

Comet A small body that generally revolves about the Sun in an elongated orbit.

Compaction A type of lithification in which the weight of overlying material compresses more deeply buried sediment. It is most important in fine-grained sedimentary rocks such as shale.

Competence A measure of the largest particle a stream can transport; a factor that is dependent on velocity.

Composite volcano A volcano composed of both lava flows and pyroclastic material. Also called a *stratovolcano*.

Compressional stress Differential stress that shortens a rock body.

Concordant A term used to describe intrusive igneous masses that form parallel to the bedding of the surrounding rock.

Condensation The change of state from a gas to a liquid.

Condensation level The height at which rising air that is cooling at the dry adiabatic rate becomes saturated and condensation begins. Also called the *lifting condensation level*.

Conditional instability Moist air with a lapse rate between the dry and wet adiabatic rates.

Conduction The transfer of heat through matter by molecular activity. Energy is transferred through collisions from one molecule to another.

Conduit A pipelike opening through which magma moves toward Earth's surface. It terminates at a surface opening called a *vent*.

Cone of depression A cone-shaped depression in the water table immediately surrounding a well.

Confined aquifer An aquifer that has impermeable layers (aquitards) both above and below.

Confining pressure Stress that is applied uniformly in all directions.

Conformable Featuring layers of rock that were deposited without interruption.

Conglomerate A sedimentary rock composed of rounded, gravel-size particles.

Constellation An apparent group of stars originally named for mythical characters. The sky is presently divided into 88 constellations.

Contact metamorphism Changes in rock caused by the heat from a nearby magma body.

Continent Large, continuous areas of land that include the adjacent continental shelf and islands that are structurally connected to the mainland.

Continental (c) air mass An air mass that forms over land; it is normally relatively dry.

Continental drift A theory which originally proposed that the continents are rafted about. It has essentially been replaced by the plate tectonics theory.

Continental margin The portion of the seafloor adjacent to the continents. It may include the continental shelf, continental slope, and continental rise.

Continental rift A linear zone along which continental lithosphere stretches and pulls apart. Its creation may mark the beginning of a new ocean basin.

Continental rise The gently sloping surface at the base of the continental slope.

Continental shelf The gently sloping submerged portion of the continental margin, extending from the shoreline to the continental slope.

Continental slope The steep gradient that leads to the deep-ocean floor and marks the seaward edge of the continental shelf.

Continental volcanic arc Mountains formed in part by igneous activity associated with the subduction of oceanic lithosphere beneath a continent.

Continuous spectrum An uninterrupted band of light emitted by an incandescent solid, liquid, or gas under pressure.

Convection The transfer of heat by the movement of a mass or substance. It can take place only in fluids.

Convection zone With respect to the Sun or another star, the radial zone in which heat is conducted outward mainly by the convective motion of hot gases.

Convective lifting Unequal surface heating that causes localized pockets of air (thermals) to rise because of their buoyancy. Also called *localized convective lifting*.

Convergence The condition that exists when the distribution of winds in a given area results in a net horizontal inflow of air into the area. Because convergence at lower levels is associated with an upward movement of air, areas of convergent winds are regions favorable to cloud formation and precipitation.

Convergent plate boundary A boundary in which two plates move together, resulting in oceanic lithosphere being thrust beneath an overriding plate, eventually to be reabsorbed into the mantle. It can also involve the collision of two continental plates to create a mountain system. Also called a *subduction zone*.

Coquina A coarse rock composed of loosely cemented shells and shell fragments.

Core The innermost layer of Earth, located beneath the mantle. The core is divided into an outer core and an inner core.

Coriolis effect The deflective force of Earth's rotation on all free-moving objects, including the atmosphere and oceans. Deflection is to the right in the Northern Hemisphere and to the left in the Southern Hemisphere.

Corona The outer, tenuous layer of the solar atmosphere.

Coronal mass ejection A huge cloud of charged particles that is ejected from the Sun's corona. When a coronal mass ejection encounters Earth, it can cause a damaging magnetic storm in the upper atmosphere.

Correlation The process of establishing the equivalence of rocks of similar age in different areas.

Cosmological redshift Changes in the spectra of galaxies which indicate that they are moving away from the Milky Way as a result of the expansion of space.

Cosmology The study of the universe.

Country rock Preexisting crustal rocks intruded by magma. Country rock may be displaced or assimilated by magmas. Also called *host rock*.

Covalent bond A chemical bond produced by the sharing of electrons.

Crater The depression at the summit of a volcano or a depression that is produced by a meteorite impact.

Craton The part of the continental crust that has attained stability; that is, it has not been affected by significant tectonic activity during the Phanerozoic eon. It consists of the shield and stable platform.

Creep The slow downhill movement of soil and regolith.

Crevasse A deep crack in the brittle surface of a glacier.

Cross bed A structure in which relatively thin layers are inclined at an angle to the main bedding. It is formed by currents of wind or water.

Crust The very thin outermost layer of Earth.

Cryosphere The portion of Earth's surface where water is in solid form, including snow, glaciers, sea ice, freshwater ice, and frozen ground. It is one of the spheres of the climate system.

Cryovolcanism A type of volcanism that results from the eruption of magmas derived from the partial melting of ice.

Crystal settling During the crystallization of magma, the settling of the earlier-formed minerals that are denser than the liquid portion to the bottom of the magma chamber.

Crystal shape Refers to the common or characteristic shape of a crystal or an aggregate of crystals. Also called *crystal habit*.

Cumulonimbus A type of towering cloud of vertical development. Cumulonimbus clouds are associated with heavy rainfall, thunder, lightning, hail, and tornadoes.

Cumulus One of three basic cloud forms; also the name given one of the clouds of vertical development. Cumulus are billowy individual cloud masses that often have flat bases.

Cup anemometer An instrument used to determine wind speed.

Curie point The temperature above which a material loses its magnetization.

Cut bank The area of active erosion on the outside of a meander.

Cutoff A short channel segment created when a river erodes through the narrow neck of land between meanders.

Cyclone A low-pressure center characterized by a counterclockwise flow of air in the Northern Hemisphere. Also called a *low*.

D

Daily mean temperature The mean temperature for a day, which is determined by averaging the hourly readings or, more commonly, by averaging the maximum and minimum temperatures for a day.

Daily range The difference between the maximum and minimum temperatures for a day.

Dark energy A hypothetical form of energy that produces a force that opposes gravity and is thought to be the cause of the accelerating expansion of the universe.

Dark matter Undetected matter that is thought to exist in great quantities in the universe.

Dark silicate mineral A silicate mineral that contains ions of iron and/or magnesium in its structure. Dark silicates are dark in color and have a higher specific gravity than light silicates.

Debris flow A relatively rapid type of mass wasting that involves a flow of soil and regolith containing a large amount of water. Also called a *mudflow*.

Decompression melting Melting that occurs as rock ascends due to a drop in confining pressure.

Deep-ocean basin The portion of seafloor that lies between the continental margin and the oceanic ridge system. This region comprises almost 30 percent of Earth's surface.

Deep-ocean trench An elongated depression in the seafloor produced by bending of oceanic crust during subduction. Also called simply a *trench*.

Deep-sea fan A cone-shaped deposit at the base of the continental slope. The sediment is transported to the fan by turbidity currents that follow submarine canyons.

Deflation The lifting and removal of loose material by wind.

Deformation A general term for the processes of folding, faulting, shearing, compression, or extension of rocks as the result of various natural forces.

Delta An accumulation of sediment formed where a stream enters a lake or an ocean.

Dendritic pattern A stream system that resembles the pattern of a branching tree.

Density Mass per unit volume of a substance, usually expressed as grams per cubic centimeter (g/cm^3).

Deposition The process by which water vapor is changed directly to a solid without passing through the liquid state.

Desert One of the two types of dry climate; the driest of the dry climates. Also called *arid*.

Desert pavement A layer of coarse pebbles and gravel created when wind removes the finer material.

Detachment fault A nearly horizontal fault that may extend hundreds of kilometers below the surface. Such a fault represents a boundary between rocks that exhibit ductile deformation and rocks that exhibit brittle deformation.

Detrital sedimentary rock Rock formed from the accumulation of material that originated and was transported in the form of solid particles derived from both mechanical and chemical weathering.

Dew point The temperature to which air has to be cooled in order to reach saturation. Also called *dew-point temperature*.

Dew-point temperature The temperature to which air has to be cooled in order to reach saturation. Also called simply *dew point*.

Diagnostic property A property of a mineral that aids in mineral identification. Taste, feel, crystal shape, and streak are examples of diagnostic properties.

Differential stress Forces that are unequal in different directions.

Differential weathering The variation in the rate and degree of weathering caused by factors such as mineral makeup, degree of jointing, and climate.

Diffused light Solar energy scattered and reflected in the atmosphere that reaches Earth's surface in the form of diffuse blue light from the sky.

Dike A tabular-shaped intrusive igneous feature that cuts through the surrounding rock.

Diorite A coarse-grained intrusive igneous rock primarily composed of plagioclase feldspar and amphibole minerals.

Dip-slip fault A fault in which the movement is parallel to the dip of the fault.

Direction The direction along the horizon to an object, usually measured in degrees clockwise from due north. Used along with the altitude above the horizon to describe the location of astronomical objects.

Discharge The quantity of water in a stream that passes a given point in a period of time.

Disconformity A type of unconformity in which the beds above and below are parallel.

Discordant A term used to describe plutons that cut across existing rock structures, such as bedding planes.

Disseminated deposit Any economic mineral deposit in which the desired mineral occurs as scattered particles in the rock but in sufficient quantity to make the deposit an ore.

Dissolved load The portion of a stream's load that is carried in solution.

Distributary A section of a stream that leaves the main flow.

Diurnal tidal pattern A tidal pattern that features one high tide and one low tide during a tidal day; a daily tide.

Divergence The condition that exists when the distribution of winds in a given area results in a net horizontal outflow of air from the region. In divergence at lower levels, the resulting deficit is compensated for by a downward movement of air from aloft; hence, areas of divergent winds are unfavorable to cloud formation and precipitation.

Divergent plate boundary A region where the rigid plates are moving apart, typified by the mid-ocean ridges. Also called a *spreading center*.

Divide An imaginary line that separates the drainage of two streams; often found along a ridge.

Dolomite Calcium/magnesium carbonate, $CaMg(CO_3)_2$, one of the two most common carbonate minerals.

Dome A roughly circular upfolded structure similar to an anticline.

Doppler effect The apparent change in wavelength of radiation caused by the relative motions of the source and the observer.

Doppler radar In addition to performing the tasks of conventional radar, a new generation of weather radar that can detect motion directly and hence greatly improve tornado and severe storm warnings.

Drainage basin The land area that contributes water to a stream. Also called a *watershed*.

Drawdown The difference in height between the bottom of a cone of depression and the original height of the water table.

Drizzle Precipitation from stratus clouds consisting of tiny droplets.

Drumlin A streamlined asymmetrical hill composed of glacial till. The steep side of the hill faces the direction from which the ice advanced.

Dry adiabatic rate The rate of adiabatic cooling or warming in unsaturated air. The rate of temperature change is 1°C per 100 meters.

Dry climate A climate in which yearly precipitation is not as great as the potential loss of water by evaporation.

Dry-summer subtropical climate A climate located on the west sides of continents between latitudes 30° and 45°. It is the only humid climate with a strong winter precipitation maximum.

Ductile deformation A type of solid state flow that produces a change in the size and shape of a rock body without fracturing. It occurs at depths where temperatures and confining pressures are high.

Dune A hill or ridge of wind-deposited sand.

Dwarf galaxy Very small galaxies, usually elliptical and lacking spiral arms.

Dwarf planets Celestial bodies that orbit stars and are massive enough to be spherical but have not cleared their neighboring regions of planetesimals.

E

Earth science The name for all the sciences that collectively seek to understand Earth. It includes geology, oceanography, meteorology, and astronomy.

Earth system Earth viewed as a dynamic system of interacting parts and processes, including the geosphere, hydrosphere, biosphere, and atmosphere.

Earth system science An interdisciplinary study that seeks to examine Earth as a system composed of numerous interacting parts or subsystems.

Earthflow The downslope movement of water-saturated, clay-rich sediment. Most characteristic of humid regions.

Earthquake Vibration of Earth produced by the rapid release of energy.

Echo sounder An instrument used to determine the depth of water by measuring the time interval between emission of a sound signal and the return of its echo from the bottom.

Ecliptic The yearly path of the Sun plotted against the background of stars.

Economic mineral A concentration of a mineral resource or reserve that can be profitably extracted from Earth.

Effusive eruption A quiescent eruption that produces mainly outpourings of fluid lava.

El Niño The name given to the periodic warming of the ocean that occurs in the central and eastern Pacific. A major El Niño episode can cause extreme weather in many parts of the world. The opposite of *La Niña*.

Elastic deformation Rock deformation in which the rock returns to nearly its original size and shape when the stress is removed.

Elastic rebound The sudden release of stored strain in rocks that results in movement along a fault.

Electromagnetic radiation Transfer of energy in the form of light and related types of radiation, including gamma rays, x-rays, ultraviolet light, infrared light, microwaves, and radio waves.

Electron A negatively charged subatomic particle that has a negligible mass and is found outside an atom's nucleus.

Elements With respect to weather and climate, quantities or properties of the atmosphere that are measured regularly and that are used to express the nature of weather and climate.

Elliptical galaxy A galaxy that is round or elliptical in outline. It contains little gas and dust, no disk or spiral arms, and few hot, bright stars.

Eluviation The washing out of fine soil components from the horizon by downward-percolating water.

Emergent coast A coast where land that was formerly below sea level has been exposed either because of crustal uplift or a drop in sea level or both.

Emission line spectrum The bright lines produced by an incandescent gas under low pressure.

End moraine A ridge of till marking a former position of the front of a glacier.

Endothermic The ability of an animal to maintain a constant body temperature through metabolic activity. Birds and mammals are endothermic.

Enhanced Fujita intensity scale (EF-scale) A scale originally developed by T. Theodore Fujita for classifying the severity of a tornado, based on the correlation of wind speed with the degree of destruction.

Environmental lapse rate The rate of temperature decrease with increasing height in the troposphere.

Eon The largest time unit on the geologic time scale, next in order of magnitude above era.

Ephemeral stream A stream that is usually dry because it carries water only in response to specific episodes of rainfall. Most desert streams are of this type.

Epicenter The location on Earth's surface that lies directly above the focus of an earthquake.

Epoch A unit of the geologic calendar that is a subdivision of a period.

Equatorial low A belt of low pressure that lies near the equator and between the subtropical highs.

Equilibrium line For a glacier, the elevation at which the accumulation and wasting of glacial ice is equal.

Era A major division on the geologic calendar; eras are divided into shorter units called *periods*.

Erosion The incorporation and transportation of material by a mobile agent, such as water, wind, or ice.

Eruption column Buoyant plumes of hot, ash-laden gases that can extend thousands of meters into the atmosphere.

Escape velocity The initial velocity an object needs to escape the surface of a celestial body.

Esker A sinuous ridge composed largely of sand and gravel deposited by a stream flowing in a tunnel beneath a glacier near its terminus.

Estuary A partially enclosed coastal water body that is connected to the ocean. Salinity here is measurably reduced by the freshwater flow of rivers.

Eukaryote An organism whose genetic material is enclosed in a nucleus; plants, animals, and fungi are eukaryotes.

Euphotic zone The portion of the photic zone near the surface where light is bright enough for photosynthesis to occur.

Evaporation The process of converting a liquid to a gas.

Evaporite deposits A sedimentary rock formed of material deposited from solution by evaporation of water.

Evapotranspiration The combined effect of evaporation and transpiration.

Exfoliation dome A large, dome-shaped structure, usually composed of granite, formed by sheeting.

Exoplanet A planet that orbits a star other than the Sun.

External process A process such as weathering, mass wasting, or erosion that is powered by the Sun and transforms solid rock into sediment.

Extrusive rock Igneous rock formed when magma solidifies at Earth's surface. Also called *volcanic rock*.

Eye A zone of scattered clouds and calm averaging about 20 kilometers (12 miles) in diameter at the center of a hurricane.

Eye wall The doughnut-shaped area of intense cumulonimbus development and very strong winds that surrounds the eye of a hurricane.

F

Fall A type of movement common to mass-wasting processes that refers to the free falling of detached individual pieces of any size.

Fall equinox The equinox that occurs on September 21–23 in the Northern Hemisphere and on March 21–22 in the Southern Hemisphere. Also called the *autumnal equinox*.

Fault A break in a rock mass along which movement has occurred.

Fault creep Displacement along a fault that is so slow and gradual that little seismic activity occurs.

Fault scarp A cliff created by movement along a fault. It represents the exposed surface of the fault prior to modification by weathering and erosion.

Fault-block mountain A mountain formed by the displacement of rock along a fault.

Felsic composition A compositional group of igneous rocks that indicates a rock is composed almost entirely of light-colored silicates. Also called *granitic composition*.

Fine-grained texture A texture of igneous rocks in which the crystals are too small for individual minerals to be distinguished with the unaided eye.

Fiord A steep-sided inlet of the sea formed when a glacial trough was partially submerged.

Fissure A crack in rock along which there is a distinct separation.

Fissure eruption An eruption in which lava is extruded from narrow fractures or cracks in the crust.

Flood The overflow of a stream channel that occurs when discharge exceeds the channel's capacity. Floods are the most common and destructive type of geologic hazard.

Flood basalts Flows of basaltic lava that issue from numerous cracks or fissures and commonly cover extensive areas to thicknesses of hundreds of meters.

Floodplain The flat, low-lying portion of a stream valley that is subject to periodic inundation.

Flow A type of movement common to mass-wasting processes in which water-saturated material moves downslope as a viscous fluid.

Focus For an earthquake, the location within Earth where slippage begins. Also called the *hypocenter*.

Fog A cloud with its base at or very near Earth's surface.

Fold A bent rock layer or series of layers that were originally horizontal and subsequently deformed.

Foliation A texture of metamorphic rocks that gives the rock a layered appearance.

Food chain A succession of organisms in an ecological community through which food energy is transferred from producers through herbivores and on to one or more carnivores.

Food web A group of interrelated food chains.

Footwall block The rock surface below a fault.

Forearc basin The region located between a volcanic arc and an accretionary wedge where shallow-water marine sediments typically accumulate.

Foreshocks Small earthquakes that often precede a major earthquake.

Foreshore The portion of the shore that lies between the normal high and low water marks. Also called the *intertidal zone*.

Fossil The remains or traces of organisms preserved from the geologic past.

Fossil assemblage The overlapping ranges of a group of fossils (assemblage) collected from a layer. Examining such an assemblage can help establish the age of the sedimentary layer.

Fossil fuel A general term for any hydrocarbon that may be used as a fuel, including coal, oil, natural gas, bitumen from tar sands, and shale oil.

Fossil magnetism The natural remnant magnetism in rock bodies; the permanent magnetization acquired by rock that can be used to determine the location of the magnetic poles and the latitude of the rock at the time it became magnetized. Also called *paleomagnetism*.

Fracture Any break or rupture in rock along which no appreciable movement has taken place.

Fracture zone Any break or rupture in rock along which no appreciable movement has taken place.

Fragmental texture An igneous rock texture resulting from the consolidation of individual rock fragments that are ejected during a violent volcanic eruption. Also called *pyroclastic texture*.

Freezing nuclei Solid particles that serve as cores for the formation of ice crystals in the atmosphere. Also called *ice nuclei*.

Freezing rain A coating of ice on objects formed when supercooled rain freezes on contact. Also called *glaze*.

Frequency For traveling waves (such as electromagnetic waves), the number of crests or troughs that pass a given point in a unit of time. It is usually expressed in hertz, the number of crests or troughs per second.

Front The boundary between two adjoining air masses having contrasting characteristics.

Frontal fog Fog formed when rain evaporates as it falls through a layer of cool air. Also called *precipitation fog*.

Frontal lifting Lifting of air that results when cool air acts as a barrier over which warmer, lighter air rises. Also called *frontal wedging*.

Frontal wedging Lifting of air that results when cool air acts as a barrier over which warmer, lighter air rises. Also called *frontal lifting*.

Frost wedging The mechanical breakup of rock caused by the expansion of freezing water in cracks and crevices.

Fumarole A vent in a volcanic area from which fumes or gases escape.

G

Gabbro A dark-green to black intrusive igneous rock composed of dark silicate minerals. Gabbro makes up a significant percentage of oceanic crust.

Galaxy A collection of interstellar matter, stars, and star remains that are gravitationally bound to one another.

Galaxy cluster A group of gravitationally bound galaxies that may contain thousands of galaxies.

Garnet A silicate mineral composed of individual silica tetrahedra. Garnet is most often brown to deep red and has a glassy luster, lacks cleavage, and exhibits conchoidal fracture.

Gas hydrate A solid, ice-like chemical structure consisting of water and natural gas that forms naturally in deep-ocean sediments, where pressure is high and temperature is low.

Geocentric The concept of an Earth-centered universe.

Geologic structure A basic geologic feature, such as a fold, fault, or rock foliation, that results from forces associated with the interaction of tectonic plates. Also called a *tectonic structure*.

Geologic time The span of time since the formation of Earth, about 4.6 billion years.

Geologic time scale The division of Earth history into blocks of time—eons, eras, periods, and epochs. The time scale was created using relative dating principles.

Geology The science that examines Earth, its form and composition, and the changes it has undergone and is undergoing.

Geosphere The solid Earth, the largest of Earth's four major spheres.

Geostrophic wind A wind, usually above a height of 600 meters (2000 feet), that blows parallel to the isobars.

Geothermal gradient The gradual increase in temperature with depth in the crust. The average is 30°C per kilometer in the upper crust.

Geyser A fountain of hot water that is periodically ejected.

Giant With reference to a star, a luminous star of large radius.

Glacial budget The balance, or lack of balance, between ice formation at the upper end of a glacier and ice loss in the zone of wastage.

Glacial drift An all-embracing term for sediments of glacial origin, no matter how, where, or in what shape they were deposited. Also known simply as *drift*.

Glacial erratic An ice-transported boulder that was not derived from bedrock near its present site.

Glacial striations Scratches and grooves on bedrock caused by glacial abrasion.

Glacial trough A mountain valley that has been widened, deepened, and straightened by a glacier.

Glacier A thick mass of ice originating on land from the compaction and recrystallization of snow that shows evidence of past or present flow.

Glassy texture A term used to describe the texture of certain igneous rocks, such as obsidian, that contain no crystals.

Glaze A coating of ice on objects formed when supercooled rain freezes on contact. Also called *freezing rain*.

Gneiss Medium- to coarse-grained banded metamorphic rocks in which granular and elongated minerals dominate.

Gondwana The southern portion of Pangaea, consisting of South America, Africa, Australia, India, and Antarctica.

Graben A valley formed by the downward displacement of a fault-bounded block.

Gradient The slope of a stream; generally measured in feet per mile.

Granite An abundant, coarse-grained igneous rock, composed of about 10 to 20 percent quartz and 50 percent potassium feldspar. Granite is used as a building material.

Granitic composition A compositional group of igneous rocks that indicates a rock is composed almost entirely of light-colored silicates. Also called *felsic composition*.

Granule A fine structure visible on the solar surface caused by convective cells below.

Gravitational collapse The gradual subsidence of mountains caused by lateral spreading of weak material located deep within these structures.

Great Oxygenation Event A time about 2.5 billion years ago, when a significant amount of oxygen appeared in the atmosphere.

Greenhouse effect The transmission of short-wave solar radiation by the atmosphere, coupled with the selective absorption of longer-wavelength terrestrial radiation, especially by water vapor and carbon dioxide.

Groin A short wall built at a right angle to the shore to trap moving sand.

Ground moraine An undulating layer of till deposited as the ice front retreats.

Groundmass The matrix of smaller crystals within an igneous rock that has porphyritic texture.

Groundwater Water in the zone of saturation.

Guyot A submerged flat-topped seamount.

Gymnosperm A group of seed-bearing plants that includes conifers and ginkgo. The term means "naked seed," a reference to the unenclosed condition of the seeds.

Gypsum A hydrated calcium sulfate mineral. It is the mineral of which plaster, drywall, and other similar building materials are composed.

Gyre The large circular surface current pattern found in each ocean.

H

Habit Refers to the common or characteristic shape of a crystal or an aggregate of crystals. Also called *crystal shape*.

Habitable zone The region around a star in which a planet with sufficient atmospheric pressure could maintain liquid water on its surface and thus might be hospitable to Earth-like life.

Hadean The name of an eon found on some versions of the geologic time scale. It refers to the earliest interval (eon) of Earth history and ended 4 billion years ago.

Hail Nearly spherical ice pellets having concentric layers and formed by the successive freezing of layers of water.

Half-graben A tilted fault block in which the higher side is associated with mountainous topography and the lower side is a basin that fills with sediment.

Half-life The time required for one-half of the atoms of a radioactive substance to decay.

Halite The mineral name for common table salt (NaCl); a nonsilicate mineral commonly found in sedimentary rocks.

Hanging valley A tributary valley that enters a glacial trough at a considerable height above its floor.

Hanging wall block The rock surface immediately above a fault.

Hard stabilization Any form of artificial structure built to protect a coast or to prevent the movement of sand along a beach. Examples include groins, jetties, breakwaters, and seawalls.

Hardness The resistance a mineral offers to scratching.

Headward erosion The extension upslope of the head of a valley due to erosion.

Heat The kinetic energy of random molecular motion.

Heliocentric The view that the Sun is at the center of the solar system.

Hertzsprung–Russell diagram A plot of stars according to their absolute magnitudes and spectral types. Also called an *H–R diagram*.

High A high-pressure center characterized by a clockwise flow of air in the Northern Hemisphere. Also called an *anticyclone*.

High cloud A cloud that normally has its base above 6000 meters (3728 miles); the base may be lower in winter and at high-latitude locations.

Highland climate A complex pattern of climate conditions associated with mountains. Highland climates are characterized by large differences that occur over short distances.

Horn A pyramid-like peak formed by glacial action in three or more cirques surrounding a mountain summit.

Hornblende A dark green to black mineral of the amphibole group, often found in igneous rocks.

Horst An elongated, uplifted block of crust bounded by faults.

Host rock Preexisting crustal rocks intruded by magma. Host rock may be displaced or assimilated by magmas. Also called *country rock*.

Hot spot A concentration of heat in the mantle capable of producing magma, which in turn extrudes onto Earth's surface. The intraplate volcanism that produced the Hawaiian islands is one example.

Hot-spot track A chain of volcanic structures produced as a lithospheric plate moves over a mantle plume.

Hot spring A spring in which the water is 6–9°C (10–15°F) warmer than the mean annual air temperature of its locality.

H–R diagram A plot of stars according to their absolute magnitudes and spectral types. Stands for *Hertzsprung–Russell diagram*.

Hubble's law A law that relates the distance to a galaxy and its velocity.

Humid continental climate A relatively severe climate characteristic of broad continents in the middle latitudes between approximately 40° and 50° north latitude. This climate is not found in the Southern Hemisphere, where the middle latitudes are dominated by the oceans.

Humid subtropical climate A climate generally located on the eastern side of a continent and characterized by hot, sultry summers and cool winters.

Humidity A general term referring to water vapor in the air but not to liquid droplets of fog, cloud, or rain.

Hurricane A tropical cyclonic storm having winds in excess of 119 kilometers (74 miles) per hour.

Hydraulic fracturing A method of opening up pore space in otherwise impermeable rocks, permitting natural gas to flow out into wells.

Hydrogen fusion The nuclear reaction in which hydrogen nuclei are fused into helium nuclei.

Hydrogenous sediment Seafloor sediments consisting of minerals that crystallize from seawater. An important example is manganese nodules.

Hydrologic cycle The unending circulation of Earth's water supply. The cycle is powered by energy from the Sun and is characterized by continuous exchanges of water among the oceans, the atmosphere, and the continents.

Hydrosphere The water portion of our planet; one of the traditional subdivisions of Earth's physical environment.

Hygrometer An instrument designed to measure relative humidity.

Hygroscopic nuclei Condensation nuclei that have a high affinity for water, such as salt particles.

Hypocenter For an earthquake, the location within Earth where slippage begins. Also called the *focus*.

Hypothesis A tentative explanation that is tested to determine whether it is valid.

I

Ice nuclei Solid particles that serve as cores for the formation of ice crystals in the atmosphere. Also called *freezing nuclei*.

Ice cap A mass of glacial ice covering a high upland or plateau and spreading out radially.

Ice cap climate A climate that has no monthly means above freezing and supports no vegetative cover except in a few scattered high mountain areas. This climate, with its perpetual ice and snow, is confined largely to the ice sheets of Greenland and Antarctica.

Ice sheet A very large, thick mass of glacial ice flowing outward in all directions from one or more accumulation centers.

Ice shelf A large, relatively flat mass of floating ice that forms where glacial ice flows into bays and extends seaward from the coast but remains attached to the land along one or more sides.

Iceberg A mass of floating ice produced by a calving glacier. Usually 20 percent or less of the iceberg protrudes above the waterline.

Igneous rock A rock formed by the crystallization of molten magma.

Impact crater A depression result from a collision with a body such as an asteroid or a comet.

Incised meander A meandering channel that flows in a steep, narrow valley. Incised meanders form either when an area is uplifted or when base level drops.

Inclination of the axis The tilt of Earth's axis from the perpendicular to the plane of Earth's orbit.

Index fossil A fossil that is associated with a particular span of geologic time.

Industrial mineral A nonmetallic mineral resource that is used in industry.

Inertia A property of matter that resists a change in its motion.

Infiltration The movement of surface water into rock or soil through cracks and pore spaces.

Infrared Radiation with a wavelength from 0.7 to 200 micrometers.

Inner core The solid innermost layer of Earth, about 1300 kilometers (800 miles) in radius.

Inner planets The four inner planets of the Solar System: Mercury, Venus, Mars, and Earth. Also called the *Terrestrial (Earth-like) planets*.

Intensity With respect to earthquakes, a measure of the degree of shaking at a given locale, based on the amount of damage.

Interferometer When applied to a set of radiotelescopes or optical telescopes, refers to a way of connecting the telescopes so that they function as a single telescope. This technique increases the resolution of the combined telescope even more than its light-gathering area.

Interior drainage A discontinuous pattern of intermittent streams that do not flow to the ocean.

Intermediate composition A compositional group of igneous rocks that contains at least 25 percent dark silicate minerals. The other dominant mineral is plagioclase feldspar. Also called *andesitic composition*.

Internal process A process such as mountain building or volcanism that derives its energy from Earth's interior and elevates Earth's surface.

Intertidal zone The portion of the shore that lies between the normal high and low water marks. Also called the *foreshore*.

Intertropical convergence zone (ITCZ) The zone of general convergence between the Northern and Southern Hemisphere trade winds.

Intraplate volcanism Igneous activity that occurs within a tectonic plate away from plate boundaries.

Intrusion A structure that results from the emplacement and crystallization of magma beneath the surface of Earth. Also called a *pluton*.

Intrusive rock Igneous rock that formed below Earth's surface. Also called *plutonic rock*.

Invertebrate An animal that does not have a vertebral column (backbone); in other words, any animal that is not a vertebrate.

Ion An atom or a molecule that possesses an electrical charge.

Ionic bond A chemical bond between two oppositely charged ions formed by the transfer of valence electrons from one atom to the other.

Irregular galaxy A galaxy that lacks symmetry.

Island arc A chain of volcanic islands generally located a few hundred kilometers from a trench where active subduction of one oceanic slab beneath another is occurring. Also called a *volcanic island arc*.

Isobar A line drawn on a map that connects points of equal atmospheric pressure, usually corrected to sea level.

Isostasy The concept that Earth's crust is floating in gravitational balance on the material of the mantle.

Isostatic adjustment Compensation of the lithosphere when weight is added or removed. When weight is added, the lithosphere responds by subsiding, and when weight is removed, there is uplift.

Isotherm A line connecting points of equal temperature.

J

Jet stream Swift (120–240 kilometer per hour), high-altitude winds.

Jetty One of a pair of structures extending into the ocean at the entrance to a harbor or river, built for the purpose of protecting against storm waves and sediment deposition.

Joint A fracture in rock along which there has been no movement.

Jovian (Jupiter-like) planets The four outer planets of the solar system: Jupiter, Saturn, Uranus, and Neptune. These planets have relatively low densities. Also called the *outer planets*.

K

Kame A steep-sided hill composed of sand and gravel that originates when sediment is collected in openings in stagnant glacial ice.

Kettle A depression created when a block of ice that became lodged in glacial deposits subsequently melted.

Köppen classification A system for classifying climates devised by Wladimir Köppen that is based on mean monthly and annual values of temperature and precipitation.

Kuiper belt A region outside the orbit of Neptune where most short-period comets are thought to originate.

L

La Niña An episode of strong trade winds and unusually low sea-surface temperatures in the central and eastern Pacific. The opposite of *El Niño*.

Laccolith A massive igneous body intruded between preexisting strata.

Lahar A mudflow on the slope of a volcano that results when unstable layers of ash and debris become saturated and flow downslope, usually following stream channels.

Lake-effect snow Snow showers associated with a cP air mass to which moisture and heat are added from below as the air mass traverses a large and relatively warm lake (such as one of the Great Lakes), rendering the air mass humid and unstable.

Laminar flow The movement of water particles in straight-line paths that are parallel to the channel. The water particles move downstream, without mixing.

Land breeze A local wind blowing from land toward the water during the night in coastal areas.

Latent heat The energy absorbed or released during a change in state.

Lateral moraine A ridge of till along the sides of an alpine glacier composed primarily of debris that fell to the glacier from the valley walls.

Laurasia The northern portion of Pangaea, consisting of North America and Eurasia.

Lava Magma that reaches Earth's surface.

Lava tube A tunnel in hardened lava that acts as a horizontal conduit for lava flowing from a volcanic vent. Lava tubes allow fluid lavas to advance great distances.

Law of conservation of angular momentum A law which states that the product of the velocity of an object around a center of rotation (axis) and the distance squared of the object from the axis is constant.

Leaching The depletion of soluble materials from the upper soil by downward-percolating water.

Leeward coast A coast where the prevailing winds blow from the land toward the ocean.

Lifting condensation level The height at which rising air that is cooling at the dry adiabatic rate becomes saturated and condensation begins. Also called simply the *condensation level*.

Light silicate mineral A silicate mineral that lacks iron and/or magnesium. Light silicates are generally lighter in color and have lower specific gravities than dark silicates.

Light-gathering area In a telescope or other optical device (such as an eye), the total area of lens or mirror that gathers the light processed by the instrument.

Light-year The distance light travels in a year; about 6 trillion miles.

Limestone A chemical sedimentary rock composed chiefly of calcite. Limestone can form by inorganic means or from biochemical processes.

Liquefaction A phenomenon, sometimes associated with earthquakes, in which soils and other unconsolidated materials containing abundant water are turned into a fluid-like mass that is not capable of supporting buildings.

Lithification The process, generally cementation and/or compaction, of converting sediments to solid rock.

Lithosphere The rigid outer layer of Earth, including the crust and upper mantle.

Lithospheric plate A coherent unit of Earth's rigid outer layer that includes the crust and upper unit. Also called simply a *plate*.

Local Group The cluster of 20 or so galaxies to which our galaxy belongs.

Local wind A small-scale wind produced by a locally generated pressure gradient. Examples include land and sea breezes and mountain and valley breezes.

Localized convective lifting Unequal surface heating that causes localized pockets of air (thermals) to rise because of their buoyancy. Also called simply *convective lifting*.

Loess Deposits of windblown silt, lacking visible layers, generally buff-colored, and capable of maintaining a nearly vertical cliff.

Longitudinal dunes Long ridges of sand oriented parallel to the prevailing wind; these dunes form where sand supplies are limited.

Longitudinal profile A cross section of a stream channel along its descending course from the head to the mouth.

Longshore current A near-shore current that flows parallel to the shore.

Low A low-pressure center characterized by a counterclockwise flow of air in the Northern Hemisphere. Also called a *cyclone*.

Low cloud A cloud that forms below a height of 2000 meters (6500 feet).

Luminosity With respect to a star, the star's total output of energy. See also *absolute magnitude*.

Lunar eclipse An eclipse of the Moon.

Lunar highlands The extensively cratered highland areas of the Moon.

Lunar regolith A thin, gray layer on the surface of the Moon, consisting of loosely compacted, fragmented material believed to have been formed by repeated meteoritic impacts.

Luster The appearance or quality of light reflected from the surface of a mineral.

M

Mafic composition A compositional group of igneous rocks indicating that the rock contains substantial dark silicate minerals and calcium-rich plagioclase feldspar. Also called *basaltic composition*.

Magma A body of molten rock found at depth, including any dissolved gases and crystals.

Magmatic differentiation The process of generating more than one rock type from a single magma.

Magnetic reversal A change in Earth's magnetic field from normal to reverse or vice versa.

Magnetic time scale A scale that shows the ages of magnetic reversals and is based on the polarity of lava flows of various ages.

Magnetometer A sensitive instrument used to measure the intensity of Earth's magnetic field at various points.

Magnitude With respect to an earthquake, the total amount of energy released during the earthquake.

Main-sequence stars A sequence of stars on the Hertzsprung–Russell diagram, containing the majority of stars, that runs diagonally from the upper left to the lower right.

Mammal A member of the vertebrate class Mammalia. Mammals are endothermic and possess hair, and females have mammary glands that secrete milk to nourish their offspring.

Mantle The 2900-kilometer- (1800-mile-) thick layer of Earth located below the crust.

Mantle plume A mass of hotter-than-normal mantle material that ascends toward the surface, where it may lead to igneous activity. These plumes of solid yet mobile material may originate as deep as the core–mantle boundary.

Maria The Latin name for the smooth areas of the Moon formerly thought to be seas.

Marine terrace A wave-cut platform that has been exposed above sea level.

Marine west coast climate A climate found on windward coasts from latitudes 40° to 65° and dominated by maritime air masses. This climate features mild winters and cool summers.

Maritime (m) air mass An air mass that originates over the ocean. These air masses are relatively humid.

Mass extinction An event in which a large percentage of species become extinct.

Mass movement The downslope movement of rock, regolith, and soil under the direct influence of gravity.

Massive Refers to an igneous pluton that is not tabular in shape.

Mean solar day The average time between two passages of the Sun across the local celestial meridian.

Meander A looplike bend in the course of a stream.

Mechanical weathering The physical disintegration of rock, resulting in smaller fragments.

Medial moraine A ridge of till formed when lateral moraines from two coalescing alpine glaciers join.

Megathrust fault The plate boundary separating a subducting slab of oceanic lithosphere and the overlying plate.

Mercury barometer A mercury-filled glass tube in which the height of the mercury column is a measure of air pressure.

Mesocyclone A vertical cylinder of cyclonically rotating air (3 to 10 kilometers in diameter) that develops in the updraft of a severe thunderstorm and that often precedes the development of damaging hail or tornadoes.

Mesosphere The layer of the atmosphere immediately above the stratosphere and characterized by decreasing temperatures with height.

Mesozoic era A span on the geologic time scale between the Paleozoic and Cenozoic eras from about 248 million to 65 million years ago.

Metallic bond A chemical bond present in all metals that may be characterized as an extreme type of electron sharing in which the electrons move freely from atom to atom.

Metamorphic rock Rocks formed by the alteration of preexisting rock deep within Earth (but still in the solid state) by heat, pressure, and/or chemically active fluids.

Metamorphism The changes in mineral composition and texture of a rock subjected to high temperature and pressure within Earth.

Meteor The luminous phenomenon observed when a meteoroid enters Earth's atmosphere and burns up; popularly called a *"shooting star."*

Meteor shower Many meteors appearing in the sky, caused by Earth intercepting a swarm of meteoritic particles.

Meteorite Any portion of a meteoroid that survives its traverse through Earth's atmosphere and strikes Earth's surface.

Meteoroid Small solid particles that have orbits in the solar system.

Meteorology The scientific study of the atmosphere and atmospheric phenomena; the study of weather and climate.

Microcontinent A relatively small fragment of continental crust that may lie above sea level, such as the island of Madagascar, or may be submerged, as exemplified by the Campbell Plateau near New Zealand.

Middle clouds Clouds that occupy the height range from 2000 to 6000 meters.

Midlatitude (middle-latitude) cyclone A large center of low pressure with an associated cold front and often a warm front. Frequently accompanied by abundant precipitation.

Mid-ocean ridge A broad, linear ridge or rise on the ocean floor. The rift at the crest of the ridge represents a divergent plate boundary, where new oceanic crust is generated. Also called an *oceanic ridge* or *rise*.

Mineral A naturally occurring, inorganic crystalline material with a unique chemical composition.

Mineral resource All discovered and undiscovered deposits of a useful mineral that can be extracted now or at some time in the future.

Mineralogy The study of minerals.

Mist A cloud of water droplets suspended in the atmosphere at or near Earth's surface.

Mixed tidal pattern A tidal pattern exhibiting two high tides and two low tides per tidal day, with a large inequality in high water heights, low water heights, or both. Coastal locations that experience such a tidal pattern may also show alternating periods of diurnal and semidiurnal tidal patterns. Also called *mixed semidiurnal*.

Mixing ratio The mass of water vapor in a unit mass of dry air; commonly expressed as grams of water vapor per kilogram of dry air.

Modified Mercalli intensity scale A 12-point scale developed to evaluate earthquake intensity based on the amount of damage to various structures.

Mohs scale A series of 10 minerals used as a standard in determining hardness.

Molecular cloud A cloud of interstellar gas and dust that is cold and dense enough for hydrogen atoms to join together into hydrogen molecules. These clouds can condense to form stars.

Moment magnitude A more precise measure of earthquake magnitude than the Richter scale that is derived from the amount of displacement that occurs along a fault zone.

Monocline A one-limbed flexure in strata. The strata are unusually flat-lying or very gently dipping on both sides of the monocline.

Monsoon Seasonal reversal of wind direction associated with large continents, especially Asia. In winter, the wind blows from land to sea; in summer, from sea to land.

Monthly mean The mean temperature for a month that is calculated by averaging the daily means.

Mountain belt A geographic area of roughly parallel and geologically connected mountain ranges developed as a result of plate tectonics.

Mountain breeze The nightly downslope winds commonly encountered in mountain valleys.

Mudflow A relatively rapid type of mass wasting that involves a flow of soil and regolith containing a large amount of water. Also called *debris flow*.

Muscovite A common member of the mica family. It is light colored, has a pearly luster, and has excellent cleavage.

N

Natural levee An elevated landform that parallels some streams and acts to confine their waters, except during flood stage.

Neap tide The lowest tidal range, which occurs near the times of the first- and third-quarter phases of the Moon.

Near-shore zone The zone of beach that extends from the low-tide shoreline seaward to where waves break at low tide.

Nebula A cloud of interstellar gas and/or dust.

Nebular theory The basic idea that the Sun and planets formed from the same cloud of gas and dust in interstellar space.

Negative-feedback mechanism A feedback mechanism that tends to maintain a system as it is—that is, maintain the status quo.

Nekton Pelagic organisms that can move independently of ocean currents by swimming or other means of propulsion.

Neritic zone The marine-life zone that extends from the low tideline out to the shelf break.

Neutron A subatomic particle found in the nucleus of an atom. A neutron is electrically neutral and has a mass approximately that of a proton.

Neutron star A star of extremely high density, composed entirely of neutrons.

Nimbostratus A solid layer of dark gray cloud that forms at low altitude (below 2000 meters [6500 feet]). Nimbostratus clouds are a major source of precipitation.

Nimbus A cloud that is a major producer of precipitation.

Nonconformity An unconformity in which older metamorphic or intrusive igneous rocks are overlain by younger sedimentary strata.

Nonfoliated Describes metamorphic rocks that do not exhibit foliation.

Nonmetallic mineral resource A mineral resource that is not a fuel and is not processed for the metals it contains.

Nonrenewable With respect to a resource, any resource that forms or accumulates over such long time spans that it must be considered as fixed in total quantity.

Nonsilicate Any mineral group that lacks silica in its structure. Nonsilicates account for less than 10 percent of Earth's crust.

Nor'easter The term used to describe the weather associated with an incursion of mP air from the North Atlantic into the Northeast and Mid-Atlantic regions. Strong northeast winds, freezing or near-freezing temperatures, and the possibility of precipitation make this an unwelcome weather event.

Normal fault A fault in which the rock above the fault plane has moved down relative to the rock below.

Normal polarity A magnetic field in which the north and south magnetic poles are at the same ends of Earth (relative to the rotational axis) as at present.

Nuclear decay The spontaneous decay of certain unstable atomic nuclei. Also called *radioactive decay*.

Nuclear fusion The source of the Sun's energy.

Nucleus The small heavy core of an atom that contains all of its positive charge and most of its mass.

Nuée ardente A highly heated mixture, largely of ash and pumice fragments, traveling down the flanks of a volcano or along the surface of the ground. Also called a *pyroclastic flow*.

Numerical date A date that specifies the actual number of years that have passed since an event occurred.

O

Obsidian A volcanic glass of felsic composition.

Occluded front A front formed when a cold front overtakes a warm front. It marks the beginning of the end of a middle-latitude cyclone.

Ocean basin A deep submarine region that lies beyond the continental margins.

Oceanic plateau An extensive region on the ocean floor composed of thick accumulations of pillow basalts and other mafic rocks that in some cases exceed 30 kilometers (20 miles) in thickness.

Oceanic ridge A broad, linear ridge or rise on the ocean floor. The rift at the crest of the ridge represents a divergent plate boundary, where new oceanic crust is generated. Also called *mid-ocean ridge* or *oceanic rise*.

Oceanic ridge system A continuous elevated zone on the floor of all the major ocean basins and varying in width from 500 to 5000 kilometers (300 to 3000 miles). The rifts at the crests of ridges represent divergent plate boundaries.

Oceanic rise A broad, linear ridge or rise on the ocean floor. The rift at the crest of the ridge represents a divergent plate boundary, where new oceanic crust is generated. Also called *oceanic ridge* or *mid-ocean ridge*.

Oceanic zone The marine-life zone beyond the continental shelf.

Oceanography The scientific study of the oceans and oceanic phenomena.

Octet rule A rule which says that atoms combine in order that each may have the electron arrangement of a noble gas; that is, the outer energy level contains eight neutrons.

Offshore zone The relatively flat submerged zone that extends from the breaker line to the edge of the continental shelf.

Oil trap A geologic structure that allows for significant amounts of oil and gas to accumulate.

Olivine A high-temperature dark silicate mineral typically found in basalt.

Oort cloud A spherical shell composed of comets that orbit the Sun at distances generally greater than 10,000 times the Earth–Sun distance.

Orbit The path of a body in revolution around a center of mass, such as Earth's orbit around the Sun.

Ore deposit A naturally occurring concentration of one or more metallic minerals that can be extracted economically.

Orogenesis The processes that collectively result in the formation of mountains.

Orogeny A specific episode of orogenesis (mountain building).

Orographic lifting Mountains acting as barriers to the flow of air, forcing the air to ascend. The air cools adiabatically, and clouds and precipitation may result.

Outer core A layer beneath the mantle about 2200 kilometers (1364 miles) thick that has the properties of a liquid.

Outer planets The four outer planets of the solar system: Jupiter, Saturn, Uranus, and Neptune. These planets have relatively low densities. Also called the *Jovian (Jupiter-like) planets*.

Outgassing The escape of gases that had been dissolved in magma.

Outlet glacier A tongue of ice that normally flows rapidly outward from an ice cap or ice sheet, usually through mountainous terrain to the sea.

Outwash plain A relatively flat, gently sloping plain consisting of materials deposited by meltwater streams in front of the margin of an ice sheet.

Overrunning Warm air gliding up a retreating cold air mass.

Oxbow lake A curved lake produced when a stream cuts off a meander.

Ozone A molecule of oxygen that contains three oxygen atoms.

P

Pahoehoe flow A lava flow with a smooth-to-ropey surface.

Paleomagnetism The natural remnant magnetism in rock bodies; the permanent magnetization acquired by rock that can be used to determine the location of the magnetic poles and the latitude of the rock at the time it became magnetized. Also called *fossil magnetism*.

Paleontology The systematic study of fossils and the history of life on Earth.

Paleoseismology The study of the timing, location, and size of prehistoric earthquakes.

Paleozoic era A span on the geologic time scale between the eons of the Precambrian and Mesozoic era from about 540 million to 248 million years ago.

Pangaea The proposed supercontinent that 200 million years ago began to break apart and form the present landmasses.

Parabolic dunes Dunes that resemble barchans, except that their tips point into the wind; they often form along coasts that have strong onshore winds, abundant sand, and vegetation that partly covers the sand.

Parasitic cone A volcanic cone that forms on the flank of a larger volcano.

Parcel An imaginary volume of air enclosed in a thin elastic cover. Typically it is considered to be a few hundred cubic meters in volume and is assumed to act independently of the surrounding air.

Parent material The material on which a soil develops.

Partial melting The process by which most igneous rocks melt. Since individual minerals have different melting points, most igneous rocks melt over a temperature range of a few hundred degrees. If the liquid is squeezed out after some melting has occurred, a melt with a higher silica content results.

Passive continental margin A margin that consists of a continental shelf, continental slope, and continental rise. These margins are *not* associated with plate boundaries and therefore experience little volcanism and few earthquakes.

Pegmatite A very coarse-grained igneous rock (typically granite) commonly found as a dike associated with a large mass of plutonic rock that has smaller crystals. Crystallization in a water-rich environment is believed to be responsible for the very large crystals.

Pelagic zone Open ocean of any depth in which animals swim or float freely.

Perched water table A localized zone of saturation above the main water table created by an impermeable layer (aquiclude).

Peridotite An igneous rock of ultramafic composition thought to be abundant in the upper mantle.

Perihelion The point in the orbit of a planet where it is closest to the Sun.

Period A basic unit of the geologic calendar that is a subdivision of an era. Periods may be divided into smaller units called epochs.

Periodic table The tabular arrangement of the elements according to atomic number.

Permafrost Any permanently frozen subsoil. Usually found in the subarctic and arctic regions.

Permeability A measure of a material's ability to transmit water.

Phanerozoic eon The part of geologic time represented by rocks containing abundant fossil evidence. The eon extending from the end of the Proterozoic eon (about 540 million years ago) to the present.

Phases of the Moon The progression of changes in the Moon's appearance during the month.

Phenocryst In an igneous rock with a porphyritic texture, a conspicuously large crystal embedded in a matrix of finer-grained crystals called the groundmass.

Photic zone The upper part of the ocean into which any sunlight penetrates.

Photon A discrete amount (quantum) of electromagnetic energy.

Photosphere The region of the Sun that radiates energy to space. The visible surface of the Sun.

Photosynthesis The process by which plants and algae produce carbohydrates from carbon dioxide and water in the presence of chlorophyll, using light energy and releasing oxygen.

Phyllite A metamorphic rock composed mainly of fine crystals of muscovite, chlorite, or both.

Phytoplankton Algal plankton, which are the most important community of primary producers in the ocean.

Piedmont glacier A glacier that forms when one or more valley glaciers emerge from the confining walls of mountain valleys and spread out to create a broad sheet in the lowlands at the base of the mountains.

Pillow lava Basaltic lava that solidifies in an underwater environment and develops a structure that resembles a pile of pillows.

Plagioclase feldspar A type of feldspar containing both sodium and calcium ions that freely substitute for one another, depending on the crystallization environment.

Plane of the ecliptic The imaginary plane that connects Earth's orbit with the celestial sphere.

Planetary nebula A shell of incandescent gas expanding from a star.

Planetesimal A solid celestial body that accumulated during the first stages of planetary formation. Planetesimals aggregated into increasingly larger bodies, ultimately forming the planets.

Plankton Passively drifting or weakly swimming organisms that cannot move independently of ocean currents. Plankton include microscopic algae, protozoa, jellyfish, and larval forms of many animals.

Plate A coherent unit of Earth's rigid outer layer that includes the crust and upper unit. Also called a *lithospheric plate*.

Playa lake A temporary lake in a playa (a flat area on the floor of an undrained desert basin).

Plucking The process by which pieces of bedrock are lifted out of place by a glacier.

Plug An isolated, steep-sided, erosional remnant consisting of lava that once occupied the vent of a volcano. Also called a *volcanic neck*.

Pluton A structure that results from the emplacement and crystallization of magma beneath the surface of Earth. Also called an *intrusion*.

Plutonic rock Igneous rock that formed below Earth's surface. Also called *intrusive rock*.

Pluvial lake A lake formed during a period of increased rainfall. During the Pleistocene epoch, this occurred in some nonglaciated regions during periods of ice advance elsewhere.

Point bar A crescent-shaped accumulation of sand and gravel deposited on the inside of a meander.

Polar (P) air mass A cold air mass that forms in a high-latitude source region.

Polar climate A climate in which the mean temperature of the warmest month is below 10°C. This climate is too cold to support the growth of trees.

Polar easterlies In the global pattern of prevailing winds, winds that blow from the polar high toward the subpolar low. These winds, however, should not be thought of as persistent winds, such as the trade winds.

Polar front The stormy frontal zone separating air masses of polar origin from air masses of tropical origin.

Polar high Anticyclones that are assumed to occupy the inner polar regions and are believed to be thermally induced, at least in part.

Porosity The volume of open spaces in rock or soil.

Porphyritic texture An igneous texture consisting of large crystals embedded in a matrix of much smaller crystals.

Positive-feedback mechanism A feedback mechanism that enhances or drives change.

Potassium feldspar An abundant, relatively hard light silicate mineral that contains potassium ions.

Pothole A circular depression in a bedrock stream channel created by the abrasive action of particles swirling in fast-moving eddies.

Precambrian All geologic time prior to the Paleozoic era.

Precipitation fog Fog formed when rain evaporates as it falls through a layer of cool air. Also called *frontal fog*.

Precursor Relative to earthquakes, an event or a change that precedes an earthquake and may provide a warning.

Pressure gradient force The amount of pressure change occurring over a given distance.

Pressure tendency The nature of the change in atmospheric pressure over the past several hours. It can be a useful aid in short-range weather prediction. Also called *barometric tendency*.

Prevailing wind A wind that consistently blows from one direction more than from another.

Primary (P) wave A type of seismic wave that involves alternating compression and expansion of the material through which it passes.

Primary productivity The amount of organic matter synthesized by organisms from inorganic substances through photosynthesis or chemosynthesis within a given volume of water or habitat in a unit of time.

Principle of cross-cutting relationships A principle of relative dating which states that a rock or fault is younger than any rock or fault through which it cuts.

Principle of fossil succession A principle by which fossil organisms succeed one another in a definite and determinable order, and any time period can be recognized by its fossil content.

Principle of inclusions A principle that uses pieces of rock contained within another to determine a relative date. According to the principle of inclusions, the rock mass that provided the inclusion is older than the rock mass containing the inclusion.

Principle of lateral continuity A principle which states that sedimentary beds originate as continuous layers that extend in all directions until they grade into a different type of sediment or thin out at the edge of a sedimentary basin.

Principle of original horizontality A principle by which layers of sediment are generally deposited in a horizontal or nearly horizontal position.

Principle of superposition A principle which states that in any undeformed sequence of sedimentary rocks, each bed is older than the one above and younger than the one below.

Proglacial lake A lake created when a glacier acts as a dam, blocking the flow of a river or trapping glacial meltwater. The term refers to the position of such lakes just beyond the outer limits of a glacier.

Prokaryote A cell or an organism such as bacteria whose genetic material is not enclosed in a nucleus.

Prominence A concentration of material above the solar surface that appears as a bright archlike structure.

Protein A class of organic molecules that provide the primary structural material for life and contribute to the functioning of cells.

Proterozoic The eon following the Archean and preceding the Phanerozoic. It extends between about 2500 million (2.5 billion) and 540 million years ago.

Proton A positively charged subatomic particle in the nucleus of an atom.

Proton–proton chain reaction A chain of thermonuclear reactions by which nuclei of hydrogen are built up into nuclei of helium.

Protoplanet A developing planetary body that grows by the accumulation of planetesimals.

Protostar A collapsing cloud of gas and dust destined to become a star.

Psychrometer A device consisting of two thermometers (wet bulb and dry bulb) that is rapidly whirled and, with the use of tables, yields the relative humidity and dew point.

Ptolemaic system An Earth-centered system of the universe.

Pulsar A variable radio source of small size that emits radio pulses in very regular periods.

Pumice A light-colored, glassy vesicular rock commonly having a granitic composition.

Pycnocline A layer of water in which there is a rapid change of density with depth.

Pyroclastic flow A highly heated mixture, largely of ash and pumice fragments, traveling down the flanks of a volcano or along the surface of the ground. Also called a *nuée ardente*.

Pyroclastic material The volcanic rock ejected during an eruption, including ash, bombs, and blocks. Also called *tephra*.

Pyroclastic texture An igneous rock texture resulting from the consolidation of individual rock fragments that are ejected during a violent volcanic eruption. Also called *fragmental texture*.

Q

Quartz A common silicate mineral consisting entirely of silicon and oxygen that resists weathering.

Quartzite A hard, nonfoliated metamorphic rock formed from quartz sandstone.

Quaternary period The most recent period on the geologic time scale. It began about 2.6 million years ago and extends to the present.

R

Radial pattern A system of streams running in all directions away from a central elevated structure, such as a volcano.

Radiation The transfer of energy (heat) through space by electromagnetic waves. Along with conduction and convection, it is a mechanism of heat transfer. The term can also refer generally to electromagnetic radiation.

Radiation fog Fog resulting from radiation heat loss by Earth.

Radiation pressure The force exerted by electromagnetic radiation from an object such as the Sun.

Radiation zone The radial zone within the Sun or another star in which energy is carried outward mainly by high-energy photons of light.

Radio telescope A telescope designed to make observations in radio wavelengths.

Radioactive decay The spontaneous decay of certain unstable atomic nuclei. Also called *nuclear decay*.

Radiocarbon dating Radiometric dating that is done using carbon-14, a radioactive isotope of carbon that is produced continuously in the atmosphere. It is used to date events from the very recent geologic past (the last few tens of thousands of years).

Radiometric dating The procedure of calculating the absolute ages of rocks and minerals that contain radioactive isotopes.

Radiosonde A lightweight package of weather instruments fitted with a radio transmitter and carried aloft by a balloon.

Rain Drops of water that fall from clouds that have a diameter of at least 0.5 millimeter (0.02 inch).

Rainshadow desert A dry area on the lee side of a mountain range. Many middle-latitude deserts are of this type.

Rectangular pattern A drainage pattern characterized by numerous right-angle bends that develops on jointed or fractured bedrock.

Red giant A large, cool star of high luminosity; a star occupying the upper-right portion of the Hertzsprung-Russell diagram.

Redshift With respect to emission or absorption line spectra, an instance in which spectral lines are shifted to longer wavelengths.

Reflecting telescope A telescope that concentrates light from distant objects by using a concave mirror.

Reflection The process whereby light bounces back from an object at the same angle at which it encounters a surface and with the same intensity.

Refracting telescope A telescope that uses a lens to bend and concentrate the light from distant objects.

Regional metamorphism Metamorphism associated with large-scale mountain-building processes.

Regolith The layer of rock and mineral fragments that nearly everywhere covers Earth's surface.

Relative date A date that results from placing rocks in their proper sequence or order to determine the chronological order of events.

Relative humidity The ratio of the air's water-vapor content to its water-vapor capacity.

Renewable With respect to a resource, a resource that is virtually inexhaustible or that can be replenished over relatively short time spans.

Reptile A group of animals that includes turtles, snakes, lizards, and crocodiles, and that traditionally also includes extinct groups such as dinosaurs, ichthyosaurs, and plesiosaurs. Modern representatives are generally scaled and ectothermic.

Reservoir rock The porous, permeable portion of an oil trap that yields oil and gas.

Resolution A measure of the ability of a telescope to separate objects that would otherwise appear as one.

Retrograde motion The apparent westward motion of the planets with respect to the stars.

Reverse fault A fault in which the material above the fault plane moves up in relation to the material below.

Reverse polarity A magnetic field opposite to that which exists at present.

Revolution The motion of one body about another, such as Earth's orbit around the Sun.

Rhyolite The fine-grained equivalent of the igneous rock granite, composed primarily of light-colored silicates.

Richter scale A scale of earthquake magnitude based on the motion of a seismograph.

Ridge push A mechanism that may contribute to plate motion, which involves the oceanic lithosphere sliding down the oceanic ridge under the pull of gravity.

Rift valley A long, narrow trough bounded by normal faults. It represents a region where divergence is taking place.

Rime A thin coating of ice on objects produced when supercooled fog droplets freeze on contact.

Ring of Fire The zone of active volcanoes surrounding the Pacific Ocean.

Rip current A strong narrow surface or near-surface current of short duration and high speed flowing seaward through the breaker zone at nearly right angles to the shore. It represents the return to the ocean of water that has been piled up on the shore by incoming waves.

Rock A consolidated mixture of minerals.

Rock avalanche Very rapid downslope movement of rock and debris. These rapid movements may be aided by a layer of air trapped beneath the debris, and they have been known to reach speeds of over 200 kilometers (125 miles) per hour.

Rock cycle A model that illustrates the origin of the three basic rock types and the interrelatedness of Earth materials and processes.

Rock flour Ground-up rock produced by the grinding effect of a glacier.

Rock-forming mineral The set of minerals that make up most of the rocks of Earth's crust.

Rockslide The rapid slide of a mass of rock downslope along planes of weakness.

Rotation The spinning of a body, such as Earth, about its axis.

Runoff Water that flows over the land rather than infiltrating into the ground.

S

Saffir–Simpson hurricane scale A scale, from 1 to 5, used to rank the relative intensities of hurricanes.

Salinity The proportion of dissolved salts to pure water, usually expressed in parts per thousand (‰).

Saltation Transportation of sediment through a series of leaps or bounces.

Sandstone An abundant, durable detrital sedimentary rock primarily composed of sand-size grains.

Santa Ana The local name given a chinook wind in southern California.

Saturation The maximum quantity of water vapor that the air can hold at any given temperature and pressure.

Savanna A tropical grassland, usually with scattered trees and shrubs.

Scattering The redirecting (in all directions) of light by small particles and gas molecules in the atmosphere. The result is diffused light.

Schist Medium- to coarse-grained metamorphic rocks having a foliated texture, in which platy minerals dominate.

Scientific method The process by which researchers raise questions, gather data, and formulate and test scientific hypotheses.

Scoria Hardened lava that has retained the vesicles produced by escaping gases.

Scoria cone A rather small volcano built primarily of pyroclastics ejected from a single vent. Also called a *cinder cone*.

Sea arch An arch formed by wave erosion when caves on opposite sides of a headland unite.

Sea breeze A local wind blowing from the sea during the afternoon in coastal areas.

Sea ice Frozen seawater that is associated with polar regions. The area covered by sea ice expands in winter and shrinks in summer.

Sea stack An isolated mass of rock standing just offshore, produced by wave erosion of a headland.

Seafloor spreading The process of producing new seafloor between two diverging plates.

Seamount An isolated volcanic peak that rises at least 1000 meters (3000 feet) above the deep-ocean floor.

Seawall A barrier constructed to prevent waves from reaching the area behind the wall. Its purpose is to defend property from the force of breaking waves.

Secondary (S) waves Seismic waves that involve oscillation perpendicular to the direction of propagation.

Sediment Unconsolidated particles created by the weathering and erosion of rock, by chemical precipitation from solution in water, or from the secretions of organisms and transported by water, wind, or glaciers.

Sedimentary rock Rock formed from the weathered products of preexisting rocks that have been transported, deposited, and lithified.

Seed An embryonic plant that is packaged with a supply of nutrients inside a protective coating.

Seismic gap A segment of an active fault zone that has not experienced a major earthquake over a span when most other segments have. Such segments are probable sites for future major earthquakes.

Seismic waves A rapidly moving ocean wave generated by earthquake activity capable of inflicting heavy damage in coastal regions.

Seismogram The record made by a seismograph.

Seismograph An instrument that records earthquake waves. Also called a *seismometer*.

Seismology The study of earthquakes and seismic waves.

Seismometer An instrument that records earthquake waves. Also called a *seismograph*.

Selective absorbers Gases that absorb and emit radiation only in certain wavelengths.

Semiarid One of the two types of dry climate. A marginal and more humid variant of the desert that separates it from bordering humid climates. Also called *steppe*.

Semidiurnal tidal pattern A tidal pattern exhibiting two high tides and two low tides per tidal day with small inequalities between successive highs and successive lows; a semidaily tide.

Sensible heat The heat we can feel and measure with a thermometer.

Settling velocity The speed at which a particle falls through a still fluid. The size, shape, and specific gravity of particles influence settling velocity.

Shale The most common sedimentary rock, consisting of silt- and clay-size particles.

Shear Stress that causes two adjacent parts of a body to slide past one another.

Sheeting A mechanical weathering process characterized by the splitting off of slablike sheets of rock.

Shield A large, relatively flat expanse of ancient metamorphic rock within the stable continental interior.

Shield volcano A broad, gently sloping volcano built from fluid basaltic lavas.

Shore Seaward of the coast, a zone that extends from the highest level of wave action during storms to the lowest tide level.

Shoreline The line that marks the contact between land and sea. It migrates up and down as the tide rises and falls.

Sidereal day The period of Earth's rotation with respect to the stars.

Sidereal month A time period based on the revolution of the Moon around Earth with respect to the stars.

Silicate Any one of numerous minerals that have the oxygen and silicon tetrahedron as their basic structure.

Silicon–oxygen tetrahedron A structure composed of four oxygen atoms surrounding a silicon atom that constitutes the basic building block of silicate minerals.

Sill A tabular igneous body that was intruded parallel to the layering of preexisting rock.

Siltstone A fine-grained sedimentary rock composed of silt-sized grains intermixed with clay-sized sediment.

Slab pull A mechanism that contributes to plate motion in which cool, dense oceanic crust sinks into the mantle and "pulls" the trailing lithosphere along.

Slate A very fine-grained metamorphic rock containing platy minerals and having excellent rock cleavage.

Sleet Frozen or semifrozen rain formed when raindrops freeze as they pass through a layer of cold air.

Slide A movement common to mass-wasting processes in which the material moving downslope remains fairly coherent and moves along a well-defined surface.

Slip face The steep, leeward slope of a sand dune; it maintains an angle of about 34 degrees.

Slump The downward slipping of a mass of rock or unconsolidated material moving as a unit along a curved surface.

Small solar system bodies Solar system objects not classified as planets or moons that include dwarf planets, asteroids, comets, and meteoroids.

Snow A solid form of precipitation produced by sublimation of water vapor.

Snowline The lower limit of perennial snow. For a glacier, equivalent to the equilibrium line.

Soil A combination of mineral and organic matter, water, and air; the portion of the regolith that supports plant growth.

Soil horizon A layer of soil that has identifiable characteristics produced by chemical weathering and other soil-forming processes.

Soil profile A vertical section through a soil, showing its succession of horizons and the underlying parent material.

Soil Taxonomy A soil classification system consisting of six hierarchical categories based on observable soil characteristics. The system recognizes 12 soil orders.

Soil texture The relative proportions of clay, silt, and sand in a soil. A soil's texture strongly influences its ability to retain and transmit water and air.

Solar eclipse An eclipse of the Sun.

Solar flare A sudden and tremendous eruption in the solar chromosphere.

Solar nebula The cloud of interstellar gas and/or dust from which the bodies of our solar system formed.

Solar winds Subatomic particles ejected at high speed from the solar corona.

Solifluction A slow, downslope flow of water-saturated materials common to permafrost areas.

Solum The O, A, and B horizons in a soil profile. Living roots and other plant and animal life are largely confined to this zone.

Sonar An instrument that uses acoustic signals (sound energy) to measure water depths. Sonar is an acronym for *so*und *na*vigation and *r*anging.

Sorting The process by which solid particles of various sizes are separated by moving water or wind. Also, the degree of similarity in particle size in sediment or sedimentary rock.

Source region The area where an air mass acquires its characteristic properties of temperature and moisture.

Source rock The sedimentary rocks in which petroleum and natural gas originate.

Southern Oscillation The seesaw pattern of atmospheric pressure change that occurs between the eastern and western Pacific. The interaction of this effect and that of El Niño can cause extreme weather events in many parts of the world.

Specific gravity The ratio of a substance's weight to the weight of an equal volume of water.

Specific heat The amount of heat needed to raise 1 gram of a substance 1°C at sea level atmospheric pressure.

Spectroscope An instrument for directly viewing the spectrum of a light source.

Spectroscopy The study of spectra.

Spheroidal weathering Any weathering process that tends to produce a spherical shape from an initially blocky shape.

Spiral galaxy A flattened, rotating galaxy with pinwheel-like arms of interstellar material and young stars winding out from its nucleus.

Spit An elongated ridge of sand that projects from the land into the mouth of an adjacent bay.

Spreading center A region where the rigid plates are moving apart, typified by the mid-ocean ridges. Also called a *divergent plate boundary*.

Spring A flow of groundwater that emerges naturally at the ground surface.

Spring equinox The equinox that occurs on March 21–22 in the Northern Hemisphere and on September 21–23 in the Southern Hemisphere.

Spring tide The highest tidal range, which occurs near the times of the new and full moons.

Stable air Air that resists vertical displacement. If it is lifted, adiabatic cooling will cause its temperature to be lower than the surrounding environment; if it is allowed to do so, it will sink to its original position.

Stable platform The part of a craton that is mantled by relatively undeformed sedimentary rocks and underlain by a basement complex of igneous and metamorphic rocks.

Standard rain gauge A gauge that has a diameter of about 20 centimeters and funnels rain into a cylinder that magnifies precipitation amounts by a factor of 10, allowing for accurate measurement of small amounts.

Star dune An isolated hill of sand that exhibits a complex form and develops where wind directions are variable.

Stationary front A situation in which the surface position of a front does not move; the flow on either side of such a boundary is nearly parallel to the position of the front.

Steam fog Fog having the appearance of steam, produced by evaporation from a warm water surface into the cool air above.

Steppe One of the two types of dry climate. A marginal and more humid variant of the desert that separates it from bordering humid climates. Also called *semiarid*.

Stock A pluton similar to but smaller than a batholith.

Storm surge The abnormal rise of the sea along a shore as a result of strong winds.

Strain An irreversible change in the shape and size of a rock body that is caused by stress.

Strata Parallel layers of sedimentary rock. Also called *beds*.

Stratified drift Sediments deposited by glacial meltwater.

Stratocumulus Low clouds with scalloped bottoms in parallel rolls.

Stratosphere The layer of the atmosphere immediately above the troposphere, characterized by increasing temperatures with height, due to the concentration of ozone.

Stratovolcano A volcano composed of both lava flows and pyroclastic material. Also called a *composite volcano*.

Stratus One of three basic cloud forms; also, the name given one of the flow clouds. They are sheets or layers that cover much or all of the sky.

Streak The color of a mineral in powdered form.

Stream terrace A flat, benchlike structure produced by a stream, which was left elevated as the stream cut downward.

Stream valley The channel, valley floor, and sloping valley walls of a stream.

Stress The force per unit area acting on any surface within a solid.

Strike-slip fault A fault along which the movement is horizontal.

Stromatolite Structures that are deposited by algae and consist of layered mounds of calcium carbonate.

Subarctic climate A climate found north of the humid continental climate and south of the polar climate and characterized by bitterly cold winters and short, cool summers. Places within this climatic realm experience the highest annual temperature ranges on Earth.

Subduction The process of thrusting oceanic lithosphere into the mantle along a convergent boundary.

Subduction erosion A process in subduction zones in which sediment and rock are scraped off the bottom of the overriding plate and transported into the mantle.

Subduction zone A boundary in which two plates move together, resulting in oceanic lithosphere being thrust beneath an overriding plate, eventually to be reabsorbed into the mantle. It can also involve the collision of two continental plates to create a mountain system. Also called a *convergent plate boundary*.

Sublimation The conversion of a solid directly to a gas without passing through the liquid state.

Submarine canyon A seaward extension of a valley that was cut on the continental shelf during a time when sea level was lower, or a canyon carved into the outer continental shelf, slope, and rise by turbidity currents.

Submergent coast A coast with a form that is largely the result of the partial drowning of a former land surface either because of a rise of sea level or subsidence of the crust or both.

Subpolar low Low pressure located at about the latitudes of the Arctic and Antarctic Circles. In the Northern Hemisphere the low takes the form of individual oceanic cells; in the Southern Hemisphere there is a deep and continuous trough of low pressure.

Subtropical high Not a continuous belt of high pressure but rather several semipermanent, anticyclonic centers characterized by subsidence and divergence located roughly between latitudes 25° and 35°.

Summer solstice The solstice that occurs on June 21–22 in the Northern Hemisphere and on December 21–22 in the Southern Hemisphere.

Sunspot A dark spot on the Sun, which is cool in contrast to the surrounding photosphere.

Supercontinent A large landmass that contains all, or nearly all, of the existing continents.

Supercontinent cycle The idea that the rifting and dispersal of one supercontinent is followed by a long period during which the fragments gradually reassemble into a new supercontinent.

Supercooled The condition of water droplets that remain in the liquid state at temperatures well below 0°C (32°F).

Supergiant A very large star of high luminosity.

Supernova An exploding star that increases in brightness many thousands of times.

Superplume A large mantle plumes. Superplumes are thought to be responsible for creating basalt plateaus.

Surf A collective term for breakers; also, the wave activity in the area between the shoreline and the outer limit of breakers.

Surface waves Seismic waves that travel along the outer layer of Earth.

Suspended load The fine sediment carried within the body of flowing water.

Suture A zone along which two crustal fragments are jointed together. For example, following a continental collision, the two continental blocks are sutured together.

Syncline A linear downfold in sedimentary strata; the opposite of *anticline*.

Synodic month The period of revolution of the Moon with respect to the Sun, or its cycle of phases.

System Any size group of interacting parts that form a complex whole.

T

Tabular Describing a feature such as an igneous pluton that has two dimensions that are much longer than the third.

Taiga The northern coniferous forest; also a name applied to the subarctic climate.

Tectonic structure A basic geologic feature, such as a fold, fault, or rock foliation, that results from forces associated with the interaction of tectonic plates. Also called a *geologic structure*.

Temperature A measure of the degree of hotness or coldness of a substance; a measure of the *average* kinetic energy of individual atoms or molecules in a substance.

Temperature control A factor that causes temperature to vary from place to place or from time to time.

Temperature gradient The amount of temperature change per unit of distance.

Tenacity A mineral's toughness or resistance to breaking or deforming.

Tensional stress The type of stress that tends to pull apart a body.

Tephra The volcanic rock ejected during an eruption, including ash, bombs, and blocks. Also called *pyroclastic material*.

Terrane A crustal block bounded by faults, whose geologic history is distinct from the histories of adjoining crustal blocks.

Terrestrial (Earth-like) planets The four inner planets of the Solar System: Mercury, Venus, Mars, and Earth. Also called the *inner planets*.

Terrigenous sediment Seafloor sediments derived from terrestrial weathering and erosion.

Texture The size, shape, and distribution of the particles that collectively constitute a rock.

Theory A well-tested and widely accepted view that explains certain observable facts.

Theory of plate tectonics A well-tested theory which proposes that Earth's outer shell consists of individual plates that interact in various ways and thereby produce earthquakes, volcanoes, mountains, and the crust itself.

Thermocline A layer of water in which there is a rapid change in temperature in the vertical dimension.

Thermohaline circulation Movements of ocean water caused by density differences brought about by variations in temperature and salinity.

Thermosphere The region of the atmosphere immediately above the mesosphere and characterized by increasing temperatures due to absorption of very shortwave solar energy by oxygen.

Thrust fault A low-angle reverse fault.

Thunderstorm A storm produced by a cumulonimbus cloud and always accompanied by lightning and thunder. It is of relatively short duration and usually accompanied by strong wind gusts, heavy rain, and sometimes hail.

Tidal current The alternating horizontal movement of water associated with the rise and fall of the tide.

Tidal flat A marshy or muddy area that is covered and uncovered by the rise and fall of the tide.

Tide Periodic change in the elevation of the ocean surface.

Till Unsorted sediment deposited directly by a glacier.

Tipping-bucket gauge A recording rain gauge consisting of two compartments ("buckets"), each capable of holding 0.025 centimeter of water. When one compartment fills, it tips, and the other compartment takes its place.

Tombolo A ridge of sand that connects an island to the mainland or to another island.

Tornado A small, very intense cyclonic storm with exceedingly high winds, most often produced along cold fronts in conjunction with severe thunderstorms.

Tornado warning A warning issued when a tornado has actually been sighted in an area or is indicated by radar.

Tornado watch A warning issued for areas of about 65,000 square kilometers (25,000 square miles), indicating that conditions are such that tornadoes may develop; it is intended to alert people to the possibility of tornadoes.

Trade winds Two belts of winds that blow almost constantly from easterly directions and are located on the equatorward sides of the subtropical highs.

Transform fault A major strike-slip fault that cuts through the lithosphere and accommodates motion between two plates. Equivalent to a transform plate boundary.

Transform plate boundary A boundary in which two plates slide past one another without creating or destroying lithosphere. Equivalent to a transform fault.

Transpiration The release of water vapor to the atmosphere by plants.

Transverse dunes A series of long ridges oriented at right angles to the prevailing wind; these dunes form where vegetation is sparse and sand is very plentiful.

Travertine A form of limestone ($CaCO_3$) that is deposited by hot springs or as a cave deposit.

Trellis pattern A system of streams in which nearly parallel tributaries occupy valleys cut in folded strata.

Trench An elongated depression in the seafloor produced by bending of oceanic crust during subduction. Also called a *deep-ocean trench*.

Trigger An event, such as an earthquake or heavy rainfall, that initiates a mass movement process.

Trophic level A nourishment level in a food chain. Plant and algae producers constitute the lowest level, followed by herbivores and a series of carnivores at progressively higher levels.

Tropic of Cancer The parallel of latitude, 23½° north latitude, marking the northern limit of the Sun's vertical rays.

Tropic of Capricorn The parallel of latitude, 23½° south latitude, marking the southern limit of the Sun's vertical rays.

Tropical (T) air mass A warm-to-hot air mass that forms in the subtropics.

Tropical depression By international agreement, a tropical cyclone with maximum winds that do not exceed 61 kilometers (38 miles) per hour.

Tropical rain forest A luxuriant broadleaf evergreen forest; also, the name given the climate associated with this vegetation.

Tropical storm By international agreement, a tropical cyclone with maximum winds between 61 and 119 kilometers (38 and 74 miles) per hour.

Tropical wet and dry A climate that is transitional between the wet tropics and the subtropical steppes.

Troposphere The lowermost layer of the atmosphere. It is generally characterized by a decrease in temperature with height.

Tsunami The Japanese word for a seismic sea wave.

Tundra climate A climate found almost exclusively in the Northern Hemisphere or at high altitudes in many mountainous regions. A treeless climatic realm of sedges, grasses, mosses, and lichens that is dominated by a long, bitterly cold winter.

Turbidity current A downslope movement of dense, sediment-laden water created when sand and mud on the continental shelf and slope are dislodged and thrown into suspension.

Turbulence In the context of astronomical observation, disturbance in the atmosphere that introduces distortions to telescopic images. Turbulence can be corrected by adaptive optics.

Turbulent flow Movement of water in an erratic fashion, often characterized by swirling, whirlpool-like eddies. Most streamflow is of this type.

U

Ultramafic Refers to igneous rocks composed mainly of iron- and magnesium-rich minerals.

Ultraviolet (UV) Radiation with a wavelength from 0.2 to 0.4 micrometer.

Unconformity A surface that represents a break in the rock record, caused by erosion or nondeposition.

Uniformitarianism The concept that the processes that have shaped Earth in the geologic past are essentially the same as those operating today.

Unsaturated zone The area above the water table where openings in soil, sediment, and rock are not saturated but filled mainly with air.

Unstable air Air that does not resist vertical displacement. If it is lifted, its temperature will not cool as rapidly as the surrounding environment, so it will continue to rise on its own.

Upslope fog Fog created when air moves up a slope and cools adiabatically.

Upwelling The rising of cold water from deeper layers to replace warmer surface water that has been moved away.

V

Valence electrons The electrons involved in the bonding process; the electrons occupying the highest principal energy level of an atom.

Valley breeze The daily upslope winds commonly encountered in a mountain valley.

Valley glacier A glacier confined to a mountain valley, which in most instances had previously been a stream valley. Also called an *alpine glacier*.

Valley train A relatively narrow body of stratified drift deposited on a valley floor by meltwater streams that issue from a valley glacier.

Vapor pressure The part of the total atmospheric pressure attributable to water-vapor content.

Vein deposit A mineral filling a fracture or fault in a host rock. Such deposits have a sheetlike, or tabular, form.

Vent The surface opening of a conduit or pipe.

Vertebrate A group of animals that are characterized by the possession of a vertebral column (backbone).

Very-long-baseline interferometer (VBLI) An interferometer that is created by linking together radiotelescopes that are located far apart on Earth's surface.

Vesicular texture A term applied to igneous rocks that contain small cavities called vesicles, which are formed when gases escape from lava.

Viscosity A measure of a fluid's resistance to flow.

Visible light Radiation with a wavelength from 0.4 to 0.7 micrometer.

Volatiles Gaseous components of magma dissolved in melt. Volatiles readily vaporize (form a gas) at surface pressures.

Volcanic cone A cone-shaped structure built by successive eruptions of lava or pyroclastic materials.

Volcanic island arc A chain of volcanic islands generally located a few hundred kilometers from a trench where active subduction of one oceanic slab beneath another is occurring. Also called simply an *island arc*.

Volcanic neck An isolated, steep-sided, erosional remnant consisting of lava that once occupied the vent of a volcano. Also called a *plug*.

Volcanic rock Igneous rock formed when magma solidifies at Earth's surface. Also called *extrusive rock*.

W

Warm front A front along which a warm air mass overrides a retreating mass of cooler air.

Watershed The land area that contributes water to a stream. Also called a *drainage basin*.

Water table The upper level of the saturated zone of groundwater.

Wave base A depth equal to one-half the wavelength measured from the still water level. Below this depth, water movement associated with a wave is negligible.

Wave height The vertical distance between the trough and crest of a wave.

Wave period The time interval between the passage of successive crests at a stationary point.

Wave refraction The process by which the portion of a wave in shallow water slows, causing the wave to bend and then to align itself with the underwater contours.

Wave-cut cliff A seaward-facing cliff along a steep shoreline formed by wave erosion at its base and mass wasting.

Wave-cut platform A bench or shelf in the bedrock at sea level, cut by wave erosion.

Wavelength For traveling waves (such as water waves or electromagnetic waves), the distance separating successive crests or troughs.

Weather The state of the atmosphere at any given time.

Weather radar Instruments that utilize transmitters to send out radio waves at wavelengths that can penetrate clouds, to produce a reflected signal called an echo that can be displayed on a monitor to show the location and intensity of precipitation.

Weathering The disintegration and decomposition of rock at or near Earth's surface.

Well An opening bored into the zone of saturation.

Westerlies The dominant west-to-east motion of the atmosphere that characterizes the regions on the poleward side of the subtropical highs.

Wet adiabatic rate The rate of adiabatic temperature change in saturated air. The wet adiabatic rate of temperature change is variable, but it is always less than the dry adiabatic rate.

White dwarf A star that has exhausted most or all of its nuclear fuel and has collapsed to a very small size. A white dwarf is believed to be near its final stage of evolution.

Wind Air flowing horizontally with respect to Earth's surface.

Wind shear A change in the speed and/or direction of wind with height. Mesocyclone formation depends on the presence of wind shear.

Wind vane An instrument used to determine wind direction.

Windward coast A coast where the prevailing winds blow from the ocean toward the land.

Winter solstice The solstice that occurs on December 21–22 in the Northern Hemisphere and on June 21–22 in the Southern Hemisphere.

Y

Yazoo tributary A tributary that flows parallel to the main stream because a natural levee is present.

Z

Zone of accumulation The part of a glacier characterized by snow accumulation and ice formation. Its outer limit is the snowline.

Zone of fracture The upper portion of a glacier, consisting of brittle ice.

Zone of saturation The zone where all open spaces in sediment and rock are completely filled with water.

Zone of wastage The part of a glacier beyond the zone of accumulation where all of the snow from the previous winter melts, as does some of the glacial ice.

Zooplankton Animal plankton.

INDEX

F

G

H

I

N

O

INTERCEPTAG
Authentic
Pearson